WILEY-BLACKWELL ENCYCLOPEDIA OF HUMAN EVOLUTION

This encyclopedia is dedicated to the life and work of Clark Howell, Glynn Isaac, Charlie Lockwood, and Elizabeth Harmon.

Wiley-Blackwell Encyclopedia of Human Evolution

Edited by

Bernard Wood

Executive Editor

Amanda Henry

WILEY Blackwell

This paperback edition first published 2013 © 2011, 2013 by Blackwell Publishing Ltd

Edition History: Blackwell Publishing Ltd (hardback, 2011)

Registered Office
John Wiley & Sons, Ltd, The Atrium, Southern Gate, Chichester, West Sussex, PO19 8SQ, UK

Editorial Offices
9600 Garsington Road, Oxford, OX4 2DQ, UK
The Atrium, Southern Gate, Chichester, West Sussex, PO19 8SQ, UK
111 River Street, Hoboken, NJ 07030-5774, USA

For details of our global editorial offices, for customer services and for information about how to apply for permission to reuse the copyright material in this book please see our website at www.wiley.com/wiley-blackwell.

Library of Congress Cataloguing-in-Publication data has been applied for

ISBN: 978-1-1186-5099-8 (paperback)

A catalogue record for this book is available from the British Library.

Wiley also publishes its books in a variety of electronic formats. Some content that appears in print may not be available in electronic books.

Cover design by Design Deluxe

Set in 9.5/11pt Ehrhardt by SPi Publisher Services, Pondicherry, India
Printed in Malaysia by Ho Printing (M) Sdn Bhd

1 2013

Contents

This encyclopedia has a companion website: www.woodhumanevolution.com

Contributors

James O'Connell
The University of Utah

Pilbeam, David
Harvard University

Roche, Hélène
CNRS, Université Paris Ouest Nanterre

Rosas, Antonio
Museo Nacional de Ciencias Naturales

Smith, Fred
Illinois State University

Smith, Richard
Washington University

Stone, Anne
Arizona State University

Stringer, Chris
Natural History Museum, London

Thackeray, Francis
University of the Witwatersrand

Ward, Carol
University of Missouri

Zilhão, João
ICREA/University of Barcelona

Section and Topic Editors

Bae, Christopher
University of Hawaii

Bailey, Shara
New York University

Collard, Mark
Simon Fraser University

Crowley, Brooke
University of Cincinnati

DeSilva, Jeremy
Boston University

Ditchfield, Peter
University of Oxford

Elton, Sarah
Hull York Medical School

Faith, Tyler
The University of Queensland

Feakins, Sarah
University of Southern California

Gordon, Adam
University at Albany-SUNY

Herries, Andy
La Trobe University

Kivell, Tracy
University of Kent

Konigsberg, Lyle
University of Illinois at Urbana-Champaign

Kramer, Andrew
The University of Tennessee, Knoxville

Moggi-Cecchi, Jacopo
Università degli Studi di Firenze

Pearson, Osbjorn
University of New Mexico

Petraglia, Michael
University of Oxford

Pickering, Robyn
University of Melbourne

Plummer, Tom
The City University of New York

Richmond, Brian
The George Washington University

Sherwood, Chet
The George Washington University

Smith, Tanya
Harvard University

Strait, David
University at Albany-SUNY

Subiaul, Francys
The George Washington University

Verna, Christine
CNRS, Dynamique de l'Evolution Humaine, UPR2147, Paris

Viola, Bence
Max Planck Institute for Evolutionary Anthropology

Wang, Steve
The City University of New York

Contributors

Adams, Justin
Grand Valley State University

Ashley, Gail
Rutgers University

Auerbach, Benjamin
The University of Tennessee,
Knoxville

Bogart, Stephanie
Agnes Scott College, Atlanta

Brooks, Alison
The George Washington University

Claxton, Alex
University of Indianapolis

Constantino, Paul
Marshall University

Crevecoeur, Isabelle
Université Bordeaux 1

Crittenden, Alyssa
University of California, San Diego

Cunningham, Andrew
Harvard University

Dechow, Paul
Texas A&M Health Science Center

Despriée, Jackie
National Museum of Natural History of Paris

de Vos, John
Netherlands Centre for Biodiversity
Naturalis

Dizon, Eusebio
National Museum of the Philippines

Domínguez-Rodrigo, Manuel
Complutense University

Drapeau, Michelle
Université de Montréal

Du, Andrew
The George Washington University

Dunsworth, Holly
University of Rhode Island

Farrell, Milly
The Royal College of Surgeons of England

Friedlaender, Jonathan
Temple University

Garcia Garriga, Joan
Rovira i Virgili University-The Human Paleoecology
and Social Evolution Institute

Green, David
The George Washington University

Grosse, Ian
University of Massachusetts, Amherst

Gunz, Philipp
Max Planck Institute for Evolutionary Anthropology

Higham, Tom
Oxford Radiocarbon Accelerator Unit
University of Oxford

Johanson, Donald
Arizona State University

Joordens, Josephine
Human Origins Group, Faculty of Archaeology,
Leiden University

Kelly, Robert
University of Wyoming

Kupczik, Kornelius
Max Planck Institute for Evolutionary Anthropology

Lestrel, Pete
7327 de Celis Place, Van Nuys,
California 91406-2853

Louchart, Antoine
CNRS, ENS de Lyon

McNulty, Kieran
University of Minnesota

McPherron, Shannon
Max Planck Institute for Evolutionary Anthropology

Martínez Molina, Kenneth
Rovira i Virgili University-The Human Paleoecology
and Social Evolution Institute

Martínez-Navarro, Bienvenido
Institut Català de Paleoecología Humana i Evolució
Social (IPHES) Universitat Rovira i Virgili,
Tarragona

Monge, Janet
University of Pennsylvania

Nelson, Emma
University of Liverpool

Patterson, David
The George Washington University

Quam, Rolf
Binghamton University (SUNY)

Pickering, Travis
University of Wisconsin-Madison

Potts, Richard
National Museum of Natural History,
The Smithsonian Institution

Reynolds, Sally
Liverpool John Moores University

Ritzman, Terry
Arizona State University

Roebroeks, Wil
Faculty of Archaeology Leiden University

Ross, Callum
The University of Chicago

Schroer, Kes
The George Washington University

Sémah, Francois
Muséum national d'histoire naturelle

Sept, Jeanne
Indiana University

Setchell, Jo
Durham University

Skinner, Matt
University College London

Smith, Holly
University of Michigan

Speth, John
University of Michigan

Sponheimer, Matt
University of Colorado at Boulder

Spoor, Fred
Max Planck Institute for Evolutionary Anthropology

Stanistreet, Ian
University of Liverpool

Steudel-Numbers, Karen
University of Wisconsin-Madison

Stewart, Kathlyn
Canadian Museum of Nature

Stout, Dietrich
Emory University

Thompson, Jennifer
University of Nevada, Las Vegas

Vogel, Erin
The George Washington University

Vonhof, Hubert
VU University Amsterdam

Willoughby, Pamela
University of Edmonton

Wynn, Jonathan
University of South Florida

Yellen, John
National Science Foundation

Zawidzki, Tadeusz
The George Washington University

Zipkin, Andrew
The George Washington University

Zonneveld, Frans
Oranjestraat 35, Middelbeers, The Netherlands

Foreword

There is at present a consensus that the 21st century will be the century of biology, just as the 20th century was the century of physics. Biology now has larger budgets and a larger workforce than physics, and it faces problems of great significance and relevance to the understanding of human nature and the conduct of human life. The core of all biological research and understanding is the theory of evolution; evolution by natural selection is the central unifying concept of biology. As the great 20th century evolutionist Theodosius Dobzhansky asserted in 1973, "Nothing in biology makes sense except in the light of evolution."

The theory of evolution has transformed our understanding of life on planet Earth. Evolution provides a scientific explanation for why there are so many different kinds of organisms and why they all share the same chemical components in similar proportions, and why all organisms share DNA as their hereditary material, and why enzymes and other proteins, which are the fundamental constituents and engines of cell processes, are all made up of the same 20 amino acids, despite hundreds of amino acids existing in organisms. Evolution demonstrates why some organisms that look quite different are in fact related, while other organisms that may look similar are only distantly related. It accounts for the origins of humans on Earth and reveals our species' biological connections with other living things in varying degrees. Evolution explains the similarities and differences among modern human groups and modern human individuals. It enables the development of effective new ways to protect ourselves against constantly evolving bacteria and viruses, and to improve the quality of our agricultural products and domestic animals.

We owe the concept of evolution by natural selection to Charles Darwin. Natural selection was proposed by Darwin primarily to account for the adaptive organization, or design, of living beings; it is a process that preserves and promotes adaptation. Evolutionary change through time and evolutionary diversification (multiplication of species) often ensue as byproducts of natural selection fostering the adaptation of organisms to their milieu. Evolutionary change is not directly promoted by natural selection and, therefore, it is not its necessary consequence. Indeed, some species remain unchanged for long periods of time, as Darwin noted. Nautilus, *Lingula*, and other so-called living fossils were used by Darwin as examples of organisms that have remained unchanged in their appearance for millions of years.

Evolution affects all aspects of an organism's life: morphology (form and structure), physiology (function), behavior, and ecology (interaction with the environment). Underlying these changes are changes in the hereditary materials. Hence, in genetic terms, evolution consists of changes in an organism's hereditary makeup and can be seen as a two-step process. First, hereditary variation arises by mutation; second, selection occurs by which useful variations increase in frequency and those that are less useful or injurious are eliminated over the generations. As Darwin (*The Origin of Species*, 1859) saw it, individuals having useful variations "would have the best chance of surviving and procreating their kind" (p. 81). As a consequence, useful variations increase in frequency over the generations, at the expense of those that are less useful or are injurious.

Natural selection is much more than a "purifying" process, for it is able to generate novelty by increasing the probability of otherwise extremely improbable genetic combinations. Natural selection in combination with mutation becomes, in this respect, a creative process. Moreover, it is a process that has been occurring for many millions of years, in many different evolutionary lineages, and in a multitude of species, each consisting of a large number of individuals. Evolution by mutation and natural selection has produced the enormous diversity of the living world with its wondrous adaptations and it accounts for our presence, *Homo sapiens*, on planet Earth.

There is a version of the history of ideas that sees a parallel between two scientific revolutions, the Copernican and the Darwinian. In this view, the Copernican Revolution consisted of displacing the Earth from its previously accepted locus as the center of the universe, moving it to a subordinate place as just one more planet revolving around the sun. Similarly, in a congruous manner, the Darwinian Revolution is viewed as consisting of the displacement of modern humans from their exalted position as the center of life on Earth, with all other species created for the service of humankind. According to this version of intellectual history, Copernicus had accomplished his revolution with the heliocentric theory of the solar system. Darwin's achievement emerged from his theory of organic evolution. (Sigmund Freud refers to these two revolutions as "outrages" inflicted upon humankind's self-image and adds a third one, his own. He sees psychoanalysis as the "third and most bitter blow upon man's craving for grandiosity," revealing that man's *ego* "is not even master in his own house.")

What the standard versions of the Copernican and Darwinian revolutions say is correct but inadequate. It misses what is most important about these two intellectual revolutions, namely that they ushered in the beginning of science in the modern sense. These two revolutions may jointly be seen as the one Scientific Revolution, with two stages, the Copernican and the Darwinian.

The Copernican Revolution was launched with the publication in 1543, the year of Nicolaus Copernicus' death, of his *De revolutionibus orbium celestium* (*On the Revolutions of the Celestial Spheres*), and it bloomed in 1687 with the publication of Isaac Newton's *Philosophiae naturalis principia mathematica* (*The Mathematical Principles of Natural Philosophy*). The discoveries by Copernicus, Kepler, Galileo, Newton, and others, in the 16th and 17th centuries, had shown that Earth is not the center of the universe, but is a small planet rotating around an average star; that the universe is immense in space and in time; and that the motions of the planets around the sun can be explained by the same simple laws that account for the motion of physical objects on our planet. These include laws such as $f = m \times a$ (force=mass×acceleration) or the inverse-square law of attraction, $f = g(m_1 m_2)/r^2$ (the force of attraction between two bodies is directly proportional to their masses, but inversely related to the square of the distance between them).

These and other discoveries greatly expanded modern human knowledge. The conceptual revolution they brought about was more fundamental yet: a commitment to the postulate that the universe obeys immanent laws that account for natural phenomena. The workings of the universe were brought into the realm of science: explanation through natural laws. Potentially, all physical phenomena could be accounted for, as long as the causes were adequately known.

The advances of physical science brought about by the Copernican Revolution had driven humankind's conception of the universe to a split-personality state of affairs, a condition that persisted well into the mid-19th century. Scientific explanations, derived from natural laws, dominated the world of nonliving matter, on the Earth as well as in the heavens. Supernatural explanations, which depended on the unfathomable deeds of the creator, were accepted as explanations of the origin and configuration of living creatures. Authors, such as William Paley in his *Natural Theology* (1802), had developed the "argument from design," the notion that the complex design of organisms could not have come about by chance, or by the mechanical laws of physics, chemistry, and astronomy, but was rather accomplished by an omnipotent deity, just as the complexity of a watch, designed to tell time, was accomplished by an intelligent watchmaker.

Darwin completed the Copernican Revolution by drawing out for biology the notion of nature as a lawful system of matter in motion that modern human reason can explain without recourse to supernatural agencies. Darwin's greatest accomplishment was to show that the complex organization and functionality of living beings can be explained as the result of a natural process: natural selection. The origin and adaptations of organisms in their profusion and wondrous variations were thus brought into the realm of science.

Darwin's theory of evolution by natural selection disposed of Paley's arguments: the adaptations of organisms are not outcomes of chance, but of a process that, over time, causes the gradual accumulation of features beneficial to organisms. There is "design" in the living world: eyes are designed for seeing, wings for flying, and kidneys for regulating the composition of the blood. But the design of organisms is not intelligent, as would be expected from an engineer, but imperfect and worse: defects, dysfunctions, oddities, waste, and cruelty pervade the living world. Darwin's focus in *The Origin of Species* (1859) was the explanation of design, with evolution playing the subsidiary role of supporting evidence.

It follows from Darwin's explanation of adaptation that evolution must necessarily occur as a consequence of organisms becoming adapted to different environments in different localities, and to the ever-changing conditions of the environment over time, and as hereditary variations become available at a particular time that improve, in that place and at that time, the organisms' chances of survival and reproduction.

Origin's evidence for biological evolution is central to Darwin's explanation of design, because this explanation implies that biological evolution occurs, which Darwin therefore seeks to demonstrate in the second half of the book.

Darwin and other 19th century biologists found compelling evidence for biological evolution in the comparative study of living organisms, in their geographic distribution, and in the fossil remains of extinct organisms. Since Darwin's time, the evidence from these sources has become stronger and more comprehensive, while biological disciplines that have emerged recently – genetics, biochemistry, ecology, animal behavior (ethology), neurobiology, and especially molecular biology – have supplied powerful additional evidence and detailed confirmation. Accordingly, evolutionists are no longer concerned with obtaining evidence to support the fact of evolution, but rather are concerned with finding out additional information of the historical process in cases of particular interest. Moreover and most importantly, evolutionists nowadays are interested in understanding further and further how the process of evolution occurs.

Nevertheless, important discoveries continue, even in traditional disciplines, such as paleontology. Skeptical contemporaries of Darwin asked about the "missing links," particularly between the extant apes and modern humans, but also between major groups of organisms, such as between fish and terrestrial tetrapods or between reptiles and birds. Evolutionists can now affirm that these missing links are no longer missing. The known fossil record has made great strides over the last century and a half. Many fossils intermediate between diverse organisms have been discovered over the years. Two examples are *Archaeopteryx*, an animal intermediate between reptiles and birds, and *Tiktaalik*, intermediate between fishes and tetrapods.

The missing link between apes and humans is not, either, missing any longer. Not one, but hundreds of fossil remains from hundreds of individual hominins have been discovered since Darwin's time and continue to be discovered at an accelerated rate. The history of hominin discoveries is narrated in this encyclopedia, as well as the anatomical and other changes that occur through time.

Darwin wrote two books dedicated to human evolution: *The Descent of Man, and Selection in Relation to Sex* (2 vols, 1871) and *The Expression of the Emotions in Man and Animals* (1872). What we now know about human evolution is immensely more than what Darwin knew. But, even concerning hominin fossil history, much remains to be discovered. Indeed, the sequence that goes from the most primitive hominins to *Homo sapiens*, our species, is not resolved. That is, in many cases we do not know whether a particular hominin fossil belongs to the line of descent that goes to our species, or whether it belongs to a lateral branch.

There are many other important issues concerning the evolutionary origin of modern human traits – anatomical, physiological, behavioral, and cultural – that remain largely unknown. I will briefly point out three great research frontiers that seem to me particularly significant: ontogenic decoding, the brain/mind puzzle, and the ape-to-human transformation. By ontogenetic decoding I refer to the problem of how the unidimensional genetic information encoded in the DNA of a single cell becomes transformed into a four-dimensional being, the individual that develops, grows, matures, and dies. Cancer, disease, and aging are epiphenomena of ontogenetic decoding. By the brain/mind puzzle I refer to the interdependent questions of (a) how the physicochemical signals that reach our sense organs become transformed into perceptions, feelings, ideas, critical arguments, aesthetic emotions, and ethical values; and (b) how, out of this diversity of experiences, there emerges a unitary reality, the mind or self. Free will and language, social and political institutions, technology, and art are all epiphenomena of the modern human mind. By the ape-to-human transformation I refer to the mystery of how a particular ape lineage became a hominin lineage, from which emerged, after only a few million years, modern humans able to think and love, who have developed complex societies and uphold ethical, aesthetic, and religious values. But the modern human genome differs little from the chimp genome.

I will refer to these three issues as the egg-to-adult transformation, the brain-to-mind transformation, and the ape-to-human transformation. The egg-to-adult transformation is essentially similar, and similarly mysterious, in modern humans and other mammals, but it has distinctive human features. The brain-to-mind transformation and the ape-to-human transformation are distinctively human. These three transformations define the *humanum*, that which makes us specifically modern human. Few other issues in human evolution are of greater consequence for understanding ourselves and our place in nature.

The instructions that guide the ontogenetic process, or the egg-to-adult transformation, are carried in the hereditary material. The theory of biological heredity was formulated by the Augustinian monk Gregor Mendel in 1866, but it became generally known by biologists only in 1900: genetic information is contained in discrete factors, or genes, which exist in pairs, one received from each parent. The next step toward understanding the nature of genes was completed during the first quarter of the twentieth century. It was established that genes are parts of the chromosomes, filamentous bodies present in the nucleus of the cell, and that they are linearly arranged along the chromosomes. It took another quarter century to determine the chemical composition of genes: deoxyribonucleic acid (DNA). DNA consists of four kinds of nucleotides organized in long, double-helical structures. The genetic information is contained in the linear sequence of the nucleotides, very much in the same way as the semantic information of an English sentence is conveyed by the particular sequence of the 26 letters of the alphabet.

The first important step toward understanding how the genetic information is decoded came in 1941 when George W. Beadle and Edward L. Tatum demonstrated that genes determine the synthesis of enzymes; enzymes are the catalysts that control all chemical reactions in living beings. Later it became known that amino acids (the components that make up enzymes and other proteins) are encoded, each by a set of three consecutive nucleotides. This relationship accounts for the linear correspondence between a particular sequence of coding nucleotides and the sequence of the amino acids that make up the encoded enzyme.

Chemical reactions in organisms must occur in an orderly manner; organisms must have ways of switching genes on and off since different sets of genes are active in different cells. The first control system was discovered in 1961 by François Jacob and Jacques Monod for a gene that encodes an enzyme that digests sugar in the bacterium *Escherichia coli*. The gene is turned on and off by a system of several switches consisting of short DNA sequences adjacent to the coding part of the gene. (The coding sequence of a gene is the part that determines the sequence of amino acids in the encoded enzyme or protein.) The switches acting on a given gene are activated or deactivated by feedback loops that involve molecules synthesized by other genes. A variety of gene control mechanisms were soon discovered, in bacteria and other micro-organisms. Two elements are typically present: feedback loops and short DNA sequences acting as switches. The feedback loops ensure that the presence of a substance in the cell induces the synthesis of the enzyme required to digest it, and that an excess of the enzyme in the cell represses its own synthesis. (For example, the gene encoding a sugar-digesting enzyme in *E. coli* is turned on or off by the presence or absence of the sugar to be digested.)

The investigation of gene-control mechanisms in mammals (and other complex organisms) became possible in the mid-1970s with the development of recombinant DNA techniques. This technology made it feasible to isolate single genes (and other DNA sequences) and to multiply them, or "clone" them, to obtain the quantities necessary for ascertaining their nucleotide sequence. One unanticipated discovery was that most genes come in pieces: the coding sequence of a gene is divided into several fragments separated one from the next by noncoding DNA segments. In addition to the alternating succession of coding and noncoding segments, mammalian genes contain short control sequences, like those in bacteria but typically more numerous and complex, that act as control switches and signal where the coding sequence begins.

Much remains to be discovered about the control mechanisms of mammalian genes. The daunting speed at which molecular biology is advancing has led to the discovery of some prototypes of mammalian gene control systems, but much remains to be unraveled. Moreover, understanding the control mechanisms of individual genes is but the first major step toward solving the mystery of ontogenetic decoding. The second major step will is the puzzle of differentiation.

A modern human being consists of one trillion cells of some 300 different kinds, all derived by sequential division from the fertilized egg, a single cell 0.1 mm in diameter. The first few cell divisions yield a spherical mass of amorphous cells. Successive divisions are accompanied by the appearance of folds and ridges in the mass of cells and, later on, of the variety of tissues, organs, and limbs characteristic of a human individual. The full complement of genes duplicates with each cell division, so that two complete genomes are present in every cell. Yet different sets of genes are active in different cells. This must be so for cells to differentiate: a nerve cell, a muscle cell, and a skin cell are vastly different in size, configuration, and function. The differential activity of genes must continue after differentiation, because different cells fulfill different functions, which are controlled by different genes. Nevertheless, experiments with other animals (and some with humans) indicate that all the genes in any cell have the potential of becoming activated. (The sheep Dolly was conceived using the genes extracted from a cell in an adult sheep.)

The information that controls cell and organ differentiation is ultimately contained in the DNA sequence, but mostly in very short segments of it. In mammals, insects, and other complex organisms there are control circuits that operate at higher levels than the control mechanisms that activate and deactivate individual genes. These higher-level circuits (such as the so-called homeobox genes) act on sets of genes rather than individual genes. The details of how these sets are controlled, how many control systems there are, and how they interact, as well as many other related questions, are what needs to be resolved to elucidate the egg-to-adult transformation. The DNA sequence of some controlling elements has been ascertained, but this is a minor effort that is only helped a little by plowing the way through the entire 3000 million nucleotide pairs that constitute the modern human genome. Experiments with stem cells are likely to provide important knowledge as scientists ascertain how stem cells become brain cells in one case, muscle cells in another, and so on.

The benefits that the elucidation of the egg-to-adult transformation will bring to humankind are enormous. This knowledge will make possible understanding the modes of action of complex genetic diseases, including cancer, and therefore their cure. It will also bring an understanding of the process of aging, which kills all those who have won the battle against other infirmities.

Cancer is an anomaly of ontogenetic decoding: cells proliferate despite the welfare of the organism demanding otherwise. Individual genes (oncogenes) have been identified that are involved in the causation of particular forms of cancer. But whether or not a cell will turn out cancerous depends on the interaction of the oncogenes with other genes and with the internal and external environment of the cell. Aging is also a failure of the process of ontogenetic decoding: cells fail to carry out the functions imprinted in their genetic code script or are no longer able to proliferate and replace dead cells.

The brain is the most complex and most distinctive modern human organ. It consists of 30 billion nerve cells, or neurons, each connected to many others through two kinds of cell extension, known as the axon and the dendrites. From the evolutionary point of view, the animal brain is a powerful biological adaptation; it allows the organism to obtain and process information about environmental conditions and then to adapt to them. This ability has been carried to the limit in modern humans, in which the extravagant hypertrophy of the brain makes possible abstract thinking, language, and technology. By these means, humankind has ushered in a new mode of adaptation far more powerful than the biological mode: adaptation by culture.

The most rudimentary ability to gather and process information about the environment is found in certain single-celled micro-organisms. The protozoan *Paramecium* swims apparently at random, ingesting the bacteria it encounters, but when it meets unsuitable acidity or salinity its advance is checked and it starts off in a new direction. The single-celled alga *Euglena* not only avoids unsuitable environments but seeks suitable ones by orienting itself according to the direction of light, which it perceives through a light-sensitive spot in the cell. Plants have not progressed much further. Except for those with tendrils that twist around any solid object and the few carnivorous plants that react to touch, they mostly react only to gradients of light, gravity, and moisture.

In animals the ability to secure and process environmental information is mediated by the nervous system. The simplest nervous systems are found in corals and jellyfishes; they lack coordination between different parts of their bodies, so any one part is able to react only when it is directly stimulated. Sea urchins and starfish possess a nerve ring and radial nerve cords that coordinate stimuli coming from different parts; hence, they respond with direct and unified actions of the whole body. They have no brain, however, and seem unable to learn from experience. Planarian flatworms have the most rudimentary brain known; their central nervous system and brain process and coordinate information gathered by sensory cells. These animals are capable of simple learning and hence of variable responses to repeatedly encountered stimuli. Insects and their relatives have much more advanced brains; they obtain precise chemical, acoustic, visual, and tactile signals from the environment and process them, making possible complex behaviors, particularly in search for food, selection of mates, and social organization.

Vertebrates are able to obtain and process much more complicated signals and to respond to the environment more variably than do insects or any other type of invertebrate. The vertebrate brain contains an enormous number of associative neurons arranged in complex patterns. In vertebrates the ability to react to environmental information is correlated with an increase in the relative size of the cerebral hemispheres and of the neopallium, an organ involved in associating and coordinating signals from all receptors and brain centers. In mammals, the neopallium has expanded and become the cerebral cortex. Modern humans have a very large brain relative to their body size, and a cerebral cortex that is disproportionately large and complex even for their brain size. Abstract thinking, symbolic language, complex social organization, values, and ethics are manifestations of the wondrous capacity of the modern human brain to gather information about the external world and to integrate that information and react flexibly to what is perceived.

With the advanced development of the modern human brain, biological evolution has transcended itself, opening up a new mode of evolution: adaptation by technological manipulation of the environment. Organisms adapt to the environment by means of natural selection, by changing their genetic constitution over the generations to suit the demands of the environment. Modern humans, and modern humans alone to any substantial degree, have developed the capacity to adapt to hostile environments by modifying the environments according to the needs of their genes. The discovery of fire and the fabrication of clothing and shelter have allowed the ancestors of modern humans to spread from the warm tropical and subtropical regions of the Old World, to which we are biologically adapted, to almost the whole Earth; it was not necessary for wandering hominins that they wait until genes would evolve providing anatomical protection against cold temperatures by changing their physiology or by means of fur or hair. Nor are modern humans biding their time in expectation of wings or gills; we have conquered the air and seas with artfully designed contrivances: airplanes and ships. It is the modern human brain (the human mind) that has made humankind the most successful, by most meaningful standards, living species.

There are not enough bits of information in the complete DNA sequence of a modern human genome to specify the trillions of connections among the 30 billion neurons of the modern human brain. Accordingly, the genetic instructions must be organized in control circuits operating at different hierarchical levels so that an instruction at one level is carried through many channels at a lower level in the hierarchy of control circuits. The development of the modern human brain is indeed one particularly intriguing component of the egg-to-adult transformation.

Within the last two decades, neurobiology has developed into one of the most exciting biological disciplines. An increased commitment of financial and human resources has brought an unprecedented rate of discovery. Much has been learned about how light, sound, temperature, resistance, and chemical impressions received in our sense organs trigger the release of chemical transmitters and electric potential differences that carry the signals through the nerves to the brain and elsewhere in the body. Much has also been learned about how neural channels for information transmission become reinforced by use or may be replaced after damage; about which neurons or groups of neurons are committed to processing information derived from a particular organ or environmental location; and about many other matters. But, for all this progress, neurobiology remains an infant discipline, at a stage of theoretical development comparable perhaps to that of genetics at the beginning of the 20th century. Those things that count most remain shrouded in mystery: how physical phenomena become mental experiences (the feelings and sensations, called "qualia" by philosophers, that contribute the elements of consciousness), and how out of the diversity of these experiences emerges the mind, a reality with unitary properties, such as free will and the awareness of self, that persist through an individual's life.

I do not believe that the mysteries of the mind are unfathomable; rather, they are puzzles that modern humans can solve with the methods of science and illuminate with philosophical analysis and reflection. And I will place my bets that, over the next half century or so, many of these puzzles will be solved. We shall then be well on our way toward answering the biblical injunction: "Know thyself."

A contemporary development that would have greatly delighted Darwin is the determination of the DNA sequence of the modern human genome, an investigation that was started under the label, the Human Genome Project, which opens up the possibility of comparing the modern human DNA sequence with that of other organisms, observing their similarities and differences, seeking to ascertain the changes in the DNA that account for distinctively modern human features. The Human Genome Project was initiated in 1989, funded through two US agencies, the National Institutes of Health (NIH) and the Department of Energy (DOE), with eventual participation of scientists outside the USA. The goal set was to obtain the complete sequence of one human genome in 15 years at an approximate cost of $3000 million, coincidentally about $1 per DNA letter. A private enterprise, Celera Genomics, started in the USA somewhat later, but joined the government-sponsored project in achieving, largely independently, similar results at about the same time. A draft of the genome sequence was completed ahead of schedule in 2001. In 2003 the Human Genome Project was finished, but the analysis of the DNA sequences chromosome by chromosome continued over the following years. Results of these detailed analyses were published on June 1, 2006. The draft DNA sequence of the chimpanzee genome was published on September 1, 2005 (an entry in the encyclopedia presents the first fossil evidence of panins ever to be discovered). In the regions of the genome that are shared by modern humans and chimpanzees the two species are about 99% identical. These differences may seem very small or quite large, depending on how one chooses to look at them: 1% is only a small fraction of the total, but it still amounts to a difference of 30 million DNA nucleotides out of the 3 billion in each genome.

Twenty-nine percent of the enzymes and other proteins encoded by the genes are identical in both species. Out of the one hundred to several hundred amino acids that make up each protein, the 71% of nonidentical proteins that differ between modern humans and chimps do so by only two amino acids, on average. If one takes into account DNA stretches found in one species but not the other, the two genomes are about 96% identical, rather than nearly 99% identical as in the case of DNA sequences shared by both species. That is, a large amount of genetic material, about 3% or some 90 million DNA nucleotides, have been inserted or deleted since ancestors went their separate evolutionary ways, about 8–6 million years ago. Most of this DNA does not contain genes coding for proteins, although it may include toolkit genes and switch genes that impact developmental processes, as the rest of the noncoding DNA surely does.

Comparison of the modern human and chimpanzee genomes provides insights into the rate of evolution of particular genes in the two species. One significant finding is that genes active in the brain have changed more in the human lineage than in the chimp lineage (Khaitovich et al. 2005). Also significant is that the fastest-evolving modern human genes are those coding for transcription factors. These are switch proteins which control the expression of other genes; that is, they determine when other genes are turned on and off. On the whole, 585 genes have been identified as evolving faster in humans than in chimps, including genes involved in resistance to malaria and tuberculosis. (It might be mentioned that malaria is a severe disease for humans but much less so for chimps.)

Genes located on the Y chromosome, found only in the male, have been much better protected by natural selection in the human than in the chimpanzee lineage, in which several genes have incorporated disabling mutations that make the genes nonfunctional. Also, there are several regions of the human genome that contain beneficial genes that have rapidly evolved within the past 250,000 years. One region contains the *FOXP2* gene, involved in the evolution of speech.

Other regions that show higher rates of evolution in humans than in chimpanzees and other animals include 49 segments, dubbed human accelerated regions or HARs. The greatest observed difference occurs in *HAR1F*, an RNA gene that is expressed specifically in Cajal-Retzius neurons in the developing human neocortex from 7 to 19 gestational weeks, a crucial period for cortical neuron specification and migration.

Extended comparisons of the human and chimpanzee genomes and experimental exploration of the functions associated with significant genes will surely advance further our understanding, over the next decade or two, of what it is that makes us distinctively human, what is it that differentiates *H. sapiens* from our closest living species, chimpanzees and bonobos, and will surely shine some light on how and when these differences may have come about during hominin evolution of the human species. Surely also, full biological understanding of the ape-to-human transformation will only come if we solve the other two conundrums: the egg-to-adult transformation and the brain-to-mind transformation. The distinctive features that make us modern human appear early in development, well before birth, as the linear information encoded in the genome gradually becomes expressed into a four-dimensional individual, an individual who changes in configuration as time goes by. In an important sense the most distinctive human features are those expressed in the brain, those that account for the human mind or for human identity.

Francisco J. Ayala
University of California, Irvine

Preface

Once upon a time, say 60 years ago, the only background required to appreciate what was known about human evolution was a familiarity with a relatively sparse fossil record, an understanding of the limited information there was about the context of the sites, some knowledge of gross anatomy, a familiarity with a few simple analytical methods, and an appreciation of general evolutionary principles.

Times have changed. The fossil record has grown exponentially, imaging techniques allow researchers to capture previously unavailable gross morphological and microstructural evidence in unimaginable quantities, analytical methods have burgeoned in scope and complexity, phylogeny reconstruction is more sophisticated, molecular biology has revolutionized our understanding of genetics, evolutionary history, modern human variation, and development, and a host of different developments in biology, chemistry, earth sciences, and physics the have enriched evidence about the biotic, climatic, and temporal context of the hominin fossil record. In short, the fossil evidence and the range of methods used to study human evolution have grown by at least one order of magnitude in the past six decades. Yet there is no single reference source where students and researchers involved in human evolution research, be they archeologists, earth scientists, molecular biologists, morphologists, paleontologists, or paleoecologists, can go to find out about topics as diverse as C_4 plants canalization, the candelabra model, canonical variates analysis, Carabelli('s) trait, carcass transport strategy, cardioid foramen magnum, carrying capacity, and Caune de l'Arago.

Antecedents

Quenstedt and Quenstedt (1936) were probably the first to compile a comprehensive list and associated bibliography of the hominin fossil record. Hue (1937) illustrated some of the hominin fossil evidence, but his bibliographic coverage was less comprehensive than Quenstedt and Quenstedt's. Vallois and Movius' *Catalogue des Hommes Fossiles* (1953) made a commendable attempt to survey the human fossil record, as did Day's *Guide to Fossil Man* (1965). The first comprehensive attempt to collate information about both the hominin fossil record and its context came in the late 1960s and thereafter with the publication of the British Museum (Natural History)'s *Catalogue of Fossil Hominids*. The first volume, "Part I: Africa," edited by Oakley and Campbell, was published in 1967. Subsequent volumes covered fossil hominin discoveries from Europe (Oakley, Campbell, and Molleson 1971) and Asia (Oakley, Campbell, and Molleson 1975). Unfortunately (but understandably given the work involved in assembling a resource such as this in a non-electronic format and on a rapidly enlarging hominin fossil record) only Part I ran to a second updated edition (Oakley, Campbell, and Molleson 1977) and the series has been discontinued. The most recent attempt to take up the challenge of providing a comprehensive catalogue of the fossil evidence for hominin evolution has been the initiative by Orban and her colleagues, entitled *Hominid Remains – an up-date*. This is distributed as part of a general series of publications called the Supplements au Societe Royale Belge d'Anthropologie et de Prehistoire. The first in the series of separate publications was published in the beginning of 1988 and publication has continued thereafter. Each slim volume covers one or more countries and retains the format of the *Catalogue of Fossil Hominids*. The early issues were organized by an Editor (Orban) and Associate Editors (Slachmuylder, Semal, and Alewaeters). The individual entries are put together by experts who are familiar with one or more temporal phases of the hominin fossil record from that country or countries. Spencer's excellent *History of Physical Anthropology*, published in 1997, is an important and authoritative source of information about the history of paleoanthropology, but although it has entries for the major fossil sites there are few entries about the fossil and archeological evidence. The four-volume series *The Human Fossil Record* edited by Jeffrey Schwartz and Ian Tatterall and colleagues, focuses on the better-preserved hominin fossils, but there is little about their context and its taxonomic interpretations are idiosyncratic. There is also a *Catalogue of Fossil Hominids* available on the internet at http://gbs.ur-plaza.osaka-cu.ac.jp/kaseki.

Scope

The long-term plan for the *Wiley-Blackwell Encyclopedia of Human Evolution* (W-BEHE) is that it will be an authoritative and accessible source of information about the hominin clade of the tree of life. Entries cover:

- general evolutionary principles,
- information about the molecular and developmental biological approaches used to help understand the pattern and process of evolution,
- methods used to investigate relationships among the living great apes and modern humans,
- methods germane to understanding the origins and evolutionary history of the hominin clade and its climatic and ecological context,
- what makes the behavior of modern humans distinctive and the evolutionary history of that distinctiveness,
- information about the modern methods used to capture and interpret data from the hominin fossil and archeological records, as well as comparative biology,
- nonhominin fossil evidence germane to the evolution of the hominin clade,
- specialist terms used to describe the hominin fossil and archeological records,
- hypotheses germane to interpreting human evolutionary history,
- biographies of individuals who have made significant contributions to the accumulation of the fossil and archeological, and other, evidence and to its interpretation,
- institutions and organizations that have contributed to our understanding of human evolutionary history,
- information about hominin fossil and comparative great ape collections and about some of the repositories that hold hominin fossil collections.

Organization

The entries for the important fossil and/or archeological sites (later editions will aim to cover *all* the relevant sites) relevant to hominin evolution are structured to make it easier for the reader to find out what they want to know. This is also the case for the entries for the more complete hominin fossils, and for some less complete fossils that have a particular significance. There are also formally structured entries for taxa that have been used to accommodate the hominin fossil record. The depth of the entries varies; the entries for fossils and sites do not aim to be exhaustive, but they are comprehensive. The entries for methods and important biological principles try to explain complex concepts in plain and simple language and where possible they include examples. As Editor I constantly needed to remind myself, the editors, and the contributors that the W-BEHE is *not* an encyclopedia of archeology, earth science, genetics, etc.; readers should go to other sources for technical information about the methods.

The entries have been prepared as a collective exercise. The drafts prepared by an editor or contributor were all screened by me as Editor for depth, style, and intelligibility and then sent to other editors, or to outside experts, for their comments. Human evolution is a controversial field, so where appropriate entries describe competing hypotheses, and if the entry comes down in favor of one of them, the entry explains why. The aim, doubtless not always met, was that the entries are self-contained and intelligible; we did not want users of the W-BEHE to look up one term they did not understand only to be faced with an entry that uses five words they are unfamiliar with. Within the text of entries, where appropriate, any term used that has its own entry are shown in bold.

Planning

The draft list of entries was drawn up by the Editor, four Section Editors (Laura Bishop, paleontology; Craig Feibel, earth sciences; Tom Plummer, archeology; and Anne Stone, molecular biology and genetics) and 17 Topic Editors (Shara Bailey, Robin Bernstein, Mark Collard, Sarah Elton, Matthew Goodrum, Adam Gordon, Katerina Harvati, Lyle Konigsberg, Andrew Kramer, Jacopo Moggi-Cecchi, Osbjorn Pearson, Brian Richmond, Chet Sherwood, Tanya Smith, David Strait, Francys Subiaul, and Christian Tryon). Important advice at this stage also came from several of the Advisory Editors.

The original plan was that these individuals would prepare the entries, but like all strategic plans this one had to be modified. It was clear that the sheer volume of the work involved could not be tackled by these editors, and as polymathic as many of them are, there were still regions of the world (e.g., China) and topics (e.g., paleoclimate, stable isotopes, zooarcheology, etc.) for which we lacked coverage. Also, because this project has been 6 years in the making, circumstances change (children are born, grants are written and awarded, relatives get sick, individuals take up additional responsibilities) so I soon learned that it was adapt or die! Thus, the original editor classification has been abandoned and in its place there are now Associate Editors, Section and Topic Editors, and Contributors. The criteria for allocating individuals to categories are inevitably subjective and I am also conscious that my memory is fallible, so I offer my abject apologies to those who feel their contributions have not been recognized at the appropriate level. The Associate Editors did all, or some, of the following. They were the lead author on a substantial number of entries, they contributed generously to the editing well beyond their immediate area of expertise, and they responded promptly and generously to my many unreasonable requests. Only I (and they) know just how hard they worked and how important their contribution was. Section Editors did all, or some, of the above, but unlike the Associate Editors their contributions were focused on their immediate area of expertise. Topic Editors were much like Section Editors, but they contributed fewer entries and their contributions were confined to their immediate area of expertise. Contributors contributed at least one entry, but were not involved in the editing. Several Advisory Editors (e.g., Rebecca Ackermann, Leslie Aiello, Chris Dean, Colin Groves, Benedikt Hallgrimsson, and Bill Kimbel) contributed entries and some of the above played a major editorial role. Many other people offered advice about entries and they are listed in the Acknowledgments; once again my apologies if my system for recording such help has failed. Please, if you do not see your name on that list and you think you should be on it, please do not be bashful about letting me know and I can rectify any omissions in the next edition.

Delivery

I urge those of you who have not read Simon Winchester's *The Meaning of Everything* to do so; those of you who have read it will know the *Oxford English Dictionary* took 70 years to complete. Much of the information in the W-BEHE is time-sensitive and the problem is that if you wait too long for that sincerely promised but much delayed entry you run the risk that many other entries will become outdated in the meantime. Yet you cannot go the press with an encyclopedia that is lacking essential entries. So we had to make a fine judgment about when to stop, and we had to close the headword list knowing we would be leaving some fossil, site, museum, and biographical entries for a later edition. You cannot have an encyclopedia without entries, but you *can* have entries without an encyclopedia. My efforts as Editor, and those of the Associate, Section, and Topic Editors and the Contributors, would have all come to naught without the hard work of the Executive Editors (Amanda Henry and Jennifer Baker) and the Editorial Assistants (Alex Claxton and Erin Mikels), who helped prepare the material for copy-editing and then checked it at various stages in the process my sincere thanks to all of them. Of course the buck stops with the Editor, so any mistakes are my responsibility alone.

The future

The plan is to follow up the hardback library edition with a more affordable edition intended for individuals, an abridged version for students, and an electronic edition that can be updated approximately every 6 months.

Errors

In a project of this scale errors will have been made. The only way to eliminate as many mistakes as humanly possible is to contact me (bernardawood@gmail.com) and I will see to it they are rectified in later editions.

Bernard Wood
The George Washington University, Washington, DC
December 2010

Acknowledgments

It goes without saying that none of this would have been possible without the editors and contributors. At many times in the process it must have seemed as if it was "all grit and no gravy" and I am conscious that contributing in various ways to the *Wiley-Blackwell Encyclopedia of Human Evolution* (W-BEHE) has deflected individuals from their core activities of teaching, grant writing, research, and writing up their research. I am more grateful to them than I can express in words. Some contributions deserve special mention. A few people really *did* read and comment on the entries I sent out for general review, and I am especially grateful to Rebecca Ackermann, Chris Dean, and Alan Bilsborough for consistently doing so. I am also grateful to Katerina Harvati for her support despite having other much more important claims on her time. Colin Groves (in Australia) would be getting up as I was going to sleep, but I cannot recall how many times I would send him complex requests and enquiries at the end of my day and when I woke up they would invariably be answered thoroughly and cheerfully. I am also immensely impressed by the graduate students who gave of their time and expertise; if Jennifer Baker, Amy Bauernfeind, Serena Bianchi, Habiba Chirchir, Andrew Du, Tyler Faith, Liz Renner, Kes Schroer, Steve Wang, and Andrew Zipkin are representative of young researchers then our discipline is in good hands. Lastly, for those who cherish such achievements, Tyler Faith and Christian Tryon tied for the shortest interval (*c*.18 minutes) between a request and a received entry.

This project could not have been completed without the generosity of the community of scholars who work in paleoanthropology and in cognate areas. People who helped prepare and edit entries include undergraduates, junior and senior graduate students, postdoctoral fellows, and junior and senior professors; some probably felt they had no choice, others I "cold-called" and some I have known for 40 years. The willingness of people, young and old, to work for the common good has been humbling. It is inadequate thanks to list them by name, but that is the currency I must, perforce, use. Some of them devoted substantial time and effort to make sure that entries, especially site entries, were correct; I hope they conclude that the product of their collective labors justifies their efforts: Zeray Alemseged, Juan Luis Arsuaga, Lucinda Backwell, Amy Bauernfeind, Hazel Beeler, John Beeler, Kay Behrensmeyer, Miriam Belmaker, Mike Benjamin, Vadim Berg, Barry Berkovitz, Serena Bianchi, Felicitas Bidlack, Robert Blumenschine, René Bobe, Silvia Bortoluzzi, Frank Brown, Judith Brown, Peter Brown, Marta Camps, Sarah Cassel, Habiba Chirchir, Rich Cifelli, Robert Cifelli, Bryanne Colby, Paul Constantino, Raymond Corbey, Robin Crompton, Daryl de Ruiter, Alexandra de Sousa, Eric Delson, Catherine Denial, Alnawaz Devani, Marietta Dindo, Michelle Drapeau, Logan Ferree, Reid Ferring, John Fleagle, Jens Franzen, Kate Freeman, Pascal Gagneux, Emmanuel Gilissen, John Gowlett, Jay Greene, Nicole Griffin, Ana Gracia, Eileen Le Guillou, Frank Guy, Brian Hall, Will Harcourt-Smith, Malcolm Harman, the late Elizabeth Harmon, Geoffrey Harrison, Terry Harrison, William Hart, Claire Heckel, Simon Hillson, Leslea Hlusko, Louise Humphreys, David Hunt, Kevin Hunt, Joel Irish, Adam Jagich, Kayla Jarvis, William Jungers, Jon Kalb, John Kappelman, Richard Klein, Chris Kocher, Steve Kuhn, Joel Kuipers, Ottmar Kulmer, Joanna Lambert, Beth Lawrie, Daniel Lieberman, Julien Louys, Shannon McFarlin, Andrew McGurn, Lindsay McHenry, Jeff McKee, Roberto Machiarelli, Daniel Miller, Ignacio Martínez, Mary Marzke, Jim Moore, Michael Morwood, Jackson Njau, Rick Potts, Bryan Pratt, Yoel Rak, Jean-Paul Raynal, Safia Razzuqi, Kaye Reed, Liz Renner, Vernon Reynolds, Philip Rightmire, Mirjana Roksandic, Lorenzo Rook, Chris Ruff, Margaret Schoeninger, Friedemann Schrenk, Geralyn Schulz, Sileshi Semaw, Brian Shea, John Shea, Courtney Snell, Katya Stansfield, Christine Steininger, Chikako Suda-King, Fred Szalay, Mark Teaford, Matt Tocheri, Peter Ungar, Sidney Vasquez, Julian Waller, Steve Ward, Randy White, Tim White, David Wilkinson, Phillip Williams, Pamela Willoughby, Milford Wolpoff, Barth Wright, Roshna Wunderlich, and Bernhard Zipfel. My thanks also to Francisco Ayala for his interest in the project and for finding the time to write his fine Foreword. I have come to enjoy Francisco's friendship through CARTA, the brainchild of Ajit Varki, and now co-directed by Fred Gage, Margaret Schoeninger, and Ajit Varki; its meetings and its members have done much to widen my intellectual horizons. All of us involved in CARTA are grateful to the G. Harold and Leila Y. Mathers Foundation and its

Executive Director, James Handelman, for their long-term support for CARTA and to Pascal Gagneux, the Associate Director, and Linda Nelson for organizing it and its meetings. Those of you who know me will realize that I am the last person who should have been entrusted with the task of making sure that thousands of entries were curated so that only the latest versions find their way into the final text; my skills as a curator of references are also sorely lacking. It is no exaggeration to say that without the help of a series of student colleagues, especially Jennifer Baker, Kaitlyn Baraldi, Alex Claxton, Gabriel Mallozzi, and Erin Mikels, that this project would still be "a work in progress;" Kayla Jarvis also helped in many ways. Alex Claxton set up the topic entry list and collated the master reference file; my thanks to them both. Sometimes serendipity comes to one's aid, and in my case it came in the form of Amanda Henry. She just happened to have defended her thesis at the time of greatest need for the W–BEHE and she was willing to pitch in and help bring the project to fruition. Without her hard work, determination, stamina, common sense, and intelligence we would still be assembling entries. Working with her has been my pleasure and privilege; it was also fun. Amanda and Jennifer Baker read and commented on every entry before they were sent for copy-editing.

My notes remind me that I met with Jane Huber (then an Editor at Blackwell Publishing) in Boston at the end of June 2005. Jane's encouragement and efficiency were the reason I decided to work with Blackwell on this project, and although she left to work for Oxfam it proved a wise decision for as many will realize Blackwell subsequently merged with Wiley, a publisher with deep connections with paleoanthropology via the *American Journal of Physical Anthropology* and *Evolutionary Anthropology*. Andy Slade has been my supervising editor and W–BEHE's champion and I thank him for his efforts on its behalf. I was fortunate that Kelvin Matthews, Project Editor for Life Sciences, has been the lead person in the editing and production process within Wiley-Blackwell. His quiet but insistent determination to make sure that W–BEHE happened and his behind-the-scenes work to keep the project on track are sincerely appreciated. Among the many reasons I am grateful to Kelvin is that it was he who arranged for Nik Prowse to be the copy-editor. Nik works from Durham, England, and he and I did not meet until after W–BEHE was published until after, but it is no exaggeration to say that without his intelligence, attention to detail (he could have been a neurosurgeon), knowledge, common sense, and good humor the project might well have foundered. The world is replete with people who create problems; Nik (and Kelvin) belong to a much smaller number who were put into this world to solve them.

My sincere thanks are also due to my colleagues at George Washington University. They have tolerated my distraction with W–BEHE and several have made important contributions as editors. I also want to give special thanks to Don Lehman, until recently George Washington University's Vice-President for Academic Affairs. He not only gave me unstinting support as a University Professor, but via the university's Signature Program his support made possible work study programs, internships, and graduate fellowships that have supported Erin Mikels, Alex Claxton, and Jennifer Baker. Bill Kimbel, Jacopo Moggi-Cecchi, David Pilbeam, and David Strait were always on hand with support and advice and to remind me there are many more important things in life than editing an encyclopedia.

Finally, my heartfelt thanks to Sally who stoically honed her already considerable Sudoku and KenKen skills while I worked on "just a few more entries."

Topic Entry List

ANATOMY

I. General terms
    1) abduction
    2) abductor
    3) adduction
    4) adductor
    5) agenesis
    6) anatomical position
    7) anatomical terminology
    8) appendicular skeleton
    9) associated skeleton
    10) atavism
    11) atavistic
    12) autopod
    13) axial digit
    14) axial skeleton
    15) axis
    16) base
    17) branchial arches
    18) calcification
    19) caudal
    20) coronal
    21) cortex
    22) cortical
    23) coxa
    24) cranial
    25) cross-section
    26) dento-gnathic
    27) distal
    28) epi-
    29) fibrocartilage
    30) fovea
    31) gnathic
    32) gomphoses
    33) histology
    34) human anatomical terminology
    35) hyaline cartilage
    36) hyoid
    37) jaw
    38) joint
    39) kyphosis
    40) ligaments
    41) lordosis
    42) mineralization
    43) muscles of mastication
    44) *Nomina Anatomica*
    45) *norma basalis*
    46) *norma basilaris*
    47) *norma frontalis*
    48) *norma lateralis*
    49) *norma occipitalis*
    50) *norma verticalis*
    51) os
    52) paleoanthropological terminology
    53) primary cartilaginous joint
    54) promontory
    55) pronation
    56) proximal
    57) reference digit
    58) rostral
    59) secondary cartilaginous joints
    60) skeleton
    61) stylopod
    62) supination
    63) symphyses
    64) symphysis
    65) synchondroses
    66) syndesmoses
    67) *Terminologia Anatomica*
    68) valgus
    69) varus
    70) volar
    71) zeugopod

II. Bone(s)
    A. Terms
        1) alveolar process
        2) alveoli
        3) alveolus
        4) bone
        5) bone density
        6) cancellous bone
        7) cartilage
        8) chondroblasts
        9) chondroclasts
        10) chondrocytes
        11) compact bone
        12) connective tissue
        13) dense bone
        14) diaphysis
        15) endochondral ossification
        16) lamellar bone
        17) osteoblast

93) r
94) r^2
95) randomization
96) reduced major axis regression
97) regression
98) resampling
99) RMA
100) Roy's largest root
101) sample
102) sample statistics
103) sampling with replacement
104) sampling without replacement
105) seriation
106) simple regression
107) three-dimensional morphometrics
108) t test
109) Type I error
110) Type II error
111) variability
112) variation
113) Wilcoxon signed rank test
114) Wilks' lambda

EARTH SCIENCES

I. Terms
1) aeolian
2) Afar Rift System
3) Australasian strewn-field tektites
4) alluvial
5) angular unconformity
6) anticline
7) basin
8) bed
9) bedrock
10) biostratigraphy
11) BPT
12) Cenozoic
13) chronostratigraphic unit
14) clastic
15) closed basin lakes
16) composite section
17) DABT
18) deflation
19) delta
20) deltaic
21) diagenesis
22) dip
23) disconformity
24) dome
25) East African Rift System
26) Eocene
27) epoch
28) era
29) erathem
30) Ethiopian Rift System
31) exposure
32) facies
33) fault
34) fluvial
35) fluviatile
36) formation
37) GATC
38) GB
39) geochronologic unit
40) glacial cycles
41) glacial period
42) graben
43) Grenzbank zone
44) group
45) Günz
46) Holocene
47) horst
48) hematite
49) interglacial
50) interstadial
51) isochron
52) isotope
53) ka
54) KAI
55) KAL
56) karst
57) karstic
58) KBS
59) KBS CC
60) kirongo
61) Koobi Fora Tuff Complex
62) k.y.
63) kya
64) lacustrine
65) lag deposit
66) Lake Bosumtwi
67) Lake Turkana Group
68) LAL
69) Last Glacial Maximum
70) Last Glacial Maximum low stand
71) Late Pleistocene
72) lithology
73) lithostratigraphy
74) Little Ice Age
75) locality
76) loess
77) LOM
78) LON
79) lower-middle tuffs

ECOLOGY

FUNCTIONAL MORPHOLOGY

14) power stroke
15) swallowing
16) working side

GENETICS AND MOLECULAR BIOLOGY

1) 3′
2) 5′
3) aDNA
4) adaptive landscapes
5) albumin
6) allele
7) Alu
8) Alu elements
9) Alu repeat elements
10) Alus
11) amino acid
12) ancient DNA
13) antibody
14) anticodon
15) antigen
16) apoptosis
17) array
18) ascertainment bias
19) aspartic acid
20) *ASPM*
21) bacteriophage
22) balanced polymorphism
23) balanced translocation
24) balancing selection
25) Barr body
26) base pair
27) bottleneck
28) bp
29) broad-sense heritability
30) cDNA
31) centimorgan
32) centric fusion
33) centromere
34) chromatin
35) chromosome
36) chromosome banding
37) chromosome number
38) chromosome painting
39) CNV
40) coalescence
41) coalescent
42) coalescent theory
43) coalescent time
44) codon
45) complementary DNA
46) complex trait

47) copy number variation
48) crossing over
49) cytogenetic
50) Darwinian fitness
51) deleterious mutation
52) deletion
53) denature
54) deoxyribonucleic acid
55) diploid
56) directional selection
57) divergence time
58) dizygotic twins
59) D-loop
60) DNA
61) DNA chip
62) DNA hybridization
63) DNA microarray
64) DNA methylation
65) DNA sequencing
66) domain
67) dominance
68) dominant allele
69) dosage compensation
70) downstream
71) draft sequence
72) drift
73) duplication
74) dynamic mutation
75) effective population size
76) electrophoresis
77) enhancer
78) enzyme
79) epigenetics
80) epigenesis
81) epistasis
82) euchromatin
83) eukaryote
84) exon
85) expressivity
86) extension
87) finished sequence
88) fitness
89) fitness landscapes
90) fixation
91) fixation indices
92) fluorescence *in situ* hybridization
93) FISH
94) founder effect
95) *FOXP2*
96) FOXP2
97) frameshift mutation

GROWTH AND DEVELOPMENT

HISTORY

37) Institute of Vertebrate Paleontology and Paleoanthropology
38) International Louis Leakey Memorial Institute for African Prehistory
39) IORE
40) ISMA
41) IVPP
42) Iziko Museums of Cape Town
43) Karonga Museum
44) Kenya National Museum
45) Koninklijk Museum voor Middenafrika
46) Laboratory for Human Evolutionary Studies
47) Laboratory of Vertebrate Paleontology
48) Leakey Foundation
49) Leibniz Institute for Evolutionary and Biodiversity Research
50) L.S.B. Leakey Foundation
51) Max Planck Institute for Evolutionary Anthropology
52) Musée Royal de l'Afrique Centrale
53) Musée de l'Homme
54) Museum für Naturkunde
55) National Museum and House of Culture, Dar es Salaam
56) National Museum of Natural History, Smithsonian Institution
57) National Natural History Museum, Arusha
58) National Museums of Kenya
59) Nationaal Natuurhistorisch Museum
60) National Natural History Museum, Arusha
61) Naturalis
62) NMK
63) Paleoanthropology Society
64) Pan-African Association of Prehistory and Related Studies
65) Pan-African Congress on Prehistory and Quaternary Studies
66) PARU
67) Peking Union Medical College
68) Powell Cotton Museum
69) PUMC
70) Rijksmuseum van Natuurlijke Historie
71) Rockefeller Foundation
72) Rockefeller Museum
73) Royal College of Surgeons of England
74) Royal Museum for Central Africa
75) RVRME
76) School of Anatomical Sciences, University of the Witwatersrand
77) Selenka Collection
78) Senckenberg Forschungsinstitut und Naturmuseum
79) Senckenberg Museum
80) Smithsonian Institution
81) South African Museum (Iziko Museums of Cape Town)
82) Staatliches Museum für Naturkunde
83) The L.S.B. Leakey Foundation
84) TILLMIAP
85) Transvaal Museum
86) University of Pennsylvania Museum of Archaeology and Anthropology
87) Viking Fund
88) Viking Fund Medal
89) Viking Summer Seminars in Physical Anthropology
90) von Koenigswald Collection
91) Wenner-Gren Foundation for Anthropological Research
92) Wenner-Gren Foundation Supper Conferences
93) Witwatersrand University Department of Anatomy
94) Zoologische Staatssammlung München
95) ZSM

IV. Publications
1) *Catalogue of Fossil Hominids*
2) *Evidence as to Man's Place in Nature*
3) Viking Fund Publications in Anthropology

V. Research Groups
1) Ancient Human Occupation of Britain Project
2) AHOB
3) Dikika Research Project
4) DRP
5) East African Geological Research Unit
6) East Rudolf Research Project
7) EP
8) Eyasi Plateau Paleontological Project
9) French-Chadian Paleoanthropological Mission
10) Gona Palaeoanthropological Research Project
11) GPRP
12) Hominid Corridor Research Project
13) International Afar Research Expedition
14) International Omo Research Expedition
15) Kohl-Larsen Expedition
16) Koobi Fora Research Project
17) Ledi-Geraru Research Project

18) Middle Awash Research Project
19) Middle Ledi Research Project
20) Mission Paléoanthropologique Franco-Tchadienne
21) Mission Scientifique de l'Omo
22) MPFT
23) National Geological Survey of China
24) OLAPP
25) Olduvai Landscape Paleoanthropology Project
26) Omo Group Research Expedition
27) Palaeo-Anthropology Research Unit
28) Paleoanthropological Inventory of Ethiopia
29) Rift Valley Research Mission in Ethiopia
30) stage 3 project
31) SPRP
32) Sterkfontein Research Unit
33) Swedish China Research Committee
34) Swartkrans Paleoanthropological Research Project
35) University of California's Africa Expedition
36) Woranso-Mille Paleontological Research Project

LIFE HISTORY

1) altricial
2) adolescence
3) adolescent growth spurt
4) adrenarche
5) age at maturity
6) age at weaning
7) androgens
8) antagonistic pleiotropy hypothesis
9) catch-up growth
10) Charnov's "dimensionless numbers"
11) childhood
12) development
13) DLNs
14) disposable soma hypothesis
15) Euler–Lotka equation
16) estrogens
17) fecundity
18) fixation
19) genetic correlation
20) genotype–environment interaction
21) gestation
22) gestation length
23) generation time
24) grandmother hypothesis
25) growth
26) growth factors
27) growth hormone
28) growth plate
29) growth spurt
30) heritability
31) interbirth interval
32) *K*-selection
33) life history
34) maternal effect
35) maternal energy hypothesis
36) maximum life span
37) menarche
38) menopause
39) mortality
40) mutation accumulation hypothesis
41) parent/offspring conflict
42) precocial
43) post-weaning dependency
44) puberty
45) *r*-selection
46) reproductive effort
47) reproductive investment
48) reaction norm
49) secondarily altricial
50) secondary sexual characteristics
51) secular trend
52) senescence
53) weaning
54) trade-off

MORPHOMETRY

I. Capturing morphology
   1) body size
   2) body-mass estimation
   3) cross-sectional geometry
   4) discrete morphological variants
   5) discrete traits
   6) epigenetic traits
   7) hyperostotic
   8) hypostotic
   9) macrostructure
   10) microstructure
   11) moments of area
   12) non-metrical traits
   13) proxy
   14) qualitative variants
   15) variability
   16) variable
   17) variance
   18) variation

II. Data capture
  A. Terms
    1) Cartesian coordinates
    2) condylar position index

PALEOANTHROPOLOGY

PALEOCLIMATE/PALEOENVIRONMENT

PALEONTOLOGY

REGIONAL GEOGRAPHY

SYSTEMATICS

Abbreviations

19thC, etc.	19th century, etc.
2D	two-dimensional
3D	three-dimensional
Afks.	Afrikaans
aka	also known as
ant.	antonym
Ar.	Arabic
BCE	before current era
BP	before present
Ca.	Catalan
cal. years BP	calibrated years before present
CE	current era
dim.	diminutive
Fr.	French
Ge.	German
Gk	Greek
He.	Hebrew
Hu.	Hungarian
It.	Italian
IUPAC-IUGS	International Union of Pure and Applied Chemistry and International Union of Geological Sciences
ka	thousand years ago
L.	Latin
Ma	million years ago
ME	Middle English
MFr.	Medieval French
N/A	not available
OD	old Dutch
OE	old English
OF	old French
OIt	old Italian
ON	old Norse
pl.	plural
SI	*Système international d'unités*, International System of Units
Sp.	Spanish
Swa.	Swahili
syn.	synonym
var.	variant
Yidd.	Yiddish

Taxonomic conventions

aff.	affinity (*see* **aff.**)
cf.	*See* **cf.**
sp. indet.	species indeterminate
sp. nov.	novel species

Informal taxonomic categories

Individual fossils are placed in one of six informal taxonomic categories, namely 'Possible hominins', 'Archaic hominins', 'Megadont archaic hominins', 'Transitional hominins', 'Pre-modern *Homo*' and 'Anatomically modern human. These are inclusive grade categories. Whereas clades reflect the *process* of evolutionary history, the grade concept is based on the *outcome* of evolutionary history. Taxa in the same grade have similar relative brain sizes, they are judged to eat foods with the same mechanical properties and they share the same posture and mode(s) of locomotion; no store is set by how they came by those behaviors. The judgment about how different two masticatory systems or two locomotor strategies have to be before the taxa concerned are considered to belong to different grades is a subjective one, but until we can be sure we are generating reliable hypotheses about the relationships among hominin taxa the grade concept enables taxa to be sorted into broad functional categories, albeit sometimes frustratingly 'fuzzy' (e.g., where to place *Homo floresiensis*) ones.

2D:4D *See* hand, relative length of the digits.

3′ One of the two ends of a single strand of DNA. It is called the 3′ (or "three prime") end because the third carbon of the **nucleotide** at that end has a hydroxyl group attached. The other end, the **5′** (or "five prime") end, is so-called because the fifth carbon of the deoxyribose sugar of the nucleotide at that end has a terminal phosphate group attached to it. The direction towards the 3′ end of a single strand of DNA is called the **downstream** direction because as you move in that direction the number of carbon molecules in the sugars in the DNA backbone decreases. The significance of the directionality of a DNA sequence is that the processes of **replication** and **transcription** only occur in the 5′ to 3′ direction. *See also* **DNA**.

5′ One of the two ends of a single strand of DNA. It is called the 5′ (or "five prime") end because the fifth carbon of the deoxyribose sugar of the **nucleotide** at that end has a terminal phosphate group attached to it. The other end, the 3′ (or "three prime") end, is so-called because the third carbon of the nucleotide at that end has a hydroxyl group attached. The direction towards the 5′ end is called the upstream direction because as you move in that direction the number of carbon molecules in the sugars in the DNA backbone increases. The significance of the directionality of a DNA sequence is that the processes of **replication** and **transcription** only occur in the 5′ to 3′ direction. *See also* **DNA**.

23 ka world *See* **astronomical time scale**.

41 ka world *See* **astronomical time scale**.

100 ka world *See* **astronomical time scale**.

1939 mandible *See* **Sangiran 5**.

1941 mandible *See* **Sangiran 6a**.

a The abbreviated form of annum. The joint IUPAC-IUGS Task Group (2006-016-1-200) in 2006 urged that the SI unit a be used for both ages and time spans (i.e., 36 ka for 36 thousand years and 2.3 Ma for 2.3 million years). The IUPAC-IUGS Task Group discourage the use of y, yr and yrs in combination with k, K, m, M, etc.

AAC *See* **alcelaphine plus antilopine criterion**.

AARD *See* **amino acid racemization dating**.

AAS *See* **Aurora archaeostratigraphic set**; **Gran Dolina**.

abduction (L. *ab* = away and *ducere* = to lead) Limbs Refers to moving a limb away from the midline (i.e., the true **sagittal** plane) when the body is in the **anatomical position**. Digits In the case of the hands and feet abduction means moving fingers or toes away from the reference digit. The reference digit of the hand is the middle finger; the reference digit of the foot is the second toe. Thus, the muscles that move the limbs away from the midline (e.g., the deltoid and the gluteus medius) and those that move a digit away from the reference digit (e.g., the dorsal interosseus muscles of the hand) are referred to as abductors. *See also* **abductor**; **reference digit**.

abductor Any muscle that contributes to moving a limb away from the midline reference plane (e.g., deltoid, gluteus medius), or a digit away from the **reference digit** (e.g., dorsal interossei of the hand).

Abdur (Location 15°09′N, 39°52′E Eritrea; etym. the name of a nearby village) History and general description Faure and Roubet (1968) had reported finding Acheulean artifacts on the surface of marine deposits on the African shore of the Red Sea. Abdur is a Pleistocene reef terrace on the western shore of the Buri Penisula, and is northeast of the Danakil Depression. The terrace, comprised of a 6.5 km/4 mile by 1 km/0.6 mile exposure of the Abdur Reef Limestone (or ARL), is divided into three sections: Abdur North (AN), Abdur Central (AC) and Abdur South (AS). Obsidian and quartz artifacts have been found in AN and AS, with most of the evidence coming from AN and the artifacts coming from the "basal cobble zone," the "lower part of the lower shell zone" and the "beach facies" (Walter et al. 2000, p. 67). Temporal span and how dated? Five of six dates based on **uranium-series dating** (uranium-thorium) performed on coral aragonite suggest the ARL was deposited between 136 and 118 ka, with a mean age of $c.125 \pm 7$ ka. Hominins found at site None. Archeological evidence found at site Bifaces made from chert and quartz and flake and blade tools mainly from obsidian. The artifacts are consistent with the early **Middle Stone Age**. Key references: Geology, dating, and fauna Walter et al. 2000; Hominins N/A; Archeology Walter et al. 2000.

"Abel" The informal name given by Michel **Brunet** to a hominin mandibular fragment recovered in 1994 by the **Mission Paléoanthropologique Franco-Tchadienne** (French-Chadian Paleoanthropological Mission, or MPFT) at a site in Chad called **Koro Toro**. Brunet and his colleagues assigned the specimen to a new species, *Australopithecus bahrelghazali*. He gave the specimen the name Abel to honor the memory of Abel Brillanceau, a colleague and close friend who died while doing field research.

abiotic (Gk *a* = not and *bios* = life) In relation to hominins this term subsumes all the "non-living" factors (e.g., climate and physical catastrophies such as massive volcanic eruptions, tsunamis, etc.) that might have influenced the outcome of human evolution.

Abri Bourgeois-Delaunay *See* **La Chaise**.

Abri d'Aurignac *See* **Aurignac**.

Abri Moula-Guercy (Location 44°52′59″N, 4°50′51″E, France; etym. Fr. *abri* = rock-shelter, Moula = the name of the discoverer, Guercy = the limestone outcrop in which the cave lies) History and general description This site, sometimes referred to as Baume Moula-Guercy (*baume* is Provençal for cave) or just Moula-Guercy, lies 80 m above the right bank of the Rhône river, in the east side of the Serre de Guercy, a small limestone outcrop in the town of Soyons, in the Ardeche region of southeast France. Discovered in 1970, the site was not excavated until 1991, when a team led by Alban Defleur began to explore the rock-shelter and deep cave beneath. They uncovered 20 layers, all of which are exclusively **Middle Paleolithic**. This site is best known for the abundant evidence of defleshing and disarticulation on bones assigned to *Homo neanderthalensis*, which the authors have suggested is due to cannibalism. It is also one of the few well-preserved sites with hominin fossils of Eemian age. Temporal span and how dated? The site represents the period between **Oxygen Isotope Stages** 6 and 4. Most of the site is dated using **biostratigraphy**, but an ash layer near the top of the site has been dated using thermo**luminescence dating** to 72 ± 12 ka. The hominin layer is estimated to be between 100 and 120 ka. Hominins found at site Neanderthals are the second most abundant taxon recovered from layer XV; 78 bone fragments were recovered, representing at least six individuals. The majority of these bones show cutmarks and intentional fractures that have been interpreted as attempts at marrow extraction; the patterns are similar to those found on deer bones found at the same site. Archeological evidence found at site The artifacts from the layer containing the fossil hominins are attributed to the Ferrassie **Mousterian**. Key references: Geology, dating, and paleoenvironment Sanzelle et al. 2000, Defleur et al. 1998, 2001; Hominins Defleur 1995, Defleur et al. 1998; Archeology Defleur 1995, Defleur et al. 1998.

Abri Pataud (Location 44°56′N, 1°00′E, France; etym. Fr. *Abri* = rock-shelter, and Pataud after the name of the family who owned the land) History and general description One of several rock-shelter sites along the Vézère River in southwest France. The site was obscured by farm buildings until 1953 when Hallam Movius began exploring the site, and it was systematically excavated between 1958 and 1964 by a joint French and American team led by Movius. Temporal span *c*.34–19 ka. How dated? Conventional **radiocarbon dating**. Hominins found at site Partial skeletons of three adult and two adolescent modern human skeletons were found in the Proto-Magdalenian layer. Archeological evidence found at site Thirteen numbered archeological layers represent an almost complete **Upper Paleolithic** sequence, from basal **Aurignacian** to a possibly lower **Solutrean** level, separated by unnamed sterile layers. Abundant fauna and some pollen suggest a cold environment consistent with the **Last Glacial Maximum**, but with local river valleys providing **refugia** for warm-climate species. Key references: Geology, dating and paleoenvironment Donner 1975, Movius Jr. 1975, Wilson 1975, Bricker and Mellars 1987; Hominins Billy 1975; Archeology Movius Jr. 1975.

Abri Peyrony   *See* **Combe-Capelle**.

Abri Suard   *See* **La Chaise**.

abridged life table   A complete **life table** gives mortality per individual year of age, while an abridged life table lumps various ages together. It is common in **paleodemography** to use age classes starting at 0, 1, 5, 10, 15, 20, 25, 30, 40, and 50, where the end of an age class is implied by the next age class. For example, the first few age classes in the above list would be 0–0.9973 (where 0.9973 is 364/365 days, 1–4.9973, and 5–9.9973. The last age class is considered to be open, so that the last age class in the above list is 50+. Any life table function which involves person-years will require that the last age interval be closed, which is usually done by continuing the age progression so that 50+ becomes 50–60. Using the notation $_nD_x$ to represent the number of deaths between age x and age $x+n$, the symbols for the abridged life table example here would be: $_1D_0$, $_4D_1$, $_5D_5$, $_5D_{10}$, $_5D_{15}$, $_5D_{20}$, $_5D_{25}$, $_{10}D_{30}$, $_{10}D_{40}$, and $_{10}D_{50}$.

absolute dating   (L. *absolutus* = free or unrestrained) In physics, "absolute" means being free, or independent, of arbitrary standards. Thus, absolute dating methods were so labeled because they were regarded as not being arbitrary in that they are based on physical or chemical systems whose dynamics are predictable. This predictability means they can be used to calibrate events or measure the passage of geological time. Very few of the current techniques can be used to determine the age of the fossils or artifacts themselves; in most cases they are used to date surrounding sediments or associated igneous rocks, which have a known stratigraphic relationship to the fossils in question. **Radiocarbon dating** is perhaps the best-known absolute dating method, but

several others (e.g., **potassium-argon dating**, thermo**luminescence dating**, **uranium-series dating**) are used to provide age estimations for fossil hominins beyond the age range of radiocarbon dating (i.e., older than approximately 50 ka). All absolute dates have various associated errors. Occasionally these are errors in the measurement of the physical parameter used [i.e. the amount of carbon-14 (^{14}C) **isotope** present in the sample]. Errors can also arise from imperfect calibration of the physical parameter to the geological time scale or from various alterations of the dated material after it was deposited. Thus, whenever the age of a tuff, cave deposit, directly dated fossil, etc. is cited it should include its error term expressed either in standard deviation units or as years, depending on the dating technique being utilized. Geochronologists are moving away from the old divisions of absolute and relative methods, and tend to refer to absolute dating methods as methods that provide a "numerical age estimate". *See also* **geochronology**.

AC   Acronym for Abdur Central. *See* **Abdur**.

acceleration   (L. *accelerationem* = a hastening) A heterochronic process that produces peramorphic results by dissociating ancestral patterns of ontogeny. Specifically, shape change is accelerated and progresses further for a given size in descendants relative to ancestors; adult size and duration of growth found in the ancestor remain unchanged in the descendant.

accelerator mass spectrometer   (AMS) A specialised mass spectrometer that can extend the range of carbon dating back to 50–60 ka. This technique uses an accelerator to produce ions with very high kinetic energies; this is done to increase the ionization state of the ions from the sample in order to facilitate the removal of potential contaminants (particularly ^{14}N and ^{13}CH⁻) that would interfere with the counting of ^{14}C ions in the detector. The technique is referred to as **AMS radiocarbon dating**, to distinguish it from conventional radiocarbon dating. *See also* **radiocarbon dating**.

accentuated line   A pronounced line within a tooth corresponding to the temporary slowing of movement of the developing **enamel** or **dentine** front, or to a change in chemical composition, that does not result from an intrinsic rhythm (*see* **incremental features**), but rather relates to a (non-specific) stressor event. The **neonatal line** found in teeth developing just prior to, during, and immediately after birth is the best-known example. Accentuated lines are sometimes found in association with enamel **hypoplasias**, but they can occur without there being any surface manifestation of developmental stress (syn. pathological lines, **Wilson bands**, or accentuated striae of Retzius).

accentuated striae of Retzius   *See* **Wilson bands**.

accessory cusp   A cusp on a maxillary (upper) or mandibular (lower) molar tooth that is not one of the main cusps (i.e., not the **proto-**, **para-**, **meta-**, or **hypocone** on a maxillary molar, and not the **proto-**, **ento-**, **meta-**, or **hypoconid**, or **hypoconulid**, on a mandibular molar). For example, the **metaconule** is an accessory cusp between the metacone and protocone on a maxillary molar, and the **tuberculum sextum** is an accessory cusp between the **entoconid** and the **hypoconulid** on the **talonid** component of the crown of a mandibular molar. *See also* **cusp**.

accessory olfactory system   *See* **olfactory bulb**.

accretion model   (L. *accrescere* = to grow) Refers to the mode of origin of the Neanderthals. The model suggests that the distinctive morphology of *Homo neanderthalensis* did not arise suddenly, but emerged gradually, or grew (i.e., accreted), over a period of several hundred thousand years (Hublin 1998). Although Jean-Jacques **Hublin** (1982, 1986, 1998, 2009) is its contemporary proponent, this type of evolutionary scenario was first proposed by Piveteau (1970), and then developed by Vandermeersch (1978b). The model proposes that initially there were relatively few distinctive features (e.g., facial recession and a wide occipital torus) and what features there were occurred at a relatively low frequency, but through time the number and frequency of distinctive features increased. Dean et al. (1998) provided the most comprehensive explanation of the model. They suggested that Neanderthal evolution occurred in four stages (*ibid*, Table 1, p. 487): Stage 1 or the early pre-Neanderthal stage, Stage 2 or the pre-Neanderthal stage, Stage 3 or the early Neanderthal stage, and Stage 4 or the classic Neanderthal stage. Examples of Stage 1 Neanderthal specimens include those from **Arago**, **Mauer**, and **Petralona**. Specimens from **Sima de los Huesos**, **Reilingen**, **Steinheim**, and **Swanscombe** have been assigned to Stage 2, with **Bilzingsleben** and **Vérteszöllös** considered as Stages 1 or 2. Examples of specimens assigned to Stage 3 include **Ehringsdorf**, **Biache**, **Saccopastore**, and **Shanidar**, with **Amud**, **Monte Circeo**, **La Chapelle-aux-Saints**, **La Ferrassie**, **La Quina**, **Neanderthal**, and **Spy** examples of specimens included in Stage 4. As for the morphological features that emerge at each stage, the Stage 2 morphology includes an incipient **suprainiac fossa**, a small **juxtamastoid eminence**, and an incipient occipital bun. The changes that occur at Stage 3 include a smaller mastoid process, a larger **juxtamastoid process**, an anterior **mastoid tubercle**, and a more exaggerated occipital bun. At Stage 4 the additions include greater midfacial **prognathism**, a more exaggerated suprainiac fossa, and a larger piriform, or nasal, aperture. However, Hublin (1998, 2009) emphasized that there is a shift in the frequencies of these features and that stages primarily result from discontinuity in the fossil record. He proposed that the taxon name *Homo neanderthalensis* should be used for *all four stages*. Other researchers support the accretion model, but they would confine the hypodigm of *H. neanderthalensis* to the later stages, and would use *Homo heidelbergensis* for the early stages (e.g., the hominins from **Sima de los Huesos**). *See also Homo heidelbergensis*; *Homo neanderthalensis*.

acculturation   The adoption by one group of a large number of cultural behaviors from another group. Thus, in the terminology of dual inheritance theory, acculturation is a form of group-level horizontal cultural transmission.

accuracy   (L. *accurare* = to do with care) Accurate refers to something that is true, or close to the truth. Thus, accurate estimates of measurements or dates are ones that are very close to the real dimensions or ages. A series of accurate measurements are not necessarily close to each other, however. Imagine a dartboard with three darts placed equidistant on a circle around the bull's eye. Each dart is close to the bull's eye, but not particularly close to the other darts. *See also* **precision**.

acetabulocristal buttress   *See* **acetabulocristal pillar**.

acetabulocristal pillar (L. *acetum* = vinegar, *abulum* = small receptacle, and *crista* = crest) Thickening of mostly the outer table of the iliac blade part of the pelvic bone between the acetabular fossa and the iliac crest about one-third of the way between the anterior and posterior superior spines. The thickening is seen in pre-modern *Homo*, but not in anatomically modern humans (Stringer 1986b). It presence is interpreted as evidence that the gluteal muscles were playing a significant abducting role (i.e., working isometrically to prevent the pelvis from dipping down on the side where the leg is no longer in contact with the ground because it is in the swing phase of the **walking cycle**). For example, Day (1971) refers to the acetabulo cristal pillar as the "vertical iliac pillar" or "vertical iliac bar" and suggests that its presence in **OH 28** is evidence that "the alternating pelvic tilt mechanism of striding bipeds was well developed" (*ibid*, p. 384). *See also* **pelvis**.

Aché A group of foragers (also known as the Guayaki) living in eastern Paraguay, an environment of mixed broadleaf evergreen forest and grassland. There are four main groups of Aché who distinguish themselves based on language, customs, and geographic range, and all contrast themselves from the neighboring sedentary farmers with whom they usually have only hostile interactions. All four Aché groups have been studied by ethnographers, and the northern group has been the particular subject of subsistence studies by behavioral ecologists. Most of the northern Aché were full-time foragers until the 1970s; now roughly 130 of this group live in an agricultural settlement. Some of these 130 people still spend a considerable amount of time in the forest on extended foraging trips, where they hunt a wide range of animals with bows, and gather plant foods. Studies of their diet have enabled several researchers to test hypotheses about evolutionary ecology in the absence of agriculture, and to use them as analogues for pre-agricultural human ancestors. Studies of the Aché have suggested that both plant and animal foods are important in an optimal diet, that men and women have different foraging goals which must be accounted for in behavioral ecology studies (Hill et al. 1987), and that hunting may be a social signal as well as a method of provisioning (Wood and Hill 2000).

Acheulean Originally proposed in the 19thC by Gabriel de Mortillet; takes its name from the French village of Saint-Acheul in the Somme River Valley where numerous **handaxe**s had been recovered. In 1872 Gabriel de Mortillet named the industry "l'Epoque de St Acheul", but in 1925 the name was changed to Acheulean. Earlier mention of handaxes was made by John Frere who in 1797 sent two examples from Hoxne, Suffolk, UK, to the Royal Academy in London claiming that "they belonged to a very ancient period indeed" and had been produced by "people who had not the use of metals." Given the general lack of knowledge of evolution in those times his claims were mostly ignored. The oldest and youngest reported Acheulean sites were in Africa from 1.6–0.16 Ma. Acheulean sites are found in a substantial portion of Eurasia (but they are generally absent from Asia east of **Movius' Line**). Most Eurasian Acheulean sites date from the Middle Pleistocene (i.e., they are *c.*500 ka). Handaxes and similar implements (such as **cleaver**s and picks) define the Acheulean; these tool types are not *confined* to the Acheulean, but when they occur at other sites they are rare and are typically outnumbered by **flake**s, **core**s, and other smaller modified tools such as **scraper**s. The Acheulean is preceded by the **Oldowan**, and many of the non-handaxe Acheulean tools are Oldowan-like, emphasizing the retention of an older but effective stone tool technology. Similarly, some later Acheulean sites, including for example those in the Somme River Valley and in the **Kapthurin Formation** of Kenya, show evidence for the use of **Levallois** technology for large flake production. This technique is seen more commonly in later **Middle Paleolithic** or **Middle Stone Age** sites, and this also suggests local technological continuity. The Acheulean site of **Gesher Benot Ya'akov** in Israel documents the earliest controlled use of fire, as well as nut fragments and processing equipment (Goren-Inbar et al. 2002a, 2002b). Meat was very likely an important component of the hominin diet, but Acheulean hunters may have been relatively ineffective compared with later populations at bringing down large game (Klein et al. 2007). Although traditionally associated with *Homo erectus*, Acheulean tools were likely made and used by more than one hominin taxon. Interpretations of the Acheulean vary widely. Its seemingly static nature is usually attributed to **social learning** and **cultural transmission**, in combination with functional constraints (cutting, scraping), repeated resharpening, and raw-material constraints (e.g., large tabular chunks versus small elongated pebbles). Some workers have speculated that the quite unchanging basic features of Acheulean implements were genetically determined, and that the variability of the final products may have signaled the genetic qualities of the maker. The Acheulean is unique in the evolution of hominins in that neither before nor since has such a singular technology dominated the activities of hominins for so long over so much of the planet.

Acheulian Alternative spelling for **Acheulean** (*which see*).

Acheulian industrial complex *See* **Acheulean**.

actualistic studies (L. *actus* = an act) Such studies, which reconstruct activities to understand the role or function of that activity, are used in archeology and paleontology. For example, in archeology, researchers use tools in controlled circumstances to help determine what paleolithic artifacts were used for. For example, if an artifact made by a skilled knapper is used to whittle wood, the wear on the modern tool can be compared with the wear on ancient artifacts to see if the wear facets show the same macro- and microwear. In paleontology, actualistic research involves studying the factors that determine the formation and nature of present-day bone assemblages and then applying that knowledge to the paleontological record. Actualistic studies are uniformitarian in the sense that they assume that processes observed in the present are same as those that took place in the past (*see* **uniformitarianism**). In turn, inferences about the past are interpreted in terms of processes operating in the present. Actualistic research methodologies range from naturalistic to experimental. In a naturalistic context, the analyst observes and records natural processes and their resulting effect on bone assemblages. In an experimental context, the analyst controls some or all of the taphonomic processes affecting bone assemblages. Actualistic research provides the basis for most taphonomic and zooarcheological analysis and interpretation and is thus central to reconstructing the behavioral and paleoenvironmental signals encoded in the fossil record.

acute margin Another term for a sharp edge on a stone tool.

adaptation (L. *adaptare* = to fit) A useful feature or trait that (a) promotes survival and reproduction and (b) is shaped by natural selection, and

formulating and testing hypotheses of adaptation is a major focus of paleoanthropological research. Adaptations are therefore heritable and perform functions. The term is also used to describe the process by which features that promote or enhance fitness evolve under **natural selection**, which can be **directional selection** or **balancing selection** (also known as stabilizing selection). Evolutionary biologists differ as to whether understanding the genesis of a trait is important in determining its adaptive status, some noting the importance of distinguishing between current function and original function(s). In some cases the function of a current trait can be assessed experimentally, but in most cases in evolutionary biology describing a phenotypic trait as a current or previous adaptation should be seen as an hypothesis. The frequency of a heritable variant, the **trait**, predominates or becomes fixed in a population because the individuals possessing it enjoy greater reproductive success (**fitness**) than those that lack it. It is common for workers to think of adaptive traits as being **derived** (i.e., **apomorphic**), but this need not be the case if a **primitive** trait has been retained in a population as a result of stabilizing selection. Adaptation can also be used as an adjective in connection with a taxon, as in "the dentition of *Paranthropus boisei* is better adapted for chewing than for slicing food." In such usage, adaptation is being used in an informal sense (i.e., "better adapted" can be read to mean "functions better"). So, what needs to be done to strengthen an hypothesis that a trait in a **hominin** taxon is an adaptation in the non-historical sense? If a trait seems to have an obvious function (e.g., the **acetabulocristal buttress** of the pelvic bone functions in bipedal hominins as a bony strut that links the proximal attachment of the hip abductors with the acetabulum of the hip joint) then all well and good. But just because our present knowledge does not allow us to assign a function to a trait, it does not follow the trait has no function; we may not yet understand what its function is. For example, if we did not know about the way the specialized receptors in the **semicircular canals** of the **membranous labyrinth** of the **inner ear** respond to motion, we would perhaps be inclined to assume that the shape of that part of the bony labyrinth of a fossil **hominid** was not adaptive. Therefore, adaptation will tend to be underrecognized (i.e., it is prone to **type I error**). It can be difficult, if not impossible, to firmly establish whether or not a given trait has been subjected to natural selection in fossil hominins, so the identification of adaptations in paleoanthropology is inevitably conjectural. *See also* **phylogenetic lag; structure–function relationships**.

adaptive landscapes *See* fitness landscapes.

adaptive radiation (L. *adaptare* = to fit and *radius* = ray) The relatively rapid diversification of a lineage into **species** that evolve a range of new **adaptive strategies** that facilitate their occupation of a range of new **adaptive zone**s. An example is the simultaneous appearance (at least in geological time) of multiple pig species at the same site(s) at several times during the **Pliocene** and **Pleistocene**. During the last 4 Ma approximately six pig species have coexisted at the same locality (i.e., **sympatric**ally) at the same time (i.e., **synchronic**ally). A number of these species exhibit M_3 hypsodonty and craniofacial elaboration (bosses, crests, tusks). This is usually interpreted as an adaptive response to more open, grassland environments.

adaptive strategy (L. *adaptare* = to fit and *strategos* = general) Sewall-Wright's term for the set of traits that enables organisms of a species to survive and reproduce. The origin of these trait sets can vary; they may have been inherited from a recent common ancestor or they may be the result of independent adaptation to similar environmental conditions (i.e., **parallel evolution** or **convergent evolution**). An adaptive strategy can be shared by more than one species (syn. organizational plan). *See also* **adaptive type; adaptive zone; grade**.

adaptive type (L. *adaptare* = to fit, *typus* = image) According to Huxley (1958) an adaptive type is a fossil taxon with a more derived **phenotype**, or organizational plan, that replaces an older fossil taxon with a less derived organizational plan. In some cases the replacement is straightforward, involving just two taxa. In others, the old organizational plan is replaced by several new organizational plans, which are then reduced in number by extinction, until only one is left. Regardless of the mode of replacement, the new taxon is called an adaptive type because, according to neo-Darwinian principles, it must have been more successful than the taxa it superseded. The rise and success of a new organizational plan is evidence that it was better adapted than

the older one, and that it was also better adapted than any potential competitor. The terms adaptive type and grade are interchangeable. Huxley's (1958) term "organizational plan" is equivalent to Sewell-Wright's adaptive strategy (syn. grade). *See also* **adaptive strategy; adaptive zone; grade**.

adaptive zone (L. *adaptare* = to fit and *zona* = girdle) Taxa that occupy the same adaptive zone survive and reproduce in a similar manner. That is, they have the same adaptive strategy. The relationship between adaptive strategies and adaptive types, or grades, is such that some adaptive zones are occupied by taxa that have the same adaptive strategy and so represent an adaptive type or grade. Occasionally, a lineage will rapidly diversify into several species that occupy a range of new adaptive zones. This is referred to as an adaptive radiation. *See also* **adaptive radiation; adaptive strategy; adaptive type; grade**.

adduction (L. *adducere* = to bring towards) Limbs Adduction refers to moving a limb towards the midline (i.e., the true **sagittal** plane) when the body is in the **anatomical position**. Digits In the case of the hands and feet, adduction means moving fingers or toes towards the reference digit. The reference digit of the hand is the middle finger; the reference digit of the foot is the second toe. Thus, the muscles that move the limbs towards the midline (e.g., teres major and obturator internus) and those that move a digit towards the reference digit (e.g., the palmar interosseus muscles of the hand) are referred to as adductors. *See also* **adductor; reference digit**.

adductor Any muscle that contributes to moving a limb towards the midline reference plane (e.g., teres major, obturator internus), or a digit towards the reference digit (e.g., palmar interossei of the hand). *See also* **reference digit**.

aDNA *See* **ancient DNA**.

adolescence (L. *adolescentia* = youth) Adolescence is the stage of modern human life history that follows the juvenile stage and the event of puberty. The length of adolescence varies across modern human populations; it ends when adult skeletal lengths, dental development, and sexual maturation are attained (e.g., the age range of its ending is from 17 to 25 years). The defining characteristic of adolescence is a spurt in height (unique to modern humans among extant higher primates; several nonhuman primate species undergo growth spurts in craniofacial dimensions) and mass (shared with many nonhuman primates). The intensity and duration of this spurt shows much variation among and within modern human populations. The onset of menstruation (**menarche**) in modern human girls usually follows within 1 year of the peak velocity of the growth spurt, and for 1–3 years they experience a period of "adolescent subfecundity" (where menstrual cycles are often irregular or anovulatory). Sexual maturation in modern human boys precedes the time of peak growth velocity, but the full development of modern human male secondary sexual characteristics takes much longer to manifest. This results in subfecund, adult-looking girls, and fecund, non-adult-looking boys. It has been proposed that this is an adaptation that enables each sex to learn as much as possible about their adult social, economic, and sexual roles in society, with as little risk as possible. It has been suggested that the adolescent stage evolved either in archaic or anatomically modern *Homo sapiens*, but it is difficult to determine, at the current time, whether extinct hominin taxa underwent a growth spurt and, if so, whether it was more similar to that of modern humans or that of nonhuman primates.

adolescent growth spurt (L. *adolescere* = to grow up, from *ad* = forward and *alescere* = to grow) Adolescence in modern humans is the period of **life history** between puberty and maturity. During the adolescent stage of modern human development rapid gains in height and weight usually occur beginning on average between ages of 10 (females) and 12 (males). This period of the highest growth velocity, second only to that experienced in infancy, usually lasts between 8 and 10 years. There is significant variation in the onset and intensity of the adolescent growth spurt within and among modern human populations, with some showing a clear and marked spurt in stature while others demonstrate no clear evidence of such a spurt. While many species of nonhuman primates have been shown to have growth spurts in mass, modern humans are the only primates that show evidence for a spurt in linear skeletal dimensions. This intense growth is seen in almost all skeletal elements, with the exception of the female pelvis (instead of experiencing a major spurt in growth, the female pelvis grows slowly until adulthood). The later onset and longer duration combined with the generally higher growth velocity experienced during the adolescent growth spurt in modern human males when compared with females is largely responsible for their greater average adult height.

adrenarche (L. *ad* = near, *renal* = kidney, and Gk *arkhe* = beginning) The onset of prepubertal adrenal androgen production (specifically of dehydroepiandrosterone, or DHEA, and dehydroepiandrosterone sulfate, or DHEAS) after a post-infancy period of very little adrenal activity. While circulating levels of adrenal androgens in nonhuman primates are generally higher than those in nonprimate mammals, none show the same pattern that is seen in modern humans and chimpanzees: a sharp postnatal decline in levels from birth with subsequent elevations before sexual maturation usually around 6–10 years of age. It has been suggested that differences in patterns of adrenal androgen secretion may be due to differences in the timing of the regression of the fetal zone and development of the *zona reticularis* (or ZR), the region of the adult adrenal gland where these hormones are produced. Specifically, these two events occur in closer temporal proximity in primates other than chimpanzees, such as baboons, whose DHEAS levels peak shortly after birth and decline steadily thereafter. It has been proposed that the decline in serum DHEA/DHEAS prior to adrenarche acts as a mechanism to delay skeletal maturation in concert with prolonged prepubertal growth (Bogin 1997). The outward effects of adrenarche (e.g., changes in body fat distribution, growth of body hair) noticeably alter an individual's physical appearance. It has been suggested that adrenarche and the slowed growth prior to it is critical to the development of an individual's cognitive function. It also aids the perception by others that an individual is transitioning from a child to a juvenile.

ADU Abbreviation of Aduma, a site in the **Middle Awash study area**, Ethiopia. *See* **Aduma**.

ADU-VP Acronym for **Adu**ma – **V**ertebrate **P**aleontology. Prefix for fossils from the site of **Aduma**.

ADU-VP-1/3 Site Aduma. Locality Aduma. Surface/*in situ* Surface collection from **lag deposits**. Date of discovery 1996. Finder Timothy Douglas **White**. Unit N/A. Horizon Bed B. Bed/member Ardu Beds. Formation **Bouri Formation**. Group Middle Awash. Nearest overlying dated horizon Bed A, Ardu Beds. Nearest underlying dated horizon Bed C, Ardu Beds, followed by the **Herto Formation**.

Geological age **Late Pleistocene**, associated with **Middle Stone Age** artifacts and sediments dated between 79 and 105 ka (Yellen et al. 2005). Developmental age Adult. Presumed sex Unknown. Brief anatomical description The most complete of five fragmentary crania from Aduma, ADU-VP-1/3 preserves part of the frontal squame and much of the parietals and occipital. The occipital is rounded and the parietals have a strong sagittal curvature, indicating a globular **neurocranium**. Announcement Haile-Selassie et al. 2004b. Initial description Haile-Selassie et al. 2004b. Photographs/line drawings and metrical data Haile-Selassie et al. 2004b. Detailed anatomical description Haile-Selassie et al. 2004. Initial taxonomic allocation *Homo sapiens*. Taxonomic revisions N/A. Current conventional taxonomic allocation *Homo sapiens*. Informal taxonomic category Anatomically modern human. Significance The Aduma crania are clearly anatomically modern and even though their dating is not precise they add support to the hypothesis that modern humans evolved in Africa. Location of original **National Museum of Ethiopia**, Addis Ababa, Ethiopia.

Aduma (Location 10°25′N, 40°31′E, Ethiopia; etym. named after a nearby Afar village) History and general description A region in the **Middle Awash study area** with **Early**, **Middle**, and **Later Stone Age** archeological sites. The locality was briefly surveyed by John Kalb in 1976, but excavations were not undertaken until John Yellen and Alison Brooks excavated there between 1993 and 1998 under the aegis of the **Middle Awash Research Project**. A variety of **lacustrine**, **fluvial**, and **tuffaceous** sediments crop out in the area; the majority of the *in situ* (Middle Stone Age) archeological sites occur within the Ardu Beds (informally subdivided into Ardu A–C from bottom to top). The archeological sites in the Ardu Beds are important for documenting temporal trends within the Middle Stone Age, and are associated with fossil fauna (including hominins) that suggest use of a range of habitats, but with a strong reliance on riverine resources. Temporal span Multiple radiometric methods provide a general estimate of the age of the Ardu Beds. An **argon-argon dating** age estimate on pumice of *c*.180 ka from the base of the sequence and a **radiocarbon dating** result of 10.50±0.07 ka on *Unio* sp. shell from sterile sediments capping the sequence provide bracketing ages. Although there are inconsistencies and stratigraphic reversals, thermoluminescence dating and optically stimulated **luminescence dating** (single and

multiple-grain) ages on silt and sand and **uranium-series** estimates on fossilized mammal and *Clarias* sp. bone and crocodile teeth suggest an age range of *c*.80–100 ka for most of the archeological sites. How dated? ^{40}Ar/^{39}Ar, ^{14}C, optically stimulated luminescence dating, thermoluminescence dating, and U-series. Hominins found at site Four adult hominins, a cranium (**ADU-VP-1/3**), parietal fragments (ADU-VP-1/1 and 1/2), and unidentified cranial fragments (ADU-VP-1/6), all attributed to *Homo sapiens*, were recovered from the surface of Aduma (likely from Ardu B sediments). Archeological evidence found at site The approximately 16,200 **lithic** artifacts are made of **lava** (including **obsidian**), **quartz**, and **chert**, and include cores for the production of **blades**, **bladelets**, **Levallois** and other flakes, diverse **retouch**ed pieces dominated by **points** and **scrapers**, and rare **grindstones**. The assemblages are notable for showing temporal variation among Middle Stone Age assemblages, particularly a decrease in core and flake/blade size with increased stratigraphic height. More than 30 taxa are recognized among the recovered fauna. Rare cutmarks implicate the role of hominins in the accumulation of at least some of this fauna. The combination of taxa that are water-dependent (e.g., *Crocodylus niloticus*, *Clarias* sp., and *Unio* sp.) and water-independent (e.g., *Oryx*) suggest the presence of steep ecological gradients away from a forested river margin to a **savanna** grassland; soil isotopic and pollen data also support the latter suggestion. Key references: Geology and dating Yellen et al. 2005; Hominins Haile-Selassie et al. 2004b; Archeology Yellen et al. 2005.

Aduma cranium *See* ADU-VP-1/3.

aeolian (Gk *Aeolus* = god of the winds) Deposited primarily by wind action. Clay-sized aeolian material is often difficult to recognize, being readily integrated into soils or sedimentary deposits. The large volumes of glacially derived silts subject to aeolian transport are responsible for **loess** deposits. Sand-size aeolian material commonly forms dunes in dry environments and coastal areas. Volcanic ash can be subjected to aeolian transport and reworking prior to deposition. At **Laetoli**, most of the sediments comprising the **Laetoli Beds** are either primary airfall tuffs, or they are formed from airfall tuffs that have been reworked by aeolian processes.

Afalou *See* Afalou-Bou-Rhummel; Iberomaurusian.

9

Afalou-Bou-Rhummel (Location 36°45′N, 5°35′E, Algeria) History and general description Site discovered by Camille **Arambourg** in April 1928. It is a rock-shelter located on the southern shore of the Gulf of Bejaïa. The hominin fossils found at this site were described as a distinctive type or race similar to those found only at other North African sites, now called the Afalou or Mechta-Afalou type. Temporal span and how dated? Late Pleistocene/Holocene *c.*15–11 ka. Hominins found at site At least 50 *Homo sapiens* individuals (26 male, 14 female, six juvenile) buried in cemeteries. Archeological evidence found at site **Iberomaurusian**, backed microlithic blades of La Mouillah type and anthropomorphic and zoomorphic clay figurines. Key references: Geology, dating, and paleoenvironment Hachi 1996; Hominins Hachi 1996; Archeology Hachi 1996, Hachi et al. 2002. *See also* **Iberomaurusian**.

Afar Rift System This is part of the **East African Rift System**, which comprises a series of river valleys and basins that extends from the Afar Rift System in the northeast, via the **Main Ethiopian Rift System**, to the **Omo Rift Zone** in the southwest. The Afar Rift System includes the **Gona**, **Middle Awash**, and the **Dikika study area**s.

Afar Triangle *See* **Awash River Basin**.

aff. Abbreviation of affinity. It is a term used in taxonomy to suggest that a specimen belongs to a hypodigm that is closely related to, but not necessarily synonymous with, a known taxon. Thus, a small piece of thick cranial vault might be assigned to "*Homo* aff. *H. erectus*."

Africanthropus Dreyer, 1935 (Gk *anthropos* = human being; this name refers to a "human-like" creature from Africa) A new **genus** established by Weinert (1939) to accommodate the species *Palaeoanthropus njarasensis* that Kohl-Larsen and Reck (1936) had proposed for the **Eyasi 1** calvaria. Type species *Africanthropus njarasensis* Weinert, 1939.

Africanthropus njarasensis Weinert, 1939 (Reck and Kohl-Larsen, 1936) Weinert (1939) transferred the **species** previously known as *Palaeoanthropus njarasensis* to a new **genus**, *Africanthropus*, as *Africanthropus njarasensis*. First discovery **Eyasi 1**. Holotype Eyasi 1. Paratypes N/A. Main sites Eyasi.

Afro-European hypothesis *See Homo heidelbergensis* Schoetensack, 1908.

Afro-European sapiens hypothesis *See* **replacement with hybridization**; **out-of-Africa hypothesis**.

age at maturity This is generally taken to mean age at reproductive maturity. It has been proposed that there is a relationship between life span and age at maturity and specifically that the ratio of adult life span to age at maturity differs among lineages but is fixed within them (Charnov and Berrigan 1990). The relationship between these two variables has been proposed to relate to the cost of reproduction (i.e., older mothers produce offspring that are more likely to survive than younger mothers).

age at weaning Weaning is most often defined as a process that begins with the introduction of supplemental foods to the nursing infant and ends with the complete termination of breast feeding. The length of this process is extremely variable, both within and among species. The end of the weaning process represents a time when offspring are, to varying degrees, responsible for their own foraging, ingestion, and food supply. However, in modern humans and many nonhuman primates, weaned juveniles forage either in different circumstances from adults, or continue to receive provisioning in some way from adults in the group. This is called the period of "postweaning dependency" and it is also seen in other nonprimate mammals, specifically the social carnivores. Among many large-bodied mammalian taxa, the timing of the end of the weaning period coincides with the attainment of a critical body weight by offspring. This weight is measured variably as a multiple of birth weight or as some proportion of adult (specifically maternal) weight.

age estimate In paleoanthropology, an age estimate indicates the number of years that are thought to have elapsed between an event (e.g., the deposition of a bone or artifact) and the present day. Paleoanthropological age estimates are expressed in thousands (**ka**) and millions (**Ma**) of years. *See also* **geochronology**.

age-heaping A common problem in reported ages, where the ages are rounded up or down to particular values, for example ages that end in zeroes or fives. This results in the histogram for ages having "heaps" (higher bars) on the particular ages. Age heaping is viewed as problematic when using "known-age" samples to work out aging methods, as the age-heaping is a clear sign that the ages are not actually known. As a consequence, some researchers will only use reference skeletal collections that have documented birth and death dates.

agenesis   (Gk *a* = absence, without and *genesis* = birth or origin) Absence or lack of development of an anatomical structure. Agenesis is one of the most common dental anomalies seen in modern humans, especially in the third molars. Agenesis of other teeth has been estimated at a frequency of 2–10% depending on the population, with most of these cases involving only one or two teeth. Agenesis also occurs in fossil hominins. For example, at least one individual of *Homo floresiensis* (**LB1**) appears to show agenesis of the lower right second premolar (RP$_4$) and upper right third molar (RM^3). It has been suggested that this may be a consequence of the small size of the individual (Brown et al. 2004), but **KNM-WT 15000** also has agenesis of the M$_3$s.

agglomerate   A coarse-grained clastic rock formed during volcanic activity. The **clasts** range in size from small pebbles to large boulders and tend to be angular in shape; they are usually enclosed in a fine-grained matrix. Examples include the fronts of lava flows where cooling lava on the surface of the flow is broken up and deposited by the underlying still molten portion of the flow; agglomerates also occur within actual volcanic vents where episodes of explosive release of volcanic gases fracture the pre-existing rocks.

aggressive scavenging   *See* **scavenging**.

Ahmado   The name used by **International Afar Research Expedition** for the **collecting area** now known as **Am-Ado** within the **Woranso-Mille study area**, Central Afar, Ethiopia.

AHOB   *See* **Ancient Human Occupation of Britain Project**.

AHS   Acronym for Awoke's hominid site at **Omo-Kibish** in the **Omo-Turkana Basin**. It was named after Awoke Amzahe, an Ethiopian scientist who found the fossil. *See also* **Omo-Kibish**.

Ailuropoda-Stegodon **fauna**   A cave fauna named after two consistent components, *Ailuropoda*,

which is the only genus in the subfamily Ailuropodinae of ursids (i.e., bears) and *Stegodon*, a genus of proboscideans within the extinct subfamily Stegodontinae. The *Ailuropoda-Stegodon* fauna is found in caves in southern China, Vietnam, and Laos, but the existence of distinct lithologies in the caves suggest different environmental settings, the faunal elements also indicate a mixed environmental signal (e.g., *Pongo* indicates a rainforest environment, whereas the presence of horses indicates a more open landscape), and many layers contain a temporal mix of genera (e.g., *Mastodon*, *Stegodon*, and *Elephas*). Most observers now consider the *Ailuropoda-Stegodon* fauna to be a faunal palimpsest (i.e., a mixed assemblage containing elements from several different faunas).

Ain Hanech   (also written as Aïn Hanech) (Location 36°16′39″N, 8°19′0″E, Algeria; etym. local settlement name) History and general description Archeological site located within the Ain Hanech Formation in the Ain Boucherit and Ain Hanech areas of northeastern Algeria. The site was discovered and initially excavated by Camille **Arambourg** in 1947 during the course of a paleontological survey conducted between 1931 and 1948; further excavations took place in 1992–3 and 1998–9. This Algerian site has yielded some of the oldest stone tools in North Africa. Temporal span A *terminus post quem* for the site is provided by the Ain Boucherit fossil-bearing stratum, which is estimated to be 2.4–2.0 Ma. The site itself has been estimated to 1.8 Ma. How dated? Biostratigraphy and magnetostratigraphy. Hominins found at site None. Archeological evidence found at site **Lithic** artifacts have been recovered from layers A (youngest), B, and C within Unit T. These artifacts are considered a North African variant of the **Oldowan** industrial complex. Artifacts recovered from overlying calcretes, paleosols, and colluvia have been attributed to the **Acheulean**. At one time the **Oldowan** and Acheulean artifacts were thought to be associated, but this is now known not to be the case. Key references: Geology and dating Sahnouni et al. 2002; Hominins N/A. Archeology Sahnouni and de Heinzelin 1998, Sahnouni et al. 2002.

A.L.   (also written as AL) Acronym for Afar Locality. Prefix for fossils recovered from the **Hadar** study area, Ethiopia.

A.L. 129-1   Site Hadar. Locality Afar Locality 129. Surface/*in situ* Surface. Date of discovery

October 30, 1973. Finder Donald **Johanson**. Unit N/A. Horizon N/A. Bed/member Sidi Hakoma. Formation **Hadar Formation**. Group N/A. Nearest overlying dated horizon Basalt. Nearest underlying dated horizon Sidi Hakoma Tuff. Geological age *c*.3.3 Ma. Developmental age Adult. Presumed sex Female, based on small size. Brief anatomical description Well-preserved right distal femur (A.L. 129-1a) and proximal tibia (A.L. 129-1b), with an associated proximal femoral fragment (A.L. 129-1c) broken inferior to the lesser trochanter (NB: A.L. 128-1, a left proximal femur fragment, may belong to the same individual). Announcement Johanson and Taieb 1976. Initial description Johanson and Coppens 1976. Photographs/line drawings and metrical data Johanson and Coppens 1976. Detailed anatomical description Johanson and Coppens 1976. Initial taxonomic allocation Specimen bears morphological similarities to proximal femora assigned to *Australopithecus* sp. Taxonomic revisions *Australopithecus afarensis*. Current convention taxonomic allocation *Au. afarensis*. Informal taxonomic category Archaic hominin. Significance One of the first hominin specimens to be recovered from Hadar, its functional morphology was interpreted as evidence of adaptations to **bipedalism**. Location of original National Museum of Ethiopia, Addis Ababa, Ethiopia.

A.L. 162-28 Site **Hadar**. Locality Afar Locality 162. Surface/*in situ* Surface. Date of discovery 1974. Finder N/A. Unit N/A. Horizon Kada Hadar Tuff. Bed/member Denen Dora. Formation **Hadar Formation**. Group N/A. Nearest overlying dated horizon BKT-2. Nearest underlying dated horizon Kada Hadar Tuff. Geological age *c*.3.15 Ma. Developmental age Adult. Presumed sex Female. Brief anatomical description Partial calvaria composed of left and right parietals and occipital. Announcement Kimbel et al. 1982. Initial description Kimbel et al. 1982. Photographs/line drawings and metrical data Kimbel et al. 1982. Detailed anatomical description Kimbel et al. 1982. Initial taxonomic allocation *Australopithecus afarensis*. Taxonomic revisions N/A. Current convention taxonomic allocation *Au. afarensis*. Informal taxonomic category Archaic hominin. Significance Compared to other larger calvariae (e.g. **A.L. 333-45, A.L. 444-2**), the small size of A.L. 162-28 suggests a high level of cranial **sexual dimorphism** in *Au. afarensis*. Location of original National Museum of Ethiopia, Addis Ababa, Ethiopia.

A.L. 199-1 Site **Hadar**. Locality Afar Locality 199. Surface/*in situ* Surface. Date of discovery October 17, 1974. Finder Ato Alemayehu Asfaw. Unit N/A. Horizon N/A. Bed/member Sidi Hakoma. Formation **Hadar Formation**. Group N/A. Nearest overlying dated horizon Basalt. Nearest underlying dated horizon Sidi Hakoma Tuff. Geological age *c*.3.3 Ma. Developmental age Adult. Presumed sex Female, based on small size. Brief anatomical description Right half of the maxilla bearing C–M^3. Announcement Johanson and Taieb 1976. Initial description Johanson and Taieb 1976. Photographs/line drawings and metrical data Kimbel et al. 1976. Detailed anatomical description Johanson and Coppens 1976. Initial taxonomic allocation Specimen bears morphological similarities to *Australopithecus* sp. and *Homo*. Taxonomic revisions *Australopithecus afarensis*. Current convention taxonomic allocation *Au. afarensis*. Informal taxonomic category Archaic hominin. Significance Found within 15 m of, and in the same horizon as, **A.L. 200-1**. Although both bear adult dentitions, A.L. 199-1 is smaller than A.L. 200-1. Location of original National Museum of Ethiopia, Addis Ababa, Ethiopia.

A.L. 200-1a Site **Hadar**. Locality Afar Locality 200. Surface/*in situ* Surface. Date of discovery October 17, 1974. Finder Ato Alemayehu Asfaw. Unit N/A. Horizon N/A. Bed/member Sidi Hakoma. Formation **Hadar Formation**. Group N/A. Nearest overlying dated horizon Basalt. Nearest underlying dated horizon Sidi Hakoma Tuff. Geological age *c*.3.4 Ma. Developmental age Adult. Presumed sex Uncertain. Brief anatomical description Complete maxilla containing both left and right I^1–M^3. The maxilla has I^2/C diastemata, extensive ribbon-like wear on the incisors and premortem enamel chipping of the C and P^3. Announcement Johanson and Taieb 1976. Initial description Johanson and Taieb 1976. Photographs/line drawings and metrical data Kimbel et al. 1976. Detailed anatomical description Johanson and Coppens 1976. Initial taxonomic allocation It was assessed as bearing morphological similarities to *Australopithecus* sp. and *Homo*. Taxonomic revisions *Australopithecus afarensis*. Current convention taxonomic allocation *Au. afarensis*. Informal taxonomic category Archaic hominin. Significance First complete australopith maxilla from Hadar. Location of original National Museum of Ethiopia, Addis Ababa, Ethiopia.

A.L. 288-1 Site Hadar. Locality Afar Locality 288. Surface/*in situ* Surface. Date of discovery 1974. Finders Donald **Johanson** and team. Unit N/A. Horizon Kada Hadar 1. Bed/member Kada Hadar. Formation **Hadar Formation**. Group N/A. Nearest overlying dated horizon BKT-2. Nearest underlying dated horizon Kada Hadar Tuff. Geological age *c.*3.2 Ma. Developmental age Adult. Presumed sex Female. Brief anatomical description A.L. 288-1 (or "Lucy") is a remarkably complete **associated skeleton** of a fossil hominin. The cranial vault remains include portions of the parietals, occipital, left zygomatic, and frontal bones. The mandible includes the right P_3–M_3, LP_3, M_3, and two M_1 fragments. The postcranial skeleton is represented by the right scapula, humerus, ulna, radius, a portion of the clavicle, the left ulna, radius, and capitate, and the axial skeleton is represented by lumbar and thoracic vertebrae and ribs. The left pelvic bone, sacrum, and left femur are well preserved. Remains of the right leg include fragments of the tibia, fibula, talus, and some foot and hand phalanges. Announcement Johanson and Taieb 1976. Initial description Johanson and Taieb 1976. Photographs/line drawings and metrical data Johanson et al. 1982. Detailed anatomical description Johanson et al. 1982. Initial taxonomic allocation Specimen bears morphological similarities to *Australopithecus* sp. Taxonomic revisions *Australopithecus afarensis*. Current conventional taxonomic allocation *Au. afarensis*. Informal taxonomic category Archaic hominin. Significance A.L. 288-1 was the first relatively complete hominin associated skeleton of this antiquity, and it remains the best preserved associated skeleton of *Au. afarensis*. Location of original National Museum of Ethiopia, Addis Ababa, Ethiopia.

A.L. 333-1 Site **Hadar**. Locality Afar Locality 333. Surface/*in situ* Surface. Date of discovery 1975. Finder N/A. Unit N/A. Horizon Denen Dora 2. Bed/member Denen Dora. Formation **Hadar Formation**. Group N/A. Nearest overlying dated horizon Kada Hadar Tuff. Nearest underlying dated horizon Triple Tuff-4. Geological age *c.*3.2 Ma. Developmental age Adult. Presumed sex Male, based on canine size. Brief anatomical description Partial facial skeleton, and maxilla containing the right P^3–P^4, left C–P^3. Announcement N/A. Initial description Kimbel et al. 1982. Photographs/line drawings and metrical data Kimbel et al. 1982. Detailed anatomical

description Kimbel et al. 1982, 2004. Initial taxonomic allocation *Australopithecus afarensis*. Taxonomic revisions N/A. Current conventional taxonomic allocation *Au. afarensis*. Informal taxonomic category Archaic hominin. Significance This specimen shares the maxillary morphology demonstrated in other *Au. afarensis* specimens (e.g., **A.L. 444-2**) and it was used in the reconstruction of a composite skull of *Au. afarensis*. Along with smaller specimens (e.g., **A.L. 200-1** and **A.L. 417-1**) it demonstrates the extent of the **sexual dimorphism** of the maxillary region and the palate. Location of original National Museum of Ethiopia, Addis Ababa, Ethiopia.

A.L. 333-45 Site **Hadar**. Locality Afar Locality 333. Surface/*in situ* Surface. Date of discovery 1975. Finder N/A. Unit N/A. Horizon Denen Dora 2. Bed/member NA. Formation **Hadar Formation**. Group N/A. Nearest overlying dated horizon Kada Hadar Tuff. Nearest underlying dated horizon Thin Tuff 4. Geological age *c.*3.2 Ma. Developmental age Adult. Presumed sex Male, based on size and robusticity. Brief anatomical description A cranium that preserves the posterior portion of the cranial vault together with a well-preserved cranial base. Announcement Johanson et al. 1982. Initial description Kimbel et al. 1982. Photographs/line drawings and metrical data Kimbel et al. 1982, 1984. Detailed anatomical description Kimbel et al. 1982. Initial taxonomic allocation *Australopithecus afarensis*. Taxonomic revisions N/A. Current convention taxonomic allocation *Au. afarensis*. Informal taxonomic category Archaic hominin. Significance First complete adult cranium of *Au. afarensis*. Location of original National Museum of Ethiopia, Addis Ababa, Ethiopia.

A.L. 333-105 Site **Hadar**. Locality Afar Locality 333. Surface/*in situ* Surface. Date of discovery 1975. Finder N/A. Unit N/A. Horizon Denen Dora 2. Bed/member Denen Dora. Formation **Hadar Formation**. Group N/A. Nearest overlying dated horizon Kada Hadar Tuff. Nearest underlying dated horizon Triple Tuff-4. Geological age *c.*3.2 Ma. Developmental age Juvenile. Presumed sex Unknown. Brief anatomical description Distorted, partial juvenile cranium. The cranial vault is represented by the frontal, the right temporal, the posterior portion of the left temporal, parts of the parietals, and fragments of the sphenoid and occipital. The face is represented by the right side of the maxilla (including the right dm^{1-2}, an unerupted M^1, and a di), plus fragments of the nasal, lacrimal, zygomatic, vomer, and palatine bones. Announcement N/A. Initial description Kimbel et al. 1982. Photographs/line drawings and metrical data Kimbel et al. 1982. Detailed anatomical description Kimbel et al. 1982. Initial taxonomic allocation *Australopithecus afarensis*. Taxonomic revisions N/A. Current conventional taxonomic allocation *Au. afarensis*. Informal taxonomic category Archaic hominin. Significance Despite being a juvenile it shows the taxonomically distinctive craniofacial morphology of *Au. afarensis*, and it is the first partially complete juvenile cranium of that taxon. Location of original National Museum of Ethiopia, Addis Ababa, Ethiopia.

A.L. 333-115 Site **Hadar**. Locality Afar Locality 333. Surface/*in situ* Surface. Date of discovery 1975. Finder N/A. Unit N/A. Horizon Denen Dora. Bed/member Denen Dora 2. Formation **Hadar Formation**. Group N/A. Nearest overlying dated horizon Kada Hadar Tuff. Nearest underlying dated horizon Triple Tuff 4. Geological age *c*.3.2 Ma. Developmental age Adult. Presumed sex Unknown. Brief anatomical description A partial left foot including portions of all five metatarsals, all five proximal phalanges, and the fourth and fifth intermediate phalanges. Announcement Johanson et al. 1982. Initial description Latimer et al. 1982. Photographs/line drawings and metrical data Latimer et al. 1982. Detailed anatomical description Latimer et al. 1982. Initial taxonomic allocation *Australopithecus afarensis*. Taxonomic revisions N/A. Current convention taxonomic allocation *Au. afarensis*. Informal taxonomic category Archaic hominin. Significance The first articulated foot skeleton of *Au. afarensis*. Location of original National Museum of Ethiopia, Addis Ababa, Ethiopia.

A.L. 400-1a&b Site **Hadar**. Locality Afar Locality 400. Surface. *in situ* Surface. Date of discovery 1976/77. Finder Dato Adan. Unit NA. Horizon Sidi Hakoma 2. Bed/member Sidi Hakoma. Formation **Hadar Formation**. Group N/A. Nearest overlying dated horizon Basalt. Nearest underlying dated horizon Sidi Hakoma Tuff. Geological age *c*.3.3 Ma. Developmental age Adult. Presumed sex Unknown. Brief anatomical description A mandible that includes the left

I_1–M_3 and the right I_2–M_3 (A.L. 400-1a) and a maxilla with the right canine (A.L. 400-1b). Announcement Johanson et al. 1982. Initial description White and Johanson 1982, Johanson et al. 1982. Photographs/line drawings and metrical data White and Johanson 1982, Johanson et al. 1982. Detailed anatomical description White and Johanson 1982, Johanson et al. 1982. Initial taxonomic allocation *Australopithecus afarensis*. Taxonomic revisions N/A. Current convention taxonomic allocation *Au. afarensis*. Informal taxonomic category Archaic hominin. Significance One of the most complete and best preserved mandibles of *Au. afarensis*. Location of original National Museum of Ethiopia, Addis Ababa, Ethiopia.

A.L. 417-1a–d Site **Hadar**. Locality Afar Locality 417. Surface/*in situ* Surface. Date of discovery 1990–3, 1999. Finders Dato Adan and team. Unit N/A. Horizon Sidi Hakoma. Bed/member 33 m above the Sidi Hakoma Tuff. Formation **Hadar Formation**. Group N/A. Nearest overlying dated horizon Basalt. Nearest underlying dated horizon Sidi Hakoma Tuff. Geological age *c*.3.3 Ma. Developmental age Adult. Presumed sex Female. Brief anatomical description Skull comprising the well-preserved left side of the mandible including C–M_3 (A.L. 417-1a), part of the right side of the mandible with M_2–M_3 (A.L. 417-1b), the basioccipital, basisphenoid, and right alisphenoid (A.L. 417-1c), and a maxilla including the right I^2–M^3 and the left C–M^3 (A.L. 417-1d). Announcement Kimbel et al. 1994. Initial description Kimbel et al. 1994. Photographs/line drawings and metrical data Kimbel et al. 1994, 2004. Detailed anatomical description Kimbel et al. 1994, 2004. Initial taxonomic allocation *Australopithecus afarensis*. Taxonomic revisions N/A. Current conventional taxonomic allocation *Au. afarensis*. Informal taxonomic category Archaic hominin. Significance This presumed female skull of *Au. afarensis* enabled researchers to assess the extent of cranial **sexual dimorphism** in *Au. afarensis*. Location of original **National Museum of Ethiopia**, Addis Ababa, Ethiopia.

A.L. 438-1a–v Site **Hadar**. Locality Afar Locality 438. Surface/*in situ* Surface and *in situ*. Date of discovery February 24, 1992. Finders Donald **Johanson** and team. Unit N/A. Horizon Kada Hadar-2. Bed/member 10–12 m below the BKT-2. Formation **Hadar Formation**. Group N/A . Nearcst overlying dated horizon BKT-2. Nearest underlying dated horizon KHT. Geological age c.3.0 Ma. Developmental age Adult. Presumed sex Male, based on skeletal size and robusticity. Brief anatomical description An **associated skeleton** consisting of a frontal fragment (A.L. 438-1b), the right side of the mandible (A.L. 438-1g) and a fragment of maxilla (A.L. 438-1s). Preserved dental elements consist of the right I_1, P_4 fragment, M_1, M_3 (A.L. 438-1h–k), upper molar root fragment (A.L. 438-1q), and a lower molar root (A.L. 438-1u). The forelimb fossils include a fragment of a clavicle (A.L. 438-1v), a right proximal humeral shaft and two shaft fragments (A.L. 438-1c, n, o), a left ulna (A.L. 438-1a), a right ulna shaft fragment (A.L. 438-1m), a right proximal radial fragment (A.L. 438-1l), and several metacarpals including the right MC2 (A.L. 438-1f) and the left MC2-3 (A.L. 438-1d, e). Announcement Kimbel et al. 1994. Initial description Kimbel et al. 1994. Photographs/line drawings and metrical data Kimbel et al. 1994, 2004, Drapeau et al. 2005. Detailed anatomical description Kimbel et al. 2004, Drapeau et al. 2005. Initial taxonomic allocation *Australopithecus afarensis*. Taxonomic revisions N/A. Current conventional taxonomic allocation *Au. afarensis*. Informal taxonomic category Archaic hominin. Significance The A.L. 438-1 associated skeleton has the most complete ulna of *Au. afarensis*, and it is the first specimen to possess metacarpals in association with forelimb remains. Location of original **National Museum of Ethiopia**, Addis Ababa, Ethiopia.

A.L. 444-2a–h Site **Hadar**. Locality Afar Locality 444. Surface/*in situ* Surface. Date of discovery February 26, 1992. Finder Yoel Rak. Unit N/A. Horizon N/A. Bed/member Kada Hadar-2. Formation **Hadar Formation**. Group N/A. Nearest overlying dated horizon BKT-2. Nearest underlying dated horizon Kada Hadar Tuff. Geological age c.3.0 Ma. Developmental age Adult. Presumed sex Male. Brief anatomical description Well-preserved (75–80% complete) hominin skull, including the frontal bone (A.L. 444-2d), portions of the right and left parietals (A.L. 444-2d, e, g), a fragment of the left squamous temporal (A.L. 444-2f), the posterior parts of both temporal bones (A.L. 444-2f), the occipital squama (A.L. 444-2f), the right zygomatic (A.L. 444-2c), the maxilla with the right I^1, C, and P^4–M^3 and left I^1, C, and P^3–M^3 (A.L. 444-2a), fragments of the nasal bones (A.L. 444-2h), and the right side of the

mandibular corpus and the symphyseal region with left and right incisors, the right canine, and damaged right P_4–M_1 (A.L. 444-2b). Announcement Kimbel et al. 1994. Initial description Kimbel et al. 1994. Photographs/line drawings and metrical data Kimbel et al. 1994, 2004. Detailed anatomical description Kimbel et al. 2004. Initial taxonomic allocation *Australopithecus afarensis*. Taxonomic revisions N/A. Current convention taxonomic allocation *Au. afarensis*. Informal taxonomic category Archaic hominin. Significance First complete adult skull of *Au. afarensis*. Location of original **National Museum of Ethiopia**, Addis Ababa, Ethiopia.

A.L. 666-1 Site **Hadar**. Locality Afar Locality 666. Surface/*in situ* Surface. Date of discovery November 2, 1994. Finders Ali Yesuf and Maumin Allahendu. Unit N/A. Horizon N/A. Bed/member N/A. Formation **Busidima Formation**. Group N/A. Nearest overlying dated horizon BKT-3. Nearest underlying dated horizon BKT-2. Geological age c.2.35 Ma. Developmental age Adult. Presumed sex Male. Brief anatomical description The maxilla, broken along the intermaxillary suture, with the left P^3–M^3, the right P^3–M^1 with M^2 and M^3 roots, and other isolated dental fragments. Announcement Kimbel et al. 1996. Initial description Kimbel et al. 1996. Photographs/line drawings and metrical data Kimbel et al. 1996, 1997. Detailed anatomical description Kimbel et al. 1996, 1997. Initial taxonomic allocation *Homo* sp. Taxonomic revisions *Homo* aff. *Homo habilis*. Current convention taxonomic allocation *H.* aff. *H. habilis*. Informal taxonomic category Transitional hominin. Significance A.L. 666-1 was found in the same layer as **Oldowan flakes** and **choppers**, and it is the first evidence of a *Homo habilis* facial morphotype in the middle **Pliocene**. Location of original **National Museum of Ethiopia**, Addis Ababa, Ethiopia.

A.L. 822-1 Site **Hadar**. Locality Afar Locality 822. Surface/*in situ* Surface. Date of discovery October 26, 2000. Finders Dato Adan and team. Unit N/A. Horizon Kada Hadar 1. Bed/member N/A. Formation **Hadar Formation**. Group N/A. Nearest overlying dated horizon BKT-2. Nearest underlying dated horizon KHT. Geological age c.3.1 Ma. Developmental age Adult. Presumed sex Female. Brief anatomical description A well-preserved skull that includes a well-preserved mandible with portions of the dentition preserved; most of the calvaria is preserved along with the zygomatics, and the maxilla including the right C–M³. Announcement Kimbel et al. 2003. Initial description Kimbel et al. 2003. Photographs/line drawings and metrical data N/A. Detailed anatomical description N/A. Initial taxonomic allocation *Australopithecus afarensis*. Taxonomic revisions N/A. Current conventional taxonomic allocation *Au. afarensis*. Informal taxonomic category Archaic hominin. Significance First relatively complete female skull of *Au. afarensis*. Its postcanine dentition is large compared to other inferred female australopith specimens (e.g., A.L. 417-1), and the specimen allows for a systematic evaluation of **sexual dimorphism** in *Au. afarensis*. Location of original **National Museum of Ethiopia**, Addis Ababa, Ethiopia.

^{26}Al/^{10}Be ratio *See* **cosmogenic nuclide dating**.

ALA Abbreviation of Alayla, and the prefix for fossils recovered from Alayla in the **Western Margin, Middle Awash study area**, Ethiopia.

ALA-VP-2/10 Site Alayla. Locality Alayla VP Locality 2. Surface/*in situ* Surface. Date of discovery July, 1999. Finder Yohannes **Haile-Selassie**. Unit N/A. Horizon N/A. Bed/member Asa Koma Member. Formation **Adu-Asa Formation**. Group N/A. Nearest overlying dated horizon Witti Mixed Magmatic Tuff. Nearest underlying dated horizon Ladina Basaltic Tuff. Geological age *c*.5.2–5.8 Ma. Developmental age Adult. Presumed sex Unknown. Brief anatomical description Right side of the mandibular corpus with M₃, together with associated teeth (left I₂, C, P₄, M₂, and part of the M₃ root). Announcement Haile-Selassie 2001. Initial description Haile-Selassie 2001. Photographs/line drawings and metrical data N/A. Detailed anatomical description N/A. Initial taxonomic allocation *Ardipithecus ramidus kadabba*. Taxonomic revisions *Ardipithecus kadabba* (Haile-Selassie 2004). Current conventional taxonomic allocation *Ar. kadabba*. Informal taxonomic category Possible primitive hominin. Significance **Holotype** of *Ar. kadabba*. Location of original **National Museum of Ethiopia**, Addis Ababa, Ethiopia.

albumin (L. *alba*=white) A **protein** given its name because it turns white when it is heated, or

coagulated (it is coagulated albumin that gives the non-yolk part of a cooked egg its white color). Albumin is produced by the liver, and is the most abundant protein in the clear, plasma, component of fresh blood. Its molecular weight is relatively small (67 kDa) and it only contains 585 **amino acid**s in its active form. Its main function is to attract and maintain water in the bloodstream, but it also transports hormones (e.g., thyroid hormone), fatty acids, and bilirubin. The importance of albumin for human evolution is that it was one of the first molecules to be used to measure the closeness of the relationships between the extant **great ape**s. Morris Goodman (1962) obtained albumin from modern humans, **chimpanzee**s, **gorilla**s, orangutans, and other primates, and injected it into woolly monkeys. It did not harm the monkeys, but because it was not their own albumin they reacted by generating **antibodies** to the injected **antigen** (i.e., the foreign albumin). Goodman then used the antiserum made when modern human albumin was injected into the woolly monkey to investigate the similarity of albumin antigens between pairs of taxa (e.g., modern human and chimpanzee albumins vs modern human antiserum; modern human and gorilla albumins vs modern human antiserum; chimpanzee and orangutan albumins vs modern human antiserum, etc.). When the modern human antiserum and the fresh albumins from the extant great apes other than *Pan* meet and react with the antiserum, they coagulate and form a white spur. Goodman showed that the reactions between the modern human antiserum and modern human and chimpanzee albumin were so similar there was no spur, yet when modern human antiserum was used to compare the albumins from modern humans and gorillas, there was a definite spur. In 1962 this was a novel use of immunochemistry, and along with Emil Zuckerkandl's (1960) use of chromatography to compare the structure of the hemoglobins of the extant great apes it represented the first use of biomolecules (e.g., albumin and hemoglobin) to compare the molecular affinities of the extant great apes. Jerold Lowenstein (1981) spearheaded attempts to see if it was possible to detect albumin in fossils, and then use this preserved albumin in a form of radioimmunoassay in much the same way that Goodman had used the albumins from living taxa. However, it was never possible to be sure that any albumin extracted from the fossils was undamaged, and this once promising technique has since fallen into disuse. *See also* **immunochemistry**; Morris **Goodman**.

Alcel+A↓C   *See* **alcelaphine plus antilopine criterion**.

Alcelaphinae   (L. *alces* = elk and Gk *elaphus* = deer) In this alternative interpretation of bovid taxonomy, the tribe Alcelaphini is elevated to a subfamilial level. It is a subfamily of antelopes, family **Bovidae**, which comprises hartebeest (the eponymous *Alcelaphus* sp.), wildebeest (*Connochaetes* sp.), and their allies. This subfamily likely arose during the late middle Miocene but it is extremely common in fossil assemblages by the later Pliocene. Modern alcelaphines prefer open grassland habitats and are restricted to Africa, although at least some examples (e.g., *Damalops*) have been recovered from the Asia. The relative proportion of alcelaphine fossils at hominin sites has been used as an important indicator of the presence of open habitats in the past as part of the **alcelaphine plus antilope criterion** developed by Elisabeth Vrba. *See also* **Alcelaphini**.

alcelaphine   Informal name for any antelope belonging to the subfamily **Alcelaphinae**.

alcelaphine plus antilopine criterion   (AAC or Alcel+A+C) Developed by Elisabeth Vrba, this is a method of **paleoenvironmental reconstruction** that relies on the relative abundance of **alcelaphine bovid**s and **antilopine bovid**s. Vrba found that in modern African game reserves characterized by closed habitats, the bovid tribes **Alcelaphini** and **Antilopini** make up a low percentage (<40%) of the bovid sample, whereas in more open habitats they make up a high percentage (>60%) of the bovid sample. The AAC has been widely used in paleoenvironmental studies at **Olduvai Gorge** in Tanzania and at sites in the **Blaauwbank valley** in South Africa.

Alcelaphini   A tribe of the family **Bovidae** that includes wildebeest, hartebeest, bonteboks, and their allies. **Alcelaphine** bovids are grazers with a preference for open grassland habitats and are characterized by hypsodont teeth and **cursorial** limb adaptations. Most alcelaphine bovids are of medium to large size (about 60–230 kg) and are highly gregarious, often forming large migratory herds. In attempts at the **paleoenvironmental reconstruction** of fossil assemblages, frequencies of alcelaphine bovids are often used to track the presence of open grasslands. For example, in the Turkana Basin, changes in the frequencies of fossil alcelaphines, together with other open habitat taxa, has been interpreted as evidence for an expansion of grassland environments associated with the origin of the genus *Homo*. *See also* **Alcelaphine plus Antilopine Criterion**.

Alemseged, Zeresenay (1969–)   Born in Axum, Ethiopia, Zeresenay Alemseged attended Addis Ababa University, graduating in 1990 with a BSc in geology. Thereafter he worked from 1991 to 1993 as a junior geologist in the Paleoanthropology Laboratory of the National Museum of Ethiopia in Addis Ababa. In 1993 he moved to France to enroll as a graduate student in the Institute of Evolutionary Sciences, University of Montpellier II, where in 1994 he obtained an MSc in paleontology. Alemseged stayed in France, but moved to the University of Paris VI for his PhD, which was awarded in 1998 for a thesis entitled "*A. aethiopicus-A. boisei* transition and paleoenvironmental changes in the Omo Ethiopia." Alemseged worked from 1998 to 2000 as research associate in the French Center for Ethiopian Studies in Addis Ababa, Ethiopia, then from 2000 to 2003 as postdoctoral research associate in the **Institute of Human Origins** at Arizona State University, and then from 2004 to 2008 as senior researcher in the Department of Human Evolution at the **Max-Planck Institute for Evolutionary Anthropology**, in Leipzig, Germany. Since 2008 he has held the Irvine Chair of Anthropology at the California Academy of Sciences. Zeresenay Alemseged began his investigation of the **Dikika study area** in 1999, and he and Denis Geraads reported **Acheulean** and **Middle Stone Age** tools from the Asbole area. During initial surveys the team identified extensive fossil deposits in the central Dikika area and in 2000 they recovered the first evidence of **DIK-1-1**, the most complete early hominin infant skeleton known and DIK-2-1, the earliest hominin from the **Hadar Formation**. In 2010 the same team also claimed to have discovered the earliest known evidence for tool use and meat consumption. *See also* **DIK-1-1**; **Dikika study area**.

alkenones   Complex molecules produced by a limited number of types of algae, used as **biomarker**s in **paleoclimate** research. Alkenones are well-preserved in marine sediments and first appeared in the Cretaceous. They are useful for **paleoclimate** reconstruction because their molecular structure varies with environmental conditions, with warmer **sea**

surface **temperature**s favoring the diunsaturated over the triunsaturated form. Sea surface temperature reconstructions are computed from a paleoclimate index (U^k37') which is based upon the abundance ratio of the diunsaturated to the triunsaturated form of the alkenones. Calibrations are based on *Emiliani huxleyii* which first appeared 250 ka (Pagani et al. 1999). In some lakes, including several in Africa, alkenones have been used to reconstruct lake surface temperature. The carbon isotopic composition of alkenones is also used to reconstruct past carbon dioxide concentrations in the atmosphere, beyond the range of the ice cores (*c*.800 ka). While there are large uncertainties, carbon dioxide levels appear to have been below 500 parts per million since the Oligocene.

allele (Gk *allos* = another) Refers to alternate forms of a **gene** at a specified site, or **locus**, in the **genome**, or alternate forms of a particular **DNA** sequence. For example, if you think of a locus in the genome as the "street address," the allele at that locus is analogous to the type of house present at that address. All houses serve the same purpose of providing shelter but the types of house can differ quite radically (i.e., one may be a luxury mansion while another is a modest bungalow). The genome is arranged into units called **chromosomes**, and with certain exceptions every chromosome is present as a pair in the cell; therefore, for each gene there is a pair of alleles (the exception is the X chromosome in males). The particular combinations of alleles at a locus can have significant effects on function. For example, in modern humans the S allele at the beta globin locus is protective against malaria if present with a wild-type allele (s) in the **heterozygous** form (i.e., Ss). However, if both copies of the beta globin allele are the S type, as in the **homozygous** dominant form (i.e., SS), a person will suffer from sickle-cell anemia.

Allen's Rule This concept (probably better described as a trend) is attributed to Joel Allen (1877). It states that animals living at lower average temperatures tend to have smaller appendages (i.e., shorter limbs or tails). **Neanderthal**s and some other high-latitude archaic *Homo* specimens have the type of body proportions (i.e., relatively shorter distal limb lengths and larger bi-iliac breadths) that would be predicted from Allen's Rule (Trinkaus 1981, Ruff 1994, 2002). These body proportions

appear, at least in part, to have resulted from climatic influences (Holliday 1999, Ruff 1994, 2002). This hypothesis is given support by research indicating that limb bone **robusticity** in modern humans might be more influenced by temperature than by habitual activity (Pearson 2000), although others have argued strongly that behavior affects the skeleton more strongly than climate (Finlayson 2004). Relatively early in hominin evolutionary history there is evidence of a relationship between temperature and postcranial morphology, for example the **KNM-WT 15000** associated skeleton has been interpreted as having tropical body proportions (Ruff and Walker 1993). However, it has been argued that KNM-WT 15000 shows signs of skeletal pathology (Ohman et al. 2002) and caution should be exercised before extrapolating from it to other members of the species.

Allia Bay (Location 3°35′4″N, 36°16′4″E, Kenya; etym. takes its name from the bay in Lake Turkana, just south of Koobi Fora) History and general description The site comprises an isolated set of exposures that forms the most southerly subregion of the **Koobi Fora** complex of sites. It includes a *c*.4.2 Ma "bone bed" that was most likely in a meander of a channel of the ancestral Omo River. The first fossil hominins were found in 1982. The collecting areas at Allia Bay are numbered in the same way as they are in other parts of the Koobi Fora site complex, with the numbering starting at 200. Fossil specimens from the site are given the **KNM-ER** prefix. All the fossil hominins from Allia Bay have been assigned to *Australopithecus anamensis*, although in some respects the fossils from this site resemble younger specimens attributed to *Australopithecus afarensis* more than they do conspecifics from the older site of **Kanapoi**. Temporal span All the hominin specimens come from *c*.3.9–4.1 Ma sediments beneath or within the Moiti Tuff which has been dated to 3.97±0.034 Ma (McDougall and Brown 2008). A hominin radius, KNM-ER 20419, was initially described as having been found at a younger locality "east of Allia Bay" (Heinrich et al. 1993, p. 139). However, this specimen's location has since been reported as being "Sibilot" (Ward et al. 1999b, p. 198). How dated? **Argon-argon dating**, **potassium-argon dating**, and **magnetostratigraphy**. Hominins found at site Thirteen hominin specimens found between 1982 and 1995 were described by Leakey et al. (1995); a further 11 found between 1995 and 1997 were announced by Leakey et al. (1998). Archeological evidence found at site None. Key references: Geology and dating Heinrich et al. 1993, Leakey et al. 1995, 1998; Hominins Heinrich et al. 1993, Leakey et al. 1995, 1998, Ward et al. 1999b.

allocortex (Gk *allos* = other and L. *cortex* = bark, thus the "outer covering" of the cerebral cortex) Refers to the parts of the cerebral cortex that have fewer than six layers of cells. Allocortical areas are concerned with the processing of olfaction (i.e., the piriform cortex, which is also called the paleocortex), as well as memory and spatial navigation (i.e., the hippocampus, which is also called the archicortex). *See also* **cerebral cortex**.

allometric *See* **allometry**.

allometry (Gk *allos* = other and *metron* = measure) Towards the end of the 19thC several researchers (e.g., Dubois 1898, 1914, Lapicque 1898, Snell 1891) recognized that there is a predictable relationship between brain size and body size. In the early part of the 20thC, Pezard (1918) and Huxley (1924) introduced the terms "heterogonic" and "constant differential growth," respectively, for this relationship. Huxley (1927, 1931) went on to show that the phenomenon was widespread, and that such a relationship could apply to dimensions of the same individual at different stages in its development, or to the same dimensions of different individuals of the same species at the same or different stages of development. Subsequently, Huxley and Tessier (1936) proposed the term allometry for the study of the growth or size of one part of an organism with respect to the growth or size of the whole (or another part which is taken as a **proxy** for the whole) of the same organism. Today, the term allometry is used in two senses, which can be a source of confusion. Allometry is often used to refer generally to the study of the "consequences of differences in size." In this sense, allometry is equivalent to the term scaling. However, allometry can also be used in a more specific sense to refer to changes in shape of part, or the whole of an organism, that are associated with changes in the overall size of the organism. When a variable increases in size more slowly than body size, this is called **negative allometry** (i.e., the variable becomes proportionally smaller as overall body size increases). The term used when a variable increases in size more quickly than body size is **positive allometry** (i.e., the variable becomes proportionally larger as overall body size increases).

In both negative and positive allometry any change in size will result in a change in **shape**. When used in this sense, the opposite of allometry is **isometry**, which refers to examples when shape is maintained as size increases. In other words, an isometric variable increases in size at the same rate as body size. *See also* **scaling**.

allopatric *See* **allopatry**; **speciation**; **vicariance biogeography**.

allopatric speciation A mode of speciation in which new species evolve as a consequence of the original species population being divided by a geographic barrier. The resulting physical isolation leads to a loss of gene flow, and the accumulation of genetic differences in the new populations is due to **genetic drift**, **natural selection**, and **mutation**. Eventually, these genetic differences lead to reproductive isolation, such that even if the physical barrier between the populations disappears they remain genetically isolated. Note that peripatric speciation can be thought of as a special case of allopatric speciation.

allopatry (Gk *allos* = other and *patris* = fatherland) When two organisms have geographic ranges that are entirely separate and distinct, and do not overlap anywhere. Given the nature of the fossil record it is difficult to be certain which hominin species were truly allopatric, but, for example, *Australopithecus africanus* (southern Africa) and *Australopithecus afarensis* (East Africa) were probably allopatric. Allopatric speciation, in which a single species splits into two separate species after a subsection of the ancestral population is geographically and reproductively isolated, is thought to be the most common cause of speciation. Geographic barriers that might promote allopatric speciation include mountain ranges, rifts formed by plate tectonics, substantial rivers, and rising sea levels that form islands out of areas that were previously connected. Forest fragmentation may also cause geographic isolation and hence allopatric speciation. It is likely that allopatric speciation was the mechanism by which many hominin species arose. *See also* **speciation**; **vicariance biogeography**.

alluvial (L. *alluere* = to wash against) Nonmarine sediments deposited by water that is flowing. If there is evidence to attribute the sediments to a more specific depositional mechanism (e.g., fluvial, lacustrine, etc.) then the term alluvial should be avoided. *See also* **riverine**.

alpha taxonomy (Gk *alpha* = first, and *taxis* = to arrange or "put in order") Mayr et al. (1953) is one of the few sources that defines the terms "alpha" (and "beta" and "gamma") taxonomy; for example, these terms do not appear in the index of Simpson's (1961) *Principles of Animal Taxonomy*. Mayr et al. (1953) suggest that "the taxonomy of a given group, therefore, passes through several stages. . .informally referred to as alpha, beta, and gamma taxonomy" (*ibid*, p. 19). The distinction made by Mayr et al. is that alpha taxonomy is about species being "characterized and named," beta taxonomy involves arranging species in "a natural system of lesser and higher categories," and gamma taxonomy involves the "analysis of intraspecific variation" (*ibid*, p. 19). If you subscribe to the philosophy that the reconstruction of phylogeny in the form of a **cladogram** should inform how higher-order taxa are arranged in a classification, then what alpha taxonomy refers to is taxonomy minus **phylogeny reconstruction**, and what Mayr et al. (1953) called beta taxonomy is the process of phylogeny reconstruction. But Mayr et al. (1953) also caution that it is "quite impossible to delimit alpha, beta and gamma taxonomy sharply from one another, since they overlap and intergrade" (*ibid*, p. 19). While it is common for reference to be made to alpha taxonomy, you very seldom read, or hear, any reference to beta or to gamma taxonomy. *See also* **systematics**; **taxonomy**.

Altamura (Location 40°52′21″N, 16°35′17″E, Italy; etym. named after Altamura, a nearby town) History and general description The hominin partial skeleton including a cranium known as Altamura 1 was found in 1993 in a cavity in the Grotta di Lamalunga, a karstic limestone cave near Altamura, Bari, Apulia, in southern Italy. The bones are covered in a variably thick coating of calcareous material. Temporal span and how dated? Biostratigraphic evidence points to a Late Pleistocene date. Hominins found at site **Altamura 1**, most likely belonging to *Homo neanderthalensis*. Archeological evidence found at site None. Key references: Geology and dating N/A; Hominins Pesce Delfino and Vacca 1993, Vacca and Pesce Delfino 2004; Archeology N/A.

Altamura 1 Site **Altamura**. Locality Grotta di Lamalunga. Surface/*in situ* *In situ*. Date of discovery October 1993. Finders Members of Centro Altamurano Ricerche Speleologiche (C.A.R.S.), the local speleological society and Eligio Vacca of the University of Bari. Unit N/A. Horizon N/A. Bed/member N/A. Formation N/A. Group N/A. Nearest overlying dated horizon N/A. Nearest underlying dated horizon N/A. Geological age Estimated to be **Middle Pleistocene**. Developmental age Adult. Presumed sex? Male. Brief anatomical description Fairly complete skeleton covered with flowstone. Announcement Pesce Delfino and Vacca 1993. Initial description Pesce Delfino and Vacca 1993. Photographs/line drawings and metrical data Pesce Delfino and Vacca 1993, 1996, Vacca and Pesce Delfino 2004. Detailed anatomical description N/A. Initial taxonomic allocation *Homo neanderthalensis*. Taxonomic revisions N/A. Current conventional taxonomic allocation *H. neanderthalensis*. Informal taxonomic category Pre-modern *Homo*. Significance The skeleton purportedly exhibits a mix of Neanderthal and anatomically modern human characteristics, but it is difficult to be sure because its detailed morphology is obscured by a thick layer of **flowstone**. Location of original The skeleton is still *in situ* in Grotta di Lamalunga except for a portion of the scapula that has been removed for DNA analysis.

altricial (L. *alere* = to nourish) The whereabouts of taxa along the altricial–**precocial** spectrum depends upon the state of the newborn. Taxa with newborn that are at a relatively early stage of development at the time of birth are called altricial. Altricial offspring usually have their eyes and ears closed, and they lack fur or feathers and they possess little to no ability to move independently. They are thus reliant on parental units for varying lengths of time after birth for temperature regulation, food, and transport. For instance, domestic cats are altricial because newborn kittens are unable to move on their own, and they rely on the mother to clean them, transport them, and direct them to the nipple. Compared to nonhuman primates, most of which are considered relatively precocial at birth, modern humans are described as secondarily altricial (Portmann 1941) for modern human babies require intensive parental care. The various manifestations of modern human newborn altriciality (e.g., poor temperature regulation, reliance on parents for feeding, poor motor control and coordination) are thought to have evolved as a consequence of a combination of a relatively large **neonate** and a relatively large neonatal brain size that caused the length of gestation to be reduced to increase the likelihood of a successful birth through the modern human birth canal. The high "catch-up" rates of brain growth during the first year of postnatal life are also likely to be consequence of this strategy.

Alu The name given to a **restriction enzyme** cut site with the sequence AGCT. This particular cut site was named "Alu" because it has a sequence recognized by an endonuclease (an enzyme that acts like scissors on specific DNA sequences) isolated from the bacterium *Arthrobacter luteus*. *See also* **Alu repeat elements**.

Alu elements *See* **Alu repeat elements**.

Alu repeat elements A family of short interspersed nucleotide elements (or **SINEs**) of **DNA**, and they are common in all primates, including the **great apes** and modern humans. Each Alu repeat element is approximately 300 **base pairs** (bp) in length, and the origin of **Alu** elements has been traced back to the 7SL RNA gene. Alu repeat elements are a class of retrotransposons (i.e., sequences that are transcribed from DNA to mRNA and then the mRNA is copied back into DNA, which is inserted elsewhere in the genome) and were originally named for the Alu restriction enzyme cut site (with the sequence AGCT) that is typically found within the element. Alu elements account for as much as 10% of the modern human genome (Smit 1996) but less than 0.5% vary (i.e., they are **polymorphic**). Alu repeat elements can help regulate the **transcription** of the DNA sequence by binding regulatory proteins, and they influence evolution via unequal **recombination** of chromosomes and **gene duplication**. They are useful for phylogenetic analyses and for studies of population history because (a) the insertion of an Alu element has a known ancestral state (i.e., no Alu insertion), (b) each Alu insertion is almost certainly homologous since the probability of two insertions at the same location within the genome is very small, (c) Alu insertions are stable, and (d) they are easy to analyze (Batzer et al. 1996, Sherry et al. 1997, Stoneking et al. 1997, Watkins et al. 2001). The most recently active Alu subfamilies in modern humans are the Ya5 and Ya8 subfamilies. Alu elements that are polymorphic within modern humans are especially useful for studies of population history while fixed Alu insertions are useful for phylogenetic studies within primates (Salem et al. 2003).

Alus *See* **Alu repeat elements**.

alveolar process (L. *alveolus* = small hollow, dim. of *alveus* = hollow, and *processus* = to go forward or advance) The inferior part of the upper jaw (i.e., the maxilla) and the superior part of the body or corpus of the lower jaw (i.e., the **mandible**). The maxillary and the mandibular alveolar processes are where the roots of the upper and lower teeth, respectively, are embedded.

alveoli Plural of **alveolus** (*which see*).

alveolus (L. *alveolus* = small hollow, dim. of *alveus* = hollow) The name for the socket in the alveolar process of the maxilla or **mandible** into which the root of a tooth is embedded. The plural of alveolus is alveoli.

AMA Prefix for fossils recovered from the Am-Ado collecting area, **Woranso-Mille study area**, **Afar Rift System**, Ethiopia. *See* **Am-Ado**.

Am-Ado Collecting area initially identified by the **International Afar Research Expedition**, but now within the **Woranso-Mille study area**, **Afar Rift System**, Ethiopia.

Amba East Collecting area within the **Central Awash Complex**, **Middle Awash study area**, Ethiopia.

Amboseli live/dead study This is an elegant use of contemporary data to validate the assumption that bone assemblages at fossil sites are an accurate reflection of the types of animals that lived in and around that site. Two researchers, David Western and Kay Behrensmeyer, collected live/dead census data (i.e., census data for live animals and for animal carcasses in the same location) in the Amboseli National Park across four periods ranging from 5 to 6 years. During each of the four periods, and across the four decades that separated the beginning of the first period and the end of the last, a period that showed "rapid ecological change" (Western and Behrensmeyer 2009, p. 1063), they found highly significant correlations for variables such as **relative taxonomic abundance** between the live and dead data sets. They also found similar values for "standard ecological measures of community structure" (*ibid*, p. 1063). Thus, the study suggests that as long as post-depositional taphonomic biases are taken into account, then at least in tropical settings fossil animal **assemblage**s can be used to make inferences about paleohabitats. *See also* **paleoenvironmental reconstruction; taphonomy**.

Ambrona (Location 41°09′34″N, 2°29′55″W, Spain; etym. named after a nearby town) History and general description This site lies along the Masegar River, and it, along with the nearby site of **Torralba**, was originally excavated by Clark **Howell** and Les Freeman in the early 1960s. They believed the sites to be contemporaneous, and presented their finding of many faunal remains, particularly from elephants, and a quantity of stone tools, as the earliest evidence of hunting. Later researchers, including Lewis **Binford** and Richard Klein, argued that the faunal profiles from these sites suggested scavenging, not hunting. The most recent analysis by Santonja and Pérez-González has shown that, first, Ambrona is significantly older than Torralba, and second, there is a mix of natural and human components to the site, and thus too little evidence of actual interaction between the hominins and fauna that it is impossible to say anything more than some butchery occurred at the site. Temporal span and how dated? Based on the **uranium-series dating** of a nearby river terrace, and on fauna and stratigraphic correlation, the site is thought to be from **Oxygen Isotope Stage** 12 (*c.*470–430 ka). Hominins found at site None. Archeological evidence found at site Large quantities of faunal remains, a very few of which have cutmarks, and a few stone tools. Key references: Geology, dating, and paleoenvironment Villa et al. 2005, Freeman 1994, Villa et al. 2005; Hominins N/A; Archeology Freeman 1994, Villa et al. 2005.

AME Prefix for fossils recovered from **Amba East, Central Awash Complex, Middle Awash study area**, Ethiopia. *See* **Amba East**.

ameloblast [obsolete *amel* = enamel (ME, ultimately from OF *esmail*) and Gk *blastos* = germ] The name given to secretory and maturational (i.e., functional) enamel-forming cells. Ameloblasts are derived from the inner enamel epithelium. As they mature, they become elongated, with their long axis at approximately right angles to the future **enamel-dentine junction** (or EDJ). During enamel formation, secretory ameloblasts move away from the EDJ and secrete enamel matrix from the Tomes' process located on the end of the cell facing the EDJ. The secreted matrix forms elongated **enamel prisms** approximately 5 μm in diameter. Think of the way a long bead of toothpaste is extruded from a toothpaste tube when the latter is squeezed. The free end of the bead is at the EDJ and the longer the tube is squeezed the further away from the EDJ the tube moves. Secretory ameloblasts cease to lay down enamel matrix when the final thickness of enamel is completed. They then switch function and become maturational ameloblasts. These alternately remove water and degraded proteins or pump in calcium to facilitate the final mineralization of enamel. Finally, ameloblasts become the reduced enamel epithelium that covers the crown and then lies dormant until the tooth erupts, after which it is shed. **Short-period** and **long-period incremental lines** produced by ameloblasts are believed to represent interruptions in the secretion or mineralization of the matrix. *See also* **enamel development; incremental features**.

amelogenesis [obsolete *amel* = enamel (ME, ultimately from OF *esmail*) and Gk *genesis* = birth or origin] The process of enamel formation by **ameloblasts**. *See also* **enamel development**.

American Institute of Human Paleontology In 1949 the Viking Fund was instrumental in the establishment of this institute. Its aim was to provide a forum to increase knowledge of early humans, and the founding members were Loren C. Eiseley (President), Joseph Birdsell, Paul Fejos, Theodore McCown, Hallam Movius, Dale Stewart, and Sherwood **Washburn**. An important accomplishment of the Institute was to acquire, with financial support from the Wenner-Gren Foundation for Anthropological Research, the Barlow/Damon collection of molds of hominin fossils. The collection was deposited with the University of Pennsylvania Museum of Archaeology and Anthropology and it provided the basis of that museum's hominin casting program. The Barlow/Damon molds were at the Penn Museum from 1952 to 1964 when they were returned to the the Wenner-Gren Foundation for Anthropological Research and put into storage. They were sent back to the University of Pennsylvania Museum of Archaeology and Anthropology in 1980 and they remain there. During the 1950s the Wenner-Gren Foundation supported the research of the plastics engineer, David Gilbert, to develop a new and highly accurate molding technique. This led to the establishment of the Anthrocast program in 1965. The first molds were made in the field in 1962 and molding continued until 1974. Between 1968 and 1976 Anthrocast provided over 16,000 replicas of 180 different fossil specimens to institutions and researchers worldwide. The Anthrocast molds are currently curated in the University of Pennsylvania Museum of Archaeology and Anthropology and casts continue to be available on a limited basis through their casting program. *See also* **Viking Fund**; **Wenner-Gren Foundation for Anthropological Research**.

American Museum of Natural History   (or AMNH) In 1869 Albert Smith Bickmore was successful in his proposal to create a natural history museum in New York City. He gained the support of William E. Dodge, Jr, Theodore Roosevelt, Sr, Joseph Choate, and J. Pierpont Morgan. On April 6, 1869 the museum became a reality when the Governor of New York, John Thompson, signed a bill creating the American Museum of Natural History. The collections relevant to paleoanthropology are the extensive collections of northwest coast Native Americans as well as great ape remains. There are 11 *Gorilla gorilla beringei* skeletal specimens all of which are from Zaire (present-day Democratic Republic of the Congo). Most of the 68 *Gorilla gorilla gorilla* skeletal specimens come from the Central African Republic or Cameroon; one comes from the Congo and the rest are unlabelled. None of the 21 skeletal specimens labeled *Gorilla gorilla* have any information about location; there is one *Gorilla* sp. The AMNH also holds the remains of 86 specimens of *Pongo pygmaeus*: 44 of these comprise skeletal remains with the remaining samples comprising soft tissues. The records simply list "Borneo" as the collection location. The great apes are part of the mammals collection. Contact Department of Mammalogy. Tel +1 202 769 5474. E-mail mammvisits@amnh.org.

AMH   Acronym for **anatomically modern humans**. *See Homo sapiens*.

amino acid   (Gk *ammoniacos* = the pungent resin that is the source of ammonia, NH_3, was first collected from near the temple of Amen in Libya) Relatively small **molecules** that are the components of proteins. There are 20 different standard amino acids. Chemically, amino acids are distinctive in having amine (NH_2) and carboxyl (COOH) groups. If they are attached to the alpha carbon they are called alpha amino acids, and these are the ones that make up the larger protein molecules. Differences in the side chains among amino acids determine their properties (i.e., whether they are acids or bases, hydrophilic or hydrophobic). Amino acids are transported by specific **transfer RNAs** (or tRNAs) and they are joined together in a sequence encoded by messenger **RNA** (mRNA) to form a **polypeptide** chain. The latter process occurs in a reaction catalyzed by **ribosome**s and is referred to as translation. Proteins consist of one or more of these polypeptide chains. *See also* **protein**; **translation**.

amino acid racemization   Amino acids exist in two **antimeric** forms, a "right-handed" or D-form, and a "left-handed" or L-form. When proteins are assembled in cells the component amino acids are all in the L-form, but they convert at a predictable rate by a process called racemization to the D-form. Racemization is also known as **epimerization**. This apparently regular and predictable process has been used as a **molecular clock** for dating. *See also* **amino acid racemization dating**.

amino acid racemization dating   (AARD) The apparently regular and predictable process of **amino acid racemization** has been used as a **molecular clock** for dating, but because the process proved to be temperature-dependent, the dates were found to be unreliable, and the method fell into disuse. Recently, the principle has been revived and applied to the epimerization of isoleucine, an amino acid preserved within the calcite crystals of ostrich eggshell to develop **ostrich egg shell dating** but the problem of temperature-dependency persists.

AMOC   *See* **Atlantic Meridional Overturning Circulation**.

AMS   *See* **accelerator mass spectrometer**.

AMS radiocarbon dating   First used in 1977, this is a direct ion-counting method of radiocarbon dating. It uses a particle accelerator to produce ions with a $^{+3}$ charge state and in this state many of the ions that interfere with the counting of ^{14}C can be easily removed, thus allowing the accurate measurement of very low concentrations of ^{14}C. Because this method, called **accelerator mass spectrometry** (or AMS), enables the direct measurement of individual ^{14}C atoms much smaller samples can be processed than with other techniques; AMS can routinely date samples of 1 mg of carbon. This means that previously undateable samples, such as single hominin teeth and individual grains of domesticated cereals, can now be dated. Typical starting weights required for AMS (e.g., 10–20 mg of seed/charcoal/wood, 500 mg of bone, 10–20 mg of shell carbonate) are about 1000 times less than the weights required by conventional counting systems. The AMS method also allows for more thorough chemical pretreatment of samples, this is particularly important for older samples (>25 ka BP) where small amounts of modern carbon contamination may have a large effect on the measured ^{14}C fraction and hence the date. *See also* **radiocarbon dating**.

Amud (Location 32°52′N, 35°30′E, Israel; etym. named after the Wadi Amud) History and general description Amud cave is approximately 5 km northwest of the Sea of Galilee, and is situated in a steep cliff face above the Wadi Amud. Excavations sponsored by the University of Tokyo began in the early 1960s under H. Suzuki. The researchers quickly found **Amud 1**, a fairly complete but poorly preserved adult, presumed male, *Homo neanderthalensis*, and further excavations by the team recovered fragments of at least three other individuals. The lithic evidence and the morphological affinities of the hominin remains suggested the site represents the transition between the **Middle** and **Upper Paleolithic**. Work was resumed in 1991 by a joint Israeli-American team led by Erella Hovers, Yoel Rak, and William Kimbel. The results of these excavations include a greatly expanded understanding of the lithic industry, the identification and study of fire places and fire-related behavior, the secure dating of the site, and more hominin remains, including an associated skeleton of a neonate, **Amud 7**, that may have been intentionally buried. Temporal span Early comparative analysis of the **Mousterian** lithic industry gave a date of *c*.30 ka, but recent thermo**luminescence dating** on a number of burned lithic artifacts for the various stratigraphic horizons indicates two occupation events, dated to 70 and 55 thousand years ago. A smaller sample of age estimates by electron spin resonance yielded comparable dates. The hominin remains are associated with the younger ages. How dated? Electron spin resonance spectroscopy dating, and thermoluminescence dating. Hominins found at site The remains of 18 hominins have been found, of which seven appear to be under the age of 2 years. They include a fairly complete skeleton of an adult male (Amud 1) and the skeleton of an infant (Amud 7). Archeological evidence found at site The Middle Paleolithic layer, Level B, has been divided into four stratigraphic units, all of which contain lithic assemblages roughly equivalent, but not identical, to those found at Tabun B. Hovers (1998) argues that the dates and appearance of the assemblages from these layers at Amud show that there was not a single sequence of technological change within the Levantine later Middle Paleolithic, as suggested from the archeological sequence at **Tabun**. **Phytolith** studies from sediment samples collected at the site provide evidence of the use of several kinds of plants and suggest that dates, figs, and grass seeds may have been part of the diet of the Amud hominins. Faunal evidence suggests hunting and transport into the site of mostly gazelle and fallow deer. Lithic production activities were differentially organized in the cave's space. Fire places were common features and ashes constitute the majority of the sediment bulk. Key references: Geology and dating Suzuki and Takai 1970, Grün and Stringer 1991, Schwarcz and Rink 1998, Valladas et al. 1999, Rink et al. 2001. Hominins Suzuki and Takai 1970, Rak et al. 1994, 1996, Hovers et al. 1995, Arensburg and Belfer-Cohen 1998; Archeology Suzuki and Takai 1970, Hovers 1998, Hovers et al. 2000, Madella et al. 2002, Rabinovich and Hovers 2004, Alperson-Afil and Hovers 2005, Hovers 2007, Shahack-Gross et al. 2008.

Amud 1 Site Amud. Locality N/A. Surface/*in situ* *In situ*. Date of discovery July 1961. Finders H. Suzuki and others. Unit B1/6. Horizon N/A. Bed/member N/A. Formation N/A. Group N/A. Nearest overlying dated horizon N/A. Nearest underlying dated horizon N/A. Geological age 55 ka. Developmental age Adult. Presumed sex Male. Brief anatomical description A nearly complete but poorly preserved adult skeleton. Announcement Vallois 1962. Initial description Suzuki and Takai 1970. Photographs/line drawings and metrical data Suzuki and Takai 1970. Detailed anatomical description Suzuki and Takai 1970. Initial taxonomic allocation Due to what was seen as its combination of Neanderthal and modern human features, Suzuki (1970) interpreted the skeleton as being intermediate between the **Tabun** and **Shanidar** specimens on the one hand, and the more clearly modern human-like **Skhul** and **Qafzeh** specimens on the other. Taxonomic revisions Hovers et al. (1995) cite a number of cranial and mandibular synapomorphies as evidence that the remains should be assigned to *Homo neanderthalensis*, but Arensburg and Belfer-Cohen (1998) rejected that proposal. Current conventional taxonomic allocation *H. neanderthalensis*. Informal taxonomic category Pre-modern *Homo*. Significance At 1740 cm³, Amud 1 appears to have the largest cranial capacity of any known hominin. Location of original Department of Anatomy, Tel Aviv University, Israel

Amud 7 Site Amud. Locality **Middle Paleolithic** layers. Surface/*in situ* In situ. Date of discovery 1992. Finders Excavation team led by Yoel Rak, William Kimbel, and Erella Hovers. Unit Layer B2/8, square K3a. Horizon N/A. Bed/member N/A. Formation N/A. Group N/A. Nearest overlying dated horizon N/A. Nearest underlying dated horizon N/A. Geological age 50–60 ka. Developmental age Neonate (*c.*10 months). Presumed sex Unknown. Brief anatomical description From the cranium, only the occipital, fragmentary parietals, nearly complete mandible, and a few maxillary teeth remain. Several postcranial elements are preserved, including thoracic and limb elements. Announcement Rak et al. 1994. Initial description Rak et al. 1994. Photographs/line drawings and metrical data Rak et al. 1994, 1996. Detailed anatomical description N/A. Initial taxonomic allocation *Homo neanderthalensis*. Taxonomic revisions N/A. Current conventional taxonomic allocation *H. neanderthalensis*. Informal taxonomic category Pre-modern *Homo*. Significance The remains of this young individual preserve several morphological features that have been put forward as *H. neanderthalensis* autapomorphies (e.g., an oval foramen magnum and a medial pterygoid tubercule). Location of original Department of Anatomy, Tel Aviv University, Israel.

Amudian (etym. After the site of **Amud**, Israel) A stone tool industry found in the Near East that appears in the late **Middle Paleolithic** or early **Upper Paleolithic**. It is characterized by backed blades and **Levallois**-style flake production.

amygdala (L. *corpus amygdaloideum* from the Gk *amygdale* = almond) A small almond-shaped complex of neurons in the medial temporal lobe of the **cerebral hemispheres**, located at the tip of the inferior horn of the lateral ventricle, near the tail of the caudate nucleus. It is part of the **limbic system**, and its major function is to attach emotional valence to learning and to consolidate **memory**. The amygdala is reciprocally connected to the **hypothalamus**, **hippocampus**, **neocortex**, and **thalamus**. It also coordinates the actions of the autonomic nervous system and the endocrine system. Barger et al. (2007) showed that the lateral nucleus of the amygdala in modern humans is larger than would be expected based on **allometric** scaling to overall **brain size** in living apes. This relative increase in the volume of the lateral nucleus of the amygdala suggests there has been some reorganization of the amygdala's connections with the orbitofrontal and temporal parts of the **cerebral cortex** in the course of human evolution.

AN Acronym for Abdur North. *See* **Abdur**.

anagenesis (Gk *ana* = up and *genesis* = birth or origin) An evolutionary pattern (or **mode**) in which an ancestral **species** evolves into a descendant species without lineage splitting. Anagenesis is the alternative to **cladogenesis**. For example, it has been claimed that *Australopithecus anamensis* and *Australopithecus afarensis* are time-successive species in the same lineage (Kimbel et al. 2006) in which case the relationship between them is anagenetic.

anagenetic *See* **anagenesis**.

anagenetic unit *See* **grade**.

analogous (Gk *analogos* = resembling, from *ana* = according to and *logos* = ratio) A trait (i.e., structure, gene, or developmental pathway) in two or more taxa that was *not* inherited from the most recent common ancestor. Analogous traits have similar functions, but not necessarily a similar structure. The eyes of vertebrates and cephalopods (e.g., octopuses) are an example of an analogy. They perform the same basic function, but in the former the retina is inverted (i.e., the receptors are towards the back of the retina, and the nervous connections are deep to them so that the light has to traverse the nervous connections before it reaches the receptors) whereas in cephalopods it is not (i.e., the receptors are towards the front of the retina, and the nervous connections are superficial, so that the light does not have to traverse the nervous connections before it reaches the receptors). *See also* **analogue**; **homologous**; **homoplasy**.

analogue (Gk *analogos* = resembling) An example that serves as an illustration of an organism without being closely related to it. For example, the differences between the masticatory systems of bears and pandas, animals that are distantly related to hominins, has been compared to the differences between the masticatory system of *Australopithecus africanus* or *Australopithecus afarensis*, on the one hand, and that of *Paranthropus robustus* or *Paranthropus boisei*, on the other. In this case pandas serve as an analogue for *P. robustus* and *P. boisei*. *See also* **homoplasy**; **homology**.

analysis of covariance (or ANCOVA) A variant of **multiple regression** in which a continuous variable is dependent on continuous and categorical variables (where the categorical variables are converted to binary dummy variables). It is typically used to determine whether the slopes and/or intercepts of scaling relationships between continuous variables differ between groups. For example, if cranial capacity and body mass is known for samples of individuals belonging to three different species, ANCOVA can be used to identify whether a significant difference exists between the three species in the scaling relationship between cranial capacity and body mass. The first step determines whether there is a significant interaction between the continuous and categorical independent variables (i.e., do the scaling slopes differ between groups). A significant result indicates the slopes differ significantly from each other, and no further analysis is conducted. A non-significant result indicates that the scaling slope cannot be distinguished statistically between the groups. In that case the next step determines whether the common scaling slope differs significantly from zero, and whether the intercepts differ between groups (e.g., whether cranial size is larger in one species than another at any given body size).

analysis of variance (or ANOVA) A statistical test most commonly used to determine whether there is a significant difference in the mean of a continuous variable between two or more groups. For example, if cranial capacity is known for samples of crania belonging to three different species, ANOVA can be used to identify whether a significant difference exists between the three species in mean cranial capacity. A non-significant result indicates that the means of all groups cannot be said to differ from each other significantly. A significant result indicates that the null hypothesis that all means are equal is incorrect; however, when more than two groups are included in the analysis, a significant result does not indicate which group or groups differ from the others. Typically this question is addressed using a series of pairwise tests in which an ANOVA is performed on each possible pair of groups from the total number of groups to identify which groups differ significantly in their means. Results from an ANOVA performed for two groups (as opposed to three or more) are equivalent to the results of a *t* **test**. ANOVA is a **parametric statisti**cal test; an equivalent **non-parametric statisti**cal test is the **Kruskal–Wallis test**. ANOVA is appropriate for a single continuous variable, but an extension to the multivariate case is also available. *See also* **multivariate analysis of variance**.

anatomical position The position of the body that is used as a reference when describing the surfaces of the body, the spatial relationships of the body parts, and the movements of the axial and postcranial skeleton. The anatomical position assumes the individual is standing, looking forward, with their legs and feet together, with their arms by their side, and with their palms facing forward. The plane that divides the body into right and left halves is the sagittal plane; it is also called the midline. All the surfaces that face towards the front when an individual is in the anatomical position are called anterior or ventral, and all the surfaces that face towards the back are called posterior or dorsal. Superior is nearer to the crown of the head; inferior is nearer to the soles of the feet. Medial is nearer to the midline of the body; lateral is further from the midline of the body. With respect to the limbs, proximal is in the direction of the root of the limb and distal is in the direction of the tips of the fingers or toes. Moving a whole limb forwards is to flex it; moving a whole limb backwards is to extend it. Moving a limb away from the body is to abduct it and moving it back towards the midline is to adduct it.

These latter terms also apply to movements of the fingers and toes, except that the reference digit of the hand is the middle finger and the reference digit of the foot is the second toe.

anatomical terminology The first serious systematic study of modern human anatomy in the Western tradition was undertaken by the Greek physicians Herophilus (335–280 BCE) and Erasistratus (304–250 BCE) and the first recorded human dissections were carried out at the Museum and Library of Alexander the Great, in what is now Alexandria, Egypt. However, by the time of Galen, physician to the Roman imperial court in the 2ndC AD, the practice of human dissection had effectively ceased. Its revival was largely due to the efforts of a young anatomy teacher called Andreas Van Wesele (who is better known by the latinized version of his family name, Andreas Vesalius). His seven-volume treatise *De Humani Corporis Fabrica Libri Septem* ("On the Fabric of the Human Body", or just the *Fabrica*) must rank among the foremost contributions to biology. In the Preface to the *Fabrica* Vesalius chronicles his dissatisfaction with his own anatomy teaching and writes that it was during a visit to Bologna in 1540, when he was given the opportunity to compare the skeletons of a monkey and a modern human, that he realized that much of the "human" anatomy described in the Galenic texts was based on monkey and not human anatomy. It was this appreciation that stimulated Vesalius to embark on the dissections on which the *Fabrica* is based. The latter comprises seven "books" totaling approximately 660 pages, or 400,000 words, plus excellent illustrations made by apprentices working in the studio of the painter Tiziano Vecellio, better known as Titian. The *Fabrica* was published in 1543, when Vesalius was still only 29; the scale and pace of Vesalius' achievement is without parallel in biology. This historical preamble explains why the original anatomical terms were Greek, and then why they were subsequently supplemented with Latin terminology. It is important that anatomical terminology transcends the barriers of language so at an early stage it was decided informally that any new anatomical terms should be based on Latin. This policy was formalized in 1895 when the first Basle edition of the *Nomina Anatomica* was produced. In 1950 responsibility for monitoring and maintaining human anatomical nomenclature passed to the International Anatomical Nomenclature Committee (IANC), and in 1989 it was transferred to the Federative Committee on Anatomical Terminology (FCAT). This body, under the aegis of the International Federation of Associations of Anatomists (IFAA), is now responsible for modern human anatomical terminology. The latest version of official modern human anatomical terminology is in a book called *Terminologia Anatomica* (1998). Two versions of each anatomical term are given in the *Terminologia Anatomica*: the Latin version and the approved English language version. English is the official language of anatomical terminology (just as it is the official language for air traffic control). Thus, the English-language version of *corpus mandibulae* is the "body of the mandible" and the English-language version of *facies articularis capitis fibulae* is the "articular facet on the head of the fibula." Some terms are the same in both Latin and English (e.g., sternum, os centrale, tibia, sustentaculum tali). Many anatomical terms were based on the everyday Latin (and sometimes Greek) vocabulary. Thus, the "cup-like" articular surface of the hip joint on the pelvis is called the acetabulum because Pliny thought it resembled a Roman vinegar (*acetum*) receptacle (*abrum*), and the condylar process of the mandible takes its name from the Greek word for a "knuckle". Researchers sometimes use the Latin versions of terms, but if you choose to use *os coxae* instead of the English language terms hip bone or pelvic bone, then logically you should use *os scaphoideum* instead of the scaphoid and *caput ulnae* instead of the head of the ulna. It is best to be consistent and use the English-language term(s) listed in the *Terminologia Anatomica* rather than continuing to use arcane terminology.

ancient DNA (aDNA) (OF *ancien* from the L. *ante*=before and DNA=deoxyribonucleic acid) **Deoxyribonucleic acid**, or DNA, that is extracted from old or poorly preserved bone, teeth, hair, tissue, or coprolites. The analysis of DNA extracted from archeological and paleontological materials is a relatively new area of research that was made possible by the technological revolution in genetics that began in the 1980s. The first experiments to determine whether DNA survived in ancient material used dried tissue, such as skin from a quagga (an extinct form of the plains zebra; Higuchi et al. 1984) and a 2400-year-old Egyptian mummy (Pääbo 1985a), and brain tissue preserved in anoxic conditions under water, such as at Windover Pond and Little Salt Spring, Florida (Doran et al. 1986, Pääbo 1986, Pääbo et al. 1988). A few years later, DNA was extracted from 300 to 5500-year-old modern human bone (Hagelberg et al. 1989). Although early work relevant to hominins primarily demonstrated the presence of DNA in ancient materials, subsequent analyses have addressed questions regarding the relationship between *Homo neanderthalensis* and modern humans, the initial colonization of the Americas, regional population history, social organization at a particular site, diet, the sex of individuals, and relationships among individuals within a cemetery (e.g., Krings et al. 1997, Stone and Stoneking 1998, Poinar et al. 2001, Krause et al. 2007a). Most ancient DNA research has targeted **mitochondrial DNA** (mtDNA) because of its high copy number in cells. Several studies have examined nuclear DNA loci, including **Y chromosome** sequences for sex identification and short tandem repeat (STR) loci for determining relatedness among individuals. Most recently, high-throughput emulsion **polymerase chain reaction** (PCR) and pyro-sequencing have been used for complete mitochondrial DNA or genome sequencing. One of the major concerns in ancient DNA research is contamination. Precautions must be taken to ensure the authenticity of the results. The so-called criteria of authenticity for ancient DNA results include a physically isolated work area, the extensive use of controls to test reagents, small size of PCR products, cloning of PCR products, assessment of preservation, and reproducibility of results (Handt et al. 1994a, 1996, Richards et al. 1995, Stoneking 1995, Cooper and Poinar 2000). In particular, because ancient DNA analyses are destructive, time-consuming, and expensive, assessments of DNA preservation such as testing the DNA from associated faunal remains or the extent of **amino acid racemization** is recommended. For the latter, high levels of **racemization** are thought to be associated with poor DNA preservation and, specifically, the lower the ratio of the D to L enantiomers of **aspartic acid**, the better DNA preservation is expected to be (Poinar et al. 1996). However, Collins et al. (2009) found no correlation between aspartic acid racemization and DNA preservation in a large sample of ancient specimens from different sites and time periods.

Ancient Human Occupation of Britain Project (AHOB) A project designed to investigate ancient human occupations of Britain and continental Europe from the **Paleolithic** to the Mesolithic. The AHOB project, which has been funded by the Leverhulme Trust, is a multi-disciplinary collaborative project directed by Christopher **Stringer** that involves members and associate members from various fields including paleoanthropology, paleontology, archeology, and the earth sciences. The first phase of the AHOB project, which was confined to Great Britain, ran from 2001 to 2006 and its successes included the excavations at **Pakefield**. The second phase extended the scope of investigation from Britain to continental Europe and ran from 2006 to 2009 and included the excavations at **Happisburgh**. The third phase, "Dispersals of Early Humans: Adaptations, frontiers and new territories," began in October 2009, and will run for 3 years, and its members come from Britain, Europe, and North America. The third phase of the AHOB project focuses on four complementary research questions: (a) what is the chronology of human dispersals over the last million years and where were the frontiers of human occupation at different times?, (b) how did the nature of occupation change through time, and what were the factors controlling dispersal into marginal environments: climate, resource availability, and changing geography?, (c) was occupation continuous or episodic, and how viable were human populations at the limits of their range?, and (d) what survival strategies were deployed in these marginal situations: seasonal migration, technological innovation, or physical adaptation? For information about the personnel involved in the AHOB 3 project see www.ahobproject.org/People.php and for the resources and publications produced by the three AHOB projects see www.ahobproject.org/Publications.php.

ANCOVA   Acronym for **analysis of covariance** (*which see*).

Andersson, Johan Gunnar (1874–1960)
Johan Gunnar Andersson was born in Knista, Sweden, and studied geology at the University of Uppsala. In 1914 Andersson became a geological and mining advisor to the government in China. In this capacity he worked with China's National Geological Survey, led by V.K. Ting (Wenjiang Ding). He became interested in the paleontology and archeology of China and he began searching for fossils in traditional Chinese pharmacies where they were sold for use in traditional

medicincs. He published a description of some **Neolithic** implements in 1920 and conducted research at a prehistoric site in Hunan in the early 1920s. It was around that time that Andersson learned of a fossiliferous cave site near Peking (now Beijing) called Choukoutien (now **Zhoukoudian**) and in 1919 he published a "Preliminary description of a bone-deposit at Chow-kou-tien in Fang-shan-hsien, Chili Province" in *Geografisker Annaler*. Andersson conducted sporadic excavations at Chou-koutien (and at other sites) over the next few years. In 1921 he was joined by the Austrian paleontologist Otto **Zdansky**, and their excavations at Choukoutien began to yield **Cenozoic** fossils, including what was then thought to be the tooth of an anthropoid ape. In 1923 Zdansky left China for Sweden and the fossils he and Andersson had unearthed from Choukoutien were sent to the University of Uppsala for analysis by Zdansky and the paleontologist Carl Wiman. In 1924–5 Wiman and Zdansky found what they interpreted to be a hominin tooth in the collection and this prompted them to suggest that a tooth found in 1921 might have belonged to a hominin and not to an ape. Andersson learned of Wiman and Zdansky's conclusions by letter and with support from the Swedish industrialist Axel Lagrelius the China Fund (which in 1921 became the Swedish China Research Committee) was established to fund and organize paleontological exploration in China, including at Choukoutien. By this time Andersson had enlisted the help of Canadian anatomist Davidson **Black** (they had first met in 1919). Black was initially the Professor of Embryology and Neurology at the newly created **Peking Union Medical College**, and only later did he hold the Chair of Anatomy. He joined forces with Weng Wenhao, who was the director of China's Geological Survey, to conduct large-scale excavations at Choukoutien. In 1926 Andersson founded the Museum of Far Eastern Antiquities in Stockholm, Sweden, to house part of the paleontological and archeological remains he collected in China; the remaining portion stayed in China. Andersson left China in 1925 to become director of the museum, a position he held until 1938, and the excavations in China were left to Black and Weng. The hominin teeth identified by Zdansky (possibly also Wiman), along with subsequent material discovered by Black and his Chinese colleagues at Choukoutien, resulted in the recognition of a new hominin species that Black named *Sinanthropus pekinensis* Black, 1927 (now *Homo erectus*). *See also* *Sinanthropus pekinensis*; *Homo erectus*.

Andresen lines The eponymous name recognizes Viggo Andresen's (1898) description of the microstructure of **dentine**. Andresen's name is used for the long-period (greater than **circadian**) incremental features in dentine, which correspond to **striae of Retzius** in enamel (Dean et al. 1993a). They are the 2D manifestation of a 3D structure, namely the surface of the developing dentine. Counts and measurement of Andresen lines have been used to infer the rate of dentine secretion and extension, as well as the age at death in developing teeth (e.g., Dean et al. 1993a, Smith et al. 2006). *See also* **dentine development; incremental features**.

androgens (Gk *andro* = male and *gennan* = to produce) A class of steroids with 19 carbon atoms ("C19 steroids") that are produced by conversion of pregnenolone (itself produced from cholesterol) in various tissues in the body but mainly in the gonads and adrenal glands. Androgen steroids include testosterone, dihydrotestosterone (or DHT), and the adrenal androgens dehydroepiandrosterone (or DHEA), and dehydroepiandrosterone sulfate (or DHEAS). Contrary to popular understanding, androgens, while responsible for many aspects of masculine phenotype and behavior, are not strictly "male" hormones. In addition to their importance in normal male fetal and postnatal development, androgens are implicated as playing key potentiating roles in various behaviors in both modern humans and nonhuman primates and their levels are themselves affected by behavior and emotion. For example, seasonal shifts in reproductive behavior and associated agonistic interactions and territorial aggression are associated with changes in gonadal androgen levels. Similarly, gonadal androgens are elevated in modern humans preceding physical or mental contests and they remain elevated in winners of these contests but decline in the losers. Levels of circulating gonadal androgens vary substantially among modern human populations, potentially related to differences in diet and body composition. Prolonged, chronically elevated levels of androgens may be associated with a number of health consequences, including prostate cancer.

Anglian *See* **glacial cycles**.

angular gyrus (L. *angulus* = angle and Gk *gyros* = ring, circle) The angular gyrus of the **cerebral cortex** corresponds to **Brodmann's area 39**, which is part of the inferior parietal cortex. It lies immediately posterior to the **supramarginal gyrus**. The cortex of the angular gyrus is involved in processing the spatial relationships among objects, the semantics of words, and mathematical problems. The cortex of the angular gyrus is also involved in the memory of meaningful gestures and the sequence of actions of the upper limb; hence the angular gyrus, like the supramarginal gyrus, has been implicated in the neural basis of the production and use of **tool**s.

angular unconformity   *See* unconformity.

anisognathous   *See* chewing.

anisotropy   (Gk *an* = not, *iso* = equal, and *tropus* = direction) Meaning directionally dependent. In paleoanthropology, anisotropy is used most commonly either in reference to the material properties of a substance (e.g., **bone** or **enamel**) or to dental microwear textures. Materials are anisotropic when their material properties, particularly their various measures of material behavior (stiffness as reflected by the **elastic modulus** or Young's modulus, shear stiffness as reflected by the **shear modulus**, and the relationship between axial and lateral strains as reflected by **Poisson's ratio**) are different in different directions and/or planes. Bone, like many other biological materials, is anisotropic, although its material properties can approach **orthotropy** or **transverse isotropy**, which can be thought of as special cases of anisotropy. With respect to dental microwear, anisotropy is a term used to describe the orientation of microwear features. Microwear is anisotropic when the features are consistently aligned in a given direction (i.e., as when microwear scratches are parallel). Anisotopic microwear textures are typically found in folivorous primates. *See also* **dental microwear**.

anorthoclase feldspar   A mineral of the **feldspar** group in which sodium (Na) and potassium (K) are the dominant cations (i.e., positively charged ions). The chemical formula is $(Na \text{ or } K)AlSi_3O_8$. Anorthoclase feldspar is common in alkaline (i.e., Na- and K-rich) **lava**s and in the **sediments** derived from them. Older anorthoclase feldspars, especially those with high potassium content, may be suitable for **potassium-argon dating**.

ANOVA   Acronym for **analysis of variance** (*which see*).

antagonistic pleiotropy theory   (Gk *pleion* = more) One of several theories put forward to explain the evolution of senescence. The antagonistic pleiotropy theory (Williams 1957) builds upon the premise that a **gene** can have both beneficial and detrimental effects on **fitness**. More specifically, it suggests that genes that confer fitness benefits early in the life cycle and/or at reproductive age may also have deleterious effects later in life. For example, in young males testosterone can benefit fitness through its effects on **body size**, coloration, secondary sexual ornamentation, and competitive behavior. However, later in the life cycle testosterone can have negative effects (e.g., depression of the immune system, or predisposing individuals to prostate cancer, etc.). *See also* **senescence**.

Antarctic Circumpolar Current   The ocean current that flows from west to east around Antarctica. It is the strongest ocean current in the world at 125 Sverdrups (or 125×10^6 m^3/s), and it is the dominant feature of the Southern Ocean circulation. It is the cause of the thermal isolation of Antarctica, which has supported a large ice sheet throughout much of the **Cenozoic**.

ante-Néandertalien   *See*   pre-Neanderthal hypothesis.

antelope   (Gk *antholpos* = a fabulous beast from the orient) An informal name for a member of any of the taxa within **Antilopini**, a tribe of the family **Bovidae** that includes the gazelles and their allies.

anterior buttress   A bony column that begins lateral to the nasal, or piriform, aperture and then runs inferiorly to the **canine eminence**.

anterior cingulate cortex   (L. *cingulum* = collar and *cortex* = shell, husk) The cingulate cortex is part of the limbic lobe, the portion of the **cerebral cortex** that wraps around the **corpus callosum**. The cingulate cortex is involved in regulating emotional and cognitive behavior. Its anterior portion is home to a class of large spindle-shaped neurons called **von Economo neurons**, which occur in high densities exclusively among large--brained and social mammals (e.g., whales, dolphins, great apes, modern humans, and elephants). The anterior cingulate cortex has been implicated in emotional self-control, focused problem solving, errii rccognition, and adaptive responses to changing conditions.

anterior condylar canal *See* hypoglossal canal.

anterior fovea *See* fovea; fovea anterior (mandibular); fovea anterior (maxillary).

anterior fovea (mandibular) *See* fovea anterior (mandibular).

anterior fovea (maxillary) *See* fovea anterior (maxillary).

anterior pillars Columns of bone described by Yoel Rak that run from the side of the nasal, or piriform, aperture down to the **alveolar process** of the upper jaw. Rak (1983) makes the point that the anterior pillars are found in many (e.g., **Sts 5, Sts 71, Sts 17, Stw 13**), but not all (e.g., **Sts 52a**), of the faces of *Australopithecus africanus* (anterior pillars are also present on **Stw 252**, a specimen found subsequent to Rak's 1983 study). They are also present in a reduced form in *Paranthropus robustus*, but they are not seen as distinct structures in *Australopithecus afarensis*, in most *Paranthropus boisei* specimens (but note that the presumed *P. boisei* female, **KNM-ER 732**, has an anterior pillar), or in **KNM-WT 17000**. Some *Homo* specimens are also said to have an anterior pillar (e.g., **Stw 53, OH 24**). The anterior pillar is often confused with, but is distinct from, the **canine jugum** that covers the root of the upper canine, and some specimens (e.g., Sts 5) display both structures.

anterior teeth Refers to the two incisors and the canine in each quadrant of the jaws; the balance of the teeth in the tooth row are referred to as the posterior, or **postcanine teeth**.

anterior transverse crest Term used by Korenhof (1960) for a structure on the crowns of the maxillary postcanine teeth that others call the **mesial marginal ridge** (*which see*).

anterior trigon crest Term used by Robinson (1956) and Tobias (1967a) for a structure on the crowns of the maxillary postcanine teeth that others call the **mesial marginal ridge** (*which see*).

anthracothere *See* Anthracotheriidae.

Anthracotheriid Unit (AU) The fossiliferous unit at **Toros-Menalla**, one of several "fossiliferous areas" discovered in the **Chad Basin** by the **Mission Paléoanthropologique Franco-Tchadienne**. The Anthracotheriid Unit was initially dated to 6–7 Ma using biochronology, but more recently **cosmogenic nuclide dating** suggests an age range of 6.8–7.2 Ma. *See also* **Anthracotheriidae; Toros-Menalla**.

Anthracotheriidae (Gk *anthrax* = coal or carbuncle, *therion* = beast) A family of **artiodactyls** that lived in the Old World and North America. They arose at the end of the middle Eocene in Asia and were extinct in Africa by the late Miocene. Anthracotheres are quite variable in their dental morphology, ranging from bunodont to selenodont in tooth form. Physically, they appear to have been ectomorphic, having relatively big bodies and small heads and limbs. Some forms (e.g., *Bothriogenys*) likely inhabited semi-aquatic habitats. Their dental and body forms, supposed habitats, and the fact that they disappeared as true hippopotamus arose lead many to consider them as stem hippos, but they are not universally accepted as ancestors of the Hippopotamidae.

Anthrocast *See* American Institute of Human Paleontology.

anthropogenic bone modification Any alteration of the completeness, structure, or surface of bone resulting from hominin activities. It subsumes bone surface modifications (e.g., cutmarks and hammerstone percussion marks) and fracture/breakage patterns associated with carcass butchery and dismemberment, marrow processing, and grease extraction. Other forms of anthropogenic bone modification include tooth marks and digestion, heating, and burning, and use-wear on bone tools. However, anthropogenic tooth marks and evidence of digestion can be difficult to distinguish from similar modifications produced by other taphonomic agents, such as carnivores. Recognition of anthropogenic bone modifications is central to demonstrating that a fossil bone assemblage has been accumulated by hominins and/or altered by hominin activities, as opposed to other taphonomic agents including carnivores, porcupines, or fluvial processes. *See also* **bone breakage patterns**.

anthropoid (Gk *anthropos* = human being) Primates that are relatively modern-human-like.

It is usually used in one of two senses, either to refer to the nonhuman higher primates (i.e., **chimpanzee**, **gorilla**, and orangutan and their immediate ancestors) as in "anthropoid apes", or to all the members of the Anthropoidea (i.e., living anthropoids include all the extant New World monkeys, Old World monkeys, and apes, plus modern humans). Strictly speaking, the latter use is the only correct one.

anthropoid apes   *See* anthropoid.

Anthropopithecus* de Blainville, 1839**   (Gk *anthropos* = human being and *pithekos* = ape-like) A **genus** created by Henri Marie Ducrotay de Blainville (1839) to accommodate chimpanzee material, but because it postdates Oken's 1816 designation of ***Pan, it is a **junior synonym** of *Pan*. *Anthropopithecus* was sometimes used as the generic name for any great ape, but for much of the 19thC it was the genus name of choice for chimpanzees (e.g., Sutton 1883). It was the genus used by Pilgrim for *Anthropopithecus sivalensis*, a fossil ape from the Siwaliks, and Eugène **Dubois** (1893) used the same genus to accommodate *Anthropopithecus erectus*, the species he introduced to accommodate the **Trinil 2** skullcap. However, Dubois subsequently abandoned *Anthropopithecus*, replacing it with *Pithecanthropus* (Dubois 1894), and later still Weidenreich (1940) subsumed *Pithecanthropus* into *Homo*. *See also Pithecanthropus*.

***Anthropopithecus erectus* Dubois, 1893** (Gk *anthropos* = human being, *pithekos* = ape-like, and L. *erectus* = to set upright) (*NB: this citation is often given the year 1892, but it should be 1893. Dubois was obliged to submit official reports of his field activities to the government of **Batavia**, and although he first refers to *Anthropopithecus erectus* in the report of his field activities for the third quarter of 1892, the report was not published as a government document until 1893) A **hominin** species established by Eugène **Dubois** (1893) to accommodate the **Trinil 2** skullcap. The sequence of events was as follows: in 1891 Dubois reported that at **Trinil** "The most important find is a molar (upper third molar) of a chimpanzee (*Anthropopithecus*). This genus of humanlike apes, now present in West- and Central Equatorial Africa, lived in Pliocene times in British India and, as is clear from the discovery, also during

Pleistocene periods in Java" (Dubois 1891, pp. 13–14) (this, and all the subsequent translations from Dutch are by kind favor of John de Vos). Later in the same year (but published in 1892) he reported that "Close to the site at the left bank of the river where the molar was found, a very nice skullcap was found, which, like the the the molar, must undoubtedly attributed to the genus *Anthropopithecus* (*Troglodytes*)" (Dubois 1892, pp. 13–14). Later in the same report Dubois wrote that "With the Pliocene *Anthropopithecus sivalensis*, of which only an incomplete upper jaw is known, only the molar is comparable, however in a limited amount. Probably one cannot exclude a close relation between both," and he continued "The above mentioned *Anthropopithecus sivalensis* [originally described under the name of *Palaeopithecus* and then as *Troglodytes*] represented by an incomplete upper jaw in the Pliocene of the Punjab in 1878 discovered by Theobald" (*ibid*, pp. 13–14). However, it was not until 1892, and the discovery of the straight-shafted **Trinil 3** femur, that Dubois decided to name the new taxon *Anthropopithecus erectus*, and in the same year (but published in 1893) he wrote that "The three discovered skeletal elements show that *Anthropopithecus erectus* Eug. Dubois is closer to Man than any other of the anthropoid apes, the most by the femur…" (Dubois 1893, p. 11). In Dubois' report for the last quarter of 1893 (Dubois 1894, p.14) the new taxon was subsequently transferred to the genus ***Pithecanthropus* Haeckel, 1868**, as *Pithecanthropus erectus*, and much later it was transferred to *Homo* by Weidenreich (1940). *See also Homo erectus*; *Homo erectus javanensis*; *Pithecanthropus erectus*.

antibody   (Gk *anti* = opposite and ME *body* = container) Antibodies (also known as immunoglobulins or Igs) are **proteins** that are part of the immune system. They are produced by lymphocytes (a type of white blood cell) when the former react with foreign particles (e.g., bacteria and viruses) collectively called **antigens**. Each antibody reacts to a specific antigen, binding with it and then tagging it for destruction by other parts of the immune system, or blocking it directly from growing or causing damage. Antibodies may be free and soluble, and found in blood and tissue fluids, or they may be bound to the surface of a type of white cell (a type of lymphocyte called B cells, so-called because they mature in the bursa of

Fabricius in birds). Antibodies consist of four chains of **amino acids**, two heavy ones and two light ones. There are five types of antibody: IgA, IgD, IgE, IgG, and IgM. These are defined by differences in the structures, functions, and properties of their heavy amino acid chains. Antibodies raised against foreign albumins are the basis of the experiments undertaken by Morris **Goodman** to investigate the relationships among the **great apes**. *See also* **albumin**.

anticline (Gk *anti* = against and *klinein* = to slope) In structural geology this is a type of fold where the oldest rocks occupy the center of the structure and the rocks become progressively younger towards the margins. The simplest form of anticline is a symmetrical arch-shaped fold, which is open downwards. Such folds are important in geological and paleontological fieldwork, as they will determine the direction in which successively older or younger strata are encountered and will cause a repetition of the **outcrop** pattern on either side of the fold's axis. They are common features in tectonically active areas.

anticodon A sequence of three nucleotides in a **transfer RNA** (or tRNA) molecule that is complementary to a **codon** (i.e., a sequence of three nucleotides) in a messenger **RNA** (or mRNA) molecule.

antigen (from *anti*body *gen*eration) Any foreign molecule capable of stimulating the production of an antibody, or of provoking other responses by the immune system. *See also* **antibody**.

antilopine *See* Antilopini.

Antilopini (Gk *antholpos* = a fabulous beast from the orient) A tribe of the family **Bovidae** that includes the gazelles and their allies. Extant members of the tribe include springbok (*Antidorcas marsupialis*), Grant's gazelle (*Gazella granti*), and Thomson's gazelle (*Gazella thomsoni*). Antilopine **bovids** are of small to medium size (approximately 15–65 kg) and they exhibit adaptations for **cursorial** behavior (i.e., running) suited for the open and arid grasslands they typically inhabit. In **paleoenvironmental reconstructions** of fossil assemblages, high frequencies of antilopine and **alcelaphine bovids** are generally interpreted as indicating open habitats. *See also* **antilopine and alcelaphine criterion**.

antimere (Gk *anti* = opposite and *meros* = a part) Refers to the version of a bilateral structure that belongs to the opposite side of the body [e.g., "the crown area of the right P_3 of **KNM-ER 992** is larger than its antimere" (i.e., the left P_3)]. Amino acids come in forms that are antimeres (e.g., D- and L-forms) and the rate of conversion from one antimere to the other has been used as a dating method. *See also* **amino acid racemization dating**; **ostrich egg shell dating**.

antimeric *See* **antimere**.

anvil A stationary object against which another object, such as a bone or **core**, can be struck to fracture it. Anvils are generally made of stone, although materials such as wood may be used when stone is not available (e.g., Tai forest nut-cracking chimpanzees). Anvils can be used in combination with a hammer, made of stone (a **hammerstone**) or wood. The object to be fractured (e.g., bone, core, nut) is rested on the anvil surface and struck from above with the **hammer**. Bones fractured with the "hammer-and-anvil" technique frequently have opposing load points and surface damage attributable to both the hammerstone and the anvil. Stone cores flaked using a hammer-and-anvil or bipolar technique usually have flakes removed from both ends. *See also* **bipolar**; **chimpanzee tool-use**.

Anyskop Blowout (Location 32°58′11.6″S, 18°06′50.66″E, South Africa; etym. in Afrikaans *anyskop* translates to "anise hill") History and general description This open-air archeological locality is located within the boundaries of the West Coast Fossil Park, approximately 1 km/0.6 miles south of the site of **Langebaanweg**. Discovered in the late 1970s, this site was extensively collected and excavated in 2001 and 2002 by researchers from the University of Tübingen, Germany. The archeological remains provide clear evidence that archaic and modern humans occupied this elevated setting during all of the southern African archeological periods, including the **Early Stone Age**, the **Middle Stone Age**, and the **Later Stone Age**. Temporal span and how dated? Allocation to the Early, Middle and Late Stone Ages is based on the types of artifact. Hominins found at site None. Archeological evidence found at site Stone tools (see above), **Late Pleistocene** and **Holocene** fauna, ceramics, and hearths. Key references: Hominins N/A; Archeology Conrad

2001, 2002, Dietl 2004, Dietl et al. 2005, Kandel 2006. *See also* **Later Stone Age**.

apatite (Gk *apate* = deceit, because of apatite's reputation for being confused with other minerals) Apatite is one of the common names (hydroxyapatite and bioapatite are others) used for the dominant mineral phase of **bone**, **cementum**, and **dentine**, as well as **enamel**. The proper name for bioapatite or hydroxyapatite is apatitic calcium phosphate, chemical formula $Ca_{10}(PO_4)_6(OH)_2$. Bioapatite or hydroxyapatite makes up approximately 96% of the mineral phase of mature enamel, and this high proportion is responsible for the latter's extreme hardness and resilience to **diagenesis**.

ape (OE *apa* = ill-bred and clumsy. Groves (2008) suggests that *apa* became *apan* (pl.), *apan* became *apen*, and then approximately 1350 years BP *apen* became apes. Before the apes had been investigated scientifically and appreciated on their own terms, they were regarded as being "clumsy" because they lacked **dexterity**) Refers to an informal taxonomic category that includes extant and fossil taxa and it is coincident with the superfamily **Hominoidea**. The extant taxa in this superfamily are chimpanzees/bonobos, gorillas, orangutans and gibbons/siamangs; the fossil taxa are all the extinct forms that are more closely related to chimpanzees/bonobos, gorillas, orangutans and the gibbons/siamangs than to any other living taxon. In the traditional, pre-molecular, taxonomy the informal term ape was equivalent to the families Pongidae and Hylobatidae. syn. hominoid.

ape hand All extant apes are capable of specialized suspensory and climbing behaviors and thus the ape hand, more so than the modern human hand, is well adapted for grasping by hook-like flexed fingers. Extant ape fingers are relatively long and the phalanges are curved to facilitate hanging and grasping. The thumb is relatively short and is not usually used during these locomotor behaviors. These adaptations, and others such as a highly mobile midcarpal joint, are particularly accentuated in orangutans, gibbons, and siamangs to facilitate suspension and, in the hylobatids, ricochetal brachiating (Lewis 1989). The gorilla, in contrast, has relatively shorter and straighter phalanges compared to other apes because of its frequent use of terrestrial locomotion. African apes differ from Asian apes in having

scaphoid-os centrale fusion, extra bony ridges, and the bevelling of articular surfaces that make the wrist and finger joints more stable. This increased stability has been interpreted as being advantageous for knuckle-walking locomotion (Tuttle 1969, Richmond et al. 2001).

apical closure The developmental condition that signifies the closing of the root canal at its distal, or apical, end. At this point the growth in length of the root (also called extension) ceases, although secondary dentine secretion continues within the pulp cavity. Apical closure, which is traditionally taken as the final stage of **tooth formation**, can be evident macroscopically, microscopically, or radiographically.

apical tuft (L. *apex* = point and OF *tof(f)e* = projection) The most distal part of the distal phalanx of a **manual digit**. The fingers of the primate **hand** each comprise three bones, called the **proximal**, intermediate, and distal phalanges. The small distal phalanx has a base, a shaft, and an apex. In higher primates the apex is rounded and takes the form of a bony excrescence called an apical tuft. The edges of this apical tuft extend more proximally than does its midpart, and they are referred to as ungual spines. The apical tuft provides bony support for the nail and for the soft tissue, or pulp, that lies beneath the nail. The apical tuft is markedly enlarged in modern humans and in *Homo neanderthalensis*, and it is relatively enlarged in chimpanzees and gorillas compared to other primates. In modern humans, the expanded apical tuft supports the elaboration of a pulpy pad (the "ungual pulp") on the palmar side of the distal phalanx. This palmar pad is functionally compartmentalized such that only the distal end is tethered by ligaments to the bone while the proximal end is more flexible, allowing for deformation in various ways during grasping of objects (Marzke and Marzke 2000) (syn. ungual process, tuberosity, or tuft) (Susman 1979).

Apidima (Location Mani peninsula, Southern Greece; etym. local dialect of Greek for "jump", probably referring to the steep sea-side cliff face where the caves are located) History and general description Discovered in 1978 and partially excavated between 1978 and 1985, the site comprises four limestone caves accessible only by sea.

Temporal span Dating investigations have proved to be inconclusive. Cave A is probably **Middle Pleistocene**, and Cave Γ possibly **Late Pleistocene**. How dated? **Biostratigraphy**, lithic typology, and morphology of the hominin remains. Hominins found at site Cave A: Apidima 1 and 2; Cave Γ: Apidima 3. Archeological evidence found at site Stone artifacts reported from Cave Γ, possibly **Upper Paleolithic**. Key references: Geology and dating Lax 1995, Liritzis and Maniatis 1995; Hominins Pitsios 1985, Harvati and Delson 1999; Archeology Darlas 1995. *See also* **Apidima 1 and 2**; **Apidima 3**.

Apidima 1 and 2 Site **Apidima**. Locality Cave A. Surface/*in situ In situ*. Date of discovery 1978. Finders Team from National Kapodistrian University. Unit N/A. Horizon N/A. Bed/member N/A. Formation N/A. Group N/A. Nearest overlying dated horizon N/A. Nearest underlying dated horizon N/A. Geological age Probably **Middle Pleistocene** on morphological grounds. Developmental age Adult. Presumed sex Apidima 2 proposed Female. Brief anatomical description Two crania; Apidima 2 is largely complete, whereas Apidima 1 lacks a face and the superior part of the cranial vault. Announcement Pitsios 1985. Initial description Koutselinis et al. 1995. Photographs/line drawings and metrical data Koutselinis et al. 1995. Initial taxonomic allocation No proper classification or description exists. Taxonomic revisions N/A. Current conventional taxonomic allocation **Pre-Neanderthal** (Harvati and Delson 1999). Significance These specimens represent two out of the total of three fossil hominin crania from the Middle Pleistocene of southeastern Europe. Location of original Anthropological Museum, National Kapodistrian University, Athens, Greece. *See also* **Apidima**.

Apidima 3 Site **Apidima**. Locality Cave Γ. Surface/*in situ In situ*. Date of discovery 1984. Finders Excavating team from the National Kapodistrian University. Nearest overlying dated horizon N/A. Nearest underlying dated horizon N/A. Geological age Possibly Late-Terminal Pleistocene. Developmental age Young adult. Presumed sex Female. Brief anatomical description Largely complete skeleton of a modern human, presumably female and possibly a burial. Announcement Pitsios 1985. Initial description Mompheratou and Pitsios 1995. Line drawings

Mompheratou and Pitsios 1995. Initial taxonomic allocation *Homo sapiens*. Significance Possible **Upper Paleolithic** burial. The brief description of the stone artifacts suggests they may be **Aurignacian**. The **Upper Paleolithic**, and the Aurignacian in particular, is almost completely absent from the fossil and archeological records of Greece. Location of original Anthropological Museum, National Kapodistrian University, Athens, Greece. *See also* **Apidima**.

Apollo 11 Cave (Location 27°45′S, 17°06′E, Namibia; etym. named after the 1969 return flight and landing of Apollo 11 spacecraft, which occurred the day excavations commenced) History and general description A large (28 m×11 m) limestone cave on the western slopes of the Huns Mountains. W.E. Wendt excavated 12 m^2 to a maximum depth of approximately 2 m, recovering more than 50,000 lithic and faunal remains. The site contains a number of **Middle Stone Age** strata including Howieson's Poort and Still Bay levels. Stone slabs with painted representations of a rhinoceros and an animal, or perhaps modern human, form are among the earliest known examples of *art mobilier*, with age estimates using all the available data ranging from *c*.20 to 59 ka. Temporal span Wendt recognized eight strata (A–H from top to bottom) separated by unconformities; Jacobs et al. (2008) further subdivided the strata, although correlations with Wendt's scheme are not provided. The sediments have been dated by single-grain optically stimulated **luminescence dating** on **sand**, conventional **radiocarbon dating** of charcoal, and **accelerator mass spectrometry** age estimates and **amino acid racemization dating** on ostrich egg shell. The pooled samples of age estimates suggests discontinuous hominin occupation of the cave from *c*.0.3 ka to more than 83 ka. Excluding conventional radiocarbon age estimates from the 1970s, the stratum containing the painted slabs (Layer E of Wendt 1976) has an estimated age of 41–59 ka. How dated? ^{14}C, amino acid racemization dating, accelerator mass spectrometry, and optically stimulated luminescence dating. Hominins found at site None. Archeological evidence found at site More than 27,000 stone artifacts and more than 23,000 faunal elements have been recovered from the Apollo 11 Cave excavations. The former have been divided by Wendt and later researchers into (from bottom to top) early Middle Stone Age, Still Bay, Howieson's Poort, late Middle Stone Age, early **Later Stone Age**,

Later Stone Age, and pottery-bearing strata. Apollo 11 Cave reflects the northernmost and westernmost limit of known Still Bay sites, and in addition to the distinctive bifacial points, the Still Bay stratum there (Layer G of Wendt 1976, Komplex 2 of Vogelsang 1998) also contains worked **hematite** and **bone tools**. Incised and perhaps painted ostrich egg shell fragments also occur in overlying levels (i.e., Layers F and E of Wendt 1976, and Komplex 3 and 4 of Vogelsang 1998). The associated fauna is interpreted as primarily the result of hominin accumulation, with the notable selective transport of high-utility elements of large **antelope**s that were subsequently discarded in the cave. Key references: Geology and dating Wendt 1976, Vogelsang 1998, Miller et al. 1999, Jacobs et al. 2008; Archeology Wendt 1976, Thackeray 1979, Vogelsang 1998, Jacobs et al. 2008.

apomorphic (Gk *apo* = different from and *morph* = form) A term used in **cladistic analysis** to refer to the state of a **character** that is different from the ancestral or **primitive** condition of that character. Hence it is a "catch-all" word that refers to any derived **character state**. For example, within the hominin clade there are two morphoclines for anterior lower premolar (P_3) root morphology. One leads from the primitive condition [i.e., two roots, one mesiobuccal and one distal (2R: MB+D)] to a derived morphology of root reduction [i.e., a single conical root (1R)], whereas the other leads to a derived morphology of further root complexity [i.e., two plate-like roots, one mesial and one distal (2R: M+D)]. Both of these character states are apomorphic. Apomorphy is one of several terms used in cladistics that is relative. The same morphology can be derived or apomorphic in one context and primitive or symplesiomorphic in another. *See also* **autapomorphy**; **synapomorphy**.

apomorphy (Gk *apo* = different from and *morph* = form) *See* **apomorphic**.

apoptosis (Gk *apo* = from, and *ptosis* = falling) A process (also called programmed cell death) whereby cells in a multicellular organism die according to a program determined by the cell. Apoptosis plays important roles in development, homeostasis, the removal of damaged cells, and the prevention of cancer. There is much apoptosis during the development of the brain, and most of the tubes of the body

(e.g., the tear duct) start as a solid core of cells that becomes tubular when the cells in the middle of the core undergo apoptosis.

appendicular skeleton (L. *appendere* = to hang upon) The hard-tissue components of fore or upper, and the hind or lower, limbs. In the upper limb it comprises the pectoral or shoulder girdle (i.e., the scapula and clavicle), the bone of the arm (humerus), and the bones of the forearm (radius and ulna) and those of the hand (carpals, metacarpals, and phalanges). In the lower limb it comprises the pelvic girdle (i.e., pelvic bone made up of the ilium, ischium, and pubic bones, but not the sacrum), the bone of the leg (femur), the patella, the bones of the lower leg (tibia and fibula), and the bones of the foot (tarsals, metatarsals, and phalanges).

appositional enamel (L. *appositus* = to put near) Strictly speaking, all enamel is appositional, but this term usually refers to the enamel formed during the initial phase of enamel formation, or cuspal enamel. *See also* **cuspal enamel**; **enamel development**.

appositional growth *See* **ossification**.

approximal wear *See* **interproximal wear**.

aptation (L. *adaptare* = to fit) There are two main categories of aptations. If a functional **trait** was fixed in a population by **natural selection**, and it still performs that function, then it is referred to as an adaptation. But if there is evidence the trait now performs a *different* function, or if what is now a functional trait was non-functional prior to its being co-opted for its current function, then the trait is referred to as an exaptation (*sensu* Gould and Vrba 1982). *See also* **adaptation**; **exaptation**.

^{40}Ar/^{39}Ar *See* **argon-argon dating**.

ARA *See* **Aramis**.

ARA-VP Acronym for Aramis – Vertebrate Paleontology. Prefix for fossils recovered from **Aramis, Middle Awash study area**, Ethiopia.

ARA-VP-1/1 Site Aramis. Locality Aramis VP Locality 1. Surface/*in situ* Surface. Date of discovery December 17, 1992. Finder Gen Suwa. Unit N/A. Horizon N/A. Bed/member Lower Aramis Member. Formation **Sagantole Formation**. Group N/A. Nearest overlying dated horizon Daam-Aatu Basaltic Tuff. Nearest underlying dated horizon Gaala Vitric Tuff Complex. Geological age *c*.4.4 Ma. Developmental age Adult. Presumed sex Unknown. Brief anatomical description Right M^3. Announcement White et al. 1994. Initial description White et al. (1994). Photographs/line drawings and metrical data White et al. 1994, Suwa et al. 2009a. Detailed anatomical description N/A. Initial taxonomic allocation *Australopithecus ramidus*. Taxonomic revisions *Ardipithecus ramidus*. Current conventional taxonomic allocation *Ar. ramidus*. Informal taxonomic category Possible primitive hominin. Significance First specimen of *Ar. ramidus* to be discovered. Location of original National Museum of Ethiopia, Addis Ababa, Ethiopia.

ARA-VP-1/300   Site **Aramis**. Locality Aramis VP Locality 1. Surface/*in situ* Surface. Date of discovery December 26, 1993. Finder Unknown. Unit N/A. Horizon N/A. Bed/member Lower Aramis Member. Formation Sagantole Formation. Group N/A. Nearest overlying dated horizon Daam-Aatu Basaltic Tuff. Nearest underlying dated horizon Gaala Vitric Tuff Complex. Geological age *c*.4.4 Ma. Developmental age Adult. Presumed sex Unknown. Brief anatomical description Associated permanent dentition that includes the crowns and roots of at least one side of all the upper teeth (ARA-VP-1/300A–M) and of all the lower teeth (ARA-VP-1/300N–ZA). Announcement White et al. 2009a. Initial description Suwa et al. 2009a. Photographs/line drawings and metrical data Suwa et al. 2009a. Detailed anatomical description N/A. Initial taxonomic allocation *Ardipithecus ramidus*. Taxonomic revisions None. Current conventional taxonomic allocation *Ar. ramidus*. Informal taxonomic category Possible primitive hominin. Significance Excellently preserved dentition of a single *Ar. ramidus* individual. Location of original National Museum of Ethiopia, Addis Ababa, Ethiopia.

ARA-VP-1/401   Site **Aramis**. Locality Aramis VP Locality 1. Surface/*in situ* Surface. Date of discovery December 27, 1993. Finder Unknown. Unit N/A. Horizon N/A. Bed/member Lower Aramis Member. Formation Sagantole Formation. Group N/A. Nearest overlying dated

horizon Daam-Aatu Basaltic Tuff. Nearest underlying dated horizon Gaala Vitric Tuff Complex. Geological age *c*.4.4 Ma. Developmental age Adult. Presumed sex Unknown. Brief anatomical description The symphysis and both sides of the corpus of a subadult mandible containing the crowns and roots of the left and right I_1–M_1, plus the left M_2. Announcement White et al. 2009a. Initial description N/A. Photographs/line drawings and metrical data Suwa et al. 2009b. Detailed anatomical description N/A. Initial taxonomic allocation *Ardipithecus ramidus*. Taxonomic revisions None. Current conventional taxonomic allocation *Ar. ramidus*. Informal taxonomic category Possible primitive hominin. Significance The best preserved mandible of *Ar. ramidus*. Location of original National Museum of Ethiopia, Addis Ababa, Ethiopia.

ARA-VP-1/500   Site **Aramis**. Locality Aramis VP Locality 1. Surface/*in situ* Surface. Date of discovery December 28, 1993. Finder Unknown. Unit N/A. Horizon N/A. Bed/member Lower Aramis Member. Formation Sagantole Formation. Group N/A. Nearest overlying dated horizon Daam-Aatu Basaltic Tuff. Nearest underlying dated horizon Gaala Vitric Tuff Complex. Geological age *c*.4.4 Ma. Developmental age Adult. Presumed sex Unknown. Brief anatomical description Three pieces (right and left temporal and occipital) of a partial adult cranium. Announcement White et al. 2009a. Initial description Suwa et al. 2009b. Photographs/line drawings and metrical data Suwa et al. 2009b. Detailed anatomical description N/A. Initial taxonomic allocation *Ardipithecus ramidus*. Taxonomic revisions None. Current conventional taxonomic allocation *Ar. ramidus*. Informal taxonomic category Possible primitive hominin. Significance The three pieces (right and left temporal and occipital) of this partial adult cranium were used in the reconstruction of the cranium of the **ARA-VP-6/500** associated skeleton. Location of original National Museum of Ethiopia, Addis Ababa, Ethiopia.

ARA-VP-1/701   Site **Aramis**. Locality Aramis VP Locality 1. Surface/*in situ* Surface. Date of discovery December 20, 1994. Finder Unknown. Unit N/A. Horizon N/A. Bed/member Lower Aramis Member. Formation Sagantole Formation. Group N/A. Nearest overlying dated horizon Daam-Aatu Basaltic Tuff. Nearest underlying dated horizon Gaala Vitric Tuff Complex. Geological age *c*.4.4 Ma. Developmental age Adult. Presumed sex Unknown. Brief anatomical description Proximal end of a left femur. Announcement White et al. 2009a. Initial description Lovejoy et al. 2009b. Photographs/line drawings and metrical data Lovejoy et al. 2009b. Detailed anatomical description N/A. Initial taxonomic allocation *Ardipithecus ramidus*. Taxonomic revisions None. Current conventional taxonomic allocation *Ar. ramidus*. Informal taxonomic category Possible primitive hominin. Significance A fragment of the proximal end of the left femur, one of only two proximal femoral fragments (the other is **ARA-VP-6/500**–5) in the *Ar. ramidus* hypodigm. Location of original National Museum of Ethiopia, Addis Ababa, Ethiopia.

ARA-VP-6/1　　Site **Aramis**. Locality Aramis VP Locality 6. Surface/*in situ* Surface. Date of discovery December 29, 1993. Finder Gada Hamed. Unit N/A. Horizon N/A. Bed/member Lower Aramis Member. Formation Sagantole Formation. Group N/A. Nearest overlying dated horizon Daam-Aatu Basaltic Tuff. Nearest underlying dated horizon Gaala Vitric Tuff Complex. Geological age *c*.4.4 Ma. Developmental age Adult. Presumed sex Unknown. Brief anatomical description Several associated teeth, including the left I^1, C, P^3, and P^4, and right, I^1, C, P^4, M^2, P_3 and P_4. Announcement White et al. 1994. Initial description White et al. 1994. Photographs/line drawings and metrical data White et al. 1994, Suwa et al. 2009a. Detailed anatomical description N/A. Initial taxonomic allocation *Australopithecus ramidus*. Taxonomic revisions *Ardipithecus ramidus*. Current conventional taxonomic allocation *Ar. ramidus*. Informal taxonomic category Possible primitive hominin. Significance **Holotype** of *Ar. ramidus*. Location of original National Museum of Ethiopia, Addis Ababa, Ethiopia.

ARA-VP-6/500　　Site **Aramis**. Locality Aramis VP Locality 6. Surface/*in situ* Surface. Date of discovery November 5, 1994 and thereafter. The first evidence of ARA-VP-6/500 were two fragments of a second right metacarpal found on the surface by Yohannes **Haile-Selassie**. Thereafter further evidence of the ARA-VP-6/500 partial skeleton was recovered by sieving and surface prospecting and from excavations. Finders Yohannes Haile-Selassie and members of the **Middle Awash Research Project**. Unit N/A. Horizon N/A. Bed/member Lower Aramis Member. Formation Sagantole Formation. Group N/A. Nearest overlying dated horizon

Daam-Aatu Basaltic Tuff. Nearest underlying dated horizon Gaala Vitric Tuff Complex. Geological age *c*.4.4 Ma. Developmental age Adult. Presumed sex Female. Brief anatomical description One hundred and thirty recognizable fragments belonging to a single individual were recovered. Most were in perilous condition and much of the cranial morphology has been recovered from micro-CT scans of unconsolidated cranial fragments still in matrix. Much of the cranial vault and the face have been reconstructed. The crowns and roots of all of the upper teeth on the right side, and the left lower canine through to the M_3, are preserved. The right forearm is intact apart from the distal end of the ulna; the partial right hand includes carpal bones and a complete **ray**. The only evidence of the left forearm is part of the radial shaft, but much of the skeleton of the left hand is preserved. All that remains of the thorax that can be identified precisely are two vertebrae (a cervical and a thoracic) and the left first rib. Much of the left innominate is preserved, but it is crushed and distorted, as is a piece of the lower part of the body of the sacrum, and part of the right ilium. Of the long bones of the lower limb, all that remains is a substantial length of the shaft of the right femur, most of the right tibia, and all but the proximal end of the right fibula. Between them the two preserved foot skeletons provide most of the bones of the tarsus and the toes. Endocranial volume: "300 ± 10 cm^3, with a larger range of 280 to 350 cm^3" (Suwa et al. 2009b, p. 68e6). Announcement White et al. 2009a. Initial description Lovejoy et al. 2009a–c, Suwa et al. 2009a, 2009b. Photographs/line drawings and metrical data Lovejoy et al. 2009a–c, Suwa et al. 2009a, 2009b. Detailed anatomical description N/A. Initial taxonomic allocation *Ardipithecus ramidus*. Taxonomic revisions None. Current conventional taxonomic allocation *Ar. ramidus*. Informal taxonomic category Possible primitive hominin. Significance An associated skeleton whose exceptional completeness and eventual preservation is a testament to the many hours of painstaking work, in both the field and laboratory, that went into its recovery, restoration, and reconstruction. This associated skeleton is the centerpiece of the **hypodigm** of *Ar. ramidus*, and functional interpretations based on it form the main evidential platform for the assumption that *Ar. ramidus* is a basal hominin. Location of original National Museum of Ethiopia, Addis Ababa, Ethiopia.

ARA-VP-7/2　　Site **Aramis**. Locality Aramis VP Locality 7. Surface/*in situ* Surface. Date of discovery December 30, 1993. Finder A. Asfaw. Unit N/A. Horizon N/A. Bed/member Lower Aramis Member. Formation Sagantole Formation. Group N/A. Nearest overlying dated horizon Daam-Aatu Basaltic Tuff. Nearest underlying dated horizon Gaala Vitric Tuff Complex. Geological age *c*.4.4 Ma. Developmental age Adult. Presumed sex Unknown. Brief anatomical description Fragments of the long bones of an arm of a single individual; the humeral fragment includes the proximal end. Announcement White et al. 1994. Initial description White et al. 1994. Photographs/line drawings and metrical data White et al. 1994. Detailed anatomical description N/A. Initial taxonomic allocation *Australopithecus ramidus*. Taxonomic revisions *Ardipithecus ramidus*. Current conventional taxonomic allocation *Ar. ramidus*. Informal taxonomic category Possible primitive hominin. Significance The first associated postcranial remains of *Ar. ramidus*, and the size of the humeral head was the source of the initial body mass estimate of approximately 40 kg. Its discovers claimed that "the arm displays a mosaic of characters usually attributed to hominids and/or great apes" and "a host of characters usually associated with modern apes" (White et al. 1994, p. 311). Location of original National Museum of Ethiopia, Addis Ababa, Ethiopia.

Aragai (Location 00°34′N, 35°50′E, Kenya; etym. local, Tugen, name for the area) History and general description Site within the **Lukeino Formation** in the foothills of the Tugen Hills, Baringo District, Kenya. Along with **Cheboit**, **Kapcheberek**, and **Kapsomin**, Aragai is one of four localities from which remains attributed to *Orrorin tugenensis* have been recovered. Temporal span *c*.6–5.7 Ma. How dated? **Argon-argon dating**, **potassium-argon dating**, and **magnetostratigraphy**. Hominins found at site The only specimen recovered to date is BAR 1215′00, a proximal right femur found by M. Pickford in 2000 which has been assigned to *O. tugenensis*. Archeological evidence found at site None. Key references: Geology and dating Hill et al. 1985, Pickford and Senut 2001, Deino et al. 2002, Sawada et al. 2002; Hominins Senut et al. 2001; Archeology N/A.

Arago *See* **Caune de l'Arago**.

Arago II Site **Caune de l'Arago**. Locality N/A. Surface/*in situ In situ*. Date of discovery July 21, 1969.

Finders Henry and Marie-Antionette de Lumley, Unit N/A. Horizon Layer G. Bed/member Middle Stratigraphic Complex. Formation N/A. Group N/A. Nearest overlying dated horizon Speleothem floor just above dates to *c*.350 ka. Nearest underlying dated horizon N/A. Geological age **Arago XXI** from the same layer was directly dated to *c*.400 ka. Developmental age 40–55 years. Presumed sex Female. Brief anatomical description A nearly complete mandible with six teeth and lacking the left ramus. Announcement de Lumley and de Lumley 1971. Initial description de Lumley and de Lumley 1971. Photographs/line drawings and metrical data de Lumley et al. 1982. Detailed anatomical description de Lumley et al. 1982. Initial taxonomic allocation Pre-Neanderthal/*Homo erectus tautavelensis*. Taxonomic revisions Rightmire 1998. Current conventional taxonomic allocation *Homo heidelbergensis*, pre-Neanderthal, or transitional form. Informal taxonomic category Premodern *Homo*. Significance The most complete mandible from Arago. Location of original Institut de Paléontologie Humaine, Paris.

Arago XXI Site **Caune de l'Arago**. Locality N/A. Surface/*in situ In situ*. Date of discovery July 1971. Finders Henry and Marie-Antionette de Lumley. Unit N/A. Horizon Layer G. Bed/member Middle Stratigraphic Complex. Formation N/A. Group N/A. Nearest overlying dated horizon Speleothem floor just above dates to *c*.350 ka. Nearest underlying dated horizon N/A. Geological age Directly dated using **uranium-series dating** to *c*.400 ka (but this is near the maximum limit of this dating method, so the fossil may be older). Developmental age Young adult. Presumed sex Male. Brief anatomical description A deformed cranial fragment with a complete frontal, sphenoid and all of the face. **Arago XLVII**, a right parietal, fits with the frontal and was recovered from the same layer in 1979. Announcement de Lumley and de Lumley 1971. Initial description de Lumley and de Lumley 1971. Photographs/line drawings and metrical data de Lumley and de Lumley 1971, 1973, Spitery 1982a, 1982b, Bouzat 1982. Detailed anatomical description Spitery 1982a, 1982b, Bouzat 1982. Initial taxonomic allocation Pre-Neanderthal/*Homo erectus tautavelensis*. Taxonomic revisions Rightmire 1998. Current conventional taxonomic allocation *Homo heidelbergensis*, pre-Neanderthal or transitional form. Informal taxonomic category Premodern *Homo*. Significance One of the most complete pre-Neanderthal or *H. heidelbergensis* fossil faces.

Location of original Institut de Paléontologie Humaine, Paris.

Arago XLVII Site **Caune de l'Arago**. Locality. Surface/*in situ In situ*. Date of discovery July 1979. Finders de Henry and Marie-Antionette de Lumley. Unit N/A. Horizon Layer G. Bed/member Middle Stratigraphic Complex. Formation N/A. Group N/A. Nearest overlying dated horizon Speleothem floor just above dates to *c*.350 ka. Nearest underlying dated horizon N/A. Geological age This piece likely joins with **Arago XXI**, which has been directly dated to *c*.400 ka. Developmental age 20 years. Presumed sex Male. Brief anatomical description Right parietal. Announcement Grimaud 1982. Initial description Grimaud 1982. Photographs/line drawings and metrical data Grimaud 1982. Detailed anatomical description Grimaud 1982. Initial taxonomic allocation Pre-Neanderthal/*Homo erectus tautavelensis*. Taxonomic revisions Rightmire 1998. Current conventional taxonomic allocation *Homo heidelbergensis*, pre-Neanderthal, or transitional form. Informal taxonomic category Pre-Neanderthal. Significance This parietal fits with Arago XXI. Location of original Institut de Paléontologie Humaine, Paris.

aragonite *See* **calcium carbonate**.

Aralee Issie A collection area initially identified by the **International Afar Research Expedition** but now within the **Woranso-Mille study area**, Central Afar, Ethiopia.

Arambourg, Camille Louis Joseph (1885–1969) Camille Arambourg was born in Paris, France. He graduated from the private school of Sainte-Croix de Neuilly in 1903, and later that year he entered the Institut National Agronomique (National Agronomic Institute) where he completed a degree in agricultural engineering in 1908. Arambourg left France after graduation to join his family in Oran, Algeria, where his father had moved to establish a vineyard. While improving irrigation around the family vineyard Arambourg found specimens of fossil Miocene fish and so began to excavate and collect fossils in the area. As Arambourg's paleontological interests grew he made contacts at the geological laboratory of the Algiers Faculty of Sciences, which allowed him to expand his knowledge of geology and paleontology. When WWI broke out Arambourg left Algeria to join the French army and fought

in the Dardanelles Campaign in 1915 and was stationed in Macedonia from 1916 to 1918. While in Macedonia, Arambourg collected Miocene mammalian fossils from the region around Salonika. At the end of the war Arambourg sent this collection to the **Muséum National d'Histoire Naturelle** (National Museum of Natural History) in Paris and he returned to Algeria. In 1920 Arambourg accepted a position as Professor of Geology at the Institut Agricole d'Alger (Agricultural Institute of Algiers) and over the next decade he collected fossil Miocene fish and Pleistocene vertebrates in Algeria and North Africa. These researches offered Arambourg opportunities to visit the renowned French paleontologist Marcellin **Boule** at the Muséum National d'Histoire Naturelle in Paris, and in 1930 Arambourg moved to Paris to become Professor of Geology at the Institut National Agronomique. Arambourg continued to conduct paleontological research in Africa, and while excavating the **Upper Paleolithic** site of Bejaia, in northern Algeria, between 1928 and 1930, he found six skeletons of early *Homo sapiens*. Soon thereafter he organized an expedition, the Mission Scientifique de l'Omo, to the Omo river valley in Ethiopia from 1932 to 1933 where he collected Pleistocene fossils and explored along portions of Lake Rudolf (now called Lake Turkana). The research conducted during this expedition on the geology, paleontology, and anthropology of the region was later published as *Mission scientifique de l'Omo, 1932–1933* (3 vols., 1935–48). Arambourg was appointed Professor of Paleontology at the **Muséum National d'Histoire Naturelle** in Paris in 1936 and during the 1938–9 field season he explored Cretaceous deposits in Lebanon and Oligocene deposits in Iran. In 1943 Arambourg published a short work on prehistory titled *La genèse de l'humanité* (*The Genesis of Humanity*), which went through eight editions by 1969. After WWII, Arambourg excavated the Villafranchian site of **Ain Hanech**, near the village of Sétif in northeastern Algeria. In 1947–8 where he discovered Lower Pleistocene mammal fossils and **Lower Paleolithic** artifacts including **spheroid** stones. Even greater success came during excavations at **Ternifine** (now called **Tighenif**) during 1954–5 where he found crude **Acheulean** tools and several mandibles that he attributed to a "pithecanthropoid" form of early human that he called *Atlanthropus mauritanicus* (now subsumed into *Homo erectus*). Arambourg's last major work, *Vertébres villafranchiens d'Afrique du nord* (*Villafranchian Vertebrates of*

North Africa) was published posthumously in 1979. Arambourg retired from the Muséum National d'Histoire Naturelle in 1955, but he continued to write and conduct research. He served as President of the Société Géologique de France (Geological Society of France) in 1950, President of the Société Préhistorique Française (French Prehistoric Society) in 1956, and he was President of the Fourth Pan-African Congress on Prehistory held in Leopoldville, the Belgian Congo, in 1959. When the **International Omo Research Expedition** was established in 1966 Arambourg became the leader of the French contingent and participated in a limited fashion in its first three field seasons from 1967 until his death in 1969. Regarding human evolution, Arambourg argued that there were long periods of relative biological stability as well as stability in the stone tool industries as seen in the archeological record, punctuated by periods of rapid change. Moreover, he believed that human biological and cultural evolution was progressive and proceeded along successive stages. As a result, Arambourg considered what he referred to as **australopithecine**s, **pithecanthropine**s, **Neanderthal**s, and *Homo sapiens* to represent successive stages of human evolution.

Aramis (Location 10°28′N, 40°26′E, Ethiopia; etym. the site is named after a tributary of the Awash River) History and general description Aramis, the type site of *Ardipithecus ramidus*, is situated between the headwaters of the Aramis and Adgantoli drainages on the west side of the Awash River in the **Middle Awash study area** in the Afar Depression in the **Afar Rift System**. All of the localities (**ARA-VP**) are on exposures of the approximately 300 m-thick Sagantole Formation. Temporal span *c.*4.2–4.1 Ma. Hominin-bearing localities in the Aramis Member are dated by **argon-argon dating** to 4.419 ± 0.058 Ma and 4.416 ± 0.031 Ma. The hominin-bearing locality in the Adgantole Member is constrained by argon-argon dates of 4.041 ± 0.060 Ma for an overlying volcanic stratum (MA 94–55C), and 4.317 ± 0.055 for the underlying Kullunta basaltic tuff (KUBT). An age of approximately 4.12 Ma has been assigned based on additional dating through **biostratigraphy** of the Aramis Member and from **magnetostratigraphy**. How dated? Argon-argon dating, biostratigraphy, magnetostratigraphy. Hominins found at site Specimens recovered from the Aramis Member attributed to *Ar. ramidus* include the **holotype ARA-VP-1/1** and the

associated skeleton, **ARA-VP 6/500**. A left maxilla, ARA-VP-14/1, has been attributed to *Australopithecus anamensis*. Archeological evidence found at site None. Key references: Geology and dating WoldeGabriel et al. 1994, 2009, Renne et al. 1999, White et al. 2006b; Hominins White et al. 1994, 2006, 2009; Archeology N/A.

arboreal (L. *arbor* = tree) The term used to describe animals that live in trees. Some of the defining features of primates (binocular vision, generalized limb structure retaining a clavicle) are important for life in the trees (e.g., binocular vision helps animals to judge distances). Although Elliott Smith and Wood Jones' arboreal theory, in which they argued that tree living was key to primate origins, has now fallen out of favor, arboreality remains an important trait within the order. Today and in the past many primates use trees for moving, foraging, and resting. Indeed, the vast majority of primates are dependent on trees, and the platyrrhines are exclusively arboreal. Nonetheless, many living and extinct primates have successfully radiated into **terrestrial** niches, or have combined life in the trees with life on the ground. The early hominins are a good example of this combination of arboreality and terrestriality, and some researchers have suggested that at least one form of hominin bipedalism may have emerged as a way of moving or foraging in trees (Thorpe et al. 2007). *See also* **locomotion**.

arboreality (L. *arbor* = tree) The tendency to live partially, or wholly, in the trees. *See* **arboreal**.

"archaeomorphous" *Homo sapiens See* **Reilingen calvaria**.

Archanthropinae (Gk *arch* = first or primary and *anthropos* = man) A subfamily apparently introduced by Franz **Weidenreich** in 1946. In the second of his five 1945 Hitchcock Lectures entitled the "The Development of the Specifically Human Form" Weidenreich refers to the introduction by Sir Arthur **Keith** (1925) of the terms "neoanthropic man" and "paleoanthropic man", for modern humans "with all its variants" and Neanderthal man, respectively (Weidenreich 1946, p. 28). Weidenreich proposes that "…only a small step is required to alter the terms in use and to designate all the groups or subfamilies represented by these types as Neoanthropinae and

Paleoanthropinae, respectively" (*ibid*, pp. 28–9). He then goes on to suggest that "…when *Sinanthropus* and *Pithecanthropus* are measured according to the characteristics of the Paleoanthropinae, they reveal new features which are not found in this group and, therefore, have to be classified as a more primitive subfamily, for which the name Archanthropinae ("primary man") seems to be adequate." (*ibid*, p. 29). In a subsequent publication Weidenreich (1951) referred *Homo soloensis* to the Archanthropinae. Robert **Broom** was presumably interpreting Archanthropinae in a different way when he used it as a higher taxon for *Australopithecus prometheus* (Broom 1950), but thereafter the subfamily Archanthropinae fell into disuse.

Archi (Location 38°08′20″N, 15°39′38″E in Calabria, Italy; etym. from the nearby village) History and general description Roadwork on the hill of San Francesco d'Archi yielded an immature hominin mandible in 1970. The stratigraphic and geological context were subsequently investigated and published along with an initial description of the specimen. Temporal span and how dated? **Uranium-series dating** on marine molluscs in overlying level suggests an age of more than 40 ka. Hominins found at site Archi 1, a mandible, was found in layer C-3, a gravelly layer containing numerous vertebrate faunal remains. The specimen was mostly likely about 3 years old at the time of death and can be confidently assigned to *Homo neanderthalensis*. Archeological evidence found at site None. Key references: Geology, dating, and paleoenvironment Ascenzi and Segre 1971; Hominins Mallegni and Trinkaus 1997; Archeology N/A.

arcuate fasciculus (L. *arcuatus* = bent like a bow and L. *fascis* = bundle) A bundle of axons in each cerebral hemisphere that connects Wernicke's area and **Broca's area** on the dominant side, as well as other areas of the temporal cortex and frontal cortex in both hemispheres. In modern humans, damage to this pathway can cause a language deficit known as conduction aphasia, which is a difficulty in repeating words that have been heard; otherwise language comprehension and production are intact. Comparative neuroimaging studies have shown that the arcuate fasciculus in modern humans is larger and makes more extensive interconnections, and thus links a greater number of areas in the temporal cortex and frontal cortex, as compared

with chimpanzees and macaque monkeys. These specializations of the modern human arcuate fasciculus may be important in conveying the stored meaning of words in the temporal cortex to the frontal cortical areas that are involved in sentence comprehension and construction.

Arcy-sur-Cure (Location 47°35′N, 3°45′E, France; etym. after a nearby town) The many caves that constitute the site of Arcy-sur-Cure lie in the limestone cliffs above the Cure river in central France. They were originally excavated for saltpeter, and then since the 1850s they have been excavated with different levels of competence by amateur and professional archeologists. The site comprises a complex of more than 10 named caves, including the **Grande Grotte**, the **Grotte du Renne**, the **Grotte du Hyène**, the **Grotte des Fées**, and the **Grotte du Loup** among others (Leroi-Gourhan 1958). The Grande Grotte is best-known for its *c*.28–33 ka cave paintings, which are the second-oldest such paintings in France. The Grotte du Renne is best known for several beads and objects of personal ornamentation found in the **Châtelperronian** level, alongside a juvenile *Homo neanderthalensis* and several isolated Neanderthal teeth. Researchers debate whether these finds reflect independent Neanderthal invention of so-called modern behaviors, or evidence of **acculturation** from interaction with modern humans, or if the finds were the product of Neanderthals at all. In the Grotte du Hyène, several hominin remains, including a nearly complete mandible with dentition, were recovered from the lower **Mousterian** levels. A mandible attributed to *Homo neanderthalensis* was found in the Grotte des Fées by de Vibraye in 1859, but researchers at the site have argued on morphological and stratigraphic grounds that this mandible likely represents a Neolithic burial. The Grotte du Loup is a minor site, preserving only a few Châtelperronian and Mousterian tools, but contains evidence of a possible Neanderthal burial. For more information, see the individual entries about each cave.

Ardipithecus **White et al., 1995** (etym. *ardi* = ground or floor in the Afar language and Gk *pithekos* = a postfix that means ape or "ape-like") A genus established in 1995 by White et al. (1995) to accommodate the species *Ardipithecus ramidus*. The authors claimed the new genus was justified because the type species, *Ar. ramidus* (*c*.4.4 Ma), was significantly more primitive than species previously assigned to *Australopithecus*. Subsequently a second more primitive and temporally older (*c*.5.8–5.2 Ma) species, *Ardipithecus kadabba*, was recognized and included in the same genus (Haile-Selassie 2001). Type species *Ardipithecus ramidus* (White et al., 1994) White et al., 1995. *See also Ardipithecus ramidus*; *Ardipithecus kadabba*.

Ardipithecus kadabba Haile-Selassie, 2001

(etym. *see Ardipithecus*, and *kadabba* = a "family ancestor" in the Afar language) A hominin subspecies with the same name was established by **Haile-Selassie** (2001), but it was subsequently elevated to species rank by Haile-Selassie et al. (2004b). The new species was established to accommodate the original cranial and postcranial remains announced in 2001 and six new dental specimens announced in 2004. All of the hypodigm was recovered from five *c*.5.8–5.2 Ma localities in the **Middle Awash study area**, Ethiopia. Four of the localities, **Saitune Dora**, Alayla, **Asa Koma**, and Digiba Dora, are in a region called the **Western Margin**, and one, **Amba East**, is in the **Central Awash Complex** of the Middle Awash study area. The main differences between *Ar. kadabba* and *Ardipithecus ramidus* are that the apical crests of the upper canine crown of the former taxon are longer and the P_3 crown outline of *Ar. kadabba* is more asymmetrical than that of *Ar. ramidus*. The morphology of the postcranial evidence is generally ape-like. Haile-Selassie et al. (2004) suggest that there is a morphocline in upper canine morphology, with *Ar. kadabba* exhibiting the most ape-like morphology, and *Ar. ramidus*, *Australopithecus anamensis*, and *Australopithecus afarensis* interpreted as becoming progressively more like the lower and more asymmetric crowns of later hominins (see Fig. 1D in Haile-Selassie et al. 2004). The proximal foot phalanx (AME-VP-1/71) combines an ape-like curvature with a proximal joint surface that is like that of *Au. afarensis* (Haile-Selassie 2001). Some researchers have suggested that because of its ape-like dental morphology the case for *Ar. kadabba* being a hominin is a relatively weak one. The fossil evidence for *Ar. kadabba* and its context are reviewed in Haile-Selassie and Wolde-Gabriel (2009). First discovery **ALA-VP-2/10** (1997). Holotype ALA-VP-2/10 (1997). Paratypes ALA-VP-2/11, -2/101; ASK-VP-3/78, -3/160; DID-VP-1/80; STD-VP-2/62,

2/63, 2/893; AMF-VP-1/71. Main sites Middle Awash study area, Ethiopia.

Ardipithecus ramidus (White et al., 1994) White et al., 1995

(etym. *see Ardipithecus* and *ramid* = "root" in the Afar language) Hominin species established in 1994 by White et al. to accommodate cranial and postcranial fossils recovered from *c*.4.5–4.4 Ma localities at **Aramis** on the northeastern flank of the **Central Awash Complex** in the **Middle Awash study area**, Ethiopia. The authors claimed *Ar. ramidus* shares some features with living species of **Pan**, has other features that are shared with the African apes in general and, crucially, has several dental and cranial features that are shared only with later hominins such as *Australopithecus afarensis*. The taxon was initially included within the genus *Australopithecus*, but it was subsequently transferred to a new genus, *Ardipithecus* (White et al. 1995). Fossils recovered from four localities (**Saitune Dora**, Alayla, **Asa Koma**, and Digiba Dora) in the **Western Margin** region of the Middle Awash study area were initially assigned to *Ar. ramidus* as a separate subspecies, *Ardipithecus ramidus kadabba* (Haile-Selassie 2001), but these and additional specimens were subsequently transferred to separate species, *Ardipithecus kadabba* (Haile-Selassie et al. 2004). The first reported additions to the *Ar. ramidus* **hypodigm** came from the **Gona Paleoanthropological study area** (Semaw et al. 2005), but subsequently more fossils, including the **ARA-VP-6/500** associated skeleton, recovered from the Aramis locality (White et al. 2009a), as well as from two other localities, Kuseralee Dora and Sagantole, in the Central Awash Complex have also been reported (White et al. 2009a). Initial estimates based on the size of the shoulder joint suggested that *Ar. ramidus* weighed about 40 kg, but researchers claim the enlarged hypodigm indicates an estimated body mass of approximately 50 kg (Lovejoy et al. 2009d). The chewing teeth of *Ar. ramidus* are relatively small and the position of the **foramen magnum**, the form of the reconstructed pelvis, and the morphology of the lateral side of the foot have all been cited as evidence that the posture and gait of *Ar. ramidus* were respectively more upright and bipedal than is the case in the living apes. The enamel covering on the teeth is not as thin as that of chimpanzees/bonobos, but it is not as thick as that seen in archaic hominins such as *Au. afarensis* (Suwa et al. 2009a).

White et al. (2009a) claim that *Ar. ramidus* is a basal hominin, but this assessment implies that "dental and locomotor specializations evolved independently in each extant great ape genus" (Suwa et al. 2009a, p. 99), and the researchers who judged *Ar. ramidus* to be a basal hominin suggest that its discovery "highlights the alacrity with which similar anatomical structures can emerge,"…"resulting in convergent adaptations" (White et al. 2009a, p. 81). However, by including *Ar. ramidus* in the hominin clade the researchers who found and described it are implicitly assuming there is little, or no, evidence of such "convergent adaptations" (i. e., convergent evolution) in the hominin clade. Yet the inclusion of *Ar. ramidus* in the hominin clade necessitates substantial amounts of convergent evolution in the closely related clades that include the extant great apes. Indeed, the hypothesis that *Ar. ramidus* is not a hominin, but instead is a member of an extinct ape clade, would, in many respects, be more parsimonious than assuming it is a basal hominin. First discovery **ARA-VP-1/1** (1993) (N.B.: but if either a mandible, **KNM-LT 329**, from Lothagam, Kenya or the mandible **KNM-TH 13150** from Tabarin, Kenya, prove to belong to the *Ar. ramidus* hypodigm, then they would be the initial discovery). Holotype **ARA-VP-6/1** (1993). Paratypes ARA-VP-1/1–4, -1/125, -1/127–9, -1/182, -1/183, -1/200, -1/300, -1/400, -1/401, -1/500, -7/2. Main sites Localities in the Gona and Middle Awash study areas, Ethiopia.

Ardipithecus ramidus kadabba **Haile-Selassie, 2001**

A new subspecies of *Ardithecus ramidus* proposed by Haile-Selassie (2001) to accommodate 11 fossils recovered from localities in the Western Margin of the **Middle Awash study area**. When additions were made to the hypodigm Haile-Selassie et al. (2004b), the subspecies was elevated to the level of a separate species as *Ardipithecus kadabba* Haile-Selassie et al., 2004. See *Ardipithecus kadabba* **Haile-Selassie et al., 2004.**

Arene Candide

(Location 44°10′N, 8°20′E, Italy) History and general description A large cave site located on the Ligurian coast of Italy, about 90 m above sea level. Excavations by Bernabo Brea and Cardini began in the early 1940s and culminated in 1942 when the burial of Arene Candide I was discovered; WWII halted work shortly thereafter and it did not resume until 1970, under the direction of Cardini. Temporal span and how dated? Using **radiocarbon dating**, Arene Candide I assigned an age of $23,440 \pm 190$ years BP. The upper layers (3–4) were radiocarbon dated to $11,750 \pm 95$ years BP. Hominins found at site The ceremonial burial of an adolescent *Homo sapiens* male, Arene Candide I, was found in layer 10. The ornate nature of the burial prompted workers to nickname the skeleton "Il Principe" or the prince. At least seventeen other burials dating from the latest Pleistocene have been recovered, many of which are also ornately decorated. Archeological evidence found at site Arene Candide I was found in a bed of red **ochre**, with a probable cap made of perforated shells and deer bones on its head. Mammoth ivory pendants, a long flint blade, and four perforated elk antler batons also adorned the skeleton. The later burials of the site similarly include ochre and perforated shells, but also have painted pebbles and bone ornaments. A **Gravettian** industry characterizes the lower layers, whereas the upper layers have a late Epigravettian microlithic industry. Key references: Geology, dating, and paleoenvironment Pettitt et al. 2003; Hominins Cardini 1980, Sergi et al. 1974; Archeology Bietti 1994.

argon-argon dating

An isotopic dating method based upon the K/Ar system, in which radioactive ^{40}K is driven to ^{40}Ar in a reactor, and used as a proxy of K content. Subsequent analyses can be done in a single experiment, by measuring isotopes of Ar in a mass spectrometer. This approach avoids the necessity of separate measurements of K and Ar (in different aliquots of a sample), thus reducing potential error. As a result, single-crystal age determinations have become a possibility. The method has been used to date materials as young as the 79 AD eruption of Vesuvius (Renne et al. 1997) but most applications are to much older rocks. The analytical methods are so sensitive they can be applied to single crystals of feldspar or volcanic glass, and this more precise version is referred to as single-crystal laser fusion $^{40}Ar/^{39}Ar$ dating (e.g., McDougall and Brown 2008).

ARI

The acronym for the Aralee Issie collection area. ARI is used as the prefix for fossils recovered from the Aralee Issie collection area, **Woranso-Mille study area**, Afar, Ethiopia. *See also* **Aralee Issie.**

Aristotle University of Thessaloniki, Department of Geology Description The School of Geology belongs to the Faculty of Sciences of the Aristotle University of Thessaloniki founded in 1925. Hominin fossil collections **Petralona cranium**. Contact information Professor Dimosthenis Mountrakis (e-mail dmountra@ geo.auth.gr; tel +30 2310 998481) or Dimitra Christou (e-mail dimitra@-geo.auth.gr; tel +30 2310 998558; fax +30 2310 998482). Website www.geo.auth.gr/en_tomeis_gewlogias.htm.

arithmetic mean A method for calculating the average of a set of numbers, it is the method commonly referred to simply as the mean. A sample of n values are summed and divided by n. For example, the arithmetic mean of X_1, X_2, X_3, and X_4 is $(X_1+X_2+X_3+X_4)/4$. There are alternative methods to calculate the mean of a set of numbers, such as the **geometric mean**.

ARL Acronym for Abdur Reef Limestone. *See* **Abdur**.

armature (L. *armatura* = armor or equipment) A term used to refer to any item used for offense or defense. In archeology the term armature refers to most **points** and any other obvious hunting equipment (e.g., the wooden spears from **Schöningen**).

array (L. *arredare* = to arrange) An array has two meanings in molecular biology. First, it can be an "orderly arrangement" of repetitive **DNA** in the genome that has a similar sequence. Such arrays include **microsatellites** and **satellite DNA**. Second, an array is also used in the sense of a microarray, which is a piece of glass that has molecules of DNA, RNA, or protein affixed to it in an orderly arrangement. These microarrays are used to capture molecules of interest. For example, they can be used to identify copy–number variants (or CNVs), DNA expression, or proteins.

Arsuaga Ferreras, Juan Luis (1959–) Born in Madrid, Spain, Juan Luis Arsuaga displayed an early interest in human evolution and prehistory and he earned both his masters and doctoral degree in biological sciences at the Complutense University of Madrid. In 1982 Arsuaga became a member of the excavation team, headed by Emiliano Aguirre, that was investigating potential hominin cave sites in the Sierra de Atapuerca, which is 14 km/8.7 miles east of Burgos in northern Spain. The Sierra de Atapuerca contains many caves and its two most famous sites, the **Sima de los Huesos** and **Gran Dolina**, have yielded some of the earliest hominin fossils and stone tools in Western Europe. Access to the Sima de los Huesos was particularly hazardous, but by 1991 the infrastructure needed to access the cave was in place and the team began excavations with Arsuaga codirecting the effort, together with José María Bermúdez de Castro and Eudald Carbonell i Roura. The Sima de los Huesos has proved to be a rich source of well-preserved pre-modern *Homo* fossils, all of which are assigned to either *Homo neanderthalensis* or *Homo heidelbergensis*. Arsuaga also helped explore and excavate the TD6 level at the Gran Dolina site where six individuals and 268 lithic artifacts were recovered; the hominins from Gran Dolina cave have been assigned to *Homo antecessor*. In 1997, the Atapuerca team won the Prince of Asturias Award for Technical and Scientific Research and the Castilla y León Prize in social sciences and humanities, and in 2000 UNESCO added the Pleistocene sites of the Sierra de Atapuerca to their list of World Heritage Sites. Currently, Arsuaga is a professor at the Complutense University of Madrid in the Faculty of Geological Sciences in the Paleontology Department where he serves as the Director of the Center of Evolution and Human Behavior (UCM-ISCIII). Arsuaga is vice-president of the Commission of Human Paleontology and Paleoecology of the International Union Quaternary Research (INQUA). He is a Foreign Member of the US National Academy of Sciences and a member of the Real Academia de Doctores de España.

art (L. *art* = art) The term art refers to the use of non-utilitarian images for symbolism or self-expression. Prehistoric art is divided into mobile (L. *mobilis* = to move) or portable art (e.g., small figurines), and parietal (L. *paries* = wall) or fixed art (e.g., wall paintings or engravings).

artifact Any portable object made, modified, or used by hominins. The earliest artifacts presently known are stone tools and their manufacturing debris from the site of Gona, Ethiopia, dating to 2.55 Ma, although indirect traces of stone tool use may be preserved as cutmarks on bones from the surface at Dikika, Ethiopia dating to 3.39 Ma, but the latter will require additional finds of comparable age for wide acceptance. Because they preserve well, stone tools form the largest part of the early archeological record, although comparison with extant primates suggests that artifacts of perishable materials (e.g., wood or fiber) remain difficult to detect and their origins may well remain in the realm of speculation. Artifacts are one of the fundamental units of analysis for archeologists.

artiodactyl The informal name for the Artiodactyla, the mammalian order that includes all the taxa with an even number of hoofed toes. *See also* **Artiodactyla**.

Artiodactyla (Gk *artios* = even and *daktulos* = toe; literally, the "even-toed") The name of the mammalian order that includes all the taxa with an even number of hoofed toes. Artiodactyl fossils are common within some hominin faunal communities. The order Artiodactyla is now often subsumed within the clade Cetartiodactyla, named to reflect the recent **DNA** studies showing that whales and artiodactyls are closely related (e.g., Ursing and Arnason 1998). There are both ruminating, called ruminants, and non-ruminating groups of artiodactyls. Artiodactyls are **terrestrial** and largely **herbivorous**, although some artiodactyl taxa (e.g., the **Suidae**) are more omnivorous. The most diverse family of artiodactyls is the **Bovidae** (antelopes and their allies), although in Africa other common families within this order are the Suidae (pigs), the Hippopotamidae (hippopotamus), and the Giraffidae (giraffes). Other, less common artiodactyls in African hominin sites are the Camelidae (camels) and the **Tragulidae** (chevrotains). In Eurasia, the **Cervidae** (deer) are common at hominin sites, as are the **Moschidae** (musk deer). There are other artiodactyl families, both extant and extinct, but these are not known from hominin sites. The earliest known artiodactyl, *Diacodexis*, is from the early Eocene of Portugal.

AS Acronym for Abdur South. *See* **Abdur**.

As Duma (Location 11°10′N, 40°20′E, Ethiopia; etym. named after a local drainage) History and general description Site on the west side of the Awash River in the **Gona Western Margin** sector of the **Gona Paleoanthropological Research Project** study area in the Afar Depression. Located in the headwaters of the Busidima and Gawis drainages. Temporal span Sediments are *c*.4.4 Ma. Hominins found at site Remains of *Ardipithecus ramidus* have been recovered from several localities. Archeological evidence found at site None. Key references: Geology, dating, and hominins Semaw et al. 2005.

Asa Issie (Location 11°10′N, 40°20′E, Ethiopia; etym. named after a local drainage) History and general description An area of fossiliferous sediments 10 km/6 miles west of **Aramis**. The hominin-bearing localities called Asa Issie and Hana Hari are in exposures of the Adgantole Member of the Sagantole Formation. Temporal span The sediments have been dated using **biochronology** to *c*.4.2–4.1 Ma. Hominins found at site The hominin fossils recovered there, that include a partial maxilla (ARA-VP-14/1), two sets of associated teeth (ASI-VP-2/2 and -2/334), and three postcranial specimens (ASI-VP-2 and -5, and ASI-VP-5/154), have all been assigned to *Australopithecus anamensis*, but are claimed to be transitional between *Ardipithecus ramidus* and *Au. anamensis*. Archeological evidence found at site None. Key references: Geology, dating, and hominins White et al. 2006b.

Asa Koma An area of fossiliferous exposures that lies against the **Western Margin** of the Afar Depression section of the **East African Rift System** in the **Middle Awash study area**, Ethiopia.

ascertainment bias (L. *ad* = near and *certus* = to determine, so certain came to mean "indisputable"; ascertainment is the process of discovering something indisputable by experimentation, so ascertainment bias is a synonym for experimental bias) Refers to the circumstances when nonrandom sampling produces false results. Such biases can lead to incorrect inferences about an entire **population** either because of distorted or non-typical sampling of the population, or because the data (i.e., a specific marker) used for the analysis were identified in a biased way. For example, in modern humans many of the first **polymorphic** markers were identified through surveys of Europeans. When these markers were examined, Europeans were found to be the most diverse of the limited numbers of modern humans included in the sample. However, when the modern human samples were expanded to better reflect modern humans across the world (i.e., the geographic representation of "modern humans" was unbiased) the results then showed that modern humans from Africa and not

Europeans are the most diverse (Rogers and Jorde 1996, Wakeley et al. 2001).

Asfaw, Berhane (1954–) Born in Gondar Province, Ethiopia, Berhane Asfaw taught chemistry and physical science in high school in Gondar before moving in 1978 to Addis Ababa to work in the Centre of Research and Conservation of Cultural Heritage in the Ministry of Culture and to attend Addis Ababa University, graduating with a BSc in geology in 1980. He then moved to the University of California, Berkeley, USA where he studied for an MA and PhD, gaining the latter in 1988 for a thesis entitled "New perspectives on the evolution of the hominid frontal bone." He returned to Addis Ababa where he became paleoanthropology coordinator at the Centre for Research and Conservation of Cultural Heritage and then in 1990 he was appointed the Director of the National Museum of Ethiopia, a post he held until 1992. Since 1997 he has managed the Rift Valley Research Service. Berhane Asfaw's paleoanthropological fieldwork began in 1980 when he worked at **Melka Kunturé**, but his important involvement with the **Middle Awash Research Project** began in 1981. In 1983 he reported the discovery of a parietal fragment from **Bodo** and since then he has been either the first author, or a co-author, of papers that have reported the discovery of hominin fossil evidence from **Olduvai Gorge** (Johanson et al. 1987), **Belohdelie** (Asfaw 1987), **Maka** (White et al. 1993), **Konso** (Suwa et al. 1997), **Bouri** (Asfaw et al. 2002), **Herto** (White et al. 2003), and **Aramis** (White et al. 2009a), and announced new taxa, including *Australopithecus ramidus* (White et al. 1994) and *Australopithecus garhi* (Asfaw et al. 1999). He was also the lead organizer of the Paleoanthropological Inventory of Ethiopia. In recognition of his contributions to paleoanthropology, in 2008 Berhane Asfaw was elected as a foreign associate of the US National Academy of Sciences and in 2010 a fellow of the Ethiopian Academy of Sciences. *See also* **Paleoanthropological Inventory of Ethiopia**.

ASI Acronym for the Asa Issie collection area. It is used as the prefix for fossils recovered from **Asa Issie, Middle Awash study area**, Ethiopia.

ASK Acronym for the Asa Koma collection area. It is used as the prefix for fossils recovered from **Asa Koma, Western Margin, Middle Awash study area**, Ethiopia.

aspartic acid (etym. a non-essential amino acid found in asparagus, hence its name) It is an empirical observation that fossil bones in which (a) the amino acid content is more than 30,000 parts per million, (b) the ratio of glycine to aspartic acid is between 2 and 10, or (c) the amount of racemized aspartic acid is low (D/L ratios <0.10) are likely to contain sufficient DNA to warrant further analysis (Poinar et al. 1996). *See also* **amino acid racemization dating**.

ASPM *See* **brain size evolution, molecular basis of**.

assemblage Generally, any collection of objects. For archeologists, an assemblage is a stratigraphically-bounded, spatially associated, set of **artifacts**. For example, a single archeological site may contain several artifact assemblages (e.g., as is the case at many of the excavations at **Olduvai Gorge**). These may derive from different **strata** at the site, or from different **facies** within the same stratum (e.g., from channel and floodplain deposits of the same river system). Assemblages form one of the basic comparative units above the level of single artifacts or artifact types, and variations within an assemblage in the range of tool types found have been used to infer past site function(s) (e.g., Binford and Binford 1969 for French **Mousterian** sites). Also used as an inclusive term to describe all the paleontological evidence from a site.

assimilation model A model for the origin of modern humans that accepts an African origin for the biologically complex characteristic of modern humans but rejects a total replacement of local archaic populations, including *Homo neanderthalensis*, by modern humans as they spread into Eurasia. This perspective grew out of morphological observations suggesting the archaic contribution to modern human populations throughout Eurasia (and possibly parts of Africa) was always relatively small, though not insignificant, and was limited to morphological details rather than fundamental morphological gestalt. First fully articulated in 1989 (Smith et al. 1989), aspects of the assimilation model were emerging by the mid-1980s. Beginning at this time, the application of thermo-**luminescence dating** and **electron spin**

resonance spectroscopy dating to fossil samples in Africa, the Near East, and Europe began to show that modern human anatomy clearly was established earlier outside of Europe than within that continent and also that Neanderthals survived until relatively late in some portions of Europe. This, along with early observations on the origin of extant human mitochondrial DNA, appeared to weaken the possibility of a classical multiregional hypothesis for modern human origins. Some "intermediate models," such as the replacement with hybridization model accept the theoretical possibility of admixture but reject any morphological evidence for it. In contrast, the assimilation model has consistently held that evidence for continuity in morphological details is demonstrable in late archaic and early modern human samples in various portions of Eurasia. The assimilation model is consistent with the mostly out-of-Africa genetic model of modern human origins and with the recent evidence from the Neanderthal genome project indicating low levels of Neanderthal contribution to modern Eurasian populations. Recent statements of the assimilation model are found in Smith et al. (2005), Trinkaus (2005, 2007), Smith (2010).

associated skeleton (L. *associare* = to join with) Refers to a fossil specimen that includes more than one skeletal element from the same individual. The term is sometimes used in a more restricted sense to refer to specimens that include evidence of *both* the skull and the postcranial skeleton, or *more than one* part of the postcranial skeleton. Associated skeletons are particularly valuable specimens for several reasons. First, most fossil hominin taxa are diagnosed and identified on the basis of skull morphology, so associated skeletons that include skull and postcranial elements help sort out which limb bones go with which skulls. For example, the lack of a securely associated skeleton of *Paranthropus boisei* that preserves both taxonomically distinctive skull evidence *and* evidence of the postcranial skeleton is hampering attempts to sort postcranial fossils from East Africa into taxa. Second, because the evidence is from a single individual well-preserved associated skeletons allow researchers to compare the size of the teeth with the rest of the body, or the relative lengths of the limbs and/or limb segments, or the relative sizes of joint surfaces. Examples of associated skeletons include **A.L. 288-1** (*Australopithecus afarensis*), **KNM-WT 15000** (*Homo ergaster*), and **Dederiyeh 1** (*Homo neanderthalensis*).

association cortex (L. *cortex* = husk, shell) This term refers to areas of the **neocortex** that are *not* concerned with the primary processing of sensory or motor information. The association cortex comprises part of the **prefrontal cortex**, the limbic cortex, and large regions of the parietal cortex, temporal cortex, and occipital cortex. The association cortices are concerned with the higher-order processing and integration of sensory modalities, abstract thinking, and the planning of movements.

asterionic notch (Gk *aster* = star) The asterion is a bilateral bony landmark on the **cranial vault** where the lambdoidal, parietomastoid, and occipitomastoid sutures meet. In modern human crania the mastoid angle of the parietal, the lateral margin of the occipital squame, and the posterior angle of the mastoid portion of the temporal bone articulate in an "edge-to-edge" fashion. But in the African apes, in which **compound temporal-nuchal crest**s frequently dominate this part of the vault, the mastoid angle of the parietal is a laterally flaring, nearly horizontal, shelf of bone that fits into a narrow, reciprocal slot in the temporal bone between the base of the squamous and the mastoid portion; the mastoid fits between the flared mastoid angle superiorly and the lateral margin of the nuchal plane (occipitomastoid suture) medially (Kimbel and Rak 1985). Kimbel referred to this morphology as the "asterionic notch" (White et al. 1981, p. 456), whereas Delattre and Fenart (1960) called the tip of the mastoid angle the "angle incisural du parietal," and the deepest (i.e., most lateral) part of the female part of the bevel in the temporal the "point incisural externe". This type of asterionic notch articulation is also seen in most crested *Australopithecus afarensis* crania (e.g., **A.L. 333-45**, **A.L. 444-2**), including some small, presumably female, individuals (e.g., A.L. 162-28). Large, presumed male, *Paranthropus boisei* crania exhibit different anatomy in this region: the posterior part of the temporal squame overlaps the non-flared mastoid angle of the parietal, which, in turn, overlaps the occipital squame in a beveled imbrication of these bones. In this species the overlap of the temporal squame on the parietal is superiorly extensive further anteriorly, as evidenced in specimens with missing or damaged temporal bones by rugose ridges running superomedially high up on the exposed parietal surface toward the midline of the vault. *See also* **striae parietalis**.

astrochronology *See* **astronomical time scale**.

astronomical theory (Gk *astron* = star and *kronos* = time) Joseph Adhemar, James Croll, and John Murphy each suggested the intensity of the sun's rays could drive glacial cycles, but it was Milutin Milanković, a Serbian astronomer, who provided the mathematical foundation for what became known as astronomical theory. The theory suggests that cyclic changes in three important aspects of the Earth's **orbital geometry (precession, obliquity,** and **eccentricity**) largely determine long-term changes in **climate**. Evidence supporting the theory of astronomical pacing of Earth's climate came from cyclical events in ocean sediments (Hays et al. 1976). *See* **astronomical time scale**.

astronomical time scale (Gk *astron* = star and *kronos* = time) A geological time scale based on regular changes involving three aspects of the Earth's orbital geometry (i.e., the way it rotates about its axis and the shape of its orbit around the sun), namely **precession, obliquity,** and **eccentricity**. Precession, or the "wobble" of the Earth's axis of rotation, has a cycle that completes every 19–23 ka. Precession controls the seasonality of the intensity of the sun's rays on the Earth's surface (called **insolation**). For example, stronger summer insolation means stronger **monsoon** intensity. Obliquity, or the tilt of the Earth's axis, has a dominant periodicity of *c*.41 ka and it controls the length of the winter polar darkness. Eccentricity, or the elliptical nature of the Earth's orbit, has *c*.100 and *c*.400 ka cycles. It is thought that obliquity determines the timing (the pacing) of the northern hemisphere glacial cycles with less summer insolation at 65°N resulting in less summer melting and more glacier growth. Note that only eccentricity changes insolation, and even then by a very small amount; the other orbital cycles change only the seasonality or latitudinal distribution of insolation. The climate has not always been uniformly sensitive to these ongoing orbital cycles. Precession has had a long-term influence on the strength of the monsoons (the "23 ka world"). Only at times of global cooling and **northern hemisphere glaciation**, as has been the case for the last 3 Ma, do strong signals of obliquity (the "41 ka world") emerge; obliquity paces the timing of glacial/interglacial cycles. For the last 1 Ma, *c*.100 ka-long cycles appear to dominate (called the "100 ka world"). Eccentricity might be driving these cycles, or ice-sheet dynamics might instead be responding to multiples of two or three obliquity cycles. The regularity of these various cycles is such that astrochronology can be used to calibrate, or "tune," other forms of age estimation. (syn. astrochronology). *See also* **orbital tuning**.

AT Abbreviation of Arala Issie Tuff.

Atapuerca (Location 42°21′N, 3°31′W, Spain) The Sierra de Atapuerca is a series of eroded limestone hills 14 km/9 miles east of Burgos in Northern Spain. It is permeated by several sediment-filled cave systems, one of which is the **Cueva Mayor-Cueva del Silo**, and within this system are several cave/fissure complexes including the Sima de los Huesos. The first hominin fossils were recovered from the Sima de los Huesos (or **SH**) in 1976 and since then more than 6500 hominin fossils have been recovered from this location. Just 1 km/0.6 miles away is another system called **Trinchera del Ferrocarril**, which includes several sites (e.g., **Galería**, Gran Dolina, and Sima del Elefante), that have yielded hominin fossils and archeological evidence. Excavations at Gran Dolina (or **TD**) began in 1978 and six levels (TD3–4, 5–7, 10–11) have produced stone artifacts. In 1994 more stone tools and the first hominin fossils were recovered from the **Aurora stratum**, a lithostratigraphic level within **TD6**. The Sima del Elefante site, which may be 1.2 Ma, has provided the oldest hominin remains from the Atapuerca hills. *See also* **Gran Dolina**; *Homo antecessor*; *Homo heidelbergensis*; *Homo neanderthalensis*; **Sima del Elefante**; **Sima de los Huesos**.

atavism (L. *atavus* = ancestor, from *atta* = father plus *avus* = grandfather) The term used for a morphological variant or anomaly that usually appears in what is assumed to be an ancestor (recent or distant). Thus, the occasional appearance as a normal variant of contrahentes muscles in the palm of the hand of modern humans (Cihak 1972) would be an example of an atavism, as would be the presence of a rhomboideus occipitalis (Aziz 1981), or an independent **os centrale** (Cihak 1972) in the carpal bones of the hand of a modern human (in modern humans the os centrale normally fuses with the main part of the scaphoid to form the scaphoid tubercle).

atavistic   *See* atavism.

ATD   Acronym for the Atapuerca – Gran Dolina. It is the prefix for fossils recovered from the **Gran Dolina** site within the Sierra de **Atapuerca**, northern Spain. The prefix ATD is followed by an Arabic number that indicates the **lithostratigraphic** level, then another Arabic number that indicates where in the sequence of **hominins** discovered in the cave that particular specimen was found (e.g., **ATD6-96** is the 96th hominin recovered from the ATD6 lithostratigraphic level).

ATD6-1 through ATD6-12 Site Gran Dolina. Locality "6 m^2 planar section" of TD6 (i.e., in the Aurora stratum). Surface/*in situ* *In situ*. Date of discovery July, 1994. Finders Team led by Eudald Carbonell and Jóse Maria Bermúdez de Castro. Unit N/A. Horizon TD6. Bed/member N/A. Formation N/A. Group N/A. Nearest overlying dated horizon N/A. Nearest underlying dated horizon N/A. Geological age Pre-780 ka (TD6 is assumed to be in the **Matuyama chron**) and uranium-series dating and **electron spin resonance spectroscopy dating** age is 731 ± 63 ka. Developmental age Young adult. Presumed sex Unknown. Brief anatomical description All of the remains, the mandible ATD6-5, plus the isolated teeth listed as ATD6-1–6-4 and ATD6-6–6-12, belong to a single individual, and they comprise the **holotype** of *Homo antecessor*. Announcement Carbonell et al. 1995. Initial description Carbonell et al. 1995. Photographs/line drawings and metrical data Carbonell et al. 1995. Detailed anatomical description Bermúdez de Castro et al. 1997, 1999, Rosas and Bermúdez de Castro 1999. Initial taxonomic allocation The holotype of *H. antecessor*. Taxonomic revisions None. Current conventional taxonomic allocation *H. antecessor*. Informal taxonomic category Pre-modern *Homo*. Significance The holotype of *H. antecessor*. Location of original Centro Nacional de Investigación sobre la Evolución Humana, Burgos, Spain.

ATD6-15 Site Gran Dolina. Locality "6 m^2 planar section" of TD6 (i.e., in the Aurora stra-

tum). Surface/*in situ* *In situ*. Date of discovery July, 1994. Finders Team led by Eudald Carbonell and Jóse Maria Bermúdez de Castro. Unit N/A. Horizon TD6. Bed/member N/A. Formation N/A. Group N/A. Nearest overlying dated horizon N/A. Nearest underlying dated horizon N/A. Geological age Pre-780 ka (TD6 is assumed to be in the **Matuyama chron**) and **uranium-series dating** and **electron spin resonance spectroscopy dating** age is 731 ± 63 ka. Developmental age Adolescent. Presumed sex Unknown. Brief anatomical description Frontal bone fragment with part of the right supraorbital torus. Announcement Carbonell et al. 1995. Initial description Carbonell et al. 1995. Photographs/line drawings and metrical data Carbonell et al. 1995. Detailed anatomical description Arsuaga et al. 1999. Initial taxonomic allocation *Homo antecessor*. Taxonomic revisions None. Current conventional taxonomic allocation *H. antecessor*. Informal taxonomic category Pre-modern *Homo*. Significance Most complete cranial vault fragment of *H. antecessor*. Location of original Centro Nacional de Investigación sobre la Evolución Humana, Burgos, Spain.

ATD6-69 Site Gran Dolina. Locality "6 m^2 planar section" of TD6 (i.e., in the Aurora stratum). Surface/*in situ* *In situ*. Date of discovery 1995. Finders Team led by Eudald Carbonell and Jóse Maria Bermúdez de Castro. Unit N/A. Horizon TD6. Bed/member N/A. Formation N/A. Group N/A. Nearest overlying dated horizon N/A. Nearest underlying dated horizon N/A. Geological age Pre-780 ka (TD6 is assumed to be in the **Matuyama chron**) and **uranium-series dating** and **electron spin resonance spectroscopy dating** age is 731 ± 63 ka. Developmental age Juvenile. Presumed sex Unknown. Brief anatomical description Much of the lower part of the face of a juvenile hominin. It includes the nasal floor and the floor of the right orbit, plus the roots and crowns of the right I^2, P$^{3-4}$, and M^1, and the left P^3 and M^1. Announcement Bermúdez de Castro et al. 1997. Initial description Bermúdez de Castro et al. 1997. Photographs/line drawings and metrical data Bermúdez de Castro et al. 1997. Detailed anatomical description Arsuaga et al. 1999. Initial taxonomic allocation *Homo antecessor*. Taxonomic revisions None. Current conventional taxonomic allocation *H. antecessor*. Informal taxonomic category Pre-modern *Homo*.

Significance The most complete specimen in the hypodigm of *H. antecessor*. Location of original Centro Nacional de Investigación sobre la Evolución Humana, Burgos, Spain.

ATD6-113 Site **Gran Dolina**. Locality Between squares G-13 and F-13. Surface/*in situ In situ*. Date of discovery 2006. Finders Team led by Eudald Carbonell and Jóse Maria Bermúdez de Castro. Unit Silt layer identified as "Pep." Horizon TD6. Bed/member N/A. Formation N/A. Group N/A. Nearest overlying dated horizon N/A. Nearest underlying dated horizon N/A. Geological age Pre-780 ka (TD6 is assumed to be in the **Matuyama chron**) and **uranium-series dating** and **electron spin resonance spectroscopy dating** age is 731 ± 63 ka. Developmental age Young adult. Presumed sex Male. Brief anatomical description The fragment consists of the left side of the mandibular corpus from the alveolus of the left M_1, the crowns and roots of the left M_{2-3}, and part of the base of the ramus including the lateral prominence. Announcement Bermúdez de Castro et al. 2008. Initial description Bermúdez de Castro et al. 2008. Photographs/line drawings and metrical data Bermúdez de Castro et al. 2008. Detailed anatomical description Bermúdez de Castro et al. 2008. Initial taxonomic allocation By inference it is assigned to *Homo antecessor*. Taxonomic revisions None. Current conventional taxonomic allocation *H. antecessor*. Informal taxonomic category Pre-modern *Homo*. Significance Resembles the morphology of the more complete ATD6-96 mandible. The combination of a reduced M_3 and a relatively primitive corpus morphology is reminiscent of the hominins from **Dmanisi**. Location of original Centro Nacional de Investigación sobre la Evolución Humana, Burgos, Spain.

ATE Acronym for Atapuerca – Sima del Elefante. It is the prefix for fossils recovered from the **Sima del Elefante** site within the Sierra de **Atapuerca**, northern Spain. The prefix ATE is followed by an Arabic number that indicates the **lithostratigraphic** level, then another Arabic number that indicates where in the sequence of hominins discovered in the cave that particular specimen was found (e.g., **ATE9-1** is the first hominin recovered from the ATE9 lithostratigraphic level).

ATE9-1 Site **Sima del Elefante**. Locality Square I-31. Surface/*in situ In situ*. Date of discovery 2007. Finders Team led by Eudald Carbonell and Jóse María Bermúdez de Castro. Unit "c". Horizon TE9. Bed/member N/A. Formation N/A. Group N/A. Nearest overlying dated horizon TE9c (see below). Nearest underlying dated horizon N/A. Geological age A combination of **biostratigraphy** using rodents and insectivores, an observed reversed magnetic polarity (consistent with the TE16 and older **lithostratigraphic** layers being in the **Matuyama chron**), and **cosmogenic nuclide dating** using ^{26}Al and ^{10}Be gives an age of 1.22 ± 0.16 Ma for TE9c suggests the ATE9-1 mandible is 1.2–1.1 Ma. Developmental age Adult. Presumed sex Unknown. Brief anatomical description The mandibular **corpus** fragment extends from the alveolus of the left P_4 to the alveolus of the right M_1. The broken crowns of the right I_2 and C and the left C and P_3 are preserved, as are the roots of the left P_3 and I_2, and the right P_3 and P_4. The crown and root of the left P_4 is also preserved. Announcement Carbonell et al. 2008. Initial description Carbonell et al. 2008. Photographs/line drawings and metrical data Carbonell et al. 2008. Detailed anatomical description N/A. Initial taxonomic allocation Provisionally assigned to *Homo antecessor*. Taxonomic revisions None. Current conventional taxonomic allocation *H. antecessor*. Informal taxonomic category Pre-modern *Homo*. Significance At the time of its discovery it was the earliest reliably dated hominin from Europe. Location of original Centro Nacional de Investigación sobre la Evolución Humana, Burgos, Spain.

Aterian (etym. the name is based on Bir el Ater, a community near the **type site** of this industrial complex) An **industrial complex** of the African **Middle Stone Age**. The type site for the Aterian is Oued Djebbana, Algeria (Morel 1974). Aterian sites occur in northern Africa, from the **Maghreb** (i.e., the northern parts of Morocco, Algeria, and Tunisia), south to Niger, throughout the Sahara and east to Egypt. **Tang**ed pieces (especially **point**s and **scrapers**) as well as **bifacially worked points** characterize Aterian sites, and these co-occur with a range of otherwise typical **Mousterian** elements such as side scrapers and diverse **Levallois** flakes and cores, as well as **blades** (Wendorf and Schild 1992). The tangs are considered an adaptation for hafting, and the tanged (or pedunculate) pieces are sufficiently distinct that Clark (1988) and McBrearty and Brooks

(2000) use the Aterian as one of the strongest cases for early regional diversity in the archeological record of the Middle Stone Age. Aterian sites may be locally contemporary with other assemblages that lack tanged pieces, and the extent to which the tanged pieces reflect temporal or group variation rather than activity-specific elements of a more generalized toolkit remains unclear (Wendorf and Schild 1992). Aterian sites were formerly considered to date to less than 40 ka, but subsequent age estimates by optically stimulated **luminescence dating**, thermoluminescence dating, and **electron spin resonance spectroscopy dating** methods suggests that this age simply represents the upper limit of the conventional **radiocarbon dating** method previously used. New results from **Mughret el 'Aliya** and other sites suggest an antiquity of approximately 35–90 ka for the Aterian (Garcea 2004). Although limited evidence from some sites suggests occupation during conditions more humid than the present, Aterian populations occupied a range of habitats, including desert areas in the present Sahara (Garcea 2004, Vanhaeren et al. 2006). A perforated *Nassarius gibbosulus* shell from Oued Djebbana was transported more than 200 km/144 miles from the Mediterranean Sea (Morel 1974, Vanhaeren et al. 2006), and green silicified tuff used to make artifacts at Adrar Bous, Algeria, were apparently transported 280 km/ 170 miles (Clark 1993). Such long-distance transport strongly suggests that Aterian populations were either highly mobile and/or had highly developed exchange networks, and it is consistent with expectations for desert-adapted populations. At **Dar es Soltane II**, Témara (**Smuggler's Cave**), and Zouhrah (El Harhoura), Aterian artifacts are associated with hominin fossils generally considered to be robust examples of *Homo sapiens*.

Atlanthropus Arambourg, 1955 (Gk *Atlas* = refers to the proximity of the site to the Atlas Mountains, a system of ranges and plateaus in northwest Africa extending from Morocco to Tunisia, and *anthropos* = human being). The taxonomic name given by Camille **Arambourg** to the genus he established to accommodate a new species he provisionally proposed for three hominin mandibles and a cranial fragment recovered by Arambourg and Hoffstetter in June, 1954 at Ternifine (now called **Tighenif**), near Mascara, Algeria, in the foothills of the Atlas mountains. Although Arambourg noted the new mandibles

were "very closely related to the Asiatic *Pithecanthropus* and *Sinanthropus*" and that "some features (are) reminiscent of the australopithecines, *Telanthropus* particularly," (Arambourg 1955, p. 195), he concluded that the fossils "cannot be identified exactly with either *Pithecanthropus* or *Sinanthropus* or *Telanthropus*" so he "assigned to them the provisional name, *Atlanthropus mauritanicus*" (*ibid*, p. 195). Arambourg's paper was published in the same year as the first edition of Wilfrid **Le Gros Clark**'s *The Fossil Evidence for Human Evolution*, but in the second, 1964, edition, when discussing the fossil record for *Homo erectus*, Le Gros Clark concludes his discussion of the remains from what is now called Tighenif by writing that "there can be little doubt…that these remains do belong to the same species (i.e., to *H. erectus*)" (Le Gros Clark 1964, p. 112). Type species *Atlanthropus mauritanicus* **Arambourg, 1955**. *See Homo erectus.*

***Atlanthropus mauritanicus* Arambourg, 1955** (Gk *Atlas* = refers to the proximity of the site to the Atlas Mountains, a system of ranges and plateaus in northwest Africa extending from Morocco to Tunisia, and *anthropos*=human being; L. *Mauritania*=the name of a province of the Roman Empire that is coincident with Morocco and Algeria). The taxonomic name given by Camille **Arambourg** to three hominin mandibles and a cranial fragment recovered by Arambourg and Hoffstetter in June, 1954 at Ternifine (now called **Tighenif**), near Mascara, Algeria, in the foothills of the Atlas mountains. Although Arambourg noted the new mandibles were "very closely related to the Asiatic *Pithecanthropus* and *Sinanthropus*" and that "some features (are) reminiscent of the australopithecines, *Telanthropus* particularly," (Arambourg 1955, p. 195), he concluded that the fossils "cannot be identified exactly with either *Pithecanthropus* or *Sinanthropus* or *Telanthropus*" so he "assigned to them the provisional name, *Atlanthropus mauritanicus*" (*ibid*, p. 195). In 1964 Wilfrid **Le Gros Clark** formally suggested *Atlanthropus mauritanicus* be sunk into *Homo erectus*. First discovery Ternifine 1, now **Tighenif 1** (1954). Holotype None formally proposed, but *de facto* it is Tighenif 1. Paratypes None proposed. Main site Ternifine (but if the Tighenif material proves to be conspecific with the hominin remains from **Gran Dolina** then *Atlanthropus mauritanicus* Arambourg, 1955 would have priority over *Homo antecessor*, but because it was just a "provisional" name according to the **International Commission on Zoological Nomenclature** (or ICZN) rules it is therefore not an available name.). *See also* **Homo antecessor**; **Homo erectus**.

Atlantic Meridional Overturning Circulation (AMOC or MOC) Describes the net northward flow of surface water in the Atlantic associated with the Gulf Stream, then it sinking in the North Atlantic to form a current known as North Atlantic deep water which flows southward until it meets with the Antarctic bottom water current, at which point it rises to the surface and either returns northward or connects via the **Antarctic Circumpolar Current** to currents in other ocean basins. The overturning circulation in the Atlantic can be conceptualized as a conveyor belt carrying warm surface water northwards and cold deep oxygenated water southwards. It is thought that changes in the AMOC could cause large changes in continental climate around the Atlantic and also globally. This is because any reduction or interruption of this "conveyor belt" would drastically reduce the heat transported by the oceans from low to high latitudes.

attrition *See* **tooth wear**.

AU *See* **Anthracotheriid Unit**; **Toros-Menalla**.

auditory cortex *See* **temporal lobe**.

auditory ossicles (L. *audire* = to hear and *ossiculum* = small bone) Three small bones in the **middle ear**, or tympanic cavity, of mammals. They connect the tympanic membrane (which is at the medial end of the external ear) with the oval window (within which is the outer of the two fluid-filled cavities of the **inner ear**). The malleus (L. *malleus* = hammer) has a head (which occupies the epitympanic recess), a handle or manubrium, and a lateral process. The handle attaches to just below the center of the tympanic membrane and the head articulates with the body of the **incus**. The incus (L. *incudere* = to beat upon, hence an anvil) has a body, and two processes, short and long. The long process articulates with the head of the stapes. The stapes (L. *stare* = to stand and *pes* = foot, hence a stirrup) has a head, neck, two limbs (or crurae), and a footplate that

fits into the oval window beyond which is the perilymph-filled cochlear part of the inner ear. The footplate is hinged anteroinferiorly and moves like a constrained piston to compress the perilymph. The ossicles articulate via synovial joints; the malleus and incus move as a unit around a horizontal axis that coincides with the anterior ligament of the malleus and the posterior ligament of the incus. Excessive movement of the ossicles is prevented by muscles in the middle ear (the tensor tympani and stapedius attach to the malleus and stapes, respectively) that contract in response to noise above approximately 90 db. Some debate exists about the precise embryological origin of the ossicles, but it is generally believed that most of the malleus and incus develop endochondrally from the Meckel's cartilage of the first **branchial arch** cartilage, while the head, neck, and crurae of the stapes develop from the second arch (hyoid) cartilage and the stapes footplate derives from the mesoderm of the otic capsule. Auditory ossicles attain their adult size prior to birth. They occasionally survive to be recognized as fossils [e.g., a *Paranthropus robustus* incus (SK 848), an *Australopithecus africanus* stapes (Stw 151)] and at the latest count approximately 25 ear ossicles had been recovered from the **Sima de los Huesos** at **Atapuerca**. *See also* **bony labyrinth**.

auditory tube (L. *audire* = to hear and *tubus* = hollow cylinder) The auditory tube connects the **middle ear** to the nasopharynx. The lateral one-third is bone (so this is the only part that fossilizes) and the medial two-thirds consists of fibrocartilage. The auditory tube is approximately 35 mm long in adult modern humans, and it is oriented 45° to the **sagittal** plane and 30° to the horizontal plane. In infants the auditory tube is shorter and more horizontally inclined; this is one of the reasons children are so prone to middle-ear infections. The auditory tube is inclined more closely to the sagittal plane in extant apes and in archaic hominins, and some specimens of *Paranthropus robustus* and *Australopithecus africanus* have club-like processes, called Eustachian processes, on the medial ends of the bony auditory tube (syn. Eustachian tube, pharyngotympanic tube, tympanic tube).

Aurignac (Location 43°13′22″N, 0°51′57″E, France; etym. Fr. after the nearby town) <u>History and general description</u> Aurignac cave (or Abri d'

Aurignac) is located the foothills north of the Pyrenees mountains. A road-worker found human bones in a shallow cave. Eight years later Édouard Lartet described its stratigraphic sequence and major artifacts, which later became the basis for establishing the "**Aurignacian**" culture. In 1938–9 Fernand Lacorre investigated Lartet's backfill and excavated the unexplored terrace outside the cave. Temporal span and how dated? The site has not been dated, but the Aurignacian in Western Europe generally dates between 36 and 26 ka. Hominins found at site Apparently 17 individuals were removed from the site by the discoverer and buried in an unmarked grave. Because these fossils were found on the surface of the deposit, it is likely they post-date the Aurignacian. Archeological evidence found at site Lartet recovered stone and bone tools, bone beads, and abundant faunal remains. Key references: Geology, dating, paleoenvironment, hominins, and archeology Lartet 1861.

Aurignacian   A material culture named after the site of Aurignac in France. Gabriel de Mortillet was the first to propose an Aurignacian industry for the unique combination of stone tool types, beads, and bone tools, but he abandoned it in his later classification schemes. The Aurignacian was reintroduced and redefined by Henri Breuil in the early 20thC. It is generally thought to be the oldest modern human (or Cro-Magnon) culture in Europe, although there are very few fossils that are securely dated to this time period. The hallmarks of this culture are the use of **blades**, **bone tools**, beads, and other objects of personal decoration, as well as figurines and other figurative art. Its appearance in Eastern Europe around 43 ka and in Western Europe between 40 and 36 ka is consistent with a migration of anatomically modern humans from the Near East. It is replaced by the **Gravettian** culture between 28 and 26 ka. *See also* **Aurignac**.

Aurora archaeostratigraphic set (AAS) Researchers now recognize six layers within the Aurora stratum, which is part of **TD6**, a lithostratigraphic level in Gran Dolina. The latter is one of the sites within the sediment-filled cave system called the **Trinchera del Ferrocarril**, which is in the Sierra de **Atapuerca**, a series of eroded limestone hills 14 km/9 miles east of Burgos in Northern Spain. *See also* **Aurora stratum**; **Gran Dolina**.

Aurora stratum   A layer within **TD6**, which is a lithostratigraphic level in **Gran Dolina**, one of the sites within the sediment-filled cave system called the **Trinchera del Ferrocarril**, one of the cave complexes in the Sierra de **Atapuerca**, a series of eroded limestone hills 14 km/9 miles east of Burgos in Northern Spain. Researchers now recognize six lithostratigraphic layers within the Aurora stratum, and they refer to these as the **Aurora archaeostratigraphic set** (AAS).

Australasian strewn-field tektites   From time to time the Earth collides with showers of small meteorites, called **tektites**. If the area of impact (the strewn-field) is widespread, and if the tektites are physically or chemically distinctive, and if they can be securely dated at least one location, strewn-field tektites can be used as a dating tool. The Australasian is the largest and the youngest of the major tektite strewn-fields. The *c.*800 ka strewn-field extends across most of Southeast Asia (Vietnam, Thailand, Southern China, Laos, and Cambodia), as well as the Philippines, Indonesia (including Java), and Malaysia. It also stretches into the Indian Ocean and south to the western side of Australia. The identification of Australasian strewn-field tektites in sediments is one of the many ways researchers have tried to date hominins from China and Southeast Asia.

australopith   Informal name for some, or all, of the fossil **hominins** not included in the genus *Homo*. This term is not used consistently in paleoanthropology. Some workers use this term to categorize *all* non-*Homo* hominins, while others include only non-*Homo* hominins that exhibit **megadontia** (i.e., species typically attributed to *Australopithecus* and/or *Paranthropus*). The term is increasingly used in place of australopithecine, because the latter term should *only* be used if the writer believes that all *Australopithecus* and *Paranthropus* taxa belong in their own hominin subfamily, the **Australopithecinae**. Examples of australopith taxa include the **type species** of *Australopithecus*, *Australopithecus africanus*, and *Paranthropus boisei*. *See also* **australopithecine**.

Australopithecinae (L. *australis* = southern and Gk *pithekos* = ape) This formal name should only be used by those who conclude that all *Australopithecus* and *Paranthropus* taxa belong in a hominin subfamily separate from the genus *Homo*. Some researchers have explicitly abandoned its use. For example, Robinson (1972a) sank *Australopithecus* into *Homo* and abandoned Australopithecinae in favor of a more encompassing Homininae. Anyone who supports all hominins being included within a single tribe cannot logically continue to refer to early hominin taxa as "australopithecines." *See also* **australopithecine**.

australopithecine Informal name for the subfamily **Australopithecinae**. Strictly speaking this term should only be employed if the user supports elevating the archaic hominin taxa included in genera such as *Australopithecus* and *Paranthropus* to the level of a subfamily. More generally, the term has been used in the past to refer to all, or to most, of the early hominins that do not belong to the genus *Homo*, but researchers are increasingly adopting the term **australopith** for this purpose.

Australopithecus Dart, 1925 (L. *australis* = southern and Gk *pithekos* = ape) Hominin **genus** established by Raymond **Dart** in 1925 to accommodate the **type species** *Australopithecus africanus*. Because it has priority as the genus name for archaic Pliocene hominins, taxa such as *Paranthropus* and *Plesianthropus*, which were not recognized until 1938, have been sunk into *Australopithecus* by those who recognize only a single genus for these hominins. In the way the taxon is used by **lumpers**, the genus *Australopithecus* includes species that are bipeds with small brains, large cheek teeth, and small, nonhoning canines. However, it almost certainly subsumes taxa representing parts of more than one **clade**. Thus, it is very likely that *Australopithecus* is a **paraphyletic group** (i.e., the taxa it contains represent some, but not all, of the descendants of their most recent common ancestor). The list of species assigned to the genus *Australopithecus* has varied over time. From its discovery up until the seminal publications of John **Robinson**, the genus subsumed three species, *Au. africanus* from the site of **Taung**, *Australopithecus transvaalensis* (later, *Plesianthropus transvaalensis*) from **Sterkfontein**, and *Australopithecus prometheus* from **Makapansgat**. Robinson sank *Au. prometheus* and *Pl. transvaalensis* into *Au. africanus*, but he did not include the fossil hominins from **Swartkrans** and **Kromdraai**, which he interpreted as belonging to the genus *Paranthropus*. In subsequent decades it became conventional to assign all of the above hominins as well as other species such as *Zinjanthropus boisei* and *Meganthropus africanus* to the genus *Australopithecus*. However, in the 1980s, the researchers who recognized the "robust" species as a monophyletic group revived the genus *Paranthropus* for *P. robustus*, *P. boisei*, and, eventually, *P. aethiopicus*, although many researchers still retain these species in *Australopithecus*. In 1994, a newly identified species, *Australopithecus ramidus*, was transferred to the genus *Ardipithecus* shortly after its initial description, and since that time several early hominin species have been discovered, described, and assigned to genera other than *Australopithecus* (e.g., *Orrorin*, *Sahelanthropus*, *Kenyanthropus*). Since 1996, four new species of *Australopithecus* have been described (*Australopithecus anamensis*, *Australopithecus bahrelghazali*, *Australopithecus garhi*, and *Australopithecus sediba*). In recent years, the evident paraphyly of *Australopithecus* has led a minority of researchers to advocate removing species from the genus until there is sound evidence that it is clearly **monophyletic**. This position has not gained widespread support, although some workers have transferred the hypodigm of *Au. afarensis* to the genus *Praeanthropus*. *See also* **Praeanthropus**.

Australopithecus aethiopicus (Arambourg and Coppens, 1968) (L. *australis* = southern, Gk *pithekos* = ape and *aethiopicus* = Ethiopia) A taxon used by researchers who recognize neither *Paraustralopithecus* nor *Paranthropus* as separate genera, but who do recognize pre-2.3 Ma **megadont** hominins from the Omo region as a species separate from *Paranthropus boisei*. No one has formally suggested that this **binomial** be adopted, but Groves (1999) warns that if they did, it would result in **homonymy** because the taxon name *Australopithecus aethiopicus* is preoccupied by Tobias's (1980b) usage of *Australopithecus africanus aethiopicus* for the **Hadar** component of the hypodigm of *Australopithecus afarensis*. *See also* *Paraustralopithecus aethiopicus*; *Australopithecus africanus aethiopicus*.

Australopithecus afarensis Johanson, 1978 (L. *australis* = southern, Gk *pithekos* = ape and *afarensis* recognizes the contributions of the local Afar people) **Hominin species** established in 1978 by Donald **Johanson** and colleagues to accommodate the *c.*3.7–3.0 Ma cranial and postcranial remains recovered from **Laetoli**, Tanzania, and **Hadar**, Ethiopia. White et al. (1981) compared the **hypodigm**s of *Australopithecus afarensis* and ***Australopithecus africanus*** and made a compelling case for recognizing *Au. afarensis* as a distinct species with generally more primitive craniodental anatomy than *Au. africanus*; Kimbel et al. (1984, 2004), Wood (1991), and Strait et al. (1997) provide supplementary information about the morphological differences between the two hypodigms. Several researchers have drawn attention to dental, facial, and mandibular differences between the early Laetoli component of the hypodigm and the geologically younger Hadar fossils assigned to *Au. afarensis*, and to similarities between the Laetoli remains and those of ***Australopithecus anamensis*** (e.g., Ward et al. 1999, Kimbel et al. 2006). It remains to be seen whether these similarities are sufficient evidence to sustain the hypothesis that *Au. anamensis* evolved via **anagenesis** into *Au. afarensis*. Rak et al. (2007), on the basis of mandibular ramus form, revived the suggestion (originally advocated for part of the Hadar sample by Olson 1981) that *Au. afarensis* may be a primitive taxon of the clade that includes ***Paranthropus boisei, Paranthropus robustus***, and ***Paranthropus aethiopicus***. The main differences between *Au. anamensis* and *Au. afarensis* relate to details of the jaws and dentition, which are generally more primitive in the former taxon. The curved, receding anterior mandibular corpus and more rectangular dental arch of *Au. anamensis* contrasts with the *Au. afarensis* anatomy. In some respects the teeth of *Au. anamensis* are more primitive than those of *Au. afarensis* (for example, the asymmetry of the premolar crowns and the relatively simple crowns of the deciduous first mandibular molars), but in others (for example, the low cross-sectional profiles and bulging sides of the molar crowns) *Au. anamensis* shows some similarities to *Paranthropus* (see below). The upper limb remains are similar to those of *Au. afarensis*, but a tibia attributed to *Au. anamensis* has features associated with **bipedalism**. Most body mass estimates for *Au. afarensis* range from approximately 30 to 45 kg and known endocranial volumes range between 385 and 550 cm^3. This is larger than the average endocranial volume of a chimpanzee, but if the estimates of the body size of *Au. afarensis* are approximately correct then relative to estimated body mass the brain of *Au. afarensis* is not substantially larger than that of *Pan*. It has smaller incisors than those of extant chimps/bonobos, but its premolars and molars are larger. The hind limbs of **A.L. 288-1** are substantially shorter than those of a modern human of similar stature. The appearance of the pelvis and the relatively short lower limb suggests that although *Au. afarensis* was capable of bipedal walking, it was not adapted for long-range bipedalism. This indirect evidence for the locomotion of *Au. afarensis* is complemented by the discovery at Laetoli of several trails of fossil footprints. These provide very graphic direct evidence that a contemporary hominin, presumably *Au. afarensis*, was capable of bipedal locomotion. The upper limb, especially the hand and the shoulder girdle, retains morphology that some workers suggest reflects a significant element of arboreal locomotion. The size of the Laetoli footprints, the length of the stride and stature estimates based on the length of the limb bones suggest that the standing height of adult individuals in this early hominin species was between 1.0 and 1.5 m. Most researchers interpret the fossil evidence for *Au. afarensis* as consistent with a substantial level of **sexual dimorphism**. A few researchers have suggested that sexual dimorphism in *Au. afarensis* is relatively poorly developed (Reno et al. 2003), but most researchers accept that this taxon shows a substantial level of sexual dimorphism. First discovery A.L. 128-1, 129-1 (1973). Holotype **LH 4** (1974). Paratypes (e.g., see Johanson et al. 1978 for a complete list) LH 1, 3 (a–t), 3/6 a, 5, 6 (a–e), 7, 8, 10, 11, 12, 13, 14 (a–h); **A.L.** 128-1, 128-23, **129-1a–c**, 129-52, 137-48a–b, 145-35, 161-40, 166-9, 198-1, 198-17a–b, 198-18, **199-1**, **200-1a**, 200-1b, 207-13, 211-1, 228-1, 241-14, 266-1, 277-1, **288-1**, 311-1, 322-1, **333-1**, 333-2, 333w-1a–e, 333x-1, 366-1, 388-1, **400-1a&b**, 411-1. Main sites **Belohdelie** (tentative), Dikika, **Hadar**, **Maka**, and **White Sands**, Ethiopia, **Koobi Fora**, Kenya, and Laetoli, Tanzania. *See also* ***Praeanthropus afarensis***.

Australopithecus africanus **Dart, 1925**

(L. *australis* = southern, Gk *pithekos* = ape and L. *africanus* = pertaining to Africa) **Hominin species** established in 1925 by Raymond **Dart** for an immature **skull** recovered from the limeworks at Taungs (now called **Taung**) in 1924. As presently interpreted the taxon *Australopithecus africanus* includes, in addition to the type specimen from Taung, fossils from Member 4 at **Sterkfontein**, fossils initially assigned to *Australopithecus prometheus* from Members 3 and 4 at

Makapansgat, and fossils recovered from lime-workers dumps and extracted *in situ* from the breccia exposed at **Gladysvale**; the hypodigm spans the period between *c*.3.0 and 2.5 Ma. It remains to be seen whether the associated skeleton **StW 573** from Sterkfontein Member 2 and 12 hominin fossils recovered from the **Jakovec Cavern** since 1995 (Partridge et al. 2003) belong to the *Au. africanus* hypodigm; Clarke (2008) suggests they belong to a second, so far unnamed, *Australopithecus* taxon (see below). The hypodigm of *Au. africanus* is one of the better fossil records of an early hominin taxon. The cranium, mandible, and dentition are well sampled. The postcranium and particularly the axial skeleton are less well represented, but there is at least one specimen of each of the long bones. However, many of the fossils have been crushed and deformed by rocks falling on the bones before they were fully fossilized. The picture emerging from morphological and functional analyses suggests that although *Au. africanus* was capable of walking bipedally it was probably more arboreal than most other archaic hominin taxa. It had relatively large chewing teeth and apart from the reduced canines the skull is relatively ape-like. Its mean endocranial volume is approximately 460 cm^3. The Sterkfontein evidence suggests that males and females of *Au. africanus* differed substantially in body size, but probably not to the degree they did in *Au. afarensis*. For a long time *Au. africanus* was regarded as the common ancestor of all later hominins, but the discovery of the even more primitive *Australopithecus afarensis* challenged that interpretation. In most cladistic analyses *Au. africanus* is either the sister taxon of **Homo** or **Paranthropus**, or the sister taxon of the common ancestor of the *Homo* and *Paranthropus* clades. Clarke (e.g., 1988, 1999, 2008) has consistently argued, and others have also suggested (Kimbel and White 1988), that the Sterkfontein Member 4 and Makapansgat hypodigm of *Au. africanus* samples a second, *Paranthopus*-like, *Australopithecus* taxon. Clarke (2008) would include **Sts** 1 and **71**, **StW** 183, **252**, 384, 498 and **505**, and MLD 2 in the second *Australopithecus* species; the differences between the second taxon and *Au. africanus* relate to dental size and craniofacial structure; details are set out by Clarke (2008, p. 448). First discovery **Taung 1** (1924). Holotype As above. Paratypes None. Main sites Gladysvale, Makapansgat (Members 3 and 4), Sterkfontein (Member 4), and Taung.

Australopithecus africanus aethiopicus Tobias, 1980

Phillip **Tobias** (1980b) made two suggestions about *Australopithecus afarensis*. First, he claimed that *Au. afarensis* was not sufficiently different from *Australopithecus africanus* to merit recognition as a distinct **species**. Second, he considered the **Hadar** and **Laetoli** parts of the **hypodigm** were subspecifically distinct, proposing that the Hadar component of the hypodigm be referred to as *Australopithecus africanus aethiopicus*. *See also Australopithecus afarensis*.

Australopithecus africanus afarensis Johanson et al., 1978

Phillip **Tobias** (1980b) proposed that the **Laetoli** part of the *Australopithecus afarensis* **hypodigm** be referred to as *Australopithecus africanus afarensis*. Even though Tobias proposed the subspecies, according to the Principle of Coordination in the **International Code of Zoological Nomenclature** (Article 46.1) a name established for a taxon at either rank in the species group (i.e., species or subspecies) retains the name of the author who proposed the species (i.e., *Australopithecus africanus afarensis* Johanson et al., 1978). The describer of *Australopithecus afarensis* is Johanson et al., 1978, and nothing can alter that, not even the transfer of the type specimen to a subspecies. *See also Australopithecus afarensis; Australopithecus africanus aethiopicus*.

Australopithecus africanus africanus Dart, 1925

John **Robinson** (1954a) proposed that the **Taung 1** skull be referred to a **subspecies** of *Australopithecus africanus* as *Australopithecus africanus africanus*. He did this to distinguish the latter from the "Sterkfontein, Makapan and East Africa" (*ibid*, p. 196) subset of *Au. africanus* which he referred to a separate subspecies, *Australopithecus africanus transvaalensis*. Bernard Campbell (Campbell 1973) subsequently used the same **nomen** to apply to the whole of the **hypodigm** while most other researchers recognize the taxon at the specific level as *Australopithecus africanus*. Even though Campbell proposed the subspecies, according to the Principle of Coordination in the **International Code of Zoological Nomenclature** (Article 46.1) a name established for a taxon at either rank in the species group (i.e., species or subspecies) retains the name of the author who proposed the species (i.e., *Australopithecus africanus* Dart, 1925). The describer of *Au. africanus* is Dart, 1925, and nothing can alter that, not even the transfer of the type specimen to a subspecies. *See also Australopithecus africanus; Australopithecus africanus transvaalensis.*

Australopithecus africanus habilis (Leakey, Tobias and Napier, 1964)

A subspecies of *Australopithecus africanus* introduced by Bernard Campbell (Campbell 1973) for the **hypodigm** most other researchers recognize at the specific level as *Homo habilis*. Even though Campbell proposed the subspecies, according to the Principle of Coordination in the **International Code of Zoological Nomenclature** (Article 46.1) a name established for a taxon at either rank in the species group (i.e., species or subspecies) retains the name of the author who proposed the species (i.e., *Homo habilis* Leakey, Tobias and Napier, 1964). The describers of *H. habilis* are Leakey, Tobias and Napier, 1964 and nothing can alter that, not even downgrading to a subspecies.

Australopithecus africanus modjokertensis (von Koenigswald, 1936)

A subspecies of *Australopithecus africanus* introduced by Bernard Campbell (Campbell 1973) for specimens from Indonesian sites Ralph **von Koenigswald** assigned to *Homo modjokertensis*. Most researchers now regard these specimens as being part of the hypodigm of *Homo erectus*. Even though Bernard Campbell proposed the subspecies, according to the Principle of Coordination in the **International Code of Zoological Nomenclature** (Article 46.1) a name established for a taxon at either rank in the species group (i.e., species or subspecies) retains the name of the author who proposed the species (i.e., *Homo modjokertensis* von Koenigswald, 1936). The describer of *H. modjokertensis* is von Koenigswald, 1936, and nothing can alter that, not even the downgrading of *H. modjokertensis* to a subspecies.

Australopithecus africanus robustus Broom, 1938 (Broom, 1938)

A subspecies of *Australopithecus africanus* introduced by Bernard Campbell (Campbell 1973) for the **hypodigm** most researchers recognize at the specific level as *Paranthropus robustus*. Even though Bernard Campbell proposed the subspecies, according to the Principle of Coordination in the **International Code of Zoological Nomenclature** (Article 46.1) a name established for a taxon at either rank in the species group (i.e., species or subspecies)

retains the name of the author who proposed the species (i.e., *Paranthropus robustus* Broom, 1938). The describer of *P. robustus* is Broom, 1938, and nothing can alter that, not even the downgrading of *P. robustus* to a subspecies.

Australopithecus africanus tanzaniensis Tobias, 1980

There is now wide acceptance that *Australopithecus afarensis* is a biologically valid species, but when Don **Johanson**, Tim **White**, and Yves **Coppens** set out the case for recognizing a new species for the hominins recovered from **Hadar** (Johanson et al. 1978) it met with spirited resistance from Phillip **Tobias**. In several publications (Tobias 1980a, 1980b) he laid out the detailed case for his claim that the authors had failed to demonstrate the new species was anything other than an East African variant of *Australopithecus africanus*, and thus should be recognized as a junior synonym of that taxon. In one of these publications, Tobias suggested that what was then called the Laetolil (now **Laetoli**) part of the *Au. afarensis* hypodigm be recognized at the level of a subspecies. He wrote "as we already have two southern African subspecies, *A. africanus africanus* and *A. africanus transvaalensis*, it would perhaps be appropriate to refer to this Tanzanian subspecies as *A. africanus tanzaniensis*" (Tobias 1980b, p. 107).

Australopithecus africanus transvaalensis (Broom, 1936)

John **Robinson** (1954a) proposed that this **subspecies** of *Australopithecus africanus* should be used to distinguish the **australopiths** from "**Sterkfontein**, Makapan and East Africa" (*ibid*, p. 196) from the **Taung 1** skull, which he referred to a separate subspecies, *Australopithecus africanus africanus*. Even though John Robinson proposed the subspecies, according to the Principle of Coordination in the **International Code of Zoological Nomenclature** (Article 46.1) a name established for a taxon at either rank in the species group (i.e., species or subspecies) retains the name of the author who proposed the species (i.e., *Plesianthropus transvaalensis* Broom, 1938). The describer of *transvaalensis* is Broom, 1936, and nothing can alter that, not even the downgrading of *transvaalensis* to a subspecies. *See also Australopithecus africanus; Australopithecus africanus africanus.*

Australopithecus anamensis Leakey et al., 1995

(etym. *see Australopithecus* and

anam — means "lake" in the Turkana language) Hominin species established in 1995 by Meave **Leakey** et al. to accommodate a left distal humeral fragment, **KNM-KP 271**, recovered by Bryan Patterson (Patterson and Howells 1967), and cranial remains recovered from *c*.3.9–4.2 Ma localities at **Allia Bay** and **Kanapoi**, Kenya [additional fossils from the two sites were described 3 years later (Leakey et al. 1998)]. The authors claim *Australopithecus anamensis* teeth are more primitive than those of *Australopithecus afarensis* (e.g., mandibular canine morphology, the asymmetry of the premolar crowns, and the relatively simple crowns of the deciduous first mandibular molars), but in other respects (e.g., the low cross-sectional profiles and bulging sides of the molar crowns) the teeth of *Au. anamensis* show similarities to *Paranthropus* (see below). Upper limb remains were also recovered and were described as being australopith-like (Leakey et al. 1998), but a tibia (KNM-KP 29283) attributed to *Au. anamensis*, but not included in the list of paratypes, has features associated with obligate bipedalism. White et al. (2006) reported 31 additional fossils from the **Middle Awash study area**. One, a maxilla, was found at **Aramis**, with the remainder coming from three localities at **Asa Issie**, a collecting area 10 km/6 miles west of Aramis. Biostratigraphic dating suggests an age of *c*.4.2–4.1 Ma for both sets of Middle Awash fossils. Craniodentally (e.g., anterior tooth and postcanine relative tooth size, crown morphology, enamel thickness, etc.) the new material from the Middle Awash is consistent with an attribution to *Au. anamensis* and ASI-VP-5/154, the proximal three-quarters of a right femur shaft, is a slightly more primitive version of the *Au. afarensis* femoral morphology. White et al. (2006) support the proposal made by others (e.g., Ward et al. 1999) that although the hypodigms of *Au. anamensis* and *Au. afarensis* are distinct there is evidence (e.g., relative canine size, upper canine morphology, enamel thickness, femoral shaft morphology, etc.) of the type of morphological continuity that one would expect to see within an evolving lineage, and Kimbel et al. (2006) provide compelling evidence that *Au. anamensis* and *Au. afarensis* are "parts of an anagenetically evolving lineage, or evolutionary species" (*ibid*, p. 134). Suggestions that *Au. anamensis* might have been a hard-object feeder (e.g., Ward et al. 1999, Teaford and Ungar 2000, Macho et al. 2005, White et al. 2006b) are apparently not supported by dental microwear evidence (Grine et al. 2006). First discovery KNM-KP 271 (1965). Holotype **KNM-KP 29281** (1994). Paratypes KNM-KP 271, 29282-8; KNM-ER 7727, 20419-23, 20427-8, 20432, 22683, 24148, 30202, 30200. Main sites Allia Bay and Kanapoi, Kenya; Aramis and Asa Issie, Ethiopia. *See also Australopithecus afarensis.*

Australopithecus bahrelghazali **Brunet et al., 1996**

(etym. *see Australopithecus*, and *bahr el ghazali* = place of discovery) Hominin species established by Michel **Brunet** and colleagues in 1996 to accommodate a tooth-bearing midline mandible fragment that includes the symphysis and an upper premolar tooth recovered from *c*.3.5–3.0 Ma sediments in the **Bahr el Ghazal** region, **Koro Toro**, Chad. The mandibular fragment had been previously assigned to *Australopithecus* aff. *Australopithecus afarensis* (Brunet et al. 1995). The authors claim that it has thicker enamel than *Ardipithecus ramidus*, a more vertically orientated and more gracile symphysis than *Australopithecus anamensis*, more complex premolar roots than *Australopithecus afarensis* and *Australopithecus africanus*, and larger incisors and canines than *Au. africanus*. However, other researchers interpret these differences as geographical variation within *Au. afarensis*. First discovery **KT 12/H1** (1995). Holotype As above. Paratypes KT 12/H2, right P^3. Main site Bahr el Ghazal, Chad. *See also Australopithecus afarensis.*

Australopithecus boisei **(Leakey, 1959)**

(etym. *see Australopithecus*, plus *boisei* to recognize the substantial help provided to Louis **Leakey** and Mary **Leakey** by Charles **Boise**) The genus *Zinjanthropus* was established by Louis Leakey in 1959 to accommodate fossil hominins recovered in 1955 and 1959 in Bed I, **Olduvai Gorge**, Tanzania. John **Robinson** (1960) suggested it was a synonym of *Paranthropus*, and Louis Leakey (1963) subsequently sank *Zinjanthropus* into *Australopithecus*. *See also Zinjanthropus boisei.*

Australopithecus garhi **Asfaw et al., 1999**

(etym. *garhi*=surprise in the Afar language) Hominin species established by Asfaw et al. (1999) to accommodate a fragmented cranium and two partial mandibles recovered from the *c*.2.5 Ma Hatayae Member of the **Bouri Formation** at the **Bouri, Gamedah**, and **Matabaietu** collection areas,

Middle Awash study area, Ethiopia. *Australopithecus garhi* combines a primitive cranium with large-crowned postcanine teeth. However, unlike *Paranthropus boisei* the incisors and canines are large and the enamel apparently lacks the extreme thickness seen in the latter taxon. A partial skeleton combining a long femur with a long forearm was found nearby, but is not associated with the type cranium (Asfaw et al. 1999) and these fossils have not been formerly assigned to *Au. garhi*. The authors claim that despite its large postcanine tooth crowns its cranium lacks the derived features of *Paranthropus*, and they suggest it may be ancestral to *Homo*, but what little evidence there is from phylogenetic analyses does not support a close link with *Homo*. The morphology of the mandibles reported in the same publication as *Au. garhi* is in some respects like that of *Paranthropus aethiopicus*. If it is demonstrated that the type specimen of *P. aethiopicus*, **Omo 18-18**, belongs to the same taxon as the mandibles that appear to match the *Au. garhi* cranium, then *P. aethiopicus* would have priority as the name for the *Au. garhi* hypodigm. First discovery GAM-VP-1/1 (1990). Holotype **BOU-VP-12/130** (1997). Paratypes None. Main sites Bouri, Ethiopia.

Australopithecus prometheus Dart, 1948

(L. *australis* = southern, Gk *pithekos* = ape and *Prometheus* = in Greek mythology the Titan who stole fire from Olympus. Raymond **Dart** chose this **species** name because organic chemists had confirmed the presence of carbon in the Makapansgat Cave; Dart 1948, p. 201). Hominin species established by Raymond Dart in 1948 to accommodate **MLD 1**, a hominin occipital recovered by the Kitching brothers at the Makapansgat Limeworks Dump. Dart (1948) considered the combination of the large size of MLD 1, the separation of **inion** and opisthocranion, and its distinctive **suture** pattern, to justify his conclusion that MLD 1 "differs vividly" (*ibid*, p. 200) from *Plesianthropus*. Most researchers interpret the Makapansgat hominin remains as being similar enough to those from **Taung** and **Sterkfontein** Member 4, etc. for *Australopithecus prometheus* to be regarded as a junior synonym of *Australopithecus africanus* Dart, 1925, and Robinson (1954a) suggested that the hominins from Makapansgat be transferred to *Australopithecus africanus* as *Australopithecus africanus transvaalensis*. However, Aguirre (1970) considered the dental morphology

of some of the Makapansgat hominins to be sufficiently like that of *Paranthropus robustus* to argue the case for a second hominin at Makapansgat. First discovery **MLD 1** (1947). Holotype As above. Paratypes None. Main site Makapansgat, South Africa. *See also* **Makapansgat**; *Australopithecus africanus transvaalensis*.

Australopithecus ramidus White et al., 1994

A new species of *Australopithecus* proposed by White et al. (1994) to accommodate 17 fossils recovered from localities at Aramis in the **Central Awash Complex** of the **Middle Awash study area**. A year later, in a corrigendum to the original paper, the same authors made a new genus name, *Ardipithecus*, available along with a diagnosis. They also noted that "in late 1994 a mandible of *A. ramidus* was found at Aramis associated with a partial postcranial skeleton, 50 m north of the holotype specimen, and at the same stratigraphic level. Analysis of this specimen has begun, and will provide further features with which to characterize *Ardipithecus*" (White et al. 1995, p. 88). Presumably White and his colleagues were referring to ARA-VP-6/500 (White et al. 2009a). *See Ardipithecus ramidus*.

Australopithecus sediba Berger et al. 2010

(L. *australis* = southern, Gk *pithekos* = ape, and Se Sotho *sediba* = fountain or wellspring) A hominin taxon established by Berger et al. (2010) to accommodate two hominin associated skeletons, **MH1** and **MH2**, recovered from **Malapa** in South Africa. These authors suggest that although the lower limb of *Australopithecus sediba* is like those of other archaic hominins, they claim that it has cranial morphology (e.g., more globular neurocranium, gracile face), mandibular morphology (e.g., more vertical symphyseal profile, a weak **true chin**), dental morphology (e.g., simple canine crown, small anterior and postcanine tooth crowns), and pelvic morphology (e.g., **acetabulocristal pillar**, expanded ilium and short ischium) that is only shared with early and later *Homo* taxa. First discovery MH1 (2008). Holotype MH1. Paratype MH2. Main site Malapa.

Australopithecus transvaalensis Broom, 1936

(L. *australis* = southern and Gk *pithekos* = ape; literally the "southern ape of the Transvaal") Hominin species established by Robert **Broom** in 1936 to accommodate **TM 1511**, the cranium and endocranial cast of a young adult recovered at **Sterkfontein** from breccia now referred to as being from Member 4. Broom considered the new cranium "agrees fairly closely with the Taungs ape," but that "the molar teeth differ in a number of important details" (Broom 1936, p. 488). The species *Australopithecus transvaalensis* was subsequently transferred to the genus *Plesianthropus* as *Plesianthropus transvaalensis* by Broom (1938), but it is now almost universally regarded as a junior synonym of *Australopithecus africanus*. In any event it is doubtful whether Broom's presentation of the new species complied with the then current (i.e., 1930 revision) version of the Rules of Zoological Nomenclature. First discovery TM 1511 (1936). Holotype As above. Paratypes None. Main site Sterkfontein, South Africa. *See also Plesianthropus; Plesianthropus transvaalensis; Australopithecus africanus.*

Australopithecus walkeri Ferguson, 1989
(L. *australis* = southern and Gk *pithekos* = ape, literally (Alan) Walker's "southern ape") Hominin species established by Walter Ferguson in 1989 to accommodate the "black skull" **KNM-WT 17000** and named "in honor of its discoverer" (*ibid*, p. 231). Ferguson (1989) argued that shape and size differences between **OH 5**, belonging to *Paranthropus boisei*, and KNM-WT 17000 were (a) greater than those between "*Australopithecus boisei*" and "*Australopithecus robustus*" (his terminology) and (b) too great to be subsumed within a single species. For example, he suggested the differences in cranial capacity (410 cm^3 in KNM-WT 17000 and 530 cm^3 in OH 5), both presumed male crania, were beyond what would be expected within one sex in a higher primate taxon. Ferguson's presentation of the new species complied with the **International Code of Zoological Nomenclature** rules, but his proposal is almost universally ignored because researchers who see the need for a second species for the pre-2.3 Ma **megadont** archaic hominins from East Africa use *Paraustralopithecus* or *Paranthropus aethiopicus* as the name for such a **taxon**. Groves (1999) has pointed out that the taxon name *Australopithecus aethiopicus* is preoccupied by Tobias (1980b), who suggested that *Au. africanus aethiopicus* could be used for the **Hadar** component of the hypodigm of *Australopithecus afarensis*. First discovery KNM-WT 17000. Holotype As above. Paratypes None. Main site **West Turkana**, East Africa. *See also Paranthropus boisei; Paraustralopithecus aethiopicus.*

Austronesia (L. *australis* = southern and Gk *nesos* = island, literally "southern islands") An inclusive term used for all the island groups in the Pacific (i.e., **Indonesia**, **Melanesia**, **Micronesia**, and **Polynesia**).

autapomorphic *See* **autapomorphy**.

autapomorphy (Gk *autos* = self and *morphe* = shape) A term used in **cladistic analysis** for a derived **character state** that is confined to one of the taxa, or **operational taxonomic units**, used in that analysis. Autapomorphies can be used for taxonomic identification, but because the morphology is by definition confined to a single taxon it cannot be used to explore how closely that taxon is related to another taxon. Examples of probable autapomorphies are the enlarged talonid of the mandibular postcanine teeth of *Paranthropus boisei* and the large globular brain case of *Homo sapiens*.

autecology (Gk *auto* = self) The branch of **ecology** that focuses on the interactions between an individual (or a single species) and its environment. Although autecology is a term used infrequently by paleoanthropologists, many ecological studies of hominins are essentially autecological, as some species are represented by a single fossil and in many cases research focuses on a single species.

autopod (Gk *autos* = self and *pod* = foot) The **distal** of the three compartments of a tetrapod, and therefore of a hominin, limb. The proximal compartment is the **stylopod** (the upper arm or the thigh), the intermediate one is the **zeugopod** (the forearm or the lower leg), and the distal one is the autopod (the hand or the foot).

availability *See* **available**.

available (L. *valere* = to be worthy) In relation to nomenclature a taxonomic name that is available is a name that has been generated according to the rules and recommendations of the **International Code of Zoological Nomenclature**. For example, the rules stipulate that an available name must not have been used in that context before, it must be formed from the 26 letters of the alphabet, and it must not have any commercial connotation. So whereas at the time it was proposed *Plesianthropus transvaalensis* was an available name, if it was proposed today it would not be available, and for different reasons neither would the taxon names *Homo 2feet* or *Homo bigmacensis*. To be a **valid** as a taxon name, the name must be available, but among many available names for a taxon, only one is valid. *See also* **International Code of Zoological Nomenclature**.

Aves (L. *avis* = bird) The name of the class that contains all bird taxa. Birds are bipedal tetrapod vertebrates; their forelimbs are specialized as wings, although the capability for powered flight has been lost in some lineages (e.g., ratites and penguins). Birds are warm blooded, lay eggs, and have feathers. Birds have a global distribution and there are both resident and migratory species. Bird bones are extremely fragile, having a lightweight structure to make flight less energetically costly. This makes them less durable in death and bird bones are disproportionally destroyed by perimortem, post mortem, and other **diagenetic** processes. Therefore bird osseous true fossils are relatively rare in the fossil record as are bird trace fossils in the form of prints and trails, but fossil egg shell, especially ostrich egg shell, is relatively common in open-air sites in Africa. Structural fragility means that bird remains are unlikely to be transported far from their place of death since they are especially prone to mechanical destruction. When bird remains are found in the fossil record they can be useful paleoenvironmental indicators (e.g., wading birds indicating the nearby presence of water). The most common bird remains found at early hominin fossil sites is probably ostrich egg shell, which can be used for ostrich egg shell dating as well as for paleoenvironmental reconstruction. *See also* **birds; ostrich egg shell dating; paleoenvironmental reconstruction**.

Awash River The Awash River runs for approximately 1100 km/700 miles within Ethiopia. It begins in the Ethiopian Highlands west of Addis Ababa, where it shares a watershed with the headwaters of the Blue Nile. It passes south of Addis Ababa and then turns north to run through the Awash National Park which is approximately 240 km/150 miles to the east of the Ethiopian capital. The Awash River then enters the Afar Triangle and runs north–northeast for 320 km/200 miles before turning east and then south–southeast for 120 km/75 miles before coming to an end in the most arid part of the Afar Triangle in Lake Abbé, some 112 km/70 miles west of the deepest part of the

western, landward, end of the Gulf of Tadjoura. The Upper, Middle, and Lower Awash regions are informal divisions of the part of the Awash River valley that runs through the Afar Triangle. *See also* **Middle Awash study area**.

Awash River Basin This refers to the approximately 440 km-/275 mile-long part of the Awash River that passes north, then in a south-easterly direction, through the Afar Triangle. Mammalian fossils have been reported in the region since the beginning of the 20thC, but it was not until geologist Maurice Taieb undertook reconnaissance work there in the 1960s that its potential as a source of fossils began to be realized. Taieb, together with Yves **Coppens**, John Kalb, and Don **Johanson**, formed the **International Afar Research Expedition** (or IARE) to explore the northern part of the basin in the early 1970s, but thereafter Kalb shifted his attention to the south and in 1975 and 1976 he formed the **Rift Valley Research Mission in Ethiopia** (or RVRME) to explore the region now known as the **Middle Awash study area**. The RVRME's activities focused on the eastern side of the Awash River, but Kalb and colleagues visited the Bouri Peninsula late in 1975, and located fossils and **Acheulean** artifacts at a locality identified as **Dakanihylo**. In the next year they discovered a hominin cranium (**Bodo 1**) in a region on the eastern side of the river at **Bodo**.

Awirs *See* **Engis**.

awl (ME *aul* = pointed tool) A small pointed tool commonly made of bone or metal for engraving or perforating. Awls are, and presumably were, used for a variety of tasks including sewing, leatherwork, basketry, woodworking, and inscribing brittle materials like ceramic and glass. Awls may be used hafted or hand-held. Bone awls, manufactured by longitudinal splitting of long bones and subsequent sharpening of the pointed tip, occur as components of early bone industries, such as found, for example, at **Blombos Cave**, South Africa.

axial digit (L. *axis* = a straight line around which a body or object rotates, or a line around which something is symmetric) The term used for the **reference digit** of the hand (the middle finger), or the reference digit of the foot (the second toe). *See also* **axis**.

axial skeleton Comprises the skull, the vertebral column, the ribs, and the sternum. It is the presence of these components of the hard tissue skeleton that defines the vertebrate subphylum. What unites the components of the axial skeleton is that it is what is left of the skeleton after the limbs and limb girdles have been removed or excluded. The components of the axial skeleton are all either in the midline, or they are connected directly, or indirectly, to midline stuctures.

axis (L. *axis* = a straight line around which a body or object rotates, or a line around which something is symmetric) Four meanings of axis are relevant to human evolution. First, it refers to the midline, or axis, of the body, as in **axial skeleton**. Second, it refers to the straight line, or axis, around which a part of the body rotates. For example, the axis of **pronation** and **supination** of the forearm is a straight line joining the center of the head of the radius and the styloid process of the ulna. Third, it refers to the reference digit of the hand (the middle finger) and foot (the second toe). The fourth meaning refers to the second (C2) of the seven cervical vertebrae. The C2 vertebra is called the axis because it incorporates the odontoid process, which forms the axis around which the atlas [the first (C1) cervical vertebra] plus the cranium, which articulates with the superior aspect of the atlas, rotate.

axon (Gk *axon* = axis) A tubular process that projects from a nerve cell, or **neuron**, and is the way neurons connect to other neurons in a nervous pathway, or with the target structure (e.g., muscle fiber, blood vessel, or sweat gland) or target organ (e.g., heart) being innervated. Axons have diameters that range from 0.2 to 20 µm and they can extend for up to 1 m. Axons are capable of conducting a nerve signal, or action potential, over great distances. Large axons are surrounded by a fatty insulating sheath called **myelin**, which is important for enabling the high-speed conduction of action potentials. The sheath is interrupted at regular intervals by nodes of Ranvier (named after the neuroanatomist Louis Antoine Ranvier who first described them towards the end of the 19thC). Near its end the axon divides into fine branches that have specialized swellings called presynaptic terminals where neurotransmitter molecules are released into the synaptic cleft that lies between the presynaptic terminals and the next neuron, or the structure or organ being innervated. During development, a neuron sends out axons by way of a growth cone. The growth cone appears as an enlargement of the shaft of the axon, and several extensions, called filopodia, project from it and facilitate the migration of the axon through tissue. Maxwell Cowan, Giorgio Innocenti, and their colleagues independently found that neurons send axons to many more targets in the developing brains than are eventually retained in the adult animal, and the projections are focused on discrete areas within the brain (e.g., the **cerebral cortex**) by pruning projections to inappropriate targets.

Azilian A technocomplex that includes small scrapers and distinctive points; the type site is Mas-d'Azil. *See* **Mas-d'Azil**.

Azokh Cave complex (Location 39°37′09″N, 46°59′19″E, Nagorno-Karabagh, an autonomous region within Azerbaijhan in the Lesser Caucasus; etym. named after a nearby village) History and general description The Azokh Cave was discovered in 1960. The initial, 15 year-long, period of excavations at Azokh 1 was brought to an end in 1989, and in that time 2430 m³ out of an estimated 3400 m³ total volume of sediment had been removed from that part of the cave complex. Since 2002 the Azokh Cave Project, an international and multidisciplinary program, has been working at the site. Four more entrances were discovered and two of these, Azokh 2 and Azokh 5, contain fossils and artifacts. A hominin mandible was found at Azokh 1 in 1968. Temporal span and how dated? Five beds have been identified at Azokh 1, with Bed V being dated at *c*.200 ka by **uranium-series dating** and *c*.300 ka by **amino-acid racemization dating** and **electron spin resonance spectroscopy dating**. Archeological evidence found at site In Beds II and II "the lithic assemblages are primarily Middle Paleolithic in character" and Bed V contains "retouched scrapers and unretouched flakes" (Fernández-Jalvo et al. 2010, p. 105). Key references: Geology and dating Fernández-Jalvo et al. 2004, 2010; Hominins Kasimova 2001; Archeology Fernández-Jalvo et al. 2004, 2010.

B

B. A prefix used by the American contingent of the **International Omo Research Expedition** for fossil localities in the Brown Sands group of exposures of the **Usno Formation** (e.g., B. 8). The prefix is also used for fossils found within one of those localities (e.g., B. 8-27).

Ba/Ca ratios *See* **strontium/calcium ratios**.

baboon The common name for a member of the **Old World monkey** genus *Papio*. The term is also sometimes used more inclusively as the term for a grade that includes the large-bodied **papionin** monkeys (including *Papio*, *Theropithecus*, and *Mandrillus* plus a range of extinct genera). After originating most likely in southern Africa and subsequently undergoing an adaptive radiation in the Pleistocene, baboons are now widespread across sub-Saharan Africa, with a small, relatively ancient population also found in Arabia (Wildman et al. 2004). Based on the **phenotype**, most modern authorities classify the *Papio* or common baboons as a single species (*Papio hamadryas*) with multiple subspecies (Jolly 1993) [i.e., the guinea baboon, *P. h. papio*, found in the far west of Africa, the sacred baboon, *Papio h. hamadryas*, found in the Horn of Africa and Arabia, the olive baboon, *P. h. anubis*, found across a wide geographic area, from west to east Africa, the yellow baboon, *P. h. cynocephalus*, found in east and south central Africa, the kinda baboon, *P. h. kindae* (but which some refer to as *Papio cynocephalus kindae*) found in south central Africa, and the chacma baboon, *P. h. ursinus*, found in southern Africa]. However, recent molecular data suggest there are seven major haplotypes that are highly complex genetically, with most likely multiple examples of divergence, reconnection, and hybridization since the origin of the genus (Zinner et al. 2009). Several authors, most notably Jolly (2001), have highlighted the ecological similarities between baboons and hominins (e.g., their **eurytopy**, preference for open or wooded habitats, terrestriality, and relative abundance in the fossil record). The baboons, as discussed in the *Theropithecus* entry, are thus a popular comparator species for hominins. *See also* ***Theropithecus***.

Bacho Kiro (Location 42°56′50″N, 25°25′47″E, Bulgaria; etym. named after a Bulgarian teacher and revolutionary who fought at the nearby Dryanovo Monastery) History and general description This site in eastern Bulgaria is a series of caves in a limestone hill that is a popular tourist attraction for its decorative **speleothem**s. It is the type site of an early Upper Paleolithic technology called the Bachokirian. The hominin fossils found at the site have been interpreted by some as a *Homo sapiens*, which has led to the association of this early industry with the very first migration of modern humans into Europe. Temporal span and how dated? **Radiocarbon dating** of the Bachokiran contexts have provided a range of >43–29 ka (uncalibrated). Hominins found at site Several remains were recovered, including a mandibular corpus with dm_1 from the Bachokiran level, and isolated teeth, a right parietal fragment, and a right mandibular corpus with dm_2 and M_1 from the **Aurignacian** level. Most of these have been assigned to *H. sapiens*. Archeological evidence found at site Stone tools assigned to both the Bachokiran and Aurignacian industries. Key references: Geology, dating, and paleoenvironment Jöris and Street 2008; Archeology Kozlowski and Otte 2000; Hominins Churchill and Smith 2000.

Bachokiran *See* **Bacho Kiro**.

backed A term used to describe the blunted or abraded edge of a **microlith**. The blunting, or backing, strengthens the edge so the microlith can be mounted to form a **composite tool**. The opposing edge is usually left unmodified for maximum sharpness.

Wiley-Blackwell Encyclopedia of Human Evolution, First Edition. Edited by Bernard Wood.
© 2013 Blackwell Publishing Ltd. Published 2013 by Blackwell Publishing Ltd.

backed bladelet A type of bladelet with a sharp margin opposite one that has been dulled, or "backed." Abrasion or percussion may be used to dull the edge to be backed, and backing is often done to facilitate holding of sharp-edge pieces or to strengthen the edge to be inserted into a slot or handle as part of a composite, or multi-part tool.

bacteriophage A virus (or virus-like agent) that infects bacteria. The use of bacteriophages during the development of molecular biology enabled researchers to understand the role DNA plays in heredity. In 1952, the Hershey–Chase experiments confirmed that DNA was the genetic material. These experiments used the T2 bacteriophage labeled with ^{32}P to make the DNA radioactive, and then with ^{35}S to make the protein component radioactive. After the bacteriophages infected the bacteria, Hershey and Chase found that it was the DNA (and not the protein) from the bacteriophage that entered the bacterial cell and infected it.

Bahr el Ghazal (Ar. *Bahr el Ghazal*=river of the gazelles) A collection area approximately 45 km/28 miles east of Koro Toro, Chad. More than 17 numbered localities have been identified where there are exposures of **Pliocene** fossiliferous sediments. At one of them, KT 12, researchers recovered the remains of two hominins, one of which is the type specimen of *Australopithecus bahrelghazali*. *See also* **Koro Toro**; **KT 12/H1**; *Australopithecus bahrelghazali*.

Baia de Fier (Location 45°11′N, 23°46′E, Romania; etym. Romanian for "cave of the old woman") History and general description The Petera cu Muierii is a **karstic** cave system comprising several galleries. It had been known of since the 1870s, but the initial archeological investigations were not undertaken until 1929, with further excavations carried out in the early 1950s. Excavations in two of the galleries yielded **paleolithic** sequences, including **Middle** and **Upper Paleolithic** remains, and mammal fossil bones. Hominin skeletal remains, thought to represent one individual (Muierii 1), were discovered in 1952 in a surface depression at the back of one of these two galleries (the Galeria Musterian). Two additional hominin remains (of unknown provenance) have been reported for this site (Muierii 2 and 3). Temporal span Middle and Upper Paleolithic, as well as the **Holocene**. The Mousterian levels of the Galeria Musterian yielded a conventional radiocarbon age of 42,560±1100–1300, and a more recent ultra-filtration **AMS radiocarbon dating** age of 40,850±450 years BP. The hominin remains themselves have yielded AMS ultrafiltration ^{14}C ages of 30,150±800, 29,930±170, and 29,110±190 years BP; all the ages are uncalibrated. How dated? Conventional and AMS ultrafiltration radiocarbon dates on hominin and faunal remains. Hominins found at site Partial cranium, mandible, scapula and tibia (Muierii 1), temporal bone (Muierii 2), and fibular diaphysis (Muierii 3). Archeological evidence found at site Eastern **Charentian** (Cave Mousterian), Upper Paleolithic (possibly evolved **Aurignacian**). Key references: Geology and dating Punescu 2000, Olariu et al. 2003, Soficaru et al. 2006; Hominins Gheorghiu and Haas 1954, Soficaru et al. 2006, Trinkaus 2007, 2008; Archeology Mertens 1996, Punescu 2000, Soficaru et al. 2006. *See also* **Muierii 1, 2, and 3.**

Baichbal valley *See* **Hunsgi-Baichbal valleys.**

Baise Basin *See* Bose Basin.

balanced polymorphism The simultaneous presence in a population of two or more distinct forms of a phenotypic character (e.g., S and A **hemoglobin**) in proportions well above those resulting solely from recurrent mutation. The determining **allele**s (in this case the S-type and wild-type alleles of the beta-globin **gene**) are maintained at elevated frequencies by **balancing selection** or frequency-dependent selection.

balanced translocation *See* **translocation.**

balancing selection A type of **natural selection** favoring more than one **allele** that, because it prevents the fixation of any allele, maintains different alleles in the population. An example in modern humans is the sickle-cell, or S, allele of the beta-globin **gene**. In the **homozygous** form this allele is deleterious since it causes sickle-cell anemia. However, in the **heterozygous** form in regions where malaria is endemic, it *protects* against malaria. In such an environment, homozygotes for the wild-type allele are at a disadvantage because they are more susceptible to malaria. Thus, both alleles are maintained in the population. This phenomenon, known as **heterozygous advantage**, is also referred to as overdominance.

balancing side   *See* chewing.

Balla Barlang   (Location 48°02′N, 20°31′E, Hungary; etym. Hu. *balla*=local family name, *barlang*=cave) History and general description This cave, in a limestone outcrop near Répáshuta village, contained a **Gravettian** occupation site with a few hominin remains. Temporal span and how dated? **Radiocarbon dating** provides an age of 20,000±200 years BP. Hominins found at site Several subadult cranial fragments, including a mandible with three isolated teeth and a partial skeleton, all attributed to *Homo sapiens*. Archeological evidence found at site A Gravettian stone tool assemblage was recovered. Key references: Geology, dating, and paleoenvironment Thoma and Vértes 1971; Archeology Vértes 1967; Hominins Thoma and Vértes 1971.

Balzi Rossi   *See* Grimaldi.

Bañolas   *See* Banyoles.

Banyoles   (Location 42°06′N, 2°45′E, Catalonia, Spain; etym. after the nearby town) History and general description Quarry near the town and lake of Banyoles, about 60 km/35 miles northeast of Barcelona. A hominin mandible was found embedded in **travertine** at a depth of 4–5 m by the stonemason Lorenzo Roura in 1887. It came into the possession of Don Pedro Alsius, a well-known local pharmacist and naturalist. Alsius himself removed much of the travertine matrix from the specimen, and the mandible is still in the hands of the Alsius family. Temporal span and how dated? The travertine matrix stuck to the mandible was dated to 45±4 ka using **uranium-series dating**. Direct uranium-series/**electron spin resonance spectroscopy dating** on the tooth enamel gave an age of 66±7 ka. Hominins found at site The Banyoles mandible was originally identified as *Homo neanderthalensis*, but later workers noted the lack of derived **Neanderthal** features (implying that the specimen was **pre-Neanderthal** in character). The more recent date of the specimen, however, makes the pre-Neanderthal attribution unlikely. Archeological evidence found at site None. Key references: Geology, dating, and paleoenvironment Julià and Bischoff 1991, Grün et al. 2006; Hominins MacCurdy 1915, de Lumley 1971/2, Stringer et al. 1984; Archeology N/A.

Baousse Rousse   *See* Grimaldi.

Bapang Formation   (etym. Named after a local village near the type section) General description The younger of the two fossil-hominin-bearing geological formations of Central and East Java, Indonesia. Formerly known as the Kabuh Formation, the Bapang Formation is largely a product of local uplift, volcanism, and sedimentary deposition from river and stream activity. Most of the hominin fossil specimens from the **Sangiran Dome** are derived from the Bapang Formation. The lower layers of the Bapang Formation were characterized by high-energy deposition so many of the mammalian fossils, including the hominins, are fragmentary and abraded. In the higher layers the sediments of the Bapang Formation are finer-grained which is indicative of low-energy deposition. The fossils from these strata are better preserved and more complete than those from the lower layers. A coarse-grained **conglomerate** layer rich with fossil bone (formerly known as the **Grenzbank zone**) marks the transition between the younger Bapang Formation and the older **Sangiran Formation**. Sites where this formation is exposed **Mojokerto**, **Ngrejeng**, **Sangiran**, and **Sendangbusik**. Geological age Argon-argon geochronological analyses have dated the lowermost Bapang hominins to 1.51–1.47 Ma, and later specimens from the same formation to 1.33–1.24 Ma. Most significant hominin specimens **Sangiran 2**, **Sangiran 17**, **Sangiran 21** ("Mandible E"). Key references: Geology and dating Watanabe and Kadar 1985, Swisher et al. 1994, Larick et al. 2001; Hominins Weidenreich 1944, von Koenigswald 1968, Jacob 1973, Sartono 1976, Pope and Cronin 1984.

Bapaume   (Location 50°05′37″N, 2°52′19″E, France; etym. named after the nearby town of Bapaume) History and general description First recognized during road construction in 1966, this large open-air **Lower** and **Middle Paleolithic** site was further exposed during excavation for a high-speed train line in northern France. Temporal span and how dated? No radiometric dating has been attempted. The site age is based on correlation to **glacial cycles**; the Middle Paleolithic deposit is thought to be early Weichselian [**Oxygen Isotope Stage** (OIS) 2 or 4], whereas the Lower Paleolithic deposit is dated to the Saalian (possibly OIS 6, 8, or 10). Hominins found at site None. Archeological evidence found at site The exposure of Lower Paleolithic **loess**, known as les Osiers, produced a variety of late **Acheulean** stone tools, made with the **Levallois** technique, and a few handaxes. The Middle Paleolithic exposures, known as Riencourt-lès-Bapaume, contains several **Mousterian** layers, as well as a late Acheulean layer. Key references: Geology, dating, and paleoenvironment Tuffreau 1976, Tuffreau et al. 1991; Hominins N/A; Archeology Tuffreau 1976, Tuffreau et al. 1991.

BAR Abbreviation for Baringo, and the prefix used for the fossils recovered at **Lukeino**, Tugen Hills, Baringo District, Kenya, from 2000 onwards (e.g., **BAR 1000'00**, the **holotype** of *Orrorin tugenensis*).

BAR 1000'00 Site Lukeino. Locality Kapsomin Surface/*in situ* Surface. Date of discovery 2000. Finder Kiptalam Cheboi. Unit N/A. Horizon N/A. Bed/member N/A. Formation **Lukeino Formation**. Group N/A. Nearest overlying dated horizon Kaparaina Basalts. Nearest underlying dated horizon Kabarnet Trachytes. Geological age *c.*6 Ma. Developmental age Adult. Presumed sex Unknown. Brief anatomical description Two fragments of a mandible, plus the left M$_2$ and left and right M$_3$: all three teeth are damaged. Enamel thickness at the apex on the **paraconid** is 3.1 mm and is comparable to other hominins with the exception of *Ardipithecus*, which has thinner enamel. Announcement Senut et al. 2001. Initial published description Senut et al. 2001. Photographs/line drawings and metrical data Senut et al. 2001. Detailed anatomical description N/A. Initial taxonomic allocation *Orrorin tugenensis*. Taxonomic revisions N/A. Current conventional taxonomic allocation *O. tugenensis*. Informal taxonomic category Possible primitive hominin. Significance Holotype of *O. tugenensis*. Location of original Kipsaramon Museum, Baringo District, Kenya.

BAR 1002'00 Site Lukeino. Locality Kapsomin. Surface/*in situ* Surface. Date of discovery 2000. Finder Kiptalam Cheboi. Unit N/A. Horizon N/A. Bed/member N/A. Formation **Lukeino Formation**. Group N/A. Nearest overlying dated horizon Kaparaina Basalts. Nearest

underlying dated horizon Kabarnet Trachytes. Geological age *c.*6 Ma. Developmental age Adult. Presumed sex Unknown. Brief anatomical description A nearly complete proximal left femur. The femoral head is preserved, but the greater trochanter is not. Announcement Senut et al. 2001. Initial published description Senut et al. 2001. Photographs/line drawings and metrical data Galik et al. 2004. Detailed anatomical description Galik et al. 2004. Initial taxonomic allocation *Orrorin tugenensis*. Taxonomic revisions None. Current conventional taxonomic allocation *O. tugenensis*. Informal taxonomic category Possible primitive hominin. Significance There has been debate about whether this specimen provides compelling evidence for a locomotor mode that involves substantial bouts of bipedal behavior. Many researchers were skeptical that the internal morphology of BAR 1002'00 justified such a conclusion, but the most recent analysis of its external morphology suggests that it does resemble femora attributed to *Australopithecus* and *Paranthropus* (Richmond and Jungers 2008), taxa that were almost certainly facultative bipeds. Location of original Kipsaramon Museum, Baringo District, Kenya.

Bar-Yosef, Ofer (1937–) Ofer Bar-Yosef was born in Jerusalem, Israel. His father's family moved to Israel from Morocco around 1850 and his mother's family moved from Latvia in 1891. Bar-Yosef developed an interest in history and archeology as a schoolboy. He conducted his first excavation of an entrance to a Byzantine cistern when he was in the fifth grade, and when he was in the eighth grade he surveyed the ridge on which the Givat Ram campus of Hebrew University would later be built. During his mandatory service in the Israeli Defense Force (1955–8) he spent time in the Negev where two members of his future kibbutz taught him the basics of lithic analysis. In 1957, Bar-Yosef volunteered to work with Moshe Stekelis in **Kebara** cave, and in 1959 Bar-Yosef surveyed an Early Bronze Age (3330–2300 years BCE) mound in the kibbutz located in the Jordan Valley. Before beginning his undergraduate education in the fall of 1959 at Hebrew University, Bar-Yosef worked in the excavation of Nahal Oren Terrace in **Mount Carmel** under the supervision of Stekelis. Bar-Yosef studied at Hebrew University, earning a BA in archeology and geography (1963), an MA in prehistoric archeology (1965), and a PhD in prehistoric archeology (1970). Bar-Yosef's interest in Paleolithic archeology led him to conduct numerous excavations in the Levant, predominantly at **'Ubeidiya** in the Jordan Valley (1960–6 under the supervision of M. Stekelis and in 1967–74 jointly with E. Tchernov) and Kebara Cave, Mount Carmel (1982–90 with Bernard **Vandermeersch**); excavations in Kebara Cave resulted in the recovery of well-preserved evidence of *Homo neanderthalensis*; he has also participated in fieldwork at sites in Turkey, China, the Czech Republic, and the Republic of Georgia. Throughout his academic career, Bar-Yosef has also maintained active field research programs involving the Neolithic, both in Israel and also, most recently, in China. Since 1988 Bar-Yosef has taught at Harvard University, where he is currently the MacCurdy Professor of Prehistoric Archaeology and head of the Peabody Museum's Stone Age Laboratory.

bare area This term, or more precisely the "bare area of the occiput," was introduced by Raymond **Dart** when he described **MLD 1**, the type specimen of *Australopithecus prometheus* (Dart 1948). He used the term to describe the area on the squamous part of the **occipital** between lambda and the **superior nuchal line**s that was not covered by either the temporalis or the nuchal muscles. Because the temporal lines or sagittal crests of extant higher primates nearly always extend posteriorly to intersect the nuchal crests, extant apes seldom have a significant bare area. This contrasts with the **megadont** archaic hominins in which the sagittal crest tends to be more developed anteriorly and as a result there is often a bare area (e.g., **OH 5**, MLD 1).

barium/calcium ratios *See* strontium/calcium ratios.

Barlow, George *See* **Taung**; **Sterkfontein**.

Barma del Caviglione (Location 43°47'00''N, 7°32'08''E, Luguria, Italy; etym. It. *barma*=cave in the local dialect) History and general description One of the caves in the **Grimaldi** complex, it was primarily excavated by E. Rivière in the early 1870s. Temporal span and how dated? The cave itself has not been directly dated, although direct **radiocarbon dating** on the biostratigraphically correlated remains from **Arene Candide** and **Barma Grande** suggest an age of *c.*24 ka. Hominins found at site BC1, or the "Homme de Menton," is a fairly complete young adult ceremonial

burial covered with ochre; perforated shells and deer canines ornament the head. However, despite its name, the skeleton appears to be female. Other hominin fossils found at the site include a right cuboid, a right hallucial phalanx, and a fragmentary radius; they and BC1 are anatomically modern *Homo sapiens* and were assigned by Verneau to the "Cro Magnon race." Archeological evidence found at site In addition to the burial, both **Gravettian** and **Aurignacian** assemblages were recovered, but the burial is associated with the Gravettian. A parietal engraving of a horse was discovered in the 1970s. Key references: Geology, dating, and paleoenvironment Alciate et al. 2005; Hominins Rivière 1872, Verneau 1906, Alciate et al. 2005; Archeology Mussi 2001.

Barma Grande (Location 43°47'00''N, 07°32'08''E, Italy; etym. It. *barma*=cave in the local dialect and *grande*=large) History and general description The activity at **Grotte des Enfants** and **Barma del Caviglione** in the early 1870s prompted many semi-professionals to explore the other caves in the **Grimaldi** complex. L. Jullien, an antiquities dealer, discovered the burial of an adult male (Barma Grande 1) in 1884, as well as a number of figurines and artifacts. The skeleton, however, was destroyed a few days after discovery due to a quarrel. Eight years later the owner of the cave, G. Abbo, carried out new excavations that resulted in the discovery of a triple burial (BG 2, 3, and 4). Two more burials were discovered in the rear of the cave in 1894 (BG 5 and 6). Temporal span and how dated? The Barma Grande 6 skeleton was directly dated using **radiocarbon dating** to 24,800±800 years BP. Hominins found at site The surviving skeletons are fairly complete (BG 6 being an exception), although BG 5 was damaged in WWII. BG 1, 2, 5, and 6 are male, whereas 3 and 4 are most likely female. Archeological evidence found at site The skeletons, like many other **Gravettian** burials, were lavishly ornamented with flint blades, ivory pendants, deer canines, and ochre. At least two "Venus" figurines were recovered by Jullien in 1884. Key references: Geology and dating Formicola et al. 2004, Hominins Verneau 1908, Formicola 1987, Churchill and Formicola 1997; Archeology Mussi 2001.

Barnfield Pit *See* **Swanscombe**.

Barr body Refers to a dense stainable structure (discovered by Murray Barr) that is the inactive **X**

chromosome in a female cell. This inactivation results in an equal expression of most genes on the X chromosome in both males and females (males have one X chromosome, but females have two). In female mammals, one X chromosome is randomly inactivated early in embryonic development through the initiation of the X-inactivation center (Xic). Since inactivation is random, different areas of the body of a female individual may have different active X chromosomes. If different alleles are present, this genetic mosaic condition can result in a patchy **phenotype** (e.g., the hair color in calico cats). Differential X-chromosome activation may play a role in evolution, including human evolution.

Barranco León One of the sites of the Orce region, this area is known for abundant fanual remains dating to the Plio-Pleistocene along with some potential stone tools. *See also* **Orce region**; **Venta Micena**.

basal ganglia A group of **nuclei** at the base of the **forebrain** in vertebrates. The basal ganglia includes the **striatum** (comprising the caudate nucleus and putamen in primates), globus pallidus, substantia nigra (which is in the midbrain), and the subthalamic nucleus. The basal ganglia are involved in a variety of functions (e.g., motor control, learning, emotional regulation, and **cognition**) via their recurrent connections through the thalamus to the **cerebral cortex** (syn. basal nuclei).

basal nuclei *See* **basal ganglia**.

basalt A hard, dense, basic (i.e., between 45 and 52% SiO_2) igneous rock, rich in the minerals plagioclase and pyroxene. It is formed when liquid magma cools either on the Earth's surface as a lava flow or at shallow depths as an intrusion (dyke or sill). Within rift valley sequences basic igneous rocks such as basalts often underly sedimentary rocks, and during the early stages of dating the sediments at **Olduvai Gorge** a layer of basalt beneath Bed I provided a maximum **potassium-argon** age for the fossil-bearing sediments that lay above it.

base *See* **base pair**; **basicranium**; **cranium**; **mandible**.

base pair The fundamental unit of a DNA molecule. The latter consists of **nucleotides** that each contain one of four bases – adenine, guanine, cytosine, and thymine – that are joined in pairs by hydrogen bonds to form one of the "rungs" that make up the double-helix "ladder" structure of DNA. The bases are divided into two classes: purines (adenine and guanine) and pyrimidines (cytosine and thymine). In the DNA double helix, guanine always binds with cytosine and adenine always binds with thymine to form what are called base pairs (NB: if you remember the registration number of the late Francis Crick's car, AT GC, then you will have remembered the appropriate pairing). The size of a DNA molecule is expressed as the number of base pairs (or bp) it contains (e.g., 1789 bp). *See also* **RNA**.

basicranium [Gk (and L.) *kranion*=brain case and *basis*=base] This term refers to the part of the bony **cranium** beneath the brain. Its inner or **endocranial** surface is divided into three hollowed areas called cranial fossae. The anterior cranial fossa lies beneath the frontal lobes of the **cerebral cortex**, the middle cranial fossa lies beneath the temporal lobes of the cerebral cortex, and the posterior cranial fossa lies beneath the **cerebellum**. The un-paired bones that contribute to the cranial base are, from front to back, a small part of the frontal, the ethmoid, sphenoid, and the **occipital**, except for the occipital squame. Only one paired bone, the temporal, contributes to the cranial base. Except for the squamous parts of the occipital and temporal bones, the rest of the cranial base forms from the **chondrocranium** via a process called **endochondral ossification**. Studies of the cranial base or basicranium usually either consider its midline morphology (i.e., in the sagittal plane; e.g., *see* **cranial base angle**), or its parasagittal morphology (i.e., the morphology that lies either side of the sagittal plane). The latter studies usually concentrate on the bilateral structures that can be seen away from the midline when the cranial base is viewed from above and below; the former view is the endocranial aspect and latter view is called the *norma basilaris*. The cranial base completes its growth and development before the **face** and the **neurocranium**, and thus may constrain their development. *See also* **normal views**. syn. Basicranium is synonymous with the more modern term cranial base.

basin Teeth A depressed area on the surface of the crown of a postcanine tooth bounded by enamel ridges or cusps (e.g., **talonid** basin) (syn. **fovea**, fossa). <u>Geology</u> Basin is used to describe large-scale structural formations that may have formed after the deposition of the strata (structural basins) or may have formed contemporaneously or prior to the deposition strata in response to tectonic processes (e.g., crustal stretching and rifting during rift valley formation). Such sedimentary basins are important as they provide the major locations for deposition and preservation of terrestrial sediments and fossils (e.g., **Turkana Basin**).

Batavia (L. *Batavi*=Germanic tribe whose homeland approximates to the region now known as The Netherlands) In antiquity the name was used for the "Low Countries", and it was also chosen as the name for the city of Jakarta on the island of Java when the former was the capital of the Dutch East Indies. It features in some of the early articles about *Homo erectus* in Java.

bats *See* **Chiroptera**.

Baume Moula-Guercy *See* **Abri Moula-Guercy**.

bauplan (Ge. *bau*=architect and *plan*=sketch or drawing) A term used to describe the general organization, or body plan, of a **taxon**. It is usually used for large taxonomic groups (e.g., the four limbs of tetrapods), but it can be applied to a species (e.g., the flat face, robust mandible, and the large postcanine tooth crowns of *Paranthropus boisei*).

Bayesian methods Bayesian statistical methods trace their roots to the Reverend Thomas Bayes (1701–61), a Presbyterian minister who formulated a rule that described how to calculate the posterior probability of a hypothesis after observing data and after considering one's prior belief in the hypothesis (i.e., at a time before the data could be observed) (Bellhouse 2004). What is known as Bayes' rule or Bayes' theorem states that $P(HD)=P(DH)*P(H)/P(D)$, where $P(HD)$ is the probability of the hypothesis given the data, $P(DH)$ is the probability of the data given the hypothesis (i.e., the "likelihood" of the data), $P(H)$ is the prior probability of the hypothesis, and $P(D)$ is the unconditional probability of the data; that is, the data probability given all possible hypotheses (Bolker 2008). The advantages of Bayesian methods for biological research include the ease with which they can fit complex models for trait evolution, and their ability to incorporate statistical uncertainty during the estimation of evolutionary models. In evolutionary biology and paleoanthropology, researchers rarely have any good prior information and thus they consider all hypotheses equally probable *a priori*; this condition, also known as "flat priors," means the conclusion about the alternative hypotheses is driven largely by the likelihood. The likelihood is the probability of the observed data given a hypothesis or model (e.g., the likelihood of the coin flip, TTTHHH, under the model of a fair coin is $0.5^6=0.016$). Because all such exact outcomes in larger problems will be exceedingly small values of probability, they are typically expressed as log-transformed and negative values. Bayes' theorem is difficult to solve exactly for most real-world problems, and posterior probabilities are often approximated through Monte Carlo Markov chain (or MCMC) procedures. MCMC procedures involve a series of proposals that randomly modify some aspect of an evolutionary model (e.g., a proposal may modify a branch position in a phylogenetic tree, or a rate of character evolution, where characters can be DNA base pairs or presence/absence and multistate morphological data). The software then calculates the likelihood of the data under the new proposal and accepts or rejects the proposal in proportion to its posterior probability. The final set of accepted proposals, usually in the thousands, represents the posterior probability distribution of tree topologies and parameter values. An MCMC for tree inference [using software such as BayesPhylogenies (Pagel and Meade 2004), BEAST (Drummond and Rambaut 2007), or MrBayes 3 (Ronquist and Huelsenbeck 2003)] will include various clades in some trees but not in others. The proportion of accepted trees with a given clade (called "clade credibility") reflects posterior belief in that evolutionary relationship given the data analyzed. Hypotheses of adaptive trait evolution can also be tested in a Bayesian context by estimating phylogenetic generalized least squares (or PGLS) and related models through an MCMC process [using software such as BayesTraits (Pagel and Meade 2007) or the R package MCMCglmm (Hadfield and Kruuk 2010)]. These models share the same goal as phylogenetically independent contrasts, but importantly they incorporate estimates of phylogenetic uncertainty by assessing trait correlations across multiple phylogenetic trees. PGLS techniques also allow for many alternative models of trait evolution (e.g., punctuational versus gradual change) that are difficult to assess with independent contrasts. Bayesian methods are gaining in popularity for phylogenetic tree inference and for examining correlated trait evolution in a phylogenetic context. *See also* **maximum likelihood**; **Monte Carlo**; **phylogenetic generalized least squares**; **phylogenetically independent contrasts**.

BBC Abbreviation for **Blombos Cave** (*which see*).

BC Abbreviation for Baringo Chemeron and Border Cave. Prefix used for fossils recovered from the Baringo Chemeron Formation, Tugen Hills, Baringo District, Kenya (e.g., **KNM-BC 1**), and for fossils from Border Cave (e.g., BC 1). *See also* **Border Cave**; **Chemeron Formation**.

^{10}Be/^{9}Be ratio *See* **cosmogenic nuclide dating**.

beam-hardening One of several technical problems involved in the data-acquisition stage of **computed tomography** (or CT). Conventional X-ray sources are polychromatic; that is, they encompass a mixture of radiation energies (called their spectrum) and if an X-ray beam passes through matter the lower-energy X-rays (i.e., soft rays) are progressively filtered out so that the average energy of the X-ray spectrum increases. To cope with this problem, clinical CT scanners are calibrated for the different muscle-like tissue thicknesses encountered in patients, but they are not calibrated for fossilized hard tissues. When clinical CT scanners are used to image fossils, inappropriate calibration can cause erroneous "lightening" of tissue boundaries (i.e., elevated gray values), an effect that increases with sample thickness and density. This results in erroneous CT numbers, and/or streak artifacts in the CT image. **Synchrotron** X-ray methods (e.g., **synchrotron radiation micro-computed tomography**, or SR-μCT) avoid these problems by using monochromatic sources.

bed (ME *Bedd*=bed) A layer, or stratum, of rock that can be distinguished from the layers, or **strata**, that lie above it and below it. The bed is the smallest stratigraphic unit that is routinely used in the formal geological nomenclature (e.g., Bed I at **Olduvai Gorge** can be distinguished from the **basalt** below it and the base of Bed II above it).

bedrock A general term for the deep rock layers that lie beneath a superficial (technically "surficial") stratigraphic sequence, usually **Quaternary** or later **Cenozoic**. The term bedrock is also used to refer to lithified or solid rock beneath unconsolidated material such as cemented sandstone beneath loose sand. For example, at **Olorgesailie** the relatively unconsolidated sedimentary rocks that make up Beds 1–4, etc., lie on a bedrock of **Pleistocene** lavas.

Beeches Pit (Location 52°18′56″N, 00°38′20″E, England, UK; etym. the name of a disused brickpit at the edge of Thetford Forest) History and general description Beeches Pit, which is northwest of Bury St Edmunds in Suffolk, is an **Acheulean** site that also preserves some of the earliest evidence of controlled fire in western Europe. In 1999 artifacts were discovered about 0.5 m beneath the principal artifact horizon. Excavation uncovered archeological sequences in two areas of the northwest part of the disused brickpit, AF to the west and AH to the east, where vertical sections of up to 5 m are preserved. The sediments in the eastern (i.e., AH) trench are stratigraphically older. Here an artifact horizon was exposed across excavations of approximately 75 m^2, and the evidence suggests there were multiple occupations, probably by the side of a pool. Several thousand knapped flints come from an approximately 30 cm-thick horizon within which around 100 **refit**ting pieces and microdebitage have been found. These artifacts are stratified within clays containing organic material including **microfauna** and molluscs, and they probably come from an occupation on the top of the channel bank in an area later erased by solifluction; only the edge of the distribution is preserved. Temporal span and how dated? The main Beeches Pit site preserves an interglacial sequence overlying glacial sediments; the western area, AF, records a later sequence within the same interglacial. On the basis of stratigraphy, environmental indicators, and **thermoluminescence dating** and **uranium–series dating** it is inferred that these sediments represent the Anglian glaciation (i.e., **Oxygen Isotope Stage** 12) and the following interglacial (Oxygen Isotope Stage 11). Hominins found at site None. Archeological evidence found at site The artifacts, which are Acheulean but not Clactonian, are associated in multiple phases with environmental evidence indicating a temperate environment, sometimes with closed vegetation. The site is also notable for evidence of fire and burning. In area AH, in addition to visible evidence of at least three large hearths, a flint refit set of around 30 pieces is consistent with selective burning. In area AF there is evidence of a similar, but later, event, with localized burning on a sloping bank adjacent to the presumed pond, and at higher levels in AF there is a widespread dark horizon that contains burnt material. The repeated nature of burning concurs with evidence from sites such as **Gesher Benot Ya'akov**, and with the presence of large hearths at **Schöningen**. Key references: Geology and dating Gowlett et al. 2005, Gowlett 2006, Preece et al. 2006; Hominins N/A; Archeology Gowlett et al. 2005, Gowlett 2006.

Behanga *See* **Nyabusosi**.

behavioral ecology Behavioral ecology focuses on the evolutionary and ecological factors that influence behavior. Behavioral ecologists might consider, for example, whether the costs of an activity or behavior outweigh its benefits (*see* **optimization**). Applying behavioral ecological principles to modern humans and to their evolution has, among other things, contributed to debates over the sexual division of labor, foraging, provisioning, and **life history** strategies (e.g., Winterhalder and Alden Smith 2000).

BEL Abbreviation of Belohdelie, Ethiopia. *See* **Belohdelie**.

BEL-VP Abbreviation of Belohdelie – Vertebrate Paleontology. Prefix used for fossils recovered from **Belohdelie**, Ethiopia (e.g., **BEL-VP-1/1**).

BEL-VP-1/1 Site Belohdelie. Locality Belohdelie Vertebrate Paleontology Locality 1 (BEL-VP-1). Surface/*in situ* Surface. Date of discovery 1981. Finder J. Krishtalka. Unit N/A. Horizon N/A. Bed/member MAK-G1. Formation Sagantole Formation. Group Awash Group. Nearest overlying dated horizon Cindery Tuff. Nearest underlying dated horizon Tuff VT-1. Geological age *c*.3.89–3.86 Ma. Developmental age Adult. Presumed sex Unknown. Brief anatomical description The specimen consists of seven vault fragments, three of which comprise most of the frontal bone. Announcement Clark et al. 1984. Initial description Clark et al. 1984. Photographs/line drawings and metrical data Clark et al. 1984, Asfaw 1987. Detailed anatomical description Asfaw 1987, Kimbel et al. 2004. Initial taxonomic allocation *Australopithecus* aff. *Au. afarensis*. Taxonomic revisions *Australopithecus afarensis* (Kimbel et al. 1994). Current conventional taxonomic allocation *Au. afarensis*. Informal taxonomic category Archaic hominin. Significance If confirmed as *Au. afarensis* the Belohdelie frontal would be among the earliest, if not the earliest, fossil evidence for that taxon. Location of original National Museum of Ethiopia, Addis Ababa, Ethiopia.

Belohdelie One of the named drainages/subdivisions within the **Bodo-Maka** fossiliferous subregion of the **Middle Awash study area**. It was first explored for its paleoanthropological potential in 1981 by a multidisciplinary team organized by the Ethiopian Ministry of Culture and Sports Affairs and led by Desmond **Clark**. A hominin frontal bone, **BEL-VP-1/1**, was recovered during the initial survey. *See also* ***Australopithecus afarensis***.

belowground food resources The terminology used by Hatley and Kappelman (1980) for what are now referred to as underground storage organs (or USOs). Hatley and Kappelman (1980) reviewed the use of belowground food resources (i.e., bulbs, rhizomes, roots, stem bases, and tubers) by bears and pigs and they also considered the data about the belowground food resources eaten by the ≠Kade San. It is estimated that belowground food resources make up just less than half of the vegetable foods that form the majority of the food intake of the ≠Kade San (Tanaka 1976). Hatley and Kappelman (1980) suggest the digging sticks used by modern humans are the analogs of the claws of the bears and the tusks and snout of pigs, and they suggest that belowground food resources would have provided "security, stability, and independence" from "grazing, fire, and drought" for hominins "seeking a living in **Pliocene** and early **Pleistocene** times" (*ibid*, p. 384). *See also* **digging stick; underground storage organs**.

bending energy *See* **transformation grids**.

benthic (Gk *benthos*=deep in the sea) Organisms that live in, on, or close to the sea floor are referred to as benthic. Deep-sea cores are extracted from ocean floors to sample the shells of ancient benthic **foraminifera** within the sediment core. The $^{18}O/^{16}O$ ratios of these shells contain a temporal record of **deep sea temperatures** and **glacial ice volume**. Any planktonic foraminifera in the cores will provide a record of **sea surface temperature**.

benthic foraminifera   *See* foraminifera.

bent-hip bent-knee walking   This form of bipedal walking, with flexed hips and knees, was suggested by John **Napier** (1967) as the type of bipedal gait that archaic hominins might have been capable of. Later Jack Stern and Randall Susman (1983) suggested this form of gait was most likely used by *Australopithecus afarensis* given the skeletal morphology and inferred muscular arrangements seen in **A.L. 288**. Since then, the nature of gait in *Australopithecus* has become one of the main debates concerning archaic hominins. Currently there is no consensus regarding whether the gait of *Australopithecus* was characterized by a bent hip and bent knee, was similar to that of modern humans, or was unique in other ways.

Berg Aukas   (Location 10°30′58″S, 18°15′10″E, Namibia; etym. named after the Berg Aukas vanadium mine) History and general description In 1965 a massive right proximal femur, Berg Aukas 1, was found with other bones in a **breccia** deposit in level 5 of the mine, 177.4 m below the surface: the source breccia was subsequently destroyed by mining operations. The specimen passed to a private collection, but was acquired by C. Begley and taken to the Museum of Man and Science in Johannesburg in 1968. It was later loaned to Phillip **Tobias** for study. Temporal span and how dated? Early to **Middle Pleistocene** based on morphological comparison with better-dated specimens. Hominins found at site Berg Aukas 1, a well-preserved and densely mineralized proximal right femur with a massive head, low neck-shaft angle, and thick cortical bone. It resembles the presumed *Homo erectus* specimen **KNM-ER 736** from **Koobi Fora**, Kenya, and the massive femoral fragment from **Castel di Guido**, Italy, attributed to *Homo heidelbergensis*. Archeological evidence found at site None. Key references: Geology and dating Grine et al. 1995, Conroy 1996; Hominins Grine et al. 1995, Trinkaus et al. 1999b; Archeology N/A.

Berg Aukas 1   *See* **Berg Aukas**.

Bergmann's Rule   This concept (probably better described as a trend rather than a rule), attributed to Carl Bergmann (1847), states that animals living at lower average temperatures tend to have larger body sizes overall. Bergmann's Rule (and its "sister" ecogeographic rule, **Allen's Rule**) have not been extensively tested in nonhuman primates, and the studies that have been undertaken show limited support. This could be because for tropical primates temperature is a less important determinant of climate than other factors (such as rainfall) and therefore it has less influence on morphology, but the body sizes and shapes of modern humans and their closest relatives show relatively strong relationships with latitude (Holliday 1999, Ruff 1994, 2002). Bergmann's Rule applies to modern humans with respect to mean body mass (Ruff 2002).

Bernard Price Institute for Palaeontological Research   Dr Bernard Price, a South African industrialist and philanthropist, was so impressed with the research carried out by Robert **Broom** that in 1945 he provided the funds to establish this institute (or BPI). James Kitching was the first employee of the BPI; he worked there full-time until 1990, remaining associated with the BPI until his death in 2003. The institute is presently part of the School of Geosciences in the Faculty of Science of the University of the Witwatersrand, Johannesburg, South Africa. Its mission is to collect, conserve, study, and interpret the fossil heritage of southern Africa. The BPI houses one of the largest collections of Karoo vertebrates in the world, and in conjunction with the **Institute for Human Evolution** it now curates the faunal fossil collections from **Makapansgat**, **Cooper's Cave** A and B, **Drimolen**, **Gladysvale**, **Sterkfontein**, Plovers Lake, etc. It also has extant comparative faunal collections. Hominin fossil collections When the repository at the BPI is completed, it will hold all of the hominins collected at Sterkfontein other than those held at the **Ditsong National Museum of Natural History** (previously the **Transvaal Museum**) (i.e., the material collected by Ron **Clarke**, Alun Hughes, and Phillip **Tobias** since 1966 from Members 2–6), and the hominins recovered from Cooper's Cave A and B, Drimolen, and Gladysvale. Comparative collections See above. Contact person Bernhard Zipfel (e-mail Bernhard Zipfel@wits.ac.za). Street address 1 Jan Smuts Avenue, Johannesburg, South Africa. Mailing address PO Wits, 2050 Wits, South Africa. Website http://web.wits.ac.za/geosciences/bpi/index.htm. Tel +27 11 717 6683. Fax +27 11 717 6694.

Bervavölgy (Location 48°00′N, 20°20′E, Hungary; etym. named after the valley in which it is found) History and general description This cave site in the Bükk mountains contains Late Magdalenian deposits. Temporal span and how dated? Comparison of the lithics suggests a Late Magdalenian age (*c.*12 ka). Hominins found at site A mandibular fragment with two teeth attributed to *Homo sapiens*. Archeological evidence found at site Late Magdalenian lithics. Key references: Geology, dating, and paleoenvironment Thoma and Vértes 1971; Hominins Thoma and Vértes 1971.

beta taxonomy (Gk *beta*=the second letter in the Greek alphabet and *taxis*=to arrange or "put in order") It is common for reference to be made to **alpha taxonomy**, but you very seldom read, or hear, any reference to beta taxonomy. Mayr et al. (1953) is one of the few sources that defines these terms [e.g., they do not appear in the index of Simpson's (1960) *Principles of Animal Taxonomy*]. Mayr et al. (1953) suggest "the taxonomy of a given group, therefore, passes through several stages...informally referred to as alpha, beta, and gamma taxonomy" (*ibid*, p. 19). The distinction made by Mayr et al. is that whereas alpha taxonomy results in species being "characterized and named," and gamma taxonomy refers to the "analysis of intraspecific variation and to evolutionary studies," beta taxonomy involves arranging species in "a natural system of lesser and higher categories" (*ibid*, p. 19). So what Mayr et al. (1953) called beta taxonomy includes the process of **phylogeny reconstruction**, which is part of systematics. But Mayr et al. (1953) caution that it is "quite impossible to delimit alpha, beta and gamma taxonomy sharply from one another, since they overlap and intergrade" (*ibid*, p. 19). *See also* **systematics**; **taxonomy**.

BHBK Acronym for bent-hip bent-knee. *See* **bent-hip bent-knee walking**.

Biache *See* **Biache Saint-Vaast**.

Biache Saint-Vaast (Location 50°18′54″N, 2°56′55″E, France; etym. named after the town in which it is located) History and general description This open-air site is located in Pas-de-Calais, on an upper terrace along the Scarpe river. The site was discovered during the expansion of a steel mill and was excavated as a salvage operation from 1976 to 1982 by Alain Tuffreau and his team. The site preserved evidence of several episodes of hominin occupation during warm interglacial periods, a large faunal assemblage, and the fragmentary remains of two *Homo neanderthalensis* crania, both from archeological layer IIA. Temporal span and how dated? The **biostratigraphy** suggests that most of the archeological layers were deposited during **Oxygen Isotope Stage** 7, or *c.*240–190 ka. Thermo**luminescence dating** of burned **flints** from the main archeological layer, IIA, suggests an age of 175±13 ka, but thermoluminescence dating using gamma spectrometry on the second hominin cranium suggests an older age of 263+53/−37 ka. **Electron spin resonance spectroscopy dating** of two animal teeth from the same layer also suggests an older age of *c.*272 ka. Hominins found at site Two fragmentary crania were recovered. One, Biache I, was found *in situ*, and includes a portions of the rear of the cranium, parts of the maxillae, and 11 upper teeth belonging to a young, presumably female, individual. The other fossil, Biache II, was found after excavation, during the study of the faunal remains, and consists of a portion of the frontal, and the left parietal and temporal bones. It is thought to be from a young male. Both fossils come from the same thin archeological layer near the base of the site, and are considered to belong to *H. neanderthalensis*, or to a **pre-Neanderthal** hominin. Archeological evidence found at site The abundant lithic assemblage has been characterized as Ferrassie **Mousterian**. Many of the faunal remains show signs of butchery and evidence of marrow extraction. Key references: Geology and dating Huxtable and Aitken 1988, Tuffreau and Sommé 1988, Yokoyama 1989, Bahain et al. 1993; Hominins Rougier 2003, Vandermeersch 1978a, 1978b; Archeology Tuffreau and Sommé 1988.

bias (Fr. *biais*=slant from the Gk *epikarsios*= slanted) A term that refers to several different concepts in statistics and measurement, all of which refer to errors in the estimation of a population parameter. Statistical bias generally refers a difference between the sample estimate of a population parameter (such as the mean) and the value of that parameter in the full population. Such differences can result from different types of bias. For example, selection (also called sampling or ascertainment bias) refers to nonrandom sampling of a population, in which case a variable measured for the sample is a poor estimate for the actual value for a population. Consider a fossil assemblage that only preserves the largest individuals in a population; in that case, the mean value of any measurement for that sample will be an overestimate of the actual value of the mean for the population from which the sample was drawn. Systematic bias is another source of error and is usually the result of an error in measurement where all measurements are off by the same amount. For example, a set of calipers may be incorrectly zeroed such that the calipers read a measurement of 10 mm when the jaws are completely closed (i.e., a real measurement of zero). In that case, all measurements will be 10 mm too high, and the estimate of the mean will reflect this error. If the systematic bias is truly equal for all measurements, then measures of absolute variation (such as variance and standard deviation) will not be statistically biased, but measure of absolute variation (such as the coefficient of variation) will be. *See also* **ascertainment bias**.

biface (L. *bis*=twice and *facies*=surface) A stone tool that has been worked twice on opposed surfaces, or faces, as opposed to those worked on one side; the latter are termed unifaces. Handaxes are a typical example of a biface.

bifacially worked (L. *bis*=twice and *facies*=surface) A term used in the analysis of stone tools that refers to a piece that is worked (i.e., shaped), on two opposed surfaces, or faces. It contrasts with a piece that is worked (i.e., shaped), on just one surface, or face; such pieces are said to be **unifacially worked**. *See also* **biface**.

Bilzingsleben (Location 51°16′52″N, 11°4′7″E, Germany; etym. named after the village of Bilzingsleben 1.5 km/0.9 miles north of the site) History and general description An open-air hominin site in west central Europe, 35 km/22 miles north of Erfurt, Germany, with controversial evidence of dwelling structures, hearths, and worked bone. The site lies near the base of a sequence of Pleistocene travertine built up by a **karst** spring that flowed into a shallow lake. The first mention of fossil bones and teeth is from 1710 by David Siegmund Büttner. Excavations began in 1922 under the direction of Adolf Spengler. Further excavations, initially directed by Dietrich Mania, began in 1969 and are ongoing. Temporal span and how dated? Biostratigraphy suggests the site is within **Oxygen Isotope**

Stage 11e, and **uranium-series dating** on calcite samples and **electron spin resonance spectroscopy dating** on teeth are consistent with a date between *c.*362 and 423 ka. Hominins found at site Thirty-seven hominin fossils, including 29 cranial fragments (occipital, parietal, and frontal), an edentulous mandible fragment, and nine isolated teeth. At least three individuals are represented, including one juvenile. Archeological evidence found at site The travertine-preserved Pleistocene fauna, molluscs, and flora, as well as hominin-modified bones and Lower Paleolithic lithic material, described as "Micro-Clactonian." In addition, Mania found a circular arrangement of stones that he interpreted as evidence of dwelling structures. The archeological horizon lies in the alluvial deposits by the ancient lake shore and has yielded around 140,000 flint artifacts, and many other tools made of stone, bone, antler, ivory, and wood. Key references: Geology, dating, and paleoenvironment Schwarcz et al. 1988, Jöris and Baales 2003; Archeology Mania 1983, Mania and Vlček 1987, Mania 1989, 1997; Hominins Vlček 1978a, Mania et al. 1993, Stringer 1981, Street et al. 2006. Location of the hominin fossils found at site Landsmuseum für Vorgeschichte, Halle/Saale, Germany (specimens discovered before 1993) and Forschungsstelle Bilzingsleben der Friedrich Schiller Universität Jena, Jena, Germany (specimens collected in 1993 and thereafter).

bimaturism (L. *bi*=two and *maturus*=ripe) Term introduced by Jarman (1983) for the process by which two organisms, or two structures, differ in size because the larger one has grown for longer. Usually it is a sex difference with the males having the longer period of growth (e.g., canine dimorphism in extant hominoids; Schwartz and Dean 2000).

Binford, Lewis (1930–) Lewis Binford was born in Norfolk, Virginia, USA. He began undergraduate studies at Virginia Polytechnic Institute, where he studied wildlife biology and thus began a lifelong scholarly interest in the environment. His education was interrupted by the Korean War, but it was during his military service when he was assigned to be an interpreter for displaced Japanese that he developed an interest in anthropology. Afterwards, he attended the University of North Carolina, where he studied anthropology and archeology. He attended graduate school at the University of Michigan, graduating in 1964 with a dissertation on the archeology and ethnohistory of coastal Virginia. Starting in 1961, Binford taught at the University of Chicago, and then at the University of California, first at Santa Barbara, and then at UCLA. His primary academic home, however, was the University of New Mexico, where he taught for 23 years before taking a position at Southern Methodist University in 1991. His magnum opus, *Constructing Frames of Reference*, which used a large database of hunter–gatherer societies and environmental data to create predictive models for prehistory, was published in 2001. In 1962, Binford published *Archaeology as Anthropology*, widely regarded as the foundation of the "New Archaeology" of the 1960s, now known as processual archeology. The call for an archeology relying on scientific methods, including a hypothesis-testing approach, was revolutionary at the time but is *de rigueur* today. Binford took a materialist approach to understanding prehistory, relying on Leslie White's concept of culture as "man's extrasomatic means of adaptation." He helped pioneer ethnoarcheology (with research among the Nunamiut Eskimo and Australian Aborigines), and made fundamental contributions to faunal analysis, taphonomy, and the use of modern ethnographic data to model past behavior. Binford was an outspoken advocate for the development of "middle-range theory," a body of information intended to allow archeologists to make reliable inferences about past dynamics from archeological remains. Binford's involvement with paleoanthropology began early in his career, through a debate with eminent French prehistorian François Bordes over the meaning of variability in stratified **Mousterian** cave assemblages. Bordes proposed that the assemblages reflected different cultural constructs about appropriate ways to make stone tools, and hence different "tribes" of *Homo neanderthalensis*, but Binford (and Binford) argued that the assemblages reflected different functional uses of a cave. Through his Nunamiut ethnoarcheological research he was later able to demonstrate that the same people did produce different assemblages at different places, and that the same place could be used for different purposes as the settlement system shifted its geographic placement. This work, which appeared first in *Nunamiut Ethnoarchaeology*, led Binford into faunal analysis. This new focus in turn led him to a concern with the inferences that archeologists were making from the faunal remains found in Lower and Middle Paleolithic sites. His efforts in this area appeared in *Faunal Remains from Klasies*

River Mouth Cave and *Bones: Ancient Men and Modern Myths* as well as in a series of articles, mostly published in the 1980s. Binford argued that most paleoanthropologists saw hominins of the Lower Paleolithic as simply modern humans stripped down to bare essentials: hunters of large game who lived in small bands that foraged in much the same manner as Bushmen and other modern hunter–gatherers, with a similar division of labor, settlement pattern (with "base camps"), and technological capacity. Binford asserted that there was no good evidence of large game hunting in the Lower and possibly the Middle Paleolithic at sites such as **Klasies River, Zhoukoudian Locality 1, Swanscombe,** Abri Vaufrey, Torralba, **Koobi Fora,** and **Olduvai Gorge.** Instead, he argued from skeletal element representation that most of the faunal remains in association with stone tools was a result of scavenging of low-utility body parts. Binford also argued that the stone tool assemblages of the Lower Paleolithic showed little evidence of the type of planning seen among modern hunter–gatherers. Binford considered the role that tools played in a foraging adaptation and he thought two associations were typical of **Oldowan** and **Acheulean** assemblages: (a) a transported toolkit consisting of heavy-duty core tools such as handaxes or choppers and (b) an expediently produced toolkit consisting of small scrapers and pebble tools; at one point Binford suggested this second toolkit might be produced by children foraging from scavenged carcasses. Binford further argued that the transported toolkits were found in sites that (based on geologic data) were **palimpsests** of artifacts that accumulated in one place slowly over time. These sites were not, he maintained, base camps, but repeatedly revisited foraging stations where hominins extracted and ate food; hence, these sites were not comparable to the residential camps of living foragers. Binford further argued that the transported tools showed little evidence of extensive use and were not "piled up" in particular locations, as if hominins did not remember or know there were already tools at one location. In brief, Binford saw little evidence for extensive "planning depth" until the Upper Paleolithic. Binford's reading of the data suggested to him that Lower and Middle Paleolithic hominins behaved in ways unlike any recent human foragers, thus opening a door onto a world inhabited by foragers that did not necessarily follow familiar modern human patterns. Binford has been criticized for his interpretations, his selection of study samples and his use of data, and he engaged in vigorous debate with researchers whose analyses and conclusions did not align with his own. His main contributions to paleoanthropology come from leading people to question easy assumptions and to work harder to justify their interpretations. Along with his contemporaries C.K. (Bob) **Brain**, Glynn **Isaac**, and others, Binford helped catalyzed renewed scrutiny of the empirical record. He inspired succeeding generations to make **taphonomy** and the study of site-formation processes central features of Paleolithic research projects. Binford's *Bones: Ancient Men and Modern Myths*, together with Brain's *The Hunters or the Hunted*, are two seminal case studies that demonstrate the value of taphonomy and **actualistic studies** to faunal analysis. Binford's approach to the study of technology as an integral part of foraging systems has become standard in the field and while many of his specific interpretations may not stand the test of time, he nonetheless helped shape paleoanthropology into the more rigorous field it is today.

binomen   *See* **Linnaean binomial**.

binomial   *See* **Linnaean binomial**.

Binshof   *See* **Binshof, Speyer**.

Binshof, Speyer   (Location 49°21′48″N, 8°27′03″E, Germany; etym. named after the village of Binshof, 30 km/19 miles west of Heidelberg) History and general description In 1974 an extremely well-preserved cranium was recovered containing most of the facial bones and maxillary teeth. The find became known as "Binshof, Speyer Woman" and was dated to the Pleistocene. The dating of the remains fell into question in 2001 when Thomas Terberger of Greifswald University had some of the fossil samples reanalyzed and the resulting ages were considerably younger. The find is now associated with the downfall of Reiner Protsch who is alleged to have falsified data. Temporal span and how dated? The original **radiocarbon dating** age for "Fra-40" was 21,300±320 years BP, but recent **AMS radiocarbon dating** (OxA-9880) suggests an age of 3090±45 years BP or 1420–1290 years BCE. Key references: Geology, dating, and paleoenvironment Henke 1980, Terberger and Street 2001, Street and Terberger 2002; Hominins Henke 1980.

biochronology   (Gk *bios*=life, *chronos*=time, and *ology*=the study of) A correlated-age method of **relative dating** that uses fossils preserved in a sedimentary rock to determine its age. Many fossil sites cannot be dated using absolute methods because they lack the raw materials needed for absolute dating (e.g., igneous rocks). These sites have to be dated using the fossils they contain. Fossils are matched with those found at other sites that cannot be dated by numerical-age and calibrated-age methods (e.g., isotopic or radiogenic) of **absolute dating**, as well as fossils from sites that can be dated using such methods. When a taxon is restricted to a particular time period or if their **first appearance datum** (or FAD) or **last appearance datum** (or LAD) is the same across several absolutely dated fossil sites then such a taxon can be used alone, or preferably in combination with other time-delineated taxa, to generate a biochronometric date for the stratum. The use of a single taxon for this purpose is exemplified by the use of the rodents *Mimomys* and *Microtus* in European sites. The vole *Mimomys* predominates in earlier sites, is rare in Cromerian sites, and by the **Middle Pleistocene** is replaced by *Arvicola* or *Microtus*. Data from one taxon, or a series of taxa, can be combined to define one, or more, **biozones**. An example of the former is the use of the extinct pig *Metridiochoerus andrewsi* as a marker taxon in East African strata, while an example of the latter is the European mammal neogene (or MN) system for dating Miocene and Pliocene sites whose ages range from 24 to 2 Ma. Recently the latter system has been extended below 2 Ma. The MN system is increasingly being used in preference to the classical European mammalian biozonations (i.e. Vallesian, Turolian, Ruscinian, Villafranchian, etc.). *See also* **biostratigraphy; European mammal neogene; geochronology**.

biogenetic law   (Gk *bios*=life, plus *genesis*=origin or generation) The biogenetic law, or "ontogeny recapitulates phylogeny," proposes that during embryonic development animals pass through the adult stages of their ancestors, thus representing a nesting sequence of orthogenesis. This idea was popularized by Ernst Haeckel in the late 19thC, and was supported by comparisons of embryonic morphology from various organisms, identifying stages that represented adult fish, birds, and mammals, moving from earlier embryonic stages to later ones. In this way, one could "read" the entire evolutionary history of an organism by observing its embryonic development. In this scheme modern humans represented the furthest point on this evolutionary scale, and embryologists endeavored to find examples of adult representatives of all lower life forms in the successive stages of modern human embryonic development. Any malformed fetus was seen as an atavism, or as the result of development that did not progress as far as it should have. The strictest interpretation of the biogenetic law is now understood to be false; that is, evolution does not occur in a linear manner wherein the developmental stages of ancestors are literally compressed into the ontogeny of descendant species. Development itself evolves, rather than just tacking on successive stages to an ancestral template. However, that being said, there are broad similarities in embryonic development among all mammals, and among mammals and reptiles, and among mammals and fish, but these mostly occur early in embryogenesis.

biogeography (Gk *bios*=life, *geo*=earth and *graphein*=to write) The scientific investigation of how and why organisms came to be distributed on the Earth in the way they are. Biogeographers document the spatial distribution of organisms and species and determine the processes that might explain such patterns. Historically, biogeography was an important focus of study for many pre-Darwinian naturalists. The geographic distribution of closely related, but adaptively distinct, species was an important factor leading Charles **Darwin** to accept the concept of evolution, and to propose **natural selection** as a mechanism to explain evolution. Darwin can reasonably be considered the father of evolutionary biogeography. Biogeographic patterns have also influenced the development of ideas about **allopatric speciation**. More recently, the modern development of biogeography as a scientific field is tied to the development and acceptance of plate-tectonic theory insofar as the movement of continental plates can have the effect of opening up or shutting off avenues for the **dispersal** (range change) or **vicariance** (range splitting) of populations (e.g., the separation of two continental plates may split what was formerly a single terrestrial habitat, but in doing so may unite two marine habitats). Some of the most important hypotheses in paleoanthropology concern biogeographic events. These include (a) the possible dispersal of the ancestor of the African ape and modern human **clade** from Europe to

Africa at the end of the **Miocene**, (b) the dispersals of various **australopith** and early *Homo* species within Africa during the Plio-Pleistocene, (c) the earliest dispersal of **hominins** out of Africa, probably comprising early *Homo* or *Homo erectus* near the **Plio-Pleistocene boundary**, (d) the colonization of Europe by hominins, (e) the colonization of Indonesia, (f) a subsequent dispersal out of Africa of modern *Homo sapiens* (or, at least, hominins possessing modern human morphology) during the **Late Pleistocene**, and (g) the colonization by hominins of the New World. Biogeography can be studied by using **phylogeny** as a tool to identify biogeographic events, because when an ancestor and its descendant are found in different regions, then obviously some type of dispersal or vicariance event must have taken place. Hominin biogeography may also be studied by comparing the biogeographic patterns of hominins to those of vegetational zones or other mammalian clades, within a tight temporal and/or climatic framework. Key biographical questions related to paleoanthropology include whether other mammals disperse in the same direction and at the same time as hominins, whether the mammalian dispersals and/or vicariance events provide an insight into how and why hominin populations disperse or split, whether hominin dispersals track the presumed changes in range of vegetational zones, and how, if at all, climate change has affected hominin biogeographic patterns.

biological species One of the many terms used to describe an **extant** species. It will be diagnosed on the basis of evidence of interbreeding, or the lack of any evidence of interbreeding between individuals belonging to that species and individuals belonging to any other species. It contrasts with a paleospecies that has to be diagnosed on evidence from the hard tissues alone. *See also* **species** (syn. **neontological species, extant species**).

biological species concept Defined by Ernst Mayr (Mayr 1942) and modified by him in 1982 (Mayr, 1982, p. 273). It is one of the "process" species definitions in which species are defined as "groups of interbreeding natural populations reproductively isolated from other such groups." Note this definition depends on there being at least one additional species. It is a "relational" and not a "free-standing" species definition. The biological species concept is of limited value in considering extinct hominins because it is impossible to observe direct evidence for interbreeding in the fossil record, and also because modern analogs of closely related primate species (as well as myriad other organisms) suggest that primate speciation can and has occurred even in the presence of gene flow. *See also* **species**.

biomarkers *See* **alkenones**.

biome (Gk *bios*=life) A distinctive regional ecological community of plants and animals adapted to the prevailing climate and soil conditions. Biomes are collections of ecosystems that have similar biotic and abiotic characteristics. Biomes can be (a) terrestrial, (b) freshwater, and (c) marine. There are several schemes for classifying terrestrial biomes (e.g., Holdridge classification scheme, Walter system, Bailey system, and Whittaker's classification scheme) on the basis of variables such as the amount of precipitation, evapotranspiration potential, mean annual biotemperature and vegetation type. It is conventional to recognize six major terrestrial biomes – **forest**, **grassland**, **woodland**, **shrubland**, semi-desert shrub, and desert – each of which can be further subdivided on the basis of climatic conditions and elevation.

biospecies A contraction of biological species; one of the many terms used to describe an extant species defined on the basis of criteria such as evidence of interbreeding, or the lack of any evidence of interbreeding between individuals belonging to that species and individuals belonging to any other species. *See also* **biological species**; **species**.

biostratigraphy (Gk. *bios*=life, *stratos*=to cover or spread, and *graphos*=to draw or write) The process of linking rock layers within and among stratigraphic sequences by comparing the fossil assemblages preserved within them. This can be via either the matching of individual taxa (e.g., the rodent *Mimomys* in Pleistocene European sites) or by using a system such as the **European Mammal Neogene** (or MN) system, which uses a combination of taxa. The use of the linked rock layers to infer the age of the strata is sometimes called biochronology. *See also* **biochronology**.

biotic (Gk *bios*=life) Describes all the living factors (e.g., competition for food, predator pressure) that affect the evolutionary history of a **taxon**, or of a **clade**.

bioturbation (Gk *bios*=life and *turbe*=turmoil) The displacement and mixing of sediment or soil by **biotic** agents or processes. Bioturbation is caused by plants or animals. In Africa, for example, both **termites** and rodents can burrow during and after the deposition of **artifacts** and **ecofacts** on a land surface. The movement of termites and rodents through the stratigraphic sequence can incidentally transport objects on and within the sediment, removing them from their original context. Their new context is often temporally incongruous, and can create or remove spatial and temporal associations. Detailed sedimentological study can identify strata that have been affected by bioturbation (i.e., they are bioturbated). Any evidence of bioturbation is important when considering the temporal and contextual integrity of a paleontological or archeological site, especially when evidence from the site might be used to develop hypotheses about hominin activity patterns and/or to reconstruct the paleoenvironment.

biozone A series of strata at one, or more, sites all of which sample the same fossil taxon, or suite of fossil taxa. An example would be all the East African strata that sample the extinct pig *Metridiochoerus andrewsi*.

biped (L. *biped*=two-footed) A hominin whose resting posture is standing on the feet of the two hind, or lower, limbs, and whose locomotor repertoire involves a significant component of bipedal travel. *See also* **bipedal**.

bipedal (L. *biped*=two-footed) Form of posture or locomotion in which the body is supported exclusively on the feet of the two hind, or lower, limbs. Prost (1980) usefully distinguishes between taxa in which adults are able to travel bipedally, which he calls "facultative bipedalism," and taxa such as modern human adults for which bipedal gait is the only efficient way for them to move around. Prost (*ibid*, p. 187) refers to the latter as "preeminent bipedalism," but others have referred to the latter category as "obligate bipeds." For example, the earliest hominins were most likely facultative bipeds, whereas *Homo neanderthalensis* was almost certainly a preeminent, or obligate, biped. Prost also makes the point that the possession of some of the morphological traits seen in known bipeds (Prost calls these **bipedal traits**) does not mean the taxon that possesses them was necessarily a biped (i.e., practices "bipedalism"). He cites some bipedal traits that are seen in both vertical climbers and bipeds; thus the possession of these traits should not be equated with bipedalism.

bipedal locomotion *See* bipedal; walking cycle.

bipedal trait A term used by Prost (1980) for a morphological feature found in known bipeds that is functionally related to bipedal posture and/or locomotion. For example, an adducted hallux is a bipedal trait established prenatally that is present in all known hominin bipeds and plays an important role in propulsion at the end of the stance phase of bipedal **gait**. Some bipedal traits, such as the **valgus** knee joint, are **epigenetic**ally sensitive and only develop in bipeds in response to bipedal walking (e.g., Tardieu and Trinkaus 1994). *See also* **bipedal**.

bipedalism The act of walking or running using only the feet of the two hind, or lower, limbs for support. *See also* **bipedal**.

bipedalism, facultative *See* bipedal.

bipedalism, obligate *See* bipedal.

bipolar (L. *bi*=two and *polus*=axis) A term used to refer to a specific technological approach for making stone flakes in which the core is rested on an **anvil** and then struck to split the core, or to remove a flake. The flakes produced by this technology are characterized by evidence of percussion from opposite ends, or "poles" (the anvil and hammer ends) of the struck core. Bipolar technology, which is used wherever stone is available in small packages (e.g., pebbles), occurs at **Oldowan** sites such as **Olduvai Gorge** (Jones 1994), and it is also common at many **Later Stone Age** sites in southern Africa (Barham 1987). Chimpanzee populations that crack nuts mostly use bipolar technology when they do so (Haslam et al. 2009).

bipolar percussion (L. *bis*=twice and Gk *polus*=axis, poles are at each end of an axis) Bipo-

lar means "two poles", so bipolar percussion is a term used for a method of fracturing stone for the production of sharp-edged **flake**s. Percussive forces are initiated on a stone from two opposed directions, or poles. Typically, a small pebble is rested on an **anvil**, and struck from above, and the percussive force splits the pebble in two. Bipolar percussion is a technique seen early in the archeological record (e.g., at **Oldowan** sites). *See also* **bipolar**.

birds All extant birds and fossil birds found at early hominin sites belong to the Neornithes, the "modern" birds, within the class Aves. Bird true fossils are relatively rare in the fossil record, but there are trace fossils of birds (e.g., prints and tracks) as well as evidence from their ontogeny, such as fossil egg shell. However, when the remains or traces of birds are found in the fossil record they can be useful paleoenvironmental indicators (e.g., wading birds indicating the nearby presence of water). The most common bird remains found at open-air early hominin fossil sites is probably ostrich egg shell, which can be used for **ostrich egg shell dating** as well as for **paleoenvironmental reconstruction**. *See also* **Aves**.

Bishop, Walter William (Bill) (1931–77) Bill Bishop graduated from Birmingham University in 1952 with a degree in geography. He stayed at Birmingham for his PhD studies, graduating in 1956, and then left the UK to work as a geologist with the Uganda Protectorate Geological Survey. After a period back in the UK as an Assistant Lecturer at Glasgow University and the Keeper of that University's Hunterian Museum he returned to Uganda in 1962 as the Director of the Uganda National Museum. From the beginning of his time in Uganda Bill Bishop began to take in interest in paleontology and human evolution and he did field work at many locations including Napak, a volcano mapped by Basil King in 1939. In 1965 Basil King recruited Bishop to join him at Bedford College in the University of London as a Lecturer in Geology, where be became part of Kings' East African Geological Research Unit (or EAGRU), which was carrying out an ambitious mapping project in the northern part of the Rift Valley in Kenya, in collaboration with the Geological Survey of Kenya and the Department of Geology at what was then University College, Nairobi. This project involved several graduate students (e.g., Greg Chapman, Andrew Hill, John Martyn, and Martin Pickford) who mapped regions (e.g., the Tugen Hills) that became the focus of several subsequent productive field research projects. For example, Bishop and Chapman (1970) and then Bishop and Pickford (1975) described the sediments and fossils from the Ngorora Formation; Bishop, Chapman, and Hill, together with Jack Miller (Bishop et al. 1971), described the c.15 Ma-long succession of vertebrate assemblages in the Tugen Hills and Carney et al. (1971) described the first hominin from the Chemoigut Formation at **Chesowanja** and later Bishop, Pickford, and Hill (1975) reported archeology and a further hominin from Chesowanja. In 1974 Bill Bishop was appointed to the Chair of Geology at Queen Mary College, also in the University of London, and at the time of his unexpected death at the age of 46 he was due to take up the Directorship of the Peabody Museum at Yale University. Bill Bishop was a great advocate of the importance of a sound chronology and he was a skilled synthesizer of information about the fossil record and geochronology. His 1973 review in *Nature*, "The tempo of human evolution," emphasized the former, as did volumes he edited or co-edited (e.g., *Background to Evolution in Africa*, 1967 with Desmond Clark; *Calibration of Hominoid Evolution*, 1972 with Jack Miller; *Geological Background to Fossil Man*, 1978). At the time of his death Bishop had just begun working as coordinator of the geological studies in David Pilbeam's project in the Potwar Plateau of Pakistan. Bill Bishop was an amateur singer and he was both noted and notorious for his own (often long) lyrics written to suit the conferences he attended. These usually featured noted conferees and were always set to tunes from Gilbert and Sullivan operettas.

bite force *See* chewing.

BK Acronym for Baringo Kapthurin. It is the prefix used for fossils recovered from the Kapthurin Formation, Tugen Hills, Baringo District, Kenya (e.g., KNM-BK 67). It is also an abbreviation of Blimbingkulon, and it is used as the prefix for fossils from that site in Java, Indonesia (e.g., **BK 7905**). It is also the acronym for Bell's Korongo, a site in the side gorge at Olduvai Gorge. Excavations at Bell's Korongo between 1952 and 1958 resulted in the recovery of many artifacts and in 1954 Mary and Louis **Leakey** recovered two hominin teeth, OH 3, which turned out to be the first evidence to be discovered of the hominin *Zinjanthropus boisei*. *See also* **Blimbingkulon; Kapthurin Formation; Olduvai Gorge**.

BK 7905 Site **Blimbingkulon**. Locality Dried tributary of the Brangkal River. Surface/*in situ* Surface. Date of discovery 1979. Finder Mr Sutanto, a local villager. Unit N/A. Horizon N/A. Bed/member Grenzbank Zone. Formation Between the overlying **Bapang Formation** and the underlying **Sangiran Formation**. Group N/A. Nearest overlying dated horizon N/A. Nearest underlying dated horizon N/A. Geological age c.1.51–1.60 Ma. Developmental age Late adolescent or young adult. Presumed sex Unknown. Brief anatomical description Fragmentary and heavily mineralized right mandibular corpus preserving the partial roots of M_1 and a complete M_2. Announcement Aziz 1983. Initial description Aziz 1983. Photographs/line drawings and metrical data Kaifu et al. 2005b. Detailed anatomical description Kaifu et al. 2005b. Initial taxonomic allocation "*Meganthropus*" (Aziz 1983). Taxonomic revisions *Homo erectus* (Kaifu et al. 2005b). Current conventional taxonomic allocation *H. erectus*. Informal taxonomic category Pre-modern *Homo*. Significance The specimens from Blimbingkulon extend the range of morphological and metrical diversity of early *H. erectus* from Java (see Kaifu et al. 2005b). Location of original Geological Research and Development Center, Bandung, Indonesia. (syn. Sangiran 33, Mandible H).

BK 8606 Site **Blimbingkulon**. Locality N/A. Surface/*in situ* Surface. Date of discovery 1986. Finder Mr. Sutanto, a local villager. Unit N/A. Horizon N/A. Bed/member **Grenzbank Zone**. Formation Between the overlying **Bapang Formation** and the underlying **Sangiran Formation**. Group N/A. Nearest overlying dated horizon N/A. Nearest underlying dated horizon N/A. Geological age c.1.51–1.60 Ma. Developmental age Adult. Presumed sex Unknown. Brief anatomical description Fragmentary and heavily mineralized right mandibular corpus preserving the distal portion of the P_3 root, the roots of M_1–M_2 and a partial M_3 crown. Announcement Aziz et al. 1994. Initial description Aziz et al. 1994. Photographs/line drawings and metrical data Kaifu et al. 2005a. Detailed anatomical description Kaifu et al. 2005a. Initial taxonomic allocation *Homo erectus* (Aziz et al. 1994). Taxonomic revisions None. Current conventional taxonomic allocation *H. erectus*. Informal taxonomic category Pre-modern *Homo*. Significance The specimens from Blimbingkulon extend the range of morphological and metrical diversity of early *H. erectus* from Java (see Kaifu et al. 2005b). Location of original Geological Research and Development Center, Bandung, Indonesia (syn. Sangiran 39).

Blaauwbank valley A shallow valley in the limestone hills that form the high veldt in the vicinity of Krugersdorp in Gauteng Province, South Africa. The hills contain many substantial breccia-filled solution cavities and some of these (e.g., Cooper's Cave, Drimolen, Kromdraai, Malapa, Sterkfontein, and Swartkrans) contain fossil hominins. *See also* **Cooper's Cave; Drimolen; Kromdraai; Malapa; Sterkfontein; Swartkrans.**

Black, Davidson (1884–1934) Davidson Black, born in Toronto, Canada, entered the University of Toronto as a medical student in 1903. After completing his medical degree in 1906 and BA degree in 1909, Black obtained an appointment as an anatomist at Western Reserve University (now Case Western Reserve University) in Cleveland, Ohio, USA. While at Western Reserve, Black was introduced to physical anthropology by Thomas Wingate Todd, who had studied with the prominent Australian neuroanatomist and physical anthropologist Grafton Elliot Smith. In 1914 Black spent a sabbatical leave in Manchester, England, where he worked with Elliot Smith and met Sir Arthur **Keith** and Sir Arthur Smith Woodward (with whom he visited **Piltdown**). Black visited the neuroanatomist Cornelius Kappers in Amsterdam (but the impending hostilities of WWI forced him to cut short the visit) before returning to America fired with enthusiasm for human paleontology and neuroanatomy. For much of the early 20thC many anthropologists believed that the first humans had evolved in Asia, not Africa, and for this reason Black became interested in the possibility of searching for hominin remains in Asia. An opportunity arose when Black learned that the newly established **Peking Union Medical College** (or PUMC) in Beijing was looking for faculty. Black obtained a position teaching embryology and neurology at the PUMC in 1918 (but he and his wife did not arrive in China for more than a year) and while his primary duties were to train

Chinese medical students, by 1921 the position offered Black enough free time to pursue his search for hominin fossils together with the Swedish geologist Johan Gunnar **Andersson**. Even before Black began his explorations, Andersson and the Austrian paleontologist Otto **Zdansky** had begun collecting **Pleistocene** fossils from sites in China. The most promising site was Choukoutien (now **Zhoukoudian**) near Beijing, a series of limestone hills where locals had long extracted the limestone and fossils ("dragon bones"), the latter of which were ground into powder for use in traditional medicines. In addition to mammalian fossils Andersson had found crude stone tools at Zhoukoudian, so there was good reason for Black to focus his efforts at this site. In 1926–7 Black made an agreement with Wenhao Weng, the director of the **National Geological Survey of China**, to collaborate on the exploration of China's Cenozoic geology and paleontology. Black obtained funds from the **Rockefeller Foundation** to initiate excavations at Zhoukoudian, which immediately produced a substantial quantity of mammal fossils and then late in 1927 a tooth Black concluded belonged to a new species of hominin. On the basis of this tooth and two others found by Zdansky in the early 1920s, Black (1927) announced the discovery and the new taxon, *Sinanthropus pekinensis* (later reclassified as *Homo erectus*). Zhongjian Yang, a paleontologist, and Wenzhong Pei, a geologist, joined the excavation team in 1928. They soon discovered two lower jaws belonging to *Sinanthropus*, and the large numbers of fossil animals being recovered and the recovery of the additional hominin remains prompted Black and Weng to found the **Cenozoic Research Laboratory** (the forerunner of the **Institute of Vertebrate Paleontology and Paleoanthropology**) in 1929. Then, in December of 1929, Wenzhong Pei found a skullcap, followed by more modern human-like teeth, all of which strengthened Black's interpretation of *Sinanthropus* as a new type of hominin. Black made a trip to Europe in 1930 to discuss the discoveries at Zhoukoudian and he found that most of his colleagues had accepted *Sinanthropus* as a human ancestor. Excavations at Zhoukoudian expanded during the 1930s and produced several skulls and other skeletal material. Researchers soon recognized that the *Sinanthropus* specimens unearthed in China bore many anatomical similarities to the *Pithecanthropus erectus* fossils discovered by the Dutch anatomist Eugène **Dubois** in Java during the 1890s, and while Black acknowledged some similarities with the Javan material he was convinced that *Sinanthropus* was more advanced than *Pithecanthropus*. Apparent evidence of *Sinanthropus*' cultural achievements was discovered in 1931 when Pei and Black found charred bones and other evidence they interpreted as showing that *Sinanthropus* used controlled fire (this hypothesis has since been refuted). Moreover, crude quartz tools were recovered, all of which suggested to Black that *Sinanthropus* was not only anatomically, but also culturally, relatively modern human-like. Black and his Chinese colleagues were assembling an impressive collection of hominin fossils, artifacts, and animal fossils, but they were also forming an institutional framework for the study of human origins in China. In the midst of these intensive investigations Davidson Black, who had a congenital heart defect, died unexpectedly in 1934. The German-American physical anthropologist Franz **Weidenreich** was appointed to replace Black in 1935 and excavations continued at Zhoukoudian until 1937 when war with Japan made excavation impossible. After the end of WWII the excavations at Zhoukoudian were resumed.

blade (OE *blaed*=sharp-edged tool) A term used for a **flake** that is at least twice as long as it is wide. A blade can also be defined as a flake that has parallel edges, or a flake that has multiple flake scars on its outer (or dorsal) surface attesting to previous attempts to remove similar parallel-sided blades from a core. Blade production was most likely invented multiple times, and perhaps by more than one hominin taxon. Blades appear intermittently in stone-tool-making industries globally from the **Middle Pleistocene** onwards, and they are particularly common in **Upper Paleolithic** assemblages (e.g., the upper strata at **Kebara**; see Bar-Yosef and Kuhn 1999).

bladelet (OE *blaed*=sharp-edged tool and OF *elet*=small) A term used to describe small **blades**. The distinction between blades and bladelets is arbitrary, although some have suggested that a bladelet must be less than 10 mm wide (Bar-Yosef and Kuhn 1999, Inizan et al. 1999). The earliest reported bladelets are those from the coastal cave site of **Pinnacle Point**, South Africa (Marean et al. 2007).

Blanc, Alberto Carlo (1906–60) The son of Gian Alberto Blanc (one of the founders of the Istituto Italiano di Paleontologia Umana), Alberto Blanc obtained his degree in geology at University of Pisa where he concentrated on prehistoric archeology and the Pleistocene phase of human evolution. He was instrumental in the discovery of the **Saccopastore 2** cranium in 1935 and of the numerous caves on the **Monte Circeo** promontory, some of which (e.g., the Grotta Guattari) were rich in Paleolithic deposits, and he directed the only excavation carried out at the site. The discovery of Saccopastore 2 prompted the exploration of the basins of the Tiber and Aniene rivers for further sites and this research led Alberto Blanc and his collaborators to identify several periods (the "Cassio," "Nomentano," "Flaminio") when the climate was much colder than at present. In the Pianura Pontina (Pontina Plain) he identified in what is now called Canale delle Acque (Ditch of the Waters) a stratigraphic sequence with fossil fauna and flora and Paleolithic artifacts. Exploration of caves along the coast of the **Monte Circeo** promontory led to the partly fortuitous discovery of the **Grotta Guatteri** in 1939. In 1954 Alberto Blanc initiated the exploration and subsequent formal excavation of the site of Torre in Pietra and in the same year he founded the journal *Quaternaria*. Blanc held a number of positions during his career, culminating in his becoming professor of Paleoethnology and Human Paleontology at the University of Rome as well as Director of the Instituto di Paletnologia in the Faculty of Letters and Philosophy at the University of Rome.

Blimbingkulon (Location 7°20′S, 110°58′E, Indonesia; etym. named after a local village) <u>History and general description</u> Within the **Sangiran Dome**, this is the site of a 1979 surface find of a hominin mandibular fragment (**BK 7905**) from a dry tributary of the Brangkal River; a second hominin mandibular fragment (**BK 8606**) was found nearby in 1986. Both specimens most likely derived from the **Grenzbank zone**, a coarse-grained **conglomerate** between the older **Sangiran Formation** and the younger **Bapang Formation**. <u>Temporal span and how dated?</u> Fluorine analyses support the derivation of these fossils from the Grenzbank zone. The most recent radiometric estimates of the age of this deposit cluster around 1.5 Ma. <u>Hominins found at site</u> BK 7905 is the fragmentary right side of a mandibular corpus with M_2 and a partially preserved root of M_3. BK 8606 is also the right side of a mandibular corpus preserving a partial M_3, the roots of M_1–M_2 and the distal portion of the P_4 root. These specimens have been attributed to *Meganthropus* and *Homo erectus*. Archeological evidence found at site None. Key references: Geology and dating Aziz 1983, Matsu'ura et al. 1992, 1995b, Kondo et al. 1993, Larick et al. 2001; Hominins Aziz et al. 1994, Kaifu et al. 2005a.

Blind River (Location 27°1′30″S, 33°0′30″E, Eastern Cape Province, South Africa; etym. unknown) History and general description The site of discovery of a hominin femur. It was found by P.W. Laidler 150 m upstream from the river mouth during an archeological survey of the area. Temporal span The site is on a raised beach likely formed during the high sea level of **Oxygen Isotope Stage** 5e, so *c*.127–122 ka. How dated? Optically stimulated **luminescence dating** and geomorphological correlation. Hominins found at site An adult left femur lacking the proximal epiphysis. Archeological evidence found at site No tools in close association. Key references: Geology and dating Laidler 1933, Wells 1935, Wang et al. 2008; Hominins Wells 1935, Wang et al. 2008; Archeology Laidler, 1933.

Blind River femur *See* **Blind River**.

Blombos Cave (Location 34°25′S, 21°13′E, South Africa; etym. Afrikaans for "flowering bush," named after the local village) History and general description Blombos Cave is situated on the coast approximately 300 km/190 miles east of Cape Town, was formed during the **Plio-Pleistocene** when waves cut away at the base of a cliff, but it is now 35 m above sea level and 100 m from the present shoreline of the Indian Ocean. The approximately 50 m^2 of sediments in the cave have been excavated since 1991. Temporal span The youngest deposits (*c*.2 ka) preserve evidence of the **Later Stone Age** (or LSA) and some of the earliest evidence of domestic animals in southern Africa. A sterile layer of *c*.70 ka dune sand divides the LSA from the more extensive **Middle Stone Age** (or MSA) deposits below, which constitute three levels, M1–3. Layers M1 and M2 have been dated to *c*.75 ka and *c*.75–105 ka, respectively; M3 is a shell **midden** that is presumed to be >100 ka. How dated? Optically stimulated **luminescence dating** and thermo**luminscence dating**. Hominins found at site Nine isolated teeth. Archeological evidence found at site Blombos is mainly important because of the recovery of the evidence

in MSA layers M1–2 of precociously early symbolic and other advanced behaviors including apparently deliberately fashioned cross-hatch patterns on pieces of **ochre**, more than 40 pierced "Dunker" shells in MSA levels M1–2 that have been interpreted as evidence of symbolic behavior (d'Errico et al. 2005), **bone tool**s, and exploitation of deep water fish. All of these behaviors are more commonly associated with the LSA rather than the MSA. Ocher has also been recovered from MSA level M3. Key references: Geology and dating Jacobs et al. 2003, 2006, 2008, Jacobs 2003, Tribolo 2006; Hominins Grine and Henshilwood 2000, 2002; Archeology Henshilwood and Sealy 1997, d'Errico et al. 2001, 2005, Henshilwood et al. 2004, Henshilwood 2007, d'Errico and Henshilwood 2007.

BOD Abbreviation of Bodo. One of the three three-letter prefixes (the others are **DAW** and **HAR**) used for the paleontological evidence collected by the **Middle Awash Research Project**. Each three-letter prefix refers to the geographical area where it was found, followed by VP (for Vertebrate Paleontology), the number of the collecting site within the area, then a final number that places the fossil in the sequence of fossils recovered from that collecting site. Thus BOD-VP-1/1 is the first fossil to be found in the first collecting site to be recognized in the **Bodo** area of the **Middle Awash study area**.

Bodo (Location 10°37.5′N, 40°32.5′E, Ethiopia; etym. named after a local stream, and also known as Bodo D'Ar) History and general description One of the named drainages/subdivisions within the **Bodo-Maka** fossiliferous subregion of the **Middle Awash study area**. Abundant vertebrate fossils and **Acheulean** artifacts attracted prospecting efforts by the **Rift Valley Research Mission in Ethiopia** and its members looked for fossils and conducted archeological excavations at Bodo between 1975 and 1978. During the course of these efforts, a portion of a hominin cranium was discovered and subsequent intensive excavation and screening yielded most of the anterior two-thirds of a hominin cranium. The site was revisited by members of the **Middle Awash Research Project** and additional hominin fragments were recovered in 1981 (BOD-VP-1/1) and 1990 (BOD-VP-1/2). Temporal span Correlations between the Bodo fauna and the fauna from sites in Kenya suggest the site dates from the **Middle**

Pleistocene, and **argon-argon dating** and stratigraphic correlation suggest an age for the site of *c*.600 ka. How dated? **Biostratigraphy** and argon-argon dating. Hominins found at site **Bodo 1**, a partial cranium with a massive supraorbital torus and face; BOD-VP-1/1, a left parietal (found roughly 400 m from Bodo 1); and BOD-VP-1/2, a small distal humerus. Archeological evidence found at site Large Acheulean **handaxe**s fashioned of locally available **basalt**. Key references: Geology and dating Kalb et al. 1980, Clark et al. 1984a, 1994; Hominins Conroy et al. 1978, 2000, Asfaw 1983, White 1986, Adefris 1992, Rightmire 1996; Archeology Kalb et al. 1980, Clark et al. 1994.

Bodo 1 Site Bodo. Locality N/A. Surface/*in situ* Surface and *in situ* in the top 4–10 cm of loose silt covering the surface. Date of discovery 1976. Finders Alemayehu Asfaw, Craig Wood, Paul Whitehead, and other members of the **Rift Valley Research Mission in Ethiopia**. Unit Upper Bodo Sand Unit (UBSU). Horizon N/A. Bed/member UBSU atop the Upper Bodo (UB) Bed. Formation Wehaietu Formation. Group N/A. Nearest overlying dated horizon N/A. Nearest underlying dated horizon A volcanic horizon dated to 0.64±0.03 Ma. Geological age **Middle Pleistocene** based on associated fauna, *c*.600 ka based on **argon-argon dating** and **biostratigraphy**. Developmental age Adult. Presumed sex Male based on its massive brow ridge and large face. Brief anatomical description A partial cranium reconstructed from many fragments. Its face and supraorbital torus are large and the nose is wide. Endocranial volume Approximately 1250 cm³ and within the range of 1200–1325 cm³. Announcement Conroy et al. 1978. Initial description Conroy et al. 1978. Photographs/line drawings and metrical data Conroy et al. 1978, 2000, White 1985, 1986, Adefris 1992, Rightmire 1996, Seidler et al. 1999. Detailed anatomical description Adefris 1992, Rightmire 1996, Conroy et al. 2000. Initial taxonomic allocation Intermediate between *Homo erectus* and *Homo sapiens*. Taxonomic revisions Early archaic *H. sapiens* or *Homo heidelbergensis*. Current conventional taxonomic allocation *H. heidelbergensis*. Informal taxonomic category Pre-modern *Homo*. Significance Bodo 1 resembles the **Petralona cranium, Arago XXI, Kabwe 1**, and **Saldanha 1**, and many regard it as consistent with the hypothesis that these specimens all belong to the hypodigm of *H. heidelbergensis*. If that is the case then, if Bodo 1 is *c*.600 ka, then it is the earliest widely accepted evidence for that taxon.

Location of original National Museum of Ethiopia, Addis Ababa, Ethiopia.

Bodo D'Ar *See* Bodo.

Bodo-Maka The only major fossiliferous subregion within the **Middle Awash study area** on the east side of the Awash River; the ones on the west side include the **Central Awash Complex, Western Margin**, and **Bouri Peninsula**. It is named to represent the narrow sedimentary exposures on the east side of the **Awash River** between the Awash floodplain and the basalt plateau of the **Afar Rift System** floor that are bounded by, and adjacent to, the **Bodo** and **Maka** drainages.

body-mass estimation A wide range of ecological, physiological, and morphological features vary among individuals and among species in association with body size. In general, as similar animals evolve to larger adult body sizes, they have larger brain volumes, larger teeth, thicker bones that support more mass, longer life spans, and occupy larger home ranges, among a large number of other size-correlated attributes. To interpret variation in these size-correlated features, most investigators attempt to use statistical methods to control or adjust for variation "due to" or correlated with body size. Body mass is the most commonly accepted measure of overall body size in paleoanthropology; stature is used less often. A number of issues have been evaluated regarding the process of estimating body mass from fragmentary fossil remains, the three most important being (a) the morphological traits employed for the estimation, (b) the form of the statistical relationship between the trait(s) and body mass, and (c) the composition of the sample used to determine the relationship. Following calculation of an estimation equation between body mass and selected traits in a sample of species or individuals with known body mass, the equation is then applied to the interpretation of the unknown body mass in the fossil specimen; simple ratios cannot be used for this purpose (Smith 2005b). Auerbach and Ruff (2004) examine mechanical criteria for selecting traits, which depend upon functional relationships between body mass and elements of the skeleton that are weight-bearing, and the morphometric methods used in attempts to reconstruct overall body size; these are mostly applicable when relatively complete specimens are available (Ruff, 2002). Smith (2002) suggests that weight-bearing is not essential to the selection of predictor traits, because predictive equations are not functional equations and others have used non-weight-bearing facial dimensions and teeth to develop equations to estimate body mass (e.g., Aiello and Wood 1994, Kappelman 1996). Most body-mass prediction equations have been simple bivariate relationships, although multiple regression would improve estimation in many cases. Konigsberg et al. (1998) and Hens et al. (2000) have called attention to the difference between an equation in which the trait to be used for estimation is the dependent variable and body mass is the independent variable (classical calibration), which is then solved so that it can be used to estimate body mass, and inverse calibration, in which body mass is the dependent variable. Smith (2009) has reviewed the considerations involved in selecting a line-fitting criterion. **Ordinary least squares regression** rather than **major axis regression** or **reduced major axis regression** is recommended for estimation equations. The species (or individuals) used to calculate the estimation equation determine the species or individuals to which it is appropriate to apply the equation. The subject must be from the same population as the subjects used to generate the equation. The body mass of a fossil hominin might be estimated from a general mammalian data set, from a primate data set, or from an ape data set; although each is valid, it is expected that estimation is improved when the data set is more specific to its intended purpose. It would be invalid to estimate an australopith body mass from a regression equation that had been calculated only with fully modern humans, since australopiths are not members of this population. Phylogenetic relationship is not the only criterion that defines population membership. Extrapolation (e.g., estimating the body mass of A.L. 288-1 from an equation based on great apes and modern humans) is problematic, because the need for extrapolation indicates that by the criterion of size the estimated specimen is not from the same population as the specimens used to determine the equation. In calculating equations specifically for juvenile specimens, Ruff (2007) has identified stage of ontogenetic development as another feature that defines population membership. All body-mass estimates for fossil species should be reported with explicit recognition of the statistical uncertainty involved (e.g., a confidence interval).

body size A linear, area, volume, or mass measurement designed to reflect the overall dimensions of an organism in a single value. Body size usually is measured with some functional and comparative purpose in mind, which determines the appropriate units. For example, wing span may be an appropriate measure of body size for some aerodynamic considerations in birds, and volume may be appropriate for some questions related to displacement in fishes. Most applications in paleoanthropology refer to body mass. Although the terms "mass" and "weight" are sometimes used interchangeably, the correct measurement is mass, which is measured in kilograms, as opposed to weight, which is mass multiplied by the acceleration due to gravity, and which is measured in Newtons ($mass \times length \times time^{-2}$). Living organisms vary over many orders of magnitude in body size, from bacteria to blue whales. Body size is causally related to a variety of physiological, morphological, and ecological features. Many of these causal interactions occur through the square-cube law originally explained by Galileo Galilei in *Discorsi e dimostrazioni matematiche intorno à due nuove scienze* (*Two New Sciences*; 1638), which states that when objects undergo proportional increases in size, volumes increase proportional to the cube of the multiplier and areas increase proportional to the square of the multiplier. Therefore, as organisms increase in size, the ratio of areas to volumes decreases. Many physiological properties take place across areas, exemplified by diffusion across cell membranes. The science of **scaling** examines and seeks to explain much of the functional variation among organisms in relationship to variation in body size (McMahon and Bonner 1983, Schmidt-Nielsen 1984, Jungers 1985, Brown and West 2000, Samaras 2007). Because body size changes during the evolution of the hominin lineage, it is necessary to take body size into account (e.g., when using brain size to judge the evolution of intelligence, tooth size to infer diet, or sexual dimorphism to infer social organization).

Bohunician *See* **Brno-Bohunice**.

Bohunice *See* **Brno-Bohunice**.

Boise, Charles Watson (1884–1964) Charles Watson Boise was the second of six children of Watson and Grace Boise. He was born and raised on a homestead in North Dakota, not far from the Saskatchewan border. He went to college at the University of North Dakota, Grand

Forks, and graduated in 1908 with the degree of Engineer of Mines from the College of Mining Engineering. After graduation Boise followed an elder brother to New Mexico where he worked in a copper mine at Santa Rita. He made the first of many trips to Africa in 1911 where he worked for a time at the Union Miniere in Katanga and he is credited on the photographs in a paper on the economic geology of the Congo published in 1914. In 1919 he led an expedition to Spitzbergen. The narrative of the next part of Boise's career are taken from his obituary in *The Times* of London (November 11, 1964): "In 1920 Boise made the first examination of diamond areas in the Gold Coast (now Ghana), which led to the formation of the Consolidated African Selection Trust, Ltd. In 1925 he was involved with the early exploration of minerals in Northern Rhodesia (now Zambia) that resulted in the formation of copper mining companies that formed part of the Rhodesian Selection Trust Group, and in 1934 he was actively involved in opening up the diamond fields in Sierra Leone. He had a long association with the Selection Trust Group of mining companies (N.B., in the 1934 University of North Dakota Alumni Magazine he is listed as a 'Director', and in the 1936 issue he is referred to as the 'General Manager and part owner') and he was one of the earliest colleagues of their founder Alfred Chester Beatty (later Sir Chester Beatty) and William Selkirk." Boise spent the rest of his career in London, where he was involved in the diamond trade. Boise received an Honorary Degree of Doctor of Science from his alma mater in 1958, and he retired from active business in 1959. In 1927 he purchased Emmetts, a country estate with an extensive garden at Idle Hill, near Sevenoaks, in Kent, England, and became a British citizen in the wake of the Great Depression. Boise continued to develop the house and garden at Emmetts, and on his death in 1964 the fine garden was transferred to The National Trust. Boise's ashes were scattered in the garden at Emmetts. It is likely that Boise first became aware of the activities of Louis and Mary **Leakey** in the 1930s, but his patronage of their research did not begin until January 1948. Boise was vacationing in Switzerland when he read a report in *The Times* about the results of the first British-Kenya Miocene Research Expedition to Rusinga Island in Lake Victoria in Western Kenya. Boise wrote from his Swiss hotel and offered his assistance to the

Leakeys to ensure the Rusinga research remained "under British auspices," and in February 1948 he mailed the Leakeys a check for £1000. This enabled Louis and Mary Leakey to buy a lorry in which they could travel (and sleep) to and fro to Western Kenya. On October 1, 1948 Mary Leakey discovered the first skull of *Proconsul* africanus, a **Miocene** ape, and less than a month later, on October 29, she left by flying boat from Lake Naivasha, Kenya, for the UK, to show the new skull to Wilfrid **Le Gros Clark**. While she was there she also showed it to Boise, who gave the Leakeys further support, which enabled them to buy a motorboat. The next recorded contact between the Leakeys and Boise was in 1950 when they accompanied him on a visit to the Dordogne, where they visited **Lascaux** and **Pech Merle Cave**. The following year Boise visited the Leakeys in Africa, and it was during this visit that he pledged to fund their fieldwork, including their renewed exploration and excavations at **Olduvai Gorge**, for a further 7 years. In 1955 Boise vested a capital sum of £25,000 in the University of Oxford with instructions that a trust fund, the **Boise Fund**, be established to support research "on the antiquity and origin of man and his precursors (with particular emphasis on the continued exploration of appropriate sites in Africa) and on the early migration of paleolithic communities." Among the first research supported by the Boise Fund was that of Louis and Mary Leakey who were given grants in 1960 and 1961 for excavations at Olduvai Gorge, and later for excavations at Fort Ternan in Western Kenya. In recognition of Charles Boise's support for their research, after Mary Leakey had discovered the **OH 5** cranium at Olduvai Gorge, in Tanzania, in 1959, they named it *Zinjanthropus boisei*.

Boise Fund A research fund based on a gift from Charles **Boise** to the University of Oxford to support research on human origins in Africa. The fund was created in 1955 when Charles Boise vested a capital sum of £25,000 in The University of Oxford with instructions that a trust fund, the Boise Fund, be established. The instructions were that the interest on the capital be used to support research "on the antiquity and origin of man and his precursors (with particular emphasis on the continued exploration of appropriate sites in Africa) and on the early migration of paleolithic communities." Charles Boise probably chose Oxford for several reasons. First, although he was by birth a US citizen by 1955 he had become a UK citizen. Second, since 1948 Boise had been helping to support the research of Louis and Mary **Leakey**, both at **Miocene** sites in Western Kenya and later at **Olduvai Gorge**, and in doing so he would have encountered Wilfrid **Le Gros Clark**. The latter, an authority on Miocene apes, had become a strong supporter of the Leakey's research, and Le Gros Clark was the Dr Lee's Professor of Anatomy at Oxford University and a Professorial Fellow of Hertford College, Oxford. Le Gros Clark was much involved with the fund when it was first established, and among the first work supported was that of Louis Leakey who was given grants in 1960 and 1961 for excavations at Olduvai Gorge, and later for excavations at Fort Ternan in Western Kenya. However, Le Gros Clark retired from the Dr Lee's Professorship in 1962, and 2 years later, in 1964, at his suggestion, the Council of the University approved a change to the statute and a committee was established with *ex officio* members, namely the Professor of Anthropology, the Professor of Geology, and the Director of the Pitt Rivers Museum (the holders of these posts were to be asked to appoint deputies when their expertise did not fit the Boise remit). From 1964 this committee met to consider applications from researchers working in appropriate subject areas. For a long time the representative from Anthropology (and the chair of the board of management of the Boise Fund) was Geoffrey Ainsworth Harrison. Enquiries at The University of Oxford suggest that support from the Boise Fund is now only available for members of the University of Oxford.

Boker Tachtit (Location 30°50'N, 34°46'E, Israel; etym. He. *Boker Tachtit*="Cowboy's Pit," probably derived from the name of a nearby kibbutz *Sde Boker*=Cowboy's Field) History and general description One of several sites found in the late 1970s during the Central Negev Project, Boker Tachtit is located along the Nahal Zin river. The site was excavated over several field seasons from 1973 to 1980 by A. Marks and his team. Marks and others (see Fox and Coinman 2004) have suggested that Boker Tachtit documents a local transition from **Middle** to **Upper Paleolithic** rather than the introduction of a foreign technology (as argued by Bar-Yosef 1999). Temporal span The site is likely beyond the age of **radiocarbon dating** (>47 ka). How dated? ^{14}C. Hominins found at site N/A. Archeological evidence found at site Abundant stone tools from the four

main occupation levels supposedly document the Middle to Upper Paleolithic transition. Many of the artifacts could be **refit**ted, allowing a precise reconstruction of the **knapping** sequence used to produce them. The older layers record a **Levallois**-based core-reduction strategy and the more recent ones a single-platform core-reduction strategy that resulted in **blade**s. Emireh points, a type of tool unique to the Levantine Middle to Upper Paleolithic transition, were abundant. Key references: Geology and dating Marks 1983; Hominins N/A; Archeology: Marks 1977, 1983.

Bølling-Allerød A warm interval (or interstadial) that followed the abrupt termination of the **Last Glacial Maximum**. The identification of a Bølling-Allerød-aged **sapropel** in the eastern Mediterranean indicates that this warm interval was associated with increased precipitation over Northeast Africa in the headwaters of the Nile. In records where there is suffcient temporal resolution two separate warming intervals can be recognized: the Bølling *c*.13–12 ka and the µAllerød *c*.11.8–11 ka. The two warm intervals are separated by the Older Dryas stadial or cold interval 12–11.8 ka and followed by the Younger Dryas stadial or cold interval *c*.11–10 ka (Mangerud *et. al.* 1974).

bone (ME *bon*=bone) Bone is one of the two connective tissues (**cartilage** is the other) that form the bony skeleton. All connective tissues are made up of cells plus tiny organic fibers, both of which are embedded in a matrix. What makes bone and cartilage harder and stiffer than, say, the fibrous connective tissue of the capsule of a synovial **joint**, is that the matrix of bone is mineralized. The cells in bone are called osteoblasts, osteocytes, and osteoclasts; osteoblasts make, deposit, and mineralize the matrix, osteoclasts remove it, and osteocytes are what osteoblasts turn into once the matrix they have made matures. The fibers in bone are collagen fibers. These are arranged in flat sheets in immature bone, or in concentric bundles in lamellar, or layered, bone, or in irregular bundles in immature woven, non-lamellar, bone. As bone matures it reorganizes into either compact bone or dense bone, or cancellous bone or trabecular bone. The shafts of a long bone are formed from an outer "skin" or cortex of compact bone, and an inner cavity or medulla of cancellous bone. There are very few spaces in compact bone, and what few there are filled either with the cells that form bone (see above) or with blood vessels; there are spaces in cancellous bone (it is built like an Aero chocolate bar). These spaces are filled either by fat (yellow bone marrow) or blood-forming tissues (red bone marrow), or a mixture of the two. The minerals in bone matrix are in the form of small (approximately 30 nm×5 nm) crystals of hydroxyapatite $[Ca_{10}(PO_4)_6(OH)_2]$. These are mostly in the matrix and among the collagen fibers; they are aligned with their long axis parallel to the long axis of the collagen fibers. It is the combination of the strong collagen fibers and the hard matrix that gives bone its unique properties. If you put a rib in acid and remove the mineral, the rib retains its shape, but it becomes so flexible you can tie a knot in it. If you take a bunsen burner and burn all the organic material out of a vertebra it will retain its shape but when you press lightly on it with your hand it will easily crush down to a small pile made up of the mineral components of hydroxyapatite. *See also* **fossilization**.

bone breakage patterns Bone breakage, or fracture, occurs when the application of an external force (loading) causes a localized mechanical failure in the bone. Loading can be either static or dynamic and can generate tensile, compressive, shearing, torsion, or bending forces. Static loading involves the application of constant compressive pressure, typically with an even distribution of force, such as the steady pressure on bone from the jaw of a large carnivore. Dynamic loading involves a concentrated and sudden impact, such as that from a hammerstone blow. Various types of bone fractures, breaks, or cracks are known to result from different types of loading on different types of bone (e.g., long bones, vertebrae) in various states (e.g., fresh, wet, dry, weathered, fossilized). On long bones, these can include spiral (radial, oblique, curvilinear) fracture, perpendicular (transverse) fracture, longitudinal fracture, and stepped or columnar fractures. For example, the breakage of fresh bone is often associated with spiral fractures characterized by curved outlines and oblique fracture edges. Conversely, the breakage of dry bone is associated with transverse fractures and perpendicular fracture angles. Thus, the study of bone breakage patterns in an archeological assemblage can inform on whether bones were broken while fresh, possibly during carcass processing, or while dry, long after discard. Bones can be fractured by numerous taphonomic agents and

processes, including hominins, carnivores (i.e., chewing), large mammals (i.e., trampling), falling rocks (i.e., within cave sites), and sediment compaction. Identifying the taphonomic agent responsible for the fractured bones involves a detailed study of bone breakage patterns. For example, in zooarcheological contexts, it has been suggested that notches along fracture surfaces show different morphologies depending on whether loading was static or dynamic. Dynamic loading associated with hammerstone percussion is thought to produce notches that are broader and shallower in cortical view than those created by the static loading of carnivore teeth. In turn, the study of these particular fracture morphologies can provide insight into whether bones were fractured by hominins or carnivores. Determining the identity of the agent responsible for bone fracture is best accompanied by a study of bone surface modifications (e.g., percussion marks, carnivore tooth marks).

bone density Measured as the average value for whole bones, or as a more accurate value at a standardized location on the same bone (e.g., the calcaneum at the heel is one of the sites clinicians use to screen for and diagnose osteoporosis). Photo densitometry gives accurate values at a standard location, but **computed tomography** provides more reliable average values for bones with complex shapes and/or cavities. Bone density is used in **zooarcheology** to study its influence on differential bone preservation at fossil and archeological sites (see Lam et al. 2003).

bone remodeling (L. *re*=in a different way and *modus*=a measure) Refers to internal reorganization within a **bone** that does *not* result in a net change in mass or in any change of external shape; also called Haversian remodeling. It involves osteoclasts resorbing existing bone, and osteoblasts laying down an equivalent mass of new secondary osteonal or Haversian bone. The surfaces of a bone where bone resorption is occurring have a distinctive morphology (e.g., Howship's lacunae) and this enables researchers to characterize parts of the surface of bones as either areas of bone resorption or of bone deposition. Bromage (1989) and Rosas et al. (2010) have utilized this distinctive morphology to explore whether there are consistent differences among taxa in the sites where bone resorption is taking place. Note that "modeling" is the equivalent term used when bone is modified so that its mass and/or its external shape *are* changed (e.g., during growth and as the result of excessive use). *See also* **ossification**.

bone strength   In material science the strength of a material is its ability to withstand an applied stress without failure. There are two common types of failure: yielding and fracture. Yielding, which is defined by the material's yield strength, occurs when the material undergoes macroscopic permanent deformation. A material with an applied stress value less than the yield strength of the material will not be permanently deformed upon unloading. Fracture is defined by the ultimate strength of the material, which is the largest absolute value of applied stress the material can withstand without fully fracturing. Since yielding occurs at an applied stress less than the applied stress that produces full fracture, the yield strength of a material is always less than its ultimate strength. The type of loading (e.g., tensile, compressive, or shear) also influences failure; thus there are corresponding types of material strength (e.g., ultimate tensile strength, ultimate compressive strength, and ultimate shear strength). However, bone is not a homogeneous structure made of single material; it is a heterogeneous composite structure. One can measure the strength of a bone (such as the femur) by loading it to failure, but this is a measurement of its structural strength and not of the bone tissue. Strictly speaking, bone tissue strength is a material property and not a structural property and therefore should be measured at the tissue level. This means one should measure the strength of compact bone, or the component individual trabeculae of cancellous bone, most likely via micro-indentation. Conversely, one can infer bone tissue strength via a coupled *in vitro* and *in silico* testing. This involves physically testing the bone structure to determine the failure load and location and also developing a finite element model of the same bone specimen loaded and constrained as in the physical experiment. The finite element model is used to provide the stress state of the tissue at the critical failure load and failure location. The appropriate measure of stress predicted by the model, such as maximum principal stress, minimum principal stress, maximum shear stress, or von Mises' or equivalent stress at the failure location is then taken as the tissue strength.

bone tool   Any osseous (i.e., bone) object used during the performance of a task. Objects made of osseous materials used by hominins to assist in carrying out tasks. The first category of bone tools is exemplified by the *c*.1–1.8 Ma bone tools found in Members 1–3 at **Swartkrans**, South Africa (Backwell and d'Errico 2001). They have been interpreted as implements used to extract **termites** from mounds and they were probably bones or bone fragments whose shape was modified through use rather by being deliberately shaped prior to their use. The second category comprises bones that are carefully shaped through cutting, grinding and/or polishing. For example, some bones at **Olduvai Gorge**, Tanzania, were deliberately modified and shaped through flaking (the technique is similar to that used for stone tool production) to make handaxe-like pieces (Backwell and d'Errico 2004a). The **Middle Stone Age** levels at **Blombos Cave**, South Africa provide another good example of deliberately fashioned bone tools (Henshilwood et al. 2001), and at **Katanda**, Demographic Republic of Congo, bone harpoons, presumably for fishing, have been dated to *c*.90 ka (Yellen 1998). Many purportedly deliberately shaped bone tools at **Lower** and **Middle Paleolithic** sites in Eurasia such as **Torralba** and **Ambrona** have been shown to be bones that were shaped by natural causes (Villa and d'Errico 2001). Bone tools are common in **Upper Paleolithic** and **Later Stone Age** sites.

Bonferroni correction   A method that addresses the problem of identifying the appropriate significance level when multiple significance tests are performed. For example, a significance level of $\alpha = 0.05$ is typically used to identify a statistically significant result. This means that we accept that 5% of the time a result that appears as significant as this could happen by random chance even if the **null hypothesis** is true (i.e., no significant difference between groups or significant relationship between variables actually exists). However, suppose one were to conduct 20 significance tests in the course of an analysis. Even if no significant relationships existed in any of those 20 tests, by random chance we would expect that 5% of the tests, or 1 out of 20, would yield a statistically significant result even though no real relationship exists. The Bonferroni correction recognizes this problem and adjusts the α value (the value at which a result is deemed statistically significant) downwards to compensate for multiple tests. In the traditional application of the Bonferroni method, α is set to α/n for all tests when n significance tests have been preformed. For example, if *P* values are calculated for four separate correlations of stature and four different skeletal variables, then to accept a result as significant at 0.05, α must be set equal to 0.05 divided by four, or 0.0125. Only those correlations which have *P* values equal to or less than 0.0125 would be considered significant. However, it has been argued that this correction is overly conservative, and Holm (1979) suggests a modification which is widely used. The *P* values for all tests are ordered from smallest to largest. If the smallest *P* value is less than α/*n* then the test associated with that *P* value is found to be significant. The next smallest *P* value is then compared to α/(*n*–1), and if *P* is smaller it is found to be significant. The next smallest *P* is compared to α/(*n*–2), and so on. If at any point the *P* value is not smaller than the adjusted α value, then that test and all others with higher *P* values are considered non-significant.

bonobo Susman (1984) and Groves (2008) review the possible etymology of "bonobo." As Susman (1984) suggests it is most likely a "mispronunciation of the word Bolobo, which is a town on the Zaire River, between Kinshasa and Lukolela" (*ibid*, p. xix). *See also* **Pan paniscus**.

bony labyrinth (Gk *labyrinthos*=complex of cavities) Complex of **periosteum**-lined cavities in the **petrous** part of the temporal bone. It has three regions, a central vestibule, the cochlea anteriorly, and the semicircular canals posterosuperiorly. It contains the **membranous labyrinth** (ML). The space between the walls of the bony labyrinth and the ML is filled with **perilymph**. It is this complex of cavities that is visualized using **computed tomography**. *See also* **semicircular canals**.

bootstrap A **resampling** procedure that can be used to calculate a confidence interval and/or the degree of bias for sample statistics (e.g., mean, variance, median, correlation coefficient, regression slope, etc.). The bootstrap is used when equations do not exist to calculate the confidence interval of interest (e.g., for the median), or when data are not normally distributed, and/or when sample sizes are small. Consider a statistic that is calculated for some sample (e.g., the median cranial capacity for 10 crania). A sample of size equal to the full sample is subjected to random **sampling with replacement**. In this example, 10 endocranial capacities are selected at random from the 10 measurements, but each time

all 10 measurements have an equal probability of being selected, so some measurements may be selected more than once and other measurements not at all. The statistic, in our example the median, would be calculated for the resulting randomly sampled 10 measurements. This procedure is repeated many times to generate a distribution of values for the statistic of interest (in this case, the median). The total number of unique samples that can be drawn from a sample of size *n* is equal to (2*n*–1)!/[*n*!(*n*–1)!], where the exclamation point is the **factorial** symbol. For our example of 10 measurements, this corresponds to 92,378 possibilities. Whereas it is theoretically possible to cycle through all possible samples (this would be an example of an **exact test**), the number of possibilities is usually prohibitively high. Instead, it is more usual to randomly choose a large number of the possibilities (typically 5000 or more) to stand in for the full set of possibilities (this is an example of a **Monte Carlo** method). In either case, the resulting distribution of values for the statistic of interest, in this case the median, is the bootstrap distribution. There are several ways to calculate a confidence interval for the statistic of interest using the bootstrap distribution, but the simplest and most common is the percentile bootstrap. In this technique, the values in the bootstrap distribution are sorted from highest to lowest. To generate a 95% confidence interval, the largest 2.5% and the smallest 2.5% of values are discarded, with the remaining maximum and minimum values giving the limits of the confidence interval. One can also check for bias in the bootstrap estimate by comparing the mean of the bootstrap distribution with the original statistic from the full sample. The difference between the two is the bias. For other methods of calculating confidence intervals for the statistic of interest and for ways to remove bias from the confidence interval generated by the bootstrap see Manly (2007).

Border Cave (Location 27°01′19″S, 31°59′24″E, Kingdom of Swaziland, South Africa, 365 m from the border with South Africa; etym. named after its location overlooking the border) History and general description Initial investigations directed by Raymond **Dart** in the 1930s and by Basil Cooke in the 1940s were followed by more extensive excavations by Peter Beaumont in the 1970s. Border Cave was formed when an **agglomerate** layer within lava flows was preferentially weathered. The more than 4.5 m sequence of sediments exposed in the >60 m^2 of excavated area consist of alternating deposits of brown windblown sandy silts and reworked white-black hearth ash. These span the **Middle Stone Age** to the Iron Age. The Stone Age sequence is divided from bottom to top into MSA 1, MSA 2 (Howieson's Poort), MSA 3a, MSA 3b, and the early **Later Stone Age** (ELSA). The finds include several hominins including a controversial early infant burial. Temporal span and how dated? Border Cave has been dated by a number of methods, including **electron spin resonance spectroscopy dating** (or ESR) of mammalian teeth and **Bayesian** estimates of age boundaries as well as **amino acid racemization dating** (or AARD) of ostrich egg shell. Ostrich egg shell and wood charcoal have been **radiocarbon dated** by traditional and **AMS** methods, and the site served as an initial test case for the acid-dichromate oxidizing solution pretreatment method. Combined AARD and ESR age estimates suggest an age range from >100 ka to 227±11 ka for MSA 1, with an age range of 82–56 ka for MSA 2 (Howieson's Poort). ESR, AARD, and radiocarbon age estimates suggest a range of 69–53 ka for MSA 3a, 56–41 ka for MSA 3b, with the ELSA dating to *c*.38 ka. Hominins found at site BC 1 (a calvaria) and BC 2 (a partial mandible) derive from commercial digging in the cave for guano, they are both possibly from the MSA 1 levels, but their **provenance** is uncertain. BC 3 was discovered in the 1941 excavations, and is an infant skeleton buried with a *Conus bairstowi* shell from coastal sources at least 80 km/130 miles away, with the lip of the grave shaft at the base of the MSA 2 levels; BC 4, a complete skeleton except for the cranium, dates to the Iron Age; BC 5 is a nearly complete lower jaw. Additional hominin remains from contexts disturbed by early excavations and thus of uncertain provenance include BC 6 (humerus), BC 7 (proximal ulna), and BC 8a and BC 8b (both metatarsals). Bone crystallinity studies suggest that BC 3 and BC 5 may be **intrusive** burials from overlying (perhaps Iron Age) strata, but ESR dating of a tooth fragment from BC5 has produced an age of 74±5 ka. Archeological evidence found at site In addition to well-preserved organic remains, the archeological sample includes more than 69,000 stone artifacts, and a substantial macromammalian fauna. As in many other southern African sites, the production of elongated **flakes** is common throughout the MSA strata, with the MSA 1 further characterized by

bifacial points, MSA 2 by the **backed** pieces characteristic of the Howieson's Poort, with the MSA 3 levels also including backed pieces and basally thinned points. The early LSA shows an informal flaked **lithic** assemblage as well as rare ground stone pieces, ostrich egg shell beads, and polished pointed pieces of bone and warthog tusk. Stone tools are predominantly of local **lavas** or **chalcedony**, the latter from an unknown source not in the immediate vicinity of the site. The frequency of the use of fine-grained chalcedony increases as one goes up the Stone Age sequence, from approximately 6% at the base to roughly 84% in the upper strata. The fauna suggest environmental conditions similar to those found at the site today for much of the occupation of the cave, and the sample includes at least two extinct species, the "giant Cape horse" (*Equus* cf. *capensis* in MSA 3) and the bovid *Antidorcas bondi* (MSA 3 and ELSA strata). Key references: Geology and dating Butzer et al. 1978, Sillen and Morris 1996, Miller et al. 1999, Grün and Beaumont 2001, Bird et al. 2003, Grün et al. 2003, Millard 2006; Hominins Cooke et al. 1945, de Villiers 1973, 1976, Rightmire 1979a, Morris 1992, Pearson and Grine 1996, Pfeiffer et al. 1996; Archeology Beaumont 1973, Klein 1977, Beaumont et al. 1978, Thackeray 1992.

Bose Basin (Location approximately 23°66'N, 106°37'E, Guangxi Zhuang Autonomous Region, China) History and general description Bose is a large basin, roughly 800 km^2/309 square miles, located north of Nanning, the capital city of Guangxi Zhuang Autonomous Region, southern China. The Mandarin transcription of the basin was "Baise," but the current spelling follows the local spelling "Bose"; researchers have used both terms. The Youjiang River cuts through the basin from northwest to southeast. Seven river terraces have been identified in the basin with Paleolithic and Neolithic archeological remains surface collected and excavated from the first four terraces (T1–T4). Bifacially worked Paleolithic stone implements were initially discovered in 1973 in T4 and many more have since been identified. **Australasian strewn-field tektites** have also been excavated and surface collected from T4. To date, about 80 Paleolithic open-air sites have been identified, the best known of which are Damei, Fengshudao, and Gaolingpo. Cave sites have been identified and excavated in the Bose Basin, but the cave deposits have been tentatively dated to the terminal Pleistocene and Holocene. Temporal span and how dated? Fission-track and ^{40}Ar/^{39}Ar dating on Australasian strewn-field tektites have produced dates of c.732±39 and c.803±3 ka, respectively. Some have questioned these dates based on the suggestion that the tektites were not found in their original context, but hundreds of tektites have been excavated from T4 and only T4, suggesting that T4 is the stratigraphic level where the tektites were originally deposited, and although some of the tektites are rounded, probably from fluvial activity, the majority of them still display sharp edges, suggesting rapid deposition and that T4 was their original context. Thus, it is generally accepted that T4 dates to the Early–Middle Pleistocene transition. Hominins found at site No hominin fossils dating to the Pleistocene have been reported, but there are Neolithic modern human burials. Archeological evidence found at site The reason why Bose receives the attention it does is because of the presence of bifacially worked stone tools in T4. In particular, 129 handaxes and similar "large cutting tools" were surface collected and excavated from the Fengshudao site, whereas lower frequencies of these large bifacially and unifacially worked implements were found at other sites in the Bose Basin. The evidence of handaxes in Bose has been used, primarily by Chinese Paleolithic archaeologists, to refute the idea of the **Movius' Line**. Detailed morphological and statistical analyses of the bifaces and other large cutting tools suggest that although the there are differences between those from Bose and those found in India and in East Africa, the Bose artifacts manifest similar flaking capability and technological efforts as Acheulean tool kits from other Early Pleistocene sites. Paleolithic artifacts have also been identified in T2 and T3, stratigraphic layers that are younger than T4, but the T2 and T3 assemblages do not include bifacially worked implements or Australasian tektites. Furthermore, variation appears to be present in the composition of the lithic assemblages and the morphology of the artifacts from T2 and T3 when compared with the materials from the handaxe-bearing T4 level. There are Neolithic deposits, including modern human burials, in at least one cave in the area that divides the Bose Basin and **Bubing Basin**, but it is geographically located on the Bose side and there may be Pleistocene deposits beneath the Neolithic layers. Repository Natural History Museum of Guangxi Zhuang Autonomous Region, China; Guangxi Institute of Archaeology, China; Bose City Museum, Guangxi, China; Institute of Vertebrate Paleontology and Paleoanthropology, Beijing, China. Key references: Geology and dating Huang et al. 1990, Hou et al. 2000, Potts et al. 2000, Zhu et al. 2001, Xie and Bodin 2007, Wang et al. 2008; Hominins N/A; Archaeology Li and You 1975, Hou et al. 2000, Xie and Bodin 2007, Wang et al. 2008, Zhang et al. 2010.

Boskop (Location 26°34′S, 27°07′E, Gauteng Province, South Africa; etym. Afrikaans for a bush- or brush-covered head or hill) History and general description The open-air site was discovered in 1914 by F.W. Fitzsimons in a field belonging to the Kolonies Plaas farm on the east bank of the Mooi River in the southwest of what is now Gauteng Province. The remains recovered included a calvaria that attracted intense interest at the time because of its large size and endocranial volume (approximately 1700–1950 cm^3). Some inferred it to have substantial antiquity, but later authors cast doubt on its antiquity and emphasized its similarity to a variety of other large-brained crania from the **Holocene** of the Cape. Temporal span **Pleistocene**. How dated? Archeological association, degree of mineralization and **fluorine dating**. Hominins found at site Boskop 1, a calvaria, a left mandibular ramus and highly fragmentary partial skeleton. The calvaria was made the holotype of *Homo capensis* Broom, 1918 Archeological evidence found at site One **Middle Stone Age** artifact. Key references: Geology and dating Fitzsimons 1915, Tobias and Wells in Oakley and Campbell 1967; Hominins Haughton 1917, Thomson 1917, Broom 1918, Galloway 1937, Dreyer 1938, Wells 1959, Tobias 1959, Singer, 1958, 1961; Archeology van Riet Lowe 1954.

Boskop 1 *See* **Boskop**.

bottleneck A bottleneck (as in a population or genetic bottleneck) is an evolutionary event during which genetic diversity is reduced due to a severe decrease in population size. A population bottleneck increases **genetic drift** (since the latter is a stronger force of evolution in smaller populations) and it also increases inbreeding. Some researchers have suggested that modern humans have undergone one, or more, population bottlenecks, possibly in connection with natural disasters, disease or migrations (e.g., Rogers and Harpending 1992, Sherry et al. 1994, Hammer 1995, Ambrose 2003, Amos and Hoffman 2009).

BOU Abbreviation of Bouri Peninsula. It is the prefix used for localities in the **Bouri Peninsula, Middle Awash study area**, Ethiopia. *See also* **Bouri**.

BOU-VP-2/66 Site Daka. Locality Dakanihylo Member, Vertebrate Paleontology Locality 2. Surface/*in situ In situ*. Date of discovery December 27, 1997. Finder Henry Gilbert. Unit N/A. Horizon Silty sand layer in the section of Vertebrate Paleontology Locality 2. Bed/member Dakanihylo "Daka" Member. Formation **Bouri Formation**. Group N/A. Nearest overlying dated horizon Paleomagnetic sample PMMA 97-5 shows reversed polarity, and is assumed to belong to the Matuyama chron. Nearest underlying dated horizon Pumiceous unit at the base of the Dakanihylo Member $c.1.04\pm0.01$ Ma. Geological age Lower Pleistocene by faunal correlations and absolute dates (see above). Developmental age Adult. Presumed sex Unknown. Brief anatomical description A well-preserved calvaria of *Homo erectus* that retains much of the basicranium except for the sphenoid and ethmoid. It bears a series of parallel scratches and may have been defleshed perimortem. Endocranial capacity 986 cm^3 (Asfaw et al. 2008). Announcement Asfaw et al. 2002. Initial description Asfaw et al. 2002. Photographs/line drawings and metrical data Asfaw et al. 2002, 2008. Detailed anatomical description Gilbert et al. 2003, Asfaw et al. 2008. Initial taxonomic allocation *Homo erectus*. Taxonomic revisions N/A. Current conventional taxonomic allocation *Homo erectus*. Informal taxonomic category Pre-modern *Homo*. Significance The Daka calvaria presents a mix of characters often associated with *Homo ergaster* and *Homo erectus*, supporting the conclusion the two are conspecific. The specimen also shows some morphological features such as more vertical parietal walls that foreshadow **Middle Pleistocene** specimens of *Homo heidelbergensis*. Location of original National Museum of Ethiopia, Addis Ababa, Ethiopia.

BOU-VP-12/130 Site Bouri. Locality Bouri Vertebrate Paleontology Locality 12. Surface/*in situ* Surface. Date of discovery November 20, 1997. Finder Yohannes **Haile-Selassie**. Unit N/A. Horizon Maoleem Vitric Tuff. Bed/member Hatayae Member. Formation **Bouri Formation**. Group N/A. Nearest overlying dated horizon See information for horizon. Nearest underlying dated horizon See information for hor-

izon. Geological age $c.2.5$ Ma. Developmental age Adult. Presumed sex Unknown. Brief anatomical description Associated cranial fragments consisting of the frontal, parietals, and maxilla with a complete and well-preserved dentition. The teeth show little occlusal wear, with dentin exposure limited to the occlusal edges of the incisors and canine, and pin-hole-sized areas of dentin exposure on the cusps of the M^1. Announcement Asfaw et al. 1999. Initial description Asfaw et al. 1999. Photographs/line drawings and metrical data Asfaw et al. 1999. Detailed anatomical description N/A. Initial taxonomic allocation *Australopithecus garhi*. Taxonomic revisions N/A. Current conventional taxonomic allocation *Au. garhi*. Informal taxonomic category Megadont archaic hominin. Significance **Holotype** of *Au. garhi*. Location of original National Museum of Ethiopia, Addis Ababa, Ethiopa.

BOU-VP-16/1 Site Herto. Locality Bouri Vertebrate Paleontology Locality 16. Surface/*in situ In situ*. Date of discovery 1997. Finder David DeGusta. Unit N/A. Horizon N/A. Bed/member Upper Herto Member. Formation **Bouri Formation**. Group N/A. Nearest overlying dated horizon Waidedo Vitric Tuff, which caps the Upper Herto Member, correlates with an unnamed tuff at **Konso** that lies 6 m below the Konso Silver Tuff, which dates to 154 ± 7 ka by **argon-argon dating**. Nearest underlying dated horizon Lower Herto Member, 260 ± 16 ka by argon-argon dating. Geological age "On the basis of the combined stratigraphic, geochemical and radioisoptopic evidence the Upper Herto archeological and paleontological remains are therefore securely constrained to be between 160 ± 2 and 154 ± 7 kyr old" (Clark et al. 2003). At the least the cranium can be considered to be **Middle Pleistocene**. Developmental age Adult. Presumed sex Male. Brief anatomical description A well-preserved, large cranium that lacks the left side of the face and part of the left side of the cranial vault. The preserved tooth crowns show substantial occlusal wear. The BOU-VP-16/1 cranium is long, the distance between the articular eminence and the occlusal plane is substantial, the upper and lower scales of the occipital meet at a more acute angle than is the case in most modern human crania, and the external occipital protuberance is exceptionally rugged compared with most modern human crania. The vault bones are described as "thick" (*ibid*, p. 742).

Endocranial volume Approximately 1450 cm^3 (by doubling the volume of the better-preserved right side of the endocranial cavity). Announcement Gilbert et al. 2000. Initial description. White et al. 2003. Photographs/line drawings and metrical data White et al. 2003. Detailed anatomical description Gilbert and Asfaw 2009. Initial taxonomic allocation *Homo sapiens idaltu*. Taxonomic revisions N/A. Current conventional taxonomic allocation *Homo sapiens*. Informal taxonomic category Anatomically modern human. Significance The BOU-VP-16/1 cranium is the **holotype** of the subspecies of *H. sapiens*, *Homo sapiens idaltu*. The cranial vault is long and high, as in modern humans, and the supraorbital torus, while robust, is modern human-like. The molars decrease in size from M^1 to M^3. Although its non-metrical morphology is closest to that of modern humans, BOU-VP-16/1 (and the other hominins recovered from Herto) can readily be distinguished metrically from living modern humans, and it is this that prompted their classification as a subspecies of *H. sapiens*. Until the redating of **Omo I** and **Omo II** from Omo-Kibish, the BOU-VP-16/1 cranium was the earliest evidence of *H. sapiens* in Africa; it is still the best preserved among the early well-dated occurrences of modern human cranial anatomy. Location of original National Museum of Ethiopia, Addis Ababa, Ethiopia.

BOU-VP-16/5 Site Bouri. Locality BOU-VP-16. Surface/*in situ* Surface. Date of discovery 1997. Finders N/A. Unit N/A. Horizon N/A. Bed/member Upper Herto Member. Formation **Bouri Formation**. Group N/A. Nearest overlying dated horizon Waidedo Vitric Tuff, which caps the Upper Herto Member, correlates with an unnamed tuff at **Konso** that lies 6 m below the Konso Silver Tuff, which dates to 154±7 ka by **argon-argon dating**. Nearest underlying dated horizon Lower Herto Member, 260±16 ka by argon-argon dating. Geological age "On the basis of the combined stratigraphic, geochemical and radioisoptopic evidence the Upper Herto archeological and paleontological remains are therefore securely constrained to be between 160±2 and 154±7 kyr old" (Clark et al. 2003). At the least the crania can be considered to be **Middle Pleistocene**. Developmental age Immature: the provisional chronological age estimate is 6–7 years. Presumed sex Unknown. Brief anatomical description The cranium, which has been restored from >180 small fragments, preserves much of the cranial vault, little of the cranial base and much of the upper face. There is also a partial dentition that is reported to include both left deciduous molars, the germs of a canine and premolars and a permanent first molar with a wear facet. Announcement White et al. 2003. Initial description White et al. 2003. Photographs/line drawings and metrical data White et al. 2003. Detailed anatomical description Gilbert and Asfaw 2009. Initial taxonomic allocation *Homo sapiens idaltu*. Taxonomic revisions N/A. Current conventional taxonomic allocation *Homo sapiens*. Informal taxonomic category Anatomically modern human. Significance Until the redating of **Omo I** and **Omo II** from Omo-Kibish, this immature cranium was part of the earliest evidence of *H. sapiens* in Africa, and it is the oldest immature *H. sapiens* specimen. Location of original National Museum of Ethiopia, Addis Ababa, Ethiopia.

Bouffia Bonneval (Fr. *bouffia*=the word for cave in the local dialect and *Bonneval* is the name of the family who own the land on which the cave is situated) The name of the cave, and the name used in the national site register, for the site better known as **La Chapelle-aux-Saints** (*which see*).

Boule, Marcellin (1861–1942) Marcellin Boule who was born in Montsalvy, France, entered the University of Toulouse in 1880 to study science where encounters with the archeologist of prehistory Émile Cartailhac and the paleontologist Louis Lartet influenced his decision to study human paleontology. Boule completed his degree in the natural and physical sciences in 1884 and in 1886 he moved to Paris where he studied geology with Ferdinand Fouqué and paleontology with Albert Gaudry. Boule became the latter's assistant at the **Muséum National d'Histoire Naturelle** (National Museum of Natural History) in 1894 and in 1902 he succeeded Gaudry as Professor of Paleontology at the museum. In addition to this position at one of France's premier scientific institutions, Boule was also appointed the Director of the **Institut de Paléontologie Humaine** (Institute of Human Paleontology), which had been established by Prince Albert I of Monaco in 1914. Boule's positions at the Museum and the Institute gave him access to substantial collections of anatomical material, both fossil and recent, and these proved invaluable in his paleoanthropological researches. His prestigious appointments also gave him a measure of scientific authority that gave added weight to his later ideas about human evolution. Besides these powerful institutional affiliations, Boule also served as an editor of the journal *L'Anthropologie* from its creation in 1893 until 1930, and he founded two other journals, the *Annales de paléontologie* (in 1906) and the *Archives de l'Institut de Paléontology Humaine* (in 1920). Boule concluded early in his career that the best way to approach the study of human prehistory was through a combination of geological, paleontological, and archeological perspectives where each was used to help interpret and inform the other. He rejected the linear progressive conception of human cultural development advocated by some prehistorians and believed that cultural development, like biological evolution, should be viewed as being like a tree with many branches rather than like a ladder. This image of biological evolution had a significant impact on Boule's theories about human evolution and especially his interpretation of the hominin remains from La Chapelle-aux-Saints. This largely complete hominin skeleton evidently resembled previously known *Homo neanderthalensis* remains and because of its significance the skeleton was sent to Boule at the Museum of Natural History. Boule's meticulous study of the La Chapelle-aux-Saints hominin skeleton was published in four parts in the *Annales de paléontology* between 1911 and 1913. Boule concluded from his analysis that Neanderthals shared more features in common with the apes than with modern humans. His influential reconstruction of the skeleton depicted Neanderthals as slouching rather than standing fully upright. He also emphasized the "bestial" appearance of the face and the low sloping skull to suggest that Neanderthals possessed minimal intelligence, a conclusion that was strengthened, he argued, by the crude **Mousterian** tools found at other Neanderthal sites in Europe. For Boule, the evidence was overwhelming that the Neanderthals had not evolved into modern humans. He concluded that the geological evidence suggesting a relatively short period separating Neanderthal deposits and early Cro-Magnon *Homo sapiens* deposits did not provide enough time for Neanderthals to have evolved into modern humans. Indeed, there seemed to be evidence the two species had co-existed in some parts of Europe, a possibility that corresponded well with Boule's conception of evolution where

there are numerous branches to the evolutionary tree, many of which are dead ends. The Neanderthals, for Boule, represented just such a dead end. Moreover, he considered the *Pithecanthropus* and *Sinanthropus* (now *Homo erectus*) fossils from Indonesia and China to possess greater similarities to the Neanderthals than to modern humans, which suggested that neither group were ancestral to modern humans. Boule's assessment of the La Chapelle-aux-Saints Neanderthal was widely accepted and had a profound impact on the way subsequent Neanderthal specimens were interpreted and upon the popular perception of Neanderthals as unintelligent brutes. Boule expanded upon these ideas is his major monograph *Les Hommes Fossiles*, published in 1921 and the English translation, *Fossil Men*, appeared in 1923. In that textbook Boule examined the available evidence of hominin evolution from the first primates to modern humans. Again the *Pithecanthropus* and Neanderthals are relegated to a side branch of human evolution, with the ancestors of modern humans evolving along a separate branch. Indeed, Boule accepted as evidence of this theory the fossils attributed to *Eoanthropus dawsoni* unearthed in Sussex, England, between 1908 and 1915. Unfortunately for Boule and many others, it was not until much later, in 1953, that Piltdown Man was demonstrated to be a hoax. Boule's interpretation of the Neanderthals was the dominant one until the middle of the 20thC when some paleoanthropologists began to reassess Boule's reconstruction of the La Chapelle-aux-Saints skeleton. It is now clear that Boule allowed his preconceptions to affect his research, and that some features of the bones that had influenced his "brutish" depiction of the skeleton were in fact the result of the fact that the La Chapelle-aux-Saints was an aged individual with severe osteoarthritis. Boule received numerous scientific honors and remained at the Muséum National d'Histoire Naturelle until his retirement in 1936. *See also* **La Chapelle-aux-Saints**; **Piltdown**.

Bourgeois-Delaunay   *See* **La Chaise**.

Bouri (Location approximately 10°15–18′N, approximately 40°30–34′E; etym. named after a village at the northern end of the Bouri Peninsula) <u>History and general description</u> The name given to the Bouri Fault Block, a peninsula in the **Middle Awash study area**. The long axis of the Bouri Fault Block trends northnorth-west–southsoutheast. It has the Awash River on its eastern margin, around the tip of the peninsula and on the southern part of its western margin, and Lake Yardi (into which the Awash River discharges) on the northern part of its western margin. Its sediments were initially explored by the **Rift Valley Research Mission in Ethiopia** (Kalb et al. 1982), and were assigned to the Wehaitu Formation (Kalb 1993). However, they were subsequently mapped in more detail in 1981 and thereafter by members of the **Middle Awash Research Project**. Three members of the **Bouri Formation** are exposed on the peninsula, from oldest to youngest these are the Hata Member (an abbreviation of Hatayae), the Daka Member (an abbreviation of Dakanihylo), and the Herto Member. The Hata Member is contemporary with the sediments exposed at **Gona** approximately 100 km/60 miles to the north. Research at Bouri has resulted in the recovery of the type specimen of *Australopithecus garhi*, hominin postcranial remains, isolated **Mode I** artifacts and cut-marked and percussed mammal bones from the Hata Member, **Acheulean** artifacts and hominins including the **BOU-VP-2/66** calvaria from the Daka Member, and the **Herto** crania from the Herto Member. Temporal span *c*.2.5 Ma for the Hata Member, *c*.1.0 Ma for the Daka Member, and *c*.150–200 ka for the Herto Member. How dated? **Argon-argon dating**, incremental laser heating, **tephrostratigraphy**, and **lithostratigraphy** Hominins found at site The type specimen of *Australopithecus garhi*, the Daka (BOU-VP-2/66) cranium, three femoral shaft fragments, one with the base of the neck preserved (BOU-VP-1/75) and one with part of the distal articular surface preserved (BOU-VP-19/63), a proximal tibia (BOU-VP-1/109), a talus (BOU-VP-2/95), and the Herto crania. Archeological evidence found at site Isolated Mode I artifacts and cut-marked and percussed mammal bones from the Hata Member, Acheulean artifacts from the Daka Member, and "Acheulian and Middle Stone Age technocomplexes" from the Herto Member. Key references: Geology and dating de Heinzelin et al. 1999; Hominins Asfaw et al. 1999, 2002, 2008, White et al. 2003; Archeology: de Heinzelin et al. 1999, 2000, White et al. 2003.

Bouri calvaria   *See* **BOU-VP-2/66**.

Bouri Formation (etym. Named after a village at the northern end of the Bouri Peninsula) General description The approximately 80 m-thick Bouri Formation is exposed in the southern two-thirds of the **Bouri** Peninsula. It consists of three members, and from oldest to youngest these are the Hata Member (an abbreviation of Hatayae), the Daka Member (an abbreviation of Dakanihylo), and the Herto Member (de Heinzelin et al. 1999). Sites where the formation is exposed The fossiliferous localities within the Bouri Formation are designated BOU-VP. Geological age *c*.2.5 Ma for the Hata Member, *c*.1.0 Ma for the Daka Member, and *c*.150–200 ka for the Herto Member. Hominin evidence The type specimen of *Australopithecus garhi* (**BOU-VP-12/130**) from the Hata Member, the Daka (**BOU-VP-2/66**) calvaria, three femoral shaft fragments, one with the base of the neck preserved (BOU-VP-1/75) and one with part of the distal articular surface preserved (BOU-VP-19/63), a proximal tibia (BOU-VP-1/109), and a talus (BOU-VP-2/95) from the Daka Member, and the Herto crania from the Herto Member. Archeological evidence Isolated **Mode I** artifacts and cut-marked and percussed mammal bones from the Hata Member, and **Acheulean** artifacts from the Daka Member. Key references: Geology and dating de Heinzelin et al. 1999, WoldeGabriel et al. 2000; Hominins Asfaw et al. 1999, 2002, 2008, White et al. 2003; Archeology de Heinzelin et al. 1999, 2000, White et al. 2003.

Bouri Peninsula   One of the major fossiliferous subregions within the **Middle Awash study area** (others include the **Central Awash Complex**, **Western Margin**, and **Bodo-Maka**). The Bouri Peninsula is in the southeastern portion of the Middle Awash study area and is situated between the **Awash River** and Yardi Lake. The peninsula is an approximately 10 km/6 miles long and 4 km/2.5 miles wide **horst** that was uplifted and tilted westward along a major NW-SE trending transverse **fault**.

bovid   The informal name for the artiodactyl family, the Bovidae, comprising the antelopes and their allies. Bovid remains are found at many early hominin sites. *See also* **Bovidae**.

bovid size classes   African faunal assemblages are often dominated by **bovid** remains. Although bovid teeth and horn cores are readily identifiable to tribe, genus, and/or species, their postcrania are extremely difficult to identify to taxon, particularly when fragmented or if multiple species are represented. For this reason, faunal analysts typically assign bovid remains to one of several size classes proposed by Bob **Brain**. They are: bovid size class I, <23 kg (e.g., duikers and dwarf antelopes); bovid size class II, 23–84 kg (e.g., springbok and Grant's gazelle); bovid size class III, 84–296 kg (e.g., wildebeest and hartebeest) and bovid size class IV, >296 kg (e.g., eland and buffalo); Richard Klein has added bovid size class V (>900 kg) to accommodate the extinct giant buffalo *Pelorovis antiquus*. Faunal analysts are increasingly utilizing these body size classes as analytical units for all mammal remains, particularly as it relates to studies of skeletal-element representation or bone-surface modifications that often include specimens that cannot be identified beyond the class Mammalia. Ethnographic and experimental observations indicate that an animal's body size can influence a number of behavioral and/or taphonomic factors, including butchery patterns, carcass transport strategies, and carnivore ravaging.

Bovidae   One of the artiodactyl families whose members are found at many hominin sites. The group is almost certainly monophyletic (Gatesy et al. 1997, Hernández-Fernández and Vrba 2005). A commonly used taxonomy recognizes the eight subfamilies listed below (the modern taxa in those subfamilies are given in parentheses): the Bovinae (cattle and buffalo), Antilopinae (dik-diks and gazelles) Cephalophinae (duikers), Reduncinae (kob), Aepycerotinae (impala), Caprinae (ibex and musk ox), Hippotraginae (antelopes and oryx), and the Alcelaphinae (wildebeest, topi, etc.). Bovids are recognized on the basis of their cranial morphology and particularly the size and shape of the bony horn cores (the horns themselves are made of keratin and do not survive, but horns have a bony core that does fossilize). *See also* **Artiodactyla**.

bovine   (L. *bos*=cow) The informal name for the artiodactyl subfamily, the Bovinae, comprising the cattle and buffalos, within the family **Bovidae**. *See also* **Bovidae**.

Boxgrove   (Location 50°51′N, 0°43′E, England, UK; etym. name of nearby village) History and general description The site is in a quarry 7 km/4 miles east of Chichester and 10 km/6 miles from the current English Channel shoreline. Excavations began in

1982 and continue to the present. A hominin tibia, **Boxgrove 1**, was found in 1993, and two hominin isolated incisors were recovered in 1996. Temporal span Oxygen Isotope Stage 13 and 12. How dated? Biostratigraphy (Cromerian mammal fauna), comparative geomorphology, optically stimulated **luminescence dating**, and **electron spin resonance spectroscopy dating**. Hominins found at site The tibial shaft of an adult and two isolated incisors, attributed to *Homo heidelbergensis*. Archeological evidence found at site Acheulean. Key references: Geology and dating Roberts et al. 1994, Roberts and Parfitt 1999; Archeology Roberts and Parfitt 1999; Hominins Roberts et al. 1994, Stringer et al. 1998, Trinkaus et al. 1999, Hillson et al. 2010.

Boxgrove 1   Site Boxgrove. Locality Quarry 1 (Q1-B). Surface/*in situ In situ*. Date of discovery 1993. Finder Roger Pedersen. Unit Approximately 40 m at the base of the "calcareous silts and gravels" (Roberts et al. 1994, p. 311). Horizon "Higher-energy correlatives of the upper part of the Slindon silts" (*ibid*). Bed/member N/A. Formation N/A. Group N/A. Nearest overlying dated horizon N/A. Nearest underlying dated horizon N/A. Geological age *c*.500 ka/Oxygen Isotope Stage 13. Developmental age Adult (Streeter et al. 2001). Presumed sex Unknown. Brief anatomical description A left hominin tibial shaft (approximately 294 mm) broken near the midshaft. Its morphology includes a "medially placed nutrient foramen, a well marked soleal line and a particularly prominent vertical line" (Roberts et al. 1994). Announcement Roberts et al. 1994. Initial description Roberts et al. 1994. Photographs/line drawings and metrical data Roberts et al. 1994. Detailed anatomical description Given the nature of the specimen a detailed anatomical description is unnecessary, but Trinkaus et al. (1999) provide a comprehensive analysis of its cross-sectional geometry. Initial taxonomic allocation *Homo* cf. *heidelbergensis*. Taxonomic revisions None. Current conventional taxonomic allocation *H. heidelbergensis*. Informal taxonomic category Pre-modern *Homo*. Significance This specimen only preserves the shaft, or diaphysis, of what is presumed to be an adult tibia, but early **Middle Pleistocene** postcranial hominin fossils are few and far between, so this specimen makes a significant contribution to what (little) we know about the limb morphology of *H. heidelbergensis*. The cortical bone area is modest, but the shaft is robust overall (i.e., externally thick for its length) and the specimen suggests that the overall

98

skeletal proportions of this individual are consistent with an adaptation for cold conditions. Location of original Natural History Museum, London, UK. *See also* **Homo heidelbergensis**.

BP Abbreviation of "before the present". Used in absolute dating techniques particularly in radiocarbon dating although it has been appled to absolute dates derived from other techniques, typically for relatively recent dates (e.g., the last few thousand years). To standardize radiocarbon dates the "Present" is formally assigned to the arbitrary date of January 1, 1950 (Stuiver and Polach 1977). Thus a date given as 5000 years BP refers to a date 5000 years before 1950.

bp Acronym for **base pair** (*which see*).

BPI Abbreviation of the **Bernard Price Institute for Palaeontological Research** (*which see*).

BPT Acronym for the Black Pumice Tuff in the **Koobi Fora Formation, West Turkana,** Kenya.

brachial index (L. *brachium*=arm) Ratio of the lengths of the radius and humerus, calculated as 100×(radius length/humerus length), that measures the relative length of the forearm. Like the **intermembral index**, the brachial index varies among primates with different locomotor strategies. In primates with long arms, the forearm tends to contribute disproportionately to the length of the arm (possibly due to developmental constraints), so that suspensory primates (e.g., orangutans) with long upper limbs also have high brachial indices. The brachial index also distinguishes arboreal and terrestrial quadrupeds because the latter have long limbs relative to their body size. Disproportionately elongated distal elements (e.g., forearm), which increases speed and efficiency on the ground, also increases the lengths of the parts of the forelimb that have the least mass. This contrasts with what happens in above-branch arboreal quadrupeds for which having their trunk closer to the branch improves balance. Climate also influences the brachial index, as part of a trend described in Allen's Rule. Limbs typically become narrower distally, and this increases their surface-area-to-volume ratio. The ability to lose or retain heat is proportional to relative surface area, and this explains why primates (including modern humans) that inhabit hot, dry environments have elongated distal limb elements (high brachial indices), while those in cold, humid habitats tend to have relatively short distal elements (low brachial indices). That is most likely why the high brachial index of *Homo ergaster* in Africa (as seen in **KNM-WT 15000**), and the low brachial index of *Homo neanderthalensis* are at, or surpass, the ends of the range of brachial index among modern human populations. *See also* **Allen's Rule**.

brachiationist model *See* climbing hypothesis.

Brachystegia *See* miombo woodland.

Brain, Charles Kimberlin (Bob) (1931–) C.K. (Bob) Brain was born in Salisbury, Southern Rhodesia (now Zimbabwe), and graduated from the Pretoria Boys High School in 1947. He entered the University of Cape Town in South Africa where he completed his BSc in geology and zoology in 1950. Brain continued at the University of Cape Town for his graduate work and in 1951 he began to study the complex stratigraphy of the cave deposits in the Transvaal where Robert **Broom** and John **Robinson** had found numerous hominin fossils. While working with Robinson at **Sterkfontein** in 1956, Brain found stone **artifacts** in **breccia** deposits where **hominin** fossils at the time assumed to belong to **australopith**s had previously been recovered. These discoveries apparently offered the first evidence that at least some australopiths may have made stone tools. Brain's research on the formation and structure of the australopith-bearing (then called australopithecine-bearing) deposits at Sterkfontein, **Swartkrans, Kromdraai,** and **Makapansgat** formed the basis for his PhD in geology from the University of Cape Town in 1957. The results were published as a monograph *The Transvaal Ape-man Bearing Cave Deposits* in the *Transvaal Museum Memoir* series in 1958. In 1963 Robinson resigned his position as the head of the Department of Vertebrate Palaeontology and Physical Anthropology at the **Transvaal Museum** in Pretoria and in 1965 Brain was appointed to succeed him. The discovery of australopith fossils along with **ungulate** fossils by Raymond **Dart** at Makapansgat in the 1950s, and Dart's assertion that some of the ungulate bones had been modified for use as **bone tools**, prompted Brain to re-examine the fossil collections in the Transvaal Museum and to begin new excavations at Swartkrans. Brain's work at Swartkans continued until 1986 and the main results were published in 1993 (see below). He collected over 200,000 bone fragments, including several hundred hominin fossils, from Swartkrans in an attempt to assemble as complete a fossil assemblage as possible. Brain developed field and laboratory techniques to try to understand how the deposits had been formed and he investigated the effects of weathering and of carnivore gnawing on bones to show that these processes could modify bones in ways that resembled human modification. He also gathered ethnographic data on animal butchering and the disposal of bones by indigenous people of the Namib Desert as a way of testing the notion that animal fossils found in cave deposits could have been the remains of game animals eaten by hominins. Brain's research made a major contribution to cave **taphonomy**. By the early 1970s Brain had come to understand how animal bones and other debris were deposited in caves. Significantly, by studying the eating habits of carnivores and other living African animals Brain concluded that many of the bones found at Swartkrans and other cave sites had been accumulated by predators such as leopards, and by scavenging animals such as porcupines. Brain presented these conclusions at a **Wenner-Gren Foundation for Anthropological Research**-sponsored conference on "Taphonomy and Vertebrate Paleoecology: with Special Reference to the Late Cenozoic of Sub-Saharan Africa" co-organized by Brain in 1976. Brain also argued that a juvenile *Paranthropus* cranium from Swartkrans that had two perforations at the top of the brain case was evidence that leopards had preyed upon australopiths, the perforations resulting from the canines of the leopard biting into the skull as it carried its prey away to be eaten. Whereas Dart had argued that the bone accumulations at Makapansgat were evidence that the australopiths were hunters, Brain argued that they were likely the prey of carnivores and had not gathered the bones found in cave deposits. Instead, Brain suggested that at least some of the hominin bones found in these deposits had been brought there by predators. To test his ideas he conducted experiments feeding carcasses to large predators to see what affect they had on the bones. He also collected animal bones left by carnivores in their lairs or by owls in caves to determine the types and quantity of bones that accumulated there. These original investigations led to many publications that culminated in *The Hunters or the Hunted? An Introduction to African Cave Taphonomy*, published in 1981. In addition to unraveling the processes that formed the bone-bearing deposits in southern African caves, Brain also discovered numerous hominin fossils and a large quantity of stone artifacts at Swartkrans as well as some fossil **antelope** bones that do appear to have been used as tools. Brain also found burnt bones at Swartkrans suggesting that campfires were maintained at the cave. Brain retired from the directorship of the Transvaal Museum in 1991. The first edition of his most recent book, *Swartkrans: A Cave's Chronicle of Early Man*, was published in 1993, and a second, revised, edition was published in 2004.

brain (OE *braegen*=brain, most likely from the Gk *bregma*=the front of the head) The brain is defined as the part(s) of the central nervous system that lie within the **endocranial cavity**. The central nervous system develops from the dorsal ectoderm germ layer (i.e., a long strip of ectoderm in the middle of the back of the embryo). This strip of ectoderm rolls up along its long axis into a tube called the neural tube that for all but a small section of its length (between the hindbrain and the midbrain) is normally sealed at the back. The caudal (i.e., towards the tail) part of the neural tube, which is the majority of its length, becomes the **spinal cord**. The cranial part of the neural tube initially develops as three swellings, or vesicles, and it is these vesicles that give rise to the brain. The **caudal** swelling (i.e., the one that connects with the spinal cord at the level of the **foramen magnum**) is called the hindbrain (or rhombencephalon), the middle swelling is called the midbrain (or mesencephalon), and the cranial of the three swellings is called the **forebrain** (or prosencephalon). The hindbrain develops into the medulla oblongata, the pons, and the cerebellum; the **cerebellum** is a dorsal outgrowth from the same embryonic brain subdivision that contains the pons. The forebrain comprises the **thalamus** (the diencephalic part of the forebrain vesicle), which lies close to the midline, and the two **cerebral hemispheres** (the telencephalic part of the forebrain vesicle), which are more laterally situated. It is the external surfaces of the cerebral hemispheres that are visible when the cranial cavity is opened by removing the roof of the cranium (i.e., the **calotte**). The **basal ganglia** (i.e., the **striatum**, globus pallidus, substantia nigra, and subthalamic nuclei) develop during ontogeny deep in the wall of the part of the expanding forebrain that comes to rest up against the thalamus. In primates, but especially in the great apes and particularly in modern humans, it is the forebrain that undergoes the most expansion during ontogeny. This is an extension of the trend seen across a range of mammals, wherein increases in **brain size** are generally accompanied by a positive **allometry** of forebrain components, particularly the **neocortex** and the striatum (Finlay and Darlington 1995). *See also* **brain size**; **brain volume**.

brain mass   The average specific mass or density of 78 adult modern human brains was 1.032 g/cm^3 (Zilles 1972). Thus, brain mass=brain volume (cm^3)×1.032. European researchers tend to refer to "brain weight," whereas US-based researchers generally refer to "brain mass." Count (1947) suggested that brain mass=**endocranial volume**/1.14. *See also* **brain size**.

brain size   The size of the brain can be expressed either in terms of its volume or its weight (or mass). Brain size is equal to the **endocranial volume** (also called cranial capacity) minus the volume (or weight) of the meninges, the extracerebral cerebrospinal fluid (i.e., the cerebrospinal fluid outside the ventricles), the intracranial (but extracerebral) vessels, and the cranial nerves within the cranial cavity. *See also* **endocranial volume**.

brain size evolution, molecular basis of
The genes involved in regulating brain size may also be among the molecular mechanisms underlying brain evolution within the hominin clade. Several genetic loci associated with primary microcephaly, a genetic disorder characterized by an extreme reduction in the size of the cerebral cortex, have undergone a high rate of nonsynonymous amino acid substitutions within the primate lineage, and especially within the hominin clade. Primary microcephaly has been mapped to six regions of the modern human genome (i.e., *MCPH1* to *MCPH6*, microcephaly, primary autosomal recessive 1–6) and null mutations have been identified at four loci: microcephalin (aka *MCPH1*), *CDK5RAP2* (CDK5 regulatory-subunit-associated protein 2; aka *MCPH3*), *ASPM* (abnormal spindle-like, microcephaly-associated; aka MCPH5), and *CENPJ* (centromeric protein J; aka *MCPH6*). The microcephaly related genes *ASPM* and *MCPH1* are involved in regulating brain size through mitotic spindle activity and cell cycle progression, respectively. All primary microcephaly genes (i.e., *MCPH1* to *MCPH6*) function in cell-cycle control, and their ability to influence brain size likely stems from their role in regulating the proliferation of neural precursor cells during embryogenesis. *ASPM* has been shown to have undergone positive selection throughout the primate lineage, including at the stem of the hominid clade (i.e., great apes and modern humans) and in the hominin clade. One variant of the *ASPM* gene seems to have undergone a recent **selective sweep** (i.e., a sharp reduction in variation as the favored variant quickly spreads) around 14 ka and one common variant of the allele might have entered the modern human gene pool via admixture with archaic *Homo sapiens*, perhaps even *Homo neanderthalensis*, *c.*37 ka. Microcephalin shows a strong indication of positive selection primarily at the base of the hominid clade, whereas both *CDK5RAP2* and *CENPJ* show higher rates of nonsynonymous substitutions in primates than rodents, and *CDK5RAP2* shows especially high rates in the hominin and panin clades. Another gene involved in neural precursor proliferation, *ADCYAP1* (adenylate cyclase-activating polypeptide 1), which helps regulate the transition from proliferative to differentiated states during neurogenesis, has also been found to bear signatures of accelerated protein-sequence evolution in the hominin clade. Examples of genes with critical roles in determining the relative size of areas within the cerebral cortex include *Emx2*, *Pax6*, and *COUP-TFI* (*Nr2f1*) and variation in some of these genes among modern humans leads to cognitive deficits.

brain specific mass   The average specific mass, or density, of the central nervous system within the cranial cavity. *See also* **brain mass**.

brain tissue volume   The volume of the brain itself (i.e., brain volume minus the volume of the cerebrospinal fluid within the cavities, or ventricles, of the brain and the volume of any meninges and cranial nerves that may adhere to the brain).

brain volume Usually defined as the sum of the volume of the brain tissue, the **cerebrospinal fluid** within the ventricles of the brain (i.e., the intracerebral cerebrospinal fluid), and any cranial nerves and **meninges** that may adhere to the brain. In practice adult brain volume is approximately 85% of the **endocranial volume** or cranial capacity.

brain weight *See* **brain mass**.

branch length The lengths of the branches of a **phylogenetic tree** are sometimes scaled according to the numbers of character-state changes or time intervals that separate the tree's nodes. Because the nodes of a phylogenetic tree represent taxonomic units, when branch lengths are employed the length of the branch that separates two nodes indicates either the amount of evolutionary change that has taken place since the ancestral taxon gave rise to the descendant taxon, or the amount of time that has elapsed since the ancestral taxon gave rise to the descendant taxon. A phylogenetic tree whose branch lengths represent the numbers of character-state changes between nodes is known as a phylogram, whereas one whose branch lengths represent the time intervals between nodes is referred to as a chronogram.

branchial arches (Gk *branchion*=gill) System of arches (also called pharyngeal arches) from which the side of the head and neck is formed. Each arch comprises a branchial arch **cartilage**, a dominant nerve and branchial muscle(s); some arches also have a branchial arch artery that persists into adulthood. The structures they give rise to are covered externally with skin or modified skin, and internally they are lined by a layer of epithelium moistened by mucus, called mucosa. On the outside the region in the embryo between two contiguous branchial arches is called a branchial cleft; on the inside it is called a branchial pouch. For example, the cleft between the first and second branchial arches (the first pharyngeal cleft) is the external ear, and the equivalent pouch (the first pharyngeal pouch) is the **auditory tube**; the tympanic membrane is all that prevents the first branchial cleft communicating with the first branchial pouch. The first branchial can serve as an example. Its arch cartilage is **Meckel's cartilage**; this gives rise, from medial to lateral

respectively, to the **incus**, the malleus, the anterior ligament of the malleus, the **sphenomandibular ligament**, and the **lingula**. The **mandible** develops in membrane *around* (not from) the lateral extension of Meckel's cartilage. The muscles of the first branchial arch comprise the **muscles of mastication**, the tensors of the palate and the tympanic membrane, the mylohyoid, and the anterior belly of the digastric. The first branchial arch motor nerve, the motor part of the mandibular division of the trigeminal, supplies all of these muscles (syn. pharyngeal arches).

Bräuer, Günter (1949–) Born in Altena, Germany, Günter Bräuer attended the University of Mainz, and was awarded a Diploma in Biology in 1975 and a Dr. rer. nat (PhD) in 1976. Following his PhD research he moved to the University of Hamburg, becoming Professor of Physical Anthropology in 1985. His PhD dissertation research focused on the **Later Stone Age** skeletal remains from **Mumba Shelter** and from other localities along the shores of Lake Eyasi in Tanzania, and his subsequent field work and anatomical studies have focused on African Middle and Late Pleistocene–Holocene hominin fossils. His Habilitation in 1983 was a detailed assessment of the evidence for the origin of modern humans in Africa. In 1982 he proposed a version of the **out-of-Africa hypothesis** of modern human origins called the Afro-European Sapiens hypothesis (Bräuer 1984), which favors a gradual, mosaic-like, process of anatomical modernization in Africa over about half a million years. He was also one of the first researchers to recognize the importance of the hominin remains from **Klasies River Mouth** in South Africa (Bräuer et al. 1992a), and his research in Europe suggested there was only minimal evidence for interbreeding between *Homo neanderthalensis* and the modern humans who eventually supplanted them. Günter Bräuer has been a major figure in the debate between those who advocate one or other version of the out-of-Africa hypothesis and those who preferred one or other versions of the **multiregional hypothesis** for the origin for modern humans. Working with Richard **Leakey** and Emma Mbua, he described the **Middle Pleistocene** cranial remains from **Eliye Springs** and the **KNM-ER 3884** cranium from **Koobi Fora**. His morphological studies on *Homo erectus*, which focused on the origin and evolution of *H. erectus* in Africa and Asia, led to his support for *Homo erectus sensu lato* and his rejection of *Homo ergaster* (Bräuer 1994). He is co-editor (with Fred Smith) of *Continuity or Replacement – Controversies in Homo sapiens evolution* (1992). He is currently Professor of Human Biology and Physical Anthropology at the University of Hamburg.

breccia (Ge. *brehhan*=to break, or It. *breccia*= rubble) A rock containing angular fragments that have been cemented together in a fine-grained **matrix** (e.g., soil or cave earth) hardened with a mineral cement such as calcium carbonate. The fragments have not been subjected to prolonged water transport and thus because their edges have not been rolled and rounded by water action they are still sharp. Nearly all the early **hominin**s from southern Africa have been extracted from breccia-filled limestone caves (e.g., **Drimolen, Sterkfontein**). The breccia and speleothem that adheres to many of these fossils can be removed mechanically (e.g., precision drills) or chemically (e.g., by immersion, under strictly controlled conditions, in 10% acetic acid).

bregma (Gk *brechein*=moist) The name given by Aristotle to the part of the **cranium** of a newborn that is moist and membranous. This is because in the newborn the frontal and parietal bones have not fused, and in that region beneath the skin there is a soft diamond-shaped patch of **dura mater** called the anterior fontanelle. It is now used as the name for the midline landmark where the sagittal suture between the two parietal bones meets the coronal suture, which separates the frontal from the parietal bones.

bristle-cone pine *See* **radiocarbon dating**.

Brno (Location 49°11′N, 16°39′E, Czech Republic; etym. named after the city of Brno, in or near which the sites are located) History and general description This name is applied to three hominin specimens discovered near Brno in 1885 (Brno 1), 1891 (Brno 2), and 1927 (Brno 3). The latter two specimens are burials. Brno 3 was destroyed by fire in 1945. Temporal span and how dated? Brno 1 has uncertain dating; Brno 2 is suggested to be Pavlovian (**Gravettian**) by cultural association, and Brno 3 was initially thought to be **Upper Paleolithic**, but this attribution has been questioned. Hominins found at site All three are *Homo sapiens*. Brno 1 consists of a calotte,

maxilla, and fragmentary postcrania. Brno 2 consists of a calvaria, partial face, partial mandible, and fragmentary postcrania. Brno 3 was an associated skeleton. Archeological evidence found at site Brno 1 was found in a horizon with **Aurignacian**-like tools, but it is unclear how they are associated with the fossil remains. Brno 2 was buried with mammoth bone and ivory, a figurine, and perforated shells, which suggest a Pavlovian (Gravettian) context. Brno 3 was not associated with any archeological remains. Key references: Geology, dating, and paleoenvironment Smith 1982; Hominins Vlček 1971, Smith 1982.

Brno-Bohunice (Location 49°10′N, 16°34′E, Czech Republic; etym. namd after the suburb of Brno in which it is located) History and general description This open-air site near the city of Brno in Southern Moravia is the type site of the Bohunician transitional technocomplex. Valloch collected stone tools exposed in bulldozer trenches in the 1970s, and performed limited excavation. The site was re-excavated in 2002. Temporal span and how dated? **Radiocarbon dating** suggests a range of 40–35 ka (calibrated). Thermo**luminescence dating** gives an age of *c*.48 ka, but optically stimulated **luminescence dating** provides a large range of dates: 30.9±3.1, 58.7±5.8, and 104.3±10 ka. Hominins found at site None. Archeological evidence found at site The recovered stone tools are from an early **Upper Paleolithic** industry that Valloch initially described as a kind of **Szeletian**, but was later renamed the Bohunician. It is marked by the use of **Levallois** with a significant proportion of blades and, like the **Châtelperronian** in France, it is a transitional industry with many similarities to both Middle and Upper Paleolithic technologies, but which is usually attributed to *Homo neanderthalensis*. Key references: Geology, dating, and paleoenvironment Svoboda 2003, Richter et al. 2009; Archeology Svoboda 2003.

broad-sense heritability (H or H^2) Defined as the genetic **variance** divided by the phenotypic variance. This value is considered the broad-sense (as opposed to the narrow-sense) heritability since it reflects *all* of the possible genetic contributions to the phenotypic variance. *See also* **heritability**.

Broca's area (etym. Named after Paul Broca) A region of the inferior frontal gyrus of the cerebral cortex that in modern humans is important for the production of **speech**, the processing of **syntax**, lexical retrieval, the organization of hierarchically ordered movements of the face and hands, and the processing of rules relating to **recursion**. This cortical region is named after the French physician and anthropologist Paul Broca (1824–80), who discovered its function by studying the post-mortem brains of patients with aphasia, a profound speech impairment. Broca found that lesions that included the left inferior frontal gyrus were often observed in these cases. Broca's area is generally thought to include **Brodmann's** cytoarchitectonic areas 44 and 45. In modern humans, different parts of the inferior frontal gyrus are typically associated with different cytoarchitectonic components of Broca's area. The ascending (vertical) ramus of the **lateral fissure** usually separates the pars opercularis (area 44) from the pars triangularis (area 45), and the anterior ramus divides the pars triangularis from the pars orbitalis (area 47). Because approximately 95% of modern humans show left-**hemispheric dominance** for language, numerous studies have investigated macrostructural asymmetry in Broca's area of modern humans using the sulcal landmarks listed above to subdivide the region. Results from these macrostructural studies, however, have differed markedly depending upon methodology and anatomical definitions and it is not clear whether the gross anatomical structure of inferior frontal gyrus in modern humans displays lateral asymmetry. Histological studies of Broca's area of modern human brains, however, have revealed several significant asymmetries. Volumetric cytoarchitecture-based studies have shown that area 44, but not area 45, is leftward dominant (Galaburda 1980, Amunts et al. 1999). In addition, leftward asymmetries have also been reported in terms of the dendritic branching of **pyramidal cells**, and the gray-level index. Based on connectivity, **cytoarchitecture**, and function, homologues of Brodmann's areas 44 and 45 have been identified in macaque monkeys and great apes. The region equivalent to Broca's area in the inferior frontal cortex of macaques has been shown to be activated during the perception of species-specific vocalizations and facial expressions, and, in addition, when this region is electrically stimulated in macaques it results in orofacial movements. When the equivalent region in the inferior frontal cortex of chimpanzees is stimulated movements of

the larynx and tongue are evoked and **positron emission tomography** neuroimaging experiments have shown activity in the left inferior frontal gyrus during the production of communicative vocalizations and hand gestures. As with modern humans it is unclear whether gross morphological asymmetries are present in the inferior frontal gyrus of chimpanzees. Cytoarchitecture-based studies, furthermore, indicate that Brodmann's areas 44 and 45 are *not* asymmetric in terms of volume or neuron numbers in chimpanzees (Schenker et al. 2010). This differs from the documented leftward asymmetry of area 44 in modern humans. When compared to the other cortical areas that have been measured in both modern humans and chimpanzees on the basis of a cytoarchitectonic definition of boundaries (including areas 10, 13, and 17), area 44 and area 45 on the left side are among the most enlarged in the modern human brain.

Brodmann's areas   Numbered regions of the **cerebral cortex**. They are named after Korbinian Brodmann who, after studying sections of the cerebral cortex, subdivided it according to what he interpreted were differences in the neuronal morphology, cell packing density, and relative thickness of the six main cortical layers. For example, the primary motor speech areas of the cerebral cortex are given the numbers 44 and 45, and the primary visual cortex is numbered 17.

Broken Hill   *See* **Kabwe**.

Broom, Robert (1866–1951)   Robert Broom was born in Paisley, Scotland, and entered the University of Glasgow in 1883. He completed his medical degree in 1889, and emigrated in 1892 to Australia where he lived for 4 years before settling in South Africa in 1897. Initially Broom practiced medicine in Cape Province, but in 1903 he was appointed Professor of Zoology and Geology at Victoria College (now Stellenbosch University). The fossil-rich deposits of the Karoo located at the outskirts of Cape Province provided Broom with the opportunity to pursue paleontology and he quickly became an expert on therodonts, the topic of his University of Glasgow DSc thesis in 1905; thus began Broom's long interest in the evolutionary origin of the mammals from mammal-like reptiles. In 1910, Broom was dismissed from his position at Victoria College, possibly because of his support of the idea of evolution,

which was a controversial subject in the conservative cultural environment of South Africa at the
time. Broom returned to Britain during WWI and
worked in a London hospital, but in 1916 he
returned to South Africa. He continued to practice
medicine and to pursue his paleontological
researches, but the emphasis of his research changed when he learned of Raymond **Dart**'s discovery
of *Australopithecus africanus*. Late in 1924
Dart, who was Professor of Anatomy at the
newly founded University of Witwatersrand in
Johannesburg, recognized the juvenile skull of a
previously unknown species intermediate between
the anthropoid apes and modern humans and an
evolutionary ancestor to modern humans among
the fossils from a lime quarry in Taungs (now
called **Taung**). Accounts of Dart's discovery were
published in *Nature* and in a local newspaper in
Johannesburg during the first week of February
1925, and within 2 weeks of learning of the discovery Broom traveled to Johannesburg to inspect
the fossil for himself. Broom immediately recognized the importance of Dart's "discovery"
and agreed with the latter's assessment that
Au. africanus represented an evolutionary link
between our ape-like ancestors and modern
humans. But Dart's discovery received a mixed
response among European anthropologists, and
few accepted Dart's conclusion that *Australopithecus* was ancestral to modern humans. Broom, however, acted quickly to defend Dart's interpretation
of the fossil by writing two papers about it. The
first, entitled "Some notes on the Taungs skull,"
was published in *Nature*, and the second paper,
entitled "On the newly discovered South African
man-ape," was published in *Natural History*; both
appeared in 1925, but Broom's support did little to
convince his European and American colleagues.
There were several reasons for this. First, most
anthropologists at that time believed that humans
had evolved in Asia, not Africa, and the discovery
of *Pithecanthropus erectus* by Eugène **Dubois** in
Java in the 1890s and of *Sinanthropus pekinensis*
in China by Davidson **Black** in the late 1920s only
strengthened this conviction. Second, the *Australopithecus* specimen discovered by Dart was of a
juvenile and it was known that apes during their
infancy can resemble the morphology of modern
humans, acquiring their more distinct ape-like features as they mature. Lastly, since the geology of
southern Africa was poorly understood, there was
skepticism about the geological antiquity that Dart

and Broom claimed for *Australopithecus africanus*.
Broom realized that *Australopithecus* would not be
accepted as an evolutionary link between early
anthropoids and modern humans until further,
more complete, adult specimens were discovered.
His experience in paleontology provided Broom
with the knowledge necessary to search for additional specimens and his appointment in 1934 as
curator of paleontology at the **Transvaal
Museum**, in Pretoria, gave Broom the time and
resources to devote all his efforts to the search.
The discovery of fossil baboon skulls in cave
deposits at **Sterkfontein**, near Johannesburg, in
1936 by Harding le Riche and Trevor Jones, who
were students working with Dart, prompted
Broom to visit the site. There he met the local
quarry manager, George Barlow, who gave Broom
several fossil baboon skulls and a partial fossilized
brain cast that clearly was from an "ape-man."
Broom recovered portions of the skull belonging
to the **endocast** several days later, and he considered the specimen to be closely related to Dart's
Australopithecus. He originally named it *Australopithecus transvaalensis*, but he later renamed it
Plesianthropus transvaalensis (it is now considered to be conspecific with *Au. africanus*). Broom
collected teeth and small pieces of bone from
Sterkfontein over the next 2 years, but his next
major discovery occurred in 1938 when Broom
received teeth and portions of a skull from a
hominin that seemed to differ from both *Australopithecus* and **Plesianthropus**. The specimen,
found at a cave site called **Kromdraai** in the same
valley as Sterkfontein, had larger postcanine teeth
and a larger mandibular body than *Australopithecus*,
so Broom referred the specimen to a new taxon,
Paranthropus robustus. Broom now had fossils
from several adult australopiths, and because the
outbreak of WWII had stopped excavations in
South Africa, this allowed Broom and his assistants to study the fossils. This resulted in the
publication in 1946 of *The South African Fossil
Ape-Men: The Australopithecinae* by Broom and
G.W.H. Schepers, the second in the *Transvaal
Museum Memoir* series. Broom resumed excavations at Sterkfontein in 1947 together with John
Robinson, who had just been appointed to
Broom's department at the Transvaal Museum.
Over the next few years they recovered more
skeletal material there, but in 1948 at a recently
discovered site called **Swartkrans** they unearthed
another robust hominin mandible, which they

105

assigned to yet another new species called *Paranthropus crassidens*. A year later, also at Swartkrans, they found several fossils of a much less robust type that they believed belonged to a species intermediate between "ape-men" and modern humans. Broom and Robinson named this new species *Telanthropus capensis* and by 1952 several additional specimens had been discovered there. In 1950 Broom, Robinson, and G.W.H. Schepers published *Sterkfontein ape-man, Plesianthropus*, the fourth volume in the *Transvaal Museum Memoir* series, and in 1952 the sixth volume in the series, *Swartkrans ape-man, Paranthropus crassidens*, was published, this time by Broom and J.T. Robinson. This was Broom's last publication. In addition to making major discoveries of hominin fossils, Broom also wrote on theoretical issues concerning the mechanism and operation of biological evolution. While discussing vertebrate evolution and the origin of mammals in *The Origin of the Human Skeleton* (1930), Broom criticized both the Neo-Lamarckian and the Darwinian conceptions of evolution. He repeated these criticisms in *The Coming of Man: Was It Accident or Design?* (1933), where he argued that mutationist, Neo-Lamarckian, and Darwinian mechanisms do not adequately account for what we know about the fossil record and the evolution of humans. Broom instead argued that an intelligent spiritual agent must operate in biological evolution, and moreover he believed that evolution had reached its completion in modern humans. He wrote a popular account of his discoveries on paleoanthropology entitled *Finding the Missing Link*, which was published in 1950, and he continued his paleontological research and remained curator of paleontology at the Transvaal Museum until his death in 1951. Broom was elected a Fellow of the Royal Society in 1920 and was awarded its Royal Medal in 1928.

brow ridge (ME *brow*=eyelash) An informal inclusive term that applies to either discrete thickening of the bone above each superior orbital margin (e.g., *Homo neanderthalensis*) or to a continuous bar of thickened bone that extends from the lateral end of one superior orbital margin across the midline to reach the lateral end of the superior orbital margin on the other side (e.g., **Trinil 2**). The term brow ridge is usually not applied to the compound structure seen above each

orbital opening in *Homo sapiens*, which consists of the **superciliary arch** medially and the **supraorbital trigon** laterally.

brown breccia   *See* Sterkfontein; Swartkrans.

Brown Sands   One of two hominin-bearing localities in the Usno Formation in the **Lower Omo Basin** (the other one is called **White Sands**). Fossils found in the Brown Sands locality have been given the prefix "B" for Brown Sands. *See also* **Usno Formation**; *Australopithecus afarensis*.

brown striae   Less common term used to refer to **long-period incremental lines** in enamel; the term is derived from the common appearance of these structures as brown lines when viewed using transmitted light microscopy. *See also* **striae of Retzius**.

brown striae of Retzius   *See* striae of Retzius.

Brown, Francis H. (1943–)   Francis Brown was born and raised in northern California, USA. In 1965 he graduated from the University of California, Berkeley with a BA in geology. He stayed there for his graduate studies, which addressed the petrology of kalsilite-bearing lavas in Toro-Ankole, Uganda; he was awarded his PhD in 1971. He began stratigraphic work on the **Shungura Formation** in 1966, and between 1967 and 1974 Brown joined the **International Omo Research Expedition** in Ethiopia where he worked with Jean de **Heinzelin** and Paul Haesaerts on the stratigraphy of the Usno and Shungura Formations, and between 1972 and 1974 he collaborated with Ralph T. Shuey to determine the geomagnetic polarity of the Shungura and Usno Formations; his study of tephra (volcanic ash layers) of the Shungura Formation also began in the 1970s. After fieldwork in southern Utah and at Sahabi, Libya, in 1980, Brown, at the invitation of Richard Leakey, began reviewing the stratigraphy at Koobi Fora and this began a long association with the **Koobi Fora Research Project** and by 1986 the stratigraphic framework of the Koobi Fora had been revised in collaboration with Craig Feibel; work on the Nachukui Formation, west of Lake Turkana, began in 1981. In conjunction with Thure Cerling, Brown pioneered **tephrostratigraphy** (i.e., the use of tuffs as lithostratigraphic markers) at hominin sites in East Africa. By 1982, the first correlations were established between the **Hadar** study area and Koobi Fora, and by 1985 links were established with the deep sea record in the Gulf of Aden. During the same period, together with Harry Merrick, he began analysis of Kenyan obsidians to determine sources of archeological material, and more recently has worked on obsidians from Ethiopia. Fieldwork with Bereket Haileab in 1985 and 1986 formed the basis for tephra correlations throughout the Turkana Basin, and also to other sites, such as **Gadeb** and the **Middle Awash study area** (also with Bereket Haileab), Baringo (with Fulbert Namwamba), and the Suguta Valley (with Benson Mboya). In 1980 Brown began a long-term association with Ian McDougall (Australian National University), and together they have provided secure ages for many Miocene sites in the region as well as ages for the many tephra in the Pliocene and Pleistocene **Omo Group**, and on two tephra in the Pleistocene Kibish Formation; Brown's long-term fieldwork on the Nachukui and Shungura Formations continued until 2010. His important role in the Koobi Fora Research Project was recognized by his lead-authorship of the paper that announced the recovery of **KNM-WT 15000** from **West Turkana** (Brown et al. 1985). Brown is currently Dean of the College of Mines and Earth Sciences at the University of Utah, where he is also a Distinguished Professor of Geology and Geophysics.

browser   An animal that feeds on above-ground vegetation (e.g., leaves, fruits, and shoots in bushes and trees). *See also* **browsing**.

browsing   (OF *broust*=shoot or twig) The consumption of leaves, fruits, and shoots of usually dicotyledonous plants (flowering plants with two embryonic seed leaves or cotyledons) that grow above ground level. This term is often used in paleoanthropology to refer to the inferred diets of the **herbivore**s whose remains are found at many archeological and paleoanthropological sites. A browsing diet in extinct animals is usually inferred through studies of tooth and jaw morphology (e.g., a narrow snout and evidence of hypsodonty) or from studies of stable isotopes. Plants that browsers feed on (i.e., shrubs, trees, and non-grass herbs) use the C_3 photosynthetic system and they are depleted in ^{13}C (i.e., have less ^{13}C) relative to plants that use the C_4 and **CAM** photosynthetic systems; thus the bones of browsing animals have bones and enamel that are depleted in ^{13}C (i.e., have less ^{13}C) relative to the bones of grazing

animals that consume plants that use the C_4 photosynthetic system. Modern giraffes and elephants are browsers and the same dietary adaptation has been suggested for many of their extinct fossil relatives. Like all forms of herbivory, browsing can be considered a form of predation in which an organism predates on plants, in which case browsers are primary consumers in the **ecosystem**. *See also* C_3 and C_4; $^{13}C/^{12}C$; **stable isotopes**.

Brunet, Michel (1940–)   Michel Brunet was born in Southern Vienne, France, in the midst of WWII and spent his early childhood in the village of Poitou, near Poitiers. After the end of the war Brunet completed his early education at Versailles and in 1962 he entered the Sorbonne in Paris to study paleontology. Brunet completed his thèse d'Université in paleontology in 1966, then he moved to the University of Poitiers to study Paleogene mammals, and it was there he completed his Natural Sciences State Doctorate in 1975. In 1989 Brunet accepted a position as Professor of Paleontology at the University of Poitiers. His early research dealt with the paleontology of mammals, but the focus of his research changed to primate and hominin paleontology in 1976 after he learned of efforts to find new specimens of *Ramapithecus* or other early hominoid fossils in Pakistan. During the late 1970s Brunet and Emile Heintz, a paleontologist at the Centre National de la Recherche Scientifique (National Center of Scientific Research) conducted excavations in Afghanistan and Iraq searching for fossil apes, but they had to abandon their search in 1980 due to the dangerous political situation in both countries. Brunet then turned his attention to Africa, but rather than focusing in the eastern or southern regions of Africa where fossil hominins had already been found Brunet decided to explore western Africa for ape and hominin fossils. Brunet, along with David Pilbeam, began explorations in Cameroon in 1984 and although research continued there throughout the late 1980s no fossil apes were found. Brunet was also hopeful about finding fossils in Chad, especially since French paleontologist Yves **Coppens** had found fossil mammals there during the 1960s. In 1993 Brunet received a research permit from the government of Chad to conduct excavations in the **Chad Basin** (also known as the Djurab Desert) in what had once been the bed of Lake Chad. He established the **Mission Paléoanthropologique Franco-Tchadienne** (French-Chadian Paleoanthropological Mission, or MPFT), a scientific collaboration between the University of N'Djamena, the University of Poitiers, and the Centre National d'Appui à la Recherche, whose purpose was to research the origin and evolution of early hominins and their environment in western-central Africa. In 1994 Brunet and his team began exploring the sediments of the Lake Chad basin for fossils and in the following year Brunet found a hominin mandibular fragment at a site called **Koro Toro** that he biochronologically dated at 3.5 Ma. He considered the specimen to be a new species of *Australopithecus* that he named *Australopithecus bahrelghazali* (nicknamed "Abel" in honor to the memory of Abel Brillanceau, a colleague and close friend who died while working in the field). This fossil was important evidence that hominins were also present in central Africa, but even more intriguing was the fact that in 1997 Brunet's team was discovering mammal fossils from the late **Miocene** deposits at **Toros-Menalla** that were biochronologically dated to between 6 and 7 Ma. Then in 2001 Ahounta Djimdoumalbaye, a Chadian student from the University of N'Djamena working on Brunet's team, found a largely complete but crushed and distorted hominin cranium from the *c.*6–7 Ma deposits. The fossil, named "Toumaï" (meaning "hope of life" in the Goran language) by the Chadian authorities, was assigned by Brunet and his colleagues to a new genus and species, *Sahelanthropus tchadensis*. Because it possesses what are interpreted to be derived hominin characters Brunet considers this species to represent the stem hominin, and it would be the earliest hominin fossil yet found. More recently Brunet has been conducting field surveys to look for fossil mammals and primates in Libya, and since 2004 he has served as the director of the **Mission Paléontologique Franco-Libyenne** (Franco-Libyan Paleontological Mission, or MPFL), a scientific collaboration between the University of Poitiers and the Al Fateh University of Tripoli. Brunet was appointed a professor in the Faculté des Sciences Fondamentales et Appliquées at the University of Poitiers in 1989, and from 1992 he was also the head of the Laboratoire de Géobiologie, Biochronologie, et Paléontologie Humaine (Laboratory of Geobiology, Biostratigraphy, and Human Paleontology), which was integrated into the Centre National de la Recherche

Scientifique in 2000 (the laboratory has since been renamed the Institut International de Paléoprimatologie et Paléontologie Humaine, Evolution et Paléoenvironnements, CNRS UMR 6046). He is also the director of the Mission Paléoanthropologique Franco-Tchadienne. In 2007 Brunet was appointed a Professor in the College de France.

Bruhnes *See* **geomagnetic polarity time scale**.

Brünn *See* **Brno**.

Brussels Natural History Museum *See* **Royal Museum for Central Africa**.

BSC Acronym for the **biological species concept** (*which see*).

BSN49/P27 Site **Gona**. Locality BSN49. Surface/*in situ* Surface. Date of discovery February 12, 2001. Finders Ali Ma'anda Dato and team. Unit N/A. Horizon Maoleem Vitric Tuff. Bed/member Hatayae Member. Formation **Busidima Formation**. Group N/A. Nearest overlying dated horizon Silbo Tuff (*c.*0.751±0.022 Ma). Nearest underlying dated horizon Base of the C1r polarity chron (*c.*1.778 Ma). Geological age Given the measured sedimentation rates in the Busidima Formation, the "likely age" is "0.9–1.4 Ma" (Simpson et al. 2008, p. 1089). Developmental age Adult. Presumed sex Female. Brief anatomical description Most of the right side and a substantial part of the left side of a pelvis, plus the associated sacrum and the last lumbar vertebra. The pelvic brim is preserved except for the medial end of the right pubic ramus. Announcement Simpson et al. 2008. Initial description Simpson et al. 2008. Photographs/line drawings and metrical data Simpson et al. 2008. Detailed anatomical description N/A. Initial taxonomic allocation *Homo erectus*. Taxonomic revisions N/A. Current conventional taxonomic allocation *H. erectus*. Informal taxonomic category Pre-modern *Homo*. Significance Exceptionally well preserved pelvis that provides evidence about the size of the birth canal of a "short-statured *H. erectus* adult female" (Simpson et al. 2008, p. 1090), but Ruff (2010) suggests that the size of the pelvis as assessed from the diameter of the acetabulum is more consistent with it belonging to an archaic hominin, possibly *Paranthropus boisei*. Location of original National Museum of Ethiopia, Addis Ababa, Ethiopia.

Bubing Basin (Location approximately 23°25′N, 107°00′E, Guangxi Zhuang Autonomous Region, China) History and general description Bubing is a small basin located to the southwest of the larger **Bose Basin**. Bubing Basin is 16 km/9.9 miles long by 2 km/1.2 miles wide and runs in a northwest to southeast direction. Bubing comprises Paleozoic limestone hills that uplifted at different stages during the **Neogene** and **Quaternary** periods. More than 50 caves have been identified in Bubing, between 140 and 215 m above sea level; the older caves are situated at higher elevations and the younger caves are located along the floor of the basin. The best-known caves in the basin are Chuifeng, Cunkong, Dingmo, Lower Pubu, Luna, Mohui, Upper Pubu, Wuyun, and Zhongshan. Evaluation of the vertebrate paleontological fossils from Bubing has resulted in a greater appreciation of variation in the *Ailuropoda-Stegodon* **fauna**, a category that has often been considered representative of the southeast Asia, including southern China. For example, *Stegodon* and *Ailuropoda* are absent from Cunkong Cave, which dates to the Holocene, but in the Late Pleistocene Lower Pubu Cave *Ailuropoda* is present, but *Stegodon* is absent. *Elephas maximus* is present in Lower Pubu and *Equus* was also identified in Chuifeng and Lower Pubu. Some link these differences with the amounts of open grassland in the region. Temporal span and how dated? **Biostratigraphy, uranium-series dating, electron spin resonance spectroscopy dating**, and **AMS radiocarbon dating** suggest that the caves in the Bubing Basin date from the Early Pleistocene to the Holocene. Biostratigraphic evidence places the caves in the following chronological order: Chuifeng and Mohui *c.*2 Ma, Wuyun and Upper Pubu *c.*280–76 ka, Luna Cave to 70–20 ka, and Zhongshan to 12–6 ka. Hominins found at site Two hominin teeth excavated from Mohui have been allocated to *Homo* cf. *erectus*; two anatomically modern human teeth were excavated from Luna Cave along with an anatomically modern human partial skull that appears to be pathological. Many nonhominin primates (e.g., *Presbytis*, *Macaca*, *Pongo*) have been recovered from the caves [e.g., Chuifeng yielded the largest collection of *Gigantopithecus* teeth (*n*=92) found in one locality except for the type site Liucheng (where more than 1000 teeth and three partial mandibles were found)]. Archeological evidence found at site Surface collections of Paleolithic artifacts plus excavated evidence from cave and open-air sites (e.g., eight core and flake tools were excavated from Mohui Cave, but six of these were excavated from disturbed deposits and

the relationship between the hominin teeth and the remaining two artifacts is not clear). Dingmo Cave has evidence of rock art, although the age of the art needs to be ascertained. Surface survey of the area in front of Dingmo resulted in the recovery of Paleolithic-like stone flakes. A systematic study of the archeological residues surface collected and excavated in Bubing is under way. Repository Natural History Museum of Guangxi Zhuang Autonomous Region, China; Tiandong Museum, Guangxi, China. Key references: Geology, dating, and paleoenvironment Chen et al. 2001, Wang et al. 2005, 2007a, 2007b, Rink et al. 2008, Wang 2009, Norton et al. 2010b; Hominins Wang et al. 2005, 2007a, 2007b, Wang 2009, Norton et al. 2010b; Archeology Wang et al. 2005.

buccal (L. *bucca*=cheek) Refers to the cheek. It is used to describe the outer, or lateral, aspect of the premolars and molars. The equivalent term for the canine and the incisors is labial, because the lateral aspect of these teeth faces the lips, not the cheek. The measurement of the breadth of a premolar or molar tooth that is taken at right angles to its long, or mesiodistal, axis is referred to as its buccolingual breadth (i.e., the measurement runs from the outer, buccal, aspect of the tooth crown, to the inner, tongue-side, or **lingual** aspect of the crown).

buccostyle A term used by Tobias (1967a) for an enamel feature more commonly referred to as a **paramolar tubercle** (*which see*).

Buia (Location approximately 14°49′N, 39°50′E, Eritrea; etym. unknown) History and general description In 1994–5 geological and paleontological exploration of the Danakil Depression by an Italian team headed by Ernesto Abbate led to the recognition of the Dandiero Group, an about 1000 m thick succession, previously attributed to the Danakil Formation or the Red Series. This group includes three main unconformity-bounded units (Abbate et al. 2004): the Maebele Synthem, the Curbelu Synthem, and the Samoti Synthem. The Maebele Synthem is divided into six formations that sample **fluvial, deltaic**, and **lacustrine** environments with late Early to **Middle Pleistocene** faunas. In the lower portion of the Maebele Synthem, the Alat Formation yielded six hominin fossils preliminarily attributed to *Homo erectus*. Temporal span The sedimentary succession covers from *c.*1.3 to 0.6 Ma. How dated? **Fission track dating** on volcanic glasses from an ash layer approximately 80 m below the *Homo*-bearing sediments (*c.*1.3±0.3 Ma). Rocks of normal polarity correlated with the Jaramillo subchron (C1r.1n) and associated late Early to Middle Pleistocene mammal fauna. Hominins found at site A nearly complete cranium (**UA 31**), two permanent upper incisors (UA 222, an upper left I², and UA 369, a lower left I¹), two conjoined pelvic fragments, a right iliac blade (UA 173 *s.s.*), and a right acetabulum and partial ischium (previously reported as UA 405), forming an incomplete adult hip bone (UA 173) and a pubic symphysis (**UA 466**). Archeological evidence found at site **Acheulean handaxe**s, cleavers, choppers, and flakes. Key references: Geology and dating Abbate et al. 1998, Albianelli and Napoleone 2004, Bigazzi et al. 2004, Martínez-Navarro et al. 2004, Ghinassi et al. 2009; Hominins Abbate et al. 1998, Macchiarelli et al. 2004a, Bondioli et al. 2006; Archeology Abbate et al. 1998, Fiore et al. 2004, Martini et al. 2004.

Buia cranium   *See* **UA 31**.

Buia pubis   *See* **UA 466**.

Bukuran   (Location 7°20′S, 110°58′E, Indonesia; etym. named after a local village) History and general description Within the **Sangiran Dome** and less than 1 km/0.6 miles west of the village of **Sendangbusik**, this is the site of a 1989 surface find of 16 hominin cranial fragments of a single individual named **Hanoman 1**. These specimens were derived from upper levels of the **Sangiran Formation** (NB: there is no **Grenzbank zone** between the older Sangiran and younger **Bapang Formation**s at Bukuran). Temporal span and how dated? The most recent radiometric estimates of the age of nearby Upper Sangiran Formation deposits cluster around 1.5 Ma. Hominins found at site Hanoman 1 includes 16 cranial vault fragments including portions of the frontal, parietal, and occipital bones. This specimen has been attributed to *Homo erectus*. Archeological evidence found at site Two heavily rolled siliceous flakes from the overlying Bapang Formation. Key references: Geology and dating Widianto et al. 1994, Larick et al. 2001; Hominins Widianto et al. 1994.

bulk strategy   *See* **economic utility index**.

Bulletin of Zoological Nomenclature   The name of the journal that carries the opinions of the International Commission on Zoological Nomenclature (or ICZN; *which see*).

bunodont   (Gk *buno*=hill and *dont*=teeth) Teeth with low, conical crowns, as opposed to hypsodont (i.e., high crowned) teeth. **Hominin** teeth are relatively bunodont, and among hominins the teeth of *Paranthropus boisei* are the most bunodont, for they have a large crown base and curved buccal and lingual surfaces.

Bura Hasuma   One of the designated subregions of Koobi Fora. It includes localities where some of the oldest sediments in the Koobi Fora and the Kubi Algi Formations are exposed. *See also* **Koobi Fora**.

Burg Wartenstein International Symposium Program   In 1958 The Wenner-Gren Foundation for Anthropological Research acquired Burg Wartenstein castle in Austria as a gift from Axel Wenner-Gren. Between 1959 and 1980 Burg Wartenstein was the Foundation's European headquarters and the primary venue for its International Symposium Program. International Symposia were (and still are) meetings of intensive discussion that aim to include international scholars with broad interests. During the 1960s and 1970s the Foundation hosted almost 2000 scholars at 86 symposia held at the castle and elsewhere during the summer months. The titles of landmark meetings that focused on paleoanthropology are given below with the year in which they were held (the publication date and the organizer(s) are given in parentheses): "Social Life of Early Man," 1959 (1961, Washburn), "Early Man and Pleistocene Stratigraphy in the Circum-Mediterranean Regions," 1960 (1962, Blanc and Howell), "African Ecology and Human Evolution," 1961 (1963, Howell and Bourlière), "Classification and Human Evolution," 1962 (1963, Washburn), "Background to Evolution in Africa," 1965 (1967, Bishop and Clark), "Man the Hunter," 1966 (1968 Lee and DeVore, held in Chicago), "Calibration of Hominoid Evolution: Recent Advances in Isotopic and Other Dating Methods as Applicable to the Origin of Man," 1971 (1972, Bishop and Miller), "Earliest Man and Environments in the Lake Rudolf Basin," 1973 (1976, Coppens, Howell, Isaac, and Richard Leakey, held in Nairobi and Lake Rudolf, Kenya), "After the Australopithecines," 1973

(1975, Butzer and Isaac), and "Early Hominids of Africa," 1974 (1978, Jolly, held at The Wenner-Gren Foundation, New York). The results of these meetings were published through the Foundation's Viking Fund Publications in Anthropology, as well as through special arrangement with publishing houses (such as Aldine Publishers) and various university presses. *See also* **Wenner-Gren Foundation for Anthropological Research.**

burin (Fr. *burin*=a term used by engravers and printmakers for an elongated metal chisel) This term was borrowed by archeologists to describe stone tools made on **flakes** that are particularly common at some European **Upper Paleolithic** sites. Burins have a distinctive chisel-like edge or tip and were originally considered to function as tools for engraving, bone, antler, wood, or similar media (e.g., de Sonneville-Bordes and Perrot 1954–6, Debénath and Dibble 1994). Burins are "flaked flakes," in which the relatively thick chisel-like edge is made by removing an elongated flake (called a burin spall) along one or more edges of a flake. Some suggest burins may have served a wider range of functions than originally conceived (see Barton et al. 1996). This includes their use as **cores**, and under this interpretation, it is the thin elongated flake (the burin spall) rather than the burin that was used as a tool.

burin spall *See* **burin.**

bush pig The vernacular name for the taxa within the genus *Potamochoerus*. *See also* **Suidae.**

bushland Biomes that are dryer than **woodland** forests, receiving 250–600 mm mean annual rainfall. Due to lower levels of rainfall and poorer-quality soils in terms of nutrients and drainage, bushlands are dominated by small trees between 3 and 9 m in height with multiple stems. By definition 40% or more of bushland is covered by bushes. Most bushlands have some degree of grass cover. More open habitats, such as bushland and shrubland biomes, have been suggested to have been the major biomes for *Paranthropus* and early *Homo* between 2.5 and 2.0 Ma.

Bushmen *See* **San.**

Busidima Formation [Location <10°55′–11°15′N, 40°20′–40°35′E, Ethiopia; etym. Busidima is the local Afar (Danakil) name for the ephemeral stream central to outcrops of the Busidima Formation] History and general description A wedge-shaped basin (half-graben) of sediments adjacent to the foothills of the western escarpment of the Ethiopian Rift near the Gona Paleoanthropological study area, the **Dikika study area,** and **Hadar.** The strata from which archeological evidence has been recovered at Gona were originally assigned to the upper Kada Hadar Member of the **Hadar Formation** (Semaw et al. 1997), but more recent stratigraphic work has recognized a significant angular unconformity between the Hadar Formation and the overlying sediments. This required redefining the Busidima Formation as those sediments that overlie the unconformity and fill the half-graben structure up to a prominent paleosol surface, the Halalalee Bed (Quade et al. 2004, Quade and Wynn 2008). The Busidima Formation is not further subdivided, but contains several prominent tuffs exposed throughout the region that can be used as stratigraphic markers. Temporal span and how dated? The lowest dated tuff in the Busidima Formation is an unnamed tuff dated to 2.69 Ma and the youngest is a correlate to the Waidedo Vitric Tuff, dated to 0.16 Ma. The sediments have been dated using the **argon-argon dating** method, **magnetostratigraphy, tephrostratigraphy,** and **biostratigraphy.** Hominins found at site Fossil hominins recovered from the Busidima Formation have been assigned to early *Homo* and *Homo erectus.* Archeological evidence found at site The Busidima Formation includes evidence of **Oldowan, Acheulean,** and **Middle Stone Age** sites. Key references: Geology, dating, and paleoenvironment Quade et al. 2004, Quade and Wynn 2008; Hominins Kimbel et al. 1996, Simpson et al. 2008; Archeology Corvinus and Roche 1976, Kimbel et al. 1996, Semaw et al. 1997. *See also* **Gona Paleoanthropological study area; Hadar Formation.**

Buxton Limeworks *See* **Taung.**

C

c. Abbreviation of circa (L. *circum*=around) meaning 'about' or 'approximately' the true value.

$^{13}C/^{12}C$ A ratio of **stable isotopes** in bone collagen used for **diet reconstruction** and to reconstruct the **paleoenvironment**. The system works on the basis that the levels of ^{13}C in the bone collagen and bioapatite and enamel bioapatite of herbivores reflect the types of plants they consumed (e.g., C_3, C_4, or **CAM**) and the levels of ^{13}C in the bones and enamel of carnivores reflects the diets of the herbivores preyed on by those carnivores. For example, animals that eat plants, such as most shrubs, trees, herbs, and cool-season grasses, that use the C_3 photosynthetic pathway should have carbon isotope ratios that are depleted in $\delta^{13}C$ (i.e., have less $\delta^{13}C$) relative to animals that consume plants such as succulents, sedges, and warm-season grasses that use the C_4 and CAM photosynthetic pathways. Carbon isotope ratios can indicate whether or not a particular environment was open (e.g., grassland or wooded savannah) or closed (e.g., forest) (van der Merwe and Medina 1989). Within closed environments, carbon isotope ratios can indicate if an animal was consuming foods close to the ground or up in the canopy due to the **canopy effect** (Cerling et al. 1999). Carbon isotope ratios have also been found to be correlated with temperature, and negatively correlated with relative humidity, rainfall, and elevation.

C_3 and C_4 Two groups of plants, C_3 **plants** and C_4 **plants**, show minimal overlap in $^{13}C/^{12}C$ ratios (commonly represented as $\delta^{13}C$ values parts per thousand, as in the ‰ notation). The C_3 plant group, which consists of trees, herbaceous plants, and cool-season grasses and constitutes the vast majority of plants, uses the C_3 photosynthetic system and has lower $\delta^{13}C$ values (global mean -27‰). The C_4 plant group, which includes warm-season grasses such as the vast majority of native grasslands in the Great Plains of the USA and many tropical grass species such as maize, uses the C_4 photosynthetic system and has higher $\delta^{13}C$ values (global mean approximately -13‰). A third photosynthetic pathway, CAM, is used by **CAM plants**, which includes cacti and other succulents. Depending on environmental conditions, CAM plants exhibit $\delta^{13}C$ values ranging from those typical of C_3 to those typical of C_4 plants. Fauna eating either C_3 and C_4 plants or some mixture of the two record those diet sources in their tissues (DeNiro and Epstein 1978). Dietary isotope ratios are passed on to consumers with some **enrichment**. For example, carbon isotope ratios in herbivore keratin tend to be about $3–4$‰ higher than those in their diets (see data in Kellner and Schoeninger 2007). **Marine** animals have $\delta^{13}C$ values in their flesh that are higher than most C_3 plants and they can overlap the values for C_4 plants (Schoeninger and DeNiro 1984). Therefore, in regions where modern humans eat maize *and* have access to marine foods, bone collagen $\delta^{13}C$ values are not useful in reconstructing the amount of marine food in diet (Schoeninger et al. 1990). *See also* $^{13}C/^{12}C$; **diet reconstruction**.

C_3 and C_4 proxies Several proxies from the geological record can be used to reconstruct the relative proportion of C_3 **plants** and C_4 **plants** in a region. These include the $^{13}C/^{12}C$ isotope ratio of soil carbonates, **flowstones**, plant leaf wax biomarkers, and indices of **phytolith** types.

C_3 foods These are either plants that use the C_3 photosynthetic pathway, such as trees, herbaceous plants, and cool-season grasses, or animals that browse or graze on C_3 **plants**. *See also* C_3 **and C_4**; C_3 **plants**.

C_3 plants Plants that use the C_3 photosynthetic pathway such as most trees, herbaceous plants, cool-season grasses, and the vast majority of other plants that have relatively low $\delta^{13}C$ values. *See also* C_3 **and C_4**.

Wiley-Blackwell Encyclopedia of Human Evolution, First Edition. Edited by Bernard Wood.
© 2013 Blackwell Publishing Ltd. Published 2013 by Blackwell Publishing Ltd.

C₁ foods These are either plants that use the C₄ photosynthetic pathway such as many tropical grass species (e.g., maize), or animals that feed or graze on **C₄ plants**. *See also* **C₃ and C₄; C₄ plants**.

C₄ plants Plants that use the C₄ photosynthetic pathway such as warm-season grass species and many tropical grass species (e.g., maize) that have relatively high $\delta^{13}C$ values. *See also* **C₃ and C₄**.

C5 A term commonly employed for an accessory cusp on the crown of a maxillary molar. The term has also occasionally been used for the mandibular molars where it refers to the fifth main cusp (i.e., the **hypoconulid**); syn. **distal accessory tubercle** (*which see*).

C6 A term introduced by Hellman (1928) for an accessory cusp on the crown of a mandibular molar that others call a **tuberculum sextum** (*which see*); syn. tuberculum sextum.

C7 A term introduced by Hellman (1928) for an accessory cusp on the crown of a mandibular molar originally described as the **tuberculum intermedium** (*which see*).

Cabezo Gordo *See* **Sima de las Palomas**.

CAC Abbreviation of the **Central Awash Complex** (of sites), Western Margin, **Middle Awash study area**, Ethiopia.

caenotelic (Gk *kainos*=recent and *telikos*=end, so "towards a purpose") *See* **palimpsest evolution**.

calcarine sulcus (L. *calx*=heel or spur and *sulcus*=fissure) The calcarine sulcus is the most conspicuous sulcus on the medial aspect of the occipital lobe of the **cerebral hemisphere** in primates. The **primary visual cortex** lies largely within this sulcus. The representation of the central, high-acuity, part of the visual field is in the posterior portion of the calcarine sulcus and the representation of the peripheral visual field is located in the more anterior portion of the sulcus.

calcification The process of calcium deposition that occurs during the development of hard tissues such as teeth and bones. Typically during this process an organic framework is replaced by calcium (and other associated minerals); this

part of the process is referred to as **mineralization**. During enamel development the calcium pumped into a developing tooth is incorporated into the crystalline structure of the **enamel prisms** by displacing the enamel matrix proteins.

calcite *See* **calcium carbonate**.

calcium carbonate Calcium carbonate ($CaCO_3$) is a solid abundant natural mineral and is a component of many sedimentary rocks but is particularly abundant in chalk and limestones. Calcium carbonate has two polymorphs aragonite and its more stable form calcite. Water seeping through cave bedrock carries dissolved calcium carbonate. When this solution reaches an air-filled cave, the de-gassing of carbon dioxide alters the water's ability to hold minerals in solution, causing its solutes to precipitate and accumulate to form **speleothem**s, of which **flowstone, stalactites**, and **stalagmites** are all examples. Stalactites are icicle-shaped mineral deposits which grown *down* from cave roofs, whereas stalagmites are conical mineral deposits fed from drip waters above that grow *up* from cave floors. Flowstones, which are formed by water flowing on the floor or walls of a cave, are generally horizontal, finely laminated layers of calcium carbonate that may found between breccia deposits in caves, and some of the fossil hominins in the southern African cave sites are preserved in flowstones (e.g., **Stw 573**). Stalagmites and flowstone can be dated with **uranium-series dating** and they are also important sources of paleoenvironmental information. The terms travertine and **tufa** are commonly used in older literature to describe finely crystalline, usually white, calcium carbonate formed by the precipitation of carbonate minerals from geothermally heated hot springs and from ambient-temperature water bodies such as lakes, springs, and rivers, respectively, but today speleothem or flowstone are the preferred terms.

calibrated-age methods *See* **biostratigraphy; geochronology**.

calibrated radiocarbon age *See* **radiocarbon dating**.

calibration (Gk *kalapous*=shoemaker's last) An adjustment made to a measurement by comparing it to a more reliable standard. For example, this is the process used to correct radiocarbon dates to allow for the changes that have occurred over time to the levels of radiocarbon in the atmosphere. *See also* **radiocarbon dating**.

calotte (Fr. *calota*=skullcap worn by a Roman Catholic priest or by an observant member of the Jewish faith) So called because it is the part of the **cranium** covered by a skullcap. The unpaired bones making up the calotte are the frontal and the squamous part of the **occipital**; the paired bones are the parietals and the squamous parts of the temporal bones. The calotte is one of the three parts of the cranium; the others are the **face** and the **cranial base**.

calvaria (L. *calvus*=bald) Literally, the part of the **cranium** that is exposed when a person loses their hair (NB: it has to be long hair). It is the cranium minus the **face**. The word calvaria is Latin feminine singular; the proper plural is **calvariae**. Note that the singular term calvarium (plural calvaria) is incorrect, but this has not prevented its widespread use. *See also* **cranium**.

calvariae (L. *calvus*=bald) The Latin feminine plural version of calvaria. *See also* **calvaria**; **cranium**.

calvarium (L. *calvus*=bald) This term is a neologism and should not be used. The correct term is calvaria. *See also* **calvaria**; **cranium**.

CAM Acronym for the crassulacean acid metabolism photosynthetic pathway. *See also* **C₃ and C₄**; **CAM plants**.

CAM plants Plants, such as cacti and other succulents, that use the crassulacean acid metabolism (or CAM) photosynthetic pathway. Depending on environmental conditions such as water availability, CAM plants exhibit $\delta^{13}C$ values ranging from those in C_3 plants to those in C_4 plants. Whereas CAM plants are a small fraction of global vegetation they may be locally important components of **flora** in some tropical seasonal forests and drylands, such as southern Africa. *See also* **C₃ and C₄**.

camelid The informal name for the artiodactyl family, **Camelidae**, comprising the camels and their allies, whose members are found at some hominin sites. *See also* **Artiodactyla**.

Camelidae One of the artiodactyl families whose members are found at some hominin sites. *See also* **Artiodactyla**.

canalization (L. *canalis*=conduit, hence "to channel") A term introduced by Conrad H. Waddington (1942) to describe the tendency for **development** to produce a similar **phenotype** despite different genetic and environmental influences. Waddington developed the concept of canalization to explain the results of his **genetic assimilation** experiment. In this experiment, Waddington exposed *Drosophila* to ether, which caused a proportion of individuals to develop the bithorax phenotype. For several generations, he selected for the bithorax phenotype in the presence of ether and eventually produced populations in which the bithorax phenotype appeared *even in the absence of the exposure to ether*. To explain this, he argued that selection for the bithorax phenotype had stabilized this state by acting on modifier **genes** to the extent that the phenotype appeared without the original stimulus; he referred to this process of stabilization as "canalization". Gunter Wagner et al. (1997) have formalized the population genetic definition of canalization. They distinguish between environmental and genetic canalization. Environmental canalization is the reduction in the genotype-specific environmental variance, whereas genetic canalization is the reduction in the average effect of mutations, or genetic variants.

cancellous bone (L. *cancellus*=lattice; ME *bon*=bone) (syn. trabecular bone). *See* **bone**.

candelabra model The term candelabra was introduced by William "Bill" **Howells** in *Mankind in the Making* (1959). He used it to describe the polycentric theory of German physical anthropologist Franz **Weidenreich**, which he suggested was representative of what he referred to as the "polyphyletic or Candelabra School" of thought. In a diagram portraying Weidenreich's theory, Howells indicated that modern Australasians, Mongoloids, Africans, and Eurasians had evolved from populations represented by *Pithecanthropus*, *Sinanthropus*, Rhodesian Man, and **Cro-Magnon** respectively. However, although Weidenreich (1946) had argued that a kind of racial continuity existed in geographically separated populations of hominins as they evolved over time into modern humans, he also made it clear that his model assumed a certain degree of interbreeding had occurred between these populations and that this genetic exchange ensured that these geographical "races" belonged to one species. In this sense, then, Howells misrepresented Weidenreich's views when he used a candelabra metaphor, although other polyphyletic theories of human evolution proposed in the late nineteenth and early twentieth centuries were strictly consistent with Howells' metaphor. For example, the candelabra metaphor was entirely appropriate for the views of Carleton **Coon**. In his *The Origin of Races* (1962) Coon had assembled fossil, ethnographic, and linguistic evidence to support the view that the five major modern human races he identified (Caucasoid, Negroid, Capoid, Mongoloid, and Australoid) had existed for more than half a million years. Coon argued that modern human races had evolved separately from geographically separate subspecies of *Homo erectus*. Coon also believed that the races of modern humans had evolved at different rates, which explained why they had achieved such different levels of cultural development. For Coon, therefore, modern humans are a **polytypic** species consisting of separate geographical populations that differ from one another, and if the geographical barriers separating these populations are large enough and the populations remain isolated for long enough periods then they can actually become distinct species. Given the political and social conditions of the 1960s, Coon's conception of modern human races generated considerable controversy and Coon himself complained that some of his ideas were misconstrued by others. Coon's extreme manifestation of the candelabra model of human evolution garnered few supporters, but since the 1990s it has been more common for paleoanthropologists to use the candelabra analogy (even though it may be inappropriate) to describe Weidenreich's polycentric theory and Milford **Wolpoff** and Alan **Thorne's** **multiregional hypothesis** (Thorne and Wolpoff 1981, Wolpoff et al. 1984, Wolpoff 1989).

canid The informal name for the **Canidae** (dogs), one of the **caniform** families whose members are found at some hominin sites. *See also* **Carnivora**.

Canidae The family that includes wild dogs and their close relatives, and it is one of the **caniform** families whose members are found at some hominin sites. *See also* **Carnivora**.

caniform Informal term for taxa within the suborder Caniformia, one of the two suborders in the order Carnivora. *See also* **Carnivora**.

Caniformia One of the two suborders in the order Carnivora. Common caniform families at hominin sites include the **Ursidae** (bears), **Mustelidae** (weasels and otters), and the **Canidae** (dogs). *See also* **Carnivora**.

canine buttress (L. *canis*=dog) *See* **anterior buttress**.

canine dimorphism *See* **sexual dimorphism**.

canine eminence (L. *canis*=dog) Rounded area of bone that covers the anterior, or facial, aspect of the root of the canine; syn. canine jugum.

canine fossa Hollowed area posterior to the **canine eminence** that covers the anterior, or facial, aspect of the root of the upper canine tooth. Its superior and anterior boundary is the **transverse buttress** that runs inferiorly from the infraorbital foramen to blend with the canine eminence.

canine jugum Rounded area of bone covering the anterior, or facial, aspect of the root of the canine; syn. canine eminence.

Cann, Rebecca L. (1951–) Rebecca Cann attended the University of California at Berkeley where she completed her PhD in anthropology in 1982. While at Berkeley Cann worked in the laboratory of Australian biochemist Allan **Wilson**. During the 1970s Wes Brown had conducted research on **mitochondrial DNA** (or mtDNA) taken from mothers of different ethnic groups and he had concluded that the surprisingly small amount of variation in their DNA, caused by the accumulation of mutations over time, indicated that modern human populations arose recently from a relatively small aboriginal population. Cann collaborated with Wilson and Mark Stoneking to collect mtDNA from 147 women from Europe, Asia, and the South Pacific, and because there were difficulties in obtaining samples from Africans, they obtained samples from African-American women instead. Cann and Stoneking used a computer program to analyze the variations in the mtDNA samples to construct a phylogenetic tree by grouping together samples with the fewest differences. The resulting tree had two main branches, one containing only African–American populations and the other containing populations from every part of the world. Since the two branches diverged close to the base of the tree Cann and her colleagues concluded that all the mtDNA in living human populations descended from a woman who had lived in Africa and Cann, Stoneking, and Wilson estimated that all modern humans were descended from a woman who lived in Africa about 200 ka (Cann et al. 1987). This hypothetical female common ancestor was soon given the nickname "**Mitochondrial Eve**" and Cann and colleagues' research generated considerable interest, but for methodological reasons it was also controversial. *See also* **mitochondrial DNA**.

canonical (L. *canon*=rule) In relation to **multivariate analysis** it refers to a standard or criterion. Canonical analysis is a form of multivariate analysis that captures the **variation** subsumed in a large number of variables and summarizes it in a smaller number of canonical axes. The axes are orientated at right angles to each other (this relationship is also called **orthogonal**), and the amount of variation represented in each axis can be inferred from the sum of the variable loadings on each axis. *See also* **canonical analysis; canonical axis; multivariate analysis**.

canonical analysis (L. *canon*=rule) A form of **multivariate analysis** also known as canonical variates analysis and canonical correlation analysis. Canonical analysis divides a data set into two groups of variables and finds a set of functions that maximizes the correlation between groups. For example, a variety of cranial and postcranial measurements may be collected for all specimens in a skeletal collection of modern humans. Canonical analysis can be used to find functions that generate new pairs of variables, or canonical variates, which maximize the correlation between cranial and postcranial variables. Similar to **principal components analysis**, the first pair of canonical variates account for the highest possible correlation between the two groups of variables, the second pair account for the highest possible correlation in the remaining variation in the two groups, and so on. The loadings of the original variables in the canonical variates can then be used to identify the relative contribution of

variables in the two groups to the overall correlation. For example, out of 10 cranial variables and 10 postcranial variables, eye orbit height and eye orbit width may contribute the most to the cranial canonical correlate, which is highly correlated with a postcranial canonical correlate mainly influenced by humeral head diameter, femoral head diameter, and tibial plateau width. In this way, canonical analysis can be thought of as a form of **multiple regression** in which there is more than one dependent variable.

canonical axis (L. *canon*=rule) Multivariate analysis summarizes the **variation** subsumed in a large number of variables in a smaller number of canonical axes that are orientated at right angles to each other. The amount of variation represented in each axis can be inferred from the sum of the loadings on each axis; the greater the loading associated with a variable, the greater the influence that variable has on a particular canonical axis. *See also* **canonical analysis**.

canonical correlation analysis *See* **canonical analysis**.

canonical variates analysis *See* **canonical analysis**.

canopy effect The canopy effect occurs when some of the CO_2 released during respiration by soil microbes is trapped beneath the forest canopy and is fixed by understorey plants (Medina and Minchin 1980, van der Merwe and Medina 1989). Because this CO_2 results from the breakdown of plant material, it has $\delta^{13}C$ values that are similar to the C_3 plants from whence it came (e.g., approximately $-27‰$) rather than atmospheric CO_2 (approximately $-8‰$). This results in lower $\delta^{13}C$ values in understorey leaves. Carbon isotope ratios in leaves and other vegetative plant parts tend to be more negative (^{13}C-depleted) than those in reproductive parts such as fruits and flowers (Cernusak et al. 2009).

Carabelli('s) anomaly *See* Carabelli('s) trait.

Carabelli('s) complex *See* Carabelli('s) trait.

Carabelli('s) cusp *See* Carabelli('s) trait.

Carabelli('s) structure *See* Carabelli('s) trait.

Carabelli('s) trait The term refers to an enamel feature on the mesiolingual face of a maxillary molar tooth crown between the tip of the **protocone** and the **cemento-enamel junction**. It ranges in expression from a pit to a well-developed accessory cusp. Its name derives from Georg Carabelli (1842) who provided the first description of the character, although he actually used a different term (**tuberculum anomalus**) for it [syn. Carabelli('s) anomaly, Carabelli('s) complex, Carabelli('s) cusp, Carabelli('s) structure, Carabelli('s) tubercle, entostyle, tuberculum anomalus, protostyle, pericone].

Carabelli('s) tubercle *See* **Carabelli('s) trait**.

carbon An element commonly used for **absolute dating** and **stable isotope biogeochemistry**. There are three naturally occuring isotopes of carbon: ^{14}C, ^{13}C, and ^{12}C. ^{13}C and ^{12}C are stable and do not decay; they are used in stable isotope biogeochemistry. The carbon isotope ^{14}C is not stable and decays to ^{14}N at a regular rate with a **half-life** of 5730 years. *See also* $^{13}C/^{12}C$; **radiocarbon dating**; **stable isotope**.

carbon dating *See* **radiocarbon dating**.

carbon isotope analysis *See* **isotope ratio mass spectrometry**.

carcass acquisition strategy Refers to the method used to acquire animal remains (e.g., **hunting, confrontational scavenging, passive scavenging**). In hominin evolution, identifying carcass acquisition strategies helps characterize the ecological niche and behavioral capabilities of our hominin ancestors. Hunting refers to the active pursuit and capture of animal prey and is generally understood to be a basic constituent of modern human behavior. Confrontational scavenging, also known as "power" or "aggressive" scavenging, is defined as the capture of a nearly complete animal carcass from nonhominin predators that are actively consuming it. Modern hunter-gatherers, such as the **Hadza** of northern Tanzania, have been observed to occasionally obtain meat through confrontational scavenging. Passive scavenging includes the culling of small amounts of meat or marrow from heavily ravaged animal carcasses abandoned by their initial predators. Ethnographic studies document opportunistic passive scavenging by some hunter-gather groups (e.g., the

Hadza), but no modern human populations have a carcass acquisition strategy characterized by obligate passive scavenging of terrestrial mammals. The emergence of hunting in the hominin behavioral record is a particularly controversial issue (i.e., the hunting-scavenging debate), much of which has focused on the **FLK** site in **Olduvai Gorge**. The timing of the appearance of hunting as a carcass acquisition strategy is particularly important because hunting falls within the realm of modern human behavior and implies the presence of relatively complex social interactions and cooperation within hominin groups.

carcass transport strategy A term that refers to decisions made by foragers when transporting animal remains from the point of carcass acquisition (e.g., the kill site) to the point of consumption. Ethnographic observations of **hunter–gatherers** in sub-Saharan Africa (e.g., the **Hadza**, the Kua [San], the !Kung), the Arctic (e.g., the Nunamiut), and elsewhere indicate that carcass transport strategies are influenced by constraints including the number and size of carcasses to be transported, transport distance, transport mode, the number of carriers, and even the time of day and season of prey acquisition. Decisions are presumably mediated by recognition of the nutritional/energetic value associated with different skeletal parts. In archeology, carcass transport strategies are often evaluated by examining the relationship between estimates of **skeletal part frequency** in a fossil assemblage and their corresponding energetic return, often calculated using an **economic utility index**. For example, at **Olduvai Gorge**, studies of skeletal part frequencies and the carcass transports strategies inferred from them have been used to evaluate the timing of hominin access to prey (i.e., primary or secondary access) and to determine whether individual fossil assemblages represent kill sites or consumption sites. Carcass transport strategy also refers to the method used to acquire animal remains, such as **hunting, confrontational scavenging**, or **passive scavenging**. In hominin evolution, identifying carcass acquisition strategies helps characterize the ecological **niche** and behavioral capabilities of our hominin ancestors. Ethnographic studies document opportunistic passive scavenging by some hunter–gatherer groups (e.g., the Hadza), but no modern human populations have a carcass acquisition strategy characterized by obligate passive scavenging of terrestrial mammals. The emergence of hunting in the hominin behavioral record is a particularly controversial issue (i.e., the hunting-scavenging debate), much of which has focused on the **FLK-*Zinjanthropus*** site in Olduvai Gorge. The timing of the appearance of hunting as a carcass acquisition strategy is particularly important because hunting falls within the realm of modern human behavior and implies the presence of relatively complex social interactions and cooperation within hominin groups.

cardioid foramen magnum (Gk *kardia*= heart) This is the expression used by Tobias (1967) to describe the shape of the foramen magnum of **OH 5**. He described the anteriorly truncated foramen magnum as "more or less heart-shaped, with a blunt point posteriorly" (*ibid*, p. 59). *See also* **foramen magnum**.

Carnegie Institution for Science *See* **Carnegie Institution of Washington**.

Carnegie Institution of Washington Founded by Andrew Carnegie in 1902 as an organization for scientific discovery, the Carnegie Institution of Washington is an endowed, independent, nonprofit institution. Its headquarters is in Washington DC, but the scientific staff work in five scientific departments (plant biology, developmental biology, Earth and planetary sciences, astronomy, and global ecology) on the west and east coasts of the USA. Although it remains officially and legally the Carnegie Institution of Washington, in 2007 it adopted a new name, the Carnegie Institution for Science. The Carnegie Institution of Washington provided crucial support for Ralph **von Koenigswald**'s work in Java. In 1930 von Koenigswald had been employed by the Geological Survey of Java as a mammalian paleontologist and despite his success at identifying actual and potential fossil sites the project was abandoned, and his salary and support from the Survey ceased, in December 1934. However, soon after Teilhard de Chardin visited Java on his way to China, and it was he who brought von Koenigswald's predicament to the attention of John Merriam, a well-known paleontologist and President of the Carnegie Institution of Washington at the time. Merriam arranged for some temporary support for von Koenigswald and he invited him to a meeting about fossil man held in Philadelphia early in 1937. Shortly thereafter, von Koenigswald was appointed a Research Associate of the Carnegie Institution of Washington, and according to von

Koenigswald at Merriam's instigation "a considerable sum was placed at my disposal to pursue the search for fossil man in Java" (von Koenigswald 1956, p. 93).

Carnivora (L. *carn*=flesh and *vora*, from stem of *vorare*=to devour) The order of mammals so named because its members principally consume meat. The proper informal term for members of this order is **carnivoran**, but the term usually used is **carnivore**. The carnivoran order is **speciose** and widely distributed. There are two suborders of Carnivora: the **Feliformia** (cat-like) and the **Caniformia** (dog-like) carnivorans. Common feliform families at hominin sites are the **Felidae** (cats), Hyenidae (**hyenas**), **Viverridae** (civets and their relatives), and **Herpestidae** (mongooses and relatives). Families within the Caniformia include **Ursidae** (bears), **Mustelidae** (weasels and otters), and the **Canidae** (dogs). The aquatic carnivorans of the superfamily Pinnipedia are also subsumed within the Caniformia. Carnivorans first appear in the fossil record in the Paleocene epoch.

carnivoran (L. *carn*=flesh and *vora*, from stem of *vorare*=to devour) The proper informal term for members of the order Carnivora, but the term usually used in its place is carnivore. *See also* **Carnivora** (syn. carnivore).

carnivore (L. *carn*=flesh and *vora*, from stem of *vorare*=to devour) The proper informal term for members of the order Carnivora is **carnivoran**, but the term usually used in its place is carnivore. *See also* **Carnivora** (syn. carnivoran).

carnivore guild A carnivore **guild** is a group of mammalian carnivores that garner faunal resources using similar predation strategies and dental adaptations. For example, at the **Miocene** hominoid site of Pasalar, in Turkey, four carnivore guilds have been identified on the basis of dental morphology: hypercarnivores, including bone crushers, main-carnivores (that may also have included some bone crushers), omnivores, and invertebratevores (including insectivores) (Viranta and Andrews 1995). More carnivore guilds, such as those comprising sabre- and dirk-toothed cats, existed in the past. It has been suggested, both on the basis of archeological data and also on the rather unusual evidence of **tapeworms** evolution, that hominins may have entered a carnivore guild during the **Plio-Pleistocene** (Hoberg et al. 2001).

carnivore modification *See* carnivore ravaging.

carnivore ravaging Also known as carnivore modification, this is the spatial disturbance and modification, or destruction, of skeletal remains by **carnivore**s. Archeological bone assemblages, particularly those from open-air sites, frequently show tooth marks or other signs of alteration by carnivores. Because carnivore ravaging can alter the post-depositional integrity of a bone assemblage, it is important that it be identified and corrected for when making informed decisions on hominin behavior or past environments. **Actualistic studies** indicate that carnivores, especially **hyena**s and **canid**s, readily destroy grease-rich, low-density skeletal elements, such as long-bone epiphyses, ribs, pelves, and vertebrae. When examining **carcass transport strategies**, the observation that carnivore ravaging can alter skeletal part frequencies, and even mimic the **schlepp effect**, should be taken into account. Evidence of carnivore ravaging can be unambiguously detected through the identification of carnivore tooth marks on bone. *See also* **skeletal element survivorship**.

carnivory (L. *carn-*=flesh plus *vorous*, from the stem of *vorare*=to devour) The consumption of flesh, or other animal tissues. Although members of the order **Carnivora** are popularly viewed as being exclusively meat eaters (i.e., **carnivores**), many of them eat plant food as well, and some (e.g., the grey wolf) are most accurately described as **eclectic feeder**s. Many primates are known to engage in at least occasional meat-eating, and many more are carnivorous in the wider sense of including some animal tissues (from throughout the animal kingdom) in their diets, either regularly or opportunistically. In paleoanthropology, the examination of carnivory as a hominin subsistence strategy is a primary research question. Chimpanzees, our closest living relatives, are known to hunt and consume small mammals, so this capability might also have been present in early hominins. With the advent of stone tool manufacture (*c.*2.6 Ma) hominin–faunal interactions can be demonstrated through **cutmarks** and the breakage patterns of fossil bones. Exactly when, and in what context, hominins began to consume more meat than contemporary chimpanzees remains a topic of lively debate in paleoanthropology. *See also* **Dikika**; **expensive tissue hypothesis**.

carrying capacity The fixed population size above which a population will decrease in size and below which it will increase in size. The carrying capacity represents a stable equilibrium population size in the **logistic population growth** model.

Cartesian coordinates A system invented in the 17thC by René Descartes that specifies a point uniquely in a plane by a pair of numbers, or **coordinate**s, which are the distances from the point to two fixed perpendicular lines, measured in the same units. The same principle can be used to specify the position of any point in three-dimensional space by three Cartesian coordinates, its signed distances (i.e., using positive and negative numbers) to three mutually perpendicular planes (or, equivalently, by its perpendicular projection onto three mutually perpendicular lines). In general, one can specify a point in a space of any dimension *n* by use of *n* Cartesian coordinates. Geometric shapes can be described by Cartesian equations that involve the coordinates of the points lying on the shape. *See also* **coordinate data**; **geometric morphometrics**.

cartilage (L. *cartilago*=gristle) Cartilage is one of the two connective tissues (**bone** is the other) that form the skeleton. All connective tissues are made up of a combination of cells and microscopic organic fibers, both of which are embedded in a cellular matrix. There are three types of cartilage: hyaline, fibro-, and elastic cartilage. Chondrocytes, the cells that form the matrix, are the predominant cell type in hyaline and elastic cartilage; fibroblasts are also found in all three types, but they are especially common in fibrocartilage. Collagen fibers (produced by fibroblasts) are found in all three types of cartilage, but only elastic cartilage contains yellow elastic fibers in any significant numbers. The first manifestation of the skeleton during ontogeny is made of hyaline cartilage and the bones that form the bulk of the adult skeleton develop within that cartilaginous template by a process called endochondral **ossification**. In adults, hyaline cartilage covers the surfaces of synovial **joints** and the tips of the ribs. It is also the basis of structures such as the xiphoid and most of the cartilages of the larynx. The matrix, also known as ground substance, consists of mainly water, together with dissolved salts, proteins, and glycoproteins, all held in a network of molecules called proteoglycans. Hyaline cartilage is particularly stiff, but it is also elastic and its elasticity helps it to dissipate stress.

CAS Acronym for the **Chinese Academy of Sciences** (*which see*).

Casablanca *See* Sidi Abderrahman; **Thomas Quarry**.

Casal de'Pazzi (Location 41°56′N, 12°44′E, Italy; etym. It., named after the neighborhood in Rome in which it is located; *Casal*=variant of farmhouse or castle; *Pazzi*=the name of a family which developed the land) History and general description This is an open-air site on the middle terrace of the Aniene river in the suburbs of Rome. The sedimentation of the lower Aniene valley has produced a number of other Pleistocene sites in fluvial deposits, such as **Sedia del Diavolo** and **Saccopastore**. The site was discovered in 1981 during surveys commissioned by the Soprintendenza Archeologica di Roma. The hominin parietal fragment was discovered in 1983. Temporal span and how dated? **Amino acid racemization dating** on *Bos* teeth suggests an age of *c*.360 ka. Hominins found at site The right parietal fragment, known as Casal de'Pazzi 1, displays affinities of both *Homo sapiens* and *Homo erectus*. Manzi et al. (1990) conclude that the chronology and morphology of the specimen indicate it being part of an archaic variant of *Homo*, "tend[ing] increasingly towards the Neanderthal form." Archeological evidence found at site A stone tool technology similar to the early - **Mousterian**, called the "proto-Pontinian," has been recovered from the site. Key references: Geology, dating, and paleoenvironment Bietti 1985, Belluomini et al. 1986; Hominins Passarello et al. 1989, Manzi et al. 1990; Archeology Bietti 1985.

Castel di Guido (Location 41°53′59″N, 12°16′33″E, Italy; etym. It., named after the neighborhood in Rome in which it is located; *castel*=variant of home or castle; *Guido*=the name of the duke of Spoleto who defeated the Saracens in a battle in this area) History and general description This open-air site on a low hill along the Aurelian Way in the outskirts of Rome was discovered in 1976 and systematically excavated from 1980 to

1990 by A.M. Radmilli from the University of Pisa. The top agricultural level of humus and silt lies above a layer of lacustrine tuffaceous sands, which itself overlays a rich paleosurface that may have been a butchering site. Temporal span and how dated? Technostratigraphic and Biostratigraphic comparisons with other Italian sites suggests that the site was deposited during **Oxygen Isotope Stage** 9. Hominins found at site Seven fragmentary hominin bones have been recovered, both from the surface (CdG-1–4) and from the tuffaceous sands (CdG-5–7). The remains appear to show a mixture of *Homo erectus* and *Homo neanderthalensis* features, and have been interpreted as being "intermediate" between the two. Archeological evidence found at site More than 5800 stone and bone artifacts were found. The assemblage is characteristic of the Italian **Acheulean**. Bifaces and choppers have been recovered, along with a microlithic pebble and flake industry. Many of the bone implements were made from fragments of elephant bone. Key references: Geology, dating, and paleoenvironment Mariani-Constantini et al. 2001; Hominins Mallegni et al. 1983, Mallegni and Radmilli 1988, Mariani-Constantini et al. 2001; Archeology Radmilli and Boschian 1996.

CAT Acronym for both computer-assisted tomography and computerized axial tomography. *See* **computed tomography**.

Catalogue of Fossil Hominids Quenstedt and Quenstedt (1936) was probably the first attempt at a comprehensive list and bibliography of the hominin fossil record. Vallois and Movius' *Catalogue des Hommes Fossiles* (Vallois and Movius 1953) was the next attempt to survey the hominin fossil record, but the first comprehensive attempt to collate information about both the hominin fossil record *and* its context came in the late 1960s with the publication of the first volume (Part I) of the British Museum (Natural History) *Catalogue of Fossil Hominids*. The first volume, *Part I: Africa*, edited by Kenneth **Oakley** and Bernard Campbell, was published in 1967. Subsequent volumes covered fossil hominin discoveries from Europe (Oakley et al. 1971), and Asia (Oakley et al. 1975), but only Part I ran to a second updated edition (Oakley et al. 1977). Unfortunately (but understandably given the work involved to assemble a resource such as this in a non-electronic format and on a rapidly enlarging hominin fossil record) the series was discontinued and it has long been out of print. In the *Catalogue of Fossil Hominids* the fossil evidence for human evolution was organized by country and then by site. Within the entry for each site the fossils are grouped according to their preserved morphology (crania, mandibles, teeth, postcranial bones) and references were provided for publications relating to the morphology, taxonomy, and dating of the regional morphological groups of specimens. A more recent attempt to take up the challenge of providing a comprehensive catalogue of the fossil evidence for hominin evolution has been the initiative by Orban and her colleagues, entitled *Hominid Remains – an up-date*. This is distributed as part of a general series of publications called the *Supplements au Bulletin de la Societe Royale Belge d'Anthropologie et de Prehistoire*. The first volume in this series of separates was published in the beginning of 1988, and publication has continued thereafter. The early issues were organized by an Editor (Orban) and Associate Editors (e.g., Slachmuylder, Semal, and Alewaeters). The individual entries are put together by experts who are familiar with one, or more, temporal phases of the hominin fossil record from that country, or countries. Each slim volume covers one, or more, countries. The *Hominid Remains – an up-date* series is in effect an updated version of the *Catalogue of Fossil Hominids*, and it retains the latter's format. In 2005, and published as a supplement to the *Journal of Anthropological Sciences*, Alciati et al. (2005a, 2005b) published a two-part inventory of fossil hominins from Italian sites. Note that the Catalogue of Fossil Hominids was current at a time when the term hominid was routinely used for the what many now refer to as the hominin clade.

catch-up growth   In cases where an infant is born prematurely or a child experiences prolonged nutritional deficiency, disease, or severe psychological stress, normal growth patterns may be disrupted and growth in various dimensions affected. Once the source inhibiting growth is removed (or after birth, in premature infants) catch-up growth facilitated by a marked increase in growth velocity can serve to return the child to age-appropriate size. Catch-up growth may occur through rapid gains in height, weight, and head circumference. Depending on when catch-up growth occurs there may or may not be permanent deficits in the growth of certain tissues, especially the brain. Whereas individuals who do not experience catch-up growth in many cases end up as smaller adults and may show compromised cognitive development, individuals who do go through catch-up growth can incur a greater risk of obesity and related metabolic problems later in life.

catfish   Catfish bones, in particular those of *Clarias*, a large, air-breathing catfish, are often associated with early and late **Pleistocene** hominin sites, especially in Africa. While many fish bones are naturally washed into sites, *Clarias* bones with cut and/or tooth marks have been recovered from early hominin sites at **Koobi Fora** (Pobiner 2007, Braun et al. 2010) and **Olduvai Gorge** (Stewart 1994, in press). The importance of *Clarias* as a food source continues into the late Pleistocene, with the occurrence of dozens of archeological sites along the Nile and other rivers that contain hundreds of thousands of *Clarias* bones (Van Neer 1986). These were apparently procured seasonally during annual spawning runs. Similarly, *Clarias* is known historically as a valued food in Egypt and elsewhere and today sustains many groups as a staple in the dry season. *Clarias* is readily caught with bare hands or with spears in shallow waters in the food-scarce dry season and is even more accessible and predictable in the rainy season, when groups spawn in shallow floodplains. Further, *Clarias* is adapted to low water and drought conditions, in that it burrows into mud (hence the informal name "mudfish"), often for long periods, until water returns. There are anecdotal reports of bonobos digging in mud for fish, probably for *Clarias*. Many other predators, including hyenas, baboons, and large cats, hunt *Clarias* seasonally. A major attraction of *Clarias* is its nutritional value: it is an excellent source of high-quality fats, and this source is critically important in the dry season when other animals are fat-depleted and vegetation is scarce.

caudal   (L. *cauda*=tail) The anatomical term used to refer to structures in the axial skeleton that are "towards the tail": in upright hominins this means it refers to structures that are inferior. Thus, the axis vertebra (or C2) is caudal to the atlas vertebra (or C1), and the first lumbar (or L1) vertebra is caudal to the twelfth thoracic (or T12) vertebra. The opposite of caudal is **cranial**.

Caune de l'Arago   (Location 42°50′22″N, 2°45′18″E, France; etym. *caune*=local dialect for cave and Arago=name of the valley over which the cave is located) <u>History and general description</u> This cave site is located at the eastern end of the Pyrenees overlooking the Tautavel plain (which has led to the fossils being called "l'Homme de Tautavel" or "Tautavel Man"). The cave has been excavated since 1964 under the direction of Henry de Lumley. It is large and deep and consists of three main stratigraphic complexes. <u>Temporal span and how dated?</u> Several dating methods, including **biostratigraphy**, **uranium-series dating**, **thermoluminescence dating**, **electron spin resonance spectroscopy dating**, **fission track dating**, and **amino acid racemization dating** have been attempted, but the results are not consistent. However, recent uranium-series dates on a stalagmitic floor that immediately overlies that hominin-bearing levels have provided a date of more than 350 ka for the hominin layers, which is consistent with the uranium-series date on the Arago XXI face of *c.*400 ka. <u>Hominins found at site</u> More than 100 human remains belonging to 26 individuals have been found in layer G of the Middle Stratigraphic Complex. Many of the fossils (about 50%) belong to adolescents. These individuals have several traits that align them with *Homo neanderthalensis*, but their generally primitive features and their apparent age have led to their being considered pre-Neanderthals or assigned to *Homo heidelbergensis*. The most important among these fossils are the **Arago XXI** face, the **Arago II** mandible, and the **Arago XLVII** parietal. <u>Archeological evidence found at site</u> Several tens of thousands of stone flakes and retouched tools were recovered, mostly from the Middle Stratigraphic Complex, made on a variety of raw material including quartz, quartzite, flint, and hornfels. Many faunal remains were also recovered. Key references: <u>Geology, dating, and paleoenvironment</u> Falguères et al. 2004, Yokoyama and Nguyen 1981; <u>Hominins</u> de Lumley and de Lumley 1973, de Lumley 1982; <u>Archeology</u> Byrne 2004.

Cave of Hearths (Location 24°08′25″S, 29°12′00″E, South Africa; etym. named after the numerous hearths and presumed hearths throughout the stratigraphic sequence) <u>History and general description</u> The Cave of Hearths, formed in limestone deposits in the **Makapansgat** Valley, was first investigated by C. **van Riet Lowe** in 1937. In 1947 G. Gardiner and J. Kitching collected fauna (including hominin remains) that had fallen into a sinkhole, and large-scale excavations direc-

ted by Revil Mason were conducted in 1953–4. Mason removed more than 1800 tons of calcite-hardened breccia and calcite blocks, and exposed a 7 m-thick sequence with archeological evidence that spanned the period between the **Early Stone Age** and the **Iron Age**. <u>Temporal span and how dated?</u> No radiometric dates are available for the Cave of Hearths, but the Acheulean-like artifacts are >780 ka and may be >500 ka reported fauna and the artifact typology are consistent with a **Middle Pleistocene** to **Holocene** age estimate. <u>Hominins found at site</u> An *in situ* juvenile hominin mandible, Cave of Hearths 1, which has a thick corpus, large teeth, and congenital absence of the M_3 and was originally attributed to *Homo sapiens rhodesiensis*, was recovered from **Acheulean** Bed 3. A proximal radius showing a mosaic of primitive and modern human features was recovered from a swallow hole and hence has poor provenance; it is assumed to be **Middle Stone Age** or Early Stone Age in age. <u>Archeological evidence found at site</u> The archeological sequence is divided into 11 beds numbered from bottom to top: beds 1–3 contain **Acheulean**-like artifacts with variably shaped **handaxes** and **cleavers**, and beds 4–9 contain Middle Stone Age artifacts that become smaller as one goes up the section. They include **flake-blades**, **Levallois** and other **cores**, rare **points**, and **ochre**. Beds 10 and 11 are **Later Stone Age** and Iron Age, respectively. It has been suggested that some of the "hearths" in the lower strata may be deposits of bat guano that were ignited by lightning strike. Key references: <u>Geology and dating</u> Latham and Herries 2004, Mason 1988, Herries and Latham 2009; <u>Hominins</u> Tobias 1971, Pearson and Grine 1997, Curnoe 2009; <u>Archeology</u> Mason 1962, 1988, McNabb et al. 2004, McNabb and Sinclair 2009.

Cave of Scladina *See* **Scladina**.

Caverna delle Fate (Location 44°11′45″N, 8°22′03″E, Italy; etym. It. Cave of the Fairies) <u>History and general description</u> This cave is located about 5 km/3 miles northeast of Finale in Liguria and was originally excavated by Gian Battista Amerano in the late 19thC. The rediscovery of three hominin bone fragments among the faunal assemblage in the early 1980s revived interest in the site and new excavations were carried out between 1984 and 1988. This work led to the recovery of more hominin remains, but they were recovered from the spoil heaps of the old excavations so their strati-graphic position cannot be established. Temporal span and how dated? Gamma-ray spectrometry dating of the remains and **electron spin resonance spectroscopy dating** on the stalagmitic floor of the cave suggest an age of between 60–80 ka. Hominins found at site The three "original" remains found by Giacobini include a fragmentary immature frontal (Fate I), a nearly complete immature mandible (Fate II) and a fragmentary adult mandible (Fate III). Archeological evidence found at site A **Mousterian** industry was recovered, along with faunal remains, including a large number of cave bear bones. Key references: Geology, dating, and paleoenvironment Falguères et al. 1990; Hominins Giacobini et al. 1984; Archeology Amerano 1889.

Caviglione *See* **Barma del Caviglione**.

CB Abbreviation for Central Basin, the term used for the persistent, central, part of the paleo-Lake Olduvai. *See also* **Olduvai Gorge**.

CBA Abbreviation for cranial base angle. *See* **basicranium**; **cranial base angle**.

cDNA *See* **complementary DNA**.

CEJ Acronym for **cemento-enamel junction** (*which see*).

cemento-enamel junction The junction between the inferior (lower teeth) or superior (upper teeth) limit of the enamel cap of the crown of a tooth and the **cementum**, the tissue that covers the external surface of its **root** (syn. cervix).

cementum [L. *caementum*=rough (pre-dressed) stone] One of the three hard tissues (along with **enamel** and **dentine**) that make up a tooth. There are two kinds of cementum: acellular and cellular. The initial cementum that forms against the root dentine is acellular but cementocytes get trapped in the later-forming dentine as it grows in an appositional fashion. Cementum contains fibers that run from the periodontal ligament into the tooth root. Cementum deposition at the apex of a tooth root is a response to axial tooth movement and new cementum forms over resorbed root surfaces (Nanci 2003). Cementum is believed to show an annual incremental feature known as cementum annulations (Kay et al. 1984) and at certain latitudes its microstucture may

respond to seasonal environmental changes. These responses have been used by zooarcheologists to infer the season at death of fauna found at hominin fossil sites (Lieberman 1994).

Cenozoic (Gk *kainos*=new and *zoe*=life) The term for the youngest (and current) **era** of the Phanerozoic eon. Cenozoic refers to a unit of geological time (i.e., a **geochronologic unit**) that spans the interval from 65.5 Ma to the present. The onset of the Cenozoic is marked by the **Cretaceous–Paleogene** event that saw the mass extinction of numerous faunas, including the non-avian dinosaurs. Compared to the previous Mesozoic era, the Cenozoic is noted for a worldwide decline in temperatures. Although mammals were numerous and diverse throughout the previous era, the extinction of the dinosaurs allowed them to flourish and become the dominant large fauna of the Cenozoic. Historically the Cenozoic was subdivided into the **Tertiary** and **Quaternary** periods, but today most geologists opt to subdivide the former Tertiary into two separate periods, the Paleogene and **Neogene**. The International Union of Geological Sciences recently ratified the Quaternary as the third period in the Cenozoic, although opinion is divided on this issue. *See also* **Plio-Pleistocene boundary**.

Cenozoic Research Laboratory The Cenozoic Research Laboratory (or CRL) was established in 1929 to facilitate the collection and study of geological and paleontological material relevant to the problem of human origins in China. In 1926 the Canadian anatomist Davidson **Black** approached Weng Wenhao, the director of the **National Geological Survey of China**, about a collaborative research program to study the **Cenozoic** deposits outside of Beijing, and the value of a formal institution became apparent when excavations at Choukou-tien (now **Zhoukoudian**) led to the discovery in 1927 of a tooth that Black claimed belonged to a hitherto unknown species of hominin he named *Sinanthropus pekinensis* (now *Homo erectus*). The Cenozoic Research Laboratory was established as a department within the Geological Survey of China, but it was also associated with the newly established **Peking Union Medical College** where Davidson Black was the Professor of Anatomy. The new laboratory received its funding from the **Rockefeller Foundation**, which also funded the excavations at Zhoukoudian, but the laboratory was a Chinese institution run by the director of the Geological

Survey of China. Reports of any geological and paleontological discoveries made under the auspices of the Cenozoic Research Laboratory were to be published in the journal of the Geological Survey of China, *Paleontologia Sinica*. Chinese scientists who became prominent members of the Cenozoic Research Laboratory's staff over the years included the paleontologist Yang Zhongjian, geologist Pei Wenzhong, and paleontologist Jia Lanpo. During the 1930s Black and his colleagues discovered a substantial quantity of *Sinanthropus* fossils as well as faunal remains and stone tools that were curated by the Cenozoic Research Laboratory. After Davidson Black's premature death in 1934 the German anatomist Franz **Weidenreich** was invited to replace him and Weidenreich and his Chinese colleagues continued excavations at Zhoukoudian until 1937 when the Japanese invasion of China made further excavation impossible. Weidenreich remained in China until 1941, but it became unsafe for him to remain so Weng Wenhao, Weidenreich, and Pei Wenzhong decided that photographs and plaster casts of all the "Peking Man" remains should be made and the fossils themselves were prepared to be taken out of the country, since it was feared that the Japanese would remove the fossils to Japan. Weidenreich left China with the photographs of the fossils, but the original fossils were entrusted to a group of US Marines who were also evacuating China, but the fossils were lost and their fate is still a mystery. Pei Wenzhong became the director of the Cenozoic Research Laboratory after Weidenreich's departure and he spent the war years at the laboratory. After the creation of the People's Republic of China in 1949 the Cenozoic Research Laboratory was reorganized and renamed the **Institute of Vertebrate Paleontology and Paleoanthropology** (or IVPP) in 1953 and became part of the **Chinese Academy of Sciences**: the IVPP has conducted archeological and paleoanthropological excavations throughout China.

census size   *See* **population size**.

center of mass   The experimentally determined, or estimated, location within an object of the center of its mass. This term is usually used in the sense of the center of mass of the whole body, but it can refer to the center of mass of a segment of part of the body, or a segment of a limb.

centimorgan   *See* **linkage**.

Central Awash Complex (or CAC) One of the major fossiliferous subregions within the **Middle Awash study area**; the others include the **Western Margin, Bouri Peninsula,** and **Bodo-Maka**. The Central Awash Complex is a dome-like structure that lies on the west side of the **Awash River** and it extends to within 25 km/15.5 miles of the western escarpment of this section of the **Afar Rift System**. It rises approximately 200 m above the surrounding rift floor and it is dissected by the Amba, Urugus, Sagantole, and Aramis drainages. Collecting areas within the CAC that have yielded hominins include **Amba East, Aramis,** and **Asa Issie**.

central facial hollow   A term used by Phillip **Tobias** (1967, p. 209) to refer to the depressed central part of the face, relative to the forwardly positioned zygomatic bones, of OH 5. Similar-shaped faces are seen in other specimens of *Paranthropus boisei*, and to a lesser extent in the faces of *Paranthropus robustus* crania (syn. dished face).

central fissure   The term used by Weidenreich (1937) for a tooth crown feature that others call the central fovea (maxillary). *See also* **central fovea (maxillary)**.

central fossa (mandibular)   The term used by **Robinson** (1956) for a tooth crown feature that others call the **central fovea (mandibular)** (*which see*).

central fossa (maxillary)   The term used by **Robinson** (1956) and Korenhof (1960) for a tooth crown feature that others call the **central fovea (maxillary)** (*which see*).

central fovea (mandibular)   A depressed triangular area on the occlusal surface of the crown of a mandibular molar that is bounded by the **protoconid, metaconid, hypoconid, entoconid,** and **hypoconulid** (syn. central fossa, central pit, fovea centralis, postfossid, talonid basin).

central fovea (maxillary)   The term used by **Tobias** (1967a) and **White** (1977) for the depressed triangular area on the occlusal surface of the crown of a maxillary molar that is bounded by the **paracone, metacone,** and **paracone** (syn. central fissure, central fossa, central pit, fovea centralis, protofossa, protocone basin, trigon basin).

central nervous system *See* brain.

central pit (mandibular) *See* central fovea (mandibular).

central pit (maxillary) The term used by Kraus et al. (1969) for a feature on the crown of a maxillary molar that others call the central fovea (maxillary). *See also* central fovea (maxillary).

central place foraging *See* home base hypothesis.

central sulcus (L. *sulcus*=furrow or groove) A prominent morphological landmark located on the lateral surface of the **cerebral hemispheres** of most anthropoid primates. Originally called the Rolandic fissure after Luigi Rolando, the central sulcus defines the border between the **parietal lobe** and **frontal lobe**. The primary motor cortex is located along the anterior wall of the central suclus and the primary somatosensory cortex is situated along the posterior wall of the sulcus.

Centre for Prehistory and Palaeontology *See* **National Museums of Kenya**.

centric fusion *See* Robertsonian translocation.

centroid The average of a multivariate data set, usually calculated as either the mean or median of all variables in the data set.

centromere *See* chromosome.

Ceprano (Location 41°31′42″N, 13°54′16″E; Italy; etym. named after the nearest town) History and general description The region in which Ceprano is located for some time has been recognized as an area of geological and archeological interest, but in March 1994 a hominin calvaria was recovered during road construction. The discovery of the calvaria was brought to the attention of the archeologist, I. Biddittu. It was recovered by a team of the Istituto Italiano di Paleontologia Umana directed by A.G. Segre. New excavations in the area of Campogrande started in 2001 under the direction of I. Biddittu e G. Manzi, which drove to a reconsideration of both the paleoenvironmental history and chronological succession at Campogrande. Temporal span and how dated? Tephrostratigraphy and biostratigraphy suggest an age of *c*.400 ka (i.e., Oxygen Isotope Stage 11). Hominins found at site **Ceprano 1**, a well-preserved (i.e., undistorted) calvaria. Archeological evidence found at site No artifacts are directly associated with the hominin, but **choppers** and **flakes** have been recovered from an underlying layer and **Acheulean** artifacts have been found in overlying deposits. Key references: Geology and dating Ascenzi *et al.* 1996, 2000, Muttoni et al. 2009, Basilone and Civetta 1975, Manzi et al. 2010; Hominins Ascenzi et al. 1996, 2000, Clarke 2000, Ascenzi *et al.* 1996, 2000, Manzi et al. 2001, Mallegni et al. 2003, Bruner and Manzi 2005, 2007, Muttoni et al. 2009. Archeology Ascenzi et al. 1996, Manzi et al. 2010.

Ceprano 1 Site Ceprano. Locality Campogrande. Surface/*in situ* In situ. Date of discovery March 13, 1994. Finders Construction workers and I. Biddittu. Unit Fourth Ensemble, upper part of Unit D. Horizon Muddy/sandy lake margin deposit. Bed/member N/A. Formation N/A. Group N/A. Nearest overlying dated horizon Local leucitic tephra layer. Nearest underlying dated horizon N/A. Geological age *c*.400 ka. Developmental age Adult. Presumed sex Male. Brief anatomical description A well-preserved (i.e., undistorted) calvaria that lacks much of the left parietal and most of the base. Announcement Ascenzi et al. 1996. Initial description Ascenzi et al. 1996. Photographs/line drawings and metrical data Ascenzi et al. 1996, Manzi et al. 2001, Bruner and Manzi 2005, 2007. Detailed anatomical description Ascenzi et al. 1996, Manzi et al. 1996, Manzi et al. 2001, Mallegni et al 2003, Bruner and Manzi 2005, 2007. Initial taxonomic allocation "Late" *Homo erectus*. Taxonomic revisions *Homo* cf. *antecessor* (Manzi et al. 2001), *Homo cepranensis* (Mallegni et al. 2003) or *Homo heidelbergensis* (Bruner and Manzi 2005, 2007). Current conventional taxonomic allocation *Homo sp. indet.* Informal taxonomic category Pre-modern *Homo*. Significance Some suggest the morphological affinities of Ceprano 1, especially after the reconstruction by Clarke (2000), are with *H. erectus*, but other researchers have suggested it should be assigned to *H. heidelbergensis*, **Homo mauritanicus**, *Homo antecessor*, or to a novel taxon, *Homo cepranensis*. Bruner and Manzi (2005) suggest Ceprano 1 "represents the best available candidate for the ancestral phenotype of the cranial variation observed among Middle Pleistocene fossil samples in Africa and Europe" (*ibid*, p. 643).

Location of original **Soprintendenza Archeologica del Lazio**, Rome, Italy.

cercopithecid An informal term for the Cercopithecidae, one of the two families within the superfamily **Cercopithecoidea**, also known as Old World monkeys. *See also* **Cercopithecidae**.

Cercopithecidae (L. *cercopithecus* from the Gk *kerkos* = tail and *pithekos* = ape; i.e., an ape with a tail) One of the two families (the extinct family the **Victoriapithecidae** is the other) that make up the **Cercopithecoidea**, the Old World monkey superfamily. The Cercopithecidae is further divided into two subfamilies, the **Cercopithicinae** or cercopithecines, and the **Colobinae** or colobines. Genetic data suggest the two subfamilies diverged around 16 Ma.

cercopithecin An informal term for the tribe Cercopithecini, one of the two tribes within the subfamily **Cercopithecinae**. *See also* **Cercopithecini**.

Cercopithecinae (L. *cercopithecus* from the Gk *kerkos* = tail, and *pithekos* = ape; i.e., an ape with a tail) A subfamily of the family **Cercopithecidae**. The cercopithecines comprise two tribes: the **Papionini** or papionins (baboons and macaques) and the **Cercopithecini** or guenons. The paucity of late Miocene fossil remains obscures the early stages of cercopithecine evolution but molecular evidence indicates that they split at 11.5±1.3 Ma (Tosi et al. 2005). These sister clades share many features, but they can be distinguished because cercopithecins lack a **hypoconulid** on their lower third molars (Szalay and Delson 1979).

cercopithecine An informal term for the subfamily Cercopithecinae. *See also* **Cercopithecinae**.

Cercopithecini (L. *cercopithecus* from the Gk *kerkos* = tail, and *pithekos* = ape; i.e., an ape with a tail) A tribe of monkeys within the subfamily **Cercopithecinae**. The Cercopithecini, informally referred to as guenons, includes the genera *Cercopithecus*, *Chlorocebus*, *Erythrocebus*, *Allenopithecus*, and *Miopithecus*. The guenons have a poor fossil record, and they are only known from **Koobi Fora** and a handful of other African **Pliocene** and **Pleistocene** sites in very small numbers, insufficient to properly determine the evolutionary history of the tribe. However, genetic data have filled

some of the gaps in knowledge. The divergence of *Allenopithecus* and *Miopithecus* is inferred to have been in the late **Miocene** (Tosi et al. 2005). The **terrestrial** guenons (the vervet, patas, and L'Hoest's monkeys) and major arboreal groups (*Cercopithecus mona*/*Cercopithecus neglectus*/*Cercopithecus diana* and *Cercopithecus mitis*/*Cercopithecus cephus*) diverged in the Pliocene (*ibid*) and speciation within the arboreal groups probably happened rapidly during the **Plio-Pleistocene** (*ibid*).

cercopithecoid An informal term for members of the primate superfamily Cercopithecoidea, also known as the Old World monkeys. *See also* **Cercopithecoidea**.

Cercopithecoidea (L. *cercopithecus* from the Gk *kerkos* = tail, and *pithekos* = ape; i.e., an ape with a tail) Members of this primate superfamily are referred to informally as cercopithecoids or Old World monkeys (or OWMs). The OWMs diverged from the **Hominoidea** *c*.23–25 Ma, and the extant members of the Cercopithecidae are widely distributed across Africa and Asia. The OWMs can be distinguished from the New World Monkeys (or NWMs) of South and Central America because of differences in their dental formulae (OWMs have two premolars per quadrant, whereas NWMs have three premolars per quadrant), molar crown morphology (OWM molars are dominated by two crests, or lophs, that run across the crown; hence the term bilophodont), and their nasal anatomy (the NWMs have broader noses than OWNs; hence the names **platyrrhine** for the NWMs and catarrhine for the latter). The OWM superfamily comprises two families: one extinct, the **Victoriapithecidae**, and one extant, the **Cercopithecidae**. The cercopithecoid fossil record is sparse but towards the end of the early **Miocene** there is evidence for the presence of two genera within Africa *c*.19 Ma, *Prohylobates* and *Victoriapithecus*, both belonging to the family Victoriapithecidae. OWMs are relevant to paleoanthropology for two main reasons. First, they are prevalent at **Plio-Pleistocene** paleoanthropological and archeological sites, so they can be used to help reconstruct the paleohabitat at those sites. Second, they have been used in comparative models (Jolly 1972, 2001, Elton 2000, 2006, Elton et al. 2001, Hughes et al. 2008) that have been devised to help explain the events and patterns observed within human evolutionary history.

cerebellum (L. diminutive of *cerebrum*=brain; so "little brain") The part of the central nervous system that occupies the most posterior part of the **endocranial cavity** called the posterior cranial fossa. The cerebellum has a central midline component called the vermis and two paired cerebellar hemispheres. Its main function is the coordination of motor activity, but the lateral hemispheres have also been shown to play a significant role in **language** and **cognition**. The volume of the lateral part of the cerebellar hemispheres in hominoids is greater than that predicted by allometric extrapolation based on the observed relative size relationships in monkeys (MacLeod et al. 2003).

cerebral cortex (L. *cerebrum*=brain and *cortex*=bark; thus the "outer covering") The outer layer of the **cerebral hemispheres** and unless the cerebral hemispheres are sectioned, either actually or virtually by imaging, it is the only part of the cerebral hemispheres that can be inspected externally and it is the only part whose surface morphology can be seen on natural and researcher-prepared endocasts. It is mainly made up of the cell bodies of **neuron**s and **glial** cells and because this region of the brain does not contain a significant amount of myelin the unstained cerebral cortex is gray-colored (hence the term gray matter). In modern humans the adult cerebral cortex is deeply folded. The crest of a fold is called a **gyrus** and a fissure between two gyri is called a **sulcus**. The sulci and gyri are named according to their location (e.g., the superior temporal gyrus) or shape (e.g., the **lunate sulcus**). The cerebral cortex is also subdivided into functional areas. Anatomical mapping of the cerebral cortex using **cytoarchitecture** (the microstructure of the cerebral cortex shows distinct layers of neuronal cell bodies that vary in size and shape) and **chemoarchitecture** also reveals distinct areas that are associated with different functions. The main subdivisions of the cerebral cortex of mammals include the **neocortex** (also called the isocortex) and the **allocortex** (comprised of the archicortex and the paleocortex). The neocortex is distinguished by a six-layered horizontal arrangement of cells, whereas allocortical areas have fewer than six layers. The allocortex includes parts of the cerebral cortex concerned with the processing of olfaction (i.e., the piriform cortex or paleocortex), as well as memory and spatial navigation (i.e., the **hippocampus**, also called the archicortex). Most regions of the cerebral cortex have a distinctive pattern of the cell layers and a widely used system of cerebral cortical areas proposed by Korbinian Brodmann (1909) is based on these patterns (e.g., the **motor speech area** comprises Brodmann's areas 44 and 45). *See also* **Brodmann's areas**.

cerebral hemispheres (L. *cerebrum*=brain, Gk *hemi*=half and *sphaira*=surround) The cerebral hemispheres are a major component of the telencephalon, which itself forms the majority of the forebrain (or prosencephalon). Each hemisphere consists of an outer surface layer, the **cerebral cortex**, which is made up of the cell bodies of neurons. Beneath the cerebral cortex is mainly white matter consisting of the myelin-covered processes (called dendrites and axons) of the cortical neurons. There are clusters of neuronal cell bodies buried in the white matter; the biggest of these are the basal nuclei. The cerebral hemispheres also contain a cerebrospinal fluid-filled cavity called the lateral ventricle. Named fissures or sulci separate the major **frontal**, **parietal**, **occipital**, and **temporal lobes** of the cerebral hemispheres. The paired hemispheres occupy the anterior and middle cranial fossae and they are connected by a substantial bundle of fibers called the **corpus callosum**.

cerebral petalia *See* petalia.

cerebrospinal fluid (or CSF) Clear and straw-colored fluid the composition of which resembles that of extracellular fluids; it is secreted by the choroid plexuses in the walls of the ventricles. The total volume of this fluid (approximately 150 cm^3 in an adult modern human) is divided into two main compartments, called intracerebral and extracerebral. The intracerebral CSF compartment consists of the ventricles (a pair of lateral ventricles and the third and fourth ventricles in the midline) plus the small channels that connect them. The extracerebral CSF compartment occupies the space between the arachnoid and pia (the subarachnoid space). Most of the extracerebral CSF is in well-defined spaces called cisterns; the rest is in the form of a thin film of fluid between the arachnoid and the pia. The volume of the extracerebral compartment of the CSF comprises most of the difference between the volume of the brain and **endocranial volume** (or ECV), which is the volume of the cranial cavity.

cervid (L. *cervus*=deer) The informal name for the artiodactyl family, the Cervidae, that comprises deer, reindeer, and moose. Mostly known from fossil and archeological sites in the northern hemisphere. *See also* **Cervidae**.

Cervidae The artiodactyl family that comprises deer, reindeer, and moose. All male (and, in a few taxa, the female) cervids have keratin-derived antlers that are shed annually. *See also* **Artiodactyla**.

cervix (L. *cervix*=neck, or a narrow region between two wider regions) On the outside of the tooth, it is the narrow part of a tooth where the crown and the root meet.

cf. A term used in taxonomy to suggest that incompleteness or damage make the allocation of a specimen to a taxon problematic. Thus, a small fragment of the crown of a tooth with evidently thick enamel might be assigned to "*Paranthropus* cf. *P. boisei*."

CF-bF Acronym for the Cromer Forest-bed Formation. *See* **Pakefield**.

CFT *See* **crown formation time**.

Chad Basin The Chad Basin is presently occupied by the Djurab Desert and it is there that the **Mission Paléoanthropologique Franco-Tchadienne** (French-Chadian Paleoanthropological Mission, or MPFT) has located several "fossiliferous areas" (e.g., **Kollé**, **Kossom Bougoudi**, **Koro Toro**, and **Toros-Menalla**). Each area has a different prefix – KL, KB, KT, and TM respectively – followed by a number for each locality within that area; further numbers are used to identify each specimen (e.g., TM 266 is the locality where **TM 266-01-060-1**, the type specimen of *Sahelanthropus tchadensis*, was discovered).

chaîne opératoire Literally translated as "operational sequence," this refers to the analysis of technological action sequences, especially in the case of stone tool production. The approach has its intellectual roots in the ethnology of Marcel Mauss (1935), who argued that bodily techniques (e.g., ways of walking or sitting) are culturally conditioned and can thus provide a window on non-discursive elements of thought and culture. André Leroi-Gourhan (1993, originally 1964)

extended these ideas and applied them in an evolutionist analysis of human technological development, coining the term *chaîne opératoire* to refer to the sequences of technical "gestures" said to be characteristic of particular technologies. The approach applies to any technology, but has been most enthusiastically adopted in **lithic analysis** where it refers to the entire sequence of behaviors involved in tool production, use, and discard (Inizan et al. 1999). In this sense, it is very close to the American concept of a lithic reduction sequence (Shott 2003). Both focus on the sequential transformation of artifacts by hominin behavior, and represent improvements over previous, typological approaches. However, the intellectual origins of the *chaîne opératoire* approach are reflected in a much greater emphasis on the cognitive and cultural interpretation of tool-making behavior. *Chaîne opératoire* analysis aims to decipher the intentions of tool makers through piece-by-piece refitting and/or mental reconstruction of action sequences based on the technological expertise of the analyst (Pelegrin 2005). Results are commonly expressed in terms of operational flow-charts intended to reflect the normative "technological choices" characteristic of a particular prehistoric social group (Bar-Yosef and Van Peer 2009). In contrast, reduction sequence analysis more typically proceeds through the use of experimentally derived quantitative models to interpret assemblage level variation in terms of production stages, intensity of re-sharpening, and evidence of transport across the landscape (e.g., papers in Andrefsky 2008). This reflects a theoretical interest in reconstructing the behavioral ecology, as opposed to cognitive capacities, of prehistoric populations.

chalcedony (Gk *khalkedon*=a mystical stone) A microcrystalline variety of silica, often white or translucent, with a fibrous, cryptocrystalline, structure that typically forms as a secondary deposit in igneous and other rocks. Chalcedony is fine-grained and percussion results in a **conchoidal fracture**. When available chalcedony is widely used in artifact production as in the many **Howieson's Poort** sites (e.g., **Klasies River**) in southern Africa.

Changyang () (Location 30°15′N, 110°50′E, central China; etym. named after the local county) *History and general description* The cave site of

Longdong (or Dragon Cave) is located on the southern slope of Guanlao Hill near the village of Xiazhongjiawan, 45 km/27.9 miles southwest of Changyang City, Changyang County, Hubei Province. A hominin maxilla was collected by non-specialists and sent to the **Institute of Vertebrate Paleontology and Paleoanthropology** (or IVPP) for analysis in the fall of 1956, and an isolated lower tooth, along with mammalian fossils, was found by an IVPP team in 1957. Temporal span and how dated? A mean **uranium-series** age of 195 ka is derived from fossil mammal teeth. Hominins found at site A fragment of a small left maxilla with C, P^4, M^2 (PA 76), and an isolated left P_4 (PA 81). Archeological evidence found at site None. Repository Institute of Vertebrate Paleontology and Paleoanthropology, Beijing, People's Republic of China. Key references: Geology and dating Yuan et al. 1986; Hominins Wu and Olsen 1985, Pope 1992, Wu and Poirier 1995.

Chaohu *See* **Chaoxian**.

Chaoxian () (Location 31°32′55″N, 117°51′51″E, eastern China; etym. named after the local county) History and general description A breccia-filled fissure in Yinshan (Silvery Hill) located southeast of the town of Yinping and west of Yuxihe (Yuxi River), 6 km/3.7 miles south of Chaohu City, Chaoxian County, Anhui Province. Acting on reports of mammalian fossil finds, researchers from the **Institute of Vertebrate Paleontology and Paleoanthropology** (or IVPP) organized two field seasons of excavation in 1982 and 1983. Two groups of fossiliferous deposits, Locus A and Locus B, were identified at the site. The hominin fossils were discovered in layer 2 (of four layers) at Locus B. Temporal span and how dated? Based on nine fossil animal teeth and bones Chen et al. (1987) report an age range of 200–160 ka based on **uranium-series dating**. Subsequent sampling of stalagmites and animal bones collected from the upper part of layer 2 produce uranium-series ages older than 310 ka. To constrain the chronological position of Chaoxian, Shen Guan Jun and colleagues collected additional calcite samples in 2006 and 2008. Their new **thermal ionization mass spectrometry** uranium-series dates suggest the fossils may be between 360 and 310 ka. Hominins found at site A cranial fragment (female?) including a large part

of the occipital squama was discovered in 1982. The occipital has been described as having a weak occipital torus, lacking an external occipital protuberance and possessing a distinct supratoral sulcus. It is also said to possess a "depression similar to fossa supratoralis" (Xu et al. 1984, p. 209), which Wu and Poirier describe as a "small fovea corresponding to the suprainionic [*sic*] fossa above the middle of the torus" (*ibid*, p. 134). And yet, the estimated large occipital angle and thin bone are said to resemble most closely that of "early" Homo sapiens (Xu et al. 1984). A maxillary fragment with lateral incisors and right P^3–M^1 and three isolated left teeth (P^4, M^1, and M^2) were discovered in 1983. Bailey and Lu (2010) concluded that the occlusal morphology of the maxillary dentition differs from that seen in *Homo neanderthalensis*. Archeological evidence found at site None. Repository Institute of Vertebrate Paleontology and Paleoanthropology, Beijing, People's Republic of China. Key references: Geology and dating Chen et al. 1987, Shen et al. 1994, 2010; Hominins Xu et al. 1984, 1986, Xu and Zhang YY 1989, Pope 1992, Wu and Poirier 1995, Bailey and Liu 2010.

character (ME *caractere*=to describe) A unit of the genotype or the phenotype used to describe and compare taxa: sometimes also called a **trait**. The condition of a character in a taxon is referred to as its **character state**. For example, the character "root system of the first mandibular premolar tooth" has four states among the hominins: (a) a single root (1R), (b) one root divided for part of its length (1/2R), (c) one main distal root and a second mesiobuccal accessory root (2R: MB+D), and (d) two main roots (2R: M+D). Characters can be qualitative (e.g., a sharp or a rounded boundary between the nasal cavity and the face) or they can be derived from continuous measurements.

character conflict (ME *caractere*=to describe) When the branching pattern of a **cladogram** generated using the states of one character differs from the branching pattern of a cladogram generated by the states of a second character or by a number of other characters.

character evolution The pattern by which characters evolve (i.e., change state between ancestors and descendants) in a **phylogeny**. It is not always appreciated that the pattern of character

129

evolution implied by a **cladogram** is logically equivalent to the tree topology of that cladogram. Thus, the most parsimonious tree implies a pattern of character evolution; therefore it is illogical to accept the phylogenetic hypothesis of the cladogram, but reject the hypothesis about the pattern of character evolution.

character state (ME *caractere*=to describe) Character states are different versions of a **character**. For example, if the character is the root system of the first mandibular premolar tooth, among archaic hominins and *Homo* the states would be a single root (1R), one root divided for part of its length (1/2R) (also called a Tomes' root), one main distal root and a second mesiobuccal accessory root (2R: MB+D), and two main roots (2R: M+D). *See also* **cladistic analysis**; **character**.

character state data matrix Takes the form of a table in which the rows represent **taxa** or **operational taxonomic units**, the columns represent **character**s and the cells of the table record the **character state**s seen in each of the taxa (e.g., the state of the character **chin** in *Homo sapiens* is "present" while the state of the same character in *Homo neanderthalensis*, *Homo habilis*, and *Paranthropus boisei* is "absent"). The construction of a character state data matrix is the first step in any form of **phylogenetic analysis**.

Charentian (etym. Named after the French department of Charente, where sites such as **La Quina** are located) A group of lithic assemblages within the **Mousterian** tradition as defined by François Bordes. Charentian assemblages are rich in **scraper**s (particularly side-scrapers) that may account for more than 50% of the **retouch**ed pieces, but there are few denticulates. The Charentian is divided into two types: Quina and Ferrassie, distinguished by a greater number of scrapers with typical steep, scaled Quina-type retouch in the former and a greater proportion of Levallois flakes and cores in the latter. The Charentian and other Mousterian variants, or facies, have figured prominently in debates about the implications of different archeological assemblages that have been interpreted as different cultures (Bordes and de Sonneville-Bordes 1970), time-successive entities (Mellars 1996), activity variants (Binford 1969), or the outcomes of different amounts of tool reduction prior to discard (Dibble 1991).

Charnov's "dimensionless numbers" (or DLNs) These measures, introduced by and named for the evolutionary ecologist Eric Charnov, are used in the analysis of life history. They simplify the description of relationships by describing the value of the ratio of variables, rather than the values of individual variables. It has been argued that the use of DLNs allows researchers to identify cases where the DLNs are invariant and then explore the possible reasons for their invariance. Examples include (a) exponents derived from the analysis of the relationship of life history traits to **body size**, (b) measures of reproductive effort, and (c) sex ratios.

Châtelperron (Location 46°23′N, 3°38′E, France; etym. named after the nearby village) History and general description This cave site, originally called the Grotte des Fées de Châtelperron, lies in the Massif Central of France. It was discovered during railway construction in the late 1800s and it was excavated by A. Poirrier in the 1840s, G. Bailleau between 1867 and 1872, and by Henri Delporte between 1951 and 1955. The relatively poor archeological material was determined by the Abbé Breuil to be distinctive, and he gave it the name **Châtelperronian**. Recent reanalyses of the site by several researchers have led to very different interpretations. Paul Mellars and colleagues (Gravina et al. 2005, Mellars 2005, Mellars et al. 2007) argue that there is a small **Aurignacian** deposit at Châtelperron that is interstratified with the Châtelperronian layers (this is based on the **radiocarbon dating** of the layers) and they claim this is evidence that modern humans were in the area and that they exposed Neanderthals to the Upper Paleolithic culture. Zilhão et al. (2006, 2008) have argued that the levels from which the dates come are actually from intermixed backfill from the excavations in the 1800s and cannot therefore be evidence of interstratification. Riel-Salvatore et al. (2008) come to a similar conclusion, concluding that the Aurignacian determination of the deposit relies on a very small number of Aurignacian-type tools, a percentage that is within the range of Aurignacian-type tools found at other Châtelperronian deposits. Both the latter research groups suggest that *Homo neanderthalensis* acquired the wherewithal to make this Upper Paleolithic technology independently. Temporal span and how dated? Radiocarbon dating of the layers results in a range of 41–34 ka (uncalibrated). Hominins found at site Although a calvaria was reported as coming from the Grotte des Fées, the hominin linked with Châtelperron is from a different open-air site in the area and is not from the Grotte des Fées; its age and taxonomic status are unknown. Archeological evidence found at site The approximately 217 retouched tools are consistent with a Châtelperronian assemblage; Gravina et al. (2005) identified 10 Aurignacian-type artifacts. Key references: Geology, dating, and paleoenvironment Gravina et al. 2005, Zilhão et al 2006, 2008, Mellars et al 2007; Hominins Lacaille and Cave 1947, Montagu 1952. Archeology Breuil 1910, Riel-Salvatore et al. 2008. *See also* **Châtelperronian**.

Châtelperronian (Fr. after the site of Châtelperron) A western European technocomplex from the early Upper Paleolithic that includes many aspects of what is usually considered modern human behavior, including bone tools, composite tools, and beads, but which is associated with sites with *Homo neanderthalensis* fossils. Some authors have suggested that the Châtelperronian and other so-called transitional technocomplexes are evidence that Neanderthals adopted the behavior of the first modern humans that had migrated into Europe; Mellars (1999, 2005) supports this acculturation model. Other researchers suggest that *H. neanderthalensis* individuals developed these techniques independently (e.g., d'Errico et al. 1998, Riel-Salvatore et al. 2008). Some see continuity between the Châtelperronian and later Upper Paleolithic cultures, and argue that this is evidence that Neanderthals either evolved into, or were assimilated into, modern human populations. *See also* **Châtelperron**.

Cheboit (Location 00°46′N, 35°52′E, Kenya; etym. Tugen name for the area) History and general description Site within the **Lukeino Formation** in the eastern foothills of the Tugen Hills, Baringo District, Kenya. Temporal span *c.*6–5.7 Ma. How dated? Magnetostratigraphy, argon-argon, and potassium-argon dating. Hominins found at site The only specimen recovered to date is **KNM-LU 335**, a left lower molar, either an M_1 or an M_2, but probably an M_1, discovered by Martin Pickford in 1973 and now assigned to *Orrorin tugenensis*. Archeological evidence found at site None. Key references: Geology and dating Sawada et al. 2002; Hominins Pickford 1975; Archeology N/A.

Chellean *See* **Chellean man**.

Chellean man This term was used by Louis **Leakey** to describe the population that included the individual **OH 9** whose calvaria was found at **Olduvai Gorge** in December 1960 (Leakey 1961). The term Chellean was introduced by the French prehistorian Gabriel de Mortillet, the curator of the Musée des Antiquités nationales (Museum of National Antiquities) at Saint-Germain-en-Laye, near Paris, France. He subdivided John Lubbock's **Paleolithic** period into a series of distinct "industries" named after sites in France. The final version of this scheme was published in his book titled *Le préhistorique: antiquité de l'homme* (C. Reinwald, Paris, 1883). Gabriel de Mortillet identified four industries: **Chellean, Mousterian, Solutrean**, and Magdalenian. He had earlier used the term Acheulean to refer to what he was now proposing to call Chellean, but for complicated reasons he changed the type site from St. Acheul (in the Somme Valley) to Chelles (near Paris). There were debates at the beginning of the 20thC over the existence of "Tertiary Man" (linked to supposed eoliths) but there were also supposed to be distinct races of early humans that corresponded to each archeological industry. Thus, there was an "Aurignacian Man' (often considered to be Cro-Magnon), a "Mousterian Man" (usually *Homo neanderthalensis*), and a hypothetical "Chellean Man." [NB: both Leakey (1961) and Tobias (1967c) use lower case for "man."]

Chemeron Formation (Location 00°47′N, 36°5′E, Kenya; etym. named after the Chemeron River that runs through and exposes part of the formation) History and general description The Chemeron Formation, previously referred to as the Chemeron Beds (Martyn and Tobias 1967), lies immediately above the Kaparaina Basalts and it is separated by an **unconformity** from the overlying **Kapthurin Formation**. It is part of the Tugen Hills succession and in the mid-1960s the **East African Geological Research Unit** was conceived to analyze and map the geology in and around the Tugen Hills. It is exposed at several locations and fossil hominins have been found at several locations (e.g., **Tabarin**, the Chemeron Formation Site, etc.). In 1965 John Martyn was mapping the area when a Kenyan field assistant, John Kimengich, discovered the **KNM-BC 1** temporal. The potential of the site was drawn to the attention of Louis and Mary **Leakey**, and subsequently Richard **Leakey** and Margaret Cropper found an elephant skeleton at site BPRP#23, and recovered evidence of two new species of fossil monkeys, *Paracolobus chemeroni* and *Papio baringensis*, at site BPRP#97. Since 1981 the **Baringo Paleontological Research Project**, headed now by Andrew Hill, has worked in the Chemeron Formation. Temporal span 5.6–1.6 Ma. How dated? **Argon-argon dating** and **biostratigraphy**. Hominin evidence The Chemeron Formation is best known for **KNM-TH 13150**, a fragment of >4.15 Ma from the right side of a mandibular corpus recovered from Tabarin, KNM-BC 1745, a >4 Ma proximal left humeral fragment from Mabaget, and KNM-BC 1, a *c.*2.4 Ma well-preserved but isolated right temporal bone. The former may, with hindsight, prove to be the earliest evidence for *Ardipithecus ramidus* and the latter is among the fossil evidence proposed as the earliest evidence for the genus *Homo* (but see Kimbel et al. in press). Archeological evidence Despite sustained survey efforts (McBrearty and Tryon, pers. comm.) no archeological evidence has been found. Key references: Geology and dating Martyn and Tobias 1967, Deino et al. 2002, Deino and Hill 2002; Hominins Martyn and Tobias 1967, Hill 1992, Wood 1999, Sherwood et al. 2002, Kimbel et al. in press.

chemoarchitecture (L. *alchymista*=alchemist and *architectus*=builder) The term used to describe the microanatomical organization of the cerebral cortex as revealed by staining for biochemical substances using techniques such as immunohistochemistry and enzyme or lectin histochemistry. These techniques allow researchers to define the borders of areas in the neocortex and nuclei in the thalamus. Additionally, chemoarchitectural information can provide insight into how cortical areas may differ in their neurochemical makeup across species. These techniques have been used to reveal several unique specializations of neocortical organization in great ape and human evolution. *See also* **cerebral cortex**; **neocortex**.

Chenjiawo *See* **Lantian Chenjiawo**.

chert A sedimentary rock consisting primarily of microcrystalline to cryptocrystalline quartz. Generally it takes the form of a light-colored variety of cryptocrystalline silica (SiO_2), but trace amounts of impurities impart color variations to a rock that is otherwise mostly white or translucent. Chert can be in the form of nodules, or it may be more widespread as in the beds of ancient lakes or seas, and it may contain trace fossils diagnostic of the age or location of its formation. Chert is globally perhaps the most abundant raw material used to make stone tools. This is because it is fine-grained, relatively hard, and homogeneous, thus it has excellent fracture properties (i.e., it shows **conchoidal fracture**s when percussed) (Luedtke 1992). Many **Middle Paleolithic** assemblages from Europe, the Levant, and Africa (e.g., **Porc-Epic Cave** in Africa and **Tabun** in Israel) are made almost exclusively from chert. **Flint** is a variety of dark-colored chert that has formed in marine environments.

Chesowanja (Location 36°12′E, 0°39′N, Kenya; etym. unknown) History and general description A Lower Pleistocene locality/site east of Lake Baringo, Kenya. Hominins, bovids, suids, and other fossils were first recovered from the site in 1970 by John Carney during mapping as part of **East African Geological Research Unit**. The geological record at Chesowanja is complex, as many of the sediments formed as fluvial cut-and-fill deposits, the fossil-bearing strata dip steeply as a result of post-depositional folding, and all of the outcrops are poorly exposed as a result of low topographic relief. Subsequent fieldwork by W.W. (Bill) **Bishop**, Andrew Hill, and Martin Pickford focused on clarifying the geological sequence, whereas excavations by John Gowlett and Jack **Harris** targeted Pleistocene-Holocene artifacts found in the area. Temporal span **Early** and **Middle Pleistocene** *c.*>1.42±0.07 to 0.78 Ma. How dated? **Potassium-argon dating** and **magnetostratigraphy**. Hominins found at site The remains of at least two individuals have been recovered from the site: **KNM-CH 1**, a partial cranium, KNM-CH 302, a partial right M^1 or M^2 crown plus other crown fragments consistent with them being from maxillary postcanine tooth crowns, and **KNM-CH 304**, five fragments of a hominin calotte. These derive from the **Chemoigut Formation**, the basal geological formation at Chesowanja. Archeological evidence found at site **Choppers**, polyhedrons, and **discoids** attributed to the **Developed Oldowan** occur in the Chemoigut Formation, as does purported evidence for the controlled use of fire. This includes a concentration of baked clay fragments washed into a stream channel. Some magnetic tests suggest that these fragments were heated at temperatures beyond that found in wildfires. However, the absence of clear evidence of a **hearth** (e.g., a ring of stones) and the unclear association of the artifacts and fossils with the burnt fragments makes Chesowanja an unreliable first appearance datum for the hominin control of fire. **Acheulean handaxes** have been recovered from the overlying sediments of the Chesowanja Formation, and the latter is overlain by a "non-biface" industry in the limited exposures of the Karau Formation. The uppermost strata in the Karau Formation show reversed magnetization, suggesting an age greater than the Middle Pleistocene (i.e., the Brunhes-Matuyama boundary at *c.*780 ka). Holocene pottery-bearing sites unconformably overlie Karau Formation sediments. Key references: Geology and dating Carney et al. 1971, Bishop et al. 1975, Bishop 1978, Hooker and Miller 1979; Hominins Carney et al. 1971, Bishop et al. 1975, Gowlett et al. 1981; Archeology Gowlett et al. 1981, Isaac 1982, Gowlett 1999.

chewing (ME *cheuan*=chew) Chewing is the cyclic application of bite forces to a food item, but the specific kind of chewing employed by mammals is called **mastication**. The primitive condition for primate mastication involves an anteriorly, superiorly, and (most distinctively) medially (i.e., lingually) directed power stroke. In animals (including many primates) in which the upper and lower jaws are not the same width (i.e., they are anisognathous), the distances between a tooth and its **antimere** in the upper and lower tooth rows are different, with the equivalent teeth in the mandible being closer together than those in the maxilla. In anisognathous animals bite forces can only be applied on one side at a time. The side where the bite force is applied is called the "working side;" the other side is referred to as the "balancing side." Mastication gape cycles are usually characterized by four phases, defined using a plot of either vertical gape distance or gape angle against time (a gape time plot). During the initial, *slow*, component of the open phase, the lower jaw is slowly depressed from a closed (or minimum gape) position while the tongue is protruded and elevated to make contact with the food item. During the ensuing fast component of the open phase the jaws are then depressed more rapidly while the

tongue is retracted and depressed in order to transport and/or reposition the food item between the teeth for the ensuing power stroke. Starting from their maximally depressed position, or the position of maximum gape, the jaws are then rapidly elevated in the *fast* component of the close phase until the teeth contact the food item. This tooth–food–tooth contact slows the jaws, and initiates the *slow* component of the close phase of the gape cycle. The slow close phase is also referred to as the "power stroke" because this is when bite force is applied as the teeth move through the food. The slow close phase of the chewing cycle ends at minimum gape. It used to be thought that the primate power stroke consisted of two phases: phase I being the last part of slow close phase, and phase II being the start of the slow open phase. However, it is now clear that in primates little or no bite force is generated after minimum gape; thus the primate power stroke is equivalent to just phase I. Tetrapod chewing, including mastication, evolved from and was integrated with the intra-oral transport behaviors. Tongue-based intra-oral transport is evolutionarily primitive and functionally fundamental for tetrapods because, once they moved out of the water, tetrapods were unable to use intra-oral fluid movements to move prey items through the oral cavity. Instead, coordinated movements of tongue and jaws move prey items from the mouth to the pharynx for swallowing. Chewing evolved as biting behavior was added on to these cyclic jaw and tongue movements. Single bites to kill prey items would not qualify as chewing, but once animals were applying multiple bites in a cyclical fashion to the food item, chewing, as we presently understand it, would have evolved. Many tetrapods chew, but only mammals employ the cyclic, rhythmic, medially directed power strokes characteristic of mastication.

chewing cycle *See* chewing.

chi-square test A statistical test used to identify whether a set of observed frequencies differs significantly from a set of expected or theoretical frequencies. This test is most commonly used to determine whether proportions differ between groups. For example, two cave sites each contain fossils, some of which have cutmarks and some of which do not. A chi-square test (or χ^2 test) can be used to answer the question "does the proportion of cutmarks differ significantly between the two sites?" Note that from a statistical perspective the answer to this question is identical to the answer of the question "does the proportion of bones at each site differ between cutmarked and non-cutmarked fossils?" Data for a chi-square test can usually be laid out in an $n \times m$ table where observations can be apportioned into one of n groups (e.g., 1 to n cave sites) and one of m categories (e.g., 1 to m types of bone modification). In the example given above this corresponds to a 2×2 table, although the table can be larger. Each cell of the table contains the number of observations that fall into that combination of group and category. For example, one cell of the table in the example above will contain the number of cutmarked fossils at site A, another cell will contain the number of cutmarked fossils at site B, a third cell contains the number of non-cutmarked fossils at site A, and the final cell contains the number of non-cutmarked fossils at site B. The counts in the data are used to calculate expected frequencies within each cell, and then squared differences between the observed and expected counts are summed to get the chi-square value. A significantly high value of chi-square indicates that observed frequencies are very different from expected values, indicating a significant difference in proportions between groups (alternatively, it can be thought of as a significant difference between categories; both interpretations are equally valid). Chi-square tests are generally robust at high sample sizes, but should not be used when the expected count in any cell of the table is less than five.

childhood   (OE *cild* = child or infant plus *-had* = condition or position) Childhood, lasting from approximately 3 to 7 years, has been proposed to be a unique life history stage in modern human ontogeny, inserted between infancy and the juvenile stage. During this time there is a leveling off of the postnatal growth rate following the precipitous decline in growth velocity that characterizes the infant stage. This stage is also characterized by post-weaning dependency; that is, while modern human children are weaned by the beginning of this stage, they are still highly dependent on other members of their social group to assist in procurement and preparation of food resources. This is because their dentition is still immature, and because of their low growth rate they are small and are not able to compete effectively at sites of food resource; the first perma-

nent tooth, usually one of the first molars or central incisors, erupts toward the end of this stage. Furthermore, children are neurologically and cognitively immature; brain growth is still rapid, but decreases in velocity toward the end of childhood. It is difficult to ascertain whether childhood is truly present in other primates (e.g., chimpanzees are weaned at approximately 4 or 5 years of age and begin foraging for themselves). However, analyses of primate growth trajectories have shown that chimpanzees, like modern humans, have much longer pre-adolescent periods of growth than would be predicted for their size. Because the presence of the childhood stage is associated with relatively shortened interbirth intervals in modern humans (e.g., children are weaned relatively early and remain dependent on other adults in the group, freeing up mothers for further reproduction) it is an adaptation that increases fitness. Estimates for the first appearance of the childhood stage in the hominin fossil record extend back to *c.*2 Ma although evidential support for this and other such claims is weak at best.

chimpanzee   *See* **Pan**.

chimpanzee archeology   This is the name proposed by Carvalho et al. (2008) to describe the study of **chimpanzee** tool manufacture and use (Boesch and Boesch 1990, McGrew 1992). However, tool manufacture and tool use in non-human primates is emphatically not confined to chimpanzees, so a more inclusive term was needed and in a multi-authored review Haslam et al. (2009) proposed the term **primate archeology** as a more inclusive term to describe the study of the "past and present material record of all members of the order Primates" (*ibid*, p. 339).

chimpanzee fossils   *See* **KNM-TH 4519-4521**.

chimpanzee tool-use   *See* **chimpanzee archeology**.

chin   *See* **true chin**.

China Fund   *See* **Johan Andersson**.

Chinese Academy of Sciences   (or CAS) Founded in 1949 and set up under the administration of the State Council as a government institution for the management of the nation's scientific research. It superseded the **Academia Sinica** and the Peiping Academy of Sciences. Currently it comprises five Academic Divisions, over 200 science and technology enterprises, more than 20 supporting units, and 108 scientific research institutes. The **Institute of Vertebrate Paleontology and Paleoanthropology** (or IVPP) is one of the CAS's scientific research institutes, and the former is the repository of many of the hominin fossils found in China prior to regional administrations insisting that hominin fossils stay in regional museums.

Chiroptera (Gk *cheir*=hand and *pteron*=wing) The order of mammals comprising the bats. Bat wings are formed by a membrane of skin stretched over modified manual digits and all bats are capable of powered flight. Bats are widespread in their distribution; however, they are relatively rare in the terrestrial fossil record of open-air hominin fossil sites, but they have a better chance of preservation within cave deposits. Chiropterans are phylogenetically close relatives of the primates.

Chiwondo Beds (etym. Named after Lake Chiwondo, northern Malawi) History and general description The ancient lake deposits of northern Malawi were first described as Chiwondo and Chitimwe Beds by Dixey (1927). The Chiwondo Beds (70 km/43 miles north–south and 10 km/6 miles east–west) are an uplifted sedimentary succession of fossil- and artifact-bearing **lacustrine** and **fluviatile** deposits derived from the paleo-Lake Malawi and river flows and they are situated in the Karonga District of northern Malawi on the northwest shore of the modern Lake Malawi. Desmond **Clark** (1995) provides a history of the mapping and exploration of the Chiwondo Beds, their sedimentology and stratigraphy are described by Kaufulu et al. (1981), Ring and Betzler (1995), and Betzler and Ring (1995), and their paleoanthropological significance is reviewed by Bromage et al. (1995). They were the focus of the fieldwork of the **Hominid Corridor Research Project** (or HCRP) that resulted in the recognition of more than 145 fossil localities, two of which, **Uraha** (U18) and **Malema** (RC11), have each yielded a single hominin specimen (**HCRP UR 501** and **HCRP RC 911**, respectively). Sites where the beds are exposed Malema, Uraha, and many other localities. Geological age Four biostratigraphic time periods (units/**biozones**) are represented in the

Chiwondo Beds, unit 2 >4 Ma, unit 3A-1 3.75–2.7 Ma, unit 3A-2 2.7–1.8 Ma, and unit 3B 1.74–0.6 Ma (Kullmer 2008). Most significant hominin specimens recovered from the beds The mandibular corpus HCRP UR 501 from Uraha, and a maxillary fragment HCRP RC 911 from Malema. Archeology Although Kaufulu and Stern (1987) reported artifacts from the Chiwondo Beds, subsequent geological work has shown these to be reworked and intrusive from younger strata. Key references: Geology and dating Ring and Betzler 1995, Betzler and Ring 1995, Kullmer 2008; Hominins Bromage et al. 1995, Kullmer et al. 1999. Archeology Stern and Kaufulu 1987, Juwayeyi and Betzler 1995.

Chokier *See* **Schmerling, Philippe-Charles (1790 or 1791–1836).**

chondroblasts *See* **cartilage.**

chondroclasts *See* **ossification.**

chondrocranium (Gk *khondros*=cartilage and *kranion*=brain case) The part of the **cranium** formed from cartilage by the process called **endochondral ossification**. In the second month of intrauterine life, symmetrical foci of hyaline **cartilage** appear beneath the developing **brain**. From posterior to anterior they are (a) four pairs of occipital cartilages that give rise to all of the adult occipital bone except for the occipital squame, (b) a pair of parachordal cartilages that lie on either side of the anterior end of the notochord that eventually contribute to the body of the sphenoid, (c) a pair of otic cartilages that surround the **inner ear** and later develop into the **petrous** component of the temporal bone, (d) a pair of hypophyseal cartilages that surround the pituitary that also eventually contribute to the body of the sphenoid, (e) two pairs of laterally situated sphenoid cartilages: the orbitosphenoids form the lesser wings of the sphenoid and the alisphenoids form the greater wings of the sphenoid, (f) two pairs of trabecular cartilages that eventually contribute to the **ethmoid** and to the nasal skeleton, and (g) a final pair of components that contribute to the presphenoid. All the cartilages anterior to and including the orbitosphenoids derive from neural crest cells, whereas all the paired cartilaginous elements posterior to and including the hypophyseal cartilages derive from mesoderm in the form of **somites**. The bony equivalent of the chondro-cranium is the **basicranium**. Bone formation, or **ossification**, begins at the caudal (i.e., the poster-ior or inferior) end of the chondrocranium and most of the occipital bone, the petrous part of the temporal bone, and most of the sphenoid, and ethmoid bones derive from the chondrocra-nium. The rest of the cranium, the **desmocra-nium**, develops from connective tissue by a process called intramembranous ossification. *See also* **ossification**.

chondrocytes *See* **cartilage**.

Chongokni (Location 38°00′42″N, 127°03′52″E, Kyunggi Province, South Korea; named after the town where the artifacts were discovered) History and gen-eral description Chongokni is an open-air site located in the Imjin/Hantan river basins (IHRB) in Kyunggi Province, South Korea, bordering the demilitarized zone separating modern-day North and South Korea. The IHRB is a region where more than 30 open-air Paleolithic sites have been identified, including Chongokni, Kumpari, Chuwolli, and Kawolli. Chongokni, the best known of the IHRB sites, came to the attention of the broader international scientific community in 1978 following the discovery of Acheu-lean-like bifacially worked stone implements. This evi-dence has been used by some researchers to refute the **Movius' Line** hypothesis. Temporal span and how dated? **Fission-track dating, potassium-argon dating,** thermo**luminescence dating,** infrared and optically stimulated **luminescence dating, AMS radiocarbon dating,** and presence/absence of AT and K-Tz tephra which originated from two separate volcanic explosions on Kyushu, Japan, have all been used to date the site. However, these analyses have resulted in a wide range of ages, with the age of the earliest Chongokni stone tools placed between 350 and 300 ka (i.e., the Middle Pleistocene). Hominins found at site None reported. Due to the acidity of the soil in Korea, bone preservation at open-air sites (except shell middens) is almost non-existent. Archeological evi-dence found at site More than 5000 stone artifacts have been collected through surface surveys and excavations at the IHRB sites. The most common raw materials are locally available poor-quality quartz and quartzite river cobbles. The Chongokni lithic assemblage comprises core and flake tools. Although most of the bifacially worked handaxes and cleavers were surface finds some came from excavations. The bifacially worked implements represent less than 5% of the total number

of lithic implements, leading one of the principal inves-tigators of the IHRB (Kidong Bae) to suggest these types of assemblages should be referred to as "Chon-goknian." Through systematic excavations at sites like Chongokni, the Korean researchers were able to collect small waste flakes (i.e., debitage), indicating at least some of these sites should be considered places where stone knapping occurred. Most of the bifacially worked artifacts differ morphologically from typical **Acheu-lean** handaxes found west of Movius' Line, although a small minority of researchers suggest a number of similarities. Repository Hanyang University Museum, Korea National Museum, Kyounghee University Museum, Seoul National University Museum, Yonsei University Museum, and Chungnam University Museum (all in South Korea). Key references: Geology, dating, and paleoenvironment Yi 1986, 1989, 1996, Bae 1988, 1994, 2002, Yi et al. 1998, Norton 2000, Danhara et al. 2002, Norton et al. 2006, Yoo 2007, Norton and Bae 2009; Hominins N/A (but see Norton 2000, Bae 2010 for list and description of hominin fossils from Korea); Archeology Yi and Clark 1983, Yi 1986, Bae 1988, 1994, 1997, 2002, Yi and Lee 1993, Bae et al. 1999, Norton 2000, Norton et al. 2006, Yoo 2007, Norton and Bae 2009.

Chonyogul *See* **Turubong**.

chopper A term used to describe flaked Mode 1 stone tools particularly common in Bed I at **Olduvai Gorge** and other early archeological sites. As defined by Mary Leakey (1971), a chop-per is a rounded cobblestone or angular stone block of stone with a unifacially or bifacially flaked edge along part of its circumference. Although the original terminology suggests their use as tools, more recent work (e.g., Toth 1985) suggests that choppers defined in this way may have functioned as **cores** for the production of sharp-edged **flakes**.

Choukoutien *See* **Zhoukoudian**.

chromatin *See* **chromosome**.

chromosome (Gk *chroma*=color and *soma*=body) A single, very long strand of **nuclear DNA**, plus its associated **protein** scaffolding. In **eukaryotes**, the chromosomes are found in the **nucleus** of the cell, they are typically linear, and they are packaged into a condensed structure referred to as **chromatin**. Chromatin can itself be packaged either tightly as heterochromatin (which stains darkly in Giemsa-stained **karyotypes**) or more loosely as euchromatin (which stains lightly in Giemsa-stained karyotypes). Heterochromatin is found in regions with limited gene **transcription** and is only present in eukaryotes, whereas euchromatin is typically rich in genes that are actively transcribed and is present in both prokaryotes and eukaryotes. Organisms differ in their **chromosome numbers**. Each chromosome is linear, with a centromere (a heterochromatic region containing the **tandemly repeated sequences** involved in cell division) that separates the long (q) arm from the short (p) arm of the chromosome. Chromosomes are classified by size and by the location of the centromere. Chromosomes with centromeres in the center of the chromosome are referred to as metacentric while those with centromeres closer to one end are referred to as submetacentric and catogorized further as acrocentric (p arm is still present) or telocentric (no real p arm is present). Chromosome replication occurs during interphase in both **mitosis** and **meiosis**. The evolution of chromosome structure is studied using karyotypes.

chromosome banding *See* **karyotype**.

chromosome painting *See* **karyotype**.

chromosome number Modern humans have 23 pairs of chromosomes (22 pairs of autosomes and one pair of sex chromosomes: XX in females and XY in males) whereas chimpanzees and gorillas have 24 pairs of chromosomes (23 pairs of autosomes and one pair of sex chromosomes). A **Robertsonian translocation** may have been responsible for producing the modern human chromosome 2 from two ape chromosomes (Yunis and Prakash 1982).

chron (Gk *kronos* = time) This term is an abbreviation of "polarity chronozone," and it refers to a substantial period in the history of the Earth's geomagnetic **polarity** (subdivisions of chrons are called **subchrons**). Chrons are numbered and these numbered chrons have replaced the eponymously named **epochs** (e.g., Gauss, Gilbert, etc.) that were used in older geomagnetic polarity timescales. The most recent chron is 1, and the numbers increase going back in time. A capital-letter postfix is used to identify chrons that have been added since the scheme was adopted (e.g., chron 3A). A chron can have both

normal and reversed components identified by a lower case 'n' or 'r' (e.g., 3An and 3Ar). Subchrons within a chron are also numbered and their magnetic direction is identified using the same abbreviation system (e.g., the first reversed subchron in chron 3An is 3A.1r). Some schemes mix the old and the new terminology (e.g., the "Matuyama chron"). *See also* **geomagnetic polarity time scale**; **magnetostratigraphy**.

chronocline *See* **cline**.

chronogram *See* **branch length**.

chronospecies (Gk *kronos* = time and *eidos* = form or idea) One of several terms used for a temporally distributed group of fossils all of which belong to the same **lineage**, or lineage plexus (*sensu* Westoll 1956). For example, all the fossils attributed to *Paranthropus boisei*, from its first appearance in the fossil record to its last appearance, would make up a chronospecies. Some, but not all researchers, claim *Paranthropus aethiopicus* should be included in the same chronospecies, whereas others would include it in a separate chronospecies in the same lineage. *See also Paranthropus boisei*; **species** (syn. **paleospecies**, paleontological species).

chronostratigraphic units (Gk *kronos* = time and L. *stratum* = cover or layer) Intervals of rock accumulated through Earth's history. The geologic time scale has a dual character in which intervals of time share terms with the sequence of rock (strata) deposited in that interval. However, chronostratigraphic unit subdivisions are sequential (e.g., thus they have the prefix Lower, Middle, and Upper), whereas **geochronologic units** are time-referenced, and thus their subdivisions are temporal (e.g., thus their prefixes have different names, as in Early, Middle, and Late). In the hierarchy of terminology for the geologic time scale, chronostratigraphic units are (from the largest to the smallest) called Eonothems, Erathems, Systems, Series, and Stages (e.g., *Australopithecus afarensis* fossils from **Laetoli** derive from strata of the **Pliocene** Series).

CHS Acronym for Chad's Hominid Site. *See* **Omo-Kibish**.

Chuwolli *See* **Chongokni**.

CI Acronym for **consistency index** (*which see*).

"Cinderella" Nickname given by Louis and Mary **Leakey** to **OH 13** from **Olduvai Gorge**.

cingulum (L. *cingulum*=girdle) A bulging of the **enamel** just above the base of the enamel cervix (**cemento-enamel junction**, or **CEJ**). In some teeth this bulging is so well-marked that it is given the name "style" with an appropriate prefix and suffix (e.g., **protostylid** is the part of the cingulum related to the **protoconid** in mandibular lower molars).

Cioclovina *See* **Petera Cioclovina Uscata**.

Cioclovina calvaria Site Petera Cioclovina Uscat. Locality Cioclovina cave. Surface/*in situ* Probably *in situ*. Date of discovery 1941. Finders Phosphate miners. Geological age Direct AMS ^{14}C uncalibrated **radiocarbon dating** suggests an age of *c*.28 ka (ultrafiltration pretreatment) for the hominin remains. Developmental age Adult. Presumed sex Disputed. Brief anatomical description A calvaria, missing much of the face and parts of the **basicranium**. Announcement Rainer and Simionescu 1942. Initial description Rainer and Simionescu 1942. Photographs/line drawings and metrical data Rainer and Simionescu 1942, Harvati et al. 2007, Soficaru et al. 2007. Detailed anatomical description Harvati et al. 2007, Soficaru et al. 2007. Initial taxonomic allocation *Homo sapiens*. Informal taxonomic category Modern human. Significance Among the earliest securely dated modern human specimens in Europe. Location of original **University of Bucharest, Department of Paleontology**. *See also* **Petera Cioclovina Uscat**.

Ciota Ciara *See* **Monte Fenera**.

circadian (L. *circum*=around and *dies*=day) An intrinsic biological rhythm with an approximately 24 hour cycle (see Smith 2006 for a review). The development of **enamel** and **dentine** is under the influence of a circadian cell rhythm, which manifests as **short-period incremental lines** that can be seen in both tissues. These take the form of fine dark **cross-striations** in enamel viewed microscopically that run at approximate right angles to the long axis of the prisms. The equivalent fine markings in predentine as well

as dentine are called **von Ebner's lines**. *See also* **incremental features**.

circaseptan (L. *circum*=around and *septem*= seven, thus "around seven") A term used to describe the time interval in days between long-period incremental markings (i.e., striae of Retzius) in enamel. The temporal repeat interval of the striae of Retzius (also known as their periodicity) is assessed by counting the number of short-period incremental markings (i.e., daily cross-striations) between consecutive striae. The range observed in modern humans is 6–12 days; the mean and median value is 8 days. The range in *Pan* is 5–9 days; the mean and median value is 7 days (Smith 2004). A long-period incremental marking analogous to striae of Retzius exists in dentine (called Andresen lines) and within the same organism its temporal repeat interval has been found to match that in enamel. *See also* **Andresen lines**; **striae of Retzius**.

CISC Acronym for the **Coimbra Identified Skeletal Collection** (*which see*).

Ciutarun *See* **Monte Fenera**.

clade (Gk *clados*=branch) This term, introduced by Julian Huxley (1958b, p. 27), refers to what Willi Hennig later called a **monophyletic group**. This is a grouping of two or more **taxa** that contains all (no more and no less) of the descendents of their most recent common ancestor. A clade is analogous to a *make* of car (all Rolls-Royce cars share a recent common ancestor not shared with any other make of car), whereas a **grade** is analogous to a *type* of car (luxury cars made by Mercedes, Jaguar, and Lexus are functionally similar, yet they have different evolutionary histories and therefore have no uniquely shared recent common ancestor). The smallest clade consists of just two taxa, called **sister taxa**; the largest includes all living organisms. It is conventional to indicate the structure of a clade by a series of parentheses, each of which represents a sister taxon relationship. So the shorthand way of expressing the relationships of the taxa in the clade that contains *Homo sapiens* and *Homo neanderthalensis* is (*H. sapiens*, *H. neanderthalensis*), and for the taxa in the clade that contains *Homo*, *Pan*, and *Gorilla* it would be ((*Homo*, *Pan*) *Gorilla*) (syn. monophyletic group).

cladistic (Gk *clados*=branch; i.e., a branching pattern) A pattern of relationships among **taxa** based on taxa sharing recently evolved (i.e., **derived**) features, called **character**s, of the **phenotype** or **genotype**. Phenotypic characters are either discrete morphological features (e.g., a sharp nasal sill), or a particular range of values of a measurement. *See also* **cladistic analysis**.

cladistic analysis (Gk *clados*=branch, so-called because it is an analysis that results in **taxa** being related in a tree-like branching pattern) A method devised by Willi Hennig (1966) that uses shared **derived** features to generate hypotheses about the relationships among taxa that conform as closely as possible to the phylogenetic history of the group. The term cladistic analysis was not used by Hennig, who instead called his approach to classification and phylogeny reconstruction **phylogenetic systematics**. The taxa used in a cladistic analysis are called **operational taxonomic unit**s (or OTUs). In the case of cladistic analysis applied to the phenotype, the nature of the taxa must be decided beforehand on the basis of **alpha taxonomy**. Cladistic analysis of the phenotype works by breaking morphology down into discrete features called **characters**. Characters can be described either non-metrically [e.g., the form of the root(s) of a mandibular premolar] or metrically (the cross-sectional area of the mandibular corpus), but the assumption is that they are inherited independently (i.e, they are not redundant sources of information about relationships). The different forms each of the characters takes (e.g., the different ways nasal bones articulate with the frontal bone, or the different sizes of a tooth crown) are called **character state**s. Several methods can be used (e.g., using an **outgroup** or **ontogenetic criterion**) to generate hypotheses about how to order the states of each character from the most **primitive** (shared by the most taxa) to the most **derived** (shared by the least number of taxa, or found in just one taxon). The taxa are then arranged in a branching diagram that is consistent with the distribution of the states of each character. The branching pattern that is supported by the greatest number of characters (called the **consensus cladogram**) becomes the null hypothesis for the relationships among those taxa. That hypothesis can be tested by finding new characters, and then comparing their distribution with the consensus cladogram. Any morphology shared by two taxa, but which was not inherited from their most recent common ancestor, will support a hypothesis of relationships that is not consistent with their phylogenetic history. This type of similarity is called **homoplasy** and any shared character states not inherited from a recent common ancestor are called homoplasies (syn. **phylogenetic analysis**). *See also* **character**; **character state**.

cladistics *See* **cladistic analysis**.

cladogenesis (Gk *clados*=branch and *gena*= birth or origin) The formation of two new **species** when one species splits, or bifurcates, into two. The splitting process is called vicariance if a geographical barrier is responsible for initiating the split. *See also* **speciation**; **vicariance**; **vicariance biogeography**.

cladogram (Gk *clados*=branch and *gramma*=a letter or something written or drawn) A tree-like diagram, in which the branching is usually dichotomous, indicating the relationships among **taxa** (i.e., one old taxon splits into two new taxa). The detail of the branches is determined by the distribution of shared **derived character** states (or **synapomorphies**) and it is usually based on the assumption of **parsimony**. The cladogram that results is conventionally the most parsimonious tree (i.e., the pattern of branching requiring the fewest evolutionary changes). *See also* **cladistic analysis**.

Clarias *See* **catfish**.

"classic" Neanderthals *See* **pre-Neanderthal hypothesis**.

Clark, John Desmond (1916–2002) Desmond Clark was born in London, England, and he claimed he became interested in archeology because as a boy he explored the hill forts and castles in nearby Buckinghamshire. He entered Christ's College, Cambridge University, in 1934 where he studied archeology and anthropology with Miles Burkitt and Grahame Clark, and history. While at Cambridge, Clark joined the excavations at the Iron Age hill camp at Maiden Castle conducted by Mortimer Wheeler, then Keeper of Archaeology at the Museum of London. Clark took his BA in anthropology and archeology in 1937, but the lack of professional positions in the UK led him to accept a position as curator of the David Livingstone Memorial Museum and secretary of the Rhodes-Livingstone Institute for Social Anthropology in the Northern Rhodesian (now Zambian) town of Livingstone. When Clark arrived in Livingstone in 1938 there were only a handful of archeologists working in Africa and little was known of Africa's prehistory. Clark began to expand the museum's archeological and ethnological collection and in 1938 he collaborated with geologist Basil **Cooke** on excavations of deposits along the Zambezi River that produced numerous stone implements and animal fossils. The following year Clark decided to re-examine the site of **Mumbwa** in the Kafue valley, which had first been excavated by a team of Italian scientists in 1930. His careful excavation of the site uncovered a remarkable archeological sequence that extended from the **Middle Stone Age** to the Iron Age. With the outbreak of WWII Clark joined the Field Ambulance Unit of the Northern Rhodesia Regiment in 1941 and was sent to Ethiopia and Somalia where the Allies were fighting the Italians. Clark used whatever opportunities he could to look for prehistoric artifacts and by the end of the war he had learned enough to allow him to publish *The Prehistoric Cultures of the Horn of Africa* in 1954. He also met Louis and Mary **Leakey** in Nairobi, where the Headquarters of the British East Africa Command was located, a meeting that formed the basis for a long professional relationship. When Clark was discharged from the army in 1946 he returned to Livingstone to resume his duties at the museum, which was now housed in a new building and renamed the Rhodes-Livingstone Museum in 1951. While exploring Africa's prehistoric ruins during the war Clark recognized the growing threat to archeological sites from farming, treasure hunting, and settlement. Thus, in order to preserve archeological sites Clark established the Northern Rhodesia National Monuments Commission and served as its first secretary. Clark also became active in African archeological institutions such as the South African Archaeological Society and in 1947 he participated in the First **Pan-African Congress on Prehistory and Quaternary Studies**, which was organized by Louis Leakey and held in Nairobi, Kenya. Clark served as the president of the prehistory section of the Second Pan-African Congress on Prehistory held in Algiers in 1952 and he organized the third congress, which met in Livingstone in 1955. Clark began

new excavations in the Zambezi River Valley near Victoria Falls in 1948, where he worked out the stratigraphic sequence of stone tool industries found there. Clark returned to England in 1950 to complete his PhD with Miles Burkitt at Cambridge using the archeological materials he had collected in the Horn of Africa and in the Zambezi gravels, but after completing his degree in 1951 Clark returned to Livingstone. In 1953 he explored the archeological potential of the Kalambo River, near Lake Tanganyika, and found an archeologically rich site at **Kalambo Falls**. He conducted a brief excavation in 1955 but large-scale work began in 1956 when Clark received financial support from the **Wenner-Gren Foundation**. Over the next decade Clark unearthed a wealth of archeological material from a succession of deposits containing **Acheulean hand-axe**s in the earliest levels to artifacts from the Middle Stone Age and ending with material from the Iron Age. Although no hominin and few animal fossils were found at Kalambo Falls, Clark did learn a great deal about the climate and environmental conditions that prevailed during the different periods represented at the site. Clark also used the newly devised **radiocarbon dating** method to obtain absolute dates for the different levels, and although these dates were later found to be inaccurate they did mark a major contribution to the understanding of the chronology of African prehistory. Excavations at Kalambo Falls continued until 1966 and Clark's painstaking investigation of the immense quantity of material recovered from the site was published in *The Kalambo Falls Prehistoric Site*, the first volume of which appeared in 1969, the second volume in 1974, and the third in 2001 just before Clark's death. Clark expanded into new territory in 1959 when a Portuguese diamond mining company invited him to excavate sites in northern Angola where their mining operations had uncovered numerous artifacts. Clark subsequently published the results of these excavations as *Prehistoric Cultures of Northeast Angola and their Significance in Tropical Africa* (1963) and *The Distribution of Prehistoric Culture in Angola* (1966). By 1960 Clark had become a world-renowned specialist in African prehistory, had dramatically expanded the collections of the Rhodes-Livingstone Museum, and had added two young British archeologists to the museum's staff, Ray Inskeep and Brian Fagan, who would go on to have distinguished careers of their own. In 1960 Clark received an invitation to join the Department of Anthropology at the University of California at Berkeley. Sherwood

Washburn, an American physical anthropologist whom Clark met at the Third Pan-African Congress on Prehistory in 1955 when Washburn was studying baboons in southern Africa, had just joined the department at Berkeley and was putting together a program in paleoanthropology. Clark joined the department in the autumn of 1961, Glynn **Isaac** was recruited in 1964, Clark **Howell** in 1970, and later still Tim **White**; this team of researchers would make Berkeley the premier center for the study of African prehistory and paleoanthropology over the next few decades. Before leaving for Berkeley, Desmond Clark participated in a symposium sponsored by the Wenner-Gren Foundation as part of the **Burg Wartenstein International Symposium Program** held in Burg Wartenstein, Austria, on African Ecology and Human Evolution. The 20 participants from varying disciplines discussed the implications of new evolutionary, primatological, and ecological discoveries for the study of human origins. The success of this symposium led Clark, William (Bill) **Bishop** and Clark Howell to organize a follow-up symposium in 1965 that resulted in the publication of *Background to Evolution in Africa* (1967), which helped to shape the developing field of paleoanthropology. Clark's fieldwork continued with excavations at a **Lower Paleolithic** site in the Orontes Valley in Syria (1964–5) and he supervised excavations of lake-bed deposits in the **Karonga** region of Malawi (1965–8) that brought together archeologists, paleontologists, and geologists. From 1970 to 1973 Clark investigated the origins of agriculture in Neolithic sites in the central Sahara and the Sudanese Nile valley. He also published his influential *Atlas of African Prehistory* (1967) that contained maps showing the distribution of archeological sites throughout Africa in different geological periods in relation to topography, vegetation, climate, and other factors. Publications such as *The Prehistory of Africa* (1970) emphasized his encyclopedic knowledge of that continent and the later *From Hunters to Farmers: The Causes and Consequences of Food Production in Africa* (Clark and Brandt 1984) further showed his interest in the full time range of the African past. In 1974 Desmond Clark led an international team of researchers from a variety of disciplines to explore the Paleolithic sites in Ethiopia. In the **Gadeb** Plain in central Ethiopia the team discovered **Oldowan** and Acheulean tools in deposits dated at 1.5 Ma. This was followed by the discovery in 1980 of Oldowan and Acheulean tools in deposits along the Middle Awash River in Ethiopia. Plans to

begin large-scale excavations the following year were interrupted by rebel fighting in the country but Clark was able to return with Tim White of the University of California at Berkeley and a team of researchers in 1990. Clark remained a member of the research team that explored the Middle Awash study area and discovered several new hominin taxa, including *Australopithecus ramidus* (later *Ardipithecus ramidus*) in 1994 and *Australopithecus garhi* between 1996 and 1997 (see de Heinzelin et al. 2000). Desmond Clark retired from the University of California at Berkeley in 1986, but he continued to be active as a field worker and as an analyst well into his retirement. Throughout his professional life Desmond Clark was supported by his wife Betty whose skills as an illustrator and editor did much to enhance the impact of his publications.

Clarke, Ronald J. (1944–) Born in Wokingham, England, Ron Clarke's longstanding interest in antiquities took him first to a course in their conservation, and in 1963 he was awarded Diploma in the Conservation of Antiquities and Works of Art from the University of London. From 1963 to 1969 Ron Clarke worked as technical officer and archeological and palaeontological assistant to Louis **Leakey** at the Centre for Prehistory and Palaeontology that was attached to what in 1963 was the Coryndon Museum and which from 1964 became known as the Kenya National Museum. His work included cleaning, reconstructing, and casting Miocene, **Pliocene**, and **Pleistocene** fossils, including early East African hominins. From 1969 to 1972 Clarke attended the University of London and he was awarded a BSc in 1972 and from 1973 onwards he was a graduate student at the University of the Witwatersrand in Johannesburg, South Africa. His thesis, entitled "The cranium of the Swartkrans hominid SK 847 and its relevance to human origins," was awarded in 1977. From 1978 to 1980 Clarke worked as a fossil conservator at the Transvaal Museum, but in 1981 he moved to the National Museum, Bloemfontein, where he was Senior Professional Officer in palaeontology and responsible for the excavations at **Florisbad**. From 1984 to 1986 he was Senior Research Officer in the **Palaeo-Anthropology Research Unit** within the Department of Anatomy at the University of the Witwatersrand Medical School; thereafter he returned to work in England as a freelance conservator. For a year in 1990–1 Clarke worked as Senior Exhibitions Assistant in the Department of Vertebrate Paleontology at the **American Museum of Natural History**, New York, before returning to South Africa in 1991 to the Palaeo-Anthropology Research Unit as its Senior Research Officer and its Deputy Director; it was at this time that Clarke took over the day-to-day running of the excavations at **Sterkfontein**. In January 1999 Clarke moved to Frankfurt am Main, Germany, to be the Director of Research at the Institut der Anthropologie und Humangenetik fur Biologen, Johann Wolfgang Goethe-Universitat, but he remained Deputy Director of the **Sterkfontein Research Unit**. In 2002 he returned to South Africa to work at the University of the Witwatersrand where he is now a professor and the Director of the Sterkfontein Research Unit. His main contributions to paleoanthropology are his management and then direction of the research at the Sterkfontein site and his recognition of the importance of the isolated hominin foot bones that were the key to the eventual discovery and recovery of the Stw 573 associated skeleton from the Silberberg Grotto. *See also* **Silberberg Grotto**; **SK 847**; **Stw 573**.

classification (L. *classis*=the name of the units used to subdivide the Roman army fleet, etc.) **Systematics** is the inclusive term used for (a) assembling individual organisms into groups, (b) formalizing those groups as Linnaean **taxa**, (c) giving formal names to the taxa, (d) allocating the taxa to taxonomic categories, and (e) assembling the taxa into a hierarchical scheme, or classification: classification subsumes the activities involved in (c) and (d). Classification schemes generally try to reflect the evolutionary history of a taxon, but they also need to be relatively stable. This explains why there was so much resistance to revising classifications based on the results of applying **cladistic** analytical methods to the hominin **clade**. Strictly speaking you cannot classify an individual fossil. It has to be assigned to an existing taxonomic group, or if its morphology is novel, to a new taxon, and then, and only then, can you proceed to classify the taxonomic group to which the fossil has been assigned. *See also* **nomenclature**; **systematics**; **taxonomy**.

clast (Gk *klastos*=broken) A piece, or fragment, of a rock that results from a larger rock fragment being broken or reduced in some way into smaller pieces. Clasts can be as large as a boulder or as small as a grain of sand. Clastic rocks are formed from rock fragments, or clasts.

clastic *See* **clast**.

clavicle, evolution in hominins An S-shaped bone lying superior to the first rib and articulating medially with the sternum and laterally with the acromion process of the scapula. When viewed from above the modern human clavicle has two main curvatures: an anterior and inferior curve at the medial end and laterally a superior and posterior curve. The conoid tubercle on the inferior aspect of the lateral curve is the attachment site for the coracoclavicular ligament (an important stabilizing ligament connecting the clavicle and the coracoid process of the scapula). Modern humans share the pronounced S shape with *Pan* and *Pongo*, but not all apes have this same sigmoid-shaped clavicle. *Gorilla* has a more pronounced posterior curvature laterally, and a relatively straight medial portion, whereas in *Hylobates* the lateral part is straight and the medial section is curved (Voisin 2006). In posterior view, modern humans are unique in having just one inverted U curvature. In contrast, apes have two curvatures that reflect the relative height of the shoulder girdle relative to the thorax; the relatively low position of the modern human shoulder has been associated with this single dorsal curve (Voisin 2006, Larson 2007). Most early hominin clavicles are fragmentary, and it is not always clear whether they display the two curves and an oblique orientation, or the presumed derived modern human condition, which would imply a more modern human-like thorax configuration (Larson 2007). The 1.8 Ma OH 48 clavicle from **Olduvai Gorge** is incomplete and there are conflicting reports as to how derived its morphology is. The clavicle of **KNM-WT 15000** is more derived in the direction of modern humans, but it is shorter than would be expected for modern humans, and this has led to the suggestion that a shorter clavicle may be the primitive condition for the great apes. Since modern humans and orangutans share relatively longer clavicles, it is possible that long clavicles were independently derived in these lineages (Larson 2007). It is important to note that pathology has not been ruled out in KNM-WT 15000 and some diseases (e.g., hypoplastic clavicle syndrome) do result in shorter clavicles. However, if KNM-WT 15000 is an accurate representation of the clavicular morphology of early *Homo*, Larson (2007) suggests the following evolutionary transformation of the shoulder. If the ape-like condition of short clavicles and a superiorly facing scapula positioned dorsally on the thorax was the primitive condition, early *Homo* would have maintained these shorter clavicles and possibly accommodated a more laterally oriented shoulder joint with a scapula that was more laterally situated on the thorax. From this configuration, elongation of the clavicle would have resulted in a more dorsal position of the scapula on the thorax while maintaining a laterally oriented glenohumeral joint (i.e., the *Homo sapiens* condition). It has been suggested that this configuration would have allowed for more range of motion behind the head, thus increasing leverage for overhand throws (Larson 2007).

clay Sedimentary materials defined primarily on the basis of particle size (i.e., a mean diameter of <4 m). Loose (unlithified) sediment in this range is referred to as clay, whereas consolidated or cemented (lithified) examples would be termed claystone. Clays are produced when deposition occurs in standing water (e.g., lakes) and they are also the active components of most soils. The term clay also designates a class of minerals (typically occurring as clay-sized particles) defined by their chemical structure of inter-layered silicate sheets. Clay minerals often form through the chemical weathering of other geological materials (e.g., the formation of smectite clay from the alteration of volcanic glass in an alkaline environment). The lacustrine deposits exposed in **Olduvai Gorge** include a wide range of clays that reflect complex fluctuations in the water chemistry of paleo-**Lake Olduvai**.

claystone *See* clay.

cleaver (OE *cleofan*=to split) A stone tool common among **Acheulean** assemblages. Cleavers are similar to handaxes in size and in some manufacturing details. The lateral margins of cleavers may be unifacially or bifacially **retouch**ed, but cleavers are distinguished from handaxes by having an unretouched distal end. This distal end (the cleaver part) forms a sharp cutting edge sub-perpendicular to the long axis of the tool. Cleavers are almost always made on large flakes; thus the form of the cleaver is determined at the time of flake removal (Inizan et al. 1999). Cleavers are particularly abundant at the site of **Isenya**, Kenya, and experimental evidence suggests they were used for butchery (Jones 1980, Schick and Toth 1993).

Cleveland Museum of Natural History In the 1830s Cleveland's first natural history collections were kept in a small, two-room wooden building on Cleveland's Public Square called the Ark. Its creators, the "Arkites," were amateur natural scientists, and in 1845 they joined Dr Jared Potter Kirtland, a doctor and natural scientist, to create the Cleveland Academy of Natural Science. The new Academy was in effect a museum with specimens donated by Kirtland and by some of the Arkites and it became a place where local natural historians could discuss scientific matters of the day. During the Civil War the Academy was dormant, but in 1869 members of the Academy created the Kirtland Society of Natural History (later called the Kirtland Society of Natural Science). The Kirtland Society, whose purposes were to study natural history and create a museum, lasted for 20 years, but it became dominated by lawyers and businessmen who supported it because it was good for the community, not because they had any personal interest in natural history. Eventually, with interest waning, the Kirtland Society turned its collections over to the Case School of Applied Science and Western Reserve College (in 1967 the two institutions merged to form Case Western Reserve University). In 1920, the Cleveland Museum of Natural History (or CMNH) was founded by a group of lawyers, businessmen, and scientists, and via Henry Wood Elliott, the only surviving member of the Kirtland Society of Natural Science, the Kirtland Society was incorporated into the new Museum. The CMNH has played an important role in paleoanthropology for two reasons: its links with field research in Ethiopia and the Hamann–Todd Osteological Collection. The collection, which consists of more than 3100 human skeletons and over 900 primate skeletons, includes the largest collection of lowland gorilla skeletons in the world as well as one of the largest collections of common chimpanzees in the western hemisphere. The collection was begun in 1893 by Carl August Hamann and his work was enlarged and vigorously expanded on by T. Wingate Todd until the latter's death in 1938. The collection was originally housed in its own museum facility in the Western Reserve College Medical School, but after Todd's death it was put into storage. Finally, after prolonged negotiations, it was transferred to the CMNH in several stages between 1950 and 1969. It was the Hamann–Todd Osteological Collection that attracted Donald

143

Johanson to the CMNH when he came as a student in 1968 to study the dentition of the collection's chimpanzees for his PhD dissertation. He was invited to join the **International Omo Research Expedition** for the 1970 and 1971 field seasons, and while working in the Omo Johanson met French geologist Maurice Taieb, who invited Johanson to join him to examine some promising deposits located in the Afar Triangle region in northeast Ethiopia. It was when Taieb, Johanson, French paleoanthropologist Yves **Coppens**, and American geologist Jon Kalb explored the region in 1972 that they realized the potential of the study area now known as **Hadar**. The first large-scale expedition to Hadar in 1973 resulted in the recovery of a large quantity of mammal fossils along with an early hominin knee joint, and in the following year Johanson and Tom Gray recovered the **A.L. 288-1** associated skeleton. In the meantime, in 1972 Johanson had accepted a position as a member of the CMNH's Department of Physical Anthropology and in 1974, the year he completed his PhD at the University of Chicago, he became Curator of Physical Anthropology. The original Hadar hominin fossils had been brought to the CMNH and it was there that Johanson and his team became convinced that all the Hadar remains belonged to one, novel, species of *Australopithecus* and in 1978 they proposed the Hadar hominin fossils should be referred to a new taxon, *Australopithecus afarensis*. In 1981 Johanson left his position at the CMNH and founded the **Institute of Human Origins** in Berkeley, California. Currently the CMNH is still involved with fieldwork in Ethiopia via Yohannes **Haile-Selassie**'s research in the Woranso-Mille study area. Hominin fossil collections N/A. Comparative collections Hamann–Todd Osteological Collection. Contact department Department of Physical Anthropology. Contact person Yohannes Haile-Selassie (yhailese@cmnh.org). Postal address 1 Wade Oval Drive, Cleveland, Ohio 44106, USA. Website www.cmnh. org. Tel +1 216 231 400. Fax +1 216 231 5919. E-mail jalpern@cmnh.org. *See* **Hamann–Todd Collection**; **Woranso-Mille study area**.

CLIMAP Acronym for the Climate: Long range Investigation, Mapping and Prediction research consortium, formed in 1971 to study past ocean climates. A major achievement was the reconstruction of **sea surface temperature**s across the world at the **Last Glacial Maximum** (CLIMAP Project Members 1976). *See also* **sea surface temperature**.

climate (Gk *klima* = which refers to the slope of the Earth's surface, is the origin of L. *clima* = referring to climate, or latitude) Climate is the time-averaged (>30 years) description of the Earth's weather or the average weather in a particular region (e.g., East Africa, Antarctica) of the globe. Weather, and therefore climate, is usually described in terms of four variables: temperature, precipitation (rainfall, snowfall), wind speed, and wind direction. There are local records of how these variables have changed over the course of the last few hundred years and before that climatic conditions have been recorded in literature and in painting, but once you get beyond the previous millennium nearly all information about past climate come from **proxies** that must take the place of direct observations of the four climate variables. These proxies can be divided into two main categories: **biotic** and geochemical. The climate experienced in any region is the result of interactions among component climate systems (e.g., the **El Niño Southern Oscillation**, **thermohaline circulation**, etc) and the global climate system is the product of interactions among these regional climate systems. The component climate systems have their own internal dynamics, as does the Earth's climate system, but these dynamics can be influenced, or "forced," by external factors such as changes in the way the Earth is inclined relative to the sun, changes in the shape of the Earth's orbit around the sun, massive volcanic eruptions, and changes in the Earth's atmosphere that affect the intensity of the sun's rays that reach the Earth (i.e., insolation). Regional climates are also influenced by the shape, or topography, of the Earth's surface (e.g., regional precipitation is influenced by rain shadow phenomena, and regional temperature by altitude). *See also* **climate forcing**; **insolation**; **paleoclimate**.

climate forcing (Gk *klima* = which refers to the slope of the Earth's surface, is the origin of L. *clima* = climate, latitude, and L. *fortis* = strong) One of the definitions of "force" is the ability to affect the rate of progression of normal processes (e.g., as in "forcing" the growth of plants by changing their environment). A number of physical processes (e.g., fluctuations in the shape of the Earth's orbit, and in the degree of tilt and amount of wobble of the Earth's axis, changes to ocean circulation, variations in atmospheric gases, etc.) can affect the dynamics of and thus force regional and global climate systems. In paleoanthropology there is considerable interest in climate forcing as a mechanism for environmental change, thereby potentially functioning as a catalyst for evolutionary change. Climate forcing has been linked to environmental change at some Pliocene East African hominin localities and to hominin speciation events (Kingston et al. 2007), including the emergence of *Homo erectus* (Lepre et al. 2007).

climate models *See* **general circulation model**; **models of intermediate complexity**; **paleoclimate**.

climate systems *See* **climate**; **El Niño Southern Oscillation**.

climbing hypothesis This hypothesis summarizes several separate but related models introduced to explain the origin and evolutionary history of upright posture and **bipedal** locomotion within the hominin clade. All the models regarding the evolutionary stages preceding the origin of hominin bipedalism have relied heavily on existing ideas about the phylogenetic relationships of modern humans and other higher primates. In this context, Sir Arthur **Keith** (1903, 1923), followed by William King Gregory (e.g., 1928, 1930), offered the first explicit hypothesis of the locomotor behaviors preceding hominin bipedalism. They proposed three stages prior to the appearance of hominin bipedalism: (a) a **pronograde** (horizontal-bodied) arboreal monkey (like modern catarrhines) stage, (b) a gibbon-like, "hylobatian" small-bodied **orthograde** (upright-bodied) ape stage, and (c) a large-bodied, "troglodytian" ape stage. It is from this stage that (d) hominin bipedalism evolved. Keith and Gregory surmised that the earliest hominin bipeds retained orthograde adaptations in the trunk and upper limbs from their large-bodied ape ancestors and maintained this upright posture when traveling on the ground. A variation of this model was proposed by Morton (e.g., 1924), who suggested that the earliest hominins evolved directly from small-bodied apes that like modern gibbons habitually employed bipedal locomotion when traveling arboreally. This model became known as the **brachiationist model** for its emphasis on the orthograde behaviors requiring considerable upper-limb

weight support, but this term has been largely abandoned. **Brachiation** now refers to the pendulum-like arm-swinging locomotion commonly practiced by gibbons. Russell Tuttle (1974) built on Morton's hylobatian model by arguing that bipedalism on large branches, like that observed in modern hylobatids, could be preadaptive for hominin bipedalism from a small-bodied ape. In the same paper Tuttle provides an excellent summary of the early hypotheses seeking to explain human origins (Richmond et al. 2001 provide a more up-to-date review). A more recent hypothesis to focus on the concept of **preadaptation** is the **vertical climbing hypothesis**. Following work by by Stern (1976), Cartmill and Milton (1977), and especially Prost (1980), and elaborated by Fleagle et al. (1981), this hypothesis is based principally on experimental studies of modern humans, apes, and other primates. Fleagle et al. (1981) noted that forelimb muscles that are large or unique in orthograde primates are more active in climbing than in pendular brachiation, suggesting that climbing more likely underlies the orthograde adaptations shared among apes and modern humans. Indeed, when apes climb vertical supports the movements of the hindlimb are similar to those observed when modern humans walk bipedally. Moreover, compared with quadrupedal walking, bipedal walking and vertical climbing are closer kinematically (Prost 1980) and involve more similar muscle recruitment patterns. Finally, the **ground reaction force**s produced by climbing primates when they walk bipedally are more similar to those produced by modern humans than they are to those produced by non-climbing primates. Thus, primates that habitually climb vertical supports when traveling arboreally might be expected to adopt bipedal locomotion when forced to travel terrestrially and would have fewer functional anatomical obstacles in an evolutionary transition to habitual bipedal walking. Thorpe et al. (2007) argue that instead of vertical climbing it was arboreal bipedalism on flexible branches that was preadaptive for hominin bipedalism because they observe that orangutans use extended knee and hip postures on flexible branches that resemble the extended limb postures used by modern humans walking on the ground. They argue that this extended-limbs arboreal bipedality was present in the great ape and modern human **most recent common ancestor** (or MRCA), and was subsequently lost independently in *Gorilla* and

Pan. The current climbing hypotheses are based on the concept of preadaptation and a key prediction implicit in all of them is that the ancestors of the earliest hominins were predominantly arboreal and they lacked adaptations for traveling on the ground in any way other than bipedally. The hypothesis that the pre-bipedal ancestor had an upper limb too specialized to evolve any terrestrial style other than bipedalism is no longer tenable under our current understanding of phylogeny because the MRCA of chimpanzees/bonobos and modern humans gave rise to both bipedal hominins and climbing/knuckle-walking panins. Recent analyses of fossil hominins and living apes by Richmond and Strait (2000) and Richmond et al. (2001) have revived the hypothesis originally proposed by Sherwood **Washburn** (e.g., 1968) that modern humans are descended from knuckle-walking ancestors. If true, hypotheses based on pre-adaptation could still explain why bipedalism evolved from an African ape-like MRCA because living African apes display the biomechanical similarities between climbing and bipedalism, but would complicate evolutionary scenarios requiring the MRCA to be exclusively arboreal because such hypotheses would need to explain why an ape that was already a terrestrial knuckle-walker would switch to travelling bipedally when on the ground. *See also* **bipedal**.

clinal variation *See* **cline**.

cline (Gk *k*linein=to lead, or bend) A gradual change in a feature as one moves across a given geographic region (geocline) or over time (chronocline). Clinal variation is found in morphology, genetics, and behavior. Clines occur either because dispersed populations may have slightly different adaptations in response to local conditions, or because two separate populations have been "merged" through **gene flow**. A stepped cline occurs when there is an abrupt change in a feature; stepped clines are often observed at **hybrid zones**. In modern humans there are multiple examples of clinal variation, many of them genetic, and some have argued that the majority of modern human variation is clinal (Lawson-Handley et al. 2007).

clock model Models introduced by Gould (1977) to address deficits encountered by other attempts to visualize **heterochronic** transformations in bivariate space. A vertical bar representing age forms the base of a semicircular clock composed of two arcs (one representing shape, and one representing size). Two clock hands are set in the midline, calibrated to the age representing the developmental stage of the ancestor of interest for a given analysis. The hands of the clock are then moved in either direction to demonstrate how shape and size in the descendant vary in relationship to the ancestor, and a marker on the vertical age bar is moved to demonstrate whether an adult morphology is achieved earlier, later, or at the same age as the ancestor.

closed basin lakes Lakes with no inflow or outflow. They are useful for **paleoenvironmental reconstruction** because they represent the equivalent of a "rain barrel" in that any change in lake level can be attributed solely to variations in the amounts of meteoric precipitation and evaporation, without the complication of variations in stream flow. Two examples are the **Tswaing Impact Crater** in southern Africa and Lake Bostumtwi in West Africa (Partridge et al. 2007, Scholz et al. 2007).

Clovis Currently the earliest well-documented culture in the Americas. The culture is named after the town of Clovis, New Mexico, which is close to the site of Blackwater Locality No. 1, where the first artifacts assigned to the Clovis culture were discovered in 1929. Clovis sites date to *c.*13,480–12,440 years BP using calibrated **radiocarbon dating** (Hamilton and Buchanan 2007, Haynes et al. 2007) and are found throughout most of the contiguous USA and parts of southern Canada. In terms of archeological visibility, the key Clovis artifact is a distinctive **projectile point**. Clovis points are bifacially flaked and usually have a flute (a flake scar that runs from the base part way towards the tip). They are frequently described as being **lanceolate** in shape, but recent work has shown that they can also be elliptic, linear, and deltoid (Buchanan and Collard 2009). It has been suggested on the basis of similarities between certain Clovis and **Solutrean** artifacts that Clovis Paleoindians were descended from people who migrated from Europe to North America via an "ice bridge" (Bradley and Stanford 2004). However, this hypothesis has not been widely accepted and the consensus is that the Clovis Paleoindians were descended from people who migrated to North America from Beringia via an ice free corridor between the Laurentide and Cordilleran ice sheets or along the northwest coast (Straus et al. 2005, Buchanan and Collard 2007). With regard to subsistence, the available evidence suggests that Clovis Paleoindians hunted a wide range of large mammals, including mammoth, mastodon, and ancient bison (Waguespack and Surovell 2003).

CMCK Acronym for the **Cultural and Museum Centre, Karonga** (*which see*).

CNV Acronym for **copy number variation** (*which see*).

CO Prefix for fossils recovered from Cooper's Cave. The prefix CO is followed by a capital letter indicating locations (A, B, C, etc.) within the cave (e.g., COB), which is then followed by the number of the fossil (e.g., COB-101). *See also* **Cooper's Cave.**

Coale and Demeny model life tables

This series of **life tables** was first published in 1966 by Ansley J. Coale and Paul Demeny and later updated in 1983. The 1966 life table categorized age as 0–1 year, 1–4 years (with an implied end of the interval at 4.997; i.e., up to 4 years and 364 days), and then progressed in 5-year intervals, ending with an open interval of 80+ years. The second edition added two more 5-year intervals at the end of the life table, bringing the final open interval to 90+ years. The addition of these older age intervals in the second edition was made so that the models would be "particularly useful in applications to populations with low mortality" (Coale and Demeny 1983, p. 1). As anthropological demography typically does not focus on low-mortality populations, the 1966 models suffice. The model life tables were constructed based on a large number of historical life tables, and were grouped into four families referred to as North, South, East, and West. The West model life tables are the ones most often used in anthropological demography. This follows from a statement in the first edition (and repeated in the second edition) that, "The four families of model tables will also provide a basis for estimating life tables in populations outside of the ones that underlie the model tables themselves. We would suggest utilizing the 'West' family in the usual circumstances of underdeveloped countries where there

is no reliable guide to the age pattern of mortality that prevails" (Coale and Demeny 1966, p. 29). Within each family the model life tables are numbered by level such that level 1 has a **life expectancy** at birth of 20 years and each higher level has the addition of 2.5 years of life expectancy at birth. Thus the highest numbered level in the first edition (level 24) has a life expectancy at birth of 77.5 years while the second edition's highest level (level 25) has a life expectancy at birth of 80 years. These round levels only apply to the "female" life tables, as the "male" tables are adjusted for differences in mortality between the sexes. Because the Coale and Demeny model tables were formed from historical census data, there has been considerable question as to whether they are applicable in **paleodemography** and in anthropological demography: **Weiss' model life tables** are probably more appropriate in paleodemography.

coalescence (L. *co*=together and *alescere*=to grow) Used in **molecular evolution** to describe the observation that all copies of a **gene** can be traced back to a single ancestral gene called the **coalescent**. Thus, the evolutionary histories of all the copies of a gene are said to coalesce at this ancestral gene. *See also* **coalescent theory.**

coalescent (L. *co*=together and *alescere*=to grow) The ancestral **gene** that gave rise to all the copies found in subsequent generations. *See also* **coalescent theory.**

coalescent theory (L. *co*=together and *alescere*=to grow) A retrospective model used in **molecular evolution** to trace all the **alleles** of a specific **DNA sequence** back to a single ancestral sequence called the **most recent common ancestor** (or MRCA). Thus, the evolutionary histories of all the lineages of a sequence are said to coalesce at this ancestral sequence. The coalescent process is affected by population size and structure, and coalescent theory can be used to infer the demographic or selective processes that produced the resulting **gene genealogy**. For example, compared with the coalescent produced under a neutral model, growing populations show a gene genealogy with long terminal branches and populations undergoing positive selection of one lineage (a **selective sweep**) or those populations that have undergone a bottleneck show a shorter **coalescent time** since older lineages are lost.

It is possible to distinguish between selection and **demography** as the cause of shortened terminal branches by looking at multiple loci: demography affects all loci while selection only affects a single locus.

coalescent time (L. *co*=together and *alescere*= to grow) The age of the ancestral sequence (called the **most recent common ancestor** or MRCA) that gave rise to all the copies in subsequent generations. Self-evidently each DNA sequence potentially has its own coalescent time, so the coalescent time of one sequence should not be interpreted as providing information about the timing of the evolutionary history of a species. For example, the coalescent time for mitochondrial DNA often differs from that of a nuclear DNA sequence (i.e., it is much shorter) because of the difference in **effective population size** (NB: the effective population size of mitochondrial DNA is one-fourth that of nuclear DNA) between these two types of DNA.

Cobb Collection Taxon *Homo sapiens*. History William Montague Cobb, a student of T. Wingate Todd, took over from Todd as Professor of Anatomy at Howard University in 1932. Cobb, emulating his mentor (*see* **Hamann–Todd Collection**), began to collect the skeletons of dissecting room cadavers. The collection now numbers around 700 modern human skeletons representative of the mid-19thC through to about 1969. The collection is located in the Cobb laboratory within the Anthropology Department at Howard University. Contact department Department of Sociology and Anthropology. Contact person Mark Mack (e-mail mmack@howard.edu). Postal address P.O. Box 987, Washington DC 20059, USA. Website www.coas. howard.edu/sociologyanthropology/. Tel +1 202 806 6853. Fax +1 202 806 4893. Relevant references Rankin-Hill and Blakey 1994, Hunt and Albanese 2005.

Cobb Mountain subchron *See* **geomagnetic polarity time scale**.

cochlea (Gk *kochlias*=snail) Snail-shaped anterior part of the membranous and bony labyrinths. The cochlea has a central bony spine (called the modiolus) and a spiral canal that consists of two and three quarter turns. The lumen of the canal is partly divided by a shelf of bone to which is attached the part of the membranous labyrinth called the **cochlear duct**. The endolymph-filled cochlear duct (called the scala media) subdivides (except at the apex of the cochlea) the perilymph-filled part of the cochlea into two channels, the scala vestibuli and the scala tympani. Compared to the semicircular canals, which form the posterior part of the labyrinth, the cochlea part of the bony labyrinth is relatively invariant among living higher primates. *See also* **bony labyrinth**.

cochlear duct One of the components of the endolymph-filled **membranous labyrinth**. Part of its wall consists of the Organ of Corti, the specialized linear arrangement of hair cells that interprets the relative movement of the hair cells as sound.

codon *See* **genetic code**.

coefficient of determination (or r^2) The square of the **correlation coefficient** (r). It is a measure of how much of the total variation in one variable, y, can be explained, or determined, by variation in a second variable, x. For example in the case of two variables, if $r=0.8$ then $r^2=0.64$, so knowledge about one of the variables provides the researcher with information that explains approximately 64% of the variation in the value of the other. This is a useful measure, especially when combined with **regression**, which describes the manner in which y changes with changes in x. More broadly, for any predictive model including one or more independent variables, the coefficient of determination indicates the proportion of variation in the dependent variable that can be explained or accounted for by variation in the independent variable(s). *See also* **correlation**.

coeval (L. *coaevus*=of the same age) The term used for fossils that are to all intents and purposes the same geological age. For example, hominin fossils recovered from different localities in the same stratigraphic interval in the Turkana Basin (e.g., **KNM-ER 406** and **KNM-ER 732**) are considered to be coeval.

cognition Refers to the mental process of "knowing:" awareness, perception, reasoning, and judgment are traditionally subsumed within cognition. Cognition also involves the ability to learn and subsequently use the knowledge that has been acquired in ways that are appropriate and adaptive. Researchers refer to different types of cognition as different types of **intelligence**. Among primatologists three types of intelligence (technical, ecological, and social) have received the most attention, and these also seem most relevant to the evolution of modern human behavior. Technical intelligence refers to knowledge about physics and causality, with a special emphasis on knowledge relevant to object and tool-use. Ecological intelligence refers to knowledge about the environment and other members of the ecological community, such as the seasonal availability of foods and the behavior of predators. The third type of intelligence of particular relevance to human evolution is **social intelligence**. It refers to an individual's knowledge of relationships with kin and non-kin, and with the behaviors and dispositions of conspecifics. *See also* **intelligence**.

Cohuna (Location 35°55′S, 144°15′E, Australia; etym. named after a local town) History and general description Cohuna 1, a young male's cranium, was recovered in the red loam of an alluvial formation during canal excavations in November 1925. Temporal span and how dated? Presumed **Late Pleistocene** age, but no radiometric dates are associated with the fossil. Cohuna's physical proximity to, and morphological similarity with, **Kow Swamp** suggest that Cohuna 1 is very likely contemporary with, and may even be a member of, the Kow Swamp population. Hominin found at site The Cohuna 1 skull is robust and morphologically similar to the Kow Swamp series found just to the east, displaying a long, flattened frontal with a well-developed supra-orbital torus. Archeological evidence found at site None. Key references: Geology and dating Thorne and Macumber 1972; Hominins MacIntosh 1972, Thorne and Macumber 1972, Thorne 1976.

Cohuna 1 *See* **Cohuna**.

COHWHS Acronym for the **Cradle of Humankind World Heritage Site** (*which see*).

collagen (Gk *kola*=glue and *genes*=born, for when collagen is boiled it forms gelatin, which is a glue) Collagen is the dominant protein in animal bone and connective tissue (e.g., tendons, ligaments). Collagen molecules form aggregates called collagen fibrils. Collagen fibrils contain three polypeptide alpha chains twisted together to form a right-handed triple helix (Orgel et al. 2006, Perumal et al. 2008) and the helices are held together by hydrogen bonds. With the aid of specialized proteins, collagen fibrils bundle together to form a collagen fiber. There are at least 29 types of collagen, but Types I, II, III, and IV are the most common types in mammals. Type I collagen is the most abundant and is the main collagenous component of bone. Collagen in bone can be isolated and used for **radiocarbon dating** or **stable isotope biogeochemistry**.

collecting areas *See* **Koobi Fora**.

Colobinae (L. *colobus* from the Gk *kolobos*= maimed, referring to the short thumbs that characterize the genus *Colobus*) One of the two subfamilies (the other is the **Cercopithecinae**) of Old World monkeys within the **Cercopithecidae**, one of the two families within the superfamily **Cercopithecoidea**. Members of this subfamily are commonly referred to as colobines. Colobines are commonly split into Asian and African forms and genetic evidence suggests the two groups diverged *c.*11 Ma (with a confidence limit of between 9.6 and 12.3 Ma). The earliest fossil colobine currently known is *Microcolobus*, recovered from Kenya and dated to between 9 and 8.5 Ma. By the late **Miocene** colobines had dispersed into Eurasia, with the extinct *Mesopithecus* recovered from sites in Greece, the Ukraine, Macedonia, Iran, Afghanistan, and Italy and the smaller colobine *Presbytis sivalensis* found in some terminal Miocene deposits in the Siwaliks. Colobines radiated in Eurasia during the **Pliocene** and **Pleistocene**. *Dolichopithecus*, a terrestrial colobine associated with wooded paleoenvironments, emerged in the Pliocene. Like *Mesopithecus*, it became extinct *c.*2.5 Ma. Compared to Africa and Europe, relatively little is known about the fossil record of colobines in Asia. However, it is clear that some modern genera were present in south and central Asia at least from the Pliocene and in southeast Asia from the latest Pliocene and Pleistocene (Jablonski 2002). The African fossil record of colobines between 7 and 9 Ma is generally poor, so little is known about colobine evolution on that continent in the late Miocene. However, by the time of the **Plio-Pleistocene**, African colobines were much more diverse than they are today. Unlike modern forms, which are relatively small (ranging in size from around 8 to 13 kg), arboreal, and restricted to primary or secondary forest, several extinct African colobines were large and terrestrial. The body masses of fossil *Paracolobus*, *Rhinocolobus*, and *Cercopithecoides* from East Africa were probably in the range of 20–50 kg (Delson et al. 2000), but even these three taxa contain a range of locomotor adaptations. For example, whereas *Cercopithecoides williamsi* was terrestrial, exploiting open environments (Birchette 1982, Elton 2006), *Paracolobus* may have been capable of a mix of terrestrial (Ting and Ward 2001) and arboreal locomotion (Elton 2000). In addition to these large-bodied colobines, smaller-bodied species, including some that are closely related to modern African forms, were also present in the Pliocene and Pleistocene fossil record of East Africa.

colobines The informal name for the Colobinae, one of the two subfamilies within the **Cercopithecidae**. *See also* **Colobinae**.

Columnata (Location 35°28′48″N, 1°16′12″E, near Sidi Hosni, Algeria) History and general description A collapsed rock-shelter/cave site in inland western Algeria best known for its "cemetery" with an extensive collection of modern human remains and the presence of a transitional Epipaleolithic tool industry known as the Columnatian. The latter, which is characterized by a high proportion of microliths and bone tools of Iberomaurusian tradition, was excavated by Cadenat (1948, 1957, 1966) and Brahimi (1972). Temporal span and how dated? **Radiocarbon dating** of charcoal and shell provided Late Pleistocene through the Early Holocene dates, specifically 10,800–5200 years BP. Hominins found at site Many modern human skeletons (minimum number of individuals, >60), classified as Mechta-Afalou, were recovered from what has been interpreted as the cemetery area at the site; one mandible reported from the site may have been used as a ritual item. Demographic markers suggest a healthy and growing population during the course of the site's habitation. Archeological evidence found at site The stratified sequence spans the Iberomaurusian (Oranian) through the Neolithic. The Iberomaurusian deposits are dated to 10,800 years BP but also appear to be laterally bedded with the Columnatian deposits (8200 years BP), rather than underlying them; this is interpreted as evidence that the Columnatian inhabitants removed the materials left behind by the previous Iberomaurusian

inhabitants. Above the Columnatian layers, deposits identified to the Capsian Mesolithic industry have been dated to 6750 years BP; the uppermost Neolithic layer dates to 5200 years BP. Key references: Geology, dating, and paleoenvironment Lubell 2001; Hominins Chamla et al. 1970; Archeology Lubell et al. 1976.

COM Acronym for **center of mass** (*which see*).

Combe-Capelle (Location 44°45′14″N, 0°51′00″E, France; etym. Fr. *combe*=valley or hollow; *Capelle*=name of the family that owned the farm) History and general description This site in the **Dordogne** region of France has several distinct archeological areas, including the Plateau de Ruffet, Roc du Combe-Capelle, Combe-Capelle haut (also called Abri Peyrony), and Combe-Capelle bas. On the Plateau de Ruffet, stone tools were uncovered by plowing. Roc du Combe-Capelle and Combe-Capelle haut are two rockshelters along the steep face of the plateau. Combe-Capelle bas is a deep and extensive deposit stretching from the base of the plateau towards the Couze River and is the main archeological area. Much of the site was first explored by amateur archeologists in the early 20thC including the Swiss Otto Hauser who in 1909 found the only hominin burial in the **Perigordian** layers of Roc de Combe-Capelle. Professional excavations were initiated in 1910 by D. Peyrony and his colleague Henri-Marc Ami in Combe-Capelle haut and bas and continued for several years under their direction. Combe-Capelle bas was revisited in the late 1980s by a team led by Harold Dibble and Michel Lenoir. Temporal span and how dated? Burned stone tools from the lower portion of Combe-Capelle bas have been dated using **thermoluminescence dating** to *c*.51 ka, but some of the stone tool technologies found in the Peyrony and Ami excavations suggest that the site may also have been inhabited much earlier, since *c*.100 ka. Hominins found at site The Perigordian modern human skeleton was buried with several grave goods, including a shell necklace and flint artifacts. It was sold by O. Hauser to the Museum für Vor- und Frühgeschichte in Berlin and the postcrania were burned during the bombing of the museum in WWII. The skull was damaged and misidentified until 2002 and only it and the shell necklace remain; it was dated using accelerator mass spectrometry radiocarbon dating to *c*.7 ka. In addition, a small fragment of a frontal bone also from *Homo* *sapiens* was recovered in 1914. Archeological evidence found at site With the exception of Roc de Combe-Capelle, the site contains predominantly **Mousterian** layers, with some late Acheulan assemblages in the lower layers. There is some debate about the kinds of Mousterian technologies present and their ordering within the site, which would influence interpretations of local cultural evolution within this area of France. Roc de Combe-Capelle has an Upper Paleolithic sequence including Perigordian (**Châtelperronian** or **Gravettian**), **Aurignacian**, and **Solutrean** layers. Key references: Geology, dating, and paleoenvironment Peyrony 1943, Valladas et al. 2003, Dibble and Lenoir 1995; Hominins Ambrose and Bouvier 1973, Hoffmann and Wegner 2002; Archeology Mellars 1969, Peyrony 1943, Valladas et al. 2003, Dibble and Lenoir 1995.

Combe-Capelle bas   *See* **Combe-Capelle**.

Combe-Capelle haut   *See* **Combe-Capelle**.

Combe-Grenal (Location 44°48′N, 1°13′E, France; etym. Fr. *combe*=small valley) History and general description This cave and talus deposit was discovered in 1817, but was extensively excavated much later by François Bordes between 1953 and 1965. It was at this site that Bordes formulated the original concept of the distinctive **Mousterian** tool typologies, and the Mousterian facies (e.g. Denticulate Mousterian, Typical Mousterian, etc.) that he believed represented individual tribes or ethnicities. Others have since debated these interpretations, but the site remains an important reference site for the Mousterian. Temporal span and how dated? Geostratigraphic correlation suggested an age range of **Oxygen Isotope Stage** (OIS) 5e or 6 through late OIS 3. Hominins found at site Several fragments were recovered from the Mousterian levels, all of which have been attributed to *Homo neanderthalensis*. Several of the recovered fragments have marks that have been interpreted as evidence of intentional defleshing. Archeological evidence found at site Bordes recognized 64 layers, 54 of which he identified as Mousterian. Key references: Geology, dating, and paleoenvironment Bordes 1955, 1972, 1973, Bordes et al. 1966, Dibble et al. 2009a; Hominins Le Mort 1989, Genet-Varcin 1982; Archeology Bordes 1955, 1972, 1973, Bordes et al. 1966, Dibble et al. 2009a.

communality analysis   A method of deciding which state of a character is primitive and which is **derived**. In communality analysis, the states exhibited by a range of **taxa** beyond those being investigated are examined. The state that occurs most frequently in the comparative sample is the one that is deemed to be primitive. This course of action is defended on the grounds of the principle of **parsimony**. *See also* **cladistic analysis**: **outgroup criterion**.

communication (L. *communicare*=to share) The term refers to the process by which information in the form of a message is passed from one individual to another. The various means of communication – visual, tactile, olfactory, gustatory, and auditory – correspond to the major sensory modalities. Communication has three components: (a) encoding, (b) transmission, and (c) decoding, or translation. According to the most liberal definitions of communication, all living things (including plants and animals) communicate. In this view, communication need not be intentional. That is, the encoding, or codification process need not be designed with a specific appreciation or understanding of the audience. Rather, codification may be the result of selection, whereby specific messages are favored because they enhance the sender's fitness (e.g., estrus swellings in many non-human primates). The decoding process may be similarly shaped by selection, so that the individuals attuned to certain messages are favored over those that are not. More conservative or anthropocentric conceptualizations of communication argue that **intentionality** is critical. In this view, the codification of messages and their perception is premised on mutual understanding between the individual sending the message (i.e., the sender) and the individual that is the focus of the message (i.e., the receiver). This is certainly the case for a uniquely modern human mode of communication, **language**.

comminution (L. *comminuere*=to reduce) A term used to describe the reduction in the particle size of food that takes place during the process of **mastication**.

community ecology (L. *communis*=common and Gk *oikos*=house) A community is a group of organisms that live together in the same area and **ecology** refers to all the interactions between an organism and its environment. Community ecology is therefore concerned with the interactions and features of co-existing populations of different types of organism. Although individual species may stay the same, the nature of the overall community (i.e., the total number and types of species) is likely to change over time or across space, influenced by the external physical environment as well as the other community members. Studying the competition between organisms is a key aspect of community ecology, as is understanding their ecological **niches**, diversity, species composition, and **predation**. Paleontologists have developed numerous qualitative and quantitative ways for measuring the influence of these factors in extinct communities. Community ecology studies help to reconstruct the environments inhabited by extinct hominins, and they can also be useful for increasing our understanding of speciation and extinction events. *See also* **ecology**.

compact bone *See* **bone**.

competitive exclusion This principle, also known as **Gause's rule**, states that two species cannot stably occupy an identical **niche** within the same **ecosystem**. Competitive exclusion leads either to local extinction of a species or to adaptation as one species moves into a different niche (i.e., character divergence). The competitive exclusion principle is best known in paleoanthropology because it underpinned the **single species hypothesis**. Wolpoff (1968) suggested that human culture is such a specialized ecological niche that "no more than one culture-bearing hominid could have arisen and been maintained" (*ibid*, p. 477). Thus, once culture had been acquired there could only ever have been one **synchronic** hominin species. In fact the competitive exclusion principle does not necessarily mean that two closely related species cannot co-exist: in guenon monkeys, for example, two or more species may live closely together but avoid direct and sustained competition by foraging on a different mix of resources or exploiting different areas of the canopy. Behavioral and morphological adaptations assist such differentiation. It has been argued (e.g., Banks et al. 2008) that competitive exclusion may have caused the extinction of *Homo neanderthalensis* in Europe.

complementary DNA (or cDNA) DNA that is synthesized using reverse transcriptase from messenger RNA. Because cDNA is synthesized from the mature mRNA, it does not contain the intron sequences present in the original DNA copy of the gene. Studies of gene expression include creation of a cDNA library so that the genes expressed in a given tissue can be identified and quantified. *See also* **gene expression**.

complex trait *See* **epistasis**; **heritability**; **quantitative trait locus**.

composite section (L. *compositus*=to put together) A diagram summarizing the **stratigraphy** of several localities within a fossil site, or several subregions within a site complex (e.g., Koobi Fora). It is assembled by using **marker bed**s that are defined lithologically or chemically to correlate shorter sections from individual localities or subregions into a longer composite section.

composite tool (L. *compositus*=to put together) Any tool made of more than one component (e.g., a spear consisting of a stone point mounted on a wooden shaft, bound together by leather, sinew, or resin). Composite tools make their first appearance in the **Middle Stone Age** and **Middle Paleolithic** at sites such as **Twin Rivers** and Königsaue (Clark 1988, Koller et al. 2001).

compound sagittal crest *See* **sagittal crest**.

compound temporal-nuchal crest A term introduced by John **Robinson** (1958) to describe the feature that results when the **temporal crest** fuses with the **superior nuchal line** or **nuchal crest**. Because the temporalis muscles of the extant higher primates, and especially the gorilla, are largest posteriorly, in many large gorilla males the temporal crest fuses with the nuchal crest to form a compound temporal-nuchal crest. Because the temporalis muscles are largest more anteriorly in the megadont archaic hominins, most of them do not display a compound temporal-nuchal crest, but a few do. Among hominins it is especially common in *Australopithecus afarensis*, in which the temporalis muscles are inferred to have been especially well developed posteriorly. It is given the prefix "compound" because what may look like a single crest actually consists of two elements, the temporal and the nuchal crests.

compound (T/N) crest Abbreviation for **compound temporal-nuchal crest** (*which see*).

computed axial tomography *See* computed tomography.

computed tomography (or CT) (Gk *tomos* = section and *graphein*=to write) An X-ray-based radiological imaging method. In paleontological and paleoanthropological applications it is mainly used to nondestructively visualize the inside and surface of modern and fossil bones and teeth. Compared to **plain film radiography**, CT (aka computed axial tomography or CAT) has three advantages: (a) no parallax distortion, (b) no superimposition of underlying structures, and (c) a higher contrast resolution. Medical CT scanners have an X-ray source and an array of detectors that rotate about a static specimen (the patient/specimen being examined). As X-rays pass through the specimen their absorption and scatter (attenuation), which depend on the local density properties, are detected and converted into an electronic signal. If this is repeated from many different angles the spatial distribution of the attenuation values in a cross-sectional slice can be computed and reconstructed as a 2D grayscale image consisting of a matrix of picture elements or pixels. Each pixel is assigned a CT number proportional to the degree that the X-ray beam is attenuated by the object (attenuation coefficient). The pixel size depends on the area to be scanned per axial section, also known as the field of view (or FOV). Since every slice has a given thickness, each pixel corresponds to a small cube known as a volume element or voxel. Medical CT systems can collect cross-sectional images slice by slice (sequential mode) or by acquiring a "corkscrew" of measurements from which slices are calculated afterwards (helical or spiral mode). The latter mode has the advantage that the acquisition time is much shorter compared to sequential CT scanning, but the reconstructed images may show small distortions. Modern CT scanners often have multiple lines of detectors that speed up the scanning process. Medical CT images have a spatial resolution typically between 0.2 and 0.5 mm (in the scan plane) and 0.5 and 1.0 mm (between slices). Industrial **micro-computed tomography** (or micro-CT or μCT) uses the same basic principles as medical CT, but the specimen is rotated on a turntable while the source/detector system remains static. Moreover, the cross-sectional images are calculated from a large series of digital radiographs obtained by projecting an X-ray cone beam onto a flat detector, rather than using a collimated X-ray fan beam and one or more lines of detectors as with medical CT systems.

The spatial resolution of μCT typically ranges between 1 and approximately 200 μm depending on the size of the specimen, but scanning of small objects at sub-micrometer resolution is possible as well. **Synchrotron radiation micro-computed tomography** (or SR-μCT) uses X-rays produced by a **synchrotron**. The monochromatic, high-energy, and parallel X-ray beam provides a better image quality than can be obtained with regular μCT scanners. Medical CT has adequate spatial resolution to reliably visualize internal skull morphology such as that of the paranasal sinuses, the endocranial surface, the inner ear, tooth roots, or the cross-sectional geometry of long-bone diaphyses. Moreover, it can provide a surface model of the external bone surface for use in the virtual reconstruction of a fossil or for morphometric analysis. With its better spatial resolution, μCT can also be used to examine structures with finer detail such as trabecular bone and the enamel-dentine junction of teeth. SR-μCT, especially in combination with propagation phase-contrast techniques, can visualize microscopic interfaces within structures, including the dental microstructure required to determine the age at death in juvenile fossil hominins. Persistent problems encountered when applying CT to fossil specimens include (a) density artifacts when scanning highly mineralized fossils because of the limited CT scale range of medical CT, (b) **beam-hardening** artifacts, (c) **lack-of-signal artifacts** in large and/or highly mineralized fossils, and (d) stitching artifacts due to the cone beam projection in μCT.

computer-assisted tomography *See* computed tomography.

conceptual modeling (L. *conceptus*=to conceive and *modus*=standard) One of two types of behavioral modeling suggested by Tooby and DeVore (1987). In conceptual modeling the model is not directly based on "real" observations, but on general principles developed from observations of a wide range of animals (e.g., bees) not just those closely related or analogous to the referent. Therefore, for example, baboon behavior could be used as a literal **referential model** or observations on baboons could be just one component of the information used to generate a conceptual model. *See also* **modeling**.

conchae (L. *concha*=shell) Used in two senses with respect to morphology. The first refers to the skin-covered expanded lateral end of the cartilaginous part of the external ear. The second refers to three (superior, middle, and inferior) scroll-like bones, also called turbinates, attached to the lateral wall of the nose. The superior and middle are parts of the ethmoid bone; the inferior concha is a separate bone. They are very delicate and are seldom preserved in the hominin fossil record.

conchoidal fracture (etym. This descriptive term refers to an analogy between the broken surface of a stone and the smooth inner surface of a shell) Describes how brittle, homogenous (glass-like) material breaks when the crack does not follow any natural plane of separation. It is a desirable feature of a lithic raw material that allows it to be made into tools, because the material fractures in a controllable fashion and can be made into a variety of shapes. Materials that have conchoidal fracture include siliceous rocks (**flint**, **chert**, **chalcedony**), bone, and some igneous rocks (**obsidian**, fine-grained **basalt**) (Whittaker 1994).

condylar position index The index introduced by Wilfrid **Le Gros Clark** (1950a) as a variation on Raymond **Dart**'s **head-balancing index**. The center of the convexity on the occipital condyles is projected onto the **Frankfurt Horizontal** and then the distance between the projected point and **opisthocranion** (CD) is divided by the maximum length of the cranium (CE) (i.e., the distance between **prosthion** and opisthocranion projected onto the Frankfurt Horizontal), multiplied by 100. The more posterior the occipital condyles, as in *Pan* compared to modern humans, the lower the value of the condylar position index.

confocal microscopy In a conventional light microscope the observer can see into a solid object as far as light can penetrate, but they see a confused image because information from multiple depths is superimposed. The optics of a confocal microscope are designed so that out-of-focus information is eliminated and only information from a specified depth, called the "plane of interest," is collected to form the image. There are various types of confocal microscopes (e.g., light, fluorescence, laser scanning), but the one that has been used in paleoanthropology is the confocal scanning light microscope (or CSLM). Alan Boyde and colleagues pioneered the use of tandem scanning reflected light microscopy (or TSRLM) to investigate dental tissues (Boyde et al. 1985, Boyde and Martin 1987) and Tim Bromage and colleagues have developed a confocal scanning light microscope that is portable enough to take to museums yet can still see **enamel prism**s in a dense solid structure such as dental enamel up to a depth of $50\,\mu m$ (Bromage et al. 2005). *See also* **portable confocal scanning optical microscope**.

confocal scanning light microscope *See* confocal microscopy.

confrontational scavenging Also referred to as power scavenging or aggressive scavenging, this is a **carcass acquisition strategy** defined as the capture of a nearly complete animal carcass from the nonhominin predators still engaged in consuming it. Modern **hunter–gatherers**, such as the **Hadza**, have been observed to occasionally practice confrontational scavenging.

congener (L. *congener*=of the same race) Strictly interpreted it refers to another **species** in the same **genus**, but some use "congener" in a more general sense to refer to members of the same taxonomic category, whatever that category's level in the Linnæan hierarchy, although they should not do so.

conglomerate (L. *glomus*=ball) A rock made of large, rounded fragments (pebbles, gravel) that may be cemented together by a fine-grained matrix or mineral cement.

Congo Museum The name of the museum now known as the **Royal Museum for Central Africa**.

Congress of the Pan-African Association of Prehistory and Related Studies The Pan-African Congress on Prehistory and Quaternary Studies was inaugurated in Nairobi in January 1947 and it met thereafter in Algiers (1952), Livingstone (1955), Leopoldville (1959), Santa Cruz de Tenerife (1963), Dakar (1967), Addis Ababa (1971), Nairobi (1977), and Jos (1983). At the Jos meeting the name of the congress was changed to the Congress of the Pan-African Association of Prehistory and Related Studies, and the **Pan-African Association of Prehistory and Related Studies** has organized the later congresses in this series in Harare (1995), Bamako (2001), Gaborone (2005), and Senegal (2010). *See also* **Pan-African Congress on Prehistory and Quaternary Studies**.

connective tissue *See* bone; cartilage.

consensus cladogram *See* cladistic analysis; cladogram.

consistency index (or CI) Developed by Kluge and Farris (1969) as a measure of the goodness of fit between a **cladogram** and a **character state data matrix**. The CI for a single **character** is calculated as m/s, where m is the minimum number of character state changes required by any possible cladogram and s is the number of character state changes required by the cladogram being investigated. To compute the CI for a group of characters used in a **cladistic analysis**, the m values for the characters are summed and then divided by the sum of the s values for the characters (M/S). The values for CI can range between 1 and 0. A CI of 1 indicates that the character state data matrix is perfectly congruent with the cladogram (i.e., the cladogram requires no changes due to **homoplasy**). Homoplasy levels increase as the CI decreases toward 0. The CI value is affected by the number of taxa or the number of characters included in a character state data matrix, whereas the retention index was designed so that it is not influenced by either the number of taxa or the number of characters included in a character state data matrix. *See also* **cladistic analysis**.

contagion *See* social enhancement.

contingency (L. *contigens*=to touch, but contigency now refers to an unintended event) An evolutionary perspective that stresses the effects of earlier factors or events (usually random or near random) in influencing subsequent evolutionary outcomes. For example, in small populations in the absence of countervailing factors, **genetic drift** will ultimately lead to the elimination or fixation of alleles, thereby reducing the population's genetic variation. But which of a pair of alleles is fixed and which lost is due to chance, but the population's subsequent evolution will be determined by (i.e., be contingent upon) the pattern of random fixation and loss. This is a micro-

evolutionary example, but in recent years, influenced by punctuational theory, and especially by the writings of Stephen J. Gould, workers have considered the possible role for contingency at the macroevolutionary level. In ascribing most evolutionary change to speciation, **punctuated equilibrium** identifies many more speciation and extinction events than **phyletic gradualism**. The isolation that promotes speciation does not of itself necessarily result in enhanced fitness or improved adaptation, and many species, especially when originating as clusters, will be of identical or near-identical fitness. Their continuation or extinction may therefore be strongly influenced by chance: those that do persist provide the starting points for subsequent evolutionary diversification, which is thus contingent on the fortuitous pattern of species survival. For example, Gould has argued that the survival of only a small number of forms from the extraordinarily diverse Burgess Shale fauna, most of which became extinct in the Middle–Late Cambrian, predetermined subsequent metazoan evolution in general and arthropod evolution in particular (Gould 1990). Any other pattern of Burgess fauna survival would have meant other starting points and so very different patterns of post-Cambrian evolution. Possible examples of contingency in hominin evolution include (a) differences of detail in the locomotor systems of East and southern African *Australopithecus*, (b) the initial appearance of several early *Homo* species (*H. habilis*, *H. rudolfensis*, and *H erectus/ergaster*) with differing morphologies and then the extinction of all save the last, (c) the evolution of *Homo neanderthalensis* morphology from later *Homo heidelbergensis* populations in higher-latitude Eurasia around the Middle/Upper Pleistocene boundary, possibly associated with population bottleneck(s) during Oxygen Isotope Stage 6, and (d) the evolution of *Homo floresiensis* in Southeast Asia. In this last case, **founder effect**, genetic drift, transilience event(s) that mitigate the adverse impacts of disadvantageous recessive alleles exposed as homozygotes, and endemic dwarfing were probably all involved, illustrating that chance factors can interact with other evolutionary processes. It is also possible that the fixation of some or all of the characteristic morphology of *Homo sapiens* (e.g., globular neurocranium, retracted face, gracile mandible with a chin, pronounced basicranial flexion, slender limb-bones, etc.) results from random factors influencing diversity in later Middle

Pleistocene African populations. The spread of *H. sapiens* would then reflect subsequent behavioral, technological, and/or reproductive or demographic innovations of some *H. sapiens* populations compared with other later-Pleistocene *Homo*, rather than any enhanced fitness resulting from the morphology *per se*, which nonetheless becomes ubiquitous. See further explanations and examples in papers by Stephen J. Gould, Chris Paul, Robert Foley, and Alan Bilsborough in *Structure and Contingency: Evolutionary Processes in Life and Human Society* (Bintliff 1999).

conule   (L. *conule*=diminutive of cone) A small accessory cusp of a maxillary (upper) postcanine tooth that is not one of the four main cusps (i.e., not the **proto-**, **para-**, **meta-**, or the **hypocone**). For example, the **metaconule** is an accessory cusp between the metacone and protocone. *See also* **accessory cusp**.

conulid   (L. *conule*=diminutive of cone) A small accessory cusp of a mandibular (lower) postcanine tooth (hence the suffix "-id") that is not one of the five main cusps (i.e., not the **proto-**, **ento-**, **meta-**, or **hypoconid**, or the **hypoconulid**). For example the **tuberculum intermedium** is an extra distal cusp between the metaconid and the entoconid. *See also* **accessory cusp**.

conventional radiocarbon dating   *See* **radiocarbon dating**.

conventional radiography   *See* **plain film radiography**.

convergence   (L. *convergere*=to incline together) When a character shared by taxa has not been inherited from their **most recent common ancestor** (i.e., the taxa have evolved the same character independently). Convergence is a common cause of homoplasy. For example, are the coronally rotated petrous bones seen in *Homo* and *Paranthropus* inherited from their most recent common ancestor, or are they an example of convergence? *See also* **homoplasy**; **convergent evolution**.

convergent evolution   Convergent evolution is the independent origin of similar **traits** in distantly related taxa (e.g., the acquisition of thick-enameled teeth in orangutans and cebus monkeys).

Parallel evolution is the independent origin of similar traits in closely related taxa. For example, the coronally rotated petrous bones seen in *Homo* and *Paranthropus* may not have been inherited from their most recent common ancestor, in which case they would be an example of convergence in the form of parallel evolution, or if bipedalism evolved independently in *Australopithecus afarensis* and *Australopithecus africanus*, then this would also be an example of parallel evolution. Simpson (1950) was most likely the originator of these terms, or, if not, he certainly was among the first to formalize them (Simpson 1960). He suggested "there is no really fundamental difference between" them (*ibid*, p. 181), and inevitably whether a trait in two taxa that are not in the same clade qualifies as an example of convergent or parallel evolution (i.e., whether the taxa are "distant" or "close") is a subjective decision. Convergent and parallel evolution both result in **homoplasy**: the appearance of similar morphology in two taxa that is not seen in the most recent common ancestor of those taxa. "Deep" convergent evolution (e.g., similar body form in whales and fishes) is relatively easy to recognize because only a minority of features is affected. Parallel evolution in closely related taxa is more difficult to distinguish from characters inherited from the most recent common ancestor because the taxa involved are more similar overall and are likely to have responded to common selection pressures in similar ways. *See also* **homoplasy**.

Coobool Creek   (Location 35°12′S, 143°45′E, Australia) <u>History and general description</u> An ancient cemetery near the Coobool Crossing of the Wakool River between Swan Hill and Deniliquin in southern New South Wales where G.M. Black collected a series of hominin skeletal remains in 1950. <u>Temporal span and how dated?</u> A preservative applied to the hominin remains upon their collection unfortunately precludes direct radiocarbon dating of the skeletons. Based on morphological similarities to the robust hominins that comprise the **Kow Swamp** sample, and on distinctions from the more gracile mid-Holocene specimens from **Green Gully** and **Lake Nitchie**, the Coobool Creek series is presumed to belong to *Homo sapiens* and to be of Late Pleistocene age. <u>Hominins found at site</u> The Coobool Creek crania are robust and morphologically similar to those from Kow Swamp, **Cohuna**, and **Naccurie**. The anterior skull vault shows evidence of artificial cranial deformation, and the faces are large, robust, and sub-nasally prognathic. The Coobool Creek mandibular sample displays the greatest average dimensions (e.g., height of the symphysis, and height and breadth of the corpus and the buccolingual dimensions of many of the teeth) among all of the hominin samples found at Australian sites. Archeological evidence found at site None. Key references: Geology, dating, and paleoenvironment Brown 1989; Hominins Brown 1989, 2010, Durband 2008.

Coobool Crossing *See* Coobool Creek.

Coon, Carleton Stevens (1904–81) Carleton Coon was born in Wakefield, Massachusetts, USA. He entered Harvard University to study Egyptology with George Reisner, but he changed to anthropology after taking classes with Earnest Hooton. Coon graduated from Harvard in 1925 and remained there for his graduate studies, which involved conducting fieldwork in North Africa studying the Rif Berbers of Morocco. Coon completed his PhD in anthropology in 1928 and his dissertation, *Tribes of the Reef*, was published in 1931. Coon accepted a position as a lecturer at Harvard in 1928 and continued to conduct anthropological research, spending a period from 1929 to 1930 studying the people of northern Albania and traveling to Ethiopia and Yemen in 1933–4. Coon was appointed an instructor at Harvard in 1935 and became a Professor of Anthropology there in 1938. In 1939 he published a reworked version of William Z. Ripley's *The Races of Europe*, a work originally published in 1899, in which Coon used craniometry to examine the physical anthropology of the people of Europe. Coon took a leave of absence from Harvard in 1941 and in 1942 he was sent to German-occupied Morocco with a mission to mobilize Riffian tribesmen in preparation for the landing of Allied troops in Africa. After the war Coon returned to teaching at Harvard, but in 1948 he accepted a position as Professor of Anthropology at the University of Pennsylvania, as well as curator of ethnology at the University of Pennsylvania Museum. In addition to conducting research into the physical anthropology of current populations, Coon also conducted archeological excavations. During one of his first excavations in Morocco in 1939 he unearthed what was then thought to be a *Homo neanderthalensis* maxillary bone from a cave in Tangier. From 1948 to 1949 he participated in excavations in Iraq and Iran, and during the mid-1950s he excavated sites in Afghanistan and Syria, where he explored an archeological record that spanned the Paleolithic and the Neolithic. In 1950 Coon published, in collaboration with anthropologist Stanley Garn and evolutionary biologist Joseph Birdsell, *Races: A Study of the Problems of Race Formation in Man*, which used Darwinian ideas of adaptation, especially in response to climate, to explain modern human racial variation. Coon also published a general overview of recent discoveries and theories in archeology and physical anthropology in *The Story of Man* (1954). However, Coon's most extensive and controversial treatment of the idea of human racial diversity was *The Origin of Races* (1962) in which he argued that the five major human races (Caucasoid, Negroid, Capoid, Mongoloid, and Australoid) have a deep evolutionary history. According to Coon, human racial diversity was already present in *Homo erectus* and he argued that the different modern human races evolved separately from geographically separate subspecies of *H. erectus* into *Homo sapiens*. Moreover, he believed these races had evolved at different rates, which explained why the modern human races have attained such different levels of cultural development. This model of human evolution is often referred to as the **candelabra model** and differs from the so-called **multiregional model** of human evolution originally espoused by Franz **Weidenreich**, and more recently by Milford **Wolpoff** and Alan **Thorne**. According to Coon, the candelabra model suggests that the various geographical races evolved independently from one another (polygenism), while the multiregional model argues that the distinct geographical populations evolved into modern humans in only relative isolation, maintaining their regional identity, but with some **gene flow** between these populations, and thus did not evolve independent of one another (polycentrism). Coon's ideas were controversial and in later years he backed away from them. Sherwood **Washburn** and Ashley Montagu, both of whom had written on the question of race from the perspective of physical anthropology, used new ideas derived from the **modern evolutionary synthesis** and population genetics to criticize Coon's interpretations. Scientific and political attitudes of the time were turning against the scientific racism of the 19thC and early 20thC, yet Coon's work did

demonstrate that human racial diversity was a subject that could be studied using evolutionary theory and the hominin fossil record. Coon published a less controversial, but valuable, study of modern human racial diversity titled *The Living Races of Man* in 1965. He later included new biological and anthropological discoveries, especially evidence from molecular biology, in *Racial Adaptations* (1982), which expanded upon his earlier work. Coon retired from the University of Pennsylvania in 1963, and in 1981 he published his autobiography *Adventures and Discoveries*.

Coopers *See* Cooper's Cave.

Cooper's Cave [Location 26°00'46"S, 27°44'45"E, South Africa; etym. a cave in the **Sterkfontein** Valley named after R.M. Cooper. **Broom** (in Broom and Schepers 1946) refers to a "Mr. R. M. Cooper" who "owns the Sterkfontein Caves" and also refers to Cooper giving Broom "every facility to collect ... on his farm" (p. 144). Some publications refer to it as "Coopers Cave", but Cooper's Cave is correct] History and general description A cluster of breccia-filled caves in Pre-Cambrian dolomite in the Blauuwbank Valley, Gauteng, South Africa, between Sterkfontein to the west and **Kromdraai** to the east. A single hominin tooth was discovered in 1939 and since then hominins have been reported in 1995, 2000, and 2009. Temporal span and how dated? **Biostratigraphy** (1.9–1.6 Ma) and **uranium-series dating** (U-Pb) (1.5–1.4 Ma) for Locality D. Hominins found at site Hominins have been recovered from three spatially distinct infills (A, B, and D) at Cooper's. From Cooper's A there is an upper incisor (COA-1), and a squashed face (COB-101). From Cooper's B there is an upper third molar (TM 1514), but the original was lost during the 1980s and only a cast remains. The hominins recovered from Cooper's D include four isolated teeth, three deciduous molars and a permanent molar (CD 1634, 1638, 5774 and 22619), two mandibular fragments (CD 6807 and 17796) and a thoracic vertebra (CD 5773). All have been assigned to *Paranthropus robustus*, except COA-1, which has been assigned to *Homo sp.*; only the hominins from Cooper's D are known to be from that location. The hominins from Cooper's A and B were found in boxes in the **Transvaal Museum**, not at the site. The fossil prefixes were assigned by staff at the museum and do not refer to the lettered infills referred to above. Archeological evidence found at site Possibly early **Acheulean**. Key references: Geology and dating Brain 1958, Berger *et al.* 2003, de Ruiter *et al.* 2009; Hominins Berger *et al.* 1995, 2003, Steininger and Berger 2000, Steininger *et al.* 2008, de Ruiter *et al.* 2009. Collections prior to 1998 are at the **Ditsong National Museum of Natural History** (formerly the **Transvaal Museum**), but the hominins recovered after 1998 at kept at the **Bernard Price Institute**.

coordinate In mathematics, a coordinate is one of a set of numbers that specifies where a point is located in space. *See also* **geometric morphometrics**; **Cartesian coordinates**.

coordinate data The **Cartesian coordinates** of **landmarks**. Bookstein (1991, p. 2) defines landmarks as loci that are homologous across individuals in a sample (i.e., the landmarks have the "same" location in every other member of the sample being investigated). These coordinate data can be one-, two-, or three-dimensional. 2D coordinates are usually captured using a digitizing tablet or by measuring an image on the computer. 3D data can be captured directly using a coordinate digitizer such as a Microscribe or Polhemus, or may be measured on surface or volumetric scans. Volumetric data are based on image slices from **computed tomography** (CT or μCT) or **magnetic resonance imaging** (MRI or μMRI) or, more recently, optical projection tomography. These slices contain grayscale values that correspond to tissue densities and are concatenated to obtain a 3D representation of an object. Surface scanners provide high-resolution 3D representations of an object's surface using either laser or more traditional optical technology and may also include texture information, and virtual surfaces can be generated from CT or MRI data. Most software packages allow landmark coordinates to be measured directly on these virtual surfaces or volumetric objects.

Cope's rule This "rule," which is probably better described as a trend, is attributed to Edward Cope (1840–97) and it suggests that lineages tend to increase in body size over the course of evolution. For example, all marine phyla, except molluscs, show a size increase between the

Cambrian and Permian (Novack-Goppottshall 2008). The *post hoc* explanation is that the trend can be explained by **directional selection** (e.g., larger body size buffers against changes in the paleoenvironment, lowers the risk of predation, etc.), but several factors that impose limits on the size (e.g., birds must be light enough to fly, changes in size may be accompanied by changes in ecological niche, and that a new niche may already be occupied) may oppose these pressures (Van Valkenburgh et al. 2004). Moreover, a complication with the "rule" is that it may be an artifact of the tendency for **adaptive radiation**s to begin with small-bodied forms. Thus, if a clade begins with small-bodied ancestors, then as it diversifies to fill available niche-spaces, the evolution of both large- and small-bodied descendents may give the appearance of a trend towards increasing body mass simply because there was a small-bodied starting point. Some workers cite hominins as an example of Cope's rule insofar as the later members of the genus *Homo* are larger-bodied than the australopiths.

Coppens, Yves (1934–) Yves Coppens was born in Vannes, in the Brittany region of France, in 1934. He participated in local excavations during his school years and after graduating from the lycée in Vannes, Coppens entered the University of Rennes, where he completed a degree in the natural sciences. He pursued graduate studies in paleontology at the University of Paris (Sorbonne) where he studied the **Proboscidea** under the direction of Jean Piveteau. Coppens accepted an appointment at the **Centre National de la Recherche Scientifique** (National Centre for Scientific Research) in 1956 where he worked on vertebrate and human paleontology during the late **Tertiary** and **Quaternary** periods. In 1960 he undertook the first of four fieldwork seasons in Chad, Central Africa, and in 1961, while collecting fossils during this expedition, Coppens discovered a hominin cranium that he named *Tchadanthropus uxoris* (now considered *Homo sapiens*). He later took over from Camille **Arambourg** as the leader of the French team that, along with an American team lead by Clark **Howell** and a Kenyan team led by Richard **Leakey**, composed the **International Omo Research Expedition** exploring the deposits in the **Lower Omo Basin** in Ethiopia between 1967 and 1976. In 1974, Coppens participated in excavations with American paleoanthropologist

Donald **Johanson** and French geologist Maurice **Taieb** at **Hadar** that resulted in the discovery of *Australopithecus afarensis*. Besides these expeditions, Coppens has organized, led, or participated in field research in Algeria, Tunisia, Mauritania, Indonesia, and the Philippines. Coppens left the Centre National de la Recherche Scientifique in 1969 to accept a joint position in the Laboratory of Anthropology at the Muséum national d'histoire naturelle (National Museum of Natural History) and the **Musée de l'Homme** (Museum of Man). Beginning in 1981 Coppens became a prominent supporter of the idea, first proposed by the Dutch primatologist Adriaan Kortlandt, that the evolutionary separation of the **hominin** and **panin** clades was caused by environmental changes, namely the formation of the East African Rift System, which cut off forest living primates of the western side of the rift from primates that were forced to adapt to savannah conditions on the eastern side of the rift. This theory, popularly called the "east side story," argues that savannah conditions prompted our hominin ancestors to walk upright while forest-dwelling primates west of the rift valley faced no such selective pressure. However, the recent discoveries of *Australopithecus bahrelghazali* and *Sahelanthropus tchadensis* in Chad have posed a challenge to this hypothesis. Coppens was appointed to the chair of paleoanthropology and prehistory at the Collège de France in 1983 where he remained until his retirement in 2005.

coprolite (Gk *copros*=dung) The term for fossil feces or dung. Coprolites are an example of a **trace fossil**. *See also* **fossil**.

coprophagy (Gk *kopros*=dung, and *phagein*= to eat) The consumption of feces or dung. Coprophagy is a common strategy in animals with simple digestive systems that consume low-quality foods. Many primates use this dietary strategy, usually temporarily. Gorillas have been observed to eat their own feces and those of conspecifics, but in modern humans coprophagy is usually considered to be a sign of significant mental disorder or disease.

copy number variation (or CNV) Refers to a segment of DNA a thousand, or more, **base pairs** (or 1 kb) long that is devoid of high-copy repetitive sequences such as **long interspersed nuclear elements** (LINEs) and which is present in variable copy number among individuals. CNVs may be inter- or intra-chromosomal, although the vast majority of CNVs are intra-chromosomal. Intra-chromosomal CNVs may be dispersed along a chromosome or they may be arranged in tandem. For molecular geneticists studying human and primate evolution, it has long been thought that the genomes of two modern human individuals are approximately 99.9% similar (Li and Sadler 1991, HapMap 2003, Hinds et al. 2005), while other research has indicated that the modern human and chimpanzee genomes are nearly 99% identical (Sibley and Ahlquist 1984, Ruvolo et al. 1994, Ebersberger et al. 2002, Shi et al. 2003, Watanabe et al. 2004). At both levels of comparison, this variation was thought to be primarily represented by single nucleotide differences. However, these basic assumptions have recently been reconsidered based on findings that copy number variants are more common than previously expected, both among modern humans (Iafrate et al. 2004, Sebat et al. 2004, Tuzun et al. 2005) and between modern humans and chimpanzees (Newman et al. 2005, Wilson et al. 2006). In fact, in both cases, it appears that the total content of genomic variation (measured in base pairs) is greater for copy number variation than for single nucleotide-level variation (Cheng et al. 2005, Consortium 2005, Redon et al. 2006). The generation of such segments can involve both gains or losses of genomic material and an immediate consequence of a genomic copy number gain or loss may be a corresponding increase or decrease in gene expression (Ohno 1970). Thus, intra-specific CNVs may be involved in intra-specific phenotypic variation including individual differences in disease susceptibility. Inter-specific CNVs may likewise account for between-species differences in gene expression.

core (ME *core*=the hard, seed-containing central part of some fruits) Any piece of stone from which flakes have been deliberately removed. Cores are present at the earliest archeological sites such as those at **Gona**, Ethiopia, and they occur at most archeological sites where stone tool production has occurred (e.g., in Africa they are not restricted to one or other of the **Early**, or **Middle**, or **Later Stone Age**s).

coronal (L. *coronalis*=a crown) Refers to a plane at right angles to the **sagittal** plane or to the cranial **suture** that is aligned at right angles to the **sagittal**

suture. The coronal suture is where the front of a small crown, or a wreath, such as those worn by Roman notables, would cross the crown of the head. Coronal is a term applied to the bony skeleton and not to the dentition.

coronoid process *See* mandible.

corpus *See* **mandible**.

corpus callosum [L. *corpus*=body and *callosum*=callous (i.e., firm)] The largest bundle of **axons** in the brain of placental mammals. It runs across the midline of the brain and it physically and functionally links the **cerebral cortex** of the left and right **cerebral hemispheres**. The corpus callosum consists of both **myelin**ated and unmyelinated axon fibers. In general, as brain size increases among primates, the axons that traverse the corpus callosum increase in diameter and develop a greater degree of myelination to minimize delay in the conduction of nerve impulses between the cerebral hemispheres, but there are no significant differences in axon diameters within the corpus callosum between modern humans and chimpanzees (Caminiti et al. 2009). This suggests that nerve impulses take longer to travel between the cerebral hemispheres in modern humans, which may relate to increased functional lateralization. In addition, several studies in modern humans have reported **sexual dimorphism** in the cross-sectional area of the corpus callosum relative to total brain size. These results have been disputed, and studies of corpus callosum size among common chimpanzees have not revealed a similar pattern of sexual dimorphism.

corpus striatum *See* **striatum**.

correlated-age methods *See* geochronology.

correlation (L. *cor*=together and *relatus*=relate) Earth sciences An attempt to relate **strata** that are exposed in two or more different locations within the same, or different, sites. Strata can be compared using the appearance of the fossils they contain or their chemical make-up. For example, **tuff** layers exposed in isolated sediment blocks within the **Koobi Fora** site have been correlated on the basis of their **lithology** and chemistry. This method for stratigraphic correlation is known as **tephrostratigraphy**. Statistics A measure of the amount of association between two variables, such that when one variable changes the other also changes. If the changes are the same then the correlation is perfect, if they are similar the correlation is strong, but if the changes are only a little better than random the relationship is weak. The **correlation coefficient** (r) measures the strength of the relationship. The **coefficient of determination** (or r^2) is a measure of how much of the total variation in y can be explained, or determined, by variation in x. For example, in the case of two variables, if $r=0.8$ then $r^2=0.64$, so knowledge about one of the variables provides the researcher with information that explains approximately 64% of the variation in the value of the other. This is a useful measure, especially when combined with **regression**, for it describes the manner in which x changes in response to any with changes in y. *See also* **correlation coefficient**.

correlation coefficient   The correlation coefficient (r) measures the strength of the relationship between two variables with $r=1$ or -1 for a perfectly positive or negative relationship, respectively, and $r=0$ for no relationship. The significance of r varies with the number of observations (N); generally the value of r needs to be high (approximately >0.8 or -0.8) if N is a small number. *See also* **correlation**.

correlation matrix   An $n \times n$ square **matrix** corresponding to a data set with n **variables**. The value in each cell of the matrix corresponds to the **correlation** between the row and column variable for that cell. The values on the diagonal of the matrix are all correlations of variables with themselves and are thus all equal to one.

cortex   [L. *cortex*=bark (i.e., outer covering)] The outer covering, or layer, of a structure or organ (e.g., the outer covering of gray matter of the **cerebral hemispheres**) or the outer layer of dense, or cortical, bone of a limb bone that lies immediately beneath the **periosteum**.

cortical   *See* **cortex**.

Coryndon Museum   *See* **National Museums of Kenya**.

cosmogenic nuclide dating   Principle At least two versions of this dating method, $^{26}Al/^{10}Be$

and $^{10}Be/^9Be$, have been applied to dating hominin sites. The principle of the $^{26}Al/^{10}Be$ system is that the isotopes ^{26}Al and ^{10}Be are produced in **quartz** when neutrons, protons, and muons penetrate the upper few meters of the Earth, but if sediments are buried at depths greater than 5–10 m (this is why this version is called cosmogenic nuclide *burial* dating) the production of these isotopes ceases. Once the burial has halted the production of new ^{26}Al and ^{10}Be the different decay characteristics of the inherited ^{26}Al and ^{10}Be (the former has a shorter half-life than the latter) cause the ratio of these elements to change over time, thus providing an estimate of how long the quartz has been buried. The principle of the $^{10}Be/^9Be$ system is analogous to that of **radiocarbon dating**. Atmospheric ^{10}Be is adsorbed on aerosols then it gets transferred to the Earth's surface via precipitation where it is incorporated into sediments. It decays into ^{10}B with a half-life of *c.*1.4 Ma, and by using ^9Be as a "normalizing species" (Lebetard et al. 2008, p. 3228) the ratio of $^{10}Be/^9Be$ in silicate frustules can be used to provide an age for **diatomite** layers. Method ^{26}Al, ^{10}Be, and ^9Be are measured using **accelerator mass spectrometry** (or AMS). Suitable materials Quartz and diatomite. Time range This is determined by the half-lives of ^{26}Al (*c.*0.7 Ma) and ^{10}Be (*c.*1.4 Ma). After 5 Ma the amount of ^{26}Al is <1% of its original value, and this is too small to measure; thus the time range of the $^{26}Al/^{10}Be$ system is *c.*1 Ma to 4 or 5 Ma. The time range of the $^{10}Be/^9Be$ system is 0.2 to 14 Ma. Problems There are two sources of uncertainty: measurement uncertainty and uncertainty about the length of the half-lives of ^{26}Al and ^{10}Be. These two sources place a limit of *c.*100 ka on the precision of ages derived using cosmogenic nuclide dating. Examples *See* Partridge et al. 2003, Gibbon et al. 2009.

Cosquer Cave   (Location 43°11′57″N, 5°27′07″E, France; etym. named after Henri Cosquer, the professional diver who found the site) History and general description Cosquer Cave is unique among **Upper Paleolithic** cave sites in that it is located underwater about 40 m off the French coast near Marseille. The large cave, discovered by divers in 1985, was recognized as an archeological site in 1991 and it has since been explored by a team led by Jean Clottes and Jean Courtin. The isolated environment has provided excellent preservation of a variety of cave art, mostly painted hands and drawn and engraved animals, including horses, bison, and an assortment of birds. Temporal span and how dated? Conventional **radiocarbon dating** on charcoal from the sediments and from the paintings themselves have suggested two periods of inhabitation, one from 27.5 to 26.5 ka and the other from 19.5 to 18.5 ka. Hominins found at site None. Archeological evidence found at site The excellent cave paintings and engravings are the main feature of this site. Because of the difficulty accessing the cave, excavations have not been attempted. Key references: Geology, dating, and paleoenvironment Clottes et al. 1992, 1997, Clottes 1992; Hominins N/A; Archeology Clottes et al. 1992, 1997, Clottes 1992.

Cossack (Location 20°40′S, 117°11′E, Australia; etym. named after a local village) History and general description In 1972 a hominin partial skeleton was found eroding from a deflating coastal sand dune situated about 3 km/1.8 miles east of Cossack in Western Australia. Temporal span and how dated? The hominin skeletal remains were not suitable for direct dating but shell **midden**s in the immediate area yielded radiocarbon ages of 2000–6500 years BP. Hominins found at site The Cossack individual was a *c.*40 year-old male *Homo sapiens* whose robusticity and cranial morphology (especially its low frontal and long overall skull length) is similar to the skulls from terminal Pleistocene sites (e.g., **Kow Swamp**, **Cohuna**, and **Coobool Crossing**) in eastern Australia. Archeological evidence found at site Surface scatters of stone artifacts, bone, and shell fragments. Key references: Geology and dating Freedman and Lofgren 1979; Hominins Freedman and Lofgren 1979; Archeology Freedman and Lofgren 1979.

COT Acronym for cost of transport. The cost of transport of an animal is the amount of oxygen it uses to transport a unit of its mass over a fixed distance. It is usually expressed as ml O_2 kg^{-1} m^{-1}.

Cova del Gegant (Location 41°13′17″N, 1°46′08″W, Catalonia, Spain; etym. Ca. Cave of the Giant) History and general description Cave site on the Mediterranean coast near the city of Sitges, about 40 km/25 miles south of Barcelona. First excavated in 1952 by the Agrupació Muntanyenca de Sitges, and excavations continued through the 1970s and 1980s, partially in an

attempt to save material threatened by the encroaching sea. Temporal span and how dated? **Biostratigraphy** suggests that the sequence in the cave spans between 128 and 40 ka. A recent study has shown that much of the archeological sequence dates between 49.3±1.8 ka, the **uranium-series date** of the overlying flowstone, and 60.0±3.9 ka, the optically stimulated **luminescence** date of the basal deposits. A uranium-thorium date of 52.3±2.3 ka has been acquired on the Neanderthal mandible. Hominins found at site A hominin mandibular corpus was recovered during the first field season in 1952 but was not recognized as such until 2001. The morphology of the mandible, its chronology, and its **Mousterian** archeological context all point to the mandible belonging to *Homo neanderthalensis*. Archeological evidence found at site Around 60 Mousterian stone tools were recovered. Key references: Geology, dating, and paleoenvironment Viñas 1972, Daura et al. 2010; Hominins Daura et al. 2005; Archeology Martínez and Mora 1985.

Cova Negra (Location 38°59′N, 0°32′W, Spain; etym. Ca. Black Cave) History and general description Cave site located on the left bank of the Albaida River in the municipality of Xàtiva in the region of Valencia, Spain, approximately 17 m above the current water level. The initial excavations were carried out by G. Viñes from 1928 to 1933, the next phase by Fusté was from 1953 to 1958 and the third phase by V. Villaverde from 1981 to 1991. Cova Negra provides one of the most complete late Pleistocene stratigraphic sequences in Spain, and it has yielded numerous remains of *Homo neanderthalensis*, including abundant juvenile specimens. Archeological evidence from Cova Negra indicates sporadic, short-term hominin occupations of the site (Arsuaga et al. 2007) that has been interpreted as indicating a high degree of mobility among Neanderthals. Temporal span and how dated? Thermo**luminescence dating** produced remarkably old ages for the oldest layers, 235±21 ka, but these ages are inconsistent with the biostratigraphic evidence, which suggests the most recent levels at the site are 50±8 ka. Hominins found at site Twenty-four hominin fragments have been recovered from the three excavation phases, although many specimens lack precise stratigraphic information due to older imprecise excavation and recovery practices. The parietal bone originally described by Fusté (1953) was attributed to *H. neanderthalensis*, and while some subsequent authors have judged it to be a "pre-Neanderthal", the combined chronological, archeological, and morphological evidence is consistent with the initial judgment. Archeological evidence found at site The recovered artifacts are **Mousterian** in character and they are surprisingly uniform throughout the different levels; the majority of the artifacts are side **scrapers**. Key references: Geology, dating, and paleoenvironment Arsuaga et al. 2007; Hominins Fusté 1953, Arsuaga et al. 1989, 2007; Archeology Villaverde and Fumenal 1990.

covariance A measure of association between two variables similar to **correlation** in which positive values indicate that an increase in one variable is associated with an increase in the other, and negative values indicate that an increase in one variable is associated with a decrease in the other, and the magnitude of the covariance indicates the strength of the association. Unlike correlation, covariance can take on any value and is not limited to the range –1 to 1. Correlation is actually a scaled version of covariance; the **correlation coefficient** for two variables is equal to their covariance divided by the standard deviation of the first variable and the standard deviation of the second variable. The covariance of a variable with itself is equal to the **variance** of that variable.

covariance matrix An $n \times n$ square **matrix** corresponding to a data set with n **variables**. The value in each cell of the matrix corresponds to the **covariance** between the row and column variable for that cell. The values on the diagonal of the matrix are all covariances of variables with themselves and are thus are each equal to the **variance** or the corresponding variable.

covariation A concept similar to **variation**, covariation refers to how two variables change in response to each other. Covariation is typically measured as **covariance** or **correlation**.

coxa (L. *coxa*=hip, or hip bone, or hip-joint) An old-fashioned term that is best avoided. The proper name for the "hip bone" is the pelvic bone. *See also* **anatomical terminology**.

CPR Abbreviation for Ceprano. *See* **Ceprano 1**.

Cradle of Humankind World Heritage Site Located in South Africa, this World Heritage Site includes important **karstic** early **hominin** sites (e.g., **Cooper's**, **Drimolen**, **Gladysvale**, **Gondolin**, **Kromdraai**, **Malapa**, **Sterkfontein**, and **Swartkrans**), which are clustered to the southwest of Pretoria and to the northwest of Johannesburg in the province of Gauteng. This area was declared a World Heritage Site in 1999 thanks in large measure to the advocacy and persistence of Phillip **Tobias** and his South African colleagues. Subsequently, the sites of **Makapansgat** and **Taung** were added as components of the World Heritage Site. There is a fine interpretative center at Maropeng, near Sterkfontein.

crania (Gk *kranion*=brain case) Plural of **cranium**.

cranial (Gk *kranion*=brain case) Used informally for fossils of the bony skull and dentition. When used formally it refers to structures that are in the direction of the head; the opposite of **caudal**.

cranial base (Gk *kranion*=brain case and L. *basis*=base) This term is synonymous with the more formal term **basicranium** (*which see*).

cranial base angle (Gk *kranion*=brain case and L. *basis*=base) (or CBA) Quantifies the angular relationship between the two major (anterior and posterior) components of the **basicranium** in the sagittal plane. Various landmarks have been used to capture this relationship, but they are all variants on Huxley's use of nasion–prosphenion–basion (Huxley 1867). Some use a different anterior landmark (e.g., the anterior cribriform point; Scott 1958) and others a different middle landmark (e.g., the "pituitary point"; Cameron 1927), but the choice of basion as the posterior landmark is standard. Lieberman and McCarthy (1999, Table 1, p. 490) provide a summary of the various combinations of landmarks that have been used to measure the CBA. The anterior portion of the basicranium is derived from the prechordal (the part anterior to the end of the notochord) portion of the embryonic cranial base and the posterior cranial base is formed by the parachordal (the part alongside the notochord) part of the embryonic cranial base. The anterior component comprises the ethmoid, the posterior part of the frontal, and the anterior part of the sphenoid body, and underlies the anterior cranial fossa. The posterior component comprises the posterior portion of the sphenoid body and the basioccipital, and underlies part of the posterior cranial fossa. The transverse "hinge" (i.e., the place where the anterior and posterior components of the basicranium meet) is the mid-sphenoidal synchondrosis, which is in the floor of the hypophyseal fossa. The region beneath and between the two components of the basicranium is occupied, from posterior to anterior, by the oropharynx, nasopharynx, and part of the nasal cavity. Numerous factors influence the CBA, including the volume of the brain relative to the area of the cranial base, the length of the face, the size of the pharynx, and the relationships that must be maintained to ensure that the eyes point forwards during normal posture. Among the extant great apes and modern humans the average CBA is most acute in modern humans (mean $\pm$ SD$=135\pm3.1°$ in adult modern human crania) compared with adult chimpanzees ($157\pm4.9°$) and adult bonobos (approximately $148°$). The CBA changes during ontogeny; so, for example, in modern humans the CBA is $143°$ in neonates and then flexes (becomes more acute) during the first 2 years of postnatal life. This differs from the ontogenetic changes in non-human primates, in which the crania base extends (becomes more acute) throughout ontogeny. Few fossil hominin crania are complete enough to be able to make a reliable estimate of the CBA, but what evidence there is suggests there is much variation in CBA among hominins depending on relative face size, brain size, and other factors. In pre-modern *Homo* taxa such as *Homo neanderthalensis* the CBA is approximately 15% more obtuse than in anatomically modern humans. Lieberman et al. (2000) provide a review of the comparative context and significance of the CBA in fossil hominins. *See also* **basicranium; chondrocranium**.

cranial base flexion *See* cranial base angle.

cranial fossa(e) *See* basicranium.

cranial vault (Gk *kranion*=brain case and L. *volvitus*=arched) The roof and side walls of the **cranium**. Formed from a pair of flat bones, the parietals, the flat or squamous parts of the two temporal bones, and the squamous parts of the frontal and occipital bones. Vault bones have an inner and an outer layer of compact bone called the inner and outer tables and an intervening layer of spongy bone filled with red bone marrow, the **diploë**; this type of "layered" structure is said to minimize deformation and maintain a "low-strain" environment (Herring and Teng 2000). The **meningeal arteries** supply the vault bones and the diploic veins drain blood from the bones of the cranial vault to **dural venous sinuses** internally and also to veins on the external surface of the cranium. Veins that pass through the cranial vault and which thus connect the endocranial veins (i.e., the veins on the inside of the cranial vault) to the pericranial veins (i.e., the veins on the outside of the cranial vault) are called **emissary veins** and they run through **emissary foramina** (syn. **calvaria**).

cranial venous drainage (Gk *kranion*= brain case and L. *vena*=vein) Two venous systems drain the head. The intracranial system drains venous blood from the tissues that lie *inside* the bones of the **cranium** (e.g., the brain and the covering meninges) and from the cranial bones themselves. The extracranial system drains venous blood from the tissues that lie *outside* the bones of the cranium (e.g., the soft tissues of the scalp and face). The **intracranial venous system** is divided into two components: an internal and an external. The internal one consists of the deep cerebral veins that do not leave any trace on the bones of the cranium. The external component of the intracranial venous system consists of superficial cerebral veins that drain the surface of the brain, and these, in turn, drain into a system of venous channels called the **dural venous sinuses** that run between the fibrous and endosteal layers of the **dura mater**, the outer of the three meningeal layers. Importantly for paleoanthropology these dural venous sinuses do leave impressions on the inner, or **endocranial**, surface of the bones of the cranium. The venous sinuses that are of most concern to paleoanthropologists are the superior sagittal sinus that drains most of the blood from the superior surface of the brain and the dural venous sinuses within the posterior cranial fossa; namely the transverse, sigmoid, occipital, marginal, and petrosquamous sinuses. The transverse sinus runs from the midline laterally in the base of the tentorium cerebelli in a groove on the internal surface of the **squamous** part of the **occipital**. In most modern humans and in the vast majority of chimpanzees and gorillas (Kimbel 1984) the venous blood draining the deep structures of the brain forms the deep cerebral veins that normally drain into the left transverse sinus and the superior sagittal sinus drains into the right transverse sinus. Each transverse sinus drains into a sigmoid sinus that runs inferiorly in a sigmoid-shaped groove from the lateral end of the transverse sinus to the superior jugular bulb, which marks the beginning of the internal jugular vein. This pattern of dural venous sinuses is known as the **transverse-sigmoid system** and individuals (e.g., **KGA 10-525**) with transverse and sigmoid venous sinuses on both sides are said to show transverse-sigmoid system dominance. In just a few percent of modern humans the venous blood from the brain takes a different route. Instead of draining into the transverse sinuses, the deep cerebral veins and the superior sagittal sinus drain into an enlarged midline occipital sinus, which runs down from the cruciate eminence towards the foramen magnum in a midline groove. The occipital sinus drains into uni-, or bilateral, marginal sinuses, which run(s) around the margin(s) of the posterior quadrants of the foramen magnum to drain into the superior jugular bulb and/or the **vertebral venous plexuses**. This is known as the **occipito-marginal system** of venous sinuses and individuals with a substantial occipital sinus and with marginal venous sinuses on both sides (e.g., **OH 5**) are said to show occipito-marginal system dominance. Some individuals (e.g., A.L. 333–45 and **KNM-ER 23000**) have a transverse-sigmoid system of venous sinuses on one side and an occipito-marginal system of venous sinuses on the other, and a few individuals (e.g., **Taung 1**) have been reported to show coexistence of the two systems on the same side (Tobias and Falk 1988). The petrosquamous sinus links the sigmoid sinus with the extracranial venous system (see below) and when present it occupies a groove that runs along the endocranial surface of the temporal bone at the junction of the squamous and **petrous** parts of the temporal bone. The extracranial venous system consists of four main external veins; from anterior to posterior they are the facial, superficial temporal, posterior auricular, and occipital veins. The facial drains into the internal jugular vein, the superficial temporal into both the internal and external jugular veins, the posterior auricular mainly into the external jugular vein, and the occipital into both the internal jugular and the vertebral veins. None of the veins of the extracranial venous system leave any trace on the cranium. A system of **emissary veins** (plus some venous sinuses such as the petrosquamous sinus) connects the intra- and extracranial venous systems by passing through small foramina in the walls of the cranium. The first researcher to suggest that the pattern of intracranial venous drainage may have some taxonomic valency was Tobias (1967), who pointed out that OH 5 showed clear evidence of occipito-marginal system dominance. Falk and Conroy (1983) followed this up with observations about the cranial venous drainage in *Australopithecus afarensis*, and subsequently Falk (1986, 1988b) suggested the incidence of occipito-marginal system dominance in *Paranthropus boisei* and *Au. afarensis* was 100%. Since then it has became clear that, while the incidence of occipito-marginal system dominance is distinctively higher in *P. boisei*, *Paranthropus robustus*, and *Au. afarensis* (Wood and Constantino 2007, Kimbel and Delezene 2009), cranial venous drainage is a **polymorphism** and the presence or absence of one system or the other cannot be used as a definitive taxonomic marker.

cranial venous sinuses *See* **cranial venous drainage**.

cranium (Gk *kranion*=brain case) The term used for a **skull** minus its **mandible**. In one scheme the **cranium** is divided into the **neurocranium** and the **viscerocranium**. The viscerocranium is the part of the cranium that covers the anterior aspect of the **brain** and it is equivalent to the **face** including the upper jaw. The viscerocranium develops from a combination of the frontonasal process and the first pharyngeal arch; all of its components derive from cells that come from the neural crest. The neurocranium consists of the bony elements that surround the brain and the special sense organs. It develops from the paraxial mesenchyme (i.e., the first five somites and the somite-like material rostral to the first somite) with a contribution from the neural crest. Another way to think of the cranium is that it consists of the **calvaria** (or cranial vault), the **face**, and the **basicranium**. *See also* **endocranial cavity**.

cribriform plate (L. *cribrum*=a sieve) Small, bilateral, perforated plates of bone that are part of the ethmoid bone. They are rectangular, with their long axis oriented anteroposteriorly, and are situated on either side of the midline *crista galli* in the depressed medial part of the floor of the anterior cranial fossa. The perforations transmit the nerves connecting the olfactory receptors in the nasal cavity with the olfactory bulbs that lie beneath the frontal lobes of the brain.

crista anterior The term used by Remane (1960) for an enamel crest on the crowns of the maxillary postcanine teeth that others call the **epicrista** (*which see*).

crista galli (L. *crista*=crest and *gallus*=a Gaul. Early anatomists thought the *crista galli* resembled the emblem on a Gallic soldier's helmet.) A midline crest on the **endocranial** surface of the ethmoid in the floor of the anterior cranial fossa. It marks the anterior attachment of the falx cerebri, a fold made up of the fibrous layer of the dura mater that lies between the parts of the cerebral hemisphere not joined by the **corpus callosum**.

crista nova The term used by Remane (1960) for a crest that others call the **mesial marginal ridge** (*which see*).

crista obliqua The term used by Remane (1960) for an enamel crest on the crowns of the maxillary postcanine teeth that others call the **distal trigon crest** (*which see*).

crista occipitomastoidea *See* **occipitomastoid crest**.

crista paramastoidea (L. *crista*=crest, Gk *para*=beside, *mastos*=breast, and *œides*=shape. Literally "a crest beside the mastoid process") In some crania the **occipitomastoid crest** is separate from another more lateral structure, which Franz **Weidenreich** (in his description of the Solo calvariae; Weidenreich 1951, p. 280) referred to as the *crista paramastoidea*. *See* **juxta-mastoid eminence**.

crista transversa The term introduced by Selenka (1898/1900) for an enamel crest on the crowns of the maxillary molar teeth that others call the **distal trigon crest** (*which see*).

crista transversa anterior The term used by Korenhof (1960) for an enamel crest on the crowns of the maxillary postcanine teeth that others call the **epicrista** (*which see*).

critical function This term was introduced by Rosenberger and Kinsey (1976) for the morphological counterpart of the ecological concept of **fallback foods**. It refers to morphology that enables an animal to process its fallback food(s) as opposed to its preferred foods. For example, it has been suggested that the distinctive **anterior pillars** of the face of *Australopithecus africanus* may be an adaptation to allow the premolars to be used for processing hard objects too large to be accommodated between the molars.

CRL Acronym for the **Cenozoic Research Laboratory** (*which see*).

Cro-Magnon (Location 44°56′N, 1°00′E, France; etym. local French name for the cliff and rock-shelter, originally written as "Cramagnon" on the public register of land boundaries) History and general description The site of Cro-Magnon was discovered in 1868 when a road crew was moving large blocks of limestone from the talus slope at the base of the steep cliffs lining the Vézère river just outside of the village of Les Eyzies. Édouard Lartet, a lawyer turned paleontologist and archeologist, excavated and described the 17 m-wide cave, which was shallow and low-ceilinged and only about 6 m from the mouth to the rear wall and 4 m from the base of the deposit to the stone ledge that made up the ceiling of the cave. Except for a small area at the back of the cave it was filled with sediment. Lartet recognized several archeological layers with animal bones, flint tools, and distinct hearths. The cave was excavated prior to the naming of the various tool technologies, so the layers were not linked with any particular stone tool industry. The modern human fossils found at the site were among the first in Europe, and thus gave rise to the term "Cro-Magnon Man." Temporal span and how dated? 25–30 ka based on correlation with **Abri Pataud**. Hominins found at site Several modern human skeletons were found at the back of the cave in the top-most archeological layer. The area contained the remains of five individuals. The most famous is the "Old man of Cro-Magnon" (**Cro-Magnon 1**), but within a 1.5 m-radius circle around this individual they also found the remains of a female (**Cro-Magnon 2**), a perinatal infant (Cro-Magnon 5), and two more remains, likely males (Cro-Magnon 3 and 4). Lartet (1868) and subsequent researchers (e.g., Vallois and Billy 1965a) have interpreted the restricted area of the fossils as evidence of either a single burial or repeated use of a burial area. The skeletons were found on and just below the surface in the back of the cave, but the skulls were filled with calcrete and covered in a stalagmitic crust that continued towards the front of the cave and was covered by calcareous deposit that filled the front of the cave. Archeological evidence found at site Lartet (1868) mentions that **scraper**s make up the majority of the stone tools in the main archeological layers (which he labels F, G, and H) and that layer H contained several **bone tool**s, including what he called punches and arrowheads (*poinçons* and *flèches*). Layer I, the layer in which the skeletons were found, was quite thin and contained no hearths, but it did yield a variety of stone tools, animal bones, bone tools, and amulets. Rivière (1897) excavated more of the site, and found what he determined were undisturbed **Magdalenian** layers, including several pieces of engraved bones. Pittard (1963) noted some tools that are typical of the Gravettian in a private collection of tools that were collected from the yard of the house next to the site. Key references: Geology and dating Movius 1969a; Hominins Broca 1868, Lartet 1868, Vallois and Billy 1965a, 1965b, Dastugue 1967, Vallois 1970; Archeology Lartet 1868.

Cro-Magnon 1 Site Cro-Magnon. Locality Layer I. Surface/*in situ In situ*. Date of discovery March 1868. Finder Édouard Lartet. Unit N/A. Horizon N/A. Bed/member N/A. Formation N/A. Group N/A. Nearest overlying dated horizon None. Nearest underlying dated horizon None. Geological age *c*.25–30 ka. Developmental age *c*.50 years. Presumed sex Male. Brief anatomical description This specimen comprises a nearly complete skull that lacks all but the right M^2; the rest of the dentition was lost post-mortem. Several postcranial elements, including robust femora and a tibia, were assigned to this individual, but the association between cranial and postcranial material from this site is much debated. Announcement Lartet 1868. Initial description Broca 1868. Photographs/line drawings and metrical data Vallois and Billy 1965a, 1965b. Detailed anatomical description Vallois and Billy 1965a, 1965b. Initial taxonomic

allocation *Homo sapiens*, "Cro-Magnon race" (Quatrefages and Hamy 1874). Taxonomic revisions None. Current conventional taxonomic allocation *Homo sapiens*. Informal taxonomic category Anatomically modern human. Significance This was among the first fossil evidence of *Homo sapiens* recovered in Europe. Location of original **Musée de l'Homme**, Paris, France.

Cro-Magnon 2 Site Cro-Magnon. Locality Layer I. Surface/*in situ In situ*. Date of discovery March 1868. Finder Édouard Lartet. Unit N/A. Horizon N/A. Bed/member N/A. Formation N/A. Group N/A. Nearest overlying dated horizon None. Nearest underlying dated horizon None. Geological age *c*.25–30 ka. Developmental age *c*.20–30 years. Presumed sex Female. Brief anatomical description This fragmentary calvaria includes parts of the face and maxilla, including the right M^1 and M^2. A few postcranial elements have been assigned to this individual, but the exact association between the cranial and the postcranial material from this site is much debated. Announcement Lartet 1868. Initial description Broca 1868. Photographs/line drawings and metrical data Vallois and Billy 1965a, 1965b. Detailed anatomical description Vallois and Billy 1965a, 1965b. Initial taxonomic allocation *Homo sapiens*, "Cro-Magnon race" (Quatrefages and Hamy 1874). Taxonomic revisions None. Current conventional taxonomic allocation *Homo sapiens*. Informal taxonomic category Anatomically modern human. Significance Much has been made of a minor scratch and fracture of the frontal bone, which Lartet and later Broca argued was made by a flint tool (Broca 1868, Lartet 1868), possibly due to violence or post-mortem defleshing. More recently researchers have suggested this was much more likely to be damage caused during excavation (Vallois and Billy 1965a, 1965b, Dastugue 1967, Vallois 1970). Location of original **Musée de l'Homme**, Paris France.

Cro-Magnon Man *See* **Cro-Magnon**.

Crocuta (L. *crocuta* = literally the "saffron-colored one." The term was said to have been used by Pliny the Elder to refer to a wild animal in northern Africa.) The **genus** name of the modern spotted **hyena**, *Crocuta crocuta*, a member of the family **Hyaenidae**. These bone-crunching **carnivoran**s are common and distinctive members of the modern African fauna and extinct forms of hyena have been implicated in the bone **comminution** seen at some archeological and hominin paleontological sites. Thus hyenas may have been an important competitor for the earliest tool-making and tool-using hominins. *Crocuta* probably diverged from the true hyenas of the genus *Hyena* and first appears in the African fossil record in the late **Pliocene**.

Cromerian *See* **glacial cycles**.

cross-section Section normally taken at right angles to the long axis of a bone or a tooth. Cross-sections of teeth can also be in any plane that can be consistently located among observers.

cross-sectional data Data collected from many subjects at a single point in time. This type of "broad-swath" sampling aims to gather information from large numbers of individuals of similar or different ages and character states. It is used to acquire information about, and is considered representative of, the larger population from which the sample is drawn. **Longitudinal data** (derived from measurements taken from a series of observations on a single individual or group of individuals) are preferable to cross-sectional data for studies that wish to examine how individuals change over time, but for obvious reasons this type of data is more difficult to collect.

cross-sectional geometry (or CSG) A method for investigating the mechanical adaptations of bones, based on mechanical engineering formulae used for modeling the bending stress experienced in straight beams under static (non-moving) loads. A section of a bone is taken at a specified location (e.g., midshaft, 66% of the length from the proximal end, etc.) and in a plane perpendicular to the long axis of the bone. Sections can be obtained either by physically sectioning the bone or by medical imagery such as computed tomography scans or magnetic resonance imaging (in living people), or by tracing the external (periosteal) surface of the bone and approximating the thickness of the cortical walls and the approximate endosteal surface by means of two radiographs of the bone taken at a right angles to each other [generally anteroposterior (AP) and mediolateral (ML)]. Once the periosteal and endosteal contours of the section have been obtained, most techniques model the space between the two contours as cortical bone of uniform mechanical properties. Using software such as SLICE (Nagurka and Hayes, 1980) or Sylvester et al.'s (2010) routine in R, a series of properties of the cross-section are then calculated including the centroid of mass [computed as the average (x, y) coordinate of each unit of area of cortical bone in the section], the total area (or TA) (i.e., the total area circumscribed by the periosteal contour), the medullary area (or MA) (i.e., the area circumscribed by the endosteal membrane), the cortical area (or CA) (i.e., the area occupied by cortical bone, calculated as TA–MA), the percentage of cortical bone in the section (%CA) [calculated as $100 \times (\text{CA/TA})$], and a series of second moments of area. Each second moment of area captures the rigidity of the section and is partially proportional to that section's ability to resist deformation under bending loads. Despite this imperfect proportionality with rigidity, second moments of area are often used in the literature as proxies for the strength of the bone. Second moments of areas are calculated around axes constrained to pass through the centroid of mass of the section; this is the case for straight beams but not necessarily the case for bones, which are rarely perfectly straight. Each second moment of area is calculated as the sum of each unit of area in the section occupied by cortical bone multiplied by that part of the area's distance to bending axis squared. This gives the bone located farthest away from the bending axis exponentially more ability to resist bending than bone located close to the axis; bone located directly in the axis of bending has no ability to resist bending. Commonly used second moments of area include I_{xx} (the second moment of area around the x axis, generally set to equal AP bending), I_{yy} (the second moment of area around the y axis, generally set to equal ML bending), I_{max} (the maximum second moment of area of the section, whether it occurs along an AP, ML, or other direction), I_{min} (the minimum second moment of area of the section, which is always positioned $90°$ from I_{max}), and J, the polar moment of area, which captures twisting around an axis perpendicular to the plane of the section and running through the centroid of mass (i.e., twisting about the long axis of the bone). J is proportional to the sum of any two perpendicular second moments of area of the section (i.e., $J = I_{xx} + I_{yy} = I_{max} + I_{min}$). Some additional properties are also occasionally calculated, including theta (Θ), the angle between the y axis and the direction of I_{max}, and additional distance measures from the centroid of mass to the most distant point of bone in each of the bending axes to be considered. One must divide a second moment of area by that distance to compute the section modulus (Z) in the direction of bending; the section modulus, rather than the second moment of area, is the key aspect of the geometry of a section that determines its resistance to deformation under a bending moment. Once the cross-sectional properties have been calculated they have to be standardized for body size and shape to compare the strength relative to size of different individuals or populations. Several adjustments have been proposed to control for size; the best currently available divides areas (CA, TA, and MA) by body mass and second moments of area by the product of body mass times the bone's length (Ruff 2008). Ideally one should use section moduli rather than raw second moments of area, but this has rarely been done in the anthropological literature. One exception was Ruff's (1995) research on a sample of femora in which he found that the polar section modulus (Z_p) of a sample of femora could be approximated as $Z_p = (J)^{0.73}$; dividing Z_p by the product of body mass times bone length then provides a better approximation of the relative rigidity of the bone, albeit with the disadvantage of ignoring individual variation in the ratio of the length of the outermost piece of bone to the second moment of area. The drawbacks of CSG analysis stem from the fact that virtually all of its assumptions can be violated (e.g., the neutral axis may not pass through the centroid of mass or the bones do not bend in the direction one might expect based on their external form), thus rendering calculations based on that assumption inaccurate (Demes et al. 2001, Lieberman et al. 2004, Demes 2007). Also the various mechanical properties of bone may vary across a section, along the axis of a bone, and between individuals, thus leading to further inaccuracies (Currey 2002, Wang et al. 2010). With respect to inferences about recent activity patterns from bones there is reason to suspect that the plasticity of bones to given levels of activity varies across the life span and by location in the limb, thus adding additional levels of complexity to inferences about activity patterns based on CSG (see Lieberman et al. 2003, Pearson and Lieberman 2004, Ducher et al. 2009). Some researchers minimize the import of these complications (e.g., Ruff et al. 2006) but little empirical evidence has been offered is to justify such a dismissal. Computer-aided studies of CSG, which became common in the mid-1980s, have produced insights into hominin functional anatomy. These include evidence of a strong pattern of gracilization from archaic through modern *Homo* (Ruff et al. 1993, 1994, 1997); evidence of diachronic changes in bilateral asymmetry of the upper limb, with ***Homo neanderthalensis*** and early **Upper Paleolithic** humans having a stronger right than left upper limb (Trinkaus et al. 1994, Churchill and Formicola 1997); evidence that the marked sexual dimorphism in I_{xx} of the femur and tibia in many hunter–gatherer societies largely disappears in industrialized societies (Ruff 1987); evidence of a marked decline in mobility from late archaic *Homo* and early modern humans to recent times (Trinkaus and Ruff 1999a, 1999b, Holt 2003); and evidence that hip proportions, especially bi-acetabular breadth and femoral neck length, may account for some of the differences in morphology between *Homo erectus* and modern human femora and perhaps other otherwise unexplained anomalies such as the round femoral midshaft sections of Neanderthals and the Tyrolean Iceman (Ruff 1995, Ruff et al. 2006).

cross-striations (L. *stria* = furrow) **Short-period incremental lines** (i.e., **circadian**) in enamel running at approximate right angles across the long axis of **enamel prisms**, best seen under polarized light and with a typical spacing of 3–6 μm in hominoids (reviewed in Boyde 1989, Smith 2006). They reflect a 24-hour cycle of ameloblast secretory activity that can be used to determine the daily secretion rate of enamel. Cross-striations are equivalent to **von Ebner's lines** in **dentine** (Dean 1998b). *See also* **enamel development**; **incremental features**.

cross-validation A form of **resampling** that is used to evaluate how well predictive models perform for known cases. In particular, cross-validation is useful for identifying when models are "overfit" with respect to a particular data set (i.e., when predictions are thought to be *more* accurate than they actually are because the efficacy of a model is highly dependent on the data set used to generate it). Cross-validation has been used heuristically in human evolutionary research, but typically it is not used for hypothesis testing. In predictive models such as **regression** or **discriminant function analysis**, portions of the data set are sequentially held back during the initial calculation and then predictions from the model are compared to the subset that was *not* included in the generation of the model. In this way the model is not influenced by the data to be predicted and the **accuracy** and **precision** of predictions derived from the model can be assessed. A common application of cross-validation leaves each observation out in turn, predicts it from the resulting model, and calculates an overall measure of success; see Smith (2002) for other variants. For example, to investigate how well a regression predicts body mass from one or more variables, an investigator could calculate the regression using all but the first specimen, then calculate the difference between the predicted value and the actual value of the first specimen. This is repeated for the second, third, etc. specimen, enabling a metric of success to be calculated (for example, the standard deviation of differences, proportion of predictions within 20% of the actual value, the proportion of predictions that fall outside of a 95% prediction interval, etc.). Similarly, cross-validation can be used to evaluate the success of categorical predictions, for example as in the use of discriminant function analysis to predict habitat categories based on bovid skeletal metrics (DeGusta and Vrba 2003, Kovarovic and Andrews 2007). In general, cross-validation is not necessary in simple (bivariate) regression or for linear discriminant functions, but it becomes important to consider its use as more variables are added to multiple regression analyses or when quadratic discriminant function analysis is used. Because in these cases the models fit more tightly to the data, the possibility that the models are overfit with respect to the data becomes a concern. *See also* **discriminant function analysis**.

crossing over *See* **recombination**.

crown (L. *corona* = crown) The part of a tooth covered by enamel and which, after eruption, projects above the surrounding alveolar bone. It is made up of an outer enamel cap and an inner core of dentine. The rest of a tooth, made of dentine and embedded in the bony alveolus, is called the root. The boundary between the crown and the root on the surface of the tooth is called the cervix or the cemento-enamel junction (or CEJ), and the boundary between the crown and the dentine within the tooth is called the enamel-dentine junction (or EDJ).

crown area *See* **crown base area**.

crown base area A measure of the surface area of a **tooth** crown mainly used for **postcanine teeth**. It has most often been expressed as the two-dimensional area of the base of the crown, rather than a three-dimensional estimate of the functional occlusal area. Crown base area has been either computed from the product of the maximum buccolingual breadth and the mesiodistal length of the crown, or measured using a device that can calculate the surface area of an area based on tracing its perimeter.

crown completion Developmental stage signified by the completion of the last-formed enamel covering the crown of a tooth so that extension of enamel formation towards the cervix and secretion of enamel matrix onto the crown surface are complete on all aspects of the tooth. It is identified radiographically when a spicule of root has formed beyond the cervix on the mesial or distal aspect of a tooth and histologically **enamel development** has completely ceased on the buccal or lingual aspects of a crown. In multicusped teeth this may be especially confusing as completion occurs in different cusps at different times and may even occur after the stage of root bifurcation in some hominid molars and premolars; the sectorial P_3 of great apes being an example (e.g., Beynon et al. 1998). It can also be assessed by scoring the distinctive developmental stages [A–H (Demirjian et al.) or 1–14 (Moorrees et al.)] from cuspal initiation through to crown completion (e.g., Moorrees et al. 1963, Figs 1 and 2; Demirjian et al. 1973). *See also* **dental development**.

crown formation time (or CFT) The time it takes to develop the whole of a tooth crown; i.e., from the onset of **amelogenesis** at the tip of the first (or only) **cusp** to be formed to the end of amelogenesis at the cervix (the future **cemento-enamel junction**) of the last cusp to form. The same principle can be applied to the time it takes to form a single cusp of multi-cusped teeth but cusp-specific or total crown formation times should not be confused. Crown formation time is typically estimated through counts (Risnes 1986, Boyde 1990) of the **cross-striations** and **striae of Retzius** in the **cuspal enamel** (i.e., **appositional enamel**) and lateral enamel (i.e., **imbricational enamel**) (e.g., Beynon et al. 1991). *See also* **enamel development**.

crown group (Gk *koronos* = curved, refers to the round wreath on the very top of the head, hence the meaning of the "highest" taxon) A term introduced by Jefferies (1979) for the smallest subset of sister **taxa** within a **clade** or **monophyletic**

group that includes the living taxon, or taxa, within that clade. For example, the crown group of the **hominin** clade is the (*Homo sapiens*, *Homo neanderthalensis*) subclade.

crown module A compound variable used in dental metrical analysis. It is half the sum of the maximum mesiodistal length and the maximum buccolingual breadth of a tooth crown (MD+BL/2). It was used by Middleton Shaw (1931) in his monograph on the **gnathic** morphology of the modern Bantu, by Robinson (1956) in his monograph on the dentition of the southern African hominins, and by Tobias (1967, 1991).

crude birth rate The number of births observed within in a year divided by the living population size at mid-year. Often written simply as b, the crude birth rate is expressed per unit time, usually a year, much as automobile speeds are given as distance traveled per time. But unlike automobile speeds (i.e., rates), b cannot be measured in terms of distance traveled, so instead it is measured in terms of the number of births per mid-year population size. In **paleodemography**, where the number of births within a year and mid-year population size cannot be directly observed, the crude birth rate is defined as the number of births per population size per year. This rate can be multiplied by 100 and given as a percentage, or by 1000 when the measure is per 1000. The use of "crude" is to distinguish this rate from the "age-specific fertility rate" as well to note that the rate is per population without specifying females. Consequently, the crude birth rate expresses the number of births per population, where much of the population is not at risk of giving birth (either because they are males or because they are pre- or post-reproductive females). Together with the **crude death rate**, the crude birth rate defines the intrinsic rate of increase of a population.

crude death rate The number of deaths observed within a year divided by the living population size at mid-year. The crude death rate, d, is expressed per unit time, usually a year, much as automobile speeds are given as distance traveled per time. But unlike automobile speeds (i.e., rates), d cannot be measured in terms of distance traveled, so instead it is measured in terms of number of deaths per mid-year population size. In

paleodemography, where the number of deaths within a year and mid-year population size cannot be directly observed, the crude death rate is defined as the proportion of a population that dies per year. This rate can be multiplied by 100 and given as a percentage, or by 1000 when the measure is per 1000. "Crude" is used to distinguish this rate from the "age-specific death rate." Together with the **crude birth rate**, the crude death rate defines the **intrinsic rate of increase** of a population. *See also* **death rate, age-specific**.

crural index (L. *cruralis*=leg) The lower-limb equivalent of the brachial index; the ratio of the lengths of the tibia and femur, calculated as 100×(tibia length/femur length), thus it measures the relative length of the leg. Among primates, the crural index varies by locomotor strategy and climate. The long-limbed terrestrial quadrupeds have disproportionately elongated **distal** elements (high crural indices). The distal elements are the narrowest and therefore the lowest-mass portions of the lower limbs, thus by elongating the distal part of the lower limb it minimizes the energetic costs of accelerating and decelerating the lower limbs during gait. Elongating the distal portion of the lower limb also increases the surface-area-to-volume ratio, which improves the ability to lose heat in hot, dry climates. Thus, short legs are more effective at retaining heat, and low crural indices are characteristic of taxa, including *Homo neanderthalensis*, that live in cold climates. *See also* **brachial index**.

crypt *See* **dental crypt**.

Crystal Tuff An **ignimbrite** eruption found over a large area in the **Ethiopian Rift System**. Dated at 3.56±0.03 Ma (WoldeGabriel et al. 1992) it is a likely source for the **Kilaytoli Tuff** that is exposed in the **Woranso-Mille study area** and the **Lokochot Tuff** that is exposed in **Koobi Fora**. It is also the more general name for any fine-grained volcanic ash deposit containing a significant proportion of crystals from the parent magma.

CSF Abbreviation of **cerebrospinal fluid** (*which see*).

CSLM Acronym for confocal scanning light microscope. *See* **confocal microscopy**.

CT Acronym for **computed tomography** (*which see*).

CT-number scale overflow One of several technical problems involved in the data-acquisition stage of **computed tomography** (or CT). The CT-number scale ranges from -1000 to $+3095$ Hounsfield units: water has the value 0 and air -1000. This means that the maximum X-ray attenuation coefficient is about four times that of water. When CT is used clinically this limit is reached by dense bone and enamel, but in fossils in which minerals of high atomic number have penetrated the bone during fossilization this can cause this limit to be surpassed. For example, the **Kabwe 1** cranium has been infiltrated with lead and zinc ore and the CT scans register 5000–6000 Hounsfield units, and the rock matrix in the **Gibraltar 1** cranium causes the CT scans to register between 2250 and 3150 Hounsfield units. When scanning small fossils such as single teeth there is a great risk of CT-number scale overflow as there is hardly any **beam-hardening** (i.e., most soft X-rays reach the detector, resulting in high attenuation coefficients). CT-number scale overflow manifests itself in different ways. In most cases CT-numbers over 3095 will be reconstructed as 3095 (the maximum value). This "white overflow" can lead to errors in distance measurements. In some CT scanners the CT scale is duplicated above 3095 (i.e., it reverts to -1000 and the sequence starts over again). This is called "black overflow" because it registers as black areas on the image.

Cueva de les Malladetes *See* **Malladetes**.

Cueva de Nerja (Location 36°45'54"N, 3°50'30"W, Spain; etym. Sp. *cueva*=cave and Nerja= the name of the nearby city) History and general description This extensive cave complex is located in the town of Maro, 3 km/1.8 miles from Nerja, in the province of Málaga, Spain. The caves lie on the slopes of the Sierra Almijara Mountains and since the cave was discovered by speleologists in 1959 the site has been under organized excavation and exploration but recent work has focused predominantly on the geological context. The cave is over 4 km/2.4 miles long but the only evidence of human habitation is near the opening where there is extensive evidence of anatomically modern human occupation including burials and artwork. Temporal span and how dated? The site preserves an intact sequence through the **Upper Paleolithic**

to the Chalcolithic. Conventional **radiocarbon dating** suggests the Upper Paleolithic is *c*.16,500 years BP and the Epipaleolithic is between 13,300 and 8770 years BP: an anatomically modern human skeleton from the Epipaleolithic level has been dated to 6800–7600 years BP. The Neolithic ranges from 7390 to 4950 years BP and the Chacolithic ranges from 3750 to 2750 years BP. Hominins found at site Many fragmentary anatomically modern human remains were found in most levels. A well-preserved skeleton of a young woman was found in the Epipaleolithic levels (García Sánchez 1986). Several intact burials were found in the Solutrean level, but these fossils were badly damaged in a fire in the laboratory in which they were stored. Archeological evidence found at site The stone and bone tools and ceramics record evidence of occupation of the cave through the Upper Paleolithic (including **Aurignacian**, Magdalenian, and Epipaleolithic industries), Neolithic, and Chalcolithic. Key references: Geology, dating, and paleoenvironment Jordí Pardo 1986, Durán et al. 1993; Hominins García Sánchez and Jiménez Brobeil 1995; Archeology Jordí Pardo 1986, Sanchidrián Torti 1994.

Cueva Mayor-Cueva del Silo One of several sediment-filled cave systems in the Sierra de Atapuerca hills, 14 km/9 miles east of Burgos in Northern Spain. In this system there are several cave/fissure complexes including the Sima de los Huesos. *See also* **Atapuerca**; **Sima de los Huesos**.

Cueva Morín (Location 43°22'N, 3°51'W, Spain; etym. Sp. *cueva*=cave and meaning of *Morín* is unknown) History and general description This limestone cave is located in a ridge near the Cantabrian coast in the town of Villanueva de Villaescusa. Discovered in 1910, it was excavated between 1917 and 1920 first by the Reverend J. Carballo and then by the Conde de la Vega del Sella. Modern excavations by González-Echegaray and Freeman took place between 1966 and 1969. The recovered lithic material is significant because the site spans the transition between the **Middle Paleolithic** and **Upper Paleolithic** periods. Temporal span and how dated? Samples for **AMS (accelerator mass spectrometry) radiocarbon dating** taken in 2001 yielded ages of *c*.39 ka for the final **Mousterian** and *c*.36 ka for the archaic **Aurignacian** layers. Hominins found at site Two graves were discovered in the Aurignacian layer, both capped by red **ochre**.

The smaller grave, Morín II, contained no recognizable hominin remains, but the larger one, Morín I, contained well-preserved mold of an adult, created by the replacement of the organic material by fine sediment that was distinguishable from the surrounding matrix (Freeman and González-Echegaray 1970). In addition to the mold of the skeleton, there is also the mold of a small ungulate placed in the grave and molds of other, unidentified, organic grave goods. The Aurignacian date of these burials suggests they are *Homo sapiens*. Archeological evidence found at site The 22 archeological layers, which represent most technologies from the Mousterian to the Azilian, include evidence of a Mousterian bone industry, the first appearance of the **Châtelperronean** in Spain, and the presence of structural features (a hut foundation and post holes). Key references: Geology, dating, and paleoenvironment Maíllo et al. 2001, Stukenrath 1978; Hominins Freeman and González-Echegaray 1970, González-Echegaray et al. 1971, 1973; Archeology Freeman and González-Echegaray 1970, González-Echegaray et al. 1971, 1973.

Cultural and Museum Centre, Karonga (or CMCK) An initiative of the Karonga community leaders and members of the **Hominid Corridor Research Project**, is a public/private partnership between the Government of Malawi represented by the Ministry of Wildlife, Tourism and Culture and the Uraha Foundation Malawi. Constructed with European Union funding it opened in December 2004 and it promotes the rich cultural and natural heritage of northern Malawi and provides facilities for research in that region. The main building in Karonga includes an exhibition area and a training and research facility; a separate paleoanthropological field station at **Malema** is also equipped with training and research facilities. Hominin fossil collections **HCRP UR 501** from **Uraha** and **HRCP RC 911** from Malema. Contact person Harrison Simfukwe (harrison.simfukwe@hotmail.com). Postal address Private Bag 16, Karonga, Malawi. Website www.palaeo.net/cmck. Tel/fax +265 1 362 574.

cultural transmission The process that gives rise to **culture**. In cultural transmission an individual learns information that is capable of affecting its behavior and/or other aspects of its phenotype from another individual (Boyd and Richerson 1985). In the past researchers debated whether the term should be restricted to cases in which individuals learn from other individuals via **imitation**. Today, the consensus is that limiting the term in this way is not helpful, and most workers do not distinguish between different forms of **social learning** when discussing cultural transmission. In line with this, the terms cultural transmission, social learning, and cultural learning are now used interchangeably. Anthropologists often talk about culture as if it is unique to hominins. However, over the last few decades it has become clear that such is not the case. Work in the laboratory and field has pointed to the existence of cultural learning in a number of nonhominin animals, including chimpanzees, orangutans, killer whales, humpback whales, a variety of passerine bird species, and guppies (e.g. Whiten et al. 1999, Yurk et al. 2002, van Schaik et al. 2003, Laland and Galef 2009, Lycett et al. 2010). Some of these cases are still being debated, but others are not. For example, the notion that the songs of many passerine bird species are socially learned is no longer controversial. Similarly, there is general agreement that guppies engage in cultural learning (Laland and Janik 2006). The adaptive significance of cultural transmission appears obvious. The ability to learn new solutions to problems from other individuals would seem to be something that natural selection can be expected to always favor. So, why is cultural transmission not more widespread? Part of the answer seems to be that it is cognitively costly and its benefits outweigh its costs only in certain circumstances. Modeling work suggests that cultural learning is only favored when the rate of environmental change is moderate (Boyd and Richerson, 1985, Whitehead, 2007). When the rate of environmental change is slow, it is outcompeted by genetic adaptation, whereas when the rate of environmental change is fast, it is outcompeted by individual learning. In addition to investigating whether nonhominin animals have culture and trying to identify the conditions that favor the evolution of cultural learning, researchers have attempted to understand the operation of cultural transmission within human populations (e.g., Cavalli-Sforza and Feldman 1981, Boyd and Richerson 1985, Hewlett and Cavalli-Sforza 1986, Guglielmino et al. 1995, Shennan and Steele 1999, Collard et al. 2006, Tehrani and Collard 2009). This body of work has identified a number of modes of cultural transmission and several cultural learning strategies (Cavalli-Sforza and Feldman 1981, Boyd and Richerson 1985). The key modes of cultural

transmission are vertical transmission, oblique transmission, and horizontal transmission. In vertical transmission, a child learns from a parent. In oblique transmission, an individual in one generation learns from an individual in the previous generation who is not their parent. In horizontal transmission, an individual learns from another individual in the same generation. Of the cultural learning strategies that have been identified, the most notable are direct bias, indirect bias, and frequency-dependent bias. In direct bias, individuals evaluate alternative behaviors and choose among them. In indirect bias, individuals use some traits, such as those connoting health or prestige, to choose a cultural model, and then copy a range of the model's behaviors. In frequency-dependent bias, an individual copies a behavior on the basis of its frequency in the population; the most common form is thought to be conformist bias, which involves copying the behavior that is most widespread in the population.

culture (L. *cultura*=cultivation) Numerous definitions of culture have been put forward, but most are problematic for researchers interested in human evolution. The reason is that they automatically preclude the possibility that nonhominin animals have culture, which affects how the issue of the origins of culture is approached. There is substantial agreement among the other definitions. Crucially, they all hold that culture involves **social learning**. The issues on which these definitions differ are whether all forms of social learning give rise to culture, and whether social learning is the only process involved. Some researchers contend that a behavior exhibited by an individual is cultural if it was learned from another individual by any form of social learning (e.g., van Schaik et al. 2003, Laland 2008). Thus, according to these researchers, culture is simply socially learned behavior. Other researchers argue that culture should be restricted to behaviors that result from two particular forms of social learning: **teaching** and **imitation** (e.g., Galef 1992). Still other researchers suggest that the crucial feature of culture is that it is cumulative (e.g., Tomasello 1999, Boyd and Richerson 2005). According to these researchers, for a behavior to be considered cultural it not only has to be socially learned, but also has to show signs of elaboration through time and across generations. The latter two groups of researchers use the term "traditions" to refer to non-cultural socially learned behaviors, whereas proponents of the first definition often use the terms culture and traditions interchangeably. Depending on which of the three definitions is employed culture may, or may not, be present in nonhominin animal species. If the first definition is employed, there is compelling evidence for the existence of culture in a number of nonhominin animal species, including chimpanzees (Whiten et al. 1999, Lycett et al. 2007), orangutans (van Schaik et al. 2003), and killer whales (Yurk et al. 2002). The term "culture" is also used in anthropology to refer to a complex of beliefs and behaviors, or to a suite of artifacts, that is diagnosably different from other such complexes or suites (e.g., American culture versus Japanese culture, or the **Aurignacian** versus the **Gravettian**) and which is associated with a group of individuals who interact socially. These uses of culture are problematic for several reasons. In the first place, in most cases we are ignorant about the mechanism(s) underlying a given belief, behavior, or artifact, and we do not know whether it is underpinned by social learning or by some other mechanism (e.g. individual learning). Second, the application of more effective dating methods has made it clear that many of the cultures that archaeologists have identified cannot be linked in any simple way to recognizable social groups. Thus, this fourth definition of culture is potentially inconsistent with the other three definitions of culture, which are all linked to social learning. Given these problems it may be more useful to refer to the complexes of beliefs, behaviors, and artifacts that we document ethnographically and archeologically as "ethnographic taxonomic units" and "archeological taxonomic units," respectively (see Gamble et al. 2005). *See also* **social learning**; **teaching**; **imitation**.

Cunkong *See* **Bubing Basin**.

cursorial (L. *cursorius*=running) Animals, quadrupedal or **bipedal**, adapted for fast terrestrial travel (i.e., running). Cheetahs and gazelles are cursorially adapted and some researchers have suggested that the body size and shape of early *Homo* may have been adapted for endurance **running**. *See also* **endurance running hypothesis**.

Curve of Spee *See* **tooth wear**.

cusp (L. *cuspis*=point) A portion of a tooth crown demarcated by primary **fissures** and with an independent apex. Cusps are divided into primary, or main, and accessory. The primary cusps on the mandibular molars are the **entoconid**, **hypoconid**, **hypoconulid**, **metaconid**, and **protoconid**. The latter two cusps are also the primary cusps on mandibular premolars. Examples of **accessory cusps** on mandibular molars are the **C6** and the **C7**. The primary cusps on the maxillary molars are the **hypocone**, **metacone**, **paracone**, and **protocone**. The latter two cusps are also the primary cusps on maxillary premolars. The **C5** is an example of an accessory cup on a maxillary molar. Enamel features that are not large and discrete enough to qualify as accessory cusps are called **conules** on maxillary teeth and **conulids** on mandibular teeth. The word "cusp" is a misnomer with respect to hominin teeth for their cusps are blunt rather than pointed. *See also* **accessory cusp**; **cuspulid**; **cusp nomenclature**; **cuspule**.

cusp 6   *See* **C6**; **tuberculum sextum**.

cusp 7   *See* **C7**; **tuberculum intermedium**.

cusp morphology   This term refers to the relative size and shape of the main and accessory cusps on the occlusal surface of a postcanine tooth crown. It refers to both metrical and non-metrical studies, with most of the latter addressing the incidence and prevalence of accessory cusps. Nearly all of the existing research in this area has focused on the taxonomic value of interspecific differences in cusp morphology; relatively few studies have looked at the functional significance of such differences. Research methods to quantify non-metric variation have often been adapted from standards used for contemporary humans (Dahlberg 1956). In modern human populations variability of expression of main and accessory cusps is usually described using the Arizona State University Dental Anthropology System (or ASUDAS; Turner et al. 1991, Scott and Turner 1997). This system, and its modifications, have also been employed to give an semiquantitative account of morphological details in fossil hominin dentition (e.g., Irish and Guatelli-Steinberg 2003, Bailey 2002, 2004, Martinón-Torres et al. 2008). Most, but not all, of the variation in cusp morphology observed in contemporary humans can be observed in fossil hominins (Bailey 2006, Irish and Guatelli-Steinberg 2003, Martinon-Torres 2007). Robinson (1956) drew attention to differences in

cusp morphology between *Paranthropus robustus* and *Australopithecus africanus,* but these differences were not quantified. The research methods used to capture metrical information about the relative size of the cusp areas of fossil hominin teeth were mostly adapted from similar comparative studies of extant modern human and higher primate populations (e.g., Hanihara 1981, Erdbrink 1965, 1967, Biggerstaff 1969, Corruccini 1977a, 1977b, 1978, Lavelle 1978). Corruccini and McHenry (1980) were the first researchers to focus on fossil teeth and this study was followed by a series of analyses of the East African early hominin fossil record (Wood and Abbott 1983, Wood et al. 1983, Wood and Uytterschaut 1987, Wood and Engleman 1988, Suwa et al. 1994, 1996). Bailey (2004, 2006) was the first to successfully apply and adapt these research methods to the fossil record of later hominins and many (e.g., Martinón-Torres et al. 2006, 2007) but not all (Moggi-Cecchi and Boccone 2007, Quam et al. 2009) of the more recent studies have addressed this later phase of hominin evolution. Relative variability in cusp morphology has also been studied in living apes (e.g., Uchida 1998a, 1998b, Pilbrow 2006, 2007). All of the above studies focused on cusp morphology at the outer enamel surface and they expressed cusp areas in two dimensions, but recently **micro-CT** imaging has been used to investigate cusp morphology in three dimensions and at the **enamel-dentine junction** as well as at the outer enamel surface (e.g., Skinner et al. 2008, 2009a).

cusp nomenclature   (L. *cuspis*=point) The names of the cusps of the postcanine teeth of the upper and lower jaw were devised by Henry Fairfield Osborn (1888, 1892). They are consistent with what we now know is almost certainly an incorrect hypothesis (called the tritubercular theory) about the evolution of tooth crowns, but Osborn's naming scheme has outlasted his theory. Osborn's scheme begins with a reference **cusp**, which is on the **lingual** side of the crown in an upper postcanine tooth (the **protocone**) and on the **buccal** side of the crown in a lower postcanine tooth (the **protoconid**). He suggested that these cusps in mammals are **homologous** with the mesiodistally elongated main cusp in a reptile tooth hence the prefix "proto". In the upper jaw he called the cusp on the mesiobuccal aspect of the reference cusp the **paracone** and the cusp situated distobuccally to it the **metacone**. The equivalent cusps in the lower postcanine teeth are the mesiolingually situated **paraconid** (which has been lost in higher primates) and the distolingually situated **metaconid**. These triangular (or tricono-dont) units made up of three cusps are called the **trigon** in the upper teeth and the **trigonid** in the lower teeth. In higher primates each postcanine tooth may have a distal addition to the trigon or trigonid. This is called the **talon** in the upper teeth and the **talonid** in the lower teeth. In the upper postcanine teeth the talon is represented by the distolingual **hypocone**. In the lower teeth it is represented by the **entoconid** lingually, the **hypoconid** buccally, and the **hypoconulid** (if there is one) distally. Biggerstaff (1968) provides a succinct review of the history of the scheme: he refers to it as a nomenclature for molar cusps but it is used for all the postcanine teeth, especially the molarized premolars of *Paranthropus*.

cuspal enamel (L. *cuspis*=point) Term used for the first-formed **enamel** over the dentine horn that makes up the occlusal aspect of a tooth and which does not show long-period growth increments at the surface. In a completed unworn tooth no portion of the cuspal enamel is visible on the surface of a tooth crown, but the imbricational (or lateral) enamel does have increments visible on the tooth surface. In anterior teeth cuspal enamel forms a smaller percentage of the crown volume than in the posterior (or postcanine) teeth. Several methods have been applied to determine the cuspal enamel formation time, many of which involve assessment of the linear **enamel thickness** and the **daily secretion rate** or the total number of **cross-striations** (e.g., Dean 1998, Smith 2006).

cuspule An accessory cusp on a maxillary postcanine tooth crown. A small **C5** is an example of a cuspule. *See also* **accessory cusps**.

cuspulid An accessory cusp on a mandibular postcanine tooth crown. A small **C7** is an example of a cuspulid. *See also* **accessory cusps**.

cutmarks A hominin-induced modification inflicted on bone surfaces by stone **artifacts** or knives during carcass processing (e.g., the deflesh-ing, disarticulation, or skinning of animal remains). Cutmarks have a deep, V-shaped cross-section, whereas the tooth marks of large carnivores tend to be shallower and U-shaped. Stone tool cutmarks frequently have fine, parallel lines within them called internal striations that are made by the undulating edge of an **artifact** as it is dragged across the bone and contextually stone tool cutmarks often appear in roughly parallel groups. When archeologists identify a cutmark on bone, they are demonstrating a temporal, spatial, and behavioral association between the hominins that formed an archeological site and the fauna. The percentages of bones showing cutmarks as opposed to those showing carnivore toothmarks have been used as a proxy for the intensity and scope of hominin activity involved in generating a given archeological assemblage and they have been used to make inferences about the relative timing of hominin access to animal remains. These indicators have been particularly relevant in the study of early archeological sites such as those at Bed I, **Olduvai Gorge**. Sharp-edged pieces of shell, bamboo, and bone may also produce cutmarks. Lyman (1994) and Dominguez-Rodrigo and Pickering (2003) provide useful introductions to cutmark identification and analysis. *See also* **Bouri**; **Dikika study area**; **taphonomy**.

cynodont (Gk *cyno*=dog and *dont*=teeth) Molar teeth that have short, dog-like, **pulp chambers** relative to the distance between the roof of the pulp chamber and the tip of the longest **root** (ant. **taurodont**).

cytoarchitecture (L. *cyto*=cell and *architectus*= builder) The cellular composition of any structure in the body, but in the **cerebral cortex** the term refers to the characteristic arrangement of **neurons** into layers according to their size, morphology, and spatial packing density. These variables allow researchers to distinguish the boundaries of **cortical areas** in the brain. For example, one of the most widely used maps of the **neocortex** by Brodmann is based on cytoarchitecture.

cytogenetic *See* **karyotype**.

D

D Prefix for fossils from **Dmanisi**, Georgia.

D211 (NB: this specimen was referred to as "Dmanisi 211" in the initial publication) Site **Dmanisi**. Locality Block 1. Surface/*in situ In situ*. Date of discovery September 24, 1991. Finder Antje Justus. Unit Stratum B1. Horizon See previous. Bed/member N/A. Formation N/A. Group N/A. Nearest overlying dated horizon None. Nearest underlying dated horizon Stratum A1 (aka "VI"), 1.81 Ma. Geological age *c*.1.77 Ma. Developmental age Adult. Presumed sex Unknown. Brief anatomical description A mandibular corpus that preserves all the tooth crowns and roots, but the base on both sides is damaged. Announcement Gabunia and Vekua 1995. Initial description Gabunia and Vekua 1995. Photographs/line drawings and metrical data Gabunia and Vekua 1995, Bräuer and Schultz 1996, Rosas and Bermúdez de Castro 1998b. Detailed anatomical description Bräuer and Schultz 1996. Initial taxonomic allocation *Homo erectus*. Taxonomic revisions *Homo sp. indet.* (aff. *Homo ergaster*). Current conventional taxonomic allocation *Homo sp. indet.* Informal taxonomic category Pre-modern *Homo*. Significance The first hominin recovered from Dmanisi. The D211 mandible combined relatively small postcanine tooth crowns and a modern human-like molar size order with a corpus that was more robust than usually seen in modern humans and Neanderthals. The mandible is consistent in size with the cranium **D2282** and because they were found "close together" it has been suggested that they "probably represent one individual" (Rightmire et al. 2006b, p. 121). Location of original **Georgian National Museum**, Tiblisi, Georgia.

D2280 Site **Dmanisi**. Locality Block 1/Excavation Unit 50/62. Surface/*in situ In situ*. Date of discovery May 1, 1999. Finder Antje Justus. Unit Stratum B1. Horizon See previous. Bed/member N/A. Formation N/A. Group N/A. Nearest overlying dated horizon None. Nearest underlying dated horizon Stratum A1 (aka "VI"), 1.81 Ma. Geological age *c*.1.77 Ma. Developmental age Adult. Presumed sex Unknown. Brief anatomical description An almost complete calvaria, plus substantial parts of the basioccipital, left temporal, and sphenoid. Endocranial volume 775 cm^3 (Rightmire et al. 2006b). Announcement Gabunia et al. 1999. Initial description Gabunia et al. 2000. Photographs/line drawings and metrical data Gabunia et al. 2000, Rightmire et al. 2006b. Detailed anatomical description Rightmire et al. 2006b, de Lumley et al. 2006. Initial taxonomic allocation *Homo* aff. *Homo ergaster*. Taxonomic revisions *Homo erectus* (Rightmire et al. 2006b), *Homo georgicus* (de Lumley et al. 2006). Current conventional taxonomic allocation *Homo sp. indet.* Informal taxonomic category Pre-modern *Homo*. Significance D2280 is one of the better-preserved hominin calvaria from Dmanisi. Location of original **Georgian National Museum**, Tiblisi, Georgia.

D2282 Site **Dmanisi**. Locality Block 1/Excavation Unit 52/62. Surface/*in situ In situ*. Date of discovery July 22, 1999. Finder Antje Justus. Unit Stratum B1. Horizon See previous. Bed/member N/A. Formation N/A. Group N/A. Nearest overlying dated horizon None. Nearest underlying dated horizon Stratum A1 (aka "VI"), 1.81 Ma. Geological age *c*.1.77 Ma. Developmental age Adult. Presumed sex Unknown. Brief anatomical description A cranium that has suffered substantial damage to the mid and lower face. In addition the cranial base and what remains of the face have been deformed. The roots and crowns of the right P^4 and M^1 and the left M^1 and M^2 are preserved in the maxilla. Endocranial volume 650–660 cm^3 (Rightmire et al. 2006b). Announcement Gabunia et al. 2000. Initial description Gabunia et al. 2000. Photographs/line drawings and metrical data Gabunia et al. 2000, Rightmire et al. 2006b, Martinón-Torres et al. 2008. Detailed anatomical

Wiley-Blackwell Encyclopedia of Human Evolution, First Edition. Edited by Bernard Wood.
© 2013 Blackwell Publishing Ltd. Published 2013 by Blackwell Publishing Ltd.

description Rightmire et al. 2006b, de Lumley et al. 2006. Initial taxonomic allocation *Homo* aff. *Homo ergaster*. Taxonomic revisions *Homo erectus* (Rightmire et al. 2006b), *Homo georgicus* (de Lumley et al. 2006). Current conventional taxonomic allocation *Homo sp. indet.* Informal taxonomic category Pre-modern *Homo*. Significance The D2282 cranium is consistent in size with the mandible **D211** and because they were found "close together" they "probably represent one individual" (Rightmire et al. 2006b, p. 121). Location of original **Georgian National Museum**, Tiblisi, Georgia.

D2600 Site **Dmanisi**. Locality Block 2/Excavation Unit 64/59. Surface/*in situ In situ*. Date of discovery September 26, 2000 Finder Gocha Kiladze. Unit Stratum B1y. Horizon See previous. Bed/member N/A. Formation N/A. Group N/A. Nearest overlying dated horizon None. Nearest underlying dated horizon Stratum A1 (aka "VI"), 1.81 Ma. Geological age *c*.1.77 Ma. Developmental age Aged adult. Presumed sex Unknown. Brief anatomical description A substantially complete aged adult mandibular corpus. Announcement Gabounia et al. 2002. Initial description Gabounia et al. 2002. Photographs/line drawings and metrical data Gabounia et al. 2002, Martinón-Torres et al. 2008, Rightmire et al. 2008. Detailed anatomical description N/A. Initial taxonomic allocation *Homo georgicus*. Taxonomic revisions *Homo sp. indet.* (aff. *Homo ergaster*). Current conventional taxonomic allocation *Homo sp. indet.* Informal taxonomic category Pre-modern *Homo*. Significance The D2600 mandible has a corpus that is larger and more robust than that the other Dmanisi mandibles (Skinner et al. 2006) and the preserved tooth root size and morphology, together with the molar size order (Martinón-Torres et al. 2008), have prompted these authors to suggest that D2600 may represent a second hominin **taxon**, or **paleodeme**, at Dmanisi. Location of original **Georgian National Museum**, Tiblisi, Georgia.

D2700/D2735/D2724/D3160,   etc. Site **Dmanisi**. Locality Block 2/Excavation Units 65/60 and 66/60. Surface/*in situ In situ*. Dates of discovery August 11 and 30, 2001. Finder Gocha Kiladze. Unit B1x. Horizon See previous. Bed/member N/A. Formation N/A. Group N/A. Nearest overlying dated horizon None. Nearest underlying dated horizon Stratum A1 (aka "VI"),

1.81 Ma. Geological age *c*.1.77 Ma. Developmental age Subadult. Presumed sex Unknown. Brief anatomical description A skull comprising a generally well-preserved adult cranium (D2700) and mandible (D2735) together with associated postcranial bones (D2724, etc.). Nearly all of the teeth are preserved. Endocranial volume 600 cm^3, with a possible correction to 612 cm^3 because of the immaturity of the specimen (Rightmire et al. 2006b). Announcement de Lumley and Lordkipanidze 2006. Initial description de Lumley and Lordkipanidze 2006. Photographs de Lumley and Lordkipanidze 2006, Rightmire et al. 2006b, de Lumley et al. 2006. Detailed anatomical description Rightmire et al. 2006b, de Lumley et al. 2006. Initial taxonomic allocation *Homo georgicus*. Taxonomic revisions *Homo erectus* (Rightmire et al. 2006b). Current conventional taxonomic allocation *Homo sp. indet.* Informal taxonomic category Pre-modern *Homo*. Significance The skull and postcranial skeletal elements are not necessarily from a single individual but several of the postcranial bones (clavicle, ribs, humeri) are juvenile and probably go with the D2700/D2735 skull. If so they would provide an important source of information about the relative size relationships of the main hominin taxon sampled at Dmanisi. Location of original **Georgian National Museum**, Tiblisi, Georgia.

D2724 *See* **D2700/D2735/D2724/D3160, etc.**

D2735 *See* **D2700/D2735/D2724/D3160, etc.**

D3444/D3900 Site **Dmanisi**. Locality Block 2/Excavation Units 64/61. Surface/*in situ In situ*. Date of discovery August 26, 2002, and thereafter. Finder Gocha Kiladze. Unit B1y. Horizon See previous. Bed/member N/A. Formation N/A. Group N/A. Nearest overlying dated horizon None. Nearest underlying dated horizon Stratum A1 (aka "VI") 1.81 Ma. Geological age *c*.1.77 Ma. Developmental age Old adult. Presumed sex Unknown. Brief anatomical description A skull comprising an aged adult cranium (D3444) together with an associated mandible (D3900) and with the exception of the left upper canine, the individual was edentulous. The remains of an adult associated skeleton (see Lordkipanidze et al. 2007) are believed to belong to the same individual. Announcement Lordkipanidze et al. 2005. Initial description Lordkipanidze et al. 2005. Photographs Lordkipanidze et al. 2005. Detailed anatomical description N/A. Initial taxonomic allocation early *Homo*. Taxonomic revisions *Homo erectus*

(Rightmire et al. 2006b). <u>Current conventional taxo-nomic allocation</u> early *Homo sp. indet.* <u>Informal taxonomic category</u> Pre-modern *Homo*. <u>Significance</u> The extent of the resorption in the alveolar processes of the maxilla and mandible suggest that this indivi-dual lived for some time without the ability to chew effectively. The paper announcing the find suggested that this specimen "raises interesting questions regarding social structure, life history and subsistence strategies of early *Homo*" (Lordkipanidze et al. 2005, p. 718). <u>Location of original</u> **Georgian National Museum**, Tiblisi, Georgia.

D3900 *See* **D3444/D3900**.

D-loop Abbreviation of displacement loop. *See* **mitochondrial DNA**.

DABT Acronym for Daam Aatu Basaltic Tuff. *See* **Aramis**.

daily enamel increment *See* **cross-striation**; **daily secretion rate**; **enamel development**.

daily secretion rate The amount of enamel matrix measured in micrometers (μm) secreted by an **ameloblast** during each 24 hour period (i.e., the distance between adjacent **cross-striations** along an enamel prism). These linear (i.e., along the long axis of the prism) daily increments con-tinue until the ameloblast ceases to secrete matrix and the end result is an elongated **enamel prism**. The average daily secretion rate in modern humans is close to 4 μm/day but inner rates are close to the enamel dentine junction (or EDJ) of approximately 2.8 μm/day and these rise as a gradient to approximately 5.5 μm/day close to the enamel surface. The secretion rate in *Pan* (as well as other great apes) is a little higher at the EDJ (approximately 3.3 μm/day) but the main difference is that rates rise more quickly to approximately 5.5 μm/day such that the cumula-tive trajectory of enamel thickness formed over time is greater than in *Homo sapiens*. This linear rate of matrix secretion may not reflect the volume of enamel matrix secreted in a day very accurately since prism diameter increases in some regions of a tooth crown (i.e., in the cusp) more than others. *See also* **enamel development**.

Daka Abbreviation of Dakanihyalo, the name of a locality in the **Bouri** region within the **Middle Awash study area** in the Ethiopia section of the East African Rift Valley. *See* **Dakanihyalo**.

Daka calvaria *See* **BOU-VP-2/66**.

Dakanihyalo Locality in the Bouri region within the **Middle Awash study area** in the Ethiopia section of the East African Rift Valley. Dakanihyalo was discovered in 1981 by the **Rift Valley Research Mission in Ethiopia** (or RVRME) and an archeological assemblage from Dakanihyalo was briefly referred to by Clark et al. (1984). It is one of 29 localities within the Bouri region and it is now given the designation BOU-A1. It is widely referred to in the literature by the first for four letters of the full name (i.e., as "Daka", as in the "Daka calvaria"). *See also* **Bouri**; **Bouri Formation**.

Dali (大荔) (Location 34°51′57.49″N, 109°43′58.58″E, northwestern China; etym. named after the local county) <u>History and general descrip-tion</u> An open-air site, discovered in 1978, located in the basal gravels of the third **loess** terrace of Luohe (Luo River) known as Tianshuigou, 1 km/0.6 miles south of Jiefang village, Dali County, Shaanxi Province. <u>Temporal span and how dated?</u> Initial thermo**luminescence** **dating** and **magnetostratigraphy** assign Dali to Late **Middle Pleistocene**. **Uranium-series dating** of fossil mammals suggests an age range of 230–180 ka. Recent correlation between magnetic susceptibility records of the loess-paleosol sequences at Dali and those at Luochuan in the central part of Chinese Loess Plateau places the hominin-bearing layer between the loess L3/paleosol S3 boundary, which gives an age of *c.*270 ka. <u>Hominins found at site</u> A well-preserved, complete, and edentulous **cra-nium** with slight distortion on the left side of face and a cranial capacity of 1120 cm³. The specimen is presumed to be a young adult male. <u>Archeological evidence found at site</u> Thirteen stratigraphic horizons were identified at the site. Animal fossils and stone tools were recovered from layers 3–5 from bottom to top; the hominin fossil was recovered from layer 3. The lithic assemblage consists of 181 artifacts: seven small and irregular **cores** with no evidence of platform preparation, 152 small **flakes** (mostly **debitage**), and 22 simple, **retouch**ed tools (mostly **scrapers**). Raw materials consist of **quartzite**, vein quartz, and **flint**. <u>Repository</u> **Institute of Vertebrate Paleontology and**

Paleoanthropology. Key references: Geology and dating Wang et al. 1979, Chen et al. 1984, Xiao et al. 2002; Hominins Wang et al. 1979, Wu 1981, Wu and Olsen 1985, Pope 1992, Wu and Poirier 1995, Holloway 2000, Brown 2001; Archeology Wu and You 1979, Wu and Olsen 1985.

Damei *See* **Bose Basin**.

Danakil cranium *See* UA 31.

Danakil pubis *See* UA 466.

Dansgaard–Oeschger cycles Refers to *c*.1.5 ka cycles of cooling and warming that have been identified in the Greenland ice core and North Atlantic records during the last glacial. **Heinrich events** occur prior to periods of abrupt warming. *See also* **Greenland Ice Core Project**; **Heinrich events**.

Dar el 'Aliya *See* **Mugharet el 'Aliya**.

Dar es Soltane 5 Site Dar es Soltane II. Locality N/A. Surface/*in situ* In situ. Date of discovery 1975. Finders A Debénath and others. Unit N/A. Horizon N/A. Bed/member N/A. Formation N/A. Group N/A. Nearest overlying dated horizon N/A. Nearest underlying dated horizon N/A. Geological age 70–40 ka. Developmental age Adult. Presumed sex Male. Brief anatomical description A robust partial skull preserving the upper face and the left side of the mandible. Announcement Debénath 1975. Initial description Ferembach 1976. Photographs/line drawings and metrical data Bräuer and Rimbach 1990. Detailed anatomical description Bräuer and Rimbach 1990, Schwartz and Tattersall 2003. Initial taxonomic allocation *Homo sapiens*. Taxonomic revisions None. Current conventional taxonomic allocation *H. sapiens*. Informal taxonomic category Modern human. Significance Similarity to the temporally earlier Jebel Irhoud (**Irhoud 1** and **Irhoud 2**) crania may suggest local continuity in this region between pre-modern *Homo* and anatomically modern humans. Location of original Musée Archéologique, Rabat Chellah, Morocco.

Dar es Soltane II (Location 33°57′36″N, 6°35′59″W, Morocco; etym. Ar. sultan's house) History and general description Several cave sites line the El Menzah range along the Atlantic coast of Morocco near Rabat. These were first explored by A. Ruhlmann before WWII, but in 1969 a group led by André Debénath excavated another cave site roughly 200 m from the first Dar es Soltane deposit. They named the second site Dar es Soltane II. The cave is known for many hominin remains, but a relative lack of faunal and artifacts. Temporal span and how dated? The Aterian layers at nearby Dar es Soltane I have **electron spin resonance spectroscopy dates** of <110 ka, although other Aterian sites are mostly between 60 and 80 ka. Hominins found at site The remains of at least five individuals were recovered; **Dar es Soltane 5** is the most complete, consisting of a nearly complete cranium and mandible. Archeological evidence found at site Seven numbered layers were recovered with the second from the bottom containing Aterian technology. Key references: Geology, dating, and paleoenvironment Barton et al. 2009; Hominins Ferembach 1976, Bräuer and Rimbach 1990; Archeology Debénath 1976.

Dart, Raymond Arthur (1893–1988) Raymond Dart, who was born in Toowong, a suburb of Brisbane, Australia, entered the University of Queensland in 1911 to study for a Bachelor of Science degree and, despite being raised in a religiously fundamentalist family, he became interested in zoology and evolutionary theory. He graduated in 1913 and in 1914 he moved to the University of Sydney to study medicine. The British Association for the Advancement of Science held their annual meeting in Sydney that same year and the young Dart was able to meet the neuroanatomist and physical anthropologist Grafton Elliot Smith and the paleontologist and archeologist William Johnson Sollas. Between 1915 and 1917 Dart worked with James T. Wilson, head of the Anatomy Department at the University of Sydney, on anatomical studies of the human brain. After Dart completed his medical degree in 1917 he enlisted in the Australian Army Medical Corps and spent the last year of WWI in France. After his demobilization in 1919, Dart studied with Grafton Elliot Smith in the Anatomy Department at University College London. Dart was given funds by the **Rockefeller Foundation** to visit the USA and between the end of 1920 and late 1921 he worked with Robert J. Terry at Washington University in St. Louis and later he served as a demonstrator of anatomy in Cincinnati. By the time Dart returned to University College London he was interested in physical anthropology and began examining the collection of brains kept at the **Royal College of Surgeons of England**. Dart's career changed course in 1922 when he accepted the Chair of Anatomy at the newly established University of the Witwatersrand in Johannesburg, South Africa. Soon after his arrival, Dart began to assemble a collection of primate fossils, including brain endocasts, unearthed in the many lime quarries located in the Transvaal. Early in 1924 Josephine Salmons, a medical student at the University of the Witwatersrand, brought Dart a fossil baboon skull that had been blasted out of the Northern Lime Company's quarries located near the town of **Taungs**. The find prompted Dart to arrange for any other interesting fossils to be sent to him and in November 1924 he received a crate of fossils from the quarry. He noticed two pieces of **breccia**, one containing a natural **endocranial cast** (the fossilized cast of a brain) and in the other was the anterior portion of a cranium. After 4 weeks of preparation Dart freed the face from the surrounding breccia, revealing the partial skull of a juvenile primate. Dart judged that many anatomical features of the skull and endocranial cast were modern human-like and from the position of the foramen magnum Dart concluded that the creature was bipedal. There were, however, also ape-like features to the skull, notably its small cranial capacity. Dart published a brief description of the fossil in *Nature* in February 1925, where he described its anatomy, the geological evidence for its antiquity and an interpretation of the fossil that suggested it represented a "man-like ape" that might be ancestral to modern humans. To reflect this interpretation Dart named the specimen *Australopithecus africanus*, meaning the "southern ape from Africa." Eugène **Dubois**' discovery of *Pithecanthropus erectus* in Java in the 1890s and the later discoveries of *Sinanthropus pekinensis* (both species are now classified as *Homo erectus*) in China during the 1920s and 1930s seemed to support Ernst Haeckel's idea that modern humans had evolved in Asia, thus few anthropologists were expecting fossil evidence of the early stages of human evolution in Africa. There were other reasons for scientists to be doubtful about Dart's interpretation of *Australopithecus*, for the small-brained but upright *Pithecanthropus* seemed to support the view that hominins had first become bipedal and that bipedalism then led to an increase in the size of the brain since the hands were free to use tools.

However, other researchers, such as Arthur **Keith** in England, believed that a large brain had evolved first and bipedalism later and the controversial discoveries at **Piltdown** of *Eoanthropus dawsonii* seemed to support this view. Most prominent physical anthropologists (e.g., Grafton Elliot Smith, Arthur Keith, and Arthur Smith Woodward) all expressed skepticism about Dart's claim that *Au. africanus* represented an intermediate form between the anthropoid apes and modern humans. However, Dart did gain one important supporter, the Scottish paleontologist Robert **Broom**, and shortly after the *Nature* article appeared Broom visited Dart and inspected the fossil. Broom immediately agreed the specimen represented a new species of primate that was intermediate between apes and modern humans. Dart's duties in the Anatomy Department and as Dean of the Medical School from 1925 to 1943 at the University of the Witwatersrand left little time for paleontological expeditions and Broom's considerable experience in searching for fossils resulted in Broom making most of the significant new hominin discoveries. In 1936 Robert Broom recovered evidence of an australopith from a quarry at Sterkfontein and in 1938 he obtained more evidence from **Kromdraai**. Dart worked and published on *Au. africanus* only sporadically during the 1920s and 1930s but WWII caused a serious interruption in research. Dart was thrust back into active research in 1947 when James Kitching of the **Bernard Price Institute** for **Palaeontological Research** found a portion of an *Australopithecus* cranium along with a large quantity of animal fossils at a limestone quarry at **Makapansgat** in the former central Transvaal, now Limpopo Province. The new fossil and the abundance of animal remains assumed a new significance when Kitching found fossil baboon skulls that had been fractured in a way that suggested they had been struck by a club. Dart concluded from the material recovered at Makapansgat that the australopiths had hunted big game animals thus explaining the vast quantities of **ungulate** bones, but more significantly Dart argued that there was evidence that some of those ungulate bones had been modified or used as tools. Dart suggested that the australopiths at Makapansgat had developed an osteodontokeratic (bone/tooth/horn) industry that preceded the use of stone tools. Between 1949 and the mid-1960s Dart studied and published extensively on the osteodontokeratic culture of the australopiths, but few other researchers accepted the idea and Dart's theory has subsequently been rejected in favor of interpretations pioneered by C.K. (Bob) **Brain**. Dart retired from the University of the Witwatersrand in 1958, but continued to conduct research into the 1960s.

Dart collection *See* Raymond A. Dart Collection.

Darwin, Charles Robert (1809–82) Charles Darwin was born in Shrewsbury, England. Darwin's grandfather, Erasmus Darwin, had been a prominent physician and naturalist who had speculated upon the possibility that species might change over time. Darwin's father, Robert Darwin, was also a physician and it seemed that Charles was set to continue the family tradition when he was admitted to Edinburgh University in 1825 to study medicine. However, while there he studied with Robert Edmond Grant, a zoologist who taught the theories of Jean Baptiste de Lamarck who, in his *Philosophie zoölogique* (*Zoological Philosophy*) published in 1809, had formulated a transmutationist theory of how biological species change over time. Darwin continued with his medical studies at Edinburgh University until 1827 after which he transferred to Cambridge University to study what is now called natural sciences, where he took botany classes with John Henslow and geology classes with Adam Sedgwick. This training would prove invaluable for after his graduation in 1831 Darwin was invited to join the HMS *Beagle*. The plan was to circumnavigate the globe, but a substantial proportion of the effort was to be devoted to surveying and mapping the coast of South America. During the 5 years of the voyage (1831–5) Darwin collected a vast quantity of geological, zoological, and botanical specimens. He also read widely and perhaps most importantly he read the *Principles of Geology* (1831–3) by Charles Lyell, who advocated the principle of **uniformitarianism**, which proposed that the geological processes operating today are the same as those that operated in the past. The detailed observations Darwin made while on the HMS *Beagle* were instrumental in the development of his seminal conclusion that species had changed over time. When Darwin returned to England he was already respected as a naturalist and he quickly became a member of Britain's scientific community. He married and eventually established a home, Down House, in Downe, Kent, where he set to work to explain how species could change and give rise to new species. By the late 1830s Darwin had formulated the idea of **natural selection** and in 1842 he wrote a brief

outline of the theory. Darwin wrote a long essay on his theory in 1844 and was working on a major treatise on his ideas when he received a letter in 1858 accompanied by a scientific paper written by the naturalist Alfred Russel Wallace. Wallace, when exploring the flora and fauna of the Malay Archipelago, had become convinced that species changed over time and in the paper he sent to Darwin he proposed a mechanism to explain how new species emerged that in many ways resembled Darwin's theory of evolution by natural selection. Darwin's and Wallace's accounts of their theories were presented at a meeting of the Linnaean Society in London in 1858 and thereafter Darwin lost no time in completing and publishing his *On the Origin of Species by Means of Natural Selection* (1859). Darwin largely avoided the question of human origins in *The Origin of Species* and the first substantial discussion of the implications of Darwin's evolution theory for human origins was Thomas Henry **Huxley**'s *Evidence as to Man's Place in Nature* (1863). In that volume Huxley demonstrated the close anatomical affinities between modern humans and the African apes and he also discussed the then known examples of human fossil remains, including the fossils found in the **Kleine Feldhofer Grotte** in Germany in 1856. Darwin finally confronted the question of human evolution in *The Descent of Man and Selection in Relation to Sex* (1871) where Darwin examined the anatomical and physiological similarities shared between modern humans and the rest of the animal kingdom, discussed how these similarities could be explained by evolution, and addressed the more difficult problem of explaining how modern human mental faculties could have evolved by comparing the instincts, mental abilities, and social habits of modern humans and animals in order to trace continuities between the two. Darwin also addressed the issue of modern human racial variation and explained this from a biological perspective, while retaining the prevailing 19thC hierarchical arrangement of races, and he discussed the question of whether all modern humans share a common origin (monogenism) or whether the different races had separate origins and represent distinct species (polygenism). In *The Descent of Man*, Darwin explained how modern humans could have evolved from an ape-like ancestor and discussed the prehensile hands of monkeys and the origins of tool use by early human ancestors, arguing that bipedalism arose because of the increasing use of the hands for manipulating tools. He argued that the use of tools led to the decrease in size of the canine teeth in modern humans and he claimed that this would have altered the morphology of the face in general. Darwin's view was that the evolution of **bipedalism** preceded the evolution of a larger brain and he argued that modern humans evolved from an Old World primate. Few fossil apes were known at this time except for the European *Pliopithecus* and *Dryopithecus* fossils discovered by French paleontologist Édouard Lartet earlier in the century, yet Darwin argued that modern humans had probably evolved in Africa. Darwin wrote *The Descent of Man* at a time when very few hominin fossil remains had been discovered and when many considered that the **Cro-Magnon** fossils from the Dordogne region in France, discovered by Louis Lartet in 1868, and the Neanderthal fossils discovered in 1856, were either ancient examples of modern humans and not truly ancestral forms, or they were frankly pathological. Darwin died in 1882, before the major hominin fossil discoveries of the late 19thC, and by the time a hominin fossil record began to be accumulated many anthropologists and biologists had turned away from Darwin's mechanism of natural selection and instead invoked non-Darwinian mechanisms to explain human evolution. Yet the description of human evolution presented by Darwin in *The Descent of Man* influenced generations of human origins researchers and he continued to examine the problem of the evolution of human behavior and mental faculties in *The Expression of the Emotions in Man and the Animals* (1872), showing that again there are continuities between modern humans and other animals even in relation to such an apparently unique human attribute as emotions. Darwin, who never held an academic position, was a prominent member of Britain's scientific establishment. He was elected a fellow of the Royal Society in 1839 and in 1864 he was awarded the Copley Medal, the Royal Society's highest scientific honor. In recognition of his scientific stature, after his death in 1882 Darwin was accorded the signal honor of being buried in Westminster Abbey in London.

Darwinian fitness   *See* **fitness**.

data matrix   Any data set that is organized into an $m \times n$ table, where m refers to the number of rows and n refers to the number of columns. Data matrices are typically used in a **multivariate analysis**. The number of rows and columns may be equal or different; when they are equal the matrix is referred to as a

square matrix. Some common examples of square matrices include **correlation matrix**, **covariance matrix**, and **distance matrix**.

date   Used as a verb it refers to the act of dating, whereas as a noun it refers to a specific point in time (e.g., 1347 years BCE). Dates measured in years were abbreviated to "yrs," those in thousands of years were abbreviated to "ky," and those measured in millions of years were abbreviated to "myrs." Now it is conventional to refer to a specimen's age rather than its date and to use the abbreviation **ka** to refer to an age in thousands of years and the abbreviation **Ma** to refer to an age in millions of years before the present.

dating   *See* **absolute dating**; **relative dating**; **geochronology**.

DAW   Abbreviation of Dawaitoli. It is one of the three three-letter prefixes (the others are **BOD** and **HAR**) used for the paleontological evidence collected by the **Middle Awash Research Project**. Each fossil is identified by a three-letter prefix that refers to the geographical area where it was found, followed by VP (for Vertebrate Paleontology), the number of the collecting site within the area, then a final number that places the fossil in the sequence of fossils recovered from that collecting site. Thus DAW-VP-1/1 is the first fossil to be found in the first collecting site to be recognized in the Dawaitoli area of the **Middle Awash study area**. *See also* **Dawaitoli**.

Dawaitoli   One of the named drainages/subdivisions within the **Bodo-Maka** fossiliferous subregion of the **Middle Awash study area** in Ethiopia. The area was first explored for its paleoanthropological potential in 1981 by a multidisciplinary team organized by the Ethiopian Ministry of Culture and Sports Affairs and led by Desmond **Clark** with more extensive excavations undertaken in 1990. The localities are in deposits that are part of the Bodo–Dawaitoli–Hargufia sedimentary sequence and several localities within Unit U of the Dawaitoli Formation have yielded archeological assemblages. Temporal span A tuff that is also exposed within Unit U of the Bodo–Dawaitoli–Hargufia sedimentary sequence suggests an age of 0.64 ± 0.03 Ma and biostratigraphic evidence suggests that all the localities are **Middle Pleistocene**. How dated?

Argon-argon dating and biostratigraphic correlation with **Olduvai** Bed IV and **Olorgesailie**. Hominins found at site None. Archeological evidence found at site Localities DAW-A6 (Lower), DAW-A7, and DAW-A11 have yielded **Mode 1 Acheulean** artifacts. Localities DAW-A1, DAW-A2, DAW-A3, DAW-A4, DAW-A5, DAW-A6 (Upper), DAW-A8, and DAW-A9 have yielded **Mode 2** Acheulean artifacts. Artifacts consistent with the **Middle Stone Age** have been recovered from locality DAW-A10. Key references: Geology and dating Clark et al. 1994, de Heinzelin et al. 2000; Hominins N/A; Archeology de Heinzelin et al. 2000.

"Dawn man" *See* **Piltdown**.

Day, Michael Herbert (1927–) Michael Day was born in London, UK. He served in the Royal Air Force from 1945 to 1948 and then entered the Royal Free Hospital School of Medicine where he earned his MB BS degree plus Membership of the Royal College of Surgeons (MRCS) and became Licentiate of the Royal College Physicians (LRCP) in 1954. In 1962 he was awarded his PhD for research on the anatomy and blood supply of the lumbosacral plexus. He spent his academic career within the University of London, first at the Royal Free Hospital School of Medicine (1957–62), then at the Middlesex Hospital Medical School (1962–72), and finally at St. Thomas's Hospital Medical School where he was the Head of the Anatomy Department (1972–89); he was appointed Reader in Physical Anthropology in 1969 and Professor of Anatomy in 1972. Although his PhD was in soft-tissue anatomy Day became involved, via John **Napier**, with the analysis of the hominin postcranial remains from **Olduvai Gorge** and this began a long collaboration with Louis and, more substantively, Mary Leakey. Initially together with Napier, but later on his own, he described and analyzed many of the key postcranial hominin fossils from Olduvai Gorge (**OH 8**, Day and Napier 1964; **OH 10**, Day and Napier 1966; **OH 20**, Day 1969; **OH 28**, Day 1971; **OH 34**, Day and Molleson 1976). He carried out the first multivariate analysis of a hominin postcranial fossil (OH 10, Day 1967), described and undertook the initial analysis of the hominin remains recovered from **Omo-Kibish** (Day 1969), and the hominin cranium from Laetoli (**LH 18**, Magori and Day 1983b) and contributed to the analysis of the postcranial fossils from Swartkrans and Kromdraai (Day and Napier 1964, Day and

Thornton 1986). Day was also one of the team of three anatomists who generated the initial descriptions of the hominin fossils recovered from **Koobi Fora**. His other important contribution to paleoanthropology was his authorship of the *Guide to Fossil Man*. This careful site-by-site treatment of the hominin fossil record was deservedly popular among researchers and students; the first edition was published in 1965 and the last, fourth, edition was published in 1986. Day also made notable contributions to professional organizations relevant to paleoanthropology, serving as President of the Royal Anthropological Institute (1979–83), and together with Peter Ucko he was instrumental in establishing the World Archeological Congress, being elected its Founding President in 1986.

day range The area exploited by an individual animal or group of animals during a day (i.e., a 24 hour period). Because it is a daily subset of movements within a home range, the day range is usually smaller than the home range. *See also* **home range**.

Daylight Cave *See* **Silberberg Grotto**.

dc Abbreviation for a deciduous canine. Upper, or maxillary, deciduous canines are identified by a line below the abbreviation (dc) and lower, or mandibular, deciduous canines are identified by a line above the abbreviation (dc).

de Heinzelin, Jean *See* **Heinzelin, Jean de (1920–98)**.

Deacon, Hilary John (1936–2010) Hilary Deacon was born in Cape Town, South Africa, and he completed his BSc in geology and archeology at the University of Cape Town in 1955. He worked as a geologist in East and West Africa before being appointed as an archeologist at the Albany Museum in Grahamstown in South Africa in 1963. In 1967, he spent a year at the University of London as a British Council Scholar and received his PhD from the University of Cape Town in 1974. Apart from the periods when he was a Visiting Professor at the University of California, Berkeley (1986), a Visiting Fellow at the Australian National University, Canberra (1984), and a Visiting Professor at the University of Chicago (1978), Hilary Deacon served as the head of the Archeology Department at

Stellenbosch University from 1971 until his retirement at the end of 1999. Deacon's special interest was the archeology of southern Africa and the origin of modern humans and arguably his most important contribution was the research he carried out at **Klasies River** from 1984 onwards. In addition, he contributed to paleoenvironmental studies and to a detailed study of the Fynbos Mediterranean System in the Cape. Deacon was a Past-President of the South African Archaeological Society, the Southern African Association of Archaeologists, and the South African Society for Quaternary Research, and was a member of the Board of Iziko Museums in Cape Town.

"Dear Boy"   The nickname used by **Louis** and **Mary Leakey** for **OH 5**.

death rate, age-specific   The instantaneous hazard of death at exact age t, and is often written as $\mu(t)$ or $h(t)$. The concept is closely related to the **age-specific probability of death**, the proportion of individuals who enter an age interval, but who die *within* that age interval. The age-specific death rate generalizes this by taking smaller and smaller age intervals, so that for example $\mu(50.29)$ is the death rate at exactly age 50.29 years. In the case where there is a constant hazard of death independent of age (also known as the exponential survival model) the hazard is the same as the instantaneous **crude death rate**. Organisms as diverse as yeasts, flies, worms, rodents, monkeys, and modern humans (Finch 2007) show age-specific death rates that do not follow an exponential survival model, but instead often follow a **Gompertz hazard model** for those who survive to adulthood.

débitage   (Fr. *débiter*=to cut up) A French term used by English-speaking lithic technologists to refer to the flakes, flake fragments, and other debris that results from the process of stone tool manufacture. In French and some French translations the term is used to describe the action of intentional **knapping** as well as its byproducts.

deciduous dentition   (L. *decidere*=to fall off and *dentes*=teeth) Refers to the teeth of the primary dentition (or milk teeth) shed during the normal course of dental development. In each jaw quadrant of all hominins and hominids it consists of two deciduous **incisors** (di1 and di2), a deciduous canine (dc), and two deciduous **molars** (dm1 and dm2, or dp3 and dp4). Deciduous molars are occasionally found as fossils, but deciduous incisors and canines are rare; one of the exceptions is the southern African site of **Drimolen**. First lower deciduous molars, dm_1s, have been shown to be particularly useful for **alpha taxonomy**. For example, Robinson (1956) showed that the dm_1s of *Paranthropus robustus* and *Paranthropus boisei* have especially complex cusp morphology on their crowns.

declarative memory   (L. *clarus*=to make clear) Memories come in two forms: those that can be explicitly recalled and those that cannot. Memories that can be articulated or consciously reckoned are referred to as declarative memories. Declarative memories stand in contrast to procedural memories, which are long-term memories for either skills or motor responses that are difficult to articulate or describe linguistically. Examples of this form of memory include most forms of motor learning such as tying one's shoes or riding a bicycle. Examples of declarative memory include **episodic memory**, having to do with the recollection of biographically relevant events, and **semantic memory**, having to do with knowledge of everyday facts or concepts that are disconnected from personal, autobiographical knowledge. Tulving (1983) has argued that while declarative semantic memory is likely to be phylogenetically ancient and shared among mammals and birds, declarative episodic memory is likely to be unique to modern humans, but researchers are unlikely to be able to determine whether this is the case.

decussation   (L. *decussis*=the crossing of two structures after the roman numeral for 10, X, in which two lines intersect) A pattern formed by **enamel prism**s seen in hominin teeth, but it is most evident in rodent enamel. Enamel prisms do not typically run in a straight line from the **enamel dentine junction** to the surface of the tooth. Some researchers believe that over the dentine horn of a tooth cusp enamel prisms turn helically (like a snail shell) and intertwine as they move towards the tooth surface (e.g., Risnes 1986); this is known as gnarled enamel because of its complex appearance under a microscope. Others report a wave-like pattern in the **lateral enamel** (e.g., Osborn 1990, Macho et al. 2003). Typically prisms are said to twist in a sinusoidal manner two or three times before running straight to the surface in the outer one-third of the enamel. When cut in cross-section, bands or tracts of prisms appear to intersect at an angle (thus the term enamel decussation) but actually they are just crossing over and past each other much as the component strands of a rope cross over one another. Christian Schreger and John Hunter both independently described the banding pattern visible on fractured or cut enamel surfaces that results from prisms moving in and out the plane of section; hence the eponymous term **Hunter–Schreger bands**, which are a manifestation of enamel decussation in enamel, bears both of their names. *See also* **diazone**; **enamel development**; **Hunter–Schreger bands**; **parazone**.

Dederiyeh (Location 36°24′N, 36°52′E, northern Syria; etym. "two entrances" in Kurdish) History and general description This cave site in the Afrin Valley, northern Syria, which is 400 km/250 miles north of Damascus and 60 km/40 miles northwest of Aleppo, was discovered in 1987 and was excavated for 5 years prior to the discovery of a child's skeleton (**Dederiyeh 1**) in 1993. A second child's skeleton (Dederiyeh 2) was identified in 1997 and as of 2000 the remains of at least 15 more individuals had been recovered. Fifteen recognizable layers are grouped into four major stratigraphic units with a combined thickness of 3 m. The artifacts from all four units correlate with those found at **Tabun** layer B, and are generally described as Levantine **Mousterian**. Each unit contains artifacts equivalent to the "Tabun-B Levantine Mousterian." Temporal span and how dated? Synchronous with Tabun layer B from the archeological evidence. Hominins found at site Dederiyeh 1 and 2, plus the remains of at least 15 other individuals. Archeological evidence found at site Tabun-B Levantine Mousterian. Key references: there is an excellent monograph on the context and morphology of the two Dederiyeh infant skeletons (Akazawa and Muhessen 2003) and several of the references below are from that publication; Geology and dating Akazawa et al. 2003; Hominins Akazawa et al. 1995, 2003; Archeology Akazawa et al. 2003.

Dederiyeh 1 Site **Dederiyeh**. Locality N/A. Surface/*in situ In situ*. Date of discovery August 1993. Finders Team led by Takeru Akazawa and Sultan Muhesen. Unit SU-III. Horizon Layer 11, but Akazawa et al. (1995) suggest it comes from Layer 8. Bed/member N/A. Formation N/A.

Group N/A. Nearest overlying dated horizon N/A. Nearest underlying dated horizon N/A. Geological age No dates, but the artifacts are similar to those from **Tabun** layer B. Developmental age *c.*2 years. Presumed sex Unknown. Brief anatomical description This fragmentary, but remarkably complete, infant skeleton was found in more than 200 pieces. It includes much of the skull, axial skeleton, and parts of all four limbs, but because of its immaturity the hands and feet are relatively poorly represented. Announcement Akazawa et al. 1993. Initial description Akazawa et al. 1995. Photographs/line drawings and metrical data Akazawa et al. 1995, 2003. Detailed anatomical description Akazawa et al. 2003. Initial taxonomic allocation *Homo neanderthalensis*. Taxonomic revisions None. Current conventional taxonomic allocation *H. neanderthalensis*. Informal taxonomic category Pre-modern *Homo*. Significance A remarkably complete skeleton that provides important information about the ontogeny of *H. neanderthalensis*. Location of original Not known.

"Deep Skull" See **Niah cave**.

deep sea temperatures The temperature of the deep sea (defined as depths of >1 km/0.6 miles) is set by conditions in the regions of sinking water masses, which today are in the North Atlantic and around the coast of Antarctica. Deep sea temperatures are remarkably homogeneous and relatively stable through time within 2–4°C; however, subtle changes did occur between glacial and interglacial conditions. The **benthic** oxygen isotope record derived from calcitic **foraminifera** is used to reconstruct deep sea temperatures and **glacial ice volumes**. The cyclicity captured in these records and their uniformity across the world's oceans provides a means of temporal correlation between ocean drilling sites. *See also* **astronomical time scale**.

definition A list of the features shared by all, or at least a large majority, of the members of a **taxon**. The features given in a definition of a taxon include all of its attributes, not just the ones that make it distinctive: the latter are the ones that are emphasized in the diagnosis of a taxon. (syn. **description**). *See also* **diagnosis**.

deflation The winnowing process whereby the surface of a body of sediment is partially eroded, particularly by wind activity This preferentially removes the finer-grained components of the sediment (sands, silts, and clay) and therefore concentrates the remaining larger/heavier **clasts**, including bones and stone tools. The sediments at sites in Chad (e.g., **Toros-Menalla**) are an example of deflation.

deflecting wrinkle In most lower molars the main ridge that runs from the metaconid cusp tip to the central fossa follows a straight path in a distobuccal direction. However, in some cases, the initial section of the ridge is orientated buccally and then it runs in a distal direction. This latter variant of the ridge running off the metaconid is described as a "deflecting wrinkle." The feature is illustrated in Morris (1970) and Scott and Turner (1997).

degrees of freedom Functional morphology Refers to the number of independent axes at which movement can take place at a moveable (i.e., synovial) **joint**. A joint like the modern human elbow joint can only flex and extend, so it is uni-axial and thus it has *one* degree of freedom. The carpometacarpal joint of the thumb (between the first metacarpal and the trapezium) has two independent axes that are determined by the shape of the joint surfaces and thus it *two* degrees of freedom. Since there is a maximum of three axes about which independent rotations can occur, three is the maximum number of degrees of freedom there can be at a joint. Statistics In statistics, degrees of freedom refers to the number of values that are free to vary when calculating a particular statistic. For example, for a sample of n measurements with a given mean, $n1$ of the measurements can take on any value. But if you know $n-1$ measurements then one can calculate the *exact value* of the nth measurement that will produce the given mean. Thus only $n-1$ of the n values are free to vary, for the nth value *must* equal whatever number will result in that particular mean value. This is why a one-sample t test that compares the mean of a set of measurements to a particular value has $n-1$ degrees of freedom. However, in **linear regression**, for a given slope and intercept, $n-2$ values can vary freely and knowing those $n-2$ measurements allows one to calculate what the $(n-1)$th and nth values *must* be in order to produce the given slope and intercept. This is why there are $n-2$ degrees of freedom for linear regression.

But when linear regression is performed with the intercept constrained to zero (as is the case **phylogenetically independent contrasts**) the intercept is *not* a function of the data, thus given a particular slope only $n-1$ values can vary, so linear regressions with constrained slopes have $n-1$ degrees of freedom. In general, statistics tend to have as many degrees of freedom as the number of observations made (n) minus the number of **parameters** estimated in the calculation of that statistic (e.g., two parameters, the slope and the intercept, are estimated in the case of linear regression). With regard to the effect of degrees of freedom on statistical tests, as the number of values that can vary independently increases, so does the confidence in the accuracy of estimated parameters (e.g., mean, slope intercept, etc.). Thus, if all else is held constant, statistical significance increases as the number of degrees of freedom increases.

deinothere The informal name for the taxa within the family Deinotheriidae, which contains all of the proboscideans with downward-curving tusks. *See also* **Proboscidea**.

Deinotheriidae (Gk deinos=terrible plus *ther*= combining form, wild beast) A family of extinct Old World Proboscidea whose main distinguishing characteristics are downward-facing tusks in the lower jaw and, in later forms, strongly lophodont molars with extremely thick enamel; deinotheres have no upper tusks. Even when fragmentary, deinothere molars are easy to identify in fossil assemblages so their temporal and spatial distribution is well documented. The African species *Deinotherium bozasi* is known from the late Miocene and was the latest surviving form, with a last appearance date of approximately 1 Ma. It is found at many archeological and hominin paleontological sites. *Deinotherium* remains found with **Developed Oldowan** stone tools at FLKN II at **Olduvai Gorge**, Tanzania, have been cited as an early example of a single carcass butchery site.

Deinotherium See Deinotheriidae.

Deir El-Fakhuri See Esna.

DEJ Acronym for dentine-enamel junction. It is one of two conventional abbreviations (**EDJ** is the other) used for the interface between the underlying dentine and the overlying enamel. *See* **enamel-dentine junction**.

Deko Weko One of the Early and Middle **Pleistocene** sites in the exposures of the **Ola Bula Formation** within the **Soa Basin** of central Flores, eastern Indonesia.

deleterious mutation (L. *deleterius*=to harm and *mutare*=to change) A **mutation** that causes a reduction in the **fitness** of an individual (e.g., it may cause disease, reduce survival in other ways, or affect reproduction). Examples of deleterious mutations affecting modern humans are mutations in the *CFTR* gene (which encodes the cystic fibrosis transmembrane conductance regulator protein) that cause cystic fibrosis, deletions in the DAZ region of the **Y chromosome** that result in low sperm count and infertility, and mutations in the *HEXA* gene that cause Tay–Sachs disease.

deletion (L. *delere*=to abolish) A mutation in the **genome** where one or more **base pairs** (i.e., pairs of **nucleotides**) have been removed or deleted from a sequence of nucleotides. These deletions can be as small as one base pair or involve many thousands of base pairs. In a coding region any deletion that does not involve a multiple of three base pairs causes a **frameshift mutation**, which is typically deleterious. Large deletions that include genes, such as the DAZ region of the **Y chromosome**, or a substantial portion of a **chromosome** are usually detrimental whereas deletions in non-coding regions are usually neutral. **Insertions** and deletions are commonly referred to as **indels**.

δ Delta, or δ, is a standardized notation used in **stable-isotope biogeochemistry**. The δ notation is used to refer to the proportion of heavy to light **isotope** within a given substance relative to an international standard, where $\delta=[(R_{sample}/R_{standard})-1]$ *1000, and $R=^{13}C/^{12}C$, $^{15}N/^{14}N$, $^{18}O/^{16}O$, etc. Values are reported as parts per thousand (‰). It is important to note that δ values are relative. International standards differ for each element; thus δ values for each element are on different scales and are not directly comparable.

delta (Gk *delta*=fourth letter of the Greek alphabet, Δ, that is shaped like a triangle) *See* **carbon isotope analysis; deltaic; stable isotopes; hydrogen isotopes**.

deltaic (Gk *delta*=fourth letter of the Greek alphabet, Δ, that is shaped like a triangle) The name given to sediments formed in the sometimes triangular-shaped region at the mouth of a river. They are deposited when the flow in the river reduces prior to the river entering a lake or the sea. Because the water energy in a delta is very low the sediments tend to be dominated by fine-grained material (e.g., **silts** and **clays**).

deme (Gk *demos*=people) An informal category used for a geographically localized population of closely related individuals. When applied to the fossil record the concept is sometimes referred to as a "p-deme," paleodeme, or paleocommunity-deme, and the closeness of the relationship among individual fossils is determined from their morphology; the more similar they are, the closer the relationship is inferred to be (but *see* **homoplasy**). Clark **Howell** (1999) recommended that it might be less controversial to use p-demes rather than species for discussions about the taxonomy of the hominin fossil record. Presumably one or more p-demes may correspond to regional variants within a single species (i.e., equivalent to a sub-species) or they could correspond to the hypodigm of a species. The use of p-demes does not solve the problem of how many species to recognize in the fossil record, but this informal category may be a useful way to discuss the different parts of the fossil record prior to making decisions about its taxonomy.

demography Hinde (1998, p. 1) defines demography as "the study of population structure and change." This definition is not particularly useful within biological anthropology, as "population structure" is often used to refer to population genetic structure. A better definition of demography for the purposes of paleoanthropology is the study of age and sex structure and/or population size.

denature (L. *de*=cessation or separation and *hatura*=born or produced) The process whereby the DNA double helix is separated (or denatured) into two single strands. If you visualize DNA as a twisted ladder, denaturing breaks the hydrogen bonds that form the rungs of the ladder. Denaturation requires either heating or chemical treatment. *See also* **base pair; deoxyribonucleic acid; DNA hybridization**.

dendrite (Gk *dendrites*=of, or pertaining to, a tree) An extension of the cell body of a **neuron** that branches in a tree-like fashion (also known as a "dendritic arbor") and that serves as the primary site for receiving and integrating information coming into that neuron from other neurons. Some dendrites have specialized outgrowths along their shaft called spines. These spines are the sites of excitatory synaptic inputs. Although some neurons lack dendrites (e.g., dorsal root ganglion cells), others have multiple dendritic trunks that arise from the neuronal cell body. For example, the **pyramidal cells** of the **cerebral cortex** have long slender sets of dendrites that emerge from the apex of the cell body, called *apical* dendrites, and other sets that emerge from the base of the cell, called *basal* dendrites. The basal dendrites of the **prefrontal cortex** in modern humans have been shown to have a much higher density of spines than in any other primate examined (Elston et al. 2006), but as yet comparable observations about ape prefrontal cortical neurons have not been made.

dendritic arbor (Gk *dendrites*=of, or pertaining to, a tree and L. *arbor*=tree) An extension of the cell body of a **neuron** that branches in a tree-like fashion and serves as the primary site for receiving and integrating information coming into that neuron from other neurons.

dendrochronology *See* **radiocarbon dating**.

Denisova Cave (Location 51°40′N, 84°68′E, Russia; etym. unknown, although Denisova is a common last name) History and general description This cave site in the Altai Republic of Siberia has been explored since 1977 and contains **Middle** and **Upper Paleolithic** assemblages. The **mitochondrial DNA** (mtDNA) from a hominin phalanx found in 2008 was found to be distinct from that of both *Homo neanderthalensis* and *Homo sapiens*, suggesting the late survival of another, yet-unidentified, hominin lineage. Furthermore, analysis of the Denisova mtDNA established that it was more different from modern human mtDNA than it was from Neanderthal mtDNA. Temporal span and how dated? The lowest levels were dated to between 171 and 282 ka using thermo**luminescence dating**, but these dates are likely too old. **Biostratigraphy** suggests a maximum age of **Oxygen Isotope Stage** 5e (*c*.125 ka). The phalanx derives from a layer with three **uncalibrated radiocarbon age**s of 48,650 ± 2380 years BP, an infinite age of more than 37,235 years BP, and 29,200 ± 360 years BP. Hominins found at site The proximal epiphysis of a juvenile manual phalanx from an unidentified hominin species was recovered. Archeological evidence found at site The site contains a sequence of Middle and Upper Paleolithic layers. The layer from which the fossil was recovered contains elements of both industries, including both bone tools and **Levallois** points. The authors argue that this is not evidence of post-depositional stratigraphic mixing, since other Early Upper Paleolithic sites in the area have similar assemblages. Key references: Geology, dating, and paleoenvironment Derevianko 1998, Krause et al. 2010; Hominins Krause et al. 2010; Archeology Derevianko 1998, Krause et al. 2010.

dense bone *See* **bone**.

dental crypt (L. *dens*=tooth and *crypta*=hidden) The space within the **alveolar process** of the maxilla or mandible within which a tooth germ develops. Bone is resorbed to make space for the developing **tooth germ**, and this space may be detected radiographically before hard-tissue formation begins.

dental development A broad term that encompasses the initiation, morphogenesis, mineralization, and eruption of the teeth and even the establishment of occlusion. Modern human dental development is a process of *c*.20 years, stretching from embryo to young adult; the same process takes *c*.12 years in chimpanzees (Nanci 2007, Kuykendall 2009). Because teeth are highly mineralized, they are conserved as fossils more often than other tissues, and because they are not remodeled their macrostructure and microstructure preserve a record of history, of their formation, and of interruptions to their growth. Some of dental development have a lower coefficient of variation than comparable measures of skeletal development (Lewis and Garn 1960) and a lower component of environmental variance (Tanner 1955). Mammalian dental development is both revealing and amenable to study because individuals mature through a series of distinctive stages to a fixed final size with many clues to developmental stage and age. Each tooth of each tooth type is a variable and within each tooth each component of the crown and root is another variable; dental development is the antithesis of a simple continuous increase in size. Recognizable dental stages are often used to define traditional life stages of infant, juvenile, and adult in both human biology and primatology (Schultz 1960, Bogin 1999b). In modern humans initial dental development is observed in the late stage embryo, with the differentiation of precursor tissues. Tooth buds form as an organic matrix that is subsequently mineralized: in modern humans mineralization of deciduous teeth begins in the second trimester and for earliest permanent teeth just before birth (see Kraus and Jordan 1965). Cell activity responds to circadian rhythms, leaving lines and bands in enamel and dentine that reflect both the passing days and also longer-period physiological rhythms (see Aiello and Dean 1990). Stressors can interrupt growth, leading to more pronounced lines; the best known of these is the neonatal line, a virtual "birth certificate" preserved in teeth that were mineralizing at birth. Eruption of teeth through the alveolar bone and gingival tissue occurs after tooth crowns are complete and roots are actively forming; roots close after eruption. At this point the enamel cap may be worn or decayed, but cannot grow, although dentine-producing cells on the interior of the tooth can respond to stimuli in maturity by gradually filling in the pulp chamber. Most chronologies of dental development are chronologies of mineralization or emergence, although chronologies based on dissection or histology and microscopy are far more sensitive to the early stages of dental development than are radiographic studies (Beynon et al. 1998). Although most primates are born with some teeth already erupted, modern humans and great apes are born with toothless gums (Smith et al. 1994). At birth, in both modern humans and chimpanzees mineralization of the crowns of deciduous teeth is well under way and first permanent molars are mineralized at the cusp tips, although the chimpanzee neonate has more advanced deciduous crowns. In slow-growing mammals like anthropoid primates, the deciduous dentition may function alone for a prolonged period before the first permanent molar emerges. Molars emerge distal to the deciduous dentition, apparently developing in time with the growing face and jaws. The small deciduous teeth, partly or wholly formed *in utero* in placental mammals, are eventually shed and succeeded by more densely mineralized "replacing" permanent incisors, canine, and premolars, emerging in limited and characteristic sequences (Schultz 1935, Smith 2000). Dental development has been both the subject

and a tool of human evolutionary research. As actual data on timing became available on a variety of primates, the age of emergence of M_1 was shown to be highly correlated with onset or duration of other events in life history, from gestation length to reproductive maturation and life span, brain size and, to a lesser degree, body size (Smith 1989). Some events in dental development are related to events in life history (e.g., weaning and emergence of first permanent molars; sexual maturity and canine emergence). On finer scales, tooth surfaces preserve evidence of growth disturbances that reveal the timing of stress episodes that characterize populations and particular ages of death (Armelagos et al. 2009). In the fossil record, tooth histology can be used to count or estimate (depending on the scale of inquiry) the time elapsed between birth and death for individual juvenile fossils. The key to opening the hominin fossil record to such inquiry has been the introduction of methods that avoid destruction of specimens, either by using surface manifestations of incremental features (Bromage and Dean 1985) or, more recently, through high-energy imaging of hidden increments (Smith and Tafforeau 2008). As work progresses, each fossil juvenile with an associated age of death builds towards a chronology of growth and development of its own species. To date, dental histologists have established that *Australopithecus* and *Paranthropus* matured on a time scale closer to apes than to modern humans (Beynon and Dean 1988); dental development within the genus *Homo* is under investigation (Smith et al. 2007, Dean and Smith 2009). In recent years, experimental studies (often with mice) manipulate dental development to discover how it is regulated on a genetic or cellular level and how a change in developmental program can alter tooth shape, size, and proportion (Jernvall and Thesleff 2000). This renewed interest in ontogeny and phylogeny (as "evo-devo") is yielding new perspectives on the historical record of dental evolution. *See also* **enamel development; dentine development; hypoplasia; long-period incremental lines; perikymata; secondary dentine; short-period incremental lines; synchrotron radiation micro-computed tomography.**

dental follicle (L. *dens*=tooth and *folliculus*= a little bag, diminutive of *follis*=a bag) The fibrous sac that surrounds the developing tooth germ in the bony crypt. The dental follicle begins as a cellular condensation of mesenchymal origin and when fully formed it encloses the developing tooth germ, both enamel organ and dental papilla (the future pulp). The follicle is maintained in constant relationship to the mucous membrane that lines the mouth, or oral cavity, through its attachment to the gubernacular cord that runs in the **gubernacular canal**. The dental follicle is involved in the formation of alveolar bone and **cementum** as well as with tooth eruption.

dental formula (L. *dens* = tooth) The numbers of teeth in each of the four quadrants of the jaws. The formula is written beginning mesially with the number of incisors (I or di), the number of canines (C or dc), the number of premolars (P), and ending with the number of molars (M or dm/dp). In all higher primates, including fossil hominins, the dental formula for the normal **deciduous dentition** is "2. 1. 2," and for the normal **permanent dentition** it is "2. 1. 2. 3."

dental macrowear (L. *dens* = tooth and Gk *macros* = large) Tooth wear you can see with the naked eye. *See also* **tooth wear**.

dental mesowear (L. *dens* = tooth and *mesos* = middle) Tooth wear you can see with the naked eye. *See also* **tooth wear**.

dental microwear Occlusal tooth wear that needs a microscope in order to see it. Dental microwear focuses on the size, number and orientation of microscopic scratches, pits, etc. on the enamel, or on the overall complexity and the degree of **isotropy** of the enamel surface. Whereas dental macrowear is a measure of the abrasiveness of the diet in the long term, dental microwear signals on fossil teeth indicate whether the food ingested in the days or weeks before death contained hard or abrasive material. Such abrasive material can be either intrinsic to the ingested foods (e.g., **phytoliths**) or extrinsic to the foods (e.g., adherent sand grains). Dental microwear can also detect evidence for softer or tougher foods (e.g., polished or finely pitted surfaces). If any striations or scratches have a predominant orientation, this can be used to reconstruct the direction of the jaw movement(s) that generated the microwear. Dental microwear studies were revolutionized when precision-molding materials were introduced, enabling positive molds of whole teeth to be examined with a scanning electron microscope (Walker et al. 1978). Fred Grine, Peter Ungar, and Mark Teaford have been in the van of applying dental

microwear analysis to fossil hominins (e.g., Ungar et al. 2006) and in the process they have helped develop new analytical methods (e.g., dental microwear textural analysis; see Scott et al. 2005).

dental microwear textural analysis *See* **dental microwear**.

dental reduction Decrease in the metric proportions of teeth between successive taxa or populations. For, example the reduction in crown size from the large postcanine teeth of *Australopithecus* to the smaller postcanine teeth of early *Homo* and the further reduction in crown size from pre-modern *Homo* to *Homo sapiens*. The term may also refer to a **metameric** trend of decreasing size along the tooth row in a tooth type (e.g., the pattern of decreasing crown dimensions from M1 to M3 seen in later fossil hominins and in modern humans).

dental wear *See* **tooth wear**.

dentine (L. *dens* = a tooth and *inus* = "of" or "relating to," hence "pertaining to a tooth") The term coined by Richard Owen for the hard tissue that forms the bulk of a tooth crown and tooth root. Dentine is one of the three dental hard tissues (the others are **enamel** and **cementum**). Dentine (or ivory, which is the same thing), unlike bone, contains no blood vessels or included cell bodies. However, fluid-filled dentine tubules containing some nerves and odontoblast processes make dentine a vital (i.e., living) and exquisitely sensitive tissue. Like enamel, dentine is biphasic [i.e., it consists of both a mineral inorganic **hydroxyapatite** phase (70%) and an organic phase made up of collagen, lipids, and noncollagenous proteins]. The dentine core of a tooth is continuous between the crown and root. Coronal dentine underlies the enamel cap and root dentine, which is covered by cementum, supports the crown and anchors the tooth within its bony **alveolus** or socket. Primary dentine forms the first-formed inner part of the coronal dentine and the root dentine. Secondary dentine is slowly and continuously laid down after primary crown and root formation are complete around the periphery of the pulp chamber, thus reducing the size of the pulp chamber with age. Tertiary dentine, a reactionary tissue that is secreted quickly and irregularly in response to exposure of dentine through attrition, abrasion, caries, or trauma, blocks off open tubules from the pulp (Nanci 2003). Dentine is not as hard or dense as enamel, but it is harder and denser than compact bone. Unlike enamel, dentine has no "grain" or planes of cleavage and it is a relatively elastic material (hence ivory was originally the material of choice for billiard or pool balls). Several classes of **incremental features** are found in dentine, including long-period **Andresen lines** and short-period **von Ebner's lines**. They correspond to long- and short-period increments in enamel (Dean et al. 1993).

dentine development **Odontoblasts** derived from the mesenchymal cells of the dental papilla (the precursor of the dental pulp) secrete a collagenous matrix called predentine which undergoes **mineralization** to form primary **dentine**. Dentine formation begins at the **dentine horn** that underlies the future cusp tip and progresses inward through secretion, thus slowly defining and reducing the size of the central pulp chamber as tooth formation proceeds. Extension of the odontoblast sheet finally reaches the apex of the **root**. The extension rate of the dentine is the rate at which new terminally differentiated odontoblasts appear along the length of the future **enamel-dentine junction**, or cemento-dentine junction. Like enamel formation the secretion of dentine has short-period (*c*.24 hour) and long-period (>24 hour) rhythms; these result in dentine increments that are preserved in the fully formed tissue as **von Ebner's lines** and **Andresen lines**, respectively. The secretory process of a single odontoblast remains behind in a canal called a **dentine tubule**, which runs from the enamel-dentine junction to the pulp. Dentine formation and **enamel formation** proceed in tandem, but with dentine formation slightly in advance of enamel formation (Witzel et al. 2008).

dentine horn The tallest point of the **dentine** underlying one of the cusps or features that make up the **enamel** cap of a tooth. The tip of the dentine horn is usually slightly posterior to the corresponding enamel cusp tip but projections of dentine that may be "horn-like" or pointed exist at the **enamel-dentine junction** (or EDJ) with little or no overlying enamel relief related to them. A signaling center, the enamel knot, which controls cell proliferation, appears initially over the tip of each of the future dentine horns. There is reciprocal induction between the epithelial and mesenchymal components of the inner enamel epithelium so that signals exchanged between the future enamel- and dentine-secreting cells initially trigger the deposition of dentine and then subsequently enamel. Dentine matrix secretion proceeds from the EDJ inwards towards the pulp and then enamel matrix secretion follows in the opposite direction from the EDJ outwards towards what will eventually be the surface of the crown.

dentine tubule Small rounded tunnels approximately 2–3 m in diameter running through the dentine from the pulp chamber towards the **enamel-dentine junction** or the cemento-dentine junction at the root surface. In life each tubule is fluid-filled and for a variable distance it is occupied by the elongated process of an **odontoblast** and a proportion of dentine tubules also contain nerve fibers. Changes in either osmotic pressure (e.g., sugar on exposed dentine) or temperature cause intertubular fluid to rush inwards or outwards and trigger dentine sensitivity. This is a major mechanism for protective feedback against tooth damage. Dentine tubules are similar in function to the canaliculi in bone that connect osteocytes and which signal changes in stresses or strains via fluid-filled tubules. See Nanci (2003) for an illustration. *See also* **dentine development**.

dentinogenesis (L. *dens*=a tooth and *inus*= "of" or "relating," and Gk *gena*=to give birth to) The cellular activity involved in the development of **dentine**. *See also* **dentine development**.

dentition (L. *dentes*=teeth) A collective term for all of the teeth in an individual. In the context of the hominin fossil record it is used to refer to the teeth preserved in a particular specimen (e.g., "the dentition of KNM-ER 3734 comprises the roots and crowns of the left C, P_3, P_4, M_1, M_2, and the root and part of the M_3 crown"). Otherwise it refers to the type of dentition [i.e., **deciduous dentition** (primary) or **permanent dentition** (secondary)].

dento-gnathic (L *dens*=tooth and Gk *gnathos*= jaw) Shorthand for "teeth and jaws."

deoxyribonucleic acid (or DNA) A nucleic acid the components of which are **bases** (the four standard versions of bases are adenine, guanine, cytosine, and thymine), and a backbone made of a phosphate and a sugar. In DNA the sugar is 2-deoxyribose and it alternates with a phosphate to form the backbone. These three components (i.e., a base+phosphate+a sugar) make up a **nucleotide**, and nucleotides linked by phosphodiester bonds make up a single strand of DNA. The phosphodiester bond joining adjacent nucleotides is formed from the phosphate group of one nucleotide and the hydroxyl group of the other. The two ends of a single strand of DNA are distinctive. The **5′** (or five prime) end is called that because the fifth carbon of the deoxyribose sugar has a terminal phosphate group attached to it. The other end, the **3′** (or three prime) end, is called that because the third carbon has a hydroxyl group attached. The significance of the directionality of a DNA strand is that the processes of **replication** and **transcription** only occur in the 5′ to 3′ direction. Typically, DNA is present as a double strand arranged in a double helix with the two strands being held together by hydrogen bonds that link a base in one of the DNA strands with a complementary base in the other strand (i.e., adenine always pairs with thymine and guanine always pairs with cytosine). In double-stranded DNA the two DNA molecules are antiparallel (i.e., one of the strands is in the 3′ to 5′ orientation, while the other is in the 5′ to 3′ orientation). The structure of DNA was solved in 1953 by Francis Crick and James Watson with the help of X-ray diffraction data from Rosalind Franklin. Some DNA (the minority) contains genetic information in the form of **genes** and is called coding sequence, while the vast majority of the DNA is called structural or non-coding DNA (e.g., the DNA in the centromeres of **chromosomes**). Coding DNA is transcribed (*see* transcription) into **ribonucleic acid** (or RNA). The RNA is then either translated (*see* **translation**) into **protein** or it forms one of many kinds of functional RNA molecule [e.g., transfer RNA (tRNA), ribosomal RNA (rRNA), or small interfering RNA (siRNA)].

dependency ratio   A ratio of those individuals "too young" or "too old" to be self-sufficient against those individuals who are between these two age thresholds, and are consequently at an age where they can be self-sufficient. In the **demography** of recent modern human populations these ages are often taken as the age at which one reaches maturity (generally taken as 15 years) and the age of retirement (generally 65 years), so

that the dependency ratio is the sum of those individuals less than 15 years old and those individuals greater than 65 years old, divided by the number of individuals between 15 and 65 years old. This is therefore the number of individuals who are "dependent" per the number who are "self-sufficient." Note that this ratio, or the related proportion of individuals who are "self-sufficient," must be calculated using the number of individuals alive in an age interval. Consequently, a **life table** or hazard model approach must be used in **paleodemography** to calculate the dependency ratio from an estimate of the **living age distribution**. Also note that the age thresholds used in the demography of recent modern human populations may be completely inappropriate for prehistoric modern humans and for other hominin taxa.

derived   (L. *derivare*=to draw off) A version or state of a **character** that is not its primitive **character state**. For example, if the character is the root system of the first mandibular premolar tooth, the primitive condition for the hominin clade is most likely two roots, a plate-like distal root, and a mesiobuccal accessory root, or in shorthand 2R: MB+D. In this example there are two derived trends within the hominin clade. One is toward root reduction and simplification; that is, a single root (1R) via one root divided for part of its length (1/2R: Tomes' root). The other is toward a more complex molar-like system of two plate-like roots (2R: M+D). *See also* **apomorphy**.

derived character   A version or state of a **character** that is not its primitive **character state**. *See also* **derived**; **apomorphy**.

dermal bones   (Gk *derma*=skin) Bones of the cranium the shape of which resembles that of the flat cells that make up the epidermis of the skin. The dermal bones form from connective tissue by a process called intramembranous **ossification** and they make up the **desmocranium**. *See also* **ossification**.

dermocranium   *See* **desmocranium**.

description   A list of features shared by all or at least a large majority of the members of a taxon. It is sometimes referred to as the **definition** of a taxon.

desmocranium   (Gk *desmos*=to connect and *kranion*=brain case) The part of the **cranium**

formed from connective tissue by the process called intramembranous **ossification**. The desmocranium consists of most of the **calvaria**, a very small part of the **basicranium**, and much of the **face** and **mandible**. The calvarial elements are the flat, or squamous, components of two unpaired bones (the frontal and the occipital), the two parietal bones, and the squamous part of the paired temporal bones. The frontal and temporal components form from neural crest cells, the parietals from mesoderm, and the occipital component has a mixed neural crest/mesodermal origin. The parts of the basicranium that form via intramembranous ossification are the tympanic component of the temporal bone and the medial pterygoid plate. The bones of the face that form via intramembranous ossification include the paired lacrimal, **maxilla**, **palatine**, and **zygomatic** bones, the unpaired **vomer**, and the mandible. *See also* **ossification**. (syn. dermocranium).

deuterium A **stable isotope** of **hydrogen** with two neutrons, also called ^{2}H. *See also* ^{2}H/^{1}H: **hydrogen**.

Developed Oldowan An African stone tool tradition defined by Mary **Leakey**. It was based on **artifact assemblages** excavated from Beds II to IV, **Olduvai Gorge**, Tanzania, that were judged to be advanced relative to the **Oldowan** and which largely overlapped the **Acheulean** industry in time. The Developed Oldowan, which was seen by Leakey as emerging directly from the Oldowan, was suggested to have three time successive phases. The oldest phase, Developed Oldowan A (*c.*1.7–1.4 Ma), was known from lower and middle Bed II and differed from the **chopper**- and **flake**-dominated Oldowan *sensu stricto* primarily in having a higher frequency of subangular to round artifacts termed "subspheroids and spheroids," as well as a higher frequency of crude **handaxe**-like artifacts termed "protobifaces." Developed Oldowan B was viewed as a continuation of Developed Oldowan A, with the addition of "poorly made," generally small, bifaces as well as a higher frequency of light duty (small flake) tools. It is known from middle and upper Bed II and possibly Bed III at Olduvai Gorge. Developed Oldowan C was described as having a higher proportion of light duty tools as well as a higher frequency of pitted **anvils** and **hammerstones** relative to the Developed Oldowan B and was best known from Bed IV. In reality, there is not much distinguishing

the Developed Oldowan B and C and some would subsume the former into the latter and if that is done they are **coeval** with the Acheulean at Olduvai Gorge for approximately a million years (*c.*1.5–0.5 Ma). Leakey argued that the Developed Oldowan and Acheulean likely represented two distinct cultural traditions, perhaps made by *Homo habilis* and *Homo erectus*, respectively. This interpretation has been called into question on several grounds. First, the protobifaces seen in the Developed Oldowan A may be bifacial cores that fortuitously attained a biface-like shape through heavy reduction rather than through a conscious attempt to make a handaxe. There is thus little distinguishing the Developed Oldowan A and the Oldowan *sensu stricto* and the former could reasonably be collapsed within the latter. Second, differences in biface form between the Developed Oldowan B and the Acheulean likely reflect differences in blank shape (cobble vs flake) and raw material, rather than "cultural" differences. Finally, the presence in the Developed Oldowan B and C of low frequencies of bifaces made on large flakes in a manner indistinguishable from the Acheulean suggests a connection between these traditions. Rather than being transitional between the Oldowan *sensu stricto* and the Acheulean, or a separate tradition from the Acheulean, the Developed Oldowan may be best regarded as a biface-poor **facies**, or variant, of the Acheulean.

development Indicating an increase in functional ability, sometimes used interchangeably with maturity. The end point of development is often considered the age at which successful procreation is functionally possible. This involves biological as well as behavioral maturity, and is complete in the late teens in modern humans. The relationship between growth and size and progress in maturation is not a linear or straightforward one, especially in modern humans. The conflation of chronological age, overall size, and degree of development is one that obscures the meaningful way in which these systems can vary among individuals or populations. The most commonly used markers of development include secondary sexual development, skeletal maturity, and dental maturity (which also mature according to dissimilar schedules). This dissociation between processes of development, as seen in KNM-WT 15000, creates problems for researchers wishing to estimate age of extinct species based on developmental criteria drawn from extant populations: both the intra- and interpopulational variation provide challenges, as do the inconsistent agreement between chronological and developmental age assessed by examination of multiple systems. *See also* **KNM-WT 15000**.

developmental stability Developmental stability refers to the tendency for **development** to produce the same **phenotype** under the same environmental conditions. One measure of developmental stability is the **variance** that remains once both genetic and environmental factors are accounted for. The closest approximation of this is the random variation that occurs across planes of symmetry in symmetrical organisms, or fluctuating asymmetry. The variance of genetically identical individuals under identical environmental conditions is also a theoretically possible measure. Variation in developmental stability has been linked to stress, sexual selection, phenotypic extremeness, and **heterozygosity** although all of these correlations are weak, difficult to detect, and controversial.

Devil's Lair (Location 34°19′S, 115°9′E, Australia; etym. a cave in which bones of an extinct Tasmanian Devil were found) History and general description Devil's Lair, a cave near Augusta in southwestern Australia, documents a long human cultural occupation during the late Pleistocene that includes hearths, artifacts, rock art, and items of personal adornment. A hominin tooth and an ilium have been recovered from the site. Temporal span and how dated? The cave was periodically occupied, presumably by modern humans, from *c*.48 to 13 ka with the modern human skeletal remains likely dating to the end of the occupation. The earliest date of occupation in the cave was corroborated by four independent dating techniques thus supporting an early occupation of the Australian continent. Hominin found at site The Devil's Lair ilium is that of an aged modern human adult male. Archeological evidence found at site Hearths, tools, rock art, and personal adornment including a pendant and beads. Key references: Geology, dating, and paleoenvironment Turney et al. 2001; Hominins Davies 1968; Archeology Dortch 1979.

Devil's Tower *See* **Gibraltar**.

dexterity (L. *dexter*=on the right side, or skillful) Having precision and/or skill when performing a function using the fingers/hands (e.g., the manufacture of an **Aurignacian** fine bone needle requires more dexterity than the manufacture of an **Oldowan** chopper). As the etymology implies it refers to most people's preference for their right hand when performing such functions.

dextra (L. *dexter*=skillfull) Refers to the D- or right-handed form of a complex molecule. *See also* **epimerization**.

DFA *See* **discriminant function analysis**.

DFT Abbreviation for **Duinefontein** (*which see*).

DHA Abbreviation of docosahexaenoic acid. *See* **essential fatty acids**.

di Abbreviation for a deciduous incisor. Upper, or maxillary, deciduous incisors are identified by a superscript number (di^1 and di^2) and lower, or mandibular, deciduous incisors by a subscript number (di_1 and di_2).

diachronic (Gk *dia*=through and *kronos*=time) Any change or process that occurs through time (e.g., evolution, radioactive decay).

diagenesis (Gk *dia*=through, across and *genesis*=birth, coming into being) The complex physical, chemical, and biological changes that occur when sediments are converted into rocks and mineralized tissues (bones and teeth) turn into fossils. During the process organic and inorganic components of the biogenic material can be altered and replaced by chemicals from the surrounding burial environment. An understanding of the processes involved, their effect, and the degree of alteration of biogenic signals is critical for the meaningful interpretation of data obtained from archeological/fossil bone and teeth for the reconstruction of diet, habitat, migration, or climate. Post-mortem diagenesis of mineralized tissues affects both the organic and mineral phases through processes of either degradation and/or exchange of single elements and molecules. Investigations of the mechanisms and effects of diagenesis include the measurement of changes in crystallinity and phase of the mineral, rare earth element uptake and distribution in the specimen, the assessment of collagen preservation, microbial degradation, and stable isotope analyses of various compounds such as oxygen from mineral derived phosphate and carbonate, and compound specific stable isotope analyses of the organic phase. See reviews by Collins et al. (2002), Hedges (2002), and Lee-Thorp (2008).

diagenetic *See* diagenesis.

diagnosis (Gk *diagnosis*=to distinguish). When a new hominin **species** or **genus** is announced one of the requirements is that the authors provide a diagnosis. This is a list of the features or character- istics shared by all (or a large majority) of the mem- bers of a taxon that enable its members to be distinguished from the those of other taxa. There is considerable confusion about the proper use of the terms diagnosis and **definition**. Diagnosis is what a medical doctor does when she/he uses their skills and experience to discriminate between the likely causes of an illness. Thus, in taxonomic terms a diagnosis is how you tell taxa apart. A definition, on the other hand, concentrates on the morphology the members of a taxon have in common. Thus, a definition is a list of the features that binds members of a taxon together, whereas a diagnosis lists the ways a taxon differs from other taxa.

diaphysis (Gk *dia*=apart and *physis*=growth) The part of a long bone, usually equivalent to the shaft, that is formed from the primary center of ossification. *See also* **ossification**.

diatomite (Gk *diatomos*=to cut in half) A **sedimentary rock** composed primarily of the siliceous skeletons (frustules) of diatoms (i.e., algae with cell walls made of two interlocking silica "valves" that belong to the class Bacillariophy- ceae). Diatomite is characteristically powdery and has a low density. Diatomite often contains accessory **biotic** components such as sponge spicules, mol- luscs, and ostracods. Diatoms occur in both marine and non-marine settings and because some types of diatoms have limited ecological ranges they can be used for **paleoenvironmental reconstruction**. For example, localities with significant diatomite sequences, such as those at **Olorgesailie** (Owen et al. 2008) can be used to reconstruct the paleoecology of that part of the landscape.

diazone (Gk *dia*=apart and *zone*=girdle) The region of a **Hunter–Schreger band** that when cut longitudinally shows **enamel prisms** running transversely (cross-cut or end on). Diazone, like parazone and perikymata, were terms first invented

to describe structural features in ungulate enamel by Gustav Preiswerk (1895), but see also later descriptions by Pickerill (1913) and Osborn (1990) (ant. **parazone**). *See also* **decussation**.

DID Abbreviation for Digiba Dora and the prefix for fossils recovered from Digiba Dora, Western Margin, **Middle Awash study area**, Ethiopia. *See* **Digiba Dora**.

Die Kelders (Location 34°32′S, 19°22′E, Western Cape Province, South Africa; etym. Afrikaans for "the cellars;" and also known as Klipgat, which means "cliff cave"). History and general description A network of caves on the shore of the Indian Ocean, roughly 120 km/74 miles south- east of Cape Town. The caves are formed in the contact between the Table Mountain Sandstone Group (Paleozoic quartzites) and the overlying Bre- dasdorp group (**Cenozoic** sediments). The caves were initially excavated in 1969–73 by archeologists from the **South African Museum** under Franz Schweitzer. Schweitzer's excavation revealed 17 layers, the topmost of which was a shell **midden** from the **Later Stone Age**. **Middle Stone Age** (or MSA) deposits begin under a roof fall in layer 4 and continue through layer 15. The MSA occupation levels consist of dense concentrations in sand of bone and lithics that alternate with layers containing only sporadic fauna and lithics. Further excavations con- ducted in 1992–5 by a team led by Graham Avery, Frederick Grine, Curtis **Marean**, and Richard **Klein** substantially expanded the horizontal exposure of MSA sediments. Temporal span and how dated? Optically stimulated **luminescence** dates from the MSA sand layers date point to an age of *c*.60–70 ka and **electron spin resonance spectroscopic** dates on teeth spanning the MSA sequence are very similar (i.e., 70±4 ka assuming early uptake of uranium) and suggest the whole MSA sequence accumulated in less than 10 ka. Hominins found at site The MSA layers contained 27 hominin specimens including 24 isolated teeth, an edentulous mandibular fragment, and two manual middle phalanges. These remains may derive from a minimum of 10 indivi- duals some of which are likely juvenile. The teeth are morphologically similar to, but tend to be larger than, than those of living Africans. Archeological evidence found at site The new studies of the strati- graphy confirmed Schweitzer's basic outline, but demonstrated that MSA sediments were more complex than originally envisioned. Some bone fragments found in layers 10–13 **refit** and there is evidence of downward movement of bones and perhaps artifacts in these levels. Of the more than 150,000 identifiable faunal remains recovered from the MSA levels, the most common are dune mole rats. Fauna accumulated by hominins include tortoises, small to medium-sized **bovid**s, Cape fur seals, and sporadic remains of many other taxa. There is no evidence of artifacts characteristic of the Howieson's Poort industry. The great majority of the artifacts are fashioned of locally available fine-grained gray **quartzite** and generally resemble the artifacts in the later MSA levels at **Klasies River**. The artifacts from the lower MSA levels contain a higher frequency of **silcrete** and other non-local raw materials. The vertebrate faunal remains largely mirror patterns observed at Klasies River and other MSA sites in southern Africa and have helped fuel debates about the hunting abilities of MSA people as well as the impact of hominin population density on the accumulation of the faunal assemblage. Key references: Geology, dating, and paleoenvironment Tankard and Schweitzer 1976, Schweitzer 1979, Schwarcz and Rink 2000, Feathers and Bush 2000, Goldberg 2000; Hominins Grine et al. 1991, Grine 2000; Archeology Schweitzer 1979, Klein and Cruz-Uribe 2000, Grine et al. 1991, Avery et al. 1997, Klein et al. 1999b, Marean 2000, Marean et al. 2000, Thackeray 2000a.

diencephalon *See* forebrain.

diet (L. *diaeta* = way of life from the Gk *diaita* = to live one's life) The food and drink consumed by an animal. The diets of the extant higher primates have proved to be surprisingly complex, in that they vary according to the abundance and availability of a wide range of possible food items. It is heuristically useful to think of possible foods in two categories, preferred foods and **fallback foods**. The former are the food items animals prefer to eat in times of plenty; the latter are the food items animals eat when their preferred foods are scarce or unavailable. Diets can be broadly characterized as either ecologically specialized (i.e., **stenotopy**), or ecologically generalized (i.e., **eurytopy**). Reconstructing the diet of extinct hominins has proved to be a challenging task for in some cases the various lines of evidence result in conflicting evidence about the nature of the food items. *See also* **diet reconstruction**; **fallback foods**.

diet breadth Refers to the total number of resources incorporated in the diet of an individual or of a taxon. An obvious measure of diet breadth is the number of taxa recovered in an archeological bone assemblage. Diet breadth has long been examined within an **optimal foraging theory** framework in an effort to explain changes in modern human and fossil hominin subsistence behavior in response to social and/or environmental changes. For example, according to the broadspectrum revolution hypothesis the emergence of the Neolithic in western Asia was preceded by an expansion of diet breadth in response to environmental and demographic stress at the end of the Paleolithic.

diet reconstruction (L. *diaeta* = way of life, from Gk *diaita* = to live one's life) This entry briefly reviews the lines of evidence that can be used to reconstruct the types of foods eaten by extinct organisms. It should be emphasized that the effectiveness of many of these lines of evidence is sensitive to the scale of the reconstruction being attempted. For example, just because tooth morphology is effective at coarse-grained levels of taxonomic and functional dietary discrimination (e.g., the carnassials of a **carnivore** vs the hypsodont molars of a **herbivore**) it does not mean tooth morphology will be as effective at discriminating at much finer-grained levels (e.g., among omnivorous extinct hominins). It is also important to consider that diet embraces possible **fallback foods** as well as preferred foods. The former may only be eaten on the relatively rare occasions when preferred foods are in short supply, but the ability to access them may determine the fate of individuals and thus potentially the fate of species. The more general, species-level, lines of evidence for reconstructing diet include information about archeology, **dexterity**, **locomotion**, masticatory morphology, **paleoenvironment**, and **paleohabitat**. If archeological evidence can be securely associated with a particular extinct hominin, the form of the **artifact**s (stone, bone, antler, wood, etc.) and evidence such as **use-wear** and **residue analysis** can be used to infer what foods were being accessed by the hominins that manufactured and used the tools. The presence of **cutmarks** is direct evidence that artifacts are being use to process carcasses. More proximate, but still species-level, lines of evidence include masticatory morphology. Do the species' parameters for the absolute and relative sizes and shapes of the jaws and teeth and the thickness and microstructure of the enamel either point towards, or preclude, particular feeding strategies? In other groups of large mammals observations about the food preferences of morphologically analogous extant taxa have proved to be useful sources of evidence for reconstructing the diets of extinct taxa. However, some extinct hominin taxa (e.g., *Paranthropus*) have no obvious extant primate analogue, so some researchers have started with observations about the masticatory morphology of hominins such as *Paranthropus* and then used information from materials science and the principles of **optimization** (e.g., Rudwick 1964) to make predictions about the physical attributes of the foods (e.g., particle size, hardness, toughness, etc.) that the observed derived morphology is best adapted to process. The paleoenvironment and the paleohabitat determine the types of foods available (e.g., lake-shore habitats vs habitats far from aquatic resources) and other lines of evidence (e.g., dexterity and locomotor mode) constrain an animal's ability to access those resources. Observations at the level of the individual (which can be combined to generate species' parameters) are both physical and chemical. The physical methods include the detection of distinctive **phytoliths** and **starch grains** from ingested plant foods embedded in the hardened dental plaque (called calculus or tartar) that accumulates around the base of tooth crowns. **Tooth wear** can be investigated at the gross and microscopic levels. Observations about gross dental wear, also called dental macrowear or dental mesowear, focus on the development of wear facets, including measurements of their size and orientation. Dental microwear focuses on the size, number, and orientation of microscopic scratches, pits, etc. on the enamel. Dental macrowear is a measure of the abrasiveness of the diet in the long term, whereas dental microwear indicates whether the food ingested in the days or weeks before death contained hard or abrasive material. Such abrasive material can be either intrinsic to the ingested foods (e.g., phytoliths) or extrinsic to the foods (e.g., adherent sand grains). Chemical methods of dietary reconstruction applied to the individual may be based on the isotopic ($^{13}C/^{12}C$, $^{15}N/^{14}N$, $^{18}O/^{16}O$) or trace-element composition (Sr/Ca, Ba/Ca) of hard tissues. The $^{13}C/^{12}C$ system is based on photosynthesis. Plants such as shrubs, trees, and non-grass herbs that use the Calvin Benson cycle (in which CO_2 is initially fixed in molecules that have *three* carbons; thus they are referred to as C_3 plants) for photosynthesis have less ^{13}C relative to plants that use a photosynthetic system that initially fixes CO_2 in molecules that contain *four* carbons (C_4 plants). Plants that use a third photosynthetic system based in crassulacean acid (CAM plants) have an intermediate proportion of ^{13}C. These carbon ratios are passed on to the animals that consume the various plant types. The $^{15}N/^{14}N$ system is based on the finding that levels of the heavier nitrogen isotope, ^{15}N, generally increase through the food chain, so **carnivores** tend to have significantly higher levels of ^{15}N than **herbivores**. As for the $^{18}O/^{16}O$ system, leaves are relatively enriched in ^{18}O compared to the water animals drink from streams, freshwater lakes (except for lakes with extensive evaporation), and water holes (this is referred to as meteoric water) so $^{18}O/^{16}O$ ratios can be used to discriminate between animals that get much of their water needs from browsing on leaves compared to those that drink mostly meteoric water. The **Sr/Ca** chemical system is based on the observation that herbivores discriminate against dietary strontium, so their bones and late-forming tooth enamel have lower Sr/Ca ratios than the plants they consume. The carnivores that consume those herbivores also discriminate against dietary strontium, so their bones and tooth enamel have even lower Sr/Ca ratios than the herbivores they consume. Among herbivores, grazers tend to have a higher Sr/Ca than browsers and animals that consume insects and underground storage organs are also generally richer in strontium, but grazers and insectivores are also rich in barium (Ba), so relatively high **Ba/Ca ratios** may enable researchers to discriminate between high Sr/Ca ratios due to grazing or insectivory from high Sr/Ca ratios due to the consumption of, say, **underground storage organs**. All of the relative differences referred to are within a single trophic system; wide geographic variation in soil concentrations of both Sr and Ba prevent other types of comparison. Reconstructing the diet of extinct hominins has proved to be a challenging task for in some cases the various lines of evidence result in conflicting evidence about the nature of the food items. For further reading see Ungar (2007). *See also* $^{13}C/^{12}C$; **diet; fallback foods;** $^{15}N/^{14}N$; $^{18}O/^{16}O$; **Sr/Ca; structure–function relationship; tooth wear**.

differential preservation The bones and teeth that make up the hominin fossil record are made up of a mixture of organic matrix and inorganic material. The dense cortical bone that forms much of the cranium and mandible and the outer cortical bone of the long bones of the limbs is comprised approximately 70% by volume of mineral, or inorganic, material, and 20% by volume of organic material. The balance (approximately 10% by volume) is made up of water. **Dentine** forms the core of mammalian teeth and when mineralized it has a composition similar to that of bone. **Enamel** is the hard, white, outer covering of teeth. It consists of approximately 95% by weight (approximately 98% by volume) inorganic material, and less than 1% by volume organic material. Enamel is the hardest tissue in the body. Thus, for the reasons set out above, teeth are denser than bones and bones that have a higher proportion of cortical bone are denser than bones with more cancellous bone. Because of these regional differences the components of the skeleton are not equally dense (Lam and Pearson 2005) and therefore not equally durable, nor do they behave in the same way if they are transported by water. Whereas the light, less dense, bones such as vertebrae are carried along by high-energy streams of water and only fall to the bottom when the water is flowing more slowly, teeth and the dense bones fall to the beds of streams and will be incorporated into the rocks that form from stream and river beds. Thus, the lighter, less dense, bones float out into the lakes where they are incorporated into the sediments that accumulate in the bed of the lake. These lake-bed (called **lacustrine**) sediments are usually ignored by researchers looking for fossils. The honeycomb construction that makes bones like vertebrae lighter also makes them more vulnerable to damage prior to burial and this together with their relative buoyancy are probably the main reasons why bones like vertebrae are so scarce in the hominin fossil record. Other factors, such as the predilection of carnivores for the hands and feet of primates (Brain 1981), explain why some parts of the skeleton are relatively rare in the hominin fossil record. The hands and feet of most early hominins were gnawed and devoured before they had a chance to become part of the fossil record. *See also* **bone**; **taphonomy**.

differentiation (L *differentia* = diversity, difference) At the cellular level, this refers to the process whereby naïve cells become differentiated into specific cell types (e.g., skin, muscle, neurons,

blood cells). During this process, several properties of the cell (e.g., size, shape, and responsiveness) alter. Although it had been previously believed that genetic material was lost during this process of specialization, it is now known that each differentiated cell remains capable of giving rise to a new animal with the genetic information contained therein (e.g., Dolly, the cloned sheep) and that the process of differentiation is guided by gene expression. Totipotent cells (such as the zygote) are able to differentiate into all cell types; pluripotent cells (such as embryonic stem cells) are able to differentiate into any cell type seen in an adult. In some organisms that possess regenerative capabilities, differentiated cells are able to undergo a process of "dedifferentiation" in which they revert back to a less-specialized form.

digging stick A tool used for excavating soil, acquiring **belowground food resources**, and/or planting seeds. Digging sticks may be made of wood, bark, other plant materials, or bone, and may also have stone components. The simplest digging sticks known are unmodified tree branches, sticks, and pieces of bark used by chimpanzees to aid in the excavation of the **underground storage organs** of edible plants (Hernandez-Aguilar et al. 2007), termites, ants, and sweat bees (Boesch et al. 2009). Digging sticks are widely used by modern humans in traditional hunter–gatherer and agricultural societies. **Hunter–gatherers** such as the **Hadza** of Tanzania, the Bushmen/San of southern Africa, Australian Aborigines, and various South American Indian groups currently use, or are known to have used in the past, sharpened digging sticks in the excavation of belowground resources (including plants and animals) or other activities such as digging post holes or planting. Digging sticks may be regularly re-sharpened depending upon how often they are used and the nature of the soil being excavated. Digging sticks are sometimes further modified by fire-hardening though this is not a universal feature of the tool type (Marlowe 2005). The most complex variation of the digging stick involves weighting the tool with a bored stone to increase the force with which the stick strikes/penetrates the ground. This type of digging stick is more commonly associated with agricultural applications such as planting. Stone digging-stick weights often survive in the archeological record while the sticks themselves more often than not decompose. A notable exception to the absence of digging tools in **Paleolithic**

archeological assemblages is the discovery of probable digging sticks in Members 1–3 at **Swartkrans**, **Sterkfontein** Member 5, and **Drimolen** in South Africa. These putative digging tools are made from medium-large mammal long bones, and recent analyses indicate that anthropogenic modification by grinding sharpened at least some of these artifacts. Further research suggests that these tools may have been employed variously in excavating underground storage organs, digging into **termite** mounds, and processing fruits (d'Errico and Backwell 2003, 2009). The 1.7–1.0 Ma putative digging sticks associated with *Paranthropus robustus* and *Homo ergaster*, taken in combination with the use of digging tools by extant chimpanzees, raise intriguing questions about the possibility of tool use by hominins prior to the advent of the flake and core-based **Oldowan** stone tool culture, with implications for our understanding of early human technological evolution, innovation, cognition, social organization, learning strategies (imitation vs emulation) and mechanisms of cultural transmission.

digit   (L. *digitus*=finger) The term applies to either fingers (i.e., manual digits) or toes (i.e., pedal digits). In clinical anatomy manual digits are normally named and pedal digits are numbered. *See also* **ray**.

digit ratios   *See* **hand, relative length of the digits**.

DIK   Abbreviation of Dikika and prefix for fossils recovered from Dikika, Ethiopia after 2000. *See* **Dikika study area**.

DIK-1-1   Site Dikika. Locality DIK-1. Surface/ *in situ* In situ. Date of discovery 2000 with further excavations and screening in 2002, 2003, and 2005. Finders Team led by Zeresenay **Alemseged**. Unit N/A. Horizon N/A. Bed/member Sidi Hakoma. Formation **Hadar Formation**. Group N/A. Nearest overlying dated horizon TT-4 tuff (3.24 Ma). Nearest underlying dated horizon SHT tuff (3.42 Ma). Geological age *c*.3.37 and 3.34 Ma. Developmental age *c*.3 years. Presumed sex Female. Brief anatomical description The skeleton includes much of the cranium, a natural endocast, a well-preserved mandible, all the deciduous teeth except the crowns of the left lower incisors, the hyoid, both scapulae and clavicles,

much of the vertebral column, sternum, many ribs, a fragment of the right humerus, a manual ray, both knee joints, both patellae, and the distal end of the left lower limb including the left foot. Announcement Alemseged et al. 2006. Initial description Alemseged et al. 2006. Photographs/ line drawings and metrical data Alemseged et al. 2006. Detailed anatomical description N/A. Initial taxonomic allocation *Australopithecus afarensis*. Taxonomic revisions None. Current conventional taxonomic allocation *Au. afarensis*. Informal taxonomic category Archaic hominin. Significance A remarkably complete skeleton that provides exquisite detail about the morphology and life history of *Au. afarensis*. Location of original National Museum of Ethiopia, Addis Ababa, Ethiopia.

Dikika Research Project   (or DRP) Intensive paleoanthropological research in the **Dikika study area** began in 1999 when Zeresenay **Alemseged** led an Ethiopian expedition to explore the deposits between the **Hadar** and **Gona** areas to the north and the Middle Awash region to the south. Alemseged and Denis Geraads studied a diverse **Middle Pleistocene** fauna and reported **Acheulean** and **Middle Stone Age** tools from the Asbole area. During initial surveys the team identified extensive fossil deposits in the central Dikika area and in 2000 they recovered the first evidence of **DIK-1-1**, the most complete early hominin infant skeleton known and DIK-2-1, the earliest hominin from the **Hadar Formation**. New scientists expanded the scope of the DRP in 2002 and 2005, including two vertebrate paleontologists, René Bobe and Denné Reed, a geologist, Jonathan Wynn, and an archeologist, Shannon McPherron. Wynn mapped the geology of the site in detail, particularly the volcanic tuffs and marker beds that provide a chronostratigraphic framework for the regional stratigraphy that spans >3.8 to <0.16 Ma. McPherron et al. (2010) announced the recovery in 2009 of two faunal bones (DIK-55-2 and DIK-55-3) from deposits that date to approximately 3.39 Ma that they claim bear stone tool cut and percussion marks made prior to fossilization.

Dikika study area   (Location 11°05′N, 40°35′E, Ethiopia; etym. Dikika is the local Afar name for a hill in the central part of the Dikika study area) History Portions of the Dikika study area adjacent to **Hadar** were explored by the

International Afar Research Expedition during expeditions from 1973 to 1977 and the **Rift Valley Research Mission in Ethiopia** (or RVRME) during expeditions from 1975 to 1978. Following a long hiatus, surveys were undertaken by the **Dikika Research Project** (or DRP) and have continued since 1999. General description Area located south of the Awash River and south and east of the **Hadar** and **Gona** paleoanthropological study areas in Ethiopia's **Afar Rift System** section of the East Africa Rift System. Temporal span and how dated? Tephrostratigraphic correlation with argon-argon dated units of the **Hadar Formation** and **Busidima Formation** suggest deposits in the area contain the complete sequences of the Hadar Formation $c.>3.8$–2.9 Ma and the Busidima Formation $c.2.7$–<0.16 Ma. Hominins found at site The first hominin specimen recovered was A.L. 277-1, a left mandible fragment with C–M_2 found in 1974. Specimens recovered from 1976 to 1977 are A.L. 400-1, a complete mandible and associated right upper canine, and A.L. 411-1, a right mandible fragment with broken M_1–M_3. **DIK-1-1**, a nearly complete juvenile skeleton, was recovered in the field seasons of 2000, 2002, and 2003 by members of the DRP. DIK-2-1, a left mandibular fragment and associated dentition, was found at locality DIK-2 in 2000. All the hominin specimens recovered from the area are attributed to *Australopithecus afarensis.* Archeological evidence found at site Oldowan, Acheulean, and Middle Stone Age technologies are found in the area but they are not associated with the reported cutmarked bones (see below). McPherron et al. (2010) announced the recovery in 2009 of two faunal bones (DIK-55-2 and DIK-55-3) found in deposits dated to approximately 3.39 Ma that they claim bear stone tool cut and percussion marks made prior to fossilization. Key references: Geology, dating, and fauna Alemseged and Geraads 2000, Geraads et al. 2004, Wynn et al. 2006, 2008; Hominins Johanson et al. 1982a, 1982b, Alemseged et al. 2005, 2006; Archeology McPherron et al. 2010.

dimension A property of a variable that describes how many of the three dimensions of space the variable occupies. **Size** variables may have a dimension equal to one (linear), two (area), or three (volume) dimensions. Measures of mass are typically treated as three-dimensional because mass is a property of objects in three-dimensional space, and an object's mass is generally proportional to its volume. **Shape** variables are typically dimensionless (dimension equals zero). Dimension is important in a **scaling** analysis because the isometric slope is determined by the ratio of the dimensions of the two variables. *See also* **scaling**.

Dingmo *See* **Bubing Basin**.

dip In structural geology this term describes the orientation, in 3D space, of a 2D geological feature such as a stratum or bed. Dip has two main components, the first being the maximum angle between the feature and the horizontal, and the second being the direction, which is recorded as a compass bearing from north in the direction of the maximum angle (this is sometimes referred to as true dip as opposed to apparent dip). These data are reported either directly as the dip direction and bearing relative to north (i.e., $45°$ to $270°$), or as a dip and **strike** (where strike refers to a bearing at $90°$ to the dip direction). Dip and strike are usually measured in the field using a compass clinometer. A consideration of both the angle of dip and its direction is important during paleontological fieldwork, as knowing these allows surveying teams to identify the directions in which the strata will be relatively older or younger. In archeology, it is important to consider whether beds dip when excavating for tectonic processes, such as **fault**ing or folding, may have altered their orientation from the horizontal plane in which they were originally lain down.

diploë (Gk *diploë*=doubling from the feminine of *diplous*=twofold or double) The cancellous or spongy bone between the dense inner and outer tables of the bones of the cranial vault. It contains endothelial-lined spaces containing venous blood. *See also* **cranial vault**.

diploid (Gk *diplous*=double and *eidos*=shape or form) The term used for organisms (e.g., all primates, including modern humans and the great apes) or cells that have two **homologous** copies of each **chromosome**. The term also refers to possessing two complete sets of chromosomes. Thus, the genetic complement of modern humans is 22 pairs of diploid chromosomes (also called autosomes) and one pair of sex chromosomes. The diploid number of chromosomes for modern humans is 46.

directional selection Natural selection that favors (i.e., increases the **fitness** of) a particular phenotype so that **allele** frequency shifts in a direction that results in an increase of that particular phenotype (and a decrease of all other phenotypes) in the population. An example of an advantageous trait subject to positive natural selection in modern humans is the null allele of the *FY* gene, which is protective against *Plasmodium vivax* **malaria**; the frequency of this allele is almost fixed in sub-Saharan Africa. Other examples of alleles likely to have been selected for during human evolution include those involved in the morphology needed to be an efficient biped, to increase brain size, etc. (Sabeti et al. 2006) (syn. positive selection).

disconformity   *See* **unconformity**.

discrete morphological variants   *See* **non-metrical traits**.

discrete traits   *See* **non-metrical traits**.

discriminant function analysis   A form of **multivariate analysis** similar to **principal components analysis** (or PCA), discriminant function analysis finds linear transformations of the variables in a data set that maximize the variation between user-identified groups as opposed to the maximizing the variation across all observations in a data set. Like PCA, the resulting new variables (discriminant functions) are independent of each other; that is, they have **correlation**s of zero with each other. However, whereas PCA produces as many principal components as there are original variables, discriminant function analysis produces a number of discriminant functions equal to the lesser of the number of original variables and the number of groups minus one. For example, a researcher may collect a set of 10 measurements from a sample of chimpanzee, gorilla, and modern human skulls where groups correspond to species. A PCA would produce 10 principal components while a discriminant function analysis would produce two discriminant functions. Each specimen would have a score for both discriminant functions, and a bivariate plot of discriminant function scores for each specimen would show whether the discriminant function does a good job of identifying differences between groups: if each group of skulls plots in their own portion of the graph with no overlap between groups, the discriminant functions provide good separation between groups, whereas if there is substantial overlap between skulls belonging to different groups, then the discriminant functions do not provide information that separates groups well. Typically researchers use discriminant function analysis to generate functions which separate specimens for known groups, then calculate discriminant function scores for specimens of unknown group membership (e.g., use skeletal metrics from extant bovids living in known habitat categories to generate functions which can be applied to extinct bovids of unknown habitat). The group **centroid** that is closest to the unknown specimen is identified as the group to which the specimen has the highest probability of belonging. Note that each unknown specimen will be classified as belonging to some group regardless of whether or not it is actually a member of any of the user-specified groups in the analysis, and it can only be classified into one of the groups identified by the user. If the discriminant functions do not separate the known groups well, it is likely that unknown specimens will be misclassified. To get a sense of how likely this is to happen, it is common to calculate prediction accuracy by predicting the group membership of every known specimen and calculating the percentage of specimens that are correctly classified into the group to which they actually belong. In any case, if an unknown specimen belongs to a group not represented by the known sample, it is impossible for the unknown specimen to be correctly classified. Some versions of discriminant function analysis also weight results by the number of specimens in each original group (i.e., unknown specimens are more likely to be classified into groups represented by many specimens in the known sample than into groups with relatively few known specimens); end users of computer programs that perform discriminant function analysis should be aware of whether this is case for their particular analysis. It is worth noting that discriminant function analysis and **multivariate analysis of variance** (MANOVA) are both based on the same **matrix** derived from the original data (which is functionally equivalent to a ratio of between-group and within-group covariances); thus a significant result from a MANOVA indicates that a discriminant function analysis should provide functions that separate groups well; conversely a non-significant result suggests that one should not waste any time conducting a discriminant function analysis.

dispersal (L. *dis*=apart and *spargere*=to scatter) A change in the location of a population of organisms brought about by the physical movement of the entire population or a subset. Three types of dispersal can be recognized. **Range expansion** occurs when the area in which a population is found increases but the increased range still includes the original, ancestral area. For example, toward the end of the **Pliocene** increasing aridity led to an expansion of grasslands in Africa and many grazing mammalian species experienced range expansion during this time period. **Range shift** occurs when the area in which a population is found changes location. For example, during aridification trends in the southern hemisphere, **mesic** vegetational zones shift their location in the direction of the equator (*see* **habitat theory hypothesis**). This should lead mammals adapted to mesic conditions to experience a northward range shift. **Jump dispersal** occurs when a subpopulation of a larger population of organisms crosses an existing barrier to dispersal. Stochastic processes may play an important role in generating jump dispersals, or novel adaptations may enable individuals to traverse regions that had previously been impassable. A particularly dramatic example of jump dispersal is the colonization of South America by African **anthropoids** during the early **Miocene**, in which the founding populations of **platyrrhines** may have crossed the Atlantic Ocean on raft-like mats of floating vegetation. Jump dispersal may be a prelude to **allopatry**. The peopling of **Oceania** is one of a number of examples of jump dispersal during human evolution.

disposable soma hypothesis (Gk *soma*= body) One of several theories put forward to explain the evolution of **senescence**. The disposable soma hypothesis (Kirkwood 1977) suggests that the deterioration of the body with age is an inevitable consequence of an imbalance between the resources devoted to reproduction and those devoted to maintenance. Since the resources available to an organism are finite, if a greater proportion of that finite resource is directed to reproduction then inevitably the proportion of the total resource devoted to repair reduces such that bodily functions decline with age. *See also* **life history**.

dissociation (L. *dissociare*=to disunite) One of the major evolutionary mechanisms by which variation in development is generated. Specifically, **modules** or processes that were previously linked become unlinked, or dissociated. Relative timing of developmental events can also become dissociated, leading to morphologies shaped by heterochronic processes (e.g., **acceleration**, **neoteny**) in the descendants that differ from those in the ancestors. Furthermore, dissociations can occur between size change and shape change during ontogeny; these kinds of dissociations can result in adult shape differences that are allometrically scaled. Dissociations between modules and processes often take place early in ontogeny, and patterns of dissociation often differ between regions of the body (e.g., dissociations among early growing tissues, such as the brain, commonly occur). For example, comparisons of bonobo and common chimpanzee ontogeny suggest that shape differences between these taxa have been produced by localized dissociations of size and shape.

distal (L. *distare*=to stand apart) In the limbs distal refers to structures that are in the direction of the extremity of that limb (i.e., towards the tips of the fingers or toes), and with respect to the teeth, or their components, distal refers to teeth or structures within teeth that are in the direction of the back of the jaw. Thus, the elbow joint is distal to the shoulder joint, the hallucial-metatarsophalangeal joint is distal to the ankle joint, the right M_2 is distal to the right P_4 of the same dentition, and the **hypoconulid** of an M_1 crown is distal to the **protoconid** on the crown of the same tooth.

distal accessory ridge An **enamel** crest that is sometimes present on the lingual face of maxillary or mandibular canines. It is located between the **median ridge** and the **distal marginal ridge** and it is variably expressed.

distal accessory tubercle An accessory cusp on the **distal marginal ridge** of a postcanine tooth crown. In the case of maxillary molars it has a number of synonyms (e.g., C5, distoconule, distoconulus, hypostyle, metaconule, and postentoconule).

distal fovea (maxillary) *See* **fovea posterior (maxillary)**.

distal marginal ridge A crest of **enamel** that delineates the distal border of the lingual face of a canine or incisor anterior tooth or the occlusal surface of a postcanine tooth crown. In a maxillary postcanine tooth crown it is an enamel crest running between the **hypocone** and the **metacone** and in a mandibular postcanine tooth crown it is an enamel crest running between the **entoconid** and the **hypoconulid** (syn. **distal margocrista (-id)**, distocrista (-id), postcristid, distal rim, posterior margin, posterior wall).

distal margocrista (-id) Term used by Schwartz and Tattersall (2002) for an **enamel** crest more commonly referred to as the **distal marginal ridge** (*which see*).

distal pit The term used by Kraus et al. (1969) for a tooth crown feature that others call the fovea posterior. *See also* **fovea posterior (maxillary)**; **fovea posterior (mandibular)**.

distal rim *See* distal marginal ridge.

distal trigon crest One of the three **enamel** crests on the **trigon** of maxillary molar teeth; the others are the **mesial marginal ridge** and the **epicrista**. It runs between the **metacone** and the **protocone** and it forms the distal boundary of the trigon. The lingual/metacone end of the ridge is sometimes called the postprotocrista and the buccal/protocone end is called the premetaconule crista (syn. *crista obliqua, crista transversa*, metaloph, **plagiocrista**, postprotocrista, transverse ridge, **trigon crest, oblique crest, oblique ridge**).

distal trigonid crest An **enamel** crest on the crown of a mandibular molar that runs between the **protoconid** and the **metaconid**, distal to the **mesial marginal ridge**. An additional crest, the **mid-trigonid crest**, can also be present, in some cases, between the mesial marginal ridge and the distal trigonid crest (syn. protolophid, protocristid, posterior or distal trigonid ridge, **trigonid crest**, tranverse ridge).

distance matrix An $n \times n$ square **matrix** corresponding either to a data set with n observations/specimens or to a single observation/specimen with n spatial landmarks. In the first case, values in each cell correspond to the distance (morphological or otherwise) between the row and column observations/specimens. For example, it may refer to the

morphological distance between two skulls, or the geographical distance between two fossil sites. This type of distance matrix is often referred to as a pairwise distance matrix. In the second case, the value in each cell corresponds to the spatial distance between two landmarks in a single specimen; for example, the distance between prosthion and inion in a single cranium. This type of distance matrix is used in **Euclidean distance matrix analysis**. In either case, the values on the diagonal of the matrix are distances of observations or landmarks from themselves and are thus all equal to zero.

distobuccal cusp A term sometimes used for the **primary cusp** on a maxillary molar called the **metacone**, and for the primary cusp on a mandibular molar called the **hypoconid**. *See also* **hypoconid; metacone**.

distocone A term introduced by Vanderbroek (1961, 1967) for what most cusp nomenclature schemes refer to as the **metacone** (*which see*).

distoconid A term sometimes used for the enamel feature on mandibular molars that is more commonly referred to as the **protostylid** (*which see*).

distoconule *See* distal accessory tubercle.

distoconulid A small accessory cuspulid on the **distal marginal ridge** of a mandibular postcanine tooth crown (syn. distoconulus).

distoconulus *See* distal accessory tubercle.

distolingual cusp Cusp on the distolingual part of the maxillary or mandibular molar tooth crown. The distolingual cusp on a maxillary molar is the hypocone and on a mandibular molar it would be the entoconid. *See also* **entoconid; hypocone**.

distosagittal ridge A rare non-metric trait on the maxillary first premolar crown occurring mostly in American Indian populations. It is described by Scott and Turner (1997) as an "exaggerated distobuccal rotation of the paracone in combination with the presence of a fossa at the intersection of the distal occlusal ridge and distal marginal ridge." *See also* **Uto-Aztecan premolar**.

distostyle A term introduced by Vanderbroek (1961, 1967) and used by Grine (1984) for a small enamel feature on the **distal marginal ridge** close to the tip of the **metacone** on the occlusal surface of a maxillary molar (syn. hypoconule, metastyle, tuberculum accessorium posterium externum). *See also* **metastyle**.

distostylid A term introduced by Vanderbroek (1961, 1967) for what most cusp nomenclature schemes refer to as the **hypoconulid** (*which see*).

Ditsong National Museum of Natural History On May 28, 2010, the Transvaal Museum was renamed the Ditsong National Museum of Natural History. *See also* **Transvaal Museum**.

diurnal (L. diurnus=daily) Referring to animals that are active in the day (i.e., between dawn and dusk). For most **platyrrhines** and for all catarrhines it is the predominant activity pattern. Dependence on daylight may have been a crucial factor in limiting most nonhuman primate ranges to the tropics. Primates spend their days undertaking four main activities: feeding, resting, traveling, and socializing. Since the cohesion of complex primate groups relies on social interactions such as grooming, time is a major constraint in the lives of primates, especially **anthropoid**s (Dunbar 1988). In temperate (i.e., high) latitudes, daylight is highly seasonal and in winter there may be insufficient time for primates to maintain social bonds and fulfill their need to feed, travel, and rest. This latitudinal effect may therefore impose limits on the geographic ranges of most primates. When hominins expanded into higher latitudes behavioral and cultural developments may have freed them from such constraints. For example, technology may have made it easier and more efficient for hominins to procure food in temperate regions where resources are generally more seasonal and less abundant. The evolution of language, hypothesized by Dunbar (1996) to be a form of grooming that is essential for group bonding, may have reduced the time needed for the maintenance of social relationships and the control of fire may have increased the time available for social interactions. These innovations may have helped diurnal hominins prosper in higher latitudes with variable day lengths.

divergence time (L. *dis*=apart and *vergere*=to bend) The time since two **lineage**s separated. So, for example, the divergence time of the lineages leading to modern humans and to *Homo neanderthalensis* has been estimated to be c.350–700 ka using data from the genome (e.g., Krings et al. 1997, Noonan et al. 2006). It is important to stress that the genetic **coalescence**, population separation, and phenotypic differentiation of two taxa represent chronologically distinct successive events. However the genetic **coalescent time** of two taxa cannot be more recent than their divergence time.

dizygotic twins When two offspring produced in the same pregnancy are derived from two separately fertilized eggs. Dizygotic twins are also known as fraternal or non-identical twins. Both dizygotic and monozygotic twins have been important in studies of **heritability** in modern humans because they can help identify the balance between the genetic and environmental influences on a phenotypic trait.

Djebel Irhoud *See* **Jebel Irhoud**.

Djebel Qafzeh *See* **Qafzeh**.

Djurab Desert *See* **Chad Basin**.

DK Acronym for the Douglas (Leakey) Korongo, one of the archeological localities at **Olduvai Gorge**. Excavations beginning in 1962 resulted in many artifacts and faunal evidence. *See also* **Olduvai Gorge**.

DLNs Acronym for dimensionless numbers. *See* Charnov's "**dimensionless numbers**".

dm Acronym for deciduous molar. Individual upper, or maxillary, deciduous molars are identified by a superscript number (dm^1 and dm^2) and lower, or mandibular, deciduous molars by a subscript number (dm_1 and dm_2). John **Robinson** (1956) showed that lower deciduous molars, especially the dm_1s, are a useful taxonomic tool for discriminating among early hominin taxa. (e.g., the dm_1s of *Paranthropus robustus* and *Paranthropus boisei* have especially complex crowns). These teeth are sometimes referred to as deciduous premolars (or dps) because the permanent premolars develop under the deciduous molars.

Dmanisi (Location 41°20′10″N, 44°20′38″E, Georgia; etym. after the medieval city of Dmanisi, a major center on the Silk Road occupied until the 18thC) History and general description This open-air site is located on a basalt and bedrock promontory formed at the confluences of the easterly flowing Masavera and northerly flowing Pinasaouri rivers. Bronze Age and medieval archeological deposits including substantial masonry architectural remains cover the promontory and these deposits overlie and sometimes intrude into the underlying Plio-Pleistocene sediments. Excavations of the medieval ruins at Dmanisi, 55 km/34 miles southwest of Tbilisi, began in the 1930s. Between 1983 and 1987 excavations in Medieval Room XI resulted in the recovery of Plio-Pleistocene fossils and this attracted paleontologists from the Georgian Academy of Sciences and the discovery that artifacts were buried in the same deposits led to the 1989 archeological excavations in Block 1 by the Archaeological Center of the Georgian Academy of Sciences. In 1991, archeologists from the Römisch-Germanisches Zentral-Museum joined the field team, and on September 24, 1991 the first hominin fossil (the **D211** mandible) was recovered. The joint Georgian/German team continued excavations from 1991 to 1999. The excavations involved expanding Block 1 (only within Strata B2–B3 deposits), five shallow test trenches (M1–M5) peripheral to Block 1, and the opening of Blocks 2 and 3, east and southeast of Block 1, respectively. In 1999 the deep part of Block 1 was reopened and the **D2280** and **D2282** crania were discovered. Since 2000, the Dmanisi investigations have been the responsibility of the **Georgian National Museum**, under the direction of David **Lordkipanidze**. Fieldwork between 2000 and 2008 included (a) expansion of Block 2 to the north, with recovery of numerous hominin fossils and associated faunal remains and artifacts, (b) excavations in Block M6, 30 m northwest of Block 1, and (c) deep testing in Unit M5 (75 m west of Block 2). Stratigraphic and paleomagnetic studies in Block 1 and Room XI by Ferring and Swisher resulted in stratigraphic revision of the 1980s VI–I scheme (Gabunia et al. 2000). The Plio-Pleistocene sediments are dominated by primary and locally reworked ashfall and they are divided into two major units: Stratum A (with subunits A1–A4), which conformably overlies the Masavera Basalt, and Stratum B (with subunits B1–B5), which is separated from Stratum A by a minor erosional disconformity (Lordkipanidze et al. 2007). Subsequent stratigraphic and paleomagnetic studies in Block 2 by Ferring and Ohms (Lordkipanidze et al. 2007) contradict the reports of the recovery of hominin remains and artifacts

from "Layers VI-IV" (de Lumley et al. 2005, de Lumley and Lordkipanidze 2006). All the hominin remains and the associated fauna and artifacts recovered from Blocks 1 and 2 are from firmly defined Stratum B contexts. Temporal span and how dated? Argon-argon ages for the Masavera Basalt average 1.848 ± 0.005 Ma (Gabunia et al. 2000) and 1.81 ± 0.05 Ma on Stratum A1 ashes (de Lumley et al. 2002). The Orozmani Basalt, which covers the Dmanisi deposits west of the locality, has been dated by ^{40}Ar/^{39}Ar to 1.759 ± 0.005 Ma (Gabunia et al. 2000). The Masavera Basalt and all Stratum A deposits exhibit normal geomagnetic polarity, suggesting correlation with the **Olduvai subchron**. The Strata A/B contact corresponds with the 1.77 Ma Olduvai–Matuyama boundary, and all Stratum B deposits exhibit reversed polarity and are correlated with the **Matuyama** chron (Lordkipanidze et al. 2007). The fauna shows paleozoogeographical similarities with the Late **Villafranchian** of Western Europe (Lordkipanidze et al. 2007). Evidence from sedimentology, soils, and bone **taphonomy** suggests that the temporal span of the deposits to be "less than 10,000 years" (Lorkipanidze et al. 2007) and the best estimate of their age is 1.77 Ma. Hominins found at site A rich collection of skulls (e.g., **D2700/D2735/D2724/D3160, etc., D2282, D211, D3444/D3900**), crania and calvaria (e.g., **D2280**), mandibles (e.g., **D2600**), and adult and subadult associated skeletons (e.g., D2724, 4166); all the components of the associated skeletons have been given different specimen numbers (see Lordkipanidze et al. 2007). There is an ongoing debate about how primitive the Dmanisi remains are, with some arguing they are more primitive than *Homo erectus sensu stricto* (e.g., Gómez-Robles et al. 2008), and others arguing that they are best interpreted as being relatively small-bodied examples of that taxon. An initial review of the dental remains suggests the dentition displays a mix of presumed primitive features seen in archaic hominins and more derived features seen in pre-modern *Homo* (Martinón-Torres et al. 2008). Archeological evidence found at site The stone artifacts are all classic **Mode 1** type. The Dmanisi industry is dominated by **debitage, cores, choppers**, and **manuports; retouch**ed pieces are rare. Key references: Geology, dating, and paleoenvironment Gabunia et al. 2000, Lordkipanidze et al. 2007; Hominins Gabunia and Vekua 1995, Gabunia et al. 2000, Vekua et al. 2002, Lordkipanidze et al. 2005, 2007,

Rightmire et al. 2006b, 2008, Martinón-Torres et al. 2008, Pontzer 2010; Archeology Gabunia et al. 2000, de Lumley et al. 2005.

Dmanisi 211 *See* **D211**.

Dmanissi The spelling used in Gabunia et al. (2001, 2002) and de Lumley et al. (2006) for the site now referred to as **Dmanisi** (*which see*).

DNA *See* **deoxyribonucleic acid**.

DNA chip *See* **DNA microarray**.

DNA hybridization (L. *hybrida*=something made of at least two different components) A method for comparing DNA fragments or whole genomes. It can also be used to assess genetic distance (as DNA–DNA hybridization) or, when very stringent conditions are applied, to identify specific DNA sequences. The method involves separating the double-stranded nuclear DNA from two samples into single strands by heating them and then cooling them to allow them to anneal (i.e., to come together again). When used to identify DNA fragments, the conditions are set such that only a sample with an identical sequence will anneal to the control sequence (often referred to as a **probe**). This is what is involved in methods such as **Southern blotting** and denaturing high-performance liquid chromatography (or DHPLC). When whole genomes are being compared, the heat that needs to be injected into the system to separate the annealed single strands is used as a **proxy** for the number of bonds joining the strands. Because DNA strands from the same individual would be linked by the maximum number of bonds this is taken as the standard: the more distant the relationship the fewer the bonds linking the annealed strands, so less heat is required to separate them. Whole-genome DNA hybridization has been superseded by **DNA sequencing**. Sibley and Ahlquist (1984) used DNA–DNA hybridization to investigate the phylogeny of the great apes and they confirmed the hypothesis that chimpanzees , bonobos, and modern humans were more closely related to each other than to gorillas.

DNA methylation (Gk *methu*=wine and *hule*=wood, from "wood alcohol" the common name for methyl alcohol) Usually the modification of the cytosine in a **DNA** sequence by the addition of a methyl group to form 5-methylcytosine. Methylation is associated with **gene expression** being silenced and it typically occurs in CpG dinucleotides or "CpG islands" (i.e., clusters of CpG dinucleotides) that are common in the regulatory regions of **genes**.

DNA microarray A technology used to (a) distinguish particular **alleles** of a **DNA** sequence (such as the detection of **single nucleotide polymorphisms**), (b) sequence a specific region of DNA (*see* **DNA sequencing**), (c) investigate patterns of **gene expression**, or (d) identify **deletions** or **insertions** (such as **copy number variations**). It is a glass, plastic, or silicon chip onto which fragments of DNA have been spotted. Fluorescent **probes** are then used to distinguish a specific **base pair**, allele, or gene expression pattern. These arrays are the basis of the rapid high-throughput analysis that allows researchers to sequence DNA more rapidly than in the past.

DNA sequencing (L. *sequi*=to follow) Several methods can be used to determine the sequence of **bases** present in a **DNA** molecule. The most common is the Sanger or chain-termination method. This method uses dideoxynucleotide triphosphates (or ddNTPs) which lack the 3'-OH group necessary for the formation of a phosphodiester bond as "chain terminators." Specifically, the DNA is **denatured** and a mixture comprising a DNA primer, polymerase, deoxynucleotide triphosphates (or dNTPs), and ddNTPs is added. The standard dNTPs (dATP, dGTP, dCTP, and dTTP) and the ddNTPs are used by the polymerase to copy the single-stranded DNA. However, whenever a ddNTP is added by the polymerase the copying stops. The newly synthesized DNA is then denatured and the DNA is separated by its size (i.e., its molecular weight) using **electrophoresis**, which can distinguish DNA strands that differ in size by as little as one **base pair** (1 bp). In order to **read** the sequence the ddNTPs are labeled either radioactively or with different fluorescent dyes.

DNH Abbreviation for Drimolen hominids and the prefix for hominins recovered from **Drimolen** (*which see*).

DNH 7 Site **Drimolen**. Locality N/A. Surface/ *in situ* *In situ*. Date of discovery October 21, 1994.

Finders R. Smith and André Keyser. Unit N/A. Horizon N/A. Bed/member N/A. Formation Monte Christo Formation. Group Chunniesport. Nearest overlying dated horizon N/A. Nearest underlying dated horizon N/A. Geological age c.2.0–1.5 Ma. Developmental age Adult. Presumed sex Female. Brief anatomical description An almost complete adult skull extensively restored over a period of 5 years by Ron **Clarke**. The face lacks only the left zygomatic arch, parts of the left orbital margin, and the nasal bones, and the base lacks substantial parts of the sphenoid and the anterior margin of the foramen magnum. The walls of the orbits, the ethmoid, and the pterygoid plates are badly damaged, but damage to the mandible is confined to the base. Other than distortion caused by bilateral compression, the skull overall is in remarkably good condition. Announcement Keyser 2000. Initial description Keyser 2000. Photographs/line drawings and metrical data Keyser 2000. Detailed anatomical description N/A. Initial taxonomic allocation *Paranthropus robustus*. Taxonomic revisions None. Current conventional taxonomic allocation *P. robustus*. Informal taxonomic category Megadont archaic hominin. Significance An exceptionally well-preserved specimen of a *P. robustus* skull that confirms the claimed distinctions between *Paranthropus boisei* and *P. robustus*. It also provides support for the hypothesis that specimens such as **KNM-ER 407** and **KNM-ER 732** represent females of *P. boisei*. Location of original **School of Anatomical Sciences, University of the Witwatersrand**, South Africa.

Dolní Věstonice I (Location 48°53′N, 16°40′E, Czech Republic; etym. name of nearby village) History and general description One of three (Dolní Věstonice I, II, and III) open-air **loess** sites along the northern slopes of the Pavlovské Hills above the Dyje River floodplain. Dolní Věstonice I (or DV I) is closely associated with the nearby Pavlov sites (Pavlov I and II) that have yielded *Homo sapiens* remains and **Upper Paleolithic** industries. Excavation (both systematic and salvage) and survey have been conducted at DV I since 1924 and intermittently until 1993. DV I is a large, complex settlement site and has yielded Upper Paleolithic (**Gravettian**) archeological and hominin skeletal remains, including burials (DV 3 and possibly DV 4), as well as scattered individual modern human skeletal remains (DV 1, 2, 5–10, 23–32, 35, 37, 38). Temporal span and how dated? Conventional and **accelerator mass spectrometer** radiocarbon dates on charcoal suggest that settlement of the site probably spans from c.29 to 27 ka uncalibrated radiocarbon years BP for the lowermost layers to c.26–25 ka uncalibrated radiocarbon years BP for the uppermost layers. A direct date on an anatomically modern human femur has yielded a late age of 22,840 radiocarbon years BP, but this has been disputed. Hominins found at site An associated skeleton (DV 3), two calvarae (DV 1 and 2), cranial fragments (DV 4–6, 23–25, 28, 30), teeth (DV 7–10, 26, 27, 29, 31, 32, 37, 38), and a femoral shaft (DV 35). Archeological evidence found at site Upper Paleolithic (Gravettian) figurines, dwelling structures. Key references: Geology and dating N/A; Hominins Jelinek 1969, Trinkaus and Svoboda 2006; Archeology Klima 1954, 1959.

dolomite The name, which derives from the French mineralogist Deodat de Dolomieu, is used for a magnesium-rich carbonate mineral [calcium magnesium carbonate or $CaMg(CO_3)_2$] and for limestone rocks which are predominantly composed of this mineral. Dolomite is often formed in warm tropical ocean environments; however, it can also be formed by the diagenetic alteration of pre-existing limestones. The early hominin cave sites in southern Africa are breccia-filled solution cavities formed within the Precambrian dolomite that makes up much of the high veld.

domain In molecular biology, a domain is part of a **protein** that has a specific structure and function. Specific domains may be found in many different protein families. For example, the alpha helix, a right-handed spiral, is one of the most common protein domains, and it is found in such proteins as hemoglobin, rhodopsins, and G-protein-coupled receptors. Specific domains may also be characteristic of a particular family of proteins. For example, *HOX* genes encode a protein domain known as the homeodomain, which is a highly conserved helix-turn-helix structure of 61 amino acids that can bind to enhancers of other genes and turn them on or off.

dome In structural geology this term has more than one meaning. Dome can refer to an anticlinal fold, which is closed (by folding) in all directions; in other words, having a dome or upside-down bowl-like shape. In its simplest symmetrical form such a feature when eroded, would give a circular bull's-eye-like **outcrop** pattern with the oldest rocks in the core of the structure. The term dome can also refer to the accumulation of magma and solid rock around and above a volcanic vent; this is sometimes referred to as a lava dome. *See* **Sangiran dome**.

dominance *See* allele; dominant allele.

dominant allele In an organism that has two **homologous** copies of each **chromosome**, the dominant allele is the one that determines the **phenotype** no matter whether two copies of that allele (i.e., as in a **homozygote**) are present or just one copy (i.e., as in a **heterozygote**) is present. An example of a dominant allele in modern humans would be the normal or wild-type allele for the *CFTR* gene that encodes the cystic fibrosis transmembrane conductance regulator (or CFTR) protein. Homozygotes and heterozygotes for this allele are both phenotypically normal, whereas those who are **homozygous** for the recessive (mutated) allele have the disease cystic fibrosis.

dosage compensation For genes that are present on the part of the **X chromosome** not found in males, the amount (or dosage) of protein produced is regulated so that it is the same in males (XY) and females (XX). To do this, in placental mammals in every female cell one X chromosome is randomly inactivated. It is this inactivated X chromosome that forms the **Barr body**.

double stance phase *See* walking cycle.

downstream Downstream is in the direction (as indicated by the numbered carbon molecules of the sugars in the DNA backbone) toward the $3'$ (three prime) end of a DNA strand. The $3'$ end is so called because the *third* carbon has the hydroxyl group attached. The other end of a single strand of **DNA** is called the $5'$ (five prime) end because the *fifth* carbon of the deoxyribose sugar has a terminal phosphate group attached to it. The significance of the directionality of a DNA sequence is that the processes of **replication** and **transcription** only occur in the $5'$ to $3'$, or downstream, direction.

Dozu Dhalu One of the Early and Middle **Pleistocene** sites in the exposures of the **Ola Bula Formation** within the **Soa Basin** of central Flores, eastern Indonesia.

dp *See* dm.

draft sequence A **genome** sequence that has either not been completed and/or checked for errors. For example, the first modern human genome sequence to be published (International Human Genome Sequencing Consortium 2001) was a draft sequence because it covered only 94% of the genome and only 25% of that sequence was "finished" (i.e., sequenced with at least fourfold coverage and with a quality score of 99% accuracy). The first chimpanzee sequence to be published (The Chimpanzee Sequencing and Analysis Consortium 2005) was also a draft sequence.

drift Geology Term used to describe surficial sediments deposited by glaciers or associated processes. In practice it is commonly used to describe a wide variety of quaternary sediments in areas that have been glaciated. Genetics *See* **genetic drift**.

Drimolen (Location 25°58'08"S, 27°45'21"E, Gauteng Province, South Africa; etym. named after van Drimolen, the owner of the farm at the time the site was discovered) History and general description Discovered by André Keyser in 1992, Drimolen is a complex of **breccia**-filled caves in Precambrian **dolomite** 5.5 km/3.4 miles northwest of **Sterkfontein**. Temporal span 2.0–1.5 Ma. How dated? Biostratigraphy. Hominins found at site Most of the many well-preserved hominins recovered from Drimolen (including **DNH 7**, a complete skull, and **DNH 8**, a mandible with an almost complete dentition) have been referred to *Paranthropus robustus* and after **Swartkrans** its sample (>80 specimens) is the next largest contributor to the *P. robustus* **hypodigm**. The Drimolen contribution to the *P. robustus* dental hypodigm is smaller in overall size than the contributions from Swartkrans and **Kromdraai**, but the range of variation is greater, and Lockwood et al. (2007) have suggested this may be due to taphonomic differences between the sites, with males dominating in the Swartkrans sample of *P. robustus*. The Drimolen contribution to the *P. robustus* hypodigm is interpreted as intermediate in non-metrical morphology between the Swartkrans and the Kromdraai samples of *P. robustus*. A minority of the hominins, mainly teeth (e.g., DNH 35), have been assigned to *Homo* sp., and Curnoe and Tobias (2006) have referred DNH 45 and 70 to *Homo* aff. *H. habilis* and *Homo* *habilis*, respectively. Curnoe (2010) includes DNH 70 in the hypodigm of *Homo gautengensis*. Archeological evidence found at site **Bone tools**. Key references: Geology, dating, and paleoenvironment Keyser et al. 2000; Hominins Keyser 2000, Curnoe and Tobias 2006, Moggi-Cecchi et al. 2010; Archeology d'Errico and Backwell 2003.

DRP Acronym for the **Dikika Research Project** (*which see*).

DSR Acronym for **daily secretion rate** (*which see*).

dual-unit foraging model A model for **Oldowan** site formation based on an analogy with social carnivores. It argues that site formation resulted from the intersection of two independent strategies used by hominins to reduce predation risk. In the first, males unencumbered by offspring transport carcass parts away from high-risk, highly competitive, death sites, and in the second hominins are attracted to relatively secure areas where **altricial** offspring and caregivers can forage safely (Oliver 1994). The fitness benefits of provisioning mothers, infants, and caregivers with this high-quality food may then have introduced and established the principle of goal-directed transport of food for sharing, thus providing the selective milieu for the further evolution of cognition and for the eventual emergence of modern human language.

Dubois Collection *See* **Rijksmuseum van Natuurlijke Historie**.

Dubois, Marie Eugène François Thomas (1858–1940) Eugène Dubois was born in Eijsden, The Netherlands, and in 1877 he entered the University of Amsterdam to study medicine. There he came under the influence of Max Fürbringer who instilled in Dubois the idea that comparative anatomy and embryology should be used to reconstruct phylogeny. Dubois completed his medical degree in 1884 and in 1886 he accepted a post as a lecturer of anatomy at the University of Amsterdam where he became interested in human evolution. Influenced by Ernst Haeckel and by the discovery of primate fossils in the Siwalik Hills of India by Pilgrim in 1878, Dubois resigned from his university post in 1887 and enlisted as a medical officer in the Royal Dutch East Indies Army so he could search in

Southeast Asia for the evolutionary link between apes and modern humans. Dubois managed to obtain financial assistance from the colonial government as well as practical assistance in the form of convict laborers to help in his search. Dubois arrived on Sumatra in 1887 and soon found animal fossils, but news of the discovery in 1888 of a human skull near the village of **Wadjak** on the island of Java prompted Dubois to request a transfer to Java in 1889. In November 1890 he found a mandibular fragment at Kedung Brubus and in 1891 he began excavating along the banks of the Solo River near the village of **Trinil**. In September of that year a molar was discovered and in October Dubois' excavators found a calotte. At first he thought it belonged to an extinct ape, but with features that were more modern human-like than those seen in any previously discovered anthropoid ape. In August 1892 a nearly complete modern human-like femur was found in the same strata and this led Dubois to reassess the affinities of the calotte. In a preliminary report prepared in 1892 but presented and published in 1893, Dubois assigned the specimen to *Anthropopithecus erectus*, but when a fuller description of the fossils appeared in 1894 the genus name of the taxon was changed to *Pithecanthropus erectus*. In 1895 Dubois returned to The Netherlands and exhibited his fossils at the Third International Congress of Zoology held in Leiden. Dubois' interpretation that the skullcap belonged to an intermediate form that linked apes and modern humans attracted the support of Ernst Haeckel, Gustav **Schwalbe**, and the American paleontologist Othniel Marsh but few others accepted his interpretations. This position was not improved by the discovery of the fossils from **Piltdown**, England between 1908 and 1915, since they supported the view that the large brain of modern humans had evolved before bipedalism. It was only when Piltdown Man was exposed as a fraud that paleoanthropologists began to reconsider the importance of *Pithecanthropus*. Soon after his return to The Netherlands Dubois was granted an honorary doctorate and accepted a position at the University of Amsterdam in 1898 to teach geology and paleontology. He also became curator of the Teylers Museum in Haarlem in 1897, a position he held until his death. Dubois continued to present his specimen at scientific meetings during the late 1890s and even prepared a life-size reconstruction of *Pithecanthropus erectus* for the World Exhibition held in Paris in 1900.

He became increasingly disappointed and discouraged with the lack of acceptance for his theories, so much so that after 1900 he stopped presenting the fossils, and eventually would not allow anyone to see them. He published little on *Pithecanthropus* between 1900 and 1912 except for a few papers on the fossil fauna of Trinil that supported his interpretation of the age of *Pithecanthropus*. However, Dubois did not abandon research and he began to work on the problem of brain evolution in vertebrates. Specifically, he investigated the mathematical relationship between the size of the brain and the size of the body and was one of the first researchers to systematically investigate the relationship between brain and body size (Dubois, 1898, 1914). In 1920 Dubois returned to human origins research by publishing papers on *Pithecanthropus* and the Wadjak skulls, but made no significant new contributions to paleoanthropology before retiring from his position at the University of Amsterdam in 1929. *See also* **allometry**.

Duinefontein (Location 33°43'S, 18°27'E; South Africa; etym. in Dutch it means "dune spring." Colonial farms were always sited near springs and many farms, including this one, were named after that local spring) History and general description The Duinefontein sites occur within the Koeberg Nature Reserve, adjacent to the Koeberg Nuclear Power Station, about 35 km/21 miles north of Cape Town on the Atlantic Coast of South Africa. Power station construction revealed numerous buried fossil and archeological sites, of which two – Duinefontein 1 (or DFT 1) and Duinefontein 2 (or DFT 2) – have been extensively investigated. DFT 1 is a fossil hyena den excavated in 1973. DFT 2 is a multilayered archeological site, excavated in 1973, 1975, and 1997–2002. Artifacts and bones at DFT 2 occur on paleosurfaces within a thick sequence of aeolian sands. Excavation exposed two paleosurfaces known as Horizons 2 and 3. Wind-polished bones and artifacts on the present surface are assumed to come from a now-deflated paleosurface designated "Horizon 1." The sediments at DFT 2 broadly resemble those at **Elandsfontein**. Temporal span and how dated? Optically stimulated **luminescence** has provided an age of 270 ka for the sands surrounding Horizon 2 and 290 ka for sands between Horizons 2 and 3. **Uranium-series dating** has provided an age of 160 ka for a capping **calcrete** and **biostratigraphy** suggests an age of 400–200 ka. Hominins found at site None. Archeological evidence found at site Lithic artifacts from Horizon 2 have been attributed to the Late **Acheulean** Industrial Complex. Sediments and fauna (e.g., buffalo, wildebeest, and kudu) suggest an environment with locally abundant water, grass, and broad-leafed brush, substantially different from the fynbos (or "fine-leafed shrub") that characterizes the area today. Although numerous large mammal bones occur at the site, many more were damaged by hyena teeth than by stone tools, suggesting perhaps limited Acheulean hominin hunting or scavenging. Key references: Geology and dating Klein et al. 1999a, Feathers 2002, Cruz-Uribe et al. 2003; Hominins N/A; Archeology Cruz-Uribe et al. 2003.

duplication A region of **DNA** that is present in more than one copy. For example, nonhomologous **recombination** or retrotransposition can result in the duplication of a region of the **genome** involving one or more genes. **Copy number variants**, **microsatellites**, and segmental duplications of chromosomes are examples of duplications. *See also* **gene duplication**.

dura mater (L. *dura* = hard and *mater* = mother) The outer of the three meningeal layers. The dura mater is made up of an outer endosteal layer, which adheres to the endocranial surface of the cranial vault and is effectively the endocranium and an inner meningeal, or fibrous, layer that forms the folds of dura mater (e.g., the **falx cerebri** and the **tentorium cerebelli**) that incompletely subdivide the endocranial cavity. The **dural venous sinuses** are endothelium-lined spaces between the endosteal and the fibrous layers of the dura mater. *See also* **meninges**.

dural venous sinuses (L. *dura* = hard) Venous channels lying between the inner, or fibrous, and the outer, or endosteal, layers of the **dura mater**, the outer layer of the three meningeal layers. They transmit blood from the veins draining the cerebral hemispheres and from the veins draining the **diploë** of the bones that form the walls of the cranial cavity. Dural venous sinuses usually leave a groove on the **endocranial** surface of the cranium, so their presence can be inferred from fossils that preserve that part of the endocranial surface, but because they do not *always* leave a groove, the lack of a groove is not unambiguous evidence of an absent dural venous sinus. Small **emissary veins**

provide a communication between the cerebral venous system and the veins of the scalp. The dominant pattern of venous drainage in modern humans is superior sagittal sinus→right transverse sinus→right sigmoid sinus→right jugular bulb→right internal jugular vein. However, in approximately 3% of modern human crania the main venous drainage route is superior sagittal sinus→occipital sinus→marginal sinuses→jugular bulbs→internal jugular veins. This is also the dominant pattern in *Paranthropus boisei* (Wood and Constantino 2007) and *Australopithecus afarensis* (Kimbel et al. 2004).

durophagy (ME *dure*, from L. *dura*=hard, plus Gk *phagein*=to eat) Durophagy, which refers to hard-object feeding, was first described in aquatic animals and fish that consume objects such as shelled invertebrates (Pregill 1984). Most major vertebrate groups feature at least one durophagous lineage [e.g., fish (Grubich 2003), sharks (Huber 2005), rays (Summers 2000), snakes (Jackson and Fritts 2004), lizards (Metzger and Herrel 2005), bats (Nogueira et al. 2005), cats (Biknevicius 1996), bears (Sacco and van Valkenburgh 2004), red pandas (Pradhan et al. 2001), and primates (Lambert et al. 2004)]. The diets of some early hominin taxa have been interpreted as being durophagous on the basis of morphological features that are considered adaptations to hard-object feeding (e.g., thick tooth enamel, chewing teeth with low, blunt cusps, and anteriorly positioned jaw musculature; Rak 1983) and more recently it has been suggested that tooth chipping in some hominin taxa is also indicative of hard-object feeding (Constantino et al. 2010). However, there is considerable debate about the relative importance of durophagy in hominins, with some researchers supporting the hypothesis that hard foods were not a dietary staple, but were primarily consumed during periods of resource scarcity (i.e., as **fallback foods**) (Ungar et al. 2008).

Duynefontein *See* **Duinefontein**.

DV Acronym for Dolní Věstonice and the prefix for hominin fossils from Dolní Věstonice. *See* **Dolní Věstonice I**.

DV I *See* **Dolní Věstonice I**.

dynamic mutation (Gk *dunamikos*=powerful and L. *mutare*=to change) An unstable region of a **DNA** sequence, such as a **microsatellite** or **short tandem repeat**, where the probability of **mutation** increases as the size of the region increases. Dynamic mutations at trinucleotide (e.g., CAG, CTG, etc.) repeats are responsible for pathologies that cause diseases such as Huntington's disease (CAG repeat), myotonic dystrophy (CTG repeat), and fragile X syndrome (CGG repeat). Such repeats are a special class of **duplication**.

dynamic modeling Most simulations of locomotion are performed using dynamic modeling, which is concerned with predicting force from motion (or vice versa) in a multibody system of given mass distribution. Inverse dynamic modeling predicts force from known motion. It is less computationally demanding than the reverse approach, forward dynamic modeling (e.g., Yamazaki et al. 1996) and hence was applied earlier to the reconstruction of hominin gait mechanics (e.g., Crompton et al. 1998, Kramer and Eck 2000) to compare the mechanical costs of locomotion of *Australopithecus afarensis* in different gaits and by Wang and Crompton (2004) to compare the mechanical costs of load transport in *Au. afarensis* and *Homo ergaster*. However, since it assumes known joint motion, inverse dynamics do not allow the identification of mechanically optimal motion for a given morphology. Thus, more recent studies have increasingly adopted forward dynamic modeling, which calculates motion from force. In practice, forces are not required to be known as they can be established on the basis of simulated muscles activated by a neuromuscular model (Ogihara and Yamazaki 2001) or another pattern generator (e.g., Sellers et al. 2003) plus a muscle model (e.g., Umberger et al. 2003). This system also allows researchers to predict metabolic, rather than mechanical, energy costs (e.g., for the gaits of *Au. afarensis* by Sellers et al. 2004 and Nagano et al. 2005). Such models are far more computationally demanding than inverse dynamics approaches in that they require extensive repetition to search for optimal solutions using optimization algorithms such as those from evolutionary robotics (Sellers et al. 2003, 2005). On the other hand, the derivation of optimal gaits for a given morphology means that they can be used together with other data to assess a much wider range of phenomena (e.g., the speed of the Laetoli trailmaker from stride lengths and predicted energetically optimal speed; Sellers et al. 2005). In the future inverse dynamic modeling is likely to be restricted to analyses where motion is experimentally determined (e.g., predicting the internal mechanics of the foot from surface-marker motion).

Dzeravá Skála (Location 48°29′N, 17°19′E, Slovakia; etym. the site is also known as Deravá Skála. Slovak *derav*=sieve, *skála*=rock) History and general description This cave site near the town of Plaveck Mikuláš in western Slovakia is important for understanding the **Szeletian** transitional lithic industry. A hominin tooth recovered from a layer with Szeletian lithics was recently classified as *Homo sapiens*, suggesting the Szeletian was not created by *Homo neanderthalensis* as previously thought. However, questions remain about the exact stratigraphic association of this tooth since it may also be associated with **Aurignacian** tools. Three main layers were found at the site, with the Szeletian overlain by an Aurignacian layer and then a **Gravettian** layer. Temporal span and how dated? Radiocarbon dating was performed on all three layers, providing dates of 35–30 ka for the Szeletian, 36–30 ka for the Aurignacian, and *c.*25 for the Gravettian layer. Optically stimulated **luminescence dating** provided slightly earlier dates for the Szeletian layer (*c.*57 ka), but similar dates for the Aurignacian layer (36–30 ka). Hominins found at site An M_2 germ attributed to *H. sapiens* was found either in the Szeletian or Aurignacian layer. Archeological evidence found at site Stone tools attributed to the Szeletian, Aurignacian, and Gravettian. Key references: Geology, dating, and paleoenvironment Kaminská et al. 2004; Archeology Kaminská et al. 2005; Hominins Churchill and Smith 2000, Bailey et al. 2009.

213

E

E686 The British Natural History Museum, London (or BMNH) accession number for the Kabwe 1 cranium. *See* **Kabwe 1**.

E691 The British Natural History Museum, London (or BMNH) accession for the Kabwe 2 tibia recovered from the site of **Kabwe** the same day that the **Kabwe 1** cranium was discovered. *See* **Kabwe 2**.

EAGRU Acronym for East African Geological Research Unit. *See* **Bishop, Walter William**.

ear *See* **external ear**; **middle ear**; **inner ear**.

early African *Homo erectus* An informal term used by some authors for a subset of *Homo erectus sensu lato* that others refer to as *Homo ergaster* Groves and Mazák, 1975. *See* ***Homo ergaster***.

Early Stone Age (or ESA) The oldest period in the tripartite division of the African archeological lithic record formalized by A.J.H. Goodwin and C. van Riet Lowe (1929) in *The Stone Age Cultures of South Africa*. Although it was developed for sequences in southern Africa, the terminology has subsequently been applied to sites across sub-Saharan Africa. There have been numerous calls to abandon the tripartite division (e.g., Clark et al. 1966b, Barham and Mitchell 2008) but the terminology persists because of its utility for describing broad stages in the general trend towards a more sophisticated lithic record (i.e., a trend towards aspects of the manufacture of stone tools that are conceptually and technically more elaborate). As used today, the ESA incorporates lithic evidence from **Oldowan** and **Acheulean** sites. The evidence at Oldowan sites consists largely of stone flakes and cores made using direct freehand percussion, whereas at Acheulean sites **handaxes** and **cleavers** are added to the mix. At some Acheulean sites there is evidence that the artifacts may have been shaped by hammers or billets made of organic (e.g., wood, bone, or antler) materials. Recognizing that first and last appearance dates vary locally and are subject to change as new sites are found, existing dating methods are improved, new methods are introduced, etc., the present evidence suggests the ESA began *c*.2.6 Ma and ended *c*.200 ka.

East Africa and Uganda Natural History Society *See* **National Museums of Kenya**.

East African Rift System (or EARS) A continental-scale tectonic feature that dominates the East African landscape and which has influenced climate, vegetation, and faunal distribution throughout the **Neogene**. The East African Rift System comprises two major branches, eastern and western. In the north, the eastern branch begins in the **Afar Depression**, a triple junction where the East African Rift System meets the Red Sea and Gulf of Aden oceanic rifts. The Afar trends into the **Main Ethiopian Rift System**, and continues southwards into the Gregory Rift that bisects Kenya. The eastern branch splays out and terminates in northern Tanzania. Throughout its extent, the eastern branch is characterized by abundant volcanism (Oligocene to recent) both within and flanking the rift valley and it is partly occupied by numerous shallow lakes commonly saline and alkaline in nature (e.g., Lakes Baringo and Turkana). The eastern branch is renowned for the rich fossiliferous sedimentary deposits of its **basins**, along with an archeological record from the earliest sites (2.6 Ma in the Afar) to recent accumulations. The western branch curves from Lake Albert through large mainly freshwater lakes (e.g., Lakes Tanganyika and Malawi) to the coast in Mozambique. Deep lakes and more localized volcanism characterize this branch and although fossiliferous and archeological sites occur along its length they are generally associated with thinner and more localized sedimentary accumulations. Although its most prominent feature is the East African Rift System the entire region has seen significant uplift, effectively blocking penetration of Atlantic-derived moisture from the Congo Basin and leaving the region in a rain shadow subject to monsoonal influence from the Indian Ocean (see Sepulchre et al. 2006). Most of the fossil hominin sites in East Africa occur either in, or close to, the eastern branch of the East African Rift System. The eastern branch consists of a series of river valleys and basins that includes in the north the **Afar Rift System**, the **Ethiopian Rift System**, and the **Omo rift zone**. *See also* **paleoclimate**; **tectonism**.

East Lake Turkana   *See* **Koobi Fora**.

East Rudolf   The name previously used for the site now known as either Koobi Fora or East Turkana. It was given this name because the lake now known as Lake Turkana was originally named Lake Rudolf and the site lay to the east of the lake. *See also* **Koobi Fora**.

East Rudolf Research Project   Paleoanthropological research along Lake Turkana (formerly Lake Rudolf) began in 1968 when Richard **Leakey** led a **National Museums of Kenya** expedition, comprising Margaret Leakey, Paul Abell, Kamoya **Kimeu**, and two graduate students, John Harris and Bernard Wood, to explore the deposits along the eastern shore of Lake Rudolf. During this initial survey the group identified extensive fossil deposits and recovered several fragmentary hominin fossils. As a result a base camp was established in the following year on a spit along the lakeshore called Koobi Fora. New scientists joined the project in 1969, including Kay Behrensmeyer, a geologist, Meave Epps (now Meave **Leakey**), a zoologist, and the paleoanthropologist Glynn **Isaac**. Behrensmeyer began the task of mapping the geology of the site, particularly the volcanic tuffs that allowed the radiometric dating of the deposits. **Oldowan** tools were soon discovered and before the end of the field season two crania belonging to *Paranthropus boisei* were found (**KNM-ER 406** and **KNM-ER 407**). In 1970 the East Rudolf Research Project was established under the co-leadership of Richard Leakey and Glynn Isaac to pursue expanded excavations at Lake Rudolf with an even larger company of scientists participating. Specialists studied the various species of fossil mammals present, including hominins, while geological mapping also continued. A mandible (**KNM-ER 992**) was discovered in 1971 that later became the type specimen of *Homo ergaster* and in 1972 Bernard Ngeneo discovered a hominin cranium (**KNM-ER 1470**) that was to become the

lectotype of *Homo rudolfensis*. Arguments over the dating of the **KBS Tuff** became a major issue during the 1970s since it had implications for the age of the KNM-ER 1470 cranium and other remains from below the KBS Tuff, and the controversy was not resolved until the early 1980s. Two additional crania discovered in 1973, the first (**KNM-ER 1805**) by Paul Abell and the second (**KNM-ER 1813**) by Kamoya Kimeu, were subsequently assigned to *Homo habilis*. In 1975 the Kenyan government renamed Lake Rudolf Lake Turkana and this prompted the East Rudolf Research Project to be renamed as well, so in the same year it became the **Koobi Fora Research Project**. *See also* **Koobi Fora Research Project**.

"east side story"   *See* **Coppens, Yves**.

East Turkana   *See* **Koobi Fora**.

eccentricity   (L. *eccentricus*=not having the same center) One of the three rhythms, or cycles, affecting the Earth's orbital geometry (i.e., the manner in which the Earth orbits the sun). The *c.*100 ka-long eccentricity cycle affects the shape of the Earth's orbit and especially affects **insolation** at high latitudes. It has been the dominant cyclical signal affecting global climate for close to a million years. *See also* **astronomical theory**; **astronomical time scale**.

eclectic feeder   One of the several terms (others include ecological generalist, **eurytope**, and omnivore) used for animals that include a diverse array of foodstuffs in the diet. Eclectic feeders might include an exceptionally wide range of plant foods into their diets (flowers, fruits, young leaves, seeds) or they might mix plant and animal foods (for example, supplementing a largely fruit-based diet with insects, small vertebrates, or eggs). Within primates, baboons are the classic eclectic feeders, ingesting a wide range of foods from cultivated crops to flamingos and juvenile antelopes. Much of the foraging for supplementary items may be done opportunistically. Eclectic feeding may have enabled Old World monkeys to inhabit a wide range of habitats, respond quickly to environmental changes (either climatic or anthropogenic), and exploit a wide geographic area (O'Regan et al. 2008). However, primates that feed eclectically do not necessarily incorporate a wide range of foods in the short term, but adjust their subsistence strategies seasonally, cycling between different foodstuffs as they become available (Chapman and Chapman

1990). The ability to feed eclectically is seen as one of the hallmarks of ecological generalists or eurytopes. Hominins were most likely eclectic feeders.

ecofact (Gk *oikos*=house and L. *facere*=to do) In the context of archeology an ecofact is a naturally occurring object that has not been handled or altered by modern humans or hominins. Examples include seeds, shells, or bones, which can become incorporated into archeological deposits through natural processes. Although ecofacts are not directly relevant to understanding hominin behavior, they can provide important information concerning the environmental context of an archeological site. For example, the identification of plant seeds in an archeological deposit could provide insight into the past vegetation cover at the time of deposition.

ecological intelligence The ability to adapt to an ecological niche by having an understanding of organisms and their ecosystems. By saying an organism has ecological intelligence, it is inferred that the organism has the capacity to learn from experience and deal effectively with the environment. This term was first introduced as "natural intelligence" by Howard Gardner, a psychologist at Harvard University, but Daniel Goleman (2009) adapted this term to ecological intelligence and applied it to how modern humans adapt to their changing environment in the face of industrialism. *See also* **cognition**.

ecology (Gk *oikos*=house and *logia*=study) First used by Haeckel (1866), this term refers to all the interactions between an organism and its environment. These can include the relationships between the organism and its physical, or **abiotic**, environment, as well as between it and its organic, or **biotic**, environment (i.e., the other organisms it comes into contact with). Three subdisciplines of ecology particularly relevant to hominin evolution are **paleoecology**, **behavioral ecology**, and **human ecology**.

economic utility index Measures the combination of the meat, marrow, fat, grease, or other energetic returns associated with a skeletal element. In zooarcheology these indices are used to evaluate skeletal part frequencies in terms of **carcass transport strategies**. The first economic utility index was developed for caribou (*Rangifer*) by Lewis Binford in 1978 and has since been developed for a number of mammalian taxa, including buffalo (*Bison*), horse (*Equus*), and Phocid seals, among others. Binford proposed a family of hypothetical utility curves, representing a bivariate scatterplot comparison of skeletal part frequencies against economic utility to illustrate three hypothetical carcass transport strategies: (a) the bulk strategy, whereby all but the lowest-utility elements are transported, (b) the gourmet strategy, in which only the highest-utility elements are transported, and (c) the unbiased strategy, in which skeletal elements are transported in direct proportion to their economic utility. Economic utility indices are also referred to as food utility indices. *See also* **carcass transport strategy**.

economy *See* efficiency.

ecosystem The biotic community and its abiotic environment functioning as a system. It consists of dynamic interactions between plants, animals, and micro-organisms and their environment working together as a functional unit. An ecosystem can be as large as the Sahara Desert, or as small as a puddle or vernal pool. Ecosystems are typically comprised of several **habitats**.

ecotone [etym. *eco*(logy) plus *tone*, from Gk *tonos*=tension; i.e., an ecology in tension] In ecology, an ecotone is the transitional zone between two adjacent plant communities (e.g., open grassland and forest). The ecotone may represent a sharp boundary between adjacent habitats or a gradual blending of habitats across a broader area. Plant and animal species found in the adjacent communities are often found in the ecotone, in addition to organisms largely restricted to the ecotone. For example, the African impala (*Aepyceros melampus*) is often regarded as an ecotone species because it favors the ecotone between open grassland and savanna woodland; early hominins may well have occupied the same ecotone.

ectocranial morphology (Gk *ectos*=outside and *kranion*=brain case) Morphological features on the outside of the cranium. They can take the form of lines, crests, or grooves and mark where **connective tissue** (either the covering of a muscle, the epimysium, or the septa within a muscle, the perimysium) attaches to the external surface of the cranium. *See also* **nuchal crest**; **parasagittal crests**; **sagittal crest**; *striae parietales*.

ECV Acronym for endocranial volume (*which see*).

edaphic (Gk *edaphos*=ground or soil) Refers to anything related to **soil** (*which see*).

EDJ Acronym for the enamel-dentine junction, the three-dimensional boundary between the dentine core and the enamel cap of a tooth. *See* **enamel-dentine junction** (syn. dento-enamel junction).

EDMA Acronym for **Euclidean distance matrix analysis** (*which see*).

effect hypothesis One of several hypotheses proposed by Elisabeth Vrba (the others are the **habitat theory hypothesis** and the **turnover-pulse hypothesis**) that seek to explain how environmental change influences macroevolutionary patterns in the fossil record. The effect hypothesis provides a mechanism to explain the evolution of morphological trends in the fossil record that does not involve **directional selection**. Vrba (1980) noted the evolutionary histories of species that are ecological generalists (also called **eurytopes**) are expected to differ from those of species that are ecological specialists (also called **stenotopes**). Eurytopic, or generalist, species are able to survive under a wide variety of environmental conditions, whereas stenotopic, or specialist, species can only survive under a narrow set of environmental conditions. Thus, as environmental conditions change, stenotopic species are expected to experience higher rates of speciation and extinction than eurytopic species because the former are less well equipped to tolerate the new environment. In contrast, eurytopic species are more likely to be able to cope with, or adapt to, the new environment, so they are expected to be less likely to speciate or to go extinct. As a result, within a clade that contains both stenotopes and eurytopes, there are expected to be more species of the former than the latter. This gives the appearance that natural selection has favored the evolution of specialized morphological characteristics when in fact the observed species pattern (i.e., more stenotopes than eurytopes) is the result of a **macroevolution**ary "effect" and not the result of selection at the microevolutionary level. *See also* **eurytope**; **habitat theory hypothesis**; **macroevolution**; **stenotope**; **turnover-pulse hypothesis**.

effective population size In an idealized population (with random mating and no selection) it is the number of breeding individuals who contribute genes to the next generation. Introduced by the population geneticist Sewall Wright, effective population size (or N_e) influences the amount of **genetic drift** experienced by a population. Specifically, genetic drift is weaker in large populations and stronger in small ones. Usually the **census size** of a population is at least three times larger than the effective population size because it includes individuals who are too young, or too old, to produce offspring as well as those who are of reproductive age, but who do not reproduce. Effective population size can be affected by age structure, overlapping generations, unequal numbers of males and females, skewed family size distributions, fluctuations in population size, and the geographical dispersion of populations, but all of these issues can be corrected for statistically. Fluctuations in population size results in effective population size being the harmonic mean of the actual population size. Since the harmonic mean is typically dominated by the smaller population sizes, one or more periods of small population sizes (often called population **bottleneck**s) have a significant effect. In modern humans, N_e has been estimated at approximately 10,000 despite our current estimated population size of over 6 million people. This value is due to the fact that modern human effective population size still reflects much smaller population sizes in the past.

efficiency The ratio of work done to energy expended. Thus a large car can be more efficient than a small car even if it requires more energy to travel a given distance, as long as the amount of energy per unit of work done is less. It is useful to make a distinction between efficiency and economy. Economy refers to the thrifty use of resources; economy cars can be less efficient than larger automobiles, but they are more economical if they travel further on less fuel. In general, larger animal species are less economical than smaller species in that they use more energy to power their larger bodies. However, they are generally more efficient because they use less energy per gram of their body mass.

efficiency hypothesis This hypothesis is one of several purporting to explain the origin of upright posture and bipedal locomotion within the **hominin clade**. The hypothesis was proposed by Peter Rodman and Henry McHenry (1980) and is based on what were then ground-breaking experimental data on the energetics of locomotion in two juvenile chimpanzees (Taylor and Rowntree 1973). When the locomotor energetics in two juvenile chimpanzees was compared with those of adult modern humans, Taylor and Rowntree (1973) showed that modern human bipedalism is more energetically efficient than chimpanzee quadrupedalism. However, the key aspect of Rodman and McHenry's efficiency hypothesis is the observation that chimpanzee quadrupedalism is approximately as energetically efficient as chimpanzee bipedalism (Taylor and Rowntree 1973). Rodman and McHenry (1980) interpreted this to mean that the transition from quadrupedalism to bipedalism would not have involved passing through an energetically inefficient mode of locomotion against which natural selection would have acted. Thus, as the earliest hominins were adapting to shrinking forests and increased fragmentation of forest patches at the end of the **Miocene**, neither terrestrial quadrupedalism nor bipedalism would have initially been favored on energetic grounds. However, over time, natural selection would act to improve the efficiency of the locomotor system and Rodman and McHenry (1980) suggest that bipedalism would have been favored over quadrupedalism because it would have allowed the earliest hominins to more effectively compete for arboreal resources by more efficient travel between forest patches. The major assumption that the energetic pattern for chimpanzee adults would be equivalent to Taylor and Rowntree's (1973) results collected on 3 year-old juveniles was confirmed by data from Sockol et al. (2007) on five adult chimpanzees. They showed that the energetic cost of chimpanzee knuckle-walking and bipedality were similar (and above average for comparably sized mammals) and modern human bipedality was about 75% less costly (and well below average for comparably sized mammals). Perhaps most importantly, in most chimpanzee individuals Sockol et al. (2007) found significant differences in efficiency between knuckle-walking and bipedal walking and they showed that this could be explained by biomechanical differences in gait. Some individuals, using more extended hip and knee postures (that more closely resemble the gait of moderns) walked more efficiently with a bipedal gait than during knuckle-walking. Sockol et al. (2007) suggest that intraspecific variation in the last common ancestor of modern humans and chimpanzees/bonobos meant that some individuals would have been able to better extend their hindlimbs and thus have longer contact

times, resulting in a bipedal gait that was more efficient than knuckle-walking; thereafter small increases in hindlimb extension or length would accentuate the energetic savings of bipedal gait in the earliest bipeds. Sockol and colleagues' version of the efficiency hypothesis posits that bipedality in the earliest apelike hominins could have been less energetically costly than knuckle-walking and could have been a driving factor in the origin of hominin bipedality (Sockol et al. 2007). *See also* **bipedal locomotion**.

Ehringsdorf (Location 50°58′0″N, 11°21′0″E, Germany; etym. named after Ehringsdorf, a suburb of Weimar) <u>History and general description</u> The Ehringsdorf fossils were discovered as a result of blasting in the Fischer and Kämpfe quarries; all come from the 15 mm-thick Lower Travertine. Most attribute the Ehringsdorf remains to *Homo neanderthalensis*, but some researchers have drawn attention to the presence of more modern human-like morphology. <u>Temporal span and how dated?</u> *c.*230 ka based on **uranium-series dating** of the **travertine**, **electron spin resonance spectroscopy dates** on tooth enamel, and **biostratigraphy** of the **microfauna**. <u>Hominins found at site</u> At least nine individuals are represented by 35 hominin fragments that include cranial fragments, two mandibles, and postcranial evidence. <u>Archeological evidence found at site</u> A variant of the **Mousterian** that includes worked **scrapers** and **points**. Several hearths have been identified in the lower and middle parts of the Lower Travertine. Key references: <u>Geology, dating, and paleoenvironment</u> Blackwell and Schwarcz 1986, Grün et al. 1988, Schüler 1994; <u>Hominins</u> Virchow 1920, Weidenreich et al. 1928, Smith 1984, Vlcek 1993; <u>Archeology</u> McBurney 1950, Behm-Blanke 1960, Feustel 1983.

Ein Gev (Location 32°46′N, 35°39′E, Israel; etym. He. *Ein Gev*=''cistern spring,'' named after the nearby Ein Gev Kibbutz) <u>History and general description</u> Ein Gev refers to a complex of four sites on the eastern shore of the Sea of Galilee, just outside of the Golan Heights. <u>Temporal span and how dated?</u> 15.7 ka based on **radiocarbon dating** of associated burnt faunal remains. <u>Hominins found at site</u> A fairly complete female skeleton (though lacking a mandible) was recovered by Ofer **Bar-Yosef** and M. Stekelis from site I during the 1964 season. The remains are consistent with other female inhabitants of the **Natufian**. <u>Archeological evidence found at site</u> The sites of Ein Gev are characterized by a primarily

Kebaran assemblage Key references: Geology, dating, and paleoenvironment Vogel and Waterbolk 1972; Hominins Arensburg and Bar-Yosef 1973; Archeology Arensburg and Bar-Yosef 1973.

El Harhoura (Location 33°57′17″N, 6°55′27″W, Morocco; etym. named after the nearby town) History and general description These two cave sites (El Harhoura 1, also known as Zouhrah Cave, and El Harhoura 2) near Témara, Morocco, were excavated in the late 1970s by André Debenath and F. Sbihi-Alaoui. They lie on the coast roughly halfway between Dar es Soltane and Smuggler's Cave, and contain predominantly Upper Paleolithic deposits. Temporal span and how dated? Upper Paleolithic based on stone tool technology. Hominins found at site Roughly 15 individuals were found in the Epipaleolithic layer of El Harhoura 1, and one canine, similar to that found at Dar es Soltane, was found in the lower layers. Two individuals were found in the Neolithic layer at El Harhoura 2. Archeological evidence found at site El Harhoura 1 had an epipaleolithic layer directly overlying two layers containing a mixed assemblage with both primitive-looking stone tools, like choppers, as well as relatively advanced Aterian points. The faunal profile and abundance of coprolites suggests that the site was primarily a carnivore den with only ephemeral human habitation. El Harhoura 2 contained Neolithic, Epipaleolithic, and Aterian layers. Key references: Geology, dating, and paleoenvironment Débénath and Sbihi-Alaoui 1979; Hominins Débénath and Sbihi-Alaoui 1979; Archeology Débénath and Sbihi-Alaoui 1979, Monchot and Aouraghe 2009.

El Kherba (Location 36°25′12″N, 5°49′12″E, near Sétif, Algeria) History and general description El Kherba is an **Oldowan** archeological site sometimes considered a sublocality of the nearby **Ain Hanech** site, which was discovered in 1947 by French paleontologist Camille **Arambourg**. El Kherba itself was discovered and first excavated by Sahnouni and **de Heinzelin** in 1992–3. El Kherba preserves the earliest stratigraphically contextualized evidence of hominins living in North Africa. Temporal span and how dated? Based on **magnetostratigraphy** (the site has normal polarity) and **biostratigraphy**, Sahnouni and de Heinzelin (1998) proposed that the site falls with the Olduvai subchron and is thus within the 1.95–1.78 Ma range. Other researchers such as

Raynal et al. (2001) dispute this and claim a date closer to 1.2 Ma. Even if the younger date is the correct one, this site is still holds the earliest primary-context evidence for hominin occupation in North Africa. Hominins found at site None. Archeological evidence found at site El Kherba has yielded a **Mode 1** tool assemblage with a low degree of standardization. As of 2002 some 952 artifacts had been recovered by Sahnouni et al. and of these around 72% were less than 2 cm in length and were classified as debitage. The primary raw materials for the tools are limestone and flint, both locally available from eroding Plio-Pleistocene conglomerates. Three artifacts consisting of two whole flakes and one denticulate fragment display use wear consistent with their use for cutting animal soft tissue and bone. Key references: Geology, dating, and paleoenvironment Sahnouni and de Heinzelin 1998, Raynal et al. 2001; Hominins N/A; Archeology Sahnouni and de Heinzelin 1998, Sahnouni et al. 2002, Lahr 2010.

El Niño Anomalously warm waters in the eastern equatorial Pacific that contrast with cool waters in normal years and the even cooler water temperatures in the opposite (cold) phase called La Niña. This phenomenon is part of the **El Niño Southern Oscillation** (or ENSO). More recent observations have uncovered and distinguished two distinct warm patterns of El Niño: the central Pacific or Modoki El Niño and the canonical eastern Pacific El Nino (Yeh et al. 2009).

El Niño Southern Oscillation (or ENSO) A quasi-periodic (2–7 year) pattern of natural climate variability in the tropical Pacific with far-reaching consequences around the globe. ENSO is preferred over the terms **El Niño** (ENSO positive or warm phase) and La Niña (ENSO negative or cold phase) because ENSO includes reference to both the ocean surface temperatures and the changes in atmospheric pressures (**Southern Oscillation Index**) that describe the full extent to which this ocean/atmosphere oscillation can alter **climate** around the globe. ENSO has been identified as the dominant mode of modern inter-annual climate variability. Paleoclimate reconstructions show that the modern sea surface temperature gradient in the Pacific Ocean between a warm western **tropical** Pacific and a cool eastern tropical Pacific was established *c*.2 Ma; prior to that warm temperatures across the tropical Pacific indicate a "super-**El Niño**" state that may not have included the periodic ENSO cyclicity and if this is correct the pattern of global climate and inter-annual variability would have been quite different. In Africa ENSO influences the amount and **seasonality** of precipitation with regional characteristics. ENSO events and related changes in the Indian Ocean most strongly influence East and southern Africa with strengthened upper westerly winds that lead to decreased rainfall in southern Africa (December–March) and increased rainfall in East Africa (October–December) (Hastenrath et al. 1993). North and West Africa are partially influenced by ENSO variability. In the **Sahel**, El Niño events tend to correlate with dry conditions (July–September). Sustained **El Niño**- or La Niña-like conditions may also explain global climate patterns during **paleoclimate** events such as the warm, wet mid-**Pliocene** (Ravelo et al. 2006) or be a possible trigger mechanism for abrupt climate events such as the **Younger Dryas** (Clement et al. 2001).

El Sidrón (Location 43°23′01″N, 5°19′44″W, Spain; etym. uncertain, but local pseudo-mythological humans are called "sidrones") History and general description This site, located in the Asturias region of Spain, was first explored in the 1970s by a team led by T. Pinto, but it was not until 1994 that the first hominin fossils were accidentally unearthed: systematic excavations have been ongoing since 2000. The fossils were found in the Osario Gallery, one of several galleries that run at right angles to the long axis of the cave system and were recovered from the exposed surface of "stratum III." Many of the modest collection of artifacts can be refitted. Temporal span and how dated? The age of the hominin fossils is c.39–49 ka and probably closer to the latter age. The ages come from calibration of the raw **AMS radiocarbon dates** obtained from two teeth and a bone fragment coming from excavations. **Amino acid racemization dating** of gastropods, **electron spin resonance spectroscopy dating** and optically stimulated **luminescence dating** have also been performed. Hominins found at site Hominins dominate the fauna at the site, and more than 1900 hominin specimens have been recovered from the El Sidrón cave. Of these 140 were unearthed by amateurs (designated with the prefix SDR) with the remainder being recovered during formal excavations (designated with the prefix DR). All parts of the skeleton are preserved. The crania are mostly fragmentary with a bias towards young individuals. There is evidence of at least 11 individuals and of these one is an infant, one a juvenile,

two are adolescents, and four are young adults. There is no evidence that the collection samples more than one hominin taxon and there is ample morphological evidence (e.g., supraorbital torus, suprainiac fossa, small mastoid, shovel-shaped, labially convex incisor crowns with strong lingual tubercles, distinctive premolar occlusal morphology, taurodont molars, large joint surfaces, etc.) that the collection samples *Homo neanderthalensis*; the El Sidrón Neanderthals have broader and shorter faces than more northerly samples. Mitochondrial and nuclear ancient DNA has been recovered from several specimens (two specimens that could belong to the same individual) and the extracted DNA provides evidence that evolutionary changes in the *FOXP2* gene seen in modern humans are also seen in Neanderthals. There is copious evidence of hominin-induced bone modification. Archeological evidence found at site A total of 333 stone artifacts have been recovered, including **side scrapers**, denticulates, a **handaxe**, and several **Levallois point**s. The raw material, which is dominated by chert with some quartzite, was available locally. Key references: Geology, dating, and paleoenvironment Rosas et al. 2006, Fortea et al. 2008, de Torres et al. 2010. Hominins Lalueza-Fox et al. 2007, Rosas et al. 2006, Krause et al. 2007a; Archeology Rosas et al. 2006.

El-Wad (Location 32°40′N, 35°05′E, Israel; etym Ar. *Mugharet el-Wad*=Cave of the Valley) History and general description El Wad is one of the "classic" **Mount Carmel** sites excavated by Dorothy **Garrod** from 1928 to 1934. It is by far the largest of the caves and is situated near the entrance to the **Wadi-el-Mughara** (Valley of the Caves). Temporal span and how dated? The early **Natufian** layers have been dated using **radiocarbon dating** from c.12.9 to 11.7 ka; the ages of the lower layers are unknown. Hominins found at site A number of burials ornamented with shells are known from the epipaleolithic Natufian layers, but no hominin remains were found in the lower **Mousterian** layers. Archeological evidence found at site There is a dense lithic sequence that spans the **Middle Paleolithic** to the Epipaleolithic, including artifacts from what Garrod called the **Levalloiso-Mousterian**, **Aurignacian**, and Natufian industries. Key references: Geology, dating, and paleoenvironment Garrod and Bate 1937, Weinstein-Evron 1991; Hominins Garrod and Bate 1937, Mastin 1964, Oakley et al. 1975; Archeology Garrod 1932, Garrod and Bate 1937.

Elandsfontein [Location 33°05'S and 18°15'E, South Africa; etym. named after the Elandsfontein (Afrikaans for "eland's spring") farm. The alternative names are taken from the nearby towns of **Hopefield** and **Saldanha** (the latter named after the Portuguese explorer António de Saldanha)] <u>History and general description</u> After the initial surface collection by G. Smit, excavations were undertaken between 1951–4 and 1965–6 by Ronald **Singer**, J. Wymer, and M. Drennan. Elandsfontein is a former dune field near South Africa's Atlantic coast where wind **deflation** exposed Pleistocene–Holocene artifacts and fossils in numerous interdunal "bays." The mammalian fauna and artifacts date mainly from the Middle Pleistocene, but later Pleistocene and Holocene artifacts and fauna, including **Middle Stone Age** lithics and Khoekhoe pottery, have also been found. The large mammalian fauna, **Acheulean biface**s, and flaking debris are associated with a calcareous duricrust formed on quartzose and shell-rich **aeolian** sands that promoted bone preservation. Most of the 13,000 faunal elements and numerous handaxes and other tools were collected unsystematically in deflation bays, but some were found in the excavation known as Cutting 10. <u>Temporal span and how dated?</u> Assuming the same pattern of faunal turnover observed in eastern Africa, the co-occurrence of the extinct long-horned buffalo *Pelorovis antiquus* and the ancestral hartebeest *Rabaticerus arambourgi* implies an age of between 1.0 and 0.6 Ma. The meticulously made handaxes suggest an age near the younger end of this range. <u>Hominins found at site</u> In 1953 K. Jolly and Ronald Singer found the **Saldanha 1** calvaria, and a mandibular fragment that may belong to the calvaria was reported separately in 1955. <u>Archeological evidence found at site</u> Most of the Acheulen artifacts associated with the duricrust likely accumulated independently of the animal bones, for stone-tool marks or other traces of hominin activity occur on less than 1% of the studied sample of 2184 **bovid** limb elements. Cutting 10 provided the only *in situ* artifact sample, comprising 37 **handaxe**s, two **cleaver**s, and several **flake**s related to their manufacture and resharpening, as well as other **core**s, flakes, and **retouch**ed pieces made of **silcrete** and quartz (n = 260). Hippopotamus (*Hippopotamus amphibius*), reedbuck (*Redunca arundinum*), and other taxa suggest the environs were much moister than they were historically, but the most striking element of the Elandsfontein fauna is its diversity and lack of a modern or historic analogue. Occasional Middle Stone Age artifacts, including **Still Bay point**s, and Later Stone Age artifacts, including sherds of Khoekhoe pottery, occur in deposits that overlie the Acheulean duricrust. Key references: <u>Geology, dating, and paleoenvironment</u> Klein et al. 2007, Butzer 2004; <u>Hominins</u> Singer 1954, Drennan and Singer 1955, Bräuer 1984, 2008, Rightmire 2008; <u>Archaeology</u> Singer and Wymer 1968, Volman 1984, Klein et al. 2007.

Elandsfontein 1 *See* **Saldanha 1**.

elastic modulus A material property defined as the ratio of **stress** to **strain** within a material being subjected to tensile forces applied along an axis, as the material is being deformed elastically (meaning that the material returns to its original shape after the applied force is removed). The modulus can be thought as a uniaxial measure of the stiffness of a material and can be visualized as the slope of the elastic portion of the stress/strain curve, which is linear. Strain is a dimensionless value, so the elastic modulus, denoted by E and also known as Young's modulus, is measured in Pascals (Newtons per square meter), which is the unit that describes stress. Typically, the elastic modulus of cortical bone is measured in terms of gigapascals (i.e., GPa, or billions of Pascals). Cortical bone is often said to have an elastic modulus of 15–18 GPa, although this is an oversimplification, as the elastic modulus varies by direction (anisotropy) in many regions. Values of 15–18 GPa or greater are usually found in the direction of greatest stiffness in the cortical plane, whereas values of approximately 5–12 GPa can often be found in one or two directions orthogonal to the direction of greatest stiffness in bone.

electromyography (or EMG) A technique that measures electric activity within a muscle and then uses that as a proxy for the work performed by that muscle. EMG research is most commonly conducted using superficial EMG sensors applied to the skin that detect electrical activity of muscles deep to the skin, but more precision can be obtained with implanted electrodes inserted directly into the muscle tissue. By back-stimulation (i.e., using a weak electrical current to stimulate the muscle of interest to become active) researchers can verify the electrode lies in the targeted muscle. Then, when the subjects recruit their muscles during natural activities, telemetry systems send the electrical signal wirelessly to a detector, allowing researchers to study anatomical movements while the subjects are unencumbered. Stern and Susman (1981) provide an example of successful EMG research. They collected EMG data on the activity of gluteus medius, gluteus superficialis (the equivalent of gluteus maximus in modern humans), and tensor fascia femoris in gibbons, chimpanzees, and an orangutan during climbing and bipedal and quadrupedal locomotion. They found that, contrary to previous suggestions, the gluteus medius in apes acts primarily to medially rotate, not extend, the thigh while in its typically flexed position. They conclude that the role of gluteus medius during bipedality is the same in apes as in modern humans, namely to provide side-to-side balance of the trunk at the hip. This is turn suggests that the evolution of the hominin lateral balance mechanism during bipedal gait was primarily osteological, thus allowing the same muscle to function in the same way, but with the thigh extended (i.e., in modern humans) instead of flexed (i.e., chimpanzees, etc.).

electron spin resonance spectroscopy dating (or ESR) A radiation-based dating method involving the measurement of the trapped electrons resulting from "molecular-level" radiation damage caused by the uptake of uranium and other radioactive elements into crystalline materials such as in **calcite** (e.g., shells and coral) or **apatite** (e.g., bones and teeth), often at the time of their burial. The same provisos as those for **luminescence dating** methods apply to ESR dating (i.e., the resetting of the "electron clock" must be unambiguously linked with hominin activity, radiation dose rate must be established, etc.). One suitable crystalline material whose formation can be linked with hominin intervention is the enamel of the teeth of animals used by the hominins for food or as a source of raw materials. The technique can provide dates over a time range of a few thousand years to more than *c.*1 Ma. The ESR technique is not without its problems and an important unknown is the timing of the uranium uptake by the sample. The "early uptake model" assumes that uranium reaches its present levels soon after burial, whereas the "linear model" assumes that the rate of uranium accumulation is linear, but neither model may accurately account for the complex history of uranium accumulation or loss in particular specimens. An example of an appropriate application of the ESR technique is the dating of **Hexian** (Grun et al. 1998).

electrophoresis [L. *electricus*=literally "like amber" (electricity is produced when amber is rubbed) and Gk *phoresis*=to bear] A method that separates biological molecules (e.g., **DNA**, **RNA**, or **proteins**) according to their mobility in a gel (commonly made of agarose, starch, or polyacrylamide) under the influence of an electric field. The size of the pores in the gel affects the speed that molecules can move through it and the distance the molecules move depends on their electric charge (which is determined by their chemical make-up) and on their mass (i.e., their molecular weight). Electrophoresis is commonly used in genetics laboratories for such things as identifying protein variants, checking **polymerase chain reaction**s to make sure that they worked, identifying size variants in **restriction fragment length polymorphisms** or insertion/deletion polymorphisms, and for **DNA sequencing**.

Elephantidae The formal name of the only extant family of **proboscideans**. It includes *Loxodonta*, the African elephant, and *Elephas*, the Asian elephant. *See also* **Proboscidea**.

Eliye Springs *See* **KNM-ES 11693**.

Eliye Springs cranium *See* **KNM-ES 11693**.

ELM Acronym for eastern lake margin, the term used for the eastern margin of the paleo-lake in the **Olduvai Basin** called Lake Olduvai. *See also* **Lake Olduvai; Olduvai Gorge**.

EMG Abbreviation of **electromyography** (*which see*).

Emireh points *See* **Boker Tachtit**.

emissary foramina (L. *emissarium*=a drain and *foro*=to pierce, the root of *foramen*=a hole, or opening) The holes in the cranium (e.g., the hypoglossal, mastoid, occipital, parietal, and posterior condylar foramina) that transmit **emissary veins** which provide a communication between the **intracranial cerebral venous system** and the extracranial venous system (e.g., the veins of the scalp or the vertebral venous plexus).

emissary veins (L. *emissarium* = a drain and *vena* = a water course, or blood vessel) Small thin-walled veins (e.g., hypoglossal, mastoid, occipital, parietal, posterior condyloid) that provide a communication between the **intracranial cerebral venous system** and the extracranial venous system. With the exception of the parietal emissary vein, which connects the superior sagittal sinus with the veins of the scalp via the parietal foramen, the remaining emissary veins transmit venous blood through foramina in the **cranial base**. From posterior to anterior, the major emissary veins of the cranial base are the occipital (occipital foramen), mastoid (mastoid foramen), condylar (condylar canal), hypoglossal (jugular foramen), and sphenoidal (emissary sphenoidal foramen). Falk (1986) suggested that in some early hominin taxa (e.g., *Paranthropus boisei* and *Australopithecus afarensis*) the emissary veins around the base of the cranium, along with the **occipitomarginal system** of intracranial venous sinuses, provided a route for significant amounts of venous blood to drain from the superficial and deep cerebral veins into the vertebral venous system.

enamel (OF *esmail* = a vitreous coating, often white, baked on metal, glass, etc.) The hard, white, outer coating of the **crown** of a tooth. One of three dental hard tissues (along with **dentine** and **cementum**), enamel is the hardest of the biological tissues and this accounts for the relative abundance of teeth in fossil assemblages. Mature enamel consists of two components or phases, inorganic and organic. Most mature enamel (approximately 96% by weight) is an inorganic mineral called hydroxyapatite, sometimes abbreviated to just **apatite**. The smallest structural unit of enamel, the enamel crystallite, is only a few nanometers wide in cross-section but it can grow to be a millimeter or more long. The cells responsible for the manufacture of what eventually becomes mature enamel, the **ameloblast**s, secrete a matrix of proteins, water, and mineral ions. The organic phase of mature enamel (approximately 4% by weight) consists of proteins that function as a "glue" to bind the crystallites together. Some of the proteins (e.g., amelogenins) provide a framework or scaffold along which crystallite growth is guided in its long axis (or C-axis). In mature aprismatic, or prismless, enamel the crystallites are bound together by remnants of the organic matrix. Most reptiles have aprismatic enamel, but although it occurs in primate teeth occasionally, especially close to the crown surface, almost all mammalian enamel is prismatic. In the same way that minute glass fiber threads can be bound together to form flexible glass fiber optic bundles, so too can enamel crystallites become associated into bundles called **enamel prism**s, or enamel rods. Prism boundaries are no more than places where crystallites meet each other at different angles and where remnants of the organic matrix get squeezed during enamel maturation. In cross-section, prisms may be rounded pencil-like structures or alternatively more complex and keyhole-like in outline. They may pack together in bundles, as flat-sided pencils would, in a "close-packed" hexagonal manner with one at the center surrounded by six prisms, or in layers that alternate one above the other like bricks in a wall. The appearance of enamel prisms viewed end-on has been classified into pattern 1, pattern 2, and pattern 3 depending on their outline, area, and packing arrangement. **Hominids** mostly have the keyhole-shaped hexagonally packed pattern 3 enamel, but with areas of round hexagonally packed pattern 1 enamel; pattern 2 enamel arranged into alternate layers of rounded prisms predominates in Old World Monkeys. Enamel is known to show several classes of **incremental features**. **Short-period incremental lines** form either daily or even occasionally every approximately 8–12 hours. These daily **cross-striations** and so-called intradian lines look like fine dark lines in transmitted light microscopy and run across the prisms in line with the face of the secretory end of the ameloblast (or Tomes' process) that formed them. Laminations, another incremental phenomenon, look like (and may actually be) daily increments in regions of aprismatic and prismatic enamel and their layered appearance suggests that many cross-striations formed at the same time may have merged. Long-period markings called **striae of Retzius** are aligned along previous positions of the whole ameloblast cell sheet. These **long-period incremental lines** have a modal value of 7 days apart in australopiths and 8 days in fossil *Homo* (Lacruz et al. 2008). All these structures in primate enamel are described, illustrated, and reviewed in Boyde (1989, 1990), Aiello and Dean (1990), and Smith (2006). The presence of incremental lines in enamel allows researchers to determine the **crown formation time** and rate, and in some cases the **mean age at death**; these applications are reviewed in Aiello and Dean (1990), Dean (2000, 2006), and Smith et al. (2006). *See also* **enamel development**; **enamel microstructure**.

enamel chipping The loss of volume that affects the dentition is mostly a gradual process (*see* **tooth wear**), but tooth volume is also lost suddenly in the form of substantial (i.e., millimeter-sized) edge chips of surface enamel. Enamel chipping has been documented in numerous mammalian species and more than 50 years ago John **Robinson** compared the frequency of chips on the teeth of *Australopithecus africanus* and *Paranthropus robustus* from southern Africa (Robinson 1954). He concluded that the latter featured more chips and he suggested that this may have been because *P. robustus* was eating **underground storage organs** covered in exogenous grit. Phillip **Tobias** also looked at enamel chipping in his 1967 monograph of **OH 5** (Tobias 1967a), the type specimen of *Paranthropus boisei*, but he did not find any difference in the incidence of chipping in the southern African hominins and Wallace (1973) reached a similar conclusion. Tobias suggested that the large size of the chips precluded them from being caused by small hard particles like grit and instead proposed that they might have been the result of chewing on bones. Enamel chipping has since been documented in a number of other East African taxa including *Australopithecus afarensis* (Johanson and Taieb 1976), *Australopithecus anamensis* (Ward et al. 2001), and early *Homo* (Constantino et al. 2010). Constantino et al. (submitted) showed how a simple fracture equation could be used to estimate peak bite forces from the sizes of enamel edge chips. A preliminary analysis revealed that bite forces calculated from chips on the posterior teeth of fossil hominins frequently exceed 1000 N and can approach 2000 N, thus supporting the idea of consumption of large hard foods like seeds; this latter hypothesis is given some support from the high frequency of similar chips seen in extant seed crushers (e.g., orangutans and peccaries). The semi-permanence of enamel chips means that hard-object consumption can be detected even if the hard foods are eaten rarely, as is the case with **fallback foods**.

enamel decussation *See* decussation; enamel development.

enamel-dentine junction (or EDJ) The interface or boundary between the outer surface of the **dentine** core of a tooth crown and the inner surface of the enamel cap that covers it. Although the EDJ looks smooth at the gross level, at the microscopic scale the microtopography of the dentine surface is sometimes scalloped to help key enamel into the dentine. Its overall shape defines the shape of the crown and the shape of the EDJ is determined by the contours of the **inner enamel epithelium** and the underlying mesenchymal cells that make up the dental papilla in developing tooth germs (see illustrations in Aiello and Dean 1990 and Nanci 2003). The shape of the EDJ is generally replicated at the outer enamel surface, except where differential rates of enamel formation modify the EDJ template (e.g., Skinner et al. 2009b). Smith et al. (2006b) have shown that within a single species (modern humans) both EDJ shape and **enamel thickness** differ among molar types, between the sexes and among regional populations; syn. dento-enamel junction.

enamel development Enamel is formed by **ameloblasts** that are derived from the **inner enamel epithelium**, the invaginated **ectodermal** layer of a developing **tooth germ**. In the early stages of the development of the tooth germ, cell division begins where the tips of the future **dentine horn** will be. At the apex of each developing horn the inner enamel epithelium differentiates into ameloblasts with secretory poles facing inwards towards the future **enamel-dentine junction** (or EDJ). The underlying mesenchymal cells of the dental papilla differentiate into **odontoblasts** whose secretory poles face outwards towards the EDJ: the interface between the two layers of cells is what will become the EDJ. Secretory odontoblasts differentiate first at the tips of the dentine horns and the wave of secretory activation moves down the sides of the tooth germ along the EDJ to the cervical margin or **cervix**. Enamel formation always lags slightly behind dentine formation. The rate at which fully differentiated ameloblasts become active secretory cells is known as the enamel **extension rate** (Shellis 1984). As each ameloblast moves outwards away from the EDJ in its wake it secretes an increasing thickness of matrix consisting of proteins, water, and mineral ions. Once the full thickness of enamel matrix is secreted, these secretory ameloblasts switch roles and become maturational ameloblasts. They alternate between pumping mineral (mostly Ca^{2+} and PO_4^{3-} ions) into the matrix and drawing water and degraded proteins out of the matrix thus enabling the crystallites (*see* **enamel**) to expand in diameter and pack increasingly tightly together. Enamel maturation can take years but by the time of its eruption the enamel of a tooth is hard and mostly mature. Rhythmic slowings of enamel matrix secretion are preserved in the form of **short-** and **long-period incremental lines**. The short-period enamel incremental features are called **cross-striations** and the longer-period enamel incremental features are called **striae of Retzius**. See the references in Humphrey et al. (2008) for more information about enamel mineralization. *See also* **extension rate**; **mineralization**.

enamel extension rate   *See* **extension rate**.

enamel formation   *See* **enamel development**.

enamel microstructure   The microstructural features of enamel reflect the cellular activity of the ameloblasts that secrete a protein-rich matrix during **enamel development**. Enamel is comprised of **enamel prism**s that are marked by various types of incremental line (e.g., **striae of Retzius, cross-striations**, laminations, and **intradian lines**). Enamel prisms weave across each other and create a regular banding pattern known as **decussation** and its manifestation is **Hunter–Schreger bands**. Counts and measurements of the incremental microstructure of enamel can be used to estimate **daily secretion rate**s and/or **crown formation time**. *See also* **enamel development**; incremental features.

enamel prism   Enamel prisms consist of bundles of hydroxyapatite crystallites formed into long rods. They are approximately 5 μm in diameter and run from the **enamel-dentine junction** to the surface of the **enamel**. The shape of the secretory tip of an **ameloblast** (the Tomes' processes) determines the cross-sectional shape of a prism and the prism path is determined by the direction of movement of the secretory ameloblast. Enamel prisms preserve microscopic **incremental features** such as daily **cross-striations** and the longer-period **striae of Retzius** (syn. enamel rod).

enamel thickness   Enamel thickness is typically quantified by linear (e.g., Gantt 1977), areal (i.e., two-dimensional; e.g., Martin 1983, Smith et al. 2006a) or volumetric (i.e., three-dimensional; e.g., Kono 2004, Olejniczak et al. 2008a) measures of the enamel cap. Martin (1983) proposed that measurements of average and relative enamel thickness be made from two-dimensional controlled section of the tooth crown. Average enamel thickness represents the average linear thickness across the tooth crown (i.e., the enamel cap area divided by the enamel-dentine junction length), while relative enamel thickness represents a measure of average enamel thickness divided by coronal dentine area. The latter approach facilitates comparisons across taxa (as enamel thickness is scaled using a proxy for body mass – the square root of the area of the dentine cap – that scales isometrically with body mass across primates; Shellis et al. 1998). Non-destructive **micro-CT** has been used by several researchers to calculate these indices of relative enamel thickness in three dimensions (e.g., Kono 2004, Olejniczak et al. 2008a), thus providing a more accurate representation of whole-crown enamel thickness. Linear measurements of enamel thickness have been used to support taxonomic distinctions among hominin taxa (e.g., Beynon and Wood 1986), generate functional models of tooth structure (e.g., Schwartz 2000), as well as to estimate cuspal or whole-crown formation times (e.g., Risnes 1986, Dean et al. 2001). Average enamel thickness in modern human molars has been shown to be greater in upper than in lower molars (Smith et al. 2006a) and it also increases along the tooth row from incisors through molars (Smith et al. 2008). Enamel thickness in hominins is known to vary from the very thick enamel of *Paranthropus boisei* to the thick enamel of *Australopithecus* and *Homo* (e.g., Beynon and Wood 1986, Grine and Martin 1988, but see Olejniczak et al. 2008b) to the reportedly thin enamel seen in *Ardipithecus kadabba* (Suwa et al. 2009a), but the use of different measurement techniques presently precludes meaningful direct comparisons. *See also* **enamel thickness, relative**.

enamel thickness, relative   A method of quantifying **enamel thickness** proposed by Martin (1983, 1985) that involves measuring the area of the **enamel** cap, the length of the **enamel-dentine junction**, and the area of the coronal **dentine** from a controlled (two-dimensional) buccolingual plane of section (i.e., across the tips of the dentine horns of the mesial or distal molar cusps). The measure, relative enamel thickness (or RET), is intended to facilitate comparisons across taxa (as it is scaled by a proxy for tooth size). Recent studies using **micro-CT** have quantified the three-dimensional relative enamel thickness using volumes and surface areas rather than areas and linear distances (e.g., Kono 2004, Olejniczak 2006) and Kono (2004) and Smith et al. (2006b) review the published two-dimensional enamel thickness data for extant hominoid molars. *See also* **enamel thickness**.

encephalization (L. *cephalon*=brain from the Gk *enkephalos*=literally "marrow in the head") A relative measure of a species' **brain size** that tries to capture the degree to which it is larger or smaller than expected for a typical animal of the same **body size**. Formulae for calculating encephalization differ according to the choice of **line-fitting technique** and what taxa are included in the reference sample. In primates, increased encephalization has been linked to a number of factors including **intelligence**, **diet**, innovation rate, the degree to which the animals use binocular vision, and the total amount of visual input to the brain. *See also* **encephalization quotient**.

encephalization quotient (L. *cephalon*= brain from the Gk *enkephalos*=literally "marrow in the head" and L. *quotiens*=how many times) (or EQ) Introduced by Harry Jerison in his 1973 book *Evolution of the Brain and Intelligence*, but the idea originated with Eugène **Dubois**' (1897) initial proposal for an equation to quantify **encephaliza-tion**. It was devised to account for the interspecific **allometric** effects of body size on brain size so that the degree of encephalization can be meaningfully compared across a range of different species. A number of studies (e.g., Bauchot and Stephan 1969, Gould 1975, Jerison 1973, Martin 1981) have shown that brain size scales with **negative allometry** with respect to body size. The exact scaling exponent depends on the species included in the reference sample and the statistical method of line fitting employed. In practice, the EQ of a particular species is deter-mined by calculating the ratio of its observed brain size over its "expected" brain size. The expected brain size is calculated from a prediction equation based on a scaling relationship, either theoretical (i.e., Jerison 1973) or empirically determined (i.e., Martin 1981, Holloway and Post 1982). Thus, a species with an EQ that is greater than 1.0 has a brain that is larger than expected for its body size and an EQ that is less than 1.0 indicates that the species has a brain that is smaller than expected. Although the exact EQ depends on the composition of the reference sample used in the analysis, without exception modern humans have been found to have the highest EQ in com-parative studies of mammals and primates.

Attempts to link variation in EQ to general cogni-tive capacities across diverse mammalian taxa have met with mixed results.

encounter-contingent prey-choice model *See* **prey-choice model**.

endemism (Gk *endmia*, *en*=in, *demos*=people) Belonging, or native to, a particular place, group, field, area, or environment. In an ecological con-text, endemism refers to the state of being unique to a particular geographic location. Typically, endemic types or species are more likely to develop on actual, or what are effectively, islands due to the geographic isolation of these popula-tions. For example, all lemur species (Lemuri-formes) are endemic to the island country of Madagascar and 50% of the primate species of Indonesia are endemic to its islands. In contrast, only 30% of the species in Brazil are endemic, and even fewer primate species in Africa are endemic to individual countries (about 11%). By default larger tropical countries may have more endemic species simply because they may have more species and greater habitat diversity (e.g., Brazil has more endemic species than Madagascar, but it covers an area around 15 times as large). Among fossil homi-nin taxa the obvious and best candidate for being an endemic species is *Homo floresiensis*.

endocasts (Gk *endo*=within and ME *casten*= to form an object from a mold) Applies to a cast made of any cavity in the body, but it is usually used as an abbreviation for an endocranial cast. *See also* **endocranial cast**; **endocranial cavity**.

endochondral ossification (Gk *endo*= within and *chondros*=cartilage) Bone formation within a framework of cartilage. Endochrondral ossification is a two-staged process. First, cells called chondroblasts form a hyaline cartilage model of the future bone. Cells in the middle of the cartilage model first divide and grow by interstitial growth (as distinct from appositional growth). This enables rapid formation of the cartilage model. Chondrocytes then swell (hypertrophy) and die, leaving only a mineralized shell of cartilage matrix. This mineralized cartilage is then resorped by osteoclasts (strictly speaking they should be called chondroclasts) and at the same time cells called osteoblasts start forming osteoid matrix on the walls of the spaces once occupied by chondroblasts. This happens first at the primary center of ossification, and that center is soon surrounded by a sleeve of subperiosteal bone that forms beneath the **periosteum**. In long bones the bone formed from the primary center of ossification is called the **diaphysis**. Primary centers usually appear between 7 weeks and 4 months of intrauterine life. Some bones (e.g., **auditory ossicles**, zygomatic, carpals, and tarsals) usually form from a single primary center but others, called complex bones (e.g., sphenoid, temporal), form from multiple primary centers. Secondary centers of ossification, or **epiphyses**, form in complex bones (e.g., vertebrae), at the ends of long bones and at the sites of bony processes (e.g., trochanters of the femur and the tubercles of the humerus). Long bones continue to grow in length because new cartilage is laid down in epiphyseal plates between the primary center in the shaft of the bone and the epiphyses at either end. *See also* **ossification**.

endocone A term introduced by Vanderbroek (1961, 1967) for what more commonly used cusp nomenclature schemes refer to as the **hypocone** (*which see*).

endoconid A term introduced by Vanderbroek (1961, 1967) for what more commonly used cusp nomenclature schemes refer to as the **entoconid** (*which see*).

endocranial (Gk *endo*=within and *kranion*= brain case) Noun The cavity formed by the walls, roof, and floor of the cranium. Adjective Refers to any structure within the cranial cavity (e.g., the dural venous sinuses are sometimes referred to as the endocranial venous sinuses).

endocranial capacity *See* endocranial volume.

endocranial cast A natural or prepared cast that uses the walls, roof, and floor of the cranial cavity as a mold. The best known natural hominin endocasts are those recovered from the southern African cave sites (e.g., **Taung**). Ralph **Holloway** developed the method that is widely used to make artificial endocranial casts. This involves stopping up the foramina and fissures then pouring just enough liquid latex into the **foramen magnum** to form a thin coating over the endocranial surface of the cranial cavity. Once the latex has cured, it is peeled off

the endocranial surface (hence it is referred to as a "peel"), extracted via the foramen magnum, and used to make the mold from which solid latex or plaster casts can be made. It is also possible to use the data collected from a computed tomography (or CT) scan to make a virtual endocranial cast that can then be used to manufacture a solid rendering of the virtual cast. Only rarely are fossil crania well enough preserved to enable a really accurate endocranial cast to be made and in most cases some guesswork is involved when reconstructing the parts of an endocast that are missing. The problem is that the volume of inaccurate endocranial casts can be measured as precisely as the volumes of accurate endocranial casts. Thus most estimates of the endocranial volume of fossil hominin taxa contain a mixture of reliable and less reliable measures of endocranial volume depending on the degree to which the walls of the endocranial cavity have been reconstructed.

endocranial cavity Refers to the cavity formed by the walls, roof, and floor of the cranium.

endocranial fossae The endocranial surface of the cranial base is conventionally divided into three endocranial fossae. The two anterior endocranial fossae house the **frontal lobes** of the **cerebral hemispheres**, the two middle endocranial fossae house the **temporal lobes** of the cerebral hemispheres, and the two posterior endocranial fossae house the cerebellar hemispheres. *See also* **endocranial cavity**.

endocranial morphology Endocranial morphology comprises whatever features of the external surface of the dura mater have been preserved in the form of negative impressions on the endocranial surface of the bones of the cranial vault. These preserved features include negative impressions of **sulci**, **gyri**, and the branches of the **meningeal vessels**.

endocranial volume The sum of **brain volume**, plus the volumes of the **meninges**, the extra-cerebral cerebrospinal fluid (or CSF) (i.e., the CSF outside the ventricles), the intracranial (but extra-cerebral) vessels, and the cranial nerves within the endocranial cavity. All of the latter (i.e., the meninges, etc.) generally amount to approximately 15% of the endocranial volume. Count (1947) suggested that endocranial volume=brain mass × 1.14. The volume of the endocranial cavity is usually measured by stopping up the foramina and fissures with cotton wool, inverting the **cranium**, pouring bird seed or another type of granule into the **foramen magnum**, and then decanting the seed or granules into a measure. Endocranial volume can also be measured by dipping a plaster or latex **endocranial cast** into water and measuring the volume of water displaced, or by using specialized software to convert data obtained from computed tomography (or CT) scans into a measure of endocranial volume (syn. endocranial capacity).

endolymph (Gk *endon*=within and L. *lymph*= water) Watery fluid contained within the **membranous labyrinth** of the **inner ear**. Chemically it resembles intracellular fluid.

endosteum The osteogenic layer lining the internal surface of the bones that make up the walls of the **endocranial cavity**. It is equivalent to the periosteum that covers the external surface of the skull and limb bones. *See also* **periosteum**.

endurance running hypothesis Posits that modern humans perform unusually well compared to other mammals at sustained running and that selection for this capacity shaped the evolution of modern human anatomy and physiology. The endurance running (ER) hypothesis was first articulated in detail by Carrier (1984), who attributed conversations with Dennis Bramble (see below) and S.C. Carrier as the motivation for his article. Acclimatized modern humans can run down prey (e.g., duicker, gemsbock, wildebeest, zebra, deer, antelope, kangaroo) by persistent pursuit until the animal collapses from exhaustion. Carrier (1984) pointed out the "energetic paradox" that modern humans have a very high energetic cost (ml O_2/kg×km) of transport (i.e. movement over distance), and yet modern humans are able to sustain their running over much longer distances than mammals of the same body size. Energetic inefficiency and ER would seem incompatible, but the ER hypothesis argues that modern humans can sustain long-distance running because they have evolved physiological mechanisms to (a) dissipate metabolic heat and (b) store and use large reserves of energy. The former include sweat glands that secrete enough moisture for evaporative cooling (horses and camels are other exemplars), but at substantial costs to water and salt reserves, and diminished body hair that

facilitates evaporative cooling. These factors support the argument that modern humans are derived in their abilities to dissipate heat, and evidence suggests that heat generated from activity is the limiting factor. Carrier hypothesized that adaptations for ER may have shaped the anatomy of *Australopithecus*, noting for example that its bipedal posture for locomotion would have decoupled breathing and running for, unlike panting, sweating is decoupled from breathing while running (Carrier 1984). However, the relative energetic inefficiency of modern human running means that ER (e.g., for predation or scavenging) would require unusually high energy stores to outlast prey. Twenty years later, Bramble and Lieberman (2004) published an influential article that built on Carrier's ER hypothesis, presenting data that had accrued since 1984, and more explicitly linked modern human musculoskeletal anatomy (e.g., long Achilles tendon compared with that in great apes) with the potential advantages of ER. Bramble and Lieberman's version of the ER hypothesis argues that selection for ER was the main driver of the musculoskeletal changes seen in the transition from *Australopithecus* to *Homo*, notably **early African *Homo erectus***. Other circumstantial evidence that modern humans are adapted for sustained running over long distances (ER) includes the observation that modern humans frequently experience an endocannabinoid "runner's high;" a similar response is seen in mammalian cursors but it is lacking in non-cursors (Raichlen et al. 2010). What remains to be established is whether ER is the only possible explanation for the adaptations seen in modern humans presently linked with ER, or whether long-distance walking or sprinting might be equally good adaptive explanations, and at what stage in human evolution ER (or walking, or sprinting) became an important selective force.

energetic returns *See* **optimal foraging theory**.

Engihoul *See* **Schmerling, Philippe-Charles (1790 or 1791–1836)**.

Engis (Location 50°34′N, 5°25′E, Belgium; etym. named for the town of Engis) History and general description This cave site (also called Awirs) has been destroyed by mining. The **hominin** fossils come from stratified cave deposits that were excavated by P.C. Schmerling in 1829 (Engis 1–3) and by

E. Dupont in 1872 (Engis 4). Although it was not recognized as such until 1936, Engis 2 was one of the first, if not the first, specimen of *Homo neanderthalensis* to be discovered. Engis 1 is recognized as a modern human buried in **Mousterian** deposits. Temporal span and how dated? Any date is based on the associated **Mousterian** and **Aurignacian** archeological evidence. Hominins found at site Nearly complete adult **neurocranium** of *Homo sapiens* (Engis 1), partial child's cranium with separate maxilla of *H. neanderthalensis* (Engis 2), cranial and postcranial fragments (Engis 3; NB: this specimen has been lost), and a taxonomically undiagnostic ulna fragment (Engis 4). Archeological evidence found at site Mousterian (Engis 2) and **Upper Paleolithic** (Engis 1). Russell and Le Mort (1986) suggested there is evidence of cutmarks on the Engis 2 cranium. White and Toth (1989) agreed that the marks were anthropogenic but claimed they are not cutmarks, but "sandpaper striae formed during restoration of the vault, moulding striae formed when mold part lines were incised into the fossil and profiling striae formed when craniograms were made with sharp steel instrument tips" (*ibid*, p. 361). Key references: Geology, dating, and paleoenvironment Fraipont 1936; Hominins Fraipont 1936, Fenart and Empereur-Buisson 1970; Archeology Fraipont 1936, Russell and Mort 1986, White and Toth 1989. Location of the hominin fossils found at site Service de Paléontologie Animale et Humaine, Université de Liège.

enhancement The presence of another individual may enhance or inhibit specific behaviors in the observing individual. These socially driven effects are considered to be basic (low-level) motivational and affective states that compel individuals to act and the mechanisms are believed to be phylogenetically ancient and shared among all mammals. But enhancement is also found in many bird species, particularly the passerines including the corvids (crows); it is not known whether reptiles show evidence of enhancement. As such it is likely that the enhancement seen in birds and mammals are the result of **convergent evolution**. In the animal behavior and comparative literature several motivational (enhancement) states are associated with **social learning**. These include **social enhancement**, **stimulus enhancement**, and **local enhancement**. These different "primitive" enhancement effects serve to orient attention and motivation learning mechanisms and they are often contrasted with **imitation**, considered by many to be a derived cognitive trait in primates (see Subiaul 2007 for a review). *See also* **social learning**.

enhancer A regulatory sequence of **DNA** that positively influences **gene expression**, often from a distance (i.e., the enhancer sequence may be a long way away – in terms of the number of intervening nucleotides – from the gene whose expression it influences). Enhancers do not themselves bind to the **promoter** region of a gene where **transcription** is initiated. Rather, they are chemically bound by activator and repressor proteins that interact with **transcription factors** and RNA polymerase II, which then transcribe the gene.

Enkapune ya Muto (Location 0°50′S, 36°09′E, Kenya; etym. unknown, but the site is also known as Twilight Cave) History and general description Stanley Ambrose directed excavation of this rockshelter located west of Lake Naivasha in the Kenyan section of the **East African Rift System**. An area of 14 m^2 was excavated from 1982 to 1987 and bedrock was reached at a maximum depth of 5.54 m below the surface and the excavations yielded **Iron Age**, **Later Stone Age**, and **Middle Stone Age** artifacts. Sediments include windblown silts and sands and at least three layers of **tephra**, gravels, and anthropogenic deposits such as **hearths** and wood ash. Temporal span and how dated? **Radiocarbon dating** estimates on charcoal, charcoal-impregnated sediment, and ostrich eggshell document presumably modern human occupation of the rockshelter from 41.4±0.7 to 0.5±0.15 ka (the former age is uncalibrated). Age estimates using **obsidian hydration dating** on artifacts from the Later and Middle Stone Age strata, corrected for Pleistocene temperatures −5°C from present, range from 32.46±1.25 to 46.41±2.76 ka. Average sedimentation rates are used to argue for an age for the earliest Later Stone Age of more than 55 ka. Hominins found at site None. Archeological evidence found at site Enkapune ya Muto is an important site for dating Iron Age and **Neolithic** sites in East Africa and these upper strata include a rich faunal assemblage. The Later Stone Age artifact assemblages are divided into the Sakutiek and older Nasampolai industries. The Sakutiek industry includes stone implements typical of the Later Stone Age (e.g., thumbnail **scrapers** and backed **microliths**) and the Middle Stone Age (e.g., discoidal **cores** and facetted platform **flakes**), as well as a range of ostrich eggshell bead fragments demonstrating on-site manufacture, directly dated to 39.9 ± 1.60 ka. The Nasampolai industry is characterized by geometric microliths and large backed blades, many of the latter retaining traces of red **ochre**, suggesting hafting parallel to the long axis of the blade. The flake-based Middle Stone Age Endingi Industry also includes flakes with traces of ochre on them, as well as an ochre-stained lower **grindstone**. Obsidian is the dominant raw material type throughout the sequence. Faunal evidence is rare in the Later and Middle Stone Age strata. Key references: Geology, dating, and paleoenvironment Ambrose 1998; Archeology Ambrose 1998, 2001.

enrichment The isotope enrichment value (ϵ) provides the exact isotopic difference between two substances and is not limited by the isotopic scale (e.g., the Pee Dee Belemnite for carbon and Standard Mean Ocean Water for oxygen) on which it is calculated. The ϵ and Δ (*see* **fractionation**) values are nearly identical when isotopic differences among tissues are approximately 1–2‰, but the difference between the two values increases as the isotopic differences among tissues increases. When tissues are approximately 10‰ different, Δ and ϵ values can differ by as much as 0.5‰ (Cerling and Harris 1999). To calculate ϵ, one first must calculate α, the fractionation factor (*see* **fractionation factor**): $\alpha_{a-b} = (\delta^H X_a + 1000)/(\delta^H X_b + 1000)$ and $\epsilon_{a-b} = (\alpha_{a-b} - 1) \times 1000$. Note that the sign of enrichment is dependent on which substance is in the numerator and which in the denominator.

ENSO Acronym for the **El Niño Southern Oscillation** (*which see*).

entoconid (Gk *ento*=within; *konos*=pine cone) One of the terms proposed by Osborn (1888) for the main cups of mammalian molar teeth. It is the more distal (hence the prefix meaning "within") of the two main lingual cusps of a mandibular (lower) (hence the suffix "-id") molar tooth crown. It forms part of the **talonid** component of a mandibular postcanine tooth crown. (syn. **endoconid, distolingual cusp**, tuberculum posterium internum)

entoconulid A term sometimes used for an accessory cusp situated mesial to the **entoconid** on the occlusal surface of a mandibular molar. The term entoconulid is used by Scott and Turner

(1997) as a synonym for **tuberculum sextum**, but the latter is normally distal and not mesial to the entoconid. *See also* **tuberculum sextum**.

entostylid A term sometimes used for an enamel feature on the occlusal surface of a mandibular molar more commonly referred to as a tuberculum intermedium. *See* **tuberculum intermedium**.

environment (OFr. *environer*=to encircle and L. *mentum*=verb-to-noun suffix) The sum of the biotic and abiotic conditions with which an organism interacts. The term environment is used to refer to processes at different geographical scales (e.g., local, regional, and global). Biotic and abiotic conditions often interact [e.g., local vegetation (biotic) is affected by the amount of groundwater, precipitation, and soil conditions (abiotic)] so the concept of environment is necessarily a complex one.

enzyme (Gk *en*=in and *zume*=leaven; i.e., the factor that causes bread to rise) A protein whose function is to change the rate (i.e., act as a catalyst for) a biochemical reaction. Enzymes are often named after the substrate they act upon or the chemical reaction that they catalyze with an addition of the ending "-ase". For example, lactase is the enzyme that breaks down the sugar lactose and it is the enzyme that is lacking in individuals who cannot digest lactose into glucose and galactose as adults. Thus, individuals with lactose intolerance cannot consume most milk products without discomfort.

***Eoanthropus dawsoni* Dawson and Woodward, 1913** (Gk *eos*=dawn and *anthropos*=human, and *dawsoni*=to acknowledge the role played by Charles Dawson in the recovery of the remains of **Piltdown I**) On December 18, 1912, Charles Dawson and Sir Arthur Smith Woodward read a communication to The Geological Society entitled *On the discovery of a Paleolithic Human Skull and Mandible* and a paper with the same title and with the same authorship appeared in the *Quarterly Journal of the Geological Society* in 1913 (Dawson and Woodward 1913). In that paper the authors proposed "that the Piltdown specimen be regarded as the type of a new genus of the family Hominidæ to be named *Eoanthropus* and defined by its ape-like mandibular symphysis, parallel molar-premolar series, and narrow lower molars which do not decrease in size backwards; to which diagnostic characters may probably be added the steep frontal eminence and slight development of the brow ridges" and they went on to suggest "The species of which the skull and mandible have now been described in detail may be named *Eoanthropus dawsoni*, in honour of its discoverer." *See also* **Piltdown**.

Eocene (Gk *eos*=dawn and *kainos*=new) The Eocene epoch is the second epoch of the **Paleogene** period within the **Cenozoic** era, spanning the interval of time from 55.8 to 33.9 Ma. Rocks of the Eocene series are the global record of Earth history through that interval of time. The Eocene was a "hothouse world" and saw significant diversification of early mammalian orders.

eocone A term introduced by Vanderbroek (1961, 1967) for what more commonly used cusp nomenclature schemes refer to as the **paracone** (*which see*).

eoconid A term introduced by Vanderbroek (1961, 1967) for what more commonly used cusp nomenclature schemes refer to as the **protoconid** (*which see*).

eolith (Gk *eos*=dawn and *lithos*=stone, thus "dawn stone") A term used in the late 19thC and early 20thC to describe minimally modified stones considered to be evidence for hominin tool use in pre-Quaternary sediments. Although most eoliths were subsequently shown to be naturally modified rocks, their study initiated a series of experimental approaches aimed at distinguishing hominin-modified stones from those made by other means (e.g., stream abrasion) that is the basis for contemporary **lithic analysis** (Grayson 1986). *See also* **Piltdown**.

EP Acronym for Eyasi Plateau Paleontological Project and the prefix for fossil hominins recovered from Laetoli, Tanzania by the **Eyasi Plateau Paleontological Project**. *See also* **Laetoli**.

epi- (Gk *epi*=upon, around or "outside of," "close to," or "at the same time") Prefix used for structures that lie on top of, or around, other structures, or for events or effects that occur around the same time (e.g., the epidural space is the space immediately outside/around the dura mater and an epidemic refers to an illness that affects many people more or less simultaneously).

epicone A term introduced by Vanderbroek (1961, 1967) for what more commonly used cusp nomenclature schemes refer to as the **protocone** (*which see*).

epiconid A term introduced by Vanderbroek (1961, 1967) for what more commonly used cusp nomenclature schemes refer to as the **metaconid** (*which see*).

epiconule A term introduced by Vanderbroek (1967) for a small enamel feature on the occlusal surface of a maxillary postcanine tooth crown at the lingual end of the **mesial marginal ridge** close to the tip of the **protocone** (syn. paraconule, **protoconule**, protoconulus, tuberculum accessorium anteriorium internum, mesial accessory tubercle).

epicrista One of the three enamel crests on the **trigon** of maxillary molar teeth; the other crests are the **distal trigon crest** and the **mesial marginal ridge**. The term was introduced by Vanderbroek (1961) for the crest-like structure that runs between the **paracone** and either the **protocone** or the **epicristid**. It is also a term introduced by Hershkovitz (1971) for an enamel structure that others call the middle trigonid crest. *See also* **middle trigonid crest**.

epicristid A term introduced by Hershkovitz (1971) for a structure on the crowns of the mandibular postcanine teeth that others call the **middle trigonid crest** (*which see*).

epicristoconule A term introduced by Grine (1984) for a small cuspule on the occlusal surface of maxillary molars that may be on, or take the place of, an **epicrista**. When present it is either between the **paracone** and the **mesial marginal ridge** or between the former and the **protocone**.

epigenesis (Gk *epi*=around and *genesis*=birth or origin; literally the "influences around development") The generation of form through a series of causal interactions between **genes** and extragenetic epigenetic factors. Epigenetic effects include heritable information (e.g., patterns of gene expression) that is not encoded in the organism's **DNA**. *See also* **epigenetics**.

epigenetic *See* **epigenetics**.

epigenetic landscape A visual metaphor used by Conrad H. Waddington (1957) to illustrate his concepts of **canalization** and **epigenetics**. In the metaphor, the development of an individual organism is likened to a ball rolling down a grooved landscape. At one level the landscape represents the developmental system, at another it represents the probability of different phenotypic outcomes. Genes are likened to guy-wires that underpin the landscape and shape its valleys and ridges. According to this metaphor mutations can shift both the mean (the locations of valleys) and also the variance (the steepness of the sides of the valleys) of the phenotypic outcomes produced by a particular **genotype** or developmental configuration.

epigenetic traits *See* **non-metrical traits**.

epigenetics (Gk *epi*=around and *genesis*= birth or origin; literally the "influences around development") This term is used in two quite distinct ways in the literature. The original Conrad H. Waddington definition of epigenetics is that it subsumes all that is involved in the process of converting the **genotype** into the **phenotype**. Thus, it would include *all* of the factors, intrinsic or external/environmental, acting on and influencing the genetic program for the **development** of an organism. Under this definition intrinsic influences include gene–gene interactions and hormones; external influences include diet and temperature. In modern genetics, the term has acquired a different meaning. In this context, it refers to the influence of various forms of chromatin modification on gene expression. This is a growing field that has relevance to both evolution and human disease and for this reason this latter and more specific meaning of the term has become dominant in the literature. Another way of expressing this contemporary and more exclusive definition of epigenetics is that it is the "study of heritable changes in gene function that *cannot* [our italics] be explained by changes in the DNA sequence" (Russo et al. 1996). Thus, under this more exclusive definition epigenetics embraces information (e.g., patterns of gene expression) that is heritable, but which is not encoded in the DNA of the organism. Thus, as Bird (2007) stresses, without epigenetic mechanisms "hard-won changes in genetic programming could be dissipated and lost" (*ibid*, p. 398). Practical examples of epigenesis include the propensity of offspring to inherit levels of energy extraction from food in the gut. If a previous generation was relatively short of food, their gut is programmed to extract more energy. If this propensity is passed on to a successive generation that has access to potentially excessive amounts of processed, calorie-rich food, and if they continue to extract the maximum amount of energy from the food they ingest, then this can result in high levels of obesity. For a review of this and other similar hypotheses see Gluckman and Hanson (2006). Both meanings of epigenetics are broadly consistent with Waddington's original intent since they both refer to mechanisms that act either directly on genes (or regulatory sequences) or are acting above the level of genes, but the parallel use of the word in these two contexts is a potential source of confusion. *See also* **epigenetic landscape**.

epimerization (Gk *epi*=upon and *meristos*= divided) Epimerization is when there is a change from one of two possible forms of a complex molecule to the other, usually from the L, left-handed, or levo form to the D, right-handed, or dextra form. For example in **ostrich egg shell dating** the change is from L-isoleucine to D-alloisoleucine. That reaction is temperature-dependent, but if average temperature can be estimated independently of sample age, the method can provide dates for material that is difficult to date using other methods.

epiphyseal plate (Gk *epi*=upon and *physis*= growth) The radiotransluscent cartilaginous plate between the **diaphysis** (formed from a primary ossification center) of a long bone and the epiphyses that are formed from the secondary ossification centers. *See also* **endochondral ossification**; **growth plate**; **ossification**.

epiphysis (Gk *epi*=upon and *physis*=growth) The part of a bone that develops from a secondary ossification center. *See also* **ossification**.

episodic memory According to the multiple memory systems model (Tulving 1983) episodic memory, also known as autonoetic or autobiographical memory, is a declarative form of memory that can be explicitly described and articulated. It is also independent of other forms of memory such as the procedural and **semantic memory** systems. Episodic memories encode and structure multiple forms of semantic knowledge or information for a given event, including: *what* (i.e., categorization), *where* (i.e., location), *who* (i.e., individual/s involved), and *when* (i.e., the time when a given event occurred or duration). This information is bound to a context and charged with emotion. These features of episodic memory are what allow us to re-visit the past as well as re-live past events. Tulving has referred to this feature of episodic memory as mental time travel. But just as one can travel to the past, one can also travel to the future by using past episodic experiences to imagine a possible, future world (Tulving 2002). Tulving has long argued that while procedural and semantic memories are phylogenetically ancient memory systems present in all mammals and birds, episodic memory is unique to modern humans. However, there is evidence that birds and apes encode information in an "episodic-like" fashion (Clayton and Dickinson 1998, Schwartz et al. 2002) and that they are capable of encoding *what*, *where*, *when*, and even *who*. *See also* **memory**.

epistasis (Gk *ephistanai* = to place upon, from *epi* = upon and *histanai* = to place) The name given to the interactions among **genes** that influence a **complex trait**. The action of a gene can be affected by one or more modifier genes. For example, the *ABO* locus on chromosome 9 encodes the blood-group antigens that determine blood type. In individuals with the *ABO* secretor allele, these antigens are also expressed in most cells of the body that produce bodily fluids interacting with the environment (e.g., saliva). This allele is encoded by the *FUT2* (fucosyltransferase 2) locus on chromosome 19. However, the antigens are not expressed in individuals called nonsecretors who are homozygous for the nonsecretor *FUT2* allele. Epistasis is also applied to the phenomenon whereby the effect of a **mutation** on the **phenotype** depends on its context (i.e., where in the **genome** it occurs). In classical Mendelian genetics, epistasis referred to non-allelic dominance relationships. Epistasis increases the complexity of adaptive, or fitness, landscapes and thus may have been involved in the evolution of complex morphological and behavioral traits during the course of human evolution.

epistemology (Gk *episte* = to understand and *logos* = to learn) The study of the nature, validity, and scope of knowledge. Discussions of what types of knowledge can, and cannot, be deduced from the hominin fossil record are epistemological discussions. Colloquially, how we know what we know or what we think we know.

epoch (Gk *epokhe* = a point in time) Epoch is used in at least two senses that are relevant to paleoanthropology; both are units of time. As a geochronologic unit an epoch is a subdivision of a **period**, and it is itself subdivided into ages. For example, the **Pliocene**, **Pleistocene**, and **Holocene** are all epochs within the **Tertiary** period. The corresponding **chronostratigraphic unit** equivalent of an epoch is a **series**. In the second sense it is a prolonged period in the past when either a normal or a reversed magnetic field predominated. Superseded by the term "polarity chronozone" which is usually shortened to the word **chron**. *See also* **magnetostratigraphy**.

equid The informal name for the perissodactyl family comprising the horses and their allies. *See* **Perissodactyla**.

Equidae One of the perissodactyl families comprising the horses and their allies, whose members are found at some hominin sites. *See* **Perissodactyla**.

equifinality (L. *aequi* = equal and *finis* = end) The principle that different processes can produce similar, if not identical, patterns or results. The term was coined in 1949 by Ludwig von Bertalanffy in the context of general systems theory. The concept became widely applied in the zooarcheological and taphonomic literature in the 1980s, often in reference to cases where it is unclear which taphonomic process created the patterns observed in a fossil assemblage. A classic case of equifinality is observed in the archeological bone assemblages from **Olduvai Gorge**. The bone assemblages from FLK-*Zinjanthropus* and FLKN levels 1–2 are characterized by an abundance of limb elements and relatively low frequencies of pelves, vertebrae, and ribs. This pattern was originally interpreted as evidence that hominins preferentially transported limbs to the site, while discarding the axial elements at the point of carcass acquisition (*see* **schlepp effect**). This was consistent with the view that early archeological sites were "home bases" or "central-place foraging sites" to which hominins supposedly transported and shared a variety of foods, including meat hunted from large game. Subsequent taphonomic research, however, demonstrated that carnivores preferentially consume and destroy vertebrae, ribs, and pelves (*see* **low-survival elements**; **carnivore ravaging**), calling into question the argument that hominins preferentially transported appendicular elements to these sites. Thus, the dominance of limb elements at FLK-*Zinjanthropus* and FLKN levels 1–2, which could be the result of hominin activities or carnivore ravaging, represents a case of equifinality.

Equus Cave (Location approximately 27°37′S, 24°38′E, northern Cape Province, South Africa; etym. after Latin for horse, named after the abundant equid fossils recovered) <u>History and general description</u> A cave in the western face of the Oxland **tufa** fan on the Gaap escarpment at Bruxton, northern Cape Province. It was initially located as part of a geological study of the long series of tufa deposits, but Peter Beaumont and Myra Shackley were the first to excavate the site in 1978 and Beaumont returned in 1982. The site contains abundant vertebrate fossils in four stratigraphic units designated 1A, 1B, 2A, and 2B. 1A contains **Later Stone Age** artifacts; the others contain **Middle Stone Age** (or MSA) tools. Eight isolated teeth were recovered from units 1B–2B and C.K. **Brain** found a mandibular fragment in 1971 in a scree slope that was the result of a road cutting. <u>Temporal span and how dated?</u> **Late Pleistocene** to **Holocene**. Unit 1A dates to 2.39–7.48 ka by radiocarbon and correlations with nearby tufa deposits suggest that the MSA-containing units 1B–2B likely date to between *c.*33 and 94 ka. <u>Hominins found at site</u> Eight isolated teeth from the MSA levels and a left mandibular corpus with M_2–M_3 that may derive from the same MSA levels. <u>Archeological evidence found at site</u> See above. Key references: <u>Geology, dating, and paleoenvironment</u> Butzer et al. 1978, Grine and Klein 1985; <u>Hominins</u> Grine and Klein 1985; <u>Archeology:</u> Beaumont et al. 1984, Grine and Klein 1985.

ER Acronym for East Rudolf and the prefix for fossils recovered from East Rudolf (now called **Koobi Fora** or East Turkana), Kenya.

era (L. *aera*=counters used for calculations) A unit of geological time (i.e., a **geochronologic unit**). It is next in order of magnitude below an eon (e.g., the Phanerozic eon is divided into three eras, the Paleozoic, Mesozoic, and **Cenozoic**, each

marked by major faunal turnovers). Eras, which usually span hundreds of millions of years, are subdivided into **period**s. The **chronostratigraphic unit** equivalent of an era is an **erathem**.

erathem The chronostratigraphic unit equivalent of an **era**; erathems are subdivided into **system**s.

error *See* **Type I error**; **Type II error**.

eruption (L. *erumpere*=to break up) The process of tooth movement through the alveolar bone, past the gumline (gingival emergence) and into functional position. The assessment of the stage of eruption is typically based on either hard or soft tissue evidence. The hard tissue evidence measures the position of a tooth from a relatively stable landmark (e.g., the bony alveolar margin) usually from radiographs. One key stage in this process is alveolar emergence, when the tip of the tooth passes the level of the margin of the tooth crypt opening in the bony alveolus. Gingival emergence, when the tip of the tooth first pierces the gingival soft-tissues, can only be observed in living subjects. The extrinsic stain on the teeth of many wild collected primates that forms when they erupt into the oral cavity is a way to judge how much of a tooth was subgingival and how much was beyond gingival emergence. Eventually, the tooth reaches the occlusal plane, where it comes into functional occlusion with the teeth in the other jaw and thus begins to play a functional role in **mastication**. Studies of modern humans show there is only *c.*10–12 months between alveolar and gingival eruption of a lower first molar and that the gingival eruption of molar teeth is very close in time to their coming into functional occlusion. It is self-evident that the progress of tooth eruption in the hominin fossil record has to be based on evidence from the hard tissues. Evidence for the age of gingival eruption (notably the first permanent **molar**) has been used to compare the age at death in developing hominin dentitions with those in modern humans and great apes at the same developmental stage and so to predict **life history** schedules in fossil great apes and hominins (e.g., Smith 1986, 1991a, Kelley and Smith 2003). The sequence of tooth eruption has also been reported to distinguish **hominin** taxa from each other and among fossil hominins, great apes, and modern humans (e.g., Dean 1985). Data on tooth eruption ages and sequences in modern humans (and other primates) are given in Smith (1991b), Smith et al. (1994), Godfrey et al. (2001), and Swindler (2002).

ESA Acronym for the **Early Stone Age** (*which see*).

ESC Acronym for the **evolutionary species concept** (*which see*).

Esna Site Deir El-Fakhuri E71K1 (Location 25°22′30″N, 32°28′50″E, Egypt). Locality Site 5 Locality A. Surface/*in situ* Subsurface. Date of discovery 1964–5. History and general description The Late Paleolithic occupation at Deir El-Fakhuri is represented by six sites spread in a rough semicircle slightly to the south of an abandoned Coptic monastery after which the site is named. The monastery is located 11 km/6.8 miles northwest of Esna. On the western slope of the E71K1 site, heavily weathered and partially disarticulated modern human remains were found eroding out on the surface. Finder Martin A. Baumhoff. Unit N/A. Horizon N/A. Bed/member N/A. Formation Dibeira-Jer silts. Group N/A. Nearest dated horizon Samples of *Unio* shell collected from the artifact-bearing levels at E71K1 gave a date of 16,070±330 years. Archeological evidence found at site The assemblage has many microliths as well as denticulates, retouched pieces, end scrapers, double-backed perforators, and bladelets with Ouchtata retouch. The presence of both chipping debris and faunal remains indicates that the site was used both as a workshop and a living site. The assemblage is considered to be part of the Fakhurian industry. Temporal span and how dated? **Radiocarbon dating** of the related artifact-bearing level suggests that the site is between *c*.24 ka and *c*.12 ka but the deposition of the Dibeira-Jer silts began before *c*.24 ka and ended shortly after *c*.18 ka. Developmental age The cranial and upper limb material belong to a young adult (about 18 years old) and the pelvis and lower limb bones belong to a more mature individual. Presumed sex The more mature individual is probably male. Brief anatomical description The skeletal material from site E71K1 represents the remains of a minimum of two individuals. The skeletal material recovered is quite fragmentary. The cranial fragments appear to be morphologically similar to the juveniles of a similar age from the Sahaba series. Initial taxonomic allocation *Homo sapiens*. Announcement Butler 1974. Initial description Butler 1974. Photographs/line drawings and metrical data Butler 1974. Geology, archeology, and dating Buckley and Willis 1969 (I-3416), Lubel 1974. Detailed anatomical description Butler 1974. Location of original Department of Ancient Egypt and Sudan, British Museum, London, England.

ESR Abbreviation for **electron spin resonance spectroscopy dating** (*which see*).

essential crest A term employed to indicate the most prominent enamel ridge running down from the incisal edge of the incisors or from the cusp tip of a canine or from a cusp on a post-canine tooth crown. (syn. essential ridge).

essential ridge *See* **essential crest**.

estrogens (New L. *estrus*=period of fertility plus *gen*=to generate) A class of steroids with 18 carbon atoms ("C18 steroids") that includes estrone, estriol, and estradiol. Estrogens are all produced from androgen precursors through enzymic action and this conversion primarily takes place in the ovaries and peripheral tissues. Estradiol is the main estrogen during premenopausal life in women and estrone, a weaker estrogen, is more plentiful in postmenopausal women. Critical for female reproductive function and key in the development of female secondary sexual characteristics, estrogens are also believed to exert a protective effect on neural and other tissues. However, excessively high levels of estrogens and so-called "environmental estrogens" have been linked to unusually high incidences of cancer.

Ethiopian rift system The Ethiopian section of the East African Rift valley comprises three main components, the **Afar Rift System** (also known as the **Afar Depression**) in the northeast, the **Main Ethiopian Rift System** in the middle, and what some call the **Omo rift zone** and others refer to a "300-km-wide rifted zone" (Renne et al. 1999, p. 869) in the southwest of the country. The Ethiopian rift system contains many productive hominin sites, including those in the **Dikika**, **Gona**, **Hadar**, and **Middle Awash study areas** in the Afar Rift, **Chorora**, **Gadeb**, **Gademotta**, **Kesem-Kebena**, **Konso-Gardula**, and **Melka Kunturé** in the Main Ethiopian rift, and **Fejej**, **Omo-Shungura**, and **Usno Formation** in the Omo Rift Zone.

ethology (Gk ethos=*character* and *logia*= study) The scientific study of animal behavior. This discipline is important to paleoanthropologists because primate ethology generates information about the behavior of living primates (and sometimes other mammals) and thus provides the context for developing hypotheses about the behavior of early hominins and the evolution of modern human behavior.

euchromatin *See* **chromosome**.

Euclidean distance A multivariate measure of distance that can be applied to the distance between objects in space or the morphological distance between objects or groups. Euclidean distance is equal to the square root of the sum of squared differences in all variables. For example, the spatial distance between two specimens is equal to the square root of $(X_1-X_2)^2+(Y_1-Y_2)^2+(Z_1-Z_2)^2$. The morphological distance between two specimens with measurements in four variables A, B, C, and D is equal to the square root of $(A_1-A_2)^2+(B_1-B_2)^2+(C_1-C_2)^2+(D_1-D_2)^2$. However, morphological distance is more typically measured using distances that take into account the standard deviation of variables and their correlation with each other (e.g., **Mahalanobis distance**). *See also* **Mahalanobis distance**.

Euclidean distance matrix analysis (or EDMA) A technique for comparing the two- or three-dimensional **shape** and **form** of objects such as crania, mandibles, and postcranial elements (Lele and Richtsmeier 1991). In EDMA a series of **landmark**s is identified for an object and the form of that object is represented by a square **matrix** of pairwise Euclidean distances between landmarks. The rows and columns of the matrix are the landmarks and the value for each cell in the matrix corresponds to the physical distance in two- or three-dimensional space between that pair of landmarks. For example, consider an analysis consisting solely of two landmarks: **prosthion** and **inion**. The upper-left-most cell (row 1, column 1) will contain the Euclidean distance between prosthion and itself (i.e., zero). Both the next cell to the right (row 1, column 2) and the cell directly below (row 2, column 1) will contain the distance between prosthion and inion. The final cell (row 2, column 2) will contain the distance between inion and itself (again, zero). Thus the matrix for any object will contain only zeros along its diagonal, and all of the values below the diagonal will be duplicated above the diagonal. As a consequence, EDMA typically only considers those cells either below or above the diagonal (i.e., a triangular matrix). The resulting matrix is typically referred to as a form matrix. Form matrices for multiple objects can be analyzed in a variety of ways. Often, the form or shape is compared between two groups of objects [e.g., between male and female articular surfaces (Lague and Jungers 1999) or between crania of two taxa (Krovitz 2003)]. An estimated form matrix is calculated for each group, which can be thought of as an average of the form matrices for each individual object within a group. Form difference matrices are calculated as the ratio of the value in each cell for the estimated form matrices in each group (e.g., in the cell corresponding to prosthion–inion distance, the form difference matrix will contain the ratio of the average prosthion–inion distance from the first group to the average prosthion–inion distance from the second group). Shape matrices can also be calculated by dividing a form matrix by a scaling factor [i.e., a measure of overall size of the object(s) represented in the matrix]. This scaling factor is typically the **geometric mean** of all pairwise distances in the matrix. Shape difference matrices are calculated as the difference between two shape matrices. Continuing the example above in the cell corresponding to prosthion–inion distance the shape difference matrix will contain the scaled average prosthion–inion distance from the first group minus the scaled average distance from the second group. A variety of statistical comparisons of shape and form (e.g., testing whether the form of one group is a scaled-up version of the form of another group, identifying influential landmarks that distinguish two shapes, etc.) are available using these matrices. Hypothesis tests in EDMA typically rely on **bootstrap** procedures to generate confidence intervals for scaling ratios, difference values, etc. It should be noted that there is a debate in the literature between proponents of EDMA and **Procrustes analysis** regarding the validity and/or appropriateness of these various procedures, focusing in particular on (a) the power and Type I error rates of the methods, (b) the consistency within a method for estimating mean form and/or shape, and (c) any bias in shape estimation (e.g., Rohlf 2000, 2003, Lele and Richtsmeier 2001).

eukaryote (Gk *eu*=good or true, *karyote*=the inside, or kernel, of a nut) A single-cell or multicellular organism where each cell contains at least one membrane-bound nucleus usually containing multiple different chromosomes and other membrane-bound organelles such as **mitochondria** and Golgi bodies. Eukaryotic organisms include animals, plants, fungi, and protists.

Euler–Lotka equation   The problem of how to relate the proportional growth rate of a population to the characteristic functions that define life history was originally solved by Euler (1760) and rediscovered in the context of modern population genetics by Lotka (1907). This equation, considered to be the central equation relating to life history evolution and demography, specifies the relationships of age at maturity (α), age at last reproduction (ω), probability of survival to a given age class (l_x), and number of offspring expected in a given age class (m_x), to the intrinsic rate of increase (r), which measures the potential for growth rate in a population. The Euler–Lotka equation is one way to estimate r; using life tables based on the Euler–Lotka equation is another. The Euler–Lotka equation is especially useful because it provides a way to weight contributions of different age classes to overall fitness, although one of the assumptions for solving the equation is that there is a stable age distribution within a population.

European hypothesis   *See Homo heidelbergensis* **Schoetensack, 1908**.

European mammal neogene   A system that uses **biozones** for dating **Miocene** and **Pliocene** sites within the age range of 24 to more than 2 Ma. The original system focused on sites in the **Neogene** (Mein 1975), but it has since been updated (Mein 1990) and is constantly being refined (e.g., Agusti 1999). The zones of the European mammal neogene (or MN) system run from MN1, the oldest, at the beginning of the Miocene, to MN17 in the **Plio-Pleistocene**. Some of the MN zones span several millions of years (e.g., MN5 between *c.*17 and *c.*14 Ma), whereas others span less than a million years (e.g., MN15 between 4 and 3.5 Ma). Each zone has one or more sample localities containing the taxa characteristic for that zone (see Steininger et al. 1989, 1996). The MN system is replacing the classical system of European mammalian biozonations (e.g., **Vallesian, Turolian, Ruscinian, Villafranchian, Cromerian**, etc.). *See also* **biostratigraphy**.

eurytope   (Gk *eurus*=broad or wide and *typos* = place) A **species** adapted to a broad range of environmental conditions or to a broad ecological **niche** (e.g., bush pig), or having **morphology** that is interpreted as being part of a eurytopic adaptation (e.g., the molars of a bush pig). The term was introduced to paleoanthropology by Elisabeth Vrba (e.g., Vrba 1980, 1985a, 1985b, 1988, 1992) in a series of hypotheses that sought to explain how environmental change influences **macroevolution**ary patterns in the fossil record. Several recent studies (Teaford and Ungar 2000, Wood and Strait 2004, Scott et al. 2005, Sponheimer et al. 2005) have concluded that most early hominin species were ecological generalists rather than specialists, particularly with respect to diet. *See also* **effect hypothesis; turnover-pulse hypothesis; habitat theory hypothesis** (syn. **generalist**).

eurytopic   *See* **eurytope**.

eurytopy   *See* **eurytope**.

Eustachian tube   The endodermally derived first pharyngeal pouch that connects the middle ear to the nasopharynx. *See also* **auditory tube**.

event   (L. *evenire*=to happen) A short period of the opposite field direction during a prolonged period when either a normal or a reversed magnetic field predominated (e.g., the **Olduvai event** is a normal event within the Matuyama reversed chron, previously referred to as an **epoch**). Superseded by the term **subchron**.

eversion   Foot movement around an anteroposterior axis in which the sole of the foot turns outward away from the midline of the body. Often incorrectly used interchangeably with pronation (see **foot movements**), eversion refers only to rotation about the long (anteroposterior) axis of the foot. Eversion occurs primarily at the subtalar joint but movement at the talocrural (ankle) joint can be incorporated in extreme eversion. Eversion is limited by the deltoid ligament, which is robust in modern humans. *See also* **foot movements**.

Evidence as to Man's Place in Nature
This book by Thomas Henry Huxley, first published in London in 1863, is arguably the first scientific

book about human origins. It consists of three essays: *On the Natural History of the Man-like Apes*, *On the Relations of Man to the Lower Animals*, and *On Some Fossil Remains of Man*. The first essay considers and evaluates reports of sightings of what we now refer to as chimpanzees and bonobos, gorillas, orangutans, and gibbons, and in that section Huxley pays a fulsome tribute to Richard Owen's 1835 memoir *On the Osteology of the Chimpanzee and Orang*. The second essay addresses what Huxley refers to as "The question of questions for mankind-the problem which underlies all others, and is more deeply interesting than any other-is the ascertainment of the place which Man occupies in nature and of his relations to the universe of things" (p. 57) and in this essay Huxley writes that he "set(s) forth...the chief facts upon which all conclusions respecting the nature and the extent of the bonds which connect man with the brute world must be based" (p. 59). He reviews the evidence that existed at the time, including observations about limb proportions and the teeth which he describes as "organs which have a peculiar classificatory value, and whose resemblances and differences of number form and succession, taken as a whole, are usually regarded as more trustworthy indicators of affinity than any others" (p. 81). As for the brain, he writes of "the impossibility of erecting any cerebral barrier between man and the apes" (p. 96). The best-known quotations that reflect Huxley's assessment of the morphological evidence are "Whatever part of the animal fabric-whatever series of muscles, whatever viscera might be selected for comparison-the result would be the same-the lower apes and the Gorilla would differ more than the Gorilla and Man" (p. 84), and "Thus, whatever system of organs be studied, the comparison of their modifications in the ape series leads to one and the same result-that the structural differences which separate Man from the Gorilla and the Chimpanzee are not so great as those which separate the Gorilla from the lower apes" (p. 103). It is as a postlude to this second essay that Huxley rounds on Richard Owen for claiming that the "posterior horn of the lateral ventricle" and the "hippocampus minor" are peculiar to the genus *Homo*. The third essay discusses the child's cranium from **Engis** and the adult cranium from the **Kleine Feldhofer Grotte**, which Huxley refers to as "the Neanderthal cavern." Huxley's analysis of the two fossil crania is perceptive and prescient. He suggests that even though the Neanderthal remains are "the most pithecoid of known human

skulls" he goes on to write that "in no sense...can the Neanderthal bones be regarded as the remains of a human being intermediate between Man and Apes" and he notes that if we want to seek "the fossilized bones of an Ape more anthropoid, or a Man more pithecoid" than the Neanderthal cranium, then researchers need to look "in still older strata" (Huxley, 1863, p. 159). Huxley is also the first writer to propose objective criteria for comparing crania, and he writes that "I have arrived at the conviction that no comparison of crania is worth very much, that is not founded upon the establishment of a relatively fixed base line, to which measurements, in all cases, must be referred. Nor do I think it is a very difficult matter to decide what that base line should be. The parts of the skull, like those of the rest of the animal framework, are developed in succession: the base of the skull is formed before its sides and roof; it is converted into cartilage earlier and more completely than the sides and roof: and the cartilaginous base ossifies, and becomes soldered into one piece long before the roof. I conceive then that the base of the skull may be demonstrated developmentally to be its relatively fixed part, the roof and sides being relatively moveable" (*ibid*, p. 148). Moreover, the "cranio-facial" angle he refers to (*ibid*, p. 150) and illustrates (*ibid*, p. 151) is effectively the same as the **cranial base angle** that is still in use today. *See also* **basicranium**.

evo-devo Abbreviation of **evolutionary developmental biology** (*which see*).

evolutionary developmental biology Often abbreviated to "evo-devo," this term was first used by Brian Hall (1992), but interest in the relationship between development and evolution has a long history. Within evolutionary biology, interest in development as the missing element of the modern synthesis reached a turning point in the 1970s and 1980s with works by Gould, Raff and Kaufman, Hall, and others (e.g., Hall 1975, Gould 1977, Raff and Kaufman 1983). To some, the modern discipline of evolutionary developmental biology is mainly concerned with the ways in which development relates to evolutionary explanation (Hendrikse et al. 2007), while others regard it more inclusively as being about the evolution of developmental mechanisms. Either way, the field of evo-devo represents a novel synthesis of evolutionary and developmental biology that has had tremendous influence on both fields and it has the potential to illuminate the molecular basis of the important morphological changes that have made the hominin clade distinctive.

evolutionary scenario   (L. *scaena*=scene) A complex hypothesis that attempts to explain why a phylogeny looks the way it does. It provides explanations for speciation and extinction events and for patterns of morphological evolution (e.g., the **endurance running hypothesis**).

evolutionary species concept   (or ESC) An attempt by Simpson (1961) to add a temporal dimension to the **biological species concept**. Simpson suggested that under the evolutionary species concept a species is "an ancestral-descendant sequence of populations evolving separately from others and with its own evolutionary role and tendencies". *See also* **species**.

evolutionary taxonomy   A major approach to **classification**. Like the rest of biology, the discipline of **taxonomy** was transformed by the work of Charles **Darwin**. During the first half of the 19thC, most naturalists held that the evidently hierarchical organization of **taxa** revealed the plan of a divine creator. In *On the Origin of Species* Darwin argued that this view was unsatisfactory because it added nothing to knowledge and he offered an alternative interpretation of the fact that **species** could be grouped into **genera** and genera into higher-level taxa on the basis of phenotypic similarities and differences. He suggested that the so-called "natural order" that previous generations of naturalists had uncovered was exactly what would be expected if species had evolved by descent with modification. Evolutionary taxonomy is one of the approaches that taxonomists have developed in response to Darwin's insight. The other is **phylogenetic systematics**. Both approaches hold that classifications should reflect descent. However, they differ over the validity of **paraphyletic** taxa and the importance of incorporating information about adaptation into classifications. Phylogenetic systematics insists that taxonomic groups include all the descendants of the most recent common ancestor of that group and therefore rejects paraphyletic taxa. In addition, information about **adaptation** plays no part in the creation of taxa in phylogenetic systematics. In contrast, evo-

lutionary taxonomy recognizes that one or more of the descendants of a given ancestor may be so adaptively different (i.e., they are in a different **grade**) from the ancestor or the other descendants that they warrant allocation to a distinct group. Thus, unlike phylogenetic systematics, evolutionary taxonomy accepts both **monophyletic** and paraphyletic taxa.

evolvability   (Ge. *Evolutionsfähigkeit*=the "ability to evolve") This term, coined by Wagner (1986), refers to the degree to which a species or structure responds to natural selection with evolutionary change. Key determinants of evolvability are the extent to which genetic variation is expressed at the phenotypic level, the rate of **mutation**, and the extent to which variation in the trait under selection is correlated with variation in other traits. Evolvability is a key concept in quantitative genetic theory. It is also key to **evolutionary developmental biology** because the determinants of evolvability themselves have a basis in development. *See also* **canalization**; **integration**; **modularity**.

exact randomization   A form of **randomization** that computes exactly the right number of data combinations to test a hypothesis. *See* **randomization**; **resampling**.

exact test   *See* **exact randomization**.

exaptation   A useful feature or **trait** that did not originate as a direct consequence of **natural selection** acting on its current function. Rather, the trait may have been co-opted to perform its current functional role *after* having been initially nonfunctional or adapted (through natural selection) to perform a different role (e.g., feathers most likely evolved in the ancestors of birds as an **adaptation** related to thermoregulation, but, once present, feathers were co-opted for flight and display and thus can be thought of as exaptations for those two functions). A complication, of course, is that once the exaptation has been co-opted, natural selection may act upon it to modify its form and function. Thus, modern birds have experienced considerable evolution in feather and, more broadly, wing form related to locomotion, and these changes can legitimately be thought of as adaptations. Thus, the presence of feathers is an exaptation to flight, but the modifications to feathers associated with different types of flight (e.g., long distance "cruising" and agile maneuvering through a dense forest canopy) are adaptations. In some cases, therefore, adaptation and exaptation can be hierarchically nested concepts. An example of an exaptation in modern humans and later hominins is the opposable thumb. A mobile, grasping thumb evolved in our earliest primate ancestors as an adaptation for grasping branches, and that basic design, plus subtle but important modifications, was co-opted for tool use by later members of the hominin clade. Exaptation explains the related term **preadaptation**, which describes the phenomenon in which a trait appears to be well designed for a functional role it does not currently perform. The term exaptation was introduced to the lexicon of evolutionary biology by Stephen J. Gould and Elisabeth Vrba (Gould and Vrba 1982). In historical context, this contribution can be seen as part of a broader critique of what Gould and Lewontin (1979) called the "adaptationist programme," namely the tendency for evolutionary biologists to ascribe to natural selection a nearly limitless ability to explain the evolution of form.

exon (L. *ex*=to move out) The portion of a **gene** that undergoes **transcription** and eventually forms part of a functional messenger **RNA** molecule (as opposed to the **intron** part of a gene, which is excised from the RNA). The mature RNA may then move into the cytoplasm and be translated into a **protein**. Portions of this mature RNA include the $5'$ and $3'$ untranslated regions, which help guide and/or regulate the process of **translation**. Other exonic sequences may be directly transcribed into one of the many different varieties of structural and/or functional RNAs (e.g., tRNA, rRNA, siRNA, miRNA, and snRNA).

expensive tissue hypothesis (or ETH) Proposed by Leslie Aiello and Peter Wheeler in 1995 to suggest that the metabolic requirements of relatively large brains are offset by a corresponding reduction of the gut. Because gut size is highly correlated with diet, relatively small guts imply high-quality, easy-to-digest food. This led to the conclusion that brain-size increase in hominin evolution could not have been achieved without incorporating increasing amounts of high-quality animal-based foods in the diet, irrespective of the selective factors resulting in large hominin brains.

Dietary change was proposed to have occurred with the appearance of *Homo ergaster* approximately 1.7 Ma. This is based on the claimed increase in brain size, the smaller teeth and jaws, and the emergence of more modern human-like body proportions implying a smaller gut and the increased evidence for scavenging and/or hunting in the archeological record. The basic premise of an energetic trade-off with brain size has stood the test of time. However, new research suggests that it might involve features other than, or in addition to, diet and gut size, such as muscle mass or locomotor efficiency. The expensive brain hypothesis allowed Isler and van Schaik (2009) to provide a unifying explanatory framework, arguing that the costs of a relatively large brain in mammals in general must be met by any combination of increased total energy turnover or reduced energy allocation to another expensive function such as digestion, locomotion, or production (growth and reproduction). Application to the fossil record has also been complicated by evidence for animal-based foods in the diet predating significant brain expansion, the suggestion that cooked food played an important role in hominin brain evolution, and new interpretations of the body proportions of early *Homo*. However, these factors do not diminish the importance of the ETH in focusing research on energetic trade-offs in hominin evolution.

exploratory analysis The process of using multiple statistical techniques to identify patterns in a data set, as distinct from **hypothesis testing** in which a specific relationship which has been identified *a priori* is tested.

exponential population growth A model of population growth in which a population increases (or decreases) in size by a constant factor per unit time. The simplest equation for this model is $N_{t+1} = N_t \exp(r)$. If $r=0$, then it is a **stationary population** (i.e., neither increasing nor decreasing in size) because $\exp(0)=1.0$. If $r>0$, then the population is increasing in size whereas if $r<0$ then the population is decreasing in size.

exposure (L. *exponere*=to set forth) A place where cross-sections of **strata** have been exposed by erosion or by faulting and thus these are places where the fossils contained in those strata are also partly or fully exposed to view (syn. **outcrop**).

expressivity When different levels of expression of a **gene** are reflected in differences in the **phenotype** and specifically when individuals with the same **genotype** vary in their phenotype. For example, two individuals who have the same genotype for a gene influencing weight may differ in their weight (i.e., their phenotypes differ) because the expression of the gene could be affected by factors such age, diet, illness (i.e., epigenetic factors), sex, or other genes.

extant (L. *ex*=out and *stare*=stand) Means literally "still standing" and refers to taxa that are alive today (syn. living, surviving).

extant species One of the many terms used to describe a contemporary **species** that is defined on the basis of criteria that can be observed, such as evidence of interbreeding, or the lack of any evidence of interbreeding between individuals belonging to that species and individuals belonging to any other species. *See also* **species** (syn. **neontological species, biological species**).

extension *See* **polymerase chain reaction**.

extension rate The rate at which fully differentiated **ameloblasts** or **odontoblasts** become active secretory cells (Shellis 1984). Extension rates are usually measured along the **enamel-dentine junction** (or EDJ) in longitudinal sections of teeth. Initial extension rates close to the cusp tip are fast (approximately 20–30 μm/day) in all tooth types, but the rate falls to approximately 3–6 μm/day in the cervical two-thirds of the crown. Extension rates in the longitudinal plane of section of a tooth relate directly to the time taken to form a tooth from dentine horn to enamel cervix or onwards through root extension rates to apex completion of the root tip. However, circumferential expansion (or reduction) in tooth crown or root diameter also occurs. The initial spread or extension of ameloblasts over the occlusal surface of a tooth as cusps form and coalesce is very fast. Newly differentiated ameloblasts in the crown must, therefore, also feed into circumferential extension (or out of it and into tooth length extension). The rate of extension is inferred from the angle between the **long-period incremental lines** or **accentuated lines** in the enamel and the dentine and the enamel-dentine junction; low angles are taken to infer a rapid extension rate

and high angles a relatively slow extension rate (Boyde 1964, Shellis 1984, Beynon and Wood 1986) or it may be calculated directly by inspection. *See also* **dentine development; enamel development**.

external auditory canal *See* **external ear**.

external auditory meatus (L. *ex*=outer, *audire*=to hear, and *meatus*=passage) The passage between the ear lobe (auricle or pinna) laterally and the tympanic membrane medially. The lateral one-third is made of a U-shaped fibrocartilaginous tube that is continuous laterally with the elastic cartilage of the pinna. The medial two-thirds is the bony part of the external auditory meatus. Two components of the temporal bone contribute to the margin of the lateral end of the bony external auditory meatus. The superior part of the meatal margin is formed from the squamous part of the temporal bone and the inferior part from the tympanic plate of the temporal bone. Some researchers have suggested that the size and shape of lateral end of the osseous part of the external auditory meatus are useful for discriminating among early hominin taxa (e.g., *Australopithecus anamensis* or *Kenyanthropus platyops*, Leakey *et. al.* 1998).

external ear (L. *ex*=outer) The external ear is made up of the earlobe (auricle or pinna) and the **external auditory meatus**. The skeleton of the lateral one-third of the external auditory meatus is fibrocartilage and the pinna is made up of elastic cartilage; neither of these structures fossilizes. The medial two-thirds of the external auditory canal is bony and is made up of contributions from the squamous and tympanic plate components of the temporal bone. The lateral end of the bony part of the external auditory meatus is called the osseous auditory meatus.

extinct (L. *ex*=out and *stinguere*=to put out, or quench) The term means literally "to extinguish" and it refers to all taxa (e.g., *Australopithecus afarensis*) that are no longer alive.

extractive foraging The name for the sequence of behaviors that enables organisms to extract embedded foodstuffs (e.g., fruits enclosed in hard shells, underground storage organs buried in the soil, marrow within the trabecular cavities of bones, or the soft tissues within the exoskeleton of insects) to remove the edible or palatable parts (King 1986). In **hominin** evolution tool use is seen as being key to this behavior (Gibson 1986) although extractive foraging can take place in nonhominins without using **tools** (e.g., the aye-aye). The importance of extractive foraging has been discussed widely in the primate **cognition** literature (e.g., Gibson 1986, King 1986) as it has been argued that it takes greater mental ability to remove or process embedded foods than foods that are not embedded. Thus, extractive foraging might act as a selective pressure for the evolution of the brain and cognition in primates and hominins (see references in Elton et al. 2001). Of the nonhuman primates, **chimpanzees**, capuchins, and the aye-aye all use extractive foraging and they are also relatively highly encephalized (Sterling and Povinelli 1999). However, there are a number of limitations to the extractive foraging model of primate, and perhaps hominin, intelligence. One of the most important of these is the disjunction between observations of extractive foraging and measured intelligence in a range of animals and birds (King 1986, Russon, 2004). Nonetheless, the hypothesis that extractive foraging was central to hominin food procurement and allowed hominins access to a wider range of foods, enabling competition with other members of their ecological communities, is an attractive one.

extrastriate cortex (L. *striatus*=furrowed, striped, ridged and *cortex*=husk, shell) Refers to parts of the cortex of the occipital and parietal lobes of the cerebral hemispheres other than the primary visual cortex (also known as the **striate cortex** in primates). These regions correspond to areas 18 and 19 in the nomenclature of **Brodmann's areas**. In modern parcellation schemes based on **cytoarchitecture** and physiology, the extrastriate cortex of primates includes a large number of distinct areas that contain both complete and partial representations of the retina, which each process different features of visual stimuli (e.g., movement, color, contour). One of the main signs of brain reorganization in hominin evolution involves the relative size of the striate and extrastriate cortex. The **lunate sulcus** forms the anterior boundary of the striate cortex (or primary visual cortex) and it has been proposed that the more posterior displacement of the lunate sulcus in some hominins compared to its location

in chimpanzees and other great apes indicates a reduction in the proportion of striate cortex (or primary visual cortex) relative to the extrastriate cortex and the **posterior parietal cortex**.

Eyasi (Location 3°32′26″S, 35°16′05″E, Tanzania; etym. after Lake Eyasi, which borders the site) History and general description Eyasi comprises at least two fossiliferous localities on the shore of Lake Eyasi. **Kohl-Larsen** collected at the locality in the 1930s, recovering fossil evidence of three partial hominin crania. Subsequently, it has been revisited several times by researchers trying to clarify the stratigraphy. Some of these studies have uncovered additional fragmentary hominin cranial fragments to make a total of seven hominins. **Eyasi 1** to Eyasi 3 were recovered by Kohl-Larsen, Eyasi 4–5 (two undescribed mandibles) by C.A. Keilland and J. Ikeda, Eyasi 6 by A. Mabulla, and Eyasi 7 by Domínguez-Rodrigo and colleagues. Temporal span and how dated? Late **Middle** or early Upper **Pleistocene**. Domínguez-Rodrigo et al. (2008) obtained **electron spin resonance spectroscopy** (104 ± 13–207 ± 26 ka) and **uranium-series** (^{230}Th/^{234}U) (92.4 ± 4.1–138.3 ± 0.7 ka) ages for a wildebeest tooth found 5 m from the hominin frontal. Hominins found at site Eyasi 1, a fragmentary cranium, Eyasi 2, a fragmentary occipital, Eyasi 3, cranial fragments and teeth, Eyasi 4 and 5, two mandibles, Eyasi 6 (NB: Eyasi IV of Bräuer and Mabulla 1996), an occipital fragment, and Eyasi 7, a frontal. All of the hominins have been assigned to archaic *Homo sapiens*. Archeological evidence found at site An early **Sangoan** or **Njarasan** late **Acheulean** industry as well as an overlying **Middle Stone Age** industry. Key references: Geology, dating, and paleoenvironment Kohl-Larsen and Reck 1936, Mehlman 1987, Manega 1993, Domínguez-Rodrigo et al. 2007b, 2008; Hominins Leakey 1936a, Weinert 1939, Kohl-Larsen 1943, Bräuer 1984, Bräuer and Mabulla 1996, Domínguez-Rodrigo et al. 2008; Archeology Kohl-Larsen 1943, Mehlman 1984, 1987, 1989, Domínguez-Rodrigo et al. 2007b.

Eyasi 1 Site Eyasi. Locality N/A. Surface/*in situ* Surface. Date of discovery November 29, 1935. Finder N/A. Unit N/A. Horizon N/A. Bed/member N/A. Formation N/A. Group N/A. Nearest overlying dated horizon N/A. Nearest underlying dated horizon N/A. Geological age *c*.100 ka. Developmental age Adult. Presumed sex Female. Brief anatomical description Most of the occipital and much of the left parietal (but many fragments are not contiguous), some of the left temporal, and parts of the frontal including the right supraorbital margin. Announcement Kohl-Larsen and Reck 1936. Initial description Leakey 1936. Photographs/line drawings and metrical data Kohl-Larsen and Reck 1936, Weinert 1939, Schwartz and Tattersall 2003. Detailed anatomical description Weinert 1939. Initial taxonomic allocation *Palaeoanthropus njarasensis*. Taxonomic revisions "*Pithecanthropus*" (Leakey 1936), *Africanthropus njarasensis* (Weinert 1939). Current conventional taxonomic allocation *Homo* sp. Informal taxonomic category Pre-modern *Homo*. Significance The holotype of *Pa. njarasensis* and *Afri. njarasensis*. Location of original Eberhard-Karls-Universität Tübingen, Institut für Ur- und Frühgeschichte, Tübingen, Germany.

Eyasi Plateau Paleontological Project A multidisciplinary field project investigating the geology and paleontology of the fossil hominin site at **Laetoli** and at other paleontological localities on the Eyasi Plateau to the north of Lake Eyasi.

The project, which was initiated in 1998, is directed by Terry Harrison and organized in collaboration with the **National Museum of Tanzania**. The major aims of the project are to recover additional fossil hominins and to obtain more detailed information on their paleoecological and geological context. In addition to new specimens of *Australopithecus afarensis* from the Upper Laetoli Beds, the project has recovered the first fossil hominins from the Upper Ndolanya Beds attributed to *Paranthropus æthiopicus*. Study of the fossil hominins has been accompanied by research on the fossil vertebrates, invertebrates, and plants (more than 22,000 specimens have been collected since 1998), as well as on the geology, geochronology, **palynology**, **stable isotopes**, **taphonomy**, and modern ecology, by a team of more than 50 specialists. Further investigations of the fossil footprints at Laetoli, initially discovered and excavated by Mary **Leakey**, have not been a primary focus of this project, but an independent team of researchers, led by Charles Musiba, has continued this effort.

F

F. One of the prefixes used by the US contingent of the **International Omo Research Expedition** for localities in the Omo **Shungura Formation** (e.g., F. 511). The prefix is also used for fossils found within that locality (e.g., F. 511-16).

F statistics *F* statistics (also known as fixation indices) were developed by the population geneticist Sewall Wright to measure the effects of various types of population subdivision. *F* statistics describe the level of **heterozygosity** in a population and they can be used as a measure of **genetic distance**. The *F* statistics are derived from *F*, called the inbreeding coefficient. The three types of *F* statistic are F_{ST} (which compares the heterozygosity of the subpopulation to that of the total population), F_{IS} (which compares the heterozygosity of an individual to that of the subpopulation), and F_{IT} (which compares the heterozygosity of an individual to that of the total population). The three types of *F* statistic are related to one another through the equation $1-F_{IT}=(1-F_{IS})(1-F_{ST})$. A high F_{ST} means that (a) homozygosity in the subpopulation is high relative to the total population, (b) heterozygosity within the subpopulation is low, (c) inbreeding within the subpopulation is high, (d) there is high genetic differentiation (or structure) between subpopulations, (e) genetic drift is high, and (f) genetic distance is large. In modern humans, F_{ST} is typically approximately 9–15%, which is a relatively low value compared with other animal species.

F test A statistical test applied to a ratio of two measures of variability to identify whether they differ from each other. *F* tests have two different degrees of freedom that are associated with the numerator and the denominator. The most common application of an *F* test is in testing for the significance of linear models such as **regression** or **analysis of variance**. The *F* test is used to identify whether the **variance** of observed values for the dependent variable around their estimated values is significantly less than the variance of the observed values about their mean; that is, does the addition of the independent variables (be they continuous, categorical, or both) significantly improve the ability to predict particular observations relative to predictions that are simply the mean of the dependent variable? In a similar way, *F* tests can be used to determine whether the addition of one or more variables to a regression model significantly improves the fit of the model by analyzing the difference in the residual error produced by the regression with and without the extra variable(s). An *F* test can also be used to test whether two variances are equal (e.g., comparing variance in femoral head size in gorillas and a sample of fossil hominins), but such comparisons are highly sensitive to non-normality in the distribution of the data. Instead, alternative tests for the equality of variance such as the **Levene test** should be applied.

face (L. *facies*=face) One of the three main components of the **cranium**; the others are the **vault**, or **calotte**, and the **basicranium**. The unpaired bones making up the face are the anterior part of the frontal (the flat, or squamous, part of the frontal belongs to the vault), the **ethmoid**, and the **vomer**; the paired bones are the **nasal, maxillary, palatine,** and **zygomatic**. Rak (1983) lamented the lack of an authoritative definition of the face, and suggests that a good working definition is that it is the part of the cranium that extends from the alveolar plane of the upper jaw superiorly and posteriorly to the point of minimum frontal breadth. Thus, in this definition, the face includes the supraorbital region of the frontal bone.

facetted platform A platform on a flake that has multiple facets resulting from previous flake removals undertaken to shape or modify the **striking platform**. Many (but not all) **Levallois** flakes have facetted platforms.

Wiley-Blackwell Encyclopedia of Human Evolution, First Edition. Edited by Bernard Wood.
© 2013 Blackwell Publishing Ltd. Published 2013 by Blackwell Publishing Ltd.

facial mask   The term used by Yoel Rak (1983) for the parts of the facial skeleton that are visible when the cranium is seen in a *norma frontalis* (i.e., when the cranium is viewed from the front or anteriorly).

facial visor   The term used by Yoel Rak (1983) for the mostly relief-free, antero-superiorly facing, plate-like infraorbital surface of the face in *Paranthropus boisei* (*ibid.* Fig. 18B, p. 137). This relief-free morphology contrasts with the morphology of the same part of the face of *Paranthropus robustus*, which displays, from anterior to posterior, an **anterior pillar**, a **maxillary trigon**, and a **zygomaticomaxillary step**.

facies   (L. *facies*=face) Earth science A lateral, different-looking, subdivision of the same stratigraphic unit that formed under different conditions. For example, at lake margins, sedimentary strata often grade from sandy clays near the lake margin to coarser sands further inland. Archeology The lateral variability implied by facies as used in geology has also been applied to archeological assemblages, notably by François Bordes (e.g., Bordes and de Sonneville-Bordes, 1970), who used the term to describe variation within the **Mousterian** that he attributed to contemporaneous but distinct cultural groups.

factorial   A mathematical operation such that the factorial of an integer is equal to the product of all positive integers equal to or less than the original value. It is denoted by an exclamation point after an integer. For example, factorial 4, or 4!, is equal to $4 \times 3 \times 2 \times 1$.

facultative biped   *See* **bipedal**.

FAD   Acronym for **first appearance datum** (*which see*).

failure   Mechanical failure is the inability of a system to perform its intended mechanical function. In an organism the failure of a tissue such as bone or enamel could mean excessive elastic deflection without yielding (the bone or enamel is too compliant to perform its intended function), yielding, or full fracture.

fallback foods   [etym. The origin of the term is lost in the "mists of time," but the concept has been used in agriculture for a long time because the *Oxford English Dictionary* cites its use in a contribution to the *Journal of the Royal Agricul-tural Society* (1851, 12, p. 402) suggesting that if you keep sheep then "it is…advisable…to provide a 'fall-back', or adjacent stubble field into which the flock may retire at pleasure" (*ibid*, p. 402)]. With respect to an animal's diet, fallback foods are not the foods an animal prefers, but the foods an animal turns to when its preferred foods are not available. Marshall and Wrangham (2007) suggest that fallback foods are food items that are abundant, available year round, and are of low nutritional quality and they suggest that their consumption is negatively correlated with the availability of preferred foods. **Optimal foraging theory** suggests that animals forage in such a way that they favor foods that are (a) abundant, (b) energy-rich, (c) easily accessed, and (d) easily processed, so that, all things being equal, an individual gets the greatest yield per unit of energy expended on feeding. But when these preferred foods are unavailable for either predictable (e.g., seasonality) or unpredictable (e.g., drought, volcanic eruption) reasons, animals have to turn to foods that are either less energy-rich or that cost more energy to procure and process. Hladik (1973) was among the first to point to an example of fallback foods among primates, namely that chimpanzees increase their consumption of leaves and stems when their preferred food item, fruit, is scarce. Altmann (1998) provided an operational definition of fallback foods as those foods consumed in negative correlation with the availability of preferred resources and suggested that grass corms are a fallback food of **baboons**. Lambert et al. (2004) suggested that hard foods such as bark and seeds are fallback foods of the grey-cheeked mangabey. Although Rosenberger and Kinzey (1976) proposed that specific aspects of morphology may be linked to the need to deal with mechanically demanding foods, Lambert et al. (2004) was the first to suggest that the critical ability to process fallback foods may have shaped the evolution of the masticatory system (especially enamel thickness) in at least some primates. Laden and Wrangham (2005) followed up on the observations of Hatley and Kappelman (1980) to suggest that the need to be able to process **underground storage organs**, a possible early hominin fallback food, may have been an important factor in the emergence of the derived masticatory morphology of megadont early hominins. Since then Grine et al. (2006) and Ungar et al. (2008) have suggested that fallback foods may explain mismatches between two lines of evidence about diet, dental microwear, and gross morphology.

Paleoanthropologists and morphologists are increasingly interested in examining the role of fallback foods in shaping hominin craniodental morphology and in developing hypotheses about the adaptive niches of early hominins. Marshall and Wrangham (2007) review their possible role in shaping evolution and a whole issue of *Journal of Human Evolution* [140 (4)], edited by Paul Constantino and Barth Wright, was devoted to fallback foods.

falx cerebelli (L. *falx*=sickle and *cerebelli*= diminutive of *cerebrum*=brain, so the "little brain") A small, usually insignificant, sickle-shaped fold of the meningeal, or fibrous, layer of the **dura mater** that partially separates the cerebellar hemispheres. If an **occipital sinus** is present it will run between the meningeal and the endosteal layers of the dura mater within the base of the falx cerebelli. *See also* **cranial venous sinuses**.

falx cerebri (L. *falx*=sickle and *cerebrum*= brain) A sickle-shaped fold of the meningeal, or fibrous, layer of the **dura mater** superior to the **corpus callosum** that partially separates the **cerebral hemispheres**. It runs from the **crista galli** anteriorly to merge with the **tentorium cerebelli** posteriorly. The superior and inferior sagittal venous sinuses run between the meningeal and the endosteal layers of the dura mater within its base and free border, respectively. In modern humans the superior sagittal venous sinus usually drains into the right **transverse sinus**, thence to the **sigmoid sinus** and **jugular bulb** of the same side. *See also* **cranial venous sinuses**.

Fate *See* **Caverna delle Fate**.

fault In structural geology this is a planar fracture through the Earth's crust such that the rocks above and below the fracture move relative to each other. If the hanging wall (i.e., the rocks above the fault plane) is displaced downwards (i.e., down the **dip**) relative to the rocks of the footwall (i.e., the rocks below the fault plane) this type of fault is referred to as having an extensional geometry and it is called a **normal fault**. However, if the hanging wall is displaced upwards (i.e., up the dip) relative to the rocks of the footwall this type of fault is referred to as having a compressional geometry, and it is called a **reverse fault**.

fauna (L. *fauna*=in Roman mythology the sister of the Faunus, the god of nature and *faunus*= a mythological creature that has the head of a man and the body of a goat) Meaning literally "animals," this term can be used to refer to the complete animal component of an **ecosystem** or, combined with a taxonomic modifier (e.g., mammal/ian, vertebrate, avian) a particular component of the animal life. The most common ways the term fauna is used in paleoanthropology is (a) to refer to all animal remains from a particular site (e.g., **Swartkrans** fauna) or (b) to refer animals known from a defined time period (e.g., **Pliocene** fauna), or from a region (e.g., British fauna). Usually at a paleoanthropological site the fauna is first described by a faunal or taxon list which enumerates, in rank taxonomic order and to the minimum determinable level, all the species to which faunal remains recovered from the site have been assigned. In the absence of other more precise and reliable dating methods, a **faunal list** can be used as an indication of the age and paleoenvironment of the fossil site.

faunal assemblage (L. *fauna*=in Roman mythology the sister of the Faunus, the god of nature and L. *faunus*=a mythological creature that has the head of a man and the body of a goat) A group of associated animal bones or fossils, generally found in a specific spatial and temporal context, which can be treated as a unit for analysis (e.g., the accumulated fauna from all of the localities in Bed I of **Olduvai Gorge** would be regarded as a "faunal assemblage"). *See also* **assemblage**.

faunal break A discontinuity that results in an apparently abrupt faunal change. Within Bed II at Olduvai Gorge there is an abrupt change in the fauna just above the Lemuta Member, which is referred to as the "faunal break." *See also* **Olduvai Gorge**.

Fauresmith A southern African stone artifact industry characterized by the variable presence of diminutive **handaxes** (Humphreys 1970 suggested that their size may reflect the material properties of the local stone raw material), elongated **flakes**, or **blades**, **points**, and **Levallois** technology. The Fauresmith has traditionally been considered a regional variant of the later or terminal **Acheulean** or intermediate between the Acheulean and the industries of the **Middle Stone Age**. Evidence from **Wonderwerk Cave** and elsewhere suggests a **Middle Pleistocene** age (i.e., >*c*.285 ka) (Porat et al. 2010).

favored place hypothesis This model for **Early Stone Age** site formation was proposed by Schick (1987; see also Schick and Toth 1993) and it suggests that archeological site formation was a byproduct of hominin transport and discard behavior. The anticipated need for stone tools led to the habitual transport of **lithic artifacts**. Over time, the occasional discard of artifacts at "favored places" (i.e., frequently visited foraging areas where hominins would find and consume foods, rest, carry out social activities, sleep, etc.) would lead to the passive accumulation of a local store of raw material. This would reduce the need for artifact transport while foraging in the immediate area and, over time, stone and debris from multiple butchery events could form dense archeological concentrations.

FEA Acronym for **finite element analysis** (*which see*).

feature Archeology Any non-portable object made, modified, or used by hominins. Houses and **hearths** are examples of features. Morphology Used in studies of dental morphology to refer to cusps, crests, ridges, and pits on the outer surface of the **enamel**.

fecundity (L. *fecundus* = fruitful) Typically a measurement of the potential reproductive capacity of an organism. Fecundity can be gauged generally by an organisms' age in relationship to the average reproductive life span for that taxon, or more precisely by the number of gametes (eggs or sperm) possessed by an individual at a given time during their life span. This more specific measure of fecundity is used to calculate **fitness** in demographic studies. In addition to gamete number, other factors such as gamete quality, nutritional status, stress, and environmental conditions can affect fecundity. In modern human females, fecundity reaches a peak in the early 20s and declines thereafter (until menopause, when it reaches zero), whereas modern human males do not show such clear age-related patterns of diminishing reproductive capacity.

feeding Feeding is an inclusive term for the ingestion and intra-oral processing (including chewing) of food items. Feeding occurs in sequences of gape cycles, in which the lower jaws are cyclically depressed and elevated relative to the upper jaws and the tongue is moved to position the food item (s) for chewing and swallowing. A feeding sequence begins with an ingestion cycle, often followed by a variable number of cycles in which the food is manipulated, investigated, and transported to the molar tooth row for chewing. A feeding sequence ends with a swallow. *See* **chewing**.

Fejej (Location 4°07'N, 36°04'E; Ethiopia; etym. from the name of a nearby police post) History and general description Site in the northeast of the **Turkana Basin** in southern Ethiopia. An extensive reconnaissance survey during the 1970s by the Canadian Geological Survey provided broad-scale geologic mapping of the region and established the area contained fossiliferous sedimentary rocks ranging in age from the Oligocene through the **Miocene** and into the early **Pliocene**. The **Paleoanthropological Inventory of Ethiopia** conducted a reconnaissance of the Fejej area in 1989 to evaluate its potential for paleoanthropological research and this led to a collaboration between the National Museum of Ethiopia and Stony Brook University. A French archeological team directed by Henry de Lumley has also worked at the site. Temporal span *c*.34.2–*c*.1.88 Ma, but the radiogenic ages of the basalts capping the FJ-4 locality plus the geomagnetic dating evidence suggest a minimum age for the FJ-4 locality of 4.0–4.18 Ma (Kappleman et al. 1996). How dated? **Argon-argon dating, potassium-argon dating**, and **biostratigraphy**. Hominins found at site Two specimens have been recovered from locality FJ-4: FJ-4-SB-1, a collection of six heavily worn mandibular teeth and tooth fragments, and FJ-4-SB-2, an unworn right P_4, have been attributed to *Australopithecus afarensis*. However, the FJ-4 dentition is in the size range of *Australopithecus anamensis* and some (e.g., White 2002) assign them to that taxon. Archeological evidence found at site Artifact assemblages attributed to the **Oldowan Industry** were found at locality FJ-1. Geology, dating, and paleoenvironment Asfaw et al. 1991, 1993, Kappelman et al. 1996; Hominins Fleagle et al. 1991; Archeology Asfaw et al. 1991, de Lumley and Beyene 2004.

Feldhofer Grotte *See* **Kleine Feldhofer Grotte**.

feldspar The most abundant group of minerals in the Earth's crust, having a basic composition of $xAlSi_3O_8$, where x is usually K, Na, or Ca. Feldspars are generally of igneous or metamorphic origin, but may form authigenically (i.e., formed in place) within sediments. The alkali (i.e., rich in sodium and potassium) feldspars are good targets for **isotopic dating**, as radioactive potassium decays to argon gas which, if it is trapped and accumulates within the crystal lattice, functions as a geochronometer. *See also* **argon-argon dating**; **potassium-argon dating**.

felid   The informal name for the **Felidae** (the cat family), one of the feliform families whose members are found at some hominin sites. *See also* **Carnivora**.

Felidae   The Felidae are carnivores and it is one of the **feliform** families whose members are found at some hominin sites. The Felidae include two sub-families: the Felinae (cheetahs and other small-medium size cats) and the Pantherinae (lions, leopards, and tigers). Felids first appear in the fossil record in the Oligocene. *See also* **Carnivora**.

feliform   Informal term for taxa within the sub-order **Feliformia**, one of the two suborders in the order **Carnivora**.

Feliformia   One of the two suborders in the order **Carnivora**. Common feliform families at hominin sites are the **Felidae** (cats), **Hyenidae** (**hyenas**), **Viverridae** (civets and their relatives), and **Herpestidae** (mongooses and relatives).

FEM   Acronym for finite element models. *See* **finite element analysis**.

Fengshudao   *See* **Bose Basin**.

fertility   Although this term is used loosely in a number of different ways in the paleoanthropological literature, in demography it is generally synonymous with the **total fertility rate**. As such, fertility as a population-level measurement is the number of births divided by the number of childbearing lives. On a per-individual basis, fertility is the number of births that a woman can be expected to have over her entire reproductive career. The measure does not include any aspects of mortality, so for an individual woman who has not yet reached menopause the assumption is that she will survive until the end of her reproductive career. The measure also does not consider the survival of her offspring, and only requires that they be live births.

fertility rate, age-specific   The age-specific fertility rate (or ASFR) in its simplest form is the number of births in a year to women age x years old divided by the number of women at age x alive at the midpoint of the year. For example, f_{20} would be the number of births in 1 year to women who were 20 years old (from 20 years old to 20 years and 364 days old) on the date of birth, divided by the number of women 20 years old (again, 20–20.9973 years old) at the midpoint of the year. As in an **abridged life table**, it is more common to see the ASFR given for broader age intervals. For example, the ASFR for women ages 20–25 (24.9973) years old could be written as $_5f_{20}$, and would be the number of births in a year to women ages 20–24.9973 years old divided by the midyear number of women in the age bracket.

fibrocartilage   *See* **cartilage**; **joint(s)**.

field theory   Model of dental morphogenesis proposed by Percy Butler (1939) that posited the existence of developmental fields within which regional (**incisor, canine, molar**) morphogenetic cues influence the development of specific tooth types. Butler suggested that the position of the germ and the concentrations of the specific morphogens led to the different tooth types seen in mammals, including hominins and the other higher primates. See Butler (1982) and Kieser (1990) for critical reviews of this theory.

finger   *See* **digit**.

finished sequence   A **genome** that has been sequenced (usually multiple times) and checked for errors. For example, when first completed only 25% of the modern human genome had a "finished" sequence and in that example there was at least fourfold coverage and with a quality score of 99% accuracy.

finite element analysis (or FEA)   An advanced computational technique used by engineers to predict the response of physical systems subjected to known loading conditions. The most common anthropological application of FEA is to biomechanical problems concerning the **stress** and **strain** experienced by bones when exposed to various types of forces. In FEA, an object of complex geometry (e.g., a bone) is typically modeled as a mesh of tetrahedra or hexahedra (these are the "elements" referred to in the name of the method) joined at nodes (e.g., the vertexes of the tetrahedra). The elements are assigned the material properties (e.g., stiffness) of a given substance (e.g., cortical bone), forces are applied to nodes, certain nodes are constrained from moving and displacements are found at each node. The displacement field within each element is interpolated from the calculated using known polynomial functions. Strains and stresses throughout each element are then obtained by differentiating the element displacement field using the known or estimated material property for that element. The result is a characterization of stress and strain patterns across the object as a whole. In general, the generated stress and strain patterns are approximations to the theoretically exact stress and strain patterns. The accuracy of the solution can be improved by "refining the mesh" (i.e., by modeling the system with more nodes and elements) and by improving the modeling of geometry, material properties, loads, and boundary conditions (e.g., constraints). Recent advances in computer power and software have made feasible the relatively rapid construction of geometrically precise finite element models. Although geometry is the most visible component of a model, the importance of incorporating realistic forces, material properties, and constraints is sometimes unappreciated. Poor assumptions with respect to any of these variables may potentially render a model unrealistic even if it accurately reflects bony architecture. For this reason, it is critical that all finite element models be subjected to a validation study in which results from the FEA are compared to experimentally derived strain data. If there is good correspondence between the experimental and the FEA data, then the model can be interpreted and used to test biological hypotheses. FEA has been widely employed in orthopedics to understand bone growth and adaptation, and it is becoming increasingly important in anthropology as a tool for testing hypotheses about functional morphology (see Richmond et al. 2005 for a review). For example, Strait et al. (2009) used FEA to test hypotheses of facial biomechanics in early hominins.

finite element method   *See* finite element analysis.

finite element modeling   *See* finite element analysis.

fire   *See* pyrotechnology.

first appearance datum   (or FAD) Refers to the date of a taxon's earliest appearance in the fossil record and in paleoanthropology it is usually used in connection with a species (e.g., *Paranthropus boisei*) or a genus (e.g., *Homo*). For various reasons the FAD of a taxon is almost certainly later than when the taxon actually originated in, or migrated into, that region. Just how much earlier the origination or migration occurred is determined by two factors. The first is any error in the date, the second is the nature of the relevant fossil record prior to the FAD. The problem is the old adage that "absence of evidence is not evidence of absence." Hominins are such a rare component of the mammalian faunal record that researchers need to find a substantial number of nonhominin mammalian fossils (at least several hundred) without finding *any* evidence of a particular hominin taxon before it can be reasonably assumed that hominin taxon was not part of the faunal assemblage being sampled. In such cases (e.g., *Homo* at Omo-Shungura in Bobe and Leakey 2009, pp. 178–9) the FAD has an acceptably low 95% confidence interval, but if, as is the case at Koobi Fora, there is a major unconformity spanning several hundred thousand years prior to the FAD of *Homo* (i.e., there is no fossil record during that time), then the FAD of *Homo* at that location has an unacceptably high 95% confidence interval (*c*.0.7 Ma).

F_{IS}   *See* F statistics.

FISH   Acronym for fluorescence *in situ* hybridization (*which see*).

fish   Despite being less dense than bones of terrestrial animals, fish bones are surprisingly common in early hominin sites (e.g., **Lothagam**, **Kanapoi**, **Kanjera**, **Koobi Fora**, Middle Awash, **Olduvai Gorge**, and **Toros-Menalla**). Most, but not all, of these bones were deposited post-mortem through natural water transport, which promotes rapid burial and preservation. Fish bones can provide information on hominin predation and other aspects of hominin behavior, as well as biogeographic and environmental information relevant for the reconstruction of hominin **paleoecology**. For example, the ecology of the fish taxa helped to reconstruct the paleoenvironment of the late **Miocene** Lothagam site as being close to a large and slow-moving river with non-brackish, well-vegetated back bays and swamps (Stewart 2003). This reconstruction indicates the availability to hominins of potable water and sources of river margin foods (e.g., **sedges** and **catfish**). A further example lies in the similarities between certain fish groups at Lothagam and Wadi Natrun (Egypt), which suggest there must have been a fluvial connection between the two regions with biogeographic implications for hominins. Evidence for fish as prey is rare in early hominin sites, due to either taphonomic processes and/or consumption of the whole fish that would leave few traces. Nevertheless fish bones with tooth and/or cutmarks have been reported from Koobi Fora (Pobiner 2007, Braun et al. 2010) and Olduvai Gorge (Stewart 1994). Fish bone scatters occur at other archeological sites in the **Turkana Basin** (Braun et al. 2010, Stewart 2010). The dominant fish taxon in these and other Pleistocene archeological sites is the large catfish *Clarias*, a common food today in Africa, which is readily caught in shallow or receding waters. Other inshore fish, including cichlids, are also easily caught when stranded in receding waters. While most **Pleistocene** hominins could only catch the most accessible fish; improved technology, especially boats, in the latest Pleistocene and **Holocene** resulted in much increased numbers and diversity of fish in later archeological sites. Such sites include the late Pleistocene site of **Cueva de Nerja** (coastal Spain) which is one of the first with a diversity of marine fish, and the early Holocene site of Esh Shaheinab, Sudan, which has large numbers of both shallow and deep freshwater fish.

Fish Hoek [Location 34°07′S, 18°24′E, Western Cape Province, South Africa; etym. *Fish Hoek*, from Afrikaans *Vishoek* = "fish corner," derived from the High Dutch meaning "Fish Bay" and the name of a nearby town. Also known as Skildergat, from the Afrikaans meaning either "Schilder's (a local farm hand's) cave" or "cave with paintings" and Peers Cave, after the original excavators] History and general description A.J.H. Goodwin excavated a trench from the front to the back of the cave in 1925 and encouraged the local father-and-son pair of naturalists and amateur archeologists, Victor and Bertie Peers, to continue exploration of the site, which they did between 1927 and 1929. Goodwin wrote an unpublished report on those excavations in 1929, and Sir Arthur **Keith** published a summary of the site in 1931. The Peers found a recent, culturally sterile level in the first 15 cm, followed by up to 1.5 m of **Later Stone Age** (or LSA) shell **midden** that contained six skeletons. The **Middle Stone Age**

(or MSA) (**Stillbay**) deposits began below the midden and contained two poorly preserved skeletons, both apparently intrusive from the LSA. A ninth skeleton (**Fish Hoek 1**) was recovered from the Howieson's Poort layers beneath the Stillbay and a tenth skeleton that was too poorly preserved to be recovered was discovered in what may have been MSA layers below the Howieson's Poort. The Peers' excavation in the cave stopped at 3 m below the top of the deposit. They also excavated a trench in the talus, encountering at a depth estimated as 6 m below surface late **Acheulean** (Stellenbosch Culture) artifacts. Keith Jolly conducted more excavations at the site in 1946–7; Jolly thought there was no post-Howieson's Poort MSA below the LSA midden. Barbara Anthony conducted a third excavation in 1963, the results of which are contained in an unpublished report to the National Monuments Council. Temporal span and how dated? Charcoal samples from Anthony's excavation of the upper MSA yielded uncalibrated conventional radiocarbon ages of >35 ka and 36±2.4 ka; Protsch later obtained a date of 35 ka from an equid bone from the level of Fish Hoek 1 but all his dates are suspect and are most likely fabricated. The MSA layers are apparently beyond the range of radiocarbon dating and optically stimulated **luminescence** dates at other sites suggest the Stillbay dates to *c*.70 ka and the Howieson's Poort to *c*.65–60 ka (Jacobs et al. 2008). Hominins found at site They include Fish Hoek 1, a well preserved cranium, mandible, and postcranial skeleton. Archeological evidence found at site A stratified sequence of deposits containing, from top to bottom, LSA, Stillbay (post-Howieson's Poort MSA), Howieson's Poort, Stillbay (MSA), and possibly **Acheulean** artifacts in the talus of the cave. Key references: Geology, dating, and paleoenvironment Jacobs et al. 2008; Hominins Keith 1931, Schwartz and Tattersall 2003; Archeology Keith 1931, Deacon and Wilson 1992, d'Errico and Henshilwood 2007.

Fish Hoek 1 Site Fish Hoek. Locality Peers Cave. Surface/*in situ In situ*. Date of discovery 1927–9. Finders V.S. and B. Peers. Unit Excavated from the Howieson's Poort levels, although it was recognized that the burial may be intrusive. Horizon N/A. Bed/member N/A. Formation N/A. Group N/A. Nearest overlying dated horizon N/A. Nearest underlying dated horizon N/A. Geological age Strata are **Late Pleistocene** through to the **Holocene** (see below). Developmental age Adult. Presumed sex Male, based on size. Brief anatomical description A well-preserved cranium and mandible that are larger in size than most Holocene Khoesan analogues, but it is otherwise reminiscent of them in morphology. Announcement Keith 1931. Initial description Keith 1931. Photographs/line drawings and metrical data Keith 1931, Schwartz and Tattersall 2003. Detailed anatomical description Keith 1931, Schwartz and Tattersall 2003. Initial taxonomic allocation *Homo sapiens*. Taxonomic revisions N/A. Current conventional taxonomic allocation *H. sapiens*. Informal taxonomic category Anatomically modern human. Significance Fish Hoek was long considered to possibly derive from the Howieson's Poort layers and thus be a representative of Middle Stone Age hominins in southern Africa. However, recent direct, calibrated AMS radiocarbon dates obtained by Stynder et al. (2009) indicate the cranium only dates to 7457–7145 cal. years BP, thus it is mid-Holocene in age and much too young to be from the Howieson's Poort levels. Location of original The cranium and mandible are in the **South African Museum (Iziko Museums of Cape Town)**; the postcranial skeleton is missing.

fission track dating (L. *fissus*=to split) Principle A radiogenic dating method that exploits the tendency of an ubiquitous isotope of uranium ^{238}U to undergo spontaneous disintegration, or fission. Method The fragments produced by the fission pass through the rock crystal leaving a trail, or "track," of damage in the crystal. Thin sections of the rock are etched with chemicals to visualize the tracks. First, the "natural" tracks – those that have accumulated since the formation of the crystal – are counted. Then the specimen is subjected to a dose of neutrons large enough to drive all the remaining ^{238}U to undergo fission, at which time the "induced" tracks are counted. The ratio of the two track counts and the neutron dose are sufficient to calculate the age of the crystal. Suitable material To be suitable for fission-track dating, rocks have to be rich in uranium. This effectively confines its use to natural glasses, zircon, sphene, and apatite. Time range Certainly older than 100 ka and preferably older than 300 ka. Problems Generally reliable, but only used on tephra if for some reason they are unsuitable for argon-argon dating. Example The Soa Basin of Flores, Indonesia (O'Sullivan et al. 2001).

fissures (L. *fissura*=cleft) Morphology: central nervous system In adult brains the outer layer or cortex of the cerebral and cerebellar hemispheres is thrown into folds. The blunt crests of the folds are called **gyri** and the fissures between the gyri are called **sulci**. The major gyri and sulci are named (e.g., temporal gyrus and the **lunate sulcus**). Morphology: teeth The enamel that covers the crowns of postcanine teeth is shaped into folds called fissures where developing enamel fronts meet one another. The primary fissures are between the main cusps and the secondary fissures mark the surfaces of the main cusps and delineate accessory cusps.

F_{IT} See **F statistics**.

fitness The capability of an individual to reproduce so that its **genes** are represented in the next generation. Fitness is measured as the proportion of genes contributed to the next generation by an individual relative to the contributions of others. It is equal to the relative probability of survival and reproduction for a **genotype**. Since fitness is a relative term, the genotype with the best fitness is usually assigned a value of 1 and other genotypes have values less than 1. The concept is also known as genetic fitness or Darwinian fitness.

fitness-generating function A numerical function that describes the genetic **fitness** of a particular combination of **traits** by taking into account all the fitness trade-offs related to that **phenotype** (Rosenzweig et al. 1987). Rosenzweig et al. (1987) suggest that **grades** or **genera** share a similar fitness-generating function (or **G-function**) and that when a new adaptive type emerges it shares a new G-function, which has better fitness trade-offs than the G-function of the group of taxa it replaced. For example, the ability to consistently and efficiently manufacture small sharp stone flakes that can be used to cut the hide of a large mammal would allow the user access to important food resources that would be denied hominins that had to wait until nonhuman predators had penetrated the carcass. This ability would be an example of a new G-function.

fitness landscapes A metaphor introduced by Sewall Wright for the three-dimensional landscape in which the peaks are regions of higher genotypic **fitness** and the valleys are regions of lower fitness. He and others make the point that these landscapes are complex and that complexity is in part due to interactions among genes, and in part due to phenomena such as **epistasis**, which is when the effect of a **mutation** on the **phenotype** depends on its context (i.e., whereabouts in the genome it occurs). Many of the possible pathways across the fitness landscape are inaccessible for one reason or another and these restrictions result in evolutionary **constraints** (e.g., additional cusps on mandibular postcanine teeth tend to occur in the **talonid**, the distal part of the crown, and not in the more mesial **trigonid**). George Gaylord Simpson later applied a similar principle to population-level dynamics in the form of adaptive landscapes.

fixation   In a population where more than one form of a gene (an allele) exists, fixation occurs when, through any number of processes (including random **genetic drift**, **natural selection**, or as a consequence of small breeding populations), variation in allelic distribution declines to the point where only one allele remains. Once this happens, mutation or the immigration of new alleles are the only ways for the population to move out of fixation for this allele.

fixation indices   These are better known as *F* statistics. *See F* **statistics**; **fixation**.

FJ   Abbreviation of Fejej and the prefix for fossils from Fejej, Ethiopia. *See* **Fejej**.

flake   An inclusive term for any fragment removed from a **core** made of stone (or much less commonly) bone. Some types of flake (e.g., **blades**) are morphologically distinctive. The term can be applied to complete flakes, as well as to the various flake-like fragments (which go under a variety of terms including debris and **débitage**) that are produced during the process of fracturing stone to make flakes (also called **knapping**). Flakes may also be produced in the absence of hominin activity as a result of post-depositional processes, but these flakes do not normally show the distinctive morphology associated with **conchoidal fracture**.

flake-blade   This term is encountered in descriptions of southern African **Middle Stone Age** sites such as **Klasies River**. Unlike the term **blade** that refers to artifacts that are at least twice as long as they are wide, a flake-blade is an elongated **flake** with dorsal scars that suggest several parallel (or subparallel or convergent) flakes have been previously removed.

flake types   A typology used to describe stone flakes produced during knapping based on the occurrence of cortex (original exterior surface of the raw material) on the surface of the flake. The distribution of types in an assemblage provides information about the knapping methods employed and the representation of different stages within a **reduction sequence**. The widely used system is that of Toth (1985), which includes seven types defined in terms of the partial or total absence of cortex from (a) the dorsal (exterior) surface of the flake and (b) the "striking platform" (surface surrounding the point of percussion). These are Flake Type I, a cortical platform and dorsal surface; Flake Type II, a cortical platform and partially cortical dorsal surface; Flake Type III, a cortical platform and noncortical dorsal surface: Flake Type IV, a noncortical platform and cortical dorsal surface; Flake Type V, a noncortical platform and partially cortical dorsal surface; Flake Type VI, a noncortical platform and noncortical dorsal surface; and Flake Type VII, indeterminate. Experimental replication generates expectations for the distribution of flake types that would be associated with early or late reduction stages, and with assemblages that include all stages of reduction versus those in which only part of the reduction sequence is represented or from which certain types have been preferentially removed (for example due to utility as tools). With respect to knapping methods, cortical platforms are associated with "unifacial" reduction, in which percussion is consistently addressed against the original raw material surface, whereas noncortical platforms are associated with "bifacial" or "polyfacial" methods in which the "scars" left by previous flake removals are used as the platforms for subsequent flake removals.

flint   *See* **chert**.

FLK   Acronym for the Frida Leakey Korongo, one of the localities at **Olduvai Gorge**. Discovered and excavated in 1931, it was the first site where artifacts were found *in situ* in Bed I. In 1959 the **OH 5** cranium was recovered "within a few yards of the 1931 excavation," OH 6 was recovered there in 1960, and **OH 16** in 1963. Two other localities, **FLKN** and **FLKNN**, are nearby. *See also* **Korongo**; **Olduvai Gorge**.

FLK-*Zinjanthropus*   *See* **FLK**.

FLKN Acronym for Frida Leakey Korongo North, one of the localities at **Olduvai Gorge**. Excavations in 1961–2 resulted in many artifacts and faunal evidence and **OH 10** was recovered there in 1961. *See also* **Korongo**; **Olduvai Gorge**.

FLKNN Acronym for Frida Leakey Korongo North North, one of the localities at **Olduvai Gorge**. The locality is some 365 m north of **FLK**. Hominin remains (**OH 7** and **OH 8**) were discovered here in 1960–1. *See also* **Korongo**; **Olduvai Gorge**.

flora (L. *flora*=Roman goddess of flowers) A collective term for the plant life that occurs in a specific area or at a particular time. The fossil records of plants at paleoanthropological sites tend to be sparser and less well known than the **faunal** records. Nonetheless, the flora at several **hominin** sites have been studied by investigating fossilized pollen (e.g., Bonnefille et al. 2004), fossilized leaves, vines, woods, and fruits (e.g., Bonnefille and Letouzey 1976, Goren-Inbar et al. 2002). The floral record can also be studied through **phytoliths** (i.e., the silica components of plants) (e.g., Alexandre et al. 1997, Bamford et al. 2006), **paleosols** (i.e., fossilised soils) (e.g., Levin et al. 2004), and plant leaf wax biomarkers (e.g., Huang et al. 1999a, Schefuß et al. 2005, Feakins et al. 2007). *See also* **palynology**; **trace fossil**.

Florisbad (Location 28°45′S, 26°00′E, Orange Free State, South Africa; etym. Afrikaans "Floris' bath," after a spa operated on the site by Floris Venter) History and general description Vertebrate fossils were discovered in an old spring eye 40 km/24 miles northnorthwest of Bloemfontein, in what was the Orange Free State (now Free State Province), by Floris Venter in 1912. Robert **Broom** described some of the fossils, but it was the subsequent excavations by T.F. Dreyer that resulted in the **Florisbad 1** cranium. A variety of more recent excavations have attempted to clarify the stratigraphy and obtain absolute dates and additional fossils. Temporal span and how dated? The Peat I (see below) levels are beyond the range of **radiocarbon dating**. **Biostratigraphy** of the associated fauna suggests a Late **Middle** to **Upper Pleistocene** age. **Electron spin resonance spectroscopy dating** applied to teeth (both faunal and the hominin M^3) produced an age of 259±35 ka, and optically stimulated **luminescence dating** suggests an age of 300–100 ka. Hominins found

at the site **Florisbad 1**, a fragmentary cranium, was made the holotype of *Homo (Africanthropus) helmei* Dreyer, 1935. Archeological evidence found at site Three levels (Peats I–III) of old land surfaces containing **Middle Stone Age** artifacts. The provenance of the cranium is uncertain, but its taphonomic state best matches the fauna from Peat I. Key references: Geology, dating, and paleoenvironment Van Zinderen Bakker 1957, Oakley 1957, Cooke 1963, Vogel 1970, Protsch 1975, Clarke 1985a, Grün et al. 1996; Hominins Dreyer 1935, Clarke 1985a, Day 1986, Curnoe and Brink 2010; Archeology Dreyer 1935, Clarke 1985a, Brink and Henderson 2001.

Florisbad 1 Site **Florisbad**. Locality N/A. Surface/*in situ* *In situ*, but from an old spring eye choked with debris and fossils. Date of discovery 1932. Finders G. Venter and T.F. Dreyer. Unit N/A. Horizon N/A. Bed/member N/A. Formation N/A. Group N/A. Nearest overlying dated horizon N/A. Nearest underlying dated horizon N/A. Geological age Late **Middle** or **Upper Pleistocene**. Developmental age Adult. Presumed sex Unknown. Brief anatomical description A fragmentary cranium comprising the frontal, both parietals, both nasals, the right zygomatic and maxilla, part of the left maxilla, and the right M^3. Endocranial volume Uncertain, but probably approximately 1200–1400 cm^3 based on vault dimensions. Announcement Dreyer 1935. Initial description Dreyer 1935. Photographs/line drawings and metrical data Dreyer 1935, Drennan 1937, Galloway 1937, Rightmire 1978, 1984, Bräuer 1984, Day 1986, Schwartz and Tattersall 2003. Detailed anatomical description Dreyer 1935, Rightmire 1978, 1984, Bräuer 1984, Tappen, 1987, Schwartz and Tattersall, 2003. Initial taxonomic allocation *Homo helmei*. Taxonomic revisions *Homo florisbadensis* (Drennan 1935), *Homo sapiens* (Rightmire 1978). Current conventional taxonomic allocation Some favor placing it in its own species, *H. helmei* (Lahr and Foley 1994, McBrearty and Brooks 2000), others advocate placing it in late archaic *H. sapiens* (Bräuer 1984), and yet others view it as an early form of modern humans, *H. sapiens* (Galloway 1937, Rightmire 1984b, 2008, Bräuer 1984, 2008). Informal taxonomic category Pre-modern *Homo* or anatomically modern human. Significance The supraorbital torus is large, yet smaller than in **Kabwe** or **Saldanha** and, according to some, shows incipient separation of its medial and lateral segments. The forehead rises more steeply than in the latter specimens, the face is broad but vertically short and bears a canine fossa. The cranial vault is wide. The thick vault bone bears two circular tooth marks, suggesting the individual was killed or scavenged by a large carnivore. The Florisbad cranium appears to be intermediate, and probably transitional, in morphology between *Homo heidelbergensis* and *Homo sapiens* in its vertical facial dimensions, brow ridge size, and inclination of the frontal bone. It retains primitive features such a large brow ridge, receding frontal, and broad vault that distinguish it from anatomically modern humans. If *H. helmei* can be satisfactorily diagnosed, which is debatable, Florisbad would serve the holotype of the taxon. Location of original **National Museum, Bloemfontein**, Free State Province, South Africa.

flowstone *See* calcium carbonate; speleothem.

fluorescence *in situ* hybridization (or FISH) A **cytogenetic** method in which fluorescently labeled **probes** are used to detect specific **DNA sequences**. The method can also be applied to entire **chromosomes**, in which case the method is called **chromosome painting**. FISH can be very useful for identifying the physical location of **genes**, or of other **loci**, on chromosomes and for identifying duplicated, deleted, or rearranged DNA sequences.

fluorine analysis *See* fluorine dating.

fluorine dating (L. *fluor*=the name of a group of minerals used to make fluxes) No longer used to date fossil hominins (see Goodrum and Olson 2009 for an excellent history of fluorine dating), it played an important role early in the history of paleoanthropology and its contribution to uncovering the Piltdown fraud was a crucial one. In the middle of the 19thC James Middleton suggested that fossil bone absorbs fluorine from the surrounding groundwater and he hinted that the process could be used to track time. However, it was a French scientist, Adolphe Carnot, who exploited the potential of the system. On the principle that bones buried at the same time in the same deposit should have taken up equal amounts of fluorine, Carnot showed that a human tibia found at the site of Billancourt, outside of Paris, was much younger in age than the many animal bones recovered from the site. Carnot's method was applied to Eugène **Dubois**' discoveries at Trinil, and the results were consistent with the proposed antiquity of those fossils because they had levels of fluorine comparable to those

of Pliocene fossils from Europe. The method was little used in the early decades of the 20thC, until it attracted the attention of Kenneth Oakley who spoke about its potential at the First Pan-African Congress on Prehistory in Nairobi in 1947. Although fluorine dating failed to resolve the controversy over the age of the **Kanjera** (the volcanic sediments contained so much fluorine that it saturated *all* of the bones from the site) and **Kanam** hominin fossils, it did show that the anatomically modern human **Galley Hill** skeleton, one of the critical pieces of evidence in favor of the **pre-sapiens hypothesis**, was an intrusive burial and not **Pleistocene** in age (Oakley and Montagu 1949) and it confirmed the antiquity of the hominin fossils recovered from **Swanscombe** (Oakley 1949). However, fluorine dating is probably best known for the role it played in helping to expose the Piltdown fraud. The animal fossils at the Piltdown site were found to have fluorine levels that varied between 3 and 1.6%, whereas the levels in the alleged hominin fossils claimed to be from Piltdown were only approximately 0.2% (Oakley and Hoskins 1950). These levels were similar to those found in contemporary bones, and the authors' suggestion that the result "requires some explanation" (*ibid*, p. 381) was undoubtedly an important element in spurring Joseph Weiner and Wilfrid **Le Gros Clark** to re-examine the fossil hominin evidence from Piltdown. They collaborated with Oakley to apply a more sensitive form of fluorine analysis to the allegedly ancient hominin fossils. These analyses showed that the fluorine levels in the cranial bones and the mandible were 0.1 and 0.03%, respectively (Weiner et al. 1953). These fluorine levels implied two things. First, the cranial bones and the mandible of **Piltdown I** were not the same age, and second, the mandible was almost certainly modern. These results, along with other lines of evidence showing the cranial bones and the mandible had been stained and that the cusps of the teeth in the mandible had been filed down, were sufficient for Weiner, Oakley, and Le Gros Clark to say with confidence that the Piltdown hominin fossils were modern and not ancient and that their association with the Piltdown site was the result of an elaborate hoax. *See also* Kenneth **Oakley**; **Piltdown**.

fluvial (L. *fluvius*=river) Pertaining to a river or stream, so a fluvial paleoenvironment is one involving a river or stream, and in geology fluviatile is the name applied to sediments laid down by rivers or streams.

fluviatile *See* fluvial.

fMRI Acronym for **functional magnetic resonance imaging** (*which see*).

foliate point (L. *foliates*=leafy) A pointed stone artifact whose shape resembles that of a leaf. It is presumed that, depending on their size, foliate points were hafted onto a wooden shaft for use either as a spear (larger points) or as arrows (smaller points). *See also* **point**.

folivory (L. *folium*=leaf and *vorous*, from the stem of *vorare*=to devour) Literally, leaf eating; animals that eat leaves are folivores. Leaves are less energy-rich than fruits or insects. This is because the complex cellulose of leaves requires specialized digestive processing and bacterial colonies in the fore- or hindgut to break it down. Folivorous primates tend to have shearing crests on their teeth that mechanically process leaves much as scissors slice through paper. Soft-tissue digestive adaptations to folivory in primates include adaptations to the foregut in the form of a large, complex, multi-chambered stomach (as found in **colobine** monkeys) and adaptations to the hindgut as either an enlarged caecum (found in some strepsirhines) or a long colon (found, for example, in howler monkeys and gorillas). Folivory is most common among larger primates, which require relatively less dietary energy per unit of body mass than smaller animals. Large animals can eat leaves because they have a long digestive tract that enables them to successfully process low-quality food. Nonetheless, some small primates are folivorous (e.g., *Lepilemur*) and they solve the problem of energy extraction by eating feces that contain high proportions of undigested plant materials. Folivory only features prominently in the diets of one extant great ape, the gorilla, and it has not been seriously proposed as a dietary specialization of extinct hominins. *See also* **coprophagy; Jarman/Bell principle**.

follicle *See* **dental follicle**.

Fond-de-Forêt *See* **Schmerling, Philippe-Charles (1790 or 1791–1836)**.

Fontana Ranuccio (Location 41°45′35″N, 13°16′03″E, Italy; etym. It. *fontana*=fountain and *Ranuccio*=a family name, possibly of the local landowner) History and general description This open-air site 60 km/40 miles southeast of Rome in the Sacco-

Liri valley was exposed during mining and has been explored since the late 1960s by the Istituto Italiano di Paleontologia Umana. An ancient volcano produced pyroclastic flows, which allowed for fairly specific **potassium-argon dating** on many of the levels in the sequence. The level in which the artifacts and hominin remains were found is also made of pyroclastic sand, allowing direct dating of this level. Temporal span The archeological levels were dated to *c*.458 ka. How dated? Potassium-argon dating. Hominins found at site Three hominin teeth were recovered: a lower left I_1, and a left and right lower M_2. They have been compared with the teeth of Homo erectus, but no formal taxonomic assignment has been made. Archeological evidence found at site The few artifacts that have been recovered include elephant bone implements and small flint handaxes and scrapers and are likely **Acheulean**. The **Levallois** technique does not appear to have been used. Key references: Geology, dating, and paleoenvironment Biddittu 1979, Segre and Ascenzi 1984, Muttoni et al. 2009; Hominins Segre and Ascenzi 1984, Ascenzi and Segre 1996; Archeology Segre and Ascenzi 1984, Ascenzi and Segre 1996, Segre-Naldini et al. 2009.

fontanelle (L. *fontanelle*=small fountain or spring) Fontanelles are places on the newborn **cranium** where the bone of the cranial vault is not yet completely ossified. The largest of the fontanelles are the rhomboid-shaped anterior fontanelle at **bregma**, where the frontal (**metopic**), **coronal**, and **sagittal** sutures meet and the triangular posterior fontanelle at **lambda**, between the sagittal and the lambdoid sutures. They were called fontanelles because they are the places on a newborn cranium that appear to pulsate. They do this because they (and especially the anterior fontanelle) transmit the pulsations of the intracranial arteries via the cerebrospinal fluid and the venous blood in the **dural venous sinuses**.

Fontéchevade (Location 45°40′43″N, 0°28′47″E, France; etym. named after a nearby village) History and general description This cave site in the Charente region was excavated by several amateurs since the 1870s, and then by Germaine Henri-Martin from 1937 to 1954. It is known as one of the reference sites for the Tayacian. The two recovered hominin fragments were thought to be quite old [approximately Oxygen Isotope Stage (OIS) 5e]; one fragment resembled *Homo neanderthalensis* traits but the other appeared more modern and was thought to represent an early modern human population, and was part of the evidence used to support the **pre-sapiens hypothesis**. However, a program of excavation and a re-dating effort have shown that these fossils are more recent; the modern human-like fossil could even be an early **Upper Paleolithic** *Homo sapiens* individual. Temporal span and how dated? The lowest level was originally thought to date to the last interglacial (OIS 5e) based on the warm-climate fauna and the crude stone tools, but recent **radiocarbon dating** and **electron spin resonance spectroscopy dating** have suggested an age of *c.*33–60 ka (i.e., OIS 3) for the top of this level. Hominins found at site A fragment of a frontal bone [Fontéchevade (or Homo) I] and a partial calotte [Fontéchevade (or Homo) II] were both recovered near the top of layer E; Fontéchevade II was clearly Neanderthal-like but Fontéchevade I lacks a supraorbital torus and appears more modern. Archeological evidence found at site The lowest layer consists of a thick Tayacian deposit (a type of the **Mousterian**), but recent work at the site has shown that this "industry" is largely the result of taphonomic processes. This is overlain by a brecciated roof-fall, and then several thin archeological layers containing Mousterian, **Châtelperronian**, **Aurignacian**, and Bronze Age artifacts. Key references: Geology, dating, and paleoenvironment Chase et al. 2007; Hominins Henri-Martin 1957, Vallois 1949, 1958; Archeology Henri-Martin 1957, Chase et al. 2009.

food utility index *See* **economic utility index**.

foot The foot consists of seven tarsal bones, five metatarsals, and 14 phalanges (three for each toe, except for the big toe, or hallux, which has two) plus any sesamoid bones. The modern human foot differs considerably from the ape foot, reflecting a shift from a grasping appendage to a propulsive one. The most apparent differences are that in apes such as the chimpanzee, the foot is positioned in an "inverted set" (i.e., the sole of the foot is positioned with the lateral aspect of the foot on the ground) and it possesses an abducted (i.e., deviated away from the longitudinal axes of the other toes) grasping hallux, both features important for arboreal locomotion. In contrast the modern human foot has evolved a more everted "plantar" set (i.e., the sole of the foot is positioned with the bottom, or

plantar, aspect of the foot towards the ground), and the hallux is adducted (i.e., it is in line with the longitudinal axes of the other digits). Adaptations for habitual bipedality include a more robust calcaneus with a derived lateral plantar process. The modern human calcaneocuboid joint is unique among living taxa in possessing an interlocking facet that stabilizes the joint during the **push-off** phase of walking, and the tarsals are elongated (proximodistally), which increases the lever arm for efficient bipedal push-off. Metatarsal robusticity differs between modern humans and chimpanzee feet with the former typically having a $1 > 5 > (3,4) > 2$ formula and chimpanzees possessing a $1 > 2 > 3 > 4 > 5$ formula. The metatarsals of the modern human foot also have a distinct torsion (i.e., the long axis of the metatarsal head is not in line with the long axis through the articular surface at the base of the metatarsal) that allows the bases to form a transverse arch while all of the metatarsal heads lie squarely on the ground during the stance phase of the walking cycle. In contrast, the metatarsal torsion seen in the apes puts the lateral toes in opposition with the grasping hallux. While variation exists in the length and degree of curvature of ape phalanges, modern humans possess relatively short and straight phalanges compared to all apes. Modern human feet have a longer Achilles tendon and a thicker plantar aponeurosis (fascia) than ape feet. *See also* **bipedalism**; **walking cycle**.

foot arches The elevated midregion of the foot that results from the architecture of the hard-tissue skeleton in combination with soft-tissue structures. Modern humans, chimpanzees, gorillas, and orangutans possess a transverse arch, a mediolaterally oriented structure formed at the tarsometatarsal junction. Modern humans differ, however, from all other extant primates in also possessing a **longitudinal arch** that runs the length of the foot and serves both to stiffen the foot to make it an effective lever and to allow ligaments and the plantar aponeurosis to store elastic energy during the stance phase of walking. *See also* **longitudinal arch**; **transverse arch**; **walking cycle**.

foot function The modern human foot is stiff and straight and thus adapted for terrestrial bipedalism, whereas the extant ape foot is more flexible and adapted for a wider range of motion. Modern humans and nonhuman apes are both plantigrade; that is, they strike the ground with their heel (unlike monkeys, which are digitigrade or semiplantigrade). During modern human walking forces are greatest at heel strike and again at **push-off**, producing a characteristic "double-humped" force/time curve in the **walking cycle**. During the stance phase of walking the modern human midfoot is normally elevated as a result of the **longitudinal arch**, maintained by ligaments and the plantar aponeurosis (fascia), whereas the ape midfoot is in contact with the ground during the stance phase. As the stance phase proceeds the foot undergoes **eversion**, due in part to the action of the fibularis muscles. Eversion helps transfer the weight-bearing function towards the robust hallux for push-off and it also keeps the transverse axes of the midtarsal joints roughly parallel and allows for some midfoot mobility, which is important for keeping the foot in contact with uneven substrates. Activity of the gastrocnemius-soleus complex at push-off plantar flexes the foot at the ankle, which causes inversion of the hindfoot. This motion twists the calcaneus against the cuboid in a close-packed position and this converts the midfoot into a structure that allows for uninterrupted weight transfer from the hindfoot to the forefoot. In apes such as the chimpanzee, the midfoot remains in contact with the ground after heel lift, the so-called "**midtarsal break**" or "**midfoot break**." As the heel lifts in modern humans, the fulcrum shifts directly to the metatarsophalangeal joints and the resulting dorsiflexion of the phalanges tightens the plantar aponeurosis (the "**windlass effect**"), which helps convert the foot into a stiff and effective lever. *See also* **midtarsal break**; **walking cycle**.

foot movements The primary motions at the foot joints are dorsiflexion (extension) and plantarflexion (flexion) at the ankle (talocrural) joint that occurs around an approximately mediolateral axis, inversion and eversion, primarily at the subtalar joint, that occurs around an approximately anteroposterior axis, and internal and external rotation of the foot that occurs around an approximately superoinferior axis. The toes can also abduct and adduct with respect to the second digit; movement of the toes (other than the second) toward the second digit is **adduction**, and movement of the toes away from the second digit (or away from its normal longitudinal axis in the case of the second digit itself) is **abduction**. These motions often occur together as a result of the close articulation of the foot bones. Triplanar motion incorporating

these different foot movements is sometimes termed supination or pronation. Pronation involves external rotation, dorsiflexion, and eversion, and supination combines internal rotation, plantarflexion, and inversion. During the stance phase of locomotion, some pronation and some supination occur. At the very beginning of stance the hindfoot pronates slightly, allowing the midfoot to move. After this the foot is said to supinate, increasing stability of the foot during midstance. After midstance, the foot pronates slightly to absorb the forces of body weight in the arch and at the very end of stance phase the foot supinates slightly to create a rigid foot. That supination combined with the **windlass effect** promotes efficient propulsion by the plantarflexors.

footprints *See* **Koobi Fora hominin footprints**; **Laetoli hominin footprints**.

forager (OF *fourrage*=to search for food) The term used to describe societies that practice a subsistence strategy based primarily on the collection of wild plant foods, fishing, and the hunting and **scavenging** of animals. Contemporary foragers are of interest to paleoanthropologists because they provide the only means to directly observe behavioral patterns, subsistence strategies, and social interactions in the absence of horticulture, agriculture, or pastoralism. Studies of contemporary forager populations can provide insights into the selective forces that shaped the aboriginal modern human adaptation. In terms of caloric return, plant foods are generally the most important component of the contemporary forager diet, but plant **productivity** decreases with increasing latitude, so high-latitude foragers rely less on plant foods and more on animal foods than those living in the tropics. A predominantly foraging (hunting and gathering, or hunter–gatherer) lifestyle is practiced by a small number of groups today, including the **Hadza** of Tanzania, the **Aché** of Paraguay, the **San** of southern Africa, the Inuit of Canada and Alaska, and the **Mbuti** of the Democratic Republic of the Congo. Hunter–gatherers tend to forage in small, relatively egalitarian groups. Forager movements depend on the spatial distribution and seasonal availability of resources. In regions with a relatively low and even resource distribution, foragers live in mobile groups that move seasonally to exploit resources as they become available. Where resources are spatially and/or temporally more clumped (e.g., in regions with seasonal fish or large mammal migrations), hunter–gatherers may reside in larger groups that do not need to move as frequently and small, special task groups may be sent out to obtain food for the group at large. Food storage is more likely in the latter context, where seasonal abundance may provide surpluses that can be stored for later consumption. Hunting and gathering was the predominant mode of subsistence before plant and animal **domestication** began *c*.10,000–12,000 years BP. Paleoanthropologists and archaeologists analyze the ways in which foragers produce and use tools in an attempt to glean how tool use developed in hominin evolution. Characteristics of certain tools used by contemporary foragers, such as wooden digging sticks and carrying devices used to transport infants and tools, remind us of the danger of interpreting the behavior of early hominins on the basis of durable (i. e., lithic) evidence, for the tools referred to above are typically constructed out of material that does not survive the archeological record. Recent hunting technology, such as iron-tipped tools, most likely replaced their bone and/or stone counterparts in most parts of the old world. The behavioral ecology of extant hunter–gatherer groups provides an indication of what the hunting and gathering adaptation might have been like for more ancient forms of *Homo*, although the extent to which this adaptive complex, including a sexual division of labor, extensive food sharing, and **central place foraging**, extends back into the **Pleistocene** is debated.

foraging (OF *fourrage*=to search for food) Foraging is an inclusive term for the search for and acquisition of food items. With respect to human evolution researchers use information about the foraging behaviors of extant organisms to develop **referential models** from which inferences can be made about the functional morphology of extinct hominin taxa. **Optimal foraging theory** involves setting the estimated energy expended in the search for food against the net energy yielded by ingesting the food. *See also* **extractive foraging**; **forager**.

foraging efficiency Refers to the efficiency with which foragers obtain calories from their food resources, often measured as the amount of energy obtained relative to the amount of time spent searching for and handling food. **Optimal foraging theory** assumes that the objective of all foragers is to maximize foraging efficiency with declines in foraging efficiency often signaled by an expansion of **diet breadth** and increased utilization of **low-ranking prey**. Foraging efficiency can be affected by numerous factors including environmental change, technological innovation, and resource depression. The detection of changes in foraging efficiency through time in archeological contexts is central to understanding the factors mediating prehistoric human subsistence strategies.

foramen magnum (L. *foramen*=an opening and *magnus*=great) An unpaired opening in the occipital bone, this is the largest opening in the **basicranium** (hence its name). It transmits the spinal cord together with the **meninges** that cover it, the spinal contributions to the accessory (XI) cranial nerves, the vertebral and spinal arteries, and the thin-walled veins that connect the internal vertebral venous plexuses with the **dural venous sinuses** within the **endocranial cavity**. Two important midline cranial landmarks involve the foramen magnum; basion is defined by its anterior margin and opisthion by its posterior margin. Important variables associated with the foramen magnum involve its *location* relative to either the long (i.e., the **sagittal**) axis of the neurocranium or to axes defined by bilateral landmarks in the basicranium (e.g., bi-carotid, bi-tympanic, etc.), its *orientation* relative to horizontal planes such as the **Frankfurt Horizontal**, its *shape* described either non-metrically or metrically, and its *size* captured either by linear dimensions or by estimating or measuring its surface area. Location Interest in quantifying the location of the foramen magnum can be traced back to Bolk (1909) (see references in Luboga and Wood 1990), but the **head-balancing index** of Dart (1925), the index devised by Schultz (1942), and the **condylar position index** of Le Gros Clark (1950a) are among the best known of many attempts to locate the foramen magnum, or a proxy (e.g., porion or the occipital condyles) for the foramen magnum, along the midline axis of the cranium. But the utility of these indices is limited by the requirement that a well-preserved cranium is available. The realization that this is an unrealistic expectation for the early hominin fossil record provided the stimulus for researchers to investigate locating the foramen magnum with respect to transverse axes defined by bilateral landmarks within the basicranium (e.g., Dean and Wood 1981, 1982, Ahern 2005). The nonhuman great ape with the most anteriorly situated foramen magnum with respect to landmarks within the basicranium is *Pan paniscus*

(Luboga and Wood 1990) and there is overlap, albeit very limited, between the ranges of *Pan paniscus* and some modern human populations. Bolk (1915b) also noted that the position of the foramen magnum early in ontogeny in the nonhuman great apes resembled its location in adult modern humans. Among adult fossil hominins, the position of basion relative to the bi-tympanic line is most anterior in *Paranthropus boisei* but it lies close to this line in all specimens where this relationship can be assessed. Orientation Bolk (1910) was also a pioneer in the quantification of the orientation of the foramen magnum; much later Luboga and Wood (1990) showed that there is modest overlap of the ranges of the orientation of the foramen magnum in *Pan paniscus* and in modern humans. Shape A **cardioid foramen magnum** (i.e., heart-shaped) has been reported in several crania belonging to *P. boisei* and to the one cranium attributed to *Paranthropus aethiopicus* (Tobias 1967a, Leakey and Walker 1988). Size When the fossil record of a taxon is limited to the cranium at least one researcher suggested that the foramen magnum may be an appropriate and convenient proxy for body weight (Radinsky 1967), but subsequent research has shown that foramen magnum area has a relationship with **endocranial volume** that is independent of its relationship via body weight. *See also* **cranial venous drainage**.

foraminifera (L. *foramen* = an opening) An order of protozoans with many holes, or foramina, in their hard calcareous (Ca_2CO_3) shells. The oxygen in the **calcium carbonate** in the shells is the source of the oxygen isotopes used for **paleoclimate** reconstruction and the method of temporal correlation known as **orbital tuning**. Deep sea cores are extracted from ocean floors to sample the shells of ancient foraminifera along the sediment core. The $^{18}O/^{16}O$ ratios of the shells of deep sea, or benthic, foraminifera contain a temporal record of **deep sea temperature** and **glacial ice volume**. Often in the same cores planktonic foraminifera can be found, whose shells record **sea surface temperature**.

Forbes' Quarry (Location 36°08′N, 5°20′W, **Gibraltar**, UK; etym. named after the owner of the quarry) History and general description In 1848, the cranium of an early hominin was recovered from a limestone quarry on the north face of the Rock of Gibraltar, and in March of that year its discovery was the subject of a presentation to the Gibraltar Scientific Society by Lieutenant Flint.

The cranium was brought to the attention of Hugh Falconer, the paleontologist, and George Busk, a surgeon and paleontologist, when they visited the excavations at Gibraltar in 1862. In 1864 a report on the cranium was made to the British Association for the Advancement of Science and Falconer (1864) proposed it be referred to *Homo calpicus* (after Mons Calpe, the name for Gibraltar). This was the same year that the fossils from **Kleine Feldhofer Grotte** were published as *Homo neanderthalensis* and the significance of the Gibraltar 1 cranium was overlooked. Temporal span and how dated? The exact provenance of the fossil is unknown, and the fossil has not been directly dated. Hominins found at site The Forbes' Quarry Gibraltar 1 cranium is a small but nearly complete cranium, showing nearly all of the distinctive suite of Neanderthal characteristics; with hindsight it was the first adult cranium of *H. neanderthalensis* to be recovered. Archeological evidence found at site None. Key references: Geology, dating, and paleoenvironment N/A; Hominins Busk 1865, Barton et al. 1999; Archeology N/A. *See also* **Gibraltar 1**; *Homo neanderthalensis*.

force (L. *fortis* = strong) The capacity to cause an object to change velocity. A force has both a magnitude and a direction, thereby making it a vector. Every force has an equal and opposite reaction force. Thus, if the foot applies a force to the ground the ground applies a reaction force of equivalent magnitude (and opposite direction) to the foot. Anatomical systems (e.g., the upper limb in tool making and the lower limb in running) must be constructed in a way that enables them to both apply and resist forces.

force plate A device that measures **force** using load cells (these are electronic devices used to convert, or transduce, a force into an electric signal). The simplest force plates measure only the force **normal** (i.e., perpendicular) to and in the center of the plate, much like a bathroom scale does. Most force plates used in biomechanical research calculate the force at the center of pressure and provide all three main directional components of force (normal, side-to-side, and fore–aft) over time. Most applications in **biomechanics** concern the force exerted by (and on) the hindlimbs (and forelimbs in nonhuman animals) during **gait**. For example, Kimura et al. (1985) used force plates to show how the **ground reaction force** (force at the foot) during **chimpanzee**

bipedal walking differs from the force profile during modern human walking. *See also* **kinetics**.

forcing   *See* **climate forcing**.

forebrain   (ME *fore*=the front and OE *braegen*=brain, most likely from the Gk *bregma*= the front of the head) The anteriormost of the three swellings or vesicles in the embryo that give rise to the brain. The forebrain is comprised of the **thalamus** (the diencephalic part of the forebrain vesicle), which lies close to the midline, and the two **cerebral hemispheres** (the telencephalic part of the forebrain vesicle), which are more laterally situated. The **basal ganglia** (i.e., the **striatum**, globus pallidus, substantia nigra, and subthalamic nuclei) develop during ontogeny deep in the wall of the part of the expanding forebrain that comes to rest up against the thalamus. In primates, but especially in the great apes and particularly in modern humans, it is the forebrain that undergoes the most expansion during development. This is an extension of the trend seen across a range of mammals, wherein increases in brain mass size are generally accompanied by the positive **allometry** of the growth of the forebrain components, particularly the **neocortex** and the striatum (Finlay and Darlington 1995).

forest   (L. *foris*=outside) An assemblage of ecosystems dominated by trees and other woody vegetation. Within this broad category, there are several distinct forest biomes relevant to hominin evolution including **tropical dry**, tropical wet, **temperate deciduous**, montane, and gallery or riparian forests. In general, a forest is defined as a stand of trees, which can vary from 10 to 50 m or more in height, with interlocking crowns. Forests can have several layers, including shrub and ground layers, and in some forest biomes a grass layer may be present. The majority of African forests are evergreen or semi-evergreen, although localized patches of deciduous forests do exist.

forest hog   The vernacular name for the taxa within the genus *Hylochoerus*. See **Suidae**.

form   Objects (such as teeth, crania, long bones, etc.) have a number of spatial properties, including their location, orientation, size, shape, etc. Form refers to those aspects of an object that remain when all but **size** and **shape** are removed (e.g., ignoring location in space, rotation, etc.). Form is often studied through the use of **geometric morphometrics** and/or **Euclidean distance matrix analysis**. Form and its subsets of size and shape are of great interest in human evolution: the analysis of similarity and variation in form, size, and/or shape are used in studies of topics such as **alpha taxonomy**, phylogenetic reconstruction, **sexual dimorphism**, developmental scaling, hormonal control of growth, and paleoecological reconstruction.

formation   An inclusive term used in stratigraphy to refer to a set of strata whose upper and lower boundaries are evident. Formations are divided into **member**s and members into **bed**s. For example, the **Koobi Fora Formation** is divided into eight members: from oldest to youngest the Lonymun, Moiti, Lokochot, Tulu Bor, Burgi, KBS, Okote, and the Chari.

Fossellone   (Location, 41°14′N, 13°05′E, Italy) <u>History and general description</u> This is one of the large cave sites in the **Monte Circeo** complex. Explorations by A.C. Blanc in 1937 revealed an **Upper Paleolithic** industry and he spent the next several seasons working there. It was not until 1954 that he found a **Mousterian** industry and hominin remains in level 4 of the Obermeier entry of the cave. <u>Temporal span and how dated?</u> Tool typology suggests a **Middle Paleolithic** date. <u>Hominins found at site</u> A fragmentary immature mandible and three teeth have been recovered from the Mousterian levels. Previously known as Circeo IV, this individual is now generally called Fossellone 3 and is identifiably *Homo neanderthalensis* in character. At least two other individuals (Fossellone 1 and 2) were found in the **Aurignacian** layers. <u>Archeological evidence found at site</u> A long Aurignacian (from the upper levels) and Mousterian (from the lower levels) sequence was recovered by Blanc, as well as a large sample of bone tools. Key references: <u>Geology, dating, and paleoenvironment</u> Blanc 1954; <u>Hominins</u> Mallegni 1992; <u>Archeology</u> Vitagliano and Piperno 1991.

fossil   (L. *fossilis*=to dig) The word fossil now means a relic, or trace, of a formerly living organism, but in the past the term fossil was also applied to inorganic materials, such as rocks and crystals. In its modern interpretation, there are two categories of fossil. The first category comprises **true fossil**s, which are remnants of the organism itself, either in the form of hard (i.e., bones and teeth) or soft (i.e., skin, muscle, etc.) tissues. The latter are seldom preserved in the fossil record and in the case of human evolution soft-tissue fossils are confined to relatively recent and rare examples of bodies preserved in bogs, or within the ice of glaciers. The second category comprises **trace fossils** (also called ichnofossils). These provide direct evidence that an organism has been at a particular place at a particular time. Sometimes sediments function like casting material, retaining details of the surface of plant (e.g., leaves) and animal soft tissues or bones long after the individual, or its hard tissues, have disappeared. Footprints (e.g., the Laetoli and Ileret footprints) are an example of a trace fossil that preserves the impression of a soft tissue (i.e., the skin of the sole of the foot), and the natural endocranial casts that faithfully reproduce the inner (i.e., endocranial) surface of the missing parts of the brain case of the southern African fossil hominins (e.g., **Taung 1**) are an example of a trace fossil of a hard tissue. Other categories of non-cast-like trace fossils include coprolites (fossilized feces).

fossilization   The process whereby organic materials (usually bones or teeth but other materials like wood) are transformed into a **fossil**. During life, the bones and teeth – the structures that make up the vast majority of the hominin fossil record – are made up of a mixture of organic **matrix** and inorganic material. The dense cortical bone that forms much of the **cranium** and **mandible** and the outer cortical bone of the long bones of the limbs are comprised approximately 70% by volume of mineral, or inorganic, material, and approximately 20% by volume of organic material. The inorganic component is mostly a complex of chemicals called hydroxyapatite. The organic material consists predominantly of **collagen** embedded within a matrix of other organic components (as well as the cells that resorb bone and secrete bone matrix). The balance (approximately 10% by volume) is made up of water. **Dentine** forms the core of mammalian teeth and has a chemical composition similar to that of bone. **Enamel** is the hard, white, outer covering of teeth; it consists of approximately 95% by weight inorganic material and less than 1% by volume organic material. Enamel is the hardest biological tissue known and is almost pure **apatite**. During fossilization the organic component of hard tissues degrades and is replaced either totally or partially by minerals from the surrounding rock. Many trabecular

spaces in bone become filled with deposits of, for example, calcite (calcium carbonate). The details of the fossilization process are site-specific and fossilization may result in the elimination of all of the original tissue and its replacement by minerals from the surrounding deposits; thus, fossils are effectively "bone-shaped" or "tooth-shaped" rocks. Plant fossils, for example, may become completely silicified but they can retain exquisite details of microstructure (e.g., Bamford 2005). Hydroxyapatite easily reduces to fluorapatite as the hydroxyl groups are replaced with fluoride from groundwater and it is this process that makes fossils extremely hard. Enamel, dentine, and bone may take up elements from the surrounding deposits (e.g., manganese and iron) which stain them brown or black. The chemicals in the inorganic component are exchanged with those in the rock to different degrees and in different ways. Some cells actually mineralize and preserve during life within bone and fossils bones can contain evidence of cells and even cell structure. Klinge et al. (2005, 2009) provide examples of the processes involved with respect to the fossilization of bone and dentine that have taken place at Koobi Fora, Kenya. The chances that the bones and teeth of a dead animal will end up in the fossil record are vanishingly small. Most of the animals that die in the wild are eaten by **scavengers**, their soft tissues are consumed or degraded by insects and bacteria and once the soft tissues that link the bones at joints are lost the bones are disarticulated and scattered. They are then reduced to small taxonomically unrecognizable fragments by a combination of the destructive effects of **diurnal** heating and cooling and by being trampled underfoot by large mammals. In all but a very few cases these processes effectively eliminate the soft and hard tissues thus removing any evidence that the animal had existed and denying it a place in the fossil record. If an animal does survive as a fossil it is usually because the remains were covered with **sediment** of some kind soon after death. These covering layers of sediment protect the skeletal remains from damage and the sediment moisture provides a medium for the exchange of chemicals between the sediments surrounding the fossil and the fossil itself. Intentional burial increases the chance that hard tissues will fossilize, which partially explains the increase in human fossils in later time periods. Higher population densities likely also contribute to this increase in numbers of fossils. Some of the organic material in bones is retained long enough to make it useful for dating (*see* **radiocarbon dating; amino acid racemization** dating) and for helping to identify what type of animal the bone comes from (e.g., protein and **DNA** analysis). Researchers have been able to recover short fragments of DNA from relatively recent (<100 ka) fossil evidence such as that for *Homo neanderthalensis* and *Homo sapiens*, thus opening up a new line of evidence for scientists to use to assess the relationship between Neanderthals and modern humans. *See* **ancient DNA**.

Foster's island rule   *See* **island rule**.

founder effect   A concept developed by Ernst **Mayr** based on the foundational research of Sewall Wright. The term refers to the reduction in **genetic diversity** that occurs when a subset of the individuals belonging to a larger population forms the basis of (or "founds") a smaller subpopulation. The new subpopulation is likely to have allele frequencies that differ from those of the parent population and since this sampling event would also most likely change the gene frequencies in the parental populations, this phenomenon is often referred to as intragenerational genetic drift. In modern humans, founder effects can be seen in subpopulations such as the Amish, the Finns, and the French-Canadians, for each of these subpopulations has an increased incidence of particular inherited diseases. The founder effect very likely contributed to the genetic distinctiveness of subpopulations of early hominin taxa that formed the basis of new species. Often, founder populations become extinct, but occasionally they may survive and evolve into a new species via the process of **peripatric speciation**. *See also Homo floresiensis*.

fovea   (L. *fovea*=a small pit) A depression on an otherwise smooth surface. Teeth The fovea anterior, or trigonid basin, is a depressed area of enamel between the **mesial marginal ridge** and the **trigonid crest**, and the fovea posterior, or **talonid basin**, is a depressed area of enamel between the ridge connecting the **entoconid** with the **hypoconulid** and the **distal marginal ridge**. Bones There is a fovea on the head of the **femur** (the fovea capitis) where the ligament of the head of the femur would have been attached. Eye There is a fovea in the central portion of the macula region of the retina of all haplorhine primates (called the fovea centralis retinae). This part of the retina has an extraordinarily high packing density of cone photoreceptors and makes high visual acuity possible.

fovea anterior (mandibular)   Term used by Selenka (1898/1900) for the depression on the occlusal surface of a mandibular molar between the **mesial marginal ridge** and the **distal trigonid crest** of a mandibular molar (syn. anterior fovea, mesial pit, precuspidal fossa, prefossid, trigonid basin). *See also* **fovea**.

fovea anterior (maxillary)   Term used by Selenka (1898/1900) for the depression on the occlusal surface of a maxillary molar between the **mesial marginal ridge** and the **epicrista** of a maxillary molar (syn. anterior fovea, mesial fovea, mesial pit). *See also* **fovea**.

fovea centralis (mandibular)   *See* **central fovea (mandibular)**.

fovea centralis (maxillary)   *See* **central fovea (maxillary)**.

fovea posterior (mandibular)   A depression (sometimes a pit or a groove) in the talonid component of a mandibular molar crown between the ridge connecting the **entoconid** with the **hypoconulid** and the **distal marginal ridge** (syn. posterior fovea, postcuspidal fossa, post-talonid basin, distal pit). *See also* **fovea**.

fovea posterior (maxillary)   Term used by Selenka (1898/1900) for the depression in the **trigon** component of a maxillary molar crown between the **hypocone**, the **distal marginal ridge**, and the **distal trigon crest** (syn. distal fovea, distal pit, distal triangular fossa, posterior fovea, oblique furrow, oblique groove, talon basin). *See also* **fovea**.

FOXP2   An abbreviation of forkhead box P2. The FOXP2 **protein** is a **transcription factor** that regulates **genes** involved in the development of the brain, lung, and gut. Evidence from songbirds and mice implicates FOXP2 expression in the brain with the production of vocalizations. In modern humans, expression of FOXP2 protein is important for the normal development of the neural circuits underlying language (Lai et al. 2001). There are two amino acid substitution differences between the FOXP2[human] protein and the equivalent protein in chimpanzees, FOXP2[chimp]. Researchers have speculated that these differences may be involved in the evolution of complex language in modern humans. *See also* *FOXP2*.

FOXP2 An abbreviation of forkhead box P2. *FOXP2* (modern human genes are always italicized) is the shorthand for a **gene** that encodes a **transcription factor** important for the development of the brain, lung, and gut. Recent evolutionary changes in the *FOXP2* gene sequence are thought to be involved in the production of modern human **language**. Lai et al. (2001) showed that the gene had undergone a **mutation** in all of the members of a family of modern humans that were affected by the same dysfunction of orofacial movement and language. The affected family members, furthermore, showed functional and anatomical abnormalities in regions of the brain that are important in language processing, such as **Broca's area**, the striatum, and the cerebellum. When Enard et al. (2002) compared the product of the *FOXP2* gene in modern humans, the FOXP2[human] **protein**, with the equivalent protein in chimpanzees, FOXP2[chimp], they found that two amino acid substitutions had occurred in the hominin clade sometime after hominins shared their most recent common ancestor with chimpanzees and bonobos, but Krause et al. (2007) showed that *Homo neanderthalensis* had the modern human-like form of *FOXP2*. Subsequently Konopka et al. (2009) showed that the "genes that are differentially regulated" by FOXP2[human] and FOXP2[chimp] include "some with functions critical to the development of the human central nervous system" and they claim that their study "reveals enrichment of differential FOXP2 targets with known involvement in cerebellar motor function, craniofacial formation, and cartilage and connective tissue formation" (*ibid*, p. 217). They also comment that the "differences exhibited by humans and chimpanzees cannot be explained by differences in DNA sequence alone, and are probably due to differences in gene expression and regulation" (*ibid*, p. 217). Congruent with this view of the differential gene regulation of FOXP2[human], Enard et al. (2009) showed that mice with insertion of FOXP2[human] display differences from wild-type littermates in their vocal production, exploratory behavior, dopamine neurotransmission, and neuronal morphology in the striatum.

fractionation (L. *frangere* = to break) When small differences in mass cause pairs of **stable** isotopes (e.g., ^{15}N and ^{14}N) to behave slightly differently during chemical and physical reactions, leading to different abundances of isotopes in the product and reactant or in multiple products. This is because isotopes of an atom have a different number of neutrons and thus different masses. There are two categories of fractionation: equilibrium and kinetic. Equilibrium fractionation leads to the concentration of isotopes in different components of a reversible system that is in equilibrium. The component, or phase, with relatively more of the heavy isotope is said to be "enriched" with respect to the heavy isotope and "depleted" with respect to the lighter isotope. Usually, the heavy isotope (e.g., ^{15}N) concentrates in the component with the strongest bonds (see International Atomic Energy Agency; www.unu.edu/unupress/food2/uid05e/uid05e00.htm). Kinetic fractionation occurs when one isotope reacts more rapidly than the other in an irreversible system, or in a system in which the products are removed from the reactants before they have an opportunity to reach equilibrium. In general, the lighter isotope (e.g., ^{14}N) reacts faster than the heavier isotope (e.g., ^{15}N), leaving more of the heavy isotope in the reactant and more of light isotope in the product(s) (see International Atomic Energy Agency). In this case fractionation will *only* occur if the reaction does not reach completion, for if all of the reactant is turned into product there cannot be an isotopic difference between the product and the reactant. The isotopic offset, or fractionation, between two substances (a and b) is often expressed using Δ (or fractionation factor) notation (Martínez del Rio et al. 2009), where $\Delta^H X_{a-b} = \delta^H X_a - \delta^H X_b$. The Δ values are trivial to calculate and are accurate as long as the differences in δ values among tissues are small. However, Δ values become less accurate as the δ value differences among tissues increase. If the Δ values between substances are 10 or greater, it is preferable to calculate **fractionation factors** (α) and **enrichment** values (ϵ) instead of δ.

fractionation factor The fractionation factor (or α) provides the exact isotopic difference between a product and a reactant. The fractionation factor is similar to δ, but unlike δ it is not limited by the isotopic scale on which it is calculated (e.g., the Pee Dee Belemnite for carbon and Standard Mean Ocean Water for oxygen). The calculation for the fractionation factor, or α,

263

between two substances (a and b) is straightforward: $\alpha_{a-b} = (\delta^H X_a + 1000)/(\delta^H X_b + 1000)$. If the calculated fractionation factor is $>1‰$, the substance a is more enriched in the heavy isotope than substance b. Conversely, if α is $<1‰$, the substance a is less enriched (also called depleted) in the heavy isotope than substance b.

fracture and deformation biomechanics

The study of the way loads applied to a structure initiate and propagate irreversible damage (e.g., cracks). It focuses on the ways the shape and the material properties of a structure influence its response to such loads. These methods have been used to predict the ways the shape and internal structure of tooth crowns affect the ability of teeth to fracture food without themselves suffering irreversible damage (e.g., Lucas 2004, Lucas et al. 2008).

frameshift mutation

A **mutation** within the **coding sequence** of a **gene** that involves changes other than the insertion or deletion of whole **codon**s (i.e., other than multiples of three **nucleotide**s). Such mutations disturb the **reading frame** of the gene and because of this they typically disrupt the **messenger RNA** (mRNA) produced by the gene, which can lead to a truncated version of the **protein** being translated, a protein with disrupted function, or no protein at all.

French-Chadian Paleoanthropological Mission

See **Mission Paléoanthropologique Franco-Tchadienne**.

frontal lobe

One of the four main subdivisions of the **cerebral cortex** of each **cerebral hemisphere**. It forms the anterior part of the cerebral hemisphere; the **parietal lobe** is immediately posterior to it. The most anterior point on the cerebral hemisphere, called the anterior pole, is formed by the frontal lobe. The frontal lobe is important for controlling movement and in planning and coordinating behavior. The frontal lobe contains the primary motor cortex, which lies in the **precentral gyrus** and is important for controlling voluntary movements of the body. Although neuroscientists traditionally believed that enlargement of the frontal lobe in modern humans was responsible for enhanced cognitive abilities in our species, recent studies have shown that the frontal lobe of modern humans comprises approximately the same proportion of the cerebral hemisphere as in the great apes (Semendeferi et al. 2002).

frontal trigon

The region, which straddles the midline between the **supraorbital tori** and the **temporal crest**s or raised temporal lines. It can be either flat or hollowed and it is especially evident in some megadont archaic hominin crania (e.g., **OH 5**).

frugivory

(L. *frug-*, *frux*=fruit, and *vorous*, from the stem of *vorare*=to devour) Literally, fruit-eating. Animals that eat fruit are frugivores. Most primates are frugivores to a greater or lesser extent. Primate frugivores may eat ripe or unripe fruits and seeds. **Hominoids** favor ripe fruits, whereas some monkeys can tolerate unripe fruits, which for them may be an important **fallback food**. Some have argued that the ability to eat unripe fruit is one important reason for the adaptive radiation of the Old World monkeys, and it may have helped them take over many of the niches previously occupied by hominoids in the late **Miocene** and **Pliocene**. Frugivory has been linked to the evolution of color vision and to the expansion of brain size through **extractive foraging** (Gibson 1986) via the need to exploit patchily distributed resources (Clutton-Brock and Harvey 1980), which may require mental maps for successful navigation (Milton 1988). In general, frugivorous primates have larger **brain** sizes and larger **home range**s relative to body mass than do those relying on **folivory**.

FS

Used in at least two senses in paleoanthropology. Koobi Fora In connection with **Koobi Fora**, Kenya, FS is used as the acronym for field surface. At the beginning of field research at Koobi Fora the recording system differentiated the fossils found on the surface (i.e., FS for field surface) from those found *in situ* (i.e., FI for field *in situ*). But because so few fossils were being found *in situ* the system was abandoned in favor of a single system (i.e., **KNM-ER** then the field catalogue number) that covered all the fossils that were recovered and catalogued (e.g., the cranium FS-158 became **KNM-ER 406**). Sangiran In connection with **Sangiran**, Java, Ralph **von Koenigswald** had given all of the isolated teeth recovered from Sangiran the prefix FS [i.e., fossils (from) Sangiran] followed by a number. Later the prefix S7 was introduced by Jacob (1975) for teeth subsequently recognized as hominin. In Jacob's scheme if teeth were recognized as hominin they carried their FS number with them (e.g., FS 89 became S7-89). *See also* **KNM-ER 406**; **Sangiran 7**.

FS-158   *See* **KNM-ER 406**.

FS-210   *See* **KNM-ER 407**.

F_{ST} *See* **F statistics**.

Fuentenueva   One of the sites of the **Orce region**. The latter is known for abundant faunal remains dating to the **Plio-Pleistocene** along with some potential stone tools. *See also* **Orce region**; **Venta Micena**.

fulcrum   (L. *fulcire*=to support) A fulcrum is the point about which a lever pivots.

fulcrumate   An antiquated term used by Morton (1924) to distinguish between "tarsi-fulcrumating" and "metatarsi-fulcrumating" primates. The former, exemplified by tarsiers and galagos, produce propulsion using an enlarged posterior tarsal region that acts as a lever about a fulcrum positioned anteriorly in the tarsus. Morton (1924) categorizes all anthropoids as "metatarsi-fulcrumating" primates that include the metatarsals as part of the propulsive lever. In these animals the fulcrum is positioned at the metatarso-phalangeal joints.

Fumane   (Location 45°35′N, 10°54′E, Veneto, Italy; etym. after the nearby town. The site is also known as Stazione della Neve or Riparo Solinas) History and general description A rock-shelter/cave site in the Venetian pre-Alps. Excavations began in 1988 under A. Broglio and M. Cremaschi; so far, only the area at the front of the cave has been systematically excavated. Temporal span and how dated? One of the **Upper Paleolithic** levels, A2, was dated using **radiocarbon dating** to 40+4/−3 ka. Level S7, a **Middle Paleolithic** level, has a thermo**luminescence dating** age of 79±13 ka. Hominins found at site Three fragmentary hominin teeth were found, two of which are from the Middle Paleolithic levels and thus most likely belong to *Homo neanderthalensis*. Archeological evidence found at site Typical **Mousterian** artifacts with **Levallois** debitage characterize the lower occupation levels. The abundant faunal assemblage in the Mousterian levels are most likely due to hominin activity. Key references: Geology, dating, and paleoenvironment Bartolomei et al 1992; Hominins Bartolomei et al. 1992; Archeology Fiore et al. 2004.

functional genomics   The study of basic functional aspects of the **genome**. Functional genomics subsumes research topics such as the mechanisms of **gene expression** during development, gene network interactions, gene–protein interactions, and the **epigenetic** processes responsible for proteome production.

functional magnetic resonance imaging   (or fMRI) A method commonly used to investigate brain activity in modern human subjects and occasionally in macaque monkeys. fMRI works by the external application of magnetic fields and electromagnetic pulses to manipulate the spins of atomic nuclei, inducing them to emit electromagnetic signals, which can be monitored to map the functional activity of living brains. This technique is based on the fact that changes in neural activity are accompanied by changes in regional blood flow and glucose utilization. Functional imaging maps these regional metabolic changes by tracking the different magnetic properties of oxygenated and deoxygenated hemoglobin present in the blood during regional activity.

furrow defects   *See* **hypoplasias**.

FwJj14E   (Location 4°18′44″N, 36°16′16″E, Koobi Fora, Kenya) A 1.5 Ma archeological and paleontological site within the **Koobi Fora Formation** in Area 1A at **Ileret** where hominin footprints have been discovered at two footprint surfaces vertically separated by 5 m and separated in time by *c*.10–20 ka (Bennett et al. 2009). The presence of animal footprints was recognized in 2005 and the first hominin footprints were exposed by excavation in 2007. In the upper footprint surface (which is less than a meter above the Ileret Tuff and about 2 m below the Northern Ileret Tuff) in addition to the prints of many nonhominin animals there are three hominin footprint trails (two trails of two footprints and one of seven footprints) plus several isolated single hominin footprints. In the lower surface (which is less than a meter above the Lower Ileret Tuff) there is one trail of two hominin footprints, plus one isolated hominin footprint. For details of their analysis and functional interpretation *see* **Koobi Fora hominin footprints**.

G

G-function *See* **fitness-generating function**.

Gadeb (Location 7°05′N, 39°21′E, Ethiopia; etym. named after the Plains of Gadeb, Ethiopia below the site) History and general description Gadeb is near the western edge of the South-East (or Somali) Plateau that borders the **Afar Rift System** near the headwaters of the Webi Shebeli River. It was initially explored in 1975–7 by international teams directed by Desmond **Clark**, but dam construction has since buried the excavations. Temporal span Artifacts occur within sediments of the Mio Goro Formation. Pumice at the base of the formation has a potassium-argon age estimate of 1.48 Ma and all sediments are normally magnetized, suggesting an age greater than the Brunhes-Matuyama boundary (i.e., >780 ka). An age range of c.1.5–0.78 Ma is supported by **tephra** correlated from the **Turkana Basin** (Silbo Tuff, Lower Nariokotome Tuff, and the Black Pumice Tuff) to stratigraphic sections near the Gadeb archeological sites, but because dam construction has buried the excavations, more precise correlation is impossible. How dated? **Potassium-argon dating, tephrostratigraphy**, and **magnetostratigraphy**. Hominins found at site None. Archeological evidence found at site Artifacts occur at a number of localities and stratigraphic levels in the Gadeb area have been provisionally attributed to the **Developed Oldowan** (but containing **handaxes**) and to the **Acheulean**. Analysis and publication focused on Locality 8 where more than 20,000 stone artifacts were recovered from a >115 m^2 excavated area. Notable are the abundant (approimately >450) handaxes and **cleavers** made from local **basalt** boulders and **ignimbrite** outcrops a few kilometers away, as well as a possible hippopotamus butchery site. The analysis of the recovered artifacts was conducted by H. Kurashina. Key references: Geology, dating, and paleoenvironment Eberz et al. 1988, Haileab and Brown 1994; Archeology Clark 1987, Clark and Kurashina 1979.

Gademotta (Location 8°03′N, 38°15′E, Ethiopia; etym. unknown) History and general description This artifact-bearing area on the flanks of a collapsed caldera approximately 5 km/3 miles west of Lake Ziway in the **Main Ethiopian Rift** includes the type section of the Gademotta Formation, which consists largely of **volcaniclastic** sediments and **paleosols** of approximately 40 m thickness. Gademotta and the slightly younger site of **Kulkuletti**, about 2 km/1.2 miles away, also within the Gademotta Formation, are correlated through shared **tephra**. Temporal span and how dated? **Argon-argon dating** of **sanidine** crystals in tephra and **tephrostratigraphy** suggest the Unit 10 tuff above the **Acheulean** and the oldest **Middle Stone Age** (or MSA) strata is c.280±8 ka and the tuffaceous Unit D that overlies it has an age estimate of c.180±10 ka. Hominins found at site None. Archeological evidence found at site Among the >40,000 recovered artifacts are **Levallois flake**s, **point**s, **core**s, and **blade**s (many of them large), unifacially and bifacially **retouch**ed points, as well as blades and blade cores. These are sufficient to attribute the four excavated strata from Gademotta to the MSA. Many of the points have **resharpening spall**s removed from the tip that are distinct from impact fractures; this, combined with **use wear** evidence, suggests the points may have been used for cutting and sawing rather than as projectiles. **Obsidian** from the c.1.3 Ma volcanics on which the Gademotta Formation sediments lie was used as the primary stone raw material and the site was likely either a **quarry**, or a near-quarry workshop. The recovered fauna are sparse and limited to hippopotamus, equid, and three varieties of bovid. The age estimates and numerous diagnostic artifact forms suggest that Gademotta is the oldest known MSA site. Key references: Geology, dating, and paleoenvironment Laury and Albritton 1975, Morgan and Renne 2008; Archeology Wendorf and Schild 1974, Schild and Wendorf 2005.

Wiley-Blackwell Encyclopedia of Human Evolution, First Edition. Edited by Bernard Wood.
© 2013 Blackwell Publishing Ltd. Published 2013 by Blackwell Publishing Ltd.

gait (ON *gata*—path) The way in which an organism (e.g., hominin, horse, etc.) uses its feet for locomotion. For example, the difference between a **bipedal** running gait and a bipedal walking gait is that in the latter one foot is in contact with the ground at all times.

GaJi10 (Location 3°44′15″N, 36°55′48″E, Koobi Fora, Kenya) A 1.43 Ma archeological site in Area 103 of the **Koobi Fora Formation**, Kenya, where a single footprint trail comprising seven hominin footprints were discovered in 1978 on a land surface (Behrensmeyer and Laporte 1981). Four of these seven footprints were recently re-excavated (Bennett et al. 2009) and an additional hominin footprint was exposed during that re-excavation. *See also* **Koobi Fora hominin footprints**.

Galeria (Location 42°21′09″N, 3°31′08″W, Spain; etym. karst gallery) History and general description One of the fossiliferous cave fillings in the **Sierra de Atapuerca**, a series of eroded, karstic, limestone hills 14 km/9 miles east of Burgos in northern Spain, which is permeated by sediment-filled cave systems. The largest is the Cueva Mayor-Cueva del Silo system, where the **Sima de los Huesos** is located, but just 1 km/0.6 miles away, in the Trinchera del Ferrocarril, is the Galeria (TG) site with 17 m of sediment. Excavations at Galeria, which is very close to **Gran Dolina** (TD), began in 1979. Originally the site of Galería was included as part of a larger complex of karst infillings, taking the general name of Galeria-Tres Simas complex. Later on, the subsystem Galeria was in turn subdivided into three areas or sectors. The middle part was called Galeria *sensu stricto*, to the south is the Covacha de los Zarpazos (TZ), and to the north a vertical infilling called TN. All three sectors are connected but they apparently received distinct sources of sediments. The early terminology distinguished 12 geologic levels (TG1–12) (Aguirre et al. 1987), but these have subsequently been rationalized into six stratigraphic units, GI–VI (Carbonell et al. 1999); of these GIIa, GIIb, and GIII have produced stone tools and Middle Pleistocene faunal remains. Temporal span and how dated? **Biostratigraphy**, thermo**luminescence**, and **uranium-series dating** suggest the sediments range from 503±95 ka at the base of the lowermost "surface-inwash" facies to 185±26 ka at the top. Hominins found at site A mandible fragment with M_{2-3} (AT76-T1H) was discovered in 1976 and a cranial vault fragment was found at the base of GIII

stratigraphic units Archeological evidence found at site **Mode 2** artifacts in layers GIIa, GIIb, and GIII, plus clear evidence of butchering on many faunal remains. Key references: Geology, dating, and paleoenvironment Aguirre et al. 1987, Rosas et al. 1998, Carbonell et al. 1999, Berger et al. 2008; Hominins Bermúdez de Castro 1992; Archeology Carbonell et al. 1999.

Galería del Osario *See* **El Sidrón**.

Galeria Musteriană *See* **Baia de Fier**.

Galili [Location, in the Mulu Basin, east of Gedameyto, 9°46′N, 40°33′E, Somali Region, Ethiopia; etym. the name of the site derives from Mount Galili, the hill in the center of the area (White Mountain in the language of the local nomads, the Issa). On official maps the hill is marked as Serkoma, but the locals refer to it as Galili (depending on transcription also Galila or Galilie)] History and general description The site is 30 km/18 miles to the west of the Mulu basin, and the latter is separated by several horsts from the Galili deposits. The Galili area drains not towards the Mulu, but into the Awash to the northwest. The Paleoanthropological Research Team began prospecting in the area in 2000 and they report a rich mammalian fauna including hominins from four members (Lasdannan, Dhidinley, Godifray, and Shabeley Laag) of the Mount Galili Formation. Two other members higher in the section (Dhagax and Caashacaado) have so far not yielded any fossil evidence except a bovid phalange from the Caashacaado Member; the total thickness of the deposits is probably close to 200 m. Temporal span and how dated? Biostratigraphic evidence suggests the fossil mammals are equivalent to those from the Kanapoi Formation, the uppermost part of the Apak Member at Lothagam, and the lower parts of the Kaiyumung Member of the Nachukui Formation (Kullmer et al. 2008, p. 461). Unpublished argonargon ages suggest that the sediments at Galili date from *c*.5 Ma to *c*.3.4 in the lower (fossiliferous) part. Hominins found at site "Six hominid teeth and a proximal femur fragment" (Kullmer et al. 2008, p. 452), plus a distal humerus, a metatarsal, and four additional teeth; some of the possible hominin fossils are >4.4 Ma. Archeological evidence found at site The archeological evidence all postdates the deposits. The few handaxes recovered are from from recent alluvium, and the **Middle Stone Age** and later lithic scatters are all on the surface; none of it is *in situ*. Key references: Geology, dating, and paleoenvironment Kullmer et al. 2008; Hominins Macchiarelli et al. 2004b; Archeology N/A.

gallery forest (L. *galeria*=covering, as in the covering, or portico, over a door) Forests running alongside rivers or bordering wetlands. These strips of evergreen forest survive in environments that otherwise may have little forest, or even woodland, because of (a) the greater fertility of soils around rivers and wetlands, (b) the reliability of the water supply, and (c) the protection provided from natural fires by rivers and wetlands. Part of the woodland and/or grassland biome, these forest ecosystems border rivers and wetlands and although occasionally or seasonally flooded by river waters, they are dry for most of the growing season. The alluvial soils of these forests are high in nutrients and have better drainage than adjacent woodland and grassland ecosystems, thus hosting unique flora and fauna that are not present in drier forest types. A key characteristic is that they project into biomes such as woodland, grasslands, or deserts; the boundary between gallery forests and the surrounding woodland or grassland is usually abrupt. Gallery forests are important for primate biodiversity as they allow arboreal primates to survive in regions that would otherwise be suitable only for open-habitat adapted animals. Most early **hominins** were associated with tropical woodlands, but it is also likely that some (particularly those in the **Miocene** and **Pliocene**) existed in areas that contained tropical, and possibly gallery, forest. However, tropical forest is seen as a less favorable habitat for hominins than woodland because of the inaccessibility of fruits in the higher canopy and the cryptic nature of forest animals.

Galley Hill *See* **fluorine dating**.

Gamedah One of the named drainages/subdivisions within the **Bodo-Maka** fossiliferous subregion of the **Middle Awash study area**, Ethiopia.

gamete (Gk *gamos*=marriage) A reproductive or germ cell with a **haploid** (i.e., they only have one copy of each **chromosome**) complement of chromosomes. The gametes (i.e., the egg and sperm) pass genetic information from one generation to the next. All the other cells in the body are **diploid** (i.e., they have two copies of each chromosome) and are called **somatic cells**. *See also* **germline**.

gametic phase The term refers to combinations of **alleles** at different **loci** linked together on the same **chromosome**.

gamma taxonomy (Gk *gamma*=the third letter in the Greek alphabet and *taxis*=to arrange or "put in order") It is common for reference to be made to **alpha taxonomy**, but you very seldom read, or hear, any reference to "**beta**" or "gamma taxonomy." Mayr et al. (1953) is one of the few sources that define these terms [e.g., they do not appear in the index of Simpson's (1961) *Principles of Animal Taxonomy*]. Mayr et al. (1953) suggest that "the taxonomy of a given group, therefore, passes through several stages...informally referred to as alpha, beta, and gamma taxonomy" (*ibid*, p. 19). The distinction made by Mayr et al. is that whereas alpha taxonomy results in species being "characterized and named" and beta taxonomy refers to phylogenetic reconstruction, gamma taxonomy refers to the "analysis of intraspecific variation and to evolutionary studies" (*ibid*, p. 19). But Mayr et al. (1953) caution that it is "quite impossible to delimit alpha, beta and gamma taxonomy sharply from one another, since they overlap and intergrade" (*ibid*, p. 19). *See also* **systematics**; **taxonomy**.

Gánovce (Location 49°10′N, 20°18′E, Slovakia, etym. named after a local town) History and general description The site is in a travertine quarry 3 km/1.8 miles southeast of Poprad below the High Tatra Mountains. The first finds, including the hominin remains, were reported in 1926 and excavations were conducted in 1955. Temporal span Oxygen Isotope Stage 5e. How dated? Uranium-series dates on travertine (130, 105, and 84 ka years BP); biostratigraphy. Hominins found at site A natural endocast, cranial fragments, and natural molds of postcranial fragments attributed to *Homo neanderthalensis*. Archeological evidence found at site Artifacts described as Taubachian (**Mousterian**). Geology, dating, and paleoenvironment Hausmann and Brunnacker 1986; Hominins Vlček 1955, Smith 1982; Archeology Vlček 1969.

Gaolingpo *See* **Bose Basin**.

gape *See* **chewing**.

Garba *See* **Melka Kunturé**.

Garrod, Dorothy Annie Elizabeth (1892–1968)

Dorothy Garrod was born in London, England, and in 1913 was admitted to Newnham College, University of Cambridge, to study history. After graduating in 1916, Garrod joined the war effort, serving in the Catholic Women's League in France from 1917 to 1919 taking care of wounded soldiers. In 1919 she joined her father in Malta and he suggested that Garrod explore Malta's prehistoric ruins, a recommendation that sparked her interest in archeology. Garrod returned to England and enrolled in classes at the University of Oxford in 1921 to study anthropology with Robert R. Marett, who had excavated at the Paleolithic site of La Cotte de St. Brelade in France. Given Garrod's interest in prehistoric archeology, Marrett suggested that she go to France to work with the eminent French prehistorian Abbé Henri Breuil. Garrod arrived in Paris in 1922 and studied with Breuil at the **Institut de Paléontologie Humaine** (Institute of Human Paleontology) until 1924. While in France, Garrod participated in excavations at the Paleolithic site of **La Quina** with Henri Martin and also worked at Les Eyzies. During this period Garrod also met George Grant MacCurdy, who taught anthropology and archeology at Yale University and who had just become the director of the newly founded American School of Prehistoric Research. With Breuil's encouragement, Garrod began working on a book that would synthesize what was known about the late Paleolithic in Britain. The book, *The Upper Paleolithic in Britain* (1926), earned Garrod an undergraduate degree from the University of Oxford. Breuil also arranged for Garrod to excavate a site in Gibraltar called the **Devil's Tower**. Excavations began in 1925 and soon Garrod found **Mousterian** tools and skull fragments belonging to a *Homo neanderthalensis* child. While Garrod was excavating Neanderthal remains in Gibraltar, fossil hominins were being unearthed by Francis Turville-Petre of the British School of Archeology in Jerusalem at Zettuphe cave in what was then Palestine. This discovery prompted George Grant MacCurdy to plan a joint Anglo-American archeological expedition to the region to be led by Garrod. Garrod arrived in Palestine in 1928 and began excavating at the cave of Skukbah where she discovered Levalloiso-Mousterian tools and a more recent Mesolithic tool type that she called **Natufian**. Also in 1928 Garrod was elected president of the Prehistoric Society of East Anglia and in her presidential address entitled "*Nova et Vetera*: a plea for a new method in Paleolithic archaeology" she argued for a more phylogenetic approach in archeology that would focus on the inter-relationship of **Paleolithic** cultures globally which then would result in the construction of family trees showing how different cultures and archeological industries influenced each other. This would replace the more geological stratigraphic and primarily European approach developed by the French prehistorian Gabriel de Mortillet. Garrod began excavations at **Mount Carmel** in Palestine in 1929 and soon found Natufian-type tools and human remains. Her careful excavations of several caves at the site, including the **Tabun** and **Skhul** caves, allowed her to work out a stratigraphy of tool types ranging from the Mousterian to the Bronze Age that spanned about 600 ka of human prehistory. In 1932, while Garrod was away in Britain, the American paleontologist Theodore McCown discovered the remains of several burials in the Skhul cave. By the time excavations ended at Mount Carmel in 1934 Garrod and her team had recovered a substantial quantity of tools, animal fossils, and Neanderthal remains that provided a remarkably clear picture of the Paleolithic in Palestine. It was also clear that the stratigraphic sequence of Paleolithic cultures and industries originally developed from excavations in Europe did not correspond to the sequence discovered in western Asia. Some of the discoveries indicated that certain industries emerged in Asia earlier than they appeared in Europe and may have been introduced into Europe from western Asia. Garrod published the results of her research in Palestine in *The Stone Age of Mount Carmel* (1937). In recognition of this achievement Garrod was granted a doctorate by the University of Oxford. Garrod accepted the position of Disney Professor of Archaeology at the University of Cambridge in 1939, but the outbreak of WWII put an end to any prospects of continuing excavations in Palestine. In addition to her teaching, Garrod also served in the Women's Auxiliary Air Force from 1942 to 1945, interpreting aerial photographs of bomb damage. Garrod established anthropology and archeology as full-degree courses at Cambridge during her tenure and at the end of the war she returned to France to conduct excavations at **Fontéchevade** with Germaine Henri-Martin and at Angles-sur-l'Anglin with Suzanne de St. Mathurin. Garrod retired from teaching in 1952 and in 1954 she returned to western Asia to continue the research that had been interrupted by the war. Between 1958 and 1962 she excavated sites in Lebanon and she worked out the sequence of cultures and tool industries for the region, comparing them with the sequence established at Mount Carmel and correlating these with the corresponding sequences of prehistoric Europe. In addition to making significant contributions to prehistoric archeology, Garrod mentored a generation of women archeologists and strengthened the teaching of archeology at Cambridge. In 1968, shortly before her death, Garrod was awarded the Gold Medal of the Society of Antiquaries.

Garusi The name used by Ludwig Kohl-Larsen for the site now known as Laetoli. *See* **Laetoli**.

Garusi 1 Site Laetoli. Locality There are photos of the find spot of this early hominin maxilla as well as a map showing its approximate location (Protsch 1981) but it is not possible to relate these to the specific localities currently known at Laetoli. The Garusi 1 maxilla was recovered from sediments exposed at the head of the Garusi River Valley and the preservation of the fossil matches fossils recovered from the Upper Laetolil Beds. Surface/*in situ* Surface. Date of discovery February 8, 1939. Finders Members of the Kohl-Larsen Expedition. Unit N/A. Horizon N/A. Bed/member Laetoli Beds. Formation N/A. Group N/A. Nearest overlying dated horizon N/A. Nearest underlying dated horizon N/A. Geological age Given the preservation similarity noted above Garusi 1 is likely to have come from a horizon that is $c.3.7$–3.59 Ma. Developmental age Adult. Presumed sex Unknown. Brief anatomical description A piece of the right maxillary bone extending from the midline to the distal aspect of the right P^4. The crowns and roots of the P^3 and P^4 are preserved, but the alveoli are all that remain of the incisors and canine. The specimen preserves the junction of the floor and the right lateral margin of the nasal aperture. Announcement Weinert 1950. Initial description Weinert 1950. Photographs/line drawings and metrical data Weinert 1950, Puech et al. 1986. Detailed anatomical description N/A. Initial taxonomic allocation *Meganthropus africanus*. Taxonomic revisions *Praeanthropus afarensis* Strait et al. 1997. Current conventional taxonomic allocation *Australopithecus afarensis*. Informal taxonomic category Archaic hominin. Significance The Garusi 1 maxilla was an important part of

the evidence that led Johanson et al. (1978) to link the hominins from **Hadar** and Laetoli and to combine them into a single hypodigm as *Au. afarensis*. Location of original **National Museum and House of Culture, Dar es Salaam**.

Garusi Hominid 1 *See* **Garusi 1**.

Garusi River Series *See* **Laetoli**.

gas chromatography isotope ratio mass spectrometry Gas chromatography-coupled isotope ratio mass spectrometry (or GC-IRMS) uses a continuous flow of carrier gas over a solid surface with an absorptive coating to separate compounds by molecular weight and these compounds are supplied to the mass spectrometer. In this way the isotopic composition of individual compounds can be analyzed. This is important since different compound classes may have very different isotopic compositions as a result of differences in synthesis pathways, making the compound-specific isotopic analysis an advance over bulk isotopic analyses. Compounds can be combusted (for **carbon isotope analysis**) or pyrolyzed (for **hydrogen isotope analysis**) and supplied via continuous flow to the mass spectrometer for determination of isotope ratios of each sequentially eluting compound. This system can be used for stable isotope measurements on semi-volatile compounds (e.g., plant **leaf waxes**). *See also* C_3 and C_4; **stable isotopes**.

GATC Abbreviation for Gàala Tuff Complex. *See* **Aramis**.

Gause's Rule Also known as the principle of competitive exclusion, this states that two species cannot stably occupy an identical niche within the same ecosystem. *See also* **competitive exclusion; niche**.

Gauss chron *See* **geomagnetic polarity time scale**.

Gaussian distribution *See* **normal distribution**.

gazelle (Ar. *ghazal*=hoofed animals belonging to the genus *Gazella*) *See* **Antilopini**.

GB Abbreviation of the Galana Boi Formation, **Turkana Basin, Kenya**. *See* **Galana Boi Formation**.

GBY Acronym for **Gesher Benot Ya'akov** (*which see*).

GCM Acronym for **general circulation model** (*which see*).

Geißenklösterle (Location 48°23′53″N, 9°46′17″E, Germany; etym. Ge. *Geißen*=women, *klöster*=cloister) History and general description This cave site near Blaubeuren, in the Schwabian Jura, Baden-Württemberg, was first investigated in 1957 by G. Riek and from 1973 onwards excavated by J. Hahn, and, after his death, by Nicholas Conard and Hans-Peter Uerpmann. It is one of the most important **Aurignacian** sites in Europe, particularly for the evidence of art and symbolic behavior. Temporal span and how dated? The site ranges from the **Middle** to **Upper Paleolithic**. **Radiocarbon dating** and **electron spin resonance spectroscopy dating** have provided dates of *c.*40–29 ka for the Aurignacian deposits. Hominins found at site An unpublished deciduous tooth from the Aurignacian, and two deciduous teeth from the Gravettian deposit. Archeological evidence found at site The site contains Middle Paleolithic, Aurignacian, and Gravettian lithic industries. The Aurignacian is best known for its abundant ivory figurines and ivory and bone flutes, and for evidence of pigment use on the walls, which is otherwise rare in Germany. Key references: Geology, dating, and paleoenvironment Conard and Bolus 2008, Street et al. 2006; Archeology Hahn 1986; Hominins Street et al. 2006.

gel electrophoresis *See* **electrophoresis**.

gen. et sp. indet. (etym. *gen. et sp. indet.* is an abbreviation of "indeterminate genus and species") When dealing with fragmentary specimens it is sometimes possible to identify the **higher taxon** a specimen belongs to, but it may not be possible to be sure what genus, and therefore, also what species it belongs to. In that case the taxonomic allocation researchers should use is *gen. et sp. indet.* (e.g., although some researchers are confident assigning the associated skeleton **KNM-ER 1500** to *Paranthropus boisei*, others are more cautious and refer it to **Hominidae** *gen. et sp. indet.*)

gene (Gk *genes*=to be born) A gene is a region of **DNA** that contains information for both the protein or RNA product it encodes and for regulating the expression of messenger RNA. Thus, a gene includes both coding and non-coding DNA. The structure of a gene typically includes a **promoter** region. This is usually upstream (at the 5′ end) of the gene sequence and it is where **transcription** is initiated and regulated. Most genes also have multiple **exons** and **introns**. Over evolutionary history genes (or parts of genes) can be highly conserved (i.e., they are resistant to change because they are subject to strong **purifying selection**) if change is highly detrimental or they may be less conserved if some change is compatible with preservation of protein function. Genes that have been subject to **directional selection** (e.g., *FOXP2*, which encodes a transcription factor linked to phenotypes of speech and language) have been used to reconstruct adaptive changes in the hominin clade. Gene regions, such as introns or third positions of codons, that are less conserved may effectively evolve neutrally and thus can be used to reconstruct population history. *See also* **neutral theory of molecular evolution**.

gene conversion The process whereby one copy of a **gene** is converted or changed such that it is identical to a **homologous** gene (either the **allele** of the same gene on the homologous chromosome or another gene from the same multigene family). This can occur during meiosis or mitosis as a result of the mismatch repair that occurs during **recombination**.

gene duplication If a **gene** is duplicated and if the additional copy of the gene is not selected against (for example, because of changes in the amount of protein produced), it may be free to mutate and that mutation may result in a change in function. This is one way many novel genes are created over deep evolutionary time. For example, the alpha and beta globin genes evolved via gene duplication from a myoglobin-like gene. Genes may also be duplicated because they are located in regions of the genome that are subject to **copy number variation**. For example, the salivary amylase gene, *AMY1*, which produces the enzyme responsible for starch hydrolysis in the mouth, has been duplicated such that five to seven copies are present on average in modern humans while only two copies are present in chimpanzees (Perry et al. 2007). It has been suggested that this is because modern humans, particularly agricultural populations, typically have diets that are high in starch, while this is not usually the case with chimpanzees. *See also* **duplication**.

gene expression   The process whereby the information encoded by a **gene** is used to make a gene product (e.g., a protein or one of the many kinds of RNA). Gene expression can be regulated in many ways including by **enhancers** and repressors as well as at different stages of **transcription** and **translation**. Gene expression may change during development, it may differ among the tissues of the body, and it may change according to environmental influences.

gene family   A group of related (i.e., paralogous) **genes** that arose over time via **gene duplication** and thus are descended from a common ancestral gene. Members of the same gene family often have similar functions. For example, all the genes in the **globin** and *HOX* **gene** families are genes involved, respectively, in heme transport and development. A gene family may also be referred to as a multigene family.

gene flow   The movement of alleles of genes from one population to another. Along with **genetic drift**, **mutation**, and **natural selection**, gene flow (also known as gene migration) is one of the four forces of evolution. High levels of gene flow cause two populations to become more similar (i.e., it decreases the **genetic distance** between them). In modern humans, culture, language, and geography are factors that may restrict gene flow. Gene flow can result when substantial numbers of people simultaneously migrate from one area to another in a single event or it can result when there are many episodes of smaller numbers of people moving from one place to another.

gene genealogy   *See* **coalescent**.

gene networks   Gene networks or gene regulatory networks are sets of genes that interact (usually via their products) to perform some physiological or developmental function. This is one of the core concepts of systems biology. The impetus behind the concept is that gene function can only be understood within the context of the larger system within which it operates.

gene pool   The gene pool of a population consists of all the **allele**s present in that population.

gene tree   A diagram that depicts the phylogenetic relationships among alleles or between homologues of the same gene. For example, Retief et al. (1993) analyzed the phylogenetic relationships of the protamine P1 gene in primates and found that the **chimpanzee** and **gorilla** protamine P1 genes are more closely related to each other than either is to modern humans or the orangutan. Note that one gene tree, such as that for protamine P1, can differ from the gene tree of another gene or from the structure of a multigene tree or from the species tree. Other gene trees, as well as phylogenies based on many genes, are consistent with the hypothesis that chimpanzees and modern humans are more closely related to each other than to any other taxon (Ruvolo 1997, Kumar and Hedges 1998).

general circulation model   (or GCM) A type of climate model in which the initial and the external conditions (i.e., **insolation**, **sea surface temperature**, etc.) are set and then the model is run until it has reached equilibrium with the external **forcing**s that influence it. An example of a GCM equilibrium experiment is the paired experiments for 105 ka BP run with the University of Victoria (UVic) Earth System Model to test the sensitivity of African climate to cooling and circulation changes in the North Atlantic as a result of freshwater inputs during **ice rafting** (sometimes called ice discharge) events known as **Heinrich events** (Carto et al. 2009).

generalist   *See* **eurytopy**.

generation length   The generation length, often symbolized as g or T, is the number of years that it takes a nonstationary population to grow or decline in size the amount given by the intrinsic rate of increase (r). The generation length is related to the **net reproductive rate** (or NRR), being $T = \frac{\ln(NRR)}{r}$. The generation length is difficult to estimate and is undefined when the population is stationary (i.e., the growth rate is zero). As a consequence, the mean age at childbearing is often used in place of generation length. For modern humans the mean age at childbearing (and hence the generation length) is around 25–7 years.

generation time   The average interval between the birth of an individual and the birth of their offspring, or the average age of mothers giving birth in a given population.

genetic assimilation *See* canalization.

genetic bottleneck *See* bottleneck.

genetic code The genetic code is a three-**nucleotide** system (in **DNA** and **RNA**) that encodes genetic information. Each sequence of three nucleotides (also referred to as a triplet codon) defines either an **amino acid** or a punctuation message [i.e., the beginning or end of a messenger RNA (or mRNA) or protein sequence] for the processes of **translation** and **transcription**. Examples of DNA codons are AAA and AAG for phenylalanine (UUU and UUC are the corresponding mRNA codons) and ACC for tryptophan (UGG is the corresponding mRNA codon). Since there are four nucleotides, there are 64 possible three-nucleotide sequences. However, there are only 20 commonly occurring amino acids and thus the genetic code is redundant with many amino acids such as serine and leucine (as well as the "stop" punctuation) encoded by several different codons. The code used in the modern human genome is commonly referred to as the standard or universal genetic code and is shared among most organisms, but the genetic code can vary (e.g., the mitochondrial genome differs very slightly from the universal code and the mitochondrial code in mammals differs slightly from that in *Drosophila*).

genetic correlation A value that indicates how much of the genetic influence on two traits is common to both. Different traits in the same organism (e.g., length and mass) are often correlated. Typically, such correlations (also called phenotypic correlations) have genetic and environmental components. Genetic correlation is reflected in the proportion of variance that two traits share due to genetic causes (i.e., pleiotropy and gametic phase disequilibrium between genes affecting different traits). If the genetic correlation is greater than zero, this suggests the two traits are influenced by common genes.

genetic distance A statistical measure of the relationships among **DNA** sequences, **protein** sequences, or other genetic materials. Genetic distances can be then used to construct a phylogeny or other visual depiction of the relationships between **alleles**, **species**, or **populations**. The relationships among the DNA sequences, protein sequences, or populations determine genetic distance, but there are circumstances in which two

different sets of relationships can result in the same genetic distance.

genetic drift A random change in **gene** frequencies between generations. It occurs because the gametes produced by each parent contain a haploid genome that is a random assortment of chromosomes (i.e., for each type of chromosome an individual receives only one of the two chromosomes of that type within each parent). Genetic drift can also occur because of the **founder effect**. Genetic drift is more likely to be a significant factor in small populations because they have a smaller **effective population size** and it may have played an important role in shaping the diversity of past hominin populations (Ackermann and Cheverud 2004, Weaver et al. 2007). Along with **gene flow**, **mutation**, and **natural selection**, genetic drift is one of the four forces of evolution (syn. random drift, drift).

genetic hitchhiking *See* selective sweep.

genetic load A measure between 1 and 0 of the cost to the population of lost **alleles** due to **mutation** or **selection** (i.e., deaths due to deleterious alleles). It is the relative difference between the fitness of the "most fit" genotype and the mean population fitness for all the genotypes at that locus. If all the alleles in the population are of optimal fitness, then the genetic load is 0.

genetic map A map of the **genome** generated using a set of markers. The distance between markers is calculated not by the physical distance (that would generate a physical map) but by observing the number of **recombination** events that occur between them. For example, if two markers on the same chromosome are typically separated after each meiosis they are likely located on opposite ends of the chromosome, whereas two markers that are tightly linked (i.e., typically inherited together) are likely to be close to each other on the same chromosome. The distance between markers in a genetic map is measured in centimorgans. One centimorgan equals the distance between markers in which there is a 1% recombination frequency (i.e., one recombination event per 100 meioses). *See also* **linkage**.

genetic variation The genetic diversity present in a species. Genetic variation is created through **mutation** and it is affected by **genetic drift, natural selection,** and **gene flow**. Patterns of genetic variation can thus inform researchers about the relative importance of these forces of evolution. They can also be used to infer demographic features of a species or population such as its **effective population size**, whether the population is expanding, stable or contracting, or hybridization patterns.

genome The complete **DNA** sequence of an organism. Traditionally the term genome is used to refer to the complete set of **chromosomes** in a **gamete** (i.e., one set of chromosomes) so somatic cells in primates, which have two sets of chromosomes, have a **diploid** genome. Organelles such as mitochondria have their own DNA complement that can also be called a genome (e.g., the **mitochondrial genome**).

genomic imprinting In **diploid** organisms, whether a gene is expressed or not depends on the parent from whom the individual inherited that gene's **allele**. So, for some genes, **gene expression** only occurs from the allele inherited from the mother (or vice versa). Genetic imprinting occurs in the **germline** cells (typically through methylation), and it determines for any particular allele whether it is the allele inherited from the mother or the one inherited from the father that is expressed in the offspring. Genomic imprinting is present in all marsupial and placental mammals. Hypotheses to explain its origins include the "parental conflict hypothesis," which states that differences in expression are related to an evolutionary conflict of interest between males and females over the growth of the fetus versus the conservation of resources by the mother for herself and for other offspring, and the "protection against foreign DNA hypothesis" which is that imprinting evolved to silence foreign DNA elements. In modern humans genomic imprinting explains why children with deletions in one region of chromosome 15 have different syndromes depending on which copy is deleted. Children who are missing the copy that is paternally imprinted have Prader–Willi syndrome while those without the maternally imprinted copy have Angelman syndrome.

genomics (Gk *genes*=to be born and *nomos*= to manage) An umbrella term that subsumes all of the activities involved in the study of the **genome** (e.g., mapping, sequencing, etc).

genotype (Gk *genes*=to be born and L. *typus*= form) The genetic composition of an organism (as opposed to its physical appearance or **phenotype**). The term is also used for the combination of allelic states of a locus (or of a set of loci) on a pair of homologous chromosomes. So, for example, a person with type A blood could have the genotype AO or AA for the ABO locus on chromosome 9.

genotype–environment interaction A technical term used in **quantitative genetics** to refer to the phenotypic effect of the interaction of variation in genotypes with variation in the environment (i.e., $G \times E$). This term encapsulates the consensus view that variation in traits is caused not by variation in genotype or by variation in the environment, but is rather the result of interactions between the two. The interactions between the genotype and the environment occur both prenatally (through the *in utero* environment) and in postnatal life. Phenylketonuria (PKU) is an example of such an interaction in modern humans. This metabolic disorder impairs proper breakdown of the amino acid phenylalanine and results in phenotypic effects such as mental retardation. However, avoidance of dietary items containing phenylalanine can prevent many of the major effects of this genetic condition.

genu valgum *See* **valgus**.

genus (Gk *genos*=race or stock) The next to lowest category in the original classificatory system introduced by Linnaeus in 1740. The most common contemporary understanding of a genus is the evolutionary systematic interpretation associated with Ernst Mayr. Mayr (1950, p. 110) suggested "a genus consists of one species, or a group of species of common ancestry, which differ in a pronounced manner from other groups of species and are separated from them by a decided morphological gap." He went on to state that the genus "has a very distinct biological meaning. Species that are united in a given genus occupy an ecological situation which is different from that occupied by the species of another genus, or, to use the terminology of Sewall Wright, they occupy a different adaptive plateau" (*ibid*). Thus, according to Mayr, a genus is as a group of **species** of **common ancestry** that is adaptively both homogeneous and distinctive. Mayr et al. (1953, p. 50) acknowledged the phylogenetic and functional evidence may be in conflict if "unrelated species acquire a superficial similarity owing to parallel adaptations to similar environments" and in such cases they recommended that the phylogenetic evidence should be given precedence. However, it is implicit in Mayr's (1950) definition that "common ancestry" subsumes both **monophyletic group**s and **paraphyletic group**s. It is one of the unfortunate aspects of taxonomy that except for the category species there are no objective criteria for taxonomic ranking. Goodman et al. (1998) suggested a time-depth criterion and proposed that a genus should be at least 4 million years old, but this idea has not attracted adherents. Wood and Collard (1999) proposed that a genus should be defined as a species or monophylum whose members occupy a single **adaptive zone**. It differs from Mayr's (1950) concept in that it excludes paraphyletic taxa. It also differs from Mayr's (1950) concept in that it does not require the adaptive zone to be unique or distinct (*contra* Leakey et al. 2001 and Cela-Conde and Altaba 2002). Rather, it simply requires the adaptive zone to be consistent and coherent across the species in the putative genus. Wood and Collard's (1999) suggested definition allows for the possibility that species assigned to different genera may occupy the same adaptive zone, but it precludes species in the same genus from occupying different adaptive zones. Wood and Collard (1999) suggested two criteria for assessing whether or not a group of species has been correctly assigned to a genus. First, the species should belong to the same monophyletic group as the type species of that genus. Second, the adaptive strategy of the species should be closer to the adaptive strategy of the type species of the genus in which it is included than it is the type species of any other genus.

geochronologic units (Gk *geo*=earth and *kronos*=time) Intervals of time in Earth history. The geologic time scale has a dualistic character in which intervals of time share terms with the sequence of rock (strata) deposited in that interval. Geochronologic units are time-referenced and thus subdivisions are temporal (e.g., Early, Middle, and Late), whereas their rock equivalents are **chronostratigraphic units**, and their relative subdivisions are physically sequential (e.g., Lower, Middle, and Upper). In the hierarchy of terminology for the geologic time scale, geochronologic units are (from larger to smaller) eons, eras, periods, **epochs** and

ages (e.g., *Australopithecus afarensis* fossils from **Lactoli** date to the **Pliocene** epoch).

geochronology (Gk *geo*=earth and *kronos*=time) The umbrella term used for methods that can provide ages for rock layers or fossils. Traditionally, dating methods have been divided into two categories, relative and absolute. **Relative dating** methods date a horizon by finding similar horizons elsewhere that are part of an ordered and independently calibrated sequence. For example, the same fossil taxa preserved in the horizon being investigated and in the reference horizon, or the same sequence of magnetic reversals. **Absolute dating** methods are dating methods that use physical and chemical "clocks" that measure the passage of time in years (e.g., **potassium-argon** and **radiocarbon dating** methods). Colman and Pierce (2000) suggested that geochronological methods be divided up on the basis of the nature of the method and how the results of those methods are expressed. What follows is an adaptation of their terminology. Methods (a) **Sidereal** dating is based on counting annual events (e.g., dendrochronology); (b) **isotopic dating** measures changes in the absolute amounts or ratios of isotopes (e.g., **argon-argon dating**, radiocarbon, etc.); (c) **radiogenic dating** relies on measuring the effects of radioactive decay (e.g., **electron spin resonance spectroscopy dating**, **fission track dating**, etc.); (d) chemical and biological dating methods rely on measuring the progress of time-dependent processes (e.g., amino acid racemization); (e) **geomorphic dating** relies on processes that affect the landscape (e.g., sedimentation rates); (f) **correlation** relies on relationships established using differences that are themselves time-independent (e. g., **biostratigraphy**, **magnetostratigraphy**, **stratigraphy**, **tephrochronology**). Results Numerical-age methods provide an age in years (syn. absolute dating methods), and the term refers to methods a–c listed above; calibrated-age methods provide an approximation of the real age and the term refers to methods d and e listed above; **relative dating** provides an age via establishing links between the site in question and an age sequence established elsewhere, and refers to methods d and e listed above; correlated-age methods provide an age by relating the fossil or horizon to another independently dated sequence, and refers to methods in section f listed above.

geocline *See* **cline**.

Geological Survey of India (or GSI) An organization that undertakes the recovery of paleontological, including paleoanthropological, evidence within India and which is also a repository for fossils.

geological time An expression usually referring to long stretches of time, often on the order of many thousands or even millions of years across which evolution occurs. These long periods are "geological" in the sense that they are long enough to be detectable in the geological record.

geomagnetic anomaly (Gk *geo*=earth, *magnes*=magnet, and *anomalos*=uneven) A boundary between two prolonged periods during which the direction of the Earth's magnetic field has been consistent. These boundaries, or magnetozones, are between what were previously referred to as epochs and events, but contemporary geomagnetic columns now use numbered chrons and subchrons. Because the direction of the Earth's magnetic field reverses at these boundaries or anomalies, they are also known as **polarity reversals** or field reversals (syn. anomaly).

geomagnetic dating (Gk *geo*=earth and *magnes*=magnet) *See* **geomagnetic polarity time scale**; **magnetostratigraphy**.

geomagnetic polarity time scale (or GPTS) Time scale based on past changes in the direction of the Earth's magnetic field. The liquid core of the Earth can be likened to a dipole magnet that generates a geomagnetic field. In the present day a magnetized needle will point to the North Pole (this is what we call the **normal** direction). However, at the beginning of the 20thC Bernard Bruhnes (1867–1910) showed that when a magnetized needle was placed in the magnetic field preserved by iron oxide in some older lavas the needle pointed to the South Pole (the **reversed** direction). Motonori Matuyama (1884–1958) demonstrated that the Earth's magnetic field had been reversed as recently as the early **Quaternary**. Scientists were initially unsure whether the lavas had been laid down during a global magnetic reversal ("field reversal"), or whether they had subsequently reversed their primary normal magnetism ("self reversal"). However, when **potassium-argon dating** began to be applied to the reversed rocks (Rutten 1959) it became clear that "field reversal" was a reality. The first published land-based paleomagnetic timescale was relatively crude (Cox et al. 1963) but within a short time the basic structure of the GPTS (i.e., long periods of either normal or reversed magnetism, then called **epochs**, punctuated by shorter periods of the opposite magnetic direction, then called **event**s) was established (Cox et al. 1964). Researchers then began the process of integrating the land-based record with magnetic anomalies detected in the ocean floor, and the fit was a good one (Vine 1966). By the late 1960s a geomagnetic reversal time scale was established with four epochs named after pioneer physicists (i.e., listed in order starting with the most recent: Brunhes, Matuyama, Gauss, and Gilbert) and events named according to the locations where they were first identified (e.g., Jaramillo, Cobb Mountain, Olduvai) (Cox 1969). Confusingly, any change of magnetic direction, either from normal to reversed or from reversed to normal, is referred to as a "reversal" of the geomagnetic field. The latest version of the GPTS has dropped the eponymous terminology (but see below) in favor of a numerical sequence adapted from the scheme used by scientists studying the magnetic anomalies associated with spreading from ridges in the deep sea floor. Epochs have been replaced by numbered **chrons** and events by **subchrons**. The most recent chron is 1 with the numbers increasing going back in time. A letter postfix is used to identify chrons that have been added since the scheme was adopted (e.g., 3A). A chron can have both normal and reversed components identified by lower case n or r (e.g., 3An and 3Ar). Subchrons within a chron are also numbered and their magnetic direction is identified using the same abbreviation system (e.g., the first reversed subchron in 3An is 3An.1r). The precision of the GPTS has been substantially increased by matching changes in geomagnetism with the increasingly detailed record of oscillations in the $^{16}O/^{18}O$ record. This in turn takes advantage of the correlations possible between such continuous geological records and the predicted ages derived from calculations using the known frequencies of the Earth's orbital components (eccentricity of the Earth's orbit, tilt, and precession of the Earth's axis), enabling the ages of geological transitions to be tuned to the astronomical signal; this is referred to as **orbital tuning**. Geomagnetic data are not by themselves sufficient evidence to provide ages for strata containing fossils. One of the other dating methods is needed to provide an absolute date so that the sequence of changes in magnetic direction in the sedimentary sequence can be matched with the GPTS. When this is done the detailed sequence of magnetic reversals preserved in a section of sediments can then be used to date fossiliferous strata within it. The GPTSs change as the evidence they are based on is refined, with each scale having a title made up of the first letters of the author's names followed by the year of publication (e.g., HDHPL68, CK92). Initially they were based on assumptions about rates of seafloor spreading, but latterly astronomical, isotopic, and other evidence has been utilized to calibrate the timing and duration of the changes in the direction of the Earth's magnetic field. One of the first events to be recognized was a brief episode of normal magnetic direction at *c*.1.9 Ma within the Matuyama reversed epoch identified in the basalt underlying the fossiliferous sediments at **Olduvai Gorge**, Tanzania (Grommé and Hay 1963), and this was the first example of geomagnetic data being used to help date a hominin fossil site. It used to be called the **Olduvai event**, but it is now called the **Olduvai subchron**. The Cenozoic epoch boundaries are now tied to the GPTS chron system. The Cretaceous/Paleocene boundary lies in chron C29 at 65.5. Ma, Paleocene/Eocene in chron C24 at 55.8 Ma, Eocene/Oligocene in chron C13 at 33.9 Ma, Oligocene/Miocene in chron C6C at 23.0 Ma, Miocene/Pliocene in chron C3 at 5.3 Ma, and the Pliocene/Pleistocene either in chron C2 at 2.6 Ma or at the chron C2/chron C1 boundary at 1.8 Ma. *See also* **geomagnetic dating** (syn. magnetic anomaly time scale, magnetic polarity timescale, **magnetostratigraphy**).

geomagnetic reversal time scale *See* **geomagnetic polarity time scale**.

geometric mean A method for calculating the average of a set of numbers, it differs from the more commonly known **arithmetic mean** as follows. Rather than summing n values and dividing by n, the geometric mean calculates the product of n values then takes the nth root of the result. For example, the geometric mean of X_1, X_2, X_3, and X_4 is $\sqrt[4]{X_1 \times X_2 \times X_3 \times X_4}$. The geometric mean may also be calculated as the antilog of the arithmetic mean of logged values; thus the geometric of the four measurements above could also be calculated as $e^{\wedge}$ ($[\ln X_1 + \ln X_2 + \ln X_3 + \ln X_4]/4$). The geometric mean of a set of values is always less than or equal to the arithmetic mean of those values. The geometric mean is often used to combine multiple measurements into a single measurement of **size**. *See also* **size**.

geometric morphometrics A family of methods for the analysis of **form** that preserve the original geometry of the measured objects during all the stages of the analysis. A second major strength of geometric morphometric methods is their ability to visualize shape differences and shape changes using deformation grids (i.e., morphing of two-dimensional images or three-dimensional surfaces). Geometric morphometrics is based on the coordinates of **landmarks**. Bookstein (1991, p. 2) defines landmarks as loci that are homologous across individuals in a sample (i.e., the landmarks have the "same" location in every other member of the sample being investigated). Coordinate data can be one-, two-, or three-dimensional. Two-dimensional coordinates are usually captured using a digitizing tablet or by measuring an image on the computer. Three-dimensional data can be captured directly using a coordinate digitizer such as a Microscribe or Polhemus or they can be measured on surface or volumetric scans. Surface scanners provide high-resolution three-dimensional representations of an object's surface using either laser or more traditional optical technology. Volumetric data are based on image slices from **computed tomography** (CT or μCT) or **magnetic resonance imaging** (MRI or μMRI) or, more recently, optical projection tomography. These slices contain gray values that correspond to tissue densities and they are concatenated to obtain a three-dimensional representation of an object; virtual surfaces can be generated from CT or MRI data. Most software packages allow landmark coordinates to be measured directly on these virtual surfaces or on volumetric objects. Statistical analysis in geometric morphometrics is performed using Procrustes shape coordinates (or their equivalent). In geometric morphometrics the shape variables all possess the same units so analyses are based on the covariance matrix and there exists a well-defined metric (the Procrustes metric) in shape space. Results of multivariate methods that preserve this metric (e.g., **principal component analysis**, multivariate regression, and partial least squares) can be visualized as actual shapes or shape deformations in the geometry of the original specimens. The algebra of the statistical methods is the same for two- and three-dimensional shape data. Principal component analysis (or PCA) is usually used to analyze the Procrustes shape coordinates and the principal components are computed in a way that preserves the Procrustes distances among the specimens.

Principal component scores are the projections of the shapes onto the low-dimensional space spanned by the eigenvectors. The weightings for linear combinations of the original variables can be visualized as actual shape deformations (called relative warps; see Bookstein 1991). However, the principal components are statistical artifacts and should not be interpreted as representing biologically meaningful factors (Mitteroecker and Bookstein 2007), with the exception that the first principal component of a species or population often represents the shape variation induced by overall size variation (i.e., allometry) (Klingenberg 1998). Partial least squares (or PLS) can be used to assess relationships among blocks of variables (Bookstein et al. 2003), and the results can be visualized as shape deformations (called singular warps). When the blocks of variables are shape coordinates, PLS can also be used to test hypotheses of **morphological integration** (Bookstein et al. 2003, Mitteroecker and Bookstein 2007, 2008) as well as functional, environmental, or behavioral influences (e.g., Bookstein et al. 2002, Frost et al. 2003, Manfreda et al. 2006). Many geometric morphometric analyses employ **randomization** tests (using permutations or bootstrapping) rather than parametric methods to assess the statistical significance level of a given hypothesis (e.g., Mitteroecker et al. 2005). See Good (2000) and Mitteroecker and Gunz (2009) for a more technical description and Mitteroecker et al. (2005) and Gunz et al. (2009a) for examples of its use.

geometric population growth

Geometric population growth refers to an increase in population size as a fixed percentage per unit time. For example, if a population consists of 10,000 individuals on January 1 of a year and by December 31 the population consists of 10,100 individuals then it has grown in size by 1%. The growth rate must be given per unit time, so in this case the growth rate would be 1% per annum (per year). It is more common to give the growth rate on a basis of 1.0 rather than on a percentage basis, so a growth rate of 1% per annum would be given as 0.01. In an analogy to interest earned, geometric growth means that interest is earned at fixed periods, so that the compounding of interest (earning interest on previous interest) only occurs at a few fixed times (quarterly or annually). This can be contrasted with the exponential growth rate, where the time interval is infinitesimally small. Carrier (1958) was the first to suggest that a geometric

population growth model could be used in paleodemography to adjust **life tables** for population growth. Note that the population "growth" rate can also be negative if the population is in decline.

geophagy (Gk *geo* = earth and *phagein* = to eat) Literally, the consumption of clays and soils, but the term may also be extended in common usage to eating other organic or inorganic materials (such as charcoal). Several species of primates practice geophagy to help neutralize ingested secondary plant compounds (chemical deterrents produced by plants as a defense mechanism). Primate species differ in their ability to break down secondary compounds and what is poisonous to one may be edible by another. Colobines commonly use geophagy, presumably because their diets include mature leaves, which contain many alkaloids. Other primates minimize the effects of secondary compounds in their diets by adopting behavioral strategies, for example by varying their foodstuffs to avoid over-consumption from a single plant species (Strier 2000). Another function of geophagy is to balance mineral (including trace element) deficiencies; some primates and other mammals preferentially eat soils (e.g., salts at saltlicks) to add minerals which would otherwise be inadequate in their diets. Geophagy is also known to occur in modern human groups, especially in pregnant women. Studies of the evolution of hominin diets have not yet focused on how hominins procure important minerals, but this requirement must have played a role in food choice.

geophyte *See* **underground storage organ**.

"George" Nickname given by Louis and Mary Leakey to **OH 16** from **Olduvai Gorge**.

George S. Huntington Collection *See* **Huntington (George S.) Collection**.

Georgian National Museum (or GNM) A network of 11 museums in the Republic of Georgia. It was established in its current form in 2004 by a Presidential decree, but the origins of the museum can be traced back to the Museum of the Caucasian Department of the Russian Geographic Society founded in 1852. The central branch, the Simon Janashia Museum of Georgia, is currently undergoing a large-scale reorganization and renovation. In addition to the hominin and faunal specimens from

Dmanisi, the Simon Janashia Museum houses a large collection of Bronze Age and Classical artifacts from around Georgia. A new branch of the GNM was recently opened in 2009 at the Dmanisi site, allowing visitors to view reconstructions of the hominins, casts of the fossils, and the excavation itself. Hominin fossil collections **D211, D2280, D2282, D2600, D2700/ D2735/D2724/D3160, etc., D3444/D3900,** and **D3900**, among others. Comparative collections N/A. Contact department N/A. Contact person Teona Jakeli (e-mail teonajakeli@museum.ge). Postal address 3 Rustaveli Avenue, 0105 Tbilisi, Georgia. Website www.museum.ge. General e-mail info@ museum.ge. Tel +995 32 99 80 22. Fax +995 32 98 21 33. *See also* **Lordkipanidze, David**.

germ cells Cells, also called gametes (i.e., egg and sperm) that have a **haploid** complement of chromosomes. *See also* **germline**.

germline (L. *germen*=offshoot) The gametes (i.e., egg and sperm) plus their immature precursors that have a **haploid** complement of chromosomes. The gametes are the cells that pass genetic information from one generation to the next. The remaining cells in the body are called **somatic cells**.

Gesher Benot Ya'akov (or GBY) (Location 33°00′28″N, 35°37′44″E, Israel; etym. He. literally "the bridge or crossing of Jacob's daughters." The name given from at least the time of the Crusaders to this point where the Jordan River can be crossed conveniently and the site of a battle called "Jacob's ford.") History and general description The GBY deposits stretch for approximately 3.5 km/2 miles along both sides of the Jordan River. They were first explored in a series of surveys in the 1980s and one area on the left bank of the Jordan River was excavated in seven seasons between 1989 and 1997 and again in the early 2000s. Temporal span **Oxygen Isotope Stages** 18–20 (i.e., at least 790 ka). How dated? **Biostratigraphy** and **magnetostratigraphy**. Hominins found at site None. Archeological evidence found at site GBY has at least 13 archeological horizons, which preserve fauna, plant remains, and **Acheulean handaxes** and **cleavers**. There is reliable evidence (perhaps the earliest reliable evidence) for the use of fire (Goren-Inbar et al. 2004), for nut-cracking (Goren-Inbar et al. 2002), and for a system of spatial organization (Alperson-Afil et al. 2009). Key references: Geology, dating, and paleoenvironment Goren-Inbar and Belitzky 1989, Goren-Inbar et al. 2000; Hominins N/A; Archeology Goren-Inbar et al. 1994, 1996, 2000, 2006, Alperson-Afil and Goren-Inbar 2010.

gestalt (Ge. *stellen*=place) A pattern that has properties or meaning(s) that are greater than the sum of its components (e.g., the distinctive overall appearance of the face of *Paranthropus boisei*).

gestation Gestation includes the period of time in which an embryo or fetus is carried inside a female in a species that gives birth to live young. During gestation, a fertilized egg develops into a neonate of varying levels of maturity (e.g., marsupials versus precocial mammals). Gestational age is typically calculated to include the time of the last menstrual period, although this is not always accounted for. Modern human pregnancy is divided into three trimesters: last menstrual period to 13 weeks, 14–27 weeks, and 28–40 weeks. Therefore, a full-term modern human birth occurs at 40 weeks' gestational age. However, birth occurring after 37 weeks is typically considered full term, and anything prior to 37 weeks is preterm. Preterm birth, together with the associated low weight and immature organ systems, comprises a large proportion of infant mortality.

gestation length The period of time between ovulation and birth. Gestation length varies widely among primate taxa. While there exists a within-order scaling of gestation length with body size in primates, the same is not true for mammals in general: bats have the longest gestation lengths for a mammal of their size. Within primates, mouse lemurs have a gestation period of *c*.60 days, lemurs *c*.133 days, macaques from 145 to 185 days (it varies across species), great apes between *c*.240 and 250 days, and modern humans average 267 days. Gestation length is an important component of reproductive investment. An active area of discussion in the past 30 years has involved the question of extended gestation length in *Homo neanderthalensis* with Erik Trinkaus proposing that Neanderthal gestation lasted 12 months based on aspects of pelvic morphology (from partial remains) that he suggested would have accommodated the birth of a larger-headed neonate. This was hotly debated, however, but evidence from subsequent finds of pelvic elements led to it eventually being discounted and recent work using reconstructions of Neanderthal pelves and neonatal head size suggest that brain size at birth was comparable in Neanderthals and *Homo sapiens.* Further, using information gleaned from virtual reconstructions of the passage of a Neanderthal head through a Neanderthal birth canal, this research also suggests that the obstetrical constraints for birthing a large-brained baby in Neanderthals were similar to those in modern *Homo sapiens.* Differences in adult brain size between Neanderthals and *Homo sapiens* are due, the authors suggest, to increased rates of brain growth in the early postnatal period in Neanderthals.

gesture A voluntary non-vocal action used to express meaning or significance in a socially specified manner. Gestures do not include various non-vocal signals such as smiling, laughing, and crying that index emotional states and serve important species-wide communication functions. Such affective or emotion signals are neither entirely voluntary nor are they socially specified. So, whereas everyone in the world expresses happiness with a smile (Ekman 1993), not everyone expresses approval with the "thumbs up" gesture. When defined this way, gesturing becomes an activity that is almost exclusive to modern humans (Tomasello and Call 1997). This is perhaps because gesture interacts with language in important ways, including the facilitation of language development (Goldin-Meadow 2007) and understanding (Skipper et al. 2009). There is some evidence that enculturated apes raised in modern human households use pointing gestures to refer to things out of sight and apparently comprehend the referential functions of another's gesture as a means of directing one's attention (Tomasello and Call 1997). However, work with non-enculturated apes supports the notion that the most basic type of gesture (e.g., pointing) is absent in chimpanzees, despite its presence in 12 month-old modern human infants (Liszkowski et al. 2009).

Geula (Location 32°40′N, 35°05′E, Israel; etym. named after the suburb in which it was found) History and general description Geula Cave, which is located on the northern slope of **Mount Carmel** in the suburbs of the city of Haifa, was excavated by E. Wreschner from 1958 to 1964. Despite the large faunal and lithic assemblages, relatively few **hominin** remains have been found. One of the more notable characteristics of the fauna from the cave is the abundance of porcupine remains. Temporal span *c.*41 ka. How dated? Conventional **radiocarbon**

dating of charred bones. Hominins found at site A total of six hominin bone fragments are known from the site. Their fragmentary nature hinders the ability to generate any reliable hypothesis about their taxonomy, but Arensburg (2002) notes that they have some affinities with hominins from both **Qafzeh** and **Tabun**. Raymond **Dart** (1967) suggested that at least one of the fragments had been fashioned into a tool. Archeological evidence found at site A typical Levantine **Mousterian** assemblage was found. Key references: Geology, dating, and paleoenvironment Vogel and Waterbolk 1963; Hominins Wreschner 1967, Arensburg 2002; Archeology Dart 1967, Wreschner 1967.

G.H. Abbreviation of Garusi Hominid and the prefix for fossil hominins recovered from **Laetoli**, Tanzania by the expedition by the 1938–9 Kohl–Larsen Expedition.

G.H. 1 *See* **Garusi 1**.

GI Acronym for **gyrification index** (*which see*).

Gibraltar A British overseas territory (a "Crown Colony") on the southern coast of Spain, on a large limestone outcrop ("the Rock") that juts into the Mediterranean. The cave sites in this outcrop preserve evidence of some of the latest-surviving *Homo neanderthalensis* in Europe, and its southern location and relative isolation on the far end of the Iberian peninsula are consistent with it being a **refugium**. Two Neanderthal fossils were found in the late 19thC and early 20thC. The first, a cranium from **Forbes' Quarry** (**Gibraltar 1**) on the north side of Gibraltar was discovered during blasting for limestone in 1848. It was the first Neanderthal fossil to be discovered, but the importance of the find was not recognized until 1862, two years before the *H. neanderthalensis* fossils from **Kleine Feldhofer Grotte** were recognized as a distinct species. This discovery led several researchers to begin more concentrated examination of other Paleolithic sites on the Rock, and in 1926 Dorothy Garrod found the second Gibraltar fossil in Devil's Tower, along with animal bones, charcoal, and other evidence of human occupation. Several other caves with **Mousterian** levels have since been discovered and documented, including Ibex Cave, Vanguard Cave, and Gorham's Cave, but to date no fossils have been recovered.

Gibraltar 1 Site Gibraltar. Locality Forbes' Quarry. Surface/*in situ* Not known. Date of discovery 1848. Finders Military prisoners. Unit N/A. Horizon N/A. Bed/member N/A. Formation N/A. Group N/A. Nearest overlying dated horizon N/A. Nearest underlying dated horizon N/A. Geological age N/A. Developmental age Adult. Presumed sex Female. Brief anatomical description Relatively well-preserved cranium. Announcement The discovery of the cranium was reported to the Gibraltar Scientific Society on March 3, 1848. Initial description The cranium was brought to the attention of Hugh Falconer, the paleontologist, and George Busk, a surgeon and paleontologist, when they visited the excavations at Gibraltar in 1862. Busk caused the cranium to be presented to the **Royal College of Surgeons of England** in 1868, where it remained until 1955 when it was transferred to what was then called the British Museum (Natural History) and is now the Natural History Museum in London. Photographs/line drawings and metrical data Morant 1927. Detailed anatomical description None. Initial taxonomic allocation *Homo calpicus* Falconer, 1864. Taxonomic revisions *Homo neanderthalensis* Keith (1911). Current conventional taxonomic allocation *H. neanderthalensis*. Informal taxonomic category Pre-modern *Homo*. Significance With hindsight, this was the first adult *H. neanderthalensis* cranium to be recovered. Location of original Natural History Museum, London, where it is on permanent loan from the Royal College of Surgeons of England.

Gilbert chron *See* **geomagnetic polarity time scale**.

giraffid The informal name for the artiodactyl family comprising the giraffes and their allies, whose members are found at some hominin sites. *See* **Artiodactyla**.

Giraffidae One of the artiodactyl families whose members are found at some hominin sites. *See* **Artiodactyla**.

GISP2 Abbreviation of the **Greenland Ice Sheet Project Two** (*which see*).

glabella (L. *glabellus*=smooth and hairless) The most anteriorly projecting point in the midline of the **cranium** between the superior orbital margins.

glacial cycles The term "ice-age" was apparently introduced by Goethe, but it became clear that it was not one monolithic event, but a series of colder glacial periods punctuated by warmer interglacial periods. These cold/warm cycles were originally identified and defined on the basis of lithological evidence, and before the use of **Oxygen Isotope Stages** the antiquity of a horizon or a site was reflected by it being linked with one of these named glacial cycles. There are several regional terminologies for glacial cycles, but the one with historical priority is the scheme developed for use in the Alps. This has named glacial periods and from recent to the oldest they are the Würm, Riss, Mindel, Günz, and Donau. The interglacial periods are named after the adjacent glacial periods, the older one first. Hence they are the Riss-Würm, the Mindel-Riss, the Günz-Mindel, and the Donau-Günz. Geologists in northern Europe used a different terminology, with separate terms for the glacial and interglacial cyles, hence the Würm glacial equivalent is the Weichselian, the Riss-Würm interglacial equivalent the Eemian, the Riss glacial equivalent the Saalian, the Mindel-Riss interglacial equivalent the Holsteinian, the Mindel glacial equivalent the Elsterian, the Günz-Mindel interglacial equivalent the Cromerian and the Bavelian, the Günz glacial equivalent the Menapian, the Günz-Mindel interglacial equivalent the Waalian, and so on. The British Isles had yet another scheme, with the Ipswichian interglacial equivalent to the Riss-Würm/Eemian, the Hoxnian interglacial equivalent to the Mindel-Riss/Holsteinian, and the Anglian glaciation equivalent to the Mindel/Elsterian. Around the time this latter scheme was introduced "cold stage" began to replace "glacial" and "temperate stage" replaced "interglacial." But *all* of these named stages or cycles have been superseded by the numbered Oxygen Isotope Stages (OISs; see Gibbard and van Kolfschoten 2004 for a useful chart that matches the various named schemes of glacial cycles with OISs; for an updated version see Gibbard et al. 2007). *See also* **Oxygen Isotope Stages**.

glacial ice volume Glacial ice volume refers to the volume of ice locked up in continental glaciers and it may be usefully measured in sea-level-rise equivalent units. As the volume of continental ice has waxed and waned during glacial (cold) and interglacial (warm) cycles, the changing volume of continental ice has dramatically altered sea levels. The volume of continental ice locked up in the East Antarctic Ice Sheet is equivalent to approximately 65 m of sea-level rise and in the West Antarctic Ice Sheet and the Greenland ice cap it is equivalent to approximately 8 and 6 m, respectively. All the other ice caps and valley glaciers are estimated to be equivalent to approximately 0.5 m of sea-level rise. At the **Last Glacial Maximum**, sea levels were about 120 m lower than they are today and this may have provided the opportunity for hominins to migrate out of Africa across the narrow, shallow, channel at the southern end of the Red Sea and out of Southeast Asia into Australasia. Lowered sea levels also provided new coastal habitats (e.g., around southern Africa) now presumably below sea level. Glacial ice volume can be roughly estimated from geomorphological evidence on land, **ice rafting** in the oceans, and most importantly the oxygen isotope ratio of **benthic foraminifera**.

glacial period *See* glacial cycles.

Gladysvale (Location 25°53′42″S, 27°46′21″E, Gauteng Province, South Africa; etym. named for Gladys, the daughter of former owner of the land in the 1920s) History and general description Cave system within Precambrian **dolomite** 13 miles northeast of **Sterkfontein** and approximately midway between **Drimolen** and **Gondolin**. Its **breccia** deposits were recognized as fossiliferous by Robert **Broom** in 1936 (in Broom and Schepers 1946, p. 43). Andre Keyser began the latest series of studies in 1988. In 1991–2, Lee Berger and Keyser analyzed breccia dumps scattered during mining of the cave that took place during the 1920s, eventually finding isolated hominin teeth. Berger took over excavations in 1993 when Keyser started looking for other sites (eventually finding Drimolen). The cave includes open de-roofed sections (called GVED), a smaller series of stratified deposits within the cave (GVID), and large series of underground chambers not yet formally described. Temporal span and how dated? **Biostratigraphy** based on faunal remains recovered from breccia dumps suggests the presence of deposits of an age comparable to Sterkfontein Mb 5 and **Kromdraai** A. **Electron spin resonance spectroscopy dating** on tooth enamel from the GVED suggests an age of *c.*578–830 ka and all the GVED sediments are normally magnetized. Uranium-thorium dating of the GVID sediments provided a broad temporal range from

*c.*570 ka to the Holocene. Herries et al cite Curnoe (1999) as suggesting that the teeth assigned to *Au. africanus* may come from "interior deposits" in the cave that may be as old as *c.*2.4 Ma. Hominins found at site The first hominins, isolated teeth from breccia dumps, were assigned to *Australopithecus africanus*; a single phalanx assigned to *Homo* sp. was recovered from later *in situ* excavations. Archeological evidence found at site A single quartzite **handaxe** was recovered from just below the **flowstone** that marks the boundary between the GVED and a separate, and as yet undated, set of sediments called the South Western Cone (SWC). A probable human hair found in a hyaena coprolite was reported from GVID sediments dated to *c.*200 ka. Key references: Geology, dating, and paleoenvironment Lacruz et al. 2002, 2003, Pickering et al. 2007, Herries et al 2010; Hominins Berger 1992, Berger et al. 1993, Berger and Tobias 1994, Schmid and Berger 1997; Archeology Hall et al. 2006, Backwell et al. 2009.

glass An amorphous solid usually formed by the rapid cooling of magma. Volcanic glass forms when magma is quenched (i.e., it is rapidly cooled). It occurs in large volumes as **obsidian** and in small, microscopic volumes it forms the matrix of most **lavas**. Fragmented glass explosively erupted from a volcano is referred to as vitric tephra. Because the chemistry of a glass closely approximates the composition of the parent magma at the time of cooling, vitric tephras record a geochemical fingerprint of an eruption. Widely dispersed products of such an eruption can thus be correlated geochemically to provide an isochronous marker and in some cases these markers can be dated (e.g., by **argon-argon dating** of associated potassium-bearing **feldspars**). *See* **tephrostratigraphy**.

glia (Gk *glia*=glue) Refers to a diverse class of cells in the nervous system that surround and ensheath **neurons** (i.e., their cell bodies, **axons**, and dendrites). The functions of glial cells include providing structural and metabolic support for neurons, maintaining homeostasis, synthesizing **myelin**, signaling, and enabling the nervous system to respond to damage. Glial cells are classified according to their morphology, function, and location and the types include microglia, astrocytes, oligodendrocytes, and Schwann cells. Most types of glial cells are derived from the neuroectoderm (one of the germ layers) but microglia are part of the immune system and enter the brain from the blood during development. Microglial cells engulf dead cells and neuronal debris and they have been implicated in the synaptic remodeling that occurs during development. Oligodendrocytes (in the central nervous system) and Schwann cells (in the peripheral nervous system) produce a lipid-rich membrane called myelin, which wraps around axons and insulates axonal segments, thereby speeding up the conduction of electrical impulses. Astrocytes provide neurons within the central nervous system with energy and substrates for neurotransmission. Their processes act as a physical barrier between neurons (the glial membrane), and between neurons and blood vessels (the blood–brain barrier), as well as removing excess neurotransmitter molecules from the extracellular space around the synapses that connect neurons. Recently, researchers have demonstrated that astrocytes are involved in the formation and modulation of synapses and that the astrocytes of the extant hominids have a distinctive morphology.

globins Globins form the protein component of two functionally important molecules: hemoglobin (the molecule that carries oxygen in blood) and myoglobin (the oxygen-storing molecule in muscle). The globin genes comprise three families: the myoglobin family (located on chromosome 22 in modern humans), the alpha globin family (on chromosome 16 in modern humans), and the beta globin family (on chromosome 11 in modern humans). Together they comprise the globin **gene family**. Mutations in the alpha globin or beta globin genes result in hemoglobinopathies such as sickle cell anemia. Deletions or regulatory changes in these genes result in thalassemias. These proteins and the genes encoding them have been intensively studied to understand modern human adaptation to **malaria**. *See also* **hemoglobin**.

gnathic (Gk *gnathos*=jaw) Anything that relates to either the upper (maxilla) or the lower (mandible) jaw.

Gomboré *See* **Melka Kunturé**.

Gompertz hazard model This model of adult **mortality** was proposed by Benjamin Gompertz (1825). Gompertz's model states that adult mortality increases exponentially with age. A parameter in Gompertz's model, an initial hazard rate, is usually symbolized by α and the variable that specifies the increasing hazard of death that accompanies increasing age is usually symbolized by β. In the Gompertz hazard model the hazard of death at exact age t is $h(t) = \alpha e^{\beta t}$ and the **survivorship** to exact age t is $s(t) = \exp(\alpha/\beta(1 - \exp(\beta t)))$. The **probability density function** for age at death is the product of the hazard and the survivorship: $f(t) = h(t)s(t)$. The logarithmic scale is particularly useful in examining the Gompertz model because the natural logarithm of the hazard is a straight line given by $\alpha + \beta t$, so that the Gompertz hazard plotted on a logarithmic scale is a straight line. In the **Siler hazard model** the Gompertz is the third (senescent) component of mortality, so it is often written with a_3 for α and b_3 for β. The Gompertz model is assumed whenever authors refer to a **mortality rate doubling time** and the model can have important implications when using age-at-death data to try to reconstruct the age distribution in the living. As such, the Gompertz model can been used to estimate the living age structure for fossil hominins based on the ratio of older to younger adults from the dead and the assumption of an initial mortality rate (α).

gomphoses *See* **joint(s)**.

gomphothere The informal name for the taxa within the family Gomphotheriidae, which contains all the proboscideans with shovel-shaped tusks. *See* **Proboscidea**.

Gomphotheriidae (Gk *gomphos*=joint, *ther*= combining form, wild beast) A diverse family of extinct Proboscidea known from every continent except Australia and Antarctica. Their cheek teeth are relatively simple and bunodont, and many forms have both lower and upper tusks. Some species (e.g. *Anancus*) lack the lower tusks whereas in others (e.g. *Amebelodon*, *Platybelodon*) these were specialized and had an elongate, shovel shape. In Africa the family is known from the late Oligocene to the late Pliocene; their later distribution overlaps with that of the **Elephantidae**, which they gave rise to and which eventually replaced the Gomphotherids. Gomphotheres lived as far south as South America. They survived in Central America and Mexico until the end of the Pleistocene, and perhaps into the

Holocene. Gomphotheriidae remains are associated with early human occupation of the New World.

Gomphotherium *See* **Gomphotheriidae**.

Gona Palaeoanthropological Research Project

This multidisciplinary field and laboratory project directed by Sileshi Semaw was established in 1997 to investigate the fossil- and artifact-rich deposits in the **Gona Paleoanthropological study area**, Afar Rift System, Ethiopia. Other researchers involved with the project include Michael Rogers, Dietrich Stout, and Dominique Cauche (palaeolithic archeology); Tesfaye Kidane, Naomi Levin, and Jay Quade (geology); Scott Simpson (hominin paleontology); Stephen Frost (non-hominin paleontology); and Manuel Dominguez-Rodrigo (zooarcheology). The local Afar contingent is led by AsAhmed Humet. Among the achievements of the Gona Palaeoanthropological Research Project are the recovery of evidence of early stone artifacts in association with deliberately fractured fauna (Semaw et al. 2003, Domínguez-Rodrigo et al. 2005), the recovery from sites in the **Western Margin** of the Gona Paleoanthropological study area of more than 30 hominin specimens (belonging to at least nine individuals) that have been assigned to *Ardipithecus ramidus* (Semaw et al. 2005), additional evidence of *Ar. ramidus* recovered during the 2003–8 field seasons, and a pelvis, BSN49/P27, whose "likely age" is "0.9–1.4 Ma" (Simpson et al. 2008, p. 1089). Simpson et al. (2008) assign it to a "short-statured *H. erectus* adult female" (*ibid*, p. 1090) but Ruff (2010) suggests that the size of the pelvis as assessed from the diameter of the acetabulum is more consistent with it belonging to an archaic hominin, possibly *Paranthropus boisei*. Further, as yet unpublished, hominin remains have been also recovered from the Late Miocene deposits exposed in the Gona Western Escarpment and in sediments containing **Acheulean** artifacts.

Gona Paleoanthropological study area

(Location 11°07′–08′N, 40°18′–19′E; Ethiopia; etym. the local name for the area) History and general description The Gona Paleoanthropological study area is located in the west-central section of the **Afar Rift System**; it is west of the **Hadar study area** and north of the **Middle Awash study area**. Initial fieldwork at Gona was carried out by Helene Roche and later by Jack Harris in the 1970s and collaborative work by Jack Harris and Sileshi Semaw

began in 1987. Systematic excavations at **Kada Gona** began in 1992 and these resulted in the recovery of the oldest known stone artifacts dated to 2.6 Ma (Semaw et al. 1997, Semaw 2000) and the **Gona Palaeoanthropological Research Project** began large-scale field investigations in 1999. The early Pliocene deposits exposed in the Gona section of the **Western Margin** have yielded hominins attributed to *Ardipithecus ramidus* (Semaw et al. 2005). The eastern and southeastern portion of the study area in the Kada Gona, Ounda Gona, and Dana Aoule drainages contain sediments of more than 80 m thick containing Plio-Pleistocene fossil fauna and **Oldowan** artifacts. In addition, several **Acheulean** occurrences estimated to be between 1.7 and 0.5 Ma have also been documented in these drainages and some of these have also yielded hominin fossils including a female pelvis (Simpson et al. 2008) and a cranium discovered in 2006. The Gawis drainage and the area the east of it contain late Acheulean, **Middle Stone Age**, and **Later Stone Age** sites. Temporal span and how dated? *c.*2.6–0.5 Ma determined by **argon-argon dating** and magnetostratigraphy. Hominins found at site More than 30 hominin specimens (belonging to at least nine individuals) that have been assigned to *Ar. ramidus* (Semaw et al. 2005), additional evidence of *Ar. ramidus* was recovered during the 2003–8 field seasons and a pelvis, BSN49/P27, whose "likely age" is "0.9–1.4 Ma" (Simpson et al. 2008, p. 1089); other hominins recovered from this time interval have yet to be published. Archeological evidence found at site Many levels in the *c.*2.6–1.5 Ma time interval contain **Oldowan** artifacts and the Gawis drainage and the area to the east of it has late Acheulean, Middle Stone Age, and Later Stone Age sites. Geology, dating, and paleoenvironment Semaw et al. 1997, 2003, 2005, Quade et al. 2004, 2008, Levin et al. 2004, 2008, Kleinsasser et al. 2008; Hominins Semaw et al. 2005, Simpson et al. 2008; Archeology Semaw et al. 1997, 2003, Semaw, 2000, Stout et al. 2005, Dominguez-Rodrigo et al. 2005.

Gona pelvis *See* **BSN49/P27**.

Gondolin

(Location 25°49.837′S, 27°51.857′E, North West Province, South Africa; etym. named by a former owner of the farm where the site is located. He combined the names of his daughters – Gonda and Lynn – into the name "Gondalynn" but the site has only ever been published on with the spelling "Gondolin") Ḥistory and general description The site is an abandoned limeworks based on a cave system within Precambrian **dolomite** near Broederstroom, 24 km/15 miles northwest of **Sterkfontein**. The potential of the site was noted in 1977 by MacKenzie and Elisabeth Vrba, who excavated there in 1979. Two *in situ* sequences are recognized at the site of the old cave, GD2 and GD1/3, along with extensive *ex situ* dumpsite deposits. Temporal span Fauna, specifically Stage III *Metridiochoerus andrewsi* recovered during the excavations at GD2, suggests an age for GD2 of *c*.1.5–1.9 Ma (equivalent to **Swartkrans** Mb 1) for the excavated breccia. The **normal** polarity of the GD2 sediments is consistent with their sampling the **Olduvai subchron** in which case the dates would be between 1.78 and 1.95 Ma but the **reversed** polarity recorded in GD1/3 suggests that other parts of the cave filling may be older or younger. How dated? Biostratigraphy and magnetostratigraphy. Hominins found at site Two hominin teeth, GA 1, a ?*Homo* left lower molar and GA 2, a large *Paranthropus robustus* left M_2, have been recovered from the site. Archeological evidence found at site None. Key references: Geology, dating, fauna, and hominins Menter et al. 1999, Adams and Conroy 2005, Herries et al. 2006, 2009, Adams et al. 2007, Adams 2010.

Gongwangling *See* **Lantian Gongwangling**.

Goodman, Morris (1925–2010) Morris Goodman was born in 1925 in Milwaukee, Wisconsin. In 1942, Goodman enrolled at the University of Wisconsin-Madison, but from mid-1943 to the fall of 1945 he served in the US Air Force flying as a navigator on bombing missions over Europe. He returned to the university as a sophomore in the winter of 1946 and became a research assistant to Harold Wolfe, Professor of Comparative Anatomy. After obtaining a BS in Zoology with a minor in Biochemistry in 1948, Goodman went on to receive his MS a year later and a PhD in Zoology with a concentration in biochemistry in 1951, all from the University of Wisconsin-Madison. Wolfe, a former student of Alan Boyden who was renowned for his work in comparative serology, influenced Goodman to concentrate on using molecular techniques to explore mammal relationships during the latter's master's degree. This influenced Morris Goodman's decision to focus on molecular

systematics and to examine the effects of salt concentration on the precipitin reaction in birds as the centerpiece of his PhD; it also helped foster Goodman's interest in immunology and protein chemistry. After his PhD, Goodman left Wisconsin to work at California Institute of Technology on a comparative study of hemoglobin with immunochemist Dan Campbell. Morris's interest in evolution was furthered during his subsequent work at the Detroit Institute of Cancer Research where he used the Ouchterlony method to make a comparative study of a range of proteins, including **albumin** (e.g., Goodman 1962). In 1962 Goodman proposed that chimpanzees and gorillas should be classified within the family Hominidae instead of the Pongidae because the comparative protein evidence suggested that chimpanzees and gorillas are more closely related to modern humans than to the other apes. On the basis of this evidence, especially that from albumin, Goodman also proposed the "hominoid slowdown hypothesis," the concept that rates of protein evolution had slowed down in the recent ancestry of modern humans and other hominoids. Goodman continued his work on serology and molecular taxonomy in the 1960s and 1970s, with work that focused on the evolutionary history of hemoglobin (Goodman et al. 1975). In the 1980s, Goodman used protein sequence data to reconstruct primate evolution and to inform primate classification. A 1988 paper published by Goodman's research group showed how important motifs in the non-coding DNA of genomes could be identified and in 1997 Goodman and Edwin McConkey lobbied for funding for an initiative to sequence the chimpanzee genome (McConkey and Goodman 1997). Goodman continued to work on evolutionary problems; until his death in 2010 his interests were placental evolution and the genetic changes responsible for enlargement of the **neocortex** and for **encephalization**. Morris Goodman, who was elected as a fellow of the American Academy of Arts and Sciences in 1996 and a member of the National Academy of Sciences in 2002, was a Distinguished Professor in the Department of Anatomy and Cell Biology and the Center for Molecular Medicine and Genetics at Wayne State University School of Medicine, where he had been for much of his career. He was also the chief editor of the journal *Molecular Phylogenetics and Evolution*.

Gorham's Cave *See* **Gibraltar**.

Gorilla The gorilla was first described by the Rev. Savage in 1847, as a new species of chimpanzee and called *Troglodytes gorilla* (at that time the generic name *Troglodytes* was used for chimpanzees), but the genus *Gorilla* was described for it in the 1860s. Paul Matschie and others in the early 20thC described a number of supposed new species of gorillas, but in 1929 Coolidge sank them all into a single species, *Gorilla gorilla*, with two subspecies (the western gorilla, *G. g. gorilla*, and the eastern gorilla, *G. g. beringei*). In the 1960s it became clear that eastern gorillas are divisible into two subspecies, the true mountain gorilla (*G. g. beringei*) and an eastern lowland or Grauer gorilla (*G. g. graueri*). Nowadays it is more usual to recognise two species of gorilla, *Gorilla gorilla* (western) and *Gorilla beringei* (eastern), as they have many absolute differences, both morphological and genetic. Each has two subspecies: the Cross River gorilla (*Gorilla gorilla diehli*) and the more widespread *Gorilla gorilla gorilla* in the west, and the Grauer gorilla (*Gorilla beringei graueri*) and the mountain gorilla (*Gorilla beringei beringei*) in the east. Pilbrow (2010) provides a useful map locating the various extant and historical gorilla populations and a comprehensive review of the molecular and morphological evidence. Her study of molar crown morphology supports a species-level distinction between *Gorilla gorilla* and *Gorilla beringei*.

gourmet strategy *See* **economic utility index**.

GPRP *See* **Gona Palaeoanthropological Research Project**.

GPTS *See* **geomagnetic polarity time scale**.

graben (Ge. *graben*=ditch) In structural geology a graben is a geological structure consisting of two (or more) inwardly directed **normal faults**, which form a trough-like feature. Because of their trough-like nature, grabens can become locations where potentially fossil-bearing sediments are preferentially deposited and preserved and thus grabens make obvious targets for geological and paleontological exploration. The rift valleys of East African are examples of grabens. *See also* **Afar Rift System**; **East African Rift System**.

grade (L. *gradus*=stage) Julian Huxley (1958, p. 27) introduced this term for what he called an "**anagenetic unit**." Whereas a **clade** is analogous

to a make of car (e.g., all Rolls-Royce cars share a recent common ancestor not shared with any other make of car), a grade is analogous to a type of car (e.g., luxury SUVs made by Mercedes, Jaguar, and Lexus are functionally similar yet they have different evolutionary histories and therefore have no uniquely shared recent common ancestor). Thus, it is a category based on what an animal looks like and does, rather than on what its phylogenetic relationships are. Thus "leaf-eating monkey" is a grade that contains both Old and New World monkeys, but it is not a **clade** because the two groups of monkeys from the Old and New Worlds are components of different larger clades. Grades may also be clades, but they do not have to be. *See also* **adaptive zone**; **genus**.

graminivory (L. *gramen*=grass and *vorous*, from stem of *vorare*=to devour) Literally, the consumption of grass. This is a highly specialized primate dietary strategy, found only among primates in the genus ***Theropithecus***. Today *Theropithecus gelada* live in the Ethiopian highlands, where their diet consists entirely of grasses of which they consume the leaves, rhizomes, stems, and seeds. *See also* **seed-eating hypothesis**.

Gran Dolina (Location 3°31′08″W, 42°21′09″N, Spain; etym. at the beginning of the investigation of this major European hominin site, it was thought the site was a large karstic doline, hence the name *Gran*=large and *Dolina*=doline, but now, even though it is a cave infill, the name persists) History and general description One of the fossiliferous cave fillings in the Sierra de **Atapuerca**, which is a series of eroded, **karstic**, limestone hills 14 km/8.5 miles east of Burgos in Northern Spain. It is permeated by sediment-filled cave systems. The largest is the **Cueva Mayor-Cueva del Silo** system where the **Sima de los Huesos** is located. Just 1 km/0.6 miles away in the **Trinchera del Ferrocarril** is the Gran Dolina (**TD**) with 18 m of sediment. Excavations began here in 1978 and seven levels (TD3–4, 5–7, and 10–11) have produced stone tools. In 1994 more stone tools and hominin fossils were recovered from the **Aurora stratum** (a conspicuous layer within the **TD6** lithostratigraphic level). More hominins were recovered in 2003, including **ATD6-96**, and 25 additional hominin specimens were recovered between 2003 and 2008. In total the hominins from the TD6 sample came from at least 10 individuals. Researchers now recognize six lithostratigraphic layers within the Aurora stratum, which they refer to as the Aurora archaeostratigraphic set (or AAS). Temporal span and how dated? **Magnetostratigraphy** points to an age >780 ka (it is in the **Matuyama** chron) and the best age estimate using **uranium series dating** and **electron spin resonance spectroscopy dating** methods is 731 ± 63 ka. Hominins found at site Between 1994 and 2008 more than 100 hominin fossils belonging to at least 10 individuals were recovered from the Gran Dolina cave. In 1997 they were assigned to a new species *Homo antecessor* on the basis of a hitherto unknown combination of a modern human-like midfacial morphology (e.g., a **canine fossa** and an acute zygomaticoalveolar angle) and a primitive *Homo ergaster*-like dental morphology. Critics suggest that the "modern" facial morphology is due to the immaturity of the ATD6-69 facial fragment. Archeological evidence found at site Mainly small, **Mode 1** artifacts, such as **flakes**, **denticulates**, **notches**, and **side scrapers**, many of which have been used for butchery and woodworking. The **cutmarks** on some of the hominin specimens shows that the hominins were being deliberately defleshed. Key references: Geology, dating, and paleoenvironment Parés and Pérez-González 1995, 1999, Cuenca-Bescós et al. 1999, Falguères et al. 1999, Berger et al. 2008; Hominins Carbonell et al. 1995, 2005, Bermúdez de Castro et al. 1997, 1999, 2008, Arsuaga et al. 1999a, Carretero et al. 1999, Lorenzo et al. 1999, Rosas and Bermúdez de Castro 1999, García-González et al. 2009, Gómez-Olivencia et al. 2010; Archeology Carbonell et al. 1995, 1999, Fernández-Jalvo et al. 1996, 1999 (syn. Trinchera Dolina).

Grande Grotte (Location 47°35′N, 3°45′E, France, etym. Fr. large cave) History and general description One of the largest caves of the **Arcy-sur-Cure** complex, the Grande Grotte is known for its many cave paintings and etchings, which include mostly animal forms, but also hand outlines and abstract symbols (points, wavy lines, etc.). The site has been extensively explored by a team led by Dominique Baffier and Michel Girard, who have painstakingly uncovered the artwork and performed a series of archeological excavations on the sediments below the artwork since they were discovered preserved under a thick layer of calcite in 1990. The site is currently open for tours. Temporal span and how dated? **Radiocarbon dating** of ochre-colored bones found at the base of some of the artwork in the cave provided dates of 26–28 ka, and radiocarbon dates from torch smears over one of the friezes are about 27 ka. Hominins found at site N/A. Archeological evidence found at site An archeological layer was found below a 30 cm sterile surface deposit. This layer contained evidence of cave-painting activity, including pestles, colorants, lamps, and hearths. Key references: Geology, dating, and paleoenvironment Girard et al. 1996, Baffier et al. 2001; Hominins N/A; Archeology Baffier and Girard 1995.

grandmother hypothesis A model of human evolution introduced by Kristen Hawkes and colleagues (Hawkes et al. 1998) which emphasizes the important role played by post-menopausal women in helping to provide food for their daughter's children. This is an attempt to explain why in modern humans female **fertility** and **mortality** are decoupled so that modern human females have a long post-reproductive lifespan. The hypothesis suggests that post-reproductive adults and particularly grandmothers can continue to indirectly influence their own reproductive success (i.e., **inclusive fitness**) by contributing to the reproductive success of their daughters. The model suggests that grandmothers achieve this by **foraging** for foods such as **tubers** that their grandchildren would find difficult if not impossible to access on their own. This assistance increases the chances of existing offspring surviving and it also allows daughters to wean infants early, move on to the next pregnancy, and thus shorten the **interbirth interval**.

granivory (L. *granum*=grain and *vorous*, from stem of *vorare*=to devour) Literally, the consumption of seeds. This term applies principally to rodents and insects that are exclusive seed-eaters, and whereas living primates consume seeds, particularly of flowering plants and sometimes grasses, as part of their diet (e.g., *Theropithecus*), no living primate has an exclusively granivorous diet. *See also* **graminivory**; **seed-eating hypothesis**.

grassland A biome with mean annual rainfall between 250 and 800 mm, a high rate of evaporation, periodic severe droughts, and a rolling to flat terrain, and where most animals are grazing or burrowing species. In Africa, grasslands can be completely devoid of woody plants, but in general grasslands are defined as those biomes with less than 10% cover of woody plants. The majority of modern African grasslands are dominated by C_4 grasses. However, prior to 1.7 Ma, most of the fossil localities in East Africa characterized as grasslands were a mix of C_4 and C_3 plants, indicating that modern C_4 grasslands are a more recent phenomenon (Cerling 1992). Most grasslands require periodic fires for maintenance, renewal, and elimination of encroaching woody growth. Production in grasslands is positively correlated with precipitation. Whereas grasslands at one time are thought to have covered about 40% of the land surface on Earth, today they cover less than 12%. Much of the secondary grasslands in Africa would revert to woodland if not for overgrazing and fire (Pratt and Gwynne 1978). Grasslands tended to dominate at many hominin sites in Africa between 2.0 and 1.0 Ma, as evidenced by carbon isotopes in soils and deposits of mammalian grazer fossils. *Homo erectus* first appeared during this period and is thought to have existed in grasslands and lightly wooded habitats. Many of these grasslands became bushlands and woodlands after 1.0 Ma.

Gravettian An early **Upper Paleolithic** technocomplex found throughout Europe, dated to between 28 and 19 ka, a particularly cold phase during the last glacial maximum. It is characterized the appearance of diagnostic stone tools including burins and straight-backed points, the exploitation of large herd mammals, particularly mammoths, the expansion of long distance trade as evidenced by shells, and the production of ivory and clay stylized figures of women (known as Venus figurines) and animal designs. Sites in central and eastern Europe (e.g., **Pavlov, Dolni Věstonice**) show further technological specialization, including meat-storage pits, bone burning for fuel, and construction of shelters out of animal bones. A similar technocomplex from the same time period found only in central and southwestern France is known as the **Perigordian**.

gray matter A term that describes regions of the central nervous system that mainly comprise the cell bodies of **neuron**s. There is gray matter in the **cerebral cortex**, the cerebellar cortex, the nuclei of the **brain** and the **brain stem**, and in the central portion of the spinal cord.

gray-scale **image** *See* computed tomography.

gray-scale **values** *See* computed tomography.

grazer An animal that feeds on ground-level vegetation (e.g., terrestrial or aquatic grasses). Grazers tend to have a relatively broad snout (e.g., extant hippopotami) so that the anterior teeth can trap as much grass as possible, whereas browsers have a narrower snout (e.g., modern giraffes) that allows them to select items in bushes and trees. In tropical environments, the vegetation that grazers feed on (i.e., grasses) uses the C_4 photosynthetic system and it has more ^{13}C relative to plants that use the C_3 photosynthetic system; thus grazing animals have bones and enamel that have more ^{13}C relative to the bones of browsing animals that consume plants using the C_3 photosynthetic system. *See also* C_3 and C_4; **grazing**.

grazing (ME *grasen*=to feed on grass) The consumption of the leaves and shoots of monocotyledonous plants (i.e., grasses and sedges having a single embryonic seed leaf or cotyledon) which grow at, or slightly above, ground level. Many **antelopes**, particularly the alcelaphines (i.e., wildebeest and allies that belong to the tribe **Alcelaphini**) and the antilopines (i.e., gazelles and allies that belong to the tribe **Antilopini**), graze on grasses. A grazing diet in extinct animals is usually inferred through studies of tooth morphology or **stable isotope**, usually carbon, geochemistry. Grazing on tropical grasses that have a C_4 photosynthetic pathway leaves a stable carbon isotopic signal that can be detected in tissues, including the enamel of fossil teeth. Grazing is important for understanding the environmental context of hominin evolution. **Adaptive radiation**s of grazing ungulates occurred along with the reduction of forest and more wooded landscapes in Africa. The spread of grazing as a dietary strategy in the faunal community thus helps document changes in the environments inhabited by early hominins. *See also* $^{13}C/^{12}C$; **stable isotopes**.

great ape (OE *great*=thick or large and *apa*= ill-bred and clumsy; literally the "large and clumsy" ones, as opposed to the "small and clumsy" ones, known as the "lesser apes". Before the apes had been investigated scientifically and appreciated on their own terms, 'apes' were regarded as being 'clumsy' because they lacked **dexterity**) An informal taxonomic category that includes extant and fossil taxa. The extant "great ape" taxa are chimpanzees/bonobos, gorillas, and orangutans; the fossil taxa are all the extinct forms that are more closely related to chimpanzees/bonobos, gorillas, and orangutans than to any other living taxon. In the traditional, pre-molecular, taxonomy the informal term "great apes" was equivalent to the family Pongidae. *See also* **ape**.

Great Cave of Niah   See **Niah cave**.

Green Gully (Location 37°44′S, 144°50′E, Victoria, Australia) History and general description A soil pit 14.5 km/9 miles northwest of Melbourne and 1.6 km/1 mile south of **Keilor**. In 1965 hominin bones were accidentally uncovered during commercial sand-extraction operations. Temporal span and how dated? The direct **radiocarbon dating** of collagen from the skeletal suggests they are *c.*6500 years BP. Hominins found at site Most researchers suggest that the modern human remains from Green Gully represent two individuals, Green Gully 1, the partial skeleton of an adult female that includes portions of the cranium, thorax, and right upper limb, and Green Gully 2, the partial skeleton of an adult male that includes most of the left upper limb, a fragmentary pelvis, and both lower limbs. Archeological evidence found at site Large and small **quartzite scrapers** and quartzite fabricators. Key references: Geology and dating Bowler 1970a; Hominins Macintosh 1967, 1970; Archeology Mulvaney 1970.

Greenland Ice Core Project (or GRIP) A European initiative that complemented the US National Science Foundation-funded **Greenland Ice Sheet Project Two** (or GISP2) that drilled approximately 30 km/18 miles west of the site of the GRIP core. Between them, the GRIP and GISP2 provided a long, high-resolution record of past climate events. For more information see www.nerc-bas.ac.uk/public/icd/grip/griplist.html.

Greenland Ice Sheet Project Two (or GISP2) In late 1988 the Office of Polar Programs of the US National Science Foundation initiated GISP2 as part of its Arctic Systems Science Program that focused on environmental change in the Arctic. The aim of the project was to use a drill to recover an ice core that captured as much as possible of the history of the Greenland ice cap. After 5 years

of drilling, on July 1, 1993 the GISP2 drill penetrated the full thickness of the ice sheet and entered the bedrock. The ice core recovered was approximately 3050 m deep and at the time it was the deepest ice core ever recovered. The GISP2 drilling program complemented the European-funded **Greenland Ice Core Project** (or GRIP) that drilled approximately 30 km/18 miles east of the site of the GISP2 core. Between them the two projects provided a long, high-resolution record of past climate events. For more information see www.gisp2.sr.unh.edu/.

Grenzbank zone (Ge. for "border bed") General description A 1 m-thick boundary layer between the underlying **Sangiran Formation** and the younger overlying **Bapang Formation** in the **Sangiran Dome** of central Java. This stratum is a calcareous **conglomerate** rich in fossil bone, including hominins. Sites where this formation is exposed **Blimbingkulon**, **Mojokerto**, and **Sangiran**. Geological age In 2001, published **argon-argon dating** suggested the lowermost Bapang hominins are 1.51–1.47 Ma and the youngest hominins from the Sangiran Formation are *c.*1.6 Ma so the Grenzbank is dated to *c.*1.51–1.6 Ma. Most significant hominin specimens **BK 7905**, **BK 8606**, **Sangiran 6** ("*Meganthropus* A"), and **Sangiran 9** ("Mandible C"). Key references: Geology, dating, and paleoenvironment Watanabe and Kadar 1985, Larick et al. 2001; Hominins Weidenreich,1944, Sartono 1961, Kaifu et al. 2005a.

Grimaldi (Location 43°47′N, 7°37′E, Italy; etym. It. after the nearby village, or after the ruling family of Monaco, due to Prince Albert I having led the excavations in the **Grotte des Enfants**) History and general description The Grimaldi complex of sites consists of a series of limestone caves and rockshelters at the base of a cliff about 1 km/0.6 miles from the Italy/France border on the Mediterranean coast; it is also known as the "Balzi Rossi" or "Baousse Rousse" caves after the reddish color of the cliff. A number of these have yielded hominin fossils. From west to east they are Grotte des Enfants, Riparo Mochi, **Barma del Caviglione**, **Riparo Bombini**, **Barma Grande**, Baousso da Torre, and **Grotta del Principe**. Some of the excavations that took place in the latter quarter of the 19thC were undertaken by amateurs and even the remains and artifacts that were collected by trained archeologists tended to be poorly looked after and were put in different museums. Most of the caves have since collapsed, making further inquiry logistically difficult. Temporal span and how dated? As many of the caves were explored before rigorous excavation techniques were developed, stratigraphic and chronologic questions remain unanswered. However, direct **radiocarbon dating** of a number of the skeletons gives an age of *c*.24 ka for the **Gravettian** layers and *c*.11 ka for the **Epigravettian**. Hominins found at site Many complete or near-complete skeletons are known from the **Paleolithic** deposits in the caves. All of the skeletons belong to modern *Homo sapiens* and appear to have been intentionally buried. Beginning with Verneau (1906), the lower burials of Grotte des Enfants 5 and 6 were regarded as "Negroid" in character. Boule and Vallois (1957) elaborate on this, stating that the Grotte des Enfants remains display affinities of South African Bushmen, even finding similarities between Grotte des Enfants 5 and 6 and the skeleton of the "Hottentot Venus," which was at their disposal. These differences are now accepted as being within the normal range of anatomically modern human cranial robusticity, but the theories of Verneau, **Boule**, and **Vallois** were influential in eroding the Eurocentism in physical anthropology. Archeological evidence found at site **Mousterian** assemblages occupy the lower **Middle Paleolithic** levels, while **Aurignacian** and Gravettian assemblages are known from the **Upper Paleolithic**. In addition to Gravettian lithic technology, a number of "Venus" figurines have been found in the Upper Paleolithic levels of the caves, but there is some confusion in a few cases as to which figurines come from which cave. Key references: Geology, dating, and paleoenvironment Formicola et al. 2004; Hominins Verneau 1906, Boule and Vallois 1957; Archeology Mussi 2001.

grindstone Stone tools used for cracking open or reducing the particle size of plant or animal foods, or of pigments. Grindstones usually appear in two forms; handheld grindstones are referred to as a hammer or mano, fixed or stationary grindstones are referred to as an anvil or metate. *See also* **ground stone tool**.

GRIP Acronym for the **Greenland Ice Core Project** (*which see*).

gross reproductive rate (or GRR) The integral of the age-specific fertility rate in a continuous age model and the sum of the age-specific fertility rates across the reproductive age intervals in a **life table**. The GRR is also sometimes referred to as the total fertility rate (or TFR) because it represents the expected total number of females born to a woman throughout her life, although some use GRR to refer to female births and TFR to refer to all births. The GRR should be contrasted with the **net reproductive rate** (or NRR), which is lower because it includes the mortality function for potential mothers.

Grotta Breuil (Location 41°14′N, 13°05′E, Italy; etym. It. *grotta*=cave and *Breuil*=for Henri Breuil, the French archeologist) History and general description One of the caves in the **Monte Circeo** complex, this site is a large karstic chamber located on the southwestern face of the promontory, just above the present sea level. The entrance opens towards the west, with a large central chamber extending to the east. Systematic excavations began in 1986 under the direction of A. Bietti. A few hominin remains and an abundance of associated stone tools and a rich faunal assemblage were found, and represent a late occurrence of *Homo neanderthalensis* in the Italian peninsula. Temporal span The hominin remains are dated to late **Oxygen Isotope Stage** 3, *c*.35–45 ka; How dated? **Electron spin resonance spectroscopy dating, biostratigraphy**. Hominins found at site Three fragments have been recovered: a portion of a left parietal, an adult left M_2, and a juvenile M_3. The remains are typical of *H. neanderthalensis*. Archeological evidence found at site A Pontian (local **Mousterian**) industry was found associated with the remains. Key references: Geology, dating, and paleoenvironment Bietti and Manzi 1991, Schwarcz et al. 1991b; Hominins Manzi and Passarello 1995; Archeology Bietti et al. 1991.

Grotta dei Fanciulli *See* **Grotte des Enfants**.

Grotta del Caviglione *See* **Barma del Caviglione**.

Grotta del Fossellone *See* **Fossellone**.

Grotta di Lamalunga (Location 40°52′21″N, 16°35′17″E, Italy; etym. It. *lama*=cave system in the local dialect and *lunga*=long) A cave system in karstic limestone near Altamura, Bari, Apulia, where a hominin skeleton covered in a variably thick coating of calcareous material, **Altamura 1**, was found in 1993 by speleologists. *See also* **Altamura 1**.

Grotta Guattari (Location 41°14′N, 13°05′E, Italy; etym. It. *grotta*=cave; *Guattari*=from A. Guattari, the name of the discoverer of the fossils, the owner of the land where the site is and the owner of what used to be an eponymous hotel near the site, now called the Hotel Neanderthal) History and general description This, one of the many caves in the **Monte Circeo** complex, contained the most iconic of the hominin remains recovered from the Monte Circeo promontory. Alberto Carlo **Blanc** spent much of the late 1930s exploring the caves of Monte Circeo (e.g., **Grotta Breuil**, Fossellone, and many others). In 1939 Guattari, a local hotelier, told Blanc that he and his workmen had discovered a narrow entrance to a cave on his property and that there appeared to be a number of animal and human remains, including a cranium. The cranium had already been removed when Blanc arrived; it had been sitting on the cave floor, surrounded by a "circle of stones." The following week one of Guattari's staff who had explored the cave handed over a mandible that she had apparently found near the cranium. Another mandible was discovered in a breccia deposit outside the cave after WWII by A. Ascenzi and G. Lacchei. The remains were then given to Sergio **Sergi** for study. Temporal span and how dated? *c*.51–7 ka by **uranium-series dating** and **electron spin resonance spectroscopy dating**. Hominins found at site Three specimens were recovered: Guattari 1 (formerly Circeo I), a fairly complete cranium, and two fragmentary mandibles, Guattari 2 and 3 (formerly Circeo II/A and III/B), all attributed to *Homo neanderthalensis*. Archeological evidence found at site A Pontian (local **Mousterian**) industry was found. It was thought that Guattari 1 was the subject of a cannibalistic ritual, because of the damaged foramen magnum and the "circle of stones" surrounding the skull and other evidence. This is most likely not the case, since more thorough study in the 1980s suggests the cave was primarily a hyena den with only secondary occupation by hominins. Key references: Geology, dating, and paleoenvironment Schwarcz et al. 1991a; Hominins Bietti and Manzi 1991, Piperno and Scichilone 1991,

Sergi 1954; Archeology Taschini 1979, Bietti and Manzi 1991, White and Toth 1991.

Grotta Paglicci *See* **Paglicci**.

Grotte Cosquer *See* **Cosquer Cave**.

Grotte d'Aurignac *See* **Aurignac**.

Grotte de Spy *See* **Spy**.

Grotte des Contrebandiers *See* **Smugglers' Cave**.

Grotte des Enfants (Location 43°47′00″N, 07°32′08″E, Italy; etym. presumably named after the infant skeletons found in the upper layers) History and general description The westernmost of the caves in the **Grimaldi** complex; it is also known as the Grotta dei Fanciulli. The cave was first excavated by Émile Rivière in the 1870s and subsequently by Prince Albert I of Monaco in the early 20thC. Temporal span and how dated? Direct **radiocarbon dating** of Grotte des Enfants 1 gave an age of 11,130±100 years BP. The older skeletons, Grotte des Enfants 5 and 6, are presumed to be *c*.24 ka, based on direct dating of remains from **Arene Candide** and **Barma Grande**, which can be **biostratigraphically** correlated. Hominins found at site Six complete or nearly complete skeletons are known from the site and were almost certainly intentionally buried. The first three can be chronologically grouped together and include two complete infant skeletons and one adult female. The second group of three are from the **Gravettian** layers (layer F) and consist of two males and one female. Ochre and marine shell ornaments were found decorating the burials. The hominins are all unquestionably *Homo sapiens*. Verneau (1906), and later Boule and Vallois (1957) thought the Gravettian remains Grotte des Enfants 5 and 6 to be "Negroid" in character. Boule and Vallois (1957) elaborate on this and stated that the Grotte des Enfants remains display affinities of South African Bushmen, finding similarities with the skeleton of the "Hottentot Venus." Their morphology is now accepted as being within the normal range of anatomically modern cranial robusticity, but the "Negroid theory" was influential in eroding the Eurocentrism in physical anthropology of that time. Archeological evidence found at site Most of the **Upper**

Paleolithic artifacts were originally identified as **Aurignacian** but recent work has definitively attributed the industry to the Gravettian. Key references: Geology and dating Formicola et al. 2004; Hominins Boule and Vallois 1957; Archeology Mussi 2001.

Grotte des Fées (Location 47°35′N, 3°45′E, France, etym. Fr. *grotte*=cave and *fées*=fairies) History and general description The first among the **Arcy-sur-Cure** caves to be excavated was the Grotte des Fées, which was explored by Marquis de Vibraye in the 1850s and 1860s and later by the Abbe Parat and others until 1905, when the site was considered exhausted. de Vibraye never formally published his results, and instead gave a series of talks that were summarized by Parat (1903). Later researchers with more modern techniques subsequently excavated several of the nearby caves (e.g., Leroi-Gourhan 1950, 1958, Movius Jr 1969b), which helped to place the Grotte des Fées in context, although they raised several questions about the conclusions drawn from the earlier excavations. The Grotte des Fées consists of several chambers, with one long gallery running mostly north–south for nearly 150 m through the limestone cliffs. It contained several archeological layers, many faunal remains, and a mandible of questionable taxonomic status. Temporal span and how dated? The cave is only very approximately dated using **technostratigraphy**, especially with nearby caves, such as the **Grotte du Renne**. The stone tools are mostly **Mousterian**, suggesting a date of more than 35 ka. Hominins found at site de Vibraye found the only hominin remains, a mandible he attributed to *Homo neanderthalensis*, and a fragmentary atlas. Several authors have argued that the mandible is not ancient (de Mortillet and de Mortillet 1900, Leroi-Gourhan 1950, 1958). Leroi-Gourhan (1950) noted that the mandible lacked both the patina that marked the **Middle Pleistocene** bones from the site and the red ochre that marked the bones from the **Upper Paleolithic** layers. Furthermore, he argued the mandible lacked many of the distinctive traits of *H. neanderthalensis* and manifested many traits found in modern humans (Leroi-Gourhan 1958) and he suggested the mandible may have been a recent, perhaps **Neolithic**, burial. Archeological evidence found at site de Vibraye described three layers based on changes in the fauna, although Parat and others noted that the bottom layer could be further subdivided. The few tools associated with de Vibraye's lower level were Mousterian, although some suggested **Acheulean handaxes** were found in the lowermost part of the excavation (Parat 1903). Key references: Geology, dating, and paleoenvironment Leroi-Gourhan 1950, 1958; Hominins Parat 1903, de Mortillet and de Mortillet 1900, Leroi-Gourhan 1950, 1958; Archeology Parat 1903.

Grotte des Fées de Châtelperron *See* Châtelperron.

Grotte des Pigeons (Location 34°48′38″N, 2°24′30″E, Morocco; etym. F. for "cave of pigeons") History and general description This large (>400 m²) cave formed in Permo-Triassic dolomitic limestone in the Beni Snassen mountains near the village of **Taforalt** was excavated discontinuously between 1944 and 1977 by Abbé J. Roche and since 2003 by N. Barton and A. Bouzzouggar. The approximately 10 m-thick archeological sequence consists of **Aterian** and **Iberomaurusian** artifacts as well as fossils within multiple layers of silt, silty loams, interstratified **speleothem**s, and anthropogenic ashes and charcoal. The Aterian levels contain early shell beads with an age estimate of 82 ka, whereas the Iberomaurusian strata include more than 180 Late Pleistocene (*c.*17–13 ka) hominin burials. Temporal span and how dated? Sediments at the Grotte des Pigeons are divided into six sediment groups (A–F, from top to bottom) bounded by erosive unconformities. Archeological strata at the Grotte des Pigeons are dated by **AMS radiocarbon, optical stimulated luminescence**, thermo**luminescence**, and **uranium-series dating** methods, with age estimates ranging from 10.94 ± 0.04 to *c.* 95.4 ± 0.90 ka. A **Bayesian age model** incorporating 13 age estimates constrains the age of the shell bead-bearing group E to between 73.4 and 91.5 ka with a most likely date of *c.*82.5 ka. Hominins found at site More than 180 modern human individuals from two necropolises have been reported for the Iberomaurusian levels. Archeological evidence found at site Sediment group E contains **Middle Paleolithic** tools such as **scrapers**, small **Levallois cores**, and thin **bifacially worked** (Aterian) **foliate point**s, well-defined **hearths**, and 13 small (<2 cm) *Nassarius gibbosulus* shells transported >40 km/25 miles to the site. Some of the shells may have been deliberately pierced, 10 show wear patterns suggestive of "string damage," and hematite is present on 10 of them. Older excavations recovered a number of typical Aterian tanged or pedunculate points and other retouched implements, although the spatial and stratigraphic relation of these to the present excavations is unclear. Diverse **Upper Paleolithic** (including Iberomaurusian) **flake**, **blade**, and **backed bladelet** industries occur in sediment groups A–B. The abundance of cedar (*Cedrus*) charcoal fluctuates throughout the Pleistocene, and dates from some of the Iberomaurusian strata suggest the presumed modern human occupations coincided with **Heinrich events**. Key references: Geology, dating, and paleoenvironment Bouzouggar et al. 2007, Raynal 1980; Hominins Ferembach 1962, Irish 2000; Archeology Roche 1967, Barton et al. 2007, Bouzouggar et al. 2007.

Grotte du Hyène (Location 47°35′N, 3°45′E, France, etym. Fr. cave of hyenas, named for the most frequent fauna found during excavation) History and general description This name is given to two adjacent caves within the **Arcy-sur-Cure** complex. One cave, Hyène I, was systematically excavated by André Leroi-Gourhan between 1947 and 1957; Hyène II, was not formally excavated. Leroi-Gourhan described 16 layers in Hyène I numbered 14–30. Temporal span and how dated? Stratigraphic **correlation** to layers at the **Grotte du Renne** that have been **radiocarbon dated** suggests that the early **Mousterian** layers are considerably older than 39 ka, and the post-Mousterian layers may be *c*.33–5 ka. Hominins found at site Several isolated hominin remains were found in layer 20, which was assigned to the lower Mousterian period. These included a mandible with nearly complete dentition, a maxilla from a different individual, several isolated teeth, a fragmentary fibula, and a fragmentary occipital. Archeological evidence found at site The lower layers at the site contain a variety of **Acheulean** and Mousterian artifacts. There is a depositional hiatus between layers 16 and 15, and the two top layers (14 and 15) are from a more recent, post-Mousterian or possibly **Châtelperronian** period. These top layers contained ochre, shells, pyrite, and bolas stones in addition to flaked stone tools. Key references: Geology, dating, and paleoenvironment Leroi-Gourhan 1961; Hominins Leroi-Gourhan 1959, 1961; Archeology Leroi-Gourhan 1961.

Grotte du Loup (Location 47°35′N, 3°45′E, France, etym. Fr. cave of wolves, named after the

taxon found most frequently during excavations) History and general description This cave, part of the **Arcy-sur-Cure** complex, was covered by a rockfall and thus was protected from looting and was first excavated by André Leroi-Gourhan in 1947. Compared to other caves in the complex, the Grotte du Loup is quite small, measuring roughly 4 m by 3 m. Leroi-Gourhan described six layers, two of which contained archeological material. One was assigned to the **Châtelperronian**, the other to the denticulate **Mousterian**. Temporal span and how dated? The archeological layers have been dated by **correlation** to those at the **Grotte du Renne**, where the Châtelperronian was dated using **radiocarbon dating** to *c*.33–5 ka, and the denticulate Mousterian to *c*.39 ka. Hominins found at site A molar and two incisors, along with a few fragmentary cranial pieces, were found in Layer IV. Leroi-Gourhan noted that the area immediately surrounding the remains had pieces of bones and flint, whereas the rest of this layer was devoid of cultural material. He suggested this was evidence of an intrusive burial from Layer III into Layer IV, but could not tell if the entire body had been buried (and subsequently decayed) or if the head only had been interred. Archeological evidence found at site A total of approximately 150 stone artifacts mostly from Layer III were assigned to the denticulate Mousterian; the rest were found in Layer I and the Upper Paleolithic layer assigned to the Châtelperronian. A few faunal remains – including reindeer, fox, cave bear, and wolf – were found, mostly in the Layer III. Key references: Geology, dating, and paleoenvironment Leroi-Gourhan 1950; Hominins Leroi-Gourhan 1950, 1959; Archeology Leroi-Gourhan 1950.

Grotte du Renne (Location 47°35′N, 3°45′E, France, etym. Fr. *grotte* = cave and *renne* = reindeer, named for the most frequent fauna found during excavation) History and general description This cave, which is part of the **Arcy-sur-Cure** cave complex, was excavated between 1949 and 1966 by André Leroi-Gourhan. The **Châtelperronian** levels are the subject of much debate, as they contain beads and other personal ornaments that are usually associated with modern human behavior (Leroi-Gourhan 1961), but the fragmentary hominin remains found in these layers are characteristic of *Homo neanderthalensis* (Hublin et al. 1996, Spoor et al. 2003, Bailey and Hublin 2006). Some argue this is evidence of Neanderthal social complexity either via acculturation from interaction with modern humans (Mellars

2005) or independent evolution of those traits (d'Errico et al. 1998). Others argue there is no stratigraphic integrity to these layers and that the tools are from the **Aurignacian** levels (White 2001). Temporal span and how dated? **Radiocarbon dating** on charcoal suggest the youngest layers at the site date to *c*.20 ka, while the oldest layers are more than 39 ka. The Châtelperronian levels date to between *c*.33 and 35 ka, with some much older and much younger dates. Hominins found at site Several teeth were found throughout the site and a partial skeleton of a juvenile Neanderthal was found in the bottom of the Châtelperronian levels (d'Errico et al. 1998). Archeological evidence found at site A total of 14 layers were found and these have been interpreted as representing a sequential progression from the Mousterian through Châtelperronian and Aurignacian to the **Gravettian** (Leroi-Gourhan 1961). Key references: Geology, dating, and paleoenvironment Farizy 1990, Hedges et al. 1994, Leroi-Gourhan 1961, 1988, Leroi-Gourhan and Leroi-Gourhan 1964; Hominins Hublin et al. 1996, d'Errico et al. 1998, Spoor et al. 2003, Bailey and Hublin 2006; Archeology Movius 1969b, Leroi-Gourhan 1961, Leroi-Gourhan and Leroi-Gourhan 1965, Farizy and Schmider 1985, Farizy 1990 and articles therein, Schmider and Perpere 1995, d'Errico et al. 1998, Mellars 2005, White 2001.

Grotte du Rhafas   *See* **Rhafas Cave**.

Grotte Scladina   *See* **Scladina**.

ground reaction force   *See* **force**.

ground stone tool   Any stone tool made by abrasion rather than by percussion. Ground stone tools include deliberately shaped implements (such as the stone adzes still used on Irian Jaya; Stout 2002) as well as those shaped by use, such as the upper and lower components (manos and metates, respectively) of **grindstones**. Deliberately shaped ground stone tools become common globally only in the **Holocene**, but the first grindstones appear in the archeological record in the **Middle Pleistocene** at sites such as **Twin Rivers**, Zambia.

group   Formal geological term for a stratigraphic package comprised of several **formations**. The group is a useful comprehensive term to encompass related subunits that are closely related genetically and temporally (e.g., in the **Turkana**

Basin, the **Omo Group** includes the **Koobi Fora Formation**, Nachukui Formation, and **Shungura Formation**).

growth   Growth is very broadly defined as an increase in size. During ontogeny, growth occurs in a number of dimensions, including mass (which represents the sum of lean and fat mass, water, organ weight, and skeletal weight), proportion, and linear dimensions. In modern humans pronounced growth in stature occurs shortly after puberty; this is unique in comparison to nonhuman primates. The end point of growth is the size attained by adulthood, and growth in various dimensions may reach this point at different times.

growth factors   Substances, usually peptides or steroid hormones, that are capable of stimulating and regulating cellular growth, proliferation, differentiation, and maturation. They also play key roles in the growth rate of some cancers. Examples of these include the insulin-like growth factors, fibroblast growth factor, and epidermal growth factor, which regulate a wide variety of processes within organisms ranging from neural development and neuroprotection, to angiogenesis and embryonic development, to wound healing.

growth hormone   (or GH) A peptide hormone released from the anterior pituitary gland. Release of GH is stimulated by growth hormone-releasing hormone (or GHRH) from the hypothalamus and it is downregulated (i.e., reduced in quantity) by somatostatin (or SST). GH exerts its potent effects on cell hypertrophy and differentiation both through direct effects at target tissue as well as through modulation of insulin-like growth factor I (or IGF-I), which is synthesized mainly in the liver. Body size dimorphism is regulated largely through different patterns of pulsatile secretion of GH in males and females, which are initiated at puberty under the influence of increasing levels of sex steroids. Abnormalities in GH secretion, or in receptors or binding proteins for GH, can result in dwarfism or gigantism.

growth plate   Also known as the epiphyseal plate, the growth plate is a cartilaginous plate located at the ends of long bones in individuals who are still undergoing growth (e.g., juveniles, individuals with estrogen insensitivity). The chondrocytes of the growth plate undergo proliferation, resulting in cells that eventually ossify and contribute to bone growth that occurs in a proximal–distal manner. Under the influence of increasing steroid hormones, the cartilage cells of the growth plate cease duplication and the whole area becomes ossified. The rate at which this occurs, and the relationship of this ossification to chronological age, can exhibit a great degree of variation among individuals. Methods that assess bone age based on degree of ossification of hands of the bone and wrist, include the Gruelich–Pyle method and the Tanner–Whitehouse method. The state of fusion of epiphyses from fossil hominins is used to estimate chronological age, although the efficacy of such methods will vary depending on the reference population used and the skeletal elements preserved.

growth spurt   *See* **adolescent growth spurt**.

GRR   Acronym for **gross reproductive rate** (*which see*).

GSI   *See* **Geological Survey of India**.

GSI-Narmada   Site Hathnora. Locality Hathnora. Surface/*in situ* Surface. Date of discovery 1982. Finder Sonakia. Unit N/A. Horizon N/A. Bed/member "ca. 4m thick boulder gravels and sandy-pebble beds" (Patnaik et al. 2009, p. 120) also referred to as the "Boulder Conglomerate" (*ibid*, p. 117). Formation Surajkund Formation. Group N/A. Nearest overlying dated horizon N/A. Nearest underlying dated horizon N/A. Geological age de Lumley and Sonakia (1985) suggested a Late Middle Pleistocene age on the basis of the presence of *Stegodon ganesa* and *Elephas hysudricus*; the combination of poorly constrained paleomagnetic and isotopic ages for equivalent sediments suggests an age range of 50 to more than 200 ka (Patnaik et al. 2009). Developmental age Adult. Presumed sex Initially male (Sonakia 1984), but subsequently revised to female (de Lumley and Sonakia 1985, Kennedy et al. 1991, Kennedy 1992). Brief anatomical description Almost complete right half of a calvaria to which part of the left parietal is attached. Announcement Sonakia 1984. Initial description Sonakia 1984. Photographs/line drawings and metrical data Kennedy 1991. Detailed anatomical description N/A. Initial taxonomic allocation *Homo erectus*. Taxonomic revisions *Homo sp. indet*. Current conventional taxonomic allocation *Homo sp. indet*. Informal taxonomic category Pre-modern *Homo*. Significance It is the only significant fossil hominin evidence from the Indian subcontinent. Location of original Geological Survey of India, Calcutta, India.

Guattari Cave   *See* **Grotta Guattari**.

gubernacular canal   *See* **gubernacular foramen; gubernaculum**.

gubernacular foramen   The opening of a gubernacular canal at the surface of the alveolar process. The opening communicates with the **dental crypt** that contains the developing **tooth germ**. *See also* **gubernaculum**.

gubernaculum   (L. *gubernaculum*=a rudder or helm) Cords of connective tissue that play a role in the ontogeny of the testis (i.e., *gubernaculum testis*) or of a tooth (i.e., *gubernaculum dentis*). The *gubernaculum dentis* is a cord of fibrous tissue that connects the outer layer of the **dental follicle** (the fibrous sac surrounding the developing tooth germ in the bony crypt) with the mucous membrane that lines the mouth or oral cavity. This fibrous cord has a central epithelial strand and lies in a bony canal within the alveolar bone. Its function is to guide or direct the erupting tooth to its proper location in the **alveolar process**. The bony **gubernacular canals** containing the gubernacular cords of the permanent **anterior teeth** open on the lingual side of the erupted deciduous teeth at a **gubernacular foramen**, but those of the permanent molars (and sometimes the premolars) rise directly upwards through the opening in the roof of the bony crypt. During eruption the gubernacular cords decrease in length, increase in thickness, and become replaced at the surface by an epithelial plug, through which the tooth cusp(s) eventually emerge into the mouth.

gubernaculum dentis   *See* **gubernaculum**.

guenon   The informal name for the tribe Cercopithecini that includes the genera *Cercopithecus*, *Chlorocebus*, *Erythrocebus*, *Allenopithecus*, and *Miopithecus*. *See also* **Cercopithecini**.

guild   (ON *gildi*=payment. Craftsmen or tradesmen had to pay dues to belong to their appropriate guild) In ecology the term guild is used to describe a group of species that exploit resources in a similar way. Overall body size and functional morphology-based interpretations of the type of resource consumed are considered when defining a guild. For example, dental morphology was used to identify four **carnivore guilds** at the Miocene hominoid site of Pasalar, Turkey.

Günz *See* **glacial cycles**.

Guomde cranium *See* **KNM-ER 3884**.

GVH Abbreviation of Gladysvale hominins and the prefix for fossils recovered from **Gladysvale** Cave, Blauuwbank Valley, Gauteng, South Africa.

GWM Acronym for the Gona Western Margin sector of the Gona Paleoanthropological study area. *See* **Gona Paleoanthropological study area**.

gyri Plural of **gyrus** (*which see*).

gyrification index (or GI) An attempt to capture in the form of an index the amount of folding of the cerebral (or cerebellar) cortex. The method uses brain slices and for each slice the gyrification index is the total length of the cortex (i.e., including the part of the cortex buried in sulci) divided by the length of the cerebral or cerebellar cortical contour that is exposed on the surface of the hemispheres.

gyrus (Gk *gyros*=circle, or ring) The visible part of a fold in the cerebral (or cerebellar) cortex that can be seen on the surface of the brain.

H

²H/¹H Two **stable isotopes** of hydrogen. The relative ratio of $^2H/^1H$ (referred to as δD values; *see* **δ**) can be measured in plant and animal tissues and is used in **stable-isotope biogeochemistry**. Plants get all of their hydrogen from water. Water sources (e.g., groundwater, surface water, and rain water) have different δD values (Kendall and Coplen 2001), thus plants using different water sources due to differences in growing season or rooting depth may have different δD values. Evapotranspiration leads to increased δD values in leaves. This is because the lighter isotope (^1H) has a higher vapor pressure and evaporates first so that the remaining leaf water is enriched in ^2H. This process is exacerbated in windy or arid regions and is reduced in humid regions. Yakir (1992) provides an excellent review of hydrogen isotopes in plants. Among animals, δD is most commonly measured in hair or fur, feathers, and in bone **collagen**. Animals get their hydrogen from the water they drink and from the food that they eat. There is some disagreement over how much of the δD values in animals reflect diet versus drinking water. Interpretations of consumer δD therefore require a good understanding of both the δD values of potential water and food sources and the physiology of the organism of interest (Hobson et al. 1999).

H or **H²** Abbreviation of **broad-sense heritability**. *See* **heritability**.

h or **h²** Abbreviation of **narrow-sense heritability**. *See* **heritability**.

habitat (L. *habitare* = to dwell) The place where an organism, population, or community lives. Almost all habitats are shared by a range of taxa that together contribute to the community ecology of that habitat. Habitat differs from **niche** in that the former stresses location and the latter the environmental parameters.

habitat mosaic *See* **mosaic habitat**.

habitat theory hypothesis An overarching theory proposed by Elisabeth Vrba that seeks to explain how environmental change influences evolutionary patterns in the fossil record. Habitat theory is complex because it embraces no fewer than seven related hypotheses: (a) most species survive periods of environmental change by passively "tracking" the changing geographical distribution of their preferred **habitat** (e.g., as **savanna** conditions spread, so do the ranges of savanna-adapted mammals); (b) areas with irregular topography (i.e., peaks and valleys) will show higher rates of species' ranges being split (this phenomenon is called **vicariance**) than will areas of gentle topography (i.e., smooth gradations in elevation) because irregular topography induces habitat fragmentation; (c) during periods in which global temperatures vary strongly by latitude, species living in equatorial latitudes should be more prone to vicariance, **speciation**, and extinction as a result of cyclical environmental change than species living at higher latitudes; (d) during extreme cooling trends, species' extinctions should be concentrated at low latitudes because warm equatorial-type habitats are fragmenting or disappearing; conversely, extinctions should be concentrated at high latitudes during extreme warming trends; (e) speciation does not occur unless it is "forced" by environmental change; thus, most of the turnover in lineages (i.e., the extinction of species and the appearance of new ones) occurs simultaneously in multiple taxa in what is known as a "turnover-pulse"; (f) major global climatic changes are responsible for the majority of mammalian speciation events; (g) species whose ecological resources persisted even as the Earth oscillated between environmental extremes experienced low vicariance, speciation, and extinction rates. Persistent species will include both ecological generalists, **eurytopes**, who can exploit a variety of resources, as well as ecological specialists, **stenotopes**, whose preferred resource did not

Wiley-Blackwell Encyclopedia of Human Evolution, First Edition. Edited by Bernard Wood.
© 2013 Blackwell Publishing Ltd. Published 2013 by Blackwell Publishing Ltd.

diminish in the face of environmental change. Conversely, species whose ecological resources disappeared during periods of environmental change will have experienced more vicariance, and thus higher speciation and extinction rates. *See also* **effect hypothesis**; **eurytope**; **macroevolution**; **stenotope**; **turnover-pulse hypothesis**.

Hadar (Location 11°10′N, 40°35′E, Ethiopia; etym. Ahdi D'ar or Adda Da'ar is the local Afar (Danakil) name for the ephemeral stream central to what is now the Hadar Research Project area. The Afar name was transcribed by Maurice Taieb as "Hadar," which is now the widely accepted name for the area and formation that outcrops here) History and general description Maurice Taieb discovered fossils at the site of Hadar, which is 300 km/180 miles northeast of Addis Ababa in the **Afar Rift System**, in the 1960s while conducting a geological survey. In 1972 Taieb invited Donald **Johanson**, Yves **Coppens**, and Jon Kalb to join him for a field season and in 1973 the group formed the **International Afar Research Expedition**. The first fossil hominins were found at the site in 1973 and specimens were found each year until this first phase of intensive fieldwork at Hadar came to an end in 1977. Sustained fieldwork resumed from 1990 until 1994 and it has been ongoing since 1999. Temporal span The **Hadar Formation** is 3.4–3.0 Ma and the overlying **Busidima Formation** is *c*.2.4 to less than 2.0 Ma. How dated? **Biostratigraphy**, zircon fission track, **magnetostratigraphy**, **argon-argon dating**, and **potassium-argon dating**. Hominins found at site The vast majority of the many hominins found at Hadar have been attributed to *Australopithecus afarensis* on the basis of their dental and gnathic similarities to specimens from **Laetoli** (including the type specimen, **LH 4**). Since 1973 the site has yielded approximately 400 *Au. afarensis* specimens and the Hadar sample constitutes approximately 90% of all known *Au. afarensis* fossils. Teeth, jaws, and crania are the three largest specimen categories, but postcranial fossils make up approximately 38% of the site sample. The Hadar hominin sample includes **A.L. 288-1** ("Lucy"), a remarkably complete *Au. afarensis* associated skeleton and **A.L. 444-2**, a 75–80% complete *Au. afarensis* adult skull. The material from A.L. 333, comprising a rich assortment of adult and juvenile skull, skeletal, and dental remains, prob-

ably represents a single biological population of *Au. afarensis*. A small minority of the hominin fossils have been attributed to *Homo*, including **A.L. 666-1**, which has been referred to *Homo* aff. *Homo habilis* (Kimbel et al. 1997). Archeological evidence found at site Oldowan artifacts. Key references: Geology, dating, and paleoenvironment Johanson et al. 1982, Taieb et al. 1976, Walter 1994, Bobe and Eck 2001, Bonnefille et al. 2004, Campisano and Feibel 2008a, 2008b, Reed 2008, Aronson et al. 2008, Wynn and Bedaso 2010; Hominins Johanson et al. 1982, Kimbel et al. 2004; Archeology Kimbel et al. 1996, Hovers 2009.

Hadar Formation [Location 10°55′–11°20′N, 40°20′–40°50′E, Ethiopia; etym. Ahdi D'ar or Adda Da'ar is the local Afar (Danakil) name for the ephemeral stream central to what is now the **Hadar study area**. The Afar name was transcribed by Maurice Taieb as "Hadar," which is now the widely accepted name for the area and formation that outcrops here] History and general description A region of badlands in an incised channel of the Awash River exposes substantial thickness of sediments, which can be mapped throughout the region surrounding Hadar. Although the expeditions of both Nesbitt (in 1928; see Nesbitt 1930) and Thesiger (in 1933–4; see Thesiger 1996) both passed through the region, Maurice Taieb was the first to recognize the fossiliferous localities (in 1970; see Taieb et al. 1972). Taieb et al. (1972) initially designated four formations: Hadar, Meschelle, Leadu, and Haouna-Lédi. At least two of these formations (Hadar and Meschelle) are coeval, and all four can now be rationalized into a single unit called the Hadar Formation. The Hadar Formation was further subdivided into four members by the later description of Taieb et al. (1976) (i.e., the Basal, Sidi Hakoma, Denen Dora, and Kada Hadar Members). Although the boundaries of the uppermost and lowermost members were not originally defined, the three stratigraphic boundaries between the four members remain in use. Each of the members above the Basal Member is defined as the sediments between the base of a designated tuff and the base of an overlying designated tuff. The three designated tuffs forming the stratigraphic divisions are the Sidi Hakoma Tuff (SHT), Triple Tuff-4 (TT-4), and the Kada Hadar Tuff (KHT). The names of the members come from ephemeral streams near the most prominent outcrop of the sediments that defines the member. The Basal Member was originally defined as a "catch-all" to include the sediments below the Sidi Hakoma Tuff, although this has been refined to the sediments between the uppermost surface of a basin-bounding weathered basalt flow (attributed to the Dahla Series Basalts; see Kalb et al. 1978a and Wynn et al. 2008). Likewise, the original definition of Kada Hadar Member had an undefined upper boundary and was a "catch-all" including all sediments above the Kada Hadar Tuff. This imprecise definition was problematic for dating sites in the upper part of the formation, and the upper boundary of the formation has since been refined to the level of a major unconformity with the overlying Busdima Formation (Quade et al. 2008). Temporal span and how dated? The oldest member, the Basal Member, includes sediments that date from an unconstrained lower age of less than 4 to 3.42 Ma, and the youngest, the Kada Hadar Member, dates from 3.20 to *c.*2.9 Ma. The sediments have been dated using the **argon-argon dating** and **potassium-argon dating** methods, as well as by **magnetostratigraphy** and **biostratigraphy**. Hominins found at site Fossil hominins recovered from the Hadar Formation, *sensu novo*, include only *Australopithecus afarensis*, of which much of the hypodigm is defined by specimens from the Hadar Formation; the AL-666 early *Homo* site was originally attributed to the Hadar Formation (Kimbel et al. 1996), although the upper formation boundary remained problematic. Archeological evidence found at site The Hadar Formation includes evidence of pre-**Oldowan** bone modifications (cutmarks and percussion marks), without accompanying stone tools; Gona archeology was originally attributed to the Hadar Formation (Corvinus and Roche 1976, Kimbel et al. 1996), although the definition of the upper formation boundary remained problematic. Key references: Geology, dating, and paleoenvironment Taieb et al. 1972, Taieb 1976, Tiercelin 1986, Walter 1994, Walter and Aronson 1993, Quade and Wynn 2008; Hominins Johanson et al. 1982a, Johanson and Coppens 1976, Alemseged et al. 2005, 2006, Kimbel and Delezene 2009; Archeology McPherron et al. 2010. *See* also **Gona Paleoanthropological study area; Busidima Formation.**

Hadley circulation (or Hadley cell) Describes the general circulation of the atmosphere in the tropics. The equator-ward air supplied by the trade winds that rises at the **Intertropical Convergence Zone** (or ITCZ) is matched by high poleward air that subsides in the subtropics. Rising air at the ITCZ accounts for high rates of precipitation there and the sinking air in the subtropics accounts for the global distribution of arid regions.

Hadza (etym. "Hadza" is short for "Hadzabe," the word that identifies the Hadza people in their own "Hadzane" language) The Hadza are one of the world's few remaining populations of hunter–gatherers. They live in a 4000 km^2/1544 square-mile region of northern Tanzania, south of the Serengeti, in a savanna woodland habitat around the shores of Lake Eyasi. Approximately 300–400 individuals, out of a total population of 1000, depend on hunting and gathering wild foods for over 90% of their diet; the remaining members of the tribe combine foraging with trading and begging or gaining income from the tourist trade. The Hadza do not practice agriculture or animal husbandry and they do not participate in a market economy. They reside in camps whose average size is about 30 people. People frequently move between camps and this has been linked with the Hadza's lack of traditional land rights or sense of individual ownership of natural resources. Camps move approximately every 2 months in response to seasonal availability of water and foods. Distinct wet and dry seasons are associated with differences in subsistence behaviors and social arrangements. During the dry season (i.e., June to November) camp size is relatively large, due to the limited availability of drinking water and thus the greater concentration of animals near watering holes. During the wet season (i.e., December to May) drinking water is more freely available in many areas and camp sizes are smaller. The Hadza diet consists of a wide variety of plant and animal foods and is characterized by a distinct sexual division of labor. Men typically go hunting alone with poisoned tipped arrows and take a wide variety of birds and mammals. They also collect baobab fruit and honey. Women forage in groups and primarily focus on gathering plant foods such as baobab, berries, and tubers (underground storage organs). Observations about the Hadza enable researchers to test hypotheses about evolutionary ecology in the absence of agriculture. The Hadza collect and consume foods in an ecosystem that may have been quite similar to that of our early hominin ancestors.

haemoglobin *See* **hemoglobin**.

Hahnöfersand (Location 53°32'50"N, 9°43'38"E, Germany; etym. name of the island) History and general description A hominin skull fragment was recovered from the bank of the river Elbe just downstream of Hamburg in 1973, supposedly associated with Pleistocene fauna. **Radiocarbon dating** by Reiner Protsch, who is alleged to have falsified data, resulted in an age of (36.3 ka±300 years), suggested this specimen was the earliest anatomically modern human in Europe, but the specimen was recently redated by more precise **AMS radiocarbon dating** to the Holocene. Temporal span and how dated? Direct AMS radiocarbon dating to 7.5 ka±55 years (uncalibrated). Hominins found at site A single cranial fragment attributed to *Homo sapiens*. Archeological evidence found at site None. Key references: Geology, dating, and paleoenvironment Street and Terberger 2002, Street et al. 2006; Archeology N/A; Hominins Bräuer 1980, Churchill and Smith 2000.

Haile-Selassie, Yohannes (1961–) Born in Adigrat, Ethiopia, Yohannes Haile-Selassie entered Addis Ababa University in Ethiopia in 1979 and graduated with a BA in history in 1982. From 1983 to 1985 he worked as a historian in the Center for Research and Conservation of Cultural Heritage in Addis Ababa and from 1986 to 1992 he was a research assistant at the Paleoanthropology Laboratory of the National Museum of Ethiopia, also in Addis Ababa. In 1992 Haile-Selassie enrolled in the University of California, Berkeley, USA where he studied for an MA and PhD. He received his MA in anthropology in 1995 and PhD in integrative biology in 2001 for a thesis entitled "Late Miocene mammalian fauna from the Middle Awash valley, Ethiopia." In 2002 he was appointed Curator and Head of Physical Anthropology at the Cleveland Museum of Natural History. Yohannes Haile-Selassie has undertaken paleoanthropological fieldwork in Ethiopia, beginning in 1988–9, at **Kesem-Kebena**, Chorora, and Melka Werer, and later at **Fejej**, Burji, **Konso**, Galili, and **Middle Awash study area**. Latterly he surveyed and explored the north central Afar region. Haile-Selassie currently conducts fieldwork research at the **Woranso-Mille study area** and directs the Woranso-Mille Paleontological Research Project. He was the sole author of the paper that proposed the new taxon *Ardipithecus kadabba* (Haile-Selassie 2001) and the senior author on the paper that reported further evidence of that taxon (Haile-Selassie et al. 2004). An edited volume on *Ar. kadabba* (edited jointly with Giday WoldeGabriel) was published in 2009. *See also **Ardipithecus kadabba*; **Woranso-Mille Paleontological Research Project**.

half-life *See* **radiometric dating**.

Hamann–Todd Collection Taxa *Homo sapiens, Pan, Gorilla*, etc. History Collection of modern human, great ape, and monkey skeletons held at the **Cleveland Museum of Natural History**. The collection of modern humans was assembled at Case Western Reserve University (but at the time the collection began that institution was known as Western Reserve University). The initial collection of approximately 100 modern human skeletons was assembled by Carl Hamann but the majority of the specimens were added by Hamann's successor as Professor of Anatomy, T. Wingate Todd. The latter began his academic career with Sir Grafton Elliot Smith at what was then called the Victoria University of Manchester (now called the University of Manchester), but in 1912 Todd moved to Case Western Reserve University. The influence of Sir Grafton Elliot Smith on Todd is evident, for Todd collected not only modern human skeletons but also the skeletons of non-human primates. Curator/contact person Yohannes Haile-Selassie (e-mail yhailese@cmnh.org). Postal address Cleveland Museum of Natural History,1 Wade Oval Drive, University Circle, Cleveland, OH 44106 USA. Relevant reference Hunt and Albanese 2005.

hammer As used in the analysis of stone tools made by **percussion**, a hammer refers to an object used to fracture rock. Hammers may be made of a variety of materials, including stone, bone, antler, or wood. *See also* **hammerstone**.

hammerstone A rock, often oval in shape, held in the hand and used for a variety of pounding tasks. In hard-hammer percussion, a hammerstone is used to strike flakes off a **core**. In combination with an anvil, a hammerstone may be used to flake rocks (**bipolar** technique) or to break open foods such as nuts, marrow bones, or shellfish. Hammerstones often develop characteristic pitting on one or both ends during use.

HAN Abbreviation of Hana Hari and the prefix for fossils recovered from Hana Hari, **Middle Awash study area**, Ethiopia.

hand The primate hand can be divided into the wrist, palm, the thumb (the first digit, also known as the pollex), and the medial four fingers (digits two through five). The primate hand differs from that of most other mammals in its prehensile or grasping ability. The opposable thumb is also generally considered to be a specialized feature of primates, although only catarrhines have a truly opposable thumb. Opposability is accentuated in the modern human hand compared to other primates, especially apes, because the thumb is relatively long and the fingers relatively short, making touching the thumb to the finger tips relatively easy. In contrast, grasping by the flexed fingers or by the thumb against the side of the index finger is accentuated in apes; the fingers are remarkably long and the thumb relatively short. Because the ape hand is specialized for suspension and climbing, modern human hand proportions are most similar to those of a terrestrial Old World monkeys, like the baboons and mandrills (Napier 1993). The modern human hand is composed of 27 bones: there are eight carpal bones in the wrist, five metacarpals that make up the palm, three phalanges (proximal, intermediate and distal) in each finger and two phalanges (proximal and distal) in the thumb. African apes and modern humans share the same number of hand bones. However, all other primates, apart from some lemurs, have an additional carpal bone, the **os centrale**. The wrist is generally divided up into two rows of bones: the proximal row including, from the radial to ulnar side, the scaphoid (and os centrale in nonhominid primates), lunate, triquetrum, and pisiform, and the distal row including the trapezium, trapezoid, capitate, and hamate. The pisiform is described in modern humans by some (Scheuer and Black 2000) as a sesamoid bone and not a "true" carpal because it is small and forms late in ontogeny within the tendon of the flexor carpi ulnaris. However, in all other primates the pisiform is a much larger bone that plays a more prominent functional role within the wrist (Lewis 1989). Both rows of carpals are connected by several interosseous ligaments, both within and across the two rows, which help maintain both mobility and stability within the wrist. The two carpal rows together form a mediolaterally concave arch on the palmar surface that creates the carpal tunnel. The "walls" of the carpal tunnel are formed on the lateral side by tubercles on the scaphoid and the trapezium and on the medial side by the pisiform and the "hook" or hamulus of the hamate, all of which

extend into the palm more than the other carpals. These anatomical "walls" serve as the lateral and medial attachment sites for the flexor retinaculum, a band of tough connective tissue that forms the palmar surface (or "roof") of the carpal tunnel and keeps the digital flexor tendons from bow-stringing and buckling during flexion of the fingers. The attachments of the carpal tunnel are also the key areas of origin for the intrinsic thenar and hypothenar muscles of the hand (see below). The eight carpal bones of the modern human wrist form multiple complex articulations, some of which are relatively mobile (e.g., between the capitate and scaphoid) and others that are highly stable (e.g., between the capitate and hamate). In general the proximal row is more mobile than the distal row. However, most of the wrist bones function to create two main synovial joints, the radiocarpal joint and the midcarpal joint (see below), that permit the large range of mobility found in the wrist. The palm of the hand is comprised of five metacarpals, the first for the thumb and the fifth on the medial side of the palm. Each metacarpal articulates with at least one carpal bone at its proximal end and the corresponding proximal phalanx at its distal end. Several intrinsic muscles that move the thumb attach to the scaphoid and trapezium and create the muscular mass on the lateral side of the palm, called the thenar eminence. On the opposing medial side, several intrinsic muscles that move the fifth (or "pinky") finger create a muscle mass called the hypothenar eminence. The **digit**s can be divided into the thumb and four fingers. Each finger is comprised of a proximal, intermediate, and distal phalanx. The thumb is unique in lacking an intermediate phalanx. The metacarpal and corresponding phalanges of any digit are called a **ray**. For example, the second metacarpal and the proximal, intermediate, and distal phalanges of the index finger are together called the "second ray." The joints between the phalanges are synovial and each ray receives at least one muscular tendon; many are independent but some have connections to the tendons of adjacent digits, making all of the digits of the hand highly mobile and capable of fine manipulation of objects. *See also* **opposable thumb; os centrale.**

hand, evolution in hominins Hominin hand fossils, especially early in hominin evolutionary history, are rare and, rarer still are hand fossils from the same individual. Much of what is known about hominin hand evolution is inferred from only a few fossils from different individuals and thus our understanding of overall hand proportions and function is limited. The hand of the last common ancestor of modern humans and *Pan* is generally most parsimoniously considered to be similar to that of an African ape (Tocheri et al. 2008). However, fossil evidence from the time of divergence is scarce. The earliest evidence of hominin hand morphology are two phalanges associated with *Orrorin tugenensis* (*c.*6 Ma), the most notable of which is an adult distal phalanx of the thumb which is quite broad at its distal end and foreshadows the morphology of much later hominins (Gommery and Senut 2006, Almécija et al. 2010). However, the researchers who wrote the preliminary descriptions of two nearly complete hands of *Ardipithecus ramidus* (4.4 Ma) suggest it represents the morphotype of the last common ancestor of panins and hominins. *Ardipithecus* is interpreted as lacking evidence of specializations for suspension, climbing, or knuckle-walking found in extant African apes and instead is said to be similar to more generalized, arboreal, quadrupedal Miocene apes (e.g., *Proconsul*). The thumb is large and robust and the metacarpals and phalanges are relatively short, such that the hand proportions are more similar to Old World monkeys than to extant African apes (Lovejoy et al. 2009a). The researchers who discovered and analyzed these fossils (Lovejoy et al. 2009a) suggest that the morphology of the chimpanzee is too derived to be a good model for the hand of the stem hominin. Numerous hand bones of australopiths, including those of *Australopithecus anamensis*, *Australopithecus afarensis*, and *Australopithecus africanus*, shed light on Pliocene evolution of the hominin hand. Several hand fossils of *Au. afarensis* from Hadar, Ethiopia (*c.*3.2 Ma), indicate these hominins had relatively short fingers compared to the length of the thumb and a more derived capitate–trapezium–second metacarpal articulation; both lines of evidence suggest the ability to use some modern human-like precision grips (Marzke 1997). However, compared to later hominins and modern humans the phalanges are relatively more curved and the thumb is gracile with a more curved trapezial articulation. Hand fossils associated with *Au. africanus* and the nearly complete hand of Stw 573 from **Sterkfontein** in southern Africa suggest a similar derived morphology to that seen in *Au. afarensis* (Tocheri et al. 2008). Several isolated

hand bones from **Swartkrans** and associated hand fossils from **Olduvai Gorge** (**OH 7**) also demonstrate the **mosaic** nature of hominin hand evolution. However, the taxonomic attribution of these fossils to *Paranthropus* or early *Homo* (or another taxon) is uncertain as both taxa are found at each site. The most discussed fossils from Swartkrans include two first metacarpals, one that is very large and robust with a flatter, more modern human-like trapezial facet (SKX 5020), and one that is gracile with a more curved and primitive trapezial facet (SK 84). Some of the phalanges show a combination of features, such as being relatively straight like those of modern humans but with relatively larger flexor sheath attachments as seen in the extant apes. Thirteen bones are attributed to the immature OH 7 hand, including carpals, metacarpals, and phalanges. Again, some features of the hand are similar to African apes (e.g., the morphology of the scaphoid and the large flexor sheath ridges on the phalanges), while other aspects are more modern human-like, including the remarkably flat and broad metacarpal facet on the trapezium and the expanded apical tuft of the thumb's (pollical) distal phalanx (Susman and Creel 1979). The hand of later *Homo* species such as *Homo antecessor*, *Homo neanderthalensis*, and early *Homo sapiens* is well represented in the fossil record and their morphology can be generally described as fully modern. However, there are remarkably few fossils attributed to *Homo erectus* and this gap in the hominin fossil record makes it unclear exactly when and how a modern human-like morphology evolved (Tocheri et al. 2008). Nearly complete hands from **Shanidar**, in Iraq, and **Kebara**, in Israel, show that the biggest differences between Neanderthal and modern human hands are the robusticity of the bones and slight variations in the orientation of the carpometacarpal joints. The Neanderthal hand morphology is interpreted as having increased mechanical advantage of certain muscles and being well adapted for stronger grip strength, especially that of the thumb. The morphology of early modern human hands from **Qafzeh** and **Skhul** is interpreted as being more similar to that of modern humans than to Neanderthals, suggesting that these hominins were capable of the full repertoire of modern manipulative abilities and tool making. The discovery of the *Homo floresiensis* challenged traditional ideas of hominin evolution and the hand morphology is no different. Although *H. floresiensis* is a late surviving hominin (0.018 Ma), its wrist bones are remarkably primitive. Five articulating wrist bones have been uncovered and all demonstrate more morphological similarities to African apes and early hominins, such as the OH 7 wrist bones, than to later *Homo* (Tocheri et al. 2007, 2008, Larson et al. 2009). Thus, this primitive morphology brings into question not only the taxonomy of *H. floresiensis* but also current ideas about what type of morphology is necessary for tool use and tool making.

hand function The modern human hand is capable of both prehensile and nonprehensile movements. Prehensile movements are often used when gripping or pinching an object and generally fall into one of two categories: **precision grip**s and **power grip**s. Nonprehensile movements include pushing, lifting, or tapping (Napier 1993). All of these movements are accomplished through complex, integrated actions of the numerous joints, muscles, and ligaments within the forearm, wrist, and hand. Muscles that move the joints of the hand can be divided into extrinsic and intrinsic muscles. Extrinsic muscles have proximal attachments outside of the wrist or hand, and most of these attach to the medial and lateral condyles of the humerus, but that insert (or distally attach) onto the digits. Intrinsic muscles have proximal attachments within the hand, most attaching to the carpals and inserting on the digits or running between two rays (e.g., the interosseous muscles). All of the articulations among the carpal bones (i.e., intercarpal joints), between the carpals and metacarpals (i.e., carpometacarpal joints), between the metacarpals and proximal phalanges (i.e, metacarpophalangeal joints), and, finally, between phalanges (i.e., interphalangeal joints) are synovial joints and are mobile, especially the radiocarpal, midcarpal, metacarpophalangeal, and interphalangeal joints. Intercarpal joints within each carpal row and some carpometacarpal joints are comparatively more stable. The functional axis of the hand runs along the middle finger (third ray), which is in line with the forearm; side-to-side movements of the digits, such as abduction and adduction, are made in reference to this axis. *See also* **abduction**; **adduction**.

hand, ray proportions Primate hands can be generally divided into three categories based on variation in ray proportions. Mesaxonic hands are characterized by a longer third ray, which is found in most haplorhines. Ectaxonic hands are characterized by a longer fourth ray and are commonly found in most strepsirrhines. Finally, paraxonic hand describes a primate hand in which the third and fourth rays are relatively equal in length. This morphology is found in colobine monkeys, as well as some strepsirrhines (e.g., dwarf lemurs) and New World monkeys (e.g., spider monkeys) (Jouffroy et al. 1991).

hand, relative length of the digits Comparisons of relative digit lengths (within individuals) in the form of ratios potentially provide information on how early developmental processes can impact behavior and health (Manning 2002). Differences in length between the index and ring finger in male and female modern humans were described over 100 years ago (Ecker 1875, Casanova 1894), but it was not until the 1990s that John Manning rediscovered the trait and began his pioneering work on digit ratios (Manning et al. 1998). Sex differences in the ratio of the index (2D) to ring finger (4D), termed 2D:4D, are fixed early in life and it has been proposed that this reflects the effect of prenatal sex hormones on the fetus. Modern human individuals exposed to high prenatal androgens (masculine hormones) have longer ring fingers relative to their index fingers resulting in a lower 2D:4D ratio (<1). In contrast, estrogens (feminine hormones) appear to increase the length of the index finger relative to the ring finger making the digits more equal in length, resulting in a ratio that more closely approximates to 1 (Manning 2002). Although other digit ratios show sex differences, 2D:4D has become the preferred ratio as it tends to be the most sexually dimorphic. The links between prenatal sex hormones, sexual differentiation, and 2D:4D appear to be via shared developmental pathways. The formation of the fetal reproductive system (that produces sex hormones) and the development of the distal parts of the limbs (digits) are influenced by the same group of *HOX* genes (Zákány et al. 1997, Kondo et al. 1997). *HOX* genes are developmental regulators implicated in the expression of DNA and the action of *HOX* genes can be altered by sex hormones (Taylor 2002). The overarching theory is that 2D:4D acts as a biological marker for the effects of prenatal sex hormones on fetal tissues and thus for their subsequent influence on the body and brain during growth and into adulthood (Manning 2002).

Consequently, 2D:4D has been employed widely in the study of sex differences in modern human developmental psychology, behavior, and disease and these kinds of analyses are being extended to other animals (see Lombardo and Thorpe 2008, McIntyre et al. 2009). In primates, differences between ray proportions and relative digit length vary between taxonomic groups and are strongly linked to patterns of locomotion and hand posture (Susman 1979, Jouffroy et al. 1993). After controlling for possible functional effects on relative digit lengths, studies of primate 2D:4D indicate that evolutionary changes in prenatal sex hormones may be linked to important differences in social behaviors across species; lower 2D:4D is linked to promiscuous social systems while higher 2D:4D is associated with pair-bonded, monogamous social systems (Nelson and Shultz, 2010). These relationships also appear to be detectable in fossil hominoids (Nelson et al. 2010). *See also* **hand ray proportions**.

hand, tool use and tool making Traditionally, past research has generally focused on the length of the thumb compared to the fingers and the potential to form one type of **precision grip**, between the end of the thumb and distal pad of the index finger, as the key to tool-use and tool-making ability in humans and hominins. However, experimental research with modern humans making or using **Oldowan** tools reveals that three types of precision grip are primarily used: (a) the pad-to-side grip between the thumb and the side of the index finger, (b) the "baseball," or three-jaw chuck, grip, and (c) a pinch grip between the thumb and the four finger pads (Markze 1997); **power grip**s are not used (see below). These three precision grips allow for effective tool use and tool making while minimizing muscle fatigue and stress on the hand joints. Thus, researchers have looked for skeletal morphologies that facilitate these grips in modern humans and in the hominin fossil record to understand how and when tool use and tool making evolved (*see* **hand, evolution in hominins**). Kinematic studies of modern humans using tools have shown that the wrist plays a particularly important role in stone-tool production and the increased mobility in the modern human wrist relative to that of chimpanzees may have played a key role in the evolution of hominin tool use (Williams et al. 2010). Electromyographic studies of modern humans making

Oldowan tools have shown that muscles on both the radial (thumb) side and ulnar (fifth digit) side are used most consistently (Markze et al. 1998). Apes trained to use crude stone tools demonstrate that modern human hand morphology is not necessary for this behavior, but perhaps a key adaptation in hominins was the ability to use forceful manipulation of stone with one hand (instead of two). Modern human hand function during tool use and tool making differs from that of nonhuman primates in two distinct ways: (a) modern humans are able to apply more force to pinch, rather than hold, an object securely with precision grips; (b) because of the increased mobility of the digits, modern humans are able to accommodate the thumb and fingers to the shape of objects. In contrast to modern humans, the relatively short thumb and long fingers of chimpanzees are not able to create a stable, firm pinch of an object or to control it by the pads of cupped thumb and fingers, but recent research suggests they can maneuver objects within their hand.

handaxe Large (typically more than approximately 10 cm in length) bifacially flaked, oval or teardrop in plan view and lenticular in cross section, stone tools that are characteristic of the Acheulean industry. Handaxes can be made on cobbles, nodules, or large **flakes** and they were probably used for a wide variety of tasks, including butchery and perhaps woodworking. Handaxes occur in Africa and parts of Eurasia from *c.*1.6 Ma to perhaps as recently as 0.16 Ma.

hanging remnant *See* **Swartkrans**.

Hanoman 1 Site Bukuran. Locality Near the village of Bukuran in the **Sangiran Dome**. Surface/*in situ* Surface. Date of discovery 1989. Finder Local villager. Unit N/A. Horizon N/A. Bed/member Within the upper levels of the **Sangiran Formation**. Formation Sangiran Formation. Group N/A. Nearest overlying dated horizon N/A. Nearest underlying dated horizon N/A. Geological age *c.*1.5 Ma. Developmental age Adult. Presumed sex Unknown. Brief anatomical description Sixteen cranial vault fragments including portions of the frontal, both parietals, and occipital bones. Announcement Widianto et al. 1994. Initial description Widianto et al. 1994. Photographs/line drawings and metrical data Widianto et al. 1994. Detailed anatomical description Widianto et al.

1994. Initial taxonomic allocation *Homo erectus* Widianto et al. 1994. Taxonomic revisions None. Current conventional taxonomic allocation *H. erectus*. Informal taxonomic category Pre-modern *Homo*. Significance This specimen documents morphological variation in *H. erectus* crania from Java (i.e., it has a typical frontal with a **supraorbital torus** combined with a gracile occipital region that lacks a marked **nuchal torus**. Location of original **Palrad Laboratory**, Bandung, Indonesia.

haplogroup (Gk *haplo*=single and It. *gruppo*=group) A group of similar **haplotypes** that share a common ancestor. Haplogroups are commonly used in studies of **mitochondrial DNA** and the **Y chromosome**. For example, A, B, C, D, and X are the mitochondrial DNA haplogroups most commonly found in Native Americans.

haploid (Gk *haplo*=single and *oid*=like or resembling) Organisms or cells having only one copy of each **chromosome**. For example, sperm and egg cells, called gametes or germline cells, are haploid. The **Y chromosome** is also referred to as haploid since there is only one present in a normal male cell.

haplotype (Gk *haplo*=single and *typos*=figure or model) An abbreviation of "haploid genotype" that refers to a specified sequence of **DNA** that is confined to one **chromosome**. The term haplotype is also used to refer to sets of **single nucleotide polymorphisms** (or SNPs) or **short tandem repeats** (or STRs) found together (or linked) on the same chromosome (or the same part of a chromosome).

Happisburgh (Location 52°49′36″N, 1°31′58″E, England; etym. name of a nearby village) History and general description Changes in the level of the North Sea have recently re-exposed sections of the Hill House Formation (or HHF) just south of Mundesley on the Norfolk coast of eastern England. Temporal span and how dated? The artifacts and the fauna from Happisburgh Site 3 come from channel deposits that were laid down in the lower reaches of the proto-Thames that at the time flowed into what is now the North Sea at this location. The paleomagnetic signal is weak, but the sediments show reversed polarity, which constrains them to either the **Matuyama** epoch or to a period of reversed magnetism in the **Brunhes** epoch.

Biostratigraphy (e.g., *Mammathus meridionalis* and *Microtus arvalis*) and paleobotanical [e.g., *Tsuga* (hemlock) and *Ostrya*-type (hop-hornbeam)] dating evidence further constrains these age estimates and suggests the site was occupied between 0.99 and 0.78 Ma. Hominins found at site None. Archeological evidence found at site **Cores**, **flakes**, and flake tools ($n=76$) made of flint were found *in situ* at several levels in the site; there were no handaxes. Key references: Geology, dating, and paleoenvironment Parfitt et al. 2010; Hominins N/A; Archeology Parfitt et al. 2010.

HAR Abbreviation of Hargufia, one of three three-letter prefixes (e.g., **BOD** and **DAW** are the others) used for the paleontological evidence collected by the **Middle Awash Research Project**. Each fossil has a three-letter prefix that refers to the subdivision/drainage area where it was found followed by VP (for Vertebrate Paleontology), then the number of the collecting site within the area, and finally a number that places the fossil in the sequence of fossils recovered from that collecting site. Thus HAR-VP-1/1 is the first fossil to be found in the first collecting site to be recognized in the Hargufia drainage of the **Middle Awash study area**.

Hardy–Weinberg equilibrium (etym. Named after the English mathematician Godfrey Hardy and the German physician Wilhelm Weinberg) A principle in population genetics. Hardy and Weinberg pointed out that under the assumption of a randomly mating (the technical term is panmictic) population **allele** frequencies do not change from generation to generation (i.e., they are in equilibrium) unless the population is acted upon by one or more of the four forces of evolution (**mutation**, **genetic drift**, **natural selection**, and **gene flow**). In the simplest case of two alleles at a **locus**, the allele frequencies are represented by the equation $p+q=1$ (where p=the allele frequency of one allele and q=the allele frequency of the other; together their frequencies must equal 100% or 1.0) and the genotype frequencies are represented by the equation $p^2+2pq+q^2=1$. These equations can be used to calculate the expected frequency of carriers ($2pq$) of a recessive trait or to examine whether a particular locus is at Hardy–Weinberg equilibrium. But the above only applies so far as genotype frequencies are concerned if mating is random.

If mating is not random (e.g., if there is assortative mating or endogamy/exogamy the genotype frequencies will change and not in the way predicted by the Hardy–Weinberg equilibrium. If the mating is assortative the effect will influence the frequencies of the genotypes involved in determining the phenotypic character causing assortative mating and in the case of endogamy/exogamy the frequencies of all of the genotypes will be affected.

Hargufia One of the named subdivisions/ drainage areas within the **Bodo-Maka** fossiliferous subregion of the **Middle Awash study area**, Ethiopia. History and general description The drainage area was first explored for its paleoanthropological potential in 1981 by a multidisciplinary team organized by the **Ethiopian Ministry of Culture** and led by Desmond **Clark**, with more extensive excavations taking place in 1990. Temporal span A **tuff** exposed within Unit U of the Bodo-Dawaitoli-Hargufia sedimentary sequence has a suggested age of 0.64±0.03 Ma and biostratigraphic evidence points to a **Middle Pleistocene** age for the sediments exposed at Hargufia. How dated? **Argon-argon dating** and **biostratigraphy** based on biostratigraphic correlation with **Olduvai Gorge** Bed IV and **Olorgesailie**. Hominins found at site None. Archeological evidence found at site. **Mode 1 Acheulean** artifacts have been recovered from localities HAR-A2, HAR-A5, and HAR-A6 and **Mode 2** Acheulean artifacts have been recovered from localities HAR-A1, HAR-A3, and HAR-A4. Key references: Geology, dating, and paleoenvironment Clark et al. 1994, de Heinzelin et al. 2000; Hominins N/A; Archeology de Heinzelin et al. 2000.

Harris lines Dense lines seen in plain radiographs of **juveniles**, subadults, and young adults that are orientated at right angles to the long axis of long bones. Insults such as serious infections apparently do not interfere with **mineralization** but they do cause **cartilage** growth to slow. Following such insults, **osteoblasts** produce horizontally orientated trabeculae that show up as lines of increased density.

HARs An acronym for **human accelerated regions** (*which see*).

Hathnora (Location 22°49′14″N, 77°51′14″E, India; etym. local name for a village close to the find site) There are several paleontological and arche- ological localities at Hathnora in the **Narmada Basin**. Four are in the Surajkund Formation (Surajkund 1, 2, and 3, and **Hathnora 1**) and one is in the Baneta Formation (Hathnora 2). *See also* **GSI-Narmada**; **Hathnora 1**.

Hathnora 1 (Location 22°49′14″N, 77°51′14″E, India; etym. local name for a village close to the find site) History and general description Hathnora is one of several localities near the village of Hathnora on the northern bank of the Narmada River in Madhya Pradesh, India. The presence of Pleistocene sediments has been recognized since the late 19thC (de Terra and de Chardin 1936). Temporal span and how dated? Late **Middle Pleistocene** based on **biostratigraphy**, but **electron spin resonance spectroscopy dating** of "three mammalian teeth stratigraphically associated with the hominin calvaria" (Patnaik et al. 2009, p. 114) suggest a minimum age of *c.*50 ka and a maximum age of *c.*160 or more than 200 ka. Hominins found at site Hathnora 1, also known as the GSI-Narmada calvaria, plus Patnaik et al. (2009) suggest there are also two hominin clavicles and a partial hominin rib. Archeological evidence found at site **Quartzite hand-axes**, **flakes**, etc. and **flint** and **chert scraper**s were reported in the area near to the find site. Key references: Geology, dating, and paleoenvironment Sonakia 1984, Agarwal et al. 1988, Tiwari and Bhai 1997, Patnaik et al. 2009; Hominins Sonakia 1984, 1985, Kennedy et al. 1991, Sankhyan 1997; Archeology Sonakia 1985, Patnaik et al. 2009. *See also* **GSI-Narmada**.

Haua Fteah (Location 32°55″12N, 22°07′48″E, Cyrenaica, Libya; etym. Berber for great cave) History and general description Haua Fteah (or Hawa Ftaih) is located 60 m above sea level on the northern slopes of Gebel el Akhdar (etym. Ar. green mountain) on the Mediterranean coast of Libya, east of the Gulf of Sirte. The site was discovered by Charles McBurney in 1948 during a geological survey and he returned to excavate the cave between 1951 and 1955. Despite excavating to 14 m depth in the deposits McBurney never reached an archeologically sterile layer. His 1967 monograph on the site was a landmark study in African archeology and incorporated some of the earliest uses of **Oxygen Isotope Stage** dating. The cave itself preserves one of the longest archeological chronologies in North Africa with deposits spanning the Historic period through the **Middle Paleolithic**. Temporal span and how dated? Dating of the site had been somewhat controversial since the original **radiocarbon dates** published by McBurney (1967) have a high margin of error. More recent dating efforts using **biostratigraphy** and geostratigraphy have established the antiquity of the oldest deposits at *c*.200 ka while also indicating that the recent Holocene deposits are the result of many brief depositional episodes separated by hiatuses (Hunt et al. 2010). Hominins found at site Two hominin mandible fragments were recovered by McBurney from early Middle Paleolithic levels (McBurney et al. 1953a). The fragments represent two individuals, one young adult and one juvenile, from the same occupation period (McBurney 1967). Phillip **Tobias**, in his analysis in McBurney's 1967 text, assigned the mandibles to *Homo sapiens rhodesiensis*. The modern taxonomic equivalent would be *Homo heidelbergensis* or *Homo rhodesiensis*. Archeological evidence found at site McBurney divided the deposits into Phases G–A from youngest to oldest. The phases correspond respectively to Historic-Protohistoric, Neolithic of Libyco-Capsian Tradition, Libyco-Capsian, Eastern Oranian (Iberomaurusian), Dabban, **Mousterian**, and Pre-**Aurignacian** (Lubell 2001, McBurney 1967). Faunal remains, both terrestrial and marine, are well represented at the site. Barbary sheep are present throughout the sequence of deposits, as are auroch and gazelle. Domesticated caprines first appear in the Libyco-Capsian levels (Lubell 2001). Sea shells in the pre-Aurignacian levels are among the earliest evidence for human exploitation of marine resources; land snail shells in the same deposits evoke the Capsian culture *escargotières* found to the west in the Maghreb. Key references: Geology, dating, and paleoenvironment Hunt et al. 2010, McBurney 1967; Hominins Tobias in McBurney 1967, McBurney et al. 1953a; Archeology Moyer 2003, Lubell 2001, McBurney 1967.

Haua Fteah I Site **Haua Fteah**. Locality N/A. Surface/*in situ In situ*. Date of discovery 1952. Finder Unnamed assistant to R. Hey. Unit Phase BI, 1st substage of the Levalloiso-Mousterian complex, 2.5 m below the earliest Upper Paleolithic deposits. Horizon Layer XXXIII, Spit 1952/32. Bed/member N/A. Group N/A. Nearest overlying dated horizon Layer XXVIII, dated to 41.45±1.3 ka. Nearest underlying dated horizon N/A. Geologic age The layer containing the fossil (Layer XXXIII) is dated to 45.05±3.2ka by **radiocarbon dating**; however, more recent efforts at dating the site based on Mousterian artifact associations suggest that an

age between 130 and 50 ka is more likely (Hublin 2000, Klein 2009). Developmental age Young adult, 18–25 years old. Presumed sex Female. Brief anatomical description Virtually the entire left ramus, except for the lateral end of the condyle, and a part of the mandibular corpus as far forwards as the P$_4$; the second and third permanent molars are in place. Announcement McBurney et al. 1953a. Initial description McBurney et al. 1953a. Photographs/line drawings and metrical data McBurney 1967 (Appendix 1B by P.V. Tobias). Detailed anatomical description McBurney et al. 1953a, 1953b, Tobias 1967b. Initial taxonomic allocation *Homo sapiens neanderthalensis*, specifically described as having affinities to the **Tabun** *Homo sapiens* hominins (McBurney et al. 1953b). Taxonomic revisions *Homo sapiens rhodesiensis* (Tobias 1967b). Tobias notes similarities to sub-Saharan African specimens and suggests that Haua Fteah I represents an "advanced Afro-Asian Neandertaloid population." Current conventional taxonomic allocation *Homo heidelbergensis* or **Homo rhodesiensis** Informal taxonomic category Anatomically modern human. Significance Used to support McBurney's hypothesis that advanced **Neanderthaloid** populations, now generally assigned to archaic *Homo sapiens*, arose in the tropics of sub-Saharan Africa and moved northward into the Maghreb and southwest Asia. Location of originals University Museum of Anthropology and Archaeology, Downing Street, Cambridge, UK (Duckworth Laboratory reg. no. Af.8.7.1).

Haua Fteah II Site **Haua Fteah**. Locality N/A. Surface/*in situ In situ*. Date of discovery 1955. Finder C.M.B. McBurney. Unit Phase BI, 1st substage of the Levalloiso-Mousterian complex, 2.5 m below the earliest Upper Paleolithic deposits. Horizon Layer XXXIII, Spit 1955/110. Bed/member N/A. Group N/A. Nearest overlying dated horizon Layer XXVIII, dated to 41.45±1.3 ka. Nearest underlying dated horizon N/A. Geologic age The layer containing the fossil (Layer XXXIII) is dated to 45.05±3.2 ka by **radiocarbon dating**; however, more recent efforts at dating the site based on Mousterian artifact associations suggest that an age between 130 and 50 ka is more likely (Klein 2009, Hublin 2000). Developmental age Juvenile, 12–14 years old. Presumed sex Female. Brief anatomical description The left mandibular ramus is present though missing the lateral end of the condyle, as in **Haua Fteah I**. A small part of the corpus is present, as far forward as the distal edge of the socket for M_2. Radiographs taken at the British Museum in 1955 indicated the presence of a developing M_3 within its follicle. This was subsequently extracted for further study. Announcement Described at the Neandertal Centenary Symposium, Dusseldorf, Germany, in 1956. Initial description McBurney 1958. Photographs /line drawings and metrical data McBurney 1967 (see Appendix 1B by P.V. Tobias), McBurney 1958. Detailed anatomical description Tobias 1967b. Initial taxonomic allocation *Homo sapiens rhodesiensis*. Taxonomic revisions None. Current conventional taxonomic allocation *Homo heidelbergensis* or *Homo rhodesiensis*. Informal taxonomic category Anatomically modern human. Significance Used to support McBurney's hypothesis that advanced **Neanderthaloid** populations, now generally assigned to archaic *Homo sapiens*, arose in the tropics of sub-Saharan Africa and moved northward into the Maghreb and southwest Asia. Location of originals University Museum of Anthropology and Archaeology, Downing Street, Cambridge, UK (Duckworth Laboratory reg. no. Af.8.7.2).

Haupt knochen schicht (Ge. for "main bone layer") *See* **Trinil**; **Trinil H.K. fauna**.

Hay, Richard L. (1929–2006) Born in Indiana, USA, Richard Hay received his BA in 1946 and his MA in 1948, both from Northwestern University. He went on to receive a PhD in 1952 from Princeton University. Hay spent 2 years in the US Army before becoming an Assistant Professor at Louisiana State University in 1955. Two years later he went to the University of California at Berkeley where he became a Professor of Geology. At Berkeley Hay joined a group of the world's most distinguished petrographers, such as Howell Williams, Francis Turner, and Charles Gilbert. He studied mineral alteration in sedimentary rocks using petrographic analysis and emphasized integrating both analytical data and field relations with petrography. The skills he developed and used at Berkeley were the basis of many of his later achievements. In 1962 Mary and Louis **Leakey** invited Hay to work on the stratigraphy of Bed I at **Olduvai Gorge**. In the event he spent a total of 12 years there working out the stratigraphy, magnetostratigraphy, chronology, sedimentology and the Pleistocene paleogeography of the area. Hay's 1976 book, *Geology of the Olduvai Gorge*, was eminently accessible to non-geologists and it has had a lasting

impact among archeologists and anthropologists. When Hay and Mary Leakey moved the focus of their activities to Laetoli, he combined field studies and petrography with chemical, isotopic, and X-ray diffraction analysis to generate hypotheses about the nature of syndepositional volcanism that resulted in the unique record of footprints, including those of early hominins. In 1983, Hay became the Ralph E. Grim Professor in the Department of Geology at the University of Illinois at Urbana-Champaign. There he discovered the low-temperature megareplacement of uppermost Precambrian and Cambrian-Ordovician rocks of the US mid-continent by potassium-feldspar, a huge geochemical alteration still not quite understood in terms of its origin in the geological history of the North American continent. Hay retired in 1997 and moved to Tucson, Arizona, with his wife, Lynn, where he became an Adjunct Professor at the University of Arizona. In 2000 Hay was awarded the Rip Rapp Archaeological Geology Award from the Geological Society of America. A year later, in 2001, he shared the Leakey Prize with Garniss Curtis. *See also* **Laetoli hominin footprints**; **Olduvai Gorge**.

Hayonim (Location 32°56′N, 35°13′E, Israel; etym. He. *hayonim*=dove or pigeon) History and general description Cave site located in western Galilee on the left bank of a tributary of the Nahal Yassof river. The excavations were carried out in three phases, 1965–1971, 1974–1979 and 1992–1999. The uppermost **Natufian** layers possibly represent a permanent multiseasonal occupation. Temporal span and how dated? Thermo**luminescence dating** on burned flints gives an age of 230–140 ka for the lower **Mousterian** layers and **electron spin resonance spectroscopy dating** on teeth gives a range of 177–182 ka. Hominins found at site A large number of complete or nearly complete skeletons have been recovered from the Natufian layers of the site, many of which are burials. The hominin remains from the **Acheulean** and Mousterian layers are far more fragmentary and are unsuitable for taxonomic assignment; Arensburg et al. (1990) provide a list of the hominins recovered. Archeological evidence found at site An extensive lithic sequence is known from the site extending from Natufian in the upper layers to the Mousterian and Acheulean in the lower layers. The Mousterian assemblage includes early **Middle Paleolithic** blade tools. Key references: Geology, dating, and paleoenvironment Weiner et al. 2002, Rink et al. 2004, Mercier et al. 2007a; Hominins

Bar-Yosef and Goren 1973, Arensburg et al. 1990; Archeology Bar-Yosef 1991, Meignen 1999.

hazard rate The limiting value for the probability of an event (typically a death) within a time interval, divided by the width of the time interval. As the time interval becomes progressively shorter this probability divided by the interval reaches the instantaneous hazard rate. The average age at death for a constant hazard (also referred to as the residual component in a **Siler hazard model**) is the inverse of the hazard rate, so that a constant hazard of, for example, 0.02 gives an average age at death of $1/0.02 = 50$ years. The first graph in a series shows the probability of death within the first year, then within the first 2 years, and so on up to the probability of death within the first 80 years. The age-specific **probability of death** can be found by assuming a constant hazard of death within the interval (see Wood et al. 1992).

Hazorea (Location 32°39′N, 35°06′E, Israel; etym. He. *hazore'a* = the sower, named after the kibbutz on which it is located) History and general description Open-air site in the middle of a field on the Hazorea Kibbutz. The **Paleolithic** faunal and lithic assemblages came to light when a field was ploughed and because none of the recovered artifacts or bones were found *in situ*, it is difficult to judge the relative age of the specimens. Further excavations revealed numerous **Neolithic** and historical artifacts. Temporal span and how dated? The phenotypically Paleolithic assemblages are undated. Hominins found at site The more complete specimens, Hazorea 1 and 2, are nearly complete occipital bones and share morphological affinities with both *Homo erectus* and the **Steinheim** cranium. **Fluorine dating** revealed that the hominin remains belong to temporally separate groups, with Hazorea 2, 4, and 5 being older than Hazorea 1 and 3. Archeological evidence found at site Most of the flint tools found appear to be "typical of the Lower Paleolithic" (Anati and Haas 1967). Key references: Geology, dating, and paleoenvironment Anati et al. 1973; Hominins Anati and Haas 1967; Archeology Anati et al. 1973.

HCRP RC Acronym for the Hominid Corridor Research Project Ruasho and Chomolo and the prefix for fossils recovered by the **Hominid Corridor Research Project** from localities between the Ruasho and Chomolo Rivers (e.g., **HCRP RC 911** from Malema). *See* **Malema**.

HCRP RC 911 Site Malema. Locality RC 11. Surface/*in situ In situ*. Date of discovery 1996. Finder Stephen Mwanyongo. Unit Middle Sand at the base of the Bone Bed. Horizon Bone Bed. Bed/member Unit 3A. Formation Chiwondo Beds. Group N/A. Nearest overlying dated horizon N/A. Nearest underlying dated horizon N/A. Geological age *c*.2.3–2.5 Ma. Developmental age Adult. Presumed sex Unknown. Brief anatomical description Alveolar process of the left side of the maxilla, including the heavily worn and badly damaged crown and the roots of M^1 and the even more damaged crown and roots of the M^2. No measurements are given, but the tooth crowns are estimated to be equivalent in size to those of KNM-ER 733 and 1804 (both belonging to *Paranthropus boisei*); they are substantially larger than the equivalent crowns of *Paranthropus robustus*. Announcement Kullmer et al. 1999. Initial description Kullmer et al. 1999. Photographs/line drawings and metrical data Kullmer et al. 1999. Detailed anatomical description N/A. Initial taxonomic allocation *Paranthropus boisei sensu lato*. Taxonomic revisions None. Current conventional taxonomic allocation *Paranthropus boisei*. Informal taxonomic category **Megadont** archaic hominin. Significance The value of this fragmentary specimen is that enough morphology is preserved to be sure that (a) it does not belong to *Homo* and (b) it belongs to an archaic megadont hominin. It is assumed to belong to *P. boisei* but the size of its tooth crowns is also consistent with it belonging to the *Australopithecus garhi*. Either way, its main significance is biogeographical, for it extends the range of either of these taxa more than 1000 km/620 miles to the south. Location of original **Cultural and Museum Centre, Karonga**.

HCRP UR Acronym for the Hominid Corridor Research Project Uraha and the prefix for fossils recovered by the **Hominid Corridor Research Project** from localities at **Uraha**, Malawi. *See* **Uraha**.

HCRP UR 501 Site Uraha. Locality U 18. Surface/*in situ* Surface, then *in situ*. Date of discovery 1991 and 1992. Finder Tyson Mskika. Unit Paleosol in the lower part of Unit 3. Horizon See above. Bed/member Unit 3A. Formation Chiwondo Beds. Group N/A. Nearest overlying dated horizon N/A. Nearest underlying dated horizon N/A. Geological age *c*.2.3–2.5 Ma.

309

Developmental age Adult. Presumed sex Unknown. Brief anatomical description Adult mandibular corpus preserved in two nearly equally complete parts, the one on the left side extending to the distal border of the M_2 whereas the fragment of the right side of the corpus extends to the mesial plate-like root of the M_3. Parts of the crowns of the P_3s are preserved and the P_4 crowns are well preserved, as are the M_1 crowns. The right M_2 crown is well preserved, the left M_2 crown less so. Announcement Schrenk et al. 1993. Initial description Schrenk et al. 1993. Photographs/line drawings and metrical data Bromage et al. 1995. Detailed anatomical description Bromage et al. 1995. Initial taxonomic allocation *Homo rudolfensis*. Taxonomic revisions None. Current conventional taxonomic allocation *H. rudolfensis*. Informal taxonomic category Transitional hominin. Significance The value of this specimen is that enough morphology is preserved to be sure that (a) it does not belong to an archaic **megadont** hominin and (b) it resembles **KNM-ER 1802**, a mandible assigned to *H. rudolfensis*. The main significance of HCRP UR 501 is biogeographical, for if it does belong to *H. rudolfensis* it extends the range of that taxon more than 1000 km/620 miles to the south. Location of original **Cultural and Museum Centre, Karonga, Malawi**.

head-balancing index An index introduced by Raymond **Dart** (1925, p. 197) to provide a quantitative way of demonstrating the relatively anterior position of the **foramen magnum** in the **Taungs** child. It is the distance between basion and inion, divided by the distance between basion and prosthion, multiplied by 100. Dart cited index values of approximately 50 in *Pan*, more than 90 in modern humans, and he estimated its value in Taung(s) to be roughly 60.

hearth (OE *hoerth*) Following standard English usage, the hearth is the floor of a fireplace, but when the term is used by paleolithic archeologists it refers to visible evidence for the controlled use of fire. Fire use may be recognized through sediment colorization caused by high heat and oxidation or by the presence of large amounts of charcoal or wood ash (e.g., Meignen et al. 2007). Often, but not always, evidence of attempts to control a fire took the form of shallow pits or stone barriers (e.g., a ring of stones) recoverable through careful excavation. The absence of such features at **Chesowanja** makes it difficult to be sure whether that site records a fire deliberately set by hominins or a natural fire (e.g., lightning setting fire to a tree resulting in a burned out tree stump).

heel strike *See* **walking cycle**.

Heidelberg mandible *See* **Mauer mandible**.

Heinrich events Heinrich events are **ice-rafting** events named after Hartmut Heinrich. They were identified by Heinrich on the evidence of "dropstones" (i.e., pebbles considerably larger than the ambient sedimentation) found in sediments on the floor of the North Atlantic Ocean (Heinrich 1988). Sea ice often forms in shallow waters and incorporates sediments from the continental shelf, and the subsequent movement of sea ice can transport these sediments (with their included dropstones) that fall to the sea floor when the sea ice melts. The morphology of ice-rafted sediments can be distinguished from other forms of transport (e.g., **fluvial** or **aeolian**), and the mineralogy, elemental, and isotopic composition of these dropstones can be used to infer their source. Six Heinrich events have been identified during the last glacial (i.e., between 60 and 17 ka), and they are numbered H1–H6, with H6 being the oldest. Some consider the **Younger Dryas** to be the most recent Heinrich event, in which case it would be H0. Climate shifts coinciding with Heinrich events have been reported in paleoclimate records well beyond the Atlantic. For example, cold **sea surface temperatures** in the North Atlantic associated with increased ice rafting may have altered patterns of precipitation in Africa (Carto et al. 2009).

Heinzelin, Jean de (1920–98) Born in Belgium to a family of noble descent, de Heinzelin attended the Université Libre Bruxelles (ULB) where he obtained his undergraduate degree in 1941, followed by a master's degree in geology in 1946 and a doctoral degree in geology and the mineralogical sciences in 1953. He was also awarded a Diplôme d'Etat from the Sorbonne in Paris, France. In 1944 de Heinzelin became a scientist at the Belgian Natural Sciences Museum, where he eventually became deputy curator and assistant director of the laboratory from 1947 until 1951. He also taught geology, successively at ULB and the Vrije Universiteit Brussel, finally becoming a full Professor at the Royal

University in Ghent in 1964. de Heinzelin's involvement in archeology and prehistory began in the 1950s. While exploring in what was then the Belgian Congo in 1957 de Heinzelin discovered the Ishango bone, a baboon fibula bearing a series of tally marks engraved along it and having a piece of quartz fixed to one end. This instrument was dated to the Upper Paleolithic and is now thought to be more than 20,000 years old. From 1961 to 1966 he participated in a UNESCO project to rescue important archeological sites in the Nile Valley, and he also undertook field research in Sahabi (Libya) between 1977 and 1982. However, it was de Heinzelin's involvement with the **International Omo Research Expedition** with Clark **Howell** between 1967 and 1974, and subsequently with the **Middle Awash Research Project** with Desmond **Clark** and Tim **White**, that resulted in his seminal contributions to paleoanthropology. His detailed assessments of the stratigraphy of the Omo-Shungura Formation enabled the paleontologists working with the International Omo Research Expedition to locate fossils with precision and thus the Omo-Shungura Formation became an important reference for biostratigraphical studies at other sites in East and southern Africa. His contributions were recognized by the award of the V. Van Straelen and Adolphe Wetrems prizes in Belgium, as well as the Paris Prix des laboratories. de Heinzelin was also a lifetime member of the scientific council for the Belgian Royal Institute for the Natural Sciences.

Helicobacter pylori (Gk *helix*=spiral, plus L. *bacter*=rod and *pylorus*=gatekeeper) Over half of all modern humans are infected by this Gram-negative bacterium, which, as its name implies, is found in the stomach. It is harmless in the short term, but its spiral structure allows it to burrow into the mucosa of the upper gastrointestinal tract, where it causes peptic ulcers in both the stomach and duodenum and it is a risk factor for gastric cancer. Its relevance for human evolution is that *Helicobacter pylori* is highly **polymorphic**, so there are more than 300 distinct strains of *H. pylori*. Linz et al. (2007) investigated 769 isolates of *H. pylori* from 51 different "ethnic" sources and used **Bayesian Markov chain Monte-Carlo cluster analysis** to investigate the relationship between the distribution of strains and the geographical locations of the samples. They found that 73% of modern human variation as assessed using **microsatellite** data could be explained by a linear relationship

with the variation among *H. pylori* strains. The geographical patterns of genetic isolation by distance (or IBD) assessed using *H. pylori* and modern human genetic diversity are a good match, both showing a continuous loss of genetic diversity with increasing geographic distance from East Africa. Simulations suggest that *H. pylori* migrated from East Africa *c.*58 ka, a time that is consistent with the estimated departure of modern humans from East Africa. The evidence suggests that modern humans have harbored *H. pylori* for at least 60 ka, and that modern humans and the micro-organism have been co-evolving ever since. More recently Moodley et al. (2009) used the distribution of *H. pylori* haplotypes to investigate the history of the peopling of the Pacific. They found that "a distinct biogeographic group" called hpSahul split from the main Asian populations of *H. pylori c.*31–7 ka and moved into **Sahul** (i.e., New Guinea and Australia) and they suggest those populations remained isolated there for *c.*23–32 ka. Much later a second biogeographic group of *H. pylori*, hpMaori, split off as a subpopulation of hpEastAsia *c.*5 ka and that subpopulation subsequently spread into Melanesia and Polynesia. The timing of the origin of hpSahul is consistent with the estimated age of the Q mtDNA **haplogroup** and the timing of the origin of hpMaori is consistent with the results of a computational linguistic analysis of the dispersal and development of the Austronesian language group (Gray et al. 2009).

helicoidal wear plane (Gk from *helix*= spiral) The pitch (i.e., the transverse orientation) of the occlusal wear plane is not the same along the tooth row. Wear is normally greater on the lingual, or tongue, side of the maxillary teeth in the anterior part of the postcanine tooth row, with the posterior part of the tooth row either showing less emphasis on the lingual side or more wear on the buccal, or cheek, side. Tobias (1980b) proposed that this twist in the pitch of the occlusal wear plane was not seen in hominins until the emergence of *Homo*, but Richards and Brown (1986) have shown that the twisting nature of the occlusal wear plane has no taxonomic value.

hematite (Gk *haimatites*=blood-like stone) Hematite (α-Fe_2O_3), the principal ore of iron, is a common component of metamorphic, igneous, and sedimentary rocks. It is the primary component of (red) **ochre**, a common colorant found as an exotic material at many **Late Pleistocene** sites from southern Africa to France (see Watts 2002, White 2007). *See also* **ochre**.

hemispheric dominance Hemispheric dominance refers to an asymmetry of the structure or function of the **cerebral hemispheres** of the brain. For example, there is strong population-wide leftward functional hemispheric dominance among modern humans in the neural representation of **language**. Similarly, the modern human brain also displays anatomical hemispheric dominance as evident by **petalias** and by the leftward asymmetry of the **planum temporale**.

hemizygous (Gk *hemi*=half and *zygon*=yoke) The possession of only one copy of an **allele** or **chromosome** in a **diploid** organism. All mammals and thus all primates are diploid organisms and males are hemizygous for both the **X chromosome** and **Y chromosome**. If a gene is deleted on just one of the two autosomal chromosomes, then the individual is hemizygous for that gene. Because males are hemizygous for the X chromosome, recessive X-linked phenotypes are more common in males. This is also the case for X-linked recessive diseases and disorders such as hemophilia and color blindness.

hemoglobin (Gk *haîma*=blood plus L. *globulus*=globe and *in*=pertaining to) Abbreviation of hematinoglobulin (i.e., hematin plus globulin). It is the major **protein** in red blood cells and it carries oxygen from the lungs to the tissues. It is made up of two alpha globin and two beta globin protein subunits, which each contain a heme group. Each heme group can bind one oxygen molecule. The significance of hemoglobin for human evolution is that hemoglobins were some of the earliest and most extensively studied proteins in regional populations of modern humans. In addition, the alpha globin and beta globin protein subunits have been subject to selection for protection against malaria. *See also* **globins**.

Heoanthropus *See* **Sergi, Giuseppe (1841–1936)**.

herbivore (L. *herba*=grass, plus *vorous*, from the stem of *vorare*=to devour) An animal that consumes plants. *See also* **herbivory**.

herbivory (L. *herba*=grass, plus *vorous*, from the stem of *vorare*=to devour) The consumption of plants. Herbivory is an inclusive term that may subsume a more specific dietary focus on a type of plant (i.e., grass, or **graminivory**) or a specific part of plants in general (i.e., fruit, or **frugivory**). Herbivory can be considered as a form of predation in which an organism consumes a plant.

heritability (L. *hereditare*=to inherit, from *heres*=heir) An individual's **phenotype** is determined by that individual's **genotype** and environment. Heritability is the proportion of phenotypic variation in a **population** that is attributable to **genetic variation** among individuals. The heritability of a particular **trait** may be different among populations because of both genetic and environmental factors. For example, the heritability of a complex trait such as stature might be high in a population that is well nourished and has access to health care and lower in a population where malnourishment and disease are common (i.e., the conditions that most likely prevailed in early hominins) even if the underlying genetic variation is the same. This is because in the latter population the environmental variance accounts for a higher proportion of the **phenotypic variance**. **Broad-sense heritability** (H or H^2) is defined as the total genetic variance divided by the phenotypic variance. It is called "broad-sense" because it reflects *all* possible genetic contributions to the phenotypic variance such as **additive genetic variance**, **dominance**, **epistasis**, and parental effects. **Narrow-sense heritability** (h or h^2) is the additive genetic variance divided by the phenotypic variance where the additive genetic variance is the weighted average of the squared values of the additive effects. In other words, the additive variance is variance due to differences among **alleles** only. Thus, narrow-sense heritability only refers to the portion of the phenotypic variation that is additive (or allelic) and does not take into account variation due to epistatic, dominance, or parental effects. However, the additive effects are important because this variation is the only variation that natural selection can act on. Narrow-sense heritability estimates are typically used in agricultural genetics to estimate how the mean of a trait of interest (such as milk production or size) will change in the next generation given how much the mean of the selected parents differs from the mean of the population.

heritable *See* heritability.

herpestid The informal name for the **Herpestidae** (mongooses and relatives), one of the **feliform** families whose members are found at some hominin sites. *See also* **Carnivora**.

Herpestidae The animal family that includes mongooses and their close relatives. It is one of the **feliform** families whose members are found at some hominin sites. *See also* **Carnivora**.

Herto (Location 10°15′55″N, 40°33′38″E, Ethiopia; etym. named after a nearby Afar village) History and general description The Herto locality was discovered as part of the **Middle Awash Project**'s research efforts on the **Bouri Peninsula**. Prospecting in 1997 led to the discovery of six hominins, including three partial crania, and this intensified efforts to clarify the geology, archeology, and paleontology of the site. Two additional hominin fragments were recovered in 1998. All the hominins derive from the base of Upper Herto Member and were found on an **erosional surface** containing fossils, artifacts, and vertebrate fossils in a pebble **conglomerate**. Vertebrate fauna from the Upper Herto Member include taxa that are indicative of both grassland and aquatic habitats (e.g., *Hippopotamus*, *Thryonomys*, *Equus*, and *Connochaetes*). Temporal span and how dated? **Biostratigraphy** suggests a late **Middle Pleistocene** age, and **argon-argon dating** of **anorthoclase** grains from **pumice clast**s from the sediments containing the hominins resulted in an age estimate of 163±3 ka. **Tephrostratigraphy** of an overlying volcanic ash (tephra) deposit has provided a correlation with an unnamed tuff at **Konso**, Ethiopia, that lies 6 m beneath the Konso Silver Tuff, which has been dated by argon-argon dating to 154±7 ka. The bounding tephra deposits suggest a very brief temporal span of *c*.160 ka. Hominins found at site **BOU-VP-16/1**, a large, well-preserved adult cranium and dentition, BOU-VP 16/2, a larger but fragmentary cranium, BOU-VP-16/3, a parietal fragment, BOU-VP-16/4, a parietal fragment, BOU-VP-16/5, a child's cranium, BOU-VP-16/6, a right M^1 or M^2, BOU-VP-16/18, vault fragments, BOU-VP-16/42, a left P^4, and BOU-VP-16/43, a parietal fragment. These specimens are large and robust but their morphology is anatomically modern human. Archeological evidence found

at site. As well as butchered faunal elements, numerous tools were found in the Upper Herto Member, including **handaxes**, **cleavers**, and other **biface**s as well as **Levallois** flake tools. The bifaces were made from fine-grained **basalt**, but the points and light-duty blades are made from **obsidian**. The technology and typology of the tools most closely resemble those from **Garba III** and **Melka Kunturé** in Ethiopia. Key references: Geology, dating, and paleoenvironment Clark et al. 2003; Hominins White et al. 2003; Archeology Clark et al. 2003.

Hesperanthropus *See* **Sergi, Giuseppe (1841–1936)**.

heterochromatin *See* **chromosome**.

heterochrony (Gk *hetero*=other and *kronos*=timing) Changes in the relative timing of the appearance or the rate of development of a feature in a descendant relative to the developmental timetable of the same feature in an ancestor. Heterochrony involves changes in ancestral developmental patterns that include shifts in rate and/or timing. Even when such shifts are apparently minor, they can produce significant alterations in morphology. Analyses of heterochronic transformations focus on the ontogeny of size and size-related changes, and explore the idea that selection may act on size through growth rates or timing, or on the rate or timing of growth itself. Since a large proportion of studies focus on adult morphologies and their presumed evolutionary transformations, and detailed studies of ontogeny permitting a comparative analysis of rate and duration of growth are relatively rare, it is important to discriminate between the results of heterochrony from the processes that generate the results. *See also* **acceleration**; **paedomorphosis**; **peramorphosis**; **hypermorphosis**; **hypomorphosis**; **neoteny**.

heteroplasmy (Gk *hetero*=other and L. *plasma*=form or shape) This term refers to the mixture of several different sequences of mitochondrial DNA in the same cell or individual. For example, individuals with mitochondrial diseases (e.g., Leber's hereditary optic neuropathy) are typically heteroplasmic since their mitochondria will contain a mixture of the wild-type DNA (i.e., mitochondrial DNA without the mutation) and some mtDNA with the deleterious mutation. The higher

313

the proportion of mitochondrial DNA with the deleterious mutation, the greater the severity of the disease. In the debate about "Mitochondrial Eve" (*see* **mitochondrial DNA**), the assumption that mitochondrial DNA was inherited strictly from the mother was questioned since occasional examples of heteroplasmy due to a paternal contribution were found in other animals. If this were true in modern humans, if would affect the coalescence time for mitochondrial DNA (i.e., the age of "Mitochondrial Eve"). However, subsequent research has failed to provide any evidence of a significant paternal contribution to mitochondrial DNA.

heterosis (Gk *heteroios*=different in kind) Heterosis can result from heterozygote advantage (overdominance) or recombinant hybrid vigor. In the skeleton heterosis is typically manifested as increased size. The role of **hybridization** in human evolution is currently poorly understood, though hotly debated (e.g., modern humans and Neanderthals) (syn. hybrid vigor). *See also* **hybridization**.

heterozygosity *See* heterozygote.

heterozygote (Gk *hetero*=other and *zygotos*=yoked) An individual organism that has two different **alleles** at a **diploid locus**. For example, an individual who has the genotype AB for the ABO locus and thus the blood type AB is referred to as being heterozygous. The significance of heterozygosity is that it is one of the sources of the variation that is necessary for evolution to occur.

heterozygous *See* heterozygote.

heterozygous advantage (Gk *hetero*=other and *zygotos*=yoked) The circumstance where it is advantageous to be heterozygous (i.e., have different versions of the allele on the two chromosomes) because the **heterozygote** has a higher fitness. An example in modern humans is the sickle-cell, or S, **allele** of the beta globin **gene**. In the **homozygous** form this allele is deleterious because it causes sickle-cell anemia, but in regions where malaria is endemic the heterozygous form protects against malaria. In such an environment, homozygotes for the wild-type allele are at a disadvantage because they are more susceptible to malaria. Thus, both alleles are maintained in the population (syn. heterozygote advantage). *See also* **balancing selection**; **hemoglobin**.

Hexian (和县) (Location 31°53′3″N, 118°12′18″E, eastern China; etym. named after the local county) History and general description The site of Longtandong (Dragon Pool Cave) is situated on the northern slope of Wangjiashan ("hill of Wang's family") just south of the town of Taodian, Hexian County, Anhui Province. Temporal span and how dated? Thermoluminescence and **uranium-series dating** methods of the fossil fauna yielded age ranges of 211–179 ka and 190–150 ka, respectively. Subsequent **electron spin resonance spectroscopy dating** and uranium-series analyses of rhinoceros and bovid teeth give a combined age estimate of 412±25 ka. This places Hexian equivalent to the base of layer 3 of **Zhoukoudian Locality 1**. Hominins found at site PA 830 (also known as **Hexian 1**), a well-preserved calvaria, PA 831, a fragmentary left side of the mandibular corpus with M_2 and M_3, and PA 832, 833, and 834(1–2), four isolated teeth, were discovered in 1980. The Hexian (PA 830) calvaria is presumed to be a young adult male individual. In 1981, additional fossil material consisting of a frontal with part of the **supraorbital region** (PA 840), a parietal fragment (PA 841), and five isolated teeth (PA 835–839) were recovered. Archeological evidence found at site None. Repository **Institute of Vertebrate Paleontology and Paleoanthropology**, Beijing, China. Key references: Geology, dating, and paleoenvironment Chen et al. 1987, Grün et al. 1998; Hominins Wu and Dong 1982, Wu and Olsen 1985, Pope 1992, Wu and Poirier 1995, Brown 2001, Durband et al. 2005, Wu et al. 2006.

Hexian 1 Site Longtandong. Locality N/A. Surface/*in situ* Surface. Date of discovery 1980. Finders Unknown. Unit N/A. Horizon Near the top of Bed II. Bed/member Bed II. Formation N/A. Group N/A. Nearest overlying dated horizon N/A. Nearest underlying dated horizon N/A. Geological age: The most recent electron spin resonance spectroscopy and uranium-series dates suggest an age of *c*.412±25 ka. Developmental age Adult. Presumed sex Unknown. Brief anatomical description The cranial vault is well preserved, but the only parts of the basicranium to be preserved are the temporal bones and the posterior border of the foramen magnum. Endocranial volume 1025 cm³. Announcement Huang et al. 1981. Initial description Huang et al. 1982, Wu and Dong 1982. Photographs/line drawings and metrical data Huang et al 1982, Wu et al. 2006. Detailed anatomical description N/A. Initial taxonomic allocation *Homo erectus*. Taxonomic revisions N/A. Current conventional taxonomic allocation *Homo erectus*. Informal taxonomic category Pre-modern *Homo*. Significance The Hexian calvaria is equivalent in age to **Zhoukoudian X**, **XI**, and **XII** and its endocranial morphology is closest to the Zhoukoudian sample (Wu et al. 2006). Location of original **Institute for Vertebrate Paleontology and Paleoanthropology**, Beijing, China.

High Cave, The *See* **Mugharet el 'Aliya**.

high-ranking prey Within an **optimal foraging theory** framework high-ranking prey are those taxa or food resources that provide higher-than-average energetic returns. The **prey-choice model** predicts that high-ranking prey should always be pursued when encountered. Accordingly their abundances in archeological contexts should approximate their abundances on past landscapes. In archeological contexts prey rankings are often determined according to body size since larger taxa tend to provide greater energetic returns than small taxa. For example, large ungulates are typically regarded as high-ranking prey. However, a growing body of ethnographic evidence suggests that other factors (e.g., prey mobility) may play an important role in explaining human foraging decisions and prey rankings. *See also* **low-ranking prey**.

high-survival elements A subset of skeletal elements known to be resistant to destructive taphonomic processes. These include elements with substantial amounts of high-density cortical bone and relatively little cancellous bone (e.g., for most ungulates, these include the long bones, cranium, and mandible). Because high-survival elements provide the most accurate representation of those bones originally deposited by hominins, their study can be particularly useful when examining **carcass transport strategies** in bone assemblages that have been subjected to destructive processes. *See also* **skeletal element survivorship**.

high-utility bones *See* **economic utility index**.

higher taxon (Gk *taxis* = to arrange or "put in order") Any **taxon** above (i.e., more inclusive than) the level of the **genus**. Thus, the subtribe Australopithecina and the tribe **Hominini** are examples of higher taxa, whereas *Paranthropus boisei*, a **species** is not.

hindbrain (OE *hinder* = behind and OE *braegen* = brain, most likely from the Gk *bregma* = the front of the head) The hindbrain (or rhombencephalon) is the posteriormost, or **caudal**, of the three swellings, or vesicles, that give rise to the brain. The hindbrain is comprised of the medulla oblongata, the pons, and the cerebellum. The **cerebellum** is a dorsal outgrowth from the same embryonic brain subdivision that contains the pons. *See also* **brain**.

hippocampus (Gk *hippos* = horse and *kampos* = sea monster) A seahorse-shaped structure located in the medial part of the **temporal lobe** of the brain, adjacent to the **amygdala**. An important component of the **limbic system** the hippocampus is involved in functions such as spatial orientation, navigation, emotions, and the storage of **declarative memory** (i.e., the parts of the memory concerned with persons, places, and objects).

hippopotamid The informal name for the artiodactyl family comprising the hippopotami. *See* **Artiodactyla**.

Hippopotamidae The family that includes hippopotami and their close relatives. It is one of the **artiodactyl** families whose members are found at some hominin sites. *See* **Artiodactyla**.

histology The study of microscopic anatomy. Histology often involves generating a section of a tissue to visualize structures with light or transmission electron microscopy. Alternatively, a replica can be made for study with scanning electron microscopy or fractured or polished surfaces can be examined with **confocal microscopy**. For example, traditional approaches to understanding **enamel development** and **dentine development** involve preparing thin sections (approximately 100 m) (or **cross-section**s) to view them under transmitted or polarized light microscopy (e.g., Schmidt and Keil 1971).

H.K. Abbreviation for Haupt knochen schicht. *See* **Trinil**; **Trinil H.K. fauna**.

HLA Acronym for human leukocyte antigens. *See* **major histocompatibility complex.**

Hoedjiespunt (Location 33°0′S, 17°57′E, Western Cape Province, South Africa; etym. named after the nearby peninsula) History and general description Fossiliferous **Middle** and **Late Pleistocene** dunes are well known in the area and they have provided evidence of shellfish collecting during **Oxygen Isotope Stage** 5 (Volman, 1978). Lee Berger collected the first hominin tooth in 1993 from the base of a 6 m-deep sequence of fossiliferous sands exposed by a road cut in a fossil dune deposit on the southern edge of Hoedjiespunt Peninsula, 6 m above sea level in Saldanha Bay. Further collection yielded additional vertebrate fossils and artifacts. The vertebrate fossils, including the hominins, derive from a brown hyena den. Temporal span and how dated? Middle Pleistocene based on associated fauna and **Middle Stone Age** (or MSA) artifacts and on geological evidence including **foraminifera** that are older than 78 ka. There is a **uranium-series date** of 300 ka for **calcrete** capping the deposit but Butzer (2004) argues the hyena burrow is intrusive and he favors a Late Pleistocene age for the site. Hominins found at site Four permanent teeth of a juvenile and a fragmentary right tibia that may belong to the same individual. The teeth are an upper left molar (HDP1-1), a right M^3 (HDP1-2) that bears three cusps, and a left I_1 and I_2 (HPD1-3 and HDP1-4). Archeological evidence found at site MSA artifacts occur in the hominin level and in a sand and shell deposit at the top of the 6 m section. Key references: Geology, dating, and paleoenvironment Volman 1978, Berger and Parkington 1995, Churchill et al. 2000, Butzer 2004; Hominins Berger and Parkington 1995, Churchill et al. 2000, Stynder et al. 2001; Archeology Volman 1978, Berger and Parkington 2005, Stynder et al. 2001, Matthews et al. 2005.

Hofmeyr (Location 25°58′E, 31°34′S, South Africa; etym. after the nearby town of Hofmeyr, Eastern Cape Province, South Africa) History and general description The skull was recovered in 1954 from a dry channel bed of the Vlekport River, but erosional controls downstream have caused the location to be covered by silt. The endocranium was filled with a hard matrix of sand cemented with carbonate. The specimen was severely damaged in 1970 and comparison with photographs taken in 1965, reveal the loss of the anterior part of the face, the anterior teeth, and part of the ramus of the mandible. Temporal span and how dated? Optically stimulated **luminescence** and **uranium-series dating** suggest an age of 36.2±3.3 ka. Hominins found at site **Hofmeyr 1**, a skull comprising a partial cranium and a fragmentary mandible. Archeological evidence found at site N/A. Key references: Geology, dating, and paleoenvironment Grine et al. 2007; Hominins Grine et al. 2007, 2010; Archeology N/A.

Hofmeyr 1 Site Hofmeyr. Locality N/A. Surface/*in situ In situ* in a channel of the Vlekport River. Date of discovery 1954. Finder C. Hattingh, Jr. Unit N/A. Horizon The skull may have derived from another sedimentary context, but it shows no evidence of rolling. Bed/member N/A. Formation N/A. Group N/A. Nearest overlying dated horizon N/A. Nearest underlying dated horizon N/A. Geological age **Late Pleistocene**. Developmental age Adult. Presumed sex Unknown, but its size and morphology suggest it may be male. Brief anatomical description A large cranium with a moderately strong supraorbital torus, a projecting glabella, square orbits, a slightly projecting, wide nose, and (in its pristine state) marked alveolar **prognathism**. Announcement Grine et al. 2007. Initial description Grine et al. 2007. Photographs/line drawings and metrical data Grine et al. 2007, 2010. Detailed anatomical description N/A. Initial taxonomic allocation *Homo sapiens*. Taxonomic revisions N/A. Current conventional taxonomic allocation *Homo sapiens*. Informal taxonomic category Anatomically modern human. Significance Hofmeyr 1 (its catalogue number at the East London Museum is among the earliest evidence of late Pleistocene *Homo sapiens* in southern Africa. The results of a **multivariate analysis** show that Hofmeyr 1 falls closest to the centroid for European crania associated with the **Upper Paleolithic**, and is at the edge of the range of variation of recent sub-Saharan Africans. Thus, Hofmeyr 1 provides more evidence that at *c*.36 ka modern human crania resemble each other rather than specimens of modern *Homo* from the same region. These results fit predictions of the **out-of-Africa hypothesis** for the origins of modern humans. Location of original East London Museum, South Africa.

Hohle Fels Venus A carved ivory figurine recovered in six fragments between September

8 and 15, 2008, from the southern German **Aurignacian** site of **Hohle Fels**. The figurine depicts a female modern human that is lifelike, with the exceptions that it lacks a head, has de-emphasized limbs and it has exaggerated secondary sexual characteristics. It is in the tradition of the Venus figures from the **Gravettian** period such as the **Willendorf Venus** discovered in 1908. The Hohle Fels Venus (*c.*35 ka calibrated radiocarbon years) is at least 5 ka older than other figurines and is the earliest known figurative image of a modern human (Conard 2009).

Höhlenstein-Stadel (Location 48°32′57″N, 10°10′21″E, Germany; etym. *Höhlen*=caves, *stein*=rock, and *stadel*=barn; presumably because it was used to house livestock) History and general description A cave site along the Lone river in the Ulm district that is just downstream from **Vögelherd** cave. It was excavated in the 1930s by Robert Wetzel, and several hominin specimens and archeological levels were recovered. Temporal span and how dated? Biostratigraphic correlation has suggested an age range from the Eemian or Early Würm to the Holocene. Hominins found at site Several hominin remains were found, including some Holocene *Homo sapiens*, but also a presumed **Upper Paleolithic** (**Aurignacian**) *H. sapiens* adult premolar, and a likely juvenile male *Homo neanderthalensis* right femoral diaphysis. Archeological evidence found at site The site is best known for a large anthropomorphic lion figurine carved from mammoth tusk that was found in the Acheulean levels. Key references: Geology, dating, and paleoenvironment Street et al. 2006; Archeology Hahn et al. 1985; Hominins Czarnetzki 1983, Kunter and Wahl 1992.

Holloway, Ralph (1935–) Born in Philadelphia, Pennsylvania, USA, Ralph Holloway began his undergraduate career in metallurgical engineering at the Drexel Institute of Technology in 1953, but after transferring to the University of New Mexico he received a BS in geology and engineering in 1959. He attended the University of California at Berkeley for his graduate studies and in 1964 he was awarded a PhD in anthropology for his thesis entitled "Some quantitative relations of the primate brain." Holloway joined Columbia University in 1964, where he is still on the Faculty; he became a Research Associate at the American Museum of Natural History in 2005. Holloway is a leading expert on hominin and comparative **endocranial morphology**. His important technical contribution

was to develop a method that uses latex to make an impression of the endocranial surface of relatively complete crania. The method involved stopping holes other than the foramen magnum, introducing a small volume of liquid latex, swirling it around so that it formed a thin coat against the endocranial surface, letting that cure, then repeating the process until the latex was thick enough to be used to make an endocranial cast and thin enough that it could be easily detached from the endocranial walls and removed via the foramen magnum. This technique gave Holloway a unique insight into hominin endocranial morphology and this led him to develop the hypothesis that early hominins underwent major neural reorganization (specifically a relative increase in the parietal association areas that resulted in the lunate sulcus being shifted posteriorly) *before* there was evidence of any significant increase in overall brain size. Ralph Holloway has also made major contributions to the interpretation of hominin endocranial morphology with regard to the evolution of cerebral asymmetry, the significance of encephalization, and the co-evolution of language and the brain. *See also* **encephalization**; **endocranial morphology**; **lunate sulcus**.

Holocene (Gk *holos*=whole and *kainos*=recent) The youngest (and current) epoch of the **Neogene** Period within the **Cenozoic** Era, spans the interval of time from 11.7 ka to the present. The onset of the Holocene follows the termination of the **Younger Dryas** stadial (i.e., cooling) event that punctuated the most recent deglaciation. The Holocene was previously thought to be a period of relatively monotonous interglacial conditions; however, much attention has recently been devoted to **proxy**-based **paleoclimate** reconstructions uncovering the climatic variability of the Holocene. For example, the early Holocene was much wetter, with a perhaps abrupt transition to drier conditions around 5 ka (deMenocal et al. 2005). More recently historical records have documented climatic swings during the Medieval Climate Anomaly (*c.* AD 1100–1300), marked by warm conditions in Europe and a northwards migration of agricultural zones, and the Little Ice Age (*c.* AD 1400–1900), a period of cooler conditions in Europe. These climate events identified as temperature excursions in mid and high latitudes may be regional or global in impact and in some low-latitude regions are instead felt as changes in aridity. For example, the Medieval Climate Anomaly, which is associated with a warming in Britain, coincides with an arid shift affecting the Anasazi in the southwestern USA.

holotype (Gk *holos*=whole and *typus*=image) According to the **International Code of Zoological Nomenclature** a holotype is the specimen listed as the **type specimen** in the original description of a **taxon** (e.g., **OH 7** is the holotype of *Homo habilis*). *See also* **type specimen**.

Holstein *See* **glacial cycles**.

Homa peninsula *See* **Kanam**; **Kanjera**.

home base hypothesis A model for **Oldowan** site formation developed by Glynn Isaac (1978a, 1981, 1984). It incorporated elements that were claimed to distinguish extant **hunter–gatherer** and African ape adaptations, including (a) the use of a central place from which hominins dispersed and returned on a daily basis, (b) a sexual division of labor whereby males hunted or scavenged for animal tissue and females gathered plant food resources, (c) delayed consumption, and (d) transport of food to the base where food sharing and social activities took place. Food sharing was a central element of the home base hypothesis and it was claimed that it could have provided the selective milieu for the further evolution of **cognition** and for the emergence of modern human **language**.

home range The total area exploited by an individual animal or a group of animals. Home range size varies between species and even between populations of the same species. It is influenced by a variety of factors including the **productivity** of the **habitat** and the dietary strategy of the animal: insectivorous and frugivorous primates tend to have relatively larger home ranges than primates that rely on more ubiquitous resources, such as grass or leaves. **Body mass** is also an important determinant of home range size; larger animals tend to need absolutely more food, so will have larger home ranges. Thus, the total population (or group) weight must also been considered. Estimates of hominin home ranges have been based on body mass within a model derived from foraging human populations (e.g., Antón and Swisher 2004).

homeobox genes (Gk *homoio* from *homos*=the same and "box" apparently because these genes have

a small enough sequence that it can be written in a "box" in a manuscript) A family of genes involved in **development**, specifically in the regulation of **morphogenesis**. *Homeobox* (also abbreviated to *Hox*) genes control spatial positioning for both the anterior–posterior and proximal–distal body axes. They are called *homeobox* genes because of a short 180 **base pair DNA** sequence that encodes part of a protein molecule called the homeodomain. Homeodomain molecules can bind DNA, so proteins that include a homeodomain can act as **transcription factors** and thus help regulate the **gene expression** of other genes. In modern humans, the 39 *HOX* genes are found in four clusters on chromosomes 2, 7, 12, and 17 (NB: the convention is that if the name of the gene is written with all capital letters it means the gene is a human gene, so *HOXA1* is a modern human gene, whereas the homologous gene in mice would be *HoxA1*).

homeodomain (Gk *homoio* from *homos*=the same and L. *dominium*=property) *See* **homeobox genes**.

hominid (L. *homin*=man or human being) Informal or vernacular name for a **species** (or an individual specimen belonging to a **taxon**) within the family **Hominidae**. *See* **Hominidae**.

Hominid Corridor Research Project A field and laboratory project that began in 1983 that addressed outstanding problems in the Plio-Pleistocene paleobiogeography of continental Africa (e.g., the origin and dispersion of Plio-Pleistocene faunas) by focusing on the region between the clusters of hominin sites in East Africa and those in southern Africa. Most of the project's fieldwork was conducted in Malawi, specifically in the Chiwondo Beds in northern Malawi. The project, which was founded and co-directed by Friedemann Schrenk and Timothy Bromage (e.g., Bromage et al. 1987, Schrenk et al. 1993, Bromage and Schrenk 1995), recovered a hominin mandible (**HCRP UR 501**) from Uraha in 1991 and a maxillary fragment (**HCRP RC 911**) from Malema in 1996.

Hominidae (L. *homin*=man or human being) In the past the family Hominidae was interpreted to include only the modern human **clade** and it would have excluded all the non-human great apes. But given the copious morphological, molecular, and genetic evidence for a close relationship between modern humans and chimpanzees/bonobos, some researchers now take the view that the former clade should be recognized at a lower level than that of the family. These researchers support a more inclusive interpretation of the family Hominidae so that it would include the three subfamilies, Ponginae, Gorillinae, and Homininae or just two subfamilies, Ponginae and Homininae. In this more inclusive interpretation the living representatives of the family Hominidae are, respectively, the orangutans, gorillas, **chimpanzees/bonobos**, and **modern humans**. Therefore, the fossil representatives of the family Hominidae are all those extinct taxa that are more closely related to the above living taxa than to any other living **taxon**. Some researchers still use the family name Hominidae for the modern human twig of the tree of life, but most are moving towards a usage that is consistent with modern biology (i.e., recognizing the modern human clade at the level of the tribe as the **Hominini**).

hominin   (L. *homin*=man or human being) The informal or vernacular term for a specimen that belongs to a **taxon** within the tribe **Hominini**. For example, *Paranthropus boisei* is a hominin taxon and **KNM-ER 406** is a hominin fossil.

hominin footprints   Footprints are **trace fossil**s (syn. **ichnofossils**) that provide a record of fossilized behavior on a relatively short time scale (on the scale of minutes to days). They provide unique information about gait in extinct hominin taxa in that they provide direct evidence of the shape of the soft tissues of the plantar surface of the foot and the manner in which the foot makes contact with the ground during the gait cycle. Modern human footprints are distinctive because they begin with a deep, rounded heel print, followed by an impression made by the lateral side of the foot, then a deep impression under the metatarsal heads as the center of pressure moves medially and the impression sequence ends with a relatively deep impression for the hallux in line with the other toes. Modern human footprints often lack an impression under the medial side of the foot where the longitudinal arch, if well developed, does not contact the substrate. The deepest part of the print often occurs below the medial metatarsal heads and hallux where the highest toe-off forces occur (Vereecke et al. 2003). However, footprint morphology shows substantial variation due in part to variation in foot anatomy, walking gait, speed, and the properties of the substrate such as its firmness, particle size, and moisture content (Bennett et al. 2009). Footprints are rare in the hominin fossil record. The only hominin footprints in the Pliocene are those made by three individuals walking in a wet ash deposited 3.75 Ma in **Laetoli**, Tanzania (Leakey and Hay 1979). The **Laetoli hominin footprints** show compelling evidence that bipedal gait was well established in the hominins that made them, but ambiguities remain in the evidence of an arch and regarding the nature of gait (Stern and Susman 1983, White and Suwa 1987, Bennett et al. 2009). In the early Pleistocene, three instances of hominin footprints are known from the **Koobi Fora** (1.4–1.53 Ma). While they are distinguishable from modern human footprints, they show distinctive characteristics of a modern human style of gait noted above. Fossilized footprints are more common in the late Pleistocene and Holocene. A set of footprints in the Roccamonfina volcanic ash, southern Italy (Mietto et al. 2003) dates to 385–325 ka and appear fully modern human-like. Footprints from *c*.117 ka sediments in South Africa (Roberts and Berger 1997) and from 19–23 ka layers at **Willandra Lakes** in Australia (Webb et al. 2006) match those made by modern humans today.

Hominina   If the tribe **Hominini** is interpreted to include both the **clade** that contains modern humans and the clade that contains extant chimpanzees/bonobos then some researchers (e.g., Andrews and Harrison 2005) discriminate between the two clades at the level of the subtribe. In which case the clade that contains modern humans would be called the Hominina and the clade that contains extant chimpanzees/bonobos would be called the **Panina**.

Homininae   (L. *homin*=man or human being) A subfamily formally introduced by John **Robinson** (1972). He suggested that it contained just one genus, *Homo*, that subsumed "*Australopithecus*, *Meganthropus* from Africa, *Telanthropus*, *Sinanthropus*, *Pithecanthropus*, *Atlanthropus*, and other supposed genera that are now generally treated as falling into either *H. erectus* or *H. sapiens*" (*ibid*, p. 6). However, this was at a time when the family **Hominidae** was interpreted more exclusively and included only modern humans and all

319

the extinct taxa more closely related to modern humans than to any other non-human great apes.

homininan The informal term for individuals or taxa within the subtribe Hominina. See **Hominina**.

hominine (L. *homin*=man or human being) The informal or vernacular term for a specimen that belongs to a **taxon** within the subfamily Homininae.

Hominini (L. *homin*=man or human being) The tribe that includes modern humans and all the fossil taxa more closely related to modern humans than to any other living **taxon**.

hominoid (L. *homin*=man or human being and *oides*=resembling) The vernacular term for a member of the superfamily **Hominoidea**. The living hominoids (i.e., modern humans, chimpanzees/bonobos, gorillas, orangutans, and gibbons) all belong to the same **clade**. Fossil hominoids are all the taxa that are more closely related to living hominoids than they are to any other living **taxon**.

Hominoidea (L. *homin*=man or human being and *oides*=resembling) The superfamily comprising the following living taxa – modern humans, chimpanzees/bonobos, gorillas, orangutans, and gibbons – and all the fossil taxa that are more closely related to these taxa than to any other living **taxon**.

Homo See *Homo* **Linnaeus, 1758**.

Homo Linnaeus, 1758 (L. *homin*=man or human being) The genus introduced by Carl Linnaeus in 1758 to accommodate modern humans and which referred exclusively to modern humans. The history of the interpretation of the genus *Homo* since its introduction in the tenth edition of Linnaeus' *Systema Naturae* has been one of episodic relaxation of the criteria used for deciding what taxa should be included within the genus, with each episode resulting in one, or more, extinct taxon being added to *Homo*. As originally conceived by Linnaeus, the genus *Homo* incorporated two species, *Homo sapiens* (i.e., modern humans) and *Homo troglodytes*. William King was the first to suggest that an extinct hominin species, namely *Homo neanderthalensis*, should be included within *Homo* (King 1864). Researchers still argue about the taxonomic significance of the differences between *H. sapiens* and *H. neanderthalensis* but the inclusion of the latter within *Homo* meant expanding the definition of *Homo* to included archaic and derived morphology (e.g., discrete and rounded supraorbital margins, midline facial projection, a distinctive parietal and occipital morphology, and robust limb bones with relatively large joint surfaces) either not seen at all, or not seen in this combination, in *H. sapiens*. The next modification of the interpretation of the genus *Homo* was in 1908 when *Homo heidelbergensis* was added to the genus (Schoetensack 1908). This meant that *Homo* now embraced at least one individual with a mandible that had a robust mandibular corpus and lacked a true chin. Then came the addition of *Homo rhodesiensis* (Woodward 1921) from the site now known as **Kabwe** and subsequently the addition of *Homo soloensis* from **Ngandong** (Oppenoorth 1937). The addition of the Ngandong fossils meant that crania substantially more robust than those of modern humans and *H. neanderthalensis* were now included in *Homo*. Nonetheless, the endocranial volumes of all the crania associated with *H. neanderthalensis*, *H. rhodesiensis*, and *H. soloensis* are still within, or close to, the modern human range. When Franz **Weidenreich** formally proposed that two existing extinct hominin hypodigms, *Pithecanthropus erectus* and *Sinanthropus pekinensis*, should be merged into a single species and transferred to *Homo* as *Homo erectus* (Weidenreich 1940) the addition of these taxa resulted in *Homo* subsuming an even wider range of morphology. Thereafter, the hypodigms of *Meganthropus*, *Telanthropus*, and *Atlanthropus* were also transferred to *Homo* by, respectively, Mayr (1944), Robinson (1961), and Le Gros Clark (1964). The addition of the *H. erectus* hypodigm at this time, even though it was well before the discovery of the small *H. erectus* crania from East Africa (Anton 2004, Leakey et al. 2003, Potts et al. 2004, Spoor et al. 2007) and **Dmanisi** (Rightmire et al. 2006) substantially increased the range of endocranial volume within the genus *Homo*. Compared with most of the pre-1940 hypodigm of *Homo*, fossils attributed to *H. erectus* have a smaller neurocranium, a lower vault, a broader base relative to the vault, and more complex premolar roots. They also have a substantial, shelf-like torus above the orbits and there are often both sagittal and angular tori, although the expression of some, or all, of this morphology may be size-related (Antón et al. 2007). The occipital sagittal profile is sharply angulated in *H. erectus*, with a well-marked supratoral sulcus and the inner and outer tables of the vault are thickened. The cortical bone of the postcranial skeleton is generally thick. The long bones are robust, and the shafts of the femur and the tibia are flattened from front to back relative to those of other *Homo* species. However, all the postcranial elements of *H. erectus* are consistent with a habitually upright posture and obligate long-range bipedalism. In 1964 the remains of *Homo habilis* from Olduvai Gorge were added to *Homo* (Leakey et al. 1964). The inclusion of this hypodigm (e.g., **OH 7, 8, 13, 16**) in *Homo* substantially expanded the range of morphology within the genus and meant that the Le Gros Clark (1955) diagnosis of *Homo* needed to be amended. To accommodate *H. habilis* in the genus *Homo*, Leakey et al. (1964) were forced to reduce the lower end of the range of brain size of fossils attributed to *Homo* to 600 cm^3, but these authors claimed that other criteria, such as dexterity, an erect posture, and a bipedal gait, did *not* need to be changed because their interpretation of the functional capabilities of the type specimen and the paratypes of *H. habilis* was consistent with these functional criteria. Ultimately, however, fresh evidence and fresh interpretations of existing evidence have led others to offer rather different functional assessments of the same material. After the announcement of *H. habilis* in 1964, the next significant addition to the *Homo* hypodigm was the recovery of **KNM-ER 1470** from the Upper Burgi Member of the **Koobi Fora Formation** (Leakey 1973) and the subsequent proposal to subsume *Pithecanthropus rudolfensis* into *Homo* as *Homo rudolfensis*. Morphologically, the latter taxon presented an apparently unique mixture of a relatively large *Homo*-like neurocranium, and a broad *Paranthropus*-like face, but it lacked the distinctive combination of small anterior and large postcanine teeth typical of *Paranthropus*, especially *Paranthropus boisei*. As a consequence, from 1972 onwards, the genus *Homo* subsumed a substantially wider range of facial morphology (Wood 1991) than it did prior to the discovery of KNM-ER 1470. Walker (1976) was alone among the early commentators to caution that KNM-ER 1470 may sample a large-brained **australopith** taxon and may not belong to *Homo*. The KNM-ER 1470 cranium was initially not allocated to a species, but instead was referred to an informal category called "early *Homo*." In due course other cranial specimens from Koobi Fora (e.g., **KNM-ER 1590, 1802, 1805, 1813, 3732**) (Wood 1991) and Olduvai (e.g., **OH 62**) (Johanson et al. 1987) were added to the early *Homo* hypodigm, as were fossils from Members G and H of the **Shungura Formation** (Howell et al. 1976, Boaz and Howell 1977, Coppens 1980), Members 4 and 5 at **Sterkfontein** (Hughes and Tobias 1977, Clarke 1985, Kimbel and Rak 1993), Member 1 at **Swartkrans** (Clarke and Howell 1972, Grine and Strait 1994, Grine et al. 1993, 1996), the **Chemeron Formation** (Hill et al. 1992), **Uraha** (Bromage et al. 1995), **Hadar** (Kimbel et al. 1996), the Nachukui Formation in **West Turkana** (Prat et al. 2005), and Dmanisi in Georgia (Gabunia and Vekua 1995, Gabunia et al. 2000, Vekua et al. 2002, Lordkipanidze et al. 2005, 2006, 2007). These additions to "early *Homo*" subsumed a wide range of cranial and dental morphology. For example, the endocranial volumes of the specimens range from just less than 500 to around 850 cm^3. The mandibles also vary in size, but all have relatively robust bodies (Wood and Aiello 1998) and premolar teeth with relatively complex crowns and roots (Bromage et al. 1995). The discovery of OH 62 provided the first unequivocal postcranial evidence for *H. habilis*, so it is significant that OH 62 has been interpreted as having limb proportions that are at least as ape-like as those of individuals attributed to *Australopithecus afarensis* (Johanson et al. 1987, Hartwig-Schrerer and Martin 1991, Richmond et al. 2002) and limb cross-sectional geometries that are unusual in a committed biped (Ruff 2008). It is also likely that the even more fragmentary KNM-ER 3735 associated skeleton belongs to *H. habilis* (Wood 1991). Some researchers have suggested that the inclusion of *H. habilis* and *H. rudolfensis* into *Homo* is unjustified on the basis that to do so weakens the case for recognizing *Homo* as a distinct **grade**. The most recent species to be added to *Homo*, *Homo floresiensis* (Brown et al. 2004) broadened the morphological scope of *Homo* even further. The specimens attributed to this species were recovered from deposits in the **Liang Bua** cave on the Indonesian island of Flores and are dated to between approximately 74 and 18 ka (Brown et al. 2004, Morwood et al. 2004, Morwood and Jungers 2009). Its small brain and body size makes *H. floresiensis* a particularly significant addition to *Homo*. The endocranial volume of the partial associated female skeleton, LB1, was initially reported to be 380 cm^3 (Brown et al. 2004), but Falk et al. (2005) increased this figure to 417 cm^3. Even at 417 cm^3, the endocranial volume of *H. floresiensis* is considerably smaller than those of the other species assigned to *Homo* for adult endocranial volume in *H. habilis* ranges between 509 and 687 cm^3 (de Sousa and Wood 2006). However, the stature estimates of 106 cm for LB1 (Brown et al. 2004) and 109 cm for LB8 (Morwood et al. 2005) are only slightly smaller than McHenry's (1991) stature estimate of 118 cm for the *H. habilis* OH 62 partial skeleton. Some researchers have interpreted *Homo* more inclusively. For example, John **Robinson** proposed that *Australopithecus* be sunk into *Homo* (Robinson 1972). Curnoe and Thorne (2003) have proposed a much more radical scheme whereby they only recognize three species in the hominin clade and they include all three in *Homo*, as *Homo ramidus*, *Homo africanus*, and *Homo sapiens*. See **genus**; *Homo africanus*.

Homo africanus (Dart, 1925)

(L. *homin*= man or human being and *africanus*=Africa) A new taxonomic combination introduced by John **Robinson** when he proposed that *Australopithecus* is a subjective junior synonym of *Homo* (Robinson 1972). Robinson (1967) suggested that *H. africanus* had been used by Broom (1918) for the **Boskop** associated skeleton, but Broom had assigned the latter to *Homo capensis* and not to *H. africanus*. Later Robinson realized that Giuseppe **Sergi** had introduced *Homo africanus* in 1908 and in deference to Sergi's prior use of the name Robinson used the next most senior synonym, *Australopithecus transvaalensis* Broom, 1936, and thus the binomen *Homo transvaalensis* in place of *Australopithecus africanus*. However, in his 1972 book Robinson judged that von Eickstedt (1937) was the first person to use *H. africanus* formally and so because *Australopithecus africanus* Dart, 1925 had priority he adopted *H. africanus* in place of *Homo transvaalensis*. Robinson (1972, p. 246) wrote that his inclusion of *Au. africanus* in *Homo* was the logical consequence of his belief that "a genus represents a clearly defined adaptive zone or way of life," and that the species within a genus should "represent no more than variations in detail on the basic adaptive theme" and he concluded that "the basic adaptation of *H. africanus* seems to me to be the same as that of man, hence the two should be in the same genus." Wolpoff (1969) used *H. africanus* in place of *Homo habilis* when writing about OH 7, presumably because he saw it as a senior synonym of *H. habilis*. See also *Homo transvaalensis*.

Homo antecessor Bermúdez de Castro et al., 1997*

(L. *homin*=man or human being and *antecessor*=forerunner) Between 1994 and 1996 more than 80 hominin fossils were recovered from the **Gran Dolina** cave in the **Sierra de Atapuerca**. Their discoverers considered them too primitive to be allocated to either *Homo neanderthalensis* or *Homo heidelbergensis* so in 1997 they assigned them to a new species, *Homo antecessor* (Bermúdez de Castro et al. 1997), claiming that *H. antecessor* combines a modern midfacial morphology (a **canine fossa**, an acute zygomaticoalveolar angle, and a modern human nasal morphology) with primitive *Homo ergaster*-like dental morphology. Critics of the decision to recognize a new taxon suggest that the "modern" facial morphology is due to the immaturity of the **ATD6-69** facial fragment. First discovery ATD6-1 (1994). Holotype The details of the holotype of *Homo antecessor* are given in note no. 5 of Bermúdez de Castro et al. (1997). They suggest that the holotype consists of *all* the remains belonging to an individual whose remains comprise the mandible **ATD6-5**, plus the isolated teeth listed as **ATD6-1**–**ATD6-4** and **ATD6-6**–**ATD6-12**. Paratypes Listed in Table 2 in Carbonell et al. (1995, p. 1393). Main site Gran Dolina, Sierra de Atapuerca.

Homo antiquus Ferguson, 1984

(L. *homin*= man or human being and *antiquus*=former) Ferguson (1984) reviewed "14 hominoid fossil jaws" (*ibid*, p. 526) from **Hadar** and **Laetoli** and concluded that they should be distributed among three taxa: one unnamed pongid taxon and two hominin taxa, namely *Australopithecus africanus* for the larger hominin jaws (see also Ferguson 1999) and a new primitive *Homo* taxon, dubbed *Homo antiquus* for the smaller jaws. In doing so he claimed he was merely restoring "the original determination of A.L. 199-1 as belonging to the genus *Homo* (Johanson, Taieb and Coppens, 1982)" (*ibid*, p. 527). In a later paper Ferguson (1987) suggests that **KNM-ER 1813** and KNM-ER 1501 should also be assigned to *H. antiquus*. Holotype **A.L. 288-1** (1974). Paratypes A.L. 128-23, 198-1, 199-1, 207-1, and 333-2. Main site Hadar, Ethiopia. See also *Australopithecus afarensis*.

Homo antiquus praegens Ferguson, 1989

(L. *homin*=man or human being, *antiquus*= former, and *praegens*=progenitor) A subspecies introduced by Walter Ferguson to accommodate the **Tabarin** mandible with **KNM-TH 13150** (Ferguson mistakenly refers to it as "KNM-ER TI 13150") as its **holotype**. No matter what anyone thinks of the logic behind the proposal for this new species the paper introducing it meets the criteria set out in the **International Code of Zoological Nomenclature**, and thus the name is available.

Homo calpicus Falconer, 1864 *See* Gibraltar 1.

Homo capensis Broom, 1918 *See* Boskop; *Homo erectus capensis* (Broom & Robinson, 1949).

Homo cepranensis Mallegni et al., 2003 *See* Ceprano 1.

Homo erectus (Dubois, 1893*)

(L. *homin*= man or human being and *erectus*=to set upright) **Hominin** species established by Eugène **Dubois** to accommodate fossil hominins recovered in 1891 at **Trinil**, Java. In his initial publication of the Trinil remains, published in 1893, Dubois referred the **skullcap** to *Anthropopithecus erectus*. The name reflected Dubois' initial conviction that he had discovered the remains of a fossil ape, for at that time *Anthropopithecus* and not *Pan* was the preferred genus name for chimpanzees. But he changed his mind and a year later in 1894 he transferred the new species to *Pithecanthropus*. The discovery of the Trinil **calotte** was significant because of its small cranial capacity (approximately 940 cm^3), its primitive shape with its low brain case, and quite sharply angulated occipital region, for in 1894 the only two hominin taxa known were *Homo sapiens* and *Homo neanderthalensis*. Discoveries made by Ralph **von Koenigswald** at **Sangiran**, also in Indonesia, were added to the hypodigm of *Pithecanthropus erectus*. Later additions came when the fossils recovered from what was then called Choukoutien (now called **Zhoukoudian**) and initially assigned to *Sinanthropus pekinensis* Black, 1927 were compared with the *P. erectus* hypodigm. The researchers responsible for analyzing the two collections suggested that the Indonesian and Chinese hypodigms were "related

to each other ... in the same way as two different races of present mankind" (von Koenigswald and Weidenreich 1939, p. 928) and a year later Weidenreich formally proposed the two hypodigms should be merged in a single genus as *Homo erectus pekinensis* and *Homo erectus javanensis*, respectively (Weidenreich 1940). Subsequently the hypodigms of *Meganthropus* (Mayr 1944, p. 14, Le Gros Clark 1955, pp. 86–7), *Atlanthropus* (Le Gros Clark 1964, p. 112), and *Telanthropus* (Robinson 1961) were transferred to *Homo* as *Homo erectus*. First discovery **Kedung Brubus 1**, Kedung Brubus, Java (1890). Holotype **Trinil 2**, adult calotte, Trinil, **Ngawi**, Java (1891). Paratypes None. Main sites **Olduvai Gorge, Sambungmacan,** Sangiran, Trinil, Zhoukoudian. *See also Homo ergaster.* (*NB: the year of publication is usually given as 1892, but this is incorrect. The publication was a report by Dubois of excavations carried out in 1892, but it did not become a public record until 1893, so this should be the year of its publication).

Homo erectus capensis (Broom & Robinson, 1949)

(L. *homin*=man or human being, *erectus*= to set upright, and *cape*=a reference to land that includes the Cape of Good Hope) A name combination introduced by Phillip **Tobias** (1968) for a subspecies of *Homo erectus* that presumably included specimens such as **SK 15** and **SK 45**. The name is preoccupied by *Homo capensis* Broom, 1918.

Homo erectus erectus (Dubois, 1893)

(L. *homin*=man or human being and *erectus*=to set upright) A name combination introduced by Campbell (1973) as the nominotypical subspecies of *Homo erectus* for what most other researchers recognize as the temporally later part of the Indonesian hypodigm of *Homo erectus*.

Homo erectus heidelbergensis Schoetensack, 1908

(L. *homin*=man or human being, *erectus*=to set upright, and *heidelbergensis*=a city near the location of the holotype of *Homo heidelbergensis*) A name combination introduced by Campbell (1973) for what most other researchers recognize as either *Homo heidelbergensis* or the temporally early part of the **hypodigm** of an inclusive interpretation of *Homo neanderthalensis*.

Homo erectus javanensis

(L. *homin*=man or human being, *erectus*=to set upright, and

323

javanensis=the name of the island where the holotype of *Homo erectus* was found) Name combination used by Franz **Weidenreich** (1940) for the **hypodigm** of what was then called *Pithecanthropus erectus*. It is derived from a poorly constructed name *Homo javanensis primigenius* Houze, 1896; the correct combination should have been *Homo erectus erectus*.

Homo erectus leakeyi (or Homo leakeyi) Heberer, 1963

(L. *homin*=man or human being, *erectus*=to set upright, and *leakeyi*=to recognize the contribution of Louis **Leakey**) A name combination introduced by Heberer (1963) for a subspecies of *Homo erectus* that accommodated **OH 9**. Campbell (1973) later used the same name combination for a subspecies that subsumed the temporally later part of the East African hypodigm of *Homo erectus*.

Homo erectus mauritanicus (Arambourg, 1954)

(L. *homin*=man or human being, *erectus*= to set upright, and *mauritanicus*=named after Mauritania, the name the Romans gave to the province that included the western part of modern Algeria and the northernmost part of Morocco. The name Mauritania come from the Gk *mauros*=black, from which the Moors took their name) A name combination introduced by Tobias (1968) and perpetuated by Campbell (1973) for what most other researchers recognize as the North African component of the hypodigm of *Homo heidelbergensis* or *Homo erectus* or the North African component of the temporally early part of the hypodigm of an inclusive interpretation of *Homo neanderthalensis*.

Homo erectus olduvaiensis Tobias, 1968

(L. *homin*=man or human being, *erectus*=to set upright, and *olduvaiensis*=named after **Olduvai Gorge**, the location of **OH 9** and **OH 12**, see below) A name combination introduced by Tobias (1968) for a subspecies of *Homo erectus* that included specimens such as OH 9 and perhaps OH 12, but because it was proposed conditionally it has no standing.

Homo (erectus or sapiens) palaeohungaricus Thoma, 1966

A subspecies suggested by Thoma (1966) to recognize a subspecies of *Homo erectus* to accommodate the hominin remains recovered from **Vérteszöllös**, Hungary. The holotype is **Vérteszöllös 2**.

Homo erectus pekinensis (Black, 1927)

(L. *homin*=man or human being, *erectus*=to set upright, and *pekinensis*=named after the site where *Sinanthropus pekinensis* had been discovered) Name combinations suggested by Weidenreich (1940) and Campbell (1973) for what most other researchers recognize as the mainland East Asia part of the hypodigm of *Homo erectus*.

Homo erectus reilingensis Czarnetzki, 1989

A subspecies of *Homo erectus* proposed by Czarnetzki (1989) for the hominin remains recovered from **Reilingen**, Germany in 1978. The holotype is the Reilingen calvaria. *See also* **Reilingen calvaria**.

Homo erectus sinensis

(L. *homin*=man or human being, *erectus*=to set upright, and *sinensis*= named after the country site where *Sinanthropus pekinensis* had been discovered) A name combination used by Weidenreich (1940) for a subspecies to recognize the hypodigm of what was then referred to as *Sinanthropus pekinensis*. The name derives from *Pithecanthropus sinensis* Weinert, 1931. *See also Homo erectus pekinensis*.

Homo ergaster Groves and Mazák, 1975

(L. *homin*=man or human being and *ergo*=work) A hominin species established by Colin Groves and Vratislav Mazák to accommodate fossil hominins recovered from **Koobi Fora** that in their judgment did not belong in the taxa that existed at the time. Wood (1994) used the taxon name *Homo ergaster* for East African hominin remains that are generally more **primitive** and lack the more extreme expressions of some of the **derived** features (e.g., thick inner and outer table, sagittal keeling, etc.) seen in Asian *Homo erectus*. Other researchers acknowledge these differences exist, but most are inclined to view them as regional variations within a single species, *Homo erectus*. First discovery KNM-ER 730 (1970). Holotype **KNM-ER 992** (1971). Paratypes **KNM-ER 730**, 731, 734, 803, 806-9, **820**, 1480, **1805**. Main sites Koobi Fora, **West Turkana**.

Homo floresiensis Brown et al., 2004

(L. *homin*=man or human being and *flores*=the name of the island, Flores, where the type site is located) Hominin species erected by Brown et al. (2004) to accommodate **LB1**, a partial adult hominin skeleton and LB2, an isolated left P_3 recovered in 2003 from the **Liang Bua** cave on the Indonesian island of Flores. More material belonging to LB1 and evidence allocated to LB4–9, including a second partial skeleton (but lacking a cranium), **LB6**, was recovered in 2004 (Morwood et al. 2005). The hypodigm now includes close to 100 individually numbered specimens that are estimated to represent fewer than 10 individuals. The taxon was immediately controversial for at least two reasons. First, its estimated geological age of between *c.*17 and *c.*74 ka substantially overlapped evidence of the presence of modern humans in the region. Second, while its discoverers and describers acknowledged its small overall size (the stature of LB1 is estimated to be 106 cm and its body mass to be roughly between 25 and 30 kg), its especially small brain (approximately 417 cm^3) for an adult hominin and its primitive morphology were most parsimoniously interpreted to be evidence of a novel endemically dwarfed pre-modern *Homo* or a transitional hominin taxon. Other researchers have suggested that no new taxon needs to be erected because they claim the "*Homo floresiensis* hypodigm" has been sampled from a population of *Homo sapiens* – most likely related to the small-statured Rampasasa people who live on Flores today – afflicted by either an endocrine disorder or one, or more, of a range of syndromes that include **microcephaly**. Both explanations are exotic, but those who espouse a pathological explanation for the individuals represented by LB1–15 need to explain what pathology results in an early *Homo*-like cranial vault, primitive mandibular, dental, carpal and pedal morphology, and a brain that while very small apparently has none of the morphological features associated with the majority of types of microcephaly. Initially it was suggested that *H. floresiensis* was a dwarfed *Homo erectus*, but the burden of subsequent analyses suggest that it may be more closely related to a transitional hominin such as *Homo habilis* (Tocheri et al. 2007, Argue et al. 2009, Brown and Maeda 2009, Morwood and Jungers 2009). Aiello (2010) provides a useful review of the evidence and the controversy surrounding *H. floresiensis*. First discovery LB1 (2003). Holotype As above. Paratype LB2 (2003). Only site **Liang Bua**, Flores, Indonesia.

Homo florisbadensis Drennan, 1935 *See* Florisbad 1.

Homo gautengensis Curnoe, 2010 (L. *homin*=man or human being and *gauteng*=the name of the South African province within which are the caves of Sterkfontein, Swartkrans, and Drimolen) A hominin species established to accommodate fossil hominins recovered from Sterkfontein, Swartkrans, and Drimolen that Curnoe judged could not be accommodated within existing species within *Australopithecus*, *Paranthropus*, and *Homo*. The diagnosis stresses dental morphology and, in particular, the unusual combination of mandibular postcanine tooth crowns that are both buccolingually narrow and have a large talonid. First discovery **SK 15** (1949). Holotype **Stw 53** (1976). Paratypes The 21 paratypes listed in Curnoe (2010) include SE 255, SK 45, and DNH 70. Main sites **Sterkfontein** and **Swartkrans**, plus one specimen at **Drimolen**.

Homo georgicus Gabounia et al., 2002 (L. *homin*=man or human being and Georgia is the name of the country within which the **Dmanisi** site is located) A hominin species established in 2002 by Gabounia et al. (the spelling of Leo Gabunia's name in the 2002 reference is not the conventional one but according to the rules of the **International Commission on Zoological Nomenclature** (or ICZN) the spelling mistake has to be retained in the formal name of the species) for an adult mandible recovered from Dmanisi in 2000. Most researchers recognize that the vast majority of the hominin fossils recovered from Dmanisi belong to a **taxon** that shares features with both *Homo habilis* and *Homo erectus* (and with *Homo ergaster* if that taxon is recognized). However, at this stage few researchers (e.g., de Lumley and Lordkipanidze 2006 are an exception) are willing to assign this material to a novel taxon, but if that need arises in the future *Homo georgicus* is the species name that has priority. The workers who introduced the new taxon saw it as intermediate between *H. habilis* and *Homo rudolfensis* on the one hand and *Homo ergaster* on the other. First discovery **D2600** (2000). Holotype D2600 (2000). Paratypes No paratypes were formally designated, but under the heading "Material," the mandible **D211**, the calvaria **D2280**, the cranium **D2282**, and the skull and **associated skeleton D2700** were referred to as evidence that "will complete the characteristics of the new species" (Gabounia et al. 2002, p. 244). Only site Dmanisi, Georgia.

Homo habilis Leakey, Tobias and Napier, 1964 (L. *homin*=man or human being and *habilis*=handy) A hominin species established by Louis **Leakey**, Phillip **Tobias**, and John **Napier** in 1964 to accommodate fossil hominins (**OH 4, 6, 7, 8,** and **13;** OH 14 and 16 were not included as paratypes, but they were "referred" to *Homo habilis*) recovered between 1959 and 1963 from Beds I and II, **Olduvai Gorge**, Tanzania. Leakey et al. (1964) claimed that brain size, cranial, and dental morphology, and inferences about **dexterity** and **locomotion**, distinguished the new taxon from *Australopithecus africanus* and justified its inclusion in *Homo*. Subsequent discoveries at Olduvai (e.g., **OH 24, 62,** and **65**) and from other sites (e.g., **Koobi Fora: KNM-ER 1470, 1802, 1805, 1813, 3735; Sterkfontein: Stw 53; Swartkrans: SK 847; Hadar: A.L. 666-1**) have also beenn assigned to this taxon. The hypodigm has a relatively wide range of cranial and dental morphology. Endocranial volume ranges from just less than 500 to approximately 800 cm^3. All the crania in this group are wider across the base of the cranium than across the vault but facial morphology varies, with KNM-ER 1470 having its greatest width across the mid-face whereas KNM-ER 1813 is broadest across the upper face. Mandibles vary in size and **robusticity**, with those from the larger individuals having **robust** mandibular bodies. **Postcanine teeth** vary in size and crown morphology. Some mandibular **premolars** and **molars** are buccolingually narrow (e.g., OH 7), whereas larger mandibular premolar teeth within the hypodigm (e.g., KNM-ER 1802) have buccolingually broader crowns, complex **talonids**, and more complex root systems. Some researchers consider the cranial variation within a single taxon to be excessive in scale and unlike the pattern of intraspecific variation seen in the *Pan/Homo* clade. They suggest *H. habilis* subsumes two taxa, *Homo habilis sensu stricto* and *Homo rudolfensis*. First discovery OH 4 (1959). Holotype OH 7 (1960). Main sites **Koobi Fora, Olduvai, Shungura Formation, ?Sterkfontein, ?Swartkrans**. *See also Homo habilis sensu lato; Homo habilis sensu stricto.*

Homo habilis sensu lato A **taxon** whose **hypodigm** includes all the specimens referred to under *Homo habilis*. This is the sense in which *H. habilis* is used by researchers (e.g., Tobias 1991a, Suwa et al. 1996) who are content that variation within that hypodigm is compatible with a single-species interpretation. *See also Homo habilis; Homo habilis sensu stricto.*

Homo habilis sensu stricto A **taxon** whose **hypodigm** includes a subset of the specimens referred to under *Homo habilis*. This is the sense in which *H. habilis* is used by researchers (e.g., Wood 1992, **Stringer** 1986a, Kramer et al. 1995, Grine et al. 1996) who suggest that the variation within that hypodigm is not compatible with a single-species interpretation. For Stringer the *H. habilis s. s.* hypodigm would consist of OH 7 and OH 24, but not specimens such as OH 13 and OH 16, but for Wood (1992) *H. habilis s. s.* comprises all the Olduvai hypodigm of *H. habilis*, plus a subset of the Koobi Fora site component of the hypodigm (e.g., **KNM-ER 1805, 1813, 3785**). *See also Homo habilis; Homo habilis sensu lato.*

Homo heidelbergensis Schoetensack, 1908 (L. *homin*=man or human being and *heidelbergensis*=because the holotype comes from the Grafenrain commercial sandpit, which is southeast of Heidelberg, Germany) This species name was created by Otto Schoetensack to accommodate a robust hominin mandible found in 1907 in a commercial sandpit at **Mauer**, near Heidelberg, Germany. The **Mauer mandible** lacks a chin and exhibits other traits Schoetensack interpreted as primitive, including a broad ramus and an anterior–posteriorly deep mandibular symphysis, but these are combined with modern human-like dental proportions, including reduced canines. In the light of this evidence Schoetensack concluded the Mauer mandible's mosaic of primitive and derived traits was sufficient to distinguish it from *Homo sapiens, Homo neanderthalensis*, and what was then called *Pithecanthropus erectus* (Schoetensack 1908). However, until relatively recently the new species name attracted little interest. Howell (1960) does not mention it in his influential review of Middle Pleistocene hominins and Le Gros Clark's (1955, 1964) only reference to *H. heidelbergensis* was to suggest that it is by no means certain the species belongs in *Homo*. Most European researchers described fossils with similar features and geological ages as belonging to a poorly defined group called "preneanderthals," but did not refer to any specific taxa. In 1995 Rightmire suggested that *H. heidelbergensis* might be the most appropriate species name for a group of Afro-European hominin fossils (e.g., **Arago, Bodo, Kabwe, Mauer, Ndutu,** and **Petralona**) that others were referring to as "archaic" *Homo sapiens* and in the same paper he considered the pros and cons of several hypotheses for the relationships among *H. heidelbergensis*, modern humans, and Neanderthals (Rightmire 1995, Fig. 4, p. 454). Mounier et al. (2009) provide the most detailed morphological and taxonomic analysis of the holotype of *H. heidelbergensis*. Their conclusion is that there are morphological grounds for recognizing *H. heidelbergensis* as a taxon separate from *H. neanderthalensis*, *H. sapiens*, and *H. erectus*, and they supply both a definition and a differential diagnosis for the taxon (*ibid*, pp. 243–4). They consider the mandibles from Arago, **Montmaurin, Sima de los Huesos**, and **Tighenif** to be conspecifics of the Mauer mandible and their conclusions about the taxonomic options for and the phylogenetic relationships of a *H. heidelbergensis* taxon include some of those reviewed by Rightmire (1995). The one Mounier et al. (2009) call the "Afro-European hypothesis" is that "*H. heidelbergensis* was a geographically widespread and diverse species that gave rise to *H. neanderthalensis* in Eurasia, and *H. sapiens* in Africa" (Stringer 2001, p. 568 and see Mounier et al. 2009, Fig. 6, p. 242). In what Mounier et al. (2009) call the "European hypothesis," *H. heidelbergensis* is restricted to Europe and is only ancestral to *H. neanderthalensis*. In this hypothesis modern humans are descended from an African taxon that would be *Homo rhodesiensis* if that taxon's hypodigm included Kabwe and Tighenif and *Homo mauritanicus* if it included the latter, but not the former. However, if, as some argue, *contra* Mounier et al. (2009), *H. heidelbergensis* represents an early stage in the **accretion model** of Neanderthal characteristics (e.g., Arsuaga et al. 1997, Dean et al. 1999, Carbonell et al. 2005) then *H. heidelbergensis* would be a junior synonym of *H. neanderthalensis*. However, if the latter view prevails then *H. heidelbergensis* would have priority as the name of the taxon that gave rise to *H. neanderthalensis* and *H. sapiens* (the "Afro-European hypothesis") or just *H. neanderthalensis* (the "European hypothesis"). Until recently *H. heidelbergensis* was considered one of the oldest hominin taxa in Europe, but if *Homo antecessor* is a good taxon, then this would antedate it. First discovery Mauer mandible

(1907). Holotype As above. Paratypes None. Main sites Arago, Mauer, Mountmaurin, Sima de los Huesos, Tighenif. *See also* **pre-Neanderthal hypothesis**.

Homo helmei Dreyer, 1935

The taxonomic name assigned to a partial skull discovered in 1932 by T.F. Dreyer in Florisbad, South Africa. Some researchers interpret the **Florisbad 1** cranium as being intermediate in morphology between *Homo heidelbergensis* and *Homo sapiens*. Its vertical facial dimensions, brow ridge size, and inclination of the frontal bone distinguish it from the former yet it retains primitive features (e.g., a large brow ridge, receding frontal and broad vault) that distinguish it from anatomically modern humans. While some suggest that Florisbad 1 could serve as the holotype of the taxon *Homo helmei* most researchers take the view that there is no satisfactory diagnosis for such a taxon. Other specimens that have been suggested as belonging to the same hypodigm as the Florisbad 1 cranium include **Jebel Irhoud**, Ngaloba (*see* **LH 18**), **Omo II**, and **Singa** (e.g., Group 2 in McBrearty and Brooks 2000, Fig. 2, p. 481).

Homo kanamensis Leakey, 1935
See **Kanam**.

Homo leakeyi Clarke, 1990

(L. *homin=* man or human being and *leakeyi=*to recognize the role of Louis and Mary **Leakey** in the recovery of **OH 9**) This name combination was suggested, but not formally proposed, as the binomial for OH 9 by Heberer (1963). Several authors (Simons et al. 1969, Harrison 1993, Groves 1999) have pointed out that Heberer's name was unavailable in nomenclature and that Clarke (1990) was the first to use *Homo leakeyi* in a way that conforms with the requirements of the **International Code of Zoological Nomenclature**, but Groves (1989) points out that the introduction of *Homo (Proanthropus) louisleakeyi* by Kretzoi (1984) means that *Homo louisleakeyi* Kretzoi, **1984** would have priority.

Homo louisleakeyi Kretzoi, 1984
See Homo (Proanthropus) louisleakeyi.

Homo (Proanthropus) louisleakeyi Kretzoi, 1984
A name combination suggested by Kretzoi (1984) for the taxon of which **OH 9** would be the type specimen. *See also Homo erectus.*

"Homo mauritanicus" (L. *homin*=man or human being and *Mauritania*=the name of a Roman province that is coincident with what is now Morocco and Algeria) Camille **Arambourg** (1955) provisionally suggested a new species and a new genus *Atlanthropus mauritanicus* to accommodate three hominin mandibles and a cranial fragment recovered by Arambourg and Hoffstetter in June 1954 at **Ternifine** (now called **Tighenif**) near Mascara, in Algeria. *Atlanthropus* was never widely adopted as a genus, and because Le Gros Clark (1964) formally suggested that *Atlanthropus mauritanicus* is a subjective junior synonym of *Homo erectus* the new species was never formally incorporated into *Homo* as *H. mauritanicus*. Some suggest that if *H. heidelbergensis* is restricted to Europe and is only ancestral to *H. neanderthalensis* and if modern humans are descended from an African taxon that includes Tighenif but not Kabwe then *H. mauritanicus* would have priority as the name of such a taxon. However, that name is not available because its introduction was not associated with a formal description "giving characters purporting to differentiate the taxon."

Homo microcranous Ferguson, 1995 (L. *homin*=man or human being, *micro*=small, and *cranion*=brain case) In his 1995 review of the taxonomic status of KNM-ER 1813 Ferguson reiterates his earlier claim that the fossil is a male (Ferguson 1987) and because, in his opinion, the small brain size of KNM-ER 1813 precludes it from being included in *Homo habilis*, he sees the need to erect a new species, *Homo microcranous*, to accommodate KNM-ER 1813. He does not discuss the implications of this proposal for other early *Homo* specimens with a similar brain size. No matter what anyone thinks of the logic behind the proposal for this new species the paper introducing it meets the criteria set out in the **International Code of Zoological Nomenclature**; thus the species name is available. If it was judged that KNM-ER 1813 does not belong to *H. habilis* and that it belongs to a species other than *Homo rudolfensis* then *Homo microcranous* is a species name that is both valid and available. Holotype KNM-ER 1813 (1973). Paratypes None. Only site **Koobi Fora**. *See also Homo habilis*; **KNM-ER 1813**.

Homo modjokertensis von Koenigswald, 1936 (L. *homin*=man or human being and *modjokertensis* from the name of a village in East Java near the site where the calvaria was discovered) The name given by Ralph **von Koenigswald** to a taxon he introduced to accommodate the juvenile **Modjokerto** calvaria, **Modjokerto 1**. Subsequently he and others transferred the taxon to *Pithecanthropus*. Harvati et al. (2010) show that some of the "characteristic morphology" of *H. neanderthalensis* is shared with African crania and they suggest that some of these features may be retained symplesiomorphies. First discovery Modjokerto 1 (1936). Holotype As above. Paratypes None. Only site Modjokerto.

Homo mousteriensis Forrer, 1908 [Homo mousteriensis hauseri Klatsch and Hauser, 1909] *See* **Le Moustier I**.

Homo neanderthalensis King, 1864 (L. *homo*=man or human being and *neanderthalensis*=literally "of Neander's valley", the Neander in question being Joachim Neander, a 17thC non-ordained Lutheran ecclesiastical poet and composer) A hominin species established in 1864 by William King for the partial skeleton recovered in 1856 from the **Kleine Feldhofer Grotte** in the part of the Düssel valley named after Joachim Neander. No faunal or archeological evidence was reported in the original publication. However, through careful archival research, Ralf Schmitz and Jürgen Thissen were able to identify what remained of the sediments from the Kleine Feldhofer Grotte at the base of the south wall of the Neander valley (Schmitz et al., 2002). Excavations of these sediments in 1997 resulted in the recovery of fauna, artifacts, and hominin bone fragments and "a small piece of human bone (NN 13) was found to fit exactly onto the lateral side of the left lateral femoral condyle of Neandertal 1" (*ibid*, p. 13342). In 2000 more fauna, archeological, and hominin skeletal fragments were recovered and "two cranial fragments…were found to fit onto the original Neandertal 1 calotte" (*ibid*, p. 13343). Some 80 fragments from at least two individuals have been recovered. Dates obtained from the sediments indicate an age of *c*.40 ka/**Oxygen Isotope Stage** (or OIS) 3 (*ibid*, p. 13345). King proposed the species name *Homo neanderthalensis* in a paper read to the British Association in 1863, but in the 1864 publication he apparently revised his assessment for he wrote that "I now feel strongly inclined to believe that it (i.e., the Kleine Feldhofer skeleton) is not only specifically but generically distinct from Man." However, King did not propose any new generic name for the specimen and the binomen *Homo neanderthalensis* is now well established. King's conclusion about the taxonomic implications of the specimen's distinctive morphology was in sharp contrast to workers such as Rudolf Virchow who considered it that of a recent individual affected by pathology, or others such as Hermann Schaaffhausen, Charles Lyell, and Thomas Henry **Huxley** who, while arguing for its antiquity and evolutionary significance, refrained from assigning it to a separate species. Discoveries made earlier, such as the infant's cranium from **Engis** (1828) and the partial cranium from Forbes' Quarry, **Gibraltar** (1848), were subsequently recognized as belonging to the same species. In the following half-century further Neanderthal remains were discovered at other European sites including **La Naulette** and **Spy** (Belgium), **Šipka** (Moravia), **Krapina** (Croatia, with over 900 specimens representing multiple individuals), and **Malarnaud, La Chapelle-aux-Saints, Le Moustier** (lower shelter), **La Ferrassie**, and **La Quina**, among others, in France. So by the early 20thC *H. neanderthalensis* was by far the best-known extinct hominin species. In 1924–6 the first *H. neanderthalensis* remains were found outside of western Europe at **Kiik-Koba** in the Crimea. Thereafter came discoveries at **Tabun** cave on **Mount Carmel** in the Levant, at **Teshik-Tash** in central Asia, and at two sites in Italy: **Saccopastore** and **Monte Circeo**. Further evidence was added after WWII, first from Iraq (**Shanidar**), then from additional Levantine sites in Israel (**Amud** and **Kebara**), and from Syria (**Dederiyeh**). New fossiliferous localities continue to be discovered in Europe and Western Asia (e.g., **Saint-Césaire** and **Moula-Guercy** in France, **Zaffaraya** in Spain, **Mezmaiskaya** in Russia, Vindija in Croatia, and **Lakonis** in Greece). To date, Neanderthal remains have been found throughout much of Europe below 55°N, as well as in the Near East and Western Asia, but no fossil evidence has been found in North Africa. Early views about Neanderthal morphology and phylogenetic status were based predominantly upon Marcellin **Boule's** highly influential description of the La Chapelle-aux-Saints skeleton, which he reconstructed as massively built, with numerous primitive features including incomplete truncal erectness, no cervical lordosis, a habitually flexed knee joint, and an abducted hallux. Boule argued that this morphology precluded Neanderthals from the ancestry of *Homo sapiens* and he concluded that the species became extinct at the end of the **Mousterian**. Even after Boule's description a few workers continued to argue for a Neanderthal ancestry for modern humans (*see* **Neanderthal phase hypothesis**) but it was not until the mid-1950s that Boule's reconstruction was shown to be anatomically incorrect and functionally impossible (Straus and Cave 1957). Many mid-20thC workers considered Neanderthals to be only subspecifically distinct from modern humans, as *Homo sapiens neanderthalensis*. This was likely due to the combination of a reaction against the "Boulean" view and the effect of the neo-Darwinian **modern evolutionary synthesis** with its emphasis on intraspecific variation and taxonomic lumping. By the 1980s the notion of conspecificity was increasingly challenged by the growing interest in **punctuated equilibrium** models and **speciation** as an evolutionary driver and by the growing evidence for an African origin for modern human (*see* **out-of-Africa hypothesis**). In the past decade or so there has been an increasing acceptance that the Neanderthals are morphologically distinctive, so much so that many consider it unlikely that such a derived form could have given rise to the morphology seen in modern humans (e.g., **Stringer** 1996, Tattersall 1986, Rak et al. 2003, Harvati et al. 2004). There is, however, another school of researchers (e.g., **Wolpoff** 1989, Frayer et al. 1993, Wolpoff and Frayer 2005) who point to, and stress, the morphological continuity between the fossil evidence for *H. sapiens* and the remains others would attribute to *H. neanderthalensis*. The inclusion of Neanderthals within *Homo sapiens* is consistent with the **multiregional hypothesis** for the evolution of modern humans. Yet another perspective is the assimilation model, first proposed in 1989 (Smith et al. 1989), which recognizes the African origin of modern humans but argues that a small amount of admixture occurred between expanding modern and indigenous archaic populations. This admixture is reflected only in certain morphological details that do not overshadow the basic differences between modern and archaic morphology (Smith, 2010). Some have argued that morphologically intermediate specimens are evidence of admixture between Neanderthals and modern humans (Duarte et al. 1999, Trinkaus

2005) but this argument has been challenged (Tattersall and Schwartz 1999, Pearson 2004, Harvati et al. 2007). Those who claim that *H. neanderthalensis* is specifically distinct from *H. sapiens* recognize morphological autapomorphies that distinguish the former from both earlier (e.g., **Homo erectus, Homo heidelbergensis**) and contemporary taxa (i.e., *H. sapiens*). These features are found in the cranium (e.g., large, rounded discrete brow ridges, projecting mid-face, inflated cheeks, small mastoid process, suprainiac fossa, and occipital bun), mandible (long corpus, retromolar space, and asymmetric mandibular notch), dentition (e.g., large shovel-shaped incisors, occlusal morphology of molars and premolars, a high incidence of taurodontism), and the postcranial skeleton (e.g., long clavicle, teres minor groove extending onto the dorsal surface of the scapula, large infraspinous fossa, long, thin pubic ramus, and large joints). The taxon *H. neanderthalensis* is currently the only extinct hominin for which there is sound ancient DNA evidence. In 1997 Mathias Krings and his colleagues (Krings et al. 1997) announced they had recovered short fragments of the hypervariable region I (HVRI) of the **mitochondrial DNA** (or mtDNA) control region from a piece of bone taken from the right humerus of the Kleine Feldhofer Neanderthal 1 type specimen and 3 years later Krings et al. (1999) sequenced a 340 bp segment of a different part (HVRII) of the mtDNA control region of the type specimen. A year later two laboratories confirmed the extraction of a 345 bp sequence of HVRI from a rib fragment belonging to a Neanderthal from Mezmaiskaya (Ovchinnikov et al. 2000) and Krings et al. (2000) reported sequences obtained from a Neanderthal individual (Vi75; now Vi 33.16) from layer G3 at **Vindija**, Croatia. Schmitz et al. (2002) extracted mtDNA from the humeral shaft (NN 1) from a second individual at the Neanderthal type site. Serre et al. (2004) extracted mtDNA from four more Neanderthal individuals, two more from Vindija (Vi77 and Vi80), and one each from Engis (Engis 2) and La Chapelle-aux-Saints, as well as from five fossil *Homo sapiens* individuals. The Vi80 sequence was identical to that of the sequence reported for Vi75 (Krings et al. 2000) and none of the modern humans yielded any amplification products with the Neanderthal primers. The latest Neanderthal fossils to yield mtDNA is the evidence of HVR1 that has been found in samples from the left femur of the Teshik-Tash Nean-

derthal from Uzbekistan and from the femur of the subadult individual from Okladnikov, a site in the Altai mountains in Western Asia (Krause et al. 2007). Subsequently the complete mtDNA of a specimen from Vindija has been sequenced (Green et al., 2008) and Briggs et al. (2009) reported the full mtDNA sequences of five individuals (Neanderthal 1 and 2, Sidron 1253, Vi33.25 from Vindija, and the Mezmaiskaya hominin). That study concluded that the genetic diversity within *H. neanderthalensis* was substantially less than that seen in modern humans. The results of the study of mtDNA by Serre et al. (2004) suggested that if Neanderthals made any contribution to modern humans then it was small one. This contribution was confirmed by the results of the analysis of the draft sequence of the Neanderthal nuclear genome (Green et al. 2010). The study focused on three individuals (Vi33.16, 25, and 26) from Vindija. Green et al. compared the draft sequence with smaller amounts of sequence data from Neanderthal specimens from **El Sidron, Kleine Feldhofer Grotte**, and Mezmaiskaya, as well as the sequenced nuclear genomes of five modern humans: two from sub-Saharan Africa, one each from Papua New Guinea and France, and a Han Chinese. The results show that whereas the modern humans from sub-Saharan Africa displayed no evidence of any Neanderthal DNA, the three modern humans from outside of Africa showed similar, low amounts (between 1 and 4%) of shared DNA with Neanderthals. These results are compatible with either a deep split within Africa between the population that gave rise to modern Africans and a second one that gave rise to present-day non-Africans plus Neanderthals, or with the hypothesis that there was hybridization between Neanderthals and modern humans soon after the latter left Africa, perhaps in western Asia. Overall, Smith's assimilation model and Bräuer's African hybridization and replacement model of modern human origins appears most compatible with the current morphological and genetic evidence on Neanderthal/modern human diversity. The combination of two distinctive morphologies with molecular evidence for gene flow – albeit partial and limited – points to a lack of congruence between *H. neanderthalensis* and *H. sapiens* defined as morphological species and biological **species**. A similar lack of correspondence has been demonstrated in other primates [e.g., baboons (*Papio*); Jolly 1993]. It has also been noted that the projected time of the divergence between the Neanderthal and modern human lineages (see below) is probably not sufficient to establish reproductive isolation between them (Holliday 2006), but this may not be critical. Molecular and morphological evidence can also contribute to the debate about the origin of *H. neanderthalensis*. Nuclear and mtDNA have been used to generate estimates of the date of divergence of *H. sapiens* and *H. neanderthalensis*. A recent estimate using mtDNA and based on an assumed divergence time of *c.*6–7 Ma for the modern human and chimpanzees/bonobo lineages suggests a divergence date of 660 ± 140 ka (Green et al. 2008). Estimates of the divergence time of the ancestral modern human and Neanderthal populations based on the nuclear genome range from 440 to 270 ka; 440 ka is based on an early (8.3 Ma) divergence date for chimpanzees/bonobos and modern humans and 270 ka on a later (5.6 Ma) divergence date (Green et al. 2010). As far as the fossil evidence is concerned, the debate is about when the distinctive morphology of *H. neanderthalensis* first appeared in the fossil record. There is a growing literature about the distinctive morphology of *H. neanderthalenis* (see Wood and Lonergan 2008 for a brief survey of this literature) but there is no comprehensive review that brings together the cranial, dental, and postcranial evidence. The earliest fossils that most researchers would accept as *H. neanderthalensis* are from OIS 5 (i.e., *c.*130 ka), but beyond that date there is contention. For example, some interpret the fossil evidence from **Swanscombe** (OIS 11) and the **Sima de los Huesos** (possibly as early as OIS 13) as showing enough *H. neanderthalensis*-like morphology to justify those remains being included in *H. neanderthalensis* (Hublin 2009), whereas others see a distinction between these specimens, which they would include in *H. heidelbergensis* and later "true" Neanderthals that they claim do not appear until OIS 6 (Rosas et al. 2006). These latter researchers see the evolution of *H. neanderthalensis* as a two-phase process and Hublin (2009) refers to this as the "two-phase model." This contrasts with Hublin's own "**accretion model**" (Hublin 1998, 2009), which suggests that after the first *H. neanderthalensis* **apomorphies** appear the process of accumulating further apomorphies is so seamless that there is no obvious species boundary after that time that can be identified using morphological evidence. First discovery **Neanderthal 1** (1856).

Holotype As above. Paratypes No paratypes were formally designated. Main sites Many sites in Europe and the Western Asia.

Homo neanderthalensis aniensis **Sergi, 1935** *See* **Saccopastore 1**.

Homo neanderthalensis soloensis Oppenoorth, 1932 *See Homo (Javanthropus) soloensis* **Oppenoorth, 1932**.

Homo nocturnus (L. *homin*=man or human being and *nocturnus*=of the night) This is not a formal species name. *See* **Homo trogolodytes**.

Homo ramidus **(White et al., 1994)** *See Homo* Linnaeus, 1758.

Homo rhodesiensis **Woodward, 1921** (L. *homin*=man or human being and *rhodesiensis*= from what was in 1921 the British protectorate of Northern Rhodesia – now Zambia – a country named after Cecil Rhodes) A pre-modern *Homo* taxon introduced by Sir Arthur Smith **Woodward** (1921) for the cranium and limb bones recovered from the site of **Broken Hill** lead mine at **Kabwe**. Woodward reasoned that a new species of *Homo* was needed for the specimen because it did not fit within any of the existing *Homo* taxa; it was not as primitive as **Homo erectus** and not as derived as either **Homo sapiens** or **Homo neanderthalensis**. Many researchers regard *Homo rhodesiensis* as a **junior synonym** of either an inclusively defined *H. sapiens* or **Homo heidelbergensis**. However, the taxon is used by researchers who see *H. heidelbergensis* as an exclusively European pre-modern *Homo* taxon. First discovery **Kabwe 1** or **E 686** (1921). Holotype As above. Paratypes None. Main sites Kabwe, **Bodo**.

Homo rudolfensis **(Alexeev, 1986) Groves, 1989 sensu Wood, 1992** (etym. L. *homin*= man or human being and *rudolfensis*=after the old name, Lake Rudolf, for Lake Turkana) Hominin species formally established by Alexeev as *Pithecanthropus rudolfensis* and transferred to *Homo* by Groves (1989) and more comprehensively differentiated from *Homo habilis sensu stricto* by Wood (1992). Kennedy (1999) claims that the species name *Homo rudolfensis* is unavailable, but Groves (1989) and Wood (1999) point out that although Alexeev did not follow all the

recommendations of the **International Commission on Zoological Nomenclature** (or ICZN), he did follow its rules and thus the name is available. The differences between *H. rudolfensis* and *H. habilis sensu stricto* include that the crania of the former taxon are wider across the midface instead of the upper face, the mandibles are larger and more robust, the crowns of the postcanine teeth are larger and broader, and the talonids and the root systems of the mandibular premolars of the latter are more complex. Leakey and Walker (1989, pp. 212–13) consider that *H. rudolfensis* makes a more likely ancestor for *Homo erectus* than *H. habilis sensu stricto*. However, not all researchers recognize *Homo rudolfensis* as a separate taxon, with many preferring to include its hypodigm within a more inclusively interpreted *Homo habilis sensu lato*. First discovery KNM-ER 819 (1971). Lectotype **KNM-ER 1470** (1972). Main sites **Koobi Fora, Uraha**. *See also Pithecanthropus rudolfensis*.

Homo saldanensis **Drennan, 1955** *See Saldanha 1*.

Homo sapiens **Linnaeus, 1758** (L. *homin* = man or human being and *sapiens* = to be wise) Hominin species established by Linnaeus to accommodate modern humans. The first discovery of a fossil modern human was made in 1822–3 in Goat's Hole Cave in Paviland, Wales, but the first widely accepted evidence that modern humans were ancient enough to have fossilized representatives came when a series of skeletal remains were discovered by workmen at the Cro-Magnon rock-shelter at Les Eyzies de Tayac, France, in 1868. A male skeleton, Cro-Magnon 1, was made the type specimen of a novel species, *Homo spelaeus* Lapouge, 1899, but it was not long before it was realized that it was not appropriate to discriminate between this material and modern *Homo sapiens* (Topinard 1890, Keith 1912) and further discoveries of *H. sapiens*-like fossils were soon made elsewhere in Europe [e.g., Mladec (1881–1922), Predmosti (1884–1928), and Brno (1885)]. In Asia and Australasia similar fossils were recovered from Wadjak, Indonesia (1889–90), the Zhoukoudian Upper Cave, China (1930), Niah Cave, Borneo (1958), Tabon, Philippines (1962), and the Willandra Lakes, Australia (1968 and thereafter). In the Near East comparable fossil hominins have been recovered from sites such as Skhul (1931–2) and Djebel Qafzeh (1933 and 1965–75). The first African fossil evidence of populations that are difficult to

distinguish from modern humans came in 1924 from Singa in the Sudan. Thereafter comparable evidence has come from Dire-Dawa, Ethiopia (1933), Dar es Soltane, Morocco (1937–8), Border Cave, Natal (1941–2 and 1974), Omo-Kibish, Ethiopia (1967), Klasies River Mouth, Cape Province (1967–8), and Herto, Middle Awash study area, Ethiopia (1997). With the exception of Omo-Kibish (*c*.190 ka, McDougall et al. 2005) and Herto (*c*.170 ka, White et al. 2003), there is no firm evidence to suggest that any of the above sites are likely to be more than 150 ka, and most are probably less than 100 ka. Stringer (2002) provides an excellent overview of the fossil evidence for *H. sapiens* and the ways that the fossil evidence might be used to investigate modern human origins. All the fossil evidence referred to above has been judged to be within, or close to, the range of variation of living regional samples of modern human populations, and thus researchers concluded it is was not appropriate to distinguish it taxonomically from *H. sapiens*. But what are the features of the cranium, jaws, dentition, and postcranial skeleton that are only found in *H. sapiens*, and what are the limits of living *H. sapiens* variation? How far beyond these limits, if at all, should we be prepared to go and still be prepared to refer the fossil evidence to *H. sapiens*? These are simple enough questions, to which one would have thought there would be ready answers, but the assembly of a set of morphological criteria for "modern humanness" has proved to be a surprisingly difficult task. With respect to the cranium, little progress has been made since William (Bill) **Howells**' seminal studies of modern human cranial variation. Howells (1973, 1989) carried out the most comprehensive sampling ($N = 28$ groups and 57 measurements) of modern human cranial measurements and so he has, as well as anybody, captured the essence of modern human cranial variation. He showed that the totality of variation, as measured in **Mahalanobis distance**, among his 28 groups is comparable to the distance that separates all modern human crania from his small sample of *Homo neanderthalensis* crania. Small-bodied modern humans tend to have smaller crania, but overall there is very little among-sample difference in the overall size of the modern human cranium. Qualitatively, Howells comments that modern human crania share a " ... universal loss of robustness," and he goes on to write that within modern humans "variation in shape seems to be largely located in the upper face, and particularly the upper nose and the borders of the orbits"

(Howells 1989, p.83). A few attempts have been made to specify acceptable ranges of morphometric variation for the cranium of *H. sapiens* (e.g., Stringer et al. 1984, Day and Stringer 1991), but even these authors conceded that a sample need only comply with about 75% of the defining characteristics to qualify for inclusion in *H. sapiens*. In a review of variation in regional samples of modern human crania Lahr (1996) emphasized that regional peculiarities should not be incorporated into criteria for inclusion in *H. sapiens*. Lieberman (1998) distilled existing cranial definitions of *H. sapiens*, and suggested that to be regarded as "anatomically modern human," crania need to have "a globular braincase, a vertical forehead, a diminutive brow ridge, a canine fossa and a pronounced chin" (*ibid*, p. 158). Others have suggested that all these features may be related in one way or another to a reduction in facial projection (Spoor et al. 1999) and Lieberman et al.'s (2002) conclusion was that all modern human crania have an unusually globular neurocranium. The most comprehensive survey of modern human dental metrics is still that of Kieser (1990). The equivalent source for non-metrical dental traits is Scott and Turner (1997) and a series of papers by Irish and colleagues (see Irish and Guatelli-Steinberg 2003). The post-canine teeth of modern humans are notable for the absolutely and relatively small size of their crowns, and for a reduction in the number of cusps and roots (Kraus et al. 1969, Hillson, 1996). There are no equivalent book-length surveys of the modern human postcranial skeleton. The postcranial skeleton of *H. sapiens* has limbs that are long relative to the trunk (Holliday 1995), elongated distal limb bones (Trinkaus 1981), a relatively narrow trunk and pelvis, and low estimated body mass relative to stature (Ruff et al. 1997). Many of these traits cause the earliest modern humans (e.g., those from Skhul and Qafzeh) to resemble extant people from hot, arid climates, and the contrasts in postcranial morphology between modern humans and Neanderthals probably have more to do with the uniqueness and distinctiveness of Neanderthal morphology than with the ability of researchers to define the distinctive characteristics of *H. sapiens* (Pearson 2000). The features of the postcranial skeleton that make extant and subrecent modern human populations distinctive have to be described in terms that encompass the full range of modern human climatic and altitudinal adaptations. In summary, compared to their more archaic immediate precursors, modern humans are characterized postcranially by their reduced body

mass (e.g., Kappelman 1996, Ruff et al. 1997), linear physique, and a distinctive pelvic shape that includes a short, stout, pubic ramus, and a relatively large pelvic inlet (Pearson 2004). In Africa there is fossil evidence of crania that are generally more robust and archaic-looking than those of anatomically modern humans, yet they are not archaic or derived enough to justify being allocated to *Homo heidelbergensis* or to *H. neanderthalensis*. Specimens in this category include Jebel Irhoud from North Africa, Laetoli 18 from East Africa, and Florisbad and the Cave of Hearths from southern Africa. There is undoubtedly a gradation in morphology that makes it difficult to set the boundary between anatomically modern humans and *H. heidelbergensis*, but the variation in the later *Homo* (i.e., post-*Homo erectus*) fossil record is too great to be accommodated in a single taxon. Researchers who wish to make a distinction between fossils such as Florisbad and Laetoli 18 and subrecent and living modern humans either do so taxonomically by referring the former specimens to a separate species, *Homo helmei* **Dreyer, 1935**, or they distinguish them informally as "archaic *Homo sapiens*." A few researchers have suggested that *H. sapiens* should be a much more inclusive taxon. For example, because they could see no obvious morphological discontinuity between *H. sapiens* and *H. erectus*, Wolpoff et al. (1994) recommended the boundary of *H. sapiens* be changed to incorporate *H. erectus*, thus echoing a similar proposal made earlier by Mayr (1950). This taxonomy has received little support, but at least the authors made an explicit statement about the scope of the morphology they were prepared to subsume within *H. sapiens*. First discovery Goat's Hole Cave, Wales, UK (1822–3). Holotype No holotype was nominated by Linnaeus. Paratypes None. Main sites Many sites in the Old World and some in the New World.

Homo sapiens idaltu White et al., 2003

A subspecies of *Homo sapiens* established "because the **Herto** hominids are morphologically just beyond the range of variation seen in [anatomically modern *Homo sapiens*], and because they differ from all other known fossil hominids, we recognize them here as *Homo sapiens idaltu*" (White et al. 2003, p. 315).

Homo sapiens shanidarensis Senyürek, 1957 *See* **Shanidar 7**.

Homo sapiens neanderthalensis King, 1864 A name combination introduced by

Campbell (1973) for what most other researchers recognize as the European part of the **hypodigm** of *Homo neanderthalensis*.

Homo sapiens palestinus McCown and Keith, 1934
A name combination introduced by Campbell (1973) for what most other researchers recognize as the Near Eastern part of the **hypodigm** of *Homo neanderthalensis*.

Homo sapiens protosapiens Montandon, 1943
A new name combination suggested by Montandon (1943). *See* **Swanscombe 1**.

Homo sapiens rhodesiensis Woodward, 1921
A name combination introduced by Campbell (1973) for what most other researchers recognize as *Homo rhodesiensis*, or as the African part of either the **hypodigm** of *Homo heidelbergensis*, or the temporally earlier part of the hypodigm of an inclusive interpretation of *Homo neanderthalensis*.

Homo sapiens soloensis
See Homo (Javanthropus) soloensis Oppenoorth, 1932.

Homo sapiens soloensis Oppenoorth, 1932
A name combination introduced by Campbell (1973) for what most other researchers recognize as the temporally later part of the Indonesian hypodigm of *Homo erectus*. *See Homo (Javanthropus) soloensis*.

Homo sapiens steinheimensis Linnaeus, 1758
A new name combination suggested by Weiner and Campbell (1964). *See* **Swanscombe 1**.

Homo-simiadae
In his 1925 paper reporting the discovery of what was then called the Taungs skull, Raymond Dart proposed "tentatively" (Dart 1925, p. 198) that the novel taxon he announced, *Australopithecus africanus*, should be placed in a new family the Homo-simiadae. There is no evidence that Dart seriously pursued the proposal.

Homo soloensis Oppenoorth, 1932
See Homo (Javanthropus) soloensis.

Homo (Javanthropus) soloensis Oppenoorth, 1932
(L. *homin*=man or human being, *Java*=from the name of the island where the fossils were found, Gk *anthropos*=human being and *solo*=refers to the river Solo in whose banks the site is located) Taxon introduced by Oppenoorth (1932) for the calvaria (**Ngandong 1, 2,** and **3**) recovered from **Ngandong** in 1931. In that paper Oppenoorth compares the new calvaria with "Neandertalers," the "Rhodesia-skull," and "**Wadjak** man." He does not specifically refer to Ngandong 1 as the holotype, but it is evident from the text that he regards it as such. In a later paper Oppenoorth (1937) jettisons the subgenus *Javanthropus* and just refers to the taxon as *Homo soloensis*. Since then researchers have either relegated the taxon to a subspecies of *Homo sapiens* as *Homo sapiens soloensis* (e.g., Dubois 1940, Tobias 1985), a subspecies of *Homo neanderthalensis* as *Homo neanderthalensis soloensis* (Weidenreich 1940), or it has been subsumed into *Homo erectus* (e.g., Santa Luca 1980) as either a **junior synonym** of *H. erectus* or as a recognized subspecies of that taxon (e.g., *Homo erectus soloensis*; Mayr 1944). Kaifu et al. (2008) support the use of separate subspecies for the Ngandong and Bapang-AG samples of *Homo erectus* and they also support Santa Luca's conclusion that the Ngandong hominins are more derived than the Bapang-AG sample. First discovery Ngandong 1/Skull 1 (1931). Holotype As above. Paratypes None. Main Sites Ngandong, also called Solo, and some researchers also include the remains from **Ngawi** and **Sambungmachan** in this hypodigm.

Homo spelaeus Lapouge, 1899
See Homo sapiens Linnaeus, 1758.

Homo spyensis Krause, 1909
See Spy.

Homo steinheimensis Berkhemer, 1936
See Steinheim 1.

Homo sylvestris
(L. *homin*=man or human being and *silva*=forest) This is not a formal species name. It was included in the discussion when the genus *Homo* was introduced in 1758 by Carl Linnaeus in the tenth edition of his *Systema Naturae*. It apparently referred to a mythical nocturnal cave-dwelling form from Java but the discovery of *Homo floresiensis* means that *Homo sylvestris* might not have been mythical after all (Collard and Wood 2006). *See Homo troglodytes*.

Homo transvaalensis (Broom, 1936)

(L. *homin*=man or human being and *transvaal*= the name of the state where the fossil site is located) Taxon used by Ernst Mayr (1950) for specimens then subsumed under the following taxa: *Australopithecus africanus*, *Plesianthropus transvaalensis*, *Australopithecus prometheus*, *Paranthropus robustus*, and *Paranthropus crassidens*. Apparently Mayr did not use the taxon with obvious priority, *Australopithecus africanus* Dart, 1925, because the type specimen belongs to a juvenile. In John **Robinson**'s discussion of the relationship between *Homo habilis* and the southern African australoliths (Robinson 1965) he suggests that *Homo* should be divided into two taxa: *H. transvaalensis* for the "tool-using phase of the lineage" and *Homo sapiens* for the "tool-makers who possessed greatly improved means of communication and comparatively complex social structure" (*ibid*, p. 123). Robinson points out that his usage of *H. transvaalensis* differs from Mayr's in the sense that he would exclude the *Paranthropus* taxa because "their basic adaptation...was quite different from that of the *Homo* lineage" (*ibid*, p. 123). For Robinson, the genus *Homo* differed from earlier taxa and from *Paranthropus* because the component taxa had shifted to an omnivorous diet that included "a significant degree of carnivorousness" (*ibid*, p. 123). *See also* **Homo**; *Homo africanus*.

Homo troglodytes Linnaeus, 1758

(L. *homin*=man or human being and *troglodyta*= cave dwellers, from the Gk *troglos*=cave and *dutai*=those who enter) The second of two species of *Homo* coined by Linnaeus in the key tenth edition of *Systema Naturae* (the first species being *Homo sapiens*). He defined it as "*Homo nocturnus*" (=nocturnal humans), and as evidence for the existence of this species he cited "Homo sylvestris Orang Outang" from Jakob de Bondt, an early 17thC Dutch physician in Java and "Kakurlacko" from Kjoeping, a Swedish traveller, and evidence from Dalin, the president of the Swedish Royal Academy. De Bondt may have been describing a real orangutan, or maybe just a hypertrichous woman; Kjoeping and Dalin were describing albinos on the Indonesian island of Ternate, who hid away from the light in caves. Thus, *H. troglodytes* Linnaeus, 1758 is, in part or in whole, a synonym of *Homo sapiens*. Some authors have proposed that the chimpanzee should be placed in the genus *Homo*, in which case the name would be *Homo troglodytes* (Blumenbach, 1785). On the face of it this name combination looks as if it should be preoccupied by *Homo troglodytes* Linnaeus, 1758, but the current (4th) edition of the *International Code of Zoological Nomenclature*, Article 23.9.1, states that "prevailing usage must be maintained" when "the senior homonym has not been used as a valid name after 1899, and the junior homonym has been used for a particular taxon, as it is presumed valid name, then at least 25 works, published by at least 10 authors in the immediately preceding 50 years." It is highly unlikely that the sporadic use of *Homo troglodytes* for the chimpanzee over the past half century qualifies, thus these criteria would appear to have been met; thus, Linnaeus' *Homo troglodytes* can be ignored.

homoiology (Gk *homoios*=similar) Despite the meaning of the Greek word on which it is based, homoiology refers to the part of the **phenotype** whose morphology is influenced by the type and level of activity undertaken (i.e., by environmental factors) (Remane 1961). Homoiology contributes to variation within a **taxon** and in theory homoiology can contribute to phylogenetically misleading morphological similarities among taxa when different genotypes express the same or similar phenotypes because of similar functional demands and not because of shared recent evolutionary history (i.e., they were present in the most recent common ancestor of the taxa concerned). Homoiologies are the result of a more inclusive phenomenon called **phenotypic plasticity**. *See also* **homoplasy**.

homologous (Gk *homologos*=agreeing, from *homo*=the same and *logos*=proportion) A structure, gene, or developmental pathway shared by more than one taxon that was inherited from the most recent common ancestor of the taxa concerned. For example, canine teeth are homologous across all primates but the eyes of vertebrates and cephalopods are not homologous. In genetics, the term homologous refers to **DNA** or protein sequences that share a common ancestry. *See also* **homology**.

homology (Gk *homologos*=agreeing, from *homo*=the same and *logos*=proportion) A term introduced by Richard Owen in 1843 that now refers to similar structures, or genes or developmental pathways found in more than one taxon whose similarity is due to their being inherited from the most recent common ancestor of the taxa concerned. Homologies are the basis for assembling taxa into **clades** or **monophyletic groups**. For example, within the hominin clade the outcome of the debate about whether *Paranthropus* is a monophyletic group rests on resolving whether the distinctive facial and dental morphologies shared by taxa currently included in *Paranthropus* are homologies or are the result of parallel evolution. There are, however, cases where the structure remains homologous but the developmental processes change (e.g., the frontal is derived from the neural crest in some mammalian and bird taxa and from cranial mesoderm in others) but most researchers would retain the homology of the frontal across taxa. The way around this is to include Van Valen's addition of "continuity of information" to the processes that ground homology. Another nuance with respect to homology is that some people recognize degrees of homology whereas others argue that structures are either homologous or not. The term "deep homology" has been introduced for cases where even though organisms do not share the same morphology they do share the same genetic regulatory cascade (e.g., common photoreceptor precursors that result in differently structured eyes in arthropods, squids and vertebrates; Shubin et al. 2009). *See also* **homoplasy**.

homonym (Gk *homonymous*=the same name, from *homo*=the same and *onuma*=name) The same name used for more than one taxon. For example, the binomial *Meganthropus africanus* was introduced by Hans Weinert (1950) for the maxilla found by Ludwig Kohl-Larsen in 1939 at the site that was then called **Garusi** and is now called **Laetoli**. As long as *M. africanus* stays in a genus other than *Australopithecus* then there is no confusion with *Australopithecus africanus* Dart, 1925, but if *M. africanus* is transferred to *Australopithecus* (Senut 1996) then it becomes a secondary homonym. *See* **homonymy**.

homonymy In nomenclature, homonymy refers to a situation when two or more available names have the same spelling but they denote different nominal taxa. The **International Code of Zoological Nomenclature** stipulates that two different species or subspecies deemed to belong to the same genus may not have the same spelling, two different genera or subgenera in the Animal Kingdom may not have the same spelling, and two different families, subfamilies, tribes, or subtribes may not have the same spelling or differ only in suffix (-idae, -inae, -ini, -ina respectively). The rules for homonymy between species or subspecies of the same genus are the most complex because some "different" spellings are deemed to be homonymous (*caeruleus*, *coeruleus*, and *ceruleus*, for example, are deemed to be "the same" spelling). Another category of problems are secondary homonyms. These are identical species or subspecies names that were initially used in different genera, but which later find themselves in the same genus because the two original genera were combined. In that case the junior homonym is dropped (and the species it represents renamed, if it is still thought to be a valid species), but if the two genera are then separated again, the two names are now no longer homonyms. Paleoanthropology provides at least two examples. The first is that, in 1950, Hans Weinert described Laetoli Hominid 1 (at that time known as "the Garusi maxilla") as a new species *Meganthropus africanus*. But most researchers consider **LH 1** to be a member of the genus *Australopithecus* in which case the specific name "*africanus* Weinert" would be a junior secondary homonym of "*africanus* Dart, 1925", for there cannot be two species called *Australopithecus africanus*. Subsequently, in 1978 Johanson and colleagues gave the Laetoli species a new name, *Australopithecus afarensis*, but some researchers preferred to keep the Laetoli species in a different genus, in which case the name *africanus* Weinert would cease to be a junior secondary homonym. The case was referred to the **International Commission on Zoological Nomenclature** (or ICZN), which suppressed *africanus* Weinert. The second example is that of the **Omo 18-1967-18** mandible, which was made the holotype of a new genus and species *Paraustralopithecus aethiopicus* by Arambourg and Coppens in 1968. However, many researchers regard Omo 18-1967-18 as belonging to either *Australopithecus* or to *Paranthropus*. But in 1980 Tobias designated the Hadar hypodigm of *Au. afarensis* species as *Australopithecus aethiopicus*. So under the ICZN rules relating to homonymy, if Omo 18-1967-18 is included in *Australopithecus* it *cannot* retain its species name as *Australopithecus aethiopicus*, but if it is included in *Paranthropus*, it can retain its species name as *Paranthropus aethiopicus*.

homoplasy (Gk *homos*=same and *plasis*=mold) Introduced by Ray Lankester in 1870. He wrote "When identical or nearly similar forces, or environments, act on two or more parts of an organism...the resulting correspondences called forth in the several parts in the two organisms will be nearly or exactly alike. I propose to call this kind of agreement *homoplasis* or *homoplasy*" (Lankester 1870, p. 39). The contemporary interpretation is that homoplasy refers to resemblances between taxa that can be ascribed to processes other than descent from the most recent common ancestor of the taxa concerned and which imply phylogenetic relationships that conflict with the best estimate of phylogeny for the taxa. Several forms of homoplasy have been identified. Analogy and **convergence** are both caused by adaptation to similar environments. Natural selection operates on different developmental processes in the case of analogies but on the same developmental processes in the case of convergences. Parallelisms result from aspects of ontogeny that limit phenotypic diversity, but have no necessary connection with the demands of the environment. Parallelisms, in other words, are by-products of development, not adaptations. A fourth type of homoplasy is reversal, in which, for example, brain size increases and then decreases. Most cases of reversal are probably due to natural selection, but recent work on silenced gene reactivation suggests that some reversals may also be neutral with regard to adaptation. The last form of homoplasy is **homoiology**. Homoiologies are homoplasies attributed to non-genetic factors (e.g., activity-induced bone remodeling). Homoplasies can be mistaken for shared derived similarities or **synapomorphies**, which are the principal evidence for phylogeny. As such, homoplasy complicates attempts to estimate phylogenetic relationships. Indeed, if homoplasies are sufficiently numerous they can prevent a reliable phylogeny from being obtained. Homoplasy is also used to refer to a single trait shared by two or more taxa that was not inherited from the most recent common ancestor of those taxa. *See also* **cladistic analysis; homoiology**.

homozygote (Gk *homos*=the same and *zygotos*=yoked) A **diploid** organism that has the same **alleles** at a particular **locus** is referred to as being homozygous at that locus. For example, an individual who has the genotype AA at the ABO locus is homozygous and has type A blood.

homozygous *See* homozygote.

honing (OE *hān*=stone, from a whetstone used to sharpen knives or tools) Honing refers to teeth that rub together resulting in a sharp edge on one or both of them. In paleoanthropology it is used to refer to the honing complex made up of the upper canine and the anterior lower premolar that is found in apes and that is assumed to be the primitive condition for hominins. The loss of a honing complex is assumed to be related to a shift in behavior where males do not need to fight with their teeth to get access to mates, and thus the presence or absence of this feature in early hominins is much discussed. *See also* **sexual selection**.

Hoogland (Location 25°48.483′S, 28°0.204′E, Gauteng Province, South Africa; etym. Named by the landowner of the farm property). History and general description The site is an abandoned limeworks based on a cave system within Precambrian **dolomite** near the Hennops River, 22.5 km west of Pretoria in the Schurveberg Mountain Range. The site is consistent with field notes and descriptions by Broom (1936) as one of the localities he sampled to generate the 'Schurveberg collection' at the **Ditsong National Museum of Natural History** (formerly the Transvaal Museum), which includes the type specimen of *Papio (Dinopithecus) ingens*. Geologic description The cave system initially formed as a series of deep, vertical shafts that allowed exogenous materials to form a significant talus slope at the base of the entrance. A continuous record of intact *in situ* deposits on the cave wall records five 'flowstone bounded units' (FBUs), of which FBU 2–5 are fossiliferous. Fauna A diverse range of Neogene species, including *Theropithecus oswaldi oswaldi* and bovids similar to those recovered from **Makapansgat** Member 3. Temporal span Paleomagnetic sampling across the Hoogland deposits recorded a series of polarity reversals, indicting that the deposits span an extremely long time period. The densest fossil deposits (FBU4) occur during a normal polarity period. Initial analysis suggests deposition of FBU4 at the end of the Gauss Chron (C2An.1n; 3.03–2.58 Ma) normal polarity period given the morphology of the recovered bovids from the site. How dated? Biostratigraphy and magnetostratigraphy. Hominins found at site None. Archeological evidence found at site None; Key references: Geology and fauna Adams et al., 2010.

Hopefield *See* **Elandsfontein**.

Hopefield 1 *See* **Saldanha 1**.

hornfels A hornfels is a fine-grained, dark-colored metamorphic rock, formed by contact metamorphism adjacent to an igneous intrusion. Hornfels typically forms from high-temperature baking and recrystallization of a pre-existing rock, a process that usually removes features such as bedding planes. Its appearance is often similar to fine-grained basalt but it is much more brittle than basalt. This can make it a poor choice as raw material for **knapping** but it is selected where better options are unavailable (e.g., **Florisbad**).

horst Refers to an area of relative uplift between a pair of normal faults; the adjacent down thrown area is known as a **graben**. These faults tend to form in conjugate pairs within heavily faulted areas (e.g., parts of the floor of the East African rift valley) develop a landscape with relatively uplifted areas adjacent to relatively low-lying areas (e.g., the faulted margins of the **Olorgesailie** paleolake basin). This can lead to substantial environmental heterogeneity over relatively short lateral distances. Horsts that form after deposition of sedimentary sequences can expose otherwise buried formations (e.g., **Bouri Formation** in the Middle Awash study area).

Hortus (Location 43°47'N, 3°50'E, France; etym. originally spelled Ortus, of unknown etymology) History and general description A cave site located high in a limestone massif in southeast France. Excavations by a series of professional and amateur scientists since 1908 focused on the superficial Late Bronze Age deposit. In the 1950s, Cazals and Audibert discovered deeper **Mousterian** layers and formal excavations of these layers were carried out by Henry de Lumley between 1960 and 1964. Temporal span and how dated? Conventional **radiocarbon dating** on bones from the Mousterian levels was unsuccessful. **Geochronology** suggests that the main deposition occurred during the Würm I and II cold phases. Hominins found at site The remains of between 20 and 36 individuals were recovered; there was no evidence of intentional burial. A high proportion of the individuals were young adults. The majority of the bones are maxillae or mandibular fragments, and de Lumley suggested they may be the result of cannibalism, but none show unambiguous cutmarks. Archeological evidence found at site Mousterian artifacts in the lower levels. Key references: Geology, dating, and paleoenvironment de Lumley 1972; Hominins de Lumley 1972, Mann and Trinkaus 1973; Archeology de Lumley 1972.

Hotelling T^2 test A generalization of the t test used to test whether a vector of means of two or more continuous variables differ between two groups. For example, it can be used to test whether the **centroid** of two sets of cranial measurements differ from each other.

housekeeping gene A gene that is expressed in connection with the basic functions of a cell. For example, the genes involved with **transcription**, **translation**, energy production, and cell division would all be regarded as housekeeping genes.

Howell, Francis Clark (1925–2007) Francis Clark Howell was born in Kansas City, Missouri, USA, and grew up on a farm in Kansas. He was introduced to human prehistory by reading Henry Fairfield Osborn's *Men of the Old Stone Age* (1916) and William W. **Howells'** *Mankind so Far* (1944). During WWII Howell served in the US Navy in the Pacific from 1944 to 1946 and after his discharge he spent a short period at the **American Museum of Natural History** in New York where he met American paleontologist George Gaylord Simpson, as well as Franz **Weidenreich** and Ralph **von Koenigswald**. Howell wrote that "he had the humbling pleasure at the age of not yet 21, to meet von Koenigwald" and that the latter "was immensely kind to me, as was Franz Weidenreich, at a time that I only hoped for a career in human evolutionary studies" (Howell, 1994, p. 18). This experience strengthened Howell's conviction to pursue a career in paleoanthropology and he entered the University of Chicago where he studied anthropology with Sherwood **Washburn** and Robert J. Braidwood. After completing his BA in 1949 Howell decided to stay at the University of Chicago for graduate school where he conducted research on *Homo neanderthalensis*. Howell completed his MA in 1951 and his PhD in 1953. Howell claimed that it was the influence of the Spring Seminar Series organized by Sherwood Washburn that convinced him of the importance of the relationship between evolutionary biology, paleontology, and ecology. After completing his PhD Howell taught anatomy as an instructor at Washington University in St. Louis from 1953 to

1955 He also began to spend a portion of each year traveling to Europe to conduct excavations at prehistoric sites, and the latter led to early publications on Paleolithic archeology and fossil mammals. He also expanded upon his earlier research on Neanderthals by integrating the principles of the **modern evolutionary synthesis** with what was known about the climate of Europe during the Pleistocene, to explain the anatomical differences between the "classic" Neanderthals of Western Europe and the Neanderthals found in **Krapina** and in the Near East. In 1955 Howell accepted a position as Professor of Anthropology at the University of Chicago, but he continued to visit important prehistoric sites and museum collections throughout the world. Howell drew upon his broad knowledge of paleoanthropology to publish *Early Man* (1965), a popular synthetic work that appeared in revised editions in 1968 and again in 1980. Howell carried out important fieldwork at the **Acheulean** sites of **Isimila** in Tanzania, **Ambrona**, and **Torralba** in Spain, and at the **Lower Paleolithic** site of Yarimburgaz in Turkey. However, his most influential fieldwork was carried out in the **Lower Omo Valley** in southern Ethiopia. The Borg de Bosas expedition in 1902 had shown that region to be fossiliferous and Camille **Arambourg** had assembled a substantial collection of mammalian fossils in 1933. Clark Howell had visited the Lower Omo Valley in 1959 and it was on the basis of this experience that in 1966 he was invited to join a team of researchers assembled by Louis **Leakey** and Camille Arambourg to re-explore the Lower Omo Valley. Howell emerged as one of the leaders of what came to be called the **International Omo Research Expedition** (or IORE), which brought together specialists from a variety of disciplines including geology, paleontology, archeology, and paleoanthropology between 1967 and 1976. The IORE was important because the **Omo** deposits, a mixture of river sediments and lava flows, were found to represent close to 3 Ma of evolutionary history during the **Plio-Pleistocene**. The presence of numerous animal fossils and volcanic layers dateable using the then newly devised **potassium-argon dating** and **magnetostratigraphy** methods meant that the Omo deposits could be used to calibrate the evolutionary and paleoenvironmental history of the past several million years. The site thus became an important focus of attempts to develop a comprehensive and reliable biostratigraphy of the Plio-Pleistocene. This careful

attention to stratigraphy and dating allowed the team to generate an absolute chronology for the evolution of many mammal groups. For example, correlation of the abundant fossil pig species with absolute dates from the volcanic deposits in the **Shungura Formation** at the Omo enabled researchers to date deposits elsewhere that contained similar fossil pigs but where reliable absolute dates were not available and this eventually helped resolve the **KBS tuff dating controversy**. Howell left his position at the University of Chicago in 1970 to move to the Department of Anthropology at the University of California at Berkeley, where he joined Sherwood Washburn, Theodore McCown, Desmond **Clark**, and Glynn **Isaac**, and together they formed an impressive human origins research group. There he co-founded with Tim **White** the Human Evolution Research Center in the Department of Integrative Biology. It is difficult to exaggerate the influence Howell had on paleoanthropology. He had an encyclopedic knowledge of both the human fossil record and the more general mammalian fossil record, especially that of carnivores and Old World monkeys. His seminal review of the Neanderthals (Howell 1951) stimulated much of the subsequent research eventually showing that Neanderthals were unlikely to have been the direct ancestors of modern Europeans and his review of hominin evolution in Africa was equally influential (Howell 1978). Finally, his insistence on scientific rigor meant that the Omo Shungura Formation became the reference for African mammalian biostratigraphy. Howell taught at Berkeley until his retirement from formal teaching in 1991. Thereafter, he was appointed emeritus professor, a position he held until his death in 2007.

Howells, William White (1908–2005)

William "Bill" Howells who was born in New York City obtained his BS in anthropology from Harvard University in 1930. He stayed on at Harvard to study with Earnest Hooton and completed his PhD in physical anthropology in 1934 with a dissertation entitled *The Peopling of Melanesia as Indicated by Cranial Evidence from the Bismarck Archipelago*. After graduating Howells accepted a research position at the **American Museum of Natural History** where he remained until 1937 when he left to become an Assistant Professor in the Department of Sociology and Anthropology at the University of Wisconsin. Howells served as an officer in Naval Intelligence from 1943 to 1945 and at the end of the war he returned to the University of Wisconsin. Howells put this time to good use and his published first book, *Mankind So Far* (1944), was an overview of human evolution accessible for a general audience, that was followed by *The Heathens* (1948), a study of "primitive" religions, and *Back of History* (1954), a popular survey of prehistoric archeology. These and his most influential popular book, *Mankind in the Making* (1959), reflect Howells' broad knowledge of cultural anthropology, physical anthropology, and archeology. When Earnest \Hooton died Howells accepted an offer to succeed his mentor as Professor of Anthropology at Harvard in 1954. Howells, assisted by his wife Muriel, expanded upon his doctoral work by collecting measurements of crania from populations in **Oceania**, Europe, Africa, and Asia to understand the pattern and extent of the metrical variation in modern human populations. The results of his study, one of the first examples of the application of **multivariate statistics** in biological anthropology, convinced Howells that all living modern human populations represented a single homogeneous species and that the **Neanderthals** represented a distinctly different species. The results of these pioneering analyses were published in *Cranial Variation in Man* (1973) and the metrical data set accumulated by Howells has been used by many researchers. Howells used the conclusions drawn from this research to criticize the **Neanderthal phase hypothesis** of human evolution advocated by Aleš **Hrdlička** and others as well as the multiregional continuity model (later called by Howells the **candelabra model**) of human origins proposed by Franz **Weidenreich**, which argued that geographically distinct populations of anatomically modern humans had evolved relatively independently throughout the Old World from an original pre-*Homo erectus* population that had left Africa. He was also very critical of the ideas of Carleton **Coon** who had argued that regional populations of *H. erectus* evolved into the modern human races of *Homo sapiens* and therefore the races had been biologically separate for a long period. Howells instead promoted a view he called the Noah's ark hypothesis where there were several migrations of hominins (only later did he specify that Africa was the most likely origin for all of these migrations) with the last migration comprising modern humans that then replaced the populations of more archaic hominins that had formed the earlier migrations. The current version of this idea is referred to as the **out-of-Africa hypothesis**. During the 1960s Howells also helped to organize the Harvard Solomon Islands Project, an interdisciplinary study of the interaction of culture, natural selection, and disease and Howell's own emphasis on the need to integrate biological anthropology, cultural anthropology, linguistics, and archeology are apparent in *The Pacific Islanders* (1973). That same year he also published *Evolution of the Genus Homo* (1973), a short overview of current theories of human origins. He was awarded the Viking Fund Medal in 1954 and was elected to the National Academy of Sciences in 1967. In 1978 he was awarded the Distinguished Service Award of the American Anthropological Association, and the American Association of Physical Anthropologists presented him with the Charles Darwin Lifetime Achievement Award in 1992. Howells retired from Harvard in 1974 but continued to conduct research and to write. In 1975 Howells was one of the first American paleoanthropologists to be allowed to visit China. He returned to the subject of craniometry and the evolution and migration of human populations in *Skull Shapes and the Map: Craniometric Analyses in the Dispersion of Modern Homo* (1989) and *Who's Who in Skulls: Ethnic Identification of Crania from Measurements* (1995). His last popular book that surveyed human origins, *Getting Here: The Story of Human Evolution*, was published in 1993, with a second edition in 1997.

HOX genes An abbreviation of *homeobox*. *See homeobox* genes.

Hoxne *See* **Acheulean**.

HP Acronym for the Howieson's Poort industry.

hpAfrica *See Helicobacter pylori*.

hpAsia *See Helicobacter pylori*.

hpMaori *See Helicobacter pylori*.

hpSahul *See Helicobacter pylori*.

Hrdlička, Aleš (1869–1943) Aleš Hrdlička was born in Humpolec, Bohemia (now the Czech Republic), but he immigrated with his parents to the USA in 1881. He entered the New York

Eclectic Medical College in 1889 to study medicine and graduated with honors in 1892. He continued his medical education at the New York Homeopathic College from 1892 to 1894 and after graduation he worked as a physician at the New York Homeopathic Hospital for the Insane and it was here that he became interested in anthropometry. In 1895 Hrdlička joined the newly founded Pathological Institute of the New York State Hospitals where he remained until 1899. To prepare himself for his new position, Hrdlička spent several months in Paris during 1896 studying anthropometry with the French physical anthropologist Léonce-Pierre Manouvrier at the École d'Anthropologie and the Laboratoire d'Anthropologie de l'École Pratique des Hautes Études. When he returned to the Pathological Institute, Hrdlička began to collect anthropometric data. In the course of these studies he met Frederic W. Putnam of the **American Museum of Natural History** who arranged for Hrdlička to join an expedition to Mexico in 1898 and this initiated his life-long interest in the native peoples of the Americas. Hrdlička resigned from the Pathological Institute in 1899 and joined Putnam at the American Museum of Natural History where between 1899 and 1902 he conducted several large anthropometric studies of the native peoples of the American Southwest and northern Mexico. Hrdlička's extensive research experience enabled his appointment as assistant curator of the Division of Physical Anthropology at the **National Museum of Natural History, Smithsonian Institution** in Washington DC in 1903, and then his appointment as curator in 1910. While at the Smithsonian, Hrdlička assembled an extensive collection of modern human skeletal material from around the world and thus made the Smithsonian a center for research in physical anthropology. He traveled widely, visiting many of the most important sites where hominin fossils had been discovered and examining many of the fossils himself and he was an ardent promoter of the science of physical anthropology, both in America and abroad. As part of this effort Hrdlička founded first the *American Journal of Physical Anthropology* in 1918, serving as its editor from 1918 to 1942, and then the American Association of Physical Anthropologists in 1930. The latter provided an institutional organization for the physical anthropology and contributed to its professionalization. For much of his life Hrdlička studied the question of how and when the first humans had arrived in the New World. He spent many years inspecting much of the evidence that suggested the first humans arrived in the Americas before or during the "Glacial Epoch" (i.e., the Pleistocene). In *The Skeletal Remains Suggesting or Attributed to Early Man in North America* (1907) and *Early Man in South America* (1912) Hrdlička concluded that this evidence was suspect and that it was more likely that the first inhabitants of the New World had not arrived until the end of the Glacial Epoch and he remained a staunch supporter of this view throughout his life. These views were reiterated in *The Quest for Ancient Man in America* (1937) where Hrdlička argued that modern humans arrived in the New World from northeast Asia via the Bering Strait from where they migrated south to populate North and South America. In 1927 Hrdlička published an article in the *Journal of the Royal Anthropological Institute* entitled "The Neanderthal Phase of Man," in which he argued that all the modern human races had a common origin and that anatomically modern *Homo sapiens* evolved from Neanderthals. The extensive investigation of the hominin fossil evidence that formed the foundation for Hrdlička's theory was reflected in the *The Skeletal Remains of Early Man*, published in 1930, which at the time was one of the most comprehensive expositions of the fossil evidence for human evolution. Despite his many contributions to modern paleoanthropology, Hrdlička's methodology and outlook were rooted more in the 19thC than in the 20thC. He had little training in biology or evolutionary theory and the revolutions that were occurring in genetics largely passed him by. Some of his theories, especially the Neanderthal origin of modern humans, were controversial and not widely accepted, although his analysis and description of individual fossil specimens were of a high quality. Through his work at the Smithsonian and through works such as *Physical Anthropology: Its Scope and Aims* (1919), *Anthropometry* (1920b), and *Practical Anthropometry* (1939) Hrdlička helped to expand the scientific scope of physical anthropology as a discipline in the USA. Hrdlička retired from his position at the Smithsonian in 1942 and spent the last years of his life studying the native peoples of Alaska and the Aleutian Islands in search of support for his theories about the peopling of the Americas.

HSBC Acronym for Hunter–Schreger band curvature, which is defined as the "number of cervically located **diazone**s and **parazone**s" crossed by the projection to the enamel-dentine junction of a line drawn "parallel to the H(unter) S(chreger) B(and)s in the outer enamel" (Beynon and Wood 1986, p. 182). *See also* **enamel microstructure**; **Hunter–Schreger bands**.

HSBW Acronym for Hunter–Schreger band width, which is defined as the average width of "at least ten alternating **parazone**s and **diazone**s in the occlusal one-third of the lateral enamel" (Beynon and Wood 1986, p. 181). *See also* **enamel microstructure**; **Hunter–Schreger bands**.

HTU Acronym for **hypothetical taxonomic unit** (*which see*).

Hublin, Jean-Jacques (1953–) Jean-Jacques Hublin was born in Mostaganem in former French Algeria. He gained a BA in Geology in 1975, an MA in Vertebrate Paleontology and Human Paleontology in 1976, and a third-cycle Doctorate in Vertebrate Paleontology and Human Paleontology in 1978, all at the University of Paris VI where his mentors were R. Hoffstetter and Bernard **Vandermeersch**. While working on his PhD Hublin focused on the anatomy of *Homo neanderthalensis* and one of his early papers was a cladistic investigation of the cranial morphology of *H. neanderthalensis* (Hublin 1978). In 1991 he was awarded a State Doctorate (Habilitation) in Anthropology from the University of Bordeaux. Hublin has conducted field work in North Africa (**Ternifine**, **Jebel Irhoud**), and Europe (**Zafarraya**). Much of his work has been on **Middle Pleistocene** and **Late Pleistocene** hominins, and his research demonstrated that *H. neanderthalensis* was older than initially thought, and that Neanderthals were unlikely to be ancestral modern humans. He is the contemporary proponent of the **accretion model** for the evolution of *H. neanderthalensis* (Hublin 1998). His research also suggests that assemblages considered as early **Upper Paleolithic** (such as the **Châtelperronian**) were actually produced by late Neanderthals, maybe as a result of an acculturation process (Hublin et al. 1996). He has also worked extensively on the origin of modern humans in North Africa and his research has provided evidence of *Homo sapiens* beyond sub-Saharan Africa and put the temporal origin of *H. sapiens* between 300 and 100 ka (Hublin and Tillier 1981,

Hublin 1992, Smith et al. 2007). He was also one of the first researchers to recognize the potential of imaging for the investigation of **life history** (e.g., Coqueugniot et al. 2004). From 1976 to 1978 he was under research contract with the French Ministry of Education and Research and then worked for the **Centre National de la Recherche Scientifique** (National Center for Scientific Research) from 1980 to 2000. In 2000 he succeeded Bernard Vandermeersch as Professor of Anthropology at the University of Bordeaux and in 2004 he was appointed Professor at the **Max-Planck Institute for Evolutionary Anthropology**, Leipzig, Germany, and the Founding Director of the Department of Human Evolution.

Hulu 1 *See* **Tangshan Huludong**.

human (L. *humanus*=people) This term is used in many ways but it should be reserved for referring to either the species taxa within the genus *Homo*, or to the species *Homo sapiens*, or to individuals within the those species.

human accelerated regions (or HARs) Regions of the **genome** that show relative conservation in vertebrates but which show changes in the lineage leading to modern humans subsequent to the most recent common ancestor of modern humans and chimpanzees and bonobos.

human anatomical terminology *See* **anatomical terminology**.

Human Genome Project An international, publically funded, effort to produce the first sequence of the modern human genome initiated in the late 1980s by the US Department of Energy and the National Institutes of Health. The first published **draft sequence** (International Human Genome Sequencing Consortium 2001) covered 94% of the modern human genome. At the time of its publication only approximately 25% was considered **finished sequence**, but at present approximately 99% of the human genome sequence is finished sequence.

human leukocyte antigens *See* **major histocompatibility complex**.

Human Origins Program *See* **National Museum of Natural History, Smithsonian Institution**.

humerofemoral index Ratio of the lengths of the humerus and femur calculated as 100×(humerus length/femur length). Like the **intermembral index** this index reflects major differences in locomotor strategies among primates. Because it requires fewer limb bones to calculate, it is more commonly used in studies of the hominin fossil record. Due to its relatively short femora *Australopithecus afarensis* (i.e., A.L. 288-1) has a humerofemoral index intermediate between Jungers (1982) and significantly different from Richmond et al. (2002), the high index of great apes and the low index of modern humans. The earliest evidence of a modern human-like humerofemoral index comes from associated limb bones from Bouri, **Middle Awash study area**, Ethiopia. By 1.8 Ma early *Homo* at Dmanisi has a modern human-like humerofemoral index, reflecting a relatively long femur that confers the locomotor benefits of speed and energetic efficiency. *See also* **A.L. 288-1**; **Dmanisi**.

Hungsugul *See* **Turubong**.

Hunsgi-Baichbal valleys (Location 16°30′N, 76°35′E, India; etym. named after two towns) History and general description Over 200 **Acheulean** archeological sites located in two separate but interconnected valleys in the Gulbarga District, Karnataka, southern India. Surveys and excavations have been performed by K. Paddayya since the late 1960s, recovering information about Acheulean behavior. Joint research since the 1980s, with Michael Petraglia, examined the formation of the archeological sites and the stone tool technology. Temporal span and how dated? **Uranium-series dating** at several localities suggests an age of more than 350 to 166 ka. **Electron spin resonance spectroscopy dating** at **Isampur Quarry** suggests an age of 1.27 Ma (linear-uptake model) or 730 ka (early-uptake model). Archeological evidence found at site Acheulean localities contain handaxes, cleavers, choppers, picks, knives, polyhedrons, scrapers, discoids, and unifaces. A variety of local materials were used for tool manufacture including limestone, quartzite, granite, dolerite, basalt, and schist. Variability in hard-hammer and soft-hammer techniques were observed between the localities. Technological variations, depositional contexts, and the range of chronometric ages suggest an earlier and later stage of the Acheulean. Specialized localities include a cluster of handaxes interpreted as a cache and dense concentrations of limestone and basalt manufacturing debris at geological outcrops, interpreted as sources for stone-tool manufacture. Based on the distribution of archeological occurrences, the inferred paleomonsoonal climate, and the identification of seasonal springs and ponds, an Acheulean settlement model was proposed by Paddayya, consisting of wet-season dispersal of groups and dry-season aggregation. One notable discovery was the Isampur Quarry, a limestone source for the procurement and manufacture of stone tools. Key references: Geology, dating, and paleoenvironment Paddayya et al. 2002, Jhaldiyal 2006; Archeology Paddayya 1982, Paddayya and Petraglia 1993, Petraglia et al. 1999, Shipton et al. 2009a, 2009b.

hunter–gatherer A group of people, or a member of such a group, whose primary subsistence strategy is based on the hunting or gathering of wild plant and animal resources. Prior to the advent of **domestication**, virtually all hominin groups were hunter–gatherers and it is presumed that our more recent hominin ancestors (i.e., members of the genus *Homo*) had a hunter–gatherer subsistence economy. Present-day hunter–gatherers are of particular interest to paleoanthropologists because they provide an analogue for understanding past hominin subsistence strategies, behavioral patterns, and social interactions. For example, modern hunter–gatherers are known to obtain the majority of animal remains through hunting although they occasionally obtain meat from **passive scavenging** or **confrontational scavenging**. In an archeological context the identification of a subsistence strategy focused on scavenging rather than hunting therefore falls outside the realm of modern human behavior, which has implications concerning hominin social interactions and behavioral complexity. Ethnographic observations indicate that plant remains comprise the majority of food resources exploited by modern hunter–gatherers. In turn, this implies that plant resources likely made up a large portion of past hominin diets despite a strong paleoanthropological research bias towards the study of animal remains. *See also* **forager**.

Hunter–Schreger bands Structural microscopic features of tooth enamel that result from contrasting patterns of enamel prism decussation and named after two of the pioneers of dental microstructure, John Hunter (1778) and Heinrich

Schreger (1800) (reviewed in Boyde 1964 and Lynch et al. 2010). These bands are formed by successive layers of enamel prisms running in different directions, a phenomenon that results in patterns of light scattering that are particularly evident under reflected or polarized light microscopy. Attempts have been made to quantify differences in the degree of enamel prism decussation among hominins (e.g., Beynon and Wood 1986) and to model their formation (e.g., Osborn 1990). The latter has proven to be difficult because of the three-dimensional nature of these features. *See also* **diazone**; **enamel microstructure**; **parazone**.

hunting The pursuit and capture of wild animals for food or trade. Hunting is the primary carcass acquisition strategy of contemporary hunter–gatherer groups and is generally considered to be a basic constituent of aboriginal modern human behavior. In turn, the extent to which our hominin ancestors practiced a subsistence economy based on hunting, as opposed to scavenging, is of particular interest to paleoanthropologists since it implies complex social interactions and cooperation within hominin groups, but cooperative hunting is not confined to hominins, or indeed to primates. *See also* **carcass transport strategy**.

hunting and gathering *See* **hunter–gatherer**; **forager**.

Huntington (George S.) Collection Taxon *Homo sapiens*. History Collection of modern human skeletons assembled by anatomist George S. Huntington (1861–1927) based on cadavers used for dissection at the College of Physician and Surgeons, New York City. The Huntington Collection comprises the remains of immigrants and US citizens dying in New York City between 1893 and 1921. Many of the approximately 4054 individuals in this collection have documented sex, age, nationality, cause of death, and in some cases occupation and post-mortem anthropometrics. However, being from dissection cadavers, the skeleton is not always complete and few skulls are complete. It should also be noted that approximately 14% of the collection is commingled due to decades-old cataloguing errors. The demographic profile of this collection from the registered database of 4054 cataloged individuals is 71% male and 23% female. European and US Whites comprise 66.5% of the series,

while US Blacks are 7.6%, and individuals of unknown ancestry are 25.9%. Approximately 64% are immigrants (Irish 15%, Germans 10%, Italians 8%), and 36% are identified as US-born. The mean age at death is around 46 years (ranging from 5 to 96 years, with very few individuals under the age of 19). The collection is unique in the numerous excellent examples of trauma and occupational stress as well as infectious and congenital diseases. In the 1920s, before Huntington's death, one of his students, Aleš **Hrdlička**, arranged for the collection to be transferred to the Department of Anthropology at the **National Museum of Natural History, Smithsonian Institution**; this transfer was completed in 1927. Curator/contact David Hunt (e-mail huntd@si.edu). Postal address Department of Anthropology, National Museum of Natural History, Washington DC, USA. Website http://anthropology.si.edu/cm/terry.htm. Relevant references Hunt and Albanese 2005, Hrdlička 1937.

Huxley, Thomas Henry (1825–95) *See Evidence as to Man's Place in Nature*.

HWK Acronym for Henrietta Wilfrida Korongo, a locality at Olduvai Gorge. *See* **Olduvai Gorge**.

Hyaenidae The mammalian family containing the hyenas. *See* **Carnivora**; **hyenas**.

hyaline cartilage (Gk *hyalos*=glass and L. *cartilago*=gristle) *See* **cartilage**; **joint(s)**.

hybrid (L. *hybrida*, var. of *ibrida*=mongrel) The offspring resulting from the interbreeding of individuals from two different recognized species (interspecific hybridization) or from the interbreeding of individuals from two different subspecies, "races," demes, or allotaxa (intraspecific hybridization). Hybridization can occur in areas where the ranges of two species overlap. For example, it has been demonstrated in the Awash region of Ethopia with respect to the gelada (*Theropithecus gelada*) and the Hamadryas baboon (*Papio hamadryas*) (Jolly et al. 1997) and in connection with the Hamadryas baboon and the olive baboon (*Papio anubis*) (Phillips-Conroy et al. 1991, Bergman and Beehner 2004). Hybridization is a feature of **reticulate evolution** and Jolly (2001) has been a proponent of the hypothesis that during the course of hominin evolution species may have arisen as the result of hybridization as well as by classical processes such as **allopatry**. *See also* **reticulate evolution**; **speciation**.

hybrid zone A geographic region of contact and hybridization between two populations that are genetically distinct. Hybrid zones can be fleeting or stable; the latter has been demonstrated for well-studied Old World primates (Jolly 2001). *See also* **hybrid**; **hybridization**.

hybridization (L. *hybrida*, var. of *ibrida*= mongrel, Gk *izein*=to make) Evolution Interbreeding between individuals from genetically differentiated lineages or between local demes. The role of hybridization (and hence **reticulate evolution**) in human evolution is not well understood although there is considerable debate over whether certain groups (e.g., modern humans and Neanderthals) hybridized. By analogy to other living primates it has been argued that hybridization among distinct hominin taxa is likely (Jolly 2001) and emerging evidence from the broader literature also supports this conclusion (e.g., Arnold and Meyer 2006, Arnold 2009), but our understanding of how to detect this in the skeleton is still poor. Recent work suggests that hybrids may be detectable by the presence of skeletal developmental anomalies – especially sutural and dental abnormalities – or other signatures of breakdown in the coordination of early development (Ackermann et al. 2006, Ackermann 2007, Ackermann and Bishop 2010), but these traits have not been shown to have a fitness advantage or disadvantage. Genetics and molecular biology A molecular biology technique where two complementary strands of **DNA** are joined together. *See also* **DNA hybridization**; **hybrid**.

hydrogen The lightest of all elements, hydrogen is being increasingly used in **stable-isotope biogeochemistry**. There are two common hydrogen isotopes: ^{1}H (more common) and ^{2}H (less common); ^{2}H is commonly referred to as **deuterium**. Both of these isotopes are stable and do not decay. *See* ^{2}H/^{1}H; **stable isotopes**; **hydrogen isotopes**.

hydrogen isotope analysis *See* **gas chromatography isotope ratio mass spectrometry**.

hydrogen isotopes Stable isotopes of hydrogen in water provide a useful tracer of changes in the hydrological cycle and they are increasingly being incorporated into general circulation models (or GCMs). Hydrogen in cellulose from tree rings can be used to reconstruct past climate variations even in tropical trees which often lack the annual rings that facilitate dendroclimatology in temperate climates. Tree ring records of past climates can be used to reconstruct climate in the **Holocene** and hydrogen isotopes in leaf wax biomarkers can be used to reconstruct climate across millions of years.

hydroxyapatite One of the two common names (the other is apatite) for the dominant salt (also called the mineral phase) in **bone**, **cementum**, **dentine**, and **enamel**. *See* **apatite**.

hyenas (Gk *huaina*=swine) Hyenas belong to the mammalian family Hyaenidae, within the order **Carnivora**. The group includes the striped (*Hyaena hyaena*), the spotted (*Crocuta crocuta*), and the brown (*Parahyaena brunnea*) hyenas and the aardwolf (*Proteles cristata*). Hyenas are of particular interest from a paleoanthropological perspective for two reasons. First, in the **Plio-Pleistocene** hyenas were a prominent member of the **carnivore guild** and if, as some researchers suggest (Walker 1984), hominins became members of that guild as **scavengers** and/or hunters, then hyenas and hominins would have been in competition. The second reason is that like modern humans and early hominins hyenas are known to accumulate large collections of bones (i.e., **faunal assemblages**). In the absence of rigorous taphonomic investigations, hyena-accumulated bone concentrations can potentially be misidentified as the product of hominin activities, leading to erroneous behavioral interpretations. However, taphonomic observations suggest that hyena-accumulated faunal assemblages can be distinguished from hominin-accumulated assemblages because the former include large numbers of juvenile hyenas, high frequencies of tooth-marked long-bone shaft fragments, and a preponderance of limb bones with complete shafts, but lacking their epiphyses. In addition to accumulating faunal assemblages, hyenas are also known to modify or destroy skeletal remains discarded by modern humans and probably by early hominins, a phenomenon known as **carnivore ravaging**.

Hylochoerus (Gk *hylo*=woodlands, forest; *choerus*=pig) The giant forest hog, *Hylochoerus*

345

meinertzhageni is the sole species of the genus. The largest of all living pigs, it is thought to be a descendent of the *Kolpochoerus* clade. The earliest known representative is from the late to middle Pleistocene. Giant forest hogs prefer closed habitats when possible, but in parts of their range where forests and woodlands have been disturbed they have been seen to form large groups and graze in the open. Their range is restricted to tropical central Africa and they are patchily distributed from Guinea to Kenya. *See also* **Suidae**.

hyoid (Gk *hyoeides*=shaped like the Greek letter hypsilon) Bone palpable in the midline of the neck, between the mandibular symphysis and the manubrium of the sternum. It consists of a central body and bilaterally the greater and lesser horns. With the mylohyoid muscle it forms the floor of the mouth and the base of the tongue. The strap muscles of the neck connect it to the thyroid cartilage and to the manubrium of the sternum. Hyoid bones occur occasionally in the hominin fossil record (e.g., **Dikika study area, Kebara, Sima de los Huesos**). Some researchers (e.g., Arensburg et al. 1989) claim that the morphology of the hyoid can be used to make inferences about the capacity for spoken language, but this claim has been disputed.

hypermorphosis (Gk *hyper*=over or beyond, *morphe*=form or shape) A **heterochronic** process that involves the extension of a shared ancestral trajectory to new size ranges in the descendants. Size and shape covariation found in the ancestors remains similarly associated in hypermorphic descendants; hypermorphosis requires a change to larger size of descendants to preserve the ancestral size/shape relationship. Hypermorphosis can occur two ways: by increasing the ancestral growth rate so that descendants grow beyond the ancestral size by the same age of maturity (rate hypermorphosis) or by extending the growth period of ancestors so that descendants maintain similar growth rates but reach maturation at later ages (time hypermorphosis).

hyperostotic *See* **non-metrical traits**.

hypervariable region *See* **mitochondrial DNA**.

hypocone (Gk *hypo*=under, below and *konos*=pine cone) The term proposed by Osborn (1888) for the main cusp on the distolingual aspect of a maxillary (upper) mammalian molar. In some teeth its tip is lower than the tips of the **meta-, para-,** and **protocone** hence the prefix meaning "below." It is part of the **talon** (syn. endocone, distolingual cusp, tuberculum posterium internum).

hypoconid (Gk *hypo*=under, below and *konos*=pine cone) The term proposed by Osborn (1888) for the main buccal cusp situated distal to the **protoconid** on a mandibular (lower) (hence the suffix "-id") molar tooth crown. Its prefix is a misnomer in hominins because its tip is not usually lower than those of the other main cusps. It forms part of the **talonid** component of a mandibular postcanine tooth crown (syn. teleconid, distobuccal cusp, tuberculum mediale externum).

hypoconule The term proposed by Osborn (1888) for an accessory cusp on the occlusal surface of a maxillary molar that others more commonly refer to as a distostyle. *See* **distostyle**.

hypoconulid (Gk *hypo*=under, below and *konos*=pine cone) The term proposed by Osborn (1888) for the main cusp at the distal end of a mandibular (lower) (hence the suffix "-id") molar crown; it is wedged between the **hypoconid** on the buccal side and the **entoconid** on the lingual side. It forms part of the **talonid** component of a mandibular postcanine tooth crown. Modern humans may not have a hypoconulid but it is found in all archaic hominins and most pre-modern *Homo* taxa (syn. distostylid, tuberculum posterium internum, mesoconid).

hypodigm (Gk *hypo*=under, below and *deik*=to show) All the fossil evidence assigned to a taxon, but the term is usually used to refer to all the fossil evidence assigned to a **species**.

hypoglossal canal (Gk *hypo*=under, below and *glossa*=tongue) A bony canal that passes through the basioccipital bone just in front of the occipital condyles (hence its other name, the anterior condylar canal). It transmits the motor axons of the twelfth cranial or hypoglossal nerve, which innervate all of the intrinsic muscles of the tongue (i.e., the muscles that make up the tongue itself) plus all but one of the muscles that move the tongue around in the mouth. It was suggested by Kay et al. (1998) that the size of the hypoglossal canal could be used as a proxy for the density of the nerve supply to the tongue muscles. The logic was that **speech** needs precise movements of the tongue, which are achieved by having smaller motor units (i.e., fewer muscle fibers supplied by each axon) thus more axons are required, and therefore the larger-diameter hypoglossal nerves needed for a hominin capable of speech compared to a hominin of the same overall mouth size that was incapable of speech would demand a commensurately larger diameter hypoglossal canal. Specifically, it was claimed that *Homo neanderthalensis* had similar speech capacities as modern humans on the basis of hypoglossal canal cross-sectional area relative to oral cavity size. However, subsequent recent studies have demonstrated that (a) the relative size of the hypoglossal canal shows too much within-species variability to be informative and (b) modern humans do not have a greater relative number of hypoglossal motor neurons relative to medulla size than do apes who do not use spoken language.

hypomorphosis (Gk *hypo*=under, *morphe*= form or shape) A **heterochronic** process that involves the truncation of ancestral growth trajectories resulting in descendant adults that resemble ancestral juveniles. As in **hypermorphosis**, size and shape covariation is maintained in descendants; hypomorphosis brings shape change through selection for reduced size growth. Hypomorphosis can operate by shortening the duration of growth in a descendant relative to an ancestor, while maintaining similar rates of growth (time hypomorphosis, also called progenesis), or by reducing the rates of ancestral growth together with similar duration of growth (rate hypomorphosis).

hypoplasia (Gk *hypo*=under, below and *plasis*=growth) Literally "arrested growth," the term refers to areas on teeth where the production of **enamel** matrix by **ameloblast**s has been disrupted by a non-specific stressor experienced during **tooth formation** (e.g., systemic infection, a deficient diet, or starvation), which results in a depression on the surface of a tooth. Witzel et al. (2008) have proposed three levels of disturbance of enamel matrix production: a reduction in volume, the formation of aprismatic enamel when one would expect prismatic enamel (this is due to

interference with the function of the distal portion of the Tomes' processes of ameloblasts), and a complete cessation of matrix production. Hillson and Bond (1997) classified areas of hypoplasia as either furrow defects or pit defects. Furrow defects are the most common type of hypoplasia and are the result of an interruption to enamel matrix production that affects many ameloblasts; they typically manifest as linear bands encircling a tooth crown (and possibly other teeth developing simultaneously). In pit defects the disturbance is more localized, appearing as an indentation or pit that does not include all simultaneously formed enamel. Because hypoplasias usually affect all the teeth in which enamel matrix formation is occurring, they can be used to compare the relative ontogeny of teeth along the tooth row. Guatelli-Steinberg (2001, 2003, 2004a, 2004b) has documented linear enamel hypoplasias in large samples of hominins and non-human primates, which show some differences in the frequencies and durations of stress.

hypostotic *See* **non-metrical traits**.

hypostyle A term introduced by Osborn (1892) for an accessory cusp others more commonly refer to as a distal accessory tubercle. *See* **distal accessory tubercle**.

hypothalamus (Gk *hypo*=under, below and *thalamus*=inner chamber, bedroom) The hypothalamus is situated inferior to the **thalamus**; together they form the **diencephalon**. The hypothalamus is involved in a diverse array of functions such as homeostasis, controlling the endocrine functions of the pituitary gland, and regulating emotional behavior and reproduction.

hypothesis testing The process of testing a specific relationship which has been identified *a priori*, as opposed to **exploratory analysis** in which multiple techniques are used to explore possible patterns within a data set.

hypothetical common ancestor When a cladistic analysis groups two taxa or two clades together as **sister taxa** or sister clades, certain assumptions can be made about the character states of the common ancestor of the two taxa or clades. These assumptions amount to a prediction about the morphology of the most recent common ancestor of the two taxa or clades. Researchers who are skeptical about the chances of ever finding an actual ancestor prefer to describe the predicted taxon as a hypothetical common ancestor.

hypothetical taxonomic unit A concept used in **phylogenetic comparative analysis** to refers to an internal branching node within a phylogeny as opposed to an actual taxon from which data have been collected. *See also* **operational taxonomic unit**.

hystricomorphs The informal name for the suborder of the Rodentia that includes the porcupines and cavies. *See* **Rodentia**.

I

I.A.R.E. Acronym for the **International Afar Research Expedition** (*which see*).

IBD Acronym for **identical by descent** (*which see*).

Iberomaurusian (A combination of *Ibero-*, as in the Iberian Peninsula, and L. *Mauritania* = the name of a Roman province that is coincident with what is now Morocco and Algeria) A term that refers to **Late Pleistocene** artifacts and to the hominins that made them found at coastal sites in Morocco, Algeria, and Tunisia that date from *c.*18 ka and 11 ka (Bouzouggar et al. 2008). Archeologically, Iberomaurusian stone artifact assemblages are dominated by **bladelets** and **microliths** with subsistence data suggesting the seasonal exploitation of coastal and interior habitats. Archeological evidence suggests there were evident technological and behavioral differences between the Iberomaurusian and its regional antecedents. In fact, Iberomaurusian strata often unconformably overly **Aterian** ones, suggesting that the area may have been temporarily abandoned by hominins until settlement by Iberomaurusian populations (Close and Wendorf 1990). This hypothesis receives further support from the distinctive, skeletally robust "Mechta-Afalou" types found at sites with abundant burials such as **Grotte des Pigeons** (Taforalt; Irish 2000) and genetic evidence linking the appearance of the U6 mitochondrial DNA clade with the spread of populations into North Africa from southwestern Asia at the time of the appearance of the Iberomaurusian (Maca-Meyer et al. 2003). High frequencies of dental evulsion and similarities in lithic technology suggest the presence of widespread social networks to maintain behavioral norms among Iberomaurusian populations (Close 2002, Humphrey and Bocaege 2008).

IBS Acronym for **identical by state** (*which see*).

ice-rafting Ice rafting refers to the process of sediment transport via sea ice. Icebergs that calve off continental glaciers carry sediments from bedrock into the ocean and sea ice that forms in shallow waters incorporates sediments from the continental shelf. Icebergs calved from glaciers and sea ice are both transported by wind and ocean currents. Eventually the ice will start to melt and release its sediment load, which sinks and is then incorporated into ocean sediments. These ice-rafted sediments are known as dropstones (i.e., pebbles considerably larger than ambient sedimentation) (Heinrich 1988). The morphology of ice-rafted sediments can be distinguished from other forms of transport (e.g., fluvial or aeolian). The mineral, elemental, and isotopic composition of these dropstones can be used to infer their source. The extent of sea ice transport increases during times of glaciations and it is the sediments derived from sea ice that have provided evidence of abrupt climate change in the form of **Heinrich events**.

ichnofossils (Gk *ikhnos* = footprint) Evidence of one sort or another (e.g., footprints, tracks, burrows, leaf impressions, etc.) that an organism had been in that place at some time in the past. The **Laetoli hominin footprints**, **coprolites**, and leaf impressions are all examples of ichnofossils (syn. trace fossil).

ichnology The study of ichnofossils (i.e., fossilized footprints, tracks, burrows, leaf impressions, etc.) An analysis of the **Laetoli hominin footprints** would be an example of ichnology.

icons *See* **symbol**; **symbolic**.

ICZN Acronym for the **International Commission on Zoological Nomenclature** (*which see*).

ID Acronym for **immunological distance**. *See* **molecular clock**; Vincent M. **Sarich**.

-id (L. *-id*=source or origin) When this suffix is used after the name of a cusp it means it is a cusp on a mandibular (lower) tooth. Thus, **metaconid** is the term for the cusp mesial to the **protoconid** for the lower teeth whereas **metacone** is the term for the equivalent cusp in an upper tooth.

identical by descent The condition where two **alleles** are identical because they share a **common ancestry**; it is the genetic equivalent of a phenotypic **homology**.

identical by state The condition where two **alleles** are identical *not* because they share a **common ancestry** but because they have mutated independently to become identical to each other; it is the genetic equivalent of a phenotypic **homoplasy**. For example, **microsatellite** alleles are often identical by state because of the stepwise **mutation** process that produces the alleles. In this process a new allele is likely to have one additional repeat, or one less repeat, than the parent allele. Since microsatellites have high mutation rates, alleles with the same number of repeats are often created independently. For example, one allele with 14 repeats may have arisen from a gain of one repeat from a parental allele with 13 repeats while another allele with 14 repeats may have arisen because of a loss of one repeat from a parental allele with 15 repeats.

identification The process of allocating individual fossil specimens to an existing **taxon** or to a new taxon. Classification only applies to taxa; you identify an individual fossil, but classify the taxon it belongs to. *See* **classification**; **systematics**.

IHE Acronym for the **Institute for Human Evolution** (*which see*).

Il Dura One of the designated subregions of the Koobi Fora site complex. *See* **Koobi Fora**.

Il Naibar Lowlands One of the designated subregions of the Koobi Fora site complex. *See* **Koobi Fora**.

Ileret One of the designated subregions of the Koobi Fora site complex. *See* **Koobi Fora**.

image stack *See* **computed tomography**.

imbrication lines *See* **imbricational enamel**; **perikymata**.

imbricational enamel (L. *imbricatus*=roof made of tiles) Enamel formed after the cuspal (or appositional) enamel has been formed. The **striae of Retzius** within the imbricational enamel emerge on the crown surface as parallel ridges. They were called imbrication lines by Pickerill (1913) because especially towards the cervical region they can look like the edges of the tiles on a tiled roof. Their other name, **perikymata**, was given to them earlier by Gustav Preiswerk in his thesis on ungulate enamel structure (NB: Preiswerk is also responsible for the names **diazone** and **parazone** in enamel microstructure). *See also* **enamel development**; **incremental features** (syn. cuspal enamel).

imitation (L. *imitari*=to copy) The comparative study of imitation can be traced back to Thorndike (1858) who defined imitation as "learning to do an act from seeing it done" (*ibid*, p. 1911). In the *Poetics*, Aristotle noted that modern humans are "the most imitative creature[s] in the world and learns at first from imitation." Modern human children imitate everything from the words used to describe new objects to actions with tools (e.g., telephones, brooms, and remote controls). As adults, modern humans from all cultures imitate everything from postures to styles of dress to conventions of affection, and most of this is done automatically. This phenomenon has been referred to as "the chameleon effect" (Chartrand and Bargh 1999). However, imitation may be divided functionally; that is, depending on the type of information that is being imitated (Subiaul 2010). For example, familiar imitation refers to the copying of actions that already exist in one's behavioral repertoire (e.g., sticking out one's tongue). Familiar imitation may be contrasted with novel imitation, which refers to the copying of novel actions or behaviors that do not already exist in the individual's behavioral repertoire (e.g., riding a bicycle). Additionally, one can specify the type of information that is imitated. For example, vocal imitation refers to the imitation of specific sounds, motor imitation refers to the copying of actions, and cognitive imitation refers to the copying of abstract rules that are neither motor nor vocal (e.g., social conventions or the order of events) (Subiaul et al. 2004). Motor imitation has been widely studied in both modern humans and non-human primates (Tomasello and Call 1997, Hermann et al. 2007). According to Call and Carpenter (2002) motor imitation has three components: (a) understanding the goals of a model or conspecific, (b) recreating the results of the event, and (c) using the model's actions rather than your own to achieve an end result. In this view, motor imitation is associated with high copying fidelity. This view of motor imitation is consistent with the notion that motor imitation is a specific form of social learning. But, as far as social learning mechanisms are concerned, motor imitation is considered to be the most cognitively sophisticated as it may require perspective taking, **self-awareness**, and **theory of mind** (or ToM). Certainly, various studies have demonstrated how ToM improves the flexibility and fidelity of imitation (Carpenter et al. 1998, Gergely et al. 2002, Meltzoff 1995). Arguably, the best examples of imitation come from animals that are believed to have a ToM (e.g., modern humans and chimpanzees). For example, chimpanzees have been shown to imitate novel actions of a modern human experimenter. In the "Do as I do" (DAID) paradigm (Hayes and Hayes 1952) chimpanzees were trained to respond to the phrase, "Do this!" followed by a range of behaviors, all of which are novel to the chimpanzee (e.g., touching an elbow, stomach, an ear, or nose, the latter of which are not visible). Chimpanzees were successful at imitating the modern human demonstrator, except they had difficulty matching behaviors that were visually opaque, such as facial gestures. In contrast, modern human children imitate touching both transparent actions (such as manipulating an object) and opaque actions (such as copying facial expressions). In the case of transparent actions, one can visually match one's own actions with those of others through a visual–visual match. But this is not possible in the case of opaque actions; for example, one does not have direct visual access to one's own face. Despite the differences of opinion regarding the nature of the heterogeneity of imitation competence in non-human apes, it is clear that from an early age the imitative skills of modern humans exceed those observed in apes. One possible scenario is that in the course of human evolution an elaboration of ToM (e.g., perspective taking, **intentionality**, reasoning about

belief) might have modulated imitation performance, eventually making modern human imitation more flexible and accurate than the predicted level of performance in the most recent common ancestor of chimpanzees/bonobos and modern humans (Tomasello 1999). Another possibility is that modern humans simply evolved more specialized imitation mechanisms than other apes. In this view, modern humans are exceptional imitators relative to other apes because we have learning mechanisms that allow our species to imitate more widely and flexibly (Subiaul 2007). There is debate about how well archeological evidence can track the evolution of imitative behavior from the level one would predict for the *Pan/Homo* common ancestor to the imitative capability of modern humans. *See also* **theory of mind**.

Imjin/Hantan river basins   *See* **Chongokni**.

immunochemistry   (L. *immunis*=an individual who is exempt from public service and *chemista*= alchemist, a person skilled in the art of alloying metals) A branch of biochemistry that investigates the components and chemical reactions of the immune system. Morris **Goodman** (1963) used the reactions between an antigen and an antibody to determine the relationships among the higher primates and showed that the patterns of agglutination produced by the reaction between specific antibodies and serum **albumin** were virtually identical in modern humans and chimpanzees. Others have tried less successfully to use the same techniques to investigate the relationships among fossil taxa. Immunochemistry has also been used to test whether a morphologically undiagnostic specimen claimed to be a hominin actually is a hominin, but these methods have not always proved to be accurate (see references in Martínez-Navarro 2002).

immunoglobulins   *See* **antibody**.

immunological distance   *See* **molecular clock**; Vincent M. **Sarich**.

imprinting   *See* **genomic imprinting**.

in silico   A term, equivalent to *in vivo* (i.e., an investigation made within a living organism) or *in vitro* (i.e., an investigation made outside of a living organism using isolated organs, tissues, or cells) for an investigation conducted by means of computer simulation. Like "Silicon Valley" in California, USA, the term has its roots in the silicon that was used in the early integrated circuits.

in situ (L. *in*=in and *situs*=place) Fossils or artifacts found in their parent horizon either during the course of an excavation or in the process of eroding out of that horizon (e.g., **KNM-ER 1813** was found eroding out of the sediments of the Upper Burgi Member of the Koobi Fora Formation and all the **Dmanisi** fossils have been recovered from excavations).

incertae sedis [L. *incertae sedis*="of uncertain (taxonomic) position"] The phrase indicates that a particular specimen is difficult to assign to any taxon, and sometimes it is used to indicate that a named taxon may be difficult to assign to a higher taxonomic group. For example, the **KNM-ER 1482** mandible from below the KBS Tuff at Koobi Fora, which when first described was not allocated to any taxon; the Ceprano 1 calvaria, which in some ways resembles *Homo erectus* and in others resembles later Lower Pleistocene African specimens such as Daka (**Bouri**) or *Homo heidelbergensis*. *See also* **taxonomy**.

increment (L. from *increscere*=to increase) The term means literally "a small increase." The term is used in paleoanthropology when a large quantity is broken up into smaller ones. So, for example, when researchers count **perikymata** on the buccal surface of anterior teeth to compare details of enamel microstructure they break up the surface into smaller "increments." If they break the height of the crown up into 10 equal increments, (e.g., Guatelli-Sternberg et al. 2005) they are called "10% increments" or deciles. Breaking the crown up into smaller increments allows researchers to investigate whether an overall increase or decrease in the rate of enamel deposition is distributed over all of the crown height or whether any change is focused on just one region of the total height of the crown.

incremental features (L. from *increscere*=to increase). Physical features in tissues such as **enamel**, **dentine**, and **cementum** that record their growth at regular, predictable time intervals. The growth of hard tissues is appositional in nature (i.e., one layer of tissue is deposited upon a previous layer) but the term "incremental" also

implies that the layers are deposited according to a periodic time scale regulated by an internal physiologic rhythm. Microscopic markings representing intrinsic temporal rhythms in hard-tissue secretion may be annual (e.g., cementum annulations, lines of arrested growth in bone), long-period (i.e., greater than one day; e.g., **striae of Retzius/ perikymata** in enamel or **Andresen lines/ periradicular bands** in dentine), short-period (i.e., 24 hours or circadian; e.g., **cross-striations** in enamel, **von Ebner's lines** in dentine), or less than 24 hours (**intradian lines** in enamel and dentine; reviewed in Smith 2006). These increments of growth are preserved throughout life in mineralized tooth tissues because they are never replaced or "turned over." For a long time it was widely believed that there were no equivalent preserved growth increments in bone. However, reports have demonstrated the presence of annual/seasonal and more recently possible long-period growth increments in the bone tissues of many vertebrates, including mammals (reviewed in Castanet 2006, Bromage et al. 2009). *See also* **enamel microstructure**.

incurvatio mandibulae A midline depression on the external aspect of the **symphysis** of the **mandible** between the **alveolar process** and the **base**. There needs to be an *incurvatio mandibulae* for a **mental protuberance** to qualify as a **true chin**.

incus (L. *incus*=anvil) One of the three **auditory ossicles** in the middle ear. The ossicles, including the incus, occasionally survive into the fossil record (e.g., SK 848, a *Paranthropus robustus* incus). *See also* **auditory ossicles**; **branchial arches**; **middle ear**.

indel Shorthand for insertion(s) and deletion(s). *See* **insertion**; **deletion**.

independent assortment The phenomenon observed by Gregor Mendel whereby the **alleles** of **genes** that are unlinked (i.e., they are on different **chromosomes** or located far apart on the same chromosome) assort independently during **meiosis** when gametes are formed. Independent assortment is one of the sources of genetic variation and therefore provides new combinations of variants upon which evolution can act.

independent contrasts *See* **phylogenetically independent contrasts**.

indexes *See* symbol; symbolic.

Indonesia (L. *Indus*=Indian and Gk *nesos*= island, literally "Indian Islands;" equivalent to the Malay Archipelago) Indonesia was a geographical entity before it became a political one. It includes the large and small islands that extend eastwards from the Malay Peninsula. The large islands include Borneo, Java, the Moluccas, Sulawesi, Sumatra, and Timor; the small islands include Bali, Flores, and Sumba.

industrial complex A higher-rank archeological taxonomic term defined initially by Clark et al. (1966). It defines an aggregate consisting of multiple related **industries** (e.g., the **Acheulean Industrial Complex**). This particular terminology is meant to both highlight the broad similarities among sites across the world that have **handaxe**s and similar **Acheulean** tools, and emphasize local variation.

industry (L. *industria*=diligent) A spatially and temporally constrained collection of archeological assemblages, considered by some to represent the range of material items produced by a prehistoric group of people. Industries, particularly as defined and codified in the 1960s (Clarke 1968, Clark et al. 1999), came closest to archeological manifestations of the ethnographic ideals of "cultures" or "culture groups," despite the temporal resolution at most sites being too coarse to establish contemporaneity (a) among different artifacts deposited at a site and (b) between assemblages at two, or more, sites. The precise boundaries of an industry are rarely defined and usage varies among authors. The term is used informally as the archeological equivalent of a Linnaean taxonomic category – the genus is probably the best analogy – to highlight inter-assemblage similarities.

infant (L. *infans*=not being able to speak) Literally a child before it can talk, but normally used to refer to children when they are relatively helpless. There is no formal, more precise, definition.

inferior nuchal line A roughened line on the occipital bone that is produced by the attachment of the fascia between the biggest nuchal muscle, the semispinalis capitis, posterosuperiorly and a smaller nuchal muscle, the obliquus capitis superior, anterorinferiorly.

infraorbital visor *See* facial visor.

inner ear A cavity medial to the middle ear within the petrous part of the temporal bone. It consists of a series of complex intercommunicating cavities referred to as the **bony labyrinth**. The bony labyrinth is filled with a fluid called perilymph and suspended in the perilymph is the membranous labyrinth, a smaller endolymph-filled facsimile of the bony labyrinth, in the walls of which are the hair cells that detect sounds, posture, and motion. The membranous labyrinth is divided into three functional components: the snail-shaped cochlea for sound, the saccule and utricle for posture, and the **semicircular canals** for motion. The relative size and shape of the semicircular canals of the bony labyrinth have proved to be taxonomically valid variables (Spoor and Zonneveld 1998). *See also* **semicircular canals**.

inner table *See* cranial vault.

innominate (L. *innominatus*=no name) The old name for what the *Terminologia Anatomica* encourages us to call the hip bone or pelvic bone.

insectivory (L. *insectum*=animal, derived from "a notched or divided body," literally "cut into," plus *vorous*, from stem of *vorare*=to devour) Literally, the consumption of insects. Insects are very high in energy and protein but can be rare, cryptic, and difficult to catch. Insectivory is therefore most common in the smallest primates, which require high-quality food because of their need for relatively more energy per unit body mass than large primates and because they are inefficient processors of low-quality foods. Generally insectivorous primates weigh less than 500 g, whereas folivorous ones tend to exceed this body mass: this is known as Kay's threshold. Insectivores often have teeth with high, sharp cusps that facilitate the penetration of the insect exoskeleton. Modern humans lack these dental specializations, but nonetheless they are known to consume insects (and insect products such as honey) opportunistically and it is likely that early hominins would have done the same.

Isernia la Pineta (Location 41°36′N, 14°14′E, Italy; etym. named after the nearby town) <u>History and general description</u> Road work outside of the town of Isernia uncovered paleosurfaces rich in both faunal and artifactual assemblage. At least four layers have been identified. <u>Temporal span and how dated?</u> **Potassium-argon dating** and **magnetostratigraphy** suggest that the sediments immediately overlaying the archeological layers are 0.73 ± 0.04 Ma, but the **biostratigraphy**, particularly the micromammals, suggest an early Late Pleistocene date. <u>Hominins found at site</u> None. <u>Archeological evidence found at site</u> Flint choppers, flakes, and cores were all recovered from the site. Some of the faunal elements show signs of butchering. Key references: <u>Geology, dating, and paleoenvironment</u> Coltorti et al 1982; <u>Archeology</u> Mussi 2001.

insertion A mutation in the **genome** where one or more **base pairs** (i.e., one or more **nucleotides**) have been added, or inserted, into a sequence of nucleotides. An insertion can be as small as one base pair or it can encompass many thousands of base pairs. In a coding region any insertion that is not a multiple of three base pairs causes a **frameshift mutation**, which is typically deleterious. The significance of insertions for human evolution is two-fold. First, insertions can add genetic material that is functionally important, including genes. For instance, the *CCL3L1* gene has been duplicated (i.e., inserted) multiple times in some individuals. This gene encodes a protein that binds to the CCR5 receptor that is used by HIV to enter cells. Thus, individuals who have more copies of this gene have a lower risk of HIV infection and if they do become infected they have slower progression to AIDS (Gonzalez et al. 2005). Second, insertions can be used to understand the relationships among groups. For example, **Alu** insertions have been used to study population history and argue for a larger population size in Africa over most of human evolutionary history (Stoneking et al. 1997). Insertions and **deletions** are commonly referred to by the abbreviation indels.

insolation (L. *insolare*=in the sun) The amount of solar radiation reaching the Earth's surface. Solar radiation is more intense at or near the equator than at or near the poles. Changes in (a) the shape of the Earth's orbit (i.e., **eccentricity**), (b) the tilt of the Earth's axis (i.e., **obliquity**), and (c) the unsteadiness of that axis (i.e., **precession**), all of which take the Earth (or parts of the Earth in the case of changes in the tilt angle) nearer to or further away from the sun, affect climate by altering insolation.

Institute for Human Evolution (or IHE) This institute, at the University of the Witwatersrand, was established in January 2004 under the interim Directorship of Professor Trefor Jenkins. The current Director, Professor Francis Thackeray, was appointed in February 2009. The objectives of the IHE include fieldwork and research focusing partly on South African Plio-Pleistocene hominin fossils and non-hominin fauna (from **Gladysvale**, **Gondolin**, **Makapansgat**, **Malapa**, **Sterkfontein**, **Swartkrans**, and **Taung**) and partly on late Pleistocene archeological sites (including **Blombos Cave**, Sibudu, and Rose Cottage Cave) associated with artifacts reflecting technological innovation in the **Middle Stone Age**. In addition to pursuing paleoanthropological research, the IHE's mission is to (a) promote South Africa's heritage to a wider audience, (b) stimulate a new generation of South African scientists with an interest in human evolution, and (c) contribute to the development of young researchers, especially from previously disadvantaged communities in South Africa, through outreach programs at the community, school, and university levels.

Institute for the Study of Man in Africa (or ISMA) Established in 1963 to honor the contributions of Raymond **Dart**, who retired from the University of the Witwatersrand in 1958. It functioned as a focus within South Africa for educating the public about human evolution and for a brief period it ran the Museum of Man in Africa, but the collections have long since been subsumed into those of the Africana Museum that in 1994 was renamed in MuseuMAfricA. Currently the most visible manifestation of ISMA is the Dart Lecture series.

Institute of Human Origins (or IHO) A research center in the School of Human Evolution and Social Change on the Tempe campus of Arizona State University (ASU), USA. The IHO was founded in 1981 by paleoanthropologist Donald **Johanson** as a non-profit research foundation in Berkeley, California, a still extant second identity, although all of its operations take place at ASU, where it has been headquartered since 1997. In 2009, Donald Johanson stepped aside as Director of IHO to become Founding Director; William Kimbel was appointed Director and Curtis Marean Associate Director. IHO's mission is three-fold: to recover and analyze the primary fossil and archeological evidence for human evolution and its ccological and geological contexts, to advance scientific understanding of human origins and its contemporary relevance through focused public outreach programs, and to provide expertise that improves teaching of human evolution in primary- and secondary-school science classrooms (primarily through its website). IHO scientists have been involved in a number of field projects that have provided key evidence for early hominin evolution. The discovery of "Lucy" and other *Australopithecus afarensis* fossils by the **International Afar Research Expedition**, co-directed by Johanson at Hadar, Ethiopia, in the mid-1970s, provided the scientific centerpiece for the founding of IHO. After 15 field seasons research at **Hadar** is ongoing (directed by William Kimbel, Kaye Reed, and Chris Campisano), since 2007 as a biennial ASU-sponsored field school. More than 400 fossils of *Au. afarensis* have been recovered from 3.0–3.4 Ma sediments at the site, along with rare remains of early *Homo* and **Oldowan** stone tools dated to *c*.2.35 Ma. IHO's research in Ethiopia's Afar region expanded in 2002 with the launch of the **Ledi-Geraru Research Project** (directed by Kaye Reed and the late Charles Lockwood). From 1986 to 1988 an IHO/University of California Berkeley team (directed by Gerry Eck, Donald Johanson, and Tim White) worked at **Olduvai Gorge**, Tanzania, where, in 1986, a 1.7 Ma partial skeleton of a female *Homo habilis* (**OH 62**), was recovered. IHO research projects also address the later human record. From 1991 to 1994 William Kimbel collaborated with Israeli researchers (Erella Hovers and Yoel Rak) on IHO-cosponsored excavations at **Amud** Cave, in the northern Galilee, which produced several fossils of *Homo neanderthalensis*, including an infant's skeleton, Amud 7, amid abundant Mousterian tools dated to *c*.57–60 ka. In 1998–9 this team conducted strategic survey across Israel for new sites, identifying several with paleoanthropological potential. Beginning in 2000 an IHO-led multinational team directed by Curtis Marean has conducted excavations at **Pinnacle Point**, near Mossel Bay, South Africa, where cave deposits dated to 72–164 ka have yielded evidence for human exploitation of marine resources, use of pigment, and heat-treatment of stone to enhance knapping ability, implying the early advent of "modern" cultural behaviors. IHO's outreach programs include publically accessible print, web, and social media outlets, as well as lectures, symposia, and school tours. IHO's website (becominghuman.org) pro-

vides news and information on human origins research and its Learning Center is a curriculum resource for primary and secondary school science teachers, a primary vehicle for IHO's outreach into the educational community.

Institute of Vertebrate Paleontology and Paleoanthropology (or IVPP) Formed in 1957 in China when the **Laboratory of Vertebrate Paleontology** was renamed. It is an independent research institution under the **Chinese Academy of Sciences**. The IVPP conducts paleontological research in collaboration with local research entities and publishes the journal *Acta Anthropologica Sinica*. It is the repository of collections of vertebrate fossils, including many, but not all, of the hominins from Chinese sites.

Institutul de Speologie "Emil Racoviță"
The "Emil Racovita" Institute of Speleology, which is part of the Romanian Academy, includes Departments of (a) Biospeleology and Karst Edaphobiology, (b) Geospeleology and Paleontology, and (c) Karstonomy, Karst Inventory and Protection. Hominin fossil collections **Oase 1**, **Oase 2**. Contacts Ioan Povara and Cristian Goran (e-mail iser_b@yahoo.com). Postal address 13 Septembrie Road, No. 13, 050711, Sector 5, Bucharest; note that the hominins are at the branch in Cluj-Napoka. Website www.iser.ro/index_en.php. Tel and fax +40 21 311 08 29.

integration *See* **morphological integration**.

interbirth interval The time between the birth of one offspring and birth of the following offspring of a different gestation. Interbirth interval is a crucial determinant of lifetime reproductive output in females. It can be affected by social elements, including group size, competition, and sex ratio. It is also influenced by ecological factors, including those affecting food availability (e.g., rainfall, altitude, seasonality), and maternal characteristics such as parity and survival of previous offspring. Three main phases have been proposed to make up the composition of primate interbirth intervals, some of which are more variable than others. The first phase consists of a period of postpartum amenorrhea, the length of which is related to the duration of lactation and patterns of nursing behavior (e.g., frequency and intensity). The second period is one in which ovulatory cycling is resumed, and the duration of this can vary especially in relationship to food availability and maternal condition. The third period is defined by gestation of the subsequent offspring. The length of the interbirth interval, playing as central a role as it does in reproductive rate, is one of the primary life history variables that factor into discussions about "fast" and "slow" life histories. *See also* **grandmother hypothesis; life history.**

interglacial *See* glacial cycles.

intermembral index Ratio of upper-limb length relative to lower-limb length, calculated as $100\times$(humerus length+radius length)/(femur length +tibia length). This index is the most common way researchers measure and compare limb proportions in primates and hominins and it reflects major differences in locomotor strategies among primates. Large-bodied primates (e.g., orangutans) tend to have high intermembral indices, their long upper limbs allowing them to reach across gaps between trees and hold onto multiple supports to distribute their weight. In contrast, many small-bodied primates (e.g., galagos) have low intermembral indices reflecting relatively long lower limbs that are mechanically advantageous for leaping. Modern humans are an exception to this trend, for although their body mass is large they have long lower limbs, which increase the speed and energetic efficiency of bipedal walking and **running**. *See also* **humerofemoral index.**

internal ear *See* inner ear.

International Afar Research Expedition (or I.A.R.E.) A collaboration among Maurice Taieb (its head), Yves **Coppens**, Donald **Johanson**, and Jon Kalb. Taieb and Kalb were to be responsible for the geology and Coppens and Johanson were to be jointly responsible for paleontology in general and hominid paleontology in particular. The arrangement persisted until the I.A.R.E. was disbanded in September 1974.

International Code of Zoological Nomenclature (current, fourth edition is Ride et al. 1999, with an online version available at www. iczn.org/iczn/index.jsp) Almost universally referred to as "the code," published by the International Trust for Zoological Nomenclature on behalf of the **International Commission on Zoological Nomenclature** (or ICZN) (for information about the latter see below or www.iczn.org/ICZN homepage.htm). The code sets out rules and recommendations that help "promote stability and universality in the scientific names of animals and to ensure that the name of each taxon is unique and distinct." If you name a new **species** you need to abide by these rules and it is also sensible to follow the recommendations.

International Commission on Zoological Nomenclature (or ICZN) Founded in 1895, its activities are supported by prominent national science societies and academies, and leading natural history museums and foundations. The ICZN is mandated by its scientific membership of the International Union of Biological Sciences (IUBS) and its members are elected by zoologists attending General Assemblies of the IUBS or other International Congresses. Its mission is "to create, publish and, periodically, to revise the **International Code of Zoological Nomenclature**. The ICZN also considers and rules on specific cases of nomenclatural uncertainty." The rulings of the ICZN are published as "Opinions" in the ***Bulletin of Zoological Nomenclature***. The code sets out rules and recommendations that help "promote stability and universality in the scientific names of animals and to ensure that the name of each taxon is unique and distinct." For more information about the ICZN see www.iczn.org/ICZNhomepage.htm. *See also* **International Code of Zoological Nomenclature.**

International Louis Leakey Memorial Institute for African Prehistory *See* Leakey, Louis Seymour Bazett (1903–72).

International Omo Research Expedition (or IORE) A 1967 international expedition organized by Louis **Leakey**, Camille **Arambourg**, and F. Clark **Howell** to explore the fossiliferous sediments exposed in the **Lower Omo Basin**, that is the region just north of where the Omo River drains into what was then called Lake Rudolf and what is now called Lake Turkana. The Kenyan contingent consisted of Richard **Leakey** and Margaret Leakey, the French contingent was led by Camille Arambourg, and the American contingent by F. Clark Howell. In its first year, the three contingents of the International Omo Research Expedition worked in different geographical areas and therefore concentrated on different parts of the strata exposed in the area.

The Kenyan contingent (which only participated in 1967, and thereafter left to work further south on the east side of what was then called Lake Rudolf, at East Rudolf, now called **Koobi Fora**) worked in the north where they concentrated on the **Kibish Formation** and **Mursi Formation**. The US contingent focused on the Kibish and **Usno Formations** and the French contingent on the **Shungura Formation**, but thereafter the US and the French contingents concentrated on the Shungura Formation, the US working in the northern part of the type area north of the so-called "Watering Road" which ran between the combined Sectors 15 and 16 and Sector 17 and the French in the south. Full field seasons were worked until the mid-70s and thereafter research was mostly restricted to refining the geology. The fossiliferous sediments exposed in the Lower Omo Basin were assigned to formations in either the older **Omo Group** or the younger **Turkana Group**. Formations in the Omo Group in the Lower Omo Valley are the Usno, Mursi, Nkalabong, and Shungura Formations and in the Turkana Group they are the Bume, Kibish, and the Errum Formations. Compared with Koobi Fora, the geology in the Lower Omo Basin generally, and in the Shungura Formation specifically, was relatively simple. This relative simplicity combined with isotopic dates obtained from the tuffs that occur at frequent intervals in the Shungura Formation enabled the French and the US contingents to amass an impressive collection of well-dated mammalian fossils. This collection formed the basis of a reference biostratigraphy that has been and still is being used to date sites where there is no immediate prospect of obtaining absolute dates. Details of the history of the exploration of the Lower Omo Basin are given in Howell and Coppens (1983).

interproximal wear Wear between tooth crowns (also called approximal wear) that occurs when adjacent teeth move slightly against each other in the mouth during the process of **chewing**. In the early stages of the wear between two teeth in the same jaw that are in contact with one another (hence interproximal wear is also called interstitial wear) most of the wear occurs in the form of a concave wear facet on the proximal (i.e., mesial) face of the crown of the distal tooth (e.g., between the left P_4 and M_1 of **KNM-ER 992**), hence the emphasis in the term on proximal wear, but if the teeth wear against each other long enough a substantial wear facet also appears on the distal face of the crown of the mesial proximal tooth (e.g., the left M_1 of KNM-ER 1509 shows a distal interproximal facet) Some taxa (e.g., **Paranthropus boisei**) show marked interproximal wear that results in laterally and vertically extensive, more or less flat, interproximal wear facets. Marked interproximal wear can reduce the mesiodistal length and the overall occlusal area of a tooth crown, thus interproximal wear needs to be taken into account when comparing the size and shape of worn and unworn tooth crowns. *See also* **tooth size**; **tooth wear**.

interstadial *See* **glacial cycles**.

interspecific allometry *See* **allometry; scaling**.

interstitial growth *See* **ossification**.

interstitial wear Wear that occurs between two teeth in the same jaw and between adjacent bones (e.g., between adjacent metatarsals in the foot). In both cases the relative movement between the adjacent teeth or bones produces wear facets (e.g., interproximal wear facets between the teeth and smooth, eburnated, areas between the bones). *See also* **interproximal wear; OH 8**.

intertropical convergence zone (or ITCZ) Refers to the region where the trade winds converge and the air rises. The ITCZ migrates between the Tropics of Capricorn and Cancer as it follows the overhead sun with the seasons. A secondary control is contrasts between land and sea temperature and these are such that the migration of the ITCZ is not uniform around the planet. Seasonal variations in the position of the ITCZ influence the pattern of seasonal precipitation maxima across much of Africa. For example, East Africa receives most of its precipitation in April to March and in September to November, in association with the biannual movement of the ITCZ and with the seasonally reversing winds.

intracranial venous drainage *See* **cranial venous drainage; intracranial venous system**.

intracranial venous system The venous system that drains venous blood from the tissues that lie inside the bones of the **cranium** (e.g., the brain and the covering meninges) and from the cranial bones themselves. The intracranial venous system consists of superficial cerebral veins that drain the surface of the brain, which, in turn, drain into a system of venous channels called the **dural venous sinuses** that run between the fibrous and endosteal layers of the outer of the three meningeal layers, the **dura mater**, and which leave impressions on the **endocranial** surface of the cranium. *See also* **cranial venous drainage**.

intradian line (L. *intra*=within and *dies*=day) An **incremental feature** found in **enamel** and **dentine** that shows a subdaily periodicity. Experimental evidence in primate enamel suggests that either one or two of these fine lines exist within a 24-hour period, and this is also supported by some studies of rodent dentine (reviewed in Dean 2000 and Smith 2006). Other rhythms, such as an 8-hour maturational rhythm, also occur in enamel formation. To date there is no experimental evidence to pin down the exact periodicity of these intradian lines but they can complicate counts and measurement of daily **cross-striations** in enamel. *See also* **enamel microstructure**.

intramembranous ossification (L. *intra*= within and *membrana*=membrane) *See* **ossification**.

intraspecific variation Variation within species due to differences in size, shape, or (almost always) a mixture of the two. The following factors may contribute to intraspecific variation in a collection of fossils assigned to a single species: (a) ontogeny, differences due to fossils coming from individuals at different stages of development, (b) geography, differences due to fossils coming from different regions of a species' range, (c) time, differences due to directional or random-walk changes through time, (d) sex, primary or secondary sexual differences in the hard tissues (this category of intraspecific variation is referred to as **sexual dimorphism** and in some extinct hominin taxa this can be a major component of intraspecific variation), and (e) intrasexual variation, differences between same-sex individuals within a species that are not due to (a)–(c). These intrasexual differences are sometimes due to **homoiology** (i.e., morphological variation that accrues due to differences in factors such as the activity levels among individuals).

introgression (L. *introgredi*=to step in) Gene flow between lineages whose individuals are hybri-

dizing. Introgression can result in **reticulate evolution**, best characterized as a "web-like" rather than a "tree-like" pattern of relationships through evolutionary time. *See* **hybrid**; **hybridization**.

intron An abbreviation of intragenic region. It refers to the non-coding regions of genes between the **exon**s. After **transcription** the introns are removed from the pre-mRNA by a process called **splicing**. Introns may contain regulatory elements such as **enhancers**. Mutations in intronic splice sites can lead to human diseases such as β^0-thalessemia.

intrusive (L. *intrudere*=to thrust in) Fossils or artifacts that have been secondarily introduced into sediments either naturally (e.g., by burrowing rodents) or deliberately (e.g., by ritual burial).

inversion A foot movement around an anteroposterior axis in which the sole of the foot turns inward to face the midline of the body. Although some of this motion can occur at the ankle (talocrural joint) during extreme inversion, inversion is primarily a motion between the calcaneus and the talus (i.e., at the subtalar joint). Often incorrectly used interchangeably with supination (see **foot movements**), inversion refers only to rotation about the long (anteroposterior) axis of the foot. This movement is limited in modern humans by joint morphology, and by tension in the fibularis muscles and in the anterior talofibular ligament, which is one of the most often-sprained ligaments in the modern human body. Nonhuman primates have a greater range of rear-foot inversion due to an inverted set to the ankle, increased mobility at the subtalar joint, and the absence of an anterior talofibular ligament.

IORE Acronym for the **International Omo Research Expedition** (*which see*).

Irhoud 1 Site Jebel Irhoud. Locality N/A. Surface/*in situ* In situ. Date of discovery 1961. Finder Mohammed Ben Fatami. Unit N/A. Horizon N/A. Bed/member N/A. Formation N/A. Group N/A. Nearest overlying dated horizon N/A. Nearest underlying dated horizon N/A. Geological age 190–90 ka. Developmental age Adult. Presumed sex Male. Brief anatomical description Cranium lacking base, nasal cavity structures, posterior portion of orbits, and all teeth; its endocranial volume is estimated to be 1305 cm^3 (Holloway et al. 2004). Announcement Ennouchi 1962a. Initial description Ennouchi 1962b . Photographs/line drawings and metrical data Ennouchi 1962b, Hublin 2001. Detailed anatomical description Hublin 2001, Rightmire 1990. Initial taxonomic allocation *Homo neanderthalensis* (Ennouchi 1962b). Taxonomic revisions *Homo sapiens* (Hublin 2001). Current conventional taxonomic allocation *H. sapiens*. Informal taxonomic category Anatomically modern human. Significance Among the earliest *H. sapiens* in North Africa. Location of original Musée Archéologique, Rabat Chellah, Morocco.

Irhoud 2 Site Jebel Irhoud. Locality N/A. Surface/*in situ* *In situ*. Date of discovery December 23, 1963. Finders Mohammed Ben Fatami and Carlton **Coon**. Unit N/A. Horizon N/A. Bed/member Ashy level C. Formation N/A. Group N/A. Nearest overlying dated horizon N/A. Nearest underlying dated horizon N/A. Geological age 190–90 ka. Developmental age Adult. Presumed sex Male. Brief anatomical description Calvaria, but all of the face below the brow ridges missing; a metopic suture divides the frontal bone. Cranial capacity estimated to be 1400 cm^3 (Holloway et al. 2004). Announcement Ennouchi 1963. Initial description Ennouchi 1968. Photographs/line drawings and metrical data Ennouchi 1968. Detailed anatomical description Hublin 2001, Rightmire 1990. Initial taxonomic allocation *Homo neanderthalensis* (Ennouchi 1963). Taxonomic revisions *Homo sapiens* (Hublin 2001). Current conventional taxonomic allocation *H. sapiens*. Informal taxonomic category Anatomically modern humans. Significance Among the earliest *H. sapiens* in North Africa. Location of original Musée Archéologique, Rabat Chellah, Morocco.

Isaac, Glynn Llywelyn (1937–85) Glynn was born in Cape Town, South Africa, to two botanists: his mother was the flowering-plant botanist at Cape Town's Bolus Herbarium and his father a marine ecologist and Professor of Botany at the University of Cape Town (UCT). Glynn's interest in archeology began at an early age as he and his identical twin, Rhys, explored ancient outcrops near the Cape, and investigated castles, cathedrals, and museums together on their first trip to Britain as 9-year-olds. When Glynn finished secondary school in South Africa he returned to England where he worked on Bronze Age and late Mesolithic digs, and in 1955 he successfully completed a 6-month

diploma course in archeological techniques at the University of London. He returned to South Africa to study geology, zoology, and archeology/ethnography at UCT and received a BSc with distinction in 1958. In the summer of 1956 he hitchhiked up to Zambia to study the collections at the Rhodes-Livingstone Museum, thus beginning his longstanding relationship with Desmond **Clark**. His mentors at UCT were John Goodwin and Monica Wilson, and when Professor Goodwin fell gravely ill, Isaac, just 21, was invited to take over his courses. Isaac's accumulated experience prepared him well and he was awarded an Elsie Ballot Scholarship to study at the University of Cambridge, England. He studied Paleolithic Archaeology with Charles McBurney from 1959 to 1961, taking the Archaeology and Anthropology honors degree examination (tripos) in 2 years and thus completing his second undergraduate degree. During his time at Cambridge he participated in a number of different field projects at sites in England and Jersey, a survey in Libya with Eric Higgs, and excavations at the site of **Abri Pataud** in France with Hallam Movius Jr. Eager to return to Africa to pursue field research, Isaac accepted Louis **Leakey**'s invitation to become the Warden of Prehistoric Sites of the Royal National Parks of Kenya. As Warden (1961–2) and subsequently as Deputy Director of the Center for Prehistory and Paleontology at the **Coryndon Museum** (1963–4) Isaac was responsible for the museum's oversight of several prehistoric sites in Kenya, including **Kariandusi** and Hyrax Hill. Louis Leakey encouraged Isaac to visit **Olduvai Gorge** and consult with Mary **Leakey** as her archeological excavation program developed. In 1964 Isaac also co-led a field expedition to the remote exposures west of Lake Natron in Tanzania, in collaboration with Richard **Leakey**. The team surveyed sediments and discovered several promising archeological localities with distinctive early **handaxe**s, as well as a massive hominin mandible from the site of **Peninj**. Isaac focused most of his scientific energies during this period on new investigations at the Acheulean site complex of **Olorgesailie**. He developed the project in close intellectual partnership with his new wife Barbara Miller and their professional collaboration grew to become a central and enduring pillar of his career. The impressive concentrations of Acheulean artifacts and faunal remains at this site had been initially discovered by Louis and Mary Leakey during WWII, yet it took Isaac's detailed and problem-oriented research to reveal their true potential for understanding Acheulean tool use during the Middle Pleistocene. Olorgesailie became Isaac's PhD dissertation project and in it he began to formulate research questions and develop methodological approaches that became hallmarks of his later research programs, including a focus on site-formation processes, artifact function and design, and hominin hunting abilities and social organization. For example, in an early example of taphonomic research, Isaac began a program of experimental layouts of bones and stone tool replicas to evaluate sedimentary processes that may have led to the unusual concentrations of large **bifaces** and bones at Olorgesailie. He also developed innovative multivariate approaches to analyze attributes of handaxes and **cleavers**. In 1965 Isaac returned to the University of Cambridge to spend a year analyzing the results of his Olorgesailie research and he was awarded his PhD in 1968. A monograph based on this work, *Olorgesailie: Archaeological Studies of a Middle Pleistocene Lake Basin*, was published in 1977. Desmond Clark recruited Isaac to the faculty of the University of California at Berkeley in 1966, where he joined a group of colleagues, Sherwood **Washburn**, Phyllis Dolhinow, Theodore McCown, Elizabeth Colson, Richard **Hay**, and Garniss Curtis, with special interests in Africa and human evolution. At Berkeley, Isaac and Clark worked closely together to establish what quickly became the preeminent program in African prehistory and the archeology of human origins. In 1970 Clark **Howell** joined the program followed by Tim **White** in 1977. Isaac returned to Kenya to begin field research in the Late Pleistocene deposits in the Naivasha-Nakuru basin of the eastern Rift Valley in 1969–70. Also during this period, Richard Leakey invited Isaac to be co-director of the **East Rudolf Research Project**, later renamed the **Koobi Fora Research Project**. Although he was focused on documenting and interpreting the archeological record in the region, Isaac's field experience and training in geological and evolutionary sciences shaped his leadership of what became a collaborative and interdisciplinary investigation. Isaac, Kay Behrensmeyer, and a small team discovered **Oldowan** artifacts associated with the remains of a variety fossil animals in relatively undisturbed channel fill and deltaic deposits. These archeological

sites, particularly **KBS** (FxJj1) and HAS (FxJj3), led Isaac to formulate his food sharing/**home base hypothesis** and began a decade of empirical investigations into the technological abilities, subsistence patterns, ranging behavior, and social strategies of early Oldowan tool makers. Isaac's archeological teams grew to involve colleagues from many other institutions and provided dissertation topics and support for a large number of students at Berkeley and other schools. Much of the research was later published as Volume 5 of the *Koobi Fora Research Project* monograph series. The KBS and HAS sites were initially thought to be close to 2.6 Ma because of their stratigraphic relationship to the **KBS Tuff**, potentially making them the oldest evidence of stone tools and animal butchery. These early dates stoked a debate, however, because they placed the sites and other hominin remains much earlier than comparable finds at the nearby Omo **Shungura Formation**. Ultimately, a comprehensive reanalysis of **chronostratigraphy** in the **Turkana Basin** resolved that the Koobi Fora sites were closer to 1.9 Ma. During this dating controversy, Isaac was a strong advocate for collegial dialogue and the importance of undertaking new research and analyses to test the divergent dating hypotheses and resolve the scientific uncertainty. Isaac moved from Berkeley to Harvard University in 1983 but his intellectual home base remained in Africa. He prioritized African scholarship, training students and expanding international scientific collaboration. One of his last publications *Ancestors for us All: Towards Broadening International Participation in Paleoanthropological Research* (Isaac 1984) focused optimistically on the ethical responsibilities and potential of integrative human origins science. Pursuing new international collaboration was one motive for his trip to China in 1985, sponsored by the National Academy of Sciences. Tragically he collapsed from an illness shortly after arriving in Beijing, an illness that resulted in his untimely death at the height of his powers.

Isampur Quarry (Location 16°30′N, 76°29′E, India; etym. after nearby village) History and general description A significant **Acheulean** quarry consisting of more than 15,000 excavated artifacts in the **Hunsgi-Baichbal valley** of southern India. Discovery of the stone-tool quarry was made in 1994 during a survey to locate stratified sites. Temporal span and how dated? Electron spin resonance spectroscopy dating ages on two herbivore teeth suggest an age of 1.27 Ma (linear-uptake model) or 730 ka (early-uptake model). Archeological evidence found at site The artifact assemblage is mostly limestone and includes giant cores, flakes, and bifacial and unifacial tools. Hammerstones are made from a variety of raw materials, some of which were imported into the site. The entire biface-manufacturing sequence was carried out at the limestone source, from the extraction of the bedrock to the creation of finished handaxes and cleavers. Two main reduction strategies were identified, consisting of the manufacture of handaxes from thin slabs and the production of cleavers from large flakes struck from prepared cores. Analysis of the stone-tool reduction sequences indicates that hominins were engaged in repeated imitative behaviors, while demonstrating a long sequence of hierarchically organized cognitive actions. Key references: Geology, dating, and paleoenvironment Paddayya et al. 2002; Archeology Paddayya and Petraglia 1997, Petraglia et al. 1999, 2005, Shipton et al. 2009a, 2009b, Shipton 2010. *See also* **imitation**.

Isenya (Location 1°40'S, 36°50'E, Kenya; etym. named after a nearby small trading village) History and general description A large open-air **Acheulean** site characterized by its dense archeological levels (particularly its abundance of **cleavers**) and a rich fossil fauna. More than 300 m² were excavated from 1983 to 1988 by a team led by Hélène Roche, revealing five archeological strata within fluvial sediments some 55 km/34 miles south of Nairobi in the Eastern Highlands and approximately 45 km/28 miles east of Acheulean sites of the **Olorgesailie Formation**, Kenya. Temporal span and how dated? Isenya is estimated to be **Early** to **Middle Pleistocene** in age. **Tephrostratigraphic** correlation to Member 10 of the Olorgesailie Formation (Roche and Texier 1995, p. 157) suggests an age greater than *c.*0.6–0.7 Ma although the data supporting this correlation have yet to be published. **Biostratigraphical** evidence suggests similarities with Bed IV at **Olduvai Gorge**, Tanzania, with extinct taxa including *Elephas recki*, two species of the genus *Metridiochoerus*, and *Hipparion* sp. Hominins found at site Brugal and Denys (1989) report a small shaft of right fibula, which they attributed to *Primates gen. et sp. indet.* that may be hominin. Archeological evidence found at site The excavated assemblage includes more than 17,000 arti-

facts. Larger tools such as the 1300 cleavers and 800 **handaxes** were made primarily of locally available lavas, with smaller tools primarily made of **chert** from sources approximately 3 km/1.8 miles upstream of the site. The more than 2000 faunal elements in the bovid-dominated fauna indicate a savanna habitat. Cut and percussion marks are rare but they suggest that hominins were involved in the accumulation of some of the faunal remains. Key references: Geology, dating, and paleoenvironment Roche et al. 1988, Roche and Texier 1995; Hominins Brugal and Denys 1989; Archeology Brugal and Denys 1989, Roche et al. 1988, Roche and Texier 1995.

Ishango (Location 0°8'S, 29°36'E, Democratic Republic of the Congo; etym. the name means "confluence" in KiNande, referring to the place where the Semliki River exits from Lake Edward) History and general description The site complex is on the right bank of the Semliki River at its exit from Lake Edward, approximately 6 km/3.5 miles south of Katanda. The terraces were explored briefly by H. Damas in 1935, in 1950–7 by a Jean **de Heinzelin**-led expedition, and again in 1983–4 by Noel Boaz. From 1985 to 1990 Alison Brooks, John Yellen, and Kanimba Misago conducted further excavations both at the original site (Ishango 11) and at a second site further downstream (Ishango 14). The main artifact- and bone-bearing levels at Ishango were deposited on an eroded bench of Plio-Pleistocene-age deposits (Lusso beds, formerly equated with the Kaiso beds of Uganda). These latter were then covered by at least two later Pleistocene to Holocene depositional episodes. An erosional unconformity separates the lower, later Pleistocene deposits, capped by a cemented layer ("terrasse tufacée" or "Tt"), from the later or Holocene deposits ("Terrasse posterieure" or "Tp"). This terrace sequence is approximately 10–12 m above the current river level and is banked up against a much higher and older Middle Pleistocene to early later Pleistocene terrace series ("Terrasses supérieures" or "Ts") that contains the Semliki and Katanda Beds. Terrace Tt, which lies 10.5 m above the present lake level, includes a fining-upward sequence consisting of a lower gravel deposit (GI), the Principal Fossiliferous Layer (NFPr) of fine sands and small gravels, and the uppermost cemented tufaceous levels (NT). A Holocene-age catastrophic flood removed most of the massive ashfall from the early

Holocene volcanic eruption of the Katwe volcano (Uganda) which underlies the modern soil elsewhere in the region. Terrace Tp, which consists of **midden** deposits in fine sands and volcanic ash, was deposited on top of the NT. It contains both the later Holocene Post-Emersion Zone (P-EZ) and several dark soil horizons of Neolithic and Iron Age materials above it. Cultural materials from the Belgian expeditions and the hominins Ishango 1–8 are curated at the Institut Royal des Sciences Naturelles de Belgique, Brussels, Belgium. Temporal span and how dated? Initial **radiocarbon** ages of 18 and 21 ka on mollusc shell were not regarded as accurate. Citing comparisons of the material with European post-Pleistocene fishing industries, it was suggested that most of the hominin remains were $c.9$ ka (de Heinzelin 1962). New chronometric dates suggest that the "Tp" and the zone post-emersion post-date 7 ka BP, based on an **AMS radiocarbon date** on charcoal, while the NFPr and other deposits of the Tt date to between 25 and 20 ka BP, based on new AMS radiocarbon dates on mollusc shell and ostrich eggshell, and amino acid racemization ages on both mollusc shell and ostrich eggshell. Hominins found at site Ishango 1, a right mandibular ramus. Ishango 2–4 derive from NFPr and comprise three partial mandibles, 10 cranial fragments, assorted teeth, dens, atlas, cervical vertebra, adult left humerus, juvenile right humerus, three left (one is juvenile) and one right ulnae, three right radii, two metacarpals, nine manual phalanges, three femoral fragments, left patella, right tibia, left talus, 20 metatarsals, and three pedal phalanges. Ishango 5–8 and unnumbered individuals derive from P-EZ and comprise six mandibular fragments, some with teeth, a fragmentary left ulna and left radius, four metacarpals, two manual phalanges, three patellae, proximal epiphyses of a right and a left tibia, left talus and calcaneus, and one pedal phalanx, plus Boaz et al. (1990) recovered 19 additional hominin fragments from NFPr. All of the hominin remains are anatomically modern, but the molar teeth are strikingly large and the mandibular symphyses are deep but the shafts of the limb bones are straight and slender. Archeological evidence found at site NFPr to NT contained **Later Stone Age** artifacts including microlithic scrapers and backed crescents, harpoons with barbs on two sides in the earlier level gradually replaced by those with barbs on only one side, grindstones, but no pottery.

Cultural material from P-EZ contained a microlithic Later Stone Age industry, with bone artifacts that include harpoons with barbs on one edge, but no pottery or grindstones. Key references: Geology and dating Damas 1940, Lepersonne 1949, 1970, 1974, de Heinzelin 1955, 1957, 1962, Helgren 1997, Brooks and Smith, 1987, Brooks et al. 1991, Verniers and de Heinzelin 1990, Peters 1990, de Heinzelin and Verniers 1996, Brooks et al. 2008a; Hominins Twiesselmann 1958, de Heinzelin 1962, Boaz et al. 1990, Brooks et al. 1991, Crevecoeur 2008b; Archeology de Heinzelin 1955, 1957, 1961, 1962, Brooks and Smith 1987, Stewart 1989, Boaz et al. 1990, Peters 1990, Yellen 1998, Mercader and Brooks 2001, Brooks et al. 2008b.

Isimila   (Location 7°53′48″S, 35°36′12″E, Tanzania; etym. meaning unknown) History and general description This rich open-air **Acheulean** site located in the Iringa Highlands (outside of the Rift Valley) was discovered in 1951, with extensive excavations beginning in 1957 under the direction of F. Clark **Howell** together with Glen Cole and Maxine Kleindienst. The excavations at Isimila are an important early example of landscape-scale excavations to look at spatial and temporal variation in the **Early Stone Age** archeological record. Temporal span and how dated? **Uranium-series dates** on bone fragments (a poor material for this method) from Sands 4 range from more than 170 ka to $c.330$ ka, but these may well be underestimates. Although fauna is sparse, the presence of the **suid** *Metridiochoerus compactus* is consistent with an older, perhaps **Early Pleistocene** age. Hominins found at site None. Archeological evidence found at site Gully erosion has exposed sediments of the >18 m-thick Isimila Beds, a series of five sand deposits (Sands 1–5 numbered from top to bottom) overlying a paleosol developed on bedrock. The Isimila Beds are exposed over an area of at least 2 km^2, and were the subject of numerous large-scale excavations (some apparently as large as 620 m^2) from laterally continuous deposits. Excavation results were important in demonstrating both the presence and absence of **handaxes** and similar implements in Acheulean stone tool assemblages and the importance of environment and hominin activity differences in explaining this variation. All of the fauna, mainly dental specimens of elephant, rhino, equids, suids, and bovids, plus some small taxa, derive from Sands 4. Key references: Geology, dating, and paleoenvironment Hansen and Keller 1971, Harris and White 1979, Howell et al. 1962, 1972; Archeology Cole and

Kleindienst 1971, Coryndon et al. 1972, Hansen and Keller 1971, Howell 1961, Howell et al. 1962.

island rule Proposed by Foster (1964), who suggested that island populations of mammals tended towards either gigantism or dwarfism depending on the original size of the organism. According to the island rule, if a mammal was on average smaller than a rabbit it will get larger, and if it is larger than a rabbit then it will tend to dwarf. The logic is that in the absence of mainland predators organisms no longer need to be small in order to be cryptic. Similarly, larger organisms, especially herbivores, would have access to less forage on an island and in the absence of mainland predators there is no need for them to resort to a large body mass to avoid predation, and so will get smaller.

ISMA Acronym for the **Institute for the Study of Man in Africa** (*which see*).

Isoberlinia *See* **miombo woodland**.

isochron (Gk. *isos*=the same and *kronos*=time) Refers to events happening at the same time, or to what a complex surface (e.g., a land surface or the outer enamel surface of a developing tooth) looks like at a point in time. It is also used in connection with potassium-argon dating.

isocortex (Gk *iso*=same or equal) The isocortex is synonymous with the **neocortex**. *See* **cerebral cortex; neocortex**.

isometric relationship (Gk *iso*=same or equal and *metron*=measure) *See* **isometry**.

isometry (Gk *iso*=same or equal and *metron*=measure) A change in size without an associated change in shape. Isometry is identified in scaling analyses when variables of equivalent **dimension** (i.e., both variables are lengths, or areas, or volumes/masses) are log transformed and plotted against each other and the first variable changes at the same rate as the second so that the slope (α) of the regression line is equal to 1. The objects on the regression line change their size, but not their shape. If the variables are of different dimensions then the isometric slope is equal to the ratio of the y-axis dimension to the x-axis dimension. For example, if a log-transformed area measurement (dimension=2) is plotted against a log-transformed

volume measurement (dimension=3) then the isometric slope is 2/3. It should be noted that if the y-axis variable is not a size variable but a shape variable (e.g., an angle or a ratio of size measurements) then the isometric slope is equal to zero (i.e., a change in size with no change in shape).

isotope (Gk *iso*=same or equal and *topos*=place) One of two or more atoms that have the same atomic number but different atomic masses. Because they have the same atomic number (i.e., the same number of protons) they occupy the same place in the periodic table (hence the name "isotope") and have the same element name, but because they differ in the number of neutrons they will have a different number superfix (atomic mass). Contrast this with what happens in non-isotopic decay. The daughter product has the same atomic mass, but one less proton than the parent isotope. This changes its atomic number and thus its place in the periodic table (e.g., when ^{40}K decays to ^{40}Ar). Many isotopes of the lighter elements such as carbon, nitrogen, and oxygen do not decay. Instead their proportions are determined by processes such as photosynthesis (e.g., $^{13}C/^{12}C$), climate or diet (e.g., $^{2}H/^{1}H$, $^{13}C/^{12}C$, $^{15}N/^{14}N$, and $^{18}O/^{16}O$). These isotopes are called **stable isotopes** and researchers use them to reconstruct diet, habitat, and migration in modern and extinct animals.

isotropy (Gk *iso*=equal and *tropus*=direction) Meaning, literally, equal in all directions. Within paleoanthropology, isotropy is used most commonly in reference to the material properties of a substance (e.g., **bone** or **enamel**) or to **dental microwear** textures. Materials are isotropic when their material properties, particularly their various measures of material behavior (stiffness as reflected by the **elastic modulus** or **Young's modulus**, shear stiffness as reflected by the **shear modulus**, and the relationship between axial and lateral strains as reflected by **Poisson's ratio**) are the same in all directions. Moreover, in isotropic structures those measures are related to each other in a mathematically predictable fashion such that the elastic modulus is equal to one plus the Poisson's ratio, multiplied by twice the shear modulus. Bone, like many other biological materials, is not isotropic, but it is often assumed to be so as a simplifying assumption in biomechanical analyses (e.g., **finite element analysis**). Regarding dental microwear, isotropy is a term used to describe the orientation of microwear features. Microwear is isotropic when the features are not consistently aligned in any given direction (i.e., feature alignment varies without patterning). The opposite of isotropy is **anisotropy**, although additional terms are used in materials science (e.g., **orthotropy, transverse isotropy**) also denote departures from isotropy. *See also* **dental microwear**.

Istállóskő (Location 48°04'18"N, 20°25'08"E, Hungary etym. inversion of Hu. *kő istálló*=stone barn, since the cave was used to house pigs in historic times) History and general description This cave site in the Bükk Mountains, near Szilvásvárad, was excavated throughout the 20thC. It preserves evidence of one of the oldest examples of the **Aurignacian** in Europe, suggesting a long period of overlap between **modern humans** and *Homo neanderthalensis*. However, some suggest there are more recent dates for the site. Temporal span and how dated? Early **radiocarbon dating** of bone provides an age of *c*.40 ka (uncalibrated); more recent dating has given a range of *c*.33-28 ka (uncalibrated). The anatomically modern human tooth has been directly dated to *c*.39 ka, although this date is questioned by some. Hominins found at site The germ of a right M_2 that is assigned to *Homo sapiens*. Archeological evidence found at site Aurignacian I and Aurignacian II lithic technology. Key references: Geology, dating, and paleoenvironment Allworth-Jones 1990, Vogel and Waterbolk 1972, Adams and Ringer 2004; Archeology Vértes and Vries 1952, Adams and Ringer 2004; Hominins Malán 1954, Thoma and Vértes 1971.

Isturitz (Location 43°21'10"N, 1°12'21"W, France; etym. Basque, name of the nearby town) History and general description This cave site located in a limestone hill near Bayonne on the Atlantic Coast of France has been known since at least the late 1700s when it was explored for potential gold deposits and then in the late 1800s it was partially excavated for phosphate. Organized amateur and professional archeological excavations in the main area of the cave (Salle d'Isturitz or Grande Salle) and in a second, smaller chamber (Salle de Saint-Martin) began in 1912 and the current excavations directed by Christian Normand have been underway since 1999. This is one of very few sites with well-dated *Homo sapiens* fossils in association with an early **Aurignacian** deposit. Temporal span and how dated? The earliest Aurignacian levels have been dated using **AMS radiocarbon**

dating to 41,355– 42,625 cal. radiocarbon years BP. Hominins found at site 119 fragments and isolated teeth were found, all from the Upper Paleolithic levels. All of material is assigned to *H. sapiens*, even those from the earliest Aurignacian. Archeological evidence found at site Most deposits are **Upper Paleolithic**, with some evidence of the **Middle Paleolithic**. The Salle d'Isturitz has thin Aurignacian and **Solutrean** layers and thick **Magdalenian**, and **Gravettian** deposits, all capped by Bronze Age sediments and burials. Perforated human teeth were found in the upper levels. Amber pendants were found in the earliest Aurignacian deposits and whale bone artifacts were found in the Magdalenian layers. The **Mousterian** deposits in the Salle de Saint-Martin are topped by Aurignacian and Magdalenian; there is no evidence of the Châtelperronian. Key references: Geology, dating, and paleoenvironment Normand 2002, Szmidt et al. 2010; Hominins Gambier 1990/1991; Archeology Normand 2002, Szmidt et al. 2010.

IVPP Acronym for the **Institute of Vertebrate Paleontology and Paleoanthropology** (*which see*).

Iwo Eleru (Location 7°26'N, 5°8'E, Nigeria; etym. unknown) History and general description A large rock-shelter in the rain forest of western Nigeria excavated by Thurstan Shaw in 1965. Temporal span and how dated? Conventional **radiocarbon dating** on charcoal suggests and age of 2.52±0.07 ka and 11.25±0.2 ka. Hominins found at site Iwo Eleru 1, a tightly contracted fragmentary skull and skeleton was excavated from the bottom of the sequence and Shaw reported he was confident from the stratigraphy that the burial was not intrusive. The cranium preserves much of the vault, which is long, low, and narrow. Its markedly receding frontal aligns it with "archaic" crania in **multivariate space** and the **chin** is weakly developed. Archeological evidence found at site A **Later Stone Age** sequence containing more than half a million artifacts. Key references: Geology, dating, and paleoenvironment Shaw 1965, 1968, Brothwell and Shaw 1971; Hominins Brothwell and Shaw 1971, Stringer 1974, Bräuer 1984; Archeology Shaw 1965, 1968, Brothwell and Shaw 1971.

Iwo Eleru 1 *See* Iwo Eleru.

Iziko Museums of Cape Town *See* **South African Museum**.

J

jackknife A **resampling** procedure similar to the **bootstrap** that can be used to calculate a confidence interval and/or the degree of bias for any sample statistic (e.g., mean, variance, median, correlation coefficient, regression slope, etc.). In its simplest application, the jackknife is conceptually more difficult than the bootstrap and it is also a less flexible technique than the bootstrap. For these reasons, and because the computing power necessary for performing bootstrap analyses is now readily available, the jackknife is not often used. For a sample of size n, the statistic of interest is calculated n times, each time leaving out one measurement. For example, for a sample of 10 endocranial capacities, the median can be calculated for all but the first measurement, all but the second measurement, and so on. These calculated values are known as partial estimates. For each partial estimate a quantity known as a pseudovalue can be calculated using the following formula: $\hat{\theta}_1^* = n\hat{\theta} - (n-1)\hat{\theta}_1$, where $\hat{\theta}$ is the statistic of interest (e.g., median) calculated for the full sample and $\hat{\theta}_1$ is the partial estimate calculated by leaving out the ith value. In our example, the first pseudovalue would be calculated as 10 times the median for the whole sample, minus nine times the median calculated for the sample without the first measurement. This equation takes into account the fact that the partial estimates are calculated on samples that are smaller than the full sample. The resulting pseudovalues are equivalent to the values that are generated in a bootstrap distribution. The mean and standard deviation of these pseudovalues can be treated as the jackknife estimate of the statistic and the standard error of the jackknife estimate can be used to generate a confidence interval. The pseudovalues can also be used to estimate bias in the jackknife estimate of the statistic. Manly (2007) provides further detail.

Jacob, Teuku (1929–2007) Teuku Jacob was born in Peureulak, Aceh, in what was then the Dutch East Indies (now Indonesia). During the late 1940s he participated in the Indonesian independence movement and remained an ardent nationalist throughout his life. He entered **Gadjah Mada University** in Yogyarkarta in 1953 to study anatomy in the university's School of Medicine and completed his medical degree in 1956. He became an assistant in anthropology at the university in 1954 and became a lecturer in anthropology at the university in 1962. He continued his studies in anthropology at Howard University in Washington DC in the USA where he completed a Master's degree in 1960. He then went to The Netherlands where he studied paleoanthropology with Ralph **von Koenigswald** at the Rijksuniversiteit in Utrecht, completing his doctorate "Some problems pertaining to the racial history of the Indonesian region" in 1967 that included a study of the small-statured modern humans from Flores, including the negritos from Liang Toge. When he returned to Gadjah Mada University after completing his studies in The Netherlands, Jacob accepted a position as Professor of Anthropology and was appointed to head the Laboratory of Bioanthropology and Paleoanthropology at the university, serving as the Dean of the School of Medicine from 1975 to 1979. He was also involved in Indonesian politics, serving as a member of the Indonesian People's Consultative Assembly from 1982 to 1987. Throughout his career at Gadjah Mada University, Jacob pursued studies of Indonesian *Homo erectus* fossils and rose to considerable prominence in Indonesian paleoanthropology. Early in his career, in 1962, he conducted excavations at **Sangiran** where he later found *Homo erectus* and early *Homo sapiens* fossils and stone tools. It was most likely Jacob's familiarity with the small-statured modern humans from Flores, including the negritos from Liang Toge, that caused him to involve himself centrally in the debate about *Homo floresiensis*, joining those who strenuously opposed the proposal that the **Liang Bua** fossils represent a novel species

Wiley-Blackwell Encyclopedia of Human Evolution, First Edition. Edited by Bernard Wood.
© 2013 Blackwell Publishing Ltd. Published 2013 by Blackwell Publishing Ltd.

of dwarfed primitive hominin. Jacob proposed a very different interpretation, namely that the Liang Bua fossils were only 1300–1800 years old, that they sample an "Australomelanesid race" of modern humans. He also suggested that the unusually small skull could be explained as being the result of a congenital abnormality called **microcephaly**.

Jacovec Cavern *See* **Jakovec Cavern**.

Jakovec Cavern (etym. Justin Wilkinson, who explored the Sterkfontein Cave system, named this particular cave after a friend whose family name was Jakovec) Part of the **Sterkfontein** cave system containing **breccia** equivalent in age to Sterkfontein Member 2. In 1995 two of the charges laid to blast the breccia in the **Silberberg Grotto** were detonated, but no breccia was removed. This indicated to one of the miners laying the charges that the force of the charge must have dissipated into a cavity on the other, eastern, side of where the charges had been laid. Ron **Clarke** knew the Jakovec Cavern was east of the Silberberg Grotto and he surmised that the breccia in the grotto might also be exposed in the nearby cavern. Meshaka Makgothokgo, Stephen Motsumi, and Nkwane Molefe were sent to search the walls of the Jakovec Cavern for signs of fossils, and within a day they had located a section of the brain case of a hominin. This was the first of the 12 hominins recovered in 1995 that were reported in Partridge et al. (2003). *See also* **Sterkfontein**.

Jaramillo *See* **geomagnetic polarity time scale**.

Jarman/Bell principle A model that relates dietary quality, metabolism, and body size. The model was initially derived from studies of antelopes (Bell 1970, Jarman 1974), but it has also been shown to be applicable to primates (Gaulin 1979). The Jarman/Bell principle suggests that because large-bodied animals require less energy per unit body mass than smaller-bodied ones, they can exist on lower-quality diets. However, although the relative energy intake of large-bodied animals is lower compared to small-bodied animals, their food consumption is still absolutely greater than that of smaller-bodied animals and if they focused on less abundant, high-quality foods (such as ripe fruits and insects, for example) large-bodied animals may not be able to meet their daily energy needs. The Jarman/Bell principle provides a useful basis

for investigating the relationship between body mass, diet, and metabolic rate in studies that examine the role energetics may have played in hominin evolution.

jasper A characteristically reddish variety of cryptocrystalline silica (SiO_2) (i.e., chert). The distinctive color is due to the presence of trace amounts of iron oxide. Nodules of jasper have the hardness and **conchoidal fracture** properties typical of siliceous materials and at some **Middle Stone Age** sites it was the selected raw material for flaking.

***Javanthropus* Oppenoorth, 1932** (*Java*= from the island and Gk *anthropos*=human being) A subgenus of the genus *Homo* introduced by Oppenoorth (1932) to accommodate the new species *Homo soloensis*. In a later paper Oppenoorth (1937) suggested the subgenus should be abandoned and nearly all researchers (e.g., Le Gros Clark 1955) took his advice.

jaw (E *jowe* or *joue*=cheek) Refers to either the maxilla (i.e., upper jaw) or the mandible (i.e., lower jaw).

Jebel Irhoud (Location 31°51'36"N, 8°52' 12"W, Morocco; etym. unknown; alternate spellings Jebel Ighoud, Djebel Irhoud) History and general description Discovered during mining, this cave site has produced several **Middle Paleolithic** hominin fossils. It was originally excavated by Ennouchi, and subsequently by Tixier and de Bayle des Hermens in 1967 and 1969. Much of the recent analysis, including new excavations beginning in 2004, have been done by an international group led by Jean-Jacques **Hublin** and Abdelwahed Ben-Ncer. Temporal span and how dated? **Electron spin resonance spectroscopy** (ESR) **dates** on horse teeth provide a range of 90–125 ka assuming an early-uptake model or 105–190 ka assuming a linear-uptake model. More recent uranium-series and ESR dates on one hominin tooth suggest an age of 160±16 ka. Hominins found at site The remains of at least four individuals have been recovered. They were originally attributed to *Homo neanderthalensis* or were at least considered to possess some Neanderthal traits, but reanalysis suggests that similarities to early European fossils may be due to shared ancestral traits and that the Irhoud material displays more similarities with early modern humans from the

Near East. **Irhoud 1** and **Irhoud 2** are mostly complete fossil crania. Irhoud 3 is a juvenile mandible with a well-preserved dentition. A study of dental microstructures suggest that this individual experienced a prolonged life history, in which case it would be among the earliest evidence of a modern human-like life history. Many authors consider these fossils to be the predecessors of the later individuals found at **Dar es Soltane II** in Aterian context. Archeological evidence found at site A Middle Paleolithic stone tool technology with use of **Levallois** technique and a high proportion of scrapers, which has been named "Irhoud Mousterian." Faunal remains are poorly preserved but dominated by gazelles. Key references: Geology, dating, and paleoenvironment Hublin 1992, Grün and Stringer 1991, Smith et al. 2007; Hominins Hublin 1992, Hublin and Tillier 1981, Hublin et al. 1987, Tixier et al. 1998, Smith et al. 2007; Archeology Hublin et al. 1987.

Jebel Qafzeh   *See* **Qafzeh**.

Jebel Sahaba (Site 117)   (Location 21°59'N, 31°20'E, Sudan; etym. Arabic, "mountain of the companion of the Prophet," a local geographical feature) History and general description This site, 3 km/1.8 miles north of Waldi Halfa, was discovered and excavated during the rescue archeological efforts that preceded the construction of the Aswan Dam. The existence of the site was first noted in 1962, but the first skeletal and archeological evidence was recovered in 1965 by a team led by Fred Wendorf. Tony Marks conducted additional excavations in 1966, recovering nine more skeletons of a total for the site of 58. The burials were in flexed positions, heads to the north, and faces turned east. Most of the skeletons were partially mineralized and 24 were associated with **microlith**ic artifacts that had apparently been attached to darts or arrows shot into their bodies. Temporal span and how dated? Associated, possibly Qadan (Nile Valley Later Stone Age, or LSA) lithics and conventional **radiocarbon dating** of the skeletons suggest an age of 12–14 ka. Hominins found at site Anderson (1968) analyzed 52 of the 58 skeletons and that subsample included 19 adult males, 17 adult females, seven adults of indeterminate sex, and nine juveniles. The crania are large and robust with strong supraorbital tori and some are prognathic; the mandibular bodies are robust with large teeth. Morphologically, the crania

closely resemble those from **Wadi Halfa**, and recall the Mechoid people from the **Ibero-Maurusian** of North Africa. The dental morphology resembles that of West Africans rather than the later agricultural populations of Nubia. Archeological evidence found at site 110 microlithic artifacts, possibly from the Qadan (Nile Valley LSA). Key references: Geology, dating, and paleoenvironment Wendorf 1968a; Hominins Anderson 1968, Turner and Markowitz 1990, Irish and Turner 1990, Irish 2005; Archeology Wendorf 1968a.

JHS   Acronym for John's Hominid Site. *See* **Omo-Kibish**.

Jinniushan   (金牛山) (Location 40°34'40"N, 122°26'38"E, northeastern China; etym. named after a local hill) History and general description The fossil-hominin-bearing locality (Site A) is one of several breccia-filled fissures located in an isolated karst tower locally known as Jinniushan (Golden Ox Hill), 1 km/0.6 miles west of Sitian Jiatun village and about 20 km/12 miles southeast of the city of Yingkou, Liaoning Province. Archeological survey and excavation began in 1974 and three potential localities, Site A (located on the eastern slope), Site B (located 20 m north of A), and Site C (located on the western slope), were discovered. In 1984, archeology field school students from Peking University, under the direction of Professor Lü Zun'E, discovered a partial hominin skeleton in layer 7 (not in layer 6 as originally reported) of Site A. Although Jinniushan is one of the most important paleoanthropological sites in China, most of the excavated materials (fauna, hominin, and lithics) have not been widely studied and results have mostly been published in Chinese. Temporal span and how dated? Chen et al. (1994) report a mean of 237 ka from **uranium-series dating** and a mean **electron spin resonance spectroscopy** (early-uptake model) date of 187 ka for Jinniushan based on five fossil mammalian teeth. Taken together, the authors suggest a conservative, minimum age of *c*.200 ka. Lü (1985) reports another uranium-series date of 280 ka for layer 7. However, both Wu Ru-kang (1988) and Geoffrey Pope (1992) have called the reliability of the older date into question on stratigraphic grounds. Hominins found at site A partial skeleton belonging to a male around 20 years of age. The cranium was reconstructed by Wu Xin Zhi and Zhao Zhong Yi of the **Institute of Vertebrate**

Paleontology and Paleoanthropology from over 100 broken fragments. It is thin-vaulted and has an endocranial volume of about 1300 cm^3. The postcranial elements include six **vertebrae** (one cervical and five thoracic), two left ribs, a complete left ulna, a complete left os coxa, a complete left patella, and numerous articulated bones of both wrists, hands, ankles, and feet. Archeological evidence found at site Only two lithic artifacts were reported from the seven layers (numbered from top to bottom) of excavated deposits at Site A. From the dense accumulations of well-stratified **microfauna**, Pope (1992) suggests that Site A was likely a natural faunal trap not an occupation site. At Site C, 15 small stone artifacts were recovered from layers 4–6 (the Lower Cultural Zone of presumed **Early Paleolithic** antiquity) of six layers in 1975. These are dominated by **scrapers** and most show **retouch**ing by simple direct percussion. Abundant mammalian fossils and evidence of fire utilization (details unclear) have also been reported from the Lower Cultural Zone. Repository Beijing University Department of Archaeology. Key references: Geology, dating, and paleoenvironment Lü 1985, 1990, Chen et al. 1994; Hominins Lü 1985, 1990, Wu 1988, Pope 1992, Wu and Poirier 1995, Brown 2001, Rosenberg et al. 2006; Archeology Jinniushan Joint Excavation Team 1976, 1978, Wu and Olsen 1985, Pope 1992.

Johanson, Donald Carl (1943–) Donald Johanson was born in Chicago, Illinois, USA. He became interested in anthropology as a child and after initially choosing chemistry as his major at the University of Illinois at Urbana–Champaign he later changed to anthropology and received his BA in 1966. Johanson entered The University of Chicago to study with paleoanthropologist Clark **Howell**, completing his MA in 1970. Johanson began his graduate studies and took part in the 1970, 1972, and 1973 fieldwork seasons of the **International Omo Research Expedition** that was exploring the deposits of the Omo River Basin in Ethiopia. While working in the Omo, French geologist Maurice Taieb invited Johanson to join him in examining some promising deposits located in the **Afar Triangle** region in northeast Ethiopia and when Taieb, Johanson, French paleoanthropologist Yves **Coppens**, and American geologist Jon Kalb explored that region in 1972 they came to realize the potential of the site now known as **Hadar**. The first formal expedition to Hadar in

1973 resulted in the recovery of a large quantity of mammal fossils along with an early hominin knee joint. The following year an even more remarkable hominin discovery was made when Johanson and Tom Gray recovered a substantially complete associated skeleton, **A.L. 288-1**, to which they gave the nickname "Lucy." Fieldwork continued at Hadar and in 1975 the team unearthed the remains of at least 13 hominin individuals, about 200 fossils in all, at locality 333. Johanson began working in the Department of Physical Anthropology at the **Cleveland Museum of Natural History** in 1972. Upon completing his PhD on variation in chimpanzee dentition at the University of Chicago in 1974, he became Curator of Physical Anthropology at the Cleveland Museum. Since it was impossible to continue fieldwork at Hadar, Johanson began to examine the Hadar fossils with the assistance of William Kimbel, Owen Lovejoy, Bruce Latimer, and Tim **White** among others. There was initially uncertainty as to whether the Hadar specimens represented *Homo* or *Australopithecus*, and whether they sampled one, or more, species. Johanson and his team became convinced that all the Hadar remains belonged to one, novel, species of *Australopithecus*, and in 1978 they proposed that the Hadar hominin fossils should be referred to new taxon *Australopithecus afarensis*. In 1981 Johanson left his position at the Cleveland Museum of Natural History (as well as the adjunct faculty positions he held at Case Western Reserve University and Kent State University) and founded the **Institute of Human Origins** (or IHO) in Berkeley, California, to promote and facilitate further multidisciplinary research into human origins. He also wrote, with Maitland A. Edey, a popular book titled *Lucy: The Beginnings of Humankind* (1981), which described the discoveries at Hadar and his interpretation of the role played by *Au. afarensis* in hominin evolution. Johanson and his team returned to Africa in 1985 when he received permission from the Tanzanian government to excavate at **Olduvai Gorge** where Louis and Mary **Leakey** had worked. In 1986, Tim White found a 1.8 Ma associated hominin skeleton (**OH 62**) that White and Johanson argued belonged to *H. habilis.* In addition to directing the IHO, conducting field research in the Hadar study area and contributing to the analysis of the Hadar hominins Johanson was a Professor of Anthropology at Stanford University from 1983 to 1989 and also continued to write books for a popular audience. Johanson collaborated with Maitland Edey to write *Blueprints: Solving the Mystery of Evolution*

(1989), with J. Shreeve to produce *Lucy's Child* (1989), and with Blake Edwards to produce *From Lucy to Langauge* (1996). Johanson and the IHO moved from Berkeley to Tempe, Arizona, in 1997, where it is affiliated with Arizona State University and where Johanson is a Professor and still occupies the Virginia M. Ullman Chair in Human Origins. In July 2008 Johanson stepped down as director of the IHO.

joint (ME *joint*=junction) A discontinuity between two bones. Joints are classified according to whether the tissues that separate the bones are solid (i.e., fibrous and cartilaginous joints) or have a cavity (i.e., synovial joints). There are three types of fibrous joints: suture joints between flat cranial bones (e.g., the sagittal **suture** between the parietal bones), gomphoses joints that keep teeth in their sockets, and syndesmoses fibrous bands of collagen that join two, or more, bones (e.g., the sacro-iliac, tibio-fibular, and interspinous ligaments). Cartilaginous joints are of two types: primary cartilaginous joints (also called synchondroses) in which hyaline cartilage covers the surfaces of the bones involved in the joints and secondary cartilaginous joints (also called symphyses) in which each bone is covered by a plate of hyaline **cartilage** with fibrocartilage between the plates. Synchondroses are what we all have before the ends of our long bones (i.e., epiphyses) fuse with the shaft (i.e., diaphysis). They are also found at the lateral ends of ribs and between some cranial bones (e.g., between the sphenoid and the ethmoid and between the sphenoid and the occipital) formed from the **chondrocranium**. All secondary cartilaginous joints are in the midline (e.g., the intervertebral discs and the pubic symphysis). The surfaces of the bones involved in synovial joints (like those involved in synchondroses) are covered with hyaline (called articular) cartilage and the bones are connected by a fibrous capsule lined with synovial membrane. The space between the membrane and the cartilage contains synovial fluid and may also include a flat meniscus made of fibrocartilage. Synovial joints are usually protected by ligaments. These may be thickenings of the capsule (e.g., the ilio-femoral ligament) or they are non-capsular. Non-capsular ligaments may be within the capsule (e.g., the cruciate ligaments of the knee joint) or outside the capsule (e.g., sacro-iliac ligament). Tension in ligaments is one of the main factors restricting movement at synovial joints.

Jones, Trevor *See* Robert **Broom**; **Sterkfontein**.

jugal (L. *iugum* or *jugum*=yolk, as in what is worn by beasts of burden) The name of a cranial bone in amphibians, reptiles, and birds; the equivalent in mammals is the zygomatic bone. The term jugal is used in old anatomy books (e.g., the 7th edition of Buchanan's *Manual of Anatomy* refers to what Martin calls the "frontomalare temporale" as the "jugal point") and some researchers refer to the postcanine teeth as the "jugal" teeth. However, all told jugal is an arcane term that is best avoided. *See also* **malar**.

"Jonny's Child" The nickname given by Louis and Mary **Leakey** to OH 7 from **Olduvai Gorge** that became the **holotype** of *Homo habilis*.

jugular bulb *See* **cranial venous drainage**.

Julbernardia *See* **miombo woodland**.

jump dispersal A type of **dispersal** that occurs when a subpopulation of a larger population of organisms crosses an existing barrier to dispersal. Stochastic processes may play an important role in generating jump dispersals, or novel adaptations may allow individuals to traverse regions that had previously been impassable. A particularly dramatic example of jump dispersal is the colonization of South America by African **anthropoid**s during the early **Miocene**, in which the founding populations of **platyrrhine**s may have crossed the Atlantic Ocean on mats of floating vegetation that functioned as rafts. Jump dispersal may be an important process leading up to **allopatric speciation** and involuntary jump dispersal may have been how hominins reached islands in South Asia. Once hominins were able to build seagoing craft they were effectively able to undergo jump dispersal either by accident or by choice.

jumping genes *See* **transposition**; **transposon**.

junior synonym A name for a taxon that has been superseded by a different name that has historical priority. The junior synonym may be an objective synonym (when two taxa have been erected with the same type specimen) or a subjective synonym (when two taxa, formerly thought to be different, have been synonymized). For example, the names *Homo neanderthalensis*

King, 1864, *Homo primigenius* Schwalbe, 1903, and *Palaeanthropus europaeus* Sergi, 1910 were all given to a taxon whose type specimen was the calvaria from the Kleine Feldhofer Cave and the latter two names are consequently objective junior synonyms of *H. neanderthalensis*. The name of *Australopithecus transvaalensis* was given by Broom, 1936 to a specimen from Sterkfontein, and later transferred to a new genus *Plesianthropus*; most authorities since then have considered that the Sterkfontein fossils are the same species as represented by the Taung skull, which was described as *Australopithecus africanus* Dart, 1925. Consequently the specific name *transvaalensis* is considered a junior subjective synonym of *africanus*, and the generic name *Plesianthropus* is considered a junior subjective synonym of *Australopithecus*.

juxtamastoid eminence A bony process that is seen in some, but not all, modern human crania. When present, it lies between the digastric fossa laterally and the groove for the occipital artery medially and it probably owes its presence to bone being pushed up between the two aforementioned structures, especially when the occipital artery takes a course lateral to the occipitomastoid crest and the occipitomastoid suture. A juxtamastoid eminence is present in early hominin crania such as **SK 27** and **SK 847** and the combination of a large juxtamastoid eminence but small mastoid process is one of the distinctive features of *Homo neanderthalensis*. The juxtamastoid eminence is probably equivalent to Franz **Weidenreich**'s *crista paramastoidea* (syn. juxtamastoid process).

juxtamastoid process *See* **juxtamastoid eminence**.

Jwalapuram (Location 15°19′20″N, 78°08′01″E, India; etym. named after a local village) History and general description A group of archeological sites located in the Jurreru River Valley, Kurnool District, southern India. Surveys and excavations were performed between 2003 and 2010, identifying rock-shelters and open-air sites, some in association with the Toba super-eruption event. Temporal span and how dated? **Radiocarbon dating** (AMS) and optically stimulated **luminescence dating** suggest an age of *c.*78–10 ka. Hominins found at site Four fragmentary and burned (calcined) cranial fragments and a tooth attributed to *Homo sapiens*, and dating to 20–12 ka at Jwalapuram Locality 9 rock-shelter. Archeological evidence found at site Jwalapuram Locality 3 contains **Middle Paleolithic** assemblages in association with Toba ash. The 511 artifacts excavated from Jwalapuram Localities 3 and 23 are made of limestone, chert, quartzite, quartz, chalcedony, and dolerite. The assemblage includes multi- and single-platform cores and **Levallois** cores for the production of flakes and blades, and diverse retouched pieces, including scrapers, notches, and points. The assemblages are notable for showing technological continuity after the Toba event. Discriminant core analysis indicates that the Jwalapuram cores are similar to sub-Saharan **Middle Stone Age** industries. The Jwalapuram Locality 9 rock-shelter dates up to 35 ka, and contains 53,000 excavated lithic artifacts, beads, and worked bones. The transition from the Middle Paleolithic to Microlithic at Jwalapuram occurs *c.*35 ka, and has been attributed to innovations arising from environmental deterioration and demographic increase. Key references: Geology, dating, and paleoenvironment Jones 2010, Petraglia et al. 2007, 2009; Hominins Clarkson et al. 2009; Archeology Petraglia et al. 2007, 2009, Haslam et al. 2010.

K

K-selection *See* r/K-selection.

K/Ar *See* potassium-argon dating.

ka (L. *kilo*=thousand and *annum*=year) The abbreviation for an age determination or **age estimate** expressed in thousands of years. For example, "the date of the anatomically modern human crania from the Omo is *c.*190 ka." The term "ky" or "kyr" should be used to describe periods of time as in "30 kyr have elapsed since anatomically modern humans first reached Europe." *See also* **age estimate**.

Kaalam *See* **Shungura Formation**.

Kabua (Location 3°27′N, 35°47′E, Kenya; etym. named after the water-hole) History and general description The site is 1143 m south-southeast of Kabua water-hole on the west side of Lake Turkana. A fossilized hominin skull was found *in situ* in Late Pleistocene lacustrine deposits that outcrop near Kabua Gorge by T. Whitworth and a research group from the University of Durham in September 1959. Temporal span and how dated? Association with an elevated lake level suggests a **Late Pleistocene** or **Holocene** age. Hominins found at site A fragmentary skull comprising a calotte, right half of the mandible with M_2–M_3 and fragments of I_2–P_4, two isolated upper molars, and a right maxillary fragment with I^1–I^2 and P^4. The calotte is long and somewhat low but anatomically modern. The large mandible has a deep corpus, a **retromolar gap**, and the signs of a strong **chin**. Archeological evidence found at site No artifacts in close association, but a large, fossilized hippopotamus skeleton was found *in situ* from the same level nearby. Key references: Geology, dating, and paleoenvironment Whitworth 1965a, 1966; Hominins Whitworth 1960, 1966; Archeology Whitworth 1965b, 1966.

Kabuh Formation *See* **Bapang Formation**.

Kabwe (Location 14°27′S, 28°26′E, central Zambia; etym. Kabwe is the name of the nearest village. The site, formerly known as **Broken Hiil**, was named after a circular depression that divided the artifact- and fossil-bearing hill or kopje and apparently also because it resembled the mines at Broken Hill, Queensland, Australia) History and general description Fossils and artifacts were discovered in 1921 during lead and zinc mining operations at Number 1 Kopje, a dolomitic outcrop containing an approximately 30 m-long, **clay**- and **breccia**-filled cavity. By 1930 Number 1 Kopje had been leveled and the precise provenance of the recovered fossils and artifacts remains uncertain, although mineralogical and chemical analyses suggest that most derive from the lower portions of the cavity. Subsequent excavations by Desmond **Clark** in 1953 targeted surficial deposits some 200 m from Number 1 Kopje. Temporal span and how dated? The more than 23 large mammal species from Kabwe include extinct large baboons, saber-toothed cat, warthog, giraffid, and giant buffalo, and show a general similarity to fauna from Bed IV at **Olduvai Gorge** and thus the assemblage is estimated by some authors to be as old as 0.78–1.3 Ma. However, the precise association of the extinct taxa with the hominin remains is unclear and such a date is at odds with the **Middle Stone Age** (or MSA) attribution of the (equally poorly provenanced) artifacts recovered with the fossils, as elsewhere the MSA dates to <300 ka. Hominins found at site **Kabwe 1** (British Museum of Natural History registration no. E686 and formerly Broken Hill 1) is a nearly complete cranium found by T. Zwigelaar in 1921. The specimen became the holotype for *Homo rhodesiensis* but is attributed by many authors to *Homo heidelbergensis*. Additional hominin remains recovered from Kabwe include a left tibia (no. E691), distal right humerus (no. E898), fragment of a femoral shaft (no. E793), cranial fragment (no. E897), right maxilla (no. E687),

Wiley-Blackwell Encyclopedia of Human Evolution, First Edition. Edited by Bernard Wood.
© 2013 Blackwell Publishing Ltd. Published 2013 by Blackwell Publishing Ltd.

fragmentary right and left innominates (nos. E719–20), sacrum (no. E688), proximal right femur (no. E907), proximal left femur and distal left femur (no. E689), and left femoral diaphysis (no. E690). Most of these specimens were also collected in 1921 and mineralogical analyses and comparisons of the chemical compositions of hominin and non-hominin fossils (as well as sediment adhering to artifacts) suggest that most derive from the lower portions of the sediment-filled cavity, but Hrdlicka (1930) reported finding the distal end of a humerus and a parietal fragment on a nearby bone pile. Archeological evidence found at site The majority of the artifacts from Kabwe have been lost, although some lithic artifacts and putative bone tools have been preserved. Artifacts associated with the hominin fossils recovered from Number 1 Kopje include **flakes** with **facetted platform**s, discoidal and perhaps **Levallois** cores, **hammerstone**s and possible **anvil**s, as well as three shaped bone tools that Desmond Clark attributed to the MSA. Clark's 1953 excavations revealed more than 2 m of stratified sediments that contained a sequence of **Acheulean**, **Sangoan**, and MSA artifacts, but unfortunately no fauna was preserved for comparison with the fauna from Number 1 Kopje. Key references: Geology, dating, and paleoenvironment Clark et al. 1947, Klein 1973, McBrearty and Brooks 2000; Hominins Woodward 1921, Pycraft 1928, Clark et al. 1968, Kennedy 1984b, Stringer 1986b, Trinkaus 2009; Archeology Clark et al. 1947, Clark 1950, Barham et al. 2002.

Kabwe 1 Site **Kabwe**. Locality Discovered during lead and zinc mining operations at Number 1 Kopje, a dolomitic outcrop containing an approximately 30 m-long cavity filled with clay and breccia. Surface/*in situ In situ*. Date of discovery June 17, 1921. Finders A.S. Armstrong and A.W. Whittington. Unit N/A. Horizon N/A. Bed/member N/A. Formation N/A. Group N/A. Nearest overlying dated horizon N/A. Nearest underlying dated horizon N/A. Geological age **Biostratigraphy** suggests a general similarity to fauna from Bed IV at **Olduvai Gorge** and thus the assemblage is estimated by some authors to be as old as 0.78–1.3 Ma. Developmental age Adult. Presumed sex Male, based on its massive brow ridge and large face. Brief anatomical description A cranium including the worn maxillary dentition except for the right I^2 and the crown of the right M^3. Endocranial volume Approxi-

mately 1280 cm^3 (Rightmire 2004). Announcement Woodward 1921. Initial description Woodward 1921, Pycraft 1928. Photographs/line drawings and metrical data Pycraft 1928, Rightmire 1990. Detailed anatomical description Rightmire 1990. Initial taxonomic allocation *Homo rhodesiensis* **Woodward, 1921**. Taxonomic revisions *Cyphanthropus* (Pycraft 1928), *Homo erectus* (Coon 1962), *Homo sapiens rhodesiensis* (Campbell 1964, Rightmire 1976). Current conventional taxonomic allocation *Homo heidelbergensis* or *H. rhodesiensis*. Informal taxonomic category Pre-modern *Homo*. Significance Kabwe 1 resembles **Bodo 1**, **Arago**, **Petralona**, and **Saldanha**, and many regard it as consistent with the hypothesis that these specimens all belong to the hypodigm of *H. heidelbergensis*, or *H. rhodesiensis* if the former taxon is interpreted as having some *Homo neanderthalensis* autapomorphies. If Kabwe 1 is the same age as Olduvai Gorge Bed IV then it is possibly the earliest evidence for *H. heidelbergensis* and *H. rhodesiensis*. Location of original **Natural History Museum**, London, UK.

KAI Abbreviation of the Kaitio Member in the Nachukui Formation, **West Turkana**, Kenya.

Kaitio [etym. Named for an ephemeral river system (laga) in whose drainage the localities are located] (or KI) Refers to a complex of fossil localities in the Nariokotome, Natoo, Kaitio, and Kalochoro Members of the Nachukui Formation within the Kaitio drainage in West Turkana, Kenya. *See also* **West Turkana**.

KAL Abbreviation of the Kalochoro Member in the Nachukui Formation, **West Turkana**, Kenya.

Kalahari Dune Expansion In southern Africa, fossil dunes in the Mega Kalahari dated by **luminescence dating** (Stokes et al. 1998) indicate arid conditions and dune expansion between 95 and 115 ka, 41 and 46 ka, 20 and 26 ka, and post-20 ka. Comparable periods of aridity are also seen in the **paleoclimate** reconstruction from the **Tswaing Impact Crater** (Partridge et al. 1997).

Kalokodo (etym. Named for an ephemeral river system (laga) in whose drainage the localities are located) (or KK) Refers to a complex of fossil localities in the Kalochoro Member of the Nachukui Formation within the Kalokodo drainage in West Turkana, Kenya. *See also* **West Turkana**.

Kalaloo One of the named drainages/subdivisions within the **Bodo-Maka** fossiliferous subregion of the **Middle Awash study area**, in Ethiopia.

Kalam *See* **Shungura Formation**.

Kalambo Falls (Location 8°30′S, 31°15E, Zambia; etym. unknown) History and general description Desmond **Clark** directed intermittent excavations between 1953 and 1966 in sediments of the Kalambo Falls Formation. Kalambo Falls Formation sediments crop out immediately upstream of the falls of the Kalambo River, which drains into Lake Tanganyika approximately 6 km/3.6 miles downstream. Although no fauna is present, the waterlogged conditions at the site preserved abundant organic remains including wood and pollen. The composite stratigraphic sequence of the different excavations provides a crucial benchmark for the cultural historical succession of southern and East Africa, from the **Early Stone Age** (e.g., **Acheulean**) to the Iron Age. Temporal span and how dated? **Uranium-series dating** on fossil wood, a material for which the uranium and thorium geochemistry remains poorly understood, may underestimate the true sample age. They suggest an age of 76 ± 10 ka for the **Sangoan** levels, 182 ± 10 ka for the Upper Acheulean level, and 182 ± 16 for the Lower Level Acheulean. **Radiocarbon dating** estimates from the 1950s suggest that that the **Later Stone Age** and Iron Age strata may be as old as *c*.25 ka, although most are **Holocene** in age. Hominins found at site None. Archeological evidence found at site Stone artifacts likely spanning much of the Pleistocene occur at Kalambo Falls, from Early Stone Age (i.e., Acheulean), **Middle Stone Age** (i.e., **Lupemban**), to the Later Stone Age as well as "intermediate" industries (i.e., Sangoan and "**Magosian**"). Most are made from locally available **quartzite**, **mudstone**, or **silcrete**. These stratified stone-tool assemblages in alluvial and colluvial sediments have provided a fundamental relative dating tool used throughout much of Africa, although many of them have undergone **winnowing** or otherwise been reworked by **fluvial** processes. The typology developed for their study by Desmond Clark and Maxine Kleindienst enjoyed widespread use beyond Kalambo Falls. Waterlogged conditions, particularly in the lowermost strata, assist in the preservation of pollen that suggests fluctuations in the mosaic of forest and swamp environments in the area, an interpretation supported by preserved seeds, tree bark, and wood charcoal. A number of wood specimens that suggest hominin modification and use occur in Acheulean and Sangoan strata. They include a number of pointed pieces that may have functioned as **digging sticks** but smoothing by stream flow has removed many of the fine traces necessary for confirmation. Key references: Geology, dating, and paleoenvironment Clark 1969, 2001; Archeology Clark 1969, 1974, 2001, Sheppard and Kleindienst 1996.

Kalochoro (etym. Named for an ephemeral river system (laga) in whose drainage the localities are) History and general description Complex of fossil localities within the Kalochoro, Kaitio, Natoo, and Nariokotome Members of the Nachukui Formation in the vicinity of the Kalochoro drainage in **West Turkana**, Kenya. The **Koobi Fora Research Project** conducted surveys between 1980 and 1983 and more intensive fieldwork began in 1984. By 1988 six vertebrate fossil localities had been identified (Harris et al. 1988) but no hominin remains have been recovered to date. Archeological excavations have been undertaken by the West Turkana Archaeological Project and have continued since 1994. Temporal span and how dated? The localities occur in two blocks separated by a major fault, and are well constrained by **argon-argon dating**. The western sequence spans from the Kalochoro Tuff (2.32 Ma) to well above the KBS Tuff (1.87 Ma), and includes most of the paleontological localities. The archeological sites lie east of the fault, and are immediately above the Lower Nariokotome Tuff that is *c*.1.3 Ma. Hominins found at site None. Archeological evidence found at site Test excavations were undertaken at Kalochoro 1 (KL 1) in 1994 and Kalochoro 2 (KL 2) in 1996. Surface lithic assemblages have been found in four localities, but no *in situ* material has been recovered to date. Key references: Geology and dating Harris et al. 1988, Kibunjia 2002; Hominins N/A; Archeology Kibunjia 2002. *See also* **West Turkana**.

Kamoya *See* **Kimeu, Kamoya**.

Kanam (Location 0°21′S, 34°30′E, Kenya; etym. named after the Kanam Beds exposed at the site) History and general description A fossilized hominin mandible was found at the site in an erosion gully on the south shore of the Kavirondo Gulf, Lake Victoria, in 1932 by Juma Gitau, a member of Louis **Leakey's**

East African Expedition. Temporal span and how dated? Inferred to be late **Pliocene** or early **Pleistocene** based on correlations (later doubted) with the Kanam Beds. Hominins found at site Kanam 1, the anterior portion of a mandible with pathological anterior surface of the symphysis (causing it to superficially resemble a chin) and the right P_3 and P_4. The mandible is the holotype of *Homo kanamensis*. Archeological evidence found at site None found with the mandible, but there are **Oldowan** tools from the Kanam Beds. Key references: Geology, dating, and paleoenvironment Leakey 1935, Cooke 1963; Hominins Woodward 1933, Leakey 1935, 1936b, Tobias 1960; Archeology Leakey 1935.

Kanapoi (Location 2°19′N, 36°04′E, Kenya; etym. the site was apparently misnamed by Bryan Patterson; the local name for the area is Kanapunyi) History and general description The site located southwest of Lake Turkana and just north of the Kakurio River was discovered in 1965 by Bryan Patterson's team. The Kanapoi sediments were deposited in a channel of the Kerio River that drained northwards into the paleolake Lonyumun. Temporal span and how dated? Three stratigraphic intervals are represented in the Kanapoi exposures. All lie beneath the *c.*3.4 Ma Kalokwanya basalt and all but one of the hominin fossils were recovered from sediments dated to between 4.17 ± 0.03 and 4.07 ± 0.02 Ma by **argon-argon dating**. Hominins found at site The first hominin fossil recovered from Kanapoi was a distal left humerus found by Bryan Patterson's expedition in 1965 (Patterson and Howells 1967). Eight more hominins were recovered in 1994 (Leakey et al. 1995) and a further 24 more hominins recovered between 1994 and 1997 were reported in Leakey et al. (1998). The initial humerus was regarded as *Australopithecus* sp. but a mandible found in 1994, **KNM-KP 29281**, was made the holotype of *Australopithecus anamensis* and this is the taxon to which the 1965 humerus and all the subsequent (i.e., 1994 and thereafter) discoveries have been allocated. All of the hominins except a mandible come from the interval between the Kanapoi Tuff and a tephra layer called the "lower pumiceous tuff." Archeological evidence found at site Mary **Leakey** reported on Oldowan-style artifacts recovered from Kanapoi, but they are not associated with the Pliocene fossils. They are believed to be of recent derivation and remain somewhat enigmatic (Leakey 1966b).

Key references: Geology and dating Leakey et al. 1995, 1998, Feibel 2003; Hominins Leakey et al. 1995, 1998, Ward et al. 1999b.

Kanapoi Hominoid 1 *See* **KNM-KP 271**.

Kangaki (etym. Named for an ephemeral river system (laga) in whose drainage the localities are located) (or KK) Refers to a complex of fossil localities in the Kalochoro Member of the Nachukui Formation within the Kangaki drainage in West Turkana, Kenya. *See also* **West Turkana**.

Kangatukuseo (etym. Named for an ephemeral river system (laga) in whose drainage the localities are located) (or KU) Refers to a complex of fossil localities in the Upper Lomekwi Member of the Nachukui Formation within the Kangatukuseo drainage in West Turkana, Kenya. The early hominin mandible **KNM-WT 16005** was recovered from Kangatukuseo III. *See also* **West Turkana**.

Kanjera (Location 0°20′S, 30°02′E, Kenya; etym. unknown) History and general description The site of Kanjera, a series of erosion gullies 500 m from the south shore of the Kavirondo Gulf, Lake Victoria, was initially explored by Louis **Leakey**'s East African Expedition, which recovered anatomically modern-looking hominins in 1932. Kanjera comprises several collection areas around Kanjera North (KN) and Kanjera South (KS). The apparent association of anatomically modern humans with Acheulean tools, along with Hans Reck's discovery of an anatomically modern skeleton (**OH 1**), erroneously associated with **Pliocene** fauna at **Olduvai Gorge**, formed the basis of Louis Leakey's view that modern humans had a long, separate, lineage in East Africa extending back to **Oldowan** times. Subsequent excavations at Kanjera have shown the hominins to be from much more recent contexts or probably intrusive burials into geologically older strata. Temporal span and how dated? Behrensmeyer et al. (1995) restudied the approximately 37 m of deposit and divided the sequence into three units, the Kanjera Formation (*c.*1.5–0.5 Ma), the **Apoko Formation** (<0.5 Ma), and the Black Cotton Soil (terminal Pleistocene-Holocene). Behrensmeyer et al. concluded that all the hominins except Kanjera 3 derived from the Black Cotton Soil, and that Kanjera 3 was probably an intrusive burial in the Kanjera Formation at KN-2t. Subsequent geological work and paleomagnetic sampling by Plummer et al.

(1999) and Ditchfield et al. (1999) demonstrated that sediments at Kanjera South extend back to 2.2 Ma. The strata exposed in the Kanam and Rawi Gullies have been divided into the Rawi Formation, Abundu Formation, and Kasibos Formation (c.3.0–1.0 Ma based on faunal and paleomagnetic correlations). Today, Kanjera is chiefly notable for its faunal sequence and for the archeological evidence that samples activities in a more open environment than at most Oldowan sites. Hominins found at site Five fragmentary hominins, mostly surface finds; Kanjera 1, fragmentary cranium; Kanjera 2, parietal and rib fragments; Kanjera 3, a partially *in situ* cranium an femoral fragments; Kanjera 4, two frontal fragments; Kanjera 5, two fragmentary femora. The cranial fragments are clearly anatomically modern. Archeological evidence found at site None in direct association with the hominins, but Leakey emphasized that the Kanjera Beds contain evolved **Acheulean** bifaces. The site of Kanjera South also preserves strata containing Oldowan artifacts. The importance of Kanjera is that it provides evidence about how hominins contemporary with the Oldowan artifacts procured raw material. Key references: Geology, dating, and paleoenvironment Leakey 1936, Kent 1942, Oakley 1962, Cooke, 1963, Plummer and Potts 1989, Behrensmeyer et al. 1995, Plummer et al. 1999, Ditchfield et al. 1999; Hominins Woodward 1933, Leakey 1935, Tobias 1962; Archeology Leakey 1935, Cooke 1963, Plummer and Potts 1989, Behrensmeyer et al. 1995, Plummer et al. 1999.

Kapcheberek (Location 00°45′N, 35°52′E, Kenya; etym. the local, Tugen, name for the area) History and general description Locality within the **Lukeino Formation** in the eastern foothills of the Tugen Hills, Baringo District, Kenya. It is one of four localities from which remains attributed to *Orrorin tugenensis* have been recovered. Temporal span and how dated? c.6–5.7 Ma based on a combination of **argon-argon dating**, **potassium-argon dating**, and **magnetostratigraphy**. Hominins found at site The only specimen recovered to date is BAR 349′00, a manual phalanx found in 2000 and assigned to *O. tugenensis*. Archeological evidence found at site None. Key references: Geology, dating, and paleoenvironment Deino et al. 2002, Hill et al. 1985, Pickford and Senut 2001, Sawada et al. 2002; Hominins Senut et al. 2001; Archeology N/A.

Kapedo Tuffs (Location 1°04′N, 36°05′E, Kenya; etym. named after the nearby trading village of Kapedo) History and general description Deposits of volcanic **tuffs** and intercalated fluvial **conglomerates** and **sands** with a maximum thickness of 20 m, discontinuously exposed over an area of approximately 50 km²/19 square miles west of Silali, the largest **Quaternary** volcano in the Kenyan section of the **East African Rift System**. Temporal span and how dated? Electron probe microanalyses of glass from several of the tuff layers suggest the sediments exposed at Kapedo Tuffs are equivalent to the Upper **Pyroclastic** Deposits exposed in the caldera of Silali and the latter have **argon-argon dating** estimates of 132 ± 3 and $135 \pm$ ka. The Arzett Tuffs, with $^{40}Ar/^{39}Ar$ age estimates of 123 ± 3 ka, are also exposed on Silali but they are not represented at Kapedo Tuffs; this suggests the latter site has an age range of c.135–123 ka. Hominins found at site None. Archeological evidence found at site The sparse archeological record includes surface and excavated **lithic** assemblages made predominantly of locally available lavas that include rare **Levallois flakes** and **cores** and retouched or deliberately shaped pieces, attributed to the **Middle Stone Age**. No fauna was recovered. Key references: Geology, dating, and paleoenvironment Dunkley et al. 1993, Macdonald et al. 1995, McCall 1999, Smith et al. 1995, Tryon et al. 2008; Archeology Tryon et al. 2008.

Kapsomin (Location 00°45′N, 35°52′E; Kenya; etym. the local, Tugen, name for the area) History and general description Locality within the **Lukeino Formation** in the eastern foothills of the Tugen Hills in the Baringo District. It is one of four localities from which remains attributed to *Orrorin tugenensis* have been recovered. Temporal span and how dated? It and other nearby sites have been bracketed c.6–5.7 Ma based on a combination of **argon-argon dating**, **potassium-argon dating**, and **magnetostratigraphy** but it is suggested that "the Kapsomin hominid specimens that are the most important" can be bracketed to 5.9–5.8 Ma (Sawada et al. 2002). Hominins found at site All specimens were recovered in 2000 and have been assigned to *O. tugenesis*. **BAR 1000′00**, comprising two mandibular fragments, is the **holotype** of *O. tugenensis*. Other important finds include BAR 1001′00, an upper central incisor, BAR 1002′00, a left femur, BAR 1003′00, a proximal left femur, BAR 1004′00, a right humeral shaft, BAR 1390′00, a P_4, BAR 1425′00, an upper right C, BAR 1426′00, a left M^3, and BAR 1900′00, a right M^3. Archeological evidence found at site None. Key references: Geology, dating, and paleoenvironment Deino et al. 2002, Hill et al. 1985, Pickford and Senut 2001, Sawada et al. 2002; Hominins Senut et al. 2001; Archeology N/A.

Kapthurin Formation (Location 00°47′N, 36°5′E, Kenya; etym. named after the Kapthurin River where sediments are best exposed) History and general description An approximately 125 m-thick sequence of **tuffaceous**, predominantly **lacustrine**, and **fluvial** sediments exposed over an area of about 150 km²/58 square miles, west of Lake Baringo, Kenya, the **Middle Pleistocene** portion of the Tugen Hills sequence. The sediments are divided into five geological members numbered K1–K5 from bottom to top. The 1966 discovery of an adult hominin mandible, postcranial remains, and **lithic** artifacts led to additional geological mapping in the area under the aegis of William (Bill) **Bishop** and Basil King of the **East African Geological Research Unit**. The **Baringo Paleontological Research Project** has conducted research in the region since 1980 and in 1982 John Kimengich found another hominin mandible during archeological investigations in the area directed by Francis Van Noten. Since 1990, the area has been studied by Sally McBrearty with notable fossil finds including the first fossil chimpanzee remains and the discovery of blades in sites that are *c*.500 ka. Temporal span and how dated? **Argon-argon dating** of lavas and tephra and tephrostratigraphic correlation by electron probe microanalysis suggest an age bracket of 780 to less than 200 ka, with the lower Kasurein Basalts within K1 dated to 610 ± 40 ka and upper portions of K4 (the Bedded Tuff Member) dated to 235 ± 20 ka. Hominin evidence An adult hominin mandible, KNM-BK 67, was discovered in 1966. Subsequently two manual phalanges, a right metatarsal, and a right ulna (KNM-BK 63–66) were recovered from excavations. A second mandible, **KNM-BK 8518**, from a second individual was found in 1982 by John Kimengich. The individual represented by KNM-BK 63–67 was referred to *Homo* cf. *erectus* by Leakey et al. (1969), while KNM-BK 8518 was referred to *Homo sp. indet.* (Wood and van Noten 1986). The **panin** remains comprise eight dental specimens (**KNM-TH 45519–45522** and KNM-TH 46437–46440), and all but one undiagnostic root fragment are maxillary. There is no duplication so the eight teeth may sample one individual. Archeological evidence Sediments of the Kapthurin Formation contain numerous archeological sites across the extent of the exposures and

throughout much of its temporal sequence. The richest and best-studied sites occur in members K3 and K4. Interbedded tephra have been dated and correlated, permitting an understanding of temporal and spatial variation among multiple surface-collected and excavated **Acheulean** and **Middle Stone Age** (or MSA) archeological sites. Particularly notable are the presence of large **Levallois flakes** in Acheulean strata at sites LHA and LHS (*c*.250–500 ka) and diverse methods of Levallois flake production at MSA strata at Koimilot (*c*.250 ka). The Kapthurin Formation sequence is unusual in showing complex temporal patterning in Acheulean and MSA artifacts, inconsistent with simple replacement scenarios during the early appearance of MSA lithic technology. The Kapthurin Formation also includes early traces of the use of **ochre** and **blade** production. Key references: Geology, dating, and paleoenvironment Leakey et al. 1969, McCall et al. 1967, Deino and McBrearty 2002; Hominins Leakey et al. 1969, Wood and van Noten 1986, Solan and Day 1992, McBrearty et al. 1996, McBrearty 1999, 2005, Wood 1999, Hill 2002; Archeology McBrearty et al. 1996; McBrearty 1999, Tryon and McBrearty 2002, 2006, Tryon et al. 2005, Tryon 2006, Johnson and McBrearty 2010.

Karari Ridge One of the subregions of the Koobi Fora region. *See* **Koobi Fora**.

Karonga Alternative name used by Richard Hay for a small valley at **Olduvai Gorge** that Louis and Mary **Leakey** refer to as a **korongo**. *See also* **Olduvai Gorge**.

Karonga Museum *See* **Cultural and Museum Centre, Karonga**.

karst (Ge. *karst*=kras, after the "kras" region in Slovenia not far from Trieste, where this geological feature is common) Limestone deposits that have been affected by water rich in carbonic acid. This mildly acidic ground water produces small (e.g., flutes), medium (e.g., sinkholes), or large-scale (e.g., limestone pavements) surface features, and underground the water action leads to fissures and solution cavities and when breccia and bones fill these solution cavities the process results in cave sites like those in the Blaauwbank valley (e.g., **Sterkfontein**, **Swartkrans**).

karstic A landscape typically dominated by small, medium, and large-scale karst features on the surface and by channels and cavities within the rock. *See also* **karst**.

karyotype (Gk *karyon*=nut, kernel and *typos*=impression, mark, figure, or original form) The complement of **chromosome**s found in an individual or normally found in a taxon. The modern human karyotype has 46 chromosomes including 22 pairs of autosomes and one pair of **sex chromosomes**. The techniques used to investigate karyotypes are called cytogenetic. The most common cytogenetic technique is Giemsa staining (also called G-banding) that reveals specific banding patterns on each chromosome, but other dyes or fluorescently labeled probes can also be used. Karyotypes are used to detect chromosomal abnormalities in fetal cells (amniocentesis) and cancer cells. Karyotypes have also been used to study the evolution of chromosome number and structure. For example, the karyotypes of modern humans and the great apes show that they have 46 and 48 chromosomes, respectively. The banding patterns reveal that sometime during hominin evolution chromosomes 2a and 2b found in the great apes fused to form modern human chromosome 2. Prior to the availability of large amounts of DNA data, differences in chromosomal number and pattern were often used to determine the relatedness between species (e.g., Yunis and Prakash 1982).

Katanda (Location 0°06'N, 29°35'E, Democratic Republic of the Congo; etym. denotes a resting place by a ford in KiNande) History and general description Katanda, which incorporates three sites exposed on the eastern bank of the Semliki River, was first explored by Jean de Heinzelin in the 1950s and was excavated by John Yellen and Alison Brooks from 1986 to 1990. The excavations at Katanda 2 (108 m^2 excavated, 21 m^2 of them in the Katanda beds), Katanda 9 (35.2 m^2), and Katanda 16 (11 m^2) focused on the Katanda Beds, alluvial sediments attributed to lateral fills along an ancient floodplain margin and the valleys of ephemeral tributary streams, 34–41 m above the current Semliki river. The Katanda beds unconformably overlie the ASB paleosol complex and are themselves overlain by the Holocene-aged Katwe Ashes whose base is dated to 6.89±0.75 ka by AMS radiocarbon dating on charcoal. Yellen (1996) argues from spatial patterning in artifacts and fauna that the Katanda 9 excavations reveal a short-term occupation or base camp. The range of radiometric age estimates by multiple methods allow for a considerably longer period of sediment accumulation and site burial. Original thermolu minscence dating estimates on sands overlying the artifact horizon at Katanda 9 yielded an estimated age of 82.8±8 ka, while a second series of optically stimulated luminescence dating determinations on the 20 sand samples suggested an age range of 65–90 ka in all but one of the results. **Electron spin resonance spectroscopy dating** age estimates on fossil *Hippopotamus* teeth from Katanda 9 range from 89±22 ka (early uptake) to 155±38 ka (late uptake) whereas **uranium-series dating** on the teeth was more-or-less consistent with the older determination (140±4 ka, 174±1 ka). None of the dating results suggested an age younger than 50 ka. Because of the broad similarity of recovered bone harpoons at the Katanda sites with younger examples from **Ishango**, and differences in Katanda 9 faunal preservation states, Klein (2008) among others contends that the true age of the sites remains unresolved. However, the Ishango sites are located on a terrace some 24–8 m below the level of the Katanda sites and the bone points there are smaller, manufactured by a different process and used a different shape for both the barbs and the base. Hominins found at site None. Archeological evidence found at site The Katanda sites contain 12 worked bone tools, including seven barbed points or harpoons, five unbarbed pointed pieces, and one "knife," plus approximately 16,000 lithic artifacts composed primarily of quartz, quartzite, and chert flaking debris, including discoidal and single and multi-platform cores. Rare rubbing or grinding stones are also present. The approximately 19,000 faunal elements include a diverse mammalian taxa that suggests a riverine setting with a gallery forest fringed by a savanna. Of the minimum number of individuals of identified fish remains, 65% are catfish (*Clarias* sp. and *Synodontis* sp.) whose size and age profiles suggest they were acquired during spawning, which today is limited to the rainy season. The spatial distribution of artifacts and fauna at the Katanda 9 excavation have been argued to reflect an anthropogenic feature or "pavement" similar to nuclear family areas ethnographically documented among the Kalahari San and other groups. Key references: Geology, dating and environment Damas 1940, Lepersonne 1949, 1970, 1974, de Heinzelin 1955, 1961, Brooks et al. 1995, Helgren 1997, Verniers and de Heinzelin 1990, de Heinzelin and Verniers 1996, Feathers and Migliorini 2001, Brooks et al. 2008a. Archeology de Heinzelin 1955, 1961, Stewart 1989, Yellen 1998, Brooks et al. 2008b.

Kawolli *See* **Chongokni**.

KB Prefix used for fossils from two locations, the Kossom Bougoudi "fossiliferous area," **Chad Basin**, Ethiopia, and Kromdraai B, South Africa.

KBS Acronym for the Kay Behrensmeyer site at **Koobi Fora** and used for the KBS Member in the **Koobi Fora Formation**, **Koobi Fora**, Kenya.

KBS CC Abbreviation of the KBS Channel Complex in the **Koobi Fora Formation**, **Koobi Fora**, Kenya. *See also* **KBS**.

Kebara (Location 32°34′25″N, 34°58′08″E; etym. Hebrew, פערת בכאַרה, *Me'arat Kebbara* or Arabic, مغارة الكبارة, *Mugharat al-Kabara* both mean "the great cave") History and general description Kebara is located on the western escarpment of Mount Carmel in Israel and has undergone several phases of excavations. The first, by Dorothy **Garrod** and Francis Turville-Petre in the early 1930s examined the **Natufian**, Kebarran, and **Upper Paleolithic** layers. Stekelis, who was the next to work at the site from 1951 to 1965, focused on the Upper and **Middle Paleolithic** layers and it was during his excavations that the fossil neonate (KMH1) was found. The latest excavations by Ofer **Bar-Yosef** and colleagues (1982–90) further explored the Middle Paleolithic and recovered more hominin remains (Bar-Yosef et al. 1992). Temporal span and how dated? The Mousterian levels have been dated by **radiocarbon dating**, **electron spin resonance spectroscopy dating**, and thermo**luminescence dating** methods (Bar-Yosef et al. 1996, Schwarcz et al. 1989, Valladas et al. 1987); the Upper Paleolithic levels have been dated by radiocarbon (Bar-Yosef et al. 1996). The Middle Paleolithic levels have been dated at 62–48 ka, the transition between the Upper and Middle Paleolithic has been assigned a date of *c*.45 ka, and the Upper Paleolithic levels have been dated at *c*.36–32 ka. Hominins found at site The excavations of Garrod and Turville-Petre recovered the remains of at least 15 individuals from the Upper Paleolithic layers, but these have not been formally described in detail. The later excavations of Stekelis and Bar-Yosef yielded the remains of approximately 29 individuals, most of which are from the Middle Paleolithic layers; all belong to *Homo neanderthalensis*. The remains are mostly craniodental frag-

ments, but include a nearly complete skeleton of an infant (KMH1) as well as the nearly complete torso of an adult (**KMH2**). Archeological evidence found at site The Upper Paleolithic levels include Bronze age, Lower Natufian, Kebarran, **Aurignacian**, and **Ahmarian** industries. The Middle Paleolithic levels contain Levantine Mousterian tools. There is fauna from all layers and several studies have suggested that fluctuations in occupation frequency may be related to environmental change as well as to human demographic change (Davis 1977, Speth and Tchernov 2002). Charred macrobotanical remains and **phytoliths** from the Middle Paleolithic levels have been interpreted as evidence of broad-spectrum plant foraging by Neanderthals, as well as for the controlled use of fire (Lev et al. 2005, Albert et al. 2000). Key references: Geology, dating, and paleoenvironment Bar-Yosef et al. 1996, Valladas et al. 1987, Schwarcz et al. 1989; Hominins Smith and Arensburg 1977, Rak and Arensburg 1987, Bar-Yosef and Vandermeersch 1991; Archeology Garrod 1954, Bar-Yosef et al. 1992, Schick and Stekelis 1977.

Kébibat (Location 34°02′N, 6°51′W, Morocco; etym. named after the nearby city) History and general description This site and its hominin cranial specimen were discovered in 1933 by workmen with M. Alenda at Mifsud-Guidice Quarry in the suburb of Rabat. It was subsequently studied by J. Marçais. The first Middle Pleistocene hominin found in North Africa; its fragmentary nature (23 fragments) is due to dynamite blasting and subsequent reburial by quarry workers. Temporal span and how dated? **Biostratigraphy** and **uranium-series dating** on overlying shell suggest 200–130 ka; it is assigned to the middle Tensiftian local Maghreb continental stage. Hominins found at site Mandible with left I_1–M_1 and right I_1, $P_{3–4}$, and M_3; left maxilla with I^2–M^2; parietals and occipital fragments of an adolescent assigned to what Saban (1977) called "evolved *Homo erectus*." Archeological evidence found at site None. Key references: Geology, dating, and paleoenvironment Hublin 2001, Stearns and Thurber 1965; Hominins Hublin 2001; Archeology N/A.

Kedung Brubus (Location 7°30′S, 111°54′E, Indonesia; etym. named after a local village) History and general description An exposed river-bed deposit 35 km/22 miles southeast of **Trinil** where Eugène **Dubois** discovered a fragmentary fossil hominin mandible on November 24, 1890. The fauna from

Kedung Brubus, including the hominin mandible, comes from the sandstones of the Trinil beds of the **Bapang Formation** (formerly called Kabuh). Temporal span and how dated? There has been debate in the literature concerning the relative age of Kedung Brubus. Leinders and colleagues contend that the Trinil mammals are older than those from Kedung Brubus, whereas Hooijer's interpretations of these faunas lead to the opposite conclusion. **Argon-argon dating** of volcanics from breccias and tuffs at and near Kedung Brubus suggest a latest Pliocene/earliest Pleistocene age for these deposits. Hominin found at site **Kedung Brubus 1** is a right mandibular fragment containing part of the canine and P$_3$ roots that most likely represents a juvenile *Homo erectus*. Dubois (1935) described what he rather confusingly referred to as "the sixth (fifth new) femur of *Pithecanthropus erectus*", and he suggested that there is "some probability that this fossil was found at Kedung Brubus" (*ibid*, p. 852). However, when Day and Molleson (1973) examined all of the Trinil femora (i.e., Femora I–VI) in detail (using radiography and microscopy) they concluded that "'Femur VI' is not hominine, hominid or even primate" (*ibid*, p. 151). Archeological evidence found at site None. Key references: Geology, dating, and paleoenvironment Hooijer 1983, Hooijer and Kurtén 1984, Leinders et al. 1985, Bandet et al. 1989; Hominins Dubois 1892, 1935, Tobias 1966b, Day and Molleson 1973.

Kedung Brubus 1

Site **Kedung Brubus**. Locality N/A. Surface/*in situ* Not known. Date of discovery November 24, 1890. Finder Eugène **Dubois**. Unit N/A. Horizon N/A. Bed/member N/A. Formation N/A. Group N/A. Nearest overlying dated horizon N/A. Nearest underlying dated horizon N/A. Geological age Latest Pliocene/earliest Pleistocene. Developmental age Probably juvenile. Presumed sex Unknown. Brief anatomical description Small fragment of a right mandibular corpus that extends from the region of the canine posteriorly to include the anterior margin of the mental foramen. It preserves the base of the crown and the root of the right P$_3$ and the tip of the root of the right canine. Announcement Dubois 1891. Initial description Dubois 1924a. Photographs/line drawings and metrical data Dubois 1924b, Tobias 1966b. Detailed anatomical description Tobias 1966. Initial taxonomic allocation *Homo sp. indet.* Taxonomic revisions *Pithecanthropus erectus* Dubois (1924a, 1924b). Current conventional taxonomic allocation *Homo erectus*. Informal taxonomic category Pre-

modern *Homo*. Significance As poorly preserved as it is, Kedung Brubus 1 was the first early hominin to be recovered from Java. Location of original **Geological Research and Development Center**, Bandung, Indonesia.

Keilor

(Location 37°42′S, 144°51′E, Australia; etym. named after a local town) History and general description An adult probable male cranium and a fragmentary femur were discovered in 1940 during quarrying operations on a river terrace about 16 km/10 miles northwest of Melbourne. The specimen was recovered from a silt deposit, below a minor disconformity. Secondary carbonates adhering to the cranium suggest it was deposited into a paleosol. Temporal span and how dated? **Radiocarbon dating** of the bone suggest an age of 12 ka while dates from the adhering carbonates cluster around 6.8 ka. Hominin found at site The Keilor cranium is anatomically modern human, but it is unlike its more robust contemporaries from **Kow Swamp** in that it is relatively orthognathic, has moderately sized teeth and a curved frontal region. Archeological evidence found at site Quartzite flakes. Key references: Geology, dating, and paleoenvironment Brown 1987; Hominins Wunderly 1943, Brown 1987; Archeology Simmons and Ossa 1978.

Keith, Arthur (1866–1955)

Arthur Keith was born in Old Machar, near Aberdeen in Scotland and entered the University of Aberdeen in 1884 to study medicine. He completed his studies in 1888 and the following year accepted a post as a medical officer for a gold-mining company in Siam (now Thailand). While there Keith dissected monkeys to see if they carried malaria and this led to a lifelong interest in the comparative anatomy of primates. Keith returned to Britain in 1892 where he studied anatomy with George D. Thane at University College London and with Robert W. Reid at the University of Aberdeen and it was there that Keith completed his MD in 1894. His thesis, entitled 'The myology of the Catarrhini: a study in evolution' was based on the numerous dissections he had conducted in Siam. Keith became a Fellow of the **Royal College of Surgeons of England** in 1894 and after spending a brief period in 1895 working with the Swiss anatomist and embryologist Wilhelm His in Leipzig, Germany, he returned to Britain to accept an appointment as Senior Anatomy Demonstrator at the London Hospital Medical

College. Keith conducted a study of primate skulls in 1895 and in 1896 he published *An Introduction to the Study of Anthropoid Apes*. Keith's professional success led to his appointment as Head of the Anatomy Department at the London Hospital Medical College in 1899 where he combined his primate research with a detailed study of the conducting tissue of the heart. The latter resulted in the recognition of one of the two foci of electrical activity in the heart and his contribution is recognized in the eponymous name for the atrio-ventricular node, which is called the node of Keith and Flack. In 1908 Keith resigned from the London Hospital Medical College to become conservator of the Hunterian Museum of the Royal College of Surgeons of England, and thereafter his research focused on physical anthropology. Keith published a paper on the evolution of the orthograde posture in 1903 and in *Ancient Types of Men* (1911) Keith first expressed his opinion that the modern human type was at least as old as extinct hominins such as *Homo neanderthalensis*. This view, later referred to as the **pre-sapiens hypothesis**, had an important influence on theories of human evolution and on the interpretation of hominin fossils during the early 20thC. Keith was influenced by discoveries such as the **Galley Hill** skeleton, an anatomically modern human discovered in 1888 in geological deposits in southern England that appeared to be very old (the skeleton was later shown to be a more recent intrusive burial). It was perhaps Keith's view that the modern human type evolved very early that led him to be an ardent supporter of Piltdown Man (*Eoanthropus dawsoni*). The controversy surrounding the competing interpretations of Piltdown Man and Keith's own investigations of the fossils prompted him to publish *The Antiquity of Man* in 1915 in which he examined the existing fossil evidence for human evolution and discussed the pre-sapiens hypothesis. Over the following decades Keith gradually modified his views, and this is reflected in the views expressed in *New Discoveries Relating to the Antiquity of Man* (1931). Keith collaborated with American anthropologist Theodore McCown between 1933 and 1937 on studies of the archeological and hominin skeletal material excavated at the **Mount Carmel** sites in Palestine by the American School of Prehistoric Research and the British School of Archaeology in Jerusalem. This resulted in the publication in 1939 of the second of the two volumes entitled *The Stone Age*

of Mount Carmel. Keith retired as Conservator at the Royal College of Surgeons of England in 1933, but in the same year he was appointed the first Master of the Royal College of Surgeons of England's Buckston Brown Research Farm located at Downe, in Kent, and he lived in Charles Darwin's old home, Down House. Keith had been instrumental in the establishment of the Buckston Brown Research Farm and he lived there until his death in 1955. Keith was elected a Fellow of the Royal Society in 1913 and he served as President of the Royal Anthropological Institute from 1914 to 1917. Keith was also the editor of the *Journal of Anatomy* from 1916 to 1936 and in recognition of his accomplishments he was knighted in 1921. During the last decades of his life Keith continued to write, focusing on the issue of race and the mechanisms of evolution. He published *A New Theory of Human Evolution* (1948), *An Autobiography* (1950), and *Darwin Revalued* (1955).

Kelsterbach (Location 50°03′N, 8°31′E, Germany; etym. named after the nearby town) History and general description A hominin calvaria found in Kelsterbach, near Frankfurt, was dated by Reiner Protsch to *c*.32 ka. The specimen is currently missing so redating is impossible, but many other fossils dated by Protsch's lab (e.g., **Paderborn**, **Hanöfersand**) have been recently redated to much more recent times, making this Upper Paleolithic date highly questionable. Temporal span and how dated? Probably Holocene. Hominins found at site A calvaria attributed to *Homo sapiens*. Archeological evidence found at site None. Key references: Geology, dating, and paleoenvironment Street et al. 2006; Archeology N/A; Hominins Protsch and Semmel 1978.

Kendeng Hills A range of hills on the south side of the Solo River in the northern part of central Java. The Kendeng Hills were explored by Eugène **Dubois** and fossil sites in that region include **Kedung Brubus**, Pati Ajam, Teguan, and **Trinil**.

Kenya National Museum *See* **National Museums of Kenya**.

Kenyanthropus platyops **Leakey et al. 2001** (Gk *anthropos*=human being, *platus*=flat, and *opsis*=face; etym. *Kenya*=Kenyan) A hominin genus and species established in 2001 by Meave

Leakey and colleagues to accommodate cranial remains recovered from the *c.*3.5 Ma Kataboi Member at **Lomekwi, West Turkana**. The initial report lists the **holotype** cranium and the paratype maxilla plus 34 specimens that include three mandible fragments, a maxilla fragment, and isolated teeth some of which may also belong to the hypodigm, but the researchers reserved their judgment about the taxonomy of most of these remains (Leakey et al. 2001), some of which had only recently been referred to *Australopithecus afarensis* (Brown et al. 2001). The main reasons Leakey et al. (2001) did not assign the Lomekwi material to *Au. afarensis* are its reduced subnasal prognathism, anteriorly situated zygomatic root, flat and vertically orientated malar region, relatively small but thick-enameled molars, and the unusually small M^1 compared to the size of the P^4 and M^3. Some of the morphology of the new genus, including the shape of the face, is *Paranthropus*-like yet it lacks the postcanine megadontia that characterizes *Paranthropus*. The authors note the face of the new material resembles that of *Homo rudolfensis* (see below), but they point out that the postcanine teeth of the latter are substantially larger than those of KNM-WT 40000. *K. platyops* apparently displays a hitherto unique combination of facial and dental morphology. White (2003) has taken the view that the new taxon is not justified because the cranium could be a distorted *Au. afarensis* cranium, but even if this explanation is correct (see Spoor et al. 2010 for a response) it would not explain the small size of the postcanine teeth. First discovery **KNM-WT 8556** (1982). Holotype **KNM-WT 40000** (1999). Main site **West Turkana**.

Kesem-Kebena (9°13′N and 39°58′E, Ethiopia; etym. from the names of two nearby rivers) History and general description Area located on the Awash River upstream from the **Middle Awash study area** in the northern sector of the Ethiopian section of the **East African Rift Valley**. It was explored between 1988 and 1989 by the **Paleoanthropological Inventory of Ethiopia**. Eleven localities were identified, two of which have yielded lithic assemblages. Temporal span and how dated? **Potassium-argon dating** of Tuff K-K2E overlying the deposits at locality K-K2 provided a date of 3.76±0.6 Ma. Tuff K-K6G is just above the deposits at locality K-K6 and **argon-argon dating** provided an age

of 1.002±0.041 Ma. Thus, the age of these Plio-Pleistocene sediments date is >3.75 1.0 Ma. Hominins found at site None. Archeological evidence found at site Localities K-K6 and K-K7 have yielded **Acheulean** assemblages and K-K5 **Later Stone Age** assemblages, but the artifacts at locality K-K3 were in a secondary context. Key references: Geology, dating, and paleoenvironment WoldeGabriel et al. 1992; Hominins N/A; Archeology WoldeGabriel et al. 1992.

key innovations Evolutionary innovations that allowed a lineage to exploit a new **adaptive zone**. Taxa that share these innovations are normally recognized as occupying the same **grade** (Miller 1949, Jensen 1990, Hunter 1998).

Keyser, André W. (1938–2010) *See* **Drimolen**.

keystone resource (ME *key(e)*=central and *stan*=stone; it refers to the central wedge-shaped stone in an arch that locks the other stones in place) A taxon that has a major impact on community structure through its consumption by other organisms, even though a keystone resource might not itself be a consumer. The effect of a keystone resource is measured by quantifying the taxon's community importance (or CI) (Power et al. 1996). Robert Paine introduced the term keystone species in 1969 to refer to species that have a disproportionate effect on their environment and on local biodiversity. Paine focused on the impact two species of starfish, *Pisaster ochraceus* of California and *Acanthaster planci* of the Great Barrier Reef, have on their local intertidal zones. Closer to paleoanthropology Terborgh (1986) applied the term to the figs produced by trees in the genus *Ficus*. During periods of low fruit availability, primates, marsupials, birds, and many other frugivorous animals rely on figs for a significant portion of their energy. Keystone resources have been documented and studied in all of the world's major ecosystems (Power et al. 1996).

KGA Abbreviation of Konso-Gardula, the original name for the site now known as Konso, Ethiopia. Thus it is the prefix used for fossils from Konso. *See* **Konso**.

KGA 10-1 Site Konso. Locality KGA 10. Surface/*in situ* Surface, then excavation from a lag deposit. Date of discovery 1991. Finder Gen Suwa. Unit N/A. Horizon N/A. Bed/member Kayle Member. Formation Konso Formation. Group N/A. Nearest overlying dated horizon By stratigraphic correlation, the Karat Tuff (or KRT) is the closest overlying dated tuff at 1.41 ± 0.02 Ma. Immediately above KRT lies the Bright White Tuff, also dated to 1.40 ± 0.02 Ma. The latter corresponds to the Chari Tuff at **Koobi Fora** in its major and minor element compositions of glass shards. Nearest underlying dated horizon Lehayte Tuff inferred to be *c*.1.42 Ma based on 1.47 ± 0.07 Ma age on plagioclase and stratigraphically underlying Trail Bottom Tuff dated to 1.43 ± 0.02. Geological age Lower Pleistocene. Developmental age Adult. Presumed sex Male, based on large size. Brief anatomical description The left side of the mandibular corpus including P_4–M_3 and the left ramus and condyle. Announcement Asfaw et al. 1992. Initial description Asfaw et al. 1992. Photographs/line drawings and metrical data Asfaw et al. 1992, Suwa et al. 2007. Detailed anatomical description Suwa et al. 2007. Initial taxonomic allocation *Homo erectus*. Taxonomic revisions N/A. Current conventional taxonomic allocation *Homo erectus* or *Homo ergaster*. Informal taxonomic category Pre-modern *Homo*. Significance The molar crown morphology of KGA 10-1 resembles that of **KNM-ER 992**. The tooth crowns have inter-proximal grooves that resemble those made by modern humans who habitually use tooth-picks or sinews and the KGA 10-1 mandible is associated with **Acheulean** artifacts. Location of original National Museum of Ethiopia, Addis Ababa, Ethiopia.

KGA 10-525 Site Konso. Locality KGA 10. Surface/*in situ* Surface. Date of discovery 1993. Finder A. Amzaye. Unit N/A. Horizon N/A. Bed/member N/A. Formation N/A. Group N/A. Nearest overlying dated horizon Karat Tuff, *c*.1.41 Ma. Nearest underlying dated horizon Trail Bottom Tuff, *c*.1.43 Ma. Geological age *c*.1.4 Ma. Developmental age Mature adult. Presumed sex Male. Brief anatomical description A skull comprising a well-preserved **calotte** lacking the frontal and the anterior part of the cranial base and much of the lower face along with the lateral parts of both sides of the upper face. The remarkably complete mandible preserves the worn crowns of all tooth types other than the incisors. Announcement Suwa et al. 1997. Initial description Suwa et al. 1997. Photographs/

line drawings and metrical data Suwa et al. 1997. Detailed anatomical description N/A. Initial taxonomic allocation *Paranthropus boisei*. Taxonomic revisions N/A. Current conventional taxonomic allocation *P. boisei*. Informal taxonomic category Megadont archaic hominin. Significance The well-preserved first skull of *P. boisei* and although its describers claimed that KGA 10-525 shows "striking deviations...from the known Olduvai/Turkana *A. boisei* condition" (Suwa et al. 1997, p. 492), other researchers have concluded that most of the morphology of KGA 10-525 can be accommodated within the morphological range of the pre-Konso hypodigm of *P. boisei*. Location of original National Museum of Ethiopia, Addis Ababa, Ethiopia.

KHS (Location 5°24.152′N, 35°55.812′E, Ethiopia; etym. acronym for "Kamoya's Hominid Site") History and general description The site was discovered by Kamoya Kimeu in 1967 during the surveys carried out by Richard **Leakey**'s Kenyan component of the **International Omo Research Expedition** to the **Lower Omo Basin**. The site is located on a hill overlooking an erosional gully leading to the Omo River. Geological work by Karl Butzer and colleagues placed the site in Member I of the **Kibish Formation**. The site provided the earliest date for anatomically modern human based on two infinite (>39.9 ka) radiocarbon dates and a uranium-thorium (or $^{240}Th/^{234}U$) date of *c*.130 ka on Nile oysters from unit f above the hominin. Skepticism among anthropologists over these dates led to renewed work led by John Fleagle and Frank **Brown** in the Kibish Formation in 1999, 2001, 2002, and 2003. The new excavations relocated KHS, clarified the site's stratigraphic position, and determined **Omo I** to be even older than previously claimed; thus if the age of Omo I is correct it is currently the earliest evidence of anatomically modern humans within *Homo sapiens*. Temporal span and how dated? **Argon-argon dating** suggests an age of 195 ± 5 ka on an underlying tuff combined with correlation with Mediterranean **sapropels**. Hominins found at site Omo I, a fragmentary skull and skeleton. Archeological evidence found at site A small sample of **Middle Stone Age** stone artifacts made of fine-grained silicate rocks including **chert**, **jasper**, and **chalcedony** as well as some made of **rhyolite** or **basalt**. Of the artifacts, un-retouched **flakes** predominate but 24 small **cores**, 20 retouched pieces, and two **hammerstones** were recovered. Scattered faunal remains were also recovered, including a complete bird. Key references: Geology, dating, and paleoenvironment Butzer 1969, Butzer et al. 1969, Butzer and Thurber 1969, McDougall et al. 2005, 2008, Brown and Fuller 2008, Feibel 2008; Hominins Day 1969a, Stringer 1974, Day and Stringer 1982, 1991, Kennedy 1984b, Day et al. 1991, Pearson et al. 2008a, Voisin 2008; Archeology Leakey 1969, Shea 2008, Assefa et al. 2008, Louchart 2008. *See also* **Omo-Kibish**.

Kibish   *See* **Omo-Kibish**.

Kibish Formation   A formation in the Lower Omo Basin, Ethiopia, that takes its name from a local village. Its main outcrops occur along the Omo river, where it skirts the west side of the Nkalabong mountains about 100 km/60 miles north of the northern end of the present Lake Turkana. It overlies, but is not continuous with, the sediments belonging to the **Omo Group**. In the north, near the village of Kibish, it overlies the **Nkalabong Formation** and **Mursi Formation** but further south it overlies the **Usno Formation** and **Shungura Formation**. Karl **Butzer** divided the Kibish Formation into four Members (Mb. I–IV). Three absolute dates are available for the Kibish Formation. Single crystal **argon-argon dating** provide ages for Mb. I and Mb. III tuff feldspars of 196 ± 2 and 104 ± 1 ka respectively and **radiocarbon dating** on molluscs from Member IV gives ages in the range of *c*.3250 to *c*.9500 years ago. The three hominin individuals, **Omo I**, an associated skeleton including a fragmentary skull of a young adult, **Omo II**, an adult calvaria, and **Omo III**, fragments of an adult cranial vault, are judged to have come from Mb. I and their ages have been estimated at *c*.195 ka (McDougall et al. 2005).

Kiik-Koba   (Location 45°03′N, 34°18′W, Ukraine; etym. Turkik phrase for mountain goat cave) History and general description This **Middle Paleolithic** cave site in the Crimea was excavated by Gleb Bonch-Osmolovskii in 1924–6, who found the skeletons of an adult and child buried close to each other. Temporal span and how dated? The site has never been absolutely dated; but the presence of cold-phase fauna and technostratigraphy has suggested an age around the end of **Oxygen Isotope Stage** (OIS) 5 or the beginning of OIS 4 for the layer containing

the hominins. Hominins found at site Kiik-Koba 1 is the remains of an adult, probably a male. Its grave was disturbed in antiquity, removing all but a few teeth, portions of both hands, the right lower leg, and both feet. Kiik-Koba 2 is the remains of a very young child, less than a year old. Most of the postcrania are preserved. Both individuals are assigned to *Homo neanderthalensis*. Archeological evidence found at site Three Middle Paleolithic layers were recovered, and the site is the type site for the Kiik-Koba **Mousterian**. Key references: Geology, dating, and paleoenvironment Bonch-Osmolovskii 1940; Hominins Trinkaus 2008b, Trinkaus et al. 2008; Archeology Bonch-Osmolovskii 1940.

Kilaytoli Tuff   A 3.57 ± 0.014 Ma (Deino et al. 2010) vitric tuff found in several collection areas in the **Woranso-Mille study area**. Its geochemistry matches that of the **Lokochot Tuff** known from the **Koobi Fora Formation**.

Kilombe   (Location: 0°6′S, 35°53′E, Kenya; etym. the site is about 4 km/2.4 miles southeast of a mountain of the same name) History and general description Kilombe is an **Acheulean** site complex which lies on the western flank of the Rift Valley in Central Kenya. The site was discovered by W.B. Jones in 1972 and its geological context was first described by William (Bill) **Bishop** (1978). Fossiliferous brown clays (with a sparse fauna including *Elephas recki*, hippopotami, **suid**s, and a range of **bovid**s) formed on top of trachyphonolite lava that is approximately contemporary with the *c*.1.8 Ma trachyte lavas of the Kilombe volcano (Jones and Lippard 1979). The main archeological horizon at Kilombe accumulated on a clay surface during a pause in deposition in what was probably a broad drainage way or lakeside environment. Temporal span and how dated? Reversed paleomagnetization in layers immediately overlying the main artifact horizon indicates a probable age of more than 0.8 Ma. The sequence resumes with deposition of weathered tuffs and airfall tuffs through a depth of about 15 m and is capped by an ashflow tuff or agglomerate several meters thick. Finds of occasional bifaces have been made up to the base of this tuff which probably derives from Menengai (*c*.0.35 Ma). Hominins found at site None. Archeological evidence found at site The extensive exposures of its main biface horizon at Kilombe can be traced horizontally for more than 200 m.

It provides exceptional opportunities for comparisons of contemporaneous local occurrences at points more than 100 m apart on the same horizon. Its geological context is very unusual for East Africa in that the artifacts are on a clay surface rather than in a sandy channel. Small flakes found with the bifaces indicate some local resharpening and are evidence that the material was not redeposited. Some biface cores have been found, but most of the most of the biface blanks were probably made at nearby lava exposures and imported. Several **obsidian** bifaces indicate transport over long distances. The bifaces are most similar to those from **Kariandusi**, **Olorgesailie**, and **Olduvai Gorge** Bed IV. **Cleavers** occur across the surface but not in high frequencies. Details of the Kilombe bifaces differ from area to area within the relatively small site. For example, their relative thickness varies markedly (the mean thickness/breadth ratio ranging from 0.30 to 0.56) whereas the breadth/length ratio is a closely controlled **allometric** relationship. The former observation calls into doubt the notion that the thickness/breadth ratio of bifaces has any chronological significance. Cluster analyses indicate that biface morphology is correlated with location within the site and in some areas of the site **Developed Oldowan**-like small bifaces are found alongside large classic Acheulean specimens. The size (approximately 60–250 mm in length) and weight (about 50–2000 g) range of the latter is striking and suggests that no single function is likely to account for the large biface accumulations of the East African Acheulean. As at Olorgesailie, some artifact occurrences at Kilombe do not include bifaces. Key references: Geology, dating, and paleoenvironment Dagley et al. 1978, Jones and Leat 1985; Hominins N/A; Archeology Gowlett 1988, 2005, 2006, Crompton and Gowlett 1993, Lycett and Gowlett 2008.

Kimeu, Kamoya (1940–) A renowned fossil hunter and field director of numerous expeditions, Kamoya Kimeu (universally referred to as "Kamoya") began his career working for Louis and Mary **Leakey** at **Olduvai Gorge** in 1960. Trained in excavation by Mary Leakey, he quickly learned to recognize hominin fossils and in 1964 made his first major discovery with the **Peninj 1** mandible. Working with the Kenyan contingent of the **International Omo Research Expedition** in 1967 he discovered the **Omo I** cranium at **Omo-Kibish**. Kamoya was a member of Richard

Leakey's first expedition to **Koobi Fora** (then East Rudolf) in 1968, and continued working there over four decades, making numerous important hominin finds. Over the years Kamoya was regularly dispatched with a small field crew to investigate potential new localities, and in the process he became familiar with sites all over Kenya. He worked extensively in Miocene localities of western Kenya as well as in the Turkana Basin. During early exploration of **West Turkana** he discovered the first fragments of the Nariokotome Boy (**KNM-WT 15000**), and directed the subsequent 5 years of excavations at that site. During the expeditions to **Kanapoi** and Allia Bay he was responsible for the discovery of important specimens attributed to *Australopithecus anamensis* including the tibia **KNM-KP 29285**. In 1985 Kamoya was awarded the LaGorce Medal of the National Geographic Society by President Ronald Reagan in a ceremony at the US White House, in recognition of his contributions to the understanding of human evolution.

kind (OE *gecynd*=race or offspring) A group made up of individuals that share one, or more, traits. Kind is also used in connection with the philosophical distinction between "degree" and "kind," as in the following quote from Charles **Darwin**'s *The Descent of Man*: "the difference in mind between man and the higher animals, great as it is, is one of degree and not kind."

kinematics (Gk *kinesis*=movement and *kinein*=to move) A branch of **biomechanics** that describes the motion of objects (kinematics sounds like, but should not to be confused with, **kinetics**, which describes the effects of forces on objects, particularly the effects of forces on the motion of objects). Kinematics includes information on the position, the **velocity** (speed) and accelerations (rate of change of velocity) of an animal or of an anatomical element (e.g., the hand) or of an object. Kinematic data are collected using **motion analysis** equipment, accelerometers, or other movement-measuring devices and these data are often collected in concert with data on kinetics. For example, building on Jenkins' (1972) landmark observations about how **chimpanzee** bipedalism involves movements that differ from those of modern humans, D'Août et al. (2002) measured the kinematics of the trunk and hind limb of nine **bonobos** during **knuckle-walking** and bipedal walking. They showed that, unlike modern humans, when bonobos walk bipedally they use flexed hips and knees throughout the **stance phase** and the knee joint rarely passes behind the hip joint. This **bent-hip bent-knee walking** gait prevents bonobos and chimpanzees from enjoying the energetic efficiency that characterizes the **gait** of modern humans (Sockol et al. 2007). It is probable that these efficiency gains were critical for the origin and/or later evolution of **obligate bipedalism** in later hominins.

kinetics (Gk *kinesis*=movement) A branch of **biomechanics** describing the effects of forces on objects, particularly the effects of forces on the motion of objects. Kinetic data are collected with **force plates**, **pressure pads**, or with other force-measuring devices, and such data are often collected in concert with data on **kinematics**. For example, Kimura (1985) used a force plate to measure the vertical force (i.e., **ground reaction force**) generated by the foot of **chimpanzees** during bipedal walking. In contrast with the double-peaked force pattern of **modern human** walking they found that in chimpanzees the vertical force remains at or slightly above body weight throughout the **stance phase**. This indicates that chimpanzees do not enjoy the energy-saving benefits of the oscillation between potential and kinetic energy that characterizes the **inverted-pendulum** type of bipedal walking in modern humans.

K-K Acronym for **Kesem-Kebana** (*which see*).

KL Abbreviation of Kollé and the prefix for fossils from the Kollé "fossiliferous area," Chad Basin, Ethiopia. *See* **Kollé**.

Klasies River (Location 34°06′S, 24°24′E, Eastern Cape Province, South Africa; etym. named after the outlet of Klasies River approximately 500 m to the east) History and general description The Klasies River Main Site on the Tsitsikamma coast of South Africa between Klasies River and Druipkelder Point comprises several caves and shelters (Caves 1 and 2, Shelters 1a, 1b, and 1c). They were initially excavated by John Wymer and Ronald **Singer** in 1967–8 and they were subsequently intensively investigated from 1984 onwards by Hilary **Deacon**. The excavations revealed a long **Middle Stone Age** (or MSA) sequence capped by a dense **Later Stone Age** shell **midden** and underlain by a culturally sterile storm beach formed during a period of high

sea level. Singer and Wymer (1982) divided the 30 m-deep sequence into 38 layers of which 13–38 contain evidence of the MSA. Singer and Wymer grouped these layers into cultural phases designated, from oldest to youngest, as MSA I, MSA II, Howieson's Poort, MSA III, and MSA IV. Deacon revised the stratigraphy, dividing the cultural deposits into members named, from lowest to highest, Light-Brown Sands (LBS), Sands-Ash-Shell (SAS), Rockfall (RF), and Upper Member; a disconformity and occupation hiatus separates the LBS from the SAS member. Temporal span and how dated? The storm beach at the base of the section likely formed in **Oxgen Isotope Stage** 5e (i.e., *c.*125–115 ka) or a previous interval of high sea level. The MSA faunal remains have absorbed a substantial amount of uranium, producing disparate age estimates for the linear and early-uptake dating models. Linear uptake age estimates from **electron spin resonance spectroscopy dating** suggest ages of around 93.5 ± 10.4 or 88.3 ± 7.8 ka for the base of the SAS member, which agree with **amino acid racemization** dating estimates and **uranium-series dating** estimates for the end of **speleothem** growth beneath the SAS. ESR dates indicate ages of 60–80 ka for overlying samples in the SAS and 40–60 ka for the Howieson's Poort-bearing strata (Grün et al. 1990). Jacobs et al. (2008) reported optically stimulated **luminescence dating** ages for the pre-Howieson's Poort of 72.1 ± 3.4 and 71.6 ± 2.9 ka, for the Howieson's Poort-bearing strata between 63.4 ± 2.6 and 65.5 ± 2.3 ka and an age for the post-Howieson's Poort MSA levels of 57.9 ± 2.3 ka. Hominins found at site There are hominins from the MSA levels but none from the Howieson's Poort strata. The vast majority of the specimens derive from the SAS Member (MSA II) with a few found in the LBS member and two in the post-Howieson's Poort MSA levels (MSA III). The MSA hominin collection comprises 10 isolated teeth, five partial mandibles, a frontal, a zygomatic, a left temporal bone, 13 vault fragments, two fragmentary maxillae, a proximal left radius, a fragmentary right ulna, a fragmentary atlas, a lumbar vertebra, a small fragment of an innominate, a left metatarsal I, a large left metatarsal II, and a right metatarsal V. The hominins are remarkable for several reasons. They exhibit a striking variability in overall size, suggesting either that a higher level of **sexual dimorphism** existed than is seen in recent modern humans or that the collection samples more than one morphologically distinctive population. Many support the claim that these fossils represent the earliest anatomically modern humans from southern Africa, but the degree to which the specimens are "anatomically modern" has been the subject of vigorous debate. Many of the hominin specimens show evidence consistent with having been defleshed and burned. <u>Archeological evidence found at site</u> Singer and Wymer (1982) organized the Klasies River strata into (from oldest to youngest) MSA I, MSA II, Howieson's Poort, MSA III, and MSA IV. The Howieson's Poort differs dramatically in frequencies and types of artifacts and in the greater proportion of exotic raw materials (e.g., **silcrete**) used. Shellfish as well as mammal and tortoise bone occur throughout the sequence. The MSA I occupies the LBS and the bottom of the SAS Member, the MSA II occurs in the rest of the SAS and RF member, and the Howieson's Poort and MSA III–IV the Upper Member. Most of the MSA hominins from the site derive from the SAS member, but a few come from the LBS as well as the Upper Member deposits above the Howieson's Poort. Stone raw materials are predominantly locally available **quartzite**s, with a shift to finer-grained rocks such as **hornfels** and **chalcedony** in the Howieson's Poort levels. Hearths, shell middens, and a highly fragmented fossil fauna with abundant cut and percussion marks attest to the use of the site as living area. Among the fauna are a number of burned and cutmarked hominin elements associated with other food remains from levels dating to c.100–110 ka, suggesting cannibalism. Analysis of the faunal remains from the MSA of Klasies and other sites in southern Africa have fueled lively debates over the hunting abilities of MSA versus LSA people, as well as the relative roles played by taphonomy and population density in shaping the faunal record. Key references: <u>Geology, dating, and paleoenvironment</u> Bada and Deems 1975, Butzer 1978, 1982, Shackleton 1982, Hendle and Volman 1986, Deacon et al. 1988, van Andel 1989, Grün et al. 1990, Miller et al. 1999, Jacobs et al. 2008; <u>Hominins</u> Singer and Wymer 1982, Bräuer 1984, White 1987, Rightmire and Deacon 1991, Smith 1992, Bräuer et al. 1992a, Churchill et al. 1996, Bräuer and Singer 1996a, 1996b, Wolpoff and Caspari 1996, Lam et al. 1996, Pearson and Grine 1997, Grine et al. 1998, Rightmire and Deacon 2001, Schwartz and Tattersall 2003, Rightmire et al. 2006a; <u>Archeology</u>

Klein 1975, 1976, 1982, 1989, Volman 1978, Singer and Wymer 1982, Voigt 1982, Binford 1984, Deacon et al. 1988, Deacon and Geleijnse 1988, Deacon 1989, 1992, 1995, 2001, Henderson 1992, Thackeray A 1992, Klein and Cruz-Uribe 1996, Klein et al. 1999, Milo 1998, Bertram and Marean 1999, Wurz 1999, 2002, 2005, d'Errico and Henshilwood 2007, Faith 2008.

Klasies River Mouth *See* **Klasies River**.

Kleine Feldhofer Grotte (Ge. *kleine*=small, *Feldhofer*=the name of a nearby farm, and *grotte*=cave) (Location 51°14′36″N, 6°57′4″E, Germany; etym. named after a local farm) <u>History and general description</u> Limestone cave now destroyed by mining. The original hominin associated skeleton (**Neanderthal 1**; an adult **calotte** plus 15 postcranial bones) was part of a collection of fossils discovered by miners in 1856. The hominins in the collection were identified by a local school teacher, C. Fuhlrott, who brought it to the attention of H. Schaaffhausen and the two of them collaborated to report the find to the local natural history society; the Neanderthal 1 associated skeleton was made the holotype of *Homo neanderthalensis* King, 1864. No faunal or archeological evidence from the Kleine Feldhofer Grotte was reported and there seemed to be no prospect that such evidence, and thus information about the context and dating, could ever be obtained. However, as a result of careful archival research, Ralf Schmitz and Jürgen Thissen (Schmitz et al. 2002) were able to identify what remained of the sediments from the Kleine Feldhofer Grotte at the base of the south wall of the Neander valley. Excavations of these sediments in 1997 resulted in the recovery of fauna, artifacts, and fragments of hominin bone and the researchers reported that "a small piece of human bone (NN 13) was found to fit exactly onto the lateral side of the left lateral femoral condyle of Neandertal 1" (*ibid*, p. 13342). In 2000 more fauna, more archeological evidence, and hominin skeletal fragments were recovered and "two cranial fragments...were found to fit onto the original Neandertal 1 calotte" (*ibid*, p. 13343). The balance of the remains recovered by Schmitz and his colleagues has been attributed to two additional individuals (Neanderthal 2 and 3). Researchers have recovered short fragments of mitochondrial DNA from the humerus of the Neanderthal 1 type specimen (Krings et al. 1997, 1999) from what might be the right tibial shaft of the same specimen (NN 4), and from the shaft of a right humerus (NN 1) of a second individual (Schmitz et al. 2002). Temporal span and how dated? Direct **radiocarbon dating** based on Neanderthal 1 and 2 suggest an age of *c*.40 ka. Hominins found at site Neanderthal 1 (1865/1997/2000), Neanderthal 2 (2000), and Neanderthal 3 (2000). Archeological evidence found at site Micoquian artifacts typical of late **Middle Paleolithic** as well as **Upper Paleolithic** artifacts from the **Gravettian**. Key references: Geology, dating, and paleoenvironment Schmitz et al. 2002; Hominins Fuhlrott and Schaaffhausen 1857, Schaafhausen 1888, Krings et al. 1997, 1999, Schmitz et al. 2002; Archeology Schmitz and Thissen 2000a, 2000b. Location of the hominin fossils found at site **Rheinisches Landesmuseum**, Bonn, Germany.

Klipgat   *See* **Die Kelders**.

KMH   Acronym for Kebara Mousterian hominids and the prefix used by Baruch Arensburg for hominin fossils from **Kebara** associated with the **Mousterian** occupation levels in order that they are distinguished from the hominin fossils excavated in the 1930s from the **Natufian** levels.

KMH1   Site **Kebara**. Locality Middle Paleolithic layer VII. Surface/*in situ* *In situ*. Date of discovery 1965. Finders Hebrew University of Jerusalem team led by Moshe Stekelis. Unit N/A. Horizon **Mousterian**. Bed/member N/A. Formation N/A. Group N/A. Nearest overlying dated horizon N/A. Nearest underlying dated horizon N/A. Geological age The Mousterian levels at Kebara have been dated by thermoluminescence dating to 61–59 ka (Valladas et al. 1987) and by electron spin resonance spectroscopy dating to 64–60 ka (Schwarcz et al. 1989). Developmental age Infant *c*.7–9 months. Presumed sex Unknown. Brief anatomical description Fragments of the skull and postcranial skeleton and a complete set of deciduous teeth. Announcement Schick and Stekelis 1977. Initial description Smith and Arensburg 1977. Photographs/line drawings and metrical data Smith and Arensburg 1977. Detailed anatomical description Smith and Arensburg 1977. Initial taxonomic allocation *Homo neanderthalensis*. Taxonomic revisions None. Current conventional taxonomic allocation *H. neanderthalensis*. Informal taxonomic category Pre-modern *Homo*. Signifi-

cance This is one of the more complete *H. neanderthalensis* juvenile fossils. Location of original Department of Anatomy and Anthropology, Tel Aviv University, Israel.

KMH2   Site **Kebara**. Locality N/A. Surface/*in situ* *In situ*. Date of discovery 1983. Finders The excavation team led by Ofer **Bar-Yosef**, Baruch **Arensburg**, and Bernard **Vandermeersch**. Unit N/A. Horizon **Mousterian**. Bed/member N/A. Formation N/A. Group N/A. Nearest overlying dated horizon N/A. Nearest underlying dated horizon N/A. Geological age The Mousterian levels at Kebara have been dated by **thermoluminescence dating** to 61–59 ka (Valladas et al. 1987) and by electron spin resonance spectroscopy dating to 64–60 ka (Schwarcz et al. 1989). Developmental age Adult. Presumed sex Male. Brief anatomical description Mandible but no cranium and the postcranial skeleton is missing most of left leg and all of the right leg. Announcement Arensburg et al. 1985. Initial description Arensburg et al. 1985. Photographs/line drawings and metrical data Bar-Yosef and Vandermeersch 1991, Bar-Yosef et al. 1986, 1992. Detailed anatomical description Bar-Yosef and Vandermeersch 1991. Initial taxonomic allocation The affinities of this individual with the Tabun, Amud, and Shanidar remains were noted early, aligning this fossil with other near-eastern *Homo neanderthalensis* rather than with the more modern human-looking hominins from **Skhul** and **Qafzeh**. Taxonomic revisions N/A. Current conventional taxonomic allocation *H. neanderthalensis*. Informal taxonomic category Pre-modern *Homo*. Significance This skeleton preserved a nearly complete Neanderthal pelvis, which helped end a debate about whether Neanderthals had different obstetric constraints than do modern humans (Rak and Arensburg 1987, Rak 1990). It also preserved the first Neanderthal hyoid, which was interpreted by the excavators to indicate that Neanderthals were potentially capable of speech (Arensburg et al. 1988, 1989, 1990, Tillier et al. 1991), although others have disagreed (Lieberman 1993). The preservation of the mandible, hyoid, and cervical vertebrae but the lack of a cranium led Bar-Yosef and colleagues (1992) to suggest that the skull was intentionally removed sometime after the ligaments that attached it to the spine had decomposed. Location of original Department of Anatomy and Anthropology, Tel Aviv University, Israel.

knapping (ME *knappen*=apparently it refers to the sound made when flakes are struck from a core) The act of intentionally fracturing stone by percussion for the production of **flake**s.

KNM Acronym for Kenya National Museum (now the **National Museums of Kenya**) and the prefix for fossils collected from sites in Kenya that are catalogued and stored in the National Museums of Kenya. The subsequent initials identify the site of origin of the fossil (e.g., **KNM-WT 15000** is an associated skeleton from West Turkana and **KNM-ER 1470** is a cranium from **Koobi Fora**, the site previously known as **East Rudolf**). The number that follows is the field number of the fossil that also functions as the museum accession number. *See also* **National Museums of Kenya**.

KNM-BC Prefix for fossils accessioned at the Kenya National Museum (now the **National Museums of Kenya**) that derive from sites in the Baringo region, Kenya, that are in the Chemeron Formation (e.g., **KNM-BC 1**). *See also* **Chemeron Formation; National Museums of Kenya**.

KNM-BC 1 Site Chemeron. Locality **EAGRU** site J.M. 85/BPRP#2. Surface/*in situ* Surface. Date of discovery 1965. Finder John Kimengich. Unit "The fossil was found weathered out on the surface of an exposure of the Upper Fish Beds" (Martyn and Tobias 1967, p. 478). Horizon Approximately 2.4 m above the Lower Tuffs (Deino and Hill 2002, p. 143 and Fig. 2). Bed/member Upper Fish Beds. Formation **Chemeron Formation**. Group N/A. Nearest overlying dated horizon KAP-3 lapilli. Nearest underlying dated horizon ALD86-1B "near the top of the Lower Tuffs" (Deino and Hill, 2002, Table 1a, p. 145). Geological age Between 2.456±0.006 and 2.393± 0.013 Ma. Developmental age Adult. Presumed sex Unknown. Brief anatomical description A well-preserved partial right temporal bone. Announcement Martyn and Tobias 1967. Initial description Martyn and Tobias 1967. Photographs/line drawings and metrical data Martyn and Tobias 1967, Hill et al. 1992, Sherwood et al. 2002, Lockwood et al. 2011. Detailed anatomical description N/A. Initial taxonomic allocation Hominidae *gen. et sp. indet.* Taxonomic revisions *Australopithecus boisei* (Tobias 1991b), *Homo sp. indet.* (Hill et al. 1992), *Homo rudolfensis* (Wood 1999), Homi-

nini *gen. et sp. indet.* (Lockwood et al. 2011). Current conventional taxonomic allocation *Homo sp. indet.* Informal taxonomic category Transitional hominin. Significance This securely dated isolated temporal includes several morphological traits said to align it with *Homo* (Sherwood et al. 2002). However, Lockwood et al.'s (2011) morphometric and phylogenetic study of the specimen found ambiguous phenetic affinities and little evidence for a cladistic relationship with *Homo*, echoing the original finding of Martyn and Tobias (1967). Location of original **National Museums of Kenya**, Nairobi, Kenya.

KNM-BK Prefix for fossils accessioned at the Kenya National Museum (now **National Museums of Kenya**) that derive from sites in the Baringo region, Kenya, that are in the Kapthurin Formation (e.g., **KNM-BK 8518**). *See also* **Kapthurin Formation; National Museums of Kenya**.

KNM-CH Prefix for fossils accessioned at the Kenya National Museum (now **National Museums of Kenya**) that derive from sites in the Baringo region, Kenya (e.g., Chesowanja), that are in the Chemoigut Formation (e.g., **KNM-CH 1**). *See also* **Chemoigut Formation; National Museums of Kenya**.

KNM-CH 1 Site **Chesowanja**. Locality Area 2. Surface/*in situ* Surface. Date of discovery 1970. Finders N/A. Unit It was described as coming from "yellow marls" that "overly the green silts in the west" in Carney et al. (1971), but Bishop et al. (1971, p. 205) suggest it was recovered from a "grey tuffaceous grit." Horizon N/A. Bed/member N/A. Formation **Chemoigut Formation**. Group N/A. Nearest overlying dated horizon Chesowanja basalt. Nearest underlying dated horizon Unknown. Geological age >1.42±0.07 Ma. Developmental age Subadult. Presumed sex Female. Brief anatomical description The specimen consists of a partial face, part of the frontal bone, and the anterior part of the base of a hominin cranium. The preserved teeth in the facial fragment include the right C–M^1. Announcement Carney et al. 1971. Initial description Carney et al. 1971. Photographs/line drawings and metrical data Carney et al. 1971. Detailed anatomical description N/A. Initial taxonomic allocation "Late robust australopithecine." Taxonomic revision *Australopithecus boisei* (Szalay 1971). Current conventional taxonomic allocation ***Paranthropus***

boisei. Informal taxonomic category Megadont archaic hominin. Significance The generally massive build of the face, the absolute and relative size of the dentition and details of facial morphology were all judged to be characteristic of *P. boisei* (Carney et al. 1971). However, when they compared KNM-CH 1 with **OH 5** Carney et al. (1971) interpreted the greater postorbital breadth and the apparent absence of temporal ridges in KNM-CH 1 as evidence that it sampled "...a population of evolved robust australopithecines, most likely descended from *A. boisei*" (*ibid*, p. 513). Szalay (1971) and others were not persuaded that the cranial capacity of KNM-CH 1 was large, nor were they convinced that its morphology departed from that of *P. boisei*. The KNM-CH 1 cranium is now more usually interpreted (e.g., Howell 1978b, Wood 1991) as a small-bodied probably female member of *P. boisei*. Location of original **National Museums of Kenya**, Nairobi, Kenya.

KNM-CH 304 Site **Chesowanja**. Locality Area 2. Surface/*in situ* Surface. Date of discovery 1978. Finders Bernard Ngeneo and Wambua Mangao. Unit The five cranial vault fragments were found in two separate scatters, 15 m apart. The fragments most likely came from the same "silting-up channel" in "clayey silts" that yielded the fauna and artifacts at the GnJi 1/6 "burnt clay" site. Horizon N/A. Bed/member N/A. Formation **Chemoigut Formation**. Group N/A. Nearest overlying dated horizon Chesowanja basalt. Nearest underlying dated horizon Unknown. Geological age $>1.42\pm0.07$ Ma. Developmental age Adult. Presumed sex Male. Brief anatomical description The specimen consists of five cranial vault fragments (A–E). They include parts of the occipital (A), the parietals (B and D), and the temporal (C) bone(s). Announcement Gowlett et al. 1981. Initial description Gowlett et al. 1981. Photographs/line drawings and metrical data N/A. Detailed anatomical description N/A. Initial taxonomic allocation *Australopithecus boisei*. Taxonomic revisions None. Current conventional taxonomic allocation *Paranthropus boisei*. Informal taxonomic category Megadont archaic hominin. Significance Although KNM-CH 304 consists of just a few fragments of a calotte the nature and location of the ectocranial cresting on the occipital bone, the presence of *striae parietalis*, and the evidence for an **occipito-marginal system** of cranial venous sinuses and the geological age of the specimen all point to it belonging to *P. boisei*. Location of original **National Museums of Kenya**, Nairobi, Kenya.

KNM-ER Prefix for fossils accessioned at the Kenya National Museum (now **National Museums of Kenya**) that derive from a series of sites that used to be known as East Rudolf, and which are now referred to informally as **Koobi Fora** (e.g., **KNM-ER 3733**). *See also* **Koobi Fora**; **National Museums of Kenya**.

KNM-ER 406 (NB: Referred to as FS-158 when announced and later given a "KNM-ER" number) Site **Koobi Fora** (Ileret). Locality Area 10. Surface/*in situ* Surface. Date of discovery 1969. Finder Richard **Leakey**. Unit N/A. Horizon N/A. Bed/member KBS Member. Formation **Koobi Fora Formation**. Group Omo Group. Nearest overlying dated horizon N/A. Nearest underlying dated horizon **KBS Tuff**. Geological age *c*.1.7 Ma. Developmental age Adult (but some cranial vault sutures are unfused). Presumed sex Male. Brief anatomical description Nearly complete but edentulous cranium. Most of the surface bone is missing from the alveolar processes and surface bone has also been lost from the malar region and the underside of the sphenoid. Endocranial volume 510 ± 10 cm^3 (Holloway 1983c). Announcement Leakey 1970. Initial description Leakey et al. 1971. Photographs/line drawings and metrical data Leakey et al. 1971, Wood 1991. Detailed anatomical description Wood 1991. Initial taxonomic allocation *Australopithecus boisei*. Taxonomic revisions *Paranthropus boisei*. Current conventional taxonomic allocation *P. boisei*. Informal taxonomic category Megadont archaic hominin. Significance Two mandibular fragments (KNM-ER 403 and 404) and a poorly preserved palate (KNM-ER 405) had suggested the presence of *P. boisei* at Koobi Fora, but the discovery of KNM-ER 406 confirmed it. It conforms to the same basic **bauplan** as **OH 5** (a modest-sized neurocranium associated with a large, wide, orthognathic face) but it also showed morphological differences in facial and ectocranial morphology. These differences, and the ones between it and **KNM-ER 407** and **732**, helped researchers establish the types of intraspecific variation to be expected within a sexually dimorphic megadont archaic early hominin taxon. Location of original **National Museums of Kenya**, Nairobi, Kenya.

KNM-ER 407 (NB: Referred to as FS-210 when announced and later given a "KNM-ER" number) Site **Koobi Fora** (Ileret). Locality Area 10. Surface/ *in situ* Surface. Date of discovery 1969. Finder Mwongela Muoka. Unit N/A. Horizon N/A. Bed/ member KBS Member. Formation **Koobi Fora Formation**. Group Omo Group. Nearest overlying dated horizon N/A. Nearest underlying dated horizon **KBS Tuff**. Geological age *c*.1.85 Ma. Developmental age Adult. Presumed sex Female. Brief anatomical description Well-preserved presumed small-bodied female adult calvaria in three pieces that preserve most of both parietals and the posterior two-thirds of the frontal, the base, and sides of the cranial vault and the left petrous temporal. Endocranial volume 506 cm^3 (Falk and Kasinga 1983), 510 cm^3 (Holloway 1983c). Announcement Leakey 1970. Initial description Day et al. 1976. Photographs/line drawings and metrical data Day et al. 1976, Wood 1991. Detailed anatomical description Wood 1991. Initial taxonomic allocation *Homo sp. indet.* Taxonomic revisions *Australopithecus boisei* (Leakey 1976a). Current conventional taxonomic allocation *Paranthropus boisei*. Informal taxonomic category Megadont archaic hominin. Significance The cranial base is smaller than that of KNM-ER 406 but otherwise it conforms to the same basic **bauplan**. Unlike **KNM-ER 406** the vault has no ectocranial crests. The overall resemblance to KNM-ER 406 along with the differences between it and KNM-ER 406 suggested an attribution to *P. boisei*, but as a small-bodied presumed female. The difference in ectocranial morphology between it and KNM-ER 406 helped researchers establish the type of sex-related intraspecific variation to be expected within a sexually dimorphic megadont archaic early hominin taxon. Location of original **National Museums of Kenya**, Nairobi, Kenya. *See also* **KNM-ER 732**.

KNM-ER 732 Site **Koobi Fora** (Ileret). Locality Area 10. Surface/*in situ* Surface. Date of discovery 1970. Finder Harrison Mutua. Unit N/A. Horizon N/A. Bed/member KBS Member. Formation **Koobi Fora Formation**. Group Omo Group. Nearest overlying dated horizon N/A. Nearest underlying dated horizon KBS Tuff. Geological age *c*.1.7 Ma. Developmental age Adult. Presumed sex Female. Brief anatomical description Well-preserved cranium of a presumed small-bodied adult female, preserving facial morphology and dental evidence. Preserved portions include the right side of the face, including the alveolar process from the right canine alveolus to the maxillary tuberosity, part of the right side of the basicranium, the frontal, and parts of both parietals. Most of the surface bone is missing from the alveolar process and the only tooth crown preserved is the lingual half of the P^4. Surface bone has also been lost from the vault and mastoid region. Endocranial volume 500 cm^3 (Holloway 1983c). Announcement Leakey 1971. Initial description Leakey et al. 1972. Photographs/line drawings and metrical data Leakey et al. 1972, Wood 1991. Detailed anatomical description Wood 1991. Initial taxonomic allocation *Australopithecus sp. indet.* Taxonomic revisions *Australopithecus boisei* (Leakey 1973b), *Australopithecus africanus* (Wolpoff 1978a), *Paranthropus boisei* (Robinson 1960). Current conventional taxonomic allocation *P. boisei*. Informal taxonomic category Megadont archaic hominin. Significance Conforms to the same frontal and facial **bauplan** as OH 5 (modest-sized neurocranium with prominent glabella associated with an orthognathic face widest in its middle part) but it shows morphological differences in facial and ectocranial morphology that helped researchers establish the types of sex-related intraspecific variation to be expected within a sexually dimorphic megadont archaic early hominin taxon. Location of original **National Museums of Kenya**, Nairobi, Kenya. *See also* **KNM-ER 407**.

KNM-ER 736 Site Koobi Fora. Locality Area 103. Surface/*in situ* Surface. Date of discovery 1970. Finder Meave **Leakey**. Unit N/A. Horizon A6+7m. Bed/member KBS Member. Formation **Koobi Fora Formation**. Group Omo Group. Nearest overlying dated horizon Lower Koobi Fora Tuff (1.48 Ma). Nearest underlying dated horizon KBS Tuff. Geological age *c*.1.6 Ma. Developmental age Adult. Presumed sex Male. Brief anatomical description The shaft of a left femur, broken proximally just distal to the lesser trochanter and distally proximal to the divergence of the supracondylar lines. Announcement Leakey 1971. Initial description Leakey 1971. Photographs/line drawings and metrical data Leakey et al. 1972, Day 1976, Leakey et al. 1978. Detailed anatomical description Leakey et al. 1972. Initial taxonomic allocation *Homo* sp. Taxonomic revisions of. *Australopithecus* (Leakey et al. 1972), *Homo* (Day 1976), *Homo erectus* (Walker 1993). Current conventional taxonomic allocation *Homo erectus* or *Homo ergaster*. Informal taxonomic category Pre-modern *Homo*. Signifi cance This massive platymeric femoral shaft has a clearly developed *linea aspera* and for these reasons it is now generally attributed to *Homo* as *H. erectus*. However, if ***Homo rudolfensis*** proves to be valid taxon, then it is conceivable it could belong to that taxon. Location of original **National Museums of Kenya**, Nairobi, Kenya.

KNM-ER 737 Site **Koobi Fora**. Locality Area 103. Surface/*in situ* Surface. Date of discovery 1970. Finder Richard **Leakey**. Unit N/A. Horizon LKF-1m. Bed/member Okote Member. Formation **Koobi Fora Formation**. Group Omo Group. Nearest overlying dated horizon Lower Koobi Fora Tuff (1.48 Ma). Nearest underlying dated horizon KBS Tuff. Geological age *c*.1.5 Ma. Developmental age Late adolescent or adult. Presumed sex Unknown. Brief anatomical description A left femur, lacking the greater and lesser trochanter, head, most of the neck, and the condyles. Announcement Leakey 1971. Initial description Leakey 1971. Photographs/line drawings and metrical data Day and Leakey 1973, Leakey et al. 1978. Detailed anatomical description Day and Leakey 1973. Initial taxonomic allocation *Homo* sp. Taxonomic revisions *Homo erectus* (Day and Leakey 1973). Current conventional taxonomic allocation *H. erectus* or ***Homo ergaster***. Informal taxonomic category Pre-modern *Homo*. Significance The shaft of the KNM-ER 737 femur is platymeric and it bears a clearly defined *linea aspera* and for these reasons it is now generally attributed to *Homo* as *H. erectus*. However, if ***Homo rudolfensis*** proves to be valid taxon, then it is conceivable that it could belong to that taxon. Location of original **National Museums of Kenya**, Nairobi, Kenya.

KNM-ER 738A Site **Koobi Fora**. Locality Area 105. Surface/*in situ* Surface. Date of discovery 1970. Finder Bernard Ngeneo. Unit N/A. Horizon KBS CC. Bed/member KBS Member. Formation **Koobi Fora Formation**. Group Omo Group. Nearest overlying dated horizon N/A. Nearest underlying dated horizon KBS Tuff. Geological age *c*.1.87 Ma. Developmental age Adult. Presumed sex Unknown. Brief anatomical description Two fragments of a proximal left femur. The proximal fragment preserves the head, neck, and lesser trochanter; the distal fragment preserves a portion of the medial half of the diaphysis. The shaft bears a trace of a fracture callus. Announcement Leakey 1971. Initial description Leakey 1971. Photographs/line drawings and metrical data Leakey

1971, Leakey et al. 1972. Detailed anatomical description Leakey et al. 1972, Walker 1973. Initial taxonomic allocation *Australopithecus*. Taxonomic revisions N/A. Current conventional taxonomic allocation ***Paranthropus boisei*** or ***Homo habilis***. Informal taxonomic category Megadont archaic or transitional hominin. Significance One of the first femora to be recovered from Koobi Fora that differed from the femoral morphology associated with later *Homo* and resembled the femora of ***Paranthropus robustus*** in that the femoral neck shows the same marked anteroposterior flattening as that seen in the two proximal femora from Swartkrans, **SK 82** and **SK 97**. This is one of several proximal femoral fragments that fuelled the suggestion that similar femora from Koobi Fora belonged to *P. boisei*. But evidence from **OH 62** suggests that much the same morphology is also seen in *H. habilis* in which case this and other similar femora could belong to *H. habilis*. It is one of the earliest examples of a healed fracture in a hominin long bone. Location of original **National Museums of Kenya**, Nairobi, Kenya.

KNM-ER 739 Site **Koobi Fora**. Locality Area 1. Surface/*in situ* Surface. Date of discovery 1970. Finder Harrison Mutua. Unit N/A. Horizon CHA-18m. Bed/member Okote Member. Formation **Koobi Fora Formation**. Group Omo Group. Nearest overlying dated horizon Chari Tuff. Nearest underlying dated horizon Lower Ileret Tuff. Geological age *c*.1.45 Ma. Developmental age Adult. Presumed sex Male. Brief anatomical description A right humerus lacking its head and the greater and lesser tubercles. The shaft is thick and the muscular attachments are highly developed. Announcement Leakey 1971. Initial description Leakey 1971. Photographs/line drawings and metrical data Leakey 1971, Leakey et al. 1978. Detailed anatomical description Leakey et al. 1972, McHenry 1973. Initial taxonomic allocation *Australopithecus*. Taxonomic revisions *Homo rudolfensis* (Lague and Jungers 1996). Current conventional taxonomic allocation ***Paranthropus boisei***. Informal taxonomic category Megadont archaic hominin. Significance KNM-ER 739 is one of the best-preserved humeri from Koobi Fora and differs strikingly from humeri of ***Homo sapiens*** in having a pronounced brachioradialis flange and a deltopectoral crest. Visual inspection (Senut and Tardieu 1985) and the results of a broad comparative analysis (Lague and Jungers 1996) of distal humerus shape variation linked this specimen with **KNM-ER 1504**, which shares morphological similarities with **KNM-ER 3735** (Leakey et al. 1989), the fragmentary partial skeleton attributed to *H. habilis*. Based on its similarity with humeri of early *Homo* and its large size, Lague and Jungers (1996) suggest that KNM-ER 739 be tentatively attributed to *H. rudolfensis* whereas most authors have considered it to be most likely attributable to *P. boisei* (Leakey 1971, McHenry 1973, Day 1976, Howell 1978b). Location of original **National Museums of Kenya**, Nairobi, Kenya.

KNM-ER 741 Site **Koobi Fora**. Locality Area 1. Surface/*in situ* Surface. Date of discovery 1970 Finder Mary **Leakey**. Unit Collection unit 5. Horizon ILN+10m. Bed/member Okote Member. Formation **Koobi Fora Formation**. Group Omo Group. Nearest overlying dated horizon Chari Tuff. Nearest underlying dated horizon Lower Ileret Tuff. Geological age *c*.1.45 Ma. Developmental age Adult. Presumed sex Unknown. Brief anatomical description Proximal end and adjacent shaft of a left tibia. Announcement Leakey 1971. Initial description Leakey 1971. Photographs/line drawings and metrical data Leakey et al. 1972, 1978. Detailed anatomical description Leakey et al. 1972. Initial taxonomic allocation *Australopithecus boisei*. Taxonomic revisions *Homo erectus* (Walker 1993, Brown et al. 2001). Current conventional taxonomic allocation *H. erectus* or Hominini *gen. et sp. indet.* Informal taxonomic category N/A. Significance One of several well-preserved tibiae from Koobi Fora, KNM-ER 741 has been attributed to *H. erectus* based on a comparison (Walker 1993) with **KNM-WT 15000** (a juvenile associated skeleton attributed to early *H. erectus* or *H. ergaster*) but because so little is known about the tibial morphology of early hominins other than *Australopithecus afarensis* it may be premature to rule out the possibility that it belongs to *Homo habilis* or to *Paranthropus boisei*. Location of original **National Museums of Kenya**, Nairobi, Kenya.

KNM-ER 803 Site **Koobi Fora**. Locality Area 8. Surface/*in situ* Surface. Date of discovery 1971. Finder Meave **Leakey**. Unit N/A. Horizon BPT + 1m. Bed/member Okote Member. Formation **Koobi Fora Formation**. Group Omo Group. Nearest overlying dated horizon Chari Tuff. Nearest underlying dated horizon N/A. Geological age *c*.1.5 Ma. Developmental age Adult. Presumed sex Unknown. Brief anatomical description A partial

associated skeleton of an adult comprising the lower left C, right I[1], a fragment of the shaft of the left ulna, a fragment of a radius shaft, an abraded base of a metacarpal, a fragment of a phalangeal shaft, a fragment of a proximal phalanx (possibly II or III), a partial shaft of the left femur, a partial shaft of the left tibia, a fragment of the medial condyle of the left tibia, fragments of the left and right fibula, a fragmentary left talus, left metatarsal III except its head, the base of left metatarsal V, a fragment of the shaft of proximal pedal phalanx I, two intermediate pedal phalanges (possibly II and III), and a terminal toe phalanx. Announcement Leakey 1971. Initial description Leakey 1971. Photographs/line drawings and metrical data Day and Leakey 1973, Leakey et al. 1978. Detailed anatomical description Day and Leakey 1973. Initial taxonomic allocation *Homo* sp. Taxonomic revisions *Homo erectus* (Wood 1976, 1991, Walker 1993, Ruff and Walker 1993). Current conventional taxonomic allocation *H. erectus* or *Homo ergaster*. Informal taxonomic category Pre-modern *Homo*. Significance KNM-ER 803 is one of the few associated skeletons that has been attributed to *H. erectus* or *H. ergaster*. However, whereas the C and I[1] are large enough to suggest that KNM-ER 803 does not represent *Paranthropus boisei*, there is not sufficient taxonomically diagnostic evidence (e.g., there is no proximal femur) to rule out its attribution to *Homo habilis*. Location of original **National Museums of Kenya**, Nairobi, Kenya.

KNM-ER 813 Site **Koobi Fora**. Locality Area 104. Surface/*in situ* Surface. Date of discovery 1971. Finder Bernard Ngeneo. Unit N/A. Horizon A2-4m. Bed/member KBS Member. Formation **Koobi Fora Formation**. Group Omo Group. Nearest overlying dated horizon N/A. Nearest underlying dated horizon KBS Tuff. Geological age *c*.1.78 Ma. Developmental age Unknown. Presumed sex Unknown. Brief anatomical description The right talus and a fragment of the distal shaft of the right tibia apparently from a single individual. Announcement Leakey 1972. Initial description Leakey and Wood 1973. Photographs/line drawings and metrical data Wood 1974, Leakey et al. 1978. Detailed anatomical description Leakey and Wood 1973, Gebo and Schwartz 2006. Initial taxonomic allocation *Homo* sp. *indet*. Taxonomic revisions *Homo erectus* or *Homo ergaster* (Tobias 1991b), *Homo rudolfensis* (Wood 1992). Current conventional taxonomic allocation *Homo sp. indet.* Informal taxonomic category Pre-modern *Homo* or transitional hominin. Significance The morphological differences between **OH 8** and KNM-ER 813 suggest these two fossil tali do not belong to the same species (Wood 1974, Gebo and Schwartz 2006). If OH 8 belongs to *Homo habilis*, then KNM-ER 813 (together with Omo 323-76-898) may belong to either *H. rudolfensis* or to *H. erectus*. Location of original **National Museums of Kenya**, Nairobi, Kenya.

KNM-ER 815 Site **Koobi Fora**. Locality Area 10. Surface/*in situ* Surface. Date of discovery 1971. Finder Andrew Hill. Unit Collection unit 4. Horizon N/A. Bed/member KBS Member. Formation **Koobi Fora Formation**. Group Omo Group. Nearest overlying dated horizon N/A. Nearest underlying dated horizon KBS Tuff. Geological age 1.77±0.10 Ma. Developmental age Adult. Presumed sex Unknown. Brief anatomical description A weathered proximal femur retaining its neck, base of the lesser trochanter, and the subtrochanteric diaphysis. Announcement Leakey 1972. Initial description Leakey and Walker 1973. Photographs/line drawings and metrical data Leakey and Walker 1973, Leakey et al. 1978. Detailed anatomical description Leakey and Walker 1973, Walker 1973. Initial taxonomic allocation *Australopithecus*. Taxonomic revisions N/A. Current conventional taxonomic allocation *Paranthropus boisei*. Informal taxonomic category Megadont archaic hominin. Significance The femoral neck shows the same marked anteroposterior flattening seen in the two proximal femora from Swartkrans, **SK 82** and **SK 97**, attributed to *Paranthropus robustus*. This is one of the proximal femur fragments that fuelled the suggestion that similar femora from Koobi Fora belonged to *P. boisei*. But evidence from **OH 62** suggests that much the same morphology is also seen in *Homo habilis*, in which case this and other similar femora could belong to *H. habilis*. Location of original **National Museums of Kenya**, Nairobi, Kenya.

KNM-ER 820 Site **Koobi Fora** (Ileret). Locality Area 1. Surface/*in situ* In situ. Date of discovery 1971. Finder Harrison Mtua. Unit N/A. Horizon N/A. Bed/member Okote Member. Formation **Koobi Fora Formation**. Group Omo Group. Nearest overlying dated horizon Chari Tuff. Nearest underlying dated horizon N/A. Geological

age *c.*1.6 Ma. Developmental age Juvenile (both dm_1 and dm_2 are heavily worn). Presumed sex Unknown. Brief anatomical description Well-preserved juvenile mandible. The right and left sides of the corpus, the base of the left ramus, the crowns and roots of the deciduous canines and molars, the crowns and roots of the permanent incisors and M_1, the germs of the M_2, and the crypts for the M_3s. Announcement Leakey 1972. Initial description Leakey and Wood 1973. Photographs/line drawings and metrical data Leakey and Wood 1973, Wood 1991. Detailed anatomical description Wood 1991. Initial taxonomic allocation *Homo sp. indet.* Taxonomic revisions *Homo ergaster* (Groves and Mazák 1975). Current conventional taxonomic allocation *Homo erectus* or *H. ergaster*. Informal taxonomic category Pre-modern *Homo*. Significance Along with **KNM-ER 992**, KNM-ER 820 provided the first evidence of a *H. erectus*-like mandibular morphology that was in several ways more primitive than that seen in the *H. erectus* hypodigm from Java and **Zhoukoudian**. Location of original **National Museums of Kenya**, Nairobi, Kenya. *See also* **KNM-ER 992**.

KNM-ER 992 Site **Koobi Fora** (Ileret). Locality Area 3. Surface/*in situ* Surface. Date of discovery 1971. Finder Bernard Ngeneo. Unit N/A. Horizon ILN+10m. Bed/member Okote Member. Formation **Koobi Fora Formation**. Group Omo Group. Nearest overlying dated horizon Chari Tuff. Nearest underlying dated horizon N/A. Geological age *c.*1.52 Ma. Developmental age Subadult (both M_3s show occlusal wear, but the roots of both the M_3s are incomplete with wide-open apices). Presumed sex Unknown. Brief anatomical description Mandible broken close to the midline but otherwise well preserved. It includes both sides of the corpus, part of the right and much of the left ramus, and the crowns and roots of the canines to the M_3s on both sides. Announcement Leakey 1972. Initial description Leakey and Wood 1973. Photographs/line drawings and metrical data Leakey and Wood 1973, Wood 1991. Detailed anatomical description Wood 1991. Initial taxonomic allocation *Homo sp. indet.* Taxonomic revisions *Homo ergaster* (Groves and Mazák 1975). Current conventional taxonomic allocation *Homo erectus* or *H. ergaster*. Informal taxonomic category Pre-modern *Homo*. Significance This mandible was made the **holotype** of *H. ergaster*. Along with **KNM-ER 820**, KNM-

ER 992 provided the first evidence of a *H. erectus*-like mandibular morphology that was in several ways more primitive than that of *H. erectus* from Java and **Zhoukoudian**. Location of original **National Museums of Kenya**, Nairobi, Kenya. *See also* **KNM-ER 820**.

KNM-ER 993   Site **Koobi Fora**. Locality Area 1. Surface/*in situ* Surface. Date of discovery 1971. Finder Kamoya **Kimeu**. Unit Collection unit 5. Horizon ILN+10m. Bed/member Okote Member. Formation **Koobi Fora Formation**. Group Omo Group. Nearest overlying dated horizon Chari Tuff. Nearest underlying dated horizon Lower Ileret Tuff. Geological age *c*.1.45 Ma. Developmental age Adult. Presumed sex Unknown. Brief anatomical description Right femur lacking its proximal end. Announcement Leakey 1972. Initial description Leakey 1972. Photographs/line drawings and metrical data Leakey 1972, Leakey and Walker 1973, Leakey et al. 1978. Detailed anatomical description Leakey and Walker 1973, Walker 1973. Initial taxonomic allocation *Australopithecus*. Taxonomic revisions N/A. Current conventional taxonomic allocation *Paranthropus boisei*. Informal taxonomic category Megadont archaic hominin. Significance The femoral shaft is moderately robust, platymeric in the proximal portion but more circular in cross-section distally owing to the development of a pilaster. Although the femoral condyles are abraded, the knee is highly **valgus** with a bicondylar angle of about 15° (from vertical) compared with mean modern human values of between 8.5° and 10.7° (Tardieu and Trinkaus 1994). This suggests a substantial inter-acetabular distance and thus wide hips for this individual. While it has been attributed to *P. boisei*, the lack of associated craniodental fossils leaves its attribution open to question. Location of original **National Museums of Kenya**, Nairobi, Kenya.

KNM-ER 999   Site **Koobi Fora** (Ileret). Locality Area 6A. Surface/*in situ* Surface. Date of discovery 1971. Finder Kamoya **Kimeu**. Unit G. Horizon Collection unit 6, approximately 9–11 m above the bottom of the Guomde Formation (Leakey et al. 1978). Bed/member The specimen likely derives from undifferentiated Middle–Late Pleistocene sediments above the Chari Member. Formation Likely unassigned strata between the **Koobi Fora Formation** and the Galana Boi Formation *sensu stricto*. Group N/A. Nearest overlying

dated horizon Galana Boi Formation. Nearest underlying dated horizon Chari tuff. Geological age Bräuer et al. (1997) reported direct **uranium-series dating** ages for the specimen of 301 ka (+infinity/−96 ka) by uranium-thorium and >180 ka by uranium-protactinium. Developmental age Adult. Presumed sex Male, based on its large size. Brief anatomical description A massive left femur lacking most of the distal epiphysis and with substantial damage to the greater trochanter and head. The femur has a high neck-shaft angle, a pronounced *linea aspera*, but a moderate pilastric index. The estimated bicondylar length of the complete bone is between 470 and 500 mm Announcement Day and Leakey 1974. Initial description Day and Leakey 1974. Photographs/line drawings and metrical data Leakey et al. 1978, Trinkaus 1993. Detailed anatomical description Day and Leakey 1974, Trinkaus 1993. Initial taxonomic allocation *Homo*. Taxonomic revisions Late archaic *Homo sapiens*, anatomically modern *Homo sapiens*, **Homo helmei**. Current conventional taxonomic allocation *Homo sapiens*. Informal taxonomic category Anatomically modern *Homo*. Significance Despite its geological antiquity, the femur is essentially modern human-like in its anatomy, supporting speculation that a modern human-like lower-limb morphology had emerged in Africa by *c*.300 ka. Location of original **National Museums of Kenya**, Nairobi, Kenya.

KNM-ER 1463   Site **Koobi Fora**. Locality Area 1A. Surface/*in situ* Surface. Date of discovery 1972. Finder Kamoya **Kimeu**. Unit Collection unit 5. Horizon MFB+1m. Bed/member Okote Member. Formation **Koobi Fora Formation**. Group Omo Group. Nearest overlying dated horizon Chari Tuff. Nearest underlying dated horizon Lower Ileret Tuff. Geological age *c*.1.4 Ma. Developmental age Adult. Presumed sex Unknown, possibly male based on size. Brief anatomical description The neck and intact shaft of a right femur; the head, the greater and lesser trochanters, and the condyles are missing. Announcement Leakey 1973. Initial description Day et al. 1976. Photographs/line drawings and metrical data Day et al. 1976, Leakey et al. 1978. Detailed anatomical description Day et al. 1976. Initial taxonomic allocation *Australopithecus*. Taxonomic revisions N/A. Current conventional taxonomic allocation *Paranthropus boisei*. Informal taxonomic category Megadont archaic hominin. Significance The femoral neck shows the same marked anteroposterior flattening as that seen in the two proximal femora from Swartkrans, **SK 82** and **SK 97**, attributed to *Paranthropus robustus*, so this is one of the proximal femur fragments that fuelled the suggestion that similar femora from Koobi Fora belonged to *P. boisei*. But evidence from **OH 62** suggests that the same morphology is also seen in *Homo habilis*, in which case this and other similar femora could belong to *H. habilis*. Location of original **National Museums of Kenya**, Nairobi, Kenya.

KNM-ER 1464 Site Koobi Fora. Locality Area 6A. Surface/*in situ* Surface from the same locality as KNM-ER 801, 802, 1171, 1816, 1823, 1824, 1825, and 3737. Date of discovery 1972. Finder Peter Nzube Mutiwa. Unit N/A. Horizon LIL-8m. Bed/member KBS Member. Formation **Koobi Fora Formation**. Group Omo Group. Nearest overlying dated horizon Lower Ileret Tuff. Nearest underlying dated horizon KBS Tuff. Geological age *c*.1.7 Ma. Developmental age Adult. Presumed sex Unknown. Brief anatomical description A well-preserved right talus. Announcement Leakey 1973. Initial description Leakey 1973. Photographs/line drawings and metrical data Day et al. 1976, Leakey et al. 1978, Gebo and Schwartz 2006. Detailed anatomical description Day et al. 1976, Gebo and Schwartz 2006. Initial taxonomic allocation *Australopithecus*. Taxonomic revisions *Incertae sedis* (Day et al. 1976, Gebo and Schwartz 2006), *Australopithecus boisei* (Grausz et al. 1988). Current conventional taxonomic allocation **Hominini** *gen. et sp. indet.* Informal taxonomic category N/A. Significance This enigmatic talus shows pronounced trochlear wedging, a short neck, and a strong lateral flange suggesting a tibiofibular articulation different from that of modern humans. It does not clearly align with either specimens attributed, rightly or wrongly, to *Paranthropus* or to *Homo* aff. *erectus*. Location of original **National Museums of Kenya**, Nairobi, Kenya.

KNM-ER 1465 Site Koobi Fora. Locality Area 11. Surface/*in situ* Surface. Date of discovery 1972. Finder Kamoya **Kimeu**. Unit Collection unit 5. Horizon Approximately MFB. Bed/member Okote Member. Formation **Koobi Fora Formation**. Group Omo Group. Nearest overlying dated horizon Chari Tuff. Nearest underlying dated horizon N/A. Geological age *c*.1.42 Ma. Developmental age Unknown. Presumed sex Unknown. Brief anatomical description Two frag-

ments of a proximal femur, a fragment of the shaft and neck missing both trochanters, and a small portion of the head. Announcement Leakey 1973. Initial description Day et al. 1976. Photographs/line drawings and metrical data Day et al. 1976. Detailed anatomical description Day et al. 1976. Initial taxonomic allocation *Australopithecus*. Taxonomic revisions N/A. Current conventional taxonomic allocation *Paranthropus boisei*. Informal taxonomic category Megadont archaic hominin. Significance The femoral neck shows the same marked anteroposterior flattening as that seen in the two proximal femora from Swartkrans, **SK 82** and **SK 97**, attributed to *Paranthropus robustus*, so this is one of the proximal femur fragments that fuelled the suggestion that similar femora from Koobi Fora belonged to *P. boisei*. But evidence from **OH 62** suggests that the same morphology is also seen in *Homo habilis*, in which case this and other similar femora could belong to *H. habilis*. Location of original **National Museums of Kenya**, Nairobi, Kenya.

KNM-ER 1470 Site Koobi Fora (Karari Ridge). Locality Area 131. Surface/*in situ* Surface. Date of discovery 1972. Finder Bernard Ngeneo. Unit N/A. Horizon KBS-36m. Bed/member upper Burgi Member. Formation **Koobi Fora Formation**. Group Omo Group. Nearest overlying dated horizon KBS Tuff. Nearest underlying dated horizon N/A. Geological age *c*.1.88–1.95 Ma. Developmental age Adult. Presumed sex Male. Brief anatomical description Cranium lacking most of the base; the base was finely fragmented when discovered. The reassembled cranium comprises separate calvarial and facial components; there is a join at nasion but it is not possible to determine the facial angle with any reliability. Bromage et al. (2008) used Enlow's 90° "PM-NHA rule" (Enlow and Azuma 1975) to suggest the face was most likely more prognathic than in the original reconstruction. Most of the surface bone of the specimen is missing and the specimen has suffered from plastic deformation which according to Walker (1981, p. 203) has depressed the parietotemporal suture on the right side so that "the right temporal and porion have been carried downwards and forwards." Endocranial volume 752 cm^3 (Holloway 1983c). Bromage et al. (2008) suggest their more prognathic reconstuction of KNM-ER 1470 is not compatible with this estimate, but Holloway's methodology is well tried and it is difficult to see how the endocranial volume could be as low as the value (625 cm^3) predicted by the regression used in Bromage et al. (2008) and in fact the latter authors suggest a more reasonable endocranial volume estimate of 700 cm^3. Announcement Leakey 1973c. Initial description Day et al. 1975. Photographs/line drawings and metrical data Day et al. 1975, Wood 1991. Detailed anatomical description Wood 1991. Initial taxonomic allocation *Homo* sp. indet. Taxonomic revisions *Homo habilis* (Leakey 1976a), *Pithecanthropus rudolfensis* (Alexeev 1986), *Homo rudolfensis* (Groves 1989, Wood 1992). Current conventional taxonomic allocation *H. habilis* or *H. rudolfensis*. Informal taxonomic category Pre-modern *Homo* or transitional hominin. Significance The cranial vault is similar in shape to that of *H. habilis*, but larger. The face is more orthognathic than that of *H. habilis* and it is widest in the midface, not in the upper face as in the case in *H. habilis*. Holloway (1983c) suggested the endocranial morphology of KNM-ER 1470 is more *Homo*-like than the endocasts of australopiths. It is either a large-bodied presumed male of an inclusively interpreted *H. habilis* (in which case *H. rudolfensis* is a junior synonym of *H. habilis*) or it is the **lectotype** of *H. rudolfensis*, a second species of early *Homo*. Walker (1976) was one of the first to point out the relatively primitive nature of the shape of the calvaria (e.g., widest part of the vault is low down). Similar cranial vault morphology is seen in **KNM-ER 1590** and similar upper facial and endocast morphology is seen in the **KNM-ER 3732** calotte. Location of original **National Museums of Kenya**, Nairobi, Kenya.

KNM-ER 1471 Site Koobi Fora. Locality Area 131. Surface/*in situ* Surface. Date of discovery 1972. Finder Bernard Ngeneo. Unit Collection unit 3. Horizon KBS-27m. Bed/member Upper Burgi Member. Formation **Koobi Fora Formation**. Group Omo Group. Nearest overlying dated horizon KBS Tuff. Nearest underlying dated horizon N/A. Geological age *c*.1.88–1.95 Ma. Developmental age Adult. Presumed sex Unknown. Brief anatomical description Proximal portion of a right tibia with abrasion of the cortex below the condyles and above the tibial tuberosity. Announcement Leakey 1973. Initial description Day et al. 1976. Photographs/line drawings and metrical data Day et al. 1976, Leakey et al. 1978. Detailed anatomical description Day et al. 1976. Initial taxonomic allocation *Australopithecus*.

Taxonomic revisions N/A. Current conventional taxonomic allocation Hominini *gen. et sp. indet.* Informal taxonomic category Archaic hominin. Significance The initial attribution to *Australopithecus* (Leakey 1973) implies that its morphology departs sufficiently from the morphology seen in modern humans to suggest it may belong to *Paranthropus boisei*. However, more detailed study is warranted before attributions to other contemporaneous hominin taxa can be ruled out. Location of original **National Museums of Kenya**, Nairobi, Kenya.

KNM-ER 1472 Site **Koobi Fora**. Locality Area 131. Surface/*in situ* Surface. Date of discovery 1972. Finder John Harris. Unit N/A. Horizon KBS-30m. Bed/member Upper Burgi Member. Formation **Koobi Fora Formation**. Group Omo Group. Nearest overlying dated horizon KBS Tuff. Nearest underlying dated horizon N/A. Geological age *c*.1.88–1.95 Ma. Developmental age Adult. Presumed sex Unknown. Brief anatomical description A nearly complete right femur. Announcement Leakey 1973b. Initial description Leakey 1973a. Photographs/line drawings and metrical data Day et al. 1975, Leakey et al. 1978. Detailed anatomical description Day et al. 1975. Initial taxonomic allocation *Homo* sp. indet. Taxonomic revisions *Homo erectus* or *Homo ergaster* (Tobias 1991b), *Homo rudolfensis* (Wood 1992). Current conventional taxonomic allocation *Homo* sp. Informal taxonomic category Pre-modern *Homo* or transitional hominin. Significance The most notable feature of KNM-ER 1472 (and **KNM-ER 1481**) is its overall length, which resembles the femora of later members of the genus *Homo* and contrasts with the relatively short femora of *Australopithecus* taxa (e.g., **A.L. 288-1**). KNM-ER 1472 (along with KNM-ER 1481) was found geographically close to the cranium **KNM-ER 1470** and is more or less the same geological age. Compared with the evidently archaic morphology (i.e., small head, and long, anteroposteriorly flattened neck) of the hominin femora previously recovered at Koobi Fora (e.g., **KNM-ER 738A, 815, and 993**), KNM-ER 1472 was clearly more modern human-like. Some have suggested that it shares diagnostic morphology with *H. erectus* (Kennedy 1983a), and others have suggested, for no particularly compelling reason other than its **sympatry** and approximate synchronicity with KNM-ER 1470, it should be attributed to

H rudolfensis (Wood 1992). Ruff's (1995) biomechanical analysis showed that the distinctively mediolaterally broad, platymeric femoral shaft with an inferiorly positioned minimum breadth seen in KNM-ER 1472, KNM-ER 1481, and **KNM-WT 15000** can be explained by elevated bending moments resulting from a long femoral neck (especially when combined with a wide inter-acetabular distance). McHenry and Corruccini (1978) used traditional linear data to compare the morphology of early hominin femora, including KNM-ER 1472. Their results showed that among early hominin femora from East and southern Africa, the relatively large femoral head and rounded femoral neck of KNM-ER 1472 (and 1481) were more like those of modern humans and distinct from the smaller femoral heads and anteroposteriorly narrow necks of australopith femora such as **KNM-ER 1503**, **SK 82**, and **SK 97** (see also Day 1969a, Walker 1973, Wood 1976). Richmond and Jungers (2008) showed that KNM-ER 1472 and 1481 bear a stronger resemblance to femora securely attributed to *Homo* (e.g., KNM-WT 15000, the **Berg Aukas I** femur) than they do to australopith femora, including A.L. 288-1. Using three-dimensional landmark data in addition to traditional measurements, Harmon (2009) confirmed that KNM-ER 1472 (and 1481) shared these anatomical features, as well as a short greater trochanter, with femora securely attributed to *Homo* (e.g., the Berg Aukas I femur) compared with femora such as KNM-ER 738A, 815, 993, 1503, and 1505. It is for these reasons that most knowledgeable researchers now consider KNM-ER 1472 likely belongs to early African *H. erectus*, or to *H. rudolfensis* (if there is some convincing independent evidence that this taxon has a less archaic femoral morphology than *Homo habilis*). Whatever its ultimate taxonomic allocation, the morphometric evidence from KNM-ER 1472 (and 1481) suggests a pattern of bipedalism virtually identical to that inferred for *H. erectus* had emerged by 1.9 Ma. Location of original **National Museums of Kenya**, Nairobi, Kenya.

KNM-ER 1476 Site Koobi Fora. Locality Area 105. Surface/*in situ* Surface. Date of discovery 1972. Finder Kamoya **Kimeu**. Unit Collection unit 4. Horizon KBS CC. Bed/member KBS Member. Formation **Koobi Fora Formation**. Group Omo Group. Nearest overlying dated horizon N/A. Nearest underlying dated horizon N/A. Geological age *c.*1.87 Ma. Developmental age Adult. Presumed

sex Unknown. Brief anatomical description Three geologically associated hominin lower limb fragments that are presumed to come from the same individual, a left talus (1476A), proximal end of a left tibia (1476B), and fragment of the **diaphysis** of a right tibia (1476C). Announcement Leakey 1973. Initial description Leakey 1973. Photographs/line drawings and metrical data Day et al. 1976, Leakey et al. 1978. Detailed anatomical description Day et al. 1976, Gebo and Schwartz 2006. Initial taxonomic allocation *Australopithecus*. Taxonomic revisions N/A. Current conventional taxonomic allocation *Paranthropus boisei*. Informal taxonomic category Megadont archaic hominin. Significance One of the better-preserved hominin tali from Koobi Fora and one of the few associated with other skeletal elements. Gebo and Schwartz (2006) concluded that KNM-ER 1476 shares distinctive morphology with tali from Kromdraai (**TM 1517**) and Olduvai Gorge (**OH 8**) and suggested that they all belong to megadont archaic hominins. However, secure attributions of isolated postcranial remains must be tentative until relevant taxonomically unambiguous associated skeletons are recovered. Location of original **National Museums of Kenya**, Nairobi, Kenya.

KNM-ER 1481 Site **Koobi Fora**. Locality Area 131. Surface/*in situ* Surface. Date of discovery 1972. Finder John Harris. Unit N/A. Horizon KBS-12m. Bed/member Upper Burgi Member. Formation **Koobi Fora Formation**. Group Omo Group. Nearest overlying dated horizon KBS Tuff. Nearest underlying dated horizon N/A. Geological age *c.*1.89 Ma. Developmental age Adult. Presumed sex Unknown. Brief anatomical description Associated elements of the left lower limb, including a nearly complete femur, proximal tibia, distal tibia, and the distal end of the fibula. Announcement Leakey 1973b. Initial description: Leakey (1973a). Photographs/line drawings and metrical data Day et al. 1975, Leakey et al. 1978, Kennedy 1983a. Detailed anatomical description Day et al. 1975, Kennedy 1983a, Trinkaus 1984. Initial taxonomic allocation *Homo sp. indet.* Taxonomic revisions *Homo erectus* or *Homo ergaster* (Kennedy 1983b, Tobias 1991b), *Homo habilis* (Trinkaus 1984), *Homo rudolfensis* (Wood 1992). Current conventional taxonomic allocation *Homo sp. indet.* Informal taxonomic category Pre-modern *Homo* or transitional hominin.

Significance The most notable feature of the KNM-ER 1481 femur (KNM-ER 1481A) and the **KNM-ER 1472** femur is their overall length, which resembles the femora of later members of the genus *Homo* and contrasts with the relatively short femoral length in *Australopithecus* (e.g., **A.L. 288-1**). KNM-ER 1481 was found geographically close to the cranium **KNM-ER 1470**, and is more or less the same geological age. Compared with the evidently archaic morphology (i.e., small head, and long, anteroposteriorly flattened neck) of the hominin femora previously recovered at Koobi Fora (e.g., **KNM-ER 738A, 815** and **993**), KNM-ER 1481A was clearly more modern-human-like. Some have suggested that it shares diagnostic morphology with *H. erectus* (Kennedy 1983a) and others have suggested, for no particularly compelling reason other than its **sympatry** and approximate synchronicity with KNM-ER 1470, it should be attributed to *H. rudolfensis* (Wood 1992). Ruff's (1995) biomechanical analysis showed that the distinctively mediolaterally broad, platymeric femoral shaft with an inferiorly positioned minimum breadth seen in KNM-ER 1472, KNM-ER 1481, and KNM-ER-WT 15000 can be explained by elevated bending moments resulting from a long femoral neck (especially when combined with a wide inter-acetabular distance). McHenry and Corruccini (1978) used traditional linear data to compare the morphology of early hominin femora, including KNM-ER 1472. Their results showed that among early hominin femora from East and southern Africa, the relatively large femoral head and rounded femoral neck of KNM-ER 1472 (and 1481A) were more like those of modern humans and distinct from the smaller femoral heads and anteroposteriorly narrow necks of australopith femora such as **KNM-ER 1503, SK 82**, and **SK 97** (see also Day 1969a, Walker 1973, Wood 1976). Richmond and Jungers (2008) recently showed that KNM-ER 1472 and 1481A bear a stronger resemblance to femora securely attributed to *Homo* (e.g., **KNM-WT 15000**, the **Berg Aukas I** femur) than they do to australopith femora, including A.L. 288-1. Using three-dimensional landmark data in addition to traditional measurements, Harmon (2009) used three-dimensional landmark data to confirm that KNM-ER 1481A and 1472 shared these anatomical features, as well as a short greater trochanter, with femora such as KNM-ER 738A, 815, 993, 1503, and 1505. It is for these reasons

that most knowledgeable researchers now consider KNM-ER 1481 most likely belongs to early African *H. erectus* or to *Homo rudolfensis* (if there is some convincing independent evidence that this taxon has a less archaic femoral morphology than *H. habilis*). Whatever its ultimate taxonomic allocation, the morphometric evidence from KNM-ER 1481 (and 1472) suggests a pattern of bipedalism virtually identical to that of *H. erectus* had emerged by 1.9 Ma. Location of original **National Museums of Kenya**, Nairobi, Kenya.

KNM-ER 1482 Site **Koobi Fora** (Karari Ridge). Locality Area 131. Surface/*in situ* Surface. Date of discovery 1972. Finder H. Muluila. Unit N/A. Horizon KBS-33m. Bed/member upper Burgi Member. Formation **Koobi Fora Formation**. Group Omo Group. Nearest overlying dated horizon KBS Tuff. Nearest underlying dated horizon N/A. Geological age *c*.1.88–1.95 Ma. Developmental age Young adult. Presumed sex Unknown. Brief anatomical description Mandibular body preserved from the distal root of the M_1 on the right side to an oblique break through the ramus on the left side, plus three dental fragments. Announcement Leakey 1973. Initial description Leakey and Wood 1974. Photographs/line drawings and metrical data Leakey and Wood 1974, Wood 1991. Detailed anatomical description Leakey and Wood 1974, Wood 1991. Initial taxonomic allocation Hominidae *gen. et sp. indet.* Taxonomic revisions aff. *Paraustralopithecus* (Leakey 1974), *Australopithecus boisei* (Howell 1978a, White 1977b), *Australopithecus robustus* (Suwa 1988), *Homo* sp. (Wolpoff 1988), *Homo aethiopicus* (Groves 1989), *Homo sp. indet.* (Wood 1991). Current conventional taxonomic allocation *Homo sp. indet.* Informal taxonomic category Pre-modern *Homo* or transitional hominin. Significance As can be seen from the range of taxa it has been assigned to, KNM-ER 1482 cannot be easily accommodated in any of the existing recognized taxa. However, Wood (1991) suggested that the form of the corpus and the size and shape of the tooth crowns and roots and especially the thickness of the enamel (Beynon and Wood 1986) effectively ruled out its allocation to *Paranthropus boisei*. It most likely belongs to the nexus of specimens that some link with the origins of *Homo*. Location of original **National Museums of Kenya**, Nairobi, Kenya.

KNM-ER 1500 Site **Koobi Fora**. Locality Area 130. Surface/*in situ* Surface. Date of discovery 1972. Finder John Kimengich. Unit N/A. Horizon N/A. Bed/member Upper Burgi Member. Formation **Koobi Fora Formation**. Group Omo Group. Nearest overlying dated horizon KBS Tuff. Nearest underlying dated horizon N/A. Geological age *c*.1.9 Ma. Developmental age Adult. Presumed sex Unknown. Brief anatomical description An associated skeleton including a clavicle fragment, the glenoid of the left scapula, fragment of left distal humerus shaft, proximal end and shaft fragment of the right radius, proximal end and shaft fragment of the right ulna, left radius shaft fragment, left proximal femur fragment, distal left femur, nearly complete left tibia (proximal half and distal third), distal end and shaft fragment of right tibia, proximal right metatarsal III, and a small fragment of the base of the mandibular corpus. Announcement Leakey 1973b. Initial description Day et al. 1976. Photographs/line drawings and metrical data Day et al. 1976, Leakey et al. 1978, Wood 1991. Detailed anatomical description Wood 1991, Grausz et al. 1988. Initial taxonomic allocation *Australopithecus sp. indet.* Taxonomic revisions *Australopithecus boisei* (Grausz et al. 1988). Current conventional taxonomic allocation *Paranthropus boisei* or Hominini *gen. et sp. indet.* Informal taxonomic category Megadont archaic or transitional hominin. Significance KNM-ER 1500 is one of the few hominin associated skeletons from Koobi Fora and many interpret it to be a small *P. boisei* individual. This species diagnosis is based on the morphology of the fragment of the base of the mandibular corpus (Grausz et al. 1988), which is robust and has a rounded inferior border. However, while this morphology is consistent with it being a small *P. boisei* individual, it is also consistent with it belonging to whatever taxon the KNM-ER 1802 mandible belongs to (i.e., *Homo rudolfensis* or *Homo habilis*). Whatever species it represents, KNM-ER 1500 does not have enlarged lower limb joints like those of modern humans; instead, it resembles **A.L. 288-1** and *Pan troglodytes* in having small lower limb joints relative to the joints of its upper limbs. Location of original **National Museums of Kenya**, Nairobi, Kenya.

KNM-ER 1503, 1504, 1505 Site **Koobi Fora**. Locality Area 123. Surface/*in situ* Surface. Date of discovery 1972. Finders Maundu Muluila (KNM-ER 1503), M. Mbithi (KNM-ER 1504), and Bernard Ngeneo (KNM-ER 1505). Unit Collection Unit 4. Horizon C4-5m. Bed/member Upper Burgi Member. Formation **Koobi Fora Formation**. Group Omo Group. Nearest overlying dated horizon N/A. Nearest underlying dated horizon N/A. Geological age *c*.1.87 Ma (Feibel et al. 2009). Developmental age Adult. Presumed sex Female, based on small size. Brief anatomical description These fossils comprise five well-preserved specimens that may derive from a single individual. KNM-ER 1503 is a proximal femur, KNM-ER 1504 is a small distal humerus, and KNM-ER 1505 comprises a left femoral head and a fragment of the distal shaft of a left femur. Announcement Leakey 1973a. Initial description Leakey 1973a. Photographs/line drawings and metrical data Leakey 1973a, Day et al. 1976, Leakey et al. 1978. Detailed anatomical description Day et al. 1976. Initial taxonomic allocation *Australopithecus*. Taxonomic revisions N/A. Current conventional taxonomic allocation *Paranthropus boisei*. Informal taxonomic category Megadont archaic hominin. Significance The KNM-ER 1503 proximal femur has a small head and a long, anteroposteriorly compressed neck, the KNM-ER 1504 distal humerus is small, but it has a broad distal shaft and strongly developed medial and lateral epicondyles, and the femoral head of KNM-ER 1505 is small. If these specimens do belong to a single individual, that individual is absolutely small and its proximal femoral morphology resembles that of the two proximal femora from Swartkrans, **SK 82** and **SK 97**, attributed to *Paranthropus robustus*. Thus, KNM-ER 1503 and 1505 are among the proximal femoral fragments that fuelled the suggestion that similar femora from Koobi Fora belonged to *P. boisei*. But evidence, albeit less complete, from **OH 62** suggests that the same morphology is also seen in *Homo habilis*, in which case this and other similar femora could belong to *H. habilis*. The distal humeral morphology (KNM-ER 1504) resembles that of the large KNM-ER 739 humerus (Leakey 1973a) and differs from other extant hominoid and fossil hominin humeri in having a shape characterized by a more projecting medial epicondyle (associated with a greater biepicondylar width), shallower olecranon fossa, narrower zona conoidea width, smaller minimum anteroposterior trochlear width, wider lateral pillar, and wider medial pillar (Lague and Jungers 1996). No matter what taxon it is eventually assigned to, if the fossils in this group do belong to one individual their size suggests that individual is most likely a female. Location of original **National Museums of Kenya**, Nairobi, Kenya.

KNM-ER 1590 Site Koobi Fora (Ileret). Locality Area 12. Surface/*in situ* Surface. Date of discovery 1972. Finder Bernard Ngeneo. Unit N/A. Horizon Approximately C4+. Bed/member KBS Member. Formation **Koobi Fora Formation**. Group Omo Group. Nearest overlying dated horizon Okote Tuff. Nearest underlying dated horizon KBS Tuff. Geological age *c.*1.85 Ma (Feibel et al. 1989), but an alternative, younger, age of 1.8 Ma has been suggested (Brown et al. 2006). Developmental age Immature. Presumed sex Male. Brief anatomical description Cranial vault fragments with associated permanent and deciduous maxillary teeth. The cranial fragments include most of the left and right parietal bones and part of the frontal bone. The associated dentition provides an estimated age at death of 4.0–8.2 years based on a modern human standard and 5.5–6.0 years based on a non-human great ape standard (Smith 1986). Endocranial volume >800 cm^3. Announcement Leakey 1973b. Initial description Day et al. 1976. Photographs/line drawings and metrical data Day et al. 1976, Wood 1991. Detailed anatomical description Wood 1991. Initial taxonomic allocation *Homo sp. indet.* Taxonomic revisions *Homo habilis* (Howell 1978b), *Homo rudolfensis* (Groves 1989, Wood 1992). Current conventional taxonomic allocation *H. habilis* or *H. rudolfensis*. Informal taxonomic category Transitional hominin. Significance The **calotte** is similar in shape and a little larger in size than that of KNM-ER 1470, but whereas KNM-ER 1470 has no tooth crowns preserved, KNM-ER 1590 provides good evidence about the size and morphology of the maxillary dentition of a KNM-ER 1470-like hominin. Location of original **National Museums of Kenya**, Nairobi, Kenya. *See also* **KNM-ER 1470**.

KNM-ER 1802 Site Koobi Fora (Karari Ridge). Locality Area 131. Surface/*in situ* Surface. Date of discovery 1973. Finder John Harris. Unit N/A. Horizon below GPC. Bed/member Upper Burgi Member. Formation **Koobi Fora Formation**. Group Omo Group. Nearest overlying dated horizon KBS Tuff. Nearest underlying dated horizon N/A. Geological age *c.*1.88–1.95 Ma. Developmental age Young adult. Presumed sex Unknown. Brief anatomical description Well-preserved mandible, including both sides of the corpus but lacking any evidence of the ramus. The crowns of three premolar and four permanent molar

teeth are well preserved. Announcement Leakey 1974. Initial description Day et al. 1976. Photographs/line drawings and metrical data Day et al. 1976, Wood 1991. Detailed anatomical description Wood 1991. Initial taxonomic allocation *Homo habilis*. Taxonomic revisions *Homo rudolfensis* (Groves 1989, Wood 1992). Current conventional taxonomic allocation *H. habilis* or *H. rudolfensis*. Informal taxonomic category Transitional hominin. Significance A mandible with a relatively robust mandibular corpus and with premolar and molar crowns and roots that differ from the morphology of the Olduvai hypodigm of *Homo habilis* (e.g., more complex premolar roots and M$_1$ crowns with C6s). Location of original **National Museums of Kenya**, Nairobi, Kenya.

KNM-ER 1805 Site Koobi Fora (Karari Ridge). Locality Area 130. Surface/*in situ In situ*. Date of discovery 1973. Finder Paul Abell. Unit N/A. Horizon KBS+11m. Bed/member KBS Member. Formation **Koobi Fora Formation**. Group Omo Group. Geological age *c.*1.85 Ma (Feibel et al. 1989). Suwa et al. (2007) suggest that the age may be "ill-constrained" and maybe be between 1.8 and 1.6 Ma. Developmental age Adult. Presumed sex Male. Brief anatomical description Incomplete skull comprising the calvaria, the lower part of the face and the mandibular corpus. The calvaria includes the posterior of the frontal, both parietals and most of the temporal bones. Although the maxillary dentition is nearly complete (only the left I^1 and the right P^3 are missing) the combination of occlusal wear, chemical erosion, and other post-mortem damage has removed most of the occlusal morphology. The crowns and roots of the mandibular molars are preserved, but only the roots of the mandibular premolars remain. Endocranial volume 582 cm^3 (Holloway 1983c). Announcement Leakey 1974. Initial description Day et al. 1976. Photographs/line drawings and metrical data Wood 1991, Day et al. 1976. Detailed anatomical description Wood 1991. Initial taxonomic allocation *Australopithecus sp. indet.* Taxonomic revisions *Homo africanus* (Olson 1978), *Homo ergaster* (Groves and Mazák 1975), *Homo erectus* (Wolpoff 1978b), *Australopithecus boisei* (Tobias 1980a), *Homo habilis sensu lato* (White et al. 1981), *Australopithecus africanus* (Falk 1986b), *Homo habilis sensu stricto* (Wood 1991). Current conventional taxonomic allocation *H. habilis*. Informal taxonomic category Transitional hominin. Significance Although KNM-ER 1805 has a compound nuchal crest that is the only morphology linking it with *Paranthropus boisei* and almost all of its morphology is inconsistent with it being assigned to *Au. africanus*. It is most similar to *H. habilis* specimens from Olduvai and along with **KNM-ER 1813** it is the best evidence for that taxon at Koobi Fora. However, Thompson (1993) cautioned researchers about the "peculiar basicranial proportions" and the "unusual features of the frontal, occipital, and dentition of KNM-ER 1805" (p. 261) and Prat (2002) referred to a "number of unique features that differentiate it from other hominids" (p. 32). Falk (1983b) suggested that "the frontal lobe of KNM-ER 1805 is clearly pongid-like and similar to frontal lobes of earlier dated South African australopithecines" (p. 1073). Location of original **National Museums of Kenya**, Nairobi, Kenya.

KNM-ER 1808 Site **Koobi Fora** (Koobi Fora Ridge). Locality Area 103. Surface/*in situ* Surface. Date of discovery 1973. Finder Kamoya **Kimeu**. Unit N/A. Horizon A6-2m. Bed/member KBS Member. Formation **Koobi Fora Formation**. Group Omo Group. Nearest overlying dated horizon Lower Koobi Fora Tuff. Nearest underlying dated horizon KBS Tuff. Geological age *c*.1.6 Ma. Developmental age Adult. Presumed sex Unknown. Brief anatomical description An associated skeleton including the right mandibular condyle, fragments of the mandibular body and associated dentition, maxilla, right temporal, right supraorbital torus, left side of the occipital, atlas vertebra, much of the right clavicle, two fragments of the left scapula, distal portion of the left humerus, most of the shaft of the right humerus, two fragments of the ilium, parts of both femora, left fibula shaft fragment, and assorted long-bone fragments. Sections through the long bones show evidence of new bone growth beneath the periosteum. Announcement Leakey 1974. Initial description Leakey and Walker 1985. Photographs/line drawings and metrical data Leakey and Walker 1985, Leakey et al. 1978, Wood 1991. Detailed anatomical description Wood 1991, Walker et al. 1982. Initial taxonomic allocation *Homo sp. indet.* Taxonomic revisions *Homo erectus* (Walker et al. 1982). Current conventional taxonomic allocation *H. erectus* or *Homo ergaster*. Informal taxonomic category Pre-modern *Homo*. Significance After **KNM-WT 15000**, KNM-ER 1808 is the most complete skeleton of East African *Homo erectus*, but the bones are affected by an exuberant deposit of subperiosteal woven bone. The etiology of this pathology is uncertain. It may be a case of hypervitaminosis-A, possibly from eating an excess amount of uncooked carnivore liver (Walker et al. 1982), or it may be the result of a treponemal disease such as yaws (Rothschild et al. 1995). Whatever its cause, it is an early example of paleopathology in a hominin. Location of original **National Museums of Kenya**, Nairobi, Kenya.

KNM-ER 1813 Site **Koobi Fora** (Bura Hasuma subregion). Locality Area 123. Surface/*in situ* In situ. Date of discovery 1973. Finder Kamoya **Kimeu**. Unit N/A. Horizon N/A. Bed/member Upper Burgi Member. Formation **Koobi Fora Formation**. Group Omo Group. Nearest overlying dated horizon N/A. Nearest underlying dated horizon N/A. Geological age *c*.1.88–1.90 Ma (Feibel et al. 1989). An alternative, younger, age of 1.65 Ma has been suggested (Gathogo and Brown 2006), but others dispute the grounds for the revised age. Suwa et al. (2007) suggest that its age is ill-constrained and that it may be anywhere from >1.8 Ma to 1.55 Ma, but Feibel et al. (2009) present a cogent argument that KNM-ER 1813 must be older that 1.87 Ma and may be older than 1.89 Ma. Developmental age Adult. Presumed sex Unknown. Brief anatomical description Nearly complete cranium missing the roof of the right mandibular fossa, part of the border of the foramen magnum, much of the sphenoid, the right zygomatic arch, the left malar region, and most of the body of the ethmoid. Aside from the I^1, there is at least one well preserved crown of each tooth. Endocranial volume 510 cm^3 (Holloway 1983c). Announcement Leakey 1974. Initial description Day et al. 1976. Photographs/line drawings and metrical data Wood 1991. Detailed anatomical description Wood 1991. Initial taxonomic allocation *Australopithecus sp. indet.* Taxonomic revisions *Homo africanus* (Olson 1978), *Homo habilis* (Howell 1978b), *Australopithecus* cf. *africanus* (Walker 1981), *Homo antiquus* (Ferguson 1987), *Homo ergaster* (Groves 1989), *Homo habilis* (Wood 1991), **holotype** of *Homo microcranous* (Ferguson 1995). Current conventional taxonomic allocation *H. habilis*. Informal taxonomic category Transitional hominin. Significance The small endocranial volume of KNM-ER 1813 prompted some researchers to allocate this specimen to *Au. africanus*, but its overall morphology and especially the morphology of the cranial base is unlike that of *Au. africanus*. Its affinities are with early *Homo* and if that taxon is to be divided into *H. habilis* and *Homo rudolfensis* the cranial vault and facial and dental morphology of KNM-ER 1813 are closer to the former than to the latter. Location of original **National Museums of Kenya**, Nairobi, Kenya.

KNM-ER 2598 Site **Koobi Fora**. Locality Area 15. Surface/*in situ* Surface. Date of discovery 1974. Finder Peter Nzube. Unit N/A. Horizon KBS-4m. Bed/member Upper Burgi Member. Formation **Koobi Fora Formation**. Group Omo Group. Nearest overlying dated horizon KBS Tuff. Nearest underlying dated horizon N/A. Geological age *c*.1.87 Ma. Developmental age Adult. Presumed sex Unknown. Brief anatomical description An occipital fragment that preserves lambda and the medial portions of the lambdoid sutures. Announcement Leakey 1976b. Initial description Leakey and Walker 1985. Photographs/line drawings and metrical data Leakey and Walker 1985, Wood 1991. Detailed anatomical description Wood 1991. Initial taxonomic allocation cf. *Homo* aff. *H. erectus*. Taxonomic revisions N/A. Current conventional taxonomic allocation **Homo ergaster** or **Homo erectus**. Informal taxonomic category Pre-modern *Homo*. Significance Although this is only a cranial fragment, the thickness of the cranial vault, the rugosity of the occipital torus, and the acuteness of the angle between the external surface of the squamous and nuchal parts of the occipital are all strong indicators that this specimen represents a pre-modern *Homo* taxon and if so it would be among the earliest evidence, if not the earliest evidence, for such a taxon in East Africa. However, Suwa et al. (2007) do *not* accept the *c*.1.88–1.9 Ma age. Instead, they cite White (1996; but it should be 1995) as the basis for their suggestion that KNM-ER 2598 "was found on a desert-pavement lag surface that potentially samples overlying deposits now eroded away" (p. 135). Location of original **National Museums of Kenya**, Nairobi, Kenya.

KNM-ER 2602 Site **Koobi Fora**. Locality Area 117. Surface/*in situ* Surface. Date of discovery 1974. Finder Wambua Mangao. Unit N/A. Horizon Just above the Tulu Bor Tuff. Bed/member Tulu Bor Member. Formation **Koobi Fora Formation**. Group Omo Group. Nearest overlying dated horizon Toroto Tuff. Nearest underlying dated horizon Tulu Bor Tuff. Geological age *c*.3.2–3.3 Ma. Developmental age Juvenile. Presumed sex Unknown. Brief anatomical description Several broken and weathered portions of a cranium including substantial portions of the occipital, the parasagittal parts of both parietals, a piece of right frontal, left zygomatic, several fragments of the deciduous dentition, and a fragment of a molar germ. Announcement Leakey 1976b. Initial description Leakey and Walker 1985. Photographs/line drawings and metrical data Kimbel 1988, Wood 1991. Detailed anatomical description Kimbel 1988, Wood 1991. Initial taxonomic allocation Hominini *sp. indet.* Taxonomic revisions *Australopithecus afarensis* (Boaz 1988, Kimbel 1988), *Australopithecus sp. indet.* Current conventional taxonomic allocation *Au. afarensis*. Informal taxonomic category Archaic hominin. Significance If KNM-ER 2602 does belong to *Au. afarensis* then it would be the first evidence of that taxon recovered from a Kenyan fossil site and, with the possible exception of the hominins from Kubi Algi, it would be the only evidence of that taxon from Koobi Fora. Location of original **National Museums of Kenya**, Nairobi, Kenya.

KNM-ER 3228 Site **Koobi Fora** (Koobi Fora Ridge). Locality Area 102. Surface/*in situ* Surface. Date of discovery 1974–5. Finder Bernard Ngeneo. Unit N/A. Horizon Level C1-10m. Bed/member Upper Burgi Member. Formation **Koobi Fora Formation**. Group Omo Group. Nearest overlying dated horizon KBS Tuff. Nearest underlying dated horizon N/A. Geological age 1.89 ± 0.05 Ma. Developmental age Adult. Presumed sex Male. Brief anatomical description A right pelvic bone missing the pubic and ischial rami. Announcement Leakey 1976b. Initial description Leakey 1976b. Photographs/line drawings and metrical data Rose 1984. Detailed anatomical description Rose 1984. Initial taxonomic allocation *Homo sp. indet.* Taxonomic revisions *Homo erectus* (Rose 1984). Current conventional taxonomic allocation *H. erectus* or **Homo habilis**. Informal taxonomic category Pre-modern *Homo*. Significance KNM-ER 3228 is presently the earliest pelvic bone that suggests a form of bipedalism similar to that of **Homo sapiens**. The KNM-ER 3228 pelvic bone is similar to the modern human morphology in many ways including functionally significant features such as the relative lengths of the ischial body and tuberosity, sagittally expanded iliac blade, large acetabulum, and the positions of the sacroiliac articular area and the acetabulosacral buttress (Rose 1984). However, KNM-ER 3228 resembles archaic hominin pelvic bones in some features including an iliac blade that (a) diverges somewhat laterally (relative to lower parts of the hip) and (b) is oriented at a relatively wide angle relative to the sagittal plane and a somewhat protuberant anterior superior iliac spine region. The KNM-ER 3228 specimen also shares some morphology with the **OH 28**, Arago XLIV, **Jinniushan**, and the Gona **BSN49/P27** pelvic bones (e.g., broad, laterally flaring ilia, pronounced acetabulocristal buttress, and a deep fossa for the origin of gluteus medius). Location of original **National Museums of Kenya,** Nairobi, Kenya.

KNM-ER 3728 Site **Koobi Fora** (Koobi Fora Ridge). Locality Area 100. Surface/*in situ* Surface. Date of discovery 1975. Finders N/A. Unit N/A. Horizon C2+5m. Bed/member Upper Burgi Member. Formation **Koobi Fora Formation**. Group Omo Group. Nearest overlying dated horizon KBS Tuff. Nearest underlying dated horizon N/A. Geological age 1.89±0.05 Ma. Developmental age Adult. Presumed sex Probably male based on size. Brief anatomical description Shaft and neck of a left femur. Announcement Leakey and Walker 1985. Initial description Leakey and Walker 1985. Photographs/line drawings and metrical data Leakey and Walker 1985. Detailed anatomical description N/A. Initial taxonomic allocation N/A. Taxonomic revisions N/A. Current conventional taxonomic allocation *Paranthropus boisei*. Informal taxonomic category Megadont archaic hominin. Significance It is one of the better preserved and oldest members of a group of femora from Koobi Fora that share a long and anteroposteriorly compressed neck (others include **KNM-ER 738A, 815, 993,** and **1503**). A similar morphology is seen in two proximal femora from Swartkrans, **SK 82** and **SK 97**, attributed to *Paranthropus robustus*, so this is one of the proximal femur fragments that continues to fuel the suggestion that similar femora from Koobi Fora belonged to *P. boisei*. But evidence from **OH 62** suggests that the same morphology is also seen in *Homo habilis*, in which case this and other similar femora could belong to *H. habilis*. Location of original **National Museums of Kenya**, Nairobi, Kenya.

KNM-ER 3732 Site **Koobi Fora**. Locality Area 115. Surface/*in situ* Surface. Date of discovery 1975. Finder M. Muluila. Unit N/A. Horizon GPC+1m.

Bed/member upper Burgi Member. Formation **Koobi Fora Formation**. Group Omo Group. Nearest overlying dated horizon KBS Tuff. Nearest underlying dated horizon N/A. Geological age *c*.1.88–1.95 Ma. Developmental age Adult. Presumed sex Unknown. Brief anatomical description A single fragment comprising most of the calotte, which is better preserved on the left than on the right side, plus the left zygoma; there is a separate natural endocast. Announcement Leakey 1976. Initial description Leakey and Walker 1985. Photographs/line drawings and metrical data Leakey and Walker 1985, Wood 1991. Detailed anatomical description Leakey and Walker 1985, Wood 1991. Initial taxonomic allocation Hominidae *gen. et sp. indet.* Taxonomic revisions aff. *Homo sp. indet.* (Wood 1991), *Homo rudolfensis* Wood (1992). Current conventional taxonomic allocation *Homo sp. indet.* Informal taxonomic category Pre-modern *Homo* or transitional hominin. Significance The lack of the distinctive frontal morphology of *Paranthropus boisei* rules out allocation to that taxon, and the robust upper face make allocation to *Homo ergaster* unlikely. The fragment looks to be a smaller version of **KNM-ER 1470** and Wood (1992) included it in the hypodigm of *H. rudolfensis*. Location of original **National Museums of Kenya,** Nairobi, Kenya.

KNM-ER 3733 Site **Koobi Fora** (Koobi Fora Ridge). Locality Area 104. Surface/*in situ* Surface. Date of discovery 1975. Finder Bernard Ngeneo. Unit N/A. Horizon A2+6m. Bed/member KBS Member. Formation **Koobi Fora Formation**. Group Omo Group. Nearest overlying dated horizon N/A. Nearest underlying dated horizon KBS Tuff. Geological age *c*.1.78 Ma (Feibel et al. 1989). An alternative, younger, age of 1.65 Ma has also been suggested (Brown et al. 2006). Developmental age Adult. Presumed sex Unknown. Brief anatomical description Well-preserved cranium missing only portions of the face and nasal cavity. Some of the anatomy of the cranial base is obscured by matrix and there is some plastic deformation of the cranial base and face, but overall the preservation allows for useful morphological detail. The maxillary dentition is represented by the crown and roots of the right M^2 and the damaged crowns of the right P^3, P^4, and M^1 and left P^4, M^1, and M^2. Endocranial volume 848 cm^3 (Holloway 1983a). Announcement Leakey and Walker 1976. Initial description Leakey and Walker 1985. Photographs/line drawings and metrical data Leakey and

Walker 1985, Wood 1991. Detailed anatomical description Wood 1991. Initial taxonomic allocation *Homo erectus* Taxonomic revisions *Homo ergaster* (Groves 1989). Current conventional taxonomic allocation *H. erectus* or *H. ergaster*. Informal taxonomic category Pre-modern *Homo*. Significance There is widespread consensus that KNM-ER 3733 resembles *H. erectus* crania from Java and China. What is at issue is whether the generally less-derived morphology displayed by KNM-ER 3733 (and also KNM-ER 3883) is just a manifestation of intraspecific geographical variation within *H. erectus*, or whether it indicates that KNM-ER 3733 and KNM-ER 3883 belong to a separate species, *H. ergaster*, that would be the sister taxon of *H. erectus*. Location of original **National Museums of Kenya**, Nairobi, Kenya. *See also* **KNM-ER 3883**.

KNM-ER 3735   Site **Koobi Fora**. Locality Area 116. Surface/*in situ* Surface, then from sieving. Date of discovery 1975. Finder Kamoya **Kimeu**. Unit N/A. Horizon F1-4m. Bed/member Upper Burgi Member. Formation **Koobi Fora Formation**. Group Omo Group. Nearest overlying dated horizon KBS Tuff. Nearest underlying dated horizon N/A. Geological age *c*.1.88–1.9 Ma. Developmental age Adult. Presumed sex If it belongs to *Homo habilis* (see below) possibly a male based on how its size compares with that of **OH 62**. Brief anatomical description An associated skeleton consisting of 14 cranial fragments plus postcranial fragments belonging to the upper limb girdle (clavicle and scapula), upper limb (radius, ulna, and two hand phalanges), axial skeleton (sacrum), and lower limb (femur and tibia). Announcement Leakey and Walker 1985. Initial description Leakey et al. 1989. Photographs/line drawings and metrical data Leakey et al. 1989, Wood 1991. Detailed anatomical description Wood 1991, Haeusler and McHenry 2004, 2007. Initial taxonomic allocation *H. habilis*. Taxonomic revisions *Homo sp. indet.* (Wood 1991). Current conventional taxonomic allocation *H. habilis*. Informal taxonomic category Pre-modern *Homo*. Significance The cranial morphology precludes allocation to *Paranthropus boisei*, some of it is consistent with an allocation to *Homo erectus*, but it is most like the morphology of *H. habilis*. The postcranial evidence has been compared preliminarily with **A.L. 288-1** and **OH 62**, and was found to be more like the latter than the former (Leakey et al. 1989). Haeusler and McHenry (2004) concluded that

the proportions of upper- to lower-limb shaft diameters in KNM-ER 3735 could be encountered among modern humans, but not chimpanzees. In a later paper (Haeusler and McHenry 2007) the same authors suggested that KNM-ER 3735 and A.L. 288-1 had relatively small sacra compared to modern humans, but that KNM-ER 3735 had a proportionately longer radial neck and, by inference, a longer forearm than *Australopithecus afarensis* [e.g., "the robusticity of the scapular spine and relative radial neck length" show "a more ape-like pattern in KNM-ER 3735 than in A.L. 288-1" (*ibid*, p. 18)]. These authors also claim that KNM-ER 3735 is consistent with the "the possible evolution of *A. africanus* into *H. habilis*" (*ibid*, p. 22). If KNM-ER 3735 does belong to *H. habilis* then it is consistent with an assessment that the relative limb strengths of OH 62 and *H. erectus* are very different (Ruff 2009). Location of original **National Museums of Kenya**, Nairobi, Kenya.

KNM-ER 3883   Site **Koobi Fora**. Locality Area 1. Surface/*in situ* Surface. Date of discovery 1976. Finders Richard **Leakey** and Maundu Muluila. Unit N/A. Horizon N/A. Bed/member Okote Member. Formation **Koobi Fora Formation**. Group Omo Group. Nearest overlying dated horizon Chari Tuff. Nearest underlying dated horizon N/A. Geological age *c*.1.50–1.65 Ma. Developmental age Adult. Presumed sex Unknown. Brief anatomical description Well-preserved calvaria, but with considerable cracking of the surface bones, insinuation of the matrix, and plastic deformation. Endocranial volume 804 cm^3 (Holloway 1983c). Announcement Leakey and Walker 1985. Initial description Leakey and Walker 1985. Photographs/line drawings and metrical data Leakey and Walker 1985, Wood 1991. Detailed anatomical description Wood 1991. Initial taxonomic allocation *Homo erectus*. Taxonomic revisions *Homo ergaster* (Groves 1989). Current conventional taxonomic allocation *H. erectus* or *H. ergaster*. Informal taxonomic category Pre-modern *Homo*. Significance There are many general similarities between KNM-ER 3883 and KNM-ER 3733 and most researchers who have studied them in detail (e.g., Picq 1983, Rightmire 1984, Wood 1991) assign these two crania to the same taxon. Location of original **National Museums of Kenya**, Nairobi, Kenya. *See also* **KNM-ER 3733**.

KNM-ER 3884   Site **Koobi Fora**. Locality Western edge of Area 5 on the Ileret Ridge.

Surface/*in situ* Some fragments found on the surface, others were excavated and thus were *in situ*. Date of discovery 1976. Finder Wambua Mangao. Unit N/A. Horizon N/A. Bed/member Galana Boi (Feibel et al. 1989), initially attributed to the Guomde Formation, most of which is now subsumed into the Chari Member. The specimen likely derives from undifferentiated sediments above the Chari Formation. Formation Likely unassigned strata between the **Koobi Fora Formation** and the **Galana Boi Formation** *sensu stricto*. Group N/A. Nearest overlying dated horizon Galana Boi Formation. Nearest underlying dated horizon Chari Tuff. Geological age Bräuer et al. (1997) reported direct **uranium-series** ages for the specimen of 272 ka (+infinity/–113 ka) by uranium-thorium and >180 ka by uranium-protactinium. Developmental age Adult. Presumed sex Unknown. Brief anatomical description A fragmentary **cranium** of which the posterior half of the cranial vault, frontal, maxillae, and the dentition are well preserved. Announcement Day 1986. Initial description Bräuer et al. 1992b, Bräuer 2001. Photographs/line drawings and metrical data Bräuer et al. 1992b, Bräuer 2001. Detailed anatomical description N/A. Initial taxonomic allocation Late archaic *Homo sapiens*. Taxonomic revisions None. Current conventional taxonomic allocation Late archaic *H. sapiens* or *Homo helmei*. Informal taxonomic category Pre-modern *Homo*. Significance The posterior of the cranial vault is anatomically modern in configuration, but the frontal bone is more receding and broader, and bears a larger supraorbital torus than most modern human crania. Its mix of primitive and modern features, in association with its likely antiquity, is consistent with the hypothesis that modern humans evolved in Africa. Location of original **National Museums of Kenya**, Nairobi.

KNM-ER 5428 Site Koobi Fora. Locality Area 119. Surface/*in situ* Surface. Date of discovery 1978. Finders Not reported. Unit N/A. Horizon BPT-1m. Bed/member KBS/Okote Member (Feibel et al. 1989). Formation **Koobi Fora Formation**. Group Omo Group. Nearest overlying dated horizon Lower Ileret Tuff. Nearest underlying dated horizon KBS Tuff. Geological age *c*.1.54 Ma (Feibel et al. 1989). Developmental age Adult. Presumed sex Unknown. Brief anatomical description A well-preserved right talus. Announcement Leakey and Walker 1985. Initial description Leakey

and Walker 1985. Photographs/line drawings and metrical data Leakey and Walker 1985. Detailed anatomical description N/A. Initial taxonomic allocation N/A. Taxonomic revisions *Homo erectus* (Walker 1993). Current conventional taxonomic allocation *H. erectus* or *Homo ergaster*. Informal taxonomic category Pre-modern *Homo*. Significance KNM-ER 5428 is the most modern human-like of all the hominin tali (including **KNM-ER 813**) from the Koobi Fora Formation. If it does belong to *H. erectus*, this hints at subtle yet perhaps important differences in gait between transitional hominins and *H. erectus*. Location of original **National Museums of Kenya**, Nairobi, Kenya.

KNM-ER 6020 Site Koobi Fora. Locality Area 8A. Surface/*in situ* Surface. Date of discovery 1980. Finders Not reported. Unit N/A. Horizon N/A. Bed/member KBS Member. Formation **Koobi Fora Formation**. Group Omo Group. Nearest overlying dated horizon N/A. Nearest underlying dated horizon KBS Tuff. Geological age 1.77± 0.10 Ma. Developmental age Adult. Presumed sex Male, based on large size. Brief anatomical description A massive but heavily weathered distal end of a right humerus. Announcement Leakey and Walker 1985. Initial description Leakey and Walker 1985. Photographs/line drawings and metrical data Leakey and Walker 1985. Detailed anatomical description N/A. Initial taxonomic allocation N/A. Taxonomic revisions None. Current conventional taxonomic allocation Hominini *gen. et sp. indet.* Informal taxonomic category N/A. Significance KNM-ER 6020 is similar in size to **KNM-ER 739**, but it differs from the latter in having only a weakly developed brachioradialis flange. A detailed comparative analysis of KNM-ER 6020 has yet to be undertaken. Location of original **National Museums of Kenya**, Nairobi, Kenya. *See also* **KNM-ER 739**.

KNM-ER 13750 Site Koobi Fora. Locality Area 105. Surface/*in situ* Surface. Date of discovery 1985/6. Finder Aila Derekitch. Unit N/A. Horizon KBS CC. Bed/member KBS Member. Formation **Koobi Fora Formation**. Group Omo group. Nearest overlying dated horizon N/A. Nearest underlying dated horizon N/A. Geological age *c*.1.87±0.05 Ma. Developmental age Adult. Presumed sex Male. Brief anatomical description A calotte, plus much of the right zygomatic process and zygomatic bone. Much of the surface bone has been removed by erosion. Endocranial volume Approximately 480–530 cm^3 (Holloway 1988, Leakey and Walker 1988, Brown et al. 1993). Announcement N/A. Initial description Leakey and Walker 1988. Photographs/line drawings and metrical data Leakey and Walker 1988. Detailed anatomical description Leakey and Walker 1988. Initial taxonomic allocation *Australopithecus boisei*. Taxonomic revisions None. Current conventional taxonomic allocation *Paranthropus boisei*. Informal taxonomic category Megadont archaic hominin. Significance Apart from some of the details of the sagittal crest, KNM-ER 13750 resembles **KNM-ER 406**, which is some 100 ka younger. Thus, KNM-ER 13750 and KNM-ER 406 are examples of the temporal stability of the cranial morphology of the presumed males of *P. boisei*. It supports the hypothesis that *P. boisei* provides an example of evolutionary **stasis**. Location of original **National Museums of Kenya**, Nairobi, Kenya. *See also* **KNM-ER 406**.

KNM-ER 23000 Site **Koobi Fora**. Locality Area 104. Surface/*in situ* "Mostly *in situ*." Date of discovery 1990. Finder Benson Kyongo then excavated by Meave and Richard **Leakey**. Unit N/A. Horizon A2-9m. Bed/member KBS Member. Formation **Koobi Fora Formation**. Group Omo Group. Nearest overlying dated horizon N/A. Nearest underlying dated horizon N/A. Geological age *c.*1.85 Ma. Developmental age Young adult. Presumed sex Male. Brief anatomical description A calvaria comprising the frontal, the temporal, and parietals of both sides, most of the squamous part of the occipital, the most lateral parts of the greater wings of the sphenoid, and a mesial fragment of the crown and root of a mandibular molar. The bone is permeated by cracks but there is little, or no, matrix-infilling. Brown et al. (1993) commented on the calvaria that there was "very mild distortion which precluded a perfect fitting together of the bones" (*ibid*, p. 149). Endocranial volume 491 cm^3 (Brown et al. 1993). Announcement N/A. Initial description Brown et al. 1993. Photographs/line drawings and metrical data Brown et al. 1993. Detailed anatomical description Brown et al. 1993. Initial taxonomic allocation *Australopithecus boisei*. Taxonomic revisions None. Current conventional taxonomic allocation *Paranthropus boisei*. Informal taxonomic category Megadont archaic hominin. Significance KNM-ER 23000 is a well-preserved calvaria similar in geological age to OH 5 and

the differences between the two approximately synchronic specimens [e.g., greatest height of the sagittal crest posterior in OH 5 whereas it is anterior in KNM-ER 23000, with a consequent lack of a compound temporonuchal crest in the latter; thin cranial vault bones that overlap at asterion in the former and thicker cranial bones and no temporo-parietal and parieto-occipito/temporal overlap at asterion in the latter (Kimbel and Rak 1985); bilateral occipito-marginal cranial venous drainage in the former and unilateral occipito-marginal cranial venous drainage in the latter] provide a good illustration of the range of phenotypes, especially of ectocranial and facial morphology, that can be expressed within a single early hominin taxon. Otherwise KNM-ER 23000 is a good example of the temporal persistence and geographical distribution of the distinctive cranial morphology of the presumed males of *P. boisei* and it supports the general hypothesis that *P. boisei* provides an example of evolutionary **stasis**. Location of original **National Museums of Kenya**, Nairobi, Kenya. *See also* **OH 5**.

KNM-ER 42700 Site **Koobi Fora**. Locality "Area 1." Surface/*in situ In situ*. Date of discovery 2000. Finder Frederick Manthi. Unit N/A. Horizon Between the IL02-043 tuff above and the Main Ileret Caliche below. Bed/member KBS Member. Formation **Koobi Fora Formation**. Group Omo Group. Nearest overlying dated horizon Lower Ileret Tuff (1.527±0.014 Ma). Nearest underlying dated horizon KBS Tuff. Estimated geological age *c.*1.55 Ma. Developmental age Young adult/late subadult. Presumed sex Unknown. Brief anatomical description Well-preserved calvaria missing only portions of the right side of the anterior cranial base and part of the right side of the cranial vault. The spheno–occipital synchrondrosis is two-thirds fused. There is some plastic deformation of the anterior part of the cranial vault. Endocranial volume 691 cm^3 (Spoor et al. 2007). Announcement Spoor et al. 2007. Initial description Spoor et al. 2007. Photographs/line drawings and metrical data Spoor et al. 2007. Detailed anatomical description N/A. Initial taxonomic allocation *Homo erectus*. Taxonomic revisions *Homo* sp. (Baab 2008). Current conventional taxonomic allocation *H. erectus* Informal taxonomic category Pre-modern *Homo*. Significance Despite the small overall size of this calvaria, the researchers involved in the recovery of KNM-ER 42700 cite its overall shape (after correction for deformation), details of the cranial base (specifically the angle between the tympanic and the petrous elements of the temporal bone), and the keeling of the frontal and parietal bones to support their recommendation that the calvaria be assigned to *H. erectus*. If this allocation survives refutation then it would be evidence of a substantial period of sympatry between *H. erectus* and *Homo habilis* and Spoor et al. (2007) suggest that this would be evidence that "an anagenetic relationship between the two taxa is implausible" (*ibid*, p. 690). Location of original **National Museums of Kenya**, Nairobi, Kenya. *See also* **KNM-ER 42703**.

KNM-ER 42703 Site Koobi Fora. Locality Area 8. Surface/*in situ* Surface. Date of discovery 2000. Finder John Kaatho. Unit N/A. Horizon N/A. Bed/member Okote Member. Formation **Koobi Fora Formation**. Group Omo Group. Nearest overlying dated horizon Chari Tuff (1.383±0.028 Ma). Nearest underlying dated horizon Lower Ileret Tuff (1.527±0.014 Ma). Estimated geological age *c*.1.44 Ma. Developmental age Young adult/late subadult. Presumed sex Unknown. Brief anatomical description Right maxilla that preserves up to the midline and the zygomatic and palatine processes, as well as the heavily worn crowns and roots of the right canine back to the M^3. Announcement Spoor et al. 2007. Initial description Spoor et al. (2007). Photographs/line drawings and metrical data Spoor et al. 2007. Detailed anatomical description N/A. Initial taxonomic allocation *Homo habilis*. Taxonomic revisions None. Current conventional taxonomic allocation *H. habilis*. Informal taxonomic category Transitional hominin. Significance The molars are larger than those of *Homo erectus* and they are within the size range of *H. habilis*. On this basis its discoverers allocated it to *H. habilis*, and unless its taxonomic allocation is refuted, it and KNM-ER 42700 suggest that *H. erectus* and *H. habilis* were most likely sympatric in East Africa. Location of original **National Museums of Kenya**, Nairobi, Kenya. *See also* **KNM-ER 42700**.

KNM-ES Prefix for fossils accessioned at the Kenya National Museum (now the **National Museums of Kenya**) that derive from Eliye Springs, Kenya (e.g., **KNM-ES 11693**). *See also* **National Museums of Kenya**; **Eliye Springs**.

KNM-ES 11693 Site West Turkana. Locality **Eliye Springs**. Surface/*in situ* Surface find on a beach on the western shore of Lake Turkana at Eliye Springs. Date of discovery 1986. Finders Dr and Mrs Till Darnhofer. Unit N/A. Horizon N/A. Bed/member Probably Pleistocene sediments beneath the Galana Boi Formation. Formation N/A. Group N/A. Nearest overlying dated horizon N/A. Nearest underlying dated horizon N/A. Geological age Uncertain, but possibly Middle to Late Pleistocene, but based only on comparisons with other hominin crania purported to be of Middle Pleistocene age. The cranium and several other fossils appear to be in a secondary context after having eroded from nearby sedimentary exposures and moved by wave action to the find spot. The other fossils recovered belong to extant taxa, hippo, crocodile, and Cape buffalo. Developmental age Adult. Presumed sex Unknown. Brief anatomical description A heavily mineralized, darkly stained, well-preserved cranium that lacks the supraorbital region, parts of the nasal bones, most of both maxillae and zygomatic bones, and all of the dentition. The endocranium is filled with matrix, but data from CT scans shows a strongly flexed endocranial base. The cranial vault bone is strikingly thick and the outer table porous; this may be due chronic anemia in childhood. Endocranial volume 1243 cm^3. Announcement Bräuer and Leakey 1986a. Initial description Bräuer and Leakey 1986a. Photographs/line drawings and metrical data Bräuer and Leakey 1986a, 1986b, Bräuer et al. 2004. Detailed anatomical description Bräuer and Leakey 1986b, Bräuer et al. 2003, 2004. Initial taxonomic allocation *Homo sapiens* "late archaic grade." Taxonomic revisions Possibly *Homo helmei* (Foley and Lahr 1997, Howell 1994). Current conventional taxonomic allocation *H. sapiens* or *H. helmei*. Informal taxonomic category Pre-modern *Homo*. Significance The cranium is long, low, and broad, but the occipital is more rounded and the frontal bone much closer to vertical than in specimens such as **Kabwe** and **Saldanha**; overall it resembles the cranium from Florisbad. KNM-ES 11693 is one of the best-preserved African fossils of a population that may have been transitional to, or closely related to, anatomically modern *Homo sapiens*. Location of original **National Museums of Kenya**, Nairobi.

KNM-KP Prefix for fossils accessioned at the Kenya National Museum (now the **National Museums of Kenya**) that derive from localities at Kanapoi, Kenya (e.g., **KNM-KP 31724**). *See also* **National Museums of Kenya**; **Kanapoi**.

KNM-KP 271 <u>Site</u> **Kanapoi**. <u>Locality</u> "On the surface of exposures at the base of the west side of Naringangoro Hill" (Patterson and Howells 1967, p. 64). <u>Surface/*in situ*</u> Surface. <u>Date of discovery</u> 1965. <u>Finder</u> Bryan Patterson. <u>Unit</u> N/A. <u>Horizon</u> "Upper fossiliferous zone"…"distributary channel associated with the Kanapoi tuff" (Leakey et al. 1995, p. 571). <u>Bed/member</u> N/A. <u>Formation</u> Kanapoi Formation. <u>Group</u> N/A. <u>Nearest overlying dated horizon</u> Kanapoi Tuff. <u>Nearest underlying dated horizon</u> Upper pumiceous tuff. <u>Geological age</u> *c.*4.1 Ma. <u>Developmental age</u> Adult. <u>Presumed sex</u> Unknown. <u>Brief anatomical description</u> Distal left humerus. <u>Announcement</u> Patterson and Howells 1967. <u>Initial description</u> Patterson and Howells 1967. <u>Photographs/line drawings and metrical data</u> Patterson and Howells 1967. <u>Detailed anatomical description</u> Hill and Ward 1988. <u>Initial taxonomic allocation</u> Hominidae cf. *Australopithecus* sp. <u>Taxonomic revisions</u> *Homo* (Senut 1981), *Australopithecus afarensis* (Hill and Ward 1988), *Australopithecus anamensis* (Leakey et al. 1995. <u>Current conventional taxonomic allocation</u> *Australopithecus anamensis*. <u>Informal taxonomic category</u> Archaic hominin. <u>Significance</u> This distal humerus, the first hominin to be recovered from Kanapoi, was a puzzle because, notwithstanding its antiquity, it was closer morphologically to modern humans than is a temporally much younger distal humerus from Kromdraai that belongs to **TM 1517**, the **holotype** of *Paranthropus robustus*. If KNM-KP 271 does belong to *Au. anamensis* then this taxon has a remarkably modern human-like distal humeral morphology. <u>Location of original</u> **National Museums of Kenya**, Nairobi, Kenya.

KNM-KP 29281 <u>Site</u> **Kanapoi**. <u>Locality</u> N/A. <u>Surface/*in situ*</u> Surface. <u>Date of discovery</u> 1994. <u>Finder</u> Peter Nzube. <u>Unit</u> N/A. <u>Horizon</u> N/A. <u>Bed/member</u> N/A. <u>Formation</u> Kanapoi Formation. <u>Group</u> N/A. <u>Nearest overlying dated horizon</u> Upper pumiceous tuff. <u>Nearest underlying dated horizon</u> Lower pumiceous tuff. <u>Geological age</u> *c.*4.15 Ma. <u>Developmental age</u> Adult. <u>Presumed sex</u> Female. <u>Brief anatomical description</u> A mandible preserving the corpus with all the tooth crowns present and undamaged. Part of the left temporal bone most likely belongs to the same

individual. <u>Announcement</u> Leakey et al. 1995. <u>Initial description</u> Leakey et al. 1995. <u>Photographs/line drawings and metrical data</u> Leakey et al. 1995, Ward et al. 1999b. <u>Detailed anatomical description</u> Ward et al. 1999b. <u>Initial taxonomic allocation</u> *Australopithecus anamensis*. <u>Taxonomic revisions</u> None. <u>Current conventional taxonomic allocation</u> *Au. anamensis*. <u>Informal taxonomic category</u> Archaic hominin. <u>Significance</u> The distinctive symphyseal morphology of this mandible was one of the reasons for creating a new species, *Au. anamensis*, for the Kanapoi hypodigm and KNM-KP 29281 is its **holotype**. <u>Location of original</u> **National Museums of Kenya**, Nairobi, Kenya.

KNM-KP 29285 <u>Site</u> **Kanapoi**. <u>Locality</u> N/A. <u>Surface/*in situ*</u> Surface. <u>Date of discovery</u> 1994. <u>Finder</u> Kamoya **Kimeu**. <u>Unit</u> N/A. <u>Horizon</u> N/A. <u>Bed/member</u> N/A. <u>Formation</u> Kanapoi Formation. <u>Group</u> N/A. <u>Nearest overlying dated horizon</u> Kanapoi Tuff. <u>Nearest underlying dated horizon</u> Upper pumiceous tuff. <u>Geological age</u> *c.*4.1 Ma. <u>Developmental age</u> Adult. <u>Presumed sex</u> Unknown. <u>Brief anatomical description</u> The proximal and distal ends of a tibia, together with the adjacent shaft; the middle third of the shaft is missing. <u>Announcement</u> Leakey et al. 1995. <u>Initial description</u> Leakey et al. 1995. <u>Photographs/line drawings and metrical data</u> Leakey et al. 1995, Ward et al. 1999b. <u>Detailed anatomical description</u> N/A. <u>Initial taxonomic allocation</u> *Australopithecus anamensis*. <u>Taxonomic revisions</u> None. <u>Current conventional taxonomic allocation</u> *Au. anamensis*. <u>Informal taxonomic category</u> Archaic hominin. <u>Significance</u> The tibia displays several features consistent with bipedalism (e.g., rectangular proximal surface, straight and robust shaft, small fibular articulation, and a horizontally orientated distal articular surface). <u>Location of original</u> **National Museums of Kenya**, Nairobi, Kenya.

KNM-KP 29287 <u>Site</u> **Kanapoi**. <u>Locality</u> N/A. <u>Surface/*in situ*</u> Surface. <u>Date of discovery</u> 1994–5. <u>Finder</u> Samuel Ngui. <u>Unit</u> N/A. <u>Horizon</u> N/A. <u>Bed/member</u> N/A. <u>Formation</u> Kanapoi Formation. <u>Group</u> N/A. <u>Nearest overlying dated horizon</u> Kalokwanya Basalt. <u>Nearest underlying dated horizon</u> Kanapoi Tuff. <u>Geological age</u> *c.*4.0 Ma; "likely to be only marginally younger than the Kanapoi Tuff" (Leakey et al. 1998, p. 63).

Developmental age Adult. Presumed sex Unknown. Brief anatomical description Mandible with P$_4$–M$_2$ of both sides, plus the left I$_1$, both I$_2$s, the right P$_3$, and both M$_3$s. Announcement Leakey et al. 1998. Initial description Leakey et al. 1998. Photographs/line drawings and metrical data N/A. Detailed anatomical description N/A. Initial taxonomic allocation *Australopithecus anamensis*. Taxonomic revisions None. Current conventional taxonomic allocation *Au. anamensis*. Informal taxonomic category Archaic hominin. Significance This mandible is larger than that of the **holotype** and the size and morphology of the canine suggests it belonged to a male individual. Location of original **National Museums of Kenya,** Nairobi, Kenya.

KNM-KP 31724 Site **Kanapoi**. Locality N/A. Surface/*in situ* Surface. Date of discovery 1996. Finder Peter Nzube. Unit N/A. Horizon N/A. Bed/member N/A. Formation Kanapoi Formation. Group N/A. Nearest overlying dated horizon Upper pumiceous tuff. Nearest underlying dated horizon Lower pumiceous tuff. Geological age *c.*4.15 Ma. Developmental age Adult. Presumed sex Unknown. Brief anatomical description A well-preserved left capitate. Announcement Leakey et al. 1998. Initial description Leakey et al. 1998. Photographs/line drawings and metrical data Leakey et al. 1998. Detailed anatomical description N/A. Initial taxonomic allocation *Australopithecus anamensis*. Taxonomic revisions None. Current conventional taxonomic allocation *Au. anamensis*. Informal taxonomic category Archaic hominin. Significance This left capitate shares at least one feature (a laterally facing facet for the second metacarpal) with the extant apes that is not seen in later hominins such as *Australopithecus afarensis*. Location of original **National Museums of Kenya,** Nairobi, Kenya.

KNM-LT Prefix for fossils accessioned at the Kenya National Museum (now **National Museums of Kenya**) that derive from localities at Lothagam, Kenya (e.g., **KNM-LT 329**). *See also* **National Museums of Kenya; Lothagam**.

KNM-LT 329 Site **Lothagam**. Locality Locality 1. Surface/*in situ* Surface. Date of discovery 1967. Finder "A.D. Lewis, assistant to B. Patterson" (Oakley et al. 1977). Unit N/A. Horizon "The lowermost part of the Apak Member, less

than 3 m above the Purple Marker" (Leakey and Walker in Leakey and Harris 2003, p. 219). Bed/member Apak Member. Formation Nachukui Formation. Group Omo Group. Nearest overlying dated horizon "Pumiceous bed that lies 17 m below the Lothagam Basalt and 35 m above the Purple Marker" (McDougall and Feibel 1999, p. 249) *c.*4.2 Ma. Nearest underlying dated horizon Purple Marker *c.*5.23 Ma. Geological age *c.*4.5–5.0 Ma. Developmental age Adult. Presumed sex Unknown. Brief anatomical description A piece of the right side of the body of the mandible. The oblique anterior fracture surface runs from mesial to the M^1 at the alveolar process to between M$_1$ and the M$_2$ at the base; the posterior fracture surface is also oblique and runs from just distal to the roots of the M$_3$ at the alveolar process to just anterior to the mandibular angle. The crown of the M$_1$ is preserved, but all that remains of the M$_2$ and M$_3$ is their root systems. Dentine is exposed on the main buccal cusps of the M$_1$. Announcement Patterson et al. 1970. Initial description Kramer 1986, White 1986b. Photographs/line drawings and metrical data Kramer 1986, White 1986b, Hill and Ward 1988, Hill et al. 1992, Leakey and Walker 2003. Detailed anatomical description N/A. Initial taxonomic allocation *Australopithecus* sp. cf. *Australopithecus africanus*. Taxonomic revisions *Australopithecus* sp. cf. *Australopithecus afarensis* (Kramer 1986, and by inference McHenry and Corruccini 1980 and White 1986b), *Australopithecus afarensis* (Hill and Ward 1988), *Australopithecus* cf. *afarensis* (Hill et al. 1992), *Australopithecus* sp. cf. *Australopithecus anamensis* (by inference Leakey and Walker 2003). Current conventional taxonomic allocation *Australopithecus* sp. or *Ardipithecus* sp. Informal taxonomic category Archaic or possible primitive hominin. Significance For a long time the KNM-LT 329 mandible was among the few tantalizing pieces of evidence (e.g., other evidence included the **Kanapoi** humerus, **KNM-KP 271**, the **Lukeino** molar, **KNM-LU 335**, and the **Tabarin** mandible **KNM-TH 13150**) of one, or more, African **hominin**, perhaps even **hominid**, taxa that predated *Au. afarensis*. Location of original **National Museums of Kenya**, Nairobi, Kenya.

KNM-LU Prefix for fossils accessioned at the Kenya National Museum (now **National Museums of Kenya**) that derive from sites in the Lukeino Formation in the Baringo region,

Kenya (e.g., **KNM-LU 335**). *See also* **National Museums of Kenya**; **Lukeino Formation**.

KNM-LU 335 <u>Site</u> Cheboit. <u>Locality</u> N/A. <u>Surface/*in situ*</u> Surface. <u>Date of discovery</u> 1973. <u>Finder</u> Martin Pickford. <u>Unit</u> N/A. <u>Horizon</u> N/A. <u>Bed/member</u> Kapgoywa Member. <u>Formation</u> **Lukeino Formation**. <u>Group</u> N/A. <u>Nearest overlying dated horizon</u> Kaparaina Basalts. <u>Nearest underlying dated horizon</u> Kabarnet Trachytes. <u>Geological age</u> *c.*6 Ma. <u>Developmental age</u> Adult. <u>Presumed sex</u> N/A. <u>Brief anatomical description</u> A left lower molar, either M_1 or M_2, but probably an M_1. <u>Announcement</u> Pickford 1975. <u>Initial published description</u> Pickford 1975. <u>Photographs/line drawings and metrical data</u> N/A. <u>Detailed anatomical description</u> There is no detailed anatomical description, but McHenry and Corruccini (1980) undertook an analysis of the diameters (not the areas) of the main cusps and concluded that their results did not support the "separation of the Lukeino molar from the genus *Pan*" (*ibid*, p. 397). Ungar et al. (1994) undertook a multivariate analysis of the crown morphology of KNM-LU 335 based on the relative locations of the cusp tips and the peripheral termini of the main fissures. Their conclusions were somewhat different from those of McHenry and Corruccini, for the latter authors found that the flaring of the buccal surface of the crown had no modern analogues in modern humans and chimpanzees, and they further concluded that their results "do not exclude KNM-LU 335 as a potential ancestral morphotype for *Pan* and *Homo*" (*ibid*, p. 165). <u>Initial taxonomic allocation</u> Hominidae *gen. et sp. indet.* <u>Taxonomic revisions</u> ***Orrorin tugenensis*** (Senut et al. 2001). <u>Current conventional taxonomic allocation</u> *O. tugenensis*. <u>Informal taxonomic category</u> Possible primitive hominin. <u>Significance</u> The first specimen of the hypodigm of *O. tugenensis* to be discovered. <u>Location of original</u> **National Museums of Kenya**, Nairobi, Kenya.

KNM-OG Prefix for fossils accessioned at the Kenya National Museum (now **National Museums of Kenya**) that derive from localities at Olorgesailie (e.g., **KNM-OG 45500**). *See also* **National Museums of Kenya**; **Olorgesailie**.

KNM-OG 45500 <u>Site</u> Olorgesailie. <u>Locality</u> OL 45500. <u>Surface/*in situ*</u> Frontal fragment was found *in situ* in June 2003 within a sediment block that had been removed from the site in 1999; the remaining fragments were found in July and August, 2003 on the surface of the slope below where the block was extracted. <u>Date of discovery</u> June, July, and August, 2003. <u>Finders</u> Messrs Kanunga, Kilonzi, Kinyua, Kioko, Mumo, Nume, and Potts. <u>Unit</u> Sandy silts containing reworked diatomite and volcanic cobbles just above the Member 5 paleosol. <u>Horizon</u> See above. <u>Bed/member</u> Member 7. <u>Formation</u> **Olorgesailie Formation**. <u>Group</u> N/A. <u>Nearest overlying dated horizon</u> Air-fall ash at the top of Member 8. <u>Nearest underlying dated horizon</u> Pumice clasts from Member 5. <u>Geological age</u> 970–900 ka. <u>Developmental age</u> Young adult. <u>Presumed sex</u> Unknown. <u>Brief anatomical description</u> Frontal and left temporal bones and nine cranial vault fragments. No endocast has been reconstructed, but Potts et al. (2004) suggest that the endocranial volume is likely to be less than 800 cm^3. <u>Announcement</u> Potts et al. 2004. <u>Initial description</u> Potts et al. 2004. <u>Photographs/line drawings and metrical data</u> Potts et al. 2004. <u>Detailed anatomical description</u> N/A. <u>Initial taxonomic allocation</u> Compared to, but not formally assigned to ***Homo erectus***. <u>Taxonomic revisions</u> None. <u>Current conventional taxonomic allocation</u> cf. *H. erectus*. <u>Informal taxonomic category</u> Pre-modern *Homo*. <u>Significance</u> Although the KNM-OG 45500 calvaria is incomplete, it comes from a time interval in which hominin fossil evidence is frustratingly scarce. Enough is preserved for it to be confidently allocated to *H. erectus*. It is one of a growing number of fossil specimens from this species that appear to be characterized by small body size (as inferred from the size of the cranium). <u>Location of original</u> **National Museums of Kenya**, Nairobi, Kenya.

KNM-OL This was cited as the prefix for fossils accessioned at the Kenya National Museum (now National Museums of Kenya) that derive from localities at Olorgesailie (e.g., **KNM-OL 45500**) (Potts et al. 2004), but the correct prefix is **KNM-OG**. *See also* **National Museums of Kenya**; **Olorgesailie**.

KNM-OL 45500 This was cited as the accession number for the hominin frontal fragment from **Olorgesailie** (Potts et al. 2004) but the correct accession number is **KNM-OG 45500**.

KNM-TH Prefix for fossils accessioned at the Kenya National Museum (now **National**

Museums of Kenya) that derive from sites in the Tugen Hills in the Baringo region, Kenya (e.g., **KNM-TH 13150**). *See also* **National Museums of Kenya**; Tugen Hills.

KNM-TH 13150 Site Tabarin. Locality BPRP#77. Surface/*in situ* Surface. Date of discovery 1984. Finder Kiptalam Cheboi. Unit N/A. Horizon N/A. Bed/member N/A. Formation **Chemeron Formation**. Group N/A. Nearest overlying dated horizon 4.3 Ma. Nearest underlying dated horizon 4.428±0.005 Ma. Geological age 4.42 Ma. Developmental age Adult. Presumed sex Unknown. Brief anatomical description Fragment of the right side of the mandibular corpus with M_1, M_2, part of the alveolus of M_3, and part of the root and alveolus of P_4. Announcement Hill 1985. Initial description Hill 1985. Photographs/line drawings and metrical data Hill 1985. Detailed anatomical description Hill 1985, Ward and Hill 1987. Initial taxonomic allocation cf. *Australopithecus afarensis*. Taxonomic revisions *Homo antiquus praegens* Ferguson, 1989. Current conventional taxonomic allocation *Ardipithecus ramidus*. Informal taxonomic category Possible primitive hominin. Significance With hindsight the KNM-TH 13150 mandible may be the first evidence of *Ar. ramidus* to be recovered and if it belongs to this taxon it would extend its known geographical range. Location of original **National Museums of Kenya**, Nairobi, Kenya.

KNM-TH 45519–45522, KNM-TH 46437–46440 Site Kapthurin. Locality Loc.99. Surface/*in situ* Surface. Date of discovery KNM-TH 45519–45522 (2004), KNM-TH 46437–46440 (2006 and thereafter). Finders B. Kimeu, N. Kanyenze, and M. Macharwas. Unit N/A. Horizon N/A. Bed/member K3'. Formation **Kapthurin Formation**. Group N/A. Nearest overlying dated horizon K2 (545±3 ka). Nearest underlying dated horizon K4 (284±12 ka). Geological age *c.*545 ka. Developmental age Adult. Presumed sex Unknown. Brief anatomical description KNM-TH 45519 and 45520 are, respectively, well-preserved right and left upper central permanent incisors with broad, spatulate, crowns, KNM-TH 45521 the crown of a left upper molar, probably the M^1, KNM-TH 45522 is an upper or lower supernumary molar, KNM-TH 46437–46438 are upper lateral incisors, KNM-TH 46439 is a lower lateral incisor, and KNM-TH 46440

is a P^4. Announcement McBrearty and Jablonski 2005 (NB: KNM-TH 45522 was not reported in this paper). Initial description McBrearty and Jablonski 2005. Photographs/line drawings and metrical data McBrearty and Jablonski 2005. Detailed anatomical description N/A. Initial taxonomic allocation *Pan sp. indet.* Taxonomic revisions None. Current conventional taxonomic allocation *Pan*. Informal taxonomic category Panin. Significance These teeth, probably from the same individual, represent the first fossil evidence of the panin clade; thus they are the first fossils to be attributed to the extant sister clade of hominins. Location of original **National Museums of Kenya**, Nairobi, Kenya.

KNM-WT Prefix for fossils accessioned at the Kenya National Museum (now **National Museums of Kenya**) that derive from localities within the West Turkana site complex (e.g., **KNM-WT 8556**). *See also* **National Museums of Kenya**; West Turkana.

KNM-WT 8556 Site West Turkana. Locality Lomekwi (LO5). Surface/*in situ* Surface. Date of discovery 1982. Finder Peter Nzube. Unit N/A. Horizon N/A. Bed/member Lomekwi Member. Formation Nachukui Formation. Group Omo Group. Nearest overlying dated horizon Lokalalei Tuff. Nearest underlying dated horizon Tulu Bor Tuff. Geological age 3.26±0.10 Ma. Developmental age Adult. Presumed sex Unknown. Brief anatomical description Mandibular corpus that extends from just distal to the left P_3 to midway along the right M_2. The crowns or partial crowns of the left and right P_3, and the right P_3 and M_3 are preserved. Announcement Brown et al. 2001. Initial description Brown et al. 2001. Photographs/line drawings and metrical data Brown et al. 2001. Detailed anatomical description N/A. Initial taxonomic allocation *Australopithecus afarensis*. Taxonomic revisions aff. *Kenyanthropus platyops* (Leakey et al. 2001). Current conventional taxonomic allocation *Au. afarensis*. Informal taxonomic category Archaic hominin. Significance Significance Leakey et al. (2001) point out how it differs from *Au. afarensis* and Kimbel and Delezene (2009) in their review of *Au. afarensis* agree, writing "Although Leakey et al. (2001) discussed these specimens in the context of their description of the species *Kenyanthropus platyops*, they did not attribute them to this taxon; they did enumerate ways in which they departed from *A. afarensis* morphology, especially in the dentition—which we can affirm from examination of the original fossils" (*ibid*, p. 13). Location of original **National Museums of Kenya**, Nairobi, Kenya.

KNM-WT 15000 Site **West Turkana**. Locality Nariokotome III (NK-III). Surface/*in situ* Surface, then *in situ*. Date of discovery August 1984, then four seasons of excavation. Finders Kamoya **Kimeu**, then excavated by a team led by Alan **Walker**. Unit N/A. Horizon "Siltstone that immediately overlies a tuff identified as a component ash of the Okote Tuff complex" (Brown et al. 1985, p. 788). Bed/member Natoo. Formation Nachukui Formation. Group Omo Group. Nearest overlying dated horizon Chari Tuff. Nearest underlying dated horizon Immediately above the Okote Tuff complex. Geological age "Very close to 1.6 Myr" (*ibid*, p. 788) [NB: according to Brown et al. 2006, NK-III is above the tuff that was correlated to a tuff in the Kaitio drainage, which places it stratigraphically above the Ebei Tuff, dated to 1.47 Ma (McDougall and Brown 2006). The implication is that KNM-WT 15000 should be closer to 1.44 Ma in age]. Developmental age Adolescent, on the basis that the second permanent molars are just in occlusion and the M^3 germs are probably only just crown complete. The most precise and accurate chronological age estimate based on tooth microstructure (Dean and Smith 2009) suggests an age of *c*.8 years of age (range 7.5–8.9 years); the modern human equivalent would be *c*.12 years old. Presumed sex Male. Brief anatomical description At the time of its recovery KNM-WT 15000 was the most complete early hominin skeleton, as well as being remarkably free from plastic deformation. It is so complete it is easier to list what major elements are *missing* rather than what is preserved. The KNM-WT 15000 cranium is complete apart from the supraorbital tori, and the axial skeleton lacks most of the cervical and some of the thoracic vertebrae, parts of the sacrum and pelvic bone and some ribs. The upper limbs lack the left scapula and humerus, both radii, and most of the hand bones. The lower limbs lack the proximal end of the right femur, most of the epiphyses of the bones of the leg, and most of the bones of the foot. Endocranial volume 909 cm³ (Walker and Leakey 1993). Estimated stature At the time of death 160 cm; 185 cm is the projected stature estimate for an equivalent adult (Walker and Leakey 1993), but see Graves et al. (2010) for a more more modest (about 163 cm) stature estimate. Estimated body mass At the time of death 48 kg; 68 kg is the projected body mass for an equivalent adult (Walker and Leakey 1993). Ruff (2007) provides an estimate of 50–53 kg at the time of death based on the size of the femoral head, and a similar estimate was provided by Aiello and Wood (1994) based on orbit size. Announcement Brown et al. 1985. Initial description Brown et al. 1985. Photographs/line drawings and metrical data Brown et al. 1985, Walker and Leakey 1993. Detailed anatomical description Walker and Leakey 1993. Initial taxonomic allocation *Homo erectus*. Taxonomic revisions *Homo ergaster* (Wood 1991, p. 270). Current conventional taxonomic allocation *H. erectus*. Informal taxonomic category Premodern *Homo*. Significance It is difficult to exaggerate the importance of the KNM-WT 15000 skeleton for before its discovery relatively little was known about the postcranial skeleton, ontogeny, and life history of *H. erectus*. It provided conclusive proof of the limb proportions and limb morphology of one of the earliest taxa within the genus *Homo* and in doing so helped researchers make more sense of the isolated limb bones recovered from sites such as Koobi Fora. Location of original **National Museums of Kenya**, Nairobi, Kenya.

KNM-WT 16002 Site **West Turkana**. Locality Lomekwi (LO3). Surface/*in situ* Surface. Date of discovery 1985. Finder John Harris. Unit N/A. Horizon N/A. Bed/member Lokalalei. Formation Nachukui Formation. Group Omo Group. Nearest overlying dated horizon Kalochoro Tuff. Nearest underlying dated horizon Tulu Bor Tuff. Geological age 2.7 ± 0.03 Ma (Brown et al. 2001). Developmental age Adult. Presumed sex Unknown. Brief anatomical description Right femur preserving the neck and shaft, but missing the head, both trochanters, and the distal end. Announcement Brown et al. 2001. Initial description Brown et al. 2001. Photographs/line drawings and metrical data Brown et al. 2001. Detailed anatomical description N/A. Initial taxonomic allocation *Australopithecus sp. indet.* Taxonomic revisions None. Current conventional taxonomic allocation *Australopithecus sp. indet.* Informal taxonomic category Archaic hominin. Significance There are very few relatively well-preserved femora from this important time range and there is a possibility that this femur may belong to *Paranthropus aethiopicus*. Location of original **National Museums of Kenya**, Nairobi, Kenya.

KNM-WT 16005 Site **West Turkana**. Locality Given as Kangatukuseo II (KU2) in Feibel et al. (1989) but listed as Kangatukuseo III (i.e., KU3) in Walker et al. 1986. Surface/*in situ* Surface. Date of discovery 1985. Finders Not known. Unit 3.8 m below the Lokalalei Tuff or 19 m above the Lokalalei Tuff. Horizon N/A. Bed/member Lomekwi – Upper. Formation Nachukui Formation. Group Omo Group. Nearest overlying dated horizon Lokalalei Tuff. Nearest underlying dated horizon Tulu Bor Tuff. Geological age *c*.2.55 Ma. Developmental age Adult. Presumed sex Unknown. Brief anatomical description The upper part of the corpus of an adult mandible that extends from just distal to the M_1 on the right, to the mid-point of the M_3 alveolus on the left. The roots of the anterior teeth are preserved except for the left I_1, as are most of the worn crowns of the left P_3 to M_2 and the right P_3 to M_2. Announcement Walker et al. 1986. Initial description Leakey and Walker 1988. Photographs/line drawings and metrical data Walker et al. 1986, Leakey and Walker 1988. Detailed anatomical description Leakey and Walker 1988. Initial taxonomic allocation *Australopithecus boisei*. Taxonomic revisions None. Current conventional taxonomic allocation *Paranthropus aethiopicus*. Informal taxonomic category Megadont archaic hominin. Significance This mandible was found in sediments that are effectively synchronic (Harris et al. 1988) with those that contained **KNM-WT 17000** and it is possible, if not probable, that this was the type of mandible that would have belonged to an individual in the same taxon as KNM-WT 17000. Location of original **National Museums of Kenya**, Nairobi, Kenya.

KNM-WT 17000 Site **West Turkana**. Locality Lomekwi (LO1). Surface/*in situ* Surface. Date of discovery 1985. Finder Alan Walker. Unit N/A. Horizon Lokalalei Tuff. Bed/member Lomekwi – Upper (Harris et al. 1988), but Feibel et al. (1989) list it as coming from the Lokalalei Member; it is the Lokalalei Member. Formation Nachukui Formation [but some publications e.g., Feibel et al. (1989) refer to the Koobi Fora Formation]. Group Omo Group. Nearest overlying dated horizon

"Within the Lokalalei Tuff" (Feibel et al. 1989, p. 610). Nearest underlying dated horizon See above. Geological age 2.53 ± 0.02 Ma. Developmental age Adult. Presumed sex Male. Brief anatomical description Cranium missing the posterior part of the frontal, the anterior parts of the parietals, the lateral orbital margins, the zygomatic arches, parts of the sphenoid and the maxillae, and the occipital squame. Parts of the alveoli of the anterior teeth are preserved as are many of the roots of the postcanine teeth, but the only well-preserved tooth crown is that of the left P^4. Endocranial volume Approximately 410 cm^3 (Walker and Leakey 1988). Announcement Walker et al. 1986. Initial description Leakey and Walker 1988. Photographs/line drawings and metrical data Walker et al. 1986, Leakey and Walker 1988. Detailed anatomical description Leakey and Walker 1988. Initial taxonomic allocation *Australopithecus boisei*. Taxonomic revisions *Australopithecus aethiopicus* (Kimbel et al. 1988). Current conventional taxonomic allocation *Paranthropus aethiopicus*. Informal taxonomic category Megadont archaic hominin. Significance There are very few well-preserved crania in this time period, and the discovery of KNM-WT 17000 prompted a re-evaluation of the relationships among the more primitive hominin taxa such as *Australopithecus africanus* and *Australopithecus afarensis*, and the later, and presumed to be more derived, megadont archaic hominin taxa such *Paranthropus robustus* and *Paranthropus boisei*. Walker et al. (1986) and Leakey and Walker (1988, pp. 21–3) set out the reasons why they chose to interpret KNM-WT 17000 as belonging to "an early *A. boisei* population" (*ibid*, p. 22), rather than assign it to the available species *Paraustralopithecus aethiopicus*. By that time one of the authors of the paper that proposed the recognition of *P. aethiopicus* had already cast doubt on the wisdom of recognizing a new species (Coppens 1979), and Walker et al. (1986) suggested that if there was future need for a new taxon then that taxon should be subsumed into the genus *Australopithecus*, as *Au. aethiopicus*. Other researchers were more impressed with the differences between KNM-WT 17000 and the main hypodigm of *P. boisei*, and were not convinced the taxon to which KNM-WT 17000 belongs was part of a *P. boisei* lineage, but instead they suggested that the taxon might be a link between *Au. afarensis* and the common ancestor of *P. robustus* and *P. boisei* (e.g., Kimbel et al.

1988). Location of original **National Museums of Kenya**, Nairobi, Kenya.

KNM-WT 17400 Site West Turkana. Locality Naiyena Engol (NY2). Surface/*in situ* Surface. Date of discovery Not known. Finders Not known. Unit Not known. Horizon Not known. Bed/member Kaitio Member. Formation Nachukui Formation. Group Omo Group. Nearest overlying dated horizon N/A. Nearest underlying dated horizon KBS Tuff. Geological age *c*.1.77±0.10 Ma. Developmental age Late juvenile. Presumed sex Probable female. Brief anatomical description A cranium that lacks the supraorbital region of the face, the zygomatic arches, some of the cranial base and cranial vault, and the occipital squame. The tooth crowns, with the exception of the medial incisors, are all present and well preserved; the M^3 germs are exposed in their crypts. The paranasal sinuses are large and the temporal bone is extensively pneumatized. Endocranial volume 390–400 cm^3 (Holloway 1988) and 500 cm^3 (Brown et al. 1993). Announcement None. Initial description Leakey and Walker 1988. Photographs/ line drawings and metrical data Leakey and Walker 1988. Detailed anatomical description Leakey and Walker 1988. Initial taxonomic allocation *Australopithecus boisei*. Taxonomic revisions None. Current conventional taxonomic allocation *Paranthropus boisei*. Informal taxonomic category Megadont archaic hominin. Significance This cranium is one of several late juvenile/young adult crania in the *P. boisei* hypodigm and it adds to our knowledge of the ontogeny of that taxon. It is the best-preserved face of *P. boisei* after **OH 5**. Location of original **National Museums of Kenya**, Nairobi, Kenya.

KNM-WT 19700 Site West Turkana. Locality Nachukui (NC1). Surface/*in situ* Surface. Date of discovery 1988. Finder Kamoya **Kimeu**. Unit "8–8.5 m below stomatolite horizon S3" (Brown et al. 2001, p. 42). Horizon N/A. Bed/member Nariokotome Member. Formation **Nachukui Formation**. Group Omo Group. Nearest overlying dated horizon N/A. Nearest underlying dated horizon Lower Nariokotome Tuff. Geological age 1.0±0.15 Ma. Developmental age Adult. Presumed sex Unknown. Brief anatomical description The proximal end of an adult tibia. Announcement Brown et al. 2001. Initial description Brown et al. 2001. Photographs/line drawings and metrical data Brown et al. 2001. Detailed anatomical description Brown et al. 2001. Initial

taxonomic allocation *Homo erectus*. Taxonomic revisions None. Current conventional taxonomic allocation *H. erectus* or *Homo ergaster*. Informal taxonomic category Pre-modern *Homo*. Significance Other than **KNM-WT 15000** few limb bones have been assigned to *H. erectus* or *H. ergaster*. Brown et al. (2001) assigned KNM-WT 19700 to *H. erectus* on the grounds that "this fragment compares closely…with KNM-WT 15000" (*ibid*, p. 43); "especially obvious is the way in which the epiphysis is set at a strong angle to the anteroposterior axis of the shaft" (*ibid*, p. 43). This compares with *Australopithecus* (i.e., *Paranthropus*) in which the "long axis of the epiphysis set at right angles to the anteroposterior axis of the shaft" (*ibid*, p. 43). Location of original **National Museums of Kenya**, Nairobi, Kenya.

KNM-WT 40000 Site West Turkana. Locality Lomekwi (LO-6N). Surface/*in situ* Surface. Date of discovery August 1999. Finder Justus Erus. Unit N/A. Horizon "Dark mudstone" (Leakey et al. 2001, p. 439). Bed/member Kataboi Member. Formation Nachukui Formation. Group Omo Group. Nearest overlying dated horizon 8 m below the β-Tulu Bor Tuff. Nearest underlying dated horizon 12 m above the Lokochot Tuff. Geological age *c*.3.5 Ma. Developmental age Adult. Presumed sex Unknown. Brief anatomical description A relatively complete but plastically deformed cranium in two parts, one preserving the neurocranium, the supraorbital region and part of the cranial base, the other preserving the rest of the face. Endocranial volume N/A. Announcement Leakey et al. 2001. Initial description Leakey et al. 2001. Photographs/line drawings and metrical data Leakey et al. 2001. Detailed anatomical description N/A. Initial taxonomic allocation *Kenyanthropus platyops*. Taxonomic revisions *Australopithecus afarensis* (White 2003). Current conventional taxonomic allocation *K. platyops*. Informal taxonomic category Archaic hominin. Significance The main reasons Leakey et al. (2001) did not assign this material to *Au. afarensis* are its reduced subnasal prognathism, anteriorly situated zygomatic root, flat and vertically orientated malar region, relatively small but thick-enameled molars, and the unusually small M^1 compared to the size of the P^4 and M^3. Some of the morphology of the new genus including the shape of the face is *Paranthropus*-like yet it lacks the postcanine **megadontia** that characterizes *Paranthropus* taxa.

The authors note the face of the new material resembles that of **Homo rudolfensis**, but they rightly point out that the postcanine teeth of the latter are substantially larger than those of KNM-WT 40000. *K. platyops* apparently displays a hitherto unique combination of facial and dental morphology. White (2003) has taken the view that the new taxon is not justified because the cranium could be a distorted *Au. afarensis* cranium, but these arguments have been challenged and/or refuted by Spoor et al. (2010). Although distortion could conceivably account for why the preserved cranial vault morphology differs from that of *Au. afarensis*, it cannot account for the much smaller postcanine teeth of KNM-WT 40000 compared to the *Au. afarensis* dentition. Location of original **National Museums of Kenya**, Nairobi, Kenya.

KNM-WT 42718 Site **West Turkana**. Locality Lokalalei (LA1α). Surface/*in situ* Surface. Date of discovery 2002. Finders Not reported. Unit "Yellow beige clayey siltstone" (Prat et al. 2005, p. 231). Horizon N/A. Bed/member Kalochoro Member. Formation Nachukui Formation. Group Omo Group. Nearest overlying dated horizon KBS Tuff. Nearest underlying dated horizon Kalochoro Tuff. Geological age 2.3–2.4 Ma. Developmental age Juvenile. Presumed sex Unknown. Brief anatomical description The unworn germ of a right M_1. Announcement Prat et al. 2005. Initial description Prat et al. 2005. Photographs/line drawings and metrical data Prat et al. 2005. Detailed anatomical description Prat et al. 2005. Initial taxonomic allocation *Homo sp. indet.* Taxonomic revisions None. Current conventional taxonomic allocation *Homo sp. indet.* Informal taxonomic category Transitional hominin. Significance The narrow crown, the absence of a C6, and mildly expressed protostylid suggest affinities with *Homo*. If this tooth does belong to *Homo*, it would be one of the earliest examples of this taxon. Location of original **National Museums of Kenya**, Nairobi, Kenya.

knock-in *See* **knockout**.

knock-out A genotype, usually of a model organism (e.g., mice), in which a particular gene has been rendered inoperative. Knockout mice are made through a targeted insertion of DNA into embryonic stem cells, and the subsequent introduction of these cells into blastocysts, which are then implanted into a pregnant female mouse. To target a specific region of DNA, the inserted sequence has two regions that are complementary to DNA sequences that flank (i.e., they are upstream and downstream) of the targeted gene. There are two crucial selection steps in the process of making a knockout mouse. The first involves making sure the DNA is incorporated correctly into the genome of the embryonic stem cells. To accomplish this, two extra genes, one coding for neomycin resistance and the other coding for sensitivity to ganciclovir, are included with the DNA insertion. The gene for neomycin resistance is placed between the two complementary sequences and the gene coding for ganciclovir sensitivity is placed outside the two complementary sequences. If the sequence is not incorporated, the embryonic stem cells will be killed by neomycin. If the sequence is incorporated randomly into the genome, then it will also include the ganciclovir-sensitivity gene, which means that such cells will be killed by ganciclovir. However, if the sequence is inserted so as to replace the desired gene (which is the objective), it will include only the regions between (and including) the two complementary sequences. Such cells will be resistant to neomycin and will also survive being treated with ganciclovir. These are the cells selected for insertion into a blastocyst. Once the injected blastocyst is implanted, a second selection step is necessary to select for mice in which the inserted gene has been incorporated into the germ line. This is done by using stem cells from a mouse with a different coat colour than the mice from which the blastocyst is obtained. The first-generation mice are chimeras but breeding can establish lines of mice in which the inserted gene is segregating in such a way that "knockout" individuals can be compared to normal littermates. The ability to knockout specific genes has revolutionized developmental biology and genetics and in 2007 Mario Capecchi, Martin Evans, and Oliver Smithies were awarded the Nobel Prize for Physiology or Medicine for their pioneering work on this technology. A disadvantage of gene knockouts, however, is that they only tell researchers what happens to the organism when the targeted gene does not perform its function. Often the result is a nonviable embryo or there is no detectable phenotype, usually because other genes have overlapping functions with the one that was targeted. Interpreting such results can be like trying to figure out how a car works by taking components out one by one and seeing what each deletion does to the car's performance. To overcome these limitations, other related strategies have been developed. One is the use of multiple knockouts of related genes. Another is to replace the targeted gene with an allele that performs an altered function rather than one that is inactive; this is known as the knockin approach. A third is to make a knockout (or a knockin) that is conditional on the expression of some other gene. Such conditional knockouts only disrupt the function of the targeted gene in specific tissues or at specific times in development and thus can provide much more precise information about gene function. A somewhat different technology can be used to knockdown gene function through the use of morpholinos, which are small synthetic RNA molecules that block or reduce the translation of a specific protein. Transgenic technologies such as these have been used in human evolution and biological anthropology-related research. Mouse models, for instance, have been used to test the predicted relationship between brain size and cranial shape in hominins (Hallgrimsson and Lieberman 2008). An early use of transgenic mice in human evolutionary studies involved the role of growth hormone in producing allometric variation (Shea et al. 1990).

knuckle-walking *See* **ape hand; hand, evolution in hominins**.

Kobatuwa One of the Early and Middle **Pleistocene** sites in the exposures of the Ola Bula Formation within the **Soa Basin**, central Flores, Indonesia.

Kocaba(s) Site Quarry of the same name in Western Turkey ($37°86'37''$N and $29°38'45''$E). Locality See above. Surface/*in situ* Found in 35 mm-thick blocks of cut travertine in a factory that processes travertine for tiles. Date of discovery Not known. Finders Not known. Unit Not known. Horizon Not known. Bed/member Not known. Formation Not known. Group Not known. Nearest dated horizon Travertines in the area of the quarry have been dated using **thermoluminescence dating**. Geological age Between 510 ± 0.05 and 490 ± 0.05 ka. Developmental age Late juvenile/early adult. Presumed sex Male. Brief anatomical description Two fragments of the frontal that articulate with parts of the right and left parietals. Announcement Kappelman et al. 2008. Initial description Kappelman et al. 2008. Photographs/line drawings and metrical data Kappelman et al. 2008. Detailed anatomical description Kappelman et al. 2008. Initial taxonomic allocation *Homo* cf. *erectus*. Taxonomic revisions None. Current conventional taxonomic allocation *H*. cf. *erectus*. Informal taxonomic category Pre-modern *Homo*. Significance Hominins of this age anywhere are rare and they are even rarer in Western Eurasia. Impressions on the endocranial surface of the frontal may be evidence of inflammatory lesions consistent with infection by *Leptomeningitis tuberculosa*. Location of original Denizli Müze Müdürlügü, Istiklal Caddesi, Deliklicinar, Denizli, Turkey.

Koenigswald, Gustav Heinrich Ralph von (1902–82) *See* **von Koenigswald, Gustav Heinrich Ralph (1902–82)**.

Kohl-Larsen Expedition *See* **Laetoli**.

Kokiselei [etym. Named for an ephemeral river system (laga) in whose drainage the localities are located] (or KS) Refers to a complex of fossil localities in the Kaitio Member of the Nachukui Formation within the Kokiselei drainage in West Turkana, Kenya. *See also* **West Turkana**.

Kollé One of the "fossiliferous areas" within the Chad Basin; others include **Kossom Bougoudi**, **Koro Toro**, and **Toros-Menalla**. *See also* **Chad Basin**.

Kom Ombo (Location $24°28'$N, $32°57'$E, Egypt) History and general description Located on the east bank of the Nile in the southern area of Upper Egypt, Kom Ombo is an agricultural town known since antiquity for its 2ndC temple and its strategic location that facilitated control of trade routes from Nubia into the Nile Valley. The area's relevance to the Paleolithic was first recognized by Edmond Vignard who collected artifacts from multiple open-air sites on the approximately 500 km²/193 square mile Kom Ombo plain in 1919 and 1924. Temporal span and how dated? Primarily by correlation of technocomplexes. Sites on the edges of the plain and on the Nile and wadi terraces are **Lower** and **Middle Paleolithic** in age while those in the silt plain proper are primarily **Upper Paleolithic**. Published and unpublished **radiocarbon dates** determined during the 1960s suggest a span of 14,000 to around 9000 years BP for the local Upper Paleolithic Sebilian tool industry. The modern human remains are well dated by radiocarbon to between 11,500 and 11,000 years BP. The rock art is estimated to date from *c.*15,000 BP. Hominins found at site Fragments of the vault of at least two crania assigned to *Homo sapiens* were recovered from a Sebilian II site, Gebel Silsila 2A, during excavations by the 1962–3 Yale University Prehistoric Expedition to Nubia. Archeological evidence found at site During his work in the early 20thC, Vignard defined a three-stage local tool industry that he termed the Sebilian after the nearby village of Sebil, and proposed, controversially, that it was a wholly indigenous outgrowth of the **Mousterian** tool industry of the Middle Paleolithic. Vignard's Sebilian I is characterized by short, broad Levallois flakes, retouched flake points, and, more rarely, scrapers and blades. The Sebilian II includes broad simple and retouched flake points with the addition of backed blades, large trapezoidal pieces triangular and crescent forms, grindstones, and a reduced frequency of Levallois flakes. The Sebilian III is characterized by many diverse geometric microliths, possibly linked to bow hunting, as well as end-of-blade scrapers, flake scrapers, stemmed points, perforators, and microburins. Also notable are the extensive examples of rock art found at Qurta I, II, and III on the Kom Ombo plain, which consist of roughly 160 mostly naturalistic depictions of animals and humans. Key references: Geology, dating, paleoenvironment, and archeology Smith 1964, 1966a, 1966b; Hominins Reed 1965.

Koneprusy   *See* **Zlat Ků**.

Koninklijk Museum voor Middenafrika
See **Royal Museum for Central Africa**.

Konso (Location 5°20′N, 37°25′E, Ethiopia; etym. Konso is the name of the region and its ethnic group) History and general description Previously known as Konso-Gardula, this site on the western flank of the **Ethiopian Rift System** was discovered in 1991 during the **Paleoanthropological Inventory of Ethiopia**. Temporal span and how dated? 1.9–0.8 Ma based on **argon-argon dating** of tuffs, tephrostratigraphy, and paleomagnetic correlation. All the hominins are from sediments dated to 1.5–1.3 Ma. Hominins found at site Field work conducted beginning in 1991 resulted in the discovery of 1.4–1.45 Ma hominin fossils that were assigned to *Homo erectus* and

subsequent discoveries (Suwa et al. 1997) include excellently preserved specimens of *Paranthropus boisei* dated to between 1.4 and 1.5 Ma, as well as other specimens attributed to *H. erectus*. Archeological evidence found at site Discoveries include <1.6 Ma **Acheulean** artifacts. Key references: Geology, dating, and paleoenvironment Asfaw et al. 1992, Katoh et al. 2000, Suwa et al. 2003, White and Suwa 2004, WoldeGabriel et al. 2005, McDougal and Brown 2006; Hominins Asfaw et al. 1992, Suwa et al. 1997, 2007; Archeology Asfaw et al. 1992, Beyene et al. 1997, Beyene 2003.

Konso-Gardula   *See* **Konso**.

Koobi Fora [Location 3°35′–4°25′N and 36°10′–36°30′E, Kenya; etym. Koobi Fora is the local Das(s)enetch name for several low hills on the northeastern shore of Lake Turkana] History and general description Koobi Fora (formerly known as **East Rudolf** and also known as East Turkana) is a substantial area of outcrops and exposures on the northeastern shore of Lake Turkana. It is one of several fossil sites (the others include **Lothagam**, **Omo**, and **West Turkana**) within the Lower Omo and the Turkana Basin and its size and complexity make it equivalent to the larger study areas in Ethiopia (e.g., **Hadar**, Middle Awash). Richard **Leakey** recognized the paleontological potential of Koobi Fora region in 1967 when he was participating in the **International Omo Research Expedition**. The first **East Rudolf Expedition** to Koobi Fora in 1968 was based at **Allia Bay** and recovered three hominin fossils. Field seasons continued annually and when in 1975 the Kenya Government decreed that the name of the lake be changed from Lake Rudolf to Lake Turkana the **Koobi Fora Research Project** (or KFRP) superseded the East Rudolf Expedition and its members continued to explore, map, and undertake intensive paleoanthropological, paleontological, and archeological fieldwork in the Koobi Fora region until 1979 when the effort at Koobi Fora would down and the KFRP began to focus on other sites within the Turkana Basin. These many cumulative months of field seasons resulted in substantial collections held in the **National Museums of Kenya** in Nairobi. The strata from which fossil hominins have been recovered at Koobi Fora were originally assigned to the Kubi Algi, Koobi Fora, and the Guomde

Formations (Bowen and Vondra 1973), all overlain by the Galana Boi Beds. The three original formations have since been rationalized into a single unit called the **Koobi Fora Formation** within which eight members are recognized. The fossiliferous exposures were initially subdivided into three named subregions: **Ileret** in the north, **Koobi Fora** midway, and Allia Bay to the south. Within each of these subregions natural features such as sand rivers were used to divide the landscape into collecting areas. These are numbered 1–18 and 40–2 in Ileret, 101–10, 112, 114–21, 125, 127–31, and 133–9 in Koobi Fora, and 200–7, 210, 212, 250–4, and 260–1 at Allia Bay. Subsequently the scheme using three main subregions was superseded by one that recognizes seven subregions (Brown and Feibel 1991). Their names, together with the numbered collecting areas within each subregion, are given below in order from north to south: Il Dura (14, 15, 41, and 42), Ileret (1–12), Il Naibar Lowlands (116–17, 136–7), Karari Ridge (105, 118, 129, 131), Koobi Fora Ridge (102–15), Bura Hasuma (107, 110, 119–21, 123–4), and Sibilot (202–4, 207, 212, 250–2). All the vertebrate fossils recovered were recorded and numbered consecutively in a field book. Until 1970 hominin fossils from Koobi Fora had the prefix *FS* (acronym for field: surface) followed by a number (e.g., what are now **KNM-ER 406** and **407** were *FS*-158 and *FS*-210, respectively). Since 1970 each hominin from Koobi Fora has the prefix **KNM** for the Kenya National Museum (now **National Museums of Kenya**) where they are curated, **ER** for East Rudolf, and then the field number (e.g., **KNM-ER 1813**). Temporal span and how dated? The Lonyumun Member, the oldest sediments in the Koobi Fora Formation, dates from 4.35 to 4.0 Ma, and the youngest, the Chari Member, dates from 1.39 to 0.7 Ma. The sediments have been dated using various radiogenic methods, **magnetostratigraphy**, and **biostratigraphy**. Hominins found at site Fossil hominins recovered from Koobi Fora have been assigned to *Paranthropus boisei* (e.g., **KNM-ER** 406, 407, and **732**), *Homo erectus* or *Homo ergaster* (e.g., **KNM-ER 992, 3733,** and **3883**), *Homo habilis* (e.g., **KNM-ER 1478,** 1501, and 1813), *Homo rudolfensis* (e.g., **KNM-ER 1470, 1590,** and **1802**), and probably also *Australopithecus afarensis* (**KNM-ER 2602**). Several hominin footprint trails have also been uncovered. Archeological evidence found at site The Koobi Fora region is best known for its abun-

dant carefully excavated early Pleistocene **Oldowan** and **Developed Oldowan** archeological sites whose investigation was directed by Glynn **Isaac**. Extensive lateral sediment exposures combined with tephrostratigraphic correlations have permitted landscape-scale studies of early archeological and environmental variation. Other Pleistocene sites, particularly those characterized by **Acheulean** and **Middle Stone Age** artifacts, are sparse and their geological contexts are in many cases poorly understood. There is a robust mid-to-late regional **Holocene** archeological record, including sites characterized by fishing economies as well as early domesticates (primarily cattle) in the region. Key references: Geology, dating, and paleoenvironment Brown and Feibel 1991, Brown et al. 2006, McDougall and Brown 2006; Hominins Leakey and Leakey 1978, Wood 1991, Bennett et al. 2009; Archeology Barthelme 1985, Rogers et al. 1994, Kelly 1996, Isaac 1997, Braun et al. 2009.

Koobi Fora Formation [Location 3°35′–4°25′N and 36°10′–36°30′E, Kenya; etym. Koobi Fora is the local Das(s)enetch name for several low hills on the northeastern shore of Lake Turkana] History and general description A substantial thickness of sediments on the northeastern shore of Lake Turkana. The strata from which fossil hominins have been recovered at Koobi Fora were originally assigned to the Kubi Algi, Koobi Fora, and the Guomde Formations (Bowen and Vondra 1973), all overlain by the Galana Boi Beds. These three formations have since been rationalized into a single unit called the Koobi Fora Formation within which eight members are recognized (Brown and Feibel 1986). Each member is defined as the sediments between the base of one designated tuff and the base of the designated tuff immediately overlying it. The members, starting at the base, are the Lonyumun, **Moiti, Lokochot, Tulu Bor, Burgi, KBS, Okote,** and the **Chari Members.** The name of the Lonyumun Member comes from a stream near its type section; the other names are taken from the name of designated tuff that forms the base of each member. The fossiliferous exposures are subdivided into seven subregions (Brown and Feibel 1991) and from north to south they are Il Dura, Ileret, Il Naibar Lowlands, Karari Ridge, Koobi Fora Ridge, Bura Hasuma, and Sibilot. Temporal span and how dated? The oldest member, the Lonyumun Member, dates from 4.35 to

4.0 Ma and the youngest, the Chari Member, dates from 1.39 to 0.7 Ma. The sediments have been dated using various radiogenic methods, **magnetostratigraphy**, and **biostratigraphy**. Hominin evidence Fossil hominins recovered from the Koobi Fora Formation have been assigned to or represent *Paranthropus boisei*, *Homo erectus* or *Homo ergaster*, *Homo habilis*, *Homo rudolfensis*, and probably also *Australopithecus afarensis*. There are also several footprint trails. Archeological evidence The Koobi Fora Formation includes evidence of **Oldowan**, **Developed Oldowan**, **Acheulean**, and **Middle Stone Age** sites and artifacts and there is a mid-to-late **Holocene** archeological record, including sites characterized by fishing economies as well as early domesticates (primarily cattle) in the region. Key references: Geology, dating, and paleoenvironment Brown and Feibel 1986, 1991, Brown et al. 2006, McDougall and Brown 2006; Hominins Leakey and Leakey 1978, Wood 1991, Bennett et al. 2009; Archeology Barthelme 1985, Rogers et al. 1994, Kelly 1996, Isaac 1997, Braun et al. 2009.

Koobi Fora hominin footprints Hominin footprints dating to *c*.1.4–1.53 Ma have been found at Koobi Fora, Kenya, on at least three footprint surfaces at two localities: FwJj14E located at Ileret and GaJi10 on the Koobi Fora Ridge. The three-dimensional shape of the hominin footprints on the three surfaces was captured using optical laser scanning, 13 landmarks were identified on each well-preserved print, and their shapes were compared using generalized **Procrustes analysis** (Bennett et al. 2009). The FwJj14E hominin footprints are better preserved than those at GaJi10 and those from the upper footprint surface at FwJj14E more numerous and include some that are better preserved than those from the lower footprint surface; thus data from the former provide most of the evidence used for morphometric analysis and functional interpretation. The hominin prints from both footprint surfaces at FwJj14E were described as having a "well-defined, deeply depressed and adducted hallux; visible lateral toe impressions that vary in depth of indentation; a well-defined ball beneath the first and second metatarsal heads; and a visible instep reflecting a medial longitudinal arch" (*ibid*, p. 1198). The average angle of hallucial abduction for the FwJj14E hominin footprints (14°) is significantly

greater than the mean for the modern human reference footprints (8°) and significantly smaller than the mean for the **Laetoli hominin footprints** (27°) (*ibid*, p. 1198). Compared to modern humans, the FwJj14E hominin footprints have a "narrower heel and ball area," a "wider instep," and "less pronounced arch elevation" (*ibid*, p. 1201), but when compared to the Laetoli hominin footprints the FwJj14E footprints have a "more contracted proximal mid-foot," a "deeper instep," and "the location of the narrowest point of the instep" is more proximal (*ibid*, p. 1201). Bennett et al. (2009) suggest that the morphology of the FwJj14E hominin footprints is consistent with modern human-style walking notably an "an adducted hallux, a medial longitudinal arch" and a "medial weight transfer before push-off" (*ibid*, p. 1201). Bennett et al. (2009) also suggest that although other taxa (*Homo habilis*, *Paranthropus boisei*) cannot be ruled out as the makers of the prints the morphology and size of the Ileret hominin footprints "are most consistent with the large size and tall stature evident in some *Homo ergaster/erectus* individuals" (*ibid*, p. 1201). *See also* *Homo erectus*; **Koobi Fora**.

Koobi Fora Research Project Paleoanthropological research along Lake Turkana (formerly Lake Rudolf) began in 1968 when Richard **Leakey** led a Kenya National Museum (now the **National Museums of Kenya**) expedition comprising Margaret Leakey, Paul Abell, Kamoya **Kimeu**, and two students, John Harris and Bernard Wood, to explore the deposits along the eastern shore of Lake Rudolf. During this initial survey the group identified extensive fossil deposits and recovered several fragmentary hominin fossils. As a result a base camp was established in the following year at a site along the lake shore called Koobi Fora. New scientists joining the project in 1969 included Kay Behrensmeyer, Meave Epps (now Meave **Leakey**), and Glynn **Isaac**. Behrensmeyer began the task of mapping the geology of the site, particularly the volcanic tuffs that facilitated radiometric dating of the deposits. Discoveries included Oldowan artifacts and two crania belonging to *Paranthropus boisei* (**KNM-ER 406** and **KNM-ER 407**). In 1970 the **East Rudolf Research Project** was established under the leadership of Richard Leakey and Glynn Isaac and the number of participating scientists studying fossil mammals, including hominins, increased. A hominin mandible discovered in 1971 (**KNM-ER 992**) later became the type specimen of *Homo ergaster* and in 1972 Bernard Ngeneo discovered a hominin cranium (**KNM-ER 1470**) that was to become the **lectotype** of *Homo rudolfensis*. Two crania discovered in 1973 (**KNM-ER 1805** and **KNM-ER 1813**) were subsequently assigned to *Homo habilis*. Arguments over the dating of the KBS Tuff became a major issue since it had implications for the age of the KNM-ER 1470 cranium and of other hominins recovered from below the KBS Tuff and the controversy continued until its resolution in the early 1980s. In 1975 Lake Rudolf was renamed Lake Turkana by the Kenyan government and this prompted the East Rudolf Research Project to be renamed as the Koobi Fora Research Project and the first (**KNM-ER 3733**) of several hominin crania resembling *Homo erectus* was recovered. In 1981 the focus of research at Lake Turkana shifted from the east side to the west side of the lake and this led to the discovery in 1984 of a remarkably complete skeleton of an adolescent male at a locality called **Nariokotome**. The specimen (**KNM-WT 15000**) represents one of the most complete and earliest skeletons of *Homo erectus*. In 1985 researchers recovered the nearly complete cranium of a megadont archaic hominin (**KNM-WT 17000**) that was dated to *c*.2.5 Ma, making it substantially older than most other East African megadont archaic hominin fossils; it has been classified as either *P. boisei* or *Australopithecus aethiopicus*. The numerous geological, archeological, and paleoecological research projects undertaken during the 1980s (e.g., Kanapoi, Lomekwi, Lothagam) expanded the number of scientists working at Lake Turkana. Richard Leakey resigned from his position as director of the National Museums of Kenya in 1989 and Meave Leakey assumed leadership of research at Koobi Fora. The significance of the research carried out at Koobi Fora is exemplified by the publications in the *Koobi Fora Research Project* series. The six volumes in this series include *The Fossil Hominids and an Introduction to their Context, 1968–1974* (1978), *The Fossil Ungulates: Proboscidea, Perissodactyla, and Suidae* (1983), *The Fossil Ungulates: Geology, Fossil Artiodactyls, and Palaeoenvironments* (1991), *Hominid Cranial Remains* (1991), *Plio-Pleistocene Archaeology* (1997), and *The Fossil Monkeys* (2008).

Koobi Fora Ridge One of the designated subregions of the Koobi Fora region. *See* **Koobi Fora**.

Koobi Fora Tuff Complex *See* Okote Tuff Complex.

Kopwatu One of the Early and Middle Pleistocene sites in the exposures of the **Ola Bula Formation** within the **Soa Basin**, central Flores, Indonesia.

Koro Toro (Location 16°00′21″N, 18°52′34″E, Chad, Central Africa; etym. named after a local village) History and general description In a region called Bahr el Ghazal approximately 45 km/28 miles east of Koro Toro researchers identified more than 17 numbered localities where Pliocene fossiliferous sediments are exposed. Temporal span and how dated? 3.58±0.27 Ma based on **biostratigraphy** and **cosmogenic nuclide dating**. Hominins found at site There are four hominins from Koro Toro, three from locality KT 12 and one (a maxillary fragment) from KT 13. The three hominins from KT 12 are a mandibular fragment (**KT 12/H1**; the holotype of *Australopithecus bahrelghazali*), an upper premolar (KT 12/H2), and one (as yet unpublished) mandible fragment from the KT40 locality (Guy et al. 2008). Archeological evidence found at site None. Key references: Geology, dating, and paleoenvironment Brunet et al. 1995, 1996, Lebatard et al. 2008; Hominins Brunet et al. 1995, 1996, Guy et al. 2008; Archeology N/A.

Kossom Bougoudi One of the "fossiliferous areas" within the Chad Basin; others include **Kollé**, **Koro Toro**, and **Toros-Menalla**. *See also* **Chad Basin**.

Kostenki (Location 51°23′N, 39°02′E, Russia; etym. named after the nearby town) History and general description The left bank of the Don River near Voronezh contains 28 open-air early Upper Paleolithic sites, known collectively as the Kostenki-Borshchevo localities. These sites have been excavated by various workers since the late 1800s. The age of these sites has been much studied, since they represent some of the oldest modern human occupations of Europe. Temporal span and how dated? Several methods including **AMS radiocarbon dating** and optically stimulated **luminescence dating** have been attempted, as well as magnetostratigraphic and geostratigraphic correlation. The sum of these methods suggests that the earliest deposits date to *c*.48 ka, and the latest to *c*.30 ka. The earliest Upper

Paleolithic deposits date to *c*.45 ka. Hominins found at site No fossils are found in the oldest layers. Two isolated teeth were found in the earliest Upper Paleolithic layer and these are tentatively assigned to *Homo sapiens*. More complete *H. sapiens* remains have been recovered from the later Upper Paleolithic/Aurignacian deposits. Archeological evidence found at site The lowest layers at the site contain a local "transitional" technocomplex, like the **Szletian**, that retains many Middle Paleolithic traits. The excavators suggest this was created by Neanderthals, but there are no fossils found in these layers. The earliest Upper Paleolithic layers contain a distinctive **Aurignacian** deposit, including bone and ivory artifacts, possible art, and shells. The later layers also contain Aurignacian assemblages, dwelling structures made of bones, and clay figurines. Key references: Geology, dating, and paleoenvironment Anikovich et al. 2007; Hominins Sinitsyn et al. 1997; Archeology Anikovich et al. 2007, Sinitsyn et al. 1997.

Kostenki-Borshchevo *See* **Kostenki**.

Kotzetang One of the three main areas within Zhoukoudian Locality 1. *See* **Zhoukoudian Locality 1**.

Kow Swamp (Location 35°55′S, 144°19′E, Australia; etym. named after the local swamp where the burials were discovered) History and general description Between 1968 and 1972 researchers recovered the primary burials of 17 robust anatomically modern human individuals from a silt and sand deposit 3.2 km/2 miles south of Leitchville, Victoria. Temporal span and how dated? Initially dated to the **Late Pleistocene** to Early **Holocene** by **radiocarbon dating** on associated charcoal (9590 years BP) and directly from the bone apatite of two specimens (KS 1, 10,070 years BP, and KS 9, 9300 years BP). More recent ages using optically stimulated **luminescence dating** on associated sediments place the Kow Swamp people during the **Last Glacial Maximum** (or LGM) (i.e., *c*.22–19 ka). Hominins found at site The 17 Kow Swamp individuals represent the largest single population of Late Pleistocene *Homo sapiens* known in the world. Their remarkable robusticity initially led researchers to link these modern humans directly to earlier *Homo erectus* specimens from Java. But another explanation for their osteological robusticity is that it represents long-term (*c*.20 ka) adaptation to a

deterioration in climatic conditions that culminated in the LGM. The Kow Swamp skeletal remains were re-interred in 1990. Archeological evidence found at site Quartz artifacts. Key references: Geology, dating, and paleoenvironment Thorne and Macumber 1972. Stone and Cupper 2003; Hominins Thorne and Macumber 1972, Kennedy 1984a, Habgood 1991, Brown 2010; Archeology Thorne and Macumber 1972.

KP Abbreviation of Kanapoi and the prefix used for fossils recovered from Kanapoi, Kenya. *See* **Kanapoi**.

Krapina (Location 46°10′N, 15°52′E, Croatia; etym. the name of a nearby town) History and general description The site is a sandstone rock-shelter in Hušnjakovo Brdo (Hill) where fossil mammal bones were discovered by local people when quarrying for building sand and stone. Dragutin **Gorjanović-Kramberger** found a fossil hominin molar when he visited Krapina in August 1899. Formal excavations, which began in September 1899 and lasted until 1905, yielded approximately 900 hominin fossils, more than 1191 lithic artifacts, and 3000 animal bones. Temporal span and how dated? **Electron spin resonance spectroscopy dating** on hominin dental enamel and **uranium-series dating** on the fauna suggest a short depositional event at the end of **Oxygen Isotope Stage** (or OIS) 6 or the beginning of OIS 5e (*c*.130 ka). Hominins found at site The roughly 900 hominin fossils represent a probable minimum of 70 individuals. They include Kr 1 (juvenile calotte) also known as the "A cranium," Kr 2 (juvenile occipital and parietals) also known as the "B cranium," Kr 3 (partial cranium) also known as the "C cranium," Kr 2/5 (parietals), Kr 6 (partial cranium) also known as the "E cranium," Kr 46 (juvenile maxilla with teeth) also known as the "B maxilla"), and Kr 59 (mandible) also known as the "J mandible." Several hundred specimens are cranial vault fragments, isolated teeth, and postcranial elements. Archeological evidence found at site **Mousterian**. The Krapina site is significant for a number of reasons. Some of these are historical in that Gorjanović-Kramberger's 1906 monograph is the earliest detailed monographic analysis of a sample of *Homo neanderthalensis* and his use of plain film radiography and fluorine dating were the first applications of these techniques to fossil hominins.

Early on the Krapina sample played an important role in debates about variation in Neanderthals. Because the sample includes many anatomical elements it has proved an important source of information about the pattern of intraspecific variation and the ontogeny of Neanderthals (Monge 2008). Krapina has also been influential in the debate over differences between early and "classic" Neanderthals in Europe (Smith 1976b), the accelerated pattern of Neanderthal growth and development (Wolpoff 1979, Mann et al. 1991), Neanderthal hunting (Miracle 2007), and the practice of cannibalism by Neanderthals (White 2001). Key references: Geology, dating, and paleoenvironment Rink et al. 1995, Simek and Smith 1997; Hominins Gorjanović-Kramberger 1906, Smith 1976b, Wolpoff 1979, Radovčić et al. 1988, Monge et al. 2008; Archeology Gorjanović-Kramberger 1913, Simek and Smith 1997, Karavanić 2007, Miracle 2007.

Kromdraai (Location 26°00′0″S, 27°45′0″E, South Africa; etym. Afrikaans for a "crooked turn," referring to a bend in the road that runs along the Blaauwbank valley) History and general description Kromdraai is a complex of breccia-filled caves in Precambrian dolomite a mile or so east of **Sterkfontein** in the Blaauwbank valley in Gauteng Province, South Africa. In 1938 a local schoolboy, Gert Terblanche, found a hominin cranium at the site. He broke off all four of the teeth and gave one to the manager of the Sterkfontein mine (the same Mr Barlow who had managed the mine at **Taungs**), who showed it to Robert **Broom**. Broom recognized the tooth was unlike those he had collected at Sterkfontein so Broom visited Gert's school and Gert took Broom to Kromdraai and showed him where he had hidden the cranium that became the type specimen (**TM 1517**) of *Paranthropus robustus*. There are two fossiliferous localities within 30 m of each other: Kromdraai A (the "faunal site" or "KA") and Kromdraai B (the "hominid site" or "KB"). Three members, Members 1–3, are recognized in Kromdraai B. Temporal span and how dated? **Biostratigraphy** and **magnetostratigraphy** suggest that KB is *c*.2.0–1.5 Ma; Member 3 is likely between 1.78 and 1.65 Ma. Hominins found at site Thus far the majority of the hominin fossils have come from KB Member 3 and with only one exception they have been assigned to *P. robustus*. The exception, KB 5223, is a collection of decid-uous and permanent isolated teeth presumed to come from a single individual. Braga and Thackeray (2003) claim that the size, shape, and morphology of the M_1 and dm_1 crowns of KB 5223 differ from the characteristic morphology of the equivalent teeth of *P. robustus* and resemble that of teeth belonging to early **Homo**. However, a study of the enamel microstructure of the right I_1 and M_1 of this specimen suggest that the teeth are not developing like those of **Homo** and in some critical ways (e.g., enamel extension and secretion rates) their development is more like that of *Paranthropus* (Lacruz 2007); Moggi-Cecchi et al. (2010) are also not convinced there is dental evidence of *Homo* at Kromdraai. A recently reported fragment of left humeral shaft (Thackeray et al. 2005) may well belong to the type specimen. Archeological evidence found at site Stone artifacts recovered from KA have been classified as **Developed Oldowan/Early Acheulean** and residue analysis suggests that the hominins were using them to process bone. Key references: Geology, dating, and paleoenvironment McKee et al. 1995, Thackeray et al. 2002b, Herries et al. 2009; Hominins Thackeray et al. 2001, 2005, Braga and Thackeray 2003, Lacruz 2007; Archeology Kuman et al. 1997, Thackeray et al. 2005.

Kruskal–Wallis test A nonparametric statistical test (i.e., it does not assume normally distributed variables) for difference in medians between two or more groups that is analogous to the **analysis of variance** (or ANOVA).

KS Abbreviation of Kow Swamp and the prefix for the hominin fossils recovered from Kow Swamp. Also an abbreviation of Kokiselei, one of several groups of localities within the site of West Turkana named after ephemeral river drainages. *See* **Kokiselei**; **Kow Swamp**; **West Turkana**.

Ksar 'Akil (Location 33°9′N, 35°6′E, Lebanon; etym. Palace of Akil) History and general description This cave site in eastern Lebanon was excavated in the 1920s, 1940s, and 1970s, and represents a type site for the Upper Paleolithic of the northern Levant. The early Upper Paleolithic deposits may be among the earliest in Eurasia, and include shell beads. Temporal span and how dated? Radiocarbon dating was attempted during the most recent excavations. The Mousterian deposits were dated to *c*.43 ka, although this is probably an infinite date. Dates from the Upper Paleolithic layers average *c*.32 ka. The early Upper Paleolithic layers were not directly dated, but they may date to between 43 and 50 ka assuming constant rates of sediment accumulation. Hominins found at site None. Archeological evidence found at site Several layers that span from the late Middle Paleolithic through the Epipaleolithic were found in the 19 m-thick deposits. Key references: Geology, dating, and paleoenvironment Mellars and Tixier 1989; Hominins N/A; Archeology Azoury et al. 1988, Kuhn et al. 2001.

KSD-VP-1/1 Site Woranso-Mille. Locality KSD-VP-1. Surface/*in situ* Surface then *in situ*. Date of discovery February 10, 2005. Finder Alemayehu Asfaw, then excavation crew. Unit/horizon Mudstone. Bed/member N/A. Formation N/A. Group N/A. Nearest overlying dated horizon Presumed C2An2/C2An3n boundary. Nearest underlying dated horizon Tuff dated by **argon-argon dating** to 3.6±0.03 Ma. Geological age *c*.3.6–3.58 Ma. Developmental age Adult. Presumed sex Male. Brief anatomical description Cervical vertebrae, four ribs, pelvic bone, scapula, shaft, and distal end of the humerus and shaft and proximal end of the ulna of the right side, clavicle, second rib, part of the adjacent shaft and the distal femur, and whole tibia of the left side. Announcement, initial description, photographs/line drawings, and metrical data Haile-Selassie et al. 2010a. Detailed anatomical description N/A. Initial taxonomic allocation *Australopithecus afarensis*. Taxonomic revisions N/A. Current conventional taxonomic allocation *Au. afarensis*. Informal taxonomic category Archaic hominin. Significance This, the third adult associated skeleton of *Au. afarensis*, is substantially larger than A.L. 288-1 and thus enables researchers to get a sense of whether inferences based on the latter were influenced by any effects of scaling. Its describers interpret the preserved morphology as indicating that the "habitual bipedality" in *Au. afarensis* "was highly evolved." Location of original **National Museum of Ethiopia**, Addis Ababa, Ethiopia.

KT Acronym for the Kilaytoli Tuff and prefix for fossils from the Koro Toro "fossiliferous area", **Chad Basin**, Ethiopia. *See* **Koro Toro**.

KT 12/H1 Site **Koro Toro**. Locality KT 12. Surface/*in situ* Surface. Date of discovery January 23, 1995. Finders Not reported. Unit N/A. Horizon N/A. Bed/member N/A. Formation N/A. Group N/A. Nearest overlying dated horizon N/A. Nearest underlying dated horizon N/A. Geological age *c*.3.58±0.27 Ma. Developmental age Adult. Presumed sex Unknown. Brief anatomical description A section of the left side of the mandibular corpus including the symphysis. It extends from the distal end of the left P_4 to the distal end of the right P_4 including the crowns of the left $C–P_4$ and the right $I_2–P_4$. Announcement Brunet et al. 1995. Initial description Brunet et al. 1996. Photographs/line drawings and metrical data Brunet et al. 1995, 1996. Detailed anatomical description N/A. Initial taxonomic allocation *Australopithecus* aff. *Australopithecus afarensis*. Taxonomic revisions *Australopithecus bahrelghazali* Brunet et al. (1996). Current conventional taxonomic allocation *Au. bahrelghazali* or *Au. afarensis*. Informal taxonomic category Archaic hominin. Significance Although KT 12/H1 is the holotype of *Au. bahrelghazali* its real significance is that the Koro Toro region is 2500 km/1550 miles west of the East African Rift Valley so it was the first sound evidence of an early hominin from a site outside of East and southern Africa. Location of original N'Djaména, Chad.

Kulkuletti (Location 8°03′N, 38°15′E, Ethiopia; etym. not known) History and general description Artifact-bearing area in the Gademotta Formation on the flanks of a collapsed caldera about 5 km/3 miles west of Lake Ziway in Ethiopia's **Main Ethiopian Rift System**. The tephra at Kulkuletti and **Gademotta** (approximately 2 km/1.2 miles away) have been correlated using electron probe microanalysis and the combination of the younger sediments at Kulkuletti with those found at Gademotta provides one of the oldest and richest samples of **Middle Stone Age** (or MSA) artifact assemblages. Temporal span and how dated? **Argon-argon dating** of sanidine crystals in tephra and correlation on the basis of field stratigraphic evidence and geochemical similarity suggest the Unit 10 tuff above the **Acheulean** and the oldest MSA strata are 280±8 ka and the tuffaceous Unit D that overlies it has an age estimate of 180±10 ka. Hominins found at site None. Archeological evidence found at site Among the more than 8000 recovered artifacts are **Levallois flake**s, **point**s, and **core**s (many of them large), unifacially and bifacially **retouch**ed points, as well as blades and blade cores. These are sufficient to attribute the artifacts from the two excavated strata at Kulkuletti to the MSA. **Obsidian** from the volcanics beneath the Gademotta Formation sediments was used as the primary stone raw material and the site was likely either a **quarry** or a near-quarry workshop. No faunal evidence has been recovered. Key references: Geology, dating, and paleoenvironment Laury and Albritton 1975, Morgan and Renne 2008; Archeology Schild and Wendorf 2005, Wendorf and Schild 1974.

Kůlna (Location 49°25′N, 16°40′E, Czech Republic; etym. Czech for "shed") History and general description The potential importance of this cave site near Sloup village 35 km/22 miles north of Brno was recognized in 1880. It was first excavated in the late 19thC and early 20thC, but it was not until 1961 that a formal program of excavation was begun. In 1965 hominin remains were recovered from layer 7a in association with stone artifacts described as **Mousterian** of the Micoquian tradition. Temporal span and how dated? **Biostratigraphy**, conventional **radiocarbon dating** (>45 ka BP uncalibrated), and **electron spin resonance spectroscopy dating** (layer 7a, c.50 ka BP) all suggest a Last Interglacial to Last Glacial date. Hominins found at site A partial maxilla (with I^1–M^1), a right parietal fragment, and isolated teeth have all been attributed to *Homo neanderthalensis*. Archeological evidence found at site Mousterian (Taubachian, Micoquian) and **Upper Paleolithic** (Magdalenian). Geology, dating, and paleoenvironment Valoch 1988, Rink et al. 1996; Hominins Jelinek 1966, Smith 1982; Archeology Valoch 1967, 1968.

Kumpari *See* **Chongokni**.

KUS Abbreviation of Kuseralee Dora. *See* **Kuseralee Dora**; **Middle Awash study area**.

KUS-VP-2/100 Site **Kuseralee Dora**. Locality Kuseralee Dora VP Locality 2. Surface/*in situ* Surface. Date of discovery November 7, 1995. Finder Unknown. Unit N/A. Horizon N/A. Bed/member Lower Aramis Member. Formation **Sagantole Formation**. Group N/A. Nearest overlying dated horizon Daam-Aatu Basaltic Tuff.

Nearest underlying dated horizon Gaala Vitric Tuff Complex. Geological age c.4.4 Ma. Developmental age Adult. Presumed sex Unknown. Brief anatomical description Associated upper dentition comprising the right C, the right and left M^1, the left M^2, plus parts of the crowns of the right P^3 and P^4. Announcement White et al. 2009a. Initial description N/A. Photographs/line drawings and metrical data N/A. Detailed anatomical description N/A. Initial taxonomic allocation *Ardipithecus*. Taxonomic revisions None. Current conventional taxonomic allocation *Ar. ramidus*. Informal taxonomic category Possible primitive hominin. Significance An associated dentition that includes both anterior and postcanine teeth. Location of original National Museum of Ethiopia, Addis Ababa.

Kuseralee Dora A concentration of fossil localities situated in the **Central Awash Complex** in the Afar Rift section of the East African Rift System in the **Middle Awash study area**, Ethiopia. Remains attributed to *Ardipithecus ramidus* have been found here.

k.y. (L. *kilo*=thousand) The abbreviation for a period of time expressed in thousands of years, as in "30 k.y. have elapsed since anatomically modern humans first reached Europe." The result of an age determination, or **age estimate**, should be expressed differently, as in "the age of the anatomically modern human crania from the Omo is c.190 ka," but many contemporary reports use the abbreviation "kyr." *See also* **age estimate**.

kyphosis (Gk *kyphos*=bent forwards) The term kyphosis and its antonym **lordosis** refer to the angular relationship within or between sections of a compound structure (e.g., the vertebral column) or between two planes (e.g., the anterior and posterior components of the cranial base). Thus, the parts of the adult modern human vertebral column that are concave anteriorly (i.e., the thoracic and sacral) are referred to as being kyphotic, whereas the parts of the adult modern human vertebral column that are concave posteriorly (i.e., the cervical and lumbar regions) are referred to as being lordotic. With respect to the cranial base, organisms, or individuals with a smaller, more acute (i.e., more modern human-like) **cranial base angle** are said to have a more kyphotic **basicranium**, whereas organisms or individuals with a larger, more obtuse (i.e., more chimpanzee-like) cranial base angle have a more lordotic basicranium.

kyr (L. *kilo*=thousand) The way many contemporary reports abbreviate a period of time expressed in thousands of years, as in "30 kyr have elapsed since anatomically modern humans first reached Europe". The result of an age determination, or **age estimate**, should be expressed differently, as in "the age of the anatomically modern human crania from the Omo is *c.*190 ka." *See also* **age estimate**.

L

L. Prefix used by the American contingent of the **International Omo Research Expedition** for localities in the **Shungura Formation** (e.g., L. 396). The prefix is also used for fossils found within one of the L. localities (e.g., **L. 40-19**).

L. 7a-125 Site **Omo-Shungura**. Locality 7-a. Surface/*in situ* Surface. Date of discovery 1968. Finder Not known. Unit N/A. Horizon Upper G 5. Bed/member Member G. Formation **Shungura Formation**. Group Omo Group. Nearest overlying dated horizon Tuff H. Nearest underlying dated horizon Tuff G. Geological age *c*.2.2 Ma. Developmental age Presumed adult. Presumed sex Unknown. Brief anatomical description A mandibular corpus that extends to just distal of the M₃ on both sides and includes the crowns of the right I₂–M₃ and the crowns of the left C–M₃ in different degrees of completeness. Announcement Howell 1969a. Initial description Howell 1969a. Photographs/line drawings and metrical data Howell 1969a. Detailed anatomical description N/A. Initial taxonomic allocation *Australopithecus* cf. *boisei*. Taxonomic revisions *Australopithecus boisei* (Howell 1978, Coppens 1980). Current conventional taxonomic allocation *Paranthropus boisei*. Informal taxonomic category Megadont archaic hominin. Significance Although the extensive matrix-filled cracking makes it difficult to be sure of what some of the pristine morphology would have been like, there is little doubt that this is one of the largest *P. boisei* mandibles in terms of the size of the corpus, yet it has one of the smallest canine crowns of known *P. boisei*. Location of original National Museum of Ethiopia, Addis Ababa, Ethiopia.

L. 40-19 Site **Omo-Shungura**. Locality L. 40. Surface/*in situ* "Found almost *in situ*, outcropping from E-5" (de Heinzelin 1983, p. 244). Date of discovery 1971. Finder Gerry Eck. Unit Sand. Horizon E-5. Bed/member Member E. Formation

Shungura Formation. Group Omo Group. Nearest overlying dated horizon Tuff F. Nearest underlying dated horizon Tuff E. Geological age *c*.2.35 Ma. Developmental age Adult. Presumed sex Unknown. Brief anatomical description Nearly complete right ulna. The edges of the proximal articular surface are broken away and there are many linear cracks that limit the extent to which the size and shape of the shaft can be reconstructed, but any length measurements are liable to be accurate. Announcement Howell and Wood 1974. Initial description Howell and Wood 1974. Photographs/line drawings and metrical data Howell and Wood 1974. Detailed anatomical description Howell and Wood 1974. Initial taxonomic allocation *Australopithecus boisei*. Taxonomic revisions *Australopithecus* cf. *aethiopicus/boisei* (Howell et al. 1987), *Homo* (Aiello et al. 1999), *Paranthropus aethiopicus* (McHenry et al. 2007). Current conventional taxonomic allocation *Paranthropus boisei*. Informal taxonomic category Megadont archaic hominin. Significance The fact that researchers have difficulty assigning this relatively complete ulna to a taxon is a reflection of how little we know about the upper limb of 2–3 Ma-old hominins. McHenry et al. (2007) accepted Howell and Wood's (1974) assignment of L. 40-19 to *P. boisei*, but they also accept Aiello et al.'s (1999) assignment of OH 36 to *P. boisei* and thus conclude that the forelimb of *P. boisei* is highly variable. Location of original National Museum of Ethiopia, Addis Ababa, Ethiopia.

L. 74a-21 Site **Omo-Shungura**. Locality 74-a (aka P 420). Surface/*in situ* Surface. Date of discovery 1968. Finder N/A. Unit N/A. Horizon G-3 to 5. Bed/member Member G. Formation **Shungura Formation**. Group Omo Group. Nearest overlying dated horizon Tuff H. Nearest underlying dated horizon Tuff G. Geological age *c*.2.2 Ma. Developmental age Presumed adult. Presumed sex Unknown. Brief anatomical description

The right side of the corpus of an adult mandible that extends from a fracture through the right I$_2$ alveolus to an oblique fracture that passes though the M$_2$ alveolus and the base of the ramus so that the inferior part of the anterior edge of the ramus is preserved but the angle is not. The crowns of the canine and P$_4$ are the only ones preserved but clean cross-sections through the roots of P$_3$ and M$_1$ are exposed. Announcement Howell 1969a. Initial description Howell 1969a. Photographs/line drawings and metrical data Howell 1969a. Detailed anatomical description N/A. Initial taxonomic allocation *Australopithecus* cf. *boisei*. Taxonomic revisions *Au. boisei* (Howell 1978a, Coppens 1980). Current conventional taxonomic allocation *Paranthropus boisei*. Informal taxonomic category Megadont archaic hominin. Significance One of the earliest known mandibles of *P. boisei*. Location of original National Museum of Ethiopia, Addis Ababa, Ethiopia.

L. 338y-6 Site **Omo-Shungura**. Locality L. 338y* [*Omo-Shungura locality letter postfixes are listed in Howell and Coppens (1974) in capital letters, but Rak and Howell (1978) use lower case]. Surface/*in situ In situ*. Date of discovery Jean **de Heinzelin** recognized the occipital in a geological section and the two parietals were recovered from excavations conducted in 1969 by Gerry Eck, but subsequent excavations in the 1970–2 field seasons recovered no more of the hominin calvaria. Unit E-3. Horizon See above. Bed/member Member E. Formation **Shungura Formation**. Group Omo Group. Nearest overlying dated horizon Tuff F. Nearest underlying dated horizon Tuff D-3. Geological age *c*.2.4 Ma. Developmental age Juvenile (the spheno-occipital synchondrosis is unfused). Presumed sex Male on the basis of the precocious development of the ectocranial crests. Brief anatomical description All that remains of this incomplete juvenile calvaria is most of an occipital, both parietals, and a small fragment of the left side of the frontal, but these few elements are close to being perfectly preserved. Endocranial volume Approximately 420 cm^3. Announcement Howell and Coppens 1974. Initial description Rak and Howell 1978. Photographs/line drawings and metrical data Rak and Howell 1978. Detailed anatomical description Rak and Howell 1978. Initial taxonomic allocation *Australopithecus boisei*. Taxonomic revisions Holloway (1981a) suggested that some features of the endocranial

morphology of the L. 338y-6 calvaria aligned it more with *Australopithecus africanus* or *Australopithecus afarensis* than with *Au. boisei*, but Leakey and Walker (1988) point out that many of these same features are seen in the **KNM-WT 17000** cranium. Current conventional taxonomic allocation *Paranthropus boisei*. Informal taxonomic category Megadont archaic hominin. Significance Despite being a juvenile the L. 338y-6 calvaria already manifests features seen in most adult *P. boisei* crania [e.g., the heart-shaped, or cardioid, foramen magnum, *striae parietalis*, temporal lines closest anteriorly, etc., but *contra* White and Falk (1999) it does *not* have an **occipitomarginal system** of endocranial venous drainage). The calvaria provides important information about the ontogeny of *P. boisei* crania and these data were put to good use by Rak and Howell (1978). One of the two round holes in the anterior third of the right parietal is consistent with it being made by the canine of a large carnivore. Location of original National Museum of Ethiopia, Addis Ababa, Ethiopia.

L. 894-76-898 Site **Omo-Shungura**. Locality L. 894. Surface/*in situ* Surface and then *in situ*. Date of discovery 1973 and then more recovered from excavations later that year and in 1974. Finders J. Mathias, J. Kithumbi, and Clark **Howell**. Unit "Silty sand". Horizon G-28. Bed/member Member G. Formation **Shungura Formation**. Group Omo Group. Nearest overlying dated horizon Tuff H$_2$. Nearest underlying dated horizon Tuff G. Geological age *c*.2.0 Ma. Developmental age Adult. Presumed sex Unknown. Brief anatomical description Fragments that make up approximately 30% of the cranium, including a fragment of the right maxilla, the right temporal that includes the mandibular fossa and the zygomatic arch, and a fragment of the vault around the left asterion. No anterior tooth crowns are preserved, but all but the left P^4 of the postcanine tooth crowns are preserved. Announcement Boaz and Howell 1977. Initial description Boaz and Howell 1977. Photographs/line drawings and metrical data Boaz and Howell 1977. Detailed anatomical description Boaz and Howell 1977. Initial taxonomic allocation Boaz and Howell (1977) suggest that Omo L. 894-76-898 is an "an early species of *Homo*" (*ibid*, p. 105). The authors claim the partial cranium "most closely parallels Old. Hom. 24 and Old. Hom. 13" (*ibid*, p. 105) and thus by implication they assign it to *Homo habilis*. Taxonomic revisions None. Current conventional taxonomic allocation *H. habilis*. Informal taxonomic category Transitional hominin. Significance If this cranium does belong to *H. habilis* then it would be among the early evidence of that taxon and one of a small number of specimens from the Shungura Formation assigned to early *Homo*. Location of original National Museum of Ethiopia, Addis Ababa, Ethiopia.

L. 996-17 Site Omo-Shungura. Locality L 996 (this was originally a "P" locality, so some references refer to this fossil as "P. 996-17"). Surface/*in situ* Surface "on Upper Member K" (de Heinzelin 1983, p. 303). Date of discovery 1971. Finder N/A. Unit N/A. Horizon N/A. Bed/member Member K or older. Formation **Shungura Formation**. Group Omo Group. Nearest overlying dated horizon N/A. Nearest underlying dated horizon Tuff K. Geological age *c*.1.4 Ma. Developmental age Presumed adult. Presumed sex Unknown. Brief anatomical description Two cranial vault fragments, one "parietal," and one near pterion that includes parts of the "frontal, sphenoid and parietal" (Coppens 1980, p. 219). Announcement Coppens 1980. Initial description Coppens 1980. Photographs/line drawings and metrical data N/A. Detailed anatomical description N/A. Initial taxonomic allocation *Homo erectus*. Taxonomic revisions None. Current conventional taxonomic allocation *H. erectus*. Informal taxonomic category Pre-modern *Homo*. Significance These are fragments from a thick cranium (approximately 11.5 mm thick at lambda) and if it does belong to *H. erectus* it would be the best evidence of this taxon in the Shungura Formation. Location of original National Museum of Ethiopia, Addis Ababa, Ethiopia.

L'Homme de Tautavel *See* **Caune de l'Arago**.

LA Abbreviation of Lokalalei, one of several groups of localities within the site of **West Turkana** named after ephemeral river drainages. *See* **Lokalalei**.

La Chaise (Location 45°40′12″N, 0°26′48″E, France; etym. named after the nearby town) History and general description Discovered in 1850, the La Chaise cave complex is near Angoulême. There are three main galleries: Abri Suard, Abri Bourgeois-Delaunay, and Grotte Duport (all named after the people who found them); the first two are the best-studied. Formal excavations were by P. David between 1930 and 1963 and he found several hominin remains in the Abri Suard. André Debénath excavated between 1967 and 1983 and found more hominin remains in both the Abri Suard and the Abri Bourgeois-Delaunay. Temporal span and how dated? **Thermoluminesence dating** and **uranium-series dating** suggest that Abri Suard dates to between 185 and 101 ka and Abri Bourgeois-Delaunay to between 146 and 106 ka. Hominins found at site Several cranial and postcranial fragments assigned to *Homo neanderthalensis* were recovered from the lower layers (Debénath's layer 11) of Abri Bourgeois-Delaunay. The fossils from the Abri Suard are older and are described as being either Neanderthal or **pre-Neanderthal**. Many of the fossils in the Abri Suard come from juvenile individuals. Archeological evidence found at site The Abri Bourgeois-Delaunay contained **Mousterian** and **Aurignacian** deposits. Artifacts in the older deposits in Abri Suard have been assigned to the late **Acheulean** and the Mousterian. Key references: Geology, dating, and paleoenvironment Debénath 1974, Schvoerer et al. 1977, Schwarcz and Debénath 1979, Blackwell et al. 1983; Hominins Debénath 1974, 1977, Genet-Varcin 1974, 1975a, 1975b, 1976; Archeology Debénath 1977, Matilla 2004, 2005.

La Chaise-de-Vouthon *See* **La Chaise**.

La Chapelle-aux-Saints (Location 44°59′N, 1°43′E, France; etym. named after a nearby village) History and general description The cave of **Bouffia Bonneval**, near to the village of La Chapelle-aux-Saints, is located along the Sourdoire river, a tributary of the Dordogne, in the Corrèze, France. It is a small, shallow cave, no more than 4 m wide. Excavations at the site by A. Bouyssonie, J. Bouyssonie, and L. Bardon began in 1905 and they extended approximately 6 m into the cave. The site is best known for the nearly complete skeleton of *Homo neanderthalensis* recovered in 1908. Temporal span and how dated? **Electron spin resonance spectroscopy dating** of the hominin skeleton suggested an age of either *c*.47 ka or 56 ka depending on the radiation-uptake model used. Hominins found at site A single aged male hominin skeleton, **La Chapelle-aux-Saints 1**, was buried at the base of the infill that contained the archeological assemblage. Archeological evidence found at site The excavators described only one archeological level in which they found a **Mousterian** assemblage. It was dominated by **scrapers**, but it also included **handaxes**, **discoids**, and **points**. The types and relative quantities of artifacts matched the Quina Mousterian assemblage (Bouyssonie et al. 1913). Although the excavators recovered many animal bones, none of them were worked. The fauna was dominated by *Rangifer tarandus*, but it also included *Bos/Bison*, *Equus*, *Capra ibex*, *Canis lupus*, *Canis vules*, *Rhinoceros*, *Hyena spelaea*, *Sus scrofa*, and *Arctomys*. Bouyssonie et al. (1913) suggested the cave sediments were laid down during a cold period. Key references: Geology, dating, and paleoenvironment Grün and Stringer 1991; Hominins Boule 1908, 1909, 1911, 1912, Trinkaus 1985, Heim 1989; Archeology Bouyssonie et al. 1908, 1913, Boule 1911.

La Chapelle-aux-Saints 1 Site La Chapelle-aux-Saints.

Locality Mousterian layer. Surface/*in situ* In situ. Date of discovery 3 August 1908. Finders A. Bouyssonie, J. Bouyssonie, P. Bouyssonie, and L. Bardon. Unit N/A. Horizon N/A. Bed/member N/A. Formation N/A. Group N/A. Nearest overlying dated horizon N/A. Nearest underlying dated horizon N/A. Geological age **Electron spin resonance spectroscopy** dating of the hominin skeleton suggested an age of either *c.*47 ka or 56 ka depending on the radiation-uptake model used. Developmental age *c.*40–45 years. Presumed sex Male. Brief anatomical description Nearly complete cranium and mandible with all of the postcanine teeth lost pre-mortem. There is significant resorption of the alveolar process of the mandible. Many postcranial elements are present, including 21 vertebrae, 20 ribs, a clavicle, both humeri, both radii (though fragmentary), both ulnae, some hand bones, pieces of the pelvic bone, both femora (though fragmentary), both patellae, both tibae, an astragalus, a calcaneus, five right metatarsals, fragments of left metatarsals, and one pedal phalanx. Announcement Bouyssonie et al. 1908. Initial description Boule 1908, 1909. Photographs/line drawings and metrical data Boule 1908, 1909, 1911, 1912, Trinkaus 1985, Heim 1989. Detailed anatomical description Boule 1908, 1909, 1911, 1912, Tappen 1985, Trinkaus 1985, Heim 1989. Initial taxonomic allocation *Homo neanderthalensis*. Taxonomic revisions None. Current conventional taxonomic allocation *Homo neanderthalensis*. Informal taxonomic category Pre-modern *Homo*. Significance Marcellin **Boule**'s original reconstruction of this fossil (1911, 1912) portrayed this individual as a stooped and hunched, which fostered the stereotype of Neanderthals as substantially less-evolved, both intellectually and physically, than modern humans. The reconstruction was corrected to a more upright posture in the 1950s (Straus Jr and Cave 1957). Other researchers, noting that this individual had lost all of its molar teeth and had widespread severe degenerative joint disease, and thus may not have been able to chew or hunt, argued that this was evidence that Neanderthals had a complex social system that involved care for less able individuals (Jolly and Plog 1982). Location of original **Musée de l'Homme**, Paris.

La Fate *See* **Caverna delle Fate**.

La Ferrassie (Location 44°57′N, 0°56′E, France; etym. named after a nearby village)

History and general description The site of La Ferrassie, which is located in the Vézère valley between the towns of Bugue and Sauvignac-de-Miremont in the Dordogne, consists of a large and deep cave with a small rock-shelter to one side and a long rock-shelter to the other, all hollowed out from a south-facing cliff, overlooking a valley with a perennial spring. It was first explored by archeologists in 1896, when M. Tabanou excavated a small area of the site. In 1902, Denis Peyrony began systematic excavations that resulted in the removal of all the sediments in the shallow rock-shelter and the cave, and much of the talus slope below the longer rock-shelter was also excavated. Delporte and colleagues returned to the site in the late 1960s and early 1970s to focus on the **Aurignacian** and later levels. Temporal span and how dated? The dating of this site is problematic due to some mixing of the levels. Current dates rely on **radiocarbon dating** of fauna, **chronostratigraphic** correlations to other sites, and some **electron spin resonance spectroscopy dating** (or ESR). The oldest, **Mousterian**, levels have been correlated with the Ferrassie Mousterian at **Combe Grenal**, suggesting a date between 68 and 74 ka, which is equivalent to the transition between **Oxygen Isotope Stage** (or OIS) 5 and OIS 4. ESR ages on bovid teeth that may be from the Mousterian levels have been difficult to interpret, but they suggest a general age of OIS 3. Radiocarbon dates from the later levels suggest the site was abandoned *c*.27 ka (i.e., during the **Gravettian** period). Hominins found at site The remains of eight hominin individuals were found at this site over the period from 1909 to 1921. There were two adults (**La Ferrassie 1** and **La Ferrassie 2**), three juveniles, one neonate, and two fetuses, at least one of which was probably intentionally buried with a possible grave good. One of the juveniles (**La Ferrassie 6**) was secondarily buried. Archeological evidence found at site The small rock-shelter and the cave contained archeological material from the Mousterian, Aurignacian, and Gravettian periods, and the talus slope below the longer rock-shelter contained Mousterian, **Châtelperronian**, Aurignacian, and Gravettian levels. Key references: Geology, dating, and paleoenvironment Capitan and Peyrony 1912a, 1912b, Mellars 1986, Texier 2001, Delpech and Rigaud 2001, Delpech 2007; Hominins Capitan and Peyrony 1912a, 1912b, Heim 1970, 1974a, 1974b, 1976, 1982a, 1982b; Archeology Capitan and Peyrony 1912a, 1912b, Peyrony 1934, Delporte 1969, 1978, Delporte and Tuffreau 1972/3, Delporte et al. 1977, 1983.

La Ferrassie 1 Site La Ferrassie. Locality **Mousterian** layer. Surface/*in situ In situ.* Date of discovery September 1909. Finders Peyrony and Capitan. Unit N/A. Horizon N/A. Bed/member N/A. Formation N/A. Group N/A. Nearest overlying dated horizon N/A. Nearest underlying dated horizon N/A. Geological age *c*.68–74 ka. Developmental age *c*.45 years. Presumed sex Male. Brief anatomical description Well-preserved cranium with most of the face and nearly complete dentition, and many well-preserved postcranial elements, including 23 vertebrae, sacrum, 20 ribs, both clavicles, scapulae, humeri, radii, ulnae, pelvis, both femora, tibae, fibulae, and some hand and foot bones. Announcement Capitan and Peyrony 1909. Initial description Boule 1921. Photographs/line drawings and metrical data Boule 1921, Heim 1970, 1974a, 1974b, 1976, 1982a. Detailed anatomical description Heim, 1970, 1974a, 1974b, 1976, 1982a. Initial taxonomic allocation *Homo neanderthalensis.* Taxonomic revisions None. Current conventional taxonomic allocation *Homo neanderthalensis.* Informal taxonomic category Pre-modern *Homo.* Significance One of the more complete Neanderthal skeletons, and, together with **La Ferrassie 2** and **La Chapelle-aux-Saints 1**, it

was used by Marcellin **Boule** to reconstruct the morphology of Neanderthals. Although the anterior teeth of this individual were incorrectly placed in the reconstruction, there is evidence they may have played a non-masticatory role (Wallace 1975, Puech 1981). A piece of animal bone with incised marks on the surface was found in association with this individual, and Capitan and Peyrony (1909) suggested this was a grave good. Location of original **Musée de l'Homme**, Paris.

La Ferrassie 2 Site La Ferrassie. Locality **Mousterian** layer. Surface/*in situ In situ.* Date of discovery September 1910. Finders Peyrony and Capitan. Unit N/A. Horizon N/A. Bed/member N/A. Formation N/A. Group N/A. Nearest overlying dated horizon N/A. Nearest underlying dated horizon N/A. Geological age *c*.68–74 ka. Developmental age *c*.25–30 years. Presumed sex Female. Brief anatomical description Fragmented cranium, though with most pieces represented, maxilla with six teeth, and three loose mandibular teeth. Well-preserved and complete postcranial skeleton, with 11 vertebrae, 19 ribs, both scapulae, humeri, one radius, one ulna, the pelvis, both femora, patellae, tibae, fibulae, and parts of both hands and both feet. Announcement Capitan and Peyrony 1911. Initial description Boule 1921. Photographs/line drawings and metrical data Boule 1921, Heim 1970, 1974a, 1974b, 1976, 1982a. Detailed anatomical description Heim 1970, 1974a, 1974b, 1976, 1982a. Initial taxonomic allocation *Homo neanderthalensis.* Taxonomic revisions None. Current conventional taxonomic allocation *Homo neanderthalensis.* Informal taxonomic category Pre-modern *Homo.* Significance This is one of the more complete Neanderthal skeletons, and, together with **La Ferrassie 1** and **La Chapelle-aux-Saints 1**, it was used by Marcellin **Boule** to make his reconstruction of Neanderthal morphology. Location of original **Musée de l'Homme**, Paris.

La Ferrassie 6 Site La Ferrassie. Locality **Mousterian** layer. Surface/*in situ In situ.* Date of discovery 1921. Finders Peyrony and Capitan. Unit N/A. Horizon N/A. Bed/member N/A. Formation N/A. Group N/A. Nearest overlying dated horizon N/A. Nearest underlying dated horizon N/A. Geological age *c*.68–74 ka. Developmental age *c*.3–5 years. Presumed sex Unknown. Brief anatomical description Well-preserved, nearly complete skeleton of a juvenile, including 16–20 vertebrae, 11 ribs, both humeri, radii, ulnae, the pelvis, both femora, tibae, one patellac, one fibula, and some hand and foot bones. The cranium is badly damaged and fragmentary, and was separated from the body. Announcement Capitan and Peyrony 1921. Initial description Capitan and Peyrony 1921. Photographs/line drawings and metrical data Capitan and Peyrony 1921, Peyrony 1934, Heim 1982b. Detailed anatomical description Heim 1982b. Initial taxonomic allocation *Homo neanderthalensis*. Taxonomic revisions None. Current conventional taxonomic allocation *Homo neanderthalensis*. Informal taxonomic category Pre-modern *Homo*. Significance The cranium of this individual appears to have been secondarily separated from the skeleton and reburied 1.25 m away. Peyrony and Capitan (1921) also found a large calcite stone with several cupules engraved on its surface overlying the grave of this individual, which they considered to be an intentional grave good. Location of original **Musée de l'Homme**, Paris.

La Madeleine (Location 44°58′05″N, 1°01′52″E, France; etym. named after a nearby village) History and general description La Madeleine is an extensive rock-shelter in a south-facing cliff along the floodplain of the Vézère river in the commune of Tursac in the Dordogne region. It was first explored by Lartet and Christy in 1863, and fully excavated by Peyrony and Capitan in 1921–2 (Capitan and Peyrony 1928). These investigations were followed up by Bouvier and others in attempt to better understand the stratigraphy of the site. The unique stone tool technology found at this site was named the **Magdalenian**, and examples of this technocomplex are found throughout western Europe. Temporal span and how dated? The Archeological layers have not been directly dated, but the Magdalenian burial of a child was dated by **radiocarbon dating** to *c.*10,190 years BP. This date is young when compared with other Magdalenian occupations, and it is unclear whether the burial is intrusive or if the dating was somehow compromised. Hominins found at site Several hominin remains were found at the site, but only two were more than small fragments. Near the middle of the deposit Lartet and Christy found the fragmentary remains of anatomically modern human, including parts of the skull, the mandible and a few long bones (Lartet and Christy 1864). They believed this was an intentional burial, even though it was not associated with any grave goods. The most famous fossil from La Madeleine is that of

a child (**LM4**) found in the uppermost layer, where several thousand pierced shells and other decorations were found with the burial (Capitan and Peyrony 1928, Peyrony 1928, Heim 1991, Vanhaeren and d'Errico 2001). Archeological evidence found at site Lartet and Christy's excavation took place over a brief period in 1863, and warranted only a two-page description of the site (Lartet and Christy 1864). They excavated a deposit that was 15 m by 5 m in area, and between 2.5 and 3 m deep, with what they described as a homogeneous fauna and tool typology. They found both bone and stone tools, including harpoons and carved pieces, and a large quantity of flint tools, including knives and scrapers. Capitan and Peyrony removed the backfill of the preceding excavations, and in the undisturbed areas they recognized three main sedimentary layers. Although later archeologists described further subdivisions within each of these layers, the overall pattern of these three layers and the archeological finds from each layer came to define the main stages within the classical Magdalenian period (Bouvier 1973). In sum, the Magdalenian is known for its bone harpoons, large quantities of stone burins, and bones and antlers decorated with engravings, mostly of animals. Key references: Geology, dating, and paleoenvironment Gambier et al. 2000, Bouvier 1973; Hominins Heim 1991; Archeology Lartet and Christy 1864, Capitan and Peyrony 1928, Bouvier 1973, Vanhaeren and d'Errico 2001.

La Naulette (Location 50°12′46″N, 4°55′53″E, Belgium) History and general description La Naulette is 1 km/0.6 miles southwest of Chaleux on the left bank of the River Lesse. Hominin remains were found in a series of campaigns by Dupont in 1866. Temporal span and how dated? The remains have not been dated, but were thought by Dupont to be older than "classic" *Homo neanderthalensis* because of their stratigraphic location. Hominins found at site An incomplete mandible with all teeth lost postmortem, an ulna, and a third metatarsal of undiagnosed species. The mandible, thought to belong to a female, possesses primitive characteristics (e.g., the symphyseal region) but it does not show any *H. neanderthalensis* characters, such as a retromolar space. Metrically the mandible falls between the variation of Neanderthals and their predecessors. The ulna and third metatarsal, on the other hand, are said to be morphologically and metrically modern (although there is overlap with Neanderthals).

There is a possibility that the postcranial remains are intrusive but only direct dating will answer this question. Archeological evidence found at site No stone tools are associated with the human remains. Key references: Hominins Dupont 1866, 1867, 1872.

La Quina (Location 45°30′N, 0°17′E, France; etym. named after a nearby town) History and general description La Quina is located in the Charente region, southeast of the city of Angoulême. It is made up of a series of rock-shelters in and at the base of limestone cliffs lining a floodplain of the Voultron River. There were two main archeological accumulations. The main one, Station Amont, consisted of a roughly 7 m-deep deposit that extended more than 100 m along the base of the cliff. This deposit contained several **Mousterian** layers and yielded many faunal remains and stone tools, and more than 50 fragments from *Homo neanderthalensis* individuals. About 200 m to the southwest, the smaller accumulation, called Station Aval, contained Mousterian, **Châtelperronian**, and **Aurignacian** levels (Henri-Martin 1961). The upper area was first explored between 1881 and 1885 by Chauvet and Ramonet, then Henri Martin more thoroughly excavated both areas from 1905 to 1935; thereafter Martin's daughter, Germaine Henri-Martin, continued the excavations from 1953 to the late 1960s (Henri-Martin 1964), but she did not publish a comprehensive summary of her findings, nor was the stratigraphic context clear. After her death, the site was given to the Museé des Antiquités Nationales of France, and Arthur Jelinek and André Debénath have continued excavations at the site since 1985. Temporal span and how dated? Débenath and Jelinek dated two of their layers using **thermoluminescence dating** on burned flints to *c*.43 ka and *c*.48 ka, but these layers correspond to Martin's upper Mousterian levels, suggesting that the majority of the Station Amont is much older. Others have suggested dates corresponding to **Oxygen Isotope Stages** 3–4. Hominins found at site Fifty-three remains of Neanderthals were recovered in the Station Amont, but most of these are very fragmentary. One fairly complete Neanderthal skeleton, **La Quina H5**, was recovered and possibly represents an intentional burial. Archeological evidence found at site In the Station Amont Henri Martin described four Mousterian layers that graded into

each other. He interpreted these finds to indicate a local development and improvement in technology, that eventually included very rare bone tools and pierced teeth and bone pieces (Martin 1910). The stone tool technocomplex found in Martin's middle Mousterian layer became known as the Quina Mousterian, and is found at sites throughout Europe. Debénath and Jelinek (1998) excavated a small witness baulk and areas to the north and south of Martin and Henri-Martin's excavations in the Station Amont. While they also found only Mousterian deposits, their stratigraphy differs from that of Martin, making correlation between the new and old excavations difficult. The excavation of the Station Aval revealed a Mousterian level, a Châtelperronian level, and a "typical" Aurignacian level. Key references: Geology, dating, and paleoenvironment Debénath and Jelinek 1998; Hominins Martin 1911a, Martin 1911b, Martin 1912a, Martin 1912b, Martin 1913, Verna 2006, Verna et al. 2010; Archeology Henri-Martin 1961, Jelinek et al. 1989, Debénath and Jelinek 1998, Chase et al. 1994, Hardy 2004.

La Quina H5 Site La Quina. Locality N/A. Surface/*in situ In situ*. Date of discovery 18 Sept 1911. Finder Germaine Henri-Martin. Unit N/A. Horizon N/A. Bed/member Quina **Mousterian** Layer, or Martin's layer 3. Formation N/A. Group N/A. Nearest overlying dated horizon N/A. Nearest underlying dated horizon N/A. Geological age >*c*.48 ka. Developmental age Adult. Presumed sex Unknown. Brief anatomical description Cranium, mandible, and fragments of cervical vertebrae, both scapulae, both clavicles, both humeri, one ulna, both femora. Announcement Martin 1911a. Initial description Martin 1911a. Photographs/line drawings and metrical data Martin 1912a. Detailed anatomical description Martin 1912a. Initial taxonomic allocation *Homo neanderthalensis*. Taxonomic revisions N/A. Current conventional taxonomic allocation *H. neanderthalensis*. Informal taxonomic category Pre-modern *Homo*. Significance The two humeri assigned to this individual varied in circumference by about 11 mm, suggesting pathology of the right arm. Location of original **Musée de l'Homme**, Paris.

La Roche à Pierrot *See* **Saint-Césaire**.

Labeko Koba (Location 43°03′N, 2°29′W, Spain, etym. Basque for "baking caves") History and general description This cave site is located in Guipúzcoa on the coast of the Basque Country of Spain. It was excavated between September 1987 and December 1988 primarily by Alvaro Arrizabalaga and Jesús Altuna as part of a salvage operation prior to the construction of a highway. Although the site was heavily affected by damage from water flow many of the discovered artifacts display signs of decoration and artistic behavior. Temporal span and how dated? Conventional radiocarbon dating of bone from the oldest and youngest layers have given dates of 29,750±740 to 34,215±1265 years BP. Hominins found at site None. Archeological evidence found at site Many bone and lithic artifacts were found in the five archeological layers. These layers record the start of the Upper Paleolithic in Cantabria, including the Châtelperronian, Proto-Aurignacian, and the Aurignacian technocomplexes. Key references: Geology, dating, and paleoenvironment Arrizabalaga and Altuna 2000, Arrizabalaga et al. 2003; Hominins N/A; Archeology Arrizabalaga and Altuna 2000, Arrizabalaga et al. 2003.

Laboratory for Human Evolutionary Studies Established in 1970 by Clark **Howell** within the Department of Integrative Biology at the University of California at Berkeley, and co-directed by Howell and Tim **White**. It is now directed by Tim White. It conducts fieldwork in Ethiopia, and has conducted fieldwork in Turkey. It is the home of the Researching Hominid Origins Initiative, which is a NSF-HOMINID program-funded initiative.

Laboratory of Vertebrate Paleontology Established in 1949 by the People's Republic of China, it was part of the Institute of Palaeontology within the **Chinese Academy of Sciences**. It was charged with continuing the study of early man in China by field and laboratory research. In 1953 it became an independent institute within the **Academia Sinica** and in 1957 changed its name again to the **Institute of Vertebrate Paleontology and Paleoanthropology** (or IVPP).

labyrinth *See* bony labyrinth; membranous labyrinth.

lack-of-signal artifacts One of several technical problems involved in the data-acquisition stage of **computed tomography**. When fossils are heavily mineralized and attenuation levels are high, the level of the X-rays reaching the detector may be quantitatively close to the level of the noise in the signal. A positive noise fluctuation is not disastrous, but a negative noise fluctuation reduces the signal to almost zero, and results in fine high-density streaks that can obscure the object of interest.

lack of spatial resolution One of several technical problems involved in the data-acquisition stage of **computed tomography**. In the scan plane resolution is limited by the beam geometry of the CT scanner (e.g., focal spot width, detector width, geometry) and it should not be limited by the pixel size if an appropriate reconstruction matrix size is chosen. Perpendicular to the scan plane, it is limited by the slice thickness and the slice-indexing geometry. The result of a limited spatial resolution in the resulting images is a blurring effect, so that an instantaneous transversal from, say, bone to air, is represented not by a sharp boundary, but by a gradual change in the density of the image. This means that when making linear measurements on CT images researchers have to make sure that the position of that transversal is halfway along the density slope that reflects the lack of spatial resolution. The researcher should not try and locate the termini of measurements by eye for this may lead to substantial errors, especially when the distances are relatively small.

lactase persistence The ability to digest lactose into adulthood. All mammals can digest milk as infants; however, this ability is typically lost after the infant is weaned. Specifically, the gene (called *LCT* in modern humans) that encodes the enzyme lactase is turned off. Lactase breaks down the sugar lactose into glucose and galactose, which can be absorbed into the bloodstream. In individuals who are lactose intolerant, this process does not take place and the lactose is metabolized by intestinal bacteria, which produce byproducts including hydrogen and methane (typically causing cramping, bloating, diarrhea, and nausea). In modern humans lactase persistence has evolved multiple times in populations with a history of herding and dairying (Bersaglieri et al. 2004, Itan et al. 2010, Tishkoff et al. 2007).

lactose intolerance *See* lactase persistence.

lacustrine (L. *lacus*=lake and *inus*=relating to) Anything that relates to lakes, so "lacustrine

433

sediments" are sediments formed in the bed of a lake.

LAD Acronym for **last appearance datum** (*which see*).

Laetoli (formerly Laetolil) [Location 3°13′27″S and 35°11′38″E, Tanzania; etym. Leakey et al. (1976) suggest that the name of the site derives from the Maasai name for a red lily plant (*Haemanthus*) that grows abundantly in the area. However, it is more likely that the name "Laetoli" is actually a corruption of the name of a local river called the Olaitole. Peter Kent, the geologist with Louis **Leakey**'s 1935 expedition to the area, recognized three stratigraphic subdivisions and named the lowermost the Laetolil Beds. However, in 1979 Mary **Leakey** reported that "the Tanzanian authorities have now asked that the term Laetolil should be dropped in favor of Laetoli" (Leakey and Hay 1979, p. 317). Hence the site is now called "Laetoli", but for reasons to do with the rules of geological nomenclature the old name for the sediments, the "Laetolil Beds," has to be retained] History and general description The site now called Laetoli is where fossiliferous sediments known variously as the Garusi River Series, the Vogel River Series, and the Laetolil Beds are exposed mainly along the headwaters of the Garusi River. The area was first explored by Louis Leakey and colleagues in 1935 and a hominin lower canine (M.42323, but published as "M.18773") was recovered. In 1938–9 an expedition led by Ludwig Kohl-Larsen recovered three hominin fossils (**Garusi I**, II, and 4). Louis and Mary Leakey made a brief visit to the area in 1959 and Mary Leakey returned briefly to visit with Richard Hay in 1964. Attention was refocused on the area when George Dove reported finding fossil vertebrate teeth in the headwaters of the Gadjingero River (that flows northwestward into Lake Masek) in 1974 and it was during that visit that Mary Leakey located hitherto unexplored exposures, from which her team recovered a hominin upper premolar and a juvenile mandible. The hominin-bearing Upper Laetolil Beds are exposed over an approximately 60 km^2/23 square mile area and the discrete localities are numbered from 1 to 24. The hominin fossils referred to as Laetoli hominids (LH) 1–14 were collected in 1974–5 and the hominins numbered LH 15–19 and 21–30 (NB: LH 20 is a monkey) were collected by Mary

Leakey's expedition between 1975 and 1979. An isolated upper molar (LH 31) was recovered by the **Institute of Human Origins'** expedition in 1987. Four more hominins (EP 1000/98, 162/00, 2400/00, and 1500/01) were collected between 1998 and 2001 by the **Eyasi Plateau Paleontological Project**. Fossil footprints of small and large animals, including the trails of three hominins (one individual is included in the footprints preserved at A in Locality 6 and at least two individuals at G in Locality 8) were first recognized in 1976. Temporal span and how dated? *Australopithecus afarensis* remains are restricted to the Upper Laetolil Beds dated to 3.85–3.63 Ma; *Paranthropus aethiopicus* is known only from the Upper Ndolanya Beds (2.66 Ma). The sediments have been dated using **potassium-argon dating** and **argon-argon dating**. Hominins found at site The 40 hominins recovered from Laetoli include a right maxillary fragment (Garusi I) the **holotype** of *Praeanthropus africanus*, a virtually complete mandibular corpus, the holotype of *Australopithecus afarensis* (**LH 4**), an associated skeleton (LH 21), a fragment of the right maxilla that has been assigned to *Paranthropus aethiopicus*, and a proximal tibia that may belong to the same taxon. A partial cranium of an archaic *Homo sapiens* (LH 18) is known from the **Late Pleistocene** Upper Ngaloba Beds. Archeological evidence found at site No tools are known from the Upper Laetolil Beds. A single stone artifact has been reported from the Upper Ndolanya Beds, but this is likely to be intrusive. **Middle Stone Age** and **Later Stone Age** artifacts are known from the Late Pleistocene Olpiro and Ngaloba Beds. Key references: Geology, dating, and paleoenvironment Drake and Curtis 1987, Hay 1987; Hominins Leakey et al. 1976, White 1977, 1980, Leakey 1987, Day et al. 1980; Archeology N/A. *See also Australopithecus afarensis*; **Laetoli hominin footprints**; *Paranthropus aethiopicus*.

Laetoli 18 *See* LH 18.

Laetoli Beds *See* **Laetoli**.

Laetoli hominin footprints In 1976, Andrew Hill, who was visiting Mary **Leakey**'s camp, discovered animal prints and raindrop impressions preserved on an exposed airfall tuff (White and Suwa 1987). In the following year,

Phillip Leakey and Peter Jones found a trail of five prints, potentially made by a hominin, at Footprint Site A and in 1978 Paul Abell discovered footprints of hominins and a variety of other animals at Footprint Site G; both sets of prints are in a series of footprint-bearing layers of petrified volcanic airfall ash (Leakey and Hay 1979) most likely emitted from the now quiescent Sadiman volcano, some 20 km/12.5 miles to the east. The age of the footprint-bearing stratum is estimated to be 3.66 Ma. Initially Ndibo Mbuika and later Tim **White** and Ron **Clarke** eventually uncovered a 41 m-long surface, of which 27.5 m exposed the surface of the footprint tuff. The excavations unearthed the trails of three hominin individuals at Site G. Two trails (G-1, G-2/3) occur side by side, suggesting that two individuals may have walked together; prints of the third individual (G3) were made in the prints of G2, obscuring some of the details of prints in both trails. A total of 38 footprints were uncovered in the G-1 trail, and 31 in the G-2/3 trail. Based on foot length estimated from the prints and limb lengths inferred from foot length and fossil anatomy, it is estimated that the hominins were moving at a leisurely 0.4–0.45 m/s walking pace (Tuttle et al. 1990). Researchers continue to debate interpretations of the prints. Virtually all researchers agree that the Laetoli hominin prints represent evidence of hominins well adapted for bipedal walking with derived foot anatomy, including relatively short toes and a hallux (big toe) more closely aligned with the other toes than is the case in the extant African apes. However, researchers disagree over interpretations regarding other aspects of foot anatomy and gait, with some arguing that the Laetoli prints were made by a hominin with a foot with a **longitudinal arch** and modern human-like hallucal adduction and with a **gait** like that of modern humans (e.g., Leakey and Hay 1979, White and Suwa 1987, Tuttle et al. 1990, Raichlen et al. 2010) and others suggesting that the anatomy of the foot and/or the gait were not modern human-like (e.g., Stern and Susman 1983, Bennett et al. 2009). Prints of other animals found along with the hominin footprints include those of (in order of abundance) members of the Bovidae (especially *Madoqua*, dik-dik), Lagomorpha, Giraffidae (including large and small species), Rhinoceratidae, Equidae, Suidae, Proboscidea, Rodentia, Carnivora, and Cercopithecidae.

lag deposit   What is left after high-energy flow (usually subaqueous) has preferentially removed the finer-grained or lower-density components of the sediment. One effect of this is to concentrate particles that require a higher energy to transport them. The same process can result in the preferential concentration and accumulation of skeletal elements particularly within fluvial channels.

Lagar Velho   (Location 39°45′20″N, 8°44′6″W, Portugal; etym. Portuguese for "old olive press") <u>History and general description</u> The site is a rock-shelter located in the Lapedo gorge, about 7 km/4.3 miles east of the city of Leiria, in central Portugal. An early Upper Paleolithic burial was identified at the site in November 1998, and excavated as a salvage operation over 4 weeks (under the direction of João Zilhão), followed by more extensive area excavation of the Upper Paleolithic occupation levels in the subsequent decade (under the direction of Francisco Almdeida). The discovery occurred after the uppermost levels of the site had been bulldozed away, which brought about extensive breakage of the skull but only marginally affected the postcranial skeleton, which was buried a few centimeters deeper in a shallow pit. <u>Temporal span and how dated?</u> The burial was covered by approximately 3 m of sterile deposits, but about 20 m to the east the rock-shelter fill contained a rich Gravettian-to-Solutrean sequence that overlay a late **Oxygen Isotope Stage** 3 non-archeological fluvial and colluvial succession. Lateral stratigraphic correlation and radiocarbon dating of associated finds place the burial event *c*.30 ka. <u>Hominins found at site</u> The burial contained the articulated postcranial skeleton of a 4–5-year-old child, with the mandible, dentition, and most of the fragments of the broken cranium having been recovered from surrounding disturbed sediments (**Lagar Velho 1**). <u>Archeological evidence found at site</u> The burial context provided detailed evidence of ritual (burning of a Scots pine branch prior to deposition of the body, body laid down with body ornaments and wrapped-up in a red-ochre-painted shroud, and deposition of grave gifts, namely parts of a deer at the head and feet, and a juvenile rabbit skeleton across the legs) and the Late Gravettian occupation surfaces were very well preserved, containing both evident features (e.g., hearths) and circumstantial evidence of activities *in situ* (e.g., about 100% by weight stone tool refitting). Key references: Duarte et al. 1999, Zilhão and Trinkaus 2002, Almeida et al. 2009, Bayle et al. 2010.

Lagar   Velho   1   <u>Site</u> Lagar Velho. <u>Locality</u> Grid unit L20. <u>Surface/*in situ*</u> *In situ*. <u>Date of</u>

discovery November 28, 1998. Finders João Maurício and Pedro Souto. Unit/horizon gs complex. Bed/member N/A. Formation N/A. Group N/A. Nearest overlying dated horizon Rabbit placed across lower legs radiocarbon dated to 23,920±220 years BP (OxA-8422). Nearest underlying dated horizon Scots pine charcoal under the body radiocarbon dated to 24,860±200 years BP (GrA-13310). Geological age c.24 ka. Developmental age Juvenile. Presumed sex Unknown. Brief anatomical description Although clearly modern human by most anatomical criteria, the child also presented a series of traits (e.g., low crural index, receding symphysis, semispinalis capitis fossae, I_2 shoveling, dental maturational pattern) usually associated with *Homo neanderthalensis*. Announcement Duarte et al. 1999. Initial description Zilhão and Trinkaus 2002, Bayle et al. 2010. Photographs/line drawings and metrical data Zilhão and Trinkaus 2002, Bayle et al. 2010. Detailed anatomical description Zilhão and Trinkaus 2002, Bayle et al. 2010. Initial taxonomic allocation *Homo sapiens*. Taxonomic revisions N/A. Current conventional taxonomic allocation *H. sapiens*. Informal taxonomic category Anatomically modern human. Significance The mosaic of anatomically modern human and archaic traits prompted the suggestion that Lagar Velho 1 provides evidence for admixture between Neanderthals and modern humans at the time of contact in Europe, but most assessments consider it to be within the range of variation of *H. sapiens*. Location of original Museu Nacional de Arqueologia, Lisbon, Portugal.

lagomorph   *See* **lagomorpha.**

Lagomorpha   (Gk *lagos*=hare, *morph*=form). The zoological order comprising rabbits, hares (Leporidae), and pikas (Ochotonidae). Lagomorpha resemble rodents superficially (e.g., they are small and appear to have only two upper incisors, which grow continually), but are classified in their own order due to other morphological differences. Originating in the late Paleocene, their natural distribution is worldwide except for Oceania, where they have latterly been introduced by modern humans. Pikas prefer colder climates and are mostly found in parts of Asia, Eastern Europe, and North America.

Lainyamok   (Location 1°47′S, 36°11′E, Kenya; etym. Lainyamok means "place of thieves" in the local, Maa, language) History and general description Lainyamok lies in **Middle Pleistocene** sediments in the East African Rift Valley in southern Kenya, 40 km/25 miles southwest of **Olorgesailie** and about 8 km/5 miles northwest of Lake Magadi. W.W. ("Bill") **Bishop** commenced geological work and mapping of the site in 1976 and a team led by Phillip Leakey and Kamoya **Kimeu** made a collection of artifacts and fossils including three associated hominin teeth. Additional fieldwork by Richard Potts, Pat Shipman, and colleagues in 1984 refined the stratigraphy, found vertebrate fossil localities possibly indicative of butchery by hominins, and a hominin femoral fragment. Temporal span and how dated? c.0.39–0.36 Ma on the basis of tephra bracketing the primary fossil layer dated by **single-crystal argon-argon dating** plus an obviously more modern fauna than that collected from Members 1–7 of the Olorgesailie Formation (c.0.9–0.7 Ma) at **Olorgesailie**. Hominins found at site A fragment of the diaphysis of a left femur (KNM-WM 13350) and associated right P^4–M^2. The femur and the teeth are consistent with attribution to either *Homo sapiens* or *Homo heidelbergensis*. Archeological evidence found at site Fossils and artifacts at Lainyamok occur in layer "Khaki 2," a 1–2.5 m-thick, poorly sorted **silt** and cobble layer near the middle of the section of Middle Pleistocene sediments. The layer appears to be the result of mudflows that redeposited material from the highlands, including stone tools and animal bones, to the lowlands around Lake Magadi. Denser concentrations of fauna appear to have accumulated within hyena burrows into these redeposited sediments. Thus, most of the dense bone accumulations at Lainyamok are likely the result of the bone-accumulating activities by hyenas rather than traces of hominin activities. Erosion and deflation of sediments at Lainyamok produced a few locally dense surface scatters of lithics (**flakes**, **discoid**al cores, **handaxes**, and **spheroids**, largely fashioned of green lava or **quartzite**), but not dense concentrations *in situ* or in association with faunal remains. Key references: Geology, dating, and paleoenvironment Shipman et al. 1983, Potts et al. 1988, Potts and Deino 1995; Hominins Shipman et al. 1983, Potts et al. 1988; Archeology Shipman et al. 1983, Potts et al. 1988.

Lake Botsumtwi   Lake Botsumtwi occupies a meteor-impact crater in Ghana. The 74 m-deep lake preserves a sedimentary record of paleoclimate for subtropical southern Africa stretching back c.1 Ma. Oxygen isotopes in carbonates preserve a detailed record of changes in African monsoon strength. Decadal- to centennial-length droughts are common over the past three millennia, suggesting that droughts similar to, or worse than, recent **Sahel**ian droughts would have been a factor affecting modern human ancestors in the region (Shanahan et al. 2009) and that on longer time-scales there is evidence for extreme aridity between 135 and 75 ka (Scholz et al. 2007). *See also* **megadroughts**.

Lake Mungo (Location 33°07′S, 143°10′E, Australia; etym. the site is named after the lake of the same name) History and general description Three gracile hominin skeletons (LM 1 and LM 2 in 1968 and 1969, respectively, and LM 3 in 1974) were recovered from this site; all were found in the core of an eroded lake-shore dune on the southern edge of the dry Lake Mungo, 100 km/62 miles northeast of Mildura. Two of the skeletons (LM 1 and LM 2) were cremated and buried in a conical pit, while the third was interred as a complete skeleton (LM 3). The specimens were buried in calcrete horizons within brownish grey sands. Temporal span and how dated? LM 1 and LM 2 were originally attributed to the **Late Pleistocene** by **radiocarbon dating** of collagen from LM 1 (*c.*25 ka) and from charcoal found with the burials (*c.*26 ka). There has been a great deal of debate surrounding the age of LM 3. Some researchers claimed this individual is the most ancient on the Australian continent, citing ages of *c.*62 ka from **uranium-series dating** and **electron spin resonance spectroscopy dating** of the skeleton and ages of *c.*61 ka from optically stimulated **luminescence dating** (or OSL) of the associated sediments. However, subsequent OSL analyses point to both LM 1 and LM 3 being *c.*40 ka. Hominins found at site LM 1 and LM 3 are remarkably gracile, especially when compared to the later but much more robust individuals from **Kow Swamp** and **Willandra Lakes**. Two competing explanations have been offered regarding the transition from the earlier gracile to the later robust crania and postcranial skeletons in Pleistocene Australia. The first interprets the fossil evidence as being consistent with a single evolving lineage that began by being relatively gracile and then over time it came to consist of more robust individuals, a change that some see as an adaptation to an increasingly harsh and arid environ-

ment. The second interprets the same fossil record as evidence for multiple migrations from Asia, with an early migration of gracile individuals from north Asia and a later migration of more robust individuals from southeast Asia. Genetic evidence from the site In 2001 it was reported that **mitochondrial DNA** (mtDNA) had been isolated from LM 3. The recovered mtDNA sequence is unknown among modern Australian Aboriginals and it was suggested that it represented one of the most ancient modern human mtDNA lineages. However, researchers who have re-examined the mtDNA evidence suggest that the "ancient mtDNA" purportedly isolated from LM 3 was most likely either a contaminant and/or was the result of degradation. Archeological evidence found at site Australian core-tool and **scraper** tradition. Key references: Geology, dating, and paleoenvironment Bowler et al. 1972, 2003, Thorne et al. 1999, Grün 2006; Hominins Bowler et al. 1970, Brown 2000, Thorne and Curnoe 2000; Genetics Adcock et al. 2001, Smith et al. 2003; Archeology Bowler et al. 1970.

Lake Nitchie (Location 33°40′S, 142°00′E, Australia; etym. the site is named after the lake of the same name) History and general description The site consists of the deflated surface of what was formerly a sand dune that had formed on the northern shore of Lake Nitchie, 90 km/56 miles northwest of Mildura, 870 km/540 miles west of Sydney, Australia, that in 1969 yielded a hominin skeleton. Temporal span and how dated? **Radiocarbon dating** of bone collagen indicates a **Holocene** age *c.*6860 years BP. Hominins found at site The robust modern human male skeleton was interred at the site in a semi-recumbent position with the body flexed and the face pointing downwards. The Lake Nitchie specimen may document the earliest evidence of ritual tooth avulsion. Archeological evidence found at site **Cores** and **scrapers** (but no microliths) were recovered from the upper stratigraphic unit of the site. The modern human individual was buried wearing a necklace of pierced canine teeth of Tasmanian devil (*Sarcophilus harrisii*). Key references: Geology, dating, and paleoenvironment Bowler 1970b; Hominins Macintosh et al. 1970; Archeology Bowler 1970b.

Lake Olduvai The name suggested by Hay and Kyser (2001) for the highly alkaline, shallow, and saline paleolake that occupied the region of the Olduvai Main Gorge during the deposition of Beds I and II. It varied in size, but its persistent central portion, known as the Central Basin (or CB) was centered on the western part of the Olduvai Main Gorge. The extensive lake margin areas to the west are known as the Western lake margin (or WLM) and to the east they are referred to as the Eastern lake margin (or ELM). The period of greatest alkalinity was in Bed I times and Lake Olduvai was most extensive during lower Bed II times. *See also* **Olduvai Gorge.**

Lake Tandou (Location 32°37′S, 142°08′E, Australia; etym. the site is named after the lake of the same name) History and general description This site, which yielded a hominin skeleton in 1967, consists of the deflated surface of what was formerly a sand dune that had formed on the eastern shore of Lake Tandou, 150 km/93 miles northwest of **Lake Mungo.** Temporal span and how dated? The hominin skeletal remains were too few, fragmentary, and fragile to allow direct dating, but **radiocarbon dating** of shell middens in the immediate area yielded ages of *c*.15 ka. Hominins found at site The extremely fragmented gracile male skeleton belongs to a modern human individual estimated to be 20–5 years old at death. It was interred at the site in a kneeling position. The Lake Tandou cranium is most similar to the gracile individuals from **Lake Mungo, Lake Nitchie,** and **Keilor** and, like them, it contrasts with the more robust modern human crania from **Kow Swamp** and **Coobool Creek.** Archeological evidence found at site Shell middens, hearths and burial features. Key references: Geology, dating, and paleoenvironment Hope et al. 1983; Hominin Merrilees 1973, Freedman and Lofgren 1983; Archeology Hope et al. 1983.

Lake Turkana This lake, previously known as Lake Rudolf, occupies the southern part of the **Omo-Turkana Basin.** The lake has fluctuated in size through time becoming larger when inflow exceeds any outflow and evaporation and smaller when it does not. The major fossiliferous formations around Lake Turkana are the **Koobi Fora, Mursi, Nachukui, Nkalabong, Shungura,** and **Usno Formations.** The major fossil and archeological sites surrounding Lake Turkana are Koobi Fora, **Lothagam,** the **Omo** region, and **West Turkana.**

Lake Turkana Basin *See* **Turkana Basin.**

Lake Turkana Group *See* **Turkana Group.**

Lakonis Location 36°46′59″N 22°34′49″E, Greece; etym. named after the province of Lakonia) History and general description The site, which was discovered by an amateur archeologist, has been excavated systematically by the Ephoreia of Paleoanthropology and Speleology (Greek Ministry of Culture) since 1999. It consists of a cave and a series of collapsed karstic formations, of which Locality I is the richest. Temporal span and how dated? Uranium-series dating on travertine, and AMS radiocarbon dating on charcoal suggest a range of *c*.100–40 ka for the cave sediments. Hominins found at site Isolated lower M_3 (LKH1). Archeological evidence found at site Middle Paleolithic (Mousterian) and the Initial Upper Paleolithic. Key references: Geology, dating, and paleoenvironment Panagopoulou et al. 2002–4; Hominins Harvati et al. 2003, Richards et al. 2008, Smith et al. 2009; Archeology Panagopoulou et al. 2002–4, Elefanti et al. 2008.

LAL Abbreviation of the Lokalalei Member in the Nachukui Formation, **West Turkana,** Kenya.

Lamalunga Cave *See* **Altamura; Altamura 1; Grotta di Lamalunga.**

Lamarckism (and neo-Lamarckism) One of the first to argue that species are not fixed (this had been accepted dogma since the time of the Greek philosopher and naturalist Aristotle) was Jean-Baptiste de Lamarck whose theory of the "transmutation" of species was published in *Philosophie zoologique* in 1809. In this work Lamarck argued that through the combination of the principles of the use and disuse of organs and the inheritance of acquired characteristics, individuals could change over time until their descendants would no longer closely resemble them. It is important to note that for Lamarck it is not the species that changes, but individuals. Lamarck suggested that there was an inherent tendency for organisms to be transformed into "higher" (i.e., more advanced) forms, and that this tendency resulted in a kind of escalator effect where, over many generations, lower organisms would progressively be transformed, or transmuted, into more advanced taxonomic categories. However, Lamarck believed that the major taxonomic categories (fishes, reptiles, birds, mammals, etc.) had always existed; there was no suggestion, as Darwinian evolutionary theory would later argue, that these categories emerged over time. In the early 19thC Lamarck's theory failed to garner adherents and was widely criticized. However, in his *Vestiges of the Natural History of Creation* (1844) the Scottish publisher Robert Chambers (1802–71) linked Lamarck's theory of transmutation with the then recent recognition by geologists of a progression in the fossil record (more advanced taxa appearing in more recent geologic strata) and with the equally recent description of the development of the embryo (which seemed to progress through stages corresponding to the succession of taxonomic categories, an idea that would later be described as phylogeny recapitulating ontogeny) to formulate a naturalistic and progressive history of life on Earth. Lamarck's transmutation theory remained an idea on the periphery of accepted science until the publication of Charles **Darwin**'s *On the Origin of Species by Means of Natural Selection* in 1859. Darwin's convincing argument, supported by immense quantities of supporting evidence, gradually drew the support of many prominent European scientists during the last half of the nineteenth century. Many naturalists in France, Germany, and America accepted the fact of biological evolution but relied upon Lamarck's notion of the inheritance of acquired characteristics to explain how evolution occurred. Rather than random variations in a population being acted upon by natural selection ("survival of the fittest" as Darwin's friend Herbert Spencer referred to it) these scientists believed that a biologically beneficial trait could be acquired by an organism and then that trait could be passed on to the next generation. Moreover, these naturalists reintroduced the Lamarckian idea that evolution was goal-directed and that the history of life on Earth was a linear progression leading from simple to more complex forms. Darwin had rejected these ideas and had argued that evolution by natural selection was random rather than progressive or teleological. The neo-Lamarckian version of evolution made the idea of biological evolution more acceptable than Darwin's version because within the neo-Lamarkian framework it could be argued that evolution proceeded according to a natural, or divinely directed, plan that was still consistent with the history of life on Earth as recorded in the fossil record. It was also consistent with a "progressive" interpretation (i.e., **orthogenesis**) that saw the appearance of modern humans in the world as the ultimate goal of evolution. Whereas for Darwin the traits shared between the extant apes and modern humans were derived from a recent common ancestor, neo-Lamarckians were more likely to argue that these traits had evolved independently. The British anatomist and physical anthropologist Frederic Wood Jones supported a neo-Lamarckian view of modern human evolution during the early decades of the 20thC as did Robert **Broom**. The German physical anthropologist Hermann **Klaatsch** relied upon a neo-Lamarckian model of evolution to explain the origin of bipedalism and invoked the same principles in his theory that modern human races had evolved independently, each from a separate early primate species. Neo-Lamarckian ideas remained appealing and influential in biology and among theories of human evolution until the importance of the role of natural selection and the Darwinian model of biological evolution was reasserted as part of the rise of the **modern evolutionary synthesis** in the 1940s.

lamellar bone   *See* **bone**.

lanceolate   (L. *lanceola*=the diminutive of *lancea*=lance) A term used to describe long and relatively thin bifacially flaked pieces such as those found at many **Lupemban** sites in Africa. An inference implicit in the term lanceolate is that such pieces may have served as lance or spear points rather than points on smaller **armature**s such as arrows.

landmark   (etym. *landmark*=a nautical term meaning a visible place onshore that sailors use as a reference point to help them fix a course into a harbor) A place on a bone or tooth that can be defined precisely enough so that independent observers are likely to agree on its location. For example, nasion is where the suture between the nasal bones meets the frontal bone and basion is the most posterior point on the anterior border of the foramen magnum in the sagittal plane. In the context of **geometric morphometrics** Bookstein (1991, p. 2) defines landmarks as loci that are homologous across individuals in a sample (i.e., the landmarks have the "same" location in every other member of the sample being investigated).

Their location can be recorded in one, two, or three dimensions. *See also* **coordinate data**; **geometric morphometrics**.

Langebaanweg (Location 32°57′41″S, 18°06′ 38″E, South Africa; etym. in Afrikaans Langebaanweg literally translates to "long path road," but here the "weg" refers to the railway stop and therefore Langebaanweg means "the railway station near the town of Langebaan") History and general description The paleontological site of Langebaanweg, located on the west coast of South Africa, approximately 120 km/74 miles north of Cape Town, is renowned for its exceptionally well-preserved and diverse faunal remains. The fossil deposits at Langebaanweg were discovered during phosphate-mining operations that began at Baard's Quarry in 1943 and later moved to "C" and "E" quarries in the 1960s. From the late 1950s until the late 1980s Langebaanweg was a major focus of research at the South African Museum (now the **South African Museum, Iziko Museums of Cape Town**) under the leadership of Brett Hendey. Hendey published extensively during this time period, on both the fossil fauna and the geology of the region. In the 1980s, following Hendey's departure from the museum, research slowed considerably, and mining of the area stopped in 1992. Following the establishment of the West Coast Fossil Park (or WCFP), the educational, research, and tourism institution that incorporates the Langebaanweg paleontological site, in 1998 research recommenced at Langebaanweg and excavations were resumed at "E" quarry. This has resulted in an influx of international researchers to the site and to study the collections, but much work remains to be undertaken as entire families of organisms are poorly understood. To date, over 230 vertebrate and invertebrate taxa have been described from these deposits that date to the terminal **Miocene/ earliest Pliocene** (*c.*5 Ma). For many taxa, including carnivores, birds, and frogs, Langebaanweg is unique in terms of species number, specimens, and quality of preservation. Although no hominin fossils have been found at the site to date, Langebaanweg is important for understanding the emergence of early hominins, because it provides essential context for interpreting and dating hominin-bearing sites in other parts of Africa with comparable faunal assemblages. At the end of 2005, the WCFP became an independent institution, managed by the WCFP Trust, and separate from Iziko

Museums, although the collections currently remain housed at the South African Museum, Iziko Museums of Cape Town. The open-air archeological locality known as the **Anyskop Blowout** is also located within the boundaries of the WCFP, approximately 1 km/0.6 miles south of the Pliocene fossil beds. Key references: Geology, dating, and paleoenvironment Hendey 1970a, 1970b, 1972, 1981, Roberts 2006; Paleontology Bernor and Kaiser 2006, Denys 1990, 1991, 1994a, 1994b, Franz-Odendaal et al. 2002, Haarhoff 1988, Hendey 1976, 1978a, 1978b, 1982, 1984, Matthews et al. 2007, Olson 1984, 1985a, 1985b, Pocock 1987, Rich and Haarhoff 1985. Contact person Ms Pippa Haarhoff (e-mail pjh@fossilpark.org.za). Postal address West Coast Fossil Park, PO Box 42, Langebaan 7375, South Africa. Website www.fossilpark.org.za. Tel +27 (0)22 766 1606. Fax +27 (0)22 766 1765.

language (ME *language* from OF *langage* from L. *lingua*=tongue or speech) Communication is ubiquitous among animals, but most language experts subscribe to the view that modern human language is a unique form of **communication**. The study of modern human languages can be broken down into the following subareas. **Phonology** is the study of how specific sounds (or signs) encode meaning, semantics is the investigation of how this meaning is generated and constituted in language, grammar is the study of how the various elements of language are structured and organized into larger units, and pragmatics focuses on the way language is used socially. According to Hauser et al. (2002), the only unique feature of modern human language is **recursion**. They argue that "though bees dance, birds sing and chimpanzees grunt, these systems of communication differ qualitatively from human language. In particular... [they] lack the rich expressive and open-ended power of human language" (*ibid* p. 1570). Various studies have demonstrated that while apes as well as other mammals (e.g., dogs) and birds (e.g., parrots) may learn, for example, to recognize specific sounds (i.e., phonology) and associate these sounds with particular objects (i.e., semantics) and even recognize that the social context may change the meaning of these sounds (i.e., pragmatics), no non-modern human animal has shown evidence of stringing different sounds to generate new meaning (i.e., grammar) into a sentence (cf., Terrace et al. 1979, Hauser et al. 2002). So, for instance, while a dog may respond appropriately when asked to *fetch* *ball* and when asked to *roll over*, it will be entirely perplexed when asked to *roll over the ball* even though such a command, while odd, would be immediately understood by a 2 year-old modern human. Researchers who focus on the evolution of modern human language try to understand how and when our ancestors acquired the facility for recursion. It is still unclear if the capacity for recursion gradually evolved or if it appeared "fully formed" relatively recently in our evolutionary history.

language, evolution of Discussion of the evolutionary origins of modern human language has focused primarily on the origins of its distinctive recursive **syntax**. There is evidence that, unlike modern humans, nonhuman primate species otherwise very good at detecting patterns of speech syllables cannot detect patterns generated by recursive rules (Fitch and Hauser 2004). Recursive syntax makes possible several important semantic features of modern human language. Modern humans can communicate about specific objects or events with which they can have had no acquaintance because they can use recursive syntax to generate descriptions of arbitrarily precise specificity (Bolender 2007). Few animal communication systems are capable of communicating information about spatiotemporally displaced referents, and none seem capable of specifying them with the precision that recursive syntax makes possible. Syntactic rules apply to parts of speech defined independently of their semantics (e.g., any noun phrase, referring to any object, combines with any verb phrase, referring to any action, to form a sentence). This makes possible two other powerful semantic properties: generality and domain independence. According to the so-called "generality constraint" (Evans 1982) to count as having mastered any word, one must be capable of understanding sentences that combine that word with *any* other word one counts as having mastered. This includes even rather unlikely word combinations (Camp 2004); thus, if you count as understanding the terms "cat," "can of tuna," and "eats" you must be able to understand not just the sentence "The cat eats the can of tuna" but also "The can of tuna eats the cat." One of the advantages of this is that we can use language to represent situations that have never been perceived and perhaps never could be perceived (e.g., a can of tuna eating a cat). This is closely related to the modern human capacity for metaphorical thought and language, including many scientific insights (e.g., the claim that light is a wave was a central insight of 19thC physics). Domain independence is a byproduct of such generality (e.g., noun phrases and verb phrases referring to members of disparate domains can be combined into sentences expressing domain-independent propositions). Given these biologically unique properties explaining the evolution of the capacity for distinctively modern human language is extremely challenging. The classic approach to this problem adopts the traditional philosophical understanding of modern human language as an expression of modern human thought (Pinker and Bloom 1990). According to this view, many of the problematic properties of language are actually properties of uniquely modern human thinking, inherited by modern human language because it functions to communicate modern human thought. Unfortunately, this strategy only kicks the problem "up a level," for now, to explain the puzzling properties of modern human language we must explain how and why the thought of our immediate, prelinguistic precursors acquired these properties. It is not clear how or why a group of hominins, not yet able to use a modern human-like language, would develop a capacity to think recursive, domain-independent, and fully general thoughts. Numerous species of primates have inhabited environments similar to those of our precursors without developing the capacity for such thought. Furthermore, archeological proxies of such thought, like geographically and temporally diverse toolkits and the integration of information from multiple domains in tools, weapons, and ornaments (Mithen 1996), date to *c.*200 ka, by which time our species with, presumably, the capacity for language, was already on the scene. This leads many to argue that the puzzling properties of modern human language are not inherited from a kind of thinking that evolved prior to it. Rather, structurally complex language evolved as a tool both for thinking *and* communication in the relatively recent past (Bickerton 1990, 1995, Hauser et al. 2002). Such views distinguish the faculty of language in the broad sense from the faculty of language in the narrow sense (Hauser et al. 2002). The faculty of language in the broad sense incorporates many capacities that modern humans share with nonhuman species, including the capacity to refer to objects and events using vocalizations, the capacity to conceptualize objects and events, and the capacity to read the communicative intentions of others. The unique component of the modern human faculty of language – the faculty of language in the narrow sense – is the capacity to detect and produce recursively structured signals. Bickerton (1990, 1995, 1998, 2000) similarly distinguishes between "proto-language," in which unstructured combinations of two to three distinct vocalizations communicate task-specific information, and fully grammatical language. According to Bickerton (1990) pidgins and the language of two-year-olds are linguistic fossils: relics of such proto-language. Fully grammatical language requires an appreciation of the various parts of speech and the recursive rules for combining them into sentences. This Chomskyian perspective is hostile to Darwinian explanations of fully grammatical language. One problem is that there does not appear to be any plausible, biological function for which such an expressively powerful communication system could have been selected. As many followers of Chomsky note, many of the functions of quotidian communication can be achieved with proto-language, as they are even today among immigrant workers with no formal instruction in the languages of their host countries. These individuals communicate successfully with short, unstructured sequences of words drawn from the local language; a kind of linguistic competence that Klein and Perdue call "The Basic Variety" (1997). If we can meet our communicative needs without employing the fully recursive structure of language, what advantage can a mutant capable of such a language have had in human prehistory? Origgi and Sperber (2000) raise and try to solve a closely related bootstrapping problem for Darwinian approaches. Namely, what advantages could accrue to a mutant individual capable of understanding and producing grammatically complex language if there were no others with this capacity in her group? Finally, Bickerton (1990) notes that many aspects of fully recursive grammar cannot be acquired piecemeal. Sentences cannot function without their component parts of speech, but these component parts of speech cannot function without playing roles in sentences. In short, fully recursive, grammatical language appears to be a highly complex structure, with many components that make no independent contributions to functionality and which have no obvious independent function. This undermines the assumption that adaptive complexity is the product of incremental accumulation and adjustment of multiple components, each of which has some adaptive justification independent of the others. These problems lead many followers of Chomsky to propose biologically implausible hypotheses about the phylogeny of grammatically complex language, like catastrophic macro-mutation (Bickerton 1990), wholesale **exaptation** of other cognitive faculties (Bickerton 2000), or some kind of emergent byproduct of anatomical constraints. Darwinian theorists hotly dispute such proposals and the claims that motivate them (Pinker and Bloom 1990, Pinker 2003, Pinker and Jackendoff 2005). If grammar is a textbook adaptation to the task of communicating propositional thought then the interdependence of grammatical components is no more problematic than the interdependence of components of other adaptations, like the eye. Furthermore, according to these theorists, there is much more that is unique to the modern human faculty of language than just recursive grammar. So, the picture of language as largely continuous with the communicative capacities of other species, except for one glaring, non-adaptive, structurally complex component – recursion – is, according to these theorists, misleading. Some theorists argue that grammatical complexity is a product of cultural, not biological evolution. For example, Wray and Grace (2007) argue that grammatical complexity arose gradually as isolated hominin populations began interacting with each other in prehistory. The need for communication among individuals sharing little background knowledge or context, drove the development of increasingly context-independent languages, capable of conveying clear, unambiguous information between interlocutors of vastly different backgrounds. This proposal is consistent with recent ethnographic evidence from an Amazonian population, the Piraha, said to indulge in self-imposed isolation. Although it is currently a matter of controversy among linguists, the researcher who has studied the Piraha language the longest claims it lacks recursive structure (Everett 2005). Charles **Darwin** first proposed another possibility, namely that modern human language is descended from sexually selected mating song (Darwin 1871). This proposal receives some support from evidence that sexually selected birdsong approaches the structural complexity of human language (Okanoya 2002) and that the *FOXP2* gene which is known from neuropathological evidence to play a role in the modern human capacity for structurally complex language (Lai et al. 2001) apparently plays a role in the capacity for sexually selected birdsong in some species (White et al. 2006a). The structural analogies between modern human language and music have long been appreciated (Lerdahl and Jackendoff 1996) and there is growing evidence of neural overlap between areas underlying modern human musical and linguistic competence (Patel 2010). Such evidence leads some contemporary theorists of human evolution to revive Darwin's proposal that modern human language is descended from a sexually selected, song-like proto-language (Miller 2000, Mithen 2006). There are other notable disputes among theorists of modern human language evolution. One of the most prominent debates concerns whether or not language had a manual origin (Corballis 2002). Voluntary control is one of the distinctive features of modern human communication. Vocalizations of other primates are emotion-triggered and involuntary, like the sighs, laughs, and groans of modern humans. In contrast, all primates have voluntary control of manual gestures. Assuming that the most recent common ancestor with chimpanzees/bonobos fits this pattern, it seems more likely that a voluntarily controlled communication system first evolved in connection with the hand. The discovery of mirror neurons in primate brains, including those of macaques (Rizzolatti et al. 1996) and modern humans (Mukamel et al. 2010), has been interpreted as evidence for the manual origins view of language evolution (Rizzolatti and Arbib 1998, Arbib, 2006). The reason is that a central feature of language is the linkage between productive and receptive competence, for to produce linguistic signals properly one must understand how they will be interpreted, and to interpret linguistic signals properly one must understand the intent with which they are produced. Mirror neurons fire both when the same action is observed and executed, suggesting a mechanism capable of implementing this Janus-faced property of modern human linguistic communication. Terrence Deacon's *The Symbolic Species* (1997) and Tecumseh Fitch's *The Evolution of Language* (2010) provide more extensive surveys of hypotheses about the origin of language.

language, hard tissue evidence for

Hard-tissue evidence that may potentially be related to spoken language can be divided into two cate-

gories: evidence related to the generation of spoken language and evidence related to its perception. With respect to the generation of spoken language only two of the four categories of potential neuroanatomical evidence suggested by Ralph **Holloway** (1983b) are relevant, namely evidence for the reorganization, or enlargement, of **Broca's area** and cerebral cortical **petalia**s. Broca's area, which is located along the inferior frontal gyrus of the anterior prefrontal cortex and which is generally enlarged in the left hemisphere of most modern humans (Amunts et al. 1999) is thought to play a role in the production of spoken language (Damasio and Damasio 1989). The lateralization of Broca's area, though it is present in apes, occurs less frequently than it does in modern humans (Holloway 1996). A recent survey of extinct and extant hominins showed that while asymmetrical enlargement does occur in transitional and pre-modern *Homo* taxa (i.e., *Homo rudolfensis*, *Homo erectus*, *Homo heidelbergensis*, *Homo neanderthalensis*) it is absent in *Australopithecus afarensis* (de Sousa and Wood 2007). However, doubt has been cast on the wisdom of using endocranial morphology as a proxy for identifying functional regions of the cortex (e.g., Holloway 2009). Petalias, or asymmetries in the cortical hemispheres of the brain, once thought to be peculiar to modern humans (and thus a sound way of imputing functions such as spoken language), have been identified in samples of higher primates other than modern humans (Hopkins and Marino 2000). Two other categories of hard-tissue evidence for speech production have been put forward. The first used the size of the vertebral canal to predict the presence of the "extra" neurons in the thoracic spinal cord that allows for the fine control of the muscles of respiration, thus facilitating spoken language in modern humans. Examination of thoracic spinal cords across a wide range of primates suggested their cross-sectional areas are correlated with the cross-sectional area of the matching part of the vertebral canal (MacLarnon 1995), and when the vertebral canal sizes were adjusted for body weight only the relative size of the thoracic canals of modern humans and Neanderthals were larger than their body sizes would predict (MacLarnon and Hewitt 1999). However, experience with other systems suggests that broad across-primate allometric relationships cannot necessarily be extrapolated to much narrower allometric contexts, such as that within the hominin clade. The second category of evidence used the size of the hypoglossal vertebral canal to predict the presence of the "extra" neurons that facilitate the fine control of the muscles of the tongue. The initial results of attempts to infer the size of the hypoglossal nerve (the motor nerve to most of the extrinsic and all of the intrinsic muscles of the tongue) from the size of the hypoglossal canal seemed promising (Kay et al. 1998), but closer examination (DeGusta et al. 1999) revealed unwarranted assumptions, and thus flaws, in this approach. Potential hard-tissue evidence related to the perception of spoken language involves the external and the middle ear. The external ears of chimpanzees are both longer and smaller in cross-section than those of modern humans [the external auditory meatus of modern humans (mean = 115 mm^2) is more than twice that of chimpanzees (mean = 45 mm^2)] and what is known of the size of the external auditory meati of early hominins prior to *Au. afarensis* suggests that they were also chimp-sized. Middle ear ossicles have been recovered at Swartkrans and Sterkfontein (an incus, SK 848, belonging to *Paranthropus robustus*, from the former, and stapes, belonging to *Australopithecus africanus* and *Homo habilis*, from the latter). Moggi-Cecchi and Collard (2002) showed there was a marked similarity in size between the footplates of the early hominin stapes and the stapes belonging to the living great apes, and that they were both substantially smaller than those of modern humans. But what, if anything, this implies about the speech perception capabilities of *Au. africanus* (and *H. habilis*) is unclear.

language vocalization *See* **spoken language**.

Lantian Chenjiawo (蓝田陈家窝子) (Location 34°14′N, 109°14′E, northwestern China; etym. named after the local village) <u>History and general description</u> An open-air site, discovered in 1963, located near the village of Chenjiawozi, 10 km/6 miles northwest of Lantian City, Shaanxi Province. Initially made the type specimen of *Sinanthropus lantianensis*, the mandible is now referred to *Homo erectus*. Together with the **Gongwangling** cranium, they are commonly referred to as "Lantian Man." <u>Temporal span and how dated?</u> On biostratigraphic grounds, the fossil mandible is estimated to be *c*.300 ka. However, **magnetostratigraphy** and paleoclimatic data derived from deep-sea cores place Chenjiawo at *c*.650–500 ka. A more detailed magnetostratigraphic and loess-paleosol lithostratigraphic study conducted by Sheng and Kun (1989) suggests the locality dates to 650 ka. <u>Hominins found at site</u> Presumed to be

an older adult female the Chenjiawo (PA 102) **mandible** is generally well preserved, missing just parts of the rami, the right P$_3$ (antemortem), and both M$_3$s (congenitally). The crowns of several teeth on the left side (C–M$_1$) were damaged during excavation. <u>Archeological evidence found at site</u> A lithic **scraper** and nine **flake**s were found scattered near the fossil site. One of these, found 70 m away, was reported to be from the same stratum as the hominin mandible. Key references: <u>Geology, dating, and paleoenvironment</u> Aigner and Laughlin 1973, An and Ho 1989; <u>Hominins</u> Woo 1964, Aigner and Laughlin 1973, Wu and Olsen 1985, Pope 1992, Wu and Poirier 1995, Brown 2001; <u>Archeology</u> Aigner and Laughlin 1973, Wu and Olsen 1985, Schick and Dong 1993.

Lantian Gongwangling (蓝田公王岭) (Location 34°11′0.73″N, 109°29′38.94″E, northwestern China; etym. named after a hill near Gongwang village) <u>History and general description</u> An open-air site, discovered in 1964, located on the northern slope of a hill south of Gongwang village, 17 km/10.5 miles southeast of Lantian City, Shaanxi Province. Together with the Chenjiawo mandible they are commonly known as "**Lantian Man**," or *Sinanthropus lantianensis* (now referred to *Homo erectus*); <u>Temporal span and how dated?</u> Gongwangling is considerably older than Chenjiawo. Earlier estimates using **biostratigraphy**, **magnetostratigraphy**, and correlation with deep-sea cores place the locality at *c*.700 ka, *c*.800–750 ka (or 1.0 Ma), and *c*.800–730 ka, respectively. Based on new **geomagnetic dating** evidence and the lithostratigraphic position of the fossil in the loess-paleosol sequence, Sheng and Kun (1989) infer an age of *c*.1.15 Ma, making Gongwangling one of the oldest hominin fossil sites in mainland East Asia. <u>Hominins found at site</u> Presumed to be a female over 30 years of age, Gongwangling (PA 105 [1-6]) is a partial **cranium** consisting of a complete frontal, much of both parietals, the right **temporal** without the **mastoid**, parts of both nasal bones, incomplete maxillae, and an isolated left M^2. The fossils are poorly preserved and heavily mineralized; very little ectocranial topography is preserved. <u>Archeological evidence found at site</u> Twenty **quartzite** and **quartz** artifacts (11 cores, five flakes, and four scrapers) made using a hard hammering technique with minimal flaking, were found in association with the fossil cranium. Since these artifacts were distributed through a 13 m depth of deposits they cannot be considered a single archeological assemblage and it is assumed that the site was only temporarily occupied. From the nearby site of Pingliang, a thick, large, and **Acheulean**-like **handaxe** (measuring roughly 20 cm×10 cm) was collected on the surface; it is known as the "Lantian **biface**." Key references: Geology, dating, and paleoenvironment Aigner and Laughlin 1973, An and Ho 1989, Zhu et al. 2003; Hominins Woo 1966, Wu and Olsen 1985, Pope 1992, Wu and Poirier 1995, Holloway 2000, Brown 2001, Shang et al. 2008; Archeology Aigner and Laughlin 1973, Wu and Olsen 1985, Schick and Dong 1993.

Lantian Man *See* Lantian Chenjiawo; Lantian Gongwangling; *Sinanthropus lantianensis.*

Laron syndrome A particular variety of dwarfism that results from a deficiency of growth hormone receptors in the liver, which, in turn leads to a deficiency of IGF-1, also known as somatomedin. The Laron phenotype includes short stature, a small and abnormally shaped skull, malformed and misaligned teeth, small extremities, and relatively as well as absolutely short limbs. Hershkovitz et al. (2007) suggested that the morphology seen in **LB1**, the **holotype** of *Homo floresiensis*, was consistent with the Laron phenotype, but Falk et al. (2007) show that LB1 departs in several significant ways from the phenotype seen in Laron syndrome. *See also* **Homo floresiensis**; LB1.

laryngopharynx *See* pharynx.

Lascaux (Location 45°03′15″N, 1°10′01″E, France; etym. unknown) History and general description This cave site was discovered by accident by several small boys in September 1940. The cave houses some of the most famous **Upper Paleolithic** cave paintings in Europe, with many scenes of animals, humans, and abstract designs. These paintings have been the subject of many interpretations, as evidence of sympathetic hunting magic, totem worship, astronomy, and/or shamanism. At one time open to the public, the cave has been closed since 1963, although an exact replica of two of the galleries, Lascaux II, is available for tours. Temporal span and how dated? **Radiocarbon dating** of charcoal and antler have provided a range of dates from 15,500 to 18,900 years BP;

stylistic interpretations of the art suggest it was created during the **Solutrean**. Hominins found at site None. Archeological evidence found at site A variety of artifacts were found at the site, including lamps, decorated bones, and more than 300 stone tools, some of which show evidence of having been used for engraving. Key references: Geology, dating, paleoenvironment, and archeology Delluc and Delluc 2003, Glory et al. 2008.

laser ablation Because enamel develops incrementally any bulk-sampling method will result in a stable isotope (e.g., $^{13}C/^{12}C$) or elemental (e.g., **strontium/calcium ratio**) signal that is subject to **time-averaging**. Laser ablation helps ameliorate time-averaging by focusing a thin laser beam on a small area of **imbricational enamel** on the outer surface of a whole tooth, or on enamel that has been exposed by the tooth having been naturally fractured or deliberately sectioned, and this much more precise sampling strategy can result in a $^{13}C/^{12}C$ or Sr/Ca signal that samples days to weeks rather than years of time. This greater precision enables researchers to examine whether the $^{13}C/^{12}C$ ratio changes during the development of a tooth, as would be the case if the diet differed from one season to another (e.g., Sponheimer et al. 2006) or to use the ratio of Sr/Ca ratio to detect when weaning occurs (Humphrey et al. 2008). However, laser ablation cannot eliminate time-averaging because we are still ignorant about the dynamics of enamel mineralization. Only approximately 15% of the mineral in enamel is laid down during enamel secretion; the rest is deposited during enamel maturation and the timing and patterning of that process is poorly understood. Enamel maturation does not even start until the full crown thickness is attained, and it may take three times as long as the process of enamel secretion. *See* **C_3 and C_4**; **enamel development**; **stable isotopes**; **strontium/calcium ratio**.

laser ablation inductively coupled plasma mass spectrometry (or LA-ICP-MS) This process uses lasers to convert a solid by ablation into a phase that can be sampled using inductively coupled plasma mass spectrometry. The energy in the laser transforms the solid into plasma and its elemental composition is measured by a mass spectrometer. The system can be used for determining stable isotope or trace element ratios. Laser ablation helps ameliorate the effects of

time-averaging by focusing a thin laser beam on a small area of **imbricational enamel** on the outer surface of a whole tooth, or on enamel that has been exposed by the tooth having been naturally fractured or deliberately sectioned, and this much more precise sampling strategy can result in a signal that samples days to weeks rather than years of time. Humphreys et al. (2008) have used this system to measure the strontium/calcium ratio in thin sections of enamel to detect when weaning occurs in extant animals. *See* **laser ablation**; **strontium/calcium ratio**.

laser ablation stable isotope analysis
See **laser ablation**.

last appearance datum
(or LAD) The term refers to the date of that taxon's last occurrence in the fossil record and in paleoanthropology it is usually used in connection with a species (e.g., *Paranthropus boisei*) or a genus (e.g., *Homo*). For various reasons the LAD of a taxon is almost certainly earlier than when the taxon actually became extinct, or emigrated from that region. Just how much later than its extinction or emigration from that region is determined by two factors. The first is any error in the date and the second is the nature of the relevant fossil record after the LAD. The problem is the old adage that "absence of evidence is not evidence of absence." Hominins are such a rare component of the mammalian faunal record that researchers need to find a substantial number of nonhominin mammalian fossils (at least several hundred) without finding any evidence of a particular hominin taxon before it can be reasonably assumed that the hominin taxon in question was not part of the faunal assemblage being sampled. In such cases the LAD has an acceptably low 95% confidence interval, but in cases where there is a major unconformity spanning several hundred thousand years after the LAD of a taxon, so there is no fossil record during that time, then the LAD of that taxon at that location has an unacceptably high 95% confidence interval. *See also* **first appearance datum**.

last common ancestor
See **most recent common ancestor**.

Last Glacial Maximum
This is defined as "the most recent interval in Earth history when global ice sheets reached their maximum volume" (Clark et al. 2009, p. 710). Clark and colleagues investigated glaciers in five regions of the world and concluded that some ice sheets reached their maxima at 33 ka, all of them reached their maxima by 26.5 ka, and then they remained at these maxima until 20–19 ka. Sea levels, which are inversely correlated with glacial maxima, fell during **Oxygen Isotope Stage** 3 and reached their lowest level, called the Last Glacial Maximum low stand, between 26.5 and 19 ka.

Last Glacial Maximum low stand
See **Last Glacial Maximum**.

Late Pleistocene
(Gk *pleistos*=most and *kainos*=new) An informal younger division of the **Pleistocene** epoch, synonymous with the Tarantian Age (126–11.7 ka). The lower boundary of the Late Pleistocene is traditionally placed at the base of **Oxygen Isotope Stage** 5e.

Late Pliocene transition
Shackleton (1967) argued that the $^{18}O/^{16}O$ ratio of **benthic foraminifera** in the oceans would fall when ^{18}O-rich water was locked up in expanded ice caps and enlarged glaciers. Glaciation intensified in the northern hemisphere, this is called the **Northern Hemisphere Glaciation** (or NHG) from the mid **Pliocene** to the late **Pleistocene**, but during this time there were two periods when the rate of reduction in the $^{18}O/^{16}O$ ratio was more pronounced, from *c.*3.2 to *c.*2.7 Ma and from *c.*1.2 to *c.*0.8 Ma. The first of these two shifts is called the late Pliocene transition (or LPT) and the second later shift in the ratio of $^{18}O/^{16}O$ is called the **Mid-Pleistocene transition** (or MPT). The end of the LPT marks the beginning of the 41 ka-long obliquity-driven climate cycles and it is when ice-rafted debris begin to appear in sediments in the North Atlantic.

Late Stone Age
See **Later Stone Age**.

Later Stone Age
The youngest period in the tripartite division of the African archeological lithic record formalized by A.J.H. Goodwin and C. van Riet Lowe (1929) in their *The Stone Age Cultures of South Africa*. Although it was developed for sequences in southern Africa, the terminology has subsequently been applied to sites across sub-Saharan Africa. There have been numerous calls

to abandon the tripartite division (e.g., Clark et al. 1966, Barham and Mitchell 2008), but it, and thus the term Later Stone Age, persists because of the latter's utility for describing a broad stage in the general trend toward a more sophisticated lithic record (i.e., a trend toward aspects of the manufacture of stone tools that are conceptually and technically more elaborate). The Later Stone Age is distinguished from the **Middle Stone Age** by the presence of small **microliths** with one blunted, or **backed** edge, small circular **scraper**s, and in later assemblages **ground stone tools** and **pottery**. Although the term Later Stone Age is frequently used to describe stone tool-equipped modern human **forager**s, the term may also encompass stone tool-using **pastoralist** populations. Recognizing that first and last appearance dates vary locally and are subject to change as new sites are found, existing dating methods are improved, or when new methods are introduced, the present evidence suggests the Later Stone Age began *c.*40 ka. Its last appearance dates are contentious, and some contemporary African populations still occasionally use stone tools.

lateral enamel   *See* imbricational enamel.

lateral fissure   *See* lateral sulcus.

lateral geniculate nucleus   (L. *geniculum*= knee, knotted, or bent and *nux*=nut, kernel) A cluster of cells situated lateral to the main body of the **thalamus** on its ventroposterior aspect. The optic nerve conveys visual information from the retinal receptors via retinal ganglion cells to the lateral geniculate nucleus, and neurons in the lateral geniculate nucleus project to (i.e., communicate with) the **primary visual cortex** (also called the **striate cortex** or **Brodmann's area 17**) of the **occipital lobe** of the **cerebral hemisphere**. The lateral geniculate nucleus is one of three subcortical regions that receive direct input from the retina and the only one that processes information ultimately resulting in the conscious perception of visual information. The lateral geniculate nucleus contains a **topographic** representation of the contralateral half of the visual field. In primates, the lateral geniculate nucleus is divided into layers, each of which receives inputs from only one eye. In modern humans and great apes, there are six main layers: two ventral magnocellular (i.e., large-celled) layers, which process motion-related information, and four dorsal parvocellular (i.e., small-celled) layers, which are concerned with processing color and fine details of form. Neurochemically distinctive koniocellular cells located between these main layers are involved in processing color information from short-wavelength "blue" cones in the retina. The lateral geniculate nucleus and the primary visual cortex are considerably smaller in modern humans than predicted by **allometry** for a typical primate of the same brain size. This has been interpreted as reflecting reorganization of the modern human cerebral cortex towards a greater emphasis on regions of the **association cortex** that focus on higher-order processing and function. *See also* **lunate sulcus**.

lateral longitudinal arch   A mainly hard-tissue structure consisting of the calcaneus, cuboid, and the lateral two rays. It is not a structural arch *per se* in modern humans as the lateral side of the foot is often in contact with the ground during the stance phase of modern human walking. The lateral longitudinal arch mainly serves to stiffen the lateral side of the midfoot to increase propulsive capacity during bipedal walking. This stiffening is produced by the dorsal inclination of the calcaneus and it is stabilized by a locking mechanism between the calcaneus and the cuboid. Additionally, the long plantar ligament and the well-developed plantar aponeurosis, unique among extant primates to the modern human foot, prevent dorsiflexion at the cuboid-metatarsal joint, thereby eliminating the midfoot flexion (the **midfoot break**) in the modern human foot. A stiff lateral side of the foot may be quite ancient given the evidence for tarsometatarsal stability in *Australopithecus* and perhaps even in *Ardipithecus*.

lateral sulcus   (L. *fissus*=split, cloven) The lateral fissure is the *Terminologia Anatomica* term for the structure some comparative neuroscientists refer to as the Sylvian fissure. This eponymous name commemorates its discover, Franz de le Boë (1614–1672), a Dutch physician and scientist whose Latinized name is Franciscus Sylvius. The lateral fissure is a prominent cleft, or sulcus, found in primates. It is located on the lateral surface of each **cerebral hemisphere** and it separates the **temporal lobe** from the **frontal lobe** and **parietal lobe**. The lateral fissure is evident early in development, appearing before most other sulci. Lateral fissure length has been demonstrated to be asymmetric (with a longer fissure in the left than on the right side) in modern humans and in the great apes. Similar asymmetry of lateral fissure length has also been reported in several Old and New World monkey species, but not all studies of monkeys are consistent in finding this asymmetry. The greater length of the lateral fissure in the left hemisphere may be correlated with anatomical **lateralization** of the posterior temporal lobe and possibly related to hemispheric specialization of auditory processing for vocal communication signals.

lateralization The term used for the parts of the **cerebral hemisphere** whose functions are asymmetric. The classical example is that of **Brodmann's areas** 44 and 45 in the frontal lobe. These areas are involved with **speech** in the dominant hemisphere (the left in the majority of people, i.e., in those who are right-handed) and with spatial cognition on the non-dominant side.

Lautsch *See* **Mladeč**.

lava (L. *labi* = to fall) The term lava refers to molten material (magma) extruded at the Earth's surface by volcanic processes and to the rocks formed when this material cools. The character of lavas varies widely with chemistry and cooling rate. Rapid quenching of the molten material can yield the volcanic glass **obsidian**, usually at the base or at the front of a lava flow. The rate of cooling determines the texture of the lava, and the size of the crystals it contains. Slower cooling allows the growth of crystals, with exterior portions of a lava flow characterized by a mix of fine crystals and a glassy groundmass, whereas the inner portions of the flow are more coarse-grained with large crystals. Classification schemes for lavas rely primarily upon mineralogy, crystallinity, and chemical composition. The most abundant lava types are **basalt**, andesite, **rhyolite**, **trachyte**, and phonolite.

Lawley–Hotelling trace *See* **multivariate analysis of variance**.

Lazaret (Location 43°41′38″N, 7°17′21″E, France; etym. Fr. quarantine area; the cave may have been used as a quarantine area for the port of Nice or it may have resembled the quarantine area of a boat) History and general description A large cave on the Mediterranean coast of France on the western slope of Mount Boron in Nice. Long

known to amateur archeologists, it was first explored professionally by Emile Rivière in 1879 and then more fully excavated by C.F.E. Octobon in the 1950s and 1960s. The most recent excavations were undertaken by Henry and Marie-Antoinette de Lumley in the mid 1960s. The site contains abundant archeological and faunal remains as well as several fragmentary hominin remains. Though the site is on the coast, there is no evidence for consumption of marine resources. Temporal span and how dated? **Biostratigraphy** suggests an age of **Oxygen Isotope Stages** 5d and 5c for the archeological layers. The underlying beach deposits and the overlying flowstone have both been the object of a variety of dating methods, including **uranium-series dating** and **electron spin resonance spectroscopy dating**. Fauna from the archeological layers have been directly dated using uranium-series dating, but none of these dating methods have proved conclusive. The overall consensus for the archeological layers is roughly the beginning of Oxygen Isotope Stage 5. Hominins found at site Nine hominins have been discovered, three of which are complete enough to tell they resemble other material that is considered to be **pre-Neanderthal**. Archeological evidence found at site The archeological layer, layer C, is divided into three units. CI and CII contain **Acheulean** artifacts and CIII contains **Mousterian** items of Acheulean tradition with many flake tools. Key references: Geology, dating, and paleoenvironment Michel et al. 2000, de Lumley H. et al. 1976; Hominins de Lumley M.-A. and Piveteau 1969, Puech and Albertini 1981; Archeology de Lumley H. et al. 1976.

LB Prefix used for fossils recovered from **Liang Bua**, Flores, Indonesia.

LB1 Site Liang Bua. Locality Sector VII, Spits 59 and 62 (possibly also 61), and Sector XI, Spits 56B, 57A, and 58A. Surface/*in situ In situ*. Date of discovery 2003 (the right humerus, both ulnae, and left fibula were found in 2004). Finders Benyamin Tarus, Emanuel Wahyu Saptomo, and Rhokus Due Awe. Unit N/A. Horizon Dark-brown silty clay (Layer R) immediately above a clayey silt matrix with reworked conglomerate (Morwood et al. 2005, 2009). Bed/member N/A. Formation N/A. Group N/A. Nearest dated horizon Dated samples, ANUA-27116 and -27117 for **radiocarbon dating**, and LBS7-42 for **luminescence dating**, were taken from the same location as the LB1 associated skeleton. The 95% confidence intervals of the former two samples were 18.7–17.1 ka and 19.0–17.9 ka, respectively, and the luminescence sample provided a minimum age of 6.8 ka (Roberts et al. 2009). Geological age *c*.18 ka. Developmental age Adult. Presumed sex Probably female based on pelvis. Brief anatomical description The associated skeleton comprises evidence of the cranium and mandible, and evidence of the clavicle, humerus, ulna, part of the hand, vertebrae, ribs, sternum, partial pelvis, femora, tibiae, fibulae, and most of the foot (Brown et al. 2004, Morwood et al. 2005). Endocranial volume Approximately 417 cm^3 (Falk et al. 2005). Announcement Brown et al. 2004. Initial description Brown et al. 2004, Morwood et al. 2005. Photographs/line drawings and metrical data Brown et al. 2004, Falk et al. 2005, 2007, 2009, Morwood et al. 2005, Tocheri et al. 2007, Baab and McNulty et al. 2009, Brown and Maeda 2009, Jungers et al. 2009a, 2009b, Larson et al. 2009; Detailed anatomical description Falk et al. 2005, 2009, Tocheri et al. 2007, Brown and Maeda 2009, Jungers et al. 2009a, 2009b, Larson et al. 2009. Initial taxonomic allocation *Homo floresiensis*. Taxonomic revisions *Homo sapiens* Jacob et al. 2006. Current conventional taxonomic allocation *H. floresiensis*. Informal taxonomic category Pre-modern *Homo*. Significance The **holotype** of *H. floresiensis*. There are two interpretations of this specimen. The first concludes that its small overall size (the stature of LB1 is estimated to be 106 cm and its body mass to be 25–30 kg), its especially small brain (approximately 417 cm^3) for an adult hominin, and its primitive, pre-modern *Homo* or transitional hominin-like morphology are most parsimoniously interpreted as evidence of a novel endemically dwarfed early *Homo* taxon. The second concludes that no new taxon needs to be erected for in this interpretation LB1 is viewed as a pathological *Homo sapiens* individual afflicted by, variously, endocrine disorders or a range of syndromes that include microcephaly. Both explanations are exotic, but the ones that support a pathological explanation for LB1 need to explain what pathology affecting a modern human results in an early African *Homo erectus*-like cranial vault, *Homo habilis*-like cranial, mandibular, dental, carpal, and pedal morphology, and a brain that while very small, apparently has none of the morphological stigmata of **microcephaly**. Location of original National Research and Development Center for Archaeology, Jakarta, Indonesia.

LB6 Site Liang Bua. Locality Sector XI, Spits 51–53. Surface/*in situ* *In situ*. Date of discovery 2004. Finders Emanuel Wahyu Saptomo, Thomas Sutikna, Rhokus Due Awe, Jatmiko, and Sri Wasisto. Unit N/A. Horizon A brownish clayey silt (Layer Q) (Morwood et al. 2005, 2009). Bed/member N/A. Formation N/A. Group N/A. Nearest dated horizon Two dated samples, ANUA-23610 and -27117 taken from Sector III and VII respectively for **radiocarbon dating**, potentially constrain the ages of the LB6 associated skeleton. The confidence intervals of these two samples are >15.7–17.1 ka and <17.1–18.7 ka, respectively. Geological age *c*.17 ka. Developmental age Adult. Presumed sex Unknown. Brief anatomical description The associated skeleton comprises 14 numbered fossils including a mandible, the right scapula, ulna and radius, bones from the digits of the hand, and three pedal phalanges. The mandible lacks the right ramus, and the left coronoid and condylar processes have suffered some damage. As for the teeth, all but the left I$_{1-2}$ and the right I$_1$ are preserved. The scapula is nearly complete, and the ulna lacks just the distal end. The radius is complete, but the distal end displays the type of deformity associated with a healed (but unreduced and unsplinted) fracture. Announcement Morwood et al. 2005. Initial description Morwood et al. 2005. Photographs/line drawings and metrical data Morwood et al. 2005, Brown and Maeda 2009. Detailed anatomical description Larson et al. 2009, Jungers et al. 2009b, Brown and Maeda 2009. Initial taxonomic allocation *Homo floresiensis*. Taxonomic revisions *Homo sapiens* (Jacob et al. 2006). Current conventional taxonomic allocation *H. floresiensis*. Informal taxonomic category Pre-modern *Homo* or transitional hominin. Significance This, the second associated skeleton of *Homo floresiensis*, has a mostly well-preserved mandible that in general matches the morphology seen in the **LB1** mandible. As for LB1, the primitive early *Homo*-like morphology of this specimen can either be interpreted as evidence of a novel endemically dwarfed early *Homo* taxon, or as the result of a second modern human individual afflicted by a congenital abnormality or endocrine disorder resulting in small stature and primitive mandibular, dental, and postcranial morphology. Location of original National Research and Development Center for Archaeology, Jakarta, Indonesia.

LCA Acronym for last common ancestor, sometimes also referred to as the hypothetical last common ancestor. In the context of hominin evolution, the last common ancestor of the hominin clade would be the hypothetical last common ancestor of the clades whose extant members are, respectively, chimpanzees/bonobos and modern humans. *See also* **most recent common ancestor**.

LCN Prefix for fossils recovered from Lincoln Cave North, at Sterkfontein, in Gauteng Province, South Africa. *See* **Lincoln Cave**; **Sterkfontein**.

LCS Prefix for fossils recovered from Lincoln Cave South, at Sterkfontein, in Gauteng Province, South Africa. *See* **Lincoln Cave**; **Sterkfontein**.

LD Acronym for **linkage disequilibrium** (*which see*).

Le Fate *See* **Caverna delle Fate**.

Le Gros Clark, Wilfrid Edward (1895–1971) Wilfrid Le Gros Clark was born in Hemel Hempstead, England, and entered St. Thomas's Hospital Medical School in London in 1912. After qualifying in medicine in 1916 he joined the Royal Army Medical Corps in 1917 and was sent to France early in 1918 to take part in WWI. He was demobilized in 1919 and he returned to St. Thomas's to study surgery. After passing his surgical qualification exams he became the Principal Medical Officer in Sarawak, part of the island of Borneo, in 1920. There he developed his longstanding interest in natural history (for details see below and Le Gros Clark's autobiography) by studying tarsiers (*Tarsius*) and tree shrews (*Tupaia*) with a particular emphasis on their evolutionary relationship to the primates. In 1923 Le Gros Clark left Sarawak to accept a position as head of the Department of Anatomy at St. Bartholomew's Hospital Medical School in London where he remained from 1924 until 1929. He moved again in 1930 to become professor of anatomy at his alma mater, St. Thomas's Hospital Medical School, and again in 1934 when he accepted the Dr Lee's Professorship of Anatomy at the University of Oxford. In the same year he was elected as a Professorial Fellow of Hertford College, Oxford. Through Le Gros Clark's studies of the tree shrew he concluded that they are the closest living example of the earliest primate to become differentiated from the mammalian/insectivore stem. New evidence has caused this interpreta-

tion to be rejected, but at the time Clark's research represented state-of-the-art comparative primatology, and it influenced many researchers. In 1934 Le Gros Clark published *Early Forerunners of Man*, which discussed human evolution and used his deep knowledge of the anatomy of living and fossil primates to review the pertinent comparative evidence. Because he believed that parallel evolution was common in primates he argued that the hominin and anthropoid lines could have separated very early. Non-Darwinian models of evolution were common throughout the early 20thC and Clark initially supported the idea of **orthogenesis** (i.e., that evolution was directional) as opposed to a strictly Darwinian conception of evolution driven solely by natural selection. Also, like many of his colleagues, Clark initially considered the australopiths discovered by Raymond **Dart** and Robert **Broom** in southern Africa to be anthropoid apes and not closely related to our own evolution. Le Gros Clark's experiences in WWI led him to be a pacifist, so that in WWII he contributed to the war effort through his research on muscle and nerve regeneration. Le Gros Clark attended the First **Pan-African Congress on Prehistory** held in Nairobi, Kenya, in 1947 and while he was in Africa he traveled to South Africa to inspect the australopith fossils that had been collected by Dart and Broom. Clark, who was one of the few Europeans to actually inspect the full collection thoroughly, became convinced that Dart and Broom had been correct in their interpretation that *Australopithecus* represented a hominin intermediate between anthropoid apes and modern humans and ancestral to modern humans. His article "Observations of the anatomy of the fossil australopithecine" published in the *Journal of Anatomy* in 1947 marked a turning point in the general acceptance of australopiths as at least potential, if not actual, human ancestors. Le Gros Clark was drawn into the controversy over **Piltdown Man** by his Oxford colleague Joseph Weiner who, together with Kenneth **Oakley**, began in 1953 to re-examine the Piltdown fossils and their context. The Piltdown (*Eoanthropus dawsoni*) fossils were found between 1908 and 1915 in Sussex, England, by Charles Dawson, an amateur archeologist. The large cranium, but apelike jaw, of the specimen differed from the other fossil hominins that were becoming known by that time, but Piltdown had many supporters as a human ancestor. While some physical anthropologists had been skeptical of the specimen, it was not until 1953 that Weiner, Oakley, and Le Gros Clark used fluorine dating to show that the fossils were for geries and thus were able to finally remove Piltdown Man from the accepted ranks of valid hominin species. The removal of Piltdown Man from modern human evolutionary history had implications for the way other controversial hominin species, especially those assigned to the genus *Australopithecus*, were viewed. Le Gros Clark re-examined human evolution in *The Fossil Evidence for Human Evolution: an Introduction to the Study of Palaeoanthropology* (1955). This influential work discussed the hominin fossil record and contemporary debates over primate taxonomy and phylogeny, and in it the australopiths are treated as important links in the human lineage. By backing away from the centrality of parallel evolution and by drawing upon the work of the American paleontologist George Gaylord Simpson, Le Gros Clark helped reform **paleoanthropology**. Simpson had revolutionized paleontology by integrating the new emphasis on Darwinian evolutionary mechanisms and population genetics that were part of the **modern evolutionary synthesis** into paleontology. Le Gros Clark's 1955 book rejected theories, common during the early 20thC, suggesting modern human races had separate evolutionary histories. In 1959, in perhaps his best-known book, *The Antecedents of Man*, Le Gros Clark discussed primate comparative anatomy, including their embryology and dental and cranial anatomy. In it he examined the known hominin fossils, including treatments of the recent discoveries made by Louis Leakey in East Africa. This work went through three editions by 1971 and was widely used by students of primatology and paleoanthropology. Le Gros Clark retired as Dr Lee's Professor of Anatomy at the University of Oxford in 1962, but he was elected an Honorary Fellow of Hertford College, Oxford, in 1962 and he retained that appointment until his death. After his retirement he continued to conduct research and write books, including a second edition of *The Fossil Evidence for Human Evolution: an Introduction to the Study of Palaeoanthropology* (1964), *Man-apes or Ape-men? The Story of Discoveries in Africa* (1967), and an autobiography, *Chant of Pleasant Exploration* (1968). Le Gros Clark was elected a Fellow of the Royal Society in 1935, he was the President of the International Anatomical Congress in 1950, and he was elected the President of the Anatomical Society of Great Britain and Ireland in 1951. In 1955 he was made a Knight Commander of the British Empire.

Le Moustier (Location 44°59′38″N, 1°03′35″E, France; etym. named after the nearby town) History and general description A site along the Vézère river consisting of two shelters. The upper shelter was excavated in the late 19thC by Edouard Lartet and Henri Christy and the material they recovered led Gabriel de Mortillet to define the **Mousterian** technocomplex. Otto Hauser explored the much deeper deposit in the lower shelter in the early 20thC and found a deliberately buried associated skeleton of a young individual (Le Moustier 1) and a skull fragment (Le Moustier 3). Denis Peyrony excavated the remains of the lower shelter and found at least 11 different archeological layers. He also discovered the remains of an infant (Le Moustier 2). Temporal span and how dated? Thermo**luminescence** and **electron spin resonance spectroscopy dating** of the lower shelter suggest an age range of c.55.8–40.3 ka for the Mousterian levels and c.42 ka for the **Châtelperronian** level. Hominins found at site Le Moustier 1, the cranium, mandible, and a few postcranial remains of a young male c.10.5–12.1 years of age, attributed to *Homo neanderthalensis*; **Le Moustier 2**, a nearly complete neonatal skeleton, likely also intentionally buried; and **Le Moustier 3**, a cranial fragment. Archeological evidence found at site The deep Mousterian deposit contained several layers that Francois Bordes attributed to different types or facies of the Mousterian (e.g., Mousterian of Acheulean Tradition, Denticulate Mousterian, Typical Mousterian, etc.). These layers were topped by thin Châtelperronian and **Aurignacian** levels. Key references: Geology, dating, and paleoenvironment Mellars and Grün 1991, Valladas et al. 1986; Hominins Hauser 1909, Klaatsch and Hauser, 1909, Archeology Peyrony 1930.

Le Moustier 1 Site Le Moustier. Locality N/A. Surface/*in situ* In situ. Date of discovery March 7, 1908. Left radius and ulna and lower extremities removed on that date. Vertebrae and part of the pelvis disintegrated when an attempt was made to excavate them (Hauser 1909); the cranium and mandible were excavated on August 12 by anatomist Hans Klaatsch at Otto Hauser's invitation. Finders Otto Hauser and crew. Unit N/A. Horizon Probably layer J (Typical **Mousterian**). Bed/member N/A. Formation N/A. Group N/A. Nearest overlying dated horizon N/A. Nearest underlying dated horizon N/A. Geological age c.40 ka. Developmental age c.10.5–12.1 years (Smith et al. 2010). Presumed sex Male (Thompson and Nelson 2005a). Brief anatomical description Nearly complete cranium, now missing parts of the midface and the upper right central incisor lost some time after 1945, and fragments of the postcranial skeleton (some bones were damaged in a fire in 1945 but these burned fragments were subsequently recovered from the rubble of the museum). Casts of the original postcranial remains made soon after discovery retain important morphological and diagnostic information. The elements cast include the left clavicle, fragment of the right scapula, left second rib, right radius and ulna (missing epiphyses), left femur and tibia, left patella, both fibulae (fragments of diaphyses), and the left first metatarsal. Burned fragments that remain include left femur diaphysis (incomplete), a fragment of the diaphysis of the right humerus, a fragment of the right scapula, a fragment of the proximal diaphysis of the right ulna, a fragment of the distal end of the left radius, fragments of the pelvic bone, a fragment of the left talus, a fragment of a rib, several diaphyseal fragments of long bones, fragments of epiphyses of several bones, and small unidentified bone fragments (Thompson and Nelson 2005b). Despite the individual's young age at death the specimen exhibits many features typical of *Homo neanderthalensis*. Announcement Hauser 1909. Initial description Klaatsch and Hauser 1909. Photographs/line drawings and metrical data Weinert 1925. Detailed anatomical description Weinert 1925, Bilsborough and Thompson 2005, Thompson 2005, Thompson and Bilsborough 1997, 2005, Thompson and Illerhaus 1998, Thompson and Nelson 2005b (see also papers in Ullrich 2005). Initial taxonomic allocation *Homo mousteriensis* (Klaatsch and Hauser 1909). Taxonomic revisions *H. neanderthalensis* (Weinert 1925). Current conventional taxonomic allocation *H. neanderthalensis*. Informal taxonomic category Pre-modern *Homo*. Significance A fairly complete cranium and mandible of a late "classic" Neanderthal. The mandible shows evidence of having been broken some time prior to the death of the individual (MacGregor 1964; see also Ullrich 2005). The individual is a juvenile and so provides important information about patterns of Neanderthal growth and development. Estimates using synchrotron virtual histology provide a dental age of 11.6–12.1 years (Smith et al. 2010), which contrasts with estimates of 15–16 years based on modern rates of dental formation. This new dental age is consistent with the postcranial age estimates (Nelson and Thompson 2005), particularly that based on the length of the femur (10.5 years) and indicate that Neanderthal growth was accelerated compared to modern humans.

Given that Le Moustier 1 had completed about 85% of his skeletal and dental growth at the time of death it is estimated that Neanderthals had a total duration of growth of 13–14 years, much shorter than seen in most modern human populations. The implication is that the modern human pattern of growth and development is recent. Location of original Although the postcranial skeleton was damaged in a fire at the end of WWII, fragments were recovered from the building rubble (Ullrich 2005). The cranium, mandible, and remaining postcranial fragments are housed at the Staatliche Museen zu Berlin, Museum für Ur- und Frügeschichte, Berlin, Germany.

Le Moustier 2 Site Le Moustier. Locality N/A. Surface/*in situ* In situ. Date of discovery May 19, 1914. Finder D. Peyrony. Unit N/A. Horizon Probably layer J (Typical **Mousterian**). Bed/member N/A. Formation N/A. Group N/A. Nearest overlying dated horizon N/A. Nearest underlying dated horizon N/A. Geological age *c*.40 ka. Developmental age Neonate *c*.4 months or less (Maureille 2002) Presumed sex N/A. Brief anatomical description Almost complete cranium and deciduous dentition (missing left temporal) and fairly complete postcranial skeleton (missing scapulae and pubes) of a neonate. Announcement Peyrony 1921, 1930. Initial description Maureille 2002a, 2002b. Photographs/line drawings and metrical data Maureille 2002a, 2002b, 2005. Detailed anatomical description Maureille 2005. Current conventional taxonomic allocation *Homo neanderthalensis*. Informal taxonomic category Pre-modern *Homo*. Significance A fairly complete neonate skeleton of a late "classic" Neanderthal. Location of original Musée National de Préhistoire (Les Eyzies-de-Tayac), France.

Le Moustier 3 Site Le Moustier. Locality N/A. Surface/*in situ* In situ. Date of discovery 1910. Finder O. Hauser. Unit N/A. Horizon N/A. Bed/member N/A. Formation N/A. Group N/A. Nearest overlying dated horizon N/A. Nearest underlying dated horizon N/A. Geological age *c*.40 ka. Developmental age N/A. Presumed sex N/A. Brief anatomical description Cranial fragment. Announcement Hauser 1911, 1925. Initial description N/A. Photographs/line drawings and metrical data N/A. Detailed anatomical description N/A. Current conventional taxonomic allocation *Homo neanderthalensis*. Informal taxonomic category Pre-modern *Homo*. Significance A cranial fragment from

Le Moustier site was mentioned in two of Hauser's publications but there is no formal description of the specimen. Hauser noted that the skull fragment was from the "lower grotto" at Le Moustier and in the travel diaries of Dr F. Krantz Le Moustier 3 is mentioned as "a small skull fragment isolated from a Moustier layer" (Rosendahl 2005, p. 75). <u>Location of original</u> Unknown.

Le Regourdou *See* **Regourdou**.

le Riche, William Harding *See* Robert Broom; Sterkfontein.

leaf waxes Leaf waxes are the waxy molecules on the surface of terrestrial plant leaves. They are resilient to degradation and thus they are preserved in the geological record and can act as a **proxy** for past vegetation and climate. The carbon and hydrogen isotope ratios in individual leaf wax molecules (such as *n*-alkanes and *n*-alkanoic acids) can be analyzed by **gas chromatography isotope ratio mass spectrometry** dating and used for **paleoclimate** reconstruction. This approach has been used to measure C_3 and C_4 vegetation changes on Mount Kenya (Huang et al. 1999b), off northwest Africa (Huang et al. 2000), in the Congo fan (Schefuß et al. 2003), in the Namib margin (Rommerskirchen et al. 2006), in the Gulf of Aden (Feakins et al. 2005), and in the Turkana Basin (Feakins et al. 2007). Based on the carbon isotopic composition of plant leaf wax biomarkers in marine sediments researchers have reported repeated shifts between forest and grassland conditions in northeast Africa over the past 4 Ma. **Hydrogen isotope analysis** in plant leaf waxes has been applied to the Congo fan (Schefuss et al. 2003), Lake Tanganyika (Tierney et al. 2008), and Lake Wandakara in Uganda (Russell et al. 2009). *See* C_3 and C_4; **stable isotopes**.

Leakey, Louis Seymour Bazett (1903–72) Louis Leakey was born at the Kabete missionary station near Nairobi, in what was then the British East Africa Protectorate (it became the Kenya Colony in 1920). The outbreak of WWI prevented Louis from attending school in Britain, but at the end of the war in 1919 he completed his education at Weymouth College in England. In 1922 he was admitted into St. John's College, Cambridge, where he studied anthropology and archeology with Alfred C. Haddon and Miles Burkitt. During a break in his studies at Cambridge in 1924, Leakey joined the British Museum East African Expedition to what was then Tanganyika Territory to look for dinosaurs with Hans Reck at Tendaguru. Leakey completed his degree in 1926 and resolved to organize "an expedition to Kenya Colony to commence the scientific investigation of the history of the country during Stone Age times" (Leakey 1937, p. 179). He was appointed to a Research Studentship, also at St. John's College, and in July 1926 he returned to East Africa as the leader of the East African Archeological Expedition and between 1926 and 1935 Louis Leakey led four archeological expeditions to investigate the prehistory of Kenya and Tanganyika. By 1927 he had recovered sufficient human skeletal material from Nakuru, Elmentaita, and Gamble's Cave to justify their description and analysis. Leakey approached Sir Arthur **Keith**, by then the Conservator of the **Royal College of Surgeons of England**, to do this; Keith declined, but offered Louis space at the College so that he could do it himself. In 1931 Leakey traveled with Hans Reck to **Olduvai Gorge**; this was the first time Reck had been to Olduvai since the outbreak of WWI had prevented him from returning in 1914. Leakey was joined on that expedition by Hopwood, Donald MacInnes, and Vivian Fuchs. Subsequent expeditions also included visits to Olduvai Gorge, where, with his first wife Frida Leakey, Louis found and mapped many sites. The results of these early expeditions were published as *The Stone Age Cultures of Kenya Colony* (1931) and *Stone Age Africa* (1936). Leakey had completed is PhD at Cambridge in 1930, and while lecturing at the Royal Anthropological Institute in 1933 he met Mary Nicol, a student of archeology and an artist who worked on the illustrations for Gertrude Caton-Thompson's book *The Desert Fayum* (1934). Impressed by her work, Leakey asked her to illustrate his *Adam's Ancestors* (1934). Louis and Mary Nicol were married in 1936 and thus began a remarkably productive professional partnership. As well as resuming work at Olduvai Gorge in 1942 Louis and Mary Leakey relocated the site at **Olorgesailie** where hundreds of naturally unearthed **handaxes** had been reported some 30 years earlier by the geologist and explorer John Walter Gregory. They also explored Miocene sites on the shores of Lake Victoria, and in 1948 they discovered on Rusinga Island the partial skull and jaw of a Miocene primate they named *Proconsul africanus*. In the meantime Louis Leakey had accepted a position in 1945 as Curator of the **Coryndon Memorial Museum** in

Nairobi, a post he held until 1961. Louis also organized the First **Pan-African Congress on Prehistory**, which was held in Nairobi in 1947. The meeting brought together archeologists, geologists, and paleontologists from across Africa as well as Europe to discuss African prehistory. Louis served as the general secretary of the Pan-African Congress on Prehistory from 1947 to 1951, and was its president from 1955 to 1959. Excavation work at Olduvai Gorge initially progressed slowly due to lack of funds and the limited amounts of time the Leakeys could spend in the field, but this changed in 1951 when Charles **Boise** provided long-term financial support for their research at Olduvai. Excavations at Olduvai Gorge produced considerable quantities of prehistoric artifacts that were described in *Olduvai Gorge: A Report on the Evolution of the Hand-Axe Culture in Beds I–IV* (1951). But despite decades of searching, and apart from some isolated hominin teeth that at the time were not considered of sufficient moment to report, no well-preserved fossils of the tool makers had been discovered. However, in July 1959 Mary **Leakey** found the cranium of a fragmented robust hominin (**OH 5**) in association with **Oldowan** tools and what was interpreted as a "living site." Leakey named the specimen *Zinjanthropus boisei* (the species name recognizes the support given by Charles Boise), and concluded that *Zinjanthropus* made the tools that were found with it. Specimens of volcanic rock from adjacent deposits were analyzed by a team of physicists, Garniss Curtis and Jack Evernden, from the University of California at Berkeley using a new technique called **potassium-argon dating**, and they established that *Zinjanthropus* had lived *c.*1.75 Ma; this was the first use of this dating method to establish the age of a fossil hominin. In May 1960 Jonathan Leakey found a hominin tooth, and this led to the recovery of the skull (**OH 7**) and the postcranial bones (**OH 8**) of a more gracile form of hominin in deposits that also contained artifacts and processed animal bones; more evidence was found in 1960 and in 1963. Louis enlisted the help of South African paleoanthropologist Phillip **Tobias** (who was already undertaking the detailed description of OH 5 for the Leakeys) and British primatologist John **Napier** to analyze the post-1959 hominin discoveries, and in 1964 the three authors published a paper that introduced and described a new hominin taxon, *Homo habilis*. *H. habilis* possessed a larger brain than *Zinjanthropus* as well as other anatomical traits more akin to modern humans, and this led Louis to suggest that it was *H. habilis* and not *Zinjanthropus* that made

the Oldowan tools found at Olduvai Gorge. In 1961 Leakey discovered an upper jaw and teeth of a Miocene primate in western Kenya. This specimen, which he named *Kenyapithecus wickeri*, was followed in 1965 by fragments of another jaw and teeth that Louis referred to *Kenyapithecus africanus*. Louis and Mary Leakey's excavations at Olduvai Gorge transformed **paleoanthropology**, through both the hominin specimens they recovered and the archeological evidence they discovered there. Moreover, because the geology and paleontology at Olduvai needed to be properly studied Louis and Mary enlisted the assistance of specialists (e.g., Shirley Coryndon, Richard **Hay**, Dick Hooijer) to conduct the work. Likewise, they relied upon the assistance of Phillip **Tobias**, John **Napier**, Michael **Day**, and Peter Davis to describe and analyze the hominin fossils; the archeological evidence was the bailiwick of Mary Leakey. The immense productivity of the team of researchers assembled by the Leakeys is evident in the publications produced during this period, which include *The Progress of Man in Africa* (1961), *Olduvai Gorge, 1951–1961. Volume 1. A Preliminary Report on the Geology and Faunas* (1965), *Fossil Vertebrates of Africa* (1969 and 1970), *Olduvai Gorge, Volume 2. The Cranium and Maxillary Dentition of Australopithecus (Zinjanthropus) boisei* (1967), *Olduvai Gorge, Volume 3. Excavations in Beds I and II* (1971), *Olduvai Gorge, Volume 4. The Skulls, Endocasts and Teeth of Homo habilis* (1991), and *Olduvai Gorge, Volume 5. Excavations in Beds III, IV and the Masek Beds* (1994). The **L.S.B. Leakey Foundation** was established in California in 1968 to support the research of Louis (and Mary) Leakey and to encourage further cognate research in Africa and elsewhere, and after Louis' death in 1972 the International Louis Leakey Memorial Institute for African Prehistory (or TILLMIAP) was established at the National Museums of Kenya in recognition of Louis Leakey's immense impact on paleoanthropology and African prehistory. In addition to his paleoanthropological contributions Louis Leakey published many seminal scholarly articles on the ethnography and culture of the Kikuyu, and he initiated, with Mary Leakey, programs that recorded Angolan String Figures and rock-shelter paintings in Tanzania. Louis Leakey was also instrumental in instigating the primate research of Jane Goodall, Dianne Fossey, and Birute Galdakis. *See also* Charles **Boise**; *Homo habilis*; **Olduvai Gorge**; *Paranthropus boisei*.

Leakey, Mary Douglas Nicol (1913–96)

Mary Nicol was born in London, England, the daughter of Erskine Nicol, a well-known painter, and Cecilia Frere, whose ancestor was the British antiquary, John Frere. In 1800 Frere had argued that a flint axe he had found several years earlier was of great antiquity, and was evidence that modern humans were more ancient than conventional wisdom suggested. The Nicol family lived for a time in the Dordogne region of France where Mary visited the prehistoric sites of the region with French prehistorian Elie Peyrony. When her father died in 1926, Mary and her mother returned to London. Mary had been unable to gain admittance to university, but in 1930 she began attending archeology and geology lectures at The University of London, as well as the archeology lectures of Mortimer Wheeler at the London Museum. She also worked with Dorothy Liddell on excavations at the Neolithic site of Hembury in Devon. Mary, who was a skilled artist, not only gained experience in archeological excavation, but also made fine drawings of the flint tools unearthed at the site. The drawings came to the notice of British archeologist Gertrude Caton-Thompson, who asked Mary to illustrate her book *The Desert Fayum* (1934). Mary and Caton-Thompson developed a close friendship and it was Caton-Thompson who, in 1933, introduced Mary to Louis **Leakey**, who was in London lecturing at the Royal Anthropological Institute. During the early 1930s Mary joined Dorothy Liddell in the excavation at Stockbridge in Hampshire and in 1934 Mary conducted excavations on her own at Jaywick Sands in Essex. Mary and Louis Leakey were married in 1936, and later in that year she joined Louis in Africa. During the early 1940s Mary along with Louis explored many prehistoric sites including **Olorgesailie**. While excavating on Rusinga Island in Lake Victoria in 1948 Mary discovered the partial skull of a **Miocene** primate later named *Proconsul africanus*. When Louis and Mary began more intensive work at **Olduvai Gorge** in 1951, they directed their efforts to the lowermost deposits of the gorge where they found numerous stone artifacts and processed animal bones. After years of excavating at Olduvai, in July 1959 in Bed I Mary found a fragmented cranium belonging to a robust looking hominin. Louis assigned the specimen to a new genus and species, *Zinjanthropus boisei*, and assumed it had made the tools found along with it. Mary mapped the

archeological sites and studied the stone artifacts found in the various deposits at Olduvai. Working with a team of Kenyan assistants and with geologist Richard **Hay** of the University of California at Berkeley, Mary devised a systematic way of classifying the sequence of stone tool industries at Olduvai, as well as suggesting the sources of the raw material for the stone tools. Mary's careful studies of what she interpreted as an apparent camp or living floor found in the *Zinjanthropus* deposits were at the time judged to be important information about the social life of these hominins. The results of Mary's archeological investigations were reported in detail in the third and fifth volumes of the *Olduvai Gorge* series, respectively: *Excavations in Beds I and II, 1960-1963*, published in 1971, and *Excavations in Beds III, IV and the Masek Beds*, published in 1994. Mary continued to conduct research at Olduvai after Louis' death in 1972. In 1974 she began work at a site called **Laetoli** and in 1975 she discovered fragmentary hominin remains including several jawbones, which have been dated to *c*.3.8–3.6 Ma. These remains were attributed to *Australopithecus afarensis*, although Mary demurred from this interpretation. Between 1974 and 1981 more hominin fossils were found at Laetoli and in 1978 trails of fossil footprints left by at least two hominin individuals were found in deposits of hardened volcanic ash; the **Laetoli footprints** indicate that hominins were walking upright at least 3.6 Ma. Mary Leakey retired from fieldwork in 1983 and moved back from Olduvai to the family home in Langata, near Nairobi. In 1984 she published an autobiography entitled *Disclosing the Past*.

Leakey, Meave (1942–)

Meave Epps was born in London, England. She attended the University of North Wales where she completed a BSc in zoology and marine zoology. In 1965 she began her PhD work in zoology at the University of North Wales and later that year she took a staff position at the Tigoni Primate Research Centre near Nairobi, Kenya. Meave completed her PhD in 1968, and early the following year, she returned to Kenya to take a temporary position as the Director of the Tigoni Primate Research Centre. Later that year she joined Richard **Leakey's** team of researchers who were searching for hominin fossils at **Koobi Fora** on the eastern shores of Lake Rudolf (now called Lake Turkana) in northern Kenya. Richard and Meave were married in

1970 and Meave became an important contributor to the **East Rudolf Research Project**, the precursor of the **Koobi Fora Research Project**. Glynn Isaac, the co-leader of the Koobi Fora Research Project, had died in 1985, and when in 1989 Richard Leakey undertook other responsibilities Meave took over the leadership of the field project. Meave accepted a position at the **National Museums of Kenya** in 1969 and from 1982 to 2001 she was the head of the museum's paleontology division. Meave Leakey's research in the Turkana Basin has focused on the evolution of East African mammals, especially monkeys, and the discovery of fossil remains of early human ancestors. She collaborated on the reconstruction of the **KNM-ER 1470** skull discovered in 1972 at East Rudolf and assisted with the recovery and reconstruction of the **KNM-ER 3733** *Homo erectus* (*Homo ergaster*) cranium discovered in 1975, as well as the cranium KNM-ER 1813, and others. Meave Leakey's role in the Koobi Fora Research Project expanded in 1989 when she became the coordinator of research. It was under her leadership between 1994 and 1997 that fragments of upper and lower jaws, a number of associated isolated teeth, and a proximal tibia, now classified as *Australopithecus anamensis*, were found at **Kanapoi** and **Allia Bay** in the Turkna Basin. This was followed in 1999 by the discovery at **West Turkana** of a mostly complete cranium (**KNM-WT 40000**) of a hominin that was assigned to a new species and genus, *Kenyanthropus platyops*. The specimen was dated at *c*.3.5 Ma, thus making it the contemporary of *Australopithecus afarensis*, and thereby challenging the notion that *Au. afarensis* is necessarily the ancestor of later hominins. In addition to publishing numerous research articles Meave Leakey also co-authored, with Richard Leakey, the first volume of the Koobi Fora Research Project series titled *The Fossil Hominids and an Introduction to Their Context, 1968-1974*, published in 1978. In 2003 John Harris and Meave Leakey published *Lothagam: The Dawn of Humanity in Eastern Africa* that surveys the geology, dating, and fossil fauna of **Lothagam**, a site on the western side of Lake Turkana, which documents the late Miocene/early Pliocene faunal turnover and the associated spread of grasslands. Meave Leakey continues to conduct field research in the Turkana Basin, and now co-leads the Koobi Fora Research Project with her daughter Louise Leakey who completed her PhD in paleontology from University College London. *See also* **Kanapoi**; **Lothagam**; **Koobi Fora**.

Leakey, Richard Erskine Frere (1944–)

Richard Leakey was born in Nairobi, Kenya, the middle of the three sons of Louis and Mary **Leakey**. Initially Richard Leakey had no interest in pursuing human origins research and after dropping out of high school he began trapping animals to supply skeletons to universities and museums and later started his own safari business. But while flying over Lake Natron on the border between Kenya and Tanganyika (now Tanzania) in 1963 he noticed geological deposits that looked promising possible locations for vertebrate fossils. Richard explored the **Peninj** region early in 1964, and was later joined by archeologist Glynn **Isaac**. In sediments exposed by the Peninj river Kamoya **Kimeu** discovered an exceptionally well-preserved hominin mandible that belongs to what is now called *Paranthropus boisei*. Leakey took a decisive step into paleoanthropology in 1967 when he represented Louis Leakey on the the **International Omo Research Expedition** that had been organized to explore the lower reaches of the Omo River valley in southern Ethiopia. During that expedition he was loaned the expedition's helicopter to inspect the sediments exposed on the eastern shore of Lake Turkana (then called Lake Rudolf) in northern Kenya. Despite recovering two early *Homo sapiens* crania that are now considered to be among the earliest fossil evidence for anatomically modern humans in sediments exposed at **Omo-Kibish**, Leakey determined that the exposures along the eastern shore of the lake were more promising than the exposures the Kenyan contingent had been allocated to the east of the Omo River, so he withdrew from the International Omo Research Expedition to concentrate on the sites in northern Kenya. In 1968, with support from the National Geographic Society, Leakey and a team of colleagues conducted their first survey of the area in 1968 and quickly began to discover artifacts, animal fossils, and some hominin cranial and postcranial fragments including a partial *P. boisei* jaw. When Richard Leakey returned to work at Lake Turkana in 1969 he focused his attention on a site called **Koobi Fora**, which together with exposures on the western side of the lake, has become one of the most productive hominin sites in the world. In 1969 Leakey invited Glynn Isaac, his partner in the Lake Natron expedition, to join him as co-director of what became known as the **East Rudolf Research Project**. Over the next two decades Leakey and a team of researchers, together with a team of experienced fossil hunters lead by Kamoya Kimeu, explored the fossil-rich exposures along the eastern shore of Lake Turkana and their efforts resulted in the recovery of a number of important hominin fossils. The first of the relatively complete hominins to be recovered was in 1969 when Leakey and Meave Epps discovered an adult *P. boisei* cranium, **KNM-ER 406**. The following year a partial cranium (**KNM-ER 732**) was found that resembled *P. boisei*, but it was smaller and lacked a sagittal crest. In 1972 in deposits originally thought to be 2.6 Ma Bernard Ngeneo discovered a relatively large-brained hominin cranium, **KNM-ER 1470**, and in 1973 Paul Abell discovered a small-brained hominin skull, **KNM-ER 1805**; that skull together with a cranium with an even smaller brain **KNM-ER 1813**, discovered by Kamoya Kimeu, contributed to the debate about the validity of the taxon *H. habilis*. In 1975, Bernard Ngeneo found yet another hominin cranium, **KNM-ER 3733**, that resembled *Homo erectus* specimens from Indonesia and China. The specimen was also significant since it was found in deposits that were of the same age as the deposits where the KNM-ER 406 *P. boisei* cranium had been found, which meant that at least two species of hominin had lived in the Turkana Basin at the same time. During the late 1970s Leakey's team began to expand beyond sites at Koobi Fora to explore the potential of sites on the west side of the lake. Important discoveries continued to be made at in the Turkana Basin and in 1984 Kamoya Kimeu discovered a skeleton of an adolescent male at West Turkana (**KNM-WT 15000**), which provided important information about the stature, body size, and limb proportions of early pre-modern *Homo*. This find was followed by the discovery in 1985 by Alan **Walker** of an almost complete *c.*2.5 Ma cranium of a megadont archaic hominin, **KNM-WT 17000**. Leakey left his position as director of the National Museums of Kenya in 1989 when he was appointed as director of the Kenya Wildlife Service. During his tenure he worked to prevent elephant and rhinoceros poaching and to reform Kenya's park service. He was forced to resign in 1994 due to political opposition and he published a book about his experience titled *Wildlife Wars: My Battle to Save Kenya's Elephants* (2001). As a result of a plane crash in 1993 both of Leakey's legs

had to be amputated below the knee, and this made continued fieldwork difficult. Leakey's interests became increasingly political during the 1990s, and he joined a group of Kenyan intellectuals in 1995 to form a new political party called the Safina Party. Kenya's President, Daniel arap Moi, appointed Leakey to Head the Civil Service in 1999, but he was forced to resign in 2001. In addition to many papers in scientific journals announcing and describing the hominin and nonhominin fossils from the Turkana Basin, and elsewhere, Leakey published two popular books about his field research and ideas about human origins, *Origins* (1977) and *People of the Lake* (1978) and together with Meave Leakey they authored the first volume in the Koobi Fora Research Project series entitled *The Fossil Hominids and an Introduction to their Context, 1968-1974* that appeared in 1978. Although Richard Leakey became less involved in fieldwork in the Turkana Basin after 1989, he continued to write on the subject. He published two general works on human origins during the early 1980s, *The Making of Mankind* (1981) and *Human Origins* (1982) and following the discoveries at Lake Turkana during the 1980s Leakey joined with Roger Lewin to write *Origins Reconsidered* (1992). He is presently a Professor at Stony Brook University, where he is working to establish and develop the Turkana Basin Institute as a focus for a wide range of research programs, including paleoanthropology, in an effort to foster sustainable development in the Turkana Basin. *See also* **Koobi Fora; Koobi Fora Research Project; Omo-Kibish; West Turkana**.

Leakey Foundation *See* **L.S.B. Leakey Foundation**.

learning *See* **communication; cooperation; imitation; social intelligence; social learning; teaching; theory of mind**.

least squares regression *See* **ordinary least squares regression**.

lectotype (Gk *lectos*=chosen and *typus*=image) If a **holotype** was not designated at the time of the original description of a **taxon**, one of the specimens (or **syntypes**) listed in that description can subsequently be designated as the type. This is then referred to as the lectotype of that taxon. For example, because **Alexeev** (1986) did not specifically name a holotype of *Pithecanthropus rudolfensis*, Wood

(1999) formally named **KNM-ER 1470** as the lecto-type of that taxon.

Ledi-Geraru Research Project   This paleo-anthropology project was begun in 2002 and was originally known as the Middle Ledi Research Project. The project was organized by the late Charles Lockwood (paleoanthropology), Ramon Arrowsmith (geology), and Kaye Reed (paleoecol-ogy/paleontology). In 2002, the researchers collected material in what Jon Kalb (see the **International Afar Research Expedition** and **Rift Valley Research Mission in Ethiopia**) referred to as the **Middle Ledi**, but research in 2002 focused on the Hawoona Dora and Leadu Basins, which are Afar names for the dry drainages in the area. Two *Australopithecus afarensis* teeth were recovered from the Kada Hadar Member of the **Hadar Formation** near the Hawoona Dora Basin. Other fauna from the western region was attribu-table to taxa also found in the **Hadar** and **Dikika** study areas of the same age. Subsequent surveys from 2003 through 2008 expanded the boundaries of the research area and the name was changed to the Ledi-Geraru Research Project to better describe the broader region. These later expeditions included a comprehensive study of the geology (DiMaggio et al. 2008, Dupont-Nivet et al. 2008) and identification of potential paleontological and archeological sites. These include regions such as Grufaitu in which there are conformable sediments less than 2.95 Ma, as well as Geraru and Asbole in which there were numerous fossils and stone tools. No fossils were collected during the surveys of 2003–8. Researchers on this expanded project include Lars Werdelin (paleontology), Guilloume Dupont-Nivet (paleo-magnetic dating), Chris Campisano (sedimentology, spatial analysis), Erin DiMaggio (geology), Mark Sier (geology), and Andy Cohen (geology, continen-tal drilling).

Ledi-Geraru study area   (Location 11°20′N, 40°45′E, Ethiopia; etym. Ledi and Geraru are Afar names for the river beds that extend to the west and east in this area) History and general descrip-tion Early research in this area was accomplished by the **International Afar Research Expedition** (or I.A.R.E.) from 1972 through 1974 and by the **Rift Valley Research Mission in Ethiopia** (or RVRME) from 1975 to 1978. Paleoanthropological and geological research was restarted in the area in 2002 by the **Ledi-Geraru Research Project**

and is ongoing. The area is between the Awash and Mille River basins in northeastern Ethiopia. Fossil exposures are directly east and somewhat north of the Hadar site, to the north of the Dikika site, and to the south of the Woranso-Mille study area. Temporal span and how dated? Tephrostra-tigraphic correlation with argon-argon dated units of the Hadar and Busidima Formation as well as direct **argon-argon dating** (Roman et al. 2008) suggest the deposits in the Ledi-Geraru range from slightly less than 3.4 to 1.7 Ma. Thus, the sediments encompass much of the **Hadar For-mation** and **Busidima Formation**. Hominins found at site Left M_3 (HD 117-1) and left M_3 fragment (HD 118-1); both teeth have been attributed to *Australopithecus afarensis*. Archeo-logical evidence found at site Evidence of **Oldowan**, **Acheulean**, and **Middle Stone Age** technologies has been found in the study area. Key references: Geology, dating, and fauna DiMaggio et al. 2008, Dupont-Nivet et al. 2008, Roman et al. 2008.

Leibniz Institute for Evolution and Biodiversity Research   *See* **Museum für Naturkunde**.

les Osiers   *See* **Bapaume**.

Lembahmenge   One of the Early and Middle **Pleistocene** sites in the exposures of the **Ola Bula Formation** within the **Soa Basin** of central Flores, Indonesia.

Lemuta Member   *See* **Olduvai Gorge**.

leptomeninges   (Gk *lepto*=thin and *meninx*= membrane) The inner pair of meninges (i.e., the pia and arachnoid). They are both thin-layered and both are derived from ectoderm. *See also* **meninges**.

Levallois   (etym. the Levallois **industry** takes its name from Levallois-Perret, a Paris suburb, the location of the archeological site where this method of stone tool manufacture was first recog-nized) Levallois refers to a distinctive strategy or approach to the manufacture of stone tools that involves the production of large, relatively thin, flakes (called Levallois flakes) from carefully pre-pared cores (Boëda 1994, 1995, Inizan et al. 1999, Van Peer 1992). Levallois cores have an upper surface from which Levallois flakes are removed and a lower surface containing the "unflaked" volume of the core that diminishes in volume as successive flakes are removed. The two surfaces are separated by a **striking platform** that may extend around the full perimeter of the core. The "preparation" involved in a prepared core consists of removing multiple small flakes to shape the various convexities on the "upper" surface of the **core**; it is these convexities that determine the size and shape of the resulting Levallois flakes. Preparation also involves carefully shaping the striking platform to ensure successful flake removal; this latter phase of the preparation results in the characteristic **facetted platform**s of Levallois flakes. The Levallois production strategy is capable of producing Levallois flakes, **blade**s, or **point**s, and it can be preferential or recurrent depending on whether one, or several, Levallois flakes, points, or blades are removed from the upper surface before further preparatory flakes are removed to reshape it. Levallois technology first appears in **Acheulean** sites in Africa and Eurasia, and it is seen commonly in many **Middle Stone Age** or **Middle Paleolithic** assemblages (e.g., the **Kapthurin Formation**; see Monnier 2006 and Tryon et al. 2005). The Levallois technology may have been invented once, but more likely it was developed independently on several occasions (cf. Foley and Lahr 1997, Brantingham and Kuhn 2001).

Levant (MFr. *levant* = the Orient from L. *lever* = to rise) Its first recorded use according to the *Oxford English Dictionary* is in 1497 for "the Mediterranean lands east of Italy," but today it refers to the lands on the shores of Mediterranean that are in the direction of the sunrise as seen from the sea. Today the Levant comprises Israel, Lebanon, and Palestine. So "Levantine Neanderthals" is an inclusive term for *Homo neanderthalensis* remains from sites in those countries.

Levantine Neanderthals *See* **Levant**.

Levene test Sometimes known as Levene's test, a statistical test for the equality of two variances that is not as sensitive to non-normality as an *F* test. Rather than determining whether a ratio of two variances differs significantly from 1 like the *F* test, the Levene test calculates the absolute value of observations from their group mean for two groups and compares the resulting absolute values of the deviations. If variances in the two groups are equal, then on average the observations should be the same distance from their group mean for both groups. If one set of absolute values is significantly greater than the other as identified by a *t* test or **analysis of variance**, then that group has a significantly greater variance. Variants of this test calculate deviations from the median rather than the mean, and may use **randomization** to determine whether average deviations differ between groups.

lever A lever is a rigid bar with a fulcrum (pivot, or point of rotation). Power (or force) arm length refers to the distance between the fulcrum and the point of application of the force; load arm length refers to the distance between the fulcrum and the load. A lever is a simple machine that can be used to multiply the mechanical force that can be applied to an object (load), based on the simple principle that the moments (M = force times distance), or torques, will be equivalent about the fulcrum. For example, if the power arm is five times longer than the load arm, then the force applied to the load will be five times greater than the applied (power) force. Levers are commonly found in musculoskeletal systems. For example, in the modern human foot, the triceps surae (calf) muscles contract and pull on the calcaneus via the Achilles tendon (the force), which rotates about the ankle joint (the fulcrum) to apply a force through the forefoot (the load). A longer calcaneus would allow the foot lever system to generate greater forces at the forefoot; however, but that enhanced force is achieved at the expense of speed.

levo (L. *laevus* = left) Refers to the L- or left-handed form of a complex molecule. *See* **epimerization**.

LGM Acronym for the **Last Glacial Maximum** (*which see*).

LH (or L.H.) Acronym for Laetoli hominid and the prefix used for hominin fossils recovered from **Laetoli**, Tanzania.

LH 2 Site **Laetoli**. Locality Locality 7. Surface/ *in situ* In situ. Date of discovery 1974. Finder Maundu Muluila. Unit N/A. Horizon Found within aolian tuff "*b*" deposits (Leakey et al. 1976, Fig. 3). Bed/ member Laetolil Beds. Formation N/A. Group N/A. Nearest overlying dated horizon Tuff *c*.

Nearest underlying dated horizon "Two dates, 3.82±0.16 Ma and 3.71±0.12 Ma, (from a 60 cm thick tuff lying approximately 50 m below tuff c) average 3.77 and give a lower limit to the age of the hominid remains" (*ibid*, p. 463). Geological age If LH 4 (which is within the tuff b, the tuff immediately below tuff c) is approximately the same age as a biotite crystal from tuff b "approximately 1–2 m below...tuff c" (*ibid*, p. 463) then it is c.3.59 Ma. Developmental age Juvenile. Presumed sex Unknown. Brief anatomical description An immature mandible unfused at the midline. It preserves most of the alveolar process on the right side, but less than half is preserved on the left side. The crown of the right dc, most of the right dm_1 crown, the crowns of the right dm_2 and M_1, and the crowns of the left dm_1, dm_2, and M_1 are preserved. Announcement Leakey et al. 1976. Initial description Leakey et al. 1976. Photographs/line drawings and metrical data Leakey et al. 1976. Detailed anatomical description White 1977. Initial taxonomic allocation *Homo* sp. Taxonomic revisions *Australopithecus afarensis* (Johanson et al. 1978). Current conventional taxonomic allocation *Au. afarensis*. Informal taxonomic category Archaic hominin. Significance One of the better preserved juvenile *Au. afarensis* mandibles. Location of original **National Museum and House of Culture, Dar es Salaam**, Tanzania.

LH 4 Site **Laetoli**. Locality Locality 7. Surface/*in situ* Surface. Date of discovery 1974. Finder Maundu Muluila. Unit N/A. Horizon Found within aolian tuff "b" deposits (Leakey et al. 1976, Fig. 3). Bed/member Laetolil Beds. Formation N/A. Group N/A. Nearest overlying dated horizon Tuff c. Nearest underlying dated horizon "Two dates, 3.82±0.16 Myr and 3.71±0.12 Myr, (from a 60 cm thick tuff lying approximately 50 m below tuff c) average 3.77 and give a lower limit to the age of the hominid remains" (*ibid*, p. 463). Geological age If LH 4 (which is within the tuff b, the tuff immediately below tuff c) is approximately the same age as a biotite crystal from tuff b "approximately 1–2 m below...tuff c" (*ibid*, p. 463) then it is c.3.59 Ma. Developmental age Adult. Presumed sex Unknown. Brief anatomical description A mandibular corpus. The crowns of the right P_3–M_3 and the left P_4–M_2 are well preserved, plus the damaged crown of the right canine crown, the damaged root of the left canine, and the root only of the left P_3. The internal mandibular contour is narrow and

the anterior surface of the symphysis slopes posteriorly. Announcement Leakey et al. 1976. Initial description Leakey et al. 1976. Photographs/line drawings and metrical data Leakey et al. 1976 [NB: the occlusal view of LH 4 (Fig. 7) is mislabeled as **LH 2**], White 1977. Detailed anatomical description White 1977. Initial taxonomic allocation *Homo* sp. Taxonomic revisions *Australopithecus afarensis* (Johanson et al. 1978). Current conventional taxonomic allocation *Au. afarensis*. Informal taxonomic category Archaic hominin. Significance The distinctive symphyseal morphology of the LH 4 mandible was central to the argument for recognizing a new species of *Australopithecus* for the hominin fossils from Laetoli and **Hadar** and it was probably because of this that LH 4 was chosen as the **holotype** of *Au. afarensis*. Location of original **National Museum and House of Culture, Dar es Salaam**, Tanzania.

LH 18 Site **Laetoli**. Locality Locality 2. Surface/*in situ* In situ. Date of discovery 1976. Finder E. Kandini. Unit "Water-worked vitric tuff" (Day et al. 1980, p. 55). Horizon N/A. Bed/member Ngaloba Beds. Formation N/A. Group N/A. Nearest overlying dated horizon Eroding out of an exposure of a horizon that includes a trachytic tuff correlated with "the marker tuff in the lower unit of the **Ndutu** Beds at **Olduvai Gorge**" (*ibid*, p. 55). Nearest underlying dated horizon Lavas overlying the Ndolanya Beds. Geological age The vitric tuff has been "tentatively correlated with the marker tuff in the lower unit of the Ndutu Beds at Olduvai Gorge," (*ibid*, p. 55) the age of which has been estimated to be 120±30 ka (*ibid*, p. 55). Manega (1993) revised the date to 200 ka based on **amino acid racemization** dating. Developmental age Adult. Presumed sex Unknown. Brief anatomical description A cranium comprising a calvaria and a face, with no contact between them. The vault is well preserved, but much of the midline of the cranial base and most of the medial wall of the left infratemporal fossa are missing. The facial fragment includes the maxillary bones, and much of the hard palate. All of the postcanine teeth are represented on at least one side. The heavily worn crowns of the left M^1 and M^3 and the right P^3, P^4, and M^1 are affected by postmortem damage and no useful occlusal morphology is preserved. It has a rounded occipital that is modern in configuration, expanded parietals, and small mastoid processes, but a strongly receding frontal bone makes the specimen recognizably archaic. Endocranial volume Approximately 1200 cm³. Announcement Day et al. 1980. Initial description Day et al. 1980. Photographs/line drawings and metrical data Day et al. 1980, Magori and Day 1983a, 1983b, Schwartz and Tattersall 2003. Detailed anatomical description Magori and Day 1983a, 1983b, Cohen 1996, Schwartz and Tattersall 2003. Initial taxonomic allocation Homo sapiens. Taxonomic revisions Late archaic Homo sapiens (Bräuer 1984). Current conventional taxonomic allocation H. sapiens. Informal taxonomic category Anatomically modern human. Significance The initial description of LH 18 said that "its anatomical features are mixed in that it has some modern characters and some that are archaic" (Day et al. 1980, p. 56). The modern part included an expanded vault, a rounded occipital, and low inion, and the archaic features included frontal flattening, an obvious supraorbital torus, small mastoid processes, an occipitomastoid crest, and a thick vault. The cranium has been likened to **Omo I** and **II** and the Ndutu cranium. Until the discovery of the **Herto** crania, and the redating of Omo I and II from **Omo-Kibish**, many regarded this cranium as the best-preserved and earliest evidence of H. sapiens in Africa. Location of original **National Museum, Dar es Salaam, Tanzania**.

Liang Bua (Location 08°31′50.4″S, 120°26′36.9″E, Flores, Indonesia; etym. Liang Bua in the local Manggarai language means "Cool Cave") History and general description The cave is 30 km/18.6 miles from the north coast of Flores and is 7 km/4.3 miles northwest of Ruteng. Some 500 m above sea level it is a **karstic** solution cavity 30 m wide and 25 m high at the entrance and extends 40 m into Miocene limestone. Its north-facing entrance was exposed by the Wae Racang river, but this feature now lies 30 m below and 200 m distant from the cave entrance. The first large-scale excavations at the cave were carried out by Father Theodor Verhoeven in 1965. The next substantial investigation of the cave was carried out by R.P. Soejono, who between 1978 and 1989 excavated 10 squares (Sectors I–X). Between 2001 and 2004 a team directed by Soejono and Michael Morwood removed the backfill from the previous excavations of Sectors I, III, IV, and VII and extended the excavations in those sectors, and in 2003 they recovered the c.18 ka

associated skeleton, **LB1**, the **holotype** of *Homo floresiensis*, from Sector VII. The following year Morwood et al. excavated deeper into Sector VII and at a depth of 11 m they had not reached bedrock. In the same year they also excavated the adjacent Sector XI and recovered more evidence of LB1 (right humerus, both ulnae, and left fibula) as well as additional hominin remains (e.g., **LB6**). Most recently, Morwood and his colleagues have investigated Sectors V, XII, XIII, and XIV. Morwood et al. (2009) provide a detailed history of the research at Liang Bua. Temporal span and how dated? The c.190 ka conglomerate at the rear, which contains stone artifacts but no faunal/hominin remains, documents the original reopening of the cave by the river. The sediments that contain the artifacts and skeletal remains attributed to *Homo floresiensis* range in age from those in the basal layer of Sector IV (95–74 ka) to around 17 ka, the approximate date attributed to the black tuffaceous sands in Sector XI. Remains of modern humans and other non-endemic fauna are found in the upper layers that lie above the white tuffaceous silts, which are dated to between 11 and 13 ka. **Radiocarbon dating**, various **luminescence dating** techniques (infrared-stimulated luminescence, optically stimulated luminescence dating, and thermoluminescence dating) and coupled **electron spin resonance spectroscopy** and **uranium-series dating** were the main methods used. Hominins found at site A formal analysis of the minimum number of *Homo floresiensis* individuals has yet to be presented, but estimates range between six and 12. The different estimates are due to uncertainties regarding particular aspects of the stratigraphy, but further excavations may help resolve these issues. Archeological evidence found at site Thirty-two artifacts were found at the same level as LB1, but the greatest concentration of artifacts is in Sector IV, where the density reaches 5500 per m³. Most are **flakes** made from volcanics and **chert**, large flakes of which were bifacially and centripetally reduced using freehand, burination, truncation, and bipolar techniques (Moore et al. 2009). Many stone artefacts are found in association with the remains of a dwarf variety of Stegodon. The Stegodon at the site, *Stegodon florensis insularis*, is a subspecies of *Stegodon florensis* known from the Soa Basin of central Flores between 880 and 680 ka and is 30% smaller than *Stegodon florensis*. Key references: Geology and dating Morwood et al. 2004, 2005, Roberts et al.

461

2009, Westaway et al. 2009; Hominins Brown et al. 2004, Morwood et al. 2005, Falk et al. 2005, 2007, 2009, Tocheri et al. 2007, Brown and Maeda 2009, Jungers et al. 2009a, 2009b, Larson et al. 2009, Morwood and Jungers 2009; Archeology Morwood et al. 2004, Moore and Brumm 2007, Moore et al. 2009.

life expectancy The average future time remaining for a person at age t to live. At age zero the life expectancy is referred to as the life expectancy at birth. For a **stationary population** (i.e., a population that is neither increasing or decreasing in size) the life expectancy at birth is equal to the **mean age at death**, and the mean age at death for those who survive to age t is equal to the life expectancy at that age plus the age (e.g., if the life expectancy at age 50 is 7.8 years, then the mean age at death for those alive at age 50 will be 57.8 years). In a stationary life table the life expectancies are found by dividing the total people-years to be lived (T_x) by the **survivorship** (l_x). This amounts to dividing people-years to be lived by the number of people to do the living, resulting in an average number of years to be lived. So, for example, in a population where 1000 individuals are born, 800 survive to age 20 years, and these 800 individuals accumulate 27,200 years of life past age 20, the life expectancy at age 20 will be $27,200/800 = 34$ years. Similarly, the mean age at death for those who reach age 20 will be $20 + 34 = 54$ years. In a hazard model the life expectancy at age t is the integral of the survivorship ($S(t)$) from age t up until the age at which survivorship equals 0.0 (i.e., up until the age by which all individuals are dead, the **maximum life span**). For nonstationary populations the life table or hazards analysis calculations must be adjusted using the population growth rate.

life history Individuals, populations, and **species** all have life histories. However, the term is usually used to describe the relative rate at which members of a species proceed through important developmental, maturational, and reproductive milestones. Life histories reflect the ways taxa have adapted to their ecological context by dividing the energy of individuals between maturation, the maintenance of the individual, and its reproduction, with the latter component being further subdivided between the production of offspring and their subsequent maintenance. An organism's

life cycle is punctuated by events that reflect "decisions" regarding the allocation of resources such as time and energy expenditure, and trade-offs such as those between growth and reproduction. It is the sum of the average of these decisions that define the life history strategy of a taxon. For mammals, this includes when to be born, when to be weaned, how many stages of development to pass through on the way to reproductive competence, when to procreate, and when to die. The life history of modern humans can be divided into six recognized life history stages: infancy, childhood, juvenile, adolescence, adult, and grandmotherhood/postmenopausal. On the basis of comparative physiological and behavioral data, childhood, adolescence, and grandmotherhood are proposed to be unique to modern humans (Bogin 1999a). In modern humans the total life span is relatively long, and the intervals between developmental and reproductive milestones are also relatively long. Modern humans are exceptional because they wean their infants early, their age at first birth is later than would be expected for a great ape of the same body mass, they have an absolutely long life span, and they decouple female fertility and mortality so that females have a long post-reproductive lifespan. Measures of life history include length of gestation, age at first molar eruption, age at weaning, age at sexual maturity, ages at first and last birth, interbirth interval, mean lifespan, and length of post-reproductive lifespan. *See also* **grandmother hypothesis**.

life table A statistical method used to describe mortality such that it can be summarized within various age categories. Shryock and Siegel (1976, p. 251) note that the "basic life table functions" are $_nq_x$, l_x, $_nd_x$, $_nL_x$, T_x, and e_x, which correspond respectively to the **probability of death** (age-specific) between age x and age $x+n$, the **survivorship** at age x, the number of deaths out of the **radix** between ages x and $x+n$, the number of person-years lived between ages x and $x+n$, the **total person-years to be lived in an age interval** at and beyond age x, and the **life expectancy** (average years of life remaining) at age x. The age categories are usually listed in the first column of a life table, while the "basic life table functions" are listed in subsequent columns. The most basic data in life tables used in **paleodemography** are the age intervals and the counts of the number of deaths observed in each age interval (often symbolized as $_nD_x$ to differentiate it from the $_nd_x$ used as a standardized count based on a radix of 100 or 1000). Note that a life table has the appearance of a spreadsheet, with the age intervals forming the rows and the life table functions forming the columns. Indeed, computer spreadsheets can easily be used to calculate life tables.

ligaments (L. *ligamen*=a bandage, or "something that binds structures together") The term ligament refers to any connective tissue that apparently connects one structure to another. These connections range from insubstantial layers of mesothelium within the body cavities (e.g., the gastro-splenic ligament runs from the stomach to the spleen) to thick bands of collagen (e.g., the ilio-femoral and sacro-iliac ligaments). Only this latter use of the term ligament is relevant to paleoanthropology. *See also* **joints**.

limbic cortex *See* **limbic system**.

limbic system (L. *limbus*=border or interface) A group of interconnected structures some of which are in the cerebral cortex and some of which are beneath the cerebral cortex (hence they are called subcortical). The limbic system includes the amygdala, hippocampus, septum, basal ganglia, and cingulate gyrus. The cingulate gyrus is the part of the cerebral cortex that wraps around the corpus callosum (also referred to as the limbic lobe). The limbic system is involved in memory, emotion, motivation, learning, and some homeostatic regulatory functions.

limestone (OE *lim*=birdlime) A sedimentary rock rich in calcium carbonate ($CaCO_3$) usually derived from the shells of microorganisms. It is easily dissolved by acidic water percolating through cracks to produce solution cavities and when this happens over a long period of time these solution cavities develop into caves. If these caves connect with the surface soil may be washed in. When this is mixed with stone fragments falling from the roof of the cave, it results in a rock called **breccia**. The early hominin cave sites in southern Africa are formed in a variety of limestone rich in magnesium called **dolomite**. Most of the fossil hominin cave sites in southern Africa (e.g., **Taung**, **Sterkfontein**) were initially mined commercially for limestone which was used as either a building material or for the local manufacture of cement.

Lincoln Cave (etym. Justin Wilkinson, who explored the Sterkfontein Cave system, named this particular cave after a friend of his whose family name was Lincoln) Part of the Lincoln-Fault cave, it is part of the **Sterkfontein** cave system. Two deposits, Lincoln Cave North (LCN) and South (LCS) are separated by a rubble ramp built by lime miners. Excavations in 1997 and 1998 suggest that the **breccia** is reworked and probably derives from the Sterkfontein Member 5 breccia via a connection between the two cave systems in or close to the L/63 grid square of the Sterkfontein excavation. There is also evidence that the cavity in the cave has been filled by **Middle Stone Age** deposits that Kuman and Clarke (2000) have referred to as "Post-Member 6." The **Acheulean** artifacts found in both LCN and LCS deposits and the *Homo* cf. *ergaster* remains found in the LCS deposit almost certainly come from Sterkfontein Member 5 West breccia, and a *Paranthropus* incisor in the LCS sample most likely eroded out of the Sterkfontein Member 5 East **Oldowan** deposit (Reynolds et al. 2003).

LINE Acronym for long interspersed nucleotide element. Long interspersed nucleotide elements are DNA sequences of more than 5000 base pairs that are dispersed throughout the genome. They are a class of mobile genetic elements or transposons that are derived from the reverse transcription of RNA molecules. Because LINEs typically contain two genes, which encode reverse transcriptase and integrase, they are able to copy themselves (as well as other elements such as short interspersed nucleotide elements, or **SINEs**) into DNA and thus insert themselves into the genome. The L1 element, the most common LINE in modern humans, is estimated to make up approximately 20% of the modern human genome. LINEs are significant for studies of evolution because, like SINEs, they can be used to study population and species relationships. In addition, when LINEs copy themselves and insert the copy into the genome, neighboring DNA which may contain regulatory elements, genes, or gene fragments may also be copied and inserted. This DNA may then affect the regulation or sequence of a gene in the new location.

lingual (L. *lingua*=tongue) The aspect of the upper and lower tooth crowns that is closest to the tongue. Therefore, the lingual side of a tooth crown is the inner side of the crown.

lingual accessory cusp *See* postmetaconulid; tuberculum intermedium.

lingual paracone tubercle A term used by Kanazawa et al. (1990) for an accessory cusp on the occlusal surface of a maxillary molar that is located mesial to the median ridge of the paracone, but which is separate from the **mesial marginal ridge**. *See also* epicristoconule.

lingual pillar *See* median ridge.

lingula (L. *lingua*=tongue) A small tongue-like spicule of bone that forms the anterior border of the mandibular foramen. It is the inferior attachment of the sphenomandibular ligament and, like the latter ligament, it is a remnant of the first branchial arch. *See* branchial arches.

linkage When two loci are physically near each other the alleles present at those loci are inherited together as a "linked" unit (i.e., they are not separated during **recombination**). Linkage strength increases as the physical distance between the alleles decreases, and because of this linkage strength can be used to calculate the relative distance between loci, and from these data genetic maps can be generated. The relative distance between two loci is calculated using the frequency of recombination between them and it is measured in centimorgans (cM). One centimorgan equals the distance between two loci for which one recombination occurs out of 100 meioses. Examples of linkage in modern humans include the Rhesus blood group and the enzyme 6-phosphogluconate dehydrogenase (an enzyme found in blood) on modern human chromosome 1. Before other polymorphic markers such as microsatellites were discovered the linkage between phenotypic traits was used to create a crude map of the modern human genome. *See also* **linkage disequilibrium**.

linkage analysis An analysis which assesses the position of a locus (or loci) relative to known loci by estimating the **recombination** frequency (or **recombination fraction**) between them. This information can then be used to create a **genetic map**. A common method of linkage analysis is the LOD score, which stands for logarithm of the odds. It compares the likelihood of obtaining the observed data if the loci are linked to the likelihood of obtaining those data by chance.

Generally, a LOD score greater than 3 is considered evidence for linkage because this corresponds to 1000-to-1 odds that the observed data are not the result of chance.

linkage disequilibrium This refers to the nonrandom association of alleles at two or more loci. The degree of association, or the **linkage**, between two loci is measured by the frequency with which recombination separates the alleles (this is also known as the crossover frequency). The farther apart two loci are the greater the chance they will be affected by **recombination**. Linkage disequilibrium occurs when a combination of alleles at different loci is found more often than expected from a random combination of alleles based on their frequencies in a population. Linkage disequilibrium is important in studies of modern humans because it can indicate the presence of positive selection in a portion of the genome (e.g., Sabeti et al. 2002). It can also be used to research population history since "young" populations show greater linkage disequilibrium in the genome than "older" populations (this is because in the latter "older" populations recombination has had more time to break up linkage disequilibrium) (e.g., Tishkoff et al. 1996, Reich et al. 2001, Kaessmann et al. 2002). For example, sub-Saharan Africans typically have less linkage disequilibrium in their genomes than do Europeans.

Linnaean binomial Refers to the combination of the generic name and the species name (e.g., *Australopithecus afarensis*) that make up the proper scientific name of a species. The genus name is always capitalized and both names are always italicized. The genus can be abbreviated by reducing it to the first letter, or to the first few letters in cases where confusion may arise (e.g., *Au. afarensis*), but the species name must always be spelt out (e.g., *Homo sapiens* is abbreviated as *H. sapiens* and not as *Homo s.*).

Linnaean taxonomy A taxonomy based on the principles introduced by Carl Linnaeus. The two names (hence it is called a binomial) used for a species are a unique combination of a genus and a species name (e.g., *Homo sapiens*, *Pan troglodytes*, etc.).

lithic (Gk *lithos*=stone) An adjective used to describe **artifacts** made of stone. Its use as a noun is common, but incorrect. Lithic artifacts are durable, and the oldest known artifacts (from the site of **Gona**, Ethiopia) are made of stone.

lithic analysis   Analyses undertaken to derive information about hominin behavior and site-formation process from lithic assemblages (i.e. collections of stone tools and their manufacturing debris). Depending on the research objectives, lithic analysis may include typological classification, metric and nominal trait analyses, quantification and/or description of flake scar patterns, refitting, microwear and residue analysis, and raw-material studies (see Odell 2004 for overview). Typological classification has a long history and remains important for description and communication; however, its classical use to infer cultural affiliations and behavior on the basis of archeologically recognized tool types (e.g., a "Mousterian point") has been vigorously questioned (e.g., Rolland and Dibble 1990, Toth 1985). In contrast, the use of technological types, such as flakes, cores, fragments, etc., to subdivide and describe lithic assemblages remains a fundamental step in virtually all lithic analyses. A special case is the system of **flake types** often used in studies of **Mode I** technology (Toth 1987). The relative representation of different technological types may be compared to experimentally derived expectations to make inferences about site-formation processes, the representation of different stages within a **reduction sequence**, and artifact transport patterns. Different technological artifact types are subjected to various different metric and nominal trait analyses. The distribution of maximum dimension and/or weight of all artifacts may be used to identify evidence of winnowing or other disturbance (Schick 1986). For cores, original form (e.g., block, cobble, large flake), linear dimensions, number and size of 'scars' left by flake removals, and amount of remaining cortex (original exterior surface) are commonly recorded and used to make inferences about knapping methods and reduction intensity. Analysis of detached pieces usually focuses on whole flakes rather than fragments and typically includes measures of length, breadth, and thickness. Often the dimensions of the "striking platform" (the surface surrounding the point of percussion) and the angles formed between the platform and the interior and exterior surfaces of the flake will be measured. On the exterior surface of the flake, the amount of cortex and number and size of flake scars may be recorded. Nominal traits might include flake shape,

platform type, and the occurrence of retouch. Artifact counts and metric and nominal data are used to make a wide range of inferences regarding knapping methods, techniques and skills, artifact transport patterns, and reduction intensity (e.g., Stout et al. 2010). The patterning and orientation of flake scars on cores and flakes may be used to infer knapping strategies. This form of "lithic reading" is particularly important in the *chaîne opératoire* tradition of lithic analysis, and where possible is complemented by actually **refit**ting of pieces. Lithic reading proceeds through qualitative "mental reconstruction" based on personal knapping expertise (Pelegrin 2005), or by quantification of the frequency of different scar pattern types ("diacritic schemes," e.g., de la Torre and Mora 2005). Characteristic microwear "polishes" and macroscopic damage patterns identified through experimental replication are used to infer artifact function (Keeley 1980), including the materials tools were used on, evidence of hafting, and use as projectiles (Shea 1988). Under favorable preservation conditions, organic residues may also provide an indication of artifact function (Wadley et al. 2004). Raw-material analyses are most commonly undertaken to identify the sources, transportation, and/or trade of lithic resources (e.g., Braun et al. 2008), but may also investigate the influence of raw-material performance characteristics on technological systems (e.g., Brantingham et al. 2000) and the preferences expressed by ancient tool makers (e.g., Stout et al. 2005).

lithic assemblage   (Gk *lithos*=stone) A collection of stone artifacts, from some temporally and spatially bounded interval, typically a stratigraphic or excavation unit. This scale varies, such that, for example at **Blombos Cave** archeologists may use the term lithic assemblage to refer to the artifacts recovered from layer CD, or to refer the **Still Bay** lithic assemblage from phase M1.

lithology   (Gk *lithos*=stone and *ology*=study) The part of earth science that deals with the physical appearance of rocks (e.g., color, grain, or particle size, etc.).

lithostratigraphy   (Gk *lithos*=stone, and see the etymology of **stratigraphy**) Recognition and correlation of sedimentary strata based on the appearance of the rock layers as seen through the naked eye, with a lens, or by using a microscope.

"Little Foot" Informal name for **Stw 573** (*which see*).

Little Ice Age *See* **Holocene**.

living age distribution (or c_x) The proportion of individuals alive in each age interval or the **probability density function** for ages in the living. The living age distribution is either observed from a census, or it is the distribution of ages among the living that is implied by the deaths analyzed in a **life table**, or in a hazards analysis. For a stationary life table the living age distribution is given by L_x divided by the life expectancy at birth. For a stationary hazard model the probability density function for age in the living is the survivorship at exact age t divided by the integral of survivorship across age. For a nonstationary life table or hazard model the estimated living age distribution needs to be adjusted by the growth rate. *See also* **stationary population**.

LK Acronym for Loruth Kaado, one of several groups of localities within the site of West Turkana named after ephemeral river drainages. *See* **Loruth Kaado**.

LKH Prefix for fossils recovered from **Lakonis**, Greece.

LKH1 Site **Lakonis**. Locality Locality 1. Surface/*in situ* *In situ*. Date of discovery 2002. Finder Maria Ntinou. Unit Unit 1a. Horizon Layer 3. Nearest underlying dated horizon Base of Unit 1a. Geological age *c*.40 ka. Developmental age Adult. Presumed sex Unknown. Brief anatomical description Left lower third molar. Announcement Harvati et al. 2003. Initial description Harvati et al. 2003. Photographs/line drawings and metrical data Harvati et al. 2003. Detailed anatomical description Harvati et al. 2003. Smith et al. 2009. Initial taxonomic allocation *Homo neanderthalensis*. Informal taxonomic category Pre-modern *Homo*. Significance The only published Neanderthal fossil from Greece, apparently associated with an initial **Upper Paleolithic** industry. Location of original Ephoreia of Palaioanthropology and Speleology, Athens, Greece.

LLK Acronym for Louis Leakey Korongo. A locality at **Olduvai Gorge**, where, in 1961, Louis Leakey discovered the **OH 9** calvaria. Matrix attached to the calvaria is consistent with it originating from Upper Bed II. *See also* **Olduvai Gorge**.

LM Acronym for La Madeleine and Lake Mungo and the prefix for fossils recovered from **La Madeleine** and **Lake Mungo**. *See* **Lake Mungo**; **La Madeleine**.

LM4 Site **La Madeleine**. Locality Base of Capitan and Peyrony's upper level (Magdalenian IV). Surface/*in situ* *In situ*. Date of discovery 11 December 1926. Finder D. Peyrony. Unit N/A. Horizon N/A. Bed/member N/A. Formation N/A. Group N/A. Nearest overlying dated horizon N/A. Nearest underlying dated horizon N/A. Geological age $10{,}190 \pm 100$ years BP. Developmental age 3–4 years. Presumed sex Unknown. Brief anatomical description Nearly complete, but very fragmented, calotte, mandible, and most postcranial elements except the right lower leg and foot. Announcement Peyrony 1928. Initial description Capitan and Peyrony 1928. Photographs/line drawings and metrical data Heim 1991. Detailed anatomical description Heim 1991. Initial taxonomic allocation *Homo sapiens*. Taxonomic revisions N/A. Current conventional taxonomic allocation *H. sapiens*. Informal taxonomic category Anatomically modern human. Significance This young child was buried with over 1000 shells, beads, and decorated animal teeth that were probably attached to clothes it wore during life, indicating a high level of social stratification (Vanhaeren and d'Errico 2001). **Radiocarbon dating** for this individual suggests it belongs in the Azilian period, but the shells and other grave goods are consistent with a late **Magdalenian** date. Location of original Musée National de Préhistoire, Les Eyzies-de-Tayac, France.

LO Abbreviation for Lomekwi one of several groups of localities within the site of **West Turkana** named after ephemeral river drainages. *See* **Lomekwi**.

local enhancement Local enhancement is a type of **enhancement** effect associated with a suite of cognitively primitive attention and motivation mechanisms that affect **learning**. Specifically, local enhancement describes learning in situations where the presence of a conspecific directs an observer's attention to a particular location such as a water hole or a termite mound (Thorpe

1963). This type of enhancement is seen in mammals and birds alike, and may result in behavior matching, which can resemble learning by **imitation**.

locality A location within a large fossil site where fossils have been found. At small sites and at excavated cave sites (e.g., the **Gran Dolina** cave at **Atapuerca**) there is no need to do other than specify within which grid square and from what level in that square a fossil came from. But in large fossil sites (e.g., **Koobi Fora**, Middle Awash, **Olduvai Gorge**) the locations where one, or more, fossils have been found, are described as localities (e.g., **FLKNN I** and **MNK I** are two localities within Olduvai Gorge, and **ARA-VP-1** is a fossiliferous locality within the **Central Awash Complex** of the **Middle Awash study area**).

Locality 1 *See* **Zhoukoudian Locality 1**.

Locality 53 *See* **Zhoukoudian Locality 1**.

locomotion (L. *loco*=place and *motivus*=to cause motion) Refers to the various ways that animals use their anatomy to move from one place to another. Locomotion allows an individual to acquire food, find mates, avoid predators, move to safe resting places, and interact socially with other members of a group. Virtually all primates use a repertoire of different types of locomotion in a given day to move from place to place. Compared to that of other mammals, primate locomotion tends to be hindlimb-dominated, meaning that peak vertical forces experienced by the hindlimbs are greater than those on the forelimbs (Demes et al. 1994). If an animal locomotes with its trunk upright and supports itself exclusively on its hindlimbs, then its locomotion is described as bipedal. The locomotion practiced by a primate species has an impact on the architecture of most of the postcranial musculoskeletal system, as well as on aspects of cranial anatomy (e.g., Fleagle 1977). Thus, the anatomy of the skeleton, especially the postcranial skeleton, provides information about the locomotor adaptations and behaviors of fossil hominins. *See also* **bipedalism**.

locus (L. *locus*=place) The physical location of a **gene** or **nucleotide** in the genome. The locus of a gene is analogous to an individual's street address. So, for example the gene for phenylthiocarbamide (PTC) tasting is called *TAS2R38* and it is on the long arm of chromosome 7 at position 141,318,900–141,320,042 according to the March 2006 assembly of the modern human genome.

LOD score A method for estimating recombination frequency in a **linkage analysis**. It is the logarithm (base 10) of the odds of linkage (i.e., the ratio of the likelihood that loci are linked to the likelihood that they are not) for a set of markers in a pedigree. Typically, a LOD score of greater than 3.0 is considered evidence for linkage.

loess (Ge. *loess*=loose) A deposit of wind-blown silt that typically forms an extensive blanket over mid-latitude landscapes during a period of cold, dry climate. The windy, dry environment of periglacial terrains and ample sediment supply from glacial meltwater streams afforded ideal conditions for silt to mobilize and form a mantle over extensive areas. About 10% of present-day continental terrain is covered by loess. Eastern and central Europe and China, in particular, have significant deposits. Cycles of alternating loess and soils, indicative of climate fluctuation between cold/dry and warm/moist conditions, have produced a long stratigraphic record of paleoclimate back to *c*.2.5 Ma over most of the Loess Plateau of north-central China, and as far back as *c*.22 Ma in the western part of this region (Kukla 1987, Guo et al. 2002). Connections between the climate history of the Loess Plateau and the Nihewan Basin have been implicated in the dispersal of early Pleistocene hominins to northeast Asia (Ao et al. 2010). Loess deposition during the Last Glacial Maximum extended in China as far south as the Bose Basin. *See also* **Bose Basin**; **Nihewan Basin**.

logistic population growth A model of population growth that, unlike exponential population growth, will bring a population to its **carrying capacity**. The carrying capacity (usually symbolized with a K) is included in the equation for logistic population growth along with the exponential rate of increase (r) found in the exponential model. Under the assumption that the exponential rate of increase must be positive, the equation for population size at time $t+1$ given the population size at time t is $N_{t+1} = N_t \times \exp\left(r(1 - \frac{N_t}{K})\right)$. Note that if $N_t=K$, then $N_{t+1}=N_t$; in other words, a population at its carrying capacity will stay at its carrying capacity.

If $N_t > K$, then $N_{t+1} < N_t$; in other words, a population above its carrying capacity will "downregulate" (decline in size) to reach its carrying capacity. If $N_t < K$, then $N_{t+1} > N_t$; in other words, a population below its carrying capacity will "upregulate" (increase in size) to reach its carrying capacity.

Lokalalei (or LA) (etym. Named for an ephemeral river system, or laga, in whose drainage the localities are located) Refers to a complex of fossil localities in the Kalochoro Member of the Nachukui Formation within the Lokalalei drainage in **West Turkana**, Kenya. Some of the earliest known archeological sites (i.e., LA1, LA2C, and LA1α) are within the Lokalalei drainage. *See also* **West Turkana**.

Lokochot Tuff A 3.61 ± 0.09 Ma **tuff** known from the **Koobi Fora Formation**. Tuffs with the same geochemistry have been located in Uganda (Pickford et al. 1991) and Ethiopia (Deino et al. 2010), and in deep-sea cores from both the Arabian Sea (deMonacal and Brown 1999) and the Gulf of Aden (Brown et al. 1992).

LOM Abbreviation of the Lomweki Member in the Nachukui Formation, **West Turkana**, Kenya.

Lomekwi (or LO) (etym. Named for an ephemeral river system, or laga, in whose drainage the localities are located) Refers to a complex of fossil localities in the Kaitio, Lomekwi, Kataboi, and Lokalalei Members of the **Nachukui Formation** within the Lomekwi and Topernawi drainages in **West Turkana**, Kenya. The early hominin crania **KNM-WT 17000** and **KNM-WT 40000** were recovered from Lomekwi localities. *See also* **West Turkana**.

LON Abbreviation of the Lonyumum Member in the **Koobi Fora Formation**, **West Turkana**, Kenya.

long distance transport *See* **carcass transport strategy**.

long interspersed nucleotide element *See* **LINE**.

long-period incremental lines Incremental features in tooth enamel or dentine that have an intrinsic temporal secretory rhythm of greater than 1 day. They are called **striae of Retzius** and **perikymata** in enamel and **Andresen lines** and **periradicular bands** in dentine. In primates, these lines have been reported to range between 2 and 12 days (reviewed in Smith et al. 2003), with a consistent **periodicity** within each individual (FitzGerald 1996). Counts and measurements of these features facilitate estimation of the **crown formation time**, root formation time, or the root **extension rate** (reviewed in Smith et al. 2006). *See also* **enamel development**; **enamel microstructure**; **dentine development**.

long-period markings *See* **long-period incremental lines**.

Longgushan *See* **Zhoukoudian Locality 1**.

longitudinal arch A structure unique among extant taxa to modern humans, the longitudinal arch runs along the length of the foot and serves both to stiffen it and to store and return elastic energy during the stance phase of walking (Ker et al. 1987). The longitudinal arch has two components, the **medial longitudinal arch** and the **lateral longitudinal arch**, formed by articulations between bones in the hind and midfoot. It is reinforced by four soft-tissue structures: the short and long plantar ligaments, the calcaneonavicular (spring) ligament, and the plantar aponeurosis. The short plantar ligament binds the cuboid to the calcaneus and the long plantar ligament anchors the calcaneus to the bases of the lateral four metatarsals. The calcaneonavicular (spring) ligament anchors the medial navicular to the sustentaculum tali and prevents plantar flexion of the talar head. Functionally, the important plantar aponeurosis tightens during dorsiflexion of the metatarsophalangeal joints and helps convert the foot into a stiff and effective lever during the final stage of toe-off. The longitudinal arch is also supported by intrinsic muscles of the foot (e.g., abductor hallucis, flexor hallucis brevis, flexor digitorum brevis, quadratus plantae, and abductor digiti minimi). It achieves its greatest height at the second metatarsal. The long flexors (flexor hallucis longus and flexor digitorum longus), fibularis longus, and tibialis posterior also help support the arch. Modern human foot arches can be excessively high (pes cavus) or low/flat (pes planus). Apes do not possess a longitudinal arch. Fossil footprints from Ileret, Kenya, indicate that the arch had evolved in its modern form by 1.53 Ma, although some suggest that the 3.66 Ma footprints from **Laetoli** provide evidence for an arch much earlier in hominin evolution. *See* **foot movements; Koobi Fora hominin footprints; Laetoli hominin footprints; walking cycle**.

longitudinal data Data derived from measurements taken from a series of observations on a single individual, or a group of individuals. Longitudinal data are preferable to cross-sectional data for studies that wish to examine how individuals change over time, although studies using longitudinal data obviously take longer to conduct and longitudinal data are methodologically more difficult to collect.

Lordkipanidze, David (1963–) Born in Tbilisi, Republic of Georgia, in 1963, David Lordkipanidze claims it was an early love of detective stories that determined his early interest in anthropology. After completing his undergraduate degree in geology and geography at Tbilisi State University in 1985, a year later Lordkipanidze registered as a graduate student in the Russian Academy of Sciences where he studied the reconstruction and distribution of early human settlements in montane regions, receiving his PhD in geography in 1992. It was during that time that Lordkipanidze was selected by the Georgian paleontologist Leo Gabunia to take part in the excavations at **Dmanisi**; Lordkipanidze was subsequently selected to succeed Gabunia as the Director of the project. In 1992 Lordkipanidze was appointed Senior Researcher in the Georgia Academy of Sciences and he stayed there until his appointment in 1997 as Head of the Geology and Paleontology Department in the Georgia State Museum. In early 2002 Lordkipanidze was awarded a Fulbright Scholarship, which enabled him to visit a number of universities in the USA. When he returned to Georgia he became the Deputy Director of the Georgian National Museum, a post he held until 2005 until his appointment as General Director of the Georgian National Museum, where he is currently overseeing a large-scale reorganization and renovation. In 1991, during Lordkipanidze's first field season as leader of the excavations at Dmanisi, the first of many hominin remains (the **D211** mandible) was recovered and the during the subsequent years under his leadership the Dmanisi research team has grown into a large multinational enterprise and its activities over many years have resulted in many more important hominin discoveries. In recognition of the importance of his longstanding and extraordinarily productive research at Dmanisi, in 2007 David Lorkipanidze was elected as a foreign associate of the United States National Academy of Sciences. *See also* **Georgian National Museum**.

lordosis (Gk *lordos*=bent backwards) The term lordosis, and its antonym, **kyphosis**, refer to the angular relationship within or between sections of a compound structure (e.g., the vertebral column), or between two planes (e.g., the anterior and posterior components of the **cranial base**). Thus, the parts of the adult modern human vertebral column that are concave posteriorly (i.e., the cervical and lumbar regions) are referred to as being lordotic, whereas the parts of the vertebral column that are concave anteriorly (i.e., the thoracic and sacral) are referred to as being kyphotic. Adult modern humans are distinctive among the extant great apes in normally having a pronounced lumbar lordosis, and it has been suggested that this is related to our habitual upright posture, and specifically to cope with anteriorward migration of the trunk's center of mass that occurs in pregnant females (Whitcombe et al. 2007). With respect to the cranial base, organisms or individuals with a larger, more obtuse (i.e., more chimpanzee-like) **cranial base angle** are described as having a more lordotic basicranium, whereas organisms or individuals with a smaller, more acute (i.e., more modern human-like) cranial base angle have a more kyphotic basicranium.

Loruth Kaado (or LK) (etym. Named for an ephemeral river system, or laga, in whose drainage the localities are located) Refers to a complex of fossil localities in the Kaitio, Kataboi, and Natoo Members of the **Nachukui Formation** within the Loruth Kaado drainage in **West Turkana**, Kenya. *See also* **West Turkana**.

Los Azules (Location 43°21′N, 5°08′W, Spain; etym. Sp. for "the blues," as in the color) History and general description This cave lies in the Llueves mountain, to the north of the village of Cangas de Onis in the province of Asturias, only about 15 km/9 miles from the Cantabrian sea. It was excavated between 1973 and 1988 by a team led by Fernández-Tresguerres. It is one of few Spanish sites that documents the end of the **Upper Paleolithic**, with layers dating to the

Magdalenian and Azilian, and includes a burial from the Azilian period. Temporal span and how dated? The Azilian burial was dated using conventional **radiocarbon dating** to between 9430 and 9540 years BP. Hominins found at site The Azilian-age burial of an adult presumed male was found close to the cave mouth, against the left wall. The skeleton of this individual is robust and tall, and it has congenitally fused foot bones and, likely, a bilateral claw foot. The grave was surrounded by **ochre**, and several artifacts were found in close association with the burial, including two bone **harpoon**s, a deer antler, shells, and stone tools. Archeological evidence found at site Several archeological layers were uncovered which record the long-term presence of hominins, presumably *Homo sapiens*, through the middle and end of the Upper Paleolithic. The Azilian layers are represented by flat bone harpoons and small "thumbnail scrapers." Key references: Geology, dating, and paleoenvironment Fernández-Tresguerras 1980, Fernández-Tresguerres and Rodríguez Fernández 1990; Hominins Fernández-Tresguerres 1976, Garralda 1986; Archeology Fernández-Tresguerras 1980, Fernández-Tresguerres and Rodríguez Fernández 1990.

Lothagam [Location 2°54′N, 36°03′E, Kenya; etym. the local Turkana people call the hill, or horst, that forms the eastern boundary of the site, Lothsagam. Leakey (2003) explains that in Turkana "lothsagam" refers to something that is "rough, varied and heterogeneous" and in the case of Lothagam "it is a reference to the many different rocks that make up the **horst**" (*ibid*, p. 2)] History and general description The first time the sediments at Lothagam were noted to be fossiliferous was in 1965 when Lawrence Robbins was excavating early to middle **Holocene** archeological sites (traces of hunter–gatherers and fishers) in the area. He recovered the remains of 29 buried individuals and many of these are stored at the Smithsonian Institution's **National Museum of Natural History**. These reports caught the attention of Bryan Patterson and he and his colleagues prospected for fossils at Lothagam in 1967–8; the Lothagam mandible, **KNM-LT 329**, was recovered by Patterson's group in 1967. A group from Princeton worked at the site in 1972–3. Personnel from the **Koobi Fora Research Project** (or KFRP) visited the site in 1980 and recovered evidence of fossil mon-

keys. A more deliberate survey conducted by Meave **Leakey** in 1989 resulted in the discovery of catarrhine and carnivore fossils, and in 1990 the same researchers spent a month at Lothagam. The KFRP focused their field research at Lothagam in 1991–3, and between 1989 and 1993 researchers recovered approximately 1700 vertebrate and other tetrapod fossils. Temporal span and how dated? Vertebrates have been recovered from exposures of the **Nachukui Formation** and Nawata Formation. The lower member of the Nawata Formation is *c*.7.4–6.5 Ma, the upper member of the Nawata Formation is *c*.6.5–5.23 Ma, and the Apak Member, the oldest member in the Nachukui Formation, is between approximately 5.0 and 4.2 Ma. The sediments have been dated using **potassium-argon dating**, **argon-argon dating**, and **magnetostratigraphy**. Hominins found at site Two teeth, a left partial M^3 (KNM-LT 22930), and a right I_1 (KNM-LT 25935), have been recovered from the upper member of the Nawata Formation, a fragment of the right side of the mandibular corpus (KNM-LT 329) from the Apak Member, and four isolated teeth [a fragment of the crown of a right dm^2 germ (KNM-LT 23181), most of the crown of a right M_3 (KNM-LT 23182), a left M_2 crown (KNM-LT 23183), and a premolar crown fragment (KNM-LT 25936) from the Kaiyumung Member of the Nachukui Formation]. The former two hominins have been assigned to Hominidae *indet.* and the four isolated teeth to *Australopithecus* cf. *Australopithecus afarensis*. Archeological evidence found at site None have been found in the sediments from which the hominin fossils were recovered. Key references: Geology, dating, and paleoenvironment Patterson et al. 1970, Feibel 2003, McDougall and Feibel 2003; Hominins Patterson et al. 1970, McHenry and Corruccini 1980, Kramer 1986, White 1986b, Hill et al. 1992, Leakey and Walker 2003; Archeology N/A.

Lothagam (LSA site) (Location approximately 2°55′N, 36°03′E, Kenya; etym. named after Lothagam Hill, around which a series of burials were discovered, but *see* **Lothagam** for the full etymology) History and general description Lawrence Robbins excavated a series of **Later Stone Age** (or LSA) sites around Lothagam Hill on the western side of Lake Turkana in 1965–6. Temporal span and how dated? Conventional **radiocarbon dating** suggests 6–9 ka. Hominins found at site Twenty-nine burials, including eight adult males, six adult females, one female subadult, two adolescents, and 12 individuals too poorly preserved or fragmentary to be identified. The burials include no infants or young children. The skeletons belong to tall people with a linear physique and long, narrow crania generally resembling modern Nilotic populations, but with larger and more robust jaws. Archeological evidence found at site Cultural traces included barbed bone harpoons, a lithic industry dominated by **scrapers** and **crescents**, and pottery. Faunal remains at the site suggest the occupants of the site exploited Nile perch, **catfish**, and other aquatic species including crocodiles and turtles; similar faunal profiles are found at sites on the eastern side of the lake, including at **Koobi Fora** and Loengalani. The remains are associated with highstands of Lake Turkana between 9.5 and 7.5 ka and between 6.6 and 4.4 ka (Butzer et al. 1972). Key references: Geology, dating, and paleoenvironment Robbins 1974, Butzer et al. 1972; Hominins Angel et al. 1980; Archeology Robbins 1974, Angel et al. 1980.

low-ranking prey Within an **optimal foraging theory** framework low-ranking prey are those taxa or food resources that provide below-average energetic returns. The **prey-choice model** predicts that low-ranking prey will only enter the diet when encounter rates with **high-ranking prey** decline and overall foraging efficiency declines, a phenomenon known as **resource depression**. In archeological contexts prey rankings are often determined according to body size since larger taxa tend to provide greater energetic returns than small taxa. For example, small mammals such as rabbits and hares are typically regarded as low-ranking prey. However, a growing body of ethnographic evidence suggests that other factors (e.g., prey mobility) may play an important role in explaining human foraging decisions and prey rankings. *See also* **high-ranking prey**.

low-survival elements A term that refers to a subset of skeletal elements known to be particularly sensitive to destructive taphonomic processes. These include elements with thin cortical walls and substantial low-density, grease-rich, cancellous portions. For most ungulates, these include the vertebrae, ribs, pelves, scapulae, and long-bone epiphyses. Phalanges, carpals, and tarsals of size 1–2 ungulates are also considered low-survival elements, given their tendency to be consumed by carnivores.

Because low-survival elements are readily destroyed following hominin discard, they are of relatively little utility when examining **carcass transport strategies** in bone assemblages that have been subjected to destructive processes. *See also* **skeletal element survivorship**.

low-utility bones *See* economic utility index.

lower bank *See* Swartkrans.

Lower Cave *See* Zhoukoudian Locality 1.

lower fissure One of the three main areas within **Zhoukoudian Locality 1**. *See* **Zhoukoudian Locality 1**.

Lower-middle tuffs *See* Okote Tuff Complex.

Lower Omo Basin The depression through which the lower reaches of the Omo River runs. The Omo River rises in the Ethiopian Highlands and ends by draining into the northern end of Lake Turkana. The last part of its course and Lake Turkana occupy the low country between the Ethiopian and Kenyan domes to the north and south, respectively, and between the western and eastern walls of the East African Rift. It was the mammalian fossils collected in 1902 by the Bourg de Bozas expedition that prompted Camille **Arambourg** to re-explore the region in 1933. Arambourg (1934) recognized that the fossils he had recovered were Plio-Pleistocene in age, but he had to wait until 1967 to return to the region as the first head of the French contingent of the **International Omo Research Expedition** (or IORE) that worked in the Lower Omo Basin from 1967 to 1976. The most recent program of fieldwork in the Lower Omo Basin, by the **Omo Group Research Expedition** led by Jean-Renaud Boisserie, began in 2006–7. The sediments in the Lower Omo Basin were mapped by Fuchs (1939) and by Butzer and Thurber (1969), but the geology of the Lower Omo Basin is described in detail by Paul Haesaerts, Jean **de Heinzelin**, and Frank **Brown** in contributions to de Heinzelin (1983). The sediments are divided into the **Omo Group** and the Lake Turkana Group (also called the **Turkana Group**). Fossils recovered from the Lower Omo Basin during the research campaigns conducted by the IORE were allocated to collecting localities under several systems (see Appendix III in de Heinzelin 1983). Within the **Shungura Formation**, fossils collected by the French team were referred to "Omo" (or "O") localities and these were given numbers (e.g., Omo 18); the American contingent also numbered localities, but used the letter "L" for locality (e.g., L. 396). Within each of these numbered localities fossils are themselves assigned a number (e.g., **L. 40-19**), and for the specimens collected by the French the year of discovery is inserted between the locality number and the fossil number (e.g., **Omo 18-1967-18**). In the **Usno Formation**, fossil hominins come from either **Brown Sands** (or "B") or **White Sands** (or "W"). Within each of these fossiliferous regions there are numbered localities, and within each numbered locality fossils are assigned a number (e.g., B 8-27 and W 7-508). Additionally, some localities were recorded as "F" (by Frank Brown) or P (for Fr. *point*) by Jean de Heinzelin. Details of the history of the exploration of the Lower Omo Basin are given in Howell and Coppens (1983) and details of the geology are found in various contributions to de Heinzelin (1983). *See also* **Omo Group**; **Omo-Turkana Basin**; **Turkana Group**.

Lower Omo Valley   *See* **Lower Omo Basin**.

Lower Paleolithic   A term used to describe a stage of the European and Near Eastern **Paleolithic** that is defined predominantly by the stone tool technology (**mode 1** and **mode 2**, including the **Acheulean**, the **Oldowan**, and those Plio-Pleistocene technocomplexes that typically do not include handaxes) that appears from *c.*2.5 Ma to 200 ka. The term is somewhat problematic since it encompasses a large period of time (at least 2 Ma) during which there were significant biological and behavioral shifts, as well as several distinct technocomplexes. The equivalent term for African sites with similar technologies is the **Early Stone Age**.

Lower Pubu   *See* **Bubing Basin**.

lower scale   Sometimes used for the lower part of the squamous part of the occipital bone (syn. **nuchal plane**).

Loyangalani   Site Loyangalani. Locality N/A. Surface/*in situ* Surface. Date of discovery 1976. Finder Olivier Braun. Unit N/A. Horizon N/A.

Bed/member N/A. Formation N/A. Group N/A. Nearest overlying dated horizon N/A. Nearest underlying dated horizon N/A. Geological age N/A. Developmental age Adult. Presumed sex Male. Brief anatomical description Mandible and fragment of maxilla of the same individual. The robust mandible points to archaic *Homo sapiens* rather than modern humans, but there is a chin. Initial taxonomic allocation *Homo sapiens*. Announcement Twiesselmann 1991. Initial description Twiesselmann 1991. Photographs/line drawings and metrical data Twiesselmann 1991. Detailed anatomical description Twiesselmann 1991. Location of original Original is lost, but there is a cast in the Institute royal des Sciences naturelles de Belgique (Royal Belgian Institute of Natural Sciences), Brussels, Belgium.

LPT   Acronym for the **Late Pliocene transition** (*which see*).

LSA   Acronym for the Late Stone Age and Later Stone Age. *See* **Later Stone Age**.

L.S.B. Leakey Foundation   Formed in 1968 to foster research into human origins. The Foundation awards more than US$600,000 annually in grants, and it is the only privately funded granting organization wholly committed to human origins research in the USA. It was established by a group who wanted to establish an eponymously named organization to support the type of research carried out and promoted by Louis **Leakey**. For a long time it was called the L.S.B. Leakey Foundation and it is only recently that it has styled itself The Leakey Foundation. The Foundation is an endowed, independent, nonprofit institution whose mission is to increase scientific knowledge, education, and public understanding of human origins, evolution, behavior, and survival. Its headquarters is in San Francisco, California, and the organization partners with other major science institutions throughout the USA to produce an annual, national, lecture series. In keeping with Louis Leakey's philosophy, the Foundation promotes a multidisciplinary approach to investigating human origins (i.e., research into the paleoenvironments, archeology, paleontology, and paleoanthropology of the Miocene, Pliocene, and Pleistocene). Research funded by the Foundation embraces the reconstruction of the behavior, biology, and ecology of hominins. The Foundation also supports projects that can shed light on the behavioral ecology of modern hunter–gatherers and the genetics and behavior of our closest living relatives, the great apes, in the belief that they provide important behavioral analogues for hominin evolution. Within the first decade of its existence the Leakey Foundation provided grants to many of the seminal studies that have informed our understanding of human prehistory, including the field research and discoveries of the Leakey family, Donald **Johanson**, Jane Goodall, Dian Fossey, and Biruté Mary Galdikas. Today, the Foundation continues to support the work of researchers including Zeresenay **Alemseged**, Jill Pruetz, Fred Grine, Sileshi Semaw, David **Lordkipanidze**, and many more. Since 1968 the Foundation has provided about $14 million to more than a thousand grantees in over 100 countries. The major funding streams are General Research Grants and Franklin Mosher Baldwin Memorial Fellowships. General Research Grants are awarded in competitions that are held twice annually, constitute the majority of the Foundation's grant program; priority for funding is given to the exploratory phases of promising new research projects. The Franklin Mosher Baldwin Memorial Fellowships are intended for scholars and students from emerging nations who wish to obtain an advanced degree or specialized training in an area of study related to human origins research. Since 1977 more than 90 Baldwin Fellowships have been awarded to scientists from developing countries to complete their graduate work and pursue careers in human origins research. Recipients include young researchers from countries including Egypt, Ethiopia, Eritrea, Iran, Kenya, Madagascar, Malawi, Nigeria, Democratic Republic of the Congo, Somalia, South Africa, Sudan, Tanzania, Togo, Uganda, Zambia, and Zimbabwe. Details of the Leakey Foundation's programs and information on how to apply can be found at www.leakeyfoundation.org.

"Lucy" Nickname for **A.L. 288–1**, the partial skeleton of *Australopithecus afarensis* discovered by Donald **Johanson** and his research team at Hadar, Ethiopia, in 1974. During a camp celebration of the discovery, The Beatles' album Sergeant Pepper's Lonely Hearts Club Band was playing and expedition member Pamela Alderman suggested calling the presumed female skeleton "Lucy" after the song "Lucy in the Sky with Diamonds."

Lukeino (Location 00°48'N, 35°52'E, in Kenya; etym. local, Tugen, name for the area) History and general description Name of a complex of localities in the Tugen Hills in the Baringo District, Kenya. More than 44 fossiliferous localities have been mapped within the **Lukeino Formation**. Temporal span and how dated? Underlain by Kabarnet Trachytes (*c*.6.2 Ma) and overlain by Kaparaina Basalts (*c*.5.7 Ma); the trachytes and basalts have both been dated using **potassium–argon dating**. Hominins found at site Fossils assigned to *Orrorin tugenensis* have been recovered from at least four localities: **Aragai**, **Cheboit**, **Kapcheberek**, and **Kapsomin**. Archeological evidence found at site None. Key references: Geology, dating, and paleoenvironment Pickford and Senut 2001; Hominins Senut et al. 2001; Archeology N/A.

Lukeino Formation (etym. Local, Tugen, name for the area) General description Late Miocene formation consisting primarily of lacustrine and fluvial sediments that outcrop extensively in the eastern foothills of the Tugen Hills, west of Lake Baringo, Kenya. Deposits are located between the underlying Kabarnet Trachyte Formation and overlying Kaparaina Basalts Formation; the sediments are divided into four members. The oldest is the Kapgoya Member. Overlying this is the Kapsomin Basalt Member, followed by the Kapsomin Member (NB: most of the *Orrorin tugenensis* specimens were recovered from sediments within the Kapsomin member). The Kapcheberek Member is the youngest member (Sawada et al. 2002). Sites where the formation is exposed **Aragai**, **Cheboit**, **Kapcheberek**, **Kapsomin**. Geological age *c*.6.0 Ma to *c*.5.7 Ma. Most significant hominin specimens recovered from the formation The current hypodigm of *Orrorin tugenensis*, including **KNM-LU 335**, BAR 349'00, **BAR 1000'00**, BAR 1001'00, BAR 1002'00, BAR 1003'00, BAR 1004'00, BAR 1215'00, BAR 1390'00, BAR 1425'00, BAR 1426'00, and BAR 1900'00, has been recovered from sediments within the Lukeino Formation. Key references: Geology, dating, and paleoenvironment Dagley et al. 1978, Deino et al. 2002, Hill et al. 1985, Pickford 1974, 1975, Pickford and Senut 2001, Sawada et al. 2002; Hominins Pickford 1975, Senut et al. 2001.

"Lukeino molar" An isolated mandibular molar found in 1975 in the **Lukeino Formation**

in the Tugen Hills. The molar is the **holotype** of *Orrorin tugenensis*. *See also* **KNM-LU 335**; *Orrorin tugenensis*.

Lukenya Hill (Location 1°29′S, 37°03′E, Kenya; etym. the local name for the hill) History and general description During a survey of archeological sites around the base of Lukenya Hill in March 1970, R.M. Gramly dug a test pit in the rock-shelter and recovered a hominin frontal bone 140 cm below the surface in association with a **Later Stone Age** (or LSA) industry. The specimen was found at the junction of two semi-concreted limestone breccias containing bones and stone tools referred to as strata IV and V. More extensive excavations in 1971 produced no additional hominin remains. Temporal span and how dated? 17.7 ka radiocarbon years for the hominin layer based on conventional **radiocarbon dating** on animal bone. Hominins found at site Partial calotte comprising two-thirds of the frontal bone and the left parietal. It has a more robust supraorbital torus and a more receding forehead than is usual in living Africans, but the calotte is clearly that of an anatomically modern human. Archeological evidence found at site A microlithic LSA industry featuring scrapers of various kinds made on blades and flakes, perforators, backed pieces, and many bladelets. The most common raw materials were fine-grained quartz followed by chalcedony, opal, obsidian, and white quartz. Strata IV and V also featured stone tools made from gneiss and sandstone and traces of ochre adhere to many of the grinding implements. No bone tools were encountered, but a few perforated ostrich eggshell beads were found. Extant taxa including zebra, eland, wildebeest, hartebeest, and impala dominated recognizable teeth in the fauna. Key references: Geology, dating, and paleoenvironment Dating of questionable reliability by Rainer Protsch is reported in Gramly and Rightmire 1973; Hominins Gramly and Rightmire 1973; Archeology Gramly and Rightmire 1973.

Luleche (Location 25°49.99′S, 27°51.36′E, North West Province, South Africa; etym. named for a nearby farm property) History and general description The site is an abandoned limeworks based on a cave system within Precambrian **dolomite** near the **Gondolin** Plio-Pleistocene locality in the Schurveberg mountain range, 15 miles northwest of **Sterkfontein**. Geologic description of the cave system is limited because of extensive lime mining obliterated all but small *in situ* remnants of the original infillings. Field collections from the site in 2005–6 concentrated on the extensive *ex situ* dumpsites that contained decalcified sediments and a small ($n=365$) sample of heavily manganese-stained fossil specimens. Temporal span No depositional age could be assigned; however, the identifiable faunal remains (exclusively Bovidae) suggest accumulation in the Pleistocene. Hominins found at site None. Archeological evidence found at site None. Key references: Geology, fauna, and taphonomy Adams et al. 2007.

lumbar lordosis (L. *lumbus*=loin and Gk *lordos*=bent backwards) Refers to the characteristic curvature in the small of the back near the base of the vertebral column seen in adult modern humans. Because the lumbar region of the vertebral column in adult modern humans is concave posteriorly it is referred to as being lordotic. The **lordosis** is due to the wedge-shaped (anteriorly tall and posteriorly short) vertebral bodies and intervertebral discs in the lumbar region. This configuration of the lumbar spine is seen in modern humans, in all extinct premodern *Homo* taxa, and in individuals assigned to *Australopithecus africanus* (e.g., **Sts 14** and **Stw 431**). In modern human females lumbar lordosis is exaggerated during pregnancy and Whitcombe et al. (2007) suggest that this helps compensate for the anteriorward migration of the center of mass that occurs in pregnant females in habitually bipedal hominins.

lumbar vertebral column (L. *lumbus*=loin) The lumbar vertebral column, which is the region between the rib-bearing thorax and the sacrum, has a considerable influence on the degree and pattern of motion between the fore- and hindlimbs. Traditionally defined, lumbar vertebrae lack ribs and do not articulate with the sacrum. An alternative functional definition is based on the orientation of the synovial (or zygapophyseal) joints between the vertebrae. In thoracic-type articulations the joint surfaces form arcs of a circle the center of which lies ventral (i.e., anterior or inferior) to the vertebral canal, whereas in lumbar type articulations the joint surfaces form arcs of a circle the center of which lies dorsal (i.e., posterior or superior) to the vertebral canal. But these are ideal types and intergradations exist. In pronograde quadrupeds, primates and many other mammals, the most caudal two or three rib-bearing vertebrae frequently bear lumbar type articulations. Traditionally defined lumbar vertebrae in most anthropoid pronograde quadrupeds generally number six or seven, whereas functionally defined lumbar vertebrae may total nine or ten. For the majority of extant hominoid individuals the two definitions coincide. The modal number of traditionally defined lumbar vertebrae is four in the living great apes, five in smaller-bodied hylobatids, and four or five in the larger siamang species. The modal number of traditionally defined lumbar vertebrae in modern humans is five, but six functionally defined lumbar vertebrae are frequent (e.g., up to 43% in some modern human populations). **Lordosis**, or posterior concavity, of the modern human lumbar region is an adaptation to **bipedal** positional behaviors, and is achieved by vertebral wedging, with dorsal vertebral margins being shorter than ventral margins. Modern humans are dimorphic, with females having three wedged vertebrae, males only two, as well as differences in relative size and orientation of lumbar zygapophyses between the sexes. These differences reflect the need to accommodate shifts in the position of the trunk's center of mass during pregnancy by increasing lumbar lordosis (Whitcome et al. 2007). The fossil records of three early hominin specimens, two belonging to *Australopithecus africanus* and one to *Homo erectus*, and several *Homo neanderthalensis* individuals, are sufficiently complete to infer the number of lumbar vertebrae. The Neanderthal specimens have five lumbar vertebrae, one *Au. africanus* specimen (**Sts 14**) has six functional lumbars, the most cranial of which bears a unilateral rib. The lumbar counts of the second *Au. africanus* and the *H. erectus* specimen (i.e., whether there are five or six lumbar vertebrae) are disputed (Haeusler et al. 2002).

luminescence dating   One of the radiogenic group of dating methods, luminescence dating relies on either a thermal (thermoluminescence dating or TL) or an optical (optically stimulated luminescence dating or OSL) signal, the strength of which is related to the numbers of electrons trapped within defects in the crystal lattice of minerals. The electron trapping results from nuclear radiation from both within, and without, the mineral. The populations of electrons in the minerals under investigation must be capable of being reset to zero. For TL this happens when temperatures exceed 400°C. In the case of OSL, trapped electrons are eliminated by prolonged exposure to daylight (i.e., when sediments are trans-ported, by wind or water, or if they lie for long periods on the surface). If these methods are to be of value for dating events in hominin evolution the "resettings" must be brought about by events that are the result of hominin activity, and not the result of natural events. For example, when TL is used on burnt flint artifacts researchers assume the flints were burned in a fire not long after they were fashioned, so the TL date provides a conservative, minimum date, or *terminus post quem* for the manufacture of the flint artifacts. These methods can be applied to minerals such as **quartz**, **flint**, and **feldspar**s that are relevant to the chronology of the later stages of hominin evolution and they can be used up to *c*.100 ka and in certain circumstances more than 500 ka. Analytical problems with respect to OSL include inadequate exposure to sunshine so that previously acquired luminescence is retained and with respect to TL the long-term instability of the signal. Examples of OSL application include its use to date quartz grains from the M1 phase of the **Middle Stone Age** levels at **Blombos Cave** (Jacobs et al. 2003) and TL has been used to date two burned stone artifacts from the M2 phase at Blombos Cave (Tribolo 2003).

lumper   (ME *lumpe*=to aggregate) A researcher who favors fewer, more inclusive, taxa [e.g., Milford **Wolpoff**, who advocated lumping all extinct *Homo* species into a single species, *Homo sapiens sensu lato* (Wolpoff et al. 1992)].

Luna   *See* **Bubing Basin**.

lunate sulcus   (L. *lunatus*=to bend like a crescent and *sulcus*=a deep furrow or groove) A prominent **sulcus** found on the lateral surface of the occipital lobe of the **cerebral hemisphere**s of some monkeys, all extant ape taxa, and a small percentage of modern humans. In general, it marks the anterior-most limit of the **primary visual cortex**. Compared to the terminus of the **Sylvian fissure**, the lunate sulcus in modern humans (when present) is in a more posterior position relative to other primates. This is thought to reflect reorganization of the cortical areas to the extent that the parietal association cortex that lies immediately in front of the lunate sulcus has expanded in size relative to the primary visual cortex. Because the endocasts of fossil hominins, both natural and those prepared by researchers, occasionally preserve evidence of the lunate sulcus, its position has historically been a focus of debate about when in hominin evolution cortical reorganization occurred. Raymond **Dart** (1925) argued for the hominin status of the **Taung 1** child, in part on the basis of his assessment of a modern human-like position of the lunate sulcus on the specimen's natural endocast. While there is general consensus that the lunate sulcus is in a modern human-like position in endocasts from early *Homo* and pre-modern *Homo*, the position of the lunate sulcus in *Australopithecus afarensis* and *Australopithecus africanus* endocasts has proven to be more difficult to determine conclusively. Ralph **Holloway** has argued that **A.L. 162-28**, **Taung**, and **Stw 505** all show the lunate sulcus in a posterior, modern human-like position, signaling evidence of brain reorganization in Pliocene hominins (Holloway 1983a, 1985a, 1991). In contrast, Dean **Falk** has claimed that the lunate sulci of Taung and A.L. 162-28 are in a more anterior, extant ape-like, position (Falk 1983a, 1986a, 1991).

Lupemban A Middle Stone Age archeological industrial complex, sites of which are widespread in central Africa, particularly the Congo River basin, from Angola to as far north as western Kenya. The type collections derive from mining operations along the Lupemba River in southern Democratic Republic of the Congo, near the Angola border, with the best-described collections and dated collections found at **Kalambo Falls** and **Twin Rivers** in Zambia, with an estimated age of *c*.200 ka (Barham 2000, Clark 2001). The diagnostic stone tools of the Lupemban are "lanceolate" **point**s, long (>20 cm in length), relatively thin **bifacially worked** tools whose function is unknown, and the more robust "core axes," tools surmised to have served heavy-duty functions such as woodworking.

L_x The total number of years (based on the **radix**) lived in an age interval is represented in a **life table** as L_x or $_nL_x$. *See* **person-years lived in an age interval**.

Lydekker's Line Richard Lydekker (1849–1915) was a naturalist and an early biogeographer. He worked for the Geological Survey of India and then at the British Museum (Natural History). In 1895 he delineated the eastern boundary of the biogeographical region known as Wallacea (i.e., the boundary between Wallacea and **Sahul**) that is now referred to as Lydekker's Line. *See also* **Wallacea**.

Lyonization *See* X-inactivation.

Lyttelton Formation (etym. Named after local town near the type section) General description. Part of the Malmani Subgroup, which collectively are dolomites of the late Archaean (2.6–2.5 Ga). The Malmani Subgroup sequence is (from bottom to top): Oaktree Formation, Monte Christo Formation (includes **Swartkrans**, **Sterkfontein**, **Kromdraai**, and **Drimolen**), Lyttelton Formation (includes **Malapa**), Eccles Formation (includes **Gondolin** and **Gladysvale**), Frisco Formation.

M

Ma Abbreviation of millions of years ago, which is an age determination, or **age estimate**, expressed in millions of years. For example, "the approximate age of **KNM-ER 3733** is 1.8 Ma." The terms "my" or "myr" should be used to describe periods of time, as in "2 myr have elapsed since hominins first appeared in Africa." *See also* **age estimate**.

MA Acronym for **major axis regression** (*which see*).

Maba (马坝) (Location 24°40′28.51″N, 113°34′48.68″E, southern China; etym. named after the local town) History and general description This limestone cave site is situated in Shizishan (Lion Hill) about 2 km/1.2 miles southwest of the town of Maba, Qujiang County, Guangdong Province. Local farmers discovered hominin and non-human mammalian fossils in 1958 while digging for natural fertilizers. Several authors (e.g., Wu Xing Zhi and Geoffrey Pope) have commented on the morphology of the right orbit (rounded orbit with medially thick supraorbital torus and no supraorbital notch) and its similarity to *Homo neanderthalensis*. Temporal span and how dated? Uranium-series dating of calcite deposit and animal teeth produced an age estimate of $129\pm c.10$ ka. Hominins found at site A **calotte** (PA 84) consisting of partial parietal bones, a frontal bone with fairly complete nasals, and most of the right orbit (frontal process of the zygomatic bone). Sutural fusion is consistent with it being an aged adult and ectocranial rugosity suggests the individual is male. Archeological evidence found at site None. Repository **Institute of Vertebrate Paleontology and Paleoanthropology**, Beijing. Key references: Geology, dating, and paleoenvironment Yuan et al. 1986; Hominins Woo and Peng 1959, Wu and Olsen 1985, Pope 1992, Wu and Poirier 1995, Brown 2001.

MacClade A piece of computer software written by Wayne and David Maddison (Maddison and Maddison 2008) that is used as an aid in **cladistic analysis**. The main function of MacClade is as a tool for exploring the consequences of tree topology on patterns of **character evolution**; it does *not* search for the **most parsimonious tree** that can be generated from a given character matrix.

Machiavellian intelligence *See* social intelligence.

macrodont (Gk *macros* = large and *odont* = tooth) Literally having large teeth, but in paleoanthropology it usually refers to the relative size of the crowns of the postcanine teeth. It was used by Phillip **Tobias** (1967) to describe the large postcanine tooth crowns of *Paranthropus robustus*. Unlike the term megadont, which has been given a quantitative definition, macrodont is only a qualitative term. *See also* **megadont**.

macrostructure *See* morphology.

MAD Acronym for **mean age at death** (*which see*).

Magdalenian (etym. Named after the site of **La Madeleine**, France) A European **Upper Paleolithic** technocomplex, characterized by unretouched stone blades, bone needles, bone harpoons, and other bone tools, as well as figurative engravings of humans and animals, jewelry, and complex burials with many grave goods. It dates to between 17 and 11.5 ka.

Maghreb (etym. "The place where the sun sets" or "western" in Arabic) The northern lowland regions of Morocco, Algeria, and Tunisia (i.e., the parts of those countries that lie between the Atlas massif and the Mediterranean Sea).

magnetic anomaly time scale *See* geomagnetic polarity time scale.

Wiley-Blackwell Encyclopedia of Human Evolution, First Edition. Edited by Bernard Wood.
© 2013 Blackwell Publishing Ltd. Published 2013 by Blackwell Publishing Ltd.

magnetic polarity time scale *See* geomagnetic polarity time scale.

magnetic resonance imaging *See* functional magnetic resonance imaging.

magnetochronology *See* geomagnetic polarity time scale.

magnetostratigraphy The use of the magnetic polarity of a stratum, or a sequence of magnetic polarities of a longer **section**, to date sediments or igneous rocks. *See* **geomagnetic polarity time scale**.

magnetozone *See* geomagnetic anomaly.

Magosi rock-shelter *See* Magosian.

Magosian The term formerly used to describe artifact assemblages attributed to the "Second Intermediate," a transitional period considered to occur between the **Middle Stone Age** (or MSA) and the **Later Stone Age** (or LSA) in Africa. The type site occurs at Magosi rock-shelter, a sediment-filled cistern at the base of a granite cliff in otherwise arid easternmost Uganda (2°35′N, 30°25′E). Early descriptions (Wayland and Burkitt 1932, Clark 1957) emphasized the co-occurrence of **points**, considered diagnostic of the MSA, and backed pieces or **microliths**, typical of the LSA. However, subsequent excavations demonstrated the presence of steeply dipping strata at Magosi. The intersection of these tilted strata during initial excavation in horizontal layers resulted in the accidental admixture of formerly discrete MSA and LSA levels (Cole 1967). As the "transitional" nature of the Magosian was demonstrably the result of excavation methods rather than past human behavior, the term "Magosian" was subsequently abandoned, and with it the concept of the Second Intermediate as part of a general movement towards a better definition of African artifact assemblages (Clark et al. 1966).

Mahalanobis distance A measure of multivariate distance between a point and the **centroid** of data set. Mahalanobis distance differs from **Euclidean distance** in that it takes into account the correlation between variables in the data set as well as the difference in standard deviations between variables, whereas Euclidean distance does not. This distinction is potentially important in calculating multivariate morphological distances. For example, consider a data set with three variables: ulna length, radius length, and radial head diameter. While all of these variables will be correlated, the two length variables will be highly correlated and they are of much greater magnitude (and standard deviation) than radial head diameter. Thus a measure of multivariate distance that treats all variables equally (i.e., Euclidean distance) will be heavily weighted towards a difference in forearm length, and the variability in radial head diameter will contribute very little to the resulting distance. However, Mahalanobis distance effectively scales variables by standard deviation and correlation such that differences in variables of small magnitude and low standard deviations (e.g., radial head diameter) will be given equal weight with differences in variables with large magnitudes and large standard deviations (e.g., ulna and radius length) in the resulting distance. Furthermore, variables which are highly correlated with each other (e.g., radius and ulna length) will be given less individual weight in the resulting distance than variables which are less correlated with the rest of the data set (e.g., radial head diameter). Because Mahalanobis distance requires knowledge of standard deviations within each variable and correlations between variables, the distance cannot be calculated between two individual points in the absence of a reference group from which the standard deviations and correlations are calculated.

Maiko Gully *See* Olduvai Gorge.

main cusp *See* cusp (syn. primary cusp).

Main Deposit One of the three main areas within **Zhoukoudian Locality 1**. *See* **Zhoukoudian Locality 1**.

Main Ethiopian rift The middle of the three components of the **Ethiopian Rift System**. To the north is the Afar rift (also known as the Afar Depression) and to the southwest is the **Omo rift zone**. The main Ethiopian rift is divided into three sectors (WoldeGabriel et al. 2000). The southern sector is divided by the Amaro Horst into two basins and sites within it include Burji and **Konso-Gardula**. The central sector contains fewer areas of exposed sediments and the sites include **Gadeb**, **Gademotta**, and **Melka Kunturé**. The main site in the northern sector is **Kesem-Kebena**.

major axis regression   A form of **regression** in which model **parameters** are fit by minimizing the sum of squared differences between observed values and the best-fit line as calculated by a perpendicular line from the best-fit line, thus taking into account error in both the independent and dependent variables. The best-fit line produced by major axis regression is identical to the axis of the first principal component in a **principal components analysis** (using a **correlation matrix**, not a **covariance matrix**). Major axis regression should only be used when independent and dependent variables are measured in the same units; otherwise, **reduced major axis regression** should be used. Because major axis regression takes into account error in all variables and not just the independent variable when fitting model parameters, it is an example of a **model II regression**.

major histocompatibility complex   (or MHC) This system (known in modern humans as the **human leucocyte antigens**, or HLA, system) is a series of antigens on the cell surface that function as a part of the antibody processing that occurs during the immune response. The HLA system also identifies host cells as "self." They are the most polymorphic system of inherited traits known in modern humans. HLA types are determined by several closely linked **loci** on chromosome 6 (i.e., HLA-A, HLA-B, HLA-C, HLA-D, and HLA-DR). The *HLA-A, -B,* and *-C* genes encode antigens (known as histocompatibility, or Class I, antigens) found on the surface of all nucleated cells, but they are *not* found on red blood cells, sperm, and some placental cells. The HLA-D region encodes Class II antigens and is composed of three subregions: HLA-DR, HLA-DQ, and HLA-DP. Class II antigens are structurally different from Class I antigens and are found only on the surface of B lymphocytes, macrophages, activated lymphocytes, and other specialized cells. The Class III genes, which are located between the HLA-B and HLA-D regions, encode the genes of the complement system, which is important in cell-mediated immunity. Each of these MHC loci has **alleles** for 8–40 antigens. Alleles are codominant, and **recombination** between loci is rare because of their close **linkage**. The set of alleles on one chromosome is referred to as a **haplotype**. The nomenclature for the alleles that code for different antigens at these loci first uses a letter (which refers to the locus) followed by a number that signifies the allele. For example A1 was the first allele described at the A locus. Matching HLA types in modern humans is important for transplantation since these antigens are what identify "self" from "non-self." Diseases such as type 1 diabetes, ankylosing spondylitis, and rheumatoid arthritis are associated with particular HLA alleles. Most of these diseases are associated with Class II alleles and are also referred to as autoimmune diseases. Many are inflammatory diseases and these disorders appear to be triggered by a combination of HLA and non-HLA effects, with environmental activation (possibly by infection by an agent that has antigens that mimic an HLA antigen; this type of mimicry is known as molecular mimicry). There is also evidence linking the MHC to mate choice. Due to its role in the immune system, mate choice for MHC genotype may provide several, non-exclusive fitness advantages, by providing offspring with increased (or optimal) MHC diversity and thus disease resistance, or by providing a "moving target" against infection by parasites that are themselves rapidly evolving. The exact patterns of mate choice vary, but they include choice for MHC-disparate partners or for a partner with an optimal set of MHC dissimilar alleles in taxa as diverse as birds, fish, and lizards, as well as nonhuman primates. Studies of whether our own species mates disassortatively with respect to MHC genes have yielded mixed results, which may be a reflection of context-dependent mate preference expression.

MAK   Abbreviation of Maka and the prefix for fossils recovered from Maka, Ethiopia. *See* **Maka**.

MAK-VP-1/1   <u>Site</u> **Maka** in the **Bodo-Maka** fossiliferous subregion of the **Middle Awash study area**. <u>Locality</u> Maka Vertebrate Paleontology Locality 1. <u>Surface/*in situ*</u> Surface. <u>Date of discovery</u> 1981. <u>Finder</u> Tim **White**. <u>Unit</u> N/A. <u>Horizon</u> "Maka sands" (see White et al. 2000, p. 46). <u>Bed/member</u> N/A. <u>Formation</u> Matabaietu Formation. <u>Group</u> N/A. <u>Nearest equivalent dated horizon</u> MA 90–16 volcanic lens (see White et al. 2000, p. 47). <u>Nearest underlying dated horizon</u> VT-3 tuff. <u>Geological age</u> *c*.3.4 Ma. <u>Developmental age</u> Subadult. <u>Presumed sex</u> N/A. <u>Brief anatomical description</u> A well-preserved left proximal femur of a subadult lacking the head. The neck is preserved, as are the greater and lesser trochanters; the epiphyseal line of the latter is clearly seen. <u>Announcement</u> Clark et al.

1984a. Initial description Clark et al. 1984a. Photographs/line drawings and metrical data Clark et al. 1984a, Lovejoy et al. 2002. Detailed anatomical description Lovejoy et al. 2002. Initial taxonomic allocation *Australopithecus afarensis*. Taxonomic revisions None. Current conventional taxonomic allocation *Au. afarensis*. Informal taxonomic category Archaic hominin. Significance It is claimed that this well-preserved proximal femur confirms that *Au. afarensis* practiced a form of "habitual terrestrial bipedality" that differed "only trivially from that of modern humans" (Lovejoy et al. 2002, p. 97). Location of original National Museum of Ethiopia, Addis Ababa, Ethiopia.

MAK-VP-1/12 Site **Maka** in the **Bodo-Maka** fossiliferous subregion of the **Middle Awash study area**. Locality Maka Vertebrate Paleontology Locality 1. Surface/*in situ* Surface. Date of discovery 1990. Finder Tim **White**. Unit N/A. Horizon "Maka sands" (see White et al. 2000, p. 46). Bed/member N/A. Formation **Matabaietu Formation**. Group N/A. Nearest equivalent dated horizon MA 90-16 volcanic lens (see White et al. 2000, p. 47). Nearest underlying dated horizon VT-3 tuff. Geological age *c*.3.4 Ma. Developmental age Subadult. Presumed sex N/A. Brief anatomical description An adult mandible assembled from many small fragments (White et al. 2000, Fig. 3). Both angles are damaged, but otherwise the bone of the mandible is well preserved, although inevitably there are minor distortions because of the need to glue the many fragments. The crowns of the left I_2–M_3 and the right P_3–M_3 are preserved in the alveolar process. Announcement White et al. 1993. Initial description White et al. 1993. Photographs/line drawings and metrical data White et al. 1993, 2000. Detailed anatomical description White et al. 2000. Initial taxonomic allocation *Australopithecus afarensis*. Taxonomic revisions None. Current conventional taxonomic allocation *Au. afarensis*. Informal taxonomic category Archaic hominin. Significance This mandible is another good example of the mandibular and dental morphology that is "typical" of *Au. afarensis* (Kimbel et al. 2004, Kimbel and Delezene 2009). The P_3s of MAK-VP-1/12 are two-rooted, but each root has a single root canal, thus they are unlike the dumbbell-shaped root cross-sections seen in many, but by no means all, specimens of *Paranthropus*. Location of original National Museum of Ethiopia, Addis Ababa, Ethiopia.

Maka (Location 3°54.03′N, 35°44.40′E, Ethiopia; etym. the name of the drainage in which the site is situated) History and general description The Maka site is located on the eastern side of the Awash River in the **Middle Awash study area** in the **Afar Rift System**; it is immediately south of **Matabaietu** and north of **Belohdelie**. Maka was first explored by Jon Kalb and his colleagues in 1975 and 1976 and then in 1981 by a multidisciplinary team led by Desmond **Clark**. Pleistocene hominin remains were recovered during a later survey and further surveying at Maka in 1990 resulted in the recovery of more hominin remains. Two localities within the Maka area have yielded archeological assemblages. Temporal span The horizon from which the hominin remains come is referred to informally as the "Maka sands." The Maka sands were deposited discomformably *c*.6 m above the 3.74±0.023 Ma VT-3 Tuff (=Wargolo Tuff). Interdigitated within the Maka sands is a volcanic lens (MA90-16) whose glass chemistry is correlated with the 3.40±0.03 Ma Sidiha Koma Tuff of the Hadar Formation and with the 3.39±0.036 Ma MA90-28 Tuff at the nearby Wee-ee fossil locality. Thus, an age of 3.4 Ma is attributed to the hominin-bearing Maka sands horizon. Currently there are no dates available for the horizons bearing archeological assemblages How dated? **Biostratigraphy** and tephrocorrelation. Hominins found at site The first hominin fossil recovered at Maka, in 1981, was **MAK-VP-1/1**, a left proximal femur. Additional specimens from the Maka area recovered in the course of surface prospecting in 1990 include the right side of a mandibular corpus with M_{1-3} (MAK-VP-1/2), a left humerus (MAK-VP-1/3), a right M_2 (MAK-VP-1/4), the left side of an endentulous mandible (MAK-VP-1/6), a nearly complete mandible with dentition (**MAK-VP-1/12**), a left M^1 (MAK-VP-1/13), a ramus from the left side of a hominin mandible found in 1990 (MAK-VP-1/83), and the proximal end of a left ulna (MAK-VP-1/111). The hominin remains from Maka are attributed to *Australopithecus afarensis* and they provided the first substantial sample of *Au. afarensis* material outside of **Hadar** and **Laetoli**. Archeological evidence found at site Two archeological localities were discovered at Maka in 1981. Locality MAK-A1 yielded evidence of an assemblage consisting of artifacts that are technologically and typologically analogous to Early **Acheulean** assemblages from East Africa. The artifacts are all surface finds and are scattered within lag gravels of an uncertain age that cap Pleistocene sediments at MAK-A1. In 1991 further investigations of a low density **Mode 2** occurrence at locality MAK-A2 yielded both surface and *in situ* lithic artifacts. The age of the artifacts at MAK-A2 is also uncertain, but they are most likely Late Acheulean. Key references: Geology, dating, and paleoenvironment Clark et al. 1984a, White et al. 1993, Renne et al. 1999; Hominins Clark et al. 1984a, White et al. 1993, 2000; Archeology Clark et al. 1984a, de Heinzelin et al. 2000.

Makah Mera Collection area within the **Woranso-Mille study area**, Central Afar, Ethiopia.

Makapansgat (Location 24°12'S, 29°12'E, South Africa; etym. in Afrikaans it means literally "Makapan's hole" and it commemorates the Ndebele Chief, Mokopane, or Makapan, who was besieged by the Boers in a cave in the valley in 1854) History and general description Makapansgat (its formal name is Makapansgat Limeworks; see below) is one of a number of cave complexes [others include Buffalo Cave, Cave of Hearths, Gwasa Cave (aka Makapan's Cave, Historic Cave), and Peppercorns] in the Makapansgat Valley, which is in the Highlands Mountains, 22.5 km/14 miles northeast of Mokopane in Limpopo Province (formerly Northern Province). The cave was mined as the "Makapansgat Limeworks" (hence the prefix "MLD" for "Makapansgat Limeworks Deposits/ Dumps") from around the beginning of the 20thC. In 1925 a local teacher, Wilfred Eitzman, in response to the publicity surrounding the **Taung**(s) discovery, sent blocks of fossiliferous breccia collected from the dumps near the limekilns to Raymond **Dart**, who confirmed their antiquity. Other caves in the valley became the focus of attention, but in 1947 the Kitching brothers (James, Ben, and Scheepers) discovered a hominin brain case in the same dumps that Eitzman had examined. The brain case was sent to Dart who referred the new fossil to *Australopithecus prometheus* (Dart 1948). Partridge (1979) divided the Makapansgat Formation into five members (1–5) and he subdivided Member 4 into two deposits (A and B); subsequently M4 B became the CDP or central debris pile. Hominins that are now attributed to *Australopithecus africanus* derive mainly from Member 3, with one specimen (**MLD 37/38**) from Member 4A. Temporal span and how dated? **Biostratigraphy** and **magnetostratigraphy** suggest Member 3 (the source of all but two hominin fossils) is between *c.*2.85 and 2.58 Ma and

Member 4 is *c.*2.58 Ma. Herries (2003) suggests that the deposits in the Makapansgat Limeworks range in age between >*c.*4.3 and *c.*2.58 Ma and that the oldest fossiliferous deposits are *c.*3.0 Ma. Hominins found at site Raymond Dart suggested that the hominins he was aware of from Makapansgat should be assigned to a single new taxon, *Au. prometheus*, which is now regarded as a junior synonym of *Au. africanus*. However, others have suggested that more than one taxon is represented in the Makapansgat hominins; Aguirre (1970) proposed that the second taxon was *Paranthropus robustus* but Clarke (2008) opts for a second unnamed species of *Australopithecus* that is also sampled in Member 4 of the main Sterkfontein cave (e.g., Sts 1, **Sts 71, Stw 252, Stw 505**) and in the **Silberberg Grotto** (i.e., **Stw 573**). Archeological evidence found at site Dart's osteodontokeratic culture has been discredited. Key references: Geology, dating, and paleoenvironment McFadden et al. 1979, Partridge 1979, 2000, Herries 2003; Hominins Dart 1948, Aguirre 1970, Tobias 2000; Archeology Dart 1957, Shipman and Phillips-Conroy 1977, Brain 1981, d'Errico and Backwell 2003.

Makapansgat Limeworks Deposit *See* Makapansgat.

MAL Acronym for **mean age in the living** (*which see*).

Malahuma One of the Early and Middle **Pleistocene** sites in the exposures of the **Ola Bula Formation** within the **Soa Basin**, central Flores, Indonesia.

Malapa (Location 25°53'39"S, 27°47'57"E, Gauteng Province, South Africa; etym. Sesotho for "homestead") History and general description The Malapa cave is *c.*15 km/9 miles northeast of a concentration of sites in the Blauuwbank Valley that includes **Kromdraai, Sterkfontein**, and **Swartkrans**. It had been mined for lime in the 1930s, but the hominins exposed by mining were not found until 2008 when they were recognized in blocks removed from the excavated area, which is 3.3 m by 4.4 m and 3.5 m deep. Temporal span and how dated? The cave filling comprises five **facies**, with A and B below being separated by a layer of **flowstone** from C, D, and E above; the fossiliferous facies are D and E. The two specimens reported so far, **MH1** and **MH2**, come from facies D, but it is suggested that there are "additional hominin remains in facies E" (Dirks et al. 2010, p. 205). Facies D is estimated to be within the range of 1.95–1.78 Ma. The dates of the fossiliferous facies are constrained by the **first appearance datum** of *Equus, Tragelephas* cf. *strepsiceros, Felis silvestris,* and *Parahyaena brunnea* (*c*.2.36 Ma), and the **last appearance datum** of *Megantereon whitei* (*c*.1.5 Ma). There is normal magnetic polarity at the base of the flowstone, but most of the flowstone is reversed polarity and the facies above it (i.e., C, D, and E) are normal. Dates of 2.024±0.062 and 2.026± 0.021 Ma using **uranium-series dating** (U-Pb) were obtained from the Bern and Melbourne labs, respectively, for specimens taken from flowstone 20 cm below MH2. Thus, the normal polarity at the base of the flowstone is interpreted as the Huckleberry Ridge subchron (2.06±0.04 Ma) and the normal polarity in facies C–E is interpreted as the 1.95–1.78 Ma **Olduvai subchron**. Hominins found at site Two hominin associated skeletons: the juvenile MH1 and the adult MH2. Archeological evidence found at site None reported. Key references: Geology, dating, and paleoenvironment Dirks et al. 2010; Hominins Berger et al. 2010; Archeology N/A.

malar (L. *mal*=cheek) Refers to the region of the **face** corresponding to the cheek. That region of the bony face is called the malar region, and the external surface of the part of the maxillary bone that lies deep to the soft tissues of the cheek is called its malar surface.

malar visor *See* **facial visor.**

Malarnaud (Location 43°01'08"N, 1°19'44"E, France) History and general description This cave complex, consisting of several long interconnected galleries, has been explored by miners, treasure hunters, and amateur and professional archeologists since at least the late 1800s. A *Homo neanderthalensis* mandible was recovered in 1888, the first evidence for Neanderthals in France. Temporal span and how dated? Radiometric dating has not been attempted, but faunal analysis and correlation to glacial cycles suggests the Riss-Würm interglacial or the beginning of the Würm (*c*.100–50 ka) for the layer that contained the fossil. Hominins found at site A nearly complete mandible of juvenile, *c*.12 years old, but which is missing the crowns of all the teeth except the right M_1 and the unerupted left M_3, which is still

visible in its crypt. Archeological evidence found at site The site was inhabited for a long period, and includes predominantly **Mousterian** deposits, but also **Aurignacian** and Magdalenian levels. Key references: Geology, dating, and paleoenvironment Heim and Granat 1995, Pales 1939; Hominins Heim and Granat 1995; Archeology Heim and Granat 1995, Pales 1939.

male-display model This model suggests that **Oldowan** sites were places on the landscape where scavenging and hunting opportunities were concentrated. It suggests that male *Homo erectus* displayed their fitness, and increased their mating success, by aggressively scavenging carcasses from carnivores in favored locations. These carcasses would then be transported to the nearest refuge, where taxonomically diverse archeologically visible assemblages would form over time (O'Connell 1997, O'Connell et al. 2002).

Malema (Location 10°1'18.59"S, 33°55'51.53"E, in Malawi; etym. named after Malema village) History and general description Discovered in 1992, Malema RC 11 is one of the many fossiliferous localities in the **Chiwondo Beds**. Its main significance is biogeographical, for it extends the range of the hominin taxon recovered from the site by more than 1000 km/620 miles to the south. Temporal span and how dated? *c*.2.5–2.3 Ma using **biostratigraphy**. Hominins found at site A maxillary fragment **HCRP RC 911** provisionally assigned to *Paranthropus boisei*. Archeological evidence found at site None. Key references: Geology, dating, and paleoenvironment Kullmer et al. 1999, Kullmer 2008; Hominins Kullmer et al. 1999; Archeology N/A.

Malladetes (Location 39°01'19"N, 0°18'01"W, Spain; etym. name of the local cave, also known as Cueva de les Malladetes) History and general description Malladetes cave lies on a hillside overlooking the town of Barx in Valencia. Excavations in the late 1940s under the direction of L. Pericot and F. Jordá uncovered a hominin occipital. Temporal span and how dated? Charcoal from the hominin level was dated by **AMS radiocarbon dating** to 25,120±240 years BP (uncalibrated age). Hominins found at site A hominin occipital was found in level 12. The age at death has been approximated to 5–7 years. It has been suggested that the Malladetes occipital supported the hypothesis of hybridization between *Homo neanderthalensis* and anatomically modern humans, but comparative studies indicate the specimen belongs to *Homo sapiens*. Archeological evidence found at site Some of the earliest tools have been characterized as **Aurignacian**. In upper layers, a **Gravettian** assemblage was recovered, including **points**, **scrapers**, and **burins**. Bone tools were also found. Key references: Geology, dating, and paleoenvironment Arsuaga et al. 2002; Hominins Arsuaga et al. 2002; Archeology Arsuaga et al. 2002.

malleus (L. *malleus*=hammer) *See* **auditory ossicles**; **branchial arches**; **middle ear**.

mammal size classes *See* **bovid size classes**.

mandible (L. *mandere*=to chew and *mandibular*= lower jaw) The bone of the lower jaw. Each side of the mandible consists of a horizontal **corpus** and a vertical ramus. In the adult the corpora are joined in the midline at the mandibular symphysis to form the body of the mandible. Each corpus has an **alveolar process** superiorly, a base inferiorly, and external and internal surfaces. The alveolar process contains the sockets, or alveoli, for the roots of the mandibular teeth. On the base, on either side of the midline, are the digastric fossae. The mental foramen perforates the external surface of the corpus and the mylohyoid line divides the internal surface into the sublingual fossa superiorly and the submandibular fossa inferiorly. The symphysis has the alveolar process superiorly, the external or mental surface anteriorly, the internal or lingual surface posteriorly, and a base. The external surface may have a **mental protuberance**, but to qualify as a **true chin** there must be a depression, called an *incurvatio mandibulae*, superior to it. The internal surface of the symphysis may have tubercles and fossae for the genial muscles. It may describe a smooth curve, or be reinforced by transverse bony buttresses. These may be separate, as superior and inferior transverse tori, or in the form of a single transverse torus. When the internal surface of the symphysis between the alveoli and the torus is flat it is called a post-incisive planum. The quadrilateral rami are attached to the corpora posteriorly. Each ramus has two surfaces, external and internal, four borders, superior and inferior, anterior and posterior, and two processes, condylar and coronoid, separated by the mandibular notch. The posteroinferior corner is called

the angle. The anterior border of the ramus can be straight or sinuous. The internal surface of the ramus is perforated by the mandibular foramen (for the inferior alveolar nerve and vessels). On the anterior border of the foramen the lingula marks the inferior attachment for the sphenomandibular ligament.

Mandible A *See* **Kedung Brubus**.

Mandible B *See* **Sangiran 1b**.

Mandible C *See* **Sangiran 9**.

Mandible E *See* **Sangiran 21**.

mandibular scaling *See* **scaling**.

Mann–Whitney U test A nonparametric **statistic**al test used to determine whether the values of a **sample** of measurements are significantly larger or smaller than the values for a second sample. Used in the same situations as one would use a two-sample *t* **test**, the Mann–Whitney U test considers the relative ranking of the values in the two groups rather than the values themselves.

MANOVA Acronym for **multivariate analysis of variance** (*which see*).

manual digit (L. *manus*=hand and *digit*= finger) The separate fingers of the primate, and therefore the hominin hand, each comprise three bones, namely the proximal, the intermediate, and the distal phalanges. The manual digit is the distal component of a **manual ray**, which comprises a manual digit plus its associated metacarpal. The thumb has just two phalanges.

manual ray (L. *manus*=hand and *radius*= radiating line) The manual ray comprises a **manual digit** (i.e., the proximal, the intermediate, and the distal phalanges) plus its associated **metacarpal**. The thumb has just two phalanges.

manuport (L. *manus*=hand and *portare*=to carry) An unmodified stone found at an archeological site that is inferred to have been transported by hominins, perhaps for future use as raw material sources for the production of sharp-edge stone tools. Manuports are usually identified as such by a process of eliminating other natural transport processes (e.g., deposition by streams) although this can be difficult in many ancient settings (e.g., de la Torre and Mora 2005).

Mapa *See* **Maba**.

marginal sinus (L. *margo*=edge) One of the **dural venous sinuses** belonging to the external component of the **intracranial venous system**. The marginal sinus, when present, runs from the inferior end of the occipital sinus around the margin of the posterior quadrant of the **foramen magnum** to drain into the superior jugular bulb and/or the **vertebral venous plexus**es. In most modern humans the superior sagittal sinus and the straight sinus (the latter drains venous blood from the deep cerebral veins) drain into the transverse sinuses. But in the majority of the crania belonging to *Australopithecus afarensis*, *Paranthropus boisei*, and *Paranthropus robustus*, the venous blood from the same sources drains into an **occipital sinus**, which in turn drains into the marginal sinuses. This pattern of intracranial venous drainage is known as the **occipito-marginal system** of venous sinuses, and individuals with a substantial occipital sinus and with marginal venous sinuses on both sides (e.g., **OH 5**) are said to show occipito-marginal system dominance. *See also* **cranial venous drainage**; **dural venous sinuses**.

MARGO *See* **sea surface temperature**.

marine isotope stages *See* **Oxygen Isotope Stages**.

marker bed A layer or stratum that is physically or chemically distinctive enough to be identified independently of its context. Marker beds are used to link isolated blocks of sediments in the same fossil site, or to provide a means of linking/correlating strata from one locality to another, or even from one site to another (e.g., the "grey tuff" and the "pumice tuff" in the **Kapthurin Formation**, Tugen Hills, Kenya; Tallon 1978). The ash spewed from each volcano has a unique mix of chemicals and even the ashes derived from different eruptions from the same volcano are chemically distinctive. The unique chemical profile of a tuff can be used to trace a tuff from one block of sediment to another and from one site to another. This technique is called **tephrostratigraphy**.

Mas-d'Azil Location 43°04′04″N, 1°21′16″E, France; etym. name of a nearby town) History and general description This late **Upper Paleolithic** site is located in the caves leading off the large tunnel created by the Arize river as it flowed through part of the Pyrenées. It is the type site of the Azilian, and was excavated in the 19thC by Edouard Piette. It contains many carvings on bone and antler of reindeer and horses, as well as humans and other animals. Temporal span and how dated? Based on the art style and stone tool technologies, as well as a **radiocarbon dating**, the site is thought to date to between 18 and 7 ka. Hominins found at site Piette discovered an incomplete modern human skeleton in 1895, and Vallois found a skull in the **Magdalenian** layers; both are attributed to *Homo sapiens*. Archeological evidence found at site Stone and bone tools and carvings attributed to the Magdalenian and Azilian; the latter is known for small scrapers and distinctive points. Key references: Geology, dating, and paleoenvironment Aleirac and Bahn 1982; Hominins Piette 1895, Vallois 1965; Archeology Cartailhac 1891, Péquart and Péquart 1941.

mastication (Gk *mastikhan*=to grind the teeth) The processes involved in **chewing** food by moving the lower teeth against the upper teeth to slice, fracture, or crush food into smaller pieces. This mechanical reduction, or **comminution**, of food in the mouth is a key feature of mammals and it was crucial for the relatively high energy consumption of mammals compared to reptiles of the same body mass. Mastication reduces the particle size of food, mixes it with saliva, and softens it so that it forms a bolus that can be transferred to the esophagus by the process called **swallowing**. *See also* **chewing**; **mandible**; **TMJ**; **mastication, muscles of**.

mastication, muscles of (Gk *mastikhan*=to grind the teeth) An informal term used for the group of striated muscles that moves the mandible during **mastication**. *Gray's Anatomy* includes the masseter, temporalis, and the lateral and medial pterygoid muscles in its section on the muscles of mastication. The main action of the masseter is to elevate the mandible, but its obliquely orientated fibers also play a minor role in side-to-side and protraction and retraction movements of the mandible. The anterior temporalis fibers are involved in elevation of the mandible and the posterior fibers with retraction. The pterygoids act as an integrated unit. When the medial and lateral pterygoids of the same side contract the condyle on that side is drawn anteriorly so that movements can then occur around an axis that passes through the opposite condyle. When the medial and lateral pterygoids of one side contract, followed by contraction of the medial and lateral pterygoids of the other side, the mandible is moved from side to side in a chewing movement. When the inferior fibers of the two lateral pterygoids contract together with the two medial pterygoids they draw both mandibular condyles and articular discs anteriorly onto the articular eminence to maximize the gape. When the superior fibers of the lateral pterygoids are involved instead of the inferior fibers, together with medial pterygoids, the mandible is protruded. The motor branch of the trigeminal nerve supplies all these muscles. Other muscles, including the buccinator supplied by the facial nerve, play an important role in mastication (e.g., the buccinator prevents masticated food collecting in the cheek) but they are not usually considered to be among the "muscles of mastication."

masticatory apparatus (Gk *mastikhan*=to grind the teeth, plus L. *ad*=to and *parare*=to make ready) The soft (i.e., muscles) and hard (i.e., bones and teeth) parts of the skull involved in **chewing** food. With respect to the hard tissues it includes the upper jaw, the maxilla, plus the upper, or maxillary, teeth, and the lower jaw, the mandible and the lower, or mandibular, teeth. The jaws and the teeth are among the more dense parts of the skeleton, and for this and other reasons they tend to resist post-mortem damage and survive long enough to be fossilized. This results in the hard tissues of the masticatory apparatus dominating the hominin fossil record.

mastoid crest (Gk *mastos*=breast, *œides*=shape, and L. *crista*=crest) A bony crest that sometimes marks the lateral boundary of the attachment of the sternocleidomastoid muscle and two deep neck muscles (the splenius and longissimus capitis) to the lateral surface of the mastoid process. If the posterior-most fibers of the temporalis muscle are especially well developed, then the bony edge of that muscle in the region of the mastoid may be raised up to form a supramastoid crest, and in *Paranthropus*, in which the supramastoid crest and the mastoid crests are usually distinct, they are

separated by a supramastoid sulcus. *See also* **supramastoid crest**.

mastoid process (Gk *mastos*=breast and *œides*=shape) A bony prominence that projects inferiorly from the posterior part of the temporal bone. It is relatively large in modern humans (you can feel it just behind your ear). The shape and size of the mastoid process shows considerable inter-individual variation in all the higher primates, including extinct hominin taxa. However, there are trends in mastoid process size and shape that are reasonably consistent. In modern humans it is typically the only substantially pneumatized part of the temporal bone and its pyramidal shape is distinctive. Among pre-modern *Homo* taxa the mastoids of *Homo neanderthalensis* are distinctively small. In archaic hominins the mastoid is less distinct because it is just one part of a more general pneumatizion of the temporal bone. The sternocleidomastoid muscle and two deep neck muscles (the splenius and longissimus capitis) are attached to its lateral surface and the bony edge of their attachment is sometimes raised up as a mastoid crest. If the posterior-most fibers of the temporalis muscle are especially well developed, then the bone may be raised up to form a supramastoid crest. In *Paranthropus* (and in most archaic hominin taxa assigned to *Australopithecus*) the mastoid normally projects further laterally than the supramastoid crest. In *Paranthropus* the supramastoid and mastoid crests are distinct and they are usually separated by a supramastoid sulcus. The posterior belly of the digastric muscle is attached in a groove (called the digastric notch) on the medial surface of the mastoid process. The sequence of the suite of structures that may be manifest posteromedially to the tip of a mastoid process are, from lateral to medial, the digastric fossa, the juxtamastoid eminence, a groove for the occipital artery, and the occipito-mastoid crest.

Mata Menge (Location 8°41′31″S, 121°5′43″E, Flores, Indonesia; etym. named after a local village) History and general description Mata Menge is the best known of a group of Early and Middle **Pleistocene** sites in the **Ola Bula Formation** located in the **Soa Basin** of central Flores. The first fieldwork was conducted between 1991 and 1994 by Paul Sondaar, John de Vos, Gert van den Bergh, and Fachroel Aziz. Michael

Morwood and colleagues began their investigations in 1997 and excavations in 1994, 1997, and 2004–5 resulted in the recovery of more than 500 artifacts, Temporal span and how dated? The age of the *c.*780 ka is based on **magnetostratigraphy** and **fission track dating**. The artifact-bearing horizon is 3 m above what is assumed to be the reversed/normal transition of the Brunhes-Matuyama boundary. Zircons for fission track dating were collected from "the pink tuff immediately below the lowest fossils and stone artefacts" (MM1) and from a "white tuff" that "overlies *in situ* artefacts" (MM2) (Morwood et al. 1998, pp. 175–6); MM1 has been dated to 0.88 ± 0.07 Ma and MM2 to 0.80 ± 0.07 Ma (*ibid*). Hominins found at site None. Archeological evidence found at site Excavations in 1994 resulted in the recovery of 14 chert artifacts, four of them with use wear consistent with being used to process plant remains. A few additional artifacts were recovered in 1997 and a more recent excavation campaign in 2004–5 recovered 487 *in situ* artifacts; roughly 50% of the combined sample of 507 artifacts are classified as "fresh/unabraded." Aside from the intrinsic value of the archeological evidence, the site is significant for two other reasons. First, because Flores is separated from the **Sunda** shelf by at least three deep-sea channels – even when sea levels were lowered at times of global cooling – hominins would have had to face a sea crossing of *c.*20 km/12.4 miles to reach Flores from the Sunda shelf. Second, the similarities between the artifacts at Mata Menge and those at **Liang Bua** (some 50 km/31 miles to the west) strengthen the case that the latter were manufactured by hominins other than modern humans. Key references: Geology, dating, and paleoenvironment Morwood et al. 1998, O'Sullivan et al. 2001; Hominins N/A; Archeology Morwood et al. 1997, 1998, 1999, Brumm et al. 2006.

Matabaietu One of the named drainages/subdivisions within the **Bodo-Maka** fossiliferous subregion of the **Middle Awash study area**, in Ethiopia.

maternal effect This occurs when an organism's phenotype is affected by the maternal genotype, environment, or condition. Sometimes an offspring will manifest aspects of the maternal phenotype, due to maternal effects resulting from a maternal supply of mRNA or proteins to the developing ovum. The maternal, or intrauterine, environment can affect offspring phenotype in ways such as intrauterine position: female fetuses located between male fetuses can experience effects of androgens on morphology, behavior, and reproduction. The condition of the mother can also affect her offspring. For example, in modern humans, uncontrolled gestational diabetes mellitus, in which the mother's normal insulin response has been altered, is related to macrosomia (large size), delayed lung maturation, or problems with blood–sugar regulation.

maternal energy hypothesis Robert Martin (1996) suggested there is a causal link between the basic metabolic rate of the mother and the brain size of her developing offspring, such that "all mammals have the largest brains that are compatible with the metabolic resources available to their mothers" (*ibid*, p.155).

matrix (L. *mater*=mother, in the sense of a substance from which something else originates) Biology The combination of water and polymers that, along with cells, fibers, and fibrils, makes up connective tissues. The more polymerized the matrix the stiffer the connective tissue. Earth sciences The rock in which a fossil is embedded. If the matrix is very mineralized it may adhere to the surface of the bone. Extreme changes in humidity and temperature can produce cracks in fossils and when matrix enters these cracks they undergo further expansion. Matrix-filled cracks can artificially increase the size of a structure like the corpus of the mandible. Statistics When raw data, or reformulations of raw data, are presented in a series of rows and columns. It is referred to as data matrix. More sophisticated matrices (e.g., correlation and covariance matrices) are used in **multivariate analysis** and in **Euclidean distance matrix analysis**.

Matupi Cave (Location $1°13'53''$N, $29°49'23''$E, in the Democratic Republic of the Congo; etym. unknown) History and general description Matupi Cave is one of approximately 40 caves formed in the Precambrian limestone and shale of Mount Hoyo in the Democratic Republic of the Congo. Archeological excavation of a 15 m^2 area in 1973–4 directed by Francis Van Noten exposed a 210 cm-deep sequence of loam formed by the combination of limestone bedrock decalcification and anthropogenic sediments; the excavation also exposed a number of **hearths**.

Later Stone Age artifacts, which occur from 25 cm below the surface to the base of the excavations, underlie strata attributed to the Iron Age. Many of the **speleothems** found in the cave complex at Matupi (but these are about 10–100 m distant from the archeological excavation) preserve a diverse pollen spectrum suggesting local shifts in the region throughout the **Late Pleistocene** between **savanna grasslands** and **montane forests**. This interpretation is also suggested by the pollen record recovered from the excavated sediments. Temporal span and how dated? Uncalibrated **radiocarbon dating** estimates (presumably on charcoal) suggest long-term presumably modern human occupation of Matupi Cave, with published values increasing with stratigraphic depth ranging from 0.72 ± 0.05 to more than 40.7 ka. A thermo**luminescence dating** age estimate of 21.35 ± 3.50 ka on burnt quartzite cobbles has also been reported. **Uranium-series dating** (^{230}Th/^{234}U) of the speleothems provide comparable Late Pleistocene estimates, ranging from 0.99 ± 0.28 to 50.33 ± 2.58 ka. Hominins found at site The Later Stone Age strata contain a single isolated lower tooth attributed to *Homo sapiens sapiens*. Archeological evidence found at site The artifact sample includes more than 8000 predominantly **quartz** artifacts attributed to the Later Stone Age, including early **microliths** in the basal layers that are associated with an infinite radiocarbon age estimate, ground pieces of **hematite** that are found throughout the sequence, and early ($c.20$ ka) evidence for the use and manufacture of "bored" stones. More than 50,000 faunal elements were recovered, including a diverse mammalian and non-mammalian fauna that suggesting proximity to a forest-savanna **ecotone**. Key references: Geology, dating, and paleoenvironment Brook et al. 1990, Van Neer 1989, Van Noten 1977; Hominins Van Neer 1989; Archeology Van Neer 1989, Van Noten 1977.

Matuyama *See* **geomagnetic polarity time scale.**

Matuyama chron *See* **geomagnetic polarity time scale.**

MAU Acronym for **minimal animal unit** (*which see*).

Mauer (Location 49°21′N, 8°48′E, Germany; etym. Ger. for "wall," the name of a nearby village,

and also of the "Mauer Sands") History and general description The site is the Grafenrain commercial sandpit 16 km/10 miles southeast of Heidelberg. The stratified fluvial sands exposed there contain mammalian fossil fauna, but the only hominin from the site is a mandible discovered in 1907. Temporal span and how dated? Between $c.450$ and $c.650$ ka (i.e., **Oxygen Isotope Stages** 11 or 15), but probably OIS 15 based on **biostratigraphy** and **magnetostratigraphy**. Recent combined **electron spin resonance spectroscopy dating** and **uranium-series dating** on mammal teeth and infrared radiofluorescence on sand grains suggests an age of 609 ± 40 ka. Hominins found at site An adult mandible, the **holotype** of *Homo heidelbergensis*. Archeological evidence found at site None. Geology, dating, and paleoenvironment Howell 1960 (pp. 199–202), von Koenigswald and Tobien 1987, von Koenigswald 1992, Hambach 1996, Wagner et al. 2010; Hominins Schoetensack 1908, Howell 1960 (pp. 211–212), Kraatz 1992, Rosas and Bermudez de Castro 1998a, Mounier et al. 2009.

Mauer mandible Site **Mauer.** Locality Grafenrain pit. Surface/*in situ In situ*. Date of discovery October 21, 1907. Finder Workmen at the sandpit. Unit N/A. Horizon N/A. Bed/member Mauer Sands. Formation N/A. Group N/A. Nearest overlying dated horizon Rhine terrace volcanic tuffs postdating Mauer Sands (>400 ka from a **potassium-argon date**). Nearest underlying dated horizon N/A. Geological age Between 450 and 650 ka (**Oxygen Isotope Stages** 11 or 15). Recent combined **electron spin resonance spectroscopy dating** and **uranium-series dating** on mammal teeth and infrared radiofluorescence on sand grains suggests an age of 609 ± 40 ka (Wagner et al. 2010). Developmental age Adult. Presumed sex Presumed male, based on large size. Brief anatomical description When the mandible was found it was in two pieces having been broken at the symphysis (in Schoetensack 1908 it is illustrated as separate right and left sides of a mandible) and some perisymphyseal bone has been lost. Most of the lower teeth are preserved in good condition. The left coronoid process is broken at the tip and the left mandibular condyle is deformed. Announcement Schoetensack 1908. Initial description Schoetensack 1908. Photographs/line drawings and metrical data Schoetensack 1908. Detailed anatomical description Schoetensack 1908, Mounier et al. 2009.

Initial taxonomic allocation *Homo heidelbergensis.* Taxonomic revisions *Homo erectus* (Hrdlička 1927), *Homo neanderthalensis* (Hublin 1998), *incertae sedis* (Hublin 2009). Current conventional taxonomic allocation *H. heidelbergensis.* Informal taxonomic category Pre-modern *Homo.* Significance The oldest hominin fossil specimen known from north-central Europe. As it is the **holotype** of *H. heidelbergensis*, any specimens allocated to that hypodigm must be compatible with being included in the same species as the Mauer mandible. Mounier et al. (2009) provide the most detailed morphological and taxonomic analysis and consider the mandibles from **Arago, Montmaurin, Sima de los Huesos**, and **Tighenif** to be conspecifics of the Mauer mandible. While some argue that the Mauer mandible represents an early stage in the **accretion model** for the origin of *Homo neanderthalensis* (e.g., Arsuaga et al. 1997, Humphrey et al. 1999, Carbonell et al. 2005), others see no evidence of the derived morphology seen in later Neanderthals (e.g., Rightmire 2008, Mounier et al. 2009). If there is sufficient evidence in favor of the former view, then *H. heidelbergensis* would be a **junior synonym** of *H. neanderthalensis.* However, if the latter view prevails, then *H. heidelbergensis* would have priority as the name of the taxon that gave rise to both *H. neanderthalensis* and *Homo sapiens*, or to just *H. neanderthalensis.* Location of original **Geologisch-Paleontologisch Institute, University of Heidelberg**, Heidelberg, Germany.

MAVT Abbreviation for the Maoleem Vitric Tuff. *See* **BOU-VP-12/130.**

Max Planck Institute for Evolutionary Anthropology

(or MPI-EVA) Founded in 1997, this research institute is one of many sponsored by the Max-Planck-Gesellschaft zur Förderung der Wissenschaften (Max Planck Society for the Advancement of Science), a nonprofit organization funded in large part by the German federal and state governments. The MPI-EVA aims to investigate the history of humankind with the help of comparative analyses of different genes, cultures, cognitive abilities, languages, and social systems of past and present human populations as well as those of primates closely related to human beings. The institute currently comprises five departments (listed here with their current heads): Primatology (Christophe Boesch), Developmental and Comparative Psychology (Michael Tomasello), Linguistics (Bernard Comrie), Evolutionary Genetics (Svante Pääbo), and Human Evolution (Jean-Jacques Hublin). The MPI-EVA funds many research projects, including field sites investigating three of the four great ape species, and studies of captive populations of all four great ape species at the Wolfgang Koehler Primate Research Center, the Neandertal Genome project, archeological excavations (e.g., Jebel Irhoud, Willensdorf II, Jonzac), and isotope analyses to reconstruct diet and environment (in collaboration with the **Ancient Human Occupation of Britain Project**). What makes the MPI-EVA unique is that these diverse perspectives – behavioral, ecological, cognitive, cultural, linguistic, and genetic – are all housed under one roof. The MPI-EVA has close connections with local research institutions including the University of Leipzig, and together with the University of Leipzig the MPI-EVA runs the Leipzig School of Human Origins, an International Max Planck Research School. Contact address Deutscher Platz 6, 04103 Leipzig, Germany. Website www.eva.mpg.de/english/index.htm.

maxillary fossula The name used by Yoel Rak (1983, Fig. 16, p. 135) to describe the small hollowed area on the face of *Paranthropus robustus* at the inferomedial corner of the **maxillary trigon**. It is what remains of the **maxillary furrow** after the anterior advance of the infraorbital region in *P. robustus*, and it contrasts with the morphology seen in *Australopithecus africanus.*

maxillary furrow A modification of the **canine fossa** seen in archaic hominin taxa (e.g., *Australopithecus africanus*) that have **anterior pillars**. It is a sulcus that extends from the infraorbital foramina inferiorly to the alveolar process. It runs parallel and posterior to the superior end of the anterior pillar (e.g., MLD 6).

maxillary sulcus The name given by Franz **Weidenreich** (1943) to the hollowed area in the face of *Homo erectus* between the **canine eminence** and the zygomatic process of the maxilla.

maxillary trigon The name used by Yoel Rak (1983, Fig. 16, p. 135) to describe the triangular area of the mid-face of *Paranthropus robustus*. Its long sides are the **anterior pillars** medially and the **zygomaticomaxillary step** laterally and its base is the **zygomaticoalveolar crest**. There is often a hollowed area, the **maxillary fossula**, at the infermedial corner of the maxillary trigon.

maxillary visor   *See* **facial visor**.

maximum life span   The greatest age to which a member of a group or species has lived. Carey and Judge (2000) is an excellent online source for the data that are available for comparative maximum life spans across many taxa. For modern humans the documented maximum life span was for Jeanne Calment, who was born on February 21, 1875 in Arles, France and died on August 4, 1997 at the age of 122 years and 164 days (Robine and Allard 1995). For nonhumans the maximum life span is much harder to obtain. While zoos keep good demographic records, the maximum life span for zoo animals can typically only be obtained if the animals were born in captivity. Further, animals that are born in captivity and live their entire lives in captivity can be subject to very different hazards of death compared to animals living in the wild. To measure maximum life span in the wild an observer must be present for both the birth and death of an individual animal and must be able to identify the animal as a unique individual so that the birth and death can be linked. A further complication is that the maximum life span represents an extreme event, so it is sample-size-dependent. In paleodemography and paleoanthropology the measurement of maximum life span is extremely difficult. This is because (a) ages at death must be estimated rather than observed by linking birth and death records, (b) ages at death for the old are particularly difficult to estimate, and (c) sample sizes are typically small, so the probability of someone dying at the maximum lifespan is extremely small. Because of these problems it is more common in paleodemographic studies to model **survivorship** (or mortality) and to determine the maximum life span from the model. Given the difficulties both in determining ages at death for fossils and the widely varying sample sizes for various paleoanthropological taxonomic units, any statements about the evolution of maximum life span are highly speculative.

maximum likelihood   A method of inferring phylogenetic relationships. Unlike **cladistic analysis**, the maximum likelihood method requires an explicit model of character evolution to be specified. It then calculates the probability or likelihood of phylogenetic trees given the observed character state data set and the chosen model of character evolution. The preferred phylogenetic tree is the one with the highest likelihood. Maximum likelihood is extensively used in DNA-based phylogenetic research but has had limited impact in paleoanthropology. This difference in popularity is due to a difference in availability of widely accepted models of character evolution for DNA and morphology. Several such models have been developed for DNA, including the Jukes–Cantor Model, which simply specifies an expected substitution rate, and the Kimura Model, which differentially weights transitions and transversions (*see* **mutation** for explanations). In contrast, so little is known about the processes of morphological evolution that a widely accepted model of morphological character evolution has yet to be developed.

MCBA   Abbreviation for measured crown base area. This is measured from an enlarged print of a photograph taken from above the tooth with the optical axis of the camera perpendicular to the plane of the tooth cervix (e.g., Wood and Abbott 1983, pp. 199–200). This provides a two-dimensional measure of the total occlusal area of a tooth crown. When this estimate is corrected for interstitial wear it is abbreviated as "MCBA(C)." These raw or corrected measurements are always less than estimates of occlusal area that are computed from linear measures of the maximum breadth (buccolingual or B/L) and maximum length (mesiodistal or M/D) of a tooth crown.

MCBA(C)   *See* **MCBA**.

Meadura   One of the named drainages/subdivisions within the **Bodo-Maka** fossiliferous subregion of the **Middle Awash study area** in Ethiopia.

mean age at death   (or MAD) The average age at which individuals die. In the analysis of a **stationary population** the mean age at death is equal to the **life expectancy** at birth. The inverse of the average age at death in a stationary population is equal to the **crude death rate** as well as the **crude birth rate**. For example, if the average age at death (life expectancy at birth) in a life table

489

is 25 years then the crude birth and death rates are 1/25, or 0.04. In a (stationary) hazards analysis the mean age at death is the integral of the product of age at death with the probability density function of death at that age. Consequently, the probability density function "averages" across all ages at death. More simply, the mean age at death can be found as the integral of the survivorship function.

mean age in the living The mean age in the living is the average age from a census or the average age among the living that is implied by the deaths analyzed in a **life table** or a hazards analysis. In the life table of a **stationary population** the average age of the living population is found by summing the product of the **living age distribution** with the mid-point of the age intervals. In a stationary hazards analysis the mean age in the living is the integral of the product of age and survivorship divided by the integral of survivorship. Nancy Howell (1982) has used arguments based on the implied age distribution in the living to argue that there may be deficits in paleodemographic analyses.

Mechta-Afalou *See* **Iberomaurusian**.

Meckel's cartilage The cartilage of the first branchial (or pharyngeal) arch. Described by Johann Meckel in 1805. It does *not* develop into the mandible; it provides the cartilaginous scaffold around which the mandible develops. *See also* **branchial arches**.

medial longitudinal arch The medial longitudinal arch is a mainly hard-tissue structure consisting of the calcaneus, talus, navicular, cuneiforms, and the medial three pedal rays. It is higher than the lateral longitudinal arch, causing the medial side of the foot to be elevated off the ground during the stance phase of walking. Although the bones produce the shape of the arch, soft-tissue structures stabilize and support it. These include the calcaneonavicular (spring) ligament, the long plantar ligament, and the plantar aponeurosis. Intrinsic muscles of the foot, fibularis longus, tibialis anterior, and tibialis posterior are all instrumental in supporting the medial longitudinal arch.

median lingual accessory cusp *See* **tuberculum intermedium**.

median ridge An enamel crest present with variable degree of expression in the central part on the lingual face of the anterior teeth, that runs from the **cingulum** towards the incisal edge or, in the case of canines, to the cusp tip. In the postcanine teeth the median ridge indicates the most prominent enamel ridge running down from the cusp tip. *See* **essential crest**.

Medieval Warm Period *See* **Holocene**.

Mediterranean climate In contrast to **monsoon**al climates, Mediterranean climates receive predominantly winter-season precipitation. Mediterranean climates around the world are characterized by high floral diversity and a large number of **endemic** species. The northern coast of Africa has a Mediterranean climate that receives winter precipitation supplied by the mid-latitude westerly winds (Nicholson 2000). Inter-annual to inter-decadal variability is influenced by regional and global atmospheric connections associated with the North Atlantic Oscillation (Hurrell 1995) and to a variable extent by the **El Niño Southern Oscillation** (Knippertz et al. 2003). Greenland ice core data indicate these patterns of variability persist for several hundred years (Dansgaard et al. 1993).

Medjez (Location approximately 36°11′N, 5°42′E, Algeria) History and general description Sometimes known as Medjez II, this open-air Mesolithic archeological site east of Sétif and 4 km/ 2.5 miles north of El Eulma contains assemblages identified to the Upper Capsian tradition. The site was first professionally excavated by Camps (1974) and has been described as a typical example of a Capsian *escargotières*, or a site characterized by large numbers of whole and crushed land snail shells and large quantities of ash and fire cracked rock (Lubell 2001). Temporal span and how dated? **Radiocarbon dating** suggests a range of 8.9–6.5 ka, but the exact age of the site is controversial due to a disparity in radiocarbon dates determined by two different laboratories (Lubell et al. 1992). However, the artifact assemblage from Medjez does match a technological transition known from other Algerian sites to date to *c*.8000 years BP (Lubell 2001, Sheppard 1987). Hominins found at site A minimum number of 15 hominin individuals, including four adult and seven neonate skeletons, were recovered by Camps-Fabrer (1975) and three adults and one infant from earlier unsystematic excavations were reported by Lubell

(2001). The hominin remains were deliberately buried in a range of flexed and unflexed positions, many in association with red ochre and other grave goods (Haverkort and Lubell 1999). The Iberomaurusian cultural practice of incisor evulsion is present in the Upper Capsian culture at Medjez, albeit with more variation in the specific teeth removed than in the earlier culture. Archeological evidence found at site Camps-Fabrer (1975) divided Medjez into four arbitrary units, all which were assigned to the Upper Capsian tool tradition. However, this interpretation has been challenged by Rahmani (2004) on the basis that two of the units lack pressure-flaked stone tools. Faunal remains include the aforementioned ubiquitous land snail shells and many vertebrate species, with larger examples including hartebeest, gazelle, and Barbary sheep. Key references: Geology, dating, and paleoenvironment Lubell et al. 1992, Rahmani 2004; Hominins Haverkort and Lubell 1999, Lubell 2001; Archeology Camps 1974, Camps-Fabrer 1975, Lubell 2001, Sheppard 1987.

megadont (Gk *megas*=great and *odont*=tooth) This term, that literally means having "very large teeth," is used in paleoanthropology to refer to the size of the crowns of the postcanine teeth (i.e., premolars plus molars). The term was probably first used by William Flower in relation to his "dental index" that related the length of the postcanine tooth row to the distance between basion and nasion. If the index was more than 44 then Flower suggested such individuals should be referred to as "megadont." Phillip **Tobias** reintroduced the term to describe the very large postcanine teeth of *Paranthropus boisei* (i.e., "a megadontic line (*A. boisei*) emerged with specialised dentition", Tobias 1967, p. 243). In 1988 Henry McHenry introduced the "megadontia quotient" (or MQ) to relate postcanine tooth area to estimated body mass. By analogy with the **encephalization quotient**, MQ= observed tooth area/12.15 (body mass), where "observed tooth area" is the sum of the areas of the P_4, M_1, and M_2 (McHenry 1988, p. 144). The specifics of this formula are derived from data used to regress postcanine tooth area against estimated body mass (McHenry 1984, Table 1, p. 299 and Fig. 1, p. 300).

megadontia *See* megadont.

megadontia quotient *See* megadont.

megadroughts Droughts are periods of below-average rainfall and when such dry intervals last decades or more they are referred to as megadroughts. Drill cores from **Lake Botsumtwi** and **Lake Malawi** indicate that periods of severe aridity between 135 and 75 ka were marked by 95% reduction in lake volume. This is notably drier than the conditions reconstructed for the **Last Glacial Maximum**, previously identified as a dry interval in the Late **Pleistocene**. Following the 135–70 ka extreme dry period, there was a shift to wetter climates around 70 ka. While the extreme arid period may have restricted the total area available for year-round habitation, the following wet interval may have facilitated the expansion and migration of early modern human populations (Scholz et al. 2007).

megafauna (Gk *megas*=large and L. *fauna*= in Roman mythology the sister of the Faunus, the god of nature) Literally "large animals," but usually interpreted as animals with an adult body mass of more than 44 kg (approximately the size of a large Labrador retriever dog). Defined in this way, in the **Plio-Pleistocene** faunas of Africa the category megafauna would include not only elephants and rhinoceros, but also lions, Grant's gazelles, and warthogs. In the more temperate faunas of Pleistocene Europe megafauna would include deer, boar, and bear. With an average body weight of around 67 kg **modern humans** are also megafauna, but they illustrate the problem of this category in an evolutionary perspective since the some of the earlier hominins almost certainly did not have a body mass over 44 kg and some populations of contemporary modern humans would not qualify, either.

Meganthropus (Gk *megas*=great and *anthropos*=human being) This genus name was introduced and used informally in the early 1940s by Ralph **von Koenigswald** for the **Sangiran** D (1941) mandible. The genus name was used by other researchers (e.g., **Weidenreich** 1942, 1944), but the correct citation for the first use of this genus name is von Koenigswald (1950) for it was in this paper that he formally introduced the species name *Meganthropus palaeojavanicus*. *Meganthropus* fell into disuse because researchers judged it to be a junior synonym of either *Homo erectus* (e.g., Le Gros Clark 1955, p. 155) or *Paranthropus* (e.g., Robinson 1954, p. 36).

491

Meganthropus A   *See* Sangiran 6a.

Meganthropus africanus Weinert, 1950

(Gk *megas*=great and *anthropos*=human being, and L. *africanus*=pertaining to Africa) The binomial introduced by Hans Weinert (1950) for the taxon represented by the maxilla found by Ludwig **Kohl-Larsen** in 1939 at the site then called **Garusi** and now called **Laetoli**. John **Robinson** (1953a, 1954) suggested that the Garusi maxilla "cannot be distinguished from the Sterkfontein apeman" (Robinson 1954, p. 195), but Remane (1954) took the view that the premolar morphology of the Garusi maxilla was more primitive than that of the Sterkfontein hominin, prompting him to "doubt whether the more pongid premolars of *Meganthropus africanus* are the same species as *Plesianthropus*" (*ibid*, p.126). Most researchers now regard *Meganthropus africanus* as a junior synonym of *Australopithecus afarensis*.

Meganthropus B   *See* Sangiran 8

Meganthropus palaeojavanicus von Koenigswald in Weidenreich, 1945   (Gk *megas*=

great, *anthropos*=human being, *palaios*=ancient, and L. *javanicus*=from Java) The taxon *Meganthropus palaeojavanicus* was mentioned by **Weidenreich** (1942, p. 62), who later referred to it as "von Koenigswald's *Meganthropus palaeojavanicus*" (Weidenreich 1944, p. 2). In his 1945 monograph Weidenreich explains that the name *M. palaeojavanicus* was used in a letter dated January 15, 1942 sent to him by W.C.B. Koolhoven, Director of the Geological Survey of the Netherlands Indies, in which it was explained that it was **von Koenigswald**'s intention that the "1939" and the "1941" mandibles should both be given the name *Meganthropus palaeojavanicus*, the former mandible belonging to a female and the latter to a male of that taxon. However, von Koenigswald did not formally propose it as a hominin taxon until 1950 (von Koenigswald 1950). Arguably, the first formal attempt to justify the new taxon was made by Weidenreich (1945) so strictly speaking his name should be attached to the taxon, but von Koenigswald's name was associated with it by Weidenreich, so *de facto* the taxon has become known as *Meganthropus palaeojavanicus* von Koenigswald, 1950. This is all theoretical for the species name is rarely, if ever, used now because most researchers (e.g., Le Gros Clark 1955, p. 155) regard *M. palaeojavanicus* as a junior synonym of *Homo*

erectus. John **Robinson** retained the species name, but suggested that it be transferred from *Meganthropus* to *Paranthropus* (e.g., Robinson 1954, p. 196), but von Koenigswald (1955) insisted that differences in the relative size of the mandibular premolars and in the details of their cusp morphology precluded *M. palaeojavanicus* being included in *Paranthropus*. First discovery Sangiran Mandible D (also known as the "1941" mandible). Holotype As above. Main site **Sangiran**. *See also* *Meganthropus*.

meiosis   (Gk *meiosis*=diminution) The process of cell division in sexually reproducing organisms that reduces the number of chromosomes in reproductive cells by half to create **haploid** gametes (i.e., the egg and sperm). It is also known as reduction division.

Melanesia   (L. *melan* from Gk *melas*=black, and Gk *nesos*=island, literally "black islands") A group of islands in the southwest Pacific that extends southwest from Fiji to the Admiralty Islands. *See also* **Oceania, peopling of.**

melanin   (L. *melan* from Gk *melas*=black) A dark organic pigment that gives brown coloration to skin, hair, feathers, scales, eyes, or other tissues. In hominins, melanin helps protect the skin from damage by ultraviolet radiation. *See also* **pigmentation.**

Melka Kontoure   *See* Melka Kunturé.

Melka Kunturé   (Location 8°42′N, 38°34′E, Awash River, Ethiopia; etym. unknown; alternate spellings: Melka Kunture, Melka-Kunturé, Melka Kontoure, and Melka Kontouré) History and general description Melka Kunturé is a complex of sites along approximately 6 km/3.7 miles of outcrops exposed along either side of the Awash River in Ethiopia. The complex cut-and-fill sequence of tuffaceous alluvial sediments was first explored and excavated by Jean Chavaillon, Marcello Piperno, and Maurice Taieb in the 1960s. Typologically, the sites range from **Early Stone Age (Oldowan)** to **Later Stone Age**. The site is currently a museum, housing a small permanent collection of artifacts with several excavations still open for viewing, and a detailed website is maintained (geoserver.itc.nl/melkakunture/index.asp). Temporal span and how dated? The numerous sites at Melka Kunturé remain poorly dated. Incompletely published **potassium-argon dating** estimates suggest it is <1.8 Ma and a reversed to normal polarity shift in the sediments is most likely the Brunhes-Matuyama boundary (i.e., *c*.780 ka). Hominins found at site Five cranial fragments from more than one individual, two mandibular fragments, a distal left humerus, and a possible hominin phalanx from three excavations at sites (Gomboré II, Garba III, and Garba IV) within the Melka Kunturé site complex have been attributed to *Homo erectus*. Archeological evidence found at site Archeological sites have been identified at more than 13 localities in the Melka Kunturé area, each locality being named after a smaller tributary that drains into the Awash River. Sites at each location are designated by roman numerals with different strata within an excavation by capital letters. The primary localities are Garba (**Developed Oldowan**, a long sequence of **Acheulean**, and perhaps **Middle Stone Age** strata), Gomboré (Oldowan and Acheulean assemblages), Karre (Oldowan), and Wofi (Middle and Later Stone Age artifacts). More than 35,000 artifacts have been recovered from more than 500 m² of excavation, although many of the assemblages have been reworked by water. The stone artifacts were made of various locally available lavas such as basalt and trachyte, as well as obsidian derived from Balchit, 7 km/4.3 miles north of Melka Kunturé. Obsidian use is unusual in the Early Stone Age sites, and is a distinguishing feature of the Melka Kunturé assemblages. The only site where fauna are abundant is Garba IV, and the Garba IV assemblage includes rare taxa such as carnivores and primates, and the presence of *Pelorovis turkanensis brachyceras* and *Metridiochoerus modestus* suggest an Early Pleistocene age. The lack of tragelaphine and near lack of reduncine bovids, combined with the dominance of alcelaphine and antilopine bovids, suggest a dry, open environment at Garba IV. Pollen preserved at Gomboré I include evidence of a thicket or scrub vegetation with a nearby forest. Key references: Geology, dating, and paleoenvironment Bonnefille 1976, Chavaillon et al. 1979, Westphal et al. 1979, Chavaillon and Piperno 2004; Hominins Condemi 2004; Archeology Hivernel-Guerre 1976, Chavaillon and Piperno 2004, Negash et al. 2005, Piperno et al. 2008.

member A physically distinct subdivision of a geological formation (e.g., the Okote Member of the **Koobi Fora Formation**).

membranous labyrinth (L. *labyrinthus*= maze and *membrana*—skin or pliable layer) Endolymph-filled connective tissue sac suspended in perilymph and lying within the **bony labyrinth**. It has three parts, anteriorly the duct of the cochlea, next the saccule and utricle, and posterosuperiorly the semicircular ducts. The cochlear duct communicates with the saccule and the three semicircular canals with the utricle. *See also* **bony labyrinth**.

meme A term first used by Richard Dawkins (1976, p. 192) in the *Selfish Gene* to describe a "unit of cultural transmission," which he viewed as "a new kind of replicator." For Dawkins, memes were unique to modern humans and "were analogous to genetic transmission in that, although basically conservative, they can give rise to a form of evolution" (*ibid*, p. 192). For Dawkins, examples of memes include, "tunes, ideas, catchphrases, clothes fashions, ways of making pots or of building arches" (*ibid*, p. 192). And so, "Just as genes propagate themselves in the gene pool by leaping from body to body via sperms or eggs, so memes propagate themselves in the meme pool by leaping from brain to brain via a process which, in the broad sense, can be called **imitation**" (*ibid*, p. 192). *See also* **imitation**.

memory (L. *memor*=to be mindful) Among contemporary cognitive scientists, memory is not viewed as a unitary psychological mechanism, but as a multi-faceted faculty that includes different memory systems (Schacter and Tulving 1994, Tulving 1972). Specifically, cognitive neuroscientists have demonstrated that there are at least two broad types of memory systems: long-term memory and short-term or **working memory**. Each of these systems has more specialized memory subsystems that include episodic or biographic memory and semantic or factual memory systems. It is widely accepted that all mammals and many birds have both long- and short-term memory. However, some have argued that certain memory subsystems such as episodic memory (Tulving 2005) and working memory (Coolidge and Wynn 2007) are unique to modern humans and are not shared with other mammals or other primate species including the great apes. However, this view is not widely accepted among comparative psychologists (Tomasello and Call 1997).

memory span This is a measure of an individual's short-term memory, generally tested by how many numbers or words an individual can remember immediately after seeing a list. A variety of animals have shown competency in delayed match-to-sample tasks involving sounds, objects, and two-dimensional images. In all these tasks, the general consensus is that the short-term memory span of both modern humans and nonhuman animals (birds and primates) is about five to seven items (Terrace 1987, 2005).

menarche (Gk *men*=month and *arkhe*=beginning) A **life history** variable that is the time of onset of the first menstrual period in females and since it represents the beginning of near-monthly cycles of egg release from the ovary it also represents the beginning of **fertility**. However, for variable lengths of time following menarche a female's cycles are usually irregular and can be anovulatory. The onset of menarche is influenced by a number of factors; nutritional status is thought to be particularly important and social factors such as the presence of a father or adult male have also been proposed to play a role. Over the past 200 years, statistics show a trend toward decline in the average age at menarche in various modern human populations across the world. The mechanism behind this apparent trend is debated. While some scholars argue that earlier attainment of a certain percentage of body fat is the main reason for the pattern, others have more recently argued that an early age at menarche is in fact a temporally old hominin trait. These scholars argue that high levels of disease and infection associated with increased population sizes worked in the past to hold off the beginning of regular menstrual cycling and that only recently, with improvements in nutrition and health care, have modern human populations started to move back to the younger age at menarche. Menarche generally occurs earlier in great apes and Old World monkeys than in modern humans, but there is some overlap among chimpanzees and modern humans (around 11 years of age). There is evidence to suggest that great apes and Old World monkeys, like modern humans, also pass through a phase of adolescent subfertility after menarche occurs.

Mendelian inheritance *See* **Mendelian laws**.

Mendelian laws These laws of inheritance were established by Johann Gregor Mendel, an Augustinian monk who experimented with simple

traits in pea plants. Mendel's work was initially published in 1865 and 1866 but was largely ignored until the early 20thC. As a result of his experiments, Mendel came to two conclusions that are now known as Mendel's laws. The first law is the law of segregation, which states that during **gamete** formation the two **alleles** of a **gene** segregate so that each gamete receives only one allele. This law is true because of the mechanism of **meiosis** where the pairs of chromosomes are separated so that each gamete receives one of each type of chromosome. The second law is the law of independent assortment, which states that alleles from different gene loci assort independently during gamete formation. Mendel examined simple traits that were located on different chromosomes so the second law was true in all of his experiments. However, the second law is not true if two genes are located close together on the same chromosome. In this case, the alleles may not sort independently because of **linkage**. In general, genes on the same chromosome that are more than 50 centimorgans apart will be unlinked, and thus for these genes the second law will apply. *See also* **linkage disequilibrium**.

meningeal arteries *See* **meningeal vessels**.

meningeal vessels (Gk *meninx*=membrane) The meningeal vessels (i.e., arteries and veins) run between the inner, or **endocranial**, surface of the **cranial vault** and the **dura mater** that lines it. They either supply blood to (the meningeal arteries) or drain blood from (the meningeal veins) both the cranial vault and the dura mater. They leave vascular impressions on the endocranial surface of the cranial vault and the pattern of these vascular markings has been explored for its taxonomic utility as well as to make inferences about the relative size of the components of the **cerebral hemispheres**. *See also* **middle meningeal vessels**.

meninges (Gk *meninx*=membrane) The term refers individually or collectively to the three membranes (the pia mater, the arachnoid mater, and the dura mater) that cover the components (i.e., the brain, the optic nerve, and the spinal cord) of the central nervous system. The innermost layer, the pia mater, adheres closely to the surface of the brain. The middle layer, the arachnoid mater, is separated from the pia by the **subarachnoid space** and from the outer dura by the subdural space.

The outer dura mater has two layers: the fibrous inner meningeal layer and the outer more vascular and osteogenic endosteal layer. The inner meningeal layer of the dura mater has two prominent folds. The **falx cerebri** and **falx cerebelli** in the sagittal plane separate the cerebral and cerebellar hemispheres, respectively, and the approximately horozontal **tentorium cerebelli** separates the posterior cranial (also called the cerebellar fossa) from the rest of the cranial cavity. The **dural venous sinuses** (e.g., superior sagittal, transverse, sigmoid, occipital, and marginal) are endothelium-lined spaces between the two layers of the dura. The endosteal layer of the dura is continuous at the foramina and sutures with the periosteum (or pericranium) that covers the outer surface of the cranium. The pia and arachnoid are derived from ectoderm and are called the **leptomeninges**; the dura mater develops from mesoderm and is also called the **pachymeninx**.

menopause (Gk *men*=month and *pausis*=pause) This **life history** variable is the time when ovulation ceases and it marks the permanent end of female reproductive fertility. In modern human females this occurs, on average, around 50 years of age, although substantial within- and between-population variation exists. Normal menopause is initiated when stores of oocytes (egg cells stored in the ovaries) built up before birth and depleted through a combination of monthly release, degeneration, and resorption diminish to the point where the hormones needed to initiate the monthly cycle are no longer stimulated. While other primates may also experience an end to fertility before death, modern humans differ in that the majority of populations now vigorously outlive their fertility by a significant margin. The extension of the life span beyond menopause in modern human women, compared to nonhuman primates and most other mammals, has been attributed to a number of factors. *See also* **grandmother hypothesis**.

mental protuberance (L. *mentum*=chin) A midline projection of bone near the base of the mandible on the external surface of the **symphysis**. For a mental protuberance to be classed as a **true chin** there must be a depression, called an *incurvatio mandibulae*, between it and the **alveolar process** superiorly. *See also* **mandible**.

mentum osseum *See* **true chin**; **mental protuberance**.

meristic (Gk *meristos*=divided) The term refers to morphology that develops from craniocaudally or mesiodistally serially arranged segmental units as well as to the number or position of those segmental structures. The **vertebral column** and the **dentition** are examples of meristic structures.

meristic variation (Gk *meristos*=divided) Variation in the number or nature of serial structures such as vertebrae or teeth. Additional vertebrae, sacralization of the fifth lumbar vertebra, additional teeth, etc., are all examples of meristic variation.

Mesgid Dora Collection area within the **Woranso-Mille study area**, Central Afar, Ethiopia.

mesial accessory cuspule *See* mesioconulid.

mesial accessory tubercle A generic term to indicate an accessory cusp on the **mesial marginal ridge** of a postcanine tooth crown; on a mandibular molar the equivalent accessory cusp is usually referred to as the **mesioconulid**. Several accessory tubercles have been identified on, or adjacent to, the **mesial marginal ridge** on the mesial portion of the occlusal surface of a maxillary molar. *See also* **mesial paracone tubercle**; **lingual paracone tubercle**; **protoconule**; **epiconule**; **mesiostyle**.

mesial fovea (maxillary) *See* fovea anterior (maxillary).

mesial marginal accessory tubercle *See* **mesial accessory tubercle**.

mesial marginal ridge A crest of enamel that delineates the mesial border of the lingual face of an anterior tooth or the mesial border of the occlusal surface of a postcanine tooth crown. In a maxillary postcanine tooth crown it is an enamel crest running between the **protocone** and the **paracone**; in a mandibular postcanine tooth crown it is an enamel crest running between the **protoconid** and the **metaconid** [syn. anterior transverse crest, anterior trigon crest, crista nova, mesial margocrista (-id), precingulum, precingulid, protoloph].

mesial margocrista (-id) A term used by Schwartz and Tattersall (2002) for an enamel crest on the anterior teeth more commonly referred to as the mesial marginal ridge. *See also* **mesial marginal ridge**.

mesial paracone tubercle A term used by Kanazawa et al. (1990) for an **accessory cusp** located on the occlusal surface of a maxillary molar on the **mesial marginal ridge** mesiolingual to the **paracone**. *See also* **mesial accessory tubercle**; **mesiostyle**.

mesial pit (mandibular) *See* **fovea anterior (mandibular)**.

mesial pit (maxillary) *See* **fovea anterior (maxillary)**.

mesic (Gk *mesos*=middle) An **environment** or **habitat** that is moist.

mesiobuccal cusp A term sometimes used for the primary cusp on a maxillary **molar** called the **paracone**, and for the primary cusp on a mandibular molar called the **protoconid**. *See also* **paracone**; **protoconid**.

mesioconulid A small **accessory cuspule** on the **mesial marginal ridge** of a mandibular postcanine tooth crown.

mesiolingual cusp A term sometimes used for the primary cusp on a maxillary molar called the **protocone**, and for the primary cusp on a mandibular molar called the **metaconid**. *See also* **protocone**; **metaconid**.

mesiostyle A term introduced by Vanderbroek (1961, 1967) and used by Grine (1984) for a small enamel feature on the **mesial marginal ridge** close to the tip of the **paracone** on the occlusal surface of a maxillary postcanine tooth. *See also* **parastyle**.

mesoconid A term used by Weidenreich (1937) for what most researchers refer to as the **hypoconulid**. *See* **hypoconulid**.

mesostyle A term used by Pilbrow (2003) for an **accessory cusp** on the buccal face of a maxillary molar, between the **paracone** and the **metacone**.

messenger RNA *See* **ribonucleic acid**.

meta-memory The ability to reflect on one's own knowledge or memory. This could include a reflection upon **semantic** facts such as "do I *know* the capital of the United States?," or episodic memory, such as "have I ever *experienced* meeting someone famous?" Studies with primates (rhesus macaques, capuchin monkeys, and the great apes) have suggested that meta-semantic-memory is shared and is, perhaps, phylogenetically ancient (Hampton 2001, Fujita 2009, Kornell et al. 2007, Call and Carpenter 2001). However, various authors have argued that meta-episodic-memory is unique to modern humans. This implies that other animals are unable to declaratively state or "re-experience" the past, an essential feature of all episodic memories (Carruthers 2006, Tulving 2005).

metacentric *See* **chromosome**.

metacone (Gk *meta*=behind or beyond; *konos*=pine cone) One of the terms proposed by Osborn (1888) for the main cusp of mammalian molar teeth. It refers to the main cusp distal to the **paracone** on a maxillary (upper) molar tooth crown. It is one of the components of the **trigon** (syn. distocone, distobuccal cusp, tuberculum posterium externum).

metaconid (Gk *meta*=behind or beyond and *konos*=pine cone) One of the terms proposed by Osborn (1888) for the main cusp of mammalian molar teeth. It refers to the main mesial cusp on the lingual aspect of a mandibular (lower) (hence the suffix "-id") molar tooth crown and to the lingual cusp of a bicuspid mandibular premolar. It was given its name because in the lower molar of a primitive mammal it is the cusp distal to, or behind, the paraconid. It forms part of the **trigonid** component of a mandibular postcanine tooth crown (syn. epiconid, mesiolingual cusp, tuberculum anterium internum).

metaconule A term introduced by Osborn (1907) for an **accessory cusp** on the crest between the metacone and the protocone of a maxillary molar crown. A synonym for this feature, a plagioconule, was introduced by Vanderbroek (1961, 1967) and has since been used by Grine (1984). Metaconule has been used by some researchers to refer to a different enamel feature most researchers refer to as the C5 (e.g., Harris and Bailit, 1980). *See* **C5**; **distal accessory tubercle**.

metaconulid *See* **tuberculum intermedium**.

metaquartzite   *See* quartzite.

metastyle   A term introduced by Osborn (1892) for an **accessory cusp** distolingual to the metacone on the crown of a maxillary molar. It is equivalent to the structure other researchers (e.g., Grine 1984) refer to as the distostyle.

metastylid   A term used by Schwartz and Tattersall (2002) for an **accessory cusp** on the occlusal surface of a mandibular molar that most other researchers refer to as **tuberculum intermedium** or **C7**. *See* **tuberculum intermedium**.

methylation   The attachment of a methyl group to a substrate. In the genome, DNA methylation typically occurs at CpG sites (an acronym for cytosine-phosphate-guanine sites) and it can have a major impact on gene expression. For example, many genes have large clusters of CpG sites called "CpG islands" in or near the promoter region that influence gene expression. Methylation (or demethylation) of these sites is catalyzed by enzymes during certain phases of development and by environmental stimuli. In general, the addition of methyl groups turns genes "off."

Metridiochoerus   (Gk *metridios*=fruitful and *choerus*=pig) A genus of African Suidae which underwent an adaptive radiation in the later Pliocene and Pleistocene, during which five species arose and then went extinct. The earliest examples of *Metridiochoerus* are relatively small-bodied and have a more generalized dentition. Later examples follow the dental pattern of their modern descendant, the warthog *Phacochoerus*, in having extremely hypsodont third molars that outlast the other cheek teeth. The more derived *Metridiochoerus* species are large-bodied, having large heads with complex hypsodont third molars and, in mature individuals, no other cheek teeth. Canines were highly developed in some forms, particularly *Metridiochoerus compactus*, which had long, relatively straight canine teeth. *Metridiochoerus* fossils found at hominin sites have been used for biostratigraphy for the radiation of *Metridiochoerus* is relatively well dated at East African hominin sites, and the presence of particular species or evolutionary stages can be used to estimate the age of sites that are otherwise undatable.

Mezmaiskaya Cave   (Location 44°10′N, 40°00′E, Northern Caucasus, Russia; etym. named after the nearby village of Mezmay) History and general description This cave site was discovered in 1987 and excavations were directed by L. Golovanova until 1997. Well-stratified Middle and Upper Paleolithic deposits were recovered, recording the transition between these two periods. An infant *Homo neanderthalensis* was found in the lowest levels of the cave. Temporal span and how dated? Radiocarbon dating has suggested that the Upper Paleolithic layers date to *c*.32 ka, whereas the electron spin resonance spectroscopy dating shows that the Middle Paleolithic layers date to *c*.36–73 ka. The Neanderthal infant was directly dated using radiocarbon to 29 ka, in contrast to the date of the level in which it was found. Hominins found at site A partial skeleton of a neonate less than 7 months old was found in the lowest Middle Paleolithic layer, and included a damaged cranium and much of the upper body. This individual was the second Neanderthal to have its mitochondrial DNA examined. The size of the cranium allowed Ponce de León and colleagues to reconstruct Neanderthal life history as being as slow as that modern humans. Cranial fragments from a 1–2-year-old infant were recovered in a pit in the upper Middle Paleolithic layer. Archeological evidence found at site Four Middle Paleolithic layers attributed to the "East European **Mousterian**," and three Upper Paleolithic levels, were recovered. Key references: Geology, dating, and paleoenvironment Ovchinnikov et al. 2000, Skinner et al 2005, Golovanova et al 1999; Hominins Ovchinnikov et al. 2000, Golovanova et al 1999, Ponce de León et al. 2008; Archeology Golovanova et al 1999.

MH   Acronym for Malapa Hominin and used as the prefix for hominin fossils from the site of **Malapa** (e.g., **MH1**). *See* **Malapa**.

MH1   Site **Malapa**. Locality Excavation 3.3 m by 4.4 m in area and 3.5 m deep. Surface/*in situ* *In situ*. Date of discovery August 2008 and thereafter. Finder The first evidence of the MH1 associated skeleton (the right clavicle, U.W.88-1) was found by Matthew Berger on August 15, 2008. Unit N/A. Horizon Facies D. Bed/member N/A. Formation N/A. Group Malmani Subgroup. Nearest overlying dated horizon N/A. Nearest underlying dated horizon Flowstone (*c*.2.025 Ma). Geological age 1.95–1.78 Ma. Developmental age Juvenile. The second molars are in occlusion, the third molar crowns are close to crown complete with no radiological evidence of root formation, and the

497

distal humeral epiphysis is the only fused long-bone epiphysis. Presumed sex Male. Brief anatomical description This associated skeleton comprises much of a cranium with the right P^3–M^3, left I^2, and P^3–M^3, separate right I^1 and C, the right side of the mandibular body with M_1 and M_2 and an unerupted M_3, and the left C with an attached section of the corpus. The axial skeleton is represented by three cervical, three thoracic, and one lumbar vertebrae, plus vertebral and rib fragments. The pectoral girdle and the upper limb are represented by part of the right clavicle, all but the proximal epiphysis of the right humerus, the right proximal ulna, the right distal radial epiphysis, and the right third metacarpal, plus the proximal part of the right humeral diaphysis. The pelvic girdle and the lower limb are represented by substantial parts of the left ilium and ischium, part of the right ilium, the proximal half of the right femur minus the proximal epiphysis, most of the right tibia (but its attribution to MH1 is said to be "uncertain") and the diaphyses of two left metatarsals. Endocranial volume c.420 cm^3 (Berger et al. 2010). Announcement Berger et al. 2010. Initial description Berger et al. 2010. Photographs/line drawings and metrical data Berger et al. 2010. Detailed anatomical description N/A. Initial taxonomic allocation *Australopithecus sediba*. Taxonomic revisions Its discoverers assign it to a new species, but the combination of its immaturity and the demonstration that it differs from samples of *Au. africanus* as opposed to the population from which those samples were drawn makes it possible that MH1 comes from a sample of *Au. africanus*. Current conventional taxonomic allocation *Au. sediba*. Informal taxonomic category Archaic hominin. Significance **Holotype** of *Au. sediba* and one of two associated skeletons attributed to that taxon. The researchers who have analysed this material suggest that despite its archaic hominin-like ankle and foot, it shares cranial (e.g., more globular neurocranium, gracile face), mandibular (e.g., more vertical symphyseal profile, a weak *mentum osseum*), dental (e.g., simple canine crown and small tooth crowns), and postcranial (e.g., **acetabulocristal buttress**, expanded ilium, and short ischium) morphology with early and later *Homo* taxa. Its discoverers assigned it to a new species, but the combination of its immaturity and the demonstration that it differs from samples of *Au. africanus* as opposed to the population from which those samples were drawn makes it possible that MH1 comes from a sample of *Au. africanus*.

498

Location of original **Bernard Price Institute for Palaeontological Research**, University of the Witwatersrand, Johannesburg, South Africa.

MH2 Site Malapa. Locality Excavation 3.5 m deep and 3.3 m by 4.4 m in area. Surface/*in situ* *In situ*. Date of discovery September 2008 and thereafter. Finder The first evidence of the MH2 associated skeleton (the right humerus, U.W.88–57) was found by Lee Berger on September 4, 2008. Unit N/A. Horizon Facies D. Bed/member N/A. Formation N/A. Group Malmani Subgroup. Nearest overlying dated horizon N/A. Nearest underlying dated horizon Flowstone (c.2.025 Ma). Geological age 1.95–1.78 Ma. Developmental age Adult. Presumed sex Female. Brief anatomical description This associated skeleton comprises two maxillary teeth (left M^{2-3}), a fragment of the right side of the mandible with M_1–M_3, AND a fragment of the left side of the mandible with M_2–M_3. The axial skeleton is represented by one cervical and three thoracic vertebrae and rib fragments. The majority of the pectoral girdle and upper limb are preserved on the right side, including the scapula, humerus, radius, and ulna. On the left side all that remains is the proximal humerus and a few hand bones. The pelvis is represented by a fragment preserving both sides of the symphysis and part of the left ischiopubic ramus. The lower limb is represented by the proximal end and distal third of the right femur, the right distal tibia, a complete talus and calcaneum and phalanges of the right foot, and on the left side the proximal and distal fibula (but the latter's attribution to MH2 is said to be "uncertain") and part of the lateral tibial condyle. Announcement Berger et al. 2010. Initial description Berger et al. 2010. Photographs/line drawings and metrical data Berger et al. 2010. Detailed anatomical description N/A. Initial taxonomic allocation *Australopithecus sediba*. Taxonomic revisions Its discoverers assign it to a new species, but the demonstration that it differs from samples of *Au. africanus* as opposed to the population from which those samples were drawn makes it possible that MH2 comes from a sample of *Au. africanus*. Current conventional taxonomic allocation *Au. sediba*. Informal taxonomic category Archaic hominin. Significance The **paratype** of *Au. sediba* and one of the two associated skeletons attributed to that taxon. The researchers who have analysed this material suggest that MH2 shares postcranial morphology (e.g., acetabulocristal buttress, expanded ilium, and short ischium) with early and later *Homo* taxa, but the foot is acknowledged to be like that of archaic hominins. Location of original **Bernard Price Institute for Palaeontological Research**, University of the Witwatersrand, Johannesburg, South Africa.

MHC Acronym for the **major histocompatibility complex** (*which see*).

MIC Acronym for **models of intermediate complexity** (*which see*).

microarray *See* **array**.

microcephaly (Gk *micro*=small and *kephale*=head) Refers in modern humans to a spectrum of pathologies, whose common denominator is that individuals have an unusually small brain for their body size. The defective growth of the brain results in a cranium with an unusually small head circumference. Some modern human microcephalic brains are relatively normally proportioned, but most have a disproportionately small cerebral cortex and hence a disproportionately large cerebellar cortex. Loss-of-function mutations at certain genetic loci have been identified that may cause primary microcephaly. Among them are *ASPM* and *microcephalin*, genes which show significant coding sequence changes in ape and human evolution, One interpretation of *Homo floresiensis* is that it samples a *Homo sapiens* population afflicted by a pathology that includes microcephaly, but if *H. floresiensis* samples a relic population of archaic or transitional hominins that at one time, or another, had undergone insular dwarfing, then its small brain would not be considered to be pathological.

micro-computed tomography (or micro-CT or μCT) Based on the same principles as **computed tomography** (or CT), but using dedicated equipment (e.g., with a micro-focus X-ray tube) designed for the nondestructive testing of materials. The limit on sample size is in the order of 7 cm in diameter, and the spatial resolution ranges between 5 and 50 μm. Micro-CT has been used to investigate the morphology of the enamel-dentine junction, enamel thickness, and enamel volume in single teeth. A special type of micro-CT, **synchrotron radiation micro-computed tomography** (or SR-μCT) enables researchers to investigate microstructure of subsurface enamel. It makes use of synchrotron radiation, which is monochromatic and therefore it does not suffer from **beam-hardening**.

micro-CT *See* **micro-computed tomography**.

microfauna (Gk *micro*=small and L. *fauna*= in Roman mythology the sister of Faunus, the god of nature) In the context of **paleoanthropology** the term microfauna is usually used loosely to refer to *all* small animals, or *all* small vertebrates, or *all* small mammals (usually rodents and insectivores) found in a fossil **faunal assemblage**. Analyses of microfauna, principally small mammals (sometimes termed micromammals, i.e., mammals that weigh less than 1000 g) have contributed significantly to studies of **taphonomy, paleoenvironments**, and **biostratigraphy** relevant to human evolution. Note that the more common definition of microfauna outside of paleoanthropology is that the term applies to microscopic animals such as protozoa, bacteria, nematodes, etc., rather than to animals visible with the naked eye.

microlith (Gk *micro*=small and *lithos*=stone) Small **blades** or **flakes** that have been modified by removing even smaller flakes to form triangular or crescent-shaped pieces. One edge is typically blunted (or **backed**) for insertion into a prepared **haft** as part of a **composite tool**. Although abundant in some <40 ka artifact assemblages (e.g., **Enkapune ya Muto** in Kenya) similarly shaped, but larger, forms dating to the **Middle Pleistocene** are seen in some African sites (e.g., Ambrose 2002, Barham 2002a).

Micronesia (Gk *micro*=small and *nesos*= island, literally "small islands") A group of small islands in the Pacific that extends east from the Phillipines; they all lie north of the equator. The islands in Micronesia include the Caroline, Gilbert, Mariana, and Marshall islands. *See also* **Oceania, peopling of**.

microsatellite (Gk *micro*=small and *satelles*= an attendant) Microsatellites, also known as short tandem repeats (or STRs) are repeated sequences that are between two and nine base pairs in length. The most common are dinucleotide, trinucleotide, and tetranucleotide repeats (i.e., two, three, and four base repeats, respectively). Some trinucleotide microsatellites have been linked to modern human diseases. These are called trinucleotide repeat disorders, and they include Huntington's disease, fragile X, and myotonic dystrophy. Trinucleotide repeat disorders commonly have a threshold number of repeats above which the microsatellite tends to become unstable and gains large numbers of repeats. A higher number of repeats above the threshhold is correlated with more severe disease and an earlier age at onset. Microsatellites occur throughout the **genome** and are often highly **polymorphic**. This makes them useful for studies of recent modern human population history, for linkage analyses, and for individual identification. For example, Bowcock et al. (1994) used 30 microsatellites to investigate relationships among 10 modern humans. They found that microsatellite diversity reflected the geographic origin of each individual and was highest in Africa. In addition, the data could be used to generate a phylogeny of the relationships among individuals and of the different populations.

microstrain   (or µε) *See* **strain**.

microstructure   *See* **enamel microstructure; incremental features**.

microtomography   *See*   **micro-computed tomography**.

microwear   *See* **dental microwear; diet reconstruction; tooth wear**.

Mid-Holocene Warm Event   *See* Holocene.

Mid-Pleistocene transition (or MPT) Shackleton (1967) argued that the $^{18}O/^{16}O$ ratio of benthic **foraminifera** in the oceans would fall when ^{18}O-rich water was locked up in expanded ice caps and enlarged glaciers. Glaciation intensifies in the northern hemisphere (this is called the **Northern Hemisphere Glaciation**, or NHG) from the mid Pliocene to the late Pleistocene, but during this time the rate of reduction in the $^{18}O/^{16}O$ ratio is more pronounced in two periods, from *c*.3.2 to 2.7 Ma and from *c*.1.2 to 0.8 Ma. The second of these two shifts is called the MPT and the first earlier shift in the ratio of $^{18}O/^{16}O$ is called the **late Pliocene transition** (or LPT). The end of the MPT *c*.0.8 Ma marks the beginning of the *c*.100 ka-long climate cycles. *See* **astronomical theory; astronomical time scale**.

midbrain   (ME *midde*=middle and OE *braegen*= brain, most likely from the Gk *bregma*=the front of the head) The midbrain (or mesencephalon) is the middle of the three swellings, or vesicles, of

the developing neural tube that give rise to the brain. The midbrain houses important processing centers for coordinating movements of the head and saccadic eye (the superior colliculi) and auditory information (inferior colliculi). *See also* **brain**.

midcarpal joint   The articulation between the proximal and distal carpal rows; it functions generally as a single ball-and-socket-like joint. The articulation between the capitate and lunate forms the crux of this joint, with the scaphoid (and os centrale if present) on the lateral side and the triquetrum-hamate articulation on the medial side. Together, these bones function to allow supination and pronation, adduction and abduction, and flexion and extension of the wrist. Each of these motions is accomplished via a complex set of movements among different combinations of carpals within the midcarpal joint (Moritomo et al. 2006).

midden   An archeologically visible accumulation of household waste, which is usually a faithful record of daily activities. In many places, middens contain large numbers of shells, which make them alkaline and thus good preservers of organic remains.

Middle Awash   *See* Middle Awash study area.

Middle Awash Research Project   The Middle Awash study area extends along both sides of the modern Awash River in the Afar Depression of Ethiopia, north of Gewane town. Geological work in the study area began in 1938 with an Italian geological mission (Gortani and Bianchi 1973), but the first formal investigation of its paleontological potential was by Maurice Taieb, a French geologist who collected the first fossils and conducted pilot stratigraphic work, and this was followed by the work of the **Rift Valley Research Mission in Ethiopia** (or RVRME). In 1976, Antiquities Officer Alemayehu Asfaw discovered the study area's first hominin fossil in the Middle Pleistocene deposits at **Bodo** (Kalb et al. 1980, Conroy et al. 1978). The RVRME project ended in 1978 and the Middle Awash research project was initiated in 1981 under the leadership of the Desmond **Clark**; Tim **White**, and Berhane Asfaw (as a graduate student) joined Desmond Clark that year and went on to co-lead the project after Desmond Clark's retirement in 1986. A permit issued by the Centre for Research and Conservation of Cultural Heritage (CRCCH) of the Ethiopian Ministry of Culture authorized the 1981 exploration of areas on both sides of the Awash River, north of Gewane and south of the Gona. Archeological excavations were undertaken at Bodo and **Hargufia**, radiometric dates for the area were determined and the first Pliocene hominins were recovered (Clark et al. 1984a). Planned Middle Awash field research was postponed from 1982 to 1990 while new Ethiopian antiquities legislation was formulated. With the new laws in place, the Middle Awash research project resumed fieldwork in 1990, concentrating on the better-known geological sequences east of the modern Awash River, but since 1992 the Middle Awash research project's primary attention has been directed to the western side of the modern Awash River. In addition to developing a comprehensive stratigraphic record for the basin the team's many significant fossil discoveries included the remains of *Ardipithecus ramidus* (White et al. 1994, 2009, WoldeGabriel et al. 1994), *Ardipithecus kadabba* (Haile-Selassie 2001, Wolde-Gabriel et al. 2001, Haile-Selassie et al. 2004), and *Australopithecus garhi* (Asfaw et al. 1999, de Heinzelin et al. 1999). The project's recent research publications include a monograph on the Acheulean of Bouri (de Heinzelin et al. 2000), monographs on *Ardipithecus kadabba* (Haile-Selassie and Wolde-Gabriel 2009) on the **Daka** *Homo erectus* calvaria (Gilbert and Asfaw, 2008), and papers on the **Asa Issie** *Australopithecus anamensis* (White et al. 2006b) and fossils from **Herto** representing some of the earliest evidence for *Homo sapiens* (White et al. 2003, Clark et al. 2003). The Middle Awash research project has arranged with the University of California Press to publish three more monographs on the Middle Awash during the next several years. Field research continues on both the eastern (e.g., Ala Kanasa, Andalee, **Gamedah**, **Maka**, and **Mata-baietu**) and western (e.g., Hatayae Member, Esa Dibo, Bahroo Koma, upper Asa Issie, Burka, Guneta) sides of the study area. Stone tools belonging to the **Oldowan**, developed Oldowan, and a range of **Acheulean, Middle Stone Age,** and **Late Stone Age** archeological technologies have been identified by the Middle Awash research project. Hominins and archeology were recovered since 2003 in the Talalak and Halibee areas in the northwestern quadrant of the study area. More than 70 scientists (from countries including Argentina, Australia, Belgium, Canada, China, England, Ethiopia, France, Greece, Guam, India, Italy, Japan, Kenya, Mexico, Spain, Tanzania, Turkey, and the USA) and hundreds of local people are involved in the research.

Middle Awash study area The Middle Awash region (Kalb et al. 1978b refer to it as the Middle Awash valley) is one of the three informal divisions of the part of the **Awash River** valley that runs through the **Afar Triangle**. The Middle Awash region is more or less coextensive with the Middle Awash study area and extends from approximately Lake Yardi in the south towards the **Hadar study area** in the north. Major fossiliferous subregions within the Middle Awash include the **Bouri Peninsula, Central Awash complex,** and **Western Margin** on the west side of the **Awash River**, and **Bodo-Maka** on the east side of the river.

middle ear The term refers to the part of the ear between the external, or outer ear, and the **inner ear**. It comprises a cavity and the hard- and soft-tissue structures contained within it. The cavity lies within the **petrous** part of the temporal bone between the medial end of the outer ear (i.e., the tympanic membrane at medial end of the external auditory meatus) and the oval window of the **bony labyrinth**. It contains three small bones or **auditory ossicles** (from lateral to medial they are the malleus, the incus, and the stapes) that connect the tympanic membrane with the oval window, two small muscles that help dampen excessive movements of the auditory ossicles, blood vessels, and several nerves. It is about the size of three stacked 10 cents (US) or 1 pence (UK) pieces stood on end.

Middle Ledi *See* Ledi-Geraru study area.

Middle Ledi Research Project *See* Ledi-Geraru Research Project.

middle meningeal vessels (Gk *meninx* = membrane) The middle meningeal vessels (i.e., arteries and veins) supply blood to (the arteries) and drain blood from (the veins) both the **cranial vault** and the **dura mater** of the middle **cranial fossa**. They leave vascular impressions on the endocranial surface of the cranial vault (the transmitted pulsations of the arteries are responsible for this, although it is the vein that is immediately adjacent to the bone of the cranial vault). The pattern of these vascular markings has been explored for its taxonomic utility as well as to make inferences

501

about the relative size of the components of the **cerebral hemispheres**.

Middle Paleolithic A term used to describe a stage of the European and Near Eastern **Paleolithic** that is defined by the dominant stone tool technology (**Mode 3**, including the **Mousterian** and its regional variants). The Middle Paleolithic is usually associated with *Homo neanderthalensis* in Europe, but it has also been associated with some anatomically modern human fossils in the Near East (e.g., **Skhul**). The term has also been applied to sites in India and the Far East, but it is not clear what hominin was responsible for the artifacts at these sites. The equivalent term for African sites with comparable technologies (but not necessarily similar ages) is **Middle Stone Age**.

Middle Pleistocene (Gk *pleistos*=most and *kainos*=new) An informal division of the **Pleistocene** epoch and series, synonymous with the Ionian age/stage (781–126 ka). The lower boundary of the Middle Pleistocene is placed at the boundary between the **Matuyama** and **Brunhes** chrons.

middle-range theory *See* Binford, Lewis (1930–).

Middle Stone Age (or MSA) The intermediate period in the tripartite division of the African archeological lithic record formalized by A.J.H. Goodwin and C. van Riet Lowe (1929) in their *The Stone Age Cultures of South Africa*. Although it was developed for sequences in southern Africa, the terminology has subsequently been applied to sites across sub-Saharan Africa. There have been numerous calls to abandon the tripartite division (e.g., Clark et al. 1966, Barham and Mitchell 2008), but it, and thus the term Middle Stone Age, persists because of the latter's utility for describing a broad stage in the general trend towards a more sophisticated lithic record (i.e., a trend towards aspects of the manufacture of stone tools that are conceptually and technically more elaborate). The MSA was initially defined by its technological and stratigraphic position between **Early Stone Age** and **Later Stone Age** sites. The distinctive features of a MSA archeological site are the absence of the **handaxes** and/or **cleavers** that distinguish Early Stone Age sites, the absence of the **microliths** that are typical of the Later Stone Age, and the presence of **points**, presumably used to make

composite tools such as spears or similar hunting implements. Since the terminology was first introduced (Goodwin 1928) it has been recognized that MSA sites show spatial and/or temporal variation in the shape of the points and of other aspects of MSA artifact assemblages (Clark 1988, McBrearty and Brooks 2000). Flake production is often by **Levallois**, or comparable methods, and **blades** or elongated flakes are present in some assemblages, particularly in southern Africa. MSA artifacts are associated with the earliest fossil remains of *Homo sapiens* (e.g., **Omo-Kibish**). Recognizing that first and last appearance dates vary locally and are subject to change as new sites are found, existing dating methods are improved, or when new methods are introduced, the present evidence suggests the MSA began *c.* 300 ka and ended *c.*30 ka.

middle trigonid crest An **enamel** crest in a mandibular molar that runs between the protoconid and the metaconid, distal to the **mesial marginal ridge** and mesial to the **distal trigonid crest**. The feature is illustrated in Zubov (1992) (syn. midtrigonid crest, epicristid, accessory trigonid crest).

midfacial prognathism *See* **prognathic**.

midfoot break Bending (or dorsiflexion) at the calcaneocuboid and cuboid-metatarsal joints during the push-off phase of walking in nonhuman primates. This establishes a temporary fulcrum at the cuboid-metatarsal joint in nonhuman primates, and helps balance the demands of propulsion and grasping in the nonhuman primate midfoot. Notably, this flexibility in the midfoot allows the forefoot to maintain its grip on arboreal supports while the hindfoot and other hind limb elements propel the body during locomotion. Originally thought to occur solely at the transverse tarsal joint (as the midtarsal break), it is now recognized to be a more complicated foot motion involving both the transverse tarsal joint and the tarsometatarsal joints. A calcaneocuboid locking mechanism and the development of strong plantar ligaments that support the longitudinal arch prevent flexion of the midfoot in modern humans, producing a rigid structure that functions as an effective lever for push-off during bipedal locomotion. *See also* **midtarsal break**.

midtarsal break Bending (or dorsiflexion) at the midtarsal region during the push-off phase of walking in nonhuman primates. First identified and studied by Elftman and Manter (1935) this flexibility in the mid foot allows the nonhuman primate hind limb to move about the forefoot while the it maintains its grasp on arboreal supports. The bending was originally argued to occur at the transverse tarsal joint (i.e., the calcaneocuboid and talonavicular joints) and it was suggested that the interlocking mechanism between the calcaneus and the cuboid prevents the midtarsal break in modern humans. More recent work (DeSilva 2009) has found that the midtarsal break involves motion at both the transverse tarsal joint and the cuboid-metatarsal joints that would more accurately be termed **midfoot break**.

migration *See* **gene flow**.

Milankovitch cycles *See* **astronomical theory**.

Miles' method This method for determining ages at death from dental wear was first put forth by A.E.W. Miles (1962). Miles suggested that ages at death for subadults could be estimated using modern information on the timing of tooth development and eruption. This "known age" archeological or paleoanthropological sample can then be used as a baseline to establish the pattern of dental wear. Because the rate of dental wear is highly dependent on diet and methods of food preparation, it is necessary to do an "internal calibration" using the actual archeological or paleoanthropological sample. The "known" age sample of subadults allows for this calibration. Because the first, second, and third molars in modern humans typically erupt at 6, 12, and 18 years of age (respectively), a 12-year-old should exhibit 6 years of dental wear on their first molar. Similarly, an 18-year-old should exhibit 12 years of wear on their first molar and 6 years of wear on their second molar. Dental development and eruption are only useful in aging individuals up until their late teens, but from the "known" age sample it is possible to estimate more advanced ages from dental wear. For example, an individual with "6 years' worth" of dental wear on their third molar should be approximately 24 years old (the age of eruption plus the years of dental wear on the erupted tooth). This 24-year-old should display 18 years (24–6) of dental wear on their first molar. In turn, an individual with "18 years' worth" of wear on their third molar should be approximately 36 years old (18+18), and should have 30 years of wear on their first molar (36–6). An individual with 30 years of wear on their third molar should then be approximately 48 years old (30+18). This method

assumes constant rates of wear across teeth. Miles suggested that for at least some populations the third molar wears more slowly than the second molar, which wears more slowly than the first. To account for this, he suggested: (a) adding 8.33% years when "reading" wear on a second molar using first molar standards, (b) adding 16.67% years when "reading" wear on a third molar using first molar standards, and (c) adding 7.69% years when "reading" wear on a third molar using second molar standards. These percentages come from a presumed wear ratio of 6:6.5:7; in other words, that 6 years of wear on a first molar is equivalent to 6.5 years of wear on a second molar, which is equivalent to 7 years of wear on a third molar. As an example, Miles (1962) gives information on a burial with "31 years of M_1 wear" on the second molar, which gives an age estimate of $31 + 0.0833 \times 31 + 12$, or about 46 years old. This same burial had "28 years of M_1 wear" on the third molar giving an age estimate of $28 + 0.1667 \times 28 + 18$, or about 51 years old, and "29 years of M_2 wear" on the third molar giving an age estimate of $29 + 0.0769 \times 29 + 18$, or about 49 years old. Miles suggested averaging these ages to arrive at a final estimate of about 49 years old for this particular example.

Millennium Man A term used by the media for **BAR 1000'00** from **Lukeino**, Kenya. *See* **BAR 1000'00**.

Minatogawa *See* **Ryukyu Islands**.

Mindel *See* **glacial cycles**.

Mindel-Riss *See* **glacial cycles**.

mineral The term is used for any naturally occurring, crystalline solid, with a fixed or limited range of chemical composition. Minerals form when a fluid cools, when chemicals precipitate from solutes, or when there is solid-state recrystallization from preexisting phases. Minerals are the fundamental building blocks of most rocks, and in the context of biology they are the chemicals that are introduced into tissues during the process of calcification and **mineralization**.

mineralization The process whereby minerals are added to a soft tissue, such as immature enamel, that results in an increase in its hardness so much so that it is classified as a hard tissue. During tooth formation, the primary organic matrix secreted by ameloblasts, cementoblasts, and odontoblasts is replaced by an inorganic crystalline lattice, which results in three of the body's hard tissues (**enamel**, **dentine**, or **cementum**). It is due to the highly mineralized nature of teeth that they are relatively common elements recovered as fossils. In effect the teeth of an organism are fossilized during its lifetime. *See also* **calcification**; **dentine development**; **enamel development**.

minicolumn Minicolumns are fundamental structural and functional units within the **cerebral cortex** made up of a single vertical row of neurons with strong vertical interconnections among the layers. The core of a minicolumn contains the majority of the neurons, their apical dendrites, and both myelinated and unmyelinated fibers, whereas the outer layer of each column has fewer neurons and more connections (e.g., dendrites, unmyelinated axons, and synapses). In modern humans there is bilateral asymmetry of minicolumn width in **Wernicke's area** (cytoarchitectonic area Tpt), with more space for neuronal interconnections in the left cerebral hemisphere. Similar analyses of minicolumns in area Tpt of chimpanzees and macaques have not revealed this asymmetry.

minimal animal unit (or MAU) A derived measurement used to quantify the frequencies of skeletal elements in a bone assemblage. The MAU is calculated as the **minimum number of elements** (or MNE) of a skeletal element divided by the number of times that element occurs in the complete skeleton (e.g., femur MAU=0.5, rib MAU= 0.042). This standardization procedure makes MAU a useful measure for examining **skeletal element survivorship**, since elements that are rare in a skeleton may show disproportional frequencies within an archeological site. The MAU measure can help show how hominins differentially dismember and transport portions of an animal carcass. In archeological contexts, for example, MAU can be used to demonstrate that hominins preferentially transported certain classes of elements from the point of prey acquisition to the point of consumption. The MAU can also help establish the nature of an archeological site (e.g., as a kill/butchery site or home base/consumption site) or suggest whether hominins had access to fully fleshed carcasses or those largely defleshed by carnivores. *See also* **carcass transport strategy**; **home base hypothesis**; **minimum number of elements**.

minimum number of elements (or MNE) A derived measurement that refers to the minimum number of skeletal elements, or portions thereof, necessary to account for the observed specimens in a bone assemblage. The MNE measurement is used to quantify skeletal part frequencies. When fossil bone assemblages are heavily fragmented, MNE counts are used to estimate the minimum number of bones originally deposited by hominins, carnivores, or any other depositional agent. MNE counts allow faunal analysts to compare the relative abundances of skeletal elements to those in a complete skeleton to address questions of **skeletal element survivorship** or differential transport by hominins. The number of fragments that represent a complete element can be used for addressing taphonomic questions concerning bone fragmentation. For example, in an archeological context, one might compare ratios of the **number of identified specimens** (or NISPs) to the MNE to determine whether those bones with higher marrow contents were fragmented to a greater extent than those bones with low marrow contents. *See also* **minimal animal unit**.

minimum number of individuals (or MNI) A derived measurement that is the minimum number of individuals of a particular taxon necessary to account for the number of observed skeletal elements, or portions thereof, in a bone assemblage. It provides an estimate of the minimum number of individual animals originally deposited by any depositional agent. The MNI measure can be used to quantify taxonomic abundances and, to a lesser extent, skeletal part frequencies. It can be a useful measure when comparing taxonomic abundances in a faunal assemblage where the bone accumulator may have deposited some species largely intact and only certain parts of another. To some extent, MNI can also correct for differential fragmentation and for the fact that some species have a greater number of taxonomically diagnostic bones than others.

minisatellite (It. *miniatura*=miniature illumination in a Medieval manuscript and L. *satelles*=an attendant) A **tandemly repeated sequence** that is 10 to over 100 **base pairs** in length. Minisatellites are located throughout the genome and are particularly common at centromeres and telomeres. They are often **polymorphic** and have been used extensively for DNA fingerprinting (i.e., individual identification) and for **linkage** analysis.

Miocene (Gk *meion*=less and *kainos*=new) One of the five **epochs** that make up the **Tertiary** period, and the first of the two epochs that make up the **Neogene** period. Miocene refers to a unit of geological time (i.e., a geochronological unit) that begins 23.5 Ma at the end of the Oligocene and ends 5.2 Ma at the beginning of the **Pliocene**. It is usually divided in to three phases, Early, Middle, and Late.

miombo woodland (Swa. *miombo*=term for the many types of trees in the genus *Brachystegia*) The term is applied to woodland that is dominated by trees belonging to the genus *Brachystegia*. Other trees growing in miombo woodland include members of the genus *Isoberlinia* and *Julbernardia*. Miombo "woodlands" can vary from areas where the trees are spaced so far apart that there are substantial areas of open tropical, or C_4, grass between them, to patches of woodland where the canopy is more or less continuous.

mirror neurons *See* **language, evolution of.**

MIS Acronym for marine isotope stages. *See* **Oxygen Isotope Stages**.

mismatch distribution A histogram of the number of pairwise **nucleotide** differences within a sample. A population that is expanding in size shows a smooth curve, while a population with a long-term constant population size shows a ragged distribution. As a population expands, the "wave" generated moves from left to right (i.e., as the differences between lineages increase with time) and the timing of the initial population expansion can be estimated. This method has been used in analyses of **mitochondrial DNA** as well as other loci and the results suggest that the population of anatomically modern humans expanded approximately 50 ka (Sherry et al. 1994).

missense mutation *See* **nonsynonymous mutation**.

Mission Paléoanthropologique Franco-Tchadienne (French-Chadian Paleoanthropological Mission, or MPFT) An international and interdisciplinary research program on the origin

and environment of early hominids in Chad. The project was founded by Michel **Brunet** in 1994 and is a collaboration among the University of Poitiers in France, the University of N'Djamena in Chad, and the Centre National d'Applui à la Recherche (C.N.A.R) of N'Djamena in Chad. It focuses on geological and paleontological research, the further education of Chadian students through PhD research and training, and it provides training in collections management. It enabled the construction of the Centre de Valorisation des Collections at the C.N.A.R. The MPFT has recovered fossil fauna and several fossil hominins, including two holotype specimens: the 3.5 Ma mandible assigned to *Australopithecus bahrelghazali* and a *c.*7 Ma cranium assigned to *Sahelanthropus tchadensis*.

Mission Scientifique de l'Omo *See* **Arambourg, Camille Louis Joseph (1885–1969)**.

mitochondria (Gk *mitos*=thread and *khondrion*=small grain) The energy-producing organelles of the cell located in the cytoplasm. Each cell typically contains several hundred mitochondria, and each mitochondrion has several copies of its own **DNA**. Similarities in the genetic code of mitochondria and an endosymbiotic type of Proteobacteria suggest that the former may be derived from the latter. *See also* **mitochondrial DNA**.

"mitochondrial Eve" (also called "African Eve") A concept that arose from research conducted by Allan **Wilson**, Rebecca **Cann**, Mark Stoneking, and a group of collaborators at Wilson's laboratory at the University of California at Berkeley in the 1980s. From studies of the mitochondrial DNA (or mtDNA) of current human populations they concluded that all current humans can trace their matrilineal ancestry back to one common source, a modern human female who lived in Africa *c.*150–200 ka. This kind of genetic evidence, when combined with fossil evidence that is consistent with the evolution and earliest appearance of *Homo sapiens* in Africa (Stringer and Andrews 1988), forms the basis of the so-called **out-of-Africa hypothesis**. Wesley Brown et al. (1979) developed techniques for comparing mtDNA samples from different individuals and he successfully compared mtDNA taken from the placentas of 21 mothers of different populations. Brown and colleagues found that the samples differed only slightly in the number of mutations that they did not share in common, which they took to indicate that contrary to the candelabra and multi-regional models, the different modern human races had diverged relatively recently. He even suggested that the amount of similarity in his samples might mean that all modern humans were descended from a single recent, small population. Meanwhile Douglas Wallace, a geneticist at Stanford University, who was the first to demonstrate that mtDNA is passed exclusively from mother to daughter, used a technique called **restriction mapping** to study mtDNA to reconstruct human phylogeny. Wallace collected mtDNA samples from over 200 individuals from diverse populations (Bantu and San from Africa, Native American, white, and east Asian). This study produced results similar to Brown et al.'s, namely that the remarkably few differences among the mtDNA of the different populations indicated a recent common origin. Furthermore, Wallace found much more variation in the mtDNA of different African populations than between African and non-African populations. By the time Wallace published the results of these studies he had joined Allan Wilson's team at Berkeley. Wilson had begun to use new techniques such as **polymerase chain reaction** (or PCR) and the **relative rate test** to build upon his earlier work on the phylogenetic relationships between mammals and his team was studying mutations in mtDNA in humans and primates (e.g., Brown et al. 1982, Wilson et al. 1951). During this time Brown met Rebecca Cann, a graduate student who had begun working in Wilson's laboratory in 1979 and was interested in mtDNA research. Cann decided to expand upon Brown's earlier research into human phylogeny for her dissertation research, but now with a larger population sample of mtDNA. Cann worked with Wilson and Mark Stoneking, who arrived at Wilson's laboratory in 1981, and began to collect mtDNA samples from placentas collected from local hospitals. By 1986 Cann and Stoneking had gathered samples from 147 individuals from Europe, Asia, and the South Pacific, although difficulties in collecting samples from Africa led them to rely upon African-American samples instead. Cann and Stoneking reconstructed a phylogenetic tree by using an algorithm to group together those samples with the greatest similarity according to their degree of difference in the quantity of mutations in their mtDNA. The result was a tree with two main branches, one a small branch containing only Africans and the other a large branch containing samples from every other geographical group, including some Africans. Since the branch containing only Africans connected with the other branch close to its base,

Cann, Stoneking, and Wilson concluded that all modern human populations, or at least their mtDNA, must have descended from a woman living in Africa. By arguing that mutations in mtDNA occur at a steady rate they had a possible means of determining approximately when this woman had lived. To estimate the mutation rate they studied the mtDNA of populations, such as Australian aborigines, whose ancestors had migrated to Australia (an event then estimated to be c.30 ka), and determined how many mutations had accumulated in this population as compared to others over that known period of time. Extrapolating from the mutation rate acquired from that research they estimated that the female mitochondrial common ancestor lived 140–290 ka, which they simplified to a figure of c.200 ka (Cann, Stoneking, and Wilson 1987). This hypothetical female common ancestor was dubbed Eve, and the research that led to the notion of the mitochondrial Eve was immediately seen as support for the out-of-Africa hypothesis, and was incompatible with the idea that *Homo sapiens* had evolved from regional populations of *Homo erectus*, as supporters of the **multiregional hypothesis** argue. Subsequently, there has been debate over mutation rates in mtDNA and about other factors that could affect the original interpretation of Cann et al.'s research, but its general conclusion is widely accepted. The paper helped to spark rapid growth in the application of molecular biology to the interpretation of recent human evolutionary history.

mitochondrial DNA (or mtDNA) Circular and approximately 16,570 **base pair**s (bp) in length. It contains 37 **gene**s, 13 that encode **protein**s involved in cellular respiration (i.e., the synthesis of adenosine triphosphate, or ATP), two that encode rRNA subunits, and 22 that encode tRNAs specific to protein synthesis in the mitochondria. The mitochondrial DNA genetic code is slightly different from the **nuclear DNA** genetic code. For instance, the almost universal nuclear mRNA stop codon UAG in mtDNA codes for tryptophan in the nuclear DNA. Most of the mtDNA genome is coding sequence except for some short sequences between genes and the displacement loop (also known as the D-loop or hypervariable region). The mitochondrial genome has been examined extensively to investigate modern human population history worldwide (e.g., Di Rienzo and Wilson 1991, Horai et al. 1996, Lum et al. 1994, Mountain et al. 1995, Richards et al. 1996, Vigilant et al. 1989, Ingman et al. 2000) as well as nonhuman primate population history (Morin et al. 1994, Wildman et al. 2004). Several important features of mtDNA make it useful for population genetic studies [e.g., it is typically maternally inherited, it does not undergo **recombination** (Giles et al. 1980), and it has a relatively fast mutation rate]. In modern humans, mtDNA analyses have largely focused on the two hypervariable segments of the control region, a noncoding "spacer" region where one of two origins of replication is located, on restriction sites located throughout the mitochondrial genome, and on a 9 bp deletion found in some populations in a small noncoding segment between the cytochrome oxidase subunit II gene and a gene coding for a lysine tRNA. Characteristic polymorphisms have been used to define mtDNA **haplogroups** (a group of similar mtDNA lineages). One limitation of mtDNA is that its lack of recombination means that it should be treated as a single locus (since everything is linked and passed from mother to child as a unit) in population genetic analyses. This reduces the statistical power of conclusions about population history that only use this locus. Although both men and women have mtDNA, only women usually pass it to their offspring, and therefore it is useful for examining female population history (including migration patterns and effective population sizes). For example, in the New World, research on mtDNA variation has included both restriction enzyme surveys and sequencing of the hypervariable regions. These data have been used as evidence for hypotheses about the timing, origins, and number of migrations from Asia into the Americas, about the size of the migrant population, and about local demographic history (e.g., Merriwether et al. 1996, Bonatto and Salzano 1997). Most analyses of **ancient DNA** focus on mtDNA because it has a much higher copy number (>700 copies) in each cell than nuclear DNA, making it much more likely to be recovered in samples where DNA quantity and quality are degraded. For example, several mtDNA hypervariable region sequences are available for Neanderthals (e.g., Krings et al. 1997, Krause et al. 2007b) as well the complete mtDNA genome (Green et al. 2008). Mitochondrial DNA has also been used extensively to generate primate phylogenies. Such analyses have primarily examined the *cytochrome B* gene and the cytochrome oxidase subunit II (*COII*) gene, but whole-genome data are becoming increasingly common (e.g., Disotell et al. 1992, Schrago and Russo 2003, Raaum et al. 2005).

mitochondrial genome *See* **mitochondrial DNA**.

mitogen (Gk *mitos*=warp thread and *genes*=born) Any substance that stimulates **mitosis**.

mitosis (Gk *mitos*=warp thread) The process of cell division that produces two identical **diploid** cells. It occurs in four stages (prophase, metaphase, anaphase, and telophase) followed by cytokinesis.

ML Acronym for **maximum likelihood** (*which see*).

Mladeč (Location 40°40′N, 17°00′E, Czech Republic, West of Litovel; etym. name of nearby village) <u>History and general description</u> Cave discovered, probably as early as 1826, while blasting for a quarry, and yielding some of the earliest *Homo sapiens* remains in Europe and evidence of **Upper Paleolithic** industries. The excavations conducted in 1881–2 yielded Mladeč 1–3 and other unnumbered bones. In 1903–4 Mladeč 4–7 and other unnumbered bones were recovered as part of salvage and/or amateur excavations. Mladeč 8–10 were recovered in 1922. All of these efforts also resulted in a rich faunal and archeological record. The process of converting the caves into a tourist attraction, starting in 1911, resulted in the disturbance and large-scale removal of sediments. Later systematic excavations conducted between 1958 and 1961 did not recover any additional hominins or Upper Paleolithic archeological remains. <u>Temporal span and how dated?</u> Uncalibrated radiocarbon dates on carbonates, faunal bone, and human bone. The modern human specimens have yielded dates between $26,330 \pm 170$ and $31,500 \pm 420$–400 years BP. <u>Hominins found at site</u> Mladeč 1 [adult cranium, Natural History Museum, Vienna, Austria (or NHMV)], Mladeč 2 (adult calvaria, NHMV), Mladeč 3 (child, calvaria, NHMV), Mladeč 4 (missing), Mladeč 5 (adult calotte, Moravian Museum, Brno, Czech Republic), Mladeč 6 (adult callotte, destroyed), Mladeč 7 (missing), and Mladeč 8–10 (missing), plus additional cranial and postcranial remains of several individuals, also missing (see Wolpoff et al. 2006 for the latest inventory). <u>Archeological evidence found at site</u> Lithic and bone industry described as Upper Paleolithic (**Aurignacian**). Bone artifacts include **point**s, **awl**s, and perforated teeth and long bones. **Ochre**

markings on the wall have also been reported. Key references: Geology, dating, and paleoenvironment Svoboda 2006, Wild et al. 2005, 2006; Hominins Szombathy 1925, Vlček 1971, Teschler-Nicola 2006, Wolpoff et al. 2006; Archeology Szombathy 1925, Oliva 2006.

MLD Acronym for Makapansgat Limeworks Deposit and prefix for fossils recovered from the Makapansgat Limeworks Deposit, Limpopo Province, South Africa.

MLD 1 Site Makapansgat. Locality N/A. Surface/*in situ* Extracted from a piece of breccia that came from one of the dumps of Lower Phase I breccia discarded by the limeworkers. Date of discovery September, 1947. Finder John Kitching. Unit N/A. Horizon N/A. Bed/member Member 3. Formation **Makapansgat Formation**. Group N/A. Nearest overlying dated horizon N/A. Nearest underlying dated horizon N/A. Geological age *c*.2.8 Ma. Developmental age Adult. Presumed sex Unknown. Brief anatomical description Posterior part of a calvaria that preserves most of the occipital bone and the posterior parts of the parietal bones. Announcement Dart 1948. Initial description Dart 1948. Photographs/line drawings and metrical data Dart 1948. Detailed anatomical description N/A. Initial taxonomic allocation *Australopithecus prometheus*. Taxonomic revisions Robinson (1954) proposed that a subspecies of *Australopithecus africanus*, *Australopithecus africanus transvaalensis*, be used to distinguish the australopiths from "Sterkfontein, Makapan and East Africa" (*ibid*, p. 196) from the australopith taxon, *Australopithecus africanus africanus*, represented by the **Taung**(s) skull. Current conventional taxonomic allocation *Au. africanus*. Informal taxonomic category Archaic hominin. Significance The **holotype** of *Au. prometheus*. Location of original School of Anatomical Sciences, University of the Witwatersrand, Johannesburg, South Africa.

MN zones Acronym for mammal neogene. *See* **European mammal neogene** system.

MNE Acronym for **minimum number of elements** (*which see*).

MNI Acronym for the **minimum number of individuals** (*which see*).

MNK Acronym for Mary Nicol Korongo, a locality in **Olduvai Gorge**. It was discovered by Mary **Leakey** (Nicol was her maiden name) in 1935. Subsequently artifacts and faunal remains have been recovered, and in 1963 the juvenile cranium **OH 13** was recovered. *See also* **Olduvai Gorge**.

mobile art *See* art.

mobile genetic element *See* transposition; transposon.

mode (L. *modus*=manner) Archeology In his global review of world prehistory, Grahame Clark (1969) divided lithic technology into five categories, or modes, that characterized the dominant artifact form and the inferred underlying hominin behavior responsible for it. The approach obscures much of the variability of the lithic technological record at any time or place. Modes emphasize only the most complex, or derived tool forms, and this feature has occasionally led to its use for comparing change in the archeological and hominin fossil records (e.g., Foley and Lahr 2003). Evolution George Gaylord Simpson's *Tempo and Mode in Evolution* (1944) helped push evolutionary biologists in the direction of recognizing these two components of the evolutionary process: rate (or **tempo**) and pattern (or mode) of evolution. The various evolutionary modes include **stasis**, **directional selection**, and **random drift**. Grey et al. (2008) provide good pictorial examples from invertebrate evolution. As one can see from the Grey et al. paper you need a lot of fossil evidence (i.e., many data points) to investigate evolutionary mode. Researchers who have attempted to investigate the mode of evolution within the hominin clade (e.g., Wood et al. 1994, Lockwood et al. 2000) suggest stasis is the dominant signal. *See also* **phyletic gradualism; punctuated equilbrium; sample statistics**.

Mode 1 The first of the five technological **modes** defined by Clark (1969) in his global review of lithic technology. Mode 1 artifacts were originally defined as "chopper-tools and flakes," but they are now widely equated with "least-effort flake production." Mode 1 assemblages are comparable to those of the **Oldowan**, characterized by the production of flakes from cobbles or similar naturally occurring rock forms by direct stone-on-stone percussion or the use of an **anvil**. Sharp flakes are the desired products, cores are generally thought to be byproducts, and there is no elaborate knapping plan or predetermination of products. Mode 1 is exemplified by Oldowan sites from Bed I of **Olduvai Gorge** (Leakey 1971), **Koobi Fora** (Isaac et al. 1997), and **Gona** (Semaw 2000); by early sites in western Eurasia such as **Dmanisi** (Baena et al. 2010) and **Atapuerca** (Carbonell et al. 1999); and by a majority of Lower Paleolithic sites east of India (Schick and Zhuan 1993).

Mode 2 The second of the five technological **mode**s defined by Clark (1969) in his global review of lithic technology. Mode 2 artifacts were originally defined as "bifacially flaked handaxes," but it is generally considered to include cleavers as well. Mode 2 assemblages are comparable to those of the **Acheulean**, characterized by the production of handaxes and cleavers through bifacial flaking using stone or organic (e.g., wood or antler) hammers. The shaped core is the desired product, and the flakes produced in shaping are considered waste. There is substantial variation, but Mode 2 forms should usually be worked on both sides (bifacial) and have a recognizable, imposed form. They may be shaped from naturally occurring clasts (e.g., cobbles, nodules, blocks) or from large (>10 cm) flake blanks detached from boulder cores. Exemplified by Acheulean sites throughout Africa and western Eurasia, including Bed II of **Olduvai Gorge** (Leakey 1971), **Kalambo Falls** in Zambia (Clark 2001), **Gesher Benot Ya'aqov** in Isreal (Saragusti and Goren-Inbar 2001), and Boxgrove in England (Roberts and Parfitt 1999).

Mode 3 The third of the five technological **mode**s defined by Clark (1969) in his global review of lithic technology. Mode 3 artifacts were originally defined as "flake tools from prepared cores." The classic example is **Levallois** technology. Cores are carefully shaped (prepared) so that one or more standardized flakes of a predetermined shape may then be produced. Cores and the flakes from shaping are considered waste. Exemplified by numerous **Middle Stone Age** sites from East (e.g., the Kapthurin Formation; Tryon et al. 2005) and North (Vermeersch 2001) Africa, and **Middle Paleolithic** sites in the Levant (e.g., Tabun, Kebara; Bar-Yosef 1998), and Europe (e.g., Le Moustier, La Quina, Combe Grenal; Mellars 1996)

Mode 4 The fourth of the five technological **mode**s defined by Clark (1969) in his global review of lithic technology. Mode 4 artifacts were originally defined as "punch-struck blades with steep retouch," but now it is often used to refer to blade production generally. Classically, a roughly cylindrical, conic, or wedge-shaped "prismatic" core is carefully prepared with long ridges around its circumference. Percussion above the ridges (often using a punch) produces multiple, uniform blades, which may then be retouched into a wide array of standardized tools. Prismatic blade production is exemplified by Mayan (Bordes and Crabtree 1969) and European **Upper Paleolithic** (e.g., Etoilles; Pigeot 1990) technologies. Other blade-production methods are more widespread (Bar-Yosef and Kuhn 1999).

Mode 5 The fifth of the five technological **mode**s defined by Clark (1969) in his global review of lithic technology. Mode 5 artifacts were originally defined as the "microlithic components of composite artifacts." Clark associated Mode 5 with the Mesolithic, but microlithic technologies are now well known from the later Paleolithic. Microlithic tools may be produced using microblades detached from diminutive cores or snapped segments of larger blades that are retouched into standardized geometric shapes and hafted to make composite tools. Classically exemplified by Mesolithic (e.g. Maglemosian) sites in Western Europe, microlithic technology is important in the late Pleistocene of East Asia (e.g., Dyuktai) and examples are found throughout the world by the mid-Holocene (see papers in vol. 12, issue 1 of *Archaeological Papers of the American Anthropological Association*, January 2002, Elston and Kuhn (eds)). Less classically, Mode 5 may include somewhat larger geometric forms from the **Middle Stone Age** Howieson's Poort from South Africa (*ibid*).

model I regression A term applied to regression techniques which minimize residuals from the best-fit line in the independent variable alone (e.g., **ordinary least squares regression**). Model I regressions should be used when the goal is to predict values rather than identify **scaling** relationships.

model II regression A term applied to regression techniques which minimize residuals from the best-fit line in both independent and dependent variables (e.g., **major axis regression** and **reduced major axis regression**). In general, model II regressions should be used when the goal is to identify **scaling** relationships rather than to predict values.

model organism   A species that is intensively studied to understand particular biological processes or traits. These species usually are easy to work with in the laboratory setting. Such organisms include bacteria (e.g., *Escherichia coli*), nematodes (e.g., *Caenorhabditis elegans*), fruit flies (*Drosophila melanogaster*), zebra fish (*Danio rerio*), and mice (*Mus musculus*). Model organisms have been important for studies of development, gene expression and function, and comparative genomics, and thus they have aided in our understanding of development, gene function, and the evolution of the modern human genome.

modeling   (L.   *modus* = standard)   Biology, general Modeling is routinely used in the experimental sciences. The aim is to build a model that best explains the observed phenomena. Well-designed experiments enable researchers to refine and validate their models by comparing what is observed with what the model predicted. The principles of modeling can also be used in the historical sciences, but there are difficulties in applying modeling in the absence of the ability to experiment. Tooby and DeVore (1987) reviewed the use of models for reconstructing the evolutionary history of hominin behavior. They distinguished between **referential** and **conceptual modeling**. In a referential model the real behavior of one, or more, living animal(s) is used to reconstruct the behavior of a fossil taxon (the **referent**). For example, if in extant forms closely related to fossil hominins substantial levels of canine-crown-height **sexual dimorphism** are always associated with a **multimale social structure**, then the principles of referential modeling suggest it is reasonable to infer that a fossil hominin with similar levels of canine-crown-height sexual dimorphism had a multimale social structure. In contrast, in conceptual modeling the model is not so directly based on "real" observations. It is based on general principles that have been developed from observations of a wide range of animals (e.g., bees), not just those closely related or analogous to the referent. In conceptual models "individual taxa should be treated as sources of data points for comparative studies,

not as models" (Moore 1996). Therefore, for example, baboon behavior could be used as a literal referential model, or observations on baboons could be just one component of the information used to generate a conceptual model. Tooby and DeVore (1986) suggest a third category of model, **strategic modeling**, based on the principle of **uniformitarianism**. The premise of strategic modeling is that "species in the past were subject to the same fundamental evolutionary laws and ecological forces as species today, so that principles derived today are applicable throughout evolutionary history" and therefore "although no present species will correspond precisely to any past species, the principles that produced the characteristics of living species will correspond exactly to the principles that produced the characteristics of past species" (*ibid*, p. 189–90). Biology, bone In bone biology modeling refers to any change in the mass and/or external shape of bone (e.g., during growth or as the result of excessive use).

models of intermediate complexity   (or MIC) Climate models that are more complex than "box models" but less complex than **general circulation model**s (or GCMs). They are useful for paleoclimate reconstruction because their reduced computational demands (compared to GCMs) mean that, whereas the latter can only be run for "equilibrium" experiments, MIC models can be used for "time-transient" experiments (i.e., simulations through time). *See also* **general circulation model**; **paleoclimate**.

modern evolutionary synthesis   The modern evolutionary synthesis emerged as the dominant model of biological evolution during the 1930s and 1940s. Charles **Darwin** unveiled his theory of evolution through natural selection in 1859 and while most biologists came to accept the idea that species evolved over time there was disagreement over the mechanisms that caused evolution. By the beginning of the 20thC a substantial number of biologists and paleontologists supported versions of evolution that differed from Darwin's. These scientists tended to reject the idea of natural selection, and instead invoked either **neo-Lamarckism**, which stressed the role of the inheritance of acquired characteristics, or **orthogenesis**, which argued that biological properties internal to organisms caused certain groups to evolve along linear progressive lines. The German biologist August Weissmann's studies of germ plasm and chromosomes during the 1880s and 1890s, and the experiments on heredity conducted by Gregor Mendel (published in 1865 but generally unknown until the turn of the 19thC) led to the emergence of the new disciplines of genetics and population genetics in the early 20thC. The relationship between evolution theory and genetics was explored by Ronald Fisher and J.B.S. Haldane in England and by Sewall Wright in the USA. Their mathematical approach to studying the frequency of genes in populations offered new ways of understanding how variation within populations originated and could be acted upon by natural selection. Fisher's *The Genetical Theory of Natural Selection* (1930) and Haldane's *The Causes of Evolution* (1932) demonstrated how Mendelian genetics and Darwin's theory of evolution by natural selection could be combined to explain how species change over time. At the same time the Russian geneticist Sergei Chetverikov was studying the impact of recessive genes in populations of organisms and their role in evolution. Chetverikov conducted this research within a Darwinian conception of how evolution operated and his research influenced a generation of Russian geneticists before they were purged by the Soviet leader Joseph Stalin. However, the ideas of this generation of Russian geneticists were introduced into America by Theodosius Dobzhansky who left Russia in 1927 to work in the genetics laboratory of Thomas Hunt Morgan at Columbia University. Dobzhansky combined population genetics and natural selection to explain the process of speciation, and his *Genetics and the Origin of Species* (1937) had a powerful impact on biological thinking during the middle of the 20thC. By the 1930s, among at least some biologists, Darwin's theory was enjoying a resurgence. The German ornithologist Ernst Mayr, who emigrated to the USA in 1931 to become the curator of the American Museum of Natural History's collection of birds, offered new insights into how new species emerge by analyzing the role of isolation. Dobzhansky nominated Mayr to present his ideas about speciation in animals at the Jesup lectures at Columbia University in March 1941 and these lectures were the basis of his *Systematics and the Origin of Species* (1942). In that work Mayr promoted a conception of evolution that was strongly Darwinian and yet integrated the latest biological knowledge, especially population genetics. This new thinking about the mechanisms of evolution was introduced into paleontology by George Gaylord Simpson,

curator of fossil vertebrates at the American Museum of Natural History. In his influential work *Tempo and Mode in Evolution* (1944) Simpson demonstrated that the kinds of small-scale evolutionary processes discussed by population geneticists could produce the kinds of large-scale evolutionary changes observed in the fossil record. He used evidence from the mammalian fossil record to show that the fossil series of many groups of organisms did not show the linear paths of evolutionary development as proponents of neo-Lamarckism or orthogenesis argued, but showed instead many diverging branches just as Darwin described. The best-known systematic survey of the new union of population genetics and Darwinian evolution theory was Julian Huxley's *Evolution: The Modern Synthesis* (1942). Reduced to its most critical components, the modern evolutionary synthesis argued that (a) evolution was the result of small genetic mutations accumulating in populations of organisms over long periods of time as well as the movement and recombination of genetic traits within populations, (b) natural selection was the primary mechanism by which species change over time, and (c) natural selection operating upon a population leads to adaptation to specific environmental conditions. Through their research, publications, and prominent institutional affiliations, Mayr at Harvard University, Dobzhansky at Columbia University, Simpson at the American Museum of Natural History, and Sewall Wright at the University of Chicago transformed modern thinking about the mechanisms underlying evolution. It was not long before the members of this group began to apply the principles of the modern evolutionary synthesis to the problem of human evolution. Dobzhansky addressed the question of human evolution and hominin species in an article titled "On species and races of living and fossil man" published in the *American Journal of Physical Anthropology* in 1944, in which he argued that **Sinanthropus** and **Pithecanthropus** are no more different than are what were then recognized as races of modern humans. As a result of this conception of a species, Dobzhansky argued that only one species of hominin had existed at any one time, and that there was often considerable variation within hominin species. He greatly expanded upon these issues later in *Mankind Evolving* (1962). The modern evolutionary synthesis was brought solidly into human origins research in 1950 at a symposium held at Cold Spring Harbor in New York. In a paper titled "Taxonomic categories in fossil hominids," Mayr sought to reform hominin taxonomy to reflect the new population thinking about species. In Mayr's opinion, anthropologists had been too ready to name new species on the basis of very slight differences between fossils, and he argued that all of these species should be collapsed into a single genus (*Homo*) consisting of only three species: *Homo transvaalensis* (australopiths), *Homo erectus* (*Sinanthropus* and *Pithecanthropus*), and *Homo sapiens* (i.e., **modern humans** and **Neanderthals**). At this same meeting George Gaylord Simpson criticized the persistence of orthogenesis in many interpretations of about human evolution and explained how the hominin fossil record could be better interpreted from the perspective of the modern evolutionary synthesis. While there was some resistance among anthropologists and anatomists to the modern evolutionary synthesis (sometimes referred to as just "the synthesis"), by the 1980s the majority of human origins researchers had fully accepted this approach to the study of human evolution.

modern human (L. *modernus*=now and *humanus*=people) When used as a noun (e.g., as in "the Herto crania more closely resemble the crania of modern humans than they do any other hominin taxon") the term modern human is equivalent to either all *Homo sapiens*, or to just **extant** *H. sapiens*. When used as an adjective (e.g., as in "modern human morphology") this term refers to the morphology seen in contemporary populations of *H. sapiens*. "Human morphology" (i.e., the morphology seen in taxa within the genus *Homo*) is sometimes used when the writer really means "modern human morphology." *See also Homo sapiens*.

modern human behavior The suite of behavioral patterns that characterize behaviorally modern *Homo sapiens* and that distinguish them from living primates and other members of the hominin lineage. There is intense debate concerning the nature, timing, and geographical origin of modern human behavior. Disagreement stems in part from the difficulty of defining modern human behavior in the first place and then being able to recognize modern human behavior through the lens of the archeological record. Some behavioral patterns that are considered to be evidence of modern human behavior include symbolic behavior and personal adornment, effective exploitation of large mammals and dietary breadth, long-distance exchange networks, standardized lithic technology, and use of composite tools, among others. But there is a danger of historicism coming into play with respect to some of these lines of evidence, because archeologists tend to assume that the links between artifacts and behavior observed in the present (e.g., ochre and body adornment) were also operating in the past. As for timing, some researchers argue that modern human behavior emerged abruptly *c.*50,000 years ago, creating a temporal lag between the origins of anatomically modern humans and the origin of behavioral modernity. Others suggest the origin of modern human behavior is the result of the gradual accumulation of behavioral patterns associated with the appearance of the **Middle Stone Age** in Africa beginning *c.*250 ka, roughly coincident with fossil evidence for the appearance of anatomically modern humans, or *H. sapiens*.

Modjokerto The old Indonesian spelling for the site now known as **Mojokerto**. Hence the fossils recovered from the site when the old spelling was in use retain that spelling (e.g., **Modjokerto 1**), as does the taxon *Homo modjokertensis*, now a junior synonym of *Homo erectus*. *See* **Mojokerto**.

Modjokerto 1 Site Mojokerto. Locality "The exact find spot remains unknown" (Anton 1997, p. 501). Surface/*in situ* Surface. Date of discovery 1936. Finder Tjokrohandojo. Unit N/A. Horizon "Tuffaceous conglomeratic sandstone" (Anton 1997, p. 500). Bed/member Djetis Beds. Formation **Sangiran Formation**. Group N/A. Nearest overlying dated horizon N/A. Nearest underlying dated horizon N/A. Geological age "From a pumice-bearing level in the region of Modjokerto" and the "youngest of these levels known from that section is 1.81 mya" (Anton 1997, p. 501). Developmental age 4–6 years (Anton 1997), *c.*1 year (Coqueugniot et al. 2004), and <4 years of age (Balzeau et al. 2005). Presumed sex Unknown. Brief anatomical description Partial calvaria that includes most of the vault and part of the floors of the middle and anterior cranial fossae. Announcement von Koenigswald 1936. Initial description von Koenigswald 1940. Photographs/line drawings and metrical data von Koenigswald 1940, Anton 1997, Coqueugniot et al. 2004, Balzeau et al. 2005. Detailed anatomical description Anton 1997. Initial taxonomic allocation *Homo modjokertensis*. Taxonomic revisions *Homo soloensis* (Dubois 1936), *Pithecanthropus modjokertensis* (Jacob 1975), *Homo erectus* (Clark 1978). Current conventional taxonomic allocation *H. erectus*. Informal taxonomic category

Pre-modern *Homo*. Significance **Holotype** of *H. modjokertensis* and the specimen used to argue for a relatively primitive pattern of growth in endocranial volume in *H. erectus* (Coqueugniot et al. 2004). Location of original **Senckenberg Forschungsinstitut und Naturmuseum**, Frankfurt, Germany.

modularity (L. *modulus*=a measure in music) Modules in biology are entities within a system that have a greater density of internal than external connections (Gass and Bolker 2003). Some regard modules as abstract entities that reflect processes, or sets of connections within networks, while others regard them as spatially contiguous physical entities within organisms. In the **phenotype** of complex organisms such as primates, there are typically many sets of integrated **trait**s and such organisms are said to be modular. Modularization can act to release evolutionary constraints, allowing different parts of the phenotype to evolve independently. Thus, modularity is a determinant of **evolvability**. Modular organization also means that changes to one part of an organism or developmental system are less likely to produce deleterious effects elsewhere. An example of the application of the modularity concept in biological anthropology is the work on integration and modularity in primate limbs (e.g., Hallgrímsson et al. 2002, Young and Hallgrímsson 2005, Rolian 2009). This work shows that the modular organization of the limb is evident in the structure of limb and that **covariation** within the limb may have influenced or constrained evolutionary changes in limb proportions. *See also* **morphological integration**.

module An internally integrated unit, such as a complex biological structure or pathway, that is comparatively independent from other surrounding modules with which it may interact. In addition to modularity another way in which the term module is used in paleoanthropology is in dental metrics. *See also* **crown module**; **modularity**.

Mohui *See* **Bubing Basin**.

MOI Abbreviation of the Moiti Tuff in the **Koobi Fora Formation**, **West Turkana**, Kenya.

Mojokerto (Location 7°22′S, 112°38′E, Indonesia; syn. Perning, Modjokerto; etym. the modern Indonesian spelling for a nearby village) History

and general description An exposed river-bed deposit, 3 km/1.8 miles north of Perning, 10 km/6 miles northeast of Mojokerto, and 35 km/22 miles west of Surabaya in East Java. On February 13, 1936 a juvenile fossil hominin calvaria was discovered *in situ* 1 m below the surface. The hominin fossil is derived from a conglomeratic sandstone stratum of the **Sangiran** (formerly Pucangan) **Formation**. Temporal span and how dated? There is debate concerning the radiometric age of the site. In 1994, **argon-argon dating** of volcanic hornblende from the deposit provided an age of 1.81±0.04 Ma. In 2003, **fission track dating** of primary zircon grains suggested a maximum age for the hominin at 1.49±0.13 Ma. Recent attempts to more reliably pinpoint the actual find site at Mojokerto have resulted in published interpretations favoring the more ancient date for the hominin partial cranium. Hominin found at site **Modjokerto 1** is the holotype of *Homo modjokertensis*, named by Ralph **von Koenigswald** in 1936. More recent interpretations place this fossil within the hypodigm of *Homo erectus*, based on its possession of a metopic eminence and an occipital torus, characters many researchers regard as synapomorphic for this hominin species. Chronological age estimates suggest that the Modjokerto child was between 4 and 6 years old at death. **Computed tomography** of the fossil's endocranial features indicates that *H. erectus*' brain growth differed from that observed in modern humans. Archeological evidence found at site None. Key references: Geology, dating, and paleoenvironment Swisher et al. 1994, Huffman 2001, Morwood et al. 2003, Huffman et al. 2005; Hominins Antón 1997, Coqueugniot et al. 2004, Balzeau et al. 2005.

mokondo A term used at the southern African cave sites for a sinkhole. This is an area of **breccia** that has been decalcified and then subsequently washed out, thus providing a cavity into which more recent breccia can accumulate. It is this type of sinkhole and new breccia accumulation that complicates the geology of sites such as **Swartkrans**. The cranium **Stw 53** from Sterkfontein was said to have been recovered from breccia that was in a mokondo/sinkhole (Hughes and Tobias 1977).

molar (L. *mola*=mill stone) A type of tooth. The name refers to the teeth distal to (i.e., behind) the **premolars** in the tooth row. Upper molars consist of a mesial **trigon** and a distal **talon**; the lower molar equivalents are **trigonid** and **talonid**. There are two **decidous** and three **permanent** molars. The upper deciduous molars are referred to as dm^1 and dm^2 (or dp^1 and dp^2) and the upper permanent molars as M^1, M^2, and M^3. The lower deciduous molars are referred to as dm_1 and dm_2 (or dp_1 and dp_2) and the lower permanent molars as M_1, M_2, and M_3. Primarily used for chewing, crushing, and grinding food.

Molare Shelter (Location 40°02′21″N, 15°28′32″E, Italy; etym. local name of the area, called "Molara," near Scario) History and general description Rock-shelter containing a large **Middle Paleolithic** deposit with at least seven successive hearth levels. An immature hominin mandible was found at the bottom of these levels in 1985. Temporal span and how dated? Technostratigraphic correlation to other sites suggested a date of Würm I (*c.*50–70 ka), but a re-examination of the evidence suggests a "pre-Wurmian" age (*c.*Oxygen Isotope Stage 5b). Hominins found at site The mandible belonged to an immature individual, probably 3–4 years old. The preserved dentition of includes only the right and left dm_1 and dm_2. Whereas the mandible is usually considered to belong to *Homo neanderthalensis*, Mallegni and Ronchitelli (1989) note that many of the mandibular and dental traits are primitive. Archeological evidence found at site **Mousterian** artifacts with some evidence of artifacts made using the **Levallois** technique. Key references: Geology, dating, and paleoenvironment Mallegni and Ronchitelli 1987, 1989; Hominins Mallegni and Ronchitelli 1987, 1989; Archeology Mallegni and Ronchitelli 1987, 1989.

molarized A term introduced by John **Robinson** (1956) to describe mandibular premolars with such a well-developed **talonid** that they come to resemble mandibular molars.

molecular anthropology Developments in biochemistry and immunology during the first half of the 20thC allowed the focus of the search for better evidence about the nature of the relationships between modern humans and the great apes to be shifted from traditional, macroscopic, morphology to the morphology of molecules. Sufficient information was known about the differences in organic molecules between organisms as early as 1944 for Marcel Florkin to assemble this knowledge in a book entitled *Biochemical Evolution*. The earliest attempts to use the proteins of primates to determine the relationships among taxa were made just after the turn of the century, but the results of the first of a new generation of analyses were reported in the early 1960s, and Linus Pauling claimed he coined the name molecular anthropology for this area of research. Zuckerkandl et al. (1960) used enzymes to break up the hemoglobin (Hb) protein into its peptide components, and showed that when the peptides were separated (using starch gel electrophoresis) the patterns of the peptides in the gel for modern humans, gorilla, and chimpanzee were indistinguishable. Morris Goodman (1963) used what was then the new technology of immunology, specifically a process called immunodiffusion, to study the affinities of the serum proteins of the apes, monkeys, and modern humans. He came to the conclusion that because the patterns produced by the albumins of modern humans and the chimpanzee in the immunodiffusion gels were effectively identical, then it is likely that the structures of the albumins were also likely to be, to all intents and purposes, identical. Proteins are made up of a string of amino acids. In many instances one amino acid may be substituted for another without changing the function of the protein. In the 1970s Vince Sarich and Allan Wilson exploited these minor variations in protein structure to determine the evolutionary history of protein molecules, and therefore, presumably, the evolutionary history of the taxa whose proteins had been sampled. They, too, concluded that modern humans and the African apes, in particular the chimpanzee, were very closely related. In a later paper Mary-Claire King and Allan Wilson (1975) suggested that 99% of the amino acid sequences of chimps and modern humans were identical. The discovery, by James Watson and Francis Crick, of the structure of DNA, and the subsequent discovery by Crick and others of the genetic code, showed that it was the sequence of bases in the DNA molecule that determined the nature of the proteins manufactured within a cell. This meant that the affinities between organisms could be pursued at the level of the genome, thus potentially eliminating the need to rely on morphological proxies, be they traditional anatomy, or the morphology of proteins, for information about evolutionary relatedness. The DNA within the cell is located either within the nucleus as nuclear DNA, or within the mitochondria as mitochondrial DNA (mtDNA). Comparisons between the DNA of organisms were initially made using a method called DNA hybridization. In the early stages of DNA research, DNA hybridization told researchers "relatively little about a lot of DNA," whereas the sequencing method told them "a lot about a little piece" of DNA. Nowadays technological advances mean that whole genomes can be sequenced. Sequencing is favored because the knowledge about the type of differences between the base sequences provides some clues about the steps that are needed to produce the observed differences. This is because the base changes called transitions (A to G and T to C) readily switch back and forth, whereas transversions (A to C and T to G), which are less common, are more stable, and thus are more reliable indicators of genetic distance. These methods, hybridization (e.g., Caccone and Powell 1989) and sequencing (e.g., Bailey et al. 1992; Horai et al. 1992), have been applied to the living great apes, and the number of sequence studies increases each year (for a review see Bradley 2008). Information from both nuclear and mtDNA suggested that modern humans and chimpanzees are more closely related to each other than either is to the gorilla (Ruvolo 1997, Li and Saunders 2005), and when these differences are calibrated using paleontological evidence for the split between the apes and the Old World monkeys, and if one assumes that the DNA differences are neutral, then this predicts that the hypothetical ancestor of modern humans and the chimpanzees and bonobos lived between about 5 and 8 Ma, and probably closer to 5 than to 8 Ma (Bradley 2008). However, when other ways of calibration are used the predicted date for the split is somewhat older (e.g., Arnason and Janke 2002 suggest >10 Ma).

molecular basis of brain size evolution

See brain size evolution, molecular basis of.

molecular biology

The study of the molecules involved with the maintenance of life. Although the term suggests it should involve the biology of all molecules, in practice molecular biologists are mainly concerned with complex molecules such as proteins, and in particular with the roles of another class of complex proteins, nucleic acids (e.g., **nuclear DNA, mitochondrial DNA, RNA**) involved in the maintenance of life.

molecular clock

A term introduced by Emile Zuckerkandl and Linus Pauling in 1962 to refer to the use of differences between molecules to generate a tree of their relationships in which branch lengths are proportional to time. Such proportional trees can then be calibrated using the fossil record and then used to estimate otherwise non-calibrated divergence times. Genetic differences between species can be estimated, among other approaches, using nucleotide sequence differences, DNA/DNA hybridization distances, amino acid sequence differences, restriction-enzyme pattern differences, and immunological distances. According to the **neutral theory of molecular evolution** some classes of mutations accumulate in a sufficiently clock-like manner over relevant time intervals as judged by the relative rate test (Sarich and Wilson 1967b), albeit in a manner that is modulated over longer time periods by factors such as generation time and effective population size. Thus, knowing the genetic distance between two species allows the estimation of the average time since the two genomes diverged. This time is calibrated using known divergence times based on the fossil record. So if D is, say, the amount of DNA sequence divergence per site and u is the mutation rate (estimated from the taxa used to calibrate the clock) then the equation $T = D/(2u)$ can be used to calculate T, the divergence time between the species of interest. For example, paleontological evidence about the timing of the divergence between Old World and New World monkeys can be used to calibrate the molecular clock in primates and from this the divergence time of hominoids from Old World monkeys, or that of modern humans from chimpanzees, can be estimated (e.g., Kumar et al. 2005). However, these times will always be older than the time of population separation, and it is the latter that the fossil record can be used to estimate. Given that population separation is recognized using derived features of one or both lineages, and that these features may not be present at the precise time of population separation, the fossil record will always underestimate the actual population separation time by some unknown amount. The molecular clock is invalid if the locus does not evolve neutrally (i.e., if it is under selection). To use a molecular clock to estimate times of divergence, sufficient external data to calibrate the clock reliably (i.e., a well-dated fossil record) as well as relative rate tests to demonstrate rate constancy at the locus or loci being used are necessary. Recently, the analysis of both the genetic data and the fossil record has frequently followed a Bayesian approach in which prior probabilities can be assigned both to a range of plausible fossil-based calibration times and to plausible variation in "local" rates of genetic change, such approaches generating best-fit, most probable hypotheses given the data. It is now clear that rates of genetic change vary between and within tree branches and it is also rarely the case that the fossil record is dense enough to demonstrate with confidence that a divergence has not occurred, even though it can show that a divergence has occurred; hence caution needs to be used in interpreting any molecular clock estimates. Neither the primate fossil record nor our understanding of the long-term fluctuations in rates of genetic evolution are currently complete enough to settle unambiguously questions such as the divergence age of crown primates, cercopithecoids, and hominoids, or *Pan* and *Homo*. *See also* **neutral theory of molecular evolution**.

molecular evolution Evolution involving changes in either **DNA** (due to one or more **bases** being substituted for another, or added, or deleted) or **proteins** (due to one or more **amino acids** being substituted for another).

molecule (L. *moles*=mass) Two, or more, atoms held together by chemical bonds, or the smallest physical unit of a substance that demonstrates the properties of that substance (e.g., a single H_2O molecule has all the properties of a larger volume of water). The term **molecular biology** suggests it should involve the biology of all molecules, in practice molecular biologists are mainly concerned with complex molecules such as proteins and nucleic acids involved in the maintenance of life.

moments of area *See* **cross-sectional geometry**.

monogenic trait (Gk *monos*=alone) A trait caused by alleles at a single gene locus, following simple patterns of inheritance from the Mendelian laws of genetics. In modern humans monogenetic traits include wet or dry earwax and attached or unattached earlobes. Such phenotypic traits are easy to measure and thus have been intensively studied; the single nucleotide polymorphism (or SNP) responsible for the variation in earwax phenotype was discovered in the *ABCC11* gene (Yoshiura et al. 2006) (NB: despite being cited as

such in many textbooks, tongue rolling is *not* a monogenetic trait).

monophyletic *See* **clade**.

monophyletic group (Gk *monos*=alone and *phylon*=race) *See* **clade**.

monophyletic species concept *See* **species**.

monosomy The condition where only one of a pair of chromosomes is present in a diploid organism. In modern humans the only monosomy that is viable is that of the X chromosome.

monozygotic twins When the two offspring produced in the same pregnancy are derived from a single zygote (i.e., a single egg fertilized by a single sperm) that then divides into two separate embryos. Monozygotic twins are generally referred to as identical twins. *See also* **dizygotic twins**.

monsoon (Ar. *mawsim*=seasonal, referring to the seasonal reversal of the winds) Monsoonal climates are characterized by summer-season precipitation. The West African monsoon is dominated by summer monsoonal precipitation associated with the northward migration of the **intertropical convergence zone**. Tropical Atlantic sea-surface temperatures are the primary variable controlling the strength of West African monsoon and Sahelian precipitation at inter-annual to inter-decadal time scales (Giannini et al. 2003). Variations in continental heating driven by **precession** also influence the land/sea temperature contrast and the strength of the monsoon on multi-millennial time scales (deMenocal et al. 2000, Liu et al. 2003). The Indian monsoon indirectly affects climate in northeast Africa, causing drying. Prior to the uplift of the Himalayas and the Ethiopian highlands, northeast Africa would also have had a substantial monsoon. Summer-season precipitation in monsoonal **tropical climates** creates favorable conditions for C_4 tropical grasses.

Monte Carlo (Fr. *Monte-Carlo* is one of the administrative areas of the state of Monaco, and the Monte Carlo quarter of Monaco is where Le Grand Casino is located) The term "Monte Carlo" is used for a family of methods that allow for statistical tests using randomly drawn samples from a larger set of possible outcomes. The term comes from the use of random–number generation in gambling. Monte Carlo methods are a subset of resampling analyses (as opposed to exact resampling analyses). *See also* **resampling**.

Monte Christo Formation (etym. Named after the local town near the type section) General description The second oldest of the five geological formations recognized in the Malmani Subgroup of the dolomite uplands near to modern-day Krugersdorp in South Africa. Cave sites within this formation **Sterkfontein**, **Swartkrans**. *See also Australopithecus africanus*; *Paranthropus robustus*; **Sterkfontein**; **Swartkrans**.

Monte Circeo (Location 41°14′N, 13°05′E, Italy; etym. from *Circe*, a sorceress from Homer's *Odyssey* who was said to dwell there) A limestone promontory along the Tyrrhenian coast of Italy about 85 km/53 km southeast of Rome that includes a number of caves known to contain **Pleistocene** deposits. There are over 30 caves within the mountain, but only three have produced hominin remains; from west to east they are **Grotta Breuil**, **Grotta del Fossellone**, and **Grotta Guattari**. In the past, the fossils found in the caves were named after the mountain itself (e.g., Circeo I, Circeo II), but more recently workers have taken to naming the remains after the individual cave (see Bietti and Manzi 1991). Hence, Circeo I would now be known as Guattari I. *See also* **Grotta Breuil**; **Grotta del Fossellone**; **Grotta Guattari**.

Monte Fenera (Location 45°42′40″N, 8°18′35″E, Italy; etym. named after the mountain on which the sites are located) History and general description Two caves, Ciota Ciara and Ciutarun, lie on the western side of Monte Fenera in northwest Italy. Excavations were carried about by C. Conti in 1955–6, but the fragments were not identified as hominin until much later. Temporal span and how dated? There does not appear to be a published chronology for the site, but the **Mousterian** artifacts and the affinities of the hominin remains with *Homo neanderthalensis* suggest a Middle Paleolithic date. Hominins found at site The hominin remains consist of two teeth and a fragmentary parietal from Ciota Ciara (known as Ciota Ciara 1–3 or Fenera 1–3), and a single tooth from Ciutarun (known as Ciutarun 1 or Fenera 4). Archeological evidence found at site The lithic assemblage in both caves is Mousterian.

Key references: Hominins Mottura 1980, Villa and Giacobini 1996.

Montgaudier (Location 45°40′N, 0°29′E, France; etym. named after a nearby village) History and general description This series of interconnected but collapsed caves in the Charente region was discovered in the late 19thC and has been excavated by a series of researchers since then. It contains both Middle and Upper Paleolithic deposits, and is known for its **Magdalenian** bone artifacts (which are of uncertain provenance) and for the discovery in 1974 of a juvenile mandible that has been attributed to *Homo neanderthalensis*. Temporal span and how dated? Radiometric dating has not been attempted; the site is dated based on technostratigraphy, faunal analysis, and geostratigraphy. The layer in which the Neanderthal fossil was found is the oldest in the cave, and could date to *c.*150 ka, but this is an extremely tentative date. Hominins found at site Several cranial and postcranial remains have been recovered, mostly from the Magdalenian levels. One juvenile Neanderthal mandible was found in the oldest levels of the cave. Archeological evidence found at site Many artifacts, predominantly from the Magdalenian and **Mousterian**. Some researchers have reported **Solutrean** and **Aurignacian** items, but this has been challenged. The complex geology of the site makes it difficult to reconstruct the archeology. Key references: Geology, dating, and paleoenvironment Duport 1967; Hominins Mann and Vandermeersch 1997; Archeology Duport 1967, Debénath and Duport 1972.

Montmaurin (Location 43°13′N, 0°36′E, France; etym. Fr. after the nearby town) History and general description This site comprises a series of caves and chimneys in a limestone cliff overlooking the Seygouade River in the Pyrenees region of southern France. The caves were long known to contain archeological materials and were first explored near the beginning of the 20thC by Marcellin **Boule** and others, but the majority of information comes from the excavations by Louis Méroc and colleagues in the late 1940s. The site is one of the oldest known in France. Temporal span and how dated? **Biostratigraphy** and palynology suggest that the earliest layers (those containing the well-preserved mandible) date to the **Mindel-Riss** interglacial (*c.*400 ka). Most of the site ranges between **Riss** I (*c.*390 ka) and **Würm** I (*c.*100 ka), although the youngest levels are from the Roman era. Hominins found at site Several hominin dental remains were recovered, including a well-preserved mandible that has been variously interpreted as belonging to a **pre-Neanderthal** or *Homo heidelbergensis*, portions of a maxilla of uncertain taxonomic status, a juvenile mandibular fragment likely *Homo neanderthalensis*, and a few isolated teeth. Archeological evidence found at site The site contains several levels, including late **Acheulean** assemblages associated with the earlier hominin, a typical **Mousterian** layer, and a **Châtelperronian** layer. Key references: Geology, dating, and paleoenvironment Méroc 1963, Girard 1973; Hominins Billy 1982, 1985, Billy and Vallois 1977a, 1977b; Archeology Méroc 1963.

morphocline (Gk *morph*=form and *klinein*= to lean or a gradient) The order of character states from the most primitive to the most derived. Morphoclines are usually established by two criteria: **ontogenetic** and **outgroup**. In the former, the earlier stages in the ontogeny of a character are assumed to be more primitive than the later stages, and in the latter closely related taxa are used as outgroups on the assumption that they are likely to display the primitive character state. For example, the ontogenetic criterion would suggest that the primitive condition for any tooth root is a single root, because all teeth with multiple roots start out as single-rooted. But the outgroup criterion suggests that the primitive condition for the root system of anterior mandibular premolar teeth within the hominin clade is the same as the most common root morphology for *Pan* (2R: MB+D; i.e., a mesiobuccal root and a plate-like distal root). Within the hominin clade there appear to be two derived morphoclines. One that trends towards root simplification, with a single root (1R) at the derived end of one morphocline, and one that trends towards complexity, with teeth with two molar-like roots (2R: M+D) at the derived end of the other morphocline.

morphogenesis (Gk *morphe*=form, shape, plus *genesis*=origin, generation) Morphogenesis is the development of morphology or form during the embryonic stage. Along with pattern formation, which is the formation of discrete tissues along a spatiotemporal pattern, and growth, morphogensis is one of the three major classes of developmental processes. Morphogenesis is technically about more than just shape, which is defined as the geometric features of an object that are left once location, orientation, and size have been removed. Morphogensis, rather, is about form, which is the combination of size and shape and it is thus not sharply distinguishable from growth. The reason is that the developmental processes that produce morphology do not allow for a clear distinction between growth and change in shape. During embryological development, the key developmental processes involved in morphogenesis are cell proliferation, cell polarity, cell adhesion, cell migration, cell hypertrophy, production of extracellular matrix, and cell death. Cell proliferation, cell hypertrophy, and the production of extracellular matrix are also primary drivers of growth. The combined activity of these processes produces changes in the form of embryos as they grow and thus results in morphology. The combination of cell proliferation and cell polarity, for example, is sufficient to produce a change in shape because cell polarity will produce preferential directions of cell division. Cell adhesion is critical for morphogenesis, particularly in early embryonic development. As has been shown in computer simulations, cells that have no properties other than proliferation and a tendency to adhere to each other on only part of their surface cell membrane will spontaneously develop into morphological structures. Cell migration can profoundly influence form by altering the location of centers of cell proliferation. The migration of neural crest cells into the branchial arches, for example, is required for morphogenesis of the face. Physical interactions between tissues and between the organism and its surroundings also play an important role in morphogenesis. During embryological development, as one region of cells grows it will displace neighboring regions. This phenomenon results in a complex interplay of physical interactions during development that is only now beginning to be understood. As the embryonic brain (neural tube) grows, for example, it will expand and stretch out the developing face. The face, which forms from prominences that grow forward from the brain around the primitive mouth (stomodeum), needs to grow at a rate that is sufficient to overcome the fact that it is being stretched out by the growing brain. Physical interactions between muscles and bones and between the musculoskeletal system and the physical environment also exert profound influences on the form of bones; these are also morphogenetic processes. Experiments that deprive bones of mechanical stimuli demonstrate the critical role that such physical factors play in the development of skeletal morphology. Evolution, which has been neatly summarized as the influence of ecology on development, mainly involves minor adjustments to morphogenesis and to growth, which between them determine the shape and size of the bones and teeth that comprise the hominin fossil record. Genes that affect morphology do so via effects on the complex developmental processes and pathways subsumed within the term morphogenesis.

morphogenetic fields (Gk *morphe*=form or shape, plus *genesis*=origin or generation) The concept of the morphogenetic field originated in the early part of the 20thC. Experiments conducted using dissection, transplantation, and grafting allowed researchers to ascertain that there were some areas (fields of cells) in the developing embryo that were under the influence of self-directed, self-contained regulation, that would eventually lead to the development of an organ, or structure (e.g., heart, limb). It was further shown that these regions had the potential to influence other cells that they were brought into contact with, by exerting an influence over the fate of these cells so that they developed according to the rest of the field. This led to the idea that it was the field itself, rather than the individual cells composing it, which was responsible for the pattern of development leading to the eventual organ or structure. After the middle part of the 20thC the concept lost popularity; there were no direct tests of the concept and the burgeoning field of studying gene expression in embryos took hold. However, from the latter part of last century to today, more sophisticated techniques have been able to demonstrate the molecular basis for the observations that led to the morphogenetic fields concept. These are, first, that there are spatially defined regions of gene expression that are related to important patterning events in diverse developmental contexts and, second, that secreted factors can produce spatial patterns of gene expression that can determine the morphogenetic fate of transplanted tissues.

morpholino *See* knockout.

morphological integration (L. *integrare*= to make complete) The tendency for development to produce **covariation** among morphological **trait**s. Under this definition, morphological inte-

gration is a property of the architecture of **development** and it is distinct from correlation or covariation observed in the **phenotype**. Patterns of phenotypic covariation or correlation result from integration when combined with genetic variation in particular populations (Hallgrímsson et al. 2009). The term was originally coined by Olson and Miller (1958) who also laid the foundation for the modern study of integration. The concept appears in Darwin's *On the Origin of Species*. In the following quotation in which he refers to "correlated variation," Darwin presages both the concepts of integration and also **modularity**. Darwin wrote, "I mean by this expression that the whole organisation is so tied together during its growth and development, that when slight variations in any one part occur, and are accumulated through natural selection, other parts become modified. … The several parts of the body which are homologous, and which, at an early embryonic period, are identical in structure, and which are necessarily exposed to similar conditions, seem eminently liable to vary in a like manner." (*On the Origin of Species*, 6th edition, 1872). The first formal study of correlated variation, however, is probably that of Sewall Wright (1932), who performed a quantitative analysis of covariation in fowl limb lengths. The population genetics framework for the study of morphological integration was developed principally by James Cheverud and Gunter Wagner in various publications in the 1980s and 1990s (e.g., Cheverud 1982, 1988, 1996, Wagner 1984, 1989, 1996). In this framework, functional integration, or shared functions among traits, leads to genetic and developmental integration through natural selection. This, in turn, produces evolutionary integration, or the tendency for structures to exhibit coordinated evolutionary changes. Much of the pioneering work on morphological integration in primates was done by James Cheverud and colleagues (e.g., Cheverud 1982, 1995, 1996, Ackermann and Cheverud 2000, Marriog and Cheverud 2001). Recent years have seen a surge in studies of integration in the literature on human evolution, where morphological integration is measured primarily via patterns of **phenotypic correlation** or covariation. For paleoanthropologists, understanding morphological integration is an important concern both at practical and theoretical levels. Practically, it is necessary for structuring cladistic studies to avoid highly correlated (integrated) phenotypic traits

(Strait 2001), or for choosing extant models that share similar patterns of integration (and thereby presumably have evolved similarly) to a fossil group being studied (Ackermann 2003, 2009). Theoretically, it provides important insights into evolutionary process, as the degree and pattern of integration can either constrain or facilitate the evolution of complex phenotypes, thereby having a profound impact on the manner in which morphological evolution proceeds. A simple way to think about this is through the following analogy: imagine three objects on your computer screen. You can "grab" each object with your cursor and move each independently; this is like selection acting independently on traits that are not integrated. But imagine using a grouping function to link the objects to each other, so that they only move as a unit (i.e., they are now integrated). In grouping the objects, you have constrained the movement of each object, as they cannot move independently. So if one object becomes fixed, the others do as well. This would be like **stabilizing selection** acting on one trait, which would as a result constrain the evolution of the other, correlated traits. But at the same time, by selecting and moving just one of the three objects, you can move the entire group. In other words, directional selection acting to change a trait would facilitate the evolution of correlated traits, even though those traits are not themselves being selected for. This is very different from the target and amount of selection necessary to move these three traits independently, if they were not integrated, although the end product might appear the same. Obviously in a complex phenotype there are many sets of integrated traits, and in such a case an organism is said to be modular. Modularization can act to release evolutionary constraints, allowing different parts of the phenotype to evolve independently. Williams (2010) provides an example of how the principles of morphological integration can be used to help answer questions related to **evolvability**. *See* **modularity**.

morphology (Gk *morph* = form and *ology* = study of) The morphology of an animal, structure, or fossil is the combination of its external and internal appearance. It subsumes both gross (i.e., what you can see with the naked eye, also called **macrostructure**) and microscopic (what you need a microscope to see, also called **microstructure**) morphology. To be useful for identification, clas-

sification, and phylogenetic analysis, morphology has to be captured in some systematic way. There are two main systems for capturing and recording morphology. One uses measurements, the other records morphology by using presence/absence criteria, or by comparing the fossil's morphology with a series of standards. The former is referred to as **metrical analysis**, **morphometrics**, or just morphometry. The latter system is called **non-metrical analysis** because it relies on categorical methods for assessing the presence/absence or the degree of development of structures whose morphology needs to be compared. Examples of non-metrical traits include the numbers of cusps or roots a tooth has, or the presence or absence of a structure such as a foramen in the cranium, or a bony tubercle on a limb bone. In the case of morphometric methods, measurements are traditionally made between homologous, standardized locations called **landmarks**. The measurements taken are usually the shortest distance between the landmarks; these are known as chord distances. If the surface between the landmarks is curved, a tape laid between the two points will record the arc distance, and the difference between the chord and arc summarizes the degree of curvature; this can also be captured by the chord distance and the subtense. It is also possible to use angles to record the orientation of a structure relative to the sagittal or coronal planes, or to a reference plane such as the Frankfurt Horizontal or the orbital plane. A family of three-dimensional morphometric techniques, also called three-dimensional morphometrics, can be used to record the shape of an object in three dimensions. The position of each reference point is recorded using a three-dimensional coordinate system, and the distances between pairs of recorded points can be recovered if needed. These are the type of three-dimensional data that are used in a family of methods called **geometric morphometrics**, and these methods of data analysis are being used with increasing frequency. Three-dimensional data can be captured using machines called "digitizers," or they can be captured from laser- or CT-generated images. Digitizers usually have a mobile arm with a fine, needle-like point at the end. The tip of the needle is placed on the reference point and then the machine automatically records the location of the reference point in three dimensions. These three-dimensional coordinate data can be manipulated using specialized software programs (e.g., Morphologika), and then converted it into a "virtual" solid object whose shape can be visualized and compared. These virtual surfaces are compared using a variety of analytical methods, some of which (e.g., thin-plate spline analysis) were developed by engineers for measuring and comparing the complex three-dimensional shapes of machinery components. Morphology is also used to refer to the scientific study of form as in "the morphological sciences" (syn. anatomy, phenotype).

morphometrics (Gk *morph*=form and *metron*=measurement) Morphometrics is literally the study of **form**, and although form comprises **size** and **shape**, most use of morphometric methods in paleoanthropology focuses on the effective capture and comparison of the shape of complex objects (e.g., teeth, cranium, pelvic bone, proximal end of the femur, carpal, and tarsal bones). Morphometric methods can be divided into traditional morphometrics (i.e., non-geometric) and **geometric morphometrics**. Traditional morphometric methods typically apply bi- and multivariate analytical statistical techniques to a range of measurements, such as distances and distance ratios, angles, areas, and volumes. These methods were and still are valuable, but a common disadvantage is that distance measurements do not preserve the geometric properties of the object, they do not allow those geometric relationships to be reconstructed, nor can such data be subjected to statistical analysis before separating shape information from overall size. In the 1980s several innovations (e.g., coordinate-based analytical methods, the introduction of more sophisticated shape statistics, and the computational power to manipulate deformation grids) meant that researchers could explore and visualize large high-dimensional data sets and take advantage of exact statistical tests based on **resampling**. This new approach to morphometry is referred to as geometric morphometrics because it preserves the original geometry of the measured objects during all phases of the analyses. Geometric approaches used in paleoanthropology include **Procrustes analysis** (Bookstein 1996, Dryden and Mardia 1998), **Euclidean distance matrix analysis** (Lele and Richtsmeier 1991, 2001), and elliptic Fourier analysis (Lestrel 1982). See Rohlf (2000a, 2000b, 2003) for a review and for comparisons of these methods.

morphospecies Cain (1954) defines a morphospecies as a species group defined on the basis

of morphology, and suggests that "all (or very nearly all) species founded upon fossils will of necessity be morphospecies" (*ibid*, p. 51). It is, in effect, the type of species most paleoanthropologists try to recognize, but nevertheless it still leaves moot how much variation can be subsumed within a single morphospecies before a researcher should contemplate establishing another morphospecies to accommodate the "overflow" variation. *See also* **species**.

mortality The number of deaths per unit of time in a population, scaled to the size of that population. Mortality rate is calculated as number of deaths per 1000 individuals in a given year. Mortality rates, and the pressures that produce these figures, are of central importance in theories that attempt to explain life history strategies in modern humans and nonhuman primates. For example, low mortality rates relax the pressure on reaching reproductive age early and allow for ontogenies to move more slowly, especially during the juvenile period. Juvenile mortality and adult mortality each exert their own influence on fitness across a wide range of life history strategies.

mortality rate doubling time The **mortality** rate doubling time is the amount of time (usually measured in years) for the hazard of death to double from its initial value in a **Gompertz hazard model**. Starting from the hazard function, we have $h(0)=\alpha$ and $h(a)=\alpha e^{\beta a}$, so that $h(a)/h(0)=2=e^{\beta a}$. Taking the natural logarithms of both sides and solving for t gives $\ln(2)/\beta$ as the mortality rate doubling time. As an example, we can use $\alpha=0.001$ and $\beta=0.08666434$, so that mortality rate doubling time is $\ln(2)/0.08666434=8$ years. This value of a mortality rate doubling time of 8 years has been noted as a general pattern for modern humans (Finch et al. 1990).

mosaic The presence of cells with two different **genotype**s in the same individual. For example, this can occur because of a mutation during development that is then found in only a subset of the cells in the body, or when more than one fertilized zygote fused early in embryonal development (chimerism). In primates, marmoset chimerism occurs as a result of frequent twinning. Twins *in utero* can exchange stem cell lines by an interconnected lattice of blood vessels attached to the same placenta (Benirshke et al. 1962). The term mosaic evolution is also applied to morphological changes that occur in stages as opposed to a synchronic package (e.g., different rates of brain, locomotor, or dental evolution).

mosaic evolution (L. *mosaicus*=of the Muses. A term applied to a decoration made of many small pieces) The term introduced by Gavin de Beer (1954) to describe what William King Gregory and Robert **Broom** had earlier described as "palimpsest evolution." Based on his study of *Archaeopteryx* and the origin of birds, de Beer pointed out that organisms typically combine a mix of evolutionarily older and more recent traits that have evolved at varying rates at different times. They thus present a combination, or mosaic, of "primitive" (plesiomorphic) and "advanced" or "derived" (apomorphic) characters. The mosaic contrasts often correspond to functionally distinct systems (masticatory, locomotor, nervous system, etc.), with those reflecting recent or current major adaptive shifts displaying the fastest rates of change and being the most derived conditions at any given time. However, the need to maintain the overall integrity of the entire phenotype constrains the pace and magnitude of change in any given trait or character complex. The hominin fossil record has been cited as a particularly clear case of mosaic evolution. *See also* **palimpsest evolution**.

mosaic habitat (Gk *mouseion*=a picture or design made up of small components, and L. *habitare*=to inhabit) A mosaic habitat (or habitat mosaic) is one in which a range of different habitat types are scattered across, and interspersed within, a given area. The paleohabitats of many Miocene, Pliocene, and Pleistocene hominin sites have been reconstructed as mosaics. In some, if not all, cases these could be accurate representations of regional habitats, since a great deal of variation in habitat can be present in a relatively small geographic area (Andrews 2006). Examples of the latter include modern Maputaland (Van Wyk 1994) and the Okavango Delta (Ramberg et al. 2006) in southern Africa. However, taphonomic processes are likely to have altered the assemblages upon which the reconstructions are based, resulting in "time-averaged" and "space-averaged" reconstructions that result in a mosaic habitat signal in the absence of a true mosaic habitat. The possibility of time and space averaging should always be considered when faced with paleoehabitat reconstructions indicating mosaics. In these cases, statistical techniques such as **ordination** and **rarefaction** may yield valuable information about assemblage biases (Andrews 2006). Modern evidence also shows that although habitats can be highly variable in a small area, they can also vary over even a short period of time, as has been seen for example at Amboseli in Kenya (Altmann et al. 2002). *See also* **time-averaging**.

Moschidae (Gk *moskhos*=mouse, testicle, musk) The family name of the musk deer, genus *Moschus*. The musk deer, which has no antlers but has a musk gland of economic importance, resembles a chevrotain, and they live in mountainous forest of southern Asia.

Mossgiel (Location 33°15′S, 144°28′E, Australia; etym. named after a nearby town) History and general description A clay pan on Turmbridge Station, near Mossgiel in New South Wales, that yielded a partial hominin skeleton in 1960. Temporal span and how dated? **Radiocarbon dating** of bone carbonate indicates a **Holocene** age *c*.6050 years before present. Hominins found at site The hominin remains were interred at the site in a sitting position and include a robust partial modern human cranium (including a mandible) and fragmentary postcrania. Archeological evidence found at site Abundant **cores**, **scrapers**, and semi-lunar **microliths** were recovered from the surface of the clay pan. Key references: Geology, dating, and paleoenvironment Macintosh 1965; Hominins Macintosh 1965, Freedman 1985; Archeology Macintosh 1965.

most parsimonious tree The most parsimonious tree, also referred as the most parsimonious **cladogram**, is the tree that requires the fewest number of evolutionary events (as recorded by **character state** changes, or steps) to account for the distribution of similarities among a group of taxa. The recovery of the most parsimonious tree (or trees) is the primary goal of **cladistic analysis**. *See* **cladistic analysis**.

most recent common ancestor The ancestor of a group of species that emerged closest to the present. Because evolution involves species splitting and then their descendents splitting in turn, any group of species will share multiple ancestors that will have emerged at different times. The most recent common ancestor is the one whose emergence date is latest in time. As an example, consider modern humans and orangutans. Even if we go no further back than the origin of the primates, modern humans and orangutans have at least three ancestors in common: the ancestor of all the haplorhine primates, that of all the hominoid primates, and that of all the great apes. Of these ancestors, the most recent was the ancestor of all the great apes. Thus, the most recent common ancestor of modern humans and orangutans is the ancestor of all the great apes. *See also* **cladistic analysis**; **coalescent time**.

motion analysis   The capture and analysis of quantitative information about movement, usually undertaken with the goal of studying functional morphology. *See also* **kinematics**.

motor imitation   *See* **imitation**.

motor speech areas   *See* **Broca's area**.

Moula-Guercy   *See* **Abri Moula-Guercy**.

Mount Carmel   A mountainous region on the Mediterranean coast of modern-day Israel contains several historic and prehistoric sites. One valley on its southwestern slope, called **Wadi el-Mughara** in Arabic and *Nahal Me'arot* in Hebrew (meaning "stream of caves" in both cases), is the location of the Middle and Late Pleistocene sites of **el-Wad**, **Tabun**, and **Skhul**, all of which were originally explored by Dorothy **Garrod** and her team in the 1930s and 1940s. Tabun and Skhul, which are, respectively, *Homo neanderthalensis* and anatomically modern human sites, are close both geographically (no more than 100 m apart on the same outcrop) and temporally (both estimated to be between 100 and 130 ka). Another site on Mount Carmel, **Kebara**, has yielded an exceptionally complete postcranial skeleton of *H. neanderthalensis*. *See also* **Garrod, Dorothy Annie Elizabeth (1892–1968)**.

Mousterian   (etym. After the site of **Le Moustier**) A technocomplex found in the European and Near Eastern record from sites dating to the **Middle Paleolithic**, which is characterized by the use of the **Levallois** technique and a high proportion of scrapers. It is associated with *Homo neanderthalensis* in Europe, but in the Near East it is found in association with both Neanderthals (e.g., **Tabun**) and with anatomically modern human fossils (e.g., **Skhul**). French archeologist François Bordes (1961a, 1961b) defined 63 different types of Mousterian stone tool and suggested that different layers within sites had different percentages of each type. He noted four distinctive variants, or facies, of the Mousterian: the Mousterian of Acheulean Tradition, which included small handaxes or backed knives; the Typical Mousterian, which was dominated by sidescrapers; the Denticulate Mousterian, which had a high proportion of notched or toothed flakes; and the Quina-Ferrassie (Charentian) Mousterian, which had steep retouch or was made predominantly on **Levallois** flakes. He argued that each of these facies represented different tribes or cultural groups, who made stone tools based on a cultural tradition. Lewis and Sally **Binford** (1966) agreed broadly with Bordes' tool typological distinctions, but argued that each tool type had a specific function, so that the different facies represented differences in behavioral demands at the various sites, not cultural differences. Harold Dibble and Nicholas Rolland (1992) have argued that Bordes' 63 tool types do not represent the final intent of the maker, and instead represent different points along the reduction sequence of tool making, from a raw flake to a heavily reused and retouched flake. They reject any explanation based on cultural traditions or function, and argue that the differences among the facies reflect only differences in raw material and intensity of reduction.

Movius' Line   The term was apparently introduced by Swartz (1980), but it refers to an observation set out by Hallam Movius in a seminal review entitled *The Lower Paleolithic Cultures of Southern and Eastern Asia* (Movius 1948). At the end of that review Movius points to the demarcation between the "Hand-Axe Culture" seen in Africa, the Near East and much of India, and the "Chopping-Tool Culture" seen in mainland and Southeast Asia (a region he refers to as Southern and Eastern Asia). The absence of **handaxes** in the latter culture struck Movius as significant enough to cause him to write that in "Southern and Eastern Asia…the tools consist for the most part of relatively monotonous and unimaginative assemblages of choppers, chopping-tools, and hand-adze" (*ibid*, p. 411), and he went on to suggest "that as early as Lower Palaeolithic times Southern and Eastern Asia as a whole was a region of cultural retardation" (*ibid*, p. 411). The line of demarcation can be clearly seen in Map 4 (*ibid*, p. 409). Evidence recovered since Movius' time has shown that hominins in eastern Asia could and occasionally did manufacture bifacially flaked, handaxe-like tools, with well-documented examples from the Bose Basin (Schick and Dong 1993, Hou et al. 2000) in China and from the Imjin/Hantan River Basins in Korea (Norton et al. 2006), but their rarity is consistent with the pattern of geographic distribution noted by Movius.

MPFT *See* **Mission Paléoanthropologique Franco-Tchadienne**.

MPT Acronym for the **Mid-Pleistocene transition** (*which see*).

MQ *See* **megadont**.

MRCA Acronym for the most recent common ancestor of a group (e.g., molecules, taxa). *See* **coalescent**; **most recent common ancestor**; **phylogeny**.

MRI Acronym for **magnetic resonance imaging**. *See* **functional magnetic resonance imaging**.

MSA Acronym for the **Middle Stone Age** (*which see*).

MSD Prefix for fossils recovered from the Mesgid Dora collection area, **Woranso-Mille study area**, Afar, Ethiopia.

μCT *See* **micro-computed tomography**.

mudstone A sedimentary rock composed of fine particles of clay and silt size. The term is used to indicate a **lithology** finer than sandstone but of generally indeterminate composition.

Mugharet el 'Aliya (Location 35°45′N, 5°56′W, Morocco; etym. the name means "cave of ascent". This cave site has also been referred to as Dar el 'Aliya, The High Cave, and Tangier) History and general description Carleton **Coon** began the first detailed excavations at the cave in the 1930s, and detailed stratigraphic investigations were directed by B. Howe in 1947. Mugharet el 'Aliya, one of the "Caves of Hercules," is located at Cap (Cape) Ashakar, 11 km/7 miles southwest of Tangier. The cave initially formed as a solution cavity within Upper Pliocene conglomeratic limestone, and during a subsequent high-sea-level phase it was breached by wave action. The Pleistocene sediments consist of five beds of well-sorted windblown fine-to-medium sand derived from beaches formed beneath the cave. The site is notable for possessing stratified **Aterian** assemblages that date to beyond the limits of radiocarbon dating. Temporal span and how dated? **Electron spin resonance spectroscopy dates** on mammalian teeth suggest that the basal Layer 10 dates to 81 ± 9 ka using a linear-uptake model, and 62 ± 6 ka using an early-uptake model, and teeth from Layers 5, 6, and 9 have provided ESR dates ranging from 39 ± 4 to 56 ± 5 ka. These results suggest an age range of 60–35 ka for the archeological sequence, including the Aterian levels in Layers 5 and 6. Hominins found at site Tangier 1, a juvenile maxilla, was recovered by Carleton Coon in 1939 and has been assigned to *Homo sapiens*. Although not recovered *in situ*, it is attributed to the Aterian Layer 5 on the basis of a similar fluorine/phosphate ratio to other fauna from that level. Two isolated teeth (an upper canine and third premolar) are attributed to the maxilla. A third tooth (a worn upper second molar) belongs to a second individual. Archeological evidence found at site The densest artifact accumulations derive from Layers 6 and 5, and are attributed to the Aterian on the basis of abundant bifacial leaf-shaped points, endscrapers, rare pieces with tangs, Levallois cores and flakes, and rare blades removed from opposed platform cores. The faunal remains are incompletely published, but include gazelle, zebra, wild cattle, *Pelorovis*, hartebeest, warthog, golden jackal, and the spotted hyena. The latter likely denned in the cave and may have contributed to the faunal accumulation. Key references: Geology, dating, and paleoenvironment Howe 1967, Wrinn and Rink 2003; Hominins Minugh-Purvis 1993; Archeology Bouzouggar et al. 2002, Howe 1967, Hencken 1948, Wrinn and Rink 2003.

Mugharet el-Kebara *See* **Kebara**.

Mugharet el-Wad *See* **El Wad**.

Mugharet el-Zuttiyeh *See* **Zuttiyeh**.

Mugharet es-Skhul *See* **Skhul**.

Mugharet et-Tabun   *See* **Tabun**.

Muierii 1   Site **Baia de Fier**, Romania. Locality Galeria Musteriană. Surface/*in situ* Surface. Date of discovery 1952. Geological age Direct **AMS radiocarbon** date $30,150 \pm 800$ years BP (on a combined sample from tibia and scapula) and AMS ^{14}C ultrafiltration $29,930 \pm 170$ years BP (on the cranium); both dates are uncalibrated. Developmental age Adult. Presumed sex Female. Brief anatomical description Partial cranium, preserving the face, most of the neurocranium, and part of the basicranium; partial mandible, preserving the right ramus and corpus, broken anteriorly at the level of the canine; partial scapula, preserving glenoid fossa, lateral spine and axillary border up to the cranial end of the teres major attachment, and a tibial fragment. Announcement Gheorghiu and Haas 1954. Initial description Gheorghiu and Haas 1954. Photographs/line drawings and metrical data Soficaru et al. 2006, Trinkaus 2008a. Detailed anatomical description Soficaru et al. 2006, Trinkaus 2008. Initial taxonomic allocation *Homo sapiens*. Informal taxonomic category Anatomically modern human. Significance Among the earliest reliably dated anatomically modern human individuals in Europe. Location of original Museum of Ethnography and Folk Art, Craiova and Speological Institut, Bucharest, Romania.

Muierii 2   Site **Baia de Fier**, Romania. Locality Unknown. Surface/*in situ* Presumed surface. Date of discovery Unknown. Geological age Ultrafiltration **AMS radiocarbon** date $29,110 \pm 190$ years BP (uncalibrated). Developmental age Adult. Presumed sex Unknown. Brief anatomical description Partial temporal bone, missing the squamous portion. Photographs/line drawings and metrical data Soficaru et al. 2006. Detailed anatomical description Soficaru et al. 2006. Initial taxonomic allocation *Homo sapiens*. Informal taxonomic category Anatomically modern human. Significance Among the earliest reliably dated anatomically modern human individuals in Europe. Location of original Museum of Ethnography and Folk Art, Craiova and Speological Institut, Bucharest, Romania.

Muierii 3   Site **Baia de Fier**, Romania. Locality Unknown. Surface/*in situ* Presumed surface. Date of discovery Unknown. Geological age Presumed to be contemporaneous with **Muierii 1** and **2**. Developmental age Adult. Presumed sex Unknown. Brief anatomical description Fibular diaphysis. Initial taxonomic allocation *Homo sapiens*. Informal taxonomic category Anatomically modern human. Significance Possibly among the earliest reliably dated anatomically modern humans in Europe. Location of original Museum of Ethnography and Folk Art, Craiova and Speological Institut, Bucharest, Romania.

multiple regression   A **regression** in which there are multiple independent variables, as opposed to simple regression in which there is a single independent variable. In multiple regression, the coefficient associated with any single independent variable is the change expected in the dependent variable associated with a unit change in the independent variable (i.e., the slope) when all other independent variables are held constant. The ***P* value** associated with each independent variable is the significance associated with adding that variable to a regression model which already contains all of the other independent variables. This can create interpretation problems when two independent variables are highly correlated (e.g., endocranial volume and brain volume). Both variables may be significantly associated with the dependent variable when analyzed by themselves, but in a multiple regression neither variable may have a significant *P* value because adding, for example, brain volume to a regression which already includes endocranial volume improves the fit of the data very little, as is also true for adding endocranial volume to a regression which already includes brain volume. It is useful to check correlations between independent variables before including them in a multiple regression analysis. *See also* **regression**.

multipoint mapping   *See* **linkage analysis**.

multiregional hypothesis   The multiregional hypothesis was formulated by Milford **Wolpoff** of the University of Michigan and Alan **Thorne** of the Australian National University during the early 1980s and elaborated subsequently (Wolpoff, Wu, and Thorne 1984, Wolpoff 1989). It is an interpretation of the human fossil record that emphasized a phenomenon known as "regional continuity," namely what appeared to be morphological continuities in the fossil record of *Homo erectus* and *Homo sapiens*. Specifically, Thorne and Wolpoff (1981) presented evidence that for much of the past million years a distinct morphological signature could be traced through the hominin fossil record in Australasia. Thorne and Wolpoff later expanded their theory to argue that a similar morphological continuity within geographical regions could be observed in the hominin fossil record over much of the Old World. The multiregional hypothesis of Thorne and Wolpoff argued that *H. erectus* was the first hominin to leave Africa, and that a million years ago these migrants dispersed to populate much of Asia and parts of Europe. It stressed that these regional populations of *H. erectus* encountered and adapted to quite different environments, and because of their relative geographic isolation, regional differences began to appear, yet there was sufficient gene exchange between adjacent groups to ensure that the populations continued to remain as one species. Thorne and Wolpoff pointed to several lines of evidence that suggested to them that the distinctive regional morphology seen in *H. erectus* remains from China could be traced via early *H. sapiens* to distinctive features visible in at least some Asian populations of modern humans. The multiregional hypothesis was similar in some ways to other models of human evolution that had emerged earlier in the 20thC. A few researchers, best represented by the German physical anthropologist Hermann Klaatsch, had argued at the turn of the century that the different modern human races had evolved from different hominin ancestors, and that they in turn had evolved from entirely different species of early anthropoid apes. These theories were shaped by the racial anthropology that was prominent during the early 20thC and they were influenced by the 19thC polygenist theories of anthropologists that considered the different modern human races to be distinct biological species with separate origins. Klaatsch's theory, and others like it that derived the different modern human races from distinct hominin ancestors, attracted few supporters. Aleš Hrdlička (1927, 1930) proposed that all humans had evolved from a "**Neanderthal** phase" and considered Neanderthals and other forms of *Homo* then known to be ancestors of living humans. Other key scientists including Sir Arthur **Keith** and Marcellin **Boule** rejected Hrdlička's hypothesis and argued for a recent common ancestor of living humans. The idea that the various modern human races emerged very early in human evolutionary history re-emerged during the 1960s in a model of human evolution proposed by the American physical anthropologist Carleton **Coon**. In 1962 Coon published *The Origin of Races* which examined the question of human evolution and drew upon evidence from the hominin fossil record to argue that the five major modern human races (Caucasoid, Negroid, Capoid, Mongoloid, and Australoid) have existed for more than half a million years. According to Coon, modern human racial diversity was already present in *H. erectus*. Coon's model of human evolution differs from the multiregional model of Thorne and Wolpoff in that Coon believed the various geographical "races" evolved independently from one another, while the multiregional model argues that the distinct geographical populations evolved into modern humans in only relative isolation. In this way Thorne and Wolpoff's theory is similar to a theory proposed in the 1940s by the German physical anthropologist Franz **Weidenreich**. Weidenreich's polycentric theory of human evolution argued that modern humans had evolved from populations of an early representative of the genus *Homo* that lived in various regions of the Old World. According to Weidenreich, populations of early hominins living in relative geographical separation from other populations evolved into modern humans and because these populations were separated geographically racial variations emerged. Thus, there is a morphological continuity through time in any given geographical region, but Weidenreich was adamant that these populations did not evolve in complete isolation from one another because there was always an important amount of interbreeding between these populations that ensured all these geographical "races" still belonged to the same species. Thorne and Wolpoff explicitly acknowledged the similarities between their own theory and the earlier theory of Weidenreich, although they had invoked more recent biological thinking to explain how this process had occurred. In 1997 Wolpoff and Rachel Caspari published a lengthy explication of the multiregional hypothesis in *Race and Human Evolution*, a text in which they explored the connections between this model of human evolution and Weidenreich's polycentric theory. The original "strong" version of the out-of-Africa hypothesis, which argued that modern *H. sapiens* evolved in Africa and that this population migrated out of Africa and replaced *Homo erectus* in Asia and other archaic hominins throughout Eurasia, is incompatible with the multiregional hypothesis. Adherents to the multiregional hypothesis expressed doubts about the genetic research conducted by Allan **Wilson**, Rebecca **Cann**, and Mark Stoneking on **mitochondrial DNA** from different human populations this and much other genetic work in the 1990s suggested that the evidence is consistent with all modern humans being descended from a small population of *H. sapiens* less than 200 ka. By the late 1990s most paleoanthropologists were interpreting the fossil and genetic evidence as incompatible with a "strong" version of the multiregional hypothesis. However, the results of recent analyses of the draft Neanderthal nuclear genome suggests that whereas modern humans from sub-Saharan Africa showed no evidence of any Neanderthal DNA, three modern humans from outside of Africa showed similar amounts (between 1 and 4%) of shared DNA with Neanderthals. These results are compatible with either a deep split within Africa between the population that gave rise to modern humans and a second one that gave rise to present-day non-Africans plus Neanderthals, or with the hypothesis that there was hybridization between Neanderthals and modern humans soon after the latter left Africa.

multivariate analysis *See* multivariate statistics.

multivariate analysis of variance (or MANOVA) A statistical test that extends **analysis of variance** (or ANOVA) to the multivariate case. MANOVA is used to determine whether there is a significant difference in **vectors** containing the means of two or more continuous variables for two or more groups. For example, if humerus length and radius length are known for skeletal samples belonging to three different species, MANOVA can be used to identify whether a significant difference exists between the three species in mean humerus and radius length. A nonsignificant result indicates that the vectors of means cannot be said to differ from each other significantly across all groups. A significant result indicates that the null hypothesis that all mean vectors are equal is incorrect; however, it does not indicate which particular variables differ in their means (e.g., humerus length or radius length in the example above) and when more than two groups are included in the analysis, a significant result does not indicate which group or groups differ from the others. As with ANOVA, pairwise tests can be used to identify where the differences

arise from. There are four widely accepted versions of the MANOVA test statistic (Wilks' lambda, Pillai–Bartlett trace, Lawley–Hotelling trace, Roy's largest root), all of which give similar results. When MANOVA is performed on two groups (as opposed to three or more) all four versions provide the same result which is equivalent to the result of a **Hotelling T^2 test**.

multivariate size Regardless of whether it is the size of a whole organism, or the size of a smaller subregion of an organism, size can be measured univariately or multivariately. Univariate measures of size are single measurements (e.g., the mass of an animal, the volume of the cranial cavity, the surface area of an articular surface, or the height of a canine). Multivariate measures of size typically reduce multiple size measurements to a single value. The four most common methods of doing so use (a) a **geometric mean**, (b) an equation for the size of a known geometric shape, (c) the first principal component of a **principal component analysis** (or PCA), or (d) centroid size. In the first case, the geometric mean of a set of measurements (usually linear) is calculated to generate a single measure of overall size (e.g., proximal femur size as in Richmond and Jungers 1995, Lockwood et al. 1996, Harmon 2006; forelimb and hindlimb size as in Green et al. 2007; overall skeletal size as in Gordon et al. 2008; and overall cranial size as in Coleman 2008). In the second case, a perfect geometric solid is used to approximate the region of interest (e.g., an articular surface is usually represented as the surface of a partial sphere) or limb segment (often represented as a cylinder or other solid of revolution). Equations for the size of interest, often surface area or volume, are used to approximate size from multiple linear measurements (e.g., femoral and humeral head size as in Ruff 1988, 1990, limb segment size as in Crompton et al. 1996 and Raichlen 2004). In the third case, multiple measurements are included in a PCA without any previous size correction, and scores on the first principal component are interpreted as representing size. It should be noted that multivariate measures of size generated through PCA are dependent on the specimens included in a sample and will change for a particular specimen depending on the other specimens that are included in the analysis. In contrast, measures of size using geometric means or size equations for known geometric shapes are sample independent (i.e., the value for a particular specimen does not change, regardless of what other specimens are included in an analysis). Furthermore, the first principal component is almost always a mix of size *and* shape information (Jungers et al. 1995). In **geometric morphometrics** analyses, size is usually calculated as centroid size (i.e., the square root of the sum of squared distances of a set of landmarks from their centroid).

multivariate statistics   A term which can be broadly applied to a wide range of statistical test as long as the test involves three or more **variables**. Multivariate statistics typically consider the interaction of multiple variables together as opposed to individual analyses of each variable. Commonly used multivariate statistical methods include **multiple regression**, **principal components analysis**, and **discriminant function analysis**.

Mumba Shelter   (Location 3°32′S, 35°19′E, northern Tanzania; etym. originally designated as Mumba-Höhle by Margit Kohl-Larsen) History and general description This rock-shelter underlies portions of a gneiss inselberg on the eastern margin of the alkaline and saline Lake Natron. Major excavations were by **Kohl-Larsen** in the 1930s and by Mehlman in the 1970s–1980s. The site remains a key regional sequence because of its *c.*11 m-thick sedimentary sequence and preserved **Middle** and **Later Stone Age** artifacts and hominin remains. Temporal span and how dated? The sequence at Mumba Shelter spans the **Late Pleistocene** to **Holocene**. Basal Middle Stone Age deposits and hominin teeth are associated with mean **uranium-series dating** age estimates of *c.*130 ka and *c.*110 ka. Overlying strata are dated by additional uranium-series analyses of bone, as well as **radiocarbon dating** assays on shell, charcoal, and ostrich eggshell; the latter material is also dated by **amino acid racemization dating**. The pre-ceramic Later Stone Age levels of Bed III are associated with a 29.96 ± 0.76 ka radiocarbon age estimate; even later deposits date to 2.01 ± 0.01 ka. Hominins found at site Three small morphologically modern human molars occur in Bed VI-B at 7.1 m depth, associated with Middle Stone Age artifacts. Archeological evidence found at site The geological sequence is divided into six beds (numbered from bottom to top), with the basal layer (Bed VI-B) overlying bedrock. Bed VI

is characterized by Middle Stone Age artifacts including rare scrapers and points, and flakes, as well as radial and **Levallois** cores. Locally available quartz and chert are the dominant raw materials throughout the sequence, but obsidian flakes and tools from Bed VI apparently derive from sources in the central Kenya Rift, more than 300 km/185 miles away. Bed V, with an estimated age of between 31 and 67 ka, contains large **backed** pieces (crescents or knives), bifacially and unifacially worked retouched points, small scrapers, and bipolar and Levallois cores, as well as **ochre** and ostrich eggshell beads. These beads have been considered intrusive from Bed III, a hypothesis supported by direct **AMS radiocarbon** age estimates of between 22 and 33 ka and one of two direct amino acid racemization assays. However, a second amino acid racemization result suggests an age of 52 ka for these artifacts, consistent with other age estimates for Bed V. Overlying deposits from the lower portion of Bed III include stone artifacts produced by **bipolar percussion** with rare radial or Levallois core types, and retouched pieces are scarce. Fauna is well preserved in Beds III and V, and include diverse **suids**, **bovids**, hippopotamus, and crocodile, as well as a fish and birds, and other taxa. Above and beneath Bed IV are dense **midden** deposits of land snail (*Achatina* sp.) shells and shell fragments. Key references: Geology, dating, and paleoenvironment Mehlman 1991, McBrearty and Brooks 2000, Conard 2007; Hominins Bräuer and Mehlman 1988; Archeology Mehlman 1979, 1991, McBrearty and Brooks 2000, Prendergast et al. 2007, Diez-Martin et al. 2009.

Mumbwa Caves   (Location 15°01′S, 26°59′E in eastern Zambia; etym. named after local village) History and general description Mumbwa Caves consists of multiple sediment-filled entrances formed in a freestanding dolomite outcrop. The site has been excavated a number of times, in the 1920s–1930s by F.B. Macrae, N del Grande, Raymond **Dart**, and Desmond **Clark**; Lawrence Barham renewed fieldwork there in the 1990s. Excavation of what in places is more than 7 m of sediment overlying bedrock has exposed an archeological sequence spanning **Holocene Later Stone Age** strata to **Later** and **Middle Pleistocene** ones including an example of a windbreak, one of the earliest evidence for hominin habitation structures. Possible **Acheulean** handaxes were reported from earlier excavations but their presence has not been confirmed by more recent work. Temporal span and how dated? The Mumbwa Caves' sedimentary and archeological sequence spans from more than 170 ka to the present, based primarily on numerous **thermoluminescence dating** estimates on burnt quartz and calcite, more limited **electron spin resonance spectroscopy dating** of teeth, and **radiocarbon dating** assays of charcoal. Hominins found at site Middle and Later Stone Age strata produced six isolated teeth or tooth fragments, and fragments of two femoral shafts, two radii, and a proximal pedal phalanx. Archeological evidence found at site The more than 200,000 stone artifacts show flake production from primarily single and multiple platform cores as well as radial cores (including **Levallois** examples); most made from stone sources less than 10 km/6 miles from the site. Retouched pieces include unifacially and bifacially flaked points as well as some of the earliest backed pieces (segments), and there are also multiple examples of polished and pointed **bone tools**. Several hundred pieces of modified specularite and hematite possibly used as a colorant were imported into the caves. Other features include multiple hearths and a roughly circular area interpreted as the remains of a windbreak. The nonhominin faunal evidence from Mumbwa Caves is sparse, highly fragmented, and bovid-dominated. Taxonomic diversity is comparable to modern or historic fauna in the region, and varies little throughout the sequence, although complementary phytolith analyses suggest some fluctuation between grasslands and wooded grasslands. The majority of the macromammal fauna was hominin- (and to a much lesser extent, hyena-) accumulated; barn owls likely contributed most of the **micromammals**. Key references: Geology, dating, and paleoenvironment Barham 2000; Hominins Barham 2000; Archeology Barham 2000.

Mursi Formation (etym. Named after the tribe occupying the area) General description The oldest of the four formations comprised in the **Omo Group, Lower Omo Valley**, Ethiopia. It was explored in the late 1960s and early 1970s by the **International Omo Research Expedition**. The exposed sedimentary sequence of the formation is located on the western side of the southern half the Nkalabong range, on the eastern side of the Omo River. The Mursi Formation is generally divided in three sedimentary members topped by the Mursi basalt; the latter is likely part of the larger Gombe Group basalts found in southern Ethiopia and northern Kenya. The sedimentary sequence, which consists of deltaic and fluvio-littoral deposits, was slightly tilted towards the west and faulted after deposition. No hominins have been found in the formation. The mammalian and non-mammalian fossils found include suids, bovids, elephantids, hippopotamids, deinotheriids, rhinocerotids, equids, turtles, crocodiles, fishes, and molluscs, with suids making more than 50% of the mammalian assemblage. Sites where this formation is exposed Yellow Sands, Cholo (discovered in 2009). Geological age **Potassium-argon dating** of the Mursi basalt that caps the sedimentary members suggests an age of *c.*4 Ma. Key references: Geology, dating, and paleoenvironment Butzer 1976a, de Heinzelin 1983, Brown 1994.

muscles of mastication *See* mastication, muscles of.

Musée de l'Homme Founded in 1937 by Paul Rivet for the Exposition International des Arts et Techniques dans la Vie Modern (International Exhibition of Arts and Techniques of Modern Life) this museum in central Paris, France, was designed to bring together all the aspects that define humanity, including evolution, present-day unity and variety, and cultural and social expression. It is part of the Muséum national d'histoire naturelle (National Museum of Natural History). In 2006, the ethnographic collection was moved to the newly created Musée du quai Branly, leaving only the physical anthropology and evolutionary history collections at the Muséum national d'histoire naturelle. This move that has garnered considerable debate, with the detractors claiming that the Musée du quai Branly is designed to display only those artifacts that are esthetically pleasing regardless of their scientific value. The Musée de l'Homme was closed for renovations in March 2009 and is expected to reopen in 2012. Hominin fossils and related collections Many of the fossils recovered from French excavations during the 19thC and 20thC including many (but not all) of the remains from **Abri Pataud, Arcy-sur-Cure, Aurignac, La Chapelle-aux-Saints, Cro-Magnon, La Ferrassie, Fontéchevade, La Madeleine, Malarnaud, Montmaurin,** and **La Quina**.

Postal address Palais de Chaillot, 17 place du Trocadéro, 75116 Paris cedex, France. Website www.museedelhomme.fr.

Musée Royal de l'Afrique Centrale *See* **Royal Museum for Central Africa**.

Museum für Naturkunde Formerly known as the Museum Alexander Humboldt, the Museum für Naturkunde in Berlin, Germany, was established in 1814 when the Humboldt University of Berlin acquired the mineralogy collections of the Berlin Mining Academy collections in 1814. During the 19thC new departments of paleontology and zoology were added, and in 1889 a new building on Invalidenstraße was constructed for the collections. The first Curator of Mammals was Paul Matschie, and his tenure coincided with the time when large collections of animals (and plants, minerals, and cultural items) were coming in from German colonies in Africa and elsewhere. The Museum für Naturkunde has large collections of skulls of gorillas and chimpanzees (100 or more of each), all of them from precisely known localities. Matschie described a large number of new species and subspecies, most of which are without substance, but the collections do include the **type specimens** of *Gorilla beringei*, *Gorilla graueri*, and *Gorilla diehli*. Contrary to rumor, the collections at the Museum für Naturkunde were not destroyed during WWII, although one wing of the museum was partly destroyed, but the museum did become part of the Eastern sector of Berlin, just to the east of the Berlin Wall. At the beginning of 2009, the museum separated from the university and is now known as the Leibniz Institute for Evolution and Biodiversity Research. Contact person Professor Frieder Mayer (e-mail frieder.mayer@mfn-berlin.de). Postal address Museum für Naturkunde, Invalidenstrasse 43, 10115 Berlin, Germany. Tel +49 30 2093-8591. Website www.naturkundemuseum-berlin.de.

mustelid The informal name for the **Mustelidae** (weasels and otters), one of the **caniform** families whose members are found at some hominin sites. *See also* **Carnivora**.

Mustelidae The family that includes weasels, otters, and their close relatives, and one of the **caniform** families whose members are found at some hominin sites. *See also* **Carnivora**.

mutation A change in the sequence of **nucleotides** in a **DNA** molecule. Mutation is one of the four forces of evolution and the only force that creates new variation. Mutations are caused by an error in DNA replication or through damage to the DNA. There are various types of mutation: (a) **point mutations** involve a change from one nucleotide to another, (b) inversions occur where a sequence is inverted by 180°, (c) translocations or transpositions are the terms used when a sequence of nucleotides is moved to another location, and (d) **insertion/deletion** (or indel) mutations are where one or more nucleotides are deleted or added to a DNA sequence. In coding regions, the latter may cause a frameshift in the reading of the nucleotide sequence and frameshift mutations are typically highly deleterious. Point mutations can be transitions, where one purine (adenine or guanine) is replaced by another purine or where one pyrimidine (cytocine or thymine) is replaced by another pyrimidine. They can also be transversions, where a purine is replaced by a pyrimidine (or vice versa); in modern humans transitions are more common than transversions. Point mutations in coding regions may be synonymous (also called silent) when they do not cause a change in the protein sequence or nonsynonymous. Nonsynonymous point mutations are called nonsense mutations when they involve a change to a stop codon, or missense mutations when a different amino acid is coded for. Another class of mutations involves the number of **chromosomes**. Mutations that result in changes in the numbers of the whole complement of chromosomes are not viable in humans, but other changes in chromosome number are. Aneuploidy, an abnormal number of chromosomes, can result from **monosomy** (e.g., monosomy X, known as Turner syndrome) or trisomy (e.g., trisomy 21, known as Down syndrome). Chromosomal aberrations such as centric fusion can lead to the evolution of chromsome number and organization. For instance, sometime during hominin evolution, the chromosomes 2a and 2b found in the other great apes fused to form the single modern human chromosome 2. *See also* **diploid**; **haploid**; **karyotype**.

mutation accumulation hypothesis One of several theories put forward to explain the evolution of **senescence**. The mutation accumulation hypothesis (Medawar 1952) suggests that senescence is due to a relaxation of the effects of selection in post-prime reproductive years. Since selection is expected to improve function in the young and in early adult life more than in the later phases of life, Medawar hypothesized that **allele**s with deleterious effects that are manifest only in the later stages of life will be balanced (between an increase in frequency due to mutation and a decrease in frequency due to selection) at higher frequencies than harmful alleles that are manifest earlier in life.

mutation rate   The occurrence of a **mutation** per generation or unit of time. Mutations are stochastic events but over longer periods of time they occur at a measurable rate. This rate differs among regions of the **genome** (for example, the mutation rate is faster in **mitochondrial DNA** than in **nuclear DNA**) and among species. Reasons for differences in rate can include the guanine/cytosine content of the DNA, differences in editing capability of the polymerase that replicates the DNA, and secondary structure of DNA in a specific region.

m.y.   *See* **myr**.

myelin   (Gk *muelos*=marrow) A fatty substance that surrounds and insulates the axons of some neurons. Myelin is arranged in concentric layers around the axon of the neuron and it is essential for the high-speed conduction of electrical currents across the cellular membrane. The myelin sheath is formed by the glia cells that are adjacent to the axon. In the central nervous system a single oligo-dendrocyte can myelinate several axons, whereas in the peripheral nervous system a single Schwann cell myelinates only one axon. The myelin sheath is interrupted at regular intervals by nodes of Ranvier (named after the neuroanatomist Louis Antoine Ranvier who first described them in the 19thC). These gaps are characterized by clusters of voltage-gated sodium and potassium ion channels, which enable the electrical signal to propagate by depolarization of the axon's plasma membrane. Nodes of Ranvier significantly increase the rate of conduction along an axon.

mylohyoid line   *See* **mandible**.

myr   The abbreviation for a period of time expressed in millions of years, as in "2 myr have elapsed since hominins first appeared in Africa." The result of an age determination, or **age estimate**, should be expressed differently, as in "the approximate age of KNM-ER 3733 is 1.8 **Ma**." *See also* **age estimate**.

N

15N/14N Two stable isotopes of nitrogen. The relative ratio of $^{15}N/^{14}N$ (commonly represented as $\delta^{15}N$ values in per mil, or ‰, notation) in an animal's bone collagen or fur keratin reflects the foods it consumed and where it consumed them. In most terrestrial ecosystems, plants obtain their nitrogen directly from soil nitrate and ammonium, resulting in nitrogen isotope ratios over 0‰ (e.g., Schmidt and Stewart 2003), but some plants, in particular many legumes, have symbiotic bacteria that fix nitrogen directly from the atmosphere. As a result, these plants have $\delta^{15}N$ values that resemble air (0‰) (e.g., Ambrose 1991, Schmidt and Stewart 2003). Among C_3 plants, factors such as rainfall, temperature, topography, plant part, rooting depth, myccorhizal associations, soil age, grain size, mineral content, chemical composition, moisture, and the available form of nitrogen can influence plant $\delta^{15}N$ values (e.g., Heaton 1987, Ambrose 1991, Schmidt and Stewart 2003). Nitrogen isotope ratios in animal tissues reflect those in the plants that they consume, with some trophic enrichment. Trophic enrichment values can vary considerably. Nevertheless, $\delta^{15}N$ values in herbivores are generally thought to be approximately 3‰ higher than those in the plants that they consume, and $\delta^{15}N$ values in secondary carnivores are about 3‰ higher than those in herbivores (6‰ higher than those in plants) (DeNiro and Epstein 1981). In addition to trophic enrichment, consumer $\delta^{15}N$ values increase with water and nutritional stress (Ambrose 1991). Thus while elevated $\delta^{15}N$ values tend to indicate higher trophic levels, they can also reflect nutrition or moisture stress (Heaton 1987, Ambrose 1991). It is less clear how accurately omnivory (i.e., some meat in the diet) is recorded although because meat contains relatively more nitrogen per unit weight than does the majority of plant material, meat intake should be reflected in higher values of bone collagen $\delta^{15}N$ than those found in plants (Phillips and Koch 2002, Schoeninger 1985). Marine animals exhibit higher $\delta^{15}N$ values in their bone collagen and other tissues than do terrestrial ones (Schoeninger and DeNiro 1984), and modern human diets that include marine foods have higher $\delta^{15}N$ values than those with only terrestrial foods. Further, marine diets high in secondary carnivores (e.g., carnivorous mammals and fish) have higher $\delta^{15}N$ values than do diets with marine invertebrates due to trophic level effects (e.g., Schoeninger et al. 1983, Little and Schoeninger 1995). In regions where prehistoric modern human populations had access to both marine and maize, which confound the use of $\delta^{13}C$ values, bone collagen $\delta^{15}N$ values identify dependence on marine foods (Schoeninger et al. 1990). Fuller et al. (2006) show how to use the $^{15}N/^{14}N$ (and the $^{13}C/^{12}C$ system) systems to investigate the timing of **weaning**. Richards et al. (2000) suggest the enriched levels of ^{15}N in *Homo neanderthalensis* indicate a diet that included a significant component of carnivory, even though some of the energy needs of *H. neanderthalensis* may have been met by low-protein foods.

Nachukui (or NC) (etym. Named for an ephemeral river system, or laga, in whose drainage the localities are located) Refers to a complex of fossil localities in the Nariokotome Member of the Nachukui Formation within the Nachukui drainage in **West Turkana**, Kenya. An early hominin proximal femur, **KNM-WT 19700**, attributed to *Homo erectus*, was recovered from locality NC1. *See also* **West Turkana**.

Nacurrie (Location 35°10'S, 143°38'E, Australia; etym. named after a local railway siding near the find site) History and general description A site near Merran Creek about 19 km/12 miles north of Swan Hill in southern New South Wales where George Murray Black collected a series of hominin skeletal remains in 1949. The better-preserved specimens were added to the G.M. Black Collection in the Department of Anatomy, Melbourne University. In 1971 the most complete skeleton was sent for cleaning, restoration, and analysis to Professor

Wiley-Blackwell Encyclopedia of Human Evolution, First Edition. Edited by Bernard Wood.
© 2013 Blackwell Publishing Ltd. Published 2013 by Blackwell Publishing Ltd.

N.W.G. Macintosh ("Black Mac") at Sydney University, Macintosh and Larnach refer to it as the "Murrabit skull." After Macintosh's death the skeleton was returned to Melbourne (now referred to as the "Nacurrie skeleton") but it was finally analyzed at the University of New England by Peter Brown. In 1984 the Nacurrie skeleton along with the rest of the G.M. Black Collection was repatriated for reburial. Temporal span and how dated? AMS radiocarbon dating on collagen from the skeleton provides a Late Pleistocene age of c.11.4 ka. Hominins found at site The "Nacurrie skeleton" was a tall and robust adult modern human male with evidence of artificial cranial deformation similar to that seen in individuals from other terminal Pleistocene sites including Kow Swamp, Cohuna, and Coobool Crossing. The Nacurrie skeleton had healed fractures affecting both forearms the right distal fibula and the right parietal, which have been interpreted as ancient evidence of the Aboriginal practice of dispute settlement with the striking of sticks. The shape of the cranial vault is consistent with manual molding. Archeological evidence found at site None. Key references: Geology, dating, and paleoenvironment Brown 1994; Hominins Brown 1994, 2010.

Nadung'a (Location 4°08′N, 35°50′E, Kenya; etym. named for the major river system (laga) near which it is located) History and general description Complex of localities within the Nariokotome Member of the Nachukui Formation in the vicinity of the Nadung'a drainage in **West Turkana**, Kenya. Initial exploration in 1988 was conducted by the West Turkana Archaeological Project (or WTAP), with more extensive excavations undertaken in 2002–4. Temporal span and how dated? The archeological sites are located within the middle of the Nariokotome Member, 17–20 m above the 1.33 Ma Lower Nariokotome Tuff. Stratigraphic correlation with the **Shungura Formation** suggests the youngest site within the Nadung'a complex is slightly above a tuff correlated with the 0.74 Ma Silbo Tuff of the Shungura Formation. Thus, an age range of 1.33–0.7 Ma has been proposed for Nadung'a. Hominins found at site None. Archeological evidence found at site It is debated whether the lithic artifacts recovered from Nadung'a 4 (NAD 4) are an atypical example of the Acheulian industrial complex, or a sample of a different technocomplex contemporaneous with the **Acheulean**. Key references: Geology, dating, and paleoenvironment Delagnes

et al. 2006; Hominins N/A; Archeology Kibunjia et al. 1992, Delagnes et al. 2006.

Nahal Ein Gev *See* **Ein Gev**.

Naiyena Engol (or NY) (etym. Named for an ephemeral river system, or laga, in whose drainage the localities are located) Refers to a complex of fossil localities in the Kaitio Member of the Nachukui Formation within the Nanyangakipi drainage in **West Turkana**, Kenya. *See also* **West Turkana**.

Nanjing Man *See* **Tangshan Huludong**.

Nanyangakipi (or NN) (etym. Named for an ephemeral river system, or laga, in whose drainage the localities are located) (or NN) Refers to a complex of fossil localities in the Kaitio and Natoo Members of the Nachukui Formation within the Nanyangakipi drainage in **West Turkana**, Kenya. *See also* **West Turkana**.

Napier, John Russell (1917–87) John Napier was born in Old Windsor near London, England, and attended Canford School. He studied medicine at the Medical College of St Bartholomew's Hospital, graduating in 1943. He specialized in orthopedic surgery, but his clinical and research interests soon focused on peripheral nerve injuries. He moved to the Anatomy Department at what was then called the London School of Medicine for Women (later known as the Royal Free Hospital School of Medicine) in 1946 and apart from a year as a Visiting Professor at Iowa State University and an interlude at the **National Museum of Natural History, Smithsonian Institution**, in Washington DC, USA, Napier spent the rest of his career within the University of London, latterly at Queen Elizabeth College, and immediately prior to his retirement as a Visiting Professor at Birkbeck College. His interest in the way neurological disorders impaired the function of the hands and feet of modern humans was soon translated into an interest in primate functional morphology, especially gait and dexterity. This interest in the hand led to Napier being invited by Sir Wilfrid **Le Gros Clark** to describe and analyze the *Proconsul africanus* forelimb remains that had been recovered by Louis and Mary **Leakey** on Rusinga Island; this he did with the help of Peter Davis (Napier and Davis 1959). It was natural therefore that when the Leakeys recovered early hominin limb bones from **Olduvai Gorge** in the 1960s they and Phillip **Tobias**

turned to John Napier for help with the analysis of the fossils. Napier focused on the hand (**OH 7**; Leakey et al. 1964) and he recruited two colleagues at the Royal Free, Michael **Day** and Peter Davis, to tackle the foot (**OH 8**; Day and Napier 1964) and the leg bones (**OH 35**; Davis 1964). John Napier was one of the first people to review the fossil evidence for the evolution of bipedal walking in early hominins (Napier 1964), and his joint authorship of the proceedings of a symposium devoted to primates (Napier and Barnicot 1963), and later the publication of *A Handbook of Living Primates* (1967), which he jointly wrote with his wife Prue Napier, were landmarks in development of primatology. *See also* **SK 82**.

Nariokotome (etym. Named for an ephemeral river system, or laga, in whose drainage the localities are located) (or NK) Refers to a complex of fossil localities in the Kaitio and Natoo Members of the **Nachukui Formation** within the Nanyangakipi drainage in **West Turkana**, Kenya. An early hominin associated skeleton, **KNM-WT 15000**, attributed to *Homo erectus*, was recovered from locality Nariokotome III. *See also* **West Turkana**.

"Nariokotome boy" *See* KNM-WT 15000.

Narmada *See* Hathnora.

Narmada Basin The Narmada Basin is the central part of the **Narmada Valley**. The sedimentary rocks exposed in the Narmada Basin are from oldest to youngest the Pilikarar, Dhansi, Surajkund, and the Baneta Formations (see Patnaik et al. 2009 for a review of the geology). There are several paleontological and archeological localities at Hathnora: four are in the Surajkund Formation (Surajkund 1, 2, and 3, and **Hathnora 1**) and one is in the Baneta Formation (Hathnora 2). *See also* **GSI-Narmada**; **Hathnora 1**.

Narmada Valley The Narmada River rises in Central India at Amarkantak and then flows westward for 1300 km/806 miles to the Arabian Sea at the Gulf of Cambay. The central part of the Narmada Valley, the **Narmada Basin**, is the location of several localities (e.g., **Hathnora**) that have yielded fossil hominin and archeological evidence (see Patnaik et al. 2009 for a review of the history of exploration in this region) *See also* **Hathnora**.

narrow-sense heritability Narrow-sense heritability (h or h^2) is defined as the **additive** genetic variance divided by the **phenotypic variance** where the additive genetic variance is the weighted average of the squared values of the additive effects. In other words, the additive variance is variance caused by differences among **alleles** only. Thus, narrow-sense heritability only refers to the portion of the phenotypic variation that is additive (or allelic) and does not take into account variation due to epistatic, dominance, or parental effects. However, the additive effects are important because this variation is the only variation that natural selection can act on. Narrow-sense heritability estimates are typically used in agricultural genetics to estimate how the mean of a trait of interest (such as milk production or size) will change in the next generation given how much the mean of the selected parents differs from the mean of the population. *See also* **heritability**.

Nasechebun (etym. Named for an ephemeral river system, or laga, in whose drainage the localities are located) (or NS) A complex of fossil localities in the Kataboi Member of the Nachukui Formation within the Nasechebun drainage in **West Turkana**, Kenya. *See also* **West Turkana**.

nasoalveolar clivus The bone between the inferior margin of the nasal, or piriform, aperture and the alveolar process in the region of the upper incisors. The clivus can be flat, concave, or convex, and vertical or inclined. The amount of demarcation between the floor of the nose and the nasoalveolar clivus varies within and among the great apes, whereas in modern humans and most pre-modern *Homo* a distinct sill separates the nasal floor from the nasoalveolar clivus. Among extinct hominins the distinctiveness of the nasal sill varies both within and between taxa. In *Australopithecus africanus* the nasoalveolar clivus and the inferior part of the **anterior pillars** are in the same plane, and they form a unitary structure, the **nasoalveolar triangular frame**, that stands out from the rest of the face.

nasoalveolar gutter Term used by Rak (1983) for the depressed upper part of the **nasoalveolar clivus** in *Paranthropus robustus* and *Paranthropus boisei*. In *P. robustus* the nasoalveolar clivus is scooped out as it passes up to the inferior margin of the anterior nasal or piriform aperture, which is set below the surface of the face between the inferior ends of the **anterior pillars**. In *P. boisei* a similar appearance is due to the plate-like infraorbital region (*see* **maxillary visor**) extending anteriorly ahead of the plane of the anterior nasal aperture.

nasoalveolar triangular frame   Term used by Rak (1983) for the combination of the **anterior pillars** and the flat **nasoalveolar clivus** in *Australopithecus africanus*. Because these structures are in the same plane they form a unitary structure that stands out from the rest of the face. In *Au. africanus* the lateral margins of the nasoalveolar triangular frame are separated from the rest of the face by the **maxillary furrows**.

nasomaxillary basin   Term used by Rak (1983) for the hollowed area between the lateral edge of the nasal bones and the medial end of the inferior margin of the orbit. It is well seen in **OH 5**.

nasopharynx   *See* pharynx.

NAT   Abbreviation of the Natoo Member in the Nachukui Formation, **West Turkana**, Kenya.

natal group   See philopatry.

Nationaal Natuurhistorisch Museum   *See* Rijksmuseum van Natuurlijke Historie.

National Geological Survey of China   (or NGS) In effect beginning in 1913 with the creation of the Geological School of Peking, but founded officially in 1916 by the Ministry of Agriculture and Mines at Peiping. Its purpose was to promote knowledge of mineral resources, carry out a general geologic survey of the country using a uniform scale, and undertake scientific geological research. It was in the process of the survey that the caves at Choukoutien were discovered. The NGS was responsible for the publication of *Palaeontologia Sinica*, the *Bulletin of the Geological Survey of China*, and the *Memoirs of the Geological Society of China*, which contain descriptions of the fossil hominins recovered from what was then called Choukoutien and which is now called **Zhoukoudian**.

National Museum and House of Culture, Dar es Salaam   One of the constituent museums of the National Museum of Tanzania system. Formerly known as the National Museum – Dar es Salaam, the National Museum and House of Culture is the largest museum in Tanzania with collections and exhibitions covering archeology, arts, biology, ethnography, and history. The original building was constructed in 1938–9 when it was known as the King George V Memorial Museum. After Tanganyika's independence in 1961 it became known as the National Museum of Tanzania, and it was extended following the presentation of **OH 5** to President Julius Nyerere by Louis **Leakey**, on January 26, 1965. The collections relevant to human evolution include some of the hominins from **Olduvai Gorge**, the hominins from **Ndutu** and **Peninj**, the non-hominin fossils collected by Mary **Leakey** and Terry Harrison at **Laetoli** and by the Wembere Manonga Paleontological Expedition (or WMPE), and the archeological evidence collected by Neville Chittick, Felix Chami, Amandus Kweka, and others. There are plans to provide suitable facilities that will allow the remaining Olduvai Gorge and the Laetoli hominins to be transferred from the **National Museums of Kenya**, as well as other collections from Tanzanian sites that are in museums outside of Tanzania. <u>Hominin fossil collections</u> Olduvai Gorge: OH 5, 7, 8, 9, 24, 62, and 63, plus isolated teeth, Peninj: **Peninj 1**; Laetoli: EP 2400/00 (mandible), EP 1500/01 (face), EP 1000/98 (proximal tibia), and isolated teeth. <u>Contact person</u> Dr Paul Msemwa, Director, or Achiles M. Bufure, Senior Curator and Head of Ethnography Department (e-mail achiles11@yahoo.co.uk). <u>Street address</u> National Museum and House of Culture of Tanzania, Shabaan Robert Street, Dar es Salaam, Tanzania. <u>Postal address</u> National Museum and House of Culture - Dar es Salaam, PO Box 511, Dar Es Salaam, Tanzania. <u>Website</u> N/A. <u>Tel</u> +255 22 2122030/+255 22 2117508. <u>Fax</u> N/A.

National Museum of Natural History, Smithsonian Institution   The Smithsonian Institution was founded in 1846 by the Congress of the United States of America through a bequest by James Smithson, a British chemist, for the purpose of creating "an establishment for the increase and diffusion of knowledge" in Washington DC. The Smithsonian's National Museum of Natural History (or NMNH), formerly the US National Museum, opened to the public March 17, 1910, and contains biological and anthropological collections, including stone artifacts, fossils, and casts related to human and primate evolution. The Smithsonian's first secretaries (executive directors) initiated the development of extensive collections of Paleolithic artifacts. These collections include approximately 22,000 objects from 332 locations in 30 countries in Europe, Asia, and

Africa (Petraglia and Potts 2004). The first NMNH curator of physical anthropology, Aleš Hrdlička, founded the American Association of Physical Anthropologists. Although Hrdlička's research focused on North American modern human skeletal biology, he also conducted exchanges with European researchers that initiated the NMNH's fossil human cast collection. Through the work of curator T. Dale Stewart, the Smithsonian's participation in excavations at **Shanidar** Cave, northern Iraq, from 1953 to 1960, led to the housing in NHMH of the Shanidar 3 *Homo neanderthalensis* skeleton, one of 10 fragmentary Neanderthal skeletons uncovered at the site. The Shanidar 3 skeleton and the NMNH's Old World Paleolithic and hominin cast collections are under the care of the Smithsonian's Human Origins Program (or HOP). The HOP was founded in 1985, when the NMNH began its formal research program in human evolutionary studies. The HOP has conducted cooperative research with scientific organizations in Kenya, Tanzania, China, Indonesia, and India. Long-term collaboration with the **National Museums of Kenya** has added extensively to the NMK's fossil and archeological collections from the sites of Olorgesailie, Kanjera, and Kanam, and associated tephra and other geological collections from these sites in the NMNH. The NMNH's National Anthropological Archives holds the research documents of Glynn **Isaac** and Richard **Hay**, who were influential in East African archeological and geological research, respectively. The NMNH also maintains physical anthropology collections of nearly 33,000 specimens, primarily modern human skeletal collections (http://anthropology. si.edu/cm/phys_intro.htm). The NMNH's permanent exhibition hall on human evolution opened to the public on the museum's 100th anniversary in 2010. Contact department Human Origins Program (http://humanorigins.si.edu/contact). Contact persons Rick Potts (Human Origins Program; e-mail pottsr@si.edu) and David Hunt (Human Skeletal collections; e-mail huntd@si.edu). Postal address Human Origins Program, National Museum of Natural History, PO Box 37012, Washington DC 20013-7012, USA. There are approximately 5000 nonhuman primate individuals, including skin, skeleton, skull, or fluid preparations of select species of cebids, cercopithecids, and hominoids. The NMNH currently holds skeletal remains of 25 *Pongo abelii* from Aru Bay in Sumatra, and 94 *Pongo pygmaeus* from the Sakaiam, Kendawangan, Sempang, Semandong, and Mambulah Rivers in Borneo. Most of the 50 *Gorilla beringei* skeletal specimens are from Rwanda, with a

few from the Democratic Republic of the Congo. Most of the 57 *Gorilla gorilla* skeletal specimens come from Cameroon or Gabon, several from Republic of the Congo, and several are marked as "locality unknown." (http://collections.nmnh.si.edu/search/mammals). Contact persons Kristofer Helgen (Curator of Mammals; e-mail helgenk@si.edu) and Darrin Lunde (for use of the primate collections; e-mail lunded@si.edu). Postal address As above. *See also* **Bose Basin**; **Kanjera**; **Liang Bua**; **Nihewan Basin**; **Olorgesailie**.

National Museums of Kenya Grew out of a desire by European colonial leaders in East Africa to establish a scientific institution devoted to the natural history of the colony. The first natural history museum in Kenya was established in Nairobi in 1910 by the East Africa and Uganda Natural History Society to house the collections generated by local naturalists, but the museum's collections quickly outgrew the original buildings and the Society began construction of a new museum in 1929. The new Coryndon Museum, named in honor of Sir Robert Coryndon, a former governor of Kenya and supporter of the East Africa and Uganda Natural History Society, opened in 1930 and housed the Society's natural history collections as well as its substantial library. The East Africa and Uganda Natural History Society ran the museum until 1939 when it became an independent institution funded by the colonial government. The Coryndon Museum became an important center of paleoanthropological research after 1941 when Louis **Leakey** accepted a position at the museum. Leakey served as the curator of the Coryndon Museum from 1945 to 1961 and he greatly expanded the museum's collections while he was conducting paleoanthropological research at **Olduvai Gorge**. During his final year as curator Leakey established the **Centre for Prehistory and Palaeontology** at the museum, which served as an important focus of research into human prehistory. The Coryndon Museum was renamed the National Museums of Kenya in 1964 and beginning in 1969 the museum grew to include institutions and sites outside of Nairobi, including the prehistoric site of **Olorgesailie**. Louis Leakey's son, Richard **Leakey**, served as director of the National Museums of Kenya from 1969 to 1989 and through his efforts the National Museums of Kenya became a major center of paleoanthropological research with a large staff of scientists and a substantial collection of hominin fossils and archeological artifacts. In 1976 the museum opened the Leakey Memorial Building named in honor of Louis Leakey, which houses research and educational facilities. The fossil collections are dominated by Kenyan sites, but there are still some hominins in the National Museums of Museum from sites in Tanzania. The National Museums of Kenya has grown from a small repository of natural history specimens into a multidisciplinary research institute housing over 20 departments and divisions including archeology and paleontology. There are also 15 regional museums and eight sites open to the public. Hominin fossil and related collections Amboseli; Chemeron; **Chesowanja**; Eliye Springs; Eshoa Kakurongori; Fort Ternan; Harrar; Hindmarsh Moshi; Homa shell mound; Hyrax Hill; Jawuoyo; Kana shell mound; **Kanam** West; Kanam East; **Kanapoi**; Kanjera; Kapthurin; Karungu; Kijabe flats; Kisese; **Kokiselei**; **Koobi Fora**; Korongo Farm; **Laetoli**: LH 1–8, 10–17, 19–22, 29; **Lainyamok**; Loboi, **Lomekwi**; **Lothagam**; **Lukeino**; Lukenya; Naivasha; Nakoret; **Nariokotome**; Narok; Ngira; Ngorora; Njoro River Cave; Ntuka river; Nyarindi rock-shelter; **Olduvai Gorge**: OH 3–4, 6, some fragments of 8, 9–17, 19–46, 48–60; Ol Kalau; Pickford's site; Sotik; **South Turkwel**; and Tugen Hills. Contact department Department of Earth Sciences, Paleontology Division. Contact person Emma Mbua. Postal address National Museums of Kenya, Museum Hill, PO Box 40658. Website www.museums.or.ke. Tel +254 20 3742131/+254 20 3742161. Fax +254 20 3741424. E-mail nmk@museums.or.ke.

National Natural History Museum, Arusha Established to house and curate the Tanzanian components of the paleontological, paleoanthropological, and archeological collections from the countries in the East African Community that had been housed in the Coryndon Museum (now the **National Museums of Kenya**) in Nairobi. Since the collapse of the East African Community in 1977, the goal was to relocate the **Olduvai Gorge, Laetoli**, and other collections from Tanzanian sites to one place. Arusha, near Mount Kilimanjaro, was identified as the best location for the new museum because of its proximity to Olduvai Gorge and Laetoli, and Lakes Natron, Manyara, and Eyasi, all important sources of evidence about human evolution in Tanzania. In the early 1980s the Tanzanian Government donated the old German Fort along Boma Road in Arusha as the site of the new museum. Dating back to 1880s, the Boma Fort, the oldest colonial building in Arusha,

served initially as a military base for the German East Africa Company, and subsequently it was the headquarters of the first provincial administration in the region. This small but fine historical building was transformed into a museum and opened in 1987, but at that time there were no adequate facilities to curate and store the fossil collections, nor were funds available to hire trained curatorial personnel. In addition, due to political tension between Tanzania and Kenya in the 1980s, it was not possible to repatriate the Tanzanian collections from Kenya. Thus, the first hominin fossils from Tanzanian sites to be housed in the National Natural History Museum in Arusha were those collected by the **Olduvai Landscape Paleoanthropology Project** (or OLAPP). The OLAPP project raised the resources needed to convert one of the buildings at the Boma Fort into a facility that was suitable for the preparation and secure storage of hominin fossils. Hominin fossil collection OH 64–74 (including the **OH 65** maxilla). In addition to the hominin fossils the museum houses a large collection of stone artifacts (>4000) and non-hominin faunal specimens (>8000 identified specimens), and a number of fossilized wood fragments from Bed I and Lower Bed II, Olduvai Gorge. Contact department Paleoanthropology. Contact person Jackson Njau (e-mail jknjau@gmail.com). Postal address National Natural History Museum, Boma Road, PO Box 2160, Arusha, Tanzania. Website N/A. Tel +255 272 507540. Fax N/A. E-mail nnhm@habari.co.tz.

Natoo (etym. Named for an ephemeral river system, or laga, in whose drainage the localities are located) (or NT) A complex of fossil localities in the Natoo Member of the Nachukui Formation within the Nanyangakipi drainage in **West Turkana**, Kenya. *See also* **West Turkana**.

natural selection The process of differential survival and reproduction of **genotypes**. Natural selection acts on the **phenotype** and in doing so affects the inherited **alleles** such that it maintains favorable alleles and genotypes in the population and removes deleterious alleles and genotypes. It promotes increased frequency of the most advantageous alleles and genotypes in that environment, and in doing will inevitably reduce the frequencies of other alleles and genotypes. It is one of the four forces of evolution (along with **mutation, genetic drift,** and **gene flow**). There are several different types of natural selection. These include **directional selection** (which can be positive or negative), balancing selection, diversifying selection, **purifying selection**, **sexual selection**, and **stabilizing selection**. Ultimately, all forms of natural selection result in differential birth rates (and some differential death rates) for individuals with different fitness levels.

Naturalis *See* **Rijksmuseum van Natuurlijke Historie**.

Nazlet Khater (Location 26°47′N, 31°21′E, Egypt; etym. named after the nearby village) History and general description Located on a remnant of a Nile terrace, the site was excavated by the Belgian Middle Egypt Prehistoric Project between 1980 and 1982. It is an underground **chert**-mining site characterized by three types of excavated structure: ditches, vertical shafts, and underground galleries. Temporal span and how dated? A combination of **AMS radiocarbon dating** on charcoal and optically stimulated **luminescence dating** on the **aeolian** sand that fills the excavations suggests the exploitation period of Nazlet Khater 4 is dated from 40 to 35 ka. Hominins associated with site Two burials found on a nearby locality known as Boulder Hill: **Nazlet Khater 1** and **Nazlet Khater 2**. Archeological evidence found at site **Upper Paleolithic** industry characterized by handaxes in association with a fully developed blade-production system. This assemblage may represent the incursion into the Nile Valley of a transformed Nubian Complex. Key references: Geology, dating, and paleoenvironment Vermeersch et al. 1984, Vermeersch 2002; Archeology Vermeersch 2002; Hominins Vermeersch et al. 1984, Thoma 1984, Pinhasi and Semal 2000, Crevecoeur 2008a.

Nazlet Khater 1 Site **Nazlet Khater**, Egypt. Locality Boulder Hill. Surface/*in situ* *In situ*. Date of discovery 1980. Finders Vermeersch and colleagues. Unit N/A. Horizon N/A. Bed/member N/A. Formation N/A. Group N/A. Nearest overlying dated horizon N/A. Nearest underlying dated horizon N/A. Geological age 37,570+350/−310 years BP (GrA-20145), an AMS radiocarbon age on charcoal associated with the burial. Developmental age Adult. Presumed sex Female. Brief anatomical description Poorly preserved adult skeleton plus the remains of a fetus. Announcement Vermeersch 2002. Initial description Vermeersch 2002. Photographs/ line drawings and metrical data Vermeersch 2002.

Detailed anatomical description N/A. Significance One of the very few modern human skeletons from this time period in this region of Africa. Location of original Bordeaux 1 University, France.

Nazlet Khater 2 Site **Nazlet Khater**, Egypt. Locality Boulder Hill. Surface/*in situ* *In situ*. Date of discovery 1980. Finders Vermeersch and colleagues. Unit N/A. Horizon N/A. Bed/member N/A. Formation N/A. Group N/A. Nearest overlying dated horizon N/A. Nearest underlying dated horizon N/A. Geological age 38±6 ka (based on an **electron spin resonance spectroscopy dating** age on an enamel fragment). Developmental age Adult. Presumed sex Male. Brief anatomical description The skeleton is complete with the exception of the distal part of the lower limb. Morphological and biometrical comparative analyses of this specimen underline the complex morphology of modern humans from this time period. Nazlet Khater 2 exhibits several retained archaic features, notably on the face and the mandible. The morphology of the bony labyrinth of Nazlet Khater 2 can be distinguished from that of contemporary modern humans. The limb bones show strongly marked muscle insertions and the cross-sectional geometric properties of the long bones suggest they were adapted to substantial loading. Furthermore, Nazlet Khater 2 vertebrae show pathological changes that are unusual for a relatively young adult and it has been suggested these lesions may be related to involvement in the intensive physical activity involved in mining. Initial taxonomic allocation *Homo sapiens*. Announcement Vermeersch et al. 1984. Initial description Thoma 1984. Photographs/line drawings and metrical data Vermeersch et al. 1984, Vermeersch 2002, Crevecoeur 2008a. Detailed anatomical description Crevecoeur 2008a. Significance One of the very well-preserved modern human skeletons from this time period in this region of Africa. Location of original Bordeaux 1 University, France.

NC Abbreviation for Nachukui, one of several groups of localities within the site of **West Turkana** named after ephemeral river drainages. *See* **Nachukui**.

Ndutu (Location 3°00′S, 35°00′E, Tanzania; etym. the site is on the north shore of Lake Ndutu, a seasonal soda-lake at the western end of the Olduvai Main Gorge) History and general description The site was first visited in 1972; surface collections were made at seven locations and a test trench established the stratigraphy. Formal excavations carried out at the site in September–October 1973 exposed "occupational floors with *in situ* lithic and faunal material" and "a hominid skull" (Mturi 1976, p. 484). Temporal span and how dated? "Later Middle Pleistocene" (Rightmire 1983, p. 246) or less than 400 ka according to **tephrostratigraphy**. Hominins found at site **Ndutu 1**. Archeological evidence found at site "270 plotted lithic and faunal materials," "94 out of the 270 plotted *in situ* materials"…"are splinters and sizeable bone pieces"…"seems to suggest that the area was probably a butchery site" (Mturi 1976, p. 485). Only 20 out of the 126 lithic pieces "are defintive tools"…"among the tools spheroids and hammerstones predominate" (*ibid*). Key references: Geology, dating, and paleoenvironment Mturi 1976; Hominins Clarke 1976, Rightmire 1983; Archeology Mturi 1976.

Ndutu 1 Site **Ndutu**, Tanzania. Locality Shoreline of Lake Ndutu. Surface/*in situ* *In situ*. Date of discovery 1973 Finder Amini Mturi. Unit N/A. Horizon Upper, sandy horizon of two, the second without much sand, in greenish clay near the shore of the lake. Bed/member N/A. Formation N/A. Group N/A. Nearest overlying dated horizon A re-worked tuff that has mineralogical similarities to the Norikilili Member, Upper Masek beds, **Olduvai Gorge**. Nearest underlying dated horizon Underlain by a sand-free unit of green clay. Geological age **Middle Pleistocene**, probably around 200–400 ka on the basis of tuff correlation with Olduvai Gorge (see above). Developmental age Adult. Presumed sex Unknown, but possibly female based on its small size. Brief anatomical description A small, fragmentary cranium that preserves the occipital, much of both parietals and temporals, some of the frontal and facial bones and the sphenoid. The occipital has a marked angulation between its upper and lower scales. The bone of the cranial vault is thick. The face is neither large nor strongly projecting. The vertically thin but projecting brow ridge is separated from the frontal squama by a sulcus. The temporal bone has an ossified styloid process. Endocranial volume Approximately 1100 cm^3. Announcement Mturi 1976. Initial description Clarke 1976. Photographs/line drawings and metrical data Clarke 1976, 1990. Detailed anatomical description Rightmire 1983, Clarke 1990, Schwartz and Tattersall 2003. Initial taxonomic allocation *Homo erectus* (Clarke 1976).

Taxonomic revisions *Homo sapiens* (Rightmire 1983, Bräuer 1984), *Homo leakeyi* (Clark 1990). Current conventional taxonomic allocation *Homo heidelbergensis* or *Homo rhodesiensis*. Informal taxonomic category Pre-modern *Homo*. Significance The Ndutu cranium provides a key morphological link between *Homo erectus* and *Homo heidelbergensis* in Africa. Location of original **National Museum and House of Culture, Dar es Salaam**, Tanzania.

Ndutu cranium *See* **Ndutu 1**.

N$_e$ *See* **effective population size**.

Neander, Joachim *See* **Homo neanderthalensis**.

Neandertal *See* **Homo neanderthalensis**; **Kleine Feldhofer Grotte**; **Neanderthal**.

Neanderthal In this encyclopedia the taxon *Homo neanderthalensis*, and any specimens belonging to that taxon, are referred to informally as "Neanderthals." In other publications this group is sometimes referred to as "Neandertals." The authors who use Neandertal presumably do so because in "new German" (German orthography following 1901) the silent "h" was dropped, so that "thal" became "tal." However, despite these changes to the German language, according to the International Code for Zoological Nomenclature the Linnaean binomial *Homo neanderthalensis* remains unchanged. This is why we use the old spelling. *See also* **Homo neanderthalensis**; **Kleine Feldhofer Grotte**.

Neanderthal 1 The holotype of **Homo neanderthalensis**. *See* **Kleine Feldhofer Grotte**.

Neanderthal phase hypothesis This hypothesis, also called the Neanderthal phase model, influenced interpretations of human evolution for much of the 20thC. It suggested that modern humans had evolved from an ape-like ancestor through several evolutionary stages, the most recent of which was represented by *Homo neanderthalensis*. Thus, supporters of a **Neanderthal** phase of human evolution believed that at least some Neanderthal populations evolved into modern *Homo sapiens*. After the discovery of *Pithecanthropus erectus* (now *Homo erectus*) by the Dutch anatomist Eugène **Dubois** in the 1890s, the German anatomist Gustav **Schwalbe**, in *Studien zur Vorgeschichte des Menschen* [*Studies on Human Prehistory*] published in 1906, suggested that modern humans had either evolved from an ape-like ancestor directly through *Pithecanthropus* and Neanderthal stages, or from other ancestors that must have been very similar to these. However, after French paleontologist Marcellin **Boule** had published his influential study of the **La Chapelle-aux-Saints** Neanderthal, Schwalbe was more willing to view Neanderthals as a side line rather than the ancestors of modern humans. The first comprehensive exposition of what we now call the Neanderthal phase hypothesis of human evolution published in English was by the Czech/American physical anthropologist Aleš Hrdlička in his 1927 Huxley Memorial Lecture of the Royal Anthropological Institution entitled "The Neanderthal phase of man." Hrdlička opposed the **pre-sapiens hypothesis** and instead interpreted the evolution of modern humans as proceeding in a unilinear fashion from an ape-like ancestor through successive stages that possessed gradually smaller numbers of primitive traits and gradually larger numbers of modern human traits. While he recognized that most of the Neanderthals of western Europe demonstrated a distinctive morphology that was interpreted to be the result of adaptation to a cold climate, he noted that the Neanderthals from **Krapina** collected by the Croatian paleontologist Dragutin **Gorjanović-Kramberger** included a surprising amount of variation with some individuals approaching much more closely the morphology of *H. sapiens*, factors noted by Gorjanović-Kramberger in his seminal 1906 monograph. Hrdlička also argued that the archeological record, rather than supporting the notion that *H. sapiens* with their **Aurignacian** industry had displaced the Neanderthals with their **Mousterian** industry, indicated instead that the Aurignacian industry may have developed from the Mousterian. Hrdlička further noted that for the Neanderthals to be replaced by a more advanced immigrant population there should be evidence of the source of such a group somewhere outside Europe, and at that time no such evidence existed. Given the widespread impact of Boule's interpretation of the Neanderthals as an extinct side branch of human evolution, as well as the acceptance of the pre-sapiens hypothesis by a large number of influential anthropologists such as Sir Arthur **Keith**, relatively few scientists accepted Hrdlička's new views about the Neanderthals. However, he did attract several important

advocates including German anatomist Franz **Weidenreich** who studied with Gustav Schwalbe and later oversaw the continuation of the excavations at Choukoutien (now **Zhoukoudian**) that ultimately produced the collection of fossils initially assigned to *Sinanthropus*. For Weidenreich different hominin groups spread throughout the Old World had evolved from an ape-like ancestor through successive *Sinanthropus/Pithecanthropus* and Neanderthal stages. While he considered it possible that the western European "classic" Neanderthals could have evolved into modern humans, he believed it more likely that European *Homo sapiens* had evolved from the Neanderthals from the Near East. The German anthropologist Hans Weinert was another strong supporter of the Neanderthal phase hypothesis during the 1940s, but his impact was limited. The excavations conducted at **Mount Carmel** in Palestine during the early 1930s by Dorothy **Garrod** and Theodore McCown had unearthed what were interpreted as Neanderthal remains that were morphologically more similar to modern humans than were the classic Neanderthals of western Europe. This prompted a number of researchers to suggest that these so-called "progressive" Neanderthals had evolved into modern humans, whereas the European classic Neanderthals had been replaced by immigrant populations of modern humans. In the 1950s, the Italian paleontologist Sergio **Sergi** suggested that a Neanderthal population like that found at Mount Carmel had evolved into both modern *Homo sapiens* and the classic Neanderthals. Similarly, Clark **Howell** argued that progressive Neanderthals (like those found at Mount Carmel and Krapina) had evolved into both modern humans and classic Neanderthals. Versions of the so-called pre-Neanderthal hypothesis proliferated during the 1950s and 1960s, but the strongest advocate of the Neanderthal phase theory during the 1960s and 1970s was the American physical anthropologist C. Loring **Brace**. Brace supported what has now become known as the **single-species hypothesis**, which was that at any one time all hominins belong to the same, morphologically diverse, species. According to Brace, in the course of their evolution modern humans had successively passed through **australopithecine, pithecanthropine**, and Neanderthal stages of evolution before becoming fully modern humans. Brace accepted Hrdlička's model of human evolution, but in addition he adopted the new biological thinking of the **modern evolutionary synthesis** to explain how Neanderthals could have evolved into *H. sapiens*. Brace suggested that stone-tool use made it less necessary for early hominins to use their teeth as a tool, thus leading to consequential morphological changes in the skull. Brace cited evidence of morphological and archeological continuity between Neanderthals and early *H. sapiens*, to support the idea that Neanderthals evolved into the immediate ancestors of modern humans. The Neanderthal phase hypothesis received several setbacks in the 1980s and 1990s as a result of new fossil discoveries, the development of radiometric dating techniques for the time span between approximately 30 and 250 ka, and molecular evidence suggesting that anatomically modern humans had evolved in Africa between 200 and 100 ka. This led to the view that members of this latter group migrated out of Africa to replace the Neanderthals in Western Asia and Europe. While at least one study (Krause et al. 2010) has demonstrated gene exchange between the Neanderthals and anatomically modern humans, the primary ancestral contribution to modern humans in Eurasia clearly derived from the out-of-Africa migrations of early modern people (*see* **out-of-Africa hypothesis**).

Neanderthaloid (Ge. *Neanderthal*=the valley of the river Neander and Gk *oeides*=resembling, as in having the shape or form of Neanderthals) Originally coined as an adjective meaning simply "Neanderthal-like" (e.g., Burkitt and Hunter 1922) this informal term subsequently came to be used in two distinct senses. A minority of workers applied it to some **Middle Pleistocene** European specimens (e. g., **Mauer, Steinheim, Montmaurin, Fontéchevade**) displaying one, or more, of the features characteristic of *Homo neanderthalensis*, and because of this they were considered potential Neanderthal ancestors (i.e., "fossils resembling Neanderthals"). Used in this sense it is broadly equivalent to the term "pre-Neanderthal" or "AnteNeanderthal" as used by some French workers. The second sense in which Neanderthaloid has been used is in connection with African and Asian fossil specimens dating from the later Middle or early Upper Pleistocene that resembled, but were distinct from, *H. neanderthalensis* known from Europe and the Near East. Specimens considered to be Neanderthaloid in this sense include fossils from **Jebel Irhoud, Kabwe, Elandsfontein, and Maba**. In both of the senses in which it has been used, Neanderthaloid refers to hominin remains that were considered to resemble *H. neanderthalensis* at a time when the latter represented the largest and best-known group of fossil hominins, so it was natural

to compare new discoveries with *H. neanderthalensis*. For example, Morant (1928) concluded that the Kabwe and Neanderthal crania resembled each other more closely (i.e., were big-faced and strongly constructed) than either did modern humans, and in the monographs about *Sinanthropus pekinensis* Franz **Weidenreich** devotes substantial effort to showing how those remains differed from *H. neanderthalensis*. However, under the influence of the **Neanderthal phase hypothesis** "Neanderthaloid" subsequently acquired a phyletic connotation, implying that specimens so described represented the African and Asian *evolutionary* equivalents of European and Near-Eastern Neanderthals. "Neanderthaloid" was in common usage from the 1940s to the 1970s. However, since then increasing awareness and adoption of **cladistic** methods, and recognition of the importance of focusing on shared, **derived**, characters when trying to delimit taxonomic groups, has led to the realization that the term "Neanderthaloid" is nothing more than an ill-defined **grade** of fossil hominins and its use has now been largely abandoned.

nearly neutral theory of molecular evolution A modification by Ohta (1992) of Kimura's **neutral theory of molecular evolution** suggesting that slightly disadvantageous and slightly advantageous mutations often evolve in a similar way as strictly neutral mutations, but that their evolution is dependent on population size. In small populations, the effects of selection are limited such that the fates of nearly neutral mutations, like those of strictly neutral mutations, are determined by random **genetic drift**. In large populations, the effects of selection are significant for nearly neutral mutations and thus it has more influence on their fates. In nearly neutral systems, the rate of substitution generally decreases as population size increases.

negative allometry (Gk *allo*=other and *metron*=measure) This term refers to a relative size relationship in which part of an organism, or a variable that functions as a proxy for part of an organism, increases in size at a slower rate than the overall size of the organism, or a variable that functions as a proxy for the whole of an organism (i.e., the variable becomes proportionally smaller as overall body size increases). *See also* **allometry; scaling**.

negative assortment This mating pattern, also known as negative assortative mating, occurs when individuals mate with individuals that are different from themselves at a higher frequency than would be expected by chance (i.e., when "like" mates with "unlike"). An example of negative assortative mating in modern humans is thought to be human leukocyte antigen (or HLA) type (as sensed via pheromones or body odor) (e.g., Ober et al. 1997, Lie et al. 2010, Chaix et al. 2008), although this has not been demonstrated in all populations of modern humans (e.g., Hedrick and Black 1997, Chaix et al. 2008). Red hair in modern humans also shows moderate negative assortative mating (Mascie Taylor and Boldsen 1988).

negative Gompertz hazard model The negative Gompertz hazard model is used as part of the **Siler hazard model** to represent premature (i.e., pre-senescent) mortality. Although it is related to the Gompertz model, the negative Gompertz hazard model was not proposed by Benjamin Gompertz and it is usually only used within the Siler model; immature mortality on its own is more typically represented with a **Weibull hazard model**. The negative Gompertz hazard model gives the hazard of death at exact age a as $h(a) = \alpha e^{-\beta a}$ and the survivorship to exact age a as $S(a) = \exp(-\alpha/\beta(1 - \exp(-\beta a)))$, where α is the initial hazard of death at age zero and β is the decrease in the hazard of death (on a logarithmic scale) with advancing age. As noted by Siler (1979), at very old ages the survivorship from the negative Gompertz model is $S(a) = \exp(\alpha/\beta)$, which is not equal to zero. This is an undesirable property of the negative Gompertz, although within the Siler model this problem does not matter as the survivorship from the (positive) Gompertz model does decrease to zero at advanced ages. As the survivorships are multiplicative across the Siler components, ultimately the survivorship for the entire model does reach zero at advanced ages. Survivorship in a Weibull model does reach zero at very old ages, which is why this mortality pattern cannot be used within the Siler model. If it were used then adult survivorship would be forced to equal zero because of the multiplicative nature of the survivorships from the immature and senescent (and residual) components.

negative selection Equivalent to **purifying selection** (*which see*).

neo-Lamarckism *See* Lamarckism.

neoanthropic man *See* Archanthropinae.

Neoanthropinae *See* Archanthropinae.

neocortex (Gk *neos* = new and L. *cortex* = bark; i.e., outer covering) Refers to the phylogenetically newest part of the **cerebral cortex** [i.e., all of the cerebral cortex apart from certain cortical areas that are linked with olfaction (the piriform cortex) and memory and spatial navigation (hippocampus), which are also referred to as the archi- and paleocortex, respectively]. The neocortex, which forms the external shell of the **cerebral hemisphere**s, is made up primarily of **neuron**s and **glial** cells. It is the largest component of the **forebrain** in mammals and it is the only part of the forebrain that is visible when viewing the whole brain. The neocortex is only present in mammals, although homologous structures, which share certain features of molecular expression and connectivity with the neocortex, have been proposed to exist in reptiles and birds. The most striking and distinguishing architectural trait of the neocortex is that its **neuron**s are arranged in several well-defined layers, numbered sequentially from the surface beneath the innermost of the three meningeal layers, the pia mater (layer I) to the layer adjacent to the underlying white matter (layer VI). Although this six-layer structure is characteristic of the entire neocortex, the thickness of individual layers, and the types of neuron within them, varies in different functional regions of the cortex. The characteristic pattern of layering in different cortical areas was shown by Korbinian Brodmann, who used evidence about **cytoarchitecture** to organize the modern human cerebral cortex into about 50 areas. The modern human neocortex is approximately four times larger than the neocortex of the living **great ape**s and it is larger than would be predicted using the observed allometry relationships within living hominoids (Rilling 2006).

neocortex ratio The ratio of the volume of the **neocortex** to the volume of the rest of the **brain**. Robin Dunbar and colleagues have proposed that this metric is correlated across primate species with variables such as social group size and grooming clique size, which might reflect the complexity of social cognition. Aiello and Dunbar (1993) have argued that the large neocortex ratio of modern humans can be used as a proxy for the size of the groups that individual modern humans are comfortable with. They also suggest that a major force driving the evolution of language was the need to track the status of the individuals in a relatively large social network. In this scenario, spoken language evolved as a replacement for grooming as a more efficient mechanism for maintaining social connections and exchanging information among a large numbers of individuals. *See also* **language, evolution of**.

Neogene (Gk *neo*=new and *genos*=race, kind) The middle of the three periods that comprise the **Cenozoic** era. Neogene refers to a unit of geological time (i.e., a **geochronological unit**) that spans the interval from *c.*23.5 to 2.58 Ma (Gelasian) or 1.8 Ma (Calabrian). The Neogene comprises two **epochs**, the **Miocene** and **Pliocene**. Some geologists disagree with the decision to elevate the **Quaternary** to period status and thus they still include the **Pleistocene** and the **Holocene** epochs in the Neogene. In this case, the Neogene would span the interval of time from *c.*23 Ma to the present. See also **Plio-Pleistocene boundary**; **Tertiary**.

neonatal line (L. *neo*=new and *natus*=to be borne) An **accentuated line** in the **enamel** and **dentine** of deciduous teeth and in the first permanent molars in higher primates caused by the stress of birth interfering with enamel and dentine development. *See also* **enamel development**.

neonate (Gk *neos*=new and L. *natus*=to be born) Literally the "newborn," but in pediatrics the term is used for infants from birth to 4 weeks old.

neontological species (Gk *neos*=new, *onto*=being, and *ology*=the study of, thus a "new," or contemporary, species) One of the many terms used to describe a contemporary species defined on the basis of criteria such as evidence of interbreeding, and lack of any evidence of interbreeding between individuals belonging to that species and individuals belonging to any other species. *See also* **species** (syn. **biological species**; **extant species**).

neoteny (Gk *neos*=young plus *teinein*=to extend) Neoteny is a **heterochronic** process that involves the dissociation of ancestral ontogenetic trajectories to produce **paedomorphosis**. In neoteny, shape change is retarded for a given size in the descendant relative to the ancestor and does not proceed as far, resulting in a juvenilized morphology. Importantly, and to distinguish this process from either form of **hypomorphosis**, neoteny does not require associated shifts in adult size or duration of growth. Neoteny has a history of being invoked as a particularly important heterochronic process for human evolution. Features including large brains, short and flat faces, and hairlessness have been attributed as results of juvenilization via uncoupling of ancestral ontogenetic trajectories. However, instead of providing an understanding of *all* of the key human adaptations as the result of a single, relatively simple, change in development, the invoking of a theory of global neoteny has resulted in other important processes being overlooked. For example, the morphology of modern human lower limbs can probably best be described in terms of **peramorphosis**. Further, modern human development is prolonged relative to that of apes, counter to the requirements of neoteny. For other comparisons, however, neoteny provides a viable explanation (e.g., juvenilized skull and face in bonobos relative to common chimpanzees, who reach similar adult sizes and grow for comparable durations of time).

neotype (L. *neo*=new and *typus*=image) If the **type specimen** of a **taxon** is destroyed for some reason, researchers can apply to the **International Commission on Zoological Nomenclature** to have another specimen, called the neotype, designated in its place.

Nerja *See* **Cueva de Nerja**.

net reproductive rate (or NRR) The integral across age of the product of the **age-specific fertility rate** and the **survivorship** in a hazards analysis. The NRR must be lower than the **total fertility rate** because the latter does not include the effect of **mortality** (measured by survivorship). For example, in the (unlikely) event that survivorship to age 13 was zero, the NRR would also be zero, regardless of the age-specific fertility rates. As NRR includes the effects of both fertility and mortality it can be used to project population size into the future. Another interpretation for NRR is as the ratio of the number of daughters in the next generation to the number of mothers in the current generation. For example, a population with a NRR equal to 1.0 is simply replacing itself but is neither growing nor declining in size over time. If T is the number of years (**generation length**), then under an **exponential population growth** model $NRR = e^{rT}$ and solving for the intrinsic growth rate gives $r = \ln(NRR)/T$. In a **life table** analysis the NRR is the average of age-interval midpoints where the averaging is weighted by the product of age-specific fertility rate and survivorship.

neurocranium (Gk *neuron*=nerve and *kranion* = brain case) The cranium is divided into the neurocranium and the **viscerocranium**. The neurocranium is the part of the cranium that surrounds all but the anterior aspect of the **brain** (this is in contact with the viscerocranium) and it contains the **inner ear** and **middle ear**. It comprises the cranial vault, or **calvaria**, which covers the top, back, and sides of the brain, and the **basicranium**, which lies beneath the brain.

neuron (Gk *neuron*=a nerve cell with an appendage) The name for a class of cells in the nervous system. Neurons are specialized for conducting electrical signals that transmit information to other cells. Neurons have long processes called **axons** that only branch close to their termination and shorter-branching projections called **dendrites**. Neurons vary widely in size, shape, and electrochemical properties. Some classes of neurons in the cerebral cortex, such as the **pyramidal cells** of the prefrontal cortex and the **von Economo neurons** of the anterior cingulate and frontoinsular cortex, display morphologies that may serve as a neuroanatomical substrate for the behavioral and cognitive specializations that are seen in the great apes and modern humans.

neuropil The portion of brain tissue that *does not* take up **Nissl stain**. It comprises mainly **dendrites**, **axons**, and **synapses**.

neurotransmitter (Gk *neuro*=nerve) A chemical substance secreted at a synaptic terminal by one **neuron** for the purpose of conveying information via receptors to other neurons or to organs. Neurotransmitters may be either neuroactive peptides or low-molecular-weight molecules. To be a neurotransmitter, a substance must (a) be synthesized in the neuron, (b) be present in the presynaptic terminal and released in quantities sufficient to exert an influence upon the post-synaptic body, (c) mimic endogenously released transmitter when applied exogenously (e.g., as a drug), and (d) be removable from the synaptic cleft by a specific mechanism. There is no evidence that modern human brains have any neurotransmitters that are not also found in the brains of chimpanzees.

neutral allele An **allele** that has a **selection coefficient** of zero ($s=0$). Any change in the frequency of such alleles is due to random **genetic drift**, **gene flow**, or **mutation**.

neutral mutation A mutation that has a **selection coefficient** of zero ($s=0$). Any change in the frequency of such mutations is due to random **genetic drift** or **gene flow**.

neutral polymorphism *See* **neutral allele**.

neutral theory of molecular evolution A theory developed by Kimura (1968) suggesting that the vast majority of **molecular evolution** involves the replacement of one **neutral mutation** by another neutral mutation.

New Archeology *See* Binford, Lewis (1930–).

NG Abbreviation of Ngrejeng and the prefix for fossils recovered from that site. *See* **Ngrejeng**.

NG 8503 Site Ngrejeng, Indonesia. Locality Near the village of Ngrejeng. Surface/*in situ* Surface. Date of discovery 1985. Finder A local villager. Unit N/A. Horizon N/A. Bed/member N/A. Formation **Bapang Formation**. Group N/A. Nearest overlying dated horizon Middle Tuff. Nearest underlying dated horizon Lower Tuff. Geological age *c.*1.02–1.51 Ma. Developmental age Juvenile. Presumed sex Unknown. Brief anatomical description Part of the right side of the mandibular corpus with M_1 and M_2 (still in its crypt); both molars preserve their occlusal morphology. Announcement Aziz et al. 1994. Initial description Aziz et al. 1994. Photographs/line drawings and metrical data Kaifu et al. 2005a. Detailed anatomical description Kaifu et al. 2005a. Initial taxonomic allocation *Homo erectus*. Taxonomic revisions None. Current conventional taxonomic allocation *H. erectus*. Informal taxonomic category Pre-modern *Homo*. Significance This specimen documents dental-size reduction in later *H. erectus* specimens from Java (Kaifu et al. 2005b). Location of original **Geological Research and Development Center**, Bandung, Indonesia (syn. Sangiran 38).

Ngaloba cranium *See* LH 18.

Ngamapa One of the Early and Middle **Pleistocene** sites in the exposures of the **Ola Bula Formation** within the **Soa Basin**, central Flores, Indonesia.

Ngandong (Location 7°18′S, 111°39′E, Java, Indonesia; etym. named after a local village)

History and general description An open-air, river terrace site on the north bank of the Solo River, about 10 km/6 miles north of **Ngawi** in East Java. C. ter Haar directed excavations at Ngandong from September 1931 through the end of November 1933. The initial finds consisted of "two fragments, one fine specimen of an adult hominid calvaria" (Oppenoorth 1932a, p. 62). The latter is referred to as **Ngandong 1** and the former as Ngandong 2 and 3; Ngandong 4 and 5 were found in 1932 (Oppenoorth 1932b). Oppenoorth suggested that the calvaria "may represent a stage in the development of man equivalent to that of the neandertaler" (*ibid*, p. 63) and he suggested it be referred to a new species and subgenus of *Homo*, as *Homo (Javanthropus) soloensis* **Oppenoorth, 1932**. The ter Haar field project produced over 25,000 mammalian fossils, including at least 11 hominin crania and two hominin tibiae. During the dry season, the fossiliferous deposits at Ngandong known as the High Terrace are 20 m above the Solo River. The High Terrace deposits are 3 m thick and have been divided variably, depending on the researcher, into three to five separate strata. Most of the vertebrate fossils were recovered from the lowermost half meter of the High Terrace. The deposit itself is mostly composed of sand and gravel with a large proportion of volcanic material included. Temporal span and how dated? The age of Ngandong has been, and still is, controversial. Using faunal correlation, the Ngandong hominins were initially attributed to the Upper Pleistocene. **Uranium-series dating** of bone samples excavated from Ngandong in the late 1980s ranged from 31 to 101 ka, thus supporting these Upper Pleistocene age estimates. In 1997, an international team analyzed additional fossil mammal teeth recovered from Ngandong and using both uranium-series dating and **electron spin resonance spectroscopy dating** techniques they determined that the fossil hominins date from 46.4–27.3 ka. These startlingly recent ages have met with considerable resistance from other scholars who have claimed that the fossil hominins were re-worked from older deposits and are therefore not contemporary with the recently dated mammal teeth, or that the very young ages of the deposit do not correspond with the well-established faunal biostratigraphy of the region. Hominins found at site Three different numbering systems have been employed in cataloguing the Ngandong hominins. Originally, the cranial specimens were referred to as "Solo I" through "Solo XI" while the tibia were referred to as "Tibia A" and "Tibia B." The modern system refers to the crania as "Ngandong 1" through "Ngandong 12" and retains the original names of the tibiae. The third system (employed by Teuku **Jacob**) names all of the specimens "Ngandong 1" through "Ngandong 14" with the tibiae included as "Ngandong 9" and "Ngandong 10." **Ngandong 1** is the **holotype** of *H. soloensis*. All of the Ngandong crania are without faces and most lack the base of the cranium. The crania have thick vaults, bar-like supraorbital tori that are thickest laterally, and an average endocranial volume of 1135 cm³. Archeological evidence found at site The High Terrace at Ngandong has yielded crude, small (generally <5 cm in length) stone implements, including cores and flakes. They are mostly water-worn, many are made from chalcedony, with a few being made from chert and **jasper**. Key references: Geology, dating, and paleoenvironment Orchiston and Siesser 1982, Bartstra et al. 1988, Swisher et al. 1996, Grün et al. 1997b; Hominins Weidenreich 1951, von Koenigswald 1975, Jacob 1981, Antón 1999, Durband 2004, Kaifu et al. 2008; Archeology Bartstra et al. 1988 (syn. Solo).

Ngandong 1 Site Ngandong, Indonesia. Locality N/A. Surface/*in situ* Surface. Date of discovery 1931. Finder Unknown. Unit N/A. Horizon N/A. Bed/member High Terrace. Formation **Bapang Formation**. Group N/A. Nearest overlying dated horizon N/A. Nearest underlying dated horizon N/A. Geological age *c*.100–27 ka. Developmental age Juvenile. Presumed sex Unknown. Brief anatomical description A thick-walled cranial vault. Announcement Oppenoorth 1932a. Initial description Oppenoorth 1932a. Photographs/line drawings and metrical data Oppenoorth 1932a, Weidenreich 1951. Detailed anatomical description Weidenreich 1951. Initial taxonomic allocation *Homo soloensis*. Taxonomic revisions *Homo erectus* (Santa Luca 1980). Current conventional taxonomic allocation *H. erectus*. Informal taxonomic category Pre-modern *Homo*. Significance The **holotype** of *H. soloensis*. Location of original **Geological Research and Development Center**, Bandung, Indonesia (syn. Solo I).

Ngandong hominins In the introduction to his unfinished monograph the "Morphology of Solo man" Weidenreich (1951) suggests that the paleontological significance of the Ngandong site was first recognized by G. Elbert, a member of the

1907–8 German Selenka Expedition to Trinil, but Elbert referred to it as "Bangoen," not Ngandong. The site was "rediscovered" in August 1931 by C. ter Haar, who was mapping the area for the Geological Survey. The initial finds were made in September of that year and consisted of "two fragments, one fine specimen of an adult hominid calvaria" (Oppenoorth 1932a, p. 62). The latter is referred to as **Ngandong 1**, or Ngandong Skull 1 and the former as Ngandong Skull 2. Oppenoorth, who was the Acting Director of the Geological Survey, went to Ngandong in October 1931 and during his visit a third skull was found, Ngandong Skull III. Four more skulls and a tibia were found in 1932 (Ngandong Skulls IV–VII and Tibia A) and more in 1933 (Ngandong Skulls VIII–XI and Tibia B). Oppenoorth suggested that the calvaria "may represent a stage in the development of man equivalent to that of the neandertaler" (*ibid*, p. 63) and he suggested it be referred to a new species and subgenus of *Homo*, as *Homo (Javanthropus) soloensis*. He later withdrew his support for *Javanthropus* (Oppenoorth 1937). The most detailed studies of the Solo crania have been carried out by Weidenreich (1951) and Santa Luca (1980). Although Weidenreich died before his analysis could be completed, he made his taxonomic philosophy clear, stating that "I believe that all the hominids now known belong morphologically to a single species" (Weidenreich 1951, p. 226). Later in the Solo monograph he writes that his earlier studies of the Solo material that were presented in his monograph on the skull of *Sinanthropus pekinensis* (Weidenreich 1943) had convinced him that "Ngandong man was not a true **Neanderthal** type but distinctly more primitive and very close to *Pithecanthropus* and *Sinanthropus*. For this reason I ranked Solo man with the same group of early hominids as the two latter forms and called the whole group Archanthropines" (Weidenreich 1951, p. 227). Santa Luca (1980) came to much the same conclusion. He also rejected any special affinity with *Homo neanderthalensis* and suggested that the Ngandong hominins be referred to as "the Ngandong group of *Homo erectus erectus*" Santa Luca (1980, p. 133). In a recent careful study Kaifu et al. (2008) make a good case that the cranial evidence suggests (a) there is continuity between the Bapang-AG hominins and the Ngandong hominins, (b) that despite the evidence for continuity there are consistent morphological differences between the two samples (e.g., larger braincase, longer basiocci-

put, more marked expression of the angular torus, and reduction in the length of the attachment of the temporal muscle, etc. in Ngandong hominins compared to the Bapang-AG sample), and (c) that the Ngandong sample displays more apparently derived features than does the collection of crania from the Bapang-AG levels. *See also Homo (Javanthropus) soloensis*; *Javanthropus*.

Ngawi (Location 7°23′S, 111°39′E, Indonesia; etym. named after a local village) History and general description This is the site of a 1987 surface find of a hominin calvaria found eroded out of the bank of the Solo River near the village of Ngawi, 10 km/6 miles south of **Ngandong**. Temporal span and how dated? Unknown, but the morphology of Ngawi matches that of the fossil hominins found at Ngandong and **Sambungmacan**. The most recent radiometric estimates of the ages of these sites range between 27 and 53 ka. Hominin found at site **Ngawi 1** is a presumed adult female calvaria that has been attributed to *Homo erectus* and to *Homo (Javanthropus) soloensis*. Archeological evidence found at site None. Key references: Geology, dating, and paleoenvironment Swisher et al. 1996. Hominins Widianto and Grimaud-Hervé 1993, Widianto and Zeitoun 2003.

Ngawi 1 Site **Ngawi**, Indonesia. Locality Near the village of Ngawi. Surface/*in situ* Surface. Date of discovery 1987. Finder A local villager. Unit N/A. Horizon N/A. Bed/member N/A. Formation N/A. Group N/A. Nearest overlying dated horizon N/A. Nearest underlying dated horizon N/A. Geological age *c*.27–53 ka based on morphological similarities to the **Ngandong** and **Sambungmacan** crania. Developmental age Adult. Presumed sex Female. Brief anatomical description Ngawi 1 is an adult calvaria and, based on its relatively low estimated cranial capacity of about 1000 cm^3, presumptively female. In its preservation and morphology it is most similar to the Ngandong and Sambungmacan crania. Announcement Sartono 1991. Initial description Sartono 1991. Photographs/line drawings and metrical data Widianto and Grimaud-Hervé 1993, Widianto and Zeitoun 2003. Detailed anatomical description Widianto and Zeitoun 2003. Initial taxonomic allocation *Homo erectus*. Taxonomic revisions cf. *Homo (Javanthropus) soloensis* (Widianto and Zeitoun 2003). Current conventional taxonomic allocation *H. erectus*. Informal taxonomic category Pre-modern *Homo*. Significance Evidence from another east Javan site of a hominin calvaria similar to those from Ngandong and Sambungmacan. Location of original **Balai Arkeologi**, Yogyakarta, Indonesia.

Ngebung (Location 7°20′S, 110°58′E, Indonesia; etym. named after a local village) A locality in the northwestern part of the Sangiran Dome where the Bapang Formation is exposed and from where stone tools, including utilized large flakes and bolas stones, have been recovered. Key references: Archeology Sémah et al. 1992. *See also* **Sangiran**.

Ngrejeng (Location 7°20′S, 110°58′E, Indonesia; etym. named after a local village) History and general description Within the **Sangiran Dome**, this is the site of a 1985 surface find of a hominin mandibular fragment. The specimen was most likely derived from a stratum between the Lower and Middle Tuffs of the **Bapang Formation**. Temporal span and how dated? Trace element analysis of a bone sample drilled from the corpus, and the mineralization and color of the specimen are consistent with its stratigraphic assignment. The most recent radiometric estimates of the age of this deposit range between 1.24 and 1.51 Ma. Hominin found at site **NG 8503** is a juvenile, consisting of the right side of the mandibular corpus with M_1 and an unerupted M_2. It has been referred to *Homo erectus*. Archeological evidence found at site None. Key references: Geology, dating, and paleoenvironment Matsu'ura et al. 1995a, 1995b, Larick et al. 2001; Hominins Aziz et al. 1994, Kaifu et al. 2005a.

NHG Acronym for **Northern Hemisphere Glaciation** (*which see*).

Niah cave (Location 3°49′09″N, 113°46′42″E, Sarawak, Borneo Island, Malaysia) History and general description The Niah cave (or the Great Cave of Niah) is located on the coastal plain of Sarawak (northern Borneo), about 15 km/9.3 miles from the South China Sea. The paucity of marine plants and animals in the Niah sediments suggests the cave was always too far away from the coast for the foraging groups to transport such foodstuffs back to the cave. The cave was excavated extensively in the 1950s and 1960s by Tom and Barbara Harrisson, and it came to the attention of the international scientific community in 1958 following the discovery of a skull of an anatomically modern human, nicknamed the "Deep Skull." The Deep Skull was excavated from an area nicknamed "Hell Trench" because the midday sun beats directly down in that area making it a very difficult area to work continuously. Niah cave contains Late Paleolithic and Neolithic deposits, the latter including as many as 1000 human burials that have been dated to between 6 and 2 ka. Temporal span and how dated? **Uranium-series dating** provides ages of 35.2±2.6 ka, ages from **radiocarbon dating** are between 41 and 39 ka, and **AMS radiocarbon dating** provides ages between 46 and 34 ka; the Neolithic remains have been radiocarbon dated to between 6 and 2 ka. One of the main questions related to the Deep Skull is the possibility that it originated from one of the overlying Neolithic burials. However, the combination of dates from charcoal samples found in close proximity to the skull, together with excavation records, supports the argument that it is Late Pleistocene and dates to between 45 and 39 ka. Hominins found at site The Deep Skull is actually a partial cranium comprising occipital, frontal and upper face, maxilla with four teeth, and part of the basicranium; no mandibular fragments or mandibular teeth were identified. The cranium is gracile in appearance, has a partially erupted M^3, and noticeable wear on the M^1; these features suggest that the cranium belongs to a young adult female, possibly between 20 and 30 years of age. Its morphology is most similar to that of modern day Tasmanian and Negroid populations. The maturity of an almost complete left femur and a right proximal tibia fragment found near the Deep Skull is consistent with those bones belonging to the same individual as the cranium. A modern human-looking talus was identified in the "nonhominin" fauna originally sent to Leiden, The Netherlands, for further study. Archeological evidence found at site Taphonomic analysis indicates that clusters of excavated burnt bone are suggestive of hearths or individual depositional events of the calcined bones. Twelve cutmarked bones identified in Hell Trench, in the area of the Deep Skull, include the bones of suids, leaf monkey/macaque, and monitor lizards. Charcoal is present in many different areas of the excavation area. Several thousand stone artifacts best described as simple core and flake tools produced on locally available raw materials were recovered from the site. The original excavator, Tom Harrisson, referred to these core tools as "choppers," which was a standard description for the time. In addition 146 bone artifacts were identified at Niah, most of them from deposits dating to the Pleistocene/Holocene transition. At least three bone tools have been identified in deposits that may be coeval with the Deep Skull. There are also suggestions that some of the pig incisors may have been used to fashion wooden implements. <u>Repository</u> Sarawak Museum, Borneo, has most of the materials. Some of the materials are also stored in the Natural History Museum, London, England, the University of Bradford, England, the University of Las Vegas, USA, and the University of Leiden, The Netherlands. Key references: <u>Geology, dating, and paleoenvironment</u> Harrisson 1958, 1970, Hooijer 1963, Gilbertson et al. 2005, Stephens et al. 2005, Barker 2010, Barker et al. 2007, 2010; <u>Hominins</u> Brothwell 1960, Birdsell 1979, Kennedy 1979, Krigbaum and Datan 1999, Barker 2010, Barker et al. 2010; <u>Archeology</u> Harrisson 1958, 1970, 1978, Barker 2005, 2010, Rabett 2005, Barker et al. 2002, 2007, 2010.

niche (L. *nidus*=nest) The physically distinctive space, or habitat, occupied by a species. It can also mean the ecological role of a species.

niche construction A term introduced by John Odling-Smee (1988) to describe **biotic** environmental modification. Organisms adapt to their environment, but they are also capable of modifying the environment to better suit their needs (e.g., beavers). This organism-driven environmental modification is seldom as dramatic or substantial as the environmental modification that is brought about by tectonism or climate change, but it is nonetheless significant. Modern humans are capable of significant environmental modification; it is unclear at what stage in their evolution they shifted from adapting to existing environments to being substantial modifiers of the environment. The principles of niche construction are set out by Odling-Smee et al. (2003) in a book of the same name.

Nihewan Basin (Location approximately 41°13′N, 114°40′E, Shanxi and Hubei Provinces and Inner Mongolia Autonomous Region, northern China) <u>History and general description</u> The Nihewan Basin is an area of some 150–200 km^2/58–77 square miles comprised of Cenozoic fluvial-lacustrine deposits.

The basin contains a paleolake and a series of tributary stream deposits. The primary river that cuts through the basin is the Sanggan. The Nihewan Basin came to the attention of geologists and paleontologists beginning in the 1920s because of the presence of thick sedimentary deposits (more than 90 m thick in some areas) and the discovery of heavily fossilized animal bones eroding out of these deposits. Beginning in the 1920s, Licent, Barbour, and Teilhard de Chardin provided much of the early geological and paleontological descriptions of the sediments and materials from the Nihewan Basin. The Nihewan Basin sites are among the oldest evidence of hominins in northern China, and currently more than 12 Early and Middle Pleistocene sites have been identified in the eastern part of the Nihewan Basin; **Xujiayao** is a major Middle–Late Pleistocene open-air site located in the western Nihewan Basin. The best-known early Pleistocene sites are Goudi/Majuangou III (they are the same site), Xiaochangliang, and Donggutuo; there are also terminal Pleistocene sites (e.g., Hutouliang). <u>Temporal span and how dated?</u> The thick sedimentary deposits have been dated to the Pliocene and Pleistocene. Traditionally, the Plio-Pleistocene vertebrate remains, referred to as the Nihewan Fauna, corresponds with the European Villafranchian Fauna. An unusual aspect of the Nihewan Fauna is the apparent co-occurrence of *Hipparion* and *Equus* at a number of sites, some of which may be *c.*1 Ma. The earliest faunas from the Nihewan Basin include *Postschizotherium*, *Hipparion*, *Zygolophodon*, and *Proboscidipparion*; what appear to be later faunas include *Palaeoloxodon*, *Equus*, *Bison*, and *Dicerorhinus*. Interpretations of the **magnetostratigraphy** have suggested Goudi/Majuangou III may date to *c.*1.66 Ma, Xiaochangliang may date to *c.*1.36 Ma, and Donggutuo to *c.*1.1 Ma; **electron spin resonance spectroscopy dating** dates are much younger and range between 0.87 and 0.21 Ma. <u>Hominins found at site</u> None reported. <u>Archeological evidence found at site</u> More than 20,000 artifacts and vertebrate fossils have been found on the surface or collected from excavations in fine-grained silt, sand, and clay fluvial deposits in the Nihewan Basin. The artifacts are primarily small core and flake tools produced on locally abundant, diverse rock types, including fine-grained silicified quartzite. One of the defining characteristics of the Nihewan Basin is the overall poor-flaking quality of the stone raw materials. While many of the Nihewan core and flake tools indicate similar flaking techniques and capabilities as Oldowan assemblages in Africa, some Nihewan lithic assemblages lack evidence of deliberate stone knapping (e.g., platform-preparation techniques, retouch, etc.), while others (e.g., Donggutuo) show careful core-reduction techniques. Some use-wear analyses indicate possible evidence of meat and/or hide processing but these studies have only been done on a small subset of the overall artifact collection. Initial taphonomic observations on the fauna from Majuanguo III have reported percussion marks, but another taphonomic analysis of a small subset of the Xiaochangliang faunal assemblage failed to identify any unequivocal evidence of hominin modification. Repository Institute of Vertebrate Paleontology and Paleoanthropology, Institute of Cultural Relics of Hebei Province, China, Donggutuo Field Research Station, China. Key references: Geology, dating, and paleoenvironment Barbour et al. 1927, Teilhard de Chardin and Piveteau 1930, Schick et al. 1991, Pope and Keates 1994, Zhu et al. 2001, 2003, 2004, Chen et al. 2003, Gao et al. 2005; Hominins N/A; Archaeology Jia 1985, Wei, 1988, Schick et al. 1991, Pope and Keates 1994, Keates 2000, Peterson et al. 2003, Shen and Chen 2003, Shen and Wei 2004, Gao et al. 2005, Keates 2010, Shen et al. 2010, Potts and Teague 2010.

NISP An acronym for **number of identified specimens** (*which see*).

Nissl stain A histological staining technique developed by the German neurologist Franz Nissl (1860–1919), used to show the **cytoarchitecture** of the nervous system. This method stains large cytoplasmic granular bodies of rough endoplasmic reticulum that are found predominantly in the cell bodies of **neuron**s; some light staining of the cell bodies of **glial** cells is also seen. The Nissl stain has been used in comparative studies of the organization of the brains of modern humans and great apes to investigate species differences in the distribution of neuronal morphology, and in neocortical area **parcellation** and organization. For example, the Nissl staining method was used to identify **von Economo neurons** in hominid primates and to examine asymmetry in the minicolumns of **Wernicke's area** of modern humans and common chimpanzees.

nitrogen An element commonly used in **stable isotope biogeochemistry**. There are two common isotopes in nitrogen, ^{15}N and ^{14}N; both are stable and do not decay. *See also* ^{15}N/^{14}N; **stable isotopes**.

NK Abbreviation of Nariokotome, one of several groups of localities within the site of **West Turkana** named after ephemeral river drainages. *See* **Nariokotome**.

NMK Acronym for the **National Museums of Kenya** (*which see*).

NN Abbreviation of Nanyanakipi one of several groups of localities within the site of **West Turkana** named after ephemeral river drainages. *See* **Nanyanakipi**.

nocturnal (L. nocturnus=belonging to the night) The term used for animals that are active at night (i.e., between dusk and dawn). Many strepsirrhine primates (lorises, galagos, and some lemurs) are nocturnal, tending to forage on their own in darkness, and they have a reflective "cat's eye" (a *tapetum lucidum*) that facilitates night-time activity. Since visual cues for things such as conspecifity are less helpful in low light, many nocturnal primates rely more on olfaction (for example, scent marking), or on calls, to distinguish species, groups, and territories. There are many fewer nocturnal haplorhines; the tarsiers and neotropical owl monkeys are the only taxa to be routinely active at night. They do not have a *tapetum lucidum*, but instead they have relatively bigger eyes to facilitate nocturnal activity. No catarrhines are nocturnal, and activity after dusk and before dawn tends to be minimal. Baboons, for example, have preferred group sleeping sites and will settle there before sundown. Most modern human activity at night is made possible because of the cultural innovation of artificial light, and is likely therefore that until fire could be controlled to provide illumination, nocturnal activity would have been severely limited for much of hominin evolutionary history.

Nojiriko *See* **Tategahana**.

nomen (L. *nomen*=name; pl. *nomina*) Commonly used in the field of biological nomenclature to refer to any scientific name, at any level (e.g., Primates, Hominidae, Hominini, *Australopithecus*, *Australopithecus africanus*). It refers to the name itself, not to any taxon that may be designated by the name. A new taxon is often published together with the phrase "*nomen novum*", meaning "new name." There are other types of nomina. A *nomen nudum* is a name that is unavailable in nomenclature (i.e., it cannot be used) because it was published without a description or definition of the taxon it denotes. A *nomen oblitum* ("forgotten name") is one which has not been used since 1899, and need not be revived as long as a younger name (a *nomen protectum*) has in the meantime become more familiar. A *nomen dubium* is a name which is of doubtful application, and therefore left in limbo.

nomenclature (L. *nomenclatura* = calling by name; e.g., calling the role) The principles and practice of providing names for taxa. The process of naming a new group is controlled by rules and recommendations set out in the **International Code of Zoological Nomenclature**, otherwise known as the "Code." The stipulations in the Code are designed to ensure that everyone who takes part in discussions about issues involving classification and nomenclature does so with a common understanding of the taxonomic categories and of the taxa in each of the categories. The purpose of the rules and recommendations set out in the Code is to make sure the names given to new taxa do not duplicate those of existing taxa (i.e., **homonymy**), and to prevent two or more names being given to the same taxon (i.e., **synonymy**). It also describes procedures that are designed to avoid new taxa being given inappropriate names.

Nomina Anatomica *See* **anatomical terminology**.

nondisjunction During **meiosis** a pair of **chromosomes** aligned at the equator of the cell may not separate properly, resulting in either no or two copies of that chromosome in all (if it occurs in meiosis I) or some (if it occurs during meiosis II) of the resulting **haploid** gametes. If such gametes join with a normal gamete and become part of a zygote, then the zygote will be aneuploid (it will have an abnormal number of chromosomes) and it will either have a **monosomy** or **trisomy** of that particular chromosome.

nonindependence *See* **independent contrasts**.

non-metrical traits Most morphology can be satisfactorily recorded using measurements. Analyses based on measurements are examples of **metrical analysis**. But some morphology is diffi-cult to measure and is more amenable to **non-metrical analysis** based on non-metrical traits. Most skeletal non-metrical traits are craniodental (like many of the examples given below), but they can also be postcranial. Non-metrical skeletal traits include additional bones called ossicles (e.g., bregmatic, asterionic), bony tubercles (e.g., marginal, pharyngeal), bony ridges or tori (e.g., angular torus, mandibular torus, palatine torus), bony processes (e.g., paracondylar process, third trochanter), bony connections called "bridges" (e.g., divided hypoglossal canal, mylohyoid bridge), fissures (e.g., squamotympanic, mastoid), sutures (e.g., infraorbital, metopic), foramina (e.g., mastoid, supraorbital), and additional cusps, ridges and fissures on the enamel of tooth crowns. Some can be recorded (also called scoring) as either present or absent – these are also called discrete variables – or they are given a nominal value by comparing the morphology observed with predefined categories of expression of that feature (e.g., small/medium/large or well-expressed, expressed, weakly expressed, absent, etc.). Bony non-metrical traits are also divided into hyperostotic and hypostotic, literally meaning traits that are the result of "more bone than usual" and traits that are due to "less bone than usual," respectively. Non-metrical traits are usually regarded as examples of the interaction between genetic information and the developmental environment, and so they are also called epigenetic traits. Hauser and De Stefano (1989), Buikstra and Ubelaker (1994), Manzi (2003), and Hallgrímson et al. (2005) provide excellent reviews of the utility of commonly used non-metric traits. *See also* **epigenetic**.

nonparametric statistics Statistical tests that do not assume that data follow any particular distribution. The most commonly used tests are of **parametric statistics** which assume particular probability distributions (e.g., the **normal distribution**). If data do not meet the assumptions of parametric tests, it is more appropriate to use nonparametric tests such as the **Mann–Whitney U test** or **Wilcoxon signed rank test**, or **resampling** techniques such as the **bootstrap** and **randomization**.

nonrandom mating When mating is affected by physical, genetic, or social preference or by a barrier (e.g., a geographic barrier such as a large river). Within a population, nonrandom mating results in increased inbreeding and random **genetic drift**. Nonrandom mating can include positive and negative assortative mating. In modern humans cultural factors (e.g., language, education, and religion) and physical characteristics (e.g., height and skin color) are known to affect mating patterns (Spuhler 1972, Lie et al. 2010).

nonrandom sampling *See* **bias**; **ascertainment bias**.

nonsense mutation A mutation that changes a codon sequence from one that encodes an amino acid to one that encodes a stop codon. This results in a truncated protein sequence and such mutations are typically deleterious. *See* **genetic code**; **mutation**.

nonsynonymous mutation A point mutation (i.e., change in one nucleotide) in a coding sequence that results in a change in the amino acid sequence of a protein. Most changes of this sort are deleterious and are subject to purifying selection, but a small fraction are adaptive and are subject to positive selection. For example, in the modern human lineage there have been two nonsynonymous mutations in the *FOXP2* gene, which encodes **FOXP2**, a **transcription factor**. This new amino acid sequence is highly conserved in evolution and in modern humans changes in the *FOXP2* gene have been linked to problems with speech and grammar. Enard et al. (2002) demonstrated that this gene has been subject to recent positive selection. *See also* **FOXP2**; *FOXP2*; **mutation**.

nonsynonymous substitution *See* **nonsynonymous mutation**.

norm of reaction *See* **reaction norm**.

norma basalis *See* **norma basilaris**.

norma basilaris (L. *normal*=made according to a pattern, or standard, and basilar is a neologism introduced by translators, that presumably comes from *basis*=base, or pedestal) One of the standard views of the cranium. It is the inferior view, the view that looks at the cranium from its underside, or base. In modern humans it mainly comprises the inferior surface of the hard palate and the undersides of the sphenoid, temporal, and occipital bones.

norma frontalis (L. *normal*=made according to a pattern, or standard, and *frons*=brow, or forehead) One of the standard views of the cranium. It is the anterior view, the view that looks at the forehead and face. In modern humans it comprises the anterior surface of the face, and the anterior surfaces of the squamous part of the frontal bone and the parietal bones.

norma lateralis (L. *normal*=made according to a pattern, or standard, and *latus*=side, or flank) One of the standard views of the cranium. It is the view from the side. In modern humans it comprises the lateral surfaces of the frontal, temporal, parietal, and occipital bones.

norma occipitalis (L. *normal*=made according to a pattern, or standard, and *occipio*=to begin or commence: the part of the head that usually appears first during a normal birth; or *ob*=against and *caput*=head: the part of the head that rests on the ground when you sleep on your back) One of the standard views of the cranium. It is the posterior view. In modern humans it mostly comprises the posterior surfaces of the parietal and **occipital** bones.

norma verticalis (L. *normal*=made according to a pattern, or standard, and *vertex*=highest point) One of the standard views of the cranium. It is the view from above (i.e., looking down onto the vertex, or the highest point, of the cranium. In modern humans it mostly comprises the superior surfaces of the frontal, parietal, and occipital bones.

normal The state of the Earth's magnetic field when the needle of a magnet points to the North Pole. *See also* **geomagnetic dating**.

normal distribution A particular probability distribution characterized by data falling symmetrically about the mean and with most data falling relatively close to the mean with observations becoming less frequent as distance from the mean increases. Most traditional statistical tests assume that data are normally distributed; these tests are referred to as **parametric statistics**. Many data sets are not normally distributed, however, and parametric tests should not be applied in these cases. For example, data sets containing counts of specimens or species at a site in which many sites have an observed value of zero are unlikely to be symmetrically distributed and thus not normally distributed. When data are not normally distributed, parametric tests are at a greater risk of **Type I error** and **nonparametric statistics** should be used instead (syn. Gaussian distribution).

normal fault In structural geology a normal fault is a fault or fracture with an extensional geometry. This type of fault is sometimes referred to as a dip-slip fault and as this term implies the dominant component of the displacement is down the **dip** of the affected strata such that the hanging wall (i.e., the rocks above the fault plane) is displaced downwards relative to the rocks of the footwall (i.e., the rocks below the fault plane)

normal views (L. *normal*=made according to a pattern, or standard) Standard views of the cranium (i.e., *norma verticalis*, superior view; *norma frontalis*, anterior view; *norma lateralis*, right or left lateral view; *norma occipitalis*, posterior view; *norma basalis*, inferior view).

normalizing selection *See* stabilizing selection.

Northern blotting A technique used in molecular biology to study gene expression. In a Northern blot **electrophoresis** is used to separate RNA by size on a gel. The RNA in the gel is transferred to a membrane that is then washed with probes that hybridize to the RNA segments of interest.

northern hemisphere glaciation Refers to the occurrence of major ice sheets over many northern landmasses including North America, Greenland, and much of Asia. The onset of the northern hemisphere glaciation (sometimes referred to as NHG or the **Late Pliocene transition**) is defined by the oxygen isotope record and it takes place in the late Pliocene between 3.2 and 2.7 Ma.

Notanthropus *See* **Sergi, Giuseppe (1841–1936)**.

NRR Acronym for **net reproductive rate** (*which see*).

NS Abbreviation of Nasechebun, one of several groups of localities within the site of **West Turkana** named after ephemeral river drainages. *See* **Nasechebun**.

NT Abbreviation of Natoo one of several groups of localities within the site of **West Turkana** named after ephemeral river drainages. *See* **Natoo**.

nuchal [ME *nucha*=nape (or back) of the neck] The area at the back of the neck where you can feel the nuchal muscles. Used also for the nuchal area on the underside of the basicranium posterior to the foramen magnum where the nuchal muscles are attached.

nuchal area height index [ME *nucha*=nape (or back) of the neck] The index introduced by Wilfrid **Le Gros Clark** (1950b) to describe the relationship between the height of the cranial vault and the uppermost limit of the nuchal area. It is the ratio between the height of the highest point of the nuchal area above the Frankfurt Horizontal (AG in the index) and the total height of the calvaria above the Frankfurt Horizontal, times 100. The higher the nuchal area the higher the value (e.g., in modern humans the average is approximately 0 to +5 and in chimpanzees it is about 50).

nuchal crest [ME *nucha*=nape (or back) of the neck] A bony crest that takes the place of the superior nuchal line, a roughened line on the squamous part of the occipital bone. It is produced by the attachment of the fascia that covers one of the biggest nuchal muscles, the semispinalis capitis, and when the muscle and the fascia are well developed the superior nuchal line takes the form of a bony crest. Such crests are common in large male gorillas, but they are also seen in some hominin taxa (e.g., *Australopithecus afarensis* and large, presumed male, megadont archaic hominins such as **OH 5**).

nuchal muscles The muscles (e.g., trapezius and sternocleidomastoid covering the semispinalis and splenius capitis) you can see and feel at the back of the neck. *See also* **nuchal**; **nuchal crest**.

nuchal plane [ME *nucha*=nape (or back) of the neck] Refers to the surface of lower or inferior part of the squamous occipital bone. The inion is the dividing point in the midline between the nuchal plane and the upper part of the squamous occipital bone that forms the occipital plane (syn. lower scale of the occipital bone).

nuclear DNA The **DNA** that is found within the **nucleus** of eukaryotic cells (i.e., it is intra-nuclear).

In most eukaryotes nuclear DNA, which in a modern human diploid cell consists of roughly 6.4 billion base pairs, forms the majority of the **genome**. The other part of the genome of eukaryotes is the **mitochondrial DNA** found within mitochondria, which are organelles in the cytoplasm (i.e., mitochondrial DNA is "extra-nuclear"). *See also* **DNA**.

nucleotide Molecules that form the basis of both **DNA** and **RNA**. They are also involved in metabolism (e.g., adenosine triphosphate is a nucleotide). Nucleotides are composed of a nitrogenous base (either a purine or a pyrimidine), a five-carbon sugar (ribose or deoxyribose), and one or more phosphate groups.

nucleus (L. *nux*=nut) The term nucleus is used in two ways relevant to human evolution. In cell biology it refers to the membrane-bound organelle in a nucleated, or eukaryotic, cell that contains DNA (known as nuclear DNA). The transcription of DNA occurs *inside* the nucleus, whereas the process of translation occurs *outside* the nucleus in the cytoplasm. In neuroscience nucleus refers to a geographically localized group of neurons that share the same morphology and/or function (e.g., the caudate nucleus, or nuclei of the cranial nerves), or to a group of nuclei (e.g., the basal ganglia, or basal nuclei at the base of the forebrain).

null hypothesis In statistics, a statement to be tested that is typically framed as the absence of a difference between groups or absence of a relationship between variables. The *P* value is the probability that the observed pattern in the data could occur if the null hypothesis were true; if the *P* value is below some threshold (usually 0.05) then the test is said to show a statistically significant result and the null hypothesis is rejected.

number of identified specimens (or NISP) A measure that reflects the number of identified specimens (i.e., a skeletal element, or portion thereof) of any taxon represented in a bone assemblage. The NISP serves as the baseline from which all other related quantitative units (e.g., **minimum number of elements, mini-** mal animal units, and **minimum number of individuals**) are derived. NISP counts are used to compare taxonomic abundances through time or across fossil localities.

numerical-age methods *See* geochronology.

NY Abbreviation of Naiyena Engol one of several groups of localities within the site of **West Turkana** named after ephemeral river drainages. *See* **Naiyena Engol**.

NY 75′86 *See* Nyabusosi.

Nyabusosi (Location approximately 1°02′N, 30°21′E, Uganda; etym. unknown) History and general description Nyabusosi is a Plio-Pleistocene site in a long series of sediments in the Western Rift about 25 km/15.5 miles southwest of Lake Albert, Uganda. The approximately 1450 m-thick sequence preserves sediments dating from the early Miocene through the Quaternary. W.W. ('Bill') **Bishop** initially explored and mapped the geology in 1963 and reported several localities with Plio-Pleistocene mammal faunas. Some sediments from Nyabusoi have yielded suids that are **Pliocene** (e.g., "Behanga-type" suids that correlate with **Omo Shungura** Member G, i.e., 1.8–2.0 Ma) and others **Middle Pleistocene** suids (e.g., *Mesochoerus majus*). Re-study of the sequence by Senut et al. yielded a badly crushed and fragmentary hominin cranium (NY 75′86) from locality NY 15. Associated fauna suggests a Lower Pleistocene date. Temporal span and how dated? Associated fauna [e.g., *Elephas recki*, *Hippopotamus imagunculus* (or *Hippopotamus aethiopicus*), and *Mesochoerus limnetes*] suggests a **Late Pliocene** to Lower Pleistocene age. Hominins found at site NY 75′86, a badly crushed and fragmentary cranium attributed to *Homo* sp. that is embedded in ferruginous sandstone matrix. Archeological evidence found at site Site NY 15 yielded several **Oldowan**-like small flakes and polyhedrons made of **quartz** and **quartzite**. Key references: Geology, dating, and paleoenvironment Bishop 1965, Gautier 1970, Harris and White 1979, Senut et al. 1987; Hominins Senut et al. 1987; Archeology Senut et al. 1987.

O

O. Prefix used by the French contingent of the **International Omo Research Expedition** for localities in the **Shungura Formation** (e.g., O. 18). The prefix is also used for fossils found within those localities (e.g., O. 18-1967-18 for **Omo 18-1967-18**).

$^{18}O/^{16}O$ ratio Oxygen isotopes can be measured in animal proteins, such as collagen, and in the carbonate or phosphate molecules present in shell, bone, or enamel apatite. The ratio of these stable oxygen isotopes reflects the isotopic ratio of the water ingested by an animal either as drinking water or as part of their **diet** (see review by Koch 1998). Oxygen isotope ratios in plants reflect their water source and the relative humidity of the environment in which they are growing. Plants growing in drier environments (and animals consuming these plants) have higher $^{18}O/^{16}O$ ratios (commonly represented as $\delta^{18}O$ values in per mil, or %, notation) than those growing in wetter, more humid environments. Among **C_3 plants**, $\delta^{18}O$ values tend to differ among plant parts. Leaves, which are subject to evapotranspiration, tend to have higher $\delta^{18}O$ values than fruits (transpiration leads to the preferential loss of ^{16}O over ^{18}O from leaf surfaces). As a result, $\delta^{18}O$ values in consumers may be able to distinguish frugivores and folivores (Cerling et al. 2004). Recent research suggests that $\delta^{18}O$ values can also be used to distinguish large herbivores that drink frequently from those that drink infrequently (Levin et al. 2006). Animals that drink infrequently (called drought-tolerant) tend to have higher $\delta^{18}O$ values than those that drink frequently (drought-intolerant). Oxygen isotope ratios can be used to interpret **paleoclimates**, for ^{16}O evaporates more readily than ^{18}O from large bodies of water. Thus rain and rain-derived terrestrial water sources tend to be ^{18}O-depleted (lower $\delta^{18}O$ values) compared to ocean water. During colder periods, water evaporated from oceans tends to be trapped on land as ice. This leaves relatively more ^{18}O in liquid water reservoirs such as oceans and lakes. Animals ingesting water during cooler glacial periods thus have higher $\delta^{18}O$ values than animals living during warmer interglacial periods. Global shifts in $\delta^{18}O$ values associated with periodic glacial/interglacial cycling have been used to create **Oxygen Isotope Stages**, which are used to calibrate sites in higher latitudes. *See also* **stable isotopes**; **Oxygen Isotope Stages**.

O/M Abbreviation for the **occipito-marginal system** of intracranial venous sinuses. Individuals with a substantial occipital sinus and with marginal venous sinuses on both sides (e.g., **OH 5**) are said to show occipito-marginal system dominance. *See also* **cranial venous drainage**.

Oakley, Kenneth Page (1911–81) Kenneth Oakley was born in Amersham, England, and entered University College London in 1930 to study geology and anthropology. After completing his BSc in 1933, Oakley joined the Geological Survey (in 1934) and in 1935 he was appointed Assistant Keeper of Paleontology at the British Museum (Natural History). He completed his PhD at University College London in 1938 with a thesis on Silurian bryozoa. Oakley's work at the British Museum was interrupted by the onset of WWII when he was attached to the Geological Survey and began work on the use of phosphate for fertilizer. This introduced Oakley to the work of Adolphe Carnot, a French mineralogist who had noted in 1893 that the absorption of fluorine in groundwater by bones might be used to determine the age of archeological or paleontological material. Prehistoric archeology had long been plagued by an inability to provide absolute dates for the artifacts or bones excavated, and at the time stratigraphy only gave a relative date. Oakley determined that the absorption of fluorine by bone increased the longer the bone had been buried, but the

Wiley-Blackwell Encyclopedia of Human Evolution, First Edition. Edited by Bernard Wood.
© 2013 Blackwell Publishing Ltd. Published 2013 by Blackwell Publishing Ltd.

amount of fluorine varied widely depending upon other factors such as the concentration of the fluorine in the groundwater and other conditions. This meant fluorine content could not be used to calculate an absolute date, but it could be used to test the relative age of objects found in the same deposit. An early opportunity to test the usefulness of this technique occurred in 1947 when Oakley attended the first **Pan-African Congress on Prehistory** organized by Louis **Leakey** in Nairobi, Kenya. In the early 1930s Leakey had unearthed portions of a cranium and jaw from two adjacent sites called **Kanam** and **Kanjera** that he considered to be from an early form of anatomically modern human, and the geological evidence suggested the fossils were extremely old. This was consistent with Leakey's belief that the australopiths were not human ancestors, and that another more modern human-like line stretched far back in time. Oakley convinced Leakey to allow a **fluorine dating** test of the bones to determine their general age, but the test was inconclusive due to the high fluorine content of the groundwater where the fossils were found. However, the fluorine dating technique proved its worth a year later when he tested the controversial Galley Hill specimen, a morphologically modern-looking skeleton found in a geologically old deposit in 1888. Galley Hill Man was used by Sir Arthur **Keith** to support his own theory that modern humans were of great antiquity, but many doubted that the skeleton was as old as had been claimed. Oakley announced at the 1948 meeting of the British Association for the Advancement of Science that the fluorine test indicated that the Galley Hill remains were quite recent when compared to the results obtained from other fossils from the same area that were known to be old. This suggested that the Galley Hill skeleton was a much later intrusive burial. The success of this test indicated that the fluorine dating method might be used to resolve doubts about the relative age of other controversial fossils. Oakley was released from his wartime duties in 1947, and he returned to the British Museum as a principal scientific officer. In addition to working on fluorine dating methods, Oakley also began work on the collection of prehistoric artifacts housed in the Museum's Department of Geology. In 1949 Oakley published *Man the Tool-Maker*, a book that examined the evidence for human antiquity, the origins and importance of tool-making, the chron-

ology of tool types during the Paleolithic, and the significance of tool use and tool-making to the evolutionary and cultural development of hominids, and contributed to debates over the role of tool use in human evolution. Oakley is perhaps best known for his contributions to exposing the Piltdown hoax. Between 1908 and 1915 Charles Dawson, a solicitor and amateur paleontologist, unearthed fragments of a cranium and mandible from a gravel pit in Sussex, England. The fossils seemed to represent a hitherto unknown hominin that possessed a large brain, but an ape-like jaw. Several prominent English anthropologists, notably Sir Arthur Smith Woodward and Sir Arthur Keith, quickly claimed **Piltdown** Man (*Eoanthropus dawsoni*) was either a direct ancestor of modern humans or on its own branch, but some researchers were uneasy about the specimen. One frequent objection was that the cranial fragments and the mandible belonged to two separate creatures. Debate over Piltdown continued over the next few decades but many anthropologists accepted the specimen at face value. As time passed, however, no more seemingly Piltdown-like hominins were discovered, while in Africa and Asia numerous specimens of *Homo erectus* and *Australopithecus* were making the Piltdown specimen look increasingly aberrant. By 1950 Piltdown Man had been pushed to the periphery of scientific discussion and critics of the fossil felt strengthened in their doubts about the validity of the specimen. The demise of Piltdown began in 1949 when Joseph Sidney Weiner, Reader in Physical Anthropology in the Department of Anatomy at Oxford University, was encouraged by Wilfrid **Le Gros Clark**, Dr Lee's Professor of Anatomy at Oxford University, to initiate an extensive investigation of the Piltdown remains. Weiner later collaborated with Oakley, whose contribution was to subject the cranium and jaw to fluorine testing. This showed conclusively that the fossils were much younger than had previously been thought. They concluded from their investigation that the Piltdown fossils were a forgery: a relatively recent cranium had been planted along with a fresh orangutan jaw that had been chemically stained and the teeth filed down to resemble human teeth. Weiner, Oakley and Le Gros Clark were the joint authors of an article entitled "The solution to the Piltdown problem," published in 1953 in the *Geological Bulletin of the British Museum (Natural History)*, the publication that finally settled the debate over Piltdown Man. Oakley continued to work on the problem of establishing reliable relative and absolute dating methods, which culminated in his publication of *Frameworks for Dating Fossil Man* in 1964. He also collaborated with Bernard Campbell and Theya Molleson in compiling the three-volume *Catalogue of Fossil Hominids* (1967, 1971, 1977). Oakley retired from his position at the British Museum (Natural History) in 1969. *See also* **fluorine dating**; **Piltdown**.

Oase 1 Site Peştera cu Oase. Locality Panta Strămoşilor. Surface/*in situ* Surface, recovered amid *Ursus spelaeus* and other large-mammal fossil bone remains. Date of discovery 2002. Finders S. Milota, A. Bîlgăr, and L. Sarcina. Nearest underlying dated horizon Stalagmite dated by **TIMS uranium-series dating** to 41,620±*c*.2400 cal. years BP. Geological age 34,950±900 years BP, an uncalibrated **AMS radiocarbon date**. Developmental age Adult. Presumed sex Male. Brief anatomical description Large and robust modern human mandible with large teeth showing modern human crown morphology. Announcement Trinkaus et al. 2003a. Initial description Trinkaus et al. 2003a. Photographs/line drawings and metrical data Trinkaus et al. 2003a. Detailed anatomical description Trinkaus et al. 2003a, 2006, Trinkaus 2007. Initial taxonomic allocation *Homo sapiens*. Informal taxonomic category Anatomically modern human. Significance Presently the oldest reliably dated modern human remains in Europe. Location of original **Institutul de Speologie "Emil Racoviţă,"** Cluj-Napoka, Romania.

Oase 2 Site Peştera cu Oase. Locality Panta Strămoşilor. Surface/*in situ* Surface, recovered amid *Ursus spelaeus* and other large-mammal fossil bone remains. Date of discovery 2003, during mapping of the site. Nearest underlying dated horizon Stalagmite dated by **TIMS uranium-series dating** to 41,620±*c*.2400 cal. years BP. Geological age 28,890±170 years BP uncalibrated **AMS radiocarbon date**. Developmental age Adolescent. Presumed sex Unknown. Brief anatomical description Modern human cranium, nearly complete. Announcement Trinkaus et al. 2003b. Initial description Trinkaus et al. 2003b. Photographs/line drawings and metrical data Trinkaus et al. 2003b, Rougier et al. 2007. Detailed anatomical description Trinkaus et al. 2003b, 2006, Rougier et al. 2007. Initial taxonomic allocation *Homo sapiens*. Informal taxonomic cate-gory Anatomically modern human. Significance Presently the oldest reliably dated modern human remains in Europe. Location of original **Institutul de Speologie "Emil Racoviţă,"** Cluj-Napoka, Romania.

Oberkassel (Location 50°43'N, 7°10'E, Germany; etym. named after a nearby town). History and general description This open-air site in a suburb of Bonn was discovered by workers in 1914. Two hominins were found associated with several arti-facts, including decorated bone objects and a dog skeleton. Temporal span and how dated? Direct **AMS radiocarbon dating** on the hominin skeleton and associated artifacts suggests an age of *c*.12 ka. Hominins found at site Two nearly complete *Homo sapiens* skeletons found buried next to each other: Oberkassel 1, associated skeleton, adult male; Oberkassel 2, associated skeleton, young adult female. Archeological evidence found at site Bone artifacts attributed to the **Magdalenian** and an associated dog skeleton. Key references: Geology, dating, and paleoenvironment Baales and Street 1998, Street et al. 2006; Archeology Bosinski 1982; Hominins Henke 1986.

objective synonym *See* **synonym**.

obligate biped *See* **bipedal**.

obligate bipedalism *See* **biped**.

oblique crest *See* **distal trigon crest**.

oblique furrow The term used by Weidenreich (1937) for a feature on a maxillary postcanine tooth crown that others call the fovea posterior. *See also* **fovea posterior (maxillary)**.

oblique ridge *See* **distal trigon crest**.

obliquity Earth science The Earth's orbital geometry is affected by three "rhythms" or "cycles." Obliquity is the *c*.41 ka cycle that relates to the tilt of the Earth's axis (i.e., the angle between the Earth's axis and the equator). It inter-acts with the *c*.23 ka precessional cycle to take the poles of the Earth either closer to or further away from the sun. It was the dominant signal affecting cyclic changes in the global climate between 3 and 1 Ma. *See also* **precession**. Morphology Obliquity is also used to refer to a methodological problem that can occur when teeth are sectioned for analysis. If the plane of section deviates from an "ideal plane" passing through enamel cusps and/or dentine horn tips, it can result in inflated, and thus erroneous, measures of **enamel thickness** and cuspal, or **crown formation time**. *See also* **precession**.

observational learning *See* **social enhancement**.

obsidian (etym. Pliny suggests it was given its name because obsidian is said to resemble a stone found in Ethiopia by Obsius) A typically black, or dark-colored, silica-rich volcanic glass. When it is subjected to percussion it fractures in a conchoidal fashion (percussion results in a so-called **conchoidal fracture**) and when it is available it is often selected for artifact manufacture at (e.g., **Enkapune ya Muto**). There is evidence from some sites that hominins traveled long distances to fetch obsidian (e.g., **Mumba Shelter**).

obsidian hydration dating Principles and method (or OHD) A chronometric dating method based on the fact that **obsidian** (volcanic glass) weathers, or hydrates, and that the amount of this hydration increases with time. In principle, accurate measurement of the hydrated area ('the hydration rim') and the rate of hydration allow calculation of the time since the obsidian surface was first exposed. In the case of stone artifacts, this exposure time is initiated during the process of flake removal, so the hydration present should be a measure of the time that has elapsed since the flake was made. Suitable materials Obsidian. Time range 200–100 ka. Problems Comparison of results obtained from obsidian hydration dating with those obtained by other methods has revealed that there is substantial variability in hydration rates due to compositional variability within obsidian sources and individual artifacts, the latter in part a consequence of substantial variation within local depositional settings (e.g., groundwater composition and past temperatures). Further difficulties arise from the precise measurement of the extent of hydration, which is typically on the scale of a few micrometers. Obsidian hydration dating remains accurate for coarse chronologies, although some advocate its use primarily as a **relative dating** tool (Hull 2001, Rogers 2008). Obsidian hydration dating, along with other

methods, has been used to provide age estimates at **Enkapune ya Muto** and other sites in East Africa.

Occam's razor William of Occam (or Ockham) was a 13th/14thC English philosopher who promulgated the rule (or "razor") that all things being equal researchers should favor the simplest hypothesis. In paleoanthopology, Occam's razor is most frequently applied to **phylogenetic** reconstruction, where the preference for the least complex explanation (i.e., the most parsimonius explanation) is a guiding principle.

occipital (L. *occipio* = to begin or commence, thus the part of the head that usually appears first during a normal birth; or from *ob* = against and *caput* = head, the occiput is the part of the head that rests on the ground when you sleep on your back) One of the unpaired cranial bones. It is a compound bone, which means that it develops from several components. They are, from front to back, a single basilar (or basioccipital) part, paired lateral (or condylar) parts, and posteriorly a single, large, flat, or squamous part. The occipital bone forms the posterior part of the basicranium and contributes to the posterior part of the cranial vault. All four components of the occipital contribute to the border of the foramen magnum. The squamous part of the occipital bone is divided in the midline by the inion into a lower part that forms the nuchal plane and an upper part that forms the occipital plane. The two parts of the occipital squame are also called the lower and the upper scales, respectively.

occipital foramen (L. *occipio* = to begin or commence, thus the part of the head that usually appears first during a normal birth; or *ob* = against and *caput* = head, the occiput is the part of the head that rests on the ground when you sleep on your back) Refers in modern anatomical terminology to a small bilateral foramen, or hole, in the occipital bone that transmits one of the venous channels that allows blood to flow back and forth between the intra- and extracranial venous systems. In times gone by (e.g., in T.H. Huxley's *Evidence as to Man's Place in Nature*, p. 76) the term occipital foramen was used for what in modern anatomical terminology is called the **foramen magnum**.

occipital index The ratio introduced by Karl Pearson to describe the curvature of the squamous part of the occipital bone in the sagittal plane. It is the ratio of the length of the radius of curvature of the occipital bone between lambda and opisthion to the chord distance between lambda and opisthion. Tobias (1967) suggested that it should be replaced by a simple chord-arc index.

occipital length index I An index introduced by Weidenreich (1943, p. 131) to express the projected distance between opisthion and opisthocranion relative to overall cranial length. Both opisthion and opisthocranion are projected onto the **Frankfurt Horizontal** and the horizontal distance between them is divided by maximum cranial length, times 100. Its value is low (approximately 10–15%) in *Pan* and *Gorilla*, and higher (about 25–35%) in modern humans.

occipital length index II An index introduced by Weidenreich (1943, p. 131) to express the distance between opisthion and opisthocranion relative to the distance between nasion and opisthion. The line between nasion and opisthion is projected posteriorly and the locations of opisthion and opisthocranion are projected onto it. The horizontal distance between them is divided by the nasion–opisthion distance, times 100. Its value is close to zero in *Pan* and *Gorilla*, and more negative (*c.*–25%) in modern humans.

occipital lobe (L. *occipio* = to begin or commence, thus the part of the head that usually appears first during a normal birth; or *ob* = against and *caput* = head, the occiput is the part of the head that rests on the ground when you sleep on your back) One of the four main subdivisions of the cerebral cortex of each cerebral hemisphere; it forms the posterior part of each cerebral hemisphere. The occipital lobe is primarily devoted to visual processing. The primary visual cortex, Brodmann's area 17 (also known as V1 and the striate cortex) is located within the calcarine sulcus, toward the rear of the occipital lobe. The most posterior point on the cerebral hemisphere, called the posterior pole, is formed by the occipital lobe.

occipital sinus (L. *occipio* = to begin or commence, thus the part of the head that usually appears first during a normal birth; or *ob* = against and *caput* = head, the occiput is the part of the head that rests on the ground when you sleep on your back) One of the dural venous sinuses that make up the external component of the intracranial venous system. The occipital sinus, when present, runs down from the cruciate eminence towards the foramen magnum in a midline groove. In most modern humans the superior sagittal sinus and the straight sinus (the latter drains venous blood from the deep cerebral veins) drain into the transverse sinuses. But in the majority of the crania belonging to *Australopithecus afarensis*, *Paranthropus boisei*, and *Paranthropus robustus*, the venous blood from the same sources drains into occipital sinus, which in turn drains into the marginal sinuses. This pattern of intracranial venous drainage is known as the **occipito-marginal system** of venous sinuses and individuals with a substantial occipital sinus and with marginal venous sinuses on both sides (e.g., **OH 5**) are said to show occipito-marginal system dominance. *See also* **cranial venous drainage**; **dural venous sinuses**; **occipito-marginal system**.

occipito-marginal system A variant of the internal component of the intracranial venous system. In most modern humans the veins draining the deep structures of the brain drain as the deep cerebral veins into the left transverse sinus, and the superior sagittal sinus drains into the right transverse sinus. Each transverse sinus drains into a sigmoid sinus, which runs inferiorly in a sigmoid-shaped groove from the lateral end of the transverse sinus to the superior jugular bulb, which marks the beginning of the internal jugular vein. This pattern of dural venous sinuses is known as the transverse-sigmoid system. However, in just a few percent of modern humans the venous blood from the brain takes a different route (Kimbel 1984). Instead of draining into the transverse sinuses, the deep cerebral veins and the superior sagittal sinus drain into an enlarged midline occipital sinus, which runs down from the cruciate eminence towards the foramen magnum in a midline groove. The occipital sinus drains into uni- or bilateral marginal sinuses, which run(s) around the margin(s) of the posterior quadrants of the foramen magnum to drain into the superior jugular bulb and/or the vertebral venous plexuses. This is known as the occipito-marginal system of venous sinuses, and individuals with a substantial occipital sinus and with marginal venous sinuses on both sides (e.g., **OH 5**) are said to show occipito-marginal system dominance. *See also* **cranial venous drainage**.

occipitomastoid crest [L. *occipio* = to begin or commence, thus the part of the head that usually appears first during a normal birth; or *ob* = against and *caput* = head, the occiput is the part of the head that rests on the ground when you sleep on your back; *mastos* = breast and *oeides* = shape and L. *crista* = crest, literally "a crest along the junction (suture) between the occipital bone and the mastoid part of the temporal bone"] The term *crista occipitomastoidea* was coined by Franz **Weidenreich** (1943, p. 64) for a crest that in the **Zhoukoudian** calvaria runs along the suture between the occipital bone and the mastoid part of the temporal bone. It may, or may not, follow the course of the occipitomastoid suture. It forms the bony lateral edge of the cranial attachment of the superior oblique (Kimbel et al. 1985), one of the short muscles of the base of the cranium. When the superior oblique muscle is confined to the occipital bone then the occipitomastoid crest and the occipitomastoid suture are likely to coincide. However, when that muscle is relatively large, or if its location is displaced laterally, then the occipitomastoid crest may run on the temporal bone instead of along the occipitomastoid suture. When the course of the occipital artery coincides with that of the occipitomastoid crest the latter may appear as a compound structure, but in effect it is a single crest divided by a groove for the occipital artery. The occipitomastoid crest is especially prominent and pneumatized in *Australopithecus afarensis*, and in that taxon it forms the lateral face of the occipitomastoid suture. In *Homo neanderthalensis* the occipitomastoid crest region is located in a relatively inferior position, resulting in the impression of a less-projecting and less-robust mastoid process and the distinctive appearance of this anatomical region in Neanderthals.

occlusal (L. *occludere* = to close) Refers to the superiorly facing surfaces of the enamel cap of the mandibular teeth and the inferiorly facing surfaces of the enamel cap of the maxillary teeth that come together in occlusion when the mandible is elevated during the chewing cycle.

occlusal morphology (L. *occludere* = to close) Refers to the size and shape of the superiorly facing surfaces of the enamel cap of the mandibular teeth and the inferiorly facing surfaces of the enamel cap of the maxillary teeth. Occlusal morphology is usually described in terms of the number and relative size of main and accessory cusps, the pattern of primary and secondary fissures, and the presence/absence of fovea (syn. cusp morphology).

occlusal wear (L. *occludere* = to close) Loss of the volume of a tooth that occurs on the functional or occlusal surface of a tooth (i.e., the surfaces of the upper and lower teeth that meet when the jaws are brought together as the mouth closes). Occlusal wear that can be seen with the naked eye is called gross wear, dental macrowear, or dental mesowear, whereas occlusal wear that can only be seen with a microscope is called **dental microwear**. *See also* **tooth wear**.

Oceania (L. *oceanus* = sea) The islands, large and small, in the central, western, and southern Pacific. The large islands are Australia and New Zealand; the three major divisions of small islands that make up Oceania are **Melanesia**, **Micronesia**, and **Polynesia**. *See also* **Sahul**.

Oceania, peopling of Past theories on the human settlement of the Pacific Islands collectively known as **Oceania** have suggested that the immigrants originated in Island Melanesia, Southeast Asia, Japan, Micronesia, or even South America. Evidence from contemporary archeology, linguistics, and genetics now points to several major migrations into Oceania over a remarkably long time period, all of them coming from the west. The initial settlement by modern humans of **Sahul** (the ancient Australia/New Guinea continent) as well as neighboring islands to the east in the Bismarck Archipelago, occurred *c*.50 ka from areas immediately to the west, mostly now submerged (as the **Sunda** shelf) (Summerhayes et al. 2010). This was most likely an extension of an early modern human expansion out of Africa that followed the shorelines of South Asia. As sea levels subsequently rose, the inhabitants of Sahul and the nearby islands were isolated for the following 20–30 ka, a unique occurrence in modern human prehistory, but by the end of the Pleistocene there is evidence of increasing interaction with peoples to the west. At *c*.3300 years BP a small number of seafaring peoples with clear linguistic roots in Taiwan some 1500 years earlier (Gray et al. 2009) entered the Bismarck Archipelago from the west. Within the following 500 years, these people developed remarkable seafaring abilities in the region of the

Bismarcks, but their impact was more linguistic and cultural than genetic (Friedlaender et al. 2008). Then it seems that they spread back along the north shore of New Guinea, along then through the off-shore islands and coastal regions of the Bismarck and Solomon Islands. By c.2800 years BP they had settled previously unoccupied islands as far east as Tonga and Samoa (i.e., the Lapita Expansion). It is in the Tonga/Samoa region during the following few hundred years that the basic elements of Polynesian language and culture developed. After a pause of a few more hundred or perhaps even 1000 years, voyagers from this area began colonizing islands far to the east, north, and south (Kirch 2000). Initial settlement dates for various Pacific archipelagos remain controversial, but the Marquesas, the Society Islands (Tahiti), Easter Island, and Hawai'i were settled by at least 1200 years BP, and New Zealand somewhat later, by about 1000–800 years BP. The Polynesians even sailed to South America, bringing back the sweet potato and bottle gourd, rather than the reverse, as had been famously proposed by Thor Heyerdahl. Because of this pattern of ancient settlement, Polynesians form a genetic, linguistic, and cultural phylogenetic entity. Micronesians are more diverse because of influences from Melanesia, Polynesia, and directly from East Asia. The inhabitants of Melanesia are the most diverse of all in genes, language, and culture, with some very ancient mitochondrial DNA genetic relationships with Australian Aborigines, as well as some relatively minor genetic similarities in particular regions with Polynesian and Micronesian groups. Some geneticists have recently even suggested that there are indications of genetic affinities of Australo-Melanesians with pre-modern human groups. *See also* **Helicobacter pylori**.

Ochtendung (Location 50°21′N, 7°23′E, Germany; etym. named after a nearby town) History and general description Hominin remains were discovered at the foot of an exposure of late **Middle Pleistocene** volcanic and loess deposits during quarrying operations in 1997. The finds were made near a known Middle Paleolithic site. Temporal span and how dated? Stratigraphic correlation suggests a date of **Oxygen Isotope Stage 6** (c.128–186 ka). Hominins found at site A few cranial fragments, assigned to *Homo neanderthalensis*. Archeological evidence found at site Three **Middle Paleolithic** stone tools were found associated with the hominin remains. Key references: Geology, dating, and paleoenvironment von Berg 1997, 2002, Street et al. 2006; Archeology von Berg 1997, 2002; Hominins von Berg 1997, 2002, von Berg et al. 2000.

Ochoz (Location 49°20′N, 16°45′E, Czech Republic; etym. after a nearby village) History and general description This cave, also known as Švédův stůl (the Swedish cave) is southwest of Ochoz village, north of Brno. A hominin partial mandible was found in 1905. Subsequent excavations were conducted in 1953–5 and in 1964, with the latter yielding some hominin cranial fragments and an isolated molar. Temporal span and how dated? **Biostratigraphy** suggests last interglacial to last glacial. Hominins found at site An adult partial mandible, cranial fragments, and isolated molar, attributed to *Homo neanderthalensis*. Archeological evidence found at site **Mousterian** and **Upper Paleolithic** industries. Key references: Geology, dating, and paleoenvironment Vlček 1969; Archeology Klima et al. 1962, Vlček 1969; Hominins Jelinek 1969, Smith 1982.

ochre (Gk *okhros* = pale yellow) A family of minerals identified by the colored streak they produce and the compounds from which those colors derive: hematite (Fe_2O_3) and goethite [α-$Fe^{3+}O(OH)$]. Ochres can, when pulverized into powder or shaped into "crayon" form, be used as pigment to decorate the body and other surfaces (Watts 2002). Other known applications include but are not limited to animal hide tanning aid (Audouin and Plisson 1982), medicine, vegetable preservative, and component of mastic used for hafting tools (Wadley 2005). Some notable archeological examples include the c.40 ka Mungo III burial from Lake Mungo in Australia which was discovered covered in powdered ochre (Bowler et al. 2003), the c.77 ka specimens of symbolically incised ochre from **Blombos Cave**, South Africa (Henshilwood et al. 2002), and the assemblage containing multiple types of ochre from between 270 and 170 ka at **Twin Rivers**, Zambia (Barham 2002b).

Ockham's razor *See* **Occam's razor**.

odontoblasts (Gk *odonto* = pertaining to dentine and *blastos* = bud or immature cell) Dentine-forming cells derived from the mesenchyme of the dental papilla that interact with the inner enamel epithelium to induce hard-tissue matrix secretion.

The cells become elongated during differentiation, with their long axis at approximate right angles to the **enamel-dentine junction** (EDJ) (see images in Nanci 2003). As they lay down matrix they move away from the EDJ and towards the future pulp chamber, leaving behind a long cell process that extends from the secretory end of the cell and into the dentine for a variable distance. This cell process is encased in a dentine tubule approximately 2 μm in diameter. The fluid-filled dentine tubule runs from the EDJ to the pulp, but the odontoblast process (that only extends a variable way along it from the pulpal aspect) responds to changes in osmotic pressure and temperature as the fluid is drawn up and down the tubule. Over a long period of time peritubular dentine, which is harder than intertubular dentine and which contains no collagen, is formed by the odontoblast cell process. This peritubular dentine slowly fills in the tubules resulting in sclerosed or transparent dentine. The extent of transparent dentine formation in roots can be used as an indicator of chronological age (see Aiello and Dean 1990). *See also* **dentine development; enamel development.**

odontogenesis (Gk *odonto* = tooth and *gena* = to give birth to) *See* **dentine development.**

OES Acronym for **ostrich egg shell dating** (*which see*).

OH Abbreviation of Olduvai Hominid and the prefix for fossil hominins recovered from **Olduvai Gorge**, Tanzania.

OH 1 Site **Olduvai Gorge**. Locality RK. Surface/*in situ In situ*. Date of discovery 1913. Finder Hans Reck. Unit N/A. Horizon Near the top of Bed II. Bed/member Bed II. Formation N/A. Group N/A. Nearest overlying dated horizon Holocene. Nearest underlying dated horizon Bed II, *c*.1.15–1.70 Ma. Geological age Initially inferred to be contemporaneous with Bed II, now thought to be intrusive and likely Holocene or terminal **Pleistocene** in age. Developmental age Adult. Presumed sex Male. Brief anatomical description An anatomically modern human skeleton found in a strongly contracted position. The cranium is long, narrow and tall and the face is orthognathic. The mandible has a deep body and a strong chin. Announcement Reck 1914. Initial description Reck 1914. Photographs/line drawings and metrical data

Reck 1914, Mollison and Gieseler 1929, Protsch 1974. Detailed anatomical description Mollison 1929, Gieseler 1929. Initial taxonomic allocation *Homo sapiens*. Taxonomic revisions N/A. Current conventional taxonomic allocation *H. sapiens*. Informal taxonomic category Anatomically modern human. Significance The erroneous association of OH 1, a clearly anatomically modern human skeleton, with the Plio-Pleistocene fauna at Olduvai Gorge, along with a similarly mistaken association at **Kanjera**, led Louis **Leakey** in the 1930s–1970s to champion the then-unorthodox idea that modern humans had probably evolved in Africa and had a very ancient evolutionary history on the continent. OH 1 remains important for studies of modern population history and microevolution in East Africa (e.g., Rightmire 1975, 1981, Bräuer 1978, Angel et al. 1980). Location of original **Staatssammlung für Anthropologie und Palaeoanatomie München**, Munich, Germany.

OH 3 See **OH 5**.

OH 4 See **OH 5**.

OH 5 Site **Olduvai Gorge**. Locality **FLK**, 45, Archeology Site no. 41a. Surface/*in situ In situ* "in the process of being eroded out on the slopes" (Leakey 1959, p. 491). Date of discovery July 17, 1959. Finder Mary **Leakey**. Unit A paleosol between Tuff IB and IC. Horizon Level 22. Bed/member Bed I. Formation N/A. Group N/A. Nearest overlying dated horizon Tuff IC. Nearest underlying dated horizon Tuff IB. Geological age Between 1.845 and 1.839 Ma. Developmental age Adult. Presumed sex Male. Brief anatomical description This well-preserved cranium is the result of painstaking assembly by Louis **Leakey**, Mary **Leakey**, and Alun Hughes. It preserves all of the upper dentition, much of the face, cranial base, and cranial vault. It is undistorted and the surfaces of the bones and teeth are remarkably well preserved. Endocranial volume 530 cm^3 (Tobias 1967). Announcement Leakey 1959. Initial description Leakey 1959. Photographs/line drawings and metrical data Tobias 1967a. Detailed anatomical description Tobias 1967a. Initial taxonomic allocation *Zinjanthropus boisei*. Taxonomic revisions Leakey et al. (1964) proposed that *Zinjanthropus* be reduced to a subgenus within *Australopithecus*, as *Australopithecus (Zinjanthropus) boisei*. Subsequently, Tobias (1967a) recommended that the subgenus *Zinjanthropus* be sunk into *Australopithecus*, so the taxon became *Australopithecus boisei*. Researchers who support the revival of *Paranthropus* as the genus name for a **megadont** australopith clade refer to the same species taxon as *Paranthropus boisei*. Current conventional taxonomic allocation *Au.* or *P. boisei*. Informal taxonomic category Megadont archaic hominin. Significance OH 5 is the **holotype** of *Au.* or *P. boisei*. Apart from two deciduous teeth (OH 3) recovered from BK in 1955 and a mandible fragment and a few isolated teeth (OH 4) recovered from MK, also in 1959, the OH 5 cranium was the first well-preserved archaic hominin to be recovered from Olduvai Gorge. Louis Leakey was initially convinced that OH 5 was the maker of the stone tools found in Bed I, but when what he judged to be more *Homo*-like specimens, such as **OH 7**, **OH 8**, and then **OH 13**, came to light they were judged to be more advanced and thus were more likely to have made the **Oldowan** artifacts. The discovery of such an iconic fossil transformed the availability of funds for research at Olduvai Gorge. In the days before televisions were ubiquitous, the *National Geographic* magazine was the main source of information about the natural world, and via that magazine the Leakey's research at Olduvai became known across the world. The discovery of OH 5 and the discovery of new sites at **Peninj**, on the shores of Lake Natron, together with the relative academic isolation of South Africa because of its apartheid policy, marked the beginning of East Africa as the major focus of activity for paleoanthropologists searching for fossil evidence of early hominins. OH 5 was also the first significant fossil hominin to be dated using the then-novel method of **potassium-argon dating**. The Leakeys used the epithet "Dear Boy" for what is still one of the most impressive and dramatic of all early hominin fossils. Location of original **National Museum and House of Culture, Dar es Salaam, Tanzania.**

OH 7 Site Olduvai Gorge. Locality FLKNN, 45b, Archeology Site no. 38b. Surface/*in situ* Some of fossils comprising this individual were found on the surface, others during the course of excavations. Date of discovery Leakey (1960, p. 1051) reported in the December 17 issue of *Nature* the recovery (presumably in October or November 1960) of "fragments of a skull and some teeth, two clavicles, a large part of a left foot, six finger bones and two ribs" at a site FLKNN I that he described as being "not far away, at what appears to be a slightly lower level of Bed I" than the FLK locality where **OH 5** was discovered. The juvenile skull and the juvenile hand bones became OH 7, the foot bones **OH 8**, and the clavicle OH 48. In the February 25, 1961 issue of *Nature* Leakey (1961a) reported that at the same site, FLKNN I, on November 2, 1960 Jonathan Leakey had recovered a hominin mandible, and subsequently "parts of two hominid parietals" were recovered. Finders Jonathan Leakey found the mandible and F. Masao found the right M_2 in 2006. Unit Clay overlying Tuff IB. Horizon Level 3. Bed/member Bed I. Formation Olduvai. Group N/A. Nearest overlying dated horizon Tuff IC. Nearest underlying dated horizon Tuff IB. Geological age *c.*1.84 Ma. Developmental age Juvenile. Presumed sex Male. Brief anatomical description OH 7 presently comprises a juvenile mandible, most of two juvenile parietals, and the balance of the 21 original hand bones that were neither non-hominin, nor obviously adult. An additional mandibular tooth (right M_2) was found in 2006; it was a surface find on an outcrop of Bed I, some 10 m from the FLKNN site. The mandibular corpus is broken at the symphysis, but all the teeth are present in the corpus except the separate right M_2 (see above), and the right and left M_3s. The left parietal is more complete than the right; both are thin. The 13 hand bones come from at least two hands. Announcement Leakey 1960, 1961a. Initial description Leakey 1960, 1961. Photographs/line drawings and metrical data Tobias 1991. Detailed anatomical description Tobias 1991, Susman and Creel 1979. Initial taxonomic allocation *Homo habilis*. Taxonomic revisions. None specific to this specimen. Current conventional taxonomic allocation *H. habilis*. Informal taxonomic category Pre-modern *Homo* or transitional hominin. Significance The specimen is the **holotype** of *H. habilis*. Most researchers regard *H. habilis* as the earliest representative of the genus *Homo*, but some are less convinced that the morphology *H. habilis* shares enough features of the *Homo* clade and grade to justify its inclusion in *Homo*. Functional morphological interpretations of the hand fossils vary from those that emphasize their similarity to modern humans, to those that stress their similarities to *Australopithecus*, including *Australopithecus afarensis*. Location of original **National**

Museum and House of Culture, Dar es Salaam, Tanzania.

OH 8 Site Olduvai Gorge. Locality FLKNN, 45b, Archeology Site no. 38b. Surface/*in situ* Found *in situ* during the course of an excavation. Date of discovery Leakey (1960, p. 1051) reported in the December 17 issue of *Nature* the recovery (presumably in October or November 1960) of "fragments of a skull and some teeth, two clavicles, a large part of a left foot, six finger bones and two ribs" at a site FLKNN I that he described as being "not far away, at what appears to be a slightly lower level of Bed I" than the FLK site where **OH 5** was discovered. The juvenile skull and the juvenile hand bones became part of **OH 7** and the adult bones OH 8. Finder Mary **Leakey**. Unit/horizon Clay overlying Tuff IB. Bed/member Bed I. Formation N/A. Group N/A. Nearest overlying dated horizon Tuff IC. Nearest underlying dated horizon Tuff IB. Geological age *c*.1.76 Ma. Developmental age Leakey initially assumed the OH 8 foot was part of the same juvenile skeleton as OH 7, but Day and Napier (1964) and Leakey et al. (1964) later judged them to be adult, and thus evidence of second individual at FLKNN I, but Susman and Stern (1982) and Susman (2008) claim that the distal ends of the OH 8 metatarsals 2 and 3 are not jagged because they were damaged by predators, but because OH 8 is a subadult and that the distal epiphyses have been lost. However, a recent careful reassessment of the specimen (De Silva et al. 2010) provides compelling evidence that the OH 8 foot belongs to an adult individual. Presumed sex Unknown. Brief anatomical description A left foot except for the phalanges, the distal ends of the metatarsals, and the calcaneal tuberosity. There are also some dental fragments. Two hand bones (both proximal phalanges) found with the foot are nonhominin. Announcement Leakey 1960. Initial description Day and Napier 1964. Photographs/line drawings and metrical data No systematic analysis of the whole foot has been undertaken, but measurements of individual tarsal bones are available in Day and Wood (1968), Kidd et al. (1996), Deloison (2004), and Gebo and Schwartz (2006). Detailed anatomical description There is no detailed description of the whole foot. Initial taxonomic allocation *Homo habilis*. Taxonomic revisions None specific to this specimen. Current conventional taxonomic allocation *H. habilis*. Informal taxonomic category Pre-modern *Homo* or transitional hominin. Significance The specimen is one of the

paratypes of *H. habilis*. Most researchers regard *H. habilis* as the first representative of the genus *Homo*, but others are less convinced that the morphology *H. habilis* shares enough features of the *Homo* clade and grade to justify its inclusion in *Homo*. The conclusions of functional morphological interpretations of the whole foot and of the component bones vary from those that emphasize similarities to modern humans to those that stress similarities to *Australopithecus*, but most of these assessments are based on the morphology of the talus, which Day and Napier (1964) suggested was the most primitive element in the foot. Some researchers claim that the OH 8 foot and the OH 35 distal tibia and fibula belong to the same individual (these reasons are reviewed in Susman 2008 and Susman et al. 2011, but see de Silva et al. 2010 for a counter argument) and it has been claimed that crocodile tooth marks on both sets of bones match up. But questions remain over whether they can be associated geologically (the OH 8 foot and the OH 35 distal tibia come from different levels in the Olduvai sequence) and an investigation of the extent of the congruence between the reciprocal joint surfaces on OH 8 and OH 35 showed that they were unlikely to belong to the same individual (Wood et al. 1998). Although OH 8 was allocated to *H. habilis* there are no taxonomically diagnostic cranial remains associated with it, so it could equally well belong to *Paranthropus boisei*. Location of original **National Museum and House of Culture, Dar es Salaam**, Tanzania.

OH 9 Site **Olduvai Gorge**. Locality LLK, 46, Archeology Site no. 64. Surface/*in situ* Surface. Date of discovery December 12, 1960. Finder Louis **Leakey**. Unit/horizon Tuff in Bed II 3–3.6 m below the base of Bed III. Bed/member Bed II. Formation N/A. Group N/A. Nearest overlying dated horizon None. Nearest underlying dated horizon Tuff IIA. Geological age *c*.1.7–1.2 Ma. Developmental age Adult. Presumed sex Male. Brief anatomical description Calvaria lacking much of the vault, but preserving the supraorbital region. Announcement Leakey 1961a. Initial description Rightmire 1979b, pp. 101–106. Photographs/line drawings and metrical data Leakey 1961a, Heberer 1963, Rightmire 1979b, Wood 1991. Detailed anatomical description The most complete systematic descriptions are by Rightmire (1979b, pp. 101–6; 1990, pp. 59–70). In one of the early applications of **computed tomography** to fossil hominins, Maier and Nkini (1984) provide information about the sagittal morphology of OH 9. Initial taxonomic allocation *Homo leakeyi*. Taxonomic revisions

Homo erectus. Current conventional taxonomic allocation *H. erectus*. Informal taxonomic category Pre-modern *Homo*. Significance The OH 9 calvaria is seen by many as an early example of *H. erectus sensu stricto* in Africa and it exemplifies the proposition that the distinctive features of *H. erectus* are best expressed in larger specimens. Location of original **National Museum and House of Culture, Dar es Salaam**, Tanzania.

OH 10 Site **Olduvai Gorge**. Locality FLKN, 45a, Archeology Site no. 40b. Surface/*in situ* Surface. Date of discovery 1960. Finders Unknown. Unit Clay horizon approximately 0.9–1.2 m below Tuff IF. Horizon Level 5. Bed/member Bed I. Formation N/A. Group N/A. Nearest overlying dated horizon Tuff IF. Nearest underlying dated horizon Tuff IE. Geological age *c*.1.79 Ma. Developmental age Adult. Presumed sex Unknown. Brief anatomical description Distal-hallucial phalanx from a right foot. Announcement It is included in a figure illustrating the approximate stratigraphic locations of hominins at Olduvai (Tobias 1965c, Fig. 1, p. 24), but it was not referred to in the text, thus its formal announcement was by Day and Napier (1966). Initial description Day and Napier 1966. Photographs/line drawings and metrical data Day and Napier 1966. Detailed anatomical description Day and Napier 1966. Initial taxonomic allocation Hominidae *gen. et sp. indet.* (Tobias 1965c). Taxonomic revisions *Homo habilis* (Day and Napier 1966). Current conventional taxonomic allocation *H. habilis*. Informal taxonomic category Pre-modern *Homo*. Significance Although just a single toe bone, OH 10 has been a critical part of the evidence used by those who suggest the postcranial skeleton of *H. habilis* is consistent with that taxon being an obligate biped. However, the results of Day's (1967) multivariate analysis showed that although OH 10 was closer to three modern human samples than to gorillas and chimpanzees, it was significantly different from *all* of the extant comparative samples. Although the OH 10 hallucial phalanx was allocated to *H. habilis* there are no taxonomically diagnostic cranial remains associated with it, so it could equally well belong to *Paranthropus boisei*. Location of original **National Museum and House of Culture, Dar es Salaam**, Tanzania.

OH 11 Site **Olduvai Gorge**. Locality **DK**. Surface/*in situ* Surface. Date of discovery 1962. Finders Mary **Leakey** and coworkers. Unit Surface of lower Ndutu Beds. Horizon N/A.

Bed/member Lower Ndutu Beds. Formation N/A. Group N/A. Nearest overlying dated horizon Upper Ndutu Beds. Nearest underlying dated horizon Masek Beds. Geological age **Middle** or Upper Pleistocene (400–32 ka, but probably closer to the earlier end of the age range). Developmental age Adult. Presumed sex Unknown. Brief anatomical description A weathered and abraded fragmentary left maxilla with roots of C–M^2. Announcement Day 1977. Initial description Rightmire 1980. Photographs/line drawings and metrical data Rightmire 1980. Detailed anatomical description Rightmire 1980. Initial taxonomic allocation *Homo* sp. Taxonomic revisions *Homo heidlebergensis* or archaic *Homo sapiens*. Current conventional taxonomic allocation *H. heidlebergensis* or archaic *H. sapiens*. Informal taxonomic category Pre-modern *Homo*. Significance One of the few hominin specimens from this time period in sub-Saharan Africa. Location of original **National Museum and House of Culture, Dar es Salaam**, Tanzania.

OH 12 Site **Olduvai Gorge**. Locality **VEK**, 86, Archeology Site no. 45b. Surface/*in situ* Surface. Date of discovery 1962. Finder Margaret Leakey. Unit Above Tuff IVA. Horizon N/A. Bed/member Bed IV. Formation N/A. Group N/A. Nearest overlying dated horizon None. Nearest underlying dated horizon Tuff IVA. Geological age *c*.>0.78 Ma. Developmental age Adult. Presumed sex Unknown. Brief anatomical description A skull consisting of posterior vault, parietal, and facial fragments plus a few fragments belonging to the coronoid process of the mandible. There is enough preserved to attempt an endocast reconstruction, and thus generate an estimate of its cranial capacity, but because of the fragmentary nature of all but the posterior part of the cranial vault, this must be regarded as a relatively crude estimate. Endocranial volume 727 cm^3 (Holloway 1983). Announcement Leakey and Leakey 1964. Initial description Rightmire 1979b, pp. 106–7. Photographs/line drawings and metrical data As above and Antón 2004. Detailed anatomical description The first systematic descriptions were by Rightmire 1979b (pp. 106–7; 1990, pp. 70–3), but the most complete is that of Antón 2004. Initial taxonomic allocation *Homo erectus*. Taxonomic revisions None. Current conventional taxonomic allocation *H. erectus*. Informal taxonomic category Pre-modern *Homo*. Significance OH 12 is one of the youngest African fossils attributed to *H. erectus*, yet its estimated endocranial volume is small. It was the first evidence that *H. erectus*-like hominins subsumed a substantial range of absolute size and endocranial volume. Location of original **National Museum and House of Culture, Dar es Salaam**, Tanzania.

OH 13 Site **Olduvai Gorge**. Locality MNK, 88, Archeology Site no. 71a. Surface/*in situ* Two fragments of the skull were found *in situ*. Date of discovery October 22, 1963. Finder N. Mbuika. Unit Clayey tuff horizon 1.8 m thick, approximately 3 m above Tuff IIA. Horizon See above. Bed/member Lower Bed II. Formation N/A. Group N/A. Nearest overlying dated horizon None. Nearest underlying dated horizon Tuff IIA. Geological age *c.*1.7 Ma. Developmental age Subadult. Presumed sex Unknown. Brief anatomical description The specimen is the skull of a subadult hominin. It comprises much of the vault, the greater part of the mandible and both maxillae, all of the mandibular teeth, and some of the maxillary cheek teeth. Announcement Leakey and Leakey 1964. Initial description Leakey et al. 1964. Photographs/line drawings and metrical data Tobias 1991a. Detailed anatomical description Tobias 1991a. Initial taxonomic allocation *Homo habilis*. Taxonomic revisions There have been no formal taxonomic revisions, but Tobias and von Koenigswald (1964) pointed out many gnathic and dental similarities between OH 13, Pithecanthropus IV, and Sangiran B, both now assigned to *Homo erectus*, and Robinson (1965) considered OH 13 to be conspecific with *H. erectus*. Current conventional taxonomic allocation *H. habilis*. Informal taxonomic category Pre-modern *Homo* or transitional hominin. Significance OH 13 was designated a **paratype** of *Homo habilis* and it provides some of the best dental and mandibular evidence for that taxon. Although its postcanine teeth are absolutely small, when they are related to its relatively small body mass it is still **megadont**. Location of original **National Museum and House of Culture, Dar es Salaam**, Tanzania.

OH 16 Site **Olduvai Gorge**. Locality Maiko Gully, FLK, 45, Archeology Site no. 43. Surface/*in situ* Surface. Date of discovery November 1963. Finder Maiko Mutumbo. Unit The specimen was found in small fragments on a slope. It is inferred to have come from a horizon in the base of Bed II, approximately 0.9–1.2 m above Tuff IF. Horizon See above. Bed/member Lower Bed II. Formation N/A. Group N/A. Nearest overlying dated horizon Tuff IIA. Nearest underlying dated horizon Tuff IF. Geological age *c.*>1.74 Ma. Developmental age Subadult. Presumed sex Unknown. Brief anatomical description This is a very fragmented skull. The cranial vault has been reconstructed from about 107 fragments and it is evident that the reconstruction featured in most illustrations and the one used by Tobias (1991a) is in need of revision. The best-preserved part of the calvaria is the supraorbital region and the region of the mandibular fossa. Fifteen maxillary and 13 mandibular permanent teeth are preserved; the M_3s were erupted, but the M^3s were not. Announcement Leakey and Leakey 1964. Initial description Leakey et al. 1964. Photographs/line drawings and metrical data Tobias 1991a. Detailed anatomical description Tobias 1991a. Initial taxonomic allocation OH 16 was not included in the list of paratypes of *Homo habilis*, but it was "provisionally referred" to that taxon (Leakey et al. 1964, p. 9). Taxonomic revisions There have been no formal taxonomic revisions. Tobias (1965c, p. 392) considered that the teeth of OH 16 "are compatible with those of an *Australopithecus*," but later formed the opinion that it belonged to *H. habilis*. Robinson (1965) considered OH 13 to be conspecific with *Homo erectus*, and some researchers have suggested OH 16 samples a second *Homo* taxon at Olduvai Gorge. Current conventional taxonomic allocation *H. habilis*. Informal taxonomic category Pre-modern *Homo* or transitional hominin. Significance OH 16 was not designated as a paratype of *H. habilis*, but has since been added to the *H. habilis* hypodigms. Location of original **National Museum and House of Culture, Dar es Salaam**, Tanzania.

OH 20 Site **Olduvai Gorge**. Locality HWK, 44a, Archeology Site no. 46c. Surface/*in situ* Surface. Date of discovery 1959 (it was not initially identified as a hominin but it was recognized as such by Mary **Leakey** in 1968 when she was sorting through the fauna from the site). Finders Unknown. Unit Surface find but "its degree of mineralization and colour agree closely with those of other bones from the same area known to be from lower Bed II" (Day 1969, p. 230). Horizon Lower augitic sandstone overlying Tuff IF, cutting out Tuff IIA. Bed/member Lower Bed II. Formation N/A. Group N/A. Nearest overlying dated horizon

None. Nearest underlying dated horizon Tuff IF. Geological age *c*.>1.725 Ma. Developmental age Adult. Presumed sex Unknown. Brief anatomical description The proximal end of a left femur preserving the greater and lesser trochanters and most of the neck. Announcement Day 1969. Initial description Day 1969. Photographs/line drawings and metrical data Day 1969. Detailed anatomical description Day 1969. Initial taxonomic allocation *Australopithecus* cf. *boisei*. Taxonomic revisions No formal revisions. Current conventional taxonomic allocation *Paranthropus boisei*. Informal taxonomic category Megadont archaic hominin. Significance Michael **Day** drew attention to the several ways (e.g., long and anteroposteriorly flattened neck, no femoral tubercle) in which OH 20 resembles two proximal femora, **SK 82** and **SK 97**, from **Swartkrans** that are attributed to *Paranthropus robustus*. Thus, he attributed OH 20 to *Australopithecus* cf. *boisei*, but there are no taxonomically diagnostic cranial remains associated with OH 20, so it could equally well belong to *Homo habilis*. Location of original **National Museum and House of Culture, Dar es Salaam,** Tanzania.

OH 22   Site **Olduvai**. Locality Unnamed location on the surface of the Beds III and IV. Surface/*in situ* Surface. Date of discovery 1968. Finders Mary **Leakey** and coworkers. Unit N/A. Horizon N/A. Bed/member Beds III/IV. Formation N/A. Group N/A. Nearest overlying dated horizon Masek Beds. Nearest underlying dated horizon Bed II. Geological age **Early** or **Middle Pleistocene** (*c*.1.15–0.6 Ma). Developmental age Adult. Presumed sex Unknown. Brief anatomical description A large right side of the mandible with roots of I_1–M_3 and crowns of P_3–M_2 and most of the ramus. The symphysis is receding, the upper and lower borders of the corpus are parallel, and the corpus is robust. Announcement Leakey 1969. Initial description Rightmire 1980. Photographs/line drawings and metrical data Rightmire 1980. Detailed anatomical description Rightmire 1980. Initial taxonomic allocation cf. *Homo erectus*. Taxonomic revisions *Homo heidelbergensis* (Rightmire 1996, 1998) or early archaic *Homo sapiens* (Bräuer 1984). Current conventional taxonomic allocation *H. heidelbergensis* or early archaic *H. sapiens*. Informal taxonomic category Pre-modern *Homo*. Significance Well-preserved post-*Homo erectus* mandibles from sub-Saharan Africa are rare. Location of original **National Museum and House of Culture, Dar es Salaam,** Tanzania.

OH 23   Site **Olduvai**. Locality FLK. Surface/*in situ In situ*. Date of discovery 1968. Finders Mary **Leakey** and coworkers. Unit N/A. Horizon N/A. Bed/member Lower Masek Beds. Formation N/A. Group N/A. Nearest overlying dated horizon Ndutu Beds. Nearest underlying dated horizon Bed IV. Geological age **Middle Pleistocene** (*c*.0.6–0.4 Ma). Developmental age Adult. Presumed sex Unknown. Brief anatomical description Fragmentary left side of the mandibular corpus with P_4–M_2 and roots or alveoli of C, P_3, and M_3. Announcement Leakey 1969. Initial description Rightmire 1980. Photographs/line drawings and metrical data Rightmire 1980. Detailed anatomical description Rightmire 1980. Initial taxonomic allocation cf. *Homo erectus* (Day 1977). Taxonomic revisions Early or archaic *Homo sapiens* (Rightmire 1980, Bräuer 1984), *Homo heidelbergensis* (Rightmire 1996, 1998). Current conventional taxonomic allocation Archaic *H. sapiens* or *H. heidelbergensis*. Informal taxonomic category Pre-modern *Homo*. Significance One of the few sub-Saharan African Middle Pleistocene hominins. Location of original **National Museum and House of Culture, Dar es Salaam,** Tanzania.

OH 24   Site **Olduvai Gorge**. Locality DK East, Archeology Site no. N/A. Surface/*in situ* The cranium was found embedded in a "mass of calcareous matrix" with only the "right supra-orbital region and the posterior part of the palate exposed" (Leakey et al. 1971, p. 308); the four teeth were found during sieving. Date of discovery October 1968. Finder Peter Nzube. Unit/horizon Tuffaceous clay between the Bed I basalt and Tuff IB. Bed/member Bed I. Formation N/A. Group N/A. Nearest overlying dated horizon Tuff IB. Nearest underlying dated horizon Bed I basalt. Geological age Between 1.845 and 1.865 Ma. Developmental age Adult. Presumed sex Female. Brief anatomical description Although this cranium is distorted and very fragmented, it does preserve substantial parts of the vault, face, palate, and base. No anterior teeth are preserved, but both upper premolars and the M^1 and M^3 are preserved; its reconstruction was undertaken by Ron **Clarke**. Announcement Leakey et al. 1971. Initial description Leakey et al. 1971. Photographs/line drawings and metrical data Leakey et al. 1971. Detailed anatomical description Tobias 1991a. Initial taxonomic allocation *Homo* aff. *habilis*. Taxonomic revisions None specific to this specimen. Current conventional taxonomic allocation *Homo habilis*. Informal taxonomic category Pre-modern *Homo* or transitional hominin. Significance The OH 24 cranium is one of the temporally oldest specimens attributed to *H. habilis* yet it has some of the more "advanced" features noted in **OH 13**. Because OH 13 is one of the youngest specimens in the *H. habilis* hypodigm it had been suggested that there was a temporal trend in *H. habilis* morphology, but the antiquity of OH 24 casts doubt on that hypothesis. Location of original **National Museum and House of Culture, Dar es Salaam**, Tanzania.

OH 28 Site **Olduvai Gorge**. Locality WK, 36, Archeology Site no. 52. Surface/*in situ* *In situ*. Date of discovery 1970. Finder Mary **Leakey**. Unit OH 28 was found "at the base of a sandstone with fragments of the (Tuff IVB) tuff" (Leakey 1971, p. 381). Horizon See above. Bed/member Bed IV. Formation N/A. Group N/A. Nearest overlying dated horizon Tuff in the base of the Masek Beds (spelled "Mesak" in Leakey 1971). Nearest underlying dated horizon Leakey (1971) suggests that the sediments that contain OH 28 postdate Tuff IVB. Geological age *c*.<780 ka. Developmental age Adult. Presumed sex Probably female. Brief anatomical description Left femur lacking the head, greater trochanter, and the condyles; much of the ischium and the ilium (the crest and the anterior superior iliac spines are lacking) of a left pelvic bone. Announcement Leakey 1971. Initial description Day 1971. Photographs/line drawings and metrical data Day 1971. Detailed anatomical description Day 1971. Initial taxonomic allocation *Homo erectus*. Taxonomic revisions None. Current conventional taxonomic allocation *H. erectus*. Informal taxonomic category Pre-modern *Homo*. Significance The apparently derived features shared among the femora recovered from **Zhoukoudian** and the OH 28 femur (e.g., anteroposterior flattening of the shaft, the convex medial profile, and the low position of the minimum breadth of the shaft, which have been related to high bending moments) link OH 28 with *H. erectus*, and provide support for the hypothesis that while *H. erectus* lower limb morphology (e.g., iliac pillars and the impressions for the ilio-femoral ligament) is consistent with that taxon being a habitual, or obligate biped, it shows consistent departures from the lower limb morphology of modern humans. Prior to the discovery of OH 28 there was no reliable information about the pelvis of *H. erectus*. Location of original **National Museum and House of Culture, Dar es Salaam**, Tanzania.

OH 34 Site **Olduvai Gorge**. Locality JK2 West. Surface/*in situ* *In situ*. Date of discovery 1962. Finder Maxine Kleindienst. Unit "Channel formation" (Day and Molleson 1976, p. 455). Horizon See above. Bed/member Bed III. Formation N/A. Group N/A. Nearest overlying dated horizon N/A. Nearest underlying dated horizon N/A. Geological age *c*.1 Ma. Developmental age Adult. Presumed sex Unknown. Brief anatomical description A left femur lacking the distal end and part of the shaft of a left tibia. Like all the mammal bones from the same excavation, these hominin bones are so highly abraded that all trace of taxonomically or functionally useful surface morphology has been lost. Announcement Day and Molleson 1976. Initial description Day and Molleson 1976. Photographs/line drawings and metrical data Day and Molleson 1976. Detailed anatomical description Day and Molleson 1976. Initial taxonomic allocation Hominini *gen. et sp. indet.* Taxonomic revisions None. Current conventional taxonomic allocation *Homo erectus* or *Homo heidelbergensis*. Informal taxonomic category Pre-modern *Homo*. Significance The preserved morphology of the femur is consistent with these associated lower limb bones belonging to either *H. erectus* or *H. heidelbergensis*, but the significance of OH 34 is that at the time of its recovery it was the *only* hominin recovered from Bed III. Location of original **National Museum and House of Culture, Dar es Salaam**, Tanzania.

OH 36 Site **Olduvai Gorge**. Locality SC, 90a, Archeology Site no. 60. Surface/*in situ* *In situ*. Date of discovery 1970. Finder M. Mutala. Unit N/A. Horizon As above. Bed/member Upper Bed II. Formation N/A. Group N/A. Nearest overlying dated horizon Tuff IVB. Nearest underlying dated horizon Tuff IID. Geological age Extrapolation based on the thickness of strata suggests an approximate age of 1.5–1.2 Ma. Developmental age Adult. Presumed sex Unknown. Brief anatomical description Nearly complete right ulna, missing only the distal end. The distal fracture surface is just distal to the pronator crest and is close to the neck. Announcement Leakey 1978. Initial description Aiello et al. 1999. Photographs/line drawings and metrical data Aiello et al. 1999. Detailed anatomical description Aiello et al. 1999. Initial taxonomic allocation *Homo erectus*. Taxonomic revisions *Paranthropus boisei* (Walker and Leakey 1993). Current conventional taxonomic allocation Hominini *gen. et sp. indet.*, or *P. boisei*. Informal taxonomic category Archaic hominin. Significance

The OH 36 ulna is one of several isolated post-cranial fossils (e.g., **OH 8, OH 10**) from Beds I and II of Olduvai Gorge. It comes from Upper Bed II, which has produced evidence of both *H. erectus* and *P. boisei*. Aiello et al. (1999) carried out a multivariate analysis in which they compared OH 36 to the ulnae of extant apes and modern humans and with fossil hominin ulnae from Omo Shungura (**L. 40–19**), Baringo (KNM-BK 66), and from southern African fossil sites. When OH 36 was paired with, respectively, KNM-BK 66 and L. 40–19, the variation between these specimens is not consistent with their coming from the same taxon. An analysis of trochlear keel shape also came to the conclusion that OH 36 and L. 40–19 most likely represent different taxa (Drapeau 2008). Thus, given it is most unlikely that KNM-BK 66 ulna belongs to *P. boisei* (it is several hundred thousand years younger than the most recent fossil evidence for *P. boisei*), Aiello et al. (1999) suggested that it is probable that OH 36 belongs to *P. boisei*. If so, it would join OH 3 and OH 38 as evidence for a *c*.1.15 Ma **last appearance datum** for *P. boisei*. Location of original **National Museum and House of Culture, Dar es Salaam**, Tanzania.

OH 51 Site Olduvai. Locality GTC. Surface/*in situ* Surface. Date of discovery 1974. Finders Mary **Leakey** and coworkers. Unit N/A. Horizon N/A. Bed/member Beds III/IV. Formation N/A. Group N/A. Nearest overlying dated horizon Masek Beds. Nearest underlying dated horizon Bed II. Geological age **Early** to **Middle Pleistocene** (*c*.1.15–0.6 Ma). Developmental age Adult. Presumed sex Unknown. Brief anatomical description A fragmentary left side of the mandibular corpus with P_4–M_1 and roots of P_3. Announcement Day 1977. Initial description Rightmire 1980. Photographs/line drawings and metrical data Rightmire 1980. Detailed anatomical description Rightmire 1980. Initial taxonomic allocation cf. *Homo erectus* (Day 1977). Taxonomic revisions *Homo heidelbergensis*. Current conventional taxonomic allocation *H. heidelbergensis* or archaic *Homo sapiens*. Informal taxonomic category Pre-modern *Homo*. Significance One of very few sub-Saharan African hominins from this time period. Location of original **National Museum and House of Culture, Dar es Salaam**, Tanzania.

OH 62 Site **Olduvai Gorge**. Locality FLK, Dik Dik Hill, adjacent to 45c. Surface/*in situ* Sur-face. Date of discovery July 21, 1986. Finder Tim **White** recognized the initial ulna fragment, and the subsequent search and screening of the deposits recovered the many fragments that comprise OH 62. Unit/horizon Most likely eroded from a sand lens below Tuff IC on Dik Dik Hill. This would make it the lateral equivalent of the horizon from which **OH 5** was recovered at FLK. Bed/member Bed I. Formation N/A. Group N/A. Nearest overlying dated horizon Tuff IC. Nearest underlying dated horizon Tuff IB. Geological age Between 1.845 and 1.839 Ma. Developmental age Adult. Presumed sex Unknown. Brief anatomical description The skull and all of the major long bones are represented in the OH 62 associated skeleton, but the specimen is *very* fragmentary (there are about 302 fragments). The palate is the most complete part of the skull, followed by the face and the mandible. The tooth crowns that are preserved are very worn. The long bones of the upper limb are better preserved than those of the lower limb, but even the relatively well-preserved humerus cannot provide a thoroughly reliable estimate of its undamaged length. Announcement Johanson et al. 1987. Initial description Johanson et al. 1987. Photographs/line drawings and metrical data Johanson et al. 1987, Johanson 1989. Detailed anatomical description N/A. Initial taxonomic allocation *Homo habilis*. Taxonomic revisions None specific to this specimen. Current conventional taxonomic allocation *H. habilis*. Informal taxonomic category Pre-modern *Homo* or transitional hominin. Significance In spite of its very fragmentary condition, OH 62 is a significant specimen for at least two reasons. First, it provides the only unambiguous evidence about the postcranial skeleton of *H. habilis*, for none of the other postcranial elements linked with *H. habilis* have associated taxonomically diagnostic cranial elements. Second, despite the fact that the shafts of the long bones are incomplete, most researchers consider enough is preserved to suggest the limb-length proportions of the OH 62 *H. habilis* individual are not significantly different from those of *Australopithecus afarensis*, but they *are* significantly different from those of modern humans. This inference is supported by an analysis of the cross-sectional geometry, demonstrating that OH 62 has a degree of humeral strength relative to femoral strength comparable to that of *Au. afarensis*, but unlike those of *Homo erectus* and modern humans (Ruff 2008). Reno et al. (2005) claim there is not enough evidence to support the conclusion that the limb-length proportions of OH 62 are *more* ape-like than those of *Au. afarensis*. Location of original **National Museum and House of Culture, Dar es Salaam,** Tanzania.

OH 65 Site Olduvai Gorge, Western lake margin. Locality 64. Surface/*in situ In situ*. Date of discovery August 10, 1995 Finders Amy Cushing and Agustino Venance. Unit The upper part of Bed I in Trench 57 from within "a zone 20 to 50 cm above the base of a 2.5-m-thick tuffaceous silty sandstone infilling" (Blumenschine et al. 2003, p. 1217) incised into Tuffs IB and IC. Horizon See above. Bed/member Bed I. Formation N/A. Group N/A. Nearest overlying dated horizon Tuff IF. Nearest underlying dated horizon Tuff IC. Geological age Between 1.839 and 1.79 Ma. Developmental age Adult. Presumed sex Unknown. Brief anatomical description A complete upper dentition in a well-preserved palate and lower face. Announcement Blumenschine et al. 2003. Initial description Blumenschine et al. 2003. Photographs/line drawings and metrical data Blumenschine et al. 2003. Detailed anatomical description N/A. Initial taxonomic allocation *Homo habilis*. Taxonomic revisions None specific to this specimen. Current conventional taxonomic allocation *H. habilis*. Informal taxonomic category Pre-modern *Homo*. Significance Blumenschine et al. (2003) stressed what they interpreted as similarities between OH 65 and **KNM-ER 1470**. They went on to argue that this supported the hypothesis that variation within the "early *Homo*" fossils from East Africa was best explained as sexual dimorphism within a single species, thus undermining the case for a second species such as *Homo rudolfensis*. Other observers accept that OH 65 strengthens the case for sexual dimorphism within the Olduvai hypodigm of *H. habilis*, but they do not necessarily accept that OH 65 and KNM-ER 1470 are conspecific. Location of original **National Natural History Museum, Arusha,** Tanzania.

Ohaba Ponor (Location 45°31′N, 23°07′E, Romania) History and general description Cave in the Streiu valley, south of Hunedoara in the Southern Carpathian mountains. Temporal span and how dated? **Biostratigraphy** and conventional **radiocarbon dates** on wood charcoal and unburnt bone suggest the following ages: Middle Paleolithic (dated to 43,600 ± 2100–2800 years BP, Level IIIb > 41,000 years BP, Level IIIa 39,200 ± 2900–4500 years BP) and

an early Upper Paleolithic horizon. Hominins found at site Right second pedal phalanx (**Ohaba Ponor 1**). Archeological evidence found at site **Mousterian** (Eastern **Charentian**), **Upper Paleolithic**. Key references: Geology, dating, and paleoenvironment Gaál 1928, Mertens 1996; Hominins Gaál 1928, Necrasov and Cristescu 1965, Necrasov 1971; Archeology Mertens 1996, Păunescu 2000.

Ohaba Ponor 1 Site Ohaba Ponor. Surface/ *in situ In situ*. Date of discovery 1923. Finder J. Mallász. Horizon Level III. Geological age *c*.40 ka based on **conventional radiocarbon dates** from Levels IIIa and IIIb. Developmental age Adult. Presumed sex Unknown. Brief anatomical description Right second pedal phalanx. Announcement Gaál 1928. Initial description Gaál 1928. Photographs/ line drawings and metrical data Gaál 1931. Detailed anatomical description Gaál 1931. Initial taxonomic allocation *Homo primigenius*. Taxonomic revisions *Homo neanderthalensis*. Current informal taxonomic category *H. neanderthalensis*. Significance The only hominin remains associated with Middle Paleolithic context in Romania Location of original Reported as the Museum of Deva, Muzeul Regional Hunedoara, Deva, Romania, but currently its whereabouts is uncertain.

Ohalo II (Location 32°43′N, 5°34′E, Israel; etym. the name of the local village) History and general description The site of Ohalo II was first surveyed in 1989 by a team led by Dani Nadel during an unprecedented drought that lowered the water level of the Sea of Galilee by 3 m. It was excavated until 1993 when heavy rains re-submerged the site; a second series of excavations were undertaken in 1999–2001 when another drought re-exposed the site. The unique geological setting caused by shallow/calm water flooding has preserved a remarkable number of materials. Temporal span and how dated? **Radiocarbon dating** on associated charcoal suggests an age of 23–19 ka. Hominins found at site The complete skeleton of an adult male was found in late 1991, apparently intentionally buried with its knees tightly flexed and its hands crossed over its chest. The skeleton is typical of other Epipaleolithic *Homo sapiens* (e.g., relatively robust jaws, rugose muscle markings, and marked asymmetry in the upper limb bones). Archeological evidence found at site The skeleton was found in direct association with a Kebaran assemblage. Apart from the fauna (Belmaker et al. 2001) the waterlogged nature of the site preserved many organic remains, including evidence of brush huts (Nadel and Werker 1999), plant foods, plant food processing and plant food preparation areas (Weiss et al. 2008, Piperno et al. 2004), and wooden implements (Nadel et al. 2006). Key references: Geology and dating Nadel et al. 1995, Carmi and Segal 1992; Hominins Hershkovitz et al. 1995; Archeology Nadel 2002, Nadel et al. 2002.

OHD   *See* **obsidian hydration dating**.

OIS   Acronym for **Oxygen Isotope Stages** (*which see*).

Okazaki fragment   A short fragment of **DNA** that is created during replication on the lagging strand of DNA (i.e., the strand of DNA that is oriented in the 5′ to 3′ direction and which is complementary to the leading strand). Since DNA replication always occurs in the 5′ to 3′ direction, it is continuous on the leading strand, but occurs in sections on the lagging strand. These fragments are then linked by DNA ligase to create a continuous strand.

Okladnik'ov Cave   (Location   54°45′50″N, 84°02′34″E, Altai Mountains, Siberia, Russia; etym. named after Alexey Okladnikov, a famous Soviet archeologist) History and general description This cave site in Siberia is best known for being the being the most easterly of the sites containing evidence of *Homo neanderthalensis*. Several hominin fossils have been recovered, but their taxonomic status is much debated, with some arguing they are *H. neanderthalensis*, others saying they evolved from Neanderthals, and still others suggesting similarities between these fossils and Asian *Homo sapiens* and *Homo erectus*. Recent studies of **mitochondrial DNA** from one adult and a juvenile suggest that the juvenile is a Neanderthal, while the adult is likely not to be. Temporal span and how dated? The remains were directly dating by radiocarbon dating; the juvenile has ages that range from *c*.30 to 38 ka (uncalibrated), while the adult remains are dated to *c*.24 ka (uncalibrated). Hominins found at site Several teeth and a few isolated postcranial bones were recovered, representing one or two adults and one juvenile. Archeological evidence found at site Three layers were identified, all with **Mousterian** stone tools. Key references: Geology, dating, and paleoenvironment Krause et al. 2007b; Hominins Krause et al. 2007; Archeology Kuzmin 2004.

Okote Tuff   Tuff exposed in the **Koobi Fora Formation** that defines the base of the Okote Member. Age 1.56±0.05 Ma (McDougall et al. 2005). Lateral equivalent H-2 in the **Shungura Formation**.

Okote Tuff Complex   Name given to the tuffs and sediments exposed along the Karari Ridge at **Koobi Fora** that occupy the interval between the KBS and the Chari Tuffs of the **Koobi Fora Formation**. The equivalent interval at **Ileret** is called the Ileret Tuff Complex or the Lower-Middle Tuffs, and at the western end of the Koobi Fora Ridge it is called the Koobi Fora Tuff Complex (Brown et al. 2006).

OKT   Abbreviation for the Okote Member in the **Koobi Fora Formation** (*which see*).

Ola Bula   One of the Early and Middle **Pleistocene** sites in the exposures of the **Ola Bula Formation** within the **Soa Basin**, central Flores, Indonesia.

Ola Bula Formation   (etym. Named after a site where the formation is exposed) General description The <80 m-thick Ola Bula Formation is exposed in the Soa Basin, which is the drainage of the Ae Sissa River in central Flores, Indonesia. It is described as comprising "white tuffs and interbedded tuffaceous sediments underlying fluvial deposits" (Morwood et al. 1998, p. 174). Sites or localities where the formation is exposed Boa Lesa, Deko Weko, Dozu Dhalu, Kobatuwa, Kopwatu, Lembahmenge, Malahuma, **Mata Menge**, Ngamapa, Ola Bula, Pauphadhi, Sagala, Tangi Talo, Wolo Milo, Wolokeo. Temporal span Early and Middle Pleistocene. How dated? **Biostratigraphy**, **geomagnetic dating**, and **fission track dating** (for details *see* **Mata Menge**). Hominins recovered from the formation None. Archeological evidence recovered from the formation Some 500 artifacts, mostly *in situ*, the majority of which are **flakes** and **cores**. Key references: Geology, dating, and paleoenvironment Morwood et al. 1998, O'Sullivan et al. 2001; Hominins N/A; Archeology Morwood et al. 1997, 1998, 1999, Brumm et al. 2006.

OLAPP   Acronym for the **Olduvai Landscape Paleoanthropology Project** (*which see*).

"Old man of Cro-Magnon"   *See* **Cro-Magnon 1**.

Old World monkeys Old World monkeys (sometimes abbreviated as OWM) is the informal name for the superfamily **Cercopithecoidea**. *See* **Cercopithecoidea**.

older to younger adults *See* **OY ratio**.

Oldowan (etym. Named after the site of **Olduvai Gorge**; Olduvai is Maasai for "the place of the wild sisal") The Oldowan, which is the oldest recognized stone tool **industry**, is known from sites that date from *c*.2.6 to 1.6 Ma, primarily in Africa. The term was first applied by Louis **Leakey** (Leakey 1935) to artifacts from Olduvai Gorge (the type site), as well as those from **Kanjera** and **Kanam** West on the Homa Peninsula in Kenya. The term Oldowan is not only used for this ancient technology, but also for the suite of artifacts produced by these technological practices, as well as for the sites where these artifacts are found. Mary **Leakey** published the first formal description of Oldowan tools from Bed I and lower Bed II Olduvai Gorge, Tanzania (Leakey 1966a, 1971) and her typology is still widely employed. **Flakes** were produced by using one stone (called a **hammerstone**) to knock flakes off of another (a **core**) in a technique termed hard-hammer percussion. Cores were sometimes set on an **anvil** and struck from above (**bipolar percussion**), others were likely fractured by being thrown against a hard substrate (Plummer 2004). Leakey (1971) classified hominin-modified Oldowan stone artifacts into three groups: **tools**, utilized pieces, and **debitage**. An artifact was only considered a "tool" if it had been intentionally **retouch**ed or flaked. Tools were further subdivided into light-duty (e.g., small, retouched flakes) and heavy-duty (e.g., **choppers**, **discoids**, polyhedrons) categories based on a mean diameter of less than 50 mm (light duty), or greater than 50 mm (heavy duty), respectively. Hammerstones, anvils, or flakes damaged through use were termed utilized pieces. Mary Leakey's debitage category included unmodified flakes, flake fragments, and other knapping debris. The term **manuport** was used for natural stones that had been transported and discarded without being modified. Leakey believed that heavy-duty tools were the most significant component of the Oldowan toolkit, that hominins shaped their tools with a clear idea of the desired end product, and that different tool forms were used for different tasks (Leakey 1971). More recently, it has been argued that the Oldowan was a simple but effective method of producing sharp flakes from stones, and that the flakes, not the cores, were the desired end product (Potts 1991, Toth 1985). Others claim that rather than following a mental template, the hominins responsible for the Oldowan industry relied on simple spatial concepts to coordinate flake production (Wynn and McGrew 1989), and that the different forms of cores were strongly influenced by the size and shape of the raw material, its flaking characteristics, and the extent to which a stone had been flaked (Toth 1985). Recent analyses (e.g., Stout et al. 2005, Braun et al. 2009, Harmand 2009) emphasize that the hominins responsible for Oldowan artifacts displayed a keen understanding of the physical properties of different raw materials and preferentially chose good-quality rock (i.e., tractable and not readily prone to shattering) to flake. Oldowan occurrences are best known from East Africa and sites in the 2.0–2.6 Ma range are found exclusively in this region. Important late Pliocene localities include those in the Omo **Shungura Formation**, the **Gona study area**, and the **Hadar study area** in Ethiopia, and **West Turkana** and **Kanjera** in Kenya. Whether the earliest usage of Oldowan tools was restricted to East Africa, or whether behaviors forming Oldowan sites were more broadly distributed across Africa, is at this point unclear. More recent (i.e., between 1.6 and 2.0 Ma) Oldowan sites have been found in North (e.g., **Ain Hanech**), southern (e.g., **Sterkfontein**), and East (e.g., **Koobi Fora**, **Olduvai Gorge**) Africa. The recovery of a comparable technology to the Oldowan at **Dmanisi**, Georgia, at approximately 1.7 Ma and from Majuangou and **Yuanmou**, China, at about the same time suggests that the earliest travelers out of Africa took with them the capacity to make the Oldowan toolkit (Gabunia et al. 2001, Zhu et al. 2004, 2008). The uses that Oldowan tools were put to, the adaptive significance of the earliest lithic technology, and the behaviors that led to archeological site formation have been controversial topics (Plummer 2004). One demonstrable use of these artifacts was animal butchery, but Oldowan artifacts may have been used to work wood and process plant foods as well. Behavioral models of Oldowan site formation include the **home base hypothesis**, **routed foraging**, stone caching, the **favored place hypothesis**, the **resource defense model**, the **dual-unit foraging model**, **refuging**, near-kill accumulation, and male display.

Olduvai *See* **Olduvai Gorge**.

Olduvai event *See* **Olduvai subchron.**

Olduvai Gorge (Location 2°56–59′S, 35°15–22′E, Tanzania; etym. Maasai for "the place of the wild sisal") <u>History and general description</u> Olduvai Gorge is an approximately 40 km/25 mile-long dry river valley that has been incised at the western margin of the eastern limb of the East African Rift System and which has eroded through sediments dating from *c*.2 Ma to the recent past. The first nonhominin fossils were recovered from Olduvai Gorge by Wilhelm Kattwinkel in 1911 and more were located by Hans Reck in 1913. Reck also recovered a hominin skeleton (**OH 1**), but this proved to belong to a recently buried modern human. Louis **Leakey** took over research at Olduvai from Reck in 1931 and for the next 40 years Louis and Mary **Leakey**, and latterly just Mary Leakey, undertook fieldwork at Olduvai. Since Mary Leakey stopped fieldwork at Olduvai Gorge several other groups have conducted research there (e.g., *see* **OH 62, Olduvai Landscape Paleoanthropology Project**); more on the 20thC history of research at Olduvai Gorge can be found in Leakey (1965, p. xi–xiv) and Hay (1976, p. 1–6). The hominin fossil and archeological localities at Olduvai Gorge are located in sedimentary rocks in either the Main Gorge or in a branch called the Side Gorge. The sediments exposed in Olduvai Gorge all lie above the basement lavas and they are divided into seven beds, collectively referred to as the Olduvai Beds, and Hay (1976) makes the point that "in modern stratigraphic usage, the various beds would be considered formations" (*ibid*, p.6). The four beds immediately above the basalt are identified using roman numerals (Beds I–IV, with Bed I at the bottom), whereas the three beds above them were given local names, the Masek immediately above Bed IV, then the Ndutu, with the Naisiusiu Bed the most recent; what used to be Bed VI of Reck and Leakey (1951) is now known as the Namarod Ash Formation. Beds I and II are readily distinguished wherever they are exposed, but there are places in the gorge where Beds III and IV cannot be readily distinguished and Hay (1976) proposed they be merged into a single unit called Beds III–IV. The sediments of Beds I and II contain several tuff layers (Bed I, Tuffs IA–IF; Bed II, Tuffs IIA–IID) and Olduvai was the first fossil hominin site to be dated using **potassium-argon dating** methods. Within Bed II, just above the Lemuta Member, there is an abrupt change in the fauna, which is referred to as the "faunal break." The sediments exposed in the gorges at Olduvai were laid down in and around a paleolake, called Lake Olduvai. The paleolake varied in size, sometimes being small and highly saline and at other times larger with fresher water. The localities at Olduvai are identified using a letter and number code. The letters refer to the names used by the Leakeys for the small valleys, or korongos (Mary Leakey uses that spelling, but Hay 1976 uses "karonga"), made by the streams that drained into the Main and Side Gorges, and to the situation of the locality within that small stream valley. The roman numeral numbers refer to the bed the locality is in. Most of the korongos were named after the Leakey's relatives and colleagues, with additional letters to denote how adjacent sites are geographically related to each other. So FLKN I is a fossil locality in Bed I in the Frida Leakey korongo that is north of locality FLK I, and FLKNN I is a fossil locality in Bed I in the Frida Leakey korongo that is north of locality FLKN I. Each locality has two other numbers associated with it, a geological locality number and an archeological number. Thus, FLKNN I is geological locality 45 and archeological site no. 31. There is even more terminological complexity for a few localities such as TK, MLK, FLK, FLKN, and FLKNN. At FLK, for example, 10 separate levels are recognized, each of which has a lower-case letter after the site number, so FLK Level 12 is archeological site no. 41g (but all the different levels have the same geological locality number). Three other sites have the FLK designation followed by some qualifier [e.g., "FLK (west of 40)," "FLK, Maiko Gully," and "FLK South"]. Just to make it even more complicated, these "extra" FLK sites all have the same geological locality number (45), but different archeological site numbers (42, 43, and 44, respectively). <u>Temporal span and how dated?</u> **Argon-argon dating, geomagnetic dating,** and **potassium-argon dating** suggest the sediments at Olduvai Gorge span the period between *c*.2 Ma to *c*.1300 years BP. <u>Hominins found at site</u> More than 60 fossil hominins have been recovered from Olduvai Gorge including **OH 5**, the type specimen of *Paranthropus boisei*, **OH 7**, the type specimen of *Homo habilis*, **OH 9**, a *Homo erectus* cranium, and **OH 62**, a partial skeleton of *H. habilis*. <u>Key references</u>: <u>Geology, dating, and paleoenvironment</u> Hay 1976, 1990, Tamrat et al. 1995, Hay and Kyser 2001;

Hominins Leakey 1959, Leakey et al. 1964, Tobias 1967, 1991a, Johanson et al. 1987, Blumenschine et al. 2003; Archeology Leakey 1971, Leakey and Roe 1995, Potts 1988, Domínguez-Rodrigo et al. 2007, Ashley et al. 2010, plus entries in a special issue of *Quaternary Research* [2010, 74(3): 301–424].

Olduvai Landscape Paleoanthropology Project Multidisciplinary field and laboratory project established in 1989 to investigate the paleoecological context of the hominin activities at **Olduvai Gorge**. It is co-directed by Robert Blumenschine (who heads the archeology subgroup) and Charles Peters (who heads the paleobotanical subgroup.); the third, geological, subgroup is headed by Ian Stanistreet. The focus of the fieldwork has been lowermost Bed II and Middle-Upper Bed I. The early field seasons focused on the margin of the paleolake, **Lake Olduvai**, and on the regions of Olduvai Gorge that had not been explored in detail by Louis and Mary **Leakey**. Its first major field season was in 2000. The project has built and maintains a field laboratory at Olduvai Gorge, where it carries out much of the analysis of the stone tools and faunal remains it has recovered from the excavations conducted, and it has helped train Tanzanian students.

Olduvai subchron One of the first events (now called **subchron**s) to be recognized in the **geomagnetic polarity time scale** (GPTS). It is an episode of normal magnetic direction within the Matuyama reversed epoch (now called the **Matuyama chron**) from *c*.1.9 to *c*.1.785 Ma that was recognized in the sediments of **Olduvai Gorge** (Grommé and Hay 1963). The lower boundary (i.e., between the R1 and N1 magnetozones) is 6 m below Tuff IA, and the upper boundary (i.e., between the N1 and R2 magnetozones) is just above Tuff IF, between it and Tuff IIA.

olfactory bulb The olfactory bulb is the part of the vertebrate forebrain involved in the detection and discrimination of odors. The olfactory bulb is the most rostral (i.e., anterior) structure of the brain in most vertebrates, but in haplorhine primates (i.e., apes, monkeys, and tarsiers) in which the olfactory bulbs are relatively small compared to the **cerebral hemispheres** they are mostly hidden beneath the overlying **frontal lobe**. The olfactory bulbs sit either side of the midline crista galli in the floor of the anterior cranial fossa

above the cribriform plate of the ethmoid, which separates it from the olfactory epithelium of the nasal cavity. Olfactory receptor neurons in the olfactory epithelium project their axon fibers through perforations in the cribriform plate to the main olfactory bulb. The main olfactory bulb in haplorhine primates is small relative to overall brain size, which may reflect a reduction in olfactory sensitivity. In addition, as judged by the accumulation of **pseudogene**s, apes have lost many olfactory receptors and modern humans have approximately twice as many olfactory receptor pseudogenes as other apes, suggesting there has been further relaxation of selection on olfactory capacity in the hominin clade. In most vertebrates, an additional structure, an accessory olfactory bulb, is located dorsal and posterior to the main olfactory bulb. The accessory olfactory bulb is part of an independent system for the processing of pheromones and other chemical substances that are important in sociosexual behavior. It receives inputs from the vomeronasal organ and sends outputs to the amygdala and hypothalamus to influence aggression and other aspects of social behavior. The vomeronasal organ and accessory olfactory bulb are highly variable among catarrhine primates (i.e., Old World monkeys and apes) and they appear to regress during fetal development in many species, including modern humans. The extent to which the accessory olfactory system is functional in modern humans is unclear.

oligonucleotide (Gk *oligos* = few or little, so literally "a few nucleotides") A short fragment of single-stranded **DNA** that is typically less than 30 **bases**, or **nucleotides**, long and which is used as a **probe** for **DNA hybridization**, or as a **primer** for the **polymerase chain reaction**. It is also possible to make oligonucleotides from RNA.

Olorgesailie (Location 1°35′S, 36°37′E, Kenya; etym. Maasai for "the place of the Giselik people," and applied to the nearby volcano) History and general description A large site complex located in the southern Kenya part of the **East African Rift System** with numerous archeological and paleontological sites in well-dated **tuffaceous** sediments near the margins of a fluctuating lake, best known for locally dense concentrations of **handaxe**s and other **Acheulean** implements, but the upper layers contain substantial evidence of the Middle Stone Age. J.W. Gregory first noted stone tools at Olorgesailie in 1919 but it was the survey and excavation undertaken by Louis and Mary **Leakey** (with the help of Italian prisoners of war) between 1942 and 1945 that led to Olorgesailie being gazetted as part of the **Coryndon Museum** in Nairobi. Further excavations were conducted from 1961 to 1965 by Glynn **Isaac** and have been ongoing since 1985 under the direction of Richard Potts. Temporal span and how dated? Sediments of the Olorgesailie basin are divided into two geological formations. The **Early** and **Middle Pleistocene**-aged **Olorgesailie Formation** has been studied in detail; the Middle and Later Pleistocene Oltulelei Formation has only been recently defined. The Olorgesailie Formation has been divided into 14 numbered members. **Argon-argon dating** of **anorthoclase feldspars** in **tephra** and magnetostratigraphic analyses of sediments have provided an age estimate of 992 ± 39 ka for a water-lain ash deposit in the middle of Member 1, and an age estimate of 493 ± 1 ka for **pumices** within Member 14. The **Brunhes-Matuyama boundary** occurs near the base of Member 8. Formal dates for the overlying Oltulelei Formation have yet to be published. Hominins found at site The diminutive partial cranium **KNM-OG 45500** (but published as KNM-OL 45500) was recovered in 2003 from 970–900 ka sediments. Archeological evidence found at site More than 100 excavations have been undertaken in Olorgesailie Formation sediments, primarily in Members 1, 6/7, and 11. The more than 34,000 recovered stone artifacts, predominantly made from local (<5 km/3 miles) Mio-Pliocene **lavas**, demonstrate substantial technological and typological variability within the Acheulean behavioral system. Reconstructions using **soil carbonate isotopic analysis**, analyses of **diatoms**, and the rich mammalian fauna suggest the paleoenvironment ranged from open grassland settings to shallow floodplains and wetlands on the margin of a variably fresh or alkaline series of lakes. The distribution of archeological traces indicates a relationship between the palenvironment and hominin behavior. Key references: Geology, dating, and paleoenvironment Isaac 1977, 1978b, Deino and Potts 1990, Potts et al. 1999, Behrensmeyer et al. 2002, Potts 2007, Owen et al. 2008; Hominins Potts et al. 2004; Archeology Isaac 1977, Potts et al. 1999.

OLS Acronym for **ordinary least squares regression** (*which see*).

OMD 1'28 *See* **Oranjemund**.

omega-3 fatty acids *See* **essential fatty acids**.

omega-6 fatty acids *See* **essential fatty acids**.

omnivore *See* **omnivory**; **eclectic feeder**.

omnivory (L. *omni*, from *omnis* = all, every, plus *vorous*, from *vorare* = to devour) The ability to eat a wide range of plants and animals. Several primates, including vervets, baboons, chimpanzees, and modern humans, are described as omnivores. However, the ability of modern humans to process foods (e.g., by cooking), means they can incorporate a particularly wide range of items into their diets. Thus, modern humans are the only habitually omnivorous extant primate and among large mammals only pigs and bears are as omnivorous. *See also* **eclectic feeder**.

Omo Prefix sometimes used in place of the "O." used by the French contingent of the **International Omo Research Expedition** for localities in the **Shungura Formation** (e.g., Omo 18) and for the fossils found within those localities (e.g., **Omo 18-1967-18**).

Omo 18-1967-18 (or Omo 18-18) Site Omo-Shungura. Locality 18. Surface/*in situ* Surface. Date of discovery 1967. Finder René Houin. Unit N/A. Horizon C-8. Bed/member Member C. Formation **Shungura Formation**. Group Omo Group. Nearest overlying dated horizon Tuff D. Nearest underlying dated horizon Tuff C. Geological age *c*.2.6 Ma. Developmental age Presumed adult. Presumed sex Unknown. Brief anatomical description An adult mandibular corpus that extends from a fracture through the M_3 alveolus on the right side through to a fracture just mesial to the left M_3 alveolus. There are no tooth crowns preserved, but the broken surfaces of all, or part, of the roots of the left $C–M_2$ and the right $I_2–M_2$ are exposed. Announcement Arambourg and Coppens 1967. Initial description Arambourg and Coppens 1967. Photographs/line drawings and metrical data Arambourg and Coppens 1967. Detailed anatomical description N/A. Initial taxonomic allocation *Paraustralopithecus aethiopicus*. Taxonomic revisions *Australopithecus africanus* (Howell 1978), *Australopithecus boisei* (Coppens 1980). Current conventional taxonomic allocation

Paranthropus aethiopicus. Informal taxonomic category Megadont archaic hominin. Significance The **holotype** of *Pa. aethiopicus*. Although almost all researchers regard the genus *Paraustralopithecus* as a junior synonym of the genus *Paranthropus*, researchers who want to maintain a distinction between the pre- and post-2.3 Ma parts of the East African **hypodigm** of *Paranthropus* do this by recognizing the species *Paranthropus aethiopicus*. Location of original National Museum of Ethiopia, Addis Ababa, Ethiopia.

Omo 323-1976-896 Site **Omo-Shungura**. Locality Omo-323. Surface/*in situ In situ*. Date of discovery 1976. Discoverer Claude Guillemot. Unit N/A. Horizon G-8. Bed/member Member G. Formation **Shungura Formation**. Group Omo Group. Nearest overlying dated horizon G-28. Nearest underlying dated horizon Tuff G. Geological age *c*.2.1 Ma. Developmental age Adult. Presumed sex Male. Brief anatomical description This partial **cranium** consists of nine substantial fragments (a–i). They include (a) a frontal fragment that preserves glabella, (b) the non-squamous parts of the left temporal, (c) a fragment that includes the posterior vault and the right mandibular fossa, and (g) part of the right maxilla including the worn crowns of P^3–M^1, and a left canine and P^4. Endocranial volume Approximately 490 cm^3 (Brown et al. 1993) [NB: Brown et al. (1993] claim that "with the recovery of **KNM-ER 23000**, we feel that these new estimates of cranial volume are more accurate" (*ibid*, p. 157), but there is very little of the cranial vault preserved and Alemseged et al. (2002) are more prudent and do not venture an endocranial volume estimate). Announcement Coppens and Sakka 1980. Initial description Coppens and Sakka 1980. Photographs/line drawings and metrical data Alemseged et al. 2002. Detailed anatomical description Alemseged et al. 2002. Initial taxonomic allocation *Australopithecus* sp. Taxonomic revisions *Australopithecus boisei* (Alemseged et al. 2002). Current conventional taxonomic allocation *Paranthropus boisei*. Informal taxonomic category Megadont archaic hominin. Significance Omo 323-1976-896 shows affinities with the hypodigms of both *Paranthropus aethiopicus* and *Paranthropus boisei*, but it is closer to the latter than to the former. Any differences between it and the existing *P. boisei* bauplan are consistent with within-species **polymorphism**, but Alemseged et al.

(2002) claim that the Omo 323-1976-896 calvaria "emphasizes the anagenetic link between *A. aethiopicus* and *A. boisei*" (*ibid*, p. 111). Location of original National Museum of Ethiopia, Addis Ababa, Ethiopia.

Omo I Site **Omo-Kibish**. Locality KHS (Kamoya's Hominid Site). Surface/*in situ* Surface. Date of discovery 1967 and 2001. Finder Kamoya **Kimeu**. Unit N/A. Horizon N/A. Bed/member Member 1. Formation **Kibish Formation**. Group N/A. Nearest overlying dated horizon Aliyo Tuff at the base of Member III. Nearest underlying dated horizon Nakaa'kire Tuff about 15 m above the base of Member I. Geological age Late **Middle Pleistocene** *c*.195 ka. Developmental age Young adult. Presumed sex Male, based on the large mastoid process and its overall size. Brief anatomical description An associated skeleton comprising a skull that includes a substantial part of the cranial vault (more complete on the right than the left) and parts of the face and mandible, and parts of the axial skeleton, the shoulder girdle, pelvic girdle, and the upper and lower limbs. The axial skeleton is represented by several well-preserved cervical and thoracic vertebra, and numerous vertebral and rib fragments. The shoulder girdle remains include the left clavicle and both coracoid processes. The right humerus is the only relatively complete upper limb long bone, and although the bones of both forearms are represented, the only carpal, metacarpal, and phalangeal remains belong to the right hand. A large and small fragment of the left pelvic bone and the distal ends of the femur and the tibia are preserved, as is the first ray of the right foot, plus the medial cuneiform and the navicular. Endocranial capacity Probably similar to **Omo II** (1435 cm^3). Announcement Day 1969. Initial description Day 1969. Photographs/line drawings and metrical data Day 1969. Day and Stringer 1991, Day et al. 1991, Pearson et al. 2008a. Detailed anatomical description Day and Stringer 1991. Day et al. 1991, Kennedy 1984, Bartsiokas 2002, Pearson et al. 2008a, Voison 2008. Initial taxonomic allocation *Homo sapiens*. Taxonomic revisions N/A. Current conventional taxonomic allocation *H. sapiens*. Informal taxonomic category Anatomically modern human. Significance The cranial vault of Omo I is high and globular, with a nearly vertical frontal bone, rounded occipital and pronounced parietal bosses that give the cranium an "*en maison*" form in posterior view; the mandible bears a slight chin. The cranium and postcranium are modern human in overall morphology but retain some primitive features. The original inferred age of *c.*130 ka, as well as the revised age of 195 ka, places this skeleton among the earliest evidence of modern humans. Location of original National Museum of Ethiopia, Addis Ababa, Ethiopia.

Omo II Site **Omo-Kibish**. Locality PHS (Paul's Hominid Site). Surface/*in situ* Surface. Date of discovery 1967. Finder Paul Abel. Unit N/A. Horizon N/A. Bed/member Member I. Formation **Kibish Formation**. Group N/A. Nearest overlying dated horizon Aliyo Tuff at the base of Member III. Nearest underlying dated horizon Nakaa'kire Tuff approximately 15 m above the base of Member I. Geological age Late, **Middle**, or early Upper Pleistocene, most likely 195 ka. Developmental age Older adult based on suture closure. Presumed sex Male, based on its robusticity. Brief anatomical description A well-preserved calvaria preserving all of the vault bones except portions of the frontal and the posterior and lateral parts of the basicranium. Endocranial volume 1435±20 cm^3. Announcement Leakey et al. 1969. Initial description Day 1969. Photographs/line drawings and metrical data Day 1969, Day and Stringer 1982, 1991, Schwartz and Tattersall 2003. Detailed anatomical description Day and Stringer 1982, 1991, Bräuer 1984, Rightmire 1984a; Schwartz and Tattersall 2003. Initial taxonomic allocation *Homo sapiens* or *Homo erectus*. Taxonomic revisions N/A. Current conventional taxonomic allocation Taxonomic proposals for Omo II have included *H. sapiens,* **Homo helmei,** **Homo heidelbergensis,** **Homo rhodesiensis,** and late archaic *H. sapiens.* Informal taxonomic category Pre-modern *Homo.* Significance The vault is high and the frontal is moderately receding, but closer to vertical than in **Kabwe** or **Saldanha**. The lateral segment of the supraorbital torus is small, suggesting affinities with modern humans. However, the occipital bone is strongly angled, maximum vault breadth is across the supramastoid tubercles, the mastoids are mediolaterally thick, and the tympanic is robust, suggesting affinities with *H. erectus* or *H. heidelbergensis.* Whereas **Omo I** is evidently anatomically modern, the angled occipital and robust temporal bone of Omo II align it with pre-modern *Homo.* Geologic evidence indicates the two

are coeval, raising questions of whether they sample a single, morphologically variable population, two morphologically distinct populations, or whether the Omo II calvaria might have eroded from an older context. Location of original National Museum of Ethiopia, Addis Ababa, Ethiopia.

Omo III Site **Omo-Kibish**. Locality N/A. Surface/*in situ* Surface. Date of discovery 1967. Finders N/A. Unit N/A. Horizon N/A. Bed/member Uncertain, probably Member I or III. Formation **Kibish Formation**. Group N/A. Nearest overlying dated horizon If from Member I, the Aliyo Tuff at the base of Member III (but if from Member III it would be Member IV). Nearest underlying dated horizon If from Member I, Nakaa'kire Tuff approximately 15 m above the base of Member I (but if from Member III, the Aliyo Tuff at the base of Member III). Geological age Either *c.*195 ka (Member I) or *c.*104 ka (Member III). Developmental age Presumed adult. Presumed sex Unknown. Brief anatomical description A fragmented cranial vault including a frontal that includes glabella. Announcement Day 1969. Initial description Day (1969) reported "what resemblance it has lies with the more modern of the first two Omo skulls" (*ibid*, p.1138). Photographs/line drawings and metrical data Day 1969. Detailed anatomical description N/A. Initial taxonomic allocation *Homo sapiens.* Taxonomic revisions N/A. Current conventional taxonomic allocation *H. sapiens.* Informal taxonomic category Anatomically modern human. Significance A further example of an **Omo I**-type cranial vault. Location of original National Museum of Ethiopia, Addis Ababa, Ethiopia.

Omo Basin *See* **Lower Omo Basin**.

Omo Group The older of the two inclusive stratigraphic sequences (the younger one is called the **Turkana Group**) exposed within the region known as the **Lower Omo Basin** (also called the Lower Omo Valley) and the **Turkana Basin**. It consists of the **Mursi, Nkalabong, Shungura,** and **Usno Formations** to the north, the **Koobi Fora Formation** to the east, and the Nachukui Formation to the west. *See also* **Omo-Turkana Basin; Turkana Basin**.

Omo Group Research Expedition A research team, led by Jean-Renaud Boisserie, that has been conducting fieldwork in the **Lower Omo Basin** since 2006–7.

Omo-Kibish (Location *c*.5°23′N, 35°56′E, Ethiopia; etym. named after the police post at Kibish) History and general description A complex of sites in the **Kibish Formation**, part of the **Turkana Group**, that outcrops along the lower Omo River in Ethiopia. The site complex, which was initially explored by Frank **Brown** and later by a team headed by Richard **Leakey**, yielded a series of hominins (**Omo I, II**, and **III**) in 1967. Geological work by Karl Butzer and colleagues established the basic organization of the sedimentary sequence that is still used today. Doubts expressed by paleoanthropologists in the 1980s over the initial minimum age for Omo I by **uranium-series dating** (U-Th) prompted renewed work on the Kibish Formation in 1999 and 2001–3 by John Fleagle, Frank Brown, Zelalem Assefa, John Shea, and colleagues. The new research discovered additional hominin fossils, including more fragments of Omo I, and help clarified the geochronology (although not to everyone's satisfaction), which indicated that both Omo I and II likely date to *c*.195 ka. Temporal span and how dated? Each member in the Kibish Formation formed during high stands of Lake Turkana that correlate with the deposition of **sapropel**s in the Nile's sediment cone in the Mediterranean. Member I correlates with S7 (i.e., sapropel 7) *c*.197 ka, Member III with S4 *c*.104 ka, and Member IV with S1 *c*.8 ka. **Argon-argon dating** on **tephra** in the Kibish Formation is consistent with those correlations. The Nakaa'kire Tuff low in Member I just below the level of Omo I and II dates to 196±2 ka, the KHS Tuff near the base of Member II chemically correlates with a tuff from **Konso** dated to more than 154 ka, and the Aliyo Tuff in Member III dates to 104±1 ka. A series of ages by **radiocarbon dating** on beds of molluscs from the top half of Member IV range from 3.6 to 13.1 ka. Hominins found at site From Member I, Omo I, an associated skeleton including a skull, Omo II, a calvaria, and Kib-163, the AHS tibia, comprising the distal three-quarters of a left distal tibia. From Members I or III, Omo III, frontal and nasal plus parietal fragments. From Member III or IV, Kib-170-1 and Kib-170-4, two cranial fragments from CHS. From Member IV, from the site of "Pelvic Corner," an unpublished pelvis, cranial fragments, and other bones recovered in 1967, and from JHS, an undescribed frontal, mandible, and tibia recovered in 2001. Archeological evidence found at site Members I–III contain scattered concentrations of small **Middle Stone Age** artifacts with preferential use of finer-grained raw materials. Most of the tools lack retouch, but some leaf-shaped bifacial points as well as sporadic, larger handaxes, picks, or lanceolates (which resemble Acheulean artifacts) were recovered. Member IV contains **Later Stone Age** artifacts and bone **harpoons**. Key references: Geology, dating, and paleoenvironment Butzer 1969, Butzer and Thurber 1969, Butzer et al. 1969, McDougall et al. 2005, 2008, Brown and Fuller 2008, Feibel 2008; Hominins Day 1969b, Day and Stringer 1982, 1991, Bartsiokas 2002, Pearson et al. 2008a, 2008b, Voison 2008; Archeology Shea 2008, Sisk and Shea 2008, Assefa et al. 2008b, Louchart 2008, Trapani 2008.

Omo Rift Zone Part of the **East African Rift System**, a series of river valleys and basins that extends from the **Afar Rift System** in the northeast, via the **Main Ethiopian Rift System** to the **Omo Rift Zone** in the southwest. Hominin fossil and archeological sites in the Omo Rift Zone include **Fejej**, **Usno Formation**, **Shungura**, and the site complexes around Lake Turkana (e.g., **Koobi Fora**, **West Turkana**, **Lothagam**, and **Kanapoi**).

Omo-Shungura *See* **Omo Basin**; **Shungura Formation**.

Omo-Turkana Basin The Lower Omo Basin and the Turkana Basin are effectively two components of an area of low country between the Ethiopian and Kenyan domes (north and south) and between the walls of the **East African Rift System** (east and west). The Omo-Turkana Basin extends for about 500 km/310 miles from north to south and up to about 100 km/62 miles from east to west. For the past few million years, or so, the larger southern component of the basin has been occupied by Lake Turkana. The latter has fluctuated in size becoming larger when inflow exceeds any outflow and evaporation and smaller when it does not. The sediments in the Omo-Turkana Basin are divided into the Omo Group and the Turkana Group. The major fossiliferous formations within the Omo-Turkana Basin are the **Koobi Fora, Mursi, Nachukui, Nkalabong, Shungura** and **Usno Formations**. The major hominin fossil and archeological site within the Omo-Turkana Basin are Koobi Fora, **Lothagam**, the **Omo region**, and **West Turkana**. *See also* **Omo Group**; **Turkana Group**.

ontogenetic (etym. *See* **ontogeny**) Any criterion, explanation, or test, that involves ontogeny. For example, the **ontogenetic criterion** is the proposal that the morphology of a character early in its ontogeny is likely to be the primitive condition of that character.

ontogenetic allometry *See* **allometry**; **scaling**.

ontogenetic criterion A criterion for deciding which state of a **character** is **primitive** and which is **derived**. When employing the ontogenetic criterion, one examines the distribution of the states of the character in the development of the species being examined, on the grounds that the morphology of a character early in its ontogeny is likely to be the primitive condition of that character. For example, the ontogenetic criterion would suggest that the primitive condition of the root of a tooth is a single root, for this is what all tooth roots, whether they eventually have one or several roots, look like early in their development (syn. the ontogeny character polarization criterion). *See also* **cladistic analysis**.

ontogenetic scaling (Gk *ontos* = being, *geneia* = origin) The process by which an ancestrally retained ontogenetic trajectory is followed to different adult endpoints, producing allometrically scaled adults. The ontogenetic trajectories for large and small animals who are ontogenetically scaled are similar; in descendant animals larger than the ancestors the trajectory is carried to a farther endpoint (**peramorphosis** via **hypermorphosis**) and in smaller descendants the similar ontogenetic trajectory is simply truncated at an earlier point (**paedomorphosis** via **hypomorphosis/progenesis**).

ontogeny (Gk *onto* = being and *gena* = to give birth to) A term that refers to all of the phases of **life history** that precede the cessation of **growth**. It is also used to refer to the unique developmental history of each individual organism.

open reading frame (or ORF) Refers to a **DNA** coding sequence that is translated into mRNA. Such a sequence runs from an initiation **codon** to a termination codon. It has the potential to encode a protein and is, for example, unbroken by a premature stop codon (hence the term "open"). If you find an open reading frame it usually indicates you have found a **gene**.

operational taxonomic unit (or OTU) Any taxon used in a **cladistic analysis** or phylogenetic analysis. To avoid circular reasoning hypotheses about the nature of operational taxonomic units must be generated using **phenetic** and not **cladistic** methods. In phylogenetic comparative analysis an operational taxonomic unit refers to an actual extant or fossil taxon for which data have been measured as opposed to an internal branching node within a phylogeny. Measurements from operational taxonomic units are often referred to as tip data. *See also* **hypothetical taxonomic unit**; **phylogenetic comparative analysis**.

opposable thumb An opposable thumb refers to the ability to bring the palmar pulp surface of the first digit squarely into contact with – or opposite to – the palmar pads of one or more of the other digits (i.e., "true" opposition) (Napier 1993). Thus, all digits are involved in true opposition; the thumb abducts and flexes (resulting in conjunct rotation) across the palm while the finger(s) flexes to meet the thumb. Napier (1961) suggested that a truly opposable thumb is found only in Old World monkeys, apes, and modern humans. In modern humans and great apes the joint surface between the trapezium and base of the first metacarpal is saddle-shaped, such that the thumb moves around a flexion/extension axis and an abduction/adduction axis with conjunct motion about these axes, facilitating opposition in the sense of pronation (or rotation) (Cooney et al. 1981, Hollister et al. 1992, Brand and Hollister 1999, Cerveri et al. 2008, Marzke et al. 2010). In modern humans, these joint surfaces are less curved than in African apes, allowing for accommodation of large axial loads on the joint associated with forceful pinch and grasp (Marzke et al. 2010). In modern humans opposition is facilitated by a broad area of contact between the pulps of the finger(s) and thumb making the modern human hand especially able to manipulate objects. In apes, opposability is hampered by the relatively short thumb compared to long fingers. In contrast, terrestrial monkeys, such as baboons and mandrills have more enhanced manipulative skills because the thumb-to-finger length ratio is more similar to that in modern humans. In modern humans, the opposability and overall grasping function of the thumb is accentuated compared to that of apes. Modern humans have an additional muscle, the flexor pollicis longus, that is either absent or only rudimentary in other primates, in addition to relatively enlarged intrinsic muscles to the thumb, such as the opponens pollicis, that together make the modern human thumb much stronger and more manipulative than that of other primates (Tocheri et al. 2008). In modern humans, the leverage or moment arm of the flexor pollicis longus muscle is enhanced by the presence of a sesamoid bone underneath the tendon. This sesamoid is accommodated by a facet on the proximal palmar surface of the thumb's distal phalanx. However, the facet is proximal to the "hollow" proximal volar fossa. This fossa, the sesamoid facet, and sesamoid bone are found in some nonhuman primates and thus the presence of a sesamoid does not necessarily reflect potential tool-use ability (Shrewsbury et al. 2003). The potential torque of the thumb muscles is estimated to be much higher in modern humans compared to that in chimpanzees (Marzke et al. 1999). Due to the shape of the carpometacarpal joint of the thumb many New World monkeys and strepsirrhines (e.g., capuchin) have a pseudo-opposable thumb. Pseudo-opposability refers to the ability to move the thumb close to the fingers (but not using the type of movement that occurs at the trapezial-first metacarpal in modern humans, the great apes and baboons), thus in pseudo-opposability there is no pulp-to-pulp contact between the thumb and finger(s). Other primates lack an opposable thumb completely (e.g., tarsiers) or the thumb is greatly reduced or absent (e.g., spider monkey).

optically stimulated luminescence dating
See luminescence dating.

optimal foraging theory
A subset of evolutionary ecology concerned with the foraging behavior of an organism. MacArthur and Pianka (1966) are credited with developing optimal foraging theory to determine the optimal diet of a predator in terms of the net amount of energy gained from the prey relative to the amount of energy expended in acquiring it (i.e., energetic returns). Their model predicts how a predator's diet breadth, or the number of prey types included in the diet, will respond to changes in its environment. For example, a predator foraging in an unproductive environment should have a greater diet breadth than one foraging in a highly productive environment, because it must target a greater number of prey types to maximize its energetic returns and satisfy its energetic requirements. The fundamental assumption of optimal foraging theory is that all foragers behave optimally, with the objective of maximizing energetic returns (i.e., obtaining more calories with minimal effort). Early archaeological applications of optimal foraging theory date to the 1970s and focused largely on changes in subsistence behavior. Today, optimal foraging theory is increasingly used to examine subsistence change through time in response to environmental change, **resource depression**, demographic shifts, or technological innovation. Inferences derived from optimal foraging theory have featured strongly in a variety of issues, including the origin and diffusion of agriculture, the material correlates of social status, early human social organization, and the evolution of human life history.

Oranjemund
(Location *c.*28°36′S, 16°26′E, Namibia; etym. from the German for "Orange mouth" the name of a nearby diamond-mining town) <u>History and general description</u> The site on the beach between the Orange River and the Atlantic Ocean where, in 1988, Daan Marais found OMD 1'28, a calvaria that most likely eroded from a black clay horizon on the riverbank 100 m from the findspot. The calvaria is long and narrow, has a thick cranial vault, a fairly low and receding frontal region, and a thin but projecting supraorbital torus. The latter has substantially thinner lateral segments, but no clear sulcus between the medial and lateral segment. <u>Temporal span and how dated?</u> No absolute dates are available, but when the morphology of OMD 1'28 is compared with other better-dated hominin crania from Africa the age of the calvaria is estimated to be *c.*50–100 ka. <u>Hominins found at site</u> OMD 1'28. <u>Archeological evidence found at site</u> None. Key references: <u>Geology, dating, and paleoenvironment</u> Senut et al. 2000; <u>Hominins</u> Senut et al. 2000; <u>Archeology</u> N/A.

orbital geometry
Three aspects of the Earth's relationship to the sun, including the tilt of the Earth's axis, the stability of that axis, and the shape of the Earth's orbit around the sun. Changes in orbital geometry are predictable over the **Neogene**, but predictions break down further

back in time (Laskar 1989). Predictable changes in (a) the tilt of the Earth's axis (called **obliquity**) follow a *c*.41 ka cycle, (b) the extent to which the Earth's axis wobbles (called **precession**) follow a *c*.23 ka cycle, and (c) the shape of the Earth's orbit round the sun (called **eccentricity**) follow a *c*.100 ka cycle. These changes in orbital geometry alter the latitudinal and seasonal distribution of **insolation** (i.e., incoming sunlight) as calculated by Laskar et al. (2004) with demonstrated influences on Earth's **climate**. The regularity of these cycles during the Neogene is such that **astrochronology** can be used to calibrate, or "tune" (as in "orbital tuning") **paleoclimate** cycles to the **astronomical time scale**. *See also* **astronomical theory**; **astronomical time scale**.

orbital tuning The process of matching a stratigraphic record to the calculated fluctuations in **insolation** (i.e., incoming sunlight) based upon **astronomical theory**. Orbitally tuned age models are dependent on absolute dating boundaries (e.g., the geomagnetic polarity time scale). Within those age constraints cyclic variations in a paleoclimate proxy may be tied to a dominant orbital frequency. The presence of other orbital frequencies provides independent corroboration of whether or not the "tuning" is valid. This approach has been very powerful for improving the age-model resolution for the Pliocene and Pleistocene using the oxygen isotope stratigraphy of benthic foraminifera as done by the SPECMAP project (Imbrie et al. 1984), and more recently updated and extended by Lisiecki and Raymo (2005). *See* **astronomical theory**; **astronomical time scale**; **Oxygen Isotope Stages**; **stable isotopes**.

Orce region (Location 37°43′N, 2°28′W, Spain; etym. after a nearby town) Located in a geographical depression of exposed Lower Pleistocene sediments (Gibert et al. 1983) in the southeastern Spanish province of Andalucía, this region has been reported to have one of the best Plio-Pleistocene mammalian fossil records in Europe (e.g., Martínez-Navarro 2002). There are three main sites in the Orce region: **Venta Micena**, Fuentenueva, and Barranco León. Venta Micena is the oldest of the sites, and is best known for the debate over whether a cranial bone from the site is from a hominin or from a nonhuman mammal. The two other sites contain mainly faunal remains and some possible stone tools. *See also* **Venta Micena**.

ordinal taxon A taxon in the taxonomic category "order" in the Linnaean system. Orders are relatively inclusive groupings (e.g., the order **Primates**). Other orders include the order **Carnivora**, which includes all the mammalian taxa that principally consume meat, and the order **Artiodactyla**, which includes all the mammalian taxa with an even number of hoofed toes.

ordinary least squares regression A form of **regression** in which model **parameters** are fit by minimizing the sum of the squared differences between observed and predicted values of the dependent variable. Least squares techniques may be applied to **simple regression** or **multiple regression**. Ordinary least squares regression, or OLS, minimizes error about the dependent variable and assumes no error in the independent variable(s) (or ignores error in the independent variables, depending on who you ask). It is a type of **Model I regression**. *See* **regression**.

ordination *See* **seriation**.

ORF Acronym for **open reading frame** (*which see*).

organ of Corti This term refers to a ribbon-like array of hair cells in the cochlea part of the membranous labyrinth of the inner ear. The organ of Corti lies on the basilar membrane within the cochlear duct part of the membranous labyrinth. The hair cells are stimulated in response to relative movement between the endolymph and the hair cells and the stimulation of hair cells is the means by which higher animals detect sound. The hair cells of the inner ear are an example of a highly conserved structure, for they are homologous with the hair cells embedded in the lateral line of fish.

oropharynx *See* **pharynx**.

Orrorin (etym. *Orrorin* = original man in the Tugen language) A genus established in 2001 by Senut et al. (2001) to accommodate a new species they introduced for fossil remains recovered from the *c*.6.0 Ma **Lukeino Formation** at **Aragai**, **Cheboit**, **Kapcheberek**, and **Kapsomin** in Baringo District, Kenya. The authors claimed that the various cranial and postcranial features that distinguished the new fossils from *Pan* could not be accommodated in any of the existing hominin genera. White et al. (2009a) suggest that the morphological differences between species assigned to *Ardipithecus* and to *Orrorin* do not justify their being assigned to different genera, in which case the former genus would have priority and *Orrorin* would be the junior synonym of *Ardipithecus*. Type species *Orrorin tugenensis* Senut et al., 2001.

Orrorin tugenensis Senut et al., 2001 (etym. *Orrorin* = original man in the Tugen language; *tugen* = Tugen Hills) The genus and species established in 2001 by Senut et al. to accommodate cranial and postcranial remains recovered from the *c.*6.0 Ma **Lukeino Formation** at **Aragai, Cheboit, Kapcheberek,** and **Kapsomin,** Baringo District, Kenya. The femoral morphology has been interpreted (Pickford et al. 2002, Richmond and Jungers 2008) as suggesting that *O. tugenensis* is at least a facultative biped, but other researchers interpret the radiographs and computed tomography scans of the femoral neck as indicating a mix of bipedal and nonbipedal locomotion (Galik et al. 2004, Ohman et al. 2005). Otherwise, the discoverers admit that much of the critical dental morphology is "ape-like" (Senut et al. 2001, p. 6). *O. tugenensis* may prove to be a hominin, but it is equally and perhaps more likely that it belongs to another part of the adaptive radiation that included the common ancestor of panins and hominins. White et al. (2009a) suggest that the morphological differences between *Ardipithecus ramidus* and *O. tugenensis* do not justify their being assigned to different genera, in which case the latter taxon would be transferred to the genus with priority (i.e., *Ardipithecus*) as *Ardipithecus tugenensis* (Senut et al., 2001) White et al., 2009. First discovery **KNM-LU 335** (1975). Holotype **BAR 1000′00** (2000). Paratypes KNM-LU-335, BAR 349′00, BAR 1002′00, BAR 1004′00, BAR 1003′00, BAR 1001′00, BAR 1215′00, BAR 139′00, BAR 1425′00, BAR 1426′00, and BAR 1900′00. Main sites Localities in the Tugen Hills, Kenya, as listed above.

orthogenesis A model of biological evolution popular in the early 20thC. Theodor Eimer, professor of zoology and comparative anatomy at the University of Tübingen, used the term orthogenesis (also called orthogenetic evolution) to refer to a model of evolution that proceeded in straight lines instead of the branching pattern of evolution promoted by Charles **Darwin**. This may explain why by the end of the 19thC most biologists had accepted the principle of biological evolution, yet many of them rejected Darwin's mechanism of natural selection in favor of neo-Lamarckian models of evolution consistent with orthogenesis. Supporters of orthogenesis believed that it was an inherent biological property of all organisms to evolve along predetermined lines of development. Thus, selection and adaptation play little, or no, role in orthogenetic models of evolution. A strong early 20thC advocate of orthogenesis was Henry Fairfield Osborn, a paleontologist at the American Museum of Natural History. Osborn's research on the extinct group of mammals called Titanotheres had suggested there were parallel lines of development in these organisms. Osborn did not invoke orthogenesis in his discussions of human evolution, but some paleoanthropologists did, foremost of whom was Arthur Smith **Woodward** who was deeply involved in the study of the **Piltdown** fossils. Smith Woodward argued that the evolution of the human brain was driven by internal biological factors and was not an adaptive response to environmental conditions or the product of changing human behavior. Arthur **Keith** also argued that the evolution of large brains in different hominin taxa was likely the result of an inherent and general tendency for brain enlargement in Pliocene hominins. Later, the British anatomist Wilfrid **Le Gros Clark** also suggested that orthogenesis helped to explain why primates in separate lineages had evolved similar traits and developed along similar evolutionary lines despite the fact they had diverged from a common ancestor at a geologically remote period. Orthogenetic theories of evolution began to wane by the middle of the 20thC as the mammalian fossil record became more complete and many examples of nonlinear evolutionary series in the fossil record were discovered. Orthogenesis finally fell out of favor in the 1950s in response to the rise of the modern evolutionary synthesis and to the general acceptance of models of evolution that were closer to that promulgated by Darwin.

orthognathic (Gk *orthos* = straight and *gnathos* = jaw) A term that literally refers to jaws that do not project, but technically it refers to skulls in which no part of the face or mandible (except the chin) projects forwards. It is usually quantified by measurements of the distance between a landmark on the base of the cranium (e.g., **basion, porion**) and a landmark on the front of the face (e.g., subnasale, **alveolare**). *Paranthropus boisei* and *Paranthropus robustus* are examples of fossil hominin taxa that have relatively flat or orthognathic faces (ant. **prognathic**).

orthogonal (Gk *orthogonion* = right-angled) In a direction at right angles to the plane of reference. The axes generated by multivariate analytical methods are usually orthogonal.

orthograde *See* **posture**.

orthologous genes (Gk *orthos* = straight and *logos* = proportion) **Genes** present in different organisms that are descended from a common ancestor (e.g., the myoglobin genes in chimpanzees/bonobos and modern humans are orthologous).

orthologue Abbreviation of **orthologous genes** (*which see*).

orthoquartzite *See* **quartzite**.

orthotropy (Gk *orthos* = straight and *tropus* = direction) In materials science, a material whose properties differ among each of three orthogonal axes, but those properties are similar within each of the three axes. Wood is an example of an orthotropic material; the stiffness of a log in its long axis differs from that in the axis perpendicular to its growth rings and from that in the axis tangential to its growth rings. The anthropoid mandible exhibits orthotropy at a histological or supraosteonal level of organization. Bone in other regions, like the shaft of the femur, can also be orthotropic, although the differences in elastic moduli in the tangential (perpendicular to the cortical shaft) and circumferential directions often differ by about 10% and thus the elastic structure is close to **transverse isotropy**. The three-dimensional elastic structure for most bones of vertebrates has not been investigated but is presumed to be orthotropic or transversely isotropic because of regular structural patterns involved in bone growth and remodeling, although this is likely not the case in regions constructed of woven bone.

os (L. *os* = bone, as in *osseus* = bone or bone-like) The Latin term for bone. In the Latin version of **anatomical terminology** "os" is used as a prefix for many of the bones. Some researchers still use the Latin form (e.g., "os coxae" instead of the official English-language term "hip bone" or "pelvic bone"), but to be consistent those who do so should refer to *all* bones in their Latin form (e.g., the scaphoid as the "os scaphoideum" and "caput ulnae" instead of the "head of the ulna"). It is more appropriate to use the English-language terms listed in the *Terminologia Anatomica* than to continue to use terminology that is arcane. However, for some bones "os" is still used as a prefix as part of the formal name of some bones (e.g., "**os centrale**" is both the Latin *and* the English-language term for a cartilaginous nodule that in modern humans usually fuses with the scaphoid, but which in the great apes more often persists as a separate carpal bone). *See also* **anatomical terminology**.

os centrale (L. *os* = bone, as in *osseus* = bone or bone-like) A cartilaginous nodule that in modern humans usually fuses with the scaphoid, but which in the non-African apes more often persists as a separate carpal bone. *See also* **scaphoid-os centrale fusion**.

os coxae (L. *os* = bone, as in *osseus* = bone or bone-like and *coax* = hip) Some researchers use the Latin terms "os coxae" instead of the English-language term "hip bone" or "pelvic bone." But to be consistent, those who do so should refer to all bones in their Latin form (e.g., the scaphoid as the "os scaphoideum" and the head of the ulna as the "caput ulnae"). *See also* **anatomical terminology**.

OSL Acronym for optically stimulated luminescence dating. *See* **luminescence dating**.

Osmundsen, Lita *See* The **Wenner-Gren Foundation for Anthropological Research**.

ossification (L. *os* = bone and *facere* = to make) The process of bone formation, or osteogenesis. There are two types of ossification: **intramembranous ossification** and **endochondral ossification**; in both types bone formation begins at a focus called an ossification center. Bone formation by intramembranous ossification (e.g., in the flat bones of the cranial vault) begins at centers of ossification within flat sheets or "membranes" (hence the term intramembranous) of collagen fibers. Bone-forming cells called osteoblasts form a featureless osteoid matrix that first calcifies (involving calcium phosphate) and then ossifies (involving the formation of hydroxyapatite). As more and more matrix is calcified and then ossified, the distance between individual bone-forming cells increases and thus the cells are gradually isolated from one another in osteocyte lacunae. Most eventually lie quiescent, some die from inadequate nutrition, and some of these mineralize within the lacuna. Groups of osteoclasts cut out microcavities in previously formed bone (this happens during both bone formation and bone remodeling). The space created is then filled in by bone formed first at the periphery and then in circles of ever decreasing size; these are the Haversian lamellae in an osteon system. The central canal in such an osteon system contains a nutrient blood vessel and the osteocytes communicate with each other via minute canaliculi. Most of the bones of a hominin skeleton are formed by the second type of ossification, endochondral ossification. Endochondrial ossification is a two-stage process. First, cells called chondroblasts form a hyaline cartilage model of the future bone. Cells in the middle of the cartilage model initially divide and grow by interstitial growth (as distinct from appositional growth). This enables rapid formation of the cartilage model. Chondrocytes then swell (hypertrophy) and die, leaving only a mineralized shell of cartilage matrix. This mineralized cartilage is then resorbed by osteoclasts (strictly speaking, chondroclasts) and at the same time osteoblasts start forming osteoid matrix on the walls of the spaces once occupied by chondroblasts. This happens first at the primary center of ossification, and this primary center is soon surrounded by a sleeve of subperiosteal bone that forms beneath the periosteum. In long bones the bone formed from the primary center of ossification is called the diaphysis. Primary centers usually appear between 7 weeks and 4 months of intrauterine life. Some bones (e.g., auditory ossicles, zygomatic, carpals, and tarsals) usually form from a single primary center; other bones (e.g., sphenoid, temporal) always form from multiple primary centers. Secondary centers of ossification, or epiphyses, form in complex bones (e.g., vertebrae), at the ends of long bones, and at the sites of bony processes (e.g., trochanters of the femur and the tubercles of the humerus). Long bones can continue to grow in length because new cartilage is laid down in epiphyseal plates between the primary center in the shaft of the bone and the epiphyses at either end. Bones stop growing in length when the epiphyseal plates cease to form new cartilage. When this happens the primary and secondary centers of ossification fuse into a single continuous bone. Epiphyseal plates are sometimes visible in radiographs as a dark line between the diaphysis and the epiphysis. *See also* **modeling; remodeling**.

ossification center *See* ossification.

Osteichthyes (Gk *ostei* = bony and *icthys* = fish) The name of the superclass of aquatic vertebrates that subsumes all the bony fish, both freshwater and marine. Bony fish can be divided into two classes: the diverse Actinopterygii (ray-finned fish) with more than 23,000 species, and the smaller and much less common Sarcopterygii (lobe- or fleshy finned fish), which include the ancestors of terrestrial vertebrates. The earliest known ray-finned fish appeared about 395 Ma in freshwater deposits and about 230 Ma in marine deposits. Most of the fish remains found in early hominin sites are freshwater taxa, although some marine taxa are found in late Pleistocene coastal sites. Africa has a diverse freshwater fish fauna, with almost 50% of the families being endemic, including the electrosensing Mormyridae (elephantfish) and the "flying" fish Pantodontidae. Other taxa are remnants from the Gondwanaland fauna, such as the Characiformes (e.g., tigerfish) and Cichlidae (included in the diverse Perciformes). Lungfish (Lepidosireniformes) and bony-tongues (Osteoglossidae) are remnants of more archaic distributions, while catfish (Siluriformes) from Asia and the minnow-like cyprinids (Cypriniformes) are later immigrants. Rifting in the Miocene and the Pleistocene was particularly significant for the evolution and ecology of African fish. It changed much of the topography of Africa from a flat landscape crossed by river systems to one with high relief and many lake basins. These new lacustrine zones were rapidly colonized by previously river-adapted fish and the well-studied cichlids in several African great lakes are examples of "explosive" adaptive radiations (Salzburger and Meyer 2004). Many early *Homo* remains are associated with lacustrine sites (e.g., **Olduvai Gorge** where there is evidence of hominin exploitation of wetland resources including fish; Ashley et al. 2009). *See also* **catfish; fish**.

osteoblasts (Gk *osteon* = bone and *blastos* = bud or immature cell) The cells that manufacture the collagen fibers and proteoglycans that comprise the osteoid matrix which eventually becomes bone. The matrix initially becomes calcified with calcium phosphate and then the calcium phosphate is converted into hydroxyapatite crystals. This matrix plus collagen, plus a few dormat osteoblasts (called osteocytes) comprise **bone** tissue. *See* **bone**; **ossification**.

osteoclasts *See* bone; ossification.

osteocytes *See* bone; ossification.

osteodontokeratic *See* Dart, Raymond; Makapansgat.

osteogenesis (Gk *osteon* = bone and *genesis* = origin) Bone formation, or ossification. *See also* **ossification**.

osteon (Gk *osteon* = bone) *See* **ossification**.

ostrich egg shell The fossilized eggs, or more often fragments of eggs, of ostrich-related species are the most common avian fossils in African hominin sites, especially open-air sites. This is doubtless because the egg shell of ostrich-related species is less than 5 mm thick. In Africa, two main lineages of ostrich-related birds run from the Oligo-Miocene to the Pleistocene. One, called aepyornithoid (i.e., "resembling *Aepyornis*"), has erroneously been related to (or even sometimes synonymized with) the elephant birds (i.e., the Aepyornithidae) of Madagascar, that have been extinct since the 19thC. But the features used to do this are probably primitive and cannot be used to infer such a relationship. The other main lineage shows marked evolutionary changes in the arrangement of the pores in the shell, together with a global size reduction through time. It is not clear which of the two lineages is ancestral to the extant ostrich genus *Struthio*. These taxa are useful for dating and for paleoenvironmental reconstruction, for ostrich-related birds are indicative of open, dry savanna-like habitats. *See also* **ostrich egg shell dating**.

ostrich egg shell dating Principle Ostrich egg shell and the shells of other large ratites (e.g., emus, moas) constitute some of the best-known materials for the analysis of protein **diagenesis** to determine age, as the proteins are tightly bound into the crystal structure of the shell and are not affected by humidity variation in the burial environment As a result, egg shell protein approximates a closed system, unlike bone or mollusc shell. Whereas racemization or epimerization of various amino acids can be detected in egg shell using liquid or gas chromatography, dating of sites in the Middle to Later Pleistocene has most often used the epimerization reaction of L-isoleucine to D-alloisoleucine, since the reaction kinetics have been shown to be linear. Method The reaction that brings about the change is temperature-dependent, but if average temperature can be estimated independently of sample age or if an ostrich egg shell sample from the Later Pleistocene of the site in question can be simultaneously dated by isoleucine epimerization and **AMS radiocarbon dating** then age estimates for at least two to three times the age of the calibration piece can be regarded as reliable. It is also possible to control for pieces that have been heated to higher than ambient temperature through exposure to fire, by measuring the relationship between L-isoleucine/D-alloisoleucine ratios and breakdown resulting from a different chemical decomposition reaction: de-carboxylation. The availability of a proxy for temperature, and lack of response to humidity variation make ostrich egg shell dating more reliable than amino acid racemization studies of bone, and ostrich egg shell dates have generally matched AMS radiocarbon dates made on the same sample. Ostrich egg shell dating can also be corroborated through analysis of stable carbon and oxygen isotopes in the samples to obtain paleoclimate information. Time range The crystal structure of ostrich egg shell is such that amino acids may be retained within it for upwards of 15–17 Ma, although isoleucine epimerization may only be valid for *c*.100–135 ka in the tropics and up to 1 Ma in China. Problems The influence of temperature and inter-laboratory variation in results. Examples Brooks et al. 1990, Miller et al. 1992, 1999, Johnson et al. 1997. *See also* **amino acid racemization dating**.

OTU Acronym for operational taxonomic unit. *See* **cladistic analysis**; **operational taxonomic unit**; **phylogenetic comparative analysis**.

Oulad Hamida 1 *See* **Thomas Quarry**.

out-of-Africa hypothesis This hypothesis on modern human origins emerged during the

1980s and drew upon paleontological, archeological, and genetic evidence to support the idea that modern humans had evolved within Africa, and then had migrated in one, or very likely more than one event, out of Africa to populate the Old World and later the New World. Thus, according to this hypothesis, all modern humans originate from a group of *Homo sapiens* that left Africa relatively recently. Importantly, the out-of-Africa hypothesis rejects the idea that *Homo erectus* populations existing in Asia and elsewhere had evolved into *H. sapiens*, or had given rise to any current modern human populations, as is argued by the supporters of the **multiregional hypothesis**. The out-of-Africa hypothesis for the origin of modern humans is now sometimes referred to as "Out of Africa 2" in recognition of the fact that the first hominin dispersal out of Africa was undertaken much earlier by *H. erectus*. Some of the basic tenets of the out-of-Africa hypothesis were proposed by William (Bill) **Howells** in 1976. Howells' "Noah's Ark hypothesis" argued that all modern humans originated from a single recent population of *H. sapiens* that was already morphologically modern. He suggested that groups of these early *H. sapiens* spread throughout the world replacing the existing more archaic hominin populations. During the early 1980s Günter **Bräuer** suggested a slightly different version of the hypothesis. This was initially called the Afro-European sapiens hypothesis, but was later changed to "**replacement with hybridization;**" this latter version allowed for interbreeding between *H. sapiens* and the archaic populations they replaced (Bräuer 1984). The most widely accepted version of the out-of-Africa hypothesis, also called the "**recent African origin**" or "replacement model" was proposed in 1988 by Christopher **Stringer** and Peter Andrews in an article published in the journal *Science* entitled "Genetic and fossil evidence for the origin of modern humans" (Stringer and Andrews 1988). They argued that modern *H. sapiens* evolved from a *Homo heidelbergensis* ancestor in Africa *c.*200 ka and that the hominin groups existing outside of Africa did not contribute in any significant way to the ancestry of current modern human populations. According to Stringer, some of these early *H. sapiens* migrated out of Africa between *c.*100 ka and spread throughout the Old World where they replaced existing populations such as *Homo neanderthalensis*. Important evidence for this hypoth-

esis came from research conducted by Rebecca **Cann**, Mark Stoneking, and Allan **Wilson** at the University of California at Berkeley and published in their influential paper "Mitochondrial DNA and human evolution" in the journal *Nature* in 1987. By studying the accumulation of mutations in the **mitochondrial DNA** of women from different geographical populations and by estimating mutation rates over time these researchers concluded that all living human populations shared a most recent female common ancestor who lived in Africa *c.*200 ka. The out-of-Africa hypothesis also draws upon evidence from the fossil and archeological records. Fossil hominins with a morphology similar to that of anatomically modern humans and dated to *c.* 70–100 ka have been recovered at **Klasies River Mouth** and **Border Cave** in southern Africa. Debates over the role of the *H. neanderthalensis* in human evolution assumed a new significance when it became clear that the *H. sapiens*-like remains found at sites such as **Qafzeh** and **Skhul**, in Israel, were likely to be *c.*100 ka, which meant that the Neanderthal remains found at sites such as **Amud** and **Kebara** almost certainly *postdated* the earliest regional evidence for *H. sapiens*, making it very unlikely that Neanderthals had evolved into modern humans. This latter scenario has also been weakened by the discovery of Neanderthal remains that may be as recent as 32 ka and of *H. sapiens* remains in Africa at Omo-Kibish and Herto that are estimated to be between 160 and 200 ka. The archeological evidence has also been interpreted (e.g., McBrearty and Brooks 2000) as being consistent with the recent African origin or replacement model. Supporters of the out-of-Africa hypothesis reject claims by advocates of the multiregional hypothesis that the China or Southeast Asia fossil record provides evidence of morphological continuity that connects *H. erectus* to *H. sapiens*. While at least one study (Krause et al. 2010) has demonstrated gene exchange between the Neanderthals and anatomically modern humans, the primary ancestral contribution to modern humans in Eurasia clearly derived from the out-of-Africa migrations of early modern people. Thus, whereas the strongest form of the out-of-Africa hypothesis has apparently been falsified, in its weaker form it has survived refutation. However, as some have pointed out (e.g., Relethford 2003), there is little to choose between the weak form of the the out-of-Africa hypothesis and the weak form of the multiregional hypothesis.

outcrop A naturally occurring exposure of rock at the Earth's surface. Geological investigation of outcrops have significant advantages over subsurface studies utilizing core or geophysical properties in the ability to collect large samples, follow lateral **facies** variations, and directly examine 3D geometry of strata. A significant drawback of outcrops relates to the rapid alteration of some rocks exposed to the atmosphere or to soil formation.

outer table *See* **cranial vault**.

outgroup The term used for a **taxon** (or taxa if it is a composite outgroup) used to help determine the polarity of the characters used in a cladistic analysis. Outgroups are normally closely related, preferably primitive, taxa. *See* **cladistic analysis**; **phylogenetic comparative analysis**.

outgroup criterion One of the criteria used to determine the **polarity** (i.e., which is the most primitive state and which the most derived) of the states of a character. The principle is that the character states seen in closely related, preferably primitive, sister taxa of the group being studied are likely to manifest the primitive states for that character. *See also* **cladistic analysis**.

overdominance *See* **balancing selection**.

Owen's lines Richard Owen was the first to describe "contour lines" in dentine. He suggested that they correspond to successive positions of the sheet of **odontoblasts** that form the dentine during tooth formation (Owen 1840–1845). Subsequent research suggests that Owen's contour lines occur because dentine tubules simultaneously all make a short bend along a line parallel to the outer contour of the tooth, and researchers have proposed that this phenomenon is due a disturbance to the regular secretion of dentine matrix. Thus clusters, or groups, of irregularly spaced Owen's lines are superimposed onto the regular underlying pattern of long-period **Andresen lines** in dentine; Owen's lines do not always coincide with Andresen lines. The equivalent of Owen's lines in enamel are **Wilson bands**.

owl pellet The regurgitated, indigestible remains of food ingested by owls. Owl pellets, particularly those attributed to *Tyto alba*, are considered to be the main way microfaunal remains enter the fossil record at archeological and paleontological sites. Owls are specialized predators of microfauna. When the prey is small, microfauna are eaten whole without preparation and the indigestible skin, hair and bone are regurgitated resulting in the deposition of a structure called an owl pellet. This digestive treatment both concentrates micromammal remains and also may protect their bones from other diagenetic processes. Small concentrated lenses of fossil rodents within archeological and paleontological sites are attributed to the repeated use of roosting sites by owls. The activities of owls thus are important for the preservation of an important part of the paleontological and paleoenvironmental record. *See also* **microfauna**.

OWM Acronym for **Old World monkeys** (*which see*).

oxygen An element commonly used in paleoclimate reconstructions and **stable-isotope biogeochemistry**. Oxygen has two common isotopes (^{18}O and ^{16}O) and one rare isotope (^{17}O); all three isotopes are stable and do not decay. *See* $^{18}O/^{16}O$; **paleoclimate**; **stable isotopes**.

Oxygen Isotope Stage 1 (or OIS 1) Time span *c.*141 ka–present. Magnetozone Brunhes Chron, normal polarity. This is the current Holocene stage. We are currently in an interglacial period. The only living *Homo* species during this stage is *Homo sapiens*.

Oxygen Isotope Stage 2 (or OIS 2) Time span *c.*29–14 ka. Magnetozone Brunhes Chron, normal polarity. Significant events The **Last Glacial Maximum** occurs during OIS 2 and it includes three **Heinrich events** (H): H1 (*c.*14.5 ka), H2 (*c.*20.5 ka), and H3 (*c.*23.4 ka). Examples of sites in OIS 2 Abri Flageolet I, **Abri Pataud**, and **La Ferrassie** (France).

Oxygen Isotope Stage 3 (or OIS 3) Time span *c.*57–29 ka. Magnetozone Brunhes Chron, normal polarity. Significant events This period is known as the interpleniglacial which is broken up into five rapidly oscillating interstadials: Oerel (*c.*58–54 ka), Glinde (*c.*51–48 ka), Moershoofd (*c.*46–44 ka), Hengelo (*c.*39–36 ka), and Denekamp (*c.*32–28 ka). **Heinrich event** (H) H4 took place at *c.*31.8 ka and H5 transpired at *c.*41.9 ka. The last appearance date of *Homo neanderthalensis* is during this stage. Examples of sites in OIS 3 **Saint-Césaire** and **Le Moustier** (France), **Banyoles** and **Zafarraya** (Spain), **Shanidar**, **Amud**, **Kebara** Units III–VI, **Boker Tachtit** Levels 1–4, and Boker A (Israel), **Grotta Breuil** (Italy), **Pavlov** (Czech Republic), **Zhoukoudian Upper Cave** (China).

Oxygen Isotope Stage 4 (or OIS 4) Time span *c*.71–57 ka. Magnetozone Brunhes Chron, normal polarity. Significant events Heinrich event 6 took place at *c*.64.2 ka. Examples of sites in OIS 4 **La Ferrassie** (France, the Mousterian levels have been dated to OIS 5 and OIS 4); Regourdou (France) may be either stage 5e or early stage 4.

Oxygen Isotope Stage 5 (or OIS 5) Time span *c*.130–71 ka. Magnetozone Brunhes Chron, normal polarity. Significant events This stage is broken up into five substages, with a series of quick oscillations designated 5a–5e. The last major interglacial correlates to OIS 5e, and this period is also known as the Eemian. The majority of *Homo neanderthalensis* sites are during stage 5. Examples of sites in OIS 5 Regourdou (France, may be either stage 5e or early stage 4), Lazaret (France), La Ferrassie (France, the Mousterian levels have been dated to OIS 5 and OIS 4), Taubach (5e), Tata (Hungary), Saccopastore 1 and 2 (Italy), Krapina (Croatia, may be either 6 or beginning of 5e), Šipka (Czech Republic), Subalyuk (Hungary), Gánovce (5e) (Slovakia). The peaks within this OIS date to approximately: 5.1 peak, *c*.82 ka; 5.2 peak, *c*.87 ka; 5.3 peak, *c*.96 ka; 5.4 peak, *c*.109 ka; 5.5 peak, *c*.123 ka.

Oxygen Isotope Stage 6 (or OIS 6) Time span *c*.191–130 ka. Magnetozone Brunhes Chron, normal polarity. Examples of sites in OIS 6 - Krapina (Croatia) (possible alternative dating OIS 5), **Biache Saint-Vaast**, **Combe Grenal**, Fontéchevade, **La Chaise-de-Vouthon** (l'Abri Suard), **Lazaret**, and **Montmaurin** (La Niche) (France), La Cotte de St. Brelade (Jersey), Pinnacle Point (South Africa), Ochtendung (Germany), Krapina (Croatia, either 6 or beginning of 5e).

Oxygen Isotope Stage 7 (or OIS 7) Time span *c*.243–191 ka. Magnetozone Brunhes Chron, normal polarity. Examples of sites in OIS 7 Pontnewydd Cave (Wales), Steinheim (Germany).

Oxygen Isotope Stage 8 (or OIS 8) Time span *c*.300–243 ka. Magnetozone Brunhes Chron, normal polarity. Examples of sites in OIS 8 Mesvin IV (Belgium), Ariendorf 1 (Germany).

Oxygen Isotope Stage 9 (or OIS 9) Time span *c*.337–300 ka. Magnetozone Brunhes Chron, normal polarity. Distinguishing whether a site is from OIS 9 or OIS 11 is often difficult because the intervening stage (OIS 10) was very short. Examples of sites in OIS 9 Hoxne (England), Pradayrol and Terra Amata (France), **Steinheim** (Germany), Castel di Guido.

Oxygen Isotope Stage 10 (or OIS 10) Time span *c*.374–337 ka. Magnetozone Brunhes Chron, normal polarity.

Oxygen Isotope Stage 11 (or OIS 11) Time span *c*.424–374 ka. Magnetozone Brunhes Chron, normal polarity. Distinguishing whether a site is from OIS 9 or OIS 11 is often difficult because the intervening stage (OIS 10) was very short. Examples of sites in OIS 11 Sidi Abderrahman (Morocco) Beeches Pit (and OIS 12), Boxgrove (or OIS 13), and Swanscombe (England), **Bilzingsleben (Germany)**.

Oxygen Isotope Stage 12 (or OIS 12) Time span *c*.478–424 ka. Magnetozone Brunhes Chron, normal polarity. Examples of sites in OIS 12 Beeches Pit (and OIS 11).

Oxygen Isotope Stage 13 (or OIS 13) Time span *c*.533–478 ka. Magnetozone Brunhes Chron, normal polarity. Examples of sites in OIS 13 Boxgrove (or OIS 11) (England), **Mauer** (possible alternative dating OIS 15).

Oxygen Isotope Stage 14 (or OIS 14) Time span *c*.563–533 ka. Magnetozone Brunhes Chron, normal polarity. Example of a site in OIS 14 **Sima de los Huesos** (Spain).

Oxygen Isotope Stage 15 (or OIS 15) Time span *c*.621–563 ka. Magnetozone Brunhes Chron, normal polarity. Examples of sites in OIS 15 **Atapuerca** (Gran Dolina) (Spain).

Oxygen Isotope Stage 16 (or OIS 16) Time span *c*.676–621 ka. Magnetozone Brunhes Chron, normal polarity. Example of a site in OIS 16 Tangshan Huludong (China).

Oxygen Isotope Stage 17 (or OIS 17) Time span *c*.712–676 ka. Magnetozone Brunhes Chron, normal polarity. Examples of sites in OIS 17 Pakefield (England) (possible alternative dating OIS 19), Zhoukoudian Lower Cave (China).

Oxygen Isotope Stage 18 (or OIS 18) Time span *c*.761–712 ka. Magnetozone Brunhes Chron, normal polarity. Examples of sites in OIS 18 **Gesher Benot Ya'aqov** (could be in OIS 18–20) (Israel).

Oxygen Isotope Stage 19 (or OIS 19) Time span *c*.790–761 ka. Magnetozones Brunhes Chron (780–760 ka), normal polarity, and Matuyama Chron (787–780 ka), reversed polarity. Examples of sites in OIS 19 Gesher Benot Ya'aqov (could be in OIS 18–20), Pakefield (England) (possible alternative dating OIS 17).

Oxygen Isotope Stage 20 (or OIS 20) Time span *c*.814–790 ka. Magnetozone Matuyama Chron, reversed polarity. Example of a site in OIS 20 Gesher Benot Ya'aqov (could be in OIS 18–20) (Israel).

Oxygen Isotope Stage 21 (or OIS 21) Time span *c*.866–814 ka. Magnetozone Matuyama Chron, reversed polarity.

Oxygen Isotope Stages A sequence of periods of relative differences in the ratio of oxygen isotopes used to provide approximate ages for archeological and fossil sites. The initial crucial discovery that led to the use of oxygen isotopes for geochronology was the demonstration by Urey (1947) of variations in the oxygen content of the shells of foraminifera (also called benthic or planktonic foraminifera, or just "forams") recovered from the sea floor. Foraminifera contain two isotopic forms of oxygen, ^{16}O and ^{18}O, and Urey demonstrated that the relative amounts of the "light" (^{16}O) and "heavy" (^{18}O) isotopes varied according to the temperature of the seawater. When the water is warm evaporation removes more of the lighter isotope, thus increasing the $^{18}O/^{16}O$ ratio, and when the water is cool the $^{18}O/^{16}O$ ratio declines. Emiliani (1955, 1966) used the link between the $^{18}O/^{16}O$ ratio and water temperature, together with information provided by recently devised absolute dating methods, to reconstruct ocean temperatures for the past 400

ka. Shackleton (1967) argued that the $^{18}O/^{16}O$ ratio could also be used as a proxy for terrestrial glaciations, suggesting that the $^{18}O/^{16}O$ ratio in the oceans would fall when ^{18}O-rich water was locked up in expanded ice caps and enlarged glaciers. Emiliani (1972) used this association to show that the history of temperature fluctuations inferred from oxygen isotope levels in the past million years was more complex than the classical interpretation of four 100 kyr-long glaciations separated by 100–300 ka interglacial periods. Emiliani (1955) was the first to propose that the maxima and minima of the inferred paleotemperature curve be referred to as "stages" (hereafter referred to as Oxygen Isotope Stages or OIS, but also and increasingly referred to as marine isotope stages or MIS) starting with OIS 1 for the most recent maximum. Thereafter odd numbers were used for peaks of maximum temperature and even numbers for troughs of minimum temperature. Shackleton and Opdyke (1973) showed that planktonic and benthic foraminifera have similar isotope records and were able to demonstrate that sea levels predicted from the oxygen isotope record were matched by terrestrial geological evidence in both Barbados and New Guinea. They also extended the 19 stages of Emiliani (1955) to 22, and proposed that these stages should be adopted as a "standard for the latter half of the Pleistocene" (ibid, p. 48). Emiliani and Shackleton (1974) combined data from the shorter deep-sea cores from the Atlantic and the Caribbean with the longer, more densely sampled record, from a new core (V28-238) from the Pacific, and since then additional cores have expanded the temporal span of the record and better sampling methods have improved its precision. An older glacial/interglacial terminology based on continental indicators of cold and warm climate phases has been replaced by the OIS numerical system, allowing more precise estimates of age. Thus the Würm glacial corresponds with OIS 2–4, the Riss with OIS 6, the Mindel with OIS 8, 10, 12, and 14, and Günz with OIS 16 (even-numbered OIS are minima). The recognition of OISs allowed the validation of the so-called astronomical theory of climate. This theory was mathematically formulated early in the 20thC by the Serbian mathematician Milutin Milankovitch, who hypothesized that the Earth's orbital components (namely, **eccentricity** of the Earth's orbit, its tilt, or **obliquity**, and the **precession** of the Earth's axis) were strongly implicated in geologically recorded climate change. However, it was not until the development of the quasi-continuous deep-sea isotopic record that there was an adequate enough geological record to test theory; Hays et al. (1976) unambiguously showed that fluctuations in the oxygen isotopes in the deep-sea cores contained the periods associated with orbital components. Once it became clear that the orbital "Milankovitch cycles" were significant drivers of climate change it became possible to use the mathematically generated time estimates to fine tune less-precise geological records, so-called "orbital tuning." The OIS (or MIS) system is increasingly being used to indicate the age of fossil sites that cannot be dated precisely using methods that measure time in years. In some cases layers of tephra at fossil sites are also found intercalated in the marine sediments sampled by deep-sea cores. Whereas the latter sediments are deposited more or less continuously, some cycles of deposition may be missing from the nonmarine sections. These correlations between terrestrial and marine sequences have enabled the dates from the terrestrial sections to be orbitally tuned using the marine sequences (deMenocal and Brown 1999). See also **astronomical theory; astronomical time scale; orbital geometry; orbital tuning**.

OY ratio This is defined by Caspari and Lee (2004, p. 10895) as "the ratio of older to younger adults (OY ratio) in the death distribution over time." They intended the measure to provide a summary of the number of "old" adults relative to "young" adults in a death assemblage. They define adulthood as the age at which the third molars come into occlusion, and they further define "old" adults as those with estimated ages twice that of the age at which the third molars come into occlusion. Although they (Caspari and Lee 2004, p. 10896) argued that "ages were used to allocate specimens to life history categories that are themselves independent of actual ages," it is not possible to have life history categories (such as "old" and "young") that are truly independent of age. To define "twice the age" of attainment of adulthood, it is necessary to specify an age by which adulthood (occlusion of the third molars) is reached. As a consequence, they suggested 15 years as the age at which individuals enter adulthood, and 30 years (2×15 years) as the age at which individuals enter "older" adulthood. Given their definitions, the OY ratio is the number of individuals estimated to have died at or over 30 years of age divided by the number of individuals estimated to have died between 15 and 30 years of age. Ages at death were estimated in Caspari and Lee's analyses by using **Miles' method**. The proportion of "old" adult deaths out of all adult deaths (D_{30+}/D_{15+}) is related to the OY ratio [it is equal to $OY/(1+OY)$] but has simpler properties as well as a more straightforward demographic interpretation. The D_{30+}/D_{15+} death proportion is the **survivorship** to age 30 years (in a **stationary population**) for those who survive to adulthood (age 15 years).

P

P. Prefix used by the French contingent of the **International Omo Research Expedition** for localities in the **Shungura Formation** (e.g., P. 996). The prefix is also used for fossils found within that locality (e.g., **P. 996–17**).

P. 996–17 A fossil hominin from the **Shungura Formation**. Its original locality was "P. 996," but in later publications it is referred to as "L. 996." *See* **L. 996–17**.

p arm *See* **chromosome**.

p-deme *See* **deme**.

P **value** Associated with a statistical test, the probability that the observed pattern in a data set could occur if the **null hypothesis** were true. If the *P* value is below some threshold (usually 0.05) then the test is said to show a statistically significant result and the null hypothesis is rejected.

PA Prefix used for hominin fossils recovered from sites in China (e.g., PA 830 from **Hexian**).

PA 830 *See* **Hexian 1**.

pachymeninx (Gk *pachys*=thick and *meninx*= membrane) Refers to the dura mater, the thick outer layer of the meninges that is formed from mesoderm. *See also* **meninges**.

Paderborn (Location 51°41′N, 8°44′E, Germany; etym. named after the nearby town) History and general description A hominin cranial fragment was recovered from gravel deposits outside of the city of Paderborn in Westfalia in 1976. Original **radiocarbon dating** by Reiner Protsch, who is alleged to have falsified data, suggested an age of 27.4±0.6 ka, but recent redating has shown it to be post-medieval. Temporal span and how dated? Direct dating of the skull fragment by **AMS radiocarbon dating** has revealed that this skull is recent (238±39 years). Hominins found at site A relatively robust skull fragment attributed to attributed to *Homo sapiens*. Archeological evidence found at site None. Key references: Geology, dating, and paleoenvironment Street and Terberger 2002, Street et al. 2006; Archeology N/A; Hominins Henke and Protch 1978.

paedomorphosis (Gk *pedo*=child plus *morphe*=form, shape) A category of resultant morphologies that follow from three different kinds of **heterochronic** processes (time **hypomorphosis**, rate hypomorphosis, or **neoteny**). All of these processes produce, in various ways, descendants with ancestral juvenile characteristics or shapes retained during later stages of ontogeny. Since the juvenile features are shifted to the adult, there are no "new" paedomorphic morphologies. *See also* **hypomorphosis**; **neoteny**.

Paglicci (Location, 41°40′N, 13°37′E, Italy; etym. unknown) History and general description The existence of sediments in this cave, near Rignano Garganico in Apulia, was first reported by R. Battaglia in 1955, but excavations did not begin until the early 1960s under F. Zorzi and F. Mezzena. Temporal span and how dated? **Radiocarbon dating** has provided a range from 34,300±800 years BP for layer 24B2 to 11,440±180 years BP for layer 7c-2. The layer containing the two hominin burials has been dated to 24,720±420 years BP. Hominins found at site In addition to at least 30 hominin fragments scattered through the sequence, two burials are known from the bottom of layer 21. These two burials, an adolescent male (Paglicci 12) and an adult female (Paglicci 25), have been referred to as "Cro-Magnon" in character. Fragmentary remains recovered from Paglicci have been submitted to mitochondrial DNA analyses and they appear to differ significantly from *Homo neanderthalensis*. Archeological evidence found at site The sequences

Wiley-Blackwell Encyclopedia of Human Evolution, First Edition. Edited by Bernard Wood.
© 2013 Blackwell Publishing Ltd. Published 2013 by Blackwell Publishing Ltd.

primarily contains **Gravettian** and Epigravettian artifacts. Key references: Geology, dating, and paleoenvironment Galiberti 1980; Hominins Corrain 1965, Caramelli et al. 2003, 2008, Mallegni 2005; Archeology Mussi 2001.

pairwise distance matrix See distance matrix.

Pakefield (Location 52°25.9′N, 1°43.8′E, England; etym. name of a nearby village) History and general description Changes in sea level have recently re-exposed sections of the Cromer Forest-bed Formation (CF-bF) between Pakefield and Kessingland on the North Sea coast of eastern England and Pakefield is one of the excavations on that section of the coast. Temporal span and how dated? The artifacts come from sediments that probably belong to Oxygen Isotope Stage 17 (but could be as old as Oxygen Isotope Stage 19) and to the early part of the Brunhes chron. The presence of *Mimomys* suggests that Pakefield is probably older than the type section of the CF-bF. Hominins found at site None. Archeological evidence found at site Thirty-two worked flints, including a simply flaked core and a crudely retouched flake, were found *in situ*. Key references: Geology, dating, and paleoenvironment Parfitt et al. 2005; Hominins N/A; Archeology Parfitt et al. 2005.

Palaeanthropi *See* **Sergi, Sergio (1878–1972)**.

Palaeo-Anthropology Research Unit (or PARU) The umbrella term for all the paleoanthropology research undertaken within the **Witwatersrand University Department of Anatomy** during the time Phillip **Tobias** was the head of that department. In 1999 the responsibility for research at **Sterkfontein** was transferred from PARU to a new entity, the **Sterkfontein Research Unit**, with Phillip Tobias as the Director and Ron **Clarke** the Deputy Director.

Palaeoanthropus A new genus established by Reck and Kohl-Larsen (1936) to accommodate the species *Palaeoanthropus njarasensis*, which they had proposed to accommodate the **Eyasi 1** calvaria. Type species *Palaeoanthropus njarasensis* Reck and Kohl-Larsen, 1936.

Palaeoanthropus njarasensis **Reck and Kohl-Larsen, 1936** First discovery **Eyasi 1**. Holotype Eyasi 1. Paratypes N/A. Main sites **Eyasi**.

Palaeoanthropus palestinus **McCown and Keith, 1939** *See* Skhul 1; Skhul IV; **Tabun 1**; **Tabun 2**.

palaeoanthropic man *See* Archanthropinae.

Palaeoanthropinae *See* Archanthropinae.

Paleoanthropological Inventory of Ethiopia One of the few attempts to locate potential hominin fossil sites on a large, regional, scale. Led by Berhane **Asfaw**, the inventory was a collaboration between the **Ethiopian Ministry of Culture and Sports Affairs**, the **Laboratory for Human Evolutionary Studies** at the University of California, Berkeley, the NASA-Goddard Space Flight Center, the Los Alamos National Laboratory, and the University of Tokyo. It used images from two space-based systems, Landsat thematic mapping (TM) and large-format camera (LFC). The former is a satellite-based system that measures the intensity of reflected sunlight in seven wavebands, and the resulting color images were used to identify vegetation and distinctive tephra. The high-resolution LFC images were taken during the Challenger shuttle mission in 1984. The two sets of data were collected for the main Ethiopian Rift Valley and for the Afar depression, and they were combined to help identify promising sedimentary basins. Thereafter, so-called "ground-truth information" was collected from 1988 onwards by researchers who explored the basins by vehicle and on foot to verify the presence of potential sites. Several sources of hominin fossils in the Ethiopian Rift Valley were identified by the Paleoanthropological Inventory of Ethiopia, one in the north and one in the south. Examples include the site complex within the **Kesem-Kebena** basin to the north of Lake Koka, in the Awash River system upstream of the Hadar and Middle Awash study areas, and the site of **Fejej** between the river Omo to the west and Lake Chew Bahir to the east. For more information see Asfaw et al. (1990).

paleoanthropological terminology The terminology used to describe hominin fossils is a mixture of the terminology used by modern human anatomists (i.e., **anatomical terminology**) and terms introduced by anthropologists and paleoanthropologists when they have encountered structures in nonhuman primates and in the hominin fossil record that are not ordinarily seen in modern humans. Some of these terms were initially described using the Latin form (e.g., Franz **Weidenreich**'s *crista occipitomastodea* and Gustav **Schwalbe**'s *planum alveolare*) in which case it is conventional to use italics for these terms. However, there is no need to italicize terms that were initially described in English (e.g., Yoel Rak's nasoalveolar gutter and Weidenreich's preglenoidal plane). *See also* **anatomical terminology**.

paleoanthropology (Gk *palaios*=ancient, *anthropos*=human being, and *ology*=study of; literally the study of ancient humans) In the present day this term is interpreted differently in Europe and North America. In Europe it is synonymous with human, or hominin, paleontology, whereas in North America it is interpreted more inclusively to include archeology as well as hominin paleontology. The meaning of paleoanthropology has changed over the past century and a quarter. The French geologist and paleontologist Louis Lartet may have been the first scientist to use the term. In a paper published in the *Annales des sciences géologiques* in 1872 he divided the science of paleontology into three subdisciplines: paleoanthropology (*paléo-anthropologie*), paleozoology, and paleophytology. However, when he refers to paleoanthropology in that and subsequent papers Lartet primarily discusses prehistoric archeological artifacts and monuments, not prehistoric human skeletal remains. However, when Lartet and his colleague Chaplain Duparc used the term *paléo-anthropologie* in 1874, they did so in the context of a discussion about Stone Age humans *and* the possible existence of human ancestors in the Miocene. Thus, although by 1874 Lartet was using paleoanthropology to refer to *any* kind of scientific study that addressed prehistoric or Stone Age humans, before long other researchers began to narrow and refine its meaning. One of the first people to use the term paleoanthropology more exclusively was Clémence Royer, a controversial French scholar who wrote on subjects ranging from economics and philosophy to anthropology and biology. It was she who translated Charles Darwin's *On the Origin of Species* into French in 1862 and she was an early member of the Société d'Anthropologie de Paris. In a paper published in the *Journal des économists* in 1879 she described *paléo-anthropologie* as the study of lost races of humans through their fossil remains. According to this formulation paleoanthropology was distinct from, but connected with, prehistoric archeology, which was the study of the artifacts manufactured by these "lost races." According to Royer, paleoanthropology and prehistoric archeology "reveal to us the successive

phases of the social evolution of human kind, of his customs, of his instincts and of his rudimentary institutions" (1879, p. 408). The person who bears most of the responsibility for defining paleoanthropology as a scientific discipline was the French anthropologist Paul Topinard. In his influential book *Éléments d'anthropologie générale* (*Elements of General Anthropology*) published in 1885, Topinard noted the recent expansion and the changing nature of research into human prehistory. He acknowledged the contributions of the French prehistorian Gabriel de Mortillet, who since the 1860s had argued that the study of prehistory is an anthropological science, and it was Gabriel de Mortillet who proposed the term *paléoethnologie* (paleoethnology) for the scientific investigation of prehistoric humans. But Topinard was among those who objected to de Mortillet's new term. One reason for Topinard's preference for paleoanthropology was that it emphasized the study of skeletal remains and physical anthropology as distinct from the study of archeological artifacts. Indeed, Topinard was critical of prehistorians who focused on simply amassing collections of flint artifacts. In a paper entitled "La paléoanthropologie" presented at the 1889 Congrès international d'anthropologie et d'archéologie préhistoriques (the precursor of the International Congress of Prehistoric Anthropology and Archaeology) and published in the congress proceedings in 1891, Topinard distinguished between paleoanthropology, which deals with prehistoric human fossils and the identification of prehistoric human races, and paleoethnography, which deals with prehistoric artifacts and material culture, prehistoric art, and the customs of different prehistoric races. Paleoethnography was the domain of archeologists and ethnologists, whereas paleoanthropology was the domain of zoologists and anatomists. Topinard expanded upon and clarified his point in *L'homme dans la nature* (*Man in Nature*) published in 1891, where he stated that the scientific study of prehistoric humans can be divided into two separate concerns: (a) the study of the fossil remains of humans and (b) the study of the society of these humans through their artifacts, dwellings, sepulchers, and artistic remains. In essence, Topinard is saying that paleoanthropology is synonymous with human paleontology. In a paper presented to the Société d'Anthropologie de Lyon another French anthropologist, Philippe Salmon, offered a very similar definition of paleoanthropology (Salmon 1888, pp. 208–9). Thus, by 1891 paleoanthropology had emerged as a distinct scientific discipline in France and largely, but not solely through the efforts of Paul Topinard, it was increasingly understood to be the anthropological study of prehistoric humans via their fossil remains. It was a paper read by Topinard at the Congrès International d'Anthropologie et d'Archéologie Préhistoriques in 1889 that helped promulgate the new discipline of paleoanthropology outside of France. Thomas Wilson, the curator of the Department of Prehistoric Anthropology at the Smithsonian Institution and one of America's prominent archeologists, attended the congress and upon his return to the USA he published a report on it in *The American Naturalist* in 1892. He discussed Topinard's paper and introduced the term "paleoanthropology" to Americans. The same year, in the *Annual Report of the Board of Regents of the Smithsonian Institution*, Wilson listed paleoanthropology (meaning the study of prehistoric skeletal material) as one of the disciplines related to prehistoric anthropology. Similarly, in a paper published in 1892 the influential German anthropologist and biologist Rudolf Virchow followed Topinard in using paleoanthropology to mean *specifically* the study of human fossil remains, as distinct from archeological artifacts. By the end of the century Europe's leading anthropologists and archeologists recognized paleoanthropology as a science that dealt specifically with human paleontology. During the first half of the 20thC few books or articles used paleoanthropology in their titles, but this changed in the 1960s. Wilfrid **Le Gros Clark**'s *The Fossil Evidence for Human Evolution: An Introduction to the Study of Paleoanthropology*, originally published in 1955, was not only one of the first books to use paleoanthropology in its title, but it also helped to redefine the emerging modern science of paleoanthropology. Clark described paleoanthropology as the study of fossil hominids and the comparative anatomy of humans and existing primates. Significantly, Clark noted that in recent years paleoanthropologists, rather than scientists trained in other disciplines, were playing a more critical role in the interpretation of the fossil evidence of human evolution. In the 1970s several books appeared bearing the term paleoanthropology in their titles, including Russell Tuttle's *Paleoanthropology: Morphology and Paleoecology* (1975), Glynn Isaac's *Perspectives on Human Evolution: East Africa and Paleoanthropology* (1976), and James A Gavan's *Paleoanthropology and Primate Evolution* (1977). It was around this time that William (Bill) **Howells** and Patricia Jones Tsuchitani offered a more inclusive interpretation of paleoanthropology. In their introduction to *Paleoanthropology in the People's Republic of China* (1977), a report arising from a visit to China by a delegation of US paleoanthropologists, they

defined paleoanthropology as a discipline that encompasses the interdisciplinary study of early humans from the perspective of such disciplines as archeology, physical anthropology, geology, geophysics, paleobotany, and paleontology, thus distinguishing modern paleoanthropology from the way it was interpreted by Topinard et al. Howells and Tsuchitani *emphasized* the connection between the study of human fossils and prehistoric archeology, rather than distinguishing the two activities. However, when in 1980 Gail Kennedy and Milford Wolpoff both published textbooks entitled *Paleoanthropology*, they were not interpreting paleoanthropology in the same way. In his textbook Wolpoff interpreted it more inclusively to include archeology, but Kennedy exluded archeology and suggested that modern paleoanthropology was a synthetic science that relied on developments in "evolutionary biology, population genetics, biochemistry, geochronology, paleontology, ecology, biomechanics, multivariate statistics, and cybernetics," all of which supplement the traditional biometrics and descriptive, functional, and comparative anatomy of earlier studies (1980, p. 2). The most up-to-date interpretation of paleoanthropology used in the North American, more inclusive, sense can be found on the website of the Paleoanthropology Society founded in 1992 (www.paleoanthro.org/thesociety.htm). To quote from its website it suggests that "paleoanthropology is multidisciplinary in nature and the organization's central goal is to bring together physical anthropologists, archeologists, paleontologists, geologists and a range of other researchers whose work has the potential to shed light on hominid behavioral and biological evolution."

Paleoanthropology Society Founded in 1991, the two goals of the Society are to foster an understanding of the processes which underlie the evolution of behaviorally and anatomically modern humans, and to create, largely through face-to-face interaction, a multidisciplinary group of researchers who share this common goal and who will work collaboratively towards it attainment. An underlying premise of the Society is that to understand how modern humans evolved it is necessary to examine the interactions between biology and behavior and to set these within an appropriate environmental and chronological context; thus membership is drawn from a variety of disciplines. The Society, which accepts all applicants for membership regardless of nationality and level of academic attainment, focuses on three activities. The first is a publicly accessible web site, www.paleoanthro.org, which, in addition to information about the Society, contains a membership directory, a student section, a series of published dissertations, and links to other relevant websites. The second is an electronic journal, *PaleoAnthro*, which may also be accessed, free of charge, through the website. *PaleoAnthro* focuses on the publication of large, data-rich articles, which can take advantage of the less stringent spatial constraints of an electronic format; letters and book reviews are also published. Finally the Society hosts an annual 2-day meeting, the first was held in St Louis, MO, USA, in 1992, which is held in conjunction, in alternate years, with the Society for American Archaeology and the American Association for Physical Anthropology. *See also* **paleoanthropology**.

paleoclimate (Gk *palaios*=ancient and *klima*= which refers to the slope of the Earth's surface, is the origin of L. *clima*=referring to climate, or latitude) Assessments of climate made for time periods where it is not possible to take direct measurements of four climate variables, namely temperature, precipitation (rainfall, snowfall), and wind speed and direction. Information about past climates must come from proxies for the four climate variables. These proxies can be divided into two main categories: biotic and chemical. The biotic proxies are mainly to do with productivity (e.g., in warmer phases the oceans have a greater biomass and tree rings are wider) and with direct evidence of the types of plants and animals that were living at the time; the chemical proxies include changes in oxygen isotope ratios recorded in foraminifera and in air bubbles in ice. Paleoclimate can be reconstructed on a global or on a regional scale. *See also* **climate**; **climate forcing**.

paleoclimate reconstruction *See* **paleoclimate**.

paleocommunity deme *See* **deme**.

paleodeme *See* **deme**.

paleodemography The demography of past populations or samples from which no records of age or sex are available. These populations or samples may be modern humans from the recent past, or they may be from different taxa where the data are the derived from paleoanthropological research.

paleoecology (Gk *palaios*=ancient, *oikos*= house, and *logia*=study) Interactions between an organism and its environment in the past. Paleoecology is of interest in hominin paleontology because it enables researchers to investigate the factors affecting past human behavior and evolution from a holistic perspective. Studies of paleoecology are focused on either the interactions centering on a particular species (paleo**autecology**) or reconstructions of community ecology at a particular site, or during a well-defined time interval in the past.

paleoenvironment (Gk *palaios*=ancient and OF *environ*=round about) Information about the climate, landscape, and the vegetation of a fossil locality in the remote geological past. It is a more inclusive concept than paleoclimate. *See also* **environment**; **paleoenvironment reconstruction**.

paleoenvironmental reconstruction The process of deducing past environmental conditions from the physical and biological evidence left behind in the geological record. Cognate evidence can be abiotic or biotic and it can take many forms (e.g., sedimentology, soil chemistry, paleobotany, and vertebrate and invertebrate paleontology). For example, biotic evidence includes palynology, a subfield of micropaleontology. Fossil pollen and spores in lacustrine or terrestrial sediments can be used to infer the plant species that occurred at the time when the environment existed, and when the means of their dispersal and taphonomy are taken into account the relative abundance of the plants which produced them. Depending on the type of plants and sediments involved, the evidence might represent plants occurring on either a regional or local level. Knowledge of the abiotic environment can be gained either directly from evidence of natural processes in the preserved sediments or indirectly from a study of the biotic components and secondary inference based on their known ecological preferences. For example, the presence of a perennial river in the local paleoenvironment could be inferred primarily from the scale of paleochannels or particular types of cross bedding in the sedimentary deposits, or the biotic evidence such as preserved freshwater oyster reefs and riverine fish, reptiles, and mammals in the associated fossil assemblage. Like much of geology and paleontology, paleoenvironmental reconstruction relies on the principles of **uniformitarianism**; present processes are the key to the past. Paleoenvironmental reconstruction is by its nature multidisciplinary. Since many of the factors that determine the paleoenvironment interact, the nature of the evidence may require complex analyses and negotiation to arrive at a consensus interpretation. Recently, paleoenvironmental reconstructions of fossil localities have relied on a multiproxy approach, which tries to reconcile different strands of evidence recovered from a site to create a fuller picture of the past. Multidisciplinary paleoanthropological projects now routinely take this approach, since past environments are seen as a key aspect of understanding evolutionary processes.

paleohabitat   A past habitat. *See* **habitat**.

Paleolithic   The "old stone age," defined by John Lubbock in 1865 as the time when humans coexisted with extinct fossils and stone implements were rough-hewn, in contrast to the Neolithic, or "new stone age," when stone implements were generally polished. The term now refers both to a time period (the sum of human history from the invention of stone tools until *c*.10 ka) and to a distinct economic, social, and technological complex, usually defined as reliance on flaked stone tools, low population sizes, and foraging. It is customarily divided into three stages (**Lower**, **Middle**, and **Upper Paleolithic**), each having unique features. In Africa, the terms Early, Middle, and Late Stone Age first introduced by A.J.H. Goodwin and C. van Riet Lowe in 1929 are preferred, although they are broadly similar to the Eurasian terms. Prior to the advent of absolute dating, most sites were simply attributed to one of these stages.

paleomagnetic dating   (Gk *palaios*=ancient and *magnes*=magnet) *See* **geomagnetic dating**.

paleomagnetic record   (Gk *palaios*=ancient and *magnes*=magnet) *See* **geomagnetic polarity time scale**.

paleosol   (Gk *palaios*=ancient and L. *solum*= ground) A soil of the geological past, typically recognized by the preservation of characteristic structures, color, mineral segregations, or chemistry that relate to soil-forming processes at the time the paleosol was active at the Earth's surface. The characteristics of paleosols reflect the controlling factors of soil formation (i.e., climate, vegetation, relief, parent material,

and time) and thus can be used to reconstruct some of these variables for the geologic past. Investigation of paleosols at **Kanapoi** (Wynn 2000) was used to characterize the habitats available to *Australopithecus anamensis* there *c*.4 Ma.

paleotelic   (Gk *palaios*=ancient and *telikos*= end, so "towards a purpose") *See* **palimpsest evolution**.

Pálffy barlang   *See* **Dzeravá Skála**.

palimpsest   (Gk *palimpseston*=scrape again. The term originally referred to a parchment, the surface of which had been scraped in places to allow for reuse, and so by extension the text written on it was likely produced by several authors on different occasions) In paleoanthropology a palimpsest refers to an archeological or fossil assemblage formed of elements deposited at different times and often by different causes or agents. Some cave deposits are evidently palimpsests. The shelter afforded by caves make them attractive to a succession of hominins and various carnivores such as bears, hyenas, and various raptors (e.g., **Die Kelders**). As such the accumulated fauna is often a combination of the dietary leftovers from hominin and nonhominin sources that must be distinguished if the research focus is hominin behavior. It is useful to distinguish between "depositional" and "behavioral" causes. For example, the abraded *Homo erectus* fossils (calotte, femur, molar tooth) recovered some distance apart from a sandbank on a bend of the Solo River at **Trinil**, central Java, very likely comprise a "depositional" palimpsest in the sense that the bones were doubtless redeposited on the sandbank at different times, having been carried down by the river. However, an archeological assemblage can also be a "behavioral" palimpsest. Glynn **Isaac** developed the idea of a behavioral palimpsest in the early 1980s in what he termed the "dynamic flow-through model" for site formation (Isaac 1983, Kroll and Isaac 1984) and the concept has recently been revisited (Domínguez-Rodrigo et al. 2007a). For example, sites such as those at **Olduvai Gorge** (e.g., FLKN) are almost certainly the time-averaged result of many instances of stone tools and butchered bones being deposited at that place by hominins, plus bone deposition at the same location by felids. A similar argument can be advanced for Dmanisi being a behavioral palimpsest. Realistically, any site that has a sample size of artifacts or fauna large enough to be detected in the archeological or fossil record is likely to be a palimpsest, whether formed over days, months, years, decades, or thousands of years.

palimpsest evolution

palimpsest evolution (Gk *palimpseston=* scrape again. The term originally referred to a parchment, the surface of which had been scraped in places to allow for reuse, and so by extension the text written on it was likely produced by several authors on different occasions) The term used by William King Gregory and Robert **Broom** to describe what is usually referred to as "**mosaic evolution**." In 1910 Gregory differentiated between what he then termed caenotelic (later "habitus") traits and paleotelic (later "heritage") traits. Caenotelic characters are of direct current adaptive relevance whereas paleotelic characters are phyletically older (comparable to symplesiomorphies in cladistic parlance). The latter may have been functionally important, but they do not relate to a specific current niche. For example, in early hominins the form of the hindlimb adapted to bipedal locomotion is a caenotelic trait, but the limb's elements (femur, tibia and fibula, tarsus, pentadactyl extremity) constitute a paleotelic trait. Gregory made the important point that the terms are relative since paleotelic traits were once caenotelic. In 1947 he described the resulting assemblage of features as a "palimpsest," but the term had been introduced by Robert Broom in 1924 when in a review of reptilian evolution he referred to "habitus" and "heritage" features. *See also* **mosaic evolution**.

palynology

palynology (A neologism created in 1944 by Hyde and Williams. Its roots are Gk *paluno=* to sprinkle and *pale=* dust, and the suffix *ology=* study of) The study of pollen and other organic microfossils recovered from sedimentary rocks and sediments (other plant microfossils; i.e., **phytoliths** and **starch grains** are usually not included in palynology, the former because they are made of silica and not strictly organic, and the latter because they have not been observed to fossilize). Palynology includes the use of pollen (a) to reconstruct past environments and climates, (b) to investigate of the effects of the activity of modern humans and extinct hominins (e.g., logging and burning) on the flora, and (c) in the creation of plant species lists to assist in reconstructing the diet of extinct hominins. Traverse (2007) provides

a useful primer for researchers contemplating exploring the potential of pollen for any of the types of analysis set out above. *See also* **pollen**.

Pan

Pan A number of different chimpanzee species have been recognized since the 1860s. It was then that Du Chaillu described *Troglodytes koolookamba* and *Troglodytes calvus* from Gabon, in addition to the ordinary chimpanzee which he, in line with the nomenclature most commonly in use at the time, called *Troglodytes niger*. Other supposed species were described throughout the 19thC and early 20thC, until Schwarz in 1932 decided that only one species was necessary, with four subspecies. Subsequently authorities have been almost unanimous in separating out one of these as a distinct species, *Pan paniscus*. It became known as the pygmy chimpanzee but is now more commonly designated the bonobo, leaving the simple name "chimpanzee" for *Pan troglodytes*. In the latter half of the 20thC authorities tended to assume, with very little scrutiny, the existence of three subspecies within *P. troglodytes*: the West African chimpanzee, *Pan troglodytes verus*, the Central African chimpanzee, *Pan troglodytes troglodytes*, and the East African chimpanzee, *Pan troglodytes schweinfurthii*. When re-examined, by both molecular and morphological means, this simple picture has turned out to be unsatisfactory. First, **mitochondrial DNA** (or mtDNA) sequencing in the 1990s showed that West African chimpanzees differ absolutely from others, with many fixed substitutions indicating a considerable time since they separated; it has therefore been proposed that they constitute a different species, *Pan verus*. Secondly, the mtDNA of chimpanzees from between the Sanaga and Niger Rivers (i.e., between *verus* and *troglodytes*) is different yet again, and it is now customary to recognise them as a further subspecies, *Pan troglodytes ellioti* (initially, but incorrectly, called *Pan troglodytes vellerosus*). Thirdly, morphological analysis pointed out that there is as much difference within *P. t. schweinfurthii* as between that and *P. t. troglodytes*, so the former "*schweinfurthii*" is often now divided into a northwestern subspecies (*schweinfurthii*) and a southern and eastern one (*marungensis*). Finally, there is a growing view that there is not much difference between Central and Eastern chimpanzees, and some lump them together.

Pan-African Association of Prehistory and Related Studies

Pan-African Association of Prehistory and Related Studies The organization that plans and organizes the successors to the **Pan-African**

Congress on Prehistory and Quaternary Studies. It was established at the 1983 Pan-African Congress on Prehistory and Quaternary Studies in Jos, Nigeria.

Pan-African Congress on Prehistory and Quaternary Studies

This series of congresses (now called the Congress of the Pan-African Association of Prehistory and Related Studies) has played a leading role in promoting research and scholarship in African prehistory. The congresses bring together researchers for sessions of delivered papers, but an important and consistent component of the congresses has been one, or more, excursions to paleontological and archeological sites. Thus far, 13 congresses have been held: Nairobi, Kenya, 1947; Algiers, Algeria, 1952; Livingstone, Zambia, 1955; Leopoldville (now Kinshasa, Demographic Republic of the Congo), 1959; Santa Cruz de Tenerife, Canary Islands, 1963; Dakar, Senegal, 1967; Addis Ababa, Ethiopia, 1971; Nairobi, Kenya, 1977; Jos, Nigeria, 1983; Harare, Zimbabwe, 1995; Bamako, Mali, 2001; and Gaborone, Botswana, 2005; and Dakar, Senegal, 2010. The first Pan-African Congress on Prehistory in Nairobi in 1947 was organized and hosted by Louis Leakey. It brought together scholars of African prehistory with the goal of sharing and disseminating new scientific knowledge and discoveries. During that meeting the Abbé Breuil was elected President and Robert Broom Vice-President. Professor Lionel Balout and the Abbé Breuil organized the second congress, held in Algiers in 1952. Desmond Clark, who was the president of the prehistory section of the Algiers congress organized the third congress in Livingstone. Its president was Louis Leakey, who gave an impromptu demonstration of skinning an antelope with stone tools he had fashioned. It was at this congress that the terms Early, Middle, and Later Stone Age were defined and formally endorsed. Another important proposal was that the stratigraphic/climatic divisions defined for East Africa (Kageran, Kamasian, Gamblian, etc.) should not be applied to other parts of the African continent unless there was firm evidence to support such a correlation. The proposal that the Pleistocene faunal sequence should be divided into four stages has now largely passed out of use, but it provided a tool for regional or even wider correlations, and it was the best chronological tool then available. The fourth congress was held in Leopoldville (now Kinshasa, Demographic Republic of the Congo) in 1959, and it was there that Louis Leakey announced Mary Leakey's discovery of OH 5, the holotype of *Zinjanthropus boisei*, at Olduvai Gorge. Politics threatened the sixth congress scheduled for Dakar in 1967. The Senegalese media and some politicians wanted to exclude scientists from South Africa, but President Senghor intervened and allowed them to attend, and it was at this congress that the new hominin discoveries from the Omo Basin were announced by Clark Howell. The Pan-African Association for Prehistory and Related Studies was established at the ninth congress, held in Jos in 1983, and Desmond Clark and Thurstan Shaw were elected to Honorary Membership because of their "outstanding contributions towards the realization of the objectives of the Association." The scope of the twelfth congress held in Gaborone in 2005 was expanded to include cultural heritage management, historical archeology, and post-colonial archeology. The Gaborone congress was important in another way, for it was the first congress at which a native African, the Ethiopian scientist Zeresenay Alemseged, announced a major hominin discovery.

Pan fossils See KNM-TH 45519–45522, KNM-TH 46437–46440.

Pan paniscus A species of extant chimpanzee known as the bonobo. It was previously referred to as the "pygmy chimpanzee," but after Schultz (1969) showed there is extensive overlap between the limb dimensions of *Pan paniscus* and *Pan troglodytes*, and given that the mean of the field-recorded body masses of seven *P. paniscus* males exceeded the mean body mass of 15 *P. troglodytes* males (Jungers and Susman 1984), the epithet "pygmy" was inappropriate. The first formal taxonomic consideration of the bonobo suggested it was a subspecies of the chimpanzee (*Pan satyrus paniscus* Schwarz, 1929), but 4 years later Coolidge (1933) raised it to the rank of species. The type specimen, a skull and skin (#9338), is in the collection of the Royal Museum for Central Africa. Randall Susman's edited volume is the primary source for historical and for most scientific information about the bonobo (e.g., Coolidge 1984, Van den Audenaerde 1984), but Johanson (1974) and Cramer (1977) also provide important detailed assessments of the dentition and cranial morphology, respectively, of the bonobo. In addition to its importance as one of the living taxa most closely related to modern humans some have taken the view that among the extant *Pan* taxa *P. paniscus* "may approach more closely to the common ancestor of chimpanzees and man than does any living chimpanzees" (Coolidge 1933, p. 56). Adrienne Zihlman has been a consistent advocate of Coolidge's proposition and a comparison of the limb proportions of *P. paniscus* and *Australopithecus afarensis* captured simply, but effectively, in a diagram (Zihlman 1984, Fig. 15) has been a powerful promotional icon for this proposal. *See* **hominid**; **Hominidae**.

panin The informal or vernacular term for a specimen that belongs to a taxon within the tribe Panini. For example, *Pan paniscus* is a panin taxon and **KNM-TH 45519** is a panin fossil.

Panini The tribe that includes chimpanzees and bonobos and all the fossil taxa more closely related to chimpanzees and bonobos than to any other living taxon.

Panina If the tribe **Hominini** is interpreted to include both the clade that contains modern humans *and* the clade that contains extant chimpanzees/bonobos then some researchers (e.g., Andrews and Harrison 2005) discriminate between the two clades at the level of the subtribe. In this case the clade that contains extant chimpanzees/bonobos would be called the **Panina** and the clade that contains modern humans would be called the **Hominina**.

paninan The informal term for individuals or taxa within the subtribe Panina. *See* **Panina**.

Panthera (L. *panthera* from Gk *panther*, but the word is probably of Oriental origin, cf. Sanskrit *pundarikam* "tiger," probably literally "the yellowish animal") The cosmopolitan genus of large cats that includes the extant lion *Panthera leo*, tiger *Panthera tigris*, panther *Panthera onca*, leopard *Panthera pardus*, and their extinct close fossil relatives. The genus *Panthera* appears in the fossil record in the middle **Pliocene** and becomes a relatively speciose taxon; it now has modern representatives on five continents.

Panxian Dadong (Location 25°37′38″N, 104° 44′E, Guizhou Province, China; etym. means "Grand Cave") History and general description Panxian Dadong (sometimes abbreviated to "Dadong") is a middle-level cave juxtaposed between lower and upper caves within an approximately 230 m hill situated on the western part of the Guizhou Plateau that comprises carboniferous and Permian limestones, cataclas-

tic rocks, basalt, and coal deposits; the hill itself is about 1630 m above sea level. Dadong is a large cave with a roughly 55 m-wide and 50 m-high entrance and it is about 220 m from the entrance to the back wall; the total area of the cave floor is approximately 8000 m^2. The cave contains an estimated 19.5 m of deposit near the entrance of the cave, but the top 1–2 m has been disturbed by local people mining potassium nitrate for use in the production of gunpowder and it was removed during the initial excavation of the site in 1992. Four excavations were conducted at Dadong between 1996 and 2000 by a joint Sino-American research team led by Sari Miller-Antonio and Lynne Scheparz on the American side and Huang Weiwen and Hou Yamei on the Chinese side. Three primary stratigraphic units were identified – (from top to bottom) Unit 1, Unit 2, and Unit 3 – with Unit 2 divided into two separate stratigraphic layers (Units 2a and 2b). The majority of the vertebrate paleontological (including hominins) and archeological remains were excavated from Unit 2. Temporal span and how dated? Middle to Late Pleistocene based on the presence of the *Ailuropoda-Stegodon* **fauna**. Unit 2a: uranium series 180–142 ka, electron spin resonance 156–137 ka; Unit 2b: electron spin resonance 294–214 ka. Hominins found at site Five hominin teeth were allocated to "*Homo sapiens*" (Jones et al. 2004, Karkanas et al. 2008), although others suggest they belong to "archaic *H. sapiens*" (Liu and Si 1997), but none of the teeth have been described in detail. If these hominin teeth turn out to be modern human, then it will have significant implications for hypotheses about the origins of modern humans because of their early date. Archeological evidence found at site Core and flake stone implements produced on locally available limestone, basalt, and chert. Analysis of the raw materials and artifact typologies suggests more reliance on basalt and chert and an increase in the number of artifacts though time. It has also been suggested that some of the *Rhinoceros* teeth were flaked and used as tools. Analysis of the faunal remains suggests that hominins were the primary accumulators of the animal bones, with carnivores, birds, and rodents playing minor roles. Repository Dadong Village Research Station, Guizhou, China. Key references: Geology, dating, and paleoenvironment Huang et al. 1995, Shen et al. 1997, Rink et al. 2003, Jones et al. 2003, Wang et al. 2004, Karkanas et al. 2008; Hominins Liu and Si 1997; Archeology Miller-Antonio et al. 2000, 2004, Scheparz et al. 2003, Bekken et al. 2004, Scheparz and Miller-Antonio 2010.

papionin The informal name for the tribe Papionini (*which see*).

Papionini (L. *papio*=baboon, which gives rise to *Papio*, the name for the genus that contains the baboons) One of two tribes within the subfamily **Cercopithecinae** (the other tribe within the subfamily is the **Cercopithecini**, or guenons), the family **Cercopithecidae**, and the superfamily **Cercopithecoidea**. The papionini comprises **baboons** and their allies. The extant genera included in the Papionini are *Papio* (common baboons), *Mandrillus* (drills and mandrills), *Theropithecus* (geladas), *Macaca* (macaques), *Cercocebus*, *Rungwecebus*, and *Lophocebus* (mangabeys). Fossil genera within the Papionini include *Parapapio*, *Dinopithecus*, and *Gorgopithecus*. Papionins tend to be relatively large-bodied and they are mainly terrestrially adapted (except for several macaque and mangabey species that are arboreal), whereas cercopithecins are smaller and generally more arboreal. Several papionin species have "long" faces, or dog-like muzzles, in comparison to the shorter faces of the guenons. The papionins are the best studied of the groups within the Old World monkeys, and a good deal is known about their evolution and past diversity, in part because their fossils are relatively common at many archeological and hominin paleontological sites. The two cercopithecine tribes, the Papionini (papionins) and Cercopithecini (cercopithecins), radiated in the late Miocene to the Pleistocene. By the late Miocene there is fossil evidence for the modern genus *Macaca* in North Africa (Delson 1974), tallying with X-chromosome data that suggest a divergence of *Macaca* from other papionins at 7.6±1.3 Ma (Tosi et al. 2005). In the terminal Miocene and earliest Pliocene, *Macaca* dispersed around the Mediterranean and radiated into Europe and Asia, reaching latitudes as far north as 53° (UK). It was in the Pliocene and Pleistocene that the major adaptive radiations of the African and Eurasian papionins occurred. Late Miocene *Macaca* probably gave rise to *Paradolichopithecus* and *Procynocephalus*, two Eurasian papionin genera that were successful in the Pliocene, but which became extinct in the Pleistocene (Jablonski 2002). Data from mitochondrial DNA and the Y chromosome suggest substantial speciation within *Macaca* during the late Pliocene and Pleistocene in Asia (Tosi et al. 2003). The Pliocene and Pleistocene papionin radiations in Africa are better documented than those that occurred in Eurasia, but many of the species that thrived in the Plio-Pleistocene did not survive into the late Pleistocene and Holocene.

The very latest Miocene or earliest Pliocene witnessed the divergence of the taxa that gave rise to the *Mandrillus/Cercocebus* and *Papio/Theropithecus/Lophocebus* clades (Tosi et al. 2005), probably from a Central African source (Bohm and Mayhew 2005). *Papio*, *Theropithecus*, and *Lophocebus* then speciated in quick succession *c*.4 Ma (Harris 2000), although, like *Mandrillus* and *Cercocebus*, the fossil record of *Lophocebus* is poor compared to those of *Theropithecus* and *Papio*. *Parapapio* and *Theropithecus* were the two dominant cercopithecine genera in the Plio-Pleistocene. *Parapapio*, although found in East African deposits, had a stronghold in southern Africa, but it was extinct by the early Pleistocene. *Theropithecus* had a wider geographic distribution than *Parapapio*, and a longer temporal span. It was most diverse in the Turkana Basin, but fossils have been recovered from southern Africa (including Angola), North Africa, and Eurasia (see Hughes et al. 2008 for a review). The genus now has only one surviving member, *Theropithecus*, the gelada, which is confined to the highlands of Ethiopia and Eritrea. The most abundant and widespread modern sub-Saharan African papionin genus, *Papio*, was poorly represented in the East African Plio-Pleistocene record, although it was more prevalent in southern Africa. Its fossil record is relatively poor compared to that of *Theropithecus*. *Papio hamadryas* fossils (in the form of *Papio robinsoni*) have been recovered from Plio-Pleistocene sites in southern Africa, including **Swartkrans**. The classification of modern baboons has veered from placing them all into a single species, *P. hamadryas*, and new genetic data suggest there may be at least five species with a complicated history. Present-day diversification apparently began when the modern chacma baboon, *Papio ursinus*, diverged in southern Africa *c*.1.8 Ma and then dispersed out and speciated during the Pleistocene (Newman et al. 2004).

paracone (Gk *para*=beside or by and *konos*=pine cone) One of the terms proposed by Osborn (1888) for the main cups of mammalian molar teeth. It is the term used for the main cusp beside (i.e., on the buccal aspect of) the **protocone** on a maxillary (upper) molar tooth crown, and it also the term used to refer to the main buccal cusp of a bicuspid upper premolar. It is part of the **trigon** component of an upper molar crown (syn. eocone, mesiobuccal cusp, tuberculum anterium externum).

paraconid (Gk *para*=beside or by and *konos*=pine cone) A term some researchers have used for an enamel feature on a mandibular molar tooth crown more commonly referred to as a **premetaconulid**. Paraconid is one of the terms proposed by Osborn (1888) for the main cups of mammalian molar teeth. It was given its name because it is the main cusp beside (i.e., on the lingual aspect of) the **protoconid** on a mandibular molar tooth crown, but in higher primates including hominins this main cusp has been lost. *See also* **premetaconulid**.

paracristid A term used by Schwartz and Tattersall (2002) for an **enamel** crest more commonly referred to as the **mesial marginal ridge** in a mandibular postcanine tooth crown. *See also* **mesial marginal ridge**.

parallel evolution Parallel evolution is the independent origin of similar traits in closely related taxa (e.g., the coronally rotated petrous bones seen in *Homo* and *Paranthropus* may not have been inherited from their most recent common ancestor, in which case coronally rotated petrous bones in *Homo* and *Paranthropus* would be an example of parallel evolution, or if bipedalism evolved independently in *Australopithecus afarensis* and *Australopithecus africanus*, then this, too, would be an example of parallel evolution). **Convergent evolution** is the independent origin of similar traits in distantly related taxa (e.g., the acquisition of thick-enameled teeth in orangutans and cebus monkeys). Simpson (1950) was most likely the originator of these terms, or if not he certainly was among the first to formalize them (Simpson 1961). He suggests "there is no really fundamental difference between" them (*ibid*, p. 181), and inevitably whether a trait in two taxa that are not in the same clade qualifies as an example of parallel or convergent evolution (i.e., whether the taxa are "close" or "distant") is a subjective decision. Convergent and parallel evolution are examples of **homoplasy**, for both concern the appearance of similar morphology in two taxa that is not seen in the most recent common ancestor of those taxa. *See also* **evolution; homoplasy**.

paralogous genes (Gk *para*=beside and *logos*=reason) Genes formed by a **gene-duplication** event. Paralogous genes often have similar functions and they may form a series of related genes known as a **gene family** (e.g., the *HOX* gene family and the **globin** gene family). Any demonstrated **homology** between paralogous genes of the same species can be used to establish gene phylogenies.

paralogue *See* **paralogous genes**.

parameter (Gk *para*=alongside and *meter*=measure) The true, generally unknowable, population values that researchers try to estimate with sample statistics. Most statistical analyses involve collecting data on a sample whose members are part of a larger population. The statistics calculated for the sample, such as the mean, standard deviation, variance, or range, are estimates, determined with some error, of the corresponding population parameters. Paleoanthropologists can sometimes measure every known example of a trait, such as the mesiodistal length of all of the well-preserved mandibular first molars belonging to an early hominin taxon such as *Australopithecus afarensis*. However, even if the sample is large the value determined it is still not a population parameter. It is still a sample statistic for the researcher is using the known specimens as a sample to make an inference about the mean of the trait for the entire species. Many treat the terms parameter and variable as synonyms, but they are not. For example, the mesiodistal length of mandibular first molars is a variable, the mean and standard deviation of the mesiodistal length of mandibular first molars for the population of all members of the species *Pan troglodytes* are parameters, and the mean and standard deviation of the mesiodistal length of mandibular first molars as calculated for the sample of *P. troglodytes* in the **Powell Cotton Collection** are sample statistics (i.e., they are estimates of the parameters of the population of *P. troglodytes* from which the sample in the Powell Cotton Collection is drawn).

parametric statistics Statistical tests which assume that data follow particular probability distributions (e.g., **normal distribution**) to estimate **parameter**s of those distributions such as the mean and **variance**. Most common statistical tests (e.g., *t* **test**, *F* **test, analysis of variance**, etc.) are parametric. Inappropriate use of parametric tests when data do not meet the assumptions of parametric tests will result in outcomes being subject to increased **Type I error**. In such cases one should use **nonparametric statistics**.

paramolar tubercle The anglicization of a term (*tuberculum paramolare*) used by Bolk (1915) for an enamel feature on the buccal face of the **paracone** between the tip of the paracone and the cemento-enamel junction of a maxillary molar (syn. anticone, buccostyle ectostyle j, paramolar cusp, paramolar tubercle of Bolk, parastyle, tuberculum paramolare). *See* **parastyle**.

paramolar tubercle of Bolk   Anglicization of the *tuberculum paramolare* of Bolk (1915). *See* **paramolar tubercle**.

Paranthropinae   (Gk *para*=beside and *anthropos*=human being) A subfamily formally introduced by Broom (1950) for *Paranthropus crassidens* and *Paranthropus robustus*. Robinson (1972) later added *Gigantopithecus* von Koenigswald, 1935 to the list of taxa, and he widened his interpretation of the Paranthropinae to include "*Zinjanthropus, Meganthropus* from Java, and *Paraustralopithecus*" (*ibid*, p. 6). *See also* **Paranthropus**.

paranthropoid   (Gk *para*=beside, *anthropos*=human being and *oeides*=form, literally "*Paranthropus*-like") A term used to suggest a specimen is *Paranthropus*-like. For example, one of the three mandibles from Tighenif (**Tighenif 1**) is substantially larger than the other two, and Briggs (1968) suggested that it belonged to a "paranthropoid" group, whereas he assigned the two smaller mandibles, Ternifine 1 and 2, to a "telanthropoid" group.

Paranthropus   (Gk *para*=beside and *anthropos*=human being) Hominin genus established by Robert **Broom** in 1938 to accommodate the species *Paranthropus robustus*. Many researchers considered the differences between the hypodigms of *Australopithecus* and *Paranthropus* insufficient to justify a second genus for the southern African archaic hominins, so it became a junior synonym of *Australopithecus*. However, the genus *Paranthropus* has been revived by researchers who consider that *P. robustus*, *Paranthropus boisei*, and *Paranthropus aethiopicus* belong to a separate hominin clade. Species in the genus *Paranthropus* are often informally called robust or "robust" australopiths. They are characterized by postcanine megadontia, thick enamel caps on their molars and premolars, a distinctive occlusal morphology in which the cusp tips of the postcanine teeth are located towards the center of the crown, and a number of highly derived features in the cranium, and especially in the facial skeleton, generally thought to be functionally related to the generation and dissipation of high masticatory loads. Type species *Paranthropus robustus* Broom, 1938.

Paranthropus aethiopicus (Arambourg and Coppens, 1967) Wood and Chamberlain, 1987   (Gk *para*=beside, *anthropos*=human being, L. *australis*=southern, Gk *pithekos*=ape and *aethiopicus*=Ethiopia) A new name combination used by researchers who do not recognize *Paraustralopithecus* as a separate genus, but who do consider pre-2.3 Ma megadont hominins from **Omo-Shungura** to belong to a species within a *Paranthropus* clade that antedates and is distinct from *Paranthropus boisei*. The hypodigm of such a species would include a well-preserved adult cranium from **West Turkana** (**KNM-WT 17000**) together with mandibles (e.g., KNM-WT 16005) and isolated teeth from the **Shungura Formation**; some also assign **L. 338y-6** to this taxon. The only postcranial fossil that is considered to be part of the hypodigm of *P. aethiopicus* is a proximal tibia from **Laetoli**. The fossil evidence for *P. aethiopicus* is similar to *P. boisei* except that the face of the former taxon is more prognathic, the cranial base is less flexed, the incisors are larger, and the postcanine teeth are not so large or morphologically specialized. But there is only one relatively complete *P. aethiopicus* cranium, thus the warnings of Smith (2005a) about making taxonomic inferences based on small samples are especially relevant. The only source of information about the endocranial volume of this taxon is KNM-ER WT 17000. When the new taxon *P. aethiopicus* taxon was introduced in 1968 it was the only megadont hominin in this time range. With the discovery of *Australopithecus garhi* it is apparent that robust mandibles with long premolar and molar tooth rows are being associated with what are claimed to be two distinct forms of cranial morphology. *See also* **Laetoli**; *Paraustralopithecus aethiopicus*.

Paranthropus boisei (Leakey, 1959) Robinson, 1960   (Gk *para*=beside and *anthropos*=human being, and *boisei* to recognize the help provided to Louis and Mary Leakey by Charles Boise) For researchers who support the hypothesis that *Australopithecus robustus* and *Australopithecus boisei* are sister taxa it makes sense to revive the genus *Paranthropus* as a way of recognizing that clade. In which case *Zinjanthropus boisei* Leakey, 1959 that became *Au. boisei* (Leakey, 1959) Tobias, 1967a would become *Paranthropus boisei* (Leakey, 1959), a combination first used by Robinson (1960). *P. boisei* has a comprehensive craniodental fossil record. There are several skulls (the one from **Konso** being especially complete and well preserved), several well-preserved crania, and many mandibles and isolated teeth. There is evidence of both large- and small-bodied individuals, and the range of the size difference suggests a substantial degree of body-size sexual dimorphism. *P. boisei* is the only hominin to combine a massive, wide, flat, face, massive premolars and molars, small anterior teeth, and a modest endocranial volume (mean *c*.480 cm³). The face of *P. boisei* is larger and wider than that of *P. robustus*, yet their brain volumes are similar. The mandible of *P. boisei* has a larger and wider body or corpus than any other known hominin, and the tooth crowns apparently grow at a faster rate than has been recorded for any other early hominin. There is, unfortunately, no postcranial evidence that can with certainty be attributed to *P. boisei*, but some of the postcranial fossils from Bed I at **Olduvai Gorge** currently attributed to *Homo habilis* may belong to *P. boisei*. The fossil record of *P. boisei* extends across about 1 million years of time during which there is little evidence of any substantial change in the size or shape of the components of the cranium, mandible, and dentition (Wood and Constantino 2007). First discovery OH 3 (1955). Holotype **OH 5** (1959). Paratypes None. Main sites **Chesowanja, Konso, Koobi Fora, Malema, Olduvai Gorge, Peninj** (Natron), **Shungura Formation, West Turkana**. *See also Zinjanthropus boisei.*

Paranthropus crassidens Broom, 1949

(Gk *para*=beside, *anthropos*=human being, and L. *crassus*=dense) Hominin species introduced by Robert **Broom** in 1949 to accommodate **SK 6**, an adult mandible recovered at **Swartkrans** from breccia now believed to be derived from Member 1. Broom opted for the new species name because he considered the Swartkrans teeth to be similar to those of "the **Kromdraai** ape-man *Paranthropus*" but he argued that the new finds deserved their own species name because "they are much larger and differ in a number of respects" (Broom 1949, p. 57). He thus assigned it to a new species, *Paranthropus crassidens*. Some researchers have supported the distinction between the Swartkrans and Kromdraai hypodigms (e.g., Grine's 1984 careful analysis of the morphology of the deciduous teeth), but most researchers do not now make a specific distinction between the hypodigms, thus *P. crassidens* is now regarded as a junior synonym of *Paranthropus robustus*. First discovery SK 2 (1948). Holotype SK 6. Main site Swartkrans (Mbs 1, 2, and 3). *See also Paranthropus robustus; Paranthropus robustus crassidens.*

Paranthropus palaeojavanicus (von Koenigswald, in Weidenreich, 1945) (Gk *para*=beside, *anthropos*=human being, *palaios*=ancient, and L. *javanicus*=from Java). A new combination first used by Robinson (1954a). He proposed that the taxon *Meganthropus palaeojavanicus* was a "good" species, but he suggested that *Meganthropus* was not a "good" genus. So he (*ibid*, p. 196) formally proposed the species *M. palaeojavanicus* be transferred to the genus *Paranthropus*. *See also Meganthropus; Meganthropus palaeojavanicus; Paranthropus.*

Paranthropus robustus Broom, 1938 (Gk *para*=beside, *anthropos*=human being, and L. *robus*=oak or strength) Hominin species established by Robert **Broom** to accommodate fossil hominins recovered in 1938 from what was then referred to as the "Phase II Breccia" (now called Member 3) at **Kromdraai B**, in Gauteng, South Africa. The fossil record is less numerous than that of *Australopithecus africanus*. The dentition is well represented, some partial crania are well preserved, but most of the mandibles are crushed or distorted. The postcranial skeleton is not well represented. Research at Drimolen was only initiated in 1992 yet already more than 80 hominin specimens have been recovered and it promises to be a rich source of evidence about *P. robustus*. The brain, face, and chewing teeth of *Paranthropus robustus* are larger than those of *Au. africanus*, yet the incisors and canines are smaller. The crania of *P. robustus* include specimens which have ectocranial crests, whereas there are no *Au. africanus* crania with unambiguous crests. What little is known about the postcranial skeleton of *P. robustus* suggests that the morphology of the pelvis and the hip joint is much like that of *Au. africanus*. It was most likely capable of bipedal walking, but most researchers subscribe to the view that *P. robustus* was not an obligate biped (but see Susman 1988a). It has been suggested that the thumb of *P. robustus* would have been capable of the type of grip necessary for stone tool manufacture, but this claim is not accepted by all researchers. Most researchers subsume *Paranthropus crassidens* Broom, 1949 recovered from **Swartkrans** into *P. robustus*, but Robinson (1954a, p. 198) suggested that it be regarded as a subspecies of *P. robustus*. First discovery **TM 1517** (1938). Holotype As above. Main sites **Cooper's, Drimolen, Gondolin**, Kromdraai (Mb 3), Swartkrans (Mbs 1, 2, and 3). *See also Paranthropus; Paranthropus robustus crassidens.*

Paranthropus robustus crassidens Broom, 1949 (Gk *para*=beside, *anthropos*=human being, and L. *robus*=oak or strength and *crassus*= dense) A new combination first used by Robinson (1954a). Robinson (1954a, p. 198) formally proposed that the Swartkrans part of the *Paranthropus robustus* hypodigm be regarded as a subspecies of that taxon. *See also Paranthropus robustus.*

Paranthropus robustus robustus Broom, 1938 (Gk *para*=beside, *anthropos*=human being, and L. *robus*=oak or strength) A new combination first used by Robinson (1954a). Robinson (1954a, p. 196) formally proposed that the Kromdraai part of the *Paranthropus robustus* hypodigm be regarded as a subspecies of *P. robustus*. *See also Paranthropus; Paranthropus robustus.*

parapatric speciation A mode of **speciation** in which a new species evolves when part of a species population colonizes a new ecological **niche** in a geographical zone adjacent to that of the "parent" species. Unlike **allopatric speciation** there is not a physical boundary separating the parent and "daughter" populations, but because these two populations are subject to different selective pressures the hybrid individuals resulting from mating between the populations experience reduced fitness. Over time, this reduces **gene flow** between the populations and favors mating within each population. This effect, in combination with **natural selection** affecting each population in different ways, can lead to the two populations becoming reproductively isolated.

parapatry (Gk para=beside and *patris*= fatherland) Two organisms with adjacent but not significantly overlapping geographic ranges (or, if their ranges do overlap, there is only a small contact zone) are described as being parapatric (e.g., most extant baboon species are parapatric). In parapatric speciation, there is usually a physical barrier between populations and new species arise from contiguous (adjacent) populations. It may also occur when a subpopulation of the parent population moves into an area with a different environment and thus different selective pressures, causing a selective advantage for traits that differ from those seen in the majority of the parent population. This can cause **clinal variation** and eventually divergence and speciation. Interbreeding between populations occurs at hybrid zones where the populations meet, but because there is no significant gene flow between the populations the populations maintain their unique characteristics. Using a baboon analogy, Jolly (2001) has suggested that hybrid zones were important in creating diversity and hence speciation within *Homo*. *See also* **reticulate evolution; speciation**.

paraphyletic An adjective used to describe a taxon that is part of a **paraphyletic group**. *See* **paraphyletic group**.

paraphyletic group (Gk *para*=beside and *phulon*=a class) A taxonomic grouping that includes only part of a **clade** or **monophyletic group**. For example, if the term "great apes" is used to refer to a group that includes chimpanzees, gorillas, and orangutans, then the term refers to a paraphyletic group. This is because there is sound evidence that the clade structure of those taxa is (((*Homo, Pan*) *Gorilla*) *Pongo*). Thus, there is no clade made up of *Pan, Gorilla*, and *Pongo* that does not include *Homo*.

paraphyly The state of a taxonomic group that is paraphyletic. *See* **paraphyletic group**.

parasagittal crests (Gk *para*=beside, L. *sagitta*=an arrow and L. *crista*=cock's comb, or a tuft of feathers, in the midline of a bird's head) A term usually used for two sharp crests of bone running either side of the inter-parietal, or sagittal, suture. The bony crests form when the medial borders of the fascia covering the temporalis muscles are raised up. In extant hominoids the interval between the crests is occupied by fat and small blood vessels. *See also* **sagittal crest; sagittal keel**.

parastyle A term introduced by Osborn (1892) for an accessory cusp located mesiolingual to the **paracone** of a maxillary molar. It has been used to describe an equivalent feature on the crowns of fossil hominin teeth (e.g., Robinson 1956); Grine (1984) refers to a similar structure as the **mesiostyle**. Parastyle has also been used by some researchers (e.g., Scott and Turner 1997) for a different enamel feature on the buccal face of the paracone more commonly referred to as the **paramolar tubercle**. *See also* **mesiostyle, mesial accessory tubercle, paramolar tubercle**.

paratype (Gk *para*=beside and *typus*=image) Specimens other than the **holotype** listed in the original description of a taxon.

***Paraustralopithecus* Arambourg and Coppens, 1968** (Gk *para*=beside, L. *australis*=southern, and Gk *pithekos*=ape) Hominin genus established by Camille **Arambourg** and Yves **Coppens** (1967) to accommodate the type species *Paraustralopithecus aethiopicus*. The genus *Paraustralopithecus* is now regarded as a junior synonym of either *Australopithecus* or *Paranthropus*. Type species *Paraustralopithecus aethiopicus* Arambourg and Coppens, 1967. *See also* ***Paraustralopithecus aethiopicus* Arambourg and Coppens, 1967**.

Paraustralopithecus aethiopicus* Arambourg and Coppens, 1968** (Gk *para*=beside, L. *australis*=southern, Gk *pithekos*=ape, and *aethiopicus*=Ethiopia) Hominin species established by Camille **Arambourg** and Yves **Coppens** to accommodate a mandible **Omo 18-1967-18** from the **Shungura Formation, Omo region**, Ethiopia. Subsequently Howell (1978a) assigned the holotype mandible to *Australopithecus africanus* (thus making *Paraustralopithecus aethiopicus* a junior synonym of *Au. africanus*), and Coppens (1980) implied the species should be sunk into *Australopithecus boisei* (*see Paranthropus boisei*), but others have assigned pre-2.3 Ma megadont fossils from the Shungura Formation and West Turkana (e.g., **KNM-WT 17000**) to either *Australopithecus aethiopicus* or *Paranthropus aethiopicus*. When *Para. aethiopicus* was introduced in 1967 it was the only megadont hominin in this time range. With the discovery of *Australopithecus garhi* it is apparent that robust mandibles with long premolar and molar tooth rows are being associated with what are claimed to be two distinct forms of cranial morphology. First discovery Omo 18-1967-18 (1967). Holotype As above. Main sites **Omo-Shungura, West Turkana**. *See also* ***Paranthropus aethiopicus.

parazone (Gk *para*=beside and *zone*=girdle) The region of a **Hunter–Schreger band** which, when a tooth is cut longitudinally, shows enamel prisms running parallel to this plane of section. Gustav Preiswerk (1895) was the first to define and name parazones, **diazones**, and **perikymata** in his thesis on the microstructure of ungulate enamel (Pickerill 1913, Osborn 1990) (ant. diazone). *See also* **decussation**.

parcellation A term used in neuroscience to refer to the subdivision of the cerebral cortex into anatomically distinct regions by virtue of cytoarchitecture, chemoarchitecture, or the pattern of connectivity. It was an early version of parcellation that Brodmann used to divide the cerebral cortex into the cytoarchitectonically-distinctive **Brodmann's areas**. *See also* **Nissl stain**.

parent/offspring conflict According to Trivers (1974), since parents share an equal genetic relationship with all of their offspring, they should divide their investment evenly among those offspring. However, it is in the offspring's interest to get the parents to focus investment on them. A branch of parent/offspring conflict theory, called maternal/fetal conflict, examines the conflicting interests of mothers and fetuses during gestation, where the placenta plays a large role in trying to extract more resources from the mother than she would otherwise be willing to give (Haig 1993). This type of conflict can be manifested in ways such as decreasing maternal sensitivity to insulin so that more blood sugar is available to the fetus, and increasing maternal blood pressure so that a greater blood supply is available to the fetus.

parental effects *See* **genomic imprinting**; **heritability**.

parental investment Any investment by a parent in an individual offspring that increases the offspring's chance of survival (and hence reproductive success) at the cost of the parent's ability to invest in other offspring (Trivers 1972). Parental investment is usually unbalanced between the sexes, and the sex that invests the most in offspring is thought to be the more discriminating sex when it comes to finding a mate. Aspects of parental investment in modern humans and nonhuman primates include gestation, lactation, carrying of offspring, food provisioning, and protection. In most primates, females are the higher-investing sex, although callitrichines and some modern human populations show significant male parental investment.

parietal art *See* **art**.

parietal lobe One of the four main subdivisions of the **cerebral cortex** of each **cerebral hemisphere**. It is located between the **frontal lobe** anteriorly and the **occipital lobe** posteriorly. The parietal lobe is involved in processing information about somatic sensation, body position, attention, visual and spatial relations, and **language**. The parietal lobe contains the primary somatosensory cortex, which lies on the postcentral gyrus.

parietal-occipital plane In some crania (e.g., **OH 5**) the external aspect of the squamous part of the occipital bone and the external aspect of the posterior part of the **parietal bone**s form a continuous flat surface which is referred to as the parietal-occipital plane, or planum.

pars mastoidea *See* **mastoid process**.

pars opercularis *See* **Broca's area**.

pars orbitalis *See* **Broca's area**.

pars triangularis *See* **Broca's area**.

parsimony (L. *parsimonia* = frugality) The principle stating that the most economical explanation is the one that should be adopted in the first instance; it is the equivalent of **Occam's razor**. For example, in **phylogenetic systematics** the parsimony principle means that the **cladogram** that involves the fewest evolutionary changes (i.e., the fewest character state changes) is the one that should be preferred.

partial volume averaging One of several technical problems involved in the data-acquisition stage of **computed tomography** (or CT). The problem arises due to the combination of "finite slice collimation" and the different beam widths used in imaging structures that consist of different tissues each of which may have its own **attenuation value**. These different values may contribute to a single measurement made during the scanning process. The effect of mixing different tissues during the scanning measurements is that the resulting CT number is not the average of the attenuation coefficients of the different tissues. The CT reconstruction process is a nonlinear one, which means that mixing tissues before the logarithm is taken will result in a value that is different from the one that would be generated if the tissues were mixed after the logarithm has been taken. Partial volume averaging can result in erroneous CT numbers, and/or streaks (called artifacts) in the CT image. *See also* **computed tomography**.

PARU Acronym for the Palaeo-Anthropology Research Unit within the **Witwatersrand University Department of Anatomy**. *See* **Palaeo-Anthropology Research Unit**.

passive scavenging A carcass transport strategy defined as the culling of small amounts of meat or marrow from animal carcasses heavily ravaged and abandoned by their initial predators. Opportunistic passive scavenging has been documented ethnographically in some hunter–gatherer groups (e.g., the Hadza), although a carcass acquisition strategy characterized by obligate passive scavenging of terrestrial mammals is considered to be outside the range of modern human behavior.

pastoralist (L. *pastor* = a shepherd) Mobile societies that rely on livestock as a food source, which are driven to move seasonally in search of forage and water. Pastoral systems based on the herding of domesticated or partially domesticated animals arose 10,000–12,000 years BP, approximately at the same time as crop domestication. Worldwide, the most common animals herded by pastoralists are sheep, goats, cattle, horses, donkeys, camels, reindeer, and yaks. Economically, pastoralism is viable in regions too arid for the planting of crops, and pastoralists frequently control their ecosystems by using fire to limit the growth of trees and to rejuvenate pasture. The degree of mobility in pastoralist societies varies. In nomadic pastoralist societies, entire households move up to several hundred kilometers with their herds, living in impermanent, easy-to-construct shelters such as tents. Transhumance pastoralists frequently move seasonally between two locations, such as a cool, high-altitude valley in the summer and a warm, lower-altitude valley in the winter. Housing may be more permanent in the case of transhumance pastoralists for only a subset of the household may be actively involved in livestock management. Milk, blood, meat, skin, and dung are common products of pastoral societies, but wool or hair can be important as well. Agriculturalists and pastoralists frequently develop extensive regional trade networks to exchange goods, and historically competition for land has often led to conflict between pastoral and agricultural groups.

pattern *See* **mode**.

PAUP Acronym for Phylogenetic Analysis Using Parsimony. PAUP is a computer software program written by David Swofford (Swofford 2003) that is widely used to perform cladistic analysis. Typically, PAUP is used to find the most parsimonious tree or cladogram implied by a given character state data matrix. It also is capable of performing statistical tests designed to assess the stability of cladograms and of the clades contained within them (e.g., the **bootstrap** and the **jackknife**); PAUP* 4.0 includes maximum likelihood and distance methods.

Pauphadhi One of the Early and Middle **Pleistocene** sites in the exposures of the **Ola Bula Formation** within the **Soa Basin**, central Flores, Indonesia.

Paviland *See Homo sapiens*; radiocarbon dating.

Pavlov (Location 48°52′N, 16°40′E, Czech Republic; etym. named after a nearby village) History and general description A series of five open-air **loess** sites on the base of the northern slopes of the Pavlovské Hills above the Dyje River floodplain, that are labeled Pavlov I–VI. These sites are closely associated with the nearby **Dolní Věstonice** sites, which are located about 500 m away; Pavlov I is the largest and most complex. Systematic excavations at Pavlov I and II were conducted from 1952 intermittently until 1972. Pavlov VI was discovered and excavated in 2007. The sites have yielded archeological and hominin skeletal remains, including a ritual burial. Temporal span and how dated? Calibrated **radiocarbon dating** of charcoal has provided a date of **Oxygen Isotope Stage 3** (*c.*26–29 ka). Hominins found at site Associated skeleton (Pavlov 1), maxilla (Pavlov 2), two mandibles (Pavlov 3 and 4), and isolated teeth (Pavlov 5–28). Archeological evidence found at site An **Upper Paleolithic** (Gravettian) stone tool assemblage, clay figurines, and remains of dwelling structures. Key references: Geology, dating, and paleoenvironment Joris and Weninger 2004; Hominins Vlček 1971, Trinkaus and Svoboda 2005; Archeology Klima 1954, 1959, Svoboda 1997, 2005. *See also* **Dolní Věstonice I**.

PCA *See* **principal components analysis**.

PCR Acronym for **polymerase chain reaction** (*which see*).

PCSA Abbreviation of the physiological cross-sectional area of a muscle. This measure provides an estimate of the **force** a muscle can generate, and it is used in models that attempt to estimate the efficiency of different modes of locomotion (e.g., Pontzer et al. 2009).

PCSOM Acronym for **portable confocal scanning optical microscope** (*which see*).

pdf Acronym for probability density function (*which see*).

PDL Acronym for periodontal ligament (*which see*).

peak growth velocity In modern humans, after a period of latency during the childhood and juvenile life history phases, the **adolescent growth spurt** is initiated. Peak growth velocity is the point at which growth velocity reaches its maximum during this period. It occurs approximately 1 year prior to onset of menarche in girls, which is distributed across a wide range of chronological ages. Following peak growth velocity, growth velocity decreases rapidly in mass, and particularly in linear dimensions since increasing levels of sex steroids act to facilitate epiphyseal closure.

Pech de l'Azé (Location 44°51′29″N, 1°15′14″E, France; etym. the name derives from Languedoc, meaning roughly Donkey Hill. Note that the accent on the final e is not pronounced). History and general description The site consists of four adjacent localities, numbered I, II, III, and IV, in a small tributary of the Dordogne River in southwest France. Pech de l'Azé I and II are opposite ends of a tunnel-shaped cave and Pech III is a very small cave 30 m west of Pech II; Pech IV is a collapsed cave 80 m east of Pech I. Pech I was first excavated in the 19thC, but it was in 1909, at the entrance of the cave, that Capitan and Peyrony recovered the remains of a *Homo neanderthalensis* child associated with a **Mousterian** of Acheulean Tradition (or MTA) assemblage. Bordes excavated the site from 1949 to 1951 and again from 1970 to 1971; Soressi most recently excavated there in 2004–5. Bordes discovered Pech II while working at Pech I, and excavated there at times between 1950 and 1969. Pech IV was also discovered by Bordes, tested by Mortureux between 1953 and 1956, excavated by Bordes between 1970 and 1977, and excavated by Dibble and McPherron between 2000 and 2003. Temporal span and how dated? Pech I, II, and IV contain **Middle Paleolithic** industries; Pech III contained a small assemblage that was attributed to the **Acheulean** and the lower portion of the Pech II sequence is attributed to the Acheulean. **Electron spin resonance spectroscopy** (ESR) dates on Pech II show two periods of deposition: one in **Oxygen Isotope Stage** (OIS) 6 and the other starting in late OIS 5 through to *c.*40 ka. The Pech I industries consist only of MTA, and recent ESR and **AMS radiocarbon** dates on Pech I show the deposits to date to *c.*51–41 ka and a reanalysis of the stratigraphic association of the *H. neanderthalensis* remains places them in this same time range. The Pech IV sequence was thought to overlap with the upper portion of the Pech II sequence and the Pech I sequence. **Thermoluminescence dating** results from the base of Pech IV suggest an age of 100 ka. Hominins found at site Pech de l'Azé 1 (from Pech I) is a hominin partial cranium and mandible of a young child; an isolated tooth from another young individual has also been published. Isolated teeth are known from Pech IV. Archeological evidence found at site Pech IV is the type site for the Asinipodian (Latin Pech de l'Azé), a Middle Paleolithic industry dominated by small flake production. Aside from their stone tool industries, fauna are well preserved at Pech I, II, and IV. Pech I and IV have evidence for the use of fire. Pech I and IV also have abundant pieces of worked manganese dioxide. Key references: Geology, dating, and paleoenvironment Grun and Mellars 1991, Soressi et al. 2007; Hominins Maureille and Soressi 2000; Archeology Bordes 1972, McPherron and Dibble 2000, Soressi et al 2008, Dibble et al. 2009b.

Pech Merle (Location 44°40′27″N, 1°38′40″E, France; etym. Occitan *pech*=hill; *merle* is unknown) History and general description This extensive cave site near the village of Cabarets, France, has been known for its paintings since at least 1926. It is one of the few sites with cave paintings that is currently open to the public. Temporal span and how dated? Stylistic features of the artwork make it likely from the **Gravettian** period. **Radiocarbon dating** of the pigments has provided an age of *c.*25 ka. Hominins found at site None. Archeological evidence found at site Many fragmentary faunal remains, and rare pieces of engraved art on bones. Key references: Geology, dating, and paleoenvironment Lorblanchet et al. 1995; Hominins N/A; Archeology Lemozi 1922, 1929, Lemozi et al. 1969, Faurie 1999.

pectoral girdle (L. *pectoralis* from *pectus*= breast and ME. *girdle*=sash) The bones by which the proximal part of the upper limb is attached to the **axial skeleton**. The pectoral girdle, which comprises the **scapula** and **clavicle**, is part of the postcranial skeleton.

pedal digit *See also* **digit**; **foot**.

pedal ray *See also* **foot**; **ray**.

pedigree A chart showing the relationships among individuals in a family. In genetics, pedigrees are used to show the **phenotype**s of related individuals. These have been important for identifying the pattern of inheritance of normal traits (i.e., whether a trait is X-linked or autosomal) and genetic diseases.

pedogenic (Gk *pedon*=Earth and *genos*=birth) Refers to the characteristic structure, color, mineralogy, and horizontal zonation characteristics of soils in the process of being formed. Pedogenic carbonate is a mineral precipitate formed in the subsurface (or B) horizon of a soil as the result of the supersaturation of the soil solution with respect to carbonate minerals (generally calcite). The mechanisms that drive precipitation of carbonate minerals (i.e., evaporation and degassing of carbon dioxide produced by respiration) are most pronounced in arid to semi-arid conditions, and thus pedogenic carbonate is most commonly found in such environments. The nodules and rhizoliths it contains preserve the carbon and oxygen stable-isotopic signatures during the time of soil formation and thus in ancient soils they can be used as a proxies to reconstruct paleovegetation, paleoclimate, and paleoenvironment. *See* C_3 and C_4.

pedomorphic (Gk *paid*=child and *morph*= form) *See* **pedomorphosis**.

pedomorphosis Gk *paedo*=child and *morph*= form) A term introduced by Walter Garstang in 1922, pedomorphosis refers to a category of resultant morphologies that follow from three different kinds of heterochronic processes (time hypomorphosis, rate hypomorphosis, or neoteny). All of these processes produce, in varying ways, descendants with ancestral juvenile characteristics or shapes retained during later stages of ontogeny. Since the juvenile features are shifted to the adult, there are no "new" pedomorphic morphologies. *See also* **hypomorphosis**; **neoteny**.

Peers Cave   *See* **Fish Hoek**.

Pei, Wenzhong [裴文中] (1904–82)

Wenzhong Pei (also spelled Wen-chung P'ei), who was born in Lauxian (now Fangnan) in Hebei Province in northern China, entered Beijing University in 1921, and joined the Geology Department in 1923 where he specialized in paleontology. After completing his degree in 1927 he was offered a position with the **Geological Survey of China** by Wenhao Weng, its director. Weng and the Geological Survey had just completed an agreement with the Canadian anatomist Davidson **Black** of the **Peking Union Medical College** in Beijing on a collaborative effort to study China's Cenozoic deposits. They were focusing their efforts at a particularly rich fossil site called Choukoutien (now called **Zhoukoudian**), where the Austrian paleontologist Otto **Zdansky** had found modern human-like (as opposed to ape-like) teeth as part of a diverse Pleistocene fauna. When Anders Bohlin found yet another tooth in 1927, Black realized its significance and announced the discovery of a hitherto unknown species of hominin he called *Sinanthropus pekinensis* (now *Homo erectus*). Pei joined the excavations at Zhoukoudian in 1928 and in 1929 he was made the field supervisor. In December 1929 Pei discovered the first complete *Sinanthropus* calotte, and in 1930 he began to find quartz artifacts in the same sediments that had yielded the *Sinanthropus* remains. A year later Pei and Black found charred bone and other evidence that was interpreted as evidence that *Sinanthropus* had used fire. Over the next several years more cranial and postcranial fossils were discovered, along with large quantities of animal fossils. After Davidson Black's death in 1934 the German physical anthropologist Franz **Weidenreich** was appointed to direct the **Cenozoic Research Laboratory**, which had been established by Black and Wenhao Weng in 1929 to oversee excavations at Zhoukoudian and other sites. In 1935 Pei had begun to study archeology in Paris with the prominent French prehistorian Henri Breuil, but he did not complete his doctorate until 1937. Little excavating could be done during the war years, so between 1937 and 1948 Pei devoted much of his time to studying the extensive collection of archeological and paleonto-

logical material housed at the Cenozoic Research Laboratory, while also lecturing on prehistoric archeology at Beijing University and Yanjing University. With Weidenreich's departure from China in 1941, Pei became the director of the Cenozoic Research Laboratory. When the People's Republic of China was established in 1949, Pei was named head of the museum's division of the Bureau of Social and Cultural Affairs within the Ministry of Culture. He remained in this post until 1953 and in the following year was appointed a researcher in the Chinese Academy of Sciences' **Institute of Vertebrate Paleontology and Paleoanthropology** (or IVPP), the successor institution to the Cenozoic Research Laboratory. At the IVPP Pei organized research programs in human paleontology and paleolithic archeology and directed excavations at a number of Chinese prehistoric sites. Pei's excavations during the mid-1950s at Dingcun and Guanyindong transformed archeologist's understanding of the Paleolithic in China, and he also made the first survey of the Mesolithic in China. In 1963 Pei became the director of the Paleoanthropological Research Laboratory, a section of the IVPP, and in 1979 he was appointed chairman of the Beijing Natural History Museum. Pei remained at the IVPP and continued to teach and conduct research until his death in 1982.

Peking Man   The informal name given to the taxon represented by the fossil hominin remains recovered from Locality 1 in the Zhoukoudian Lower Cave, a site not far from Peking. The remains were initially assigned to *Sinanthropus pekinensis*, but early in 1940 Wilfrid **Le Gros Clark** transferred them to *Pithecanthropus erectus* and later that year Franz **Weidenreich** formally suggested that taxon be a junior synonym of *Homo erectus*. *See also* *Homo erectus*; **Zhoukoudian**.

Peking Union Medical College   (or PUMC) The college that grew out of the activities of the London Missionary Society. The China Medical Board of the **Rockefeller Foundation** began its support of PUMC in 1915 and carried on doing so until 1936. The **Cenozoic Research Laboratory** was established at the PUMC in 1929, and in a joint effort with the **National Geological Survey of China**, it organized the excavations at Choukoutien (now called **Zhoukoudian**). Both Davidson **Black** and after him Franz **Weidenreich** headed the Department of Anatomy at the PUMC and during their tenures the college became a focus for paleoanthropological research in China. During the Cultural Revolution the PUMC was re-named the "Capital University of Medical Sciences," but in 1979 it resumed activity under its old name and it is still in operation.

pelvic girdle (L. *pelvis*=basin and ME *girdle*= sash) The pelvic girdle, which is part of the postcranial skeleton, comprises the bones that form the **pelvis** (i.e., the two pelvic bones and the sacrum).

pelvis The pelvis surrounds the birth canal and provides attachments for the propulsive muscles of the hindlimb, the muscles of the anterior abdominal wall, as well as the muscles that move and support the spine. Its shape reflects the overall shape of the torso. The pelvis is formed by the sacrum and two pelvic, or hip, bones and because it forms a complete bony ring the three components are called the pelvic girdle. The pelvic bone is still sometimes referred to as the innominate bone, but this is obsolete terminology. Each adult pelvic bone is formed from three component bones – the ilium, ischium, and pubis – which fuse during early childhood to form the single pelvic bone. The ilium is the broad, cranial portion, the pubis the ventral portion that meets in the midline, and the ischium the dorsal and caudal portion; all three contribute to the acetabulum, or hip joint socket. All three also border the large obturator foramen, which is covered in life by a fibrous membrane except for a small gap that transmits a nerve and blood vessels. The hip joint is a large, synovial ball-and-socket joint capable of flexion/ extension, abduction/adduction, rotation, and circumduction. The articular surface is c-shaped and it is referred to as the lunate surface. The sacroiliac joints are synovial anteriorly but posteriorly each pelvic bone is bound to the sacrum by strong ligaments. There is little movement at these joints, but what slight movement there is (known as nutation) is important for dissipating stresses during bipedal locomotion (Lovejoy 2005). The other joint between the pelvic bones is ventrally between the two pubes. This is a symphyseal joint, with the articulating surfaces covered by hyaline cartilage with interposed fibrocartilage. This joint is normally not mobile, although it may expand slightly due to the influence of hormones during parturition. Locomotor adaptation is the chief determinant of variation in pelvic form among primate species. Monkeys tend to have narrow pelves that are short craniocaudally, which is directly related to their long, flexible lumbar spine that they use during pronograde walking and running. Apes, and especially great apes, have craniocaudally elongated pelves consistent with their shortened lumbar vertebral column and stiff torso, which provides stability for hoisting the body up during arm-hanging and climbing (Waterman 1929). The orientation of the iliac blades of the great apes generally tracks that of the rib cage. Apes have a broad rib cage and shoulder joints facing laterally, and correspondingly reoriented, dorsally-rotated, iliac blades. Monkeys, and basal hominoids such as *Proconsul* (Ward 1993), retain narrow iliac blades that face more laterally, which is consistent with their narrow thoracic cage and ventrally facing shoulder joints. The modern human pelvis is shorter and broader than in any other extant primate. The wide modern human sacrum accommodates the widely set vertebral zygapophyses in the lumbosacral region, which are necessary for upright posture and human lumbar lordosis. The expanded iliac blades of the modern human pelvis are rotated so that the external surface faces partly laterally; it is on this surface that the lesser gluteal musculature (gluteus minimus and medius) originates. These muscles abduct the femur at the hip joint and keep the pelvis level during the single support phase of modern human bipedal gait; only modern humans among the great apes have this abductor mechanism, and it is a critical component of our bipedal adaptation (Lovejoy 1988). This iliac form also places the abdominal oblique muscles in a position to wrap around the abdomen and assist in the lateral pelvic rotation that is part of the bipedal gait of modern humans. The iliac blades of *Australopithecus* tend to be more horizontal and more widely flared than those of *Homo*. With the expansion of the size of the pelvic inlet in *Homo* to accommodate larger-brained neonates, the acetabulae became more widely set and the ilia more vertical. The **acetabulocristal buttress** has become more pronounced and the portion of the ilium anterior to this buttress became expanded as well. The part of the ilium dorsal to the sacroiliac joints is expanded in modern humans and provides a large area of attachment for the erector spinae musclature. The large acetabulum of modern humans provides increased an articular area for load transmission through the hip, something not found in *Australopithecus* and perhaps some early *Homo* individuals probably because of the altered geometry of their pelves (Ruff 1995). However, in all hominins the acetabulum is inclined anteriorly and inferiorly, thus maximizing the articular area for axial load transmission in bipedal posture, but this orientation also limits the range of hip abduction. Hominins all also have short, retroflexed ischia, which allow the hamstring muscles the leverage to assist in extending the thigh in bipedal postures (McHenry 1975). This contrasts with the ischia of great apes and New World monkeys, which are elongated and dorsally directed, thus providing maximum leverage when the thigh is flexed during quadrupedal locomotion and in climbing postures. In modern humans, because of the tight relationship between fetal head size and the pelvic inlet, male and female pelves are sexually dimorphic. As such, the sex of a skeleton can be determined with a fair degree of accuracy from the form of the pelvic bone. Because of the wider set of the hip joints and the wider pelvic inlet in modern human females, the angle between the caudal rami of the pubes is wider than in males, and their internal margins are everted, forming a lip that flares ventrally. Another region that is broader in females than males is the greater sciatic notch along the dorsal margin of the pelvic bone between the sacroiliac joint area and the spine of the ischium. Sometimes, due to stretching of ligaments during pregnancy, small bony scars develop adjacent to the sacroiliac joints anteriorly and caudally. In modern humans, the pelvic inlet is broader mediolaterally than anteroposteriorly (i.e., it is platypelloid), whereas the opposite proportions are found at the outlet. As a consequence, in modern humans the neonatal head rotates during the birth process; *Australopithecus* and early *Homo* pelves appear to be platypelloid at both the inlet and the outlet and would not have had a rotational birth mechanism (Tague and Lovejoy 1986, Simpson et al. 2008). Climate influences pelvic shape in modern humans because body breadth heavily influences surface-area-to-volume ratios, thus affecting a body's capacity for radiating heat. Tropical peoples tend to have narrower pelves than those in more northern latitudes, with members of *Homo neanderthalensis* having the broadest pelves of all (Ruff 1991). Neanderthals also have particularly long, thin pubic rami, which also would have projected further ventrally than those of modern humans; this may be related to the shape of their torso (Rak and Arensburg 1987). *See also* **lumbar vertebral col-** **umn; scapula, evolution in hominins; sexual dimorphism**.

penetrance The proportion of individuals with a specific genotype that express a specific phenotype. For example, people who have a faulty copy of the *Huntington* gene have a 100% chance of being affected by Huntington's disease (i.e., the penetrance is 100%), while those who have faulty copy of the *HMBS* (or hydroxymethylbilane synthase) gene have only a 10–20% chance of being affected by acute intermittent porphyria (i.e., 10–20% penetrance). Penetrance may be affected by the environment or by the effects of other genes.

Peninj (Location 2°50′S, 35°40′E, Tanzania; etym. refers to the Maasai name for the local river, also referred to as Binini or Pinyini) History and general description The site located west of Lake Natron and near the boundary between Kenya and Tanzania was first noticed from the air but it was initially explored on foot in January 1964 by Richard **Leakey** and then later that year by an expedition jointly led by Leakey and Glynn **Isaac**. The Peninj sediments were deposited in three different alluvial environments: one was within a delta not far from a lake (the Type section), the second at the beginning of an alluvial fan (the Southern escarpment), and the third was along the course of the Peninj River at the foot of the Sambu volcano (the Northern escarpment). Temporal span and how dated? Several stratigraphic intervals are represented in the Peninj exposures. The basement is made up of the Sambu lavas, overlaid by the Hajaro Formation, which is divided into two members: the Hajaro Beds and the Hajaro lavas. Overlying the Hajaro Formation is the Humbu Formation. This 40 m unit is divided into three members: the Basal Sands with Clays (BSC), the Main Tuff Limestone basaltic unit, and the Upper Sands with Clays (USC). Overlying the Humbu Formation is the Moinik Formation, with the Clay and Trachytic Tuff member and capped by the Upper Tuff member. **Potassium-argon dating** and **argon-argon dating** of the sequence suggest the following dates (their uncertainty is due to anagenetic modification of minerals): Sambu lavas (2.5–2 Ma), the base of BSC (1.7 Ma), Main Tuff (1.9–1 Ma, with an intermediate range of 1.6–1.4 Ma), base of Moinik Formation (1.3 Ma). Hominins found at site The only hominin fossil, discovered in 1964, consists of a well-preserved and fairly complete mandible (**Peninj 1**). Archeological evidence found at site Two Acheulean sites (PEEN1 and PEES2, formerly RHS and MHS, respectively) were initially discovered and excavated by Isaac (Isaac 1967). Some Oldowan localities were later excavated, among which the most important is ST4 (Domínguez-Rodrigo et al. 2009). Key references: Geology, dating, and paleoenvironment Isaac 1967, Curtis 1974, Manega 1993, Domínguez-Rodrigo et al. 2009; Hominins Leakey and Leakey 1964; Archeology Isaac 1967, Domínguez-Rodrigo et al. 2009.

Peninj 1 Site Peninj. Locality Type section (Isaac 1967, Map 2). Surface/*in situ* In situ. Date of discovery January 11, 1964. Finder Kamoya **Kimeu**. Unit/horizon Initially reported as "clayey sands and sandy grit" layer beneath "ostracod and gastropod limestone" (Isaac 1967, Fig. 1), but now Basal Sands with Clays (BSC). Bed/member N/A. Formation Humbu Formation. Group N/A. Nearest overlying dated horizon Main tuff, 1.6–1.4 Ma (Isaac and Curtis 1964), but recent research suggests this may be *c*.1.3 Ma. Nearest underlying dated horizon TBS-1 basalt from the BSC, 1.7 Ma. Geological age Between *c*.1.5 and 1.3 Ma. Developmental age Adult. Presumed sex Unknown. Brief anatomical description Apart from many cracks in the surface bone and the lack of the coronoid process on the right side and the condyle and the coronoid process on the left side the mandible is complete. All of the teeth are in place in the mandible. Announcement Leakey and Leakey 1964. Initial description Tobias 1965a. Photographs/line drawings and metrical data Leakey and Leakey 1964, Tobias 1965a, 1965b, Wood 1991. Detailed anatomical description None available. Initial taxonomic allocation "An unmistakable australopithecine" (Leakey and Leakey 1964, p. 7). Taxonomic revisions "Probably a late member of the Olduvai species, *A. boisei*" (Tobias 1965b, p. 27). Current conventional taxonomic allocation *Paranthropus boisei*. Informal taxonomic category Megadont archaic hominin. Significance This exceptionally complete and mostly well-preserved mandible provides insights into the morphology of the mandible of *P. boisei*. The exceptional degree of wear on the postcanine tooth crowns deprives researchers of the opportunity to study the detailed crown morphology of Peninj 1, but the excessive wear seen in the postcanine tooth row (all of the enamel has been removed from the metaconid and protoconid of the right P_4, from the metaconid, protoconid, and the hypoconid of the right M_1 and much of it from

these cusps on the left M_1, and all of the enamel has been lost from the protoconid and hypoconid of the left M_2) suggests that at least this individual, if not the other members of the taxon, was consuming abrasive preferred foods. It is difficult to imagine that this degree of wear could have been caused by the consumption of **fallback foods**. Location of original **National Museums of Kenya**, Nairobi, Kenya.

peramorphosis (Skt *para*=beyond plus Gk *morphe*=form, shape) Like **paedomorphosis**, peramorphosis is a category of resultant morphologies that follow from three different kinds of **heterochronic** processes (time **hypermorphosis**, rate hypermorphosis, or **acceleration**). Unlike paedomorphosis, however, peramorphosis includes new shapes since the processes that contribute to these morphologies extend the ancestral ontogenetic trajectory beyond its original endpoint. *See also* **hypermorphosis**; **acceleration**.

percussion The action of one object striking another, used in the study of stone tools to refer to the technique by which rock is fractured. In direct percussion, the stone, bone, antler, or wood **hammer** strikes the rock directly, whereas with indirect percussion an intermediate object, typically a chisel-like piece of bone, wood, or antler is placed on the rock and struck, directing the impact force for the more reliable production of elongated **flakes** and **blades**.

perforator (L. *perforare*=to bore) A flake or blade with one (or rarely multiple) small, pointed tip(s) shaped by finely controlled retouch (see Debénath and Dibble 1994). Sometimes referred to as a "borer," perforators are inferred to have been used in the scraping or drilling of holes in materials such as wood, bone, shell, and hide.

pericone A term used by Szalay and Delson (1979) for an accessory cusp on the occlusal surface of a maxillary molar that others refer to as a Carabelli('s) trait. *See* **Carabelli('s) trait**.

pericranium *See* **periosteum**.

Perigordian A term that refers to early **Upper Paleolithic** technologies predominantly found at sites in central and southwestern France, and named after the Périgord region. Denis Peyrony introduced the term Perigordian to describe a technology that appeared at the same time as, but was distinctive from, the **Aurignacian**. The results of more recent excavations suggest that what was originally called the Perigordian I should be subsumed within the **Châtelperronian**, and the Perigoridan II and III should be subsumed within the Aurignacian. The Upper Perigordian (Perigoridian IV, V, and VI) is still seen as a separate technology, characterized by backed points made on blades, rare bone tools, carved, stylized figurines of women (called Venus figurines), parietal art, and use of shelters, and dated to 28–21 ka. Similar technocomplexes found at sites outside of this area of France are usually called **Gravettian**.

perihelion (Gk *peri*=around and *helios*=sun) The point on the Earth's axis when it is closest to the sun. It is around this time that insolation (i.e., the intensity of the sun's rays on the Earth's surface) is at its maximum.

perikymata (Gk *peri*=around and *kymata*= waves) External ridges that encircle the crown of all permanent teeth; they are especially pronounced on the labial surface of the crowns of the anterior teeth. Perikymata were named by Gustav Preiswerk (1895) in his thesis on ungulate enamel structure. The "troughs" between perikymata are coincident with **striae of Retzius** (reviewed in Risnes 1985). Perikymata counts can be used to estimate the time it takes to form the **imbricational enamel** (or lateral enamel), but it is also necessary to know the **periodicity** of the perikymata. *See also* **enamel development; striae of Retzius**.

perilymph (Gk *peri*=around and *lymph*= watery) Watery fluid between the **membranous labyrinth** and the walls of the **bony labyrinth**. It is rich in sodium, resembles **cerebrospinal fluid** (or CSF), and communicates with the CSF via a narrow canal called the cochlear canaliculus.

period A unit of geological time (i.e., a **geochronologic unit**). It is a subdivision of an era, and it is itself subdivided into **epochs**. For example, the **Neogene** period is within the **Cenozoic** era and is itself divided into the **Miocene** and **Pliocene** epochs. The corresponding **chronostratigraphic unit** equivalent of an epoch is a **system**.

periodicity (Gk *periodos*=interval of time) The number of daily **cross-striations** between **striae of Retzius** in **enamel**, and/or the number of

von **Ebner's lines** between **Andresen lines** in dentine. The periodicity of these **long-period incremental lines** is consistent within all teeth belonging to the dentition of an individual (FitzGerald 1996, Smith 2006), but it may vary among individuals of a taxon. The modal value for the periodicity of long-period incremental lines in fossil hominins is 7 days in australopiths and 8 days in *Homo* (Lacruz et al. 2008). The presence of incremental lines in enamel allows researchers to determine the rate and duration of crown formation, and in some cases it is possible to infer the age at death of individual specimens (reviewed in Aiello and Dean 1990, Dean 2000, 2006, Smith et al. 2006). *See* **Andresen lines; cross-striations; striae of Retzius; von Ebner's lines**.

periodontal ligament (or PDL) (Gk *peri*= around *odonto*=pertaining to dentine) A fibrous connective tissue ligament that exists in the 300–500 μm-wide space between the tooth root surface and the inner aspect of the alveolar bone socket. The PDL is rich in blood vessels and proprioceptive nerve endings that relay information about **bite force** to the brain and trigger protective reflexes. The PDL contains bundles of collagen fibers that run between the alveolar bone and the cementum layer covering the tooth root. These fibers suspend the tooth in the socket and cushion the impact of the bite force.

periosteum (Gk *peri*=around and *osteum*= bone) A tough fibrous and cellular covering that tightly adheres to the external surface of bones. It is an important source of blood supply to bone and its undifferentiated (i.e., unspecialized) cells are capable of converting to **osteoblasts**, which can then form new, initially lamellar, bone. On the external surface of the bones of the **cranial vault** the periosteum is called the **pericranium** and on the internal surface of the vault it is called the **endosteum**, which is equivalent to the outer endosteal layer of the **dura mater**. When a bone fractures the cells in the periosteum differentiate into the osteoblasts, which are responsible for the bone formation that eventually heals the fracture. *See also* **ossification**.

peripatric speciation A mode of **speciation** in which a new species evolves as a consequence of a small subpopulation of the original species population colonizing a new geographic area (e.g., animals trapped on a large mat of vegetation that separates from a much larger island). These periph-eral populations are geographically isolated from the parent population, and because their population size is small, **genetic drift** becomes a powerful factor, driving the new populations to become genetically distinct and, eventually, reproductively isolated from the parent population. This process is also known as the **founder effect**. Peripatric speciation can be thought of as a special case of **allopatric speciation** insofar as reproductive isolation (and, hence, a new species) evolves in geographically isolated populations. *See also **Homo floresiensis**.*

Perissodactyla (Gk *perisso*=odd-numbered and *daktulos*=toe) The name of the mammalian order in which the axis of the leg goes through a single-hoofed toe (cf. **Artiodactyla**, in which the axis passes between *two* hoofed toes). Perissodactyls were common and underwent an adaptive radiation earlier in the **Cenozoic**, particularly in North America, but during the **Neogene** times that are relevant for human evolution this order was relatively impoverished, with low species diversity. Nonetheless, perissodactyl fossils are common within some early hominin faunal communities. The perissodactyl taxa found at hominin sites include members of the **Rhinocerotidae** (rhinoceros) and the **Equidae** (zebras, horses, and their allies). Examples of another extant perissodactyl family, the **Tapiridae** (or tapirs), have been recovered from some archeological sites in Asia, particularly in Malaysia.

permanent dentition (L. *per*=throughout, *manere*=to remain, and *dentes*=teeth) In modern humans and other catarrhine primates, the perma-nent dentition consists of the two **incisors** (I), one **canine** (C), two **premolars** (P), and three **molars** (M) in each quadrant of the jaw, which either replace the **deciduous dentition** [i.e., the permanent incisors (I) replace the deciduous incisors (i), the permanent canines (C) replace the deciduous canines (c), and the premolars (P) replace the deciduous molars (dm)] or are formed in the alveolar process distal to the deciduous dentition (i.e., the permanent molars, M).

permutation test *See* **randomization**.

Perning *See* **Mojokerto**.

person-years lived in an age interval The total number of years (based on the **radix**) lived in an age interval, represented in a **life table** as L_x or

$_nL_X$. For example, consider a life table where there are 250 deaths. In the first age interval (from birth to 1 year of age) there are 10 deaths. Taking the actual number of deaths (250) as the radix, the number of individuals who survive the first year of life was 250–10, or 240. Each of these individuals lived 1 year, accounting for 240 people-years. Of the 10 indivi-duals who died in the first year of life, we assume that their ages at death were uniformly distributed across the age interval. Consequently, their average age at death would be 0.5 years, and they would on average live for half of the age interval. They consequently contributed 10×0.5 years, or 5 people-years. The total people-years lived during the first year of life was therefore 245. $_nL_X$ can be found directly from **survivorship**, such that $_nL_X = n\left(\frac{l_x+l_{x+1}}{2}\right) = \frac{n}{2}l_x + \frac{n}{2}l_{x+1}$, where n is the age interval width. Person-years lived in an age interval is consequently equal to the average of the survivorship at the opening of the interval and at the opening of the next interval multiplied by the width of the interval. The assump-tion of a uniform distribution of deaths within an age interval can be problematic, particularly for the earliest age intervals in a life table. As a conse-quence, in **Weiss' model life tables**, Weiss used $_1L_0 = 0.35l_0 + 0.65l_1$. The concept of "person years lived in an age interval" is not typically used in hazards analysis because there are no defined age intervals.

Peştera Cioclovina Uscata (Location 45°35′N, 23°07′E, Romania; etym. in Romanian literally the "dry Cioclovina cave") History and general description The Cioclovina cave is part of a large **karstic** system, known since the 1880s. Several episodes of paleontological research have been conducted, including excavation that pro-duced **Mousterian** and **Aurignacian lithic** mate-rial. In the early 1940s the site was used also for phosphate mining, and a **hominin calotte** was discovered during the mining operations. Tem-poral span and how dated? Typology of artifacts and **AMS radiocarbon dating** on hominin and faunal remains suggest the hominin activity at the Cioclovina cave was during the Mousterian and during the early **Upper Paleolithic**. Hominins found at site The Cioclovinia calotte. Archeological evidence found at site Evidence of Mousterian and Aurignacian lithic industries. Key references: Geology, dating, and paleoenvironment Olariu et al. 2003, Soficaru et al. 2007; Hominins Rainer and Simionescu 1942, Harvati et al. 2007, Soficaru et al. 2007; Archeology Păunescu 2001.

Peştera cu Muierii   A **karstic** cave system in Romania that comprises several galleries. *See* **Baia de Fier**.

Peştera cu Oase   (Location 45°01′N, 21°50′E, Romania; etym. Romanian for "cave with bones") History and general description This site, discovered in 2002, consists of previously sealed galleries that form part of a karstic complex in southwestern Romania. Paleontological fieldwork, including mapping and excavation of the fossil-bearing sediments, was undertaken in 2003–4 and in 2006. Skeletal remains of two hominin individuals were recovered among bones of *Ursus spelaeus* and other large mammals in the surface accumulation of the side gallery (Panta Strămoşilor); no cultural remains were found. Temporal span and how dated? Hominin presence in the Peştera cu Oase seems to have been ephemeral and was possibly limited to one episode. The hominin remains are dated to between 34,950±*c*.900 years BP (AMS ultrafiltration radiocarbon uncalibrated age for **Oase 1**) and 28,890±*c*.170 years BP (AMS ultrafiltration radiocarbon uncalibrated age for **Oase 2**), with a suggested oldest calibrated age of *c*.40.5 ka. Hominins found at site Oase 1 (mandible) and Oase 2 (cranium); a temporal bone originally assigned to a third individual (Oase 3) is now thought to belong with the Oase 2 cranium. Archeological evidence found at site None. *See also* **Oase 1**; **Oase 2**.

PET   Acronym for positron emission tomography, a neuroimaging modality. *See* **positron emission tomography**.

petalia   (Gk *petalos*=leaf, outspread) A cerebral petalia is a modest relative expansion of part of one **cerebral hemisphere** in relation to its size in the opposite, or contralateral, cerebral hemisphere. Petalias are considered by some to be a manifestation of a more general anatomical asymmetry of the brain. The most typical configuration in modern humans is the combination of a right frontal lobe petalia with a left occipital lobe petalia. Some authors have argued that this petalia torque pattern is unique to modern humans and can be linked to the evolution of language abilities (Crow 2000). Fossil evidence attributed to *Homo erectus*, however, suggests that endocasts of at least some earlier hominin species show similar asymmetrical cerebral petalias (Holloway and De La Coste-Larymondie et al. 1982) and recent comparative studies, using measurements from magnetic resonance imaging scans, suggest that left-occipital and right-frontal petalias are also seen in chimpanzees (Hopkins and Marino 2000, Hopkins et al. 2008). There is no evidence of the left-right petalia torque in any primate species outside of the great apes.

Petralona   (Location 40°22′N, 23°10′E, Chalkidiki peninsula, Northern Greece; etym. Greek for "stone threshing place") History and general description The Petralona Cave was discovered in 1959 by Philippos Chatzarides, a local villager, and the Petralona fossil hominin cranium was discovered the following year. Excavations were conducted in 1968 and 1974–5 by Aris Poulianos. Temporal span and how dated? **Biostratigraphy, electron spin resonance spectroscopy dating, and uranium-series dating** has been used to date the cranium, with age estimates ranging from the lower **Middle** to the late **Middle Pleistocene**. Hominins found at site **Petralona cranium**. Archeological evidence found at site Proposed lithic evidence from the cave are probably not artifacts. Key references: Geology, dating, and paleoenvironment Hennig et al. 1981, Ikeya 1980, Kurtén and Poulianos 1977, Poulianos 1982; Hominins Breitinger 1964, Kokkoros and Kanellis 1960, Rightmire 2001, Stringer et al. 1979; Archeology none relevant. *See also* **Petralona cranium**.

Petralona cranium   Site Petralona. Locality Petralona Cave. Surface/*in situ* Surface. Date of discovery 1960. Finder Christos Sarigiannides. Nearest overlying dated horizon N/A. Nearest underlying dated horizon N/A. Geological age Middle to late Middle Pleistocene. Developmental age Adult. Presumed sex Male. Brief anatomical description Complete cranium, including dentition with the exception of the incisors and canines. Announcement Kokkoros and Kanellis 1960. Initial description Kokkoros and Kanellis 1960. Photographs/line drawings and metrical data N/A. Detailed anatomical description N/A. Initial taxonomic allocation *Homo neanderthalensis*. Taxonomic revisions *Homo erectus* (e.g., Breitinger 1964), archaic *Homo sapiens* (e.g., Stringer 1974), *Homo heidelbergensis* (e.g., Rightmire 2001). Current conventional taxonomic allocation *H. heidelbergensis*. Informal taxonomic category Pre-modern *Homo*. Significance One of the most complete Middle Pleistocene crania from Europe, this specimen has featured in attempts to envisage the **pre-Neanderthal** population of Europe. Location of original Department of Geology, Aristotle University of Thessaloniki, Greece.

petro-median angle (L. *petrous*=rock-like and *medius*=middle) The term used by Phillip **Tobias** (1967a) for the angle between long axis of the petrous part of the temporal bone as seen from below (i.e., in norma basilaris), and the median, or sagittal, axis of the cranium. For example, the petromedian angle is more acute (i.e., smaller) in *Pan*, whereas in later *Homo* and in *Paranthropus boisei* it is closer to the coronal plane and is thus more obtuse (i.e., larger). For more detail see Tobias (1967a, pp. 33–4). *See also* **basicranium**.

petrous (L. *petrous*=rock-like; the same root as "petroleum" which is literally "rock oil") Refers to the hard (hence "rock-like") wedge-shaped part of the temporal bone.

petrous angle (L. *petrous*=rock-like) The angle between the long axis of the petrous part of the temporal bone and the coronal plane. Equivalent to angle α in Dean and Wood (1981) which is defined as the angle formed by a line from the lowest point in the most lateral part of the tympanic plate to the center of the carotid canal and then drawn on from that point to the medial point on the apex of the petrous pyramid.

PGLS Acronym for **phylogenetic generalized least squares** (*which see*).

Phacochoerus The name of the genus that contains the extant and extinct warthogs. *See* **Suidae**.

phage This term (an abbreviated form of bacteriophage) refers to a virus that infects bacteria. The use of phages during the development of molecular biology enabled researchers to understand the role DNA plays in heredity. In 1952, the Hershey–Chase experiments confirmed that DNA was the genetic material. These experiments used a T2 phage labeled with phosphorus-32 to make its DNA radioactive and sulfur-35 to make its protein component radioactive. After the phages infected bacteria, Hershey and Chase found that it was the DNA, and not the protein, from the phage that entered the bacterial cell and infected it.

Phaneranthropi *See* Sergi, Sergio (1878–1972).

pharyngeal arches *See* branchial arches.

pharyngotympanic tube *See* auditory tube.

pharynx (Gk *pharunx*=a chasm) The hollow soft-tissue structure that connects the mouth and nose with the esophagus and the larynx. It is in three parts. The superior part, the nasopharynx, connects anteriorly with the nasal cavities via the posterior nares. The middle part, the oropharynx, communicates anteriorly with the mouth via an opening that is bounded superiorly by the soft palate, laterally by the palatoglossal folds and inferiorly by the junction of the anterior two-thirds and the posterior third of the tongue. The inferior part, the laryngopharynx, lies posterior to the larynx and extends laterally around it as far as the attachments of the inferior constrictor muscle. The pharynx is made of soft tissues so there is no direct evidence of it in the fossil record. However, it is one of the structures that differ substantially in shape in chimpanzees and modern humans. The **basicranium** and the pharynx in the modern human **neonate** are much the same as they are in an adult chimpanzee; this configuration allows modern human neonates to suckle and breathe at the same time. However, once the cranial base flexes the soft palate no longer seals off the mouth from the respiratory tract. Instead, in adult modern humans the soft palate seals off the nose and nasopharynx from the mouth and oropharynx.

phase The term refers to combinations of **alleles** at different **loci** linked together on the same **chromosome**. It also refers to cycles of **chewing**.

phase I *See* chewing.

phase II *See* chewing.

phenetic (Gk *phainen*=to show) A form of analysis that uses information about all of the phenotype. Contrast this with a cladistic analysis that just focuses on shared-derived, or **apomorphic**, aspects of the phenotype. Phenetic methods should be used to identify taxa (also called **alpha taxonomy**), and cladistic methods should be used to investigate the relationships among taxa. *See also* **morphology**; **phenotype**.

phenetic species concept *See* **species**.

phenogenomics Phenogenomics is the use of large shared databases and data standardization to enable large-scale analyses of phenotypic data in combination with genetic data. Examples of this include the development of large-scale databases for imaging data that can be mined using bioinformatic techniques. Within biological anthropology, the move towards sharing **computed tomography** scans of primate skeletal material combined with genetic data are examples of phenogenomic approaches. Although no genetic material is usually available for fossil specimens, the move towards sharing scans of fossil material is motivated by similar principles and ideals.

phenotype (Gk *phainen*=to show and *typus*= image) All the observable characteristics of a living thing, from its molecular structure up to its overall size and shape. The phenotype is determined by complex interactions between the genotype and the environment. *See also* **epigenetic**; **phenotypic plasticity**.

phenotypic plasticity (Gk *phainen*=to show, *typus*=image, and *plaissen*=to mold) The tendency for the same genotype, or genetic program for development, to generate a range of phenotypes (called **reaction norm**s) in response to different environmental settings. Although the study of the relationship between environment and phenotype has a long history (e.g., Wolterek 1909), the concept was developed in its modern form by Schmalhausen (1949). The phenotypic differences may be initiated during development or changes to the phenotype may take place during adulthood. These phenotypic differences and changes can be behavioral as well as morphological. Ultimately, the capacity for plasticity, and the degree to which an organism's phenotype is malleable, is dependent on the fitness benefit of this plasticity. The results of this capacity for flexibility in response to environmental pressures can be described using a reaction norm, but in some cases the direction of plasticity is similar in all organisms. It is often difficult to tease apart whether phenotypic plasticity itself is an adaptation, whether variation in the trait under investigation has been under selective pressure, or whether environmentally related variation in a trait is plastic but not necessarily adaptive. In modern humans, the relationship between age at menarche in girls and the family environment has been taken as an example of phenotypic plasticity: family instability (as a proxy for environmental instability) is thought to lead to earlier age at menarche and at reproduction. The relationship between body condition and reproductive capacity in women, where very low body fat resulting from undernutrition or extreme physical exercise results in irregularities or cessation in menstrual cycling, is another example of phenotypic plasticity.

phenotypic variance *See* **heritability**.

philopatry (Gk *philo*=loving and *patris*= fatherland) The term used when an individual remains within their natal group (i.e., the group they were born into). Primate social groups tend to be structured such that once either males or females reach sexual maturity they move away from the group in which they were born, presumably to minimize inbreeding. In Old World monkeys it is common for the males to move away from their natal group, but females stay in theirs; hence this is known as female philopatry. This social structure contrasts with that for chimpanzees, in which it is the males that tend to remain in their natal group (i.e., male philopatry). Joining another group is not always straightforward for emigrants. In species that have female philopatry and a high level of male–male competition, some incoming males may not be able to successfully challenge the resident males for access to mature females. In these cases, nonresident males may form separate "bachelor" groups. Incoming chimpanzee females can face aggression from resident females, as they represent competition. Hominin social groups may have been more similar to those seen in the closest extant great ape than in monkeys, so using a chimpanzee analogy, early hominins may have exhibited male philopatry, with attendant female–female aggression.

phonation (Gk *phone*=sound) Phonation is the use of the respiratory, or vocal, tract to manufacture sounds. Phonations are normally made when the vocal folds vibrate during exhalation. Phonations range from the type of sound units that make up **spoken language** to unspoken exclamations such as a sustained "aaaagrh" sound and an aperiodic noise like a hiss.

phoneme *See* **spoken language**.

phonolite *See* **lava**.

phonology (Gk *phone*=sound and *ology*=study of) Literally the "sound system" of a language. Phonology is a more inclusive category than phonetics. For in addition to phonetics, which concerns the specific mechanics of sound production, it also encompasses the mechanisms involved in the transmission and the perception of speech sounds. A central feature of spoken languages is the way specific sounds function within a linguistic or communicative system to encode meaning. According to Hauser et al. (2002) the perception and articulation of sounds is a primitive feature of the modern human language faculty, for many animals manipulate speech sounds to communicate various social (e.g., dominance) as well as affective (e.g., fear) states. Pinker and Jackendoff (2005) argue that modern human phonology is qualitatively different from the sound system of other animals, and they suggest the "major characteristics of phonology are specific to language, uniquely human, discretely infinite, and not recursive" (*ibid*, p. 212). In this interpretation, the evolution of modern humans favored two phonological systems, both of which are rule-governed. One system combines meaningless sounds into morphemes and a second system combines morphemes into words and phrases. The first phonological system is present in other animals, such as birds, but it is absent in nonhuman primates. This discontinuity suggests that this more basic sound-combination system evolved independently in modern humans. The second system has no parallel in the animal kingdom yet it is a characteristic of all known modern human languages.

photon starvation *See* **computed tomography; lack-of-signal artifacts.**

PHS (Location 5°24.55′N, 35°54.07′E, Ethiopia) Acronym of Paul's hominid site, where Paul Abell found the fragments that were reassembled to form the **Omo II** cranium. The strata exposed at the site are near the top of Member I of the **Kibish Formation** in the **Omo Rift Zone**, Ethiopia. Note that the site is not at the location given in Butzer (1969). A map made by Paul Abell at the time of the discovery and a contemporary photograph by Karl Butzer place PHS locality 3.3 km/1.8 miles west by north of KHS (McDougall et al. 2005). *See also* **Omo-Kibish.**

phyletic gradualism The phenomenon in which an ancestral species is gradually transformed through many incremental stages into a new descendant species. This mode of evolution, which is typically associated with **anagenesis** (but can also be associated with **cladogenesis**) is the one envisioned by Charles **Darwin** to have been the end result of natural selection. There are a growing number of paleoanthropologists who believe that *Australopithecus anamensis* was transformed into *Australopithecus afarensis* through this process, and many researchers believe that *Homo neanderthalensis* evolved from earlier, even more archaic pre-modern *Homo* populations using the same mode. A taxonomic implication of phyletic gradualism is that the boundaries of species are likely to be difficult to locate and define, and because of this its adherents are likely to recognize a smaller rather than a larger number of species in the fossil record. *See also* **anagenesis; lumper.**

phylogenetic (Gk *phylon*=race, *genesis*=origin, and *etikos*=from) Adjective formed from the noun **phylogeny.** For example, a **phylogenetic tree** is a branching diagram that depicts a hypothesis about the shape of part of the tree of life. A phylogenetic analysis (also known as **cladistic analysis**) is a form of analysis that is designed to recover information about relationships that are the product of phylogeny. Confusingly phylogenetic analysis results in a branching diagram called a **cladogram** and not a phylogeny or a phylogenetic tree. A phylogenetic tree is a more complicated hypothesis than a cladogram, for it includes time and specifies ancestors and descendants whereas a cladogram does not.

phylogenetic analysis *See* cladistic analysis.

Phylogenetic Analysis Using Parsimony *See* **PAUP.**

phylogenetic comparative analysis A type of analysis which recognizes that species data are not statistically independent, and which takes into account phylogenetic information when identifying relationships across multiple taxa. The two most commonly used types of phylogenetic comparative analysis are **phylogenetically independent contrasts** and **phylogenetic generalized least squares** (*which see*).

phylogenetic constraint A cause of **phylogenetic inertia**, which is the persistence of a trait despite changes in aspects of the environment that can be expected to be selectively important for that trait. The most commonly invoked phylogenetic constraints are limited genetic variation, **pleiotropy**, and functional interdependency. *See also* **phylogenetic inertia**.

phylogenetic generalized least squares (or PGLS) A type of **phylogenetic comparative analysis** devised by Emilia Martins and Thomas Hansen (1997) that takes into account phylogenetic relatedness when comparing continuous variables among three or more taxa. This technique differs from the earlier-developed technique of **phylogenetically independent contrasts** in that rather than using the branching sequence and **branch lengths** within a phylogeny to calculate standardized differences between continuous variables for various taxa, PGLS uses the same phylogenetic information to adjust the error matrix that is used in a generalized least squares model. For example, regression typically assumes that the error term has equal variance for all data points, and that the covariance between the data points is zero (i.e., data points are independent). Phylogenetic generalized least squares adjusts the error term using a covariance matrix in which more closely related taxa have relatively high covariance, and more distantly related taxa have relatively low covariance. Bivariate regression results are identical for phylogenetically independent contrasts and PGLS under the standard regression model which assumes a Brownian motion model for the evolution of continuous values in which values are expected to diverge randomly from an original state. However, PGLS can be further modified to reflect other models of character evolution. A common variant imposes an Ornstein–Uhlenbeck process, which assumes that continuous values tend to revert to a mean value over time. This process has been likened to stretching a rubber band; the more the rubber band is stretched, the stronger the force attempting to pull the rubber band back to its original state. As with phylogenetically independent contrasts, concerns arise when PGLS slopes differ from traditional regression slopes. *See also* **phylogenetically independent contrasts**.

phylogenetic inertia (Gk *phylon*=race, *genesis*= origin, and L. *inert*=idleness) The persistence of a trait despite changes in aspects of the environment that can be expected to be selectively important for that trait (syn. phylogenetic effect, phylogenetic lag). Thus, phylogenetic inertia applies to two situations. One is where a trait persists after the selective force that produced and/or maintained it stops operating. The other is where a trait is unaffected by new environmental conditions that there is reason to think should have resulted in selection acting on the trait. A number of potential causes of phylogenetic inertia have been proposed. The most commonly invoked are limited genetic variation, **pleiotropy**, and functional interdependency. These are often referred to as **phylogenetic constraints**. Behavioral plasticity is among the other potential causes of phylogenetic inertia. In paleoanthropology, phylogenetic inertia has played an important role in the debate about the locomotor behavior of the australopiths. A number of those who contend that australopiths were striding bipeds have invoked phylogenetic inertia to explain the australopiths' primitive, chimpanzee-like, traits (e.g., their curved fingers and toes).

phylogenetic lag A synonym for **phylogenetic inertia** (*which see*).

phylogenetic species concept (or PySC) Introduced by Cracraft (1983). Nixon and Wheeler (1990) describe the PySC as "the smallest aggregation of populations diagnosable by a unique combination of character states." *See also* **species**.

phylogenetic systematics An integrated approach to phylogeny reconstruction and classification devised by Willi Hennig (1966). The phylogeny reconstruction aspect of phylogenetic systematics is commonly called **cladistic analysis**, or just cladistics, and is discussed in the entry for the former. This entry focuses on Hennig's approach to classification. According to Hennig, classifications should encode only information about phylogeny such that taxonomic groups should include all of the descendants of the most recent common ancestor (or MRCA) of that group. In other words, Hennig argued that taxonomic groups should always be **monophyletic**. This emphasis on monophyly distinguishes phylogenetic systematics from the other major approach to classification, **evolutionary taxonomy**, which allows taxa to be paraphyletic if members of a clade exhibit significant adaptive differences.

phylogenetic tree A phylogenetic tree is a branching diagram that tries to capture as accurately as possible the evolutionary history of a taxon. The evolutionary history of a taxon is the path taken as one traces its ancestors (initially recent, but subsequently increasingly remote) back into the tree of life. It is a more complex hypothesis than a cladogram because it specifies ancestors and descendants. Several different phylogenetic trees may be compatible with a single cladogram. In a phylogenetic tree, time is on the vertical axis and morphology on the horizontal axis.

phylogenetically independent contrasts (or PIC) A type of **phylogenetic comparative analysis**. Devised by Joseph Felsenstein (1985), PIC takes into account phylogenetic relatedness when comparing continuous variables among three or more taxa. Most statistical analyses assume statistical independence of data points, but Felsenstein recognized that in phylogenetic comparative analyses species are *not* statistically independent data points because some taxa share a more recent common ancestry than others, and thus would be expected to be more likely to share similar values for any variable because they have been phylogenetically independent for a shorter period of time (i.e., less time has elapsed since they were a single species). Phylogenetically independent contrasts works by converting species data into differences between pairs of species (contrasts between two operational taxonomic units), between a species and an internal node within the phylogeny (contrasts between an operational taxonomic unit and a hypothetical taxonomic unit), or between pairs of internal nodes (contrasts between two hypothetical taxonomic units). These contrasts are then standardized by a function of the expected covariance between taxa, which is determined by the branching sequence and the branch lengths implied by the phylogeny of the taxa in an analysis. If branch lengths represent time of divergence between taxa, then standardized contrasts are similar to rates of accumulated difference between taxa; these "rates" are independent of each other although the raw data are not. Thus bivariate PIC analyses test whether difference in one variable tends to be associated with difference in another variable throughout the whole phylogeny represented by the sample. This approach differs from analyses of species data that investigate the relationship between species values themselves, rather than values for differences between sister taxa.

Nunn and Barton (2001) provide an accessible discussion of PIC along with worked examples and some variants on the main technique. One early concern regarding PIC regressions was that because there are only $n-1$ contrasts for a data set of n taxa, PIC analyses would have less power than traditional regressions. However, PIC regressions have the same number of degrees of freedom as traditional regressions because PIC regressions must be forced to pass through the origin (i.e., have a Y intercept of zero), and thus only one degree of freedom is lost in PIC regression analyses instead of two. It should be noted, however, that some researchers have suggested that degrees of freedom should be adjusted downward in PIC analyses when soft polytomies are present in the phylogeny (Garland and Díaz-Uriarte 1999). A further criticism leveled against PIC is that the results are not directly comparable to traditional regressions. However, it has been demonstrated that PIC regression slopes can be placed in raw data space (i.e., superimposed on a bivariate plot of operational taxonomic unit data rather than contrast data) by passing the PIC slope through the phylogenetically weighted mean of the X and Y variables (Rohlf 2001). Traditional regression is thus shown to be a special case of PIC in which all taxa inhabit a star phylogeny (i.e., a single clade with a massive polytomy at its base) where all operational taxonomic units are equally closely related to each other. It has been argued that since comparative data sets in most analyses do not conform to star phylogenies, it is incorrect to use traditional regression when phylogenetic information is available. A further issue arises when PIC slopes are compared to traditional regression slopes. PIC slopes are often (although not always) shallower than traditional slopes, and it is not unusual for hypothesis tests for a single data set to produce a significant result using traditional regression and a non-significant result using PIC. Questions of interpretation become more difficult in this scenario. A strictly statistical interpretation would observe that the traditional regression model is flawed because the error variable is incorrectly partitioned; once the error is correctly partitioned using phylogenetic information the significance of the relationship disappears, so the relationship is not significant. A more applied biological interpretation would observe that the traditional regression indicates a real relationship among the operational taxonomic unit data, and that the lack of significance for the PIC regression would suggest that most of that relationship is due to correlated differences in the two variables between clades, with less patterned covariance between variables within adaptive radiations, implying that the relationship between the operational taxonomic units is driven primarily by differences deep in the phylogeny. A third interpretation may be that the phylogeny is incorrect. This is an important consideration, particularly when incomplete fossil taxa are included in an analysis. There is some work suggesting that phylogenetic comparative analyses are robust to some error in branch lengths, and somewhat less so to errors in branching sequence, although it is unclear how much error in the phylogeny is "acceptable" (Díaz-Uriarte and Garland 1998, Symonds 2002). Finally, even among practitioners of PIC, there is some debate as to how the technique can be extended. For example, Desdevises et al. (2003) present a method that combines traditional and PIC regression to partition the relationship between two variables into a phylogenetic component, an error component, and the component of interest (usually a functional or ecological relationship). Thus researchers can potentially identify how much of a relationship is functional, how much is phylogenetic, and how much is due to phylogenetic niche conservatism. However, Rohlf (2006) has argued that accounting for phylogeny simply allows for a more accurate estimation of the error component, implying that it is not meaningful to discuss a "phylogenetic component" of the regression. The proper name for the method is phylogenetically independent contrasts, but it is often referred to in its shortened form, "independent contrasts." *See also* **phylogenetic generalized least squares**.

phylogeny (Gk *phulon* = tribe or race and *gena* = to give birth to) The phylogeny of a taxon is the same as its evolutionary history. The evolutionary history of a taxon is the path taken as one traces ancestors (initially recent, but subsequently more remote) back into the tree of life. Evolutionary history is usually represented visually as a branching diagram called a **phylogenetic tree**, also called simply a phylogeny. A phylogeny is a more complex hypothesis than the hypothesis about relationships set out in a **cladogram**; the former includes specific hypotheses about ancestors and descendants, whereas the latter does not. Thus, a single cladogram may be consistent with several different phylogenetic trees. Strictly speaking, only taxa have phylogenies; individual organisms have

an **ontogeny**, but not a phylogeny (syn. genealogy, ancestry).

phylogeny reconstruction The generation of an hypothesis or set of hypotheses concerning the phylogenetic or descent relationships among a group of taxa. Phylogeny reconstruction can be carried out in a number of different ways. In paleoanthropology the main methods are **phenetic** and **cladistic analysis**. See these entries for these methods for further information about how phylogenetic hypotheses are generated with them. *See also* **cladistic analysis**.

phylogram *See* **branch length**.

physical map A map of the whole genome, or a region of the genome, where the relative positions of markers are defined by the number of base pairs between them rather than by the recombinational distance or **linkage**.

phytolith (Gk *phyton* = plant and *lithos* = rock) One of several kinds of plant microfossils, phytoliths are small noncrystalline silica bodies found within the tissues of plants. They function as a form of structural support and/or as a physical defense mechanism against herbivory. They are thought to be one of the major causes of tooth wear. The shapes and sizes of phytoliths are usually unique to the plant taxon (e.g., at the level of the family, tribe, or genus and sometimes even at level of the species) and/or plant organ (e.g., leaf, bark, or fruit) that produced them. They are most commonly produced in the surface organs of plants (e.g., leaves, husks, and rinds of fruits). Phytoliths (unlike pollen) usually remain where the plant died and they are stable in nonalkaline environments, so can be useful markers of ancient plants. They can be recovered from archeological sediments, stone tools, ceramics, and dental calculus, and can provide information on which plants were present in the environment or in the diet of individuals and groups. Piperno (2006) provides a useful primer for those planning to exploit plant microfossils for diet and/or paleoenvironmental reconstruction and Bamford et al. (2006) provide an example of how these methods can be successfully applied. *See also* **use wear**.

PIC Acronym for **phylogenetically independent contrasts** (*which see*).

pig Informal inclusive term to describe the taxa within the family **Suidae** (syn. suid, swine). *See* **Suidae**.

pigmentation Colored material in plant or animal cells that is not due to structural color (such as iridescence). Examples of pigmentation variation in modern humans and other primates include the color of the skin, hair, and eyes. In the skin, color is the result of three features: melanin, blood, and keratin. Over 100 genes have been identified as being involved in pigmentation in mammals (e.g., Jackson 1997, Rees 2003, Sturm et al. 2001). Many of these are not involved in normal pigment variation within a species, but were discovered because they are mutated in a genetic disease or condition (e.g., albinism). It was initially estimated that approximately three genes had major effects on normal skin pigmentation variation in modern humans (Harrison and Owen 1964), but more recent analyses have identified close to 20 genes involved in normal skin, hair, and eye color variation in different populations (for a review see Sturm 2009). Among these are the melanocortin 1 receptor gene (or *MC1R*) and *OCA1*, and variants of *MC1R* have been linked to red/fair hair, fair skin, and freckles (e.g., Flanagan et al. 2000, Valverde et al. 1995) while variants of *OCA2* have been linked to lighter iris color in the eyes (Duffy et al. 2007, Frudakis et al. 2003, Sulem et al. 2007). Rogers et al. (2004) studied variation in alleles of *MC1R*, whose protein product affects skin and hair color by modifying production of one of two forms of the pigment melanin, either eumelanin or pheomelanin. Eumelanin is protective against ultraviolet radiation while pheomelanin is not. Earlier research by Harding et al. (2000) had found that the melanocortin gene is variable in non-African population while in Africa all the individuals studied possessed an allele specifying the production of eumelanin. Rogers et al. used silent mutations in an African version of the gene to calculate that a **selective sweep** for that allele occurred approximately 1.2 Ma. Using *Pan troglodytes* as a **proxy** for the earliest human ancestors, Rogers et al. suggest that hominins, like chimpanzees, originally possessed fair skin covered by protective fur. The 1.2 Ma sweep date for *MC1R* has been interpreted by Rogers' team as suggesting that modern human ancestors evolved dark skin around the time that they lost their body hair to reduce damage from solar radiation. Several hypotheses (not necessarily mutually exclusive) exist to explain normal human pigmentation variation.

These include **sexual selection, purifying selection** because of ultraviolet intensity and folate photolysis, selection for sufficient vitamin D production, and **genetic drift** (Aoki 2002, Harding et al. 2000, Jablonski and Chaplin 2000, Madrigal and Kelly 2007, Rana et al. 1999).

Pillai-Bartlett trace *See* **multivariate analysis of variance**.

Piltdown (etym. The name of the closest village to Barkham Manor) The saga of the discovery of the remains of "Piltdown Man," their validation by the British Museum (Natural History) in London, and their later exposure as an elaborate fraud, began before 1908. What follows is extracted from Joe Weiner's fine description of his first-hand work on the Piltdown material (Weiner 1955) and from information painstakingly assembled by Frank Spencer and Ian Langham and presented in book form (Spencer 1990). Charles Dawson, a solicitor (the equivalent of a US attorney) who practiced in Uckfield, Sussex, apparently was aware of the Barkham gravel beds and their exposure in a trench alongside a lane leading to part of Barkham Manor, but it was in 1908 that a laborer reported striking and shattering a hollow object in what later became known as the Piltdown Gravel Pit. A fragment belonging to the "hollow object" was shown to Dawson, who reported that it was "an unusually thick human parietal bone" and he also remarked that "peculiar brown flints" had been recovered from the same gravel pit. Dawson paid periodic visits to the site (he was the steward, or the agent, of Barkham Manor) and sometime in the fall of 1911 he wrote that he had "lighted on a larger piece of the same skull which included a portion of the left supra-orbital border" and not long after he reported finding a piece of a hippopotamus tooth (Spencer 1990, p. 31). Dawson had been an Honorary Collector for the British Museum (Natural History) since 1884, so it was not surprising that he contacted Arthur Smith Woodward at the museum about his finds. Dawson wrote to Smith Woodward in February 1912, but he could not visit the site until June of the same year. On June 2, 1912, Smith Woodward, accompanied by Fr. Pierre Teilhard de Chardin, traveled to Uckfield where they met Dawson who took them to Piltdown, and it was during that visit that Dawson recovered a third cranial fragment. Smith Woodward and Dawson visited the site several times that summer, and near the end of June Dawson (in Smith Woodward's presence) exposed part of the right side of a mandible, and thereafter they recovered "three pieces of the parietal bone" (*ibid*, p. 33) from a spoil heap, and a further piece of cranium that fitted the broken edge of the occipital fragment found by Dawson on June 2. Someone "leaked" the news of the discoveries and the *Manchester Guardian* ran the story of November 21, 1912, but the first formal presentation of the finds (i.e., the "Pre-Chellean paleoliths," the **eolith**s, fossil mammal bones, and "nine pieces" of the cranium, a mandible with M_1 and M_2, plus a reconstruction of the Piltdown I skull by Smith Woodward) was made by Dawson, Smith Woodward, and Grafton Elliot Smith at the December 18 Meeting of the Geological Society of London held in Burlington House; the December 28 issue of the *Illustrated London News* included a long article about the finds. The specimen was named *Eoanthropus dawsoni* (literally "Dawson's dawn man") and it immediately attracted considerable attention. While many accepted Smith Woodard's interpretation that the cranium and jaw belonged to a single individual, and that while "the skull is essentially human"…"the mandible appears to be that of an ape, with nothing human except the molar teeth" (Smith Woodward quoted in Weiner 1955, p. 3), other observers were more skeptical. For example, David Waterston, Professor of Anatomy at King's College, London, thought the cranium, which he considered was "in all essentials human," and the mandible, which was "chimpanzee-like," came from different individuals. In July 1913, Dawson recovered a pair of nasal bones and on August 30, 1913, Fr. Teilhard de Chardin recovered a canine tooth (the two nasal bones and the canine are illustrated in Dawson and Woodward 1914) of Piltdown I. No further evidence of the skull was recovered from the trench at Barkham, but in January 1915, Dawson, who had widened the search to neighboring fields, reported the recovery of a piece of hominin frontal bone at Sheffield Park, and later that year in the same location he recovered a piece of occipital and a molar tooth of what would become Piltdown II; he also found a rhinoceros molar. Dawson became sick towards the end of 1915, and he died on August 10, 1916. Thereafter Smith Woodward visited the sites of Piltdown I and II many times but found nothing, and he even retired to Hayward's Heath to be close to the site, but despite continued searching no more evidence was ever found; no more hominin fossils, no more "worked" flints, and no more bones of

extinct animals. Smith Woodward arranged for a memorial stone to be erected at the Piltdown I site and it was unveiled by Sir Arthur Keith on July 22, 1938. In the foreword of Smith Woodward's 1948 book about Piltdown, *The Earliest Englishman*, Sir Arthur Keith wrote that "the Piltdown enigma is still far from a final solution." Meantime, discoveries in China and southern Africa were forcing researchers to concede that if Piltdown was to be accepted as evidence of a Pliocene hominin, then all the other well-authenticated hominin evidence must belong to one lineage, with the Piltdown evidence for "dawn man" being the only evidence for a second lineage. Two important developments in 1950 helped resolve "the Piltdown enigma." In that year vegetation around the Piltdown I site was cleared by the Nature Conservancy, and despite much excavation and sieving, nothing of any interest was recovered (Toombs 1952). The second development involved the **fluorine dating** method. This method was little used in the early decades of the 20thC, but it attracted the attention of Kenneth **Oakley** who spoke about its potential at the First **Pan-African Congress on Prehistory** in Nairobi in 1947. When Oakley applied the fluorine method to the Piltdown fossils the large-mammal fossils were found to have fluorine levels that varied between 3 and 1.6%, whereas the levels in the alleged Piltdown I hominin were only about 0.2% (Oakley and Hoskins 1950). These levels were similar to those found in contemporary bones, and the author suggested that the result "requires some explanation" (*ibid*, p. 381). These results were undoubtedly an important element in spurring Joseph Weiner and Wilfrid **Le Gros Clark** to collaborate with Kenneth Oakley to re-examine the alleged fossil hominin evidence from Piltdown. The morphological analyses showed that the nature of the flat occlusal surfaces of the molar teeth in the jaw was inconsistent with the flattening having been caused by natural wear; the edges of the cusps were sharp and not rounded, and the wear was greater on the lingual, or tongue, side of the teeth, rather than on the buccal side as is the case for naturally worn teeth. There was also evidence the teeth had been artificially abraded, for while the teeth on the mandible had been filed on the occlusal surface only, the Piltdown I canine and the Piltdown II molar had also been filed down on the sides. In addition, the dentine was flush with the enamel, and not below the level of the enamel as one would expect in naturally worn teeth. Radiography showed that the roots of the molars in the jaw were much more ape-like than Underwood had contended, and there was no secondary dentine in the canine. Furthermore, the sand grains in the pulp space were not consolidated, but were loose, consistent with being recently introduced into the cavity. Weiner et al. were convinced the morphology of the mandible was consistent with that seen in modern orangutans, and when they broke a modern orangutan jaw in the same place as the Piltdown I jaw the resemblances between the two were very obvious. As part of the same detailed examination Oakley applied a more sensitive form of fluorine analysis to the Piltdown remains. The results of these analyses showed that the fluorine levels in the cranial bones and the mandible were 0.1 and 0.3%, respectively (Weiner et al. 1953). These fluorine levels implied two things. First, the cranial bones and the mandible were not the same age, and second, the mandible was almost certainly modern. This was consistent with Oakley's observation that when he drilled the teeth to extract dentine for analysis he noticed a smell of burning; this was evidence that there was still a substantial amount of organic material in the dentine. In addition, the three scientists showed that while the brown color of the eoliths was natural, the "paleolith" flint found by Teilhard de Chardin was "deliberately stained" with chromium, an element that does not occur naturally in the Piltdown gravels, and when they deliberately chipped the flint it was white beneath the surface. Most likely all of the four or five "paleolith" flints found at the Piltdown I site were Neolithic flints stained to look older. Between 1911 and 1914 18 fossil mammal bones and teeth had been recovered from the Piltdown I site; four were found *in situ*, two on the surface of an adjoining field, and the rest on spoil heaps; *none* were ever found anywhere nearby thereafter. Apart from one very dubious exception, no other evidence of such animals had ever been found in southern England outside of East Anglia; thus an isolated pocket of Villafranchian fossils at Piltdown is literally incredible. Other researchers showed that the level of radioactivity in the stegodon (*Elephas* cf. *planifrons*) teeth (and it is probable that the four tooth fragments came from two molars) could *only* be matched at a site in Tunisia, and it and a *Hippopotamus* tooth (that could only have come from Malta) had been stained with chromate. The rest of the mammal bones were almost certainly from "the Red Crag of East Anglia" (Weiner and Stringer 2003, p. 66), and as might be surmised from the name of the site, these fossils did not need staining. Finally, Kenneth Oakley showed

that the "club-like bone implement" found at the Piltdown I site was a piece of silicized bone that had been worked with a metal knife. Piltdown II was also highly suspect: the fluorine analyses showed that the tooth does not belong to the cranium, the frontal and occipital were stained with both iron and chromate, the mammal fossils came from Red Crag, and the frontal fragment proved to be part of the same cranium that provided the pieces of Piltdown I. All this was more than sufficient evidence for Weiner, Oakley, and Le Gros Clark to say with confidence that the Piltdown hominin fossils were modern and not ancient, and that their association with the Piltdown site was the result of an elaborate hoax (Weiner et al. 1953). Joe Weiner's excellent book about the Piltdown fraud was reprinted with a new introduction and afterword contributed by Chris **Stringer** (Weiner and Stringer 2003). In the afterword Stringer reviews the "runners and riders" in the Piltdown fraudster stakes. In addition to the person who Weiner considered the chief suspect, Charles Dawson, the other potential fraudsters can be divided into those who have been proposed as his potential accomplices, those who have been proposed as the sole perpetrator of the fraud, and finally people who might have combined their energies to perpetrate the fraud. The accomplice category is small, and includes just Lewis Abbott (a local jeweler and amateur paleontologist). Those considered as potential sole perpetrators of the fraud include Frank Barlow [a preparator at the British Museum (Natural History)], William Butterfield (the Curator of the Hastings Museum), Sir Arthur Conan Doyle, Venus Hargreaves (the labourer who assisted Dawson), Martin Hinton (a junior colleague of Sir Arthur Smith Woodward's), Sir Arthur Keith, Sir Grafton Elliot Smith, William Johnson Sollas (professor of geology at Oxford), Fr. Teilhard de Chardin, and Sir Arthur Smith Woodward. The only two pairs of people proposed as co-conspirators are Martin Hinton and Fr. Teilhard de Chardin, and Samuel Woodhead (a public analyst) and John Hewitt (a professor of chemistry). References for these accusations are given in Weiner and Stringer (2003). Notwithstanding the suggestions given above, there is both circumstantial and direct evidence in favor of Dawson's involvement. The circumstantial evidence includes several examples of Dawson's dishonesty in connection with other antiquarian curiosities (Russell 2003), and Weiner sets out the direct evidence in his account of the fraud (Weiner 1955). For example, Weiner recounts that as part of his investigation he had been advised to contact A.P. Pollard, the Assistant Surveyor of the Sussex County Council, to ascertain his views about the Piltdown controversy. Weiner reports that Pollard was forthright and said that "I believe it (Piltdown) is a fraud. At least, that is what my old friend Harry Morris used to say" (*ibid*, p. 154). This refers to an assessment of one of the Piltdown flints by a noted flint expert, Harry Morris, who was an acquaintance of Pollard's. Weiner managed to track down Morris' "flint cabinet" and in it was a stained flint (resembling the three "Pre-Chellean paleolith" flints from Piltdown) with two notes. On one was written "Stained by C. Dawson with intent to defraud (all) – H.M." and on the other "Stained with permanganate of potash and exchanged by D. for my most valuable specimen." Furthermore, when dilute hydrochloric acid was applied to the three Piltdown "paleoliths," plus the flint Morris had received in exchange from Dawson, the dark stain on the surface quickly dissolved to expose a white surface. Finally, Stringer is of the opinion that the Piltdown II molar comes from the same individual as the mandible of Piltdown I. If this is the case then this is further evidence that Dawson is linked with *both* Piltdown II and Piltdown I. However, numerous other people besides Dawson were, and continue to be, suspected as the perpetrator(s) and the authorship of the Piltdown fraud has still not definitively been resolved.

Piltdown I   *See* **Piltdown**.

Piltdown II   *See* **Piltdown**.

pink breccia   *See* **Swartkrans**.

Pinnacle Point   (Location 34°12′S, 22°05′E, South Africa) History and general description This series of at least 15 caves and rockshelters overlooking the Indian Ocean is exposed in quartzite cliffs near Mossel Bay, South Africa, has been under intensive investigation by Curtis Marean and his colleagues since 2000. Site PP13B, a 30 m×8 m cave, is the most extensively studied and excavated and it is one of the few **Middle Stone Age** (or MSA) archeological sites to sample **Oxygen Isotope Stage** (OIS) 6. Unlike most coastal caves the sediments in PP13B were not scoured by the rising sea levels of the last interglacial. Temporal span and how dated? Published dates for the OIS 6 sediments (the Lightly Consolidated or LC-MSA deposit) have bounding

single-grain optically stimulated **luminescence dating** age estimates of 164±12 and 90±6 ka, with overlying **uranium-series dating** estimates on **speleothems** that formed during cave closure ranging from 39 to 92 ka. Hominins found at site Two hominin fragments were found from disturbed sediments in PP13B, but these likely derive from MSA strata. They include a left parietal fragment of a young adult (specimen 4500) and a central incisor (specimen 4501). Archeological evidence found at site Multiple hearths, abundant shellfish remains, **Levallois** and other flakes, cores, and bladelets made predominantly of quartzite, and more than 50 pieces of red ochre, many of them ground or scraped. Faunal remains from the LC-MSA deposit include tortoise and a variety of bovids, particularly eland and fewer grysbok/steenbok and mountain reedbuck. Key references: Geology, dating, and paleoenvironment Marean et al. 2004, 2007, 2010, Jacobs 2010, Karkanas and Goldberg 2010, Rector and Reed 2010; Hominins Marean et al. 2004, Marean 2010; Archeology Marean et al. 2004, 2007, Jerardino and Marean 2010, Schoville 2010, Thompson et al. 2010, Watts 2010.

Pinza-abu Cave   *See* **Ryukyu Islands**.

piriform aperture   (L. *pirium*=pear)   The opening into the nose from the face. The size, shape, and especially the form of the floor of the opening (e.g., smooth or sharp) have been used by paleoanthropologists as taxonomic and functional indicators. *See also* **Homo neanderthalensis**.

piriform cortex   (L. *pyrum*=pear, it refers to the "pear-shape" of the nasal aperture)   Part of the **allocortex** component of the **cerebral cortex** and concerned with the processing of olfaction (syn. paleocortex). *See also* **cerebral cortex**.

Pisces   (L. *piscis*=fish)   A discarded taxonomic term for the vertebrate group colloquially referred to as fish. The group of aquatic vertebrates formerly referred to as the Class Pisces is now thought to be **paraphyletic**, which is why in fish circles it has been abandoned. However, the term is still found in the literature and many faunal lists from fossil sites use it to refer to fish. The proper formal taxonomic umbrella term for the types of fish commonly recovered from the hominin fossil and archeological record should be the superclass **Osteichthyes**, which includes all bony fish.

piscivore (L. *piscis*=fish, plus *vorous*, from *vorare*=to devour) Literally, a fish eater. An organism that habitually feeds upon fish is called a piscivore, whereas the practice of eating fish is called piscivory.

Pithecanthropidae A family introduced by Eugène **Dubois** in 1894 to accommodate the taxon *Pithecanthropus erectus*, which just a year earlier he had introduced as *Anthropopithecus erectus*.

pithecanthropine Informal term for a member of the **Pithecanthropidae** (*which see*).

***Pithecanthropus* II** *See* **Sangiran 2**.

***Pithecanthropus* III** *See* **Sangiran 3**.

***Pithecanthropus* IV** *See* **Sangiran 4**.

***Pithecanthropus* B** *See* **Sangiran 1b**.

***Pithecanthropus* Haeckel, 1868** (Gk *pithekos*= ape and *anthropos*=human being) The genus name *Pithecanthropus* was introduced by Ernst Haeckel in his 1868 *Natürliche Schöpfungsgeschichte* (*The History of Creation*) for *Pithecanthropus alalus*, a taxon that Haeckel predicted would be found to provide a link between *Prohylobates* and modern humans. This, the penultimate and 29th stage in Haeckel's evolutionary scheme, was described in the 1906 English translation of his 1874 *Anthropogenie* (*The Evolution of Man*) as the "speechless primitive men" (*Alali*), which were men as far as their general structure is concerned (especially in the differentiation of the limbs), "but men who lacked one of the chief human characteristics, articulate speech and the higher intelligence that goes with it, and so had a less developed brain" (Haeckel 1906). The first researcher to use the genus *Pithecanthropus* in the conventional way was Eugène **Dubois**, who in 1894 used it to accommodate the taxon just 1 year earlier he had named *Anthropopithecus erectus*. The genus name was used until relatively recently for new species (e.g., *Pithecanthropus rudolfensis*) but Weidenreich (1940) was the first among many to recommend that *Pithecanthropus* (and *Sinanthropus*) should be transferred to *Homo*, so *Pithecanthropus* is now one of the many junior synonyms of *Homo*. *See also Pithecanthropus rudolfensis*.

***Pithecanthropus* mandible B** *See* **Sangiran 1b**.

***Pithecanthropus dubius* von Koenigswald, 1950** (Gk *pithekos*=ape, *anthropos*—human being, and L. *dubious*=doubtful) A taxon introduced by von Koenigswald (1950) to accommodate the mandible **Sangiran 5**. First discovery **Sangiran 5** (1939). Holotype As above. Main site **Sangiran**. *See also Pithecanthropus erectus*; *Homo erectus*.

***Pithecanthropus erectus* (Dubois, 1892), Dubois, 1894** (Gk *pithekos*=ape, *anthropos*=human being, and L. *erectus*=upright) A new combination used by Eugène **Dubois** (Dubois 1894) for the hominin species just a year earlier he had named *Anthropopithecus erectus* to accommodate an upper molar (**Trinil 1**) and a skullcap (**Trinil 2**) from Trinil. Wilfrid **Le Gros Clark** (1940) proposed that the *Sinanthropus* fossil hominins recovered in 1923 and thereafter from Choukoutien (now **Zhoukoudian**) should be transferred to *Pithecanthropus erectus*. Later in the same year Franz **Weidenreich** (1940) transferred the *Pithecanthropus erectus* hypodigm to *Homo*, so *Pithecanthropus erectus* is a junior synonym of *Homo erectus*. Howell (1994, Table 1) provides a useful review of how the new taxon was viewed by researchers in the two decades following its announcement. First discovery Trinil 1 (1891). Holotype Trinil 2 (1891). Main sites **Trinil**, **Sangiran**. *See also Homo erectus*.

***Pithecanthropus modjokertensis* (von Koenigswald, 1945)** (Gk *pithekos*=ape, *anthropos*=human being, and L. *modjokertenensis*=after the name of the village close to where Modjokerto 1 was found) A new combination introduced by Ralph **von Koenigswald** who was no longer willing to accommodate *Homo modjokertensis* within *Homo*.

***Pithecanthropus robustus* Weidenreich, 1945** (Gk *pithekos*=ape and L. *robus*=oak or strength) Taxon introduced by Franz **Weidenreich** (1945) to accomodate the *Pithecanthropus* IV partial cranium (**Sangiran 4**) that was recovered from **Sangiran** in December 1938 and January 1939. von Koenigswald and Weidenreich (1939) attributed the cranium to *Pithecanthropus*, but Weidenreich subsequently argued that differences between the *Pithecanthropus* IV cranium and the *Pithecanthropus* crania I–III, which he accepted as *Pithecanthropus erectus*, justified the new species and that for "this more robust and primitive group I propose the name *Pithecanthropus robustus*"

(Weidenreich 1945, p. 33). <u>First discovery</u> **Sangiran 4** (1938–9). <u>Holotype</u> As above. <u>Main site</u> Sangiran. *See also* ***Pithecanthropus erectus.***

Pithecanthropus rudolfensis Alexeev, 1986

(Gk *pithekos*=ape, *anthropos*=human being, and *rudolfensis*=after the old name, Lake Rudolf, for Lake Turkana) Hominin species established by **Alexeev** in his book *The Origin of the Human Race*. Although Alexeev did not formally designate **KNM-ER 1470** as the **holotype** of the new species, he implied as much. He also complied with the rules if not the recommendations of the **International Code for Zoological Nomenclature** in the sense that the name *rudolfensis* was both **available** and **valid** and he compared the new species with *Pithecanthropus erectus*. Colin Groves (1989) proposed that *Pithecanthropus rudolfensis* be transferred to *Homo* as *Homo rudolfensis*; Wood (1992) provided a more comprehensive differential diagnosis for *H. rudolfensis* and later still it was formally proposed that KNM-ER 1470 be the **lectotype** of *H. rudolfensis* (Wood 1999). If *Homo habilis sensu lato* does subsume more than one species, then *H. rudolfensis* would be the second species. <u>First discovery</u> KNM-ER 819 (1971). <u>Lectotype</u> KNM-ER 1470 (1972). <u>Main sites</u> **Koobi Fora, Uraha.** *See also* ***Homo rudolfensis.***

-pithecus (Gk *pithekos*=ape) Postfix meaning ape, or "ape-like."

pixel *See* **computed tomography.**

plagioconule *See* **metaconule.**

plagiocrista A term introduced by Vanderbroek (1967) for an enamel feature on a maxillary molar tooth crown more commonly referred to as a **distal trigon crest.** *See also* **distal trigon crest.**

plain film radiography Also known as conventional radiography, this is the technology used to take standard radiographs of the chest or fractured bones in a clinical setting. Because the image is made from a single source of X rays, all you see on a conventional radiographic image is an outline of the densest structure between the source and the X ray film. Franz **Weidenreich** was one of the first paleoanthropologists to make intensive use of plain film radiography in his monographs on the **Zhoukoudian Lower Cave** fossils.

plane of interest *See* **computed tomography; confocal microscopy.**

planktonic foraminifera *See* **foraminifera.**

plant microfossils (Gk *micros*=small) Plant microfossils are plant remains that cannot be seen with the naked eye. They consist of **phytoliths, pollen,** and **starch grains.** All three types of microscopic body have been identified at archeological sites, and all three can be used to reconstruct diet and/or the **paleoenvironment.** *See also* **phytoliths; pollen; starch grains.**

plantar aponeurosis *See* **foot function; longitudinal arch; push-off; windlass effect.**

planum nuchale *See* **nuchal plane.**

planum temporale (L. *planus*=flat and *tempus*=time, thus it is literally "the flat part of the surface of the brain that lies beneath the temple"; NB: the part of the hair that tends to go gray first) A feature of the surface of the superior part of the temporal lobe of the **cerebral hemisphere.** It lies immediately posterior to Heschl's gyrus (which corresponds to the location of primary auditory cortex) and anterior to the termination of the **Sylvian fissure.** The planum temporale can be defined only in species that possess a clear Heschl's gyrus (i.e., hominoids). In line with the predominant left-sided language dominance seen in modern humans, the surface area of the planum temporale is greater on the left than on the right side in modern humans. However, the planum temporale also displays left-hemisphere dominance in great ape brains. Because the planum temporale is not evident on the surface of the cerebral hemispheres, and thus its presence and size are not accessible using natural or prepared endocasts, we have no access to evidence about its evolution within the hominin clade.

plasmid (Gk *plassein*=to mold, plus L. *id*=the same) A circular DNA chromosome that is independent of the primary set of chromosomes. Plasmids are typically found in bacterial cells (but they can also be found in Archaea and Eukaryotes) and may carry **genes** for antibiotic resistance or other phenotypic products. In the laboratory, plasmids can be constructed that contain specific genes or DNA fragments of interest and these can be introduced into a cell. In this way, for example, bacteria (or yeast) can be used to replicate DNA fragments, to investigate expression patterns, or to produce a protein product.

plastic *See* phenotypic plasticity; plastic deformation.

plastic deformation (Gk *plastein*=to mold and L. *deformare*=undo form) Bones can be distorted by pressure either before they are fossilized or before fossilization has substantially hardened and stiffened them. These distortions result in permanent plastic deformation. This is difficult to correct, but researchers have recently used computer software to generate information about the "average" shape of other fossils belonging to the same fossil taxon, or the "average" shape of chimpanzee and modern human crania in order to determine the most likely shape of a deformed possible early hominin cranium (Zollikofer et al. 2005).

plasticity *See* phenotypic plasticity.

Plateau de Ruffet *See* Combe-Capelle.

platform The portion of a flake that includes the part of the **striking platform** removed from the core when a flake is detached (syn. butt).

pleiotropy (Gk *pleiōn*=more and *tropos*= towards) Describes the circumstances under which a single gene has been shown to influence multiple phenotypic traits. These pleiotropic properties may be manifested concurrently, or they may be separated temporally, across the lifetime of an individual. For example, high levels of testosterone in young adult modern human males may contribute to reproductive success, but later in life they are correlated with an increased risk of prostate cancer, and Marfan's syndrome, a defect in the *FBN1* gene and abnormal production of the protein fibrillin, is characterized by abnormally long and slender fingers, extremely mobile joints, disproportionately long limbs, as well as by heart and eye defects. Pleiotropy may be one of the reasons why phenotypic characters are found to co-vary for reasons other than shared evolutionary history.

Pleistocene (Gk *pleistos*=most and *kainos*= new) This term, introduced by Charles Lyell in 1839, refers first of the two **epochs** that comprise the **Quaternary** period. Pleistocene refers to a unit of geological time (i.e., a **geochronologic unit**) It spans the interval of time between 2.58 Ma (Gelasian) or 1.8 Ma (Calabrian) and 11.5 ka. In this encyclopedia Pleistocene is used in the sense meant before the 2009 International Union of Geological Sciences decision (*see* **Plio-Pleistocene boundary** for details).

***Plesianthropus* Broom, 1938** (Gk *plesios*= near to and *anthropos*=human being) A genus established by Robert **Broom** (1938, p. 377) to accommodate the early hominin species previously referred to as *Australopithecus transvaalensis*. *Plesianthropus* was formally sunk into *Australopithecus* by Robinson (1954a, p. 196) and it is now almost universally regarded as a junior synonym of *Australopithecus*. *See also* **Australopithecus**; *Australopithecus africanus*.

***Plesianthropus transvaalensis* (Broom, 1936) Broom, 1938** [Gk *plesios*=near to and *anthropos*=human being and *transvaal*=refers to the old name for the province (now called Gauteng) where Sterkfontein is located] A new combination employed by Broom (1938). This hominin species was originally established by Robert **Broom** 1936 as *Australopithecus transvaalensis* to accommodate **TM 1511** and similar fossils recovered from **Sterkfontein**. In 1938 Broom transferred the species to *Plesianthropus*. John **Robinson** (1954a, p. 198) formally proposed that the species *Plesianthropus transvaalensis* be relegated to a subspecies of *Australopithecus africanus* as *Au. africanus transvaalensis* suggesting that among the dental differences between *Au. africanus transvaalensis* and *Au. africanus africanus* were that the M_1 of the former lacked a C6 and that the deciduous lower canine of the latter lacked a mesial cusplet. Most researchers consider *Plesianthropus transvaalensis* to be a junior synonym of *Australopithecus africanus*. First discovery **TM 1511** (1936). Holotype As above. Main site Sterkfontein. *See also* **Australopithecus africanus**; *Australopithecus transvaalensis*.

plesioconulid A term sometimes used for an accessory cuspulid on the occlusal surface of a mandibular postcanine tooth; it is more commonly referred to as a **mesioconulid**. *See also* **mesioconulid**.

plesiomorphic (Gk *plesio*=near and *morph*= form) The primitive condition, or state, of a character used in a phylogenetic or **cladistic analysis**. *See also* **cladistic analysis**.

plesiomorphy (Gk *plesio*=near and *morph*= form) The state of a character in the hypothetical most recent common ancestor of a clade, or the state of a character in an outgroup (syn. **symplesiomorphy**). *See also* **cladistic analysis**.

Plio-Pleistocene (Gk *pleion*=more, *pleistos*= most, and *kainos*=new) The period of geological time since *c*.5 Ma (i.e., that includes the Pliocene and Pleistocene epochs). It is most often used because it covers the interval during which most of hominin evolution occurred and it almost certainly includes the origin and subsequent evolution of *Homo*. Some authors use Plio-Pleistocene because it approximates the period of time during which the northern hemisphere has been glaciated. *See also* **Northern Hemisphere Glaciation; Plio-Pleistocene boundary; Pleistocene; Pliocene**.

Plio-Pleistocene boundary In 2009 the International Union of Geological Sciences (or IUGS) ratified a proposal that the boundary between the **Pliocene** and **Pleistocene epoch**s be lowered from the Calabrian-Gelasian boundary at 1.8 Ma (near the base of the **Olduvai subchron**) to the Gelasian-Piacenzian at 2.58 Ma (near the base of the **Matuyama chron**). More than a half-century of debate has surrounded this issue and it is unlikely that this decree will fully resolve the question. The International Geological Congress of 1948 made an attempt to define the Plio-Pleistocene boundary under the stipulation that it should be based on changes in marine faunas. They determined it to be at the base of the Calabrian **stage** in Italy, later dated to the end of the Olduvai subchron (i.e., 1.8 Ma). In 1983, the International Commission on Stratigraphy agreed with this decision, and used sediments at Vrica in Calabria dated to the top of the Olduvai subchron to identify the beginning of the Pleistocene. However, as interest in global climate change grew in the ensuing decades, the problem of the **Quaternary** resurfaced. The historical association of the Quaternary with a global cooling event had many workers pushing for its official recognition (*see* **Tertiary**) arguing that the present global climate is distinct from that of the **Neogene**, and that the climate changes involved happened from 2.8 to 2.4 Ma (near the base of the Gelasian). To maintain the geochronological hierarchy, any effort to for-

mally ratify the Quaternary as the third **period** of the **Cenozoic** must necessarily be correlated with an effort to also lower the base of the Pleistocene, and this is what happened in 2009. In this encyclopedia Pleistocene is used in the sense meant before the 2009 IUGS decision. The last major published time scale, in which the base of the Pleistocene is defined by the reference section (i.e., the Global Boundary Stratotype Section and Point, or GSSP) of the Calabrian Stage at 1.806 (*c*.1.8) Ma is Gradstein et al. (2004); the revised time scale approved by IUGS, in which the base of the Pleistocene is defined by the GSSP of the Gelasian Stage at 2.588 (*c*.2.6) Ma is Gibbard et al. (2010). Opponents of the 2009 decision cite the lack of faunal turnover at the Calabrian-Gelasian boundary and stress that the latest glacial episode may not be fundamentally different from the rest of the Neogene. The resulting 44% expansion of the Pleistocene is opposed by a large body of researchers who work on late Cenozoic (post-Miocene) subjects, on the grounds that this radical shift was not adequately justified under chronostratigraphic guidelines and that representatives of the affected disciplines were not consulted; a formal move to reverse the action is being prepared for presentation to the IUGS. For more detail on the debate see Gibbard et al. (2010), Aubry et al. (2009), Head et al. (2008) and McGowran et al. (2009).

Pliocene (Gk *pleion*=more and *kainos*=new) The second and the most recent **epoch** of the **Neogene** period. Pliocene refers to a unit of geological time (i.e., it is a **geochronologic unit**). It previously spanned the interval between *c*.5.3 and 1.8 Ma (Calabrian), but in 2009 the International Union of Geological Sciences allocated the Gelasian stage to the **Pleistocene** so until and unless their recommendation is overturned the Pliocene now spans the interval of time between *c*.5.3 and 2.58 Ma (Gelasian). In this encyclopedia Pliocene is used in the sense meant before the 2009 International Union of Geological Sciences decision (*see* **Plio-Pleistocene boundary** for details). The Pliocene is typically divided into Late and Early subdivisions.

ploidy (Gk *ploos*=fold) A multiple of the basic number of **chromosomes** in a cell. Normal modern human somatic cells are diploid (i.e., two sets), while the gametes are haploid (i.e., a single set). In modern humans, any abnormally polyploid cells produced during reproduction are not viable, but polyploidy is important for the production of new plant species.

plunge   In structural geology, plunge refers to the **dip** and dip direction of linear features such as the axial trace of a fold or the orientation of the long axis of sedimentary **clast**s or fossils. It is usually given as the angle from the horizontal followed by the bearing from north (i.e., 34° to 270°). It can be useful to analyse the plunge directions of fossils to test for evidence of any **reworking** of that fossil assemblage by natural agencies such as stream action or trampling. In this type of analysis, nonrandom plunge values (i.e., orientation) can be taken as evidence of reworking of fossils by the action of streams, rivers, or slope wash.

PM   Acronym for the posterior maxillary plane of Enlow. *See* **posterior maxillary plane**.

PM plane   *See* **posterior maxillary plane**.

Podbaba   (Location approximately 59°06′N, 14°23′E, Czech Republic; etym. named after the neighborhood of Prague in which it is found) History and general description A partial calvaria, discovered in 1883 and thought to derive from an **Aurignacian** layer, was destroyed during blasting in 1921. Temporal span and how dated? Undated. Hominins found at site A partial calvaria attributed to *Homo sapiens*. Archeological evidence found at site None. Key references: Geology, dating, and paleoenvironment Churchill and Smith 2000; Hominins Churchill and Smith 2000.

point   (OF *pointe*=sharp end) In the analysis of stone artifacts, the term point refers to pieces that are pointed in shape. Typically (but not always), points are inferred to represent the tips of hunting weapons such as spears or arrows. Stone points are characteristic of many **Middle Stone Age** sites in Africa (e.g., **Blombos Cave**), and they also occur at some similarly aged Middle Paleolithic sites in Eurasia. Ethnographic, historic, and archeological data demonstrate that points were also made of other materials such as bone and wood, although for obvious reasons these rarely preserve in the archeological record.

point mutation   *See* **single nucleotide polymorphism**; **mutation**.

Poisson's ratio   A material property, denoted by v, that measures the degree to which a material contracts (or, rarely, expands) in the directions perpendicular to the axis of the direction in which it is being placed under tension. Conversely, it can represent the degree to which a material expands in the directions perpendicular to the direction in which it is being compressed. Mathematically, it is defined as the negative of the ratio of transverse **strain** divided by axial strain. Rubber has a Poisson's ratio of nearly 0.5 (which is the maximum possible value of v), while cork has a ratio that is nearly 0. Cortical bone is often said to have a Poisson's ratio of 0.3, although this is an oversimplification.

polarity   (Gk *polus*=axis) Earth sciences Refers to each end of an axis passing through a sphere and in that sense it is used to refer to the direction, either normal or reversed, of the Earth's magnetic field. *See also* **geomagnetic polarity timescale**. Systematics In **cladistic analysis** it is used to refer to the alignment of a sequence of character states, with the most primitive character state at one end and the most derived at the other. *See also* **cladistic analysis**; **morphocline**.

polarity chronozone   *See* **chron**.

polarity reversal   *See* **geomagnetic polarity timescale**.

polarization   Earth sciences In earth science polarization refers to rocks that have magnetized particles so that at the time of deposition of sediments or when a lava cools, the direction of the Earth's magnetic field is preserved. At present the Earth's field is polarized so that the needle of a compass points to the north; this is called the normal direction. This has not always been the case and the Earth's history has seen a series of polarity reversals related to changes in pattern of flow in the fluid Earth core. *See also* **geomagnetic polarity timescale**. Systematics In systematics, polarization is the process of deciding which state of a character is primitive and which is derived. Several techniques have been developed to facilitate polarization. These include the stratigraphic criterion, the ontogenetic criterion, and communality analysis, but the most widely used technique is outgroup analysis. *See also* **cladistic analysis**.

pollen   (L. *pollen*=fine flour, mill dust) One of several kinds of **plant microfossils** (the others are **phytoliths** and **starch grains**). Pollen is the male gamete of flowering plants and it is either wind-borne or carried by insects. The shapes and sizes of pollen grains are usually unique to the plant taxon (e.g., at the level of the family, tribe, or genus and sometimes even at the level of the species) that produced it, and can be used to identify the presence of these plants in the archeological record. Pollen is particularly hardy and fossilizes well, but because pollen can travel long distances it is more useful for reconstructing the paleoenvironment rather for reconstructing diet. The study of pollen is called palynology and Traverse (2007) provides a useful primer for anyone exploring the use of pollen for reconstructing the paleoenvironment. *See also* **palynology; paleoenvironmental reconstruction**.

poly-A tail   The term refers to the addition of adenosine monophosphates to the end of a **messenger RNA** sequence after the mRNA has been generated by **transcription**, but before it leaves the nucleus. The addition of adenosine monophosphates (also called polyadenylation) is important for mRNA transport from the nucleus to the cytoplasm and for proper **translation**.

polycentric   *See* **Weidenreich, Franz**.

polygenic trait   A phenotypic trait that results from the combined effect (action of) alleles at more than one locus and the environment, which does not follow simple Mendelian patterns of inheritance. Because they depend on the simultaneous presence of several alleles as well as an environmental influence, polygenic traits have more complex hereditary patterns than simple monogenic traits (e.g., skin color, hair color, height, weight, and blood pressure). The variation in polygenic traits in a population is often described as continuous, and it can be depicted by a bell curve.

polygenist theories   *See* **multiregional hypothesis**.

polymerase   An enzyme that catalyzes the synthesis of DNA or RNA using an existing DNA or RNA template. *See also* **polymerase chain reaction**.

polymerase chain reaction   (or PCR) This reaction serves to copy, or amplify, a fragment of DNA. The process has three steps: DNA denatura-

tion, primer annealing, and DNA extension. First, the DNA is heated to approximately 94°C to separate (or denature) the double-helix structure into single strands. The temperature is then lowered to enable primers to stick, or anneal, to the DNA. This temperature is dependent on the GC base composition [i.e., the guanine (G) and cytosine (C) content] of the primers and the DNA sequence of interest. Primers, which define the segment of DNA to be amplified, are short single-stranded fragments of DNA (usually 15–25 base pairs long) that are complementary to a portion of the DNA sequence of interest. In the third and last step the temperature is increased again, this time to approximately 72°C, to enhance the activity of the *Taq* polymerase, an enzyme that adds complementary bases to extend the single-stranded DNA molecule. This cycle of denaturation, annealing, and extension is repeated 25–40 times during PCR, so that although initially only a small amount (maybe only a few molecules) of DNA is present; the repeated PCR process increases the amount of the specified fragment of DNA exponentially so that following PCR millions of copies are present. There are several variants of PCR. In reverse transcription PCR (RT-PCR) RNA is first reverse transcribed into DNA and then that DNA is amplified, and in quantitative PCR (qPCR), or real-time PCR, the specified DNA is both amplified and quantified.

polymorphic   *See* **polymorphism**.

polymorphism   (Gk *poly*=much and *morphe*= form, literally "many forms") A morphological or genetic feature or character with alternative specifiable states within a biological population. For the genotype, for example, polymorphism includes having more than one type of allele at a locus (e.g., sickle cell). Examples from morphology relevant to human evolution include the several discrete forms the root system of a tooth may take or the sex-specific variation seen in the occurrence of sagittal crests. Not to be confused with **polytypism**, which describes a species with multiple phenotypically and/or genotypically distinct populations that may be classified as subspecies (e.g., chimpanzees, orangutans).

Polynesia   (Gk *polus*=much or many and *nesos*= island, literally "many islands") One of the three major divisions of small islands that make up **Oceania** (the others are **Melanesia** and **Micronesia**). The islands within Polynesia (e.g., Samoa, Tahiti) extend north from New Zealand to Hawai'i, and east to Easter Island. For obvious reasons these islands were among the last places to be reached by modern humans. *See* **Oceania, peopling of**.

polypeptide (Gk *poly*=much and peptide, a compound containing two, or more, amino acids) A string of amino acids forming all, or part of, a **protein**.

polyphyletic An adjective used to describe a taxon that is part of a **polyphyletic group**. *See* **polyphyletic group**.

polyphyletic group (Gk *poly*=many and *phylon*=races) A polyphyletic group is a taxonomic grouping that includes taxa from more than one **clade** or monophyletic group. The "baboon" group, if it includes savanna (*Papio*), forest (*Mandrillus*), and gelada (*Theropithecus*) baboons, is an example of a polyphyletic group because there is sound molecular evidence that mandrills belong in a separate clade. *Papio* and *Theropithecus* form a clade with a group of smaller-bodied monkeys that are assigned to *Lophocebus*, whereas *Mandrillus* forms a clade with a different group of smaller-bodied monkeys. The latter are assigned to the genus *Cercocebus*, which is a different group of smaller-bodied monkeys. Recent analyses suggest that the morphological similarities between mandrills on the one hand, and savanna baboons and gelada baboons on the other, are the result of convergent evolution or parallel evolution and are therefore examples of **homoplasy** rather than **homology**.

polyphyly *See* **polyphyletic group**.

polytypic *See* **polytypism**.

polytypism (Gk *poly*=much and *tupos*=impression) A species with multiple phenotypically and/or genotypically distinct populations that may be classified as subspecies (e.g., chimpanzees or orangutans). Not to be confused with **polymorphism**, which refers to a feature or character of the phenotype or genotype with alternative specifiable states (e.g., number of cusps on a molar tooth or the pattern of intracranial venous sinuses) within a biological population.

ponderal index (L. *pondus*=weight) Ratio of body mass relative to stature, calculated as [(body mass)$^{1/3}$/stature]×100. Modern humans have a high ponderal index compared to the extant great apes. Archaic hominins such as *Australopithecus afarensis* appear to have had a low, relatively ape-like, ponderal index.

Pongo At first it was thought that the flanged adult male orangutans were a different species from the juveniles that were usually seen in Europe, and they were even placed in a separate genus, *Lophotes*, but gradually it was realized that substantial change occurs during the ontogeny of male orangutans. Until the late 20thC most researchers recognized just one orangutan species, but since then fixed genetic and absolute morphological differences have been elucidated between Bornean and Sumatran orangutans, and they are now universally distinguished as different species: *Pongo pygmaeus* from Borneo and *Pongo abelii* from Sumatra. There are also differences between different populations within Borneo, and three subspecies (*Pongo pygmaeus pygmaeus* from the northwest, *Pongo pygmaeus wurmbii* from the southwest, and *Pongo pygmaeus morio* from the northeast) are recognized, and there is growing evidence that they may be nearly as distinct from one another as each of them is from the Sumatran form.

Pontnnewydd (Location 53°13′N, 3°28′W, Wales, UK; etym. named after the nearby town of Bontnewydd) History and general description Known since at least the late 1800s, this cave site was excavated between 1978 and 1995 by the National Museums and Galleries of Wales. Researchers found an Acheulean assemblage and several hominin remains, including a well-dated tooth, which is the second-oldest hominin fossil in Britain, after **Swanscombe**. Temporal span and how dated? **Uranium-series dating** and **thermoluminescence dating** suggest an age of *c*.200 ka for the oldest tooth. Hominins found at site A molar that is similar to the fossils from **Krapina**, and two other hominin fragments that are likely but not necessarily from the Acheulean layers. Several other Holocene hominin remains were also recovered. Archeological evidence found at site An Acheulean stone tool assemblage, as well as later, Middle Paleolithic deposits. Key references: Geology, dating, and paleoenvironment Green et al. 1981, Green 1993; Hominins Green et al. 1981, Green 1993, Aldhouse-Green et al. 1996; Archeology Green et al. 1981, Green 1993.

population Evolution The group of interest for a study. A population may include the entire species, a subset, or a small sample taken from a species. Statistics The full set of possible values from which a smaller set of values known as a **sample** is drawn for use in statistical tests. For example, a sample of cranial capacities can be drawn from the population of all chimpanzee crania which have ever existed. Statistics based on a sample are used to make inferences regarding the population from which the sample is drawn.

population bottleneck *See* bottleneck.

population genetics The study of allele frequencies in populations. Specifically, it focuses on the effects of the four forces of evolution (**natural selection, genetic drift, gene flow**, and **mutation**).

population size Population size is literally the number of individuals alive at any one time for a given population. Typically, in modern human demography the population size is specified at mid-year, although if the population is stationary the population size is unchanging and could be measured at any time. Population size is used synonymously with census size, as the number of individuals alive at any one time. Population size is usually greater than effective population size, which is the size of an idealized population that would produce the same amount of genetic drift as observed in the actual population. *See* **effective population size** for examples of demographic processes (such as variance in completed family sizes) that lower the effective population size relative to the (census) population size.

population structure A population is described as structured if mating is not random. For example, in modern humans, geography (including distance or barriers), language, and other cultural factors influence mate choice and thus gene flow.

Porc-Épic Cave (Location 9°34′N, 41°53′E, Ethiopia; etym. Fr. for "porcupine," presumably because porcupines frequented the cave, but the reason the name was attached to the cave is obscure) History and general description The investigation of the cave, which was formed in Mesozoic limestone in the Ethiopian highlands, has a complex history. It was first excavated by a French team in the 1920s and 1930s that included

Fr. Teilhard de Chardin and the Abbé Breuil. Desmond **Clark** excavated there in 1974, as did K.D. Williamson in 1975–6, and the material from the latter excavations has only been recently discovered and analyzed. Temporal span and how dated? The age of the Porc-Épic sequence remains poorly constrained. Three **obsidian hydration dating** age estimates of *c*.61–77 ka from artifacts from the 1930s excavations stored at the Field Museum in Chicago lack stratigraphic provenance. Three **AMS radiocarbon dating** assays on gastropod shell fragments range from *c*.33 ka to more than 43 ka. Overall, these results suggest a **Late Pleistocene** age for the Porc-Épic artifacts and fossils. Hominins found at site Mandibular fragment including two premolars and three molars. Archeological evidence found at site **Middle Stone Age** artifacts primarily made of locally available **chert**, with lesser amounts of other materials including **basalt** and **obsidian** from unknown sources occur in **breccia** interstratified with **speleothem**. Small Levallois, discoidal, and single and double platform cores produced flakes, blades, and bladelets, many retouched into a wide range of point forms (approximately 40% of the formal tools) and scrapers. The well-preserved faunal assemblage is dominated by bovids, lagomorphs, and hyrax, and cutmarks and bone breakage patterns suggest that it is primarily hominin-accumulated. The dominance of high-utility bones suggests effective hunting by hominins and selective food transport to the cave. Porc-Épic Cave is situated in a difficult to access topographic high point that affords excellent visibility of the surrounding area, possibly resulting in its use as a hunting station. The material recovered within the Middle Stone Age layers includes 419 complete perforated opercula of the terrestrial gastropod *Revoilia guillainnopsis*. Their spatial distribution, evidence of polish within the perforations, and the lack of any evidence that they were used as food suggests a possible symbolic use, perhaps as beads. Later Stone Age material from overlying loamy sands has not been published in detail. Key references: Geology, dating, and paleoenvironment Clark et al. 1984b, Assefa 2006; Hominins Vallois 1951; Archeology Clark et al. 1984, Pleurdeau 2005, Assefa 2006, Assefa et al. 2008a.

porcupines The informal name for part of the suborder Hystricomorpha of the order **Rodentia**. Porcupines are of interest from a taphonomic perspective, as they are well known to accumulate large bone assemblages that can include remains of small (e.g., rabbit-sized) and large (e.g., wildebeest-sized) mammals. Such assemblages are characterized by high frequencies (60–100%) of specimens with evidence of porcupine-gnawing. *See also* **Rodentia**.

portable confocal scanning optical microscope A portable version of a **confocal scanning light microscope** developed by Tim Bromage and colleagues. The portable confocal scanning optical microscope is based on the Nipkow disc technology. The microscope provides images in real time in a very thin optical plane, thus with the minimum of superimposition. It can generate images less than 50 μm beneath the surface of an object such as a tooth or a long bone (Bromage et al. 2009). *See also* **confocal scanning light microscope**.

positive allometry (Gk *allo*=other and *metron*=measure) This term refers to a relative size relationship in which part of an organism, or a variable that functions as a proxy for part of an organism, increases in size at a faster rate than the overall size of the organism, or a variable that functions as a proxy for the whole of an organism (i.e., the variable becomes proportionally larger as overall body size increases). *See also* **allometry**; **scaling**.

positive assortative mating *See* **positive assortment**.

positive assortment In simple terms, this occurs when individuals choose to mate with those who are similar to themselves (i.e., when "like" mates with "like"). Examples of positive assortative mating in modern humans include matings between individuals who share religious beliefs, education, height, and skin color (e.g., Spuhler 1972, Mascie-Taylor 1987). When positive assortative mating is based on heritable traits, it tends to reduce overall variation.

positive selection *See* **directional selection**.

positron emission tomography [etym. A positron, which is the antimatter equivalent of an electron, is emitted by radioactive material; emission refers to emitted positrons and photons and Gk *tomos*=a cutting (from *temnein*=to cut) and

graphein=to write refers to the name of the detection system that detect photons in virtual slices through the brain] (or PET) A noninvasive neuroimaging modality that enables researchers to visualize and measure specific biochemical reactions *in vivo* (i.e., in the case of PET the subjects are alive *and* conscious). It involves the injection or consumption of radioactive compounds (called radiotracers), which are then differentially distributed, and their concentration is captured by detectors that respond to photons generated by radioactive decay. Much like autoradiography, PET can also be used to image the distribution and density of neurotransmitter receptors and transporters in the brain by employing radioactively labeled ligands. Oxygen and glucose are commonly used as radiotracers because they accumulate in tissues that are metabolically active. In the brain, this heightened metabolic signal is correlated with activity at synapses. The radiotracers accumulate in cells proportional to the latter's rate of metabolism. When the radioactive material decays, it emits a positron. After traveling a short distance in the brain, the positron collides with an electron. The positron and the electron annihilate each other and in the process emit two photons that travel in opposite directions (i.e., at 180 degrees to each other). The photons are detected as "pairs in coincidence" by a series of specialized detectors arranged in a ring around the subject. Subsequently, these signals can be assembled into three-dimensional images by a computer. The image generated by detecting the emitted photons shows the distribution of radioactivity in the brain, which is a reflection of regional metabolism during the period when the tracer was taken up. Applications of this method include (a) the description of changes in regional brain glucose metabolism during development in modern humans and Old World monkeys, (b) the comparison of resting brain glucose consumption in modern humans, chimpanzees, and macaque monkeys, and (c) the visualization of the brain areas that are activated during communicative and cognitive tasks in chimpanzees. PET is currently the only feasible method for imaging brain function in living apes.

post-incisive planum A feature that is found when the mandibular symphysis and the perisymphyseal part of the mandibular corpus are thick. In that event the bone immediately posterior to the incisors is not vertically oriented as it is in modern humans, but instead it slopes more horizontally and posteriorly.

post-reproductive life span *See* grand-mother hypothesis; life history.

post-talonid basin The term used by Hersh-kovitz (1971) for the area others refer to as fovea posterior (mandibular). *See also* **fovea posterior (mandibular)**.

post-weaning dependency See age at weaning.

postcanine teeth A tooth that is distal to the canine in the tooth row. In the deciduous dentition it refers to the first and second deciduous molars, and in the permanent dentition to the two premolars and the three permanent molars.

postcanine tooth *See* postcanine teeth.

postcanine tooth row In the deciduous dentition it refers to the first and second deciduous molars, and in the permanent dentition the two premolars and the three permanent molars.

postcentral gyrus (Gk *gyros*=circle, ring) This area of the cerebral hemispheres of the brain is located in the parietal lobe, posterior to the central sulcus and anterior to interparietal sulcus. The post-central gyrus corresponds to Brodmann's areas 1, 2, and 3 and is the primary somatic sensory cortex. The cortex of the postcentral gyrus contains a sensory map (or topographic representation) of the surface of the body, although not all body parts are represented equally. The representation of the hands and orofacial region are disproportionately large in modern humans and other primates, reflecting the important role of sensory information from these body parts in primate behavior.

postcingulid A term used by Schwartz and Tattersall (2002) for an enamel crest on the postcanine mandibular teeth more commonly referred to as the distal marginal ridge. *See* **distal marginal ridge**.

postcingulum A term used by Schwartz and Tattersall (2002) for an enamel crest on the postcanine maxillary teeth more commonly referred to as the distal marginal ridge. *See* **distal marginal ridge**.

postcristid *See* **distal marginal ridge**.

postcuspidal fossa The term used by Hrdlička (1924) for a tooth crown feature that others call fovea posterior (mandibular). *See* **fovea posterior (mandibular)**.

postentoconid cristid A term used by Pilbrow (2006) for an enamel crest that runs distally from the tip of the entoconid on a mandibular post-canine tooth crown.

postentoconule A term introduced by Hershkovitz (1971) for an accessory cusp on the lingual end of the distal enamel ridge of a maxillary molar tooth crown (syn. distoconule, distoconulus, hypostyle, tuberculum accessorium posterium internum). *See also* **distal accessory tubercle**.

postentoconulid A term introduced by Hershkovitz (1971) for an **accessory cusp** on the crown of a mandibular molar; others more commonly refer to this feature as a **tuberculum sextum**. *See also* **tuberculum sextum**.

postentocristid A term introduced by Hershkovitz (1971) and used by Grine (1984) for a ridge of enamel in a mandibular molar tooth crown that runs between the entoconid and the hypoconulid, mesial to the distal marginal ridge.

posterior fovea *See* fovea.

posterior fovea (mandibular) *See* fovea posterior (mandibular).

posterior fovea (maxillary) *See* fovea posterior (maxillary).

posterior margin A term sometimes used for an enamel feature more commonly referred to as a distal marginal ridge on a maxillary postcanine tooth crown. *See also* **distal marginal ridge**.

posterior maxillary plane (or PM plane) Called a plane, but Enlow and Azuma (1975) defined it as a line, one terminus of which is the location in the mid-sagittal plane of a line joining the most inferior and posterior points on both maxillary tuberosities (called the pterygomaxillary point), the other terminus being the location in the mid-sagittal plane of a line joining the most anterior points on the lamina of the greater wings of the sphenoid (called the posterior maxillary point). Strictly a plane cannot be defined by fewer than three points, so to be a "proper" plane it needs to be defined by the locations of the two pairs of bilateral reference points listed above. The superior terminus of the PM plane marks the boundary between the middle and anterior cranial fossae (*see* **basicranium**), and the inferior termini mark the posterior extent of the **face**. Enlow and Azuma (1975) suggest that one of the architectural constraints in the cranium is that the angle between the posterior maxillary plane and the neutral horizontal axis of the orbit (i.e., a line joining the midpoint of the orbital opening with the midpoint of the optic canal) is 90°, and because of this some researchers have suggested that the PM plane should be used more widely as a reference for functional studies of the cranium. Bromage et al. (2008) used this relationship to reconstruct the face of **KNM-ER 1470** as being more prognathic than in the original reconstruction.

posterior parietal cortex (L. *parietalis*=of, or belonging to, walls and *cortex*=husk, shell) This region of the neocortex, which contains Brodmann's areas 5, 7, 39, and 40, is located between the occipital lobe and the primary somatosensory cortex. It is involved in the coordination of somatic sensation, auditory information, and vision. It is a crucial region for integrating sensory perception of the spatial location of objects in the external world and information about body position. By integrating information about the state of the animal with that of potential targets for behavior, areas within the posterior parietal cortex are thought to create a context or frame of reference for guiding movement. Because this region is densely connected with the frontal lobe, it may contribute to the ability to voluntarily inhibit behavior, as well as adapt movements according to novel sensory stimuli. The posterior parietal cortex is functionally lateralized, such that the posterior parietal cortex of the left hemisphere is specialized for processing linguistic information whereas in the right hemisphere it is specialized for spatial information. Some functional regions within the posterior parietal cortex of modern humans do not have homologues in macaque monkeys. These regions provide additional central visual field representations and greater sensitivity to extract three-dimensional form related to motion. These regions in the dorsal interparietal sulcus are activated in positron emission tomography (PET) imaging studies of modern humans learning how to fashion Oldowan-style stone tools (Stout et al. 2008). In the absence of comparable data from apes, however, it is not clear whether these posterior parietal areas are specific to *Homo sapiens* or whether they are shared with our close relatives. In addition, it has been argued by Raymond **Dart**, Ralph **Holloway**, and others that the posterior shift of the lunate sulcus and the relative reduction of the adjacent primary visual cortex in the modern human brain indicate that areas of the posterior parietal cortex have disproportionately expanded within the hominin clade.

posterior teeth   *See* **postcanine teeth**.

posthypocone crista   A term used by Szalay and Delson (1979) for an enamel crest on a maxillary molar tooth crown that runs distally from the tip of the hypocone (syn. posthypocrista).

posthypoconid cristid   A term used by Pilbrow (2006) for an enamel crest on a mandibular molar tooth crown that runs distally from the tip of the hypoconid.

posthypoconulid cristid   A term used by Pilbrow (2006) for an enamel crest on a mandibular molar tooth crown that runs distally from the tip of the hypoconulid.

postmetaconid cristid   A term used by Pilbrow (2006) for an enamel crest on a mandibular molar tooth crown that runs distally from the tip of the metaconid.

postmetaconulid   A discrete area of elevated enamel on the occlusal surface of a mandibular molar distolingual to the metaconid. Some regard this as the weakest expression of a tuberculum intermedium (also called a C7). *See also* **tuberculum intermedium**.

postmetacrista   A term used by Szalay and Delson (1979) for an enamel crest on a maxillary molar tooth crown that runs distally from the tip of the metacone.

postparacrista   A term used by Szalay and Delson (1979) for an enamel crest on a maxillary molar tooth crown that runs distally from the tip of the paracone.

postprotocone crista   A term used by Szalay and Delson (1979) for an enamel crest on a maxillary molar tooth crown that runs distally from the tip of the protocone (syn. postprotocrista).

postprotoconid cristid   A term used by Pilbrow (2006) for an enamel crest on a mandibular molar tooth crown that runs distally from the tip of the protoconid (syn. postprotocristid).

postprotocrista *See* distal trigon crest; postprotocone crista.

postural feeding hypothesis   This is one of several hypotheses put forward to explain the origin of upright posture and bipedal locomotion in the hominin clade. Like the **seed-eating hypothesis**, it is based on a set of careful observations of both primate behavior and anatomy. The hypothesis was proposed by Hunt (1996), who suggested that among chimpanzees and baboons it was more likely that a bipedal posture was linked to feeding rather than a bipedal mode of locomotion. He showed that bipedal postures were employed during both arboreal and terrestrial feeding, and typically involved reaching into higher branches to gather fruits. While standing bipedally, chimpanzees frequently used one of their upper limbs to steady themselves by simultaneously hanging from a branch. Hunt further noted that the postcranial skeleton of australopiths retains several ape-like features associated with arm-hanging, but lacks features that would have made bipedal locomotion energetically efficient. Given that this postcranial morphology appears to have persisted for several million years, Hunt doubts that selection for bipedal locomotion played a major role in hominin origins. Rather, he suggests that the earliest hominins may have been dependent on arboreal fruit resources, and evolved anatomical adaptations to bipedal posture as a result of habitually employing an upright posture during feeding. These adaptations allowed facultative bipedal locomotion, but selection favoring the evolution of energetically efficient obligate, or endurance, bipedalism did not occur until later in hominin evolution. *See also* **bipedal**.

posture   (L. *positura* = position) The position, or attitude, of the whole body or of a part of the body. When the long axis of the trunk is orien-tated vertically this is referred to as an upright or orthograde posture; when the long axis of the trunk is orientated horizontally this is referred to as a horizontal or pronograde posture. Musculoskeletal adaptations for posture and locomotion are not always the same. For example, in chimpanzees, suspensory arm-hanging under a branch is the only behavior requiring complete abduction of the arm (Hunt 1991), suggesting that some morphology related to shoulder mobility (e.g., the superiorly narrow "cone-shaped" rib cage) may be adaptations to an arm-hanging posture rather than to a climbing form of locomotion.

Potamochoerus   The name of the genus that contains the extant and extinct bush pigs. *See* **Suidae**.

potassium-argon dating   (or ^{40}K/^{39}Ar dating) A **radioisotopic dating** method that uses potassium-bearing minerals often from layers of volcanic **tephra**, usually in the form of ash, in which case they are called **tuffs**. It compares the amounts of ^{40}K and its daughter product ^{39}Ar. Because in this method the potassium is measured as a solid and the argon daughter product as a gas, these measurements are made on two different samples of rock and from many mineral crystals from each sample. The error introduced tends to limit its application for hominin evolution to sediments that are more than 1 Ma. Because relatively large amounts of material are needed the method is only suitable for samples that are not likely to be contaminated with older or younger crystals. The potassium-argon technique had early and important implications for the understanding of the time depth of human evolution, particularly in East Africa. In 1965 Evernden and Curtis published K-Ar ages for the Olduvai Gorge sequence that placed the base of Bed I at 1.85 Ma and the top of Bed II as >500ka. These had important implications especially for the age of what was then called ***Zinjanthropus boisei*** (**OH 5**), which was associated with the 1.85 Ma date, as well as for *Homo habilis* (**OH 7**) and pre-Chellean man (**OH 9**) from Bed II. Subsequent work has shifted the ages of the Olduvai stratigraphy earlier, especially those for Bed II (see the Olduvai Gorge entry). Nonetheless, this first set of dates had immediate consequences for understanding human evolution by providing sound evidence of a greater antiquity of hominins than had been previously accepted. The method has been largely superseded by **argon-argon dating**.

Powell-Cotton Collection The "Powell-Cotton" of the Powell-Cotton Collection refers to Major P.H.G. Powell-Cotton (1866–1940), who was an avid collector with a special interest in natural history and in particular with African "big game." The specimens in the Powell-Cotton Collection were collected by two associates, Fred Merfield and Kurt Zenker, and by local villagers in Cameroon, the former French and Belgian Congos, Guinea, and Ethiopia. The relevance of the Powell-Cotton Collection to human evolution is that it includes the world's largest collections of both Western gorillas (*Gorilla gorilla*) and Central African chimpanzees (*Pan troglodytes troglodytes*). The former are represented by 204 skulls (139 of them with skeletons, many also with skins) and the latter by 173 skulls (130 of them with skeletons, many also with skins). In addition there are six skulls and skeletons of chimpanzees from the Ituri Forest (*Pan troglodytes schweinfurthii*). The greater part of the great ape skeletal collection is backed by field notes that include the date of collection, the name of the locality, the latitude and longitude, and the weight and the linear measurements of the cadaver. Contact person Malcolm Harman (e-mail harman-malcolm@btconnect.com). Postal address The Powell-Cotton Museum, Quex House & Gardens, Quex Park, Birchington, Kent CT7 0BH, UK. Website www.quexmuseum.org. Tel +44 (0)1843 842168.

Powell-Cotton Museum The Powell-Cotton Museum in Quex Park, Birchington, Kent, England, was founded in 1898 to provide a home for the collections generated by Major P.H.G. Powell-Cotton (1866–1940) during his expeditions to Africa and Asia. He was a keen naturalist, collector, and hunter who traveled extensively in Africa and parts of Asia. His total number of collecting trips spanned from 1887 to 1939, with 22 out of the total of 28 covering the African continent. Powell-Cotton set up a business relationship with two professional collectors, Kurt Zenker and Fred Merfield (both of whom operated out of the Cameroon). The collection at the Powell-Cotton Museum grew out of the arrangement with the two collectors that the museum would retain a specimen for every specimen it sold to other institutes or researchers. The purpose-built Museum is adjacent to the fine Regency country house that was formerly the home of the Powell-Cotton family. *See also* **Powell-Cotton Collection**.

power grip A grip in which the object is held by the fingers and palm, with the thumb as a buttress (Napier 1993). When gripping large objects, the thumb may flex around the object in opposition to the fingers. A distinctive modern human form of the power grip is the squeeze grip of cylindrical tools (e.g., hammer handles), held obliquely along the palm. This type of grip is facilitated in modern humans by shorter fourth and fifth metacarpals compared to the rest of the palm, increased mobility in the ulnar direction of both the wrist and the fingers, and enhanced robusticity of the ulnar side of the hand. Chimpanzees and other apes are not capable of using this grip, and instead grasp the object transversely or diagonally across the fingers, usually without active involvement of the palm or the (relatively short) thumb (Marzke et al. 1992).

power scavenging *See* confrontational scavenging.

power stroke Refers to the slow close phase of the **chewing** cycle, which ends at minimum gape. It used to be thought that the power stroke consisted of two phases, phase I being the last part of slow close phase and phase II being the start of the slow open phase. However, it is now clear that among primates little or no bite force is generated after minimum gape, so that the power stroke is equivalent only to phase I. *See also* **chewing**.

Praeanthropus (L. *prae*=before and Gk *anthropos*=human being) The genus name introduced by Musaffer Şenyürek (1955) to accommodate the maxilla found at Garusi (now called **Laetoli**) by **Kohl-Larsen** in 1939 that was attributed by Hans Weinert (1950) to *Meganthropus africanus*. *Praeanthropus* is the genus name that should be used by those who support removing *Australopithecus afarensis* from *Australopithecus*. *See also Praeanthropus afarensis*.

***Praeanthropus afarensis* (Johanson, 1978)** (L. *prae*=before, Gk *anthropos*=human being, and *afarensis* recognizes the contributions of the local Afar people) A new combination suggested by Strait et al. (1997). The results of a cladistic analysis carried out by Strait et al. (1997) suggested that *Australopithecus afarensis* was the sister taxon of *Australopithecus africanus* and other later hominins. Thus, they argued that to continue to use the same genus name for *Au. afarensis* and *Au. africanus* made it inevitable that *Australopithecus* would be a **paraphyletic** taxon and they suggested the hypodigm of *Au. afarensis* be transferred to a different genus. They argued correctly that the genus name *Praeanthropus* was **available**, but if *Praeanthropus* was used as the genus name, then the species name *afarensis* no longer had priority, for *Meganthropus africanus* Weinert, 1950 obviously had priority over *Au. afarensis* Johanson et al., 1978. However, this latter change in nomenclature would result in two hominin species called "*africanus*;" *Australopithecus africanus* Dart, 1925 and *Praeanthropus africanus* (Weinert, 1950) Şenyürek, 1955. Application was made to the **International Commission on Zoological Nomenclature** (or ICZN) to have the specific name "*africanus* Weinert, 1950" suppressed, and in "Opinion 1941" published in the *Bulletin of Zoological Nomenclature* (1999) the ICZN upheld the application. The ICZN also ruled that a news report in the *New Scientist* magazine should have priority over Johanson et al., 1978. So, if the *Au. afarensis* hypodigm is to be removed from *Australopithecus*, then the taxon should be referred to formally as *Praeanthropus afarensis* (Johanson, 1978).

Praeanthropus africanus **Weinert, 1950**

(L. *prae*=before, Gk *anthropos*=human being, and L. *africanus*=pertaining to Africa) A new combination introduced by Musaffer Şenyürek (1955) for the maxilla found at what was then called Garusi (now called **Laetoli**) by Kohl-Larsen in 1939 and attributed by Hans Weinert (1950) to *Meganthropus africanus*. Ferguson (1986) claimed that if the Garusi maxilla belongs to the same taxon as **LH 4**, then *Praeanthropus africanus* has priority and should be used in place of *Australopithecus afarensis*. However, an application was made to the **International Commission of Zoological Nomenclature** (ICZN) to have the specific name "*africanus* Weinert, 1950" suppressed, and in "Opinion 1941" published in the *Bulletin of Zoological Nomenclature* (1999) the ICZN upheld the application. So the taxon that was known as *Pr. africanus* should now be referred to as *Praeanthropus afarensis* (Johanson 1978). *See also* *Praeanthropus*; *Praeanthropus afarensis*.

pre-eminent bipedalism *See* bipedalism.

pre-Neanderthal theory *See* pre-Neanderthal hypothesis.

pre-Neanderthal *See* pre-Neanderthal hypothesis.

pre-Neanderthal hypothesis A hypothesis that asserted that both the modern human and Neanderthal lineages evolved from populations that succeeded *Homo erectus*, but which pre-dated the emergence of "classic" Neanderthals in Europe. Also known as the pre-Neanderthal theory, it was accepted by a number of leading human paleontologists during the mid-to-late 20thC, including Clark Howell, Sergio **Sergi**, Wilhelm Gieseler, William (Bill) **Howells**, and Emil Breitinger. The term pre-Neanderthal (Fr. *anté-Néandertalien*) is also used to refer to hominin remains from Europe that date from about **Oxygen Isotope Stages** 8–13. The "pre-Neanderthal theory" was formally named by Henri-Victor Vallois in his overview of modern human origins, published as the concluding chapter to his 1958 monograph about the hominins from **Fontéchevade**. Vallois specifically contrasted the pre-Neanderthal theory with the **pre-sapiens hypothesis**, which suggested that the modern human branch extended back even earlier, and the **Neanderthal phase hypothesis**, in which all Neanderthals were seen as ancestral to modern humans. The first scholar to promote the pre-Neanderthal perspective was Sergio Sergi in the 1940s. He argued that the pre-Neanderthals (including the specimens from **Swanscombe** and Fontéchevade), which he named the **Prophaneranthropi**, exhibited characteristics of both Neanderthals and modern humans (the **Phaneranthropi** and **Palaeanthropi**, respectively). He held that this pre-Neanderthal stock was the probable ancestor of both the Neanderthal and modern human lineages, with early members of the Neanderthal lineage (e.g., **Saccopastore 1**) possessing some "phaneranthropine" features (Sergi 1953). Howell (1951) differentiated "progressive" Neanderthals from "classic" Neanderthals. Progressive Neanderthals included early Neanderthals from Europe (e.g., Saccopastore, **Krapina**) as well as specimens such as **Teshik-Tash, Zuttiyeh**, and **Skhul/Tabun**; classic Neanderthals were the **Würm**-age specimens from Europe (e.g., **La Chapelle-aux-Saints, La Ferrassie, La Quina**). Howell maintained that the progressive Neanderthals gave rise to both the classic Neanderthals in Europe and modern humans via the **Mount Carmel** hominins from the Near East. Theoretically, the pre-Neanderthal hypothesis can be seen as the intellectual precursor of the **out-of-Africa hypothesis** for modern human origins in which modern humans are seen as evolving from **Middle Pleistocene** ancestors outside of Europe (Smith 1997).

pre-sapiens hypothesis This hypothesis (also called the pre-sapiens theory) was a model of human evolution advocated by a number of prominent scientists during the early 20thC. According to the pre-sapiens hypothesis, fossil hominins morphologically similar to modern humans had existed since at least the early **Pleistocene**, if not earlier. This meant that other more recent forms of hominin (i.e., *Homo neanderthalensis* in Europe, *Sinanthropus* in China, and *Pithecanthropus* in Java) could not be ancestral to modern humans. At the time of the heyday of the pre-sapiens hypothesis there were archeological, paleontological, and theoretical reasons for believing that anatomically modern-looking humans evolved very early. In 1895 a modern human-like skeleton (the skull was discovered in 1888) was found at **Galley Hill** in Kent, England, in geological deposits that at the time were considered to be very old. In addition, geologists throughout northern Europe were finding what were considered to be the earliest evidence of simple stone tools, called **eoliths**, in Lower Pleistocene and even Pliocene deposits. Although there was some debate over whether these pieces of flint were really tools, or had been chipped by natural processes, these implements were taken as evidence of hominins of one sort, or another, occupying Europe in the Pliocene. When a modern human skeleton was found in Ipswich, East Anglia, in 1911, in Pliocene deposits where eoliths had been discovered, this was seen as further evidence of the great antiquity of anatomically modern humans. One of the earliest and most influential advocates of the pre-sapiens hypothesis was the British anatomist Arthur **Keith**. In his *Antiquity of Man* (1915) Keith noted that the archeological and fossil evidence hinted that the **Cro-Magnon** peoples with their **Aurignacian** industry had abruptly replaced the Neanderthals with their **Mousterian** industry, suggesting that modern *Homo sapiens* had not evolved from the Neanderthals, but already existed elsewhere. He also interpreted the recently unearthed **Piltdown** fossils in light of the pre-sapiens hypothesis, and suggested that among the general phylogenetic consequences of the pre-

sapiens hypothesis was the implication that the divergence between the hominin line and that leading to the great apes may have occurred as early as the late Oligocene. He also suggested that the races of modern humans had emerged during the late Pliocene or early Pleistocene. The discovery in 1935 at **Swanscombe**, Kent, England, of a partial cranium from deposits in the Thames river valley that also contained **Acheulean** implements and fossil mammals from the Lower or Middle Pleistocene appeared to provide additional support for the pre-sapiens hypothesis. A little earlier Louis **Leakey** claimed he had found evidence of *H. sapiens* at two sites in East Africa. In 1932 at **Kanjera** in western Kenya Leakey found two modern human-like skulls in association with fossils he considered to date from the Middle Pleistocene, while at **Kanam** (also in western Kenya) Leakey found a mandible in deposits he considered to be Lower Pleistocene (Leakey 1936). These discoveries convinced Leakey that he had found the oldest remains of *H. sapiens* then known. A controversy soon erupted over the accuracy of the dating of these discoveries, and the debate was fuelled by Leakey's inability to locate exactly where the bones had been found, and by disagreements over the age of the deposits. However, these factors did not deter Louis Leakey, who remained an ardent supporter of the idea that modern human-like hominins had emerged relatively early in human evolution. In *Adam's Ancestors* (1934) Leakey argued that the line leading to *Pithecanthropus*, *Sinanthropus*, and Neanderthals diverged from the line leading to modern humans during the Miocene and that the human races had appeared by the Middle Pleistocene. During the 1940s and 1950s a number of developments all tended to undermine the evidence for the pre-sapiens hypothesis. Evidence began to accumulate suggesting that the supposedly ancient skeletons found at Galley Hill and Ipswich were in fact intrusive burials of late prehistoric skeletons into earlier deposits. Then between 1953 and 1955 investigations of the Piltdown remains supported by Wilfrid **Le Gros Clark** and carried out by Kenneth **Oakley** and Joseph Weiner were published that showed that Piltdown was a hoax. Thus, this large-brained early hominin with an ape-like dentition was removed from human phylogeny. Likewise, skepticism about the Kanam and Kanjera fossils led many paleoanthropologists to question their validity as exemplars of anatomically modern humans. Meanwhile, discoveries in southern Africa of australopiths and the recovery of numerous *Sinanthropus* fossils in China was strengthening the view that modern humans had evolved from an ape-like ancestor through an australopith stage followed by a "pithecanthropine" stage, the latter being represented by *Sinanthropus* in China and by *Pithecanthropus* in Java. Support for the pre-sapiens hypothesis did not necessarily disappear in the face of this new evidence. Instead, its supporters simply adjusted the theory to fit the new evidence. During the 1940s the American physical anthropologist William (Bill) **Howells** promoted the view that anatomically modern humans arose as early as the Lower Pleistocene and that the other known hominins were side branches of the hominin phylogenetic tree. In the 1950s Howells altered this model, suggesting that modern humans had emerged in the Middle Pleistocene, and that they may well have evolved from a pithecanthropine form. Howells backed even further away from the pre-sapiens hypothesis in later years. Perhaps the most prominent advocate of the pre-sapiens hypothesis during the 1940s and 1950s was the French paleoanthropologist Henri **Vallois**. Vallois originally believed that anatomically modern humans had evolved by the Lower Pleistocene, with *Sinanthropus*, *Pithecanthropus*, and the Neanderthals representing parallel lines of hominin evolution. While his general conception of hominin evolution remained the same, in the 1950s he modified his view of when anatomically modern humans appeared, arguing that they emerged in the Middle Pleistocene. Vallois had been a strong supporter of Piltdown Man and when it was shown that the fossils were a hoax, he had to rely on the Swanscombe cranium and the hominin cranial fragments discovered at **Fontéchevade** in France in 1947 as evidence for the existence of pre-sapiens man. The pre-sapiens hypothesis became untenable as discoveries in Africa during the 1960s and 1970s brought new hominin species to light and radiometric dating methods provided a more reliable chronology for hominin evolution. The growing number of australopith species found in East Africa along with the discovery of *Homo habilis*, *Homo erectus*, and *Homo ergaster* in Africa strengthened the view that modern humans had evolved from an australopith-like ancestor through *Homo habilis* and *Homo erectus* stages, with anato-

mically modern humans emerging *c*.200 ka. The primary impact of the pre-sapiens hypothesis during the early 20thC was to push other hominin species out of the evolutionary lineage leading to modern humans and to encourage researchers to search for the fossil remains of our pre-sapiens ancestors. However, the popularity of the pre-sapiens hypothesis helps to explain why during the early 20thC so many scientists rejected the australopiths and the Neanderthals as modern human ancestors, and why there was so much influential support for the fraudulent Piltdown fossils.

pre-sapiens theory *See* pre-sapiens hypothesis.

pre-*Zinjanthropus* The term Louis **Leakey** used to refer to the hominin remains that were subsequently included in the species *Homo habilis* (Leakey et al. 1964).

preadaptation A feature or **trait** that is well designed to perform a functional role it does not currently perform. Preadaptations may be co-opted to perform the new functional role, and in the process become **exaptation**s, a term introduced in 1982 by Gould and Vrba. The term may also be used as a verb in much the same way as the term **adaptation**. For example, the **vertical climbing hypothesis** suggests that the manner in which the lesser gluteal muscles are recruited by apes when climbing vertical supports makes those primates preadapted for terrestrial bipedalism. In other words, when our ape ancestors were forced to travel on the ground, bipedalism presented itself as a mode of locomotion that could be easily performed. In historical context, the concept of preadaptation was difficult for evolutionary biologists to explain because the term often seemed to infer an orthogenetic preordination that was inconsistent with the principle of **evolution** by means of **natural selection**.

precentral gyrus (Gk *gyros*=circle, ring) The precentral gyrus of the cerebral hemispheres of the brain is located in the frontal lobe, anterior to the central sulcus and posterior to the precentral sulcus. The cortex of the precentral gyrus mostly corresponds to Brodmann's area 4, which is the primary motor cortex. The anteriormost portion of the precentral gyrus also contains

Brodmann's area 6, which is the premotor cortex. The cortex of the precentral gyrus contains a motor map (or topographic representation) of the entire body, although not all body parts are represented equally (e.g., regions used in tasks requiring precision and fine control, such as the face and hands, are disproportionately large in modern humans and other primates). The primary motor cortex forms part of the agranular cortex, because it lacks a clear granular layer IV, and its layer V contains the giant Betz pyramidal cells. The relative size of the primary motor cortex remains relatively constant across primates in proportion to overall body size.

precession (L. *praecedere*=to go before) The Earth's orbital geometry is affected by three "rhythms," or "cycles." The *c*.23 ka precessional cycle is due to changes in the gravitational pull exerted on the Earth by the sun and moon, and these affect the orientation of the Earth's axis so that it "wobbles" in a predictable way. Until 3 Ma the *c*.23 ka precessional cycle was the dominant influence on global climate. *See also* **astronomical time scale**.

precingulid A term used by Schwartz and Tattersall (2002) for an enamel crest on postcanine mandibular tooth crowns more commonly referred to as the mesial marginal ridge. *See also* **mesial marginal ridge**.

precingulum A term used by Schwartz and Tattersall (2002) for an enamel crest on postcanine maxillary tooth crowns more commonly referred to as the mesial marginal ridge. *See also* **mesial marginal ridge**.

precision (L. *praecision*=to cut off) The quality of being able to do something within well-defined limits. A precise value is one that has very little error. A series of precise measurements are not necessarily close to the actual value being measured, however. Imagine a tightly clustered group of darts on a dartboard that is far from the bullseye, the intended target. The placement of these darts is precise, but not accurate. *See also* **accuracy**.

precision grip Any grip that uses the thumb and one or more of the fingers, with or without the palm serving passively as a prop for the object being grasped (Marzke 1997). Although historically the

term precision has often been used to refer specifically to grasping a small object between the thumb and the tip of the finger, there are several different types of precision grip, including two-jaw (i.e., between the thumb and index finger), three-jaw (i.e., between the thumb, index finger, and middle finger), and four- to five-jaw (i.e., between the thumb and three or more fingers) grips. Two-jaw grips include pad-to-pad (e.g., when pinching a coin), tip-to-tip (e.g., when threading a needle), or pad-to-side (e.g., when using a key) grips between distal end of the thumb and the index finger. A three-jaw thumb/full-finger grip is used when holding spherical objects such as a baseball whereas a four- or five-jaw grip involving just the distal pads may be used when unscrewing the lid of a jar. Nonhuman primates are not capable of many of the precision grips used by modern humans. Instead, chimpanzees more often use a pad-to-side grip between the thumb and the side of the index finger. Among extant primates baboons come closest to being able to use the pad-to-pad grip because their hand proportions are most similar to that of modern humans.

precocial (L. *praecox*=early ripening) Compared to **altricial** taxa, the offspring of precocial organisms are born at a relatively advanced stage of development. Precocial offspring are usually fully furred or feathered, are able to move about on their own shortly after birth, and are born with their eyes and ears open. For example, horses are a precocial taxon, for newborn foals are able to stand up and start walking around shortly after birth. While modern humans are described as **secondarily altricial** by most, they have also been described as semi-precocial. This is because although they are altricial in many respects, modern human neonates come into the world with their eyes open, and with hair. However, they fall short in many criteria for independence when compared to truly precocial animals.

precuspidal fossa The term used by Hrdlicka (1924) for a feature on the tooth crown of mandibular molars that others call the fovea anterior. *See* **fovea anterior (mandibular)**.

predation (L. *praedari*=to rob, or plunder) Predation is when one organism, the predator, hunts and kills another organism, the prey, for food. Most extant primate species are at risk from predators of one sort or another, and this has probably been true throughout primate evolutionary history. A cranium of *Notharctus* (a North American adapiform, dated to the early Eocene) exhibits puncture marks that correspond to the teeth of the arboreal predator *Vulpavus* (Hart and Sussman 2005), and evidence of puncture marks and of the deliberate disarticulation of early anthropoid skeletons from the Fayum strongly suggest carnivore predation (Gebo and Simons 1984). Patterns of damage in the orbits and on the cranial vault of the **Taung** cranium are compatible with it being killed by a **raptor** (Berger 2006), and leopard tooth marks have been found on hominin fossils from **Swartkrans** (Brain 1981). In response to the risk of predation, primates have evolved many anti-predator behaviors. Vervet monkeys, for example, have a range of calls to warn conspecifics against different key predators, including eagles, snakes, and large cats (Struhsaker 1967). Predation pressure has also been linked to changes in primate group composition, shifts in activity patterns and sexual dimorphism. Others speculate it may have been part of a feedback mechanism that drove the tendency for large brains in primates (Shultz and Dunbar 2006). Some primates are also highly effective predators themselves. Baboons have been observed to capture and eat vervet monkeys, small antelopes, and birds. Chimpanzee hunting pressure appears to have altered group composition in red colobus monkeys (Stanford 1995). There is abundant evidence from Plio-Pleistocene archeological and paleontological sites in southern and East Africa of hominins butchering animals (including ungulates and possibly primates). Although the discovery that hominins were prey animals (*sensu* Brain 1981) contradicted the views of some early paleoanthropologists who believed that hominins were "bloodthirsty killers" (Dart 1953), observations of modern primate behaviour alongside taphonomic evidence from fossils and stone tools suggest that in early hominins there was no clear division between "the hunter" and "the hunted."

predator *See* predation.

predentine *See* dentine development.

Předmostí (Location 49°30′N, 17°25′E, Czech Republic; etym. named after a nearby village) History and general description This open-air site is located in northeastern Moravia, 1 km/0.6 miles north of Přerov. It was first excavated in 1880

by J. Wankel. The first hominin remains were recovered in 1884. Excavations continued intermittently throughout the 20thC and later, with the most recent one in 2006. In 1894 a communal grave was discovered by Maška with the remains of 18 individuals, covered by limestone slabs and mammoth bones, and associated with Pavlovian (**Gravettian**) lithic artifacts and symbolic and decorative objects. Additional hominin remains (*Homo sapiens*) were recovered in later excavations, for a total number of up to 29 individuals. Subadult, juvenile, and adult specimens are represented, as are both sexes. Most remains were destroyed by fire in 1945. Temporal span and how dated? **Radiocarbon dating** of the Pavlovian occupation provided dates between 27 and 29 ka BP. Hominins found at site Předomstí 1–29; several crania and associated skeletons, as well as individual elements are represented; all are assigned to *H. sapiens*. Archeological evidence found at site **Mousterian**, **Aurignacian**, and Pavlovian (Gravettian) artifacts have been recovered from different layers at the site. The hominin remains are associated with the Gravettian cultural layer. Key references: Geology, dating, and paleoenvironment Svoboda 2008; Archeology Svoboda 2008; Hominins Vlček 1971, Smith 1982.

preentoconid A term sometimes used for an accessory cusp on the occlusal surface of a mandibular molar that is more commonly referred to as a tuberculum intermedium. *See also* **tuberculum intermedium**.

preentoconid cristid A term used by Szalay and Delson (1979) for an enamel crest on a mandibular molar tooth crown that runs mesially from the tip of the entoconid (syn. pre-entoconid cristid).

prefossid The term used by van Valen (1966) for a feature on a mandibular molar tooth crown others call the fovea anterior. *See* **fovea anterior (mandibular)**.

prefrontal cortex (L. *cortex*=husk, shell) The prefrontal cortex is the part of the cerebral cortex that lies anterior to the premotor cortex, which, in turn, is anterior to the primary motor cortex. The prefrontal cortex is defined as the part of the frontal cortex that has a relatively thick granular layer IV, which receives inputs from the mediodorsal nucleus of the thalamus. In contrast to other parts of the frontal cortex, electrical stimulation of this region does not evoke movements. The prefrontal cortex is involved in so-called executive functions, which include differentiating among conflicting goals, evaluating different behavioral strategies, and the inhibition of immediate desires for the maximization of long-term gains. As such, the prefrontal cortex is important for decision-making, moderating social behavior, and the planning of complex sequences of actions. The prefrontal cortex can be subdivided according to anatomy, connectivity, and function. For example, the dorsolateral prefrontal is associated with tasks that require an animal to inhibit certain motor responses at appropriate times, regardless of spatial elements. This region is densely interconnected with the posterior parietal cortex and they both send connections to numerous common cortical and subcortical structures. The orbitofrontal cortex is part of the limbic cortex, and has been specifically implicated in regulating aggressiveness and emotional responsiveness, as well as general arousal and some autonomic responses. Because of its involvement in higher-order cognition, one might expect the prefrontal cortex to be selectively modified in the evolution of modern humans. Accordingly, there have been many studies of gene expression, cellular composition, and the neurochemical architecture of the modern human prefrontal cortex in comparison to other primates, including great apes. In sum, most of these studies have revealed striking similarities in the gene expression and histology of the prefrontal cortex of modern humans and our close relatives. The most consistent differences indicate a higher level of gene expression associated with heightened metabolism and synaptic activity in modern humans (Preuss et al. 2004). Additionally, while it is clear that the prefrontal cortex of modern humans comprises a relatively larger proportion of the frontal lobe than in great apes (Rilling 2006), there is debate (e.g., Semendeferi et al. 2001, Holloway 2002, Schoenemann et al. 2005, Sherwood et al. 2005) about whether the entire prefrontal cortex or any of its subdivisions have undergone disproportionate expansion in modern humans beyond what would be predicted by the known allometric relationships within higher primates. *See also* **allometry**.

prehensile (L. *prehensus*=to grasp) A hand (or foot, or lips) capable of isolating, pinching, and lifting an object. *See also* **hand function**.

prehominine A term used by, among others, John Robinson to refer to australopithecines or "early members of the family of man" (Robinson 1961, p. 3).

prehypocone crista A term used by Szalay and Delson (1979) for an enamel crest on a maxillary molar tooth crown that runs mesially from the tip of the hypocone (syn. prehypocrista).

prehypoconid cristid A term used by Pilbrow (2006) for an enamel crest on a mandibular molar tooth crown that runs mesially from the tip of the hypoconid.

prehypoconulid cristid A term used by Pilbrow (2006) for an enamel crest on a mandibular molar tooth crown that runs mesially from the tip of the hypoconulid.

premetaconid cristid A term used by Szalay and Delson (1979) and by Pilbrow (2006) for an enamel crest on a mandibular molar tooth crown that runs mesially from the tip of the metaconid.

premetaconulid An accessory cusp on a mandibular molar crown situated mesial to the metaconid, either on the mesial crest of the metaconid or on the lingual portion of the mesial marginal ridge. This cusp has also been sometimes described as paraconid, although the homology with this cusp, which is absent in hominoids, is questionable (syn. paraconid, mesial accessory cusp).

premetacrista A term used by Szalay and Delson (1979) for an enamel crest on a maxillary molar tooth crown that runs mesially from the tip of the metacone.

premutation In diseases such as Huntington's disease or fragile X syndrome, which are caused by an expansion of trinucleotide repeats, unaffected carriers have a higher number of repeats than is typical and these repeats are unstable during replication, resulting in many more repeats in the allele found in the affected offspring. For example, most people have fewer than 28 copies of the CAG repeat in the *HTT* gene whereas people with Huntington's disease have more than 36 copies. Individuals with 28–35 copies are unaffected but are described as having the unstable premutation. *See also* **micro-satellite**.

preoccupied   This term is not formally defined in the **International Code of Zoological Nomenclature**, but it is commonly used to denote a name that is a junior homonym. *See* **homonymy**.

preparacrista   A term used by Szalay and Delson (1979) for an enamel crest on a maxillary molar tooth crown that runs mesially from the tip of the paracone.

prepared core   Refers to a stone core that shows signs that the **striking platform** or flake release surface has been abraded or chipped to increase the likelihood of successful flake removal. Some authors treat the terms "prepared core" and "Levallois core" as synonymous. *See also* **Levallois**.

preprotocone crista   A term used by Szalay and Delson (1979) for an enamel crest on a maxillary molar tooth crown that runs mesially from the tip of the protocone (syn. preprotocrista).

preprotoconid cristid   A term used by Pilbrow (2006) for an enamel crest on a mandibular molar tooth crown that runs mesially from the tip of the protoconid (syn. preprotocristid).

preprotocrista   *See* **mesial marginal ridge**; **preprotocone crista**.

pressure pad   Devices (also known as pressure-distribution systems) that use capacitive (i.e., electric) sensors to measure forces. They are typically arranged in a matrix within a pad (rigid or flexible) that allows normal forces (those perpendicular to the surface) to be measured across an area over time. Pressure pads have been used most commonly to examine the forces exerted by the foot from heel contact through toe-off, but they are also used to measure the force distributions during hand grips, across seats or saddles, as well as other interactions between the body and a substrate. For example, Wunderlich and Jungers (2009) presented the first quantitative data on pressures on ape fingers during **knuckle-walking**. Their results, which show that pressure is highest on digits 2–4, but that pressure varies with support type (branch or ground) and hand posture, are broadly consistent with the hypothesis that the fusion of the **os centrale** with the scaphoid is an adaptation to knuckle-walking. *See also* **kinetics**.

prey-choice model   A model derived from **optimal foraging theory** designed to predict whether or not a forager will pursue or ignore a prey item when it is encountered. The prey-choice model (also known as the "encounter-contingent prey-choice model") assumes that the objective of the forager is to maximize energetic returns, that the forager is aware of the energetic returns associated with different types of prey, and that the forager searches for prey in a patch where encounter rates with different prey types are random. The model suggests that **high-ranking prey** will always be preferentially incorporated in the diet because they provide higher-than-average energetic returns. Energetically less profitable **low-ranking prey** will only be added to the diet when encounter rates with high-ranking prey decline and overall **foraging efficiency** declines.

prey rankings   *See* **high-ranking prey**; **low-ranking prey**.

Přezletice   (Location 50°09′N, 14°34′E, Czech Republic; etym. named after a nearby town) History and general description Extinct fauna was found in a small quarry in 1938, and subsequent excavations in the late 1960s led by Oldrich Fejfar employed fine screening, which resulted in the recovery of a small, possibly hominin, tooth fragment. Temporal span and how dated? No radiometric dating has been attempted, but stratigraphic correlation has suggested an age that spans the Günz and Mindel glacial cycles (i.e., *c*.600 ka). If the deposits are in fact that old, they represent some of the oldest human inhabitaiton in Europe. Hominins found at site A single fragment of a tooth that was originally interpreted as a hominin molar fragment has since been determined to belong to a bear. Archeological evidence found at site The approximately 250 stone tools belong to the proto-Acheulean. Key references: Geology, dating, and paleoenvironment Fejfar 1969; Hominins Fejfar 1969, Vlček 1978b; Archeology Fejfar 1969.

primary cartilaginous joint   *See* **joint**.

primary centers of ossification   *See* **ossification**.

primary consumers *See* trophic level.

primary cusp (syn. main cusp) *See* cusp.

primary dentine *See* dentine.

primary producers *See* trophic level.

primary visual cortex *See* striate cortex.

primate archeology This was the name proposed in a multi-authored review (Haslam et al. 2009) as an inclusive term to describe the study of the "past and present material record of all members of the order Primates" (*ibid*, p. 339). The early reports of tool use in nonhuman primates focused on the way chimpanzees made and used tools (Boesch and Boesch 1990, McGrew 1992) and this lead to various suggestions of terms such as "cultural panthropology" (Whiten et al. 2003) and "chimpanzee archeology" (Carvalho et al. 2008) for the study of this phenomenon. However, tool manufacture and tool use in primates is emphatically not confined to chimpanzees (see the references in Haslam et al. 2009), so a more inclusive term was needed. The goals of primate archeology are to provide a "primatocentric" perspective to balance the "anthropocentric" interpretations of the early hominin archeological record, and to investigate the spatial, environmental, and other circumstances under which modern humans, early hominins, and nonhuman primates use probing, pounding, and digging tools.

primer Primers are short single-stranded fragments of DNA (usually 15–25 base pairs long) that are complementary to a portion of the DNA sequence of interest and which define the segment of DNA to be amplified. *See* **polymerase chain reaction**.

primitive (L. *primitivus*=first of its kind) A morphological feature or character is described as being primitive if it is present in the outgroup, or in the hypothetical common ancestor. A primitive or **symplesiomorphic** character state is the character state at the primitive end, or pole, of a morphocline (syn. primitive condition).

principal components analysis (or PCA) A form of **multivariate analysis** in which variation is analyzed for a data set of *n* variables, and then a new set of *n* variables is produced which are linear transformations of the original variables in which the first variable (or first principal component) contains the most variation in the data set, the second variable contains the most variation that is independent of the first variable, the third variable contains the most variation that is independent of the first two variables, and so on. For a more intuitive explanation, consider a cloud of points in space shaped like a baguette. A PCA finds the axis which describes the most variation in the data set; in this case, an axis running from end to end lengthwise along the loaf. Each data point can be projected into this axis and thus has a value or score for the first axis or first principal component. However, most points in the baguette do not fall exactly on this axis, so there is still more variation in the data set. The remaining variation is independent of the first axis, so it is at right angles to the first principal component; this can be visualized by looking at the baguette end-on or by slicing the baguette perpendicular to the first axis (as if you were cutting slices for bruschetta rather than slicing the loaf lengthwise to make one giant sandwich). The PCA now fits a second axis (principal component) which describes the most variation in the portion of the data visible from this angle; in this case, an axis running from side to side through the loaf. The remaining variation can all be described by a third axis running from top to bottom in the loaf. In effect, the new principal components are new axes that result from rotating the baguette in space. In practice all of the new axes or principal components are fit simultaneously and any number of original dimensions or variables can be accommodated (not just three). The resulting principal components have the following properties. First, there are as many principal components as there were original variables. Second, all principal components are independent of each other (i.e., the **correlation** between any pair of principal components from a single analysis is equal to zero). Third, the first principal component accounts for the most variation in the data set, the second accounts for a proportion of variance equal to or less than the first, and so on. Many researchers use PCA to reduce a data set with many variables to a data set with relatively few principal components that account for the majority of the variation in the original data set; this procedure makes it easier to visualize patterns in data. It can also minimize the number of variables in an analysis such as multiple regression, thus reducing the number of degrees of freedom lost by having a large number of variables. This reduction is most effective when the original variables are highly correlated with each other, and will be ineffective when the original variables are mostly independent of each other. When all of the original variables are highly correlated, the first principal component will account for the overwhelming majority of the variation in the data set. For example, if 12 cranial measurements are collected for a sample of skulls spanning a size range including mouse lemurs and gorillas, all measurements are expected to be highly correlated (large species have absolutely larger skulls than small species, and thus absolutely larger cranial measurements), and the first principal component will typically account for 90% or more of the variation in the data set in such a scenario. This first principal component is clearly driven by the size difference between skulls, and indeed researchers have occasionally used the first principal component of such an analysis as a measure of **multivariate size**. However, it is important to note that the first principal component also incorporates shape differences that are correlated with size differences, and thus is not a measure of size alone. PCA is performed on standardized data (i.e., for each variable, every measurement has the mean of that variable subtracted and is divided by the standard deviation of the variable). Many computer programs provide the loadings or coefficients for the various principal components produced by aPCA; these loadings refer to the relative contribution of each standardized variable to the principal component (high magnitude loadings indicate a strong contribution of the original variable, positive loadings indicate that increases in the original variable are associated with increases in the principal component, negative loadings indicate that increases in the original variable are associated with decreases in the principal component). These loadings can be used to identify those variables in the original data set which are most responsible for the overall variation present in a data set.

priority (L. *prior*=first) A principle used in **nomenclature**. It holds that the taxon name with the earliest publication date should be regarded as the senior name and thus should have priority over any other name for that taxon published thereafter. *See also* **synonym**; **synonymy**.

prism (L. *prisma*=something that has been sawn off) *See* **enamel prism**.

probability density function The concept of a probability density function (sometimes abbreviated as pdf) generalizes the idea of a probability function for a discrete variable to the case of a continuous variable. The probability density for any given value within the defined range of the function cannot be negative, but unlike a probability function the density can be above 1.0. A proper probability density function must integrate to 1.0, which means that, for example, a uniform density on the interval 0 and 0.4 will have a density value of 2.5 at any point within this range. This is true because the area of the rectangle is then $0.4 \times 2.5 = 1.0$. In hazards analysis the probability density function is often used to represent the probability density of death at exactly age a, and is usually written as $f(a)$. This is equal to the product of survivorship to age a ($S(a)$) and the hazard of death at that age ($h(a)$).

probability of death, age-specific The probability that an individual who enters an age interval in a life table will die within that interval. Although there are a number of equivalent ways to express this probability, the most straightforward uses survivorship. As an example, consider a case where 252 individuals survive to 1 year of age but by 5 years of age there are only 215 individuals still surviving. This gives $(252-215)/252 = 0.1468$ as the probability of death between ages 1 and 5 years. Note that the difference in survivorship at age 1 and at age 5 (252–215) gives the number of deaths in the 4-year interval (equal to 37 deaths), while dividing by the number of individuals surviving to age 1 year gives the number of deaths per the number of individuals who enter the age interval. If no 1 year-olds had died before reaching age 5 years then the age-specific probability of death would be $(252-252)/252 = 0.0$ while if all had died in the interval the age specific probability of death would be $(252-0)/252 = 1.0$. The usual symbol for the age specific probability of death is q_x, where x is the age at the beginning of the age category. In the example just given q_x corresponds to q_1 in the series q_0, q_1, q_5, $q_{10}, \ldots$, where q_0 is the age-specific probability of death between birth and age 1 year, q_1 is the age-specific probability of death between 1 and 5 years, q_5 is the age-specific probability of death between 5 and 10 years, and so on.

probe A fragment of single-stranded DNA used to identify and/or quantify a specific DNA sequence in a sample. For example, an allele-specific probe can be used to identify the presence and quantity of a specific allele in either a DNA sample (for genotyping purposes) or in the cDNA (i.e., complementary DNA) obtained from a specific tissue (to assess gene expression). Some types of quantitative polymerase chain reaction (or PCR) use a probe for quantifying the amount of a specific PCR product that is produced. *See also* **complementary DNA; oligonucleotide; DNA hybridization**.

problem-solving The ability to overcome various kinds of obstacles to achieve a desired goal; the manufacture of a stone artifact requires problem-solving skills. In the early 20thC the psychologist Wolfgang Kohler conducted an experiment in which he gave chimpanzees a variety of tasks, all of which required the animals to obtain a directly inaccessible food reward (Köhler 1925). In one experiment, the apes were presented with a banana hanging from the ceiling of their enclosure, where several kinds of objects, such as crates and poles, were available. After failing attempts to jump at the food reward, the apes stopped, looked around the enclosure, and came up with various ways to reach the otherwise inaccessible food, such as stacking a couple of crates underneath and climbing on top of them. Based on such observation, Kohler argued that the chimpanzees observed and used the behavior of other apes to inform their own strategies. More recently, researchers have examined seemingly insightful problem-solving abilities in orangutans (Mendes et al. 2007). They presented the apes with a vertical Plexiglas tube that was securely attached inside their cage mesh. The tubes were filled one quarter full with water and then baited with floating peanuts. All of the tested orangutans were able to retrieve the food on the first trial by collecting mouthfuls of water from a water source and then spitting the water into the tube to raise the water level. Whether or not one is a good problem-solver depends on cognitive skills such as flexibility, better inhibitory control, planning, and causal reasoning. For instance, suppose that the subject is presented with a horizontally secured Plexiglas tube with a trap mid-way, which makes the apparatus' shape look like letter "T." The tube is baited with a food reward that is out of reach of the subject. In this task, the subject has to be flexible enough to look for any potential tool (e.g., stick) to extract the food and make tools if necessary out of available objects (e.g., making a stick out of branches or combining short sticks to make a long one). She/he also has to inhibit impulsive, fruitless responses such as poking fingers through the tube or blindly pushing/raking the food with the tool without considering the position of the trap. She/he should be skillful at planning and should have a good understanding of causation when judging whether the potential tool would be appropriate for the task (whether the tool is thin enough to go through the tube, long enough to reach the food, strong enough to rake or push the reward, etc.). Moreover, she/he should anticipate the outcome of this action and try to avoid the reward falling into the trap. Great apes have demonstrated successful problem-solving skills in the aforementioned trap-tube task, but with substantial individual variations (Mulcahy and Call 2006). Furthermore, they were more successful when they were allowed to rake the reward instead of being given an option of only pushing it, showing some limitation in their ability to effectively solve the task (Limongelli et al. 1995).

Proboscidea (Gk *proboscis*=elephant's trunk, from *pro*=forward and *boskein*=to nourish or feed) The order that includes hoofless mammals with elongated trunks. The modern examples of this order are the African elephant, *Loxodonta*, and the Asian elephant, *Elephas*, both of which are subsumed within the only remaining extant family of proboscideans, the **Elephantidae**. During the early stages of human evolution, the Proboscidea were a more diverse group than they are today. Members of the family **Deinotheriidae** (proboscideans with downward curving tusks) are known from African early hominin sites, and the family **Gomphotheriidae** (proboscideans with shovel-shaped tusks) is known from South American paleoindian sites. Numerous other families of proboscideans were extant during the course of earlier hominoid and hominin evolution. Most proboscideans are regarded as **megafauna**, although some very early and dwarfed lineages are exceptions to this. The earliest known proboscidean is the 15 kg *Phosphatherium* known from the Paleocene of northern Africa.

proboscidean (Gk *proboscis*=elephant's trunk, from *pro*=forward and *boskein*=to nourish or feed) The informal term for the order that includes hoofless mammals with elongated trunks. *See also* **Proboscidea**.

processual archeology *See* **Binford, Lewis (1930–)**.

Procrustes analysis (etym. In Greek mythology, Procrustes, a son of Poseidon, was an innkeeper who insisted his guests were an exact fit for the only bed provided for them. Short guests were stretched and tall ones had their legs amputated!) A method for comparing the shape of objects. This procedure, which is based on the least-squares principle, involves three steps. First, the landmark configurations of the objects are translated so that they share the same centroid. Second, they are scaled so that they all have the same centroid size (i.e., the square root of the summed squared deviations of the coordinates from the centroid is the same). Third, they are rotated around the centroid until the sum of the squared Euclidian distances between the homologous landmarks is minimized. For more than two objects, this algorithm has been extended to generalized Procrustes analysis (Gower 1975, Rohlf and Slice 1990). In practice, all centered and scaled landmark configurations are rotated to one configuration (e.g., the first specimen in the dataset) to minimize the distances between homologous landmarks. The resulting coordinates are averaged and the procedure is repeated until the distances are minimized. The coordinates of the resulting centered, scaled, and rotated landmarks are called Procrustes shape coordinates. Their average (the consensus configuration) is the shape whose sum of squared distances to the other shapes is minimized (Dryden and Mardia 1998). The differences between any individual and the average shape are called Procrustes residuals. The Euclidian distance between two sets of Procrustes shape coordinates (i.e., the square root of summed squared coordinate-wise differences) is referred to as the Procrustes distance, which is a summary of the similarity or dissimilarity in shape between two landmark configurations.

Procrustes distance *See* **Procrustes analysis**.

Procrustes metric *See* **Procrustes analysis**.

Procrustes shape coordinates *See* **Procrustes analysis**.

progenesis Equivalent to time hypomorphosis. *See* **hypomorphosis**.

prognathic (L. *pro*=forward and *gnathos*=jaws) Literally the degree to which the jaws project forwards, but also used for mid- and upper facial projection. Prognathism can be midline, or lateral, or both, and can affect the upper, middle, lower, or all parts of the face. It is usually quantified by measurements of the distance between a landmark on the base of the cranium (e.g., basion, porion) and a landmark on the front of the face (e.g., subnasale, alveolare), but it can also be measured using landmarks confined to the face (e.g., sellion, prosthion). For example, *Australopithecus afarensis* faces are more prognathic than those of *Homo erectus*. Prognathism usually increases during ontogeny, and it is usually more pronounced in males than in females. The projection of the nasoalveolar clivus beyond the nasal aperture is known as subnasal prognathism. As a generalization, hominins have experienced a long-term trend towards the reduction of prognathism. Flat or weakly projecting faces are called **orthognathic** (ant. orthognathic).

prognathism (L. *pro*=forward and *gnathos*=jaws) The tendency for a face to be prognathic. *See* **prognathic**.

programmed cell death *See* **apoptosis**.

"progressive" Neanderthals *See* **pre-Neanderthal hypothesis**.

projectile point In archeology this term refers to a pointed stone or bone artifact that is thought to have been attached to the tip of a flight weapon, such as a javelin, a spearthrower dart, or an arrow. The functions of projectile points are inferred from wear-pattern analysis, residues, and metric comparisons with ethnohistoric and experimental projectile points. Current evidence suggests projectile point technology developed in Africa between 50 and 100 ka and that it spread to Eurasia after 40–50 ka. Examples of African **Middle Stone Age** projectile points include bone specimens from **Blombos** and **Katanda**. Likely stone projectile points include tanged **Aterian** points from North Africa, and foliate bifaces from all over the continent. Eurasian **Upper Paleolithic** projectile points take a wide range of forms, including various bone/antler types and many local variants made on blades (El Wad points, Font Robert points, Gravette points) as well as foliate bifaces and shouldered points. After 40 ka in sub-Saharan Africa and 30 ka in Eurasia, stone and bone projectile points were either augmented or supplanted by backed microlithic weapon armatures. Knecht (1997), Shea (2006), and Shea and Sisk (2010) provide more detailed treatments.

prokaryote (Gk *pro*=before and *karyote*=nut or having nuts) An organism (e.g., bacteria and Archaea) comprised of a single cell(s) without a nucleus or any membrane-bound organelles.

promontory (L. *promontorium*=mountain) The term is used in anatomy in two ways that are relevant for human evolution. The promontory of the middle ear is the elevated part of the medial wall of the middle ear that covers the basal turn of the cochlea part of the bony labyrinth. Behind and above the promontory is the oval window (fenestra ovalis) that accommodates the footplate of the stapes, and behind and below the promontory is the round window (fenestra rotundum) that accommodates a membrane that seals the scala typmpani of the inner ear off from the middle ear. The sacral promontory is the upper part of the anterior surface of the sacrum that projects into the pelvic cavity.

promoter (L. *promotus*=to move forward, or advance) The part of the gene where **transcription** is initiated and often where regulation of gene expression occurs. Promoters are typically located at the 5′ flanking end (i.e., upstream) of the gene.

pronation (L. *pronare*=to turn face downwards) Internally rotating the forearm so that it moves from the anatomical position, with the palm facing anteriorly, to the pronated position, with the palm facing posteriorly. Also applies to the foot when the leg is internally rotated. When the foot is in contact with the ground it is pronated during the part of the walking cycle when the leg moves forward over the foot and the ankle joint is dorsiflexed.

pronograde *See* posture.

Prophaneranthropi *See* Sergi, Sergio (1878–1972).

prosencephalon *See* forebrain.

protein (Gk *protos*=first) A large molecule made up of one, or more, chains of amino acids. Proteins are fundamental components and products of cells (e.g., enzymes, hormones, and antibodies are all proteins) that are specified by the triplet **DNA** code. DNA is transcribed into messenger RNA, and then the nucleotide sequence of the RNA is translated into sequences of amino acids. *See also* **DNA**.

protein clock *See* molecular clock.

protein domain (L. *domininus*=a lord, hence the territory under the control of a lord) A portion of a protein with a specific structure or function. The same domain may be present in multiple members of a protein family or related groups of proteins. For example, the *HOX* genes all contain a protein domain known as the **homeodomain** that can bind **DNA**. It is encoded by a 180 bp DNA sequence known as the homeobox (*see homeobox* genes).

proteome All of the **proteins** in a cell, a tissue or an organism.

proto-language *See* language, evolution of.

protobifaces *See* Developed Oldowan.

protocone (Gk *protos*=first and *konos*=pine cone) One of the terms proposed by Osborn (1888) for the main cups of mammalian molar teeth. It refers to the main mesial cusp on the lingual aspect of a maxillary (upper) molar tooth crown, and also to the main lingual cusp of a bicuspid upper premolar. It forms part of the trigon component of an upper molar crown (syn. deuterocone, epicone, mesiolingual cusp, tuberculum anterium internum).

protocone basin The term introduced by MacIntyre (1966) for a tooth crown feature that others call the central fovea (maxillary). *See* **central fovea (maxillary)**.

protoconid (Gk *proto*=first and *konos*=pine cone) One of the terms proposed by Osborn (1888) for the main cups of mammalian molar teeth. It refers to the main mesial cusp on the buccal aspect of a mandibular (lower) (hence the suffix "-id") molar tooth crown, and also to the buccal cusp of a bicuspid mandibular premolar. It forms part of the trigonid component of a mandibular postcanine tooth crown (syn. eoconid, mesiobuccal cusp, tuberculum anterium externum).

protoconule A term introduced by Osborn (1907) for an accessory cusp on the occlusal surface of a maxillary molar located on the mesial marginal ridge mesiobuccal to the protocone. *See also* **epiconule**; **mesial accessory tubercle**; **mesial marginal ridge**.

protofossa The term introduced by van Valen (1966) for a tooth crown feature that others call the central fovea (maxillary). *See also* **central fovea (maxillary)**.

protostyle A term used by Schwartz and Tattersall (2002) for an accessory cusp on the occlusal surface of a maxillary molar that other researchers refer to as Carabelli('s) trait. *See* **Carabelli('s) trait**.

protostylid A term introduced by Dahlberg (1945) for an accessory cusp on the crown of a mandibular molar located on the buccal face of the protoconid. It ranges in its expression from a pit to a well-developed cuspulid; it is sometimes associated with the buccal groove (syn. tuberculum paramolare, paramolar tubercle, distoconid, eoconulid).

provenance (L. *provenire*=to come forth, from *pro*=forth and *venire*=to come) The provenance of a fossil or an artifact is the horizon from which it was excavated, or, if it is a surface find, the horizon from which it originated. This includes the exact find context of an object (e.g., spatial location, lithostratigraphic, and chronostratigraphic contexts) and for artifacts it may include sourcing an object to its point of origin (e.g., determining the primary source for the obsidian used to make tools; Carter et al. 2006, Carter and Shackley 2007). If a fossil or an artifact is excavated from its horizon of origin then its provenance is not in doubt. However, most fossils and artifacts are not found *in situ*, but found already exposed of on the surface of the landscape. In such cases geologists try to trace the horizon of origin by looking for evidence of likely overlying fossiliferous strata nearby, or by matching the rock

matrix adhering to the fossil to the appearance (called **lithology**) of the overlying strata exposed nearby. In some circumstances the provenance of a surface find can be determined precisely; in others the provenance, and thus inferences about the age of the paleoenvironment, of a fossil are much less precise. The term provenience is sometimes used synonymously with provenance, particularly by North American researchers (syn. provenience).

provenience *See* **provenance**.

provisioning hypothesis This hypothesis is one of several devised to explain the origin of upright posture and bipedal locomotion within the hominin clade. Proposed by Lovejoy (1981) the provisioning hypothesis models human origins within a demographic and evolutionary socioecological framework. Lovejoy notes that primates exhibit a trend towards longer life spans, and thus prolonged life histories, which reaches its zenith in African apes and modern humans. As part of this trend, modern humans and chimpanzees exhibit a later age of sexual maturity, meaning that individuals must survive for a relatively long period before reproducing. This delayed reproduction in turn implies fewer offspring, and as a consequence individuals in these species must invest heavily in the few offspring they have to ensure their survival. However, primate mothers must typically balance the need to care for their infants with the need to care for themselves and accidents when infants accompany mothers during maternal foraging are a major source of infant mortality. To offset accidental infant death, one solution is for males to provision both the mother and child. Males forage for themselves, but also bring food back to their mate and offspring. Provisioning behavior should favor the evolution of bipedalism because this mode of locomotion frees the hand for carrying. Provisioning reduces the need for female mobility and in turn should minimize the incidence of accidental infant death. Males will only be selected for this behavior if (a) they can be assured that they are the biological father of the infant and (b) they have exclusive reproductive privileges with the female in question. This, in turn, should favor the evolution of monogamy. Lovejoy expanded on the provisioning hypothesis (Lovejoy 2009), but it is difficult to evaluate because it makes few predictions that can be tested using the fossil record. *See also* **bipedalism**.

proximal (L. *proxime* = nearest) In the direction of the root of a limb (i.e., where it is attached to the body). Thus, the shoulder joint is proximal to the elbow joint and the ankle joint is proximal to the hallucial metatarsophalangeal joint.

proximal humerus, evolution in hominins

Apes and modern humans share a relatively rounded humeral head with a relatively low greater tubercle that does not project superiorly; quadrupedal monkeys have a relatively larger greater tubercle (Larson 1995). Modern human and African ape humeri display a high degree of torsion, which means that the humeral head is "twisted" relative to the plane of the distal humerus. Most cursorial mammals have humeral heads that face posteriorly, in the same plane as the distal humerus. Torsion appears to be determined by scapula position, which in modern humans and African apes is situated on the dorsal aspect of the thorax. Thus, for the distal humerus (elbow) to remain in the anteroposterior plane, the proximal humerus is "twisted" to articulate with the glenoid. In animals where the scapula is situated more laterally, the humeral head faces more posteriorly to articulate with the glenoid cavity. Early hominin humeri appear to have lower humeral torsion values than modern humans and African apes, and it has been suggested that high humeral torsion may have evolved independently in modern humans and African apes (Larson 2007). *See also* **scapula, evolution in hominins**.

proxy (L. *procurare* = to take care of) The term proxy used to refer to a person authorized to act for someone else, but it now refers to any sort of substitute. So a variable (e.g., $^{18}O/^{16}O$ ratio) that tracks climate is called a proxy, as is a variable (e.g., femoral head size, or orbital area) that is correlated with body size or body mass.

pseudogene (Gk *pseudes* = false) A term for a **gene** that is not functional. For example, pseudogenes include genes that have arisen by **duplications**, **frameshift mutations**, or **deletions**. Pseudogenes are of interest to evolutionary biologists because they share ancestry with functional genes (for example, the beta globin gene cluster includes several pseudogenes) and they can be used to understand the evolution of a gene family. Once a gene becomes nonfunctional it evolves neutrally, so pseudogenes can be used to construct the phylogeny of a gene.

pterygoid (Gk *pteryx* = wing and *eidos* = resembling) Thin, wing-like, vertical plates of bone, medial and lateral, that project inferiorly from the body of the sphenoid bone.

puberty An event of short duration that takes place in the central nervous system. Puberty is generally characterized by pulsatile secretion of gonadotropin-releasing hormone (GnRH), which stimulates the production and release of gonadotropins [e.g., luteinizing hormone (LH) and follicle-stimulating hormone (FSH)] from the pituitary. These, in turn, stimulate the production of gonadal steroid hormones (estrogen, testosterone), and production and maturation of gametes, from the gonads (ovaries, testes). The increased levels of gonadal steroids act to hasten bone ossification, setting a limit on growth in stature. However, the starting point for this complex feedback network is debated: some see decreased hypothalamic sensitivity to circulating steroid hormones (which acts to initiate the pattern of pulsatile GnRH secretion from the hypothalamus) as the key event; others point to an increased production of gonadotropins in response to GnRH as the beginning of the cascade, and still others point to increased gonadal steroid production as initiating the process. In any case, the event of puberty "reawakes" a process initially begun during fetal and early postnatal life, but held in check during the juvenile phase of development. Once this process is begun again, full sexual maturation and reproduction follow at varying lengths of time. For example, there is a delay of approximately 4–5 years between puberty and reproduction in chimpanzees, and a delay ranging from 5 to 10 years in modern humans. Puberty typically commences at an earlier age in females relative to males. In modern humans, puberty is initiated on average 2 years earlier in girls relative to boys, although there is substantial variation within and among populations. *See also* **life history**.

Pucangan Formation *See* **Sangiran Formation**.

pulp (L. *pulpa* = fleshy part of an animal or fruit) Soft dental tissue found deep within the tooth crown and root that supports the vitality of the **dentine**. During dentine development, the dental papilla supplies the mesenchymal cells that eventually become odontoblasts, and this papilla subsequently becomes the pulp as the dentine grows and encapsulates it. The pulp contains the blood vessels and nerves that supply the tooth. *See* **dentine development**.

pulp chamber (L. *pulpa*=fleshy part of an animal or fruit) Enlarged end of the pulp cavity (root canal) within the dentine. It has been argued that pulp chamber proportions may distinguish *Homo neanderthalensis* from modern humans, with the former having larger (taurodont) molar pulp chambers (but see Constant and Grine 2001). However, because the pulp chamber decreases in size with advancing age (due to secondary and tertiary dentine development) it is important that comparisons are made between teeth of approximately the same age and degree of wear (Constant and Grine 2001). *See also* **taurodont**.

PUMC Acronym for **Peking Union Medical College** (*which see*).

pumice A pyroclastic rock, a common component of **tephra**, which is formed by the rapid cooling (or quenching) of molten rock (magma) which has a high proportion of volatile material (gas). Because of the latter, solidified bubbles or vesicles make up the majority of the rock volume, and they give pumice its characteristic buoyancy that allows it to float on water. In addition to the vesicles and the glass that forms the vesicle walls, pumices commonly contain primary volcanic minerals including feldspar, pyroxene, quartz, and zircon. Pumice can occur in any size grade, ranging from small size particles (i.e., ash and lapilli) to large-size boulders. Potassium-rich feldspars extracted from **pumice clast**s have long been a mainstay of isotopic dating efforts, as they can be unequivocally related to the eruptive event that generated the tephra deposit.

pumice clast (Gk *clastos*=to break down into pieces) A clast is a single particle, so a pumice clast is a single particle of **pumice**. A single grain of pumice of any size (i.e., small as in ash or lapilli, or large as in a boulder) would be a pumice clast.

punctuated equilibrium The name given to the phenomenon frequently observed in the fossil record in which an ancestral species persists more or less unchanged over long stretches of geological time before suddenly speciating into one or more descendant species. The term was coined by Eldredge and Gould (1972) as an alternative to the paradigm of **phyletic gradualism** that had previously been used to explain the evolutionary transformation of one species into another. Eldredge and Gould argued that the periods of morphological stasis commonly observed in fossil species over stretches of geological time were, in fact, meaningful data and not merely an artifact of poor sampling in the fossil record. According to them, these long periods of stasis (or equilibrium) were periodically punctuated by short periods of rapid evolutionary change, hence the name. They further argued that punctuated equilibrium was a logical consequence of evolutionary theory, insofar as species are often thought to originate through the process of **allopatric speciation**. In this mode of speciation a small-to-medium sized subpopulation becomes geographically isolated from a much large parent population. In such modest-sized populations natural selection and genetic drift can be relatively powerful, and can result in relatively rapid evolutionary change. This change would not be saltational (i.e., truly sudden) but might appear to be instantaneous in the fossil record, which may not preserve information at a sufficiently fine temporal scale and thus enable researchers to detect the evolutionary transformation of the subpopulation. Such evolutionary changes are more likely to result in extinction, but in some cases they may result in speciation. In contrast, in the larger parent population, the influence of selection and drift is relatively weak, and although its morphological parameters may oscillate over time these changes would only rarely result in a directional change. In theory, it is possible to test whether or not a given species may have evolved due to punctuated equilibrium by determining whether or not that species exhibits a significant temporal trend in some aspect of morphology. In practice, such analyses require large numbers of specimens whose stratigraphic and temporal context is well understood and very few fossil hominin taxa have a good enough contextual record. The two studies that have looked at the fossil records of hominin taxa across substantial periods of geological time [i.e., *Paranthropus boisei* (Wood et al. 1994) and *Australopithecus afarensis* (Lockwood et al. 2000)] both concluded that stasis rather than gradual change was the predominant signal.

purifying selection A type of natural selection whereby alleles that are deleterious are removed from the gene pool (syn. negative selection). *See also* **negative selection**.

push-off The propulsive phase of the modern human walking cycle in which the center of mass travels over the forefoot and the foot lifts off the ground and generates forward momentum; it is also known as "toe-off." This part of the gait cycle occurs at approximately the 60% point of the walking cycle in modern humans. It is initiated by activity of the soleus and gastrocnemius plantarflexing the foot at the ankle joint. During the push-off phase inversion of the hindfoot locks the calcaneocuboid joint, limiting mobility in the midtarsal region. As the heel lifts, and the metatarsophalangeal joints dorsiflex, the plantar aponeurosis (fascia) tightens in what is called the "**windlass effect**," causing the foot to stiffen and thus be an effective lever. Activity of the plantar-flexing muscles pushing the forefoot against the substrate causes the ground reaction forces on the foot to be as high, or higher, during this push-off phase of walking as during the heel-strike phase. *See* **foot arches**; **walking cycle**.

Putjangan Formation *See* **Sangiran Formation**.

pygmy chimpanzee *See Pan paniscus.*

pyramidal cells Pyramidal cells are a type of neuron in the cerebral cortex of the cerebral hemispheres of the brain that are characterized by a cell body with a triangular, or pyramidal, shape, with the apex giving rise to an apical dendrite, which runs toward the outermost layers of the cortex. The comparatively large base of the cell body gives rise to several basal dendrites that course horizontally within the layer containing the cell body and those closely adjacent. Pyramidal neurons typically project the main trunk of their axon into the white matter and it terminates either in another area of the cerebral cortex or at a more distant site in the central nervous system. Pyramidal neurons express the excitatory neurotransmitter glutamate. Elston et al. (2006) have shown that pyramidal neurons in the granular prefrontal cortex have relatively more complex dendritic arbors and a higher density of spines in association with brain-size enlargement among primates. In contrast the pyramidal neurons of primary visual cortex do not show such dramatic differences. As a consequence, the pyramidal neurons of the modern human prefrontal cortex appear to be the most morphologically complex neurons of the cerebral cortex among all primates. Such modifications of pyramidal neurons in modern humans may serve as an underlying neuroanatomical substrate that contributes to enhancements of higher-order cognitive functions.

pyriform aperture *See* **piriform aperture**.

pyriform cortex *See* **piriform cortex**.

pyroclastic (Gk *puro*=fire and *klastos*=broken) A term used to describe rock fragments that have been ejected explosively from a volcanic source. Ash is an example of a pyroclastic material.

pyrotechnology (Gk *puro*=fire and *teckne*=skill) This term, which literally means "fire-skills," applies to the skills used in the making, curation, and use of controlled fire. The first sound evidence for controlled fire takes the form of discrete scatters of burned flint chips surrounded by unburned materials at the Israeli site of **Gesher Benot Ya'aqov** at just under 780 ka (Goren-Inbar et al. 2004). Other claims for controlled fire prior to that time are difficult to substantiate because of the potential to confuse a burned-out tree stump with a hearth. Attempts to reconstruct the maximum temperature in a paleofire to discriminate between controlled fires and natural fires yielded inconclusive results. Most of the explanations for the use of controlled fire have focused on its utility for cooking and protection, but Brown et al. (2009) suggest that as early as 164 ka at **Pinnacle Point** fire was being used to pre-treat silcrete to make it easier to work to produce flakes.

PySC Acronym for the phylogenetic species concept. *See* **phylogenetic species concept**; **species**.

Q

q_x The term used to express the age-specific probability of death in a life table. *See also* **life table; probability of death, age-specific**.

q arm *See* **chromosome**.

Qafzeh (Location 32°41′N, 35°18′E, Israel; etym. Ar. الجبل *djebel*=mountain and قفزه *qafzeh*=jump or leap) History and general description Qafzeh cave is situated about 2.5 km/1.5 miles southeast of Nazareth in the Wadi el-Hadj. The first excavations from 1933–5 were undertaken by R. Neuville and M. Stekelis. A description of the stratigraphy was published in 1951, but Neuville died before a proper description of the remains could be completed. After an extended hiatus the site was excavated again from 1965 to 1979 by Bernard **Vandermeersch**. The two excavation programs focused on different areas of the cave (Neuville and Stekelis focused on the floor of the interior of the cave whereas Vandermeersch primarily worked in the narrow cave entrance, or "vestibule") and thus the levels are not named under the same scheme (the earlier work uses letters and the more recent work uses roman numerals). Temporal span and how dated? The stratigraphic levels were dated using thermo**luminescence dating** on burned flints, **electron spin resonance spectroscopy dating** (or ESR) using both an early-uptake model and a linear-uptake model was applied to mammalian teeth, and the skull was dated using gamma spectrometric **uranium-series dating**. Thermoluminescence dates on the Middle Paleolithic levels XXIII–XVII were *c.*92 ka. ESR dates on levels XXI–XV vary based on the uptake model: late-uptake estimates average to 115 ka, whereas early-uptake estimates average to 96 ka. Hominins found at site The remains of 14 individual hominins are known from the site. Two (Qafzeh 1 and 2) are from the **Upper Paleolithic** levels excavated by Neuville and the rest are from Middle Paleolithic levels (Qafzeh 3–7 were excavated by Neuville and

Stekelis, and Qafzeh 8–18 by Vandermeersch). The Middle Paleolithic Qafzeh remains are generally interpreted as sampling a robust population of anatomically modern humans, and Howell (1958) commented on the affinities between the Qafzeh and the hominin remains from **Skhul**. Because the Skhul and Qafzeh remains represent the earliest known modern human remains outside of Africa, they have played a crucial role in understanding the radiation of modern humans from Africa. Archeological evidence found at site There are very few artifacts in the Upper Paleolithic levels, and this makes a precise identification of the industry difficult. The assemblage in the Middle Paleolithic levels is broadly defined as Levantine **Mousterian**. Key references: Geology, dating, and paleoenvironment Neuville 1951, Valladas et al. 1988, Schwarcz and Rink 1998, Grün and Stringer 1991, Yokoyama et al. 1997; Hominins Howell 1958, Vandermeersch 1971, Vandermeersch 1981; Archeology Vandermeersch 1971.

Qafzeh 6 Site Qafzeh, Israel. Locality Cave interior. Surface/*in situ In situ*. Date of discovery September 3–6, 1934. Finders R. Neuville and M. Stekelis. Unit N/A. Horizon Layer L. Bed/member N/A. Formation N/A. Group N/A. Nearest overlying dated horizon N/A. Nearest underlying dated horizon N/A. Geological age 94–80 ka. Developmental age Adult. Presumed sex Male. Brief anatomical description Well-preserved, but fragmented cranium that lacks the upper right I^2, P^{3-4}, M^3, and left M^3. Announcement Köppel 1935. Initial description Vandermeersch 1971. Photographs/line drawings and metrical data Vandermeersch 1981. Detailed anatomical description Vallois and Vandermeersch 1972, Vandermeersch 1981. Initial taxonomic allocation *Homo sapiens*. Taxonomic revisions None. Current conventional taxonomic allocation Early modern *Homo sapiens*. Informal taxonomic category Anatomically modern human. Significance Qafzeh 6 is the most complete cranial specimen from Qafzeh, and outside of Africa it is one of the oldest specimens of anatomically modern humans. Like many other early modern human specimens from Africa and the near East, Qafzeh 6 exhibits some primitive morphology, particularly the presence of a supraorbital torus. Location of original **Institut de Paléontologie Humaine**, Paris, France.

Qafzeh 9 Site Qafzeh, Israel. Locality Vestibule. Surface/*in situ In situ*. Date of discovery 1969. Finder Bernard **Vandermeersch**. Unit N/A. Horizon Levels XVII–XXIV. Bed/member N/A. Formation N/A. Group N/A. Nearest overlying dated horizon N/A. Nearest underlying dated horizon N/A. Geological age 80–100 ka. Developmental age Adult. Presumed sex Female. Brief anatomical description Fragmentary skull with incomplete postcranial skeleton. Announcement Vandermeersch 1969. Initial description Vandermeersch 1971. Photographs/line drawings and metrical data Vandermeersch 1981. Detailed anatomical description Vandermeersch 1981. Initial taxonomic allocation *Homo sapiens*. Taxonomic revisions None. Current conventional taxonomic allocation early modern *Homo sapiens*. Informal taxonomic category Anatomically modern human. Significance Qafzeh 9, which is the most complete postcranial skeleton of any Qafzeh hominin, appears to have been part of a double burial. Location of original Department of Anatomy and Anthropology of the Sackler Faculty of Medicine, Tel Aviv University, Israel.

Qafzeh 11 Site Qafzeh, Israel. Locality Vestibule. Surface/*in situ In situ*. Date of discovery 1969. Finder Bernard **Vandermeersch**. Unit N/A. Horizon Levels XVII–XXIV. Bed/member N/A. Formation N/A. Group N/A. Nearest overlying dated horizon N/A. Nearest underlying dated horizon N/A. Geological age 80–100 ka. Developmental age Immature, *c.*10 years old. Presumed sex Male. Brief anatomical description Fragmentary skeleton. Announcement Vandermeersch 1970. Initial description Vandermeersch 1971. Photographs/line drawings and metrical data Vandermeersch 1981. Detailed anatomical description Vandermeersch 1981. Initial taxonomic allocation *Homo sapiens*. Taxonomic revisions None. Current conventional taxonomic allocation Early modern *Homo sapiens*. Informal taxonomic category Anatomically modern human. Significance Qafzeh 11 was almost certainly buried intentionally, as the pit in which it was located was carved into the bedrock. The child also seemed to have been arranged so that it appeared to be holding the antler

of a deer (*Dama mesopotamica*) to its chest. Location of original Department of Anatomy and Anthropology of the Sackler Faculty of Medicine, Tel Aviv University, Israel.

Qesem Cave (Location 32°11′N, 34°98′E, Israel) History and general description Discovered during construction in 2000, this cave site preserves well-dated sequence of Acheulo-Yabrudian (terminal Lower Paleolithic) technology. The dates suggest that the Acheulo-Yabrudian represents a distinct and quite long cultural phase between the Acheulean and Mousterian phases in the Near East. Temporal span and how dated? **Uranium-series dating** on **spelcothem**s provide a range of dates between 382 and 207 ka. Hominins found at site None. Archeological evidence found at site Several distinct archeological layers, all attributed to the Acheulo-Yabrudian. Key references: Geology, dating, and paleoenvironment Barkai et al. 2003; Hominins Mann and Vandermeersch 1997, Hershkovitz et al. 2010; Archeology Barkai et al. 2003, Stiner et al. 2011.

QTL Acronym for quantitative trait locus. *See* **quantitative trait linkage analysis**; **quantitative trait locus**.

qualitative variants *See* **non-metrical traits**.

quantitative genetics The genetic basis of traits that do not vary in a discrete manner. Quantitative traits such as height plot out in a continuous distribution, hence they are also referred to as continuous traits; discrete traits, such as yellow or green coat color in peas, do not follow this statistical pattern. The variation in such traits is controlled by several genes (e.g., they are polygenic) combined with environmental effects. Quantitative genetics measures changes in the frequency of these quantitative/continuous traits across generations and within and among populations. Overall, the main thrust of quantitative genetics is to implement methodologies that allow measurement of variation in quantitative traits to be teased apart: calculating the proportion of variation due to effects from several genes, each perhaps contributing a small amount of influence, and that variation due to effects of the environment. This discipline is important because much of the morphological variation within and among human populations can be attributed to variation in quantitative traits. *See also* **polygenic trait**.

quantitative trait locus (or QTL) A genetic variant that segregates with a quantitatively measured phenotype (often with a normal distribution) (e.g., height or levels of low-density lipoproteins). These loci (QTLs) are typically regions of DNA (genetic markers) that have significant variation within and between populations, but which do not necessarily have an effect themselves on the phenotype. Their statistical association with phenotypic variation indicates that a genomic region close to that locus significantly influences the phenotype. This may be a known gene (called a "candidate gene") or one that has yet to be identified. Quantitative traits, like height, are often influenced by more than one gene and are influenced by the effects of environment, sex, and age, as well as interactions between them; this is why quantitative traits are often referred to as "complex" traits. The goal of QTL research is to discover genes and gene variants that influence quantitative, or complex traits. QTL analyses have yielded important information about the genes that underlie many genetic disorders and are frequently used in medical research. QTL analyses have also been employed to elucidate how genes influence normal variation in traits that are important to evolutionary biology (such as skull shape, pigmentation patterns, and behavior). Marker number and type, as well as sample size and structure of the population can influence the results of a QTL analysis.

quantitative traits *See* **polygenic trait**.

quarry A term used for any area repeatedly visited by hominins to obtain stone raw material. Stone raw material may be freely available on the surface but in other instances quarries provide evidence that raw material has been extracted after hominins have used simple lever systems to remove subsurface boulders. Quarry sites typically include large amounts of debris associated with stone tool production (particularly in the early stages of their manufacture) as well as evidence of the tools (e.g., hammers) needed to make stone tools (Doelman 2008, Ericson and Purdy 1984). The MNK "chert factory site" at **Olduvai Gorge** (Stiles et al. 1974) is a good example of a quarry site. See **Olduvai Gorge**.

quartz Crystalline silica (SiO_2) is one of the most common rock-forming minerals in the continental crust and is found in many different types of rock. Although quartz is widely used in the production of stone tools (e.g., **Matupi Cave**), its texture is often poorly suited to the production of stone tools as it is prone to shatter, rendering the manufacture of large pieces difficult. Agate, chert, and flint are all varieties of chalcedony. Chalcedony is a variety of silica composed of microscopic quartz crystals, micro-porosity, and variable amounts of water. Because it lacks the regular crystal structure of more crystalline forms of quartz chalcedony has better flaking characteristics and thus is used to make stone tools. *See also* **conchoidal fracture; percussion**.

quartzite A typically coarse granular to crystalline rock composed primarily of **quartz**, which may be either sedimentary or metamorphic in origin. Metaquartzites are varieties of quartzite formed by recrystallization due to metamorphism; orthoquartzites formed as the result of secondary silica cementation of quartz sand following sedimentation. Although its material properties are dependent upon the size of the individual quartz grains, quartzite generally fractures in a rough, hackly fashion, but percussion does result in a conchoidal fracture and quartzite artifacts retain a durable sharp edge. When it is readily available it is widely used at archeological sites (e.g., **Olduvai Gorge** and **Klasies River**).

Quaternary (L. *quarter* = four times) This term, introduced by Desnoyers in 1829, refers to a unit of geological time (i.e., a **geochronologic unit**). As of 2009, the International Union of Geological Sciences redefined the Quaternary as the most recent of the three **period**s that comprise the **Cenozoic** era. Thus, it spans the time interval from 2.58 Ma (Gelasian) to the present and is divided into the **Pleistocene** and **Holocene** epochs.

Quina Mousterian *See* **La Quina**.

R

r *See* **correlation coefficient**.

R A statistical and graphical programming language widely used in quantitative biology and which is rapidly being adopted by physical anthropologists. Developed by Robert Gentleman and Ross Ihaka of the Statistics Department of the University of Auckland in the mid-1990s (Ihaka and Gentleman 1996) it is a free, open source, platform-independent language that is particularly good at manipulating large matrices (www.r-project.org/). User-contributed "packages" contain functions that allow for specific types of analyses common in human evolutionary studies such as discriminant function analysis, cluster analysis, and phylogenetic comparative analysis. In addition, the programming language is flexible and can be used to create customized analyses such as resampling analyses which are often used in studies of shape and size variation in fossil samples. The basic interface is command-line-driven, which presents a steep learning curve to some users but free menu-driven windows interfaces have also been developed (e.g., R Commander; http://cran.r-project.org/web/packages/Rcmdr). There also exist a number of text editors which facilitate programming in R by highlighting functions and arguments in different colors, highlighting matched parentheses, quotation marks, and brackets, and allow direct communication between the text editor and R (e.g., Tinn-R; www.sciviews.org/Tinn-R).

r-selection *See* **r/K-selection**.

r/K-selection theory This theory captures the idea that there is a systematic relationship between the type of environment experienced by a taxon, and the type of reproductive strategy it adopts to maximize its likely success in that environment (MacArthur and Wilson 1967, Pianka 1970). In *r*-selected species the individuals are small-bodied, they mature early, they tend to have many widely dispersed offspring, and their generation times and lifespans are short. This contrasts with K-selected species in which the individuals are larger-bodied, they mature later, they tend to have fewer offspring that require intensive investment, and their generation times and lifespans are relatively long. Modern humans are usually categorized as a K-selected species. However, rather than falling neatly into either category, organisms instead often show a mixture of traits from opposing ends of the r/K continuum.

r^2 *See* **coefficient of determination**.

Rabat *See* **Kébibat**.

race In the field of genetics, "race" has been used historically to describe strains, groups, or subspecies within a species. In modern humans, the term has become tied to political and sociocultural concepts that do not correspond to meaningful biological units or biological realities, and thus race is not appropriate as a biological term or as a classificatory unit for modern humans. Compared with other species, including most hominoids, modern humans have relatively little genetic variation. Modern human variation is typically clinal (i.e., subject to geographic patterning affected by selection and migration) and cannot be pigeonholed into the boxes corresponding to socially defined racial categories. Despite all of this, race is often used (however inappropriately) as a proxy for ancestry or environment in medical genetics and forensics research [see Bamshad et al. 2004, Duster 2005, and a whole Symposium Issue (139:1–102) of the *American Journal of Physical Anthropology* devoted to the topic of race].

radial unit hypothesis (L. *radius* = ray) Proposed by Pasko Rakic in the 1980s to account for the expansion of the cerebral cortex in ontogeny and evolution. Each "radial unit" is initially generated by the symmetrical mitotic division of founder cells in the ventricular zone of the fetal telencephalon.

Wiley-Blackwell Encyclopedia of Human Evolution, First Edition. Edited by Bernard Wood.
© 2013 Blackwell Publishing Ltd. Published 2013 by Blackwell Publishing Ltd.

Each progenitor cell divides to produce two daughter progenitor cells, so with each round of symmetrical divisions the number of progenitor cells doubles. In the next phase, progenitor cells in the ventricular zone begin to divide asymmetrically so that each round of mitosis yields one daughter cell that no longer divides (called postmitotic) and one daughter cell that remains in the ventricular zone and continues mitosis. The postmitotic cell that is destined to be a cortical neuron detaches from the ventricular surface and begins to migrate along a transient glial fiber that extends towards the surface beneath the pia mater, eventually resting in the cortical plate, a developmental precursor of the cerebral cortex. After arriving at the cortical plate, the more recently generated neurons bypass neurons formed at an earlier stage, so that the neurons come to rest along an "inside-out" gradient. Each proliferative unit in the ventricular zone produces approximately 100 neurons that come to occupy a vertical **minicolumn** in the cerebral cortex. The radial unit hypothesis is an attempt to account for the precise, appropriate migration of neurons from their birthplace to their final functional location in the cerebral cortex. This hypothesis also provides a mechanism by which evolutionary variation in the size of the cerebral cortex can occur by modifying the number of founder cells in the ventricular zone prior to the onset of neuronal proliferation and migration within each radial unit. Accordingly, it is proposed that genetic mutations affecting either the dynamics of **mitosis** (i.e., cell division), or **apoptosis** (i.e., cell death) of the founder cell population, or both, result in differences in the surface area of the cerebral cortex. In confirmation of some of the predictions of the radial unit hypothesis, there is evidence that several genes (e.g., *ASPM*, *microcephalin*, and several caspases) that play a role in cell-cycle regulation and programmed cell death of cortical progenitors have been positively selected for during the course of hominoid and modern human evolution. *See also* **brain size evolution, molecular basis of**.

radiocarbon dating A radioisotope-based dating method that can be applied to the remains of a wide range of organisms (i.e., bones, teeth, seeds, charred wood/charcoal) and some inorganic precipitates (i.e., carbonates) that contain sufficient amounts of ^{14}C (also called radiocarbon). Principle Radiocarbon dating is based on the regular decay of the radiocarbon isotope ^{14}C to ^{14}N. All living organisms incorporate some ^{14}C into their body tissues, either during photosynthesis (plants) or from dietary car-

bon (animal consumers). When an animal or plant dies, it no longer adds new ^{14}C to its tissues. Because ^{14}C decays to ^{14}N at a constant rate, the relative amount of ^{14}C left in organic material can be used to estimate an age for the material. However, because the half-life of the breakdown of ^{14}C to ^{14}N is 5730 ± 40 years, the method can only be applied to material that is less than $c.50–60$ ka (i.e., less than 1% of the temporal span of the hominin fossil record); samples older than this do not contain enough ^{14}C to date. Suitable materials Carbon dating can be applied to bone, tooth, shell, charcoal, and other plant remains. Three methods of measuring radiocarbon are currently used. Two of the methods are "conventional" methods, which measure the radioactive decay of ^{14}C via the emission of beta particles (electrons). This is done by either liquid scintillation counting (or LSC) or by gas proportional counting (or GPC). The third method, first used in 1977, is a direct ion-counting technique. It uses a particle accelerator to produce ions with a $^{+3}$ charge state and in this state many of the ions that interfere with the counting of ^{14}C can be easily removed, thus allowing the accurate measurement of very low concentrations of ^{14}C. Because this third method, called accelerator mass spectrometry (or AMS), enables the direct measurement of individual ^{14}C atoms much smaller samples can be processed; AMS can routinely date samples of 1 mg of carbon. This means that previously undateable samples, such as single hominin teeth and individual grains of domesticated cereals, can now be dated. Typical starting weights required for AMS (e.g., 10–20 mg of seed/charcoal/wood, 500 mg of bone, 10–20 mg of shell carbonate) are about 1000 times less than the weights required by conventional counting systems. The AMS method also allows for more thorough chemical pretreatment of samples, this is particularly important for older samples (>25 ka BP) where small amounts of modern carbon contamination may have a large effect on the measured ^{14}C fraction and hence the date. Potential limitations. Differences in the atmospheric production of ^{14}C through time due to changes in the Earth's geomagnetic field and solar output mean that ^{14}C ages (also known as uncalibrated radiocarbon ages) need to be calibrated to correct for these potentially substantial errors. The calibration method needs to be based on a system whose own chronology is independent of radiocarbon dating and ideally it should faithfully preserve evidence of annual, or **sidereal**, events. For the period between the present back to $c.10–12$ ka researchers successfully used the annual rings preserved in long-lived trees such as the bristle-cone pine to calibrate ^{14}C ages in the northern hemisphere. Between the older end of the range of high-resolution tree-ring calibration and *c*.21 ka researchers have managed to identify precise global climatic events in coral growing in Tahiti and Mururoa. These coral calibrations were the basis of the calibration curve INTCAL98 and this curve was subsequently updated and extended to 26 ka as INTCAL04 (Reimer et al. 2004). However, between *c*.21 and 26 ka the two most precise calibrations, one using annual sediment layers (called varves) in Lake Suigetsu in Japan, the other based on stratigraphy preserved in carbonates (called **speleothems**) in a submerged cave in the Bahamas that can be linked to independently dated events in the Greenland Summit ice cores, did not agree. This difference has now been explained (see Hofmann et al. 2010 for a full explanation of the discrepancy). The Intcal calibration curve has been subsequently updated to INTCAL09 (Reimer et al. 2009) and this extends the calibration curve back to *c*.50 ka cal. years BP. Although it is provisional for the region of the curve >26 ka cal. years BP, currently it provides the best estimate. Ongoing work on new cores from the varved sequence in Lake Suigetsu (Staff et al. 2010) and marine cores from the Cariaco Basin off the coast of Venezuela and an Atlantic/Iberian Margin deep sea core (MD952042) hold the promise of a better-validated calibration curve for the period between 10 and >50 ka. Due to ocean circulation patterns, carbon in the southern hemisphere tends to be *c*.50 years younger than carbon in the northern hemisphere. As a result, calibration curves for the northern hemisphere cannot be used in the southern hemisphere. Instead, an independent curve has been developed for the southern hemisphere (McCormac et al. 2004; SHCal04), but SHCal04 can only be used back to 11 ka. For samples older than this, INTCAL04 calibrations can be combined with a hemispherical time-lag correction (i.e., 56±24) (McCormac et al. 2004). The discrepancies between raw ^{14}C ages and the calibrated radiocarbon ages can be substantial. Around 6–8 ka the ^{14}C ages are 750–1000 years too young, but *c*.20 ka the ^{14}C ages may be *c*.3 ka too young and between 35–40 ka the ^{14}C ages may be *c*.5 ka too young. When researchers provide new dates in these time ranges, they should always quote both the "raw" uncalibrated date in radiocarbon years and the calibrated date. Mellars (2006) reviewed the impact these recent efforts at calibration have had on radiocarbon

dates for sites in Europe, but his use of the NotCal calibration curve, which was not designed for this purpose, was controversial (e.g., Turney et al. 2006). A further complication is the "reservoir effect," where an apparent age is obtained from a sample that contains carbon from a source or reservoir other than atmospheric carbon dioxide. The marine reservoir effect is one example. Because there is a delay in the exchange of atmospheric carbon dioxide with ocean bicarbonate and there is mixing of older upwelled deep water with the surface waters in the ocean there is a variable offset in apparent ages obtained from marine samples. Thus a reservoir correction must be made to account for the incorporation of "old carbon" in the sample. Human bone samples may sometimes be prone to this problem if there has been significant consumption of marine resources in the diet (e.g., for coastal Mesolithic populations). Recent developments In addition to the improvements in calibration of ^{14}C ages recent developments in the chemical pretreatment of samples has led to dramatic changes in dates for some events of interest in human evolution, particularly for the timing of the earliest anatomical modern humans in Europe. The use of techniques such as ultrafiltration to remove more of the exogenous carbon contamination from bone collagen has led to the re-dating of fossils [e.g., the so-called Red Lady of Paviland (Jacobi and Higham 2008) and the reassessment of chronologies for sites such as **Abri Pataud** (Higham et al. 2011)]. Improvements in the pretreatment of charcoal samples, particularly the development of the acid base oxidation technique (or ABOX) has allowed the re-dating of key sites where bone collagen does not survive [e.g., **Border Cave** in South Africa (Bird et al. 2003) and Devil's Lair in southwestern Australia (Turney et al. 2001)]. A further development that is making significant improvements in the construction of ^{14}C-based chronologies is the availability of online calibration coupled with sophisticated statistical techniques such as Bayesian analysis (e.g., OxCal 4.1). *See also* **radiometric dating**.

radiocarpal joint (aka antebrachiocarpal joint) In modern humans this is the articulation between the distal radius and the scaphoid and lunate. Modern humans and the great apes have lost the articulation between the ulnar styloid process and the triquetrum and pisiform seen in other primates so that the radius is the only connection between the hand and the rest of the forelimb. In hominoids, the ulnar styloid process is reduced in size and an interarticular disc takes its place between it and the triquetrum, probably allowing for a greater range of pronation and supination of the hand (Lewis 1989). This mobility is advantageous for suspensory and climbing behaviors and modern humans have retained this mobility. In most other nonhominoid primates the ulnar–triquetrum–pisiform articulation is extensive and provides additional stability on the medial side of the wrist during quadrupedal locomotion.

radiogenic dating *See* **geochronology**; **radiometric dating**.

radioisotopic dating *See* **geochronology**; **radiometric dating**.

radiometric dating (L. *radiare*=to emit beams, from *radius*=ray and *metricus*=relating to measurement) A family of dating methods based on measuring radiation. The methods measure the amount of parent material that remains, and/or the amount of daughter product accumulated, or the amount of radiation received (e.g., optically stimulated **luminescence dating** and **electron spin resonance spectroscopy dating**). One or more of these two measurements, in combination with the decay constant, allow scientists to determine how long the decay process has been running. Some radiometric methods (e.g., **potassium-argon dating**, **argon-argon dating**, **radiocarbon dating**, and **fission track dating**) work independently of local levels of cosmic and other background radiation. Other radiometric methods are sensitive to local burial conditions. To be useful for dating fossil hominins a radiometric system must: have an appropriate length **half-life** (i.e., one that is long enough for there to be more than a small number of atoms to measure – this is why standard radiocarbon dating is unreliable above *c*.40 ka – and short enough that there are perceptible changes in the ratio of the parent to the daughter atoms within the past 5–8 Ma); have a stable daughter product; be a "closed" system, which means that there is no other source of daughter product, nor can the accumulated daughter product be removed; and be such that the beginning of the process of the decay of the parent atom must coincide approximately with either the death of the animal, or the formation of the rock.

radix The number of individuals chosen as the basis for a **life table**. The radix is usually chosen so that it is a power of 10. Common radix values

for life tables are 100 (e.g., a survivorship of 50 represents the point where half of the population has died), 1000, and 100,000.

rain forest Most early **hominins** were associated with tropical woodlands, but it is also likely that some hominids and perhaps even hominins (particularly those in the **Miocene** and **Pliocene**) existed in areas that contained tropical rain forest. However, tropical forest is seen as a less favorable habitat for hominins than woodland because of the inaccessibility of fruits in the higher canopy and the cryptic nature of forest animals. Nonetheless, there is increasing archeological evidence for forest exploitation by hominins in Africa during the **Pleistocene** (Mercader 2002). Archeological and paleontological data also strongly suggest that modern humans colonizing Southeast Asia 46–34 ka exploited a range of habitats, including rain forest (Barker et al. 2007).

Rancah (Location 7°15′S, 108°12′E, West Java, Indonesia; etym. named after a local village) History and general description This is the site of a controlled excavation carried out in 1999 on the banks of the Cisanca River that produced an isolated hominin right I$_2$. Temporal span and how dated? **Electron spin resonance spectroscopy dating** applied to bovid teeth found below RH 1 yielded an average age of *c*.606 ka, but this age estimate is likely younger than the actual age for there is evidence the bovid teeth had been heated during their depositional history. Hominin found at site **RH 1** is a right I$_2$ crown that has been referred to *Homo erectus*. Archeological evidence found at site None. Key references: Geology, dating, and paleoenvironment Weeks et al. 2003, Bogard et al. 2004; Hominins Kramer et al. 2005.

Rancah hominid 1 *See* **RH 1**.

random assortment Occurs in unlinked genes, so that alleles of different genes assort independently during gamete formation. During metaphase I in eukaryotic organisms pairs of homologous chromosomes align randomly at the metaphase plate (i.e., which of a pair of chromosomes is from the male parent, and which is from the female, is due to chance). This process creates an unpredictable mixture of maternally and paternally inherited alleles in the newly formed gamete. This phenomenon, along with chromosomal **crossing over** in prophase

1 and fertilization, helps increase genetic diversity by producing new genetic combinations. *See also* **Mendelian laws**.

random drift *See* **genetic drift**.

random mating A mating "system" in which every individual has an equal chance of interbreeding with every other individual of the opposite sex, regardless of genetic, physical, or social preferences. This type of mating system contrasts with a situation such as that described for positive assortment. *See also* **positive assortment**.

randomization This term refers to a type of resampling analysis used in hypothesis testing. It is also known as a permutation test (although "combination test" would be more accurate, as will be explained further below). Randomization uses an iterative process of sampling without replacement to generate P values for a test statistic. For example, suppose a researcher has a sample of 10 fossil crania from site A and 12 fossil crania from site B, and wishes to know whether the median endocranial volume differs significantly between the two sites. There is no equation that allows the P value to be calculated directly for a test of difference in medians (as opposed to means), but a randomization procedure can be developed to do this. To test the hypothesis of no difference using randomization, the researcher would first calculate the actual difference in median values between sites A and B. Under a null hypothesis of no difference in median endocranial volume between the two sites, crania at sites A and B can be considered to belong to a single population, in which case the difference in medians between the crania should be similar to a difference in medians calculated for any arbitrary grouping of all 22 crania into one group of 10 and one group of 12. For this particular example there are 646,646 unique ways to split the sample into groups of 10 and 12 if order does not matter (order is unimportant since regardless of whether a group consists of specimen X followed by specimen Y or specimen Y followed by specimen X, the median will be the same). In most randomization tests the order will not matter, but since permutations calculate the number of possible ways to sample data when order *does* matter and combinations calculate the number of possible ways when order *does not* matter, this is why "combination test" would be a more appropriate name. If one were to calculate the difference in medians for all 646,646 ways to split the

data in this example, one would perform an "**exact randomization**." However, in this case (and in many cases where this technique is likely to be used) the number of total possible combinations is too high to reasonably work through every single possibility. Thus, a large number (typically $n \geq 5000$ or more) of randomly selected possibilities is used as a proxy for the full set of possible combinations (i.e., this is a version of the **Monte Carlo** methods). In the example, the difference in medians between groups A and B is calculated for n total resampled data sets, each of which randomly selects 10 crania from the full data set of 22 crania to place in group A, and the remainder of which are placed in group B. As stated above, if there is *no* difference between groups then the actual difference in medians should be similar in value to those generated from the randomly rearranged data set and should fall somewhere in the middle of the range of resampled differences in median. However, if there *is* a difference in the populations from which site A and site B crania are drawn, then the actual difference in medians should be at the extreme of values from the resampled data. To generate a P value, the number of resampled differences in median that are equal to, or greater than, the actual differences are counted (call this value x); the P value is calculated as $(x+1)/(n+1)$. The additional 1 in the numerator represents the actual value for the difference in medians, and the denominator represents the total number of resampled values plus the original observation from the actual data set; see Manly (2007) for a fuller discussion of the method. A variant on standard randomization is often used in paleoanthropology. In many cases some measure is calculated for a fossil sample (e.g., a measure of size or shape variation) based on a small number of specimens, and researchers wish to know whether that value differs significantly from the value of that measure calculated for a sample drawn from a comparative taxon with a larger sample size (e.g., comparative samples of extant great apes or modern humans). In this case the extinct taxon is considered to fall into two samples: those individuals recovered as fossils and those individuals that have not been recovered as fossils (e.g., individuals that never fossilized, fossils that were destroyed by taphonomic processes, or fossils that simply have not yet been discovered). Thus, the fossil sample represents only one of the two possible groups for that extinct taxon. The samples of the extant taxa are thus resampled without replacement into two subsamples: one whose sample size is equal to that of the fossil sample, and one that contains the balance of the comparative sample. Since the variable of interest is calculated *only* for those extinct individuals that are recovered as fossils, it is similarly calculated *only* for the extant individuals that are randomly selected to simulate the "fossil" group (i.e., the group whose sample size equals that of the fossil sample). This resampling procedure is repeated *n* times for the comparative sample, and the fossil value is compared to the values generated for the comparative sample. This type of procedure has been used to assess size and shape variation in hominin cranial, dental, mandibular and postcranial samples (e.g., Richmond and Jungers 1995, Lockwood et al. 1996, Reno et al. 2003, Harmon 2006, Skinner et al. 2006, Gordon et al 2008). *See also* **resampling**.

range expansion A type of dispersal that occurs when the area in which a population is found increases but still encompasses the original ancestral area. For example, towards the end of the Pliocene a trend toward environmental aridity led to an expansion of grasslands in Africa. Many grazing mammalian species, and the carnivores that preyed on them, might be expected to have experienced range expansion during this time period.

range shift A type of dispersal that occurs when the area in which a population is found changes geographic position. For example, during aridification trends in the southern hemisphere, **mesic** vegetational zones shifted their geographic positions towards the equator (e.g., Vrba 1992). This resulted in mammals that were adapted to mesic conditions experiencing a northward range shift.

raptor (L. *raptor*=thief, from *rapere*=to carry off) An informal but common term for birds of prey, primarily referring to those which hunt vertebrates during daylight; the term can also include the nocturnal birds of prey, such as owls, but this use is less common. Raptors are of significance in paleoanthropology because several taphonomic studies have demonstrated that accumulations of micromammal remains at fossil sites are the result of raptor behaviour: notably feces or pellets resulting from feeding activities, sourcing prey which would have been gathered over a wider area than rodents would travel in life, unaided. Therefore paleoenvironmental reconstructions of a site that rely on micromammalian paleoecology must consider the taphonomy of the microfaunal assemblage. *See also* **Taung 1**.

rarefaction A statistical technique developed by ecologists and often employed by paleontologists and zooarcheologists to compare **species richness** (i.e., the number of species) across fossil samples of different sample sizes. Because the number of species observed in an assemblage is influenced by the number of specimens that are sampled, rarefaction is used to correct for sampling effects. This is accomplished by estimating species richness from a subsample of the total assemblage. For example, consider fossil assemblage A with 1000 specimens belonging to 15 species and fossil assemblage B with 500 specimens sampling 10 species. The greater number of species in fossil assemblage A could be indicative of a more species-rich faunal community or its larger sample size, or both. Rarefaction can be used to determine how many species one would expect to observe in fossil assemblage A at a sample size equivalent to fossil assemblage B (500 specimens), facilitating a more meaningful comparison of species richness between fossil assemblages. Rarefaction is also used to determine whether a faunal community has been sufficiently sampled. This is accomplished by generating rarefaction curves, which are graphs plotting the number of species relative to the number of specimens sampled. Rarefaction curves typically have steep slopes at first, as the most common species are sampled, and slowly plateau as rare species are recovered at increasingly large sample sizes. When the number of species no longer increases at larger samples, it can be concluded that further collecting is unlikely to yield additional species.

ray (L. *radius*=radiating line) A manual digit plus its metacarpal or a pedal digit plus its metatarsal. Rays are normally numbered (e.g., second manual ray or fourth pedal ray), whereas manual digits are normal named (e.g., ring finger, thumb).

Raymond A. Dart Collection Taxon *Homo sapiens*. History Raymond **Dart** was appointed the Chair of Anatomy at the University of the Witwatersrand in 1923. While a Senior Demonstrator with Sir Grafton Elliot Smith in the Department of Anatomy at University College London, Dart had paid a short visit to T. Wingate Todd at Case Western Reserve University, Cleveland, Ohio, USA and a more extended visit to Richard J. Terry at Washington University Medical School, St. Louis, Missouri. Inspired by the collections of modern human skeletons at Cleveland and St. Louis,

Dart began his own collection at the University of the Witwatersrand. This involved devising a sequence of dissection that did not involve damage to the skeleton. When Phillip **Tobias** took over from Dart as Professor of Anatomy at the University of the Witwatersrand there were over 2000 individuals in the collection, and in 2009 it was reported that there were 2605 cadaver-derived skeletons in the collection. An excellent account of the history and the nature of the Raymond A. Dart Collection can be found in Dayal et al. (2009). Curator/contact Brendon Billings (e-mail brendon.billings@wits.ac.za). Postal address 7 York Road, Parktown, Johannesburg, 2193, South Africa. Relevant references Tobias 1987, Rankin-Hill and Blakey 1994, Hunt and Albanese 2005, Dayal et al. 2009.

RC 911 *See* **HCRP RC 911**.

reaction norm (Ge. *reaktionsnorm*) A term introduced by Richard Wolterek to describe the range of phenotypes an individual could exhibit in response to an infinite range of environments. A reaction norm is usually illustrated by plotting the curvilinear relationship between a measure of the phenotype and a measure of the environment (e.g., body size and average annual rainfall). This method is used to show the contribution of environmental variation to phenotypic variation in different species, or how differing genotypes within a species respond to varying environments. The relationships between genetic, phenotypic, and environmental factors can be complex, and the numbers of possible norms of reaction for each genotype underlines this fact. The shape of the curve(s) can range from a narrow distribution, indicating a highly predictable phenotype from any given genotype (little to no contribution from the environment), to a broad distribution, indicating that phenotypic variation is mostly a consequence of variation in the environment. *See also* **phenotypic plasticity**.

reading frame *See* **open reading frame**.

reassembly Bones and teeth that are fragmented, but otherwise undamaged and undistorted, can be reassembled. The process of reassembly is painstaking work, and researchers used to have to sit down with the original pieces and try and fit them together by hand (e.g., Alan **Walker**'s work on KNM-ER 3733). However, recently researchers have been using computer software to create a "virtual fossil" the fragments of which can be moved and rotated on the screen. This means that researchers can work at reassembling hominin fossils without risking damage to the originals (e.g., Zollikofer et al. 1998, Ponce de León 2002).

recent out-of-Africa hypothesis *See* **out-of-Africa hypothesis**.

recessive trait A recessive trait is one that will only manifest phenotypically if both alleles at a diploid locus are the same recessive alleles. For example, an individual with blood type O has the genotype OO at the ABO blood group locus, whereas people with genotypes AO or BO have blood types A and B, respectively. This is because the A and B alleles are dominant to the O allele, which is said to be recessive. Thus, recessive alleles are not expressed in the heterozygote form.

recognition species concept (or RSC) This is a process definition of a species that tries to overcome the relational aspect of the biological species concept. The recognition species concept was defined by Hugh Paterson (1985) as "the most inclusive population of individual, biparental organisms which shares a common fertilization system." In other words species are composed of animals that share a specific mate-recognition system (or SMRS). *See also* **species**.

recombination The exchange of DNA between individual chromosomes in a pair during prophase I of meiosis. Recombination, also known as crossing over, results in the shuffling of genetic material between homologous chromosomes so that alleles at different loci that are located far apart will not be linked. The frequency of recombination varies across physical distance on a chromosome and there are recombination hotspots which are short regions (just a few thousand base pairs) of the genome at which recombination is substantially elevated relative to the genome average. The frequency of recombination between two loci can be used to calculate relative distance in the genome (known as a genetic map and measured in centimorgans). One centimorgan equals the distance between two loci for which one recombination occurs out of 100 meioses. *See also* **linkage**.

recombination fraction The proportion of meioses in which a given set of loci will be separated by **recombination**. The recombination fraction is usually denoted by the symbol θ, which can range from 0.0 (e.g., for two loci that are located within a block of loci in the same segment of a chromosome – i.e., in a single **haplotype** – and that are rarely, if ever, separated during **crossing over**) to 0.5 (e.g., for two loci that are located on different chromosomes or very far apart on the same chromosome and are, thus, assorted independently). The recombination fraction is generally higher for loci that are more distant from each other on a single chromosome because the probability that they will be separated by crossing over is greater. *See also* **recombination**.

recombination hotspot   *See* **recombination**.

reconstruction   If only part of an undistorted fossil bone or tooth has been preserved, the missing parts can be reconstructed to provide estimates of variables such as measured crown area for postcanine teeth, or endocranial volume for crania. This may involve duplication if the missing piece is a bilateral structure and the other side (called the **antimere**) is preserved, or it may involve extrapolation if there is no antimere, or if only part of a structure is preserved. In general, the more complete a tooth or bone is, the more reliable the reconstruction. The most difficult problems occur when a fossil has been affected by plastic deformation. If the deformation has affected only one side of a bilateral structure then the undeformed side can be used to reconstruct the deformed side. If both sides are deformed the only recourse for researchers is to use computed tomography (or CT) to image the fossil, and then use software programs based on D'Arcy Thompson-type transformation grids to try and estimate the undeformed shape of the fossil (e.g., Zollikofer et al. 2005). The problem with these methods is that in most cases extant taxa have to be used to generate the reconstruction algorithms, and it is difficult to determine which parts of a virtual fossil are "real" and which are reconstructed, albeit on the basis of a careful and scientifically based reconstruction process.

reconstruction algorithm   *See* **reconstruction**.

recursion   (L. *recurre* = to return or recure) A term used in language studies for the computational mechanism by which a finite set of basic elements are combined to create an infinite set of discrete expressions. Certain grammatical rules can be repeatedly applied *ad infinitum* as with "and" or "that." Both can be used to string an infinite number of sentences together. The same can be said of numerical rules where adding 1 to a product leads to an infinite number. Hauser et al. (2002) argue that recursion is a unique feature of modern human communication and forms the basis of *all* modern human languages. Thus, recursion consists of basic elements (e.g., signs) that can be combined and recombined to generate an infinite number of larger linguistic elements.

reduced major axis regression   (or RMA) A form of **regression** in which model **parameters** are fitted by minimizing the sum of products of differences between observed values and the best-fit line in both the independent and dependent variables. The principles of RMA are similar to those of **major axis regression**, but RMA is preferred in cases where the independent and dependent variables are not measured in the same units. Because RMA takes into account error in all variables and not just the independent variable when fitting model parameters it is an example of a **model II regression**.

reduction sequence   *See chaîne opératoire*.

reduction strategy   *See chaîne opératoire*.

reduncine   *See* **Reduncini**.

Reduncini   (L. *reduncus* = curved back) A tribe of antelopes (family **Bovidae**) comprising the reedbucks (*Redunca*), kobs (*Kobus*), and allies. Their modern distribution is limited to Africa but they originated in Eurasia during the late Miocene. Modern examples prefer well-watered environments and many live close to permanent sources of water. Fossil reduncines are frequently found at hominin sites, where they can be a good indicator of the local availability of marshy habitats.

reference   Reference refers to an organism establishing a relationship between a given sign (e.g., visual, auditory, gustatory, olfactory, or tactile) and something that exists in the world, the **referent**. Signs may be linked with internal states such as emotions (e.g., fear, submission, sexual interest), external entities such as predators (e.g., alarm or affiliative/contact calls), or abstract concepts (e.g., home, yesterday, beauty). Some of the most controversial questions in the comparative sciences have to do with whether or not animal signs are open or closed, bound or displaced. Open/closed reference has to do with the rigidity between a given sign and its referent and whether or not the referential relationship is biologically or genetically specified as opposed to being a socially learned convention. Another controversy has to do with whether or not reference is dependent upon the availability of the referent to the senses (i.e., a bound reference) or whether a sign can refer to a given referent in its absence (i.e., a displaced reference). Generally, non-human animal communication is considered to be largely closed and bound (Deacon 1997, Pinker 1994, but see Cheney and Seyfarth 2007). The distinctiveness of modern human language is that it is largely open and unbounded.

reference digit The digit of the hand or foot that determines the names that are given to the sideways movements of the other digits and thus the names of the muscles responsible for those movements. Therefore, because the middle finger is the reference digit of the hand the act of moving the index finger away from it is called abduction, and any muscle that moves the index finger away from it is called an abductor. The movement of any other digit away from the reference digit is also called abduction. Likewise, the movement of any other digit towards the reference digit is called adduction. *See also* **abduction; adduction**.

referent Referent is used in connection with the principle of reference. It refers to something that exists in the world for which an organism has established some sort of sign (e.g., visual, auditory, gustatory, olfactory, or tactile) to represent it. It is also the term used for the source of the model in referential modeling. *See also* **reference; referential modeling**.

referential modeling (L. *referre*=to carry and *modus*=standard) One of the two types of behavioral modeling suggested by Tooby and DeVore (1987). In referential modeling the real behavior of one, or more, living animal(s) is used to reconstruct the behavior of a fossil taxon (the referent). For example, if in extant forms closely related to fossil hominins substantial levels of canine-crown-height sexual dimorphism are always associated with a multimale social structure, then the principles of referential modeling suggest it is reasonable to infer that a fossil hominin with similar

levels of canine-crown-height sexual dimorphism most likely had a multimale social structure. *See also* **modeling**.

refit A term used in the analysis of stone tools that refers to flakes and flake fragments that have been "put back together." Refits are generally divided into two types: refitted breaks and reduction refits (Conrad and Adler 1997). Refitted breaks conjoin broken artifacts, and reduction refits conjoin two or more artifacts within a reduction sequence. The extent of refitted breaks can be important in revealing post-depositional processes such as trampling, whereas reduction refits allow the precise reconstruction of how ancient hominins knapped stone, literally in a "blow-by-blow" fashion (e.g., Roche et al. 1999). Reduction refits of long sequences of removals have been critical in demonstrating the varied ways in which a single block of stone may be modified over the course of its reduction; for example, the production of **Levallois** points as well as blades (Marks and Volkman 1987). *See also* **reduction sequence**.

refugia Plural of **refugium** (*which see*).

refuging This model of **Oldowan** site formation proposes that hominins transported carcasses relatively short distances to stands of trees where food would have been consumed in relative safety (Isaac 1983, Blumenschine 1991). Recurrent visits would lead to the accumulation of artifacts and processed bones that would eventually be manifest as an archeological site.

refugium (L. *refugium*=a place of safety) The discrete location of a small residual population of a species that was once much more widespread. Refugial populations arise when inhabited areas get cut off from the rest of the range occupied by a species. For example, during the Pleistocene glaciations, it is likely the land bordering the Mediterranean acted as a refugium for hominins and other animals that were not well adapted to the extreme cold conditions that prevailed in the higher latitudes. The populations in these refugia would then have re-expanded into higher latitudes during interglacial periods. Climatic fluctuations that reduced the total area of tropical African forest in the Pleistocene may also have created small pockets of forest refugia that would have been conducive to speciation within arboreal primates such as guenons. Any isolated population separated from their parent population in refugia offers the potential for allopatric speciation. Refuge populations can also lead to **endemism** (i.e., a species that is unique to a particular, often small, area) (pl. refugia).

Regourdou (Location 45°03′N, 1°11′E, France; etym. of unknown origin. The name is often written le Régourdou, but according to the current excavators this is a misnomer and should be avoided) History and general description A collapsed cave site, located along the Vézère River in the Dordogne area of France, on the same hillside as **Lascaux** cave. The owner of the land discovered a human skeleton in 1957 and the site was subsequently excavated by Eugène Bonifay and Bernard **Vandermeersch**. Temporal span and how dated? Chronostratigraphic comparison to other sites and **biostratigraphy** have suggested an age of late **Oxygen Isotope Stage** (OIS) 5 or early OIS 4 (*c.*70 ka) for the layer containing the burial. No radiometric dating has been attempted. Hominins found at site **Regourdou 1**, an intentional burial in the **Mousterian** layers, consisting of a mandible and several postcranial elements including nearly complete left and right upper limbs, has been determined to be a young adult *Homo neanderthalensis*, sex unknown. If the dating is correct, this individual is one of the oldest known Neanderthal skeletons. Vandermeersch and Trinkaus (1995) suggest that a second individual (Regourdou 2) may be represented by one or more foot bones. Archeological evidence found at site The two main archeological layers both contain stone tools assigned to the Quina Mousterian (*see* **La Quina**), although the assemblage is surprisingly poor in artifacts. Key references: Geology, dating, and paleoenvironment Bonifay and Vandermeersch 1962, Bonifay 1964; Hominins Piveteau 1959, 1963, 1964, 1966, Vandermeersch and Trinkaus 1995; Archeology Bonifay and Vandermeersch 1962, Bonifay 1964.

Regourdou 1 Site Regourdou, France. Locality N/A. Surface/*in situ* In situ. Date of discovery September 1957. Finder Roger Constant. Unit Layer 4 (**Quina Mousterian**). Horizon N/A. Bed/member N/A. Formation N/A. Group N/A. Nearest overlying dated horizon None. Nearest underlying dated horizon None. Geological age End of Oxygen Isotope Stage (OIS) 5 or beginning of OIS 4 (*c.*70 ka). Developmental age Young adult. Presumed sex Unknown. Brief anatomical description Mandible,

eight vertebrae, sacrum, os coxae, sternum, ribs, both humeri, both radii, both ulnae, both clavicles, one patella, and several hand and foot bones. Announcement Piveteau 1959. Initial description Piveteau 1959. Photographs/line drawings and metrical data Piveteau 1959, 1963, 1964, 1966, Maureille et al. 2001. Detailed anatomical description Piveteau 1959, 1963, 1964, 1966, Vallois 1965, Senut 1985, Maureille et al. 2001. Initial taxonomic allocation *Homo neanderthalensis*. Taxonomic revisions None. Current conventional taxonomic allocation *H. neanderthalensis*. Informal taxonomic category Pre-modern *Homo*. Significance This individual is one of the most complete Neanderthal postcranial skeletons, and was intentionally buried with several bear bones. The teeth are unusually small. Location of original Musée d'Art et d'Archaeologie de Périgord, Périgueux, France.

regression (L *regress*=to step back) Earth science Contraction of a body of water (lake or sea) to expose as part of the land surface an area that used to be part of the lake or sea bed. For example, Lake Turkana is smaller than it was when the Koobi Fora Research Camp was built in the 1970s. Then the beach was within 30 m of the camp, now the beach is more than 500 m away. The additional land exposed by the contraction of the lake is the product of lake regression. Statistics A statistical approach used to identify the relationship between two or more **variable**s in which values for one variable (the dependent variable) are predicted based on the values of one or more other variables (the independent variables). In **simple regression**, a line is fit to a plot of two sets of continuous measurements. The slope, the intercept, and the scatter about the line are all **parameter**s that can be used to describe a data set. **Multiple regression** predicts values of the dependent variable based on multiple independent variables with associated parameters identifying the effect size and significance of each independent variable. Logistic regression is a technique used to predict binary outcomes in a dependent variable which is not commonly used in paleoanthropology but sees use in paleodemography (e.g., given a set of factors, does a certain outcome occur, such as death). Regression techniques can be split into **model I regression** and **model II regression** on the basis of whether models attempt to minimize errors in independent and dependent variables or the independent variable alone. *See also* **model I regression**; **model II regression**.

Reilingen (Location 49°17′N, 8°33′E, Germany; etym. the name of a nearby town) History and general description A hominin partial calvaria together with 38 non-hominin mammalian fossils were discovered early in 1978 during a commercial dredging operation in gravel pit. Temporal span and how dated? **Biostratigraphy** from the dredged mammalian fauna suggests the **Holstein** interglacial/Late **Wurm** glacial. Hominins found at site Partial calvaria recovered in four pieces. Archeological evidence found at site None. Geology, dating, and paleoenvironment Ziegler and Dean 1998. Hominins Czarnetzki 1989, Dean et al. 1998.

Reilingen calvaria Site Reilingen. Locality N/A. Surface/*in situ* It was recovered by workers using a mechanical dredger in a commercial gravel pit. Date of discovery 1978. Finder Workmen at the gravel pit. Unit N/A. Horizon N/A. Bed/member N/A. Formation N/A. Group N/A. Nearest overlying dated horizon N/A. Nearest underlying dated horizon N/A. Geological age Holstein interglacial/Late **Würm** glacial. Developmental age Adult. Presumed sex Unknown. Brief anatomical description Most of the posterior part of an adult **calvaria**, including both parietals, much of the squamous part of the occipital bone and an almost complete right temporal bone. Announcement Czarnetzki 1989. Initial description Czarnetzki 1989. Photographs/line drawings and metrical data Dean et al. 1998. Detailed anatomical description Dean et al. 1998. Initial taxonomic allocation *Homo erectus reilingensis* Czarnetzki, 1989. Taxonomic revisions "Archaeomorphous" *Homo sapiens* (Adam 1989), early *Homo neanderthalensis* (Dean 1993). Current conventional taxonomic allocation *H. neanderthalensis*. Informal taxonomic category Pre-modern *Homo*. Significance Researchers who interpret *H. neanderthalensis* as an evolving lineage would interpret the Reilingen calvaria (plus e.g., hominins from **Sima de los Huesos, Steinheim,** and **Swanscombe**) as examples of "Stage 2" in this lineage, a stage also referred to as the pre-Neanderthal stage (e.g., Dean et al. 1998). Location of original **Staatliches Museum für Naturkunde**, Stuttgart, Germany.

relative dating (L. *relatus*=to relate or compare with something else) A dating method that relates a horizon, or an assembly of fossils or artifacts, to an externally validated sequence of change of some kind (e.g., changes in climate, changes in the direction of the Earth's magnetic field, or morphological changes within an evolutionary lineage, or in artifact design). Fossils or artifacts incorporated within a horizon are used to link that horizon to an absolutely dated horizon elsewhere. For example, mammalian fossils within the caves in southern Africa are used to date their parent horizon by matching them with the same type of fossils found in absolutely dated horizons at East African sites.

relative rate test *See* **molecular clock**.

relative risk The risk for a particular individual (e.g., for developing a genetic disorder) relative to the risk of another individual (e.g., a control, such as someone not exposed to a particular environment or someone who is not taking a specific drug). Estimates of relative risk are used in epidemiology and medical genetics to assess low-probability events, but they can also be applied to models that have one group of hominins involved in one set of behaviors and a second group doing nothing, or undertaking a different set of behaviors.

relative taxonomic abundance In a fossil assemblage, relative taxonomic abundance quantifies the numerical abundance of a given taxon relative to the total number of all taxonomic groups in that assemblage. For example, the relative abundance of wildebeest in an assemblage comprising 25 wildebeest, 15 impala, and 10 buffalo is 0.50 (=25/50). Relative taxonomic abundances are typically quantified using the **number of identified specimens** (or NISP) or **minimum number of individuals** (or MNI), although any quantitative abundance index will suffice. Relative taxonomic abundances are particularly useful for making paleoenvironmental reconstructions. For example, if the relative taxonomic abundance of **alcelaphine bovids** increases through time at a paleontological locality, one might infer an expansion of grassland environments through time. In zooarcheological assemblages, relative taxonomic abundance is used to examine how human subsistence strategies varied through time and space, particularly in response to environmental change, the development of new technology, or demographic shifts.

relative tooth size The absolute sizes of tooth crowns provide valuable taxonomic and functional information, but there is a long tradition of investigating the size of the whole, or parts, of the dentition in relation to body mass, or to variables that serve as a proxy for body mass. Researchers have also investigated the taxonomic utility of relating various components of the dentition (e.g., the size of individual tooth crowns and incisor, premolar, and molar tooth row chords, etc.). Attempts to relate the size of the whole, or parts, of the dentition to body mass, or to various proxies for body mass (e.g., the size of the cranium) have a long history. For example, Flower's dental index (Flower 1879) related the premolar-molar chord (i.e., the length of the postcanine tooth row) to basion-nasion length, a proxy for the size of the cranium. More recently, Henry McHenry introduced the **megadontia quotient** (MQ) that relates the area of three postcanine teeth to estimated body mass. By analogy with the better-known **encephalization quotient**, MQ = observed tooth area / 12.15 (body mass)$^{0.86}$, where "observed tooth area" is the sum of the areas of the P_4, M_1, and M_2 (McHenry 1988, p. 144).

relative warps *See* geometric morphometrics.

remodeling *See* bone remodeling.

Rensch's rule *See* sexual dimorphism.

replacement substitution *See* nonsynonymous mutation.

replacement with hybridization A variant of the **out-of-Africa hypothesis**. Replacement with hybridization, originally called the Afro-European sapiens hypothesis, was proposed by Günter **Bräuer** (1984), who suggested that the earliest evidence of modern human morphology appeared in East and southern Africa and that members of these populations migrated out of Africa to populate Eurasia, eventually replacing archaic hominins in those regions. These early African populations of *Homo sapiens* would thus be the direct ancestors of all modern human populations, but unlike the adherents of the strong version of the out-of-Africa hypothesis, Bräuer also accepted that these early modern humans may have interbred with pre-modern *Homo* groups such as *Homo neanderthalensis* and aboriginal East Asians, presumably *Homo erectus*.

replication The process whereby **DNA** is copied into two identical daughter sequences by DNA polymerase. Replication occurs during interphase in mitosis and meiosis. Replication errors lead to point mutations and they occur at a rate of about 10^{-9}–10^{-10} in modern humans (i.e., one replication error occurs for every 1 in 10 billion base pairs).

reproductive effort The proportion of energy or materials an organism devotes to reproduction rather than to growth and maintenance. Reproductive effort in female modern humans and nonhuman primates has been described as being higher than in males, due to anisogamy, internal gestation, and lactation. Relative to nonhuman primates, it might be expected that total lifetime reproductive effort in modern humans is comparatively high due to aspects of our life history (e.g., cooperative foraging and breeding, long juvenile periods). However, a recent study suggests that actual human lifetime reproductive effort is not unusually high (Burger et al. 2010). *See also* **life history**.

reproductive investment The energy invested by females or males in sexual maturation, sexual behavior (including competition, copulation, and mate guarding), gestation, and postnatal support of offspring. In most mammals, including primates, where reproduction is a costly event, reproductive investment is higher per reproductive event for females compared to males. Various factors have the potential to impact maternal effects on offspring development, as well as mate quality. Life history theory predicts that reproductive investment should be high when high fitness returns are expected. This, combined with the cost/benefit calculation regarding current versus future reproduction (which is affected by offspring number, sex ratio, etc.), should determine an individual's level of reproductive investment throughout their reproductive lifetime.

reproductive success The ability of an individual to reproduce, thereby passing on his or her genes to the offspring. Reproductive success is often quantified as the number of offspring that an individual has who survive or who survive and reproduce.

resampling A set of statistical techniques that rely on sampling from a data set to produce confidence intervals for a sample parameter (e.g., mean, median, standard deviation, etc.) or significance tests for some test statistic (e.g., difference in the mean between two groups, difference in a correlation coefficient from some pre-specified value such as zero, etc.), or to validate predictive models (e.g., to assess the accuracy and precision of predictions from regression or discriminant function analysis). Resampling techniques are sometimes preferred over more traditional techniques because they can find confidence intervals and *P* values for test statistics that do not have standard analytical solutions (e.g., there is no equation for calculating a confidence interval for the median), and/or because resampling methods are nonparametric in that they do not assume the data follow a particular distribution. Resampling techniques used in studies of human evolution include the bootstrap, jackknife, randomization (also known as permutation tests), and cross-validation. In each of these techniques a subset of data is sampled to generate either a parameter, test statistic, or predictive model and then this procedure is repeated multiple times to generate a confidence interval, *P* value, or measure of model validation. If all possible versions of a subset of the data are used then it is an exact test (this is required for the jack-knife and is usually the case in cross-validation), but in many cases there are more possible ways to sample the data than can feasibly be performed (as is often the case in bootstrapping and randomization). In these cases a large number (e.g., 10,000) of randomly selected subsets of data are used; these versions are known as **Monte Carlo** methods. *See also* **cross-validation**; **bootstrap**; **jackknife**; **randomization**.

resharpening spall A small flake removed from a stone or bone tool to sharpen an edge dulled by use.

residue analysis The study of biotic and abiotic residues left on the surfaces of stone, bone, or metal tools or on pottery to understand what the tool or pottery was used for. These residues can be identified by microscopy (in the case of large particles, like fur, feathers, or plant residues), by spectroscopy methods (e.g. infrared spectroscopy) that allow the identification of chemical signatures (for more amorphous or invisible residues, like amino acids, peptides, and lipids), or by a variety of other tests designed to identify specific residues (like immunology assays to identify blood types). Residue analysis must take into account the possibility that the residue did not result from an intentional contact with the substance, but rather by precipitation, secondary deposition, or other forms of contamination. Odell (2004) describes four ways by which the authenticity of the residue can be established: (a) if

similar residues are not found on non-utilitarian or non cultural items from the same site, (b) if similar residues are not found in the sediments from that site, (c) if the distribution of the residue on the surface of the tool is "aggregated and coincides with areas of use" (*ibid*, p. 156), and (d) if the residue is found in the same area as and matches the **use wear** on the tool (i.e. wood fibers and starch grains are found in the same area as wood striations). Odell (2004) and Evershed (2008) provide summaries of residue analysis.

resorption (L. *resorbere* = to suck back) The process of resorbing bone tissue. Resorption occurs when bone is remodeled during growth or after a fracture, or when the overall volume of a bone is reduced because it no longer has a function (e.g., the alveolar process of the mandible resorbs after the teeth have been lost) or because of a pathology. Bone resorption is carried out by cells with many nuclei called osteoclasts, and their activity leaves distinctive areas of localized erosion called Howship's lacunae. These have been used to identify taxonomically distinctive areas of bone resorption on surface of the face (e.g., Bromage 1990). Resorption also refers to the process whereby the roots of deciduous teeth are cleared out of the path of the developing permanent teeth during the passage of the latter through the mandible or maxilla. *See also* **bone remodeling**; **eruption**; **ossification**.

resource defense model A model of **Oldowan** site formation that rests on the hypothesis that the hominins responsible for the activities that resulted in such an archeological site were entering the predator guild. The model suggests that hominins would have responded to predation risk by increasing their social interactions. Cooperating would (a) provide protection against predation and (b) would enable them to defend resources more effectively (Rose and Marshall 1996). In this view meat from hunted and scavenged carcasses was transported to places with fixed, defendable resources (e.g., trees, water, plant foods, sleeping sites). Group defense allowed these focal sites to be used regularly for activities that would lead to the gradual accumulation of archeological debris and thus their eventual recognition as an archeological site. The model resembles Glynn **Isaac**'s **home base hypothesis** (e.g., delayed consumption, food transport to a central place, and potentially extensive food-sharing), but it lacks the latter's emphasis on a sexual division of labor.

resource depression   A reduction in the abundance or availability of prey that results directly from the foraging behavior of the predator. In archeological contexts resource depression is associated with diminished access to **high-ranking prey** and may be reflected an expansion of **diet breadth** and increased utilization of **low-ranking prey**. There is much interest in understanding prehistoric resource depression in an effort to understand how hominin foragers have impacted and structured the ecosystems that they inhabited.

restriction enzyme   An enzyme that cleaves a specific sequence (usually 4–5 base pairs) of double-stranded DNA. Restriction enzymes are named based on the organism in which they were discovered. For example, the enzyme *Eco*RI was discovered in *Escherichia coli*. It recognizes the sequence GAATTC and cuts between the G and the A. Since the complementary sequence to GAATTC is the same when read from the opposite direction, this enzyme produces cuts in the double-stranded sequence that result in a single-stranded overhang on each end. Other restriction enzymes can cut symmetrically, resulting in blunt ends. For example, the restriction enzyme *Alu*I (for *Arthrobacter luteus*) recognizes the sequence AGCT and cuts between the G and C. Restriction enzymes are used to assess DNA sequence **polymorphism** and to prepare DNA for cloning or other laboratory methods. *See also* **restriction fragment length polymorphism**.

restriction fragment length polymorphisms   (or RFLP) A **polymorphism** characterized by differences in the lengths of restriction fragments (i.e., a segment of DNA cut by a restriction enzyme). RFLP analyses can be used to assess polymorphism in a DNA sequence. It has traditionally been faster and cheaper than DNA sequencing but not all polymorphisms may be identified (depending on the enzymes used and their recognition sites) and two individuals that have lost the same cut site may not be identical by descent (i.e., the cut site may have been lost because of different changes to the recognition sequence). A classic example of RFLP analysis in a study of modern human variation is that by Cann et al. (1987) who used 12 restriction enzymes

to identify polymorphisms in the mitochondrial genomes of five geographic modern human populations. They concluded that all of the mitochondrial diversity originated in the last 200 ka thus supporting the **out-of-Africa hypothesis** of human evolution.

retention index   A measure of the goodness of fit between a **cladogram** and a **character state data matrix**. The retention index was developed by Farris (1989a, 1989b) and is equivalent to Archie's (1989) homoplasy excess ratio maximum index. In contrast to the **consistency index**, the retention index is *not* sensitive to the number of taxa or the number of characters included in a character state data matrix. The retention index for a single character (ri) is calculated as $g{-}s/g{-}m$, where g is the maximum possible amount of change required by a completely unresolved cladogram (i.e., a cladogram in which all the taxa are equally closely related), s is the number of character state changes required by the cladogram being investigated, and m is the minimum amount of change required by any conceivable cladogram. The retention index of a group of characters (RI) is calculated as $(G{-}S)/(G{-}M)$, where G, S, and M are the sums of the g, s, and m values for the individual characters. In principle, RI values can range between 1 and 0. An RI of 1 indicates a cladogram requires no **homoplasy**, with the level of homoplasy increasing as the index approaches 0. *See also* **cladistic analysis**.

reticulate evolution   (L. *reticulum*, dim. of *rete*=net) In reticulate evolution species are seen as components of a complex genetic network, and the mechanism of speciation is interpreted very differently than it is in more traditional models of speciation (i.e., species form by the **hybridization** of two existing species rather than by bifurcation). This model of evolution is close to how some researchers interpret evolution in geographically widespread groups such as contemporary baboons. In this interpretation, peaks of morphological distinctiveness are equivalent to what in other taxa are interpreted as species differences (Jolly, 2001). Yet at the boundaries between these peaks of morphological distinctiveness there are evident **hybrid zones** where baboons are less morphologically distinct. In the reticulate evolution model it is likely that the location and nature of the peaks of morphologically distinctiveness and of the hybrid zones will have changed over time.

retouch (Fr. *retouch*=to alter or touch up) The removal of one, or more, small flakes to shape a stone or bone tool. The amount of retouch and the number and shape of edges that have been retouched are key elements in many formal stone tool classification schemes (e.g., Bordes 1961a, Debénath and Dibble 1994).

retrotransposition A process used by a mobile DNA element or retrovirus to insert itself into the genome. It uses reverse transcriptase to copy an RNA version of the element or virus into DNA that can then be inserted into the genome. For example, the modern human immunodeficiency virus (HIV) is a retrovirus that uses retrotransposition to insert itself into the genome.

Retzius, Anders Olaf (1796–1860) Educated in both medicine and veterinary science at Lund University, Sweden, and in Copenhagen, Denmark, Anders Retzius was appointed Professor of Anatomy and Supervisor at the Karolinska Institute in Stockholm, Sweden, in 1824. He was both an anatomist and an anthropologist, and early in his career he coined the terms dolichocephaly and brachycephaly. He was first to describe many comparative histological structures of enamel and dentine, including the primary and secondary curvatures of dentine tubules, and the **striae of Retzius** (also referred to as **long-period incremental lines**) in enamel. *See also* **striae of Retzius**.

Retzius' line *See* **striae of Retzius**.

reverse fault In structural geology a reverse fault is a **fault** or fracture with a compressional geometry. The dominant component of the displacement is up the **dip** of the affected strata so that the hanging wall rocks (i.e., the rocks above the fault plane) is displaced upwards relative to the rocks of the footwall (i.e., the rocks below the fault plane).

reverse transcription polymerase chain reaction (or RT-PCR) A variant of the **polymerase chain reaction** in which **ribonucleic acid (or RNA)** is reversed transcribed into its **DNA** complement and then the DNA is amplified the same way as in a normal PCR.

RFLP Acronym for **restriction fragment length polymorphism** (*which see*).

RH Prefix for fossils recovered from **Rancah**, Java, Indonesia.

RH 1 Site Rancah, Indonesia. Locality Near the village of Rancah. Surface/*in situ In situ*. Date of discovery 1999. Finder A local landowner employed as part of the excavation team. Unit N/A. Horizon N/A. Bed/member N/A. Formation N/A. Group N/A. Nearest overlying dated horizon N/A. Nearest underlying dated horizon N/A. Geological age **Electron spin resonance spectroscopy dating** methods applied to bovid teeth found below RH 1 yielded an average age of 606 ka, but this age estimate is likely younger than the actual age for there is evidence the bovid teeth had been heated during their depositional history. Developmental age Adult. Presumed sex Unknown. Brief anatomical description RH 1 is a faintly shoveled right I_2 crown with a weakly developed basal tubercle. Announcement Kramer et al. 2000. Initial description Kramer et al. 2000. Photographs/line drawings and metrical data Kramer et al. 2005. Detailed anatomical description Kramer et al. 2005. Initial taxonomic allocation *Homo erectus*. Taxonomic revisions None. Current conventional taxonomic allocation *H. erectus*. Informal taxonomic category Pre-modern *Homo*. Significance RH 1 is the first hominin specimen recovered from West Java and one of very few hominins recovered as a result of controlled excavations on the island of Java. Location of original **Archaeological Research and Development Center**, Bandung, Indonesia.

Rhafas Cave (Location 34°48′N, 1°53′W, Morocco; etym. unknown) The cave has also been referred to as the Grotte du Rhafas) History and general description This small (10 m × 6 m) cave located in the Oujda Mountains was discovered in 1979 by Luc Wengler, and excavations began in the 1980s. In 2007 a joint team from the Max Plank Institute of Evolutionary Anthropology and the Institut National des Sciences de l'Archéologie et du Patrimoine (INSAP) reinvestigated the **Middle Paleolithic-to-Aterian** transition at the site. Wengler had identified 72 sedimentary strata within a depth of *c*.4.4 m. These are divided from top to bottom into four major units (I–IV). Thirty-nine of the levels have archeological evidence and they span the period between the **Middle Stone Age** and the **Neolithic**. The site is notable for containing a transitional **Mousterian–Aterian** level and for dates of the Aterian that are beyond the limits of radiocarbon dating. <u>Temporal span and how dated?</u> Twelve chalcedony artifacts from the strata of Unit II at the Mousterian–Aterian "boundary" have been dated by thermo**luminescence dating** to *c.*60–90 ka. Sediment from Mousterian levels in Unit III have been dated with optically stimulated **luminescence dating** to 107±12 ka. <u>Hominins found at site</u> None. <u>Archeological evidence found at site</u> The archeological evidence recovered from Unit III (**Mousterian**) and the overlying Unit II (which includes the shift from Mousterian to Aterian) are the only ones to be published. The Aterian from Rhafas Cave is distinguished by the appearance of tanged points. Throughout the sequence, most stone tools are made of materials available in the vicinity of the cave, but rare chalcedony tools from sources more than 60 km away are also present. **Flake**s are produced by a variety of **Levallois** methods and they are found in multiple refitted sets; the retouched tool component is dominated by **scraper**s. Unit III preserves several **hearth**s that include charcoal from cypress, olive, and juniper trees, as well as a diverse fauna that includes various bovids, equids, and rhinoceros, and **ostrich egg shell** fragments. Key references: <u>Geology, dating, and paleoenvironment</u> Mercier et al. 2007b; <u>Archeology</u> Wengler 1997, 2001.

rhinocerotid Informal name for the family **Perissodactyla**, comprising the rhinoceroses and their allies, whose members are found at some hominin sites. *See also* **Artiodactyla**.

Rhinocerotidae One of the perissodactyl families comprising the rhinoceroses and their allies, whose members are found at some hominin sites. Members of the Rhinocerotidae family are referred to informally as rhinocerotids. *See also* **Perissodactyla**.

rhyolite Gk *rhyax* = a stream of lava) A term used for fine-grained volcanic rocks, often rich in phenocrysts of alkali feldspar and quartz. The proportions of silica, sodium, and potassium are used to define lava types, with rhyolite having SiO_2 abundances of greater than 69% and total alkali ($Na_2O + K_2O$) abundances ranging from about 1 to 15% (Le Bas et al. 1986). Where available rhyolite is used for artifact production (e.g., archeological sites in the **Kibish Formation**, Ethiopia).

RI Acronym for **retention index** (*which see*).

ribonucleic acid (or RNA) A nucleic acid whose components are bases (i.e., the four standard RNA bases are adenine, guanine, cytosine, and uracil), and a phosphate and sugar backbone. In RNA, the sugar is ribose, and it alternates with phosphate (derived from phosphoric acid) in the backbone. These are the three components of each nucleotide (i.e., sugar plus phosphate plus one of the four bases), and the nucleotides are joined together by phosphodiester bonds to form a single-stranded RNA molecule. RNA is produced from **DNA** during the process of transcription via the enzyme RNA polymerase. The many different kinds of RNA are important components of protein synthesis and gene regulation. They include messenger RNA (or mRNA), ribosomal RNA (or rRNA), transfer RNA (or tRNA), microRNA (or miRNA), small interfering RNA (or siRNA), and small nuclear RNA (or snRNA). RNA may be processed before it reaches its final form. For example, mRNA is produced in the nucleus from a gene that undergoes transcription. However, before it leaves the nucleus and goes to the cytoplasm for protein translation, it is spliced to remove the intron sequences, and it may undergo further modifications (e.g., polyadenylation on the 3′ end and capping on the 5′ end). The mature RNA may then move into the cytoplasm and be translated into protein, or it may become one of the varieties of functional RNA (e.g., tRNA).

ribosomal ribonucleic acid (or rRNA) A structural component of the ribosome. In modern humans four different rRNAs can be distinguished by size and they interact with a network of 82 proteins to make up a ribosome.

ribosome The component of a cell that uses the information encoded in messenger RNA (or mRNA; *see* **ribonucleic acid**) to combine amino acids to form proteins. Ribosomes are made of protein and ribosomal RNA (or rRNA) and are the organelles that are responsible for the translation process. They interact with transfer RNA (or tRNA) by supplying peptidyl transferase, an enzyme that creates peptide bonds between adjacent amino acids activity to form the growing polypeptide chain.

ridge (OE *hrycg* = spine) Refers to a crest of enamel on the surface of a tooth crown; the term ridge is usually used for the mesial or distal border of a tooth. It may also be used to refer to a linear prominence on a bone that is less marked than a crest. *See also* **distal accessory ridge**; **distal marginal ridge**; **mesial marginal ridge**.

Riencourt-lès-Bapaume *See* **Bapaume**.

Rietputs Formation General description Younger river gravels in the lower reaches of the Vaal River, South Africa, near Windsorton. Sites or localities where the formation is exposed The type site is on the east side of the Vaal River near Windsorton. The Rietputs Formation has been exposed by diamond mining on both sides of the river, but the most extensive exposures are on the east side. Temporal span and how dated? Formerly dated using **biostratigraphy** to the middle Pleistocene, now **cosmogenic nuclide dating** ages suggest a temporal span of $1.89 \pm 0.19–1.34 \pm 0.22$ Ma. Hominins recovered from the formation None. Archeological evidence recovered from the formation Bifacial tools corresponding to "an early phase of the **Acheulean**" (Gibbon et al. 2009, p. 152) have been recovered *in situ* from the lowermost part of the formation in a stratum referred to as the "coarse gravel and sand unit" (*ibid*, p. 152) and the "lower coarse alluvium" (*ibid*, p. 157). Key references: Geology, dating, and paleoenvironment Helgren 1978, de Wit et al. 2000, Gibbon et al. 2010; Hominins N/A; Archeology Klein, 2000, Gibbon et al. 2010.

Rift Valley Research Mission in Ethiopia (or RVRME) An organization established in 1975 by geologist Jon **Kalb** [other US participants included Clifford Jolly, Glenn Conroy, and Douglas Cramer (paleoanthropology) and Fred Wendorf (archeology)] to conduct paleoanthropology research in the **Afar Triangle**, Ethiopia. The RVRME was affiliated with Addis Ababa University and in 1977 it established the country's first vertebrate paleontology laboratory in collaboration with that University's Department of Biology. Students trained by the RVRME include Tsrha Adefris, who was the first Ethiopian to receive a PhD in paleoanthropology. Following aerial surveys over the **Middle Awash** by Kalb in mid-1974 that revealed extensive exposures, the RVRME undertook fossil surveys in the area in early 1975. Ground surveys revealed fossil- and artifact-bearing sediments both older and younger than *Australopithecus afarensis* at **Hadar**. Archeological remains included sites that provided evidence of both the **Late** and **Middle Stone Age**, with *in situ* tools and faunas, including the forest-adapted **Sangoan** culture, to multiple levels and

stages of **Acheulean** artifacts. The Middle Pleistocene hominin site of **Bodo** produced a hominin cranium, **Bodo 1**, and many artifacts, and older archeological sites (e.g., **Dakanihylo** and **Matabaietu**) yielded evidence of both early Acheulean and **Oldowan** industries. Exposures of Middle Pliocene age in the eastern and western of what is now the **Middle Awash study area** were located by the RVRME and have since been assigned to the **Sagantole Formation**, and exposures of late Miocene age further west have since been assigned to the **Adu-Asa Formation**. The activities of the RVRME ceased involuntarily in 1978 and authority to search for fossils in the same region was transferred to the **Middle Awash Research Project** led by Desmond **Clark** and later by Tim **White**. In 1993 *Australopithecus ramidus* fossils (they were subsequently assigned to *Ardipithecus ramidus*) were found in the Sagantole Formation and a year later, in 1994, fossil evidence initially assigned to *Ardipithecus ramidus kadabba* and later to *Ardipithecus kadabba* was recovered from the Adu-Asa Formation.

Rijksmuseum van Natuurlijke Historie The forerunner of the present museum was founded in 1820 in Leiden in The Netherlands to house the part of the collections that made up the "Cabinet du Roi" of Louis Napoleon that belonged to Leiden University, and the private bird and mammal collection of Coenraad Jacob Temminck. In 1878 the mineralogical and paleontological part of the collections were split off and were established in a new museum in the Hooglandse Kerkgracht, also in Leiden, called the "Rijksmuseum voor Geologie en Mineralogie." At this time, whereas the "Rijksmuseum van Natuurlijke Historie" was a scientific institute without exhibitions, the "Rijksmuseum voor Geologie en Mineralogie" embraced both a scientific mission and exhibitions. In 1990 the two museums were reunited as the Nationaal Natuurhistorisch Museum, which translated into English means the "National Museum of Natural History," now known as "Naturalis." A new building in the Darwinweg 2 with seven exhibition halls was opened in April 1998, and Naturalis is now the largest national history museum in the Netherlands with about 10 million specimens. The paleoanthropological collections in Naturalis come from the fieldwork of Eugène **Dubois**. Between 1887 and 1900 Dubois, working as a medical officer in the army, collected fossils in Java, then part of the Dutch Indies colony. Dubois worked first in Sumatra, where he collected fossil rain forest mammals, including orangutan fossils. When a fossil skull was found at **Wajak** (old spelling Wadjak) in Java he turned his attentions to Java and at Wajak he found fragments of a second skull and then at **Trinil** he found a hominin molar, calotte, and femur, together with some 40,000 specimens of mammals and reptiles. Between 1895 and 1900 the collection was transferred to the Rijksmuseum van Natuurlijke Historie and Dubois became the first director of what was later called the Dubois Collection. The Dubois Collection was catalogued by Father J.J.A. Bernsen and much of it was described by D.A. (Dick) Hooijer, and it is presently housed in the Naturalis museum. Hominin fossil collection The hominin fossils collected by Eugène Dubois in 1887–1900: e.g., **Trinil 1**, the third molar found in September 1891, the calotte designated **Trinil 2** found in October 1891 and assigned as the **holotype** of *Anthropopithecus erectus*, and a femur, **Trinil 3**, found in August 1892, that was presumed to be associated with the calotte. Contact department Geology. Contact person John de Vos (e-mail vos@naturalis.nl). Postal address P.O. Box 9517, 2300 RA Leiden, The Netherlands. Street address Darwinweg 2, 2333 CR, Leiden, The Netherlands. Website N/A. Tel +31 71 5687597. Fax +31 71 5687600.

riparian　(L. *ripa*＝bank or shore) Refers to the bank of a river or lake, so a "riparian" paleoenvironment would be one that related to the bank of a stream, river or lake (NB: "riparian rights" are what you pay for when you buy a fishing license).

Riparo di Fumane　*See* **Fumane**.

Riss　*See* **glacial cycles**.

riverine　(ME *rivere*＝river, or riverbank via from the OF, from L. *ripa*＝bank or shore) Refers to a river, or to the bank of a river. A "riverine" paleoenvironment would be one that focused on a river, if it was aquatic, or on the bank of a river, if it was terrestrial. Thus a terrestrial riverine paleoenvironment is one where the flora and fauna are like those found on the banks of contemporary ephemeral or permanent rivers. In the latter case this usually means a forest-like environment.

RMA　Acronym for **reduced major axis regression** (*which see*).

RNA　Acronym for **ribonucleic acid** (*which see*).

Robertsonian fusion　*See* **Robertsonian translocation**.

Robertsonian translocation　A chromosomal rearrangement where two nonhomologous acrocentric chromosomes (where the centromere is located near one end of the chromosome) break at the centromere, and the long arms join together to form one chromosome. The short arms also fuse to form a new chromosome, but that is usually lost after a few rounds of mitosis; short arms seldom contain essential genes. Thus, a person with a Robertsonian translocation only has 45 chromosomes, but may be phenotypically normal. In modern humans, the acrocentric chromosomes that are commonly involved in Robertsonian translocations include chromosomes 13, 14, 15, 21, and 22. It is thought that Robertsonian translocation may have been the process responsible for producing the modern human chromosome 2 from two ape chromosomes (thus explaining the difference between the 48 chromosomes of the great apes and the 46 chromosomes of modern humans) during our evolutionary past (Yunis and Prakash 1982).

Robinson, John Talbot (1923–2001)　John Robinson was born in Elliot, South Africa, and grew up on a farm where he acquired an interest in natural history. He studied zoology at the University of Cape Town where he completed a BSc in 1943 and a MSc in 1944. The following year he became a research fellow in marine zoology at the University of Cape Town and enrolled in the PhD program in marine biology, but he did not complete his degree because he accepted a position in 1946 as Assistant Professional Officer at the **Transvaal Museum** in Pretoria, in the museum's Division of Physical Anthropology and Vertebrate Palaeontology. In this capacity, Robinson worked as an assistant to the Scottish-born paleontologist Robert **Broom**, who had spent much of the previous decade collecting and studying australopith fossils from southern Africa. Broom and Robinson embarked on an extensive series of excavations at the sites of **Sterkfontein** and **Swartkrans**. In 1947 at Sterkfontein they discovered **Sts 5**, the cranium of an archaic hominin, and later that year Broom and Robinson found the associated skeleton **Sts 14**. Their excavations at Swartkrans between 1948 and 1951 were equally productive, with the discovery of two crania and a mandible belonging to *Paranthropus crassidens*, a more robust australopith than that found at Sterkfontein. Robinson also unearthed a mandible (**SK 15**) at Swartkrans in 1949 that appeared to be more modern human-like than the *Paranthropus* fossils found at the site. Broom and Robinson argued it belonged to a hitherto unknown species of hominin they named *Telanthropus capensis*, although it was later reclassified by Robinson and others as *Homo erectus*. After Broom's death in 1951, Robinson became Professional Officer in the Department of Vertebrate Palaeontology and Physical Anthropology at the Transvaal Museum, where he remained until 1963. He completed a seminal study of australopith teeth that was published in the Transvaal Museum Memoir series under the title *The Dentition of the Australopithecinae* (1956) and for this work the University of Cape Town awarded Robinson a DSc degree in 1955. He also continued his research at Sterkfontein and Swartkrans focusing on the process of the formation of the fossil deposits, occasionally in collaboration with C.K. (Bob) **Brain**, who as a graduate student in the mid-1950s worked with Robinson on establishing the sequence of strata in the deposits. In 1956 Brain discovered the first stone artifacts at Sterkfontein in deposits that had produced australopith remains, and Robinson found additional stone artifacts in 1957 and 1958 along with some fossil teeth and a jaw fragment that suggested to Robinson that at least some australopiths were capable of making tools from stone. During the 1950s and 1960s Robinson devoted considerable time to investigating the problems of australopith taxonomy and anatomy. He had published, with Robert Broom and Gert. W.H. **Schepers**, a monograph on the Sterkfontein fossils entitled *Sterkfontein Ape-man Plesianthropus* (1950) and he published together with Broom a volume called *Swartkrans Ape-Man, Paranthropus crassidens* (1952), both of which appeared in the Transvaal Museum Memoir series. The inclination early in the 20thC to create new species and frequently new genera with the discovery of each new tranche of hominin fossils began to shift during the 1950s. In an important paper published in the *American Journal of Physical Anthropology* in 1954 entitled "The genera and species of the Australopithecinae" Robinson classified the hominins from Taung, Sterkfontein, and Makapansgat as *Australopithecus africanus*, whereas the hominins from Kromdraai and Swartkrans were assigned to *Paranthropus robustus*. Moreover, in a paper published in 1956 Robinson argued that the morphology of the teeth suggested that the robust australopiths were herbi-

vores whereas the gracile australopiths were omnivores that ate a significant quantity of meat. During this period Robinson accepted a position as a part-time lecturer at the University of Witwatersrand where he taught human evolution from 1954 to 1963. He was also appointed to be the assistant director of the Transvaal Museum in 1960, but controversies over the presentation of evolution at the museum contributed to his decision to resign from the museum in 1963. In the same year Robinson left South Africa to accept a position at the University of Wisconsin at Madison as Professor of Anthropology and Zoology; besides the difficulties at the Transvaal Museum it has been suggested that his opposition to the apartheid system in South Africa influenced his decision to leave and go to the USA. Although the position at the University of Wisconsin was originally a joint appointment to the Departments of Anthropology and Zoology, in 1967 Robinson left the Anthropology Department to serve full time in the Department of Zoology. Robinson continued his work on hominin evolution, publishing *Early Hominid Posture and Locomotion* in 1972 in which he argued that the australopiths were efficient bipeds. He served as the director of the University of Wisconsin Zoological Museum from 1979 to 1981, but persistent health problems led him to stop teaching in 1983 and he retired in 1985.

robust (L. *robustus*=strength, from *robus*=oak) Morphology A morphological variant that is stronger than other variants. In the mandible this means mandibular bodies (corpora) that are relatively wide, and in the limbs long bones with a relatively thicker layer of cortical bone. Taxonomy Used to describe the archaic hominin taxa assigned to either *Australopithecus* or to *Paranthropus* (i.e., *P. robustus*, *P. boisei*, and *P. aethiopicus*) that are distinguished by postcanine **megadontia** and thick (i.e., robust) mandibular bodies. *See also* **robust australopithecine**.

robust australopithecine Informal term introduced by John Robinson for an archaic hominin species with postcanine megadontia and thick (i.e., robust) mandibular bodies (i.e., *Paranthropus robustus*). The term has subsequently been applied to *Paranthropus boisei* and *Paranthropus aethiopicus*. Many researchers now avoid the term because several non-*Paranthropus* archaic hominin taxa exhibit postcanine megadontia and relatively robust mandibular bodies, but lack the derived morphology of *Paranthropus*.

"robust" australopithecine Fred Grine's solution to his disapproval of the use of "robust" in the informal term "robust australopithecine" was to put double quotes around "robust." This is why he insisted his influential edited volume about the taxa included in the genus *Paranthropus* was entitled the *Evolutionary History of the "Robust" Australopithecines.*

robusticity (L. *robustus*=strength, from *robus*= oak) The tendency to be **robust**. For example, for the mandible this means having a high value for the mandibular robusticity index (corpus width/corpus height), and for a long bone a high value for an index that relates the minimum width of its shaft to its maximum length.

Roc de Combe-Capelle *See* **Combe-Capelle**.

Roc de Marsal (Location 44°54′21″N, 0°58′45″E) History and general description This small cave overlooks a tributary valley of the Vézère River not far from its confluence with the Dordogne River in southwest France. The site was first excavated in 1953 by an amateur, Jean Lafille, who continued to work there until his death in 1971. In 1961 he discovered a nearly complete skeleton of a *Homo neanderthalensis* child reportedly placed in a pit near the back of the cave. The site was re-excavated by Dibble, Goldberg, McPherron, Sandgathe, and Turq from 2004 to 2010. Temporal span and how dated? The site contains only **Middle Paleolithic** industries including a Denticulate Mousterian with small flake production similar to the Asinipodian of Pech de l'Azé IV, a Levallois Mousterian, and a Quina Mousterian. Published thermoluminescence dates on sediment and heated flints taken from the Lafille sequence provide an age range of *c.*90–60 ka for the base of the sequence. The fauna show a gradual shift from a more temperate climate at the base of the sequence to a cold climate at the top. Hominins found at site Roc de Marsal 1 is a nearly complete skeleton of a 2–3-year-old individual. It was found in anatomical contiguity, face down, on its right side, and with its legs bent backward. The stratigraphic association of the skeleton is unclear, although from the descriptions it likely comes from the lower portion of the sequence. Isolated teeth of additional individuals have been recovered from the more recent excavations. Archeological evidence found at site Two of the lower levels at Roc de Marsal contain multiple, well-preserved, and sometimes stacked hearths. The fauna, particularly in the Quina levels, are also well preserved. The stone tool industries show significant temporal technological and typological variability. Key references: Geology, dating, and paleoenvironment Guibert et al. 2008; Archeology Lafille 1961, Turq 1989, Turq et al. 2008, Bordes and Lafille 1962, Madre-Dupouy 1991.

rock A rock is a naturally occurring solid, composed of one or more mineral phases, native elements, or organic accumulations. There are three fundamental types of rock: igneous (solidified from a molten phase, or melt), metamorphic (formed by solid-state recrystallization, usually under high pressure and/or temperature), and **sedimentary rock** (otherwise formed under Earth-surface conditions). A rock is the basic unit of **lithology**, or rock type. With very few exceptions hominin fossils are to be found in sedimentary rocks.

Rockefeller Foundation The Rockefeller Foundation was founded in 1913 by John D. Rockefeller to "promote the well-being of mankind throughout the world." It supported the **Peking Union Medical College**, financed its **Cenozoic Research Laboratory**, funded excavations at Choukoutien (now called **Zhoukoudian**), and provided support for other paleoanthropological research in China.

Rockefeller Museum Description The Rockefeller Archaeological Museum in Jerusalem, Israel, opened in 1938 and houses antiquities unearthed in excavations conducted in the country mainly during the time of the British Mandate (1919–48). It contains thousands of artifacts ranging from prehistoric times to the Ottoman period. Its collections include a 9000-year-old statue from Jericho, gold jewelry from the Bronze Age, and the Dead Sea Scrolls. The latter date from the 2ndC BCE to the 1stC CE and include books of the Hebrew Bible as well as other noncanonical texts. Hominin fossil collections Skhul I (mainly post crania), Skhul IV, **Zuttiyeh**, **Qafzeh** 1, and Qafzeh 2. Contact information James S. Snyder, Director (e-mail jsnyder@imj.org.il; tel +972 2 670 8801) and Yigal Zalmona, Chief Curator (e-mail zalmonay@imj.org.il; tel +972 2 670 8972). Postal address The Israel Museum, Jerusalem POB 71117 Jerusalem, 91710 Israel. Street address No official street address. The museum is located off Sultan

Suleiman Street, adjacent to the Flower Gate of Jerusalem's Old City walls. Website www.english. imjnet.org.il/HTMLs/Home.aspx.

Rodentia (L. *rodere*=to gnaw, or eat, away) A diverse group of mammals that have continuously growing central incisors. Although the largest rodent, the capybara, can reach 50 kg, most rodents are both small-bodied and are fast breeders. In the most commonly used pre-molecular taxonomy of rodents five suborders of rodents are recognized, each characterized by a particular masticatory apparatus and named for their main group; the Anomaluromorpha (scaly tailed flying squirrels and spring hares), Castorimorpha (beavers and allies), Hystricomorpha (porcupines and cavies), Myomorpha (mice and allies), and Scuriomorpha (squirrels). Unsurprisingly, molecular phylogenies differ significantly from this classification. The primary importance of rodents in the study of human evolution is as paleoecological indicators, for their relatively species-specific habitat preferences and diets make them very useful in the reconstruction of past ecosystems. Rodents make up much of the **microfaunal** evidence found at early hominin sites. In recent prehistory the significance of rodents is that they are notoriously destructive of both crops in the fields and also of stored resources, particularly grain, so from the standpoint of an agriculturalist they are pests. Rodents have also been vectors of disease in historic times and presumably this was also true in the past.

root (OE *rot*=root) When used with respect to the dentition it refers to the part of a tooth embedded in the bony **alveolus**. It is made up of an inner core of **dentine** and the enclosed pulp that is covered with **cementum**. The boundary between the crown and the root on the outside of the tooth is called the **cemento-enamel junction** (or CEJ), or the **cervix**, and the boundary between the crown and the root within the tooth is called the **enamel-dentine junction** (or EDJ or DEJ). Incisors and canines usually have a single root, premolars may have one or more roots, and molars usually have two or more roots. Root is also used to describe the place where one structure is attached to another (e.g., the root of the zygomatic process of the maxilla). It is also used to refer to the base of a **cladogram**.

rostra The plural of **rostrum** (*which see*).

rostral (L. *rostrum*=beak, muzzle, or snout) The rostral direction is in the direction of where the beak (or its equivalent) is. Thus, the term rostral can be equivalent to ventral (i.e., anterior) or to cranial.

rostrum (L. *rostrum*=bill of a bird, beak) In vertebrate anatomy, the rostrum is an informal term for the snout or muzzle. As a practical matter, in primates, the rostrum refers to the premaxilla and that portion of the **maxilla** that projects anterior to the orbits and to the roots of the **zygomatic arch**es. The rostra (pl. of rostrum) of cercopithecines and strepsirrhines (Old and New World monkeys, respectively) are particularly pronounced, and it determines the very strong **prognathism** exhibited by these taxa. In hominins, and particularly in *Homo*, the trend towards facial **orthognathism** has resulted in extremely reduced rostra.

routed foraging A model for **Oldowan** site formation that proposes hominins were recurrently drawn to fixed resources (e.g., stone outcrops, stands of trees acting as midday resting sites, water sources). At such locations carcass parts would have accumulated over time and because of these accumulations the location would subsequently have become visible as an archeological site (Binford 1984).

Royal College of Surgeons of England
The Royal College of Surgeons of England in Lincoln's Inn Fields, London, was, until the development of human origins research at the Natural History Museum, London, one of the main foci of interest in human evolution in the UK. Until they were transferred on permanent loan to the Natural History Museum, it housed important collections of modern human skeletons, and until 1955 it was also the repository of the **Gibraltar 1** cranium and of one of the crania recovered from **Mughares-Skhul** in Israel. The Gibraltar 1 cranium came to the College because of the eclectic scientific interests of one of its Presidents, George Busk. In addition to being a surgeon, Busk had wide interests in zoology and a genus of colonial aquatic animals (the *Buskia*) carries his name. In 1862 Busk traveled with the paleontologist Hugh Falconer to visit the excavations at **Gibraltar**, and they were shown a hominin cranium, Gibraltar 1, that had been recovered from **Forbes' Quarry** in 1848. Falconer assigned the cranium to a new species, *Homo calpicus* (Falconer 1864). However, when

Busk reported on the cranium to the annual meeting of the British Association for the Advancement of Science in Bath in 1864, he commented on its affinities with the specimen recovered from the **Kleine Feldhofer Grotte** in 1856. Busk caused the cranium to be presented to the Royal College of Surgeons of England in 1868, where it remained until 1955 when it was transferred to what was then called the British Museum (Natural History), and which is now called the Natural History Museum, London. The second connection between the College and human evolution came when Sir Arthur Keith was invited by Theodore McCown to participate in the analysis of the hominins that had been recovered by Dorothy Garrod from the site of Mughâret es-Skhul. The first material from Skhul was transferred to the College in 1931, and between 1933 and 1936 Keith and McCown worked on the remains, first at Lincoln's Inn Fields, and then at a laboratory in the College's research facility at Downe, Kent, near to Down House, Charles Darwin's former home which at the time was Keith's home. Some of the Skhul remains were returned to Israel (Skhul 1 and 4), and the bulk of them went to the Peabody Museum at Harvard University (Skhul 2, 3, and 5–8), but one cranium (Skhul 9) was deposited with the College and was part of its Hunterian Museum until 1955 when it was transferred on permanent loan to what is now the Natural History Museum, London. See also Arthur Keith.

Royal Museum for Central Africa (Belgium

is a bilingual country, with both French and Dutch being spoken, depending on the part of the country. Notwithstanding the location of Tervuren in a Dutch-speaking region of Belgium, most researchers know the museum by its French name, the Musée Royal de l'Afrique Centrale. The Dutch name is Koninklijk Museum voor Midden-Afrika) The museum had its origin in 1897 when King Leopold II organized an exposition of what was then his African fiefdom, the Congo Free State. Subsequently the King was stripped of his African possessions and they were made a colony of Belgium, under the name Belgian Congo, which in 1960 was given independence and since then has become known as, successively, Zaire and the Democratic Republic of the Congo. The items in the expo quickly outgrew the building to which it was assigned and in 1904 construction of a new building was commenced in the Royal Gardens. The present enormous neoclassical building was

opened in 1910 under the name "Museum of the Belgian Congo." The name was later changed to the "Royal Museum of the Belgian Congo," and after the Congo was granted independence it was changed once again to its present name, which in English means "Royal Museum for Central Africa." In its first years, the museum acquired large collections of Congolese fauna and flora, and later fauna and flora from other African countries. The mammal collection comprises more than 120,000 specimens, including approximately 75,000 rodents, 18,000 bats, and 10,000 primates, including cercopithecus, mangabey, and colobus monkeys, and gorillas, chimpanzees, and bonobos. The collection includes the world's largest collections of Eastern gorillas (*Gorilla beringei*) and bonobos (***Pan paniscus***), the latter being assembled by Drs Schouteden and Vandebroek between 1927 and 1960 (Thys van den Audenaerade 1984). The gorilla and bonobo collections consist mainly of skulls, but there are also postcranial skeletons, skins, and some whole specimens preserved in alcohol. Almost all of the specimens have precise contextual and other information. Comparative collections *Pan paniscus*: Whole skeletons Adults=11 (females=7, males=4, sex uncertain=N/A); juveniles=3 (females=1, males=1, sex uncertain=1); infants=3 (females=0, males=2, sex uncertain=1). Crania Adults=62 (females=25, males=24, sex uncertain=13); juveniles=86 (females=19, males=17, sex uncertain=50); infants=38 (females=4, males=11, sex uncertain=23). *Pan troglodytes schweinfurthii*: Whole skeletons Adults=9 (females=5, males=1, sex uncertain=3); juveniles=4 (females=1, males=2, sex uncertain=1); infants=3 (females=0, males=0, sex uncertain=3). Crania Adults=172 (females=40, males=47, sex uncertain=85); juveniles=106 (females=21, males=16, sex uncertain=69); infants=45 (females=7, males=4, sex uncertain=34). *Pan troglodytes troglodytes*: Whole skeletons Adults=4 (females=0, males=3, sex uncertain=1); juveniles=N/A; infants=N/A. Crania Adults=46 (females=6, males=7, sex uncertain=33); juveniles=10 (females=0, males=0, sex uncertain=10); infants=N/A. *Pan troglodytes verus*: Whole skeletons N/A. Crania Adults=3 (females=0, males=2, sex uncertain=1); juveniles=1 (females=0, males=0, sex uncertain=1); infants=1 (females=0, males=0, sex uncertain=1). *Pan troglodytes subsp. indet.*: Whole skeletons Adults=5 (females=0, males=3, sex uncertain=2); juveniles=2 (females=

2, males=0, sex uncertain=0); infants=1 (females= 0, males=0, sex uncertain=1). Crania Adults=10 (females=0, males=3, sex uncertain=7); juveniles= 7 (females=0, males=1, sex uncertain=6); infants= 16 (females=0, males=0, sex uncertain=16). *Gorilla sp. indet.*: Crania Infants=1 (females=0, males=0, sex uncertain=1). *Gorilla gorilla gorilla*: Whole skeletons Adults=2 (females=0, males=1, sex uncertain=1); juveniles=N/A; infants=N/A. Crania Adults=75 (females=5, males=6, sex uncertain=64); juveniles= 14 (females=0, males=0, sex uncertain=14); infants thinsp;=3 (females=0, males=1, sex uncertain=2). *Gorilla beringei graueri*: Whole skeletons Adults=10 (females=4, males=4, sex uncertain=2); juveniles=5 (females=4, males=0, sex uncertain= 1); infants=N/A. Crania Adults=102 (females=39, males=47, sex uncertain=16); juveniles=21 (females=9, males=6, sex uncertain=6); infants= 16 (females=0, males=1, sex uncertain=15). *Gorilla beringei beringei*: Whole skeletons Adults=2 (females= 1, males=1, sex uncertain=0); juveniles=N/A; infants=N/A. Crania Adults=7 (females=3, males=4, sex uncertain=0); juveniles=4 (females= 2, males=0, sex uncertain=2); infants=1 (females= 0, males=0, sex uncertain=1). Contact department Biology – Vertebrates. Contact person Emmanuel Gilissen (e-mail emmanuel.gilissen@africamuseum. be). Postal address Leuvensesteenweg 13, 3080 Tervuren – Belgium. Website www.africamuseum.be. Tel +32 02 769 5622. Fax +32 02 769 5642 (NB: there is also a smaller collection of Congo primates and other mammals in the Brussels Natural History Museum, in the centre of Brussels. This museum is world famous mainly for its dinosaurs. For example, there is a display of a dozen complete skeletons of Iguanadon, subject of a late-19thC monograph by Louis Dollo, of Dollo's Law of the irreversibility of evolution).

Roy's largest root   *See* **multivariate analysis of variance**.

RSC   *See* **recognition species concept**.

RT-PCR   Acronym for **reverse transcription polymerase chain reaction** (*which see*).

running   *See* **endurance running hypothesis**; **gait**.

Ruscinian   *See* **European mammal neogene**; **biochronology**.

RVRME   *See* **Rift Valley Research Mission in Ethiopia**.

Ryukyu Islands   (Location approximately 26° 19′N, 127°44′E, Japan; etym. also referred to as the Nansei Islands and "Nansei" in Japanese means "southwest") History and general description The Ryukyu Islands form a chain off the southernmost part of the larger Japanese archipelago. The Ryukyu islands are themselves divided into the Satsunan Islands in the north and Ryukyu Shoto in the south; the largest and best known of the Ryukyu Islands is Okinawa. The southernmost island in the chain is approximately 120 km/74 miles from Taiwan and similarities in flora and fauna suggest that during glacial periods a land bridge likely connected the Ryukyus to the Asian mainland, although some suggest Okinawa could only have been colonized by boat. Beginning in the late 1960s, Pleistocene human fossils were identified in deposits being mined for limestone. These localities included caves, collapsed caves, and fissures from several islands, although most of the major finds were identified on Okinawa. The primary sites are Yamashita-cho and Minatogawa (Okinawa), Pinza-abu Cave (Miyako Island), Shimojibaru Cave (Kume Island), and Shiraho-Saonetabaru (Ishigaki Island). Temporal span and how dated? Evidence from **biostratigraphy** suggests that all of these sites date to the latter half of the Late Pleistocene. Currently, the oldest hominin fossils on the Japanese archipelago are from Yamashita-cho, which has been **radiocarbon dated** to *c.*32 ka. Radiocarbon dates for the other sites are Pinza-abu *c.*26 ka and Shimojibaru *c.*15 ka, and the hominins from Shiraho-Saonetabaru were **AMS radiocarbon dated** to between 20 and 16 ka. Hominins found at site The Yamashita-cho specimens comprise femoral and tibial proximal metaphyses and diaphyses of a juvenile (*c.*6 years old). Based on diaphyseal robusticity the juvenile lower limbs are considered to be robust, within the range of archaic *Homo sapiens*, but the presence of an incipient pilaster and a high neck-shaft angle more closely align the specimens with early modern humans. The most complete and best-studied collection of hominin fossils is from Minatogawa (*c.*18 and 16 ka). This skeletal assemblage includes four almost complete skeletons and other fragments. Morphologically, the Minatogawa hominins more closely resemble Late Pleistocene humans from **Niah cave** (Borneo), **Liujiang** (south China), and **Wadjak** (Java), rather than populations from northeast Asia (e.g., **Zhoukoudian Upper Cave**). That

679

the Minatogawa anatomically modern humans originated from Southeast Asia should not be that surprising given that during glacial periods (e.g., Oxygen Isotope Stage 2: **Last Glacial Maximum**) the Ryukyus were connected to Taiwan, which was connected to the Asian mainland. Detailed comparative analysis of the mandibles from Minatogawa with those of the later Jomon people (Japanese Neolithic) suggests that the Minatogawa hominins were absolutely and relatively smaller and more prognathic than those of the later Jomon people and they lacked the typical "squarish" chin of the Jomon mandibles. Nine hominin fossils identified in the Shiraho-Saonetabaru Cave on Ishigaki Island were directly dated using AMS ^{14}C dating. The Pinza-abu hominin assemblage comprises a parietal, a occipital, three isolated teeth, a lumbar vertebra, and a few hand bones. The Shimojibaru hominin infant comprises a mandible, right humerus, and right femur. Archeological evidence found at site Paleolithic artifacts have been reported from Yama-shita-cho, but some consider them to be geofacts. Unambiguous artifacts have been reported from other sites in the Ryukyu Islands (e.g., Tanegashima Island); thus the possibility of discovering hominin fossils and paleolithic artifacts in the same context remains high in the region, particularly because of the good bone preservation in these limestone caves and fissures. Repository Okinawa Prefectural Museum and Art Museum, Naha, Japan; University of Tokyo, Japan. Key references: Geology, dating, and paleoenvironment Kobayashi et al. 1971, Hasegawa 1980, Matsu'ura 1982, 1999, Sakura 1985; Hominins Baba and Endo 1982, Suzuki 1982, Suzuki and Hanihara 1982, Baba and Narasaki 1991, Trinkaus and Ruff 1996, Kodera 2006, Kaifu 2007; Archeology Takamiya et al. 1975, Oda 2003, 2007, Miyata 2005.

S

S7a *See* **Sangiran 7**.

S7b *See* **Sangiran 7**.

Saccopastore (Location 41°57′N, 12°32′E, Italy; etym. from Sacco Pastore, the name of a meander in the Aniene River) History and general description The site was an active gravel quarry on the left bank of the Aniene River, a tributary of the Tiber, within the boundaries of modern Rome. Workmen at the quarry found the first cranium (**Saccopastore 1**) in May 1929 and it was passed to Sergio **Sergi** of the University of Rome who identified is as belonging to *Homo neanderthalensis*. The second cranium, Saccopastore 2, was found 6 years later when Alberto **Blanc** and H. Breuil were examining the now-abandoned quarry. Further excavations were carried out, but no additional hominin remains have been discovered. Temporal span and how dated? **Biostratigraphy** suggests an age *c*.130–100 ka. Hominins found at site Two adult crania unearthed in 1929 and 1935. Archeological evidence found at site Some **Mousterian** artifacts were found in the same horizon as Saccopastore 2. Key references: Geology, dating, and paleoenvironment Sergi 1944, Segre 1983, Caloi et al. 1998; Hominins Sergi 1929, 1944, 1948, Condemi 1992, Bruner and Manzi 2006; Archeology Blanc 1948.

Saccopastore 1 Site Saccopastore. Locality N/A. Surface/*in situ* Probably *in situ*. Date of discovery May 1929. Finder A. Giovannini and V. Casorri. Unit N/A. Horizon N/A. Bed/member N/A. Formation N/A. Group N/A. Nearest overlying dated horizon N/A. Nearest underlying dated horizon N/A. Geological age *c*.130–120 ka. Developmental age Adult. Presumed sex Female. Brief anatomical description Cranium lacking part of the supraorbital, the zygomatic arch region, and some teeth. Announcement and Initial description Sergi 1929. Photographs/line drawings and metrical data Sergi 1944, Condemi

1992, Bruner and Manzi 2006. Detailed anatomical description Sergi 1944, Condemi 1992, Bruner and Manzi 2006. Initial taxonomic allocation *Homo neanderthalensis* as *Homo neanderthalensis aniensis* 1935. Taxonomic revisions N/A. Current conventional taxonomic allocation *H. neanderthalensis*. Informal taxonomic category Pre-modern *Homo*. Significance The presumed geological age (Riss-Würm interglacial, or **Oxygen Isotope Stage** 5) and the combination of morphological features have suggested to some that the Saccopastore crania could possibly belong to an intermediate stage between *Homo heidelbergensis* and *H. neanderthalensis*. Location of original Sapienza – Universitá di Roma, Rome, Italy.

Saccopastore 2 Site Saccopastore. Locality N/A. Surface/*in situ In situ*. Date of discovery July 1935. Finders A.C. Blanc and H. Breuil. Unit/horizon N/A. Bed/member Layer of fluvial gravels, intermediate in the stratigraphic sequence and associated with large mammal remains, just above a thick lens of silt. Formation N/A. Group N/A. Nearest overlying dated horizon N/A. Nearest underlying dated horizon N/A. Geological age *c*.130–100 ka. Developmental age Adult. Presumed sex Male. Brief anatomical description Partial cranium lacking most of the vault, the left side of the basicranium, and some teeth. Announcement Breuil and Blanc 1935. Initial description Sergi 1929. Photographs/line drawings, metrical data, and detailed anatomical description Sergi 1944, Condemi 1992, Bruner and Manzi 2006. Initial taxonomic allocation *Homo neanderthalensis* as the paratype of *Homo neanderthalensis aniensis* Sergi, 1944. Taxonomic revisions N/A. Current conventional taxonomic allocation *H. neanderthalensis*. Informal taxonomic category Pre-modern *Homo*. Significance The presumed geological age (Riss-Würm interglacial, or **Oxygen Isotope Stage** 5) and the combination of morphological features have suggested to some that the Saccopastore crania could possibly belong to an intermediate stage between *Homo heidelbergensis* and

Wiley-Blackwell Encyclopedia of Human Evolution, First Edition. Edited by Bernard Wood.
© 2013 Blackwell Publishing Ltd. Published 2013 by Blackwell Publishing Ltd.

H. neanderthalensis. <u>Location of original</u> Sapienza – Universitá di Roma, Rome, Italy.

Saegul *See* **Turubong**.

SAG Abbreviation for **Sagantole** (*which see*).

Sagala One of the Early and Middle **Pleistocene** sites in the exposures of the **Ola Bula Formation** within the **Soa Basin**, central Flores, Indonesia.

Sagantole A concentration of fossil localities situated in the **Central Awash Complex** in the Afar Rift section of the East African Rift System in the **Middle Awash study area**, Ethiopia. Fossils attributed to *Ardipithecus ramidus* have been found there.

sagittal (L. *sagitta*=an arrow) Refers to a plane corresponding to the midline of the body. It takes its name from the arrow-shaped interparietal cranial suture in the cranium of a neonate/infant that runs in that plane from bregma (the tip of the arrow) to lambda (the feathers of the arrow).

sagittal crest (L. *sagitta*=an arrow and *crista*= cock's comb, or a tuft of feathers, in the midline of a bird's head) A sharp crest of bone running along the course of the inter-parietal, or sagittal, suture. The bony crest forms when the medial borders of the fascia covering the temporalis muscles fuse. When the fusion is complete there is just one crest. If the fusion is incomplete it is referred to as a **compound sagittal crest**, and if the separate crests are either side of the sagittal suture strictly they should be referred to as **parasagittal crests**. A sagittal crest is seen in male gorillas and orangutans, and in larger-bodied, presumed male, *Paranthropus aethiopicus* and *Paranthropus boisei* crania such as **KNM-ER 406**. *See also* **sagittal keel**.

sagittal keel (L. *sagitta*=an arrow and ME *kele*=the main longitudinal timber in a wooden boat, now used for the underside of a boat, or any projection designed to prevent a boat from "keeling over.") Refers to a blunt bony prominence running longitudinally along the course of the inter-parietal, or sagittal, suture. Most, but not all, *Homo erectus* crania have a sagittal keel, but it is better expressed in the larger, presumed male, crania.

Sahel (etym. *Sahel*=region to the south of the Sahara) Semiarid region of north-central Africa to

the south of the Sahara. The fossil evidence for *Sahelanthropus tchadensis* comes from sites in the Sahel.

***Sahelanthropus* Brunet et al., 2002** (etym. *Sahel*=region to the south of the Sahara; Gk *anthropos*=human) Hominin genus established in 2002 by Brunet et al. (2002) to accommodate a new species they introduced for cranial remains recovered from the *c.*7 Ma Anthracotheriid unit at locality TM 266 at **Toros-Menalla** in the Chad Basin. The authors claimed that various features of the cranium either distinguished the new fossils from *Pan*, or could not be accommodated in any of the existing hominin genera. White et al. (2009a) suggest the morphological differences between species assigned to *Ardipithecus* and to *Sahelanthropus* do not justify their being assigned to different genera, in which case the former genus would have priority, and *Sahelanthropus* would be the junior synonym of *Ardipithecus*. Type species *Sahelanthropus tchadensis* Brunet *et al.*, 2002.

***Sahelanthropus tchadensis* Brunet et al., 2002** (etym. *Sahel*=region to the south of the Sahara; Gk *anthropos*=human; *tchadensis*=Chad) Hominin species established by Brunet et al. (2002) to accommodate cranial remains recovered from the Anthracotheriid unit at **Toros-Menalla** in the Chad Basin; the former has been dated by **biostratigraphy** and **cosmogenic nuclide dating** to *c.*7 Ma. The authors believe that, despite primitive features in the cranium and dentition, this taxon is a primitive hominin and not a panin because of its small, apically worn canines and the intermediate thickness of the postcanine enamel. It was virtually reconstructed by Zollikofer et al. (2005), who claimed that the reconstruction of the cranial base demonstrated that *S. tchadensis* is bipedal. A further, phenetic, analysis of 3D landmark data obtained from the virtually reconstructed cranium by Guy et al. (2005) concluded that whereas the reconstructed cranium shared some basicranial features with archaic hominins, for example "long, flat (more horizontally orientated) nuchal plane," "shortened occipital," and "anteriorly-positioned foramen magnum" (*ibid*, p. 18838), other features on other regions of the cranium (e.g., the face) were either primitive or unique among hominins. If the hominin classification is upheld, and the age of the specimen is confirmed, then *S. tchadensis* would be the oldest known hominin and it would be a refutation of the hypothesis that hominins were confined to the area either within, or to the east of, the East African Rift System. There are no postcranial remains. White et al. (2009a) suggest that the morphological differences between *Ardipithecus ramidus* and *S. tchadensis* do not justify their being assigned to different genera, in which case the latter taxon would be transfered to the genus with priority (i.e., *Ardipithecus*) as a new name combination *Ardipithecus tchadensis* (**Brunet** *et al.*, **2002**) **White** *et al.*, **2009**. First discovery **TM 266-01-060-1** (2001). Holotype TM 266-01-060-1. Paratypes TM 266-01-060-2, TM 266-01-447, TM 266-01-448, TM 266-02-154-1, TM 266-02-154-2. Main site **Toros-Menalla**, Chad.

Sahul (etym. Ballard 1993 suggests that it may be a word used by the Macassan people for a sandbank or shoal on the northwestern continental shelf of Australia to the southwest of the Aru Islands and separated from Timor by the deep waters of the Timor Trough). The name "Sahull" or "Sahoel" for this shoal appears in Dutch maps as early as 1598. The first time the name Sahul was used in its contemporary sense was in a report of a Dutch marine life and bathymetric survey known as the Siboga Expedition (organized by the Society for the Advancement of Scientific Research in the Netherlands' Colonies in 1889, now called Treub Maatschappij). The terms "Greater Australian Bank" and "Great Asiatic Bank" had been introduced by Earl(e) in 1845, but in their 1919 summary of the Siboga Expedition, Molengraaff and Weber suggest that those terms be replaced by the "Sahul Shelf" and the "Sunda Shelf," respectively. The name Sahul is used today for the continental landmass, also known as Australasia, that comprises present-day Australia, Tasmania, and New Guinea. During particularly cold periods when substantial volumes of water were locked up in ice accumulations at high latitudes, sea levels would have fallen low enough for Sahul to have been one continuous land mass. However, even during the coldest periods hominins would have had to make a deep-water crossing, accidental or deliberate, to reach Sahul from **Sunda**. The northwestern margin of Sahul marks the eastern boundary of **Wallacea**. Ballard (1993) suggests that Rhys Jones introduced the term Sahul into archeology when he used it in the title of a session called "Sunda and Sahul: hunters in the tropics" at the 13th Pacific Science Congress in Vancouver in 1975. Its usage was consolidated in a 1977 book called *Sunda and Sahul: Prehistoric Studies in Southeast Asia, Melanesia and Australia*.

St. Acheul *See* **Saint-Acheul**; **Acheulean**.

Saint-Acheul (Location 49°52′N, 2°19′E, France; etym. named after the neighborhood in which it is located) History and general description This open-air site in the suburbs of Amiens is the type site of the **Acheulean** technocomplex. Handaxes were found in the gravel pits in the mid 19thC, and as unequivocally hominin-made artifacts they helped prove the antiquity of hominins (contrary to the creationist view prevalent at the time). Temporal span and how dated? Recent electron spin resonance spectroscopy dating on quartz from near the artifact-bearing layers provides an age of 403±73 ka. Hominins found at site None. Archeological evidence found at site Handaxes, mollusc shells, and other fauna were recovered,. Key references: Geology, dating, and paleoenvironment Antoine et al. 2003; Hominins N/A; Archeology de Mortillet 1872, Rigollot 1854.

Saint-Césaire (Location 45°44′57″N, 0°30′11″E, France; etym. named after the nearby village) History and general description This site, near a local cliff called "La Roche à Pierrot," was discovered in 1978, and excavated by François Lévêque. The *Homo neanderthalensis* fossil was found in a **Châtelperronian** layer and its discovery confirmed that the Châtelperronian was, in fact, an **Upper Paleolithic** culture made by Neanderthals, not by modern humans. When discovered, the Saint-Césaire hominin was the subject of considerable debate, for some (ApSimon 1980, Wolpoff 1981) argued it was evidence of a unilinear model of evolution through Neanderthals to Upper Paleolithic modern humans, whereas others (Stringer et al. 1981, Stringer 1982) argued that the appearance of a late Neanderthal overlapping significantly with anatomically modern humans elsewhere essentially disproved the **multiregional hypothesis**, at least in Europe. Temporal span and how dated? Thermo**luminescence dating** of burned flints has provided dates of *c.*40 ka for the **Mousterian**. Flints near the fossil had thermoluminescence dates of 36.3±2.7 ka and overlying Proto-Aurignacian deposits were dated at 32.1±3 ka for the Châtelperronian deposits. Hominins found at site A very fragmentary and compressed skeleton was recovered in the upper Châtelperronian layer. Most body elements are represented, except the feet. Aspects of the cranium securely identified this individual as *H. neanderthalensis*. Although the postcranium is robust and similar in overall form to that of other Neanderthals, the cross-sectional anatomy of the long bones of this individual are similar to later those of Upper Paleolithic modern humans, suggesting that this individual was engaging in locomotor patterns comparable to those of the modern humans. Studies of the carbon, nitrogen, strontium, barium, and calcium isotopes in the collagen of this individual compared to those of the fauna from the site suggest that it consumed almost no plant foods. Fish may have made up a portion of its diet, but there is no evidence of this from the faunal remains. A recent computerized reconstruction of the cranium showed evidence of a healed injury from a blade-like object, which has been interpreted as evidence of interpersonal (probably intragroup) violence (Zollikofer et al. 2002). However, the fact that this individual did not die from this injury suggests that it received a level of care and support from others. Archeological evidence found at site Seventeen archeological layers that document the transition from the Mousterian through the Châtelperronian to the **Aurignacian** have been identified. The hominin fossil was found associated with *Dentalium* shells; some have interpreted these as grave goods associated with an intentional burial. Key references: Geology, dating, and paleoenvironment Lévêque et al. 1993, Mercier et al. 1991; Hominins ApSimon 1980, Wolpoff 1981, Stringer et al. 1981, Stringer 1982, Lévêque et al. 1993, Trinkaus et al. 1998, 1999a, Balter and Simon 2006, Bocherens et al. 2005, Zollikofer et al. 2002; Archeology Guilbaud et al. 1994, Lévêque et al. 1993, Morin et al. 2005, Morin 2008.

Saitune Dora A concentration of fossil localities situated up against the Western Margin of the Afar Rift section of the East African Rift System in the **Middle Awash study area**, Ethiopia. Remains attributed to *Ardipithecus kadabba* have been found there.

Šala (Location 48°15′N, 17°05′E, Western Slovakia; etym. name of nearby town) History and general description Šala, also known as Šal'a, is a site composed of secondary gravel deposits in a sandbar of the river Váh, near the town of Šala, 50 km/31 miles northeast of Bratislava. Hominins were found 1961 (Šala 1) and in 1993 and 1995 (Šala 2). Temporal span and how dated? Evidence from **biostratigraphy** suggests possibly the last interglacial. Hominins found at site A complete, presumed female, adult frontal bone (Šala 1), and isolated cranial vault bones, presumed to be from one individual (Šala 2). Both are attributed to *Homo neanderthalensis*. Arche-

ological evidence found at site None. Key references: Geology, dating, and paleoenvironment Vlček 1969; Archeology N/A; Hominins Vlček 1969, Jelinek 1969, Smith 1982, Jakab 1996, Sládek et al. 2002.

Saldanha *See* **Elandsfontein**.

Saldanha **1** Site Elandsfontein. Locality Hopefield. Surface/*in situ* Surface find on a deflated dune surface. Date of discovery 1953. Finders K. Jolly and Ronald Singer. Unit N/A. Horizon N/A. Bed/member N/A. Formation N/A. Group N/A. Nearest overlying dated horizon N/A. Nearest underlying dated horizon N/A. Geological age **Middle Pleistocene** on the basis of associated fauna. Developmental age Adult. Presumed sex Male based on the basis of the large **supraorbital torus**. Brief anatomical description A rugged, long, low, and broad calvaria with a massive, swept-back, supraorbital torus and strong angulation between the upper and lower scales of the occipital bone. Announcement Drennan 1953a, 1953b. Initial description Drennan 1953b. Photographs/line drawings and metrical data Singer 1954, Schwartz and Tattersall 2003. Detailed anatomical description Drennan 1955, Drennan and Singer 1955. Initial taxonomic allocation *Homo saldanensis* Drennan, 1955. Taxonomic revisions *Homo sapiens rhodesiensis* (Campbell 1964), *Homo heidelbergensis* (Rightmire 1996), *Homo rhodesiensis* (Bermúdez de Castro et al. 1997). Current conventional taxonomic allocation *H. heidelbergensis* or *H. rhodesiensis*. Informal taxonomic category Pre-modern *Homo*. Significance The preserved portions of the calvaria closely resemble the analogous parts of the **Bodo**, **Kabwe**, **Petralona**, and **Arago** crania. Like Kabwe, the Saldanha cranium is one of the key specimens in the hypodigm of *H. heidelbergensis* (Rightmire 2008). Location of original **South African Museum (Iziko Museums of Cape Town)**, Cape Town, Cape Province, South Africa.

Salé (Location 34°02′N, 6°48′E, Morocco; etym. named after the nearby city) History and general description An open-air site located 5 km/3 miles northeast of Salé city on the Atlantic coast. A partial hominin cranium was discovered in a fossil dune in 1971 during sandstone quarrying. Temporal span and how dated? **Biostratigraphy** and **electron spin resonance spectroscopy dating** suggest a date of *c*.400 ka and other evidence placing it at the end of the Amirian, or the beginning of the Tensiftian, local Maghreb continental stage

(Geraads et al. 1980), is consistent with this assessment. Hominins found at site Cranium tentatively assigned to *Homo erectus* or "archaic *Homo sapiens*." Archeological evidence found at site None. Key references: Geology, dating, and paleoenvironment Geraads et al. 1980, Debénath 2000, Hublin 1985; Hominins Rightmire 1990, Hublin 2001; Archeology N/A .

Salé 1 Site Salé. Locality N/A. Surface/*in situ* *In situ*. Date of discovery 1971. Finders Local quarrymen, but collected by Jean-Jacques Jaeger. Unit Fossil dune (aeolianite). Horizon N/A. Bed/member N/A. Formation N/A. Group N/A. Nearest overlying dated horizon Paleosol with microfauna of "Presolatanian" age. Nearest underlying dated horizon N/A. Geological age *c*.400 ka. Developmental age Adult. Presumed sex Female. Brief anatomical description Occipital, parietals, frontal fragments, partial left maxilla with heavily worn I^2–M^2, and a natural sandstone endocast, all from the same individual. Endocranial volume Approximately 880 cm^3 (Holloway et al. 2004). Cranial distortion and muscular trauma indicate craniosynostosis and congenital torticollis, respectively (Hublin 1991). Announcement Jaeger 1975. Initial description Jaeger 1975. Photographs/line drawings and metrical data Jaeger 1975, Hublin 1991. Detailed anatomical description Hublin 1991, 2001. Initial taxonomic allocation *Homo erectus*. Taxonomic revisions N/A. Current conventional taxonomic allocation *H. erectus* or archaic *Homo sapiens*. Informal taxonomic category Pre-modern *Homo*. Significance Cranium exhibits a mosaic of *H. erectus* and *H. sapiens* features. Location of original Laboratoire d'Anthropologie Biologique, Universite de Bordeaux 1, Bordeaux, France.

"Sally-Anne" task *See* theory of mind.

Salzgitter-Lebenstedt (Location 52°09′N, 10°19′E, Germany; etym. name of a nearby town) History and general description This open-air site was first excavated in 1952 and again in 1977. The abundant faunal remains have been used as evidence of **Middle Paleolithic** hunting strategies (notably of reindeer). Temporal span and how dated? **Radiocarbon dating** suggests an age of more than *c*.50 ka (the site is at or beyond the limit of radiocarbon dating). Stratigraphic correlation and palynological analysis suggest a date somewhere in **Oxygen Isotope Stages** (OISs)

5–3, most likely at the beginning of OIS 3 (*c*.53 ka). Hominins found at site A fragment of occipital and right parietal, labeled Lebenstedt 1 and assigned to *Homo neanderthalensis*. Archeological evidence found at site A Middle Paleolithic lithic assemblage was recovered, along with rare Middle Paleolithic bone tools. The large faunal assemblage was overwhelmingly of reindeer, suggesting specialization in Middle Paleolithic hunting patterns. Key references: Geology, dating, and paleoenvironment: Gaudzinski and Roebroeks 2000, Street et al. 2006; Archeology Tode 1953, 1982; Hominins Hublin 1984.

Sambungmacan (Location 7°21′S, 111°07′E, Indonesia; etym. named after a local village) History and general description An open site on the south bank of the Solo River in Central Java, 12 km/7.4 miles east of Sragen. The site is an exposure of calciferous sandstone above the black clay in the lowermost portion of the **Bapang Formation**. An adult calotte (Sm 1) was discovered at the site in 1973. Temporal span and how dated? Initially assigned a **Lower** to **Middle Pleistocene** age, but recent **electron spin resonance spectroscopy dating** and **uranium-series dating** of the deposits suggest a young age of 27–53 ka for both Sambungmacan and **Ngandong**. Hominins found at site In addition to the adult calotte recovered in 1973, two other crania from this site were recently announced. The first (Sm 3) was "discovered" in a New York City curio shop and the second (Sm 4) was recovered from the site in 2001. These three hominin fossils have been variably attributed to *Homo erectus* and *Homo soloensis*. In 1977, a partial tibial shaft, generally attributed to *H. erectus*, was discovered at Sambungmacan. Archeological evidence found at site There is a brief reference to "stone tools" in a 1978 report (Jacob et al. 1978). Key references: Geology, dating, and paleoenvironment Swisher et al. 1996, Anton et al. 2002; Hominins Jacob 1973, Matsu'ura et al. 1990, Márquez et al. 2001, Baba et al. 2003; Archeology Jacob et al. 1978.

Sambungmachan *See* **Sambungmacan**.

sample A set of measurements used in the calculation of a test statistic. A sample is a subset of a larger **population** of possible measurements. For example, a sample of cranial capacities can be drawn from the population of all chimpanzee crania that have ever existed. Statistics based on a sample (i.e., sample statistics) are used to make inferences regarding the population from which the sample is drawn.

sample statistics   *See* **parameter**.

sampling with replacement
In statistics, a distinction is made between sampling with replacement and sampling without replacement. When generating a random sample of size n where $n > 1$, one typically chooses a randomly selected sample from a population or larger sample in one of two ways. Consider the following examples, both of which involve drawing two tiles from a bag of nine tiles, each marked with a different digit from 1 to 9. A person could draw a tile at random, record the number on it, replace it in the bag, and draw again. This is an example of sampling with replacement. Alternatively, one could draw a tile at random, record the number on it, leave the tile out of the bag, and draw a second tile from the bag. This would be sampling without replacement. In the first case it is possible to draw the same tile twice, but that is not possible in the second case. The difference between the two sampling methods is relatively unimportant when choosing a small sample from a large population because in that case the probability of selecting the same individual twice is relatively low (to put this in the context of the population of the USA, imagine randomly pulling the same tile twice in a row from a bag of 300 million tiles). However, the difference between the two types of sampling is important for **resampling** analyses. In the case of the **bootstrap** a sample of size n is drawn from a population of size n; in the case of **randomization** two or more samples are generated whose combined sample size is equal to that of the total population. Consider again the bag of tiles, and let us suppose that the order in which tiles are drawn out of the bag does not matter. Draw a sample of nine tiles. If sampled without replacement, there is only one possible sample that can be drawn: the sample that includes a single instance of each digit from 1 to 9 (i.e., all of the tiles). However, when sampling with replacement tiles are replaced in the bag between draws, and so that technique allows over 24,000 possible samples in this example because each digit may be represented multiple times, or not at all, in a given sample. This latter example demonstrates how sampling occurs in bootstrapping. Consider another example using the same bag of tiles. We know the bag contains four even-numbered tiles (2, 4, 6, 8) and five odd-numbered tiles (1, 3, 5, 7, 9). Let us call these groups E and O. In randomization some mea-

sure comparing the two groups (e.g., the mean of group E minus the mean of group O) is compared for the actual data set to a set of measures from randomly resampled data sets of equal size. In this case, group E consists of four tiles and group O consists of five. If four tiles are drawn randomly from the bag without replacing them between draws, then the data set has randomly been split into two groups of equal size to the first; four tiles outside of the bag, and five tiles inside of the bag. If the order of tiles in a group does not matter, there are 126 possible ways to split nine tiles into a group of four and a group of five (this is an example of a combination). If the tiles were sampled with replacement to produce the group of four tiles, then the remaining tiles would not equal the sample size for the second group. Consequently the difference in these two types of sampling is very important in resampling analyses: the bootstrap uses sampling with replacement, whereas randomization uses sampling without replacement.

sampling without replacement   *See* **sampling with replacement**.

San
A term used to refer to the hunter–gatherer people of southern Africa, who live in Botswana, Zambia, Zimbabwe, and South Africa. Other terms, including Khwe, Sho, Bushmen, and Basarwa, have also been used to refer to this group, but all are problematic as they are or have historically been used by outsiders in a perjorative manner. All the San groups speak Khoisan language variants, and as recently as 1990 some practiced an entirely foraging-based economy; today, almost all of the San rely heavily or entirely on farming. This group has been the subject of many ethnographic studies, which have allowed researchers to better understand subsistence strategies and test hypotheses derived from behavioral ecology.

sand
Sedimentary materials defined primarily on the basis of particle size (i.e., a mean diameter between 2 and 0.063 mm). Loose (unlithified) sediment in this range is referred to as sand, whereas consolidated or cemented (lithified) examples would be termed sandstone. Two essential characteristics of sand and sandstone make them a common medium for the preservation of fossils. First, the physical processes responsible for accumulation of sands are likely to associate bones and teeth in the same deposits; second, the porosity inherent in sands facilitates mineralization of the fossil material and hence its

long-term preservation. The rich Plio-Pleistocene fossil assemblages of the Turkana Basin are derived primarily from the sands of the ancient river channels that flowed through the basin, or into the lake when there was a lake in the basin.

sandstone   *See* **sand**.

Sangiran   (Location 7°20′S, 110°58′E, Indonesia; etym. named after a local village) History and general description The most productive fossil hominin locality in Indonesia, it is now recognized as a UNESCO **World Heritage Site**. It is located near the village of Sangiran approximately 12 km/7.4 miles north of Surakarta, Central Java, Indonesia. Almost 80 hominin fossils including 10 well-preserved crania, 14 mandibular and maxillary specimens, and dozens of isolated teeth, have been recovered from Sangiran. The first hominin fossils were collected by local individuals on behalf of Ralph **von Koenigswald** from 1936 to 1941. Following a hiatus during the war years, from 1950 to the present day local farmers and other individuals have continued to collect hominin fossils for researchers including P. Marks, Teuku **Jacob**, and S. Sartono. The hominin fossils from Sangiran are currently housed in the following Indonesian and German institutions: Gadjah Mada University (Yogyakarta, Central Java), **Geological Research and Development Center** (Bandung, Indonesia), Institute of Technology (Bandung), and the **Senckenberg Forschungsinstitut und Naturmuseum** (Frankfurt, Germany). Today, the **Sangiran Dome** is a relatively low-lying massif surrounded by large volcanoes. The fossil hominin-bearing geological formations of the dome are the **Sangiran Formation** with the **Bapang Formation** overlying it; these were formerly known as the Pucangan Formation and the Kabuh Formation, respectively. The two formations are the result of different geological formation processes, and are separated by a fossil-rich layer (formerly known as the **Grenzbank Zone**). The oldest hominin specimens derive from the upper levels of the Sangiran Formation, but most of the fossils are derived from the Bapang Formation. The lower layers of the Bapang were characterized by high-energy deposition so many of the mammalian fossils, including the hominins, are abraded and fragmentary. Higher in the Bapang Formation, the sediments are finer-grained (indicating lower-energy deposition) and the fossils from those strata are better preserved and more complete. Temporal span and how dated? Through the 1980s it was thought that none of the

Sangiran Dome hominins were older than 1.0 Ma, but in 1994 **argon-argon dating** suggested age of 1.66 Ma for a Sangiran hominin fossil originally discovered during canal-cutting operations in the early 1970s with ages derived from stratigraphic evidence indicating that even the Bapang Formation hominins were older than 1.0 Ma. In 2001, an argon-argon geochronology generated by a different team confirmed these ages. The Sangiran Formation has argon-argon dates of 1.92–1.58 Ma, with only the uppermost section of the formation being hominin-bearing. The Bapang Formation has argon-argon dates of 1.58–1.0 Ma. The Notopuro Formation, which caps the section but has yet to yield hominins, has an argon-argon date of 780 ka. Thus, today, the geologically youngest (not the oldest) fossil hominins from Sangiran are dated to 1.0 Ma. Hominins found at site Many hominin fossils have been discovered eroding from the Sangiran Dome deposits over the years, but most are without provenance because they were surface collections made by local farmers and other non-scientists. **Sangiran 2** (discovered in 1937) was the first partial cranium discovered at the site. **Sangiran 4** (discovered in 1939), a partial cranium and palate with much of the upper dentition preserved, is the holotype of *Pithecanthropus dubius*. **Sangiran 6** (discovered in 1941) is the holotype of *Meganthropus palaeojavanicus*, a fossil taxon once thought to belong to the type of megadont archaic hominin that is found in East Africa. The **Sangiran 17** cranium (discovered in 1969) is the most complete fossil hominin cranium to be discovered in Indonesia. Archeological evidence found at site In 1992 at **Ngebung**, a locality found in the Bapang Formation in the northwestern part of the Sangiran Dome, stone tools including utilized large flakes and bolas stones were reported. Key references: Geology, dating, and paleoenvironment Watanabe and Kadar 1985, Swisher et al. 1994, Larick et al. 2001, Antón and Swisher 2004; Hominins Weidenreich 1944, von Koenigswald 1968, Jacob 1972, Sartono 1976, Pope and Cronin 1984, Kramer 1994, Kaifu et al. 2005a; Archeology Sémah et al. 1992.

Sangiran 1a   Site Sangiran. Locality There is no information about where it was found but the preservation is similar to that of Sangiran 6b. Surface/*in situ* Surface. Date of discovery 1936. Finder N/A. Unit N/A. Horizon N/A. Bed/member N/A. Formation **von Koenigswald**, in an unpublished manuscript (entitled "Early Man in Java IV. *Meganthropus palaeojavanicus*" dated 1949, in the von Koenigswald Archive at the **Senckenberg**

Forschungsinstitut und Naturmuseum), implies that Sangiran 6b comes from the **Sangiran Formation**, so by implication Sangiran 1a may also have come from the Sangiran Formation. Group N/A. Nearest overlying dated horizon N/A. Nearest underlying dated horizon N/A. Geological age *c*.1.60 Ma. Developmental age Adult. Presumed sex Unknown. Brief anatomical description Heavily mineralized posterior part of the left maxilla including the distal half of the broken crown and root of the left M^1, the broken crown and root of the left M^2, and the well-preserved crown and root of the left M^3. Announcement None, but listed by Jacob in Oakley et al. (1975). Initial description None. Photographs/line drawings and metrical data None. Detailed anatomical description None. Initial taxonomic allocation None. Taxonomic revisions N/A. Current conventional taxonomic allocation *Homo erectus*. Informal taxonomic category Pre-modern *Homo*. Significance According to the unpublished manuscript referred to above, von Koenigswald originally regarded Sangiran 6b as belonging to a fossil ape, but he later added it to the collection of hominin fossils from Sangiran, and Jens Franzen (pers. comm.) suggests that Sangiran 1a was transferred into the hominin collection at the same time. That would make it among the earliest, if not the first, hominin to be recovered from the Sangiran region after the discoveries at **Trinil**. Location of original Senckenberg Forschungsinstitut und Naturmuseum, Frankfurt, Germany.

Sangiran 1b Site **Sangiran**. Locality The location is described as being south of **Bukuran** and west of Kertosobo, along the right bank of the western tributary of the Cemoro river (**von Koenigswald** 1940, Itihara et al. 1985). Surface/*in situ* Surface. Date of discovery 1936. Finder Atmowidjojo. Unit N/A. Horizon N/A. Bed/member N/A. Formation **Sangiran Formation**. Group N/A. Nearest overlying dated horizon N/A. Nearest underlying dated horizon N/A. Geological age *c*.1.60 Ma. Developmental age Adult. Presumed sex Unknown. Brief anatomical description Right side of the mandibular corpus extending from a section through the right I_1 alveolus to posterior to the right M_3. The lateral incisor and canine, and P_3 alveoli are empty, but the roots and crowns of the P_4–M_3 are preserved. Announcement von Koenigswald 1937. Initial description von Koenigswald 1940. Photographs/line drawings and metrical data von Koenigswald 1940. Detailed anatomical description von Koenigswald 1940. Initial taxo-

nomic allocation *Pithecanthropus*. Taxonomic revisions *Pithecanthropus modjokertensis* (von Koenigswald 1950). Current conventional taxonomic allocation *Homo erectus*. Informal taxonomic category Pre-modern *Homo*. Significance The first fossil hominin to be recognized from the Sangiran Formation. According to von Koenigswald (1940) it was called Mandible B because he considered the mandible from Kedungbrubus to be Mandible A. Location of original **Senckenberg Forschungsinstitut und Naturmuseum**, Frankfurt, Germany (syn. Mandible B, *Pithecanthropus* B, *Pithecanthropus* Mandible B).

Sangiran 2 Site **Sangiran**. Locality Approximately 0.5 km/0.3 miles downstream from Bapang village along the left bank of the Cemoro River (von Koenigswald 1940, p. 78, Itihara et al. 1985). Surface/*in situ* Surface. Date of discovery 1937. Finder N/A. Unit N/A. Horizon N/A. Bed/member N/A. Formation **Bapang Formation**. Group N/A. Nearest overlying dated horizon N/A. Nearest underlying dated horizon N/A. Geological age *c*.1.51–1.47 Ma. Developmental age Adult. Presumed sex Female. Brief anatomical description Calotte that lacks the right side and the medial part of the left side of the frontal region, and much of the base except for the mandibular fossae. Announcement **von Koenigswald** 1938. Initial description von Koenigswald 1938. Photographs/line drawings and metrical data von Koenigswald 1940. Detailed anatomical description von Koenigswald 1940. Initial taxonomic allocation *Pithecanthropus*. Taxonomic revisions *Pithecanthropus erectus* (Weidenreich 1945). Current conventional taxonomic allocation *Homo erectus*. Informal taxonomic category Pre-modern *Homo*. Significance The first relatively well-preserved evidence of the cranium from Sangiran. Location of original **Senckenberg Forschungsinstitut und Naturmuseum**, Frankfurt, Germany (syn. *Pithecanthropus* II, Skull II).

Sangiran 3 Site **Sangiran**. Locality Southern end of the top of a high hill behind the village Tandjoeng (**von Koenigswald** 1940, p. 102, Itihara et al. 1985). Surface/*in situ* Surface. Date of discovery 1938. Finder N/A. Unit N/A. Horizon N/A. Bed/member N/A. Formation **Bapang Formation**. Group N/A. Nearest overlying dated horizon N/A. Nearest underlying dated horizon N/A. Geological age *c*.1.51–1.47 Ma. Developmental age Young adult. Presumed sex Unknown. Brief anatomical description Calotte that preserves most of the right parietal, the medial part of the left parietal, and part of the right side of the occipital squame. Announcement von Koenigswald and Weidenreich 1938. Initial description von Koenigswald 1940. Photographs/line drawings and metrical data von Koenigswald 1940. Detailed anatomical description von Koenigswald 1940. Initial taxonomic allocation *Pithecanthropus*. Taxonomic revisions *Pithecanthropus erectus* (Weidenreich 1945). Current conventional taxonomic allocation *Homo erectus*. Informal taxonomic category Pre-modern *Homo*. Significance The first evidence of the calotte of a young adult from Sangiran. Location of original **Senckenberg Forschungsinstitut und Naturmuseum**, Frankfurt, Germany (syn. *Pithecanthropus* III, Skull III).

Sangiran 4 Site **Sangiran**. Locality North of Glagahombo village. Surface/*in situ* Surface. Date of discovery 1938–9. Finder N/A. Unit N/A. Horizon N/A. Bed/member N/A. Formation **Sangiran Formation**, but close to the boundary with the **Bapang Formation**. Group N/A. Nearest overlying dated horizon N/A. Nearest underlying dated horizon N/A. Geological age *c*.1.60 Ma. Developmental age Adult. Presumed sex Unknown. Brief anatomical description Cranium that lacks the upper part of the face, the frontal bone and the body, and parts of the wings of the sphenoid. Announcement von Koenigswald and Weidenreich 1939. Initial description Weidenreich 1945. Photographs/line drawings and metrical data **Weidenreich** 1945. Detailed anatomical description Weidenreich 1945. Initial taxonomic allocation *Pithecanthropus*. Taxonomic revisions *Pithecanthropus robustus* (Weidenreich 1945). Current conventional taxonomic allocation *Homo erectus*. Informal taxonomic category Pre-modern *Homo*. Significance The first evidence of a cranium, albeit incomplete, from the Sangiran Formation, and although Weidenreich (1945) attributed it to a separate species of *Pithecanthropus*, it is now widely accepted as part of the hypodigm of *Homo erectus*. Location of original **Senckenberg Forschungsinstitut und Naturmuseum**, Frankfurt, Germany (syn. *Pithecanthropus* IV, Skull IV).

Sangiran 5 Site **Sangiran**. Locality "South of the small path that connects Ngebung with Blimbing" (unpublished manuscript by **von Koenigswald** entitled "Early Man in Java IV. *Meganthropus palaeojavanicus*" dated 1949, in the von Koenigswald Archive at the **Senckenberg Forschungsinstitut und Naturmuseum**). Surface/*in situ* Surface. Date of discovery 1939. Finder N/A. Unit N/A. Horizon N/A. Bed/member N/A. Formation **Sangiran Formation**. Group N/A. Nearest overlying dated horizon N/A. Nearest underlying dated horizon N/A. Geological age *c*.1.60 Ma. Developmental age Adult. Presumed sex Unknown. Brief anatomical description Fragment of the right side of the corpus of an adult mandible that extends from the alveolus of I_1 medially to the distal end of the M_2. The anterior fracture surface is oblique and exposes the alveolus of I_2, the tip of the canine root, the alveolus of P_3, and the broken crown and root of the P_4. Announcement von Koenigswald 1938, 1940. Initial description **Weidenreich** 1945. Photographs/line drawings and metrical data Weidenreich 1945. Detailed anatomical description Weidenreich 1945. Initial taxonomic allocation In 1945 Weidenreich wrote that "I deem it best to put aside the Sangiran mandible of 1939 for the present and to return to it when we have more reliable information" (*ibid*, p. 62), so von Koenigswald (1950) was the first to allocate it to a taxon as *Pithecanthropus dubius*. Taxonomic revisions *Homo erectus* (Le Gros Clark 1955). Current conventional taxonomic allocation *H. erectus*. Informal taxonomic category Pre-modern *Homo*. Significance Its relatively primitive premolar root morphology is a reminder of the range of morphological variation subsumed within a single taxon solution to the Javan hominins. Kaifu et al. (2005b) suggest that it may be evidence of a second "older" (*ibid*, p. 517) and "more robust" (*ibid*, p. 518) morph; Tyler (2004) claims it belongs to a *Pongo*-like ape. Location of original Senckenberg Forschungsinstitut und Naturmuseum, Frankfurt, Germany (syn. Sangiran 1939 mandible).

Sangiran 6a [NB: the original Sangiran 6 has *de facto* become 6a to avoid confusion after Grine and Franzen 1994 introduced a hitherto undescribed hominin mandibular fragment from Sangiran as Sangiran 6b (*ibid*, p. 91)] Site **Sangiran**. Locality **von Koenigswald**, in an unpublished manuscript (entitled "Early Man in Java IV. *Meganthropus palaeojavanicus*" dated 1949, in the von Koenigswald Archive at the **Senckenberg Forschungsinstitut und Naturmuseum**), describes the location of its discovery as "near Glagahombo, north of Sangiran" and he writes that "the place is only a few hundred m. away from the spot where the skull of Pithecanthropus Modjokertensis has been discovered in 1939" (*ibid*, first manuscript page of the section on *Meganthropus palaeojavanicus*). According to the map in Itihara et al. (1985) the location is to the south of the site of the discovery of Sangiran 4. Surface/*in situ* Surface. Date of discovery 1941. Finder Kromopawiro. Unit N/A. Horizon N/A. Bed/member N/A. Formation **Sangiran Formation**. Group N/A. Nearest overlying dated horizon N/A. Nearest underlying dated horizon N/A. Geological age *c*.1.60 Ma. Developmental age Adult. Presumed sex Unknown. Brief anatomical description Fragment of the right side of the corpus of an adult mandible that extends from an anterior oblique break that passes through the distal wall of the alveolus of I_2 anteriorly and the symphysis posteriorly, to the posterior fracture surface that has exposed the distal aspect of the distal root plate of M_1. The posterior fracture surface is also oblique, extending more posteriorly at the base than at the alveolar margin. Announcement Weidenreich 1945. Initial description Weidenreich 1945. Photographs/line drawings and metrical data Weidenreich 1945. Detailed anatomical description Weidenreich 1945. Initial taxonomic allocation *Meganthropus palaeojavanicus*. Taxonomic revisions *Homo erectus* (Le Gros Clark 1955). Current conventional taxonomic allocation *H. erectus*. Informal taxonomic category Pre-modern *Homo*. Significance The overall size of the corpus and the inferred relatively primitive P_3 root morphology caused Franz **Weidenreich** to support von Koenigswald's contention that this mandible belonged to a large-bodied hominin taxon other than *H. erectus*, but several authors (e.g., Kramer 1989) have shown that the size of the mandible is not exceptional. Location of original Senckenberg Forschungsinstitut und Naturmuseum, Frankfurt, Germany (syn. 1941 mandible, *Meganthropus* A).

Sangiran 6b Site **Sangiran**. Locality N/A. Surface/*in situ* Surface. Date of discovery 1936. Finder N/A. Unit N/A. Horizon N/A. Bed/member N/A. Formation **von Koenigswald**, in an unpublished manuscript (entitled "Early Man in Java IV. *Meganthropus palaeojavanicus*" dated 1949, in the von Koenigswald Archive at the Senkenberg Forschungsinstitut und Naturmuseum), implies that Sangiran 6b comes from the **Sangiran Formation**, for he describes it as coming from "a deeper layer in the Black Clay than the Pithecanthropus mandible" (i.e., Sangiran 1b). Group N/A. Nearest overlying dated horizon N/A. Nearest underlying dated horizon N/A. Geological age *c*.1.60 Ma. Developmental age Adult. Presumed sex Unknown. Brief anatomical description Heavily mineralized posterior part of

the left mandibular corpus. The anterior fracture surface exposes the mesial aspect of the broken crown of the M_1; the oblique posterior fracture surface exposes the distal aspect of the distal root plate of the M_3. The crowns of the M_2 and M_3 are preserved. Announcement Grine and Franzen 1994. Initial description Grine and Franzen 1994. Photographs/line drawings and metrical data Grine and Franzen 1994. Detailed anatomical description None. Initial taxonomic allocation None. Taxonomic revisions N/A. Current conventional taxonomic allocation *Homo erectus*. Informal taxonomic category Pre-modern *Homo*. Significance In an unpublished manuscript (see Formation, above) von Koenigswald refers this mandibular specimen to *Meganthropus palaeojavanicus*. Grine and Franzen (1994) make no comment about its taxonomy. Location of original Senckenberg Forschungsinstitut und Naturmuseum, Frankfurt, Germany.

Sangiran 7 The prefix "S7" was introduced by Jacob (1975). Prior to that **von Koenigswald** had given *all* of the 52 isolated teeth recovered from **Sangiran** the prefix "FS" [i.e., an acronym for "fossils (from) Sangiran"] followed by a number, but only those teeth subsequently determined to be hominin were given the prefix S7. These teeth carried their FS number with them (e.g., FS 89 became S7-89). Jacob (1975) divided the Sangiran 7 collection of isolated teeth into Sangiran 7a (or S7a), for those from the **Sangiran Formation**, and Sangiran 7b (or S7b), for those from the **Bapang Formation**. Site Sangiran. Locality N/A. Surface/*in situ* Surface. Date of discovery 1937–41. Finder N/A. Unit N/A. Horizon N/A. Bed/member N/A. Formation See above. Group N/A. Nearest overlying dated horizon N/A. Nearest underlying dated horizon N/A. Geological age *c*.1.51–1.60 Ma. Developmental age All but four of the teeth (the three, FS 67, 72, and 83, described in Grine 1984, plus S7-13) are permanent. Presumed sex Unknown. Brief anatomical description Three isolated deciduous teeth [a left upper dc (FS 83), and a right dm_1 (FS 67), and a right dm_2 (FS 72)] were described in detail by Grine (1984); the remainder were described in detail by Grine and Franzen (1994). The teeth described by Grine and Franzen (1994) are all given "S7" numbers (e.g., S7-1 is a right I^1; S7-10 is a right M^1), but the three deciduous teeth described by Grine (1984) do not have S7 numbers. Franzen (pers. comm.) suggests that when he and Fred Grine reviewed all the S7 plus the FS 67, 72, and 83 teeth in the early 1990s they both came to the conclusion that the three deciduous "FS" teeth described by Grine (1984) were most likely pongid and not hominin. Announcement FS 67, von Koenigswald 1953; FS 72, von Koenigswald 1950; FS 83, Grine 1984; the remainder Grine and Franzen 1994. Initial description, photographs/line drawings and metrical data, and detailed anatomical description FS 67, 72, and 83, Grine 1984; the remainder Grine and Franzen 1994. Initial taxonomic allocation FS 67, *Homo modjokertensis*; FS 72 and 83, *Meganthropus paleojavanicus*, with all of the other 52 teeth by implication being referred to *Homo*. Taxonomic revisions FS 67, 72, and 83, *Pongo* sp. Current conventional taxonomic allocation *Homo erectus*. Informal taxonomic category Pre-modern *Homo*. Significance The collection is a rich source of information about dental variation in the Javan part of the hypodigm of *H. erectus*. Location of the originals **Senckenberg Forschungsinstitut und Naturmuseum**, Frankfurt, Germany.

Sangiran 8 Site **Sangiran**. Locality Near the village of Glagahombo. Surface/*in situ* Surface. Date of discovery 1952. Finder A local villager. Unit N/A. Horizon N/A. Bed/member Between underlying top of **Sangiran Formation** and the bottom of the **Grenzbank Zone**. Formation **Sangiran Formation**. Group N/A. Nearest overlying dated horizon N/A. Nearest underlying dated horizon N/A. Geological age *c*.1.60 Ma. Developmental age Adult. Presumed sex Unknown. Brief anatomical description Heavily damaged mandibular corpus preserving root fragments of the left I_1–M_1, the right I_1, and the right P_4–M_2. The complete crown of the right M_3 is present. The right side of the corpus is broken such that it appears more robust than it was in its pristine state; the damage has been interpreted as evidence of a perimortem crocodile bite (**von Koenigswald** 1968). Announcement Marks 1953. Initial description Marks 1953. Photographs/line drawings and metrical data Kaifu et al. 2005a. Detailed anatomical description von Koenigswald 1968, Kaifu et al. 2005a. Initial taxonomic allocation "*Meganthropus*" (Marks 1953). Taxonomic revisions *Pithecanthropus erectus* (Le Gros Clark 1955), *Homo erectus* (Le Gros Clark 1964). Current conventional taxonomic allocation *H. erectus*. Informal taxonomic category Pre-modern *Homo*. Significance Initially interpreted as representing an indigenous, hyper-robust lineage referred to as "*Meganthropus*" (Marks 1953, von Koenigswald 1968), or a southeast Asian form of australopith or archaic hominin (Robinson 1953). Later, it was inter-

preted as being within the range of variation of *H. erectus* (Le Gros Clark 1964, Lovejoy 1970, Kramer 1989). Location of original **Geological Research and Development Center**, Bandung, Indonesia (syn. *Meganthropus* B).

Sangiran 9 Site **Sangiran**. Locality from the slope of a hill near the village of Mandingan or Bojong. Surface/*in situ* Surface. Date of discovery 1960. Finder A local villager. Unit N/A. Horizon N/A. Bed/member Between the top of **Sangiran Formation** and the bottom of the **Grenzbank Zone**. Formation Sangiran Formation. Group N/A. Nearest overlying dated horizon N/A. Nearest underlying dated horizon N/A. Geological age *c*.1.60 Ma. Developmental age Adult. Presumed sex Unknown. Brief anatomical description Heavily mineralized right side of the mandibular corpus preserving the symphysis distal to just beyond the M_3. The alveoli of the left I_1, and on the right side the I_{1-2} alveoli, the broken crown of the canine, the complete P_{3-4} crowns, the alveolus of M_1, and the complete M_{2-3} crowns are preserved. Announcement Sartono 1961. Initial description Sartono 1961. Photographs/line drawings and metrical data Kaifu et al. 2005a. Detailed anatomical description Kaifu et al. 2005a. Initial taxonomic allocation "*Pithecanthropus*" (Sartono 1961). Taxonomic revisions *Pithecanthropus dubius* (**von Koenigswald** 1968), *Homo erectus* (Kramer 1994). Current conventional taxonomic allocation *H. erectus*. Informal taxonomic category Pre-modern *Homo*. Significance Sangiran 9 extended the range of morphological and metrical dento-gnathic diversity of early *H. erectus* from Java (Kaifu et al. 2005b). Location of original **Geological Research and Development Center**, Bandung, Indonesia (syn. Mandible C).

Sangiran 10 Site **Sangiran**. Locality Near the village of Tanjung. Surface/*in situ* Surface. Date of discovery 1963. Finders Sartono, Jacob, and local villagers. Unit N/A. Horizon N/A. Bed/member Within the middle to upper levels of the **Bapang Formation**. Formation Bapang Formation. Group N/A. Nearest overlying dated horizon N/A. Nearest underlying dated horizon N/A. Geological age *c*.1.33–1.24 Ma. Developmental age Adult. Presumed sex Male. Brief anatomical description A young adult calotte, presumably male based on the size of the supraorbital torus; a left zygomatic is also attributed to this individual. Endocranial volume Approximately 975 cm^3. Announcement Jacob 1964. Initial description Jacob 1966. Photographs/line drawings and metrical data Jacob 1966. Detailed anatomical description Jacob 1966. Initial taxonomic allocation "*Pithecanthropus erectus*" (Jacob 1966). Taxonomic revisions *Homo erectus* (Le Gros Clark 1978). Current conventional taxonomic allocation *H. erectus*. Informal taxonomic category Pre-modern *Homo*. Significance Sangiran 10 exemplifies the **total morphological pattern** of *H. erectus* crania from Java, including the thick cranial wall, robust cranial superstructures, and an endocranial volume of approximately 1000 cm^3. Location of original **Gadjah Mada University**, Yogyakarta, Indonesia (syn. Skull VI).

Sangiran 17 Site **Sangiran**. Locality near the village of Pucung. Surface/*in situ* Surface. Date of discovery September 13, 1969. Finder Mr. Tukimin, a local farmer working for S. Sartono. Unit N/A. Horizon N/A. Bed/member Between the Lower and Middle Tuffs within the **Bapang Formation**. Formation Bapang Formation. Group N/A. Nearest overlying dated horizon Middle Tuff (1.02 Ma). Nearest underlying dated horizon Grenzbank boundary layer between the Bapang and **Sangiran Formation**s (1.51 Ma). Geological age *c*.1.25 Ma. Developmental age Adult. Presumed sex Male. Brief anatomical description. The most complete Javanese *Homo erectus* cranium including most of the face, base, and cranial vault. Endocranial volume 1004 cm^3 (Holloway 1981). Announcement Sartono 1971. Initial description Sartono 1971. Photographs/line drawings and metrical data Sartono 1971, Kaifu et al. 2008. Detailed anatomical description Baba et al. 1998. Initial taxonomic allocation "*Pithecanthropus erectus*" (Sartono 1971). Taxonomic revisions *H. erectus* (Le Gros Clark 1978). Current conventional taxonomic allocation *H. erectus*. Informal taxonomic category Pre-modern *Homo*. Significance Documents the total morphological pattern of *H. erectus* crania from Java, including a continuous, bar-like, supraorbital torus, thickened cranial vault, pronounced angular torus, and broad and robust face. Its cranial morphology indicates likely phylogenetic continuity with succeeding **Ngandong** and **Sambungmacan** hominins in Java (Kaifu et al. 2008). Location of original Quaternary Geology Laboratory of the **Geological Research and Development Center**, Bandung, Indonesia (syn. Skull VIII).

Sangiran 21 Site **Sangiran**. Locality Near the village of Ngebung. Surface/*in situ* Surface. Date

of discovery 1973. Finders Local collectors. Unit N/A. Horizon N/A. Bed/member Within the lowermost levels of the **Bapang Formation**, just above the underlying top of the **Grenzbank Zone**. Formation Bapang Formation. Group N/A. Nearest overlying dated horizon N/A. Nearest underlying dated horizon N/A. Geological age *c*.1.51–1.47 Ma. Developmental age Adult. Presumed sex Unknown. Brief anatomical description A fragmentary right mandibular ramus preserving the condyle and the most distal portion of the right side of the corpus preserving the crown of M$_3$. Announcement Sartono 1974. Initial description Sartono 1974. Photographs/line drawings and metrical data Kaifu et al. 2005a. Detailed anatomical description Kaifu et al. 2005a. Initial taxonomic allocation "*Pithecanthropus*" (Sartono 1974). Taxonomic revisions *Homo erectus* (Tyler et al. 1995). Current conventional taxonomic allocation *H. erectus*. Informal taxonomic category Pre-modern *Homo*. Significance Sangiran 21 documents dentognathic size reduction in later *H. erectus* from Java (Kaifu et al. 2005b). Location of original **Geological Research and Development Center**, Bandung, Indonesia (syn. Mandible E).

Sangiran 27 Site **Sangiran**. Locality Near the village of Sangiran, north of the Chemoro River. Surface/*in situ* Surface. Date of discovery 1978. Finders Local workers during the construction of a dam. Unit N/A. Horizon N/A. Bed/member Within the "Black Clays" of the upper levels of the **Sangiran Formation**, above Tuff 10 and below the overlying **Grenzbank Zone**. Formation Sangiran Formation. Group N/A. Nearest overlying dated horizon Grenzbank Zone. Nearest underlying dated horizon Tuff 10. Geological age *c*.1.66–1.58 Ma. Developmental age Adult. Presumed sex Unknown. Brief anatomical description A partial cranium preserving portions of the face, the anterior vault, and maxillary dentition including the right P^3–M^3, and on the left side a partial P^4 crown and M$^{1–2}$; the specimen is highly mineralized and underwent significant post-mortem crushing and deformation. Announcement Jacob 1980. Initial description Jacob 1980. Photographs/line drawings and metrical data Indriati and Antón 2008. Detailed anatomical description Indriati and Antón 2008. Initial taxonomic allocation "*Meganthropus*" (Jacob 1980). Taxonomic revisions *Homo erectus* (Indriati and Antón 2008). Current conventional taxonomic allocation *H. erectus*. Informal taxonomic category Pre-modern *Homo*. Significance Sangiran 27 documents the "hyper-robust" cranio-facial morphology seen in the earliest Javan *H. erectus* and it is consistent with the derived nature of these hominins (especially the dentition) in comparison to their contemporaries from **Dmanisi** and the **Turkana Basin** of East Africa (Indriati and Antón 2008). Location of original **Gadjah Mada University**, Yogyakarta, Indonesia.

Sangiran 38   Site **Sangiran**. Locality Near the village of Sendangbusik. Surface/*in situ* Surface. Date of discovery 1980. Finders Local workers. Unit N/A. Horizon N/A. Bed/member Lower levels of the Bapang Formation. Formation Bapang Formation. Group N/A. Nearest overlying dated horizon Lowest tuff. Nearest underlying dated horizon **Grenzbank Zone**. Geological age *c*.1.58–1.47 Ma. Developmental age Adult. Presumed sex Unknown. Brief anatomical description A partial cranium preserving portions of the frontal, parietal, and occipital. Announcement Jacob 1980. Initial description Jacob 1983, Grimaud-Herve and Widianto 2001, Indriati and Antón 2010. Photographs/line drawings and metrical data Indriati and Antón 2010. Detailed anatomical description Indriati and Antón 2008. Initial taxonomic allocation *Homo erectus* Jacob, 1983. Taxonomic revisions None. Current conventional taxonomic allocation *H. erectus*. Informal taxonomic category Pre-modern *Homo*. Significance Sangiran 38 shows depressed lesions of the vault consistent with a scalp or systemic infection (Indriati and Antón 2010). Location of original Gadjah Mada University, Yogyakarta, Indonesia.

Sangiran 1939 mandible   *See* **Sangiran 5**.

Sangiran 1941 mandible   *See* **Sangiran 6a**.

Sangiran Dome   A relatively low-lying massif surrounded by large volcanoes in Central Java, Indonesia. The fossil hominin-bearing geological formations of the dome are the **Sangiran Formation** and the **Bapang Formation** overlying it. The older Sangiran Formation was primarily deposited from freshwater lakes and swamps and the sediments include dense clay and organic components. The oldest hominin specimens from Sangiran are attributed to the upper levels of the Sangiran Formation. A coarse-grained conglomeratic layer rich with fossil bone (formerly known as the **Grenzbank Zone**) lies between the older Sangiran and younger Bapang Formations. In contrast to the Sangiran Formation, the Bapang Formation is largely a product of local uplift, volcanism, and sedimentary deposition from river and stream activity. Most of the hominin fossil specimens from the Sangiran Dome are derived from the Bapang Formation. The lower layers of the Bapang were characterized by high-energy deposition so many of the mammalian fossils, including the hominins, are abraded and fragmentary. Higher in the Bapang Formation, the sediments are finer-grained (indicating lower-energy deposition) and the fossils from those strata are better preserved and more complete. The fossil hominin-bearing sites of the Sangiran Dome include **Blimbingkulon, Bukuran, Ngrejeng, Sangiran,** and **Sendangbusik**. Stone tools have been report from **Ngebung**.

Sangiran Formation   (etym. named after the site of **Sangiran** where it is exposed) General description The older of the two fossil-hominin-bearing geological formations of Central and East Java, Indonesia. Formerly known as the Pucangan Formation, the Sangiran Formation includes sediments primarily deposited from freshwater lakes and swamps that include dense clay and organic components. The oldest hominin specimens from Sangiran are attributed to the upper levels of the Sangiran Formation. A coarse-grained conglomeratic layer rich with fossil bone (formerly known as the **Grenzbank Zone**) marks the transition between the older Sangiran and younger **Bapang Formation**s. Sites where this formation is exposed Hanoman, **Mojokerto**, and sites within the Sangiran Dome. Geological age **Argon-argon dating** suggests the formation dates between 1.92 and 1.58 Ma, and that the hominins at the Sangiran Dome localities are from the upper part of this formation between Tuff 10 (1.66 Ma) and the Grenzbank Zone (1.58 Ma); those from sites elsewhere (e.g., Mojokerto) may be as old as 1.8 Ma. Most significant hominin specimens For example Mojokerto 1, **Sangiran 4**, and **Sangiran 27**. Key references: Geology, dating, and paleoenvironment Watanabe and Kadar 1985, Swisher et al. 1994, Larick et al. 2001; Hominins Weidenreich 1944, von Koenigswald 1968, Jacob 1972, Sartono 1976, Pope and Cronin 1984.

Sangoan   (etym. Named after the artifacts recovered from Sango Bay, Uganda, on the western shore of Lake Victoria) The Sangoan is an artifact **industry** found across Central Africa and portions of East Africa. The Sangoan is defined typologically by the presence of heavy-duty tools such as **picks**, choppers, and core-axes as well as a number of light-duty tools such as scrapers (see McBrearty 1987, 1991). At **Kalambo Falls**, Zambia, Sangoan assemblages overlie **Acheulean** ones, and are in turn overlain by strata containing evidence of the **Middle Stone Age**, a pattern that is repeated at other sites, particularly in the Lake Victoria basin. Many of the elements found at Sangoan sites (e.g., picks, **Levallois** cores) also are found in Acheulean or Middle Stone Age ones, and thus the Sangoan has figured in some discussions of the relationship between these two industries (e.g., Tryon and McBrearty 2002). Sangoan tools co-occur with **Lupemban** tools at several sites, and the distribution of both Sangoan and Lupemban assemblages broadly coincides with the past and present distribution of equatorial forests. This coincidence lead Clark (1970) to suggest that the Sangoan and Lupemban reflect forest adaptations, and may signal the initial hominin exploitation of this habitat type, but subsequent site-specific paleoenvironmental reconstructions (McBrearty 1992) have not always supported this hypothesis. The Sangoan industry is poorly dated, but it is most likely **Middle Pleistocene** in age.

sanidine A mineral of the **feldspar** group, with the formula $KAlSi_3O_8$. A common potassium-bearing mineral, sanidine is especially suitable for **potassium-argon dating**. The pumices encountered in **tuffs** of the **Koobi Fora Formation** have proved to be a rich source of the sanidine crystals that are the basis for dating that fossiliferous sequence.

sapropel (Gk *sapros*=rotten and *pelagos*=sea) A sedimentary rock (usually a mud or marl) that contains more than 5% organic material. Sapropels occur in **lacustrine** and marine deposits. The **Neogene** strata of the eastern Mediterranean Sea are characterized by sapropel beds, generated by precessionally driven precipitation delivered via the Nile River. These sapropels provide **proxy** records of monsoonal rainfall in East Africa (particularly in the Blue Nile headwaters in the Ethiopian Highlands) and they are also used for the **orbital tuning** of the associated marine sequences and in the construction of age models for the **cores** and outcrops in which they are recognized.

Sarich, Vincent M. (1934–) Vincent Sarich received his BS in chemistry from the Illinois Institute of Technology and his MS and PhD in

anthropology from the University of California at Berkeley. From 1966 to 1994 Sarich was a faculty member in the Department of Anthropology at the University of California at Berkeley, after which became an Emeritus Professor. He then served as a Lecturer in Anthropology at the University of Auckland in New Zealand. Sarich is best known for his work on **molecular clocks**. Working with the biochemist Allan **Wilson**, Sarich co-published two important papers in 1967. Using microcomplement fixation, Sarich and Wilson (1967a) examined the pattern of immunological differences among primate species with respect to the protein **albumin**. The degree of difference in the reaction of target albumin antibody to antisera of other primates was expressed as the index of dissimilarity; log(index of dissimilarity)×100 equals the immunological distance (or ID). Sarich and Wilson were able to demonstrate that immunological distance between the proteins of two species is proportional to the number of **amino acid** differences, which in turn is proportional to the time of species separation. But the most important insight was the description of the so-called rate test (Sarich and Wilson 1967b) in a paper that sadly is seldom cited. Given three taxa A, B, and C in which A and B are **sister taxa** and C is the outgroup, comparison of the ID of A and C with the ID of B and C will demonstrate the extent to which immunological (=amino acid=evolutionary) change is more or less constant. To the surprise of Sarich and Wilson, broad surveys of primate albumin immunological distances showed, via the rate test, that molecular change was regular, or clock-like [in the hypothetical case above, this conclusion would follow if ID_{AC} was equal or subequal (nearly equal) to ID_{BC}]. Having demonstrated the evolutionary clock-like behavior of primate albumins, Sarich and Wilson could therefore generate a so-called proportional tree. This proportional tree could then be calibrated using the fossil record; Sarich and Wilson assumed a date of 45 Ma for the divergence of New and Old World anthropoids and 30 Ma for the divergence of hominoids and cercopithecoids. The latter calibration yielded a divergence age of 5 million years for modern humans, chimpanzees, and gorillas (the relationship of these taxa then being an unresolved trichotomy), a surprisingly young date given the then widely held fossil-based belief that modern humans diverged from other primates at least three times as long ago as suggested by Sarich and Wilson. For a variety of reasons, these results were taken seriously by far too few in the relevant scientific communities. In particular the refusal or inability to understand the logic of the rate test meant that Sarich and Wilson were widely accused of *assuming* rate constancy, when in fact they were always *inferring* it. In an interesting "retrospective" Sarich (1983) generously noted that the germ of the rate test idea might have lain in a 1963 paper by Margoliash. The general approach and conclusions of Sarich and Wilson have been widely validated and are now accepted by all competent evolutionary biologists. Indeed, the articulation of the rate test could be counted as one of the most important insights of later 20thC systematics and evolutionary biology. The later phases of Sarich's career have involved more controversial issues.

Sarstedt (Location 52°14′N, 9°51′E, Germany; etym. named after a nearby town) History and general description Amateur collectors have recovered several *Homo neanderthalensis* cranial fragments, Pleistocene fauna and **Middle Paleolithic** tools in the gravel dumps from quarries near Sarstedt, Leine valley, Lower Saxony, Germany. Temporal span and how dated? Due to their lack of context, these finds are hard to date. Direct **radiocarbon dating** on the fossils was unsuccessful, so the best estimates for the age range are between 117 and 25 ka based on geological and stratigraphic analysis. Hominins found at site Three cranial fragments have been reported, although others are known to have been recovered. Sarstedt I is a temporal bone, Sarstedt II is an occipital bone, and Sarstedt III is a parietal bone; all have been attributed to *Homo neanderthalensis*. Archeological evidence found at site Middle Paleolithic assemblage attributed to the Keilmessergruppen technocomplex, along with several faunal pieces. Key references: Geology, dating, and paleoenvironment Street et al. 2006; Archeology Czarnetzki et al. 2002; Hominins Czarnetzki et al. 2001, 2002, Street et al. 2006.

SASES Acronym for the Standardized African Site Enumeration System, which was developed in the 1960s as a way of recording African archeological sites. *See also* **Standardized African Site Enumeration System**.

satellite DNA Satellite DNA comprises large arrays of **tandemly repeated sequences** of DNA that are the primary components of constitutive heterochromatin. This type of DNA does not undergo **transcription** (i.e., its message does not get converted into an **RNA** version) and it is commonly found at the centromeres and telomeres (i.e., the ends) of chromosomes. The composition of the bases in satellite DNA is dictated by the composition of the tandemly repeated units (e.g., there may be a preponderance of As or Cs) and thus it may differ from the average base composition of the rest of the genome.

satellite imagery The traditional ways of identifying potential hominin fossil sites are to use conventional geological maps, or to comb the records of local geological or natural history societies for records of fossil finds by amateurs. When aerial photographs were available (these were flown for military or some other nonpaleontological purpose, or they were commissioned by research groups) they could also be used to locate possible sites. However, the availability of satellite images (e.g., NASA's Advanced Spaceborne Thermal Emission and Reflection radiometer, ASTER, http://asterweb.jpl.nasa.gov; Google Earth, http://earth.google.com) means that these images can now be used to locate previously unrecognized locations whose geological structures and reflectance match those of existing hominin fossil sites. The first time satellite imagery was used on a large scale to locate sites was as part of the systematic inventory of paleontological and archeological sites in Ethiopia in the 1980s (Asfaw et al. 1990). This resulted in the identification of productive new localities such as **Aramis, Bouri, Fejej, Kesem-Kebena,** and **Konso-Gardula**. *See also* **Paleoanthropological Inventory of Ethiopia**.

savanna (Taino *zabana*=grassland) The inclusive term savanna covers a large number of African biomes including **woodland**, **bushland**, **shrubland**, and **grassland**. Because of this, it has been suggested that the term is not helpful in discussions of early hominin habitats (e.g., Reed 2007). Savanna biomes are generally characterized as a landscape with grass and with either scattered trees or open canopy trees. Savannas tend to occur on land surfaces of little relief, the soils tend to be low in nutrients, they have distinct wet and dry seasons, poor vertical structure root system and extensive horizontal structure root system, and they are often associated with precipitation ranging between 500 and 2000 mm per year. The current evidence suggests that what are claimed to be the earliest hominins are found in a range of "savanna" biomes, ranging from closed woodlands to open grasslands.

"savanna" chimpanzees A term used to describe chimpanzees living in habitats where the annual rainfall is around 1000 mm or less and where the dominant ground cover is tropical (i.e., C_4) grass (Moore 1992). These mosaic environments vary from wooded grassland to miombo (i.e., **Brachystegia**-dominated) woodland. Moore and others claim that chimpanzees at "savanna" sites such as Mt Assirik in Senegal, and at Kasakati and Ugalla in Tanzania, make useful behavioral models for the most recent (i.e., last) common ancestor of the hominin and panin clades (Moore 1996). Preutz and Bertolani (2007) reported that savanna chimpanzees from Fongoli, Senegal, were observed to fashion pointed wooden sticks using up to five steps, and then use them to extract prosimians from hollow tree trunks.

SB Abbreviation of Sendangbusik that is used as a prefix for fossils recovered from that site. It is also an abbreviation used to refer to the Stillbay industry. *See also* **Sendangbusik**; **Stillbay**.

SB 8103 Site Sendangbusik. Locality The base of an outcrop of the **Bapang Formation** near the village of Sendangbusik. Surface/*in situ* Surface. Date of discovery 1981. Finder Mr Djoko, a local villager. Unit N/A. Horizon N/A. Bed/member N/A. Formation **Bapang Formation**. Group N/A. Nearest overlying dated horizon Middle Tuff, Bapang Formation. Nearest underlying dated horizon Lower Tuff, Bapang Formation. Geological age *c.*1.02–1.51 Ma. Developmental age Adult. Presumed sex Unknown. Brief anatomical description Moderately mineralized fragmentary right side of the mandibular corpus together with the well-preserved crowns of P_4–M_3. Announcement Aziz 1981. Initial description Aziz 1981. Photographs/line drawings and metrical data Kaifu et al. 2005a. Detailed anatomical description Kaifu et al. 2005a. Initial taxonomic allocation "*Pithecanthropus*" (Aziz 1981). Taxonomic revisions *Homo erectus* (Aziz et al. 1994). Current conventional taxonomic allocation *H. erectus*. Informal taxonomic category Pre-modern *Homo*. Significance The SB 8103 mandible is consistent with the proposition that later *H. erectus* from Java underwent **dento-gnathic** size reduction (Kaifu et al. 2005b). Location of

original **Geological Research and Development Center**, Bandung, Indonesia.

scaling (L. *scalae*=ladder) The relationship between a variable and **size** (however it is defined). Scaling subsumes **isometry** (no change in a variable with changes in size) and **allometry** (changes in a variable accompanying changes in size). For example, brain size scales with negative allometry relative to body mass. Thus, although larger-bodied taxa generally have larger brains than smaller-bodied taxa, because their brain size increases at a slower rate than their body mass, their brains are *relatively smaller* when scaled to body mass than those of smaller taxa. Scaling relationships are typically examined between species as interspecific allometry, within species as intraspecific allometry, or in a growth series as ontogenetic allometry. The scaling relationship may differ in all of these approaches; thus these different sorts of scaling should not be conflated, or confused, with one another. Scaling relationships can also be used for functional analysis, and may reflect functional equivalence or **adaptation** (Fleagle 1985). Functional equivalence refers to underlying structural relationships between a variable and size. For example, long bones must be wider in large taxa to maintain the same strength relationship relative to body size, because mass is a function of the volume (i.e., length cubed) of an organism whereas bone strength is a function of the cross-sectional area (i.e., length squared). Thus, a species that is twice the body length of another (geometrically similar) species may have eight times (2^3) the mass, and would need bones with greater cross-sectional area than expected by isometry (twice the width of the small species) to maintain the functionally equivalent relationship. Often, however, scaling relationships do not reflect functional equivalence and may instead describe differences in adaptations at different sizes. For example, large primates tend to have relatively long forelimbs and short hindlimbs (a high **intermembral index**) compared to small primates. This does not reflect functional equivalence, but rather the different strategies large and small primates have taken to traverse gaps between trees during travel. Small primates tend to use their hindlimbs to leap across gaps, whereas large primates tend to use their forelimbs to reach and clamber across gaps. *See also* **allometry**.

scaphoid-os centrale fusion Modern humans and African apes share fusion of two wrist bones, the scaphoid and the os centrale. This condition is not typically found in most other primates, including the Asian apes. Scaphoid-os centrale fusion is considered by some as a stabilizing adaptation to knuckle-walking locomotion and evidence that modern human bipedalism may have passed through a knuckle-walking phase (Richmond et al. 2001). Fusion of these two bones occurs early in development (e.g., *in utero*) in modern humans and African apes (Kivell and Begun 2007). In contrast, fusion in orangutans and other monkeys is rare and if it does occur, it is usually during adulthood. The only other primates to show consistent fusion of these two bones are a few species of vertical-clinging and leaping strepsirrhines (e.g., *Lepilemur*, *Indri*) and the suspensory subfossil lemur *Babakotia*. The highly suspensory subfossil lemur *Palaeopropithecus* displays fusion in roughly half of the specimens. Because regular scaphoid-os centrale fusion is found only in two primate lineages (Hominidae and Lemuroidea) and it is associated with a variety of different positional behaviors, some consider its functional significance to be unclear (Kivell and Begun 2007).

scapula, evolution in hominins The scapula, which is a complex, flat bone that develops from multiple ossification centers, lies against the dorsolateral portion of the ribcage, spanning ribs 1–7 in modern humans. The modern human glenohumeral joint faces laterally, whereas the joints of the great apes are situated more superiorly, producing a permanent "shrugged" shoulder appearance that provides strength and stability for the arm when raised above the head, as in suspensory climbing behaviors. Modern humans have a mediolaterally relatively broad scapula; *Gorilla* and *Pongo* approach the modern human condition, while the scapulae of *Pan* and *Hylobates* are narrower (Shea 1986, Green 2010). The relative size and shape of the dorsal scapula fossae also set modern humans apart from the great apes. Modern humans have the relatively tallest infraspinous region, but it is also relatively narrow mediolaterally. In contrast, *Gorilla*, *Pan*, and *Hylobates* all have superoinferiorly short and relatively wider infraspinous fossae; their supraspinous fossae are broader than those of modern humans. *Pongo* has an ape-like infraspinous fossa, but it is unique in having a narrower supraspinous fossa than modern humans (Young 2008, Green 2010). These differences in the proportions

of the dorsal scapula fossae are reflected in the obliquity of the scapula spine. Most apes have a markedly oblique scapula spine, with *Hylobates* the most oblique (its spine is nearly parallel to the axillary border) whereas modern humans have a more transversely oriented spine. Modern humans are unique in having an infraspinous region that is shaped like a right-angled triangle, with the angle between the spine and medial border approaching 90°; this angle is significantly more obtuse in other hominoid taxa (Green 2010). The hominin scapula fossil record is sparse and most of the specimens are fragmentary; this is not unexpected given that the scapula blade can be so thin as to be nearly transparent. Most of the preserved scapulae consist primarily of glenoid fragments with more or less of the scapula spine and axillary border attached. All the scapula specimens attributed to *Australopithecus* (i.e., **A.L. 288-1**, Sts 7, Stw 162, and **DIK-1-1**) display the ape-like condition of a relatively cranially oriented glenohumeral joint, while **KNM-WT 15000** has a more laterally facing joint (Vrba 1979, Stern and Susman 1983, Alemseged et al. 2006, Green 2010). The functional relevance of the cranial orientation of the very small A.L. 288-1 individual has been dismissed by some as a consequence of allometry (Inouye and Shea, 1997), but the larger size of Sts 7 and Stw 162, which have cranially oriented joints, does not support this explanation. Furthermore, the diminutive *H. floresiensis* scapula specimen (LB6/4) displays a laterally oriented glenohumeral joint, despite it being roughly the same size as A.L. 288-1 (Larson 2007, Larson et al. 2007). The Dikika infant (DIK-1-1) is the first archaic hominin scapula to preserve most of the blade of the scapula. It has an obliquely oriented scapula spine like modern apes and a dorsal scapula fossa configuration that is intermediate between extant apes and modern humans (Alemseged et al. 2006, Green 2010). This is in contrast to KNM-WT 15000, which in many respects is "hyper-human" (Larson 2007, Green 2010). Two other relatively well-preserved archaic hominin scapulae are the adult female scapula (MH2) from Malapa in South Africa and the KSD-VP-1/1g specimen from Woranso-Mille, Ethiopia. The former specimen was described as possessing a cranially oriented shoulder joint (Berger et al. 2010), and the latter specimen has a narrow scapula blade relative to modern humans, and is reported to have a more transversely oriented, modern human-like scapula spine (Haile-Selassie et al., 2010), but it was not clear how the authors took into account the preservation of the spine, which was described as being bent post-mortem. The supraspinous region in KSD-VP-1/1g is not preserved, thus limiting what can be said about the relative size of the dorsal scapula fossae. *See also* **proximal humerus, evolution in hominins**.

scavenger *See* scavenging.

scavenging (ME *skawager*=a collector of tolls, from the OF *escauver*=to inspect; thus, someone who lives off another) Scavenging is the acquisition of meat or marrow from the carcasses of animals that were killed by predators, or from an animal that died due to other natural causes. Perhaps the best-known contemporary scavengers are vultures, although some carnivores that primarily hunt, such as lions or spotted hyenas, are known to scavenge on an opportunistic basis (i.e., **passive scavenging**) or engage in **confrontational scavenging**. Modern human **hunter–gatherer** groups obtain the majority of meat through **hunting**, although some forms of scavenging are occasionally practiced. The extent to which early hominins hunted versus scavenged is of particular interest, since a **carcass acquisition strategy** based primarily on scavenging is considered to be outside the range of **modern human behavior** (i.e., there are no extant modern human groups that use scavenging as a major component of their subsistence strategy).

Schepers, G. (Gerritt) W.H. *See* Robert Broom; Sterkfontein.

schlepp effect (Yidd. *shlepn*=to drag or haul) A model of carcass transport behavior predicated on the assumption that **hunter–gatherers** are more likely to transport appendicular rather than axial skeletal parts from kill sites to base camps, and that the number of elements transported varies inversely with carcass size and/or transport distance. According to the schlepp effect, sites that are dominated by appendicular elements represent residential sites and sites that are dominated by axial elements represent kill or primary butchery sites. For example, a preponderance of appendicular elements at **FLK-*Zinjanthropus*** has been interpreted as evidence that hominins had access to fully fleshed animal carcasses and preferentially transported limbs to the site while discarding vertebrae, pelves, and ribs at the point of acquisition. Ethnographic support for the schlepp effect remains mixed, particularly as it relates to the preferential transport of appendicular

skeletal parts. Experimental observations also indicate that **carnivore ravaging** is capable of mimicking the schlepp effect.

Schmerling, Philippe-Charles (1790 or 1791–1836) Philippe-Charles Schmerling was born in Delft, The Netherlands, and began studying medicine at the University of Leiden, The Netherlands, but completed his studies at the University of Liège, Belgium, in 1825. Schmerling began excavating and collecting fossils in 1829 after seeing fossil bones in a patient's home. He focused on caves along the Meuse River and at one of the caves in the **Engis** complex, Schmerling found flint implements in deposits containing the bones of hyenas, rhinoceros, and large carnivores. In a different cave in the Engis complex he found two hominin crania and some postcranial bones. Almost all scientists in the early 19thC were convinced that the diluvium was deposited very long ago, and that the extinct mammals found in the diluvium therefore were also of great antiquity. Moreover, it was agreed that modern humans certainly could not have existed at the time of the extinct mammals, especially since it was accepted dogma that human history was very short in comparison with the history of the Earth. The few discoveries of human artifacts or bones in association with extinct mammals had been explained as relatively recent intrusive burials that accidentally mixed modern human remains with those of much older animal fossils. But Schmerling found more examples of flint artifacts among extinct mammal fossils at Engihoul, Chokier, and Fond-de-Forêt, and he began to suspect that these were not accidental associations, for there was no evidence that the soil had been disturbed by a burial and the nature of the cave deposits strongly indicated that the extinct mammals and the human artifacts and remains had been deposited at the same time and by the same processes. These discoveries convinced Schmerling that human ancestors were present on the Earth at a geologically remote period in time. Schmerling recorded his discoveries and his interpretation of them in *Recherches sur les ossemens fossils découverts dans les caverns de la province de Liège* [*Researches on the Fossil Bones Discovered in the Caverns of the Province of Liège*] published in two volumes in 1833 and 1834. While few scientists were willing to accept Schmerling's interpretation of his discoveries as evidence for the great antiquity of modern humans, it did contribute to a growing debate during the 1830s and 1840s over similar evidence discovered elsewhere in Europe that eventually led in the 1860s to the acceptance by geologists of the coexistence of humans and extinct mammals. At the end of the century, and after the discovery of the first recognized *Homo neanderthalensis* remains in Germany in 1856, one of the Engis crania was recognized as a juvenile *Homo neanderthalensis*. Unfortunately Schmerling died before the ideas he espoused gained wide acceptance.

Schöningen (Location 52°08′N, 10°59′E, Germany; etym. named after the nearby town) History and general description Pleistocene sediments exposed in a brown coal mine have been monitored and excavated by archeologists since 1983. In 1995 several well-preserved wooden spears were recovered, and at the present time these are the oldest-known organic artifacts. Temporal span and how dated? Correlation to glacial cycles has suggested a range of Elster-Saale glacial periods [i.e., approximately **Oxygen Isotope Stages** (OISs) 12–10]. The wooden spears come from a layer at the middle of the sequence, during interglacial OIS stage 11, which dates to *c*.400 ka. Hominins found at site None. Archeological evidence found at site Two kinds of wooden spears were recovered. The older spears from OIS 11 are merely pointed, but were found in close association with flint tools and numerous horse bones that show signs of butchery. The more recent spears, from approximately OIS 10 show grooves that might have been slots for flint spearheads. These spears are the oldest known evidence for both spears and composite tools. Key references: Geology, dating, and paleoenvironment Thieme 1997; Hominins N/A; Archeology Thieme 1997.

School of Anatomical Sciences, University of the Witwatersrand *See* **Witwatersrand University Department of Anatomy.**

Schwalbe, Gustav Albert (1844–1916)

Gustav Schwalbe was born in Quedlinburg, Germany, and received his MD from the University of Berlin in 1866, having studied also in Zürich and Bonn. Schwalbe passed his medical boards in 1867, and he turned immediately to research. He first worked under his major professor in Bonn, where he made his first significant contribution to science, the biological identification of taste buds. Schwalbe also held academic posts in Amsterdam (1868–70), Halle (1870), Freiburg (1871), Leipzig (1871–3), Jena (1873–81), Königsburg

(1881–3), and finally Strasbourg; his appointment at Jena was his first as *ordentlicher* (full) Professor of Anatomy. Schwalbe's early research focused on a variety of topics in human anatomy and physiology. In addition to his discovery of taste buds, Schwalbe made significant contributions to the understanding of the lymphatic system, the central nervous system, and additional sense organs. He published an influential textbook on neurology in 1881 and another on sense organs published in sections between 1883 and 1887. A characteristic feature of all Schwalbe's work was his insistence on precision in methodology and comparative analysis. By the late 1880s he had begun to focus more of his attention to the discipline of anthropology, which in the European tradition was limited to biological or physical anthropology. He stressed that accurate comparative anatomy of mammals, and particularly of the order Primates, was a necessary background to understanding human biology. Schwalbe began his anthropological work by focusing on various aspects of living human "races" as he felt that such studies were crucial to unraveling human evolutionary history. Schwalbe recognized that physical anthropology was in need of improved theoretical focus and methodological precision, so in 1899 he founded the journal *Zeitschrift für Morphologie und Anthropologie* with the goal of applying detailed comparative and developmental studies to the understanding of human evolution. He developed a methodology for comparative analysis of the emerging human fossil record which he referred to as *Formanalyse* (the analysis of form) and began to apply this precise, metrical technique to major questions then prominent in human evolution. Using this approach, Schwalbe correctly recognized that the so-called "Canstatt race" was made up of a mixture of *Homo neanderthalensis* individuals and a group of late Diluvial (Pleistocene) specimens. Similarly, Schwalbe held that his analysis of form clearly established the early Diluvial (Pleistocene) Neanderthals as a distinct species of humans, which he named *Homo primigenius*, as opposed to simply a primitive "race" as many other scholars claimed. From this basis, he argued strongly against Rudolf Virchow's interpretation that the original Neanderthal 1 specimen represented a pathological individual from a recent human race. By applying his analysis of form to extant species, Schwalbe concluded that anatomical comparisons supported a strong phylogenetic connection between great apes and modern humans. A devoté of both Charles **Darwin** and Ernst Haeckel, Schwalbe argued that both Neanderthals (for Schwalbe *H. primigenius)* and Eugéne **Dubois'** original *Pithecanthropus erectus* from Java represented stages that bridged the gap between modern humans and apes. Because, based on his analysis, the latter had a smaller brain and a more ape-like form, Schwalbe considered it to be a more primitive stage than his *H. primigenius.* Further, Schwalbe argued that early modern humans (e.g., the "late Diluvial" members of the "Cannstatt race") provided an additional link between his *H. primigenius* (i.e., Neanderthals) and living people. Because of this linear "stage" view, Schwalbe is widely recognized as one of the founders of the unilineal interpretation of human evolutionary history. True to this perspective, Schwalbe denied the existence of a pre-sapiens lineage and presented detailed objections to specimens (e.g. Galley Hill, Ipswich, and **Piltdown**) used to argue for such a lineage in early 20thC Europe and elsewhere (e.g., Ameghino's claims for Pliocene humans in Argentina). It was Aleš **Hrdlička's** misunderstanding of Schwalbe's reaction to Marcellin **Boule's** analysis of **La Chapelle-aux-Saints** that led him to label Schwalbe a **pre-sapiens hypothesis** supporter. This likely stemmed from confusion surrounding Schwalbe's view that the known *Pithecanthropus* and Neanderthal specimens might not represent the exact ancestral populations but could be side branches of the ancestral modern human lineage. Schwalbe felt that regardless of whether the known fossils were actually ancestral to recent humans, those ancestors would have had to be essentially identical to the known *Pithecanthropus* and Neanderthal specimens; he did not retreat from a unilineal view of human evolution as has been claimed. Schwalbe's focus from 1899 on was almost entirely on topics related to physical anthropology. Among his most influential works are: 'Studien über *Pithecanthropus erectus* Dubois' (*Zeitschrift für Morphologie und Anthropologie* 1: 1–240; 1899), 'Der Neanderthalschädel' (*Bonner Jahrbuch* 106: 1–72; 1901), 'Die Vorgeschichte des Menschen' (Braunschweig, Vieweg; 1904), 'Studien zur Vorgeschichte des Memschen' (Stuttgart, Schweitzerbartsche Buchhandlung; 1906), and 'Kritische Besprechung von Boule's Werk: "L'homme fossile de la Chapelle-aux-Saints" mit eigenen Untersuchungen' (*Zeitschrift für Morphologie und Anthropologie* 16: 527–610; 1914), and his final (posthumous)

work 'Die Abstammung des Menschen und die ältesten Menschenformen' (In: P. Hinneburg, ed. *Die Kultur der Gegenwart.* Leipzig/Berlin: Teubner; 1923). Schwalbe's one paper in English was 'The descent of man' (In: A.C. Stewart, ed. *Darwin and Modern Science.* London: Cambridge University Press; 1910). Schwalbe served as University Rector (1893–4) at Strasbourg and as chairman of the medical examination commission; he retired from all his Strasbourg posts in 1914.

sciurids The informal name for the suborder of the **Rodentia** that includes the squirrels. *See* **Rodentia**.

Scladina (Location 50°29′9.64″N, 5°1′36.27″E, Belgium; etym. named the Grotte Scladina or the Cave of Scladina after the nearby village of Sclayn) History and general description A site in a limestone cave in the north wall of the Ri de Pontainne Valley, near the town of Andenne, in Namur Province, Belgium. Intensive excavations have been conducted at the site since 1978, but it was not until 1993 that a **Late Pleistocene** hominin, subsequently assigned to *Homo neanderthalensis*, was recovered *in situ*. **Mitochondrial DNA** extracted from a molar formed part of the evidence that suggested a high level of genetic diversity among Neanderthals. Temporal span and how dated? The dates suggest an age for the hominins of *c*.100 ka. Hominins found at site Both sides of the corpus of a *c*.8-year-old juvenile mandible with both right and left M_{1-2} and the right M_3 in its crypt, a right maxillary fragment with dm^{1-2}, plus associated isolated teeth. Key references: Geology, dating, and paleoenvironment Toussaint et al. 1998, Toussaint and Pirson 2006; Hominins Otte et al. 1993, Smith et al. 2007.

Sclayn *See* **Scladina**.

sclerotome (Gk *skleros*=hard and *temnein*=to cut or slice) Sclerotomes are one of the expressions of the segmental nature of the body. They develop from somites and all but the first four or so sclerotomes (which are incorporated into the basicranium) contribute to the development of the vertebrae, the intervertebral discs, and ribs. Each vertebra is formed from the caudal half of its equivalent sclerotome, plus the cranial half of the sclerotome below it (e.g., the C4 vertebra is formed from the caudal half of the C4 sclerotome and the cranial half the C5 sclerotome). *See also* **vertebral number**.

SCLF Acronym for single-crystal laser fusion. *See* **argon-argon dating**.

scraper A term used to describe stone tools with **retouch**ed edges that meet at an angle of approximately 60–90°. The function of scrapers has been inferred by ethnographic analogy, actualistic studies, and subsequent use-wear analysis. Although scrapers are reported from **Oldowan** sites in Africa, they are particularly common at **Middle Paleolithic** sites in Western Europe (Bordes 1961a, Debénath and Dibble 1994).

SE Prefix used for fossils recovered from the Sterkfontein "Extension Site." *See* **Sterkfontein**.

Sea Harvest (Location 33°01′23″S, 17°57′0″E, Cape Province, South Africa; etym. unknown) History and general description This site consists of a series of fossil-rich pockets of sandy sediment overlain with sand and shell deposits and **Middle Stone Age** (or MSA) shell **midden**s on the north side of Saldanha Bay. Fossil collection was begun by Brett Hendy in 1969, and has been continued by Graham Avery and Richard **Klein**, and it was they who recovered the Sea Harvest hominins. Temporal span and how dated? The archeological, biochronological, and geomorphological evidence are consistent with the infinite (>40 ka) conventional **radiocarbon dating** on associated ostrich egg shell, and suggest the site spans 128–40 ka. Hominins found at site A manual distal phalanx and a left P⁴. Archeological evidence found at site None in direct association with the hominins, but MSA artifacts occur in midden deposits in slightly higher strata. The hominin fossils and many of the other vertebrates likely derive from hyena dens that were intrusive into underlying sediments. Key references: Geology, dating, and paleoenvironment Hendey 1984, Fairall et al. 1976, Butzer 1984, 2004; Hominins Grine and Klein 1993; Archeology Volman 1978, Butzer 2004.

sea surface temperature (or SST) This climate indicator can be reconstructed from a variety of paleoclimate proxies including species assemblages, oxygen isotopes, Mg/Ca, alkenones, and TEX86. SSTs vary widely between -1.8°C (the freezing point of seawater) in polar waters to more than 30°C in some **tropical** waters in restricted basins (e.g., the Gulf of Aden). SSTs generally decrease from the tropics to the poles, although in the tropical Pacific a strong longitudinal temperature gradient occurs and is associated with the **El Niño Southern Oscillation** phenomenon. SST reconstructions are available for many marine sediment sites; however, the most complete spatial coverage has been achieved for **Last Glacial Maximum** comparison projects (e.g., CLIMAP, which used oxygen isotopes, and most recently MARGO, which used multiple SST proxies).

seasonality Regular changes in weather pattern (in both tropical and temperate regions) and day length (in temperate regions) that occur within an annual cycle. These changes are driven by changes in the **orbital geometry** of the Earth that varies the amount of sunlight reaching the Earth's surface (i.e., **insolation**) received during the year. At higher latitudes, weather is determined primarily by temperature, because the tilt of the Earth's axis means that the intensity of the insolation varies throughout the year. Thus, there are warm (summer) and cool (winter) seasons. In the tropics, rainfall is an important determinant of weather with wet and dry seasons within the annual cycle. Seasonality influences habitat and the availability of food. Primates cope with seasonal shifts in resources and habitat quality by altering their behavior (e.g., by switching to **fallback foods**). Seasonality was almost certainly an important influence on evolution within the hominin clade, and it likewise probably affected hominin behavior. It is also possible that some characteristics of modern humans, such as the propensity to gain body fat very quickly, was an adaptation to highly seasonal environments in which it was beneficial to gain weight during seasons of plenty to provide a buffer in times of scarcity. Modern humans use an array of cultural tools to buffer against seasonality (e.g., food storage), thus allowing them to inhabit regions normally inaccessible to primates. Initially, the transition within *Homo* to greater dependence on meat is likely to have helped hominins exploit high latitudes, as meat is available all year round whereas many plant foods are only available seasonally.

seasons *See* **seasonality**.

secondarily altricial *See* **altricial**.

secondary cartilaginous joints *See* **joint**.

secondary centers of ossification *See* **ossification**.

701

secondary dentine *See* dentine.

secondary homonymy This refers to the circumstance when identical species/subspecies names that were initially used in different genera were to later be placed in the same genus (e.g., if the two original genera are combined). *See also* **homonymy**.

secondary sexual characteristics Physical characteristics that are associated with reproductive maturation in males and females of sexually dimorphic species, and are differentiated from primary sexual characteristics (e.g., gonads). They include differences in skin, feather, and fur coloration, distribution and thickness of hair, body size differences, differences in the size of teeth, tusks, and antlers, musculature, fat deposition, and limb proportions. The development of secondary sexual characteristics may occur at different rates, and depending on several factors (e.g., health, dominance rank) they may be more or less developed in different individuals within a population. Since the appearance of these traits is often dependent on circulating levels of gonadal steroids, inter-individual differences in hormone levels probably also explain variation in manifestations of these features. Secondary sexual characteristics are hypothesized to have evolved through mate choice and intrasexual competition. Several traits have been described as secondary sexual characteristics in modern humans. For males, these include growth of facial hair, increased muscle mass and bone robusticity, broad shoulders relative to hips, and deepening voice. In females, some secondary sexual characteristics include enlarged breasts, lower waist/hip ratio, longer upper limbs, and more subcutaneous fat. In both sexes, growth of body hair (including axillary and pubic hair) and increased secretions of sebaceous and sweat glands are considered to be secondary sexual characteristics. The Tanner scale is regularly employed by medical professionals to assess the degree of maturity of boys and girls, based on the development of the genitalia, axillary and pubic hair, and breasts. However, the relationship of the stages used in assessment to chronological age varies widely. The fossil record is mute about the soft-tissue aspects of important topics such as this.

section Geology A term for the place where **strata** are exposed. They may have been exposed naturally or researchers may have deliberately dug a trench to expose the strata. Morphology It refers to the surface of a bone or tooth exposed by either a natural break or a deliberate saw cut. Sections provide information about the internal structure of bones and teeth. *See also* **cross-section**.

sectorial (L. *secare* = to cut) In those primates in which the upper canine and the anterior lower premolar form a **honing** complex, the latter tooth develops a "blade-like" edge such that it is called a sectorial premolar. The loss of a honing complex and therefore also the sectorial premolar is assumed to be related to a shift in behavior where males do not need to fight with their teeth to get access to mates, and thus the presence or absence of this feature in early hominins is much discussed. *See also* **sexual selection**.

secular trend (L. *seculum* = generation) The use of this term in physical anthropology is usually in reference to the marked changes that have occurred in the growth and development of certain modern human populations over the past 100–200 years. "Secular" refers to the fact that these changes have occurred over successive periods of time. Specifically, there have been significant changes in height, with children especially from developed countries being of taller stature at every age from infancy onward. There have also been documented changes towards increased hand, foot, and facial dimensions. Similarly, girls from several countries have demonstrated a secular trend in earlier age at menarche (first menstrual period), with a decrease in this age by 1–3 years. It is important to note that while this trend has been towards increased stature, or earlier age at menarche, in many populations, so-called "reverse secular trends" have also been documented in some populations (e.g., size has decreased). While the mechanistic cause of these secular trends remains unclear, it has been hypothesized that the trends toward increased height and earlier ages at maturation are due to improved nutrition, hygiene, and decreased disease burden. A recent hypothesis explains these trends as a return to the ancestral age at maturation and exploitation of full potential for size, which has been altered to this point by nutritional inadequacy and chronic infections. *See also* **stature**.

sedge The common name for a family of monocotyledonous flowering plants, the Cyperaceae, which superficially resemble grasses or rushes (Govaerts et al. 2007). Many sedges are found in wetlands and they belong to a group of plants, including warm-season grasses, some herbs, and a variety of shrubs which use a photosynthetic system that initially fixes carbon dioxide in molecules containing four carbon atoms; thus they are C_4 **plants**. Some researchers have suggested that sedges may have been a source of forage for early hominins. *See also* C_3 and C_4; C_4 **plants**.

Sedia del Diavolo (Location 41°57'N, 12°32'E, Italy; etym. It. the devil's chair) History and general description Small quarry in the hills overlooking the Aniene river in the suburbs of Rome. Original surveying work was done in 1882 by R. Meli, but excavation did not begin until 1956 under A. Blanc. Level 4 contained the hominin and archeological remains. Increasing urbanization in the second half of the 20thC has made further study of the relevant deposits impossible. Temporal span and how dated? **Biostratigraphy** suggests the site formed during the Riss glacial (*c*.240–130 ka). Hominins found at site Two specimens, a femoral diaphysis and a right II metatarsal, were recovered, but they have not been assigned to any taxon. The metatarsal shows traces of microtrauma that occurred during the individual's life. Archeological evidence found at site Fifteen lithic implements were recovered from level 4, most of which are side scrapers. This assemblage has been defined as pre-**Mousterian**. Key references: Geology, dating, and paleoenvironment Blanc 1956; Hominins Mallegni 1986; Archeology Mussi 2001.

sediment (L. *sedimentum* = the act of settling) Material that settles in the bottom of a liquid. The rocks in which most hominin fossils are preserved are rocks that have formed from sediment. The sediment can be coarse, like gravel, or finer, like sand, or much finer, like silt.

sedimentary rock Rocks that are formed from **sediment**. There are two categories of sedimentary rock. Clastic sedimentary rocks are formed from the fragments of other rocks transported from their source and deposited in water. The second category includes all the rocks formed by precipitation. For example, **limestone** is formed by the precipitation of calcium carbonate that comes from marine organisms.

seed-eating hypothesis One of several hypotheses put forward to explain the origin of upright posture and eventually bipedal locomotion within the hominin clade. Proposed by Clifford Jolly (1970), the seed-eating hypothesis has had a profound influence on paleoanthropology on many levels. At the time it was proposed it presented a formidable challenge to the prevailing wisdom of the time that bipedalism had evolved in hominins in association with tool use. It also provided a compelling synthesis of careful anatomical and ecological observations that served as a model for other workers, and it led to the consensus that small hard objects were a critical component of the diet of the **australopiths** (i.e., archaic hominins). The core of the seed-eating hypothesis rests on the observation that australopiths apparently share many derived characteristics with gelada baboons (*Theropithecus*). Such inferred shared traits include those related to manual dexterity, efficient and powerful **mastication**, a reduced reliance on incision, and an upright trunk (employed by *Theropithecus* while sitting and eating). These shared features were evidently not inherited from the most recent common ancestor of these taxa, but rather represent examples of **convergent evolution** in hominins and **papionins**. Jolly's (1970) insight was to suggest functional and/or ecological explanations for these **homoplasies**. He noted that the suite of features seen in gelada baboons is related to seed-eating, and he suggested that a similar diet may have spurred the evolution of those features in the hominin clade. However, because of their different evolutionary histories (for unlike gelada baboons, which are descended from quadrupedal ancestors, hominins were more likely to have descended from suspensory apes) the routine adoption of truncal erectness in proto-hominins apparently led eventually to the evolution of an upright posture and bipedalism. Jolly (1970) predicted that the earliest hominins evolved in **edaphic** (wet) grasslands in which grain resources would have been plentiful, but most recent fossil discoveries suggest that the earliest hominins lived in wooded or mosaic environments (Pickford and Senut 2001, Vignaud et al. 2002, WoldeGabriel et al. 2004), so a diet based heavily on grain seeds seems unlikely. However, the hypothesized link between diet and bipedalism remains relevant insofar as some researchers have suggested that the dental morphology of the earliest known bipeds may indicate a novel diet (e.g., Haile-Selassie et al. 2004a). *See also* **bipedal locomotion; tool-use hypothesis**.

segregation A process that occurs during anaphase I of **meiosis** (i.e., during gametogenesis). During segregation, pairs of **homologous chromosomes** are separated, with one member of each pair going to the opposite pole and eventually contributing to a different haploid daughter gamete.

segregational distortion This term describes the situation when one particular allele of a gene is present in more than half of the gametes (NB: during a normal meiosis, an allele should be present in half of all gametes).

selection *See* **natural selection**.

selection coefficient A measure of the fitness reduction of a given **phenotype** relative to another fitter phenotype. The selection coefficient is usually denoted by *s*, and it is calculated as the proportionate difference in offspring number of a given phenotype relative to another more favorable phenotype. An *s* value of 0 represents neutrality of the given phenotype relative to the other phenotype (i.e., no selection against the phenotype), whereas an *s* value of 1 equates to a total lack of surviving offspring or to complete lethality. The selection coefficient is equivalent to the **Darwinian fitness** of a phenotype.

selective sweep The relatively rapid **fixation** (i.e., an increase to an incidence of 100% in a given population) of an **allele** and all the alleles linked to it; the latter phenomenon is known as **genetic hitchhiking**. When a selective sweep occurs, nucleotide diversity is reduced in the region around the selected allele. This property of reduced diversity is useful for identifying regions of the genome that have been recently subjected to **directional selection**. For example, in modern Europeans there is evidence of a selective sweep around the lactase gene because of the spread of an allele conferring lactose tolerance into adulthood (Bersaglieri et al. 2004)

Selenka Collection *See* **Zoologische Staatssammlung München**.

selfish DNA DNA that copies itself and inserts itself within a genome, but does not contribute to the reproductive success of the host. Endogenous retroviruses and transposable elements such as **SINEs** or **LINEs** are examples of selfish DNA.

sellion The deepest point on the sagittal contour of the infraglabellar area of the upper face.

semantic memory (Gk *semantikos* = meaning) The form of memory that relates to knowledge of everyday facts or concepts. Semantic memory is declarative and as such it can be explicitly articulated. In contrast to episodic memory, semantic memory is impersonal and it is detached from autobiographical knowledge or experiences. For example, our knowledge (or memory) of everyday objects such as paper, computer, and banana, as well as concepts such as planet, animal, and President are all forms of semantic memory. Semantic memory is not uniquely human. Many authors have reported important homologies in the semantic memories of birds and primates, including modern humans (Terrace 2005, Wright 2007). However, there is evidence to suggest that semantic memory systems may be shaped by selection to serve unique ecological demands such as seed caching in birds (Sherry and Schacter 1987).

semicircular canals The semicircular canals of the ear form one of the three main subdivisions of the bony and membranous labyrinths. The three canals each comprise two-thirds of a circle; the separate lateral canal is in the horizontal plane, whereas the conjoined anterior and posterior canals are in the vertical plane. The anterior and posterior canals are at right angles to each other and each is oriented approximately 45° to the sagittal plane. They are arranged so that the anterior canal on one side and the posterior on the opposite side are "coplanar" (i.e., the long axis of the anterior canal on one side is parallel to the long axis of the posterior canal on the opposite side). The lateral membranous semicircular canal has a swelling, or ampulla, at both ends; the anterior and posterior each have a single swelling at their free ends. Motion is detected by the relative movement that occurs between the hairs of the hair cells embedded in the walls of the part of the membranous labyrinth that lines the ampulla, and the endolymph that fills the membranous labyrinth. Perilymph separates the membranous semicircular canals from the bony labyrinth that encloses them. Spoor et al. (1994, 1996) showed that modern humans differed from all four of the living great apes in that the "arc sizes" of the two vertically oriented canals (i.e., the anterior and posterior) are larger (i.e., they are taller) than those of the great apes, whereas the "arc size" of the single, separate, lateral canal is smaller (i.e., it is narrower) than that of the great apes. Thus far, investigations of the semicircular canals of fossil hominins suggest that the modern human-like semicircular canal morphology is first seen in crania such as **KNM-ER 3733** assigned to early African *Homo erectus* or *Homo ergaster* (Spoor et al. 1994), but Hublin et al. (1996) claim that the vertical canals of *Homo neanderthalensis* are smaller than those of modern humans. It should be noted that Graf and Vidal (1995) are skeptical about the functional inferences that have been made on the basis of relatively small differences in the size and shape of the semicircular canals of modern humans and the great apes. *See also* **bony labyrinth**; **membranous labyrinth**.

semilandmarks The landmarks used in **geometric morphometrics** must be homologous across specimens (i.e., they must represent the same biological locations in every individual). However, for some regions and structures (e.g., cranial vault, tooth crowns) there are only a limited number of possible landmarks (e.g., suture intersections and cusp tips, respectively) and substantial amounts of morphology between the landmarks would be lost unless there was some other way of capturing it. Bookstein (1991) introduced the concept of semilandmarks (also called sliding landmarks) as a way of extending landmark-based statistics to smooth curves and surfaces. Semilandmarks were initially applied to 2D outlines (Bookstein 1997), but subsequently Gunz (2001, 2005) and Gunz et al. (2005) developed the algebra needed to capture curves and surfaces in 3D. The same number of semilandmarks is placed in roughly homologous locations on every specimen of the sample. What that number is depends on the complexity of the curves or surfaces to be captured and on the level of detail being investigated. These semilandmarks are then allowed to slide along the curves and surfaces so as to remove the effect of the arbitrary initial placement. The two main semilandmark algorithms differ in their optimization criteria: in one the bending energy of the **thin-plate spline interpolation** between each specimen and the sample mean shape is minimized and in the other algorithm the Procrustes distances are minimized. It is through the sliding algorithm that semilandmarks acquire their geometric "**homology**," so that in the subsequent statistical analyses they can be treated as if they were formal landmarks. See Gunz et al. (2005, 2009b) for a more technical description and Gunz et al. (2009a), Skinner et al. (2009a, 2009b), and Neubauer et al. (2009) for examples of the use of semilandmarks. *See also* **geometric morphometrics**; **landmark** (syn. sliding landmarks).

Senckenberg Forschungsinstitut und Naturmuseum The institution now called the Senckenberg Forschungsinstitut und Naturmuseum was founded as the Senckenberg Natural History Society in 1817. It is named after Johann Christian Senckenberg, a Frankfurt physician who died in 1772 and whose endowment laid the foundation for the Natural History Society, the Anatomical Institute, and the University Library in Frankfurt, Germany. The present main building was opened 1907 and houses a research institute and a Natural History Museum. Today the Senckenberg Forschungsinstitut comprises a cluster of seven research institutions and museums in Germany. While a Professor of Anthropology at Johann Wolfgang Goethe University between 1928 and 1933, Franz **Weidenreich** prepared an exhibition outlining the stages of human evolution at the Senckenberg Natural History Museum. But Weidenreich's book on race and his own Jewish background made it increasingly difficult to remain in Germany and he was dismissed from his position at Frankfurt in 1933. The close cooperation between the Johann Wolfgang Goethe University, Frankfurt University, and the Senckenberg Natural History Museum was renewed in the 1960s when the University re-established a center for palaeoanthropological research. In 1986 the Werner Reimers Foundation provided facilities for enable Ralph **von Koenigswald**, who had retired from the Chair of Palaeontology at the Rijksuniversiteit, to establish the Palaeonthropology Department at the Senckenberg Natural History Museum. von Koenigswald brought with him the part of the **Sangiran** Collection that he collected after the Dutch Geological Survey in Java had stopped supporting his research. After von Koenigswald's death in 1982 the Department was headed by Jens Franzen, followed in 2001 by Friedemann Schrenk. <u>Hominin fossil collections</u> The von Koenigswald Collection at the Senckenberg includes three fragmentary crania (**Sangiran 2, 3, and 4**), six mandibular fragments (**Sangiran 1a, 4, 5, 6a, 6b,** and **6c**) and 52 isolated teeth (**Sangiran 7**). The von Koenigswald Collection includes the holotypes of *Pithecanthropus dubius, Meganthropus palaeojavanicus, Hemanthropus peii, Sinanthropus officinalis, Gigantopithecus blacki,* and *Chinjipithecus atavus.* <u>Comparative collections</u> The Department of

705

Palaeoanthropology also houses a collection of around 280 skulls or crania of *Pan troglodytes verus* from Liberia. They were purchased by Hans Himmelheber from the Dan and Kran people in the "northeast region of the central province of Liberia" (Protsch and Eckhardt 1988, p. 13). All stages of ontogeny are represented in the collection, but some specimens are very severely damaged, including damage to the basicranium inflicted when the brain was removed. Contact department For the hominin fossils and the comparative collections contact the Department of Paleoanthropology. Contact person Birgit Denkel (e-mail bdenkel@senckenberg.de). Postal address Senckenberganlage 25, 60325 Frankfurt am Main, Germany. Website www.senckenberg.de. Tel +49 69 7542 1260. Fax +49 69 7542 1558.

Senckenberg Museum *See* **Senckenberg Forschungsinstitut und Naturmuseum.**

Sendangbusik (Location 7°20′S, 110°58′E, Indonesia; etym. named after a local village) History and general description In 1981 a hominin mandibular fragment was found on the surface at this site, which lies within the **Sangiran Dome**. The specimen was most likely derived from a stratum between the Lower and Middle Tuffs of the **Bapang Formation**. Temporal span and how dated? Trace element analysis of a bone sample drilled from the corpus, the degree of mineralization, the adhering sandstone matrix, and the color of the specimen are consistent with the proposed stratigraphic assignment. The most recent **radiometric dating** estimates of this deposit range between 1.24 and 1.51 Ma. Hominin found at site **SB 8103** is a fragment of the right side of the mandibular corpus preserving the crowns of P_4–M_2 and the distal root of M_3. It has been referred to *Homo erectus*. Archeological evidence found at site None. Key references: Geology, dating, and paleoenvironment Matsu'ura et al. 1995a, 1995b, Larick et al. 2001; Hominins Aziz et al. 1994, Kaifu et al. 2005a.

senescence The biological process of dysfunctional change with age in organisms. This is in contrast to the concept of "aging" in which objects, both biological and nonbiological, show the effects or characteristics of increasing age. Through the process of senescence, which is universal among sexually reproducing organisms, individuals become less proficient at maintaining homeostasis, and mortality rates increase due to

intrinsic causes. Several evolutionary theories have been proposed to explain senescence, which can occur at widely varying rates among individuals and between species. The senescent processes that occur in modern humans are not unique. However, due to a relatively extended life span and medical interventions, they may proceed somewhat more slowly. Similarly, patterns of biocultural evolution may provide selection for longer life spans and slower senescence in modern humans, due to contributions of older individuals to measures of inclusive fitness. *See also* **antagonistic pleiotropy theory**; **mutation accumulation hypothesis**; **grandmother hypothesis**.

senior synonym The name for a taxon that has historical priority. *See also* **synonymy**.

sensu lato (L. *sensu*=sense and *lato*=lax) A more inclusive, less restrictive, interpretation of a taxon (e.g., *Homo habilis sensu lato* includes both *Homo habilis sensu stricto* and *Homo rudolfensis*, and *Paranthropus boisei sensu lato* includes both *Paranthropus boisei sensu stricto* and **Paranthropus aethiopicus**).

sensu stricto (L. *sensu*=sense and *stricto*=strict) A more restrictive, less inclusive, interpretation of a taxon (e.g., *Homo habilis sensu stricto* does *not* include *Homo rudolfensis*, and *Paranthropus boisei sensu stricto* does *not* include *Paranthropus aethiopicus*).

Sergi, Giuseppe (1841–1936) Giuseppe Sergi was born in Messina, Sicily, and as a young man joined Giuseppe Garibaldi's military forces in their unsuccessful attempt to unify the Kingdom of Italy. After the end of the campaign, Sergi completed his studies at the University of Messina, where he studied comparative philology and philosophy, and at the University of Bologna. After completing his degree, Giuseppe Sergi became interested in anatomy, Charles **Darwin**'s theory of evolution, and the racial anthropology of Cesare Lombroso. Giuseppe Sergi developed an early interest in human psychology in the 1870s and began to investigate it in relation to physiology, leading to the publication of *Elementi di psicologia* (*Elements of Psychology*; 1873–4). Sergi accepted a position as Professor of Anthropology at the University of Bologna in 1880 and played an important role in establishing anthropology as a science at the university. In 1884 Giuseppe Sergi left

Bologna to become Professor of Anthropology at the University of Rome. He continued to study psychology and anthropology there and in 1889 he established and directed the Laboratorio di Psicologia Sperimentale and founded the Societa Romana di Antropologia (the Roman Society of Anthropology), which changed its name to the Istituto Italiano di Antropologia (the Italian Institute of Anthropology) in 1937. His scientific interests remained broad, encompassing psychology, anthropology, biology, and philosophy. In anthropology he shared his contemporaries' focus on racial classification and the use of craniometry. However, he opposed the use of the cephalic index and argued that the general cranial morphology of human skulls was a better means of classifying crania. Giuseppe Sergi wrote extensively on the question of human and primate evolution. Like some other prominent anthropologists and biologists, Sergi adopted a view of evolution where many phylogenetic lineages had evolved parallel to one another, sharing similar traits not because they shared a recent common ancestor but because similar traits could evolve in distinct lineages independently under similar conditions. Thus, in primate evolution, Sergi argued that from a distant unknown primate ancestor three primate groups emerged – Cercopithecidae, Simiidae, and Hominidae – each a lineage that developed independently from each other. He viewed human evolution in similar terms. In his book *Varietà umane: Principio e metodo di classificazione* (*Human Varieties: Principle and Method of Classification*; 1893) Giuseppe Sergi suggested that the earliest Mediterranean Europeans derived from populations from the Horn of Africa and were therefore related to so-called Hamitic peoples. This was a controversial hypothesis, but it attracted some attention among anthropologists at the time. Equally controversial were the claims made in *Origine e diffusione della stirpe mediterranea* (1895; translated into English in 1901 as *The Mediterranean Race: a Study of the Origin of European Peoples*) where Sergi argued for a polygenist origin of the different human races. He maintained that the first modern humans to enter Europe migrated from Africa and that from this original African group three racial groups emerged: Africans, Mediterraneans, and Nordic. These three races constituted what Sergi called a Eurafrican group that is distinct from Aryan peoples, whom he considered to be of an Asiatic origin and were part of what he called aEurasiatic group of humans. These ideas were directly opposed to some of the prevailing Aryan theories promulgated

by German and British anthropologists. In works published in 1912–14 Sergi continued to promote a polygenist origin of human races, arguing that the different races diverged from one another early in the evolution of the primates and in fact represent distinct genera represented by existing groups such as African races (*Notanthropus*), Asian races (*Heoanthropus*), American races (*Hesperanthropus*), as well as extinct groups such as the **Neanderthals** (*Palaeanthropus*), and the now defunct **Piltdown Man** (*Eoanthropus*). So, not only did the different hominins not share a recent common ancestor, they also did not share a common ancestry with monkeys or apes, which also diverged from the human line at some very early period and evolved along independent lines parallel with the hominins. Given the popularity of **neo-Lamarckism** and orthogenetic theories of evolution, such **parallelism** in evolution was acceptable during the early 20thC. Giuseppe Sergi retired from teaching in 1916 but continued to publish until his death. His son, Sergio **Sergi**, followed his father into anthropology.

Sergi, Sergio (1878–1972)

Sergio Sergi was born in Messina, Sicily, son of the prominent Italian anthropologist Giuseppe **Sergi**. Sergio Sergi studied medicine at the University of Rome and after completing his degree in 1902 he entered the University of Berlin where he studied physiology and the anatomy of the brain with Wilhelm von Waldeyer and anthropology with Felix von Luschan. Sergio Sergi returned to Rome in 1908 and in 1916 he accepted a position in the Faculty of Sciences at the University of Rome where he taught anthropology. His early research dealt with the neuroanatomy of apes and humans, craniometry, and the biomechanics of the human skeleton. In 1912 Sergio Sergi published *Crania habessinica: contributo all'antropologia dell'Africa orientale*, a craniometric study of peoples of Ethiopia, and through much of his career the anatomy of the skull in modern humans and other primates remained a major subject of inquiry. This is reflected in his study of gibbon neuroanatomy, *Le variazioni dei solchi cerebrali e la loro origine segmentale nell'hylobates* (*Variations of the Cerebral Sulci and their Segmental Origin in Hylobates*), published in 1940. Sergio Sergi became the director of the Institute of Anthropology at the University of Rome in 1926, although he had effectively assumed those duties the previous year, and he remained the Institute's director until 1951. Sergio Sergi was a member of the anthropological and archeological mission to the Fezzan region of central Libya sponsored by Reale Societa Geografica Italiana from 1933 to 1934, which excavated an ancient necropolis containing modern human remains from the 5thC BCE to the 4thC CE. Soon thereafter he began a series of investigations of two crania found in **Saccopastore**, Italy, in 1929 and 1935, and housed at the Institute of Anthropology. These two *Homo neanderthalensis* crania had a significant influence on Sergio Sergi's thinking about hominin phylogeny. Sergio Sergi also examined Neanderthal fossils discovered in 1939 (additional finds were made in 1950) at the site of San Felice, or **Monte Circeo**. From his investigation of these Neanderthal remains and comparisons with the anatomy of modern humans Sergio Sergi became convinced that there might be a direct hereditary link between *some* groups of Neanderthals and early *Homo sapiens*, but he considered that the classic Neanderthals were a distinct group with no close relationship to modern humans. As a result of his work on the Saccopastore and Monte Circeo fossils, in a series of papers published during the 1940s and 1950s Sergio Sergi developed a novel conception of modern human phylogeny. He used the term **Palaeanthropi** to refer to the wide range of different Neanderthal specimens from Europe, Asia, and Africa, and according to Sergio Sergi these Palaeanthropi interbred with groups of early humans from the Middle Pleistocene (e.g., **Swanscombe** and **Fontéchevade**) that he called the **Prophaneranthropi**. These Prophaneranthropi were the common ancestors for the more specialized classic Neanderthals and modern *Homo sapiens*, which he called **Phaneranthropi**. Although he never satisfactorily explained the origin of the Prophaneranthropi and despite other difficulties with his phylogenetic hypotheses Sergio Sergi's ideas influenced later paleoanthropologists (e.g., Clark **Howell**). During WWII Sergio Sergi remained in Italy and is credited with protecting the Saccopastore fossils by hiding them from German officers who were plundering fossil treasures. From 1913 to 1943 he served as secretary of the Istituto Italiano di Antropologia (the Italian Institute of Anthropology), and after the fall of the fascist government in Italy Sergio Sergi once again acted as secretary of the institute from 1947 to 1967 and he was responsible for its reorganization after the war. His monograph on the Circeo I Neanderthal skull, *Il cranio neandertaliano del Monte Circeo (Circeo I)*, was published posthumously in 1974.

seriation (L. *serere* = to join) The process by which archeological or fossil evidence is assembled into ordered series. For example, in archeology pots are sorted into a temporal series on the basis of their shape, or on the basis of the design of the decorations on the pottery. Hominin fossils, or collections of crania of higher primates, could also be sorted, or seriated, on the basis of one, or more, attributes such as cranial capacity. In this case the collection would be listed in order of the value of that attribute (e.g., a list of fossil hominin crania based not on geological age or taxonomy, but on endocranial volume) (syn. ordination).

series The **chronostratigraphic unit** equivalent of an **epoch**; series are subdivided into ages.

sex allocation In sexually reproducing species, the allocation of resources to male versus female reproduction is referred to as sex allocation. This division of resources is dependent upon reproductive strategies and behaviors within a given species (Hamilton 1967). Trivers and Willard (1973) presented data supporting their hypothesis (the Trivers–Willard hypothesis) that environmental and maternal condition could play a role in biasing offspring sex. Depending on whether a mother is in "good" or "poor" condition, they hypothesized that there will be a greater reproductive payoff for having male ("good" condition) or female ("poor" condition) offspring. Evidence supporting the Trivers–Willard hypothesis in modern human populations is lacking, although it is often discussed in the context of families with higher social status and more resources having more sons, or investing more in their sons. The application of this hypothesis to nonhuman primates has produced mixed results. According to some studies, in matrifocal groups, where rank is inherited through the mother and most males leave at sexual maturity, mothers of higher rank and in better condition produce more female than male offspring (e.g., Silk et al. 1981). Others suggest that sex ratio is affected more by local environmental conditions, in both high- and low-ranking females; when there is increased competition over resources, higher-ranking females will produce more daughters and fewer sons do than lower-ranking females (van Schaik and Hrdy 1991). When resource competition is relaxed, higher-ranking females would be expected to produce more sons than lower-ranking females. However, recent meta-analyses provide no consistent support for either the Trivers–Willard or local resource competition hypothesis in nonhuman primates (Silk et al. 2005).

sex chromosomes The pair of chromosomes that define the sex of an individual. In modern humans and most mammals the sex chromosomes are the **X chromosome** and the **Y chromosome**; males are XY and females are XX. The Y chromosome is substantially smaller than the X chromosome and only the tips of the Y chromosome (known as the pseudoautosomal regions) recombine with the X chromosome during meiosis. The larger, non-recombining, portion of the Y chromosome contains the *SRY* gene, which is the male sex-determining gene. In modern humans, if the *SRY* gene is not functional the fetus develops phenotypically as a female, but she is infertile. Although sex chromosomes are contrasted with the autosomal chromosomes, evolutionarily the sex chromosomes in most mammals were once autosomes.

sexual dimorphism (Gk *dis*=twice plus *morphe*=form, shape) Sexual dimorphism describes difference(s) between males and females of a species. Behavior (e.g., bird song), morphology (e.g., horns, teeth, and body size), and color can be sexually dimorphic, although sex differences in color are usually referred to as sexual dichromatism. Darwin (1871) identified the root causes of sexual dimorphism as different types of sexual selection acting through the channels of intrasexual competition and female choice. Sexual dimorphism in body size is common among anthropoids, and sexual dimorphism in general among primate species has been attributed to a wide range of influences in addition to sexual selection such as female body mass, diet, substrate use, and phylogeny. For example, Rensch's rule notes that in birds and mammals sexual size dimorphism (SSD) is more pronounced in larger species (Rensch 1959); further refinements of this rule state that SSD increases with size when males are the larger sex (male-biased SSD) and decreases with size when females are the larger sex (female-biased SSD). The link between male and female body size within a species is usually explained as being the result of genetic correlations between the sexes; therefore, any change in size of one sex usually results in some change in size in the other sex, although not as extreme. Thus Rensch's rule probably results from sexual selection, with intense selection on body size in the larger sex result-

ing in increases in both body mass in the smaller sex and sexual dimorphism (Gordon 2006, Dale et al. 2007). Because sexual dimorphism in body size or of "weaponry" (e.g., canine teeth) has been linked to levels of intrasexual competition, these traits are often utilized in analyses of fossil taxa in an attempt to reconstruct social structure. Evidence from fossil hominins suggests a general decrease in sexual dimorphism of body and canine size through time, with a notable decrease occurring during the time of *Homo erectus* likely brought about by an increase in female body size. However, direct inferences of social structure and mating behavior based on dimorphism in the fossil record are complicated by the fact that multiple types of group-living social structures overlap in the degree of sexual dimorphism produced in living primates (Plavcan 2000, 2002). In addition, archaic hominins, and especially megadont archaic hominins (e.g., *Paranthropus boisei*) are unusual in their pattern of dimorphism in that they exhibit high levels of body size dimorphism, but low levels of canine size dimorphism; in other primates, dimorphism levels are usually similar between body mass and canine size. The unusual hominin pattern may reflect the loss of the canine-premolar honing complex and a change in canine shape in the hominin lineage, shifting the canine away from a weapon shape (Kimbel and Delezene, 2009) and thus removing the canine as a target of sexual selection.

sexual selection A type of natural selection in which individual selection is influenced by mate choice (i.e., some individuals reproduce more because the traits they possess are more attractive to members of the opposite sex and/or they are successful in competing with individuals of the same sex for mates). Charles **Darwin** considered this differential access to mates to constitute a different evolutionary mechanism than his natural selection.

SH Acronym for Sima de los Huesos and the prefix for fossil hominins recovered from **Sima de los Huesos**, Spain. *See* **Sima de los Huesos**.

Shanidar (Location 36°50′N, 44°13′E, Iraq; etym. after the nearby village of Shanidar) History and general description Shanidar Cave is located about 645 km/400 miles north of Baghdad in the Zagros Mountains of Iraqi Kurdistan. The cave site lies just outside of the Greater Zab River valley, about two-thirds of the way up the neighboring Shanidar Valley. Ralph Solecki initially explored the cave in 1951 as part of a general archeological survey of northeastern Iraq. Having found encouraging material in his test pit, Solecki returned in 1953 for 10 weeks of excavation. In June of 1953 a partial skeleton of an infant, now known as **Shanidar 7**, was found. Another two seasons of excavation were undertaken in 1956–7 and 1960, during which the remains of six other individuals were recovered. The detailed study of the fossils was carried out by Trinkaus (1983). One individual (**Shanidar 4**) was found buried in soil that included highly elevated pollen counts, which Solecki interpreted as the remains of flowers that were interred as grave goods. Another individual (**Shanidar 1**) had suffered trauma and injury that would likely have prevented him from procuring his own food. Solecki (1963) interpreted these remains to mean that *Homo neanderthalensis* had complex social behavior that included burial and symbolism, and that they cared for their elderly and invalids. A recent reassessment of **Shanidar 3** by Churchill et al. (2009) revealed that this individual had cutmarks on its ribs that are consistent with damage from a projectile, which suggests that Neanderthals may have engaged in interpersonal conflict. The remains of three other individuals were identified upon closer examination of the collection, the last as recently as 2007. It is worth pointing out that during the excavation the cave was occupied by a group of Kurds, limiting the area that Solecki was able to dig. Temporal span and how dated? **Radiocarbon dating** suggests the **Mousterian** levels are at least *c*.45 ka and the **Upper Paleolithic** levels are *c*.27–33 ka. Hominins found at site The remains of approximately 10 individuals have been recovered and identified, at least four of which appear to have been intentionally buried. The remains are generally grouped with those from **Amud**, **Kebara**, and **Tabun** into a Near-Eastern "late archaic" Neanderthal population. Archeological evidence found at site The four main layers of Shanidar are characterized by their respective industries, and include a deep Mousterian level (Level D), which contained material consistent with the Zagros Mousterian, an Upper Paleolithic level (Level C) with material from Baradostian industry, and two more recent levels. Key references: Geology, dating, and paleoenvironment Vogel and Waterbolk 1963; Hominins Trinkaus 1983, Cowgill et al. 2007, Trinkaus and Zimmerman 1982; Archeology Solecki 1955 and 1963, Leroi-Gourhan 1975.

Shanidar 1 Site Shanidar. Locality N/A. Surface/*in situ In situ*. Date of discovery April 27, 1957. Finders Ralph Solecki and team. Unit N/A. Horizon At the top of level D (**Mousterian**). Bed/member N/A. Formation N/A. Group N/A. Nearest overlying dated horizon N/A. Nearest underlying dated horizon N/A. Geological age 45–50 ka. Developmental age Adult. Presumed sex Male. Brief anatomical description Nearly complete skeleton, with all teeth except for the left and right I_1. Announcement Stewart 1958. Initial description Stewart 1959. Photographs/line drawings and metrical data Trinkaus 1983. Detailed anatomical description Stewart 1959, Trinkaus and Zimmerman 1982, Trinkaus 1983. Initial taxonomic allocation *Homo sapiens neanderthalensis*. Taxonomic revisions N/A. Current conventional taxonomic allocation *Homo neanderthalensis*. Informal taxonomic category Pre-modern *Homo*. Significance Shanidar 1 is the most complete specimen from the site and it has been used as a reference in reconstructing other western Asian Neanderthals (e.g., **Amud 1**). This individual suffered massive crushing injuries to the right side of the body which resulted in the amputation of the right forearm and degenerative joint disease throughout the right leg, and cranial trauma that resulted in damage to the left eye. Despite the severity of these injuries, the level of healing suggests that this individual survived for a period of time after the injuries occurred. This has been interpreted as evidence that Neanderthal social behavior was complex enough to include care for the infirm. Location of original Presumably at the National Museum of Iraq, Baghdad, Iraq, but this has not been verified since the looting that took place in 2003 as a result of the second Iraq War.

Shanidar 3 Site Shanidar. Locality N/A. Surface/*in situ In situ*. Date of discovery April 16, 1957. Finders Ralph Solecki and team. Unit N/A. Horizon layer D (**Mousterian**). Bed/member N/A. Formation N/A. Group N/A. Nearest overlying dated horizon N/A. Nearest underlying dated horizon N/A. Geological age 45–50 ka. Developmental age Adult. Presumed sex Male. Brief anatomical description Four teeth, the major portions of the thoracic and lumbar vertebrae, sacrum, ribs, clavicles, scapulae, humeri, hand bones, pelvic and foot bones, plus fragments of the ulnae, radii, femora, tibiae, and fibulae. Announcement Solecki 1957. Initial description Solecki 1960. Photographs/line drawings and metrical data Trinkaus 1981. Detailed anatomical description Trinkaus 1981, Churchill et al. 2009. Initial taxonomic allocation *Homo sapiens neanderthalensis*. Taxonomic revisions N/A. Current conventional taxonomic allocation *Homo neanderthalensis*. Informal taxonomic category Pre-modern *Homo*. Significance Shanidar 3 has cutmarks on its ribs that Churchill and colleagues have suggested is due to damage from a weapon. Location of original **National Museum of Natural History, Smithsonian Institution**, USA.

Shanidar 4 Site Shanidar Locality N/A. Surface/*in situ In situ*. Date of discovery August 3, 1960. Finders Ralph Solecki and team. Unit N/A. Horizon Middle of Layer D (**Mousterian**). Bed/member N/A. Formation N/A. Group N/A. Nearest overlying dated horizon N/A. Nearest underlying dated horizon N/A. Geological age Older than *c*.45 ka. Developmental age Adult. Presumed sex Male. Brief anatomical description Fragmentary associated skeleton. Announcement Stewart 1963. Initial description Stewart 1963. Photographs/line drawings and metrical data Trinkaus 1983. Detailed anatomical description Trinkaus 1983. Initial taxonomic allocation *Homo sapiens neanderthalensis*. Taxonomic revisions N/A. Current conventional taxonomic allocation *Homo neanderthalensis*. Informal taxonomic category Pre-modern *Homo*. Significance The remains of Shanidar 4 were intermingled with the remains of Shanidar 6, 8, and 9. This assembly is interpreted as a series of multiple sequential burials, with Shanidar 4 being the most recent. The grave has several pollen clusters within it, suggesting that the body was intentionally buried with a variety of flowers in a symbolic nature (Leroi-Gourhan 1975). Location of original Presumably at the National Museum of Iraq, Baghdad, Iraq, but this has not been verified since the looting that took place in 2003 as a result of the second Iraq War.

Shanidar 5 Site Shanidar Locality N/A. Surface/*in situ In situ*. Date of discovery August 7, 1960. Finders Ralph Solecki and team. Unit N/A. Horizon Middle of Layer D (**Mousterian**). Bed/member N/A. Formation N/A. Group N/A. Nearest overlying dated horizon N/A. Nearest underlying dated horizon N/A. Geological age Older than *c*.45 ka. Developmental age Adult. Presumed sex Male. Brief anatomical description Cranial fragments, including mostly complete facial skeleton and partial vault; fragmentary associated skeleton. Announcement Stewart 1963. Initial description Stewart 1963. Photographs/line drawings and metri-

cal data Trinkaus 1983. Detailed anatomical description Trinkaus 1983. Initial taxonomic allocation *Homo sapiens neanderthalensis*. Taxonomic revisions N/A. Current conventional taxonomic allocation *Homo neanderthalensis*. Informal taxonomic category Pre-modern *Homo*. Significance As originally reconstructed, the cranial vault appeared to be extremely long and low, such that Trinkaus (1983) proposed that the individual had been the subject of intentional cranial deformation in infancy, a "first" for Neanderthals. A later reassessment (Chech et al. 2000) concluded that a piece of parietal bone had been wrongly located and oriented, and that the original shape of the cranial vault was incorrect; consequently, the hypothesis of artificial cranial deformation can no longer be maintained. Location of original Presumably at the National Museum of Iraq, Baghdad, Iraq, but this has not been verified since the looting that took place in 2003 as a result of the second Iraq War.

Shanidar 7 Site **Shanidar**. Locality N/A. Surface/*in situ In situ*. Date of discovery June 27, 1953. Finders Ralph Solecki and team, Unit N/A. Horizon middle of Layer D (**Mousterian**). Bed/member N/A. Formation N/A. Group N/A. Nearest overlying dated horizon N/A. Nearest underlying dated horizon N/A. Geological age Older than 45 ka; Shanidar 7 was stratigraphically the lowest specimen, so presumably it was the earliest of the Shanidar hominins. Developmental age Between 6 and 9 months. Presumed sex Female. Brief anatomical description The crushed skeleton of an infant. Announcement Solecki 1955. Initial description Senyürek 1957. Photographs/line drawings and metrical data Trinkaus 1983. Detailed anatomical description Senyürek 1959, Trinkaus 1983. Initial taxonomic allocation Senyürek originally proposed that Shanidar 7 be given its own subspecies, *Homo sapiens shanidarensis*. Taxonomic revisions N/A. Current conventional taxonomic allocation *Homo neanderthalensis*. Informal taxonomic category Pre-modern *Homo*. Significance The Shanidar 7 infant was the first hominin found at the site, and prompted Solecki to return for two more seasons. Although the specimen is badly damaged, much has been learned from it concerning the ontogeny of *H. neanderthalensis*. Location of original Presumably at the National Museum of Iraq, Baghdad, Iraq, but this has not been verified since the looting that took place in 2003 as a result of the second Iraq War.

shape A spatial property of objects that is independent of location (translation and rotation) and **size**. The **form** of an object is composed of its shape and size. At its most basic, shape can be thought of as the relationship between two or more size variables (e.g., humerus length expressed as a percentage of radius length, or a ratio of first molar occlusal area to total postcanine occlusal area). Shape has been measured in a variety of ways in human evolution studies. For example, shape is often quantified as a measurement which has been adjusted using another measure of size, either through a criterion of subtraction using a **regression** analysis, or by dividing it by another measure of size such as **body mass** or a **geometric mean** of multiple measurements. In **geometric morphometrics**, shape is described using a collection of landmarks which have been adjusted for size by the division of the object's centroid size. Angles are another type of shape measurement. The study of size-related shape change is known as **scaling**, and involves the principles of **isometry** and **allometry**. Shape is used in analyses of **functional morphology** (e.g., relating limb proportions to locomotion, relative tooth size to diet), **alpha taxonomy, phylogenetic reconstruction**, and paleoecological reconstruction, among others.

shape space A configuration of shape variables (i.e., the measured shape of a specimen) corresponds to a point in shape space. For triangles the shape space is spherical; for more than three landmarks the geometry of shape space can become very complex (Kendall 1981, 1984, Slice 2001). For most practical applications this complex underlying geometry can be ignored and the analyses are carried out in a linear tangent space instead.

shape variables Variables describing an object that are independent of the overall scale, orientation and location of the object. For example, the distances between the apices of a triangle (i.e., interlandmark distances) describe the **form** of that triangle (i.e., those distances are invariant to rotation and translation) and if they are scaled to overall size they become shape variables (Bookstein 1991) (syn. two-point shape coordinates, Bookstein shape coordinates).

shared-derived character *See* **synapomorphy; apomorphy**.

shear modulus A material property defined as the ratio of shear **stress** to shear **strain** within a material that is being deformed elastically (i.e., the material returns to its original shape after the applied force is removed) and is being subjected to forces acting parallel to either a surface of the material or to a plane running through it. The shear modulus, denoted by G, can be thought as the shear stiffness of a material.

Shimojibaru Cave *See* **Ryukyu Islands**.

Shiraho-Saonetabaru *See* **Ryukyu Islands**.

short interspersed nuclear element *See* **SINE**.

short-period incremental lines Features in tooth enamel or dentine that have an intrinsic temporal secretory rhythm equal to, or less than, 1 day. **Cross-striations** in enamel and **von Ebner's lines** in dentine have been shown to have a **circadian** (*c*.24 hour) rhythm (e.g., Bromage 1991). **Intradian lines** in both enamel and dentine appear between two daily increments. Counts and measurements of the circadian features help estimate the **daily secretion rate, crown formation time**, dentine formation time, or the root extension rate (reviewed in Dean 2009). Short-period lines in enamel are much easier to visualize than those in dentine. *See also* **dentine development; enamel development; enamel microstructure; long-period incremental lines**.

short-period markings *See* **short-period incremental lines**.

short tandem repeat *See* **microsatellite**.

short-term memory *See* **working memory**.

shoulder girdle *See* **pectoral girdle**.

shovel-shaped incisors Upper incisors that show "a pronounced hollow on the lingual surface…surrounded by a well-defined elevated enamel border" (Hrdlicka 1920a, p. 429). Hrdlicka is usually credited with this description, but he points out that although a shovel-shaped incisor was illustrated in an early dental atlas (Carabelli 1844), it was probably Mühlreiter who was the first to liken the crown morphology of upper incisors to "a chisel or a shovel" (Mühlreiter 1870).

The highest incidences of shovel-shaped incisors among modern human populations are to be found in samples from northeast Asia and the Americas.

shrubland Characterized by dwarfed forms of trees that range between 1 and 3 m in height; a more extreme version of bushlands. Shrublands have lower mean annual rainfall than woodlands or bushlands, and have even poorer soil quality. They occur where taller trees cannot thrive because of low rainfall, low temperatures, drought periods, high winds, salinity, or extreme topography. Thus, they tend to be dominated by shrubs, grasses, herbs, and tubers. It has been suggested that the habitat sampled by **Sterkfontein** Member 4, where most of the *Australopithecus africanus* hypodigm comes from, was similar to shrubland or bushland biomes.

Shungura Formation (etym. The name of a local village) General description A formation of sedimentary rocks that outcrops as an elongated belt (about 60 km/37 miles north–south and less than 7 km/4.3 miles east–west) in the **Lower Omo Basin**, Ethiopia. Emil Brumpt, the naturalist on an expedition led by Comte Richard de Bourg de Bozas, reported the presence of fossils in what is now called the Shungura Formation in 1902. The paleontological potential of the area was further explored by Camille **Arambourg** in 1933, and Clark **Howell** carried out a survey of the Lower Omo Basin in 1959. It was Howell's assessment of the paleontological potential of the Lower Omo Basin that led in 1966 to the detailed plans for an **International Omo Research Expedition**. It began surveys of the region in 1967, and intensive field research continued there until 1976, and in 2006 the **Omo Group Research Expedition** was created to renew paleoanthropology research in the Lower Omo Basin. The Shungura Formation is divided into a shorter (approximately 20 km/12 miles) northern component called the "Type Area" and a longer southern component called the "Kalam (or Kaalam) Area." It is divided into 13 members named according to the lettered tuff (A–L, from oldest to youngest) that underlies each member, so Member A extends from the base of Tuff A to the base of Tuff B; the sediments below Tuff A are called the Basal Member. The sediments within the members are broken up into numbered units. In those members that contain many units (e.g., Member G) the units are subdivided into "Upper" and "Lower" groups. The sediments are tilted (down to the west and up to the east), and because the tuffs are generally more resistant to erosion they stand out as a series of lighter-colored strata, with the oldest tuff to the east close to the present river Omo and getting younger as one moves westwards. The depositional environment changes through time. In the Basal Member the sediments were deposited in the bed of a lake; thereafter in Member A to mid-Member G the dominant sedimentary environment is the floodplain of a large meandering river. This was followed by another short **lacustrine** phase in upper Member G, a delta/floodplain environment in upper Member G to Member L, and then back to a lacustrine sedimentary environment. Sites where the formation is exposed The **International Omo Research Expedition** initially comprised three contingents – US, French, and Kenyan – but only the US and the French contingents worked in the Shungura Formation. The localities where fossils were collected by the French contingent were referred to as "Omo" (or "O") localities and the localities were given numbers (e.g., Omo 18); the American contingent also numbered their localities, but instead of the prefix "Omo" (or "O") they referred to localities by using the prefix "L" for locality (e.g., L. 396). Within each of these locality systems each significant fossil is assigned a number (e.g., L. 40-19), and in the case of the specimens collected by the French the year of discovery is inserted between the locality number and the fossil number (e.g., **Omo 18-1967-18**). Geological age The oldest sediments are *c*.3.5 Ma, and the youngest *c*.0.8 Ma. Most significant hominin specimens recovered from the formation With respect to hominin fossils the Shungura Formation is best known because of the dental evidence from Member G (i.e., *c*.2.3 Ma ago) (Suwa 1988) that appears to document the shift from *Paranthropus aethiopicus* to *Paranthropus boisei*, and it is the results of Gen **Suwa**'s detailed analysis of the crown morphology of the mandibular premolars (Suwa 1990) that is still the best guide to what hominins are present in the Shungura Formation. Suwa suggests that Members C through F sample *P. aethiopicus*, and, in addition to the holotype mandible, Omo 18-1967-18, other specimens assigned to that taxon include the juvenile calvaria, L. 338y-6. The Member G and above megadont postcanine tooth crowns resemble those of *P. boisei* found at other sites, and the partial cranium **Omo 323-1976-896**, and the mandibles L. **7a-125** and L. **74a-21** are good examples of the Omo hypodigm of that taxon. Non-megadont postcanine teeth in the Shungura Formation are more difficult to assign to taxa. Those from the lowermost members (i.e., Members A and B) may sample an *Australopithecus afarensis*-like hominin, whereas the non-megadont postcanine teeth in Members G and H may, like **L. 894-76-898**, are said to sample early *Homo*. Key references: Geology, dating, and paleoenvironment Butzer and Thurber 1969, de Heinzelin and Haesaerts 1983, Feibel et al. 1989; Hominins Howell and Coppens 1973, 1974, Rak and Howell 1978, Suwa 1978, Alemseged et al. 2002.

Sibilot   One of the current designated subregions of the Koobi Fora region. *See also* **Koobi Fora**.

sidereal   (L. sidereus from sidus = star) Refers to anything that relates to stars or constellations, but particularly to measurements based on events related to stars [e.g., their daily motion, the time it takes for the moon to revolve around the Earth (c.27 days), or the Earth around the sun (c.365 days)]. Sidereal dating methods are methods that utilize such events for dating (e.g., tree ring dating, or dendrochronology). See also **geochronology**.

Sidi Abderrahman   (Location 33°35′N, 7°40′W, Morocco; etym. Sidi Abderrahmane is the name of the patron saint of Algiers. The cave site has also been referred to as Casablanca and Sidi Abderrahmanne) History and general description One of several sites discovered as a result of limestone quarrying activities that began in 1907 during harbor construction near Casablanca, Morocco. R. Neuville and A. Ruhlmann noted the site in their 1941 survey of the sequence of Miocene to Holocene sea-level transgressions and regressions on the Atlantic coast and Pierre Biberson established the basic stratigraphic and archeological framework in the 1950–60s. Since 1978 further research has been undertaken by a joint French-Moroccan research team. Temporal span and how dated? The Casablanca sequence has been dated using **biostratigraphy** and **lithostratigraphy** correlation, **magnetostratigraphy**, **amino acid racemization dating**, optically stimulated **luminescence dating** (or OSL), **electron spin resonance spectroscopy dating**, and **uranium-series dating** methods, but the destruction of the Grottes des Littorines by quarrying complicates any further attempts to date this material. Current age estimates for Sidi Abderrahman suggest a **Middle** to **Late Pleistocene** age, with the earliest deposits likely forming before **Oxygen Isotope Stage** 11 (i.e., 420–360 ka). Hominins found

at site Two mandibular fragments recovered in 1954 by P. Biberson are attributed to *Homo erectus* or *Homo erectus mauritanicus*. They derive from strata with an estimated age of *c*.400 ka based on overlying OSL age estimates and correlation with OIS 11 sediments in the nearby site of Cap Chatelier. Archeological evidence found at site The Sidi Abderrahman sequence preserves handaxes, cleavers, and other **Acheulean** implements including large **Levallois** peripheral cores, made primarily from local quartzite cobbles, with finer-grained chert becoming more abundant in the Late Pleistocene strata where Levallois flaking develops. Key references: Geology, dating, and paleoenvironment Raynal et al. 2001, Lefevre and Raynal 2002, Rhodes et al. 2006. Hominins Antón 2003, Arambourg and Biberson 1956; Archeology Balout 1967, Biberson 1961, 1967, Raynal et al. 2001, 2002.

Sidi Abderrahmanne   *See* Sidi Abderrahman.

Sidrón   *See* El Sidrón.

Sierra de Atapuerca   *See* Atapuerca.

siglecs   (etym. Acronym for sialic acid-binding Ig-like lectins) These cell-surface receptors recognize carbohydrates and have immune-modulatory functions. They are members of the immunoglobulin superfamily. There are 16 different genes coding for siglecs in primates. Many of these genes have been found to have undergone modern human-specific changes such as gene inactivation (pseudogenization, siglecs 13, 14, and 16), change of sialic acid-binding preference (siglecs 5, 7, 9,11, and 12), gene conversion (siglec 11), and change of site of expression (siglecs 1, 5, 6, and 11) (Varki 2010). Some of these changes are hypothesized to have followed the uniquely modern human change in sialic acid biology, namely the loss of the common mammalian sialic acid variant Neu5Gc (*N*-glycolyneuraminic acid) due to an **Alu repeat element**-mediated mutation dated at *c*.2–3 Ma (Chou et al. 2002).

sigmoid sinus   (Gk *sigma* = in the older form of the Greek alphabet what is now was written as a C, plus *oides* = shape) The sigmoid sinus is one of the **dural venous sinuses** that belong to the outer of the two components of the intracranial venous system. In most modern humans the veins draining the deep structures of the brain drain as the deep cerebral veins into the left transverse sinus, whereas the superior sagittal sinus drains into the right transverse sinus. Each transverse sinus drains into a sigmoid sinus, which runs inferiorly in a sigmoid-shaped groove from the lateral end of the transverse sinus to the superior jugular bulb, which marks the beginning of the internal jugular vein. This pattern of dural venous sinuses is known as the **transverse-sigmoid system**, and individuals (e.g., **KGA 10-525**) with transverse and sigmoid venous sinus connections on both sides are said to show transverse-sigmoid system dominance. But in some African apes and in the majority of the crania belonging to *Australopithecus afarensis*, *Paranthropus boisei*, and *Paranthropus robustus* the sigmoid sinuses receive venous blood from the **marginal sinus**es instead of the transverse sinuses. *See also* **cranial venous drainage**.

Silberberg Grotto (etym. named after Dr H.K. Silberberg because in c.1942 he had collected the upper and lower jaw of a fossil hyena, *Chasmoporthetes*. Also known as the "Daylight Cave") Part of the **Sterkfontein** cave complex that contains exposures of Sterkfontein Members 1, 2, and 3. Systematic exploration, facilitated by the Randfontein Estates Gold Mine, began in 1978. The exploration was intensified in 1992 when some of the *in situ* Member 2 breccia was blasted. Two years later, in 1994, when Ron **Clarke** was reviewing a bag of "animal" bones that had been recovered in 1980 from some of the breccia blocks removed from the Silberberg Grotto by the lime miners, he recognized four conjoined bones from the left foot of what was to become the **Stw 573** hominin (Clarke and **Tobias** 1995). Three years later, Clarke recognized more lower-limb bones, apparently from the same individual, to make a total of 12 bones (Clarke 1998). The breaks in the shafts of the bones of the lower leg seemed to be fresh, so Clarke surmised that the blasting had removed the distal elements of a hominin skeleton that must have remained behind in the breccia, and that it should be possible to identify the other side of the break on the surface of the remaining breccia. Stephen Motsumi and Nkwane Molefe were challenged to locate the breaks, and after only a day and a half of searching they located the complementary surfaces of the tibia and fibula. Since then painstaking excavations by Clarke and his team have recovered much of the rest of the skeleton of the the Stw 573 hominin (Clarke 1999, 2002, 2008). See also **Sterkfontein**.

silent substitution *See* **synonymous mutation**.

Siler hazard model A model of animal mortality proposed by William Siler (1979) that Gage (1998) has applied to problems in primate **demography** (including extant and prehistoric humans) and Gurven and Kaplan (2007) have applied to extant **hunter–gatherer**s, forager–horticulturalists, and to acculturated hunter–gatherers. The model combines a **negative Gompertz hazard model** of mortality (to represent immature deaths with the parameters a_1 and b_1) with a Makeham model (to represent a "baseline" risk of death independent of age with the parameter a_2 and a "senescent" component with the parameters a_3 and b_3). The total hazard of death at age t in the Siler model is $h(t) = a_1 e^{-b_1 t} + a_2 + a_3 e^{b_3 t}$. **Survivorship** to exact age t can be found by multiplying the survivorship from the negative Gompertz model with the survivorship from the Makeham model. In paleodemography it is common to ignore the "baseline" hazard (a_2) and assume that it is equal to zero. The four-parameter variant of the Siler model combines a negative Gompertz with a positive **Gompertz hazard model**.

silt Sedimentary materials defined primarily on the basis of particle size (i.e., a mean diameter between 0.063 mm and 4 m). Loose (unlithified) sediment in this range is referred to as silt, whereas consolidated or cemented (lithified) examples would be termed siltstone. Silts are common components of upward-fining **fluvial** deposits and they accumulate on levees when flow energy is lost during overbank flooding. Silt is the dominant particle in **loess**, the aeolian deposit associated with glacial activity at sites such as Dolní Vstonice.

siltstone *See* **silt**.

Sima de las Palomas (Location 37°47′59″N, 0°53′45″W, Murcia, Spain; etym. Sp. chasm of doves) <u>History and general description</u> The site is a **karstic** shaft in a hill (Cabezo Gordo). The discovery of a crushed hominin facial skeleton in the uppermost breccia in 1991 prompted Michael Walker and J. Gibert to examine the sediment column and fossiliferous rubble left by miners in the 19thC. <u>Temporal span and how dated?</u> **Radiocarbon dating** of burnt bones from Level 2f has provided dates of 34,450±600 years BP . <u>Hominins</u>

found at site The remains of at least 63 individuals have been recovered from the excavated deposits, and other isolated remains have been found in the rubble left by the miners. Although the hominin fossils have been interpreted as possessing some "modern" features, the overall morphological evidence is consistent with them being assigned to *Homo neanderthalensis*. Archeological evidence found at site A typical **Middle Paleolithic** assemblage made up of retouched flakes of flint and quartz. Key references: Geology, dating, and paleoenvironment Walker et al. 2008; Hominins Walker et al. 2008, 2010; Archeology Walker et al. 2008.

Sima de los Huesos (Location 42°20′57″N, 3°30′51″W, Spain; etym. Sp. *sima* = pit or pothole and *huesos* = bones, hence "the pit of the bones") History and general description This site is one of the many breccia-filled cave systems that make up the **Cueva Mayor-Cueva del Silo** within a range of limestone hills called the Sierra de **Atapuerca**, near Burgos in northern Spain. The cave system was well-known as a source of ancient cave bear (*Ursus deningeri*) teeth, but it was not until 1976 that the first hominin fossils were found. The first 4 years of the systematic excavations that began in 1984 under the direction of Juan Luis **Arsuaga** were devoted to removing the **breccia** disturbed by cavers, but from 1988 onwards researchers worked on undisturbed sediments (Arsuaga et al. 1997b). To date, more than 6500 hominin specimens belonging to at least 28 individuals of diverse ages and both sexes (Bermúdez de Castro et al. 2004) and one **Acheulean** handaxe have been recovered from what appears to be a single unit of breccia (Bischoff et al. 1997). Temporal span and how dated? The sediments are overlain by a **speleothem** which has yielded a minimum **uranium-series dating** age of more than 530 ka for the hominin-bearing breccia (Bischoff et al. 2007). Hominins found at site The site is one of the richest sources of hominin fossils and has yielded at least one example of every bone in the body. The hominin remains include numerous crania, mandibles, hundreds of teeth, a nearly complete pelvis, vertebrae, ribs, hand and foot bones, and multiple specimens of long bones, in various states of preservation (Arsuaga et al. 1997b). Among the cranial remains, Cranium 5 (AT-700) is exceptionally complete and Cranium 4 (AT-600) has one of the largest cranial capacities among European Middle Pleistocene hominins.

The cranial and mandibular sample as a whole shows a number of clearly derived features of **Homo neanderthalensis** including pronounced midfacial prognathism, the form of the brow ridge, a flat articular eminence of the glenoid fossa, a retromolar space, and an asymmetrical configuration of the ramus of the mandible. In contrast, the cranial vault is generally more plesiomorphic (e.g., large, projecting, mastoid processes, rounded neurocranium) with some incipient derived traits of *H. neanderthalensis* (e.g., weak expression of the suprainiac fossa). The dental remains combine a pronounced buccolingual expansion of the anterior dentition, similar to that seen in Neanderthals, with a strong reduction in the size of the postcanine tooth crowns. The most complete pelvis from the site (Pelvis 1) is a large, male specimen that shows generally plesiomorphic features for the genus *Homo*, including a wide bi-iliac breadth and long superior pubic ramus, and it has yielded a body-mass estimate of over 90 kg (Arsuaga et al. 1999b). The degree of sexual dimorphism within the sample appears to be similar to that which characterizes present-day modern humans. The age distribution of the hominin sample shows an overabundance of adolescent individuals and it has been suggested that the accumulation of the hominin bones at the site has an anthropic origin (Arsuaga et al. 1997b). The hominin sample from the Sima de los Huesos has been assigned to *Homo heidelbergensis* (Arsuaga et al. 1997a) and the researchers involved suggest that the Sima de los Huesos hominins sample the population that is ancestral to *H. neanderthalensis*. However, there is considerable debate about the systematics of Middle Pleistocene hominins (e.g., Hublin 2009, Rightmire 2008). Archeological evidence found at site One large Acheulean handaxe. Key references: Geology, dating, and paleoenvironment Bischoff et al. 1997, 2007; Hominins Arsuaga et al. 1993, 1997b, 1999b; Archeology Carbonell et al. 2003.

Sima del Elefante (Location 42°20′60″N, 3°31′09″W, Spain; etym. "elephant pit") History and general description This site is one of the many breccia-filled caves that make up the **Cueva Mayor-Cueva del Silo** within a range of limestone hills called the Sierra de **Atapuerca**, near Burgos in northern Spain. The Sima del Elefante is 18 m deep and up to 15 m wide. It contains evidence of at least one **hominin** individual (**ATE9-1**), and the **lithic** assemblage of 32 artifacts is consistent with hominin stone tool manufacture within the cave. Temporal span and how dated? A combination of **biostratigraphic dating** using rodents and insectivores, an observed reversed magnetic polarity (consistent with the TE16 and older lithostratigraphic layers being in the **Matuyama**), and the results of **cosmic radionuclide dating** with ^{26}Al and ^{10}Be giving an age of 1.22 ± 0.16 Ma for TE9c, all point to an age for the sediments in the Sima del Elefante of $c.1.2–1.1$ Ma. Hominins found at site The only hominin recovered from the Sima del Elefante, ATE9-1, a partial mandible, has been provisionally assigned to *Homo antecessor*. Archeological evidence found at site Mode 1, simple flakes and debitage. Key references: Geology, dating, and paleoenvironment Rosas et al. 2006, Carbonell et al. 2008; Hominins Carbonell et al. 2008; Archeology Carbonell et al. 2008.

simple regression A **regression** with one independent variable. The coefficient associated with the independent variable is the slope of the best-fit line for the data, and indicates the change in the dependent variable associated with a unit change in the independent variable. The **P value** for the independent variable identifies the significance of the difference between that slope and zero (a slope of zero indicating a flat line). The intercept of the best-fit line with the y-axis is also given, as well as an associated P value describing the significance of the difference between the intercept and zero. Simple regression generally refers to the **ordinary least squares regression** method for identifying regression parameters; *see* **regression** for alternative techniques.

***Sinanthropus* Black, 1927** Hominin genus introduced by Davidson Black in 1927. Type species **Sinanthropus pekinensis Black, 1927.**

***Sinanthropus lantianensis* Woo, 1964** A species of *Sinanthropus* provisionally proposed by Woo Ju-kang (aka Wu Ru-kang) in 1964 for the mandible recovered in 1963 from the site of Lantian. When he described the calotte and facial fragments from the same individual (Woo 1966) Wu Ru-kang reaffirmed his suggestion that these remains sampled a novel species, but by the mid-1960s most researchers had subsumed *Sinanthropus* into the genus *Homo* and few outside of China recognized the new taxon.

Sinanthropus pekinensis* Black, 1927** (Gk *Sinai* = Chinese and *anthropos* = human being) Hominin species established by Davidson **Black** in 1927 [NB: Black (1927, p. 21) suggested the species be referred to as "*S. pekinensis* (Black and Zdansky)" but Black is the sole author of the 1927 paper so his name alone should be used] to accommodate three fossil hominin teeth recovered in 1921, 1923, and 1927 at Choukoutien (now called **Zhoukoudian**). Wilfrid **Le Gros Clark** (1940) proposed that *Sinanthropus pekinensis* be transferred to ***Pithecanthropus as either ***Pithecanthropus erectus*** or *Pithecanthropus pekinensis*. Later that year Franz **Weidenreich** (1940) formally transferred *S. pekinensis* to *Homo* as *Homo pekinensis*. More recently Antón (2002) and Kaifu et al. (2008) have once again raised the possibility that there are significant differences between the two main fossil samples from Java, namely **Ngandong** and Bapang-AG. They also suggest that the *Sinanthropus* part of *Homo erectus* is morphologically distinct from the part of the hypodigm that comes from Java. First discovery Ckn. A 1 Upp 1 (recovered in 1923, but not recognized until 1926). Holotype Ckn. A 1. (1927). Paratypes None. Main site Zhoukoudian. *See also* **Homo erectus**.

SINE Acronym for short interspersed nuclear element. Short sequences (i.e., fewer than 500 base pairs) of **DNA** dispersed throughout the genome. They make up a class of mobile elements or **transposon**s that are derived from reverse-transcribed **RNA** molecules. In modern humans, the most common SINEs are the **Alu elements**, which comprise almost 13% of our genome.

Singa (Location $13°00'$N, $33°55'$E, Sudan; etym. named after the British District Commissioner's post at the nearby town of Singa) History and general description The **Singa 1** calvaria was recovered, encrusted with **calcrete**, from the bank of the Blue Nile about 320 km/200 miles south of Khartoum. Arthur **Smith Woodward**'s (1938) initial study noted its modern aspect and portrayed it as an ancestral Bushman, largely because of its strongly developed parietal bossing. Later studies have supported its modern affinities, but emphasized that the shape of the cranial vault may be the result of pathology and that claims of a special affinity to Khoesan populations are overstated. It may be an early anatomically modern cranium, but it retains some archaic features such as a large supraorbital torus and a somewhat low and broad vault. Temporal span and how dated? 145.5±7.5 to 133±2 ka (by **uranium-series dating** on the calcrete matrix encasing the teeth and calvaria), 89±9.3 to 159±12 ka (based on **electron spin resonance spectroscopy dating**, early and late uptake, respectively on an associated *Equus* tooth). Hominins found at site Singa 1, a well-preserved calvaria. Archeological evidence found at site **Middle Stone Age** ("Advanced Levalloisian" or "Proto-Stillbay") tools. Key references: Geology, dating, and paleoenvironment Woodward 1938, Bate 1951, Grün and Stringer 1991, McDermott et al. 1996; Hominins Woodward 1938, Wells 1951, Tobias 1968, Brothwell 1974, Stringer 1979, Stringer et al. 1985; Archeology Lacaille 1951.

Singa 1 Site **Singa**. Locality N/A. Surface/*in situ In situ*, but likely eroded from a higher fossiliferous stratum. Date of discovery February 1924. Finder W.R.G. Bond. Unit N/A. Horizon Caliche deposit on the west bank of the Blue Nile. Bed/member N/A. Formation N/A. Group N/A. Nearest overlying dated horizon N/A. Nearest underlying dated horizon N/A. Geological age Late **Middle** to early **Upper Pleistocene**, 145±7.5 to 89±9.3 ka. Developmental age Adult. Presumed sex Indeterminate. Brief anatomical description A heavily mineralized cranium preserving all of the bones of the vault as well as portions of the left zygomatic, both nasals, and the frontal processes of both maxillae. The forehead recedes only slightly more than in recent crania and the occipital is rounded. The parietal bosses are strongly developed, possibly pathologically, because the vault is also shorter than is common in crania of this antiquity and it is also markedly asymmetrical. The temporal bone shows a rare pathology (Spoor et al. 1998). Announcement Smith Woodward 1938. Initial description Smith Woodward 1938. Photographs/line drawings and metrical data Wells 1951, Stringer et al. 1985. Detailed anatomical description Wells 1951, Stringer et al. 1985. Initial taxonomic allocation *Homo sapiens*. Taxonomic revisions *Homo sapiens*, "Pre-Bushman" (Wells 1951), early anatomically modern *Homo sapiens* (Bräuer 1984, 1998), with the unusual parietal morphology possibly due to pathology (Brothwell 1974, Stringer et al. 1985). Current conventional taxonomic allocation *Homo sapiens*. Informal taxonomic category Anatomically modern human. Significance Possibly an early example of anatomically modern humans at *c.*130–150 ka. Location of original Natural History Museum, London, UK.

single-crystal argon-argon dating *See* **argon-argon dating**.

single-crystal laser fusion ⁴⁰Ar/³⁹Ar dating *See* **argon-argon dating**.

single nucleotide polymorphism A difference between two individuals at a single **nucleotide**. For example, the difference between whether an individual has dry or wet earwax is due to a single nucleotide difference (Yoshiura et al. 2006) (syn. point mutation). *See also* **mutation**.

single species hypothesis (or SSH) The single species hypothesis was formulated by Loring Brace and Milford **Wolpoff** of the University of Michigan during the 1960s and 1970s. During the mid-1960s Brace began to argue for the proposal put forward by Kenneth **Oakley** that evidence of tool use by early **hominins** formed the basis for considering them to be members of the genus *Homo*. Moreover, Brace relied upon the widely accepted principle of **competitive exclusion** (see below), which he interpreted as suggesting that no more than one hominin species could inhabit a given environment at any given time since multiple hominin species would have to compete for the same resources. As a result, Brace argued that all hominins must belong to the same species. He proposed that hominins had passed through four stages (**australopithecine, pithecanthropine, Neanderthal**, and modern) during the course of their evolution, but he argued that the stages should not be considered separate species of hominin. The term single species hypothesis came to the fore when it was used by Milford Wolpoff in 1968 as part of the title of a paper called "'*Telanthropus*" and the single species hypothesis'. In that paper Wolpoff (1968) suggested the single species hypothesis was an interpretive "framework" that was derived from the principle of competitive exclusion (i.e., that only one organism at a time can occupy a specific ecological niche). Milford Wolpoff promoted the single species hypothesis in the late 1960s, and in 1971 he published an article in the journal *Man* entitled "Competitive exclusion among Lower Pleistocene hominids: the single species hypothesis" outlining the arguments in support of the hypothesis

(Wolpoff 1971b). The gist of Brace and Wolpoff's argument was that hominin culture constituted an ecological niche and thus any hominin possessing culture, as shown by evidence of tool use, must belong to the same species. The single species hypothesis as interpreted by Wolpoff is a restatement of an earlier proposal set out by Ernst Mayr. Mayr (1950) applied the principle of the "biologically defined polytypic species" (*ibid*, p. 110) to the hominin fossil record. After "due consideration of the many differences between Modern man, Java man, and the South African apeman" Mayr stated that he "did not find any morphological characters that would necessitate separating them into several genera" and the suggested that within the single hominin genus, *Homo*, the only speciation that has occurred "has been phyletic speciation resulting in *Homo sapiens*" (*ibid*, p. 110). Within the genus *Homo* Mayr (1950) recognized three time-successive species: **Homo transvaalensis, Homo erectus,** and **Homo sapiens.** Having accepted that throughout hominin evolution "the known diversity of fossil man can be interpreted as being the result of geographic variation within a single species of *Homo*" (*ibid*, p. 116) (i.e., there has been no speciation within hominin evolution involving reproductive isolation) Mayr then speculates about the reasons for "this puzzling trait of the hominid stock to stop speciating" (*ibid*, p. 116). Mayr concludes that the reason is "man's great ecological diversity" (*ibid*, p. 116) which means that if "man occupies successfully all the (ecological) niches that are open for a *Homo*-like creature, it is obvious that he cannot speciate. This conforms to **Gause's rule**" (*ibid*, p. 116), but it is noteworthy that the reference Mayr and others cite for the principle of competitive exclusion [Gause (but sometimes also mis-spelled "Gauss") 1934] deals exclusively with protozoa and yeast. However, although Wolpoff also used Gause's rule, he did so in a different way than Mayr (1950). Wolpoff (1968) suggested that human *culture* is such a specialized ecological **niche** that "no more than one culture-bearing hominid could have arisen and been maintained" (*ibid*, p. 477). Thus, once culture had been acquired there could only ever have been one **synchronic** hominin species. Wolpoff suggests that tool use could only have occurred if a hominin was bipedal, and he credits the "hominid equals bipedalism" hypothesis to Sherwood **Washburn** (1951), but a year earlier Mayr (1950) had also used bipedalism to define *Homo* when he suggested that "when the *Homo*-line acquired upright posture it entered a

completely different adaptive zone" (*ibid*, p. 111). The single species hypothesis was taken up by Bartholomew and Birdsell (1953) and Wolpoff suggests that these authors "amplified" it (*ibid*, p. 477). Indeed, they did suggest that the "bipedal tool-using mode of life" was a "level of adaptedness" that "had previously been inaccessible" (Bartholomew and Birdsell 1953, p. 492), but instead of concluding that this would have resulted in a "uniform long-persistent type" they suggested the exact opposite, namely that the australopiths were "a variable group of related forms" (*ibid*, p. 492). Only when the australopiths later came into contact with "culturally advanced hominids" did Bartholomew and Birdsell suggest that the australopiths would have been subject to "rapid replacement" (*ibid*, p. 492). As more australopith fossils were discovered during the late 1960s and 1970s, especially in East Africa, evidence for two forms, called "gracile" and "robust," emerged, and Wolpoff argued that the differences between the "gracile" and "robust" australopiths probably represented **sexual dimorphism** in a single species rather than the existence of two separate species, australopith or otherwise. Wolpoff primarily made these arguments on the basis of the southern African australopith evidence, but he later expanded it to the early East African discoveries. Critics, however, noted that the "gracile" and "robust" australopiths in southern Africa were found in separate geographical locations and were separated by approximately half a million years. *Telanthropus* had been put forward as refuting the single species hypothesis because it provided at one site, **Swartkrans,** evidence of a *Homo erectus*-like hominin alongside *Paranthropus robustus*, but Wolpoff rejected the refutation for two reasons. First he doubted the taxa were synchronic, and second he rejected the taxonomic distinctiveness of *Telanthropus*. By dint of persistent advocacy the single species hypothesis was kept alive for nearly another decade. Its demise came with the demonstration that an undoubted early African *Homo erectus* cranium (**KNM-ER 3733**) and an undoubted *Paranthropus boisei* cranium (**KNM-ER 406**), so different that they were evidently from different taxa, had both been found *in situ* in the Upper Member of the **Koobi Fora Formation** (Leakey and Walker 1976). The term "single species" has also been applied to discussions about whether *Homo sapiens* and *Homo neanderthalensis* are distinct enough to be assigned to different species and whether *Homo erectus sensu lato* is one, or more than one, species (e.g., Villmoare 2005). The demise of the single species hypothesis, along with the proliferation of the hominin fossil record and better dating methods, has led to the general acceptance that the hominin phylogenetic tree consists of several branches, with hominin species coexisting at various times in the past. *See also* **competitive exclusion**.

sinkhole *See* mokondo.

Šipka (Location 49°35′N, 18°10′E, Czech Republic; etym. Czech for "arrow") History and general description This cave site, situated on the northern slope of Kotouč Hill, north of the Morava river valley, was excavated in the 1880s and again in 1950. Temporal span and how dated? Evidence from **biostratigraphy** suggests the hominin remains were associated with a **Mousterian** level, probably dating to **Oxygen Isotope Stage** 5a–c, and the **Upper Paleolithic** lithic remains are linked to the last glacial. Hominins found at site A juvenile mandibular fragment, comprising the perisymphyseal region with three erupted permanent incisors, and the anterior part of the right side of the corpus with the germs of the right canine and premolars in their crypts, was found during the course of the 1886 excavation. It has been attributed to *Homo neanderthalensis*. As one of the earliest associations with the Mousterian, the Šipka specimen was prominent in early debates on the significance of Neanderthals. Initial descriptions of the specimen by Hermann Schaaffhausen and H. Wankel emphasized the mandible's archaic character, but Rudolf Virchow claimed it was a pathological modern human: the same argument he used to explain the **Neanderthal** skeleton (Smith 1997). Archeological evidence found at site Mousterian and Upper Paleolithic (**Magdalenian**) lithic industries. Key references: Geology, dating, and paleoenvironment Valoch 1965; Archeology Valoch 1965; Hominins Vlček 1969, Jelinek 1969, Smith 1982.

sister taxa Two taxa that share **synapomorphies**, or a unique combination of **symplesiomorphies**. A pair of sister taxa is the minimum size for a **clade**, or **monophyletic group**, and pairs of sister taxa are the basic units of clades.

site (L. *situs*=location) A location, large or small, where fossils are found. It can vary from a cave the size of a room (e.g., **Tabun**) to an area of hundreds of square kilometers (e.g., **Hadar study area, Koobi Fora, Middle Awash study area**). Large sites may be subdivided into areas, localities, or collecting regions identified with names or numbers (e.g., Area 103, A.L. 333, **Aramis**), but if the site is small there is no need to specify separate localities.

size The form of an object is composed of its size and **shape**. Size is a spatial property of an object that is independent of its location (translation and rotation) and shape. More generally, size refers to the extent of a physical object's presence in one or more spatial dimensions. Size can be measured in a variety of ways. Size variables have two distinct properties: dimension and scope (e.g., all, or part, of the body). Size measures may be one-dimensional (i.e., linear), two-dimensional (i.e., area), three-dimensional (i.e., volume), or size can be measured by means of the mass (a property of objects in three-dimensional space) of the object. In human evolution, size measurements in common use include canine height, molar crown area, endocranial volume, and body mass. With regard to scope, measurements may reflect overall organismal size (e.g., body mass, a **geometric mean** of measurements from throughout the skeleton), the size of an organ or organ system (e.g., **brain mass**, gut surface area, testicular volume), or the size of a particular anatomical region of interest (e.g., a geometric mean of measurements from the proximal femur). Note that dimension and scope are independent of one another. For example, an overall measure of **body size** can be a measure of length (e.g., stature), an area (e.g., surface area of the skin), a volume (e.g., total body volume), or a mass (e.g., body mass). Furthermore, size can be measured as a single variable or calculated from multiple variables. For example, the volume of a brain can be measured directly by placing it in a beaker of liquid and measuring the volume of liquid displaced, or it can be estimated using formula based on the height, width, and length of the brain (*see* **multivariate size** for more detail). Size is important in studies of **alpha taxonomy, phylogeny reconstruction**, and **sexual dimorphism**. Body size in particular is related to **life history** variables, **biogeography**, and substrate use. *See also* **multivariate size**.

size classes *See* bovid size classes.

SK Abbreviation of Swartkrans and the prefix used for all the hominin fossils recovered from **Swartkrans** cave in the Blauuwbank Valley,

Transvaal (now Gauteng Province, South Africa), during the excavations by Robert **Broom** and John **Robinson** from 1948 to 1953. All of the SK fossils are curated by the **Ditsong National Museum of Natural History** (formerly the **Transvaal Museum**), Northern Flagship Institution, Pretoria, South Africa.

SK 6 Site **Swartkrans**. Locality N/A. Surface/ *in situ* *In situ*. Date of discovery November 1948. Finders Robert **Broom** and John **Robinson**. Unit/ horizon Pink breccia. Bed/member Member 1. Formation **Swartkrans Formation**. Group N/A. Nearest overlying dated horizon N/A. Nearest underlying dated horizon N/A. Geological age *c*.2.0–1.5 Ma. Developmental age Young adult. Presumed sex Unknown. Brief anatomical description Left side of the corpus of an adolescent mandible with P_3–M_3, plus a fragment of the right side of the corpus containing P_4–M_1 and the crowns and roots of the right M_2 and M_3. Announcement Broom 1949. Initial description Broom and Robinson 1952. Photographs/line drawings and metrical data Broom and Robinson 1952. Robinson 1956. Detailed anatomical description Robinson 1956. Initial taxonomic allocation *Paranthropus crassidens*. Taxonomic revisions Robinson (1954a) suggested that *P. crassidens* be transferred to, and would thus be a junior synonym of, *Paranthropus robustus*. Current conventional taxonomic allocation *P. robustus*. Informal taxonomic category Megadont archaic hominin. Significance The **holotype** of *P. crassidens*. Note that researchers have suggested that SK 100 and SK 13/14 may belong to the same individual. Location of original **Ditsong National Museum of Natural History** (formerly the **Transvaal Museum**), Northern Flagship Institution, Pretoria, South Africa.

SK 15 Site **Swartkrans**. Locality N/A. Surface/ *in situ* *In situ*. Date of discovery April 1949. Finder John **Robinson**. Unit/horizon Boundary between Member 1 Hanging Remnant and Member 3. Bed/ member Member 1/3. Formation **Swartkrans Formation**. Group N/A. Nearest overlying dated horizon N/A. Nearest underlying dated horizon N/A. Geological age *c*.1.5–1.0 Ma. Developmental age Adult. Presumed sex Unknown. Brief anatomical description A crushed and distorted mandibular corpus with the left M_{1-3} and the right M_{2-3}. Announcement Broom and Robinson 1949b. Initial description Broom and Robinson 1952. Photographs/line drawings and metrical data Broom and

Robinson 1952. Detailed anatomical description N/A. Initial taxonomic allocation *Telanthropus capensis*. Taxonomic revisions *Homo erectus* (Robinson 1961, Howell 1978a), *Paranthropus crassidens* (Wolpoff 1968), *Homo habilis* (Blumenberg and Lloyd 1983, Strait et al. 1997, Curnoe 2006), *Homo sp. indet.* (Grine 2001). Current conventional taxonomic allocation *H. erectus*. Informal taxonomic category Pre-modern *Homo*. Significance The hypodigm of *Paranthropus* was relatively meager in 1949, yet even then Broom and Robinson were convinced that the corpus of SK 15 was too small, and the teeth too like those of modern humans, for it to be even an extreme variant of *P. crassidens*, so they opted to erect a new taxon, *T. capensis*, for SK 15. A year later they announced the recovery of a second "non-*Paranthropus*" mandible, **SK 45** (Broom and Robinson 1950), and although they did not immediately assign SK 45 to *T. capensis*, its gracile corpus and small molars pointed in that direction. Wolpoff (1968) is willing to subsume SK 15 within the main *Paranthropus* hypodigm, but he is a lone voice, and others (see above) have pointed to affinities between its dentition and that of *H. habilis*. Location of original **Ditsong National Museum of Natural History** (formerly the **Transvaal Museum**), Northern Flagship Institution, Pretoria, South Africa.

SK 27 Site **Swartkrans**. Locality N/A. Surface/ *in situ* *In situ*. Date of discovery 1949. Finders Robert **Broom** and John **Robinson**. Unit/horizon Pink breccia. Bed/member Member 1. Formation **Swartkrans Formation**. Group N/A. Nearest overlying dated horizon N/A. Nearest underlying dated horizon N/A. Geological age *c*.2.0–1.5 Ma. Developmental age Subadult. Presumed sex Male. Brief anatomical description Crushed juvenile cranium with partial dentition. Announcement Broom and Robinson 1952. Initial description Broom and Robinson 1952. Photographs/line drawings and metrical data Robinson 1956. Detailed anatomical description N/A. Initial taxonomic allocation *Paranthropus crassidens*. Taxonomic revisions *Homo* (Clarke 1977, Howell 1978a, Strait et al. 1997), *Homo habilis* (Curnoe and Tobias 2006). Current conventional taxonomic allocation *Homo sp. indet.* Informal taxonomic category Pre-modern *Homo*. Significance Features of the vault, base, and the dentition of the SK 27 cranium have suggested to some researchers (e.g., Clarke 1977, Howell 1978a, Strait et al. 1997) that this specimen belongs to

Homo. Clarke (1977) claimed that SK 27 was evidence of a third *Homo* individual (the other evidence included the **SK 847** composite cranium and SK 2635, two maxillary premolars) in Swartkrans Member 1. Location of original **Ditsong National Museum of Natural History** (formerly the **Transvaal Museum**), Northern Flagship Institution, Pretoria, South Africa.

SK 45 Site Swartkrans. Locality N/A. Surface/ *in situ* In situ. Date of discovery September 1949. Finders Robert **Broom** and John **Robinson**. Unit/ horizon Equivalent to the pink breccia. Bed/member Member 1. Formation **Swartkrans Formation**. Group N/A. Nearest overlying dated horizon N/A. Nearest underlying dated horizon N/A. Geological age *c*.2.0–1.5 Ma. Developmental age Adult. Presumed sex Unknown. Brief anatomical description Fragment of the right side of the mandibular corpus with the M_{1-2} and part of the M_3 alveolus. Announcement Broom and Robinson 1950. Initial description Broom and Robinson 1950. Photographs/line drawings and metrical data Broom and Robinson 1952. Detailed anatomical description N/A. Initial taxonomic allocation aff. *Telanthropus capensis*. Taxonomic revisions *Homo erectus* (Robinson 1961), *Paranthropus crassidens* (Wolpoff 1968), *Homo habilis* (Howell 1978a, Strait et al. 1997), *Homo sp. indet.* (Grine 2001). Current conventional taxonomic allocation *H. erectus*. Informal taxonomic category Pre-modern *Homo*. Significance Robinson (1961) transferred SK 45 to *H. erectus*. Wolpoff (1968) is willing to subsume SK 45 within the main *Paranthropus* hypodigm, but he is a lone voice. Others have pointed to affinities between its dentition and that of *H. habilis*. Location of original **Ditsong National Museum of Natural History** (formerly the **Transvaal Museum**), Northern Flagship Institution, Pretoria, South Africa. *See also* **SK 15**.

SK 48 Site Swartkrans. Locality N/A. Surface/ *in situ* In situ. Date of discovery 1949. Finders Robert **Broom** and John **Robinson**. Unit/horizon Pink breccia. Bed/member Member 1. Formation **Swartkrans Formation**. Group N/A. Nearest overlying dated horizon N/A. Nearest underlying dated horizon N/A. Geological age *c*.2.0–1.5 Ma. Developmental age Young adult. Presumed sex Male. Brief anatomical description Partly crushed cranium with the left M^{1-3} and the roots or crowns of the right $\underline{C}$ to M^3. Announcement Broom and Robinson

1952. Initial description Broom and Robinson 1952. Photographs/line drawings and metrical data Robinson 1956. Detailed anatomical description N/A. Initial taxonomic allocation *Paranthropus crassidens*. Taxonomic revisions Robinson (1954a) suggested that *P. crassidens* be transferred to, and would thus be a junior synonym of, *Paranthropus robustus*. Current conventional taxonomic allocation *P. robustus*. Informal taxonomic category Megadont archaic hominin. Significance For a long time SK 48 was the best-preserved evidence of a *P. robustus* cranium, but the **DNH 7** skull from **Drimolen** is more complete and better preserved. Location of original **Ditsong National Museum of Natural History** (formerly the **Transvaal Museum**), Northern Flagship Institution, Pretoria, South Africa.

SK 50 Site Swartkrans. Locality N/A. Surface/ *in situ* In situ. Date of discovery 1949. Finders Robert **Broom** and John **Robinson**. Unit/horizon Brown breccia. Bed/member Member 1. Formation **Swartkrans Formation**. Group N/A. Nearest overlying dated horizon N/A. Nearest underlying dated horizon N/A. Geological age *c*.2.0–1.5 Ma. Developmental age Adult. Presumed sex Male. Brief anatomical description A damaged and partly crushed right pelvic bone (missing most of the iliac blade) together with a distal humerus. Announcement Broom and Robinson 1950. Initial description Broom and Robinson 1950. Photographs/line drawings and metrical data Broom and Robinson 1952, Robinson 1972b. Detailed anatomical description Broom and Robinson 1952, Robinson 1972b. Initial taxonomic allocation *Paranthropus crassidens*. Taxonomic revisions Robinson (1954a) suggested that *P. crassidens* be transferred to, and would thus be a junior synonym of, *Paranthropus robustus*. Current conventional taxonomic allocation *P. robustus*. Informal taxonomic category Megadont archaic hominin. Significance This was the first (TM 1605 from **Kromdraai** was recovered in 1954) evidence of the pelvic bone of *P. robustus*, and it formed an important part of the evidence assembled by Robinson (1972b) to justify his proposal that the locomotion of *P. robustus* was substantially different from that of *Australopithecus africanus*, with the former (i.e., SK 50) showing climbing adaptations not seen in *Au. africanus*. Location of original **Ditsong National Museum of Natural History** (formerly the **Transvaal Museum**), Northern Flagship Institution, Pretoria, South Africa.

SK 52/SKW 18 Site Swartkrans. Locality N/A. Surface/*in situ In situ.* Date of discovery 1950. Finders Robert **Broom** and John **Robinson** and (see below) C.K. (Bob) **Brain.** Unit/horizon Brown breccia. Bed/member Member 1. Formation **Swartkrans Formation.** Group N/A. Nearest overlying dated horizon N/A. Nearest underlying dated horizon N/A. Geological age *c*.2.0–1.5 Ma. Developmental age Young adult (the spheno-occipital synchondrosis is unfused). Presumed sex Male. Brief anatomical description SK 52 is a cranial fragment that includes the lower part of the face. The maxillary dentition is represented by both I^{2}s, all four premolars, both M^{1}s, plus an unerupted right M^3. A second fragment, **SKW 18**, was recovered by Bob Brain in 1968 from breccia that came from the hanging remnant part of Member 1. Clarke (1977) recognized that SKW 18 articulated with SK 52 and described some of its occipital and temporal morphology, but it was not until Daryl de Ruiter used acetic acid to remove the remaining breccia that the extent of the morphology preserved on the SKW 18 fragment was appreciated. Announcement Clarke 1977 (SKW 18). Initial description Clarke 1977 (SKW 18). Photographs/line drawings and metrical data Robinson 1956, de Ruiter et al. 2006. Detailed anatomical description Robinson (1956) for the dentition of SK 52 and de Ruiter et al. (2006) for SKW 18. Initial taxonomic allocation *Paranthropus crassidens* (SK 52). Taxonomic revisions Robinson (1954a) suggested that *P. crassidens* be transferred to, and would thus be a junior synonym of, *Paranthropus robustus.* Current conventional taxonomic allocation *P. robustus.* Informal taxonomic category Megadont archaic hominin. Significance This is the third, and in some ways the best-preserved, cranial base of *P. robustus.* Its morphology confirms the *P. robustus* pattern of basicranial morphology reported by Dean and Wood (1981, 1982) (e.g., including the presence of an **occipito-marginal system** of endocranial venous drainage). The **foramen magnum** of SK 52/SKW 18 is oval in shape, thus suggesting that the cardioid-shaped foramen magnum seen in the few available *Paranthropus boisei* specimens and in the only *Paranthropus aethiopicus* specimen is not a *Paranthropus* apomorphy. Location of original **Ditsong National Museum of Natural History** (formerly the **Transvaal Museum**), Northern Flagship Institution, Pretoria, South Africa.

SK 63 Site **Swartkrans.** Locality Unknown. Surface/*in situ* Recovered from breccia removed in 1949. Date of discovery 1949. Finders Robert

Broom and John **Robinson.** Unit N/A. Horizon N/A. Bed/member Member 1. Formation **Swartkrans Formation.** Group N/A. Nearest overlying dated horizon N/A. Nearest underlying dated horizon N/A. Geological age *c*.2.0–1.5 Ma. Developmental age Juvenile. Presumed sex Unknown. Brief anatomical description Most of the corpus, much of the left ramus and the anterior part of the right ramus, the crowns and roots of the M$_1$s, dm$_2$s, dm$_1$s, dc$s$, and the right I$_1$ crown, plus the germs of the permanent lateral incisors, the permanent canines, the permanent premolars, and the second molars. Announcement N/A. Initial description Robinson 1956. Photographs/line drawings and metrical data Robinson 1956, Conroy and Vannier 1991, Dean et al. 1993. Detailed anatomical description Robinson 1956, Dean et al. 1993. Initial taxonomic allocation *Paranthropus robustus.* Taxonomic revisions N/A. Current conventional taxonomic allocation *P. robustus.* Informal taxonomic category Archaic hominin. Significance This well-preserved juvenile mandible is notable for several reasons. First, it preserves in a single specimen much information about the dental development of *P. robustus.* Second, the microstructure of the developing dentition was investigated (Dean et al. 1993) and this allowed the researchers who conducted that study to make detailed estimates of the time it took to form the various crowns and roots. These data showed unambiguously that "the data for age at death, for age of M$_1$ emergence, and for root extension rates...accord with those known for modern great apes and fall beyond the known ranges for modern humans" (*ibid*, p. 401). Location of original **Ditsong National Museum of Natural History** (formerly the **Transvaal Museum**), Northern Flagship Institution, Pretoria, South Africa. *See also* **dental development.**

SK 80 *See* SK 847.

SK 82 Site Swartkrans. Locality N/A. Surface/*in situ In situ.* Date of discovery 1949. Finders Robert **Broom** and John **Robinson.** Unit/horizon Pink breccia. Bed/member Member 1. Formation **Swartkrans Formation.** Group N/A. Nearest overlying dated horizon N/A. Nearest underlying dated horizon N/A. Geological age *c*.2.0–1.5 Ma. Developmental age Adult. Presumed sex Unknown. Brief anatomical description A well-preserved proximal end of a right femur. Announcement Napier 1964 (but SK 82 is referred to as SK 83). Initial description Robinson 1972b. Photographs/line drawings and metrical data Robinson 1972b. Detailed anatomical description Robinson 1972b. Initial taxonomic allocation *Paranthropus*. Taxonomic revisions Howell (1978) includes SK 82 as part of the hypodigm of *Australopithecus robustus*. Current conventional taxonomic allocation *Paranthropus robustus*. Informal taxonomic category Megadont archaic hominin. Significance The two proximal right femora, SK 82 and SK 97, from Swartkrans were initially not recognized as **hominin** [e.g., they were not included in Broom and Robinson's (1970) discussion of the postcranial evidence for *Paranthropus crassidens*]. It was John **Napier** who pointed out their significance (Napier 1964) and it was the similarities between SK 82 and 97 and **OH 20** that prompted Day (1969a) to assign the latter to *Australopithecus boisei* and not to *Homo habilis*. SK 82 and 97 were part of the evidence used by Robinson (1972b) to justify his case that the locomotion of *P. robustus* was substantially different from that of *Australopithecus africanus*, with the former showing substantial adaptations for climbing that were not seen in *Au. africanus*. Location of original **Ditsong National Museum of Natural History** (formerly the **Transvaal Museum**), Northern Flagship Institution, Pretoria, South Africa.

SK 84 (NB: This SK number was originally used by Robinson 1956 for an upper canine, but now SK 84 refers to a first metacarpal) Site **Swartkrans**. Locality N/A. Surface/*in situ In situ*. Date of discovery 1949. Finders Robert **Broom** and John **Robinson**. Unit/horizon Brown breccia. Bed/member Member 1. Formation **Swartkrans Formation**. Group N/A. Nearest overlying dated horizon N/A. Nearest underlying dated horizon N/A. Geological age *c*.2.0–1.5 Ma. Developmental age Adult. Presumed sex Unknown. Brief anatomical description Well-preserved complete left first metacarpal (MC-1). Announcement Broom and Robinson 1949b. Initial description Broom and Robinson 1949b. Photographs/line drawings and metrical data Napier 1959. Detailed anatomical description Susman 1988a. Initial taxonomic allocation *Paranthropus robustus*. Taxonomic revisions *Homo erectus* (Susman 1988a). Current conventional taxonomic allocation *P. robustus* or *H. erectus*. Informal taxonomic category Megadont archaic hominin. Significance Susman (1988a) suggested that SK 84 resembled a thumb metacarpal from **KNM-WT 15000**, assigned by many to *Homo erectus*,

so he concluded that SK 84 also belonged to *H. erectus*. He also suggested that SK 84 did not belong to the same hominin taxon as another hominin first metacarpal, **SKX 5020**, so he concluded that the latter specimen must sample *P. robustus*. A recent comparative study of the shape of the proximal joint surface of the first metacarpal (Marzke et al. 2010) corroborates the suggestion that SK 84 is more modern human-like than SKX 5020, but others have concluded that the differences between the two Swartkans Member 1 first metacarpals are evidence of intra- and not inter-specific variation (Trinkaus and Long 1990). If it is confirmed that this specimen is more likely to sample *H. erectus* than *P. robustus* then it would be one of the few hand bones assigned to the former taxon. Location of original **Ditsong National Museum of Natural History** (formerly the **Transvaal Museum**), Northern Flagship Institution, Pretoria, South Africa.

SK 846b *See* **SK 847**.

SK 97 *See* **SK 82**.

SK 847 (i.e., SK 80, 846b, and 847, aka SK 847 or "the SK 847 composite cranium") Site **Swartkrans**. Locality N/A. Surface/*in situ In situ*. Date of discovery September 1949 (SK 80) and 1952 (SK 847). Finders Robert **Broom** and John **Robinson** (SK 80) and John Robinson (SK 847). Robinson recognized the SK 80 palate as representing *Telanthropus capensis* (i.e., *Homo*) but regarded the SK 847 upper facial fragment as a specimen of *Paranthropus robustus*. It was not until July 1969, when Ron **Clarke** realized that SK 847, 846, and 80 came from the same individual, that the separately numbered specimens were assembled into the partial cranium known as the "SK 847 composite cranium." Unit/horizon Member 1 "Hanging Remnant", pink breccia. Bed/member Member 1. Formation **Swartkrans Formation**. Group N/A. Nearest overlying dated horizon N/A. Nearest underlying dated horizon N/A. Geological age *c*.2.0–1.5 Ma. Developmental age Adult. Presumed sex Unknown. Brief anatomical description Part of the left side of the face (SK 847), the temporal component of the left side of the cranial base (SK 846b), and a maxillary fragment (SK 80). Announcement Clarke et al. 1970. Initial description Clarke et al. 1970. Photographs/line drawings and metrical data Clarke and Howell 1972. Detailed anatomical description Clarke and Howell 1972, Clarke 1977. Initial taxonomic allocation Prior to Clarke realizing that SK 847, 846, and 80 came from the same individual, and then observing the ways its facial morphology differed from that of *P. robustus*, SK 80 was attributed to *Homo* (i.e., *T. capensis*) but SK 847 and probably SK 846b were referred to *P. robustus*. Taxonomic revisions *Homo* sp. (Clarke and Howell 1972), *Homo habilis* (Howell 1978a, Strait et al. 1997, Curnoe and Tobias 2006), *Homo erectus* (Tobias 1991), *Homo* aff. *H. habilis.* (Grine et al. 1993, 1996, Grine 2001), paratype of *Homo gautengensis* (Curnoe 2010). Current conventional taxonomic allocation *Homo* sp. Informal taxonomic category Pre-modern *Homo*. Significance While there is near unanimity that SK 847 is not part of the *P. robustus* hypodigm, there is much less agreement about what taxon SK 847 does sample. The superficial similarities between **KNM-ER 3733** and SK 847 have led some researchers to support an allocation to *Homo erectus sensu lato*, but the more detailed analyses of Grine et al. (1993, 1996) and Smith and Grine (2008) suggest that SK 847 has both similarities with, and differences from, *Homo habilis sensu stricto*. Grine expressed it well when he wrote that "the taxonomic affinities of the early *Homo* fossils from South Africa await a comprehensive comparative analysis of all the relevant characters shovel shaped incisors, including details of dental morphology" (Grine 2001, p. 112). Location of original **Ditsong National Museum of Natural History** (formerly the **Transvaal Museum**), Northern Flagship Institution, Pretoria, South Africa.

SK 848 Site Swartkrans. Locality N/A. Surface/*in situ In situ*. Date of discovery 1949–52. Finder Unknown. Unit/horizon N/A. Bed/member ? Member 1. Formation **Swartkrans Formation**. Group N/A. Nearest overlying dated horizon N/A. Nearest underlying dated horizon N/A. Geological age *c.*2.0–1.5 Ma. Developmental age Adult. Presumed sex Unknown. Brief anatomical description A temporal bone including an **incus** bone that is complete except for the long process. Announcement The existence of the temporal bone was first mentioned in Tobias et al. in Oakley et al. (1977), but the existence of the incus was announced by Rak and Clarke (1979). Initial description, photographs/line drawings, metrical data, and detailed anatomical description Rak and Clarke 1979 re. the incus. Initial taxonomic allocation *Paranthropus robustus*. Taxonomic revisions N/A. Current conventional taxonomic allocation *P. robustus*. Informal taxonomic category Megadont archaic hominin. Significance This incus, recovered from the epitympanic recess of the middle ear cavity of SK 848, was the first middle ear ossicle to be recovered from an early hominin. The acutely angled articular surface and the small short process set this incus apart from those typical of modern humans and extant higher primates. Location of original **Ditsong National Museum of Natural History** (formerly the **Transvaal Museum**), Northern Flagship Institution, Pretoria, South Africa. *See also* **middle ear**.

SK 3155b Site Swartkrans. Locality N/A. Surface/*in situ* Recovered from breccia removed by miners in 1948/9. Date of discovery 1970 (but the breccia in which it was embedded was recovered by Robert **Broom** and John **Robinson** in 1949). Finder Elisabeth Vrba recognized the pelvis in a piece of undeveloped breccia. Unit/horizon Member 1, pink breccia. Bed/member Member 1. Formation **Swartkrans Formation**. Group N/A. Nearest overlying dated horizon N/A. Nearest underlying dated horizon N/A. Geological age *c.*2.0–1.5 Ma. Developmental age Subadult. Presumed sex Unknown. Brief anatomical description Most of the ilium, except for the anterior one-sixth and the posterior one-third of the iliac crest, the unfused acetabulum, and the acetabular portions of the ischium and pubis. Unlike **SK 50**, there is minimal plastic deformation. Announcement, initial description, photographs/line drawings, metrical data, and detailed anatomical description McHenry 1975. Initial taxonomic allocation "Probably a member of the robust australopithecine group," and "its relatively small acetabulum and sacral articular surface are enough to exclude it from [*Homo erectus*]" (McHenry 1975, p. 249); thus, by exclusion, *Paranthropus robustus*. Current conventional taxonomic allocation *P. robustus*. Informal taxonomic category Megadont archaic hominin. Significance Given its location, this well-preserved pelvic bone very likely belongs to *P. robustus* and thus it provides important information about the hip joint and swing-phase balancing mechanism of *P. robustus*. This specimen was said by Henry McHenry to share the traits common to other australopith pelvic bones (i.e., "a low broad ilium, a relatively small acetabulum and sacral articular surface, a wide flaring of the iliac blades, a well-developed anterior iliac spine with a groove for the iliopsoas muscle, and a relatively deep acetabulum," McHenry 1975, p. 260), and instead of emphasizing the differences

725

between the pelves of *Australopithecus africanus* and *P. robustus*, the evidence from SK 3155b stresses just how much pelvic morphology is likely shared by these taxa. The suite of morphological features set out above differs from the much more modern human-like configuration of the fossil hominin pelvic bones from East Africa such as **KNM-ER 3228** and **OH 28**. Location of original **Ditsong National Museum of Natural History** (formerly the **Transvaal Museum**), Northern Flagship Institution, Pretoria, South Africa.

skeletal element survivorship The probability that a particular skeletal element has survived the destructive taphonomic processes that can alter a bone **assemblage** following its deposition by hominins, carnivore, or other agents. Destructive taphonomic processes include **carnivore ravaging**, trampling, and dissolution. Skeletal element survivorship is influenced largely by the density of a skeletal element, with high-density elements more likely to survive than their low-density counterparts. Differential survivorship can radically alter the frequencies of skeletal parts in a bone assemblage, which can be especially problematic when making inferences concerning hominin behavioral patterns (e.g., **carcass transport strategies**) in archeological contexts. Skeletal elements can be classified as **high-survival elements** or **low-survival elements**. The high-survival subset is known to be resistant to destructive processes and includes elements mostly characterized by dense cortical bone with relatively little, or no, cancellous bone, such as the long bones, crania, and mandibles of most ungulates. This subset is the best candidate for making hominin behavioral inferences because its high survivorship provides the most accurate representation of the bones originally discarded by hominins. The low-survivorship subset, which is particularly sensitive to destructive processes, includes elements with thin cortical walls and much low-density, grease-rich cancellous bone. This subset includes vertebrae, ribs, pelvis, scapula, and long-bone epiphyses. The phalanges, carpals, and tarsals are also low-survival elements, because they tend to be swallowed whole by carnivores. Because low-survivorship elements are readily destroyed following hominin discard, they are of relatively little utility when examining carcass transport strategies in bone assemblages subjected to destructive processes.

skeletal part frequency The frequency of different skeletal elements or fragments thereof in a fossil bone **assemblage**, typically quantified according to calculated indices such as **minimal animal unit**, **minimum number of elements**, **minimum number of individuals**, or **number of identified specimens**. The study of skeletal part frequencies is used to evaluate the taphonomic history of a bone assemblage, with a particular focus on identifying bone-accumulating agents and depositional processes. In zooarcheological contexts, observed skeletal part frequencies are the result of a complex interaction between the hominin utilization and transportation of animal carcasses (i.e., **carcass transport strategies**) and destructive taphonomic processes operating following human discard, such as **carnivore ravaging**.

skeleton (Gk *skeletos*=to dry up) In vertebrates it refers to the hard tissue (bone and cartilage) elements that make up the endoskeleton that protects and supports the soft tissues. It is divided into the **axial skeleton** (comprising the **skull**, vertebral column, and thorax) and **appendicular skeleton** (limb girdles and limbs) components.

Skhul (Location 32°40′N, 34°58′E, Israel; etym. from Ar. (مغارة *mugharet*=cave, and (سخول *skhul*= kids) History and general description Skhul cave was first excavated in 1931 by Theodore D. McCown under the auspices of Dorothy **Garrod**'s exploration of the **Mount Carmel** caves. It consists of two parts: a small cave and a larger external rock-shelter/terrace, where most of the archeological and hominin remains were found. Despite somewhat rushed excavation techniques (the nearly 1300 m^2 area was removed over a period of 8 months), McCown recognized three main levels, including a thick **Mousterian** level (Layer B). This site is one of the oldest early modern *Homo sapiens* sites found outside of Africa, and Shea (2005) argues that it represents an early modern human migration into the Near East that disappeared *c*.75 ka when the environment changed. Temporal span and how dated? Although originally aligned with **Tabun** Level C and believed to be *c*.45 ka based on faunal and lithic similarities assemblages, Skhul Level B has recently been shown to be much more ancient. **Thermoluminescence dating** on burned flints from Level B produced a wide range of 167–99 ka, with a central tendency of 119 ka, and the most recent **uranium-series dating** and **electron spin resonance spectroscopy dating** on some of the hominin and faunal remains suggests a date of 100–130 ka. Hominins found at site Ten numbered individuals (including two infants) and 16 isolated fragments have been recovered from Skhul. McCown (in Garrod and Bate 1937) believed these were all intentionally buried, and one (**Skhul V**) may have been interred with a pig's mandible as a grave good. McCown and Keith (1939) originally argued that the fossils from Skhul were from the same population as those from Tabun, which they called *Palaeoanthropus palestinensis*. They argued that this group was undergoing evolution from a **Neanderthal**-type to a **Cro-Magnon**-type, with the Tabun fossils being older and more primitive and the Skhul fossils younger and more advanced. More recent re-evaluations of these fossils have shown that the Skhul fossils are closest to and belong with modern *Homo sapiens* (Howell 1957, Stewart 1960, Higgs 1961, Brothwell 1961) and that they predate the fossils from Tabun (Mercier et al. 2003, Grün et al. 2005). Because the Skhul and **Qafzeh** remains represent the earliest known modern human remains outside of Africa, they have played a crucial role in understanding the radiation of modern humans from Africa. Archeological evidence found at site The archeological material from Level B was thought to be equivalent to Tabun Levels C and D. The technology is a relatively primitive Mousterian with **Levallois** technique, very similar in type to that found at Tabun. Key references: Geology, dating, and paleoenvironment Garrod and Bate 1937, Higgs 1961, Grün and Stringer 1991, Mercier et al. 1993, 2003, Grün et al. 2005; Hominins Garrod and Bate 1937, McCown and Keith 1939, Howell 1958, Brothwell 1961; Archeology Garrod and Bate 1937.

Skhul I Site Skhul. Locality Terrace. Surface/*in situ* In situ. Date of discovery 1931. Finders T. McCown and team. Unit Layer B. Horizon N/A. Bed/member N/A. Formation N/A. Group N/A. Nearest overlying dated horizon N/A. Nearest underlying dated horizon N/A. Geological age 100–130 ka. Developmental age *c*.4–4.5 years. Presumed sex Unknown. Brief anatomical description Fragmentary remains of cranium, mandible, and several milk teeth, many fragments of postcranial elements. Announcement Garrod and Bate 1937. Initial description Garrod and Bate 1937.

Photographs/line drawings and metrical data McCown and Keith 1939. Detailed anatomical description McCown and Keith 1939. Initial taxonomic allocation *Paleoanthropus palestinensis* McCown and Keith, 1939. Taxonomic revisions *Homo sapiens* (e.g., Howell 1957, Stewart 1960, Higgs 1961). Current conventional taxonomic allocation *H. sapiens*. Informal taxonomic category Anatomically modern *Homo*. Significance The first of the fossils to be recovered from Skhul, this young child is one of the few early modern human juvenile specimens. Location of original Museum of the Department of Antiquities, Jerusalem, Israel.

Skhul IV Site Skhul. Locality Terrace. Surface/*in situ* In situ. Date of discovery April 30, 1932. Finders Theodore McCown and team. Unit Layer B. Horizon N/A. Bed/member. N/A. Formation N/A. Group N/A. Nearest overlying dated horizon N/A. Nearest underlying dated horizon N/A. Geological age 100–130 ka. Developmental age Adult. Presumed sex Male. Brief anatomical description A nearly complete skull and postcranial skeleton. Announcement Garrod and Bate 1937. Initial description Garrod and Bate 1937. Photographs/line drawings and metrical data McCown and Keith 1939. Detailed anatomical description McCown and Keith 1939. Initial taxonomic allocation *Paleoanthropus palestinensis* McCown and Keith, 1939. Taxonomic revisions *Homo sapiens* (e.g., Howell 1957, Stewart 1960, Higgs 1961). Current conventional taxonomic allocation *H. sapiens*. Informal taxonomic category Anatomically modern *Homo*. Significance The postcranial skeleton of Skhul IV is exceptionally complete, and the position of the skeleton when recovered suggests this individual was intentionally buried. Location of original Museum of the Department of Antiquities, Jerusalem, Israel.

Skhul V Site Skhul. Locality Terrace. Surface/*in situ* In situ. Date of discovery May 2, 1932. Finders T. McCown and team. Unit/horizon Layer B. Bed/member N/A. Formation N/A. Group N/A. Nearest overlying dated horizon N/A. Nearest underlying dated horizon N/A. Geological age 100–130 ka. Developmental age 30–40 years. Presumed sex Male. Brief anatomical description Fairly complete, though heavily reconstructed mandible and calvaria. The calvaria is missing the midfacial and most of the subnasal regions. Nearly all of the postcranial elements are represented but some have minor damage. Announcement McCown 1932. Initial description **Garrod** and Bate 1937. Photographs/line drawings and metrical data McCown and Keith 1939. Detailed anatomical description McCown and Keith (1939). Initial taxonomic allocation *Palaeoanthropus palestinensis*. Taxonomic revisions Howell 1957, Stewart 1960, Higgs 1961, Brothwell 1961. Current conventional taxonomic allocation **Homo sapiens**. Informal taxonomic category Anatomically modern human. Significance Skhul V is among the best-preserved remains from the Mount Carmel region. The remains were almost certainly intentionally buried, and the mandible of a wild boar has been associated with the find, possibly having been intentionally placed in the arms of the buried skeleton. Location of original Peabody Museum at Harvard University, Cambridge, MA, USA.

Skildergat   *See* **Fish Hoek**.

skull   (ME *skulle*=skull) The bony component of the head, made up of 28 bones, a few of which are unpaired (e.g., the frontal and the occipital), but most of which are in symmetrical pairs (e.g., the parietal and temporal). It consists of the **cranium** plus the **mandible**. *See* **cranium** and **mandible**, and their components, for more detailed descriptions.

Skull II   *See* **Sangiran 2**.

Skull III   *See* **Sangiran 3**.

Skull IV   *See* **Sangiran 4**.

Skull VI   *See* **Sangiran 10**.

skullcap   *See* **calotte**.

SKW   Abbreviation of Swartkrans Cave, University of the Witwatersrand, and the prefix for all fossils recovered from **Swartkrans** Cave, Blauuwbank Valley, Gauteng, South Africa, between 1968 (when the University of the Witwatersrand purchased the property) until 1979 when the prefix was changed to **SKX**. All of the SKW fossils are curated by the **Ditsong National Museum of Natural History** (formerly the **Transvaal Museum**), Northern Flagship Institution, Pretoria, South Africa.

SKW 18   *See* **SK 52/SKW 18**.

SKX   The prefix used for all fossils recovered from **Swartkrans** during excavations *in situ* from 1979 to 1986. All of the fossils with the SKX prefix are curated by the **Ditsong National Museum of Natural History** (formerly the **Transvaal Museum**), Northern Flagship Institution, Pretoria, South Africa.

SKX 5020   Site Swartkrans. Locality Excavation coordinates E3/N3, NE Quadrant, 630–640 cm. Surface/*in situ* *In situ*. Date of discovery Sometime between 1979 and 1983. Finder Recovered during ongoing excavations. Bed/member Member 1. Formation **Swartkrans Formation**. Group N/A. Nearest overlying dated horizon N/A. Nearest underlying dated horizon N/A. Geological age *c*.2.0–1.5 Ma. Developmental age Adult. Presumed sex Unknown. Brief anatomical description Right metacarpal that lacks part of the proximal and much of the distal articular surface. Announcement N/A. Initial description Susman 1988a. Photographs/line drawings and metrical data Susman 1988a, 1988b, 1993. Detailed anatomical description N/A. Initial taxonomic allocation *Paranthropus robustus*. Taxonomic revisions None. Current conventional taxonomic allocation *P. robustus*. Informal taxonomic category Megadont archaic hominin. Significance Susman (1988) suggested that the **SK 84** hominin first metacarpal resembled the thumb metacarpal from **KNM-WT 15000**, assigned by many to *Homo erectus*, so he concluded that SK 84 also belonged to *H. erectus*. He also judged the differences between SK 84 and SKX 5020 were too great for them to belong to the same taxon, so he concluded that the latter specimen must sample *P. robustus*. A recent comparative study of the shape of the proximal joint surface of the first metacarpal (Marzke et al. 2010) corroborates the suggestion that SKX 5020 is more primitive than SK 84. However, others have concluded that the differences between the two Swartkans Member 1 first metacarpals are evidence of intra- and not inter-specific variation (Trinkaus and Long 1990). If it is confirmed that SKX 5020 is more likely to sample *P. robustus* than *H. erectus* then it would be one of the few hand bones assigned to the former taxon. Location of original **Ditsong National Museum of Natural History** (formerly the **Transvaal Museum**), Northern Flagship Institution, Pretoria, South Africa.

sliding landmarks   *See* **semilandmarks**.

Sm Abbreviation for Sambungmacan and the prefix used for fossils recovered from that site. *See also* **Sambungmacan**.

Smith Woodward, Sir Arthur *See* **Piltdown**.

Smithsonian Institution *See* **National Museum of Natural History, Smithsonian Institution**.

SMRS Acronym for the specific mate-recognition system, the system that has been proposed as the mechanism for species defined on the basis of the individuals (usually females) identifying potential mates. *See also* **species**; **recognition species concept**.

Smugglers' Cave (Location 33°55′N, 7°00′W, Morocco; aka the Grotte des Contrebandiers in French, and sometimes refered to Témara, after the nearby town) <u>History and general description</u> This site was discovered in 1955 by Abbé Jean Roche. It is a cave site in a coastal sandstone cliff located 14 m above sea level on the Atlantic coast, 17 km/10.5 miles southwest of Rabat. M. Henrion led the first sounding from 1955 to 1957. Excavations were conducted between 1967 and 1975 by Roche and Texier and the site was reopened and excavated in 1994 by A. Bouzouggar. <u>Temporal span and how dated?</u> **Uranium-series dating** on shell and **radiocarbon dating** on bone have provided dates between 125 and 10 ka. <u>Hominins found at site</u> "Témara Man," occipital, parietal, frontal fragments, and a mandible. <u>Archeological evidence found at site</u> Aterian artifacts, Iberomaurusian and Neolithic artifacts in the upper layers, shell beads, hearths, ochre use, limpet and mussel shell middens associated with Neolithic deposits. Key references: <u>Geology, dating, and paleoenvironment</u> Debénath et al 1986, Roche 1976; <u>Hominins</u> Debénath 2000, Ferembach 1998; <u>Archeology</u> Bouzouggar 1997.

SNP Acronym for **single nucleotide polymorphism** (*which see*).

Soa Basin This geographic region of central Flores, Indonesia, which is the drainage of the Ae Sissa River, is the location of exposures of the **Ola Bula Formation**. Several Early and Middle Pleistocene sites, of which **Mata Menge** is the best known, are located in the Soa Basin. *See also* **Ola Bula Formation**.

social enhancement This term describes the phenomenon when an innate, species-typical behavior is more likely when conspecifics are engaged in the same behavior. Examples of social enhancement (also referred to as social facilitation or contagion) in modern humans include reactions to other people laughing, crying, or yawning. In some cases such behaviors are inhibited, and this converse phenomenon is referred to as social inhibition. Both social enhancement and inhibition are defined as motivational (behavioral) states. Often, social enhancement may resemble behavior matching or **imitation** because the mere presence of another individual has the potential to lead to synchronization of behaviors over space and time. Animal behaviorists believe that social enhancement is crucial in group-living species for group cohesion, behavioral coordination, foraging efficiency, and predator avoidance. While social facilitation alone does not lead to complex behavior matching or imitation, it has the potential to increase the opportunities in which observational learning can occur, and observational learning has been implicated in the development of the ability to manufacture **tools**. When coordination of behavior occurs among two or more individuals, it is referred to as social contagion, because the mere act of seeing another individual engaging in a species-typical behavior elicits that same response in others. For example, satiated brown capuchin monkeys will resume feeding when a familiar conspecific is seen eating nearby, and will also consume more food in the presence of another monkey than if alone.

social facilitation *See* social enhancement.

social inhibition *See* social enhancement.

social intelligence Social intelligence encompasses a range of skills including **social learning, imitation, teaching, communication, theory of mind**, and **cooperation**. Social intelligence is believed to have evolved in response to the specific and unique pressures associated with group living. The term has its origin in the social intelligence hypothesis originally proposed by Nicholas Humphrey (1976), but it is also known as "Machiavellian intelligence" after the book

729

Machiavellian Intelligence: Social Experience and the Evolution of Intellect in Monkeys, Apes, and Humans (Byrne and Whiten 1988). The book is named after the Italian political theorist, Niccolo Machiavelli, who argued that in order to be a good Prince, one must employ deceit and engage in various forms of social manipulation (Machiavelli 1532, translated in 1966). Byrne and Whiten (1988) and others have argued that some form of Machiavellian, or social, intelligence is a requirement of living in groups when important resources (e.g., food and potential mates) are limited. Skilled field researchers working with various kinds of primate species have reported a substantial number of anecdotal cases of tactical deception, and Byrne and Whiten cite these in support of the social intelligence hypothesis. The social intelligence hypothesis proposes that group living in primates and other animals has favored social manipulation by adding to cognitive power through **natural selection**, and it has received support from a significant amount of empirical data, examples of which include the positive correlation between group size and **neocortex** ratio (i.e., the volume of neocortex relative to that of the other parts of brain) (e.g., Dunbar 1993, 1995) and superior performance of transitive inference in species with larger group size as compared to that of closely related counterparts with smaller group size (e.g., Bond et al. 2003). As such individuals living in large groups can infer from limited social interactions that if A is dominant to B, and B is dominant to C, then A must be dominant to C.

social intelligence hypothesis *See* **social intelligence**.

social learning (L. *socius* = companion and OE *lernen* = to gain knowledge) Mammalian development and certainly modern human development is very much dependent on learning that is influenced by the actions of others who provide cues and signals that guide an observer's attention to specific behaviors (e.g., social relationships) or to objects (e.g., food). The learning that results from observation or interactions with others falls under the general category of social learning (Bonnie and Earley 2007, Zentall 2007), and it is likely that learning to make stone and other types of tools involves social learning. Social learning is different from (a) individual learning, which is learning achieved through trial and error, (b) the learning

that arises in isolation without the benefit of interacting with a conspecific or a model, and (c) instinctual behaviors. Using this broad definition, social learning is widespread in the animal kingdom. In fact, for mammals and birds, social learning has been shown to play a critical role in the development and maintenance of species-specific behaviors (e.g., the acceptance or avoidance of novel foods, mate choices, and predator-avoidance strategies). Among primates, social learning may require reasoning about other individuals, or even reasoning about one's own behavior. Examples of socially learned primate behaviors include specialized grooming, communicative signals, and tool-use techniques. These types of activities may require that an individual understand their own relationship to the demonstrator (e.g., for copying motor patterns), or understand the goal of the demonstrator's actions, both of which involve relatively sophisticated cognitive abilities that may be unique to the great apes, or perhaps only to modern humans. Determining what cognitive abilities are involved in the learning process has been the primary objective of social learning researchers for over a century, with a great amount of attention focused on nonhuman primates. **Imitation** represents a specific type of social learning, initially considered one of the few traits that animals had in common with modern humans, and many languages even have terms that involve the words "ape" or "monkey" to describe imitation (e.g., "aping" the professor). However, there is debate over whether or not monkeys, apes, or any animals other than modern humans are able to truly imitate, with a particular emphasis being placed on what the definition of imitation should be. Sometimes imitation can be confused with more subtle forms of social influence that lead to learning the same behaviors as others. For this reason, the literature on social learning distinguishes terms for the different ways in which learning can take place, such as social facilitation, contagion, stimulus enhancement, local enhancement, emulation, imitation, and **teaching**. *See also* **imitation**.

soil carbonates *See* **paleosol**.

Solo *See* **Ngandong**.

Solo Man This is the informal way Franz **Weidenreich** referred to the **Ngandong hominins**, and he entitled his unfinished monograph about these remains the *Morphology of Solo Man* (Weidenreich 1951). *See also* **Ngandong hominins**.

solution cavities   *See* **limestone**.

Solutré   (Location 46°18′N, 4°42′E, department of Saône-et-Loire, approximately 10 km/6 miles west of Mâcon, France; etym. unknown) History and general description Located in the scree slope at the base of a limestone outcrop, this is the type site of the **Solutrean** technocomplex. It was first excavated in 1866 by Adrien Arcelin and Henri Testot-Ferry, and has been under occasional excavation since then, most recently by Jean Combier. The site preserves one of the best examples of a mass kill site from the **Upper Paleolithic**. Temporal span and how dated? **Biostratigraphy** and archeological correlation, as well as **radiocarbon dating**, suggest an age span of 30.4–12ka. Hominins found at site Four hominin skeletons (a female with a neonate, and two males) were recovered from the **Aurignacian** level in the 1920s, but no detailed study of these fossils has been made. A juvenile mandible from one of the less well-documented excavations was discovered in 2002 among the collections at the Field Museum of Natural History, Chicago, USA, and was thought to be from the Solutrean, but radiocarbon dating revealed that it was from a recent intrusive burial. Archeological evidence found at site Three main stratigraphic layers were uncovered, from youngest to oldest they are **Magdalenian**, Solutrean, **Gravettian**, Aurignacian, and **Mousterian**. The overwhelming numbers of horse bones have suggested that the Solutrean hunters were regularly using a natural *cul-de-sac* to trap and kill migrating herds of horses. Key references: Geology, dating, and paleoenvironment Combier and Montet-White 2002; Hominins Bossavy 1923, Pestle et al. 2006; Archeology Combier and Montet-White 2002.

Solutrean   (etym. After the site of **Solutré**) An Upper Paleolithic technocomplex that dates to *c.*21–17 ka or during the Last Glacial Maximum. It is characterized by large, flat, and thin leaf-shaped points and it is associated with an increased site density, greater prevalence of art (particularly parietal art like cave paintings and engravings), and the hunting of reindeer.

somatic cells   (Gk *soma* = body) Cells of the body other than reproductive cells (**gametes**).

Somatic cells typically have a diploid complement of **chromosomes**. For example, the interstitial tissue of the testis is made up of diploid somatic cells, whereas the secondary spermatocytes or "spermatids" are haploid gametic cells, or **germ cells**.

somite   (Gk *soma* = body) In development, somites are the first expression of the segmental organization of the body. They form from so-called "presomitic mesenchyme," and to reflect that they lie on either side of the midline, or axis, of the developing body, these cells are referred to as paraxial (literally "either side" of the "axis") mesenchyme. Somites then form **sclerotome**s and all but the first four or so of the sclerotomes contribute to the development of the vertebrae, intervertebral discs, and ribs. The first four sclerotomes contribute to the basicranium. Each vertebra is formed from the caudal half of its equivalent sclerotome, plus the cranial half of the sclerotome below it (e.g., the C4 vertebrae is formed from the caudal half of the C4 sclerotome and the cranial half the C5 sclerotome).

South African Museum (Iziko Museums of Cape Town)   The South African Museum (or SAM) was founded by the Governor of the Cape of Good Hope, Lord Charles Somerset, in 1825, and as such it is the country's second oldest scientific institute. Dr Andrew Smith was appointed the first Director (Superintendent) of the SAM, which was originally housed in a flat at what was then the Public Library (now the Slave Lodge, or South African Cultural History Museum) located in Cape Town. Early collections focused primarily on minerals, shells, fish, reptiles, birds, mammals, and ethnographic material. Between 1825 and 1855 the collections became neglected and dispersed across a number of locations in Cape Town, and in 1855 the SAM was reconstituted by government notice, at which time the first Curator and Board of Trustees were appointed. A new building for the SAM and Library was opened in 1860, but the building was too small and plans were made for another new building, which was opened on April 6, 1897, and it is this building, with later additions including the adjacent Planetarium, that houses the current collections. On April 1, 1999, the SAM was amalgamated, along with a number of other museums and heritage sites, into the Southern Flagship Institution, which was subsequently renamed the Iziko Museums of Cape Town in July 2000. *Iziko* is an isiXhosa word meaning "hearth," which in this context symbolizes the importance of this institution as a central gathering place for South African heritage. The SAM is important from a paleoanthropological point of view for its collections that include the remains of early modern humans from coastal sites around the Western Cape of South Africa, and it also houses extensive archeological collections including artifacts and the skeletal remains of Holocene hunter–gatherers. These Holocene Khoesan collections are studied frequently by researchers interested in the emergence of the modern human phenotype. Hominin fossil collections **Klasies River, Fish Hoek** (Peers' Cave), **Blombos Cave,** and **Elandsfontein** (Saldanha, Hopefield). Contact department Archeology Department. Contact person Dr Sarah Wurz (e-mail swurz@iziko.org.za). Postal address Box 61, Cape Town 8000, South Africa. Website www.iziko.org.za/sam/index.html. Tel +27 (0)21 481 3800. Fax +27 (0)21 481 3993. E-mail info@iziko.org.za.

South Turkwel (Location 2°19′N, 36°04′E, Kenya; etym. named after a local major river) History and general description This site is between Lodwar and **Lothagam** on the west side of **Lake Turkana**. A minimum of 140 m of sediments are exposed; the lower sediments are predominantly **lacustrine** and the upper ones are mainly fluvial. Temporal span and how dated? **Tephrostratigraphy** suggests that the two tuffs in the section are a vitric tuff near the base that is the equivalent of a tuff within Lonyumun Member of the Nachukui Formation (West Turkana) and the **Koobi Fora Formation** (Allia Bay), and a lens of the Lokochot Tuff that has been dated at 3.58 Ma. Thus, the upper fluvial sediments are equivalent to the Kataboi Member of the Nachukui Formation in West Turkana and Member A of the **Shungura Formation**, and the lower lacustrine sequence is equivalent to the Lonyumun Member of the Nachukui and Koobi Fora Formations. Hominins found at site The two hominin fossils from South Turkwel are KNM-WT 22936, part of a right mandibular corpus, and KNM-WT 22944, an associated skeleton that includes a piece of weathered mandible and tooth crowns, four hand bones, and a proximal foot phalanx. These were surface finds in a part of the section where the Lokochot Tuff is not exposed. A reasonable age for the two hominin fossils is the age of the Lokochot Tuff (i.e., 3.58 Ma) with a minimum age of c.3.2 Ma. Morphologically the two hominin specimens are consistent with an attribution to *Australopithecus afarensis*, but the researchers who found them attribute them to *Australopithecus sp. indet*. Archeological evidence found at site None. Key references: Geology, dating, and paleoenvironment Ward et al. 1999a; Hominins Ward et al. 1999a.

Southern Afar Rift System The term used for the southern part of the **Afar Rift System** (also known as the **Afar Depression**) (WoldeGabriel et al. 2000). This section of the **Ethiopian Rift System** includes the **Dikika study area, Gona study area, Hadar, Middle Awash study area,** and the **Woranso-Mille study area**.

Southern blotting (etym. Named after Sir Edwin Southern who devised and developed this technique) A method used to detect particular **DNA** sequences. Fragments of DNA are transferred from an electrophoretic gel to a filter membrane, and the fragments are subsequently detected by a **hybridization** assay using probes for specific nucleic acids.

Southern Oscillation Index (or SOI) This climate variable is the barometric pressure difference between Tahiti, in French Polynesia, and Darwin, in Northern Australia. It was first suggested and used by Sir Gilbert Walker. Along with **sea surface temperature** changes the SOI is a key indicator of the **El Niño Southern Oscillation**. In the **tropics** these pressure differences account for the anomalous longitudinal **Walker circulation** as opposed to the global average **Hadley circulation**. *See also* **El Niño Southern Oscillation**.

sp. indet. (etym. *sp. indet.* is an abbreviation of "indeterminate species") When dealing with fragmentary specimens it is sometimes possible to identify the genus a specimen belongs to, but it may not be possible to be sure which species it belongs to. In that case the taxonomic allocation researchers should use is "*sp. indet*." (e.g., the mandible **KNM-ER 1506** has been referred to *Homo sp. indet*.).

spatial packing hypothesis This hypothesis is an attempt to explain variation in the overall shape of the cranium in terms of physical interactions between the growth and development of the **brain,** the **chondrocranium**, and the **face**. It is based on the premise that the craniofacial complex consists of independent and partially independent **modules**

that interact as the cranium develops because they are physically adjacent or have developmental connections. In particular, the hypothesis holds that the growth of the brain relative to the growth of the basicranium is a major determinant of the **cranial base angle** and thus of the position of the face relative to the brain case. Thus, flexion of the basicranium enables an increase in brain size to be accommodated without any significant changes in its width and/or length. The model posits that interactions among the brain, chondrocranium, and face lead to correlated variation that results in integrated evolutionary change. In particular, the highly flexed **basicranium** of modern humans, which results in the face being underneath the anterior-most part of the brain case, evolved as a byproduct of the enlargement of the brain without elongation of the cranial base. The hypothesis has a long history (Virchow 1857, Bolk 1910, Weidenreich 1941a, Hofer 1960) but it was formulated most explicitly by Biegert (1957, 1963), and expanded upon with different degrees of formality by DuBrul and Laskin (1961), Vogel (1964), Gould (1977), and Enlow (1990). *See also* **basicranium**.

spear   A weapon consisting of a long shaft and a sharp tip that can be thrust into, or thrown at, an animal. Many stone **points** recovered in the archeological record are inferred to have been manufactured in order to be mounted onto a wooden shaft and to serve as the tips of spears, but sharpened wooden tips are also effective. Chance preservation of some wooden spears suggests that the latter types date to at least 400 ka (Thieme 1997).

specialist   *See* **stenotopy**.

speciation   In the most general sense, the process by which descendant **species** evolve from ancestral ones. More specifically, the term is often used to refer to the evolution of new species via a process of "splitting" off from a pre-existing species (i.e., **cladogenesis**). Speciation is often discussed in terms of the ecological and/or biogeographic factors that may have led to the evolution of new species and a terminology has been devised to distinguish among these various factors (e.g., **allopatric speciation**, **peripatric speciation**, **parapatric speciation**, **sympatric speciation**). Ultimately, however, although it is almost certainly true that at least some hominin species evolved as a result of cladogenesis, it is very difficult to assert with any confidence the precise type of specia-

tion involved with the emergence of any particular hominin taxon.

species   (L. *specere* = to look; NB: *spectaculum*, *specimen*, and *spectio*, the source of "inspection," come from the same verb. The connection with Gk *eidos* = form or idea, is a semantic and not an etymological one) Species is the next-to-least inclusive category in the Linnaean taxonomic system. There is disagreement about how species are recognized, and de Queiroz (2007) makes the point that there has been consistent confusion between how we *conceptualize* species and how in practice we *recognize* them (he refers to the latter process as "delimitation"). Andrew Smith (1994) usefully divides species concepts into process-related and pattern-related. The former emphasize the processes involved in the generation and maintenance of species, whereas the latter emphasize the methods used for recognizing them. The three main concepts in the process category are the **biological species concept** (or BSC), the **evolutionary species concept** (or ESC), and the **recognition species concept** (or RSC). The BSC definition given below is the one provided by Ernst Mayr in 1942, and subsequently modified by him in 1982. He suggested that species are "groups of interbreeding natural populations reproductively isolated from other such groups." Note that this definition emphasizes isolation; it is a "relational" and not a "free-standing" definition. The ESC was an attempt by Simpson (1961) to add a temporal dimension to the BSC. Thus, he suggested that under the ESC a species is "an ancestral-descendant sequence of populations evolving separately from others and with its own evolutionary role and tendencies." The third concept in this category, the RSC, instead of emphasizing reproductive isolation emphasizes the process that promotes interbreeding. Hugh Paterson (1985) suggested that under the RSC a species is "the most inclusive population of individual, biparental organisms which shares a common fertilization system." In practice, species are recognized by one of three main pattern-based species concepts, the **phenetic species concept** (or PeSC), the **phylogenetic species concept** (or PySC), and the **monophyletic species concept** (or MSC). In the context of paleoanthropology, all are "morphospecies" concepts in that they each emphasize different aspects of an organism's morphology. The PeSC as interpreted by Sokal and Crovello (1970) gives equal weight to *all* aspects of the phenotype, assembles a matrix of these characters, and then uses multivariate analysis to detect clusters of individual specimens that share the same, or similar, phenotypes. Under the PySC introduced by Cracraft (1983) the emphasis is on those aspects of the phenotype that are diagnostic. According to Nixon and Wheeler (1990) in this scheme a species is "the smallest aggregation of populations diagnosable by a unique combination of character states." Finally, under the MSC the morphological emphasis is narrower still, with species defined not on the basis of unique combinations of characters, but only on unique characters, or apomorphies. The problem with the MSC and is that it assumes the observer knows which characters are apomorphies. But one needs to do a cladistic analysis to determine which characters are apomorphic, and to perform a cladistic analysis you need to have **operational taxonomic unit**s, and to have these you need a **taxonomy**; the MSC has been criticized as the product of circular reasoning. de Queiroz (2007) includes a useful figure (*ibid*, Fig. 1, p. 882) in which he makes the point that the various properties researchers have suggested as criteria for recognizing species may arise at different stages in the complex process of speciation. In this way he reconciles the various suggestions for the species definitions he lists in his Table 1 (*ibid*, p. 880). In practice most researchers use one, or other, version of the PySC. They search for the smallest cluster of individual organisms that is "diagnosable" on the basis of the preserved morphology. Because in the hominin fossil record most of that morphology is craniodental, most diagnoses of early hominin taxa inevitably emphasize craniodental morphology. A useful review of the species concept with particular reference to paleoanthropology is provided by Holliday (2003).

species concept(s)   *See* **species**.

species diversity   An ecological diversity measurement that takes into account richness and evenness of species. *See also* **species evenness**; **species richness**.

species evenness   An ecological diversity measurement that quantifies the evenness of the distribution of individuals across taxa.

species richness   An ecological diversity measurement that refers to the number of species in a given community.

specific mate-recognition system   *See* **recognition species concept**.

speciose   A taxonomic hypothesis that emphasizes a larger rather than a smaller number of species. A researcher who favors speciose taxonomies is called a **splitter** (syn. **taxic**).

SPECMAP   *See* **orbital tuning**.

specularite   (L. *specere*=mirror) This variety of **hematite** is black or gray and it has a metallic luster. It has been found as an exotic material at sites such as **Twin Rivers**, Zambia. *See also* **hematite**.

speech   (ME *speche*=speech)   *See* **spoken language**.

speleothem   (L. *speleon*=cave and *them*=deposit) The inclusive term for a range of rock types that are formed when water containing dissolved **calcium carbonate** reaches an air-filled cave. Within the cave, carbon dioxide de-gassing alters the water's ability to retain minerals in solution, and over periods of tens of thousands of years its solutes precipitate and form **flowstone**s, **stalactite**s, and **stalagmite**s; these are all forms of speleothem. Flowstones are formed when water flows on the floor or walls of a cave. They are generally horizontal, finely laminated layers of calcium carbonate that occur in sheets against walls or floor of the cave, or in thinner sheets between blocks of sediments. Stalactites are icicle-shaped mineral deposits that grow *down* from cave roofs, whereas stalagmites are conical mineral deposits fed from drip waters above that grow *up* from cave floors. *See also* **calcium carbonate**; **uranium-series dating**.

sphenomandibular ligament   (Gk *sphen*=a wedge and *oeides*=shape) Fibrous ligament derived from the first branchial (Meckel's) cartilage connecting the spine of the sphenoid to the lingula of the mandible. A line joining the inferior attachments of the two sphenomandibular ligaments is the axis around which depression/protrusion and elevation/retraction occur at the **temporomandibular joint**. *See also* **branchial arches**.

spheroid   *See* **Developed Oldowan**.

splicing The process whereby **introns** are removed from messenger RNA to form a mature **transcript** containing either an unbroken **open reading frame** for **translation** into a **polypeptide** chain or a complete RNA sequence (e.g., for a transfer RNA, or ribosomal RNA molecule).

splitter (OD *splitten*=to divide) A label for researchers who favor dividing the fossil record into a larger number of exclusive taxa (i.e., with more stringent membership criteria) rather than a smaller number of more inclusive (i.e., with more relaxed membership criteria) taxa. Some researchers who are trying to reconstruct aspects of the biology of extinct taxa such as life history have suggested that splitting taxonomies is to be preferred because lumping taxonomies may "jumble characteristics from very different subgroups, producing chimeric life histories" (Smith et al. 1994, p. 181). Others claim splitting makes the hominin fossil record unnecessarily confusing.

spoken language (ME *speche*=speech and *language* from OF *langage* from L. *lingua*=tongue or speech) Spoken language, or speech, is a ubiquitous behavior among living modern human populations. It involves the use of a learned repertoire of sound units assembled in a distinctive sequence in rapid succession to convey information and meaning in an energetically efficient way. Among the advantages of spoken language are that the individuals involved in its transmission and reception need not be in visual contact with one another, and spoken language is an exceptionally efficient way of communicating with more than one person at a time. The "ingredients" of normal spoken language are a flow of exhaled air and a structure, normally the vocal cords, that is capable of vibration. The vocal cords break the emerging air into a series of separate emitted "puffs," the fundamental frequency of which is determined by the length and the degree of tension of the vocal cords. Harmonic frequencies are added to the fundamental frequency as a result of changes in the shape and size ("gestural articulations") of the various elements that make up the **supralaryngeal vocal tract**. Modern human languages use around 100 acoustically distinctive sound units, or **phonemes**. Phonemes are broken up into the consonants and vowels that make up syllables. Vowels involve phonation, whereas consonants involve the blocking and subsequently releasing the air flow. From within to without, the flow of air can be obstructed at the glottis, the epiglottis, in the oropharynx, in the nasopharynx, in the nose, by the soft palate, by the tongue at various places along the hard palate, by the incisor teeth and by the lips (Ladefoged and Maddieson 1996). Subtle differences in the shape of the tongue can produce different formant frequencies and therefore different vowels. Compared with nonhuman primates modern human spoken language involves sound sequences that are an order of magnitude more rapid, and more than twice as long as those seen in nonhuman primates. The supralaryngeal vocal tract of modern humans is distinctive in that its two components – the horizontal one (SVT_H) and the vertical one (SVT_V) – are subequal (nearly equal) in length. In nonhuman primates and in other animals the SVT_H is usually longer than the SVT_V, and it is this 1:1 SVT_H/SVT_V ratio and the approximately 90° angle between the SVT_H and the SVT_V that makes it possible for modern humans to produce vowels such as [a], [i], and [u]. Researchers interested in language origins have been searching for a hard-tissue proxy for the shape and proportions of the supralaryngeal part of the vocal tract. They thought they had found it in the form of the **cranial base angle**, but Lieberman and McCarthy (1999) showed that flexion of the basicranium and laryngeal descent are not well enough related for the cranial base angle to be used a proxy for the height of SVT_V.

Springbok Flats *See* **Tuinplaas**.

Springbok Flats skeleton *See* **Tuinplaas 1**.

SPRP Acronym for the **Swartkrans Paleoanthropological Research Project** (*which see*).

Spy (Location 50°28′21.64″N, 4°40′50.52″E, Belgium; etym. named after the nearby village of Spy) History and general description The Grotte de Spy (also known as the Béche-aux-Roches cave) is a limestone cave about 15 km/9 miles east of Namur near the village of Spy. In 1885 Max Lohest and M. De Puydt discovered two hominin skeletons in the Grotte de Spy and they found **Late Pleistocene** fauna *in situ* low in the sequence of stratified terrace deposits in front of the cave entrance. The discoveries at Grotte de Spy were among the first attributed to *Homo neanderthalensis*, and even though they were subsequently transferred to *Homo spyensis* by Krause in 1909 their current taxonomic allocation is to *H. neanderthalensis*. Hominins found at site The discoveries include two adult presumed male partial skeletons (Spy 1 and 2), two teeth, and a tibia of a juvenile (Spy 3). Between 2004 and 2006 researchers systematically reviewed the dispersed fauna from the cave and in doing so they identified 24 new Neanderthal fragments, six of which comprise part of the mandible and some of the mandibular dentition of a Neanderthal infant (the Spy VI child). Temporal span and how dated? Radiocarbon dating suggests the hominins are *c*.36 ka. Key references: Geology, dating, and paleoenvironment Semal et al. 2009; Hominins Fraipont and Lohest 1886, Holloway 1985b, Semal et al. 2009, Crevecoeur et al. 2010.

squirrels *See* Rodentia; sciurids.

square-cube law *See* body size.

Sr/Ca ratios *See* strontium/calcium ratios.

SR-μCT Abbreviation of synchrotron radiation micro-computed tomography (*which see*).

SSH Acronym for single species hypothesis (*which see*).

SST Acronym for sea surface temperature (*which see*).

Staatliches Museum für Naturkunde (or National Museum of Natural History) This museum was founded in 1791 in Stuttgart, Germany, and has some of the most important scientific biological collections in Europe. Hominin fossil collections Reilingen calvaria, Steinheim 1. Contact information Professor Dr Johanna Eder (e-mail eder.smns@naturturkundemusum-bw.de). Address Nordbahnhofstraße 177, 70191 Stuttgart, Germany. Website www.naturkundemuseum-bw.de/stuttgart/. Tel +49 (0)711 8936 115.

stabilizing selection A type of natural selection in which alleles that produce a divergence from the average phenotype are eliminated. For example, babies that are too small or too large have a lower probability of survival, so any alleles linked with those phenotypes will be removed. Stabilizing selection results in the reduction of genetic variation in a population (syn. normalizing selection).

stable isotope biogeochemistry The science that involves determining the stable isotope composition of biological, atmospheric, and geologic samples using isotope ratio mass spectrometers, and then interpreting the ecological, chemical, or geological processes driving stable isotope ratios in different substances such as plant or animal tissues. Stable isotope biogeochemistry is a systems science, in that it aims to determine how nutrients are cycled at scales ranging from the global down to an individual organism.

stable isotopes An isotope is one of two or more atomic forms of the same element that have the same atomic number, but different atomic masses (i.e., different numbers of neutrons). Unlike the isotopes that undergo decay such as the ones used in isotopic dating, the proportions of stable isotopes do not change with the passage of time. Instead their proportions are determined by processes such as photosynthesis (e.g., $^{13}C/^{12}C$), climatic variables, or diet ($^{2}H/^{1}H$, $^{13}C/^{12}C$, $^{15}N/^{14}N$, $^{18}O/^{16}O$). Thus, the stable isotope ratios in collagen recovered from fossils can be used for diet reconstruction and to reconstruct paleoclimate. Stable isotope ratios are frequently referred to using delta (or δ) notation, where the isotope ratio in a sample is measured relative to an international standard. When there is more of one of the two isotopes the system is said to be enriched with respect to that particular isotope, and when there is less of one of the two isotopes the system is said to be depleted with respect to that particular isotope. Combining data from two different stable isotope pairs is frequently more meaningful than only using one isotope system (e.g., Fuller et al. 2006 use both the $^{13}C/^{12}C$ and the $^{15}N/^{14}N$ systems to investigate the timing of weaning). In regions where prehistoric modern human populations had access to both marine foods and C_4 foods such as maize, which confound the use of $\delta^{13}C$ values, bone collagen $\delta^{15}N$ values can help identify dependence on marine foods (Schoeninger et al. 1990). Clementz and Koch (2001) also describe how $\delta^{13}C$ and $\delta^{18}O$ values from tooth enamel can together discriminate among mammals feeding in purely terrestrial systems, animals feeding on a combination of terrestrial and near-shore (estuary) foods, and animals feeding on marine foods. Schwarcz and Schoeninger (1991), Schoeninger (1995), Koch (1998), McKechnie (2004), and Sponheimer et al. (2007) all provide useful reviews of the use of stable isotopes. *See also* **C₃ and C₄; laser ablation stable isotope analysis.**

stable population If the age-specific fertility and **mortality** do not change across time for a group living in a particular region then the living age structure (proportions of individuals at each age) for that group will remain unchanged. Similarly, the age-at-death structure will not change. In the special case where fertility and mortality are equal the population will remain a constant size. While it would be difficult to assume that past populations followed unchanging fertility and mortality schedules for any appreciable amount of time, it is usually assumed that for short periods of time such an assumption may be reasonable. Further, even if past populations may not have had fixed fertility and mortality, any fluctuations are likely to "average out" over time. *See also* **stationary population.**

stage The **chronostratigraphic unit** equivalent of an age; not to be confused with **Oxygen Isotope Stages.**

Stage 3 project A multidisciplinary initiative to place the spatial and temporal distribution patterns of European Middle and Upper Paleolithic human sites within the context of **Oxygen Isotope Stage 3** (OIS 3) climate and landscape simulations (60–24 ka). The two major goals of the program were to determine (a) what the climate of Europe was like during OIS 3 and the extent to which the changes observed in the Greenland ice cores influenced the European landscape and its flora and fauna, and (b) in what way and to what degree *Homo neanderthalensis* and anatomically modern human populations may have been influenced by OIS 3 climate and environmental history. The Stage 3 project was conceived in 1995 by Nicholas J. Shackleton and Tjeerd H. van Andel of the Godwin Institute for Quaternary Research at the University of Cambridge and launched in 1996 at the Godwin Conference on Oxygen Isotope Stage 3. It concluded in 2003 with the publication of a major integrative monograph, "Neanderthals and modern humans in the European landscape during the last glaciation" (van Andel, Tjeerd, and Davies, 2003, McDonald Institute for Archeological Research, Cambridge). At its peak, the project had 34 members including Quaternary geologists, paleoclimatologists, paleo-

ceanographers, archeologists, paleoanthropologists, and human physiologists. Innovative at the time, modeling of climate and vegetation at a resolution of 60 km²/23 square miles across Europe was done at the Earth Systems Research Center of Pennsylvania State University by Eric Barron and David Pollard. One major achievement of the Stage 3 project was the documentation of two instances of major environmental stress resulting in most part from high-frequency **Dansgaard–Oeschger cycles**, one in OIS 4 and the other around 30 ka BP, a time of major mammalian extinction, including the demise of the Neanderthals. The Stage 3 climate and plant-cover simulations and the faunal and archeological databases of published data through 2000 are freely available at https://www.esc.cam.ac.uk/research/research-groups/oistage3.

stalactite *See* **calcium carbonate; speleothem.**

stalagmite *See* **calcium carbonate; speleothem.**

stance phase *See* **walking cycle.**

Standardized African Site Enumeration System (or SASES) This system of naming sites was developed in the 1960s as a way of recording the location of African archeological sites. Its creator, the American archeologist Charles Nelson, adapted the method from Charles Borden's (1954) system of site classification for Canada, where it is still employed. Nelson's article was first published in a bulletin of nomenclature of the **Pan-African Congress on Prehistory and Quaternary Studies** in 1971 (Nelson 1971). Since this report is hard to find, the Society of Africanist Archaeologists reprinted it in *Nyame Akuma*, their research bulletin (Nelson 1993). The SASES divides the African continent into squares that measure 6° of longitude or latitude; these are identified by capital letters. Then each square is subdivided into units that correspond to 15′ of longitude or latitude; these are assigned lower-case letters. If there is more than one site in a square, the sites are assigned an Arabic numeral based on the order in which they were recorded. For example, the SASES designation of the **KBS** site at **Koobi Fora**, east of Lake Turkana in northern Kenya, is FxJj-1. The first two letters represent the latitude, the second two the longitude, and the Arabic numeral tells the user that this was the first site recorded at that location. Although the SASES was designed for the whole of the African continent, the system has only been regularly adopted in Kenya and in parts of southern Ethiopia (e.g., in the **Lower Omo Basin**), but starting in 1988 the SASES was adopted by the Department of Antiquities in Tanzania.

stapes (L. *stapes*=stirrup) *See* **auditory ossicles**; **branchial arches**; **middle ear**.

starch grains (ME *starche*=substance used to stiffen cloth, from *sterchen*=to stiffen, from OE *stercan*) One of several kinds of **plant microfossils** (the others are **phytoliths** and **pollen**), starch grains are the energy storage bodies of plants. If they can be identified precisely enough they have the potential to provide, depending on the context, information about the plants that were present in the environment or about the **diet** of individuals and groups. Starch grains consist of alternating amorphous and crystalline layers of two kinds of long-chain carbohydrate, amylose, and amylopectin. Their shapes and sizes are usually unique to plant higher taxa, and sometimes down to the level of genus and species. Within the same taxon the morphology of starch grains may differ among the plant organ (e.g., fruit, seed, leaf) that produced them. The morphologically most diagnostic forms of starch grains are found within the areas of the plant that are designed for long-term energy storage (e.g., seeds, fruits, and **underground storage organs**) and this is also where starch grains are most abundant. Starch grains are best preserved in protected archeological environments, and have been recovered from stone tools, ceramics, and dental calculus. Cooking and other types of food processing causes predictable damage to starch grains, and thus the presence of such damage can be used as evidence that food has been subject to cooking or to other types of food processing. Torrence and Barton (2006) review the archeological uses of starch grains and provide several examples of their utility.

Starosel'e (Location 44°44′N, 33°53′E, Ukraine; etym. named after the nearby village) History and general description This site in a box canyon near the village of Starosel'e, an eastern suburb of Bachchisaraj in the Crimea, was excavated from 1952 to 1956 by A. Formozov, who found a thick **Middle Paleolithic** deposit and the burial of a young child. The taxonomic status and primary context of this individual was highly debated, with some researchers suggesting it has a mixture of *Homo sapiens* and *Homo neanderthalensis* traits, while others suggesting it is a later intrusive *H. sapiens* burial. Excavations in 1993–5 by a American-Ukranian team recovered two other burials in the same level, and both are unquestionably modern, supporting the interpretation that all three skeletons are intrusive burials from the 18thC. Temporal span and how dated? AMS radiocarbon, uranium-series, and electron spin resonance spectroscopy dating have provided a range of dates for the site, with the oldest layers dating to *c*.80 ka, and the youngest to *c*.40 ka, although the excavators note that all of the layers show mixing and the exact dates are therefore questionable. Hominins found at site Three intrusive 18thC burials, and a few fragmentary remains. Archeological evidence found at site Four main cultural horizons have been reported in the new excavations, all of which contain a local variant of the **Mousterian**. Use-wear and residue analysis of the stone tools shows that the inhabitants were regularly processing plants and animals. A study of the fauna suggests that the inhabitants were engaging in both directed hunting and opportunistic scavenging. Key references: Geology, dating, and paleoenvironment Marks et al. 1997, Formozov 1958; Hominins Marks et al. 1997, Formozov 1958; Archeology Marks et al. 1997, Burke 2000, Hardy et al. 2001.

Starosele *See* **Starosel'e**.

stasis (Gk *stasis*=standstill) Randomly directed change is constantly occurring within living **species**, but the term stasis is used for a period of time (it could be the temporal span of a species) during which there is no evidence of any directional change (i.e., trend) in morphology. For example, there are very few temporal trends evident within the time span of *Paranthropus boisei sensu stricto* (Wood et al. 1994) or within the time span of *Australopithecus afarensis* (Lockwood et al. 2000). Thus, the lack of any significant trend between the first and last appearance dates of those taxa would be examples of stasis within the hominin fossil record. *See also* **punctuated equilibrium**.

Station Amont *See* **La Quina**.

Station Aval *See* **La Quina**.

stationary population A special case of a **stable population** in which the crude birth and crude death rates are equal. As the intrinsic growth rate (r) is equal to the **crude birth rate** (b) minus the **crude death rate** (d), a stationary population has a growth rate of zero. Consequently, a stationary population remains the same size through time. It is often assumed in paleodemographic analyses that the samples being analyzed came from stationary populations. If the growth rate is known for a population then the **life table** or hazards analysis can be adjusted for the known growth rate. On the other hand, if the population was not stationary, but one assumes that it was, this leads to estimation errors. Sattenspiel and Harpending (1983) proved that for nonstationary populations the inverse of the mean age at death was a better estimator of the crude birth rate than of the crude death rate.

stature (L. *statura* = height) A linear measure of standing height. Stature varies within and between populations and has been used in modern humans as an indicator of diet, environmental stress, and gene flow. Stature is also one of several ways of summarizing the overall size of an individual hominin and it has been used as a scaling factor. Standing height has increased over the course of hominin evolution. Initial estimates of stature in *Ardipithecus ramidus* place its stature between 1.17 and 1.24 m (Lovejoy et al. 2009d) and **A.L. 288-1**, the most complete *Australopithecus afarensis* skeleton recovered, has an estimated stature of 1.05 m (Schmid 1983). In contrast, estimations for **KNM-ER 15000** (the Nariokotome *Homo erectus* skeleton) suggest a maximum stature of 1.60 m for this individual (Ruff 2007); there is no evidence of modern human-like statures prior to *Homo erectus*. Among modern humans, mean stature varies among populations and to varying degrees between males and females within populations. Mean statures for modern human adult populations range between 1.3 m and 1.8 m (Eveleth and Tanner 1976). Stature is known to have increased over the 20thC so that average heights in the USA (1999–2002) are 1.62 m for women and 1.76 m for men (Ogden et al 2004). Increases in mean stature that have been documented among modern human populations are examples of a **secular trend**. For example, mean statures of adult males in populations in Oaxaca, Mexico, have increased on average by 3 cm during the latter half of the 20thC (Malina et al. 2004). When only skeletal elements are available, stature is most commonly estimated from linear dimensions of long bones, but it is important to note that all stature estimates (like all mathematical predictions based on regression analysis), especially those for fossil hominins for which there are no reference populations, are subject to varying amounts of error, and therefore may not be accurate. *See also* **stature estimation**.

stature estimation (L. *statura* = height) Stature is usually estimated by relating skeletal dimensions, usually the lengths of the long bones of the lower limb, to living stature. Early attempts to predict stature from a section of the skeleton (e.g., Dwight 1894) were limited by small sample sizes and limited population samples. Most standard contemporary methods of stature estimation regress the lengths of the femur, the tibia, or both bones against known statures to yield predictive equations for that reference sample. Bayesian methods also have their advocates and Hens et al. (2000) review the merits of various methods for estimating the stature of fossil hominins. Relatively few modern human reference samples exist for which both living stature and directly observed limb-bone lengths are available and obviously there are no directly relevant reference samples that can be used the estimate the stature of extinct hominin taxa. As in all estimations, it is statistically conservative to match the sample to be estimated with a reference sample whose range encompasses the expected range of the test sample. Estimating beyond the range of the sample from which the formulae were developed introduces unknown amounts of error, in part because a small error in the estimate of a regression slope becomes magnified the farther a predicted stature is from the mean of the reference population. Skeletal elements that contribute a greater amount to total stature (e.g., the femur versus the talus) will yield smaller estimation errors and therefore more accurate predictions of stature, but stature estimations from long bones assume that lengths of the skeletal elements are proportionally similar to total stature in the reference and estimated samples (Auerbach and Ruff 2010). This has far-reaching implications for stature estimations of extinct hominins, many of which had overall body and internal limb proportions that differ from those of modern humans. A family of stature-estimation methods, collectively known as anatomical techniques, created by Georges Fully in the 1950s (Fully 1956) and revised by Raxter and colleagues (Raxter et al. 2006) circumvents these limitations. In anatomical stature estimation methods the constituent skeletal elements that directly comprise stature [i.e., the length of the lower limb (including tarsal height), the length of the vertebral column, and the height of the cranium] are measured and added together. Corrections are needed because (a) some dimensions overlap (e.g., the medial malleolus and trochlea of the talus), (b) there are "gaps" (e.g., between the first sacral element and the femoral head), and (c) no account is taken of soft tissues. Fully's original correction was discrete, but the correction developed by Raxter et al. is continuous and based on regression analyses. An advantage of the anatomical technique for stature estimation is that it does not incorporate assumptions about body proportions and it is not dependent on specific reference samples for its calculation (assuming the correction factors employed are universal among populations). However, the technique requires nearly complete skeletons, and although the dimensions of some missing elements may be estimated, these increase the potential error of resulting stature estimates. Few early hominin associated skeletons are complete enough to justify the use of anatomical techniques but they have been used to estimate the stature of the **A.L. 288-1** *Australopithecus afarensis* associated skeleton (Schmid 1983). Stature estimation of juveniles has received much less attention (Ruff 2007).

STD Abbreviation of Saitune Dora and the prefix used for fossils recovered from that site. *See* **Saitune Dora.**

Steinheim (Location 48°58′N, 9°17′E, Germany; etym. named after nearby town of Steinheim an der Murr) History and general description The gravel deposits along the river Murr had been known to contain fossils for quite some time before the discovery of a hominin cranium in 1933. There are four principal layers of deposit, the Older Mammoth Gravels, the Straight-Tusked Elephant Gravels, the Main Mammoth Gravels, and the Younger Mammoth Gravels. The Steinheim cranium was found in the Straight-Tusked Elephant Gravels. Temporal span and how dated? **Biostratigraphy** suggests that the Straight-Tusked Elephant Gravels are from the penultimate interglacial (i.e., the European Riss-Würm, which is equivalent to **Oxygen Isotope Stage** 7). Hominins found at site Only one hominin specimen, **Steinheim 1**, is known from the site. The location of the cranium is known, for its finder, Karl Sigrist,

did not collect the fragments, instead he waited for F. Berckhemer to excavate them. The cranium is lightly built, and because of this it has been widely interpreted as belonging to a female. Archeological evidence found at site N/A. Key references: Geology, dating, and paleoenvironment Adam 1954a, 1954b, Howell 1960; Hominins Howell 1960; Archeology N/A.

Steinheim 1 Site Steinheim. Locality Sigrist Gravel Pit. Surface/*in situ In situ*. Date of discovery July 24, 1933. Finder Karl Sigrist Jr. Unit N/A. Horizon Straight-Tusked Elephant Gravels. Bed/member N/A. Formation N/A. Group N/A. Nearest overlying dated horizon N/A. Nearest underlying dated horizon N/A. Geological age *c.*225 ka. Developmental age Young adult. Presumed sex Female. Brief anatomical description Distorted cranium, missing much of the left temporal and maxilla. Announcement Berckhemer 1933. Initial description Berkhemer 1934. Photographs/line drawings and metrical data Berckhemer 1934, Weinert 1936. Detailed anatomical description No monograph exists, but Weinert (1936) and Howell (1960) provide some details. Initial taxonomic allocation *Homo steinheimensis*. Taxonomic revisions *Homo heidelbergensis* Campbell (1964). Current conventional taxonomic allocation *H. heidelbergensis*. Informal taxonomic category Pre-modern *Homo*. Significance Due to its combination of archaic and modern features and plastic deformation, the Steinheim cranium has proved to be somewhat difficult to classify, and it was one of the specimens in the informal category known as the **pre-Neanderthals**. Location of original **Staatliches Museum für Naturkunde**, Stuttgart, Germany.

stem group (OE *stemn* = supporting stalk) A term introduced by Jefferies (1979) for a **clade** (or **total group**) minus its **crown group**. For example, for hominins it would be the hominin clade minus the *Homo sapiens*/*Homo neanderthalensis* subclade.

stenotope (Gk *steno* = narrow and *typos* = place) Refers to a species adapted to a restricted or narrow range of environmental conditions, or to a narrow ecological **niche** (e.g., an ant-eater). *See also* **stenotopy** (syn. specialist).

stenotopic Refers to a species adapted to a restricted or narrow range of environmental conditions, or to a narrow ecological niche (e.g., an ant-eater). It is also used to refer to morphology interpreted as being part of a stenotopic adaptation (e.g., the tongue of an ant-eater). *See* **stenotopy**.

stenotopy (Gk *steno* = narrow and *topos* = place) The condition of being ecologically specialized. A stenotopic (or specialist) species is one that can use or consume only a narrow range of ecological resources, and that can live in only a limited set of habitats. The term was introduced to paleoanthropology by Elisabeth Vrba (e.g., Vrba 1980, 1985a, 1985b, 1988, 1992) in a series of hypotheses seeking to explain how environmental change influences evolutionary patterns in the fossil record. Several recent studies (Teaford and Ungar 2000, Wood and Strait 2004, Scott et al. 2005, Sponheimer et al. 2005) have inferred that most early hominin species were not ecological specialists, but rather were generalists, particularly with respect to diet. *See also* **effect hypothesis**; **eurytopy**; **habitat theory hypothesis**; **turnover-pulse hypothesis**.

stenotypic *See* **stenotopy**.

Sterkfontein (Location 25°58′08″S, 27°45′21″E, South Africa; etym. Afrikaans for a "strong spring" or "source of water" referring to the underground spring and lake in the cave complex) History and general description Sterkfontein is the umbrella term for a system of breccia-filled caves [e.g., Extension site, **Jakovec Cavern**, **Silberberg Grotto**, Tourist Cave(s), Type site, etc.] formed from solution cavities within pre-Cambrian dolomite in the Blauuwbank valley near Krugersdorp, Gauteng Province, South Africa. The caves at Sterkfontein had been explored for gold in the 1880s, but they were later mined for lime by Hans Paul Thomasset in, or shortly before, 1895, and it was in that year that fossil bones from Sterkfontein were sent to the British Museum. Shortly thereafter the lease on the mine at Sterkfontein was taken over by Guglielmo Martinaglia, and it was he who exposed what are now known as the Tourist Cave(s). In the 1920s the Glencairn Lime Company was mining lime from the caves, and one of its employees was George Barlow, who had worked for a time at the Taungs mine. In the wake of the Taungs discovery, fossils from Sterkfontein, including the skull of a large baboon, were sent to Raymond

Dart, but he thought the fossil baboon was so similar to extant baboons that the Sterkfontein cave was unlikely to contain a hominin as primitive as *Australopithecus africanus*. By the 1930s Sterkfontein was being exploited for bat guano, and in 1936 one of Dart's students, Trevor Jones, visited Cooper's hardware store in Krugersdorp and persuaded the owner to let him borrow some of the fossils from the display advertising the Sterkfontein guano. A week or so later, Jones visited Sterkfontein where George Barlow was selling fossils from a table. Jones reports that Barlow gave him 43 specimens, mostly baboon skulls and endocranial casts. Jones took them back to **Witwatersrand University Department of Anatomy**, where he showed them to his teachers, including Raymond Dart. Later another of Jones' teachers, Gerrit Schepers, and a fellow student, William Harding Le Riche, went once again to Sterkfontein and were given a baboon mandible by Barlow. They reported this to Robert **Broom**, who wasted no time in going to see Jones' collection in the Department of Anatomy. During Broom's first visit to Sterkfontein in August 1936 he asked George Barlow to tell him, Broom, if he ever saw any Taung-like fossils. A week later, Barlow showed Broom the first Sterkfontein hominin, **TM 1511**. Later that month Broom assigned it to *Australopithecus transvaalensis*, but 2 years later he transferred that species to a new genus, *Plesianthropus*. In the early 1960s the Sterkfontein site was donated to the University of the Witwatersrand and research at the site has been carried on by first Alun Hughes then by Ron **Clarke** who is still working at the site. The classical description of the geology of the infill at Sterkfontein is based on interpretations set out in Partridge (1978, 2000) and Partridge and Watt (1991). These recognize six Members (1–6) and they have been summarized in the reviews of Martini et al. (2003) and Reynolds et al. (2007). Pickering and Kramers (2010) present a different interpretation that focuses on sedimentary facies and which also attempts to relate the geology of Sterkfontein to events recorded in nearby caves. More details of the history of the site are given in Tobias (1973a). Hominins found at site Most hominins recovered from Sterkfontein have come from Member 4 and the majority of these have been assigned to *Au. africanus*. Some researchers think that the Member 4 hominins sample more than one taxon, but opinions are divided about which specimens are exotic, and what any second taxon should be called. Other taxa possibly represented in the Sterkfontein hominin fossil record are *Homo* sp. and ***Paranthropus robustus*** from Member 5A, and the remains of an archaic hominin *Australopithecus sp. indet.* from Member 4 breccia exposed in the main cave (e.g., **Sts 71, Stw 252, Stw 505**) and from Member 2 breccia exposed in the Silberberg Grotto (e.g., **Stw 573**) and in the Jakovec Cavern. Clark (2008) describes this latter hominin as a "*Paranthropus*-like species of *Australopithecus*" (*ibid*, p. 448), but he suggests that among the features that distinguish this species from *Paranthropus* are its "large canines and incisors" (*ibid*, p. 448). The hominins collected by Broom and John **Robinson** at the so-called "type site" were given Sts and TM catalogue designations and those collected by John Robinson at the "extension site" excavation in 1957–8 were given SE prefixes. Hominins recovered between 1966 and 1996 through the efforts of Alun Hughes, Ron Clarke, and Phillip **Tobias** were given a Stw catalogue designation. Archeological evidence found at site Archeological evidence includes stone artifacts (**Oldowan** from Member 5; **Acheulean** from the Extension site; **MSA** from the **Lincoln Cave**) and **digging sticks** fashioned from bone (Member 5). Temporal span and how dated? **Biostratigraphy, magnetostratigraphy,** and **cosmogenic nuclide dating** have provided approximate dates for the main fossil-bearing strata: Member 2 is *c*.4 Ma (Partridge et al. 2003), but others claim its age is *c*.2.8–2.6 Ma based on uranium-lead dating of the deposits (Pickering and Kramers 2010); Member 3 is not yet excavated or dated; Member 4 was interpreted to be between 2.6 and 3.0 Ma but Pickering and Kramers (2010) suggest it lies between 2.6 and 2.0 Ma. Partridge has divided Member 4 into four beds, A–D. Lockwood and Tobias (1999) report that most *Au. africanus* fossils come from Beds B–D, with Bed 4B the most prolific of the beds; Member 5 is *c*.1.4–1.9 Ma. Key references: Geology, dating, and paleoenvironment Tobias and Hughes 1969, Partridge 1978, 2000, 2005, Partridge and Watt 1991, Schwarz et al. 1994, Martini et al. 2003, Partridge et al. 2003, Herries et al. 2009, Pickering and Kramers 2010; Hominins Hughes and Tobias 1977, Clarke 1998, 1999, 2002, 2008, Lockwood 1999, Lockwood and Tobias 1999, 2002, Moggi-Cecchi et al. 2006, Partridge et al. 2003,

Toussaint et al. 2003; Archeology Kuman and Clarke 2000, Kuman and Field 2009.

Sterkfontein Formation (etym. Afrikaans for a "strong spring" or "source of water" referring to the underground spring and lake in the cave complex where the type section is located) General description The Sterkfontein Formation was established by Partridge (1978). He divided the breccias in the **Sterkfontein** cave into six members, the oldest and deepest being Member 1. The breccia that contains most of the ***Australopithecus africanus*** fossils, and which was referred to as the Lower Breccia by Robinson (1962), was termed Member 4, Robinson's stone tool-bearing Middle Breccia became Member 5, and his Upper Breccia became Member 6. Members 1, 2, and 3 (Members 2 and 3 are fossiliferous) have been identified in the **Silberberg Grotto**, which is connected to the Tourist Cave. This classical description of the geology of the infill at Sterkfontein is summarized in the reviews of Martini et al. (2003) and Reynolds et al. (2007). Pickering and Kramers (2010) present a different interpretation that focuses on sedimentary facies, on uranium-lead dating of the flowstones and which also attempts to relate the geology of Sterkfontein to events recorded in nearby caves. Sites where the formation is exposed Sterkfontein. Geological age Perhaps more than 4 Ma to maybe less than 1 Ma, but the ages suggested by Pickering and Kramers (2010) make it unlikely that any of the presently explored fossil-bearing breccias are more than 3 Ma. Most significant hominin specimens recovered from the formation Most of the hypodigm of *Au. africanus* from Member 4, and important individual fossils such as **Stw 53, Stw 573**, and the collection of fossils from the **Jakovec Cavern**; Key references: Geology and dating Robinson 1962, Tobias and Hughes 1969, Partridge 1978, 2000, 2005, Partridge and Watt 1991, Clarke 1994, 1998, 2006, 2008, Schwarz et al. 1994, Kuman and Clarke 2000, Martini et al. 2003, Partridge et al. 2003, Herries et al. 2009, 2010, Pickering and Kramers 2010; Hominins Broom 1936, Hughes and Tobias 1977; Archeology Kuman 1994.

Sterkfontein Research Unit In 1999 the responsibility for research at **Sterkfontein** was transferred from the **Palaeo-Anthropology Research Unit** in the **Witwatersrand University Department of Anatomy** to a new entity, the

Sterkfontein Research Unit, with Phillip Sambungmacan as the Director and Ron Clarke the Deputy Director.

Stetten   *See* **Vögelherd**.

stimulus enhancement   A type of **enhancement** effect that is associated with a suite of cognitively primitive attention and motivation mechanisms that affect **learning**. Specifically, stimulus enhancement describes learning in situations where the presence of a conspecific directs an observer's attention to a particular object (such as a fruit or a tool) that the conspecific is interacting with (Spence 1937). This type of enhancement is seen in mammals and birds alike, and may result in behavior-matching, which resembles **imitation** learning.

stone cache hypothesis   A hypothesis that suggests **Oldowan** sites were processing areas where artifacts and **manuport**s were deposited at various points in the foraging range of hominins. These caches were either established consciously or they developed as an unconscious byproduct of hominin discard behavior, and then they became secondary sources of raw material. According to this view, sites served not as home bases, but as recurrently visited stockpiles of stone (Potts 1988, 1991).

STR   Acronym for short tandem repeat. *See* **short tandem repeat**; **microsatellite**.

strain   Strain, or ϵ, is a measure of the relative change in length that an object undergoes in response to an applied load, and it is technically defined as the difference between the loaded and the unloaded (i.e., the original) length relative to the original length of a structure ($\epsilon = \Delta L / L$). Because strain is expressed as a change in length divided by the original length, it is a dimensionless quantity. Any change in length is usually very small, so strain is normally expressed as microstrain ($\mu\epsilon$), indicating a change in length on the order of a thousandth of the original length. Experimental evidence has shown that during normal activities (e.g., running, biting hard objects, etc.) most mammalian bones experience a peak of about 2000–3000 $\mu\epsilon$ (which indicates a 0.2–0.3% change in length). Mammalian bones typically break at around 6000 $\mu\epsilon$, indicating that most bones are overbuilt relative to their peak normal strains; thus they have a substantial safety factor of

between 2 and 3 (Rubin and Lanyon 1984). Some amount of dynamic strain is required for normal bone growth, but excessive strains can result in bone modeling and **bone remodeling**.

strain gage   A device consisting of a very fine metal grid, or foil, or semiconductor material, which measures **strain**. Lord Kelvin reported in 1856 that metallic conductors change their electrical resistance when subjected to mechanical strain (e.g., when they are stretched). Strain gages take advantage of this property by calibrating the changes in resistance against known strains, allowing users to measure very small strains (changes in length) that occur when loads are applied to stiff materials like bone or metal. The most common strain gages consist of a small piece of metal foil supported by a flexible backing. These simple gages measure the strain that occurs within the plane it occupies and in the longitudinal direction of the piece of metal foil; a single-element gage thus has just a single foil and it can only measure strain in that plane. Studies of bone strain, in which the gage is attached directly to the bone surface, typically use rosette strain gages. These consist of three elements at 45° to each other and this configuration makes it possible to calculate the strain in all directions within the plane of the gage, and to determine the directions in which the maximum and minimum principal strain lie. Multiple gages around the circumference of a long bone can be used to calculate (using certain assumptions) the strain distribution throughout the bone's cross-section. Strain gages in **force plates** are also used to measure strains, and these strain measurements are used to calculate forces. Strain studies are important for validating models based on **finite element analysis**. William (Bill) Hylander, a pioneer of *in vivo* strain studies, found that when macaques and baboons chew strain in the brow ridge was much lower than in the zygomatic arch or in the mandible (Hylander et al. 1991). These results provided compelling evidence that refutes the hypothesis that brow ridge development in higher primates, including fossil hominins, is structurally related to countering stresses resulting from biting and chewing (*see* **chewing cycle**).

strain gauge   *See* **strain gage**.

Stránská Skála   (Location 49°11′N, 16°40′E, Czech Republic; etym. named after a small hill outside of Brno) History and general description

This open-air site, roughly 7 km/4 miles east of **Brno-Bohunice**, also preserved artifacts of the **Bohunician** transitional lithic industry. It was excavated in the 1980s and 1990s. Temporal span and how dated? **Radiocarbon dating** provided a range of 41–34.5 ka for the Bohunian contexts and 33–30 ka for the **Aurignacian** contexts. Hominins found at site None. Archeological evidence found at site Stone tools documenting the early **Upper Paleolithic**. Key references: Geology, dating, and paleoenvironment Svoboda 2003, 2006; Archeology Svoboda 2003, 2006.

strata (Gk *stratos* = to cover or spread) More than one layer of **sedimentary rock**.

stratigraphic criterion A controversial criterion for deciding which state of a **character** is **primitive** and which is **derived**. When employing the stratigraphic criterion, one examines the distribution of the states of the character in the fossil record. The state that occurs in the oldest deposit is the one that is deemed to be primitive. *See also* **cladistic analysis**.

stratigraphy (Gk *stratos* = to cover or spread and *graphos* = to draw or write) The study of sedimentary rocks, especially their formation, composition, sequence, and correlation. It also refers to the work of identifying and tracing layers of **sedimentary rock**. The "stratigraphy" of a site is the formal description of the sequence of strata found at that site.

stratum (Gk *stratos* = to cover or spread) A single layer of **sedimentary rock**.

strength In material science the strength of a material (e.g., bone, enamel) is its ability to withstand an applied stress without failure. There are two common types of failure: yielding and fracture. Yielding, which is defined by the material's yield strength, occurs when the material undergoes macroscopic permanent deformation. A material with an applied stress value less than the yield strength of the material will not be permanently deformed upon unloading. Fracture is defined by the ultimate strength of the material, which is the largest absolute value of applied stress the material can withstand without fully fracturing. Since yielding occurs at an applied stress less than the applied stress that produces full fracture, the yield strength of a material is always less than its ultimate strength. The type of loading (e.g., tensile, compressive, or shear) also influences the likelihood of failure, and therefore there are corresponding types of material strength (e.g., ultimate tensile strength, ultimate compressive strength, and ultimate shear strength).

stress Stress is a measure of the intensity of the internal forces within a loaded object that is assumed to behave as a continuum. Stress, which is expressed in units of force per unit area, correlates with the failure of many materials. Stress is specific to the hypothetical planes upon which it acts. If the stress acts perpendicular to a plane, it is called a normal stress component; a negative normal stress implies a compressive stress and a positive normal stress implies the normal stress is tensile. Stress that acts tangential to a plane is called a shear stress component. At a given material point there are up to six nonzero components of stress (i.e., three normal stresses corresponding to directions of the three orthogonal coordinate axes and three shear stresses corresponding to shear existing in three orthogonal planes of the coordinate axes). It turns out that at any point in the material there always exists an orientation of three orthogonal planes such that they only have normal stresses acting upon them. These three normal stresses are called principal stresses and provide the largest absolute value of normal stress that can exist on any plane at that point. Similarly, there is another orientation of the three orthogonal planes that maximizes shear stress, resulting in the maximum value of shear stress that can exist on any given plane at that point. Shear stresses tend to distort the material without changing the volume of the material, whereas normal stresses tend to elongate or compress the material in the directions in which they act. If the three principal stresses at a material point all have the same negative (positive) value, then the material at this point is in a state of pure hydrostatic compression (tension). In such a state shear does not exist on any plane at this point and therefore there is no distortion of the material. For ductile materials distortion is found to correlate closely to yielding and fracture, and von Mises stress (also called equivalent stress) measures this material distortion. Unlike the other measures of stress, von Mises stress is not associated with a particular orientation. Instead, it is a scalar function of the stress components and is directly proportional to the square root of the energy of distortion at a given material point. The location of maximum distortion energy (i.e., maximum von Mises stress) is deemed most likely to yield or fracture. Since no distortion exists at a material point of pure hydrostatic compression or tension, the value of von Mises stress at such a point is zero.

strewn-field (OE *streowian* = to scatter) The area of the Earth's surface covered by a particular tektite shower. *See also* **tektites**.

striae of Retzius (L. *stria* = a furrow-like phenomenon described by Anders Retzius in 1837) **Long-period incremental lines** in **enamel** described by Anders Retzius of Lund in 1836–7 (see reviews in Boyde 1990 and Risnes 1990) that represent the position of the developing enamel front at successive points in time. They are manifest on the lateral surface of teeth as **perikymata** (Preiswerk 1895) or **imbrication lines** (Pickerill 1913). Although the cause of striae of Retzius is not known, they result from a regular slowing of enamel matrix secretion that keeps an intrinsic consistent temporal rhythm (see reviews in Risnes 1998, Dean 2000, Smith 2006b). One hypothesis is that they result from two **circadian** rhythms, one 24 hour and the other *c*.27 hour, that interact to give a "beat" every 7 or 8 days, but this hypothesis is untested and some suggest it is oversimplistic. The temporal repeat interval of striae of Retzius, known as their **periodicity**, is assessed by counting the number of daily **cross-striations** between consecutive striae. A long-period incremental marking analogous to striae of Retzius exists in **dentine** (called **Andresen lines**). Previous studies have examined aspects of striae of Retzius curvature, length, and the angle of intersection between the striae of Retzius and the **enamel-dentine junction** in an attempt (e.g., Beynon and Wood 1986, Grine and Martin 1988; Dean and Shellis 1998) to discrimate among primate taxa. In reflected light striae look blue but in transmitted light they look brown (hence the original description of "brown striae" by Anders Retzius) (syn. brown striae, Retzius' lines). *See also* **enamel development**.

striae parietalis (L. *striae* = scores or grooves and *parietalis* = wall) Descriptive term introduced by Phillip **Tobias** to describe fine ridges and grooves on the external, or ectocranial, surface of the parietal bones that run postero-superiorly away from the border with the squamous part of the temporal bone. This distinctive morphology is probably related to the unusual degree of overlap at the temporoparietal suture and among hominins it is apparently a derived trait confined to *Paranthropus boisei* and *Paranthropus aethiopicus*.

striate cortex (L. *striatus* = furrowed, striped, ridged and *cortex* = husk, shell) The striate cortex, also known as the primary visual cortex, V1, and **Brodmann's area 17**, is located at the posterior, or caudal, pole of the **occipital lobe** of the **cerebral hemisphere** of the brain. Most of it is on the medial aspect of the occipital lobe and much of it is buried within the walls of the **calcarine sulcus**. The name striate derives from the prominent "striped" appearance of the stria of Gennari in layer IVB (which is made up of myelinated axons) of primates and because of this distinctive myelination pattern the striate cortex is readily seen in sections of the brain, even with the naked eye. The striate cortex in each hemisphere receives inputs from the ipsilateral (i.e., the same side) **lateral geniculate nucleus** and contains a complete retinotopic map of the contralateral (i.e., the opposite side) visual field, with anthropoid primates having a magnified representation of the central portion of the visual field. Neurons in the striate cortex are tuned to relatively simple features of color and spatial frequency and information from the striate cortex is transmitted via two main pathways: the ventral and the dorsal visual streams. The ventral visual stream encompasses parts of the **temporal lobe** and contains areas that are specialized for the recognition of form and object identification. The dorsal visual stream encompasses parts of the **parietal lobe** and contains areas that are specialized for the processing of motion information, the representation of object locations, and the coordination of visually guided movements of the eyes and arms. The volume of the striate cortex in modern humans is considerably smaller than that predicted by the observed **allometry** for a generalized primate of the same brain size. This relative reduction of primary visual cortex in modern humans has been argued to reflect a reorganization of cortical area allocation, with a greater proportion of neocortical tissue devoted to the adjacent **posterior parietal cortex**. In addition, reorganization of the striate cortex in modern humans has also been demonstrated at the histological level, with a unique distribution of apical dendrite bundles and interneurons in layer IVA. Because layer IVA of the striate cortex receives inputs from the motion-sensitive magnocellular (i.e., large-celled) layers of the **lateral geniculate nucleus**, this suggests that there have been modifications of the modern human brain's ability to process motion information, possibly in relation to tracking the movement of lips during the perception of **spoken language**.

striatum (aka corpus striatum) Refers to the entity made up of the caudate nucleus and the putamen, which are some of the largest subcortical (i.e., beneath the **cerebral cortex**) components of the **forebrain**. The caudate nucleus and putamen are anatomically distinct in primates (they are separated by the internal capsule) but they are fused in other mammalian taxa. The name striatum derives from the striated appearance of these **basal ganglia** nuclei in sections of unstained brain tissue because of the axon fiber bundles that traverse it. The striatum receives inputs from the cerebral cortex and sends outputs to other basal ganglia nuclei. In addition, the striatum receives substantial dopaminergic innervation from the substantia nigra pars compacta. Because of its central role in motor and cognitive circuits of the basal ganglia, the striatum has been implicated in modern human **language** function. Notably, the striatum shows volumetric reduction and under-activation in patients with loss-of-function mutations to the *FOXP2* gene. In addition, the medium spiny neurons of the striatum in mice with a "humanized" version of the *FOXP2* gene display a greater extent of dendritic branching.

stride *See* **walking cycle.**

strike Refers, in structural geology, to a bearing 90° from the direction of **dip**. In areas where the nature of the rock outcrop is determined by the intersection of the dip with the topography, strike can be a useful descriptor of the direction of the grain of the landscape and will define a direction in which strata of equivalent age may be encountered. For example, when surveying in the field, researchers will often try to control their fossil collections temporally by following along the strike of dipping beds.

striking platform A morphologically distinctive part of a stone artifact. It is the place on the

core that is struck during the act of **percussion** (freehand or otherwise) to remove a **flake.**

Stringer, Christopher Brian (1947–) Chris Stringer, who was born in East Ham, London, England, became interested in human evolution as a boy visiting what was then called the British Museum (Natural History) in London. He entered University College London where he was exposed to the research of physical anthropologist Don Brothwell and anatomist Michael **Day.** Stringer completed his BSc in anthropology in 1969 and after working briefly part-time at the British Museum (Natural History) he began his graduate studies at the University of Bristol where he worked with Jonathan Musgrave, who studied *Homo neanderthalensis.* At a time when Loring Brace was advocating a Neanderthal ancestry for modern humans and Henri **Vallois** was supporting the **pre-sapiens theory** (which held that *H. neanderthalensis* was not the ancestor of modern humans) Stringer examined Neanderthal crania from collections across Europe and used **multivariate analysis** to judge the degree of similarity or difference between Neanderthals and early modern humans. The results of his studies convinced Stringer that the anatomical differences separating Neanderthals and *Homo sapiens* indicated that Neanderthals had *not* evolved into modern humans. Therefore, Stringer concluded, the Neanderthals must have been replaced by populations of early *H. sapiens* that had migrated into Europe. While working on his graduate research Stringer had an opportunity to work with Michael Day on two partial skulls found at **Omo-Kibish,** Ethiopia, in 1967. Stringer completed his PhD at the University of Bristol in 1974 and by this time (in 1973) he had been appointed as research fellow at the British Museum (Natural History). In 1977 Stringer was appointed senior scientific officer at the museum, and he was appointed principal scientific officer in 1987. During this period Stringer continued to study Neanderthal specimens from Europe and the Middle East and to work on the problem of the origin of modern humans. He became a supporter of a version of the **out-of-Africa hypothesis** of modern human origins, variants of which were separately formulated by the German anthropologist Günter **Bräuer,** and by the South African archeologist Peter Beaumont. During the late 1970s Bräuer had argued that anatomically modern humans had originally evolved in Africa, and then had migrated to populate Europe (where there was modest interbreeding with Neanderthals). Stringer opposed the **multiregional hypothesis** formulated and expanded by Milford **Wolpoff** in the early 1980s and 1990s. Alan Thorne and Wolpoff argued that *Homo erectus* migrated from Africa to populate Asia and parts of Europe, and that it was these regional populations of *H. erectus* that had evolved into *Homo sapiens*. Stringer could see that the results of the **mitochondrial DNA** studies of Allan **Wilson** and his colleagues were consistent with the out-of-Africa hypothesis, and together with his colleague Peter Andrews he published an influential article in *Science* in 1988 entitled "Genetic and fossil evidence for the origin of modern humans. "It summarized the molecular and fossil evidence that supported a recent migration of *H. sapiens* out of Africa that then replaced existing populations of hominins throughout Europe and Asia. During the 1980s Stringer collaborated with a number of colleagues in studies, some of which involved establishing new dates for specimens from the Near East that were consistent with the hypothesis that Neanderthals had been replaced by populations of anatomically modern humans. Stringer summarized his views on Neanderthals in *In Search of the Neanderthals* (1993), written with Clive Gamble of the University of London. In 1996 Stringer, in collaboration with Robin McKie, published *African Exodus* where they present the fossil, archeological, and genetic evidence that supports the out-of-Africa hypothesis. Stringer has served as head of the Human Origins Programme at the Natural History Museum, London, first between 1990 and 1993, and later from 1999 to the present (as of 2010). Since 2001 Stringer has also been the director of the **Ancient Human Occupation of Britain Project** (or AHOB project). This Leverhulme Trust-funded project brought together archeologists, paleontologists, and geologists from various institutions to investigate the early occupation of the British Isles and the environmental conditions at the time the first humans arrived in Britain. The discovery of the so-called "Boxgrove Man" (considered to be a specimen of *Homo heidelbergensis*) in West Sussex in 1993 had been the earliest evidence of hominins in Britain, but researchers with the Ancient Human Occupation of Britain Project have since re-examined evidence of a hominin settlement found at Westbury-sub-Mendip in Somerset in the 1980s, and they found evidence of worked flints *in situ* at **Pakefield** in East Anglia. Pakefield is likely to be at least from **Oxygen**

Isotope Stage 17 (>680 ka), thus providing, along with evidence from **Happisburgh**, the earliest definite evidence so far of human occupation north of the Alps (Parfitt et al. 2005, 2010).

stromatolite   A laminated sedimentary structure formed by the activity of a bacterial and blue-green algal community. The community mediates the precipitation of calcite and this calcite, together with additional trapped sedimentary particles, results in a spheroidal or columnar structure that can reach up to a meter in diameter. Stromatolites are rare today, and presumably require highly specific environments that both foster their growth and exclude the grazing and burrowing organisms that would otherwise destroy them. The **Koobi Fora Formation** has extensive beds of stromatolites in its early Pleistocene lake margin facies.

strontium/calcium ratios   This system of **stable isotopes** (abbreviated Sr/Ca) is used for **diet reconstruction**. **Herbivore**s discriminate against dietary strontium, so their bones and tooth enamel have lower Sr/Ca ratios than the plants they consume. The **carnivore**s that consume those herbivores also discriminate against dietary strontium, so their bones and tooth enamel have lower Sr/Ca ratios than the herbivores they consume. But leaves of broad-leaved plants have less Sr than grass, so grazing herbivores have a higher Sr/Ca ration than browsing herbivores. Dietary items such as insects and **underground storage organs** are particularly rich in strontium, but grazers and insectivores are also rich in barium (Ba), so relatively high Ba/Ca ratios should discriminate between high Sr/Ca ratios due to grazing or insectivory from high Sr/Ca ratios due to the consumption of, say, underground storage organs. Enamel **apatite** is the most suitable material for Sr/Ca analysis. **Bone** apatite can be used, but the chemical changes that occur during the process of fossilization (i.e., effects associated with **diagenesis**) are more of a problem with respect to bone than they are with enamel. The method can be applied throughout hominin fossil record. Sponheimer et al. (2005) provides more details. *See also* **stable isotope biogeochemistry**.

structural paradigm   *See* **structure–function relationship**.

structure   *See* **population structure**.

structure–function relationship The relationship between the structure of an object or body part and its function; function is usually inferred from an analysis of structure. It is relatively easy to infer function from structure when a trait seen in an extinct taxon is also seen in the same context in an extant taxon in which structure–function relationships can be observed directly. Inferring function is much more difficult when the trait is found only in one or more extinct taxa. Rudwick (1964) suggested the trait should be compared with structures that are well suited to their function. For example, a pestle and mortar is well suited to fracturing relatively small pieces of brittle material, so Rudwick (*ibid*) would consider a pestle and mortar to be a "structural paradigm" for that particular function. The fossil trait (e.g., the shape of postcanine tooth crowns) is then compared to the various relevant structural paradigms, and its function is taken to be the same as the structural paradigm it most closely resembles.

Sts Prefix used for fossil hominins recovered from the **Sterkfontein** cave site, Blaauwbank Valley, Gauteng, South Africa between 1947 and 1949.

Sts 5 Site Sterkfontein. Locality Type site. Surface/*in situ In situ*. Date of discovery April 18, 1947. Finders Robert **Broom** and John **Robinson**. Horizon N/A. Bed N/A. Member Member 4. Formation Sterkfontein Formation. Nearest overlying dated horizon N/A. Nearest underlying dated horizon N/A. Geological age Previously 3.0–2.6 Ma, but see a new U-Pb age of *c*.2.0 Ma (Pickering and Kramers 2010). Developmental age Subadult; the third molar roots are still open (Thackeray et al. 2002a). Presumed sex As its informal name "Mrs Ples" implies, Robert Broom considered Sts 5 to be a female. However, as part of a comprehensive analysis of *Australopithecus africanus* crania Lockwood (1999) judged the sex of Sts 5 to be "indeterminate" and Thackeray et al. (2002a) and Gommery and Thackeray (2006) suggest there are reasons to think that it might be a subadult male and not a female. Brief anatomical description A well-preserved subadult cranium that lacks all the tooth crowns, some of the alveolar process, and some detailed cranial base morphology. A wafer-thin layer of the surface bone of the cranial vault is preserved on five blocks of breccia (Sts 5 i–v)

removed at the time of discovery (Potze and Thackeray 2010), and after 25 cycles of consolidation and careful etching with acetic acid it is possible to see on blocks Sts 5 i and iii temporal lines that are approximately 10 mm from bregma. Endocranial volume 485 cm^3 (Conroy et al. 1998). Announcement Broom 1947. Initial description, photographs/line drawings, and metrical data Broom, Robinson and Schepers 1950. Detailed anatomical description None. Initial taxonomic allocation *Plesianthropus transvaalensis*. Taxonomic revisions Post WWII there was a growing consensus that a separate genus and species for Sts 5 and its ilk was unjustified and *Plesianthropus transvaalensis* was formally incorporated into *Australopithecus africanus* by Robinson (1954). Current conventional taxonomic allocation *Au. africanus*. Significance Although not the type specimen of *Au. africanus* (which is **TM 1511**), Sts 5 has come to represent the modal cranial morphology of *Au. africanus*. This is unfortunate because it is one of the more **prognathic**, if not the most prognathic, specimen in the *Au. africanus* hypodigm. It was found 0.4 m above and 1.9 m northeast of **Sts 14** and Thackeray et al. (2002b) and Thackeray and Klopper (2004) have raised the possibility that Sts 5 and Sts 14 may belong to the same individual. Location of original **Ditsong National Museum of Natural History** (formerly the **Transvaal Museum**), Northern Flagship Institution, Pretoria, South Africa.

Sts 14 Site Sterkfontein. Locality Type site. Surface/*in situ In situ*. Date of discovery August 1, 1947. Finders Robert **Broom** and John **Robinson**. Horizon N/A. Bed N/A. Member Member 4. Formation **Sterkfontein Formation**. Nearest overlying dated horizon N/A. Nearest underlying dated horizon N/A. Geological age Previously 3.0–2.6 Ma, but U-Pb dating suggests an age of *c*.2.6–2.0 Ma (Pickering and Kramers 2010). Developmental age Subadult. Presumed sex Gommery and Thackeray (2006) conclude that the morphology of the anterior inferior iliac spine suggests that both Sts 14 and **Stw 431** are males. Brief anatomical description Seven thoracic, six lumbar, and two sacral vertebrae, plus fragments of other vertebrae, rib fragments, both pelvic bones, and parts of the left femur and tibia. The pelvic bones and the femur have suffered significant plastic deformation. Announcement Broom and Robinson 1950. Initial description, photographs, and metrical data

Broom, Robinson and Schepers 1950. Detailed anatomical description None. Initial taxonomic allocation *Plesianthropus transvaalensis*. Taxonomic revisions Post WWII there was a growing consensus that a separate genus and species for Sts 14 and its ilk was unjustified and *Pl. transvaalensis* was sunk into *Australopithecus africanus* (e.g., Robinson 1954, Le Gros Clark 1955). Current conventional taxonomic allocation *Au. africanus*. Significance For a long time this was the most complete early hominin associated skeleton and although substantially affected by plastic deformation it was the only source of information about the femoro-pelvic morphology of *Au. africanus*. Lovejoy (1979) reconstructed the Sts 14 pelvis to resemble the **A.L. 288-1** pelvis, but Schmid (1983) preferred a reconstruction that was narrower with a more posteriorly tilted sacrum. The most recent attempt at a reconstruction by Berge and Goularas (2010) uses virtual reconstruction techniques, and they interpret the shape of the pelvis as consistent with a modern-human type of rotational birth mechanism; this runs counter to the conclusions of Tague and Lovejoy (1986). This fossil was found 0.4 m below and 1.9 m southwest of **Sts 5** and Thackeray et al. (2002b) and Thackeray and Klopper (2004) have raised the possibility that Sts 14 and Sts 5 may belong to the same individual. Location of original **Ditsong National Museum of Natural History** (formerly the **Transvaal Museum**), Northern Flagship Institution, Pretoria, South Africa.

Sts 19 Site **Sterkfontein**. Locality Type site. Surface/*in situ In situ*. Date of discovery October 1947. Finders Robert **Broom** and John **Robinson**. Unit N/A. Horizon N/A. Bed/member Member 4. Formation **Sterkfontein Formation**. Nearest overlying dated horizon N/A. Nearest underlying dated horizon N/A. Geological age Previously 3.0–2.6 Ma, but U-Pb dating suggests an age of *c*.2.6–2.0 Ma (Pickering and Kramers 2010). Developmental age Adult. Presumed sex Unknown. Brief anatomical description Most of the basicranium of an individual that some researchers link with Sts 58. Announcement Broom et al. 1950. Initial description Broom et al. 1950. Photographs/line drawings and metrical data It is illustrated as "Skull No. 8" in Broom et al. (1950, pp. 27–33), Dean and Wood (1982), and Kimbel et al. (2004). Detailed anatomical description N/A. Initial taxonomic allocation *Plesianthropus tranvaalensis*. Taxonomic revisions *Australopithecus africanus*

(Robinson 1954), *Homo* (Kimbel and Rak 1993, Kimbel et al. 2004). Current conventional taxonomic allocation *Au. africanus*. Informal taxonomic category Archaic hominin. Significance In their initial description of this cranium Broom et al. (1950) noted that "the whole glenoid region is essentially similar to that of man," and "the whole occipital fossa is almost human" (*ibid*, p. 30), and that "the manner in which the cerebellum has come to be shifted forward and below the cerebral occiput is most striking" such that the "general arrangement is the same as that found in the human brain, especially in *Homo sapiens*" (*ibid*, p. 102). Others have also noted the ways in which the cranial base of Sts 19 is more *Homo*-like than the cranial bases of other *Au. africanus* crania from Sterkfontein and Makapansgat (Clarke 1977, Kimbel and White 1988b, Kimbel et al. 2004), but although Dean and Wood (1982) refer to it as "the most *Homo*-like of the (*Au. africanus*) sample" they concluded that the morphology of Sts 19 is consistent with intra-specific variation in *Au. africanus* (*ibid*, p. 171), and Spoor (1993) concluded that both the bony labyrinth and the endocranial aspect of the basicranium give no grounds for the exclusion of Sts 19 from *Au. africanus* as represented by the adult crania **Sts 5** and MLD 37/38 (*ibid*, p. 129). Location of original **Ditsong National Museum of Natural History** (formerly the **Transvaal Museum**), Northern Flagship Institution, Pretoria, South Africa.

Sts 71 Site **Sterkfontein**. Locality Type site. Surface/*in situ In situ*. Date of discovery November 1947. Finders Robert **Broom** and John **Robinson**. Unit N/A. Horizon N/A. Bed/member Member 4. Formation **Sterkfontein Formation**. Nearest overlying dated horizon N/A. Nearest underlying dated horizon N/A. Geological age Previously 3.0–2.6 Ma, but U-Pb dating suggests an age of *c*.2.6–2.0 Ma (Pickering and Kramers 2010). Developmental age Adult. Presumed sex Lockwood (1999) suggests that it is a female. Brief anatomical description A damaged calvaria that preserves much of the right side of the face. Announcement Broom et al. 1950. Initial description Broom et al. 1950. Photographs/line drawings and metrical data Lockwood 1999. Detailed anatomical description N/A. Initial taxonomic allocation *Plesianthropus tranvaalensis*. Taxonomic revisions *Australopithecus africanus* (Robinson 1954). Current conventional taxonomic allocation *Au. africanus*. Informal taxonomic category Archaic hominin. Significance The relatively prognathic Sts 5 is often illustrated in reviews of *Au. africanus*, but one should remember that other specimens that make up the cranial hypodigm of *A. africanus*, such as Sts 71, have faces that are more orthognathic. Location of original **Ditsong National Museum of Natural History** (formerly the **Transvaal Museum**), Northern Flagship Institution, Pretoria, South Africa.

Stw 53 Site Sterkfontein. Locality From south of the Extension Site/West Pit, and 3 m south of the locality in Sterkfontein where in 1956–8 Bob **Brain** and John **Robinson** recovered stone tools (but see Bed/member below). Surface/ *in situ In situ*. Date of discovery Between June and August, 1976. Finders Alun Hughes and Phillip **Tobias**. Unit N/A. Horizon N/A. Bed/member The specimen was initially linked with Member 5, Unit A by Partridge, but Clark (1994) claims that it derives from Member 4. Kuman and Clarke (2000) suggest it is late Member 4/early Member 5 breccia and Clarke (2008) suggests that it came from a hanging remnant of the Member 4 breccia largely on the observation that the decalcified breccia that derives from the erosion "swallow hole" from which the cranium was extracted is devoid of artifacts. However, the cranium itself has been observed to exhibit stone-tool cutmarks on the right zygomatic bone (Pickering et al. 2000), rendering this rationale for stratigraphic attribution less than secure. Formation **Sterkfontein Formation**. Nearest overlying dated horizon N/A. Nearest underlying dated horizon N/A. Geological age 2.0–1.5 Ma, but Pickering et al. (2000) suggest it may be as old as 2.4 Ma and U–Pb dating suggests an age of *c*.2.6–2.0 Ma (Pickering and Kramers 2010). Developmental age Adult. Presumed sex Unknown. Brief anatomical description Most of the remains belonging to this skull belong to the cranium. These include the right frontal and parts of the parietals of both sides, the midline of the occipital, important parts of the basicranium including the basioccipital, temporal and sphenoid, much of the face, much of the palate, and the right postcanine teeth. The mandibular remains are restricted to a fragment of the right ramus that includes the mandibular notch and the coronoid process. Announcement and initial description Hughes and Tobias 1977. Photographs/line drawings and metrical data

Hughes and Tobias 1977, Curnoe and Tobias 2006. Detailed anatomical description Curnoe and Tobias 2006, but see Clarke (2008) for a critique of their reconstruction. Initial taxonomic allocation *Homo* aff. *habilis*. Taxonomic revisions *Homo habilis* (Tobias 1991b), *Homo sp. nov.* Grine et al. (1996), *Australopithecus africanus* (Kuman and Clarke 2000), holotype of *Homo gautengensis* (Curnoe 2010). Current conventional taxonomic allocation *H. habilis*. Informal taxonomic category Pre-modern *Homo*. Significance If Stw 53 is part of the southern African hypodigm of *H. habilis* as Curnoe and Tobias (2006) suggest (but see the conclusions of Grine et al. 1996), then it would have substantial biogeographic significance, and it would encourage those who claim to see dental evidence of that taxon at Sterkfontein, and at other sites such as **Drimolen**. Curnoe and Tobias (2006) make the point that the initial attempt to restore Stw 53 by Hughes and Tobias (1977) has been followed by two reconstructions, one by Clarke (1985b) and one by Curnoe and Tobias (2006). The latter claim that the two reconstructions differ in important details (*ibid*, pp. 43–4), including a larger (but unreported) endocranial volume in the latter reconstruction. Clarke (2008) interprets the history of reconstructing Stw 53 very differently and is adamant that the Curnoe and Tobias (2006) reconstruction is not valid and "does not reflect the original form of the cranium" (*ibid*, p. 447); he takes the view that Stw 53 is an unexceptional example of a male *Au. africanus*. The cutmarks reported on the right maxilla are interpreted by Pickering et al. (2000) as being "the earliest unambiguous evidence that hominids disarticulated the remains of one another" (*ibid*, p. 579). Location of original School of Anatomical Sciences, University of the Witwatersrand, Johannesburg, South Africa.

Stw 151 Site **Sterkfontein**. Locality All but one of the more than 40 fragments were found in a single solution pocket in square Aa/49 (the exception was found in X/49) between 1 and 2.2 m below datum. Surface/ *in situ In situ*. Date of discovery April/May 1983. Finders Alun Hughes and team. Unit/horizon N/A. Bed/member "late Member 4 breccia deposit" (Moggi-Cecchi et al. 1998, p. 426). Formation **Sterkfontein Formation**. Nearest overlying dated horizon N/A. Nearest underlying dated horizon N/A. Geological age Previously 3.0–2.6 Ma, but U–Pb dating suggests an age of *c*.2.6–2.0 Ma (Pickering and Kramers 2010). Developmental age

Juvenile c.5.2–5.3 year. Presumed sex Unknown. Brief anatomical description Initially the fragments were given separate specimen numbers (Stw 151–178), but Jacopo Moggi-Cecchi suggested that all but three of the specimens (Stw 162, left scapula fragment, Stw 164, fragment of an adult cranial vault, and Stw 169, a worn left I^2) likely came from the same immature individual that is now referred to as Stw 151. The cranium is represented by parts of the temporals, the occipital, and the palate; the only evidence of the mandible is part of the right side of the body and the right ramus. However, most of the dentition is preserved including the upper right deciduous molars, all of the lower deciduous molars and the crowns of at least one of each of the permanent teeth. The left stapes is also preserved. Announcement The specimen is mentioned in Smith (1986). Initial description Moggi-Cecchi et al. 1998. Photographs/line drawings and metrical data Moggi-Cecchi et al. 1998, Spoor 1993, Moggi-Cecchi and Collard 2002. Detailed anatomical description Moggi-Cecchi et al. 1998, Moggi-Cecchi and Collard 2002. Initial taxonomic allocation *Australopithecus africanus*. Taxonomic revisions Moggi-Cecchi et al. (1998) point to the many ways, both metrical and non-metrical, Stw 151 resembles *Au. africanus*, but they also suggest that its relatively narrow upper and lower permanent molar crowns and the entoglenoid process are *Homo*-like and in relation to the bony labyrinth Spoor (1993) suggested that "it is feasible that Stw 158 [Stw 151] is affiliated with a more derived form of hominid than *A. africanus*" (*ibid*, p. 131). Curnoe (2010) lists it as one of the paratypes of *Homo gautengensis*. Current conventional taxonomic allocation *Au. africanus*. Informal taxonomic category Archaic hominin. Significance Stw 151 is "the most complete set of jaws and teeth of an early hominid child since the Taung child" (Moggi-Cecchi et al. 1998, p. 425), and the same authors suggest that Stw 151 "may represent a hominid more derived towards an early *Homo* condition than the rest of the *A. africanus* sample from Member 4" (*ibid*, p. 425). Thus, Stw 151 is one of the specimens from Sterkfontein that some researchers thank may represent a second, more *Homo*-like, taxon. Location of original School of Anatomical Sciences, University of the Witwatersrand, Johannesburg, South Africa.

Stw 252 Site **Sterkfontein**. Locality N/A. Surface/*in situ* *In situ*. Date of discovery June 1984. Finders Alun Hughes and Phillip **Tobias**. Unit/horizon N/A. Bed/member Decalcified "talus cone" of Member 4. Formation **Sterkfontein Formation**. Nearest overlying dated horizon N/A. Nearest underlying dated horizon N/A. Geological age Previously 3.0–2.6 Ma, but U-Pb dating suggests an age of c.2.6–2.0 Ma (Pickering and Kramers 2010). Developmental age Adult. Presumed sex Lockwood (1999) suggests that it is a female. Brief anatomical description Clarke (1988) has made a reconstruction of this cranium that preserves the upper dentition, the anterior part of the palate, part of the face and frontal bone, the left parietal, and part of the occipital. Announcement Broom et al. 1950. Initial description, photographs/line drawings, metrical data, and detailed anatomical description Clarke 1988. Initial taxonomic allocation *Australopithecus africanus*. Taxonomic revisions Clarke (1988) has suggested that Stw 252 (plus Sts 36 and 71) represents a second, more postcanine megadont, taxon from Sterkfontein, Member 4. Current conventional taxonomic allocation *Au. africanus*. Informal taxonomic category Archaic hominin. Significance Clarke (1988, 2008) has suggested that Stw 252 is part of the male sample of a second *Paranthropus*-like *Australopithecus* taxon in Member 4 at Sterkfontein. Location of original School of Anatomical Sciences, University of the Witwatersrand, Johannesburg, South Africa.

Stw 431 Site **Sterkfontein**. Locality Grid Squares 45–6. Surface/*in situ* *In situ*. Date of discovery February 9 to March 23, 1987. Finders Alun Hughes, S. Sekowe, M. Makgothokgo, and J. Kibii. Unit Bed 4B. Horizon "Several meters below the Members 4/5 junction" and "well above the junction between Members 3 and 4" (Lockwood and Tobias 1999, p. 638–9). Bed/member Member 4. Formation **Sterkfontein Formation**. Nearest overlying dated horizon N/A. Nearest underlying dated horizon N/A. Geological age Previously 3.0–2.6 Ma, but U-Pb dating suggests an age of c.2.6–2.0 Ma (Pickering and Kramers 2010). Developmental age Adult. Presumed sex Male. Brief anatomical description Stw 431 comprises 48 postcranial fragments of an associated skeleton that includes the right humerus, ulna, and radius that articulate, the vertebral column from mid-thoracic to the sacrum, and the right sacro-iliac joint. Several isolated teeth, a vault fragment, a left first metacarpal, and a possible right metatarsal may also belong to the same individual. Announcement, initial description, photographs/line drawings, and metrical data Toussaint et al. 2003. Detailed anatomical description Toussaint et al. 2003, Kibii and Clarke 2003. Initial taxonomic allocation *Australopithecus africanus*. Taxonomic revisions None specific to this specimen. Current conventional taxonomic allocation *Au. africanus*. Informal taxonomic category Archaic hominin. Significance This is the third associated skeleton from this part of the Sterkfontein cave and it formed part of the evidence that McHenry and Berger (1998) put forward to support their contention that there are significant differences in the limb proportions of *Au. africanus* and *Australopithecus afarensis*. Location of original School of Anatomical Sciences, the University of the Witwatersrand, Johannesburg, South Africa.

Stw 505 Site **Sterkfontein**. Locality Grid Square H/42. Surface/*in situ In situ*. Date of discovery 1989. Finders Alun Hughes and Phillip **Tobias**. Unit Bed 4B. Horizon "Several meters below the Members 4/5 junction" and "well above the junction between Members 3 and 4" (Lockwood and Tobias 1999, p. 638–9). Bed/member Member 4. Formation **Sterkfontein Formation**. Nearest overlying dated horizon N/A. Nearest underlying dated horizon N/A. Geological age Previously 3.0–2.6 Ma, but U-Pb dating suggests an age of *c*.2.6–2.0 Ma (Pickering and Kramers 2010). Developmental age Adult. Presumed sex Lockwood (1999) suggests that it is a male. Brief anatomical description Stw 505a comprises most of the face and the left half of the cranial vault, the left anterior cranial fossa, and part of the right middle cranial fossa; Stw 505b comprises most of the petrous and tympanic parts of the right temporal Announcement Clarke 1988. Initial description, photographs/line drawings, and metrical data Lockwood and Tobias 1999. Detailed anatomical description Lockwood and Tobias 1999. Initial taxonomic allocation *Australopithecus africanus*. Taxonomic revisions Clarke (1988, 1994) has suggested that Stw 505 is part of the hypodigm of a second hominin taxon in Member 4 at Sterkfontein. Current conventional taxonomic allocation *Au. africanus*. Informal taxonomic category Archaic hominin. Significance Clarke (1988, 1994, 2008) has suggested that Stw 505 (plus Sts 1, 36, 71, and Stw 183, 252, 384, 498) is part of a sample of a second

Paranthropus-like *Australopithecus* taxon in Member 4 at Sterkfontein, but Lockwood and Tobias' (1999) detailed analysis suggest that Stw 505 can be accommodated within a range of variation for *Au. africanus* that is reasonable for a single taxon. Location of original School of Anatomical Sciences, University of the Witwatersrand, Johannesburg, South Africa.

Stw 573 Site **Sterkfontein**. Locality N/A. Surface/*in situ* The first and second tranche of bones belonging to this associated skeleton were recovered from Dumps 18 and 20, but the bulk of the skeleton of Stw 573, the *in situ* part, was discovered embedded in **flowstone** at the western end of the **Silberberg Grotto**. Date of discovery The initial recognition of four bones from the left foot of an early hominin took place in 1994. More bones of the left foot were recognized in May 1997, along with the distal ends of what were assumed to be the fibula and tibia of the left leg, and a right distal tibia. In June 1997 it became clear that there was a good chance that the rest of the skeleton was still *in situ* in the Silberberg Grotto. A few rib fragments were exposed between 1997 and 2002, but in 2002 Ron **Clarke** and his colleagues found a substantial parts of the pelvis and evidence of the vertebral column (Clarke 2002), and since then all of the skeleton of a single hominin individual (apart from some foot bones) has been located (Clarke 2008). Finders Clarke recognized the initial four bones as hominin, and it was also Clarke who recognized the additional foot and lower leg bones in boxes of bones marked "D18 (i.e., Dump 18)" and "D20 Cercopithecoids." However, the first signs of what proved to be the bulk of the skeleton of Stw 573, the *in situ* part, was discovered by Stephan Mostsumi and Nkwane Molefe when Ron Clarke gave them the distal ends of the left tibia and fibula and asked them to search the walls of the Grotto to look for the exposed, matching, proximal surfaces of the broken leg bones. Unit Found embedded in flowstone on a talus slope at the western end of the Silberberg Grotto. Horizon N/A. Bed/member Member 2. Formation **Sterkfontein Formation**. Group N/A. Nearest overlying dated horizon N/A. Nearest underlying dated horizon N/A. Geological age Previously estimated to be *c*.2.6–3 Ma, but recent U-Pb ages suggest an age for the associated flowstone of *c*.2.2 Ma (Walker et al. 2006, Pickering and Kramers 2010). Developmental age Adult. Presumed sex Unknown.

Brief anatomical description Clarke has reported the discovery of the left lower limb (Clarke and Tobias 1995, Clarke 1998), the skull (Clarke 1998), the left arm and hand (Clarke 1999), the pelvis and part of the axial skeleton (Clarke 2002), and all but some of the foot bones of a hominin (Clarke 2008). Announcement and initial description Clarke and **Tobias** 1995. Photographs/line drawings and metrical data Some photographs, but no metrical data, in Clarke and Tobias (1995) and in Clarke (1998, 1999, 2002, 2008). Detailed anatomical description N/A. Initial taxonomic allocation *Australopithecus* sp. Taxonomic revisions N/A. Current conventional taxonomic allocation *Australopithecus* sp. Informal taxonomic category Archaic hominin. Significance Potentially the most complete early hominin skeleton, and certainly the best-preserved archaic hominin from southern Africa. Clarke (2008) claims that Stw 573 is part of the hypodigm (along with **Sts 1**, 36, **71**, **Stw 183**, **252**, 384, 498, **505**) of a second *Paranthropus*-like *Australopithecus* taxon in Member 4 at Sterkfontein. Location of original School of Anatomical Sciences, the University of the Witwatersrand, Johannesburg, South Africa.

StW, Stw/H, StW/H Various abbreviations of "Sterkfontein Witwatersrand" and "Sterkfontein Witwatersrand Hominin" that are used as prefixes for fossil hominins recovered after 1949 from the **Sterkfontein** cave site, Blaauwbank Valley, Gauteng, South Africa.

stylopod (L. *stylus* = stem and Gk *pod* = foot) The proximal of the three compartments of a tetrapod limb. The proximal compartment (i.e., the upper arm or the thigh) is called the stylopod, the intermediate one (i.e., the forearm or the lower leg) is called the **zeugopod**, and the distal one (i.e., the hand or the foot) is called the **autopod**.

Subalyuk [Location 48°00′N, 20°30′E, Hungary; etym. Hu. Suba-lyuk = Suba's (a local landowner) cave) History and general description This cave near Cserépfalu village, in the Bükk Mountains, was excavated in 1932. Temporal span and how dated? Biostratigraphy suggests **Oxygen Isotope Stage** 5a–c. Hominins found at site Subalyuk 1 consists of a partial adult mandible that includes the symphysis, part of the left ramus, the dentition from the right P_3 to the left canine, the

left P_4 to the M_3. There are also isolated right molars, as well as postcranial fragments. Subalyuk 2 consists of a partial cranium of a child, including the right di^1–dm^2, left dc–dm^1, isolated left dm^2, and the crowns of four adult incisors and both first permanent molars. Archeological evidence found at site Late **Mousterian**. Key references: Geology, dating, and paleoenvironment Kadic 1940, Vértes 1964; Hominins Thoma 1963, Vlček 1970, Smith 1982; Archeology Bartucz et al. 1940, Mester 1990.

subarachnoid space (L. *sub* = below) The space between the arachnoid mater (the middle of the three meningeal layers) and the underlying pia mater (the innermost of the three meningeal layers) that is the most intimate covering of the surface of the brain, brain stem, and spinal cord. In adult modern humans the subarachnoid space contains approximately 150 cm^3 of **cerebrospinal fluid**. The fluid-filled intracranial part of the subarachnoid space makes a major contribution to the difference between the volume of the brain and the brain stem and the volume of the endocranial cavity. *See also* **cerebrospinal fluid**; **meninges**.

subchron (Gk *kronos* = time) This term is an abbreviation of the term "polarity subchronozone," which refers to a relatively short period of consistent geomagnetic polarity. Subchrons are subdivisions of **chron**s, and they (like chrons) are numbered (NB: the magnetic direction of a subchron is identified by a lower-case 'n' for normal and 'r' for reversed; thus the first reversed subchron in the normal chron 3An is 3A.1r). Numbered subchrons have replaced the named events (e.g., Jaramillo, Olduvai, etc.) that were used in the older geomagnetic polarity time scales. *See also* **geomagnetic dating**; **geomagnetic polarity time scale**.

subdivision *See* **population structure**.

subdural space (L. *sub* = below) The space between the dura mater (the outermost of the three meningeal layers) and the underlying arachnoid mater (the middle of the three meningeal layers). It contains the thin-walled cerebral veins that drain venous blood from the brain tissue to the endocranial venous sinuses. It makes a minor contribution to the difference between the volume of the brain and the brain stem and the volume of the **endocranial cavity**. *See also* **dural venous sinuses; meninges.**

subjective synonym   *See* synonym.

subnasal prognathism   *See* prognathism.

subspheroid   *See* Developed Oldowan.

substitution   (L. *substituere* = in place of) Used in genetics to refer to one allele replacing another or in molecular evolution to refer to a single nucleotide or base change in DNA. *See also* **mutation; single nucleotide polymorphism.**

suid   Informal term for taxa within the family Suidae (syn. pig, swine). *See* **Suidae.**

Suidae   (L. *sus* = swine) The formal name for the mammalian family that includes the pigs. Pigs are omnivorous **artiodactyls**, with simple non-ruminating stomachs, that first appear in the fossil record in the Oligocene of Asia, but successive adaptive radiations are important elements of later African faunas. In the main they are large-bodied, although the smallest highly endangered modern species *Sus* (*Porcula*) *salvania* can be as small as 10 kg. In addition to the ubiquitous domestic pig, *Sus scrofa*, the suid family contains many wild species, including the three indigenous extant African wild hogs, *Phacochoerus aethiopicus*, the warthog, *Hylochoerus meinertzageni*, the giant forest hog, and *Potamochoerus porcus*, the bush pig. While paleoanthropologists are interested in all members of the ecological communities in which hominins evolved, there at least two reasons why the Suidae have a special bearing on human evolution. First, the Suidae underwent an **adaptive radiation** during the **Pliocene** and **Pleistocene** of Africa. This radiation is characterized by relatively rapid speciation, resulting in the evolution of taxa with a larger body size than the ancestral forms and with taller tooth crowns (i.e., hypsodonty). The speed at which this radiation occurred has made the Suidae a useful biochronological tool for dating hominin sites (e.g., **Koobi Fora, Sterkfontein**, etc.). Second, because they are found alongside hominins at fossil sites it is assumed that they are ecologically similar to hominins (i.e., large-bodied, terrestrial, omnivores). Thus, their evolutionary history provides a useful comparative heuristic device for examining our own evolution.

sulcus   (L. *sulcus* = furrow)   A deep furrow within an organ. Most named sulci are in the parts of the central nervous system, such as the gray matter of the **cerebral cortex**, that are thrown into folds. In the central nervous system the crest of a folds is called a **gyrus** and the deepest part of a fold is a sulcus. *See also* **calcarine sulcus; central sulcus; lunate sulcus.**

sulcus supratoralis   *See* supratoral sulcus.

Sunda   (etym. In Hindu lore, variants of the word Sunda were used as names by ancient creatures. "Sunda," "ni-Sunda," and "Upa-Sunda" were the names of giants, or "Daityas," mentioned in the Upanishads. They fought with the gods because of their jealousy with their human half-brothers). Blavatsky suggests that Daitya was the name of an island-continent said to have been inhabited by giants such as Sunda during the time of Atlantis. The term "Great Asiatic Bank" had been introduced by Earl(e) in 1845 to refer to the "south-eastern extreme" of the Asiatic landmass that extends "to within 50 miles of Celebes." In their 1919 summary of the Dutch Siboga Expedition, Molengraaff and Weber suggest that Great Asiatic Bank be replaced by the term "Sunda Shelf." The term Sunda is now used to refer to both the Sunda Shelf (or Sundaland) and to the Sunda Islands. Sundaland is the name given to the part of the continental landmass that includes mainland Southeast Asia and the Greater Sunda Islands. The Sunda Islands are divided into the Greater (Borneo, Sumatra, Java) and the Lesser (Lombok, Sumbawa, Flores, Sumba, Timor) Sunda Islands; the former are part of the continental landmass, but the latter are not. During particularly cold periods when substantial volumes of water were locked up in ice accumulations at high latitudes, sea levels would have fallen low enough for much of Sundaland to have been one continuous landmass. However, the water channels between the Greater and Lesser Sunda Islands, and between Timor and the smaller Lesser Sunda Islands (i.e., Flores, Sumba, etc.), are deep, as is the channel between Timor and **Sahul** (i.e., Greater Australasia). *See also* **Sahul.**

Sunda Islands   *See* Sunda.

Sunda Shelf   *See* Sunda.

Sundaland   *See* Sunda.

Sunghir *See* Sungir.

Sungir (Location 56°08′N, 40°24′E, near the city of Vladimir, Russia; etym. unknown) History and general description Excavations of this very large, open-air Upper Paleolithic site were first conducted by Otto Bader in the 1950s and have been continued by the Russian Archaeological Society. The site is notable for its elaborate multiple burials, with grave goods including thousands of beads, jewelry, and pendants. Temporal span and how dated? Radiocarbon dates suggest a range of 20–28 ka. Hominins found at site Five individuals, including at two adolescents buried head to head, were recovered. Archeological evidence found at site Upper Paleolithic artifacts attributed to the "Eastern **Gravettian**," house structures, hearths, storage pits, and tool-production evidence. Key references: Geology, dating, and paleoenvironment Kuzmin et al 2004; Hominins Kuzmin et al 2004, Alexeeva et al. 2000; Archeology Alexeeva et al. 2000.

superciliary arch (L. *cilium* = eyelid and *supercilium* = eyebrow. The term "ciliary" refers to anything – i.e., lids, muscles, nerves, vessels – that pertains to the eye) The superciliary arch is a bilateral unitary bony arch that runs laterally and superiorly from the medial end of the superior orbital margin in some pre-modern *Homo* crania, and in most *Homo sapiens* crania. In such specimens the supraorbital region takes the form of a composite structure, made up of the superciliary arch medially and the **supraorbital trigon** laterally.

superciliary eminence *See* **superciliary arch**.

superior nuchal line A roughened line on the occipital bone that is produced by the attachment of the fascia that covers one of the biggest **nuchal** muscles, the semispinalis capitis. If the muscle and the fascia are well developed the line may take the form of a crest of bone called the **nuchal crest**.

supernumerary tooth (L. *super* = above and *numerus* = number) When the number of teeth exceeds the typical, or modal, number for that tooth class in each quadrant of a jaw (e.g., for Catarrhines, including hominoids, incisors = 2, canine = 1, deciduous molars = 2, premolars = 2, molars = 3). Supernumerary teeth may resemble neighboring teeth, or show less morphological complexity (crown and/or roots) than the other teeth in the respective tooth class. Mouse mutants have demonstrated that supernumerary teeth are associated with genetic disturbance of developmental pathways, especially those initiated by SHH (Sonic hedgehog), FGFs (fibroblast growth factors), and BMPs (bone morphogenic proteins) (reviewed in D'Souza and Klein 2007). Alternate hypotheses exist for how these mutations ultimately lead to the development of supernumerary teeth. Supernumerary teeth may be associated with the splitting of normal **tooth germ**s during development (reviewed in Schwartz 1984, Miles and Grigson 1990) or they may be the product of continued proliferation of cells in the dental lamina (Ackermann et al. 2006, D'Souza and Klein 2007). There are many documented cases of fourth molars among hominoid primates (e.g., Schwartz 1984, Ackermann and Bishop 2009), and in mammals supernumerary teeth are found in higher frequency in hybrids (Goodwin 1998, Ackermann et al. 2006), and may relate to the breakdown of normal genetic regulation in these individuals (Kangas et al. 2004).

superposition, law of (L. *super* = above) Principle, or law, that within a sequence of strata the strata at the top are younger than those at the bottom.

supination (L. *supinum* = lying on its back) In the forelimb, supination refers to the action involved in returning the forearm from the pronated position (i.e., with the palm of the hand facing backwards or downwards) to the supine position (i.e., with the palm of the hand facing forwards or upwards). In the hindlimb, supination refers to the action of moving the foot from the out-turned, pronated, position (i.e., with the sole of the foot facing outwards), to one in which the sole is maximally in-turned (i.e., with the sole of the foot facing inwards). *See also* **anatomical position**; **foot movements**.

suprainiac fossa (L. *supra* = above, and *inion* = which is a midline bony landmark that represents the superior attachment of the ligamentum nuchae) A **trait** consisting of a small depression, or fossa, above the inion. It has been proposed as an **autapomorphy**

of *Homo neanderthalensis* (Hublin 1978), but others have suggested that its expression is not restricted to that taxon and that it is also seen in fossil crania belonging to *Homo sapiens* (e.g., Trinkaus 2004). Balzeau and Rougier (2010) investigated the detailed morphology of the trait as it is expressed in Neanderthal crania (e.g., **Gibraltar 1**, Spy 1, etc.), and they compared it to the suprainiac fossae in *H. sapiens* crania (e.g., **Afalou Bou-Rhummel** and **Taforalt**). These authors showed that what appears superficially to be the same phenotypic trait shows detailed differences in the two taxa. In *H. neanderthalensis* the depression is due to loss of the **diploë** between the inner and outer tables of the cranial vault, whereas in *H. sapiens* the depression is due to a thinning of the outer table. This latter phenomenon has been referred to as a "**supranuchal fossa**" (Sládek 2000, Caspari 2005).

supralaryngeal vocal tract (L. *supra* = above) The part of the vocal tract above the vocal folds. The supralaryngeal part of the vocal tract (or SVT) comprises all of the structures between the vocal folds and the lips [i.e., the "aditus" (the part of the laryngeal cavity between the vocal folds and the opening into the pharynx), the oropharynx, and the mouth, made up of the oral cavity proper and the vestibule]. From within to without, it is divided into a vertical (or SVT$_V$) and a horizontal (or SVT$_H$) component. In modern humans these components are at right angles to each other and they are subequal (nearly equal) in length. This geometry is one of the reasons modern humans can make such a wide range of sound units, or **phoneme**s, especially the vowels [a], [i], and [u]. *See also* **spoken language**.

supramarginal gyrus (Gk *gyros* = ring, circle) The supramarginal gyrus of the cerebral cortex corresponds to **Brodmann's area 40**, which is part of the inferior parietal cortex. The supramarginal gyrus curves around the posterior end of the **lateral sulcus**. The cortex of the supramarginal gyrus is involved in spatial orientation, the phonological and articulatory processing of words, the memory of meaningful gestures, and the sequence of actions of the upper limb. Because of these associations, the supramarginal gyrus, like the **angular gyrus**, has been implicated in the neural basis of tool production and tool use.

supramastoid crest (L. *supra* = above, Gk *mastos* = breast, *oeides* = shape, and L. *crista* = crest)

If the posterior-most fibers of the temporalis muscle are especially well developed, then the edge of that muscle attachment in the region of the mastoid may be raised up to form a supramastoid crest. The supramastoid crest and the **mastoid crest** are usually well developed in *Paranthropus*, and are usually separated by a **supramastoid sulcus**.

supramastoid sulcus *See* **mastoid crest**; **mastoid process**.

supranuchal fossa (L. *supra* = above and ME *nucha* = nape, or back, of the neck) The term supranuchal fossa has been used to describe a depression in the midline of the occipital bone in *Homo sapiens* crania that is due to a thinning of the outer table (Sládek 2000, Caspari 2005). *See also* **suprainiac fossa**.

supraorbital bar (L. *supra* = above and *orbis* = wheel or hoop) The name given to a strut of bone that extends across the frontal bone, above the orbits and the nose, in taxa that manifest substantial postorbital constriction, or postorbital narrowing (e.g., chimpanzees, *Australopithecus afarensis*). The postorbital constriction means that the supraorbital bar is effectively separated from the rest of the cranial vault. A supraorbital bar has a midline, or glabellar, component and two lateral components. In megadont archaic hominin crania such as **OH 5** the glabellar component projects ahead of the lateral components.

supraorbital height index The index introduced by Wilfrid **Le Gros Clark** (1950) to describe the relationship between the cranial vault and the upper part of the face. The index, which is taken away from the sagittal plane, refers to the ratio between the height of the calotte above the upper margin of the orbit (FB) and the total height of the **calvaria** above the Frankfurt Horizontal (AB), multiplied by 100. The higher the cranial vault the higher the ratio (e.g., in modern humans the average is about 70 and in *Pan* it is about 50).

supraorbital region (L. *supra* = above and *orbis* = wheel or hoop) The supraorbital region refers to the region of the cranial vault that is situated above the superior orbital margins.

supraorbital torus (L. *supra* = above, *orbis* = wheel or hoop, and *torus* = bulge) Name given to a relatively straight and continuous morphological feature, essentially a bar of bone, that extends across the frontal bone above the orbits and the nose. It has a midline, or glabellar, component and two lateral components. The central portion encloses the frontal sinus. In taxa that manifest substantial postorbital constriction, or postorbital narrowing (e.g., chimpanzees, *Australopithecus afarensis*) such that the supraorbital region is effectively separated from the rest of the cranial vault, the analogous structure is referred to as a **supraorbital bar**.

supraorbital trigon (L. *supra* = above, *orbis-* = wheel or hoop, and *trigonum* = triangle) Name given to the lateral of the two components of the compound structure that forms the supraorbital margin in most *Homo sapiens* crania.

supratoral sulcus In *Homo erectus* crania there is often a hollowed area in the midline between lambda and inion, and in his description of the **Zhoukoudian** crania Franz **Weidenreich** (1943, p. 39) introduced the term *sulcus supratoralis*, or supratoral sulcus, to describe it.

survivorship The probability that an individual will live until a specified age. Note that if the survivorship at age 45 years is 0.4 then this simply indicates that at birth an individual will have a 40:60 chance of surviving until age 45 years, but could die at any age past 45 if they survive to that age. Also note that the starting point for survivorship need not be at birth, but can instead refer to any point in an individual's **life history**. For example, adult survivorships could be defined as starting at age 20 years, whereas fetal survivorship could be defined as starting at conception. The survivorship curve across age is determined by the **hazard rate** of death. Because life histories vary so greatly between modern humans and the extant apes, survivorship is one of the most intensely studied variables in **paleodemography**.

survivorship analysis A formal method for studying the "time to failure," which in **paleodemography** is typically death. Survivorship analysis usually assumes a particular form for the hazard of death and then estimates the parameters that fully describe **mortality**. Although paleodemographic data are typically too sparse for an in-depth survivorship analysis, survivorship models can sometimes be simplified and applied even to paleoanthropological samples. Survivorship analysis produces estimates of the parameter or parameters describing a hazard function, which gives the **hazard rate** at every age ($h(t)$). Survivorship can then be found from the hazard function by exponentiating the negative integral, which gives: $S(t) = \exp\left[-\int_0^t h(y)\mathrm{d}y\right]$ (Wood et al. 1992). The probability density function for age at death is then the product of survivorship and the hazard.

susceptibility The state of being at risk from a disease or condition. It is the probability that an individual will suffer from a condition because of environmental exposure (such as the infective agent of an infectious disease or exposure to a chemical) or from a condition that is influenced by both genotype and environment (i.e., genetic susceptibility to a disease such as type II diabetes or malaria). Scenarios for human evolution seldom take into account the possibility that disease may have played an important role in our evolution.

suture (L. *sutura* = a seam) A fibrous joint between two bones of the cranial vault. The fibrous tissue in these joints gradually disappears as the bone fuse together.

Svaty Prokop (Location 50°05′N, 14°20′E, Czech Republic; etym. Czech "Saint Procopius" = name of the valley in which it is found) History and general description Cave site southwest of Prague in the middle of the Svaty Prokop valley. Hominin remains found with Pleistocene fauna and lithic artifacts were initially thought to be Upper Paleolithic in age, but they have been recently redated and they have an age that is post-Paleolithic. Temporal span and how dated? **Radiocarbon dating** on the human bones provided two dates, 4–5 ka BP and 1.8 ka BP, showing that these remains are Holocene intrusions. Hominins found at site Occipital bone and femoral head attributed to *Homo sapiens*. Archeological evidence found at site **Middle Paleolithic** stone tools and a bone industry were found at the site, but they are not associated with the hominin remains. Key references: Geology, dating, and paleoenvironment Svoboda et al. 2004; Hominins Vlček 1971, Churchill and Smith 2000.

757

Švédův stůl *See* Ochoz.

Svitavka (Location 50°05′N, 14°20′E, Czech Replubic; etym. named after the valley in which it is located) <u>History and general description</u> Skeletal fragments discovered during geological surveying in 1962 were initially thought to be Upper Paleolithic in age, but they were recently redated to the Middle Ages. <u>Temporal span and how dated?</u> **Radiocarbon dating** of the fragments has provided a recent date. <u>Hominins found at site</u> A partial skeleton assigned to *Homo sapiens*. <u>Archeological evidence found at site</u> Some lithic artifacts. Key references: <u>Geology, dating, and paleoenvironment</u> Svoboda et al. 2002; <u>Hominins</u> Vlček 1971.

SVT Acronym for the supralaryngeal part of the vocal tract. *See* **supralaryngeal vocal tract**.

SVT$_H$ Acronym for the supralaryngeal part of the vocal tract, horizontal component. *See* **supralaryngeal vocal tract**.

SVT$_V$ Acronym for the supralaryngeal part of the vocal tract, vertical component. *See* **supralaryngeal vocal tract**.

swallowing (OE *swillian* = to wash out from Ge. *swelgan*) Swallowing is the act of moving a bolus of food or a volume of liquid from the mouth through the oropharynx and the laryngopharynx to the esophagus. Muscles help swallowing by changing the shape of the oral cavity, elevating the larynx, and by closing the oropharynx behind the food or liquid. Swallowing is aided by gravity, but it can take place against it. During swallowing striated muscle sphincters contract by reflex to close off the nose and the larynx to prevent the entry of food and liquid. The initial phase of swallowing involving striated muscles can be initiated voluntarily, but once the food or liquid is in the oropharynx swallowing becomes involuntary. Swallowing during normal eating and drinking is an unconscious reflex activity. You do not have to make a conscious decision to swallow; you swallow automatically when receptors in the mouth judge the food to be ready, or before you take another mouthful. In modern humans the soft palate closes off the nasopharynx, whereas in nonhuman primates with an unflexed cranial base the soft palate meets the epiglottis, effectively sealing off the mouth from the airway so animals can swallow and breathe at the same time. Researchers have made several, but as yet unsuccessful, attempts to work out when in human evolution this switch from the nonhuman primate-type to the modern human-type arrangement took place. Modern human infants have a relatively unflexed cranial base and newborn babies and neonates have the nonhuman relationship in which the soft palate meets with the epiglottis. This allows them to breathe and suckle at the same time, and this explains why babies and adults with blocked noses due to upper respiratory tract infections have to stop feeding to take a breath. *See also* **pharynx**.

Swanscombe (Location 51°26′44.12″N, 0°17′56.80″E, England; etym. the site is within the boundary of the village of Swanscombe, Kent) <u>History and general description</u> The site, called **Barnfield Pit**, is a gravel pit in the 100-ft (30-m) terrace just inland from the south bank of the estuary of the river Thames. The site was known for its **Clactonian** stone artifacts recovered *in situ* from the middle gravel beds. A single hominin calvaria (**Swanscombe 1**) is represented by three cranial vault fragments. The occipital and left parietal were found by Alvan Marston (a dentist from Clapham, south London) in 1935 and 1936, and the right parietal was found by John Wymer and A. Gibson in 1955. The remains were originally allocated to 'archaic' *Homo sapiens*, then to *Homo sapiens protosapiens*, and then to *Homo sapiens steinheimensis*. The current conventional taxonomic allocation is to *Homo neanderthalensis*. <u>Temporal span and how dated?</u> The fauna from the site is consistent with the interglacial that preceded the Anglian glaciation (i.e., **Oxygen Isotope Stage 1i**, *c.*400 ka). <u>Hominins found at site</u> Hominin adult calvaria represented by the occipital and both parietals. <u>Archeological evidence found at site</u> Excavations led by John d'Arcy Waechter between 1968 and 1972 yielded predominantly **Acheulean** handaxes. Key references: <u>Geology, dating, and paleoenvironment</u> Stringer and Hublin 1999; <u>Hominins</u> Marston 1936, Montandon 1943, Wymer 1955, Weiner and Campbell 1964, Stringer and Hublin 1999; <u>Archeology</u> Hawkes et al. 1938, Wymer 1964, Waechter 1971.

Swanscombe 1 <u>Site</u> Swanscombe. <u>Locality</u> N/A. <u>Surface/*in situ*</u> *In situ*. <u>Date of discovery</u>

June–October 29, 1935 (occipital), March 15, 1936 (left parietal), July 30, 1955 (right parietal). Finders Alvan Marston (1935–6) and John Wymer and A. Gibson (1955). Unit/horizon Near the base of the Upper Middle Gravels. Bed/member 100-ft Thames terrace. Formation N/A. Group N/A. Nearest overlying dated horizon N/A. Nearest underlying dated horizon N/A. Geological age *c*.400 ka. Developmental age Adult. Presumed sex Unknown. Brief anatomical description Almost complete occipital, well-preserved left parietal, and less well-preserved right parietal. Endocranial volume *c*.1325 cm^3 (Olivier and Tissier 1975). Announcement Marston 1936, Wymer 1955. Initial description Le Gros Clark 1938, Morant 1938. Photographs/line drawings and metrical data Ovey 1964. Detailed anatomical description Ovey 1964. Initial taxonomic allocation *Homo sapiens*. Taxonomic revisions *Homo sapiens protosapiens* (Montandon 1943), *Homo sapiens steinheimensis* (Weiner and Campbell 1964). Current conventional taxonomic allocation *Homo neanderthalensis*. Informal taxonomic category Pre-modern *Homo*. Significance The Swanscombe calvaria is one of a group of specimens (others include the hominins from **Reilingen**, the **Sima de los Huesos**, and **Steinheim**) referred to *H. neanderthalensis*. Researchers who interpret *H. neanderthalensis* as an evolving lineage would interpret the Swanscombe 1 calvaria as an example of "Stage 2," a stage that has also been referred to as the "pre-Neanderthal" (e.g., Dean et al. 1998) stage. Location of original **Natural History Museum, London, UK.**

Swanscombe skull *See* **Swanscombe 1.**

Swartkrans (Location 25°58'08"S, 27°45'21"E, South Africa; etym. Afrikaans for "black hillside," so-called because the rocks are stained black by manganese) History and general description The cave had been mined for lime in the 1930s, but the first paleontological exploration of the Swartkrans breccias was instigated by Robert **Broom** and John **Robinson** in 1948. This initial exploration, partly funded by the University of California's Africa Expedition, began in November 1948 and ran until November 1949. The first hominins, SK 2–**SK 6**, were recovered in November 1948, from what was called at the time called "pink breccia," and they were assigned to *Paranthropus crassidens* (Broom 1949); *P. crassidens* is now recognized by all authorities as a subjective junior synonym of

Paranthropus robustus. Broom and Robinson excavated together at Swartkrans until Broom's death in 1951, and Robinson continued for 2 years after that. C.K. (Bob) **Brain** reopened scientific exploration of the caves in 1965 and spent the first 7 years sorting through the breccia dumps the lime miners had left behind. In 1973 he began clearing the surface of the site of all remaining rubble, and starting in 1979 he conducted *in situ* excavations and these lasted until 1986. Since 1965 Brain was responsible for the research at the cave and for the interpretation of the discoveries; research at Swartkrans is now under the joint direction of Brain and Travis Pickering. The Swartkrans breccias were initially formally designated as the **Swartkrans Formation** (Butzer 1976b), with two Members (1 and 2), but researchers now recognize five Members (1–5). Member 1 (the old "pink breccia") is the most extensive and is divided into the "Hanging Remnant" (HR) and the "lower bank" (LB) components. Most of the Swartkrans *P. robustus* hypodigm comes from Member 1, but both Member 2 (this includes the old "brown breccia") and Member 3 also contain fossil evidence of *P. robustus*. There are two newly recognized sedimentary units, the Talus Cone Deposit (TCD) and the Lower Bank East Extension. The former contains *P. robustus* fossils and the latter, which is presently the lowest sedimentary unit in the cave, is equivalent to the "lower bank" of Member 1. Temporal span Approximate dates of the main fossil-bearing strata: Members 1–3 are *c*.1.8–1.0 Ma. A recent review suggests that "the currently best accepted dating of Members 1–3 relies on **biostratigraphy**, which, on balance, places each unit between c. 1.8–1.0 Ma. Parsing of the bovid and equid data leads some (e.g., Vrba 1985, Churcher and Watson 1993) to assign dates more specifically, with Member 1 at 1.7 Ma, Member 2 at 1.5 Ma, and Member 3 at 1.0 Ma" (Sutton et al. 2009, p. 689). Direct uranium-lead dating of the flowstone layers above and below the Hanging Remnant and Lower Bank deposits gives an age of 2.25–1.8 Ma for Member 1 (Pickering et al. 2010). Member 4 is Middle Stone Age (*c*.<100 ka); Member 5 is *c*.11 ka. How dated? Biostratigraphy, with some recent ages from **uranium-series dating, electron spin resonance spectroscopy dating**, and **radiocarbon dating.** Hominins found at site More than 400 hominin fossil specimens, probably representing fewer than 150 individuals, have been recovered from dumps of Swartkans breccia, or have been found *in situ* in the cave. In April 1949 a mandible **SK 15** was recovered from a pocket of "brown breccia" and was assigned to a second hominin, *Telanthropus capensis* (Broom and Robinson 1949c). Subsequent excavation has shown that all the Broom/Robinson hominin fossils come from what is now referred to as the "hanging remnant." In addition small numbers of fossils assigned to *Homo* sp. have been recovered from Members 1 and 2, and some researchers have suggested that one, or more, hominin postcranial fossils from Member 3 may belong to *Homo*. *Homo* remains make up about 5% of the hominins in Member 1, and some 20% of the hominins in Member 2. In addition to SK 15 these presumed *Homo* fossils include the composite cranium, **SK 847**, the mandible fragment, SK 45, and several isolated teeth. Archeological evidence found at site Stone artifacts, including cores and choppers resembling the **Developed Oldowan** from East Africa, have been recovered from Members 1–3, as have bones that appear to have been used as **digging sticks**. Assemblages from Member 1 (Lower Bank) and Members 2 and 3 have been interpreted as evidence for systematic butchery by early hominins, and it has been suggested that burned bone recovered in substantial quantities from Member 3 is evidence of the deliberate use of fire. Key references: Geology, dating, and paleoenvironment Curnoe et al. 2001, Brain 2004, Herries et al. 2009, Sutton et al. 2009, Pickering et al. 2010; Hominins Robinson 1956, Grine 2004, Susman 2004, Sutton et al. 2009; Archeology Clark 2004, Backwell and d'Errico 2004b, Pickering et al. 2008, Sutton et al. 2009.

Swartkrans I *See* **Swartkrans**.

Swartkrans II (Location 25°58′08″S, 27°45′21″E, South Africa; etym. Afrikaans for "black hillside," so-called because the rocks are stained black by manganese) A set of breccia outcrops a short distance northeast of the Swartkrans Main Site excavation. The outcrops were exposed by miners early in the 20thC and C.K. (Bob) **Brain** collected a few samples in the late 1950s. The deposit was mapped, a grid was constructed, and systematic excavations began in 2005. *See also* **Swartkrans**.

Swartkrans Formation The Swartkrans breccias were formally designated as the Swartkrans Formation (Butzer 1976b). Researchers now recognize five **members** (1–5). *See also* **Swartkrans**.

Swartkrans Main Site *See* **Swartkrans**.

Swartkrans Paleoanthropological Research Project (or SPRP) Initiated in 2005, this is the umbrella organization for the latest phase of research at **Swartkrans**. It is co-directed by Travis Pickering and C.K. (Bob) **Brain**. Other members include Morris Sutton (site supervisor and artifacts), Jason Heaton (nonhominin primate paleontology), Ron **Clarke** (hominin paleontology), and Kathleen Kuman (artifacts). The SPRP is currently focusing on excavations in (a) an artifact-rich **Middle Stone Age** (or MSA) deposit in the northeastern portion of the site and (b) in two deposits of early **Pleistocene** age that underlie the MSA unit, including the newly recognized and fossiliferous Talus Cone Deposit, and an eastward extension of the well-known Lower Bank deposit of Swartkrans Member 1. Other current SPRP excavations include (a) the reopening of Member 3, a depositional unit from which abundant burned bones – claimed to be the earliest evidence of hominin-controlled fire at *c*.1.0 Ma – have been previously recovered, and (b) the systematic excavation of Member 5, the site's most recent surviving depositional unit (*c*.11 ka), with a well-preserved fauna inferred to have been accumulated primarily by leopards and that is dominated by the fossils of the extinct springbok *Antidorcas bondi*.

Swedish China Research Committee In 1924/5 Carl Wiman and Otto **Zdansky** had found what they interpreted to be a hominin tooth in the collection from what was then called Choukoutien (now called **Zhoukoudian**) and this prompted them to suggest that a tooth found in 1921 might also belonged to a hominin and not to an ape. Johan **Andersson** learned of Wiman and Zdansky's conclusions by letter in 1926 and with support from the Swedish industrialist Axel Lagrelius the Swedish China Research Committee was established to fund and organize paleontological exploration in China, especially at Choukoutien. *See also* ***Homo erectus*; **Zhoukoudian**.

swine Informal term to describe the taxa within the family **Suidae** (syn. pig, suid). *See* **Suidae**.

swing phase *See* **walking cycle**.

Sylvian fissure *See* lateral fissure.

symbol One of three categories of possible signs used by communication systems to refer to things in the world (icons and indexes are the others) (Deacon 1997). Icons are signs that encode meaning in a transparent fashion; photographs of individuals, maps, and animals sounds are examples of iconic signs because they bear a direct relationship to the things they refer to (i.e., the referent). Indexes are signs that are causally related to the things they refer to; weathervanes index the direction of wind and speedometers index driving speed. Scents are also indexical, in that they are causally related a particular individuals or category of animal. According to Deacon (1997), symbols are arbitrarily related to their referent, but the relationship between a given symbol and its referent is derived by an explicit and learned social convention. This definition of symbol differs from standard definitions that emphasize the arbitrary relationship between a sign and a given referent. In addition to being arbitrary, symbols are generative, since they are composed of indexes, icons, and other symbols, and they are part of a "system of conventional relationships" (Deacon 1997, p. 71), or social conventions. Deacon has argued that modern human brains are uniquely specialized for creating and manipulating symbols. As such, while indexical and iconic signs are common in animal communication systems, symbols are not. Archeologists have emphasized the use of symbols (e.g., the curation of **ochre** presumably to use to color the skin or as a non-utilitarian object; McBrearty 2001) as an important element of what makes the behavior of modern humans different from that of early hominins. *See also* **word**; **symbolic**.

symbolic (L. *symbolus* = a token or sign) An adjective that describes anything that serves to effectively reference a socially shared idea, object, concept, or phenomenon. The presence of symbolic behavior (and its apparent absence from some hominin populations) is a central component of many debates on the origin of modern human behavior (e.g., Henshilwood and Marean 2003). Among extant populations symbolism takes many forms, including material representations as well as social performances such as dance, myth, folklore, song, gesture, and speech. Archeologists are necessarily most concerned with symbolic material culture because it alone is preserved in the archeological record. An object, image, or other item of material culture is not inherently symbolic, but rather its symbolic nature is socially ascribed or constructed. It is through collective agreement on the meaning of the symbol that it comes to be symbolic and the symbol itself may or may not bear a direct resemblance to that which it symbolizes. This renders interpretation of symbolism in the archeological record difficult, as we lack anyone to inform us of the cultural context necessary to interpret such items. Wedding rings provide a good example. Populations living in the USA, Canada, and the UK (for example) recognize that a person wearing a ring on the fourth finger of the left hand is probably married. However, from the perspective of the archeologist, there is nothing about the physical characteristics of (most) wedding rings that would link them to marriage; rather, it is the understanding of their meaning shared by members of a culture or community that lend them significance. Early artifacts (e.g., pierced shells from **Grotte des Pigeons**) that may have served as pendants may carry similarly complex meanings that present unique archeological challenges in their interpretation.

sympatric (Gk *sym* from *sumbion* = to live together and *patris* = fatherland) Two organisms with significantly overlapping geographic ranges. It is possible that several hominin species (e.g., *Paranthropus boisei* and *Homo habilis*) were sympatric, although co-occurrence in the fossil record does not necessarily imply the animals lived at the precisely the same time or in precisely the same place. There are very few examples of sympatric speciation (the genetic divergence of populations that remain geographically associated) in animals; it is likely to be a much more common mechanism of speciation in plants. *See also* **synecology**; **taphonomy**.

sympatric speciation A mode of **speciation** in which new **species** are formed when two or more populations within a single species diverge ecologically within the same geographic range. The ecological divergence results in disruptive selection that drives the populations apart anatomically and/or behaviorally, resulting eventually in genetic divergence and reproductive isolation between the groups. These is a considerable debate within evolutionary biology as to how frequently sympatric speciation occurs, if at all.

sympatry The state when two or more organisms are **sympatric** (*which see*).

symphyses *See* joint; symphysis.

symphysis (Gk *sym* from *sumbion* = to live together and *phyein* = to grow) A midline **secondary cartilaginous joint** such as that between the bodies of the right and left sides of the developing **mandible** and between the paired pubic components of the pelvic bones. Pl. symphyses. *See also* **joint**.

symplesiomorphic *See* symplesiomorphy.

symplesiomorphy (Gk *sym* from *sumbion* = to live together, *plesios* = near to, and *morph* = form) A term used in cladistics to refer to the primitive condition of a character (i.e., the character state of the **hypothetical common ancestor** of a **clade**, or of the **outgroup**). For example, a small canine crown is probably a symplesiomorphy, or symplesiomorphic, for the **hominin** clade, and a slender mandibular corpus is probably a symplesiomorphy for the *Homo* clade (syn. plesiomorphy).

synapomorphy (Gk *syn* from *sumbion* = to live together, *apo* = different from, and *morph* = form) A term used in cladistics for a character state that is neither symplesiomorphic (primitive or shared by many taxa) nor autapomorphic (confined to one taxon; *see* **autapomorphy**). Synapomorphies are character states shared by at least two taxa. For example, extreme postcanine **megadontia** is probably a synapomorphy of *Paranthropus boisei* and *Paranthropus aethiopicus*. Extreme postcanine megadontia is also seen in *Australopithecus garhi*, but it may be a homoplasy in that taxon and not a synapomorphy shared with the two aforementioned *Paranthropus* taxa. (syn. apomorphy).

synapse (Gk *syn* from *sumbion* = to live together and *haptein* = to fasten) The places on the cell surface of a **neuron** where the neuron touches its target cell. A neuron can form a synapse with another neuron, with a gland, or with a muscle. Synapses, depending upon the kind of impulse conveyed, can be divided into two categories, electrical and chemical. At electrical synapses, ion channels connect the cytoplasm of the presynaptic (sending) and postsynaptic (receiving) cells. At chemical synapses, there is no cytoplasmic continuity between the cells, and the neurons are separated by a synaptic cleft. In a chemical synapse it is through the synaptic cleft that **neurotransmitters** pass from the presynaptic axon terminal to the postsynaptic receptor membrane.

synchondroses *See* joint.

synchronic (Gk *sunkhronos* = contemporaneous) Term implying that events took place together, or fossils are of the same age. But "together" and the "same" are being used in the sense of geological time, so the events could be hundreds or thousands of years apart, and the organisms represented by the fossils could have been living at times that were hundreds or thousands of years apart.

synchrotron A machine that sends accelerated particles (electrons) through a cyclical course determined by magnetic fields. When the electrons are deviated by the magnets, synchrotron light is emitted, which can range from radio frequencies to high-energy X-rays. Synchrotron X-rays have properties (e.g., high flux, monochromaticity, parallel beam geometry, spatial coherence, etc.) that are superior to laboratory or medical **computed tomography**. *See also* **synchrotron radiation micro-computed tomography**.

synchrotron radiation micro-computed tomography A **synchrotron** imaging technique (abbreviated SR-μCT) that facilitates submicron imaging and phase-contrast techniques, which reveal structures invisible with conventional **computed tomography** (or CT). The small size and focused high energy of the source beam means that, compared to CT, micro-computed tomography can deliver an advantageous signal-to-noise ratio on even highly mineralized bone and tooth samples (Mazurier et al. 2006). Propagation phase-contrast techniques reveal interfaces within structures, including microscopic variations in the prismatic structure of enamel and the tubular structure of dentine. It has enabled researchers to image the incremental development of intact teeth (Tafforeau and Smith 2008), reveal the fine details of **enamel-dentine junction** in heavily mineralized fossil teeth (Smith et al. 2009), and determine the age at death in juvenile fossil hominins from dental microstructure (Smith et al. 2007).

syncline (Gk *syn* = together or with and *klinein* = to slope) In structural geology a syncline is a type of fold where the youngest rocks occupy the center of the structure and the rocks become progressively older towards the margins. The simplest form of syncline is a symmetrical U-shaped fold (open upwards). Such folds are important in geological and paleontological fieldwork, as they will determine the direction in which successively older or younger strata are encountered and will cause a repetition of the **outcrop** pattern on either side of the fold's axis.

syndesmoses *See* joint.

synecology (Gk *syn* = together, *auto* = self, and *logos* = knowledge) The branch of **ecology** concerned with the interactions of communities of organisms. There is great potential for synecological studies in paleoanthropology, as hominins lived in evolving ecological communities. Some hominins may have interacted or lived alongside other hominin species. At **Koobi Fora**, for example, as many as four **contemporaneous** hominin species have been identified: *Homo ergaster*, *Homo habilis*, *Homo rudolfensis*, and *Paranthropus boisei*. Although co-occurrence in the fossil record does not necessarily imply that the animals lived at precisely the same time or in precisely the same place, the fact that these species are found together in the same geological unit and in the same region suggests they may well have been **sympatric**. Although several arboreal primate species live in mixed-species groups, analogy with modern terrestrial primates suggest that sympatric hominins may have avoided competition by having (a) different home ranges, (b) exploiting different foodstuffs, or (c) by interacting only where their ranges met (Elton 2006). *See also* **community ecology**; **taphonomy**.

synonym (Gk *sun* = with and *onuma* = name) Any available formal name that denotes the same **taxon** as any other formal name. The earliest ("prior") available name is the **senior synonym**; any other name denoting the same taxon is a **junior synonym**. For example, the names *Paranthropus robustus* **Broom, 1938** and *Paranthropus crassidens* **Broom, 1949** are considered to denote the same taxon, so are deemed to be synonyms, but *P. robustus* has 11 years' priority, so it is the senior synonym, and *P. crassidens* is a junior synonym. Franz **Weidenreich** (1940) suggested that the species included in *Pithecanthropus* are actually better regarded as species of *Homo*, so *Pithecanthropus* is a junior synonym of *Homo*, the genus name that has priority.

If it is someone's opinion that two or more named taxa actually denote the same taxon, they are subjective synonyms. Occasionally, however, two or more taxa are established on the same type specimen: and then the names given to them are called objective synonyms. For example, when *Anthropopithecus erectus* **Dubois, 1893** was described, with the Trinil calotte as its type specimen, several other scientists felt themselves at liberty, for whatever reason, to give their own names to the taxon denoted by this specimen (e.g., *Homo javanensis primigenius* Houzé, 1896, *Homo pithecanthropus* Manouvrier, 1896, *Hylobates giganteus* Bumuller, 1899, *Pithecanthropus duboisii* Morselli, 1901, *Hylobates gigas* Krause, 1909, and *Homo trinilis* Alsberg, 1922); they are all examples of objective synonyms. A **synonymy** is the condition that exists when more than one name has been given to the same taxon.

synonymous mutation (Gk *sun* = with and *onuma* = name) A point mutation (i.e., change in one **nucleotide**) in a coding sequence that does not result in a change in the amino acid sequence of a protein. Such changes are typically neutral. For example, if the **DNA** sequence CTA changes to CTG, because of the redundancy in the genetic code it still codes for the same amino acid, aspartic acid (syn. silent mutation).

synonymous substitution *See* **synonymous mutation**.

synonymy (Gk *synonymos* = the same meaning) More than one Linnaean name given to the same taxon. Also used for a list, in chronological order, of the names previously used for a taxon (i.e., a list of synonyms for that taxon). The earliest available name (also known as the **senior synonym**) has priority; any names for the taxon other than the earliest available name are called **junior synonyms**. Thus *Australopithecus africanus* is the senior synonym for the taxon for which **Taung 1** is the **holotype**. *Plesianthropus transvaalensis* is one of several junior synonyms of *Au. africanus*; it is an available taxon name, but because it is not the senior synonym it is not a valid name.

syntactic rules *See* **syntax**.

syntax Order and structure of language (as a grammar). Language is not just about using signs (e.g,. sounds or gestures) with specific meanings to refer to things in the world. Languages are also characterized by syntax (Chomsky 2007). That is, all languages have grammars that orders and structures signs in a sentence or sequence of sounds. Consider the phrases *the boy hit the girl* versus *the girl hit the boy*. While the words in each sentence are identical, their meaning is radically different; what governs this shift in meaning is syntax. Since Chomsky (1977) stressed the centrality of syntax to modern human communication and language, many language scientists (e.g., Hauser et al. 2002, Pinker 1994) have noted the absence of syntax in nonhuman communication. However, Hauser and colleagues (e.g., Fitch and Hauser 2004) have argued that while some aspects of grammar may be present in nonhuman animals, modern humans are unique in their ability to engage in recursive syntactic processing. In one study Fitch and Hauser (2004) demonstrated that whereas cotton-top tamarins (*Saguinus oedipus*) could recognize violations to a simple grammar of the form AB, AB (i.e., $[AB]^n$), they could not recognize violations to a hierarchical grammar of the form AAA, BBB (i.e., $[A^nB^n]$) where the number of the first sound (e.g., A) constrains the number of the following sound (e.g, B). College students, however, given this same task, quickly mastered these two syntactic rules and performed equally well in both. These results have been widely interpreted to mean that there are cognitive constraints that limit syntactic processing in nonhuman primates. These constraints may explain the universality of modern human languages and the failure of the various ape language projects (Terrace 1979). *See also* **recursion**.

synteny (Gk *syn* = along with and *tainia* = band, thus "on the same ribbon") A term used to describe **loci** on the same **chromosome**. These loci are not necessarily linked; their probability of linkage is based on the physical distance between them. Shared synteny between species refers to the preserved order of genes on a chromosome; it can be lost because of translocations or inversions.

syntype (Gk *syn* = together and *typus* = image) If a **holotype** was not designated at the time of the original description of a taxon then one of the "syntypes" (i.e., all the specimens known to the authors at the time of the initial publication) can subsequently be designated as the **lectotype**. For example, when the taxon *Pithecanthropus*

rudolfensis was introduced by Alexeev, he did not formally designate a holotype. Subsequently **KNM-ER 1470**, one of the syntypes of the original taxon, was formally designated as the lectotype of *Homo rudolfensis*.

system The chronostratigraphic unit equivalent of a **period**; periods are subdivided into **series**.

systematic biology The part of biology that focuses on the principles and practice of **systematics** as applied to animals. *See also* **systematics**.

systematics (Gk *systema* = a whole made of several parts) Systematics is an inclusive term that subsumes all of the activities involved in the study of the diversity and origins of living and extinct organisms. Thus systematics includes (a) **identification** (i.e., identifying individual living, or fossil, organisms, and assembling them into groups), (b) **classification** (i.e., the formalization of those groups as taxa, giving formal names to the taxa, allocating the taxa to taxonomic categories, and then assembling the taxonomic categories into a hierarchical scheme), and (c) **phylogeny reconstruction** (i.e., generating hypotheses about the branching pattern of the tree of life). **Taxonomy** is the study of the principles and theory of classification. **Nomenclature**, which combines the principles used in the allocation of formal names to taxa and the rules (e.g., **priority**, **synonomy**, etc.) governing how those names may be used, is a subcomponent of classification, and thus of taxonomy.

Szeleta (Location 28°06′N, 20°37′E, Hungary) <u>History and general description</u> This cave site overlooks the town of Lillafüred in the Bükk mountains. It was first excavated in 1906–13 and then again in the 1920s, 1940s, and 1960s. It is the type site for the **Szeletian** transitional lithic industry, which appears between the **Middle Paleolithic** and **Aurignacian** deposits at the site. <u>Temporal span and how dated?</u> Early **radiocarbon dating** suggested an age of more than 40 ka (uncalibrated) for the early Szeletian layers. However, newer radiocarbon dates point to an age of 30–20 ka (uncalibrated) for this layer. <u>Hominins found at site</u> None. <u>Archeological evidence found at site</u> The site preserves Middle Paleolithic, Szeletian, Aurignacian, and **Gravettian** layers.

The Szeletian is an early **Upper Paleolithic** transitional technocomplex that is problematically associated with early migrations of modern humans. Key references: Geology, dating, and paleoenvironment Adams and Ringer 2004; Archeology Vértes 1968, Adams and Ringer 2004.

Szeletian (etym. Named after the site of **Szeleta**) An eastern European technocomplex that is one of several so-called transitional lithic industries (e.g., **Châtelperronian**), which appear in the late **Middle Paleolithic** or early **Upper Paleolithic** across Europe, and show aspects of both lithic traditions. These transitional technocomplexes are put forward as evidence of either a pre-Aurignacian expansion of modern humans into Europe, or the acquisition of advanced, so-called "modern human behaviors" by *Homo neanderthalensis*. Few fossils are associated with the Szeletian; one tooth found at **Dzeravá Skála** has been identified as belonging to *Homo sapiens*, but its stratigraphic context is questioned and therefore it cannot be taken as evidence that the Szeletian was associated with, and certainly not only with, modern humans.

T

t test A statistical test used to determine whether the mean of a **sample** of measurements differs significantly from some hypothetical value (often zero), or whether the means of two different samples of measurements differ significantly from each other (e.g., whether mean cranial capacity size differs significantly between a sample of modern humans and a sample of Neanderthals). For a two-sample *t* test, two versions of the test exist in which the variances within the two samples are assumed to be either equal or unequal. A paired *t* test determines whether two sets of observations on the same sample differ significantly from each other (for example, whether mean endocranial volume is significantly different than mean brain volume within a sample of modern human cadavers). All *t* tests assume that data are normally distributed; if not, a **nonparametric statistical** test such as the **Mann–Whitney U test** or the **Wilcoxon ranked sign test** should be used.

Tabarin (Location $00°75'$N, $35°86'$E, Kenya; etym. the name for the area in the local Tugen language) History and general description Tabarin is a locality within the lowest part of the **Chemeron Formation** within the Tugen Hills sequence to the west of Lake Baringo, Kenya. The Kaparaina Basalt at the base of Tabarin separates the Chemeron and **Lukeino Formation**s. It has produced abundant fossil fauna and one hominin, **KNM-TH 13150**, a partial adult hominin mandible discovered by Kiptalam Cheboi in 1984 while he was working with Baringo Paleontological Research Project. At the time of its discovery, KNM-TH 13150 was suggested to be the oldest known hominin and it may prove to belong to *Ardipithecus ramidus*. Temporal span and how dated? Single-crystal **argon-argon dating** have given a median age of 5.31 ± 0.03 Ma for the ignimbrite and the associated basal air-fall tuff that are in the lowest part of the Tabarin section. This evidence suggests a temporal span for the Chemeron succession at the site from about 5.3 Ma

to 3.8 Ma, with KNM-TH 13150 being *c*.4.42 Ma based on the dating methods above plus **magnetostratigraphy** and a calculated sedimentation rate of 23 cm/ka. Hominins found at site KNM-TH 13150. Archeology found at site None. Key references: Geology, dating, and paleoenvironment Hill 1985, 1994, 1999, 2002, Deino et al. 2002, Hill and Ward 1988; Hominins Hill 1985, 1994, 1999, Hill and Ward 1988, Ferguson 1989, Ward and Hill 1987, Wood 1999, Deino et al. 2002; Archeology None.

Tabon Cave (Location $9°16'$N, $117°58'$E, Palawan Island, The Philippines; etym. the name refers to the Tabon bird, *Megapodius cumingii*, that sheltered in the caves and deposited the guano layers) History and general description The site is located at an altitude of about 35 m on the Lipuun point, facing the small town of Quezon of the western coast of Palawan Island, some 100 km/62 miles south of Puerto Princesa. It is part of the so-called Tabon caves complex, which was explored for its archeological potential in the 1960s by Robert B. Fox, at the time chief of the Anthropology Division of the National Museum of the Philippines. Archeological surveys and excavations were made of more than 20 caves, fissures, and rock-shelters in the karstic massif of Lipuun, one of them having yielded the famous (late Neolithic or early Metal age) secondary burial Manunggul Jar, figuring a boat. Temporal span and how dated? The stratigraphy of the cave is complex and somewhat disturbed, especially by the activities involved with the secondary burial jars, the activity of birds, as well as the usual bioturbation. Recent work carried out at the site since 2000 under the authority of the National Museum of the Philippines has shown from uranium-lead dating of **speleothems** that the cave is more than 0.45 Ma and **uranium-series dating** suggests the fossil hominin remains are between *c*.15 and 50 ka. Recent **radiocarbon dating** of *in situ* charcoals from an occupation layer in the upper part of the stratigraphy confirms the presence of hominins in the cave *c.* 34 ka. Hominins found at site Fox's excavations in Tabon Cave yielded, beyond numerous Neolithic modern human remains (found as jar burials on the surface layers), several homininn fossils, including a quite complete *Homo sapiens* frontal bone. Two fragmentary mandibles were also found, one of them, quite mineralized, being especially robust for and anatomically modern human. Further excavations by the National Museum of the Philippines yielded more hominin cranial and postcranial remains, including a fragmentary tibia. Archeological evidence found at site Lithic implements (mostly flakes) considered by Fox to be associated with the anatomically modern human fossils were found in a layer dated in the 1960s by radiocarbon dating to be *c.*22 ka. The Upper Pleistocene age attributed to the hominin remains makes Tabon one of the few Southeast Asian island sites that document early colonization by modern humans. In the case of Tabon this probably involved crossing the quite narrow sea straits that would have separated Palawan from the Borneo Islands during glacial periods of low sea level. Location of originals The National Museum of the Philippines, Manila, The Philippines. Key references: Geology, dating, and paleoenvironment Fox 1970, Bellwood 1997; Hominins and archeology Bellwood 1997, Dizon et al. 2002, Detroit et al. 2004.

Tabun (Location 32°40′N, 35°57′E, Israel; etym. Ar. مغارة *mugharet*=cave, and تنور *tabun*=oven) History and general description Tabun is one of the three sites on the western slope of **Mount Carmel** excavated by Dorothy **Garrod** between 1928 and 1934. Garrod defined several levels, with Level C containing the most of the hominin remains. Subsequently Arthur Jelinek supervised work from 1967 to 1972 that sought to refine and revise Garrod's stratigraphic system and to better understand the chronological relationship among the sites of Mount Carmel, but no further hominin remains were recovered. Temporal span and how dated? Garrod originally dated Level C to *c.*40 ka. The most recent **electron spin resonance spectroscopy dating** gives a date of *c.*104 ka for Level B and *c.*208 ka for Level E. Hominins found at site The remains of up to 14 individuals have been recovered from Levels E, C, and B. A fragmentary skeleton and a dissociated mandible from Level C (**Tabun 1** and **Tabun 2**) are the most iconic. McCown and Keith saw these as belonging to the same pre-modern population as those from **Skhul**, but more recent interpretations emphasize

their affinities with *Homo neanderthalensis*. A number of teeth that were thought to have come from Level A have recently been re-examined and dated. They most closely correspond in morphology and preservation with remains from Level B. Archeological evidence found at site Many Levantine **Mousterian** and **Levallois** artifacts have been recovered. Because of the long and presumably continuous record at Tabun, and in the absence of absolute dating, other archeological sites have been dated based on how their stone tool assemblages matched those from the various levels at Tabun. Key references: Geology, dating, and paleoenvironment Garrod and Bate 1937, Jelinek 1982, Grün et al. 1991, Schwarcz et al. 1998, Grün and Stringer 2000; Hominins McCown and Keith 1939, Trinkaus 1995, Quam and Smith 1998, Stefan and Trinkaus 1998, Coppa et al. 2005; Archeology Garrod and Bate 1937.

Tabun 1 Site Tabun. Locality West sector of the terrace. Surface/*in situ* *In situ*. Date of discovery 1932. Finders Team led by **Dorothy Garrod**. Unit Level C, but Garrod noted that it "lay so near the surface of [Layer] C that the question must arise whether it does not represent a burial from Layer B" (Garrod and Bate 1937), an idea supported by Bar-Yosef and Callander (1999). Horizon N/A. Bed/member N/A. Formation N/A. Group N/A. Nearest overlying dated horizon N/A. Nearest underlying dated horizon N/A. Geological age Grün and Stringer (2000) place the remains at *c.*122 ka. Developmental age Adult. Presumed sex Female. Brief anatomical description Fragmentary cranium and skeleton; the cranium includes all teeth except for upper right M^3. Announcement Garrod 1934. Initial description McCown 1936. Photographs/line drawings and metrical data McCown 1958. Detailed anatomical description McCown and Keith 1939. Initial taxonomic allocation *Palaeoanthropus palestinus* McCown and Keith, 1939. Taxonomic revisions Howell (1958) compared the specimens to the **Krapina** remains, implying that they belong to *Homo neanderthalensis*. While this is the generally accepted view, some (Arensburg and Belfer-Cohen 1998) disagree. Current conventional taxonomic allocation *H. neanderthalensis*. Informal taxonomic category Pre-modern *Homo*. Significance The Tabun 1 skeleton helped confirm the presence of *H. neanderthalensis* in the Near East. Location of original Natural History Museum, London, UK.

Tabun 2 <u>Site</u> **Tabun**. <u>Locality</u> East sector of the terrace. <u>Surface/*in situ*</u> *In situ*. <u>Date of discovery</u> 1932–4. <u>Finders</u> Team led by Dorothy **Garrod**. <u>Unit</u> Level C. <u>Horizon</u> N/A. Bed/member N/A. Formation N/A. Group N/A. <u>Nearest overlying dated horizon</u> Level C. <u>Nearest underlying dated horizon</u> Level C. <u>Geological age</u> *c*.122 ka. <u>Developmental age</u> Adult. <u>Presumed sex</u> Male. <u>Brief anatomical description</u> A mandible that includes all lower teeth except the left I$_1$. <u>Announcement</u> Garrod 1934. <u>Initial description</u> McCown 1936. <u>Photographs/line drawings and metrical data</u> McCown 1958. <u>Detailed anatomical description</u> McCown and Keith 1939. <u>Initial taxonomic allocation</u> *Palaeoanthropus palestinus* McCown and Keith, 1939. <u>Taxonomic revisions</u> Howell (1958) compares both Tabun specimens to those from **Krapina**, implying that they belong to *Homo neanderthalensis*. Stefan and Trinkaus (1998) also conclude that *H. neanderthalensis* is the most sensible assignment for Tabun 2, but Quam and Smith (1998) note some more modern human-like features, particularly in the mandibular symphysis. <u>Current conventional taxonomic allocation</u> *H. neanderthalensis*. <u>Informal taxonomic category</u> Pre-modern *Homo*. <u>Significance</u> Like **Tabun 1**, the Tabun 2 mandible points to the Tabun material being *H. neanderthalensis*. <u>Location of original</u> **Rockefeller Museum**, Jerusalem, Israel.

Taenia *See* taeniid tapeworms.

taeniid tapeworms Flatworms belonging to the class Cestoda. In their adult form they are parasitic in carnivorous mammals, and they are debilitating because they predispose the host to chronic anemia. The life history of these cestodes involves an intermediate and a definitive host. Intermediate hosts ingest eggs that then develop into larvae. These larvae are destroyed by cooking, but if modern humans ingest undercooked meat or offal from an infected intermediate host, then the ingested larvae mature into adults, and these adults lodge in the gut, and produce the eggs which are found in human feces, thus completing the cestode life cycle. Modern humans are the definitive host for three different species of taeniid tapeworms, *Taenia saginata*, *Taenia asiatica*, and *Taenia solium*. Domesticated cattle are the intermediate hosts for *T. saginata*, whereas domestic swine are the intermediate hosts for *T. asiatica* and *T. solium*. For a long time it was assumed that

modern humans had acquired their taeniid parasites when cattle and swine were domesticated, but evidence from molecular biology suggests that cattle and swine acquired their infestations from modern humans, not the other way round (Hoberg et al. 2001). Moreover, the results of a phylogenetic analysis suggest that the three taeniid tapeworms that infest modern humans belong to two different subclades. Their relationship takes the form (*T. solium* (*T. saginata*, *T. asiatica*)), which suggests that there were at least two, and probably three, separate episodes of infestation. Furthermore, the same analyses suggest that the definitive hosts of the immediate precursors of modern human taeniid tapeworms were **carnivore**s (**canid**s, **felid**s, or hyaenids) and their intermediate hosts were **bovid**s. The results also suggest that the host-switching occurred in sub-Saharan Africa prior to the domestication of **ungulate**s, and that it most likely occurred when early hominins were interacting with carnivores in the quest for meat. As Hoberg et al. (2001) state at the conclusion of their article "this work highlights the utility of parasitological data in elucidating the history and behaviour of human ancestors" (*ibid*, p. 786).

Taforalt (<u>Location</u> 34°48′38″N, 2°24′30″W, southwest of Berkane and northwest of Oujda, Morocco) <u>History and general description</u> Also referred to as the Taforalt Caves or the Grotte des Pigeons, the site is a large, east-facing cave situated in a limestone cliff, 40 km/25 miles inland and approximately 720 m above sea level in the Atlas Mountains of eastern Morocco. The cave, which was first reported in 1908, was test excavated by Ruhlmann during the 1940s and extensively dug by Roche during the 1950s and the late 1960s. Since 2003 the site has been under excavation by a joint British/Moroccan team (Bouzzougar et al. 2007). <u>Temporal span</u> *c*.95,000–12,000 years BP. <u>How dated</u> Most recently using **AMS radiocarbon dating** on eight charcoal samples from Epipaleolithic deposits, optically stimulated **luminescence dating** on 15 sediment samples, thermoluminescence dating on five burnt lithics from the Middle Paleolithic deposits, and **uranium series dating** on two samples of the flowstone which underlies the Middle Paleolithic layers (Bouzzougar et al. 2007). One of the two burial zones has been **radiocarbon** dated from a single sample to 11,900 years BP. (Lubell et al. 1992). <u>Hominins found at site</u>

About 180 hominin individuals have been recovered from two burial zones or "cemeteries" at the site. Burials are sometimes marked with grave goods such as the horn cores or entire cranium of a Barbary sheep; around 100 individuals are subadult (Humphrey and Bocaege 2008, Lubell 2001, Ferembach et al. 1962). Dental evulsion of one or most often both maxillary central incisors is seen on all the adult skeletons; evidence of post-evulsion bone formation indicates the tooth removal occurred during life. This was a common practice of the Iberomaurusian culture and evidence of it is found at numerous contemporary sites in the region (Humphrey and Bocaege 2008). Archeological evidence found at site Artifact assemblages from different parts of Taforalt Cave have been assigned to the **Middle Stone Age/Middle Paleolithic** Aterian tool industry, known for its tanged points, and to the Epipaleolithic Iberomaurusian industry. The Aterian deposits contain numerous superimposed hearths, pedunculate points, and bifacially worked foliates. The recently discovered perforated Nassarius gibbosulus shells are an early example of human production of goods for personal adornment; these are broadly comparable in age and form to the shell beads found at Blombos Cave in South Africa (Bouzzougar et al. 2007). The Iberomaurusian deposits feature cutmarked and modified bones from Barbary sheep, ostrich eggshells, many Aterian microlithic points, and large numbers of burnt land snail shells. Key references: Geology, dating, and paleoenvironment Bouzzougar et al. 2007, Lubell et al. 1992; Hominins Humphrey and Bocaege 2008, Lubell 2001, Ferembach et al. 1962; Archeology Bouzzougar et al. 2007, Lubell 2001, Lubell et al. 1992. *See also* **Grotte des Pigeons**.

Tagliente (Location 45°35′N, 11°00′E, Italy; etym. It. sharp or pointed) History and general description Rock-shelter at the base of Mt. Tregano just north of Verona, with a long stratigraphic sequence. Excavations began in 1964 under F. Zorzi, and continued on-and-off under different leaders for the next 35 years. Temporal span and how dated? **Biostratigraphy** suggests the earliest levels date to the Early Würm. **Radiocarbon dating** on the **Upper Paleolithic** levels provide dates in the range of 13,430±180 to 12,040±170 years BP. Hominins found at site Levels 36 and 37, which are part of the **Middle Paleolithic** sequences, yielded two deciduous teeth: a right dm^2 (Tagliente 3) and upper right dc (Tagliente 4). The Upper Paleolithic levels yielded an incomplete mandible (Tagliente 1) and the fragmentary skeleton of a young adult burial (Tagliente 2). Archeological evidence found at site The **Mousterian** sequences contain primarily scrapers and Levallois flakes. The ungulates in the faunal assemblage from these levels bear evidence of human modification. The Upper Paleolithic levels are notable for their stratigraphically secure examples of art. Two engravings of bison were recovered from levels 13a and 10, and the stone block covering the young adult burial is emblazoned with a lion. Key references: Geology, dating, and paleoenvironment Bartolomei et al. 1982; Hominins Corrain 1966, Bartolomei 1974; Archeology: Mussi 2001, Fiore et al 2004.

Talgai (Location 28°01′S, 151°51′E, Australia; etym. named after a local outpost) History and general description Talgai 1 is a cranium of a young male recovered by workmen from the wall of Dalrymple Creek southwest of Brisbane in 1886. The cranium was encrusted with carbonate and derived from a reddish-brown clay stratum. Temporal span and how dated? Late Pleistocene, on the basis of dates from **radiocarbon dating** on carbonate and organic soil associated with the cranium that are 11,650 and 12,400 years BP, respectively. Hominin found at site The vault of the Talgai 1 cranium was crushed and distorted upon recovery. Its robusticity and long, retreating, frontal resemble the morphology seen in the **Kow Swamp** and **Cohuna** group, suggesting to some researchers continuity with Javan *Homo erectus*. Others claim, and this is now the consensus view, that there are no demonstrable differences between Talgai and modern Australian Aboriginals and that the apparent similarities to Kow Swamp frontal morphologies are due to post-mortem deformation. Archeological evidence found at site None. Key references: Geology, dating, and paleoenvironment Gill 1978; Hominins Macintosh 1967, Brown 1987.

talon (L. *talus*=heel) The distal (hence heel) component of a maxillary (upper) postcanine tooth crown in the dental terminology devised by Osborn (1892). In hominin maxillary molars it comprises the **hypocone**, which is on the distolingual aspect of the **trigon**, plus any accessory cups or cuspules associated with the hypocone.

talonid (L. *talus*=heel) The distal (hence heel) component of a mandibular (lower, hence the post-fix "-id") postcanine tooth crown in the dental terminology devised by Osborn (1892). In hominin lower molars it comprises the **hypoconid** on the buccal aspect, the **entoconid** on the lingual aspect, together with a **hypoconulid** (if present) and any C6s. Modern human lower premolars do not normally have a talonid. Premolars with a well-developed talonid (e.g., *Paranthropus boisei*) are referred to as "**molarized**" or "molariform." *See also* **molarized**.

talonid basin *See* **posterior fovea**.

talus cone (OF *talu*=the sloping sides of an earthwork) When soil washed into openings in the roofs of caves reaches the floor of the cave it forms a cone-shaped heap of debris. Some of the material newly washed into the cave adds to the height of the talus cone, but the remainder runs down the sides of the cone so that its base is ever increasing in size. There is evidence of talus cones in the southern African early hominin cave sites (e.g., **Silberberg Grotto, Sterkfontein; Swartkrans**), and they are one of the reasons why some of the sediments in these caves violate the law of **superposition**.

tandem repeat array *See* **tandemly repeated sequences**.

tandem repeats *See* **tandemly repeated sequences**.

tandemly repeated sequences (L. *tandem*= one after the other) An inclusive term for **DNA** sequences, two or more **bases** long, that are repeated and adjacent. Tandemly repeated sequences include **microsatellites, minisatellites, variable number of tandem repeats**, and **satellite DNA**.

tang (etym. ME of Scandinavian origin, similar to the Norse *tangi*, or point of land) The narrow elongated region at the base of a **point**, or point-like stone tool that is designed to facilitate hafting. The tang projects downward from the base of the piece, resembling the vertical segment of the letter "T." Tangs are most often seen on arrowheads, and tanged points are one element considered diagnostic of **Aterian** sites in North Africa.

Tangi Talo One of the sites in the exposures of the Early and Middle **Pleistocene Ola Bula Formation** within the **Soa Basin**, central Flores, Indonesia.

Tangier *See* **Mugharet el 'Aliya**.

Tangshan Huludong (汤山葫芦洞)(Location 32°03′24.94″N, 119°02′40.41″E, eastern China; etym. named after the local town and the karst cave) History and general description The Tangshan locality, which is approximately 80 km/50 miles east of **Hexian**, across the Yangtze River, consists of the **karst** cave site of Huludong (Hulu Cave). It is just west of the town of Tangshan, some 23 km/14 miles east of Nanjing City, Jiangsu Province. The fossil hominins discovered in 1993 are collectively known as "**Nanjing Man**." Temporal span and how dated? High-precision **thermal ionization mass spectrometry uranium-series dating** of the overlying **stalagmite** and **flowstone** provides a minimum age of 500 ka, and a possible age of 580 ka, for the hominin fossils. Further comparison with the SPECMAP deep-sea oxygen isotopic record suggests that deposition of the crania likely occurred during **Oxygen Isotope Stage 16**, a *c*.620 ka cold stage. This agrees with the biostratigraphy, which noted that the Tangshan fauna is composed of cold-adapted mammals typical of **Zhoukoudian Locality 1**. Hominins found at site Skull 1 (or Hulu 1), which belonged to a young adult female, consists of a partial **cranium** with well-preserved (and highly protruding) nasal bones, a fairly complete left upper and mid-face, and a large antemortem lesion (caused by either trauma or burning) that covers most of an otherwise complete frontal bone. Skull 2, which belonged to an older adult male, consists of a distorted calotte (a partial frontal, both parietals, a partial left temporal, and a partial occipital). There is also an isolated right M^2 (initially thought to be an M^3). Archeological evidence found at site None. Repository Nanjing Municipal Museum, Nanjing. Key references: Geology, dating, and paleoenvironment Zhou et al. 1999, Zhao et al. 2001, Wu et al. 2002; Hominins Wu and Poirier 1995, Brown 2001, Wu et al. 2002, Liu et al. 2005, Shang and Trinkaus 2008.

tapeworms *See* **taeniid tapeworms**.

taphonomy (Gk *taphos*=grave and L. *nomos*= law) Russian paleontologist I.A. Efremov coined the term taphonomy, which refers to the branch of science that investigates the factors involved in the death, decay, preservation, and **fossilization** of organisms. In other words, taphonomy is the study of the transition of organic materials from the biosphere to being part of the geological record. Taphonomy is obviously important to paleontologists but it is also of critical importance to archeologists who focus on the organic component of the archeological record (e.g., zooarcheologists, paleoethnobotanists). For a long time it was assumed that bones found at a fossil site were an unbiased sample of the animals living at that place many thousands, or even millions, of years ago. Only relatively recently have scientists taken an interest in the range of processes that convert a once living community into its fossil record. Taphonomy includes the consideration of why some organisms are more likely to fossilize than others (also called **differential preservation**, or differential survivorship). For example, animals that go to water-holes to drink and to the lakeshore to graze are vulnerable to predators, and in both places their bones are likely to be protected by sediments deposited by streams and rising lake levels. These two factors increase the likelihood that such animals are more likely to become part of the fossil record than animals that live away from sources of standing water in more arid habitats. Therefore, just because the remains of mammals adapted to more arid habitats are few and far between in the fossil record does not mean these animals were few and far between in the mammalian paleocommunity sampled by that fossil record. In zooarcheological contexts, taphonomy encompasses all the processes involved from the death of an animal (e.g., its butchery, transport, and discard, to the deposition, preservation, recovery, and analysis of that animal in the form of faunal remains). By understanding the factors that bias the fossil record, allowance can be made for those biases when researchers try and reconstruct the structure of a fossil paleocommunity or when they use the archeological record to interpret hominin behavioral patterns. Some general biases are well known. In general, bones with more dense cortical bone (e.g., the **mandible**) survive better than bones that are mostly made up of more fragile **cancellous bone** (e.g., vertebrae). Small animals tend to be systematically under-represented and large ones over-represented in the mammalian fossil record (Soligo and Andrews 2005). However, size

is not the sole determinant in all groups, for objects such as vole teeth that are small but which are also dense and durable are well represented in the fossil record. *See also* **differential preservation; carcass transport strategy; carcass acquisition strategy**.

Tapiridae The perissodactyl family that comprise the tapirs. Fossil tapirs are found at archeological sites in southeast Asia, particularly in Malaysia. *See* **Perissodactyla**.

Taramsa 1 Site Taramsa Hill. Locality N/A. Surface/*in situ In situ*. Date of discovery 1994. Finders Vermeersch and colleagues. Unit N/A. Horizon Activity Phase (or AP) III. Bed/member N/A. Formation N/A. Group N/A. Nearest overlying dated horizon AP IV (56.2±5.5 ka). Nearest underlying dated horizon AP II (102.5±7.5 ka). Geological age Optically stimulated **luminescence dating** (or OSL) provides mean ages for AP III of 55.3±3.7 ka and sand above the burial has an OSL age of 68.6±8 ka; Van Peer et al. (2010) discount an OSL age of 24.3±2.1 ka from sandy fill inside the cranium. Developmental age Immature: dental evidence suggests an age at death of *c*.8–10 years. Presumed sex Unknown. Brief anatomical description The rounded forehead and occipital, the absence of a supraorbital torus, and the slenderness of the long bones of the associated skeleton are all features of anatomically modern humans. Many detailed morphological features recall the robust Epipaleolithic populations of North Africa and the early anatomically modern humans from sites in the Levant. Initial taxonomic allocation *Homo sapiens*. Announcement Vermeersch et al. 1998. Initial description Vermeersch et al. 1998. Photographs/line drawings and metrical data Vermeersch et al. 1998, Van Peer et al. 2010. Detailed anatomical description Van Peer et al. 2010. Location of original National Museum of Egyptian Civilization, Cairo, Egypt.

Taramsa Hill (Location 26°6′N, 32°42′E, Egypt; etym. named after a nearby hill) History and general description Taramsa Hill, near Qena, is an isolated landform that has been excavated by the Belgian Middle Egypt Prehistoric Project since 1989. The numerous pits and trenches suggest the site was used for the systematic quarrying of chert cobbles during the entire northeast African **Middle Stone Age**. Temporal span There are six major activity phases (or APs) at the site that span the period the Middle to the Upper Pleistocene (i.e., from <207±52.7 to 33.2±2.6 ka) based on optically stimulated **luminescence dating** (or OSL) on the sandy pit fills and one accelerator mass spectrometery date on charcoal. Archeological evidence found at site AP I (>117±10.5 ka) is a **Lupemban**-type non-Levallois production system (i.e., blades and lanceolate foliate tools). The lithic assemblage of AP II (117±10.5–102.5±7.5 ka) is a production system based on **Levallois** cores and is part of Nubian Complex *sensu lato*, but the assemblages of AP III whose average OSL age is 78.5±5.6 ka can be assigned to the Nubian Complex *sensu stricto*. The average of the OSL ages for AP IV is 56.2±5.5 and this large **Upper Paleolithic** lithic assemblage is Taramsan. The AP V technology is similar to the older Safahan industry in terms of its efficiency features and morphological aspects; this phase has covered up the AP IV debris. The Upper Paleolithic technology of AP VI dates to between 45,348 and 40,711 years BP in calibrated radiocarbon years and an OSL date of 33.2±2.6 ka; bifacial tools of the Nazlet Khater axe type in the western part of the AP VI exploitation pit were probably used for extraction. Hominins associated with the site The skeleton of a child was found on a 10 cm layer of AP III extraction debris. Key references: Geology, dating, and paleoenvironment Vermeersch et al. 1997, Van Peer et al. 2010; Archeology Vermeersch et al. 1997, Van Peer et al. 2010; Hominins Vermeersch et al. 1998, Van Peer et al. 2010.

Tata (Location 47°39′N, 18°19′E, Hungary; etym. named after the nearby city) History and general description An open-air site in a travertine deposit excavated by L. Vertes in 1958–9. It contains an unusually advanced **Middle Paleolithic** stone tool industry, with artifacts that have been used as evidence for Neanderthal symbolic behavior. Temporal span and how dated? **Radiocarbon dating** and uranium-thorium dating provide dates between 116±1.6 and 70±2 ka, or roughly **Oxygen Isotope Stage 5**. Hominins found at site None. Archeological evidence found at site The stone tools from this site represent a microlithic **Mousterian** industry. The artifact that is most often used as evidence of symbolic behavior is a polished oval plaque carved from a mammoth molar lamella and incised at right angels to a natural crack to make a "cross." Key references: Geology, dating, and paleoenvironment Schwarcz

and Skoflek 1982; Archeology Vértes et al. 1964, Marshack 1990.

Tategahana (Location 36°49′N, 138°12′E, Nagano Prefecture, Japan) History and general description An open-air site on the shoreline of Nojiriko (*ko* is "lake" in Japanese), which is located outside Nagano City on the western side of Honshu Island. During a period of low water level, a *Palaeoloxodon naumanni* (Naumann's elephant) molar was discovered by a local resident in 1949, but it was not reported until 1958. Excavations began at Tategahana site in 1962 and to date there have been 16 seasons of fieldwork, some of which have encouraged the participation of amateurs. The Tategahana site can be divided into five geologic units: J-retsu Formation, Upper Nojiriko Member (three sublevels), Middle Nojiriko Member, Lower Nojiriko Member (three sublevels), and the Kannoki Formation. The majority of the artifacts and faunal remains are derived from the Upper, Middle, and Lower Nojiriko Member levels. The Tategahana faunal assemblage is one of the few examples in Japan (Hanaizumi is another) where Paleolithic artifacts and vertebrate fauna have been found in direct association; acidic soil conditions mean that bone preservation in Pleistocene open-air sites in Japan is generally poor. Temporal span and how dated? The presence of extinct megafauna (e.g., *Palaeoloxodon*, *Sinomegaceros*) indicates a Pleistocene age for the locality and uncalibrated **radiocarbon dating** of the first four stratigraphic levels has indicated a chronological range between 49 and 8 ka, with the majority of the materials thought to date to between 49 and 11 ka. The basal Kannoki Formation is estimated to date to between 70 and 50 ka. Hominins found at site None reported. Archeological evidence found at site Cultural modification of some of the *Palaeoloxodon* and *Sinomegaceros* bones indicates Tategahana was probably a megafaunal kill-butchery site. Three clusters of modified megafauna, scrapers, knives, and points manufactured from flakes of elephant bone and stone artifacts are interpreted to be butchery areas and in one such cluster located in the Middle Nojiriko Member, a fractured *Palaeoloxodon* rib has a stone flake deeply embedded in it. Repository Nojiriko Museum, Nagano, Japan. Key references: Geology, dating, and paleoenvironment Sawada et al. 1992, Nojiriko Excavation Research Group 1994, Kondo et al. 2001, 2007; Hominins N/A; Archeology Kondo et al. 2001, 2007, Ono 2001, Norton et al. 2010a.

Taubach (Location 50°57′N, 11°23′E, Germany; etym. named after local village) History and general description This **travertine** quarry is located 5 km/3 miles southeast of Weimar, in eastern Germany. Human remains, lithics, and fauna were recovered in the 19thC during the course of quarrying. Temporal span and how dated? **Biostratigraphic** correlation suggests a date of **Oxygen Isotope Stage** 5e (c.122 ka). Hominins found at site Two hominin teeth were recovered: Taubach 1, a left M_1, and Taubach 2, a left m_1. Archeological evidence found at site A **Middle Paleolithic** microlithic industry named Taubachian after the site. Several faunal remains were also recovered, including mammoth, rhinoceros, and other large mammals. Key references: Geology, dating, and paleoenvironment Street et al. 2006; Archeology Schäfer 1981; Hominins Gieseler 1971.

Taung (Location 27°32′S, 24°48′E, South Africa; etym. Taung is derived from two words from the language of the Tswana: *tau* = lion and *ng* = place of; therefore it means "the place of the lion." The site was formerly called Taungs) History and general description The site consists of an outcrop of Thabaseek **tufa** 9.6 km/6 miles southwest of the village of Buxton, in the Northern Cape Province. At the beginning of the 20thC it was being mined commercially by the Northern Lime Company as the Buxton Limeworks. Solution cavities filled with pink breccia occur within the lime-rich tufa and fossil **baboons** were first noted in one of these breccia-filled cavities in 1919. Raymond **Dart**, the Professor of Anatomy at the University of the Witwatersrand, learned about the presence of fossil baboons at Taung in the middle of 1924. E.G. Izod, a director of Rand Mines Limited, had visited Taung in the late summer of 1924 and had brought back to Johannesburg a fossil cranium the mine manager had kept in his office as a souvenir. Izod's son, who was a student at the University of the Witwatersrand, showed it to his fellow students, among whom was Josephine Salmons, one of Dart's students. She borrowed the cranium to show to her Professor, and Dart recognized it as a fossil baboon. Dart then asked R.B. (Robert) Young, Professor of Geology at the University of the Witwatersrand, to use his connections with the Northern Lime Company to make sure that if any similar crania and skulls were to be found that

they should be saved and shipped to Dart. Young promised to do this when he visited Taung, which he was scheduled to do in November 1924. Shortly before Young's arrival at Taung, the quarry-master, de Bruyn, noticed the front of a skull in the rubble after a blast, but the skull looked different from the now familiar baboons. de Bruyn gave the specimen to the mine manager, A.E. Spiers, who showed it to Young. There are conflicting accounts about whether Young brought the skull to Johannesburg and delivered it personally to Dart (with the boxes containing the other fossils following by rail), or whether Young made arrangements for the new fossil, together with more fossil baboon skulls, to be sent by rail to Dart. The former is the more likely, and it is reasonably certain that Young and the unusual skull arrived at Dart's house in Melrose, Johannesburg, on November 28. Dart recognized that the specimen was the skull of an infant, and after cleaning it he also concluded that it was not a baboon skull, but that it was the skull of a hitherto unknown man-like ape. Dart made what has become known as the "Taung child" the type specimen of a new genus and species, *Australopithecus africanus* (Dart 1925). No formal investigations were made at the find site, but in 1925 Ales Hrdlicka excavated fossil baboons from an adjacent cave, and between 1947 and 1950 Frank Peabody undertook excavations there. Jeff McKee, Phillip **Tobias**, and Michel Toussaint (in 1988 and 1989) worked at Taung between 1988 and 1993. They recovered fauna from breccia that is very likely the same as that adhering to the Taung child's skull and found it to be comparable to the fauna from Mb 4 at **Sterkfontein**. Temporal span and how dated? **Biostratigraphy** points to an age for Taung of c.2.6–2.8 Ma. Hominins found at site The only hominin recovered from the site, **Taung 1**, is the holotype of *Australopithecus africanus*. Berger and Clarke (1995) have suggested that the Taung child was taken to the cave by a large **raptor**, citing the otherwise unusual areas of damage in the orbit, and Berger and McGraw (2007) report similar damage to monkey crania taken by raptors in the Tai Forest, Ivory Coast. Archeological evidence found at site None. Key references: Geology, dating, and paleoenvironment McKee 1993, McKee and Tobias 1994, Partridge 2000; Hominins Dart 1925, 1926, Falk 1980, 1983, Holloway 1981b, 1984, Dean and Wood 1984, Falk and Clarke 2007.

Taung 1 Site Taung. Locality Unknown. Surface/*in situ* Recovered from pink calcified sand that formed part of what was described as the "solidified floor" (**Dart** 1926, p. 319) of the cave illustrated in that publication (*ibid*, p. 319). Date of discovery 1924. Finder A miner, probably M. de Bruyn, but the breccia fragment that included Taung 1 was selected to be sent to Johannesburg by Professor R.D. Young. Unit/horizon Pink calcified sand. Bed/member Member 1. Formation N/A. Group N/A. Nearest overlying dated horizon N/A. Nearest underlying dated horizon N/A. Geological age *c*.2.6–2.8 Ma. Developmental age Juvenile. Presumed sex Unknown. Brief anatomical description A mostly undistorted skull of a juvenile hominin. It preserves most of the right half of a natural **endocast**, the face, the upper jaw, part of the cranial base, and the mandible. The natural endocast preserves some endocranial morphology in exquisite detail, but frustratingly the area where the **lunate sulcus** is most likely located is poorly preserved (see below). The dentition of Taung 1 includes the maxillary and mandibular permanent molars, all of the deciduous teeth on at least one side, and the germs of the deciduous premolars, canines, and incisors. Dart (1926, p. 319) reported the breccia also contained "fragments of the distal ends of the forearm bones and of the phalanges," but he also wrote that "these proved too fragmentary and too friable to develop" (*ibid*). Announcement and initial description Dart 1925. Photographs/line drawings and metrical data Dart 1925, 1934. Detailed anatomical description Dart (1934) provides the closest there is to a detailed description. It focuses on the dentition and includes excellent plain radiographs with exquisite images of the germs of the permanent teeth. Endocranial volume 382 cm^3 (estimated to be equivalent to an adult value of 406 cm^3) (Falk and Clarke 2007). Initial taxonomic allocation *Australopithecus africanus*. Taxonomic revisions Partridge (1973) used geomorphological principles to investigate the timing of the opening of the southern African hominin cave sites, and suggested that the cave where Taung 1 was found may have opened as recently as 1 Ma. Tobias (1973b, p. 82) responded to these proposals by proposing the hypothesis that because the Taung site may be ">1.0 m.y., and probably about 2.0 m.y. more recent than the latest of the other African fossils assigned to *A. africanus*" then the Taung child may represent "a member of *A. robus-*

tus" (*ibid*, p. 82). However, a raft of studies (e.g., Grine 1985, Wood and Dean 1986, McNulty et al. 2006) have since demonstrated that the morphology of Taung 1 is consistent with it belonging to the same hominin taxon as that sampled at **Sterkfontein** and **Makapansgat**. Current conventional taxonomic allocation *Au. africanus*. Informal taxonomic category Archaic hominin. Significance The Taung 1 cranium is still iconic for many reasons. It was the first evidence of an early hominin to be recovered from Africa, it is the **holotype** of *Au. africanus*, it was the first hominin skull to be imaged using **computed tomography** (Conroy and Vannier 1987), and its natural endocast has been the focus of one of the longest-running disputes in paleoanthropology, namely the placement of the lunate sulcus. Whereas Ralph **Holloway** (e. g., Holloway 1984) is convinced the lunate sulcus is situated in a posterior, more modern human-like, position, Dean Falk (e.g., Falk 1989) is just as strongly convinced the lunate sulcus is in a more anterior, extant great ape-like, location. Berger and Clarke (1995) suggested the Taung child was taken to the cave by a large **raptor**, citing the otherwise unusual areas of damage in the orbit, and Berger and McGraw (2007) report similar damage to monkey crania taken by raptors in the Tai Forest, Ivory Coast. Location of original School of Anatomical Sciences, University of the Witwatersrand, Johannesburg, South Africa. *See also* **Taung**; **lunate sulcus**.

Taungs *See* **Taung**.

taurodont (Gk *tauros*=bull and *dont*=teeth) Molar teeth that have tall **pulp chamber**s relative to the distance between the roof of the pulp chamber and the tip of the longest root. In other words, the bi- or trifurcation of multirooted teeth occurs further from the cervical margin of the tooth and the common root trunk makes up a greater proportion of total root length. High incidences of taurodontism are claimed for the molars of *Homo neanderthalensis* and for African populations, although the degree of wear and age of the individual also influences this condition (Constant and Grine 2001). Quantified using a "Taurodont Index" (Keene 1966) (ant. **cynodont**). *See also* **pulp**; **pulp chamber**.

taurodontism The state of having taurodont teeth. *See* **taurodont**.

Tautavel Cave *See* **Caune de l'Arago**.

taxa Pl. of **taxon** (*which see*).

taxic (Gk *taxis*=arrangement) A term used by Eldredge (1979) for one of the two main theoretical components of evolutionary theory. He suggested that in order to be comprehensive all evolutionary theories must have at least two theoretical components. One theoretical component is concerned with explaining how evolutionary change occurs, be it at the level of the genome or the phenotype. He refers to this as the transformational component, or approach. The second theoretical component is concerned with how species originate and thus it implicitly acknowledges that species exist. He refers to this as the taxic approach. Eldredge (1979) suggested that overdominance of one or other of these emphases makes for "poor" science. He argues that whereas the taxic approach has to acknowledge the transformational approach, the converse is not always the case, claiming that some paleontologists have effectively abandoned the search for the equivalent of extant species in the fossil record. He cited George Gaylord Simpson's (1951) concept of an "evolutionary" species as an example of this tendency. A contemporary example is paleontologists who are more concerned with the recognition and identification of lineages in the fossil record than they are with the recognition and identification of species. *See also* **species**.

taxon (Gk *taxis*=to arrange, or 'put in order') A group recognized at any level, or category, in the Linnaean hierarchy. So, the tribe **Hominini**, the genus *Paranthropus*, and the species *Paranthropus boisei* are all examples of a taxon. Pl. taxa.

taxonomy (Gk *taxis*=arrange, or "put in order") The principles involved in assembling individual organisms into groups, the formalization of those groups as **taxa**, giving formal names to the taxa, allocating the taxa to taxonomic categories, and then assembling the taxonomic categories into a hierarchical **classification**. The term is also sometimes used to refer to the taxonomic hypothesis that results when Linnaean taxonomic principles are applied to a particular group of organisms; many researchers would refer to this as a hominin classification, but some refer to it as a hominin taxonomy. It is common for reference to be made to "alpha taxonomy," but very seldom do you read, or hear, any reference to "beta" or "gamma" taxonomy. Mayr et al. (1953) is one of the few sources that define these terms [e.g., these terms do not appear in the index of George Gaylord Simpson's (1960) *Principles of Animal Taxonomy*]. Mayr et al. (1953) suggests that "the taxonomy of a given group, therefore, passes through several stages...informally referred to as alpha, beta, and **gamma taxonomy**" (*ibid*, p. 19). **Alpha taxonomy** results in species being "characterized and named," **beta taxonomy** involves arranging species in "a natural system of lesser and higher categories," and gamma taxonomy to the "analysis of intraspecific variation and to evolutionary studies" (*ibid*, p. 19). *See also* **alpha taxonomy**.

***Tchadanthropus uxoris* Coppens, 1965. Unavailable (proposed conditionally)** *See* **Coppens, Yves (1934–)**.

TD Abbreviation of Trinchera Dolina and the prefix used for the fossils recovered from the **Gran Dolina** at **Atapuerca**, Spain.

TD6 One of the lithostratigraphic levels within the Gran Dolina, a site that is part of the sediment-filled cave system called the **Trinchera del Ferrocarril**, which itself is within the Sierra de Atapuerca, a series of eroded limestone hills 14 km/8.5 miles east of Burgos in northern Spain. The TD6 level includes the **Aurora stratum**. *See also* **Gran Dolina**; **lithostratigraphy**.

TDS Abbreviation for Trinchera Dolina Sondeo Sur, the name given to the focus of the excavations at **Gran Dolina**, in Atapuerca, Spain, carried out between 1993 and 1999.

teaching Caro and Hauser (1992) have proposed a working definition for teaching that takes into consideration empirical work and an appreciation of evolutionary theory. Thus, "An individual actor A can be said to teach if it modifies its behavior only in the presence of a naive observer, B, at some cost or at least without obtaining an immediate benefit for itself. A's behavior thereby encourages or punishes B's behavior, or provides B with experience, or sets an example for B. As a result, B acquires knowledge or learns a skill earlier in life or more rapidly or efficiently than it might otherwise do, or that it would not learn at all" (*ibid*, p. 153). An important feature of Caro and Hauser's conceptualization of teaching is that teachers must incur a cost to their own productivity, and may even encourage or discourage the observer. According to this influential paper, teaching need not require an explicit understanding of another's mind or knowledge, a feature that may be unique to the forms of teaching practiced by modern humans. Teaching by nonhuman animals is not limited to a specific taxonomic group, nor is it only seen in mammals or primates. The most convincing examples of teaching come from animals that breed cooperatively (e.g., ants, pied babblers, and meerkats). Meerkats, for example, take turns monitoring for predators while others forage for food; and in some cases, a group member will present young offspring with prey that is partially maimed in order to provide them with direct hunting experience. By doing this, a meerkat may lose the prey if the offspring are not successful in killing it, making this behavior a costly act for the teacher. However, it is clear that modern human teaching differs from the teaching observed among cooperative breeders such as meerkats. Csibra and Gergely (2009) have recently argued that modern human teaching is marked by the evolution of a unique communicative system that evolved to facilitate the exchange of information from one individual to another. This "natural pedagogy" facilitates the social learning of cognitively opaque cultural information that would be difficult to learn by observational learning alone. They argue that modern human infants have been shaped by natural selection to become attuned to a variety of communicative cues emitted by adults, allowing infants to learn rapidly and make specific as well as accurate generalizations from these cues.

technostratigraphy In archeology, using the recovered lithic technology (e.g. **Mousterian**, **Oldowan**) to provide a relative or correlated date for a layer or site.

teeth (OE *toth* = tooth) In each quadrant of the upper and lower jaw there are normally five deciduous teeth [mandible: di_1, di_2, dc, dm_1, dm_2 (or dp_1 and dp_2); maxilla: di^1, di^2, dc, dm^1, dm^2 (or dp^1 and dp^2)] and eight permanent teeth [mandible: I_1, I_2, C, P_3, P_4, M_1, M_2, M_3; maxilla: I^1, I^2, C, P^3, P^4, M^1, M^2, M^3]. Note that, because anthropoids retain only the distal two of the four premolars seen in primitive mammals, the two retained premolars are numbered P3 and P4. *See also* **deciduous dentition**; **permanent dentition**.

tektites (Gk *tektos* = molten) Approximately spherical natural glass rocks, usually black or olive-green, ranging up to a few centimeters in size. Showers of tektites are formed when large meteorites hit the Earth's surface. The area of the Earth's surface a particular tektite shower covers is called its strewn-field.

telanthropoid (Gk *tele* = distant, *anthropos* = human being, and *oeides* = form) A term used to suggest a specimen is ***Telanthropus***-like. For instance, one of the three mandibles from **Tighenif** (Tighenif 1) is substantially larger than the other two, and Briggs (1968) suggested that whereas it belonged to a "paranthropoid" group, he assigned the smaller Tighenif 1–2 mandibles to a "telanthropoid" group.

Telanthropus (Gk *tele* = distant and *anthropos* = human being) Genus established by Robert **Broom** and John **Robinson** in 1949 to accommodate the new species they established to accommodate **SK 15**, an adult mandible recovered at **Swartkrans** from "brown breccia" now referred to as being from Member 2. Robinson (1961) formally sank the taxon into *Homo*. Type species *Telanthropus capensis* **Broom and Robinson, 1949**. *See Telanthropus capensis*.

Telanthropus capensis **Broom and Robinson, 1949** (Gk *tele* = distant, *anthropos* = human being, and cape = reference to land that includes the Cape of Good Hope) Hominin species established by Robert **Broom** and John **Robinson** in 1949 to accommodate **SK 15**, an adult mandible recovered at **Swartkrans** from brown breccia now referred to as being from Member 2. They considered SK 15 to be "somewhat allied to Heidelberg man" (Broom and Robinson 1949, p. 322) and "intermediate between *P. crassidens* and *Homo*" (*ibid*, p. 323). Robinson (1961) transferred the mandible to *Homo erectus* and *T. capensis* is now widely regarded as a **junior synonym** of *Homo erectus* (e.g., Howell 1978, p. 198). Note that, if transferred to the genus *Homo*, the name *capensis* is a junior **homonym** of *Homo capensis* Broom, **1918** (type specimen: the Boskop calvaria, usually assigned to *Homo sapiens*), so *Homo capensis* is not an available **Linnaean binomial** for the Swartkrans mandible. First discovery and holotype SK 15 (1949). Main site Swartkrans, southern Africa.

teleconid A term introduced by Vanderbroek (1961, 1967) for what more commonly used schemes of cusp nomenclature refer to as the **hypoconid**. *See* **hypoconid**.

telencephalon *See* **forebrain**.

telocentric *See* **chomosome**.

Témara A city in Morocco. The name has been used in the literature to refer to two different cave sites: Smugglers' Cave and El Harhoura. *See* **Smugglers' Cave**; **El Harhoura**.

temperate (L. temperatus = restrained or regulated) The term for the regions (and the organisms occupying that region) between the Tropic of Cancer (approximately 23°N) and the northern polar circle, and between the Tropic of Capricorn (approximately 23°S) and the southern polar circle (i.e., between latitude 23° and 66° both north and south). The sun is never directly overhead in temperate regions. Compared to the **tropical** region its biodiversity is lower (this is known as Rapoport's rule). Very few contemporary primates live in temperate regions, although in the past, when the Earth's climate was warmer, the latitudinal range of primates was greater. Although living baboons are found below the Tropic of Capricorn in southern Africa, macaques are the only living nonhuman primates to have radiated extensively into temperate zones. Anatomically modern humans and their recent extinct relatives used cultural and behavioral adaptations to exploit temperate regions, and extant modern humans have extended their range even further, into the Antarctic and Arctic circles.

temperate deciduous forest (Latin dciduus = falling off) A biome that consists of trees, both in the upper and lower canopy, that lose their leaves simultaneously and may remain bare for several months. These forests are found in latitudes between 35 and 60°N, are in highly seasonal climates and have a mean annual rainfall of between 500–1500 mm per year and average monthly temperatures of 0–20°C. Most animals in these forests have unique behavioral and physiological adaptations to cope with the extreme fluctuations in climatic variables. Once covering large areas of Europe, Asia, and South and North America, temperate deciduous forests are now largely confined to eastern North America, eastern Australia, central Europe, Japan, and eastern China.

tempo (L. tempus = time) When tempo is used in connection with evolution it refers to the time it takes for evolution to occur (i.e., the rate at which evolution occurs). The other main variable is the **mode**, or the pattern, of evolution. George Gaylord Simpson's *Tempo and Mode in Evolution* (1944) helped push evolutionary biologists in the direction of recognizing these two critical components of the evolutionary process.

temporal lobe One of the four main subdivisions of the cerebral cortex of each cerebral hemisphere. It is located ventral to (i.e., beneath) the Sylvian fissure in both hemispheres. The temporal lobe houses the primary auditory cortex, which is located along the posterior part of its superior surface. Other parts, or areas, within the temporal lobe are involved in language, higher-order auditory processing, and the processing of visual information important for object perception and recognition. The medial part of the temporal lobe contains the hippocampus and the amygdala.

temporalis *See* **mastication, muscles of**.

temporomandibular joint (or TMJ) (L. tempus = time. The temple of the forehead that overlies the temporal bone is the first area to show grey hairs, signifying the passage of time) A modified synovial joint between the condyle of the mandible and the mandibular fossa and articular tubercle of the temporal bone. The bones are covered with white fibrocartilage not yellow hyaline cartilage. A meniscus partly divides the joint cavity into superior and inferior compartments. Gliding mostly occurs in the upper compartment and rotation in the lower. *See also* **cartilage**.

tentorium cerebelli (L. tentorium = a tent and cerebelli = diminutive of cerebrum, also called the "small brain") A fold of the fibrous, or meningeal, layer of the **dura mater** that covers the cerebellum. It transmits the brain stem, but otherwise separates the posterior cranial fossa from the rest of the cranial cavity. *See also* **endocranial cavity**.

tephra (Gk *tephra*=ash) A collective term for rock fragments that have been ejected explosively from a volcanic source (i.e., **pyroclastic** material) that forms unconsolidated (i.e., loosely bonded) deposits. As Lowe and Hunt (2001) note, strictly speaking tephra is singular and is used as a collective with singular verb forms, although appending an "s" to form the plural "tephras" may serve to reduce ambiguity. Tephra range in size from fine ash (particles of approximately 0.063 mm) to larger "bombs" greater than 64 mm in size (Schmid 1981). A consolidated (i.e., firm or hardened) tephra deposit is called a **tuff**.

tephrochronology (Gk *tephra*=ash and *kronos*=time) The use of dated **tuffs** to provide a regional chronology.

tephrostratigraphy (Gk *tephra*=ash, *stratos*= to cover or spread, and *graphos*=to draw or write) The use of chemically or lithologically distinctive tuffs to correlate strata from one **locality** to another, or from one **site** to another.

Terminologia Anatomica *See* **anatomical terminology**.

termites Eusocial insects in the order Isoptera, termites are an important source of protein and lipids for some chimpanzee communities (Bogart and Pruetz 2008). The extractive foraging techniques that have been observed involve the use of sticks or grass, sometimes extensively modified, to remove termites from the mounds that are conspicuous and often abundant features of many African landscapes (McGrew 1992). Females chimpanzees are said to be up to three times more likely to spend time fishing for termites than males (McGrew 1979) and there is evidence that chimpanzee termite consumption increases in open environments (e.g., Mt. Assirik and Fongoli; see McGrew 1983; Bogart and Pruetz 2008). Potential evidence for the consumption of termites by early hominins came in the form of use-wear patterns observed on bone tools from the sites **Swartkrans** and **Drimolen** in South Africa (Backwell and d'Errico 2001, 2008). Use was made of **actualistic**

studies to show that microscopic wear on the bone tools resembled the fine and highly oriented wear seen when experimental bone tools were used to dig termite mounds. Other potential support for termite consumption comes from evidence for the consumption of ^{13}C-rich C_4 **foods** by australopiths (Sponheimer et al. 2005, van der Merwe. 2008). Many termites, including the mound-building termites belonging to the genus *Macrotermes*, a favorite of many chimpanzee populations (McGrew 1992), consume C_4 vegetation. The diets of *Macrotermes* are subject to considerable variation and their carbon isotope compositions tend to track the relative abundance of C_3 and C_4 vegetation in their habitats (Schuurman 2006). Thus, they (and many other termite taxa) tend towards mixed C_3/C_4 consumption in wooded savanna environments (Sponheimer et al. 2005). Consequently, it is unlikely that termite consumption could account for the high proportion (roughly 80%) of C_4 foods consumed by some australopiths (van der Merwe et al. 2008). Harvester termites of the genera *Trinervitermes* and *Hodotermes* do consume principally C_4 vegetation and in theory their consumption could explain the C_4 signal of australopiths, but because they do not all build conspicuous nests they would be unlikely targets for foraging hominins (Sponheimer et al. 2005) and large-bodied insect dietary specialists are rare among mammals.

Ternifine The old name for the site of **Tighenif** (or Tighennif). *See* **Tighenif**.

Terra Amata (Location 43°41′51″N, 7°17′20″E, France; etym. It. for beloved land) History and general description This site was located on an ancient beach, about 20 m above current sea level, on Mount Boron in Nice, France. It was discovered and excavated during construction in 1966; today a museum stands on the site (Musée de paléontologie humaine de Terra-Amata, or Museum of Human Paleontology of Terra Amata). Henry de Lumley, who excavated the site, claims that the site preserves evidence of early use of shelters and fires, in a closely constrained series of living floors. Paolo Villa has argued that many of the stones de Lumley suggested were the outline of shelters were instead deposited by soil creep or other natural forces. She also noted that artifacts found in different "living floors" actually refit, implying a significant amount of vertical displacement. Temporal span and how dated?

Thermoluminescence dating of flint has suggested an age of 380 ka, but Villa argues for an age of *c*.230 ka. Hominins found at site None. Archeological evidence found at site Stones in circular arrangements centered around patches of dark sediment, that de Lumley interpreted as huts surrounding fire pits, and a variety of **Acheulean** stone tools. Coprolites were found in a few layers, and these preserved plant remains that suggested a seasonal occupation of the site. Key references: Geology, dating, and paleoenvironment Wintle and Aitken 1977, Villa 1983, de Lumley 2007; Hominins N/A; Archeology Villa 1983, de Lumley 2007.

terrestrial   (L. *terrestris*=earthly) Animals that live on the ground. The vast majority of primates are dependent on trees for living, sleeping and foraging, but some nonetheless spend a significant proportion of their time on the ground. Terrestrial primates tend to be relatively large and have more **robust** limb bones than do arboreal primates, although there are several exceptions to these trends. The most terrestrial living nonhuman primate is *Theropithecus gelada*, but many other taxa, including ring-tailed lemurs, common baboons, patas monkeys, vervets, mandrills, rhesus macaques, and gorillas also spend a significant amount of time on the ground. Extinct terrestrial primates include *Victoriapithecus*, *Gigantopithecus*, and **hominins**. Terrestriality is an important adaptation in areas where tree cover is broken, which could partially explain why hominins eventually became habitually terrestrial. Terrestrial primates often have relatively large **home range**s and **day range**s. Increased susceptibility to **predation** is an important cost of terrestriality and terrestrial primates therefore tend to live in relatively large groups and may seek refuge from predators in trees, which are often used as sleeping sites. Terrestial and bipedal locomotion are usually treated synonymously in debates about early hominins, but Kevin Hunt (1996) has argued that bipedalism might have been selected in part as a postural rather than a locomotor adaptation. *See also* **seed-eating hypothesis**; **postural feeding hypothesis**.

terrestriality   (L. *terrestris*=earthly) The tendency to live partially, or wholly, on the ground. *See* **terrestrial**.

Terry (Robert J.) Collection   Taxon *Homo sapiens*. History Collection of modern human skeletons assembled by Robert J Terry (1871–1966). The Terry Collection comprises the remains of citizens predominantly from the St Louis, Missouri, region of the USA collected between 1917 and 1966. Two experiences prompted Terry to make an anatomical skeletal collection: his time as a student of George S. Huntington and the year he spent in Edinburgh where he studied under Sir William Turner. Terry became the Professor of Anatomy at the Missouri Medical College (which later became the Washington University Medical School) and began to collect skeletal remains from the cadavers starting in 1898. The Washington University Collection was actually the third attempt by Terry to establish a collection of modern human cadavers with detailed supporting documentation; one previous start became mixed by inaccurate curation, and the other was destroyed in a fire. Terry retired in 1941 but the collection continued to accumulate under the direction of Mildred Trotter. Trotter's initial title was "Coursemaster of Human Anatomy" but she became a full Professor of Anatomy in 1946; it was Trotter who changed the name of the Washington University Collection to the Robert J. Terry Collection. More than 80% of the collection was assembled by the time Trotter took over but at the time of her succession the collection was dominated by aged males. Trotter's main contribution was to increase the numbers of females and younger individuals. Around the time of her retirement Trotter began negotiations with T. Dale Stewart with a view to transferring the collection to the Department of Anthropology at the **National Museum of Natural History, Smithsonian Institution** (or NMNH). The transfer took place in 1967, the year Trotter retired, and the collection is now part of the permanent collection of modern human skeletons at the NMNH. The Robert J. Terry Collection comprises 1728 individuals, 1608 of which have morgue-identified ages and 68% (1182) of which have accompanying anthropometric data. Curator/contact David Hunt (e-mail huntd@si.edu). Postal address Department of Anthropology, National Museum of Natural History, PO Box 37012, Smithsonian Institution, Washington DC 20013-7012, USA. Website http://anthropology.si.edu/cm/terry.htm. Relevant references Hunt and Albanese 2005.*See also* **Huntington (George S.) collection**.

Tertiary   (L. *tertius*=the third in a series) A unit of geological time (i.e., a **geochronologic unit**). Although the Tertiary is no longer recognized as a formal geochronologic or chronostratigraphic unit, many still use the term to refer to the pre-**Pleistocene** part of the **Cenozoic** era. The term is a remnant of the old classification of geologic **strata** (and time) into four units: Primary, Secondary, Tertiary, and **Quaternary**. The decision was made to split up the Tertiary into the Paleogene and **Neogene** periods in order for the Cenozoic era to more closely match the duration of the Mesozoic and Paleozoic eras.

tertiary dentine   *See* dentine.

Teshik-Tash [Location approximately 38°17′ 24″N, 67°02′46″E, Baisan Tau Mountains, near Pas-Machai (Pas-Machay), 18 km/11 miles from Baisun (Boysun), Uzbekistan; etym. Uzbek stone with an opening] History and general description In 1938, A. Okladnikov discovered a skeleton of a juvenile *Homo neanderthalensis* in this cave site. The skeleton was surrounded by five pairs of goat horns, which the discoverer and others have interpreted to be evidence of intentional burial with symbolic behavior. This interpretation is questioned by others who argue that the fossil was damaged by carnivores and may not have been buried, and that goat remains are common throughout the site and therefore may not be intentionally associated with the skeleton. The site is also one of the easternmost Neanderthal inhabitations. Temporal span and how dated? Based on the archeology, fauna, and fossil hominin remains, it has been assigned to the **Middle Paleolithic**. Hominins found at site The remains of a juvenile, possibly 8–11 years old. Most of the skeleton was poorly preserved, but the skull and mandible, although fragmentary, were sufficiently preserved to be reconstructed. Genetic analysis of the **mitochondrial DNA** from the left femur confirms that this specimen is a Neanderthal. Archeological evidence found at site Okladnikov recovered five layers "with remains of ancient culture," at least one of which has been attributed to the **Mousterian**. Key references: Geology, dating, and paleoenvironment Hrdlička 1939, Movius 1953; Hominins Hrdlička 1939, Krause et al. 2007, Movius 1953; Archeology Hrdlička 1939, Movius 1953.

TEX86 An abbreviation of "tetraether index of 86 carbon atoms," which refers to a relatively new paleoclimate index that is based upon the abundance ratio of a group of organic molecules all of which have 86 carbons. These molecules are produced by members of the Archaea, a major group of prokaryotes, and they are found in marine and lake sediments. Because their molecular structure varies with temperature they can be used as a **proxy** for **sea surface temperature** (or SST) and thus they can be used in **paleoclimate** reconstruction. For example a *c*.60 ka-long record from Lake Tanganyika implies that lake surface temperature dropped by more than 4°C during the **Last Glacial Maximum** (Tierney et al. 2008).

thalamus (Gk *thalamus*=inner chamber or bedroom) Part of the **diencephalon**, one of the two major components of the **forebrain** (the **cerebral hemispheres** are the other). It is a bilateral subcortical structure consisting of a set of nuclei and it is situated superior to the hypothalamus. The thalamus has many functions, including synthesizing and relaying incoming sensory information from lower processing centers to the cerebral cortex from all the sensory modalities except for olfaction. It is also thought to regulate levels of awareness and emotional aspects of sensory experience through a wide variety of effects on the cerebral cortex. Some evidence suggests that the nuclei in the thalamus, with interconnections with the **limbic system** in the brain, have undergone disproportionate enlargement in modern humans in comparison with the apes (Armstrong 1980).

The High Cave   *See* Mugharet el 'Aliya.

The L.S.B. Leakey Foundation   *See* L.S.B. Leakey Foundation.

theory of mind   The ability to attribute perceptions, knowledge, intentions, goals, and beliefs to oneself and others. The term was first used by Premack and Woodruff (1978) in their pioneering paper, "Does the chimpanzee have a theory of mind?" Their studies generated considerable excitement and controversy and stimulated research into the "mindreading" skills of both modern human children and nonhuman primates. Theory of mind tests may be divided into two categories: tests of true beliefs and tests of false beliefs. Tests of true beliefs assess second-order intentionality; that is, an individual's ability to attribute beliefs, intentions, goals, etc. to another individual. Tests of false beliefs assess third-order intentionality, which involves beliefs about (second-order) beliefs.

In a classic study assessing true beliefs in modern human children and chimpanzees, Povinelli and Eddy (1996) had participants gesture to one of two experimenters: one who was blindfolded (and could not see) or another who wore a bandana over their forehead (and could see). In this study, "seeing" represents a psychological property that we infer from shared experiences. I look at you and you look at me, so I see you and you see me. Seeing, in this case, is a true belief because my psychological state matches your own. However, such is not the case in false beliefs (e.g., lies), where what you believe is not the same as what others believe. In a classic false-belief study, Wimmer and Perner (1983) used the "Sally-Anne" task to assess when children understand that others may have "false beliefs" or beliefs that differ from their own. In this task, a child sees a puppet, Sally, hiding candy in a basket while a second puppet, Anne, watches. Anne then leaves the room, but while she is away Sally retrieves the candy and moves it to another location in a box instead of its original hiding place in the basket. Anne reappears and the child is asked: "Where will Anne look for the candy?" Three-year olds typically say that Anne will look for the candy in the box, where they know the candy is *now*; that is, they fail to attribute a different (false) belief to Anne. However, 4–year olds typically answer that Anne will look for the candy in the basket, which is where Anne saw the candy placed before she left the room. At the moment there is no consensus as to whether or not nonhuman primates understand either true beliefs or false beliefs. Some authors have argued that chimpanzees (Hare et al. 2000, Hare et al. 2001) and rhesus monkeys (Flombaum and Santos 2005) understand true beliefs such as "seeing" and "hearing." Another study has argued that whereas chimpanzees understand true beliefs, they do not understand false beliefs (Krachun et al. 2009). From this, one may infer that reasoning about true beliefs is a more primitive cognitive trait, whereas reasoning about false beliefs is a more derived trait and likely appeared later in hominin evolution. However, one must assess these claims cautiously. Povinelli and colleagues (Povinelli and Vonk 2004, Penn and Povinelli 2007), for example, have argued that true belief tasks cannot discriminate between an understanding of mental states and an understanding of behavioral states, because to understand mind you have to understand behavior (e.g., eyes are necessary for "seeing"). Consequently, it is possible to reason about a person's behavior without

necessarily being able to reason about their mental states. If this is true, then an individual who makes inferences about minds (and behavior) and one who makes inferences about behaviors (but not minds) will respond similarly in true belief tasks. *See also* **intentionality**.

thermal ionization mass spectrometry *See* **uranium-series dating**.

thermoluminescence dating *See* **luminescence dating**.

***Theropithecus* Geoffroy, 1843** (Gk *ther*= wild beast and *pithekos*=ape) Background A genus of Old World monkey with only one extant species, the highly terrestrial *Theropithecus gelada*, whose habitat is presently restricted to the Ethiopian highlands. The modern *T. gelada* is unusual among primates in having a diet restricted to grass. *Theropithecus* was identified in the fossil record as a distinct genus by Andrews (1916), who assigned the species we now know as *Theropithecus oswaldi* to *Simopithecus*. This classification was revised when *Simopithecus* was made a junior synonym of *Theropithecus* (Jolly 1972). Six fossil species of *Theropithecus* are currently recognized: *T. oswaldi*, *T. darti*, *T. atlanticus*, *T. brumpti*, *T. baringensis*, and *T. quadratirostris*, but some authors recognize fewer taxa (e.g., Frost and Delson 2002 classify *T. atlanticus* and *T. darti* in *T. oswaldi*). Relevance for human evolution Although *Theropithecus* is only distantly related to the hominin clade it is relevant to hominin evolution because (a) the genus apparently underwent an adaptive radiation in the Plio-Pleistocene (i.e., at much the same time as the hominin clade), (b) at most African early hominin sites *Theropithecus* is usually the most closely related primate to be found along with the hominin fossils, (c) of this any success at reconstructing the habitats of fossil theropiths might help in the reconstruction of early hominin habitats, and (d) the unique status of fossil *Theropithecus* as large-bodied, open habitat primates led to their use as an analogue for hominin behavioural evolution (e.g., the **seed-eating hypothesis**; Jolly 1970). Many of the features that distinguish *Theropithecus* from other **papionins**, including its high-crowned molars and opposable digits, are related to a diet of grass and seeds. However, palaeobiological studies indicate that fossil theropith diets might have included a broader range of food items, including leaves and fruits (Teaford 1993, Benefit 1999, Codron et al. 2005). The one extant species, *T. gelada*, is highly terrestrial, and it is likely that some extinct members of the genus, particularly *T. oswaldi*, also preferred an open woodland or a grassland habitat (Elton 2002), although *T. brumpti* has been interpreted as preferring closed, forested habitats (Krentz 1993, Ciochon 1993, Elton 2000, Jablonski et al. 2002). *T. oswaldi* is also notable because some populations exhibited a marked increase in body size, with the largest individuals weighing as much as 60 kg (Delson et al. 2000). There were two partially contemporaneous *Theropithecus* lineages in the **Turkana Basin** during the Plio-Pleistocene, one leading to *T. oswaldi* and the other to *T. brumpti* (Leakey 1993). In line with their apparently different habitat preferences, *T. oswaldi* and *T. brumpti* also had different adaptive strategies (Teaford 1993, Krentz 1993, Elton 2002, Jablonski et al. 2002). As well as dispersing from East Africa into southern Africa on at least two occasions, *Theropithecus* also colonized north Africa and parts of Eurasia (Pickford 1993). By 0.5 Ma, *Theropithecus* had all but disappeared in Africa, and the only extant member of the genus, *T. gelada*, is found in isolated populations in the highlands of Ethiopia and Eritrea. *See also* **seed-eating hypothesis**.

thin-plate spline interpolation *See* transformation **grids**.

Thomas Quarry (Location 33°34′N, 07°42′W, Morocco; etym. Thomas is the name of the family who rented the land for quarrying) History and general description One of several sites discovered as a result of sandstone quarrying activities that began in 1907 during harbor construction near Casablanca, Morocco. R. Neuville and A. Ruhlmann noted it in their 1941 survey of the sequence of Miocene to Holocene sea-level transgressions and regressions on the Atlantic coast and Pierre Biberson visited it during the 1950–60s and referred to it briefly in his 1961 synthesis. Detailed stratigraphic studies began in 1985 and modern controlled excavations, which began in 1988, were undertaken by a joint French-Moroccan research team led by Jean-Paul Raynal and Fatima-Zohra Sbihi-Alaouiled. Since 1978 further research has been undertaken by a joint Moroccan–French research team led by Jean-Paul Reynal. The Thomas Quarry site was initially divided into three quarries: I, II, and III. Caves in Thomas Quarry I and Thomas Quarry III have yielded hominin remains, but the Thomas III (also known as Oulad Hamida 1) cave has subsequently been destroyed. In 1969 P. Beriro discovered a mandible in a cave of Thomas I, and in 1972 skull fragments were discovered in Thomas Quarry III. Between 1994 and 2005 Raynal's excavations recovered four hominin teeth in the GH cave of Thomas Quarry I. In May 2008 a complete mandible was discovered in the same level and in May of the following year a further mandible belonging to a young individual was recovered. Along with the fossils from Salé and Sidi Abderrahman, the Thomas Quarry fossils represent a distinctive north African group that falls somewhere between *Homo erectus* and *Homo sapiens*. Temporal span and how dated? Lithostratigraphy suggests an upper age range of 700–600 ka for these sediments based on data from the Rhinoceros cave (GDR) at Oulad Hamida 1 Quarry (former Thomas Quarry III); its electron spin resonance age lies between 400 and 700 ka and its macrofauna, microfauna, and lithic artifacts are broadly similar. The smaller size of the micrommal fossils at GH (i.e., *Eliomys* and *Meriones*) as well as the morphology of the *Ellobius* fist mandibular molar, suggests an age earlier than the GDR cave, and probably younger than Tighenif. This latter range is consistent with optically stimulated **luminescence dating** estimates between 470 and 360 ka reported for the hominin layer of GH and with a direct laser ablation inductively coupled plasma mass spectrometry date of *c*.500 ka for a hominin premolar. Hominins found at site Several remains attributed to *Homo* have been discovered. These include the 1969 finds from Thomas Quarry I: a left mandibular ramus with four teeth; the 1972 finds from Thomas Quarry III: several calotte fragments and a left partial maxilla, and a well-preserved mandible; and the recent finds from Thomas Quarry I: several isolated teeth, a complete mandible and several other unpublished fragments. Archeological evidence found at site The lithic artifacts from Thomas Quarry include **Acheulean handaxes**, **cleavers**, **cores**, **flakes**, and other lithic debris associated with a modest faunal assemblage. Key references: Geology, dating, and paleoenvironment Geraads 2002, Geraads et al. 2004, Raynal et al. 2001, 2002, 2010, Rhodes et al. 1994, 2006; Hominins Ennouchi 1969, 1972, Raynal et al. 2010; Archeology Geraads et al. 1980, Raynal et al. 2002, 2010.

thoracic cage, comparative anatomy and evolution The thoracic cage consists of the ribs and the thoracic vertebrae to which they are attached, along with the sternum and costal cartilages. These structures provide protection for the thoracic contents, most notably the heart and lungs. The thoracic skeleton also anchors the proximal muscles of the upper limb, the muscles that move the head and neck, and those that support the walls of the abdomen; internally the thoracic cage anchors much of the diaphragm. Except for the caudalmost two ribs, each of the other ribs articulate with the vertebral column at two synovial joints; the head with one, or two, vertebral bodies at the costovertebral joints and the tubercle with with the vertebral transverse process of the same (or the lower of the two) vertebrae at the costotransverse joint. In modern humans the ventral ends of the first six ribs articulate with the sternum via independent costal cartilages and ventrally ribs 7–10 share a joint costal cartilage; the caudal two ribs do not articulate with the sternum, nor do they have costotransverse articulations. The thoracic cage must move when an animal breathes to allow expansion of the lungs. All the details of this motion within the thoracic cage are incompletely understood, but during respiration the anterior ends of especially the upper ribs move cranially with a "pump-handle" motion, while the shafts of the lower ribs tend to be elevated in a "bucket-handle" motion. The zygapophyseal facets are almost planar and coronally oriented, so the thoracic vertebral column is also capable of rotation, flexion-extension, and lateral flexion (Kapandji 1974). The overall shape of the thoracic cage differs among groups of primates in ways that reflect locomotor adaptation. The most significant differences are those between most monkeys and the apes. Monkeys, which stand and move in pronograde posture both in the trees and on the ground, have ventrally facing shoulder joints (Preuschoft et al. 2010, Schultz 1961). The thoracic cage is mediolaterally narrow and dorsoventrally deep, and the scapula sits in a parasagittal plane so that the glenohumeral (shoulder) joint is oriented normal (perpendicular) to the direction of load in this posture. The muscles that run between the thoracic cage and scapula are also in roughly parasagittal planes, so their contraction moves the limb primarily in flexion and extension. In contrast, the thoracic cage of extant apes is mediolaterally broader, thus shifting the scapulae into a more coronal plane and changing the orientation of the glenohumeral joint so that it is more laterally facing (Schultz 1961). The thoracohumeral musculature, most notably latissimus dorsi, is also more mediolaterally oriented, making it more effective at adducting the humerus, which is how an animal pulls itself up from hanging beneath a support (Ward 1993). Related to this morphology, the ribs of apes, especially the cranial ribs, are more tightly curved than those of monkeys (Kagaya et al. 2008, 2009); ape ribs are also more declined from dorsal to ventral. Modern humans also differ from great apes in the form of the thoracic cage. Extant great apes have thoracic cages that are narrow at the top and wide at the bottom, and the long lower ribs sit adjacent to the iliac crest due to their elongated pelves and short, stiff lumbar spines; monkeys, gibbons and modern humans have longer lumbar regions, and their lower rib cage dimensions do not have to match the pelvis as tightly. Modern humans resemble gibbons in having a barrel-shaped thoracic cage, whereas that of great apes is aptly described as cone-shaped (Schmid 1991). Modern humans also have their thoracic vertebral column invaginated into the rib cage, so that the neck of the rib and thoracic vertebral transverse processes flare dorsally relative to the vertebral body, before swinging laterally and anteriorly (Jellema et al. 1993). Early *Homo* appears to have a thoracic cage most like that of modern humans (Jellema et al. 1993). The thoracic cage of *Homo neanderthalensis* is similar overall to modern humans, but it is barrel-shaped (i.e., it is expanded both anteriorly and laterally), reflecting their overall stocky, presumably cold-adapted, body form (Trinkaus 1983). The initial reconstruction of the thoracic cage of *Australopithecus* suggested it was cone-shaped, like that of apes (Schmid 1991), but the new *Australopithecus afarensis* skeleton at Korsi Dora in the **Woranso-Mille study area**, Ethiopia (KSD-VP-1/1), has a nearly complete second rib that is clearly unlike that of apes but instead is similar in curvature to those of modern humans and recent premodern *Homo* (Haile-Selassie et al. 2010a). Other *Au. afarensis* thoracic vertebrae reflect a modern human-like pattern and *Australopithecus sediba* also appears to have had a modern human-like thoracic cage. Another feature of the *Australopithecus* thoracic cage that is unlike that of great apes is the first rib. The head of their first rib of modern humans articulates with the vertebral body of the first thoracic vertebrae, whereas that of great apes sits more cranially, articulating in two distinct facets with the caudal margin of the last cervical and cranial margin of the first thoracic vertebrae. The head of its first rib of *Au. afarensis* has only one facet, which points to it having a lower position of the first rib on the vertebral column, like that of modern humans (Ohman 1986). The cross-section of the ribs of *Au. afarensis* is rounder, and thus more chimp-like, compared to that those of modern humans. But this may not be a reliable indication of an ape-like thoracic cage, for the cross-sectional profile of gorilla ribs is modern human-like; rib cross-sectional shape is apparently not a reliable predictor of the overall shape of the thoracic cage. It is most likely that the thoracic cage of *Australopithecus* was broad cranially, like that of modern humans. This calls into question reconstructions of a large ape-like gut made under the assumption of a caudally flaring, cone-shaped rib cage, that would have contrasted with a more modern human-like smaller gut; reconstructing gut size from fossil evidence is necessarily highly speculative. *See also* **expensive tissue hypothesis; scapula, evolution in hominins; KSD-VP-1/1; lumbar vertebral column**.

three-dimensional morphometrics *See* **geometric morphometrics; morphology**.

Tighenif (Location 35°30′N, 0°20′E, Algeria; etym. named after a nearby plain) History and general description The site is a commercial sand-pit 17 km/10 miles southeast of Mascara, Algeria, in the foothills of the Atlas Mountains. Fossils were reported in the 1880s, but in 1931 Camille **Arambourg** realized the sediments exposed in the sandpit were laid down in the basin of an ancient paleolake. In 1954 he arranged for water to be pumped from the sandpit and he was able to excavate the deeper sediments. Temporal span and how dated? **Biostratigraphy** points to a **Middle Pleistocene** age. Hominins found at site Arambourg and Hoffstetter unearthed three hominin mandibles (Tighenif 1–3) and a parietal fragment (Tighenif 4). Although Arambourg noted the new mandibles were "very closely related to the Asiatic *Pithecanthropus* and *Sinanthropus*" and that "some features (are) reminiscent of the **australopithecine**s, *Telanthropus* particularly" (Arambourg 1955, p. 195), he concluded that the fossils "cannot be identified exactly with either *Pithecanthropus* or *Sinanthropus* or *Telanthropus*" so he "assigned to them the provisional name, *Atlanthropus mauritanicus*" (*ibid*, p. 195). Arambourg's paper was published in the same year as the first edition of Wilfrid **Le Gros Clark**'s *The Fossil Evidence for Human Evolution*, but in the second, 1964, edition, when discussing the fossil record for *Homo erectus*, Le Gros Clark concludes his discussion of the remains from what is now

called Tighenif that "there can be little doubt … that these remains do belong to the same species (i.e., to *H. erectus*)" (Le Gros Clark 1964, p. 112). The most detailed taxonomic assessment of the Tighenif mandibles was that of Mounier et al. (2009), and their conclusion was that all the mandibles should be assigned to *Homo heidelbergensis*, even though they noted that one of the mandibles, Tighenif 2, had "an incipient modern-human-like chin" (*ibid*, p. 243). One of the three mandibles (Tighenif 1) is substantially larger than the other two, and Briggs (1968) suggested that it belonged to a "paranthropoid" group, whereas he assigned Tighenif 1–2 to a "telanthropoid" group. Archeological evidence found at site Handaxes. Key references: Geology, dating, and paleoenvironment Arambourg and Hoffstetter 1955, Howell 1960 (pp. 208–9); Hominins Arambourg 1954, 1955, Howell 1960 (pp. 212–16); Archeology Arambourg and Hoffstetter 1954.

Tighenif 1 Site Tighenif. Locality N/A. Surface/*in situ* In situ. Date of discovery 1954. Finders Camille **Arambourg** and R. Hoffstetter. Unit Sandpit. Horizon N/A. Bed/member N/A. Formation N/A. Group N/A. Nearest overlying dated horizon N/A. Nearest underlying dated horizon N/A. Geological age *c*.780 ka. Developmental age Adult. Presumed sex Male. Brief anatomical description Large mandible lacking most of left and right rami, but preserving the left P_1–M_3 and right P_1–M_3 and fragmentary left I_2 and right I_2. Announcement Arambourg and Hoffstetter 1954. Initial description Arambourg 1954. Photographs/line drawings and metrical data Rightmire 1990. Detailed anatomical description Hublin 2001, Rightmire 1990. Initial taxonomic allocation *Atlanthropus mauritanicus* Arambourg, 1955. Taxonomic revisions *Homo erectus* (Le Gros Clark 1964). Current conventional taxonomic allocation *H. erectus*. Informal taxonomic category Pre-modern *Homo*. Significance Thus far the oldest hominin discovered in north Africa. Location of original Laboratoire de Paléontologie, Muséum National d'Histoire Naturelle, Paris, France.

Tighenif 3 Site Tighenif. Locality N/A. Surface/*in situ* In situ. Date of discovery 1955. Finders Camille **Arambourg** and R. Hoffstetter. Unit Sandpit. Horizon N/A. Bed/member N/A. Formation N/A. Group N/A. Nearest overlying dated horizon N/A. Nearest underlying dated horizon N/A. Geological age *c*.780 ka. Developmental age Adult. Presumed sex Male. Brief anatomical description Essentially complete mandible with the right P$_1$–M$_3$ and the left P$_2$–M$_3$. Announcement Arambourg and Hoffstetter 1955. Initial description Arambourg and Hoffstetter 1955. Photographs/line drawings and metrical data Rightmire 1990. Detailed anatomical description Hublin 2001, Rightmire 1990. Initial taxonomic allocation *Atlanthropus mauritanicus* **Arambourg, 1955**. Taxonomic revisions *Homo erectus* (Le Gros Clark 1964). Current conventional taxonomic allocation *H. erectus*. Informal taxonomic category Pre-modern *Homo*. Significance Thus far the oldest hominin discovered in north Africa, and substantially larger than **Tighenif 1**. Location of original Laboratoire de Paléontologie, Muséum National d'Histoire Naturelle, Paris, France.

Tighennif *See* **Tighenif**.

TILLMIAP Acronym for the International Louis Leakey Memorial Institute for African Prehistory. *See* **Leakey, Louis Seymour Bazett (1903–72)**.

time-averaging Substantial phenotypic change may have occurred during the time covered by many of the fossil collections researchers hypothesize may sample a single species. This increases the variance of the "time-averaged" sample, and may invalidate attempts to use museum collections of extant taxa, which are inevitably accumulated over a much shorter period of time than paleontological samples, as comparative analogues for testing whether a fossil sample subsumes more than one taxon.

time scale Framework for assigning ages to geological deposits or to events in the geological record. For example, the **astronomical time scale** is based on Earth/sun orbital systems and the **geomagnetic polarity time scale** uses calibrated changes in the direction of the Earth's magnetic field.

time units The IUPAC-IUGS Task Group (2006-016-1-200) urge that the SI unit "a" be used for both ages and time spans, where a is the abbreviated form of *annum*. The correct SI usage of units must follow algebraic rules, such as the distributive law: 100–90 Ma = (100–90) Ma = 10 Ma, and so on (Bureau International des Poids et Mesures 2006). Similarly, half-lives should be expressed in ka (thousands), Ma (millions), or Ga (billions), and rates and decay constants in (ka)$^{-1}$, (Ma)$^{-1}$, or (Ga)$^{-1}$. The IUPAC-IUGS Task Group discourage the use of "y," "yr," "yrs" along with any combination of "k," "K," "m," "M," etc. All ages of fossil material or associated deposits should be quoted with their full error margins. *See also* **a**.

timing (L. *tempus*=time) *See* **heterochrony**; **tempo**.

TIMS *See* **uranium-series dating**.

TL *See* **luminescence dating**.

TM Abbreviation of Transvaal Museum and the prefix used for fossils recovered from the **Sterkfontein** Type site between 1936 and 1938, and from **Kromdraai** between 1938 and 1955. The abbreviation is also used as a prefix for fossils recovered from Toros-Menalla, **Chad Basin**, Ethiopia. *See also* **Kromdraai**; **Toros-Menalla**.

TM 1511 Site Sterkfontein. Locality Type site. Surface/*in situ* In situ. Date of discovery August 17, 1936. Finder Robert **Broom**. Unit N/A. Horizon N/A. Bed/member Member 4. Formation **Sterkfontein Formation**. Group N/A. Nearest overlying dated horizon N/A. Nearest underlying dated horizon N/A. Geological age *c*.2.6–3.0 Ma. Developmental age Young adult. Presumed sex Male. Brief anatomical description Cranium with palate (including right P$_4$–M$_2$, and left P$_3$–M$_2$), right side of the face and occipital region, and an isolated right M$_3$. Announcement Broom 1936a. Initial description Broom 1936a, 1936b. Photographs/line drawings and metrical data Broom and Schepers 1946. Detailed anatomical description Broom and Schepers 1946. Initial taxonomic allocation *Plesianthropus tranvaalensis*. Taxonomic revisions *Australopithecus africanus* (Robinson 1954). Current conventional taxonomic allocation *Australopithecus africanus*. Informal taxonomic category Archaic hominin. Significance The **holotype** of *Plesianthropus tranvaalensis*. Location of original **Ditsong National Museum of Natural History** (formerly the **Transvaal Museum**), Northern Flagship Institution, Pretoria, South Africa. *See also* **Sterkfontein**; *Australopithecus africanus*.

TM 1517 Site **Kromdraai**. Locality Kromdraai B. Surface/*in situ* This specimen was removed from the breccia deposit by a schoolboy, Gert Terblanche, who then showed Robert **Broom** where he had found it at Kromdraai. Date of discovery June 8, 1938. Finder Gert Terblanche. Unit N/A. Horizon N/A. Bed/member Member 3. Formation Kromdraai Formation. Group N/A. Nearest overlying dated horizon N/A. Nearest underlying dated horizon N/A. Geological age *c*.2.0–1.5 Ma. Developmental age Adult. Presumed sex Male. Brief anatomical description An associated skeleton, made up of a skull, isolated teeth, the distal end of the right humerus, the proximal end of the right ulna, hand bones, a right talus, and other foot bones. The cranium of the skull is crushed, and the face, palate (with the left P^3–M^2), and the cranial base are the better-preserved regions. The right side of the mandibular corpus (with the right P_3–M_3) is damaged anteriorly. Announcement and initial description Broom 1938a, b. Photographs/line drawings, metrical data, and detailed anatomical description Gregory and Hellman 1939, Robinson 1956 (teeth), Broom and Schepers 1946 (skull), Broom 1942 (hand), Broom 1943 (foot). Initial taxonomic allocation *Paranthropus robustus*. Taxonomic revisions None specific to this specimen. Current conventional taxonomic allocation *P. robustus*. Informal taxonomic category Megadont archaic hominin. Significance The **holotype** of *P. robustus* and one of the few associated skeletons of that taxon. The talus and the arm bones have featured in attempts to reconstruct *Paranthropus* limb morphology. Location of original **Ditsong National Museum of Natural History** (formerly the **Transvaal Museum**), Northern Flagship Institution, Pretoria, South Africa. *See also* **Kromdraai**; *Paranthropus robustus*.

TM 266-01-060-1 Site **Toros-Menalla**. Locality TM 266. Surface/*in situ* Surface. Date of discovery 2001. Finder Djimdoumalbaye Ahounta. Unit **Anthracotheriid Unit**. Horizon N/A. Bed/member N/A. Formation N/A. Group N/A. Nearest overlying dated horizon N/A. Nearest underlying dated horizon N/A. Geological age *c*.7 Ma. Developmental age Adult. Presumed sex Male. Brief anatomical description A nearly complete cranium with, on the right, the I^2 alveolus, distal part of the canine crown, P^3–P^4 roots, and fragmentary M^1–M^2 crowns, and on the left, the I^2 alveolus, C–P^4 roots, and fragmentary M^1–M^3 crowns. Announcement **Brunet** et al. 2002. Initial published description Brunet et al. 2002. Photographs/line drawings and metrical data Zollikofer et al. 2005. Detailed anatomical description Zollikofer et al. 2005. Initial taxonomic allocation *Sahelanthropus tchadensis*. Taxonomic revisions **White** et al. 2009 suggest that *Sahelanthropus* may be a junior synonym of *Ardipithecus*. Current conventional taxonomic allocation *S. tchadensis*. Informal taxonomic category Possible primitive hominin. Significance **Holotype** of *S. tchadensis*. The authors believe that despite primitive features the cranium belongs to a primitive hominin and not to a panin because of its small, apically worn canines and the intermediate thickness of the postcanine enamel. It was virtually reconstructed by Zollikofer et al. (2005), who claimed that the reconstructed morphology of the cranial base demonstrated that *S. tchadensis* is bipedal. A further, phenetic, analysis of 3D landmark data obtained from the virtually reconstructed cranium by Guy et al. (2005) concluded that while the reconstructed cranium shared some basicranial features with archaic hominins (e.g., "long, flat [or horizontally orientated] nuchal plane," "shortened occipital," and "anteriorly-positioned foramen magnum"; *ibid*, p. 18838), other features on other regions of the cranium (e.g., the face) were either primitive or unique among hominins. Location of original Département de Conservation des Collections, Centre National d'Appui á la Recherche (CNAR), Ndjame´na, Chad.

TMJ *See* **temporomandibular joint**.

Toba super-eruption The Youngest Toba Tuff (YTT) is evidence of a volcanic eruption that occurred 74,000 years ago in Indonesia. This is one of Earth's largest known volcanic events and the largest eruption in the last 2 million years (Rose and Chesner 1990). It was two orders of magnitude larger in erupted mass than the largest known historic eruption in Tambora, Indonesia. The YTT involved the eruption of a minimum of 2800 km^3 of magma, of which at least 800 km^3 was transported in atmospheric ash plumes that blanketed an area from the South China Sea to the Arabian Sea (*ibid*). Its impact on Earth's atmosphere and climate (Robock et al. 2009) and on local and animal populations remains a matter of contention (Ambrose 1998, Oppenheimer 2002). The Indian subcontinent contains extensive YTT deposits (Jones 2007). Pollen and carbon isotope studies have suggested that the Toba eruption caused climatic cooling and prolonged deforestation in South Asia, leading to severe impacts on mammalian communities (Williams et al. 2009). The causative link between the YTT, global cooling, deforestation in India, and genetic bottlenecks and extinctions in mammals has been questioned (Haslam and Petraglia 2010). At **Jwalapuram**, in the Jurreru River Valley of southern India, Middle Paleolithic artifact assemblages have been found before and after the YTT, indicating that hominins survived the eruption (Petraglia et al. 2007, Haslam et al. 2010).

Tobias, Phillip Vallentine (1925–2012)

Phillip V. Tobias was born in Durban, in the former Natal province in South Africa. In the first volume of his autobiography, *Into the Past* (2005), Tobias claims the premature death, at age 21, of his sister Val from inherited diabetes, and visits to the Durban Natural History Museum to see the Campbell Collection of southern African vertebrates and an exhibit of artifacts from KwaZulu-Natal during his teenage years, were the reasons for his subsequent scientific interest in genetics, zoology, and archeology. Tobias entered the University of the Witwatersrand, in Johannesburg, and completed a BSc in physiology and histology and embryology in 1946, and earned an honors BSc in histology in 1947. It was as an undergraduate student at Wits that Tobias first met Raymond **Dart**, who not only taught anatomy at the medical school, but was also famous for his recognition of *Australopithecus africanus*. In 1945 Tobias and some of his fellow students organized a student expedition to the **Makapansgat** Limeworks. They located and recovered evidence of fossil **baboon**s, and this led to an intensive exploration of the site by Dart and colleagues over the next several decades, which produced new fossil remains of what were later referred to *Au. africanus*. After this initial experience in paleontology Tobias was soon exploring other hominin sites such as **Sterkfontein** and **Kromdraai**, and the former site came to be the focus of much of Tobias' field efforts in later years. After completing his undergraduate science studies Tobias continued his medical course at Wits, where he conducted research in genetics. He completed his medical degree in 1950 (but never practiced clinical medicine) and his PhD in 1953, with a dissertation on *Chromosomes, Sex-Cells, and Evolution in the Gerbil.* In 1951 Tobias was invited to join the French

Panhard-Capricorn Expedition, led by French explorer François Balsan, which collected anthropological and anthropometric data on the San Bushmen of the Kalahari Desert. This experience drew Tobias further into physical anthropology and this interest was further strengthened when in 1955 he was awarded a Nuffield Dominion Senior Travelling Fellowship and left for Cambridge University in England to conduct postdoctoral research on biometrics in physical anthropology. He used this opportunity to examine the hominin fossils curated in the British Museum (Natural History) in London as well as at the **Musée de l'Homme** (Museum of Man), the **Muséum National d'Histoire Naturelle** (National Museum of Natural History), and the **Institut de Paléontologie Humaine** (Institute of Human Paleontology) in Paris. It was at the latter institution that he spent time with Henri **Vallois** and Camille **Arambourg**. By the end of his time in England and Europe Tobias was turning away from a career in genetics and toward a career in paleoanthropology, and he spent 1956 in the USA as a Rockefeller Traveling Scholar. He spent half of 1956 visiting various US universities and institutions, including the University of Chicago and the University of Michigan, studying physical anthropology, human genetics, and dental anatomy. Tobias returned to South Africa in late 1956 and during 1957–8 he worked with an interdisciplinary team of researchers conducting studies of the Tonga of Northern Rhodesia (now Zambia). In 1959, shortly after Raymond Dart's retirement from the Chair of Anatomy, Tobias was appointed Professor of Anatomy and Head of the Department of Anatomy at the University of the Witwatersrand, a position he held until 1990. Tobias was drawn even further into paleoanthropology in 1959 when Louis and Mary **Leakey** invited him to undertake the detailed study of the recently discovered *Zinjanthropus boisei* cranium (**OH 5**) found at **Olduvai Gorge** in Tanzania. This led to Tobias's long-standing collaboration with Louis and Mary Leakey, which included his collaboration with the British anatomist John **Napier**, to describe and name the fossils discovered by Louis and Mary Leakey at Olduvai that were referred to *Homo habilis*. Detailed monographic treatments of OH 5 and of the cranial remains attributed to *H. habilis* were published in the Cambridge University Press series on Olduvai Gorge. The first of these, *Olduvai Gorge* volume 2, *The Cranium and Maxillary Dentition of*

Australopithecus (Zinjanthropus) boisei, was published in 1967, and the second, *Olduvai Gorge* volume 4, *The Skulls, Endocasts and Teeth of Homo habilis*, was published in 1991. For much of Tobias' career in paleoanthropology he was concerned in one way or another with the excavations at Sterkfontein in South Africa. Robert **Broom** and John **Robinson** conducted excavations there from 1947 until Broom's death in 1951, and throughout the 1950s Robinson continued to work periodically at Sterkfontein. The Sterkfontein site became the property of the University of the Witwatersrand in 1958 and Tobias initiated extensive excavations of the site that have continued ever since. Assisted by Alun Hughes, who had worked closely with Raymond Dart at Makapansgat, and later by paleontologist Ronald **Clarke**, who had worked with Louis and Mary Leakey as a research assistant/conservator, Tobias and his team amassed a substantial collection of animal fossils and hominin remains, as well as data on the paleoecology of the site. By the early 1990s Tobias' team had collected over 500 hominin fossils, most of which belonged to *Au. africanus*, but there were some fossils that Hughes and Tobias suggested belong to an early variety of *Homo* (e.g., **Stw 53**) while Clarke argued that some other specimens belong to *Paranthropus robustus*. Tobias was appointed Honorary Professor of Palaeoanthropology at the **Bernard Price Institute for Palaeontological Research** at the University of the Witwatersrand in 1977, and he served as director of the university's **Palaeo-Anthropology Research Unit** (or PARU). Tobias retired as head of the Department of Anatomy in 1990 but he continued as the director of the **Sterkfontein Research Unit**. In 1994 Ronald Clarke discovered four australopith foot bones in a box of fossils that had originally been collected in 1977. Tobias and Clarke published an account of the fossils (**Stw 573**), nicknamed "Little Foot," and subsequent excavations have recovered much of the skeleton, which has tentatively been dated to *c*.3 Ma. Tobias was elected a Fellow of the Royal Society of South Africa in 1951 and served as its president from 1970 to 1972. He was involved in the establishment of the **Institute for the Study of Man in Africa** and he became its first president in 1961. In 1987, Tobias was named a foreign associate of the US National Academy of Sciences. The Royal Anthropological Institute of Great Britain and Ireland elected him an Honorary Fellow in 1989 and he was elected a Fellow of the Royal Society of London in 1996.

toe-off   *See* **push-off**.

ToM   *See* **theory of mind**.

Tomes, Sir Charles Sissmore (1846–1928)   The eldest son of Sir John **Tomes**, Charles Tomes read (i.e., studied) natural science at Christ Church, Oxford, and then medicine at the Middlesex Hospital Medical School, London, before practicing as a dental surgeon. He published a series of important papers on the structure of **dentine** and the enamel organ and he edited the last four editions of his father's book, *Tomes's Dental Surgery*. He was author of a *Manual of Dental Anatomy, Human and Comparative*, which reached its sixth edition in 1904. Charles Tomes' name has been given to a root form variant most commonly seen in mandibular first premolars. *See also* **Tomes' root**.

Tomes, Sir John (1815–95)   The son of a West Country farmer, he was apprenticed to an apothecary before studying dental surgery at King's College, London, and medicine at the Middlesex Hospital Medical School, also in London. Inspired by lectures give by Richard Owen he presented a paper to the Royal Society aged 23, "On the structure of teeth," unaware that others in Europe had already described identical histological structures. Tomes' name is associated with Tomes' fibers (the cell processes of an **odontoblast** contained within the dentine tubules), the Tomes' process at the secretory end of an **ameloblast**, and the granular layer of Tomes, an incompletely mineralized layer of **dentine** close to the root surface.

Tomes' process   The eponymous name of the process at the secretory end of an ameloblast. The process was described by Sir John Tomes. *See also* **ameloblast**; **Tomes, Sir John**.

Tomes' root   This term refers to a distinctive root form variant that is most commonly seen in mandibular first premolars. It consists of a mesiolingual groove in the mandibular first premolar root that may deepen to a "C-shaped" cleft when seen in cross-section. Occasionally it may be deep enough to divide the apical portion of the root into a distinct mesiobuccal rootlet. Sir Charles **Tomes** drew attention to this root form, thus his name (and not that of his father Sir John **Tomes**) is commemorated in this eponymous term. *See also* Sir Charles **Tomes**.

tool (OE *tool* = to "prepare for use") Any material object (modified or unmodified) used to accomplish a task. Evidence from **Gona**, in Ethiopia, and dated to *c*.2.55 Ma, consisting of **flakes** and the **cores** from which they were struck, currently marks the beginnings of the archeological record, but if the cut-marks from **Dikika study area** (*c*.3.39 Ma) are anthropogenic then they would mark the first use of stone flakes, manufactured or natural, as tools. The Gona **first appearance datum** almost certainly underestimates the onset of habitual tool use, but because evidence of scattered unmodified stones or perishable items of the sort used by other primates is difficult to detect (Panger et al. 2002) we may never know when habitual tool use began.

tool-use hypothesis This hypothesis is one of several purporting to explain the origin of upright posture and bipedal locomotion within the hominin clade. It derives originally from no less of an authority than Charles **Darwin**, who proposed it in *The Descent of Man, and Selection in Relation to Sex* (1871). The basic premise of the tool-use hypothesis is that the advantage of **bipedalism** is that it frees the hands from locomotion, thereby allowing them to be used for tool use, which in turn stimulates the evolution of many other derived characteristics of later hominins. The hypothesis remained salient throughout much of the mid-20thC (e.g., Bartholomew and Birdsell 1953) and reached its height of influence in the writings of Sherwood **Washburn**. Washburn (e.g., Washburn 1960) proposed a sophisticated system of selective feedback mechanisms in which bipedalism allowed and/or was caused by tool use, which in turn favored canine reduction, hunting, and the evolution of a large brain. Brain enlargement and the concomitant increase in intelligence led to the evolution of ever more sophisticated tools, which re-energized the feedback loop (i.e., better tools led to enhanced bipedalism and a larger brain, etc.). Ultimately, increased brain size and hunting behavior led to language and complex human social behavior. The tool-use hypothesis is compelling and logical, and was influential in paleoanthropology, to the point that fossils of an extinct ape then known as *Ramapithecus* that possessed small canines were thought to be bipedal hominins (see review in Pilbeam 1978) despite the then lack of postcranial evidence for this taxon. Subsequently it was suggested the fossils in question were females of an ape genus (*Sivapithecus*) now thought

by most workers to be closely related to the modern orangutan (Pilbeam 1982, Andrews and Cronin 1982; but see Pilbeam et al. 1990). The ultimate demise of the tool-use hypothesis was brought about by the application of **radiometric dating** techniques (e.g., Leakey et al. 1961, Dalrymple and Lanphere 1969) and new fossil discoveries (e.g., Johanson and White 1979), which between them demonstrated that the earliest possible fossil hominins (*c*.6–7 Ma) and inferred bipedal locomotion (Richmond and Jungers 2008) substantially pre-dated the earliest archeologically visible stone tools (*c*.2.6 Ma) by several millions of years. *See also* **artifacts**; **bipedalism**.

tooth crypt *See* **alveolus**.

tooth emergence *See* **eruption**.

tooth eruption *See* **eruption**.

tooth formation Process of crown and root development that begins with soft tissue differentiation inside crypts in jaws, and ends with apical closure of the root(s). It is often categorized as discrete categories or stages that may be assessed from dental radiographs (reviewed in Hillson 1996). *See also* **enamel development**; **dentine development**; **eruption**.

tooth germ Tooth germ refers to the immature tooth bud within the mandible or maxilla that is undergoing **tooth formation**. At the earliest stages tooth germs are small and comprise only soft tissues, but as mineralization begins tooth germs are found in fully grown crypts in the jaws. Because they are so fragile, the very early mineralizing stages of tooth germs are relatively rare in the hominin fossil record. *See also* **enamel development**; **dentine development**.

tooth marks *See* **carnivore ravaging**; **cutmarks**.

tooth material An attempt by researchers to combine the size of several teeth as the unit of analysis of tooth size. Howes (1954) introduced the term tooth material to refer to the sum of the mesiodistal lengths of the maxillary and mandibular incisors, canines, premolars, and first molars. Advances in methods for image capture and in analytical methods are enabling researchers to investigate the absolute and relative sizes of the component cusps of multicuspid tooth crowns, and for taxonomic purposes these studies are beginning to replace more simplistic assessments of overall tooth size.

tooth nomenclature *See* teeth.

tooth size When researchers write and talk about hominin tooth size, unless they specify to the contrary, they are almost certainly referring to the maximum size of the tooth crown as reflected in its horizontal dimensions. For the incisors and canines this is usually the part of the tooth crown closest to the occlusal plane, and for the postcanine teeth the size of the junction at the junction between the crown and the root. Tooth crown size can be expressed in terms of one or more linear dimensions or as an area measurement, and this can be in two or three dimensions. When a single linear dimension is used as a proxy for tooth size it is almost always the mesiodistal length of the crown (e.g., premolar and molar chords, see below), but mesiodistal measurements need to take into account loss of enamel due to the **interproximal wear** that occurs between adjacent when they move against each other during **chewing**. The area of a tooth crown can be computed by multiplying the mesiodistal and labio- or buccolingual dimensions of the crown. This works well when the shape of the crown is close to being a rectangle, but it is unacceptably inaccurate when the outline of the crown is irregular. Crown area can also be measured from occlusal photographs, but care must be taken to orient the crown in the way the technique prescribes. Researchers often combine the size of several teeth as units of analysis. Examples include the chord lengths of rows of morphologically or functionally similar teeth (e.g., incisor, premolar, and molar row lengths), and even more inclusive units, such as Howes' (1954) global measure of **tooth material** (i.e., the sum of the mesiodistal lengths of the maxillary and mandibular incisors, canines, premolars, and first molars), or the combination of the maxillary or mandibular incisors and canine as the **anterior teeth**, and the combination of the maxillary or mandibular premolars and molars as the **posterior teeth**, or **postcanine teeth** (e.g., Curnoe and Tobias 2006). Advances in methods for image capture that allow for 3D measurements, as well as advances in analytical methods, are enabling researchers to investigate the absolute and relative sizes of the component cusps of multicuspid tooth

crowns, and for taxonomic purposes these studies are beginning to replace more simplistic assessments of overall tooth size. For the great apes, possible primitive hominins, and early hominins the operative measure of canine size is the height of the crown.

tooth wear Wear involves the worn object losing some of its volume, and the process of tooth wear (the loss of enamel volume, and then later the loss of dentine volume) begins as soon as a tooth loses its gingival covering and is exposed in the mouth. Most research on tooth wear in hominins is concentrated on the wear that occurs on the functional or occlusal surface of a tooth (so-called **occlusal wear**), but wear also takes place on the non-occlusal surfaces of the teeth (due to the action of acid) and between tooth crowns (called approximal or **interproximal wear**) when adjacent teeth move slightly against each other in the mouth during the process of **chewing**. Occlusal wear is often divided into the wear that occurs between the occlusal surfaces of teeth (tooth–tooth wear is also called attrition), the wear that occurs between teeth and food (tooth–food wear is also called abrasion), and the wear that results from chemical erosion. However, these processes can also interact in a complex fashion. Occlusal tooth wear that can be seen with the naked eye is called gross wear, dental macrowear, or dental mesowear, while tooth wear that needs a microscope to see it is called **dental microwear**. Observations about gross dental wear focus on the development of **wear facets**, including measurements of their size and orientation. Dental microwear focuses on the size, number, and orientation of microscopic scratches, pits, etc. on the enamel, or on overall complexity and **isotropy** of the enamel surface. Dental macrowear is a measure of the abrasiveness of the diet in the long term, whereas dental microwear indicates whether the food ingested in the days or weeks before death contained hard or abrasive material. Such abrasive material can be either intrinsic to the ingested foods (e.g., **phytoliths**) or extrinsic to the foods (e.g., adherent sand grains). Dental microwear can also detect evidence for softer or tougher foods (e.g., polished or finely pitted surfaces). If any striations or scratches have a predominant orientation, this can be used to reconstruct the direction of the jaw movement(s) that generated the microwear. When occlusal dental macrowear is examined at the level of the whole dentition, researchers have assessed it by examining the shape of the occlusal wear plane. The occlusal wear plane is rarely perfectly flat. In fact, the pitch (i.e., the transverse orientation) of the occlusal wear plane is not the same along the tooth row, for wear is normally greater on the lingual side of the maxillary teeth in the anterior part of the postcanine tooth row, with the posterior part of the tooth row either showing less emphasis on the lingual side or more wear on the buccal side (the so-called Curve of Spee). Tobias (1980b) suggested that this twist in the pitch of the occlusal wear plane was not seen in hominins until the emergence of *Homo*, but Richards and Brown (1986) suggest that the twisting nature of the occlusal wear plane has no taxonomic value. Dental microwear studies were revolutionized when precision molding materials were introduced, enabling positive molds of whole teeth to be examined with a scanning electron microscope (Walker et al. 1978). Fred Grine, Peter Ungar, and Mark Teaford have been in the vanguard of applying dental microwear analysis to fossil hominins (e.g., Ungar et al. 2006) and in the process they have helped develop new analytical methods (e.g., dental microwear texture analysis; see Scott et al. 2005). Enamel volume can also be lost when substantial-sized chips of surface enamel are removed when teeth are used intentionally, or unintentionally, to bite on hard objects. *See also* **enamel chipping**.

topographic (Gk *topos*=a place and *graphein*= to write) Topography refers to a detailed description of a place or region, usually in the form of a map. Topographic representation in the brain refers to the manner in which parts of the body are represented in a region of the brain. The brain's topographic representations of sensory systems, except olfaction, are due to the orderly preservation of information regarding the adjacency of peripheral sensory receptor arrays along the various stages of neural processing. For example, the spatial distribution of retinal photoreceptors is represented accurately in the lateral geniculate nucleus and the primary **visual cortex**. Similarly, the hair cells in the organ of Corti are tuned to different frequencies depending on their location along the basilar membrane of the cochlea, and the brain's representation of "tonotopy" is actually a topographic representation of the hair cells in the cochlea. The cortical representation of body parts may be magnified relative to their original size.

For example, the somatosensory representation of the bill of the platypus in its **neocortex** is greatly enlarged and certain acoustic frequencies are magnified in the auditory cortex of echolocating bats. Among primates, when the body surface is mapped onto the somatosensory cortex, the lips, hands, and the genitalia are "overrepresented," so the resulting map, or "homunculus," looks bizarre. Similarly, the visual cortex representation of the central visual field is highly magnified in haplorrhine primates in association with the presence of a retinal fovea.

Toros-Menalla (Location 16°14′–15′N, 17°28′–30′E, Chad; etym. Toros-Menalla is the name given by paleontologists to a region about 150 km/90 miles long in the western end of the **Chad Basin**, also known as the Djurab Desert) History and general description Toros-Menalla is a fossiliferous area in the Chad Basin of Northern Chad that was explored by the **Mission Paléoanthropologique Franco-Tchadienne** (French-Chadian Paleoanthropological Mission, or MPFT) beginning in 2001. Three fossiliferous localities within Toros-Menalla, TM 247, TM 266, and TM 292, have yielded hominin remains. All three localities are within the **Anthracotheriid Unit** (or AU) and are located within an area of 0.73 km²/0.3 square miles. Temporal span and how dated? Biostratigraphy suggests that the fauna associated with the possible hominins is older than fossils from **Lukeino**, Kenya (*c*.6 Ma), and more closely resembles material from the Nawata Formation at **Lothagam**, Kenya, which is radioisotopically dated to 5.2–7.4 Ma, so the TM 266 fauna can be tentatively dated to between 6 and 7 Ma. The results of ^{10}Be/^{9}Be **cosmogenic nuclide dating** of 28 samples from the AU point to the age of the AU being between 7.2 and 6.8 Ma. Hominins found at site In 2001 D. Ahounta recovered **TM 266-01-060-1**, a nearly complete cranium that became the **holotype** of *Sahelanthropus tchadensis*. The paratypes were a symphyseal fragment with incisor and canine alveoli (TM 266-01-060-2), a right M³ (TM 266-01-447), and a right I¹ (TM 266-01-448) recovered in 2001, and the right side of a mandible (TM 266-02-154-1) and a right c (TM 266-02-154-2) recovered in 2002. Three additional specimens, a right P³ (TM 266-01-462) from TM 266, a fragment of the right side of a mandibular corpus (TM 247-01-02) from TM 247, and a mandibular fragment (TM 292-02-01) from TM 292, were reported in Brunet et al. (2005). All these specimens have been attributed to *S. tchadensis*. It is understood there are also unpublished postcranial remains from the AU. Archeological evidence found at site None. Key references: Geology, dating, and paleoenvironment Vignaud et al. 2002, Lebetard et al. 2008; Hominins Brunet et al. 2002, 2005, Zollikofer et al. 2005, Guy et al. 2005; Archeology N/A.

torque *See* petalia.

Torralba (Location 41°08′19″N, 2°29′51″W, Spain; etym. named after the nearby town, Torralba del Moral) History and general description This site lies along the Masegar River, and it, along with the nearby site of **Ambrona**, was originally excavated by Clark **Howell** and Les Freeman in the early 1960s. They believed both sites to be contemporaneous, and presented their finding of many faunal remains, particularly from elephants, and a quantity of stone tools, as the earliest evidence of hunting. Furthermore, Freeman argued that at Torralba there were associations between the types of tools and the types of bones recovered, suggesting specialization in tool use. Later researchers, including Lewis **Binford** and Richard Klein, argued that the faunal profiles from these sites suggested scavenging, not hunting. The most recent analysis by Santonja and Pérez-González has shown that, first, Ambrona is significantly older than Torralba, and second, there is a mix of natural and hominin components to both sites with so little evidence of interaction between the hominins and the fauna that it is impossible to say anything more than some butchery occurred at the site. Temporal span and how dated? Based on the **uranium-series date** of a nearby river terrace, Torralba is thought to be more than 240 ka. Hominins found at site None. Archeological evidence found at site A large number of animal bones, mostly elephants, and stone tools. Torralba also has at least two pieces of elephant bones that have been intentionally flaked and shaped into points; these are among the oldest evidence of flaked bone tools. Key references: Geology, dating, and paleoenvironment Villa et al. 2005, Freeman 1994, Howell et al. 1963; Hominins N/A; Archeology Villa et al. 2005, Freeman 1994, Howell et al. 1963.

torus (L. *torus*=bulge) An area of projecting bone with a rounded profile (e.g., the occipital and frontal tori on the **cranium** and the transverse tori on the **mandible**).

total fertility rate (or TFR) The sum of the **fertility rate** (age-specific or ASFR) across the female reproductive years. If the ASFRs are given by yearly age intervals, then the TFR is the simple sum of all the ASFR values. If the ASFR values are given across broader age intervals, then each ASFR value must be multiplied by the width of its respective age interval. This is the case because ASFR values are given per year of life, and a woman who has, for example, lived to her 25th birthday will have spent 5 years in the 20–25-year-old age interval for which the $_5f_{20}$ fertility rate applies. Ultimately, the TFR represents the total number of births a woman can be expected to experience during her lifespan. The TFR makes the assumption that a woman will survive to the cessation of ovulation at menopause, but it makes no assumption about the mortality of her offspring. The TFR also does not specify the sex ratio of a woman's offspring, and as such the TFR is less useful than the **gross reproductive rate** (or GRR) in making statements about population growth. In the case where the sex ratio of a woman's births is 1:1 the GRR is equal to half of the TFR.

total group (L. *totus* =whole) A term intoduced by Jefferies (1979) for a **clade** or monophyletic group that includes all of the taxa more closely related to a living **taxon** in that clade than they are to any other living taxon. For example, the **hominin** and **panin** clades are both total groups. All the taxa in the former group are more closely related to *Homo sapiens* than they are to any other living taxon, and all the taxa in the latter group are more closely related to *Pan troglodytes* and *Pan paniscus* than they are to any other living taxon. A total group is made up of a **crown group** (equivalent to the "group" of Willi Hennig; Hennig 1981) plus the **stem group**. *See also* clade; crown group; stem group.

total morphological pattern This term (and the concept associated with it) was introduced by Wilfrid **Le Gros Clark**. He first used the term in a letter to *Nature* written in response to Solly Zuckerman (1950). Zuckerman had reiterated the claim made by Ashton and Zuckerman (1950) that they had successfully used dental morphometric methods to demonstrate that the australopith dentition was ape-like and not modern human-like as claimed by **Dart** and **Broom**, and supported by Le Gros Clark. Le Gros Clark (1950) wrote that "it is open to serious question whether the major dimensions and indices of individual teeth can by themselves provide adequate information on which to base statements regarding affinities of primitive hominids and anthropoid apes" (*ibid*, pp. 893–4). He suggested that the new fossil evidence then recently reported by Broom and others "has now made it possible to take a comprehensive view of the *total morphological pattern* presented by the dentition" (*ibid*, p. 894). He went on to report that "according to my own comparative studies based on more than four hundred ape skulls, the **Australopithecine** dentition displays certain hominid features which severally find no precise parallel in any of the known apes (recent or extinct), and others which may very occasionally be found singly if diligently sought in a large series. But conclusions regarding the hominid nature of the dentition must ultimately be based on the total pattern determined by all these features in a particular combination" (*ibid*, p. 894). Just in case the reader had failed to get the point he continues "So far as I am aware no similar combination of *all* these hominid features together has been found in the dentition of any of the anthropoid apes" (*ibid*, p. 894). In an earlier paper about evolutionary parallelism Le Gros Clark (1936) warns that "the systematic position of the animal can only be established by a complete anatomical survey, and the systematist is liable to fall into serious error if he confines his attention to one part of the body only" (*ibid*, p. 4). This suggests the concept of the total morphological pattern, but Le Gros Clark did not use the term in this earlier paper. *See also* Wilfrid **Le Gros Clark**.

total person-years to be lived in an age interval   The total number of years (based on the **radix**) lived in an age interval and all older age intervals, represented in a **life table** as T_x or $_nT_x$. The calculation is very closely related to **person-years lived in an age interval** (L_x). T_0 is just the sum of L_x across all age intervals; T_1 is the same sum but without L_0. It is simplest to think of T_x as a backward-running cumulative sum, where the addition starts at the bottom (oldest age interval) of the life table.

"Toumaï"   (etym. Toumaï means "hope of life" in the Goran language) According to Michel **Brunet** this was informal name given by "the highest Chadian authorities" to a hominin cranium recovered in 2001 by the **Mission Paléoanthropologique Franco-Tchadienne** (French-Chadian Paleoanthropological Mission, or MPFT) at a site called **Toros-Menalla**. Brunet and his colleagues assigned the specimen **TM 266-01-060-1** to a new genus and species, *Sahelanthropus tchadensis*.

TPS   Acronym for thin-plate spline interpolation. *See* transformation **grids**.

Tpt   Acronym for the temporal parietal transition area. *See* **Wernicke's area**.

trabecular bone   *See* **bone**.

trace element   A chemical (e.g., aluminum, zinc) that is only required in small, or "trace," amounts in the diet for normal function. Trace elements are not nutrients, but they may be essential for nutrients to be effective.

trace fossil   (L. *tractus*=a track and *fossile*=something that is dug up) Trace fossils (also called **ichnofossils**) make up a small, but important, fraction of the fossil evidence for human evolution. They provide direct evidence that an organism has been at a particular place at a particular time, but no remnant of the organism itself, either in the form of hard (i.e., bones and teeth) or soft (i.e., skin, muscle, etc.) tissues, is preserved. There are two categories of hominin trace fossils. In the first, sediments have functioned like casting material, and they have retained details of the surface of soft-tissues long after any evidence of the individual has disappeared. Examples include footprints (*see* **Laetoli hominin footprints**; **Koobi Fora hominin footprints**), which preserve the impression of a soft tissue, the skin of the foot, and natural **endocranial cast**s that faithfully reproduce the inner (i.e., endocranial) surface of the brain case of the hominins preserved. The second, much smaller category of trace fossils, comprises fossilized solid excreta in the form as **coprolites** (i.e., fossilized feces).

trachyte   (Gk *trachys*=rough) A family of fine-grained volcanic rocks, mostly rich in phenocrysts of alkali feldspar and other minerals, with the former often aligned approximately parallel to the direction of lava flow. The proportions of silica, sodium, and potassium can be used to define trachyte and other lava types, with trachyte having SiO_2 abundances of approximately 57–72% and a total alkali (Na_2O+K_2O) abundance of approximately 9–15% (Le Bas et al. 1986). Where available, trachyte is used for artifact production, as in the production of blades at archeological sites in the **Kapthurin Formation** in the Tugen Hills of Kenya.

trade-off A mutually exclusive interaction between two **life history** traits (i.e., traits that concern growth, maintenance, and/or reproduction). For each pair, you can have the benefit of one, or the benefit of the other, but you cannot have the benefit of both. Examples of trade-offs include those that occur between the following pairs of variables: extended life span versus early reproduction; egg number versus egg size; reproductive effort versus energy storage, etc.

tragelaphine See **Tragelaphini**.

Tragelaphini (Gk *tragos*=male goat and *elaphus*=deer) The antelope tribe (family **Bovidae**) comprising the genus *Tragelaphus*, kudu, and allies. Aside from one kudu from Arabia, this tribe, which arose in the late Miocene, is exclusively African. Their horn cores have keels and they are generally large bodied. Many tend to be found in relatively dry habitats, with the main exception of the Bongo, *Tragelaphus euryceros*, which is a large forest antelope.

tragulid The informal name for the **artiodactyl** family comprising the chevrotains and their allies, whose members are found at some hominin sites. *See also* **Artiodactyla**.

Tragulidae One of the **artiodactyl** families. Its members, which include the chevrotains and their allies, are found at some hominin sites. *See also* **Artiodactyla**.

trait (L. *tractus*=drawing. It refers to the distinctive pencil or brush strokes in a drawing or painting) Any measurable feature or non-metrical component of the **phenotype** (e.g., the characters used in a **phylogenetic analysis** are referred to as traits, as are the features used in a species diagnosis). *See also* **non-metrical trait**.

transcript (L. *transcriberere*=to copy) The **RNA** that is formed when the information in a **DNA** template is transcribed into RNA. *See also* **transcription**.

transcription (L. *trans*=across and *scribere*= to write, thus to transcribe is to "write across", or to copy) The process whereby the information encoded by **DNA** is transferred (or copied) into messenger **RNA** (or mRNA): the RNA product of this process is called a transcript. The information encoded in mRNA then takes part in a second process, called **translation**, to form a polypeptide chain. In the case of other types of RNA (e.g., transfer RNA and ribosomal RNA, etc.) the RNA molecule is not translated into polypeptides; instead, it functions as an enzyme or it becomes associated with proteins and performs specific functions within the cell. The process is facilitated by an enzyme called RNA polymerase. The process of DNA transcription is initiated or regulated by **protein**s called **transcription factors** (e.g., in modern humans the *FOXP2* gene codes for the **FOXP2** protein, which is a transcription factor that influences a number of target genes involved with the cognitive and the motor aspects of language).

transcription factor A protein complex that initiates or regulates the process of **DNA transcription** (e.g., in modern humans the *FOXP2* gene codes for the **FOXP2** [human] protein which influences a number of target genes involved with the cognitive and the motor aspects of language). Transcription factors may bind directly to the **promoter** or other regulatory regions of genes to influence transcription, or they may bind to other transcription factors to affect their activity. The process mediated by transcription factors is one of many ways **gene expression** is controlled.

transcriptome All of the messenger **RNA** molecules present in one, or more, cells. It includes all of the transcripts produced by genes that are expressed at a particular time.

transduction (L. *transduce*=transfer) The transfer of energy from one form or state to another (e.g., photons to electrical impulses in the retina; motion to electrical impulses in the semicircular canals and the organ of Corti).

transfer RNA A type of RNA molecule (abbreviated tRNA) that has a 3'-terminal site for an **amino acid** attachment and a three-base "anticodon" sequence. The tRNA "transfers" (hence its name) an amino acid (specified by the anticodon) to the site of messenger **RNA** (or mRNA) **translation** (the **ribosome**) so that it can be added to the growing polypeptide chain. The anticodon sequence is complementary to that of the mRNA **codon**.

transformation grid (L. *transformāre*=to change shape, and grid is a back-formation of gridiron, which is of obscure origin, perhaps from *gredire* which likely derives from OF *gredil*= griddle plus ME *ire*=iron) A device for visualizing shape differences between two objects whereby one (the reference shape) is typically mapped onto a 2D square grid and the other (the target shape) is represented as a warping of that grid to fit its corresponding features. In some representations the reference configuration is mapped onto the transformed grid rather than shown separately, or eliminated altogether since its shape can be intuited from the deformation of the square grid cells. Artist Albrecht Dürer (1471–1528) employed hand-drawn transformation grids to study variations in modern human form that arose through differences in perspective (Dürer 1557). It was D'Arcy Thompson (1860–1948), however, who more generally established the biological application of transformation grids as a mechanism to depict shape differences between organisms (Thompson 1917). Despite the apparent simplicity of this style of representation, it was another seven decades before an appropriate mathematical foundation was attached to Thompsonian transformation grids, through the apparatus associated with thin-plate splines (Bookstein 1991). The use of thin-plate splines in landmark-based geometric morphometrics has popularized transformation grids in modern anthropological and biological research. Here, differences between two sets of corresponding (presumably homologous) landmarks are visualized by warping the square grid of one set to fit the differences in the second set, with the spline function smoothly interpolating the spaces between landmarks through bending and stretching of the grid lines. Visually, the warped lines provide an intuitive and appealing model of how the target shape differs from the reference. Mathematically, however, the bending and stretching of the 2D mesh in a thin-plate spline is actually computed as vertical displacement of the target's landmarks above and below the plane of the reference configuration. Thus, the visual interpretation of grid lines bending within a flat surface actually derives from topographic undulations of a once-flat surface (this is the "thin plate" of thin-plate splines) warped through a new set of points offset from this plane. Whereas transformation grids can be applied to 3D data as well, trying to meaningfully visualize deformations using stacks and depths of grid cubes is nearly impossible. For 3D visualizations, authors typically rely on a single 2D grid characterizing a plane of anatomical interest, multiple stacked grids sampling the volume in increments, or video representation of a 2D grid passing through the entire volume with corresponding shape differences animated by changes to its gridlines. *See also* **thin-plate spline interpolation; geometric morphometrics; landmark; warp**.

transgenic mice   *See* **knockout**.

transgression (L. *transgress*=to step across) When a body of water (lake or sea) increases in volume and enlarges to cover, or transgress, the landscape (ant. **regression**).

translation (L. *translatus*=to carry across) Cell biology The process whereby a polypeptide chain is created from a messenger **RNA** (mRNA) template. During this process, which occurs in the cytoplasm, the mRNA is bound by a **ribosome** that catalyzes the transfer of amino acids from **transfer RNA** (tRNA) molecules to the gradually elongating polypeptide chain. Behavior The term also used in connection with communication, which has three components: (a) encoding, (b) transmission, and (c) decoding, or translation.

translocation (L. *trans*=to move across and *locus*=place, so to move from one place to another) The transfer of a section of one **chromosome** to another, usually nonhomologous, chromosome. Translocations are common in cancer cells. In reproductive cells, three types of translocation are possible: reciprocal translocation, non-reciprocal translocation, and **Robertsonian translocation**. A reciprocal (or balanced) translocation is the exchange of the equivalent portions of two nonhomologous chromosomes. Such exchanges are also referred to as balanced translocations because the normal diploid chromosome complement is still present (though rearranged). Individuals with reciprocal translocations are normal (unless the breakpoint is in the middle of a gene that affects the phenotype), but they have a 50% chance of having a normal child and a 50% chance of either having a child with a genetic condition or a miscarriage (depending on the amount and location of the genetic material involved). If the affected chromosomes are not "balanced" during production of the gamete, the latter may not have a complete haploid complement of some chromosomal regions and some regions may be present in diploid number resulting in a zygote that has monosomy for part of the genome and trisomy for another part (these are referred to as "unbalanced" translocations). Unbalanced translocations also occur when the initial translocation is non-reciprocal (causing an "unbalanced non-reciprocal translocation"). In this case, one segment of a chromosome attaches to another chromosome while the rest is lost. A Robertsonian translocation (also known as Robertsonian, or centric, fusion) is a chromosomal rearrangement where two nonhomologous acrocentric chromosomes (chromosomes where the centromere is located near one end of the chromosome) break at the centromere and the long arms join together to form one chromosome. The short arms are also fused to form a new chromosome. The short arms often do not contain essential genes, and this new chromosome is usually lost after a few rounds of mitosis. Thus, a person with a Robertsonian translocation only has 45 chromosomes but may be phenotypically normal. In modern humans, the acrocentric chromosomes commonly involved in Robertsonian translocations are chromosomes 13, 14, 15, 21, and 22. Individuals with Robertsonian translocations involving chromosomes 21 and 14 (or 15) run a greater risk of having children with Down syndrome because of an unbalanced trisomy 21. Over time, translocations can result in loss of shared **synteny** on chromosomes between species. A Robertsonian translocation may have been responsible for producing the modern human chromosome 2 from two ape chromosomes (Yunis and Prakash 1982). *See also* **chromosome; chromosome number.**

transmutation *See* **Lamarckism.**

transposition (L. *trans* = to move across and *poser* = to place, so to move from one place to another) The transfer of a segment of genetic material from one location to another. There are two basic methods of transposition. Retrotransposition involves **DNA transcription** into messenger **RNA**, the messenger RNA is then reverse transcribed into DNA, and finally, the DNA is inserted back into the genome in a new location. DNA transposition involves the excision of a segment of DNA from the genome and then this is reinserted at a new location. *See also* **transposon.**

transposon DNA segments that are capable of undergoing **transposition**. They are also known as **mobile genetic element**s or jumping genes. *See also* **transposition.**

Transvaal Museum The institution now called the **Ditsong National Museum of Natural History**, but formerly known as the Transvaal Museum, was founded in December 1892 and was originally the Staatsmuseum ("State Museum" in Afrikaans) of the Zuid-Afrikaansche Republiek (South African Republic). The idea of establishing the museum originated with Willem Johannes Leyds, the Secretary of State of the republic, and the museum was originally housed in the Raadsaal (Parliament House) in Pretoria. The museum's collections consisted primarily of zoological, botanical, geological, and paleontological specimens as well as ethnographic artifacts of the indigenous population and historical artifacts from the early European occupation of South Africa. Jan Willem Bowdewyn Gunning, a zoologist, served as the first director of the Staatsmuseum from 1897 to 1912 during a difficult time in its history. Construction of a new building to house the museum's collections began in 1899 but later that year the Anglo Boer War began and after the victory of the British the new museum opened (in 1900) and was renamed the Transvaal Museum in 1901. The museum moved to its current location on Paul Kruger Street in 1912, although the natural history collections were not relocated to the new buildings until 1925. The Transvaal Museum became an important center of paleoanthropological research during the 1930s when the Scottish-born paleontologist Robert **Broom** became curator of paleontology at the museum. Between 1936 and his death in 1951 Broom conducted excavations at **Sterkfontein, Kromdraai,** and **Swartkrans** that led to the discovery of the numerous **australopith** and animal fossils and some artifacts, now housed at the museum. John **Robinson** and C.K. (Bob) **Brain** continued the tradition of paleoanthropological research at the Transvaal Museum after Broom's death and expanded the museum's collection of australopith fossils and archeological artifacts, and Francis Thackeray continued that tradition. In 1999, the Transvaal Museum was amalgamated with the Pretoria-based National Cultural History Museum (also called the African Window) and the South African National Museum for Military History (situated in Johannesburg) to form the Northern Flagship Institution. On the May 28, 2010, its name was changed to the Ditsong National Museum of Natural History. Hominin fossil collections Sterkfontein (the material collected by Broom and Robinson from Members 4 and 5 between 1936 and 1958; the material collected by Clarke, Hughes, and Tobias since 1966 from Members 2–6 is curated by the School of Anatomical Sciences, University of the Witwatersrand), **Swartk-rans, Kromdraai, Cooper's Cave**. Contact department Vertebrate Department. Contact person Stephany Potze (e-mail potze@nfi.museum). Postal address Paul Kruger Street, PO Box 413, Pretoria 0001, South Africa. Website Under reconstruction. Tel +27 12 322 7632. Fax +27 12 322 7939.

transverse arch A mediolaterally oriented arch in the midfoot formed by the cuneiform, cuboid, and metatarsal bases. It is present in the foot of modern humans as well as nonhuman primates. All the heads of the metatarsal are weight-bearing so there is no distal transverse arch. However, because the bases of the metatarsals are not in the same plane, there is a transverse arch in the midfoot. The arch is also evident where the hindfoot articulates with the midfoot at the calcaneocuboid joint and at the joint between the navicular and the cuneiforms. The transverse arch is actively supported by the adductor hallucis, fibularis longus, and tibialis posterior. Loss of the transverse arch results in a splayfoot (pes transversoplanus).

transverse buttress A crest-like structure that in *Pan* and *Gorilla* runs from above the infra-orbital foramen to blend inferiorly with the **canine eminence**.

transverse crest *See* **epicrista**.

transverse isotropy In materials science, a material whose properties (such as stiffness) are similar within a cross-sectional plane but which are different from those along an axis orthogonal (i.e., at right angles) to that plane. Some bones (e.g., portions of long bones like the femoral shaft) exhibit material properties that approximate transverse isotropy.

transverse ridge *See* **distal trigon crest**.

transverse-sigmoid system A variant of the **intracranial venous system**. In most modern humans the veins draining the deep structures of the brain drain as the deep cerebral veins into the left **transverse sinus**, and the superior sagittal sinus drains into the right transverse sinus. Each transverse sinus drains into a sigmoid sinus, which runs inferiorly in a sigmoid-shaped groove from the lateral end of the transverse sinus to the superior jugular bulb, which marks the beginning of the internal jugular vein. This pattern of dural venous sinuses is known as the transverse-sigmoid system, and individuals (e.g., **KGA 10–525**) with transverse and sigmoid venous sinuses on both sides are said to show transverse-sigmoid system dominance. In just a few percent of modern humans and in some chimpanzees and gorillas the venous blood from the brain takes a different route (Kimbel 1984). Instead of draining into the transverse sinuses, the deep cerebral veins and the superior sagittal sinus drain into an enlarged midline occipital sinus, which courses from the cruciate eminence towards the foramen magnum in a midline groove. The occipital sinus drains into uni-, or bilateral, marginal sinuses, which run(s) around the margin(s) of the posterior quadrants of the foramen magnum to drain into the superior jugular bulb and/or the vertebral venous plexuses. This is known as the **occipito-marginal system** of intracranial venous sinuses, and individuals with a substantial occipital sinus and with marginal venous sinuses on both sides (e.g., **OH 5**) are said to show occipito-marginal system dominance. Some individuals (e.g., A.L. 333-45 and **KNM-ER 23000**) have a mix of the two systems, with a transverse-sigmoid system of venous sinuses on one side, and an occipito-marginal system of venous sinuses on the other, and a few individuals (e.g., **Taung 1**) have been reported to show coexistence of the two systems on the same side (Tobias and Falk 1988). *See also* **cranial venous drainage**.

transverse sinus One of the **dural venous sinuses** that belong to the **intracranial venous system**. In most modern humans the right and left transverse sinuses receive venous blood from, respectively, the **superior sagittal sinus** and the straight sinus (the latter drains venous blood from the deep cerebral veins), and they drain into the sigmoid sinuses. This **transverse-sigmoid system** of venous sinuses is the dominant system for intracranial venous drainage in modern humans. *See also* **cranial venous drainage**.

transverse tori   *See* **mandible**.

travertine   *See* **calcium carbonate**.

tree   In **systematics**, a tree is a diagram that depicts the descent relationships among a group of **taxa**. The term tree is used interchangeably with the term **cladogram** (syn. cladogram).

tree ring dating   *See* **radiocarbon dating**.

tree topology   (Gk *topos*=a place and *logos*= word) The branching pattern of a phylogenetic tree or **cladogram**.

tribe   The category in the Linnaean hierarchy between a **genus** and a **family**. It is the category that some researchers prefer for the *Pan* and *Homo* clades. There is no formal definition of a tribe.

tribosphenic   (Gk *tribein*=to rub and *sphen*= wedge) The name given to the triangular configuration of the three main cusps that forms the basic structure of the upper and lower postcanine tooth crowns. Hence the etymology "wedge-shaped structures that rub together." *See also* **cusp morphology**.

trigon   (Gk *trigonon*=triangle) The basic triangular three-cusped (or triconodont) structure of a maxillary (upper) postcanine tooth crown in the dental terminology devised by Osborn (1892). The three cusps are the **protocone** at the apex of the triangle on the lingual aspect, and the **paracone** mesially and the **metacone** distally on the buccal aspect.

trigon basin   The term used by Szalay (1969) and Hershkovitz (1971) for the area others refer to as the **central fovea**. *See* **central fovea (maxillary)**.

trigon crest   A term used by **Robinson** (1956) for an **enamel** structure on a maxillary molar tooth crown that others call the **distal trigon crest**. *See* **distal trigon crest**.

trigonid   (Gk *trigonon*=triangle) The basic triangular three-cusped (or triconodont) structure of a mandibular (lower) postcanine tooth crown in the dental terminology devised by Osborn (1892). The three cusps are the **protoconid** at the apex of the triangle on the buccal aspect, and the **paraconid** mesially (this cusp has been lost in all higher primates including hominins) and the **metaconid** distally on the lingual aspect.

trigonid basin   The term used by Szalay (1969) and Hershkovitz (1971) for the area others refer to as the **anterior fovea (mandibular)**. *See* **anterior fovea (mandibular)**.

trigonid crest   Term used by **Robinson** (1956) for an **enamel** structure on a mandibular molar tooth crown that others call the **distal trigonid crest**. *See* **distal trigonid crest**.

trigonum frontale   *See* **frontal trigon**.

Trinchera del Ferrocarril   One of several sediment-filled cave systems exposed in a railway cutting in the Sierra de **Atapuerca**, 14 km/9 miles east of Burgos in northern Spain. The fossiliferous cave/fissure complexes within this system include the sites of **Galería**, **Gran Dolina**, and **Sima del Elefante**. *See also* **Atapuerca**; **Galería**; **Gran Dolina**; **Sima del Elefante**.

Trinchera Dolina   *See* **Gran Dolina**.

Trinil   (Location   7°22′S,   11°34′E, Indonesia; etym. named after a local village) History and general description The type site of *Pithecanthropus erectus* (now referred to as *Homo erectus*) takes its name from the nearby village of Trinil in East Java, and is the best known of all the hominin-bearing sites in Indonesia. The Dutchman Eugène **Dubois** excavated the site from 1891 to 1894, and after he returned to Europe in 1895 his assistants Sergeants Kriele and De Winter continued systematic excavations until 1900. Subsequent field projects at Trinil were conducted by Selenka (1906–8) and by a joint Indonesian–Japanese team from 1976 to 1977. Hundreds of vertebrate fossils (including mammals, crocodiles, and turtles) were excavated, and the faunal collection is housed in the **Rijksmuseum van Natuurlijke Historie**, Leiden, The Netherlands. These fossils are known as the **Trinil H.K. Fauna** (H.K. is an abbreviation of *Hauptknochenschicht* meaning "main bone layer" in German). A recently completed ecomorphological study of this fauna concluded that the paleoenvironment of Trinil was characterized by open savannas, densely covered river valleys, and upland forests. The site is an exposure of river-formed deposits on the banks of the Solo River and it is derived from the Trinil Beds of the **Bapang Formation** (formerly Kabuh), which overlie the **Sangiran Formation** (formerly Pucangan). Temporal span and how dated? Initially dated by the **potassium-argon dating** method to 500 ka, most scholars now date the hominin fossil-bearing sediments to between 700 ka and 1 Ma. Hominins found at site Nine hominin specimens (Trinil 1–9), attributed today to *H. erectus*, were recovered by Dubois and his collectors between 1891 and 1900. These specimens include Trinil 1, the third molar found in September 1891, the **calotte** designated **Trinil 2** found in October 1891 and assigned the **holotype** of *Pithecanthropus erectus*, and a femur, **Trinil 3** (found in August 1892), that was presumed to be associated with the calotte. It was the latter that led Dubois to conclude that the individual walked bipedally. Trinil 4 and 5 are isolated teeth (a left M^2 and P_3, respectively) found in 1897. Much later in time Dubois' attention was drawn to four femoral fragments known as the **Trinil femora** (a proximal femur designated Trinil 6 and pieces of the femoral shaft numbered Trinil 7–9) that were also assigned to *P. erectus* (Dubois 1932). Archeological evidence found at site None. Key references: Geology, dating, and paleoenvironment Selenka and Blanckenhorn 1911, von Koenigswald 1964, Pope 1983, Weinand 2005; Hominins Dubois 1893, 1894, 1924, 1932.

Trinil 2 Site Trinil. Locality N/A. Surface/*in situ* In situ. Date of discovery October 1891. Finder N/A. Unit N/A. Horizon N/A. Bed/member Trinil Beds. Formation **Bapang Formation**. Group N/A. Nearest overlying dated horizon N/A. Nearest underlying dated horizon N/A. Geological age *c.*700 ka to 1 Ma. Developmental age Adult. Presumed sex Unknown, but its small size suggest it might be female. Brief anatomical description Calotte made up of most of the frontal, the occipital squame, and both parietals. Announcement Dubois 1892. Initial description Dubois 1894, 1924a. Photographs/line drawings and metrical data Dubois 1924b. Detailed anato-

mical description N/A. Initial taxonomic allocation *Anthropopithecus erectus*. Taxonomic revisions *Pithecanthropus erectus* (Dubois 1894), *Homo erectus* (Weidenreich 1940). Current conventional taxonomic allocation *Homo erectus*. Informal taxonomic category Pre-modern *Homo*. Significance The first good evidence for an early hominin outside of Europe and the **holotype** of *Homo erectus*. Location of original **Rijksmuseum van Natuurlijke Historie**, Leiden, The Netherlands.

Trinil 3 Site Trinil. Locality N/A. Surface/*in situ* In situ. Date of discovery August 1892. Finder N/A. Unit/horizon N/A. Bed/member Trinil Beds. Formation **Bapang Formation**. Group N/A. Nearest overlying dated horizon N/A. Nearest underlying dated horizon N/A. Geological age *c.*700 ka to 1 Ma. Developmental age Adult. Presumed sex Unknown, but the small size of the head led Kennedy (1983) to suggest that the femur "is that of a female" (*ibid*, p. 607). Brief anatomical description The specimen comprises a complete adult hominin left femur. An irregular bony excrescence that extends from the posteromedial surface of the proximal one third of the femoral shaft is most likely due to the formation of woven bone within a traumatically induced hematoma (i.e., myositis ossifans) in the adductor muscle compartment (Day and Molleson 1973). Other than that the specimen is free of significant pathology. Announcement Dubois 1893. Initial description Dubois 1926a. Photographs/line drawings and metrical data Dubois 1926a, Day and Molleson 1973. Detailed anatomical description Dubois 1926a. Initial taxonomic allocation *Anthropopithecus erectus* (Dubois 1893). Taxonomic revisions *Pithecanthropus erectus* (Dubois 1894), *Homo erectus* (Weidenreich 1940). Current conventional taxonomic allocation *H. erectus*. Informal taxonomic category Pre-modern *Homo*. Significance The resemblance between Trinil 3 and the femora of upright-walking modern humans was the reason Dubois used the species name he did for this taxon, but doubts have been expressed about whether Trinil 3 is contemporary with the type specimen of *H. erectus* (Day and Molleson 1973). Location of original **Rijksmuseum van Natuurlijke Historie**, Leiden, The Netherlands.

Trinil fauna The inclusive name given by Ralph **von Koenigswald** for the fauna found at **Trinil** and at Trinil-aged sites elsewhere in Java. Not to be confused with the more exclusive **Trinil H.K. fauna**, a term introduced by de Vos and

Sondaar (1982) for the fauna found in the same horizon as **Trinil 2**, the type specimen of *Homo erectus*.

Trinil femora The relatively complete and well-preserved **Trinil 3** femur (also referred to as "Femur I") was recovered in August 1892. It was the first of five hominin femora from **Trinil** recognized by Eugène **Dubois**. The remaining four hominin femora (i.e., Femora II–V) consist of one specimen that preserves the damaged proximal end (the greater trochanter is missing) and the shaft (Femur II), two substantial mid-shaft fragments (Femurs III and IV), and one shorter fragment from the mid-shaft region (Femur V). The four femora were excavated in 1900, but Dubois did not recognize them as hominins until 1932, exactly 40 years after the recovery of Trinil 3 (Dubois 1932). Three years later a sixth femur (Femur VI in Dubois' scheme) was recognized by Dubois as belonging to *Pithecanthropus erectus* (Dubois 1935). Its exact provenance is uncertain, but Dubois wrote that it is "certainly not from Trinil but from another part of the Kendeng region," and he suggested there is "some probability that this fossil was found at **Kedung Brubus**" (*ibid*, p. 3). However, when Day and Molleson (1973) examined all of the Trinil femora (i.e., Femora I–VI) in detail (using radiography and microscopy) they concluded that Femur VI "is not hominine, hominid or even primate" (*ibid*, p. 151). Doubts have been expressed about whether the femoral remains from Trinil are the same age as the calotte and the teeth (Day and Molleson 1973). Kennedy (1983) addressed the traditional morphometric affinities of the Trinil femora in a multivariate analysis and her conclusion was that "although the Trinil femora demonstrate a few characters, particularly in the lower shaft, similar to those in *Homo erectus*, their overall pattern unequivocally allies them with the sapient comparative group" (*ibid*, p. 614). *See also* **Trinil**; **Trinil 3**; **Kedung Brubus**.

Trinil H.K. fauna (etym. Combination of the name of the site, **Trinil**, and *Hauptknochenschicht* (Ge. for "main bone layer") The name given by de Vos and Sondaar (1982) to the fauna found in the same horizon as **Trinil 2**, the **type specimen** of *Homo erectus*. The Trinil H.K. fauna is not to be confused with the more inclusive **Trinil fauna**, a term introduced by Ralph von **Koenigswald** for the fauna found at Trinil, and at Trinil-aged sites elsewhere in Java.

trinucleotide repeat *See* **microsatellite**.

triplet codon *See* **genetic code**.

trisomy The presence of three copies of a particular **chromosome** in a cell rather than the normal diploid number (i.e., two). In modern humans, the only trisomies in which the individuals survive to adulthood are trisomy 21 (Down syndrome) and trisomy X (triple X syndrome).

tRNA Abbreviation of **transfer RNA** (*which see*).

trophic (Gk *trephein*=to nourish) A term which refers to anything having to do with the activities that promote nutrition. Any behaviors involved with the acquisition and the processing of food are examples of trophic behavior.

trophic level (Gk *trephein*=to nourish) The position an organism occupies in a food chain. The sun is the ultimate source of the vast majority of the energy that enters the Earth's food chain. Primary producers, typically photosynthetic plants, use sunlight as their energy source. **Herbivores** obtain their energy by consuming live plants, and **carnivores**, called primary consumers, obtain their energy by eating herbivores. Ultimately insects and micro-organisms degrade the carcasses of carnivores, and whatever energy is left becomes dead biomass. Substantial amounts of energy and/or biomass are lost at each stage of the food chain as (a) waste (e.g., feces and urine) and the energy expended on waste excretion, (b) energy used for locomotion, mastication, etc., and (c) energy consumed to heat and cool the body (especially by warm-blooded creatures). It is unlikely that any member of the hominin clade was either an obligate herbivore or an obligate carnivore; they were most likely **eclectic feeders** (e.g., omnivores).

tropical (Gk tropikos=pertaining to the solstice) The term for the region (and the organisms occupying that region) between the Tropics of Cancer (approximately 23° latitude in the northern hemisphere) and Capricorn (approximately 23° latitude in the southern hemisphere) where at some point in the year the sun is directly overhead. Day length in the tropics is less variable than in **temperate** regions, and climate tends to be determined by rainfall rather than temperature. The tropics have high levels of biodiversity, with the latitudinal range of each species generally being less than in the species found in temperate areas (known as Rapoport's rule). Primates are primarily a tropical order, and the hominin clade almost certainly emerged in the tropics.

tropical dry forest   A biome located within the tropical (between 23°N and 23°S) and subtropical (between 20° and 40° of latitude in both hemispheres) regions. Tropical dry forests are characterized by long periods of drought during which the abundant deciduous trees drop their leaves. The length of the dry period depends on latitude, with sites further from the equator having longer dry periods. During the dry periods, trees and shrubs drop their leaves to conserve water, and leaf flushing resumes with the start of the rainy season. Dry forests have trees that are shorter and simpler in structure than those in rain forests. The dry forests of Africa, Central and South America, and those of tropical islands, have a diverse wildlife, although not as highly diverse as tropical rain forests. African tropical dry forests tend to have a greater diversity of mammals than most rain forests in Central and South America.

tropical rain forest   A biome restricted mostly to the equatorial zone between latitudes 23°N and 23°S and which is primarily defined by climate and quantity of rainfall. Tropical rain forests have climates in which the mean temperature of the coldest month is at least 18°C and every month has at least 100 mm of rainfall, although short dry periods lasting a few days to a few weeks may exist. Overall, they are the least seasonal of all forest biomes. Some sources define tropical rain forests as those forests with a minimum annual rainfall of 1750–2000 mm of rainfall and located within the above equatorial zone. Tropical forests with several consecutive dry months (less than 60 mm rainfall) are referred to as tropical monsoon or seasonal forests. Both tropical rain forests and tropical monsoon forests are categorized under the inclusive term tropical moist forests. Tropical rain forests occur in all three tropical land areas (Americas, Africa, and Indo–Malaysia). The neotropical rain forests of the Americas are the most extensive, covering about half of the global total rain forests (4×10^4 km^2), followed by Indo-Malaysia (2.5×10^4 km^2), followed by Africa (1.8×10^4 km^2), but these numbers are in decline because of deforestation. Tropical rain forests have the highest net primary productivity of any forest biome, with approximately 9000 kilocalories per m^2 per year; to put this into perspective, woodland biomes have an average net primary productivity of 3000 kilocalories per m^2 per year (Archibold 1995). This translates into tropical rain forests containing approximately 80% of the world's biodiversity, thus they are a critical reserve for global biodiversity. Most African tropical rain forests experience wet and dry seasons, although the number may vary depending on their location; dryer seasons tend to be associated with lower fruit production, a preferred resource for many arboreal primates. During these lean periods, primates may switch to abundant, low-quality food items that tend to be more mechanically challenging. It is these **fallback food** items that have been suggested to have played an important role in the evolution of primate craniodental and digestive morphology.

Trotter, Mildred   *See* **Terry (Robert J.) Collection**.

true chin   A **mental protuberance** separated from the **alveolar process** of the **mandible** by a hollowed area called an *incurvatio mandibulae*. *See also* **mandible**.

true fossils   (OE *triewe* = steadfast, characterized by good faith and L. *fossile* = something that is dug up) True fossils are remnants of the organism itself, in the form of either hard (i.e., bones and teeth) or soft (i.e., skin, muscle, etc.) tissues. True hard-tissue fossils (i.e., bones and teeth) make up the vast majority of the fossil evidence for human evolution.

Tswaing Impact Crater   (also known as the Pretoria Salt Pan) A lake formed in the meteor impact crater at Tswaing, South Africa. The site preserves a sedimentary record of paleoclimate for subtropical southern Africa stretching back 200 ka (Partridge et al. 1997). **Proxy** records indicate that summer rainfall amounts varied with the amount of summer **insolation** in the southern hemisphere paced by the timing of orbital **precession**. *See also* **closed basin lakes**.

tuber   *See* underground storage organ; **belowground food resources**.

tubercle A small rounded projection on a bone (e.g., scaphoid tubercle) or tooth. *See also* **tuberculum**.

tuberculum (L. *tuberculum*=a small rounded projection, diminutive of L. *tuber*=lump) An **accessory cusp** on a tooth crown.

tuberculum anomalus A term used by Georg Carabelli (1842) for the distinctive feature he described on the mesiolingual face of a maxillary molar tooth crown. *See* **Carabelli('s) trait**.

tuberculum intermedium A term introduced by Gustav **Schwalbe** for a distinct **accessory cusp** located on the lingual edge of a mandibular molar crown between the **metaconid** and the **entoconid**. Some authors refer to weaker expressions of this feature as a **metaconulid** and others as a **postmetaconulid**, but Grine (1984) has reported examples of the possible co-occurrence of a tuberculum intermedium *and* a **postmetaconulid** (syn. **C7**, entostylid, lingual accessory cusp, median lingual accessory cusp, metaconulid, **metastylid**, postmetaconulid, **preentoconid**, seventh cusp). *See also* **postmetaconulid**.

tuberculum paramolare *See* **paramolar tubercle**; **protostylid**.

tuberculum sextum (L. *tuberculum*=a small rounded projection and *sextus*=sixth) One or more **accessory cusps** in the distal part of a mandibular molar crown between the **entoconid** and the **hypoconulid** (syn. **C6**, cusp 6, sixth cusp, postentoconulid).

tufa (OIt *tufo*=volcanic ash) A variety of **speleothem**, a tufa is a sedimentary rock predominantly composed of **calcium carbonate** made up of sediments deposited from water percolating through a porous rock in a spring or a lake. The **Taung** cave was formed as a solution cavity within a flow made up of tufa. *See also* **calcium carbonate**; **speleothem**.

tuff (OIt *tufo*=volcanic ash) A rock made of small-sized (approximately <4 mm) volcanic debris, often in the form of an ash. Most fragments contain enough glass crystals to enable their age to be determined using the **argon-argon dating** or **potassium-argon dating** methods (e.g., **KBS Tuff** in the **Koobi Fora Formation**).

tuffaceous Any sediment that has a significant **tephra** (or **tuff**) component (e.g., reworked volcanic ash deposits). *See also* **tephra**; **tuff**.

Tuinplaas (Location 25°00′S, 28°36′E, in Limpopo Province, South Africa; etym. Afrikaans for "Garden City," the name of a nearby town) History and general description The site is a quarry, in which a human skeleton was found, approximately 130 km/80 miles north of Pretoria. Temporal span and how dated? It was originally thought to be associated with **Middle Stone Age** artifacts, but it has since been directly dated by **uranium-series dating** to between 11.0±0.7 and 20±3 ka (i.e., it is within the time span of the **Late Stone Age**). Hominins found at site **Tuinplaas 1**, a fragmentary skull and postcranial skeleton. Archeological evidence found at site MSA (Pietersburg) artifacts. Key references: Geology, dating, and paleoenvironment Lang 1929, Wells 1959, Pike et al. 2004; Hominins Broom 1929, Schepers 1941, Toerien and Hughes 1955, Hughes 1990; Archeology van Riet Lowe 1929, Mason 1962.

Tuinplaas 1 Site Tuinplaas. Locality N/A. Surface/*in situ In situ*. Date of discovery 1929. Finders C.J. Swierstra and H. Lang. Unit N/A. Horizon N/A. Bed/member Lateritic deposit atop calcareous **tufa** in the Tuinplaas Quarry. Formation N/A. Group N/A. Nearest overlying /underlying dated horizon N/A. Geological age **Late Pleistocene**. Developmental age. Adult. Presumed sex Male, based on overall size. Brief anatomical description A large, fragmentary cranium, plus a mandible with a weakly modeled chin and many postcranial fragments, including portions of most of the major long bones. Announcement Lang 1929. Initial description Broom 1929. Photographs/line drawings and metrical data Schepers 1941, Hughes 1990. Detailed anatomical description Schepers 1941, Toerien and Hughes 1955, Hughes 1990. Initial taxonomic allocation *Homo sapiens*. Taxonomic revisions N/A. Current conventional taxonomic allocation *H. sapiens*. Informal taxonomic category Anatomically modern human. Significance Although it does not derive from the **Middle Stone Age**, Tuinplaas 1 is nonetheless a rare specimen of terminal Pleistocene *H. sapiens* from southern Africa. Location of original **Ditsong National Museum of Natural History** (formerly the **Transvaal Museum**), Northern Flagship Institution, Pretoria, South Africa.

turbinates   *See* conchae.

Turkana Basin   A hydrological and sedimentary basin located in northern Kenya and southern Ethiopia. The Turkana Basin, which is part of the **East African Rift System**, comprises several structural elements. Portions of the southern part of the basin integrate components of a transverse Mesozoic rift system, the Anza Rift, while the more northerly parts reflect a series of half-graben developed on the flank of the modern rift trend. The basin today is a closed system; the Omo, Turkwell, and Kerio Rivers drain into Lake Turkana and there is no outlet. However, in the geological past at times the basin has been an open system, with connections to both the Nile River and the Indian Ocean. The Turkana Basin is well known for its extensive fossils sites and the many vertebrate fossils that have been recovered. The oldest fossiliferous sediments, which extend back to the Mesozoic (<100 Ma) when dinosaurs were numerous, are found in the Labur Hills to the north west of the modern lake. Late Oligocene, and earliest Miocene outcrops are found to the west of Lake Turkana at Nakwai and Losodok, and early Miocene sites are found at Kalodirr, Moruorot, and Loncherangan on the western shores and at Buluk and Kajong (Mwiti) to the east. The lake basin is best known for its Pliocene and early Pleistocene record of human evolution and much of the fieldwork in Kenya and in Ethiopia over the last 40–50 years has been focused on sediments of this age. The Turkana Basin developed as an integrated sedimentary basin in the Early Pliocene (*c.*4.2 Ma). For much of ensuing times, the ancestral Omo River occupied most of the basin and extensive fluvial deposits accumulated, hosting much of the vertebrate fossil record the basin is well known for (e.g., the **Omo Group** deposits, particularly the **Koobi Fora**, **Nachukui**, and **Shungura** Formations). Tectonic and climatic developments fostered lake phases in the basin, as increased subsidence created room and the strengthened monsoonal rains supplied the necessary water. Present-day Lake Turkana has a stratigraphic record reaching back 200,000 years in the **Turkana Group** strata, during which time it has shifted between a freshwater mega-lake linked to the Nile and the current closed-basin alkaline water body seen today. Outcrops of the Late Pleistocene are known in the Omo Valley (the Kibish Formation) and they are also found on the western shores of the modern lake (at Eliye Springs), where they are less extensive. Numerous Holocene sites are present in the Lake Turkana Basin, but to date these have not been worked as extensively as those of the Cenozoic, although interest is increasing and currently several field teams have begun to survey these more recent outcrops. *See also* **Omo Group**; **Turkana Group**.

"Turkana boy"   *See* **KNM-WT 15000**.

Turkana Group   The younger of the two inclusive stratigraphic sequences (the older one is known as the **Omo Group**) exposed on the **Lower Omo Basin** (also called the Lower Omo Valley). It consists of the Bume Formation, **Kibish Formation**, and Errum Formation.

turnover-pulse hypothesis   One of several hypotheses proposed by Elisabeth Vrba that seek to explain how environmental change influences evolutionary patterns in the fossil record. In the absence of environmental change, species should experience evolutionary **stasis**. Sudden, large-scale environmental change should be associated with a "pulse" or concentration of extinction and speciation events within a relatively narrow time range. The turnover-pulse hypothesis (Vrba 1985a) states that most of the turnover within lineages (i.e., the extinction of old species and the appearance of new ones) is driven by environmental change, and that if environmental changes are strong enough to induce turnover within one **lineage** then they should be strong enough to induce turnover in other lineages. Vrba suggested that stenotopic (i.e., specialist) species are more affected by environmental change than eurytopic (i.e., generalist) species, so she predicted there will be a temporal ordering of extinction and speciation effects, with **stenotopes** being affected first and **eurytopes** later. With respect to paleoanthropology, she (Vrba 1985b, 1988, 1995) stated that as African habitats became cooler and dryer at *c.*2.5 Ma, the diversification of mammal lineages, including hominins, which also occurred around that time may be a manifestation of a turnover pulse. She furthermore suggested that evidence of this pulse can be found in the fossil record of bovids (i.e., antelopes and their close relatives), but a detailed investigation of fossil mammals from the Turkana Basin (Behrensmeyer et al. 1997) failed to find conclusive evidence of such a pulse at the predicted time. A caveat, however, is that the presence of a permanent water source (the Omo river) in the **Turkana Basin** may have buffered the species living there from the effects of environmental change. *See also* **effect hypothesis; habitat theory hypothesis; punctuated equilibrium**.

Turolian *See* **European mammal neogene; biochronology**.

Turubong (Location 36°30′16″N, 127°32′15″E, North Chungchong Province, South Korea; etym. named after the nearby Turubong Hill) History and general description Turubong is a cave complex located on Turubong Hill in the central part of South Korea. The caves were discovered in the 1970s during mining for limestone. The primary localities are Turubong Cave No. 2, No. 9, Saegul, Chonyogul, and Hungsugul. The importance of the Turubong localities is the presence of a purported burial of a child in Hungsugul and the argument that Turubong Cave No. 2 is evidence of an **osteodontokeratic**-type culture. Temporal span and how dated? Recent **AMS radiocarbon dating** suggests Hungsugul may be recent in age, possibly early **Holocene**, but the presence of *Crocuta ultima*, *Hyaena* sp., and *Coelodonta antiquitatis* in the Turubong Cave No. 2 faunal assemblage suggests it may date to the **Late Pleistocene**. Hominins found at site The nearly complete skeleton of a *c*.5–6 year-old modern human child was excavated from Hungsugul. Archeological evidence found at site There are few stone tools in the Turubong Cave localities, particularly Cave No. 2, and although the excavators have argued that some of the bones are evidence of an osteodontokeratic culture, they are almost certainly bones that have been modified by carnivores. Repository The Turubong materials are currently stored in the Chungbuk University Museum in Cheongju and the Yonsei University Museum in Seoul, both in South Korea. Key references: Geology, dating, and paleoenvironment Lee 1983, 1984, 1994, Sohn 1983, Park and Lee 1990, Park 1996, Norton 2000; Hominins Park and Lee 1990, Park 1992, Norton 2000, Norton and Jin 2009; Archeology Lee 1983, 1984, 1994, Kwon 1996, 1998, Norton 2000.

"Twiggy" Nickname given by Louis and Mary Leakey to the **OH 24** cranium from **Olduvai Gorge**.(NB: "Twiggy," also known as Lesley Hornby, was the first "supermodel." She was unusually "skinny" for a model and by 1968, when OH 24 was found, she was internationally known and had been on the covers of *Vogue* and *Tatler*).

Twin Rivers (Location 15°31′S, 28°11′E, eastern Zambia; etym. located between the Chikupi and Nangombi Rivers, also known as Twin Rivers Kopje) History and general description A hilltop site on a 52 m-high isolated outcrop of limestone, Twin Rivers consists of several exposed fissures filled with fossil fauna and artifact-bearing **speleothem**-cemented **breccia**. A series of excavations (Blocks A–G) were directed by Desmond **Clark** in the 1950s and by Lawrence Barham in the 1990s. The site represents the only well-dated stratified example of Lupemban (**Middle Stone Age**) assemblages, and contains a large amount of imported and worked potential pigments such as **specularite** and **hematite**. Temporal span and how dated? A, F, and G blocks have been radiometrically dated. **Thermal ionization mass spectrometry** and **uranium-series dating** estimates on speleothem suggests a **Middle Pleistocene** age range of 226–170 ka for A and F blocks prior to cave roof collapse. Based on **AMS radiocarbon dating** estimates on bone and **thermoluminescence dating** estimates on burnt **quartz** and **calcite**, G block was likely occupied during the **Late Pleistocene** to the **Holocene** (*c*.60 ka to the present). Hominins found at site One probable humeral shaft fragment. Archeological evidence found at site Flake production from local raw materials (especially quartz) was from single, multiple, and radial cores (including prepared/**Levallois** examples). Retouched tools include some of the earliest known examples of backed pieces and Lupemban lanceolates and lanceolate fragments. The fauna is well preserved but highly fragmented; a hominin accumulation agent is possible, but not proven. The A block excavation yielded more than 180 pieces of possible pigments (e.g., hematite and specularite), including pieces with wear facets, striations, and incised surfaces, as well as a quartzite pestle with iron-oxide-stained worn surfaces. Key references: Geology, dating, and paleoenvironment Barham 2000; Hominins Barham 2000; Archeology Barham 2000; Clark and Brown 2001.

tympanic tube *See* **auditory tube**.

type (L. *typus*=image) A specimen to which the name of a taxon is irrevocably attached. If it is designated in the original description it is called the holotype, but if it is designated subsequently it is called the **lectotype**. *See also* **type specimen**.

Type I error In statistics, this type of error occurs when a **null hypothesis** is rejected incorrectly (i.e., when a result is accepted as significant when it is actually *not* significant). For example, when the difference in cranial capacity between two samples is accepted as significant when it actually is not, or when two variables are thought to be significantly associated when there is not actually a significant relationship.

Type II error In statistics, this type of error occurs when a **null hypothesis** fails to be rejected incorrectly (i.e., when a result is judged to be non-significant when it actually *is* significant). For example, when the cranial capacities in two samples are thought to be equal when they are actually different, or when two variables are thought to be independent of each other when there is actually a significant relationship between them.

type locality The locality at which the **type specimen** of a species was collected.

type section The geological **section** specified in the first formal description of a **stratum**.

type site In archeology, a type site is considered to be the model of a particular technocomplex (**La Madeleine** is the type site for the **Magdalenian**). In geology, a type site is considered to be typical of a particular rock formation.

type species A species of a given genus that has been chosen (by the describer of the genus, or, failing that, by a subsequent reviser) to represent the genus. The generic name is thereafter irrevocably attached to the type species, just as a species name is irrevocably attached to its type specimen. *See also* **type**.

type specimen The specimen to which a species' name is irrevocably attached (e.g., Neanderthal 1 with *Homo neanderthalensis*; **OH 7** with *Homo habilis*; **Trinil 2** with *Homo erectus*). Whatever species the specimen belongs to, it carries its **Linnaean binomial** with it, and if its binomial has priority (i.e., it has the earliest publication date) then it provides

the name of the species. Thus, if someone decided that the type specimen of *H. neanderthalensis* was in fact a modern human, but that all other Neanderthals were still Neanderthals, then *H. neanderthalensis* would be sunk into *Homo sapiens* (*H. sapiens*, 1758 has priority over *H. neanderthalensis*, 1864), and the next available name for a presumed Neanderthal specimen would have to be substituted (the next name given to a Neanderthal specimen would appear to be *Homo neanderthalensis* var. *krapinensis* Gorjanovic-Kramberger, 1902, the type specimen being Krapina skull D, so the "Neanderthal" species with have to be called *Homo krapinensis*). If there had been no such name, a new name would have to found for the Neanderthal **hypodigm** and a **lectotype** would need to be designated and attached to that new species binomial. But, if someone decided that the type specimen of *H. habilis* really belonged to *Homo rudolfensis*, but that all other members of the *H. habilis* hypodigm were still *H. habilis*, then *H. rudolfensis* would be sunk into *H. habilis* (*H. habilis*, 1964 has priority over *H. rudolfensis*, 1986), a new Linnaean binomial would have to be found for the "old" *H. habilis* hypodigm (minus the transferred type specimen) and in this case it is *Homo microcranous* Ferguson, 1995. If the type specimen (**holotype** or lectotype) is destroyed then a new type specimen, called a **neotype**, can be designated from one of the **syntypes**.

typological A typological definition of a species is one that allows for relatively little variation, for, by definition, all the individual members of the species would closely conform to the specifications set out in the definition. It is an outmoded (but alas still extant) way of thinking about species.

Tyto alba (Gk *tuto*=night owl and L. *alba*= white) The formal name for the common barn owl, a **predatory** bird found throughout the Old and New Worlds. This animal is significant in paleoanthropology because it is largely responsible for the preservation of **microfauna**, particularly **micromammal**s, at paleoanthropological and paleontological sites, and these remains are important for the **biostratigraphy** and for the **paleoenvironmental reconstruction** of these sites. The common barn owl feeds on microfauna, principally small rodents of the Muridae and Cricetidae families, in addition to invertebrates.

U

U-series dating *See* **uranium-series dating**.

UA 31 Site Buia. Locality Uadi Aalad. Surface/*in situ* *In situ* (details in Ghinassi et al. 2009). Date of discovery December 1995. Finders Lorenzo Rook and team. Unit N/A. Horizon FL2a/FL2b, low-sinuosity fluvial channels sand. Bed/member N/A. Formation Alat Formation. Group Dandiero Group, Maebele Synthem. Nearest overlying dated horizon Upper part of the Alat Formation comprising sediments with reversed polarity, and correlated with the post-Jaramillo Matuyama Chron. Nearest underlying dated horizon Bukra Formation, sediments with reversed polarity correlated with the pre-Jaramillo Matuyama Chron. Geological age *c*.1 Ma. Developmental age Adult. Presumed sex Unknown. Brief anatomical description An almost complete cranium lacking part of the basicranium and the right side of the face. The vault is long, low, and narrow, and the occiput is angled; an angular torus is present. The supraorbital torus is thick and strongly projecting, and it is separated from the frontal squama by a sulcus. The maximum cranial breadth falls across the supramastoid crests, but the maximum biparietal breath is higher on the vault than is usual in *Homo erectus*. The face displays marked subnasal prognathism. All of the tooth crowns are broken off, but the roots of right P^3–M^3 and left P^4–M^2 are preserved. Endocranial capacity 950 cm^3 (directly measured by teff seed; Macchiarelli et al. 2008). Announcement Abbate et al. 1998. Initial description Abbate et al. 1998, Macchiarelli et al. 2004. Photographs/line drawings and metrical data Abbate et al. 1998. Detailed anatomical description Macchiarelli et al. 2004. Initial taxonomic allocation *H. erectus*. Taxonomic revisions N/A. Current conventional taxonomic allocation *H. erectus*. Informal taxonomic category Pre-modern *Homo*. Significance In many ways the UA 31 cranium resembles *H. erectus* yet it lacks an occipital torus and it possesses some "progressive" features such as a high position of

maximum parietal breath. Location of original Eritrea National Museum, Asmara, Eritrea.

UA 466 Site Buia. Locality Uadi Aalad. Surface/*in situ* Surface. Date of discovery 2003. Finder Luca Bondioli. Unit N/A. Horizon FL2a/FL2b, low-sinuosity fluvial channel sand. Bed/member N/A. Formation Alat Formation. Group Dandiero Group, Maebele Synthem. Nearest overlying dated horizon Upper part of the Alat Formation comprising sediments with reversed polarity that are correlated with the post-Jaramillo. Nearest underlying dated horizon Bukra Formation, sediments with reversed polarity correlated with the pre-Jaramillo Matuyama Chron. Geological age *c*.1 Ma. Developmental age Adult. Presumed sex Male, based on a very narrow subpubic angle and lack of ventral arc. Brief anatomical description A left pubic symphysis preserving the entire face of the symphysis and most of the bone medial to the obturator foramen. Announcement and initial description Bondioli et al. 2006. Photographs/line drawings and metrical data Bondioli et al. 2006. Detailed anatomical description Bondioli et al. 2006. Initial taxonomic allocation None. Taxonomic revisions N/A. Current conventional taxonomic allocation *Homo erectus*. Informal taxonomic category Pre-modern *Homo*. Significance The symphysis is tall relative to modern humans. If this fragment proves to be from *H. erectus* it would be the second pubic fragment belonging to that taxon. Location of original Eritrea National Museum, Asmara, Eritrea.

Uadi Aalad *See* **Buia**.

UB 335 Site 'Ubeidiya. Locality Square E 78. Surface/*in situ* *In situ*. Date of discovery It was excavated in 1966, but it was not recognized as a hominin until much later. Finder Excavated by M. Stekelis and colleagues, but identified by M. Belmaker. Unit I-26a. Horizon "Shoreline deposit". Bed/member Member FI. Formation 'Ubeidiya Formation. Group N/A. Nearest

Wiley-Blackwell Encyclopedia of Human Evolution, First Edition. Edited by Bernard Wood.
© 2013 Blackwell Publishing Ltd. Published 2013 by Blackwell Publishing Ltd.

overlying dated horizon Yarkouk Basalt 0.79 Ma. Nearest underlying dated horizon Erg el Ahmar Formation 1.96–1.78 Ma. Geological age *c*.1.5–1.2 Ma. Developmental age Adult. Presumed sex Unknown. Brief anatomical description Heavily worn right I_2 whose morphology matches that of UB 1700. Announcement Belmaker et al. 2002. Initial description Belmaker et al. 2002. Photographs/line drawings and metrical data Belmaker et al. 2002. Detailed anatomical description Belmaker et al. 2002. Initial taxonomic allocation *Homo* sp. *indet.* but "tentatively identified as *H. ergaster*" (*ibid*, p. 53). Taxonomic revisions N/A. Current conventional taxonomic allocation ***Homo erectus*** or *Homo ergaster*. Informal taxonomic category Pre-modern *Homo*. Significance Until the recovery of the hominins from **Dmanisi** the hominins from 'Ubeidiya constituted what little evidence there was of early hominins in the Caucasus/Near East. Location of original Department of Anatomy and Anthropology, Sackler Faculty of Medicine, Tel Aviv University, Israel.

'Ubeidiya (Location 32°41′N, 35°33′E, Israel; etym. name of a nearby settlement) History and general description The site, just south of the sea of Galilee, is located on the western flank of the Jordanian Rift and is unusual because the strata are steeply tilted. It was systematically excavated between 1960 and 1974, and then for two shorter periods (1989–94, 1997–9) thereafter (see Belmaker et al. 2002 for more details of the personnel involved and the results of the three campaigns). The 'Ubeidiya Formation comprises four members: from below they are the Li (Limnic Inferior), Fi (Fluviatile Inferior), Lu (Limnic Upper), and Fu (Fluviatile Upper). Each member represents a cycle of deposition, but it is not known how long these cycles lasted, nor how much time elapsed between the cycles. The strata within the members are numbered (1– <92 from below upwards) within each of the main trenches, but in all of the members, but particularly in the upper three members, the number of the same stratum differs from trench to trench. Temporal span and how dated? Biostratigraphic dating suggests a range of *c*.1.5–1.4 Ma (Martinez-Navarro et al. 2009), which is consistent with paleomagnetic studies indicating a sequence spanning *c*.1.53–1.2 Ma (Sagi 2005, Rink et al. 2007) for the Fi member. Hominins found at site Fragments of two parietals and one temporal (UB 1703-6), a left I_2 (UB 1700), and a right M^3 (UB 1701) were assigned to *Homo* sp. by Tobias (1966a) and later to cf. *Homo erectus* (Tchernov and Volokita 1986). However, these hominin remains are most likely intrusive. A right I_2 (**UB 335**) is the only *in situ* Lower Pleistocene hominin remain from the site (Belmaker et al 2002). Archeological evidence found at site Early **Acheulean** and later **Oldowan** tools are well known from the site, The assemblages most resemble the **Developed Oldowan** B and the Early Achulean from **Olduvai Gorge**. While taphonomic evidence indicates some hominin involvements in carcass processing (Gaudzinski 2004), the majority of the faunal assemblage was probably a natural accumulation (Belmaker 2006). Key references: Geology, dating, and paleoenvironment Tchernov 1988; Hominins Tobias 1966a, Tchernov 1986, Belmaker et al. 2002; Archeology Martinez-Navarro et al. 2009, Bar-Yosef and Goren-Inbar 1993, Shea 1999.

Uluzzo (Location 40°09′23″N, 17°57′38″E, Puglia, Italy; etym. named after a nearby tower and inlet) History and general description A cluster of caves on the edge of the Bay of Uluzzo, on the Ionian coast of Puglia. They include Serra Ciocora, Grotta del Cavallo, Grotta di Uluzzo, and Riparo di Uluzzo. Grotta del Cavallo is the only one to have yielded hominin remains; three teeth. Some of the caves have a relatively advanced stone tool technology named the Uluzzian by Palma di Cesnola (1989). The Uluzzian, like the **Châtelperronian** from French sites, is considered a transitional technocomplex that is associated with late **Neanderthals**. Temporal span and how dated? Layer IV of Grotta de Cavallo was dated using **radiocarbon dating** to more than 31 ka. Hominins found at site Three molars (left M_2 from lower levels; left M^1 and right M^2 from the upper levels) were found in Grotta del Cavallo. Archeological evidence found at site A typical **Mousterian** industry is found in the lower layers of the caves on the Bay of Uluzzo, but the upper layers contain evidence of the Uluzzian industry. The caves are also notable for evidence of shellfish consumption. Key references: Geology, dating, and paleoenvironment Messeri and Palma di Cesnola 1976; Hominins Palma di Cesnola and Messeri 1967, Alciati et al. 2005; Archeology Palma di Cesnola 1989, Bietti 1997.

unbalanced translocation   *See* **translocation**.

unbiased strategy   *See* **economic utility index**.

uncalibrated radiocarbon age *See* radiocarbon dating.

unconformity A break in the stratigraphic record that may be due to a period of non-deposition or to a period of active erosion. When there is an angular difference in the amount and/or direction of dip in the rocks on either side of the stratigraphic break the term angular unconformity is used. When there is no such angular difference the terms **disconformity** or non-sequence are sometimes used. Such non-angular unconformities may be very difficult to spot within the stratigraphic record yet within a given sequence they can account for substantial amounts of time.

underdominance A type of natural selection in which homozygous genotypes are favored and heterozygotes have lower fitness.

underground storage organ (or USO) The belowground or below-water part of a plant that has been modified and enlarged to store energy and/or water, typically found in dry or frequently burned environments. Depending on what part of the plant has been modified, these storage bodies can be called tubers, corms, bulbs, or rhizomes. USOs can be cryptic and are sometimes deeply buried, and therefore difficult to acquire. Due to their high caloric value, USOs are an important food of modern-day and historic hunter–gatherer groups (e.g., the Hadza of Tanzania). Several researchers have suggested USOs may have played an important role in hominin evolutionary history, including as **fallback foods** for the megadont archaic hominins (Laden and Wrangham 2005). This is at least in part because many USOs use the C_4 photosynthetic pathway and their consumption by hominins could explain the elevated $\delta^{13}C$ levels seen in hominins such as *Paranthropus robustus* and *Australopithecus africanus* (Yeakel et al. 2007). Estimates of their mechanical properties reveal that certain rhizomes, corms, and tubers can be harder and/or tougher than foods eaten by extant great apes (Dominy et al. 2008); their consumption could therefore possibly explain the derived craniofacial morphology of *Paranthropus*. USOs have also been proposed as a key component that allowed the evolution of the modern human-like body plan and life history of *Homo erectus* (O'Connell et al. 1999) (syn. geophyte). *See also* **belowground food resources**; **grandmother hypothesis**.

UNESCO Cradle of Humankind World Heritage Site *See* Cradle of Humankind World Heritage Site.

ungual process *See* apical tuft.

ungual spines (L. *unguis*=claw and *spina* = thorn) The fingers of the primate hand each comprise three bones; the proximal, the intermediate and the distal phalanges. The small distal phalanx has a base, a shaft, and an apex. In higher primates the apex is rounded and takes the form of a bony excrescence called an **apical tuft**. The edges of this apical tuft extend more proximally than does its mid-part, and it these edges that are referred to as ungual spines.

ungual tuberosity *See* apical tuft.

ungual tuft *See* apical tuft.

Ungulata (L. *ungulātus*=hoof, from *unguis*= nail) The formal term for the group of mammals that possesses at least one hoofed toe. The group contains common domesticates such as horses, sheep, goats, and cows, as well as their wild relatives contained within the orders **Perissodactyla** and **Artiodactyla**. Ungulates are generally large-bodied and relatively long-lived, and with a few exceptions they give birth to single live young. They are either vegetarians or omnivores (e.g., the **Suidae**). Recent genetic studies suggest that ungulates are a **polyphyletic group**; however, the use of the term to describe the ubiquitous hoofed animals remains a useful, even if cladistically inaccurate, shorthand.

ungulate (L. *ungulātus*=hoof, from *unguis* = nail) The informal term for the group of mammals that possesses hoofs. *See* **Ungulata**.

unifacial *See* unifacially worked.

unifacially worked (L. *unus*=one) A term used in the analysis of stone tools that refers to a piece that is worked (i.e., shaped), on one surface, or face. It contrasts with a piece that is worked (i.e., shaped), on two surfaces, or faces; such pieces are said to be **bifacially worked**.

uniformitarianism A fundamental assumption of all sciences that posits the spatial and temporal invariance of natural (materialistic) laws. The term was coined in 1832 by the geologist William Whewell when reviewing the second volume of Lyell's *Principles of Geology*. In the third volume of that work, Lyell effectively articulated the concept because he argued that natural law invariability was necessary to explain past geological changes. The concept has since been generalized to all sciences. Uniformitarianism is an example of inductive logic in that it assumes that modern processes operated in the past. It also asserts parsimony in that invoking different processes in the past is unnecessary (and it is simpler not to do so) as long as current processes provide sufficient explanations. A number of examples in human evolutionary studies can be cited where the concept of uniformitarianism is implicitly invoked (e.g., reconstructing the biomechanics of australopith chewing using data derived from observational and experimental studies of living species, dating tuffaceous rocks using the assumption that decay rates of potassium have been invariant through time, estimating body mass of past species using relationships between, for example, femoral head diameter and body mass deduced for living species, defining species in the fossil record using patterns of variation within and differences among living species, reconstructing aspects of paleoclimate, or diet, using the stable isotopic behavior of oxygen or carbon measured in living systems).

unit In geological terminology, a vaguely defined or user-defined interval of rock strata. Units are the basic components by which a stratigraphic sequence is described and measured in the field, in which case they are commonly bounded by natural breaks or discontinuities, but the scale of the chosen units is determined primarily by the field geologist according to the scale of analysis and type of investigation being undertaken.

University of California's Africa Expedition
See **Swartkrans**.

University of Pennsylvania Museum of Archaeology and Anthropology *See* **Wenner-Gren Foundation for Anthropological Research**.

University of the Witwatersrand site numbering system Zipfel and Berger (2009) set out a proposal for a numbering system to be used by the University of the Witwatersrand for "Cenozoic fossil-bearing sites" (*ibid*, p. 77). The system allocates each site a "U.W." number beginning with the earliest site to be recorded (e.g., **Taung** type site is U.W. 1, **Gladysvale** is U.W. 3, **Swartkrans** is U.W. 22, and **Malapa** is U.W. 88). The recommendation is that within these U.W. numbered sites fossils should themselves be given numbers, so the first evidence of the **MH1** associated skeleton to be found, the right clavicle, is U.W. 88–1.

untranslated region The parts, or regions, of **DNA** or messenger **RNA** molecules that are not translated into protein.

Upper cave *See* **Zhoukoudian Upper Cave**.

Upper Paleolithic A term used to describe the stage of the European and Near Eastern **Paleolithic** that is defined by the use of **Mode 4** stone tool technology including the use of blade and burin tools, and is characterized by an increase in number and variety of tool types made from stone and organic materials and the proliferation of artwork and symbolic behavior including personal ornaments (beads, jewelry), portable art (figurines), and parietal art (cave paintings and engravings). It first appears in eastern Europe around 40 ka (e.g., at **Istállósk** and **Bacho Kiro**) and spreads westward over the course of 5000–10,000 years and, depending on the area, it is replaced by **Mode 5** microlithic technologies between 20 and 10 ka. It is usually linked with *Homo sapiens*, but some early Upper Paleolithic or so-called transitional technocomplexes, such as the **Châtelperronian**, were apparently created by *Homo neanderthalensis*. The equivalent term for African sites with similar technologies (but not necessarily similar ages) is **Late Stone Age**.

Upper Pubu *See* **Bubing Basin**.

upper scale The name sometimes used for the upper part (i.e., above the superior nuchal line) of the squamous part of the occipital bone (syn. occipital plane).

upstream The term upstream refers to any relative position in a **DNA** or **RNA** molecule that is toward the 5′ end of that molecule.

UR 501 *See* **HCRP UR 501**.

uracil *See* **RNA**.

Uraha (Location 10°21′0.6″S, 34°09′23.3″E, Malawi; etym. named after the village of Uraha) History and general description In the 1920s F. Dixey discovered artifacts on the surface of Uraha Hill that were subsequently attributed to the **Middle Stone Age** (Clark et al. 1966). In 1963, Desmond **Clark** led a pioneering expedition to the Malawi Rift that collected various mammalian faunal remains, but no hominins. The older fossiliferous sediments around Uraha Hill were first recognized in Mawby (1970). The southernmost area of the **Chiwondo Beds**, including those at Uraha, were explored by the **Hominid Corridor Research Project** beginning in 1983. The main significance of the fossil assemblage is biogeographical, for it extends the range of the hominin taxon recovered from the site by more than 1000 km/620 miles to the south. Temporal span and how dated? Biostratigraphy suggests an age of *c*.2.5–2.3 Ma. Hominins found at site A well-preserved fragment of an adult hominin mandibular corpus, **HCRP UR 501** assigned to *Homo rudolfensis*. Archeological evidence found at site None. Key references: Geology, dating, and paleoenvironment Betzler and Ring 1995, Kullmer 2008; Hominins Schrenk et al. 1993, Bromage et al. 1995; Archeology N/A.

uranium-series dating Principle The uranium-thorium (^{230}Th/^{234}U, U-Th) system is an isotopic dating method that measures the daughter products resulting from the decay of uranium (this is referred to as the "daughter deficiency" variety of uranium-series dating). It is widely used for dating Quaternary deposits and the hominins therein, but U-Th dating only extends back to *c*.0.5 Ma because by this time the ^{230}Th has decayed away; material older than this can be dated with uranium-lead dating (^{206}Pb/^{238}U, U-Pb). This technique is relatively new and it has only recently been used to provide direct radiometric ages for the southern African hominin-bearing caves deposits. Uranium-lead dating could prove as useful for dating southern African hominins as **argon-argon dating** has proved to be for dating East African hominins. The advantages of uranium-series dating are its precision (e.g., errors are typically 0.1–1% for U-Th and 1–10% for U-Pb) and its independence from ambient conditions such as temperature. Method Material for dating is prepared in a clean labora-

tory environment to isolate and concentrate the isotopes of interest, which are then measured using a mass spectrometer. Inductively coupled plasma mass spectrometry (ICP-MS) and **thermal ionization mass spectrometry** (TIMS) are sensitive new counting methods that are up to 10 times more precise than alpha counting. For very high-precision work TIMS is the best, although ICP-MS has the advantage of simpler sample preparation and a higher sample throughput. The relationship between uranium-series dating and ICP-MS is much the same as that of accelerator mass spectrometry (or AMS) to regular ^{14}C estimations. For U-Th, a single analysis of the U and Th isotopes can produce an age, whereas U-Pb dating requires multiple analyses to construct **isochrons** from which the age is calculated. Suitable material The most suitable materials for uranium-series dating are inorganic carbonates, both marine (coral) and terrestrial (stalagmites, **flowstones**, **tufa**). Organic material, such as fossil bone, teeth, and ostrich egg shell, can be dated by U-Th but these are open systems and the uptake or loss of uranium must be modeled. Simple "early-uptake" or "late-uptake" models can be employed but most bone and mollusc shell samples have a much more complex uranium-uptake history. Mollusc shells are notorious for giving misleadingly young ages. Time range ^{230}Th/^{234}U $\geq$500 ka and ^{206}Pb/^{238}U $\geq$500 ka–billions of years. U-Pb can be applied to material within the same time range as U-Th, but given the extra analysis and sample volume, this is seldom worth the effort. Problems The **speleothem** material best suited to uranium-series dating is often not directly associated with fossil hominins or archeological remains, thus it is important to have precise and accurate site stratigraphy. Methods for dealing with calcite contaminated with detritus ("dirty calcite") and material with so-called open-system behavior, such as fossil bone and teeth, are still relatively experimental and currently ages have large error margins.

ursid The informal name for the **Ursidae** (bears), one of the **caniform** families whose members are found at some hominin sites. *See also* **Carnivora**.

Ursidae The family that includes bears and their close relatives. It is one of the **caniform** families whose members are found at some hominin sites. *See also* **Carnivora**.

use wear Polishes, pits, and striations on the surface of a stone, bone, or metal tool that are the result of the use of that tool. Archeologists are able to examine use wear and identify the tool motion and contact material. *See* **use-wear analysis**.

use-wear analysis The study of polishes, pits, fractures, and striations on the surface of a stone, bone, or metal tool to understand how the tool was used. The field of use-wear analysis began in the late 1950s with Sergei Semenov's publication of *Prehistoric Technology* (published in English in 1964), in which he documented the processes that caused wear on the surface of stone tools, and became popular throughout the late 1970s and 1980s. Questions about the accuracy and reliability of the method arose in the 1980s, and though several blind tests using both high-power (high magnification) and low-power (low-magnification) analysis have shown that the method is robust, although many researchers in the archeological community are still wary of use-wear analysis. After years of methodological improvements (e.g., Anderson-Gerfaud 1981, Plisson 1985), use-wear analyses are increasingly being used to answer questions about past life and behaviour (e.g., Clemente et al. 2008, Shea, 2006). Keeley (1980), Odell (2004), and Grace (1996), among others, provide summaries of use-wear analysis.

Usno Formation (etym. The name is taken from the Usno River that joins the Omo River just north of the exposures of the Usno Formation) General description The Usno Formation is a group of outcrops *c.*30 km/19 miles north of the northernmost extent of the **Shungura Formation**. They lie immediately to the west of the Omo River in southern Ethiopia, just south of where the Usno River joins it, and the exposures extend for *c.*7 km/4.3 miles from their most northeasterly extent to their southwestern limit.

The Usno Formation is equivalent to Member A and most of Member B of the **Shungura Formation**, and its total area (not all of which contains sediments) is only about 20% of the area occupied by the "Type area" of the Shungura Formation. The total thickness of the Usno Formation is approximately 175 m. Sites where the formation is exposed Hominins have been found in two localities, known as **Brown Sands** and **White Sands**, within the exposures of the Usno Formation. Fossils found in the two localities have been given the prefix "B" for Brown Sands or "W" for White Sands. Geological age The oldest sediments of the Usno Formation are *c.*3.6 Ma and the youngest *c.*2.7 Ma. Most significant hominin specimens recovered from the formation The only hominins recovered from the Usno Formation are isolated teeth. Few studies have addressed these teeth, but those that have suggest that they are similar to those from **Hadar** and elsewhere that have been assigned to *Australopithecus afarensis*. Key references: Geology, dating, and paleoenvironment Butzer and Thurber 1969, de Heinzelin and Haesaerts 1983b, Feibel et al. 1989; Hominins Howell 1969b, Howell and Coppens 1974, 1976.

USO Acronym for **underground storage organ** (*which see*).

Uto-Aztecan premolar *See* **distosagittal ridge**.

U.W. system Abbreviation of **University of the Witwatersrand site numbering system** (*which see*).

UW88-1 *See* **MH1**.

UW88-2 *See* **MH2**.

V1 *See* **striate cortex**.

Vaal River *See* **Rietputs Formation**.

Vaal River Gravels *See* **Rietputs Formation**.

Valdegoba Cave (Location 42°32′30″N, 0°05′ 10″W, Spain; etym. unknown) History and general description Valdegoba Cave is 28 km/17 miles northwest of Burgos, in northern Spain, at the southern end of the Cantabrian Cordillera. Four excavations took place between 1987 through 1991 by a team led by Rolf Quam, during which eight stratigraphic levels were uncovered. Of the six areas of the cave excavated, only one area in the western interior section contained *in situ* remains. Temporal span and how dated? Biostratigraphy suggests that the oldest layers of the site are more than 350 ka, whereas uranium-series dating of level 7 gives a date of 73.2 ± 5 ka. Hominins founds at the site Five hominin specimens were recovered from levels 5, 6, and 8. Mandibular fragments from an adolescent individual, VB1, were recovered from level 8, but that level is disturbed. The other fragments, with more precise stratigraphic context, consist of deciduous teeth (VB2), a proximal phalanx (VB3), and two metatarsals (VB4 and 5). The material is presumed to belong to *Homo neanderthalensis*. Archeological evidence An abundant Middle Paleolithic industry is known from levels 4–8, produced from local quartzite and flint. The presence of numerous first-order flakes implies that tool production was carried out at the site. Faunal evidence suggests at least two possible climatic regimes. Key references: Geology, dating, and paleoenvironment, hominins, and archeology Quam et al. 2001.

valgus (L. *valgus* = bent out) Refers to a joint, or an angle within a bone (e.g., the neck-shaft angle of the femur), being "bent" so that the limb segment beyond the joint, or the shaft of a long bone beyond the angulation, are inclined further away from the midline than normal. Thus, when children have knees that are abnormally close together, because the long axis of the lower leg appears to be inclined away from the midline, that "knock-kneed" deformity is called genu valgum. But because valgus (and its antonym **varus**) are terms that are often misunderstood (Fairbank and Fairbank 1984), their use should probably be avoided. *See also* **varus**.

valid (L. *validus* = strong) In relation to taxonomy a valid taxon name is the correct taxon name. Among what may be several **available** names for a taxon only one, called the **senior synonym**, is valid in the sense that it is the correct name according to the rules and recommendations of the **International Code of Zoological Nomenclature**. For example, *Australopithecus africanus* is the senior synonym for the taxon for which **Taung 1** is the **holotype**. *Plesianthropus transvaalensis* is one of several junior synonyms of *Au. africanus*; it is an available taxon name, but because it is not the senior synonym it is not the valid name.

Vallesian *See* **European mammal neogene; biochronology**.

Vallois, Henri Victor (1889–1981) Henri Vallois, who was born in Nancy, France, studied medicine at the University of Montpellier where he defended his thesis about the primate knee in 1914. He then entered the University of Paris where he studied natural history and completed his degree in 1922. Vallois became Professor of Anatomy and Zoology in the Faculty of Medicine at the University of Toulouse in 1922 and soon began to conduct research in physical anthropology and hominin paleontology. He became a close associate of anatomist and anthropologist Raoul Anthony, director of the Laboratoire d'Anthropologie

(Laboratory of Anthropology) at the École Pratique des Hautes Études (Practical School of Higher Studies) in Paris and of physical anthropologist Léonce-Pierre Manouvrier, who had written on the *Pithecanthropus erectus* fossils discovered by Eugène **Dubois**, and on the *Homo neanderthalensis* fossils found at **Spy** in Belgium. Vallois succeeded Anthony as director of the Laboratoire d'Anthropologie in 1938, and in 1941 he joined the prestigious **Museum National d'Histoire Naturelle** (National Museum of Natural History) as Chaire d'ethnologie et d'anthropologie des hommes actuels et des hommes fossiles (Chair of Ethnology and Anthropology of Current and Fossil Humans). At the Museum Vallois became a colleague and close associate of paleontologist Marcellin **Boule**, who had conducted the influential study of the **La Chapelle-aux-Saints** Neanderthal fossils discovered in 1908. Vallois investigated a substantial number of Paleolithic fossil humans during his career. During the 1930s and 1940s he wrote extensively on the paleopathology of prehistoric humans and estimated their life span, and during the 1940s and 1950s he analyzed the Upper Paleolithic hominin skeleton found at Chancelade and the mandible found at **Montmaurin**. Vallois and Germaine Henri-Martin published an extensive account of the hominin fossils discovered at **Fontéchevade** in 1947 in *La grotte de Fontéchevade* (1957–8). Vallois is best known for his interpretation of *Homo neanderthalensis* and for his opposition to the idea that Neanderthals were the ancestors of anatomically modern humans. He published a paper in 1929 in the journal *L'Anthropologie* where he argued for a monophyletic origin of modern humans and he was a supporter of the **pre-sapiens hypothesis**, which argued that anatomically modern humans were geologically very old and therefore the other known species of fossil hominins were not direct ancestors of modern humans. The discovery of the Fontéchevade specimen prompted Vallois to address what he saw as the three basic competing views regarding human evolution: (a) that modern humans had evolved through a Neanderthal phase, (b) that modern humans had evolved from a pre-Neanderthal hominin that gave rise to two lineages, one leading to *Homo sapiens* and the other to the Neanderthals, and (c) the pre-sapiens hypothesis that used evidence of the apparently ancient, but relatively anatomically modern-looking skeletons, found at Swanscombe, Fontéchevade, and **Piltdown**, to argue for an even greater antiquity for modern humans. Vallois, who was heavily influenced by Marcellin Boule, became a co-author on the third edition (1946) of Boule's classic

Les hommes fossiles (*Fossil Men*), first published in 1921. Boule and Vallois published a fourth edition of the book in 1952. Vallois continued to support the pre-sapiens theory even after 1955 when Piltdown was conclusively exposed as a hoax. In addition to his research on human paleontology Vallois also studied modern human racial variation. In *Les races humaines* (*Human Races*; 1944) Vallois identified four major racial groups (White, Black, Yellow, Australoid), which he subdivided into 27 separate human races. Vallois succeeded Boule as director of the Institut de Paléontology Humaine (Institute of Human Paleontology) and served in that position from 1942 to 1971. He served as director of the **Musée de l'Homme** (Museum of Man) from 1941 to 1945 and again from 1950 to 1960, and was president of the committee in charge of anthropology at the Centre National de la Recherche Scientifique (CNRS) from 1949 to 1966. He also served as the general secretary of the Société d'Anthropologie de Paris (Anthropological Society of Paris) from 1938 to 1969. Vallois was at one time or another also the editor of three prominent journals: *L'Anthropologie, Bulletins et Mémoires de la Société d'Anthropologie de Paris*, and the *Archives de l'Institut de Paléontology Humaine*. Vallois also published several important works on European prehistory, an anthropometric study of the French in *Anthropologie de la population française* (1943), and in 1953, together with Harvard geologist Hallam Movius, he compiled the *Catalogue des hommes fossiles* (*Catalogue of Fossil Men*). He also collaborated with Suzanne de Félice on an important study of the French Mesolithic titled *Les mésolithiques de France* (1977).

Van Wesele, Andreas *See* anatomical terminology.

Vandermeersch, Bernard (1937–) Bernard Vandermeersch was born in Wervicq in northern France. He studied zoology, geology, paleontology, and ethnology at the Sorbonne (Paris) and earned his 3rd Cycle Doctorate in 1963 under the direction of J. Piveteau. He was then hired as Assistant at the Laboratoire de Paléontologie des Vertébrés et Paléontologie of the University of Paris. He became Maître Assistant in 1964 and then full Professor at the University of Bordeaux 1 in 1983, a position he held until his retirement in 2001. After participating in various Paleolithic excavations, including those at **Arcy-sur-Cure**, **Regourdou**, Petit-Puymoyen, and Rigabe, Vandermeersch undertook his first field trip to Israel in 1964 and directed the excavation of

Qafzeh, initially by himself and then in collaboration with Ofer **Bar-Yosef**. He later co-directed excavations in Israel at **Kebara** and **Hayonim**. In France, he excavated the site of Marillac. He is best known for his studies of the exceptional series of Mousterian early modern humans of Qafzeh (Vandermeersch 1981). He established the clear association of modern humans and **Mousterian** assemblages *c*.95–100 ka. This work was seminal for the hypothesis that Neanderthals were not the immediate ancestors of modern humans and later in the development of the **out-of-Africa hypothesis**. Another important achievement at Qafzeh was the discovery of the oldest modern human burials, and the recovery of evidence that pigments were being curated and shells were being deliberately perforated well before the beginning of the Upper Paleolithic (Bar-Yosef Mayer et al. 2009). In Europe, Vandermeersch focused his work on several Mousterian sites (Petit-Puymoyen, Marillac, Regourdou, and **Saint-Césaire**) and he has been involved in the description of many *Homo neanderthalensis* fossils. His involvement in the study of some of the earliest evidence of modern humans (at Qafzeh), as well as some of the most recent evidence of Neanderthals (at Saint-Césaire), promoted his interest in the relationships between the two groups as well as the question of the extinction of the Neanderthals (Lévêque and Vandermeersch 1980). The Saint-Césaire specimen was crucial to the demonstration that the **Châtelperronnian** culture was produced by late Neanderthals and it prompted a re-examination of the significance of other assemblages underlying the **Aurignacian** in Europe. Besides his career at the University, Vandermeersch was also Directeur des Antiquités Préhistorique de Pointou-Charentes (director of prehistoric antiquities from Pointou-Charentes) from 1972 to 1984 and an active member of the Comité des Travaux Historiques et Scientifiques au Ministère de l'Education National et de la Recherche (the Ministry of National Education and Research Committee on Historic and Scientific Work).

variability (L. *varius* = various) Refers to the tendency to exhibit **variation** (Wagner 1996). The terms variation and variability are often used interchangeably in the literature, but this is not strictly correct. Variability may be thought of as referring to how different a new observation from a population is likely to be compared to earlier observations, whereas variation refers to the amount of difference in an existing set of observations. Like solubility, variability is a dispositional concept and refers not to the observation of variation but rather to the underlying property of having the potential to exhibit variation. This distinction is important in biology because the determinants of variability apply to the developmental or genetic architecture of a species (or the shared architecture of a larger taxonomic group) whereas variation is the result of the confluence of many factors in a particular population. Just as a solution is the result of the particular combination of the solvent, a physical environment, and a substance with a given solubility, variation is the result of the epigenetic variability characteristics of a species combined with a particular level of genetic **variance**. Variability cannot be observed directly, it can only be inferred from observations about variation in samples of the parent population. Halgrímsson (2002) argued that **integration**, **canalization**, and developmental stability are crucial developmental factors that contribute to determining variability, but, like variability, they are dispositional concepts that are not observed directly. Their influence can only be inferred (e.g., integration by observations about covariation, canalization by observations about variance, and developmental stability by studying fluctuating asymmetry). To illustrate the distinction between variation and variability, imagine you are interested in the integration of dental non-metrical traits because you suspect that the covariation among them is relevant to their evolution. You obtain observations about the correlations among the traits in two samples from the same species. Your samples differ, however, in that one sample consists of identical clones raised under identical environmental conditions, whereas the other sample manifests high genetic and environmental variance. When you calculate the respective correlations, you find that in the sample made up of clones, the correlations are extremely low, whereas they are high in the sample with high variance. In fact, in the absence of genetic and environmental variance, the correlations (or covariances) among the traits will be 0. However, in both of these samples, the underlying developmental architecture is the same. If you were to introduce genetic variation into the population of clones, that sample would also exhibit covariation just as it does in the non-clone population. In both cases, therefore, the tendency to exhibit covariation is the same while the observed covariation (or correlations) is radically different. Variability refers to the tendency for developmental systems to exhibit variation in the face of particular combinations of genetic and environmental variance.

variability selection A mode of **natural selection** (*see also* **directional selection**; **disruptive selection**; **stabilizing selection**) that is hypothesized (Potts 1998) to occur when environmental conditions are highly unstable (i.e., they are prone to both large- and small-scale unpredictable changes that may occur over geologically short periods of time). The variability selection hypothesis suggests that under such conditions **alleles** that confer behavioral flexibility would be favored over alleles that experience strong positive selection during a given climatic extreme (e.g., dry habitats) but negative selection during the opposite extreme (wet habitats). Many modern human adaptations (e.g., intelligence, language, manual dexterity, tool use) confer extraordinary behavioral flexibility and are likely to have evolved during the **Pleistocene** epoch, when oscillations in global climates increased in both amplitude and frequency. Thus, it is suggested that variability selection may explain the evolution of these traits, but while variability selection is an intuitively appealing hypothesis, it is not immediately apparent how it can be tested in the context of human evolution.

variable (L. *varius* = various) When used as a noun, the term variable refers to any trait, feature, characteristic, or property of interest, or symbol representing any of the above, that differs in some way among the members of a population and that can be measured, described, or categorized (e.g., the mesiodistal length of mandibular first molars, shape of the orbit, or type of mating system) in a sample of that population. The variables in a statistical study can be organized by their properties, including (a) continuous versus discrete, (b) nominal (also referred to as categorical), ordinal, interval, or ratio, (c) dependent versus independent, and/or (d) controlled, uncontrolled, and confounding. Variables can be computed (derived) from other variables. By definition, variables must be able to take on more than one value (e.g., more than one tooth crown length, more than one orbit shape, or more than one type of mating system); otherwise they are constants, not variables. Because variables can take on more than one value, sample statistics are typically used to describe the values for a set of observations of a variable for a sample drawn from a population (e.g., the mean, range, or standard deviation for continuous variables; frequencies for categorical variables, etc.). When used as an adjective, the term variable refers to how much observations within a sample differ from

each other. Compared to other samples of the same taxon, or a sample of a different taxon, a sample can be characterized as showing little, average, or substantial variation. *See also* **variability**.

variable number of tandem repeats (or VNTRs) Different numbers of **tandemly repeated sequences** at a location in the genome. Subcategories include satellite sequences, **minisatellites**, and **microsatellites**, but VNTRs usually refers to minisatellites.

variance (L. *varius* = various) A specific statistical quantity formulated by Ronald Fisher (1918) to express the variability (or dispersion) in a sample. To compute the variance, the deviation of each value from the sample mean is squared. The variance of the population is equal to the sum of these values (the "sum of squares") divided by N, but extreme values tend to be underrepresented in samples. Because of this, the sample variance is a biased estimate of the population variance, but this is corrected by dividing the sample sum of squares by $N–1$ rather than N. The variance is also the mean of the squared deviations of the variable from their mean, and is also known as the "mean squared error." The square root of the variance is the standard deviation.

variation (L. *varius* = various) Refers to the observation that the values for a **variable** or **trait** differ from each other among the individuals in a **sample**. The quality of being subject to variation is known as **variability**. Variation within a **population** can be estimated and quantified with sample statistics such as the standard deviation, variance, or range. A homogeneous sample exhibits little variation; a heterogeneous sample exhibits substantial amounts of variation. An emphasis on the study of variation is a key feature of evolutionary biology. *See also* **variability**.

varus (L. *varus* = bent or turned in.) The term varus refers to a joint, or an angle within a bone (e.g., the neck-shaft angle of the femur), being "bent" so that the limb segment beyond the joint, or the shaft of a long bone beyond the angulation, are inclined further towards the midline than is normal. In some children with an unusually obtuse (i.e., large) neck-shaft angle, such that the femur shaft is directed away from the midline, because the knees are so far apart the lower leg appears to be inclined towards the midline. Thus, that deformity is called genu varum. But because varus (and its antonym **valgus**) are terms that are often misunderstood (Fairbank and Fairbank 1984) their use should probably be avoided. *See also* **valgus**.

vault   *See* **cranial vault**.

vector   Biology A vector refers to an organism that permits the transfer of a pathogen from reservoir to host. For example, *Anopheles* mosquitoes are the vectors that transfer the protozoan parasites of the genus *Plasmodium*, which cause malaria, from reservoir to host, including modern humans. Statistical analysis A vector is a sequence of numbers written out horizontally (a row vector) or vertically (a column vector). Vectors of length two or three (i.e., vectors containing two or three numbers) are often used to represent spatial position as XY or XYZ coordinates. In kinematics, vectors of length three are used to represent the X, Y, and Z components of velocity, acceleration, force, etc. In statistics, vectors are used to contain a series of observations (for example, 12 measurements of body mass would be represented as a vector of length 12, with each element in the vector corresponding to one measurement of mass). One or more of these vectors may then be manipulated to produce single values (such as the mean or standard deviation) of a **matrix** (e.g., **covariance matrix, distance matrix**, etc.).

VEK   Acronym for Vivian Evelyn (Fuchs) korongo. A locality at Olduvai Gorge where artifacts were collected in 1931. Vivian Fuchs (or Bunny as he was always called) was a Cambridge colleague of Louis **Leakey** who joined Louis and Freda Leakey on the 1931 expedition to Olduvai to look at the geology and to check on the provenance of a hominin skeleton found by Hans Reck. In 1958 then Sir Vivian Fuchs made history when he successfully completed the first crossing of the Antarctic via the South Pole. In 1962 Margaret Cropper discovered the remains of a fragmented hominin calotte, **OH 12**. *See also* **Olduvai Gorge**.

Velika Pećina   (Location 46°17′N, 16°02′E, Croatia; etym. the words mean "reflecting cave") History and general description A cave site near Goranec on the Ravna Gora, northwest of Ivanec. Excavations were undertaken in 1948 and again from 1957 to 1979. Temporal span and how dated?

From the last interglacial to the **Holocene**. The results of **AMS radiocarbon dating** on the hominin frontal suggest these remains are Holocene (*c*.5 ka BP). Hominins found at site The frontal bone of *Homo sapiens* originally thought to be **Aurignacian**, but more recently it has been found to be Holocene. Archeological evidence found at site Artifacts described as **Mousterian**. Key references: Geology, dating, and paleoenvironment Smith et al. 1999, Karavanić 2007; Hominins Smith 1976a, 1984; Archeology Karavanić and Smith 1998, Karavanić 2007.

velocity curve   A device used to study patterns of individual growth, with an advantage not provided by a distance curve. That is, it illustrates the growth dynamic over the course of ontogeny. The specific velocity of growth will vary according to measurement, age, and individual. Growth velocities of various parts and organ systems of the body have been shown to exhibit asynchrony. To calculate velocities, researchers employ a range of techniques, from logistic to shape invariant models. The particular method utilized will depend on the nature of the sample, the age range that is being examined, and the frequency of measurement. The overall pattern for modern human postnatal weight and length growth includes initial rapid decreases in growth velocity throughout the first three years of life, followed by a slowing and plateauing of growth rate through to the adolescent growth spurt. A "mid-growth spurt" may or may not occur between the childhood and juvenile stages of growth. The growth velocities experienced *in utero* surpass any seen in postnatal life. Calculations of arithmetic velocity curves using cross-sectional samples of nonhuman primate mass indicate a wide range of growth velocity patterns across taxa, which vary in conjunction with socioecological factors.

Venta Micena   (Location 37°43′N, 2°23′W, Spain; etym. after a nearby town) History and general description This site, one of several in the **Orce region**, comprises an 80–120 cm-thick horizontal layer, which appears in outcrops for over 2.5 km/1.5 miles. Venta Micena has provided one of the best faunal records for the **Plio-Pleistocene** (Martínez-Navarro 2002) but in relation to paleoanthropology it is best known for the debate about the identification of one specimen, VM-0. Initially VM-0 was published as a hominin cranial vault fragment, making it the oldest hominin fossil in Europe (Gibert et al. 1983). But after the specimen was cleaned two of the original team (Agustí and Moyá-Solá 1987) argued that it was not a hominin but part of equid skull and other researchers suggested it belonged to a "juvenile female of a large ruminant species" (Martínez-Navarro 2002). Nonetheless, until his death in 2007 Gibert maintained it was hominin (Gibert et al. 1998). Temporal span and how dated? Biostratigraphic correlation to other sites in the area that have been dated using **magnetostratigraphy** indicates an early Pleistocene age (i.e., younger than 1.78 Ma). Hominins found at site No confirmed hominin fossils. Archeological evidence found at site There are several apparent **manuports**, associated with bones fractured by blows (Gibert et al. 1987), which also have microstriations probably caused by defleshing (Gibert and Jiménez 1989). Key references: Geology, dating, and paleoenvironment Gibert et al. 1983, Martínez-Navarro 2002, Oms et al. 2000; Hominins Martínez-Navarro 2002, Gibert et al. 1983, 1998, Agustí and Moyá-Solá 1987; Archeology Gibert et al. 1987, Gibert and Jiménez 1989.

Venus figurines *See* **Gravettian**; **Perigordian**; **Hohle Fels Venus**.

vertebral number (L. *verter* = to turn) The count of vertebrae is recorded as the total for the whole vertebral column [i.e., cervical (Cx), thoracic (T), lumbar (L), sacral (S), and caudal (Ca)], usually written out as the modal numbers for each of the above five regions, in the order given above (e.g., 7, 13, 4, 6, 3 for *Pan troglodytes*), or as the modal number within just one of the five regions of the vertebral column [e.g., for *P. troglodytes* lumbar (L) = 4 (range 2–5)]. Total vertebral number is precisely related to original **somite** number. All but the most **rostral** (i.e., anterior) somites (i.e., the 4.5 incorporated into the cranium) give rise to **sclerotomes** and the sclerotomes give rise to vertebrae and ribs. Each vertebra receives a contribution from two sclerotomes, its own and the one **caudal** to it (e.g., the C4 vertebra is formed from the caudal part of the C4 sclerotome and the cranial part of the C5 sclerotome). Comparisons of developing chick and mouse along with **knock-out** and **knock-in** experiments show that among the many gene systems involved in the generation, specification, and determination of somites, the *Hox* system plays a major role. The timing and duration of *Hox* **gene expression** apparently controls the number of vertebrae within

each region. Regional modal numbers of vertebrae can be combined as species-specific vertebral formulae or patterns (e.g., 7, 12, 5, 5, 4 for *Homo sapiens*). Regional vertebral number varies within as well as between species, especially in hominoids; a within-species sequence of patterns ordered by frequency is called a vertebral profile. For example, there are approximately 40 patterns in reasonably sized samples of *Pan troglodytes* (e.g., 7, 12, 5, 7, 2) and *Pongo pygmaeus* and over 20 in *Homo sapiens*, *Gorilla gorilla*, and *Hylobates lar*. In addition to different patterns, the total number of vertebrae varies within species, usually with a range of 4 or 5. In hominoids, as in almost all mammals, cervical number is effectively invariant at 7, vertebral number varies within more caudal regions, and vertebral number variation tends to be greater the further caudally you go. Of both phylogenetic and functional interest are the total numbers of thoracic plus lumbar vertebrae (documenting trunk length and degree of fore- and hindlimb separation) and their relative contributions to trunk length. For *H. sapiens*, the dominant modal pattern for thoracic and lumbar vertebrae is 12:5, whereas in the African apes it is 13:4. *P. troglodytes* and *Pan paniscus* differ on average by one vertebra (somite), half of this difference falling within the trunk region; *P. troglodytes* averages 13.1 thoracic and 3.6 lumbar vertebrae, whereas *P. paniscus* averages 13.4 thoracic and 3.8 lumbar vertebrae. Reasonably complete vertebral columns are rare in the early hominin fossil record, and they are only seen in useful numbers in *Homo neanderthalensis*; its vertebral column resembles that of *H. sapiens*. There is no consensus about whether the one *Homo erectus* and two *Australopithecus africanus* individuals that preserve at least some evidence of the vertebral column had five, or six, lumbar vertebrae (Haeusler et al. 2002). Prior to *c*.3 Ma the hominin fossil record is mute about the organization of the vertebral column. *See also* **lumbar vertebral column**.

vertebrate (L. *vertebratus* = having joints) As a noun the term refers to animals with a backbone, or spinal column, that protects their spinal cord, and as an adjective it means having a backbone and the characteristics associated with this condition. Fossil vertebrates are often preserved at archeological sites and at paleontological sites relevant to the study of human evolution. They are studied to provide an environmental context for human evolution as well as giving additional information on the age and taphonomic processes that affected the site. Vertebrate fossils can also provide insights into hominin behavior and ecology when they can be directly related to hominin activity by virtue of cutmarks, breakage patterns, or skeletal part representation that can only be attributed to human agency.

Vértesszöllős (Location 47°41′N, 18°20′E, Hungary; etym. the name of a nearby town) History and general description A **travertine** quarry 50 km/31 miles west of Budapest and 15 km/9 miles south of the Danube. Excavations began in 1964 and the Vértesszöllős 1 hominin remains were uncovered in 1965 during sieving of the sediments excavated in 1964 from lowest cultural horizon; Vértesszöllős 2 was excavated in 1965. Temporal span and how dated? **Biostratigraphy** suggests an Upper Biharian age, and **uranium-series dating** of the travertine suggests ages of >350 ka and 225–185 ± 25 ka BP; equivalent to **Oxygen Isotope Stages** 11–6. Hominins found at site Left lower dc, dm$_2$, and M (**Vértesszöllős 1**) and much of the squamous part of an adult occipital (**Vértesszöllős 2**). Archeological evidence found at site **Lower Paleolithic** small-sized pebble tool industry (described as the Buda industry). Key references: Geology, dating, and paleoenvironment Schwarcz 1982, Hennig et al. 1983, Schwarcz and Latham 1984, van der Muelen 1973; Archeology Kretzoi and Vértes 1965, Vértes 1965, Svoboda 1987; Hominins Kretzoi and Vértes 1965, Vértes 1965, Thoma 1967 and 1969.

Vértesszöllős 2 Site Vértesszöllős. Locality N/A. Surface/*in situ* Probably *in situ*. Date of discovery August 21, 1965. Finder Excavations under the direction of L. Vértes. Unit N/A. Horizon N/A. Bed/member N/A. Formation N/A. Group N/A. Nearest overlying dated horizon N/A. Nearest underlying dated horizon N/A. Geological age **Oxygen Isotope Stages** 7–6. Developmental age Adult. Presumed sex Male. Brief anatomical description Much of the squamous part of an adult occipital bone. Announcement and initial description Vértes 1965. Photographs/line drawings and metrical data Thoma 1966, 1969. Detailed anatomical description Thoma 1969. Initial taxonomic allocation *Homo* (*erectus* or *sapiens*) *palaeohungaricus*. Taxonomic revisions *Homo incertae sedis* (Hublin 1978), *Homo heidelbergensis* (Stringer 1983). Current conventional taxonomic allocation *Homo neanderthalensis*.

Informal taxonomic category Pre-modern *Homo*. Significance The Vértesszöllős occipital is one of a group of specimens (others include the hominins from **Reilingen**, the **Sima de los Huesos**, **Steinheim**, and **Swanscombe**) that is referred to *H. neanderthalensis*. Researchers who interpret *H. neanderthalensis* as an evolving lineage would interpret the Vértesszöllős occipital as an example of "Stage 2" (or perhaps "Stage 1") in this lineage. These stages are also referred to as pre-Neanderthal and "early pre-Neanderthal" (e.g., Dean et al. 1998), respectively (*see* **pre-Neanderthal hypothesis**). Location of original Magyar Nemzeti Muzeum, Budapest, Hungary.

vertical climbing hypothesis *See* **arboreal climbing hypothesis**.

Vesalius, Andreas *See* **anatomical terminology**.

Veternica (Location 45°51′N, 15°53′E, Croatia) History and general description Excavations at this cave site near Gornji Stenjevec, 9 km/5.5 miles west of Zagreb, started in 1951 and lasted until 1971. Temporal span and how dated? Conventional **radiocarbon dating** and **biostratigraphy** suggest the site includes artifacts from the last interglacial to the **Holocene**. Hominins found at site A calotte belonging to *Homo sapiens* is considered to be an intrusive Mesolithic burial. Archeological evidence found at site **Mousterian**. Key references: Geology, dating, and paleoenvironment Smith 1977, Karavanić 2007; Hominins Churchill and Smith 2000; Archeology Karavanić 2007.

vicariance (L. *vicarious* = a substitute, from *vicis* = to change) Refers to the subdivision of a population or species range that results in two descendant populations separated by a biogeographic barrier. Vicariance may lead to **speciation** or extinction in one or more of the descendant populations. Vicariance is caused by fragmentation of the environment, either because a geographic barrier has developed, such as an enlarged river or a recently formed mountain chain, or because of a change in climate. The study of the role played by vicariance events in evolution is called vicariance biogeography. *See also* **allopatry**; **vicariance biogeography**.

vicariance biogeography (L. *vicarious* = a substitute, from *vicis* = to change, and Gk *bios* = life, *geo* = earth, and *graphein* = to write) A mode of biogeographic analysis that uses **cladistic** methods to determine whether or not **vicariance** best explains the distributions of closely related taxa within multiple clades. The principle underlying the method is that because a single ecological/geological change might be expected to affect several different clades in similar ways, vicariance due to environmental or geophysical factors is more parsimonious than dispersal as an explanation of the same geographic distribution applying to several taxa. For in such circumstances in the absence of evidence for vicariance, researchers must invoke multiple independent dispersal events to explain the distribution of taxa.

vicariance event *See* **vicariance**.

victoriapithecid An informal term for the Victoriapithecidae, an extinct family of Old World monkeys. *See also* **Victoriapithecidae**.

Victoriapithecidae (Gk *pithekos* = ape and Victoria from the East African Lake named after Queen Victoria) This extinct family of Old World monkeys is the **sister taxon** of the family **Cercopithecidae**; the two families comprise the superfamily **Cercopithecoidea**. The earliest known **victoriapithecid**s are *Prohylobates* (*c*.19–15 Ma) known from sites in North and East Africa and *Victoriapithecus* (*c*.19–12.5 Ma) known from the sites of Maboko and Fort Ternan in East Africa. Victoriapithecids and cercopithecids share a number of derived features, including the absence of a maxillary sinus, bilophodont lower molars, and ischial callosities (Benefit and McCrossin 2002). Compared to cercopithecids, victoriapithecids have a relatively lower neurocranium, a steeper and more linear facial profile, and taller orbits.

Viking Fund In 1941 Axel Wenner-Gren established the Viking Fund, Inc. (which he endowed with US$2 million of Electrolux stock) as a means of supporting anthropology. Ten years later, in 1951, the name of the fund was changed to the Wenner-Gren Foundation for Anthropological Research, Inc. *See also* **Wenner-Gren Foundation for Anthropological Research**.

Viking Fund Medal This was introduced in 1946 by the **Viking Fund** but was then taken over by the **Wenner-Gren Foundation for Anthropological Research**. Medals were awarded annually

until 1961 and five awards have been made since that time. Leading paleoanthropologists who have received the medal include Franz **Weidenreich** (1946), Adolph Schultz (1948), William Gregory (1949), Hallam Movius (1949), William Straus (1952), T. Dale Stewart (1953), William **Howells** (1954), Wilfrid **LeGros Clark** (1955), Raymond **Dart** (1957), Henri **Vallois** (1958), Sherwood **Washburn** (1960), and Louis **Leakey** (1961). *See also* **Viking Fund**; **Wenner-Gren Foundation for Anthropological Research**.

Viking Fund Publications in Anthropology *See* **Wenner-Gren Foundation for Anthropological Research**; **Burg Wartenstein International Symposium Program**.

Viking Summer Seminars in Physical Anthropology In 1946 the **Viking Fund** introduced the Viking Summer Seminars in Physical Anthropology program. Under the leadership of Sherwood **Washburn** these seminars ran annually from 1946 to 1951, and then there was gap until the last one in 1953. The meetings helped introduce genetics and evolutionary theory to biological anthropology, and this in large measure helped determine the course of the modern subdiscipline of human origins research. Human evolution was a central theme in the summer seminars and Wilfrid **Le Gros Clark**, Raymond **Dart**, Kenneth **Oakley**, John **Robinson**, Ralph **von Koenigswald**, and Franz **Weidenreich** were among those who participated. *See also* **Viking Fund**; **Wenner-Gren Foundation for Anthropological Research**.

Villafranchian *See* **European mammal neogene**; **biochronology**.

Vindija (Location 46°17′N, 16°6′E, Croatia) History and general description A rock-shelter on the southwest side of Križnjak Peak, near Ravna Gora. The existence of the site was first noted in 1878, and excavations were conducted from 1928 to 1958 and again from 1974 to 1986. Temporal span and how dated? **Uranium-series dates** on the lower-level deposits are consistent with an age of *c*.114 ka BP. The hominin G3–G1 layers have been dated by a variety of methods: G3, **amino acid racemization** *c*.42 ka BP, U-Th *c*.41 ka BP, **AMS radiocarbon** > 42 ka radiocarbon years BP; G1, conventional ^{14}C on cave bear bone (18 ka ± 400 radiocarbon years BP, AMS ^{14}C on cave bear bone 33 ka ± 400 to 46 ka ± 2000 radiocarbon years BP, direct ultrafiltration AMS ^{14}C on hominin bone *c*.32.5 ka radiocarbon years BP. <u>Hominins found at site</u> More than 60 hominin specimens associated with **Mousterian** artefacts and identified as *Homo neanderthalensis* (levels G3 and G1). Specimen Vi-80 (33.16) from level G3 has yielded extremely well-preserved **mitochondrial DNA** and nuclear **DNA**; the 45 specimens associated with Epigravettian artefacts are identified as *Homo sapiens*. <u>Archeological evidence found at site</u> The artefacts are Late Mousterian or Mousterian in levels G3 and G2. The stratigraphic integrity of the G1 assemblage is disputed: it preserves Mousterian as well as some **Aurignacian**-like stone and bone tools, but some argue that the latter are intrusive from higher levels. <u>Geology and dating</u> Ahern et al. 2004, Higham et al. 2006, Karavanić 2007; <u>Hominins</u> Wolpoff et al. 1981, Smith et al. 1985, Ahern et al. 2004, Green et al. 2006, 2008; <u>Archaeology</u> Karavanić and Smith 1998, Karavanić 2007. <u>Significance</u> The Neanderthal fossil remains from levels G1 and G3 have consistently been argued to show cranial features [e.g., reduced facial (especially nasal and alveolar process) size, reduced supraorbital tori (including greater mid-orbital thinning), more vertical mandibular symphyses and an incipient mentum osseum] that set them apart from most other Neanderthals. Originally these were interpreted as evidence of an ancestral relationship between Neanderthals and early modern Europeans (Wolpoff et al. 1981, Smith and Ranyard 1980, Smith 1984), but others suggested the differences may be due to the unusually high proportion of subadult and generally small individuals in the Vindija sample, but apparently the Vindija people are of average body size for Neandertals (Trinkaus and Smith 1995). More recently, it has been claimed that the Vindija fossils sample a population that is an admixture of Neanderthals and early modern human populations entering Europe (Smith et al. 2005, Cartmill and Smith 2009).

virulence   Refers to the ability of a pathogen to cause disease.

viscerocranium   (L. *viscus* = organ and Gk *kranion* = brain case) More or less equivalent to the **face**, this is the part of the **cranium** that covers the anterior aspect of the **brain**. One scheme for subdividing the cranium draws a distinction between the **neurocranium** and the viscerocranium. The viscerocranium develops from a combination of the frontonasal process and the first pharyngeal arch; all of its components include cells that derive from the **neural crest**.

Visogliano   (Location 13°38′53″N, 45°43′29″E, Italy; etym. named after the nearby town) <u>History and general description</u> This rock-shelter in northeast Italy, near Trieste, was first explored in the early 1970s and excavations have continued off and on to the present day. **Paleolithic** sequences that contain some of the oldest hominin remains in Europe have been found in two distinct areas at the site: a rock-shelter (Unit A) and a nearby, chronologically older, breccia (Unit B). <u>Temporal span and how dated?</u> **Electron spin resonance spectroscopy dating** and **uranium-series dating** suggest that Unit B is *c*.350–500 ka; Unit A is slightly younger. <u>Hominins found at site</u> Six hominin specimens have been recovered from the site: Unit A yielded Visogliano 3, 4, 5, and 6, whereas Unit B yielded Visogliano 1 and 2. The specimens consist of isolated dental material, except for Visogliano 2, which is a fragmentary mandible. The remains have been assessed as showing characteristics of both *Homo erectus* and archaic *Homo heidelbergensis*. <u>Archeological evidence found at site</u> The upper layers in the rock-shelter have a flake-based tool technology similar to the Tayacian. The lower rock-shelter layers and the breccia have a variant of the **Acheulean** known as the "Italian Pebble Culture" that includes a high percentage of **choppers** and chopping tools but few **handaxes**. Key references: <u>Geology, dating, and paleoenvironment</u> Falguéres et al. 2008; <u>Hominins</u> Cattani et al. 1991, Abbazzi et al. 2000; <u>Archeology</u> Cattani et al. 1991, Abbazzi et al. 2000.

visor   *See* **facial visor**.

visual cortex   *See* **striate cortex**.

vitric tuff   (L. *vitrium* = glass) A deposit of **pyroclastic** material erupted from a volcano, in which the majority of the particles are of the non-crystalline glass form of ash. Vitric tuffs are central to the geochemical correlation of individual volcanic eruptions, and their correlation potential has been used to relate the products of the same eruption in volcanic-rich deposits (e.g., the **Omo Group** of the Turkana Basin). A distinctive "geochemical fingerprint" is preserved in vitric tuffs because the glass phase of the ash is not constrained by the rigid chemical formulae of the mineral phases; thus glass preserves a unique subsample of magma chemistry at the time of an eruption.

Viverridae The family that includes civets and their close relatives is one of the carnivore families whose members are found at some hominin sites. *See also* **Carnivora**.

viverrids Informal name for the **Viverridae** (civets and their relatives), one of the carnivore families whose members are found at some hominin sites. *See also* **Carnivora**.

VNTR Acronym for **variable number of tandem repeats** (*which see*).

vocal (L. *vocalis* = voice) A sound that has to be made by an organism (aka vocalization). A beaten drum makes a sound, but it is not a vocal sound. *See also* **language**; **speech**.

vocal imitation *See* **imitation**.

Vogel River Series *See* **Laetoli**.

Vögelherd Location 48°33′31N, 10°11′37E, Germany; etym. Ge. *vögel* = birds, *herd* = hearth or oven) History and general description This cave site along the Lone River is located roughtly 1 km/0.6 miles north of Stetten ob Lonetal, just upstream of **Höhlenstein-Stadel**. The site was excavated in the 1930s by G. Riek, who found several layers of archaeological material and some human fossils. The anatomically modern human remains were thought to have been found in an **Aurignacian** context, therefore making them some of the oldest in Europe, but direct dating has suggested a more recent date. The site is best known for several ivory and bone sculptures from the Aurignacian levels, mostly of animals, including a horse, a mammoth, and a feline. Temporal span and how dated? The site contains layers ranging from the **Middle Pleistocene** to historic period based on stratigraphic analysis. The hominin remains were directly dated using **AMS radiocarbon dating** to 3.9–5 ka (uncalibrated), indicating that they are intrusive. Hominins found at site Several numbered human remains attributed to *Homo sapiens* were recovered: Vögelherd 1 (Stetten 1), an incomplete cranium, partial mandible, two lumbar vertebrae; Vögelherd 2 (Stetten 2), a partial cranium, and Vögelherd 3 (Stetten 3), a partial right humerus. Archeological evidence found at site Stone tools from most periods from the **Mousterian** to the Neolithic were found in the several layers at the site. Key references: Geology, dating, and paleoenvironment Müller-Beck 1983, Conard et al. 2004; Archeology Riek 1934, Müller-Beck 1983; Hominins Gieseler 1971, Churchill and Smith 2000.

voice (L. *vox* = voice) In modern humans it is the sound made when expired air causes the vocal cords to vibrate as in the production of spoken language.

volar (L. *vola* = the sole of the foot, or the palm of the hand) The volar surface refers to the inferior surface, or sole, of the foot and the anterior surface, or palm, of the hand. A movement in the volar direction is a movement that in modern humans takes a finger or thumb towards the palm of the hand. The direction of one of the two main curvatures on the distal articular surface of the trapezium that articulates with the proximal articular surface at the base of the first metacarpal is described as being "dorsovolar" (Marzke et al. 2009).

volcanic ash correlation *See* **tephrochronology**; **tephrostratigraphy**.

von Ebner's lines Short-period incremental lines in **dentine** first described by Victor von Ebner (1902). They appear in predentine as well as dentine and may be spherical (calcospheretic), crescentic, or laminar in appearance. Close to the root surface, where there are few dentine tubules, they appear to intersect tubules at all angles but they become aligned at approximate right angles to the tubules deeper into dentine. von Ebner's lines reflect the 24 hour (or **circadian**) cycle of **odontoblast** secretory activity (Bromage 1991) and they are equivalent to **cross-striations** in **enamel**. These are illustrated in Bromage (1991), Dean (1998), and Smith (2006). *See also* **dentine development**; **incremental features**.

von Economo neurons (etym. Named after Constantin von Economo, the neuroanatomist who described them in 1926) A specialized class of projection neuron found in high densities within the **anterior cingulate cortex** and frontoinsular cerebral cortex of great apes and modern humans, but not in other primates. These neurons have a large, spindle-shaped cell body, with a distinctive thick, single basal dendrite in addition to its apical dendrite. von Economo neurons are found in brain areas involved in monitoring social information and feedback regarding internal body states and therefore might play a role in the evolution of specializations for social cognition, particularly self-awareness. von Economo neurons have also been documented in similar cortical areas in elephants and certain whales, suggesting that there may have been convergent evolution of neural circuits that appear to be implicated in complex social cognition.

von Koenigswald Collection

The von Koenigswald Collection comprises the hominin and other fossils from **Sangiran** that Ralph **von Koenigswald** collected after the Dutch Geological Survey in Java stopped supporting his research. The collection includes three fragmentary hominin crania (**Sangiran 2, 3,** and **4**), six hominin mandibular fragments (**Sangiran 1a, 4, 5, 6a, 6b,** and **6c**), and 52 hominin isolated teeth (**Sangiran 7**). It includes the holotypes of *Pithecanthropus dubius*, *Meganthropus palaeojavanicus*, *Hemanthropus peii*, *Sinanthropus officinalis*, *Gigantopithecus blacki*, and *Chinjipithecus atavus*. *See also* **Senckenberg Forschungsinstitut und Naturmuseum**.

von Koenigswald, Gustav Heinrich Ralph (1902–82)

Ralph von Koenigswald was born in Berlin, Germany, and studied geology and paleontology at the Universities of Berlin, Tübingen, and Cologne before completing his doctorate at the University of Munich in 1928. After his graduation he worked at the Bayerische Staatssammlung für Paläontologie (Bavarian State Collection for Paleontology) in Munich, but in 1930 he accepted a position in the Dutch East Indies as a paleontologist with the Dutch Geological Survey in Java. His tasks in Java were to produce a stratigraphic map and to work on the **Pleistocene** mammals, but what drew von Koenigswald to Java was the prospect of discovering more hominins like *Pithecanthropus erectus* (now called *Homo erectus*) recovered by the Dutch anatomist Eugène **Dubois** in the 1890s. In 1931–2 Cornelius ter Haar and a

team of surveyors recovered 11 hominin calvaria from **Ngandong** in central Java, and von Koenigswald published a description of the fossils in 1933. A year later the Geological Survey's budget was cut and von Koenigswald lost his salary but he opted to remain in Java to continue his search for fossil hominins. The French prehistorian Teilhard de Chardin (who had participated in the excavations at **Zhoukoudian** in China where *Sinanthropus pekinensis* had been discovered) visited Java in 1935 and recognizing the potential of von Koenigswald's field program he arranged funding for it through the Carnegie Institution in Washington DC. A year later von Koenigswald found a child's calvaria at Modjokerto (now called **Mojokerto**) in eastern Java and he assigned the specimen to a novel species *Homo modjokertensis*. In 1937 a hominin upper jaw fragment was discovered near the village of **Sangiran** and this was soon followed by the recovery of part of the calvaria of the same cranium; von Koenigswald assigned these fossils to *Pithecanthropus*. Fragments of another cranium were found at Sangiran in 1938 while the German anatomist Franz **Weidenreich** was visiting. Weidenreich had taken over the primary responsibility for the excavations at Zhoukoudian in China after the death of Davidson **Black**. In 1938 von Koenigswald traveled to Beijing to compare the *Pithecanthropus* specimens from Java with the *Sinanthropus* specimens. Weidenreich and von Koenigswald concluded the *Pithecanthropus* and *Sinanthropus* were so similar that they represented geographical variants of a hominin "at the same stage of evolutionary development." Another mandible belonging to *Pithecanthropus* was discovered at Sangiran in 1939, followed in 1941 by a fragment of a large hominin mandible that von Koenigswald assigned to a new genus and species, *Meganthropus palaeojavanicus*. The Japanese invasion of Java brought von Koenigswald's excavations to an end and he was interned in a prisoner of war camp by the Japanese from 1942 to 1945, but he and his wife managed to hide the fossils from his captors. In 1946 von Koenigswald, together with the fossils and his family, traveled to New York where he joined Franz Weidenreich at the **American Museum of Natural History**. The two men continued their work on the Chinese and Javan hominins until 1948 when von Koenigswald left to become Professor of Paleontology at the Rijksuniversiteit in Utrecht, The Netherlands. In the 1950s von Koenigswald published on

Gigantopithecus, a giant Pleistocene ape whose remains were found in China by Davidson Black in 1935, and on *Meganthropus*, both of which figured prominently in Franz Weidenreich's theories about how modern humans had evolved from giant apes. von Koenigswald also traveled to southern Africa to inspect the australopith fossils found by Raymond **Dart** and Robert **Broom** and he collaborated then and later with Phillip **Tobias** to compare the hominin fossils discovered in **Olduvai Gorge** with the fossils from Java. During the last decades of his life von Koenigswald worked to encourage the Indonesian researchers and insti-

tutions (Teuku **Jacob** was one of his students) and in 1979 he returned the hominins found at Modjokerto and Ngandong to Indonesia. von Koenigswald retired from his position at the Rijksuniversiteit in 1968 and returned to Germany (along with that part of the Sangiran Collection that was collected after his support from the Dutch Geological Survey in Java had lapsed) to re-establish a center for paleoanthropological research at the **Senckenberg Forschungsinstitut und Naturmuseum** in Frankfurt.

voxel   *See* **computed tomography**.

W

W. Prefix used by the American contingent of the **International Omo Research Expedition** for localities in the **White Sands** group of exposures in the **Usno Formation** (e.g., W. 7). The prefix is also used for fossils found in one of those localities (e.g., W. 7-23).

Wadi Dagadlé (Location not reported, Djibouti; etym. unknown) History and general description Wadi Dagaldé (Oued Dagaldé) is one of several **Plio-Pleistocene** vertebrate and archeological localities prospected by de Bonis and colleagues in 1983. Temporal span and how dated? **Biostratigraphy** suggests the fauna is **Early** to late **Middle Pleistocene**; the **uranium-series dating** ages are problematic but suggest an age of less than 250 ka. Hominins found at site Left maxilla with a fragment P^4–M^3 and a fragment of P^3. The tooth crowns are large but display a marked diminuition in size from M^1–M^3. The maxilla shows a slight degree of prognathism. Archeological evidence found at site None. Key references: Geology, dating, paleoenvironment, and hominins de Bonis et al. 1984; Archeology N/A.

Wadi el-Mughara (etym. Ar. for the "valley of caves") The Arabic name for a river valley in Israel, where Dorothy **Garrod** and her team excavated the sites of **el-Wad**, **Skhul**, and **Tabun**. The area is known in Hebrew as *Nahal Me'arot*. *See also* **Mount Carmel**.

Wadi Halfa (Location 21°57′N, 31°20′E, Sudan; etym. Ar. "valley of Halfa," named after the nearby town) History and general description A site excavated in 1963–4 by the University of Colorado's Nubian Expedition during rescue archeology preceding the construction of the Aswan Dam, Wadi Halfa is a terminal **Pleistocene** cemetery containing 36 burials. Temporal span and how dated? Associated Qadan cultural material suggests an age of *c*.10–14 ka. Hominins found at site Thirty-seven skeletons comprising 13 adult males, 15 adult females, and three juveniles.

The remains closely resemble those from **Jebel Sahaba (Site 117)**. Archeological evidence found at site Qadan (i.e., **Late Stone Age**) lithics. Key references: Geology, dating, and paleoenvironment Hewes et al. 1964, Saxe 1966, 1971; Hominins Armelagos 1964, Greene 1967, Greene and Armelagos 1972, Carlson 1976, Carlson and Van Gerven 1977, Armelagos et al. 1984, Calcagno 1986, Turner and Markowitz 1990, Irish and Turner 1990, Irish 2005; Archeology Saxe 1971.

Wadi Kubbaniya (Location 24°13′N, 32°51′E, Egypt; etym. unknown) History and general description Fred Wendorf and colleagues found a human fossil in an area with several rich archeological sites along the left bank of the Nile. Temporal span and how dated? The skeleton probably dates to some time prior to the deposition of the Nile silts which cover the coarse laminated sands in the area (i.e., prior to *c*.20 ka), but subsequent to the development of **bladelet** technology. Hominins found at site The skeleton of a young male (*c*.20–25 years) was found embedded in a calcareous matrix. The sum of the traits exhibited by the Wadi Kubbaniya skeleton place it in the **Jebel Sahaba (Site 117)** group. There is evidence of healed injuries and a lesion in the vertebral column. It is curated at the Aswan Museum, Egypt. Archeological evidence found at site The several archeological sites preserved evidence of a heavy use of Nilotic resources, including a reliance on fish and waterfowl, and on starchy plant foods like sedge **underground storage organs**. Key references: Geology, dating, and paleoenvironment Wendorf et al. 1989; Hominins Wendorf and Schild 1986, Stewart and Tiffany 1986, Angel and Kelley 1986; Archeology Wendorf et al. 1989.

Wadi Tushka (Location 22°30′N, 31°45′E, Egypt; etym. Ar. *wadi* = valley, *Tushka* = after nearby village) History and general description This site was discovered by J. Hester and

---

P. Hoebler under the direction of Fred Wendorf in 1965; excavations were conducted between 1965 and 1966. It is located 3.5 km/2 miles northwest of the abandoned village of Tushka on the western side of the Nile River; elevation is 140 m above sea level. The site comprises two main rock units: the Ballana Formation and the overlying Sahaba Formation; the Ballana consists of lacustrine and eolian silt and sand and the Sahaba of fluvial silt. Temporal span and how dated? Upper Pleistocene (i.e., 18–10 ka) on the basis of **radiocarbon dating** of charcoal and shell. Hominins found at site Nineteen *Homo sapiens* burials (one infant and 18 adults); two of the latter were identified as males and three as females. Archeological evidence found at site Stone tools belong to the Qadan Industry (around 11,000 artifacts), more than 100 hearths delineated by fire-cracked rocks with associated fish and ungulate bones, grinding stones, and a few examples of worked bone and egg shell. Key references: Geology, dating, and paleoenvironment Wendorf 1968b; Hominins Wendorf 1968b; Archeology Wendorf 1968b.

Wadjak (Location 8°06′S, 112°02′E, Indonesia; etym. named after a local village) History and general description Two modern human crania and various fragmentary specimens were discovered 30 km/19 miles south of Kediri, East Java, in 1889–90 by Eugène **Dubois** and B.D. van Rietschoten. Temporal span and how dated? Late Upper Pleistocene or **Holocene** faunal association. Hominin found at site Wajak 1 is an adult female cranium; Wajak 2 is a partial adult male cranium. Both are attributed to modern *Homo sapiens*. Archeological evidence found at site None. Key references: Geology, dating, and paleoenvironment von Koensigwald 1956; Hominins Dubois 1921.

Wajak *See* Wadjak.

Waki-Mille confluence Collection area in the **Woranso-Mille study area**, Central Afar, Ethiopia.

Walker, Alan (1938–) Alan Walker was born in Leicester, England, and although he considered attending art school, he eventually decided to study zoology, geology, and paleontology at Cambridge University. He received his BA with honors in the natural sciences from Cambridge in 1962 and then entered the University of London to study anatomy and paleontology as one of John **Napier**'s graduate students. Walker completed his PhD at the University

of London in 1967 and his dissertation, *Locomotor Adaptations in Living and Extinct Madagascan Lemurs*, reflected his developing interest in primate evolution and primate locomotion. Walker accepted a position at Makerere University College in Kampala, Uganda, where he was a lecturer in anatomy from 1965 to 1969, and from 1967 to 1969 he also served as Honorary Keeper of Palaeontology at the Uganda Museum. In 1969 Walker left Uganda for Nairobi where he was a lecturer in anatomy at the University of Nairobi until 1973. During these years in Africa Walker continued to study living primates and also conducted research on fossil primates. In 1970 he was brought into the research team being assembled by Richard **Leakey** and Glynn **Isaac** as part of the **East Rudolf Research Project** (renamed the **Koobi Fora Research Project** in 1975), which led to a long association between Walker and Leakey. Walker worked with Leakey and anatomists Michael **Day** and Bernard Wood to investigate the hominin remains that were being discovered at **East Rudolf** (now **East Turkana**) during the 1970s. Walker, along with Meave **Leakey**, reconstructed the **KNM-ER 1470** cranium that was discovered in 1972 by Bernard Ngeneo at **Koobi Fora**. In 1976 Leakey and Walker announced the discovery of a nearly complete cranium of a hominin (designated **KNM-ER 3733**) that they suggested represented evidence of *Homo erectus* at *c.*1.8 Ma. Since fossils belonging to *Paranthropus boisei* had been found in deposits of about the same age, Leakey and Walker suggested the discovery of KNM-ER 3733 falsified the **single-species hypothesis**. In 1974 Walker was appointed as a Professor of Anatomy at the Harvard University Medical School and Professor of Anthropology at Harvard University, positions he held until 1978. In 1978 Walker accepted a position as Professor of Cell Biology and Anatomy at the Johns Hopkins University School of Medicine, where he remained until 1995, when he was appointed Professor of Anthropology and Biology at Pennsylvania State University. Throughout the 1980s Walker remained an active member of the Koobi Fora Research Project, although he also worked on other projects relating to primate and human evolution. He published a number of papers re-examining the *Proconsul africanus* fossils discovered by Mary **Leakey** on Rusinga Island in Lake Victoria in 1948, as well as describing new *Proconsul* fossils found in Kenya. This research was described in *The Ape in the Tree: An Intellectual & Natural History of Proconsul* (2005), which Walker coauthored with his wife Pat Shipman. When Kamoya **Kimeu**, a member of the

Koobi Fora research team, found the nearly complete skeleton of a hominin at **Nariokotome, West Turkana**, in 1984, Walker was one of the lead investigators that studied the skeleton. Leakey and Walker concluded that **KNM-WT 15000** was one of the oldest and most complete *H. erectus* fossils known, and the results of an extensive study of the remains entitled *The Nariokotome Homo erectus Skeleton* was published in 1993. Walker was also involved in the discovery and analysis of the **KNM-WT 17000** cranium and between 1994 and 1997 Walker worked with Meave Leakey and others on the *Australopithecus anamensis* fossils found at **Kanapoi** in northern Kenya. Walker was elected a Fellow of the Royal Society in 1999 and retired from teaching in 2010.

Walker circulation In the tropical Pacific, Sir Gilbert Walker first noted the anomalous longitudinal (or zonal) circulation that is associated with the **Southern Oscillation Index**. Normal **tropical** circulation is a cross-latitudinal (or meridional) equator-ward flow of the trade winds feeding the **intertropical convergence zone**, and is part of the **Hadley circulation**. The Walker circulation (also known as the Walker cell) sets up the sea-surface temperature gradient in the tropical Pacific associated with the La Niña phase of the El Niño Southern Oscillation. *See also* **El Niño Southern Oscillation**.

walking *See* **foot function**; **gait**; **walking cycle**.

walking cycle A walking (or gait) cycle describes the movements, distance covered, and time elapsed during one stride [from the initial substrate contact of one foot through to substrate contact of that same (or ipsilateral) foot]. The beginning of the walking cycle (conventionally 0%) is initiated with heel strike of a single foot, at which time that foot is in contact with the ground for the roughly 60% of the gait cycle, known as the stance phase. Toe-off, or **push-off**, initiates the swing phase, which comprises the remaining 40% of the walking cycle. During the stance phase the center of gravity starts low, rises to its highest point in the middle of stance phase, and falls again at toe-off. At two points during the walking cycle, the opposite (contralateral) foot is in stance phase at the same time as the ipsilateral foot; this is known as the double stance phase. The first occurs from 0% to approximately 10% of the walking cycle, and the second occurs prior to toe-off, in which heel contact has

occurred with the opposite (contralateral) foot (approximately 50–60% of the walking cycle).

Wallace's Line The western boundary of Wallacea. *See* **Wallacea**.

Wallacea The region in Southeast Asia/Australasia where Alfred Russel Wallace showed that the fauna is transitional (i.e., it is neither Southeast Asian nor Australian). Its western boundary is Wallace's Line, which runs between Bali and Lombok in the south then northwards between the eastern edge of the **Sunda** continental land mass (i.e., Borneo) to the west and Kalimantan and Sulawesi to the east. The eastern boundary of Wallacea, also called Lydekker's Line, is the northwestern edge of **Sahul** (i.e., Wallacea excludes the New Guinea and the Aru Islands). Wallacea includes islands such as Sulawesi and the Lesser Sunda Islands (e.g., Flores, Lombok).

Warendorf (Location 51°57'21"N, 7°57'08"E, Germany; etym. named after a nearby town) <u>History and general description</u> Hominin remains uncovered in layers of silt and gravels in the 1990s have been attributed to *Homo neanderthalensis*. <u>Temporal span and how dated?</u> The finds have no secure context, but the excavators believe the hominin remains date a Weichselian interstadial (*c.*70–40 ka) based on geological correlation, lithic typology, and faunal analysis. <u>Hominins found at site</u> Incomplete right parietal that has been suggested to be identical to **La Chapelle-aux-Saints** and **Neanderthal 1**. <u>Archeological evidence found at site</u> **Middle Paleolithic** assemblage attributed to the Keilmessergruppen technocomplex. Key references: <u>Geology, dating, and paleoenvironment</u> Street et al 2006; <u>Archeology</u> Czarnetzki and Trelliso-Carreno 1999; <u>Hominins</u> Czarnetzki and Trelliso-Carreno 1999.

warp (ME *warpen* = to twist, so to twist or bend out of shape) A warp is a representation of the differences in the coordinates of landmarks or semilandmarks between two objects of interest. *See also* **transformation grids**.

warthog The vernacular name for the taxa within the genus *Phacochoerus*. *See also* **Suidae**.

Washburn, Sherwood Larned (1911–2002) Sherwood Washburn, who was born in Cambridge, Massachusetts, USA, attended Groton

School from 1926 to 1931, where he took a special interest in the natural history collections housed in the Groton Museum, and during school vacations he worked at the Museum of Comparative Anatomy at Harvard University. Washburn entered Harvard University in 1931 intending to study zoology, but, after taking a course in anthropology with Alfred Tozzer, he switched to anthropology. He completed his BA in anthropology in 1935 and continued his graduate studies in anthropology at Harvard with Ernest Hooton. Washburn joined the Asiatic Primate Expedition in 1937, where he assisted Johns Hopkins University primate morphologist Adolph Schultz preparing primate skeletons, and was also influenced by another member of the expedition, Ray Carpenter of Columbia University, who was observing gibbon behavior. While in graduate school Washburn studied comparative anatomy and vertebrate paleontology, including a term spent at Oxford University with anatomist Wilfrid **Le Gros Clark**. Washburn became increasingly interested in primate evolution, and this was reflected in his Harvard PhD dissertation entitled *A Preliminary Metrical Study of the Skeleton of Langurs and Macaques*, which he completed in 1940. The year before he had been appointed a Professor of Anatomy at Columbia University and had begun to conduct experimental research into primate, including human, evolution. At Columbia Washburn became associated with a group of prominent scientists, including the Russian geneticist Theodosius Dobzhansky, who also taught at Columbia, William King Gregory and George Gaylord Simpson, both of whom were paleontologists at the American Museum of Natural History, and Paul Fejos of the **Viking Fund** (later the **Wenner-Gren Foundation for Anthropological Research**). Washburn left Columbia University in 1947 to accept a position as Professor of Anthropology at the University of Chicago. In 1948 he received a grant from the Viking Fund to travel to Africa where he studied monkeys in Uganda, and in South Africa he inspected the **australopith** fossils discovered by Raymond **Dart** and Robert **Broom**. In 1950 Washburn and Theodosius Dobzhansky organized a symposium on the "Origin and Evolution of Man" held at the Cold Spring Harbor Biological Laboratory, which served to bring new ideas from the **modern evolutionary synthesis** into the study of human evolution. Washburn was also involved, with Sol Tax and Alfred Kroeber, in organizing a conference held in New York City in 1952 on the topic of "Anthropology Today," also sponsored by the

Wenner-Gren Foundation. At both conferences Washburn presented papers intended to bring about the modernization of physical anthropology. Washburn had published his manifesto in a groundbreaking paper entitled "The new physical anthropology" in the *Transactions of the New York Academy of Sciences* in 1951. In it he urged that physical anthropologists abandon the use of anthropometry and craniometry and their focus on racial classification, and instead stress the importance of evolutionary biology, population genetics, and an appreciation of variation among modern human and extant higher primate populations. In 1955 Washburn made his second trip to Africa to attend the Third **Pan-African Congress on Prehistory and Quaternary Studies**, held in Livingstone, Northern Rhodesia (now Zambia). While there he took the opportunity to study baboon behavior at a game reserve near Lake Victoria. This strengthened Washburn's interest in the study of living primates as a means of understanding early hominin behavior, and he contributed a paper on this subject at a conference on "Evolution and Behavior" held in 1955. In 1959 he organized a conference on "The Social Life of Early Man" sponsored by the Wenner-Gren Foundation. It brought together scientists from various disciplines to discuss the relevance of studies of primate behavior, archeology, and cultural anthropology to the problem of the reconstruction of early hominin behavior. In the midst of all of this activity Washburn left the University of Chicago in 1958 to accept a position as Professor of Anthropology at the University of California at Berkeley when its anthropology department was beginning to develop into an influential center of paleoanthropological research. He continued to promote the study of primate behavior and issues related to human evolution, and he became interested in the role of hunting in the evolution of human culture. In 1961 Washburn organized yet another conference sponsored by the Wenner-Gren Foundation, this time on the subject of "Classification and Human Evolution," which not only discussed primate studies and human evolution but also introduced the new developments in molecular anthropology. Washburn wrote widely on the subjects of primate behavior, human evolution, and the concept of "race" during the 1960s and 1970s, and his courses at the University of California, Berkeley, influenced a large number of students. He was instrumental in encouraging primate studies, which have contributed significantly to contemporary theories of human evolution, and in establishing primate research facilities in the USA. Washburn retired from teaching in 1979.

WCFP Acronym for the West Coast Fossil Park. *See* **Langebaanweg**; **West Coast Fossil Park**.

weaning *See* **age at weaning**.

wear facets *See* **tooth wear**.

Weibull hazard model A general "failure" model that has a number of applications in demography. The model was first put forward in the late 1930s by Waloddi Weibull and published in its fullest form in 1951. In the Weibull hazard model the hazard of death at exact age t is $h(t) = \frac{\beta}{\alpha}\left(\frac{t}{\alpha}\right)^{(\beta-1)}$ and the **survivorship** to exact age t is $s(t) = \exp\left(-\left(\frac{t}{\alpha}\right)^{\beta}\right)$. The probability density function for age at death is, as in all hazard models, the product of the hazard and the survivorship. The log/log scale (both the hazard and age measured on a logarithmic scale) is particularly useful because the produced plot of log hazard against log age forms a straight line (see Weiss 1990). Although the Weibull is often used to represent immature mortality, in the **Siler hazard model** the Weibull's potential role is taken by a **negative Gompertz hazard model**. This is necessary because the Weibull survivorship function reaches zero at very advanced ages, whereas the negative Gompertz model's survivorship does not. The Weibull has also been used to represent other distributions in human biology, such as the post-partum time interval to initiation of breast-feeding (Holman and Grimes 2003) and the length of post-partum amenorrhea (Holman et al. 2006).

Weidenreich, Franz (1873–1948) Franz Weidenreich was born in Edenkoben in the Palatinate, Germany. He entered university in 1893 to study medicine, first in Munich and then at Kiel and Berlin; Weidenreich's doctoral thesis on the central nucleus of the human cerebellum was completed in 1899 at Strasbourg. For the next 2 years he worked with the German anatomist and anthropologist Gustav **Schwalbe** on human anatomy and physiology, focusing on the spleen and milk secretion. After working briefly with the German immunologist Paul Ehrlich in Frankfurt, Weidenreich returned to Strasbourg as Professor of Anatomy. There, between 1902 and 1914, Weidenreich's research interests included hematology

and histology, the human chin, and the morphology of the pelvis in primates and its relationship to upright posture and bipedal locomotion. Weidenreich's scientific work stopped during WWI and after the German defeat in 1918 Weidenreich was dismissed from his Professorship at Strasbourg. In 1919 he left for Heidelberg and in 1921 he was appointed Professor of Anatomy at the University of Heidelberg. The move to Heidelberg marked a shift in Weidenreich's research. He published a lengthy paper on evolution and development in 1920, and another substantial paper on the anatomy of the human foot in 1921. His interests turned to the human skeleton, and in 1925 he published a report on a human skull excavated in Weimar-Ehringsdorf. In *Rasse und Körperbau* (*Race and Bodily Structure*) (1927) Weidenreich analyzed human racial variation from the perspective of evolutionary theory and he later criticized the politicization of race by the Nazi party. These studies resulted in Weidenreich being appointed Professor of Physical Anthropology at the Johann Wolfgang Goethe University in Frankfurt in 1928. Thereafter, Weidenreich increasingly focused his research upon human paleontology and prepared an exhibition at the Senckenberg Natural History Museum outlining the stages of human evolution. However, Weidenreich's book on race and his own Jewish background made it increasingly difficult to remain in Germany and he was dismissed from his position at Frankfurt in 1933. The University of Chicago offered him a position as Professor of Anatomy soon thereafter and Weidenreich taught briefly at the university but in 1935 he accepted an invitation from the Rockefeller Foundation to replace the recently deceased Davidson **Black** as Professor of Anatomy at the **Peking Union Medical College** and Honorary Director of the **Cenozoic Research Laboratory**. This placed Weidenreich at the center of the remarkably productive excavations at **Zhoukoudian** (formerly Choukoutien) that had already produced several *Sinanthropus pekinensis* (now called *Homo erectus*) specimens. The quantity and variety of skeletal material belonging to *Sinanthropus* meant that in addition to describing individual fossils Weidenreich could also conduct comparative studies of the material. He published an extensive account of *Sinanthropus* in an article titled "The *Sinanthropus* population of Choukoutien (Locality 1) with a preliminary report on new discoveries" in the *Bulletin of the Geological Society of China* in 1935. In 1936 Weidenreich published a comparative study of endocranial casts of *Sinanthropus*, comparing them to other hominins and the great apes. This was followed by a comparative study of *Sinanthropus* mandibles in 1936, and similarly comprehensive monographic studies of the dentition (1937), the extremity bones (1941), and the crania (1943) followed. Between 1935 and 1937 excavations at Zhoukoudian, conducted primarily by Wenzhong **Pei** and a team of Chinese colleagues, resulted in the recovery of several more calvaria and other skeletal material that expanded the already impressive quantity of material housed at the Cenozoic Research Laboratory. The Japanese invasion of China made excavations at Zhoukoudian impossible by 1937 and the following year Weidenreich accepted invitations to travel in Europe and America. After presenting the evidence for *Sinanthropus* at an anthropological congress held in Copenhagen, Weidenreich traveled to The Netherlands to visit Eugéne **Dubois**, who in the 1890s had discovered *Pithecanthropus erectus* (now *Homo erectus*) on the island of Java, and to inspect the *Pithecanthropus* specimen. From Europe, Weidenreich then traveled to Java where he met with the German paleontologist Ralph **von Koenigswald**, who had found further *Pithecanthropus* fossils. This meeting led to von Koenigswald traveling to Beijing in 1939 where he and Weidenreich compared the *Pithecanthropus* remains from Java with the *Sinanthropus* material from China. They concluded that the two hominins were so similar in morphology as to belong to the same stage of human evolution, the differences between them indicating no more than geographical variation. Weidenreich remained in China until 1941, but as conditions worsened he made plans to leave the country and return to the USA. Before leaving China Weidenreich and the Chinese staff of the Cenozoic Research Laboratory prepared casts of the major *Sinanthropus* fossils and all the material was photographed. It was believed that the fossils themselves were in danger of being captured by the Japanese so arrangements were made for them to be packed in crates and sent to America with a contingent of American Marines leaving China, but along the way the fossils were lost and have never been recovered. Weidenreich obtained a position at the **American Museum of Natural History** in New York City, where he remained for the rest of his career. It was during this period that Weidenreich proposed his polycentric theory of human evolution. In this theory, based in part upon his comparisons of *Sinanthropus* and *Pithecanthropus* and upon modern geographical variation in modern humans, Weidenreich suggested that *Sinanthropus* and *Pithecanthropus*

had evolved in parallel, one group in China and the other in Java. In 1945 Weidenreich proposed a theory of human origins that suggested hominins and modern humans had evolved from *Gigantopithecus*, a giant ape discovered in 1935 but not described by von Koenigswald until 1952. Weidenreich elaborated on his theory in *Apes, Giants, and Humans*, published in 1946, and embarked on a monograph the *Morphology of Solo Man*; this remained unfinished at his death and it was published posthumously in 1951. *See also* **multiregional hypothesis**.

Weimar-Ehringsdorf   *See* **Ehringsdorf**.

Weiner, Joseph   *See* **Piltdown**.

Weiss' model life tables   A series of model **life tables** published in 1973 by Kenneth M. Weiss and based on census data from "anthropological" samples as well as life tables from archeologically recovered skeletal samples. These model life tables are often seen as a useful contrast to the **Coale and Demeny model life tables** which were based on large historical census data sets. Weiss' models are keyed by **life expectancy** at age 15 (a measure of adult mortality) and **survivorship** to age 15 (a measure of juvenile mortality). He combined life expectancies at age 15 of 15.0, 20.0, 22.5, 25.0, 27.5, 30.0, 32.5, and 35.0 years with survivorships to age 15 (based on a **radix** of 100) of 30, 35, 40, 45, 50, 55, 60, 65, and 70 to obtain 72 model life tables. The highest-mortality model [using Weiss' notation where "MT" stands for model tables; e.g., MT:22.5-30 refers to a model life table with a life expectancy of 22.5 years at age 15 and a survivorship of 30 (out of a radix of 100) at age 15] is MT:15.0-30.0, which has a life expectancy at age 15 of only 15 more years and survivorship to age 15 of only 30%. The lowest-mortality model is MT:35.0-70.0, which has a life expectancy at age 15 of 35 more years and survivorship to age 15 of 70%. The adult mortality for Weiss' models is generally higher than that for the Coale and Demeny models and there is very little overlap in mortality between the two sets of models. Unfortunately, the Coale and Demeny models may not be appropriate for comparison with prehistoric samples, whereas the Weiss models are based, in part, on skeletal data from which the age estimates may not be reliable. Gurven and Kaplan (2007) have presented parameter estimates for **Siler hazard models** based on demographic studies of extant **hunter–gatherer**s, forager–horticulturalists, and acculturated hunter–gatherers. These could be taken as a basis for comparison to paleodemographic analyses.

welded tuff A sedimentary deposit consisting primarily of pyroclastic material erupted from a volcano, in which the heat of the eruptive cloud results in the partial melting of glass phases, effectively welding the particles together. Although not as dense and homogeneous as **obsidian**, welded tuffs are typically massive and glassy, and are occasionally utilized as raw material in stone tool manufacture.

Wenner-Gren Foundation for Anthropological Research In 1951, the name of the Viking Fund, Inc., which had been established by Axel Wenner-Gren 10 years earlier in 1941, was changed to the Wenner-Gren Foundation for Anthropological Research, Inc. Axel Wenner-Gren was a Swedish entrepreneur, the owner of the Electrolux Company, and in the 1930s one of the richest men in the world. He was persuaded by Paul Fejos, a Hungarian medical doctor and ethnographic film maker, that well-directed giving could make a significant contribution to the young and relatively neglected field of anthropology. Wenner-Gren endowed the Foundation with US $ 2 million of Electrolux stock, and because the Foundation does not fundraise the residue of that stock is the sole source of the Foundation's endowment. The original intention was to develop the Foundation into a broad-based, worldwide anthropological research institute. In the period between 1945 and 1980, in addition to hosting conferences and seminars and providing funds for research and publication, the Foundation housed nine research laboratories, had a library of over 28,000 volumes, and provided office space for visiting anthropologists in its New York City premises. Paul Fejos ran the Foundation from its inception until his death in 1963 and was succeeded by his young widow, Lita Binns Fejos (later Lita Osmundsen), who headed the Foundation until her retirement in 1986. Since that time there have been three presidents of the Foundation (Sydel Silverman, Richard Fox, and Leslie **Aiello**). The mission of the Foundation is to "support significant and innovative anthropological research into humanity's biological and cultural origins, development and variation and to foster the creation of an international community of research scholars in anthropology." From the outset the Foundation has made significant contributions to the field of human origins. In the 1940s and early 1950s human origins-related initiatives included support for Willard Libby to develop **radiocarbon dating** when many other funding sources considered the research too risky, as well as funding for Ralph **von Koenigswald** to spend 2 years at the **American Museum of Natural History** to carry out collaborative research with Franz **Weidenreich** on hominin fossils from Southeast Asia. The Foundation also funded a French Upper Paleolithic Research Program, which over a number of years supported excavations at Roc-aux-Sorciers in Angles-sur-l'Anglin (Vienne) (by Suzanne de Saint-Mathurin and Dorothy Garrod) and at Le Colombière (Aïn) (by Hallam Movius). The Foundation also supported a number of other excavations, for example at Ksar 'Akil (Lebanon), Cap Ashakar (Morocco), and Ouad Akarit (Tunisia), as well as publishing projects (e.g., the preparation of Weidenreich's monograph on the hominin remains from **Ngandong**). In 1946 the Foundation introduced the **Viking Summer Seminars in Physical Anthropology** (they ran from 1946 to 1951, and in 1953). Under the leadership of Sherwood **Washburn**, these seminars helped introduce genetics and evolutionary theory to biological anthropology, thereby in large measure determining the course of the modern subdiscipline of human origins research. The Early Man in Africa program (1947–55) was initiated by Fr. Teilhard de Chardin (a Fellow of the Foundation) to call attention to the extraordinary significance of the human origins in southern Africa, to date the southern Africa cave deposits, and to facilitate multidisciplinary team research. By the end of the 1950s 75 grants has been awarded under this program, including funds for field research at **Olduvai Gorge** (Louis **Leakey**), the **Cave of Hearths** (C. van Riet Lowe), and **Hopefield** (Ronald Singer), for the purchase of equipment for research on the southern African australopiths (John Robinson), and for their morphological analysis (Clark **Howell**). In 1946 the Foundation introduced the **Viking Fund Medal**. Medals were awarded annually until 1961 and five awards have been made since that time. The Supper Conferences began in 1944 under the aegis of the Viking Fund and they were continued by the Foundation for three and a half decades. These evening meetings provided a forum for discussion and debate for the growing field. In addition to many of the grantees and the awardees of the Viking Fund Medal, the roster of paleoanthropological speakers over the years includes Emiliano **Aguirre**, Karl Butzer, Leslie Freeman, Alexander Marshack, Phillip **Tobias**, Richard Lee, Glynn **Isaac**, and Richard **Leakey**. In 1949 the Foundation was instrumental in the establishment of the **American Institute of Human Paleontology**. The aim of the Institute was to provide a forum to increase knowledge of early humans and the founding members were Loren C. Eiseley (President), Joseph Birdsell, Paul Fejos, Theodore McCown, Hallam Movius, Dale Stewart, and Sherwood Washburn. The main accomplishment of the Institute was to acquire, with Wenner-Gren financial support, the Barlow/Damon collection of molds of hominin fossils. The collection of molds was deposited with the University of Pennsylvania Museum of Archaeology and Anthropology and it provided the basis of their hominin casting program. During the 1950s the Foundation supported the research of the plastics engineer, David Gilbert, to develop a new and highly accurate molding technique. This led to the establishment of the **Anthrocast** program in 1965. Between 1968 and 1976 Anthrocast provided over 16,000 replicas of 180 different fossil specimens to institutions and researchers worldwide. The Anthrocast molds are currently curated at the University of Pennsylvania Museum of Archaeology and Anthropology and casts continue to be available on a limited basis through their casting program. In 1958 the Foundation acquired Burg Wartenstein castle in Austria as a gift from Axel Wenner-Gren. Between 1959 and 1980 Burg Wartenstein was the Foundation's European headquarters and the primary venue for its International Symposium Program. International Symposia were (and still are) meetings of intensive discussion that aim to include international scholars with broad interests. During the 1960s and 1970s the Foundation hosted almost 2000 scholars at 86 symposia held at the castle and elsewhere during the summer months. The titles of landmark meetings that focused on paleoanthropology and the year they were held are given below (the publication date and the organizers/editors are given in parentheses): *Social Life of Early Man*, 1959 (1961, Washburn), *Early Man and Pleistocene Stratigraphy in the Circum-Mediterranean Regions*, 1960 (1962, Blanc and Howell), *African Ecology and Human Evolution*, 1961 (1963, Howell and Bourlière), *Classification and Human Evolution*, 1962 (1963, Washburn), *Background to Evolution in Africa*, 1965 (1967, Bishop and Clark), *Man the Hunter*, 1966 (1968, Lee and DeVore held in Chicago), *Calibration of Hominoid*

Evolution: Recent Advances in Isotopic and Other Dating Methods as Applicable to the Origin of Man, 1971 (1972, Bishop and Miller), *Earliest Man and Environments in the Lake Rudolf Basin*, 1973 (1976, Coppens, Howell, Isaac, and R. Leakey, held in Nairobi and Lake Rudolf, Kenya), *After the Australopithecines*, 1973 (1975, Butzer and Isaac), and *Early Hominids of Africa*, 1974 (1978, Jolly). The results of these meetings were published through the Foundation's **Viking Fund Publications in Anthropology**, as well as through special arrangement with publishing houses (such as Aldine Publishers) and various university presses. Throughout this time the Foundation continued to fund research in human origins. In 1965 it established the Origins of Man program (1965–72) under the guidance of Walter William (Bill) **Bishop**, C. K. **Brain**, J Desmond **Clark**, Francis Clark Howell, Louis Leakey, and Sherwood Washburn. In the 1960s and 1970s over 300 grants were made in paleoanthropology. These grants continued to support excavations in southern African (e.g., **Swartkrans**, Brain; **Elandsfontein**, Singer) as well as survey, excavation, geology, dating, Plio-Pleistocene vertebrate paleontology, and hominin analyses throughout East and central Africa [e.g., Ethiopia (Omo; Howell), Chad (Yves **Coppens**), Olduvai Gorge (Louis Leakey), Baringo (Phillip Tobias), and Olorgesailie (Bishop and Leakey)]. Foundation grants also funded Middle and Late Stone Age research (Desmond Clark), stratigraphy in the Albert Rift (Bishop), suid paleontology (Basil Cook and Shirley Coryndon), and the Miocene primates recovered from sites in Ugandan and Kenya (David Pilbeam, Peter Andrews, and Louis Leakey), as well as pioneering work in primate behavior (Jane Goodall, Diane Fossey, Robin Dunbar, etc.). Outside of Africa grants funded research in Europe and Asia (e.g., Clacton, Singer; Torralba and Ambrona, Aguirre, Howell, and Butzer; *Homo erectus* in Indonesia, Jacob, Sartono, and von Koenigswald; the **Mousterian**, Bindford and Klein; the Perigordian, Movius; the Paleolithic of Turkey and Syria, Solecki; Upper Paleolithic notations and art, Marshack, etc.). Funding was also provided for publication projects including among many others *Catalogue of Fossil Hominids* (Oakley), *Nomenclature of the Hominidae* (Campbell), the **Shanidar** reports (Solecki), and the description of the Pech de l'Aze child (Vallois). The late 1970s were a time of major change for the Foundation. Financial exigencies forced the termination of the Anthrocast program, and the sale of Burg Wartenstein and the Foundation's New York offices.

Since 1980 the majority of the Foundation's support has gone to graduate students. Many leading researchers in the field of human origins have benefited from support under the Foundation's doctoral research program (Dissertation Fieldwork Grant) or the Developing Country Training Fellowship program (now the Wadsworth International Fellowship program), which provides the opportunity for students from developing countries (e.g., Jackson Njau and Yohannes **Haile-Selassie**) to earn doctorates at world-class universities. The Foundation currently provides up to $5 million per year to support the field of anthropology, and this money is divided among the subfields, including paleoanthropology, in rough proportion to the numbers of applications received. Details of the programs and how to apply can be found at www.wennergren.org.

Wenner-Gren Foundation Supper Conferences *See* **Wenner-Gren Foundation for Anthropological Research**.

West Coast Fossil Park (or WCFP) This educational, research, and tourism institution in South Africa, which incorporates the Langebaanweg paleontological site, was established in 1998. At the end of 2005, the WCFP became an independent institution, managed by the WCFP Trust, and separate from Iziko Museums, although the collections from Langebaanweg currently remain housed at the South African Museum, Iziko Museums of Cape Town. The open-air archeological locality known as the **Anyskop Blowout** is also located inside the boundaries of the WCFP, approximately 1 km/0.6 miles south of the Pliocene fossil beds. *See* **Langebaanweg**.

West Turkana (Location 3°35′–4°30′N, 35° 40′–35°55′E, Kenya; etym. refers to a region to the west of Lake Turkana, a lake named after the Turkana, a Nilotic tribe that live around the lake) History and general description West Turkana refers to a substantial area of exposures on the northeastern shore of Lake Turkana whose size and complexity makes it equivalent to the larger study areas in Ethiopia (e.g., the **Hadar study area** and **Middle Awash study area**). In 1980, the Koobi Fora Research Project, led by Richard **Leakey** and in affiliation with the **National Museums of Kenya**, surveyed the exposures between Kalakol (formerly Ferguson's Gulf) to the south and Kokuro to the north. Subsequent field reconnaissance was carried out in 1981 and 1982 and

aerial photographs were taken in 1983. These confirmed the presence of Plio-Pleistocene sediments in a *c.*10 km-/6 mile-wide strip between Lake Turkana to the east and the Labur and Murua Rith hills to the west. Collection was begun in 1984 and the search for fossil and archeological evidence has been carried out subsequently by various teams. Between 1980 and 1989, the main paleontological effort was led by Richard Leakey and after 1989 by Meave **Leakey**; archeological investigations have always been led by Helene Roche. The many cumulative months of field seasons resulted in substantial collections that are held in the National Museums of Kenya in Nairobi. The sediments exposed in West Turkana belong to the Nachukui Formation. This ranges in age from 4.3–0.7 Ma, and it includes many of the tuffs recognized in the **Koobi Fora Formation** to the east and the **Shungura Formation** exposed in the lower reaches of the Omo River. The Nachukui Formation comprises eight members, all named after lagas, or ephemeral streams. From oldest to youngest they are the Lonyumun, Kataboi, Lomekwi, Lokalalei, Kalochoro, Kaitio, Natoo, and Nariokotome. The main fossil locality complexes are given below in alphabetical order (each full name is followed by its two-letter abbreviation): Kaitio (KI), Kalochoro (KL), Kalakodo (KK), Kangaki (KG), Kangatukuseo (KU), Kokiselei (KS), Lomekwi (LO), Lokalalei (LA), Loruth Kaado (LK), Nachukui (NC), Nasechebun (NS), Nanyangakipi (NN), Nariokotome (NK), Natoo (NT), and Naiyena Engol (NY). Localities within each complex are identified using roman numerals (e.g., Nariokotome III). Each hominin from West Turkana has the prefix "**KNM**" for the Kenya National Museum (the old name for the National Museums of Kenya), where they are curated, "**WT**" for West Turkana, and then the field number (e.g., KNM-WT 15000). Temporal span and dating methods The Nachukui Formation ranges in age from 4.3 to 0.7 Ma, based on various **radiogenic dating** methods, plus **magnetostratigraphy** and **biostratigraphy**. Hominins found at site Fossil hominins recovered from West Turkana have been assigned to *Australopithecus afarensis* (e.g., **KNM-WT 8556**), *Kenyanthropus platyops* (e.g., **KNM-WT 40000**), *Paranthropus aethiopicus* (e.g., **KNM-WT 16002, KNM-WT 17000**), *Paranthropus boisei* (e.g., **KNM-WT 17400**), and *Homo erectus* (e.g., **KNM-WT 15000**). Archeological evidence found at site More than 60 **Oldowan** and **Acheulean** sites have been recorded in West Turkana, grouped into eight major complexes and ranging from 2.3 to

0.7 Ma. Organized debitage sequences have been demonstrated at Lokalalei 2C site (2.3 Ma) using technological analysis and reconstruction of the cores. Key references: Geology, dating, and fauna Harris et al. 1988; Hominins Brown et al. 1985, Walker et al. 1986, Leakey and Walker 1988, Walker and Leakey 1993; Leakey et al. 2001; Archeology Roche et al. 1999, 2003, Delagnes and Roche 2005.

Western Margin One of the major fossiliferous subregions within the **Middle Awash study area**; the others include the **Central Awash Complex**, **Bouri Peninsula**, and **Bodo-Maka**. The Western Margin consists of a more than 30 km-/19 mile-wide strip of sediments that lie against the western escarpment of the **Afar Rift System** between approximately $10°00'$ and $10°10'$N. Localities within the Western Margin include Adu Dora, Alayla, **Asa Koma**, Digiba Dora, and **Saitune Dora**.

white matter A term that describes regions of the central nervous system that contain large bundles, or tracts, of **axons**. The myelin sheath that encircles axons in the white matter gives it a whitish appearance in fresh brain tissue, hence its name.

White Sands One of two hominin-bearing localities in the **Usno Formation** in the **Lower Omo Valley** (the other one is called **Brown Sands**). Fossils found in the White Sands locality have been given the prefix "W" for **White Sands**. *See also Australopithecus afarensis*; **Usno Formation**.

White, Timothy Douglas (1950–) Tim White was born in Los Angeles, and he completed BS degrees in biology and anthropology at the University of California at Riverside in 1972. After graduating, White entered the University of Michigan where he studied physical anthropology with Milford **Wolpoff**, Philip Gingerich, Loring Brace, and William Farrand. He completed his MA in 1974 and during that year he also became a member of the **East Rudolf Research Project** where he was part of the team of researchers led by Richard **Leakey** and Glynn **Isaac** at **Koobi Fora**, in Kenya. In 1976 White conducted anatomical studies of the hominin fossils discovered by Mary **Leakey** at **Laetoli** and he began a collaboration with Donald **Johanson** on the interpretation of the fossils Johanson had discovered at **Hadar**, in Ethiopia, between 1973 and 1975. White completed his PhD at the University of

Michigan in 1977 with a dissertation titled *The Anterior Mandibular Corpus of Early African Hominidae: Functional Significance of Shape and Size*. In the same year he became a Professor of Anthropology at the University of California at Berkeley and a year later, in 1978, he joined Mary Leakey's research team at Laetoli, where he assisted in the excavation of the **Laetoli hominin footprints**. In 1981 White joined his Berkeley colleague Desmond **Clark** along with Ethiopian graduate student Berhane **Asfaw** in the **Middle Awash Research Project**. The project was designed to be a multidisciplinary investigation of the geology, paleontology, and archeology of the hominin-bearing deposits of the Middle Awash region of Ethiopia. Over the following decades the Middle Awash Research Project expanded to include scientists who specialized in a variety of disciplines including archeology, geochemistry, geochronology, invertebrate and vertebrate paleontology, paleobotany, sedimentology, and structural geology, from more than 19 countries. Under joint leadership the Middle Awash Research Project has continued to unearth a wealth of hominin and animal fossils as well as archeological material. During the 1992–3 season White and two colleagues, Gen **Suwa** and Berhane Asfaw, discovered the fossil hominin remains at **Aramis**, in the Middle Awash study area. White and his colleagues concluded the hominin was an entirely new species, which they named *Australopithecus ramidus* in 1994; a year later the new species was transferred to a new genus as *Ardipithecus ramidus*. The fossils were dated c.4.4 Ma, making them among the oldest known possible hominin fossils at the time. Further evidence of that taxon came in the form of a fragile, poorly preserved but remarkably complete associated skeleton. That specimen, **ARA-VP-6/500**, other fossil evidence of that taxon, and a substantial body of contextual evidence, were published in 2009. The team made another remarkable discovery in 1997 when then-graduate student Yohannes **Haile-Selassie** found several hominin cranial fragments and a maxilla in the 2.5 Ma **Bouri Formation**. White and Berhane Asfaw, by then a paleoanthropologist with the Rift Valley Research Service in Ethiopia, assigned these fossils, as well as several other fragments discovered between 1996 and 1998 in the Middle Awash, to a new species they named *Australopithecus garhi* in a paper published in *Science* in 1999. These fossils not only filled in an important gap in the temporal record of hominin evolution, but they were also associated with some of the earliest evidence of animal butchery. The

Middle Awash Project team made yet another significant discovery near the village of **Herto** near the Awash River in 1997. This consisted of two partial adult *Homo sapiens* crania and one child's cranium that are between 154 and 160 ka. These fossils were among the earliest known modern humans and White's team assigned them to a new subspecies of *H. sapiens, Homo sapiens idaltu*. The discovery of a *c.*1 Ma *Homo erectus* skull (**BOU-VP-2/66**) from the Bouri Formation in 1997 led White and Berhane **Asfaw** to argue that the African and Asian varieties of *H. erectus* represent a single species and not two different species, as has been suggested by some. In addition, at the sites of Aramis in 1994 and at **Asa Issie** between 2000 and 2005 White's team recovered a 4.1 Ma collection of teeth and jaw fragments they assigned to *Australopithecus anamensis*; this was the first evidence of *Au. anamensis* from Ethiopia and it significantly enlarged the hypodigm of this taxon. Tim White is currently Professor of Integrative Biology at the University of California at Berkeley and served with the late F. Clark Howell as co-director of the **Laboratory for Human Evolutionary Studies** at the university from 1995 to 2003 when the laboratory was renamed the Human Evolution Research Center. Along with Clark **Howell**, White was also the co-leader of the Revealing Hominid Origins Initiative. He is also curator of biological anthropology at the Phoebe A. Hearst Museum of Anthropology and a research paleoanthropologist at the university's Museum of Vertebrate Zoology.

Wilcoxon signed rank test A nonparametric statistical test used to determine whether the values of a **sample** of measurements are significantly larger or smaller than a second set of values observed for the same sample. Used in the same situations as one would use a paired *t* **test**, the Wilcoxon signed rank test considers the relative ranking of the differences between pairs of values in the two groups rather than differences themselves.

Wildscheuer (Location 50°25′N, 8°08′E, Germany. etym. Ge. *wild* = wild and *scheuer* = barn, perhaps from being used as a place to house livestock) History and general description This cave site in the Lahn valley, Hess region of Germany was the site of active limestone quarrying and since 1953 it has been completely destroyed. Skull fragments were recovered that were thought to come from *Homo neanderthalen-*

sis but a recent analysis has shown them to be from a cave bear. Temporal span and how dated? No absolute dating methods have been attempted. Hominins found at site None. Archaeological evidence found at site Several lithic remains attributed to the **Middle Paleolithic, Aurignacian, Gravettian**, and **Madgalenian**, as well as abundant faunal remains. Key references Geology, dating, and paleoenvironment Terberger 1993, Fiedler 1994; Hominins Street et al. 2006; Archeology Terberger 1993, Fiedler 1994.

Wilks' lambda *See* multivariate analysis of variance.

Willandra Lakes (Location 33°S, 143°E, Australia; etym. named after the lake region of the same name) History and general description The earliest remains to be recovered from what is now referred to as the Willandra Lakes were discovered at **Lake Mungo**. When fossils began to be discovered at other lakes in the same complex of dried-up lakes it was decided that the name should be broadened from "Lake Mungo" to "Willandra Lakes." Thus the former Mungo 1 skeleton is now called WLH 1, etc. The Willandra Lakes Hominid 50 (**WLH 50**) partial cranium was discovered in 1982 in a deflating lake-shore dune in between the dry lakes of Garnpung and Leaghur in the Willandra Lakes region of New South Wales. Modern human footprints ($n = 124$) were discovered in 2006 about 10 km/6 miles north of the WLH 50 find site. Temporal span and how dated? Radiometric dates for WLH 50 range from 13 to 15 ka (**uranium-series dating**) to 29 ka (**electron spin resonance spectroscopy dating** on bone). Optically stimulated **luminescence dating** estimates derived from the associated sediments provide age estimates for the modern human trackways of 19–23 ka. Hominins found at site WLH 50 is the focus of an ongoing debate regarding its relevance to modern human origins. One group of researchers claim that this specimen's robusticity and morphology suggest descent from archaic southeast Asian hominids such as **Ngandong**, thereby supporting an evolutionary explanation in line with the **multiregional hypothesis**. Others claim these same features can be explained by pathology and/or intentional cranial-deformation practices. The modern human footprints were made by adults, adolescents, and children and may indicate collective activities, such as the hunting of water birds, at ephemeral water sources. Archeological evidence found at site None. Key references: Geology, dating, and paleoenvironment Caddie et al. 1987, Simpson and Grün 1998, Webb et al. 2006; Hominins Stuart-Macadam 1992, Webb 1992, Stringer 1998, Hawks et al. 2000, Webb et al. 2006.

Wilson, Allan Charles (1934–91) Allan Wilson was born in Ngaruawahia, New Zealand, and was raised on his family's farm at Helvetia, Pukekohe. He studied mathematics and chemistry at King's College in Auckland and then entered Otago University where he completed his BSc degree in 1955. Wilson left for the USA in 1955 and completed his MS in zoology in 1957 at Washington State University and in 1961 he completed his PhD at the University of California at Berkeley in biochemistry, where he worked with biochemist Arthur Pardee. Wilson took a postdoctoral fellowship at Brandeis University in 1961 and remained there until 1964 when he left to become Professor of Biochemistry at the University of California at Berkeley, where he remained until his death. At Berkeley, Wilson established a laboratory where he conducted research in molecular biology, developing novel methods for investigating evolutionary change and using genetic information to reconstruct phylogenies. Wilson, working with Vincent **Sarich**, then a graduate student, used a new technique called micro-complement fixation to compare the degree of similarity of **albumin**, a blood protein, among different species of primates. The results of their research supported conclusions reached earlier by Morris **Goodman** who, using an immunological method, had shown that modern humans were more closely related to the African apes, and particularly to chimpanzees, than to the Asian apes. Significantly, however, Wilson and Sarich also concluded from their research that mutations in the albumin molecule occurred at a constant rate, and thus the degree to which these molecules differed among species could be used to calculate how much time had elapsed since those species had shared a common ancestor, Thus, they believed they had found a "**molecular clock**" that could be used to date the origin of extant higher primate lineages. Wilson and Sarich published their findings in an article entitled "Immunological time-scale for human evolution" in *Science* in 1967. In that paper Wilson and Sarich argued that the remarkable degree of similarity of the albumins of modern humans and

chimpanzees (nearly 99%) meant that the two lineages must have shared a common ancestor *c*.5 Ma. Few paleoanthropologists were familiar with the methods used in the paper and most accepted the conventional wisdom of the time which was that the hominin and other hominoid lineages had diverged at least 15 Ma. During the 1970s, Wilson continued to develop and employ new and existing molecular biology techniques (e.g., **polymerase chain reaction** and the relative rate test) to expand the role of molecular biology in providing information about **phylogeny** and the timing of events in evolution. During the 1970s and early 1980s, Wilson and his team began to focus their attention on collecting evidence for mutations in **mitochondrial DNA** (mtDNA) in extant higher primate taxa. On the basis of this research, Wilson, along with S.D. Ferris and Wesley M. Brown, published an article in the *Proceedings of the National Academy of Sciences USA* in 1981 titled "Evolutionary tree for apes and humans based on cleavage maps of mitochondrial DNA." mtDNA seemed an ideal subject for genomic research of relatively recent events in evolutionary history because there are many copies within a single cell, it is passed directly from mother to offspring, and mutations accumulate at a relatively rapid rate. Wilson's new collaborators, Rebecca Cann and Mark Stoneking, collected samples from individuals representing different populations of modern humans from across the world, and compared the number of mutations present in the mtDNA from these groups. The results of this study were published in 1987 in *Nature* in an article titled "Mitochondrial DNA and human evolution." In it Cann, Stoneking, and Wilson argued that the relatively small number of mutations found across all the modern human populations they sampled indicated that all modern humans were derived from a single *c*.200 ka founder population. They also argued that their research suggested that this ancestral population lived in Africa, and that this evidence was incompatible with the tenets of the **multiregional hypothesis**, namely that *Homo sapiens* in Asia had evolved from populations of *Homo erectus* living in that region. Wilson and his colleagues suggested instead that the ancestral group of *Homo sapiens* in Africa they identified gave rise to populations that had migrated out of Africa and replaced pre-sapiens hominins in all the major regions of the Old World. The significance and controversial nature of these conclusions was picked up by the media, and because mtDNA is transmitted down the female line, and because this implied that all modern humans could, theoretically, be traced back to a single female, this hypothetical founding female was dubbed the "African Eve" or the "**mitochondrial Eve**." Wilson and his team continued to expand upon and refine this research over the next few years and their results have provided important support for the **out-of-Africa hypothesis**. Wilson died of leukemia in 1991. *See also* **Sarich, Vincent**.

Wilson bands Wilson bands are abnormal accentuated markings in enamel that follow the contour or alignment of the normal, regular, **striae of Retzius**. Wilson and Shroff (1970) drew attention to the fact that some striae of Retzius may be accentuated in their breadth, color, and because they show a pronounced change in prism orientation along their length. They are sometimes, but not always, associated with surface enamel **hypoplasia**. The disruptive physiological or pathological events that cause accentuated striae may not coincide exactly with the formation of a normal stria of Retzius, so clusters, or groups, of irregularly spaced Wilson bands may be superimposed onto the regular underlying pattern of long-period striae of Retzius. Counts and measurements of Wilson bands are an important way of assessing the degree of stress experienced by an individual during development. The equivalent of Wilson bands in dentine are called **Owen's lines**.

windlass effect First described by J.H. Hicks (1954) the windlass effect or windlass mechanism converts the foot into a solid lever during the **push-off** phase of walking. The plantar aponeurosis attaches to the calcaneus, runs across the plantar aspect of the foot, wraps around the metatarsal heads, and inserts into the base of the proximal phalanges. During the push-off phase of walking the metatarsophalangeal joints dorsiflex, tightening the plantar aponeurosis, raising the **longitudinal arch**, and shortening the foot. The hindfoot inverts relative to the forefoot, locking the calcaneocuboid joint, and converting the midfoot into a stiff lever that facilitates efficient propulsion during bipedal locomotion.

windlass mechanism *See* **windlass effect**.

winnowing (OE *windwian* = the use of wind to separate the lighter chaff from the denser, heavier, seed) The removal by flowing water of the smallest subset of flaked stone artifacts or bone from an arche-

ological or paleontological site. The effects of winnowing are typically recognized in archeological contexts by comparing the size distribution of artifacts in an assemblage to those from experimentally derived **lithic assemblage**s. The degree to which an assemblage has been winnowed is important to consider when making behavioral inferences since winnowing can alter the spatial integrity of an archeological site and remove certain classes of evidence. For example, by preferentially removing the smaller byproducts of flint knapping, particularly small flakes, flake fragments, and angular waste, winnowing can make it difficult to examine patterns of artifact manufacture and curation or make it difficult to determine whether artifacts were produced at, or away from, a site.

Witwatersrand University Department of Anatomy The Department of Anatomy of the Witwatersrand University was established in 1919. The first head of the department, Edward Stibbe, was replaced by Raymond **Dart** in 1923, and it was during his long tenure that the Department's involvement in human evolution research began. After Dart's involvement with **Taung**, the next fossil site that Wits Anatomy Department researchers, particularly Phillip **Tobias**, were involved in was **Makapansgat**. Then, just a few years after Tobias replaced Raymond Dart as Head of Department in 1959, Witwatersrand University took over the land that included the site at **Sterkfontein** and the **Palaeo-Anthropology Research Unit** within the department restarted excavations at Sterkfontein in 1966. The new project was headed by Tobias, with Alun Hughes, and later Ronald **Clarke**, in day-to-day charge of the excavation. Other field projects that in one form, or another, have come under the purview of the Witwatersrand University Department of Anatomy include those at **Drimolen**, **Gladysvale**, and **Gondolin**.

WK Acronym of Wayland's Korongo, a locality at **Olduvai Gorge**. Handaxes and cleavers were found on the scree below Bed IV at this site in 1932, but it was not formally excavated until 1970. Handaxes, cleavers, and scrapers were then found *in situ*, as well as **OH 28**, which comprises the shaft of a left femur and much of a left pelvic bone. *See also* **Olduvai Gorge**.

WLH Acronym for Willandra Lakes hominids and the prefix for fossils recovered from the Willandra Lakes region in Australia. *See* **Willandra Lakes**.

WLH 50 [NB: Webb (1989, 1990) does not use a hyphen, but Stringer (1998) and Hawks et al. (2000) do] Site **Willandra Lakes**. Locality Garnpung/Leaghur 20. Surface/*in situ* Surface. Date of discovery On, or before, 1982. Finders Unknown. Unit N/A. Horizon A deflating lake-shore dune in between the dry lakes of Garnpung and Leaghur. Bed/member N/A. Formation N/A. Group N/A. Nearest overlying dated horizon N/A. Nearest underlying dated horizon N/A. Geological age **Electron spin resonance spectroscopy dating** points to an age of 29±5 ka (Caddie et al. 1987) and **uranium-series dating** suggests a younger age of 15–13 ka (Simpson and Grun 1998). Developmental age Adult. Presumed sex Male. Brief anatomical description The frontal region and the posterior part of the calvaria. Announcement Webb 1989. Initial description Webb 1990. Photographs/line drawings and metrical data Webb 1989, 1990, Stringer 1998, Hawks et al. 2000 (these latter authors also cite measurements taken from a website maintained by Peter Brown). Detailed anatomical description Webb 1990. Initial taxonomic allocation The presumption is that the early authors (e.g., Webb 1989) interpreted WLH 50 as an example of "archaic *H. sapiens.*" Taxonomic revisions Stringer (1998) recognizes it as "anatomically modern" (*ibid*, p. 332) and by implication regards it as belonging to *Homo sapiens* and Hawks et al. 2000 refer to it as "modern human" (*ibid*, p. 21). Current conventional taxonomic allocation *H. sapiens*. Informal taxonomic category Anatomically modern human. Significance The WLH 50 calvaria was a key item of evidence cited by researchers (e.g., Thorne and Wolpoff 1992, Hawks et al. 2000) who interpreted its morphology as evidence of morphological continuity between the pre-modern *Homo* populations in Indonesia (e.g., **Ngandong**) and the early occupants of Australia. However, others interpret it as being "no closer metrically to the Ngandong than to the African archaic sample" (Stringer 1998, p. 331). Location of original WLH 50 has been reburied.

WLM Acronym for the western lake margin and the term used for the western margin of the paleolake some call **Lake Olduvai**. *See* **Lake Olduvai**; **Olduvai Gorge**.

WMC Acronym for Waki-Mille confluence, a collection area within the **Woranso-Mille study area**, Central Afar, Ethiopia.

Wolo Milo One of the Early and Middle **Pleistocene** sites in the exposures of the **Ola Bula Formation** within the **Soa Basin**, central Flores, Indonesia.

Wolokeo One of the Early and Middle **Pleistocene** sites in the exposures of the **Ola Bula Formation** within the **Soa Basin**, central Flores, Indonesia.

Wolpoff, Milford H. (1942–) Milford Wolpoff was born in Chicago, Illinois, USA, and he traces his interest in human origins to his reading Roy Chapman Andrew's *Meet Your Ancestors* (1945). Nonetheless, when he entered the University of Illinois at Urbana-Champaign it was to study physics, but after taking anthropology and archeology courses with Donald Lathrap he changed his major and received his BA in anthropology in 1964. He remained at his alma mater to pursue his PhD in physical anthropology, where he focused his studies on hominin dental evolution under the mentorship of Eugene Giles. During the course of this research in 1966 Wolpoff had the opportunity to work with John **Robinson**, who by then had left South Africa for the University of Wisconsin. Wolpoff completed his PhD in 1969, taught anthropology at Case Western Reserve University from 1968 to 1971 and then in 1971 he accepted a position in the Department of Anthropology at the University of Michigan where he has remained for the last four decades. In 1971 Wolpoff published his dissertation, entitled *Metric Trends in Hominid Dental Evolution*, and with David Brose, an archeologist at Case Western, he published an influential article in *American Anthropologist* entitled "Early Upper Paleolithic man and Late Middle Paleolithic tools." In it the authors argued that because the changes in stone tool industries from the **Middle Paleolithic** to the **Upper Paleolithic** did not correspond chronologically with the fossil evidence showing the transition from Neanderthals to anatomically modern humans, this suggested that the new tool types were not necessarily introduced by Cro-Magnon newcomers that replaced the Neanderthals, but rather were the product of behavioral evolution within *Homo neanderthalensis*. Wolpoff participated in a significant re-evaluation of *H. neanderthalensis* during the 1970s and 1980s when he and several colleagues began to examine anew the large collection of Neanderthal remains discovered at **Krapina**, in Croatia, early in the 20thC by the Croatian paleontologist Dragutin Gorjanović-Kramberger. Fred Smith, a graduate student who worked with Wolpoff, studied the bones and Wolpoff followed this work by investigating almost 300 teeth from the Krapina collection. This research expanded into a substantial new study of the Krapina remains and the publication by Jakov Radovčić, Fred Smith, Erik Trinkaus, and Wolpoff of *The Krapina Hominids* (1988). This strengthened Wolpoff's conviction that Neanderthals had not been replaced by a population of anatomically modern humans that had evolved elsewhere, but had themselves evolved *in situ* into modern humans. Wolpoff was also a trenchant advocate of the **single-species hypothesis**, which argued that only one hominin species should exist at any one time in the course of human evolution. As a result, Wolpoff argued that the gracile and robust australopiths in southern Africa belonged to the same species; human phylogeny was more like the straight trunk of a tree with a few minor branches, rather than a bush branching off into many different species. By the early 1980s Wolpoff began to promote, along with Australian anthropologist Alan Thorne, a view of recent human evolution that is referred to as the **multiregional hypothesis**. Thorne and Wolpoff published a paper in the *American Journal of Physical Anthropology* entitled "Regional continuity in Australasian Pleistocene hominid evolution" in 1981 where they presented evidence from the fossil record of the region that a distinct geographical population of hominins had persisted in the Far East and Australasia for the last million years. Wolpoff and Thorne later expanded their theory to argue that a similar morphological continuity within geographical regions could be observed in the hominin fossil record over much of the Old World. The multiregional hypothesis as developed by Wolpoff, Thorne, and Wu (1984) argues that the first hominins to leave Africa were *Homo erectus* about a million years ago and they dispersed to populate much of Asia and parts of Europe. Wolpoff and colleagues argued that these regional populations of *H. erectus* evolved semi-independently over time into modern *Homo sapiens*. They claimed that *H. erectus* remains in China resembled morphologically early *H. sapiens* remains in China that already possess some of the distinctive Asian racial features seen in modern populations. Wolpoff published, with his wife Rachel Caspari, a fuller account of the multiregional hypothesis in *Race and Human Evolution* (1997), where they explore the connections between his model of human evolution

and the polycentric theory of human evolution proposed by the German physical anthropologist Franz **Weidenreich** in the 1940s. Wolpoff's multiregional hypothesis not only draws upon paleontological evidence showing racial continuity over long periods for different geographical areas, but it also employed ideas borrowed from population genetics to show how geographical varieties of humans could persist over time while limited gene flow between these populations allowed these geographically distinct populations to remain part of the same species. The multiregional hypothesis was challenged by the **out-of-Africa hypothesis** that argued that modern *H. sapiens* evolved in Africa and then migrated out of Africa and into Asia where it to replaced *H. erectus*.

Wonderwerk Cave (Location 27°50′45″S, 29°33′19″E, Northern Cape province, South Africa; etym. Afrikaans, possibly referring to the many paintings formerly preserved on the cave wall). History and general description Formal excavation at this large (approximately 2400 m^2) solution cavity in dolomitic **limestone** began in 1943 by Malan and Wells. Karl Butzer worked there from 1974 to 1977 and research was continued in 1978 and thereafter by Peter Beaumont and his colleagues with Anne Thackeray focusing on the later occupation layers. Multiple discontinuous excavations in the cave, some to bedrock at about 6 m in depth, sample an archeological sequence that features excellent organic preservation, dated from the early **Pleistocene** to the **Holocene**. Temporal span and how dated? **Magnetostratigraphy** suggests the lower-most strata date to the **Olduvai subchron**, a hypothesis supported by a single **cosmogenic nuclide dating** estimate of *c.*2 Ma from these lower-most sediments. **Uranium-series dating** on **speleothems**, a single result from optically stimulated **luminescence dating**, and **radiocarbon dating** on charcoal constrain the ages of overlying strata of more than 350 ka to 1.21±0.05 ka. Hominins found at site None. Archeological evidence found at site The artifact assemblages span the **Early**, **Middle**, and **Later Stone Ages**. The site's preservation of multiple ($n = 6$) Early Stone Age strata in a cave setting is unusual, and these levels are characterized by **handaxes**, rare prepared **cores**, and **blades**. They are variably attributed to the **Acheulean** and **Fauresmith** stone industries, and on the basis of uranium-series age estimates on speleothem, are more than 276 ka. Middle Stone Age levels date to between *c.*70 to more than 220 ka and include

Levallois flakes, cores, blades, and a variety of retouched points. The Later Stone Age strata are attributed to the "Wilton Complex" of southern Africa, and are notable for a series of engraved stones, many depicting mammals, and extinct **megafauna** (*Equus capensis* and *Megalotragus* sp.) in layers radiocarbon dated to between 3.99±0.06 ka and 10.00±0.07 ka. All strata preserve **specularite** or **hematite** introduced into the cave by hominins and the Early Stone Age strata include evidence for hearths, and remarkably, vegetation mats interpreted as bedding material. Macrofauna are highly fragmented, but the large **micromammal**ian samples are well preserved. Key references: Geology, dating, and paleoenvironment Butzer 1984, Beaumont and Vogel 2006, Chazan et al. 2008; Archeology Thackeray et al. 1981, Beaumont 1990, Beaumont and Vogel 2006.

woodland The most widespread biome in tropical Africa, woodlands are areas of clumped trees that do not have a continuous canopy, and thus there is a grassy understory (NB: forests do not have a grass understory). In general, woodlands are 8–20 m in height and the tree canopies cover at least 40% of the surface. Like other biomes, woodlands are made up of a number of ecosystems that form a continuum from open to closed, with more closed woodland ecosystems associated with high mean annual rainfall. Average mean annual rainfall in woodland habitats ranges from 600 to 1000 mm (Bourliere and Hadley 1983). Woodlands in Africa are confined to tropical and subtropical regions. Most trees are deciduous or semi-deciduous but many woodlands contain some evergreen species. Woodland trees tend to be smaller than forest trees, and the density of coverage is often related to annual precipitation (Sikes 1999, Jacobs 2004) or seasonal fluctuations in rainfall (Vincens et al. 2007). Woodlands of different types (such as *Acacia* and **miombo woodland**) are a significant component of modern sub-Saharan African landscapes, and were likely to have been so throughout the whole of **hominin** evolutionary history. Early hominin sites, such as **Hadar**, are described as mosaic habitats that included open and closed woodlands, shrublands, **edaphic** grasslands, and gallery forests. While *Australopithecus afarensis* lived in several different types of biome, the most common of these were woodland habitats. Based on faunal evidence from **Olduvai Gorge** Bed I, it is likely that some African early Pleistocene woodland habitats were more species rich than modern woodland in the same region, suggesting there were differences between past and present **ecosystems** (Fernandez-Jalvo et al. 1998).

Woranso-Mille Paleontological Research Project (or WORMILPRP) A multidisciplinary project conducting field and laboratory research on the geology and paleontology of the Mio-Pliocene deposits bound by the towns of Mille, Chifra, and Kasagita, in the central Afar region of Ethiopia that make up the **Woranso-Mille study area**. The WORMILPRP started with an intensive paleontological survey of the north and central Afar region in 2002, but annual fieldwork concentrating on its current study area did not begin until 2004. The major objectives of the project are to recover early hominin fossils from the late Miocene and Pliocene deposits of the study area to understand the earlier phases of human evolution, to better understand the phylogenetic relationships among early *Australopithecus* species, and to collect new data to help understand the paleoecological and geological context for the early phases of human evolution. Thus far the WORMILPRP has been successful in collecting fossil hominins dated to between 3.6 and 3.8 Ma, a time period poorly sampled in the hominin fossil record. The finds include a 3.6 Ma partial skeleton of *Australopithecus afarensis* (KSD-VP-1/1) from the Korsi Dora locality. The project has also collected more than 4800 fossil specimens representing many vertebrate taxa; project geologists have established refined stratigraphic sections for much of the study area and geochronologists have generated radiometric ages for most of the fossiliferous horizons within the study area. The WORMILPRP, which is directed by Ethiopian paleoanthropologist Yohannes **Haile-Selassie**, involves many specialists from various academic institutions around the world. The project is also actively engaged in training graduate and undergraduate students from Ethiopia and the USA.

Woranso-Mille study area (Location 11°30′N, 40°30′E, Central Afar region, Ethiopia; etym. name of nearby rivers) History and general description What is now referred to as the Woranso-Mille study area was identified as a potential source of hominin fossils by the **International Afar Research Expedition** (or I.A.R.E.)

in the early 1970s, but it was first intensively investigated in 2003 by the **Woranso-Mille Paleontological Research Project**. Four collection areas [Am-Ado (AMA), Aralee Issie (ARI), Mesgid Dora (MSD), and Makah Mera (MKM)] extending *c.*5 km/3 miles along the north bank of the Mille River in the northwest of the study area have yielded more than 4300 vertebrate fossils including 26 dentognathic hominin specimens; the **KSD-VP-1/1** associated skeleton comes from a locality in another part of the study area. Temporal span and how dated? **Potassium-argon dating** and **magnetostratigraphy**. Most of the hominin fossils are bracketed by the overlying 3.57±0.014 Ma Kilaytoli Tuff (KT) and an underlying 3.82±0.18 Ma **basalt**. Other hominins are just above, within, or just below the 3.72±0.03 Ma Arala Issie Tuff (AT). Hominins found at site The initial collection of 26 dentognathic hominin specimens included fragments of mandibular corpus and maxillae some with teeth, an associated partial mandibular and maxillary dentition, and isolated teeth. The mandibular and dental morphology and the dental metrics are mostly in the zone of overlap between the **Allia Bay** part of the hypodigm of *Australopithecus anamensis* and the **Laetoli** part of the hypodigm of *Australopithecus afarensis*. An associated skeleton assigned to *Au. afarensis* was reported separately. Archeological evidence found at site None. Key references: Geology, dating, and paleoenvironment Haile-Selassie et al. 2007, Deino et al. 2010; Hominins Haile-Selassie et al. 2010a, 2010b; Archeology N/A.

word In any given modern human language a word consists of two components: phonemes, the basic units of sounds, and morphemes, the basic units of meaning, composed of phonemes. In written languages phonemes are represented by letters and morphemes are the combinations of these letters into a meaningful unit called a word (e.g., the sounds or phonemes /k/a/t/ = the morpheme *cat*). There is a long-standing debate as to whether or not words are unique to modern human language and whether or not they exist in the communication systems of other primates (Cheney and Seyfarth 1992, Deacon 1997, Hauser et al. 2002, Pinker and Jakendoff 2005). For example Cheney and Seyfarth (1992) have argued that alarm calls in vervet monkeys function much like words in human language. They note that vervet alarm calls are referential; that is, like words, they point to specific referents (i.e., things out in the world) in their absence. However, Deacon (1997) and others (e.g., Pinker and Jakendoff 2005) have argued that words do more than point to things in the world such as predators, they also point to symbols or to other words in the lexicon. Furthermore, Pinker and Jakendoff (2005) noted that words are compositional and generative. That is, elements of words (phonemes) can be combined recursively to generate new words and meanings. A word's meaning may change not only depending on the audience, but its linguistic function may change with respect to the same audience (e.g., from a noun as in "wave the *flag*" to a verb as in "*flag* the taxi"). None of these features characterize the calls or songs reported in nonhuman animals. *See also* **language**.

working memory A common way of referring to short-term, or temporary, **memory**. In research, working memory has been studied in relation to a modern human's or nonhuman animal's **memory span**. Working memory is regarded as a fundamental component of the cognitive system for one's ability to recall immediately what just happened is required to adaptively respond to any situation. For this reason, short-term memory is believed play a vital role in everyday decision-making. Coolidge and Wynn (2007) have argued that working memory is unique to modern humans. However, there is ample evidence from cognitive science and neuroscience that this perspective should be rejected (Tomasello and Call 1997). Nonetheless, there is evidence to suggest there are important differences in the working memory systems of modern human and nonhuman primates. For example, a recent study demonstrated that young chimpanzees outperformed modern human subjects in a working memory visual-spatial task (Inoue and Matsuzawa 2007). The authors argued that language, not any uniquely modern human or ape memory system, modulated performance and contributed to the species difference in performance.

working side *See* **chewing**.

WORMIL Acronym for the Woranso-Mille study area, Central Afar, Ethiopia. *See* **Woranso-Mille study area**.

woven bone *See* **bone**.

Wu, Xinzhi [吴新智] **(1928–)** Wu Xinzhi was born in Hefei, Anhui Province, in eastern China. He graduated from the senior high school affiliated with Wuhan University, in Hubei Province, in 1946. He briefly studied at Tongji University in Shanghai from 1946 to 1947, but in 1947 he entered the Shanghai Medical College to study medicine. While there Wu also studied anatomy at the Ministry of Health from 1952 to 1953. Immediately after completing his medical degree from the Shanghai Medical College in 1953, Wu accepted a position as a research assistant at the Dalian Medical College, in Dalian, Liaoning Province, where he stayed until 1957. Later that year, however, Wu entered the graduate school of the **Institute of Vertebrate Paleontology and Paleoanthropology** (or IVPP) at the Chinese Academy of Sciences in Beijing. By the time Wu arrived at the IVPP it had become an important institutional center for paleontological, archeological, and paleoanthropological research in China. Wu completed his graduate degree in 1961 and later that year became an assistant researcher at the IVPP. During the 1960s Wu conducted research on the *Homo erectus* fossils of China, and he also became interested in the differences among fossil humans in Asia. This interest in geographical variation in early *Homo sapiens* and *H. erectus* in Asia can be linked to the polycentric theory of human evolution proposed by Franz **Weidenreich**, who had served as the director of the Cenozoic Research Laboratory from 1935 until 1941. Wu was among the scientists who studied the so-called **Dali** cranium, which was discovered in 1978 near Jiefang Village, Shaanxi Province. Wu came to consider this specimen represented an archaic *H. sapiens* with cranial dimensions and morphological features intermediate between *H. erectus* and *H. sapiens*. In 1984 Wu joined Milford **Wolpoff** and Alan Thorne in advocating a theory they called the **multiregional hypothesis**, which built upon Franz Weidenreich's earlier suggestion that there had been a long period of continuous evolutionary development in Asia where *H. erectus* populations had developed geographical morphological characteristics and that these populations had evolved into modern *H. sapiens* populations that displayed similar regional morphological characteristics. Throughout the 1980 and 1990s Wu conducted research into human evolution in China that focused on the relationship of *H. erectus* in Asia to later populations of *H. sapiens*. Wu collaborated with American physical anthropologist Frank Poirier on *Human Evolution in China: a Metric Description of the Fossils and a Review of the Sites* (1995), which offered a detailed and comprehensive review of the hominin fossils and archeological remains from China. This work argued for the indigenous evolution of modern humans in Chinese from Asian populations of *H. erectus*, and rejected the **out-of-Africa hypothesis**. In 1998 Wu published a paper, in Chinese, in *Acta Anthropologia Sinica* that argued for continuities in cranial and dental traits from archaic *H. sapiens* in China to modern *H. sapiens* in China, as well as from archeological evidence, for a long period of continuous evolutionary change within China, with limited gene flow between this population and other human populations outside of China (NB: he did concede that in other parts of the world there may have been more replacement than continuity). He characterized this theory as "continuity with hybridization," which he presented as a submodel of the multiregional hypothesis. Wu reiterated his support for the multiregional hypothesis in a number of articles, in both English and Chinese, including "On the origins of modern humans in China" published in *Quaternary International* in 2003. He has also published criticisms of the evidence from molecular anthropology that appears to support the out-of-Africa hypothesis as it relates to human populations in China. Besides being a strong advocate of, and an important contributor to, the literature supporting the multiregional hypothesis, Wu continued to conduct basic research on newly discovered hominin fossils in China and has published broadly on paleoanthropology. He published, in collaboration with Weiwen Huang and Guoqin Qi, an important book on Chinese prehistoric sites titled 中国古人类遗址 (*Paleolithic Sites in China*) in 1999. Wu has spent his entire career at the Institute of Vertebrate Paleontology and Paleoanthropology where he served as a researcher and later as the IVPP's vice director. In 1986 he was appointed a board member of the Chinese Society for Anatomical Sciences, later becoming its Vice Chairman and the Director of the Society's professional committee on anthropology. He was awarded a prize by the Chinese Academy of Sciences in 1991 for his research into Chinese paleoanthropology and in 1999 he was elected an Academician of the Chinese Academy of Sciences.

Würm *See* **glacial cycles**.

Wuyun *See* **Bubing Basin**.

X

X chromosome One of the sex chromosomes in modern humans, the great apes, and in most mammals. Normal males possess one X chromosome; normal females possess two X chromosomes. *See also* **Y chromosome**.

X chromosome inactivation X chromosome inactivation (also known as Lyonization) is the process where one (i.e., one copy) of the X chromosomes in females is packaged into heterochromatin such that all, or most, of its gene expression is silenced. This results in dosage compensation so that females do not have twice as much X chromosome gene expression as males. In female mammals, one X chromosome is randomly inactivated early in embryonic development through the initiation of the X inactivation center (Xic). Since inactivation is random, different areas of the body may have different active X chromosomes. If different alleles are present on each X chromosome in a female, this can result in a "patchy phenotype" (e.g., hair color in calico cats).

X-linked X-linked refers to any **trait**, the alleles for which are located on the X chromosome. These traits can be dominant (in which case, the trait will be present in the phenotype whether one or two alleles for that trait are present in females) or recessive (in which case, the trait will only be present in females if two alleles for the trait are present). Because males normally have only one X chromosome (i.e., they are **hemizygous**) their phenotype will be influenced by whichever allele is present at the locus for a trait.

xeric (Gk *xeros* = dry) An **environment** or **habitat** that is dry.

Xujiayao (许家窑) (Location 40°06′N, 113° 59′E, northern China; etym. named for the local village) History and general description This open-air site, discovered in 1974, is located on the west bank of Liyigou, a small tributary of the Sanggan River about 1 km/0.6 miles southeast of Xujiayao village and 4 km/2.5 miles south of the town of Gucheng, Yanggao County, Shanxi Province. The site is represented by two localities (74093 and 73113), with the majority of the archeological materials recovered from Locality 74093. Excavations in 1976, 1977, and 1979 by the **Institute of Vertebrate Paleontology and Paleoanthropology** under the direction of Professor Jia Lan Po produced large assemblages of hominin fossils, stone artifacts, and vertebrate fossils. These collections make Xujiayao one of the most productive and important (but least-discussed) paleoanthropological sites in mainland East Asia. Temporal span and how dated? According to the original stratigraphic profile, fossils and lithic artifacts are concentrated 8–12 m below the topsoil. **Uranium-series dating** on rhinoceros teeth from a depth of 8 m yields an age range of 125–104 ka. However, a recent geomagnetic study (Løvlie et al. 2001) suggests that the fossil deposits may correspond with the early **Brunhes**, indicating that Xujiayao is coeval with **Zhoukoudian Locality 1**. A program of optically stimulated **luminescence dating** is currently being planned to determine a more reliable age range. Hominins found at site The Xujiayao hominin collection consists of a juvenile maxillary fragment with I^1 (erupting), C (unerupted), M^1 (newly erupted), and M^2 (erupting), 12 parietal fragments from individuals of different ages, two occipital bones, a temporal bone, a mandibular fragment, and two isolated molars; at least 11 individuals may be represented in the collection. Archeological evidence found at site More than 13,500 stone artifacts have been reported. These include polyhedral cores, **scrapers**, **points**, **anvils**, gravers, **choppers**, **spheroids** (potential bolas balls), and other multi-functional implements made from vein **quartz** (65%), **chert**, agate, **quartzite**, metamorphic limestone, and other siliceous and igneous rocks. Most tools are small – it is hypothesized that Xujiayao may constitute one of the principal forerunners of the North Chinese **microlith**ic tradition – and over half show evidence of secondary retouch. A taphonomic study conducted by Norton and Gao (2008) on percussion and tooth-mark frequency and **cutmark** patterns indicate that the Xujiayao hominins had regular primary access to intact **artiodactyl**s and **equid** long bones, and thereby procured a regular supply of protein and fat. Repository Institute of Vertebrate Paleontology and Paleoanthropology (original), Beijing, China. Key references: Geology, dating, and paleoenvironment Chen et al. 1984, Løvlie et al. 2001; Hominins Jia et al. 1979, Wu 1980, 1986, Wu and Olsen 1985, Pope 1992, Wu and Poirier 1995, Brown 2001; Archeology Jia et al. 1979, Wu and Olsen 1985, Norton and Gao 2008.

Y

Y chromosome One of the sex chromosomes in modern humans, the great apes, and in most mammals. Normal males possess one Y chromosome, whereas normal females possess no Y chromosomes. The Y chromosome is much smaller than the **X chromosome** and only the tips of the Y chromosome (known as the pseudoautosomal regions) undergo **recombination** with the X chromosome during **meiosis**. The non-recombining portion of the Y chromosome contains the *SRY* gene which is the sex-determining gene. The non-recombining portion of the Y chromosome has been examined extensively to investigate modern human population history worldwide (e.g., Hammer et al. 1997, 1998, 2001, Underhill et al. 2000, 2001, Karafet et al. 2008) as well as nonhuman primate population history (e.g., Burrows and Ryder 1997, Stone et al. 2002, Tosi et al. 2002). Hundreds of studies of the modern human Y chromosome have dealt with the geographic origin (Africa) and subsequent global dispersal of modern human males, thereby providing a male-based scenario complementary to the micro-evolutionary historical reconstruction yielded by female-based data (e.g., **mitochondrial DNA**). *See also* **X chromosome**.

Y-linked The term Y-linked refers to any **trait**, the alleles for which are located on the Y chromosome. The Y chromosome is small and contains few genes, thus Y-linked traits are rare and because only males normally possess Y chromosomes they are passed from fathers to sons.

Ya5 One of the subfamilies of Alu repeat elements that are most active in modern humans. *See* **Alu repeat elements**.

Ya8 One of the subfamilies of Alu repeat elements that are most active in modern humans. *See* **Alu repeat elements**.

Yamashita-cho *See* **Ryukyu Islands**.

YDS Acronym for Younger Dryas stadial. *See* **Younger Dryas**.

Young's modulus *See* **elastic modulus**.

Younger Dryas The Younger Dryas was a cold interval in the North Atlantic between 12.8 and 11.5 ka that briefly reversed the warming that took place after the termination of the **Last Glacial Maximum**. It is named for the occurrence of a cold-tolerant plant species in Scandinavia, the alpine/tundra wildflower *Dryas octopetala*. Ice rafting in the North Atlantic suggests that it may be considered as the most recent **Heinrich event**. Cold **sea surface temperatures** in the North Atlantic and a brief resumption of a glacial-type climate had implications for African precipitation patterns, generally indicating drying. Such is the complexity of the global climate systems that the Younger Dryas in the northern hemisphere coincided with a warm period in the southern hemisphere such that glaciers retreated in what is now New Zealand.

Yuanmou (Location 25°40′N, 101°53′E, China; etym. unknown) History and general description This site is in the Yuanmou Basin, 110 km/66 miles northwest of Kunming. The two original hominin incisors were recovered in 1965 from the Yuanmou Formation. Temporal span and how dated? Early estimates of the antiquity of this site suggested it may be as old as *c.*1.7 Ma, but a careful assessment of the magnetic polarity stratigraphy suggests that the predominant normal polarity of the Yuanmou Formation most likely corresponds to early in the Brunhes chron, *c.*0.7 Ma, and equivalent to **Oxygen Isotope Stage** 17. This is also consistent with the **electron spin resonance spectroscopy dating** estimates of Huang and Grün (1998). Hominins found at site Two incisors that broadly resemble those from *Homo erectus* sites. Archeological evidence found at site None. Key references: Geology, dating, and paleoenvironment Li et al. 1976, Cheng

Wiley-Blackwell Encyclopedia of Human Evolution, First Edition. Edited by Bernard Wood.
© 2013 Blackwell Publishing Ltd. Published 2013 by Blackwell Publishing Ltd.

et al. 1977, Jiang et al. 1989, Huang and Grün 1998, Hyodo et al. 2002, Hominins Hu 1973, Zhou and Hu 1979, Qian and Zhou 1991; Archeology N/A.

Yunxian Quyuanhekou (郧县曲远河口)
(Location 32°50′23.97″N, 110°34′42.35″E, central China; etym. named after the local county and the confluence of local rivers) History and general description An open-air site located at Xuetang Liangzi (a known **Paleolithic** locality) just northwest of Mituosi village and the confluence of Quyuan and Han Rivers, or Quyuanhekou (Quyuan River Mouth), Yunxian County, Hubei Province. Collaboration between Chinese and French researchers has thus far resulted in the completion of a **computed tomography** (or CT) prototype of Yunxian II in 2004. Although the reconstruction was principally based on EV 9002, morphometric data from nine Asian *Homo erectus* casts (**Zhoukoudian** III, X, XI, and XII, **Hexian**, **Sangiran 17**, and **Ngandong** 6, 7, and 12) were also referenced. The new cast suggests that (a) the cranial capacity of Yunxian II is within the range of *H. erectus* variability and (b) the mid-facial morphology is likely not anatomically modern-like. Temporal span and how dated? **Electron spin resonance spectroscopy dating** of nine stratigraphically associated animal teeth produces a mean early-uptake model age of 581±93 ka and a mean late-uptake model age of 1051±150 ka. These dates contradict the evidence from **magnetostratigraphy** that points to an age of 870–830 ka. At present, a mid-**Middle Pleistocene** age for the Yunxian hominins it the most reasonable estimate. Hominins found at site Two large, presumably male, and badly crushed adult crania (EV 9001 and EV 9002), discovered in 1989 and 1990, respectively. Archeological evidence found at site Some 300 artifacts have been recovered from the site. These include large cores, uni- and bifacial tools worked on pebbles and large cobbles (e.g., **choppers**), **flakes**, and flake fragments. The raw material is predominately (70%) **quartz** and **quartzite**. Repository Hubei Institute of Cultural Relics and Archaeology (original and CT prototype), Wuhan, China. Key references: Geology, dating, and paleoenvironment Chen et al. 1997, Li and Feng 2001; Hominins Li and Etler 1992, Pope 1992, Wu and Poirier 1995, Zhang 1998, Brown 2001, Li and Feng 2001, Vialet et al. 2005; Archeology Li et al. 1998, Li and Feng 2001.

Z

Zafarraya (Location 36°57′05″N, 4°07′36″W, Spain; etym. local name for the basin) <u>History and general description</u> A **karstic** cave located in the northeast of the Málaga province. The cave opening is located about 1100 m above sea level on the southern side of the Sierra de Alhama, and has two distinct areas, the entrance and the cave. The true cave is inclined and is divided into three parts: the entrance hall, the back hall, and the diverticulum at the rear. The excavations of Barroso-Luiz and Medina-Lara from 1980 to 1983 defined five main layers (A–E) near the entrance to the cave. Subsequent work in the early 1990s concentrated on extending the excavations further into the cave. However, the sediments in the interior proved to be highly homogenous, forcing the workers to create artificial levels at 5 cm intervals (I1, I2, I3, etc.). <u>Temporal span and how dated?</u> A **uranium-series dating** age of 33.4±2 ka was given for an ibex tooth associated with the hominin remains in level D. A tooth from level I8 was found to have a uranium-series date of 31.7±3.6 ka, and a **radiocarbon dating** age of 31.8±0.5 ka. The mean of three uranium-thorium dates of other samples was *c*.26.0 ka BP, whereas its radiocarbon equivalent was 29.8±0.6 ka BP. <u>Hominins found at site</u> The first hominin remains from this site were discovered in 1986 by García Sánchez in the lower layers D and E. Most were concentrated near the entrance to the cave, largely in the same level near the bottom of the excavation and centralized around an apparent fireplace. Both adult and juvenile specimens were found, including a well-preserved *Homo neanderthalensis* mandible. <u>Archeological evidence found at site</u> A typical **Mousterian** lithic assemblage was recovered. However, the scarcity of secondary products and debitage in the assemblage imply that little tool production was done within the cave. The hominin remains were found around a well-defined **hearth**. *Capra pyraineica* (i.e., ibex) makes up more than 90% of the identified faunal assemblage. Key references: <u>Geology, dating, and paleoenvironment</u> Hublin et al.

1995; <u>Hominins</u> Garcia Sánchez 1986, Hublin et al. 1995, Sanchez 1999; <u>Archeology</u> Barroso-Ruiz et al. 1984, Hublin et al. 1995.

Zdansky, Otto (1894–1988) A colleague of John Gunnar **Andersson** who carried out excavations at Locality 1 at Choukoutien (now called **Zhoukoudian**) in 1921 and 1923. In 1926, while working in Professor Wiman's laboratory in Uppsala, Sweden, examining material excavated from what is now called the Lower Cave at Zhoukoudian, Zdansky recognized a hominin lower premolar and an upper molar tooth, the first of what were to be many fossil hominins recovered from Zhoukoudian. *See also Homo erectus*; **Zhoukoudian**; **Zhoukoudian Locality 1**.

zeugopod (Gk *zeug* = join and *pod* = foot) The intermediate of the three compartments of a tetrapod limb. The proximal compartment is the **stylopod** (the upper arm or the thigh), the intermediate one is the zeugopod (the forearm or the lower leg), and the **distal** one is the **autopod** (the hand or the foot).

Zhiren 3 <u>Site</u> Zhirendong. <u>Locality</u> N/A. <u>Surface/in situ</u> In situ. <u>Date of discovery</u> May 2008. <u>Finder</u> N/A. <u>Unit</u> B. <u>Horizon</u> Layer 2. <u>Bed/member</u> N/A. <u>Formation</u> N/A. <u>Group</u> N/A. <u>Nearest overlying dated horizon</u> Flowstone level above, dated to 106.2±6.7 ka. <u>Nearest underlying dated horizon</u> N/A. <u>Geological age</u> *c*.110 ka. <u>Developmental age</u> Late adolescent/young adult. <u>Presumed sex</u> Unknown. <u>Brief anatomical description</u> Edentulous mandibular fragment extending from middle of the right P_4 alveolus to the distal portion of the left P_4 alveolus; there is damage to both P_4 alveoli and enlargement of both P_3 alveoli. While the morphology of the chin appears similar to a modern human, aspects of the mandibular corpus are more similar to Late Pleistocene archaic hominins. <u>Announcement</u> Jin et al.

Wiley-Blackwell Encyclopedia of Human Evolution, First Edition. Edited by Bernard Wood.
© 2013 Blackwell Publishing Ltd. Published 2013 by Blackwell Publishing Ltd.

2009. Initial description Jin et al. 2009. Photographs/line drawings and metrical data Liu et al. 2010, Jin et al. 2009. Detailed anatomical description Liu et al. 2010. Initial taxonomic allocation *Homo sapiens*. Taxonomic revisions N/A. Current convention taxonomic allocation *H. sapiens*. Informal taxonomic category Modern human. Significance If dates are secure, hominin remains at this site represent the earliest evidence of modern humans in East Asia. The so-called "mosaic" features of the gnathic remains have cited as evidence of admixture between anatomically modern and archaic *H. sapiens* populations. Location of original Institute of Vertebrate Paleontology and Paleoanthropology, Chinese Academy of Sciences, Beijing, China.

Zhiren Cave   *See* Zhirendong.

Zhirendong   (Location 22°17′13.6″N, 107°30′45.1″E, 2 km/1.2 miles northwest of Chongzuo Eco-Park within the Mulanshan, Chongzuo City, Guangxi Zhuang Autonomous Region, southern China) History and general description In November 2007 researchers from the Institute of Vertebrate Paleontology and Paleoanthropology and the Chongzuo Biodiversity Research Center of Peking University recovered a collection of mammalian fossils that included evidence of anatomically modern *Homo sapiens*. If dates at this site are secure, they represent the oldest evidence of modern humans in East Asia. Temporal span and how dated? Biostratigraphy, uranium-series dating, and stratigraphic correlation to Oxygen Isotope Stages suggest an age of *c*.110 ka. Hominins found at site The fragmentary hominin remains consist of gnathic and dental evidence, most notably Zhiren 3. Archeological evidence found at site None reported. Key references: Geology, dating, and paleoenvironment Liu et al. 2010, Jin et al. 2009; Hominins Liu et al. 2010; Archeology N/A.

Zhongshan   *See* Bubing Basin.

Zhoukoudian   History and general description The most productive fossil hominin locality in China, Zhoukoudian (former transliteration: Choukoutien) was recognized as a UNESCO World Heritage Site in 1987. The cave complex that produced both the "Peking Man" (originally attributed to the new hominin genus and species *Sinanthropus pekinensis*, now referred to *Homo erectus*) and the

"Upper Cave" fossils is located approximately 42 km/26 miles south of Beijing. In 1918, the Swedish geologist John Gunnar Andersson discovered the site having been led to the nearby "Chicken Bone Hill" by villagers who knew he was looking for fossils. In 1921, Andersson and Otto Zdansky, an Austrian paleontologist, commenced excavations on the neighboring "Dragon Bone Hill" after quarrymen showed them exotic quartz found during mining operations. Among the fossils that were excavated from there in 1923 Zdansky recognized two isolated teeth as potentially hominin and in 1926 an isolated lower first molar was sent to Davidson Black for analysis. Black, a Canadian chairing the anatomy department at the Peking Union Medical College, appreciated the specimen's importance and uniqueness and within a year Black had formally proposed a new taxon name *Sinanthropus pekinensis* for what he judged to be a new species. Large-scale excavations overseen by Black started in 1927 and continued after his death in 1934. In 1928 the first hominin cranium was recovered from Zhoukoudian, followed in the next 2 years by discoveries of the first stone tools and ash layers containing burned bone that were interpreted as evidence of the controlled use of fire. After Black's death, excavations were directed by his successor, German anthropologist Franz Weidenreich, until they were halted in 1937 by the outbreak of the Second Sino-Japanese War. Nearly 200 *H. erectus* fossils representing up to 45 individuals were recovered from Zhoukoudian from 1921 to 1966. In addition, tens of thousands of stone artifacts and fossils representing at least 98 nonhuman mammalian species have been produced by these excavations. All of the *H. erectus* fossils derive from Zhoukoudian Locality 1 in the Lower Cave, an approximately 40 m-thick depositional sequence that has been divided into layers. These are numbered 1–17 with Layer 1 being the highest and the youngest stratum and 17 the deepest and the oldest; most of the hominin fossils derive from Layers 8–9 and 3–4. The Upper Cave at Zhoukoudian was discovered in 1930 and excavations from 1933 to 1934 produced three skulls; the remains from at least eight individuals have been attributed to modern *Homo sapiens* and date to the Upper Pleistocene. Temporal span and how dated? *See* Zhoukoudian Locality 1; Zhoukoudian Upper Cave. Hominins found at site The history of discovery of the Zhoukoudian *H. erectus* fossils is fascinating, but fraught with tragedy and loss. Fortunately, Weidenreich was an extraordinarily meticulous and conscientious scientist who photographed, casted, and described in fine detail all of the hominin fossils discovered at Zhoukoudian. He produced four monographs on the *Sinanthropus* fossils and they remain the definitive descriptions of these fossils. Paleoanthropology owes Weidenreich a substantial debt because all of the "Peking Man" fossils became yet another casualty of WWII. Entrusted to a US Marine contingent tasked with escorting the fossils safely out of occupied China, the American servicemen had the bad luck to arrive at the Port of Beijing on December 7, 1941. With the US now at war with Japan following that day's attack on Pearl Harbor, the Marines were imprisoned as prisoners of war and the "Peking Man" fossils were never to be seen again. Archeological evidence found at site Tens of thousands of quartz, flint, and sandstone tools have been found at Zhoukoudian Locality 1. Most can be assigned to one of four primary tool types (choppers, scrapers, points, and burins) that were manufactured by a variety of hard-hammer percussion techniques. Initial interpretations of *H. erectus*' behavior at Zhoukoudian suggested that the site was a home base where hunted animals were consumed, tools were made, and fire was controlled. However, a more recent, taphonomically informed study suggests that giant cave hyenas were the primary resident of the Lower Cave and that *H. erectus* could be best characterized as a transient scavenger who used (but did not necessarily control) fire during their more sporadic and ephemeral occupations of the site. Key references: *see* **Zhoukoudian Locality 1; Zhoukoudian Upper Cave**. *See also* Davidson **Black**; John Gunnar **Andersson**; Otto **Zdansky**.

Zhoukoudian Locality 1 (周口店第一地点)

(Location 39°41′18.77″N, 115°55′32.50″E, northern China; etym. named after the local village) History and general description Although it is referred to as "Locality 1" it is effectively a fossil site in its own right. The site is situated on the northeastern slope of Longgushan (Dragon Bone Hill) near Zhoukoudian village, 50 km / 31 miles southwest of Beijing Municipality. The site, which is referred to as "Locality 53" in John Gunnar **Andersson**'s field notes, is subdivided into three, areas, the "Main Deposit," the "Lower Fissure," and the "Kotzetang." Sporadic excavations began in 1921, but large-scale excavations did not commence at Locality 1 until 1927 and they continued until the Second Sino-Japanese war broke out in 1937. Locality 1 has been made unnecessarily confusing because of the proliferation of systems used to describe it (e.g., loci, layers, and levels). The loci identified by a capital letter are effectively spatially bound "localities" within Locality 1 where at least one hominin specimen has been found (e.g., two calvariae have been found at Locus L, which in Davidson **Black**'s terminology are "Skull I" and "Skull II"). Both layers and levels use arabic numerals (e.g., Layer 8 is equivalent to Level 22). The layers are numbered 1–17 with Layer 1 being the highest and the youngest stratum and 17 the deepest and the oldest; most of the hominin fossils derive from Layers 8–9 and 3–4. Temporal span and how dated? Locality 1 is divided into 17 stratigraphic horizons. Layers 1–13 are considered to be culture- and fossil-bearing layers. Until very recently it was thought that Locality 1 in the Lower Cave was entirely **Middle Pleistocene** in age. An age range of *c*.500–230 ka for hominin occupation at Zhoukoudian's Lower Cave was supported by a number of studies published from the late 1970s through the late 1990s employing **fission-track dating, uranium-series dating**, and **electron spin resonance spectroscopy dating**. A review by Zhou Ming Zhen and Ho Chuan Kun (1990) suggests an age range between 520±61 ka (for Layer 11) and 230+30/−23 ka (for Layer 4). However, because of the large margins of error in the methods used, most dates are likely not very reliable. Renewed efforts in the last decade using higher-precision techniques and refined correlations to deep-sea **Oxygen Isotopic Stages** and **loess** sequences indicate that the layers of interest are older than previously thought. **Thermal ionization mass spectrometry uranium-series dating** of Layer 3, which yielded Skull V, are *c*.500–400 ka and the dates for Layers 8–11, which yielded Skulls II, III, X, XI, and XII, are more than 600 ka. Likewise, a recent **cosmogenic nuclide dating** study suggests an age of 770±80 ka for Layers 7–10. Together with climatic correlations, the hominin-bearing layers are now bracketed between 780 and 400 ka. This means that **Hexian** (437–387 ka) and **Tangshan Huludong** (*c*.620 ka) are **coeval** with some of the later Zhoukoudian hominins. Hominins found at site Before full-scale excavation commenced at Locality 1, Otto **Zdansky** and Davidson Black recovered three teeth. Between 1927 and 1937, five calvariae (Skulls II, III, X, XI, and XII) and other cranial fragments belonging to 13 individuals, 15 mandibular fragments belonging to 14 individuals, two humeral fragments, seven femoral fragments, one left clavicular fragment, a lunate, and

147 teeth belonging to at least six individuals were unearthed. Between 1949 and 1966, additional specimens were found either during excavations (teeth and craniognathic fragments) or in older, pre-WWII collections (a humeral fragment, a tibial fragment, and a right lower molar). Altogether, up to 40 individuals may be represented in the Zhoukoudian Locality 1 hominin (aka "Peking Man") collection. The *H. erectus* fossils recovered from the Lower Cave at Zhoukoudian were each assigned alphanumeric designations reflecting the location (= "Locus") and sequence of their discovery. In addition, Weidenreich proposed a new numbering scheme for the cranial (Weidenreich 1943) and postcranial (Weidenreich 1941b) remains. For the cranial (i.e., the calvariae and facial fragments) he replaced the informal specimen numbers used by the people who discovered and initially described the fossils (see Weidenreich 1943, Table 1, p. 5). For example, the first calvaria discovered in 1929 was informally known as "D I" (having been the first hominin recovered from Locus D in Layers 8–9), but in Weidenreich's scheme it is referred to as "Skull II." The other five most complete calvaria are designated as follows: "E I" (from Locus E, Layer 11, discovered in 1929) is "Skull III"; "H III" (from Locus H, Layer 3, discovered in 1934 and previously referred to by Weidenreich as "Skull III") is "Skull V"; "L I" (from Locus L, Layers 8–9, discovered in 1936 and previously referred to as "Skull I Locus L") is "Skull X"; "L II" (from Locus L, Layers 8–9, discovered in 1936 and previously referred to as "Skull II Locus L") is "Skull XI;" and "L III" (from Locus L, Layers 8–9, discovered in 1936 and previously referred to as "Skull III Locus L") is "Skull XII" (NB: in the section on "Choukoutien, Locality 1" in Oakley et al. 1975 Susan Limbrey uses arabic instead of roman numerals, as in "D 1," "E 1," etc., and uses the prefix "Upp." for the two teeth sent back to Sweden and "Ckn." for the fossils that stayed in China). Weidenreich used the same type of scheme for the postcranial remains, with separate sequences of Roman numerals used for the femora, the two humeral remains, and for the clavicle and the lunate (see Weidenreich 1941b, Table 1, p. 6). Archeological evidence found at site Large numbers of artifacts and debitage have been collected from Layers 1–13. **Quartz** (89%) or vein quartz, is the primary raw material used. Most of the tools are of **Mode 1** type and standardization is poor. Zooarcheological analyses and the presence of hyena coprolites suggest that *H. erectus* and carnivores may have competed

with each other for food and shelter in the cave. As early as the 1930s, evidence of hearth and ash deposits (e.g., from Layer 10) has been noted; however, recent investigation failed to find features of *in situ* burning or direct evidence for use of fire by *H. erectus*. Key references: Geology, dating, and paleoenvironment Wu and Wang 1985, Zhou and Ho 1990, Grün et al. 1997a, Zhou et al. 2000, Goldberg et al. 2001, Shen et al. 2001, 2009; Hominins Black 1927, Weidenreich 1936a, 1936b, 1937, 1939c, 1941b, 1943, Wu and Dong 1985, Wu and Olsen 1985, Pope 1992, Wu and Poirier 1995, Tattersall and Sawyer 1996, Holloway 2000, Brown 2001, Wu et al. 2010; Archeology Teilhard de Chardin and Pei 1932, Pei and Zhang 1985, Wu and Olsen 1985, Zhang 1985, Binford and Stone 1986, Wu and Poirier 1995, Weiner 1998, Wu et al. 1999, Boaz et al. 2004. A concise history of the initial phase of fieldwork (i.e., exploration and subsequent excavations) at Zhoukoudian Locality 1 can be found in Black (1934). *See also* **Zhoukoudian Locality 1 hominins.**

Zhoukoudian Locality 1 hominins Site

Zhoukoudian Locality Locality 1. Surface/*in situ* *In situ.* Date of discovery 1923–66. Finders The hominins were recovered by unnamed excavators. Unit N/A. Horizon Most hominins come from Layers 8–9 and 3–4. Bed/member N/A. Formation N/A. Group N/A. Geological age A review by Zhou Ming Zhen and Ho Chuan Kun (1990) suggests an age range between 520 ± 61 ka (for Layer 11) and $230 + 30/-23$ ka (for Layer 4). However, because of the large margins of error in the methods used, most dates are likely not very reliable. The combination of higher precision dating techniques and correlations with deep-sea **Oxygen Isotopic Stages** and **loess** sequences suggests the hominin-bearing layers are now bracketed between 780 and 400 ka. Developmental age Both adults and juveniles are represented in the collection. Presumed sex Sex is largely assigned on the basis of the overall size of the specimens. Brief anatomical description There are several well-preserved calvariae (e.g., Skulls II, III, X, XI, and XII), isolated cranial bones, maxillary and mandibular fragments (some with teeth) (e.g., Maxillae III, V, and VI; Mandibles B I, G I), isolated teeth, and some postcranial fossils (e.g., Femora I and IV and Humerus II). The endocranial volumes of the calvariae range from 915 cm^3 (Skull III) to 1225 cm^3 (Skull V). Announcements and initial descriptions Most of the fossils recovered in the 1920s were announced by Davidson **Black** (see Black 1934 for references). Photographs/line drawings, metrical data, and detailed anatomical description The main Locality 1 collection is well illustrated and meticulously described in the monographs published by Franz **Weidenreich** (1936a, 1936b, 1937, 1939c, 1941b, 1943). Initial taxonomic allocation *Sinanthropus pekinensis*. Taxonomic revisions *Homo erectus* (Weidenreich 1940). Current conventional taxonomic allocation *H. erectus*. Informal taxonomic category Pre-modern *Homo*. Significance There has never been any serious suggestion that the hominins recovered from Locality 1 sample a taxon other than *H. erectus* and as such they comprise the single largest site collection of that taxon. Location of originals or surviving casts University of Uppsala, Sweden (two teeth recovered by Otto Zdansky); **American Museum of Natural History**, New York, USA (casts of the 1927–37 materials); **Institute of Vertebrate Paleontology and Paleoanthropology**, Beijing, China (originals of the hominins recovered post-WWII).

Zhoukoudian Upper Cave (Location 39°41′ 18.77″N, 115°55′32.50″E, Beijing, China; etym. named after the local village) History and general description The second, and less well-known, locality at the **Zhoukoudian** site is called Upper Cave (or UC) because it is literally situated above Locality 1 (the latter is sometimes also referred to as Lower Cave). The UC was discovered in 1930 during fieldwork at Locality 1, with excavations taking place in 1933 and 1934. The UC, which is 4 m wide and 8 m deep, comprises an Entrance, Upper Room, Lower Room, and Lower Recess. Five cultural layers were identified: Layers 1–3 in the Entrance and Upper Room and Layers 4 and 5 in the Lower Room. Taphonomic analysis of the UC faunal assemblage suggests that hunter–gatherers were involved with the formation of the animal bones found in the cultural layers. Pei Wenchung suggested that the Lower Recess most likely served as a natural trap, and this hypothesis is supported by the results of taphonomic analyses, the presence of multiple semi-articulated deer skeletons, and the absence of *Celtis* seeds. The reason why UC receives so much attention is the presence of multiple modern human fossils, thought to have been interred, and the archeological evidence of modern human behavior. The evidence for modern human burials is strong, but examination of the casts of the modern human postcranial remains suggests that at least some of the limb bones appeared to have been chewed (epiphyses are missing and the midshafts display multiple toothmarks) by carnivores. A likely scenario is that the modern human remains were interred, but not deeply enough, and carnivores subsequently dug up the burials and gnawed on some of the bones. Temporal span and how dated? The UC deposits have been dated using conventional **radiocarbon dating**, thermo**luminescence dating**, **uranium-series dating**, and more recently **AMS radiocarbon dating**. The chronometric dates from Layer 4 where the modern human burials are reportedly derived are between *c*.34 and 10 ka BP. Although some researchers have argued for an Early **Holocene** age for the modern human burials, the faunal composition, particularly the presence of extinct species (e.g., *Hyaena ultima*), suggests a **Late Pleistocene** age. Assuming this is true, then the presence of warm-adapted species (e.g., *Acinonyx jubatus*) would suggest an **Oxygen Isotope Stage** 3 age of occupation, rather than a warm period during the Holocene, as proponents of a more recent, Neolithic, age have suggested. Hominins found at site The hominin fossils, which represent at least eight individuals, include three relatively intact crania (UC 101, 102, 103), four mandibles, dozens of loose teeth, and an assortment of postcranial fragments. UC 101 has been interpreted to be an aged male, while UC 102 a middle-aged female, and UC 103 a teenage female; all are considered modern human. The UC modern human fossils, particularly the three crania, have been used as evidence, primarily by Chinese paleoanthropologists, of morphological continuity between *Homo erectus* from Zhoukoudian Locality 1 and modern Chinese. However, the results of multivariate analyses (Cunningham and Jantz 2003) suggest that none of the UC crania have particularly strong phenetic ties to modern-day Chinese and instead, their affinities are with European Caucasoid (UC 101), Melanesian (UC 102), and Eskimo (UC 103) populations. Archeological evidence found at site Excavations at UC have yielded small stone flake tools and evidence of what are assumed to be personal adornments (e.g., perforated shell, bone, teeth, and stone). Some of these perforated artifacts were found near the necks of some of the interred humans and they are most likely necklaces. A perforated bone needle suggests a reliance on sewn clothing, and several bones and artifacts were covered with red hematite. Most of the artifacts are associated with the modern human burials from Layer 4, but some were excavated from the upper layers. The presence of sea shells (*Areca* sp.) attests to either a regional trade network or to long-range mobility. Repository Most of the UC materials are stored in the **Institute of Vertebrate Paleontology and Paleoanthropology**, the Zhoukoudian Museum, and the Geological Sciences Museum, Beijing, China. However, as with the Zhoukoudian Locality 1 *H. erectus* fossils, most of the original UC *Homo sapiens* fossils were lost during WWII. Casts of the three modern human crania (UC 101, 102, 103) are housed in many repositories around the world. Key references: Geology, dating, and paleoenvironment Pei 1934, 1940, Wu and Wang 1985, Kamminga and Wright 1988, Chen et al. 1989, Wu and Poirier 1995, Norton and Gao 2008; Hominins: Weidenreich 1939a, 1939b, Howells 1983, Wolpoff et al. 1984, Kamminga and Wright 1988, Wright 1995, Wolpoff 1995, Wu and Poirier 1995, Brown 1998, Turner et al. 2000, Cunningham and Wescott 2002, Cunningham and Jantz 2003, Liu et al. 2006, Harvati 2009; Archeology Pei 1934, 1939, Kamminga and Wright 1988, Kamminga 1992, Wu and Poirier 1995, Norton and Gao 2008, Norton and Jin 2009.

Zinjanthropus Leakey, 1959 (Swa. *Zinj* = East Africa and Gk *anthropos* = human being) Genus established by Louis **Leakey** in 1959 to accommodate fossil hominins recovered in 1955 and 1959 in Bed I, **Olduvai Gorge**, Tanzania. John **Robinson** (1960) suggested it was a junior synonym of *Paranthropus* and Leakey (1963) subsequently sank *Zinjanthropus* into *Australopithecus*.

Zinjanthropus boisei Leakey, 1959 (etym. Swa. *Zinj* = East Africa, Gk *anthropos* = human being, and *boisei* to recognize the substantial help provided to Louis and Mary **Leakey** by Charles **Boise**) Hominin species established by Louis **Leakey** in 1959 to accommodate fossil hominins recovered in 1955 and 1959 in Bed I, **Olduvai Gorge**, Tanzania. Arguably the most distinctive early hominin taxon. The skull resembles an exaggerated version of *Paranthropus robustus*, with ectocranial crests, a massive, flat, broad face, large, robust mandibular corpora, small anterior teeth, **molarized** premolar crowns and roots, and large molar tooth crowns, and there is substantial cranial **sexual dimorphism**. The genus *Zinjanthropus* is now regarded as a junior synonym of either *Australopithecus* or *Paranthropus*. Until the discovery of *Homo habilis* Louis Leakey assumed that *Z. boisei* was the manufacturer of the "Oldowan pre-Chelles-Acheul culture" (Leakey 1959, p. 491). First discovery OH 3 (1955). Holotype OH 5 (1959). Main sites Chesowanja, Konso, Koobi Fora, Malema, Olduvai Gorge, Peninj (Natron),

Omo-Shungura Formation, West Turkana. *See also Zinjanthropus.*

Zlatý Kůň (Location 49°55′N, 14°05′E, Czech Republic; etym. Czech for golden horse) History and general description Cave site in the Bohemian Karst, Central Bohemia, in the Zlatý Kůň Hill, 500 m from Koneprusy village. Hominin remains were found together with Pleistocene fauna and lithic artifacts between 1950 and 1953. Initially thought to represent the early Upper Paleolithic, this specimen has now been redated to the Late Paleolithic (**Magdalenian**). Temporal span and how dated? **Radiocarbon dating** of human bone provided a date of 12,900 years BP. Hominins found at site Zlatý Kůň 1 and 2 are two pieces from the same *Homo sapiens* calvaria. Additional specimens include a zygomatic bone, a mandible, five vertebrae, and three ribs. Archeological evidence found at site Non-diagnostic **Upper Paleolithic** tools were recovered. Key references: Geology, dating, and paleoenvironment Svoboda et al. 2002; Archeology Svoboda 2000; Hominins Vlček 1971, Smith 1982.

zooarcheology (Gk *zoio* = a living being and *arkhaiologia* = the study of antiquity) Zooarcheology is a subdiscipline of archeology concerned with the study of the animal remains from archeological sites, including bone, teeth, antlers, and shells. Zooarcheological data are used to reconstruct hominin diets and subsistence strategies, past environments, and the interactions between people and animal communities. As a result, zooarcheology provides information critical to understanding both hominin evolution and how environments have changed through time. For example, zooarcheological evidence concerning the foraging behavior of **Middle Stone Age** hominins plays a central role in attempts to understand the origins of **modern human behavior**. Some argue that Middle Stone Age hominins exploited animal resources less effectively than their **Later Stone Age** successors, although a growing body of evidence suggests that Middle Stone Age subsistence was essentially similar to that of Later Stone Age hominins and modern human **hunter–gatherers**. *See also* **diet reconstruction**.

Zoologische Staatssammlung München (or ZSM) This museum contains the Zoological Collection of the State of Bavaria, Germany. The collection was begun when the Elector of Bavaria in the mid-18thC decided to start collecting natural history objects. From the beginning of the 19thC the collection was housed in the Wilhelminum, a building owned by the Academy of Science, and its first curator was Johann von Spix, who added his own important South American material to the collection. During WWII the Wilhelminum was bombed, but most of the collection had been transferred to safety in the countryside. After the war a temporary home was found for the collection in the Nymphenburg Castle until a new purpose-built facility was constructed in its present location on Münchhausenstrasse. Parts of a separate collection, the Anthropologische Staatssammlung, including the Selenka Collection, which consists of the skulls of 59 Sumatran and 328 Bornean orangutans of all ages and sexes, were transferred to the Zoologische Staatssammlung in 2000. Contact information E-mail zsm@zsm.mwn.de. Postal address Münchhausenstr. 21, 81247 München, Germany. Website www.zsm. mwn.de. Tel +49 (0)89 8107 0.

Zouhrah *See* **El Harhoura**.

Zuttiyeh (Location 32°51′N, 35°30′E, Israel; etym from Ar. *Mugharet el-Zuttiyeh* = Cave of the Robbers) History and general description F. Turville Petre spent the summer of 1925 excavating this cave site, which is located in the Wadi Amud about 3.5 km/2 miles northwest of the Sea of Galilee. Temporal span and how dated? The nature of the lithic assemblage suggests an age of more than 200 ka. Hominins found at site A nearly complete frontal with part of the left zygomatic. Keith (1927) originally referred to the specimen as **Neanderthaloid**, but more recently authors have moved away from this interpretation, suggesting that the hominin from Zuttiyeh belongs to the population that may have given rise to the later population sampled at sites such as **Skhul**. Archeological evidence found at site The assemblage was originally regarded as being Lower Levantine **Mousterian**, but it is now recognized that the industry at Zuttiyeh is closer to the much older Acheulo-Yabrudian. Key references: Geology, dating, and paleoenvironment N/A; Hominins Turville-Petre 1927, Keith 1927, Sohn and Wolpoff 1993, Zeitoun 2001; Archeology Garrod and Bate 1937, Bar-Yosef 1989.

zygomatic prominence A bulbous protuberance at the anterior end of the zygomatic arch in the region of the zygomaticomaxillary suture. It is one of the features Rak (1983) linked with *Australopithecus africanus* (e.g., **Sts 5**, **Sts 71**) and similar morphology is also seen in *Paranthropus robustus* and in **KNM-WT 17000**; zygomatic prominences contribute to the wide mid-faces of these taxa. The zygomatic prominence marks the more-or-less abrupt transition from the forward-facing facial surface of the zygomatics to the laterally facing surface of the their temporal process. When the zygomatic prominences project anteriorly they contribute to the "dishing" of the face seen in some archaic hominins.

zygomaticoalveolar crest The anatomical term for the inferior margin of the maxilla's zygomatic process, which arises from the alveolar process to join the zygomatic process of the temporal bone, thus completing the zygomatic arch. In the great apes and modern humans the zygomaticoalveolar crest usually arches downward as it leaves the maxilla, but in some archaic hominins (e.g., *Australopithecus africanus*, *Paranthropus robustus*, some *Paranthropus boisei*) it is straight so that, when seen in *norma frontalis*, it angles superolaterally to meet the zygomatic process of the temporal.

zygomaticomaxillary fossa A term used by Clarke (1977) for a hollowed area that spans the inferolateral third of the zygomaxillary suture and interrupts the **zygomaticomaxillary step** in the lateral mid-face of *Paranthropus robustus*.

zygomaticomaxillary step A term used by Rak (1983, p. 32) to describe the morphology of the lateral upper mid-face of *Paranthropus robustus* (it is also seen in **KNM-WT 17000**). The anterior surface of the zygomatic bone of *P. robustus* is forward-facing and projects further anteriorly than the medial part of the mid-face. This results in a "step" that coincides with the zygomaxillary suture. In its lower one-third the zygomaticomaxillary step is interrupted by a transverse groove at right angles to its long axis. In his thesis Ron Clarke (1977) refers to this as the **zygomaticomaxillary fossa**. The zygomaticomaxillary step forms the superolateral boundary of the **maxillary trigon** (Rak 1983, Fig. 16, p. 135).

References

Abbate, E., Albianelli, A., Azzaroli, A., Benvenuti, M., Tesfamariam, B., Bruni, P., et al. (1998) A one-million-year-old *Homo* cranium from the Danakil (Afar) Depression of Eritrea. *Nature* 393: 458–60.

Abbate, E., Woldehaimanot, B., Bruni, P., Falorni, P., Papini, M., Sagri, M., et al. (2004) Geology of the *Homo*-bearing Dandiero Basin (Buia Region, Eritrea Danakil Depression). *Riv. It. Paleont. Strat.* 110: 5–34.

Abbazzi, L., Fanfani, F., Ferretti, M., Rook, L., Cattani, L., Masini, F., et al. (2000) New human remains of archaic *Homo sapiens* and Lower Palaeolithic industries from Visogliano (Duino Aurisina, Trieste, Italy). *Journal of Archaeological Science* 27(12): 1173–86.

Ackermann, R. (2007) Craniofacial variation and developmental divergence in primate and human evolution. In: *Tinkering: the Micro-evolution of Development*, Novartis Foundation Symposium 284, pp. 262–79. Wiley, Chichester.

Ackermann, R. (2009) Morphological integration and the interpretation of fossil hominin diversity. *Evolutionary Biology* 36(1): 149–56.

Ackermann, R., and Bishop, J. (2010) Morphological and molecular evidence reveals recent hybridization between gorilla taxa. *Evolution* 64(1): 271–90.

Ackermann, R., and Cheverud, J. (2000) Phenotypic covariance structure in tamarins (genus *Saguinus*): a comparison of variation patterns using matrix correlation and common principal component analysis. *American Journal of Physical Anthropology* 111(4): 489–501.

Ackermann, R., Rogers, J., and Cheverud, J. (2006) Identifying the morphological signatures of hybridization in primate and human evolution. *Journal of Human Evolution* 51(6): 632–45.

Ackermann, R.R. (2003) Using extant morphological variation to understand fossil relationships: a cautionary tale. *South African Journal of Science* 99: 255–8.

Ackermann, R.R., and Cheverud, J.M. (2004) Detecting genetic drift versus selection in human evolution. *Proceedings of the National Academy of Sciences USA* 101(5): 17946–51.

Adam, K.D. (1954a) Die zeitliche Stellung des Urmenschen Fundschicht von Steinheim an der Murr Innerhalb des Pleistozans. *Eiszeitalter v. Gegenwart* 4/5: 18–21.

Adam, K.D. (1954b) Die Mittelpleistozanen Faunen von Steinheim an der Murr (Wrttemberg). *Quaternaria* 1: 131–44.

Adam, K.D. (1989) Alte und neue Urmenschen-Funde in Südwest-deutschland—eine kritische Würdigung. *Quartär* 39–40: 177–90.

Adams, B., and Ringer, A. (2004) New ^{14}C dates for the Hungarian Early Upper Palaeolithic. *Current Anthropology* 45(4): 541–51.

Adams, J.W., and Conroy, G.C. (2005) Plio-Pleistocene faunal remains from the Gondolin GD 2 *in situ* assemblage, Northwest Province, South Africa. In: *Interpreting the Past: Essays on Human, Primate and Mammal Evolution in Honor of David Pilbeam*, Lieberman, D., Smith, R.J., and Kelley, J., eds, pp. 243–61. Brill Academic Publishers, Boston, MA.

Adams, J.W., Hemingway, J., Kegley, A., and Thackeray, F. (2007) Luleche, a new paleontological site in the Cradle of Humankind, North West Province, South Africa. *Journal of Human Evolution* 53: 751–4.

Adams, J.W., Herries, A.I.R., Hemingway, J., Kegley, A.D.T., Kgasi, L., Potze, S., et al. (2010) Initial fossil discoveries from Hoogland, a new primate-bearing karstic system in Gauteng Province, South Africa. *Journal of Human Evolution* 59: 685–91.

Adcock, G.J., Dennis, E.S., Easteal, S., Huttley, G.A., Jermiin, L.S., Peacock, W.J., et al. (2001) Mitochondrial DNA sequences in ancient Australians: implications for modern human origins. *Proceedings of the National Academy of Sciences USA* 98(2): 537–42.

Adefris, T. (1992) *A Description of the Bodo Cranium: An Archaic* Homo sapiens *Cranium from Ethiopia*. PhD thesis, New York University.

Agarwal, D.P., Kotlia, B.S., and Kusumgar, S. (1988) Chronology and significance of the Narmada formations. *Proceedings of the Indian National Science Academy, Part A* 54(3): 418–24.

Aguirre, E. (1970) Identification de "*Paranthropus*" en Makapansgat. *Crónica del XI Congreso Nacional de Arqueología, Mérida 1969*: 18–124.

Aguirre, E., Carbonell, E., and Bermúdez de Castro, J.M., eds (1987) *El Hombre Fósil de Ibeas y el Pleistoceno de la Sierra de Atapuerca*. Consejería de Cultura y Bienestar Social de la Junta de Castilla y León, Valladolid.

Agustí, J. (1999) A critical re-evaluation of the Miocene Mammal units in Western Europe: dispersal events and problems of correlation. In: *The Evolution of Neogene Terrestrial Ecosystems in Europe*, Agustí, J., Rook, L., and Andrews, P., eds. Cambridge University Press, Cambridge.

Agustí, J., and Moyá-Solá, S. (1987) Sobre la identidad del fragmento craneal atribuido a *Homo* sp. en Venta Micena (Orce Granada). *Estudios geologicos* 43(5–6): 535–8.

Ahern, J.C. (2005) Foramen magnum position variation in *Pan troglodytes*, Plio-Pleistocene hominids, and recent *Homo sapiens*: implications for recognizing the earliest hominids. *American Journal of Physical Anthropology* 127(3): 267–76.

Ahern, J.C.M., Karavanić, I., Paunović, M., Janković, I., and Smith, F.H. (2004) New discoveries and interpretations of hominid fossils and artifacts from Vindija Cave, Croatia. *Journal of Human Evolution* 46(1): 27–67.

Aiello, L., and Dean, C. (1990) *An Introduction to Human Evolutionary Anatomy*. Academic Press, London.

Aiello, L., and Wood, B.A. (1994) Cranial variables as predictors of hominine body mass. *American Journal of Physical Anthropology* 95: 409–26.

Aiello, L.C. (2010) Five years of *Homo floresiensis*. *American Journal of Physical Anthropology* 142(2): 167–79.

Aiello, L.C., and Dunbar, R.I.M. (1993) Neocortex size, group size and the evolution of language. *Current Anthropology* 34(2): 184–93.

Aiello, L.C., Wood, B.A., Key, C., and Lewis, M. (1999) Morphological and taxonomic affinities of the Olduvai ulna (OH 36). *American Journal of Physical Anthropology* 109(1): 89–110.

Aigner, J.S., and Laughlin, W.S. (1973) The dating of Lantian man and his significance for analysing trends in human evolution. *American Journal of Physical Anthropology* 39(1): 97–110.

Akazawa, T., Dodo, Y., Muhesen, S., Abdul-Salam, A., Abe, Y., Kondo, O., et al. (1993) The Neanderthal remains from Dederiyeh cave, Syria: interim report. *Anthropological Science* 101(4): 361–87.

Akazawa, T., and Muhesen, S., eds (2003) *Neanderthal Burials: Excavations of the Dederiyeh Cave, Afrin, Syria*. KW Publications, Auckland.

Akazawa, T., Muhesen, S., Dodo, Y., Dodo, O., and Mizoguchi, Y. (1995) Neanderthal infant burial. *Nature* 377: 585–6.

Albert, R.M., Weiner, S., Bar-Yosef, O., and Meignen, L. (2000) Phytoliths in the Middle Palaeolithic deposits of Kebara Cave, Mt Carmel, Israel: study of the plant materials used for fuel and other purposes. *Journal of Archaeological Science* 27(10): 931–47.

Albianelli, A., and Napoleone, G. (2004) Magnetostratigraphy of the *Homo*-bearing Pleistocene Dandiero Basin (Danakil Depression, Eritrea). *Rivista Italiana de Paleontologia e Stratigrafia* 110: 35–44.

Alciate, G., Ascenzi, A., Borgognini Tarli, S.M., Canci, A., Formicola, V., Giacobini, G., et al. (2005) *Catalogue of Italian Fossil Human Remains from the Palaeolithic to the Mesolithic*. Istituto Italiano di Antropologia, Rome.

Aldhouse-Green, S., Pettitt, P., and Stringer, C. (1996) Holocene humans at Pontnewydd and Cae Gronw caves. *Antiquity* 70(268): 444–7.

Alemseged, Z., and Geraads, D. (2000) A new Middle Pleistocene fauna from the Busidima-Telalak region of the Afar, Ethiopia. *Comptes Rendus de l'Académie des Sciences* 331: 549–56.

Alemseged, Z., Coppens, Y., and Geraads, D. (2002) Hominid Cranium from Omo: description and taxonomy of Omo-323-1976-896. *American Journal of Physical Anthropology* 117: 103–12.

Alemseged, Z., Wynn, J.G., Kimbel, W., Reed, D., Geraads, D., and Bobe, R. (2005) A new hominin from the Basal Member of the Hadar Formation, Ethiopia, and its geological context. *Journal of Human Evolution* 49: 499–514.

Alemseged, Z., Spoor, F., Kimbel, W.H., Bobe, R., Geraads, D., Reed, D., et al. (2006) A juvenile early hominin skeleton from Dikika, Ethiopia. *Nature* 443: 296–301.

Alexandre, A., Meunier, J., Lezine, A., Vincens, A., and Schwartz, D. (1997) Phytoliths: indicators of grassland dynamics during the Late Holocene in intertropical Africa. *Palaeogeography, Palaeoclimatology, Palaeoecology* 136: 213–29.

Alexeev, V. (1986) *The Origin of the Human Race*. Progress Publishers, Moscow.

Alexeeva, T.I., Bader, N.O., Buzhilova, A.P., Kozlovskaya, M.V., and Mednikova, M.B., eds (2000) Homo sungirensis: *Upper Palaeolithic Man: Ecological and Evolutionary Aspects of the Investigation*. Scientific World, Moscow.

Allen, J.J. (1877) The influence of physical conditions in the genesis of species. *Radical Review* 1: 108–40.

Almécija, S., Moyà-Solà, S., and Alba, D.M. (2010) Early origin for human-like precision grasping: a comparative study of pollical distal phalanges in fossil hominins. *Public Library of Science* 5(7): e11727.

Almeida, F., Moreno-Garcia, M., and Angelucci, D.E. (2009) From under the bulldozer's claws: the EE15 Late Gravettian occupation surface of the Lagar Velho rock-shelter. *World Archaeology* 41(2): 242–61.

Alperson-Afil, N., and Goren-Inbar, N. (2010) *The Acheulian Site of Gesher Benot Ya'aqov: Ancient Flames and Controlled Use of Fire*. Springer, Dordrecht.

Alperson-Afil, N., Sharon, G., Kislev, M., Melamed, Y., Zohar, I., Ashkenazi, S., et al. (2009) Spatial organization of hominin activities at Gesher Benot Ya'aqov, Israel. *Science* 326(5960): 1677–80.

Alteirac, A., and Bahn, P. (1982) Premières datations radiocarbone du Magdalénien moyen de la grotte du Mas-d'Azil (Ariège). *Préhistoire Ariégeoise. Bulletin de la Société Préhistorique de l'Ariège Tarascon-sur-Ariège* 37: 107–10.

Altmann, J., Alberts, S.C., Altmann, S.A., and Roy, S.B. (2002) Dramatic change in local climate patterns in the Amboseli basin, Kenya. *African Journal of Ecology* 40(3): 248–51.

Altmann, S.A. (1998) *Foraging for Survival: Yearling Baboons in Africa*. University of Chicago Press, Chicago, IL.

Ambrose, S. (1998) Chronology of the Later Stone Age and food production in East Africa. *Journal of Archaeological Science* 25(4): 377–92.

Ambrose, S. (2001) Middle and Later Stone Age settlement patterns in the Central Rift Valley, Kenya:

comparisons and contrasts. *Settlement Dynamics of the Middle Paleolithic and Middle Stone Age* 1: 21–41.

Ambrose, S.H. (1991) Effects of diet, climate and physiology on nitrogen isotope abundances in terrestrial foodwebs. *Journal of Archaeological Science* 18(3): 293–317.

Ambrose, S.H. (1998) Late Pleistocene human populations bottlenecks, volcanic winter, and differentiation of modern humans. *Journal of Human Evolution* 34: 623–51.

Ambrose, S.H. (2002) Small things remembered: origins of early microlithic industries in sub-Saharan Africa. In: *Thinking Small: Global Perspectives on Microlithization*, Elston, R.G., and Kuhn, S.L., eds, pp. 9–29. American Anthropological Association, Arlington, VA.

Ambrose, S.H. (2003) Did the super-eruption of Toba cause a human population bottleneck? Reply to Gathorne-Hardy and Harcourt-Smith. *Journal of Human Evolution* 45: 231–7.

Amerano, G.B. (1889) *La caverna delle Fate (Ligurie)*. Congrès International d'Anthropologie et d'Archéologie Préhistorique, Paris, Compte rendu de la dixième session.

Amos, W., and Hoffman, J.I. (2009) Evidence that two main bottleneck events shaped modern human genetic diversity. *Proceedings of the Royal Society of London, Series B* 277: 131–7.

Amunts, K., Schleicher, A., Bürgel, U., Mohlberg, H., Uylings, H.B.M., and Zilles, K. (1999) Broca's region revisited: cytoarchitecture and intersubject variability. *Journal of Comparative Neurology* 412: 319–41.

An, Z., and Ho, C.K. (1989) New magnetostratigraphic dates of Lantian *Homo erectus*. *Quaternary Research* 32: 213–21.

Anati, E., and Haas, N. (1967) The Hazorea Pleistocene site: a preliminary report. *Man* 2(3): 454–6.

Anati, E., Avnimelech, M., Haas, N., and Meyerhof, E. (1973) *Hazorea I*. Centro Camuno di Studi Preistorici, Brescia.

Anderson, J.E. (1968) Late palaeolithic skeletal remains from Nubia. In: *The Prehistory of Nubia*, Wendorf, F., ed., pp. 996–1040. Ft Burgwin Research Center and South Methodist University Press, Dallas, TX.

Anderson-Gerfaud, P. (1981) *Contribution méthodologique à l'analyse des microtraces d'utilisation sur les outils préhistoriques*. Thèse de 3ème cycle de l'Université de Bordeaux I.

Andrefsky, W., ed. (2008) *Lithic Technology*. Cambridge University Press, Cambridge.

Andresen, V. (1989) Die Querstreifung des Dentins. *Deutsche Monatsschrift für Zahnheilkunde* 16: 386–9.

Andrews, C.W. (1916) Note on a new baboon (*Simopithecus oswaldi*, gen. et sp. n.) from the (?) Pliocene of British East Africa. *The Annals of Magazine of Natural History, including Zoology, Botany, and Geology* 18: 410–19.

Andrews, P. (2006) Taphonomic effects of faunal impovershment and faunal mixing. *Palaeogeography, Palaeoclimatology, Palaeoecology* 241(3–4): 572–89.

Andrews, P., and Cronin, J.E. (1982) The relationships of *Sivapithecus* and *Ramapithecus* and the evolution of the orang-utan. *Nature* 297: 541–6.

Andrews, P., and Harrison, T. (2005) The last common ancestor of apes and humans. In: *Interpreting the Past: Essays on Human, Primate and Mammal Evolution in Honor of David Pilbeam*, Lieberman, D.E., Smith, R.J., and Kelley, J., eds, pp. 103–21. Brill Academic Publishers, Boston, MA.

Angel, J.L., and Kelley, J.O. (1986) Description of the human skeleton: description and comparison of the skeleton. In: *The Prehistory of Wadi Kubbaniya*, Wendorf, F., and Schild, R., eds, pp. 53–70. Southern Methodist University Press, Dallas, TX.

Angel, J.L., Phenice, T.W., Robbins, L.H., and Lynch, B.M. (1980) *Late Stone-Age Fisherman of Lothagam, Kenya*. Michigan State University Anthropological Series, East Lansing, MI.

Antón, S. (1997) Developmental age and taxonomic affinity of the Mojokerto child, Java, Indonesia. *American Journal of Physical Anthropology* 102(4): 497–514.

Antón, S. (1999) Cranial growth in *Homo erectus*: how credible are the Ngandong juveniles? *American Journal of Physical Anthropology* 108(2): 223–36.

Antón, S. (2002) Evolutionary significance of cranial variation in Asian *Homo erectus*. *American Journal of Physical Anthropology* 118: 301–23.

Antón, S. (2003) Natural History of *Homo erectus*. *Yearbook of Physical Anthropology* 46: 126–70.

Antón, S., and Swisher, C.C. (2004) Early dispersals of *Homo* from Africa. *Annual Review of Anthropology* 33: 271–96.

Antón, S.C., Spoor, F., Fellmann, C.D., and Swisher, C.C.I. (2007) Defining *Homo erectus*: size considered. In: *Handbook of Paleoanthropology*, Henke, W., Rothe, H., and Tattersall, I., eds, pp. 1655–93. Springer-Verlag, Berlin.

Ao, H., Deng, C.L., Dekkers, M.J., Sun, Y.B., Liu, Q.S., and Zhu, R.X. (2010) Pleistocene environmental evolution in the Nihewan Basin and implication for early human colonization of North China. *Quaternary International* 223–4.

Aoki, K. (2002) Sexual selection as a cause of human skin colour variation: Darwin's hypothesis revisited. *Annals of Human Biology* 29(6): 589–608.

ApSimon, A.M. (1980) The last neanderthal in France. *Nature* 287: 271–2.

Arambourg, C. (1934) Sur la présence d'un crocodilien du genre *Crocodilus* dans les gisements de phosphates du Maroc. *Comptes Rendus Sommaires des Scéances de la Société Géologique de France* 9: 108–10.

Arambourg, C. (1954) L'hominien fossile de Ternifine (Algérie). *Comptes Rendus de l'Académie des Sciences* 239: 893–5.

Arambourg, C. (1955) A recent discovery in human paleontology: *Atlanthropus* of Ternifine (Algeria). *American Journal of Physical Anthropology* 13(2): 191–201.

Arambourg, C., and Hoffstetter, R. (1954) Découverte en Afrique du Nord de restes humains du Paléolithique inférieur. *Comptes Rendus de l'Académie des Sciences* **239**: 72–74.

Arambourg, C., and Hoffstetter, R. (1955) Le gisement de Ternifine: Résultats des fouilles de 1955 et découvertes du noveaux restes d'Atlanthropus. *Comptes Rendus de l'Académie des Sciences* **241**: 431–3.

Arambourg, C., and Biberson, P. (1956) The fossil human remains from the Paleolithic site of Sidi Abderrahman (Morocco). *American Journal of Physical Anthropology* **14**(3): 467–89.

Arambourg, C., and Coppens, Y. (1967) Sur la decouverte, dans le Pleistocene inferieur de la vallee de l'Omo (Ethiopie), d'une mandibule d'Australopithecien. *Comptes Rendus de l'Académie des Sciences* **265**: 589–90.

Arambourg, C., and Coppens, Y. (1968) Découverte d'un australopithecien nouveau dans les Gisements de L'Omo (Ethiopie). *South African Journal of Science* **64**(2): 58–59.

Arbib, M. (2006) A sentence is to speech as what is to action? *Cortex* **42**(4): 507–14.

Archibold, W. (1995) *Ecology of World Vegetation*. Chapman and Hall, London.

Archie, J.W. (1989) Homoplasy excess ratios: new indices for measuring levels of homoplasy in phylogenetic systematics and a critique of the consistency index. *Systematic Zoology* **38**(3): 253–69.

Arensburg, B. (1989) A Middle Paleolithic human hyoid bone. *Nature* **338**: 758–60.

Arensburg, B. (2002) Human remains from Geula Cave, Haifa. *Bulletins et Mémoires de la Société d'Anthropologie de Paris* **14**(1–2): 141–8.

Arensburg, B., and Bar-Yosef, O. (1973) Human remains from Ein Gev I, Jordan Valley, Israel. *Paléorient* **1**(2): 201–6.

Arensburg, B., and Belfer-Cohen, A. (1998) Sapiens and Neandertals: rethinking the Levantine Middle Paleolithic hominids. In: *Neandertals and Modern Humans in Western Asia*, Akazawa, T., Aoki, K., and Bar-Yosef, O., eds, pp. 311–22. Plenum Press, New York.

Arensburg, B., Bar Yosef, O., Chech, M., Goldberg, P., Laville, Meignen, L., et al. (1985) Une sepulture neandertalienne dans la grotte de Kebara (Israel). *Comptes Rendus de l'Académie des Sciences* **300**: 227–30.

Arensburg, B., Tillier, A.M., Vandermeersch, B., Duday, H., Schepartz, L.A., and Rak, Y. (1989) A Middle Palaeolithic human hyoid bone. *Nature* **338**: 758–60.

Arensburg, B., Bar-Yosef, O., Belfer-Cohen, A., and Rak, Y. (1990) Mousterian and Aurignacian human remains from Hayonim Cave, Israel. *Paléorient* **16**(1): 107–9.

Arensburg, B., Schepartz, L.A., Tillier, A.M., Vandermeersch, B., and Rak, Y. (1990) A reappraisal of the anatomical basis for speech in Middle Palaeolithic hominids. *American Journal of Physical Anthropology* **83**: 137–46.

Argue, D., Morwood, M.J., Sutikna, T., Jatmiko, and Saptomo, E.W. (2009) *Homo floresiensis*: a cladistic analysis. *Journal of Human Evolution* **57**(5): 623–39.

Armelagos, G., Van Gerven, D., Martin, D., and Huss-Ashmore, R. (1984) Effects of nutritional change on the skeletal biology of northeast African (Sudanese Nubian) populations. In: *From Hunters to Farmers: The Causes and Consequences of Food Production in Africa, University of California Press, Berkeley*, Clark, J., and Brandt, S., eds, pp. 132–46. University of California Press, Berkeley, CA.

Armelagos, G., Goodman, A., Harper, K., and Blakey, M. (2009) Enamel hypoplasia and early mortality: bioarcheological support for the Barker hypothesis. *Evolutionary Anthropology* **18**(6): 261–71.

Armelagos, G.J. (1964) A fossilized mandible from near Wadi Halfa, Sudan. *Man* **64**: 12–13.

Arnason, U., and Janke, A. (2002) Mitogenomic analyses of eutherian relationships. *Cytogenetic and Genome Research* **96**: 20–32.

Arnold, M.L. (2009) *Reticulate Evolution and Humans*. Oxford University Press, New York.

Arnold, M.L., and Meyer, A. (2006) Natural hybridization in primates: one evolutionary mechanism. *Zoology* **109**(4): 261–76.

Aronson, J., Hailemichael, M., and Savin, S. (2008) Hominid environments at Hadar from paleosol studies in a framework of Ethiopian climate change. *Journal of Human Evolution* **55**: 532–50.

Arrizabalaga, A., and Altuna, J., eds (2000) *Labeko Koba (País Vasco). Hienas y Humanos en los albores del Paleolítico superior*. Sociedad de Ciencias Naturales Aranzadi, San Sebastián.

Arrizabalaga, A., Altuna, J., Areso, P., Elorza, M., García, M., Iriarte, M., et al. (2003) The initial Upper Paleolithic in Northern Iberia: new evidence from Labeko Koba1. *Current Anthropology* **44**(3): 413–21.

Arsuaga, J., Gracia, A., Martínez, I., Bermúdez de Castro, J., Rosas, A., Villaverde, V., et al. (1989) The human remains from Cova Negra (Valencia, Spain) and their place in European Pleistocene human evolution. *Journal of Human Evolution* **18**(1): 55–92.

Arsuaga, J., Martínez, I., Gracia, A., Carretero, J., and Carbonell, E. (1993) Three new human skulls from the Sima de los Huesos Middle Pleistocene site in Sierra de Atapuerca, Spain. *Nature* **362**: 534–7.

Arsuaga, J., Martínez, I., Gracia, A., Carretero, J., Lorenzo, C., García, N., et al. (1997a) Sima de los Huesos (Sierra de Atapuerca, Spain). The site. *Journal of Human Evolution* **33**(2–3): 109–27.

Arsuaga, J., Villaverde, V., Quam, R., Gracia, A., Lorenzo, C., Martínez, I., et al. (2002) The Gravettian occipital bone from the site of Malladetes (Barx, Valencia, Spain). *Journal of Human Evolution* **43**(3): 381–93.

Arsuaga, J., Villaverde, V., Quam, R., Martínez, I., Carretero, J., Lorenzo, C., et al. (2007) New Neandertal remains from Cova Negra (Valencia, Spain). *Journal of Human Evolution* **52**(1): 31–58.

Arsuaga, J.L., Martinez, I., Gracia, A., and Lorenzo, C. (1997) The Sima de los Huesos crania (Sierra de Atapuerca, Spain). A comparative study. *Journal of Human Evolution* 33: 219–81.

Arsuaga, J.L., Martinez, I., Garcia, A., and Lorenzo, C. (1997b) The Sima de los Huesos crania (Sierra de Atapuerca, Spain). A comparative study. *Journal of Human Evolution* 33: 219–81.

Arsuaga, J.L., Martínez, I., Lorenzo, C., Gracia, A., Muñoz, A., Alonso, O., et al. (1999a) The human cranial remains from Gran Dolina Lower Pleistocene site (Sierra de Atapuerca, Spain). *Journal of Human Evolution* 37(3–4): 431–57.

Arsuaga, J.L., Lorenzo, C., Carretero, J.M., Gracia, A., Martinez, I., Garcia, N., et al. (1999b) A complete human pelvis from the Middle Pleistocene of Spain. *Nature* 399: 255–8.

Ascenzi, A., and Segre, A.G. (1971) A new Neandertal child mandible from an Upper Pleistocene site in southern Italy. *Nature* 233: 280–3.

Ascenzi, A., and Segre, A.G. (1996) Artefacts and human teeth at the Fontana Ranuccio Middle Pleistocene site (Central Italy). *Anthropologie* 3: 39–46.

Ascenzi, A., Biddittu, I., Cassoli, P.F., Segre, A.G., and Segre-Naldini, E. (1996) A calvarium of late *Homo erectus* from Ceprano, Italy. *Journal of Human Evolution* 31: 409–23.

Ascenzi, A., Mallegni, F., Manzi, G., Segre, A.G., and Segre-Naldini, E. (2000) A re-appraisial of Ceprano calvaria affinities with *Homo erectus*, after the new reconstruction. *Journal of Human Evolution* 39: 443–50.

Asfaw, B. (1983) A new hominid parietal from Bodo, Middle Awash Valley, Ethiopia. *American Journal of Physical Anthropology* 61: 367–71.

Asfaw, B. (1987) The Belohdelie frontal: new evidence of early hominid cranial morphology from the Afar of Ethiopia. *Journal of Human Evolution* 16(7–8): 611–24.

Asfaw, B., Ebinger, C., Harding, D., White, T., and WoldeGabriel, G. (1990) Space-based imagery in paleoanthropological research: an Ethiopian example. *National Geographic Research* 6(4): 418–34.

Asfaw, B., Beyene, Y., Semaw, Suwa, G., White, T., and WoldeGabriel, G. (1991) Fejej: a new paleoanthropological research area in Ethiopia. *Journal of Human Evolution* 21(2): 137–43.

Asfaw, B., Beyene, Y., Suwa, G., Walter, R.C., White, T.D., WoldeGabriel, G., et al. (1992) The earliest Acheulean from Konso-Gardula. *Nature* 360: 732–5.

Asfaw, B., Beyene, Y., Semaw, S., Suwa, G., White, T., and WoldeGabriel, G. (1993) Tephra from Fejej, Ethiopia - a reply. *Journal of Human Evolution* 25: 519–21.

Asfaw, B., White, T., Lovejoy, O., Latimer, B., Simpson, S., and Suwa, G. (1999) *Australopithecus garhi*: a new species of early hominid from Ethiopia. *Science* 284(629): 629–35.

Asfaw, B., Gilbert, B.M., Beyene, Y., Hart, W.K., Renne, P.R., WoldeGabriel, G., et al. (2002) Remains of *Homo erectus* from Bouri, Middle Awash, Ethiopia. *Nature* 416: 317–20.

Asfaw, B., Gilbert, W.H., and Richards, G.D. (2008) *Homo erectus* cranial anatomy. In: Homo erectus: *Pleistocene Evidence from the Middle Awash, Ethiopia*, Gilbert, W.H., and Asfaw, B., eds, pp. 265–328. University of California Press, Berkeley, CA.

Ashley, G.M., Tactikos, J.C., and Owen, R.B. (2009) Hominin use of springs and wetlands: Paleoclimate and archaeological records from Olduvai Gorge (~1.79–1.74 Ma). *Palaeogeography, Palaeoclimatology and Palaeoecology* 272: 1–16.

Ashton, E.H., and Zuckerman, S. (1950) Some quantitative dental characters of fossil anthropoids. *Philosophical Transactions of the Royal Society of London. Series B, Biological Sciences* 234: 485–520.

Assefa, Z. (2006) Faunal remains from Porc-Epic: paleoecological and zooarchaeological investigations from a Middle Stone Age site in southeastern Ethiopia. *Journal of Human Evolution* 51(1): 50–75.

Assefa, Z., Lam, Y.M., and Mienis, H.K. (2008a) Symbolic use of terrestrial gastropod opercula during the Middle Stone Age at Porc-Epic Cave, Ethiopia. *Current Anthropology* 49(4): 746–56.

Assefa, Z., Yirga, S., and Reed, K.E. (2008b) The large-mammal fauna from the Kibish Formation. *Journal of Human Evolution* 55: 501–12.

Aubry, M.-P., Berggren, W.A., Van Couvering, J., McGowran, B., Hilgen, F., Steininger, F., et al. (2009) The Neogene and Quaternary: chronostratigraphic compromise or non-overlapping magisteria? *Stratigraphy* 6(1): 1–16.

Audouin, F., and Plisson, H. (1982) Les ochres et leurs témoins au Paléolithique en France: Enquête et expériences sur leur validité archéologique. *Cahiers du Centre de Recherches Préhistoriques* 8: 33–80.

Auerbach, B.M., and Ruff, C.B. (2004) Human body mass estimation: a comparison of "morphometric" and "mechanical" methods. *American Journal of Physical Anthropology* 125: 331–42.

Auerbach, B.M., and Ruff, C.B. (2010) Stature estimation formulae for indigenous populations from North America. *American Journal of Physical Anthropology* 141(2): 190–207.

Avery, G., Cruz-Uribe, K., Goldberg, P., Grine, F.E., Klein, R.G., Lenardi, M.J., et al. (1997) The 1992–1993 excavations at the Die Kelders Middle and Later Stone Age cave site, South, Africa. *Journal of Field Archaeology* 24(3): 263–91.

Aziz, F. (1983) Notes on a new *Meganthropus* S.33 from the Sangiran Dome area, Central Java. *Publication of the Geological Research and Development Centre, Paleontology Series* 4: 56–60.

Aziz, F., Baba, H., and Narasaki, S. (1994) Preliminary report on recent discoveries of fossil hominids from the Sangiran area, Java. *Journal of Geology and Mineral Resources* **4**: 11–16.

Aziz, M.A. (1981) Possible "atavistic" structures in human aneuploids. *American Journal of Physical Anthropology* **54**(3): 347–53.

Azoury, I., Bergman, C., and Copeland, L. (1988) *Ksar Akil, Lebanon: A Technological and Typological Analysis of the Transitional and Early Upper Palaeolithic Levels of Ksar Akil and Abu Halka*. B.A.R., Oxford.

Baab, K.L. (2008) A re-evaluation of the taxonomic affinities of the early *Homo* cranium KNM-ER 42700. *Journal of Human Evolution* **55**(4): 741–6.

Baab, K.L., and McNulty, K.P. (2008) Size, shape, and asymmetry in fossil hominins: the status of the LB1 cranium based on 3D morphometric analyses. *Journal of Human Evolution* **57**(5): 608–22.

Baales, M., and Street, M. (1998) Late Palaeolithic backed-point assemblages in the northern Rheineland: current research and changing views. *Notae Praehistoricae* **18**: 77–92.

Baba, H., and Endo, B. (1982) Postcranial skeleton of Minatogawa Man. In: *The Minatogawa Man: the Upper Pleistocene man from the Island of Okinawa*, Suzuki, H., and Hanihara, K., eds, pp. 61–195. University of Tokyo, Tokyo.

Baba, H., and Narasaki, S. (1991) Minatogawa Man, the oldest type of modern *Homo sapiens* in East Asia. *Quaternary Research* **30**(2): 221–30.

Baba, H., Aziz, F., and Narasaki, S. (1998) Restoration of head and face in Javanese *Homo erectus* Sangiran 17. *Bull. Natl. Sci. Mus., Tokyo, Series D* **24**: 1–8.

Backwell, L., and d'Errico, F. (2008) Early hominid bone tools from Drimolen, South Africa. *Journal of Archaeological Science* **35**(11): 2880–94.

Backwell, L., Pickering, R., Brothwell, D., Berger, L., Witcomb, M., Martill, D., et al. (2009) Probable human hair found in a fossil hyaena coprolite from Gladysvale cave, South Africa. *Journal of Archaeological Science* **36**: 1269–76.

Backwell, L.R., and d'Errico, F. (2001) Evidence of termite foraging by Swartkrans early hominids. *Proceedings of the National Academy of Sciences USA* **98**(4): 1358–63.

Backwell, L.R., and d'Errico, F. (2004a) A reassessment of the Olduvai Gorge 'bone tools'. *Palaeontologia Africana* **40**: 95–158.

Backwell, L.R., and d'Errico, F. (2004b) Additional evidence of early hominid bone tools from Swartkrans. In: *Swartkrans: A Cave's Chronicle of Early Man*, Brain, C.K., ed., pp. 279–95. Transvaal Museum, Pretoria.

Bada, J.L., and Deems, l. (1975) Accuracy of dates beyond the ^{14}C dating limit using the aspartic acid racemisaton reaction. *Nature* **255**: 218–19.

Bae, K.D. (1988) *The Significance of the Chongokni Stone Industry in the Tradition of Paleolithic Culture in East Asia*. PhD thesis, University of California.

Bae, K.D. (1994) Paleolithic tradition of East Asia: Chongoknian, presence of few bifaces in Paleolithic stone industries. In: *Paleolithic Culture of East Asia*, Bae, K.D., ed., pp. 193–211. National Research Institute of Cultural Properties, Seoul.

Bae, K.D. (1997) A study of stone industries of Korea. *Korean Journal of Quaternary Research* **11**: 1–26.

Bae, K.D. (2002) Chongokni Paleolithic site: current understandings. In: *Paleolithic Archaeology in Northeast Asia*, Bae, K.D., and Lee, J.C., eds, pp. 55–75. Yeoncheon County and The Institute of Cultural Properties, Seoul & Yeoncheon County.

Bae, K.D. (2010) Peopling in the Korean Peninsula. In: *Asian Paleoanthropology: From Africa to China and Beyond*, Norton, C.J., and Braun, D., eds, pp. 181–90. Springer Press, Dordrecht.

Bae, K.D., Hwang, S.H., and Im, D.I. (1999) *The Kumpari Paleolithic Site Report of Excavation in 1989–1992*. National Research Institute of Cultural Properties, Seoul.

Baena, J., Lordkipanidze, D., Cuartero, F., Ferring, R., Zhvania, D., Martn, D., et al. (2010) Technical and technological complexity in the beginning: the study of Dmanisi lithic assemblage. *Quaternary International* **223**: 45–53.

Baffier, D., and Girard, M. (1995) La Grande Grotte d'Arcy-sur-Cure (Yonne). Second sanctuaire paléolithique bourguignon. *L'Anthropologie* **99**(2–3): 212–20.

Baffier, D., Girard, M., Brunet, J., Guillamet, E., Chillida, J., Hardy, M., et al. (2001) Du nouveau à la Grande Grotte d'Arcy-sur-Cure, Yonne, France. *International Newsletter on Rock Art* **28**: 1–3.

Bahain, J., Sarcia, M., Falgueres, C., and Yokoyama, Y. (1993) Attempt at ESR dating of tooth enamel of French Middle Pleistocene sites. *Applied Radiation and Isotopes* **44**(1–2): 267–72.

Bailey, S. (2002) *Neandertal Dental Morphology: Implications for Modern Human Origins*. PhD thesis, Arizona State University.

Bailey, S. (2004) A morphometric analysis of maxillary molar crowns of Middle-Late Pleistocene hominins. *Journal of Human Evolution* **47**(3): 183–98.

Bailey, S., and Liu, W. (2010) A comparative dental metrical and morphological analysis of a Middle Pleistocene hominin maxilla from Chaoxian (Chaohu), China. *Quaternary International* **211**(1–2): 14–23.

Bailey, S., Weaver, T., and Hublin, J. (2009) Who made the Aurignacian and other early Upper Paleolithic industries? *Journal of Human Evolution* **57**(1): 11–26.

Bailey, S.E., and Hublin, J.-J. (2006) Dental remains from the Grotte de Renne at Arcy-sur-Cure (Yonne). *Journal of Human Evolution* **50**: 485–508.

Bailey, W.J., Hayasak, K., Skinner, C.G., Kehoe, S., Sieu, L.U., Slighthom, J.L., et al. (1992) Reexamination of the African hominoid trichotomy with additional sequences from the primate beta-globin gene. *Molecular Phylogenetics and Evolution* **1**: 97–135.

Ballard, C. (1993) Stimulating minds to fantasy? A critical etymology of Sahul. In: *Sahul in Review*.

Pleistocene Archaeology in Australia, New Guinea and Island Melanesia, Smith, M.A., Spriggs, M., and Fankhauser, B., eds, pp. 17–23. Department of Prehistory, Research School of Pacific Studies, The Australian National University, Canberra.

Balout, L. (1967) Procèdes d'analyse et questions de terminologie dans l'étude des ensembles indutriels du Paléolithique Inférieur en Afrique du Nord. In: *Background to Evolution in Africa*, Bishop, W.W., and Clarke, J.D., eds, pp. 701–35. University of Chicago Press, Chicago, IL.

Balter, V., and Simon, L. (2006) Diet and behavior of the Saint-Cesaire Neanderthal inferred from biogeochemical data inversion. *Journal of Human Evolution* 51(4): 329–38.

Balzeau, A., and Rougier, H. (2010) Is the suprainiac fossa a Neandertal autapomorphy? A complementary external and internal investigation. *Journal of Human Evolution* 58(1): 1–22.

Balzeau, A., Grimaud-Hervé, D., and Jacob, T. (2005) Internal cranial features of the Mojokerto child fossil (East Java, Indonesia). *Journal of Human Evolution* 48(6): 535–53.

Bamford, M. (2005) Early Pleistocene fossil wood from Olduvai Gorge, Tanzania. *Quaternary International* 129(1): 15–22.

Bamford, M.K., Albert, R.M., and Cabanes, D. (2006) Plio-Pleistocene macroplant fossil remains and phytoliths from Lowermost Bed II in the eastern palaeolake margin of Olduvai Gorge, Tanzania. *Quaternary International* 148(1): 95–112.

Bamshad, M., Wooding, S., Salisbury, B., and Stephens, J. (2004) Deconstructing the relationship between genetics and race. *Nature Reviews Genetics* 5(8): 598–609.

Bandet, Y., Sémah, F., Sartono, S., and Djubiantono, T. (1989) Premier peuplement par les mammifères d'une région de Java Est, à la fin du Pliocène: âge de la faune du Gunung Butak, près de Kedungbrubus (Indonésie). *Comptes Rendus de l'Académie des Sciences* 308(9).

Banks, W.E., d'Errico, F., Peterson, A.T., Kageyama, M., Sima, A., and Sánchez-Goñi, M.-F. (2008) Neanderthal extinction by competitive exclusion. *Public Library of Science* 3(12): e3972.

Bar-Yosef, O. (1989) Upper Pleistocene cultural stratigraphy in southwest Asia. In: *The Emergence of Modern Humans*, Trinkaus, E., ed., pp. 154–80. Cambridge University Press, Cambridge.

Bar-Yosef, O. (1991) The archaeology of the Natufian layer at Hayonim cave. In: *The Natufian Culture in the Levant*, Bar-Yosef, O., and Valla, F., eds, pp. 81–92. International Monographs in Prehistory, Ann Arbor, MI.

Bar-Yosef, O. (1998) The Chronology of the Middle Paleolithic of the Levant. In: *Neandertals and Modern Humans in Western Asia*, Akazawa, T., Aoki, K., and Bar-Yosef, O., eds, pp. 39–56. Plenum, New York.

Bar-Yosef, O., and Goren, N. (1973) Natufian remains in Hayonim Cave. *Paléorient* 1: 49–68.

Bar-Yosef, O., and Vandermeersch, B., eds (1991) *Le Squelette Moustérien de Kébara 2*. CNRS, Paris.

Bar-Yosef, O., and Goren-Inbar, N. (1993) *The Lithic Assemblages of 'Ubeidiya, a Lower Paleolithic Site in the Jordan Valley*. Qedem, Jerusalem.

Bar-Yosef, O., and Callander, J. (1999) The woman from Tabun: Garrod's doubts in historical perspective. *Journal of Human Evolution* 37(6): 879–85.

Bar-Yosef, O., and Kuhn, S.L. (1999) The big deal about blades: laminar technologies and human evolution. *American Anthropologist* 101(2): 322–38.

Bar-Yosef, O., and Van Peer, P. (2009) The Chaine Operatoire Approach in Middle Paleolithic archaeology. *Current Anthropology* 50(1): 103–31.

Bar-Yosef, O., Vandermeersch, B., Arensburg, B., Goldberg, P., Laville, H., Meignen, L., et al. (1986) New data on the origin of modern man in Levant. *Current Anthropology* 27: 63–64.

Bar-Yosef, O., Vandermeersch, B., Arensburg, B., Belfer-Cohen, A., Goldberg, P., Laville, H., et al. (1992) The excavations in Kebara Cave, Mt. Carmel. *Current Anthropology* 33(5): 497–550.

Bar-Yosef, O., Arnold, M., Mercier, N., Belfer-Cohen, A., Goldberg, P., Housley, R., et al. (1996) The dating of the Upper Paleolithic layers in Kebara Cave, Mt. Carmel. *Journal of Archaeological Science* 23: 297–306.

Bar-Yosef Mayer, D., Vandermeersch, B., and Bar-Yosef, O. (2009) Shells and ochre in Middle Paleolithic Qafzeh Cave, Israel: indications for modern behavior. *Journal of Human Evolution* 56(3): 307–14.

Barbour, G.B., Licent, E., and Teilhard de Chardin, P. (1927) Geological study of the deposits of the Sangkanho Basin. *Bulletin of the Geological Society of China* 6: 263–78.

Barger, N., Stefanacci, L., and Semendefari, K. (2007) A comparative volumetric analysis of the amygdaloid complex and basolateral division in the human and ape brain. *American Journal of Physical Anthropology* 134: 392–403.

Barham, L. (2000) *The Middle Stone Age of Zambia, South Central Africa*. Western Academic and Specialist Press, Bristol.

Barham, L.S. (1987) The bipolar technique in Southern Africa: a replication experiment. *The South African Archaeological Bulletin* 42(145): 45–50.

Barham, L.S. (2002a) Backed tools in Middle Pleistocene central Africa and their evolutionary significance. *Journal of Human Evolution* 43(5): 585–603.

Barham, L.S. (2002b) Systematic pigment use in the Middle Pleistocene of South-Central Africa. *Current Anthropology* 43: 181–90.

Barham, L.S., and Mitchell, P. (2008) *The First Africans: African Archaeology from the Earliest Toolmakers to the Most Recent Foragers*. Cambridge University Press, Cambridge.

Barham, L.S., Pinto Llona, A.C., and Stringer, C.B. (2002) Bone tools from Broken Hill (Kabwe) cave,

Zambia, and their evolutionary significance. *Before Farming* (3): 1–12.

Barkai, R., Gopher, A., Lauritzen, S.E., and Frumkin, A. (2003) Uranium series dates from Qesem Cave, Israel, and the end of the Lower Paleolithic. *Nature* **423**: 977–9.

Barker, G. (2005) The archaeology of foraging and farming at Niah Cave, Sarawak. *Asian Perspectives* **44**: 90–106.

Barker, G. (2010) *Rainforest Foraging and Farming in Island Southeast Asia: the archaeology of the Niah Caves, Sarawak*. McDonald Institute Monographs and Sarawak Museums.

Barker, G., Barton, H., Bird, M., Daly, P., Datan, I., Dykes, A., et al. (2007) The 'human revolution' in lowland tropical Southeast Asia: the antiquity and behavior of anatomically modern humans at Niah Cave (Sarawak, Borneo). *Journal of Human Evolution* **52**(3): 243–61.

Barker, G., Gilbertson, D., and Reynolds, T. (2010) *Archaeological Investigations in the Niah Caves, Sarawak, 1954–2003*. McDonald Institute Monographs and Sarawak Museums.

Barroso Ruiz, C. (1984) Le gisement moustérien de la grotte du Boquete de Zafarraya (Alcaucin-Andalousie). *L'Anthropologie* **88**(1): 133–4.

Barthelme, J.W. (1985) *Fisher-Hunters and Neolithic Pastoralists in East Turkana, Kenya*. B.A.R. International Series 254 (Cambridge Monographs in African Archaeology 13). B.A.R., Oxford.

Bartholomew, G.A., and Birdsell, J.B. (1953) Ecology and the protohominids. *American Anthropologist* **55**(4): 481–98.

Bartolomei, G. (1974) I talus detritici e la stabilizzazione del versante destro della valle dell'Adige nella zona di Trento. *Studi Trentini di Scienze Naturali* **51**: 213–28.

Bartolomei, G., Broglio, A., Cattani, L., Cremaschi, M., Guerreschi, A., Mantovani, E., et al. (1982) I depositi Würmiani del Riparo Tagliente. *Ann. Univ. Ferrara* **3**: 61–105.

Bartolomei, G., Broglio, A., Cassoli, P., Castelletti, L., Cremaschi, M., Giacobini, G., et al. (1992) La Grotte-Abri de Fumane. Un site Aurignacien au Sud des Alps. *Preistoria Alpina* **28**: 131–79.

Barton, C.M., Olszewski, D.I., and Coinman, N.R. (1996) Beyond the grave: reconsidering burin function. *Journal of Field Archaeology* **23**(1): 111–25.

Barton, R., Currant, A., Fernández-Jalvo, Y., Finlayson, J., Goldberg, P., Macphail, R., et al. (1999) Gibraltar Neanderthals and results of recent excavations in Gorham's, Vanguard, and Ibex Caves. *Antiquity* **73**: 13–23.

Barton, R.N.E., Bouzouggar, A., and Bronk-Ramsey, C. (2007) Abrupt climatic change and chronology of the Upper Palaeolithic in northern and eastern Morocco. In: *Rethinking the Human Revolution: New Behavioural & Biological Perspectives on the Origins and Dispersal of Modern Humans*, Mellars, P., Boyle, K., Bar-Yosef, O., and Stringer, C., eds, pp. 177–86. McDonald Institute for Archaeological Research, Cambridge.

Barton, R.N.E., Bouzouggar, A., Collcutt, S.N., Schwenninger, J.L., and Clark-Balzan, L. (2009) OSL dating of the Aterian levels at Dar es-Soltan I (Rabat, Morocco) and implications for the dispersal of modern *Homo sapiens*. *Quaternary Science Reviews* **28** (19–20): 1914–31.

Bartram, L.E., and Marean, C.W. (1999) Explaining the "Klasies Pattern": Kua ethnoarchaeology, the Die Kelders Middle Stone Age archaeofauna, long bone fragmentation and carnivore ravaging. *Journal of Archaeological Science* **26**: 9–20.

Bartsiokas, A. (2002) Hominid cranial bone structure: A histological study of Omo 1 specimens from Ethiopia using different microscopic techniques. *Anatomical Record* **267**: 52–59.

Bartstra, G.-J., Soegondho, S., and van der Wijk, A. (1988) Ngandong man: age and artefacts. *Journal of Human Evolution* **17**(3): 325–37.

Bartucz, L., Dancza, J., Hollendonner, F., Kadić, O., Mottl, M., Pataki, V., et al. (1940) *Die Mussolini-Höhle (Subalyuk) bei Cserépfalu. Geologica Hungarica Series Palaeontologica*. Editio Instituti Regii Hungarici Geologici, Budapest.

Bate, D.M.A. (1951) The mammals from Singa and Abu Hugar. In: *Fossil Mammals of Africa*, Arkell, A.J., Bate, D.M.A., Wells, L.H., and Lacaille, A.D., eds, pp. 1–28. British Museum Natural History, London.

Batzer, M., Deininger, P., Hellmann-Blumberg, U., Jurka, J., Labuda, D., Rubin, C., et al. (1996) Standardized nomenclature for Alu repeats. *Journal of Molecular Evolution* **42**(1): 3–6.

Bauchot, R., and Stephan, H. (1969) Encephalisation et niveau evolutif chez les simiens. *Mammalia* **33**: 225–75.

Bayle, P., Macchiarelli, R., Trinkaus, E., Duarte, C., Mazurier, A., and Zilhão, J. (2010) Dental maturational sequence and dental tissue proportions in the early Upper Paleolithic child from Abrigo do Lagar Velho, Portugal. *Proceedings of the National Academy of Sciences USA* **107**(4): 1338–42.

Beaumont, P. (1990) Wonderwerk Cave. In: *Guide to Archaeological Sites in the Northern Cape*, Beaumont, P., and Morris, D., eds, pp. 101–34. McGregor Museum, Kimberly.

Beaumont, P.B. (1973) Border Cave: a progress report. *South African Journal of Science* **69**: 41–46.

Beaumont, P.B., and Vogel, J.C. (2006) On a timescale for the past million years of human history in central South Africa. *South African Journal of Science* **102** (5–6): 217–28.

Beaumont, P.B., de Villiers, H., and Vogel, J.C. (1978) Modern man in Sub-Saharan Africa prior to 49,000 years BP: A review and evaluation with particular reference to Border Cave. *South African Journal of Science* **74**: 409–19.

Beaumont, P.B., van Zinderen Bakker, E.M., and Vogel, J.C. (1984) Environmental changes since 32 kyrs at

Kathu Pan, northern Cape, South Africa. In: *Late Cainozoic Palaeoclimates of the Southern Hemisphere*, Vogel, J.C., ed., pp. 324–38. Balkema, Rotterdam.

Behm-Blanke, G. (1960) *Altsteinzeitliche Rastplätze im Travertingebiet von Taubach, Weimar, Ehringsdorf*, Weimar.

Behrensmeyer, A.K., and Laporte, L.F. (1981) Footprints of a Pleistocene hominid in northern Kenya. *Nature* 289: 167–9.

Behrensmeyer, A.K., Potts, R., Plummer, T., Tauxe, L., Opdyke, N., and Jorstad, T. (1995) The Pleistocene locality of Kanjera, Western Kenya: stratigraphy, chronology and palaeoenvironments. *Journal of Human Evolution* 29: 247–74.

Behrensmeyer, A.K., Todd, N.E., Potts, R., and McBrinn, G.E. (1997) Late Pliocene faunal turnover in the Turkana Basin, Kenya and Ethiopia. *Science* 278: 1589–94.

Behrensmeyer, A.K., Potts, R., Deino, A., and Ditchfield, P.W. (2002) Olorgesailie, Kenya: A million years in the life of a rift basin. In: *Sedimentation in Continental Rifts*, Renaut, R.W., and Ashley, G.M., eds. Society for Sedimentary Geology, Tulsa, OK.

Bekken, D., Schepartz, L.A., Miller-Antonio, S., Hou, Y.M., and Huang, W.W. (2004) Taxonomic abundance at Panxian Dadong, a Middle Pleistocene cave in South China. *Asian Perspectives* 43: 333–59.

Bell, R.H.V. (1970) The use of herb layer by grazing ungulates in the Serengeti. In: *Animal Populations in Relation to their Food Resources*, Watson, A., ed. Blackwell Scientific Publications, Oxford.

Bellhouse, D. (2004) The Reverend Thomas Bayes, FRS: a biography to celebrate the tercentenary of his birth. *Statistical Science* 19(1): 3–32.

Belluomini, G., Branca, M., Delitala, L., Malatesta, A., and Zarlenga, F. (1986) Isoleucine epimerization ages of some Pleistocene sites near Roma. *Human Evolution* 1: 209–13.

Bellwood, P. (1997) *Prehistory of the Indo-Malaysian Archipelago*. University of Hawai'i Press, Honolulu.

Belmaker, M., Tchernov, E., Condemi, S., and Bar-Yosef, O. (2002) New evidence for hominid presence in the Lower Pleistocene of the Southern Levant. *Journal of Human Evolution* 43(1): 43–56.

Benefit, B.R. (1999) Biogeography, dietary specialization, and the diversification of African Plio-Pleistocene monkeys. In: *African Biogeography, Climate Change and Human Evolution*, Bromage, T.G., and Schrenk, F., eds, pp. 172–88. Oxford University Press, New York.

Benefit, B.R., and McCrossin, M.L. (2002) The Victoriapithecidae, Cercopithecoidea. *Cambridge Studies in Biological and Evolutionary Anthropology* 33: 241–54.

Benirschke, K., Anderson, J.M., and Brownhill, L.E. (1962) Marrow chimerism in marmosets. *Science* 138 (3539): 513–15.

Bennett, M.R., Harris, J.W.K., Richmond, B.G., Braun, D.R., Mbua, E., Kiura, P., et al. (2009) Early hominin foot morphology based on 1.5-million-year-old footprints from Ileret, Kenya. *Science* 323(5918): 1197–201.

Berckhemer, F. (1933) Ein Menschen-Schädel aus den diluvialen Schottern von Steinheim a.d. Murr. *Anthropologischer Anzeiger* 10: 318–21.

Berckhemer, F. (1934) Der Steinheimer Urmensch und die Tierwelt seines Lebensgebietes. *Natur. Mschr. "Aus der Hemiat"* 47: 101–15.

Berge, C., and Goularas, D. (2010) A new reconstruction of Sts 14 pelvis (*Australopithecus africanus*) from computed tomography and three-dimensional modeling techniques. *Journal of Human Evolution* 58(3): 262–72.

Berger, G.W., Pérez-González, A., Carbonell, E., Arsuaga, J.L., Bermúdez de Castro, J.M., and Ku, T.-L. (2008) Luminescence chronology of cave sediments at the Atapuerca paleoanthropological site, Spain. *Journal of Human Evolution* 55(2): 300–11.

Berger, L., De Ruiter, D., Churchill, S., Schmid, P., Carlson, K., Dirks, P., et al. (2010) *Australopithecus sediba*: a new species of *Homo*-like Australopith from South Africa. *Science* 328(5975): 195–204.

Berger, L.R. (1992) Early hominid fossils discovered at Gladysvale Cave, South Africa. *South African Journal of Science* 88: 362.

Berger, L.R., and Tobias, P.V. (1994) New discoveries at the early hominid site of Gladysvale, South Africa. *South African Journal of Science* 90: 223–6.

Berger, L.R., and Clarke, R.J. (1995) Eagle involvement in accumulation of the Taung child fauna. *Journal of Human Evolution* 29(3): 275–99.

Berger, L.R., and Parkington, J.E. (1995) A new Pleistocene hominid-bearing locality at Hoedjiespunt, South Africa. *American Journal of Physical Anthropology* 98(4): 601–9.

Berger, L.R., and McGraw, W.S. (2007) Further evidence for eagle predation of, and feeding damage on, the Taung child. *South African Journal of Science* 103: 496–9.

Berger, L.R., Keyser, A.W., and Tobias, P.V. (1993) Gladysvale: First Early Hominid Site Discovered in South Africa Since 1948. *American Journal of Physical Anthropology* 92(1): 107–11.

Berger, L.R., Pickford, M., and Thackeray, F. (1995) A Plio-Pleistocene hominid upper central incisor from the Cooper's site, South Africa. *South African Journal of Science* 91: 541–2.

Berger, L.R., de Ruiter, D.J., Steininger, C.M., and Hancox, J. (2003) Preliminary results of excavations at the newly investigated Coopers D deposit, Gauteng, South Africa. *South African Journal of Science* 99: 276–8.

Bergmann, C. (1847) Über die Verhältnisse Wärmeckönomie der Tiere zu ihrer Grösse. *Goettingen Studien* 3(1): 595–708.

Bermúdez de Castro, J.M., and Rosas, A. (1992) A human mandibular fragment from the Atapuerca Trench (Burgos, Spain). *Journal of Human Evolution* 22(1): 41–46.

Bermúdez de Castro, J.M., Arsuaga, J.L., Carbonell, E., Rosas, A., Martínez, I., and Mosquera, M. (1997) A hominid from the Lower Pleistocene of Atapuerca, Spain: possible ancestor to Neandertals and modern humans. *Science* **276**(1392): 1392–95.

Bermúdez de Castro, J.M., Rosas, A., and Nicolás, M.E. (1999) Dental remains from Atapuerca -TD6 (Gran Dolina site, Burgos, Spain). *Journal of Human Evolution* **37**: 523–66.

Bermúdez de Castro, J.M., Martinón-Torres, M., Lozano, M., Sarmiento, S., and Muela, A. (2004) Paleodemography of the Atapuerca-Sima de Los Huesos hominin sample: a revision and new approaches to the paleodemography of the European Middle Pleistocene population. *Journal of Anthropological Research* **60**(1): 5–26.

Bermúdez de Castro, J.M., Pérez-González, A., Mártinon-Torres, M., Gómez-Robles, A., Rosell, J., Prado, L., et al. (2008) A new early Pleistocene hominin mandible from Atapuerca-TD6, Spain. *Journal of Human Evolution* **55**(4): 729–35.

Bernor, R.L., and Kaiser, T.M. (2006) *Systemics and Paleoecology of the Earliest Pliocene Equid*, Eurygnathohippus hooijeri *n. sp.* Mitt. Ham. Zool. Mus. Inst., Band 103, Langebaanweg, South Africa.

Bersaglieri, T., Sabeti, P.C., Patterson, N., Vanderploeg, T., Schaffner, S.F., Drake, J.A., et al. (2004) Genetic signatures of strong recent positive selection at the lactase gene. *American Journal of Human Genetics* **74** (6): 1111–20.

Betzler, C., and Ring, U. (1995) Sedimentology of the Malawi Rift: Facies and stratigraphy of the Chiwondo Beds, northern Malawi. *Journal of Human Evolution* **28**: 23–35.

Beyene, Y. (2003) The emergence and development of the Acheulean at Konso. *Anthropological Science* **111**: 58.

Beyene, Y., Zeleke, Y., and Uzawa, K. (1997) The Acheulean at Konso-Gardula: results from locality KGA4-A2. In: *Ethiopia in Broader Perspective*, Fukui, K., Kurimoto, E., and Shigeta, M., eds, pp. 376–81. Shokado Book Sellers, Kyoto.

Beynon, A.D., and Wood, B.A. (1986) Variation in enamel thickness and structure in East African hominids. *American Journal of Physical Anthropology* **70**: 177–93.

Beynon, A., and Dean, M.C. (1988) Distinct dental development patterns in early fossil hominids. *Nature* **335**: 509–14.

Beynon, A.D., Clayton, C.B., Ramirez-Rozzi, F.V., and Reid, D.J. (1998) Radiographic and histological methodologies in estimating the chronology of crown development in modern humans and great apes: a review with some applications for studies of juvenile hominids. *Journal of Human Evolution* **35**: 351–70.

Biberson, P. (1961) Le gisement de l'Atlanthrope de Sidi Abderrahmane. *Bulletin d'Archééologie marocaine* **1**: 38–92.

Biberson, P. (1967) Some aspects of the Lower Palaeolithic of Northwest Africa. In: *Background to Evolution in Africa*, Bishop, W.W., and Desmond Clark, J., eds, pp. 447–75. University of Chicago Press, Chicago, IL.

Bickerton, D. (1990) *Language and Species*. University of Chicago Press, Chicago, IL.

Bickerton, D. (1995) *Language and Human Behavior*. University of Washington Press, Seattle, WA.

Bickerton, D. (1998) Catastrophic evolution: the case for a single step from protolanguage to full human language. In: *Approaches to the Evolution of Language: Social and Cognitive Bases*, Hurford, J., Knight, C., and Studdert-Kennedy, M., eds, pp. 341–58. Cambridge University Press, Cambridge.

Bickerton, D. (2000) How protolanguage became language. In: *The Evolutionary Emergence of Language: Social Function and the Origins of Linguistic Form*, Knight, C., Studdert-Kennedy, M., and Hurford, J., eds, pp. 264–84. Cambridge University Press, Cambridge.

Biegert, J. (1957) Der Formwandel des Primatenschadels und seine Beziehungen zur onto-genetischen Entwicklung und den phylogenetischen Spezialisationen der Kopforgane. *Morphol. Jahrb.* **98**: 77–199.

Biegert, J. (1963) The evaluation of characteristics of the skull, hands and feet for primate taxonomy. In: *Classification and Human Evolution*, Washburn, S.L., ed., pp. 116–45. Aldine Publishing Company, Chicago, IL.

Bietti, A. (1985) A Late Rissian Deposit in Rome: Rebibbia-Casal de'Pazzi. In: *Ancestors: The Hard Evidence*, Delson, E., ed., pp. 277–82. Alan R. Liss, New York.

Bietti, A. (1994) A re-examination of the lithic industry of the P layers (1940–42 excavations) of the Arene Candide Cave (Savona, Italy). Discussion and general conclusions. *Quaternaria Nova* **4**: 341–70.

Bietti, A. (1997) The transition to anatomically modern humans: The case of peninsular italy. In: *Conceptual Issues in Modern Human Origins Research*, Clark, G.A., and Willermet, C.M., eds, pp. 132–47. Aldine De Gruyter, New York.

Bietti, A., and Manzi, G. (1991) The Fossil Man of Monte Circeo. Fifty years of studies on the Neandertals in Latium. *Quaternaria Nova* **1**: 9–678.

Bietti, A., Kuhn, S., Segre, A.G., and Stiner, M.C. (1991) Grotta Breuil: a general introduction and stratigraphy. *Quaternaria Nova* **1**: 305–24.

Bigazzi, G., Balestrieri, M.L., Norelli, P., Oddone, M., and Tecle, T.M. (2004) Fission-Track Dating of a Tephra Layer in the Alat Formation of the Dandiero Group (Danakil Depression, Eritrea). *Rivista Italiana de Paleontologia e Stratigrafia* **110**: 45–49.

Biggerstaff, R.H. (1968) On the groove configuration of mandibular molars: the unreliability of the "dryopithecus pattern" and a new method for classifying mandibular molars. *American Journal of Physical Anthropology* **29**(3): 441–4.

Biggerstaff, R. (1969) The basal area of posterior tooth crown components: the assessment of within tooth variations of premolars and molars. *American Journal of Physical Anthropology* 31(2): 163–70.

Biknevicius, A., Van Valkenburgh, B., and Walker, J. (1996) Incisor Size and Shape: Implications for Feeding Behaviors in Saber-Toothed" Cats". *Journal of Vertebrate Paleontology* 16(3): 510–21.

Billy, G. (1975) Etude anthropologique des restes humains de l'Abri Pataud. In: *Excavation of the Abri Pataud, Les Eyzies (Dordogne)*, Movius, H.L., ed., pp. 201–61. Peabody Museum of Archaeology and Ethnology, Harvard University, Cambridge, MA.

Billy, G. (1982) Les dents humaines de la grotte du Coupe-Gorge a Montmaurin. *Bulletins et Mémoires de la Société d'Anthropologie de Paris* 9(3): 211–25.

Billy, G. (1985) Les restes humains de la grotte du Coupe-Gorge à Montmaurin (Haute-Garonne). *Zeitschrift für Morphologie und Anthropologie* 75(2): 223–37.

Billy, G., and Vallois, H.-V. (1977a) La mandibule pré-Rissienne de Montmaurin. *L'Anthropologie* 81(2): 273–312.

Billy, G., and Vallois, H.-V. (1977b) La mandibule pré-Rissienne de Montmaurin (suite). *L'Anthropologie* 81(3): 411–58.

Bilsborough, A., and Thompson, J. (2005) The dentition of the Le Moustier 1 Neanderthal. In: *The Neanderthal Adolescent Le Moustier 1: new aspects, new results*, Ulrich, H., ed. Staatliche Museen Zu Berlin-Preussicher Kulturbesitz, Berlin.

Binford, L. (1962) Archaeology as anthropology. *American Antiquity* 28(2): 217–25.

Binford, L. (1978) *Nunamiut Ethnoarchaeology*. Academic Press, New York.

Binford, L. (1984) *Faunal remains from the Klasies River Mouth*. Academic Press, New York.

Binford, L. (2001) *Constructing Frames of Reference: an analytical method for archaeological theory building using ethnographic and environmental data sets*. University of California Press, Berkeley, CA.

Binford, L.R., and Binford, S. (1966) A preliminary analysis of functional variability in the Mousterian and Upper Paleolithic. *American Anthropologist* 68: 236–95.

Binford, L.R., and Binford, S.R. (1969) Stones Tools and Human Behavior. *Scientific American* 220(4): 70–84.

Binford, L., and Clark Howell, F. (1981) *Bones. Ancient Men and Modern Myths*. Academic Press, New York.

Bintliff, J., ed. (1999) *Structure and Contingency: Evolutionary Processes in Life and Human Society*. Leicester University Press, London.

Birchette, M.G. (1982) *The Postcranial Skeleton of Paracolobus chemeroni*. PhD thesis, Harvard University.

Bird, A. (2007) Perceptions of epigenesis. *Nature* 447: 396–8.

Bird, M.I., Fifield, L.K., Santos, G.M., Beaumont, P.B., Zhou, Y., di Tada, M.L., et al. (2003) Radiocarbon dating from 40 to 60 ka BP at Border Cave, Stouth Africa. *Quaternary Science Reviews* 22(8–9): 943–7.

Birdsell, J.B. (1979) A reassessment of the age, sex, and population affinities of the Niah cranium. *American Journal of Physical Anthropology* 50: 419.

Bischoff, J.L., Fitzpatrick, J.A., Léon, L., Arsuaga, J.L., Falguéres, C., Bahain, J.J., et al. (1997) Geology and preliminary dating of the hominid-bearing sedimentary fill of the Sima de los Huesos Chamber, Cueva Mayor of the Sierra de Atapuerca, Burgos, Spain. *Journal of Human Evolution* 33(2–3): 129–54.

Bischoff, J., Williams, R., Rosenbauer, R., Aramburu, A., Arsuaga, J., García, N., et al. (2007) High-resolution U-series dates from the Sima de los Huesos hominids yields: implications for the evolution of the early Neanderthal lineage. *Journal of Archaeological Science* 34(5): 763–70.

Bishop, W., and Miller, J., eds (1972) *Calibration of Hominoid Evolution: Recent Advances in Isotopic and Other Dating Methods Applicable to the Origin of Man*. Scottish Academic Press, Edinburgh.

Bishop, W., Chapman, G., and Hill, A. (1971) Succession of Cainozoic vertebrate assemblages from the northern Kenya Rift Valley. *Nature* 233: 389–94.

Bishop, W.W. (1965) Quaternary geology and geomorphology in the Albertine Rift Valley and Frey Valley, Uganda. In: *International Studies on the Quaternary*, Wright Jr, H.E., and Frey, D.G., eds, pp. 293–321. Geological Society of America, Boulder, CO.

Bishop, W.W., ed. (1978) *Geological Background to Fossil Man: Recent research in the Gregory Rift Valley, East Africa*. Scottish Academic Press, Edinburgh.

Bishop, W.W., and Clark, J.D., eds (1967) *Background to Evolution in Africa*. University of Chicago Press, Chicago, IL.

Bishop, W.W., and Chapman, G.R. (1970) Early Pliocene sediments and fossils from the Northern Rift Valley, Kenya. *Nature* 226: 914–18.

Bishop, W.W., and Miller, J.A., eds (1972) *Calibration of Hominoid Evolution*. Proceedings of the Symposium held at Burg Wartenstein, Austria. Scottish Academic Press, Edinburgh.

Bishop, W.W., and Pickford, M.H.L. (1975) Geology, fauna and paleoenvironments of the Ngorora Formation, Kenya Rift Valley. *Nature* 254: 185–92.

Bishop, W.W., Pickford, M., and Hill, A. (1975) New evidence regarding the Quaternary geology, archaeology and hominids of Chesowanja, Kenya. *Nature* 258: 204–8.

Black, D. (1927) On a lower molar hominid tooth from Chou-kou-tien deposit. *Palaeontologia Sinica* 7: 1–28.

Black, D. (1934) The Croonian Lecture: on the discovery, morphology, and environment of *Sinanthropus pekinensis*. *Philosophical Transactions of the Royal Society of London, Series B. Containing Papers of a Biological Character* 223: 57–120.

Blackwell, B., and Schwarcz, H. (1986) Uranium-series analyses of the lower Travertine at Ehringsdorf, DDR (East Germany). *Quaternary Research* 25(2): 215–22.

Blackwell, B., Schwarcz, H.P., and Debénath, A. (1983) Absolute dating of hominids and Palaeolithic artifacts of La Chaise-de-Vouthon (Charente), France. *Journal of Archaeological Science* **10**: 493–513.

Blanc, A.C. (1948) Notizie sui ritrovamenti e sul giacimento di Saccopastore e sulla sua posizione nel Pleistocene laziale. *Palaeontogica Italiana* **47**: 3–23.

Blanc, A.C. (1954) Reperti fossili neandertaliani nella Grotta del Fossellone al Monte Circeo: Circeo IV. *Quaternaria* **1**: 171–5.

Blanc, A.C. (1956) Il più antico reperto fossile umano del Lazio rinvenuto a Roma: un secondo metatarsale destro umano e industria paleolitica nelle ghiaie superiori della Sedia del Diavolo (Roma). *Quaternaria* **3**: 1–259.

Blumenbach, J. (1795) *De Generis Humani Varietate Nativa*. apud Vandenhoek et Ruprecht, Gottingen.

Blumenberg, B., and Lloyd, A.T. (1983) *Australopithecus* and the origin of the genus *Homo*: aspects of biometry and systematics with accompanying catalog of tooth metric data. *Biosystems* **16**: 127–67.

Blumenschine, R.J. (1991) Hominid carnivory and foraging strategies, and the socio-economic function of early archaeological sites. *Philosophical Transactions of the Royal Society of London. Series B, Biological Sciences* **334**(1270): 211–21.

Blumenschine, R.J., Peters, C.R., Masao, F.T., Clarke, R.J., Deino, A., Hay, R.L., et al. (2003) Late Pliocene *Homo* and hominid land use from western Olduvai Gorge, Tanzania. *Science* **299**: 1217–21.

Boaz, N.T. (1988) Status of *Australopithecus afarensis*. *Yearbook of Physical Anthropology* **31**: 85–113.

Boaz, N.T., and Howell, F.C. (1977) A gracile hominid cranium from upper member G of the Shungura Formation, Ethiopia. *American Journal of Physical Anthropology* **46**(1): 93–108.

Boaz, N.T., Pavlakis, P., and Brooks, A.S. (1990) Late Pleistocene-Holocene human remains from the Upper Semliki, Zaire. In: *Evolution of Environments and Hominidae in the African Western Rift Valley*, Boaz, N.T., ed., pp. 273–99. Memoir No. 1. Virginia Museum of Natural History, Martinsville, VA.

Boaz, N.T., Ciochon, R.L., Xu, Q., and Liu, J. (2004) Mapping and taphonomic analysis of the *Homo erectus* loci at Locality 1 Zhoukoudian, China. *Journal of Human Evolution* **46**(5): 519–49.

Bobe, R., and Eck, G. (2001) Responses of African bovids to Pliocene climatic change. *Paleobiology* **27**: 1–48.

Bobe, R., and Leakey, M. (2009) Ecology of Plio-Pleistocene mammals in the Omo-Turkana Basin and the emergence of *Homo*. In: *The First Humans: Origin and Early Evolution of the Genus* Homo, Grine, F.E., Fleagle, J.G., and Leakey, R.E., eds, pp. 173–84. Springer, Dordrecht.

Bocherens, H., Drucker, D., Billiou, D., Patou-Mathis, M., and Vandermeersch, B. (2005) Isotopic evidence for diet and subsistence pattern of the Saint-Cesaire I Neanderthal: review and use of a multi-source mixing model. *Journal of Human Evolution* **49**(1): 71–87.

Boëda, E. (1994) *Le concept Levallois: variabilite des methodes*. Centre nationale de la recherche scientifique, Paris.

Boëda, E. (1995) Levallois: Volumetric Construction, Methods, a Technique. In: *Monographs in World Archaeology*, Dibble, H., and Bar-Yosef, O., eds. Prehistory Press, Madison, WI.

Boesch, C., and Boesch, H. (1990) Tool use and tool making in wild chimpanzees. *Folia Primatologica* **54**: 86–99.

Boesch, C., Head, J., and Robbins, M. (2009) Complex tool sets for honey extraction among chimpanzees in Loango National Park, Gabon. *Journal of Human Evolution* **56**(6): 560–9.

Bogard, J.S., Murray, M.E., Weeks, R.A., Weinand, D.C., Elam, J.M., and Kramer, A. (2004) *Age of an Indonesian Fossil Tooth Determined by Electron Paramagnetic Resonance*. Oak Ridge National Laboratory, Department of Energy, Oak Ridge, TN.

Bogart, S., and Pruetz, J. (2008) Ecological context of savanna chimpanzee (*Pan troglodytes verus*) termite fishing at Fongoli, Senegal. *American Journal of Primatology* **70**(6): 605–12.

Bogin, B. (1997) Evolutionary hypotheses for human childhood. *Yearbook of Physical Anthropology* **40**: 63–90.

Bogin, B. (1999a) *Patterns of Human Growth*. Cambridge University Press, Cambridge.

Bogin, B. (1999b) Evolutionary perspective on human growth. *Annual Review of Anthropology* **28**: 109–53.

Böhm, M., and Mayhew, P.J. (2005) Historical biogeography and the evolution of the latitudinal gradient of species richness in the Papionini (Primata: Cercopithecidae). *Biological Journal of the Linnean Society* **85**(2): 235–46.

Bolender, J. (2007) Prehistoric cognition by description: a Russellian approach to the Upper Paleolithic. *Biology and Philosophy* **22**(3): 383–99.

Bolk, L. (1909) On the position and displacement of the foramen magnum in the primates. *Proceedings of the Koninklijke Nederlandsche Akademie van Wetenschappen* **12**: 362–77.

Bolk, L. (1910) On the slope of the foramen magnum in primates. *Proceedings of the Koninklijke Nederlandsche Akademie van Wetenschappen* **12**: 252–534.

Bolk, L. (1915) Über Lagerung, Verschiebung und Neigung des Foramen magnum am Schadel der Primaten. *Z. Morph. Anthropol.* **17**: 611–92.

Bolker, B. (2008) *Ecological Models and Data in R*. Princeton University Press, Princeton, NJ.

Bonatto, S.L., and Salzano, F.M. (1997) Diversity and age of the four major mtDNA haplogroups, and their implications for the peopling of the New World. *American Journal of Human Genetics* **61**(6): 1413–23.

Bond, A.B., Kamil, A.C., and Balda, R.P. (2003) Social complexity and transitive inference in corvids. *Animal Behavior* **65**: 479–87.

Bondioli, L., Coppa, A., Frayer, D.W., Libsekal, Y., Rook, L., and Macchiarelli, R. (2006) A one-million-year-old human pubic symphysis. *Journal of Human Evolution* **50**(4): 479–83.

Bonifay, E. (1964) La grotte du Regourdou (Montignac, Dordogne): Stratigraphie et industrie lithique mousterienne. *Anthropologie* **68**(1–2): 49–64.

Bonifay, E., and Vandermeersch, B. (1962) Dépôts rituels d'ossements d'ours dans le gisement moustérien du Régourdou (Montignac, Dordogne). *Comptes Rendus de l'Académie des Sciences* **255**: 1635–6.

Bonnefille, R. (1976) Palynological evidence for an important change in the vegetation of the Omo basin between 2.5 and 2 million years ago. In: *Earliest Man and Environments in the Lake Rudolf Basin: Stratigraphy, Paleoecology and Evolution*, Coppens, Y., Isaac, G.L., and Leakey, R.E.F., eds, pp. 421–31. University of Chicago Press, Chicago, IL.

Bonnefille, R., and Letouzey, R. (1976) Fruits fossiles d'Antrocaryon dans la vallée de l'Omo (Ethiopie). *Adansonia* **16**: 65–82.

Bonnefille, R., Potts, R., Chalie, F., Jolly, D., and Peyron, Q. (2004) High-resolution vegetation and climate change associated with Pliocene *Australopithecus afarensis*. *Proceedings of the National Academy of Sciences USA* **101**(33): 12125–39.

Bonnie, K., and Earley, R. (2007) Expanding the scope for social information use. *Animal Behaviour* **74**(2): 171–81.

Bookstein, F. (1996) Biometrics, biomathematics and the morphometric synthesis. *Bulletin of Mathematical Biology* **58**(2): 313–65.

Bookstein, F.L. (1991) *Morphometric Tools for Landmark Data: Geometry and Biology*. Cambridge University Press, Cambridge.

Bookstein, F.L. (1997) Landmark methods for forms without landmarks: morphometrics of group differences in outline shape. *Medical Image Analysis* **1**(3): 225–43.

Bookstein, F.L., Streissguth, A.P., Sampson, P.D., Connor, P.D., and Barr, H.M. (2002) Corpus callosum shape and neuropsychological deficits in adult males with heavy fetal alcohol exposure. *NeuroImage* **15**(1): 233–51.

Bookstein, F.L., Gunz, P., Mitteroecker, P., Prossinger, H., Schaefer, M.S., and Seidler, H. (2003) Cranial integration in *Homo*: singular warps analysis of the mid saggital plane in ontogeny and evolution. *Journal of Human Evolution* **44**: 167–87.

Borden, C.E. (1954) A uniform site designation scheme for Canada. *Anthropology in British Columbia* **4**: 44–48.

Bordes, F. (1955) La stratigraphie de la grotte de Combe-Grenal, commune de Domme (Dordogne). *Bull. Soc. préhist. Fr.* **52**: 426–9.

Bordes, F. (1961a) *Typologie du Paléolithique Ancien et Moyen*. Centre National de la Recherche Scientifique, Paris.

Bordes, F. (1961b) Mousterian cultures in France. *Science* **134**: 803–10.

Bordes, F. (1966) Observationes sur la Pleistocene supériur de gisement de Combe-Grenal (Dordogne). *Act. Soc. Linn. Bordeaux. ser B No. 10*(103): 3–19.

Bordes, F. (1972) *A Tale of Two Caves*. Harper & Row, New York.

Bordes, F. (1973) On the chronology and contemporaneity of different palaeolithic cultures in France. In: *The Explanation of Culture Change*, Renfrew, C., ed., pp. 217–26. Duckworth, Surrey.

Bordes, F.H., and Lafille, J. (1962) Découverte d'une squelette d'enfant moustérien dans le gisement du Roc de Marsal, commune de Campagne-du-Bugue (Dordogne). *Comptes Rendus de l'Académie des Sciences* **254**: 714–15.

Bordes, F., and Sonneville-Bordes, D.d. (1970) The Significance of Variability in Palaeolithic Assemblages. *World Archaeology* **2**(1): 61–73.

Bosinski, G. (1982) *Die Kunst der Eiszeit in Deutschland und in der Schweiz: Kataloge Vor- und Frühgeschichte Altertümer, Band 20*. Rudolf Habelt GMBH, Bonn.

Bossavy, J. (1923) Découverte de trous squelettes Aurignaciens à Solutré. *Bulletin de la Société préhistorique française* **20**(10): 295–300.

Boule, M. (1908) L'Homme fossile de la Chapelle-aux-Saints. *Anthropologie* **19**: 519–25.

Boule, M. (1909) L'homme fossile de la Chapelle-aux-Saints (Corrèze) deuxième article. *L'Anthropologie* **20**: 257–71.

Boule, M. (1911) L'homme fossile de la Chapelle-aux-Saints. *Annales de Paléontologie* **6**: 106–72.

Boule, M. (1912) L'homme fossile de la Chapelle-aux-Saints. *Annales de Paléontologie* **7**: 21–56; 85–192.

Boule, M. (1921) *Les Hommes Fossiles*. Mason et Cie, Paris.

Boule, M., and Vallois, H.-V. (1946) *Fossil Men*. Thames and Hudson, London.

Boule, M., and Vallois, H.-V. (1957) *Fossil Men*. Thames and Hudson, London.

Bourlière, F., and Hadley, M. (1983) Present day savannas: an overview. In: *Ecosystems of the World: Tropical Savannas*, Bourlière, F., ed., pp. 1–17. UNESCO, Paris.

Bouvier, J.M. (1973) Nouvelle diagnose stratigraphique du gisement éponyme de La Madeleine (Tursac, Dordogne). *Comptes Rendus de l'Académie des Sciences* 2625–8.

Bouyssonie, A., Bouyssonie, J., and Bardon, L. (1908) Découverte d'un squelette humain moustériena la bouffia de la Chapelle-aux-Saints (Corrèze). *L'Anthropologie* **19**: 513–18.

Bouyssonie, A., Bouyssonie, J., and Bardon, L. (1913) La station moustérienne de la "bouffia" bonneval a la Chapelle-aux-Saints. *L'Anthropologie* (24): 609–40.

Bouzat, J.-L. (1982) *Le malaire de l'Homme de Tautavel in L'Homo erectus et la place de l'Homme de Tautavel parmi les Hominidés fossiles*. 1er Congrès International de Paléontologie Humaine, Nice, France.

Bouzouggar, A. (1997) Économie des matières premières et du débitage dans la séquence atérienne de la grotte d'El Mnasra I (Ancienne grotte des Contrebandiers - Maroc). *Préhistoire anthropologie méditerranéennes* **6**: 35–52.

Bouzouggar, A., Koziowski, J.K., and Otte, M. (2002) Étude des ensembles lithiques atériens de la grotte d'El Aliya à Tanger (Maroc) Study of the Aterian lithic assemblages from El Aliya cave in Tangier (Morocco). *L'Anthropologie* **106**(2): 207–48.

Bouzouggar, A., Barton, N., Vanhaeren, M., d'Errico, F., Collcutt, S., Higham, T., et al. (2007) 82,000-year-old shell beads from North Africa and implications for the origins of modern human behavior. *Proceedings of the National Academy of Sciences USA* **104**(24): 9964–69.

Bouzouggar, A., Barton, R.N.E., Blockley, S., Bronk-Ramsey, C., Collcutt, S.N., Gale, R., et al. (2008) Reevaluating the Age of the Iberomaurusian in Morocco. *African Archaeological Review* **25**: 3–19.

Bowcock, A.M., Ruiz-Linares, A., Tomfohrde, J., Minch, E., Kidd, J.R., and Cavalli-Sforza, L.L. (1994) High resolution of human evolutionary trees with polymorphic microsatellites. *Nature* **368**: 455–7.

Bowen, B., and Vondra, C. (1973) Stratigraphical relationships of the Plio-Pleistocene deposits, East Rudolf, Kenya. *Nature* **242**: 391–3.

Bowler, J.M. (1970a) Alluvial Terraces in the Maribyrnong Valley, near Keilor, Victoria. *Memoirs of the National Museum of Victoria* **27**: 19–67.

Bowler, J.M. (1970b) Lake Nitchie skeleton - stratigraphy of burial site. *Archaeology & Physical Anthropology in Oceania* **5**: 104–13.

Bowler, J.M., Thorne, A.G., and Polach, H.A. (1972) Pleistocene man in Australia: age and significance of the Mungo skeleton. *Nature* **240**: 48–50.

Bowler, J.M., Johnston, H., Olley, J.M., Prescott, J.R., Roberts, R.G., Shawcross, W., et al. (2003) New ages for human occupation and climatic change at Lake Mungo, Australia. *Nature* **421**: 837–40.

Boyd, R., and Richerson, P. (1985) *Culture and the Evolutionary Process*. University of Chicago Press, Chicago, IL.

Boyd, R., and Richerson, P.J. (2005) *The Origin and Evolution of Cultures*. Oxford University Press, Oxford.

Boyde, A. (1964) *The Structure and Development of Mammalian Enamel*. Thesis, University of London.

Boyde, A. (1989) Enamel. In: *Handbook of Microscopic Anatomy*, Oksche, A., and Vollrath, L., eds, pp. 309–473. Springer-Verlag, Berlin.

Boyde, A. (1990) Developmental interpretations of dental microstructure. In: *Primate Life History and Evolution*, De Rousseau, C.J., ed., pp. 229–67. Wiley-Liss, New York.

Boyde, A., and Jones, S. (1985) Bone modelling in the implantation bed. *Journal of biomedical materials research* **19**(3): 199–224.

Boyde, A., and Martin, L. (1987) Tandem scanning reflected light microscopy of primate enamel. *Scanning Microscopy* **1**(4): 265–78.

Bradley, B., and Stanford, D. (2004) The North Atlantic ice-edge corridor: a possible palaeolithic route to the New World. *World Archaeology* **36**(4): 459–78.

Bradley, B.J. (2008) Reconstructing phylogenies and phenotypes: a molecular view of human evolution. *Journal of Anatomy* **212**(4): 337–53.

Braga, J., and Thackeray, J.F. (2003) Early *Homo* at Kromdraai B: probabilistic and morphological analysis of the lower dentition. *Comptes Rendus Palevol* **2**(4): 269–79.

Brahimi, C. (1972) *Initiation a la préhistoire de l'Algérie*. SNED, Paris.

Brain, C.K. (1958) The Transvaal ape-man-bearing cave deposits. *Transvaal Museum Memoirs* (11): 1–131.

Brain, C.K. (1981) *The Hunters or the Hunted? An Introduction to African Cave Taphonomy*. University of Chicago Press, Chicago, IL.

Brain, C.K., ed. (2004) *Swartkrans: A Cave's Chronicle of Early Man*. Transvaal Museum Monograph. Transvaal Museum, Pretoria.

Bramble, D.M., and Lieberman, D.E. (2004) Endurance running and the evolution of *Homo*. *Nature* **432**: 345–52.

Brand, P., and Hollister, A. (1999) *Clinical Mechanics of the Hand*. Mosby-Year Book, St. Louis, MO.

Brantingham, P.J., and Kuhn, S.L. (2001) Constraints on Levallois Core Technology: A Mathematical Model. *Journal of Archaeological Science* **28**: 747–61.

Brantingham, P.J., Olsen, J.W., Rech, J.A., and Krivoshapkin, A.I. (2000) Raw material quality and prepared core technologies in Northeast Asia. *Journal of Archaeological Science* **27**(3): 255–71.

Bräuer, G. (1978) The morphological differentiation of anatomically modern man in Africa, with special regard to recent finds from East Africa. *Zeitschrift für Morphologie und Anthropologie* **69**: 266–92.

Bräuer, G. (1980) Die morphologischen affinitaten des jungpleistozanen stirnbeins aus dem Elbmundungsgebeit bei Hahnofersand. *Zeitschrift für Morphologie und Anthropologie* **71**: 1–42.

Bräuer, G. (1984) A craniological approach to the origin of anatomically modern *Homo sapiens* in Africa and implications for the appearance of modern Europeans. In: *The Origins of Modern Humans: A World Survey of the Fossil Evidence*, Smith, F.H., and Spencer, F., eds, pp. 327–410. Alan R. Liss, New York.

Bräuer, G. (1985) The "Afro-European *sapiens*-hypothesis" and hominid evolution in East Asia during the late Middle and Upper Pleistocene. *Courier Forsch Senckenberg* **69**: 145–65.

Bräuer, G. (2001) The KNM-ER 3884 hominid and the emergence of modern anatomy in Africa. In: *Humanity from the African Naissance to the Coming Millennia*, Tobias, P.V., Raath, M.A., Moggi-Cecchi, J., and Doyle, G.A., eds, pp. 191–7. Florence University Press, Florence.

Bräuer, G. (2008) The origin of modern anatomy: by speciation or intraspecific evolution? *Evolutionary Anthropology* **17**(1): 22–37.

Bräuer, G., and Leakey, R.E.F. (1986a) The ES-11693 cranium from Eliye Springs, West Turkana, Kenya. *Journal of Human Evolution* **15**: 289–312.

Bräuer, G., and Mehlman, M.J. (1988) Hominid molars from a Middle Stone Age level at the Mumba Rock shelter, Tanzania. *American Journal of Physical Anthropology* **75**: 69–76.

Bräuer, G., and Mabulla, Z.P. (1996) New hominid fossil from Lake Eyasi, Tanzania. *Anthropologie* **34** (1–2): 47–53.

Bräuer, G., and Leakey, R.E.F. (1986b) A new archaic *Homo sapiens* cranium from Eliye Springs, West Turkana, Kenya. *Zeitschrift für Morphologie und Anthropologie* **76**: 245–52.

Bräuer, G., and Rimbach, K. (1990) Late archaic and modern *Homo sapiens* from Europe, Africa and southwest Asia: craniometric comparisons and phylogenetic implications. *Journal of Human Evolution* **19**: 789–807.

Bräuer, G., and Smith, F.H. (1992) *Continuity or Replacement? Controversies in* Homo sapiens *Evolution*. Balkema, Rotterdam.

Bräuer, G., and Schultz, M. (1996) The morphological affinities of the Plio-Pleistocene mandible from Dmanisi, Georgia. *Journal of Human Evolution* **30**(5): 445–81.

Bräuer, G., and Singer, R. (1996a) The Klasies zygomatic bone: archaic or modern? *Journal of Human Evolution* **30**: 161–5.

Bräuer, G., and Singer, R. (1996b) Not outside the modern range. *Journal of Human Evolution* **30**: 173–4.

Bräuer, G., Zipfel, F., and Deacon, H.J. (1992) Comment on the new maxillary finds from Klasies River, South Africa. *Journal of Human Evolution* **23**: 419–22.

Bräuer, G., Deacon, H., and Zipfel, F. (1992a) Comment on the new maxillary finds from Klasies River, South Africa. *Journal of Human Evolution* **23**(5): 419–22.

Bräuer, G., Leakey, R.E.F., and Mbua, E. (1992b) A first report on the ER-3884 cranial remains from Ileret/East Turkana, Kenya. In: *Continuity or Replacement? Controversies in* Homo sapiens *Evolution*, Bräuer, G., and Smith, F.H., eds, pp. 111–20. Balkema, Rotterdam.

Bräuer, G., Yokoyama, Y., Falguères, C., and Mbua, E. (1997) Modern human origins backdated. *Nature* **386**: 337–8.

Bräuer, G., Groden, C., Delling, G., Kupczik, K., Mbua, E., and Schultz, M. (2003) Pathological Alterations in the Archaic *Homo sapiens* Cranium from Eliye Springs, Kenya. *American Journal of Physical Anthropology* **120**: 200–4.

Bräuer, G., Groden, C., Gröning, F., Kroll, A., Kupczik, K., Mbua, E., et al. (2004) Virtual study of the endocranial morphology of the matrix-filled cranium from Eliye Springs, Kenya. *Anatomical Record* **276**(2): 113–33.

Braun, D., Harris, J., Levin, N., McCoy, J., Herries, A., Bamford, M., et al. (2010) Early hominin diet included diverse terrestrial and aquatic animals 1.95 Ma in East Turkana, Kenya. *Proceedings of the National Academy of Sciences USA* **107**(22): 10002–7.

Braun, D.R., Plummer, T., Ditchfield, P., Ferraro, J.V., Maina, D., Bishop, L.C., et al. (2008) Oldowan Behavior and Raw Material Transport: Perspectives from the Kanjera Formation. *Journal of Archaeological Science* **35**(8): 2329–45.

Braun, D.R., Harris, J.W.K., and Maina, D.N. (2009) Oldowan raw material procurement and use: evidence from the Koobi Fora Formation. *Archaeometry* **51**(1): 26–42.

Breitinger, E. (1964) *Der Neanderthaler von Petralona*. 7th International Congress of Anthropological and Ethnological Sciences, Moscow.

Breuil, H. (1910) Etudes de morphologie palèolithique: l'industrie de la grotte de Chatelperron (Allier) et d'autres gisements similaires. *Revue Ecole Anthropol. Paris* **29**: 29–40.

Breuil, H., and Blanc, A.C. (1935) Rinvenimento "in situ" di un nuovo cranio di *Homo neanderthalensis* nel giacimento di Saccopastore (Roma). *Atti Accademia Nazionale dei Lincei* **22**: 166–9.

Bricker, H., and Mellars, P. (1987) Datations 14 C de l'Abri Pataud (Les Eyzies, Dordogne) par le procédé accélérateur-spectromètre de masse. *L'Anthropologie* **91**(1): 227–34.

Briggs, A.W., Good, J.M., Green, R.E., Krause, J., Maricic, T., Stenzel, U., et al. (2009) Targeted retrieval and analysis of five Neandertal mtDNA genomes. *Science* **325**(5938): 318–21.

Briggs, L.C. (1968) Hominid evolution in northwest Africa and the question of the north African "Neanderthaloids". *American Journal of Physical Anthropology* **29**(3): 377–85.

Brink, J.S., and Henderson, Z.L. (2001) A high-resolution last interglacial MSA horizon at Florisbad in the context of other open-air occurences in the central interior of southern Africa: an interim statement. In: *Settlement Dynamics of the Middle Paleolithic and Middle Stone Age*, Conard, E., ed. Kerns Verlag, Tübingen.

Broca, P. (1868) Sur le crânes et ossements des Eyzies. *Bulletins de la Société d'Anthropologie de Paris* **3**: 350–92.

Brodmann, K. (1909) *Vergleichende Lokalisationslehre der Grosshirnrinde*. Barth, Leipzig.

Bromage, T., and Dean, M. (1985) Re-evaluation of the age at death of immature fossil hominids. *Nature* **317**: 525–7.

Bromage, T., and Schrenk, F. (1995) Biogeographic and climatic basis for a narrative of early hominid evolution. *Journal of Human Evolution* **28**(1): 109–14.

Bromage, T.G. (1989) Ontogeny of the early hominid face. *Journal of Human Evolution* **18**(8): 751–73.

Bromage, T.G. (1990) Early hominid development and life history. In: *Primate Life History and Evolution*, DeRousseau, C.J., ed., pp. 105–13. Wiley-Liss, New York.

Bromage, T.G. (1991) Enamel incremental periodicity in the pig-tailed macaque: a polychrome fluorescent labelling study of dental hard tissues. *American Journal of Physical Anthropology* **86**: 205–14.

Bromage, T.G., Schrenk, F., and Juwayeyi, Y.M. (1995) Paleobiogeography of the Malawi Rift: Age and vertebrate paleontology of the Chiwondo Beds, northern Malawi. *Journal of Human Evolution* **28**(1): 37–57.

Bromage, T.G., Perez-Ochoa, A., and Boyde, A. (2005) Portable confocal microscope reveals fossil hominid microstructure. *Microsocopy and Analysis* **19**(3): 5–7.

Bromage, T.G., McMahon, J.M., Thackeray, J.F., Kullmer, O., Hogg, R., Rosenberger, A.L., et al. (2008) Craniofacial architectural constraints and their importance for reconstructing the early *Homo* skull KNM-ER 1470. *Journal of Clinical Paediatric Dentistry* **33**(1): 43–54.

Bromage, T.G., Lacruz, R.S., Hogg, R., Goldman, H.M., McFarlin, S.C., Warshaw, J., et al. (2009) Lamellar bone is an incremental tissue reconciling enamel rhythms, body size, and organismal life history. *Calcified Tissue International* **84**: 388–404.

Brook, G.A., Burney, D.A., and Cowart, J.B. (1990) Paleoenvironmental data for Ituri, Zaire, from sediments in Matupi Cave, Mt. Hoyo. In: *Evolution of Environments and Hominidae in the African Western Rift Valley*, Boaz, N.T., ed. Memoir No. 1. Virginia Museum of Natural History, Martinsville, VA.

Brooks, A.S. (2008) Katanda and the Development of complex technologies and economic practices by early modern humans in Africa. In: *Ishango, 22000 and 50 Years Later: The Cradle of Mathematics?*, Huylebrouck, D., ed., pp. 39–54. Koninklijke Vlaamse Academie van Belgie voor Wetenschappen en Kunsten, Brussels.

Brooks, A.S., and Smith, C.C. (1987) Ishango revisited: new age determinations and cultural interpretations. *African Archaeological Review* **5**(1): 65–78.

Brooks, A.S., Smith, C.C., and Boaz, N.T. (1991) New human remains from Ishango, Zaire, in relation to Later Pleistocene human evolution. *American Journal of Physical Anthropology* **S12**: 54–5.

Brooks, A.S., Helgren, D.M., Cramer, J.S., Franklin, A., Hornyak, W., Keating, J.M., et al. (1995) Dating and context of three Middle Stone Age sites with bone points in the Upper Semliki valley, Zaire. *Science* **268** (5210): 548–53.

Brooks, A.S., Yellen, J.E., Cornelissen, E., Verniers, J., de Heinzelin, J., Helgren, D., et al. (2008a) Archaeology, dating and paleoenvironments of the sites at Ishango and Katanda, D.R. Congo. In: *Ishango, 22000 and 50 Years Later: The Cradle of Mathematics?*, Huylebrouck, D., ed., pp. 55–66. Koninklijke Vlaamse Academie van Belgie voor Wetenschappen en Kunsten, Brussels.

Broom, R. (1918) The evidence afforded by the Boskop skull of a new species of primitive man (*Homo capensis*). *Anthropological Papers of the American Museum of Natural History*.

Broom, R. (1924) On the classification of the reptiles. *Bulletin of the American Museum of Natural History* **51**: 39–65.

Broom, R. (1929) The Transvaal fossil human skeleton. *Nature* **123**: 415–16.

Broom, R. (1930) *The Origin of the Human Skeleton*. Witherby, London.

Broom, R. (1933) *The Coming of Man: Was it Accident or Design?* Witherby, London.

Broom, R. (1936a) A new fossil anthropoid skull from South Africa. *Nature* **138**: 486–8.

Broom, R. (1936b) The dentition of *Australopithecus*. *Nature* **138**: 719–19.

Broom, R. (1938a) The Pleistocene anthropoid apes of South Africa. *Nature* **142**: 377–9.

Broom, R. (1938b) Further evidence on the structure of the South African Pleistocene anthropoids. *Nature* **142**: 897–9.

Broom, R. (1942) The hand of the ape-man, *Paranthropus robustus*. *Nature* **149**: 513–14.

Broom, R. (1943) An ankle-bone of the ape-man, *Paranthropus robustus*. *Nature* **152**: 689–90.

Broom, R. (1949) Another new type of fossil ape-man. *Nature* **163**: 57.

Broom, R. (1950) *Finding the Missing Link*. Watts & Co., London.

Broom, R., and Schepers, G.W.H. (1946) *The South African Fossil Ape-men: the Australopithecinae*. Transvaal Museum, Pretoria.

Broom, R., and Robinson, J.T. (1947) Further remains of the Sterkfontein ape-man, Plesianthropus. *Nature* **160**: 430–1.

Broom, R., and Robinson, J. (1949a) Thumb of the Swartkrans ape-man. *Nature* **164**(4176): 841–2.

Broom, R., and Robinson, J.T. (1949b) The lower end of the femur of *Plesianthropus*. *Annals of the Transvaal Museum* **21**: 181–2.

Broom, R., and Robinson, J.T. (1949c) A new type of fossil man (from Swartkrans). *Nature* **164**: 322–3.

Broom, R., and Robinson, J.T. (1950) Notes on the pelves of the fossil ape-man. *American Journal of Physical Anthropology* **8**: 489–94.

Broom, R., and Robinson, J.T. (1952) *Swartkrans Ape-man*: Paranthropus crassidens. Transvaal Museum, Pretoria.

Broom, R., Robinson, J.T., and Schepers, G.W.H. (1950) *Sterkfontein Ape-man* Plesianthropus. Transvaal Museum, Pretoria.

Brose, D.S., and Wolpoff, M.H. (1971) Early Upper Paleolithic man and Late Paleolithic tools. *American Anthropology* **73**: 1156–94.

Brothwell, D., and Shaw, T. (1971) A late Upper Pleistocene Proto-West African Negro from Nigeria. *Man* **6**(2): 221–7.

Brothwell, D.R. (1960) Upper Pleistocene human skull from Niah caves, Sarawak. *Sarawak Museum Journal* **15–16**: 323–49.

Brothwell, D.R. (1961) The People of Mount Carmel: a reconsideration of their position in human evolution. *Proceedings of the Prehistoric Society* 27: 155–9.

Brothwell, D.R. (1974) The Upper Pleistocene Singa skull: a problem in palaeontological interpretation. In: *Bevolkerungsbiologie*, Bernhard, W., and Kandler, A., eds, pp. 534–45. Gustav Fischer, Stuttgart.

Brown, B., Walker, A., Ward, C.V., and Leakey, R.E.F. (1993) New *Australopithecus boisei* calvaria from East Lake Turkana, Kenya. *American Journal of Physical Anthropology* 91: 137–59.

Brown, B., Brown, F.H., and Walker, A. (2001) New hominids from the Lake Turkana Basin, Kenya. *Journal of Human Evolution* 41(1): 29–44.

Brown, F. (1994) Development of Pliocene and Pleistocene chronology of the Turkana Basin, East Africa, and its relation to other sites. In: *Integrative Paths to the Past: Paleoanthropological Advances in Honor of F. Clark Howell*, Corruccini, R.S., and Ciochon, R.L., eds, pp. 285–312. Prentice Hall, Englewood Cliffs, NJ.

Brown, F., Harris, J., Leakey, R.E.F., and Walker, A.C. (1985) Early *Homo erectus* skeleton from west Lake Turkana, Kenya. *Nature* 316: 788–92.

Brown, F.H., and Feibel, C.S. (1986) Revision of the lithostratigraphic nomenclature in the Koobi Fora region, Kenya. *Journal of the Geological Society* 143: 297–310.

Brown, F.H., and Feibel, C.S. (1991) Stratigraphy, depositional environments, and palaeogeography of the Koobi Fora Formation. In: *Koobi Fora Research Project. Volume 3. The Fossil Ungulates: Geology, Fossil Artiodactyls, and Palaeoenvironments*, Harris, J.M., ed., pp. 1–30. Clarendon Press, Oxford.

Brown, F.H., and Fuller, C.R. (2008) Stratigraphy and tephra of the Kibish Formation, southwestern Ethiopia. *Journal of Human Evolution* 55: 366–403.

Brown, F.H., Sarna-Wojcicki, A.M., Meyer, C.E., and Haileab, B. (1992) Correlation of Pliocene and Pleistocene tephra layers between the Turkana Basin of East Africa and the Gulf of Aden. *Quaternary International* 13–14: 55–67.

Brown, F.H., Haileab, B., and McDougall, I. (2006) Sequence of tuffs between the KBS Tuff and the Chari Tuff in the Turkana Basin, Kenya and Ethiopia. *Journal of the Geological Society, London* 163: 185–204.

Brown, J.H., and West, G.B. (2000) *Scaling in Biology*. Oxford University Press, Oxford.

Brown, P. (1987) Pleistocene homogeneity and Holocene size reduction: the Australian skeletal evidence. *Archaeology Oceania* 22: 41–71.

Brown, P. (1989) *Coobool Creek: A Morphological and Metrical Analysis of the Crania, Mandibles and Dentitions of a Prehistoric Australoan Human Population*. Department of Prehistory, Australian National University, Canberra.

Brown, P. (1994) Human skeletons. In: *The Encyclopedia of Aboriginal Australia*, Horton, D., ed., pp. 990–1. Australian Aboriginal Studies Press, Canberra.

Brown, P. (1994) Cranial vault thickness in Asian *Homo erectus* and *Homo sapiens*. *Courier Forsch Senckenberg* 171: 33–46.

Brown, P. (1998) The first Mongoloids? Another look at Upper Cave 101, Liujiang and Minatogawa 1. *Acta Anthropologica Sinica* 17: 255–75.

Brown, P. (2000) Australian Pleistocene variation and the sex of Lake Mungo 3. *Journal of Human Evolution* 38: 743–9.

Brown, P. (2001) 10 Chinese Middle Pleistocene hominids and modern human origins in east Asia. In: *Human roots: Africa and Asia in the Middle Pleistocene*, Barham, L., and Robson-Brown, K., eds. Western Academic and Specialist Press, Bristol.

Brown, P. (2010) Nacurrie 1: Mark of ancient Java, or a caring mother's hands, in terminal Pleistocene Australia? *Journal of Human Evolution* 59(2): 168–87.

Brown, P., and Maeda, T. (2009) Liang Bua *Homo floresiensis* mandibles and mandibular teeth: a contribution to the comparative morphology of a hominin species. *Journal of Human Evolution* 57(5): 571–96.

Brown, P., Sutikna, T., Morwood, M.J., Soejono, R.P., Jatmiko, Wayhu Saptomo, E., et al. (2004) A new small-bodied hominin from the Late Pleistocene of Flores, Indonesia. *Nature* 431: 1055–61.

Brown, W., George, M., and Wilson, A.C. (1979) Rapid evolution of animal mitochondrial DNA. *Proceedings of the National Academy of Sciences USA* 76(4): 1967–1971.

Brumm, A., Aziz, F., van der Bergh, G.D., Morwood, M.J., Moore, M.J., Kurniawan, I., et al. (2006) Early stone technology on Flores and its implications for *Homo floresiensis*. *Nature* 441: 624–8.

Bruner, E., and Manzi, G. (2005) CT-based description and phyletic evaluation of the archaic human calvarium from Ceprano, Italy. *Anatomical Record* 285: 643–58.

Bruner, E., and Manzi, G. (2006) Saccopastore 1: the earliest Neanderthal? A new look at an old cranium. In: *Neanderthals Revisited: New Approaches and Perspectives*, Hublin, J.-J., Harvati, K., and Harrison, T., eds. Springer, Dordrecht.

Brunet, M., Beauvilain, A., Coppens, Y., Heintz, E., Moutaye, A.H.E., and Pilbeam, D. (1995) The first australopithecine 2,500 kilometers west of the Rift Valley (Chad). *Nature* 376: 273–5.

Brunet, M., Beauvilain, A., Coppens, Y., Heintz, E., Moutaye, A.H.E., and Pilbeam, D. (1996) *Australopithecus bahrelghazali*, une nouvelle espèce d'Hominidé ancien de la région de Koro Toro (Tchad). *Comptes Rendus de l'Académie des Sciences* 322: 907–13.

Brunet, M., Guy, F., Pilbeam, D., Mackaye, H.T., Likius, A., Ahounta, D., et al. (2002) A new hominid from the Upper Miocene of Chad, Central Africa. *Nature* 418: 145–51.

Buchanan, B., and Collard, M. (2009) A geometric morphometrics-based assessment of blade shape differences among Paleoindian projectile point types from western North America. *Journal of Archaeological Science* (in press).

Buckley, J.D., and Willis, E.H. (1969) Isotopes' radiocarbon measurements VII. *Radiocarbon* **11**: 53–105.

Buikstra, J.E., and Ubelaker, D.H. (1994) *Standards for data collection from human skeletal remains: proceedings of a seminar at the Field Museum of Natural History*. Arkansas Archeological Survey, Fayetteville, AK.

Burger, O., Walker, R., and Hamilton, M.J. (2010) Lifetime reproductive effort in humans. *Proceedings of the Royal Society of London, Series B* **277**(1682): 773–7.

Burke, A. (2000) The view from Starosele: Faunal exploitation at a Middle Palaeolithic site in Western Crimea. *International Journal of Osteoarchaeology* **10**(5): 325–35.

Burkitt, A.N., and Hunter, J.I. (1922) The description of a Neanderthaloid Australian skull, with remarks on the production of facial characteristics of Australian skulls in general. *Journal of Anatomy* **57**: 31–54.

Burrows, W., and Ryder, O.A. (1997) Y-chromosome variation in great apes. *Nature* **385**: 125–6.

Busk, G. (1865) *On a Very Ancient Human Cranium from Gibraltar*. 34th Meeting of the British Association for the Advancement of Science. Bath. pp. 91–2.

Butler, B. (1974) Skeletal remains from a Late Paleolithic site near Esna, Egypt. In: *The Fakhurian, a Late Paleolithic industry from Upper Egypt*, Lubell, D., ed., pp. 176–83. Ministry of Industry, Petroleum, and Mineral Wealth, Cairo.

Butler, P. (1982) Directions of evolution in the mammalian dentition. In: *Problems of phylogenetic reconstruction*, Joysey, K., and Friday, A., eds, pp. 235–44. Academic Press, London.

Butler, P.M. (1939) Studies of the mammalian dentition - differentiation of the postcanine dentition. *Proceedings of the Zoological Society of London Series B* **109**: 1–36.

Butzer, K. (1969) Geological interpretation of two Pleistocene hominid sites in the lower Omo Basin. *Nature* **222**: 1133–5.

Butzer, K. (1976a) The Mursi, Nkalabong, and Kibish Formations, Lower Omo Basin, Ethiopia. In: *Earliest Man and Environments in the Lake Rudolf Basin*, Howell, F.C., Isaac, G.Ll., and Leakey, R.E.F., eds, pp. 12–23. University of Chicago Press, Chicago, IL.

Butzer, K., and Isaac, G.Ll., eds (1975) *After the Australopithecines*. de Gruyter Mouton, The Hague.

Butzer, K., Stuckenrath, R., Bruzewicz, A., and Helgren, D. (1978) Late Cenozoic paleoclimates of the Gaap Escarpment, Kalahari margin, South Africa. *Quaternary Research* **10**(3): 310–39.

Butzer, K.W. (1976b) Lithostratigraphy of the Swartskrans formation. *South African Journal of Science* **72**: 136–41.

Butzer, K.W. (1978) Geoecological perspectives on early hominid evolution. In: *Early Hominids of Africa*, Jolly, C., ed., pp. 191–217. Duckworth, London.

Butzer, K.W. (1982) Geomorphology and sediment stratigraphy. In: *The Middle Stone Age at Klasies River Mouth in South Africa*, Singer, R., and Wymer, J., eds. Chicago University Press, Chicago, IL.

Butzer, K.W. (1984) Late Quaternary environments in South Africa. In: *Late Cenozoic Palaeoclimates of the Southern Hemisphere*, Vogel, J.C., ed., pp. 235–64. Balkema, Rotterdam.

Butzer, K.W. (2004) Coastal eolian sands, paleosols, and Pleistocene geoarchaeology of the Southwestern Cape, South Africa. *Journal of Archaeological Science* **31**(12): 1743–81.

Butzer, K.W., and Thurber, D.L. (1969) Some Late Cenozoic Sedimentary Formations of the Lower Omo Basin. *Nature* **222**: 1138–43.

Butzer, K.W., Brown, F.H., and Thurber, D.L. (1969) Horizontal sediments of the Lower Omo valley: the Kibish formation. *Quaternaria* **110**: 15–29.

Butzer, K.W., Isaac, G.L., Richardson, J.L., and Washbourn-Kamau, C. (1972) Radiocarbon dating of East Africa Lake levels. *Science* **175**(4026): 1069–76.

Byrne, L. (2004) Lithic tools from Arago cave, Tautavel (Pyrénées-Orientales, France): behavioural continuity or raw material determinism during the Middle Pleistocene? *Journal of Archaeological Science* **31**: 351–64.

Byrne, R., and Whiten, A. (1988) *Machiavellian Intelligence: Social Expertise and the Evolution of Intellect in Monkeys, Apes, and Humans*. Clarendon Press, Oxford.

Caccone, A., and Powell, J.R. (1989) DNA divergence among hominoids. *Evolution* **43**: 925–42.

Caddie, D.A., Hunter, D.S., Pomery, P.J., and Hall, H.J. (1987) The ageing chemist - can electron spin resonance (ESR) help. In: *Archaeometry: Further Australasian Studies*, Ambrose, W.R., and Mummery, J.M.J., eds, pp. 167–76. Department of Prehistory, Australian National University, Canberra.

Cain, A.J. (1954) *Animal Species and Evolution*. Princeton University Press, Princeton, NJ.

Calcagno, J.M. (1986) Dental reduction in post-Pleistocene Nubia. *American Journal of Physical Anthropology* **70**: 349–63.

Call, J., and Carpenter, M. (2001) Do apes and children know what they have seen? *Animal Cognition* **4**: 207–20.

Call, J., and Carpenter, M. (2002) Three sources of information in social learning. In: *Imitation in Animals and Artifacts*, pp. 211–28. MIT Press, Cambridge, MA.

Caloi, L., Manzi, G., and Palombo, M.R. (1998) *Saccopastore, a stage-5 site within the city of Rome*. SEQS Symposium "The Eemian-local sequences, global perspectives". Kerkrade.

Cameron, J. (1927) The main angle of cranial flexion (the nasion-pituitary-basion angle). *American Journal of Physical Anthropology* **10**(2): 275–9.

Caminiti, R., Ghaziri, H., Galuske, R., Hof, P.R., and Innocent, G.M. (2009) Evolution amplified processing with temporally dispersed slow neuronal connectivity in primates. *Proceedings of the National Academy of Sciences USA* **106**(46): 19551–6.

Camp, E. (2004) The generality constraint and categorial restrictions. *Philosophical Quarterly* **54**(215): 209–31.

Campbell, B. (1964) Just another 'man-ape'? *Discovery*: 37–38.

Campbell, B. (1973) A new taxonomy of fossil man. *Yearbook of Physical Anthropology* **17**: 194–201.

Campisano, C.J., and Feibel, C.S. (2008a) Tephrostratigraphy of the Hadar and Busidima Formations at Hadar, Afar Depression, Ethiopia. In: *The Geology of Early Humans in the Horn of Africa*, Quade, J., and Wynn, J.G., eds, pp. 135–62. Geological Society of America, Boulder, CO.

Campisano, C.J., and Feibel, C.S. (2008b) Depositional environments and stratigraphic summary of the Pliocene Hadar Formation, Afar Depression, Ethiopia. In: *The Geology of Early Humans in the Horn of Africa*, Quade, J., and Wynn, J.G., eds, pp. 179–201. Geological Society of America, Boulder, CO.

Camps, G. (1971) A propos du Néolithique ancien de la Méditerranée Occidentale. *Bulletin de la Société préhistorique française* **68**(2): 48–50.

Camps-Fabrer, H. (1975) *Un gisement capsien de faciès sétifien Medjez II, El-Eulma (Algérie)*. Editions du Centre national de la recherche scientifique, Paris.

Cann, R.L., Stoneking, M., and Wilson, A.C. (1987) Mitochondrial DNA and human evolution. *Nature* **325**: 31–36.

Capitan, L., and Peyrony, D. (1909) Deux squelettes humains au milieu de foyers de l'époque moustérienne. *Revue d'École d'Anthropologie de Paris* 402–9.

Capitan, L., and Peyrony, D. (1911) Un nouveau squelette humain fossil. *Revue Anthropologique* **21**(5): 148–50.

Capitan, L., and Peyrony, D. (1912a) Station préhistorique de La Ferrassie, commune de Savignac-du-Bugue (Dordogne). *Revue Anthropologique* **22**(1): 29–50.

Capitan, L., and Peyrony, D. (1912b) Station préhistorique de La Ferrassie, commune de Savignac-du-Bugue (Dordogne) (suite). *Revue Anthropologique* **22**(2): 76–99.

Capitan, L., and Peyrony, D. (1921) Découverte d'un sixième squelette moustérien à La Ferrassie. *Revue Anthropologique* **31**: 382–8.

Capitan, L., and Peyrony, D. (1928) La Madeleine, son gisement, son industrie, ses oeuvres d'art. *Publications of the International Institute of Anthropology* **2**: 63.

Carabelli, G. (1842) *Systematisches Handbuch der Zahnheilkunde*. Braumüller und Seidel, Wein.

Caramelli, D., Lalueza-Fox, C., Vernesi, C., Lari, M., Casoli, A., Mallegni, F., et al. (2003) Evidence for a genetic discontinuity between Neandertals and 24,000-year-old anatomically modern Europeans. *Proceedings of the National Academy of Sciences USA* **100**(11): 6593–97.

Caramelli, D., Milani, L., Vai, S., Modi, A., Pecchioli, E., Girardi, M., et al. (2008) A 28,000 years old Cro-Magnon mtDNA sequence differs from all potentially contaminating modern sequences. *Public Library of Science* **3**: e2700.

Carbonell, E., Bermúdez de Castro, J.M., Arsuaga, J.-L., Díez, J.C., Rosas, A., Cuenca Bescós, G., et al. (1995) Lower Pleistocene hominids and artifacts from Atapuerca-TD6 (Spain). *Science* **269**(5225): 826–32.

Carbonell, E., Esteban, M., Nájera, A.M., Mosquera, M., Rodríegez, X.P., Ollé, A., et al. (1999) The Pleistocene site of Gran Dolina, Sierra de Atapuerca, Spain: a history of the archaeological investigations. *Journal of Human Evolution* **37**: 313–24.

Carbonell, E., Mosquera, M., Olle, A., Rodriguez, X., Sala, R., Verges, J., et al. (2003) Did the earliest mortuary practices take place more than 350 000 years ago at Atapuerca? *L'anthropologie* **107**(1): 1–14.

Carbonell, E., Bermúdez de Castro, J.M., Arsuaga, J.L., Allue, E., Bastir, M., Benito, A., et al. (2005) An Early Pleistocene hominin mandible from Atapuerca-TD6, Spain. *Proceedings of the National Academy of Sciences USA* **102**(16): 5674–78.

Carbonell, E., Bermúdez de Castro, J.M., Parés, J.M., Peréz-González, A., Cuenca-Bescós, G., Ollé, A., et al. (2008) The first hominin of Europe. *Nature* **452**: 465–9.

Cardini, L. (1980) La necropoli mesolitica delle Arene Candide (Liguria). *Memorie dell'Istituto Italiano di Paleontologia Umana* **3**: 9–31.

Carey, J.R., and Judge, D.S. (2000) *Longevity Records: Life Spans of Mammals, Birds, Amphibians, Reptiles, and Fish*. Odense University Press, Odense.

Carlson, D.S. (1976) Temporal Variation in Prehistoric Nubian Crania. *American Journal of Physical Anthropology* **45**(3): 467–84.

Carlson, D.S., and Van Gerven, D.P. (1977) Masticatory function and post-Pleistocene evolution in Nubia. *American Journal of Physical Anthropology* **46**(3): 495–506.

Carmi, I., and Segal, D. (1992) Rehovot radiocarbon measurements IV. *Radiocarbon* **34**(1): 115–32.

Carney, J., Hill, A., Miller, J.A., and Walker, A.C. (1971) Late australopithecine from Baringo District, Kenya. *Nature* **230**: 509–14.

Caro, T.M., and Hauser, M.D. (1992) Is there teaching in nonhuman animals? *Quarterly Review of Biology* **67**(2): 151–74.

Carpenter, M., Nagell, K., Tomasello, M., Butterworth, G., and Moore, C. (1998) Social cognition, joint attention, and communicative competence from 9 to 15 months of age. *Monographs of the Society for Research in Child Development* **63**(4): i-174.

Carretero, J.M., Lorenzo, C., and Arsuaga, J.-L. (1999) Axial and appendicular skeleton of *Homo antecessor*. *Journal of Human Evolution* **37**: 459–99.

Carrier, D.R. (1984) The energetic paradox of human running and hominid evolution. *Current Anthropology* **25**: 483–95.

Carrier, N.H. (1958) A note on the estimation of mortality and other population characteristics given deaths by age. *Population Studies* **12**(2): 149–63.

Carruthers, P. (2006) *The Architecture of the Mind: Massive Modularity and Flexibility of Thought*. Oxford University Press, Oxford.

Carter, T., and Shackley, M.S. (2007) Sourcing obsidian from Neolithic Çatalhöyük (Turkey) using energy dispersive X-ray fluorescence. *Archaeometry* 49(3): 437–54.

Carter, T., Poupeau, G., Bressy, C., and Pearce, N.J.G. (2006) A new programme of obsidian characterization at Çatalhöyük, Turkey. *Journal of Archaeological Science* 33(7): 893–909.

Cartmill, M., and Milton, K. (1977) The lorisiform wrist joint and the evolution of "brachiating" adaptations in the Hominoidea. *American Journal of Physical Anthropology* 47: 249–72.

Cartmill, M., Smith, F., and Brown, K. (2009) *The Human Lineage*. Wiley-Blackwell, Oxford.

Carvalho, S., Cunha, E., Sousa, C., and Matsuzawa, T. (2008) Chaînes opératoires and resource-exploitation stategies in chimpanzee (*Pan troglodytes*) nut cracking. *Journal of Human Evolution* 55(1): 148–63.

Casanova, J. (1894) *The memoirs of Jacques Casanova de Seingalt*. G.P. Putnam's Sons, New York.

Caspari, R. (2005) The suprainiac fossa: the question of homology. *Anthropologie* 43(2–3): 229–39.

Caspari, R., and Lee, S.-H. (2004) Older age becomes common late in human evolution. *Proceedings of the National Academy of Sciences USA* 101(30): 10895–900.

Castanet, J. (2006) Time recording in bone microstructures of endothermic animals; functional relationships. *Comptes Rendus Palevol* 5: 629–36.

Caton-Thompson, G., and Gardner, E. (1934) *The Desert Fayum*. The Royal Anthropological Institute of Great Britain and Ireland, London.

Cattani, L., Cremaschi, M., Ferraris, M., Mallegni, F., Masini, F., Scola, V., et al. (1991) Les gisements du PlÈistocÈne moyen di Visogliano (Trieste): restes humains, industries, environnement. *L'Anthropologie* 95(1): 9–35.

Cavalli-Sforza, L.L., and Feldman, M.W. (1981) *Cultural Transmission and Evolution: A Quantitative Approach*. Princeton University Press, Princeton, NJ.

Cela-Conde, C.J., and Altaba, C.R. (2002) Multiplying genera versus moving species: a new taxonomic proposal for the familty Hominidae. *South African Journal of Science* 98: 229–32.

Cerling, T.E. (1992) Development of grasslands and savannas in East Africa during the Neogene. *Palaeogeography, Palaeoclimatology, Palaeoecology* 97: 241–7.

Cerling, T.E., and Harris, J.M. (1999) Carbon isotope fractionation between diet and bioapatite in ungulate mammals and implications for ecological and paleoecological studies. *Oecologia* 120(3): 347–63.

Cerling, T.E., Hart, J.A., and Hart, T.B. (2004) Stable isotope ecology in the Ituri Forest. *Oecologia* 138: 5–12.

Cerveri, P., De Momi, E., Marchente, M., Lopomo, N., Baud-Bovy, G., Barros, R.M.L., et al. (2008) In vivo validation of a realistic kinematic model for the trapezio-metacarpal joint using an optoelectronic system. *Annals of Biomedical Engineering* 36: 1268–80.

Chaix, R., Cao, C., and Donnelly, P. (2008) Is mate choice in humans MHC-dependent? *Public Library of Science Genetics* 4(9): 1–5.

Chambers, R. (1844) *Vestiges of the Natural History of Creation*. W & R Chambers, London.

Chamla, M., Biraben, J., and Dastugue, J. (1970) Les hommes épipaléolithiques de Columnata (Algérie occidentale). *Étude anthropologique: Mém Centr Rech Anthropol Préhist Ethnograph* 15.

Chapman, C.A., and Chapman, L.J. (1990) Dietary variability in primate populations. *Primates* 31(1): 121–8.

Charnov, E.L., and Berrigan, D. (1990) Dimensionless numbers and life history evolution: age of maturity versus the adult lifespan. *Evolutionary Ecology* 4(3): 273–5.

Chartrand, T.L., and Bargh, J.A. (1999) The Chameleon Effect: the perception-behavior link and social interaction. *Journal of Personality and Social Psychology* 76(6): 893–910.

Chase, P.G., Armand, D., Debénath, A., Dibble, H., and Jelinek, A.J. (1994) Taphonomy and zooarchaeology of a Mousterian faunal assemblage from La Quina, Charente, France. *Journal of Field Archaeology* 21: 289–305.

Chase, P.G., Debénath, A., Dibble, H.L., McPherron, S.P., Schwarcz, H.P., Stafford Jr, T.W., et al. (2007) New dates for the Fontéchevade (Charente, France) *Homo* remains. *Journal of Human Evolution* 52: 217–21.

Chase, P.G., Debenath, A., Dibble, H.L., and McPherron, S.P. (2009) *The Cave of Fontéchevade: Recent Excavations and their Paleoanthropological Implications*. Cambridge University Press, New York.

Chavaillon, J., and Piperno, M., eds (2004) *Studies on the Early Paleolithic site of Melka Kunture, Ethiopia*. Istituto Italiano di Preistoria e Protostoria, Florence.

Chavaillon, J., Chavaillon, N., Hours, F., and Piperno, M. (1979) From the Oldowan to the Middle Stone Age at Melka-Kunture (Ethiopia). Understanding cultural changes. *Quaternaria* 21: 87–114.

Chazan, M., Ron, H., Matmon, A., Porat, N., Goldberg, P., Yates, R., et al. (2008) Radiometric dating of the Earlier Stone Age sequence in Excavation I at Wonderwerk Cave, South Africa: preliminary results. *Journal of Human Evolution* 55(1): 1–11.

Chech, M., Groves, C.P., Thorne, A., and Trinkaus, E. (2000) A new reconstruction of the Shanidar 5 cranium. *Paléorient* 25(2): 143–6.

Chen, G.J., Wang, W., Mo, J.Y., Huang, Z.T., Tian, F., and Weiwen, H. (2001) Pleistocene vertebrate fauna from Wuyun Cave of Tiandong county, Guangxi. *Vertebrata PalAsiatica* 40: 42–51.

Chen, Q., Chen, T.M., and Yang, Q. (2003) ESR dating of Early Pleistocene archaeological sites. In: *Current*

Research in Chinese Pleistocene Archaeology, Shen, C., and Keates, S.G., eds, pp. 119–23. Archaeopress, Oxford.

Chen, T., Yuan, S., and Guo, S. (1984) The study of uranium series dating of fossil bones and an absolute age sequence for the main Paleolithic sites of north China. *Acta Anthropologica Sinica* 3: 259–68.

Chen, T., Yuan, S., Gao, S., and Hu, Y. (1987) Uranium series dating of fossil bones from Hexian and Chaoxian fossil human sites. *Acta Anthropologica Sinica* 6: 249–54.

Chen, T., Hedges, R.E.M., and Yuan, Z. (1989) Accelerator radiocarbon dating for Upper Cave of Zhoukoudian. *Acta Anthropologica Sinica* 8: 216–21.

Chen, T., Yang, Q., and Wu, E. (1994) Antiquity of *Homo sapiens* in China. *Nature* 368: 55–56.

Chen, T.-M., Yang, Q., Hu, Y.-Q., Bao, W.-B., and Li, T.-Y. (1997) ESR dating of tooth enamel from the Yunxian *Homo erectus* site, China. *Quaternary Science Reviews* 16(3–5): 455–8.

Cheney, D.L., and Seyfarth, R.M. (1992) *How Monkeys See the World: Inside the Mind of Another Species*. University of Chicago Press, Chicago, IL.

Cheney, D.L., and Seyfarth, R.M. (2007) *Baboon Metaphysics: The Evolution of a Social Mind*. University of Chicago Press, Chicago, IL.

Cheng, Z., Ventura, M., She, X., Khaitovich, P., Graves, T., Osoegawa, K., et al. (2005) A genome-wide comparison of recent chimpanzee and human segmental duplications. *Nature* 437: 88–93.

Cheverud, J.M. (1982) Phenotypic, genetic and environmental morphological integration in the cranium. *Evolution* 36(3): 499–516.

Cheverud, J.M. (1988) A comparison of genetic and phenotypic correlations. *Evolution* 42(5): 958–68.

Cheverud, J.M. (1996) Developmental integration and the evolution of pleiotropy. *American Zoologist* 36(1): 44–50.

Chimpanzee Sequencing and Analysis Consortium (2005) Initial sequence of the chimpanzee genome and comparison with the human genome. *Nature* 437: 69–87.

Chomsky, N. (1977) *Essays on Form and Interpretation*. North-Holland, New York.

Chomsky, N. (2007) Categories and Transformations. In: *Minimalist Syntax: the Essential Readings*, Bošković, Ž., and Lasnik, H., eds. Blackwell Publishing, Oxford.

Chou, H., Hayakawa, T., Diaz, S., Krings, M., Indriati, E., Leakey, M., et al. (2002) Inactivation of CMP-N-acetylneuraminic acid hydroxylase occurred prior to brain expansion during human evolution. *Proceedings of the National Academy of Sciences USA* 99(18): 11736–41.

Churchill, S.E., and Formicola, V. (1997) A case of marked bilateral asymmetry in the upper limbs of an Upper Palaeolithic male from Barma Grande (Liguria), Italy. *International Journal of Osteoarcheology* 7: 18–38.

Churchill, S.E., and Smith, F.H. (2000) Makers of the Early Aurignacian of Europe. *Yearbook of Physical Anthropology* 43: 61–115.

Churchill, S.E., Pearson, O.M., Grine, F.E., Trinkaus, E., and Holliday, T.W. (1996) Morphological affinities of the proximal ulna from Klasies River main site: archaic or modern? *Journal of Human Evolution* 31: 213–37.

Churchill, S.E., Berger, L.R., and Parkington, J.E. (2000) A Middle Pleistocene human tibia from Hoedjiespunt, Western Cape, South Africa. *South African Journal of Science* 96: 367–8.

Churchill, S.E., Franciscus, R.G., McKean-Peraza, H.A., Daniel, J.A., and Warren, B.R. (2009) Shanidar 3 Neandertal rib puncture wound and paleolithic weaponry *Journal of Human Evolution* 57(2): 163–78.

Cihak, R. (1972) Connections of the abductor pollicis longus and brevis in the ontogenesis of the human hand. *Folia Morphologica* 20(2): 102–5.

Ciochon, R.L. (1993) *Evolution of the Cercopithecoid Forelimb*. University of California Press, Berkeley, CA.

Clark, G. (1969) *World Prehistory*. Cambridge University Press, Cambridge.

Clark, G. (1978) *World Prehistory in New Perspective*. Cambridge University Press, Cambridge.

Clark, J., and Brandt, S., eds (1984) *From Hunters to Farmers: the Causes and Consequences of Food Production in Africa*. University of California Press, Berkeley, CA.

Clark, J., Stephens, E., and Coryndon, S. (1966) Pleistocene fossiliferous lake beds of the Malawi (Nyasa) Rift: a preliminary report. *American Anthropologist* 68(2): 46–87.

Clark, J., Beyene, Y., WoldeGabriel, G., Hart, W., Renne, P., Gilbert, H., et al. (2003) Stratigraphic, chronological and behavioural contexts of Pleistocene *Homo sapiens* from Middle Awash, Ethiopia. *Nature* 423(6941): 747–52.

Clark, J.D. (1950) *The Stone Age Cultures of Northern Rhodesia, with particular reference to the Curtural and Climatic Succession of the Upper Zambezi Valley and its Tributaries*. South African Archaeological Society, Cape Town.

Clark, J.D. (1957) A re-examination of the industry from the type site of Magosi, Uganda. In: *Third Pan-African Congress on Prehistory, Livingstone, 1955*, Clark, J.D., ed., pp. 228–41. Chatto & Windus, London.

Clark, J.D. (1963) *Prehistoric Cultures of Northeast Angola and their Significance in Tropical Africa*. Museo do Dundo Publicacoes Culturais, Lisbon.

Clark, J.D. (1966) *The Distribution of Prehistoric Culture in Angola* Museo do Dundo Publicacoes Culturais, Lisbon.

Clark, J.D. (1967) *Atlas of African Prehistory*. University of Chicago Press, Chicago, IL.

Clark, J.D. (1969) *Kalambo Falls Prehistoric Site, I., The Geology, Paleoecology and Detailed Stratigrapghy of the Excavations*. Cambridge University Press, Cambridge.

Clark, J.D. (1970) *The Prehistory of Africa*. Praeger, New York.

Clark, J.D. (1974) *Kalambo Falls Prehistoric Site, II., The Later Prehistoric Cultures*. Cambridge University Press, Cambridge.

Clark, J.D. (1987) Transitions: *Homo erectus* and the Acheulian: the Ethiopian sites of Gadeb and the Middle Awash. *Journal of Human Evolution* **16**: 809–26.

Clark, J.D. (1988) The Middle Stone Age of East Africa and the beginning of regional identity. *Journal of World Prehistory* **2**: 236–305.

Clark, J.D. (1993) The Aterian of the Central Sahara. In: *Environmental Change and Human Culture in the Nile Basin and Northern Africa until the Second Millennium B. C.*, Krzyzaniak, L., Kobusiewicz, M., and Alexander, J., eds, pp. 49–67. Poznan Archaeological Museum, Poznan.

Clark, J.D. (1995) Introduction to research on the Chiwondo Beds, northern Malawi. *Journal of Human Evolution* **28**: 3–5.

Clark, J.D. (2001) *Kalambo Falls Prehistoric Site, III., The Earlier Cultures: Middle and Earlier Stone Age.* Cambridge University Press, Cambridge.

Clark, J.D. (2004) Stone artefact assemblages from Members 1–3, Swartkrans Cave. In: *Swartkrans: A Cave's Chronicle of Early Man*, Brain, C.K., ed., pp. 167–94. Transvaal Museum, Pretoria.

Clark, J.D., and Kurashina, H. (1979) An analysis of Earlier Stone Age Bifaces from Gadeb (Locality 8E), Northern Bale Highlands, Ethiopia. *The South African Archaeological Bulletin* **34**(130): 93–109.

Clark, J.D., and Brown, K.S. (2001) The Twin Rivers, Kopje, Zambia: Stratigraphy, Fauna, and Artefact Assemblages from the 1954 and 1956 Excavations. *Journal of Archaeological Science* **28**(3): 305–30.

Clark, J.D., Oakley, K.P., Wells, L.H., and McCleland, J.A.C. (1947) New Studies on Rhodesian Man. *Journal of the Royal Anthropological Institute of Great Britain and Ireland* **77**(1): 7–32.

Clark, J.D., Cole, G.H., Isaac, G.L., and Kleindienst, M.R. (1966) Precision and Definition in African Archaeology. *The South African Archaeological Bulletin* **21**(83): 114–21.

Clark, J.D., Brothwell, D.R., Powers, R., and Oakley, K.P. (1968) Rhodesian man: notes on a new femur fragment. *Man* **3**(1): 105–11.

Clark, J.D., Asfaw, B., Assefa, G., Harris, J.W.K., Kurashina, H., Walter, R.C., et al. (1984a) Palaeoanthropological discoveries in the Middle Awash Valley, Ethiopia. *Nature* **307**: 423–8.

Clark, J.D., Williamson, K.D., Michels, J.W., and Marean, C.W. (1984b) A Middle Stone Age occupation site at Porc-Epic Cave, Dire Dawa (east-central Ethiopia). *African Archaeological Review* **2**(1): 37–71.

Clark, J.D., de Heinzelin, J., Schick, K.D., Hart, W.K., White, T., WoldeGabriel, G., et al. (1994) African *Homo erectus*: old radiometric ages and young Oldowan assemblages in the Middle Awash Valley, Ethiopia. *Science* **264**(5167): 1907–10.

Clark, J.D., Beyene, Y., WoldeGabriel, G., Hart, W.K., Renne, P., Gilbert, H., et al. (2003) Stratigraphic, chronological and behavioural contexts of Pleistocene *Homo sapiens* from Middle Awash, Ethiopia. *Nature* **423**: 747–52.

Clark, P.U., Dyke, A.S., Shakun, J.D., Carlson, A.E., Clark, J., Wohlfarth, B., et al. (2009) The Last Glacial Maximum. *Science* **325**(5941): 710–14.

Clarke, D.L. (1968) *Analytical Archaeology*. Columbia University Press, New York.

Clarke, R. (1988) A new *Australopithecus* cranium from Sterkfontein and its bearing on the ancestry of *Paranthropus*. In: *Evolutionary History of the "Robust" Australopithecines*, Grine, F.E., ed., pp. 285–92. Aldine de Gruyter, New York.

Clarke, R. (1999) Discovery of complete arm and hand of the 3.3 million-year-old *Australopithecus* skeleton from Sterkfontein. *South African Journal of Science* **95**(11/12): 477–80.

Clarke, R., Howell, F., and Brain, C. (1970) New finds at the Swartkrans Australopithecine site (contd): more evidence of an advanced hominid at Swartkrans. *Nature* **225**: 1219–22.

Clarke, R.J. (1976) New cranium of *Homo erectus* from Lake Ndutu, Tanzania. *Nature* **262**: 485–7.

Clarke, R.J. (1977) *The Cranium of the Swartkrans hominid SK 847 and its Relevance To Human Origins*. PhD thesis, University of the Witwatersrand.

Clarke, R.J. (1985a) A new reconstruction of the Florisbad cranium, with notes on the site. In: *Ancestors: The Hard Evidence*, Delson, E., ed. Alan R. Liss, New York.

Clarke, R.J. (1985b) A new *Australopithecus* cranium from Sterkfontein and its bearing on the ancestry of *Paranthropus*. In: *Evolutionary History of the "Robust" Australopithecines*, Grine, F.E., ed., pp. 285–92. Aldine de Gruyter, New York.

Clarke, R.J. (1990a) The Ndutu cranium and the origin of *Homo sapiens*. *Journal of Human Evolution* **19**: 699–736.

Clarke, R.J. (1990b) Observations on some restored hominid specimens in the Transvaal Museum, Pretoria. In: *From apes to angels: essays in anthropology in honour of Phillip V Tobias*, Sperber, G., ed., pp. 135–51. Wiley-Liss, New York.

Clarke, R.J. (1994) On some new interpretations of Sterkfontein stratigraphy. *South African Journal of Science* **90**(4): 211–14.

Clarke, R.J. (1998) First ever discovery of a well-preserved skull and associated skeleton of *Australopithecus*. *South African Journal of Science* **94**: 460–3.

Clarke, R.J. (1999) Discovery of complete arm and hand of the 3.3 million-year-old *Australopithecus* skeleton from Sterkfontein. *South African Journal of Science* **95**: 477–80.

Clarke, R.J. (2000) A corrected reconstruction and interpretation of the *Homo erectus* calvaria from Ceprano, Italy. *Journal of Human Evolution* **39**: 433–42.

Clarke, R.J. (2002) Newly revealed information on the Sterkfontein Member 2 *Australopithecus* skeleton. *South African Journal of Science* **98**: 523–6.

Clarke, R.J. (2006) A deeper understanding of the stratigraphy of Sterkfontein hominid site. *Transactions of the Royal Society of South Africa* **61**: 111–20.

Clarke, R.J. (2008) Latest information on Sterkfontein's *Australopithecus* skeleton and a new look at *Australopithecus*. *South African Journal of Science* **104**(11–12): 443–9.

Clarke, R.J., and Howell, F.C. (1972) Affinities of the Swartkrans 847 Hominid Cranium. *American Journal of Physical Anthropology* **37**: 319–36.

Clarke, R.J., and Tobias, P.V. (1995) Sterkfontein Member 2 foot bones of the oldest South African hominid. *Science* **269**: 521–4.

Clarkson, C., Petraglia, M., Korisettar, R., Haslam, M., Boivin, N., Crowther, A., et al. (2009) The oldest and longest enduring microlithic sequence in India: 35 000 years of modern human occupation and change at the Jwalapuram Locality 9 rockshelter. *Antiquity* **83**: 326–48.

Clayton, N.S., and Dickinson, A. (1998) Episodic-like memory during cache recovery by scrub jays. *Nature* **395**: 272–4.

Clement, A.C., Cane, M.A., and Seager, R. (2001) An orbitally driven tropical source for abrupt climate change. *Journal of Climate* **14**(11): 2369–75.

Clemente, C.I., Gassiot Ballbè, E., and Terradas Batlle, X. (2008) Manufacture and use of stone tools in the Caribbean Coast of Nicaragua. The analysis of the last phase of the shell midden KH-4 at Karoline (250–350 cal AD). In: *Prehistoric Technology 40 Years Later: Functional Studies and the Russian Legacy*, Longo, L., and Skakun, N., eds. BAR International Series, Oxford.

CLIMAP Project Members (1976) The surface of the ice-age Earth. *Science* **191**: 1131–7.

Close, A.E. (2002) Backed bladelets are a foreign country. *Archeological Papers of the American Anthropological Association* **12**(1): 31–44.

Clottes, J. (1992) L'archeologie des grottes ornees. *La Recherche* **239**(23): 52–61.

Clottes, J., Courtin, J., Valladas, H., Cachier, H., Mercier, N., and Arnold, M. (1992) La grotte Cosquer datée. *Bulletin de la Société Préhistorique Fançaise* **89**(8): 230–3.

Clottes, J., Courtin, J., Collina-Girard, J., Arnold, M., and Valladas, H. (1997) News from Cosquer Cave: climatic studies, recording, sampling, dates. *Antiquity* **71**(272): 321–6.

Clutton-Brock, T.H., and Harvey, P.H. (1980) Primates, brains and ecology. *Journal of Zoology* **190**: 309–23.

Coale, J.P., and Demeny, P. (1966) *Regional Model Life Tables and Stable Populations*. Princeton University Press, Princeton, NJ.

Coale, J.P., and Demeny, P. (1983) *Regional Model Life Tables and Stable Populations*. Academic Press, New York.

Codron, D., Luyt, J., Lee-Thorp, J.A., Sponheimer, M., de Ruiter, D., and Codron, J. (2005) Utilization of savanna-based resources by Plio-Pleistocene baboons. *South African Jouranl of Science* **101**: 245–8.

Cohen, P. (1996) Fitting a face to Ngaloba. *Journal of Human Evolution* **30**: 373–9.

Cole, G. (1967) A reinvestigation of Magosi and the Magosian. *Quaternaria* **9**. 153–68.

Cole, G.H., and Kleindienst, M.R. (1974) Futher reflections on the Isimila Acheulian. *Quaternary Research* **4**(3): 346–55.

Coleman, M.N. (2008) What does geometric mean, mean geometrically? Assessing the utility of geometric mean and other size variables in studies of skull allometry. *American Journal of Physical Anthropology* **135**(4): 404–15.

Collard, M., and Wood, B.A. (2006) Defining the genus *Homo*. In: *Handbook of Paleoanthropology*, Henke, W., Rothe, H., and Tattersall, I., eds, pp. 1575–610. Springer, Berlin.

Collard, M., Shennan, S.J., and Tehrani, J.J. (2006) Branching, blending and the evolution of cultural similarities and differences among human populations. *Evolution and Human Behavior* **27**: 169–84.

Collins, M., Penkman, K., Rohland, N., Shapiro, B., Dobberstein, R., Ritz-Timme, S., et al. (2009) Is amino acid racemization a useful tool for screening for ancient DNA in bone? *Proceedings of the Royal Society of London, Series B* **276**(1669): 2971–77.

Colman, S.M., and Pierce, K.L. (2000) Classifications of Quaternary geochronologic methods. In: *Quaternary Geochronology, Methods and Applications*, Noller, J.S., Sowers, J.M., and Lettis, W.R., eds, pp. 2–5. American Geophysical Union, Washington DC.

Coltorti, M., Cremaschi, M., Delitala, M.C., Esu, D., Fornaseri, M., McPherron, A., et al. (1982) Reversed magnetic polarity at Isernia La Pineta, a new lower paleolithic site in Central Italy. *Nature* **300**: 173–6.

Combier, J., and Montet-White, A., eds (2002) *Solutré: 1968–1998*. Mémoire de la Société Préhistorique Française no. 30. Société Préhistorique Française, Paris.

Conard, N.J. (2001) Stone Age Research at the Anyskop Blowout, Langebaanweg (Western Cape Province, RSA): Report on the 2001 Field Season. *Annual Report to the South African Heritage Resources Agency*.

Conard, N.J. (2002) *Stone Age Research at the Anyskop Blowout, Langebaanweg (Western Cape Province, RSA): Report on the 2002 Field Season*. Annual Report to the South African Heritage Resources Agency.

Conard, N.J. (2007) Cultural Evolution in Africa and Eurasia During the Middle and Late Pleistocene. In: *Handbook of Paleoanthropology*, pp. 2001–37. Springer, Berlin.

Conard, N.J. (2009) A female figurine from the basal Aurignacian of the Hohle Fels Cave in southwestern Germany. *Nature* **459**: 248–52.

Conard, N.J., and Adler, D.S. (1997) Lithic reduction and hominid behavior in the Middle Palaeolithic of the Rhineland. *Journal of Anthropological Research* **53**: 147–76.

Conard, N.J., and Bolus, M. (2008) Radiocarbon dating the late Middle Paleolithic and the Aurignacian of the Swabian Jura. *Journal of Human Evolution* **55**: 886–97.

Conard, N.J., Grootes, P.M., and Smith, F.H. (2004) Unexpectedly recent dates for human remains from Vogelherd. *Nature* **430**: 198–201.

Condemi, S. (1992) *Les Hommes Fossils de Saccopastore et leurs Relations Phylénétiques*. CNRS, Paris.

Condemi, S. (2004) The Garba IV E mandible. In: *Studies on the Early Paleolithic site of Melka Kunture, Ethiopia*, Chavaillon, J., and Piperno, M., eds, pp. 687–701. Istituto Italiano di Preistoria e Protostoria, Florence.

Conroy, G. (1996) The cave breccias of Berg Aukas, Namibia: a clustering approach to mine dump paleontology. *Journal of Human Evolution* **30**(4): 349–55.

Conroy, G., and Vannier, M.W. (1991) Dental development in South African Australopithecines. Part II: Dental Stage Assessment. *American Journal of Physical Anthropology* **86**: 137–56.

Conroy, G., Weber, G., Seidler, H., Tobias, P., Kane, A., and Brunsden, B. (1998) Endocranial capacity in an early hominid cranium from Sterkfontein, South Africa. *Science* **280**(5370): 1730.

Conroy, G.C., and Vannier, M.W. (1987) Dental development of the Taung skull from computerized tomography. *Nature* **329**: 625–7.

Conroy, G.C., Jolly, C.J., Cramer, D., and Kalb, J.E. (1978) Newly discovered fossil hominid skull from the Afar depression, Ethiopia. *Nature* **276**: 67–70.

Conroy, G.C., Weber, G.W., Seidler, H., Recheis, W., Zur Nedden, D., and Mariam, J.H. (2000) Endocranial capacity of the Bodo cranium determined from three-dimensional computed tomography. *American Journal of Physical Anthropology* **113**: 111–18.

Constant, D.A., and Grine, F.E. (2001) A review of taurodontism with new data on indigenous southern African populations. *Archives of Oral Biology* **46**: 1021–9.

Constantino, P., Lee, J., Chai, H., Zipfel, B., Ziscovici, C., Lawn, B., et al. (2010) Tooth chipping can reveal the diet and bite forces of fossil hominins. *Biology Letters* **6**: 826–9.

Cooke, H.B.S. (1963) Pleistocene mammal faunas of Africa, with particular reference to Southern Africa. In: *African Ecology and Human Evolution*, Howell, F.C., and Bouliere, F., eds, pp. 78–84. Aldine, Chicago, IL.

Coolidge, F.L., and Wynn, T. (2007) The working memory account of Neandertal cognition - How phonological storage capacity may be related to recursion and the pragmatics of modern speech. *Journal of Human Evolution* **52**(6): 707–10.

Coolidge, H.J. (1929) A revision of the genus *Gorilla*. *Memoirs of the Museum of Comparative Zoology Harvard* **50**: 291–381.

Coolidge, H.J. (1933) *Pan paniscus*: pygmy chimpanzee from south of the Congo River. *American Journal of Physical Anthropology* **18**(1): 1–59.

Coolidge, H.J. (1984) Historical remarks bearing on the discovery of *Pan paniscus*. In: *The Pygmy Chimpanzee: Evolutionary Biology and Behavior*, Susman, R.L., ed., pp. ix–xiii. Plenum Press, New York.

Coon, C. (1954) *The Story of Man*. Knopf, New York.

Coon, C. (1962) *The Origin of Races*. Knopf, New York.

Coon, C. (1982) *Racial Adaptations*. Burnham, New York.

Cooney, W.P., Lucca, M.J., Chao, E.Y., and Linscheid, R.L. (1981) The kinesiology of the thumb trapeziometacarpal joint. *Journal of Bone and Joint Surgery* **63**: 1371–81.

Cooper, A., and Poinar, H. (2000) Ancient DNA: do it right or not at all. *Science* **289**(5482): 1139–9.

Coppa, A., Grün, R., Stringer, C.B., Eggins, S., and Vargiu, R. (2005) Newly recognized Pleistocene human teeth from Tabun Cave, Israel. *Journal of Human Evolution* **49**(3): 301–15.

Coppens, Y. (1979) Les hominides du Pliocene et de Pleistocene de la Rift Valley. *Bulletin de la Societe Geologique de France* **21**: 313–20.

Coppens, Y. (1980) The differences between *Australopithecus* and *Homo*; preliminary conclusions from the Omo Research Expedition's studies. In: *Current Argument on Early Man*, Königsson, L.K., ed., pp. 207–25. Pergamon, Oxford.

Coppens, Y., and Sakka, M. (1980) Un nouveau crane d'australopitheque. In: *Morphologie evolutive: morphogenese du crane et origine de l'homme. Pre-congress symposium du VIII Congres Internatonal de Primatologie*, Sakka, M., ed., pp. 85–194. CNRS, Florence.

Coppens, Y., Howell, F., Isaac, G., and Leakey, R., eds (1976) *Earliest man and environments in the Lake Rudolf Basin*. University of Chicago Press, Chicago, IL.

Coqueugniot, H., Hublin, J.-J., Veillon, F., Houët, F., and Jacob, T. (2004) Early brain growth in *Homo erectus* and implications for cognitive ability. *Nature* **431**: 299–302.

Corballis, M. (2002) *From hand to mouth: The origins of language*. Princeton University Press, New York.

Corrain, C. (1965) Resti scheletrici umani di Grotta Paglicci. *Atti X Riunione Istituto Italiano di Preistoria e Protostoria*: 281–300.

Corrain, C. (1966) Un frammento di mandibola umana, rinvenuto al Riparo Tagliente, in Valpantena (Verona). *Atti Dell Istituto Veneto Di Scienze Lettere Ed Arti* **124**: 23–25.

Corruccini, R. (1977a) Crown component variation in the hominoid lower second premolar. *Journal of Dental Research* **56**(9): 1093.

Corruccini, R. (1977b) Cartesian coordinate analysis of the hominoid second lower deciduous molar. *Journal of Dental Research* **56**(6): 699.

Corruccini, R. (1978) Crown component variation in hominoid upper first premolars. *Archives of Oral Biology* **23**(6): 491–4.

Corruccini, R., and McHenry, H. (1980) Cladometric analysis of Pliocene hominids. *Journal of Human Evolution* **9**(3): 209–21.

Corvinus, G., and Roche, H. (1976) La préhistoire dans la région de Hadar (Bassin de l'Awash, Afar, Ethiopie): premiers résultats. *L'Anthropologie* **80**(2): 315–24.

Coryndon, S.C., Gentry, A.W., Harris, J.M., Hooijer, D.A., Maglio, V.J., and Howell, F.C. (1972) Mammalian Remains from the Isimila Prehistoric Site, Tanzania. *Nature* **237**: 292.

Count, E.W. (1947) Brain and body weight in man: their antecedents in growth and evolution. *Annals of the New York Academy of Sciences* **46**: 993–1122.

Cowgill, L.W., Trinkaus, E., and Zeder, M.A. (2007) Shanidar 10: A Middle Paleolithic immature distal lower limb from Shanidar Cave, Iraqi Kurdistan. *Journal of Human Evolution* **53**(2): 213–23.

Cox, A. (1969) Geomagnetic reversals. *Science* **163**: 237–45.

Cox, A., Doell, R.R., and Dalrymple, G.B. (1963) Geomagnetic polarity epochs and Pleistocene geochronometry. *Nature* **198**: 1049–51.

Cox, A., Doell, R.R., and Dalrymple, G.B. (1964) Reversals of the Earth's magnetic field. *Science* **144**: 1537–43.

Crabtree, D.E. (1968) Mesoamerican polyhedral cores and prismatic blades. *American Antiquity* **33**: 446–78.

Cracraft, J. (1983) Species concepts and speciation analysis. In: *Current Ornithology*, Johnson, R.F., ed., pp. 159–87. Plenum Press, New York.

Cramer, D.L. (1977) Craniofacial morphology of *Pan paniscus*. A morphometric and evolutionary appraisal. *Contributions to Primatology* **10**: 1–64.

Crevecoeur, I. (2008a) *Étude anthropologique du squelette du Paléolithique supérieur de Nazlet Khater 2 (Égypte). Apport à la compréhension de la variabilité passée des hommes modernes.* Leuven University Press, Leuven.

Crevecoeur, I. (2008b) Variability of Palaeolithic modern humans in Africa. Future Prospects of the Ishango human remain (re-)study. In: *Ishango, 22000 and 50 Years Later: The Cradle of Mathematics?*, Huylebrouck, D., ed., pp. 87–97. Koninklijke Vlaamse Academie van Belgie voor Wetenschappen en Kunsten, Brussels.

Crompton, R.H., and Gowlett, J.A.J. (1993) Allometry and multidimensional form in Acheulean bifaces from Kilombe, Kenya. *Journal of Human Evolution* **25**(3): 175–99.

Crompton, R.H., Li, Y., McN Alexander, R., Wang, W., and Gunther, M.A. (1996) Segment inertial properties of primates: new techniques for laboratory and field studies of locomotion. *American Journal of Physical Anthropology* **99**(4): 547–70.

Crompton, R.H., Li, Y., Wang, W., Günther, M., and Savage, R. (1998) The mechanical effectiveness of erect and 'bent-hip, bent-knee' bipedal walking in *Australopithecus afarensis*. *Journal of Human Evolution* **35**: 55–74.

Crow, T.J. (2000) Schizophrenia as the price that *Homo sapiens* pays for language: a resolution of the central paradox in the origin of the species. *Brain Research Reviews* **31**(2–3): 118–29.

Cruz-Uribe, K., Klein, R.G., Avery, G., Avery, M., Halkett, D., Hart, T., et al. (2003) Excavation of buried Late Acheulean (Mid-Quaternary) land surfaces at Duinefontein 2, Western Cape Province, South Africa. *Journal of Archaeological Science* **30**(5): 559–75.

Csibra, G., and Gergely, G. (2009) Natural pedagogy. *Trends in Cognitive Science* **13**(4): 148–53.

Cuenca-Bescós, G., Laplana, C., and Canudo, J.I. (1999) Biochronological implications of the Arvicolidae (Rodentia, Mammalia) from the Lower Pleistocene hominid-bearing level of Trinchera Dolina 6 (TD6, Atapuerca, Spain). *Journal of Human Evolution* **37**(3–4): 353–73.

Cunningham, D.L., and Jantz, R.L. (2003) The morphometric relationship of Upper Cave 101 and 103 to modern *Homo sapiens*. *Journal of Human Evolution* **45**: 1–18.

Cunningham, D.L., and Westcott, D.J. (2002) Within-group human variation in the Asian Pleistocene: the three Upper Cave crania. *Journal of Human Evolution* **47**: 627–38.

Curnoe, D. (2003) Problems with the use of cladistic analysis in palaeoanthropology. *HOMO-Journal of Comparative Human Biology* **53**(3): 225–34.

Curnoe, D. (2006) Odontometric systematic assessment of the Swartkrans SK 15 mandible. *Journal of Comparative Human Biology* **57**: 263–94.

Curnoe, D. (2010) A review of early *Homo* in southern Africa focusing on cranial, mandibular and dental remains, with the description of a new species (*Homo gautengensis* sp. nov.). *Homo* **62**: 151–77.

Curnoe, D., and Tobias, P.V. (2006) Description, new reconstruction, comparative anatomy, and classification of the Sterkfontein Stw 53 cranium, with discussions about the taxonomy of other southern African early *Homo* remains. *Journal of Human Evolution* **50**(1): 36–77.

Curnoe, D., and Brink, J. (2010) Evidence of pathological conditions in the Florisbad cranium. *Journal of Human Evolution* **59**: 504–13.

Curnoe, D., Grün, R., Taylor, L., and Thackeray, F. (2001) Direct ESR dating of a Pliocene hominin from Swartkrans. *Journal of Human Evolution* **40**: 379–91.

Currey, J.D. (2002) *Bones: Structure and Mechanics.* Princeton University Press, Princeton, NJ.

Curtis, G. (1974) Age of early Acheulian industries from the Peninj Group, Tanzania. *Nature* **249**: 624.

Czarnetzki, A. (1983) Zur Entwicklung des Menschen in Südwestdeutschland. In: *Urgeschichte in Baden-Württemberg*, Müller-Beck, H., ed., pp. 217–40. Konrad Theiss Verlag, Stuttgart.

Czarnetzki, A. (1989) Ein archaischer hominidencalvariarest aus einer kiesgrube in Reilingen, Rhein-Neckar-Kreis. *Quatär* **39–40**: 191–201.

Czarnetzki, A., and Trelliso-Carreno, L. (1999) Le fragment d'un os parietal du Neanderthalien classique de Warendorf-Neuwarendorf. *L'Anthropologie* **103**: 237–48.

REFERENCES

Czarnetzki, A., Gaudzinski, S., and Pusch, C.M. (2001) Hominid skull fragment from Late Pleistocene layers in Leine Valley (Sarstedt, District of Hildesheim, Germany). *Journal of Anatomy* **41**: 133–40.

Czarnetzki, A., Fangenberg, O.u.K.-W., Gaudzinski, S., and Rohde, P. (2002) Die Neandertaler-Fundstätte im Leinetal bei Sarstedt, Landkreis Hildesheim. *Geologie, Archäologie und Paläontologie* **53**: 23–46.

Dagley, P., Mussett, A.E., and Palmer, H.C. (1978) Preliminary observations on the palaeomagnetic stratigraphy of the area west of Lake Baringo, Kenya. In: *Geological Background to Fossil Man: Recent research in the Gregory Rift Valley, East Africa*, Bishop, W.W., ed., pp. 225–35. Scottish Academic Press, Edinburgh.

Dahlberg, A.A. (1945) The changing dentition of Man. *Journal of the American Dental Association* **32**: 676–90.

Dale, J., Dunn, P.O., Figuerola, J., Lislevand, T., Székely, T., Linda, A., et al. (2007) Sexual selection explains Rensch's rule of allometry for sexual size dimorphism. *Proceedings of the Royal Society of London, Series B* **274**: 2971–79.

Dalrymple, G.B., and Lanphere, M.A. (1969) *Potassium-argon dating: principles, techniques and applications to geochronology*. Freeman, San Francisco, CA.

Damas, H. (1940) Observations sur les couches fossiliferes bordant la Semliki. *Revue de Zoologie et de Botanique Africaines* **23**: 265–72.

Damasio, H., and Damasio, A.R. (1989) *Lesion Analysis in Neuropsychology*. Oxford University Press, New York.

Danhara, T., Bae, K.D., Okada, T., Matsufuji, K., and Hwang, S.H. (2002) What is the real age of the Chongokni Paleolithic site? In: *Paleolithic Archaeology in Northeast Asia*, Bae, K.D., and Lee, J.C., eds, pp. 77–116. Yeoncheon County and The Institute of Cultural Properties, Seoul & Yeoncheon County.

Dansgaard, W., Johnsen, S.J., Clausen, H.B., Dahl-Jensen, D., Gundestrup, N.S., Hammer, C.U., et al. (1993) Evidence for general instability of past climate from a 250-kyr ice-core record. *Nature* **364**: 218–20.

D'Août, K., Aerts, P., De Clercq, D., De Meester, K., and Van Elsacker, L. (2002) Segment and joint angles of the hind limb during bipedal and quadrupedal walking of the bonobo (*Pan paniscus*). *American Journal of Physical Anthropology* **119**(1): 37–51.

Darlas, A. (1995) Τα λίθινα εργαλεία του σκελετού ΛΑΟ 1/Σ 3 (Απήδημα – Μάνη). *Acta Anthropologica* **1**: 59–62.

Dart, R. (1925) *Australopithicus africanus*: the man-ape from South Africa. *Nature* **115**: 195–9.

Dart, R. (1926) Taung and its significance. *Natural History* **26**: 315–27.

Dart, R. (1934) The dentition of *Australopithecus africanus*. *Folia Anat. Japon.* **12**: 207–21.

Dart, R.A. (1948) The Makapansgat proto-human *Australopithecus prometheus*. *American Journal of Physical Anthropology* **6**(3): 259–84.

Dart, R.A. (1957) An Australopithecine Object from Makapansgat. *Nature* **179**: 693–5.

Dart, R.A. (1967) Mousterian osteodontokeratic objects from Geulah Cave (Haifa, Israel). *Quaternaria* **9**: 105–40.

Darwin, C. (1859) *On the Origin of Species by Means of Natural Selection, or the Preservation of Favoured Races in the Struggle for Life*. John Murray, London.

Darwin, C. (1871) *The Descent of Man, and Selection in Relation to Sex*. John Murray, London.

Darwin, C. (1872) *The Expression of the Emotions in Man and the Animals*. John Murray, London.

Dastugue, J. (1967) Pathologie des Hommes fossiles e l'abri de Cro-Magnon. *Anthropologie* **71**: 79–492.

Daura, J., Sanz, M., Subira, M., Quam, R., Fullola, J., and Arsuaga, J. (2005) A Neandertal mandible from the Cova del Gegant (Sitges, Barcelona, Spain). *Journal of Human Evolution* **49**(1): 56–70.

Daura, J., Sanz, M., Pike, A., Subir, M., Fornós, J., Fullola, J., et al. (2010) Stratigraphic context and direct dating of the Neandertal mandible from Cova del Gegant (Sitges, Barcelona). *Journal of Human Evolution*.

Davies, P. (1968) An 8,000 to 12,000 years old human tooth from Western Australia. *Archaeology & Physical Anthropology in Oceania* **3**(1): 33–40.

Davis, P.R. (1964) Hominid fossils from Bed I. Olduvai Gorge, Tanganyika. *Nature* **201**: 967–8.

Davis, S.J.M. (1977) The ungulate remains from Kebara cave. In: *Moshe Stekelis Memorial* Bar-Yosef, O., and Arensburg, B., eds, pp. 150–63. Israel Exploration Society, Jerusalem.

Dawkins, R. (1976) *The Selfish Gene*. Oxford University Press, Oxford.

Dawson, C., and Woodward, A.S. (1913) On the discovery of a palaeolithic human skull and mandible in a flint-bearing gravel overlying the Wealden (Hastings Beds) at Piltdown, Fletching (Sussex). *Quarterly Journal of the Geological Society* **69**(14): 117–23.

Day, M.H. (1967) Who were the first hominids? *Eugenics Review* **59**: 150–2.

Day, M.H. (1967) Olduvai Hominid 10: a multivariate analysis. *Nature* **215**: 323–4.

Day, M.H. (1969a) Femoral fragment of a robust australopithecine from the Olduvai Gorge, Tanzania. *Nature* **221**: 230–3.

Day, M.H. (1969b) Omo human skeletal remains. *Nature* **222**: 1135–8.

Day, M.H. (1971) Post-cranial remains of *Homo erectus* from Bed IV, Olduvai Gorge, Tanzania. *Nature* **232**: 383–7.

Day, M.H. (1971) Olduvai Hominid 28. *Journal of Anatomy* **109**: 354–5.

Day, M.H. (1976) Hominid postcranial remains from the East Rudolf succession: a review. In: *Earliest Man and Environments in the Lake Rudolf Basin: Stratigraphy, Paleoecology and Evolution*, Coppens, Y., Howell, F.C., Isaac, G.L., and Leakey, R.E.F., eds, pp. 507–21. University of Chicago Press, Chicago, IL.

Day, M.H. (1977) *Guide to Fossil Man*. Cassell, London.

Day, M.H. (1986) *Guide to Fossil Man*. Cassell, London.

Day, M.H., and Napier, J.R. (1964) Hominid fossils from Bed I, Olduvai Gorge, Tanganyika: fossil foot bones. *Nature* 201: 967–70.

Day, M.H., and Napier, J.R. (1966) A hominid toe bone from Bed I, Olduvai Gorge, Tanzania. *Nature* 211: 929–30.

Day, M.H., and Wood, B.A. (1968) Functional affinities of the Olduvai Hominid 8 talus. *Man* 3: 440–55.

Day, M.H., and Leakey, R.E.F. (1973) New evidence of the genus *Homo* from East Rudolf, Kenya. I. *American Journal of Physical Anthropology* 39(3): 341–54.

Day, M.H., and Molleson, T.I. (1973) The Trinil femora. In: *Human Evolution. Symposia of the Society for the Study of Human Biology*, Day, M.H., ed., pp. 127–54. Taylor & Francis, London.

Day, M.H., and Leakey, R.E.F. (1974) New evidence of the genus *Homo* from East Rudolf, Kenya (III). *American Journal of Physical Anthropology* 41: 367–80.

Day, M.H., and Molleson, T.I. (1976) The puzzle from JK2 - a femur and a tibial fragment (OH 34) from Olduvai Gorge, Tanzania. *Journal of Human Evolution* 5: 455–65.

Day, M.H., and Stringer, C. (1982) A reconstruction of the Omo Kibish remains and the erectus-sapiens transition. In: *L'Homo erectus et la Place de L'Homme de Tautavel Pami les Hominides Fossiles*, de Lumley, M.-A., ed., pp. 814–46. Louis-Jean Scientific and Literary Publications, Nice.

Day, M.H., and Thornton, C.M.B. (1986) The extremity bones of *Paranthropus robustus* from Kromdraai B, East Formation Member 3, Republic of South Africa - a reappraisal. *Anthropos (Brno)* 23: 91–99.

Day, M.H., and Stringer, C. (1991) The Omo Kibish cranial remains and classification within the genus *Homo*. *L'Anthropologie* 95(2–3): 573–94.

Day, M.H., Leakey, R.E.F., Walker, A.C., and Wood, B.A. (1975) New hominids from East Rudolf, Kenya 1. *American Journal of Physical Anthropology* 42: 461–75.

Day, M.H., Leakey, R.E.F., Walker, A.C., and Wood, B.A. (1976) New hominids from East Turkana, Kenya. *American Journal of Physical Anthropology* 45: 369–435.

Day, M.H., Leakey, M.D., and Magori, C. (1980) A new hominid fossil skull (LH18) from the Ngaloba Beds, Laetoli, Northern Tanzania. *Nature* 284: 55–56.

Day, M.H., Twist, M.H.C., and Ward, S. (1991) The Omo I (Kibish) postcranial remains. *L'Anthropologie* 95(2–3): 595–609.

Dayal, M., Kegley, A., Ätrkalj, G., Bidmos, M., and Kuykendall, K. (2009) The history and composition of the Raymond A. Dart Collection of Human Skeletons at the University of the Witwatersrand, Johannesburg, South Africa. *American Journal of Physical Anthropology* 140(2): 324–35.

Deacon, H.J. (1989) Late Pleistocene palaeoecology and archaeology in the Southern Cape, South Africa. In: *The Human Revolution: Behavioral and Biological Perspectives on the Origins of Modern Humans*, Mellars, P., and Stringer, C., eds, pp. 547–64. Edinburgh University Press, Edinburgh.

Deacon, H.J. (1992) Southern Africa and modern human origins. *Philosophical Transactions of the Royal Society of London. Series B, Biological Sciences* 337: 177–83.

Deacon, H.J. (2001) Modern human emergence: an African archaeological perspective. In: *Humanity from African Naissance to Coming Millennia-Colloquia in Human Biology and Palaeoanthropology*, Tobias, P.V., Raath, M.A., Moggi-Cecchi, J., and Doyle, G.A., eds, pp. 217–26. Florence University Press, Florence.

Deacon, H.J., and Geleijinse, V.B. (1988) The Stratigraphy and Sedimentology of the Main Site Sequence, Klasies River, South Africa. *The South African Archaeological Bulletin* 43(147): 5–14.

Deacon, H.J., Talma, A.S., and Vogel, J.C. (1988) Biological and cultural developments of Pleistocene people in an old world southern continent. In: *Early man in the Southern Hemisphere. Supplement to Archaeometry: Australian Studies 1988*, Prescott, J.R., ed., pp. 23–31. University of Adelaide, Adelaide.

Deacon, J., and Wilson, M. (1992) Peers Cave, 'the cave the world forgot'. *The Digging Stick* 9: 2–5.

Deacon, T. (1997) *The Symbolic Species: The Co-Evolution of Language and the Human Brain*. W.W. Norton & Company, London.

Dean, C. (1999) Hominoid tooth growth: using incremental lines in dentine as markers of growth in modern human and fossil primate teeth. In: *Human Growth in the Past*, Hoppa, R.D., and FitzGerald, C.M., eds, pp. 111–27. Cambridge University Press, Cambridge.

Dean, C., Leakey, M.G., Reid, D., Schrenk, F., Schwartz, G.T., Stringer, C., et al. (2001) Growth processes in teeth distinguish modern humans from *Homo erectus* and earlier hominins. *Nature* 414: 628–31.

Dean, D. (1993) *The Middle Pleistocene* Homo erectus/Homo sapiens *Transition: New Evidence from Space Curve Statistics*. PhD thesis, City University of New York.

Dean, D., Hublin, J.-J., Holloway, R., and Ziegler, R. (1998) On the phylogenetic position of the pre-Neandertal specimen from Reilingen, Germany. *Journal of Human Evolution* 34(5): 485–508.

Dean, M. (2009) Extension rates and growth in tooth height of modern human and fossil hominin canines and molars. In: *Comparative Dental Morphology*, Koppe, T., Meyer, G., and Alt, K.W., eds, pp. 68–73. Karger, Basel.

Dean, M., Beynon, A., Reid, D., and Whittaker, D. (1993) A longitudinal study of tooth growth in a single individual based on long and short period incremental markings in dentine and enamel. *International Journal of Osteoarchaeology* 3(4): 249–64.

Dean, M.C. (1998) Comparative observation on the spacing of short-period (von Ebners) lines in dentine. *Archives of Oral Biology* 43: 1009–21.

Dean, M.C. (2000) Progress in understanding hominoid dental development. *Journal of Anatomy* **197**(1): 77–101.

Dean, M.C. (2006) Tooth microstructure tracks the pace of human life history evolution. *Proceedings of the Royal Society of London, Series B* **273**(1603): 2799–808.

Dean, M.C., and Wood, B.A. (1981) Developing pongid dentition and its use for ageing individual crania in comparative cross-sectional growth studies. *Folia Primatologica* **36**: 111–27.

Dean, M.C., and Wood, B.A. (1982) Basicranial anatomy of Plio-Pleistocene hominids from East and South Africa. *American Journal of Physical Anthropology* **59**(2): 157–74.

Dean, M.C., and Shellis, R.P. (1998) Observations on stria morphology in the lateral enamel of *Pongo, Hylobates,* and *Proconsul* teeth. *Journal of Human Evolution* **35**(4–5): 401–10.

Dean, M.C., and Smith, B.H. (2009) Growth and development in the Nariokotome Youth, KNM-WT 15000: A dental perspective on the origins and early evolution of the genus *Homo*. In: *The First Humans: origins of the genus* Homo, Grine, F., Fleagle, J., and Leakey, R., eds, pp. 101–20. Springer, New York.

de Beer, G.R. (1954) Archaeopteryx and evolution. *Advancement of Science* **11**: 160–70.

Debénath, A. (1974) Position stratigraphique des restes humains antewurmiens de Charente. *Bulletins et Mémoires de la Société d'Anthropologie de Paris* **1**(13): 417–26.

Debénath, A. (1975) Découverte de restes humains probablement atériens à Dar es Soltane (Maroc). *Comptes Rendus de l'Académie des Sciences* **281**: 875–6.

Debénath, A. (1976) Le Site de Dar-es-Soltane 2, á Rabat (Maroc). *Bulletins et Mémoires de la Société d'Anthropologie de Paris* **3**(2): 181–2.

Debénath, A. (1977) The latest finds of antewürmien human remains in Charente (France). *Journal of Human Evolution* **6**: 297–302.

Debénath, A. (2000) Le peuplement préhistorique du Maroc: données récentes et problèmes. *L'Anthropologie* **104**: 131–45.

Debénath, A., and Duport, L. (1972) Y a-t-il du Solutréen à Montgaudier (Charente). *Bulletin et Mémoires de la Société Archéologique et. Historique de la Charente*: 193–205.

Debénath, A., and Sbihi-Alaoui, F. (1979) Découverte de deux nouveaux gisements préhistoriques près de Rabat (Maroc). *Bulletin de la Société préhistorique française* **79**(1): 11–14.

Debénath, A., and Dibble, H.L. (1994) *Handbook of Paleolithic Typology: Lower and middle paleolithic of Europe*. University Museum, University of Pennsylvania, Philadelphia, PA.

Debénath, A., and Jelinek, A.J. (1998) Nouvelles fouilles à La Quina (Charente): Résultats préliminaires. *Gallia Préhistoire* **40**: 29–74.

Debénath, A., Raynal, J.-P., Roche, J., Texier, J.-P., and Ferembach, D. (1986) Stratigraphie, habitat, typologie

et devenir de l'Atérien marocain: données récentes. *L'Anthropologie* **90**: 233–46.

de Blainville, H. (1839) *Ostéographie, ou description iconographique comparde du squelette et du système dentaire des mammifères recents et fossiles pour servir de base à la zoologie et à la géologie*. J.B. Balliere, Paris.

de Bonis, L., Geraads, D., Guerin, G., Abderrhamane, H., Jaeger, J.-J., and Sen, S. (1984) Découverte d'un Hominidé fossile dans le Pléistocène de la République de Djibouti. *Comptes Rendus de l'Académie des Sciences* **299**(15): 1097–100.

Defleur, A. (1995) Nouvelles découvertes de restes humains Moustériens dans les dépûts de la Baume Moula-Guercy (Soyons, Ardèche). *Bulletins et Mémoires de la Société d'Anthropologie de Paris* **7**(3): 185–90.

Defleur, A., Crégut-Bonnoure, E., and Desclaux, E. (1998) Première mise en évidence d'une séquence éémienne à restes humains dans le remplissage de la Baume Moula-Guercy (Soyons, Ardèche). *Comptes Rendus de l'Académie des Sciences* **326**(6): 453–8.

Defleur, A., Crégut-Bonnoure, Desclaux, E., and Thinon, M. (2001) Présentation paléo-environnementale du remplissage de la Baume Moula-Guercy à Soyons (Ardèche): implications paléoclimatiques et chronologiques. *L'Anthropologie* **105**(3): 369–408.

DeGusta, D., and Vrba, E.S. (2003) A method for inferring paleohabitats from the functional morphology of bovid astragali. *Journal of Archaeological Science* **30**(8): 1009–22.

DeGusta, D., Gilbert, W.H., and Turner, S.P. (1999) Hypoglossal canal size and hominid speech. *Proceedings of the National Academy of Sciences USA* **96**: 1800–4.

de Heinzelin, J. (1955) *Le Fossé tectonique sous le parallèle d'Ishango. Exploration du Parc National Albert (Fasc. 1)*. Institut des Parcs Nationaux du Congo Belge, Brussels.

de Heinzelin, J. (1957) *Les Fouilles d'Ishango. Exploration de Parc National Albert (Fasc. 2)*. Institut des Parcs Nationaux du Congo Belge, Brussels.

de Heinzelin, J. (1961) *Le Paléolithique aux abords d'Ishango. Exploration de Parc National Albert (Fasc. 6)*. Institut des Parcs Nationaux du Congo Belge, Brussels.

de Heinzelin, J. (1962) Ishango. *Scientific American* **206**(6): 105–11,13–14,16.

de Heinzelin, J. (1983) *The Omo Group*. Musée Royal de l'Afrique Centrale, Tervuren.

de Heinzelin, J., and Haesaerts, P. (1983) The Shungura Formation. In: *The Omo Group*, de Heinzelin, J., ed., pp. 25–128. Musée Royal de l'Afrique Centrale, Tervuren, Belgium.

de Heinzelin, J., and Verniers, J. (1996) Realm of the Upper Semliki (Eastern Zaire): An Essay on Historical Geology. *Sciences Géologiques* **102**: 1–87.

de Heinzelin, J., Clark, J.D., White, T., Hart, W., Renne, P., WoldeGabriel, G., et al. (1999) Environment and behavior of 2.5-million-year-old Bouri hominids. *Science* **284**(625): 625–9.

de Heinzelin, J., Clark, J.D., Schick, K.D., and Gilbert, W.H. (2000) The Acheulean and the Plio-Pleistocene deposits of the Middle Awash Valley, Ethiopia. In: *Annales Sciences Geologiques*, de Heinzelin, J., Clark, J.D., Schick, K.D., and Gilbert, W.H., eds, pp. 235. Royal Museum of Central Africa, Tervuren.

Deino, A., and Potts, R. (1990) Single-crystal ^{40}Ar/^{39}Ar Dating of the Olorgesailie Formation, Southern Kenya Rift. *Journal of Geophysical Research* **95**(B6): 8453–79.

Deino, A., and Hill, A. (2002) ^{40}Ar/^{39}Ar dating of Chemeron Formation strata encompassing the site of hominid KNM-BC 1, Tugen Hills, Kenya. *Journal of Human Evolution* **42**(1–2): 141–51.

Deino, A.L., and McBrearty, S. (2002) ^{40}Ar/^{39}Ar dating of the Kapthurin Formation, Baringo, Kenya. *Journal of Human Evolution* **42**(1–2): 185–210.

Deino, A.L., Tauxe, L., Monaghan, M., and Hill, A. (2002) ^{40}Ar/^{39}Ar geochronology and paleomagnetic stratigraphy of the Lukeino and lower Chemeron Formations at Tabarin and Kapcheberek, Tugen Hills, Kenya. *Journal of Human Evolution* **42**(1–2): 117–40.

Deino, A.L., Scott, G.R., Saylor, B., Alene, M., Angelini, J.D., and Haile-Selassie, Y. (2010) ^{40}Ar/^{39}Ar dating, paleomagnetism, and tephrochemistry of Pliocene strata of the hominid-bearing Woranso-Mille area, west-central Afar Rift, Ethiopia. *Journal of Human Evolution* **58**(2): 111–26.

Delagnes, A., and Roche, H. (2005) Late Pliocene hominid knapping skills: the case of Lokalalei 2C, West Turkana, Kenya. *Journal of Human Evolution* **48**: 435–72.

Delagnes, A., Lenoble, A., Harmand, S., Brugal, J.-P., Prat, S., Tiercelin, J.-J., et al. (2006) Interpreting pachyderm single carcass sites in the African Lower and Early Middle Pleistocene record: A multidisciplinary approach to the site of Nadung'a 4 (Kenya). *Journal of Anthropological Archaeology* **25**: 448–65.

de la Torre, I., and Mora, R. (2005) Unmodified lithic material at Olduvai Bed I: manuports or ecofacts? *Journal of Archaeological Science* **32**: 273–85.

Delattre, A., and Fenart, R. (1960) *L'hominisation du crâne: étudiée par la méthode vestibulaire*. Centre National de la Recherche Scientifique, Paris.

Delluc, B., and Delluc, G. (2003) *Lascaux retrouvé. Les recherches de l'abbé André Glory*.

Deloison, Y. (2004) Les empreintes de pas de Laetoli, Tanzanie. *Biométrie humaine et anthropologie* **22**: 61–65.

Delpech, F. (2007) Le grand abri de La Ferrassie, source de réflexion sur la biostratigraphie d'un court moment du Pléistocène. In: *Arts et Culture de la Préhistoire. Hommage à Henri Delporte*, Desbosse, R., and Thévenin, eds, pp. 303–4.

Delpech, F., and Rigaud, J.-P. (2001) Nouveaux apports des datations en Archéologie préhistorique. In: *Datation. XXIe Rencontres internationales d'Archéologie et d'histoire d'Antibes*, Barrandon, J.-N., Guibert, P., and Michel, V., eds, pp. 315–31.

Delporte, H. (1969) Les fouilles du Musée des Antiquités Nationales à la Ferrassie. *Antiquités Nationales* **1**: 15–28.

Delporte, H. (1978) Etat actuel de l'analyse de l'aurignacien de La Ferrassie. *Bulletin de la Société Préhistorique Française* **75**(6): 165.

Delporte, H., and Tuffreau, A. (1972/3) Les industries du Perigordien superieur de la Ferrassie. *Quartär* **23/4**: 93–123.

Delporte, H., Djindjian, F., and Maziere, G. (1977) L'aurignacien de La Ferrassie: Observations préliminaires a la suite de fouilles recentes. *Bulletin de la Société Préhistorique Française* **74**(1): 343–61.

Delporte, H., Djindjian, F., and Maziere, G. (1983) Etudes sur L'Aurignacien de la Ferrassie. *Etudes et recherches archéologiques de l'Université de Liège* **13**(1): 13–26.

Delson, E. (1974) The oldest known fossil Cercopithecidae. *American Journal of Physical Anthropology* **41**: 474–5.

Delson, E., Terranova, C.J., Jungers, W.L., Sargis, E.J., Jablonski, N.G., and Dechow, P.C. (2000) Body mass in Cercopithecidae (Primates, Mammalia): Estimation and scaling in extinct and extant taxa. *Anthropological papers of the American Museum of Natural History* **83**: 1–159.

de Lumley, H., ed. (1972) *La grotte moustérienne de l'Hortus*. Université de Provence, Marsailles.

de Lumley, H. (1982) *L'Homo erectus et la place de l'Homme de Tautavel parmi les Hominides fossiles*. 1er Congrès International de Paléontologie Humaine, Nice, France.

de Lumley, H. (2007) *La Grande Histoire des Premiers Hommes Européens*. O. Jacob, Paris.

de Lumley, H., and de Lumley, M.-A. (1971) Découverte de restes humains anténéandertaliens datés du début de Riss á la Caune de l'Arago (Tautavel, Pyénées-Orientales). *Comptes Rendus de l'Académie des Sciences* **272**: 1729–42.

de Lumley, H., and de Lumley, M. (1973) Preneandertal human remains from Arago cave in southeastern France. *Yearbook of Physical Anthropology* **17**: 162–8.

de Lumley, H., and Sonakia, A. (1985) Première decouverte d'un *Homo erectus* sur le continent indien A Hathnora, dans la moyenne vallée de la Narmada. *Anthropologie* **89**: 13–61.

de Lumley, H., and Beyene, Y. (2004) *Les sites préhistoriques de la région de Fejej, sud-Omo, Éthiopie, dans leur contexte stratigraphique et paléontlogique*. Éditions Rechereches sur les Civilisasions, Paris.

de Lumley, H., de Lumley, M.-A., Miskovsky, J.-C., and Renault-Miscovsky, J. (1976) Grotte du Lazaret. In: *Sites paleolithiques de la Region de Nice et grottes de Grimaldi*, pp. 53–74. IXth U.I.S.P.P. Meeting, Nice.

de Lumley, H., de Lumley, M., and Fournier, A. (1982) *La mandibule de l'Homme de Tautavel in L'*Homo erectus *et la place de l'Homme de Tautavel parmi les Hominidés fossiles*. Congrès International de Paléontologie Humaine. Nice.

de Lumley, H., Lordkipanidze, D., Féraud, G., Garcia, T., Perrenoud, C., Falguéres, C., et al. (2002) Datation par le méthode ^{40}Ar/^{39}Ar de la couche de cendres volcaniques (couche VI) de Dmanissi (Géorgie) qui a livré des restes d'hominidés fossiles de 1,81 Ma. *Comptes Rendus Palevol* **1**(3): 181–9.

de Lumley, H., Nioradzé, M., Barsky, D., Cauche, D., Celiberti, V., Nioradzé, G., et al. (2005) Les industries lithiques préoldowayennes du début du Pléistoc ne inférieur du site de Dmanissi en Géorgie. *L'anthropologie* **109**(1): 1–182.

de Lumley, M.-A. (1971/72) La Mandibula de Bañolas. *Ampurias* **33–34**: 1–92.

de Lumley, M.-A., and Piveteau, J. (1969) Les restes humains de la frotte du Lazaret (Nice, Alpes-Maritimes). *Memoire Societe Prehistorique Francaise* **7**: 223–32.

de Lumley, M.-A., and Lordkipanidze, D. (2006) L'Homme de Dmanissi (*Homo georgicus*), il y a 1 810 000 ans. *Comptes Rendus Palevol* **5**(1–2): 273–81.

de Lumley, M.-A., Gabounia, L., Vekua, A., and Lordkipanidze, D. (2006) Les restes humains du Pliocène final et du début du Pléistocène inférieur de Dmanissi, Géorgie (1991–2000). I - Les crânes, D 2280, D 2282, D 2700. *L'Anthropologie* **110**: 1–110.

deMenocal, P., and Brown, F.H. (1999) Pliocene tephra correlations between East African hominid localities, the Gulf of Aden, and the Arabian Sea. In: *The Evolution of Neogene Terrestrial Ecosystems in Europe*, Agustí, J., Rook, L., and Andrews, P., eds, pp. 23–54. Cambridge University Press, Cambridge.

deMenocal, P., Ortiz, J., Guilderson, T., Adkins, J., Sarnthein, M., Baker, L., et al. (2000) Abrupt onset and termination of the African Humid Period: rapid climate responses to gradual insolation forcing *Quaternary Science Reviews* **19**(1–5): 347–61.

Demes, B. (2007) *In vivo* bone strain and bone functional adaptation. *American Journal of Physical Anthropology* **133**: 717–22.

Demes, B., Qin, Y.-X., Stern, J.T.J., Larson, S.G., and Rubin, C.T. (2001) Patterns of strain in the macaque tibia during functional activity. *American Journal of Physical Anthropology* **116**: 257–65.

Demirjian, A. (1973) A New System of Dental Age Assessment. *Human Biology* **45**(2): 211–27.

de Mortillet, G. (1883) *Le préhistorique antiquité de l'homme*. C. Reinwald, Paris.

de Mortillet, G., and de Mortillet, A. (1900) *Le préhistorique: origine et antiquité de l'homme*, Paris.

DeNiro, M.J., and Epstein, S. (1978) Influence of diet on the distribution of carbon isotopes in animals. *Geochimica et Cosmochimica Acta* **42**: 495–506.

DeNiro, M.J., and Epstein, S. (1981) Influence of diet on the distribution of nitrogen isotopes in animals. *Geochimica et Cosmochimica Acta* **45**: 341–51.

Denys, C. (1990) Deux nouvelles especes d'*aethomys* (Rodentia, Muridae) a Langebaanweg (Pliocene, Afrique du Sud); Implications phylogenetiques et paleoecologiques. *Annales de Paleontologie* **76**(1): 41–69.

Denys, C. (1991) Un nouveau rongeur *Mystromys pocockei* sp. nov. (Crietinae) du Pliocene inferieur de Langebaanweg (Region du Cap, Afrique du Sud). *Comptes Rendus de l'Académie des Sciences* **313**: 1335–41.

Denys, C. (1994a) Nouvelles especes de *Dendromus* (Rongeurs, Muroidea) a Langebaanweg (Pliocene, Afrique du Sud) consequences stratigraphiques et paleoecologiques. *Palaeovertebrata* **23**: 153–76.

Denys, C. (1994b) Affinites systematiques de Stenodontomys (Mammalia, Rodentia) rongeur du Pliocene Langebaanweg (Afrique du Sud). *Comptes Rendus de l'Académie des Sciences* **318**: 411–16.

de Queiroz, K., and Gauthier, J. (1994) Towards a phylogenetic system of biological nomenclature. *Tree* **9**(1): 27–31.

Derevianko, A.P. (1998) *The Paleolithic of Siberia: New Discoveries and Interpretation.* University of Illinois Press.

d'Errico, F., and Backwell, L. (2003) Possile evidence of bone tool shaping by Swartkrans early hominids. *Journal of Archaeological Science* **30**(12): 1559–76.

d'Errico, F., and Henshilwood, C.S. (2007) Additional evidence for bone technology in the southern African Middle Stone Age. *Journal of Human Evolution* **52**(2): 142–63.

d'Errico, F., and Backwell, L. (2009) Assessing the function of early hominin bone tools. *Journal of Archaeological Science* **36**(8): 1764–73.

d'Errico, F., Zilho, J., Julien, M., Baffier, D., and Pelegrin, J. (1998) Neanderthal acculturation in Western Europe? A critical review of the evidence and its interpretation. *Current Anthropology* **39**(S1): 1–44.

d'Errico, F., Henshilwood, C., and Nilssen, P. (2001) An engraved bone fragment from ca. 75 Kya Middle Stone Age levels at Blombos Cave, South Africa: Implications for the origin of symbolism. *Antiquity* **75**: 309–18.

d'Errico, F., Henshilwood, C.S., Vanhaeren, M., and van Niekerk, K. (2005) *Nassarius kraussianus* shell beads from Blombos Cave: evidence for symbolic behaviour in the Middle Stone Age. *Journal of Human Evolution* **48**: 3–24.

de Ruiter, D.J., Steininger, C.M., and Berger, L.R. (2006) A cranial base of *Australopithcus robustus* from the hanging remnant of Swartkrans, South Africa. *American Journal of Physical Anthropology* **130**: 435–44.

de Ruiter, D.J., Pickering, R., Steininger, C.M., Kramers, J.D., Hancox, P.J., Churchill, S.E., et al. (2009) New *Australopithecus robustus* fossils and associated U-Pb dates from Cooper's cave (Gauteng, South Africa). *Journal of Human Evolution* **56**(5): 497–513.

Desdevises, Y., Legendre, P., Azouzi, L., and Morand, S. (2003) Quantifying phylogenetically structured environmental variation. *Evolution* **57**(11): 2647–52.

DeSilva, J.M. (2010) Revisiting the 'midtarsal break'. *American Journal of Physical Anthropology* **141**: 245–58.

de Sonneville-Bordes, D., and Perrot, J. (1954) Lexique typologique du Paléolithique supérieur. Outillage lithique: I Grattoirs—II Outils solutréens. *Bulletin de la Société Préhistorique Française* **51**: 327 35.

de Sonneville-Bordes, D., and Perrot, J. (1955) Lexique typologique du Paléolithique supérieur. Outillage lithique—III Outils composites—Peçoirs. *Bulletin de la Société Préhistorique Française* **52**: 76–79.

de Sonneville-Bordes, D., and Perrot, J. (1956a) Lexique typologique du Paléolithique supérieur. Outillage lithique— IV Burins. *Bulletin de la Société Préhistorique Française* **53**: 408–12.

de Sonneville-Bordes, D., and Perrot, J. (1956b) Lexique typologique du Paléolithique supérieur. Outillage lithique (suite et fin)—V Outillage à bord abattu—VI Pièces tronquées—VII Lames retouchées—VIII Pièces variées—IX Outillage lamellaire. Pointe azilienne. *Bulletin de la Société Préhistorique Française* **53**: 547–59.

de Sousa, A., and Wood, B.A. (2006) The hominin fossil record and the emergence of the modern human nervous system. In: *Evolution of Nervous Systems*, Kaas, J.H., ed., pp. 291–336. Elsevier, Amsterdam.

de Terra, H., and Teilhard de Chardin, P. (1936) Observations on the Upper Siwalik Formation and Later Pleistocene Deposits in India. *Proceedings of the American Philosophical Society* **76**(6): 791–822.

de Torres, T., Ortiz, J., Grün, R., Eggins, S., Valladas, H., Mercier, N., et al. (2010) Dating of the hominid (*Homo neanderthalensis*) remains accumulation from El Sidrón Cave (Piloña, Asturias North Spain): An example of a multi-methodological approach to the dating of Upper Pleistocene sites. *Archaeometry* **52**(4): 680–705.

Detroit, F., Dizon, E., Falguères, C., Hameau, S., Ronquillo, W., and Sémah, F. (2004) Upper Pleistocene *Homo sapiens* from the Tabon cave (Palawan, The Philippines): description and dating of new discoveries. *Comptes Rendus Palevol* **3**(8): 705–12.

de Villiers, H. (1973) Human skeletal remains from Border Cave, Ingwavuma District, Kwazulu, South Africa. *Annals of the Transvaal Museum* **28**(13): 229–56.

de Villiers, H. (1976) A second adult human mandible from Border Cave, Ingwavuma District, Kwazulu, South Africa. *South African Journal of Science* **72**: 212–15.

de Vos, J., and Sondaar, P.Y. (1982) The importance of the 'Dubois Collection' reconsidered. *Modern Quaternary Research in Southeast Asia* **7**: 35–63.

de Wit, M.J.C., Marshall, T.R., and Partridge, T.C. (2000) Fluvial deposits and drainage evolution. In: *The Cenozoic of Southern Africa*, Partridge, T.C., and Maud, R.R., eds, pp. 55–72. Oxford University Press, New York.

Di Rienzo, A., and Wilson, A.C. (1991) Branching pattern in the evolutionary tree for human mitochondrial DNA. *Proceedings of the National Academy of Sciences USA* **88**(5): 1597–601.

Diaz-Uriarte, R., and Garland, T. (1998) Effects of branch length errors on the performance of phylogenetically independent contrasts. *Systematic Biology* **47**: 654–72.

Dibble, H. (1991) Local raw material exploitation and its effects on Lower and Middle Paleolithic assemblage variability. *Raw Material Economies among Prehistoric Hunter-Gatherers, University of Kansas, Publications in Anthropology* **19**: 33–47.

Dibble, H., and Rolland, N. (1992) On assemblage variability in the Middle Paleolithic of Western Europe: History, Perspectives and a New Synthesis. In: *The Middle Paleolithic: adaptation, behavior, and variability*, Dibble, H., and Mellars, P., eds, pp. 1–28. University of Pennsylvania Press, Philadelphia, PA.

Dibble, H., and Lenoir, M., eds (1995) *The Middle Paleolithic Site of Combe-Capelle Bas (France)*. University of Pennsylvania Press, Philadelphia, PA.

Dibble, H., McPherron, S., Sandgathe, D., Goldberg, P., Turq, A., and Lenoir, M. (2009a) Context, curation, and bias: an evaluation of the Middle Paleolithic collections of Combe-Grenal (France). *Journal of Archaeological Science* **36**(11): 2540–50.

Dibble, H.L., Berna, F., Goldberg, P., McPherron, S.P., Mentzer, S., Niven, L., et al. (2009b) A Preliminary Report on Pech de l'Azé IV, Layer 8 (Middle Paleolithic, France). *PaleoAnthropology*: 182–219.

Dietl, H. (2004) *Die Freilandfundstellan von Geelebek un Anyskop in Sudafrika: das Siedlungsverhalten wahrend de Middle Stone Age*. Master's thesis, University of Tubingen.

Dietl, H., Kandel, A.W., and Conard, N.J. (2005) Middle Stone Age settlement and land use at the open-air sites of Geelbek and Anyskop, South Africa. *Journal of African Archaeology* **3**(2): 233–44.

DiMaggio, E.N., Campisano, C.J., Arrowsmith, J.R., Reed, K.E., Swisher, C.C., and Lockwood, C.A. (2008) Correlation and stratigraphy of the BKT-2 volcanic complex in west-central Afar, Ethiopia. In: *The Geology of Early Humans in the Horn of Africa*, Quade, J., and Wynn, J.G., eds, pp. 163–78. Geological Society of America, Boulder, CO.

Dirks, P.H.G.M., Kibii, J.M., Kuhn, B.F., Steininger, C., Churchill, S.E., Kramers, J.D., et al. (2010) Geological setting and age of *Australopithecus sediba* from Southern Africa. *Science* **328**(5975): 205–8.

Disotell, T.R., Honeycutt, R.L., and Ruvolo, M. (1992) Mitochondrial DNA phylogeny of the Old-World monkey tribe Papionini. *Molecular Biology and Evolution* **9**(1): 1–13.

Ditchfield, P.W., Hicks, J., Plummer, T.W., Bishop, L.C., and Potts, R. (1999) Current research on the late Pliocene and Pleistocene deposits north of Homa Mountain, southwestern Kenya. *Journal of Human Evolution* **36**: 123–50.

Dixey, F. (1927) The Tertiary and post-Tertiary lacustrine sediments of the Nyasan Rift-valley. *Quarterly Journal of the Geological Society* **83**(1–5): 432.

Dizon, E., Détroit, F., Sémah, F., Falguères, C., Hameau, S., Ronquillo, W., et al. (2002) Notes on the morphology and age of the Tabon Cave fossil *Homo sapiens*. *Current Anthropology* **43**: 660–6.

Dobzhansky, T. (1937) *Genetics and the Origin of Species*. Columbia University Press, New York.

Dobzhansky, T. (1962) *Mankind Evolving: the Evolution of the Human Species*. Bantam Books, New York.

Doelman, T. (2008) Flexibility and creativity in microblade core manufacture in Southern Primorye, Far East Russia. *Asian Perspectives* **47**(2): 352–70.

Domínguez-Rodrigo, M., and Pickering, T. (2003) Early hominid hunting and scavenging: A zooarcheological review. *Evolutionary Anthropology: Issues, News, and Reviews* **12**(6): 275–82.

Domínguez-Rodrigo, M., Pickering, T.R., Semaw, S., and Rogers, M.J. (2005) Cutmarked bones from Pliocene archaeological sites at Gona, Afar, Ethiopia: implications for the function of the world's oldest stone tools. *Journal of Human Evolution* **48**(2): 109–21.

Domínguez-Rodrigo, M., Barba, R., and Egeland, C.P. (2007a) *Deconstructing Olduvai: A Taphonomic Study of the Bed I Sites*. Springer, New York.

Domínguez-Rodrigo, M., Diez-Martin, F., Mabulla, A., Luque, L., Alcala, L., Tarrino, A., et al. (2007b) The archaeology of the Middle Pleistocene deposits of Lake Eyasi. *Tanzania. Journal of African Archaeology* **5**(1): 47–87.

Dominguez-Rodrigo, M., Mabulla, A., Luque, L., Thompson, J.W., Rink, J., Bushozi, P., et al. (2008) A new archaic *Homo sapiens* fossil from Lake Eyasi, Tanzania. *Journal of Human Evolution* **54**(6): 899–903.

Domínguez-Rodrigo, M., Alcalá, L., and Luque, L. (2009) *Peninj. A Research Project on Human Origins (1995–2005)*. Oxbow Books, Oxford.

Dominy, N., Vogel, E., Yeakel, J., Constantino, P., and Lucas, P. (2008) Mechanical properties of plant underground storage organs and implications for dietary models of early hominins. *Evolutionary Biology* **35**(3): 159–75.

Donner, J. (1975) Pollen composition of the Abri Pataud sediments. In: *Excavation of the Abri Pataud, Les Eyzies (Dordogne)*, Movius, H.L., ed., pp. 160–73. Peabody Museum of Archaeology and Ethnology, Harvard University, Cambridge, MA.

Doran, G., Dickel, D., Ballinger, W., Agee, O., Laipis, P., and Hauswirth, W. (1986) Anatomical, cellular and molecular analysis of 8,000-yr-old human brain tissue from the Windover archaeological site. *Nature* **323**: 803–6.

Dortch, C. (1979) 33,000 year old stone and bone artifacts from Devil's Lair, Western Australia. *Records of the Western Australian Museum* **7**(4): 329–67.

Drake, R., and Curtis, G.H. (1987) K-Ar geochronology of the Laetoli fossil localities. In: *Laetoli: A Pliocene Site in Northern Tanzani*, Leakey, M.D., and Harris, J.M., eds, pp. 48–52. Clarendon Press, Oxford.

Drapeau, M.S.M. (2008) Articular morphology of the proximal ulna in extant and fossil hominoids and hominins. *Journal of Human Evolution* **55**(1): 86–102.

Drapeau, M.S.M., Ward, C.V., Kimbel, W.H., Johanson, D.C., and Rak, Y. (2005) Associated cranial and forelimb remains attributed to *Australopithecus afarensis* from Hadar, Ethiopia. *Journal of Human Evolution* **48**: 593–642.

Drennan, M. (1953a) A preliminary note on the Saldanha skull. *South African Journal of Science* **50**: 7–11.

Drennan, M. (1953b) The Saldanha skull and its associations. *Nature* **172**: 791–3.

Drennan, M. (1955) The special features and status of the Saldanha skull. *American Journal of Physical Anthropology* **13**(4): 625–34.

Drennan, M.R. (1937) The Florisbad skull and brain cast. *Transactions of the Royal Society of South Africa* **25**: 103–14.

Drennan, M.R., and Singer, R. (1955) A mandibular fragment, probably of the Saldanha skull. *Nature* **175**: 364.

Dreyer, T.F. (1935) A human skull from Florisbad, Orange Free State, with a note on the endocranial cast by C.U. Ariens Kappers. *Proceedings of the Royal Academy of Science, Amsterdam* **38**: 119–28.

Dreyer, T.J., Meiring, A.J.H., and Hoffman, A.C. (1938) A comparison of the Boskop with other abnormal skull-forms from South Africa. *Zeitschrift für Rassenkunde* **7**(3): 289–96.

Drummond, A., and Rambaut, A. (2007) BEAST: Bayesian evolutionary analysis by sampling trees. *BMC Evolutionary Biology* **7**(1): 214.

Dryden, I., and Mardia, K. (1998) *Statistical Shape Analysis*. John Wiley & Sons.

D'Souza, R., and Klein, O. (2007) Unraveling the molecular mechanisms that lead to supernumerary teeth in mice and men: current concepts and novel approaches. *Cells Tissues Organs* **186**(1): 60–69.

Duarte, C., Mauricio, J., Pettitt, P., Souto, P., Trinkaus, E., Van Der Plicht, H., et al. (1999) The early Upper Paleolithic human skeleton from the Abrigo do Lagar Velho (Portugal) and modern human emergence in Iberia. *Proceedings of the National Academy of Sciences USA* **96**(13): 7604.

Dubois, E. (1891) Voorloopig bericht omtrent het onderzoek naar de pleistocene en tertiaire Vertebraten-Fauna va Sumatra en Java, gedurende het jaar 1890. *Natuurkundig Tijdschrift voor Nederlandsch-Indië* **51**: 93–100.

Dubois, E. (1892) Palaeontologische andrezoekingen op Java. *Verslag Mijnwezen (Batavia)* **3**: 10–14.

Dubois, E. (1893) Palaeontologische onderzoekingen op Java. *Verslag Mijnwezen (Batavia)* **10**: 10–14.

Dubois, E. (1893) *Pithecanthropus erectus, eine menschenaehnliche Ubergangsform aus Java*. Landesdruckerei, Batavia.

Dubois, E. (1894) Palaeontologische onderzoekingen op Java. *Verslag Mijnwezen (Batavia)* **81**: 12–15.

Dubois, E. (1894) *Pithecanthropus erectus, eine menschenaehnliche Ubergangsform aus Java*. Landesdruckerei, Batavia.

Dubois, E. (1897) Sur le rapport du poids de l'encephale avec la grandeur du corps chez les mammiferes. *Bulletin of the Society of Anthropology* **8**: 337–76.

Dubois, E. (1898) Ueber die Abhängigkeit des Hirnge-wichtes von der Körpergrosse beim Menschen. *Archiv für Anthropologie* **25**: 423–41.

Dubois, E. (1914) Die gesetzmäbige Beziehung von Gehirmasse zu Körpergrösse bei den Wirbeltieren *Zeitschrift für Morphologie und Anthropologie* **18**: 323–50.

Dubois, E. (1921) The Proto-Australian Fossil Man of Wadjak, Java. *Proceedings of the Koninklijke Neder-landsche Akademie van Wetenschappen* **23**: 1013–51.

Dubois, E. (1924a) On the principal characters of the cranium and the brain, the mandible and the teeth of Pithecanthropus erectus. *Proceedings of the Royal Academy of Science, Amsterdam* **27**: 265–78.

Dubois, E. (1924b) Figures of the calvarium and endo-cranial cast, a fragment of the mandible and three teeth of Pithecanthropus erectus. *Proceedings of the Koninklijke Nederlandsche Akademie van Wetenschappen* **27**(5–6): 459–64.

Dubois, E. (1926) On the principal characters of the femur of Pithecanthropus erectus. *Proceedings of the Koninklijke Nederlandsche Akademie van Wetenschappen* **29**: 730–43.

Dubois, E. (1932) The distinct organization of Pithecan-thropus, of which the femur bears evidence, now confirmed from other individuals of the described species. *Proceedings of the Koninklijke Nederlandsche Akademie van Wetenschappen* **35**: 716.

Dubois, E. (1935) The sixth (fifth new) femur of *Pithecanthropus erectus*. *Proceedings of the Koninklijke Nederlandsche Akademie van Wetenschappen* **38**(8): 1–3.

Dubois, E. (1936) Racial identity of *Homo soloensis* Oppenoorth (including *Homo modjokertensis* von Koenigswald) and *Sinanthropus pekinensis* Davidson Black. *Proceedings of the Koninklijke Nederlandsche Akademie van Wetenschappen* **39**(10): 1–6.

Dubois, E. (1940) The fossil human remains discovered in Java by Dr. G.H.R. von Koenigsvald and attributed by him to Pithecanthropus erectus, in reality remains of *Homo wadjakensis* (syn. *Homo soloensis*). *Proceedings of the Koninklijke Akademie van Wetenschappen* **43**: 494–6.

Ducher, G., Daly, R.M., and Bass, S.L. (2009) Effects of repetitive loading on bone mass and geometry in young male tennis players: a quantitative study using MRI. *Journal of Bone and Mineral Research* **24**: 1686–92.

Duffy, D., Montgomery, G., Chen, W., Zhao, Z., Le, L., James, M., et al. (2007) A three-single-nucleotide poly-morphism haplotype in intron 1 of OCA2 explains most human eye-color variation. *American Journal of Human Genetics* **80**(2): 241–52.

Dunbar, R.I.M. (1988) *Primate Social Systems*. Croom Helm, London.

Dunbar, R.I.M. (1993) Co-evolution of neocortical size, group size and language in humans. *Behavioral and Brain Sciences* **16**(4): 681–735.

Dunbar, R.I.M. (1995) Neocortex size and group size in primates: a test of the hypothesis. *Journal of Human Evolution* **28**(3): 287–96.

Dunbar, R.I.M. (1996) *Gossip, Grooming, and the Evolu-tion of Language*. Faber and Faber, London.

Dunkley, P.N., Smith, M., Allen, D.J., and Darling, W.G. (1993) The geothermal activity and geology of the Northern Sector of the Kenya Rift Valley. Research Report SC/93/1. Keyworth, British Geological Survey.

Dupont, E. (1866) Étude sur les fouilles scientifiques exécutées pendant l'hiver de 1865–1866 dans les cavernes des bordes de la Lesse. *Bull. Acad. r. Belg. Cl. Sci.* **22**: 44–54.

Dupont, E. (1867) Étude sur cinq cavernes explorées dans la vallée de la Lesse et le ravin de Falmignoul pendant l'été de 1866. *Bull. Acad. r. Belg. Cl. Sci.* **23**: 245–55.

Dupont, M. (1872) *Les temps préhistoriques en Belgique*. Baillière, Paris.

Dupont-Nivet, G., Sier, M., Campisano, C.J., Arrow-smith, J.R., DiMaggio, E., Reed, K., et al. (2008) Magnetostratigraphy of the eastern Hadar Basin (Ledi-Geraru research area, Ethiopia) and implications for hominin paleoenvironments. In: *The Geology of Early Humans in the Horn of Africa*, Quade, J., and Wynn, J.G., eds, pp. 67–85. Geological Society of America, Boulder, CO.

Duport, L. (1967) Le gisement préhistorique de mon-tgaudier. Premiers résultats des fouilles (1966–1967). *Bulletin et Mémoires de la Société Archéologique et. Historique de la Charente*: 73–80.

Durán, J.J., Grün, R., and Ford, D. (1993) Dataciones geocronológicas (métodos ESR y series de uranio) en la Cueva de Nerja. Implicaciones evolutivas, paleocli-máticas y neotectónicas. In: *Trabajos sobre la Cueva de Nerja*, Carrasco, F., ed., pp. 233–48. Patronato de Cueva de Nerja.

Durband, A.C. (2004) *A Test of the Multiregional Hypothesis of Modern Human Origins using Basicranial Evidence from Indonesia and Australia*. PhD thesis, University of Tennessee.

Durband, A.C. (2008) Artificial cranial deformation in Pleistocene Australians: the Coobool Creek sample. *Journal of Human Evolution* **54**(6): 795–813.

Durband, A.C., Kidder, J.H., and Jantz, R.L. (2005) A multivariate examination of the Hexian calvaria. *Anthropological Science* **113**(2): 147–54.

Duster, T. (2005) Medicine Enhanced: Race and Reifi-cation in Science. *Science* **307**(5712): 1050–51.

Dwight, T. (1894) Methods of estimating the height from parts of the skeleton. *Medical Records* **46**: 293–6.

Ebersberger, I., Metzler, D., Schwarz, C., and Pääbo, S. (2002) Genomewide comparison of DNA sequences between humans and chimpanzees. *American Journal of Human Genetics* **70**(6): 1490–97.

Eberz, G.W., Williams, F.M., and Williams, M.A.J. (1988) Plio-Pleistocene volcanism and sedimentary facies changes at Gadeb prehistoric site, Ethiopia. *International Journal of Earth Sciences* **77**(2): 513–27.

Ecker, A. (1875) Einige Bemerkungen über einen Schwankenden Charakter in den Hand des Menschen (Some remarks about a varying character in the hand of humans). *Archives fur Anthropologie* 8: 68–74.

Edey, M.A., and Johanson, D.C. (1989) *Blueprints: Solving the Mystery of Evolution*. Oxford University Press.

Eickstedt, E. (1937) Geschichte der anthropologischen Namengabung und Klassification. *Zeits. f. Rassenk* 5: 209–82.

Eickstedt, E.F. (1932) Hominiden und Simioiden Über den derzeitigen Stand der Abstammungsfrage. *Zeitschrift für ärztliche Fortbildung* 29: 608–13.

Ekman, P. (1993) Facial expression and emotion. *American Psychologist* 48(4): 384–92.

Eldredge, N. (1979) Alternative approaches to evolutionary theory. *Bulletin of the Carnegie Museum of Natural History* 13: 7–19.

Eldredge, N., and Gould, S. (1972) Punctuated equilibria: an alternative to phyletic gradualism. In: *Models in Paleobiology*, Schopf, T.J.M., ed., pp. 82–115. Freeman, Cooper, and Co., San Francisco, CA.

Elefanti, P., Panagopoulou, E., and Karkanas, P. (2008) The transition from the Middle to the Upper Palaeolithic in the southern Balkans: the evidence from the Lakonis I Cave, Greece. *European Prehistory* 5: 85–95.

Elftman, H., and Manter, J. (1935) Chimpanzee and human feet in bipedal walking. *American Journal of Physical Anthropology* 20: 69–79.

Elston, G.N., Benavides-Piccione, R., Elston, A., Zietsch, B., Defelipe, J., Manger, P., et al. (2006) Specializations of the granular prefrontal cortex of primates: Implications for cognitive processing. *Anatomical Record* 288A(1): 26–35.

Elton, S. (2000) Habitat preference and locomotion in Plio-Pleistocene *Theropithecus* species. *American Journal of Physical Anthropology (Supp.)* 30: 145.

Elton, S. (2002) A reappraisal of the locomotion and habitat preference of *Theropithecus oswaldi*. *Folia primatologica* 73: 252–80.

Elton, S. (2006) 40 years on and still going strong: the use of the hominin-cercopithecid comparison in human evolution. *Journal of the Royal Anthropological Institute* 12: 19–38.

Elton, S., Bishop, L.C., and Wood, B.A. (2001) Comparative context of Plio-Pleistocene hominin brain evolution. *Journal of Human Evolution* 41: 1–27.

Emiliani, C. (1955) Pleistocene temperatures. *Journal of Geology* 63: 538–78.

Emiliani, C. (1966) Paleotemperature analysis of Caribbean Cores P6304-8 and P6304-9 and a generalized temperature curve for the past 425,000 years. *Journal of Geology* 74(2): 109–26.

Emiliani, C. (1972) Quaternary paleotemperatures and the duration of the high temperature intervals. *Science* 178: 398–407.

Emiliani, C., and Shackleton, N.J. (1974) The Bruhnes Epoch: isotopic paleotemperature and geochronology. *Science* 183: 511–14.

Enard, W., Przeworski, M., Fisher, S.E., Lai, C.S., Wiebe, V., Kitano, T., et al. (2002) Molecular evolution of *FOXP2*, a gene involved in speech and language. *Nature* 418: 869–72.

Enlow, D. (1990) *Facial Growth*. Saunders and Co, Philadelphia, PA.

Enlow, D.H., and Azuma, M. (1975) Functional growth boundaries in the human and mammalian face. In: *Morphogenesis and Malformation of Face and Brain*, Bergsma, D., ed., pp. 217–30. Alan R. Liss, New York.

Ennouchi, E. (1962a) Un crâne d'Homme ancien au Jebel Irhoud (Maroc). *Comptes Rendus de l'Académie des Sciences* 254: 4330–2.

Ennouchi, E. (1962b) Un neandertalien: L'Homme du Jebel Irhoud (Maroc). *Anthropologie* 66: 279–99.

Ennouchi, E. (1963) Les néandertaliens du Jebel Irhoud (Maroc). *Comptes Rendus de l'Académie des Sciences* 256: 2459–60.

Ennouchi, E. (1969) Découverte d'un Pithecanthropien au Maroc. *Comptes Rendus de l'Académie des Sciences* 269: 763–5.

Ennouchi, E. (1972) Nouvelle découverte d'un Archanthropien au Maroc. *Comptes Rendus de l'Académie des Sciences* 274D: 3088–90.

Erdbrink, D. (1965) A quantification of the *Dryopithecus*- and other lower molar patterns in man and some of the apes. *Zeitschrift für Morphologie und Anthropologie* 57(1): 70.

Erdbrink, D. (1967) A quantification of lower molar patterns in deutero-Malayans. *Zeitschrift für Morphologie und Anthropologie* 59(1): 40.

Ericson, J.E., and Purdy, B.A. (1984) *Prehistoric quarries and lithic production*. Cambridge University Press, Cambridge.

Euler, L. (1760) Recherches sur la courbure des surfaces. *Memoires de l'academie des sciences de Berlin* 16: 119–43.

Evans, G. (1982) *The Varieties of Reference*. Oxford University Press, Oxford.

Eveleth, P., and Tanner, J.M. (1976) *Worldwide Variation in Human Growth*. Cambridge University Press, London.

Everett, D. (2005) Cultural constraints on grammar and cognition in Piraha. *Current Anthropology* 46(4): 621–46.

Evershed, R. (2008) Organic residue analysis in archaeology: the archaeological biomarker revolution. *Archaeometry* 50(6): 895–924.

Fairbank, T.J., and Fairbank, J.C.T. (1984) The crooked semantics of valgus and varus. *Clinical Orthopaedics and Related Research* 185: 6–8.

Faith, J.T. (2008) Eland, buffalo, and wild pigs: were Middle Stone Age humans ineffective hunters? *Journal of Human Evolution* 55: 24–36.

Falconer, H. (1864) *Palaeontological Notes and Memoirs* 2: 561.

Falguères, C., Bahain, J.J., Tozzi, C., Boschian, G., Dolo, J.-M., Mercier, N., et al. (2008) ESR/U-series chronology of the Lower Palaeolithic palaeoanthropological site of Visogliano, Trieste, Italy. *Quaternary Geochronology* 3(4): 390–8.

Falguères, C., Bahain, J.J., Yokoyama, Y., Arsuaga, J.L., Burmúdez de Castro, J.M., Carbonell, E., et al. (1999) Earliest humans in Europe: the age of TD6 Gran Dolina, Atapuerca, Spain. *Journal of Human Evolution* 37(3–4): 343–52.

Falguères, C., Yokoyama, Y., Shen, G., Bischoff, J.L., Ku, T.-L., and de Lumley, H. (2004) New U-series dates at the Caune de l'Arago, France. *Journal of Archaeological Science* 31(7): 941–52.

Falk, D. (1980) A reanalysis of the South African australopithecine natural endocasts. *American Journal of Physical Anthropology* 53: 525–39.

Falk, D. (1983a) The Taung endocast: a reply to Holloway. *American Journal of Physical Anthropology* 60(4): 429–89.

Falk, D. (1983b) Cerebral cortices of East African early hominids. *Science* 221: 1072–74.

Falk, D. (1986a) Endocast morphology of Hadar hominid AL 162-28. *Nature* 321: 536–7.

Falk, D. (1986b) Evolution of cranial blood drainage in hominids: enlarged occipital/marginal sinuses and emissary foramina. *American Journal of Physical Anthropology* 70: 311–24.

Falk, D. (1988) Enlarged occipital/marginal sinuses and emissary foramina: their significance in hominid evolution. In: *Evolutionary History of the "Robust" Australopithecines*, Grine, F.E., ed., pp. 85–96. Aldine de Gruyter, Hawthorne, New York.

Falk, D. (1989) Ape-like endocast of 'Ape-Man' Taung. *American Journal of Physical Anthropology* 80(3): 335–9.

Falk, D. (1991) Reply to Dr. Holloway: Shifting positions on the lunate sulcus. *American Journal of Physical Anthropology* 84(1): 89–91.

Falk, D., and Conroy, G.C. (1983) The cranial venous sinus system in *Australopithecus afarensis*. *Nature* 306: 779–81.

Falk, D., and Kasinga, S. (1983) Cranial capacity of a female robust australopithecine (KNM-ER 407) from Kenya. *Journal of Human Evolution* 12: 515–18.

Falk, D., and Clarke, R. (2007) Brief communication: New reconstruction of the Taung endocast. *American Journal of Physical Anthropology* 134(4): 529–34.

Falk, D., Hildebolt, C., Smith, K., Morwood, M.J., Sutkina, T., Brown, P., et al. (2005) The brain of LB1, *Homo floresiensis*. *Science* 308(5719): 242–5.

Falk, D., Hildebolt, C., Smith, K., Morwood, M.J., Sutkina, T., Jatmiko, et al. (2007) Brain shape in human microcephalics and *Homo floresiensis*. *Proceedings of the National Academy of Sciences USA* 104(7): 2513–18.

Falk, D., Hildebolt, C., Smith, K., Morwood, M.J., Sutikna, T., Jatmiko, et al. (2009) LB1's virtual endocast, microcephaly, and hominin brain evolution. *Journal of Human Evolution* 57(5): 597–607.

Farizy, C. (1990) The transition from Middle to Upper Palaeolithic at Arcy-sur-Cure (Yonne, France): technological, economic and social aspects. In: *The Emergence of Modern Humans: an archaeological perspective*, Mellars, P., ed., pp. 303–26. Edinburgh University Press, Edinburgh.

Farizy, C., and Schmider, B. (1985) Contribution à l'identification culturelle du Châtelperronien: les données de l'industrie lithique de la couche X de la grotte du Renne à Arcy-sur-Cure in La signification culturelle des industries lithiques. In: *La signification culturelle des industries lithiques*, Otte, M., ed., pp. 149–69. Studia Prehistorica Belgica.

Farmer, E.C., deMenocal, P.B., and Marchitto, T.M. (2005) Holocene and deglacial ocean temperature variability in the Benguela upwelling region: Implications for low-latitude atmospheric circulation. *Paleoceanography* 20(2): 2018.

Farris, J.S. (1989a) The retention index and homoplasy excess. *Systematic Zoology* 38(4): 406–7.

Farris, J.S. (1989b) The retention index and the rescaled Consistency Index. *Cladistics* 5(4): 417–19.

Faurie, J. (1999) Historique des recherches et réflexions sur la frise noire de la grotte préhistorique de Pech Merle (Lot). *Bulletin de la Société préhistorique de l'Ariège* 54: 43–82.

Feakins, S.J., deMenocal, P.B., and Eglinton, T.I. (2005) Biomarker records of late Neogene changes in northeast African vegetation. *Geology* 33(12): 977–80.

Feakins, S.J., Eglinton, T.I., and deMenocal, P.B. (2007) A comparison of biomarker records of northeast African vegetation from lacustrine and marine sediments (ca. 3.40 Ma). *Organic Geochemistry* 38(10): 1607–24.

Feathers, J. (2002) Luminescence dating in less than ideal conditions: case studies from Klasies River main site and Duinefontein, South Africa. *Journal of Archaeological Science* 29(2): 177–94.

Feathers, J.K., and Bush, D.A. (2000) Luminescence dating of Middle Stone Age deposits at Die Kelders. *Journal of Human Evolution* 38: 91–119.

Feathers, J.K., and Migliorini, E. (2001) Luminescence dating at Katanda - a reassessment. *Quaternary Science Reviews* 20: 961–6.

Feibel, C. (2003) Stratigraphy and depositional setting of the Pliocene Kanapoi Formation, lower Kerio Valley, Kenya. *Contributions in Science* 498: 9–20.

Feibel, C.S. (2008) Microstratigraphy of the Kibish hominin sites KHS and PHS, Lower Omo Valley, Ethiopia. *Journal of Human Evolution* 55: 404–8.

Feibel, C.S., Brown, F.H., and McDougall, I. (1989) Stratigraphic context of fossil hominids from the Omo group deposits: Northern Turkana Basin, Kenya and Ethiopia. *American Journal of Physical Anthropology* **78**(4): 595–622.

Feibel, C.S., Lepre, C.J., and Quinn, R.L. (2009) Stratigraphy, correlation, and age estimates for fossils from Area 123, Koobi Fora. *Journal of Human Evolution* **57**(2): 112–22.

Fejfar, O. (1969) Human remains from the early Pleistocene in Czechoslovakia. *Current Anthropology* **10**(2–3): 170–3.

Felsenstein, J. (1985) Phylogenies and the comparative method. *American Naturalist* **125**(1): 1–15.

Fenart, R., and Empereur-Buisson, R. (1970) Application de la méthode "vestibulaire" d'orientation au crâne de l'enfant du Pech-de-l'Azé et comparison avec d'autres crânes neandertaliens. *Arch. Inst. Paleontol. Hum.* **33**: 89–104.

Ferembach, D. (1962) La Nécropole épipaléolithique de Taforalt (Maroc Oriental). In: *Etude des Squelettes Humains*, pp. 123. CNRS, Paris.

Ferembach, D. (1976) Le restes humains de la grotte de Dar es Soltane 2 (Maroc): Campagne 1975. *Bulletins et Mémoires de la Société d'Anthropologie de Paris* **13**: 183–93.

Ferembach, D. (1998) La crâne Atérien de Témara (Maroc Atlantique). *Bulletin d'Archéologie marocaine* **18**: 19–66.

Ferguson, W. (1989) A new species of the genus *Australopithecus* (primates: hominidae) from Plio/Pleistocene deposits west of Lake Turkana in Kenya. *Primates* **30**(2): 223–32.

Ferguson, W.W. (1984) Revision of fossil hominid jaws from the Plio/Pleistocene of Hadar, in Ethiopia including a new species of the genus *Homo* (Hominoidea: Homininae). *Primates* **25**: 519–29.

Ferguson, W.W. (1986) The taxonomic status of *Praeanthropus africanus* (Primates: Pongidae) from the late Pliocene of eastern Africa. *Primates* **27**(4): 485–92.

Ferguson, W.W. (1987a) Revision of the subspecies of *Australopithecus africanus* (Primates: Hominidae) including a new subspecies from the late Pliocene of Ethiopia. *Primates* **28**: 258–65.

Ferguson, W.W. (1987b) Taxonomic status of the Hominine cranium KNM-ER1813 (Primates:Hominidae) from the late Plio/Pleistocene of Koobi Fora. *Primates* **28**: 423–38.

Ferguson, W.W. (1989) Taxonomic status of the hominid mandible KNM-ER TI 13150 from the Middle Pliocene of Tabarin, in Kenya. *Primates* **30**(3): 383–7.

Ferguson, W.W. (1995) A New Species of the Genus *Homo* (Primates: Hominidae) from the Plio/Pleistocene of Koobi Fora, in Kenya. *Primates* **36**(1): 69–89.

Ferguson, W.W. (1999) Taxonomic status of the skull A.L.444-2 from the Pliocene of Hadar, Ethiopia. *Palaeontologica africana* **35**: 119–29.

Fernández-Jalvo, Y., Diez, J.C., Bermúdez de Castro, J.M., Carbonell, E., and Arsuaga, J.L. (1996) Evidence of early cannibalism. *Science* **271**(5247): 277–8.

Fernández-Jalvo, Y., Denys, C., Andrews, P., Williams, T., Dauphin, Y., and Humphrey, L. (1998) Taphonomy and palaeoecology of the Olduvai Bed-I (Pleistocene, Tanzania). *Journal of Human Evolution* **34**(2): 137–72.

Fernández-Jalvo, Y., Diez, J.C., Cáceres, I., and Rosell, J. (1999) Human cannibalism in the early Pleistocene of Europe (Gran Dolina, Sierra de Atapuerca, Burgos, Spain). *Journal of Human Evolution* **37**: 591–622.

Fernández-Jalvo, Y., King, T., Andrews, P., Moloney, N., Ditchfield, P., Yepiskoposyan, L., et al. (2004) Azokh Cave and Northern Armenia. In: *Homenaje a Emiliano Aguirre*, Baquedano, E., and Rubio, S., eds, pp. 158–68. Museo Arqueológico Regional de Madrid, Madrid.

Fernández-Jalvo, Y., King, T., Andrews, P., Yepiskoposyan, L., Moloney, N., Murray, J., et al. (2010) The Azokh Cave complex: Middle Pleistocene to Holocene human occupation in the Caucasus. *Journal of Human Evolution* **58**(1): 103–9.

Fernández-Tresguerres, J. (1976) Azilian burial from Los Azules I, Asturias, Spain. *Current Anthropology* **17**: 769.

Fernández-Tresguerres, J. (1980) *El Aziliense en las provincias de Asturias y Santander*. Centro de Investigación y Museo de Altomira, Santander.

Fernández-Tresguerres, J., and Rodríguez Fernández, J.F. (1990) La cueva de Los Azules (Cangas de Onis) In: *Excavaciones arqueologicas en Asturias, 1983–86*. Servicio de publicacions del Principado de Asturias, Oviedo, Spain.

Ferris, S., Wilson, A., and Brown, W. (1981) Evolutionary tree for apes and humans based on cleavage maps of mitochondrial DNA. *Proceedings of the National Academy of Sciences USA* **78**(4): 2432–6.

Feustel, R. (1983) Zur zeitlichen und kulturellen Stellung des Paläolithikums von Weimar-Ehringsdorf. *Alt-Thüringen* **19**: 16–42.

Fiedler, L. (1994) *Alt- und mittelsteinzeitliche Funde in Hessen – Führer zur hessischen Vor- und Früh-geschichte, Band 2*. Theiss, Stuttgart.

Finch, C. (2007) *Biology of Human Longevity: Inflammation, Nutrition, and Aging in the Evolution of Life Spans*. Academic Press, Burlington, MA.

Finch, C., Pike, M., and Witten, M. (1990) Slow mortality rate accelerations during aging in some animals approximate that of humans. *Science* **249**(4971): 902–5.

Finlay, B., and Darlington, R. (1995) Linked regularities in the development and evolution of mammalian brains. *Science* **268**(5217): 1578–84.

Finlayson, C. (2004) *Neanderthals and Modern Humans: an Ecological and Evolutionary Perspective*. Cambridge University Press, Cambridge.

Fiore, I., Bondioli, L., Coppa, A., Macchiarelli, R., Russom, R., Kashay, H., et al. (2004) Taphonomic analysis of the late Early Pleistocene bone remains from Buia (Dandiero Basin, Danakil Depression, Eritrea): evidence for large mammal and reptile butchering. *Rivista Italiana di Paleontologia e Stratigrafia* **110**: 89–97.

Fiore, I., Gala, M., and Tagliacozzo, A. (2004) Ecology and subsistence strategies in the Eastern Italian Alps during the Middle Palaeolithic. *International Journal of Osteoarchaeology* **14**: 273–86.

Fisher, R. (1918) The correlation between relatives on the supposition of Mendelian inheritance. *Transactions of the Royal Society of Edinburgh* **52**: 399–433.

Fisher, R.A. (1930) *The Genetical Theory of Natural Selection*. Clarendon Press, Oxford.

Fitch, W.T., and Hauser, M.D. (2004) Computational constraints on syntactic processing in a nonhuman primate. *Science* **303**(5656): 377–80.

FitzGerald, C.M. (1996) *Tooth Crown Formation and the Variation of Enamel Microstructural Growth Markers in Modern Humans*. PhD thesis, University of Cambridge.

Flanagan, N., Healy, E., Ray, A., Philips, S., Todd, C., Jackson, I., et al. (2000) Pleiotropic effects of the melanocortin 1 receptor (MC1R) gene on human pigmentation. *Human Molecular Genetics* **9**(17): 2531–7.

Fleagle, J.G. (1977) Locomotor behaviour and muscular anatomy of sympatric Malaysian leaf monkeys (*Presbytis obscura* and *Presbytis melalophos*). *American Journal of Physical Anthropology* **46**: 297–308.

Fleagle, J.G. (1985) Size and adaptation in primates. In: *Size and Scaling in Primate Biology*, Jungers, W.L., ed., pp. 1–19. Plenum Press, New York.

Fleagle, J.G., Stern Jr, J., Jungers, W.L., Susman, R.L., Vangor, A.K., and Wells, J.P. (1981) Climbing: a biomechanical link with brachiation and bipedalism. *Symposium of the Royal Zoological Society of London* **48**: 359–75.

Fleagle, J.G., Rasmussen, D.T., Yirga, S., Brown, T.M., and Grine, F.E. (1991) New hominid fossils from Fejej, Southern Ethiopia. *Journal of Human Evolution* **21**: 145–52.

Flombaum, J., and Santos, L. (2005) Rhesus monkeys attribute perceptions to others. *Curent Biology* **15**(5): 447–52.

Florkin, M., and Morgulis, S. (1949) *Biochemical Evolution*. Academic Press, New York.

Flower, W.H. (1879) Catalogue of the Specimens of Osteology and Dentition of Vertebrated Animals in the Museum of the Royal College of Surgeons. *Man* **1**.

Foley, R., and Lahr, M.M. (1997) Mode 3 technologies and the evolution of modern humans. *Cambridge Archaeological Journal* **7**(1): 3–36.

Foley, R., and Lahr, M. (2003) On stony ground: lithic technology, human evolution, and the emergence of culture. *Evolutionary Anthropology: Issues, News, and Reviews* **12**(3): 109–22.

Formicola, V. (1987) The Upper Paleolithic burials of Barma Grande. In: *Hominidae: proceedings of the 2nd International Congress of Human Paleontology*, Giacobini, G., ed., pp. 483–6. Jaca Book, Milan.

Formicola, V., Pettitt, P.B., and del Lucchese, A. (2004) A direct AMS radiocarbon date on the Barma Grande 6 Upper Paleolithic Skeleton. *Current Anthropology* **45**(1): 114–18.

Fortea, J., de la Rasilla, M., Garcia-Tabernero, A., Gigli, E., Rosas, A., and Lalueza-Fox, C. (2008) Excavation protocol of bone remains for Neandertal DNA analysis in El Sidrón Cave (Asturias, Spain). *Journal of Human Evolution* **55**(2): 353–7.

Foster, J.B. (1964) The evolution of mammals on islands. *Nature* **202**: 234 35.

Fox, J., and Coinman, N. (2004) Emergence of the Levantine Upper Paleolithic: Evidence from the Wadi al-Hasa. In: *The Early Upper Paleolithic East of the Danube*, Brantingham, P., Kerry, K., and Kuhn, S., eds. University of California Press, Los Angeles, CA.

Fox, R.B. (1970) *The Tabon Caves*. Monograph of the National Museum. Manila.

Fraipont, C. (1936) Les Hommes Fossiles d'Engis. *Arcives Institute Paléontologie Humanine* **16**: 1–52.

Fraipont, J., and Lohest, M. (1886) La race humaine de Néanderthal ou de Sandstadt, en Belgique. *Bull. Adac. r. Belg. Cl. Sci.* **12**: 741–84.

Franz-Odendaal, T.A., Lee-Thorp, J.A., and Chinsamy, A. (2002) New evidence for the lack of C_4 grassland expansions during the early Pliocene at Langebaanweg, South Africa. *Paleobiology* **28**(3): 378–88.

Frayer, D.W., Wolpoff, M.H., Thorne, A.G., Smith, F.H., and Pope, G.G. (1993) Theories of modern human origins: the paleontological test. *American Anthropologist* **95**(1): 14–50.

Freedman, L. (1985) Human skeletal remains from Mossgiel, New South Wales. *Archaeology in Oceania* **20**: 21–31.

Freedman, L., and Lofgren, M. (1979) Human skeletal remains from Cossack, Western Australia. *Journal of Human Evolution* **8**: 283–99.

Freedman, L., and Lofgren, M. (1983) Human skeletal remains from Lake Tandou, New South Wales. *Archaeology in Oceania* **18**: 98–105.

Freeman, L. (1994) The many faces of Altamira. *Complutum* **5**: 331–42.

Frost, S.R., and Delson, E. (2002) Fossil cercopithecidae from the Hadar Formation and surrounding areas of the Afar Depression, Ethiopia. *Journal of Human Evolution* **43**: 687–748.

Frost, S.R., Marcus, L.F., Bookstein, F.L., Reddy, D.P., and Delson, E. (2003) Cranial allometry, phylogeography, and systematics of large-bodied papionins (Primates: Cercopithecinae) inferred from geometric morphometric analysis of landmark data. *Anatomical Record* **275**(2): 1048–72.

Frudakis, T., Thomas, M., Gaskin, Z., Venkateswarlu, K., Chandra, K., Ginjupalli, S., et al. (2003) Sequences associated with human iris pigmentation. *Genetics* **165**(4): 2071–83.

Fuchs, V.E. (1939) The geological history of the Lake Rudolf basin, Kenya Colony. *Philosophical Transactions of the Royal Society* **229**: 219–74.

Fuhlrott, C., and Schaaffhausen, H. (1857) Correspondenzblatt des naturhistorischen Veriens der preussischen Rheilande und Westphalens. *Verhandlungen des naturhistorischen Vereins der preussischen Rheinlande* **14**: 50–2.

Fujita, K. (2009) Metamemory in tufted capuchin monkeys (*Cebus apella*). *Animal Cognition* **12**(4): 575–85.

Fuller, B.T., Fuller, J.L., Harris, D.A., and Hedges, R.E.M. (2006) Detection of breastfeeding and weaning in modern human infants with carbon and nitrogen stable isotope ratios. *American Journal of Physical Anthropology* **129**: 279–83.

Fully, G. (1956) Une nouvelle méthode de détermination de la taille. *Annales de médecine légale* **35**: 266–73.

Fusté, M. (1953) Parietal neandertalense de Cova Negra (Jativa). *Museu de Prehistria de València* **17**: 31.

Gaál, I. (1928) Der erste mitteldiluviale Menschenknochen aus Siebenbürgen. *Pub. Muzeului Judetean Hunedoara, Deva* **3–4**: 61–112.

Gaál, I. (1931) A neandervölgyi ösumber elsö erdélyi csontamaradványa. *A Termes-settudomanyi Kozlnyhosz* **63**(suppl. 181): 23–31.

Gabounia, L., de Lumley, M.-A., Vekua, A., Lordkipanidze, D., and de Lumley, H. (2002) Découverte d'un nouvel hominidé à Dmanissi (Transcaucasie, Géorgie). *Comptes Rendus Palevol* **1**(4): 243–53.

Gabunia, L., and Vekua, A. (1995) A Plio-Pleistocene hominid from Dmanisi, East Georgia, Caucasus. *Nature* **373**: 509–12.

Gabunia, L., Vekua, A., Lordkipanidze, D., Swisher III, C., Ferring, R., Justus, A., et al. (2000) Earliest Pleistocene hominid cranial remains from Dmanisi, Republic of Georgia: taxonomy, geological setting, and age. *Science* **288**(5468): 1019–25.

Gabunia, L., Antón, S.C., Lordkipanidze, D., Vekua, A., Justus, A., and Swisher, C.C. (2001) Dmanisi and Dispersal. *Evolutionary Anthropology* **10**(5): 158–70.

Gage, T.B. (1998) The comparative demography of primates: with some comments on the evolution of life histories. *Annual Review of Anthropology* **27**: 197–221.

Galaburda, A.M. (1980) La region de Broca: observations anatomiques faites un siecle apres la mort de son decoveur. *Revista de Neurología* **136**: 609–16.

Galef, B.G. (1992) The question of animal culture. *Human Nature* **3**: 157–78.

Galiberti, A. (1980) La Grotta di Paglicci e il Paleolitico del Gargano meridionale. In: *Civilta e culture antiche tra Gargano e Tavoliere*, Galiberti, A., di Palma, A., and Cesnola, A., eds, pp. 33–39. Atti del Convego Archeologico.

Galik, K., Senut, B., Pickford, M., Gommery, D., Treil, J., Kuperavage, A.J., et al. (2004) External and internal morphology of the BAR 1002'00 *Orrorin tugenensis* femur. *Science* **305**: 1450–53.

Galilei, G. (1638) *Discorsi e dimonstrazioni matematiche intorno a due nuove scienze.* University of Wisconsin Press, Madison.

Galloway, A. (1937) The characteristics of the skull of the Boskop physical type. *American Journal of Physical Anthropology* **23**: 31–46.

Gambier, D. (1990/1991) Les vestiges humains du gisement d'Isturitz (Pyrénées-Atlantiques). Étude anthropologique et analyse des traces d'action humaine intentionnelle. *Antiquités nationales* **22–3**: 9–26.

Gambier, D., Valladas, H., Tisnérat-Laborde, N., Arnold, M., and Bresson, F. (2000) Datation de vestiges humains présumés du Paléolithique supérieur par la méthode du Carbone 14 en spectrométrie de masse par accélérateur. *Paléo* **12**(1): 201–12.

Gamble, C., Davies, W., Pettitt, P., and Richards, M. (2005) The archaeological and genetic Foundations of the European Population during the Late Glacial: implications for 'Agricultural Thinking'. *Cambridge Archaeological Journal* **15**(2): 193–223.

Gantt, D. (1977) *Enamel of Primate Teeth: Its Thickness and Structure with Reference to Functional and Phyletic Implications.* PhD thesis, Washington University.

Gao, X., Wei, Q., Shen, C., and Keates, S. (2005) New lights on the earliest hominids in East Asia. *Current Anthropology* **46**: 115–20.

Garcea, E.A.A. (2004) Crossing deserts and avoiding seas: Aterian North African-European relations. *Journal of Anthropological Research* **60**(1): 27–53.

Garcia Sánchez, M. (1986) Estudio preliminar de los restos neandertalenses del Boquete de Zafarraya (AlcaucÌn, Málaga). In: *Homenaje a L Siret (1934–1984)* pp. 49–56. Junta de Andalucia, Seville.

Garland, T., and Díaz-Uriarte, R. (1999) Polytomies and phylogenetically independent contrasts: examination of the bounded degrees of freedom approach. *Systematic Biology* **48**: 547–58.

Garralda, M. (1986) The Azilian man from Los Azules Cave I (Cangas de Onis, Oviedo, Spain). *Human Evolution* **1**(5): 431–47.

Garrod, D. (1932) A New Mesolithic Industry: The Natufian of Palestine. *Journal of the Royal Anthropological Institute of Great Britain and Ireland* **62**: 257–69.

Garrod, D., and Bate, D. (1937) *The Stone Age of Mount Carmel.* Clarendon Press, Oxford.

Garrod, D.A.E. (1926) *The Upper Paleolthic Age in Britain.* Clarendon Press, Oxford.

Garrod, D.A.E. (1934) Excavation at the Wady el-Mughara (Palestine) 1932–3. *Bulletin American School of Prehistoric Research* **10**: 7–11.

Garrod, D.A.E. (1954) Excavations at the Mugharet Kebara, Mount Carmel, 1931: The Aurignacian industries. *Proceedings of the Prehistoric Society* **20**: 155–92.

Gass, G.L., and Bolker, J.A. (2003) Modularity. In: *Keywords and Concepts in Evolutionary Developmental Biology*, Hall, B.K., and Olson, W., eds, pp. 260–7. Harvard University Press, Cambridge, MA.

Gatesy, J., Amato, G., Vrba, E., Schaller, G., and de Salle, R. (1997) A cladistic analysis of mitochondrial ribosomal DNA from the *Bovidea*. *Molecular Phylogenetics and Evolution* 7(3). 303–19.

Gathogo, P.N., and Brown, F.H. (2006) Stratigraphy of the Koobi Fora Formation (Pliocene and Pleistocene) in the Ileret region of northern Kenya. *Journal of African Earth Sciences* 45: 369–90.

Gathogo, P.N., and Brown, F.H. (2006) Revised stratigraphy of Area 123, Koobi Fora, Kenya, and new age estimates of its fossil mammals, including hominins. *Journal of Human Evolution* 51: 471–9.

Gaudzinski, S., and Roebroeks, W. (2000) Adults only: reindeer hunting at the Middle Paleolithic site Salzgitter Lebenstedt, Northern Germany. *Journal of Human Evolution* 38: 497–521.

Gaulin, S.J.C. (1979) A Jarman/Bell model of primate feeding niches. *Human Ecology* 7(1): 1–20.

Gause, G. (1934) The struggle for existence. *Nature* 227: 89.

Gautier, A. (1970) Fossil fresh water mollusca of the Lake Albert-Lake Edward rift. *Annales du Musee royal de l'Afrique Centrale* 67: 1–144.

Gavan, J.A. (1977) *Paleoanthropology and Primate Evolution* W. C. Brown, Dubuque, IO.

Gebo, D.L., and Schwartz, G.T. (2006) Foot bones from Omo: Implications for hominid evolution. *American Journal of Physical Anthropology* 129(4): 499–511.

Genet-Varcin, E. (1974) Etude des dents humaines isolées provenant des grottes de la Chaisede-Vouthon (Charente). *Bulletins et Mémoires de la Société d'Anthropologie de Paris* 1: 373–84.

Genet-Varcin, E. (1975a) Etude des dents humaines isolées provenant des grottes de la Chaisede-Vouthon (Charente). *Bulletins et Mémoires de la Société d'Anthropologie de Paris* 2: 129–41.

Genet-Varcin, E. (1975b) Etude des dents humaines isolées provenant des grottes de la Chaisede-Vouthon (Charente). *Bulletins et Mémoires de la Société d'Anthropologie de Paris* 2: 277–86.

Genet-Varcin, E. (1976) Etude des dents humaines isolées provenant des grottes de la Chaisede-Vouthon (Charente). *Bulletins et Mémoires de la Société d'Anthropologie de Paris* 3: 243–59.

Genet-Varcin, E. (1982) Vestiges humains du Wurmien inférieur de Combe-Grenal Commune de Domme (Dordogne). *Ann. Paléont. (Vertébrés)* 68: 133–69.

Geoffroy, S.-H.I. (1843) Description des mammiferes nouveaux ou imparfaitement connus. Famille des Singes. *Archives du Museum d'Histoire Naturelle* 2: 486–592.

Geraads, D. (2002) Plio–Pleistocene mammalian biostratigraphy of Atlantic Morocco. *Quaternaire* 13: 43–53.

Geraads, D., Beriro, P., and Roche, H. (1980) La faune et l'industrie des sites a *Homo erectus* des carrieres Thomas (Maroc). Precisions sur l'age de ces Hominids. *Comptes Rendus de l'Académie des Sciences* 291: 195–8.

Geraads, D., Raynal, J.P., and Eisenmann, V. (2004) The earliest human occupation of North Africa:

a reply to Sahnouni *et al.* (2002) *Journal of Human Evolution* 46(6): 751 61.

Gergely, G., Bekkering, H., and Kiraly, I. (2002) Developmental psychology: Rational imitation in preverbal infants. *Nature* 415(6873): 755–5.

Gheorghiu, A., and Haas, N. (1954) Date privind omul primitiv de la Baia de Fier: Consideraţii paleoantropologice. *Sesiunea Secţiunii de Ştiinţe Medicale a Academiei R.P.R* 41: 641–63.

Gheorghiu, A., Nicolăescu-Plopşor, C.S., Haas, N., Comşa, E., Preda, C., Bombiţă, G., et al. (1954) Raport preliminar asupra cercetărilor de paleontologie umană de la Baia de Fier. *Probleme Antropologie* 1: 73–86.

Ghinassi, M., Libsekal, Y., Papini, M., and Rook, L. (2009) Palaeoenvironments of the Buia *Homo* site: High-resolution facies analysis and non-marine sequence stratigraphy in the Alat formation (Pleistocene Dandiero Basin, Danakil depression, Eritrea. *Palaeogeography, Palaeoclimatology, Palaeoecology* 280: 415–31.

Giacobini, G., de Lumley, M.A., Yokohama, Y., and Nguyen, H.V. (1984) Neandertal child and adult remains from a Mousterian deposit in Northern Italy (Caverna delle Fate, Finale Ligure). *Journal of Human Evolution* 13: 687–707.

Giannini, A., Saravanan, R., and Chang, P. (2003) Oceanic forcing of Sahel rainfall on interannual to interdecadal time scales. *Science* 302(5647): 1027–30.

Gibbard, P.L., and van Kolfschoten, T. (2004) The Pleistocene and Holocene epochs. In: *A Geologic Time Scale*, Gradstein, F.M., Ogg, J.G., and Smith, A.G., eds, pp. 441–52. Cambridge University Press, Cambridge.

Gibbard, P.L., Boreham, S., Cohen, K.M., and Moscariello, A. (2007) Global chronostratigraphal correlation table for the last 2.7 million years. *Cambridge, Subcommission on Quaternary Stratigraphy, Department of Geography, University of Cambridge.*

Gibbard, P.L., Head, M.J., Walker, M.J.C., and the Subcommission on Quaternary Stratigraphy (2010) Formal ratification of the Quaternary System/Period and the Pleistocene Series/Epoch with a base at 2.58 Ma. *Journal of Quaternary Sciences* 25: 96–102.

Gibbon, R.J., Granger, D.E., Kuman, K., and Partridge, T.C. (2009) Early Acheulean technology in the Rietputs Formation, South Africa, dated with cosmogenic nuclides. *Journal of Human Evolution* 56(2): 152–60.

Gibert, J., Agusti, J., and Moyà-Solà, S. (1983) Presencia de *Homo* sp. en el yacimiento del Pleistoceno inferior de Venta Micena (Orce, Granada). *Paleontologia y Evolució*: 12.

Gibert, J., Campillo, D., Arques, J., Garcia-Olivares, E., Borja, C., and Lowenstein, J. (1998) Hominid status of the Orce cranial fragment reasserted. *Journal of Human Evolution* 34: 203–17.

Gibson, K.R. (1986) Cognition, brain size and the extraction of embedded food resources. In: *Primate*

Ontogeny, Cognition and Social Behaviour Else, J.G., and Lee, P.C., eds, pp. 93–103. Cambridge University Press, Cambridge.

Gieseler, W. (1971) Germany. In: *Catalogue of fossil hominids*, Campbell, B.G., Oakley, K.P., and Molleson, T.I., eds, pp. 189–215. British Museum (Natural History), London.

Gieseler, W., and Mollison, T. (1929) Untersuchungen uber den Oldoway fund. *Verhandlungen der Gesellschaft far Physische Anthropologic* 3: 50–67.

Gilbert, W., and Richards, G. (2000) Digital imaging of bone and tooth modification. *Anatomical Record* 261(6): 237–46.

Gilbert, W., and Asfaw, B. (2009) Homo erectus: *Pleistocene evidence from the Middle Awash, Ethiopia*. University of California Press, Berkeley, CA.

Gilbert, W., White, T., and Asfaw, B. (2003) *Homo erectus, Homo ergaster, Homo "cepranensis," and the Daka cranium. Journal of Human Evolution* 45(3): 255.

Gilbertson, D., Bird, M., Hunt, C., McLaren, S., Mani Banda, R., Pyatt, B., et al. (2005) Past human activity and geomorphological change in a guano-rich tropical cave mouth: initial interpretations of the Late Quaternary succession in the Great Cave of Niah, Sarawak. *Asian Perspectives* 44: 16–41.

Giles, R.E., Blanc, H., Cann, H.M., and Wallace, D.C. (1980) Maternal inheritance of human mitochondrial DNA. *Proceedings of the National Academy of Sciences USA* 77(1): 6715–19.

Gill, E.D., Neil, W.G., MacIntosh, N.W.G., Thorne, A., and Bowler, J.M. (1975) Australia. In: *Catalogue of Fossil Hominids. Part III: Americas, Asia, and Australasia*, Oakley, K.P., Campbell, B.G., and Molleson, T.I., eds, pp. 195–205. British Museum (Natural History), London.

Girard, M. (1973) La breche a "Machairodus" de Montmaurin (Pyrénées centrales). *Bulletin de l'Association française pour l'étude du Quaternaire* 3: 193–207.

Girard, M., Baffier, D., Tisnerat, N., Valladas, H., Arnold, M., and Hedges, R. (1996) Dates ^{14}C en spectrométrie de masse par accélérateur à la Grande Grotte d'Arcy-sur-Cure (Yonne). *Cahiers Archéologiques de Bourgogne* 6: 17–23.

Glory, A. (2008) *Les recherches à Lascaux (1952–1963). Documents recueillis et présentés par B. et G. Delluc.* CNRS, Paris.

Gluckman, P.D., and Hanson, M.A. (2006) The developmental origins of health and disease. In: *Early Life Origins of Health and Disease*, Wintour, E.M., and Owens, J.A., eds, pp. 1–7. Springer, New York.

Godfrey, L., Samonds, K., Jungers, W., and Sutherland, M. (2001) Teeth, brains, and primate life histories. *American Journal of Physical Anthropology* 114(3): 192–214.

Goldberg, P. (2000) Micromorphology and site formation at Die Kelders Cave I, South Africa *Journal of Human Evolution* 38: 43–90.

Goldberg, P., Weiner, S., Bar-Yosef, O., Xu, Q., and Liu, J. (2001) Site formation processes at Zhoukoudian, China. *Journal of Human Evolution* 41(5): 483–530.

Goldin-Meadow, S., Goodrich, W., Sauer, E., and Iverson, J.M. (2007) Young children use their hands to tell their mothers what to say. *Developmental Science* 10: 778–85.

Goleman, G. (2009) *Ecological Intelligence: How Knowing the Hidden Impact of What we Buy can Change Everything*. Broadway Books, New York.

Golovanova, L.V., Hoffecker, J.F., Kharitonov, V.M., and Romanova, G.P. (1999) Mezmaiskaya Cave: a Neanderthal occupation in the Northern Causacus. *Current Anthropology* 40(1): 77–86.

Gómez-Robles, A., Martinón-Torres, M., Bermúdez de Castro, J.M., Prado, L., Sarmiento, S., and Arsuaga, J.-L. (2008) Geometric morphometric analysis of the crown morphology of the lower first premolars of hominins, with special attention to Pleistocene *Homo. Journal of Human Evolution* 55(4): 627–38.

Gommery, D., and Senut, B. (2006) The terminal thumb phalanx of *Orrorin tugenensis* (Upper Miocene of Kenya). *Geobios* 39: 372–84.

Gommery, D., and Thackeray, F. (2006) Sts 14, a male subadult partial skeleton of *Australopithecus africanus? South African Journal of Science* 102: 91–92.

Gompertz, B. (1825) On the nature of the function expressive of the law of human mortality, and on a new method of determining the value of life contingencies. *Philosophical Transactions of the Royal Society* 115(1825): 513–83.

Gonzalez, E., Kulkarni, H., Bolivar, H., Mangano, A., Sanchez, R., Catano, G., et al. (2005) The influence of CCL3L1 gene-containing segmental duplications on HIV-1/AIDS susceptibility. *Science* 307(5714): 1434–40.

Good, P. (2000) *Permutation Tests: a Practical Guide to Resampling Methods for Testing Hypotheses*. Springer, New York.

Goodman, M. (1962) Immunocytochemistry of the primates and primate evolution. *Annals of the New York Academy of Sciences* 102: 219–34.

Goodman, M. (1963) Man's place in the phylogeny of the primates as reflected in serum proteins. In: *Classification and Human Evolution*, Washburn, S., ed., pp. 204–34. Aldine Publishing Company, Chicago, IL.

Goodman, M., Wimoore, G., and Matsuda, G. (1975) Darwinian evolution in the genealogy of haemoglobin. *Nature* 253: 603–8.

Goodman, M., Page, S.L., Meereles, C.M., and Czelusniak, J. (1998) Primate phylogeny and classification elucidated at the molecular level. In: *Evolutionary Theory and Processes: Modern Perspectives, Papers in Honour of Eviatar Nevo*, Wasser, S.P., ed., pp. 193. Kluwer Academic Publishers, Dordrecht.

Goodrum, M.R., and Olson, C. (2009) The quest for an absolute chronology in human prehistory: anthropologists, chemists and the fluorine dating method in palaeoanthropology. *British Journal for the History of Science* **42**(1): 95–114.

Goodwin, A.J.H. (1928) An introduction to the Middle Stone Age of South Africa. *South African Journal of Science* **25**: 410–18.

Goodwin, A.J.H., and van Riet Lowe, C. (1929) The Stone Age Cultures of South Africa. *Annals of the South African Museum* **27**: 1–289.

Goodwin, H. (1998) Supernumerary teeth in Pleistocene, recent, and hybrid individuals of the *Spermophilus richardsonii* complex (Sciuridae). *Journal of Mammalogy* **79**: 1161–69.

Gordon, A.D. (2006) Scaling of size and dimorphism in primates II: Macroevolution. *International Journal of Primatology* **27**: 63–105.

Gordon, A.D., Green, D.J., and Richmond, B.R. (2008) Strong postcranial size dimorphism in *Australopithecus afarensis*: results from two new resampling methods for multivariate data sets with missing data. *American Journal of Physical Anthropology* **135**: 311–28.

Goren-Inbar, N., and Belitzky, S. (1989) Structural position of the Pleistocene Gesher Benot Ya'aqov site in the Dead Sea Rift zone. *Quaternary Research* **31**(3): 371–6.

Goren-Inbar, N., and Saragusti, I. (1996) An Acheulian biface assemblage from Gesher Benot Ya'aqov, Israel: indications of African affinities. *Journal of Field Archaeology* **23**(1): 15–30.

Goren-Inbar, N., Lister, A., Werker, E., and Chech, M. (1994) A butchered elephant skull and associated artifacts from the Acheulian site of Gesher Benot Ya'Aqov, Israel. *Paléorient* **20**(1): 99–112.

Goren-Inbar, N., Feibel, C.S., Verosub, K.L., Melmed, Y., Kislen, M.E., Tcherns, E., et al. (2000) Pleistocene milestones on the Out-of-Africa corridor at Gesher Benot Ya'agov, Israel. *Science* **289**(5481): 944–7.

Goren-Inbar, N., Sharon, G., Melamed, Y., and Kislev, M.E. (2002a) Nuts, nut cracking, and pitted stones at Gesher Benot Ya'aqov, Israel. *Proceedings of the National Academy of Sciences USA* **99**: 2455–60.

Goren-Inbar, N., Werker, E., and Feibel, C.S. (2002b) *The Acheulian site of Gesher Benot Ya'aqov, Israel. The Wood Assemblage.* Oxbow Books, Oxford.

Goren-Inbar, N., Alperson, N., Kislev, M.E., Simchoni, O., Melamed, Y., Ben-Nun, A., et al. (2004) Evidence of hominin control of fire at Gesher Benot Ya'aqov, Israel. *Science* **304**(5671): 725–7.

Goren-Inbar, N., Belitzky, S., Goren, Y., Rabinovich, R., and Saragusti, I. (2006) Gesher Benot Ya'aqov - the 'Bar': An acheulian assemblage. *Geoarchaeology* **7**(1): 27–40.

Gorjanovic-Kranmberger, D. (1906) *Der Diluviale Mensch von Krapina in Kroatien: Ein Beitrag zur Paläoanthropologie.* Kriedel, Wiesbaden.

Gorjanovic-Kranmberger, D. (1913) Život i kultura diluvijalnog čovjeka iz Krapine u Hrvatskoj (with German summary). In: *Djela Jugoslavenske Akademije Znanosti i Umjetnosti.* Jugoslavenska Akademija Znanosti i Umjetnosti, Zagreb.

Gould, S.J. (1977) *Ontogeny and Phylogeny.* Harvard University Press, London.

Gould, S.J. (1975) Allometry in primates, with emphasis on scaling and the evolution of the brain. *Contributions to Primatology* **5**: 244–92.

Gould, S.J. (1990) *Wonderful Life: The Burgess Shale and the Nature of History.* Hutchinson Radius, London.

Gould, S.J., and Lewontin, R.C. (1979) The spandrels of San Marco and the Panglossian paradigm: a critique of the adaptationist programme. *Proceedings of the Royal Society of London, Series B* **205**: 581–98.

Gould, S.J., and Vrba, E.S. (1982) Exaptation - a missing term in the science of form. *Paleobiology* **8**(1): 4–15.

Govaerts, R., Simpson, D.A., Goetghebeur, P., Wilson, K.L., Egorova, T., and Bruhl, J. (2007) *World checklist of Cyperaceae.* The Board of Trustees of the Royal Botanic Gardens, Kew.

Gower, J.C. (1975) Generalized Procrustes analysis. *Psychometrika* **40**(1): 33–51.

Gowlett, J.A.J. (1988) A case of Developed Oldowan in the Acheulean? *World Archaeology* **20**(1): 13–26.

Gowlett, J.A.J. (1999) Lower and Middle Pleistocene archaeology of the Baringo Basin. In: *Late Cenozoic Environments and Human Evolution: A Tribute to Bill Bishop*, Andrews, P., and Banham, P., eds, pp. 123–41. Geological Society, London.

Gowlett, J.A.J. (2005) Beeches Pit – Archaeology, assemblage dynamics and early fire history of a Middle Pleistocene site in East Anglia, UK. *Eurasian Prehistory* **3**(2): 3–38.

Gowlett, J.A.J. (2006) The early settlement of northern Europe: Fire history in the context of climate change and the social brain. *Comptes Rendus Palevol* **5**(1–2): 299–310.

Gowlett, J.A.J., Harris, J.W.K., Walton, D., and Wood, B.A. (1981) Early archaeological sites, hominid remains and traces of fire from Chesowanja, Kenya. *Nature* **294**: 125–9.

Grace, R. (1996) Use-wear analysis: the state of the art. *Archaeometry* **38**(2): 209–29.

Gradstein, F.M., Ogg, J.G., and Smith, A.G., eds (2004) *A Geologic Time Scale 2004.* Cambridge University Press, Cambridge.

Gramly, R.M., and Rightmire, G.P. (1973) A fragmentary cranium and dated Later Stone Age assemblage from Lukenya Hill, Kenya. *Man* **8**(4): 571–9.

Grausz, H.M., Leakey, R.E.F., Walker, A.C., and Ward, C.V. (1988) Associated cranial and postcranial bones of *Australopithecus boisei*. In: *Evolutionary History of the "Robust" Australopithecines*, Grine, F.E., ed., pp. 127–32. Aldine de Gruyter, New York.

Graves, R., Lupo, A., McCarthy, R., Wescott, D., and Cunningham, D. (2010) Just how strapping was KNM-WT 15000? *Journal of Human Evolution* **59**: 542–54.

Gravina, B., Mellars, P., and Ramsey, C.B. (2005) Radiocarbon dating of interstratified Neanderthal and early modern human occupaations at the Chatelperronian type-site. *Nature* **438**: 51–56.

Gray, R.D., Drummond, A.J., and Greenhill, S.J. (2009) Language phylogenies reveal expansion pulses and pauses in Pacific settlement. *Science* **323**(5913): 479–83.

Grayson, D.K. (1986) Eoliths, archaeological ambiguity, and the generation of "middle-range" research. In: *American Archaeology Past and Future*, Meltzer, D.J., Fowler, D.D., and Sabloff, J.A., eds, pp. 77–133. Smithsonian Institution Press, Washington DC.

Green, D.J., Gordon, A.D., and Richmond, B.G. (2007) Limb-size proportions in *Australopithecus afarensis* and *Australopithecus africanus*. *Journal of Human Evolution* **52**(2): 187–200.

Green, H.S. (1993) *Pontnewydd Cave: A lower Palaeolithic hominid site in Wales*. Quaternary Studies Monographs. National Museum of Wales, Cardiff.

Green, H.S., Stringer, C.B., Collcutt, S.N., Currant, A.P., Huxtable, J., Schwarcz, H.P., et al. (1981) Pontnewydd Cave in Wales - a new Middle Pleistocene hominid site. *Nature* **294**(5843): 707–13.

Green, R., Krause, J., Briggs, A., Maricic, T., Stenzel, U., Kircher, M., et al. (2010) A draft sequence of the Neandertal genome. *Science* **328**(5979): 710.

Green, R.E., Krause, J., Ptak, S.E., Briggs, A.W., Ronan, M.T., Simons, J.F., et al. (2006) Analysis of one million base pairs of Neanderthal DNA. *Nature* **444**: 330–6.

Green, R.E., Malaspinas, A.S., Krause, J., Briggs, A.W., Johnson, P.L.F., Uhler, C., et al. (2008) A Complete Neandertal Mitochondrial Genome Sequence Determined by High-Throughput Sequencing. *Cell* **134**(3): 416–26.

Greene, D.L. (1967) Dentition of Meroitic, X-Group, and Christian Populations From Wadi Halfa, Sudan. In: *Sudan. Anthropol. Paper 85, Nubian Ser. 1* pp. University of Utah Press, Salt Lake City, UT.

Greene, D.L., and Armelagos, G.J. (1972) *The Wadi Halfa Mesolithic Population*. Research Report No. 11. University of Massachusetts, Amherst, MA.

Gregory, W.K. (1910) The orders of mammals. *Bulletin of the American Museum of Natural History* **27**: 1–524.

Gregory, W.K. (1928) The upright posture of man: a review of its origin and evolution. *Proceedings of the American Philosophical Society* **67**(4): 339–77.

Gregory, W.K. (1930) The origin of man from a brachiating anthropoid stock. *Science* **71**(1852): 645–50.

Gregory, W.K. (1947) The monotremes and the palimpsest theory. *Bulletin of the American Museum of Natural History* **88**: 1–52.

Gregory, W.K., and Hellman, M. (1939) The dentition of the extinct South African man-ape *Australopithecus* (*Plesianthropus*) *transvaalensis* Broom. A comparative and phylogenetic study. *Annals of the Transvaal Museum* **19**: 339–73.

Grey, M., Haggart, J.W., and Smith, P.L. (2008) Variation in evolutionary patterns across the geographic range of a fossil bivalve. *Science* **322**(5905): 1238–41.

Grimaud, D. (1982) *Evolution du parietal de l'homme fossile. Position de l'homme de Tautavel parmi les Hominides*. PhD thesis, Museum National d'Histoire Naturelle.

Grimaud-Herve, D., and Widianto, H. (2001) Les fossiles humains decouvert a Java depuis les annees, 1980. In: *Origine Despeuplements et Chronologie des Cultures Paléolithiques dans le Sud-Est Asiatique, Semenanjung*, Sémah, F., Falguères, C., Grimaud-Herve, D., and Sémah, A.-M., eds, pp. 331–57. Semenanjung, Paris.

Grine, F. (1996) Phenetic affinities among early *Homo* crania from East and South Africa. *Journal of Human Evolution* **30**(3): 189–225.

Grine, F., and Klein, R. (1985) Pleistocene and Holocene human remains from Equus cave, South Africa. *Anthropology* **8**: 55–98.

Grine, F., and Franzen, J.L. (1994) Fossil hominid teeth from the Sangiran Dome (Java, Indonesia). *Courier Forschungsinstitut Senckenberg* **171**: 75–103.

Grine, F., and Henshilwood, C. (2002) Additional human remains from Blombos Cave, South Africa: (1999–2000 excavations). *Journal of Human Evolution* **42**(3): 293–302.

Grine, F., Jungers, W., Tobias, P., and Pearson, O. (1995) Fossil *Homo* femur from Berg Aukas, northern Namibia. *American Journal of Physical Anthropology* **97**(2): 151–85.

Grine, F., Ungar, P., Teaford, M., and El-Zaatari, S. (2006) Molar microwear in *Praeanthropus afarensis*: evidence for dietary stasis through time and under diverse paleoecological conditions. *Journal of Human Evolution* **51**(3): 297–319.

Grine, F.E. (1984) Comparison of the deciduous dentitions of African and Asian hominids. *Courier Forschungsinstitut Senckenberg* **69**: 69–82.

Grine, F.E. (1984) *The Deciduous Dentition of the Kalahari San, the South African Negro and the South African Plio-Pleistocene Hominids*. PhD thesis, University of the Witwatersrand.

Grine, F.E. (1985) Dental morphology and systematic affinities of the Taung fossil hominid. In: *Hominid Evolution: Past, Present and Future*, Tobias, P.V., ed., pp. 247–53. Alan R. Liss, New York.

Grine, F.E. (1993) Australopithecine taxonomy and phylogeny: historical background and recent interpretation. In: *The Human Evolution Source Book*, Ciochon, R.L., and Fleagle, J.G., eds, pp. 198–209. Prentice Hall, Englewood Cliffs.

Grine, F.E. (2000) Middle Stone Age human fossils from Die Kelders Cave 1, Western Cape Province, South Africa. *Journal of Human Evolution* **38**: 129–45.

Grine, F.E. (2001) Implications of morphological diversity in early *Homo* crania from eastern and southern

Africa In: *Humanity from the African Naissance to the Coming Millennia*, Tobias, P.V., Raath, M.A., Moggi-Cecchi, J., and Doyle, G.A., eds, pp. 107–15. Florence University Press, Florence.

Grine, F.E. (2004) Description and preliminary analysis of new hominid craniodental fossils from the Swartkrans Formation. In: *Swartkrans: A Cave's Chronicle of Early Man*, Brain, C.K., ed., pp. 75–116. Transvaal Museum, Pretoria.

Grine, F.E., and Martin, L.B. (1988) Enamel thickness and development in *Australopithecus* and *Paranthropus*. In: *Evolutionary History of the "Robust" Australopithecines*, Grine, F.E., ed., pp. 3–42. Aldine de Gruyter, New York.

Grine, F.E., and Klein, R.G. (1993) Late Pleistocene human remains from the Sea Harvest site, Saldanha Bay, South Africa. *South African Journal of Science* **89**: 145–52.

Grine, F.E., and Strait, D.S. (1994) New hominid fossils from Member 1 "Hanging Remnant", Swartkrans Formation, South Africa. *Journal of Human Evolution* **26**: 57–75.

Grine, F.E., Klein, R.G., and Volman, T.P. (1991) Dating, archaeology and human fossils from the Middle Stone Age levels of Die Kelders, South Africa. *Journal of Human Evolution* **21**(5): 325–63.

Grine, F.E., Demes, B., Jungers, W.L., and Cole, T.M. (1993) Taxonomic affinity of the early *Homo* cranium from Swartkrans, South Africa. *American Journal of Physical Anthropology* **92**(4): 411–26.

Grine, F.E., Jungers, W.L., and Schultz, J. (1996) Phenetic affinities among early *Homo* crania from East and South Africa. *Journal of Human Evolution* **30**: 189–225.

Grine, F.E., Pearson, O.M., Klein, R.G., and Rightmire, G.P. (1998) Additional human fossils from Klasies River Mouth, South Africa. *Journal of Human Evolution* **35**: 95–107.

Grine, F.E., Henshilwood, C.S., and Sealy, J.C. (2000) Human remains from Blombos Cave, South Africa: (1997–1998 excavations). *Journal of Human Evolution* **38**: 755–65.

Grine, F.E., Bailey, R.M., Harvati, K., Nathan, R.P., Morris, A.G., Henderson, G.M., et al. (2007) Late Pleistocene human skull from Hofmeyr, South Africa, and modern human origins. *Science* **315**(5809): 226–9.

Grine, F.E., Gunz, P., Betti-Nash, L., Neubauer, S., and Morris, A.G. (2010) Reconstruction of the late Pleistocene human skull from Hofmeyr, South Africa. *Journal of Human Evolution* **59**(1): 1–15.

Grommé, C.S., and Hay, R.L. (1963) Magnetization of basalt of Bed I, Olduvai Gorge, Taganyika. *Nature* **200**: 560–1.

Groves, C.P. (1989) *A Theory of Human and Primate Evolution*. Clarendon Press, Oxford.

Groves, C.P. (1999) Nomenclature of African Plio-Pleistocene hominins. *Journal of Human Evolution* **37**: 869–72.

Groves, C.P. (2008) *Extended Family: Long Lost Cousins. A Personal Look at the History of Primatology*. Conservation International, Arlington, VA.

Groves, C.P., and Mazák, V. (1975) An approach to the taxonomy of the Hominidae: gracile Villafranchian hominids of Africa. *Casopis pro mineralogii a geologii* **20**(3): 225–47.

Grubich, J. (2003) Morphological convergence of pharyngeal jaw structure in durophagous perciform fish. *Biological Journal of the Linnean Society* **80**(1): 147–65.

Grün, R. (2006) Direct dating of human fossils. *American Journal of Physical Anthropology* **131**: 2–48.

Grün, R., and Mellars, P. (1991) ESR chronology of a 100,000-year archaeological sequence at Pech de l'Azé II, France. *Antiquity* **65**: 544–51.

Grün, R., and Stringer, C.B. (1991) Electron spin resonance dating and the evolution of modern humans. *Archaeometry* **33**(2): 153–99.

Grün, R., and Stringer, C.B. (2000) Tabun revisited: Revised ER chronology and new ESR and U-series analyses of dental material from Tabun C1. *Journal of Human Evolution* **39**(6): 601–12.

Grün, R., and Beaumont, P. (2001) Border Cave revisited: a revised ESR chronology. *Journal of Human Evolution* **40**: 467–82.

Grün, R., Schwarcz, H., Ford, D., and Hentzsch, B. (1988) ESR dating of spring deposited travertines. *Quaternary Science Reviews* **7**(3–4): 429–32.

Grün, R., Shackleton, N.J., and Deacon, H.J. (1990) Electron spin resonance dating of tooth enamel from Klasies River Mouth cave. *Current Anthropology* **31**: 427–32.

Grün, R., Stringer, C.B., and Schwarcz, H. (1991) ESR dating of teeth from Garrod's Tabun cave collection. *Journal of Human Evolution* **20**(3): 231–48.

Grün, R., Brink, J.S., Spooner, N.A., Taylor, L., Stringer, C.B., Franciscus, R.G., et al. (1996) Direct dating of Florisbad hominid. *Nature* **382**: 500–1.

Grün, R., Huang, P.-H., Wu, X., Stringer, C., Thorne, A., and McCullogh, M. (1997a) ESR analysis of teeth from the paleoanthropological site of Zhoukoudian, China. *Journal of Human Evolution* **32**(1): 83–91.

Grün, R., Thorne, A., Swisher, C.C., Rink, W.J., Schwarcz, H.P., and Antón, S. (1997b) Dating the Ngandong humans. *Science* **276**(5318): 1575–76.

Grün, R., Huang, P.-H., Huang, W., McDermott, F., Thorne, A., Stringer, C.B., et al. (1998) ESR and U-series analyses of teeth from the palaeoanthropological site of Hexian, Anhui Province, China. *Journal of Human Evolution* **34**(6): 555–64.

Grün, R., Beaumont, P., Tobias, P.V., and Eggins, S. (2003) On the age of the Border Cave 5 human mandible. *Journal of Human Evolution* **45**(2): 155–67.

Grün, R., Stringer, C., McDermott, F., Nathan, R., Porat, N., Robertson, S., et al. (2005) U-series and ESR analyses of bones and teeth relating to the human burials from Skhul. *Journal of Human Evolution* **49**(3): 316–34.

Grün, R., Maroto, J., Eggins, S., Stringer, C., Robertson, S., Taylor, L., et al. (2006) ESR and U-series analyses of enamel and dentine fragments of the Banyoles mandible. *Journal of Human Evolution* 50(3): 347–58.

Guatelli-Steinberg, D. (2001) What can developmental defects of enamel reveal about physiological stress in nonhuman primates. *Evolutionary Anthropology* 10: 138–51.

Guatelli-Steinberg, D. (2003) Macroscopic and microscopic analyses of linear enamel hypoplasia in Plio-Pleistocene South African hominins with respect to aspects of enamel development and morphology. *American Journal of Physical Anthropology* 120(4): 309–22.

Guatelli-Steinberg, D. (2004a) Analysis and significance of linear enamel hypoplasia in Plio-Pleistocene hominins. *American Journal of Physical Anthropology* 123: 199–215.

Guatelli-Steinberg, D., Larsen, C.S., and Hutchinson, D.L. (2004b) Prevalance and the duration of linear enamel hypoplasia: a comparative study of Neandertals and Inuit foragers. *Journal of Human Evolution* 47: 65–84.

Guatelli-Steinberg, D., Reid, D.J., Bishop, T.A., and Larsen, C.S. (2005) Anterior tooth growth periods in Neandertals were comparable to those of modern humans. *Proceedings of the National Academy of Sciences USA* 102(40): 14197–202.

Guglielmino, C.R., Viganotti, C., Hewlett, B., and Cavalli-Sforza, L.L. (1995) Cultural variation in Africa: role of mechanisms of transmission and adaptation. *Proceedings of the National Academy of Sciences USA* 92: 7585–89.

Guibert, P., Bechtel, F., Bourguignon, L., Brenet, M., Couchoud, I., Delagnes, A., et al. (2008) Une base de données pour la chronologie du Paléolithique moyen dans le Sud-ouest de la France. In: *Les sociétes du Paléolithique dans un Grand Sud-Ouest de la France: nouveaux gisements, nouveaux résultats, nouvelles méthodes*, Jaubert, J., Bordes, J.-G., and Ortega, I., eds, pp. 19–40. Mémoire de la Société Préhistorique Française.

Guilbaud, M., Backer, A., and Leveàque, F. (1994) Technological differentiation associated with the Saint-Césaire Neandertal. *Préhistoire Européenne* 6: 187–96.

Gunz, P. (2001) *Using Semilandmarks in Three Dimensions to Model Human Neurocranial Shape*. Master's thesis, University of Vienna.

Gunz, P. (2005) *Statistical and Geometric Reconstruction of Hominid Crania: Reconstructing Australopithecine Ontogeny*. PhD thesis, University of Vienna.

Gunz, P., Mitteroecker, P., and Bookstein, F. (2005) Semilandmarks in Three Dimensions. In: *Modern Morphometrics in Physical Anthropology*, Slice, D.E., ed., pp. 73–98. Kluwer Academic/Plenum Publishers, New York.

Gunz, P., Mitteroecker, P., Neubauer, S., Weber, G.W., and Bookstein, F.L. (2009a) Principles for the virtual reconstruction of hominid crania. *Journal of Human Evolution* 57: 48–62.

Gunz, P., Bookstein, F., Mitteroecker, P., Stadlmayr, A., Seidler, H., and Weber, G.W. (2009b) Early modern human diversity suggests subdivided population structure and a complex out-of-Africa scenario. *Proceedings of the National Academy of Sciences USA* 106: 6094–98.

Guo, Z.T., Ruddiman, W.F., Hao, Q.Z., Wu, H.B., Qiao, Y.S., Zhu, R.X., et al. (2002) Onset of Asian desertification by 22 Myr ago inferred from loess deposits in China. *Nature* 416: 159–63.

Guoliang, C., Suling, L., and Jinlu, L. (1977) Discussion on the age of *Homo erectus yuanmoensis* and the event of early Matuyama. *Chinese Journal of Geology* 1: 34–43.

Gurven, M., and Kaplan, H. (2007) Longevity among hunter-gatherers: A cross-cultural examination. *Population and Development Review* 33(2): 321–65.

Guy, F., Lieberman, D.E., Pilbeam, D., Ponce de León, M.S., Likius, A., Mackaye, H.T., et al. (2005) Morphological affinities of the *Sahelanthropus tchadensis* (Late Miocene hominid from Chad) cranium. *Proceedings of the National Academy of Sciences USA* 102(no. 52): 18836–41.

Guy, F., Mackaye, H., Likius, A., Vignaud, P., Schmittbuhl, M., and Brunet, M. (2008) Symphyseal shape variation in extant and fossil hominoids, and the symphysis of *Australopithecus bahrelghazali*. *Journal of Human Evolution* 55(1): 37–47.

Haarhoff, P.J. (1988) A new fossil stork (Aves, Ciconiidae) from the Late Tertiary of Langebaanweg, South Africa. *Annals of the South African Museum* 97: 297–313.

Habgood, P.J. (1991) Aboriginal fossil hominids: evolution and migrations. In: *The Origins of Human Behaviour*, Foley, R., ed., pp. 97–113. Routledge, Cambridge.

Hachi, H., Frölich, F., Gendron-Badou, A., de Lumley, H., Roubet, C., and Abdessadok, S. (2002) Figurines du Paléolithique supérieur en matière minérale plastique cuite d'Afalou Bou Rhummel (Babors, Algérie). Premières analyses par spectroscopie d'absorption infrarouge. *L'Anthropologie* 106: 57–97.

Hachi, S. (1996) L'Iberomaurisian, decouverte des fouilees d'Afalou (Bedjaia, Algerie). *L'Anthropologie* 100: 55–76.

Hadfield, J., and Kruuk, L. (2010) MCMC methods for multi-response generalised linear mixed models: The MCMCglmm R package. *Journal of Statistical Software* 33(2): 1–22.

Haeckel, E. (1866) *Generelle morphologie der organismen: Allgemeine grundz, ge der organischen formen-wissenschaft, mechanisch begr, ndet durch die von Charles Darwin reformirte descendenztheorie*. G. Reimer, Berlin.

Haeckel, E. (1906) *Anthropogenie*. W. Engelmann, Leipzig.

Haeusler, M., and McHenry, H.M. (2004) Body proportions of *Homo habilis*. *Journal of Human Evolution* 46: 433–65.

Haeusler, M., and McHenry, H.M. (2007) Evolutionary reversals of limb proportions in early hominids: evidence from KNM-ER 3735. *Journal of Human Evolution* 53(4): 2007.

Haeusler, M., Martelli, S., and Boeni, T. (2002) Vertebrae numbers of the early hominid lumbar spine. *Journal of Human Evolution* 43(5): 621–43.

Hagelberg, E., Sykes, B., and Hedges, R. (1989) Ancient bone DNA amplified. *Nature* 342(6249): 485–5.

Hahn, J. (1986) *Kraft und Aggression: die Botschaft der Eiszeitkunst im Aurignacien Süddeutschlands?* Archæologica Venatoria, Tübingen.

Hahn, J., Müller-Beck, H., and Taute, W. (1985) *Eiszeithöhlen im Lonetal. Archäologie einer Landschaft auf der Schwäbischen Alb.* Konrad Theiss Verlag, Stuttgart.

Haig, D. (1993) Genetic conflicts in human pregnancy. *Quarterly Review of Biology* 68(4): 495–532.

Haile-Selassie, Y. (2001) Late Miocene hominids from the Middle Awash, Ethiopia. *Nature* 412: 178–81.

Haile-Selassie, Y., Suwa, G., and White, T.D. (2004a) Late Miocene teeth from Middle Awash, Ethiopia, and early hominid dental evolution. *Science* 303: 1503–5.

Haile-Selassie, Y., Asfaw, B., and White, T.D. (2004b) Hominid cranial remains from Upper Pleistocene deposits at Aduma, Middle Awash, Ethiopia. *American Journal of Physical Anthropology* 123(1): 1–10.

Haile-Selassie, Y., Deino, A., Saylor, B., Umer, M., and Latimer, B. (2007) Preliminary geology and paleontology of new hominid-bearing Pliocene localities in the central Afar region of Ethiopia. *Anthropological Science* 115(3): 215–22.

Haile-Selassie, Y., and WoldeGabriel, G., eds (2009) Ardipithecus kadabba: *Late Miocene evidence from the Middle Awash, Ethiopia.* University of California Press, Berkeley, CA.

Haile-Selassie, Y., Latimer, B., Alene, M., Deino, A., Gibert, L., Melillo, S., et al. (2010a) An early *Australopithecus afarensis* postcranium from Woranso-Mille, Ethiopia. *Proceedings of the National Academy of Sciences USA* 107(27): 12121–6.

Haile-Selassie, Y., Saylor, B., Deino, A., Alene, M., and Latimer, B. (2010b) New hominid fossils from Woranso-Mille (Central Afar, Ethiopia) and taxonomy of early *Australopithecus. American Journal of Physical Anthropology* 141(3): 406–17.

Haileab, B., and Brown, F.H. (1994) Tephra correlations between the Gadeb prehistoric site and the Turkana Basin. *Journal of Human Evolution* 26: 167–73.

Haldane, J. (1932) *The Causes of Evolution.* Longmans, Green & Co, London.

Hall, B.K. (1975) Evolutionary consequences of skeletal differentiation. *Journal of American Zoology* 15: 329–50.

Hall, B.K. (1992) *Evolutionary Developmental Biology.* Chapman & Hall, London.

Hall, G., Pickering, R., Lacruz, R., Hancox, J., Berger, L., and Schmid, P. (2006) An Acheulean handaxe from Gladysvale Cave Site, Gauteng, South Africa. *South African Journal of Science* 102(3–4): 103–5.

Hallgrímsson, B., and Lieberman, D.E. (2008) Mouse models and the evolutionary developmental biology of the skull. *Integrative and Comparative Biology* 48(3): 373–84.

Hallgrímsson, B., Willmore, K., and Hall, B.K. (2002) Canalization, developmental stability, and morphological integration in primate limbs. *American Journal of Physical Anthropology* 119(S35): 131–58.

Hallgrímsson, B., Donnabháin, B.Ó., Blom, D.E., Lozada, M.C., and Willmore, K.T. (2005) Why are rare traits unilaterally expressed?: trait frequency and unilateral expression for cranial non-metric traits in humans. *American Journal of Physical Anthropology* 128(1): 14–25.

Hallgrímsson, B., Jamniczky, H., Young, N., Rolian, C., Parsons, T., Boughner, J., et al. (2009) Deciphering the palimpsest: studying the relationship between morphological integration and phenotypic covariation. *Evolutionary Biology* 36(4): 355–76.

Hambach, U. (1996) Paläo-und gesteinsmagnetische Untersuchungen im Quartär der Grube Grafenrain: Fundplatz des *Homo erectus heidelbergensis. Mannheimer Geschichtsblätter* 1: 41–46.

Hamilton, M.J., and Buchanan, B. (2007) Spatial gradients in Clovis-age radiocarbon dates across North America suggest rapid colonization from the north. *Proceedings of the National Academy of Sciences USA* 104(40): 15625–30.

Hamilton, W.D. (1967) Extraordinary sex ratios. *Science* 156: 477–88.

Hammer, M.F. (1995) A recent common ancestry for human Y chromosomes. *Nature* 378: 376–8.

Hammer, M.F., Spurdle, A.B., Karafet, T., Bonner, M.R., and Wood, E.T. (1997) The geographic distribution of human Y chromosome variation. *Genetics* 145: 787–805.

Hammer, M.F., Karafet, T., Rasanayagam, A., Wood, E. T., Altheide, T.K., Jenkins, T., et al. (1998) Out of Africa and back again: nested cladistic analysis of human Y chromosome variation. *Molecular Biology and Evolution* 15(4): 427–41.

Hammer, M.F., Karafet, T., Redd, A.J., Jarjanazi, H., and Santachiara-Benerecetti, S. (2001) Hierarchical patterns of global human Y-chromosome diversity. *Molecular Biology and Evolution* 18: 1189–203.

Hampton, R.R. (2001) Rhesus monkeys know when they remember. *Proceedings of the National Academy of Sciences USA* 98: 5359–62.

Handt, O., Höss, M., Krings, M., and Pääbo, S. (1994) Ancient DNA: methodological challenges. *Cellular and Molecular Life Sciences* 50(6): 524–9.

Handt, O., Krings, M., Ward, R., and Pääbo, S. (1996) The retrieval of ancient human DNA sequences. *American Journal of Human Genetics* 59(2): 368–76.

Hanihara, K. (1981) Sex determination using discriminant analysis on teeth. *Journal of the Anthropologial Society of Nippon* 89: 401–17.

Hansen, C.L., and Keller, C.M. (1971) Environment and Activity Patterning at Isimila Korongo, Iringa District, Tanzania: A Preliminary Report. *American Anthropologist* **73**(5): 1201–11.

HapMap (2003) The International HapMap Project. *Nature* **426**: 789–96.

Harding, R., Healy, E., Ray, A., Ellis, N., Flanagan, N., Todd, C., et al. (2000) Evidence for variable selective pressures at MC1R. *American Journal of Human Genetics* **66**(4): 1351–61.

Hardy, B. (2004) Neanderthal behaviour and stone tool function at the Middle Palaeolithic site of La Quina, France. *Antiquity* **78**: 547–65.

Hardy, B., Kay, M., Marks, A., and Monigal, K. (2001) Stone tool function at the paleolithic sites of Starosele and Buran Kaya III, Crimea: behavioral implications. *Proceedings of the National Academy of Sciences USA* **98**(19): 10972–77.

Hare, B., Call, J., Agnetta, B., and Tomasello, M. (2000) Chimpanzees know what conspecifics do and do not see. *Animal Behaviour* **59**: 771–85.

Hare, B., Call, J., and Tomasello, M. (2001) Do chimpanzees know what conspecifics know? *Animal Behaviour* **61**(1): 139–51.

Harmand, S. (2009) Variability in raw material selectivity at the Late Pliocene sites of Lokalalei, West Turkana, Kenya. In: *Interdisciplinary Approaches to the Oldowan*, Hovers, E., and Braun, D.R., eds, pp. 85–97. Springer, Amsterdam.

Harmon, E.H. (2006) Size and shape variation in *Australopithecus afarensis* proximal femora. *Journal of Human Evolution* **51**(3): 217–27.

Harmon, E.H. (2009) The shape of the early hominin proximal femur. *American Journal of Physical Anthropology* **139**(2): 154–71.

Harris, E.F. (2000) Molecular systematics of the Old World monkey tribe Papionini: analysis of the total available genetic sequences. *Journal of Human Evolution* **38**(2): 235–56.

Harris, E.F., and Bailit, H.L. (1980) The Metaconule: a morphologic and familial analysis of a molar cusp in humans. *American Journal of Physical Anthropology* **53**(3): 349–58.

Harris, J.M., Brown, F.H., Leakey, M.G., Walker, A.C., and Leakey, R.E.F. (1988) Pliocene and Pleistocene hominid-bearing sites from West of Lake Turkana, Kenya. *Science* **239**: 27–33.

Harris, J.M., and White, T.D. (1979) Evolution of the Plio-Pleistocene African Suidae. *Transactions of the American Philosophical Society* **69**(2): 1–128.

Harrison, G., and Owen, J. (1964) Studies on the inheritance of human skin colour. *Annals of Human Genetics* **28**(1–3): 27–37.

Harrison, T. (1993) Cladistic concepts and the species problem in hominoid evolution. In: *Species, Species Concepts, and Primate Evolution*, Kimbel, W.H., and Martin, L.B., eds, pp. 345–71. Plenum Press, New York.

Harrisson, T. (1958) The caves at Niah: a history of prehistory. *Sarawak Museum Journal* **12**: 542–95.

Harrisson, T. (1970) The prehistory of Borneo. *Asian Perspectives* **13**: 17–45.

Harrisson, T. (1978) Present status and problems for Paleolithic studies in Borneo and adjacent islands. In: *Early Paleolithic in South and East Asia*, Ikawa-Smith, F., ed., pp. 37–57. The Hague, Mouton.

Hartwig-Scherer, S., and Martin, R.D. (1991) Was 'Lucy' more human than the 'child'? Observations on early hominid postcranial skeletons. *Journal of Human Evolution* **21**: 439–49.

Harvati, K. (2004) 3-D geometric morphometric analysis of temporal bone landmarks in Neanderthals and modern humans. In: *Morphometrics, Applications in Biology and Paleontology*, Elewa, A.M.T., ed., pp. 245–58. Springer, Verlag.

Harvati, K., and Delson, E. (1999) Conference report: paleoanthropology of the Mani Peninsula (Greece). *Journal of Human Evolution* **36**(3): 343–8.

Harvati, K., Panagopoulou, E., and Karkanas, P. (2003) First Neanderthal remains from Greece: the evidence from Lakonis. *Journal of Human Evolution* **45**: 465–73.

Harvati, K., Gunz, P., and Grigorescu, D. (2007) Cioclovina (Romania): morphological affinities of an early modern European. *Journal of Human Evolution* **53**(6): 732–46.

Harvati, K., Hublin, J., and Gunz, P. (2010) Evolution of middle-late Pleistocene human cranio-facial form: A 3-D approach. *Journal of Human Evolution* **59**: 445–64.

Hasegawa, Y. (1980) Notes on vertebrate fossils from late Pleistocene to Holocene of Ryukyu Islands, Japan. *Quaternary Research* **18**: 263–7.

Haslam, M., and Petraglia, M. (2010) Comment on "Environmental impact of the 73 ka Toba super-eruption in South Asia" by M.A.J. Williams, S.H. Ambrose, S. van der Kaars, et al. *Palaeogeography, Palaeoclimatology, Palaeoecology* **296**: 199–203.

Haslam, M., Hernandez-Aguilar, A., Ling, V., Carvalho, S., de la Torre, I., DeStefano, A., et al. (2009) Primate archaeology. *Nature* **460**: 339–44.

Haslam, M., Clarkson, C., Petraglia, M., Korisettar, R., Jones, S., Shipton, C., et al. (2010) The 74 ka Toba super-eruption and southern Indian hominins: archaeology, lithic technology and environments at Jwalapuram Locality 3. *Journal of Archaeological Science* **37**: 3370–84.

Hastenrath, S., Nicklis, A., and Greischar, L. (1993) Atmospheric-hydrospheric mechanisms of climate anomalies in the western equatorial Indian Ocean. *Journal of Geophysical Research* **98**(C11): 20219–35.

Hatley, T., and Kappelman, J. (1980) Bears, pigs, and Plio-Pleistocene hominids: a case for the exploitation of below ground resources. *Human Ecology* **8**(4): 371–87.

Haughton, S.H. (1917) Preliminary note on the ancient human skull remains from the Transvaal. *Transactions of the Royal Society of South Africa* **6**: 1–14.

Hauser, G., and De Stefano, G.F. (1989) *Epigenetic variants of the human skull*. Schwizerbartsche, Stuttgart.

Hauser, M.D., Chomsky, N., and Fitch, W.T. (2002) The Faculty of Language: What Is It, Who Has It, and How Did It Evolve? *Science* **298**(5598): 1569–79.

Hauser, O. (1909) Découverte d'un squelette du type du Néandertal sous l'abri inférieur du Moustier. *L'homme Préhistorique* **7**: 1–9.

Hauser, O. (1911) *Le Périgord préhistorique*. Réjou impr., Périgueux.

Hausmann, R., and Brunnacker, K. (1986) U-series dating of Middle European travertines. In: *L'Homme de Néandertal*, Otte, M., ed., pp. 20–27. Université de Liège, Liège.

Haverkort, C., and Lubell, D. (1999) Cutmarks on Capsian human remains: implications for Maghreb Holocene social organization and palaeoeconomy. *International Journal of Osteoarchaeology* **9**(3): 147–69.

Hawkes, K., O'Connell, J.F., Alvarez, H., and Charnov, E.L. (1998) Grandmothering, menopause, and the evolution of human life histories. *Proceedings of the National Academy of Sciences USA* **95**: 1336–9.

Hawks, J., Oh, S., Hunley, K., Dobson, S., Cabana, G., Dayall, P., et al. (2000) An Australasian test of the recent African origin theory using the WLH-50 calvarium. *Journal of Human Evolution* **39**: 1–22.

Hay, R. (1987) Geology of the Laetoli area. In: *Laetoli: A Pliocene Site in Northern Tanzania*, Leakey, M.D., and Harris, J.M., eds, pp. 23–47. Clarendon Press, Oxford.

Hay, R.L. (1976) *Geology of the Olduvai Gorge: a Study of Sedimentation in a Semiarid Basin*. University of California Press, Berkeley, CA.

Hay, R.L. (1990) Olduvai Gorge: A case history of the interpretation of hominid paleoenvironments in East Africa. In: *Establishment of the Geological Framework of Paleoanthropology*, Laporte, L., ed., pp. 23–37. Special Paper 242. Geological Society of America, Boulder, CO.

Hay, R.L., and Kyser, T.K. (2001) Chemical sedimentology and paleoenvironmental history of Lake Olduvai, a Pliocene lake in Northern Tanzania. *Geological Society of America Bulletin* **113**(12): 1505–21.

Hayes, K.J., and Hayes, C. (1952) Imitation in a home-raised chimpanzee. *Journal of Comparative and Physiological Psychology* **45**(5): 450–9.

Haynes, G., Anderson, D.G., Ferring, C.R., Fiedel, S.J., Grayson, D.K., Haynes Jr, C.V., et al. (2007) Comment on "Redefining the Age of Clovis: Implications for the Peopling of the Americas". *Science* **317**(5836): 320.

Hays, J.D., Imbrie, J., and Shackleton, N.J. (1976) Variations in the earth's orbit: pacemaker of the ice-ages. *Science* **194**: 1121–32.

Head, M.J., Gibbard, P.L., and Amos, S. (2008) The Quaternary: its character and definition. *Episodes* **31**(2): 234–8.

Heaton, T.H.E. (1987) The $^{15}N/^{14}N$ ratios of plants in South Africa and Namibia: relationship to climate and coastal/saline environments. *Oecologia* **74**: 236–46.

Heberer, G. (1963) Uber einen neuen archanthropien Typus aus der Oldoway-Schlucht. *Zeitschrift für Morphologie und Anthropologie* **53**: 171–7.

Hedges, R.E.M., Housley, R.A., Ramsey, C.B., and van Klinken, G.J. (1994) Radiocarbon dates from the Oxford AMS system: Archaeometry datelist 18. *Archaeometry* **36**: 337–74.

Hedrick, P.W., and Black, F.L. (1997) HLA and mate selection: no evidence in South Amerindians. *American Journal of Human Genetics* **61**(3): 505–11.

Heim, J. (1970) L'encéphale néandertalien de l'homme de La Ferrassie. *L'Anthropologie* **74**: 527–72.

Heim, J. (1974a) Les hommes fossiles de La Ferrassie (Dordogne) et le problème de la définition des néandertaliens classiques. *L'Anthropologie* **78**: 81–112.

Heim, J. (1974b) Les hommes fossiles de La Ferrassie (Dordogne) et le problème de la définition des néandertaliens classiques (suite). *L'Anthropologie* **78**: 321–78.

Heim, J. (1991) L'enfant magdalénien de La Madeleine. *L'Anthropologie* **95**(2–3): 611–38.

Heim, J.-L. (1976) *Le Hommes Fossiles de La Ferrassie vol. 1*. Masson, Paris.

Heim, J.-L. (1982a) *Les Hommes Fossiles de La Ferrassie*. Masson, Paris.

Heim, J.-L. (1982b) *Les Enfants Néandertalien de la Ferrassie: Étude Anthropologique et Analyse Ontogénique des Hommes de Néandertal*. Masson, Paris.

Heim, J.-L. (1989) La nouvelle reconstitution du crâne néandertalien de la Chapelle-aux-Saints. Méthodes et résultats. *Bulletins et Mémoires de la Société d'anthropologie de Paris* **1**: 95–117.

Heim, J.-L., and Granat, J. (1995) La mandibule de l'enfant néanderthalien de Malarnaud (Ariège). *Anthropologie et Préhistoire* **106**: 75–96.

Heinrich, H. (1988) Origin and consequences of cyclic ice rafting in the Northeast Atlantic Ocean during the past 130,000 years. *Quaternary Research* **29**: 142–52.

Heinrich, R.E., Rose, M.D., Leakey, R.E., and Walker, A.C. (1993) Hominid radius from the middle Pliocene of Lake Turkana, Kenya. *American Journal of Physical Anthropology* **92**(2): 139–48.

Helgren, D. (1978) Acheulian settlement along the Lower Vaal river, South Africa. *Journal of Archaeological Science* **5**(1): 39–60.

Helgren, D. (1997) Locations and landscapes of paleolithic sites in the Semliki Rift, Zaire. *Geoarchaeology* **12**: 337–61.

Hellman, M. (1928) Racial characters in human dentition Part I. A racial distribution of the dryopithecus pattern and its modifications in the lower molar teeth of man. *Proceedings of the American Philosophical Society* **67**(2): 157–74.

Hencken, H. (1948) The Prehistoric Archaeology of Tangier Zone, Morocco. *Proceedings of the American Philosophical Society* **92**(4): 282–8.

Henderson, Z. (1992) The context of some Middle Stone Age hearths at Klasies River Shelter 1B:

implications for understanding Human behaviour. *Southern African Field Archaeology* 1: 14–26.

Hendey, Q. (1974) Faunal dating of the late Cenozoic of Southern Africa, with special reference to the Carnivora. *Quaternary Research* **4**(2): 149–61.

Hendey, Q.B. (1970a) A review of the geology and palaeontology of the Plio/Pliocene deposits at Langebaanweg, Cape Province. *Annals of the South African Museum* **56**: 75–117.

Hendey, Q.B. (1970b) The age of fossiliferous deposits at Langebaanweg, Cape Province. *Annals of the South African Museum* **56**: 119–31.

Hendey, Q.B. (1972) Further observations on the age of the Mammalian fauna from Langebaanweg, Cape Province. *Palaeoecology* **6**: 172–5.

Hendey, Q.B. (1976) The Pliocene occurrences in 'E' Quarry, Langebaanweg, South Africa. *Annals of the South African Museum* **69**: 215–47.

Hendey, Q.B. (1978a) Late Tertiary Hyaenidae from Langebaanweg, South Africa, and their relevance to the phylogeny of the family. *Annals of the South African Museum* **76**: 265–97.

Hendey, Q.B. (1978b) Late Tertiary Mustelidae (Mammalia, Carnivora) from Langebaanweg, South Africa. *Annals of the South African Museum* **76**: 329–57.

Hendey, Q.B. (1981) Geological succession at Langebaanweg, Cape Province, and global changes of the Late Tertiary. *South African Journal of Science* **77**(1): 33–38.

Hendey, Q.B. (1982) *Langebaanweg: A record of past life*. South African Museum, Cape Town.

Hendey, Q.B. (1984) Southern African Late Tertiary vertebrates. In: *Southern African Prehistory and Palaeoenvironments*, Klein, R.G., ed., pp. 81–106. A.A. Balkema, Rotterdam.

Hendley, Q.P., and Volman, T.P. (1986) Last interglacial sea levels and coastal caves on Cape Province, South Africa. *Quaternary Research* **25**: 189–98.

Henke, H. (1986) Die magdalénienzeitlichen Menschenfunde von Oberkassel bei Bonn. *Bonner Jahrbücher* **186**: 317–66.

Henke, W. (1980) Das Calvarium von Binshof (Speyer) im Vergleich mit anderen Jungpaläolithikern. *Zeitschrift für Morphologie und Anthropologie* **70**: 275–94.

Henke, W., and Protsch, R. (1978) Die Paderborner Calvaria – ein diluvialer *Homo sapiens*. *Anthropologischer Anzeiger* **36**: 85–108.

Hennig, G.J., Herr, W., Weber, E., and Xirotiris, N.I. (1981) ESR-dating of the fossil hominid cranium from Petralona Cave, Greece. *Nature* **292**: 533–6.

Hennig, W. (1966) *Phylogenetic Systematics*. University of Illinois Press, Chicago, IL.

Hennig, W. (1981) *Insect Phylogeny*. Wiley, New York.

Henri-Martin, G. (1957) *La Grotte de Fontechevade. Premiere Partie: Historique, Fouilles, Stratigraphie, Archeologie*. Masson et Compagnie, Paris.

Henri-Martin, G. (1961) Le niveau de Châtelperron à La Quina (Charente). *Bulletin de la Société Préhistorique Française* **58**(11): 796–808.

Henri-Martin, G. (1964) La derniere occupation Mousterienne de La Quina (Charente). Datation par le radiocarbone. *Comptes Rendues Hebdomadaires de l'Academie des Sciences* **258**: 3533–5.

Hens, S.M., Konigsberg, L., and Jungers, W.L. (2000) Estimating stature in fossil hominids: which regression model and sample to use? *Journal of Human Evolution* **38**: 767–84.

Henshilwood, C., D'errico, F., Marean, C., Milo, R., and Yates, R. (2001) An early bone tool industry from the Middle Stone Age at Blombos Cave, South Africa: implications for the origins of modern human behaviour, symbolism and language. *Journal of Human Evolution* **41**(6): 631–78.

Henshilwood, C., d'Errico, F., Vanhaeren, M., van Niekerk, K., and Jacobs, Z. (2004) Middle Stone Age shell beads from South Africa. *Science* **304**(5669): 404.

Henshilwood, C.S. (2007) Fully symbolic sapiens behaviour: innovation in the Middle Stone Age at Blombos Cave, South Africa. In: *Rethinking the Human Revolution*, Mellars, P., Boyle, K., Bar-Yosef, O., and Stringer, C., eds, pp. 123–32. McDonald Institute for Anthropological Research, Cambridge.

Henshilwood, C.S., and Sealy, J. (1997) Bone artifacts from the Middle Stone Age at Blombos Cave, Southern Cape, South Africa. *Current Anthropology* **38**(5): 890–5.

Henshilwood, C.S., and Marean, C.W. (2003) The origin of modern human behavior. *Current Anthropology* **44**(5): 627–51.

Henshilwood, C.S., d'Errico, F., Yates, R., Jacobs, Z., Tribolo, C., Duller, G.A.T., et al. (2002) Emergence of modern human behavior: Middle Stone Age engravings from South Africa. *Science* **295**: 1278–80.

Hernandez-Aguilar, R.A., Moore, J., and Pickering, T.R. (2007) Savanna chimpanzees use tools to harvest the underground storage organs of plants. *Proceedings of the National Academy of Sciences USA* **104**(49): 19210–13.

Hernández-Fernández, M., and Vrba, E. (2005) A complete estimate of the phylogenetic relationships in Ruminantia: a dated species-level supertree of the extant ruminants. *Biological Reviews* **80**(2): 269–302.

Herries, A.I.R. (2003) *Magnetostratigraphic Seriation of South African Hominin Paleocaves*. PhD thesis, University of Liverpool.

Herries, A.I.R., Adams, J.W., Kuykendall, K.L., and Shaw, J. (2006) Speleology and magnetobiostratigraphic chronology of the GD 2 locality of the Gondolin hominin-bearing paleocave deposits, North West Province, South Africa. *Journal of Human Evolution* **51**: 617–31.

Herries, A.I.R., Curnoe, D., and Adams, J.W. (2009) A multi-disciplinary seriation of early *Homo* and

Paranthropus bearing palaeocaves in southern Africa. *Quaternary International* **202**(1–2): 14–28.

Herries, A.I.R., Hopley, P.J., Adams, J.W., Curnoe, D., and Maslin, M.A. (2010) Letter to the editor: Geochronology and palaeoenvironments of Southern African hominin-bearing localities—A reply to Wrangham et al., 2009. "Shallow-water habitats as sources of fallback foods for hominins". *American Journal of Physical Anthropology* **143**(4): 640–6.

Herring, S.W., and Teng, S. (2000) Strain in the braincase and its sutures during function. *American Journal of Physical Anthropology* **112**: 575–93.

Hershkovitz, I., Speirs, M., Frayer, D., Nadel, D., Wish-Baratz, S., and Arensburg, B. (1995) Ohalo II H2: A 19,000-year-old skeleton from a water-logged site at the Sea of Galilee, Israel. *American Journal of Physical Anthropology* **96**(3): 215–34.

Hershkovitz, I., Kornreich, L., and Laron, Z. (2007) Comparative skeletal features between *Homo floresiensis* and patients with primary growth hormone insensitivity (Laron Syndrome). *American Journal of Physical Anthropology* **134**(2): 198–208.

Hershkovitz, I., Smith, P., Sarig, R., Quam, R., Rodríguez, L., García, R., et al. (2010) Middle Pleistocene dental remains from Qesem Cave (Israel). *American Journal of Physical Anthropology*.

Hershkovitz, P. (1971) Basic crown patterns and cups homologies of mammalian teeth. In: *Dental Morphology and Evolution*, Dahlberg, A.A., ed., pp. 95–150. University of Chicago Press, Chicago, IL.

Hewes, G.W., Irwin, H., Papworth, M., and Saxe, A. (1964) A new fossil human population from the Wadi Halfa area, Sudan. *Nature* **203**: 341–3.

Hewlett, B.S., and Cavalli-Sforza, L.L. (1986) Cultural transmission among Aka pygmies. *American Anthropologist* **88**: 922–34.

Hicks, J.H. (1954) The mechanics of the foot. II. The plantar aponeurosis and the arch. *Journal of Anatomy* **88**(1): 25–30.

Higgs, E.S. (1961) Some Pleistocene faunas of the Mediterranean coastal areas. *Proceedings of the Prehistoric Society* **27**: 144–54.

Higham, T., Ramsey, C.B., Karavanić, I., Smith, F.H., and Trinkaus, E. (2006) Revised direct radiocarbon dating of the Vindija G_1 Upper Paleolithic Neandertals. *Proceedings of the National Academy of Sciences USA* **103**(3): 553–7.

Higham, T.F.G., Jacobi, R.M., Julien, M., David, F., Wood, R., Basell, L.S., et al. (2011) New dates for the Grotte du Renne at Arcy-sur-Cure and their implications for the evolution of symbolic behaviour. *Proceedings of the National Academy of Sciences USA* **107**(47): 20234–9.

Higuchi, R., Bowman, B., Freiberger, M., Ryder, O., and Wilson, A. (1984) DNA sequences from the quagga, an extinct member of the horse family. *Nature* **312**: 282–4.

Hill, A. (1985) Early hominid from Baringo, Kenya. *Nature* **315**: 222–4.

Hill, A. (1994) Late Miocene and early Pliocene Hominoids from Africa. In: *Integrative Paths to the Past: Paleoanthropological Advances in Honor of F. Clark Howell*, Corruccini, R.S., and Ciochon, R.L., eds, pp. 123–45. Prentice Hall, New Jersey.

Hill, A. (1999) The Baringo Basin, Kenya: from Bill Bishop to BPRP. In: *Late Cenozoic Environments and Hominid Evolution: a Tribute to Bill Bishop*, Andrews, P., and Banham, P., eds, pp. 85–97. Geological Society of London, London.

Hill, A. (2002) Paleoanthropological research in the Tugen Hills, Kenya. *Journal of Human Evolution* **42**(1–2): 1–10.

Hill, A., and Ward, S. (1988) Origin of the Hominidae: the record of African large hominoid evolution between 14 my and 4 my. *Yearbook of Physical Anthropology* **31**: 49–83.

Hill, A., Drake, R., Tauxe, L., Monaghan, M., Barry, J.C., Behrensmeyer, A.K., et al. (1985) Neogene palaeontology and geochronology of the Baringo Basin, Kenya. *Journal of Human Evolution* **14**: 759–73.

Hill, A., Ward, S., and Brown, B. (1992) Anatomy and age of the Lothagam mandible. *Journal of Human Evolution* **22**: 439–51.

Hill, A., Ward, S., Deino, A., Curtis, G., and Drake, R. (1992) Earliest *Homo*. *Nature* **355**: 719–22.

Hill, K., Kaplan, H., Hawkes, K., and Hurtado, A.M. (1987) Foraging decisions among the Aché hunter-gatherers: New data and implications for optimal foraging models. *Ethology and Sociobiology* **8**(1): 1–36.

Hillson, S. (1996) *Dental Anthropology*. Cambridge University Press, New York.

Hillson, S., and Bond, S. (1997) Relationship of enamel hypoplasia to the pattern of tooth crown growth: A discussion. *American Journal of Physical Anthropology* **104**(1): 89–103.

Hillson, S., Parfitt, S., Bello, S., Roberts, M., and Stringer, C. (2010) Two hominin incisor teeth from the middle Pleistocene site of Boxgrove, Sussex, England. *Journal of Human Evolution* **59**: 493–503.

Hinde, A. (1998) *Demographic Methods*. Arnold Publishers, London.

Hinds, D.A., Stuve, L.L., Nilsen, G.B., Halperin, E., Eskin, E., Ballinger, D.G., et al. (2005) Whole-genome patterns of common DNA variation in three human populations. *Science* **307**(5712): 1072–79.

Hivernel-Guerre, F. (1976) *Les industries du Late Stone Age dans la région de Melka-Konture*. In: *Actes du VIII Congrès Panafricain de Préhistoire et d'Etudes du Quaternaire, 1971*. Berhanou Abebe, Chavaillon, J., Sutton, R. eds. pp. 93–98. Addis-Adaba, Ethiopia.

Hoberg, E.P., Alkire, N.L., de Queiroz, A., and Jones, A. (2001) Out of Africa: origins of the *Taenia* tapeworms in humans. *Proceedings of the Royal Society of London, Series B* **268**: 781–7.

Hobson, K.A., Atwell, L., and Wassenaar, L.I. (1999) Influence of drinking water and diet on the stable-hydrogen isotope ratios of animal tissues. *Proceedings of the National Academy of Sciences USA* **96**(14): 8003–6.

Hoffmann, A., and Wegner, D. (2002) The rediscovery of the Combe Capelle skull. *Journal of Human Evolution* **43**(4): 577–81.

Hoffmann, D.L., Beck, J.W., Richards, D.A., Smart, P.L., Singarayer, J.S., Ketchmark, T., et al. (2010) Towards radiocarbon calibration beyond 28 ka using speleothems from the Bahamas. *Earth and Planetary Science Letters* **289**: 1–10.

Holliday, T. (2006) Neanderthals and modern humans: an example of a mammalian syngameon? In: *Neanderthals Revisited: New approaches and perspectives*, Bailey, S.E., and Hublin, J.-J., eds, pp. 281–97. Springer, Dordrecht.

Holliday, T.W. (1999) Brachial and crural indices of European Late Upper Paleolithic and Mesolithic humans. *Journal of Human Evolution* **36**(5): 549–66.

Holliday, T.W. (2003) Species concepts, reticulation, and human evolution. *Current Anthropology* **44**(5): 653–73.

Holliday, T.W., and Falsetti, A.B. (1995) Lower limb length of European early modern humans in relation to mobility and climate. *Journal of Human Evolution* **29**: 141–53.

Hollister, A., Buford, W.L., Myers, L.M., Giurintano, D.J., and Novick, A. (1992) The axes of rotation of the thumb carpometacarpal joint. *Journal of Orthopaedic Research* **10**: 454–60.

Holloway, R. (1988) "Robust" Australopithecine brain endocasts: some preliminary observations. In: *Evolutionary History Of The "Robust" Australopithecines*, Grine, F. E., ed., pp. 97–105. Aldine de Gruyter, New York.

Holloway, R. (2002) Brief communication: How much larger is the relative volume of area 10 of the prefrontal cortex in humans? *American Journal of Physical Anthropology* **118**(4): 399–401.

Holloway, R.L. (1981a) The endocast of the Omo L338y-6 juvenile hominid: gracile or robust *Australopithecus*? *American Journal of Physical Anthropology* **54**(1): 109–18.

Holloway, R.L. (1981b) Revisiting the South African Taung australopithecine endocast: the position of the lunate sulcus as determined by the stereoplotting technique. *American Journal of Physical Anthropology* **56**: 43–58.

Holloway, R.L. (1983a) Cerebral brain endocast pattern of *Australopithecus afarensis* hominid. *Nature* **303**: 420–2.

Holloway, R.L. (1983b) Human paleontological evidence relevant to language behaviour. *Human Neurobiology* **2**: 105–14.

Holloway, R.L. (1983c) Human brain evolution: a search for units, models and synthesis. *Canadian Journal of Anthropology* **3**(2): 215–30.

Holloway, R.L. (1984) The Taung endocast and the lunate sulcus: a rejection of the hypothesis of its anterior position. *American Journal of Physical Anthropology* **64**(3): 285–7.

Holloway, R.L. (1985a) The past, present, and future significance of the lunate sulcus in early hominid evolution. In: *Hominid Evolution: Past, Present and Future*, Tobias, P.V., ed., pp. 47–62. Alan R. Liss, New York.

Holloway, R.L. (1985b) The poor brain of *Homo sapiens neanderthalensis*: see what you please. In: *Ancestors: The Hard Evidence*, Delson, E., ed., pp. 319–24. Alan R. Liss, New York.

Holloway, R.L. (1988) Some additional morphological and metrical observations on *Pan* brain casts and their relevance to the Taung endocast. *American Journal of Physical Anthropology* **77**(1): 27–33.

Holloway, R.L. (1991) On Falk's 1989 accusations regarding Holloway's study of the Taung endocast: a reply. *American Journal of Physical Anthropology* **84**(1): 87–91.

Holloway, R.L. (2000) Brain. In: *Encyclopedia of Human Evolution and Prehistory*, E. Delson, Tattersall, I., Van Couvering, J.A., and Brooks, A.S., eds, pp. 141–9. Garland, New York.

Holloway, R.L., and De La Coste-Larymondie, M.C. (1982) Brain endocast asymmetry in pongids and hominids: some preliminary findings on the paleontology of cerebral dominance. *American Journal of Physical Anthropology* **58**: 101–10.

Holloway, R.L., and Post, D.G. (1982) The relativity of relative brain measures and hominid mosaic evolution. In: *Brain evolution: methods and concepts*, Armstrong, E., and Falk, D., eds, pp. 57–76. Plenum Press, New York.

Holloway, R.L., Schwartz, J.H., Broadfield, D.C., Tattersall, I., and Yuan, M.S. (2004) *The Human Fossil Record*. Wiley-Liss, Hoboken, NJ.

Holm, S. (1979) A simple sequentially rejective multiple test procedure. *Scandinavian Journal of Statistics* **6**: 65–70.

Holman, D.J., and Grimes, M.A. (2003) Patterns for the initiation of breastfeeding in humans. *American Journal of Human Biology* **15**(6): 765–80.

Holman, D.J., Grimes, M.A., Achterberg, J.T., Brindle, E., and O'Connor, K.A. (2006) Distribution of postpartum amenorrhea in rural Bangladeshi women. *American Journal of Physical Anthropology* **129**(4): 609–19.

Holt, B.M. (2003) Mobility in Upper Paleolithic and Mesolithic Europe: evidence from the lower limb. *American Journal of Physical Anthropology* **122**: 200–15.

Hooijer, D.A. (1963) Further 'Hell' mammals from Niah. *Sarawak Museum Journal* **21–22**: 201–13.

Hooijer, D.A. (1983) Remarks upon the Dubois Collection of fossil mammals from Java and Kedungburbus in Java. *Geologie en Mijnbouw* **62**: 337–8.

Hooijer, D.A., and Kurtén, B. (1984) Trinil and Kedungbrubus: the *Pithecanthropus*-bearing fossil faunas of Java and their relative age. *Annales Zoologici Fennici* **21**: 135–41.

Hope, J.H., Dare-Edwards, A., and McIntyre, M.L. (1983) Middens and megafauna: stratigraphy and dating of Lake Tandou lunette, western New South Wales. *Archaeology in Oceania* **18**: 45–53.

Hopkins, W.D., and Marino, L. (2000) Asymmetries in cerebral width in nonhuman primate brains as revealed by magnetic resonance imaging (MRI) *Neuropsychologia* **38**(4): 493–9.

Hopkins, W.D., Taglialatela, J.P., Meguerditchian, A., Nir, T., Schenker, N.M., and Sherwood, C.C. (2008) Gray matter asymmetries in chimpanzees as revealed by voxel-based morphometry. *NeuroImage* **42**(2): 491–7.

Horai, S., Murayama, K., Hayasaka, K., Matsubayashi, S., Hattori, Y., Fucharoen, G., et al. (1996) mtDNA polymorphism in East Asian Populations, with special reference to the peopling of Japan. *American Journal of Human Genetics* **59**(3): 579–90.

Horai, S., Satta, Y., Hayasaka, K., Kondo, R., Inoue, T., Ishida, T., et al. (1992) Man's place in Hominoidea revealed by mitochondrial DNA genealogy. *Journal of Molecular Evolution* **35**: 32–43.

Hou, Y., Potts, R., Yuan, B., Guo, Z., Deino, A., Wang, W., et al. (2000) Mid-Pleistocene Acheulean-like stone technology of the Bose Basin, south China. *Science* **287**(5458): 1622–6.

Hovers, E. (2009) Learning from mistakes: flaking accidents and knapping skills in the assemblage of A. L. 894 (Hadar, Ethiopia). In: *The Cutting Edge: New Approaches to the Archaeology of Human Origins*, Schick, K., and Toth, N., eds, pp. 137–50. Stone Age Institute, Gosport.

Hovers, E., Rak, Y., Lavi, R., and Kimbel, W. (1995) Hominid remains from Amud Cave in the context of the Levantine Middle Paleolithic. *Paléorient* **21**(2): 47–61.

Howe, B. (1967) *The Paleolithic of Tangier, Morocco. Excavations at Cape Ashlar, 1939–1947*. American School of Prehistoric Research, Peabody Museum, Harvard University.

Howell, F. (1951) The place of Neanderthal man in human evolution. *American Journal of Physical Anthropology* **9**(4): 379–416.

Howell, F., Butzer, K., and Aguirre, E. (1963) *Noticia preliminar sobre el amplaxamento Achelense de Torralba (Soria)*. Ministerio de Educación Nacional, Madrid.

Howell, F.C. (1957) The evolutionary significance of variation and varieties of 'Neanderthal Man'. *Quarterly Review of Biology* **32**: 330–47.

Howell, F.C. (1958) Upper Pleistocene Men of the Southwest Asian Mousterian. In: *Hundert Jahre Neanderthaler*, von Koenigswald, G.H.R., ed., pp. 185–98. Keminick en Zoon, Utrecht.

Howell, F.C. (1960) European and Northwest African Middle Pleistocene Hominids. *Current Anthropology* **1**(3): 195–232.

Howell, F.C. (1961) Isimila: a paleolithic site in Africa. *Scientific American* **205**: 118–29.

Howell, F.C. (1965) *Early Man*. Time-Life Books, New York.

Howell, F.C. (1969a) Hominid teeth from White Sands and Brown Sands localities, Lower Omo Basin (Ethiopia). *Quaternaria* **11**: 47–64.

Howell, F.C. (1969b) Remains of Hominidae from Pliocene/Pleistocene formations in the Lower Omo basin, Ethiopia. *Nature* **223**: 1234–9.

Howell, F.C. (1972) Uranium-series dating of bone from the Isimila prehistoric site, Tanzania. *Nature* **237**: 51–52.

Howell, F.C. (1978a) Hominidae. In: *Evolution of African Mammals*, Maglio, V.J., and Cooke, H.B.S., eds, pp. 154–248. Harvard University Press, Cambridge.

Howell, F.C. (1978b) Overview of the Pliocene and earlier Pleistocene of the lower Omo basin, southern Ethiopia. In: *Early Hominids of Africa*, Jolly, C., ed., pp. 85–130. Duckworth, London.

Howell, F.C. (1994) Thoughts on Eugene Dubois and the "*Pithecanthropus*" saga. In: *100 years of* Pithecanthropus. *The* Homo erectus *problem*, Franzen, J.L., ed., pp. 11–20. Courier Forschungs-Institut Senckenberg, Frankfurt am Main.

Howell, F.C. (1999) Paleo-demes, species, clades, and extinctions in the Pleistocene hominin record. *Journal of Anthropological Research* **55**: 127–51.

Howell, F.C., and Coppens, Y. (1973) Deciduous teeth of Hominidae from the Pliocene/Pleistocene of the Lower Omo Basin, Ethiopia. *Journal of Human Evolution* **2**: 461–72.

Howell, F.C., and Coppens, Y. (1974) Inventory of remains of Hominidae from Pliocene/Pleistocene formation of the Lower Omo Basin, Ethiopia (1967–1972). *American Journal of Physical Anthropology* **40**(1): 1–16.

Howell, F.C., and Wood, B.A. (1974) Early hominid ulna from the Omo basin, Ethiopia. *Nature* **249**: 174–6.

Howell, F.C., and Coppens, Y. (1976) An overview of Hominidae from the Omo succession, Ethiopia. In: *Earliest Man and Environments in the Lake Rudolf Basin: Stratigraphy, Paleoecology and Evolution*, Coppens, Y., Howell, F.C., Isaac, G.L., and Leakey, R.E.F., eds, pp. 522–32. University of Chicago Press, Chicago, IL.

Howell, F.C., and Coppens, Y. (1983) Introduction. In: *The Omo Group*, de Heinzelin, J., ed., pp. 1–5. Musée Royal de l'Afrique Centrale, Tervuren.

Howell, F.C., Cole, G.H., and Kleindienst, M.R. (1962) Isimila, an Acheulian occupation site in the Iringa highlands. In: *Actes Du IV Congres Panafricain de Préhistoire et de l'Étude de Quaternaire*, Mortelmans, J., and Nenquin, J., eds, pp. 209–17. Musée Royale de l'Afrique Centrale, Tervuren.

Howell, F.C., Haesaerts, P., and de Heinzelin, J. (1987) Depositional environments, archaeological occurrences and hominids from Members E and F of the Shungura Formation (Omo Basin, Ethiopia). *Journal of Human Evolution* **16**: 665–700.

Howell, N. (1982) Village composition implied by a paleodemographic life table: The Libben site. *American Journal of Physical Anthropology* **59**(3): 263–9.

Howells, W.W. (1944) *Mankind so Far*. Doubleday, Doran and Company, Inc, New York.

Howells, W.W. (1948) *The Heathens: Primitive Man and his Religions*. Doubleday, Garden City, NY.

Howells, W.W. (1954) *Back of History: the Story of our Own Origins*. Doubleday, Garden City, NY.

Howells, W.W. (1959) *Mankind in the Making: the Story of Human Evolution*. Doubleday, Garden City, NY.

Howells, W.W. (1973) *Cranial Variation in Man*. Harvard University, Cambridge, MA.

Howells, W.W. (1973) *Evolution of the Genus Homo*. Addison-Wesley Publishing Company, Reading, MA.

Howells, W.W. (1973) *The Pacific Islanders*. Scribners, New York.

Howells, W.W. (1976) Explaining modern man: evolutionists versus migrationists. *Journal of Human Evolution* 5: 477–95.

Howells, W.W. (1983) Origins of the Chinese people: Interpretations of the recent evidence. In: *The Origins of Chinese Civilization*, Keightley, D., ed., pp. 297–319. University of California Press, Berkeley, CA.

Howells, W.W. (1989) *Skull Shapes and the Map: Craniometric Analyses in the Dispersion of Modern Homo*. Harvard University, Cambridge.

Howells, W.W. (1993) *Getting Here: The Story of Human Evolution*. The Compass Press, Washington DC.

Howells, W.W. (1995) *Who's Who in Skulls: Ethnic Identification of Crania from Measurements*. Harvard University, Cambridge, MA.

Howells, W.W. (1997) *Getting Here: The Story of Human Evolution*. The Compass Press, Washington DC.

Howells, W.W., and Tsuchitani, P.J., eds (1977) *Paleoanthropology in the People's Republic of China: A Trip Report of the American Paleoanthropology Delegation*. CSCPRC Report. National Academy of Science, Washington DC.

Howes, A.E. (1954) A polygon portrayal of coronal and basal arch dimensions in the horizontal plane. *American Journal of Orthodontics & Dentofacial Orthopedics* 40(11): 811–31.

Hrdlička, A. (1907) *Skeletal remains suggesting or attributed to early man in North America, Bulletin 33*. Bureau of American Ethnology, Washington DC.

Hrdlička, A. (1912) *Early man in South America, Bulletin 52*. Bureau of American Ethnology, Washington DC.

Hrdlička, A. (1919) *Physical Anthropology: Its Scope and Aims; Its History and Present Status in the United States*. Wistar Institute of Anatomy and Biology, Philadelphia, PA.

Hrdlička, A. (1920a) Shovel-shaped teeth. *American Journal of Physical Anthropology* 3(4): 429–65.

Hrdlička, A. (1920b) *Anthropometry*. Wistar Institute of Anatomy and Biology, Philadelphia, PA.

Hrdlička, A. (1924) New Data on the Teeth of Early Man and Certain European Apes. *American Journal of Physical Anthropology* 7: 109–32.

Hrdlička, A. (1927) The Neanderthal Phase of Man. *Journal of the Royal Anthropological Institute of Great Britain and Ireland* 57: 249–74.

Hrdlička, A. (1930) *The skeletal remains of early man*. Smithsonian Institution, Washington DC.

Hrdlička, A. (1937) *Biographical Memoir of George Sumner Huntington, 1861–1927*. The National Academy of Sciences, Washington, DC.

Hrdlička, A. (1937) *The question of ancient man in America*. Ledger Syndicate, Philadelphia, PA.

Hrdlička, A. (1939) Important Paleolithic find in central Asia. *Science* 90: 296–8.

Hrdlička, A. (1939) *Practical Anthropometry*. Wistar Institute of Anatomy and Biology, Philadelphia, PA.

Hu, C. (1973) Ape-man teeth from Yuanmou, Yunnan. *Acta Geol. Sin* 1: 65–71.

Huang, P., and Grün, R. (1998) Study on burying ages of fossil teeth from Yuanmou Man site, Yunnan Province, China. *Acta Anthropologica Sinica* 17: 165–70.

Huang, W., Fang, N.S., and Ye, Y.X. (1981) Observations on the *Homo erectus* skull that was found from Longtandong, Hexian, Anhui. *China Science Bulletin* 26: 1508–10.

Huang, W., Fang, N.S., and Ye, Y.X. (1982) Prelimary study on the fossil hominid skull and fauna of Hexian, Anhui. *Vertebrata Paleontologica Asiatica* 20: 248–56.

Huang, W., Si, X., Hou, Y., Miller-Antonio, S., and Schepartz, L. (1995) Excavations at Panxian Dadong, Guizhou province, southern China. *Current Anthropology* 36: 844–4.

Huang, Y., Freeman, K., Eglinton, T., and Alayne, S. (1999) ^{13}C analyses of individual lignin phenols in Quaternary lake sediments: A novel proxy for deciphering past terrestrial vegetation changes. *Geology* 27(5): 471.

Huang, Y., Street-Perrott, F.A., Perrott, R.A., Metzger, P., and Eglinton, G. (1999) Glacial–interglacial environmental changes inferred from molecular and compound-specific δ^{13}C analyses of sediments from Sacred Lake, Mt. Kenya. *Geochimica et Cosmochimica Acta* 63(9): 1383–404.

Huang, Y., Dupont, L., Sarnthein, M., Hayes, J.M., and Eglinton, G. (2000) Mapping of C_4 plant input from North West Africa into North East Atlantic sediments. *Geochemica et Cosmochemica Acta* 64(20): 3505–13.

Huber, D., Eason, T., Hueter, R., and Motta, P. (2005) Analysis of the bite force and mechanical design of the feeding mechanism of the durophagous horn shark *Heterodontus francisci*. *Journal of Experimental Biology* 208(18): 3553–71.

Hublin, J. (2009) The origin of Neandertals. *Proceedings of the National Academy of Sciences USA* 106(38): 16022.

Hublin, J., Tillier, A., and Tixier, J. (1987) L'humérus d'enfant moustérien (Homo 4) du Djebel Irhoud (Maroc) dans son contexte archéologique. *Bulletins et Mémoires de la Société d'Anthropologie de Paris* 4(2): 115–41.

Hublin, J.-J. (1985) Human Fossils of the North African Middle Pleistocene and the origin of *Homo sapiens*. *Ancestors: the hard evidence*. Delson, E., ed., pp. 282–8. Alan R. Liss, New York.

Hublin, J.-J. (1992) Recent Human Evolution in Northwestern Africa. *Philosophical Transactions of the Royal Society of London. Series B, Biological Sciences* 337: 185–91.

Hublin, J.-J. (2000) Modern–nonmodern hominid interactions: A Mediterranean perspective. In: *The geography of Neandertals and modern humans in Europe and the greater Mediterranean*, Bar-Yosef, O., and Pilbeam, D.R., eds, pp. 157–82. Peabody Museum of Archaeology and Ethnology, Cambridge, MA.

Hublin, J.-J. (2001) Northwestern African Middle Pleistocene hominids and their bearing on the emergence of *Homo sapiens*. In: *Human Roots. Africa and Asia in the Middle Pleistocene*, Barham, L., and Robson-Brown, K., eds, pp. 99–121. CHERUB, Western Academic & Specialist Press, Bristol.

Hublin, J.-J., and Tillier, A.-M. (1981) The Mousterian juvenile mandible from Irhoud (Morocco): A phylogenetic interpretation. In: *Aspects of Human Evolution. Symposia for the Study of Human Biology*, Stringer, C.B., ed., pp. 167–85. Taylor & Francis, London.

Hublin, J.-J. (1978) Quelques caracteres apomorphes du crane neandertalien et leur interpretation phylogenique. *Comptes Rendus de l'Académie des Sciences* 287: 923–6.

Hublin, J.-J. (1982) The myth of the European Praesapiens. *Anthropos* 9: 16.

Hublin, J.-J. (1984) The fossil man from Salzgitter-Lebenstedt (FRG) and its place in human evolution during the pleistocene in Europe. *Zeitschrift für Morphologie und Anthropologie* 75: 45–56.

Hublin, J.-J. (1986) Some comments on the diagnostic features of *Homo erectus*. *Anthropos* 23: 175–87.

Hublin, J.-J. (1998) Climatic changes, paleogeography, and the evolution of the Neandertals. In: *Neanderthals and modern humans in western Asia*, Akazawa, T., Aoki, K., and Bar-Yosef, O., eds, pp. 295–310. Plenum Press, New York.

Hublin, J.-J. (2009) The origin of Neandertals. *Proceedings of the National Academy of Sciences USA* 106(38): 16022–7.

Hublin, J.-J., Barroso Ruiz, C., Medina Lara, P., Fontugne, M., and Reyss, J.L. (1995) The Mousterian site of Zafarraya (Andalucia, Spain): dating and implications on the palaeolithic peopling processes of Western Europe. *Comptes Rendus de l'Académie des Sciences* 321: 931–7.

Hublin, J.-J., Spoor, F., Braun, M., Zonneveld, F.W., and Condemi, S. (1996) A late Neanderthal associated with Upper Palaeolithic artefacts. *Nature* 381: 224–6.

Huffman, O.F. (2001) Geologic context and the age of the Perning/Mojokerto *Homo erectus*, East Java. *Journal of Human Evolution* 40: 353–62.

Huffman, O.F., Shipman, P., Hertler, C., de Vos, J., and Aziz, F. (2005) Historical evidence of the 1936 Mojokerto skull discovery, East Java. *Journal of Human Evolution* 48: 321–63.

Hughes, A. (1990) The Tuinplaas human skeleton from the Sprinbok Flats, Transvaal. In: *From Apes to Angels: Essays in Honour of Phillip V. Tobias*, Sperber, G.H., ed., pp. 197–214. Wiley-Liss, New York.

Hughes, A.R., and Tobias, P.V. (1977) A fossil skull probably of the genus *Homo* from Sterkfontein, Transvaal. *Nature* 265: 310–12.

Hughes, J.K., Elton, S., and O'Regan, H.J. (2008) *Theropithecus* and 'Out of Africa' dispersal in the Plio-Pleistocene. *Journal of Human Evolution* 54(1): 43–77.

Hull, K.L. (2001) Reasserting the utility of obsidian hydration dating: a temperature-dependent empirical approach to practical temporal resolution with archaeological obsidians. *Journal of Archaeological Science* 28(10): 1025–40.

Humphrey, L., and Bocaege, E. (2008) Tooth evulsion in the Maghreb: chronological and geographical patterns. *African Archaeological Review* 25(1): 109–23.

Humphrey, L., Dean, M., and Stringer, C. (1999) Morphological variation in great ape and modern human mandibles. *Journal of Anatomy* 195(4): 491–513.

Humphrey, L.T., Dirks, W., Dean, M.C., and Jeffries, T.E. (2008) Tracking dietary transitions in weanling baboons (*Papio hamadryas anubis*) esing Strontium/Calcium ratios in enamel. *Folia primatologica* 79(4): 197–212.

Humphrey, N.K. (1976) The social function of intellect. In: *Growing Points in Ethology*, Bateson, P.P.G., and Hinde, R.A., eds, pp. 303–17. Cambridge University Press, Cambridge.

Humphreys, A.J.B. (1970) The role of raw material and the concept of the Fauresmith. *The South African Archaeological Bulletin* 25(99/100): 139–44.

Hunt, C., Davison, A., Inglis, R., Farr, L., Reynolds, T., Simpson, D., et al. (2010) Site formation processes in caves: The Holocene sediments of the Haua Fteah, Cyrenaica, Libya. *Journal of Archaeological Science* 37(7): 1600–11.

Hunt, D.R., and Albanese, J. (2005) History and demographic compostion of the Robert J. Terry Anatomical Collection. *American Journal of Physical Anthropology* 127(4): 406–17.

Hunt, K.D. (1996) Mechanical implications of chimpanzee positional behavior. *American Journal of Physical Anthropology* 86(4): 521–36.

Hunter, J. (1778) *The Natural History of the Human Teeth: Explaining their Structure, Use, Formation, Growth, and Disease including a Practical Treatise on the Diseases of the Teeth: Intended as a Supplement to the Natural History of these Parts, 2nd edn.* The Classics of Medicine Library, Birmingham, AL.

Hunter, J.P. (1998) Key innovations and the ecology of macroevolution. *Trends in Ecology & Evolution* 13(1): 31–36.

Hurrell, J.W. (1995) Decadal trends in the North Atlantic Oscillation: regional temperatures and precipitation. *Science* 269(5224): 676–9.

Huxley, J. (1924) The variation in the width of the abdomen in immature fiddler crabs considered in relation to its relative growth-rate. *American Naturalist* 58(658): 468–75.

Huxley, J. (1927) Further work on heterogonic growth. *Biologisches Zentralblatt* **47**: 151–63.

Huxley, J. (1931) Notes on differential growth. *American Naturalist* **65**(699): 289–315.

Huxley, J. (1942) *Evolution: The Modern Synthesis.* Harper & Brothers Publishers, New York.

Huxley, J. (1958) Cultural process and evolution. In: *Behavior and evolution*, Roe, A., and Simpson, G.G., eds. Yale University Press, New Haven, CT.

Huxley, J., and Teissier, G. (1936) Terminologie et notation dans la description de la croissance relative. *Comptes Rendus des Séances de la Société de Biologie et ses filiales* **121**: 934–6.

Huxley, J.S. (1958) Evolutionary process and taxonomy with special reference to grades. *Uppsala universitets årsskrift* **6**: 21–38.

Huxley, T.H. (1863) *Evidence as to Man's Place in Nature.* Williams and Norgate, London.

Huxtable, J., and Aitken, M.L. (1988) Datation par thermoluminescence. In: *Le Gisement Paléolithique Moyen de Biache-Saint-Vaast (Pas de Calais), Stratigraphie, Environnement, Études Archéologiques, Vol. I.*, Tuffreau, A., and Sommé, J., eds, pp. 107–8. Société Préhistorique Française, Paris.

Hyde, H.A., and Williams, D.A. (1944) The right word. *Pollen Analysis Circular* **8**: 6.

Hylander, W.L., Picq, P.Q., and Johnson, K.R. (1991) Masticatory-stress hypotheses and the supraorbital region of primates. *American Journal of Physical Anthropology* **86**: 1–36.

Hyodo, M., Nakaya, H., Urabe, A., Saegusa, H., Shunrong, X., Jiyun, Y., et al. (2002) Paleomagnetic dates of hominid remains from Yuanmou, China, and other Asian sites. *Journal of Human Evolution* **43**(1): 27–41.

Iafrate, A.J., Feuk, L., Rivera, M.N., Listewnik, M.L., Donahoe, P.K., Qi, Y., et al. (2004) Detection of large-scale variation in the human genome. *Nature Genetics* **36**: 949–51.

Ikeya, M. (1980) ESR dating of carbonates at Petralona Cave. *Anthropos* **7**: 143–51.

Imbrie, J., Hays, J.D., Martinson, D.G., McIntyre, A., Mix, A.C., Morley, J.J., et al. (1984) The orbital theory of Pleistocene climate: support from a revised chronology of the marine δ18O record. In: *Milankovitch and Climate*, A. Berger, J.I., J. Hays, G. Kukla, and B. Saltzman, eds, pp. 269–305. D. Reidel Publishing Company, Dordrecht.

Indriati, E., and Antón, S.C. (2008) Earliest Indonesian facial and dental remains from Sangiran, Java: a description of Sangiran 27. *Anthropological Science* **116**(3): 219–29.

Indriati, E., and Antón, S.C. (2010) The calvaria of Sangiran 38, Sendangbusik, Sangiran Dome, Java. *HOMO-Journal of Comparative Human Biology* **61**(4): 225–43.

Ingman, M., Kaessmann, H., Paabo, S., and Gyllensten, U. (2000) Mitochondrial genome variation and the origin of modern humans. *Nature* **408**: 708–13.

Inizan, M.L., Reduron-Ballinger, M., Roche, H., and Tixier, J. (1999) *Technology and terminology of knapped stone.* CREP, Nanterre.

Inoue, S., and Matsuzawa, T. (2007) Working memory of numerals in chimpanzees. *Current Biology* **17**(23): R1001–5.

Inouye, S., and Shea, B. (1997) What's your angle? Size correction and bar-glenoid orientation in "Lucy" (AL 288-1). *International Journal of Primatology* **18**(4): 629–50.

International Commission on Zoological Nomenclature (1999) Opinion 1941: *Australopithecus afarensis* Johanson, 1978 (Mammalia, Primates): specific name conserved. *Bulletin of Zoological Nomenclature* **56**(3): 223–4.

International Human Genome Sequencing Consortium (2001) Initial sequencing and analysis of the human genome. *Nature* **409**(6822): 860–921.

Irish, J. (2000) The Iberomaurusian enigma: North Africa progenitor or dead end? *Journal of Human Evolution* **39**: 393–410.

Irish, J.D. (2005) Population continuity vs. discontinuity revisited: dental affinities among late Paleolithic through Christian-era Nubians. *American Journal of Physical Anthropology* **128**(3): 520–35.

Irish, J.D., and Turner II, C.G. (1990) West African dental affinity of late Pleistocene Nubians: peopling of the Eurafrican-South Asian triangle, 2. *Homo* **41**: 42–53.

Irish, J.D., and Guatelli- Steinberg, D. (2003) Ancient teeth and modern human origins: an expanded comparison of African Plio-Pleistocene and recent world dental samples. *Journal of Human Evolution* **45**: 113–44.

Isaac, G.Ll. (1967) The stratigraphy of the Peninj Group - early Middle Pleistocene formations west of Lake Natron, Tanzania. In: *Background to Evolution in Africa*, Bishop, W.W., and Clark, J.D., eds, pp. 229–57. University of Chicago Press, Chicago, IL.

Isaac, G.Ll. (1982) Early hominids and fire at Chesowanja, Kenya. *Nature* **296**: 870.

Isaac, G.Ll. (1977) *Olorgesailie: Archeological Studies of a Middle Pleistocene Lake Basin In Kenya.* University of Chicago Press, Chicago, IL.

Isaac, G.Ll. (1978) The food-sharing behavior of protohuman hominids. *Scientific American* **238**: 99–108.

Isaac, G.Ll. (1978) The Olorgesailie Formation: stratigraphy, tectonics and the palaeogeographic context of the Middle Pleistocene archaeological sites. *Geological Society, London, Special Publications* **6**: 173–206.

Isaac, G.Ll. (1981) Archaeological tests of alternative models of early hominid behaviour: excavation and experiments. *Philosophical Transactions of the Royal Society of London. Series B, Biological Sciences* **292**(1057): 177–88.

Isaac, G.Ll. (1983) Bones in contention: competing explanations for the juxtaposition of early Pleistocene artifacts and faunal remains. In: *Animals and Archaeology*, Clutton-Brock, J., and Grigson, C., eds, pp. 3–19. British Archaeological Reports International Series 163, Oxford.

Isaac, G.Ll. (1984) The archaeology of human origins: Studies of the Lower Pleistocene in East Africa 1971–1981. In: *Advances in World Archaeology*, Wendorf, F., ed., pp. 1–87. Academic Press, New York.

Isaac, G.Ll. (1984) The Archaeology of Human Origins: Studies of the Lower Pleistocene in East Africa 1971–1987. *Advances in World Archaeology* **3**: 1–87.

Isaac, G.Ll., ed. (1997) *Koobi Fora Research Project*. Clarendon Press, Oxford.

Isler, K., and van Schaik, C.P. (2009) The expensive brain: a framework for explaining evolutionary changes in brain size. *Journal of Human Evolution* **57**: 392–400.

Itan, Y., Jones, B.L., Ingram, C.J., Swallow, D.M., and Thomas, M.G. (2010) A worldwide correlation of lactase persistence phenotype and genotypes. *BMC Evolutionary Biology* **10**: 36.

Itihara, M., Kadar, D., and Watanabe, N. (1985) Concluding remarks. In: *Quaternary Geology of the Hominid Fossil Bearing Formations in Java*, Watanabe, N., and Kadar, D., eds, pp. 367–78. Geological Research and Development Centre, Bandung.

Jablonski, N., and Chaplin, G. (2000) The evolution of human skin coloration. *Journal of Human Evolution* **39**(1): 57–106.

Jablonski, N.G. (2002) Fossil Old World monkeys: the late Neogene radiation. In: *The Primate Fossil Record*, Hartwig, W., ed., pp. 255–300. Cambridge University Press, Cambridge.

Jablonski, N.G., Leakey, M.G., Kiarie, C., and Antón, M. (2002) A new skeleton of *Theropithecus brumpti* (Primates: Cercopithecidae) from Lomekwi, West Turkana, Kenya. *Journal of Human Evolution* **43**: 887–923.

Jackson, I. (1997) Homologous pigmentation mutations in human, mouse and other model organisms. *Human Molecular Genetics* **6**(10): 1613–24.

Jacob, T. (1964) A new hominid skull cap from Pleistocene Sangiran. *Anthropologica* **6**: 97–104.

Jacob, T. (1966) The sixth skull cap of *Pithecanthropus erectus*. *American Journal of Physical Anthropology* **25**(3): 243–53.

Jacob, T. (1972) New hominid finds in Indonesia, and their affinities. *Mankind* **8**: 176–81.

Jacob, T. (1973) Palaeoanthropological discoveries in Indonesia with special reference to the finds of the last two decades. *Journal of Human Evolution* **2**(6): 473–85.

Jacob, T. (1975) The Pithecanthropines of Indonesia. *Bulletins et Mémoires de la Société d'Anthropologie de Paris* **2**(3): 243–56.

Jacob, T. (1980) "*Pithecanthropus*" *of Indonesia: The Phenotype, Genetics, and Ecology*. Current Arguments on Early Man, Proceedings of a Nobel Symposium. Karlskoga, Sweden.

Jacob, T. (1981) Solo man and Peking man. In: Homo erectus: *Papers in Honor of Davidson Black*, Sigmon, B.A., and Cybulski, J.S., eds, pp. 87–104. University of Toronto Press, Toronto.

Jacob, T., Soejono, R.P., Freeman, L.G., and Brown, F. H. (1978) Stone tools from Mid-Pleistocene sediments in Java. *Science* **202**(4370): 885–7.

Jacob, T., Indriati, E., Soejono, R.P., Hsü, K., Frayer, D.W., Eckhardt, R.B., et al. (2006) Pygmoid Australomelanesian *Homo sapiens* skeletal remains from Lian Bua, Flores: population affinities and pathological abnormalities. *Proceedings of the National Academy of Sciences USA* **103**(36): 13421–6.

Jacobi, R.M., and Higham, T.F.G. (2008) The "Red Lady" ages gracefully: new ultrafiltration AMS determinations from Paviland. *Journal of Human Evolution* **55**(5): 898–907.

Jacobs, B.F. (2004) Palaeobotanical studies from tropical Africa: relevance to the evolution of forest, woodland, and savannah biomes. *Philosophical Transactions of the Royal Society of London. Series B, Biological Sciences* **359**: 1573–83.

Jacobs, Z. (2010) An OSL chronology for the sedimentary deposits from Pinnacle Point Cave 13B–A punctuated presence. *Journal of Human Evolution* **59**: 289–305.

Jacobs, Z., Wintle, A.G., and Duller, G.A.T. (2003) Optical dating of dune sand from Blombos Cave, South Africa: I-multiple grain data. *Journal of Human Evolution* **44**: 599–612.

Jacobs, Z., Duller, G., Wintle, A., and Henshilwood, C. (2006) Extending the chronology of deposits at Blombos Cave, South Africa, back to 140 ka using optical dating of single and multiple grains of quartz. *Journal of Human Evolution* **51**(3): 255–73.

Jacobs, Z., Roberts, R.G., Galbraith, R.F., Deacon, H.J., Grün, R., Mackay, A., et al. (2008) Ages for the Middle Stone Age of southern Africa: implications for human behavior and dispersal. *Science* **322**(5902): 733–5.

Jarman, P. (1974) The social organisation of the antelope in relation to their ecology. *Behaviour* **48**(3/4): 215–67.

Jarman, P. (1983) Mating system and sexual dimorphism in large terrestrial mammalian herbivores. *Biological Reviews* **58**(4): 485–520.

Jefferies, R.P.S. (1979) The origin of chordates: a methodological essay. In: *The Origin of Major Invertebrate Groups. Systematics Association Special Volume 12*, House, M.R., ed., pp. 443–7. Academic Press, London.

Jelinek, A.J. (1982) The Tabun Cave and Paleolithic Man in the Levant. *Science* **216**(4553): 1369–75.

Jelinek, A.J., Debenath, A., and Dibble, H. (1989) A preliminary report on evidence related to the interpretation of economic and social activities at the site of La Quina (Charente), France. In: *L'Homme de Neandertal*, Otte, M., ed., pp. 99–106. Etudes et Recherches Archeologiques de l'Universite de Liege, Liege.

Jelinek, J. (1966) Jaw of an intermediate type of Neanderthal Man from Czechoslovakia. *Nature* **212**: 701–2.

Jelinek, J. (1969) Neanderthal man and *Homo sapiens* in central and eastern Europe. *Current Anthropology* **10**(5): 475–503.

Jellema, L., Latimer, B., and Walker, A. (1993) The rib cage. In: *The Nariokotome* Homo erectus *Skeleton*, Walker, A., and Leakey, R., eds, pp. 294–325. Harvard University Press, Cambridge, MA.

Jenkins, F.A. (1972) Chimpanzee bipedalism cineradiographic analysis and implications for the evolution of gait. *Science* **178**: 877–9.

Jensen, J.S. (1990) Plausibility and testability: assessing the consequences of evolutionary innovation. In: *Evolutionary innovations*, Nitecki, M.H., and Nitecki, D.V., eds, pp. 172–90. University of Chicago Press, Chicago, IL.

Jerardino, A., and Marean, C. (2010) Shellfish gathering, marine paleoecology and modern human behavior: perspectives from cave PP13B, Pinnacle Point, South Africa. *Journal of Human Evolution* **59**: 412–24.

Jerison, H.J. (1973) *Evolution of the Brain and Intelligence*. Academic Press, New York.

Jernvall, J., and Thesleff, I. (2000) Reiterative signaling and patterning during mammalian tooth morphogenesis. *Mechanisms of Development* **92**(1): 19–29.

Jhaldiyal, R. (2006) *Formation Processes of the Lower Palaeolithic Record in the Hunsgi and Baichbal Basins, Gulbarga District, Karnataka*. Centre for Archaeological Studies and Training, Eastern India, Kolkata.

Jia, L.P. (1985) China's earliest Palaeolithic assemblages. In: *Palaeoanthropology and Palaeolithic Archaeology in the People's Republic of China*, Wu, R., and Olsen, J.W., eds, pp. 135–46. Academic Press, Orlando, FL.

Jia, L.P., Wei, Q., and Li, C.R. (1979) Report on the excavation of Hsuchiayao Man Site in 1976. *Vertebrata PalAsiatica* **17**: 277–93.

Jiang, N., Sun, R., and Liang, Q. (1989) The late Cenozoic stratigraphy and paleontology in the Yuanmou basin, Yunnan, China. *Yunnan Geology*: 1–107.

Jin, C.Z., Pan, W.S., Zhang, Y.Q., Cai, Y.J., Xu, Q., Tang, Z.L., et al. (2009) The *Homo sapiens* cave hominin site of Mulan Mountain, Jiangzhou District, Chongzuo, Guangxi with emphasis on its age. *Chinese Scientific Bulletin* **45**: 3848–56.

Jinniushan Excavation Team (1976) Quaternary mammalian fauna from Jinniushan, Yingkou of Liaoning Province (in Chinese with English summary). *Vertebrata Palasiatica* **14**: 120–7.

Johanson, D., and Shreeve, J. (1989) *Lucy's Child: The Discovery of a Human Ancestor*. William Morrow and Company, New York.

Johanson, D., and Edgar, B. (1996) *From Lucy to Language*. Simon and Schuster, New York.

Johanson, D., Masao, F., Eck, G., White, T., Walter, R., Kimbel, W., et al. (1987) New partial skeleton of *Homo habilis* from Olduvai Gorge, Tanzania. *Nature* **327**: 205–9.

Johanson, D.C. (1974) Some metric aspects of the permanent and deciduous dentition of the pygmy chimpanzee (*Pan paniscus*). *American Journal of Physical Anthropology* **41**(1): 39–48.

Johanson, D.C. (1989) A partial *Homo habilis* skeleton from Olduvai Gorge, Tanzania: a summary of preliminary results. In: *Hominidae: Proceedings of the 2nd International Congress of Human Paleontology, Turin, September 28-October 3, 1987*, Giacobini, G., ed., pp. 155–66. Jaca Book, Milan.

Johanson, D.C., and Coppens, Y. (1976) A preliminary anatomical diagnosis of the first Plio/Pleistocene hominid discoveries in the Central Afar, Ethiopia. *American Journal of Physical Anthropology* **45**: 217–34.

Johanson, D.C., and Taieb, M. (1976) Plio-Pleistocene hominid discoveries in Hadar, Ethiopia. *Nature* **260**: 293–7.

Johanson, D.C., and White, T.D. (1979) A systematic assessment of early African hominids. *Science* **202**: 321–30.

Johanson, D.C., and Edey, M.A. (1981) *Lucy: The Beginnings of Humankind*. Simon and Schuster, New York.

Johanson, D.C., White, T.D., and Coppens, Y. (1978) A new species of the genus *Australopithecus* (Primates: Hominidae) from the Pliocene of East Africa. *Kirtlandia* **28**: 1–14.

Johanson, D.C., Taieb, M., and Coppens, Y. (1982a) Pliocene hominids from the Hadar Formation, Ethiopia (1973–1977): stratigraphic, chronologic, and paleoenvironmental contexts, with notes on hominid morphology and systematics. *American Journal of Physical Anthropology* **57**: 373–402.

Johanson, D.C., White, T.D., and Coppens, Y. (1982b) Dental remains from the Hadar Formation, Ethiopia: 1974–1977 Collections. *American Journal of Physical Anthropology* **57**: 545–603.

Johnson, C.R., and McBrearty, S. (2010) 500,000 year old blades from the Kapthurin Formation, Kenya. *Journal of Human Evolution* **58**(2): 193–200.

Jolly, C., ed. (1978) *Early Hominids of Africa*. St. Martin's Press, New York.

Jolly, C. (1993) Species, subspecies, and baboon systematics. In: *Species, Species Concepts, and Primate Evolution*, Kimbel, W.H., and Martin, L., eds, pp. 67–107. Plenum Press, New York.

Jolly, C.J. (1970) The seed-eaters: A new model of hominid differentiation based on a baboon analogy. *Man* **5**(1): 5–26.

Jolly, C.J. (1972) The classification and natural history of *Theropithecus* (*Simopithecus*) (Andrews 1916), baboons of the African Plio-Pleistocene. *Bulletin from the British Museum of Natural History: Geology* **22**(1): 1–123.

Jolly, C.J. (2001) A proper study for Mankind: analogies from the Papionin monkeys and their implications for human evolution. *American Journal of Physical Anthropology* **116**(S33): 177–204.

Jolly, C.J., and Plog, F. (1982) *Physical Anthropology and Archaeology*. Knopf, New York.

Jolly, C.J., Woolley-Barker, T., Beyene, S., Disotell, T.R., and Phillips-Conroy, J.E. (1997) Intergeneric hybrid baboons. *International Journal of Primatology* **18**(4): 597–627.

Jones, H.L., Rink, W.J., Schepartz, L.A., Miller-Antonio, S., Weiwen, H., Yamei, H., et al. (2004) Coupled electron spin resonance (ESR)/uranium-series dating of mammalian tooth enamel at Panxian Dadong, Guizhou Province, China. *Journal of Archaeological Science* **31**: 965–77.

Jones, P.R. (1994) Results of experimental work in relation to the stone industries of Olduvai Gorge. In: *Olduvai Gorge*, Leakey, M.D., and Roe, D.A., eds, pp. 254–98. Cambridge University Press, Cambridge.

Jones, S. (2007) The Toba Supervolcanic Eruption: Tephra-Fall Deposits in India and Paleoanthropological Implications. In: *The Evolution and History of Human Populations in South Asia*, Petraglia, M.D., and Allchin, B., eds, pp. 173–200. Springer, Dordrecht.

Jones, S. (2010) Palaeoenvironmental response to the 74 ka Toba ash-fall in the Jurreru and Middle Son valleys in southern and north-central India. *Quaternary Research* **73**: 336–50.

Jones, W., and Leat, P. (1985) Discussion on the geological evolution of the trachytic caldera volcano Menengai, Kenya rift valley. *Journal of the Geological Society* **142**(4): 711–12.

Jones, W.B., and Lippard, S.J. (1979) New age determinations and the geology of the Kenya Rift-Kavirondo Rift junction, W Kenya. *Journal of the Geological Society* **136**(6): 693–704.

Jordí Pardo, J. (1986) Estratigrafía y Sedimentología de la Cueva de Nerja (Salas de la Mina y del Vestíbulo). *La Prehistoria de la Cueva de Nerja* 39–97.

Jöris, O., and Baales, M. (2003) Zur Altersstellung der Schöninger Speere. In: *Erkenntnisjäger. Kultur und Umwelt des frühen Menschen. Festschrift für Dietrich Mania*, Burdukiewicz, J.M., Fiedler, L., Heinrich, W.-D., Justus, A., and Brühl, E., eds, pp. 281–8. Veröffentlichungen des Landesamt für Archäologie Sachsen-Anhalt - Landesmuseum für Vorgeschichte, Halle (Saale).

Jöris, O., and Street, M. (2008) At the end of the ^{14}C time scale - the Middle to Upper Paleolithic record of western Eurasia. *Journal of Human Evolution* **55**: 782–802.

Jouffroy, F.K., Godinot, M., and Nakano, Y. (1991) Biometrical characteristics of primate hands. *Human Evolution* **6**: 269–306.

Jouffroy, F.K., Godinot, M., and Nakano, Y. (1993) Biometrical characteristics of primate hands. In: *Hands of Primates*, Preuschoft, H., and Chivers, D.J., eds, pp. 133–72. Springer-Verlag Wien, New York.

Jungers, W., Larson, S., Harcourt-Smith, W., Morwood, M., Sutikna, T., Due Awe, R., et al. (2009a) Descriptions of the lower limb skeleton of *Homo floresiensis*. *Journal of Human Evolution* **57**(5): 538–54.

Jungers, W.L. (1982) Lucy's limbs: skeletal allometry and locomotion in *Australopithecus afarensis*. *Nature* **297**: 676–8.

Jungers, W.L. (1985) *Size and Scaling in Primate Biology*. Plenum Press, New York.

Jungers, W.L., and Susman, R.L. (1984) Body size and skeletal allometry in African apes. In: *The Pygmy Chimpanzee: Evolutionary Biology and Behavior*, Susman, R.L., ed., pp. 131–77. Plenum Press, New York.

Jungers, W.L., Falsetti, A.B., and Wall, C.E. (1995) Shape, relative size and size-adjustment in morphometrics. *American Journal of Physical Anthropology* **38**(21): 137–61.

Jungers, W.L., Harcourt-Smith, W.E.H., Wunderlich, R.E., Tocheri, M.W., Larson, S.G., Sutikna, T., et al. (2009b) The foot of *Homo floresiensis*. *Nature* **459**: 81–4.

Juwayeyi, Y., and Betzler, C. (1995) Archaeology of the Malawi rift: the search continues for early stone age occurrences in the Chiwodo beds, northern Malawi. *Journal of Human Evolution* **28**(1): 115–16.

Kadic, O. (1940) *Die Mussolini-Höhle (Subalyuk) bei Cserépfalu*, Budapest.

Kaessmann, H., Zöllner, S., Gustafsson, A.C., Wiebe, V., Laan, M., Lundeberg, J., et al. (2002) Extensive linkage disequilibrium in small human populations in Eurasia. *American Journal of Human Genetics* **70**(3): 673–85.

Kagaya, M., Ogihara, N., and Nakatsukasa, M. (2008) Morphological study of the anthropoid thoracic cage: scaling of thoracic width and an analysis of rib curvature. *Primates* **49**(2): 89–99.

Kagaya, M., Ogihara, N., and Nakatsukasa, M. (2009) Rib orientation and implications for orthograde positional behavior in nonhuman anthropoids. *Primates* **50**(4): 305–10.

Kaifu, Y. (2007) The cranium and mandible of Minatogawa 1 belong to the same individual: a response to recent claims to the contrary. *Anthropological Science* **115**: 159–62.

Kaifu, Y., Aziz, F., and Baba, H. (2005a) Hominid mandibular remains from Sangiran: 1952–1986 collection. *American Journal of Physical Anthropology* **128**: 497–519.

Kaifu, Y., Baba, H., Aziz, F., Indriati, E., Schrenk, F., and Jacob, T. (2005b) Taxonomic affinities and evolutionary history of the early Pleistocene hominids of Java: dentognathic evidence. *American Journal of Physical Anthropology* **128**(4): 709–26.

Kaifu, Y., Aziz, F., Indriati, E., Jacob, T., Kurniawan, I., and Baba, H. (2008) Cranial morphology of Javanese *Homo erectus*: new evidence for continuous evolution, specialization, and terminal extinction. *Journal of Human Evolution* **55**(4): 551–80.

Kalb, J.E., Wood, C.B., Smart, C., Oswald, E.B., Mebrate, A., Tebedge, S., et al. (1980) Preliminary geology and paleontology of the Bodo D'Ar hominid site, Afar, Ethiopia. *Palaeogeography, Palaeoclimatology, Palaeoecology* **30**: 107–30.

Kalb, J., Jaegar, M., Jolly, C., and Kana, B. (1982) Preliminary geology, paleontology and paleoecology of a Sangoan site at Andalee, Middle Awash Valley, Ethiopia. *Journal of Archaeological Science* **9**(4): 349–63.

Kalb, J.E., Jolly, C.J., and Mebrane, A. (1982) Fossil mammals and artefacts from Middle Awash Valley, Ethiopia. *Nature* **298**: 25–29.

Kaminska, L., Kozlowski, J., and Svoboda, J. (2004) The 2002–2003 excavation in the Dzeravá Skala cave, west Slovakia. *Anthropologie* **42**(3): 311–22.

Kamminga, J. (1992) New interpretations of the Upper Cave, Zhoukoudian. In: *The Evolution and Dispersal of Modern Humans in Asia*, Akazawa, T., Aoki, K., and Kimura, T., eds, pp. 379–400. Hokusen-sha, Tokyo.

Kamminga, J., and Wright, R.V.S. (1988) The Upper Cave at Zhoukoudian and the origins of the Mongoloids. *Journal of Human Evolution* **17**: 739–67.

Kanazawa, E., Sekikawa, M., and Ozaki, T. (1990) A quantitative investigation of irregular cuspules in human maxillary permanent molars. *American Journal of Physical Anthropology* **83**(2): 173–80.

Kandel, A.W., Conard, N.J., and Walker, S.J. (2006) Near-coastal settlement dynamics at the Anyskop Blowout, an archaeological locality at Langebaanweg, South Africa. *African Natural History* **2**: 186–7.

Kangas, A., Evans, A., Thesleff, I., and Jernvall, J. (2004) Nonindependence of mammalian dental characters. *Nature* **432**(7014): 211–14.

Kapandji, I. (1974) *The Physiology of the Joints: The Trunk and the Vertebral Column*. Elsevier Health Sciences, Philadelphia, PA.

Kappelman, J. (1996) The evolution of body mass and relative brain size in fossil hominids. *Journal of Human Evolution* **30**: 243–76.

Kappelman, J., Swisher, C.G., Fleagle, J.G., Yirga, S., Brown, T.M., and Feseha, M. (1996) Age of *Australopithecus afarensis* from Fejej, Ethiopia. *Journal of Human Evolution* **30**(2): 139–46.

Kappelman, J., Alçiçek, M.C., Kazancı, N., Schultz, M., Özkul, M., and Şen, Ş. (2008) Brief Communication: First *Homo erectus* from Turkey and implications for migrations into temperate Eurasia. *American Journal of Physical Anthropology* **135**(1): 110–16.

Karafet, T., Mendez, F., Meilerman, M., Underhill, P., Zegura, S., and Hammer, M. (2008) New binary polymorphisms reshape and increase resolution of the human Y chromosomal haplogroup tree. *Genome Research* **18**(5): 830.

Karavanić, I. (2007) Le Mousterien en Croatie. *L'Anthropologie* **111**: 321–45.

Karavanić, I., and Smith, F.H. (1998) The Middle/ Upper Paleolithic interface and the relationship of Neanderthals and early modern humans in the Hrvatsko Zagorje, Croatia. *Journal of Human Evolution* **34**(3): 223–48.

Karkanas, P., and Goldberg, P. (2010) Site formation processes at Pinnacle Point Cave 13B (Mossel Bay, Western Cape Province, South Africa): resolving stratigraphic and depositional complexities with micromorphology. *Journal of Human Evolution* **59**(3–4): 256–73.

Karkanas, P., Schepartz, L.A., Miller-Antonio, S., Wang, W., and Weiwen, H. (2008) Late Middle Pleistocene climate in southwestern China: inferences from the stratigraphic record of Panxian Dadong Cave, Guizhou. *Quaternary Science Reviews* **27**: 1555–70.

Kasimova, R.M. (2001) Anthropological research of Azykh Man osseous remains. *Human Evolution* **16**(1): 37–44.

Katoh, S., Nagaoka, S., WoldeGabriel, G., Renne, P., Snow, M.G., Beyene, Y., et al. (2000) Age and correlation of the Plio-Pleistocene tephra layers of the Konso Formation, southern Main Ethiopian Rift. *Quaternary Science Reviews* **19**: 1305–17.

Kaufulu, Z., and Stern, N. (1987) The first stone artefacts to be found *in situ* within the Plio-Pleistocene Chiwondo Beds in northern Malawi. *Journal of Human Evolution* **16**(7–8): 729–40.

Kaufulu, Z., Vrba, E.S., and White, T.D. (1981) Age of the Chiwondo Beds, northern Malawi. *Annals of the Transvaal Museum* **33**: 2–8.

Kay, R.F., Rasmussen, D.T., and Beard, K.C. (1984) Cementum annulus counts provide a means for age determination in *Macaca mulatta* (Primates, Anthropoidea). *Folia Primatologica* **42**: 85–95.

Kay, R.F., Cartmill, M., and Balow, M. (1998) The hypoglossal canal and the origin of human vocal behavior. *Proceedings of the National Academy of Sciences USA* **95**(9): 5417–19.

Keates, S.G. (2000) *Early and Middle Pleistocene Hominid behavior in Northern China*. B.A.R., Oxford.

Keates, S.G. (2010) Evidence for the earliest Pleistocene hominid activity in the Nihewan Basin of northern China. *Quaternary International* **223–24**: 408–17.

Keeley, L.H. (1980) *Experimental Determination of Stone Tool Uses: A Microwear Analysis, Prehistoric Archeology and Ecology*. University of Chicago Press, Chicago, IL.

Keith, A. (1896a) An introduction to the study of anthropoid apes I. The gorilla. *Natural Science* **9**: 26–37.

Keith, A. (1896b) An introduction to the study of anthropoid Apes II. The chimpanzee. *Natural Science* **9**: 250–65.

Keith, A. (1896c) An introduction to the study of anthropoid Apes III. The orangutan. *Natural Science* **9**: 316–26.

Keith, A. (1896d) An introduction to the study of anthropoid Apes IV. The gibbon. *Natural Science* **9**: 372–9.

Keith, A. (1911) The early history of the Gibraltar cranium. *Nature* **87**: 313–14.

Keith, A. (1912) *Ancient Types of Man*. Harper & Brothers, London.

Keith, A. (1915) *The Antiquity of Man*. Williams and Norgate, London.

Keith, A. (1923) Man's posture: its evolution and disorders. *British Medical Journal* **1**(3247): 499–502.

Keith, A. (1927) A report on the Galilee skull. In: *Researches in Prehistoric Galilee*, Turville-Petre, F., ed., pp. 53–106. British School of Archaeology in Jerusalem, London.

Keith, A. (1931) *New Discoveries Relating to the Antiquity of Man*. Williams and Norgate, London.

Keith, A. (1948) *A New Theory of Human Evolution*. Watts and Co., London.

Keith, A. (1950) *An Autobiography*. Watts and Co., London.

Keith, A. (1955) *Darwin Revalued*. Watts and Co., London.

Kelley, J., and Smith, T.M. (2003) Age at first molar emergence in early Miocene *Afropithecus turkanensis* and life-history evolution in Hominoidea. *Journal of Human Evolution* **44**(3): 307–29.

Kellner, C.M., and Schoeninger, M.J. (2007) A simple carbon isotope model for reconstructing prehistoric human diet. *American Journal of Physical Anthropology* **133**(4): 1112–27.

Kelly, A.J. (1996) Recently recovered Middle Stone Age assemblages from East Turkana, northern Kenya: their implications for understanding technological adaptations during the late Pleistocene. *Kaupia* **6**: 47–55.

Kendall, C., and Coplen, T.B. (2001) Distribution of oxygen-18 and deuterium in river waters across the United States. *Hydrological Processes* **15**(7): 1363–93.

Kendall, D.G. (1981) The statistics of shape. In: *Interpreting Multivariate Data*, Barnett, V., ed., pp. 75–80. Wiley, New York.

Kendall, D.G. (1984) Shape manifolds, Procrustean metrics, and complex projective spaces. *Bulletin of the London Mathematical Society* **16**(2): 81–121.

Kennedy, G.E. (1980) *Paleoanthropology*. McGraw-Hill, New York.

Kennedy, G.E. (1983a) Femoral morphology in *Homo erectus*. *Journal of Human Evolution* **12**: 587–616.

Kennedy, G.E. (1983b) A morphometric and taxonomic assessment of a hominine femur from the lower member, Koobi Fora, Lake Turkana. *American Journal of Physical Anthropology* **61**: 429–36.

Kennedy, G.E. (1984a) Are the Kow Swamp hominids "archaic"? *American Journal of Physical Anthropology* **65**(2): 163–8.

Kennedy, G.E. (1984b) The emergence of *Homo sapiens*: the post cranial evidence. *Man, New Series* **19**(1): 94–110.

Kennedy, G.E. (1999) Is "*Homo rudolfensis*" a valid species? *Journal of Human Evolution* **36**(1): 119–21.

Kennedy, K.A.R. (1979) The deep skull of Niah: an assessment of twenty years of speculation concerning its evolutionary significance. *Asian Perspectives* **20**: 32–50.

Kennedy, K.A.R. (1992) The fossil hominid skull from the Narmada Valley: *Homo erectus* or *Homo sapiens*? In: *Monographs in World Archaeology*, pp. 145–52. Prehistory Press, Madison, WI.

Kennedy, K.A.R., Sonakia, A., Chiment, J., and Verma, K.K. (1991) Is the Narmada hominid an Indian *Homo erectus*? *American Journal of Physical Anthropology* **86**(4): 475–96.

Kent, P.E. (1942) The Pleistocene beds of Kanam and Kanjera, Kavirondo, Kenya. *Geological Magazine* **79**: 117–32.

Ker, R.F., Bennett, M.B., Bibby, S.R., Kester, R.C., and Alexander, R.M. (1987) The spring in the arch of the human foot. *Nature* **325**: 147–9.

Keyser, A.W. (2000) The Drimolen skull: the most complete australopithecine cranium and mandible to date. *South African Journal of Science* **96**: 189–97.

Kibunjia, M. (2002) *Archaeological Investigations of Loka-lalei 1 (GaJh 5): A Late Pliocene Site, West of Lake Turkana, Kenya*. PhD thesis, Rutgers University.

Kibunjia, M., Roche, H., Brown, F.H., and Leakey, R.E.F. (1992) Pliocene and Pleistocene archaeological sites west of Lake Turkana, Kenya. *Journal of Human Evolution* **23**: 431–8.

Kidd, R.S., O'Higgins, P., and Oxnard, C.E. (1996) The OH8 foot: a reappraisal of the functional morphology of the hindfoot utilizing a multivariate analysis. *Journal of Human Evolution* **31**: 269–91.

Kieser, J. (1990) *Human Adult Odontometrics: the Study of Variation in Adult Tooth Size*. Cambridge University Press, Cambridge.

Kimbel, W. (1984) Variation in the pattern of cranial venous sinuses and hominid phylogeny. *American Journal of Physical Anthropology* **63**(3): 243–63.

Kimbel, W., White, T., and Johanson, D. (1984) Cranial morphology of *Australopithecus afarensis*: a comparative study based on a composite reconstruction of the adult skull. *American Journal of Physical Anthropology* **64**(4): 337–88.

Kimbel, W., Rak, Y., and Johanson, D.C. (2004) *The Skull of* Australopithecus afarensis. Oxford University Press, New York.

Kimbel, W., Lockwood, C., Ward, C., Leakey, M., Rak, Y., and Johanson, D. (2006) Was *Australopithecus anamensis* ancestral to *A. afarensis*? A case of anagenesis in the hominin fossil record. *Journal of Human Evolution* **51**(2): 134–52.

Kimbel, W.H. (1988) Identification of a partial cranium of *Australopithecus afarensis*, from the Koobi Fora Formation, Kenya. *Journal of Human Evolution* **17**: 647–56.

Kimbel, W.H., and Rak, Y. (1985) Functional morphology of the asterionic region in extant hominoids and fossil hominids. *American Journal of Physical Anthropology* **66**(1): 31–54.

Kimbel, W.H., and White, T.D. (1988a) A revised reconstruction of the adult skull of *Australopithecus afarensis*. *Journal of Human Evolution* **17**: 545–50.

Kimbel, W.H., and White, T.D. (1988b) Variation, sexual dimorphism and the taxonomy of *Australopithecus*. In: *Evolutionary History of the "Robust" Australopithecines*, Grine, F.E., ed., pp. 175–92. Aldine de Gruyter, New York.

Kimbel, W.H., and Rak, Y. (1993) The importance of species taxa in paleoanthropology and an argument for the phylogenetic concept of the species. In: *Species, Species Concepts, and Primate Evolution*, Kimbel, W.H., and Martin, L., eds, pp. 461–84. Plenum Press, New York.

Kimbel, W.H., and Delezene, L.K. (2009) "Lucy" Redux: a review of research on *Australopithecus afarensis*. *American Journal of Physical Anthropology* **140**(S49): 2–48.

Kimbel, W.H., Johanson, D.C., and Coppens, Y. (1982) Pliocene Hominid Cranial Remains from the Hadar Formation, Ethiopia. *American Journal of Physical Anthropology* **57**: 453–99.

Kimbel, W.H., White, T.D., and Johanson, D.C. (1985) Craniodental morphology of the hominids from Hadar and Laetoli: evidence of *Paranthropus* and *Homo* in the Mid-Pliocene of eastern Africa. In: *Ancestors: The Hard Evidence*, Delson, E., ed., pp. 120–37. Alan R. Liss, New York.

Kimbel, W.H., White, T.D., and Johanson, D.C. (1988) Implications of KNM-WT 17000 for the evolution of "robust" *Australopithecus*. In: *Evolutionary History of the "Robust" Australopithecines*, Grine, F.E., ed., pp. 259–68. Aldine de Gruyter, New York.

Kimbel, W.H., Johanson, D.C., and Rak, Y. (1994) The first skull and other new discoveries of *Australopithecus afarensis* at Hadar, Ethiopia. *Nature* **368**: 449–51.

Kimbel, W.H., Walter, R.C., Johanson, D.C., Reed, K.E., Aronson, J.L., Assefa, Z., et al. (1996) Late Pliocene *Homo* and Oldowan tools from the Hadar Formation (Kadar Hadar Member), Ethiopia. *Journal of Human Evolution* **31**: 549–61.

Kimbel, W.H., Johanson, D.C., and Rak, Y. (1997) Systematic assessment of a maxilla of *Homo* from Hadar, Ethiopia. *American Journal of Physical Anthropology* **103**: 235–62.

Kimbel, W.H., Rak, Y., and Johanson, D.C. (2003) A new hominin skull from Hadar: implications for cranial sexual dimorphism in *Australopithecus afarensis*. *American Journal of Physical Anthropology* **S36**: 129.

Kimura, M. (1968) Evolutionary rate at the molecular level. *Nature* **217**: 624–6.

Kimura, T. (1985) Bipedal and quadrupedal walking of primates, comparitive dynamics. In: *Primate Morphophysiology, Locomotor Analyses and Human Bipedalism*, Kondo, S., Ishida, H., and Kimura, T., eds, pp. 81–104. University of Tokyo, Tokyo.

King, B.J. (1986) Extractive foraging and the evolution of primate intelligence. *Human Evolution* **1**: 361–72.

King, M.C., and Wilson, A.C. (1975) Evolution at two levels in humans and chimpanzees. *Science* **188**: 107–16.

King, W. (1864) The reputed fossil man of the Neanderthal. *Quarterly Journal of Science* **1**: 88–97.

Kingston, J., Deino, A., Edgar, R., and Hill, A. (2007) Astronomically forced climate change in the Kenyan Rift Valley 2.7–2.55 Ma: implications for the evolution of early hominin ecosystems. *Journal of Human Evolution* **53**(5): 487–503.

Kirch, P.X. (2000) *On the Road of the Winds: An Archaeological History of the Pacific Islands before European Contact*. University of California Press, Berkeley, CA.

Kirkwood, T.B.L. (1977) Evolution of ageing. *Nature* **270**: 301–4.

Kivell, T.L., and Begun, D.R. (2007) Frequency and timing of scaphoid-centrale fusion in hominoids. *Journal of Human Evolution* **52**: 321–40.

Klaatsch, H., and Hauser, O. (1909) *Homo mousteriensis Hauseri*. *Archiv für Anthropologie* **35**: 287–97.

Klein, R. (2000) The Earlier Stone Age of Southern Africa. *The South African Archaeological Bulletin* **55** (172): 107–22.

Klein, R.G. (1973) Geological antiquity of Rhodesian Man. *Nature* **244**: 311–12.

Klein, R.G. (1975) Middle Stone Age man-animal relationships in Southern Africa: evidence from Die Kelders and Klasies River Mouth. *Science* **190**(4211): 265–7.

Klein, R.G. (1976) The mammalian fauna of the Klasies River Mouth sites, southern Cape Province. *South African Archaeological Bulletin* **31**: 75–98.

Klein, R.G. (1977) The mammalian fauna from the Middle and Later Stone Age (Later Pleistocene) levels of Border Cave, Natal Province, South Africa. *South African Archaeological Bulletin* **32**(125): 14–27.

Klein, R.G. (1982) Age (Mortality) Profiles as a means of distinguishing hunted species from scavenged ones in Stone Age Archeological Sites. *Paleobiology* **8**(2): 151–8.

Klein, R.G. (1989) Why does skeletal part representation differ between smaller and larger bovids at Klasies River Mouth and other archeological sites? *Journal of Archaeological Science* **16**(4): 363–81.

Klein, R.G. (2008) Out of Africa and the evolution of human behavior. *Evolutionary Anthropology* **17**: 267–81.

Klein, R.G. (2009) *The Human Career: Human Biological and Cultural Origins*. The University of Chicago Press, Chicago, IL.

Klein, R.G., and Cruz-Uribe, K. (1996) Exploitation of large bovids and seals at Middle and Later Stone Age sites in South Africa. *Journal of Human Evolution* **31**(4): 315–34.

Klein, R.G., and Cruz-Uribe, K. (2000) Middle and Later Stone Age large mammal and tortoise remains from Die Kelders Cave 1, Western Cape Province, South Africa. *Journal of Human Evolution* 38. 169 95.

Klein, R.G., Avery, G., Cruz-Uribe, K., Halkett, D., Hart, T., Milo, R.G., et al. (1999) Duinefontein 2: an Acheulean site in the Western Cape Province of South Africa. *Journal of Human Evolution* 37(2): 153–90.

Klein, R.G., Cruz-Uribe, K., and Milo, R.G. (1999) Skeletal part representation in archaeofaunas: comments on "Explaining the 'Klasies Pattern': Kua Ethnoarchaeology, the Die Kelders Middle Stone Age Archaeofauna, Long Bone Fragmentation and Carnivore Ravaging" by Bartram and Marean. *Journal of Archaeological Science* 26(9): 1225–34.

Klein, R.G., Avery, G., Cruz-Uribe, K., and Steele, T.E. (2007) The mammalian fauna associated with an archaic hominin skullcap and later Acheulean artifacts at Elandsfontein, Western Cape Province, South Africa. *Journal of Human Evolution* 52(2): 164–86.

Klein, W., and Perdue, C. (1997) The basic variety (or: couldn't natural languages be much simpler?). *Second Language Research* 13(4): 301.

Kleinsasser, L., Quade, J., Levin, N., Simpson, N., McIntosh, W.C., and Semaw, S. (2008) Geochronology of the Adu-Asa Formation, Gona, Ethiopia. *Geological Society of America Special Paper* 446: 33–65.

Klima, B. (1954) Palaeolithic huts at Dolni Vestonice, Czechoslovakia. *Antiquity* 28(109): 4–14.

Klima, B. (1959) Zur Problematik des Aurignaciens und Gravettiens in Mittle-Europa. *Archaeologia Austriaca* 26: 35–51.

Klima, B., Musil, R., Pelisek, J., and Jelinek, J. (1962) Die Erforschung der Hohle Sveduv stul 1953 55. *Anthropos* 13: 1–297.

Klinge, F., Dean, M., Gunnæs, A., and Leakey, M. (2005) *Microscopic Structure and Mineral Distribution in Tooth and Periodontal Tissues in a Robust Australopithecine Fossil Hominid from Koobi Fora, Kenya.* 13th International Symposium on Dental Morphology, Wydawnictwo Uniwersytetu Lodzkiego Lodz.

Klinge, R., Dean, M., Risnes, S., Erambert, M., and Gunnês, A. (2009) Preserved microstructure and mineral distribution in tooth and periodontal tissues in early fossil hominin material from Koobi Fora, Kenya. In: *Comparative Dental Morphology*, Koppe, T., Meyer, G., and Alt, K., eds, pp. 30–35. Karger Publishers, Basel.

Klingenberg, C.P. (1998) Heterochrony and allometry: the analysis of evolutionary change in ontogeny. *Biological Reviews* 73: 79–123.

Kluge, A.G., and Farris, J.S. (1969) Quantitative phyletics and the evolution of anurans. *Systematic Zoology* 18(1): 1–32.

Knecht, H. (1997) Projectile points of bone, antler, and stone. *Projectile Technology*: 191–212.

Knippertz, P., Ulbrich, U., Marques, F., and Corte-Real, J. (2003) Decadal changes in the link between El Niño and springtime North Atlantic oscillation and European-North African rainfall. *International Journal of Climatology* 23(11): 1293–311

Kobayashi, H., Matsui, Y., and Suzuki, H. (1971) University of Tokyo radiocarbon measurements IV. *Radiocarbon* 13: 97–102.

Koch, P.L. (1998) Isotopic reconstruction of past continental environments. *Annual Review of Earth and Planetary Science* 26: 573–613.

Kodera, H. (2006) Inconsistency of the maxilla and mandible in the Minatogawa Man No. 1 hominid fossil evaluated from dental occlusion. *Anatomical Science International* 81: 57–61.

Kohl-Larsen, L. (1943) *Auf den Spuren des Vörmenschen. Forschungen, Fahrten und Erlebnisse in Deutsch-Ostafrika.* Strecker and Schröder Verlag, Stuttgart.

Kohl-Larsen, L., and Reck, H. (1936) Erster Ueberblick über die Jungdiluvialen Tier un Menschenfunde Dr Kohl-Larsen's im Nordöstlichen Teil des Njarasa-Grabens (Ostafrika). *Geologische Rundshau* 27: 401–41.

Köhler, W. (1925) *The Mentality of Apes.* Kegan Paul, Trench, Trubner & Co, London.

Kokkoros, P., and Kanellis, A. (1960) Découverte d'un crâne d'homme paléolithique dans la péninsule Chalchidique. *L'Anthropologie* 64: 438–46.

Koller, J., Baumer, U., and Mania, D. (2001) Pitch in the Palaeolithic - Investigations of the Middle Palaeolithic "resin remains" from Königsaue. In: *Frühe Menschen in Mitteleuropa-Chronologie, Kultur & Umwelt*, Wagner, G.A., and Mania, D., eds, pp. 99–112. Shaker, Aachen.

Kondo, M., Matsu'ura, S., and Aziz, F. (1993) Probable chronological position of a mandible specimen of Java Man found from Sangiran 1986. *Journal of Anthropological Sciences* 101: 23.

Kondo, T., Zakany, J., Innis, J.W., and Duboule, D. (1997) Of fingers, toes and penises. *Nature* 390: 29.

Kondo, Y., Mazima, N., and Nojiri-ko Research Group (2001) Palaeoloxodon naumanni and its environment at the Paleolithic site of Lake Nojiri, Nagano Prefecture, Central Japan. In: *The World of Elephants: Proceedings of the 1st International Congress*, Cavaretta, G., Gioia, P., Mussi, M., and Palombo, M.R., eds, pp. 284–8. Consiglio Nazionale delle Ricerche, Rome.

Kondo, Y., Mazima, N., and Fossil Mammal Research Group for Nojiri-ko Excavation (2007) Vertebrate fossils and paleoenvironment from the Nojiri-ko Formation. *Jubilee Publication Commemorating Prof. Kamei's 80th Birthday* 117–26.

Konigsberg, L.W., Hens, S.M., Jantz, L.M., and Jungers, W.L. (1998) Stature estimation and calibration: Bayesian and maximum likelihood perspectives in physical anthropology. *Yearbook of Physical Anthropology* 41: 65–92.

Kono, R. (2004) Molar enamel thickness and distribution patterns in extant great apes and humans: new insights based on a 3-dimensional whole crown perspective. *Anthropological Science* 112(2): 121–46.

Konopka, G., Bomar, J.M., Winden, K., Coppola, G., Jonsson, Z.O., Gao, F., et al. (2009) Human-specific transcriptional regulation of CNS development genes by FOXP2. *Nature* **462**: 213–17.

Köppel, R. (1935) Das Alter der neuentdeckten Schädel von Nazareth. *Biblica* **16**: 58–73.

Korenhof, C.A.W. (1960) *Morphogenetical Aspects of the Human Upper Molar.* Uitgeversmaatschappij Neerlandia, Utrecht.

Kornell, N., Son, L.K., and Terrace, H.S. (2007) Transfer of metacognitive skills and hint seeking in monkeys. *Psychological Science* **18**: 64–71.

Kovarovic, K., and Andrews, P. (2007) Bovid postcranial ecomorphological survey of the Laetoli paleoenvironment. *Journal of Human Evolution* **52**(6): 663–80.

Kozlowski, J., and Otte, M. (2000) La formation de l'Aurignacien en Europe. *L'Anthropologie* **104**(1): 3–15.

Kraatz, R. (1992) La mandibule de Mauer, *Homo erectus heidelbergensis*. In: *Cinq Millions d'Années, l'Aventure Humaine*, Toussaint, M., ed., pp. 95–109. Université de Liège, Liège.

Krachun, C., Carpenter, M., Call, J., and Tomasello, M. (2009) A competitive nonverbal false belief task for children and apes. *Developmental Science* **12**(4): 521–35.

Kramer, A. (1986) Hominid-pongid distinctiveness in the Miocene-Pliocene fossil record: the Lothagam mandible. *American Journal of Physical Anthropology* **70**: 457–73.

Kramer, A. (1989) *The Evolutionary and Taxonomic Affinities of the Sangiran Mandibles of Central Java, Indonesia.* PhD thesis, University of Michigan.

Kramer, A. (1994) A critical analysis of claims for the existence of Southeast Asian australopithecines. *Journal of Human Evolution* **26**: 3–21.

Kramer, A., Donnelly, S.M., Kidder, J.H., Ousley, S.D., and Olah, S.M. (1995) Craniometric variation in large-bodied hominoids: testing the single-species hypothesis for *Homo habilis*. *Journal of Human Evolution* **29**(5): 443–62.

Kramer, A., Elam, J.M., Djubiantono, T., Aziz, F., and Hames, W.E. (2000) 1999 excavations in the Rancah District, West Java, Indonesia. *American Journal of Physical Anthropology* **30**(S1): 201.

Kramer, A., Djubiantono, T., Aziz, F., Bogard, J.S., Weeks, R.A., Weinand, D.C., et al. (2005) The first hominid fossil recovered from West Java, Indonesia. *Journal of Human Evolution* **48**(6): 661–7.

Kramer, P.A., and Eck, G.G. (2000) Locomotor energetics and leg length in hominid bipedality. *Journal of Human Evolution* **38**: 651–66.

Kraus, B., and Jordan, R. (1965) *The Human Dentition Before Birth.* Lea & Febiger, Philadelphia, PA.

Kraus, B.S., Jordon, R.E., and Abrams, L. (1969) *A Study of the Masticatory System Dental Anatomy and Occlusion.* Williams and Wilkins, Baltimore, MD.

Krause, J., Lalueza-Fox, C., Orlando, L., Enard, W., Green, R.E., Burbano, H.A., et al. (2007a) The derived *FOXP2* variant of modern humans was shared with Neandertals. *Current Biology* **17**(21): 1908–12.

Krause, J., Orlando, L., Serre, D., Viola, B., Prüfert, K., Richards, M., et al. (2007b) Neanderthals in Central Asia and Siberia. *Nature* **449**: 902–4.

Krause, J., Fu, Q., Good, J.M., Viola, B., Shunkov, M., Derevianko, A.P., et al. (2010) The complete mitochondrial DNA genome of an unknown hominin from southern Siberia. *Nature* **464**: 894–7.

Krentz, H.B. (1993) *The Forelimb Anatomy of* Theropithecus brumpti *and* Theropithecus oswaldi *from the Shungura Formation, Ethiopia.* Thesis, University of Washington.

Kretzoi, M. (1984) Note on *Homo leakeyi* Heberer. *Anthropologiai Közlemenyek* **28**: 189–90.

Kretzoi, M., and Vértes, L. (1965) Lower Paleolithic hominid and pebble industry in Hungary. *Nature* **206**: 205.

Krigbaum, J., and Datan, I. (1999) The Deep Skull of Niah: Borneo's first people. *Borneo* **5**: 13–17.

Krings, M., Stone, A., Schmitz, R.W., Krainitzki, H., Stoneking, M., and Pääbo, S. (1997) Neandertal DNA sequences and the origin of modern humans. *Cell* **90**(1): 19–30.

Krings, M., Geisert, H., Schmitz, R.W., Krainitzki, H., and Pääbo, S. (1999) DNA sequence of the mitochondrial hypervariable region II from the Neandertal type specimen. *Proceedings of the National Academy of Sciences USA* **96**(10): 5581–85.

Krings, M., Capelli, C., Tschentscher, F., Geisert, H., Meyer, S., von Haeseler, A., et al. (2000) A view of Neandertal genetic diversity. *Nature Genetics* **26**: 144–6.

Kroll, E.M., and Isaac, G.Ll. (1984) Configurations of artifacts and bones at early Pleistocene sites in East Africa. In: *Intrasite Spatial Analysis in Archaeology*, Hietala, H.J., ed., pp. 4–31. Cambridge University Press, Cambridge.

Krovitz, G.E. (2003) Shape and growth differences between Neandertals and modern humans: grounds for a species level distinction? In: *Patterns of Growth and Development in Genus* Homo, Thompson, J.L., Krovitz, G.E., and Nelson, A.J., eds, pp. 320–42. Cambridge University Press, Cambridge.

Kuhn, S.L., and Elston, R.G. (2002) Introduction: thinking small globally. *Archeological Papers of the American Anthropological Association* **12**: 1–7.

Kuhn, S.L., Stiner, M.C., Reese, D.S., and Güleç, E. (2001) Ornaments of the earliest Upper Paleolithic: new insights from the Levant. *Proceedings of the National Academy of Sciences USA* **98**(13): 7641–46.

Kukla, G.J. (1987) Loess Stratigraphy in Central China. *Quaternary Science Reviews* **6**: 191–219.

Kullmer, O. (2008) The Fossil Suidae from the Plio-Pleistocene Chiwondo Beds of Northern Malawi, Africa. *Journal of Vertebrate Paleontology* **28**(1): 208–16.

Kullmer, O., Sandrock, O., Abel, R., Schrenk, F., Bromage, T.G., and Juwayeyi, Y.M. (1999) The first

Paranthropus from the Malawi Rift. *Journal of Human Evolution* 37: 121–7.

Kullmer, O., Sandrock, O., Viola, T., Hujer, W., Said, H., and Seidler, H. (2008) Suids, elephantoids, paleochronology, and paleoecology of the Pliocene hominid site Galili, Somali Region, Ethiopia. *Palaios* 23(7): 452.

Kuman, K., and Clarke, R.J. (2000) Stratigraphy, artefact industries and hominid associations for Sterkfontein, Member 5. *Journal of Human Evolution* 38(6): 827–47.

Kuman, K., and Field, A.S. (2009) The Oldowan Industry from Sterkfontein Caves, South Africa. In: *The Cutting Edge: New Approaches to the Archaeology of Human Origins*, Schick, K., and Toth, T., eds, pp. 151–70. Stone Age Institute Press, Gosport, IN.

Kuman, K., Field, A.S., and Thackeray, J.F. (1997) Discovery of New Artifacts at Kromdraai. *South African Journal of Science* 93(4): 187–93.

Kumar, S., and Hedges, S.B. (1998) A molecular timescale for vertebrate evolution. *Nature* 392: 917–20.

Kumar, S., Filipski, A., Swarna, V., Walker, A., and Hedges, S.B. (2005) Placing confidence limits on the molecular age of the human–chimpanzee divergence. *Proceedings of the National Academy of Sciences USA* 102(52): 18842–47.

Kunter, M., and Wahl, J. (1992) Das Femurfragment eines Neandertalers aus der Stadelhöhle des Hohlensteins im Lonetal. *Fundberichte aus Baden-Württemberg* 17(1): 111–24.

Kurtén, B., and Poulianos, A.N. (1977) New stratigraphic and faunal material from Petralona Cave, with special reference to the Carnivora. *Anthropos* 4: 47–130.

Kuykendall, K.L. (2009) Reconstructing australopithecine growth and development: What do we think we know? In: *Patterns of Growth and Development in the Genus* Homo, Nelson, A.J., and Thompson, J.L., eds, pp. 191–218. Cambridge University Press, Cambridge.

Kuzmin, Y.V. (2004) Origin of the Upper Paleolithic in Siberia: A geoarchaeological perspective. In: *The early Upper Paleolithic beyond Western Europe*, Brantigham, J., Kuhn, S., and Kerry, K., eds, pp. 196–206. University of California Press, Berkeley, CA.

Lacaille, A., and Cave, A. (1947) Châtelperron: a new survey of its Palaeolithic industry. *Archaeologia* 92: 95–119.

Lacruz, R. (2007) Enamel microstructure of the hominid KB 5223 from Kromdraai, South Africa. *American Journal of Physical Anthropology* 132(2): 175–82.

Lacruz, R.S., Brink, J.S., Hancox, P.J., Skinner, A.R., Herries, A., Schmid, P., et al. (2002) Palaeontology and geological context of a Middle Pleistocene faunal assemblage from the Gladysvale Cave, South Africa. *Palaeontologia Africana* 38: 99–114.

Lacruz, R., Ungar, P.S., Hancox, P.J., Brink, J.S., and Berger, L.R. (2003) Gladysvale: fossils, strata, and GIS analysis. *South African Journal of Science* 99(5–6): 283–5.

Lacruz, R.S., Dean, M.C., Ramirez-Rozzi, F., and Bromage, T.G. (2008) Megadontia, striae periodicity and patterns of enamel secretion in Plio-Pleistocene fossil hominins. *Journal of Anatomy* 213(2): 148–58.

Ladefoged, P., and Maddieson, I. (1996) *The Sounds of the World's Languages*. Blackwell Publishing, Oxford.

Laden, G., and Wrangham, R. (2005) The rise of hominids as an adaptive shift in fallback foods: plant underground storage organs (USOs) and australopith origins. *Journal of Human Evolution* 49: 482–98.

Lafille, J. (1961) Le gisement dit "Roc de Marsal", commune de de Campagne du Bugue (Dordogne): Note préliminaire. *Bulletin de la Société Préhistorique Française* 58(11–12): 712–13.

Lague, M.R., and Jungers, W.L. (1996) Morphometric variation in Plio-Pleistocene hominid distal humeri. *American Journal of Physical Anthropology* 101(3): 401–27.

Lague, M.R., and Jungers, W.L. (1999) Patterns of sexual dimorphism in the hominoid distal humerus. *Journal of Human Evolution* 36(4): 379–99.

Lahr, M. (2010) Saharan corridors and their role in the evolutionary geography of "Out of Africa I". In: *Out of Africa I*, Fleagle, J.G., Shea, J.J., Grine, F.E., Baden, A.L., and Leakey, R.E., eds, pp. 27–46. Springer, Dordrecht.

Lahr, M.M. (1996) *The Evolution of Modern Human Diversity: A Study of Cranial Variation*. Cambridge University Press, Cambridge.

Lahr, M.M., and Foley, R. (1994) Multiple dispersal and modern human origins. *Evolutionary Anthropology* 3(2): 48–60.

Lai, C.S.L., Fisher, S.E., Hurst, J.A., Vargha-Khadem, F., and Monaco, A.P. (2001) A forkhead-domain gene is mutated in a severe speech and language disorder. *Nature* 413: 519–23.

Laidler, P.W. (1933) Dating evidence concerning the Middle Stone Age and a Capsio-Wilton culture, in the South-East Cape. *South African Journal of Science* 30: 530–42.

Laland, K.N. (2008) Animal cultures. *Current Biology* 18: R366–70.

Laland, K.N., and Janik, V.M. (2006) The animal cultures debate. *Trends in Ecology and Evolution* 21: 542–7.

Laland, K.N., and Galef Jr, B.G. (2009) *The Question of Animal Culture*. Harvard University Press, Cambridge, MA.

Lam, Y., Pearson, O., Marean, C., and Chen, X. (2003) Bone density studies in zooarchaeology. *Journal of Archaeological Science* 30(12): 1701–8.

Lam, Y.M., and Pearson, O.M. (2005) Bone density studies and the interpretation of the faunal record. *Evolutionary Anthropology* 14(3): 99–108.

Lam, Y.M., Pearson, O.M., and Smith, C.M. (1996) Chin morphology and sexual dimorphism in the fossil hominid mandible sample from Klasies River Mouth. *American Journal of Physical Anthropology* 100: 545–57.

Lambert, J.E., Chapman, C.A., Wrangham, R.W., and Conklin-Brittain, N.L. (2004) Hardness of cercopithecine foods: Implications for the critical function of enamel thickness in exploiting fallback foods. *American Journal of Physical Anthropology* **125**(4): 363–8.

Lang, H. (1929) The discovery of the 'Springbok' Man, III. *London News*: 427–8.

Lankester, E.R. (1870) On the use of the term homology in modern zoology. *Annals and Magazine of Natural History* **6**(4): 34–43.

Lapicque, L. (1898) Sur la relation du poids de l'encephale au poids du corps. *Comptes Rendus des Seances de la Societe de Biologie et de ses Filiales* **50**: 62–63.

Larick, R., Ciochon, R.L., Zaim, Y., Sudijono, Suminto Rizal, Y., Aziz, F., et al. (2001) Early Pleistocene Ar/Ar ages for Bapang Formation hominids, Central Java, Indonesia. *Proceedings of the National Academy of Sciences USA* **98**: 4866–71.

Larson, S. (1995) New characters for the functional interpretation of primate scapulae and proximal humeri. *American Journal of Physical Anthropology* **98**(1): 13–35.

Larson, S. (2007) Evolutionary transformation of the hominin shoulder. *Evolutionary Anthropology: Issues, News, and Reviews* **16**(5): 172–87.

Larson, S.G., Jungers, W.L., Morwood, M.J., Sutikna, T., Jatmiko, Saptomo, E.W., et al. (2007) *Homo floresiensis* and the evolution of the human shoulder. *Journal of Human Evolution* **53**(6): 718–31.

Larson, S.G., Jungers, W.L., Tocheri, M.W., Orr, C.M., Morwood, M.J., Sutikna, T., et al. (2009) Descriptions of the upper limb skeleton of *Homo floresiensis*. *Journal of Human Evolution* **57**(5): 555–70.

Lartet, E. (1861) Nouvelles recherches sur la coexistence de l'homme et des grands mammifères fossiles réputés caractéristiques de la dernière période géologique. *Annales des Sciences Naturelles* **15**: 177–261.

Lartet, E., and Christy, H. (1864) Cavernes du Périgord. Objets gravés et sculptés des temps pré-historiques. *Revue archéol.* **9**: 253.

Lartet, L. (1868) Une sépultre des Troglodytes du Périgord (crânes des Eyzies). *Bulletins de la Société d'Anthropologie de Paris* **3**: 335–49.

Lartet, L. (1872) Essai sur la géologie de la Palestine et des contrées avoisinantes telles que l'Egypte et l'Arabie. 2. Paléontologie. *Annales des sciences géologiques* **3**: 98p.

Lartet, L., and Duparc, M. (1874) Une sépulture des anciens Troglodytes des Pyrénées. *Matériaux pour l'histoire primitive et naturelle de l'homme, tenth year, second series, V, Paris*: 139.

Laskar, J. (1989) A numerical experiment on the chaotic behaviour of the solar system. *Nature* **338**: 237–8.

Laskar, J., Robutal, P., Joutel, F., Gastineau, M., Correia, A., and Levrard, B. (2004) A long term numerical solution for the insolation quantities of the Earth. *Astronomy and Astrophysics* **428**: 261–85.

Latham, A.G., and Herries, A.I.R. (2004) The formation of sedimentary infilling of the Cave of Hearths and Historic Cave complex, Makapansgat, South Africa. *Geoarchaeology* **19**(4): 323–42.

Latimer, B., Lovejoy, C.O., Johanson, D.C., and Coppens, Y. (1982) Hominid tarsal, metatarsal, and phalangeal bones recovered from the Hadar Formation: 1974–1977 Collections. *American Journal of Physical Anthropology* **57**: 701–19.

Laury, R.L., and Albritton, C.C. (1975) Geology and Middle Stone Age Archaeological Sites in the Main Ethiopian Rift Valley. *Geological Society of America Bulletin* **86**(7): 999–1011.

Lawson Handley, L.J., Manica, A., Goudet, J., and Balloux, F. (2007) Going the distance: human population genetics in a clinal world. *Trends in Genetics* **23**(9): 432–9.

Lax, E. (1995) Quaternary faunal remains from the cave site of Apidima (Laconia, Greece). *Acta Anthropologica* **1**: 127–56.

Leakey, L. (1931) *The Stone Age Cultures of Kenya Colony*. Cambridge University Press, Cambridge.

Leakey, L. (1936) *Stone Age Africa*. Oxford University Press, London.

Leakey, L. (1961) *The Progress of Man in Africa*. Oxford University Press, London.

Leakey, L. (1963) East African fossil Hominoidea and the classification within this super-family. In: *Classification and Human Evolution*, Washburn, S., ed., pp. 32–49. Aldine De Gruyter.

Leakey, L. (1965) *Olduvai Gorge 1951–61. Vol. 1*. Cambridge University Press, London.

Leakey, L.S.B. (1934) *Adams's Ancestors. An Up-to-date Outline of What is Known about the Origin of Man*. Methuen & Co., London.

Leakey, L.S.B. (1935) *The Stone Age Races of Kenya*. Oxford University Press, London.

Leakey, L.S.B. (1936a) A new fossil skull from Eyasi, East Africa: Discovery by a German Expedition. *Nature* **138**(3504): 1082–84.

Leakey, L.S.B. (1936b) Fossil human remains from Kanam and Kanjera, Kenya Colony. *Nature* **138**: 643.

Leakey, L.S.B. (1937) *White African*. Hodder & Stoughton, London.

Leakey, L.S.B., ed. (1951) *Olduvai Gorge: a Report on the Evolution of the Hand-Axe Culture in Beds I–IV*. Cambridge University Press, Cambridge.

Leakey, L.S.B. (1959) A new fossil skull from Olduvai. *Nature* **184**: 491–3.

Leakey, L.S.B. (1960) Recent discoveries at Olduvai Gorge. *Nature* **188**: 1050–52.

Leakey, L.S.B. (1961) New finds at Olduvai Gorge. *Nature* **189**: 649–50.

Leakey, L.S.B. (1961) The juvenile mandible from Olduvai. *Nature* **191**: 417–18.

Leakey, L.S.B. (1965) *Olduvai Gorge, 1951–1961*. Cambridge University Press, London.

Leakey, L.S.B., ed. (1969) *Fossil Vertebrates of Africa II*. Academic Press, London.

Leakey, L.S.B., and Leakey, M.D. (1964) Recent discoveries of fossil hominids in Tanganyika, at Olduvai and near Lake Natron. *Nature* **202**: 5–7.

Leakey, L.S.B., Evernden, J.F., and Curtis, G.H. (1961) Age of Bed I, Olduvai Gorge, Tanganyika. *Nature* **191**: 478–9.

Leakey, L.S.B., Tobias, P.V., and Napier, J.R. (1964) A new species of the genus *Homo* from Olduvai Gorge. *Nature* **202**: 7–9.

Leakey, M. (1984) *Disclosing the Past*. Doubleday, New York.

Leakey, M., Tobias, P.V., Martyn, J.E., and Leakey, R.E.F. (1969) An Acheulian industry with prepared core technique and the discovery of a contemporary hominid manidible at lake Baringo, Kenya. *Proceedings of the Prehistoric Society* **3**: 48–76.

Leakey, M.D. (1966a) A review of the Oldowan Culture from Olduvai Gorge, Tanzania. *Nature* **210**: 462–6.

Leakey, M.D. (1966b) Primitive artefacts from Kanapoi Valley. *Nature* **212**: 579–81.

Leakey, M.D. (1969) Recent discoveries of hominid remains at Olduvai Gorge, Tanzania. *Nature* **223**: 756.

Leakey, M.D. (1971) Discovery of postcranial remains of *Homo erectus* and associated artefacts in Bed IV at Olduvai Gorge, Tanzania. *Nature* **232**: 380–3.

Leakey, M.D. (1971) *Olduvai Gorge Volume 3: Excavations in Beds I and II, 1960–1963*. Cambridge University Press, Cambridge.

Leakey, M.D. (1987) The Laetoli hominid remains. In: *Laetoli: a Pliocene Site in Northern Tanzania*, Leakey, M.D., and Harris, J.M., eds, pp. 108–17. Clarendon Press, Oxford.

Leakey, M.D. (1994) *Olduvai Gorge Vol 5: Excavations in Beds III, IV and the Masek beds, 1968–1971*. Cambridge University Press, London.

Leakey, M.D. (2003) Introduction. In: *Lothagam: The Dawn of Humanity in Eastern Africa*, Leakey, M.D., and Harris, J.M., eds, pp. 1–2. Columbia University Press, New York.

Leakey, M.D., and Hay, R.L. (1979) Pliocene footprints in the Laetolil Beds at Laetoli, northern Tanzania. *Nature* **278**: 317–23.

Leakey, M.D., Clarke, R.J., and Leakey, L.S.B. (1971) New hominid skull from Bed 1, Olduvai Gorge, Tanzania. *Nature* **232**: 308–12.

Leakey, M.D., Hay, R.L., Curtis, G.H., Drake, R.E., Jackes, M.K., and White, T.D. (1976) Fossil hominids from the Laetoli Beds. *Nature* **262**: 460–6.

Leakey, M.G. (1993) Evolution of *Theropithecus* in the Turkana Basin. In: *Theropithecus. The Rise and Fall of a Primate Genus*, Jablonski, N.G., ed., pp. 85–1213. Cambridge University Press, Cambridge.

Leakey, M.G., and Leakey, R.E. (1978) *Koobi Fora Research Project. 1. The Fossil Hominids and an Introduction to their Context*. Clarendon Press, New York.

Leakey, M.G., and Harris, J.M., eds (2003) *Lothagam: The Dawn of Humanity in Eastern Africa*. Columbia University Press, New York.

Leakey, M.G., and Walker, A.C. (2003) The Lothagam Hominids. In: *Lothagam: The Dawn of Humanity in Eastern Africa*, Leakey, M.G., and Harris, J.M., eds, pp. 249–57. Columbia University Press, New York.

Leakey, M.G., Feibel, C.S., McDougall, I., and Walker, A. (1995) New four-million-year-old hominid species from Kanapoi and Allia Bay, Kenya. *Nature* **376**: 565–71.

Leakey, M.G., Feibel, C.S., McDougall, I., Ward, C., and Walker, A. (1998) New specimens and confirmation of an early age for *Australopithecus anamensis*. *Nature* **393**: 62–66.

Leakey, M.G., Spoor, F., Brown, F.H., Gathogo, P.N., Kiarie, C., Leakey, L.N., et al. (2001) New hominin genus from eastern Africa shows diverse middle Pliocene lineages. *Nature* **410**: 433–40.

Leakey, M.G., Spoor, F., Brown, F.H., Gathogo, P.N., and Leakey, L.N. (2003) A new hominin calvaria from Ileret (Kenya). *American Journal of Physical Anthropology* Suppl. **36**: 136.

Leakey, R. (1981) *The Making of Mankind*. Sphere Books, London.

Leakey, R., and Wood, B. (1974) A hominid mandible from East Rudolf, Kenya. *American Journal of Physical Anthropology* **41**(2): 245–9.

Leakey, R., and Lewin, R. (1977) *Origins*. Dutton, New York.

Leakey, R., and Lewin, R. (1978) *People of the Lake: Mankind and its Beginnings*. Anchor Press, Garden City, NJ.

Leakey, R., and Morell, V. (2001) *Wildlife Wars: My Battle to save Kenya's Elephants*. MacMillan, London.

Leakey, R.E. (1982) *Human Origins*. Penguin Putnam, New York.

Leakey, R.E.F. (1969) Faunal remains from the Omo Valley. *Nature* **222**: 1132–3.

Leakey, R.E.F. (1970) New hominid remains and early artefacts from North Kenya. *Nature* **226**: 223–4.

Leakey, R.E.F. (1971) Further evidence of Lower Pleistocene hominids from East Rudolf, North Kenya. *Nature* **231**: 241–5.

Leakey, R.E.F. (1972) Further evidence of Lower Pleistocene hominids from East Rudolf, North Kenya '71. *Nature* **237**: 264–9.

Leakey, R.E.F. (1973a) Further evidence of Lower Pleistocene hominids from East Rudolf, North Kenya 1972. *Nature* **242**: 170–3.

Leakey, R.E.F. (1973b) Evidence for an advanced Plio-Pleistocene hominid from East Rudolf, Kenya. *Nature* **242**: 447–50.

Leakey, R.E.F. (1973c) Australopithecines and hominines: a summary of the evidence from the early Pleistocene of eastern Africa. *Symposium of the Royal Zoological Society of London* **33**: 53–69.

Leakey, R.E.F. (1974) Further evidence of Lower Pleistocene hominids from East Rudolf, North Kenya 1973. *Nature* **248**: 653–6.

Leakey, R.E.F. (1976a) An overview of the hominidae from East Rudolf, Kenya. In: *Earliest Man and Environments in the Lake Rudolf Basin: Stratigraphy, Paleoecology and Evolution*, Coppens, Y., Howell, F.C., Isaac, G.Ll., and Leakey, R.E.F., eds, pp. 476–89. University of Chicago Press, Chicago, IL.

Leakey, R.E.F. (1976b) New hominid fossils from the Koobi Fora Formation in North Kenya. *Nature* **261**: 574–6.

Leakey, R.E.F., and Walker, A.C. (1973) New australopithecines from East Rudolf, Kenya (III). *American Journal of Physical Anthropology* **39**: 205–22.

Leakey, R.E.F., and Wood, B.A. (1973) New evidence of the genus *Homo* from East Rudolf, Kenya. II. *American Journal of Physical Anthropology* **39**(3): 355–68.

Leakey, R.E.F., and Walker, A.C. (1976) *Australopithecus*, *Homo erectus* and the single species hypothesis. *Nature* **261**: 572–4.

Leakey, R.E.F., and Walker, A.C. (1985) Further hominids from the Plio-Pleistocene of Koobi Fora, Kenya. *American Journal of Physical Anthropology* **67**: 135–63.

Leakey, R.E.F., and Walker, A. (1988) New *Australopithecus boisei* specimens from East and West Lake Turkana, Kenya. *American Journal of Physical Anthropology* **76**(1): 1–24.

Leakey, R.E.F., and Walker, A. (1989) Early *Homo erectus* from West Lake Turkana, Kenya. In: *Hominidae: Proceedings of the 2nd International Congress of Human Paleontology, Turin, September 28-October 3, 1987*, Giacobini, G., ed., pp. 209–15. Jaca Book, Milan.

Leakey, R.E.F., and Lewin, R. (1993) *Origins Reconsidered*. Anchor, New York.

Leakey, R.E.F., Butzer, K.W., and Day, M.H. (1969) Early *Homo sapiens* remains from the Omo River region of Southwest Ethiopia. *Nature* **222**: 1132–8.

Leakey, R.E.F., Mungai, J.M., and Walker, A.C. (1971) New australopithecines from East Rudolf, Kenya. *American Journal of Physical Anthropology* **35**: 175–86.

Leakey, R.E.F., Mungai, J.M., and Walker, A.C. (1972) New australopithecines from East Rudolf, Kenya (II). *American Journal of Physical Anthropology* **36**: 235–52.

Leakey, R.E.F., Leakey, M.G., and Behrensmeyer, A.K. (1978) The hominid catalogue. In: *Koobi Fora Research Project, vol. 1: The fossil hominids and an introduction to their context 1968–1974*, Leakey, M.G., and Leakey, R.E.F., eds, pp. 86–182. Clarendon Press, Oxford.

Leakey, R.E.F., Walker, A., Ward, C.V., and Grausz, H.M. (1989) A partial skeleton of a gracile Hominid from the Upper Burgi Member of the Koobi Fora Formation, East Late Turkana, Kenya. In: *Hominidae: Proceedings of the 2nd International Congress of Human Paleontology, Turin, September 28-October 3, 1987*, Giacobini, G., ed., pp. 167–74. Jaca Book, Milan.

Le Bas, M.J., Le Maitre, R.W., Streckeisen, A., and Zanettin, B. (1986) A chemical classification of volcanic rocks based on the total alkali-silica diagram. *Journal of Petrology* **27**(3): 745–50.

Lebetard, A.-E., Bourlès, D.L., Duringer, P., Jolivet, M., Braucher, R., Carcaillet, J., et al. (2008) Cosmogenic nuclide dating of *Sahelanthropus tchadensis* and *Australopithecus bahrelghazali*: Mio-Pliocene hominids from Chad *Proceedings of the National Academy of Sciences USA* **105**(9): 3226–31.

Lee, R.B., and DeVore, I., eds (1968) *Man the Hunter*. Wenner-Gren Foundation for Anthropological Research, New York.

Lee, Y.J. (1983) *Paleolithic research of Chongwon Turubong Cave No. 2 site*. PhD thesis, Yonsei University.

Lee, Y.J. (1984) *Early Man in Korea (II)*. Tamgu-Dang Publishing, Seoul.

Lee, Y.J. (1994) Paleontological and archeological remains from Turubong Cave Complex in Korea. In: *Paleolithic Culture of East Asia*, Bae, K.D., ed., pp. 91–130. National Research Institute of Cultural Properties, Seoul.

Le Gros Clark, W.E. (1936) Evolutionary parallelism and human phylogeny. *Man* **36**(2): 4–8.

Le Gros Clark, W.E. (1938) General features of the Swanscombe skull bones. *Journal of the Royal Anthropological Institute* **68**: 58–60.

Le Gros Clark, W.E. (1940) The relationship between *Pithecanthropus* and *Sinanthropus*. *Nature* **145**: 70–71.

Le Gros Clark, W.E. (1950) South African fossil hominoids. *Nature* **166**: 791–2.

Le Gros Clark, W.E. (1950) New palaeontological evidence bearing on the evolution of the Hominoidea. *Quarterly Journal of the Geological Society* **105**: 225–64.

Le Gros Clark, W.E. (1955) *The Fossil Evidence for Human Evolution: An Introduction to the Study of Paleoanthropology*. University of Chicago Press, Chicago, IL.

Le Gros Clark, W.E. (1959) *The Antecedents of Man*. University of Edinburgh Press, Edinburgh.

Le Gros Clark, W.E. (1964) *The Fossil Evidence for Human Evolution: an Introduction to the Study of Palaeoanthropology*. 2nd. Ed. University of Chicago Press, Chicago, IL.

Le Gros Clark, W.E. (1964) The evolution of man. *Discovery* **27**: 49.

Le Gros Clark, W.E. (1967) *Man-Apes or Ape-Men? The Story of Discoveries in Africa*. Holt, Rinehart and Winston, New York.

Le Gros Clark, W.E. (1968) *Chant of Pleasant Exploration*. E. & S. Livingstone, Edinburgh.

Le Gros Clark, W.E. (1978) *The Fossil Evidence for Human Evolution*. 3rd. Ed. University of Chicago Press, Chicago, IL.

Leinders, J.J.M., Aziz, F., Sondaar, P.Y., and de Vos, J. (1985) The age of the hominid bearing deposits of Java: state of the art. *Geologie en Mijnbouw* **64**: 167–73.

Lele, S., and Richtsmeier, J.T. (1991) Euclidean distance matrix analysis: a coordinate-free approach for comparing biological shapes using landmark data. *American Journal of Physical Anthropology* **86**: 415–27.

Lele, S., and Richtsmeier, J.T. (2001) *An Invariant Approach to Statistical Analysis of Shapes*. Chapel and Hall, New York.

Le Mort, F. (1989) Traces de décharnement sur les ossements néandertaliens de Combe-Grenal (Dordogne). *Bulletin de la Société préhistorique française* 86(3): 79–87.

Lemozi, A. (1929) La grotte-temple du Pech-Merle. *Un nouveau sanctuaire préhistorique. Picard, Paris* 183.

Lemozi, A., Renault, P., and David, A. (1969) *Pech Merle, Le Combel, Marcenac. Monographien und Dokumentationen*. Die Europäischen Felsbilder, Graz.

Lepersonne, J. (1949) Le Fossé tectonique du Lac Albert-Semliki-Lac Edouard. Résumé des observations géologiques effectués en 1938, 1939, 1940. *Annales du Société Géologique Belge* 72: M3–92.

Lepersonne, J. (1970) Revision of the fauna and the stratigraphy of the fossiliferous localities of the Lake Albert-Lake Edward Rift (Congo). *Annales du Musée Royal de l'Afrique Centrale* 67: 169–207.

Lepersonne, J. (1974) *Carte géologique de Zaire*. Echelle 1/2,000,000, République du Zaire, Département des Mines, Service Géologique.

Lepre, C., Quinn, R., Joordens, J., Swisher III, C., and Feibel, C. (2007) Plio-Pleistocene facies environments from the KBS Member, Koobi Fora Formation: implications for climate controls on the development of lake-margin hominin habitats in the northeast Turkana Basin (northwest Kenya). *Journal of Human Evolution* 53(5): 504–14.

Lerdahl, F., and Jackendoff, R. (1996) *A Generative Theory of Tonal Music*. MIT Press, Boston, MA.

Leroi-Gourhan, A. (1950) La grotte du Loup, Arcy-sur-Cure (Yonne). *Bulletin de la Société Préhistorique Français* 47(5): 268–80.

Leroi-Gourhan, A. (1958) Étude des restes humains fossiles provenant des grottes d'Arcy-sur-Cure (Yonne). *Annales de Paléontologie* 44: 87–148.

Leroi-Gourhan, A. (1959) Résultats de l'analyse pollinique de la grotte d'Isturitz. *Bulletin de la Société Préhistorique Française* 56(9–10): 619–24.

Leroi-Gourhan, A. (1961) Les fouilles d'Arcy-sur-Cure (Yonne). *Gallia Préhistoire* 4: 3–16.

Leroi-Gourhan, A. (1975) The flowers found with Shanidar IV, a Neanderthal burial in Iraq. *Science* 190(4214): 562–4.

Leroi-Gourhan, A. (1988) Le passage Mousterien-Chatelperronien a Arcy-sur-Cure. *Bulletin de la Société Préhistorique Fançaise* 85(4): 102–4.

Leroi-Gourhan, A. (1993) *Gesture and Speech*. MIT Press, Cambridge, MA.

Leroi-Gourhan, A., and Leroi-Gourhan, A. (1964) Chronologie des grottes d'Arcy-sur-Cure (Yonne). *Gallia Prehistoire* 7: 1–64.

Lestrel, P.E. (1982) A Fourier analytical procedure to describe complex morphological shapes. In: *Factors and Mechanisms Influencing Bone Growth*, Dixon, A.D., and Sarnat, B.G., eds, pp. 393–409. Alan R. Liss, New York.

Lev, E., Kislev, M.E., and Bar-Yosef, O. (2005) Mousterian vegetal food in Kebara Cave, Mt. Carmel. *Journal of Archaeological Science* 32(3): 475–84.

Lévêque, F., and Vandermeersch, B. (1980) Découverte de restes humains dans un niveau castelperronien a Saint-Cesaire (Charente-Maritime). *Comptes Rendus de l'Académie des Sciences* 291: 187–9.

Levin, N., Quade, J., Simpson, S.W., Semaw, S., and Rogers, M. (2004) Isotopic evidence for Plio-Pleistocene environmental change at Gona, Ethiopia. *Earth and Planetary Science Letters* 219(1–2): 93–110.

Levin, N.E., Cerling, T.E., Passey, B.H., Harris, J.M., and Ehleringer, J.R. (2006) A stable isotope aridity index for terrestrial environments. *Proceedings of the National Academy of Sciences USA* 103: 11201–5.

Levin, N.E., Simpson, S.W., Quade, J., Cerling, T.E., and Frost, S.R. (2008) Herbivore enamel carbon isotopic composition and the environmental context of *Ardipithecus* at Gona, Ethiopia. *Geological Society of America Special Paper* 446: 215–34.

Lewis, A., and Garn, S. (1960) The relationship between tooth formation and other maturational factors. *Angle Orthod* 30(1): 70–7.

Lewis, O.J. (1989) *Functional Morphology of the Evolving Hand and Foot*. Clarendon Press, Oxford.

Li, P., Qian, F., Ma, X., Pu, Q., Xing, L., and Ju, S. (1976) Preliminary study on the age of the Yuanmou man by paleomagnetic technique. *Sci. China* 6: 579–91.

Li, T., and Etler, D.A. (1992) New Middle Pleistocene hominid crania from Yunxian in China. *Nature* 357: 404–7.

Li, T.-Y., and Feng, X.B. (2001) *The Yunxian Man*. Scientific and Technological Publisher of Hubei, Wuhan.

Li, W.-H., and Sadler, L.A. (1991) Low nucleotide diversity in man. *Genetics* 129(2): 513–23.

Li, W.-H., and Saunders, M.A. (2005) The chimpanzee and us. *Nature* 437: 50–1.

Li, Y., Ji, H., Li, T., Feng, X., and Li, W. (1998) The stone artefacts from the Yunxian man site. *Acta Anthropologica Sinica* 17: 94–120.

Lie, H.C., Simmons, L.W., and Rhodesa, G. (2010) Genetic dissimilarity, genetic diversity, and mate preferences in humans. *Evolution and Human Behavior* 31(1): 48–58.

Lieberman, D. (1994) The biological basis for seasonal increments in dental cementum and their application to archaeological research. *Journal of Archaeological Science* 21(4): 525–39.

Lieberman, D.E. (1998) Sphenoid shortening and the evolution of modern human cranial shape. *Nature* 393: 158–62.

Lieberman, D.E., and McCarthy, R.C. (1999) The ontogeny of cranial base angulation in humans and chimpanzees and its implications for reconstructing pharyngeal dimensions. *Journal of Human Evolution* 36(5): 487–517.

Lieberman, D.E., McBratney, B.M., and Krovitz, G. (2002) The evolution and development of cranial form in *Homo sapiens*. *Proceedings of the National Academy of Sciences USA* **99**: 1134–9.

Lieberman, D.E., Pearson, O.M., Polk, J.D., Demes, B., and Crompton, A.W. (2003) Optimization of bone growth and remodeling in response to loading in tapered mammalian limbs. *Journal of Experimental Biology* **206**: 3125–38.

Lieberman, D.E., Polk, J.D., and Demes, B. (2004) Predicting long bone loading from cross-sectional geometry. *American Journal of Physical Anthropology* **123**: 156–71.

Lieberman, P. (1993) On the Kebara KMH 2 hyoid and Neanderthal speech. *Current Anthropology* **34**(2): 172–5.

Linnaeus, C. (1758) *Systema Naturae*. Laurentii Salvii, Stockholm.

Linz, B., Balloux, F., Moodley, Y., Manica, A., Liu, H., Roumagnac, P., et al. (2007) An African origin for the intimate association between humans and *Helicobacter pylori*. *Nature* **445**: 915–18.

Liritzis, Y., and Maniatis, Y. (1995) Μελέτες χρονολόγησης σε ασβεστίτες και οστά του Τεταρτογενούς. *Acta Anthropologica* **1**: 65–92.

Lisiecki, L.E., and Raymo, M.E. (2005) A Pliocene-Pleistocene stack of 57 globally distributed benthic δ^{18}O records. *Paleoceanography* **20**: PA1003.

Liszkowski, U., Schäfer, M., Carpenter, M., and Tomasello, M. (2009) Prelinguistic infants, but not chimpanzees, communicate about absent entities. *Psychological Science* **20**(5): 654–60.

Little, E.A., and Schoeninger, M.J. (1995) The late woodland diet on Nantucket Island and the problem of maize in coastal New England. *American Antiquity* **60**: 351–68.

Liu, W., and Si, X.Q. (1997) The human teeth discovered in Dadong, Panxian County, Guizhou Province. *Acta Anthropologica Sinica* **16**: 193–200.

Liu, W., Zhang, Y., and Wu, X. (2005) A Middle Pleistocene human cranium from Tangshan (Nanjing), Southeast China: A new reconstruction and comparison with *Homo erectus* from Eurasia and Africa. *American Journal of Physical Anthropology* **127**(3): 253–62.

Liu, W., Vialet, A., Wu, X., He, J., and Lu, J. (2006) Comparaison de l'expression de certains caractères crâniens sur les hominidés chinois du Pléistocène récent et de l'Holocène (grotte supérieure de Zhoukoudian, sites de Longxian et de Yanqing). *L'Anthropologie* **110**: 258–76.

Liu, W., Jin, C.Z., Zhang, Y.Q., Cai, Y.J., Xing, S., Wu, X.J., et al. (2010) Human remains from Shirendong, South China, and modern human emergence in East Asia. *Proceedings of the National Academy of Sciences USA* **107**(45): 19201–6.

Liu, Z., Trentesaux, A., Clemens, S.C., Colin, C., Wang, P., Huang, B., et al. (2003) Clay mineral assemblages in the northern South China Sea: implications for East Asian monsoon evolution over the past 2 million years. *Marine Geology* **201**(1–3): 133–46.

Lockwood, C.A., and Tobias, P.V. (1999) A large male hominin cranium from Sterkfontein, South Africa and the status of *Australopithecus africanus*. *Journal of Human Evolution* **36**: 637–85.

Lockwood, C.A., and Tobias, P.V. (2002) Morphology and affinities of new hominin cranial remains from Member 4 of the Sterkfontein Formation, Gauteng Province, South Africa. *Journal of Human Evolution* **42**: 389–450.

Lockwood, C.A., Richmond, B.G., Jungers, W.L., and Kimbel, W.H. (1996) Randomization procedures and sexual dimorphism in *Australopithecus afarensis*. *Journal of Human Evolution* **31**(6): 537–48.

Lockwood, C.A., Kimbel, W.H., and Johanson, D.C. (2000) Temporal trends and metric variation in the mandibles and dentition of *Australopithecus afarensis*. *Journal of Human Evolution* **39**(1): 23–55.

Lockwood, C.A., Menter, C.G., Moggi-Cecchi, J., and Keyser, A.W. (2007) Extended male growth in a fossil hominin species. *Science* **318**(5855): 1443–46.

Lombardo, M.P., and Thorpe, P.A. (2008) Digit ratio in green anolis lizards (Anolis carolinensis). *Anatomical Record* **291**: 433–40.

Lorblanchet, M., Cachier, H., and Valladas, H. (1995) Direct date for one of the Pech-Merle spotted horses. *International Newsletter on Rock Art* **12**: 2–3.

Lordkipanidze, D., Vekua, A., Ferring, R., Rightmire, G.P., Agusti, J., Kiladze, G., et al. (2005) The earliest toothless hominin skull. *Nature* **434**: 717–18.

Lordkipanidze, D., Vekua, A., Ferring, R., Rightmire, G.P., Zollikofer, C.P.E., Ponce de León, M.S., et al. (2006) A fourth hominin skull from Dmanisi, Georgia. *Anatomical Record* **288**(11): 1146–57.

Lordkipanidze, D., Jashashvili, T., Vekua, A., Ponce de León, M.S., Zollikofer, C.P.E., Rightmire, G.P., et al. (2007) Postcranial evidence from early *Homo* from Dmanisi, Georgia. *Nature* **449**: 305–10.

Lorenzo, C., Arsuaga, J.L., and Carretero, J.M. (1999) Hand and foot remains from the Gran Dolina early Pleistocene site (Sierra de Atapuerca, Spain). *Journal of Human Evolution* **37**: 501–22.

Lotka, A.J. (1907) Relation between birth rates and death rates. *Science* **26**: 121–30.

Louchart, A. (2008) Fossil birds of the Kibish Formation. *Journal of Human Evolution* **55**(3): 513–20.

Lovejoy, C.O. (1970) The taxonomic status of the *Meganthropus* mandibular fragments from the Djetis beds of Java. *Man* **5**(2): 228–36.

Lovejoy, C.O. (1979) A reconstruction of the pelvis of AL 288 (Hadar Formation, Ethiopia). *American Journal of Physical Anthropology* **50**: 460.

Lovejoy, C.O. (1981) The origin of man. *Science* **211**(4480): 341–50.

Lovejoy, C.O. (1988) Evolution of human walking. *Scientific American* **259**(5): 118–25.

Lovejoy, C.O. (2005) The natural history of human gait and posture: Part 1. Spine and pelvis. *Gait & Posture* **21**(1): 95–112.

Lovejoy, C.O. (2009) Reexamining human origins in light of *Ardipithecus ramidus*. *Science* **326**(5949). 74.

Lovejoy, C.O., Meindl, R.S., Ohman, J.C., Heiple, K.G., and White, T.D. (2002) The Maka femur and its bearing on the antiquity of human walking: applying contemporary concepts of morphogenesis to the human fossil record. *American Journal of Physical Anthropology* **119**(2): 97–133.

Lovejoy, C.O., Simpson, S.W., White, T.D., Asfaw, B., and Suwa, G. (2009a) Careful climbing in the Miocene: the forelimbs of *Ardipithecus ramidus* and humans are primitive. *Science* **326**(5949): 70e1–8.

Lovejoy, C.O., Suwa, G., Spurlock, L., Asfaw, B., and White, T.D. (2009b) The pelvis and femur of *Ardipithecus ramidus*: the emergence of upright walking. *Science* **326**(5949): 71e1–6.

Lovejoy, C.O., Latimer, B., Suwa, G., Asfaw, B., and White, T.D. (2009c) Combining prehension and propulsion: the foot of *Ardipithecus ramidus*. *Science* **326**(5949): 72e1–8.

Lovejoy, C.O., Suwa, G., Simpson, S.W., Matternes, J.H., and White, T.D. (2009d) The great divides: *Ardipithecus ramidus* reveals the postcrania of our last common ancestors with african apes. *Science* **326**(5949): 100–6.

Løvlie, R., Pu, S., Xingzhao, F., Zengjian, Z., and Chun, L. (2001) A revised paleomagnetic age of the Nihewan Group at the Xujiayao Palaeolithic Site, China. *Quaternary Science Reviews* **20**(12): 1341–53.

Lowe, D.J., and Hunt, J.B. (2001) A summary of terminology used in tephra-related studies. In: *Tephras, chronologie, archeologie*, Juvigne, E., and Raynal, J.P., eds, pp. 17–22. CDERAD, Goudet.

Lowenstein, J. (1981) Immunological reactions from fossil material. *Philosophical Transactions of the Royal Society of London. Series B, Biological Sciences* **292**(1057): 143–9.

Lü, Z. (1985) *The Excavation and Significance of the Jinniushan Site*. Peking University Archaeology Department, Beijing.

Lü, Z. (1990) La découverte de l'homme fossile de Jing-Niu-Shan. Première étude. *L'Anthropologie* **94**(4): 899–902.

Lubell, D. (1974) *The Fakhurian, a Late Paleolithic Industry from Upper Egypt*. The Geological Survey of Egypt Paper No. 58, Cairo.

Lubell, D. (2001) Late Pleistocene-Early Holocene Maghreb. In: *Encyclopedia of Prehistory*, Peregrine, P.N., and Ember, M., eds, pp. 129–49. Kluwer Academic/Plenum Publishers, New York.

Lubell, D., Sheppard, P., and Gilman, A. (1992) The Maghreb. In: *Chronologies in Old World Archaeology*, Ehrich, R.W., ed., pp. 257–76. University of Chicago Press, Chicago, IL.

Luboga, S.A., and Wood, B.A. (1990) Position and orientation of the foramen magnum in higher primates. *American Journal of Physical Anthropology* **81**(1): 67–76.

Lucas, P., Constantino, P., Wood, B.A., and Lawn, B. (2008) Dental enamel as a dietary indicator in mammals. *BioEssays* **30**(4): 374–85.

Lucas, P.W. (2004) *Dental Functional Morphology: How Teeth Work*. Cambridge University Press, Cambridge.

Luedtke, B.E. (1992) *An Archaeologist's Guide to Chert and Flint*. University of California, Los Angeles, CA.

Lum, J.K., Rickards, O., Ching, C., and Cann, R.L. (1994) Polynesian mitochondrial DNAs reveal three deep maternal lineage clusters. *Human Biology* **66**(4): 567–90.

Lycett, S.J., and Gowlett, J.A.J. (2008) On questions surrounding the Acheulean 'tradition'. *World Archaeology* **40**(3): 295–315.

Lycett, S.J., Collard, M., and McGrew, W.C. (2007) Phylogenetic analyses of behavior support existence of culture among wild chimpanzees. *Proceedings of the National Academy of Sciences USA* **104**(45): 17588–92.

Lycett, S.J., Collard, M., and McGrew, W.C. (2010) Are behavioral differences among wild chimpanzee communities genetic or cultural? An assessment using tool-use data and phylogenetic methods. *American Journal of Physical Anthropology* **142**: 461–7.

Lyell, C. (1831–3) *The Principles of Geology*. John Murray, London.

Lyman, R.L. (1994) *Vertebrate Taphonomy*. Cambridge University Press, Cambridge.

Maca-Meyer, N., González, N.M., Pestano, J., Flores, C., Larruga, J.M., and Cabrera, V.M. (2003) Mitochondrial DNA transit between West Asia and North Africa inferred from U6 phylogeography. *BMC Genetics* **4**(1): 15.

MacArthur, R.H., and Wilson, E.O. (1967) *The Theory of Island Biogeography*. Princeton University Press, Princeton, NJ.

Macchiarelli, R., Bondioli, L., Chech, M., Coppa, A., Fiore, I., Russom, R., et al. (2004) The early Pliestocene human remains from Buia, Danakil Depression, Eritrea. *Rivista Italiana de Paleontologia e Stratigrafia* **110**: 133–44.

MacCurdy, G. (1915) Neandertal Man in Spain: the Lower Jaw of Bañolas. *American Anthropologist* **17**(4): 759–62.

Macdonald, R., Davies, G.R., Upton, B.G.J., Dunkley, P.N., Smith, M., and Leat, P.T. (1995) Petrogenesis of Silali volcano, Gregory Rift, Kenya. *Journal of the Geological Society* **152**(4): 703–20.

MacGregor, A. (1964) The Le Moustier mandible: An explanation for the deformation of the bone and failure of eruption of a permanent canine tooth. *Man* **64**(4): 151–2.

Macho, G., Shimizu, D., Jiang, Y., and Spears, I. (2005) *Australopithecus anamensis*: a finite element approach to studying the functional adaptations of extinct hominins. *Anatomical Record Part A: Discoveries in Molecular, Cellular, and Evolutionary Biology* **283**(2): 310–18.

REFERENCES

Macho, G.A., Jiang, Y., and Spears, I.R. (2003) Enamel microstructure - a truly three-dimensional structure. *Journal of Human Evolution* **45**: 81–90.

Macintosh, N.W.G. (1952) The Cohuna cranium: teeth and palate. *Oceania* **23**(2): 95–105.

Macintosh, N.W.G. (1965) The physical aspect of man in Australia. In: *Aboriginal Man in Australia*, Berndt, R.M., ed., pp. 51–2. Sydney.

Macintosh, N.W.G. (1967) Fossil man in Australia: With particular reference to 1965 discovery at Green Gully near Keilor, Victoria. *Australian Journal of Science* **30**(3): 86–98.

Macintosh, N.W.G. (1970) The Green Gully remains. *Memoirs of the National Museum of Victoria* **10**: 93–100.

Macintosh, N.W.G., Smith, K.N., and Bailey, A.B. (1970) Lake Nitchie Skeleton - Unique Aboriginal Burial. *Archaeology & Physical Anthropology in Oceania* **5**: 85–101.

MacIntyre, G.T. (1966) The Miacidae (Mammalia, Carnivora) Part. 1. *Bulletin of the American Museum of Natural History* **131**: 115–210.

MacLarnon, A. (1995) The distribution of spinal cord tissues and locomotor adaptation in primates. *Journal of Human Evolution* **29**: 463–82.

MacLarnon, A.M., and Hewitt, G.P. (1999) The evolution of human speech: the role of enhanced breathing control. *American Journal of Physical Anthropology* **109**: 341–63.

MacLeod, C.E., Zilles, K., Schleicher, A., Rilling, J.K., and Gibson, K.R. (2003) Expansion of the neocerebellum in the Hominoidea. *Journal of Human Evolution* **44**: 401–29.

Madella, M., Jones, M., Goldberg, P., Goren, Y., and Hovers, E. (2002) The exploitation of plant resources by Neanderthals in Amud Cave (Israel): the evidence from phytolith studies. *Journal of Archaeological Science* **29**(7): 703–19.

Madre-Dupouy, M. (1991) Principaux caractères de l'enfant néandertalien du Roc de Marsal, Dordogne (France). *L'Anthropologie* **2–3**: 523–34.

Madrigal, L., and Kelly, W. (2007) Human skin-color sexual dimorphism: a test of the sexual selection hypothesis. *American Journal of Physical Anthropology* **132**(3): 470–82.

Magori, C.C., and Day, M.H. (1983a) Laetoli Hominid 18: an early *Homo sapiens* skull. *Journal of Human Evolution* **12**: 747–53.

Magori, C.C., and Day, M.H. (1983b) An early *Homo sapiens* skull from the Ngaloaba Beds, Laetoli, northern Tanzania. *Anthropus* **10**: 143–83.

Maier, W., and Nkini, A. (1984) Olduvai Hominid 9, New results of investigation. *Courier Forschungsinstitut Senckenberg* **69**: 123–30.

Malán, M. (1954) Zahnkeim aus der zweiten Aurignacien Schicht der Höhle von Istállósko. *Acta Archaeologica Academiae Scientiarum Hungaricae* **5**: 145–8.

Malina, R.M., Peña Reyes, M.E., Tan, S.K., Buschang, P.H., Little, B.B., and Koziel, S. (2004) Secular change in height, sitting height, and leg length in rural Oaxaca, southern Mexico, 1968–2000. *Annals of Human Biology* **31**: 615–33.

Mallegni, F. (1986) Les restes humains du gisement de Sedia del Diavolo (Rome) rémontant au Riss finaI. *L'Anthropologie* **90**: 539–53.

Mallegni, F. (1992) Quelques restes humains immatures, des niveaux moustériens de la grotte du Fossellone (Monte Circeo, Italie): Fossellone 3 (Olim Circeo IV). *Bulletins et Mémoires de la Société d'Anthropologie de Paris* **4**: 21–32.

Mallegni, F. (2005) Catalogue of Italian fossil human remains from the Palaeolithic to the Mesolithic. *Journal of Anthropological Sciences* **84**(S110).

Mallegni, F., and Ronchitelli, A.T. (1987) Découverte d'une mandibule néandertalienne a l'Abri du Molare pres de Scario (Salerno, Italie): observations stratigraphiques et palethnologies, étude anthropologique. *L'Anthropologie* **91**: 163–74.

Mallegni, F., and Radmilli, A.M. (1988) Human temporal bone from the Lower Paleolithic site of Castel di Guido, near Rome, Italy. *American Journal of Physical Anthropology* **76**: 175–82.

Mallegni, F., and Ronchitelli, A.T. (1989) Deciduous teeth of the Neandertal mandible from Molare Shelter, near Scario (Salerno, Italy). *American Journal of Physical Anthropology* **79**: 475–82.

Mallegni, F., and Trinkaus, E. (1997) A reconsideration of the Archi 1 Neandertal mandible. *Journal of Human Evolution* **33**: 651–68.

Mallegni, F., Carnieri, E., Bisconti, M., Tartarelli, G., Ricci, S., Biddittu, I., et al. (2003) *Homo cepranensis sp. nov.* and the evolution of African-European Middle Pleistocene hominids. *Comptes Rendus Palevol* **2**(2): 153–9.

Manega, P.C. (1993) *Geochronology, Geochemistry, and Isotopic Study of the Plio-Pleistocene Hominid Sites and the Ngorongoro Volcanic Highlands in Northern Tanzania.* thesis, University of Colorado, Boulder, CO.

Manfreda, E., Mitteroecker, P., Bookstein, F.L., and Schaefer, K. (2006) Functional morphology of the first cervical vertebra in humans and non-human primates. *Anatomical Record* **289**(5): 184–94.

Mangerud, J., Andersen, S., Berglund, B., and Donner, J. (1974) Quaternary stratigraphy of Norden, a proposal for terminology and classification. *Boreas* **3**(3): 109–26.

Mania, D. (1989) Die Geologie des Travertins von Bilzingsleben und ihre Bedeutung für die Gliederung des Quartärs. *Ethnographisch-archaologische Zeitschrift* **30**(2): 214–21.

Mania, D. (1993) *Homo erectus* von Bilzingsleben: seine Kultur und Umwelt. Forschungsergebnisse seit 1987. *Ethnographisch-archaologische Zeitschrift* **34**(4): 478–510.

Mania, D. (1997) Das Quartar des Saalegebiets und des Harzvorlandes unter besonderer Beriicksichtung der Travertine von Bilzingsleben-Ein Beitrag zur zyklischen

Gliederung des eurasischen Quartars. *Bilzingsleben V: Homo erectus seine Kultur und Umwelt* 23–103.

Mania, D., and Vlček, E. (1987) *Homo erectus* from Bilzingsleben (GDR) - his culture and his environment. *Anthropologie* **25**: 1–45.

Manly, B.F.J. (2007) *Randomization, Bootstrap and Monte Carlo Methods in Biology*. Chapman & Hall/CRC, Boca Raton, FL.

Mann, A., Monge, J.M., and Lampl, M. (1991) Investigation into the relationship between perikymata counts and crown formation times. *American Journal of Physical Anthropology* **86**: 175–88.

Mann, A., and Trinkhaus, E. (1973) Neandertal and Neandertal-like fossils from the Upper Pleistocene. *Yearbook of Physical Anthropology* **17**: 169–93.

Mann, A., and Vandermeersch, B. (1997) An adolescent female Neandertal mandible from Mountgaudier Cave, Charente, France. *American Journal of Physical Anthropology* **103**: 507–27.

Manning, J. (2002) *Digit Ratio: A Pointer to Fertility, Behavior, and Health*. Rutgers University Press, Pitscataway, NJ.

Manning, J.T., Scutt, D., and Lewis-Jones, D.I. (1998) The ratio of the 2nd to 4th digit length: a predictor of sperm numbers and the concentration of testosterone, luteinizing hormones and oestrogen. *Human Reproduction* **13**: 3000–4.

Manzi, G. (2003) "Epigenetic" cranial traits, Neandertals and the origin of *Homo sapiens*. *Journal of Anthropological Sciences* **81**: 57–68.

Manzi, G., and Passarello, P. (1995) At the Archaic/Modern boundary of the genus *Homo*: the Neandertals from the Grotta Breuil. *Current Anthropology* **36**(2): 355–66.

Manzi, G., Salvadei, L., and Passarello, P. (1990) The Casal de' Pazzi archaic parietal: comparative analysis of new fossil evidence from the late Middle Pleistocene of Rome. *Journal of Human Evolution* **19**: 751–9.

Manzi, G., Magri, D., Milli, S., Palombo, M.R., Margari, V., Celiberti, V., et al. (2010) The new chronology of the Ceprano calvarium (Italy). *Journal of Human Evolution* **59**(5): 580–5.

Marean, C. (2010) Pinnacle Point Cave 13B (Western Cape Province, South Africa) in context: The Cape Floral kingdom, shellfish, and modern human origins. *Journal of Human Evolution* **59**: 425–43.

Marean, C., Bar-Matthews, M., Fisher, E., Goldberg, P., Herries, A., Karkanas, P., et al. (2010) The stratigraphy of the Middle Stone Age sediments at Pinnacle Point Cave 13B (Mossel Bay, Western Cape Province, South Africa). *Journal of Human Evolution* **59**: 234–55.

Marean, C.W. (2000) The Middle Stone Age at Die Kelders Cave 1, South Africa. *Journal of Human Evolution* **38**: 3–5.

Marean, C.W., Abe, Y., Frey, C.J., and Randall, R.C. (2000) Zooarchaeological and taphonomic analysis of the Die Kelders Cave 1 Layers 10 and 11 Middle Stone Age larger mammal fauna. *Journal of Human Evolution* **38**: 197–233.

Marean, C.W., Nilssen, P.J., Brown, K., Jerardino, A., and Stynder, D. (2004) Paleoanthropological investigations of Middle Stone Age sites at Pinnacle Point, Mossel Bay (South Africa): Archaeology and hominid remains from the 2000 field season. *PaleoAnthropology* **2004**: 14–83.

Marean, C.W., Bar-Matthews, M., Bernatchez, J., Fisher, E., Goldberg, P., Herries, A.I.R., et al. (2007) Early human use of marine resources and pigmant in South Africa during the Middle Pleistocene. *Nature* **449**(7164): 905–8.

Marks, A., and Volkman, P. (1987) Technological variability and change seen through core reconstruction. In: *The Human Uses of Flint and Chert*, Sieveking, G., and Newcomer, M.H., eds, pp. 11–20. Cambridge University Press, Cambridge.

Marks, A., Demidenko, Y., Monigal, K., Usik, V., Ferring, C., Burke, A., et al. (1997) Starosele and the Starosele child: New excavations, new results. *Current Anthropology* **38**(1): 112–23.

Marks, A.E. (1977) The Upper Paleolithic sites of Boker Tachtit and Boker: a preliminary report. In: *Prehistory and Paleoenvironments in the Central Negev, Israel. Vol II*, Marks, A.E., ed., pp. 61–79. Southern Methodist University Press, Dallas, TX.

Marks, A.E. (1983) The sites of Boker Tachtit and Boker: a brief introduction. In: *Prehistory and Paleoenvironments in the Central Negev, Israel. Vol III*, Marks, A.E., ed., pp. 15–37. Southern Methodist University Press, Dallas, TX.

Marks, P. (1953) Preliminary note on the discovery of a new jaw of "*Meganthropus* von Koenigswald" in the Lower Plio-Pleistocene of Sangiran, Java. *Indonesian Journal of Natural Science* **109**: 26–33.

Marlowe, F. (2005) Hunter gatherers and human evolution. *Evolutionary Anthropology* **14**(2): 54–67.

Márquez, S., Mowbray, K., Sawyer, G.J., Jacob, T., and Silvers, A. (2001) New fossil hominid calvaria from Indonesia - Sambungmacan 3. *Anatomical Record, Part A* **262**(4): 344–68.

Marroig, G., and Cheverud, J.M. (2004) Did natural selection or genetic drift produce the cranial diversification of neotropical monkeys? *American Naturalist* **163**(3): 417–28.

Marshack, A. (1990) Early hominid symbol and evolution of the human capacity. In: *The Emergence of Modern Humans*, Mellars, P., ed., pp. 457–98. Edinburgh University Press, Edinburgh.

Marshall, A., and Wrangham, R. (2007) Evolutionary consequences of fallback foods. *International Journal of Primatology* **28**(6): 1219–35.

Marston, A.T. (1936) Preliminary Note on a New Fossil Human Skull from Swanscombe, Kent. *Nature* **138**: 200–1.

Martin, H. (1911a) Sur un squelette humain trouvé en Charente. *Comptes Rendus de l'Academie des Sciences* **153**: 728–30.

Martin, H. (1911b) Année 1911. *Bulletin de la Société Préhistorique Française* 8(10): 615–26.

Martin, H. (1912a) Position stratigraphie des ossements humains recueilles dans le moustérien de la Quina de 1908 à 1219. *Bulletin de la Société Préhistorique Française* 9: 700–9.

Martin, H. (1912b) L'Homme fossile mousterien de la Quina. *Bulletin de la Société Préhistorique Française* 9(6): 389–424.

Martin, H. (1913) Nouvelle série de débris humains disséminés trouvés en 1913 dans le gisement moustérien de La Quina. *Bulletin de la Société Préhistorique Française* 10: 540–3.

Martin, L. (1985) Significance of enamel thickness in hominoid evolution. *Nature* **314**: 260–3.

Martin, L.B. (1983) *The Relationships of the Later Miocene Hominoidea*. PhD thesis, University of London.

Martin, R.D. (1981) Relative brain size and basal metabolic rate in terrestrial vertebrates. *Nature* **293**: 57–60.

Martin, R.D. (1996) Scaling of the mammalian brain: the maternal energy hypothesis. *News in Physiological Sciences* 11: 149–56.

Martínez, J., and Mora, R. (1985) Excavaciones a la Cova del Gegant (Sitges). *Butlletí del Grup d'Estudis Sitgetants* IX: 32–33.

Martínez-Navarro, B. (2002) The skull of Orce: parietal bones or frontal bones? *Journal of Human Evolution* **43**(2): 265–70.

Martínez-Navarro, B., Rook, L., Segid, A., Yosieph, D., Ferretti, M., Shoshani, J., et al. (2004) The large fossil mammals from Buia (Eritrea): systematics, biochronology and paleoenvironments. *Rivista Italiana di Paleontologia e Stratigrafia* 110: 61–88.

Martínez-Navarro, B., Belmaker, M., and Bar-Yosef, O. (2009) The large carnivores from 'Ubeidiya (early Pleistocene, Israel): biochronological and biogeographical implications. *Journal of Human Evolution* **56**(5): 514–24.

Martini, F., Libsekal, Y., Filippi, O., Ghebre/her, A., Kashay, H., Kiros, A., et al. (2004) Characterization of lithic complexes from Buia (Dandiero Basin, Danakil Depression, Eritrea). *Rivista Italiana di Paleontologia e Stratigrafia* 110: 99–132.

Martini, J.E.J., Wipplinger, P.E., Moen, H.F.G., and Keyser, A. (2003) Contribution to the speleology of Sterkfontein Cave, Gauteng Province, South Africa. *International Journal of Speleology* 32(1–4): 43–69.

Martinón-Torres, M., Bastir, M., Bermúdez de Castro, J., Gómez, A., Sarmiento, S., Muela, A., et al. (2006) Hominin lower second premolar morphology: evolutionary inferences through geometric morphometric analysis. *Journal of Human Evolution* 50(5): 523–33.

Martinón-Torres, M., Bermúdez de Castro, J., Gómez-Robles, A., Margvelashvili, A., Prado, L., Lordkipanidze, D., et al. (2008) Dental remains from Dmanisi (Republic of Georgia): morphological analysis and comparative study. *Journal of Human Evolution* **55**(2): 249–73.

Martins, E.P., and Hansen, T.F. (1997) Phylogenies and the comparative method: A general approach to incorporating phylogenetic information into the analysis of interspecific data. *American Naturalist* **149**(4): 646–67.

Martyn, J., and Tobias, P.V. (1967) Pleistocene deposits and new fossil localities in Kenya. *Nature* **215**: 476–9.

Marzke, M.W. (1997) Precision grips, hand morphology and tools. *American Journal of Physical Anthropology* **102**(1): 91–110.

Marzke, M.W. (2009) Upper-Limb evolution and development. *Journal of Bone and Joint Surgery* **91**(4): 26–30.

Marzke, M.W., Wullstein, K.L., and Viegas, S.F. (1992) Evolution of the power ("squeeze") grip and its morphological correlates in hominids. *American Journal of Physical Anthropology* 89: 283–98.

Marzke, M.W., Toth, N., Schick, K., Reece, S., Steinberg, B., Hunt, K., et al. (1998) EMG study of hand muscle recruitment during hard hammer percussion manufacture of Oldowan tools. *American Journal of Physical Anthropology* 105(3): 315–32.

Marzke, M.W., Marzke, R.L., Smutz, P., Steinburg, B., Reece, S., and An, K.N. (1999) Chimpanzee thumb muscle cross sections moment arms and potential torques and comparisons with humans. *American Journal of Physical Anthropology* 110: 163–78.

Marzke, M.W., Tocheri, M.W., Steinberg, B., Femiani, J.D., Reece, S.P., Linscheid, R.L., et al. (2010) Comparative 3D quantitative analyses of trapeziometacarpal joint surface curvatures among living catarrhines and fossil hominins. *American Journal of Physical Anthropology* 141: 38–51.

Mascie-Taylor, C. (1987) Assortative mating in a contemporary British population. *Annals of human biology* **14**(1): 59–68.

Mascie-Taylor, C.G., and Boldsen, J.L. (1988) Assortative mating, differential fertility and abnormal pregnancy outcome. *Annals of Human Biology* **15**(3): 223–8.

Mason, R.J. (1962) *Cave of Hearths in Prehistory*, Johannesburg.

Mason, R.J., ed. (1988) *Cave of Hearths, Makapansgat, Transvaal*. Occasional Paper 21. University of Witwatersrand Archaeological Research Unit.

Mastin, B.A. (1964) The Extended Burials at the Mugharet el-Wad. *Journal of the Royal Anthropological Institute of Great Britain and Ireland* **94**: 44–51.

Matilla, K. (2004) Technotypologie du matériel sur galet de la Chaise-de-Vouthon (Charente). Présentation préliminaire à partir d'un échantillon provenant de l'abri Suard. *Bulletin de la Société Préhistorique Française* **101**(4): 771–9.

Matilla, K. (2005) L'industrie sur galet de La Chaise-de-Vouthon, Charente: synthèse des résultats. *L'Anthropologie* **109**(3): 481–98.

Matsu'ura, S. (1982) A chronological framing for the Sangiran hominids. *Bulletin National Science Museum Tokyo Series D* **8**: 1–53.

Matsu'ura, S. (1999) A chronological review of Pleistocene human remains from the Japanese Archipelago. In: *Interdisciplinary Perspectives on the Origins of the Japanese*, Omoto, K., ed., pp. 181–97. International Research Center for Japanese Studies, Kyoto.

Matsu'ura, S., Watanabe, N., Aziz, F., Shibasaki, T., and Kondo, M. (1990) Preliminary dating of a hominid fossil tibia from Sambungmacan by fluorine analysis. *Bulletin National Science Museum Tokyo Series D* **16**: 19–29.

Matsu'ura, S., Aziz, F., and Watanabe, N. (1992) Probable chronological positions of two additional mandible specimens of the Sangiran hominids. *Journal of the Anthropological Society of Nippon* **100**: 262.

Matsu'ura, S., Kondo, M., and Aziz, F. (1995a) A preliminary report on stratigraphic allocation of Sangiran hominid remains by multielement analyses of bone. *Anthropological Science* **103**: 161.

Matsu'ura, S., Kondo, S., Aziz, F., and Watanabe, N. (1995b) Chronology of four hominid mandibular fossils newly found from Sangiran and their possible evolutionary implications. In: *Report of Research Project Supported by Grant-in-Aid for Project No. 04454034* pp. 37–50. Japanese Ministry of Education, Science, Sports and Culture, Tokyo.

Matthews, T., Denys, C., and Parkington, J.E. (2005) The palaeoecology of the micromammals from the late middle Pleistocene site of Hoedjiespunt 1 (Cape Province, South Africa). *Journal of Human Evolution* **49**: 432–51.

Matthews, T., Denys, C., and Parkington, J.E. (2007) Community evolution of Neogene micromammals from Langebaanweg 'E' Quarry and other west coast fossil sites, south-western Cape, South Africa. *Palaeogeography, Palaeoclimatology, Palaeoecology* **245**(3–4): 332–52.

Maureille, B. (2002a) La redécouverte du nouveau-né néandertalien Le Moustier 2. *Paléo. Revue d'archéologie préhistorique* (14): 221–38.

Maureille, B. (2002b) Anthropology: A lost Neanderthal neonate found. *Nature* **419**(6902): 33–4.

Maureille, B., and Soressi, M. (2000) A propos de la position chrono-stratigraphique de l'enfant du Pech-de-l'Azé (commune de Carsac, Dordogne): la résurrection du fantôme. *Paléo* **12**: 339–52.

Maureille, B., Rougier, H., Houët, F., and Vandermeersch, B. (2001) Les dents inférieures du néandertalien Regourdou I (site de Regourdou, commune de Montignac, Dordogne): analyses métriques et comparatives. *Paléo* **13**: 183–200.

Mauss, M. (1935) Les techniques du corps. *Journal de psychologie* **32**: 271–193.

Mawby, J. (1970) Fossil vertebrates from northern Malawi: preliminary report *Quaternaria* **13**: 319–23.

Mayr, E. (1942) *Systematics and the Origin of Species.* Columbia University Press, New York.

Mayr, E. (1944) On the concepts and terminology of vertical subspecies and species. *National Research Council Committee on Common Problems of Genetics, Paleontology, and Systematics Bulletin* **2**: 11–16.

Mayr, E. (1950) Taxonomic categories in fossil hominids. *Cold Spring Harbor Symposia on Quantitative Biology* **15**: 109–18.

Mayr, E. (1982) *The Growth of Biological Thought: Diversity, Evolution and Inheritance* Harvard University Press, Cambridge, MA.

Mayr, E., Linsley, E.G., and Usinger, R.L. (1953) *Methods and Principles of Systematic Zoology.* McGraw-Hill Book Co., New York.

Mazurier, A., Volpato, V., and Macchiarelli, R. (2006) Improved noninvasive microstructural analysis of fossil tissues by means of SR-microtomography. *Applied Physics A: Materials Science & Processing* **83**(2): 229–33.

McBrearty, S. (1987) Une evaluation du Sangoen: son age, son environnement et son rapport avec l'origine de *l'Homo sapiens. L'Anthropologie* **91**: 127–40.

McBrearty, S. (1991) Recent research in western Kenya and its implications for the status of the Sangoan industry. In: *Cultural Beginnings: Approaches to Understanding Early Hominid Lifeways in the African Savanna*, Clark, J.D., ed., pp. 159–76. Romisch-Germanisches Zentralmuseum, Forschunginstitut fur Vor- und Fruhgeschichte, Bonn.

McBrearty, S. (1992) Sangoan technology and habitat at Simbi, Kenya. *Nyame Akume* **38**: 29–33.

McBrearty, S. (1999) The archaeology of the Kapthurin Formation. In: *Late Cenozoic Environments and Human Evolution: A Tribute to Bill Bishop*, Andrews, P., and Banham, P., eds, pp. 143–56. Geological Society, London.

McBrearty, S. (2001) The Middle Pleistocene of East Africa. In: *Human Roots: Africa and Asia in the Middle Pleistocene*, Barham, L., and Robson-Brown, K., eds, pp. 81–92. Western Academic and Specialist Press, Bristol.

McBrearty, S. (2005) The Kapthurin Formation: What we know now that we didn't know then. In: *Interpreting the Past: Essays on Human Primate, and Mammal Evolution in Honor of David Pilbeam*, Lieberman, D.E., Smith, R.J., and Kelley, J., eds, pp. 263–74. Brill, Boston, MA.

McBrearty, S., and Brooks, A.S. (2000) The revolution that wasn't: a new interpretation of the origin of modern human behavior. *Journal of Human Evolution* **39**: 453–563.

McBrearty, S., and Jablonski, N.G. (2005) First fossil chimpanzee. *Nature* **437**: 105–8.

McBrearty, S., Bishop, L., and Kingston, J. (1996) Variability in traces of Middle Pleistocene hominid behaviour in the Kapthurin Formation, Baringo, Kenya. *Journal of Human Evolution* **30**: 563–80.

McBurney, C.B.M. (1950) The geographical study of the older palaeolithic stages in Europe. *Proceedings of the Prehistorical Society* **16**: 163–83.

McBurney, C.B.M. (1967) *The Haua Fteah (Cyrenaica) and the Stone Age of the South-East Mediterranean.* Cambridge University Press, Cambridge.

McBurney, C.B.M., Trevor, J.C., and Wells, L.H. (1953a) A fossil human mandible from a Levalloiso-Mousterian horizon in Cyrenaica. *Nature* **172**: 889–92.

McBurney, C.B.M., Trevor, J.C., and Wells, L.H. (1953b) The Haua Fteah fossil jaw. *Journal of the Royal Anthropological Instiiute* **83**(1): 71–85.

McCall, G.J.H. (1999) Silali volcano, Baringo, Kenya: sedimentary structures at the western fringe. In: *Late Cenozoic Environments and Hominid Evolution: A Tribute to Bill Bishop*, Andrews, P., and Banham, P., eds, pp. 59–68. The Geological Society, London.

McCall, G.J.H., Baker, B.H., and Walsh, J. (1967) Late Tertiary and Quaternary sediments of the Kenya Rift Valley. In: *Background to Evolution in Africa*, Bishop, W.W., and Clark, J.D., eds, pp. 191–220. University of Chicago Press, Chicago, IL.

McConkey, E.H., and Goodman, M. (1997) A Human Genome Evolution Project is needed. *Trends in Genetics* **13**(9): 350–1.

McCormac, F.G., Hogg, A.G., Blackwell, P.G., Buck, C.E., Higham, T.F.G., and Reimer, P.J. (2004) SHCal04 Southern Hemisphere calibration, 0-11.0 Cal Kyr BP. *Radiocarbon* **46**: 1087–92.

McCown, T. (1932) A note on the excavation and the human remains from the Mughharet es-Skhul (Cave of the Kids), season of 1931. *Bull. Am. Sch. prehist. Res.* **8**: 12–15.

McCown, T., and Keith, A. (1939) *The Stone Age of Mount Carmel.* Clarendon Press, Oxford.

McCown, T.D. (1936) Mount Carmel Man. *Bulletin American School of Prehistoric Research* **12**: 131–9.

McCown, T.D. (1958) Upper Pleistocene men of the southwest Asian Mousterian. In: *Hundert Jahre Neanderthaler 1856–1956*, von Koenigswald, G.H.R., ed., pp. 185–98. Kemink en Zoon, Utrecht.

McCown, T.D., and Keith, A. (1934) *Palaeanthropus palestinus.* In: *Proc. 1st Int. Congr. Prehist. Protohist. Soc.* London.

McDermott, F., Stringer, C., Grün, R., Williams, C.T., Din, V.K., and Hawkesworth, C.J. (1996) New Late-Pleistocene uranium-thorium and ESR dates for the Singa hominid (Sudan). *Journal of Human Evolution* **31**: 507–16.

McDougall, I., and Feibel, C.S. (1999) Numerical age control for the Miocene-Pliocene succession at Lothagam, a hominoid-bearing sequence in the Northern Kenya Rift. *Journal of the Geological Society, London* **156**(4): 731–45.

McDougall, I., and Brown, F.H. (2006) Precise ^{40}Ar/^{39}Ar geochronology for the upper Koobi Fora Formation, Turkana Basin, northern Kenya. *Journal of the Geological Society* **163**(1): 205–20.

McDougall, I., Brown, F.H., and Fleagle, J.G. (2005) Stratigraphic placement and age of modern humans from Kibish, Ethiopia. *Nature* **433**(7027): 733–6.

McDougall, I., Brown, F.H., and Fleagle, J.G. (2008) Sapropels and the age of hominins Omo I and II, Kibish, Ethiopia. *Journal of Human Evolution* **55**: 409–20.

McFadden, P.L., Brock, A., and Partridge, T.C. (1979) Palaeomagnetism and the age of the Makapansgat hominid site. *Earth and Planetary Science Letters* **44**(3): 373–82.

McGowran, B., Berggren, W.A., Hilgen, F., Steininger, F., Aubry, M.-P., Lourens, L., et al. (2009) Neogene and Quaternary coexisting in the geological time scale: the inclusive compromise. *Earth-Science Reviews* **96**(4): 249–62.

McGrew, W. (1983) Animal foods in the diets of wild chimpanzees (*Pan troglodytes*): why cross-cultural variation? *Journal of Ethology* **1**(1): 46–61.

McGrew, W.C. (1992) *Chimpanzee Material Culture: Implications for Human Evolution.* Cambridge University Press, Cambridge.

McGrew, W.C., Tutin, C.E.G., and Baldwin, P.J. (1979) Chimpanzees, tools and termites: cross-cultural comparisons of Senegal, Tanzania and Rio Muni. *Man* **14**(2): 185–214.

McHenry, H.M. (1973) Early hominid humerus from East Rudolf, Kenya. *Science* **180**: 739–40.

McHenry, H.M. (1975) A new pelvic fragment from Swartkrans and the relationship between the robust and gracile australopithecines. *American Journal of Physical Anthropology* **43**(2): 245–61.

McHenry, H.M. (1984) Relative cheek-tooth size in *Australopithecus. American Journal of Physical Anthropology* **64**: 297–306.

McHenry, H.M. (1988) New estimates of body weight in early hominids and their significance to encephalization and megadontia in "robust" australopithecines. In: *Evolutionary History of the "Robust" Australopithecines*, Grine, F.E., ed., pp. 133–48. Aldine de Gruyter, New York.

McHenry, H.M. (1991) Petite bodies of the "robust" australopithecines. *American Journal of Physical Anthropology* **86**(4): 445–54.

McHenry, H.M., and Corruccini, R.S. (1978) The femur in early human evolution. *American Journal of Physical Anthropology* **49**: 473–88.

McHenry, H.M., and Corruccini, R.S. (1980) On the status of *Australopithecus afarensis. Science* **207**: 1103–4.

McHenry, H.M., and Berger, L.R. (1998) Body proportions in *A. afarensis* and *A. africanus* and the origin of the genus *Homo. Journal of Human Evolution* **35**: 1–22.

McHenry, H.M., Brown, C.B., and McHenry, L.J. (2007) Fossil hominin ulnae and the forelimb of *Paranthropus. American Journal of Physical Anthropology* **134**(2): 209–18.

McIntyre, M.H., Herrmann, E., Wobber, V., Halbwax, M., Mohamba, C., de Sousa, N., et al. (2009) Bonobos have a more human-like second-to-fourth finger length ratio (2D:4D) than chimpanzees: a hypothesized indication of lower prenatal androgens. *Journal of Human Evolution* **56**: 361–5.

McKechnie, A.E. (2004) Stable isotopes: powerful new tools for animal ecologists. *South African Journal of Science* **100**: 131–4.

McKee, J.K. (1993) Faunal dating of the Taung hominid fossil deposit. *Journal of Human Evolution* **25**: 363–76.

McKee, J.K., and Tobias, P.V. (1994) Taung stratigraphy and taphonomy: preliminary results based on the 1988–1993 excavations. *South African Journal of Science* **90**: 233–5.

McKee, J.K., Thackeray, J.F., and Berger, L.R. (1995) Faunal assemblage seriation of southern African Pliocene and Pleistocene fossil deposits. *American Journal of Physical Anthropology* **96**(3): 235–50.

McMahon, T.A., and Bonner, J.T. (1983) *On Size and Life*. W.H. Freeman and Co, New York.

McNabb, J., Binyon, F., Hazelwood, L., Machin, A., Mithen, S., Petraglia, M., et al. (2004) The large cutting tools from the South African Acheulean and the question of social traditions. *Current Anthropology* **45**(5): 653–77.

McNulty, K., Frost, S., and Strait, D. (2006) Examining affinities of the Taung child by developmental simulation. *Journal of Human Evolution* **51**(3): 274–96.

McPherron, S.P., and Dibble, H.L. (2000) The lithic assemblages of Pech de L'Azé IV (Dordogne, France). *Prehistoire Europeene* **15**: 9–43.

McPherron, S.P., Alemseged, Z., Marean, C.W., Wynn, J.G., Reed, D., Geraads, D., et al. (2010) Evidence for some stone-tool assisted consumption of animal tissues prior to 3.39 Ma at Dikika, Ethiopia. *Nature* **466**: 857–60.

Medawar, P.B. (1952) *An Unsolved Problem of Biology*. H.K. Lewis & Co., London.

Mehlman, M.J. (1979) Mumba-Hohle Revisited: The relevance of a forgotten excavation to some current issues in East African prehistory. *World Archaeology* **11**(1): 80–94.

Mehlman, M.J. (1984) Archaic *Homo sapiens* at Lake Eyasi, Tanzania: recent misrepresentations. *Journal of Human Evolution* **13**: 487–501.

Mehlman, M.J. (1987) Provenience age and associations of archaic *Homo sapiens* crania from Lake Eyasi, Tanzania. *Journal of Archaeological Science* **14**: 133–62.

Mehlman, M.J. (1989) *Late Quaternary Archaeological Sequences in Northern Tanzania*. PhD thesis, University of Illinois.

Mehlman, M.J. (1991) Context for the emergence of modern man in Eastern Africa: some new Tanzanian evidence. In: *Cultural Beginnings: Approaches to Understanding Early Hominid Lifeways in the African Savanna*, Clark, J.D., ed., pp. 177–96. Forschung-

institut für Vor- und Frügeschichte, Römisch-Germanischen Zentralmuseum, Bonn.

Meignen, L. (1999) Hayonim cave lithic assemblages in the context of the Near-Eastern Middle Paleolithic: a preliminary report. In: *Neandertals and Modern Humans in Western Asia*, Akazawa, T., Aoki, K., and Bar-Yosef, O., eds, pp. 165–80. Plenum Press, New York.

Meignen, L., Goldberg, P., and Bar-Yosef, O. (2007) The hearths at Kebara Cave and their role in site formation processes. In: *The Middle and Upper Paleolithic Archaeology of the Kebara Cave, Mt. Carmel, Israel*, Bar-Yosef, O., Meignen, L., ed., pp. 91–122. Peabody Museum, Harvard University, Cambridge, MA.

Mein, P. (1975) *Biozonation du neogène Mediterranée a partir des mammifères*. Proceedings of the VI Congress of the Regional Committee on Mediterranean Neogene Stratigraphy, Bratislava.

Mein, P. (1990) Updating of MN zones. In: *European Neogene Mammal Chronology*, Lindsay, E., Fahlbusch, V., and Mein, P., eds, pp. 73–90. Plenum Press, New York.

Mellars, P. (1969) The chronology of Mousterian industries in the Périgord region of southwest France. *Proceedings of the Prehistorical Society* **35**: 134–71.

Mellars, P. (1986) A new chronology for the French Mousterian period. *Nature* **322**: 410–11.

Mellars, P. (1996) *The Neanderthal Legacy: An Archaeological Perspective from Western Europe*. Princeton University Press, Princeton, NJ.

Mellars, P. (1999) CA Forum on Theory of Anthropology: the Neanderthal problem continued. *Current Anthropology* **40**(3): 341–64.

Mellars, P. (2005) The impossible coincidence: a single-species model for the origins of modern human behavior in Europe. *Evolutionary Anthropology* **14**: 12–27.

Mellars, P. (2006) A new radiocarbon revolution and the dispersal of modern humans in Eurasia. *Nature* **439**: 931–5.

Mellars, P., and Tixier, J. (1989) Radiocarbon-accelerator dating of Ksar 'Aqil (Lebanon) and the chronology fo the Upper Palaeolithic sequence in the Middle East. *Antiquity* **63**(241): 761–8.

Mellars, P., and Grün, R. (1991) A comparison of the electron spin resonance and thermoluminescence dating methods: the results of ESR dating at Le Moustier (France). *Cambridge Archaeological Journal* **1**(2): 269–76.

Mellars, P., Gravina, B., and Ramsey, C.B. (2007) Confirmation of Neanderthal/modern human interstratification at the Chatelperronian type-site. *Proceedings of the National Academy of Sciences USA* **104**(9): 3657–62.

Meltzoff, A.N. (1995) What infant memory tells us about infantile amnesia: long-term recall and deferred imitation. *Journal of Experimental Child Psychology* **59**(3): 497–515.

Mendes, N., Hanus, D., and Call, J. (2007) Raising the level: orangutans use water as a tool. *Biology Letters* **3**(5): 453–5.

Menter, C.G., Kuykendall, K.L., Keyser, A.W., and Conroy, G.C. (1999) First record of hominid teeth from the Plio-Pleistocene site of Gondolin, South Africa. *Journal of Human Evolution* 37: 299–307.

Mercader, J. (2002) Forest People: the role of African rainforests in human evolution and dispersal. *Evolutionary Anthropology* 11: 117–24.

Mercader, J., and Brooks, A.S. (2001) Across forests and savannas: Later Stone Age assemblages from Ituri and Semliki, northeast Democratic Republic of Congo. *Journal of Anthropological Research* 57(2): 197–217.

Mercer, N., and Valladas, H. (2003) Reassessment of TL age estimates of burnt flints from the Paleolithic site of Tabun Cave, Israel. *Journal of Human Evolution* 45: 401–9.

Mercier, N., Valladas, H., Joron, J., Reyss, J., Lévêque, F., and Vandermeersch, B. (1991) Thermoluminescence dating of the late Neanderthal remains from Saint-Césaire. *Nature* 351: 737–9.

Mercier, N., Valladas, H., Bar-Yosef, O., Vandermeersch, B., Stringer, C., and Joron, J.-L. (1993) Thermoluminescence date for the Mousterian Burial Site of Es-Skhul, Mt. Carmel. *Journal of Archaeological Science* 20(2): 169–74.

Mercier, N., Valladas, H., Froget, L., Joron, J.L., Reyss, J.L., Weiner, S., et al. (2007a) Hayonim Cave: a TL-based chronology for this Levantine Mousterian sequence. *Journal of Archaeological Science* 34(7): 1064–77.

Mercier, N., Wengler, L., Valladas, H., Joron, J.L., Froget, L., and Reyss, J.L. (2007b) The Rhafas Cave (Morocco): Chronology of the Mousterian and Aterian archaeological occupations and their implications for quaternary geochronology based on luminescence (TL/OSL) age determinations. *Quaternary Geochronology* 2(1–4): 309–13.

Méroc, L. (1963) Les éléments de datation de la mandibule humains de Montmaurin (Haute-Garonne). *Bulletin de la Société Géologique de France* 5(7): 508–15.

Merrilees, D. (1973) Fossiliferous deposits at Lake Tandou, New South Wales, Australia. *Memoirs of the National Museum of Victoria* 34: 177–82.

Merriwether, D.A., Hall, W.W., Vahlne, A., and Ferrell, R.E. (1996) mtDNA variation indicates Mongolia may have been the source for the founding population for the New World. *American Journal of Human Genetics* 59(1): 204–12.

Mertens, S. (1996) The Middle Paleolithic in Romania. *Current Anthropology* 37(3): 515–21.

Messeri, P., and Palma di Cesnola, A. (1976) Contemporaneità di paleantropi e fanerantropi sulle coste dell'talia meridionale. *Zephyrus* 26–7: 7–30.

Mester, Z. (1990) La transition vers le Paléolithique supérieur des industries moustériennes de la montagne de Bükk (Hongrie). In: *Paléolithique moyen récent et Paléolithique supérieur ancien en Europe*, Farizy, C., ed., pp. 111–13. Musee de Prehistoire d'Ile de France, Paris.

Metzger, K., and Herrel, A. (2005) Correlations between lizard cranial shape and diet: a quantitative, phylogenetically informed analysis. *Biological Journal of the Linnean Society* 86(4): 433–66.

Michel, V., Yokoyama, Y., Falguès, C., and Ivanovich, M. (2000) Problems encountered in the U-Th dating of fossil red deer jaws (bone, dentine, enamel) from Lazaret Cave: A comparative study with early chronological data. *Journal of Archaeological Science* 27: 327–40.

Mietto, P., Avanzini, M., and Rolandi, G. (2003) Human footprints in Pleistocene volcanic ash. *Nature* 422: 133.

Miles, A., and Grigson, C. (1990) *Colyer's Variations and Disease of the Teeth of Animals, Revised Edition.* Cambridge University Press, Cambridge,.

Miles, A.E.W. (1962) Assessment of the ages of a population of Anglo-Saxons from their dentitions. *Proceedings of the Royal Society of Medicine* 55: 881–6.

Millard, A.R. (2006) Bayesian analysis of ESR dates, with application to Border Cave. *Quaternary Geochronology* 1(2): 159–66.

Miller, A.H. (1949) Some ecological and morphological considerations in the evolution of higher taxonomic categories. In: *Ornithologie Als Biologische Wissenschaft*, Schuz, E., and Mayr, E., eds, pp. 84–8. Carl Winter Universitasverlag, Heidelberg.

Miller, G. (2000) *The Mating Mind: How Sexual Choice Shaped the Evolution of Human Nature.* Doubleday Books, New York.

Miller, G.H., Beaumont, P.B., Deacon, H.J., Brooks, A.S., Hare, P.E., and Jull, A.J.T. (1999) Earliest modern humans in southern Africa dated by isoleucine epimerization in ostrich eggshell. *Quaternary Science Reviews* 18(13): 1537–48.

Miller-Antonio, S., Schepartz, L.A., and Bakken, D. (2000) Raw material selection and evidence for rhinoceros tooth tools at Dadong Cave, southern China. *Antiquity* 74: 372–9.

Miller-Antonio, S., Schepartz, L.A., Karkanas, P., Yamei, H., Weiwen, H., and Bekken, D. (2004) Lithic raw material use at the Late Middle Pleistocene site of Panxian Dadong. *Asian Perspectives* 43: 314–32.

Milo, R.G. (1998) Evidence for hominid predation at Klasies River Mouth, South Africa, and its implications for the behaviour of early modern humans. *Journal of Archaeological Science* 25(2): 99–133.

Milton, K. (1988) Foraging behavior and the evolution of primate intelligence. In: *Machiavellian Intelligence*, Byrne, R.W., and Whiten, A., eds, pp. 285–305. Clarendon Press, Oxford.

Minugh-Purvis, N. (1993) Reexamination of the immature hominid maxilla from Tangier, Morocco. *American Journal of Physical Anthropology* 92(4): 449–61.

Miracle, P. (2007) *The Krapina Palaeolithic Site: Zooarchaeology, Taphonomy, and Catalog of the Faunal Remains.* Croatian Natural History Museum, Zagreb.

Mithen, S. (1996) *The Prehistory of the Mind*. Thames and Hudson, London.

Mithen, S. (2006) *The Singing Neanderthals*. Phoenix, New York.

Mitteroecker, P., and Bookstein, F.L. (2007) The conceptual and statistical relationship between modularity and morphological integration. *Systematic Biology* **56**(5): 818–36.

Mitteroecker, P., and Bookstein, F.L. (2008) The evolutionary role of modularity and integration in the hominoid cranium. *Evolution* **62**(4): 943–58.

Mitteroecker, P., and Gunz, P. (2009) Advances in geometric morphometrics. *Evolutionary Biology* **36**(2): 235–47.

Mitteroecker, P., Gunz, P., and Bookstein, F.L. (2005) Heterochrony and geometric morphometrics: A comparison of cranial growth in *Pan paniscus* versus *Pan troglodytes*. *Evolution & Development* 7: 244–58.

Miyata, E. (2005) Clusters of stones in Yokomine C site and Tachikiri site, and the process of their appearance. *Archaeological Journal (Japan)* 531: 13–16.

Moggi-Cecchi, J., and Collard, M. (2002) A fossil stapes from Sterkfontein, South Africa, and the hearing capabilities of early hominids. *Journal of Human Evolution* 42: 259–65.

Moggi-Cecchi, J., and Boccone, S. (2007) Maxillary molars cusp morphology of South African australopithecines. In: *Dental Perspectives on Human Evolution: State of the Art Research in Dental Paleoanthropology*, Bailey, S., and Hublin, J.-J., eds, pp. 53–64. Springer, Dordrecht.

Moggi-Cecchi, J., Tobias, P., and Beynon, A. (1998) The mixed dentition and associated skull fragments of a juvenile fossil hominid from Sterkfontein, South Africa. *American Journal of Physical Anthropology* **106**(4): 425–65.

Moggi-Cecchi, J., Grine, F.E., and Tobias, P.V. (2006) Early hominid dental remains from Members 4 and 5 of the Sterkfontein Formation (1966–1996 excavations): Catalogue, individual associations, morphological descriptions, and initial metrical analysis. *Journal of Human Evolution* **50**(3): 239–328.

Moggi-Cecchi, J., Menter, C., Boccone, S., and Keyser, A. (2010) Early hominin dental remains from the Plio-Pleistocene site of Drimolen, South Africa. *Journal of Human Evolution* **58**(5): 374–405.

Monchot, H., and Aouraghe, H. (2009) Deciphering the taphonomic history of an Upper Paleolithic faunal assemblage from Zouhrah Cave/El Harhoura 1, Morocco. *Quaternaire* **20**(2): 239–53.

Monnier, G.F. (2006) The Lower/Middle Paleolithic periodization in Western Europe: An evaluation. *Current Anthropology* **47**(5): 709–44.

Montagu, M.F.A. (1952) The Châtelperron skull. *American Journal of Physical Anthropology* 10: 125–7.

Montandon, G. (1943) *L'homme préhistorique et les préhumains*. Payot, Paris.

Moodley, Y., Linz, B., Yamaoka, Y., Windsor, H.M., Breurec, S., Wu, J. Y., et al. (2009) The peopling of the Pacific from a bacterial perspective. *Science* **323**(5913): 527–30.

Moore, J. (1992) "Savanna" chimpanzees. In: *Topics in Primatology. Volume I. Human Origins*, Nishida, T., McGrew, W.C., Marler, P., Pickford, M., and de Waal, F.B.M., eds, pp. 98–118. University of Tokyo Press, Tokyo.

Moore, J. (1996) Savanna chimpanzees, referential models and the last common ancestor. In: *Great Ape Societies*, McGrew, W.C., Marchant, L.F., and Nishida, T., eds, pp. 275–92. Cambridge University Press, Cambridge.

Moore, M., and Brumm, A. (2007) Stone artifacts and hominins in island Southeast Asia: new insights from Flores, eastern Indonesia. *Journal of Human Evolution* **52**(1): 85–102.

Moore, M.W., Sutikna, T., Jatmiko, Morwood, M.J., and Brumm, A. (2009) Continuities in stone flaking technology at Liang Bua, Flores, Indonesia. *Journal of Human Evolution* **57**(5): 503–26.

Moorrees, C.F.A., Fanning, E.A., and Hunt, E.E. (1963) Age variation of formation stages for ten permanent teeth. *Journal of Dental Research* 42: 1490–1502.

Morant, G.M. (1927) Studies of Palaeolithic Man II: A biometric study of Neanderthaloid skulls and of their relationships to modern racial types. *Annals of Eugenics* **2**(3–4): 318–80.

Morant, G.M. (1928) Studies of Palaeolithic Man III: The Rhodesian skull and its relation to Neanderthaloid and modern types. *Annals of Eugenics* 3: 337–60.

Morant, G.M. (1938) The form of the Swanscombe skull. *Journal of the Royal Anthropological Institute* 68: 67–97.

Morel, J. (1974) La Station eponyme de l'Oued Djebbana a Bir-el-Ater (Est algerien), contribution a la comaissance de son industrie et de sa faune. *L'Anthropologie* 78: 53–80.

Morgan, L.E., and Renne, P.R. (2008) Diachronous dawn of Africa's Middle Stone Age: New ^{40}Ar/^{39}Ar ages from the Ethiopian Rift. *Geology* **36**(12): 967–70.

Morin, E. (2008) Evidence for declines in human population densities during the early Upper Paleolithic in western Europe. *Proceedings of the National Academy of Sciences USA* **105**(1): 48–53.

Morin, E., Tsanova, T., Sirakov, N., Rendu, W., Mallye, J., and Lévêque, F. (2005) Bone refits in stratified deposits: testing the chronological grain at Saint-Césaire. *Journal of Archaeological Science* **32**(7): 1083–98.

Morin, P.A., Moore, J.J., Chakraborty, R., Jin, L., Goodall, J., and Woodruff, D.S. (1994) Kin selection, social structure, gene flow, and the evolution of chimpanzees. *Science* **265**(5176): 1193–1201.

Moritomo, H., Murase, T., Goto, A., Oka, K., Sugamoto, K., and Yoshikawa, H. (2006) *In vivo* three-dimensional kinematics of the midcarpal joint of the wrist. *Journal of Bone and Joint Surgery* **88**: 611–20.

Morris, D.H. (1970) On deflecting wrinkles and the *Dryopithecus* pattern in human mandibular molars. *American Journal of Physical Anthropology* **32**(1): 97–104.

Morton, D. (1924) Evolution of the human foot II. *American Journal of Physical Anthropology* **7**(1): 1–52.

Morwood, M., O'Sullivan, P., Susanto, E.E., and Aziz, F. (2003) Revised age for Mojokerto 1, an early *Homo erectus* cranium from East Java, Indonesia. *Australian Archaeology* **57**: 1–4.

Morwood, M.J., and Jungers, W.L. (2009) Conclusions: implications of the Liang Bua excavations for hominin evolution and biogeography. *Journal of Human Evolution* **57**(5): 640–8.

Morwood, M.J., Aziz, F., van den Bergh, G., Sondaar, P., and de Vos, J. (1997) Stone artefacts from the 1994 excavations at Mata Menge, west central Flores, Indonesia. *Australian Archaeology* **44**: 26–34.

Morwood, M.J., O'Sullivan, P.B., Aziz, F., and Raza, A. (1998) Fission-track ages of stone tools and fossils on the east Indonesian island of Flores. *Nature* **392**: 173–6.

Morwood, M.J., Aziz, F., O'Sullivan, P., Nasruddin, Hobbs, D.R., and Raza, A. (1999) Archaeological and palaeontological research in central Flores, east Indonesia: results of fieldwork 1997–98. *Antiquity* **73**: 273–86.

Morwood, M.J., Soejono, R.P., Roberts, R.G., Sutikna, T., Turney, C.S.M., Westaway, K.E., et al. (2004) Archaeology and age of a new hominin from Flores in eastern Indonesia. *Nature* **431**: 1087–91.

Morwood, M.J., Brown, P., Jatmiko, Sutikna, T., Saptomo, E.W., Westaway, K.E., et al. (2005) Further evidence for small-bodied hominins from the Late Pleistocene of Flores, Indonesia. *Nature* **437**: 1012–17.

Morwood, M.J., Sutikna, T., Saptomo, E.W., Jatmiko, Hobbs, D.R., and Westaway, K.E. (2009) Preface: research at Liang Bua, Flores, Indonesia. *Journal of Human Evolution* **57**: 437–49.

Mottura, A. (1980) Un frammento di osso temporale di tipo neanderlaliano del Monfenera, Vercelli (Piemonte). *Antropologia contemporanea* **3**: 373–9.

Mounier, A., Marchal, F., and Condemi, S. (2009) Is *Homo heidelbergensis* a distinct species? New insight on the Mauer mandible. *Journal of Human Evolution* **56**(3): 219–46.

Mountain, J.L., Hebert, J.M., Bhattacharyya, S., Underhill, P.A., Ottolenghi, C., Gadgil, M., et al. (1995) Demographic history of India and mtDNA-sequence diversity. *American Journal of Human Genetics* **56**(4): 979–92.

Movius, H. (1969a) The abri de Cro-Magnon, Les Eyzies (Dordogne) and the probable age of the contained burials on the basis of the evidence of the nearby Abri Pataud. *Anuario de Estudios Atlanticos* **15**: 323–44.

Movius, H. (1969b) The Châtelperronian in French archaeology: the evidence of Arcy-sur-Cure. *Antiquity* **43**: 111–23.

Movius, H.L. (1948) The Lower Palaeolithic Cultures of Southern and Eastern Asia. *Transactions of the American Philosophical Society* **38**: 329–426.

Movius, H.L. (1953) The Mousterian Cave of Teshik-Tash, southeastern Uzbekistan, Central Asia. *Bulletin of the American School of Prehistoric Research* **17**: 11–71.

Movius, H.L. (1975) *Excavation of the Abri Pataud, Les Eyzies (Dordogne)*. Peabody Museum, Cambridge, MA.

Moyer, C.C. (2003) *The Organization of Lithic Technology in the Middle and Early Upper Palaeolithic Industries at the Haua Fteah, Libya*. PhD thesis, Corpus Christi College.

Mturi, A.A. (1976) New hominid from Lake Ndutu, Tanzania. *Nature* **262**: 484–5.

Mühlreiter, E. (1870) *Anatomie des menschlichen Gebisses*, Leipzig.

Mukamel, R., Ekstrom, A., Kaplan, J., Iacoboni, M., and Fried, I. (2010) Single-neuron responses in humans during execution and observation of actions. *Current Biology* **20**(8): 750–6.

Mulcahy, N.J., and Call, J. (2006) How great apes perform on a modified trap-tube task. *Animal Cognition* **9**(3): 193–9.

Müller-Beck, H. (1983) *Urgeschichte in Baden-Württemberg*. Konrad Theiss Verlag, Stuttgart.

Mulvaney, D.J. (1970) Green Gully revisited: the later excavations. *Memoirs of the National Museum of Victoria* **30**: 59–77.

Mussi, M. (2001) *Earliest Italy*. Plenum Press, New York.

Nadel, D., ed. (2002) *Ohalo II – a 23,000 Year-Old Fisher-Hunter-Gatherers' Camp on the Shore of the Sea of Galilee*. Hecht Museum, Haifa.

Nadel, D., and Werker, E. (1999) The oldest ever brush hut plant remains from Ohalo II, Jordan Valley, Israel (19 ka BP). *Antiquity* **73**(282): 755–64.

Nadel, D., Carmi, I., and Segal, D. (1995) Radiocarbon dating of Ohalo II: archaeological and methodological implications. *Journal of Archaeological Science* **22**(6): 811–22.

Nadel, D., Tsatskin, A., Mayer, D., Belmaker, M., Boaretto, E., Kislev, M., et al. (2002) The Ohalo II 1999–2000 seasons of excavations: a preliminary report. *Mitekufat Haeven, Journal of the Israel Prehistoric Society* **32**: S17–48.

Nagurka, M.L., and Hayes, W.C. (1980) An interactive graphics package for calculating cross-sectional properties of complex shapes. *Journal of Biomechanics* **13**: 59–64.

Nanci, A. (2003) *Oral Histology*. Mosby, St. Louis, MO.

Nanci, A. (2007) *Ten Cate's Oral Histology: Development, Structure, and Function*, 7th edn. C.V. Mosby, St. Louis, MO.

Napier, J. (1961) Prehensibility and opposability in the hands of primates. *Symposia of the Zoological Society of London* **5**: 115–32.

Napier, J.R. (1959) Fossil metacarpals from Swartkrans. *British Museum (Natural History)* **17**.

Napier, J.R. (1964) The evolution of bipedal walking in the hominids. *Archives of Biology (Liege)* **75**: 673–708.

Napier, J.R. (1993) *Hands*. Revised by R.H. Tuttle. Princeton University Press, Princeton, NJ.

Napier, J.R., and Davies, P.R. (1959) The forelimb skeleton and associated remains of *Proconsul africanus*. In: *Fossil Mammals of Africa*, pp. 1–70. British Museum Natural History, London.

Napier, J.R., and Barnicot, N.A. (1963) *The Primates*. Symposia of the Zoological Society of London, London.

Napier, J.R., and Napier, P.H. (1967) *A Handbook of Living Primates: Morphology, Ecology and Behavior of Nonhuman Primates*. Academic Press, London.

Negash, A., and Shackley, M.S. (2006) Geochemical provenance of obsidian artefacts from the MSA site of Porc Epic, Ethiopia. *Archaeometry* **48**(1): 1–12.

Nelson, C.M. (1971) A standardized site enumeration system for the continent of Africa. *The Pan-African Congress on Prehistory and the Study of the Quaternary Commission on Nomenclature and Terminology Bulletin* **4**: 6–12.

Nelson, E., and Shultz, S. (2010) Finger length ratios (2D: 4D) in anthropoids implicate reduced prenatal androgens in social bonding. *American Journal of Physical Anthropology* **141**(3): 395–405.

Nesbitt, L.M. (1930) Danakil traversed from south to north in 1928. *Geographical Journal* **76**(4): 298–315.

Neubauer, S., Gunz, P., and Hublin, J.-J. (2009) The pattern of endocranial ontogenetic shape changes in humans. *Journal of Anatomy* **215**(3): 240–55.

Neuville, R. (1951) Le Paléolithique et le Mésolithique dans le désert de Judée. *Archives de l'Institute de Paléontologie Humaine* **24**: 179–84.

Newman, T.K., Jolly, C.J., and Rogers, J. (2004) Mitochondrial phylogeny and systematics of baboons (Papio). *American Journal of Physical Anthropology* **124**(1): 17–27.

Newman, T.L., Tuzun, E., Morrison, V.A., Hayden, K.E., Ventura, M., McGrath, S.D., et al. (2005) A genome-wide survey of structural variation between human and chimpanzee. *Genome Research* **15**: 1344–56.

Nicholson, S.E. (2000) The nature of rainfall variability over Africa on time scales of decades to millenia. *Global and Planetary Change* **26**(1–3): 137–58.

Nixon, K.C., and Wheeler, Q.D. (1990) An amplification of the phylogenetic species concept. *Cladistics* **6**(3): 211–33.

Nogueira, M., Monteiro, L., Peracchi, A., and de Ara´jo, A. (2005) Ecomorphological analysis of the masticatory apparatus in the seed-eating bats, genus Chiroderma (Chiroptera: Phyllostomidae). *Journal of Zoology* **266**(04): 355–64.

Nojiriko Excavation Research Group (1994) *The Late Quaternary Environment Around Lake Nojiri in Central Japan*. 29th International Geological Congress, Part B.

Noonan, J.P., Coop, G., Kudaravalli, S., Smith, D., Krause, J., Alessi, J., et al. (2006) Sequencing and analysis of Neanderthal genomic DNA. *Science* **314**(5802): 1113–18.

Normand, C. (2002) L'Aurignacien de la Salle de Saint-Martin (Grotte d'Isturitz; commune de Saint-Martin-d'Aberoue; Pyrénées-Atlantiques): Données préliminaires sur l'industrie lithique recueillie lors des campagnes 2000–2002. *Espacio, Tiempo y Forma, Serie I, Prehistoria y Arqueología* **15**: 145–74.

Norton, C.J. (2000) The current state of Korean paleoanthropology. *Journal of Human Evolution* **38**: 803–25.

Norton, C.J., and Gao, X. (2008) Hominin–carnivore interactions during the Chinese Early Paleolithic: Taphonomic perspectives from Xujiayao. *Journal of Human Evolution* **55**(1): 164–78.

Norton, C.J., and Bae, K.D. (2009) Erratum to "The Movius Line *sensu lato* (Norton et al., 2006) further assessed and defined" J.H. Evol. 55 (2008) 1148–1150. *Journal of Human Evolution* **57**: 331–4.

Norton, C.J., and Jin, J. (2009) The evolution of modern humans in East Asia: current perspectives. *Evolutionary Anthropology* **18**: 247–60.

Norton, C.J., Bae, K., Harris, J.W.K., and Lee, H. (2006) Middle Pleistocene handaxes from the Korean Peninsula. *Journal of Human Evolution* **51**(5): 527–36.

Norton, C.J., Kondo, Y., Ono, A., Zhang, Y.Q., and Diab, M.C. (2010a) The nature of megafaunal extinctions during the MIS 3–2 transition in Japan. *Quaternary International* **211**: 113- 22.

Norton, C.J., Wang, W., and Jin, J.H. (2010b) Modern human origins revisited: Perspectives from Luna Cave (Guangxi, China). *American Journal of Physical Anthropology* **S50**: 178–9.

Novack-Gottshall, P. (2008) Ecosystem-wide body-size trends in Cambrian-Devonian marine invertebrate lineages. *Paleobiology* **34**(2): 210.

Oakley, K. (1957) The dating of the Broken Hill, Florisbad and Saldanha skulls. In: *Proceedings of the 3rd Pan-African Congress of Prehistory, Livingston 1955*, pp. 76–79.

Oakley, K.P. (1949) The fluorine-dating method. *Yearbook of Physical Anthropology* 44–52.

Oakley, K.P. (1949) *Man the Tool-maker*. Trustees of the British Museum, London.

Oakley, K.P. (1964) *Frameworks for Dating Fossil Man*. Aldine Publishing Co., Chicago, IL.

Oakley, K.P., and Montagu, M.F.A. (1949) A reconsideration of the Galley Hill skeleton. *Bulletin of the British Museum (Natural History)* **1**: 25–48.

Oakley, K.P., and Hoskins, C.R. (1950) New Evidence on the Antiquity of Piltdown Man. *Nature* **165**: 379–82.

Oakley, K.P., and Muir-Wood, H.M. (1962) *The Succession of Life Through Geological Time*. British Museum (Natural History), London.

Oakley, K.P., and Campbell, B.G., eds (1967) *Catalogue of Fossil Hominids. Part I: Africa.* Trustees of the British Museum (Natural History), London.

Oakley, K.P., Campbell, B.G., and Molleson, T.I., eds (1971) *Catalogue of Fossil Hominids. Part II: Europe.* Trustees of the British Museum (Natural History), London.

Oakley, K.P., Campbell, B.G., and Molleson, T.I., eds (1975) *Catalogue of Fossil Hominids. Part III: Americas, Asia, Australia.* Trustees of the British Museum (Natural History), London.

Oakley, K.P., Campbell, B.G., and Molleson, T.I., eds (1977) *Catalogue of Fossil Hominids. Part I: Africa.* 2nd. Ed. Trustees of the British Museum (Natural History), London.

Ober, C., Weitkamp, L.R., Cox, N., Dytch, H., Kostyu, D., and Elias, S. (1997) HLA and mate choice in humans. *American Journal of Human Genetics* 61(3): 497–504.

O'Connell, J., Hawkes, K., and Blurton Jones, N.G. (1999) Grandmothering and the evolution of *Homo erectus. Journal of Human Evolution* 36: 461–85.

O'Connell, J.F. (1997) On Plio/Pleistocene archaeological sites and central places. *Current Anthropology* 38: 86–88.

O'Connell, J.F., Hawkes, K., Lupo, K.D., and Blurton Jones, N.G. (2002) Male strategies and Plio-Pleistocene archaeology. *Journal of Human Evolution* 43: 831–72.

Oda, S. (2003) Paleolithic artifacts from Yamashita-cho Cave One. *Nanto Koko (Journal of the Okinawa Archaeological Society)* 22: 1–19.

Oda, S. (2007) A research history of Paleolithic culture in Okinawa based on the findings in Ie Island. *Nanto Koko (Journal of the Okinawa Archaeological Society)* 26: 37–48.

Odell, G.H. (2004) Lithic analysis. In: *Manuals in Archaeological Method, Theory, and Technique.* Kluwer Academic/Plenum Publishers, New York.

Odling-Smee, F.J. (1988) Niche construction phenotypes. In: *The Role of Behavior in Evolution*, Plotkin, H.C., ed., pp. 73–132. MIT Press, Cambridge, MA.

Odling-Smee, F.J., Laland, K.N., and Feldman, M.W. (2003) *Niche Construction.* Princeton University Press, Princeton, NJ.

Ogden, C.L., Fryar, C.D., Carroll, M.D., and Feigal, K.M. (2004) Mean body weight, height, and body mass index: United States 1960–2002. *Advance Data From Vital and Health Statistics*, no. 347, pp. 1–17. DHHS publication no. (PHS) 2005-125004-0467 (10/04).

Ogihara, N., and Yamazaki, N. (2001) Generation of human bipedal locomotion by a bio-mimetic neuro-musculo-skeletal model. *Biological Cybernetics* 84: 1–11.

Ohman, J.C. (1986) The first rib of hominoids. *American Journal of Physical Anthropology* 70: 209–29.

Ohman, J.C., Wood, C., Wood, B.A., Crompton, R.H., Gunther, M.M., Savage, R., et al. (2002) Stature-at-death of KNM-WT 15000. *Human Evolution* 17(3–4): 129–41.

Ohman, J.C., Lovejoy, C.O., and White, T. (2005) Questions about *Orrorin tugenensis. Science* 307: 845.

Ohno, S. (1970) *Evolution by Gene Duplication.* Springer-Verlag, Berlin.

Ohta, T. (1992) The Nearly Neutral Theory of Molecular Evolution. *Annual Review of Ecology and Systematics* 23: 263–86.

Okanoya, K. (2002) Sexual display as a syntactic vehicle: The evolution of syntax in birdsong and human language through sexual selection. In: *The Transition to Language*, Wray, A., ed., pp. 46–64. Oxford University Press, New York.

Olariu, A., Skog, G., Hellborg, R., Stenström, K., Faarinen, M., Persson, P., et al. (2003) *Dating of Two Paleolithic Human Fossil Bones from Romania by Accelerator Mass Spectrometry.* Proceedings of the International Conference on Applications of High Precision Atomic & Nuclear Methods, Neptun, Romania, Editura Academiei Romane.

Olejniczak, A. (2006) *Micro-computed Tomography of Primate Molars.* PhD thesis, State University of New York at Stony Brook.

Olejniczak, A.J., Smith, T.M., Feeney, R.N.M., Macchiarelli, R., Mazurier, A., Bondioli, L., et al. (2008a) Dental tissue proportions and enamel thickness in Neandertal and modern human molars. *Journal of Human Evolution* 55(1): 12–23.

Olejniczak, A.J., Smith, T.M., Wang, W., Potts, R., Ciochon, R., Kullmer, O., et al. (2008b) Molar enamel thickness and dentin horn height in *Gigantopithecus blacki. American Journal of Physical Anthropology* 135(1): 85–91.

Oliva, M. (2006) The Upper Paleolithic find from the Mladec Cave. In: *Early Modern Humans at the Moravian Gate*, pp. 41–74. Springer Vienna, Vienna.

Oliver, J.S. (1994) Estimates of hominid and carnivore involvement in the FLK *Zinjanthropus* fossil assemblage: some sociological implications. *Journal of Human Evolution* 27(1–3): 267–94.

Olivier, G., and Tissier, H. (1975) Determination of cranial capacity in fossil men. *American Journal of Physical Anthropology* 43(3): 353–62.

Olson, E.C., and Miller, R.L. (1958) *Morphological Integration.* University of Chicago Press, Chicago, IL.

Olson, S.L. (1984) A hamerkop from the Early Pliocene of South Africa (Aves: Scopidae). *Proceedings of the Biological Society of Washington* 97: 736–40.

Olson, S.L. (1985a) Early Pliocene Procellariiformes (Aves) from Langebaanweg, south-western Cape Province, South Africa. *Annals of the South African Museum* 95: 123–45.

Olson, S.L. (1985b) Early Pliocene ibises (Aves, Plataleidae) from the south-western Cape Province, South Africa. *Annals of the South African Museum* 97: 57–69.

Olson, T.R. (1978) Hominid phylogenetics and the existence of *Homo* in Member I of the Swartkrans Formation, South Africa. *Journal of Human Evolution* 7: 159–78.

Ono, A. (2001) *Flaked Bone Tools: an Alternative Perspective on the Palaeolithic.* University of Tokyo, Tokyo.

Oppenheimer, C. (2007) Limited global change due to the largest known Quaternary eruption, Toba 74 kyr BP? *Quaternary Science Reviews* 21: 1593–609.

Oppenoorth, W.F.F. (1932a) *Homo (Javanthropus) soloensis een* Plistoceene mensch van Java. *Wetenschappelijike Mededeelingen* 20: 49–74.

Oppenoorth, W.F.F. (1932b) Ein neuer diluvialer Urmensch von Java. *Natur und Museum* 62: 269–79.

Oppenoorth, W.F.F. (1937) The place of *Homo soloensis* among fossil men. In: *Early Man*, MacCurdy, G.G., ed., pp. 349–60. J.B. Lippincott Company, Philadelphia and New York.

Orchiston, D.W., and Siesser, W.G. (1982) Chronostratigraphy of the Plio-Pleistocene fossil hominids of Java. *Modern Quaternary Research in Southeast Asia* 7: 131–49.

O'Regan, H., Chenery, C., Lamb, A., Stevens, R., Rook, L., and Elton, S. (2008) Modern macaque dietary heterogeneity assessed using stable isotope analysis of hair and bone. *Journal of Human Evolution* 55(4): 617–26.

Orgel, J.P.R.O., Irving, T.C., Miller, A., and Wess, T.J. (2006) Microfibrillar structure of type I collagen in situ. *Proceedings of the National Academy of Sciences USA* 103(24): 9001–5.

Origgi, G., and Sperber, D. (2000) Evolution, communication and the proper function of language. In: *Evolution and the Human Mind: Modularity, Language, and Metacognition*, Carruthers, P., and Chamberlain, A., eds, pp. 140–69. Cambridge University Press, Cambridge.

Osborn, H.F. (1888) The evolution of mammalian molars to and from the tritubercular type. *American Naturalist* 22(264): 1067–79.

Osborn, H.F. (1892) Taxonomy and morphology of the primates, creodonts, and ungulates. Fossil mammals of the Wahsatch and Wind River beds. Collection of 1891. Part IV. *Bulletin of the American Museum of Natural History of Science* 4: 101–34.

Osborn, H.F. (1907) *Evolution of Mammalian Molar Teeth: To and From the Triangular Type*. Macmillan and Co, London.

Osborn, H.F. (1916) *Men of the Old Stone Age: Their Environment, Life, and Art*. Charles Scribner's Sons, New York.

Osborn, J.W. (1990) A 3-dimensional model to describe the relation between prism directions, parazones and diazones, and the Hunter-Schreger bands in human tooth enamel. *Archives of Oral Biology* 35(11): 869–78.

O'Sullivan, P.B., Morwood, M., Hobbs, D., Suminto, F.A., Situmorang, M., Raza, A., et al. (2001) Archaeological implications of the geology and chronology of the Soa Basin, Flores, Indonesia. *Geology* 29(7): 607–10.

Otte, M., Toussaint, M., and Bonjean, D. (1993) Découverte de restes humains immatures dans les niveaux moustériens de la grotte Scladina à Andenne (Belgique). *Bulletins et Mémoires de la Société d'anthropologie de Paris* 5(1): 327–32.

Ovchinnikov, I.V., Gotherstrom, A., Romanova, G.P., Kharitonov, V.M., Liden, K., and Goodwin, W. (2000) Molecular analysis of Neanderthal DNA from the northern Caucasus. *Nature* 404: 490–3.

Ovey, C.D., ed. (1964) *The Swanscombe Skull*. Occasional Papers of the Royal Anthropological Institute. Royal Anthropological Institute, London.

Owen, R. (1835) On the osteology of the chimpanzee and orang utan. *Transactions of the Zoological Society of London* 1: 343–79.

Owen, R.B., Potts, R., Behrensmeyer, A.K., and Ditchfield, P.W. (2008) Diatomaceous sediments and environmental change in the Pleistocene Olorgesailie Formation, southern Kenya Rift Valley. *Palaeogeography, Paelaeoclimatology, Palaeoecology* 269(1–2): 17–37.

Pääbo, S. (1985) Preservation of DNA in ancient Egyptian mummies. *Journal of Archaeological Science* 12(6): 411–17.

Pääbo, S. (1986) *Molecular Genetic Investigations of Ancient Human Remains*, Cold Spring Harbor Laboratory Press, Cold Spring Harbor, NY.

Pääbo, S., Gifford, J., and Wilson, A. (1988) Mitochondrial DNA sequences from a 7000-year old brain. *Nucleic Acids Research* 16(20): 9775.

Paddayya, K. (1982) *The Acheulean Culture of the Hunsgi Valley (Peninsular India): A Settlement System Perspective*. Deccan College Postgraduate and Research Institute, Poona.

Paddayya, K., and Petraglia, M.D. (1993) Formation processes of Acheulean localities in the Hunsgi and Baichbal Valleys, Peninsular India. In: *Formation Processes in Archaeological Context*, Goldberg, P., Nash, D.T., and Petraglia, M.D., eds, pp. 61–82. Prehistory Press, Madison, WI.

Paddayya, K., and Petraglia, M.D. (1997) Isampur: an Acheulean workshop site in the Hunsgi Valley, Gulbarga District, Karnataka. *Man and Environment* 22(2): 95–100.

Paddayya, K., Blackwell, B.A.B., Jhaldiyal, R., Petraglia, M.D., Fevrier, S., Chaderton II, D.A., et al. (2002) Recent findings on the Acheulian of the Hunsgi and Baichbal valleys, Karnataka, with special reference to the Isampur excavation and its dating. *Current Science* 83: 5.

Pagani, M., Freeman, K.H., and Arthur, M.A. (1999) Late Miocene atmospheric CO_2 concentrations and the expansion of C_4 grasses. *Science* 285(5429): 876–9.

Pagel, M., and Meade, A. (2007) *BayesTraits*, University of Reading.

Pagel, M., Meade, A., and Barker, D. (2004) Bayesian estimation of ancestral character states on phylogenies. *Systematic Biology* 53(5): 673–84.

Pales, L. (1939) Les grottes de Malarnaud. In: *Mélanges de préhistoire et d'anthropologie offerts par ses collégues, amis et disciples au Professeur Comte H. Begouèn*. Edition du Muséum, Tolouse.

Palma di Cesnola, A. (1989) L'Uluzzien: Facies Italien du Leptolithique Archaque. *L'Anthropologie* **93**: 783–1.

Palma di Cesnola, A., and Messeri, P. (1967) Quatre dents humaines paleolithiques trouvees dans des cavernes de l'Italie meridionale. *L'Anthropologie* **71**: 249–61.

Panagopoulou, E., Karkanas, P., Kotjabopoulou, E., Tsarsidou, G., Harvati, K., and Ntinou, M. (2002–2004) Late Pleistocene archaeological and fossil human evidence from Lakonis cave, southern Greece. *Journal of Field Archaeology* **29**: 323–49.

Panger, M.A., Brooks, A.S., Richmond, B.G., and Wood, B.A. (2002) Older than the Oldowan? Rethinking the emergence of hominin tool use. *Evolutionary Anthropology* **11**(6): 235–45.

Parat, A. (1903) Les grottes de la Cure (cote d'Arcy) XXV La grotte des Fees et les petites grottes de l'anse. *Bulletin de la Société des sciences historiques et naturelles de l'Yonne* **2**: 55.

Parés, J.M., and Pérez-González, A. (1995) Paleomagnetic age from hominid fossils at Atapuerca archaeological site, Spain. *Science* **269**: 830–2.

Parés, J.M., and Pérez-González, A. (1999) Magnetochronology and stratigraphy at Gran Dolina section, Atapuerca (Burgos, Spain). *Journal of Human Evolution* **37**: 325–42.

Parfitt, S.A., Barendregt, R.W., Breda, M., Candy, I., Collins, M.J., Coope, G.R., et al. (2005) The earliest record of human activity in Northern Europe. *Nature* **438**: 1008–12.

Parfitt, S.A., Ashton, N.M., Lewis, S.G., Abel, R.L., Coope, G.R., Field, M.H., et al. (2010) Early Pleistocene human occupation at the edge of the boreal zone in northwest Europe. *Nature* **466**(7303): 229–33.

Park, S.J. (1992) The Pleistocene hominid fossils in Korea. In: *Korea–China Quaternary–Prehistory Symposium*, Sohn, P.K., ed., pp. 112–14. Korea Anthropological Institute, Dankook University, Seoul.

Park, S.J. (1996) Pleistocene faunal remains from Seakul/Chonyokul at Turubong Cave Complex with special emphasis on the large mammalian fossils. In: *Paleolithic Cave Sites and Culture in Northeast Asia*, Lee, S.J., and Lee, Y.J., eds, pp. 85–94. Chungbuk National University Museum, Chongju.

Park, S.J., and Lee, Y.J. (1990) A new discovery of the Upper Pleistocene child's skeleton from Hungsu Cave (Turubong Cave Complex) Ch'ongwon, Korea. *Korean Journal of Quaternary Research* **4**: 1–14.

Partridge, T.C. (1973) Geomorphological dating of cave openings at Makapansgat, Sterkfontein, Swartkrans and Taung. *Nature* **246**: 75–79.

Partridge, T.C. (1978) *Geomorphological Dating of Cave Openings at Makapansgat, Sterkfontein, Swartkrans and Taung*. South African Archaeological Society.

Partridge, T.C. (1979) Re-appraisal of lithostratigraphy of Makapansgat Limeworks hominid site. *Nature* **279**: 484–8.

Partridge, T.C. (1982) *The Ages and Paleo-Environments of the South African Hominid Sites in Actas*. Union Internacional de Ciencias Prehistoricas y Protohistoricas.

Partridge, T.C. (2000) Hominid-bearing cave and tufa deposits. In: *The Cenozoic of Southern Africa*, Partridge, T.C., and Maud, R.R., eds, pp. 100–30. Oxford University Press, Oxford.

Partridge, T.C. (2005) Dating the Sterkfontein hominids: progess and possibilities. *Transactions of the Royal Society of South Africa* **60**(2): 107–10.

Partridge, T.C., and Watt, I.B. (1991) The stratigraphy of the Sterkfontein hominid deposit and its relationship to the underground cave system. *Palaeontologica Africana* **28**: 35–40.

Partridge, T.C., deMenocal, P.B., Lorentz, S.A., Paiker, M.J., and Vogel, J.C. (1997) Orbital forcing of climate over South Africa: a 200,000-year rainfall record from the Pretoria saltpan. *Quaternary Science Reviews* **16**(10): 1125–33.

Partridge, T.C., Granger, D.E., Caffee, D.E., and Clarke, R.J. (2003) Lower Pliocene hominid remains from Sterkfontein. *Science* **300**(5619): 607–12.

Passarello, P., Salvadei, L., and Manzi, G. (1989) New evidence of archaic *Homo sapiens* from central Italy. The human parietal from Casal de'Pazzi (Rome) in the mid-to-late Pleistocene fossil record. In: *Hominidae: Proceedings of the 2nd International Congress of Human Paleontology, Turin, September 28-October 3, 1987*, Giacobini, G., ed., pp. 283–6. Jaca Book, Milan.

Patel, A. (2010) *Music, Language, and the Brain*. Oxford University Press, New York.

Paterson, H.E.H. (1985) The recognition concept of species. In: *Species and Speciation*, Vrba, E.S., ed., pp. 21–29. Transvaal Museum, Pretoria.

Patnaik, R., Chauhan, P.R., Rao, M.R., Blackwell, B.A.B., Skinner, A.R., Sahni, A., et al. (2009) New geochronological, paleoclimatological, and archaeological data from the Narmada Valley hominin locality, central India *Journal of Human Evolution* **56**(2): 114–33.

Patterson, B., and Howells, W.W. (1967) Hominid humeral fragment from early Pleistocene of northwest Kenya. *Science* **156**: 64–66.

Patterson, B., Behrensmeyer, A.K., and Sill, W.D. (1970) Geology and fauna of a new Pliocene locality in northwestern Kenya. *Nature* **226**: 918–21.

Păunescu, A. (2001) *Paleoliticul și mezoliticul din spațiul transilvan*. Asociatia Generala a Inginerilor din Romania, Bucharest.

Pearson, O., and Grine, F. (1996) Morphology of the Border Cave hominid ulna and humerus. *South African Journal of Science* **92**(5): 231–6.

Pearson, O., Fleagle, J.G., Grine, F.E., and Royer, D.F. (2008b) Further new hominin fossils from the Kibish Formation, southwestern Ethiopia. *Journal of Human Evolution* **55**(3): 444–7.

Pearson, O.M. (2000) Activity, climate, and postcranial robusticity: implications for modern human origins and scenarios of adaptive change. *Current Anthropology* **41**(4): 569–607.

Pearson, O.M. (2004) Has the combination of genetic and fossil evidence solved the riddle of modern human origins? *Evolutionary Anthropology* **13**: 145–59.

Pearson, O.M., and Grine, F.E. (1997) Re-analysis of the hominid radii from Cave of Hearths and Klasies River Mouth, South Africa. *Journal of Human Evolution* **32**: 577–92.

Pearson, O.M., and Lieberman, D.E. (2004) The aging of Wolff's "Law:" Ontogeny and responses to mechanical loading in cortical bone. *Yearbook of Physical Anthropology* **47**: 63–99.

Pearson, O.M., Royer, D.F., Grine, F.E., and Fleagle, J.G. (2008a) A description of the Omo I postcranial skeleton, including newly discovered fossils *Journal of Human Evolution* **55**: 421–37.

Pei, W., and Zhang, S. (1985) *A Study of the Lithic Artifacts of Sinanthropus*. Science Press, Beijing.

Pei, W.C. (1934) A preliminary report on the Late Palaeolithic cave of Choukoutien. *Bulletin of the Geological Society of China* **13**: 327–58.

Pei, W.C. (1939) On the Upper Cave industry. *Palaeontologica Sinica* **9**: 1–58.

Pei, W.C. (1940) The Upper Cave fauna of Choukoutien. *Palaeontologia Sinica* **125**.

Pelegrin, J. (2005) Remarks about archaeological techniques and methods of knapping: elements of a cognitive approach to stone knapping. In: *Stone Knapping: the Necessary Conditions for a Uniquely Hominin Behaviour*, Roux, V., and Bril, B., eds, pp. 23–34. Oxbow Books, Oxford.

Penn, D.C., and Povinelli, D.J. (2007) On the lack of evidence that non-human animals possess anything remotely resembling a 'theory of mind'. *Philosophical Transactions of the Royal Society of London. Series B, Biological Sciences* **362**(1480): 731–44.

Péquart, M., and Péquart, S. (1941) Nouvelles découvertes a la grotte du Mas d'Azil. *Bulletins et Mémoires de la Société d'anthropologie de Paris* **2**(1): 128–30.

Perry, G.H., Tito, R.Y., and Verrelli, B.C. (2007) The evolutionary history of human and chimpanzee Y-Chromosome gene loss. *Molecular Biology and Evolution* **24**(3): 853–9.

Perumal, S., Antipova, O., and Orgel, J.P.R.O. (2008) Collagen fibril architecture, domain organization, and triple-helical conformation govern its proteolysis. *Proceedings of the National Academy of Sciences USA* **105**(8): 2824–9.

Pesce Delfino, V., and Vacca, E. (1993) An archaic human skeleton discovered at Altamura (Bari, Italy). *Riv. Antrop.* **71**: 249–57.

Pesce Delfino, V., and Vacca, E. (1995) *The Altamura Human Skeleton: Discovery and in situ Examination*. Homo erectus heidelbergensis *von Mauer - Kolloquium I*. Beinhauer, K.W., Kraatz, R., and Wagner, G.A., eds. Jan Thorbecke Verlag, Sigmaringen.

Pestle, W., Colvard, M., and Pettitt, P. (2006) AMS dating of a recently rediscovered juvenile human mandible from Solutré (Saône-et-Loire, France). *Paleo* **18**: 285–92.

Peters, J. (1990) Late Pleistocene Hunter-gatherers at Ishango (Eastern-Zaire). *Revue de paleobiologie* 73–112.

Peterson, C.E., Shen, C., Chen, C., Chen, W., and Tang, Y. (2003) Taphonomy of an early Pleistocene archaeofauna from Xiaochangliang, Nihewan Basin, northern China. In: *Current Research in Chinese Pleistocene Archaeology*, Shen, C., and Keates, S.G., eds, pp. 83–97. Archaeopress, Oxford.

Petraglia, M., and Potts, R. (2004) The Old World Paleolithic and the development of a national collection. *Smithsonian Contributions to Anthropology* **48**.

Petraglia, M., LaPorta, P., and Paddayya, K. (1999) The first Acheulean quarry in India: stone tool manufacture, biface morphology, and behaviors. *Journal of Anthropological Research* **55**: 39–70.

Petraglia, M., Korisettar, R., Boivin, N., Clarkson, C., Ditchfield, P., Jones, S., et al. (2007) Middle Palaeolithic assemblages from the Indian subcontinent before and after the Toba super-eruption. *Science* **317**: 114–16.

Petraglia, M., Clarkson, C., Boivin, N., Haslam, M., Korisettar, R., Chaubey, G., et al. (2009) Population increase and environmental deterioration correspond with microlithic innovations in South Asia ca. 35,000 years ago. *Proceedings of the National Academy of Sciences USA* **106**: 12261–6.

Petraglia, M.D., Shipton, C., and Paddayya, K. (2005) Life and mind in the Acheulean: a case study from India. In: *The Hominid Individual in Context*, Gamble, C., and Porr, M., eds, pp. 197–219. Routledge, London.

Pettitt, P.B., Richards, M., Maggi, R., and Formicola, V. (2003) The Gravettian burial known as the Prince ("Il Principe"): new evidence for his age and diet. *Antiquity* **77**(295): 15–19.

Peyrony, D. (1928) *Pieces a languette de l'Aurignacien moyen*. Association Francaise pour l'Avancement des Sciences, La Rochelle.

Peyrony, D. (1930) Le Moustier, ses gisements, ses industries, ses couches géologiques. *Revue Anthropologique* **40**: 48–76, 155–76.

Peyrony, D. (1934) La Ferrassie: Moustérien - périgordien - aurignacien. *Préhistoire* **3**(1): 410–11.

Pézard, A. (1918) Le conditionnement physiologique des caractères sexuels secondaires chez les oiseaux. *Bulletin biologique de la France et de la Belgique* **52**: 1–176.

Pfeiffer, S., and Zehr, M.K. (1996) A morphological and histological study of the human humerus from Border Cave. *Journal of Human Evolution* **31**: 49–59.

Phillips, D.L., and Koch, P.L. (2002) Incorporating concentration dependence in stable isotope mixing models. *Oecologia* **130**: 114–25.

Phillips-Conroy, J.E., Jolly, C.J., and Brett, F.L. (1991) The characteristics of hamadryas-like males living in anubis baboon groups. *American Journal of Physical Anthropology* **86**: 353–68.

Pianka, E.R. (1970) On r- and K- selection. *American Naturalist* **104**(940): 592–7.

Pickerill, H.P. (1913) The structure of enamel. *Dental Cosmos* **55**: 969–88.

Pickering, R., and Kramers, J.D. (2010) Re-appraisal of the stratigraphy and determination of new U-Pb dates for the Sterkfontein hominin site, South Africa. *Journal of Human Evolution* **59**: 70–86.

Pickering, R., Hancox, P.J., Lee-Thorp, J.A., Grün, R., Mortimer, G.E., McCulloch, M., et al. (2007) Stratigraphy, U-Th chronology, and paleoenvironments at Gladysvale Cave: insights into the climatic control of South African hominin-bearing cave deposits. *Journal of Human Evolution* **53**(5): 602–19.

Pickering, T.R., White, T.D., and Toth, N. (2000) Cutmarks on a Plio-Pleistocene hominid from Sterkfontein, South Africa. *American Journal of Physical Anthropology* **111**: 579–84.

Pickering, T.R., Clarke, R.J., and Moggi-Cecchi, J. (2004) Role of carnivores in the accumulation of the Sterkfontein Member 4 hominid assemblage: a taphonomic reassessment of the complete hominid fossil sample (1936–1999). *American Journal of Physical Anthropology* **125**(1): 1–15.

Pickford, M. (1974) *Stratigraphy and Palaeoecology of Five Late Cainozoic Formations in the Kenya Rift Valley.* PhD thesis, University of London.

Pickford, M. (1975) Late Miocene sediments and fossils from the Northern Kenya Rift valley. *Nature* **256**: 279–84.

Pickford, M. (1993) Climate change, biogeography and *Theropithecus*. In: *Theropithecus. The Rise and Fall of a Primate Genus*, Jablonski, N.G., ed., pp. 227–43. Cambridge University Press, Cambridge.

Pickford, M., and Senut, B. (2001) The geological and faunal context of Late Miocene remains from Lukeino, Kenya. *Comptes Rendus de l'Académie des Sciences* **332**: 145–52.

Pickford, M., Senut, B., Poupeau, G., Brown, F.H., and Haileab, B. (1991) Correlation of tephra layers from the western Rift Valley (Uganda) to the Turkana Basin (Ethiopia/Kenya) and the Gulf of Aden. *Comptes Rendus de l'Académie des Sciences* **313**: 223–9.

Pickford, M., Senut, B., Gommery, D., and Treil, J. (2002) Bipedalism in *Orrorin tugenensis* revealed by its femora. *Comptes Rendus Palevol* **1**: 1–13.

Picq, P. (1983) *L'articulation temporo-mandibulaire des Hominidés fossiles: anatomie comparée, biomechanique, evolution, biométrie.* Thesis, L'Université Pierre et Marie Curie.

Piette, E. (1895) Hiatus et lacune. Vestiges de la période de transition dans la grotte du Mas-d'Azil. *Bulletins de la Société d'anthropologie de Paris* **6**(1): 235–67.

Pigeot, N. (1990) Technical and social actors: flintknapping specialists and apprentices at Magdelinian Etoilles. *Archaeological Review from Cambridge* **9**: 127–41.

Pike, A.W.G., Eggins, S., Grün, R., and Thackeray, F. (2004) U-series dating of TP1, an almost complete human skeleton from Tuinplaas (Springbok Flats), South Africa. *South African Journal of Science* **100**: 381–3.

Pilbeam, D. (1978) Rearranging our family tree. *Human Nature* 38–45.

Pilbeam, D. (1982) New hominoid skull material from the Miocene of Pakistan. *Nature* **295**: 232–4.

Pilbeam, D., Rose, M.D., Barry, J.C., and Shah, S.M.I. (1990) New *Sivapithecus* humeri from Pakistan and the relationship of *Sivapithecus* and *Pongo*. *Nature* **348**: 237–9.

Pilbrow, V. (2006) Population systematics of chimpanzees using molar morphometrics. *Journal of Human Evolution* **51**(6): 646–62.

Pilbrow, V. (2007) Patterns of molar variation in great apes and their implications for hominin taxonomy. In: *Dental Perspectives on Human Evolution: State of the Art Research in Dental Paleoanthropology*, Bailey, S., and Hublin, J.-J., eds, pp. 9–32. Springer, Dordrecht.

Pilbrow, V. (2010) Dental and phylogeographic patterns of variation in gorillas. *Journal of Human Evolution* **59**: 16–34.

Pinhasi, R., and Semal, P. (2000) The position of the Nazlet Khater specimen among prehistoric and modern African and Levantine populations. *Journal of Human Evolution* **39**(3): 269–88.

Pinker, S. (1994) *The Language Instinct*. William Morrow and Company, New York.

Pinker, S. (2003) Language as an adaptation to the cognitive niche. In: *Language Evolution: States of the Art*, Christiansen, M., and Kirby, S., eds, pp. 16–37. Oxford University Press, New York.

Pinker, S., and Bloom, P. (1990) Natural selection and natural language. *Behavioral and Brain Sciences* **13**(4): 707–84.

Pinker, S., and Jackendoff, R. (2005) The faculty of language: what's special about it? *Cognition* **95**(2): 201–36.

Piperno, D., Weiss, E., Holst, I., and Nadel, D. (2004) Processing of wild cereal grains in the Upper Palaeolithic revealed by starch grain analysis. *Nature* **430**(7000): 670–3.

Piperno, D.R. (2006) *Phytoliths: a Comprehensive Guide for Archaeologists and Paleoecologists*. AltaMira Press, New York.

Piperno, M., and Schchilone, G., eds (1991) *The Circeo 1 Neandertal Skull: Studies and Documentation*. Istituto Poligrafico e Zecca Dello Stato, Rome.

Piperno, M., Collina, C., Gallotti, R., Raynal, J.-P., Kieffer, G., le Bourdonnec, F.-X., et al. (2008) Obsidian exploitation and utilization during the Oldowan at Melka Kunture (Ethiopia). In: *Interdisciplinary Approaches to the Oldowan*, Hovers, E., and Braun, D.R., eds, pp. 111–28. Springer Netherlands, Dordrecht.

Pitsios, T.K. (1985) Παλαιοανθρωπολογικές έρευνες στη θέση «Απήδημα» της Μέσα Μάνης. *Αρχαιολογία* **15**: 26–33.

Pittard, E. (1963) Une gravure de Cro-Magnon exilée à Neufchâtel. *Bulletin de la Société d'Etudes et de Recherches Préhistoriques, Les Eyzies de Tayac* **12**: 36–39.

Piveteau, J. (1959) Les restes humains de la grotte de Regourdou (Dordogne). *Comptes Rendus de l'Académie des Sciences* **248**: 40–44.

Piveteau, J. (1963) La grotte de Regourdou (Dordogne). Paléontologie humaine. *Annales de Paléontologie* **49**: 285–304.

Piveteau, J. (1964) La grotte de Regourdou (Dordogne). Paléontologie humaine. *Annales de Paléontologie* **50**: 155–94.

Piveteau, J. (1966) La grotte de Regourdou (Dordogne). Paléontologie humaine. *Annales de Paléontologie* **52**: 163–94.

Plavcan, J.M. (2000) Inferring social behavior from sexual dimorphism in the fossil record. *Journal of Human Evolution* **39**: 327–44.

Plavcan, J.M. (2002) Reconstructing social behavior from dimorphism in the fossil record. In: *Reconstructing Behavior in the Primate Fossil Record*, Plavcan, J.M., Kay, R.F., Jungers, W.L., and van Schaik, C.P., eds, pp. 297–332. Kluwer Academic/Plenum Publications, New York.

Pleurdeau, D. (2005) Human technical behavior in the African Middle Stone Age: the lithic sssemblage of Porc-Epic Cave (Dire Dawa, Ethiopia). *African Archaeological Review* **22**(4): 177–97.

Plisson, H. (1985) Quels soins prendre des outillages lithiques pour l'analyse fonctionnelle? *Bulletin de la Société Préhistorique Française* **82**: 99–101.

Plummer, T. (2004) Flaked stones and old bones: Biological and cultural evolution at the dawn of the dawn of technology. *Yearbook of Physical Anthropology* **47**: 118–64.

Plummer, T.W., and Potts, R. (1989) Excavations and new findings at Kanjera, Kenya. *Journal of Human Evolution* **18**: 269–76.

Plummer, T.W., Bishop, L.C., Ditchfield, P.W., and Hicks, J. (1999) Research on Late Pliocene Oldowan Sites at Kanjera South, Kenya. *Journal of Human Evolution* **36**: 151–70.

Pobiner, B.L. (2007) *Hominin-carnivore Interactions: Evidence from Modern Carnivore Bone Modification and Early Pleistocene Archaeofaunas (Koobi Fora, Kenya; Olduvai Gorge, Tanzania)*. PhD thesis, Rutgers, The State University of New Jersey.

Pocock, T.N. (1987) Plio-pleistocene fossil mammalian microfauna of southern Africa: a preliminary report including description of two fossil muroid genera (Mammalia: Rodentia). *Palaeontologia Africana* **26**: 69–91.

Poinar, H., Hoss, M., Bada, J., and Paabo, S. (1996) Amino acid racemization and the preservation of ancient DNA. *Science* **272**(5263): 864.

Poinar, H., Kuch, M., Sobolik, K., Barnes, I., Stankiewicz, A., Kuder, T., et al. (2001) A molecular analysis of dietary diversity for three archaic Native Americans. *Proceedings of the National Academy of Sciences USA* **98**(8): 4317.

Ponce de León, M.S., Golovanova, L., Doronichev, V., Romanova, G., Akazawa, T., Kondo, O., et al. (2008) Neanderthal brain size at birth provides insights into the evolution of human life history. *Proceedings of the National Academy of Sciences USA* **105**(37): 13764–8.

Pontzer, H., Raichlen, D.A., and Sockol, M.D. (2009) The metabolic cost of walking in humans, chimapnzees, and early hominins. *Journal of Human Evolution* **56**(1): 43–54.

Pontzer, H., Rolian, C., Rightmire, G., Jashashvili, T., Ponce de León, M., Lordkipanidze, D., et al. (2010) Locomotor anatomy and biomechanics of the Dmanisi hominins. *Journal of Human Evolution* **58**: 492–504.

Pope, G.G. (1983) Evidence on the age of the Asian Hominidae. *Proceedings of the National Academy of Sciences USA* **80**: 4988–92.

Pope, G.G. (1992) Paleoenvironment and paleoanthropology in China. *Journal of Human Evolution* **22**(6): 519–21.

Pope, G.G., and Cronin, J.E. (1984) The Asian hominidae. *Journal of Human Evolution* **13**: 377–96.

Pope, G.G., and Keates, S.G. (1994) The evolution of human cognition and cultural capacity: a view from the Far East. In: *Integrative Paths to the Past: Paleoanthropological Advances in Honor of F. Clark Howell*, Corruccini, R.S., and Ciochon, R.L., eds, pp. 531–67. Prentice Hall, Englewood Cliffs, NJ.

Porat, N., Chazan, M., Grün, R., Aubert, M., Eisenmann, V., & Horwitz, L. K. (2010) New radiometric ages for the Fauresmith industry from Kathu Pan, southern Africa: Implications for the Earlier to Middle Stone Age transition. *Journal of Archaeological Science* **37**(2): 269–83.

Portmann, A. (1941) Die Tragzeiten der Primaten und die Dauer Schwangerschaft beim Menschen: Ein Problem der vergleichenden Biologie. *Revue Suisse de Zoologie* **48**: 511–18.

Potts, R. (1988) *Early Hominid Activities at Olduvai*. Transaction Publishers, New Brunswick, NJ.

Potts, R. (1991) Why the Oldowan? Plio-Pleistocene toolmaking and the transport of resources. *Journal of Anthropological Research* **47**(2): 153–76.

Potts, R. (1998) Variability selection in hominid evolution. *Evolutionary Anthropology* **7**: 81–96.

Potts, R. (2007) Environmental hypotheses of Pliocene human evolution. In: *Hominin Environments in the East African Pliocene: An Assessment of the Faunal Evidence*, Bobe, R., Alemseged, Z., and Behrensmeyer, A.K., eds, pp. 25–49. Springer Netherlands, Dordrecht.

Potts, R., and Deino, A. (1995) Mid-Pleistocene change in large mammal faunas of the southern Kenya rift. *Quaternary Research* **43**: 106–13.

Potts, R., and Teague, R.L. (2010) Behavioral and environmental background to 'Out of Africa I' and the arrival of *Homo erectus* in East Asia. In: *Out of Africa I: the First Hominin Colonization of Eurasia*, Fleagle, J.G., Shea, J.J.,

Grine, F.E., Baden, A.L., and Leakey, R.E., eds, pp. 67–85. Springer, Dordrecht.

Potts, R., Shipman, P., and Ingall, E. (1988) Taphonomy, paleoecology, and hominids of Lainyamok, Kenya. *Journal of Human Evolution* 17(6): 597–614.

Potts, R., Behrensmeyer, A.K., and Ditchfield, P.W. (1999) Paleolandscape variation and Early Pleistocene hominid activities: Members 1 and 7, Olorgesailie Formation, Kenya. *Journal of Human Evolution* 37: 747–88.

Potts, R., Weiwen, H., Yamei, H., Deino, A., Baoyin, Y., Zhentang, G., et al. (2000) Tektites and the age paradox in Mid-Pleistocene China: Response to Koeberl and Glass and Keates. *Science* 289: 507a.

Potts, R., Behrensmeyer, A.K., Deino, A., Ditchfield, P., and Clark, J. (2004) Small Mid-Pleistocene hominin associated with East African Acheulean technology. *Science* 305(5680): 75–8.

Potze, S., and Thackeray, J. (2010) Temporal lines and open sutures revealed on cranial bone adhering to matrix associated with Sts 5 (Mrs Ples), Sterkfontein, South Africa. *Journal of Human Evolution* 58(6): 533–5.

Poulianos, A.N. (1982) Petralona cave dating controversy. *Nature* 299: 280.

Povinelli, D.J., and Eddy, T.J. (1996) What young chimpanzees know about seeing. *Monographs of the Society for Research in Child Development* 61(3): 1–189.

Povinelli, D.J., and Vonk, J. (2004) We don't need a microscope to explore the chimpanzee's mind. *Mind & Language* 19(1): 1–28.

Power, M.E., Tilman, D., Estes, J.A., Menge, B.A., Bond, W.J., Mills, L.S. et al. (1996) Challenges in the quest for keystones. *Biosience* 46: 609–20.

Pradhan, S., Saha, G., and Khan, J. (2001) Ecology of the red panda *Ailurus fulgens* in the Singhalila National Park, Darjeeling, India. *Biological Conservation* 98(1): 11–18.

Prat, S. (2002) Anatomical study of the skull of the Kenya specimen KNM-ER 1805: a re-evaluation of its taxonomic allocation? *Comptes Rendus Palevol* 1: 27–33.

Prat, S., Brugal, J.-P., Tiercelin, J.-J., Barrat, J.-A., Bohn, M., Delagnes, A., et al. (2005) First occurrence of early *Homo* in the Nachukui Formation (West Turkana, Kenya) at 2.3–2.4 Myr. *Journal of Human Evolution* 49(2): 230–40.

Pratt, D.J., and Gwynne, M.D. (1978) *Rangeland Management and Ecology in East Africa*. Hodder and Stoughton, London.

Preece, R.C., Gowlett, J.A.J., Parfitt, S.A., Bridgland, D.R., and Lewis, S.G. (2006) Humans in the Hoxnian: habitat, context and fire use at Beeches Pit, West Stow, Suffolk, UK. *Journal of Quaternary Science* 21(5): 485–96.

Pregill, G. (1984) Durophagous feeding adaptations in an amphisbaenid. *Journal of Herpetology*: 186–91.

Preiswerk, G. (1895) *Beiträge zur Kenntniss der Schmelzstructur bei Säugethieren,–mit besonderer Berücksichtigung der Ungulaten*. PhD thesis, Universität Basel.

Premack, D., and Woodruff, G. (1978) Does the chimpanzee have a theory of mind? *Behavioral and Brain Sciences* 1(4): 515–26.

Preuschoft, H., Hohn, B., Scherf, H., Schmidt, M., Krause, C., and Witzel, U. (2010) Functional Analysis of the Primate Shoulder. *International journal of primatology* 31(2): 301–20.

Preuss, T., Cáceres, M., Oldham, M., and Geschwind, D. (2004) Human brain evolution: insights from microarrays. *Nature Reviews Genetics* 5(11): 850–60.

Prost, J.H. (1980) Origin of bipedalism. *American Journal of Physical Anthropology* 52: 175–89.

Protsch, R. (1974) The age and stratigraphic position of Olduvai Hominid I. *Journal of Human Evolution* 3(5): 379–85.

Protsch, R., and Semmel, A. (1978) Zur Chronologie des Kelsterbach-Hominiden, des ältesten Vertreters des *Homo sapiens sapiens* in Europa. *Eiszeitalter und Gegenwart* 28: 200–10.

Protsch, R.R. (1975) The absolute dating of Upper Pleistocene sub-Saharan fossil hominids and their place in human evolution. *Journal of Human Evolution* 4: 297–322.

Protsch, R.R.R. (1981) Die Archäologischen und Anthropologischen Ergebnisse der Kohl-Larsen-Expeditionen in Nord-Tanzania 1933–1939. In: *Band 3: The Palaeoanthropological Finds of the Pliocene and Pleistocene*, pp. Verlag Archaeologica Venatoria, Institute für Urgeschichte der Universität Tübingen, Tübingen.

Pruetz, J., and Bertolani, P. (2007) Savanna chimpanzees, *Pan troglodytes verus*, hunt with tools. *Current Biology* 17(5): 412–17.

Puech, P. (1981) Tooth wear in La Ferrassie man. *Current Anthropology* 22(4): 424–30.

Puech, P.F., Cianfarani, F., and Roth, H. (1986) Reconstruction of the maxillary dental arcade of Garusi Hominid I. *Journal of Human Evolution* 15: 325–32.

Puech, P.-F., and Albertini, H. (1981) Enamel pits of the Lazaret Man. *Journal of Human Evolution* 10(6): 449–52.

Pycraft, W.P. (1928) Description of the human remains. In: *Rhodesian Man and Associated Remains*, Bather, F.A., ed., pp. 1–51. British Museum (Natural History), London.

Pycraft, W.P., Smith, G.E., Yearsly, M., Carter, J.T., Smith, R.A., Hopwood, A.T., et al. (1928) *Rhodesian Man and Associated Remains*, Bather, F.A., ed., British Museum (Natural History), London.

Qian, F., and Zhou, G. (1991) *Quaternary Geology and Paleoanthropology of Yuanmou, Yunnan, China*, Beijing.

Quade, J., and Wynn, J., eds (2008) *The Geology of Early Humans in the Horn of Africa*. Geological Society of America, Boulder, CO.

Quade, J., Levin, N., Semaw, S., Stout, D., Renne, P., Rogers, M.J., et al. (2004) Paleoenvironments of the earliest stone toolmakers, Gona, Ethiopia. *Geological Society of America Bulletin* **116**(11–12): 1529–44.

Quade, J., Levin, N.E., Simpson, S.W., Butler, R., McIntosh, W.C., Semaw, S., et al. (2008) The Geology of Gona, Ethiopia. *Geological Society of America Special Paper* **446**: 1–31.

Quam, R., Arsuaga, J., Bermúdez de Castro, J., Dez, C., Lorenzo, C., Carretero, M., et al. (2001) Human remains from Valdegoba Cave (Huérmeces, Burgos, Spain). *Journal of Human Evolution* **41**(5): 385–435.

Quam, R.M., and Smith, F.H. (1998) A reassessment of the Tabun C2 mandible. In: *Neandertals and modern humans in Western Asia*, Akazawa, T., Aoki, K., and Bar-Yosef, O., eds, pp. 405–21. Plenum Press, New York.

Quenstedt, W., and Quenstedt, A. (1936) *Hominidae fossiles*. Dr. W. Junk, 's-Gravenhage.

Raaum, R.L., Sterner, K.N., Noviello, C.M., Stewart, C.-B., and Disotell, T.R. (2005) Catarrhine primate divergence dates estimated from complete mitochondrial genomes: concordance with fossil and nuclear DNA evidence. *Journal of Human Evolution* **48**(3): 237–57.

Radinsky, L. (1967) Relative brain size: a new measure. *Science* **155**(3764): 836–8.

Radmilli, A.M., and Boschian, G. (1996) *Gli scavi a Castel di Guido*. Istituto Italiano di Preistoria e Protostoria, Florence.

Radovčić, J., Smith, F.H., Trinkaus, E., and Wolpoff, M.H. (1988) *The Krapina Hominids: An Illustrated Catalog of Skeletal Collection*. Mladost Publishing House, Zagreb.

Raff, R.A., and Kaufman, T.C. (1983) *Embryos, Genes, and Evolution*. Macmillan, New York.

Rahmani, N. (2004) Technological and cultural change among the last hunter-gatherers of the Maghreb: the Capsian (10,000–6000 BP). *Journal of World Prehistory* **18**(1): 57–105.

Raichlen, D., Gordon, A., Harcourt-Smith, W., Foster, A., Haas Jr, W., and Rosenberg, K. (2010) Laetoli footprints preserve earliest direct evidence of human-like bipedal biomechanics. *PLoS One* **5**(3): e9769.

Raichlen, D.A. (2004) Convergence of forelimb and hindlimb natural pendular period in baboons (*Papio cynocephalus*) and its implications for the evolution of primate quadrupedalism. *Journal of Human Evolution* **46**(6): 719–38.

Rainer, F., and Simionescu, I. (1942) Sur le premier crâne d'homme Paléolithique trouvé en Roumanie. *Analele Academiei Romane, Memoriile Secţiunii Ştiinţifice* **Series III**(17): 489–503.

Rak, Y. (1983) *The Australopithecine Face*. Academic Press, New York.

Rak, Y. (1990) On the differences between two pelvises of Mousterian context from the Qafzeh and Kebara caves, Israel. *American Journal of Physical Anthropology* **81**(3): 323–32.

Rak, Y., and Howell, F.C. (1978) Cranium of a juvenile *Australopithecus boisei* from the lower Omo Basin, Ethiopia. *American Journal of Physical Anthropology* **48**(3): 345–66.

Rak, Y., and Clarke, R.J. (1979) Ear ossicle of *Australopithecus robustus*. *Nature* **279**: 62–3.

Rak, Y., and Arensburg, B. (1987) Kebara 2 Neanderthal pelvis: first look at a complete inlet. *American Journal of Physical Anthropology* **73**: 227–31.

Rak, Y., Kimbel, W.H., and Hovers, E. (1994) A Neandertal infant from Amud Cave, Israel. *Journal of Human Evolution* **26**: 313–24.

Rak, Y., Ginzburg, A., and Geffen, E. (2003) Does *Homo neanderthalensis* play a role in modern human ancestry? The mandibular evidence. *Am J Phys Anthropol* **119**: 199–204.

Rak, Y., Ginzburg, A., and Geffen, E. (2007) Gorilla-like anatomy on *Australopithecus afarensis* mandibles suggests *Au. afarensis* link to robust australopiths. *Proceedings of the National Academy of Sciences USA* **104**(16): 6568.

Ramberg, L., Hancock, P., Lindholm, M., Meyer, T., Ringrose, S., Sliva, J., et al. (2006) Species diversity of the Okavango Delta, Botswana. *Aquatic Sciences-Research Across Boundaries* **68**(3): 310–37.

Rana, B., Hewett-Emmett, D., Jin, L., Chang, B., Sambuughin, N., Lin, M., et al. (1999) High polymorphism at the human melanocortin 1 receptor locus. *Genetics* **151**(4): 1547–57.

Rankin-Hill, L., and Blakey, M. (1994) W. Montague Cobb (1904–1990): physical anthropologist, anatomist, and activist. *American Anthropologist* **96**(1): 74–96.

Ravelo, A.C., Dekens, P.S., and McCarthy, M. (2006) Evidence for El Niño-like conditions during the Pliocene. *GSA Today* **16**(3): 4–11.

Raxter, M.H., Auerbach, B.M., and Ruff, C.B. (2006) Revision of the Fully Technique for estimating statures. *American Journal of Physical Anthropology* **130**: 374–84.

Raynal, J.-P. (1980) Taforalt. Mission Prehistorique et paleontologique francaise au Maroc: Rapport d'activite pour l'annee. *Bulletin d'Archéologie Marocaine* **12**: 69–72.

Raynal, J.-P., Sbihi-Alaoui, F.-Z., Geraads, D., Magoga, L., and Mohib, A. (2001) The earliest occupation of the North-Africa: the Moroccan perspective. *Quaternary International* **75**(1): 65–75.

Raynal, J.-P., Sbihi-Alaoui, F.-Z., Magoga, L., Mohib, A., and Zouak, M. (2002) Casablanca and the earliest occupation of north Atlantic Morocco. *Quaternaire* **13**(1): 65–77.

Reck, H. (1914) Erste vorlaufige Mitteilung uber den Fund eines fossilen Menschenskelets aus Zentralafrika. *Sitzungsberichten der Gesellschaft naturforschender Freunde zu Berlin* **3**: 81–95.

Reck, H. (1951) A preliminary survey of the tectonics and stratigraphy of Olduvai. In: *Olduvai Gorge: a*

Report on the Evolution of the Hand-Axe Culture in Beds I–IV, Leakey, L.S.B., ed., pp. 5–19. Cambridge University Press, Cambridge.

Reck, H., and Kohl-Larsen, L. (1936) Erster Überblick über die jungdiluvialen Tier- und Menschenfunde Dr. Kohl-Larsen's im nordöstlichen Teil des Njarasa-Grabens (Ostafrika) und die geologischen Yerhältnisse des Fundgebietes. *Geologische Rundschau* 27(5): 401–41.

Rector, A., and Reed, K. (2010) Middle and late Pleistocene faunas of Pinnacle Point and their paleoecological implications. *Journal of Human Evolution* 59: 340–57.

Redon, R., Ishikawa, S., Fitch, K.R., Feuk, L., Perry, G.H., Andrews, T.D., et al. (2006) Global variation in copy number in the human genome. *Nature* 444(7118): 444–54.

Reed, C.A. (1965) A human frontal bone from the Late Pleistocene of the Kom Ombo Plain, Upper Egypt. *Man* 65: 101–4.

Reed, K. (2008) Paleoecological patterns at the Hadar hominin site, Afar regional state, Ethiopia. *Journal of Human Evolution* 54(6): 743–68.

Rees, J. (2003) Genetics of hair and skin color. *Annual Review of Genetics* 37(1): 67–90.

Reich, D.E., Cargill, M., Bolk, S., Ireland, J., Sabeti, P.C., Richter, D.J., et al. (2001) Linkage disequilibrium in the human genome. *Nature* 411: 199–204.

Reimer, P.J., Baillie, M.G.L., Bard, E., Bayliss, A., Beck, J.W., Bertrand, C.J.H., et al. (2004) IntCal04 Terrestrial Radiocarbon Age Calibration, 0–26 Cal kyr BP. *Radiocarbon* 46: 1029–58.

Reimer, P.J., Baillie, M.G.L., Bard, E., Bayliss, A., Beck, J.W., Blackwell, P.G., et al. (2009) IntCal09 and Marine09 radiocarbon age calibration curves, 0–50,000 years cal BP. *Radiocarbon* 51(4): 1111–50.

Relethford, J.H. (2003) *Reflections of Our Past: How Human History is Revealed in Our Genes*. Westview Press, Boulder, CO.

Remane, A. (1954) Structure and relationships of *Meganthropus africanus*. *American Journal of Physical Anthropology* 12: 123–6.

Remane, A. (1960) Zähne und Gebiss. In: *Primatologica, Handbook of Primatology*, Hofer, H., Schultz, A.H., and Starck, D., eds, pp. 637–846. S. Karger, Basel.

Remane, A. (1961) Gedanken zum Problem: Homologie und Analogie, Praeadaption und Parallelitat. *Zool. Anz.* 166: 447–65.

Renne, P.R., WoldeGabriel, G., Hart, W.K., Heiken, G., and White, T. (1999) Chronostratigraphy of the Miocene-Pliocene Sagantole Formation, Middle Awash Valley, Afar Rift, Ethiopia. *Geological Society of America Bulletin* 111(6): 869–85.

Reno, P.L., Meindl, R.S., McCollum, M.A., and Lovejoy, C.O. (2003) Sexual dimorphism in *Australopithecus afarensis* was similar to that of modern humans. *Proceedings of the National Academy of Sciences USA* 100(16): 9404–9.

Reno, P.L., DeGusta, D., Serrat, M.A., Meindl, R.S., White, T.D., Eckhardt, R.B., et al. (2005) Plio-Pleistocene hominid limb proportions: evolutionary reversals or estimation errors? *Current Anthropology* 46(4): 575–88.

Rensch, B. (1959) *Evolution Above the Species Level*. Columbia University Press, New York.

Retief, J.D., Winkfein, R.J., and Dixon, G.H. (1993) Evolution of the monotremes: The sequence of the protamine P1 genes of platypus and echidna. *European Journal of Biochemistry* 2(8): 457–61.

Reynolds, S.C., Vogel, J.C., Clarke, R.J., and Kuman, K.A. (2003) Preliminary results of excavations at Lincoln Cave, Sterkfontein, South Africa. *South African Journal of Science* 99: 286–8.

Rhodes, E.J., Singarayer, J.S., Raynal, J.-P., Westaway, K.E., and Sbihi-Alaoui, F.-Z. (2006) New age estimates for the Palaeolithic assemblages and Pleistocene succession of Casablanca, Morocco. *Quaternary Science Reviews* 25(19–20): 2569–85.

Rich, P.V., and Haarhoff, P.J. (1985) Early Pliocene Coliidae (Aves, Coliiformes) from Langebaanweg (Cape Province), South Africa. *Ostrich* 56: 20–41.

Richards, L.C., and Brown, T. (1986) Development of the helicoidal plane. *Human Evolution* 1(5): 385–98.

Richards, M., Sykes, B., and Hedges, R. (1995) Authenticating DNA extracted from ancient skeletal remains. *Journal of Archaeological Science* 22(2): 291–9.

Richards, M., Côrte-Real, H., Forster, P., Macaulay, V., Wilkinson-Herbots, H., Demaine, A., et al. (1996) Paleolithic and neolithic lineages in the European mitochondrial gene pool. *American Journal of Human Genetics* 59(1): 185–203.

Richards, M.P., Pettitt, P.B., Trinkaus, E., Smith, F.H., Paunovic, M., and Karavanic, I. (2000) Neanderthal diet at Vindija and Neanderthal predation: the evidence from stable isotopes. *Proceedings of the National Academy of Sciences USA* 97: 7663–6.

Richmond, B., and Strait, D.S. (2000) Evidence that humans evolved from a knuckle-walking ancestor. *Nature* 404: 382–5.

Richmond, B.G., and Jungers, W.L. (1995) Size variation and sexual dimorphism in *Australopithecus afarensis* and living hominoids. *Journal of Human Evolution* 29(3): 229–45.

Richmond, B.G., and Jungers, W.L. (2008) *Orrorin tugenensis* femoral morphology and the evolution of hominin bipedalism. *Science* 319(5870): 1662–5.

Richmond, B.R., Begun, D.R., and Strait, D.S. (2001) Origin of human bipedalism: the knuckle-walking hypothesis revisited. *Yearbook of Physical Anthropology* 44: 70–105.

Richmond, B.G., Aiello, L.C., and Wood, B.A. (2002) Early hominin limb proportions. *Journal of Human Evolution* 43: 529–48.

Richmond, B.R., Wright, B.W., Grosse, I., Dechow, P.C., Ross, C.F., Spencer, M.A., et al. (2005) Finite Element

Analysis in Functional Morphology. *Anatomical Record, Part A* **283A**: 259–74.

Richter, D., Tostevin, G., Skrdla, P., and Davies, W. (2009) New radiometric ages for the Early Upper Palaeolithic type locality of Brno-Bohunice (Czech Republic): comparison of OSL, IRSL, TL and 14C dating results. *Journal of Archaeological Science* **36**: 708–20.

Ride, W.D.L., Cogger, H.G., Dupuis, C., Kraus, O., Minelli, A., Thompson, F.C., et al., eds (1999) *International Code of Zoological Nomenclature*. International Trust for Zoological Nomenclature, London.

Riek, G. (1934) *Die Eiszeitjägerstation am Vogelherd*. Curt Kabitzsch Verlag, Leipzig.

Riel-Salvatore, J., Miller, A.E., and Clark, G.A. (2008) An empirical evaluation of the case for a Châtelperronian-Aurignacian interstratification at Grotte des Fées de Châtelperron. *World Archaeology* **40**: 480–92.

Rightmire, G. (2008) *Homo* in the Middle Pleistocene: Hypodigms, variation, and species recognition. *Evolutionary Anthropology* **17**(1): 8–21.

Rightmire, G., Van Arsdale, A., and Lordkipanidze, D. (2008) Variation in the mandibles from Dmanisi, Georgia. *Journal of Human Evolution* **54**(6): 904–8.

Rightmire, G.P. (1975) Problems in the study of Later Pleistocene Man in Africa. *American Anthropologist* **77**(1): 28–52.

Rightmire, G.P. (1976) Relationship of Middle and Upper Pleistocene hominids from subsaharan Africa. *Nature* **260**: 238–40.

Rightmire, G.P. (1978) Florisbad and human population succession in Southern Africa. *American Journal of Physical Anthropology* **48**(4): 475–86.

Rightmire, G.P. (1979) Implication of Border Cave skeletal remains for later Pleistocene Human Evolution. *Current Anthropology* **20**: 23–35.

Rightmire, G.P. (1979) Cranial remains of *Homo erectus* from Beds II and IV, Olduvai Gorge, Tanzania. *American Journal of Physical Anthropology* **51**: 99–116.

Rightmire, G.P. (1980) Middle Pleistocene hominids from Olduvai Gorge, Northern Tanzania. *American Journal of Physical Anthropology* **53**: 225–41.

Rightmire, G.P. (1981) Late Pleistocene hominids of eastern and southern Africa. *Anthropologie* **19**(1): 15–26.

Rightmire, G.P. (1983) The Lake Ndutu cranium and early *Homo sapiens* in Africa. *American Journal of Physical Anthropology* **61**(2): 245–54.

Rightmire, G.P. (1984a) Comparison of *Homo erectus* from Africa and Southeast Asia. *Courier Forschungs-Institut Senckenberg* **69**: 83–98.

Rightmire, G.P. (1984b) *Homo sapiens* in sub-Saharan Africa. In: *The Origins of Modern Humans*, Smith, F.H., and Spencer, F., eds, pp. 295–325. Alan R. Liss, New York.

Rightmire, G.P. (1990) *The Evolution of* Homo erectus*: Comparative Anatomical Studies of an Extinct Human Species*. Cambridge University Press, Cambridge.

Rightmire, G.P. (1995) Geography, time and speciation in Pleistocene *Homo*. *South African Journal of Science* **91**: 450–4.

Rightmire, G.P. (1996) The human cranium from Bodo, Ethiopia: evidence for speciation in the Middle Pleistocene. *Journal of Human Evolution* **31**: 21–39.

Rightmire, G.P. (1998) Human evolution in the Middle Pleistocene: the role of *Homo heidelbergensis*. *Evolutionary Anthropology* **6**(6): 218–27.

Rightmire, G.P. (2001) Patterns of hominid evolution and dispersal in the Middle Pleistocene. *Quaternary International* **75**(1): 77–84.

Rightmire, G.P. (2004) Brain size and encephalization in Early to Mid-Pleistocene *Homo*. *American Journal of Physical Anthropology* **124**: 109–23.

Rightmire, G.P., and Deacon, H.J. (1991) Comparative studies of Late Pleistocene human remains from Klasies River Mouth, South Africa. *Journal of Human Evolution* **20**: 131–56.

Rightmire, G.P., and Deacon, H.J. (2001) New human teeth from Middle Stone Age deposits at Klasies River, South Africa. *Journal of Human Evolution* **41**: 535–44.

Rightmire, G.P., Deacon, H.J., Schwartz, J.H., and Tattersall, I. (2006a) Human foot bones from Klasies River main site, South Africa. *Journal of Human Evolution* **50**(1): 96–103.

Rightmire, G.P., Lordkipanidze, D., and Vekua, A. (2006b) Anatomical descriptions, comparative studies and evolutionary significance of the hominin skulls from Dmanisi, Republic of Georgia. *Journal of Human Evolution* **50**(2): 115–41.

Rigollot, M. (1854) *Mémoire sur des instruments en silex trouvés à St-Acheul, près d'Amiens, et considérés sous les rapports géologique et archéologique, par le Dr Rigollot*. Duval et Herment.

Rilling, J.K. (2006) Human and nonhuman primate brains: Are they allometrically scaled versions of the same design? *Evolutionary Anthropology* **15**(2): 65–77.

Ring, U., and Betzler, C. (1995) Geology of the Malawi Rift: kinematic and tectonosedimentary background to the Chiwondo Beds, northern Malawi. *Journal of Human Evolution* **28**: 7–21.

Rink, W., Bartoll, J., Schwarcz, H., Shane, P., and Bar-Yosef, O. (2007) Testing the reliability of ESR dating of optically exposed buried quartz sediments. *Radiation Measurements* **42**(10): 1618–26.

Rink, W.J., Schwarcz, H.P., Smith, F.H., and Radovcic, J. (1995) ESR ages for Krapina hominids. *Nature* **378** (6552): 24.

Rink, W.J., Schepartz, L.A., Miller-Antonio, S., Weiwen, H., Yamei, H., Bekken, D., et al. (2003) Electron spin resonance (ESR) dating of tooth enamel at Panxian Dadong Cave, Guizhou, China. In: *Current Research in Chinese Pleistocene Archaeology*, Shen, C., and Keates, S.G., eds, pp. 111–18. Archaeopress, Oxford.

Rink, W.J., Schwarcz, H., Weiner, S., Goldberg, P., Meignen, L., and Bar-Yosef, O. (2004) Age of the Mousterian industry at Hayonim Cave, Northern Israel, using electron spin resonance and ^{230}Th/^{234}U methods. *Journal of Archaeological Science* 31(7): 953–64.

Rink, W.J., Wang, W., Bekken, D., and Jones, H.L. (2008) Geochronology of Ailuropoda-Stegodon fauna and *Gigantopithecus blacki* in Guangxi Province, Southern China. *Quaternary Research* 69: 377–87.

Rink, W.R., Schwarcz, H.P., Valoch, K., Seitl, L., and Stringer, C.B. (1996) ESR Dating of Micoquian Industry and Neanderthal Remains in Kůlna Cave, Czech Republic. *Journal of Archaeological Science* 23(6): 889–901.

Risnes, S. (1985) A scanning electron microscope study of the three-dimensional extent of Retzius lines in human dental enamel. *European Journal of Oral Sciences* 93(2): 145–52.

Risnes, S. (1986) Enamel apposition rate and prism periodicity in human teeth. *Scandinavian Journal of Dental Research* 94: 394–404.

Risnes, S. (1990) Structural characteristics of staircase-type retzius lines in human dental enamel analyzed by scanning electron microscopy. *Anatomical Record* 226(2): 135–46.

Risnes, S. (1998) Growth tracks in dental enamel. *Journal of Human Evolution* 35(4–5): 331–50.

Rivière, E. (1872) Sur le squelette humain trouve dans les cavernes des Baousse-Rousse (Italie), dites Grottes de Menton, le 26 mars 1872. *Comptes Rendus de l'Academie des Sciences* 74: 1204–7.

Rivière, E. (1897) Nouvelles recherches à Cro-Magnon. *Bulletins de la Société d'Anthropologie de Paris* 8(1): 503–8.

Rizzolatti, G., and Arbib, M. (1998) Language within our grasp. *Trends in Neurosciences* 21(5): 188–94.

Rizzolatti, G., Fadiga, L., Matelli, M., Bettinardi, V., Paulesu, E., Perani, D., et al. (1996) Localization of grasp representations in humans by PET: 1. Observation versus execution. *Experimental Brain Research* 111(2): 246–52.

Robbins, L.H. (1974) The Lothagam site: a Late Stone Age fishing settlement in the Lake Rudolf basin, Kenya. *Michigan State University Museum, Anthropological Series 1* 2: 151–221.

Roberts, D., and Berger, L.R. (1997) Last interglacial (c. 117 kyr) human footprints from South Africa. *South African Journal of Science* 93(8): 349–50.

Roberts, D.L. (2006) Varswater Formation (including the Langeenheid Clayey Sand, Konings Vlei Gravel, Langeberg Quartz Sand and Muishond Fontein Phosphatic Sand Members). In: *The Catalogue of South African Lithostratigraphic Unit*, Johnson, M.R., ed., pp. 27–31. SA Committee for Stratigraphy.

Roberts, M.B., Stringer, C.B., and Parfitt, S.A. (1994) A hominid tibia from Middle Pleistocene sediments at Boxgrove, UK. *Nature* 369: 311–13.

Roberts, M.R., and Parfitt, S.A. (1999) *Boxgrove: A Middle Pleistocene Hominid Site at Eartham Quarry, Boxgrove, West Sussex.* English Heritage, London.

Roberts, R.G., Westaway, K.E., Zhao, J.-x., Turney, C.S.M., Bird, M.I., Rink, W.J., et al. (2009) Geochronology of cave deposits at Liang Bua and of adjacent river terraces in the Wae Racang valley, western Flores, Indonesia: a synthesis of age estimates for the type locality of *Homo floresiensis*. *Journal of Human Evolution* 57(5): 484–502.

Robine, J.M., and Allard, M. (1995) Validation of the exceptional longevity case of a 120 year old woman. *Facts and Research in Gerontology* 363–8.

Robinson, J. (1958) Cranial cresting patterns and their significance in the Hominoidea. *American Journal of Physical Anthropology* 16(4): 397–428.

Robinson, J. (1972a) The bearing of East Rudolf fossils on early hominid systematics. *Nature* 240: 239–40.

Robinson, J.T. (1953) *Meganthropus*, australopithecines and hominids. *American Journal of Physical Anthropology* 11(1): 1–38.

Robinson, J.T. (1954) The genera and species of the Australopithecinae. *American Journal of Physical Anthropology* 12(2): 181–200.

Robinson, J.T. (1956) The dentition of the Australopithecinae. In: *Transvaal Museum Memoir no. 9.* Transvaal Museum, Pretoria.

Robinson, J.T. (1960) An alternative interpretation of the supposed giant deciduous hominid tooth from Olduvai. *Nature* 185: 407–8.

Robinson, J.T. (1961) The australopithecines and their bearing on the origin of man and of stone tool making. *South African Journal of Science* 57: 3–13.

Robinson, J.T. (1962) Australopithecines and artefacts at Sterkfontein: part I: Sterkfontein stratigraphy and the significance of the extension site. *South African Archaeological Bulletin* 17: 87–107.

Robinson, J.T. (1965) *Homo habilis* and the australopithecines. *Nature* 205: 121–4.

Robinson, J.T. (1967) Variation and the taxonomy of the early hominids. *Evolutionary Biology* 1: 69–100.

Robinson, J.T. (1972b) *Early Hominid Posture and Locomotion.* University of Chicago Press, Chicago, IL.

Robock, A., Ammann, C., Oman, L., Shindell, D., Levis, S., and Stenchikov, G. (2009) Did the Toba volcanic eruption of 74 ka B.P. produce widespread glaciation? *Journal of Geophysical Research* 114: D10107.

Roche, H., and Texier, P.J. (1995) Evaluation of technical competence of *Homo erectus* in East Africa during the Middle Pleistocene. In: *Human Evolution in its Ecological Context*, Bower, J.R.F., and Sartono, S., eds, pp. 153–67. *Pithecanthropus* Centennial Foundation, Leiden.

Roche, H., Brugal, J.-P., Lefevre, D., Ploux, S., and Texier, P.-J. (1988) Isenya: état des recherches sur un nouveau site acheuléen d'Afrique orientale. *African Archaeological Review* 6(1): 27–55.

Roche, H., Delagnes, A., Brugal, J.P., Feibel, C., Kibunjia, M., Mourre, V., et al. (1999) Early hominid stone tool production and technical skill 2.34 Myr ago in West Turkana, Kenya. *Nature* **399**: 57–60.

Roche, J. (1967) L'Aterian de la grotte de Taforalt (Maroc oriental). *Bulletin d'Archéologie Marocaine* **7**: 11–56.

Roche, J. (1976) Chronostratigraphie des restes atériens de la grotte des contrebandiers a Témara (Province de Rabat). *Bulletins et Mémoires de la Société d'Anthropologie de Paris* **3**: 165–73.

Rodman, P.S., and McHenry, H.M. (1980) Bioenergetics and the origin of hominid bipedalism. *American Journal of Physical Anthropology* **52**(1): 103–6.

Rogers, A., and Harpending, H. (1992) Population growth makes waves in the distribution of pairwise differences. *Molecular Biology and Evolution* **9**(3): 552–69.

Rogers, A., and Jorde, L. (1996) Ascertainment bias in estimates of average heterozygosity. *American Journal of Human Genetics* **58**(5): 1033.

Rogers, A., Iltis, D., and Wooding, S. (2004) Genetic variation at the MC1R locus and the time since loss of human body hair. *Current Anthropology* **45**(1): 105–24.

Rogers, A.K. (2008) Field data validation of an algorithm for computing obsidian effective hydration temperature. *Journal of Archaeological Science* **35**: 441–7.

Rogers, M.J., Harris, J.W.K., and Feibel, C.S. (1994) Changing patterns of land use by the Plio-Pleistocene hominids in the Lake Turkana basin. *Journal of Human Evolution* **27**: 139–58.

Rohlf, F. (2000a) On the use of shape spaces to compare morphometric methods. *Hystrix-the Italian Journal of Mammalogy* **11**(1): 9–25.

Rohlf, F.J. (2000b) Statistical power comparisons among alternative morphometric methods. *American Journal of Physical Anthropology* **111**(4): 463–78.

Rohlf, F.J. (2001) Comparative methods for the analysis of continuous variables: geometric interpretations. *Evolution* **55**: 2143–60.

Rohlf, F.J. (2003) Bias and error in estimates of mean shape in geometric morphometrics. *Journal of Human Evolution* **44**(6): 665–83.

Rohlf, F.J. (2006) A comment on phylogenetic correction. *Evolution* **60**(7): 1509–15.

Rohlf, F.J., and Slice, D. (1990) Extensions of the Procrustes method for the optimal superimposition of landmarks. *Systematic Zoology* **39**(1): 40–59.

Rolian, C. (2009) Integration and evolvability in primate hands and feet. *Evolutionary Biology* **36**: 100–17.

Rolland, N., and Dibble, H. (1990) A new synthesis of Middle Paleolithic variability. *American Antiquity* **55**(3): 480–99.

Roman, D.C., Campisano, C., Quade, J., DiMaggio, E., Arrowsmith, J.R., and Feibel, C. (2008) Composite tephrostratigraphy of the Dikika, Gona, Hadar, and Ledi-Geraru project areas, northern Awash, Ethiopia. In: *The Geology of Early Humans in the Horn of Africa*, Quade, J., and Wynn, J.G., eds, pp. 119–34. Geological Society of America, Boulder, CO.

Rommerskirchen, F., Plader, A., Eglinton, G., Chikaraishi, Y., and Rullkötter, J. (2006) Chemotaxonomic significance of distribution and stable carbon isotopic composition of long-chain alkanes and alkan-1-ols in C_4 grass waxes. *Organic Geochemistry* **37**(10): 1303–32.

Ronquist, F., and Huelsenbeck, J. (2003) MrBayes 3: Bayesian phylogenetic inference under mixed models. *Bioinformatics* **19**(12): 1572–4.

Rosas, A., and Bermúdez de Castro, J.M. (1998a) The Mauer mandible and the evolutionary significance of *Homo heidelbergensis*. *Geobios* **31**(5): 687–97.

Rosas, A., and Burmúdez de Castro, J.M. (1998a) On the taxonomic affinities of the Dmanisi Mandible (Georgia). *American Journal of Physical Anthropology* **107**: 145–62.

Rosas, A., and Bermúdez de Castro, J.M. (1999) The ATD6–5 mandibular specimen from Gran Dolina (Atapuerca, Spain). Morphological study and phylogenetic implications. *Journal of Human Evolution* **37**(3–4): 567–90.

Rosas, A., and Martinez-Maza, C. (2010) Bone remodeling of the *Homo heidelbergensis* mandible; the Atapuerca-SH sample. *Journal of Human Evolution* **58**(2): 127–37.

Rosas, A., Carbonell, E., Cuenca, N., Fernndez Jalvo, Y., Made, J., Ollé, A., et al. (1998) Cronología, bioestratigrafia y paleoecología del Pleistoceno Medio de Galería (Sierra de Atapuerca, España). *Revista Española de Paleontologia* **13**(1): 71–80.

Rosas, A., Huguet, R., Pérez-González, A., Carbonell, E., Bermúdez de Castro, J., Vallverd´, J., et al. (2006) The 'Sima del Elefante' cave site at Atapuerca (Spain). *Estudios Geológicos* **62**(1): 327–48.

Rosas, A., Martínez-Maza, C., Bastir, M., García-Tabernero, A., Lalueza-Fox, C., Huguet, R., et al. (2006) Paleobiology and comparative morphology of a late Neandertal sample from El Sidrón, Asturias, Spain. *Proceedings of the National Academy of Sciences USA* **103**(51): 19266–71.

Rose, L., and Marshall, F. (1996) Meat eating, hominid sociality, and home bases revisited. *Current Anthropology* **37**(2): 307–38.

Rose, M.D. (1984) A hominine hip bone, KNM-ER 3228, from east Lake Turkana, Kenya. *American Journal of Physical Anthropology* **63**: 371–8.

Rose, W.I., and Chesner, C.A. (1990) Worldwide dispersal of ash and gases from earth's largest known eruption: Toba, Sumatra, 75 ka. *Palaeogeography, Palaeoclimatology, Palaeoecology* **89**: 269–75.

Rosenberg, K.R., Lü, Z., and Ruff, C.B. (2006) Body size, body proportions, and encephalization in a Middle Pleistocene archaic human from northern China. *Proceedings of the National Academy of Sciences USA* **103**(10): 3552–6.

Rosenberger, A., and Kinzey, W. (1976) Functional patterns of molar occlusion in platyrrhine primates. *American Journal of Physical Anthropology* **45**(2): 281–97.

Rosenzweig, M.L., Brown, J.S., and Vincent, T.L. (1987) Red Queens and EES: the coevolution of evolutionary rates. *Evolutionary Ecology* **1**: 59–94.

Rothschild, B.M., Hershkovitz, I., and Rothschild, C. (1995) Origin of yaws in the Pleistocene. *Nature* **378**: 343–4.

Rougier, H. (2003) *Étude Descriptive et Comparative de Biache-Saint-Vaast 1 (Biache-Saint-Vaast, Pas-de-Calais, France*. Thesis, L'Université Bordeaux.

Rougier, H., Milota, Ş., Rodrigo, R., Gherase, M., Sarcină, L., Moldovan, O., et al. (2007) Peştera cu Oase 2 and the cranial morphology of early modern Europeans. *Proceedings of the National Academy of Sciences USA* **104**(4): 1165–70.

Royer, C. (1879) Les Congrès d'anthropologie, de démographie et d'ethnographie au Trocadéro en 1878. *Journal des éconimists ser.* **4, 5**: 405–20.

Rubin, C.T., and Lanyon, L.E. (1984) Regulation of bone formation by applied dynamic loads. *Journal of Bone Joint Surgery* **66-A**: 397–402.

Rudwick, M.J.S. (1964) The inference of function from structure in fossils. *British Journal for the Philosophy of Science* **15**(57): 27–40.

Ruff, C. (2007) Body size prediction from juvenile skeletal remains. *American Journal of Physical Anthropology* **133**: 698–716.

Ruff, C. (2010) Body size and body shape in early hominins – implications of the Gona pelvis. *Journal of Human Evolution* **58**(2): 166–78.

Ruff, C.B. (1987) Sexual dimorphism in human lower limb bone structure: relationship to subsistence strategy and sexual division of labour. *Journal of Human Evolution* **16**: 391–416.

Ruff, C.B. (1988) Hindlimb articular surface allometry in Hominoidea and *Macaca*, with comparisons to diaphyseal scaling. *Journal of Human Evolution* **17**: 687–714.

Ruff, C.B. (1990) Body mass and hindlimb bone cross-sectional and articular dimensions in anthropoid primates. In: *Body Size in Mammalian Palaeobiology: Estimation and Biological Implications*, Damuth, J., and MacFadden, B.J., eds, pp. 119–49. Cambridge University Press, Cambridge.

Ruff, C.B. (1991) Climate and body shape in hominid evolution. *Journal of Human Evolution* **21**: 81–105.

Ruff, C.B. (1994) Morphological adaptation to climate in modern and fossil hominids. *American Journal of Physical Anthropology* **37**(S19): 65–107.

Ruff, C.B. (1995) Biomechanics of the hip, and birth in early *Homo*. *American Journal of Physical Anthropology* **98**(4): 527–74.

Ruff, C.B. (2002) Variation in human body size and shape. *Annual Review of Anthropology* **31**: 211–32.

Ruff, C.B. (2008) Biomechanical analyses of archaeological human skeletons. In: *Biological Anthropology of the Human Skeleton*, Katzenberg, M.A., and Saunders, S.R., eds, pp. 183–206. Wiley-Liss, New York.

Ruff, C.B., and Walker, A. (1993) Body size and body shape. In: *The Nariokotome* Homo erectus *Skeleton*, Walker, A., and Leakey, R., eds, pp. 234–65. Harvard University Press, Cambridge, MA.

Ruff, C.B., Trinkaus, E., Walker, A., and Larsen, C.S. (1993) Postcranial robusticity in *Homo* I: Temporal trends and mechanical interpretation. *American Journal of Physical Anthropology* **91**: 21–53.

Ruff, C.B., Trinkaus, E., and Holliday, T.W. (1997) Body mass and encephalization in Pleistocene *Homo*. *Nature* **387**: 173–6.

Ruff, C.B., Holt, B.M., Sládek, V., Berner, M., Murphy Jr, W.A., zur Nedden, D., et al. (2006) Body size, body shape, and mobility of the Tyrolean "Iceman". *Journal of Human Evolution* **51**: 91–101.

Russell, J.M., McCoy, S.J., Verschuren, D., Bessems, I., and Huang, Y. (2009) Human impacts, climate change, and aquatic ecosystem response during the past 2000 yr at Lake Wandakara, Uganda. *Quaternary Research* **72**(3): 315–24.

Russo, V.E.A., Martienssen, R.A., and Riggs, A.D. (1996) *Epigenetic Mechanisms of Gene Regulation*. Cold Spring Harbor Laboratory Press, Cold Spring Harbor, NY.

Russon, A.E. (2004) Introduction. In: *The Evolution of Thought: Evolutionary Origins of Great Ape Intelligence*, Russon, A.E., and Begun, D.R., eds, pp. 1–14. Cambridge University Press, Cambridge.

Rutten, M.G. (1959) Paleomagnetic reconnaissance of Mid-Italian volcanoes. *Geologie en Mijnbouw* **21**: 373–4.

Ruvolo, M. (1994) Molecular evolutionary processes and conflicting gene trees: the hominoid case. *American journal of Physical Anthropology* **94**: 89–113.

Ruvolo, M. (1997) Molecular phylogeny of the hominoids: inferences from multiple independent DNA sequence data sets. *Molecular Biology and Evolution* **14**(3): 248–65.

Saban, R. (1977) The place of Rabat Man (Kebibat, Morocco) in human evolution. *Current Anthropology* **18**(3): 518–24.

Sabeti, P.C., Reich, D.E., Higgins, J.M., Levine, H.Z.P., Richter, D.J., Schaffner, S.F., et al. (2002) Detecting recent positive selection in the human genome from haplotype structure. *Nature* **419**: 832–1.

Sabeti, P.C., Schaffner, S.F., Fry, B., Lohmueller, J., Varilly, P., Shamovsky, O., et al. (2006) Positive natural selection in the human lineage. *Science* **312**(5780): 1614–20.

Sacco, T., and Van Valkenburgh, B. (2004) Ecomorphological indicators of feeding behaviour in the bears (Carnivora: Ursidae). *Journal of Zoology* **263**(1): 41–54.

Sagi, A. (2005) *Magnetostratigraphy of 'Ubeidiya Formation, Northern Dead Sea Transform, Israel*. MSc thesis, The Hebrew University of Jerusalem.

Sahnouni, M., and De Heinzelin, J. (1998) The site of Ain Hanech revisited: new investigations at this Lower Pleistocene site in northern Algeria. *Journal of Archaeological Science* 25: 1083–1101.

Sahnouni, M., Hadjouis, D., van der Made, J., Derradji, A.-E.-K., Canals, A., Medig, M., et al. (2002) Further research at the Oldowan site of Ain Hanech, North-eastern Algera. *Journal of Human Evolution* 43(6): 925–37.

Sakura, H. (1985) *Pleistocene Human Fossil Remains from Pinza-Abu (Goat Cave), Miyako Island, Okinawa, Japan.* Pinza-Abu: Reports on Excavation of the Pinza-Abu Cave. Department of Education Okinawa Prefectural Government. Naha, Department of Education, Okinawa Prefectural Government, pp. 161–76.

Salem, A., Ray, D., Xing, J., Callinan, P., Myers, J., Hedges, D., et al. (2003) Alu elements and hominid phylogenetics. *Proceedings of the National Academy of Sciences USA* 100(22): 12787.

Salmon, P. (1888) *Les races humaines préhistoriques.* Wattier et cie, Paris.

Salzburger, W., and Meyer, A. (2004) The species flocks of East African cichlid fishes: recent advances in molecular phylogenetics and population genetics. *Naturwissenschaften* 91: 277–90.

Samaras, T.T. (2007) *Human Body Size and the Laws of Scaling.* Nova Science Publishers, New York.

Sanchez, F. (1999) Comparative biometrical study of the Mousterian mandible from Cueva del Boquete de Zafarraya (Málaga, Spain). *Human Evolution* 14(1): 125–38.

Sanchidrián Torti, J. (1994) *Arte Rupestre de la Cueva de Nerja.* Trabajos sobre la Cueva de Nerja, Num. 4. Patronato de la Cueva de Nerja, Málaga.

Sankhyan, A.R. (1997) Fossil clavicle of a Middle Pleistocene hominid from the Central Narmada Valley, India. *Journal of Human Biology* 32(1): 3–16.

Santa Luca, A.P. (1980) The Ngandong fossil hominids: a comparative study of a Far Eastern *Homo erectus* group. *Yale University Publications in Anthropology* 78: 1–175.

Sanzelle, S., Pilleyre, T., Montret, M., Faïn, J., Miallier, D., Camus, G., et al. (2000) Thermoluminescence dating: study of a possible chronological correlation between the maar of La Vestide-du-Pal and a tephra layer from La Baume-Moula-Guercy (Ardéche, France). *Comptes Rendus de l'Académie des Sciences* 330(8): 541–6.

Saragusti, I., and Goren-Inbar, N. (2001) The biface assemblage from Gesher Benot Ya'aqov, Israel: illuminating patterns in. *Quaternary International* 75(1): 85–9.

Sarich, V. (1983) Retrospective on hominoid macromolecular systematics. In: *New Interpretations of Ape and Human Ancestry, Plenum Press, New York,* Ciochon, R.L., and Corruccini, R.S., eds, pp. 137–50. Plenum Press, New York.

Sarich, V., and Wilson, A. (1967a) Immunological time scale for hominid evolution. *Science* 158(3805): 1200.

Sarich, V.M., and Wilson, A.C. (1967b) Rates of albumin evolution in primates. *Proceedings of the National Academy of Sciences USA* 58: 142–8.

Sartono, S. (1961) Notes on a new find of a *Pithecanthropus* mandible. *Publikasi Teknik Seri Paleontologi* 2: 1–51.

Sartono, S. (1971) Observations on a new skull of *Pithecanthropus erectus* (VIII) from Sangiran, Central Java. *Proceedings of the Koninklijke Nederlandsche Akademie van Wetenschappen* 74: 185–94.

Sartono, S. (1974) Observations on a newly discovered jaw of *Pithecanthropus modjokertensis* from the Lower Pleistocene of Sangiran, Central Java. *Proceedings of the Koninklijke Nederlandsche Akademie van Wetenschappen* 77: 26–31.

Sartono, S. (1976) On the Javanese Pleistocene Hominids: a Reappraisal. *Union International Science Prehistorique et Protohistorique, September 1976, Nice.*

Sartono, S. (1991) A new *Homo erectus* skull from Ngawi, East Java. *Bulletin of the Indo-Pacific Prehistory Association* 11: 14–22.

Sattenspiel, L., and Harpending, H. (1983) Stable populations and skeletal age. *American Antiquity* 48(3): 489–98.

Sawada, K., Arita, Y., Nakamura, T., Akiyama, M., Kamei, T., and Nakai, N. (1992) ^{14}C dating of the Nojiriko Formation using accelerator mass spectrometry. *Earth Science* 46: 133–42.

Sawada, Y., Pickford, M., Senut, B., Itaya, T., Hyodoe, M., Miura, T., et al. (2002) The age of *Orrorin tugenensis*, an early hominid from the Tugen Hills, Kenya. *Comptes Rendus Palevol* 1: 293–303.

Saxe, A.A. (1966) *Social Dimensions of Mortuary Practices in the Mesolithic Population from Wadi Halfa, Sudan.* American Anthropological Association Meetings. Pittsburgh, PA.

Saxe, A.A. (1971) Social dimensions of mortuary practices in a Mesolithic population from Wadi Halfa, Sudan. In: *Memoirs of the Society for American Archaeology,* Brown, J.A., ed., pp. 39–57.

Schaaffhausen, H. (1888) *Der Neanderthaler Fund,* Bonn.

Schacter, D.L., and Tulving, E. (1994) *Memory Systems.* MIT Press, Cambridge, MA.

Schäfer, D. (1981) Taubach: zur Merkmalanalyse von Feuersteinartefakten der mittelpaläolithischen Travertinfundstell sowie zu ihrem Verhältnis zur Technologie anderer alt- und mittelpaläolithischer Fundplätze. *Ethnographisches Archäologisches Zeitschrift* 22: 369–96.

Schefuß, E., Ratmeyer, V., Stuut, J.-B.W., Jansen, J.H.F., and Damsté, J.S.S. (2003) Carbon isotope analyses of n-alkanes in dust from the lower atmosphere over the central eastern Atlantic *Geochemica et Cosmochemica Acta* 67(10): 1757–67.

Schefuß, E., Schouten, S., and Schneider, R.R. (2005) Climatic controls on central African hydrology during the past 20,000 years. *Nature* 437: 1003–6.

Schenker, N.M., Hopkins, W.D., Spocter, M.A., Garrison, A.R., Stimpson, C.D., Erwin, J.M., et al. (2010) Broca's area homologue in chimpanzees (*Pan troglodytes*): probabilistic mapping, asymmetry, and comparison to humans. *Cerebral Cortex* **20**(3): 730–42.

Schepartz, L.A., and Miller-Antonio, S. (2010) Taphonomy, life history, and human exploitation of Rhinoceros sinensis at the Middle Pleistocene site of Panxian Dadong, Guizhou, China. *International Journal of Osteoarchaeology* **20**: 253–68.

Schepartz, L.A., Bekken, D.A., Miller-Antonio, S., Paraso, C.K., and Karkanas, P. (2003) Faunal approaches to site formation processes at Panxian Dadong. In: *Current Research in Chinese Pleistocene Archaeology*, Shen, C., and Keates, S.G., eds, pp. 99–110. Archaeopress, Oxford.

Schepers, G. (1941) The mandible of the Transvaal fossil human skeleton from Sprinbok Flats. *Annals of the Transvaal Museum* **20**: 253–71.

Scheuer, L., and Black, S. (2000) *Developmental Juvenile Osteology*. Academic Press, San Diego.

Schick, K., Toth, N., Qi, W., Clark, J.D., and Etler, D. (1991) Archaeological perspectives in the Nihewan Basin, China. *Journal of Human Evolution* **21**: 13–26.

Schick, K.D. (1986) *Stone Age Sites in the Making: Experiments in the Formation and Transformation of Archaeological Occurrences*. B.A.R., Oxford.

Schick, K.D., and Dong, Z. (1993) Early Paleolithic of China and Eastern Asia. *Evolutionary Anthropology* **2**(1): 22–53.

Schick, T., and Stekelis, M. (1977) Mousterian assemblages in the Kebara Cave. In: *Moshe Stekelis Memorial*, Bar-Yosef, O., and Arensburg, B., eds, pp. 97–149. Israel Exploration Society, Jerusalem.

Schild, R., and Wendorf, F. (2005) Gademotta and Kulketti and the ages for the beginning of the Middle Paleolithic in Africa. *Israel Prehistoric Society Journal* **35**: 117–42.

Schmalhausen, I.I. (1949) *Factors of Evolution*. Blakiston, Philadelphia, PA.

Schmerling, P.-C. (1833) *Recherches sur les ossements fossiles découverts dans les cavernes de la province de Liège*, Liège, pp. 1–66.

Schmid, P. (1983) Eine rekonstruktion des skelettes von A.L. 288-1 (Hadar) und deren Konsequenzen. *Folia Primatologica* **40**: 283–306.

Schmid, P., and Berger, L.R. (1997) Middle Pleistocene hominid carpal proximal phalanx from the Gladysvale site, South Africa. *South African Journal of Science* **93**(10): 430–1.

Schmid, R. (1981) Descriptive nomenclature and classification of pyroclastic deposits and fragments. *Geologische Rundschau* **70**(2): 794–9.

Schmider, B., and Perpere, M. (1995) Production et Utilisation de Lamelles dans l'Aurignacien de la grotte du Renne a Arey-sur-Cure. In: *Palaéolithique Supér-ieur et Epipalaéolithique dans Ie Nord-est de la France*, Pautrat, Y., and Thevenin, A., eds, pp. 4–10. Actes de la Table Ronde, Dijon.

Schmidt, S., and Stewart, G.R. (2003) Δ N-15 values of tropical savanna and monsoon forest species reflect root specialisations and soil nitrogen status. *Oecologia* **134**: 569–77.

Schmidt, W.J., and Keil, A. (1971) *Polarizing Microscopy of Dental Tissues: Theory, Methods, and Results from the Structural Analysis of Normal and Diseased Hard, Dental Tissues and Tissues Associated with Them in Man and Other Vertebrates*. Pergamon Press, Oxford.

Schmidt-Nielsen, K. (1984) *Scaling: Why is Animal Size so Important?* Cambridge University Press, Cambridge.

Schmitz, R.W., and Thissen, J. (2000a) *Neandertal: Die Geschichte geht weiter*. Spektrum, Heidelberg.

Schmitz, R.W., and Thissen, J. (2000b) First archaeological finds and new human remains at the rediscovered site of the Neanderthal type specimen: a preliminary report. In: *Neanderthals and modern humans–discussing the transition: Central and Eastern Europe from 50, 000 to 30, 000 B.P*, Orschiedt, J., and Weniger, G.-C., eds, pp. 267–73. Neanderthal Museum, Mettmann.

Schmitz, R.W., Serre, D., Bonani, G., Feine, S., Hillgruber, F., Krainitzki, H., et al. (2002) The Neandertal type site revisited: interdisciplinary investigations of skeletal remains from the Neander Valley, Germany. *Proceedings of the National Academy of Sciences USA* **99**(20): 13342–7.

Schoenemann, P., Sheehan, M., and Glotzer, L. (2005) Prefrontal white matter volume is disproportionately larger in humans than in other primates. *Nature Neuroscience* **8**(2): 242–52.

Schoeninger, M.J. (1985) Trophic level effects on $^{15}N/^{14}N$ and $^{13}C/^{12}C$ ratios in bone collagen and strontium levels in bone mineral. *Journal of Human Evolution* **14**(5): 515–25.

Schoeninger, M.J. (1995) Stable isotope studies in human evolution. *Evolutionary Anthropology* **4**: 83–98.

Schoeninger, M.J., and DeNiro, M.J. (1984) Nitrogen and carbon isotope composition of bone collagen from marine and terrestrial animals. *Geochemica et Cosmochemica Acta* **48**(4): 625–39.

Schoeninger, M.J., DeNiro, M.J., and Tauber, H. (1983) $^{15}N/^{14}N$ ratios of bone collagen reflect marine and terrestrial components of prehistoric human diet. *Science* **220**: 1381–3.

Schoeninger, M.J., van der Merwe, N.J., Moore, K., Lee-Thorp, J., and Larsen, C.S. (1990) Decrease in diet quality between the prehistoric and contact periods. In: *The Archaeology of Mission Santa Catalina de Gaule*, Larsen, C.S., ed., pp. 78–93.

Schoetensack, O. (1908) *Der Unterkiefer des Homo heidelbergensis aus den Sanden von Mauer bei Heidelberg*. W. Engelmann, Leipzig.

Scholz, C.A., Karp, T., and Lyons, R.P. (2007) Structure and morphology of the Bosumtwi impact structure from seismic reflection data. *Meteoritics and Planetary Science* **42**(4–5): 549–60.

Schoville, B. (2010) Frequency and distribution of edge damage on Middle Stone Age lithic points, Pinnacle Point 13B, South Africa. *Journal of Human Evolution* **59**: 378–91.

Schrago, C.G., and Russo, C.A.M. (2003) Timing the origin of New World Monkeys. *Molecular Biology and Evolution* **20**(10): 1620–5.

Schreger, D. (1800) Beitrag zur Geschichte der Zähne. *Beitr Zergliederungskunst* **1**: 1–7.

Schrenk, F., Bromage, T.G., Betzler, C.G., Ring, U., and Juwayei, Y.M. (1993) Oldest *Homo* and Pliocene biogeography of the Malawi Rift. *Nature* **365**: 833–6.

Schüler, T. (1994) ESR dating of tooth-enamel from the lower Travertin at Weimar-Ehringsdorf. *Alt-Thüringen* **28**: 9–23.

Schultz, A.H. (1935) Eruption and decay of the permanent teeth in primates. *American Journal of Physical Anthropology* **11**: 489–581.

Schultz, A.H. (1942) Conditions for balancing the head in primates. *American Journal of Physical Anthropology* **29**: 483–97.

Schultz, A.H. (1960) Age changes in primates and their modification in man. In: *Human Growth*, Tanner, J.M., ed., pp. 1–20. Pergamon, Oxford.

Schuurman, G. (2005) Decomposition rates and termite assemblage composition in semiarid Africa. *Ecology* **86**(5): 1236–49.

Schvoerer, M., Rouanet, J.F., Navailles, H., and Debénath, A. (1977) Datation absolue par thermoluminescence de restes humains antewurmiens de l'abri Suard a La Chaise de Vouthon (Charente). *Comptes Rendus de l'Académie des Sciences* **287**: 1979–82.

Schwalbe, G.A. (1899) Studien über *Pithecanthropus erectus* Dubois. *Zeitschrift für Morphologie und Anthropologie* **1**: 1–240.

Schwalbe, G.A. (1901) Der Neanderthalschädel. *Bonner Jahrbuch* **106**: 1–72.

Schwalbe, G.A. (1904) *Die Vorgeschichte des Menschen.* Vieweg, Braunschweig.

Schwalbe, G.A. (1906) *Studien zur Vorgeschichte des Memschen.* Schweitzerbartsche Buchhandlung, Stuttgart.

Schwalbe, G.A. (1910) The descent of man. In: *Darwin and Modern Science*, Stewart, A.C., ed., pp. 112–36. Cambridge University Press, London.

Schwalbe, G.A. (1914) Kritische Besprechung von Boule's Werk: "L'homme fossile de la Chapelle-aux-Saints" mit eigenen Untersuchungen. *Zeitschrift für Morphologie und Anthropologie* **16**: 527–610.

Schwalbe, G.A. (1923) Die Abstammung des Menschen und die ältesten Menschenformen. In: *Die Kultur der Gegenwart*, Hinneburg, P., ed. Teubner, Leipzig.

Schwarcz, H., and Latham, A. (1984) Uranium-series age determination of travertines from the site of Vértesszöllös, Hungary. *Journal of Archaeological Science* **11**(4): 327–36.

Schwarcz, H.P., and Debénath, A. (1979) Datation absolue des restes humains de la Chaise-de-Vouthon (Chavente) au moyen du déséquilibre des séries d'uranium. *Comptes Rendus de l'Académie des Sciences* 1155–7.

Schwarcz, H.P., and Skoflek, I. (1982) New data for the Tata archaeological site, Hungary. *Nature* **295**: 590–1.

Schwarcz, H.P., and Schoeninger, M.J. (1991) Stable isotope analyses in human nutritional ecology. *Yearbook of Physical Anthropology* **34**(S13): 283–321.

Schwarcz, H.P., and Rink, W.J. (1998) Progress in ESR and U-series chronology of the Levantine Paleolithic. In: *Neandertals and Modern Humans in Western Asia*, Akazawa, T., Aoki, K., and Bar-Yosef, O., eds. Plenum Press, New York.

Schwarcz, H.P., and Rink, W.J. (2000) ESR dating of the Die Kelders Cave 1 Site, South Africa. *Journal of Human Evolution* **38**: 121–8.

Schwarcz, H.P., Grün, R., Latham, A.G., Mania, D., and Brunnacker, K. (1988) The Bilzingsleben archaeological site: new dating evidence. *Archaeometry* **30**: 5–17.

Schwarcz, H.P., Buhay, W.M., Grün, R., Valladas, H., Tchernov, E., Bar-Yosef, O., et al. (1989) ESR dating of the Neanderthal site, Kebara Cave, Israel *Journal of Archaeological Science* **16**(6): 652–9.

Schwarcz, H.P., Bietti, A., Buhay, W., Stiner, M.C., Grün, R., and Segre, A. (1991a) On the reexamination of Grotta Guattari: uranium-series and election-spin-resonance dates. *Current Anthropology* **32**(3): 313–16.

Schwarcz, H.P., Buhay, W., Grün, R., Stiner, M.C., Kuhn, S., and Miller, G.H. (1991b) Absolute dating of sites in coastal Lazio. *Quaternaria Nova* **1**: 51–67.

Schwarcz, H.P., Simpson, J.J., and Stringer, C.B. (1998) Neanderthal skeleton from Tabun: U-series data by gamma-ray spectrometry. *Journal of Human Evolution* **35**: 635–45.

Schwartz, B.L., Colon, M.R., Sanchez, I.C., Rodriguez, I., and Evans, S. (2002) Single-trial learning of "what" and "who" information in a gorilla (*Gorilla gorilla gorilla*): implications for episodic memory. *Animal Cognition* **5**(2): 85–90.

Schwartz, G. (2000) Taxonomic and functional aspects of the patterning of enamel thickness distribution in extant large-bodied hominoids. *American Journal of Physical Anthropology* **111**(2): 221–44.

Schwartz, G., and Dean, C. (2000) Interpreting the hominid dentition: ontogenetic and phylogenetic aspects. In: *Development, Growth and Evolution: Implications for the Study of the Hominid Skeleton*, O'Higgins, P., and Cohn, M.J., eds, pp. 207–33. Academic Press, London.

Schwartz, J. (1984) Supernumerary teeth in anthropoid primates and models of tooth development. *Archives of Oral Biology* **29**(10): 833–42.

Schwartz, J., and Tattersall, I. (2002) *The Human Fossil Record, Volume One: Terminology and craniodental morphology of genus Homo (Europe)*. Wiley-Liss, New York.

Schwartz, J.H., and Tattersall, I. (2003) *The Human Fossil Record, Volume Two: Craniodental Morphology of Genus Homo (Africa and Asia)*. John Wiley & Sons, Hoboken, NJ.

Schwarz, E. (1929) Das Vorkommen des Schimpansen auf den linken Kongo-Ufer. *Revue de zoologie et de botanique africaines* **16**: 425–6.

Schweitzer, F.R. (1979) Excavations at Die Kelders, Cape Province, South Africa: the Holocene deposits. *Annals of the South African Museum* **78**: 101–233.

Scott, G.R., and Turner, C.G. (1997) *The Anthropology of Modern Human Teeth: Dental Morphology and its Variation in Recent Human Populations*. Cambridge University Press, Cambridge.

Scott, J.H. (1958) The cranial base. *American Journal of Physical Anthropology* **16**: 319–48.

Scott, R.S., Ungar, P.S., Bergstrom, T.S., Brown, C.A., Grine, F.E., Teaford, M.F., et al. (2005) Dental microwear texture analysis shows within-species diet variability in fossil hominins. *Nature* **436**: 693–5.

Sebat, J., Lakshmi, B., Troge, J., Alexander, J., Young, J., Lundin, P., et al. (2004) Large-scale copy number polymorphism in the human genome. *Science* **305**(5683): 525–8.

Segre, A., and Ascenzi, A. (1984) Fontana Ranuccio: Italy's earliest Middle Pleistocene hominid site. *Current Anthropology* **25**(2): 230–3.

Segre, A.G. (1983) Geologia quaternaria e Paleolitico nella bassa valle dell'Aniene, Roma. *Rivista di Antropologia* **62**: 87–98.

Seidler, H., Weber, G., Mariam, A., zur Nedden, D., and Recheis, W. (1999) *The Skull of Bodo. CD-ROM Edition-Fossil Hominids Vol. 1*. Institute of Anthropology, University of Vienna, Vienna.

Selenka, E. (1898) Rassen, Schadel and Bezahnung des Orangutan. In: *Menschenaffen (Anthropomorphae). Studien uber Entwicklung und Schadelbau*. CW Kreidels Verlag, Wiesbaden.

Selenka, M.L., and Blanckenhorn, M. (1911) *Die Pithecanthropus - Schichten auf Java Geologische un Paleonto lo gische Ergebnisse der Trinil*. Wilhelm Engelmann, Liepzig.

Sellers, W., Dennis, L., W J, W., and Crompton, R. (2004) Evaluating alternative gait strategies using evolutionary robotics. *Journal of anatomy* **204**(5): 343–51.

Sellers, W.I., Dennis, L.A., and Crompton, R.H. (2003) Predicting the metabolic energy costs of bipedalism using evolutionary robotics. *Journal of Experimental Biology* **206**: 1127–36.

Sellers, W.I., Cain, G.M., Wang, W.J., and Crompton, R.H. (2005) Stride lengths, speed and energy costs in walking of *Australopithecus afarensis*: using evolutionary robotics to predict locomotion of early human ancestors. *Journal of the Royal Society Interface* **2**: 431–41.

Sémah, F., Sémah, A.-M., Djubiantono, T., and Simanjuntak, H.T. (1992) Did they also make stone tools? *Journal of Human Evolution* **23**(5): 439–46.

Semaw, S. (2000) The world's oldest stone artifacts from Gona, Ethiopia: their implications for understanding stone technology and patterns of human evolution between 2.6–1.5 million years ago. *Journal of Archaeological Science* **27**(12): 1197–1214.

Semaw, S., Renne, P., Harris, J.W.K., Feibel, C., Bernor, R.L., Fessaha, N., et al. (1997) 2.5-million-year-old stone tools from Gona, Ethiopia. *Nature* **385**: 333–6.

Semaw, S., Rogers, M.J., Quade, J., Renne, P.R., Butler, R.F., Domínguez-Rodrigo, M., et al. (2003) 2.6-Million-year-old stone tools and associated bones from OGS-6 and OGS-7, Gona, Afar, Ethiopia. *Journal of Human Evolution* **45**(2): 169–77.

Semaw, S., Simpson, S.W., Quade, J., Renne, P.R., Butler, R.F., McIntosh, W.C., et al. (2005) Early Pliocene hominids from Gona, Ethiopia. *Nature* **433**: 301–5.

Semendeferi, K., Armstrong, E., Schleicher, A., Zilles, K., and Van Hoesen, G. (2001) Prefrontal cortex in humans and apes: a comparative study of area 10. *American Journal of Physical Anthropology* **114**(3): 224–41.

Semendeferi, K., Lu, A., Schenker, N., and Damasio, H. (2002) Humans and great apes share a large frontal cortex. *Nature Neuroscience* **5**(3): 272–6.

Semenov, S. (1964) *Prehistoric Technology*. Cory, Adams & Mackay, London.

Senut, B. (1981) L'humérus et ses articulations chez les hominidés plio-pléistocènes. In: *Cahiers de Paléontologie (Paléoanthropologie)* pp. 1–281. CNRS, Paris.

Senut, B. (1985) Computerized tomography of a Neanderthal humerus from Le Regourdou (Dordogne, France): Comparisons with modern man. *Journal of Human Evolution* **14**(8): 717–23.

Senut, B. (1996) Pliocene hominid systematics and phylogeny. *South African Journal of Science* **92**: 165–6.

Senut, B., and Tardieu, C. (1985) Functional aspects of Plio-Pleistocene hominid limb bones: Implications for taxonomy and phylogeny. In: *Ancestors: the Hand Evidence*, Delson, E., ed., pp. 193–201. Alan R. Liss, New York.

Senut, B., Pickford, M., Ssemmanda, I., Elepu, D., and Obwona, P. (1987) Découverte du premier Homininae (*Homo* sp.) dans le Pleistocene de Nyabusosi (Ouganda occidental). *Comptes Rendus de l'Académie des Sciences* **305**(9): 819–22.

Senut, B., Pickford, M., Braga, J., Marais, D., and Coppens, Y. (2000) Découverte d'un *Homo sapiens* archaïque à Oranjemund, Namibie. *Comptes Rendus de l'Académie des Sciences* **330**(11): 813–19.

Senut, B., Pickford, M., Gommery, D., Mein, P., Cheboi, K., and Coppens, Y. (2001) First hominid from the Miocene (Lukeino Formation, Kenya). *Comptes Rendus de l'Académie des Sciences* **332**(2): 137–44.

Senyürek, M. (1955) A note on the teeth of *Meganthropus africanus* Weinert from Tanganyika Territory. *Belletin* **19**: 1–57.

Senyürek, M.S. (1959) *A Study of the Deciduous Teeth of the Fossil Shanidar Infant*. Division of Paleoanthropology, University of Ankara.

Sergi, G. (1893) *Le varieta umane. Principi e metodo di classicazione*. P. Bruno, Torino.

Sergi, G. (1895) *Origine e diffusione della stirpe mediterranea*. D. Alghieri, Rome.

Sergi, S. (1912) *Crania habessinica: contributo all'antropologia dell'Africa orientale*. Ermanno Loescher, Rome.

Sergi, S. (1929) La scoperta di un cranio del tipo di Neandertal presso Roma. *Rivista di Antropologia* **28**: 457–62.

Sergi, S. (1934) Sulla stratigrafia di Saccopastore. *Rivista di Antropologia* **30**: 477–8.

Sergi, S. (1944) Craniometria e craniografia del primo paleantropo di Saccopastore. *Ricerche di Morfologia* **20–21**: 733–91.

Sergi, S. (1948) Il cranio del secondo Paleantropo di Saccopastore. *Palaeontographia Italica* **42**: 25–164.

Sergi, S. (1953) Morphological position of the "Prophaneranthropi" (Swanscombe and Fontéchevade). *Actes du IV Congres de l'association internationale pour l'Etude du Quanternaire*. Rome, pp. 651–65.

Sergi, S. (1954) The mandibola neandertalien Circeo II. *Rivista di Antropologia* **41**: 305–44.

Sergi, S., Parenti, R., and Paoli, G. (1974) Il giovane Paleolitico della caverna delle Arene Candide. *Memorie dell'Istituto Italiano di Paleontologia Umana* **2**: 13–38.

Serre, D., Langaney, A., Chech, M., Tescher-Nicola, M., Paunovic, M., Mennecier, P., et al. (2004) No evidence of Neandertal mtDNA contribution to early modern humans. *PLoS Biology* **2**(3): 313–17.

Shackleton, N. (1967) Oxygen isotope analysis and Pleistocene temperatures re-assessed. *Nature* **215**: 15–17.

Shackleton, N.C. (1982) Stratigraphy and chronology of the KRM deposits: oxygen isotope evidence. In: *The Middle Stone Age at Klasies River Mouth in South Africa*, Singer, R., and Wymer, J., eds, pp. 194–9. Chicago University Press, Chicago, IL.

Shackleton, N.J., and Opdyke, N.D. (1973) Oxygen isotope and paleomagnetic stratigraphy of equatorial Pacific core V28–238: oxygen isotope temperatures and ice volumes on a 10^5 and 10^6 year scale. *Quaternary Research* **3**: 39–55.

Shanahan, T.M., Overpeck, J.T., Anchukaitis, K.J., Beck, J.W., Cole, J.E., Dettman, D.L., et al. (2009) Atlantic forcing of persistent drought in West Africa. *Science* **324**(5925): 377–80.

Shang, H., and Trinkaus, E. (2008) An ectocranial lesion on the middle Pleistocene human cranium from Hulu Cave, Nanjing, China. *American Journal of Physical Anthropology* **135**(1): 131–7.

Shaw, J.C.M. (1931) *The Teeth, the Bony Palate and the Mandible in Bantu Races of South Africa*. John Bale, Sons and Danielsson, London.

Shaw, T. (1965) Akure excavations: Stone Age skeleton 9000 B.C.? *African Notes* **3**(5–6).

Shaw, T. (1968) Radiocarbon dating in Nigeria. *Journal of the Historical Society of Nigeria* **4**: 453–65.

Shea, B.T. (1986) Scapula form and locomotion in chimpanzee evolution. *American Journal of Physical Anthropology* **70**: 475–88.

Shea, B.T., Hammer, R.E., Brinster, R.L., and Ravosa, M.R. (1990) Relative growth of the skull and postcranium in giant transgenic mice. *Genetical Research* **56**(1): 21–34.

Shea, J. (1988) Spear points from the Middle Paleolithic of the Levant. *Journal of Field Archaeology* **15**(4): 441–50.

Shea, J. (1999) Artifact abrasion, fluvial processes, and "living floors" from the Early Paleolithic site of 'Ubeidiya (Jordan Valley, Israel). *Geoarchaeology* **14**(2): 191–207.

Shea, J. (2005) Bleeding or Breeding: Neandertals vs. Early Modern Humans in the Middle Paleolithic Levant. In: *Archaeologies of the Middle East: Critical Perspectives*, Pollock, S., and Bernbeck, R., eds, pp. 129–51. Blackwell Publishing, Oxford.

Shea, J. (2006) The origins of lithic projectile point technology: evidence from Africa, the Levant, and Europe. *Journal of Archaeological Science* **33**(6): 823–46.

Shea, J., and Sisk, M. (2010) Complex projectile technology and *Homo sapiens* dispersal into Western Eurasia. *PaleoAnthropology*: 100–22.

Shea, J.J. (2008) The Middle Stone Age archaeology of the Lower Omo Valley Kibish Formation: Excavations, lithic assemblages, and inferred patterns of early *Homo sapiens* behavior. *Journal of Human Evolution* **55**(3): 448–85.

Shellis, R.P. (1984) Variations in growth of the enamel crown in human teeth and a possible relationship between growth and enamel structure. *Archives of Oral Biology* **29**(9): 697–705.

Shellis, R.P., Beynon, A.D., Reid, D.J., and Hiiemae, K.M. (1998) Variation in molar enamel thickness among primates. *Journal of Human Evolution* **35**: 507–22.

Shen, C., and Chen, C. (2003) New evidence of hominid behavior from Xiaochangliang, Northern China: site formation and lithic technology. In: *Current Research in Chinese Pleistocene Archaeology*, Shen, C., and Keates, S.G., eds, pp. 67–82. Archaeopress, Oxford.

Shen, C., and Wei, Q. (2004) Lithic technological variability of the Middle Pleistocene in the Eastern Nihewan Basin, northern China. *Asian Perspectives* **43**: 281–301.

Shen, C., Gao, X., and Wei, Q. (2010) The earliest hominin occupations in the Nihewan Basin of northern China: Recent progress in field investigations.

In: *Asian Paleoanthropology: From Africa to China and Beyond*, Norton, C.J., and Braun, D.R., eds, pp. 169–80. Springer Press, Dordrecht.

Shen, G., Ku, T.-L., Cheng, H., Edwards, R.L., Yuan, Z., and Wang, Q. (2001) High-precision U-series dating of Locality 1 at Zhoukoudian, China. *Journal of Human Evolution* **41**(6): 679–88.

Shen, G., Gao, X., Gao, B., and Granger, D.E. (2009) Age of Zhoukoudian *Homo erectus* determined with ^{26}Al/^{10}Be burial dating. *Nature* **458**: 198–200.

Shen, G.J., Fang, Y.S., and Jin, L.H. (1994) Re-examination of the chronological position of Chaoxian Man. *Acta Anthropologica Sinica* **13**: 249–56.

Shen, G.J., Liu, J., and Jin, L.H. (1997) Preliminary results on U-series dating of Panxian Dadong in Guizhou Province, S-W China. *Acta Anthropological Sinica* **16**: 221–30.

Shennan, S.J., and Steele, J. (1999) Cultural learning in hominids: a behavioural ecological approach. In: *Mammalian Social Learning: Comparative and Ecological Perspectives*, Box, H.O., and Gibson, K.R., eds, pp. 367–88. Cambridge University Press, Cambridge.

Sheppard, P. (1987) *The Caspian of North Africa: Stylistic Variation in Stone Tool Assemblages*. International Series 353. B.A.R., Oxford.

Sheppard, P.J., and Kleindienst, M.R. (1996) Technological change in the Earlier and Middle Stone Age of Kalambo Falls (Zambia). *African Archaeological Reviews* **13**: 171–96.

Sherry, D.F., and Schacter, D.L. (1987) The evolution of multiple memory systems. *Psychological Review* **94**(4): 439–54.

Sherry, S., Harpending, H., Batzer, M., and Stoneking, M. (1997) Alu evolution in human populations: using the coalescent to estimate effective population size. *Genetics* **147**(4): 1977.

Sherry, S.T., Rogers, A., Harpending, H., Soodyall, H., Jenkins, T., and Stoneking, M. (1994) Mismatch distributions of mtDNA reveal recent human population expansions. *Human Biology* **66**(5): 761–75.

Sherwood, C., Holloway, R., Semendeferi, K., and Hof, P. (2005) Is prefrontal white matter enlargement a human evolutionary specialization? *Nature Neuroscience* **8**(5): 537–8.

Sherwood, R.J., Ward, S.C., and Hill, A. (2002) The taxonomic status of the Chemeron temporal (KNM-BC1). *Journal of Human Evolution* **42**(1–2): 153–84.

Shi, J., H., X., Wang, Y., Zhang, C., Jiang, Z., Zhang, K., et al. (2003) Divergence of the genes on human chromosome 21 between human and other hominoids and variation of substitution rates among transcription units. *Proceedings of the National Academy of Sciences USA* **100**(14): 8331–6.

Shipman, P., and Phillips-Conroy, J. (1977) Hominid tool-making versus carnivore scavenging. *American Journal of Physical Anthropology* **46**(1): 77–86.

Shipman, P., Potts, R., and Pickford, M. (1983) Lainyamok, a new middle Pleistocene hominid site. *Nature* **306**: 365–8.

Shipton, C. (2010) Imitation and shared intentionality in the Acheulean. *Cambridge Archaeological Journal* **20**: 197–210.

Shipton, C., Petraglia, M.D., and Paddayya, K. (2009a) Inferring aspects of Acheulean sociality and cognition from biface technology in the Hunsgi-Baichbal Valley, India. In: *Lithic Materials and Paleolithic Societies*, Blades, B., and Adams, B., eds, pp. 219–31. Wiley-Blackwell, New York.

Shipton, C.B.K., Petraglia, M.D., and Paddayya, K. (2009b) Stone tool experiments and reduction methods at the Acheulean site of Isampur Quarry, India. *Antiquity* **83**: 769–85.

Shott, M. (2003) Chaîne opératoire and reduction sequence. *Lithic Technology* **28**: 95–105.

Shrewsbury, M.M., Marzke, M.W., Linscheid, R.L., and Reece, S.P. (2003) Comparative morphology of the pollical distal phalanx. *American Journal of Physical Anthropology* **121**: 30–47.

Shryock, H.S., and Siegel, J.S. (1976) *The Methods and Materials of Demography*. Academic Press, New York.

Shubin, N., Tabin, C., and Carroll, S. (2009) Deep homology and the origins of evolutionary novelty. *Nature* **457**: 818–23.

Sibley, C., and Ahlquist, J.E. (1984) The phylogeny of the hominoid primates as indicated by DNA-DNA hybridization. *Journal of Molecular Evolution* **20**(1): 2–15.

Sikes, N.E., Potts, R., and Behrensmeyer, A.K. (1999) Early Pleistocene habitat in Member 1 Olorgesailie based on paleosol stable isotopes. *Journal of Human Evolution* **37**(5): 721–46.

Siler, W. (1979) A competing-risk model for animal mortality. *Ecology* **60**(4): 750–7.

Silk, J.B., Clark-Wheatley, C.B., Rodman, P.S., and Samuels, A. (1981) Differential reproductive success and facultative adjustment of sex ratios among captive female bonnet macaques (*Macaca radiata*) *Animal Behavior* **29**(4): 1106–20.

Silk, J.B., Willoughby, E., and Brown, G.R. (2005) Maternal rank and local resource competition do not predict birth sex ratios in wild baboons. *Proceedings of the National Academy of Sciences USA* **272**(1565): 859–64.

Sillen, A., and Morris, A. (1996) Diagenesis of bone from Border Cave: implications for the age of the Border Cave hominids. *Journal of Human Evolution* **31**(6): 499–506.

Simek, J.F., and Smith, F.H. (1997) Chronological changes in stone tool assemblages from Krapina (Croatia). *Journal of Human Evolution* **32**(6): 561–75.

Simmons, S., and Ossa, P.P. (1978) Interim report on the Keilor excavation, May 1978. *Records of the Victorian Archaeological Survey* **8**: 63–6.

Simons, E.L., Pilbeam, D., and Ettel, P.C. (1969) Controversial taxonomy of fossil hominids. *Science* **166**: 258–9.

Simpson, G.G. (1944) *Tempo and Mode in Evolution*. Columbia University Press, New York.

Simpson, G.G. (1950) *The Meaning of Evolution*. Oxford University Press, London.

Simpson, G.G. (1951) The species concept. *Evolution* **5**: 285–98.

Simpson, G.G. (1961) *Principles of Animal Taxonomy*. Columbia University Press, New York.

Simpson, J.J., and Grün, R. (1998) Non-destructive gamma spectrometric U-series dating. *Quaternary Science Reviews* **17**(11): 1009–22.

Simpson, S.W., Quade, J., Levin, N.E., Butler, B., Dupont-Nivet, G., Everett, M., et al. (2008) A female *Homo erectus* pelvis from Gona, Ethiopia. *Science* **322**(5904): 1089–92.

Singer, R. (1954) The Saldanha skull from Hopefield, South Africa. *American Journal of Physical Anthropology* **12**: 345.

Singer, R. (1958) The Boskop 'race' problem. *Man* **1958**: 173–8.

Singer, R. (1961) Pathology in the temporal bone of the Boskop skull. *The South African Archaeological Bulletin* **16**: 103–4.

Singer, R., and Wymer, J. (1968) Archaeological investigations at the Saldanha skull site in South Africa. *The South African Archaeological Bulletin* **23**(91): 63–74.

Singer, R., and Wymer, J. (1982) *The Middle Stone Age at Klasies River Mouth in South Africa*. University of Chicago Press, Chicago, IL.

Sisk, M.L., and Shea, J.J. (2008) Intrasite spatial variation of the Omo Kibish Middle Stone Age assemblages: artifact refitting and distribution patterns. *Journal of Human Evolution* **55**(3): 486–500.

Skinner, A.R., Blackwell, B.A., Martin, S., Ortega, A., Blickstein, J., Golavanova, L.V., et al. (2005) ESR dating at Mezmaiskaya Cave, Russia. *Applied Radiation and Isotopes* **62**(2): 219–24.

Skinner, M., Gunz, P., Wood, B., and Hublin, J. (2008) Enamel-dentine junction (EDJ) morphology distinguishes the lower molars of *Australopithecus africanus* and *Paranthropus robustus*. *Journal of Human Evolution* **55**(6): 979–88.

Skinner, M.M., Gordon, A.D., and Collard, N.J. (2006) Mandibular size and shape variation in the hominins at Dmanisi, Republic of Georgia. *Journal of Human Evolution* **51**(1): 36–49.

Skinner, M.M., Gunz, P., Wood, B.A., and Hublin, J.-J. (2009a) How many landmarks? Assessing the classification accuracy of *Pan* lower molars using a geometric morphometric analysis of the occlusal basin as seen at the enamel-dentine junction. *Frontiers of Oral Biology* **13**: 23–9.

Skinner, M.M., Gunz, P., Wood, B.A., Boesch, C., and Hublin, J.-J. (2009b) Taxonomic discrimination of species and sub-species of *Pan* using enamel-dentine junction morphology of lower molars. *American Journal of Physical Anthropology* **140**: 234–43.

Skipper, J.I., Goldin-Meadow, S., Nusbaum, H.C., and Small, S.L. (2009) Gestures orchestrate brain networks for language understanding *Current Biology* **19**(8): 661–7.

Sládek, V. (2000) *Évolution des Hominidés en Europe Centrale durant le Pleistocène Superieur: origine des hommes anatomiquement modernes*. PhD thesis, Université de Bordeaux 1.

Sládek, V., Trinkaus, E., Šefčáková, A., and Halouzka, R. (2002) Morphological affinities of the Šal'a 1 frontal bone. *Journal of Human Evolution* **43**(6): 787–815.

Slice, D. (2001) Landmark coordinates aligned by Procrustes analysis do not lie in Kendall's shape space. *Systematic Biology* **50**: 141–9.

Smit, A. (1996) The origin of interspersed repeats in the human genome. *Current Opinion in Genetics & Development* **6**(6): 743–8.

Smith, A.B. (1994) *Systematics and the Fossil Record: Documenting Evolutionary Patterns*. Blackwell, Oxford.

Smith, B. (1989) Dental development as a measure of life history in primates. *Evolution* **43**(3): 683–8.

Smith, B. (1991) Dental development and the evolution of life history in Hominidae. *American Journal of Physical Anthropology* **86**(2): 157–74.

Smith, B. (1991) Standards of human tooth formation and dental age assessment. In: *Advances in Dental Anthropology*, Kelley, M., and Larsen, C., eds, pp. 143–68. Wiley-Liss, New York.

Smith, B., Crummett, T., and Brandt, K. (1994) Ages of eruption of primate teeth: a compendium for aging individuals and comparing life histories. *American Journal of Physical Anthropology* **37**(S19): 177–231.

Smith, B.H. (1986) Dental development in *Australopithecus* and early *Homo*. *Nature* **323**: 327–30.

Smith, C.I., Chamberlain, A.T., Riley, M.S., Stringer, C., and Collins, M.J. (2003) The thermal history of human fossils and the likelihood of successful DNA amplification. *Journal of Human Evolution* **45**(3): 203–17.

Smith, F., Falsetti, A., and Donnelly, S. (1989) Modern human origins. *Yearbook of Physical Anthropology* **32**: 35–68.

Smith, F., Jankovic, I., and Karavanic, I. (2005) The assimilation model, modern human origins in Europe, and the extinction of Neandertals. *Quaternary international* **137**(1): 7–19.

Smith, F.H. (1976a) A fossil hominid frontal from Velika Pećina (Croatia) and a consideration of Upper Pleistocene hominids from Yugoslavia. *American Journal of Physical Anthropology* **44**(1): 127–34.

Smith, F.H. (1976b) *The Neandertal remains from Krapina: A Descriptive and Comparative Study*. Department of Anthropology, University of Tennessee, Knoxville, TN.

REFERENCES

Smith, F.H. (1977) On the application of morphological "dating" to the hominid fossil record. *Journal of Anthropological Research* **33**(3): 302–16.

Smith, F.H. (1982) Upper Pleistocene hominid evolution in South-Central Europe: a review of the evidence and analysis of trends. *Current Anthropology* **23**(6): 667–703.

Smith, F.H. (1984) Fossil hominids from the Upper Pleistocene of Central Europe and the origins of modern Europeans. In: *The Origins of Modern Humans: A World Survey of the Fossil Evidence*, Smith, F.H., and Spencer, F., eds, pp. 137–209. Alan R. Liss, New York.

Smith, F.H. (1992) Models and realities in modern human origins: the African fossil evidence. *Philosophical Transactions of the Royal Society of London. Series B, Biological Sciences* **337**: 243–50.

Smith, F.H. (2010) Species, populations, and assimilation in later human evolution. In: *A Companion to Biological Anthropology*, Larsen, C.S., ed., pp. 357–78. Wiley-Blackwell, Oxford.

Smith, F.H., and Ranyard, G.C. (1980) Evolution of the supraorbital region in Upper Pleistocene fossil hominids from South-Central Europe. *American Journal of Physical Anthropology* **53**: 589–610.

Smith, F.H., Boyd, D.C., and Malez, M. (1985) Additional Upper Pleistocene human remains from Vindija Cave, Croatia, Yugoslavia. *American Journal of Physical Anthropology* **68**(3): 375–83.

Smith, F.H., Trinkaus, E., Pettitt, P.B., Karavanić, I., and Paunovic, M. (1999) Direct radiocarbon dates for Vindija G$_1$ and Velika Pećina Late Pleistocene hominid remains. *Proceedings of the National Academy of Sciences USA* **96**(22): 12281–6.

Smith, F.H., Janković, I., and Karavanić, I. (2005) The assimilation model, modern human origins in Europe, and the extinction of Neandertals. *Quaternary International* **137**(1): 7–19.

Smith, H., and Grine, F. (2008) Cladistic analysis of early *Homo* crania from Swartkrans and Sterkfontein, South Africa. *Journal of Human Evolution* **54**(5): 684–704.

Smith, M., Dunkley, P.N., Deino, A., Williams, L.A.J., and McCall, G.J.H. (1995) Geochronology, stratigraphy and structural evolution of Silali volcano, Gregory Rift, Kenya. *Journal of the Geological Society* **152**: 297–310.

Smith, P., and Arensburg, B. (1977) The Mousterian infant from Kebara. In: *Moshe Stekelis Memorial*, Bar-Yosef, O., and Arensburg, B., eds, pp. 164–76. Israel Exploration Society, Jerusalem.

Smith, P.E.L. (1964) Expedition to Kom Ombo. *Archaeology* **17**: 209–10.

Smith, P.E.L. (1966a) New Prehistoric Investigations at Kom Ombo (Upper Egypt). *Zephyrus* **17**: 31–45.

Smith, P.E.L. (1966b) The Late Paleolithic of Northeast Africa in the Light of Recent Research. *American Anthropologist* **68**(2).

Smith, R. (2009) Use and misuse of the reduced major axis for line-fitting. *American Journal of Physical Anthropology* **140**(3): 476–86.

Smith, R.J. (2002) Estimation of body mass in paleontology. *Journal of Human Evolution* **43**(2): 271–87.

Smith, R.J. (2005a) Species recognition in paleoanthropology; implications of small sample sizes. In: *Interpreting the Past: Essays on Human, Primate, and Mammal Evolution in honor of David Pilbeam*, Lieberman, D.E., Smith, R.J., and Kelley, J., eds, pp. 207–19. Brill Academic Publishers, Boston, MA.

Smith, R.J. (2005b) Relative size versus controlling for size. *Current Anthropology* **46**: 249–73.

Smith, T., and Tafforeau, P. (2008) New visions of dental tissue research: Tooth development, chemistry, and structure. *Evolutionary Anthropology* **17**(5): 213–26.

Smith, T., Olejniczak, A., Reid, D., Ferrell, R., and Hublin, J. (2006a) Modern human molar enamel thickness and enamel-dentine junction shape. *Archives of Oral Biology* **51**: 974–95.

Smith, T., Tafforeau, P., Reid, D., Grun, R., Eggins, S., Boutakiout, M., et al. (2007) Earliest evidence of modern human life history in North African early *Homo sapiens*. *Proceedings of the National Academy of Sciences USA* **104**(15): 6128–33.

Smith, T., Tafforeau, P., Reid, D., Pouech, J., Lazzari, V., Zermeno, J., et al. (2010) Dental evidence for ontogenetic differences between modern humans and Neanderthals. *Proceedings of the National Academy of Sciences USA* **107**(49): 20923–8.

Smith, T.M. (2006b) Experimental determination of the periodicity of incremental features in enamel. *Journal of Anatomy* **208**(1): 99–113.

Smith, T.M., Reid, D.J., and Sirianni, J.E. (2006b) The accuracy of histological assessments of dental development and age at death. *Journal of Anatomy* **208**(1): 125–38.

Smith, T.M., Olejniczak, A.J., Reh, S., Reid, D.J., and Hublin, J.-J. (2008) Brief communication: Enamel thickness trends in the dental arcade of humans and chimpanzees. *American Journal of Physical Anthropology* **136**(2): 237–41.

Smith, T.M., Harvati, K., Olejniczak, A.J., Reid, D.J., Hublin, J.-J., and Panagopoulou, E. (2009) Dental development and enamel thickness in the Lakonis Neanderthal molar. *American Journal of Physical Anthropology* **138**(1): 112–18.

Smith Woodward, A. (1948) *The Earliest Englishman*. Watts, London.

Snell, O. (1891) Das gehricht des gehirnes und des himmantels der saugetiere in beziehung zu deren geistigen fahig keiten. *Sitzungsber. Ges. Morph. Physiol. (Munchen)* **7**: 90–94.

Sockol, M.D., Raichlen, D.A., and Pontzer, H. (2007) Chimpanzee locomotor energetics and the origin of human bipedalism. *Proceedings of the National Academy of Sciences USA* **104**(30): 12265–9.

C

Soficaru, A., Dobo, A., and Trinkaus, E. (2006) Early modern humans from the Peştera Muierii, Baia de Fier, Romania. *Proceedings of the National Academy of Sciences USA* **103**(46): 17196–202.

Soficaru, A., Petrea, C., Dobos, A., and Trinkaus, E. (2007) The human cranium from the Peştera Cioclovina Uscata *Current Anthropology* **48**(4): 611–19.

Sohn, P.K. (1983) *Turubong Cave No. 9 Excavation Report.* Yonsei University Museum, Seoul.

Sohn, S., and Wolpoff, M.H. (1993) Zuttiyeh face: a view from the East. *American Journal of Physical Anthropology* **91**: 325–47.

Sokal, R.R., and Crovello, T.J. (1970) The biological species concept: a critical evaluation. *American Naturalist* **104**: 127–53.

Solan, M., and Day, M.H. (1992) The Baringo (Kapthurin) ulna. *Journal of Human Evolution* **22**: 307–14.

Solecki, R. (1960) Three adult Neanderthal skeletons from Shanidar Cave, northern Iraq. In: *Annual Report of the Smithsonian Institution*, pp. 603–35.

Solecki, R.S. (1955) Shanidar Cave, a Paleolithic site in northern Iraq. *Annual Report of the Smithsonian Institution* **3**: 389–425.

Solecki, R.S. (1963) Prehistory in Shanidar Valley, Northern Iraq: Fresh insights into Near Eastern prehistory from the Middle Paleolithic to the Proto-Neolithic are obtained. *Science* **139**(3551): 179–93.

Soligo, C., and Andrews, P. (2005) Taphonomic bias, taxonomic bias and historical non-equivalence of faunal structure in early hominin localities. *Journal of Human Evolution* **49**: 206–9.

Sonakia, A. (1984) The skull-cap of early man and associated mammalian fauna from Narmada Valley alluvium, Hoshangabad area, Madhya Pradesh (India). *Geological Survey of India* **113**(6): 159–72.

Sonakia, A. (1985) Early *Homo* from Narmada Valley, India. In: *Ancestors: The Hard Evidence*, Delson, E., ed., pp. 334–8. Alan R. Liss, New York.

Soressi, M., Jones, H., Rink, J., Maureille, B., and Tillier, A.-m. (2007) The Pech-de-l'Azé I Neandertal child: ESR, Uranium Series and AMS ^{14}C dating of MTA type B context. *Journal of Human Evolution* **52**: 455–66.

Soressi, M., Rendu, W., Texier, J.-P., Claud, E., Daulny, L., d'Errico, F., et al. (2008) Pech-de-l'Azé I (Dordogne, France): nouveau regard sur un gisement moustérien de tradition acheuléenne connu depuis le XIXe siècle. In: *Les sociétés Paléolithiques d'un grand Sud-Ouest: nouveaux gisements, nouvelles méthodes, nouveaux résultats*, Jaubert, J., Bordes, J.-G., and Ortega, I., eds, pp. 95–132. Mémoire de la Société Préhistorique Française.

Spence, K.W. (1937) Experimental studies of learning and higher mental processes in infra-human primates. *Psychological Bulletin* **34**(10): 806–50.

Spencer, F. (1990) *Piltdown. A Scientific Forgery*. Oxford University Press, Oxford.

Speth, J.D., and Tchernov, E. (2002) Middle Paleolithic tortoise use at Kebara Cave (Israel). *Journal of Archaeological Science* **29**(5): 471–83.

Spitery, E. (1982a). Le frontal de l'Homme de Tautavel. *1er Congrès International de Paléontologie Humaine, Nice, France*

Spitery, E. (1982b). L'Occipital de l'Homme de Tautavel. Essai de reconstitution. *1er Congrès International de Paléontologie Humaine, Nice, France.*

Sponheimer, M., de Ruiter, D., Lee-Thorp, J., and Spâth, A. (2005) Sr/Ca and early hominin diets revisited: new data from modern and fossil tooth enamel. *Journal of Human Evolution* **48**: 147–56.

Sponheimer, M., Passey, B.H., de Ruiter, D.J., Guatelli-Steinberg, D., Cerling, T.E., and Lee-Thorp, J. (2006) Isotopic evidence for dietary variability in the early hominin *Paranthropus robustus*. *Science* **314**(5801): 980–2.

Sponheimer, M., Grant, C.C., de Ruiter, D.J., Lee-Thorp, J.A., Codron, D.M., and Codron, J. (2007) Diets of impala from Kruger National Park: evidence from stable carbon isotopes. *Koedoe* **46**(1): 101–6.

Spoor, C.F., Wood, B.A., and Zonneveld, F.W. (1994) Implications of early hominid labyrinthine morphology for evolution of human bipedal locomotion. *Nature* **369**: 645–8.

Spoor, F. (1993) *The Comparative Morphology and Phylogeny of the Human Bony Labyrinth*. PhD thesis, Utrecht University.

Spoor, F., and Zonneveld, F. (1998) Comparative review of the human bony labyrinth. *Yearbook of Physical Anthropology* **41**: 211–51.

Spoor, F., Wood, B.A., and Zonneveld, F. (1996) Evidence for a link between human semicircular canal size and bipedal behaviour. *Journal of Human Evolution* **30**: 183–7.

Spoor, F., Hublin, J.-J., Braun, M., and Zonneveld, F. (2003) The bony labyrinth of Neanderthals. *Journal of Human Evolution* **44**: 141–65.

Spoor, F., Leakey, M.G., Gathogo, P.N., Brown, F.H., Antón, S., McDougall, I., et al. (2007) Implications of new early *Homo* fossils from Ileret, east of Lake Turkana, Kenya. *Nature* **448**: 688–91.

Spoor, F., Leakey, M., and Leakey, L. (2010) Hominin diversity in the Middle Pliocene of eastern Africa: the maxilla of KNM-WT 40000. *Philosophical Transactions of the Royal Society B, Biological Sciences* **365**(1556): 3377–88.

Spoor, F.S., O'Higgins, P., Dean, C., and Lieberman, D.E. (1999) Anterior sphenoid in modern humans. *Nature* **397**: 572.

Spuhler, J.N. (1972) Behavior and human mating patterns in human populations. In: *The Structure of Human Populations*, Harrison, G.A., and Boyce, A.J., eds, pp. 165–81. Oxford University Press, Oxford.

Staff, R.A., Bronk Ramsey, C., and Nakagawa, T. (2010) A re-analysis of the Lake Suigetsu terrestrial radiocarbon calibration dataset. *Nuclear Instruments and Methods in Physics Research Section B: Beam Interactions with Materials and Atoms* **268**(7–8): 960–5.

Stearns, C.E., and Thurber, D.L. (1965) Th^{230}/U^{234} dates of late Pleistocene marine fossils from the Mediterranean and Moroccan littorals. *Progress in Oceanography* **4**: 293–305.

Stefan, V.H., and Trinkaus, E. (1998) Discrete trait and dental morphometric affinities of the Tabun 2 mandible. *Journal of Human Evolution* **34**(5): 443–68.

Steininger, C.M., and Berger, L.R. (2000) Taxonomic affinity of a new specimen from Cooper's, South Africa. *American Journal of Physical Anthropology* Suppl. **30**: 291.

Steininger, C.M., Berger, L.R., and Kuhn, B.F. (2008) A partial skull of *Paranthropus robustus* from Cooper's Cave, South Africa. *South African Journal of Science* **104**(3–4): 143–6.

Steininger, F.F., Berggren, W.A., Kent, D., Bernor, R.L., Sen, Ş., and Agusti, J. (1996) Circum-Mediterranean Neogene (Miocene and Pliocene) marine-continental chronologic correlations of European mammal units. In: *The Evolution of Western Eurasian Neogene Mammal Faunas*, Bernor, R.L., Fahlbusch, V., and Mittman, H.-W., eds, pp. 7–46. Columbia University Press, New York.

Steininger, F.F., Bernor, R.L., and Fahlbusch, V. (1989) European Neogene marine/continental chronologic correlations. In: *European Neogene Mammal Chronology*, Lindsay, E.H., Fahlsbusch, V., and Mein, P., eds, pp. 15–46. Plenum Press, New York.

Stephens, M., Rose, J., Gilbertson, D., and Canti, M. (2005) Micromorphology of cave sediments in the humid tropics: Niah Cave, Sarawak. *Asian Perspectives* **44**: 42–55.

Sterling, E.J., and Povinelli, D.J. (1999) Tool use, aye-ayes, and sensorimotor intelligence. *Folia Primatologica* **70**: 8–16.

Stern, J. (1976) Before bipedality. *Yearbook of Physical Anthropology* **52**: 59–68.

Stern, J.T., and Susman, R.L. (1981) Electromyography of the gluteal muscles in *Hylobates*, *Pongo* and *Pan*: implications for the evolution of hominid bipedality. *American Journal of Physical Anthropology* **55**: 153–66.

Stern, J.T., and Susman, R.L. (1983) Locomotor anatomy of *Australopithecus afarensis*. *American Journal of Physical Anthropology* **60**: 279–317.

Stewart, K.M. (1989) *Fishing Sites of North and East Africa in the Late Pleistocene and Holocene.* International Series 521. Cambridge Monographs in African Archaeology 34. B.A.R., Oxford.

Stewart, K.M. (1994) Early hominid utilisation of fish resources and implications for seasonality and behaviour. *Journal of Human Evolution* **27**(1–3): 229–45.

Stewart, K.M. (2003) Fossil fish remains from Mio-Pliocene deposits at Lothagam, Kenya. In: *Lothagam: The Dawn of Humanity in Eastern Africa*, Leakey, M.G.,

and Harris, J.M., eds, pp. 75–111. Columbia University Press, New York.

Stewart, K.M. (2010) The case for exploitation of wetlands environments and foods by pre-*sapiens* hominins. In: *Human Brain Evolution: the Influence of Freshwater and Marine Food Resources*, Cunnane, S., and Stewart, K., eds, pp. 137–73. Wiley, New York.

Stewart, T.D. (1958) First views of the restored Shanidar I skull. *Sumer* **14**: 90–96.

Stewart, T.D. (1959) The restored Shanidar I skull. In: *Smithsonian Report for 1958*, pp. 473–80. Smithsonian Institution, Washington DC.

Stewart, T.D. (1960) Form of the pubic bone in Neanderthal man. *Science* **131**(3411): 1437–38.

Stewart, T.D. (1963) Shanidar skeletons IV and VI. *Sumer* **19**: 8–25.

Stewart, T.D., and Tiffany, M. (1986) Description of the human skeleton. Cleaning and casting of the skeleton. In: *The Prehistory of Wadi Kubbaniya*, Wendorf, F., and Schild, R., eds, pp. 49–52. Southern Methodist University Press, Dallas, TX.

Stiles, D.N., Hay, R.L., and O'Neil, J.R. (1974) The MNK chert factory site, Olduvai Gorge, Tanzania *World Archaeology* **5**(3): 285–308.

Stiner, M., Gopher, A., and Barkai, R. (2011) Hearthside socioeconomics, hunting, and paleoecology during the Lower Paleolithic and Qesem Cave, Israel. *Journal of Human Evolution* **60**(2): 213–33.

Stokes, S., Maxwell, T.A., Haynes, C.V., and Horrocks, J. (1998) Latest Pleistocene and Holocene sand sheet construction in the Selima Sand Sheet, Eastern Sahara. In: *Quaternary Deserts and Climatic Change*, Alsharan, A.S., Glennie, K.W., Whittle, G.L., and Kendall, C.G.S.C., eds, pp. 175–84. Rotterdam/Brookfield, Balkema.

Stone, A., and Stoneking, M. (1998) mtDNA analysis of a prehistoric Oneota population: implications for the peopling of the New World. *American Journal of Human Genetics* **62**(5): 1153–70.

Stone, A.C., Griffiths, R.C., Zegura, S.L., and Hammer, M.F. (2002) High levels of Y-chromosome nucleotide diversity in the genus *Pan*. *Proceedings of the National Academy of Sciences USA* **99**(1): 43–8.

Stone, T., and Cupper, M.L. (2003) Last Glacial Maximum ages for robust humans at Kow Swamp, southern Australia. *Journal of Human Evolution* **45**(2): 99–111.

Stoneking, M. (1995) Ancient DNA: how do you know when you have it and what can you do with it? *American Journal of Human Genetics* **57**(6): 1259.

Stoneking, M., Fontius, J.J., Clifford, S.L., Soodyall, H., Arcot, S.S., Saha, N., et al. (1997) Alu insertion polymorphisms and human evolution: evidence for a larger population size in Africa. *Genome Research* **7**: 1061–71.

Stout, D. (2002) Skill and cognition in stone tool production: an ethnographic case study from Irian Jaya. *Current Anthropology* **43**(5): 693–722.

Stout, D., Quade, J., Semaw, S., Rogers, M.J., and Levin, N.E. (2005) Raw material selectivity of the earliest stone toolmakers at Gona, Afar, Ethiopia. *Journal of Human Evolution* 48(4): 365–80.

Stout, D., Toth, N., Schick, K., and Chaminade, T. (2008) Neural correlates of Early Stone Age toolmaking: technology, language and cognition in human evolution. *Philosophical Transactions of the Royal Society of London. Series B, Biological Sciences* 363(1499): 1939–49.

Stout, D., Semaw, S., Rogers, M.J., and Cauche, D. (2010) Technological variation in the earliest Oldowan from Gona, Afar, Ethiopia. *Journal of Human Evolution* 58(6): 474–91.

Strait, D.S. (2001) Integration, phylogeny, and the hominid cranial base. *American Journal of Physical Anthropology* **114**: 273–97.

Strait, D.S., Grine, F.E., and Moniz, M.A. (1997) A reappraisal of early hominid phylogeny. *Journal of Human Evolution* 32: 17–82.

Strait, D.S., Weber, G.W., Neubauer, S., Chalk, J., Richmond, B.R., Lucas, P.W., et al. (2009) The feeding biomechanics and dietary ecology of *Australopithecus africanus*. *Proceedings of the National Academy of Sciences USA* **106**(7): 2124–9.

Straus, L.G., Meltzer, D.J., and Goebel, T. (2005) Ice Age Atlantis? Exploring the Solutrean-Clovis 'Connection'. *World Archaeology* 37(4): 507–32.

Straus, W.L., and Cave, A.J.E. (1957) Pathology and posture of Neanderthal man. *Quarterly Review of Biology* 32: 348–63.

Street, M., and Terberger, T. (2002) Hiatus or continuity? New results for the question of pleniglacial settlement in Central Europe. *Antiquity* 76: 691–8.

Street, M., Terberger, T., and Orschiedt, J. (2006) A critical review of the German Paleolithic hominin record. *Journal of Human Evolution* 51(6): 551–79.

Streeter, M., Stout, S., Trinkaus, E., Stringer, C., Roberts, M., and Parfitt, S. (2001) Histomorphometric age assessment of the Boxgrove 1 tibial diaphysis. *Journal of Human Evolution* 40(4): 331–8.

Strier, K.B. (2000) *Primate Behavioral Ecology*. Allyn and Bacon, Boston, MA.

Stringer, C. (1998) A metrical study of the WLH-50 calvaria. *Journal of Human Evolution* 34(3): 327–32.

Stringer, C. (2002) Modern human origins: progress and prospects. *Philosophical Transactions of the Royal Society of London. Series B, Biological Sciences* 357(1420): 563–79.

Stringer, C., and Andrews, P. (1988) Genetic and fossil evidence for the origin of modern humans. *Science* 239: 1263–8.

Stringer, C., and Gamble, C. (1993) *In Search of the Neanderthals: Solving the Puzzle of Human Origins*. Thames and Hudson, London.

Stringer, C., and McKie, R. (1996) *African Exodus: the Origins of Modern Humanity*. Pimlico, London.

Stringer, C., Hublin, J., and Vandermeersch, B. (1984) The origin of anatomically modern humans in Western Europe. In: *The Origins of Modern Humans: a World Survey of the Fossil Evidence*, Smith, F., and Spencer, F., eds, pp. 51–135. Alan R. Liss, New York.

Stringer, C., Trinkaus, E., Roberts, M.B., Parfitt, S.A., and MacPhail, R.I. (1998) The Middle Pleistocene human tibia from Boxgrove. *Journal of Human Evolution* 34(5): 509–47.

Stringer, C.B. (1974) Population relationships of later Pleistocene hominids: a multivariate study of available crania. *Journal of Archaeological Science* 1: 317–42.

Stringer, C.B. (1979) A re-evaluation of the fossil human calvaria from Singa, Sudan. *Bulletin of the British Museum Natural History (Geology)* 32(1): 77–83.

Stringer, C.B. (1981) The dating of European Middle Pleistocene hominids and the existence of *Homo erectus* in Europe. *Anthropologie* 19: 3–14.

Stringer, C.B. (1982) Towards a solution to the Neanderthal problem. *Journal of Human Evolution* 11: 431–8.

Stringer, C.B. (1986) The credibility of *Homo habilis*. In: *Major Topics in Primate and Human Evolution*, Wood, B.A., Martin, L., and Andrews, P., eds, pp. 266–94. Cambridge University Press, Cambridge.

Stringer, C.B. (1996) Current issues in modern human origins. In: *Contemporary Issues in Human Evolution*, Meikle, W.E., Howell, F.C., and Jablonski, N.G., eds, pp. 115–34. California Academy of Science, San Francisco, CA.

Stringer, C.B., and Andrews, P. (1988) Genetic and fossil evidence for the origin of modern humans. *Science* 239: 1263–8.

Stringer, C.B., Kruszynski, R.G., and Jacobi, R.M. (1981) Allez Neanderthal. *Nature* **289**: 823–4.

Stringer, C.B., Cornish, L., and Stuart-Macadam, P. (1985) Preparation and further study of the Singa skull from Sudan. *Bulletin of the British Museum Natural History (Geology)* 38(5): 347–58.

Stuart-Macadam, P. (1992) Cranial thickening and anemia: Reply to Dr. Webb. *American Journal of Physical Anthropology* 88(1): 109–10.

Sturm, R., Teasdale, R., and Box, N. (2001) Human pigmentation genes: identification, structure and consequences of polymorphic variation. *Gene* 277(1–2): 49–62.

Stuvier, M., and Polach, H. (1977) Discussion: Reporting of ^{14}C Data. *Radiocarbon* 19: 355–63.

Stynder, D., Brock, F., Sealy, J., Wurz, S., Morris, A., and Volman, T. (2009) A mid-Holocene AMS 14C date for the presumed upper Pleistocene human skeleton from Peers Cave, South Africa. *Journal of Human Evolution* 56: 431–4.

Stynder, D.D., Moggi-Cecchi, J., Berger, L.R., and Parkington, J.E. (2001) Human mandibular incisors from the late Middle Pleistocene locality of Hoedjiespunt 1, South Africa. *Journal of Human Evolution* 41(5): 369–83.

Subiaul, F. (2007) The imitation faculty in monkeys: evaluating its features, distribution and evolution. *Journal of Anthropological Sciences* **85**: 35–62.

Subiaul, F. (2010) Dissecting the imitation faculty: the multiple imitation mechanisms (MIM) hypothesis. *Behavioural Processes* **83**(2): 222–34.

Subiaul, F., Cantlon, J.F., Holloway, R.L., and Terrace, H.S. (2004) Cognitive imitation in rhesus macaques. *Science* **305**(5682): 407–10.

Sulem, P., Gudbjartsson, D., Stacey, S., Helgason, A., Rafnar, T., Magnusson, K., et al. (2007) Genetic determinants of hair, eye and skin pigmentation in Europeans. *Nature genetics* **39**(12): 1443–52.

Summerhayes, G.R., Leavesley, M., Fairbairn, A., Mandui, H., Field, J., Ford, A., et al. (2010) Human adaptation and plant use in highland New Guinea 49,000 to 44,000 years ago. *Science* **330**(6000): 78–81.

Summers, A. (2000) Stiffening the stingray skeleton-an investigation of durophagy in myliobatid stingrays (Chondrichthyes, Batoidea, Myliobatidae). *Journal of Morphology* **243**(2): 113–26.

Susman, R.L. (1979) Comparative and functional morphology of hominoid fingers. *American Journal of Physical Anthropology* **50**: 215–36.

Susman, R.L. (1984) *The Pygmy Chimpanzee: Evolutionary Biology and Behavior*. Plenum Press, New York.

Susman, R.L. (1988a) New postcranial remains from Swartkrans and their bearing on the functional morphology and behavior of *Paranthropus robustus*. In: *Evolutionary History of the "Robust" Australopithecines*, Grine, F.E., ed., pp. 149–72. Aldine de Gruyter, New York.

Susman, R.L. (1988b) Hand of *Paranthropus robustus* from Member 1, Swartkrans: Fossil evidence for tool behaviour. *Science* **240**(4853): 781–4.

Susman, R.L. (1993) Hominid postcranial remains from Swartkrans. In: *Swartkrans: A Cave's Chronicle of Early Man*, Brain, C.K., ed., pp. 117–36. Transvaal Museum, Pretoria.

Susman, R.L. (2004) Hominid postcranial remains from Swartkrans. In: *Swartkrans: A Cave's Chronicle of Early Man*, 2nd edn, Brain, C.K., ed., pp. 117–36. Transvaal Museum, Pretoria.

Susman, R.L. (2008) Brief communication: Evidence bearing on the status of *Homo habilis* at Olduvai Gorge. *American Journal of Physical Anthropology* **137**(3): 356–61.

Susman, R.L., and Creel, N. (1979) Functional and morphological affinities of the subadult hand (O.H. 7) from Olduvai Gorge. *American Journal of Physical Anthropology* **51**(3): 311–31.

Susman, R.L., and Stern, J.T. (1982) Functional morphology of *Homo habilis*. *Science* **217**: 931–4.

Susman, R.L., Patel, B.A., Francis, M.J., and Cardoso, H.F.V. (2011) Metatarsal fusion pattern and developmental morphology of the Olduvai Hominid 8 foot: Evidence of adolescence. *Journal of Human Evolution* **60**: 58–69.

Sutton, J.B. (1883) On some points in the anatomy of the chimpanzee. *Journal of Anatomy and Physiology* (1): 66–85.

Sutton, M., Pickering, T., Pickering, R., Brain, C.K., Clarke, R., Heaton, J., et al. (2009) Newly discovered fossil- and artifact-bearing deposits, uranium-series ages, and Plio-Pleistocene hominids at Swartkrans Cave, South Africa. *Journal of Human Evolution* **57**: 688–96.

Suwa, G. (1988) Evolution of the "robust" australopithecines in the Omo succession: evidence from mandibular premolar morphology. In: *Evolutionary History of the "Robust" Australopithecines*, Grine, F.E., ed., pp. 199–222. Aldine de Gruyter, New York.

Suwa, G. (1990) *A Comparative Analysis of Hominid Dental Remains from the Shungura and Usno Formations, Omo Valley, Ethiopia*. PhD thesis, University of California at Berkeley.

Suwa, G., Wood, B.A., and White, T.D. (1994) Further analysis of mandibular molar crown and cusp areas in Pliocene and Early Pleistocene hominids. *American Journal of Physical Anthropology* **93**: 407–26.

Suwa, G., White, T.D., and Howell, F.C. (1996) Mandibular postcanine dentition from the Shungura Formation, Ethiopia: crown morphology, taxonomic allocations and Plio-Pleistocene hominid evolution. *American Journal of Physical Anthropology* **101**: 247–82.

Suwa, G., Asfaw, B., Beyene, Y., White, T., Katoh, S., Nagaoka, S., et al. (1997) The first skull of *Australopithecus boisei*. *Nature* **389**: 489–92.

Suwa, G., Nakaya, H., Asfaw, B., Saegusa, H., Amzaye, A., Kono, R.T., et al. (2003) Plio-Pleistocene terrestrial mammals assemblage from Konso, southern Ethiopia. *Journal of Vertebrate Paleontology* **23**(4): 901–16.

Suwa, G., Asfaw, B., Haile-Selassie, Y., White, T., Katoh, S., WoldeGabriel, G., et al. (2007) Early Pleistocene *Homo erectus* fossils from Konso, southern Ethiopia. *Anthropological Science* **115**(2): 133–51.

Suwa, G., Kono, R.T., Simpson, S.W., Asfaw, B., Lovejoy, C.O., and White, T.D. (2009a) Paleobiological implications of the *Ardipithecus ramidus* dentition. *Science* **326**(5949): 69, 94–9.

Suwa, G., Asfaw, B., Kono, R.T., Kubo, D., Lovejoy, C.O., and White, T.D. (2009b) The *Ardipithecus ramidus* skull and its implications for hominid origins. *Science* **326**(5949): 68e1–7.

Suzuki, H. (1982) Skulls of the Minatogawa man. In: *The Minatogawa Man*, Suzuki, H., and Hanihara, K., eds, pp. 7–49. University of Tokyo, Tokyo.

Suzuki, H., and Takai, F., eds (1970) *The Amud Man and his Cave Site*. Academic Press of Japan, Tokyo.

Suzuki, H., and Hanihara, K., eds (1982) *The Minatogawa Man*. University of Tokyo, Tokyo.

Svoboda, J. (1987) Lithic industries of the Arago, Vértesszöllös, and Bilzingsleben hominids: comparison and evolutionary interpretation. *Current Anthropology* **28**(2): 219–27.

Svoboda, J. (2000) The depositional context of the Early Upper Paleolithic human fossils from the Koněprusy (Zlatý kůň) and Mladeč Caves, Czech Republic. *Journal of Human Evolution* **38**: 523–36.

Svoboda, J., and Bar-Yosef, O. (2003) *Stranska skala: Origins of the Upper Paleolithic in the Brno Basin.* Peabody Museum of Archaeology and Ethnology, Harvard University, Cambridge, MA.

Svoboda, J., van der Plicht, J., Vlček, E., and Kuzelka, V. (2004) New radiocarbon datings of human fossils from caves and rockshelters in Bohemia (Czech Republic). *Anthropol. Brno* **42**: 161–6.

Svoboda, J.A. (2006) The structure of the cave, stratigraphy, and depositional context. In: *Early Modern Humans at Moravian Gate*, Teschler-Nicola, M., ed., pp. 27–40. Springer Vienna, Vienna.

Svoboda, J.A. (2008) The Upper Paleolithic burial area at Předmostí: ritual and taphonomy. *Journal of Human Evolution* **54**(1): 15–33.

Svoboda, J.A., van der Plicht, J., and Kuzelka, V. (2002) Upper Palaeolithic and Mesolithic human fossils from Moravia and Bohemia (Czech Republic): some new ^{14}C dates. *Antiquity* **76**: 957–62.

Swartz, B.K. (1980) Continental line-making: a re-examination of basic Palaeolithic classification. In: *Proceedings of the 8th Congress of Prehistory and Quaternary Studies*, Leakey, R.E.F., and Ogot, B.A., eds, pp. 33–5. International Louis Leakey Memorial Institute for African Prehistory, Nairobi.

Swindler, D. (2002) *Primate Dentition: an Introduction to the Teeth of Non-human Primates.* Cambridge University Press, Cambridge.

Swisher, C.C., Curtis, G.H., Jacob, T., Getty, A.G., Suprijo, A., and Widasmoro (1994) Age of the earliest known hominids in Java, Indonesia. *Science* **263**: 1118–21.

Swisher, C.C., Rink, W.J., Antón, S.C., Schwarcz, H.P., Curtis, G.H., Suprijo, A., et al. (1996) Later *Homo erectus* of Java: potential contemporaneity with *Homo sapiens* in Southeast Asia. *Science* **274**(294): 1870–74.

Swofford, D.L. (2003) *PAUP*. Phylogenetic Analysis Using Parsimony (*and Other Methods).* Sinauer Associates, Sunderland, MA.

Sylvester, A., Garofalo, E., and Ruff, C. (2010) Technical note: An R program for automating bone cross section reconstruction. *American Journal of Physical Anthropology* **142**(4): 665–9.

Symonds, M.R.E. (2002) The effects of topological inaccuracy in evolutionary trees on the phyogenetic comparative method of independent contrasts. *Systematic Biology* **51**: 542–53.

Szalay, F., and Delson, E. (1979) *Evolutionary History of the Primates.* Academic Press, New York.

Szalay, F.S. (1969) Mixodectidae, Microsyopidae, and the insectivore-primate transition. *Bulletin of the American Museum of Natural History* **140**: 193–330.

Szalay, F.S. (1971) Biological level of organization of the Chesowanja robust australopithecine. *Nature* **234**: 229–30.

Szmidt, C.C., Normand, C., Burr, G., Hodgins, G., and LaMotta, S. (2010) AMS ^{14}C dating the Protoaurigna-cian/Early Aurignacian of Isturitz, France. Implications for Neanderthal-modern human interaction and the timing of technical and cultural innovations in Europe. *Journal of Archaeological Science* **37**(4): 758–68.

Szombathy, J. (1925) Die diluvialen Menschenreste aus der Fuerst-Johanns-Hoehle bei Lautschin Maehren. *Die Eiszeit* **2**(1–34): 77–95.

Tafforeau, P., and Smith, T. (2008) Nondestructive imaging of hominoid dental microstructure using phase contrast X-ray synchrotron microtomography. *Journal of Human Evolution* **54**: 272–8.

Tague, R.G., and Lovejoy, C.O. (1986) The obstetric pelvis of AL 288-1 (Lucy). *Journal of Human Evolution* **15**: 237–55.

Taieb, M., Coppens, Y., Johanson, D.C., and Kalb, J. (1972) Dépôts sédimentaires et faunes du Plio-Pléistocène de la basse vallée de l'Awash (Afar central, Ethiopie). *Comptes Rendus de l'Académie des Sciences* **275**: 819–22.

Taieb, M., Johanson, D.C., Coppens, Y., and Aronson, J.L. (1976) Geological and paleontological background of Hadar hominid site, Afar, Ethiopia. *Nature* **260**: 289–93.

Takamiya, H., Tamaki, M., and Kin, M. (1975) Artifacts of the Yamashita-cho Cave site. *Journal of the Anthropological Society of Nippon* **83**: 137–50.

Tallon, P.W.J. (1978) Geological setting of the hominid fossils and Acheulian artifacts from the Kapthurin Formation, Baringo District, Kenya. In: *Geological Background to Fossil Man: Recent research in the Gregory Rift Valley, East Africa*, Bishop, W.W., ed., pp. 361–73. Scottish Academic Press, Edinburgh.

Tamrat, E., Thouveny, N., Taieb, M., and Opdyke, N.D. (1995) Revised magnetostratigraphy of the Plio-Pleistocene sedimentary sequence of the Olduvai Formation (Tanzania). *Paleogeography, Paleoclimatology and Paleoecology* **114**: 273–83.

Tanaka, J. (1976) Subsistence ecology of the Central Kalahari San. In: *Kalahari Hunter-Gatherers: Studies on the !Kung San and their Neighbors*, Lee, R.B., and Devore, I., eds, pp. 98–109. Harvard University Press, Cambridge, MA.

Tankard, A.J., and Schweitzer, F.R. (1976) Textural analysis of cave sediments: Die Kelders, Cape Province, South Africa. In: *Geoarchaeology*, Davidson, D.A., and Shackley, M.L., eds, pp. 289–316. Duckworth, London.

Tanner, J.M. (1955) *Growth at Adolescence.* Blackwell Scientific, Oxford.

Tappen, N.C. (1985) The dentition of the "old man" of La Chapelle-aux-Saints and inferences concerning Neandertal behavior. *American Journal of Physical Anthropology* **67**(1): 43–50.

Tappen, N.G. (1987) Circum-mortem damage to some ancient African hominid crania: a taphonomic and evolutionary essay. *African Archaeological Review* **5**(1): 39–47.

Tardieu, C., and Trinkaus, E. (1994) Early ontogeny of the human femoral bicondylar angle. *American Journal of Physical Anthropology* **95**(2): 183–95.

Taschini, M. (1979) L'industrie lithique de Grotta Guattari au Mont Circé (Latium): définition culturelle, typologique et chronologique du Pontinien. *Quaternaria* **21**: 179–247.

Tattersall, I. (1986) Species recognition in human paleontology. *Journal of Human Evolution* **15**: 165–75.

Tattersall, I., and Sawyer, G.J. (1996) The skull of *Sinanthropus* from Zhoukoudian, China: a new reconstruction. *Journal of Human Evolution* **31**(4): 311–14.

Tattersall, I., and Schwartz, J.H. (1999) Hominids and hybrids: the place of Neanderthals in human evolution. *Proceedings of the National Academy of Sciences USA* **96**: 7117–19.

Taylor, C.R., and Rowntree, V.J. (1973) Running on two or on four legs: which consumes more energy? *Science* **179**(4069): 186–7.

Taylor, H.S. (2002) Transcription regulation of implantation by HOX genes. *Reviews in Endocrine and Metabolic Disorders* **3**: 127–32.

Tchernov, E. (1986) The Lower Pleistocene Mammals of 'Ubeidiya (Jordan Valley). In: *Mémoires et Travaux du Centre de Recherche Français de Jérusalem 5*. Association Paléorient, Paris.

Tchernov, E. (1988) The age of 'Ubeidiya Formation (Jordan Valley, Israel) and the earliest hominids in the Levant. *Paléorient* **14**(2): 63–5.

Teaford, M., and Ungar, P. (2000) Diet and the evolution of the earliest human ancestors. *Proceedings of the National Academy of Sciences USA* **97**(25): 13506–11.

Teaford, M.F. (1993) Dental microwear and diet in extant and extinct *Theropithecus*: preliminary analyses. In: *Theropithecus: Rise and Fall of a Primate Genus*, Jablonski, N.G., ed., pp. 331–49. Cambridge University Press, Cambridge.

Tehrani, J.J., and Collard, M. (2009) An integrated analysis of inter-individual and inter-group cultural transmission in Iranian tribal populations. *Evolution and Human Behavior* **30**: 286–300.

Teilhard de Chardin, P., and Piveteau, J. (1930) Les mammiferes fossils de Nihewan (Chine). *Annales de Paleontologie* **19**: 1–154.

Teilhard de Chardin, P., and Pei, W.G. (1932) The lithic industry of the *Sinanthropus* deposits in Choukoutien. *Bulletin of the Geological Society of China* **11**(4): 315–64.

Terberger, K. (1993) *Das Lahntal-Paläolithikum. Materialien zur Vor- und Frühgeschichte von Hessen 11*. Landesamt für Denkmalpflege, Wiesbaden.

Terberger, T., and Street, M. (2001) Neue Forschungen zum "jungpaläolithischen" Menschenschädel von Binshof bei Speyer, Rheinland-Pfalz. *Archäologisches Korrespondenzblatt* **31**: 33–7.

Terborgh, J. (1986) Keystone plant resources in the tropical forest. In: *Conservation biology: the science of scarcity and diversity*, Soulé, M., ed., pp. 330–44. Sinauer, Sunderland, MA.

Terrace, H. (1979) Is problem-solving language? A review of Premack's Intelligence in Apes and Man. *Journal of the Experimental Analysis of Behavior* **31**(1): 161.

Terrace, H.S. (1987) Chunking by a pigeon in a serial learning task. *Nature* **325**: 149–51.

Terrace, H.S. (2005) The simultaneous chain: a new approach to serial learning. *Trends in Cognitive Science* **9**(4): 202–10.

Terrace, H.S., Petitto, L.A., Sanders, R.J., and Bever, T.G. (1979) Can an ape create a sentence? *Science* **206**(4412): 891–902.

Teschler-Nicola, M. (2006) Taphonomic aspects of the human remains from the Mladec Cave. In: *Early Modern Humans and the Moravian Gate*, Teschler-Nicola, M., ed., pp. 75–98. Springer Vienna, Vienna.

Texier, J.-P. (2001) Sédimentogénèse des sites préhistoriques et représentativité des datations numériques. In: *Datation. XXIe Rencontres internationales d'Archéologie et d'histoire d'Antibes*, Barrandon, J.-N., Guibert, P., and Michel, V., eds, pp. 159–75. Colloque "Datation" éd. APDCA, Antibes.

Thackeray, A.I. (1992) The Middle Stone Age south of the Limpopo River. *Journal of World Prehistory* **6**(4): 385–440.

Thackeray, A.I. (2000) Middle Stone Age artefacts from the 1993 and 1995 excavations of Die Kelders Cave 1, South Africa. *Journal of Human Evolution* **38**: 147–68.

Thackeray, A.I., Thackeray, J.F., Beaumont, P.B., and Vogel, J.C. (1981) Dated rock engravings from Wonderwerk Cave, South Africa. *Science* **214**(4516): 64–7.

Thackeray, A.J. (1992) The Middle Stone Age south of the Limpopo river. *Journal of World Prehistory* **6**(4): 385–440.

Thackeray, F., and Braga, J. (2005) Early *Homo*, 'robust' australopithecines and stone tools at Kromdraai, South Africa. In: *From Tools to Symbols: from Early Hominids to Modern Humans*, d'Errico, F., and Backwell, L., eds, pp. 229–35. Witwatersrand University Press, Johannesburg.

Thackeray, F., Braga, J., Treil, J., Niksch, N., and Labuschagne, J.H. (2002a) 'Mrs Ples' (Sts 5) from Sterkfontein: an adolescent male? *South African Journal of Science* **98**: 21–2.

Thackeray, F., Kirschvink, J.L., and Raub, T.D. (2002b) Paleomagnetic analyses of calcified deposits from the Plio-Pleistocene hominid site of Kromdraai, South Africa. *South African Journal of Science* **98**: 537–40.

Thackeray, J.F. (1979) An analysis of faunal remains from archaeological sites in South West Africa (Namibia). *The South African Archaeological Bulletin* **34**(129): 18–33.

Thackeray, J.F., and Klopper, F. (2004) Relationship between orbit and femur dimensions in extant primates: application to Sts 5 and Sts 14 (*Australopithecus africanus*) from Sterkfontein, South Africa. *Annals of the Transvaal Museum* **41**: 83–4.

Thackeray, J.F., de Ruiter, D.J., Berger, L.R., and van der Merwe, N.J. (2001) Hominid fossils from Kromdraai: a revised list of specimens discovered since 1938. *Annals of the Transvaal Museum* **38**: 43–56.

Thesiger, W. (1996) *The Danakil Diary: Journeys through Abyssinia, 1930–4.* Harper Collins, London.

Thieme, H. (1997) Lower Palaeolithic hunting spears from Germany. *Nature* **385**: 807–10.

Thoma, A. (1963) The dentition of the Subalyuk Neanderthal child. *Zeitschrift für Morphologie und Anthropologie* **54**: 127–50.

Thoma, A. (1966) L'occipital de l'homme mindelien de Vertesszollos. *L'Anthropologie* **70**: 495–534.

Thoma, A. (1967) Human teeth from the Lower Palaeolithic of Hungary. *Zeitschrift für Morphologie und Anthropologie* **58**: 152–80.

Thoma, A. (1969) Biometrische studien uber das occipitale von Vertesszollos. *Zeitschrift für Morphologie und Anthropologie* **60**: 229–41.

Thoma, A. (1984) Morphology and affinities of the Nazlet Khater man. *Journal of Human Evolution* **13**: 287–96.

Thompson, E., Williams, H., and Minichillo, T. (2010) Middle and late Pleistocene Middle Stone Age lithic technology from Pinnacle Point 13B (Mossel Bay, Western Cape Province, South Africa). *Journal of Human Evolution* **59**: 358–77.

Thompson, J., and Bilsborough, A. (1997) The current state of the Le Moustier 1 skull. *Acta Praehistorica et Archaeologica* **29**: 17–38.

Thompson, J., and Nelson, A. (2005) The postcranial skeleton of Le Moustier 1. In: *The Neanderthal Adolescent Le Moustier 1: New Aspects, New Results*, Ulrich, H., ed., pp. 265–81. Staatliche Museen Zu Berlin-Preussicher Kulturbesitz, Berlin.

Thompson, J.L. (1993) The unusual cranial attributes of KNM-ER 1805 and their implication for studies of sexual dimorphism in *Homo habilis*. *Human Evolution* **8**(4): 255–63.

Thomson, R.B. (1917) Note upon the fragments of the limb-bones of the Boskop remains, appendix to S.H. Haughton 1917. *Transactions of the Royal Society of South Africa* **6**.

Thorne, A., and Curnoe, D. (2000) Sex and significance of Lake Mungo 3: reply to Brown "Australian pleistocene variation and the sex of Lake Mungo 3". *Journal of Human Evolution* **39**(6): 587–600.

Thorne, A., and Wolpoff, M. (1981) Regional continuity in Australasian Pleistocene hominid evolution. *American Journal of Physical Anthropology* **55**(3): 337–49.

Thorne, A., and Wolpoff, M. (1992) The multiregion evolution of humans. *Scientific American* **266**: 28–33.

Thorne, A., Grün, R., Mortimer, G., Spooner, N.A., Simpson, J.J., McCulloch, M., et al. (1999) Australia's oldest human remains: age of the Lake Mungo 3 skeleton. *Journal of Human Evolution* **36**(6). 591–612.

Thorne, A.G. (1976) Morphological contrasts in Pleistocene Australians. In: *The Origin of the Australians*, Kirk, R.L., ed., pp. 95–112. Australian Institute of Aboriginal Studies, Canberra.

Thorne, A.G., and Macumber, P.E. (1972) Discoveries of Late Pleistocene Man at Kow Swamp, Australia. *Nature* **238**: 316.

Thorpe, S., Holder, R., and Crompton, R. (2007) Origin of human bipedalism as an adaptation for locomotion on flexible branches. *Science* **316**(5829): 1328–31.

Thorpe, W.H. (1963) *Learning and Instinct in Animals.* Menthuen, London.

Thys van den Audenaerde, D.F.E. (1984) The Tervuren museum and the pygmy chimpanzee. In: *The Pygmy Chimpanzee: Evolutionary Biology and Behavior*, Susman, R.L., ed., pp. 3–11. Plenum Press, New York.

Tiercelin, J. (1986) The Pliocene Hadar Formation, Afar depression of Ethiopia. *Geological Society London Special Publications* **25**(1): 221–40.

Tierney, J.E., Russell, J.M., Huang, Y., Damsté, J.S.S., Hopmans, E.C., and Cohen, A.S. (2008) Northern Hemisphere controls on tropical southeast African climate during the past 60,000 years. *Science* **322**(5899): 252–5.

Tillier, A.-M., Arensburg, B., Vandermeersch, B., and Rak, Y. (1991) L'apport de Kebara a la palethnologie funeraire des Neandertaliens du Proche-Orient. In: *Le squelette mousterien de Kebara 2*, Bar-Yosef, O., and Vandermeersch, B., eds, pp. 89–95. CNRS, Paris.

Ting, N., and Ward, C. (2001) Functional analysis of the hip and thigh of *Paracolobus* and other large-bodied fossil cercopithecids. *American Journal of Physical Anthropology Supplement* **32**.

Tishkoff, S.A., Diezsch, E., Speed, W., Pakstis, A.J., Kidd, J.R., Cheung, K., et al. (1996) Global patterns of linkage disequilibrium at the CD4 locus and modern human origins. *Science* **271**(5254): 1380–7.

Tishkoff, S.A., Reed, F.A., Ranciaro, A., Voight, B.F., Babbitt, C.C., Silverman, J.S., et al. (2007) Convergent adaptation of human lactase persistence in Africa and Europe. *Nature Genetics* **39**(1): 31–40.

Tiwari, M.P., and Bhai, H.Y. (1997) Quaternary stratigraphy of the Narmada Valley. *Geological Survey of India* **46**: 33–63.

Tixier, J., Brugal, J.-P., Tillier, A.-M., Bruzek, J., and Hublin, J.-J. (1998) Irhoud 5, un fragment d'os coxal non adulte des niveaux mousteriens marocains. *Actes des lères Journe'es Nationales d'Arché ologie et du Patrimoine. Volume 1: Pre'histoire, Rabat, Socie'te' Marocaine d'Arché ologie et du Patrimoine.*

Tobias, P.V. (1959) The history and metamorphosis of the Boskop concept. In: *The Skeletal Remains of Bambandyanalo*, Galloway, A., ed., pp. 137–46. Witwatersand University Press, Johannesburg.

Tobias, P.V. (1960) The Kanam jaw. *Nature* **185**: 946–7.

Tobias, P.V. (1962) Early members of the genus *Homo* in Africa. In: *Evolution und Hominisation*, Kurth, G., ed., pp. 191–204. Gustav Fischer, Stuttgart.

Tobias, P.V. (1965a) The early *Australopithecus* and *Homo* from Tanzania. *Anthropologie Prague* 3: 43–48.

Tobias, P.V. (1965b) *Australopithecus, Homo habilis*, tool using and tool making. *The South African Archaeological Bulletin* 20: 167–92.

Tobias, P.V. (1965c) New discoveries in Tanganyika: their bearing on hominid evolution. *Current Anthropology* 6(4): 391–411.

Tobias, P.V. (1966a) Fossil hominid remains from Ubeidiya, Israel. *Nature* 211(5045): 130–3.

Tobias, P.V. (1966b) A re-examination of the Kedung Brubus mandible. *Zoologische Mededelingen* 41(22): 307–20.

Tobias, P.V. (1967) *Olduvai Gorge Vol 2. The Cranium and Maxillary Dentition of Australopithecus (Zinjanthropus) boisei*. Cambridge University Press, Cambridge.

Tobias, P.V. (1967) The hominid skeletal remains of Haua Fteah. In: *The Haua Fteah (Cyrenaica) and the Stone Age of the South-East Mediterranean*, McBurney, C.B.M., ed., pp. 338–52. Cambridge University Press, Cambridge.

Tobias, P.V. (1967) General questions arising from some Lower and Middle Pleistocene hominids of the Olduvai Gorge. *South African Journal of Science* 63(2): 41–48.

Tobias, P.V. (1968) Middle and early Upper Pleistocene member of genus *Homo* in Africa. In: *Evolution und hominisation*, Kurth, G., ed., pp. 176–94. Gustav Fischer, Stuttgart.

Tobias, P.V. (1971) Human skeletal remains from the Cave of Hearths, Makapansgat. *American Journal of Physical Anthropology* 34(3): 335–68.

Tobias, P.V. (1973a) A new chapter in the history of the Sterkfontein early hominid site. *S. Afr. Biol. Soc.* 14: 30–44.

Tobias, P.V. (1973b) Implications of the new age estimates of the early South African hominids. *Nature* 246: 79–84.

Tobias, P.V. (1980a) *Australopithecus afarensis* and *A. africanus*: critique and an alternative hypothesis. *Palaeontologica Africana* 23: 1–17.

Tobias, P.V. (1980b) The natural history of the helicoidal occlusal plane and its evolution in early *Homo*. *American Journal of Physical Anthropology* 53(2): 173–87.

Tobias, P.V., ed. (1985) *Hominid Evolution: Past, Present, and Future*. Alan R. Liss, New York.

Tobias, P.V. (1991a) *Olduvai Gorge Vol 4: The Skulls, Endocasts and Teeth of* Homo habilis. Cambirdge University Press, Cambridge.

Tobias, P.V. (1991b) The environmental background of hominid emergence and the appearance of the genus *Homo*. *Human Evolution* 6(2): 129–42.

Tobias, P.V. (2000) The fossil hominids. In: *The Cenozoic of Southern Africa*, Partridge, T.C., and Maud, R.R., eds, pp. 252–76. Oxford University Press, Oxford.

Tobias, P.V., and von Koenigswald, G.H.R. (1964) A comparison between the Olduvai hominids and those of Java and some implications for hominid phylogeny. *Nature* 204: 515–18.

Tobias, P.V., and Wells, L.H. (1967) South Africa. In: *Catalogue of Fossil Hominids Part I: Africa*, Oakley, K.P., and Campbell, B., eds, pp. 49–100. Trustees of the British Museum (Natural History), London.

Tobias, P.V., and Hughes, A.R. (1969) The new Witswatersrand University. Excavation at Sterkfontein. *South African Archaeological Bulletin* 24: 158–69.

Tobias, P.V., and Falk, D. (1988) Evidence for a dual pattern of cranial venous sinuses on the endocranial cast of Taung (*Australopithecus africanus*). *American Journal of Physical Anthropology* 76(3): 309–12.

Tocheri, M.W., Orr, C.M., Larson, S.G., Sutikna, T., Jatmiko, Saptomo, E.W., et al. (2007) The primitive wrist of *Homo floresiensis* and its implications for hominin evolution. *Science* 317(5845): 1743–45.

Tocheri, M.W., Orr, C.M., Jacofsky, M.C., and Marzke, M.W. (2008) The evolutionary history of the hominin hand since the last common ancestor of *Pan* and *Homo*. *Journal of Anatomy* 212: 544–62.

Tode, A. (1953) Die Untersuchung der palaolithischen Freilandstation von Salzgitter-Lebenstedt. 8. Einige archaologische Erkenntnisse aus der palaolithischen Freilandstation von Salzgitter- Lebenstedt. *Eiszeitalter und Gegenwart* 3: 192–215.

Tode, A. (1982) *Der altsteinzeitliche Fundplatz Salzgitter-Lebenstedt. Teil 1. Archaologischer Teil*. Bohlau Verlag, Koln.

Toerien, M.J., and Hughes, A.R. (1955) The limb bones of Springbok Flats man. *South African Journal of Science* 52(5): 125–8.

Tomasello, M. (1999) Emulation learning and cultural learning. *Behavioural and Brain Sciences* 21: 703–4.

Tomasello, M., and Call, J. (1997) *Primate Cognition*. Oxford University Press, Oxford.

Tooby, J., and DeVore, I. (1987) The reconstruction of hominid behavioural evolution through strategic modelling. In: *The Evolution of Human Behaviour*, Kinzey, W., ed., pp. 183–237. SUNY Press, New York.

Topinard, P. (1885) *Éléments d'anthropologie générale*. A Delahaye et E Legrosivier, Paris.

Topinard, P. (1890) *Anthropology*. Chapman and Hall, London.

Topinard, P. (1891) *L'homme dans la nature*. Ancienne Librairie Germer Baillière et Cie, Paris.

Topinard, P. (1891) *La paléoanthropologie*. Congrès international d'anthropologie et d'archéologie préhistoriques Compte Rendu [1889], pp. 383–93.

Torrence, R., and Barton, H. (2006) *Ancient Starch Research*. Left Coast Press, Walnut Creek, CA.

Tosi, A., Morales, J., and Melnick, D. (2002) Y-chromosome and mitochondrial markers in *Macaca fascicularis* indicate Introgression with Indochinese *M. mulatta* and a biogeographic barrier in the Isthmus of Kra. *International Journal of Primatology* **23**(1): 161–78.

Tosi, A.J., Disotell, T.R., Morales, J.C., and Melnick, D.J. (2003) Cercopithecine Y-chromosome data provide a test of competing morphological evolutionary hypotheses. *Molecular Phylogenetics and Evolution* **27**(3): 510–21.

Tosi, A.J., Detwiler, K.M., and Disotell, T.R. (2005) X-chromosomal window into the evolutionary history of the guenons (Primates: Cercopithecini). *Molecular Phylogenetics and Evolution* **36**(1): 58–66.

Toth, N. (1985) The Oldowan reassessed: a close look at early stone age artifacts. *Journal of Archaeological Science* **12**(2): 101–20.

Toth, N. (1987) Behavioural inferences from Early Stone Age artifact assemblage: an experimental model. *Journal of Human Evolution* **16**: 763–87.

Toussaint, M., and Pirson, S. (2006) Neandertal studies in Belgium: 2000–2005. *Periodicum Biologorum* **108**(3): 373.

Toussaint, M., Otte, M., Bonjean, D., Bocherens, H., Falguéres, C., and Yokoyama, Y. (1998) Les restes humains néandertaliens immatures de la couche 4A de la grotte Scladina (Andenne, Belgique). *Comptes Rendus de l'Académie des Sciences-Series IIA-Earth and Planetary Science* **326**(10): 737–42.

Toussaint, M., Macho, G.A., Tobias, P.V., Partridge, T.C., and Hughes, A.R. (2003) The third partial skeleton of a late Pliocene hominin (Stw 431) from Sterkfontein, South Africa. *South African Journal of Science* **99**(5–6): 215–23.

Trapani, J. (2008) Quaternary fossil fish from the Kibish Formation, Omo Valley, Ethiopia. *Journal of Human Evolution* **55**(3): 521–30.

Traverse, A. (2007) *Paleopalynology*. Springer, Dordrecht.

Tribolo, C. (2003) *Apport des methods de la luminescence à la chronologie des techno-facies du Middle Stone Age associés aux premiers homes modernes du Sud de l'Afrique*. PhD thesis, University of Bordeaux 1.

Tribolo, C., Mercier, N., Selo, M., Valladas, H., Joron, J., Reyss, J., et al. (2006) TL dating of burnt lithics from Blombos Cave (South Africa): Furthur evidence for the antiquity of modern human behavior. *Archaeometry* **48**(2): 341–57.

Trinkaus, E. (1981) Neanderthal limb proportions and cold adaptation. In: *Aspects of Human Evolution*, Stringer, C.B., ed., pp. 187–224. Taylor & Francis, London.

Trinkaus, E. (1983) *The Shanidar Neandertals*. Academic Press, New York.

Trinkaus, E. (1984) Does KNM-ER 1481A establish *Homo erectus* at 2.0 myr BP? *American Journal of Physical Anthropology* **64**(2): 137–9.

Trinkaus, E. (1985) Pathology and posture of the La Chapelle-aux-Saints Neandertal. *American Journal of Physical Anthropology* **67**: 19–41.

Trinkaus, E. (1993) A note on the KNM-ER 999 hominid femur. *Journal of Human Evolution* **24**: 493–504.

Trinkaus, E. (1995) Near Eastern Late Archaic Humans. *Paléorient* **21**(2): 9–24.

Trinkaus, E. (2004) Eyasi 1 and the suprainiac fossa. *American Journal of Physical Anthropology* **124**(1): 28–32.

Trinkaus, E. (2005) Early modern humans. *Annual Reviews in Anthropology* **34**: 207–30.

Trinkaus, E. (2007) European early modern humans and the fate of the Neandertals. *Proceedings of the National Academy of Sciences USA* **104**: 7367–72.

Trinkaus, E. (2008a) Behavioral implications of the Muierii 1 early modern human scapula. *Annuaire Roumain d'Anthropologie* **45**: 27–41.

Trinkaus, E. (2008b) Kiik-Koba 2 and Neandertal axillary border ontogeny. *Anthropological Science (Nippon)* **116**(3): 231–6.

Trinkaus, E. (2009) The human tibia from Broken Hill, Kabwe, Zambia. *PaleoAnthropology* **2009**: 145–65.

Trinkaus, E., and Zimmerman, M. (1982) Trauma among the Shanidar Neandertals. *American Journal of Physical Anthropology* **57**(1): 61–76.

Trinkaus, E., and Long, J.C. (1990) Species attribution of the Swartkrans Member I first metacarpals: SK 84 and SKX 5020. *American Journal of Physical Anthropology* **83**(4): 419–24.

Trinkaus, E., and Smith, F.H. (1995) Body size of the Vindija Neandertals. *Journal of Human Evolution* **28**: 201–8.

Trinkaus, E., and Ruff, C.B. (1996) Early modern human remains from eastern Asia: the Yamashita-cho 1 immature postcrania. *Journal of Human Evolution* **30**: 299–314.

Trinkaus, E., and Ruff, C.B. (1999a) Diaphyseal cross-sectional geometry of Near Eastern Middle Palaeolithic humans: the femur. *Journal of Archaeological Science* **26**: 409–24.

Trinkaus, E., and Ruff, C.B. (1999b) Diaphyseal cross-sectional geometry of Near Eastern Middle Palaeolithic humans: the tibia. *Journal of Archaeological Science* **26**: 1289–1300.

Trinkaus, E., and Svoboda, J., eds (2006) *Early Modern Human Evolution in Central Europe: the People of Dolní Věstonice and Pavlov*. Oxford University Press, New York.

Trinkaus, E., Churchill, S.E., and Ruff, C.B. (1994) Postcranial robusticity in *Homo* II: Humeral bilateral assymmetry and bone plasticity. *American Journal of Physical Anthropology* **93**: 1–34.

Trinkaus, E., Ruff, C., Churchill, S., and Vandermeersch, B. (1998) Locomotion and body proportions of the Saint-Césaire 1 Châtelperronian Neandertal. *Proceedings of the National Academy of Sciences USA* **95**(10): 5836–40.

Trinkaus, E., Churchill, S., Ruff, C., and Vandermeersch, B. (1999) Long bone shaft robusticity and body proportions of the Saint-Césaire 1 Châtelperronian Neanderthal. *Journal of Archaeological Science* **26**(7): 753–73.

Trinkaus, E., Ruff, C., and Conroy, G. (1999) The anomalous archaic *Homo* femur from Berg Aukas, Namibia: a biomechanical assessment. *American Journal of Physical Anthropology* **110**(3): 379–91.

Trinkaus, E., Stringer, C., Ruff, C.B., Hennessy, R., Roberts, M., and Parfitt, S.A. (1999) Diaphyseal cross-sectional geometry of the Boxgrove 1 Middle Pleistocene human tibia. *Journal of Human Evolution* **37**(1): 1–25.

Trinkaus, E., Moldovan, O., Milota, Ş., Bîlgăr, A., Sarcina, L., Athreya, S., et al. (2003a) An early modern human from the Peştera cu Oase, Romania. *Proceedings of the National Academy of Sciences USA* **100**(20): 11221–6.

Trinkaus, E., Milota, Ş., Rodrigo, R., Mircea, G., and Moldovan, O. (2003b) Early modern human cranial remains from the Peştera cu Oase, Romania. *Journal of Human Evolution* **45**: 245–53.

Trinkaus, E., Zilhão, J., Rougier, H., Rodrigo, R., Milota, Ş., Gherase, M., et al. (2006) The Peştera cu Oase and early modern humans in southeastern Europe. In: *When Neanderthals and Modern Humans Met*, Conard, N.J., ed., pp. 145–64. Kerns Verlag, Tübingen.

Trinkaus, E., Maley, B., and Buzhilova, A. (2008) Paleopathology of the Kiik-Koba 1 Neandertal. *American Journal of Physical Anthropology* **137**: 106–12.

Trivers, R.L. (1972) Parental investment and sexual selection. In: *Sexual Selection and the Descent of Man*, Campbell, B., ed., pp. 136–79. Heinemann, London.

Trivers, R.L. (1974) Parent–offspring conflict. *American Zoologist* **14**(1): 249–64.

Trivers, R.L., and Willard, D.E. (1973) Natural selection of parental ability to vary the sex ratio of offspring. *Science* **179**: 90–2.

Tryon, C., and McBrearty, S. (2006) Tephrostratigraphy of the Bedded Tuff Member (Kapthurin Formation, Kenya) and the nature of archaeological change in the later Middle Pleistocene. *Quaternary Research* **65**: 492–507.

Tryon, C.A. (2006) "Early" Middle Stone Age lithic technology of the Kapthurin Formation (Kenya). *Current Anthropology* **47**(2): 367–75.

Tryon, C.A., and McBrearty, S. (2002) Tephrostratigraphy and the Acheulian to Middle Stone Age transition in the Kapthurin Formation, Kenya. *Journal of Human Evolution* **42**(1–2): 211–35.

Tryon, C.A., McBrearty, S., and Texier, P.-J. (2005) Levallois lithic technology from the Kapthurin Formation, Kenya: Acheulian origin and Middle Stone Age diversity. *African Archaeological Review* **22**(4): 199–229.

Tryon, C.A., Roach, N.T., and Logan, M.A.V. (2008) The Middle Stone Age of the northern Kenyan Rift: age and context of new archaeological sites from the Kapedo Tuffs. *Journal of Human Evolution* **55**(4): 652–64.

Tuffreau, A. (1976) Les fouilles du gisement Acheuléen supérieur des osiers à Bapaume (Pas-de-Calais). *Bulletin de la Société Préhistorique Française* **73**(8): 231–43.

Tuffreau, A., Ameloot-Van Der Heijden, N., and Ducrocq, T. (1991) La fouille de sauvetage du gisement paléolithique moyen de Riencourt-lès-Bapaume (Pas-de-Calais): premiers résultats. *Bulletin de la Société Préhistorique Française* **88**(7): 202–9.

Tuffreau, A., and Sommé, J. (1988) *Le Gisement Paléolithique Moyen de Biache-Saint-Vaast (Pas-de-Calais) Stratigraphie, Environnement, Études Archéologiques, Vol. I*. Société Préhistorique Française, Paris.

Tulving, E. (1972) Episodic and semantic memory. In: *Organization of Memory*, Tulving, E., and Donaldson, W., eds, pp. 381–403. Academic, New York.

Tulving, E. (1983) *Elements of episodic memory*. Oxford University Press, New York.

Tulving, E. (2002) Episodic memory: from mind to brain. *Annual Review of Psychology* **53**: 1–25.

Tulving, E. (2005) Episodic memory and autonoesis: uniquely human? In: *The Missing Link in Cognition: Origins of Self-Reflective Consciousness*, Terrace, H.S., and Metcalf, J., eds, pp. 3–56. Oxford University Press, Oxford.

Turner, C., Nichol, C., and Scott, G. (1991) Scoring procedures for key morphological traits of the permanent dentition: The Arizona State University Dental Anthropology System. In: *Advances in Dental Anthropology*, pp. 13–31. Wiley-Liss, New York.

Turner, C.G., and Markowitz, M.A. (1990) Dental discontinuity between late Pleistocene and recent Nubians: peopling of the Eurafrican-south Asian triangle, 1. *Homo* **41**: 32–41.

Turner, C.G., Manabe, Y., and Hawkey, D.E. (2000) The Zhoukoudian Upper Cave dentition. *Acta Anthropologica Sinica* **19**(4): 253–68.

Turney, C.S.M., Bird, M.I., Fifield, L.K., Roberts, R.G., Smith, M., Dortch, C.E., et al. (2001) Early human occupation at Devil's Lair, Southwestern Australia 50,000 years ago. *Quaternary Research* **55**(1): 3–13.

Turney, C.S.M., Roberts, R.G., and Jacobs, Z. (2006) Archaeology: progress and pitfalls in radiocarbon dating. *Nature* **443**(7108): E3.

Turq, A. (1989) Le squelette de l'enfant néandertalien du Roc de Marsal: les données de fouilles. *Paléo* **1**: 47–54.

Turq, A., Dibble, H., Faivre, J.P., Goldberg, P., McPherron, S.J.P., and Sandgathe, D. (2008) Le Moustérien Récent Du Périgord Noir: Quoi De Neuf? In: *Les Sociétés du Paléolithique dans un Grand Sud-ouest de la France: nouveaux gisements, nouveaux résultats, nouvelles méthodes*, Jaubert, J., Bordes, J.-G., and Ortega, I., eds, pp. 83–94. Mémoire de la Société Préhistorique Française.

Turville-Petre, F., ed. (1927) *Researches in Prehistoric Galilee*. British School of Archaeology in Jerusalem, London.

Tuttle, R. (1969) Quantitative and functional studies on the hands of Anthropoidea. I: The Hominoidea. *Journal of Morphology* **128**: 309–64.

Tuttle, R., and Basmajian, J. (1974) Electromyography of brachial muscles in *Pan gorilla* and hominoid evolution. *American Journal of Physical Anthropology* **41**(1): 71–90.

Tuttle, R., Webb, D., Weidl, E., and Baksh, M. (1990) Further progress on the Laetoli trails. *Journal of Archaeological Science* **17**: 347–62.

Tuttle, R.H., ed. (1975) *Paleoanthropology: Morphology and Paleoecology*. World Anthropology. Mouton Publishers, The Hague.

Tuzun, E., Sharp, A.J., Bailey, J.A., Kaul, R., Morrison, V.A., Pertz, L.M., et al. (2005) Fine-scale structural variation of the human genome. *Nature Genetics* **37**: 727–32.

Twiesselmann, F. (1958) *Les ossements humains du gîte mésolithique d'Ishango. Exploration de Parc National Albert (Fasc. 5)*. Institut des Parcs Nationaux du Congo Belge, Brussels.

Twiesselmann, F. (1991) La mandibule et le fragment de maxillaire supérieur de Loyangalani (rive est du lac Turkana, Kenya). *Anthropologie et préhistoire* **102**: 77–95.

Tyler, D.E. (2004) An examination of the taxonomic status of the fragmentary mandible Sangiran 5, (*Pithecanthropus dubius*), *Homo erectus*, "*Meganthropus*", or *Pongo*? *Quaternary International* **117**: 125–30.

Tyler, D.E., Sartono, S., and Kranz, G.S. (1995) *Homo erectus* Mandible F (Sangiran 22) from Java: an announcement. In: *Evolution and ecology of* Homo erectus, Bower, J.R.F., and Sartono, S., eds, pp. 203–6. *Pithecanthropus* Centennial Foundation, Leiden University, Leiden.

Uchida, A. (1998) Variation in tooth morphology of *Pongo pygmaeus*. *Journal of Human Evolution* **34**(1): 71–9.

Uchida, A. (1998) Variation in tooth morphology of *Gorilla gorilla*. *Journal of Human Evolution* **34**(1): 55–70.

Ulrich, H., ed. (2005) *The Neanderthal Adolescent Le Moustier 1: new aspects, new results*. Staatliche Museen Zu Berlin-Preussicher Kulturbesitz, Berlin.

Umberger, B.R., Gerritsen, K.G.M., and Martin, P.E. (2003) A model of human muscle energy expenditure. *Computer Methods in Biomechanics and Biomedical Engineering* **6**: 99–111.

Underhill, P., Passarino, G., Lin, A., Shen, P., and Mirazón, L. (2001) The phylogeography of Y chromosome binary haplotypes and the origins of modern human populations. *Annals of human genetics* **65**(1): 43–62.

Underhill, P.A. (2000) Y-chromosome sequence variation and the history of human populations. *Nature Genetics* **26**: 358–61.

Ungar, P., Grine, F., and Teaford, M. (2008) Dental microwear and diet of the Plio-Pleistocene hominin *Paranthropus boisei*. *PLoS One* **3**(4).

Ungar, P., Walker, A., and Coffing, K. (1994) Reanalysis of the Lukeino molar (KNM LU 335). *American Journal of Physical Anthropology* **94**(2): 165–73.

Ungar, P.S., Grine, F.E., Teaford, M.F., and Zaatari, S.E. (2006) Dental microwear and diets of African early *Homo*. *Journal of Human Evolution* **50**(1): 78–95.

Valladas, H., Geneste, J.M., Joron, J.L., and Chadelle, J.P. (1986) Thermoluminescence dating of Le Moustier (Dordogne, France). *Nature* **322**: 452–4.

Valladas, H., Joron, J.L., Valladas, G., Arensburg, B., Bar-Yosef, O., Belfer-Cohen, A., et al. (1987) Thermoluminescence dates for the Neanderthal burial site at Kebara in Israel. *Nature* **330**: 159–60.

Valladas, H., Reyss, J.L., Joron, J.L., Valladas, G., Bar-Yosef, O., and Vandermeersch, B. (1988) Thermoluminescence dating of Mousterian "Proto-Cro-Magnon" remains from Israel and the origin of modern man. *Nature* **331**: 614–16.

Valladas, H., Mercier, N., Hovers, E., Froget, L., Joron, J., Kimbel, W., et al. (1999) TL dates for the Neanderthal site of the Amud Cave, Israel. *Journal of Archaeological Science* **26**: 259–68.

Valladas, H., Mercier, N., Joron, J., McPherron, S., Dibble, H., and Lenoir, M. (2003) TL dates for the Middle Paleolithic site of Combe-Capelle Bas, France. *Journal of Archaeological Science* **30**(11): 1443–50.

Vallois, H. (1929) Les preuves anatomiques de l'origine monophyletique de l'homme. *L'Anthropologie* **39**: 77–101.

Vallois, H. (1943) *Anthropologie de la population française*. Didier, Paris.

Vallois, H. (1958) La grotte de Fontéchevade. *Archives de l'Institut de Paleóntologie Humaine* **29**: 6–164.

Vallois, H., and de Félice, S. (1977) *Les mésolithiques de France*. Masson, Paris.

Vallois, H.-V. (1944) *Les Races Humaines*. Presses Universitaires, Paris.

Vallois, H.-V. (1949) The Fontéchevade fossil men. *American Journal of Physical Anthropology* **7**: 339–62.

Vallois, H.-V. (1951) La mandibule humaine fossile de la grotte du Porc-Épic, près Diré-Daoua (Abyssinie). *Anthropologie (Paris)* **55**: 231–8.

Vallois, H.-V. (1962) Un nouveau Neandertaloide en Palestine. *Anthropologie* **66**: 405–6.

Vallois, H.-V. (1965) Le sternum neandertalien du Regourdou. *Anthropologischer Anzeiger* **29**(1–4): 273–89.

Vallois, H.-V. (1970) La découverte des hommes de Cro-Magnon, son importance anthropologique. In: *L'Homme de Cro-Magnon: Anthropologie et Archéologie.*, Camps, G., and Olivier, G., eds, pp. 11–20. Arts et Métiers graphiques, Paris.

Vallois, H.-V., and Billy, G. (1965a) Nouvelles recherches sur les Hommes fossiles de l'abri de Cro-Magnon. *Anthropologie* **69**: 47–74.

Vallois, H.-V., and Billy, G. (1965b) Nouvelles recherches sur les hommes fossiles de l'abri de Cro-Magnon. *L'Anthropologie* **69**(3–4): 249–72.

Vallois, H.-V., and Movius, H.L. (1953) *Catalogue des hommes fossiles.* Congres Geologique International, Macon, France.

Vallois, H.-V., and Vandermeersch, B. (1972) Le crâne mousterien de Qafzeh (Homo VI). *L'Anthropologie* **76**: 71–96.

Valoch, K. (1965) Jeskynë Sipka a Certova dira u Stramberku. *Anthropos.*

Valoch, K. (1967) Die Steinindustrie von der Fundstelle des menschlichen Skelettrestes I aus der Höhle Kůlna bei Sloup (Mähren). *Anthropologie* **5**: 21–31.

Valoch, K. (1968) Le Remplissage et les Industries du Paléolithique Moyen de la Grotte Kůlna en Moravie. *L'Anthropologie* **72**(5–6): 453–66.

Valoch, K. (1988) Le Taubachien et le Micoquien de la grotte Kůlna en Moravie. In: *L'Homme de Néanderthal,* Otte, M., ed., pp. 205–17. La Technique, Liège.

Valverde, P., Healy, E., Jackson, I., Rees, J., and Thody, A. (1995) Variants of the melanocyte-stimulating hormone receptor gene are associated with red hair and fair skin in humans. *Nature genetics* **11**(3): 328–30.

van Andel, T.H. (1989) Late Pleistocene sea levels and the human exploitation of the shore and shelf of southern Africa. *Journal of Field Arcaeology* **16**(2): 133–55.

Van den Audenaerde, D. (1984) The Tervuren Museum and the pygmy chimpanzee. In: *The Pygmy Chimpanzee; Evolutionary Biology and Behavior,* Susman, R.L., ed., pp. 3–11. Plenum Press, New York.

Vanderbroek, G. (1961) The Comparative Anatomy of the Teeth of Lower and Non-specialized Mammals. *International colloquium on the evolution of lower and nonspecialized mammals. Brussels, Koninklijke Vlaamse Academie voor Wetenschappen, Letteren en Schone Kunsten van Belgie.* **1**: 215.

Vanderbroek, G. (1967) Origin of the cusps and crests of the tribosphenic molar. *Journal of Dental Research* **46**(5): 796–804.

Vandermeersch, B. (1970) Une sépulture moustérienne avec offrandes découverte dans la grotte de Qafzeh. *Comptes Rendus de l'Académie des Sciences* **268**: 298–301.

Vandermeersch, B. (1971) Récentes découvertes de squelettes humains á Qafzeh (Israel). In: *Origines de l'Homme Moderne,* Bordes, F., ed., pp. 49–54. UNESCO, Paris.

Vandermeersch, B. (1978a) Le crâne pré-würmien de Biache-St-Vaast. In: *Les Origines Humaines et les Epoques de l'Intelligence, Colloque int,* pp. 153–7. Fondation Singer-Polignac, Masson, Paris.

Vandermeersch, B. (1978b) Etude préliminaire du crâne humain du gisement paléolithique de Biache-Saint-Vaast (Pas-de-Calais). *Bulletin de l'Association française pour l'étude du Quaternaire* **54–6**: 65–7.

Vandermeersch, B. (1981) *Les Hommes Fossiles de Qafzeh (Israel).* CNRS, Paris.

Vandermeersch, B., and Trinkaus, E. (1995) Postcranial remains of the Régourdou 1 Neandertal: the shoulder and arm remains. *Journal of Human Evolution* **28**(5): 439–76.

van der Merwe, N., Masao, F., and Bamford, M. (2008) Isotopic evidence for contrasting diets of early hominins *Homo habilis* and *Australopithecus boisei* of Tanzania. *South African Journal of Science* **104**: 153–5.

van der Merwe, N.J., and Medina, E. (1989) Photosynthesis and $^{13}C/^{12}C$ ratios in Amazonian rain forests. *Geochimica et Cosmochimica Acta* **53**(5): 1091–4.

Vanhaeren, M., and d'Errico, F. (2001) La parure de l'enfant de La Madeleine (fouilles Peyrony). Un nouveau regard sur lienfance au paléolithique supérieur. *Paléo* **13**: 201–40.

Vanhaeren, M., d'Errico, F., Stringer, C., James, S.L., Todd, J.A., and Mienis, H.K. (2006) Middle Paleolithic shell beads in Israel and Algeria. *Science* **312**(5781): 1785–8.

Van Neer, W. (1986) Some notes on the fish remains from Wadi Kubbaniya (Upper Egypt, late Paleolithic). In: *Fish and Archaeology: Studies in Osteometry, Taphonomy, Seasonality, and Fishing Methods,* Brinkhuizen, D.C., and Clason, A.T., eds, pp. 103–13. Oxford University Press, Oxford.

Van Neer, W. (1989) *Contribution to the Archaeozoology of Central Africa.* Annales du Musee Royal de l'Afrique Centrale, Tervuren.

Van Noten, F. (1977) Excavations at Matupi Cave. *Antiquity* **51**: 35–40.

Van Peer, P. (1992) The Levallois Reduction Strategy. In: *Monographs in World Archaeology.* Prehistory Press, Madison, WI.

Van Peer, P., Vermeersch, P.M., and Paulissen, E. (2010) *Chert Quarrying, Lithic Technology, and a Modern Human Burial at the Palaeolithic Site of Taramsa 1, Upper Egypt.* Leuven University Press, Leuven.

van Riet Lowe, C. (1929) Notes on some stone implements from Tuinplaas, Springbok Flats. *South African Jouranl of Science* **26**: 623–30.

van Riet Lowe, C. (1954) An interesting Bushman arrowhead. *The South African Archaeological Bulletin* **9**(35): 88.

van Schaik, C.P., and Hrdy, S.B. (1991) Intensity of local resource competition shapes the relationship between maternal rank and sex ratios at birth in Cercopithecine primates. *American Naturalist* **138**(6): 1555–62.

van Schaik, C.P., Ancrenaz, M., Borgen, G., Galdikas, B., Knott, C.D., Singleton, I., et al. (2003) Orangutan cultures and the evolution of material culture. *Science* **299**: 102–5.

van Valen, L. (1966) Deltatheridia, a new order of mammals. *Bulletin of the American Museum of Natural History* **132**: 1–126.

Van Valkenburgh, B., Wang, K., and Damuth, J. (2004) Cope's rule, hypercarnivory, and extinction in North American canids. *Science* 306: 101–4.

Van Wyk, A.E. (1994) Maputaland-Pondoland region. In: *Centres of Plant Diversity. A Guide and Strategy for their Conservation. Volume 1*, Davis, S.D., Heywood, V.H., and Hamilton, A.C., eds, pp. 227–35. IUCN Publications Unit, Cambridge.

van Zinderen Bakker, I.E.M. (1957) A pollen analytical investigation of the Florisbad deposits. In: *Proceedings of the 3rd Pan-African Congress of Prehistory, Livingston 1955* pp. 56–57.

Varki, A. (2010) Uniquely human evolution of sialic acid genetics and biology. *Proceedings of the National Academy of Sciences USA* 107(Suppl. 2): 8939–46.

Vekua, A., Lordkipanidze, D., Rightmire, G.P., Agustí, J., Ferring, R., Maisuradze, G., et al. (2002) A new skull of early *Homo* from Dmanisi, Georgia. *Science* 297(5578): 85–9.

Vereeke, E., D'Août, K., De Clercq, D., Van Elsaker, L., and Aerts, P. (2003) Dynamic plantar pressure distribution during terrestrial locomotion of Bonobos (*Pan paniscus*). *American Journal of Physical Anthropology* 120: 373–83.

Vermeersch, P. (2001) "Out of Africa" from an Egyptian point of view. *Quaternary International* 75(1): 103–12.

Vermeersch, P.M., ed. (2002) *Palaeolithic Quarrying Sites in Upper and Middle Egypt*. Leuven University Press, Leuven.

Vermeersch, P.M., Gijselings, G., and Paulissen, E. (1984) Discovery of the Nazlet Khater Man, Upper Egypt. *Journal of Human Evolution* 13: 281–6.

Vermeersch, P.M., Pailissen, E., Stokes, S., Van Peer, P., de Bie, M., Steenhoudt, F., et al. (1997) Middle Palaeolithic chert mining in Egypt. In: *Siliceous Rocks and Culture*, Ramos-Milan, A., and Bustillo, M.-A., eds, pp. 173–94. Universidad de Granada, Granada.

Vermeersch, P.M., Paulissen, E., Stokes, S., Charlier, C., van Peer, P., Stringer, C., et al. (1998) A Middle Palaeolithic burial of a modern human at Taramsa Hill, Egypt. *Antiquity* 72: 475–84.

Verna, C. (2006) *The temporal bones from La Quina (Charente, France). Some evidence of a local population among neandertals?* Congress "150 years of Neanderthal discoveries". Bonn.

Verna, C., Hublin, J.-J. Debenath, A., Jelinek, A., and Vandermeersch, B. (2010) Two new hominin cranial fragments from the Mousterian levels at La Quina (Charente, France). *Journal of Human Evolution* 58(3): 273–8.

Verneau, R. (1906) *Les Grottes de Grimaldi*. Imprimerie de Monaco, Monaco.

Verneau, R. (1908) *L'Homme de la Barma Grande*. Imprimerie Colombani, Menton.

Verniers, J., and de Heinzelin, J. (1990) Stratigraphy and geological history of the Upper Semliki: a preliminary report. In: *Evolution of Environments and Hominidae in the African Western Rift Valley, Memoirs of the Virginia Museum of Natural History*, Boaz, N.T., ed., pp. 17–40. Memoir No. 1. Virginia Museum of Natural History, Martinsville, VA.

Vértes, L. (1964) Tata, eine mittelpaläolithische Travertin-Siedlung in Ungarn. *Archaeologia Hungarica* 43: 1–285.

Vértes, L. (1965) Discovery of *Homo erectus* in Hungary. *Antiquity* 39: 303.

Vértes, L. (1968) Szeleta-Symposium in Ungarn. *Quartär* 19: 381–90.

Vértes, L., and Vries, H. (1952) Az Istállósköi barlang aurignaci II. Kulturá jának rádiokarbon kormeghatá rozá sa. *Archaeol Ert* 2: 195.

Vialet, A., Li, T., Grimaud-Hervé, D., de Lumley, M.-A., Liao, M., and Feng, X. (2005) Proposition de reconstitution du deuxième crâne d'*Homo erectus* de Yunxian (Chine). *Comptes Rendus Palevol* 4(3): 265–74.

Vigilant, L., Pennington, R., Harpending, H., Kocher, T.D., and Wilson, A.C. (1989) Mitochondrial DNA sequences in single hairs from a southern African population. *Proceedings of the National Academy of Sciences USA* 86: 9350–4.

Vignaud, P., Duringer, P., Mackaye, H.T., Likius, A., Blondel, C., Boisserie, J.-R., et al. (2002) Geology and paleontology of the Upper Miocene Toros-Menalla hominid locality, Chad. *Nature* 418: 152–5.

Villa, G., and Giacobini, G. (1996) Neandertal teeth from Alpine Caves of Monte Fenera (Piedmont, Northern italy). Description of the remains and microwear analysis. *Anthropologie* 34: 225–38.

Villa, P. (1983) Terra Amata and the Middle Pleistocene archaeological record of southern France. *University of California Publications in Anthropology* 13: 1–16.

Villa, P., and d'Errico, F. (2001) Bone and ivory points in the Lower and Middle Paleolithic of Europe. *Journal of Human Evolution* 41(2): 69–112.

Villa, P., Soto, E., Santonja, M., Pérez-Gonzales, A., Mora, R., Parcerisas, J., et al. (2005) New data from Ambrona: closing the hunting versus scavenging debate. *Quaternary International* 126–128: 223–50.

Villaverde, V., and Fumanal, M. (1990) Relations entre le Paléolithique Moyen et le Paléolithique Supérieur dans le versant méditerranéen espagnol. In: *Paléolithique Moyen récent et Paléolithique Supérieur ancien en Europe*, Fairzy, C., ed., pp. 177–83. Mémoires du Musée de Préhistoire de l'Île de France 3, Nemours.

Villmoare, B. (2005) Metric and non-metric randomization methods, geographic variation, and the single-species hypothesis for Asian and African *Homo erectus*. *Journal of Human Evolution* 49: 680–701.

Viñas, A. (1972) Observaciones sobre los depósitos cuaternarious de la Cova del Gegant, Sitges. *Speleon* 19: 115–26.

Vincens, A., Garcin, Y., and Buchet, G. (2007) Influence of rainfall seasonality on African lowland vegetation during the Late Quaternary: pollen evidence from

Lake Masoko, Tanzania. *Journal of Biogeography* **34**(7): 1274–88.

Vine, F.J. (1966) Spreading of the ocean floor: new evidence. *Science* **154**(3755): 1405–15.

Viranta, S., and Andrews, P. (1995) Carnivore guild structure in the Paşalar Miocene fauna. *Journal of Human Evolution* **28**(4): 359–72.

Virchow, H. (1920) *Die menschlichen Skeletreste aus dem Kämpfeschen Bruch im Travertin von Ehringsdorf bei Weimar.* G. Fischer, Jena.

Virchow, R. (1857) *Untersuchungen uber die Entwicklung des Schadelgrundes im gesunden und frankhaften Zustande.* Gesichissiblung und Gehirnbru, Berlin.

Vitagliano, S., and Piperno, M. (1991) Lithic Industry of level 27 beta of the Fosselone Cave (S. Felice Circeo, Latina). *Quaternaria Nova* **1**: 289–304.

Vlček, E. (1955) The Fossil Man of Gánovce Czechoslovakia. *Journal of the Royal Anthropological Institute* **85**: 163–71.

Vlček, E. (1969) *Neandertaler der Tschechoslowakei.* Verlag Tschcchoslowakischen Akademie Wissenschaften, Prague.

Vlček, E. (1971) Czechoslovakia. In: *Catalogue of Fossil Hominids I. Europe,* Oakley, K.P., Campbell, B., and Molleson, T.I., eds, pp. 47–64. British Museum Natural History, London.

Vlček, E. (1978a) A new discovery of *Homo erectus* in central Europe. *Journal of Human Evolution* **7**: 239–51.

Vlček, E. (1978b) Diagnosis of a fragment of the "hominid molar" from Přezletice, Czechoslovakia. *Current Anthropology* **19**(1): 145–6.

Vlček, E. (1993) *Fossile Menschenfunde von Weimar-Ehringsdorf.* Konrad Theiss Verlag, Stuttgart.

Vogel, C. (1964) Über eine Schädelbasisanomalie bei einem in freier Wildbahn geschossen *Cercopithecus torquatus atys. Zeitschrift für Morphologie und Anthropologie* **55**: 262–76.

Vogel, J. (1970) Groningen radiocarbon dates IX. *Radiocarbon* **12**(2): 444–71.

Vogel, J.C., and Waterbolk, H.T. (1963) Groningen radiocarbon dates IV. *Radiocarbon* **5**: 163–202.

Vogel, J.C., and Waterbolk, H.T. (1972) Groningen radiocarbon dates X. *Radiocarbon* **14**: 6–110.

Vogelsang, R. (1998) *Middle Stone Age Fundstellen in Südwest-Namibia.* Heinrich - Barth Institut, Köln.

Voigt, E.A. (1982) The molluscan fauna. In: *The Middle Stone Age at Klasies River Mouth in South Africa,* Singer, R., and Wymer, J., eds, pp. 155–86. University of Chicago Press, Chicago, IL.

Voisin, J. (2006) Clavicle, a neglected bone: Morphology and relation to arm movements and shoulder architecture in primates. *Anatomical Record Part A: Discoveries in Molecular, Cellular, and Evolutionary Biology* **288**(9): 944–53.

Voison, J.L. (2008) The Omo I hominin clavicle: Archaic or modern? *Journal of Human Evolution* **55**: 438–43.

Volman, R.P. (1984) Early prehistory of southern Africa. In: *Southern African Prehistory and Palaoenvironments,* Klein, R.G., ed., pp. 169–220. Balkema, Rotterdam.

Volman, T.P. (1978) Early archaeological evidence for shellfish collecting. *Science* **201**: 911–13.

von Berg, A. (1997) Ein Homininenrest aus dem Wannenvulkan bei Ochtendung, Kreis Mayen-Koblenz. Ein Vorbericht. *Archäologisches Korrespondenzblatt* **27**: 531–8.

von Berg, A. (2002) Der Neandertaler aus dem Eifelkrater. *Archäologie in Deutschland* **6**: 18–19.

von Berg, A., Condemi, S., and Frechen, M. (2000) Die Schädelkalotte des Neandertalers von Ochtendung/Osteifel. *Archäologie, Paläoanthropologie und Geologie, Eiszeitalter und Gegenwart* **50**: 56–8.

von Ebner, V. (1902) Histologie der Zahne mit Einschluss der Histogenese. In: *Handbuch der Zahnheilkunde,* Scheff, J., ed., pp. 243–99. A. Holder, Wien.

von Ebner, V. (1906) Überdie Entwicklungderleimgebenden Fibrillen, insbesondereim Zahnbein. *Sitzungsberichte der Mathematisch-Naturwissenschaftlichen Klasse der kaiser-lichen Akademie der Wissenschaften in Wien. CXV Band. Abteilung* **115**: 281–347.

von Koenigswald, G.H.R. (1935) Eine fossile saogetierfauna mit Simia aus Sudchina. *Proceedings of the Koninklijke Nederlandsche Akademie van Wetenschappen* **38**: 872–9.

von Koenigswald, G.H.R. (1936) Ein fossiler hominide aus dem Altpleistocän Ostjavas. *De Ingenieur in Nederlands-Indië* **8**: 149–58.

von Koenigswald, G.H.R. (1938) Preliminary note on new remains of Pithecanthropus from Central Java. *Third Congress on the Prehistory of the Far East, Singapore.*

von Koenigswald, G.H.R. (1940) *Neue Pithecanthropus-Funde 1936–1938. Ein beitrag zur kenntnis der praehominiden.* Landsdrukkerij, Batavia.

von Koenigswald, G.H.R. (1949) Early Man in Java IV. *Meganthropus palaeojavanicus,* Senckenberg Forschungsinstitut und Naturmuseum.

von Koenigswald, G.H.R. (1950) Fossil hominids from the Lower Pleistocene of Java. *Proceedings of the International Geological Congress 9, London.*

von Koenigswald, G.H.R. (1952) *Gigantopithecus blacki,* a giant fossil hominoid from the Pleistocene of South China. *Anthropological Paper, American Museum of Natural History* **43**(4): 295–325.

von Koenigswald, G.H.R. (1955) *Meganthropus* and the Australopithecinae. *Proceedings of the Third Pan-African Congress on Prehistory, Livingstone, Northern Rhodesia.*

von Koenigswald, G.H.R. (1956) The geological age of Wadjak Man from Java. *Proceedings of the Koninklijke Nederlandsche Akademie van Wetenschappen* **59**: 455.

von Koenigswald, G.H.R. (1964) Potassium Argon Dates and Early Man: Trinil. *6th INQUA Congress. Warsaw* **4**: 325–7.

von Koenigswald, G.H.R. (1968) Observation upon two *Pithecanthropus* mandibles from Sangiran, Central Java. *Proceedings of the Koninklijke Nederlandsche Akademie van Wetenschappen* **71**: 99–107.

von Koenigswald, G.H.R. (1975) Early man in Java: catalogue and problems. In: *Paleoanthropology, Morphology and Paleoecology*, Tuttle, R.H., ed., pp. 303–9. Mouton Publishers, Den Haag.

von Koenigswald, G.H.R., and Weidenreich, F.H. (1939) The relationship between *Pithecanthropus* and *Sinanthropus*. *Nature* **144**: 926–9.

von Koenigswald, W. (1992) Zur Ökologie und Biostratigraphie der beiden pleistozänen Faunen von Mauer bei Heidelberg. In: *Schichten von Mauer, Hrsg*, Beinhauer, K.W., and Wagner, G.A., eds, pp. 101–10. Museum der Stadt Mannheim, Reiss.

von Koenigswald, W., and Tobien, H. (1987) Bemerkungen zur Altersstellung der pleistozänen Mosbach-Sande bei Wiesbaden. *Geologisches Jahrbuch Hessen* **115**: 227–37.

Vrba, E.S. (1979) A new study of the scapula of *Australopithecus africanus* from Sterkfontein. *American Journal of Physical Anthropology* **51**: 117–30.

Vrba, E.S. (1980) Evolution, species and fossils: how does life evolve? *South African Journal of Science* **76**: 61–84.

Vrba, E.S. (1985a) Environment and evolution: alternative causes of the temporal distribution of evolutionary events. *South African Journal of Science* **81**(5): 229–36.

Vrba, E.S. (1985b) Ecological and adaptive changes associated with early hominid evolution, with particular reference to the southern African evidence. In: *Ancestors: The Hard Evidence*, Delson, E., ed., pp. 63–71. Alan R. Liss, New York.

Vrba, E.S. (1988) Late Pliocene climatic events and hominid evolution. In: *Evolutionary History of the "Robust" Australopithecines*, Grine, F.E., ed., pp. 405–26. Aldine de Gruyter, New York.

Vrba, E.S. (1992) Mammals as a key to evolutionary theory. *Journal of Mammalogy* **73**(1): 1–28.

Vrba, E.S. (1995) On the connections between paleoclimate and evolution. In: *Paleoclimate and Evolution, with Special Emphasis on Human Origins*, Vrba, E.S., Denton, G.H., Partridge, T.C., and Burckle, L.H., eds, pp. 24–45. Yale University Press, New Haven, CT.

Waddington, C.H. (1942) Canalization of development and the inheritance of aquired characters. *Nature* **150**(3811): 563–5.

Waddington, C.H. (1957) *The Strategy of the Genes: a Discussion of Some Aspects of Theoretical Biology*. Allen & Unwin, London.

Wadley, L. (2005) Putting ochre to the test: replication studies of adhesives that may have been used for hafting tools in the Middle Stone Age. *Journal of Human Evolution* **49**: 587–601.

Wadley, L., Lombard, M., and Williamson, B. (2004) The first residue analysis blind tests: results and lessons learnt. *Journal of Archaeological Science* **31**(11): 1491–1501.

Wagner, G. (1986) Evolution der Evolutionsfähigkiet. In: *Selbstorganisation: die Entstehung von Ordnung in Natur und Gesellschaft*, Dress, A., Henrichs, H., and Küppers, G., eds, pp. 121–48. Piper, München.

Wagner, G.P. (1984) On the eigenvalue distribution of genetic and phenotypic dispersion matrices: Evidence for a nonrandom organization of quantitative character variation *Journal of Mathematical Biology* **21**(1): 77–95.

Wagner, G.P. (1989) A comparative study of morphological integration in *Apis mellifera* (Insecta, Hymenoptera). *Zeitschrift für Zoologische Systematik und Evolutionsforschung* **28**: 48–61.

Wagner, G.P. (1996) Homologues, natural kinds and the evolution of modularity. *American Zoologist* **36**(1): 36–43.

Wagner, G.P., Booth, G., and Bagheri-Chaichian, H. (1997) A population genetic theory of canalization. *Evolution* **51**(2): 329–47.

Waguespack, N.M., and Surovell, T.A. (2003) Clovis hunting strategies, or How to make out on plentiful resources. *American Antiquity* **68**(2): 333–52.

Wakeley, J., Nielsen, R., Liu-Cordero, S., and Ardlie, K. (2001) The discovery of single-nucleotide polymorphisms–and inferences about human demographic history. *American Journal of Human Genetics* **69**(6): 1332–47.

Walker, A. (1967) *Locomotor Adaptations in Living and Extinct Madagascan Lemurs*. PhD thesis, University of London.

Walker, A. (1993) The origin of the genus *Homo*. In: *The Origin and Evolution of Humans and Humanness*, Rasmussen, D.T., ed., pp. 29–48. Jones & Bartlett, Boston, MA.

Walker, A., and Leakey, R.E.F. (1993) *The Nariokotome* Homo erectus *Skeleton*. Harvard University Press, Cambridge, MA.

Walker, A., and Shipman, P. (2005) *The Ape in the Tree: An Intellectual & Natural History of Proconsul*. Belknap Press, Cambridge, MA.

Walker, A.C. (1973) New *Australopithecus* femora from East Rudolf, Kenya. *Journal of Human Evolution* **2**(6): 545–55.

Walker, A.C. (1976) Remains attributable to *Australopithecus* in the East Rudolf succession. In: *Earliest Man and Environments in the Lake Rudolf Basin: Stratigraphy, Paleoecology and Evolution*, Coppens, Y., Howell, F.C., Isaac, G.Ll., and Leakey, R.E.F., eds, pp. 484–9. University of Chicago Press, Chicago, IL.

Walker, A.C. (1981) The Koobi Fora hominids and their bearing on the origin of the genus *Homo*. In: Homo erectus: *Papers in honor of Davidson Black*, Sigmon, B.A., and Cybulski, J.S., eds, pp. 193–215. University of Toronto, Toronto.

Walker, A.C. (1984) Extinction in hominid evolution. In: *Extinctions*, Nitecki, M.H., ed., pp. 119–52. University of Chicago Press, Chicago, IL.

Walker, A.C., and Leakey, R.E.F. (1988) The evolution of *Australopithecus boisei*. In: *Evolutionary History of the "Robust" Australopithecines*, Grine, F.E., ed., pp. 247–58. Aldine de Gruyter, New York.

Walker, A.C., Hoeck, H., and Perez, L.M. (1978) Microwear of mammalian teeth as in indicator of diet. *Science* 201: 908–10.

Walker, A.C., Zimmerman, M.R., and Leakey, R.E.F. (1982) A possible case of hypervitaminosis A in *Homo erectus*. *Nature* 296: 248–50.

Walker, A.C., Leakey, R.E.F., Harris, J.M., and Brown, F.H. (1986) 2.5 Myr *Australopithecus boisei* from west of Lake Turkana, Kenya. *Nature* 322: 517–22.

Walker, M., Gibert, J., López, M., Lombardi, A., Pérez-Pérez, A., Zapata, J., et al. (2008) Late Neandertals in Southeastern Iberia: Sima de las Palomas del Cabezo Gordo, Murcia, Spain. *Proceedings of the National Academy of Sciences USA* 105(52): 20631–6.

Walker, M., Lombardi, A., Zapata, J., and Trinkaus, E. (2010) Neandertal mandibles from the Sima de las Palomas del Cebezo Gordo, Murcia, southeastern Spain. *American Journal of Physical Anthropology* 142(2): 261–72.

Wallace, J.A. (1973) Tooth chipping in the australopithecines. *Nature* 244: 117–18.

Wallace, J.A. (1975) Did La Ferrassie I use his teeth as a tool? *Current Anthropology* 16(3): 393–401.

Walter, R.C. (1994) Age of Lucy and the First Family: Single-crystal $^{40}Ar/^{39}Ar$ dating of the Denen Dora and lower Kada Hadar members of the Hadar Formation, Ethiopia. *Geology* 22(1): 6–10.

Walter, R.C., and Aronson, J.L. (1993) Age and source of the Sidi Hakoma Tuff, Hadar Formation, Ethiopia. *Journal of Human Evolution* 25: 229–40.

Walter, R.C., Buffler, R.T., Bruggemann, J.H., Guillaume, M.M.M., Berhe, S.M., Negassi, B., et al. (2000) Early human occupation of the Red Sea coast of Eritrea during the last interglacial. *Nature* 405: 65–9.

Wang, Q., Tobias, P., Roberts, D., and Jacobs, Z. (2008) A re-examination of a human femur found at the Blind River Site, East London, South Africa: Its age, morphology, and breakage pattern. *Anthropological Review* 71(1): 43–61.

Wang, Q., Ashley, D.W., and Dechow, P.C. (2010) Regional, ontogenetic, and sex-related variations in elastic properties of cortical bone in baboon mandibles. *American Journal of Physical Anthropolgy* 141: 526–49.

Wang, W. (2009) New discoveries of *Gigantopithecus blacki* teeth from Chuifeng Cave in the Bubing Basin, Guangxi, south China. *Journal of Human Evolution* 57: 229–40.

Wang, W., Liu, J., Yamei, H., Si, X.Q., Weiwen, H., Schepartz, L.A., et al. (2004) Panxian Dadong, South China: Establishing a record of Middle Pleistocene climatic changes. *Asian Perspectives* 43: 302–13.

Wang, W., Potts, R., Yamei, H., Chen, Y.F., Wu, H.Y., Baoyin, Y., et al. (2005) Early Pleistocene hominid teeth recovered in Mohui Cave in Bubing Basin, Guangxi, South China. *Chinese Science Bulletin* 50: 2777–82.

Wang, W., Potts, R., Baoyin, Y., Weiwen, H., Chen, H., Edwards, L., et al. (2007a) Sequence of mammalian fossils, including hominoid teeth, from the Bubing Basin caves, South China. *Journal of Human Evolution* 52: 370–9.

Wang, W., Tian, F., and Mo, J.Y. (2007b) Recovery of *Gigantopithecus blacki* fossils from the Mohui Cave in the Bubing Basin, Guangxi, south China. *Acta Anthropologica Sinica* 26: 329–43.

Wang, W., Mo, J.Y., and Huang, Z.T. (2008) Recent discovery of handaxes associated with tektites in the Nanbanshan locality of the Damei site, Bose basin, Guangxi, South China. *Chinese Science Bulletin* 53: 878–83.

Wang, Y., Xue, X., Yue, L., Zhao, J., and Liu, S. (1979) Discovery of Dali fossil man and its preliminary study. *Kexue Tongbao* 24(7): 303–6.

Ward, C., Leakey, M., and Walker, A. (1999b) The new hominid species *Australopithecus anamensis*. *Evolutionary Anthropology* 7: 197–205.

Ward, C.V. (1993) Torso morphology and locomotion in *Proconsul nyanzae*. *American Journal of Physical Anthropology* 92: 291–328.

Ward, C.V., Leakey, M.G., Brown, B., Brown, F., Harris, J., and Walker, A. (1999a) South Turkwel: a new Pliocene hominid site in Kenya. *Journal of Human Evolution* 36: 69–95.

Ward, C.V., Leakey, M.G., and Walker, A. (2001) Morphology of *Australopithecus anamensis* from Kanapoi and Allia Bay, Kenya. *Journal of Human Evolution* 41(4): 255–368.

Ward, S., and Hill, A. (1987) Pliocene hominid partial mandible from Tabarin, Baringo, Kenya. *American Journal of Physical Anthropology* 72: 21–37.

Ward, S., Brown, B., Hill, A.P., Kelley, J., and Downs, W. (1999) *Equatorius*: a new hominoid genus from the Middle Miocene of Kenya. *Science* 285: 1382–6.

Washburn, S. (1961) *Social Life of Early Man*. Aldine, Chicago, IL.

Washburn, S.L. (1940) *Preliminary Metrical Study of the Skeleton of Langurs and Macaques*. PhD thesis, Harvard University.

Washburn, S.L. (1951) The New Physical Anthropology. *Transactions of the New York Academy of Sciences* 2(13): 258–304.

Washburn, S.L. (1960) Tools and human evolution. *Scientific American* 203: 62–75.

Washburn, S.L. (1963) *Classification and Human Evolution*. Aldine Publishing Company, Chicago, IL.

Washburn, S.L. (1968) Speculation on the problem of man's coming to the ground. In: *Changing Perspectives on Man*, Rothblatt, B., ed., pp. 191–206. University of Chicago Press, Chicago, IL.

Watanabc, H., Fujiyama, A., Hattori, M., Taylor, T D., Toyoda, A., Kurioki, Y., et al. (2004) DNA sequence and comparitive analysis of chimpanzee chromosome 22. *Nature* **429**: 382–8.

Watanabe, N., and Kadar, D., eds (1985) *Quaternary Geology of the Hominid Bearing Formations in Java: Report of the Indonesian-Japan Joint Research Project. CTA-41, 1976–1979.* Special Bulletins. Geological Research and Development Centre, Bandug, Indonesia.

Waterman, H.C. (1929) Studies on the evolution of the pelvis of man and other primates. *Bulletin of the American Museum of Natural History* **58**: 585–641.

Watkins, W., Ricker, C., Bamshad, M., Carroll, M., Nguyen, S., Batzer, M., et al. (2001) Patterns of ancestral human diversity: an analysis of Alu-insertion and restriction-site polymorphisms. *American Journal of Human Genetics* **68**(3): 738–52.

Watts, I. (2002) Ochre in the Middle Stone Age of Southern Africa: ritualised display or hide preservative? *The South African Archaeological Bulletin* **57**(175): 1–14.

Watts, I. (2010) The pigments from Pinnacle Point Cave 13B, Western Cape, South Africa. *Journal of Human Evolution* **59**: 392–411.

Wayland, E., and Burkitt, M. (1932) The Magosian Culture of Uganda. *Journal of the Anthropological Institute of Great Britain and Ireland*: 369–90.

Weaver, T.D., Roseman, C.C., and Stringer, C.B. (2007) Were Neandertal and modern human cranial differences produced by natural selection or genetic drift? *Journal of Human Evolution* **53**: 135–45.

Webb, S. (1990) Cranial thickening in an Australian hominid as a possible palaeoepidemiological indicator. *American Journal of Physical Anthropology* **82**: 403–11.

Webb, S., Cupper, M.L., and Robins, R. (2006) Pleistocene human footprints from the Willandra Lakes, southeastern Australia. *Journal of Human Evolution* **50**(4): 405–13.

Weeks, R.A., Bogard, J.S., Elam, J.M., Weinand, D.C., and Kramer, A. (2003) Effects of thermal annealing on the radiation produced electron paramagentic resonance spectra of bovine and equine tooth enamel: fossil and modern. *Journal of Applied Physics* **93**(12): 74–6.

Wei, Q. (1988) The stratigraphic, geochronological, and biostratigraphic background of the earliest known Paleolithic sites in China. *L'Anthropologie* **92**: 931–8.

Weidenreich, F. (1927) *Rasse und Körperbau (Race and Bodily Structure).* Julius, Berlin.

Weidenreich, F., ed. (1928) *Der Schädelfund von Weimar-Ehringsdorf,* Jena.

Weidenreich, F. (1935) *Sinanthropus* population of Choukoutien (Locality1) with a preliminary report on new discoveries. *Bulletin of the Geological Society of China* **XIV**: 427–68.

Weidenreich, F. (1936a) The mandibles of *Sinanthropus pekinensis*: a comparative study. *Palaeontologia Sinica, New Series D* **72**: 1–162.

Weidenreich, F. (1936b) The new discoveries of *Sinanthropus pekinensis* and their bearing on the *Sinanthropus* and *Pithecanthropus* problems. *Bulletin of the Geological Society of China* **16**: 439–66.

Weidenreich, F. (1937) The dentition of *Sinanthropus pekinensis*: A comparative odontography of the hominids. *Palaeontologia Sinica* **1**: 1–180.

Weidenreich, F. (1939a) The duration of life of fossil man in China and the pathological lesions found in his skeleton. *Chinese Medical Journal* **45**: 33–44.

Weidenreich, F. (1939b) On the earliest representatives of modern mankind recovered on the soil of East Asia. *Bulletin of the Natural Historical Society of Peking* **13**: 161–74.

Weidenreich, F. (1939c) Six lectures on *Sinanthropus pekinensis* and related problems. *Bulletin of the Geological Society of China* **19**: 1–110.

Weidenreich, F. (1940) Some problems dealing with ancient man. *American Anthropologist* **42**(3): 375–83.

Weidenreich, F. (1941a) The brain and its role in the phylogenetic transformation of the human skull. *Transactions of the American Philosophical Society* **31**(5): 320–442.

Weidenreich, F. (1941b) The extremity bones of *Sinanthropus pekinensis*. *Palaeontologia Sinica* **5**: 1–150.

Weidenreich, F. (1942) Early Man in Indonesia. *The Far Eastern Quarterly*: 58–65.

Weidenreich, F. (1943) The skull of *Sinanthropus pekinensis*: a comparative study of a primitive hominid skull. *Palaeontologia Sinica* **10**: 1–291.

Weidenreich, F. (1944) Giant Early Man from Java and South China. *Science* **99**: 479–82.

Weidenreich, F. (1945) The puzzle of *Pithecanthropus*. *Science and Scientists in the Netherland Indies* **1**: 380–90.

Weidenreich, F. (1946) *Apes, Giants and Man.* University of Chicago Press, Chicago, IL.

Weidenreich, F. (1951) Morphology of Solo Man. *Anthropology Papers of the American Museum of Natural History* **43**(Part 3): 205–90.

Weinand, D.C. (2005) *A Reevaluation of the Paleoenvironmental Reconstructions Associated with* Homo erectus *from Java, Indonesia, based on the Functional Morphology of Fossil Bovid Astragali.* PhD thesis, University of Tennessee.

Weiner, J.S. (1955) *The Piltdown Forgery.* Oxford University Press, Oxford.

Weiner, J.S., and Campbell, B.G. (1964) The taxonomic status of the Swanscombe skull. *Occasional Paper/ Royal Anthropological Institute of Great Britain & Ireland* **20**: 175–209.

Weiner, J.S., and Stringer, C. (2003) *The Piltdown Forgery: Fiftieth Anniversary Edition.* Oxford University Press, Oxford.

Weiner, J.S., Oakley, K.P., and Le Gros Clark, W.E. (1953) The solution of the Piltdown problem. *Bulletin of the British Museum (Natural History), Geology* **2**(3): 139–46.

Weiner, S., Xu, Q., Goldberg, P., Liu, J., and Bar-Yosef, O. (1998) Evidence for the use of fire at Zhoukoudian, China. *Science* **281**(5374): 251–3.

Weiner, S., Goldberg, P., and Bar-Yosef, O. (2002) Three-dimensional distribution of minerals in the sediments of Hayonim Cave, Israel: diagenetic processes and archaeological implications. *Journal of Archaeological Science* **29**(11): 1289–1308.

Weinert, H. (1925) *Der Schädel des eiszeitlicher Menschen von Le Moustier in neuer Zusammensetzung*. Springer, Berlin.

Weinert, H. (1931) Der "Sinanthropus pekinensis" als Bestatigung des *Pithecanthropus erectus*. *Zeitschrift für Morphologie und Anthropologie* **9**: 159–87.

Weinert, H. (1936) Der Urmenschenschädel von Steinheim. *Zeitschrift für Morphologie und Anthropologie* **35**: 463–518.

Weinert, H. (1939) *Africanthropus njarasensis*, Beschreibung und phyletische Einordung des ersten Affenmenschen aus Ostafrika. *Zeitschrift für Morphologie und Anthropologie* **38**: 252–307.

Weinert, H. (1950) Uber die neuen Vor- und Fruhmenschenfunde aus Afrika, Java, China und Frankreich. *Zeitschrift für Morphologie und Anthropologie* **42**: 113–48.

Weinstein-Evron, M. (1991) New radiocarbon dates for the Early Natufian of El-Wad Cave, Mt. Carmel, Israel. *Paléorient* **17**(1): 95–8.

Weiss, E., Kislev, M., Simchoni, O., Nadel, D., and Tschauner, H. (2008) Plant-food preparation area on an Upper Paleolithic brush hut floor at Ohalo II, Israel. *Journal of Archaeological Science* **35**(8): 2400–14.

Wells, L.H. (1935) A fossilized human femur from East London, C.P. *South African Journal of Science* **32**: 596–600.

Wells, L.H. (1951) The fossil human skull from Singa. In: *Fossil Mammals of Africa*, pp. 29–42. British Museum Natural History, London.

Wells, L.H. (1959) The problem of Middle Stone Age Man in South Africa. *Man* **59**: 158–60.

Wendorf, F. (1968a) Site 177 A Nubian palaeolithic graveyard near Jebel Sahaba, Sudan. In: *The Prehistory of Nubia*, Wendorf, F., ed. pp. 954–995. Ft Burgwin Research Center and Southern Methodist University Press, Dallas, TX.

Wendorf, F. (1968b) Late Paleolithic sites in Egyptian Nubia. In: *The Prehistory of Nubia*, Wendorf, F., ed., pp. 753–953. Ft Burgwin Research Center and Southern Methodist University Press, Dallas, TX.

Wendorf, F., and Schild, R. (1974) *A Middle Stone Age Sequence from the Central Rift Valley, Ethiopia*. Institute of the History of Material Culture, Polish Academy of Sciences, Warsaw.

Wendorf, F., and Schild, R. (1986) *The Wadi Kubbaniya Skeleton: A Late Paleolithic Burial from Southern Egypt*. Southern Methodist University Press, Dallas, TX.

Wendorf, F., and Schild, R. (1992) The Middle Paleolithic of Northeast Africa: A status report. In: *New Light on the Northeast African Past*, Klees, F., and Kuper, R., eds, pp. 39–78. Heinrich-Barth-Institut, Koln.

Wendorf, F., Schild, R., and Close, A. (1989) *The Prehistory of Wadi Kubbaniya*. Southern Methodist University Press, Dallas, TX.

Wendt, W.E. (1976) 'Art Mobilier' from the Apollo 11 Cave, South West Africa: Africa's Oldest Dated Works of Art. *The South African Archaeological Bulletin* **31**(121–2): 5–11.

Wengler, L. (1997) La transition du Moustérien à L'Atérien. *L'Anthropologie* **101**(3): 448–81.

Wengler, L. (2001) Settlements "during the Middle Paleolithic of the Maghreb. In: *Settlement dynamics of the Middle Paleolithic and Middle Stone Age*, Conard, E., ed., pp. 65–89. Kerns Verlag, Tübingen.

Westaway, K.E., Sutikna, T., Saptomo, W.E., Jatmiko, Morwood, M.J., Roberts, R.G., et al. (2009) Reconstructing the geomorphic history of Liang Bua, Flores, Indonesia: a stratigraphic interpretation of the occupational environment. *Journal of Human Evolution* **57**(5): 465–83.

Western, D., and Behrensmeyer, A. (2009) Bone assemblages track animal community structure over 40 years in an African savanna ecosystem. *Science* **324**(5930): 1061–4.

Westoll, T.S. (1956) The nature of fossil species. In: *The Species Concept in Palaeontology*, Sylvester-Bradley, P.C., ed., pp. 53–62. Systematics Association Publishing, London.

Whitcome, K., Shapiro, L., and Lieberman, D. (2007) Fetal load and the evolution of lumbar lordosis in bipedal hominins. *Nature* **450**(7172): 1075–8.

White, T.D., and Falk, D. (1999) A quantitative and qualitative reanalysis of the endocast from the juvenile *Paranthropus* specimen L338y-6 from Omo, Ethiopia. *American Journal of Physical Anthropology* **110**: 399–406.

White, R. (2001) Personal ornaments from the Grotte du Renne at Arcy-sur-Cure. *Athena Reviews* **2**: 41–6.

White, S., Fisher, S., Geschwind, D., Scharff, C., and Holy, T. (2006a) Singing mice, songbirds, and more: models for FOXP2 function and dysfunction in human speech and language. *Journal of Neuroscience* **26**(41): 10376–9.

White, T., WoldeGabriel, G., Asfaw, B., Ambrose, S., Beyene, Y., Bernor, R., et al. (2006b) Asa Issie, Aramis and the origin of *Australopithecus*. *Nature* **440**(7086): 883–9.

White, T.D. (1977a) New fossil hominids from Laetolil, Tanzania. *American Journal of Physical Anthropology* **46**(2): 197–230.

White, T.D. (1977b) *The Anterior Mandibular Corpus of Early African Hominidae: Functional Significance of Shape and Size*. PhD thesis, University of Michigan.

White, T.D. (1980) Additional fossil hominids from Laetoli, Tanzania: 1976–1979 specimens. *American Journal of Physical Anthropology* **53**(4): 487–504.

White, T.D. (1985) Acheulian man in Ethiopia's Middle Awash Valley: the implications of cutmarks on the Bodo cranium. In: *Achtste Kroon-Voordracht*. Joh. Enschede En Zonen, Haarlem.

White, T.D. (1986a) Cut marks on the Bodo cranium: a case of prehistoric defleshing. *American Journal of Physical Anthropology* **69**: 503–9.

White, T.D. (1986b) *Australopithecus afarensis* and the Lothagam mandible. *Anthropos* **23**: 79–90.

White, T.D. (1987) Cannibals at Klasies? *Sagittarius* **2**: 6–9.

White, T.D. (1995) African omnivores: global climatic change and Plio-Pleistocene hominids and suids. In: *Paleoclimate and Evolution, with Special Emphasis on Human Origins*, Vrba, E.S., Denton, G.H., Partridge, T.C., and Burckle, L.H., eds, pp. 369–84. Yale University Press, New Haven, CT.

White, T.D. (2002) Earliest hominids. In: *The Primate Fossil Record*, Hartwig, W., ed., pp. 407–17. Cambridge University Press, Cambridge.

White, T.D. (2003) Early hominids – diversity or distortion? *Science* **299**(5615): 1994–7.

White, TD., Asfaw, B., DeGusta, D., Gilbert, H., Richards, G.D., Suwa, G., et al. (2003) Pleistocene *Homo sapiens* from Middle Awash, Ethiopia. *Nature* **423**: 742–7.

White, T.D., and Johanson, D.C. (1982) Pliocene hominid mandibles from the Hadar Formation, Ethiopia: 1974–1977 Collections. *American Journal of Physical Anthropology* **57**: 501–44.

White, T.D., and Suwa, G. (1987) Hominid footprints at Laetoli: facts and interpretations. *American Journal of Physical Anthropology* **72**: 485–514.

White, T.D., and Toth, N. (1991) The question of ritual cannibalism at Grotta Guattari. *Current Anthropology* **32**: 118–38.

White, T.D., and Suwa, G. (2004) A new species of *Notochoerus* (Artiodactyla, Suidae) from the Pliocene of Ethiopia. *Journal of Vertebrate Paleontology* **24**(2): 474–80.

White, T.D., Johanson, D.C., and Kimbel, W.H. (1981) *Australopithecus africanus*: its phyletic position reconsidered. *South African Journal of Science* **77**: 445–70.

White, T.D., Suwa, G., Hart, W.K., Walter, R., WoldeGabriel, G., de Heinzelin, J., et al. (1993) New discoveries of *Australopithecus* at Maka in Ethiopia. *Nature* **366**: 261–5.

White, T.D., Suwa, G., and Asfaw, B. (1994) *Australopithecus ramidus*, a new species of early hominid from Aramis, Ethiopia. *Nature* **371**: 306–12.

White, T.D., Suwa, G., and Asfaw, B. (1995) *Australopithecus ramidus*, a new species of early hominid from Aramis, Ethiopia – a corrigendum. *Nature* **375**: 88.

White, T.D., Suwa, G., Simpson, S., and Asfaw, B. (2000) Jaws and teeth of *Australopithecus afarensis* from Maka, Middle Awash, Ethiopia. *American Journal of Physical Anthropology* **111**: 45–68.

White, T.D., Asfaw, B., Beyene, Y., Haile-Selassie, Y., Lovejoy, C.O., Suwa, G., et al. (2009) *Ardipithecus ramidus* and the paleobiology of early hominids. *Science* **326**(5949): 64, 75–86.

White, T.D., WoldeGabriel, G., Asfaw, B., Ambrose, S., Beyene, Y., Bernor, R., et al. (2006b) Asa Issie, Aramis and the origin of *Australopithecus*. *Nature* **440** (7086): 883–9.

Whitehead, H. (2007) Learning, climate and the evolution of cultural capacity. *Journal of Theoretical Biology* **245**: 341–50.

Whiten, A., Goodall, J., McGrew, W.C., Nishida, T., Reynolds, V., Sugiyama, Y., et al. (1999) Culture in chimpanzees. *Nature* **399**: 682–5.

Whiten, A., Horner, V., and Marshall-Pescini, S. (2003) Cultural panthropology. *Evolutionary Anthropology* **12**: 92–105.

Whittaker, J.C. (1994) *Flintknapping: Making and Understanding Stone Tools*. University of Texas Press, Austin, TX.

Whitworth, T. (1960) Fossilized human remains from Northern Kenya. *Nature* **185**: 947–8.

Whitworth, T. (1965a) The Pleistocene Lake Beds of Kabua, Northern Kenya. *Durham University Journal* **57**: 88–100.

Whitworth, T. (1965b) Artifacts from Turkana, Northern Kenya. *South African Archaeological Bulletin* **20**: 75–8.

Whitworth, T. (1966) A fossil hominid from Rudolf. *South African Archaeological Bulletin* **21**: 138–50.

Widianto, H., and Grimaud-Hervé, D. (1993) Le crâne de Ngawi. *Les Dossiers d'Archéologie* **184**(36).

Widianto, H., and Zeitoun, V. (2003) Morphological description, biometry and phylogenetic position of the skull of Ngawi 1 (east Java, Indonesia). *International Journal of Osteoarchaeology* **13**(6): 339–51.

Widianto, H., Sémah, A.-M., Djubiantono, T., and Sémah, F. (1994) A tentative reconstruction of the cranial human remains of Hanoman 1 from Bukuran, Sangiran (Central Java). *Courier Forsch Senckenberg* **171**: 47–59.

Wild, E.M., Teschler-Nicola, M., Kutschera, W., Steier, P., Trinkaus, E., and Wanek, W. (2005) Direct dating of Early Upper Palaeolithic human remains from Mladec. *Nature* **435**: 332–5.

Wild, E.M., Teschler-Nicola, M., Kutschera, W., Steier, P., and Wanek, W. (2006) [14]C Dating of Early Upper Palaeolithic Human and Faunal Remains from Mladec. In: *Early Modern Humans at the Moravian Gate*, Teschler-Nicola, M., ed., pp. 149–58. Springer Vienna, Vienna.

Wildman, D.E., Bergman, T.J., al-Aghbari, A., Sterner, K.N., Newman, T.K., Phillips-Conroy, J.E., et al. (2004) Mitochondrial evidence for the origin of hamadryas baboons. *Molecular Phylogenetics and Evolution* **32**(1): 287–96.

Williams, E.M., Gordon, A.D., and Richmond, B.G. (2010) Upper limb kinematics and the role of the wrist during stone tool production. *American Journal of Physical Anthropology* **143**: 134–45.

Williams, G.C. (1957) Selection and the evolution of senescence. *Evolution* **11**: 398–411.

Williams, M., Ambrose, S.H., van der Kaars, S., Ruehlemann, C., Chattopadhyaya, U.C., Pal, J.N., et al. (2009) Environmental impact of the 73 ka Toba super-eruption in South Asia. *Palaeogeography, Palaeoclimatology, Palaeoecology* **284**: 295–314.

Wilson, A., Cann, R.L., Carr, E.M., George, M., Gyllensten, U.B., Helm-Bychowski, K.M., et al. (1985) Mitochrondrial DNA and two perspectives on evolutionary genetics. *Biological Journal of the Linnean Society* **26**: 375–400.

Wilson, D., and Shroff, F. (1970) The nature of the striae of Retzius as seen with the optical microscope. *Australian Dental Journal* **15**(3): 162–71.

Wilson, G.M., Flibotte, S., Missirlis, P.I., Marra, M.A., Jones, S., Thornton, K., et al. (2006) Identification by full-coverage array CGH of human DNA copy number increases relative to chimpanzee and gorilla. *Genome Research* **16**: 173–81.

Wilson, J. (1975) The last glacial environment at the Abri Pataud: A possible comparison. In: *Excavation of the Abri Pataud, Les Eyzies (Dordogne)*, Movius, H.L., ed., pp. 175–86. Peabody Museum, Cambridge, MA.

Wilson, T. (1892) Importance of the Science and of the Department of Prehistoric Anthropology. *American Naturalist* **26**(308): 681–9.

Wimmer, H., and Perner, J. (1983) Beliefs about beliefs: representation and constraining function of wrong beliefs in young children's understanding of deception. *Cognition* **13**(1): 103–28.

Winterhalder, B., and Alden Smith, E. (2000) Analyzing adaptive strategies: Human behavioral ecology at twenty-five. *Evolutionary Anthropology: Issues, News, and Reviews* **9**(2): 51–72.

Wintle, A., and Aitken, M. (1977) Thermoluminescence dating of burnt flint: application to a Lower Palaeolithic site, Terra Amata. *Archaeometry* **19**(2): 111–30.

Witzel, C., Kierdorf, U., Schultz, M., and Kierdorf, H. (2008) Insights from the inside: Histological analysis of abnormal enamel microstructure associated with hypoplastic enamel defects in human teeth. *American Journal of Physical Anthropology* **136**(4): 400–14.

WoldeGabriel, G., Walter, R.C., Aronson, J.L., and Hart, W.K. (1992) Geochronology and distribution of silicic volcanic rocks of Plio-Pleistocene age from the central sector of the Main Ethiopian Rift. *Quaternary International* **13–14**: 69–76.

WoldeGabriel, G., White, T., Suwa, G., Renne, P., de Heinzelin, J., Hart, W., et al. (1994) Ecological and temporal placement of early Pliocene hominids at Aramis, Ethiopia. *Nature* **371**(6495): 330.

WoldeGabriel, G., Heiken, G., White, T.D., Asfaw, B., Hart, W.K., and Renne, P. (2000) Volcanism, tectonism, sedimentation, and the paleoanthropological record in the Ethiopian Rift System. In: *Volcanic Hazards and Disasters in Human Antiquity*, McCoy, F.W., and Heiken, G., eds, pp. 83–99. Special Paper. Geological Society of America, Boulder, CO.

WoldeGabriel, G., Haile-Selassie, Y., Renne, P.R., Hart, W.K., Ambrose, S., Asfaw, B., et al. (2001) Geology and paleontology of the Late Miocene Middle Awash valley, Afar rift, Ethiopia. *Nature* **412**: 175–8.

WoldeGabriel, G., Renne, P., Hart, W., Ambrose, S., Asfaw, B., and White, T. (2004) Geoscience methods lead to paleo-anthropological discoveries in Afar Rift, Ethiopia. *EOS, Transactions of the American Geophysical Union.* **85**: 273–277.

WoldeGabriel, G., Hart, W., Katoh, S., Beyene, Y., and Suwa, G. (2005) Correlation of Plio-Pleistocene tephra in Ethiopian and Kenyan rift basins: Temporal calibration of geological features and hominid fossil records. *Journal of Volcanology and Geothermal Research* **147**(1–2): 81–108.

WoldeGabriel, G., Ambrose, S., Barboni, D., Bonnefille, R., Bremond, L., Currie, B., et al. (2009) The geological, isotopic, botanical, invertebrate, and lower vertebrate surroundings of *Ardipithecus ramidus*. *Science* **326**(5949): 65.

Wolpoff, M. (1971a) *Metric Trends in Hominid Dental Evolution*. PhD Thesis, University of Illinois.

Wolpoff, M. (1971b) Competitive exclusion among lower Pleistocene hominids: the single species hypothesis. *Man* **6**(4): 601–14.

Wolpoff, M. (1989) Multiregional evolution: the fossil alternative to Eden. In: *The Human Revolution*, Mellars, P., and Stringer, C., eds, pp. 62–108. Edinburgh University Press, Edinburgh.

Wolpoff, M.H. (1968) "Telanthropus" and the single species hypothesis. *American Anthropologist* **70**(3): 477–93.

Wolpoff, M.H. (1978a) Some aspects of canine size in the australopithecines. *Journal of Human Evolution* **7**: 115–26.

Wolpoff, M.H. (1978b) The Australopithecines: A Stage in Human Evolution. In: *Krapinski Pračovjek I Evolucija Hominida*, Malez, M., ed., pp. 269–91. Yugoslav Academy of Sciences and Arts, Zagreb.

Wolpoff, M.H. (1979) The Krapina dental remains. *American Journal of Physical Anthropology* **50**(1): 67–113.

Wolpoff, M.H. (1980) *Paleoanthropology*. Alfred E. Knopf, New York.

Wolpoff, M.H. (1981) Allez Neanderthal. *Nature* **289**: 823.

Wolpoff, M.H. (1988) Divergence between hominid lineages: the roles of competition and culture. In: *Evolutionary History of the "Robust" Australopithecines*, Grine, F.E., ed., pp. 485–97. Aldine de Gruyter, New York.

Wolpoff, M.H. (1989) The place of the Neandertals in human evolution. In: *The Emergence of modern humans: biocultural adaptations in the later Pleistocene*, Trinkaus, E., ed., pp. 97–141. Cambridge Univerity Press, Cambridge.

Wolpoff, M.H. (1995) Wright for the wrong reasons. *Journal of Human Evolution* **29**: 185–8.

Wolpoff, M.H., and Caspari, R. (1996) An unparalleled parallelism. *Anthropologie* **34**(3): 215–23.

Wolpoff, M.H., and Caspari, R. (1997) *Race and Human Evolution*. Simon & Schuster, New York.

Wolpoff, M.H., and Frayer, D.W. (2005) Unique ramus anatomy for Neandertals? *American Journal of Physical Anthropology* **128**(2): 245–51.

Wolpoff, M.H., Smith, F.H., Malez, M., Radovčić, J., and Rukavina, D. (1981) Upper Pleistocene human remains from Vindija Cave, Croatia, Yugoslavia. *American Journal of Physical Anthropology* **54**: 499–545.

Wolpoff, M.H., Wu, X.Z., and Thorne, A.G. (1984) Modern *Homo sapiens* origins: a general theory of hominid evolution involving the fossil evidence from East Asia. In: *The Origin of Modern Humans: A World Survey of the Fossil Evidence*, Smith, F.H., and Spencer, F., eds, pp. 411–83. Alan R. Liss, New York.

Wolpoff, M.H., Thorne, A.G., Jelinek, J., and Yinyun, Z. (1994) The case for sinking *Homo erectus*: 100 years of *Pithecanthropus* is enough! *Courier Forschungs-Institut Senckenberg* **171**: 341–61.

Wolpoff, M.H., Frayer, D., and Jelinek, J. (2006) Aurignacian Female Crania and Teeth from the Mladec Caves, Moravia, Czech Republic. In: *Early Modern Humans at the Moravian Gate*, Teschler-Nicola, M., ed., pp. 273–340. Springer Vienna, Vienna.

Wolpoff, W.H. (1992) Theories of modern human origins. In: *Continuity or replacement: controversies in* Homo sapiens *evolution*, Bräuer, G., and Smith, F.H., eds, pp. 25–63. A.A. Balkema, Rotterdam.

Woltereck, R. (1909) Weiterer experimentelle Untersuchungen uber Art veranderung, Speziell uber das Wesen Quantitativer Artunterschiede bei Daphniden. *Verhandlungen des Deutschen Zoologischen Gesellschaft* **19**: 110–92.

Woo, J.-K. (1964) Mandible of *Sinanthropus lantianenesis*. *Current Anthropology* **5**(2): 98–101.

Woo, J.-K., and Peng, R.-C. (1959) Fossil human skull of early Paleoanthropic stage found at Mapa, Shaoguan, Karatung Province. *Vertebrata Paleontologica Asiatica* **3**(4): 176–82.

Wood, B., and Hill, K. (2000) A test of the "showing-off" hypothesis with Aché hunters. *Current Anthropology* **41**: 124–5.

Wood, B., and Lonergan, N. (2008) The hominin fossil record: taxa, grades and clades. *Journal of Anatomy* **212**(4): 354–76.

Wood, B.A. (1974) A *Homo* talus from East Rudolf, Kenya. *Journal of Anatomy* **117**: 203–4.

Wood, B.A. (1976) Remains attributable to *Homo* in the East Rudolf succession. In: *Earliest Man and Environments in Lake Rudolf Basin*, Coppens, Y., Howell, F.C., Issac, G.L., and Leakey, R.E.F., eds, pp. 490–506. University of Chicago Press, Chicago, IL.

Wood, B.A. (1991) *Koobi Fora Research Project*. Clarendon Press, Oxford.

Wood, B.A. (1992) Origin and evolution of the genus *Homo*. *Nature* **355**: 783–90.

Wood, B.A. (1994) Taxonomy and evolutionary relationships of *Homo erectus*. *Courier Forsch Senckenberg* **171**: 159–65.

Wood, B.A. (1999) Plio-Pleistocene hominins from the Baringo Region, Kenya. In: *Late Cenozoic Environments and Human Evolution: A Tribute to Bill Bishop*, Andrews, P., and Banham, P., eds, pp. 113–22. Geological Society, London.

Wood, B.A. (1999) Homoplasy: foe and friend? *Evolutionary Anthropology* **8**(3): 79–80.

Wood, B.A., and Abbott, S. (1983) Analysis of the dental morphology of Plio-Pleistocene hominids. I. Mandibular molars: crown area measurements and morphological traits. *Journal of Anatomy* **136**: 197–219.

Wood, B.A., and van Noten, F.L. (1986) Preliminary observations on the BK 8518 mandible from Baringo, Kenya. *American Journal of Physical Anthropology* **69**: 117–27.

Wood, B.A., and Chamberlain, A.T. (1987) The nature and affinities of the "robust" australopithecines: a review. *Journal of Human Evolution* **16**(7–8): 625–41.

Wood, B.A., and Uytterschaut, B.A. (1987) Analysis of the dental morphology of Plio-Pleistocene hominids. III. Mandibular premolar crowns. *Journal of Anatomy* **154**: 121–56.

Wood, B.A., and Engleman, C.A. (1988) Analysis of the dental morphology of Plio-Pleistocene hominids. V. Maxillary postcanine tooth morphology. *Journal of Anatomy* **161**: 1–35.

Wood, B.A., and Aiello, L.C. (1998) Taxonomic and functional implications of mandibular scaling in early hominins. *American Journal of Physical Anthropology* **105**: 523–38.

Wood, B.A., and Collard, M. (1999) The changing face of genus *Homo*. *Evolutionary Anthropology* **8**(6): 195–207.

Wood, B.A., and Strait, D. (2004) Patterns of resource use in early *Homo* and *Paranthropus*. *Journal of Human Evolution* **46**(2): 119–62.

Wood, B.A., and Constantino, P. (2007) *Paranthropus boisei*: Fifty years of evidence and analysis. *Yearbook of Physical Anthropology* **50**: 106–32.

Wood, B.A., Wood, C.W., and Konigsberg, L.W. (1994) *Paranthropus boisei* – an example of evolutionary stasis? *American Journal of Physical Anthropology* **95**: 117–36.

Wood, B.A., Aiello, A., Wood, C., and Key, C. (1998) A technique for establishing the identity of 'isolated' fossil hominin limb bones. *Journal of Anatomy* **193**: 61–72.

Wood, D.S. (1983) Character transformations in phenetic studies using continuous morphometric variables. *Systematic Zoology* **32**: 125–31.

Wood, J.W., Holman, D.J., Weiss, K.M., Buchanan, A.V., and Lefor, B. (1992) Hazards models for human population biology. *Yearbook of Physical Anthropology* **35**: 43–87.

Wood, J.W., Milner, G.R., Harpending, H.C., Weiss, K.M., Cohen, M.N., Eisenberg, L.E., et al. (1992) The osteological paradox: problems of inferring prehistoric health from skeletal samples [and comments and reply]. *Current Anthropology* **33**(4): 343–70.

Woodward, A.S. (1921) A new cave man from Rhodesia, South Africa. *Nature* **108**: 371–2.

Woodward, A.S. (1933) Early man in East Africa. *Nature* **131**: 477–8.

Woodward, A.S. (1938) A fossil skull of an ancestral Bushman from the Anglo-Egyptian Sudan. *Antiquity* **12**: 193–5.

Wray, A., and Grace, G. (2007) The consequences of talking to strangers: Evolutionary corollaries of sociocultural influences on linguistic form. *Lingua* **117**(3): 543–78.

Wreschner, E. (1967) The Guela Caves – Mount Carmel. *Quaternaria* **9**: 69–89.

Wright, A.A. (2007) An experimental analysis of memory processing. *Journal of the Experimental Analysis of Behavior* **88**(3): 405–33.

Wright, R.V.S. (1995) The Zhoukoudian Upper Cave skull 101 and multiregionalism. *Journal of Human Evolution* **29**: 181–3.

Wright, S. (1932) The roles of mutation, inbreeding, crossbreeding and selection in evolution. *Proceedings of the 6th International Congress of Genetics* **1**: 356–66.

Wrinn, P.J., and Rink, W.J. (2003) ESR Dating of tooth enamel from Aterian Levels at Mughoret el 'Aliya (Tangier, Morocco). *Journal of Archaeological Science* **30**(1): 123–33.

Wu, M. (1980) Human fossils discovered at Xujiayao site in 1977. *Vertebrata PalAsiatica* **18**: 227–38.

Wu, R. (1986) Chinese human fossils and the origin of Mongoloid racial group. *Anthropos (Brno)* **23**: 151–5.

Wu, R., and Dong, X.R. (1982) Preliminary study of *Homo erectus* remains from Hexian, Anhui. *Acta Anthropologica Sinica* **1**: 2–13.

Wu, R., and Olsen, J.W., eds (1985) *Palaeoanthropology and Palaeolithic Archaeology in the People's Republic of China*. Academic Press, Orlando.

Wu, R.-k. (1988) The reconstruction of the fossil human skull from Jinniushan, Yinkou, Liaoning Province and its main features. *Acta Anthropologica Sinica* **7**: 97–101.

Wu, X. (1981) A well-preserved cranium of an archaic type of early *Homo sapiens* from Dali, China. *Scienta Sinica* **24**(4): 530–9.

Wu, X. (1999) Chinese human paleontological study in the 20th century and prospects. *Acta Anthropologica Sinica* **18**(3): 164.

Wu, X. (2004) On the origin of modern humans in China. *Quaternary International* **117**: 131–40.

Wu, X., and You, Y. (1979) A preliminary observation of the Dali man site. *Vertebrata PalAsiatica* **17**: 294–303.

Wu, X., and Wang, L. (1985) Chronology in Chinese paleoanthropology. In: *Paleoanthropology and Paleolithic Archaeology in the People's Republic of China*, Wu, R., and Olsen, J.W., eds, pp. 29–51. Academic Press, Orlando, FL.

Wu, X., and Poirier, F.E. (1995) *Human Evolution in China: A Metric Description of the Fossils and a Review of the Sites*. Oxford University Press, New York.

Wu, X., Huang, W., and Qi, G. (1999) *Paleolithic Sites in China*. Shanghai Scientific and Technological Education Publishing House, Shanghai.

Wu, X., Huang, W., and Qi, G. (2002) *Zhongguo Gu Renlei Yizhi (Paleolithic Sites in China)*. Shanghai Scientific and Technological Education Publishing House, Shanghai.

Wu, X., Schepartz, L.A., Falk, D., and Liu, W. (2006) Endocranial cast of Hexian *Homo erectus* from South China. *American Journal of Physical Anthropology* **130**(4): 445–54.

Wu, X., Schepartz, L., and Liu, W. (2010) A new *Homo erectus* (Zhoukoudian V) brain endocast from China. *Proceedings of the Royal Society of London, Series B* **277**: 337–4.

Wunderlich, R.E., and Jungers, W.L. (2009) Manual digital pressures during knuckle-walking in chimpanzees (*Pan troglodytes*). *American Journal of Physical Anthropology* **139**(3): 394–403.

Wunderly, J. (1943) The Keilor Fossil Skull: anatomical description. Memoirs of the National Museum of Victoria, Melbourne **13**: 57–70.

Wurz, S. (1999) The Howiesons Poort at Klasies River – an argument for symbolic behaviour. *South African Archaeological Bulletin* **54**(69): 38–50.

Wurz, S. (2002) Variability in the Middle Stone Age lithic sequence, 115,000–60,000 years ago at Klasies River, South Africa. *Journal of Archaeological Science* **29**(9): 1001–15.

Wurz, S. (2005) Exploring and quantifying technological differences between the MSA I, MSA II and Howiesons Poort at Klasies River. *Proceedings of the International Round Table Conference; From tools to symbolsfrom early hominids to modern humans, Johannesburg, South Africa*.

Wymer, B.O. (1955) The discovery of the right parietal bone at Swanscombe, Kent. *Man* **55**: 124.

Wynn, J., and Bedaso, Z. (2010) Is the Pliocene Ethiopian monsoon extinct? A comment on Aronson et al. (2008) *Journal of Human Evolution* **59**: 133–8.

Wynn, J., Roman, D., Alemseged, Z., Reed, D., Geraads, D., and Munro, S. (2008) Stratigraphy, depositional environments, and basin structure of the Hadar and Busidima Formations at Dikika, Ethiopia. *Geological Society of America Special Paper* **446**: 87–118.

Wynn, J.G., Alemseged, Z., Bobe, R., Geraads, D., Reed, D., and Roman, D.C. (2006) Geological and palaeontological context of a Pliocene juvenile hominin at Dikika, Ethiopia. *Nature* **443**: 332–6.

Wynn, T., and McGrew, W.C. (1989) An ape's view of the Oldowan. *Man* **24**(3): 383–98.

Xiao, J., Jin, C., and Zhu, Y. (2002) Age of the fossil Dali Man in north central China deduced from chronostratigraphy of the loess–paleosol sequence *Quaternary Science Reviews* 21(20–2): 2191–8.

Xie, G., and Bodin (2007) Les industries paléolithiques du bassin de Bose (Chine du Sud). *L'Anthropologie* 111(2): 182–206.

Xu, C., and Zhang, Y. (1989) Human fossils discovered at Chaoxian, Anhui, China. *Human Evolution* 4(1): 95–6.

Xu, C.H., Zhang, Y.Y., Chen, C.D., and Fang, D.S. (1984) Human occipital bone and mammalian fossils from Chaoxian, Anhui. *Acta Anthropologica Sinica* 3: 202–9.

Xu, C.H., Zhang, Y.Y., and Fang, D.S. (1986) Human fossil newly discovered at Choaxian, Anhui. *Acta Anthropologica Sinica* 5: 305–10.

Yakir, D. (1992) Variations in the natural abundance of oxygen-18 and deuterium in plant carbohydrates. *Plant, Cell and Environment* 15(9): 1005–20.

Yamazaki, N., Hase, K., Ogihara, N., and Hayamizu, N. (1996) Biomechanical analysis of the development of human bipedal walking by a neuro-musculo-skeletal model. *Folia Primatologica* 66: 253–71.

Yeakel, J., Bennett, N., Koch, P., and Dominy, N. (2007) The isotopic ecology of African mole rats informs hypotheses on the evolution of human diet. *Proceedings of the Royal Society B: Biological Sciences* 274(1619): 1723–30.

Yeh, S.-W., Kug, J.-S., Dewitte, B., Kwon, M.-H., Kirtman, B.P., and Jin, F.-F. (2009) El Niño in a changing climate. *Nature* 461: 511–14.

Yellen, J., Brooks, A., Helgren, D., Tappen, M., Ambrose, S., Bonnefille, R., et al. (2005) The archaeology of Aduma Middle Stone Age Sites in the Awash Valley, Ethiopia. *PaleoAnthropology* 10: 25–100.

Yellen, J.E. (1996) Behavioural and taphonomic patterning at Katanda 9: a Middle Stone Age Site, Kivu Province, Zaire. *Journal of Archaeological Science* 23: 915–32.

Yellen, J.E. (1998) Barbed bone points: tradition and continuity in Saharan and Sub-Saharan Africa. *African Archaeological Review* 15(3): 173–98.

Yi, S.B. (1986) *Lower and Middle Paleolithic of Northeast Asia: a Geoarchaeological Review*. PhD thesis, Arizona State University.

Yi, S.B. (1989) *Northeast Asian Paleolithic Research*. Seoul National University Press, Seoul.

Yi, S.B. (1996) On the age of the Paleolithic sites from the Imjingang Basin. *Journal of the Korean Archaeological Society* 34: 135–59.

Yi, S.B., and Lee, K.D. (1993) *Chuwoli and Kawoli Paleolithic Sites, Imjin River Basin, Korea*. Seoul National University Press, Seoul.

Yi, S.B., and Clark, G.A. (1983) Observations on the Lower Palaeolithic of NorthEast Asia. *Current Anthropology* 24(2): 181–202.

Yi, S.B., Arai, F., and Soda, T. (1998) New discovery of Aira-Tn Ash (AT) in Korea. *Journal of the Korean Geographical Society* 33: 447–54.

Yokoyama, Y. (1989) Direct gamma-ray spectometric dating of Anteneandertalian and Neandertalian human remains. In: *Hominidae, Proceedings of the 2nd International Congress on Human Paleontology*, Giacobini, G., ed., pp. 387–90. Jaca Book, Turin.

Yokoyama, Y., Falguères, C., and de Lumley, M.-A. (1997) Datation directe d'un crâne Proto-Cro-Magnon de Qafzeh par la spectrométrie gamma nondestructive. *Comptes Rendus de l'Académie des Sciences* 324: 773–9.

Yoo, Y. (2007) *Long-term Changes in the Organization of Lithic Technology: a Case Study from the Imjin-Hantan River Area, Korea*. PhD thesis, McGill University.

Yoshiura, K.-i., Kinoshita, A., Ishida, T., Ninokata, A., Ishikawa, T., Kaname, T., et al. (2006) A SNP in the ABCC11 gene is the determinant of human earwax type. *Nature Genetics* 38: 324–30.

Young, N. (2008) A comparison of the ontogeny of shape variation in the anthropoid scapula: functional and phylogenetic signal. *American Journal of Physical Anthropology* 136(3): 247–64.

Young, N.M., and Hallgrimsson, B. (2005) Serial homology and the evolution of mammalian limb covariation structure. *Evolution* 59(12): 2691–704.

Yuan, S.X., Chen, T.M., and Gao, S.J. (1986) Uranium-series dating of Miaohoushan Site. In: *Miaohoushan*, pp. 86–9. Wenwu Press, Beijing.

Yunis, J.J., and Prakash, O. (1982) The origin of man: a chromosomal pictorial legacy. *Science* 215(4539): 1525–30.

Yurk, H., Barrett-Lennard, L., Ford, J.K.B., and Matkin, C.O. (2002) Cultural transmission within maternal lineages: vocal clans in resident killer whales in southern Alaska. *Animal Behavior* 63: 1103–19.

Zákány, J., Fromental-Ramian, C., Warot, X., and Duboule, D. (1997) Regulation of the number and size of the digits by posterior Hox genes: A dose-dependent mechanism with potential evolutionary implications. *Proceedings of the National Academy of Sciences USA* 94: 13695–700.

Zeitoun, V. (2001) The taxonomical position of the skull of Zuttiyeh. *Comptes Rendus de l'Académie des Sciences-Series IIA-Earth and Planetary Science* 332(8): 521–5.

Zentall, T. (2007) Temporal discrimination learning by pigeons. *Behavioural processes* 74(2): 286–92.

Zhang, P., Huang, W., and Wang, W. (2010) Acheulean handaxes from Fengshudao, Bose sites of South China. *Quaternary International* 223–4: 440–3.

Zhang, S. (1985) The early Paleolithic of China. In: *Paleoanthropology and Paleolithic Archaeology in the People's Republic of China*, Wu, R., and Olsen, J.W., eds, pp. 147–86. Academic Press, Orlando, FL.

Zhang, Y. (1998) Fossil human crania from Yunxian, China: morphological comparison with *Homo erectus* crania from Zhoukoudian. *Human Evolution* 13(1): 45–48.

Zhao, J.-x., Hu, K., Collerson, K.D., and Xu, H.-k. (2001) Thermal ionization mass spectrometry U-series dating of a hominid site near Nanjing, China. *Geology* **29**(1): 27–30.

Zhou, C., Wang, Y., Chen, H., and Liu, Z. (1999) Discussion on Nanjing Man's age. *Acta Anthropologica Sinica* **18**(4): 255–62.

Zhou, C., Liu, Z., Wang, Y., and Huang, Q. (2000) Climatic cycles investigated by sediment analysis in Peking Man's Cave, Zhoukoudian, China. *Journal of Archaeological Sciences* **27**(2): 101–9.

Zhou, G., and Hu, C. (1979) Supplementary notes on the teeth of Yuanmou Man with a discussion on morphological evolution of the mesial incisors in hominoids. *Vertebrata PalAsiatica* **17**: 149–62.

Zhou, M.Z., and Ho, C.K. (1990) History of the dating of *Homo erectus* at Zhoukoudian. In: *Establishment of a Geologic Framework for Paleoanthropology*, Laporte, L.F., ed., pp. 69–74. Geological Society of America, Boulder, CO.

Zhu, R., Zhisheng, A., Potts, R., and Hoffman, K.A. (2003) Magnetostratigraphic dating of early humans in China. *Earth-Science Reviews* **61**(3–4): 341–59.

Zhu, R.X., Potts, R., Xie, F., Hoffman, K.A., Deng, C.L., Shi, C.D., et al. (2004) New evidence on the earliest human presence at high northern latitudes in northeast Asia. *Nature* **431**: 559–62.

Zhu, R.X., Potts, R., Pan, Y.X., Yao, H.T., Lu, L.Q., Zhao, X., et al. (2008) Early evidence of the genus *Homo* in East Asia. *Journal of Human Evolution* **55**(6): 1075–85.

Zhu, Z.-y., Zhou, H.-y., Qiao, Y.-l., Zhang, H.-x., and Liang, J.-p. (2001) Initial strata occurrence of the south China tektite in strata and its implication for event stratigraphy. *Journal of Geomechanics* **7**: 296–302.

Ziegler, R., and Dean, D. (1998) Mammalian fauna and biostratigraphy of the pre-Neandertal site of Reilingen, Germany. *Journal of Human Evolution* **34**(5): 469–84.

Zihlman, A.L. (1984) Body build and tissue composition in *Pan paniscus* and *Pan troglodytes*, with comparisons to other hominoids. In: *The Pygmy Chimpanzee*, Susman, R.L., ed., pp. 179–200. Plenum Press, New York.

Zilhão, J., and Trinkaus, E., eds (2002) *Portrait of the Artist as a Child. The Gravettian Human Skeleton from the Abrigo do Lagar Velho and its Archeological Context*. Trabalhos de Arqueologia. Instituto Português de Arqueologia, Lisbon.

Zilhão, J., d'Errico, F., Bordes, J.-G., Lenoble, A., Texier, J.-P., and Rigaud, J.-P. (2006) Analysis of Aurignacian interstratification at the Châtelperronian-type site and implications for the behavioral modernity of Neandertals. *Proceedings of the National Academy of Sciences USA* **103**: 12643–8.

Zilhão, J., d'Errico, F., Bordes, J.-G., Lenoble, A., Texier, J.-P., and Rigaud, J.-P. (2008) Grotte des Fées (Châtelperron): history of research, stratigraphy, dating, and archaeology of the Châtelperronian type-site. *PaleoAnthropology* 1–42.

Zilles, K. (1972) Biometric analysis of fresh volumes of various prosencephalic brain regions in 78 human adult brains. *Gegenbaurs morphologisches Jahrbuch* **118**(2): 234–73.

Zinner, D., Groeneveld, L., Keller, C., and Roos, C. (2009) Mitochondrial phylogeography of baboons (*Papio* sp.) – Indication for introgressive hybridization? *BMC Evolutionary Biology* **9**(1): 83.

Zipfel, B., and Berger, L.R. (2009) New Cenozoic fossil bearing site abbreviations for the collections of the University of the Witwatersrand. *Paleontologica africana* **45**: 77–81.

Zollikofer, C., Ponce de León, M., Vandermeersch, B., and Lévêque, F. (2002) Evidence for interpersonal violence in the St. Césaire Neanderthal. *Proceedings of the National Academy of Sciences USA* **99**(9): 6444–8.

Zollikofer, C.P.E., Ponce de León, M.S., Lieberman, D.E., Guy, F., Pilbeam, D., Likius, A., et al. (2005) Virtual cranial reconstruction of *Sahelanthropus tchadensis*. *Nature* **434**: 755–9.

Zubov, A. (1992) The epicristid or middle trigonid crest defined. *Dental Anthropology Newsletter* **6**: 9–10.

Zuckerkandl, E., Jones, R.T., and Pauling, L. (1960) A comparison of animal hemoglobins by tryptic peptide pattern analysis. *Proceedings of the National Academy of Sciences USA* **46**: 1349–60.

Zuckerman, S. (1950) Taxonomy and human evolution. *Biological Reviews* **25**(4): 435–85.